# Discovering BUSINESS STATISTICS

SECOND EDITION

Quinton J. Nottingham | James S. Hawkes

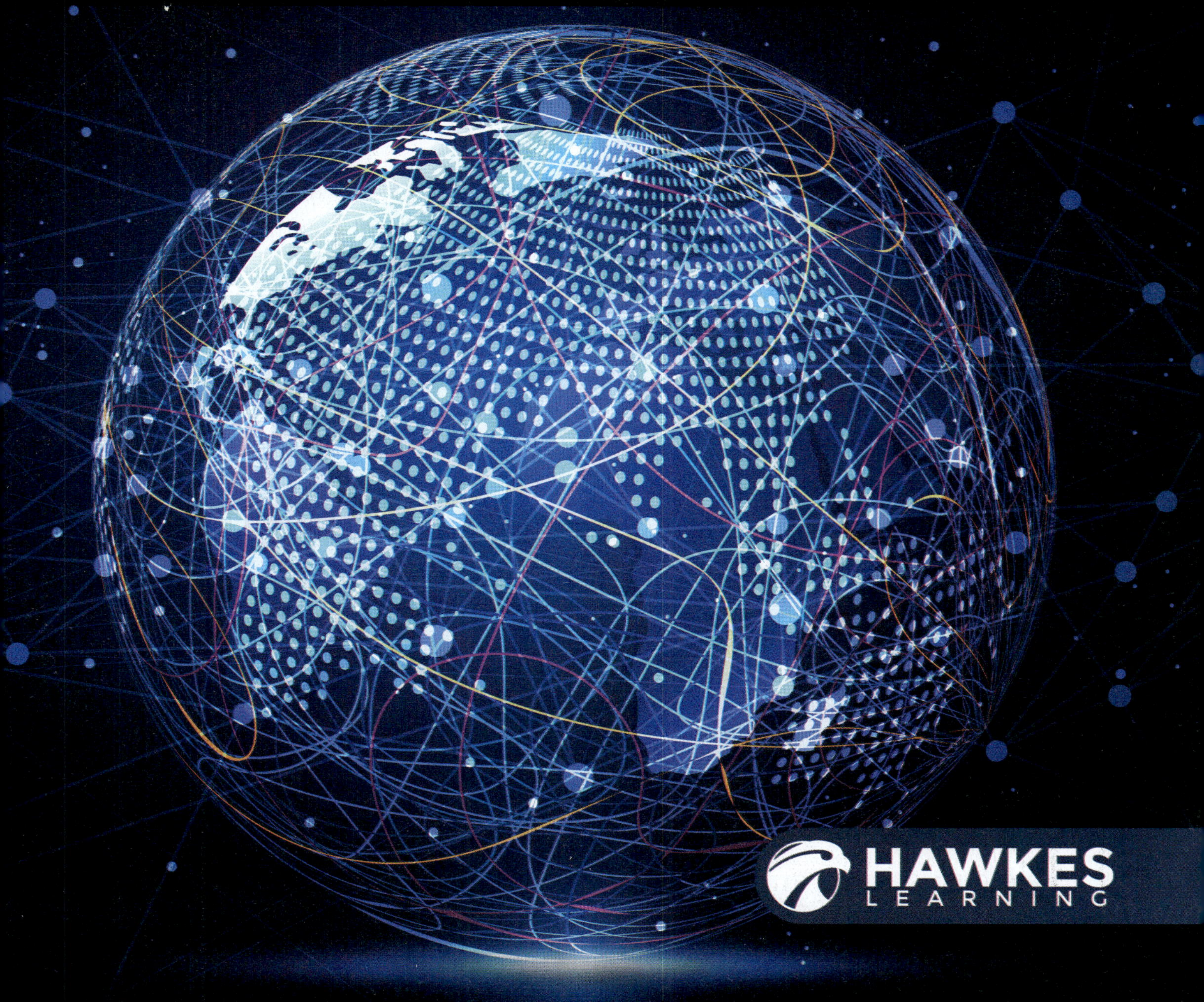

HAWKES LEARNING

**Editor:**
Robin Hendrix

**Assistant Editors:**
Vicente Muñoz,
Lauren Tubbs

**Editorial Assistants:**
Danielle Bess,
Adam Flaherty,
Marvin Glover

**Manager of Math Content Development:**
Blair Dunivan

**Creative Services Manager:**
Trudy Tronco

**Designers:**
Lizbeth Mendoza,
Patrick Thompson,
Joel Travis

**Cover Design:**
Joel Travis

**Composition and Answer Key Assistance:**
Quant Systems India Pvt. Ltd.

**A division of Quant Systems, Inc.**

546 Long Point Road
Mount Pleasant, SC 29464

Library of Congress Control Number   2022901162

Printed in the United States of America 🇺🇸

10 9 8 7 6 5 4 3 2 1

**ISBN:** 978-1-64277-511-2

# Table of Contents

## Chapter 1: Decision Making Using Statistics

## Chapter 2: Data, Big Data, and Analytics

## Chapter 3: Organizing, Displaying, and Interpreting Data

## Chapter 4: Numerical Descriptive Statistics

# Chapter 5: Probability, Randomness, and Uncertainty

# Chapter 6: Discrete Probability Distributions: About the Future

# Chapter 7: Continuous Random Variables

## Chapter 8: Samples and Sampling Distributions

## Chapter 9: Estimation with Confidence Intervals: Single Sample

## Chapter 10: Hypothesis Testing: Single Sample

## Chapter 11: Inferences about Two Samples

# Chapter 12: Analysis of Variance (ANOVA)

# Chapter 13: Regression, Inference, and Model Building

# Chapter 14: Multiple Regression

# Preface

We have taught Business Statistics for roughly thirty years. We have always wanted to write a textbook that portrays our teaching style – one that introduces the topic being discussed as it relates to the business world and then presents the methodologies in a manner that helps the students know when, where, and how these methodologies are used in the business world. This textbook does just that. As you work your way through the book, you'll see that in each chapter, the topics are introduced with examples of real-world applications. Then, as subsequent sections are covered in each chapter, the examples are presented and worked through step-by-step in a manner that will keep the students interested. When we set out to write this text, our goal was to discuss topics at a level that would engage the students, present examples that would interest them, and write the book in a relaxed, conversational style—we hope we have accomplished those goals with this text.

We have also integrated the TI-84 Plus calculator, Microsoft Excel, Minitab 21, and JMP 16 support in the form of Discovering Technology sections at the end of each chapter, along with Getting Started guides for Excel, Minitab, and JMP. Although the text includes quite a few manual computations with many of its examples, learning to use technology is helpful in analyzing the book's many business-related data sets and cases (presented as Discovery Projects at the ends of the chapters). In addition, we have included solutions to the odd-numbered exercises and a tear-out formula and table reference that we hope students will find helpful. Further, the material in this text is integrated with the *Hawkes Learning Systems: Discovering Business Statistics* software. Through the use of this text along with the accompanying software, it is our hope that students will discover the many advantages of statistical thinking and understanding, both in business environments and everyday situations.

Quinton J. Nottingham

James S. Hawkes

# Acknowledgements

We would like to thank our reviewers for giving us their time and feedback throughout the editorial process.

Scott Bailey, Troy University

Sangit Chatterjee, Northeastern University

Dr. Gregory Colman, Pace University

Dr. Jerry Dake, University of North Texas

Kim Gilbert, University of Georgia

Dr. R. Martin Jones, Virginia Tech

Dr. Lara Khansa, Virginia Tech

Dr. Onur Seref, Virginia Tech

Dr. Rick Simmons, Texas A&M University – Central Texas

Dr. Mark Skean, Mayville State University

Dr. Judith Woerner Mills, Southern Connecticut State University

Dr. Zhiwei Zhu, University of Louisiana – Lafayette

There are quite a few people at Hawkes Learning Systems who have made significant contributions to the creation of this book. First and foremost, we would like to thank our editor, Robin Hendrix, who has worked tirelessly on the development of this book with her editorial expertise. Without her thorough proofreading, we are sure it would have taken significantly longer to complete this project. We would also like to thank Vicente Muñoz, Lauren Tubbs, Danielle C. Bess, Adam O'Flaherty, and Marvin Glover for their editorial assistance.

# About the Authors

## Dr. Quinton J. Nottingham

Quinton J. Nottingham is the Associate Professor of Business Information Technology (BIT) in the Pamplin College of Business at Virginia Tech. He earned his B.S. (1989), M.S. (1991), and Ph.D. (1995) degrees in Statistics from Virginia Tech. Professor Nottingham has served on the faculty at Virginia Tech since 1995. He primarily teaches the Quantitative Methods courses and the MBA-level Managerial Statistics. As a Master Online Instructor, he has also delivered the Quantitative Methods courses online with the use of the *Hawkes Learning Systems: Discovering Business Statistics* software.

Professor Nottingham has published numerous articles in the areas of applied statistics, regression, nonparametric regression, logistic regression, time series analysis, artificial intelligence, and security. He has served on the board of the Southeastern Decision Sciences Institute (SEDSI) as the VP of Student Liaison, VP Finance, Program Chair, as well as the President (2011-2012). He is a member of or has held membership in the Institute for Operations Research and Management Science (INFORMS), Decision Sciences Institute (DSI), American Statistical Association (ASA), Institute of Mathematical Sciences, and the Virginia Academy of Sciences.

Professor Nottingham enjoys spending time with his family and friends, vacationing at various beaches in the Southeast, playing golf and basketball, and lifting weights. As a former Virginia Tech basketball player, he is an avid supporter of Virginia Tech Athletics and can be found in attendance at many Virginia Tech sporting events.

## Dr. James S. Hawkes

James S. Hawkes earned his Bachelor's degree at the University of Richmond (1969), a Master's of Business Administration at New York University (1972), and his Ph.D. in Management Science at Clemson University (1978). As a faculty member at the College of Charleston from 1977 to 1998, he primarily taught Business Statistics. In 1978, he created Hawkes Learning Systems (HLS) where he is currently the CEO. In 1986, Dr. Hawkes designed a breakthrough teaching technology called *Adventures in Statistics*. This software was the first to use expert system methodology to teach problem-solving in statistics. Successive generations of the software expanded the content and improved the expert system. The software was renamed *Hawkes Learning Systems: Statistics* and is the basis for the software that accompanies this textbook, *Discovering Business Statistics*.

Other accomplishments include the founding of Quant Systems India (QSI), a software development company located in Vishakaputnam, India, in 1994. QSI focuses on software development and quality assurance of HLS software. Dr. Hawkes also co-founded Automated Trading Desk in 1988, a proprietary stock trading company that traded approximately six percent of the NYSE and NASDAQ marketplaces before it was sold to Citigroup in 2008. Most recently, he created The Writer's Muse which was an outgrowth of his hobby of collecting interesting phrases. This company has produced a phrase thesaurus which is available for smart phones, tablets and personal computers. Dr. Hawkes enjoys composing music, playing bridge, and collecting hats.

# Discovering Business Statistics
## SECOND EDITION

# Features and Pedagogy

This text includes a number of special features developed to engage and challenge students while ensuring that their knowledge of statistics is growing. The following list outlines a number of these features.

## ▼ Discovering the Real World

A Discovering the Real World section introduces the main topic in each chapter in the context of a real-world application of the topic. These applications are designed to pique students' interest and get them excited about learning the material in the chapter. Some topics from earlier chapters are re-examined in later chapters from a different statistical perspective.

### Discovering the Real World

**Do Americans Show Support for Basic Research?**

An online survey was conducted in the United States in January 2017 and again in January of 2021. The question presented to the survey participants was, *Do you agree or disagree with the following statement? Even if it brings no immediate benefits, basic scientific research that advances the frontiers of knowledge is necessary and should be supported by the federal government.*

The results of the nationwide poll are presented in the table below.

| Opinion Results on Support for Basic Research Survey | | | | | |
|---|---|---|---|---|---|
| | Strongly Agree | Somewhat Agree | Somewhat Disagree | Strongly Disagree | Don't Know |
| January 2017 | 24% | 39% | 16% | 5% | 16% |
| January 2021 | 46% | 39% | 8% | 2% | 5% |

## ▼ An Example-Driven Text

Examples are relevant and engaging for students, and are presented in a step-by-step manner that is easy to follow. Real-world contexts are used within examples to illustrate the advantages of the statistical methods being taught. The textbook is very example-driven, and most of the concepts are taught by example. Examples are clearly presented within the layout of the textbook, making them easy for students to recognize and learn from.

**Example 14.1.1**

**Modeling Pizza Delivery Time**

A pizza delivery manager is analyzing the delivery routes in her system. She is interested in predicting the amount of time required for the driver to deliver the pizzas on a specific route. The driver has to make several stops on the route because the manager doesn't want to send several vehicles/drivers to cover the same area/route. Fit a multiple linear regression model using the data in Table 14.1.1 to predict delivery time.

| Table 14.1.1 – Pizza Delivery Time Data | | | |
|---|---|---|---|
| Observation | Delivery Time (Minutes) | Number of Pizzas | Distance (Miles) |
| 1 | 16.68 | 7 | 5.60 |
| 2 | 11.50 | 3 | 2.20 |
| 3 | 12.03 | 3 | 3.40 |
| 4 | 14.88 | 8 | 0.80 |
| 5 | 13.75 | 6 | 1.50 |
| 6 | 18.11 | 7 | 3.30 |
| 7 | 8.00 | 2 | 1.10 |
| 8 | 17.83 | 7 | 2.10 |
| 9 | 79.24 | 30 | 14.60 |
| 10 | 21.50 | 5 | 6.05 |
| 11 | 40.33 | 16 | 6.88 |
| 12 | 21.00 | 10 | 2.15 |

**⚙ Data**
This data set can be found at stat.hawkeslearning.com by navigating to **Discovering Business Statistics, Second Edition > Data Sets > Pizza Delivery Time.**

## ▼ Important Definitions, Formulas, Properties, Procedures, and Theorems Highlighted

Definitions, formulas, properties, procedures, and theorems are clearly presented and easy to reference for students. The color coding will help students focus on the most important components of the chapters. Definition boxes are blue, formula boxes are orange, and properties, procedures, and theorem boxes are outlined in green.

**Procedure**

**The Scientific Method**

1. Gather information about the phenomenon being studied.
2. On the basis of the data, formulate a preliminary generalization or hypothesis.
3. Collect further data to test the hypothesis.
4. If the data and other subsequent experiments support the hypothesis, it becomes a law. If the experiment does not support the hypothesis, construct a new hypothesis and start the method over again. Sometimes, even if the hypothesis is supported, you may want to test it again but in a different way.

## ▼ Technology and Data Links

Technology links next to examples in the text direct students to the Hawkes Learning website, where they can find detailed instructions for making a calculation using the TI-84 calculator, Microsoft Excel, Minitab, or JMP. Data links point students to where they can download large data sets used in examples and exercises.

**⚙ Data**
This data set can be found on stat.hawkeslearning.com under **Discovering Business Statistics, Second Edition > Data Sets > Tuition Consumer Price Index.**

**☁ Technology**
Confidence intervals can be calculated using technology. For instructions, please visit stat.hawkeslearning.com and navigate to Discovering Business Statistics, Second Edition > Technology Instructions > Confidence Intervals > z-Interval.

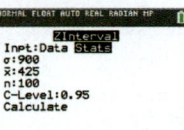
```
NORMAL FLOAT AUTO REAL RADIAN MP
      ZInterval
Inpt:Data Stats
σ:900
x̄:425
n:100
C-Level:0.95
Calculate
```

## ▼ Informational Sidebars

Interesting and engaging sidebars are presented in the margins to stimulate student interest in statistical topics. The sidebars highlight the origins of the statistical topics being covered, as well as real-world applications of statistical concepts.

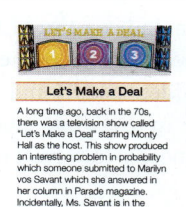

**Let's Make a Deal**

A long time ago, back in the 70s, there was a television show called "Let's Make a Deal" starring Monty Hall as the host. This show produced an interesting problem in probability which someone submitted to Marilyn vos Savant which she answered in her column in Parade magazine. Incidentally, Ms. Savant is in the Guinness Book of World Records as having the highest recorded IQ (228). Here's the problem that was posed to Ms. Savant.

**Struck by Lightning**

Many people have a greater chance of meeting someone who survived a lightning strike than someone who won the lottery. There are an estimated 1800 new thunderstorms being created around the world every

## ▼ Basic Concepts Questions

Each section has a set of exercises that begins with a list of questions geared towards testing students' knowledge of the basic concepts being taught. Students benefit from expressing the ideas they are learning about through writing. These questions both review the key definitions from the section and help students examine the parameters and limitations of the statistical methods they are applying.

### 📝 2.4 Exercises

**Basic Concepts**

1. What are time series measurements?
2. What problems are associated with the concept of population when studying time series data?
3. What is a stationary process?
4. What is a nonstationary process?
5. What is a trend? If a time series has an 'upward trend' what does this mean?
6. What are cross-sectional data?
7. What is the difference between cross-sectional data and time series data?

## ▼ Exercise Sets

Exercise sets are given at the end of each section, which range in difficulty and cover a wide variety of subject matter. Utilizing the exercises in this text, students can easily get the practice they need to master the topics being taught. Care has been taken to present students with a variety of exercise types to encourage a broad understanding of the material presented. Additionally, exercises have been created using data that is relevant to students' lives and their entry into the business world.

**Exercises**

11. A consumer magazine uses bar charts to compare four popular brands of automobiles. This particular bar chart represents a comparison of the miles per gallon (mpg) for the four brands.

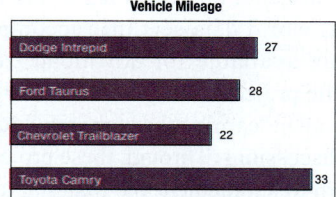

**Vehicle Mileage**

Dodge Intrepid — 27
Ford Taurus — 28
Chevrolet Trailblazer — 22
Toyota Camry — 33

a. What is wrong with this picture?
b. Evaluate the bar chart using the guidelines suggested in the section on the aesthetics of bar chart construction.

## ▼ Discovering Technology

Discovering Technology sections at the end of each chapter highlight easy-to-follow directions for using a TI-84 Plus calculator, Microsoft Excel, Minitab 21, or JMP 16 to apply the concepts learned in the chapter. Though the directions were written to accompany Minitab 21 and JMP 16, many of the steps can also be followed when using earlier versions of these software products. Additionally, many of the calculator instructions can be applied to graphing calculators other than the TI-84 Plus. Screenshots are presented alongside the step-by-step instructions to aid in student learning. Although the text includes quite a few manual computations with many of its examples, learning to use technology is helpful in analyzing the book's many business-related data sets and cases (presented as Discovery Projects at the ends of chapters). In addition, Getting Started guides are included in Appendices B, C, and D for Excel, Minitab, and JMP, respectively, to get students started with using technology and data analysis.

### T　Discovering Technology

**Note:** If you are not familiar with Excel, Appendix B is a tutorial section that will familiarize you with the controls in Excel.

**Using Excel**

Line Chart

For this exercise, use the information from Example 2.6.

1. Enter the time period data into Column A – label as "Period".
2. Enter the data for the cash operations into Column B – label as "Cash Operations".
3. Highlight the data in Column A and Column B and under the "Insert" tab click on the **Line** button. Select **Line with Markers** (the first graph in the second row under "2-D line") so that you can see each data point that is plotted.
4. You now see a time series plot of the cash operations data. To change the interval of the x-axis, select the values on the x-axis, right click and select **Format Axis**. In the "Axis Options" dialog box, next to "Major unit" select the radio button for **Fixed**. Enter 6 in the dialog box and make sure that **Months** is selected. Click **Close**.
5. To format the y-axis, select the values on the y-axis, right click and select **Format Axis**. Since there are negative values in this data set, the chart may be easier to read if the horizontal axis crosses at a negative value, say −400.0 rather than 0. Under the heading "Horizontal axis crosses," select the radio button next to **Axis value** and enter −400.0 in the dialog box. Click **Close**. Now the entire line chart is above the x-axis.

## ▾ Chapter Reviews

This end-of-chapter feature includes a list of the key terms and ideas from the chapter as well as a list of the chapter's key formulas. By gathering the key ideas and formulas in one place, students have a great resource to use while studying, as they can more easily identify the concepts they may need to review.

**R    Chapter 2 Review**

**Key Terms and Ideas**

- The Scientific Method
- Confounding Variable
- Descriptive Statistics
- Inferential Statistics
- Big Data
- Structured Data
- Unstructured Data
- Semi-Structured Data

## ▾ Additional Exercises

At the end of each chapter, additional exercises have been included to provide students with additional practice on the skills covered. Also, by providing students with a compilation of problems from the whole chapter, the guidance given in the section exercises is removed, and students are forced to evaluate which method to use to solve each of the different question types.

**AE    Additional Exercises**

1. Suppose you were the administrator of a public school system. What kinds of variables would you measure and how would you collect the measurements on the following subjects:

   a. Student learning

   b. School discipline

   c. Teacher preparation

   d. Absenteeism (pupil and teacher)

   e. Cafeteria food quality

2. The head of the Veterans Administration has been receiving complaints from a Vietnam Veterans organization concerning disability checks. The organization claims that checks are continually late. The checks are to arrive no later than the tenth of each month.

## ▾ Discovery Projects

Discovery Projects at the ends of chapters engage students with more in-depth questions regarding the chapter content. Large, real-world data sets that accompany the Discovery Projects are available for download from the companion website. The projects also give instructors the opportunity to enhance student learning through individual or group projects and class discussions. Through these projects, instructors have the freedom to supplement the material as they see fit with a longer-term, more difficult assignment.

**P    Discovery Project**

**Bachelor's Degrees Conferred by Race/Ethnicity**

Review the following graph using data from IPEDS (Integrated Postsecondary Education Data) collected from Fall 2000 through Fall 2016 and answer the questions below.

| Bachelor's Degrees Conferred by Postsecondary Institutions, by Race/ Ethnicity of Student Selected Years: 2000 through 2016 | | | | | |
|---|---|---|---|---|---|
| Year | White | Black | Hispanic | Asian/Pacific Islander | American Indian/ Alaskan Native |
| 2000 | 929,102 | 108,018 | 75,063 | 77,909 | 8,717 |
| 2001 | 927,357 | 111,307 | 77,745 | 78,902 | 9,049 |
| 2002 | 958,597 | 116,623 | 82,966 | 83,093 | 9,165 |
| 2003 | 994,616 | 124,253 | 89,029 | 87,964 | 9,875 |
| 2004 | 1,026,114 | 131,241 | 94,644 | 92,073 | 10,638 |
| 2005 | 1,049,141 | 136,122 | 101,124 | 97,209 | 10,307 |

## ▾ Answer Key and Formula Sheets

The answer key contains the solutions for the odd-numbered exercises for student reference. In addition, this text contains a tear-out formula sheet that students may use throughout the course.

# Major Changes for the Second Edition

## New topics have been added in this edition.

This edition adds a much-requested chapter on Time Series Analysis and Forecasting, which introduces students to the concepts of:

- Moving Averages
- Exponential Smoothing Techniques

- Forecast Accuracy
- Seasonality in Time Series

In addition, there are new sections on:

- Big Data and Business Analytics
- Analyzing Graphs
- Assessing Normality Graphically
- Estimating the Population Standard Deviation or Variance

- Comparing Two Population Variances
- Multiple Comparisons Procedures
- Residual Analysis

## New and more relevant examples have been added.

There have been 40 new examples created for this edition and about 70 examples were updated to reflect new methodology, more current data, and expansion of topics.

## Every chapter now has a chapter project and Discovering the Real World section.

Chapter projects and Discovering the Real World chapter openers have been updated with new scenarios and data from current events, including the COVID-19 pandemic.

## Increased emphasis has been placed on technology.

More emphasis has been placed on using technology to solve problems within examples and exercises, including expanded TI-84 Plus instructions and inclusion of Microsoft Excel, JMP, and Minitab instructions for many concepts. Solutions to exercises are given using manual computations as well as using technology, so that instructors have the freedom to assign the method of their choice. Directions on how to use additional technology options can be found by going to stat.hawkeslearning.com and navigating to Technology Instructions.

## Content has been expanded and presentation has been improved.

Numerous vignettes have been integrated to demonstrate the uses of statistics in business, technology, and science. The Table of Contents has been streamlined to contain fewer individual sections in a chapter, while at the same time expanding the content to make each section more robust and complete. Basic concept questions have been added to help students understand and reflect on what they have learned in each section (a GAISE guideline). The solution content in our examples has been expanded to better model the statistical thinking process for students and aid them in making valid conclusions from data (also a GAISE guideline). Finally, important content has been summarized in procedure, formula, or definition boxes to enhance student learning and to aid students in reviewing the content for assessments.

## Additional graphical displays were added to Chapters 3 and 7.

Choropleth maps (or heat maps) were added to the section on graphical displays of quantitative data. Normal probability plots were added to the section on assessing normality graphically.

## Hypothesis testing has been modernized.

The hypothesis testing procedure has been streamlined from ten steps to six. The biggest change is that the null hypothesis is written in this edition using only an equal sign instead of being the mathematical opposite of the alternative hypothesis. Technology directions with expanded descriptions of how to use either rejection regions or $P$-values are included throughout the hypothesis testing chapters.

## New exercises have been added.

Over 100 new exercises have been added to this edition. The majority are for the new chapter on time series and the other new sections, such as residual analysis and multiple comparisons procedures. The remaining new exercises have been added to reflect the expansion of topics within the existing content.

## New and updated technology sections have been added to the end of each chapter.

Instructions for using SAS JMP have been added to this edition. The technology instructions for Excel and Minitab have been updated to reflect the latest versions of the software.

# Companion Site

## Overview

Our companion site contains data sets, technology instructions, and other resources that may be of interest or assistance to the student. Please visit

# stat.hawkeslearning.com

and navigate to *Discovering Business Statistics, Second Edition* from the homepage.

### ⠂ Data

Throughout the book you will find marginal annotations that direct the reader to the companion site to access a data set.

You will find all these data sets on the companion site available to download in various file formats.

You will also find these annotations in the Exercise sections. We have incorporated data exercises to allow the student to practice the concepts learned in the section on real-world data. To obtain the data to complete the problem, you will visit the companion site and download the required data set.

Here are a couple of examples of what you will see.

### ⠂ Data

stat.hawkeslearning.com
**Discovering Business Statistics, Second Edition > Data > California DDS Expenditures**

### ⠂ Data

The full Amazon stock price data set can be found on stat.hawkeslearning.com under **Discovering Business Statistics, Second Edition > Data Sets > Amazon Stock Price.**

### ∽ Technology

Next to many of our examples you will find technology links to instructions that can help you find the answer(s) to the example using technology. Navigate to the technology area of the companion website and locate the topic indicated in the margin. We provide directions for many technologies including the TI-84 Plus calculator, Microsoft Excel, Minitab® Statistical Software, JMP, R, and more. As technology changes and grows, so will our resources!

Here are a couple of examples of what you will see.

### ∽ Technology

Calculation of the variance of a discrete random variable can be done using technology. For instructions go to stat.hawkeslearning. com and navigate to **Discovering Business Statistics, Second Edition >Technology Instructions > Descriptive Statistics > Two Variable.**

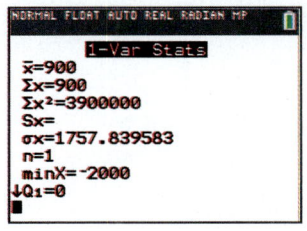

### ∽ Technology

The margin of error can be found using Excel's **CONFIDENCE.T** function. For instructions on calculating the confidence interval using Excel, please visit stat.hawkeslearning.com and navigate to **Discovering Business Statistics, Second Edition > Technology Instructions > Confidence Intervals > *t*-Interval.**

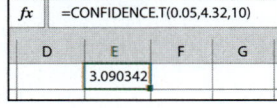

## Other Resources

Other resources available on the companion site include downloadable formula pages and statistical tables, as well as a curated collection of websites and tools that may be of interest to or aid someone who is learning statistics.

## Getting Started

Below are two scenarios from the textbook where a student would utilize stat.hawkeslearning.com.

### Accessing Data Sets Use Case

**Description**: A student is working on the exercises in Chapter 4 and needs to access a data set on the companion website in order to complete an exercise.

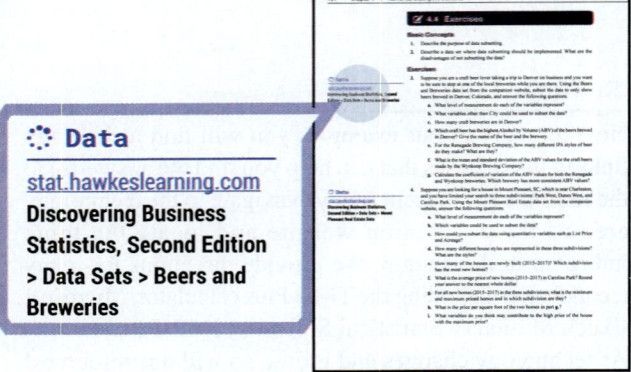

He/she navigates to the companion website using the following steps.

1. Go to stat.hawkeslearning.com.

2. Navigate to *Discovering Business Statistics, Second Edition*.

3. Select the tab at the top of the screen called Data Sets.

4. Click on "Excel", "CSV", or "DOC" next to "Beers and Breweries."

5. Open the Beers and Breweries data file you chose in the correct application.

### Accessing Technology Instructions Use Case

**Description**: A student is reading through Chapter 10 for an assignment and needs help to calculate the *P*-value for a *t*-test statistic using technology.

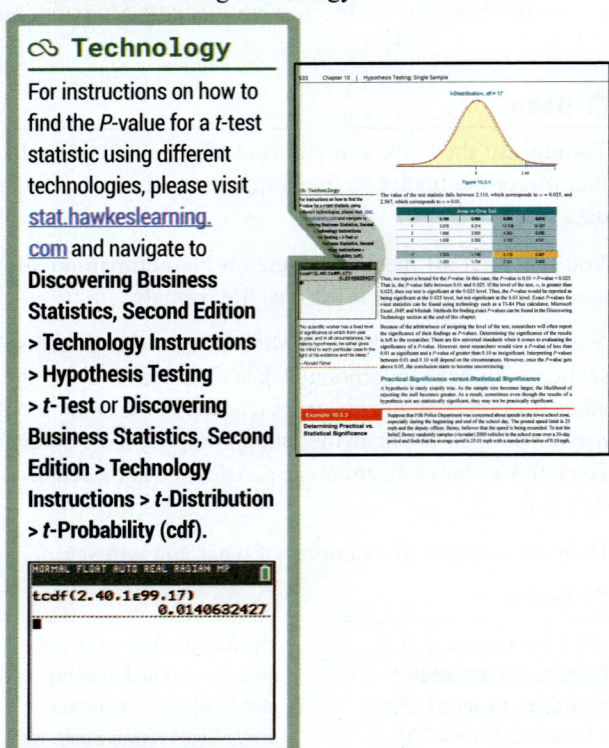

He/she navigates to the companion website using the following steps.

1. Go to stat.hawkeslearning.com.

2. Navigate to *Discovering Business Statistics, Second Edition*.

3. Select the tab at the top of the screen called Technology Instructions.

4. Click on the Topic tab.

5. Type "*t*-Distribution" within the search bar.

6. Select the button corresponding to the sub-topic and technology method you want to use (for instance, **TI Calculator** next to *t*-**Distribution** > *t*-**Probability (cdf)**).

# Hawkes Learning: A Clear Path to Mastery

Hawkes' software employs an adaptive, competency-based approach to knowledge mastery supported by a user-friendly interface. The student-centric platform promotes positive active learning by adapting to each student's needs through algorithmically generated questions based on an individual learner's pace, skill, and knowledge level. The real-time adaptive feedback addresses errors immediately, so that students learn from their mistakes when they make them. For each topic, the Hawkes Learning path to content mastery engages students through three simple modes: Learn, Practice, and Certify.

## Competency-based learning made simple in three steps:

 **LEARN** offers a multimedia-rich presentation of the lesson content. It includes instructional videos, interactive examples, and more.

 **PRACTICE** engages students with algorithmically generated questions and intelligent tutoring in an ungraded, penalty-free environment.

 **CERTIFY** requires students to demonstrate mastery of the material at a defined proficiency level without access to tutoring aids.

## Support

If you have questions or comments, we can be contacted as follows:

**24/7 Chat:** chat.hawkeslearning.com

**Phone:** 1-800-426-9538

**Email:** support@hawkeslearning.com

**Web:** support.hawkeslearning.com

# Discovering

## SECOND EDITION

# BUSINESS STATISTICS

Quinton J. Nottingham | James S. Hawkes

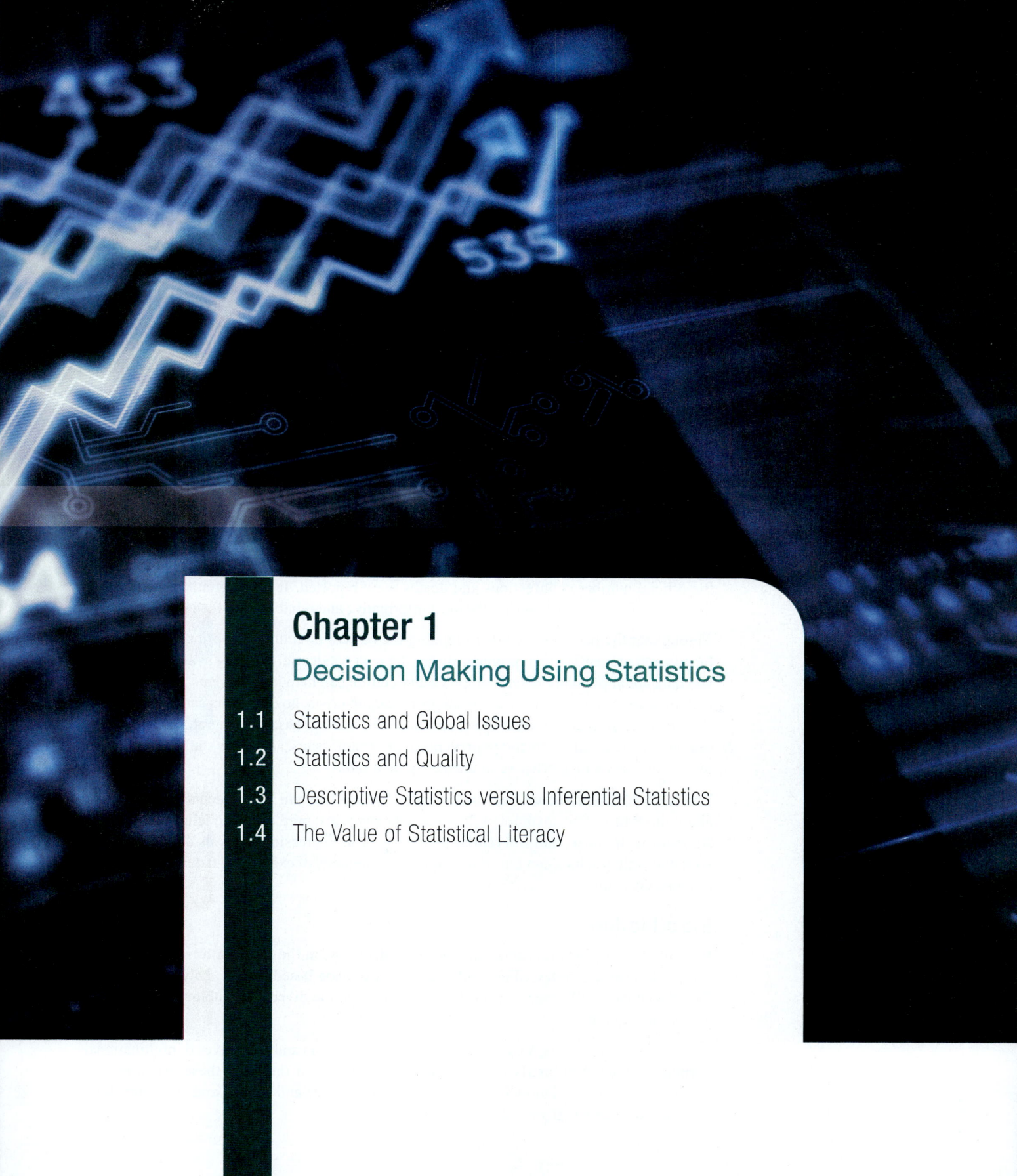

# Chapter 1

## Decision Making Using Statistics

## Discovering the Real World

### Statistics and the COVID-19 Pandemic

On March 11, 2020, the World Health Organization (WHO) declared COVID-19, the disease caused by SARS-CoV-2, a pandemic. The announcement followed a rising sense of alarm in the preceding months over a new, potentially lethal virus that was swiftly spreading around the world. From that day in March when the announcement was made to more than a year later, the United States and countries around the world have been pivoting to mitigate the effects of the pandemic—reduce the number of cases, prevent business shutdowns, and limit physical and mental health issues, just to name a few.

In addition to the many health-related issues associated with the pandemic, we must also consider the impact from a business perspective. There were countless supply-chain issues with shortages of toilet paper, used cars, computer chips, and even school supplies. The housing market was also not immune to the pandemic with a shortage of lumber and appliances. Lastly, due to COVID-19, the unemployment rate skyrocketed. As the number of infections increased, businesses (restaurants, retailers, schools, etc.) were forced to either shut down or reduce staff, which was necessary to follow CDC guidelines. The shutdowns and reduction of staff led to many businesses laying off employees, thus causing the unemployment rate to rise to 14.8% , which is the highest rate since data collection began in 1948.

According to the U.S. Centers for Disease Control and Prevention (CDC), from 2019 to 2020 the estimated age-adjusted U.S. death rate increased from 715.2 to 828.7 deaths per 100,000 population, which was an increase in mortality of 0.114 percentage points. As the infections and mortality rates increased, it became more and more critical to use data to make decisions to fight the pandemic. Many decisions surrounding the pandemic (shutdowns, social distancing, mask mandates, travel restrictions, etc.) were made as the data regarding the official number of infections and deaths were reported. It is important to realize that these decisions must be based on the most informative and reliable COVID-19 data possible.

Throughout the pandemic, while mitigating the current circumstance, priority should also be given to preventing and/or preparing for the next pandemic. Publicly available statistics and data about population demographics and culture can help governments prepare for the next pandemic. The next time there is a global health crisis, governments can use statistical techniques to determine the hot spots, understand how the disease will migrate from region to region, how to predict demand to prevent supply chain issues, as well as staffing businesses to prevent such a meteoric rise in unemployment rate.

This textbook will introduce many statistical techniques that will help students learn and think about the future. This textbook will cover data analytics, predictive modeling, descriptive statistics techniques, and inferential models such as regression analysis and analysis of variance—all of which can introduce students to methods of examining data and using data to make decisions.

### Introduction

How we conduct business and make business decisions in the real world is intensely data driven. In business today, almost all decisions are made based on the analysis of data. In fact, decisions in all aspects of our lives are based on the diverse and ubiquitous data that are available to us.

CEOs, CFOs, Presidents, Vice Presidents, and many leaders and executives of multinational corporations need to justify their decisions. Almost all of the time, these decisions are supported by statistical models, summary statistics (graphs and tables), and other statistical tools and concepts that we will discuss in this text.

Decision makers use statistics to collect data, analyze the data, and make interpretations—all relevant to their profession. They build models with data and then make decisions using those models. They describe data using graphs and summary measurements. They develop methods of designing experiments and gathering data that are cost effective and diminish bias.

## Business and Statistics

Simply put, statistics will help managers make better business decisions. Managers use data to help them oversee multiple divisions of their company, countless products, and thousands of employees. Even for small businesses, understanding data via the use of statistics is critical in today's competitive environment. Even though many business decisions are made based on expert intuition and experience, ignoring all relevant information available (such as the use of statistical techniques) would be senseless.

Thus, the use of statistical techniques must be viewed as an important part of the decision-making process. Combining intuition, experience, and statistical analysis arms the manager with a wealth of information giving him or her an advantage over the competition. If not an advantage, then this combination of tools certainly allows the manager, company, or corporation to maintain its competitive status in its respective industry.

For example, the development of Six Sigma can be traced back to the eighteenth century when Carl Frederick Gauss introduced it as a metric for the normal curve. Later Walter Shewhart began showing how three sigma deviations from the mean required a process correction. In the 1980s, a Motorola engineer coined the term Six Sigma (and copyrighted it) for improving quality management. Today Six Sigma is viewed as a concept that incorporates many statistical techniques to improve processes and make business decisions.

# 1.1 Statistics and Global Issues

We have mentioned several aspects of business and economy that were greatly impacted by the COVID-19 pandemic, such as shortages of supplies (electronic chips, furniture, paper products, used and new vehicles, etc.), unemployment, and in general, the global supply chain. While social distancing, mask mandates, and business shutdowns were ways in which governments believed they could "flatten the curve" (i.e., cases of infections would not increase and eventually plateau), the best response to mitigate COVID-19 would be the development of a vaccine and ultimately, get people vaccinated.

A number of pharmaceutical companies (Pfizer and BioNTech, Moderna, Johnson & Johnson, AstraZeneca, Novavax, Sanofi and GlaxoSmithKline) in the United States and around the world began the process of developing a COVID-19 vaccine. Without going into great scientific detail, in the United States, bringing a vaccine to the public involves many steps including vaccine development, clinical trials, United States Food and Drug Administration (FDA) authorization and approval, manufacturing, and finally distribution to hospitals, clinics, pharmacies, and other locations so that they can be given to the public.

The development of a vaccine is conducted under the conditions of a **controlled experiment**. That is, in a controlled experiment, the researcher collects data that will hopefully clarify whether the vaccine protects people from contracting the virus. A simple method of accomplishing this task is to create two groups sampled from the **population**, those receiving the vaccine and those not receiving the vaccine, and compare them. Those receiving the vaccine would be the **treatment group** (sometimes called the **experimental group**) and those not receiving the vaccine would be the **control group**. Each person is randomly assigned to one of the two groups to avoid bias. The groups would be exposed to the virus and at the end of a certain time period, researchers would compare the number of people that contracted the virus from each group. By randomly choosing members from the population,

## Definition

### Controlled Experiment, Control Group, Treatment Group, and Treatment

A **controlled experiment** is an experiment that is conducted under controlled conditions in which just one or two factors are changed at a time to determine if a relationship exists between variables.

The **treatment group** is the group that receives the treatment in the experiment.

The **control group** is the group that does not receive the treatment. Oftentimes, the control group provides a baseline to let us know if the treatment has an effect.

A **treatment** is something that is applied or administered to one or more groups in a controlled experiment.

any differences in the effect of the virus would be attributed to the **treatment** (vaccine) that was applied to one group. The drawback to this experimentation is getting people to volunteer for the experiment. Fortunately, for the development of the vaccine for COVID-19, each of the aforementioned pharmaceutical companies had tens of thousands of volunteers.

Having successfully undergone all of the clinical trials (there are three phases of clinical trials when developing a vaccine) to ensure safety and effectiveness, the data collected from the controlled experiments are submitted to the FDA to assess their findings and to determine if the vaccines can be granted Emergency Use Authorization (EUA). The EUA allows for the vaccines to be quickly distributed for use while maintaining high safety standards required of all vaccines.

What is the role of statistics in the development, creation, and deployment of these vaccines? Statisticians evaluate all of the data collected on the volunteers in each of the groups and their responses to the vaccine to determine its safety and effectiveness. Statisticians will use a variety of statistical procedures and techniques, comparing the control group and the treatment group to ensure that the vaccine is safe for the general public.

| Table 1.1.1 Vaccine Efficacy – Predicted and Actual through March 31, 2021 | | |
|---|---|---|
| **Vaccine Manufacturer** | **Predicted Percentage Effectiveness** | **Actual Percentage Effectiveness** |
| Pfizer-BioNTech | 95% | 91.3% |
| Moderna | 94.1% | 90% |
| Johnson & Johnson | 66.3% | 72% |
| Astrazeneca | 76% | 75.9% |
| Novavax | 93% | 90% |

> **Definition**
>
> Population
>
> A **population** is the set of all subjects or elements about which we are interested in making inferences.

Populations are defined by what a researcher is studying and can come in all shapes and sizes. If someone is studying houses foreclosed upon in 2020, then all the houses foreclosed upon in 2020 would constitute the population. If someone is studying banks in the United States, then all banks in the United States would represent the population. Citizens who are registered to vote in a presidential election constitute a population of considerable interest to presidential candidates. The parties spend hundreds of millions of dollars finding out what people think (sampling) and adopt or reject political positions consistent with the preponderance of the data. In the world of political science this "data-driven" approach to politics is called *populism*.

> **Definition**
>
> Frame
>
> A list containing all members of the population is referred to as a **frame**.

According to the Census Bureau there are about 328 million people in the United States and about 7.7 billion people around the world. The **frame** for the population of the U.S. would be a rather long list containing about 328 million names. Although a previous census would be a good start in developing a frame for the U.S. population, it is doubtful that an exact frame could ever be developed at a given point in time since there is one new birth every 8 seconds, one death every 12 seconds, and one new immigrant every 670 seconds (approximately 11 minutes). There are just too many people being born, dying, and immigrating over a 10-year period to get an exact frame for the U.S. population. But for problems that deal with smaller populations, frames are easily developed. For example, if your business statistics class were the population under consideration, the class roll would be the frame for the population.

> ✎ **NOTE**
>
> A strict definition of a census is a survey that includes all the elements or units in the frame.

For a presidential election, some **population parameters** in which candidates and pollsters will be interested are:

> **Definition**
>
> Population parameters
>
> **Population parameters** are facts about the population. Since parameters are descriptions of the population, a population can have many parameters.

- The percentage of eligible voters who will vote on Election Day.
- The percentage of voters who will vote for a specific candidate.
- The percentage of men who favor a candidate.
- The percentage of women who favor a candidate.
- The percentage of people in the 18–25 age group who favor a candidate.
- The average income of voters who favor a candidate.

The parameters mentioned in this instance are either averages or percentages. However, other measures such as the maximum or minimum value of the population measurement as well as other characteristics would also be considered population parameters. For a specific population at a specific point in time, population parameters do not change; they are fixed numbers. But seldom will the value of a population parameter be known since the value involves all the population measurements which are usually too expensive or time consuming to collect. It is the statistician's job to discover these values. This is done by taking a sample and using the sample measurements to estimate the desired population measurement.

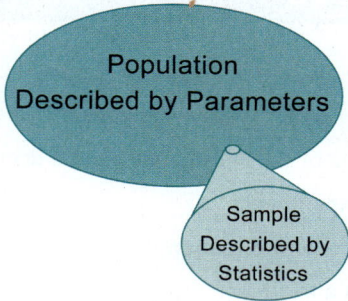

In Table 1.1.1 the vaccine's predicted percentage efficacy is a statistic since it is computed from a sample. The vaccine's actual percentage efficacy is a parameter, since it is computed using the data from the millions of people who have been vaccinated through March 31, 2021 (the population in this instance).

For any given sample a statistic is a fixed number. Because there are lots of different samples that could be drawn from a population, statistics will vary depending on the sample selected. Statistics are used as estimates of the population parameters. See Figure 1.1.1.

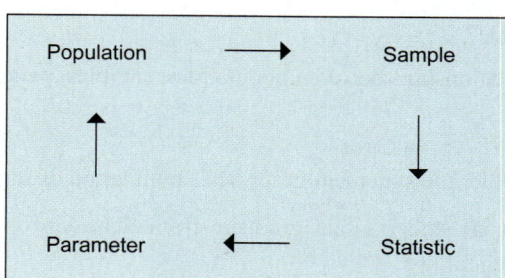

Figure 1.1.1

Studying an entire population can be an expensive proposition. In 2000 the census cost the government $39.98 per household. Total expenditure during the 2000 census cycle was approximately 4.5 billion dollars. The estimated cost of the 2010 census was around 14.5 billion dollars, or $117.33 per household. Interestingly, the estimated cost of the 2020 census was approximately 6.3 billion dollars, or approximately $96 per household. Because of the enormous expense, even the United States government with all its resources does not undertake a census of its citizens but once every ten years. Yet, amazingly accurate information about the population, even large populations like the United States, can be found by using a small sample. If the sample is a good representation of the population, then the conclusions reached using sample data will likely be reasonable for the population as a whole. A statistician faces the interesting problem of developing a representative sample without spending an inordinate amount of time or money.

It is obvious that many resources are utilized in developing a vaccine. Once the vaccine has been developed, the challenge is then to find volunteers to participate in the clinical trials to determine vaccine effectiveness. After the estimates of the vaccine effectiveness are calculated, one of the interesting statistical questions is, *how good are the estimates?*

**Definition**

**Sample**

A **sample** is a subset of the population which is used to gain insight about the population. Samples are used to represent a larger group, the population.

**Definition**

**Statistic**

A **statistic** is a fact or characteristic about a sample.

As Table 1.1.1 demonstrates, none of the estimates of the percentage of effectiveness were exactly correct, but most were very close. If we cannot determine how much faith we should place in our estimates, it will be difficult to use the estimates to make decisions. The process of selecting samples and determining the reliability of our estimates is a large part of what statistics is about.

# ✎ 1.1 Exercises

## Basic Concepts

1. What are three objectives of statistical methods?

2. Briefly describe the role of statistics in managerial decision making.

3. What is a controlled experiment?

4. Explain the purpose of a control group.

5. Explain the purpose of a treatment.

6. How is it possible to know the results of a presidential election before Election Day?

7. What is a population?

8. What is a frame?

9. What is a population parameter?

10. Why is it often difficult to determine the exact value of a population parameter?

11. What is a sample?

12. What is a statistic?

13. Describe the relationships between populations, samples, parameters, and statistics.

## Exercises

14. Determine whether the statement describes a population or sample.

    The salaries of all students that graduate from State University with degrees in Information Technology.

15. Determine whether the statement describes a population or sample.

    The number of billable man-hours logged per week by employees at Deloitte.

16. Determine whether the statement describes a population or sample.

    The number of times 6 out of 50 CEOs of technology companies in the Fortune 500 meet with their board of directors during a calendar year.

17. Determine whether the statement describes a population or sample.

    The final rankings of 5 candidates out of the 22 who applied for the open CFO position in your organization.

18. Identify the population being studied.

    The salaries of 13 out of the 30 employees who work during the night shift at a manufacturing plant.

19. Identify the sample chosen for the study.

    The price of homes of a sample of 35 employees who work at a company in Silicon Valley.

**20.** Identify the population being studied and the sample chosen.

The annual tuition paid by students in a sample of 34 from your class.

**21.** Determine if the numerical value describes a population parameter or a sample statistic.

The average number of hours students in your statistics class study per week is 15.8.

**22.** Determine if the numerical value describes a parameter or a statistic.

A survey of 1910 people in the U.S. revealed that 73% of those surveyed work a full-time job.

## 1.2 Statistics and Quality

"Statistical thinking is critical to improvement of a system."

—Mary Walton, *The Deming Management Method*

> **Definition**
>
> Process
>
> A **process** is a series of actions that changes inputs to outputs.

The idea of a **process** is closely tied to quality control. We encounter processes in all facets of our lives. A simple credit card transaction is a process—the customer inserts or swipes the card, the number is digitally read from the card, there is a credit authorization procedure, and then finally the credit card is approved or rejected for the amount of money that the customer intended to spend. In a business context, a process is a series of steps that produces a product or service. Closely monitoring and continuously improving processes produce high quality products. Monitoring the process means taking measurements of key variables over time. Improving processes means reducing process variation by finding the causes of variation and eliminating them.

> **Definition**
>
> Statistical Process Control
>
> **Statistical Process Control** (SPC) is a group of statistical methods designed to monitor and control processes.

In order to improve a process there must be an understanding of how the process is currently performing. This requires definition and monitoring of the process. Statistics helps with decisions about how the data will be collected, what data will be needed, and the analysis of the data. In addition to ferreting out production problems, **Statistical Process Control** (SPC) is a group of statistical methods designed to monitor and control processes. SPC is helpful in detecting problems in a process before they create a defective product or service. We will study this subject more extensively in chapter 18.

## 1.2 Exercises

### Basic Concepts

1. Describe the role of statistics in the quality movement.

2. What is a process?

3. How are processes improved?

4. What is SPC?

### Exercises

5. a. Describe a process at your school or place of employment.

   b. In your opinion, how could this process be improved?

   c. What type of data could you collect to use in analyzing this process?

## 1.3 Descriptive Statistics versus Inferential Statistics

The science of statistics is divided into two categories, **descriptive** and **inferential**. Descriptive methods describe and summarize data, while inferential methods aid in drawing conclusions and making decisions and predictions about populations and processes for which it is impractical to obtain measurements on each member.

### Descriptive Statistics

The emphasis in **descriptive statistics** is analyzing observed measurements, usually from a sample. With descriptive statistics we try to answer questions such as:

- What is a typical value for the measurements?
- How much variation do the measurements possess?
- What is the shape or distribution of the measurements?
- Are there any extreme values in the measurements and, if so, what does that tell us?
- What is the relative position of a particular measurement in the group of data?
- What kind of relationship exists, if any, when there are two variables and how strong is the relationship?

The application of descriptive statistical tools is usually *ad hoc*, that is, the exact method of analysis changes from one problem to the next. Sometimes the application of descriptive statistics can raise as many questions as it answers. And when that happens, statistics is working at its best as a problem-solving or process-improvement tool.

The importance of descriptive statistics as an information-producing tool relates to the amount of data to be comprehended. If there are only two observations, say 6 and 4, then comprehending the data in its entirety is not difficult and descriptive statistical aids are of little value. However, the 100 observations of the revenues of the top 100 U.S. companies in Table 1.3.1 suggest a different story. Individually inspecting 100 revenues would produce very little useful knowledge. To comprehend a large set of data, it must be summarized. That is the function of descriptive statistical techniques. Descriptive techniques are the most common statistical applications.

### Examples of Descriptive Statistics

- Frequency Distribution
- Measures of Central Tendency:
  - Mean
  - Median
  - Mode
- Measures of Dispersion:
  - Range
  - Variance
  - Standard Deviation

### Definition

**Descriptive Statistics**
**Descriptive statistics** is the collection, organization, analysis, and presentation of data.

| Table 1.3.1 – Profits of the Top 100 Companies by Revenues in the Fortune 500 (Millions of Dollars) for 2021 | | | |
|---|---|---|---|
| $13,510 | $5,580.4 | $21,331 | $4,575.2 |
| $6,205.2 | $57,411 | $7,179 | $15,403 |
| $21,180.1 | $10,103.5 | $42,521 | −$4,539 |
| $3,578.4 | $49,286.8 | $22,116.4 | $20,738.9 |
| −$3,408.7 | −$20,305 | −$21,680 | $45,783.4 |
| $40,269 | $3,456.7 | −$22,440 | $4,132.8 |
| $39,282.5 | −$5,176 | $4,002 | $8,458 |
| $31,293.4 | −$3,696 | $1,699.2 | $4,648.1 |
| $44,281 | −$1,903 | $1,638.8 | $456 |
| −$34.2 | $7,756.2 | $27,952.1 | $2,585 |
| $12,866 | $1,485.5 | $29,131 | $9,361.6 |
| $17,801 | $3,605.2 | −$1,279 | $6,201.6 |
| $6,427 | $4,572 | $1,627.7 | −$7,242 |
| $4,737.8 | $4,301.4 | $8,642.6 | $12,920.1 |
| $1,808 | $3,945.3 | $7,160.2 | $2,961 |
| $1,165 | $11,805 | $22,224 | $10,534 |
| $491.1 | $2,866.5 | $2,372.7 | $1,851.7 |
| $4,330.2 | $1,161.9 | $3,786.8 | $3,628.7 |
| $1,987.2 | $4,698.4 | −$5,543 | $3,250 |
| $17,894 | $4,368 | $13,031.1 | $5,835 |
| −$9,826 | $11,047 | $1,207.5 | $1,872 |
| $1,115.5 | $29,146 | $1,591.8 | $11,053.6 |
| $1,343 | $2,903.8 | $998.6 | $4,802.4 |
| $3,067.4 | $14,714 | $4,731.8 | $730.4 |
| $8,052.5 | $360.1 | $7,948.4 | −$670.5 |

## Inferential Statistics

**Definition**

**Inferential Statistics**

The objective of **inferential statistics** is to make reasonable estimates about population characteristics using sample data.

It would be preferable to have measurements of the entire population, but in most cases these data are either not obtainable or would be much too costly to obtain. For example, to be absolutely certain that all car air bags will inflate in head-on collisions would require each new car to be crash tested in a head-on crash. If 100 percent inspection were a requirement, cars would be a scarce commodity. Fortunately for automobile manufacturers, statistical sampling techniques can reliably estimate, with a relatively small sample, the fraction of air bags that will inflate.

**Use Sample Data to Make Inferences about Unknown Population Characteristics**

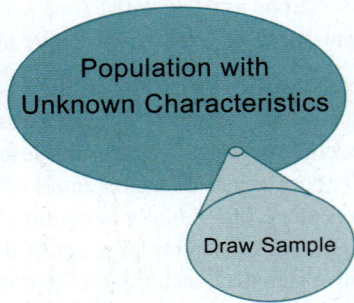

**Example 1.3.1**

**Differentiating Between Descriptive and Inferential Statistics**

The Michelin tire company has a feature called "Track Connect" that will assist racecar drivers in getting the maximum performance out of their tires when on a racetrack. Michelin has an app that will give personalized advice before, during, and after a driver takes laps around a track. The app will make suggestions for optimal tire pressure and temperature so that the car is handled efficiently as they navigate the track (road track or oval). To evaluate the app, Michelin randomly collected data from 30 drivers that took laps around various track surfaces using several car models, different tires (those with and without the sensors), and under different weather conditions. Using the data from the drivers' experiences on the tracks, Michelin has concluded that the tires lasted longer (i.e., less wear) and the cars had improved gas mileage. Were the results of this experiment an example of descriptive or inferential statistics?

**Source:** michelinman.com/trackconnect.html

### SOLUTION

The Michelin tire company has collected data on tire performance from a random sample of 30 racecar drivers who took laps around various racetracks, under various track conditions. The primary data collected from the tires were air pressure and temperature. They also collected gas mileage (miles per gallon) data for each of the racecars. Using the collected data, Michelin was able to conclude that the tires lasted longer and had better gas mileage than tires without the sensors. This is a case of inferential statistics.

# ✎ 1.3 Exercises

## Basic Concepts

1.  What is the difference between descriptive and inferential statistics?

2.  Name three questions that a descriptive statistic can be used to answer.

## Exercises

3.  Determine whether the statement describes a descriptive or inferential statistic.

    The average price of a car at the new car dealership in town is $28,200.

4.  Determine whether the statement describes a descriptive or inferential statistic.

    A survey of 885 people revealed that 51% have a college degree; therefore, it can be assumed that 51% of the U.S. population has a college degree.

# 1.4 The Value of Statistical Literacy

Part of being an intelligent human being is the desire to learn the truth about the world we live in. But as Oscar Wilde said:

> "The truth is rarely pure and never simple."
>
> —Oscar Wilde, *The Importance of Being Earnest*

Being statistically illiterate puts one (or one's organization) at a competitive disadvantage compared to companies that possess and use statistical knowledge and analytical tools. Statistics and its uses cannot be avoided. Therefore, learning and using statistical tools will give you and your organization more flexibility when making decisions.

To intelligently appreciate or produce statistical information, you must be statistically literate to defend yourself from a persuasive but fallacious statistical argument, to decrease your vulnerability to pseudo-sciences, and to diminish the chances of making poor and sometimes injurious business decisions.

A statistically literate person understands the language of statistics and understands statistical concepts and reasoning. To become statistically literate, one should be able to think "statistically". This will involve asking questions like:

- Where did the data come from?
- How was the sample taken and is the sample large enough?
- How reliable or accurate were the measures used to generate the reported data?
- Are the reported statistics appropriate for this kind of data?
- Is a graph drawn appropriately?
- How was this probabilistic statement calculated?
- Do the claims make sense?
- Should there be additional information?
- Are there alternative interpretations?

## 1.4 Exercises

### Basic Concepts

1.  What are the consequences of being statistically illiterate? How could this put you at a disadvantage in business?

2.  What kinds of questions would a statistically literate person ask?

### Liar or Statistician?

In his book *How to Tell the Liars from the Statisticians*, Robert Hooke sheds light on our exposure to misleading statistics in everyday life. In the preface he writes, "The science of statistics has made great progress in this century, but progress has been accompanied by a corresponding increase in the misuse of statistics. The public, whether it gets its information from television, newspapers, or news magazines, is not well prepared to defend itself against those who would manipulate it with statistical arguments. Many people either believe everything they hear or come to believe in nothing statistical, which is even worse." Throughout the remaining chapters, Hooke uses examples from politics, economics, entertainment, and the medical community to illustrate the dangers of being statistically illiterate. You might be surprised to learn the ways in which the misuse of statistics affects you every day. In order to digest the plethora of statistical information you encounter, you must first become statistically literate.

**Source:** Hooke, Robert. *How to Tell the Liars from the Statisticians.* New York, New York: Marcel Dekker Inc., 1983. Print.

## Exercises

3.  Do some research on the internet and locate an advertisement for a product or service that you suspect may be making a false claim.

    **a.** What leads you to suspect the claim is false?

    **b.** Does the ad include data or statistics in its claim? If so, do the data or statistics reported seem accurate?

    **c.** Does the ad reveal the source of the data and how it was collected?

    **d.** Are there any figures or graphs included in the ad? If so, is the graph appropriate and does it make sense?

# R    Chapter 1 Review

## Key Terms and Ideas

- Controlled Experiment
- Control Group
- Treatment Group
- Treatment
- Population
- Frame
- Census
- Parameter

- Sample
- Statistic
- Process
- Statistical Process Control
- Descriptive Statistics
- Inferential Statistics
- Statistical Literacy

## AE    Additional Exercises

### Exercises

1.  Jakob Nielsen, a website consultant, recently conducted an eye-tracking survey to determine whether people ignore purely decorative images when viewing web pages. An aspect of the study compared a set of products on Pottery Barn's furniture website and a page of televisions on Amazon.com. The study found that consumers tended to ignore the televisions on Amazon.com because they were generic, making the product image less inviting. When consumers viewed Pottery Barn's website, they were more inclined to view the photos of the bookcases for longer periods of time because they were images of the actual products for sale.

    **Source:** New York Times

    a.  Identify the population of interest.

    b.  What characteristic of the population is being measured?

    c.  Is the purpose of the data collection to perform descriptive or inferential statistics?

2.  A recent study at Britain's Oxford and Exeter Universities explored whether a woman's diet before conception affects the gender of her child. Researchers studied the eating habits of 740 women during their first-time pregnancies and found that higher caloric intake prior to conception can significantly increase the chances of having a son while more restricted diets are more likely to produce daughters. They found that high potassium diets (eating bananas) and calcium rich diets (cereal and milk) were associated with having a baby boy. Researchers concluded that eating a bowl of cereal for breakfast can increase the chances of a male birth. "Of women eating cereals daily, 59 percent had boys, compared with only 43 percent who bore boys in the group eating less than a bowlful per week."

    **Source:** CNNHealth.com

    a.  Identify the population of interest.

    b.  What characteristic of the population is being measured?

    c.  Identify the sample.

    d.  Is the purpose of the data collection to perform descriptive or inferential statistics?

    e.  What are some problems that could be associated with collecting data in this study?

3.  Researchers at The Ohio State University and Zeppelin University Friedrichshafen, in Germany, recently conducted a study regarding the elderly and negative news coverage. The researchers presented 276 subjects with several stories (with photos) about either old or young people. Participants were presented with one of two versions of each story. In one version the main character was painted in a positive light and in the other the same character was described negatively. After the participants finished reading their self-esteem was measured. The study found that older readers were more inclined to read the negative stories about youth. In addition, they found that the more negative stories older people read about younger individuals, the higher their self-esteem tended to be. This could explain the prominence of negative media coverage on networks with an older audience such as Fox News and MSNBC.

    **Source:** Huffpost Media

    a.  Identify the population of interest.

    b.  What characteristics of the population are being measured?

    c.  Identify the sample.

    d.  What are some problems that could be associated with this study?

    e.  Is the purpose of this study to perform descriptive or inferential statistics?

4. Researchers at Pepperdine University's Graziadio School of Business and Management recently conducted a study to determine whether the capital crunch is affecting small businesses' abilities to expand. The companies studied included alternative lenders, venture capitalists, and private equity firms, among others. The project surveyed 559 privately held businesses and 1430 lenders and investors nationwide. The study found that 78 percent of businesses had solid growth strategies but only 40 percent had access to the resources needed to grow. Lenders and investors rejected 90 percent of loan applications or investment proposals that would be secured by a business's real estate holdings and 73 percent of loan applications or investment proposals that are based on a business's cash flow. According to the survey author John Paglia, "The study shows private business owners feel they are being constrained by access to financial capital. Owners currently expect a 10 percent revenue growth over the next 12 months. If they were to receive additional capital, they estimate their revenue growth rate to jump to 25 percent."

**Source:** smallbiztrends.com

   a. Identify the population of interest.

   b. What characteristics of the population are being measured?

   c. Identify the sample.

   d. Is the purpose of this study to perform descriptive or inferential statistics?

5. Ruder Finn, one of the world's largest independent public relations agencies, recently announced the *Mobile Intent Index* which studies mobile phone user habits and explores the underlying reasons that people have for accessing the internet on mobile devices. The *Mobile Intent Index* asked 500 American adults 18 years of age and older how often they use their mobile phones to access the internet for 295 reasons. The study found that 91% of mobile phone users go online to socialize compared to 79% of desktop users. In addition, 60% of mobile internet users go online to manage finances compared to only 45% of desktop users. They also found that mobile users were less likely to use the internet for educational purposes, only 42% compared to 92% of desktop users. Finally, unsurprisingly, mobile phones are not used for creative purposes; only 42% of mobile users personally express themselves online compared to 54% of desktop users.

**Source:** PR Newswire

   a. Identify the population of interest.

   b. What characteristic of the population is being measured?

   c. Identify the sample.

   d. What are some problems that could be associated with this study?

   e. Is the purpose of this study to perform descriptive or inferential statistics?

6. A personnel director is interested in determining how effective a new reading course will be in improving the reading comprehension of her company's employees. The director randomly selects twenty employees and determines the average reading comprehension both before and after instruction in the reading course.

   a. Identify the population.

   b. What characteristic of the population is being measured?

   c. Identify the sample.

   d. Is the purpose of the data collection to perform descriptive or inferential statistics?

7.  In 2009 and 2010 the market research firm Chadwick Martin Bailey conducted several studies to provide insights into recent dating behavior in the United States. The data were collected through research via an online Consumer Research Panel. In the "Marriage Survey" (a survey of recently married people), 7000 adults 18 years of age and older who were married in the past 5 years were polled. The study found that 17% of couples married within the last 3 years met each other on an online dating site. This is compared to 26% that met their significant other through a friend or family member, 36% who met through work or school, 4% who met through a church or place of worship, 11% who met through bars, clubs, or social events, and 7% who met in some other manner.

Source: Chadwick Martin Bailey/Match.com

   a.  Identify the population of interest.

   b.  What characteristics of the population are being measured?

   c.  Identify the sample.

   d.  What are some problems that could be associated with this study?

8.  In a recent study, seat belt users were found to have 20% fewer fatalities than those who do not wear seat belts. Do these results prove that seat belts reduce the chances of a fatality?

9.  States having an abundance of coastline have an obvious advantage over landlocked states, or states with little coastline, in that their economies may profit from an extensive fishing industry, tourism, shipping, or other water related activities. Alaska, the leader by far in miles of coastline, has a total of 6640 miles, of which 5580 miles border the Pacific Ocean and 1060 miles border the Arctic. Florida, the leader in the continental United States, has a total of 1350 miles with 580 miles on the Atlantic Ocean and 770 miles on the Gulf of Mexico. Of all states with some coastline, New Hampshire, with 13 miles of coastline, is in last place.

   a.  Identify the population.

   b.  What characteristic is being measured?

10. A young actuary (statistician usually working in the insurance industry) has been asked to summarize the number of automobile accident claims by region for his company. He randomly selects 50 automobile accident claims which his company has settled in the last year and counts the number of accidents in each region: North, South, East, and West. He summarizes the counts by region in a chart and gives the results to his supervisor.

   a.  Identify the population.

   b.  What characteristic of the population is being measured?

   c.  Identify the sample.

   d.  Is the purpose of the data collection to perform descriptive or inferential statistics?

| P | **Discovery Project** |
|---|---|

## The Performance of the S&P 500 Companies During the COVID-19 Pandemic

An article published by the Wall Street Journal in August of 2021 summarized the effects of COVID-19 on the performance of the S&P 500 companies. More than three-fourths of the companies reported higher revenues than pre-pandemic levels. Of those reporting higher levels of revenue, 213 reported revenues in the second quarter of 2021 above 2019 levels after undergoing a drop in revenues in 2020. Of the remaining 153 companies that had higher revenues, revenues in the second quarter for the past two years exceeded 2019 levels. There were 101 companies in the S&P 500 that had revenues below 2019 levels and ten companies experienced a drop in revenue in 2021 after having a rise in income in 2020.

The revenue figures are based on FactSet data for the 477 S&P companies that reported their revenue for the second quarter of 2021. Approximately one-third of the S&P 500 have seen steady or rapid growth throughout the pandemic. The companies that have fared the best are the pharmaceutical, retail, and semiconductor companies. Moderna Inc. experienced the largest increase in revenue of all the S&P companies. Moderna's revenue increased 33,187% from the second quarter of 2019 to the second quarter of 2021, a value too large to include in the figure below. The consumer services sector experienced the largest decline in second quarter 2021 revenues, largely due to companies related to travel and tourism.

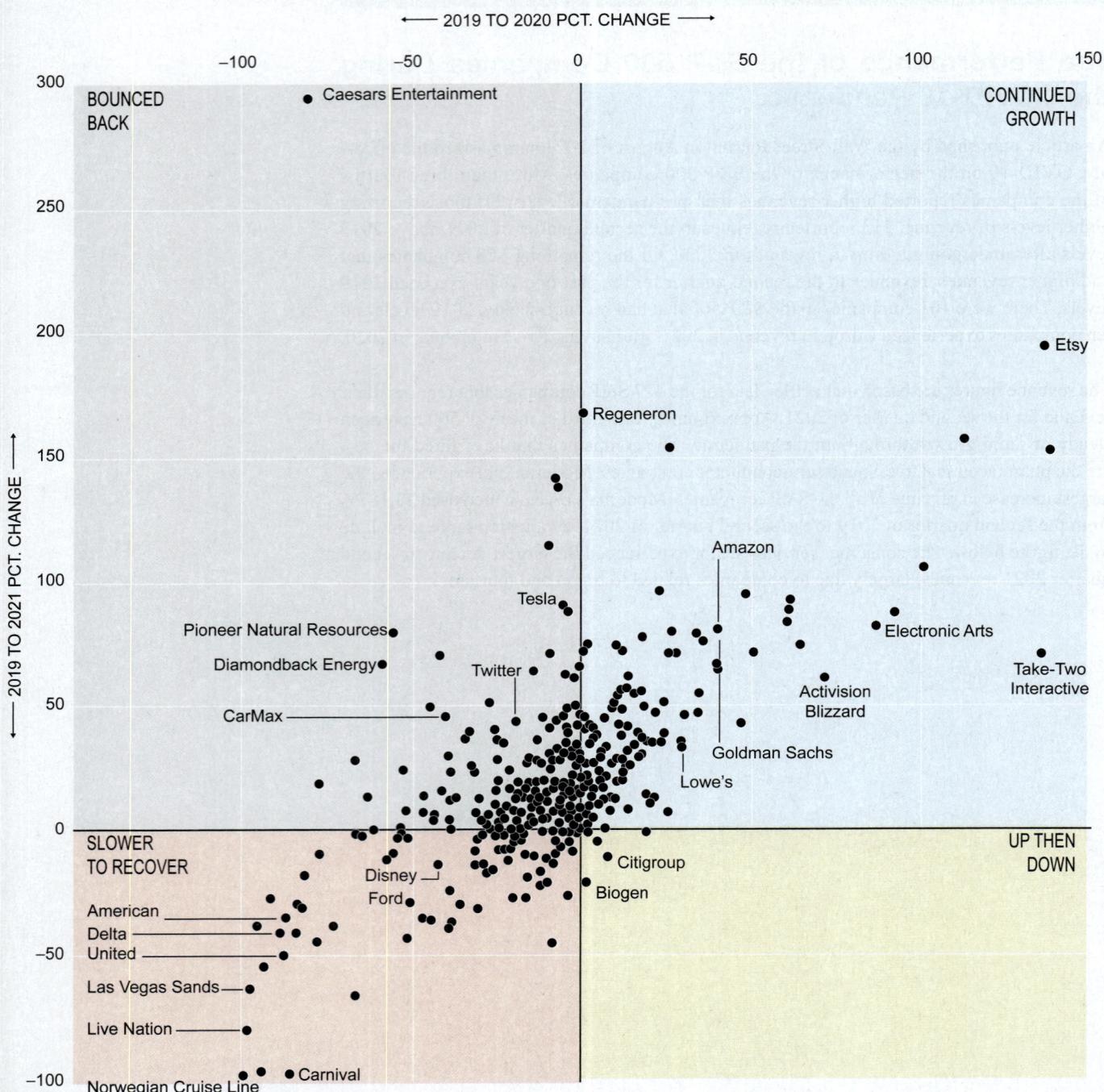

**Second-quarter Revenues Compared to Pre-pandemic Levels**

Note: Moderna is excluded. Its second quarter revenue percentage change since 2019 was 407.3% in 2020 and 33,187% in 2021.

**Source:** FactSet

Based on the article summary and the figure, answer the following questions.

1. What is the population of interest?

2. What is/are the variable(s) of interest?

3. Which company experienced the largest percentage change from 2019 to 2021?

4. Based on the figure, which company experienced the largest percentage change from 2019 to 2021? Can you think of an explanation of why this is the case?

5. Based on the figure, which company experienced the smallest percentage change from 2019 to 2021? Can you think of an explanation of why this is the case?

6. Based on the figure, which company did a major recovery from 2020 to 2021 based on revenues? Do some research on the internet to find out what caused the company to make such a major recovery.

# Chapter 2
## Data, Big Data, and Analytics

# Discovering the Real World

## Management and Measurement

The main objective of any business is to earn a profit and consequently companies depend heavily on measurement, data, and statistical thinking. In fact, there is an old management adage, *You can't manage what you can't measure*. Perhaps this adage would be better stated as *You can't manage unless you know what "reality" is*. There are many important measurement systems in a business. The accounting system is designed to measure profitability and to inform management of potential problems. The inventory system is a measurement tool designed to measure the status of inventory, indicate when orders should be placed, and spot potential inventory theft or shortfall. The cash flow system is a forecasting system that measures the company's need for cash.

Your career will essentially be a choice of the kinds of problems you desire to solve. The more difficult the problems you decide to solve, the more you will depend upon data and measurement to solve them.

## Introduction

For most people, the words "data" and "measurement" are words that generate about as much enthusiasm as watching dust settle. So why should you be interested in measurement and data?

The poem *Under Ben Bulben*, written in 1939, is one of the last poems of the Irish poet W. B. Yeats. The poem contains the insightful line "*Measurement began our might*." Long ago, our species learned that if you do not know what "reality" is, it is difficult to predictably change it to a more desirable state. Measurement is the first step in understanding the "reality" of any circumstance you wish to change in a predictable way. Thus, measurement is a fundamentally important link in controlling our environment. That is why measurement is so pervasive in our culture. We seem to want to measure just about everything in the physical world: temperature, weight, distance, pressure, hardness (Mohs scale of hardness), wind speed (Beaufort wind scale), earthquakes (Richter scale), and so on. We even try to measure feelings, like love.

Type "measurement" into an internet search engine and you will be surprised by the number of organizations that are devoted to measurement. One of those groups, The International Society for Measurement and Control (ISA), plays a prominent role in setting worldwide standards. Without measurement, standards would not be possible. Without complex measurements and standards, just about all of the conveniences that we take for granted (telephones, automobiles, refrigerators, televisions, and computers) would not exist.

Measurements can be quite costly. A company looking for oil invests substantial sums of money to obtain seismic measurements, the principal discovery tool in the hunt for oil. For companies that do oil exploration, seismic data are the company's crown jewels. Data can be just as important on the personal level. For example, when one meets with their financial advisor, they want to understand the past performance of a mutual fund to help them invest for their future.

# 2.1 Data and Decision Making

Measurement and selling the resulting data is big business. Did you know…

- …that most prescriptions are recorded in a database at the local pharmacy and then sold (without names attached) to firms that collect and summarize this information?

The data are then resold to pharmaceutical companies, enabling them to measure how often specific doctors prescribe their drugs. As a consequence, they are able to measure the effectiveness of their salespersons.

- …that when you are late on a car payment, house payment, or credit card payment, the information may wind up in a national credit database? This database is used by companies in the credit business to assess credit worthiness. It is also used by many businesses in employment screening.

- …that banks and other institutions that issue credit cards keep data on your spending habits? The data is used to build models to help prevent credit card fraud. If the bank's model determines that you have used your credit card to make "unusual" purchases, it's possible that someone from the credit card company will call and confirm that you did make the purchases. This proactive use of data has prevented an enormous amount of credit card fraud.

- …that when you order something from a mail order company, your name and what you ordered are recorded and frequently sold to other businesses? Even your grocery store gives you a card so it can collect and sell data on what items you buy and how often you buy them.

## Measurement

A large part of using statistics to make good business decisions is developing an ability to appraise the quality of measurements. For many problems, what you measure and how you measure it is more important than how you analyze the data. Thus, it is not surprising that the science of statistics is just as concerned with producing good data as it is with interpreting it.

When you encounter data, ask yourself the following questions.

---

**Procedure**

Do we have good measurements?

1. Is the concept under study adequately reflected by the proposed measurements?
2. Are the data measured accurately?
3. Is there a sufficient quantity of the data to draw a reasonable conclusion?

---

If each of these questions can be answered affirmatively, the data possess good properties.

The development of a suitable measurement involves two essential questions: *What should be measured?* and *How should the concept be measured?* Measures (sometimes called metrics) are developed from the field of study, not statistics. Suppose you decide to measure the speed of an object. The concept of speed is well-defined. It is measured in distance per unit of time (for example, 60 miles per hour). Therefore, determining the speed of an object will require two variables to be measured, one distance and the other time. There are well-defined standard measures (such as feet, yards, miles) to use for distance traveled, as well as time, so the measurement is relatively easy to obtain. However, measuring someone's intelligence is another story. How do you define intelligence? Are there standard measures that can be applied? While IQ tests exist, what do they really measure? The more well-defined a concept, the easier it is to develop measurements for it. Our ability to comprehend the world we live in is rooted in good measurement.

Let's look at some examples of variables to see if they possess good properties.

| Example 2.1.1 |
| --- |

**Identifying Well-Defined Variables**

Specify whether the following variables are well-defined or not. Justify your answer.

**a.** Interest rate on a mortgage

**b.** Sales price of a textbook

**c.** Tall

**SOLUTION**

**a.** The first variable, interest rate on a mortgage, can be adequately studied if an analyst wants to draw a conclusion about mortgage interest rates. Additionally, interest rate on a mortgage can be measured accurately and can certainly be collected in a large enough quantity to allow an analyst to draw a reasonable conclusion. Thus, if one collects data for interest rate on a mortgage, these data will possess good properties. This variable is well-defined.

**b.** Like part **a.**, sales price of a textbook can be well-defined such that it can be measured in US dollars, certainly can be collected in large quantities to allow an analyst to draw conclusions, and can be measured accurately. The sales price of a textbook is a variable that possesses good properties. This variable is well-defined.

**c.** The variable, tall, is an interesting variable. We should ask ourselves, *How do we define tall?* Is there a standard measure that can be used to define tall? Tall is a relative term from one person's perspective to the next. With that said, tall is not a well-defined variable.

## Science and Data

Measurement and data are an integral part of science. Methods for exploring research problems have been developed over a long period of time and have become standards in the scientific community. These methods are collectively known as the **scientific method**.

| Procedure |
| --- |

**The Scientific Method**

1. Gather information about the phenomenon being studied.
2. On the basis of the data, formulate a preliminary generalization or hypothesis.
3. Collect further data to test the hypothesis.
4. If the data and other subsequent experiments support the hypothesis, it becomes a law. If the experiment does not support the hypothesis, construct a new hypothesis and start the method over again. Sometimes, even if the hypothesis is supported, you may want to test it again but in a different way.

Statistics and data are fundamental to the scientific method. Data from carefully designed experiments are the ultimate evidence that support or discredit new theories.

| Definition |
| --- |

**Confounding Variable**

A **confounding variable** is a variable that was not controlled or accounted for by the researcher and thus damaged the integrity of the experiment.

The data collection process in steps one and three of the scientific method can be quite different. The first step of the scientific method is exploratory, finding out what "reality" is about the subject under consideration. Since the data in this phase need not produce convincing evidence, whatever data are available are used to generate ideas. However, the third step begins the validation of a hypothesis. Scientists are trained to be critical thinkers. If a new idea is to be accepted by the scientific community, convincing evidence must be developed at the third stage. The manner in which the data are collected is an important part of that evidence. If the evidence is to be persuasive, a data gathering strategy (an experimental design) that will produce data without the unwelcome influences of **confounding variables** (i.e., an uncontrollable or unaccounted-for variable that damages the integrity of the experiment or the other data) is required. For example, a manager wants to know if a new marketing plan has increased sales. In the analysis of the data, the manager must ensure that any increase in sales is due to the marketing plan and not other factors such as an increase in the sales force or a reduction in the cost of the product, for example. These "other factors" would be considered confounding variables.

Because of science's emphasis on data, statistics has become inextricably linked with the scientific method. Experiments are designed to yield data with maximum information. There are two branches of statistics: descriptive and inferential. The branch of **descriptive statistics** focuses on exploratory methods for examining data that yield the hypothesis mentioned in step two of the scientific method. The branch of **inferential statistics** develops theories to test the hypothesis using data collected from an experiment to make formal conclusions about a population or parameter.

## Measurement, Data, and Problem Definition

Collecting data is a natural part of our lives. For example, consider the everyday question, *What will I have for dinner?* Although virtually no one formally applies the scientific method to such a problem, most people perform experiments and collect data (by eating). This leads to generalizations such as *I hate asparagus* or *I like ice cream*. After sufficient experimentation, these generalizations become personal preference laws.

Selecting the evening meal does not have incredible consequences, yet important problems, by definition, do.

> **Procedure**
>
> ### The Decision-Making Method
>
> 1. Clearly define the problem and any influential variables.
> 2. Decide upon objectives and decision criteria for choosing a solution.
> 3. Create alternative solutions.
> 4. Compare alternatives using the criteria established in the second step.
> 5. Implement the chosen alternative.
> 6. Check the results to make sure the desired results are achieved.

Notice the first step in the decision-making method is to define the problem. This is important because almost *any solution to the right problem is better than the best solution to the wrong problem.*

When you go to a physician what is the first thing the physician does? Most physicians measure your weight, temperature, and blood pressure before he or she meets with you. During the examination the doctor will elicit more data. If there are insufficient data to make a diagnosis, more data (perhaps the dreaded blood test) will be ordered. In the medical world it is extremely important to know what reality is. That is why medical professionals gather so much data. Data are just as important in solving business problems.

Consider a trucking company manager who is beginning to hear a few complaints about freight being delivered late. If the operations manager of the trucking company keeps data on the number of shipments delivered late each week, then statistics can be used to help assess the magnitude of the problem. For example, if the number of late deliveries is plotted (see Figure 2.1.1), the data in the graph confirms a disturbing trend. Late deliveries are on the rise. Left unsolved, this problem can jeopardize jobs and the existence of the business.

> **Definition**
>
> ### Descriptive Statistics
>
> The branch of statistics that focuses on exploratory methods for examining data that help to formulate a hypothesis is called **descriptive statistics**.

> **Definition**
>
> ### Inferential Statistics
>
> The branch of statistics that develops theories to test the hypothesis using data collected from an experiment and makes formal conclusions about a population or parameter is called **inferential statistics**.

### Get Out of Here Aristotle

"Aristotle maintained that women have fewer teeth than men; although he was twice married it never occurred to him to verify this statement by examining his wives' mouths." –*Bertrand Russell*

The Greek philosopher Aristotle held that one could develop all the laws that govern the universe by pure thought and that it was unnecessary to obtain measurements that would confirm the validity of these laws. Perhaps this explains why he never bothered to count his wives' teeth. The goal of Aristotelian science was to explain why things happen. Modern science was born when Galileo began trying to explain how things happen. Galileo's approach originated the method of controlled experiments which form the basis of scientific investigation. Controlled experiments yield data. Statistics allows people to think with data. If Aristotle had been right, there wouldn't be a very large demand for statistics.

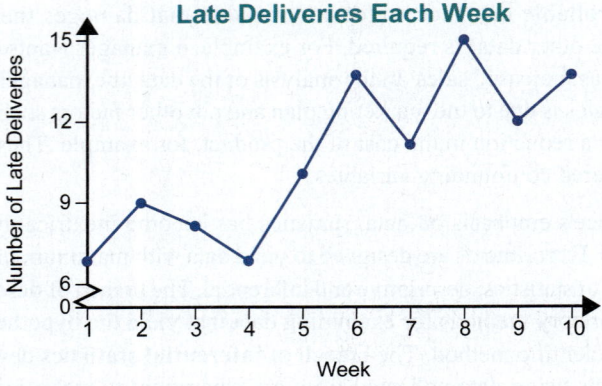

Figure 2.1.1

Suppose a plastics manufacturer takes samples from the process every hour and checks those samples for defects. After recording the number of defects for each sample, the data are plotted (see Figure 2.1.2). The data reveal a potential problem. Undoubtedly, more data will be needed, and the process may need to be carefully examined.

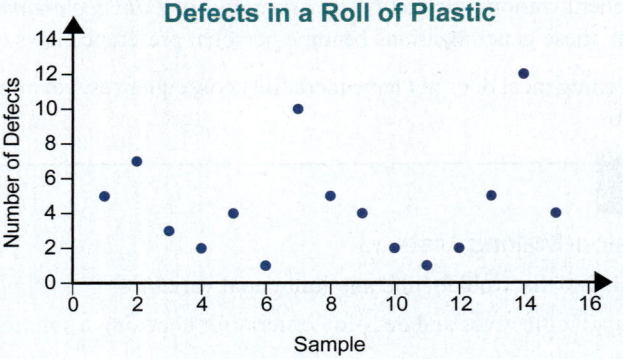

Figure 2.1.2

Graphical and numerical summaries are frequently useful in discovering the existence of a problem as well as in shedding light on what some of the potential causes may be. In many instances, problems are caused by systems that do not operate as they are designed. Collecting data (finding out system "reality") and using simple statistical tools to monitor a system are the most common ways of ensuring a system performs properly.

Problems are not always the result of a diagnosis of some complaint or system malfunction. A "problem" may well be the result of an inquiry into unexpectedly good system performance. That is, oftentimes, improving a process does not imply that something was wrong with the process. In this context, a problem presents itself as an opportunity for improvement. Suppose, for example, a finance instructor develops a new method for teaching introductory finance. Measurements are kept on his students as well as students taking subsequent finance courses. Using statistical methods to compare students using the new method to those using the old method can be valuable in pointing out potential educational improvements that may be used at other institutions. As another example, consider the development of the Intel chip. Improving the speed of its processors/chipsets is a constant goal for research and development, regardless of the current speed of the chip. It is not a matter of the chips being too slow, but as technology and hardware advance, the chipsets need to "keep up" so that computers using the chips maintain (or improve) their processing speeds.

## Statistics as Criteria

The second step in the decision-making method suggests defining objectives and developing criteria in order to evaluate various alternative decisions. Not all statistics are simple means, proportions, or standard deviations. Managers and researchers often develop their own statistical measures for summarizing some aspect of a phenomenon. Amazon.com is a web retailer that tracks a customer's "movements" on its website. For example, Amazon tracks the number of times a customer clicks on a specific item, say, a camera or many cameras. This statistic, number of clicks on cameras, allows Amazon to send emails to the customer noting price drops for cameras or highlighting cameras when the customer next visits the website. The mean time between failure (MTBF) is a statistic that is used to compare the reliability of various equipment or components. The Consumer Price Index (CPI) is a summary statistic that describes the overall price level in the United States. This statistic is an economic measure of inflation and is used in labor contracts to escalate wages as well as to calculate cost of living increases in social security payments.

## What Should Be Measured and How Should It Be Measured?

What should be measured depends on the problem to be solved. Sometimes what should be measured is obvious and relatively easy. If the problem is to maintain or improve a system, key variables are monitored and decisions are made based on the level of these variables. For example, if you are responsible for a machine that manufactures pistons, then some of the variables that should be measured and controlled are the diameter of the piston, the length of the piston, the width of the piston wall, and the number of defects on the piston's surface. However, if you are trying to design a system to perform automated stock trading, deciding what variables are important could take years to discover.

A precisely defined concept is usually easy to measure. The less precise (the fuzzier) the concept, the more difficult the measurement becomes. There is a vast difference in measuring the height of a person and measuring his or her intelligence. Defining height is relatively simple. It is nothing more than the vertical length of an object. There are standard scales, such as inches and feet, that everyone agrees upon, that can be used to measure height. The National Institute of Standards and Technology maintains rods which define a government standard for distance measures (feet and inches). Because these standards are widely accepted, if ten different people measure a person's height, there should not be large differences in their measurements. This is not true when measuring intelligence. The National Institute of Standards and Technology does not have a measuring rod for intelligence. It is unlikely that ten randomly selected people could agree on a definition of intelligence, much less on how it should be measured. Intelligence is a fuzzy concept because there is no universally accepted definition and hence there can be no universally accepted standard of measure. If a concept cannot be precisely defined, it cannot be precisely measured. How do you measure fuzzy concepts?

## Fuzzy Data and Concepts

Fuzzy concept definitions produce fuzzy measurements. One could devote an enormous amount of time developing measuring instruments for fuzzy concepts such as love, rivalry, and prejudice, and still have a poor measurement. These concepts are fuzzy because they are perceptions. No person can be sure that their perception of love, rivalry, or prejudice is the same as someone else's. Science that relies on fuzzy measurements usually makes the assumption that everyone's interpretation of the concept is more or less alike.

Often when measuring fuzzy concepts, the instrument used to measure the concept ends up defining the concept. The Wechsler Adult Intelligence Scale (WAIS) is a test that is often used to measure intelligence—an IQ test. There could be a long debate over what constitutes intelligence. For most researchers studying intelligence, developing a new instrument to measure intelligence, however it is defined, is simply not a practical alternative. If a measuring device is needed for a fuzzy concept, using an established instrument like the WAIS is usually the method of choice. However, the necessity for using a measuring instrument does not validate the instrument.

There has been a great deal of controversy in recent years over whether the SAT accurately measures scholastic aptitude. Still the SAT is an important part of the college admission process. Why? Because college admission committees apparently believe there is no better alternative. When measurements purport to represent what you believe is a fuzzy concept, it is important to think critically about the measurements. Are they valid measures of the intended concept? This issue is important and will again be addressed later in the chapter in the levels of measurement section.

In a statistical analysis, it is usually not possible to recover from poorly measured concepts or badly collected measurements. Unfortunately, during your lifetime you will be bombarded with statistics derived from poorly measured concepts, confounded measurements, and simply fictitious data. When confronted with statistical evidence of any kind, regardless of whether or not the statistical analysis is done in good faith, it is ultimately up to you to ask reasonable questions about the data. The conclusions suggested by statistics can be no stronger than the quality of the measurements which produced the statistical evidence. Fuzzy or confounded measurements must produce fragile conclusions.

## Collecting Data

Essentially there are two ways to obtain data: **observation** and **controlled experiments**. The data collection method is related to the nature of the problem to be solved and the ethical and practical constraints of collecting data in some particular environment. There are many instances in which controlled experiments that would produce particularly appropriate data are not practical or would be unethical. For example, scientists cannot exclude data in an analysis for the purpose of validating their theory or viewpoint.

There are many instances in which data that have been collected provide either no information or misleading information about the effect under study.

Suppose a finance instructor wants to determine if there is any beneficial effect of studying statistics, particularly regression, before taking finance. The instructor obtains records of her finance students and compares the group of students that had statistics or are taking it concurrently with those who have not. The average finance grade for those having had statistics is a great deal higher than those that had not. The conclusion reached from these data is: *The study of statistics improves one's understanding of finance.* Do the data and the manner in which they were collected support such a conclusion?

Students who elect to take statistics generally have above average skills in mathematics. Are the higher finance grades of the statistics students due to their mathematical skills or to the content of the statistics course? Because of the data collection method, the data are not of sufficient quality to reach a conclusion as to the benefits of taking statistics prior to taking the finance course. A confounding variable (the exceptional mathematical ability of students electing to take statistics) makes it impossible to distinguish the effects we wish to study.

## Designed Experiments and Observational Studies

Suppose you wanted to know the effect of a price increase on demand. This kind of question would be ideal for a **controlled experiment**. The purpose of controlled experiments is to reveal the response of one variable (sales, the response variable) to changes in another variable (amount of price increase, the explanatory variable). In a controlled experiment the researcher attempts to control the environment of the experiment so that the effect of one variable on another can be isolated and measured. In these studies there is a **control group** and an **experimental group**. Ideally there is no initial difference between the two groups. During the experiment a **treatment** is applied to the experimental group. The exact form of the treatment will depend on the particular experiment. If the experimental group was a particular product, a treatment might be raising the price a fixed amount. The treatment changes the level of the **explanatory variable** in the experiment. The effect of the treatment can be measured by comparing the **response variable** in the control and experimental groups.

**Definition**

### Terms Used in Experiments

In a **controlled experiment**, a researcher attempts to control the environment of the experiment so that the effect of one variable on another can be isolated and measured.

In an experiment,

- the **control group** is the group of subjects that does not receive the treatment;
- the **experimental group** is the group of subjects that receives the treatment;
- the **treatment** is the factor that changes the level of the explanatory variable.

**Definition**

### Types of Variables

The **response variable** is the variable of interest in an experiment.

An **explanatory variable** is a variable that affects the variable of interest (response variable) in an experiment.

## Comparative Experiments

Isolating the effects of one variable on another means anticipating potentially confounding variables and designing a controlled experiment to produce data in which the values of the confounding variables are regulated. In the finance example, the confounding variable is the mathematical ability of those students taking statistics before finance. To control for this *bias*, students could be randomly assigned to take the statistics course. By randomly assigning students to take the statistics course before the finance course, the mathematical ability of the two groups should be equalized. Since mathematical ability is controlled, any difference in finance scores could be attributed to a beneficial effect of the statistics course. An experimental design in which experimental units (students in this case) are randomly assigned to two different treatments is called a **completely randomized design**.

For this example of a completely randomized design:

**Response Variable:** Students' grades in the finance course.

**Explanatory Variable:** Whether students take statistics before finance.

Randomization is often used as a method of controlling bias and is an important principle in the design of experiments.

**Definition**

Completely Randomized Design

A **completely randomized design** is an experimental design in which experimental units are randomly assigned to two or more different treatments.

**Flowchart for the Statistics Before Finance Experiment**

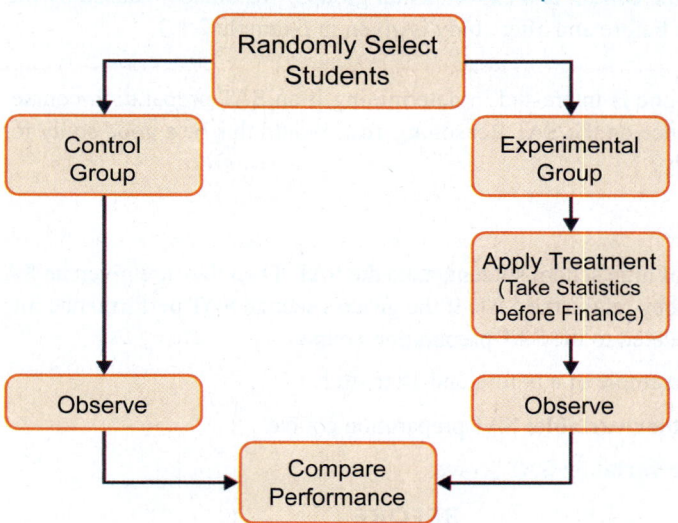

Suppose the director of dining facilities at a major university has implemented a new meal plan. The director wants to determine if the amount of money spent by students who subscribed to the old meal plan differs from the amount of money spent by students who subscribed to the new meal plan.

**Example 2.1.2**

**Identify the Control Group and the Experimental Group**

### SOLUTION

In this experimental design, the students will be divided into two groups, one that contains the students that subscribed to the old meal plan and one that contains the students that subscribed to the new meal plan. The group containing the students who subscribed to the new meal plan will be called the experimental group, and the group containing the students who subscribed to the old meal plan will be called the control group. If the experiment is properly performed, any change in the response variable (amount of money spent) can be attributed to the explanatory variable (type of meal plan) and not to other variables that are controlled. The untangling of variables at the data-gathering stage makes the analysis of the data much easier. There is no better way to establish a causal relationship.

Variables like living situation (on- or off-campus), the student's class year (freshman, sophomore, junior, senior, or graduate), or their parents' income level can affect the amount of money students spend in a dining facility. These variables must be controlled in the experiment so that differences in the amount of money spent in the dining facility between meal plans can be attributed just to the specific meal plans. One possible method of controlling for these variables is to create small groups of students in the same dormitory. The meal plan usage (i.e., the amount of money spent using each meal plan) is monitored, assuming that each student has the same opportunity and access to the dining halls to use their meal plans. Thus, any difference in the amount of money spent by the students will be due to the different plans and not any other confounding variables.

## The Before and After Study

The before and after study also contains a comparative experiment. The control group and the experimental group are initially identical. The response variable is measured in the control group at the beginning of the study, and then a treatment is applied to the control group. After the treatment is applied, the control group becomes the experimental group. The response variable is again measured after the treatment has been applied. If the treatment affects the response variable then there should be a difference between the value of the response variable for the control and experimental groups, presumably caused by the treatment. An example of a before and after study is given in Example 2.1.3.

| Example 2.1.3 |
|---|
| **A Before and After Study** |

Suppose one is interested in determining if an SAT preparation course improves the performance on the SAT Reasoning Test. Would this be a good study for a before and after study?

### SOLUTION

A group of high school students take the SAT. Then they are given an SAT preparation course. They retake the SAT. If the group's second SAT performance improves, then it may be related to the SAT preparation course.

For this example of a before and after study:

**Explanatory variable:** SAT preparation course.

**Response variable:** SAT scores.

**BEFORE**

Control Group: High school students that have taken the SAT exactly one time. The values of the response variable are the students' scores on their first SAT test.

**TREATMENT** ↓

The high school students are given an SAT preparation course.

**AFTER** ↓

Now, the same students are the experimental group and they take the SAT again. The values of the response variable are the students' scores on their second SAT test.

**COMPARE** ↓

Compare students' performances on the SAT **before** and **after** the course.

### Placebos

One might think that the effectiveness of placebos should be close to zero; this, however, is not the case. Studies have shown that up to 62% of headache sufferers, 58% of those suffering from sea sickness, and even 39% of those suffering from postoperative wound pain showed symptom relief given a placebo. This is a stunning revelation of the effect of psychology on our mind and body.

An intriguing but ethical question for you to ponder is this: If placebos can be 62% effective for curing headaches, would or should a physician treat a headache with a placebo only?

In this experiment there is only one group of students. Suppose there were two groups of students, one that had taken the course (experimental group) and one that had not (the control group). Could you assign any difference in results solely to the SAT preparation course? Because of potential differences in the cognitive abilities of the two groups of students, this experiment would be much more vulnerable to the justifiable criticism that the differences in student cognitive ability caused the difference in the group's SAT performance rather than the SAT preparation course. However, by using a before and after study we have only one group of students and cognitive ability is controlled.

### The Placebo Effect

A difficult problem arises in experiments involving people. In clinical trials of some drug or medical treatment, patients often respond favorably to any treatment, including a "dummy" or "fake" treatment. These fake treatments are called **placebos**. In medical research, a placebo is a pill that contains none of the drug that is being tested. It has been shown in several pain studies that placebos relieve pain in 30 to 40 percent of patients, even though the placebo has no active ingredients. The **placebo effect** is not confined to medicine. Similar effects have been noticed in psychological research in which the subjects seem to try to help the researcher prove some conclusion.

One of the more interesting experiments contaminated by this effect was the Hawthorne Study conducted at the Western Electric Company's Hawthorne Works in Chicago between 1927 and 1932. The studies were initiated to determine the effect of lighting on worker productivity. Lighting was increased in stages, and the investigators found that each time lighting increased worker output increased. The investigators were suspicious that another effect might be causing worker productivity to improve. So, workers were told lighting was to be increased when, in fact, it was decreased. Despite the decrease, worker output increased again. Clearly there was some other variable affecting worker output.

The workers wanted the study to be successful, and their desire was confounding the experiment. Instead of discovering the expected relationship between worker output and lighting, the investigators found that the social system and the employees' roles within that system had a great deal more to do with worker productivity than lighting. The Hawthorne Study has been credited with introducing psychology to the workplace. It also points out the hazards of measurement, even in a controlled experiment.

### Double Blind Studies

The placebo effect is prevalent in medical studies. **Double blind studies** are used to counteract this effect. In a double blind study the subjects are not told whether they are members of the experimental group or the control group. The evaluators (the persons that measure the response variable) are also not told whether their subjects are members of the experimental or control group.

The following scenario is an example of a double blind study.

Until recently, ulcers in the upper intestine were a rather common illness. A new treatment for ulcers was proposed. This treatment involved gastric freezing and required the patient to swallow a deflated balloon with tubes attached. A refrigerated solution was then pumped through the balloon for an hour. The idea behind this therapy was to cool the stomach wall and reduce the amount of acid produced. Initially, the results looked promising and the treatment was used for several years. However, none of the initial studies were double blind.

In the double blind study half of the patients were given the procedure and the other half were given a placebo treatment, which included swallowing the balloon, but no cooling solution was injected into the balloon. Patients that received the placebo treatment actually did better than the ones receiving the refrigerated solution. Gastric freezing was eventually abandoned as a treatment for upper intestinal ulcers.

---

**Definition**

**Placebo**

A **placebo** is a fake treatment that contains none of the drug being tested.

**Definition**

**Placebo Effect**

The **placebo effect** is the belief that the subject improves (or has a reaction to the placebo) when they haven't received the treatment.

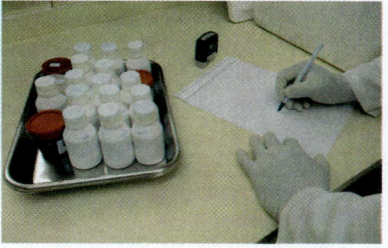

**Double Blind Studies**

Double blind studies are the gold standard in design of experiments for new drug therapies. The Food and Drug Administration has a gauntlet of "phases" that a new drug must pass through before it can be offered to the market. The last of these phases, Phase III, consists of clinical trials. Only about 1 in 3 drugs makes it to Phase III clinical trials. Phase III trials have between 300 and 3000 participants. All the participants have the condition the drug is intended to treat. If it is possible, Phase III trials are conducted as double blind studies.

**Definition**

**Double Blind Study**

A study in which the subjects are not told whether they are members of the experimental group or the control group and the evaluators are also not told whether their subjects are members of the experimental or control group is called a **double blind study**.

### Observational Studies

**Observational data** comes about by measuring "what is." If you are trading stocks, the market data you receive are observational. The data show simply what is happening in the marketplace at the time. Census data are observational; they are a measure of how things are in a specific geographic area at a given point in time. There is no experimentation to see how manipulating one variable will affect another variable or variables. Virtually all of the data we routinely encounter are observational. Examples regularly appearing in the newspaper include:

- Stock, commodity, bond, option, and currency market data
- Almost all federal government data, including census, economic, and educational data
- Virtually all local and state government data
- Sports data (scores, outcomes, etc.)

The data described are often collected to satisfy state and government regulation as well as for business purposes, such as to examine trends in a particular stock price or interest rates. These data values are not the result of a designed experiment.

Observational studies can be extremely valuable. For example, many marketing agencies use observational studies to assist them in decision making. For example, the behavior of children shopping with their parents was examined using observational studies (Rust, Langbourne, "How to reach children in stores: marketing tactics grounded in observational research," Journal of Advertising Research, vol. 33, no. 6, p. 67, 1993). Through observational studies, researchers found several general patterns of behavior of children when shopping with their parents. Researchers found that parents and children shopping together had significant implications in marketing their products. For example, researchers found that it was beneficial if packaging and materials were made to stand out while children were riding in carts so that the children were attracted to the items portrayed in the display.

### Untangling Variables in an Observational Study

In 1973, the Graduate Division at the University of California, Berkeley, carried out an observational study on gender bias in admissions to the Graduate School. There were 8442 men and 4321 women who applied for admission. Of the men that applied, 3738 were subsequently accepted, and 1494 of the women were accepted. Given that almost twice as many men applied to the graduate program, it is reasonable to expect that more men would be accepted. To make a reasonable comparison, the first thing that is needed is to adjust for the difference in application rates between men and women. This is achieved by comparing the percentage of each group that was accepted. Approximately 44% of the men and nearly 35% of the women were admitted. These admission statistics suggested there was rather substantial support for the idea of discrimination against women.

The admission process at Berkeley was done by major. If there was discrimination against women, those departments that were discriminating would stand out when the admissions data were examined separately. But when the data were examined, the investigator did not find what was expected.

Suppose the data in Table 2.1.1 represent the six largest majors on the Berkeley campus.

> **Definition**
>
> **Observational Data**
>
> Data that we routinely encounter and do not involve experimentation are called **observational data**. A distinguishable fact about observational data is that the data are collected without any influence or interaction with the experimenter.

| Table 2.1.1 – College Acceptance Rates | | | | |
|---|---|---|---|---|
| | Men | | Women | |
| Major | Number of Applicants | Percent Admitted | Number of Applicants | Percent Admitted |
| I | 825 | 62 | 108 | 82 |
| II | 560 | 63 | 25 | 68 |
| III | 325 | 37 | 593 | 34 |
| IV | 417 | 33 | 375 | 35 |
| V | 191 | 28 | 393 | 24 |
| VI | 373 | 6 | 341 | 7 |

In four of the six majors, women were admitted more frequently than men. Not only were women not being discriminated against, but there appears to be potential discrimination against men in Major I. How could this completely opposite conclusion be true? A close examination of the data shows that in the majors with the largest percentage admitted (the easiest to get into) there were a large number of male applicants and very few female applicants. The majors that had very low acceptance rates (difficult to be admitted) had relatively very few men and a large number of women applying. Thus, the variable major field of study was *confounding* the variable gender in the original analysis and *biasing* the original conclusion.

The original conclusion was the result of the fact that women were applying to the most difficult departments for admission, not because of sex discrimination. By separating the data by major, the analyst was able to *control* for the confounding variable, choice of major, and remove the bias. When it is possible to remove the effect of one variable, we are said to be *controlling for* that variable. The Berkeley data illustrates a subtle problem in the comparison of two or more proportions.

In this instance the analyst was able to untangle the two variables, but this is not always the case. Unless the data are gathered with a controlled experiment, it may not be possible to untangle the effects of the causal factors. This is an example of **Simpson's Paradox** where the results are counterintuitive. This paradox is often seen with reports based on frequency data when making inferences about relationships between two or more variables. For the Berkeley example, the relationship between the admissions data for men and women is reversed when the majors are combined and when they are considered separately.

## Surveys

A great deal of the statistical information presented to us is the result of surveys. Often we will see in the news that one of the major polling organizations, Gallup, Harris, ABC-Washington Post, or NBC-New York Times, is reporting findings on various topics, from the approval rating of the President to the popularity of gun control.

In some instances, the purpose of a survey is purely descriptive, as those described above. However, in many cases the researcher is interested in discovering a relationship. Because virtually all surveys produce observational data, survey research belongs to steps one and two of the scientific method. Sometimes a plausible relationship is discovered and a designed experiment is undertaken to more convincingly demonstrate the relationship.

A famous observational study, known as the Framingham Study, recorded various data on 4500 middle-aged men. The men were followed for many years with the hope of uncovering what factors relate to the development of heart disease. It was discovered that the development of heart disease seemed to be associated with obesity, heavy smoking, and high blood pressure. Because of the large number of participants and the researcher's ability to control for potentially confounding variables after the data were collected, this research influenced many physicians to work with their patients to control the three **causal factors** found in the study.

Even Hollywood studios do survey research. Suppose the survey shown in Figure 2.1.3 is used by a major studio. Looking at the questions that are posed, you can see they want data on *why you came to the movie and what you liked about it*. Collecting data on a wide variety of movies gives the studio insight into what makes a movie successful, the importance of stars, and what other factors have universal appeal.

---

**Definition**

**Simpson's Paradox**

**Simpson's Paradox** is a phenomenon in statistics in which an effect or trend appears in several different groups of data when considered separately but disappears or reverses when the groups are combined.

**Steps to Consider When Conducting a Survey**

1. Have specific goals.
2. Consider alternatives for collecting data.
3. Select samples to represent the population.
4. Match question wording to the concepts being measured.
5. Pretest questionnaires.
6. Construct quality checks.
7. Use statistical analysis and reporting techniques.
8. Disclose all methods used to conduct the survey.

**Source:** American Association for Public Opinion Research

**Definition**

**Causal Factors**

**Causal factors** are factors or variables that influence the response variable.

# SAMPLE MOVIE SURVEY

*Please complete the first 9 questions of this survey before the movie begins and questions 10 through 14 after the movie. Please give the completed questionnaire to the persons collecting them at the exits.*

1. Indicate the <u>one item</u> in the list below that was most influential in your decision to see the movie. (Check one)

| Newspaper: | Ad................................................( )1 | Radio: | Ad.........................................................( )1 |
|---|---|---|---|
| | Review...........................................( )2 | | Review..................................................( )2 |
| | Article............................................( )3 | Around Town: | |
| Magazine: | Ad................................................( )4 | | Received a Complimentary Pass...............( )3 |
| | Review...........................................( )5 | | Coming Attractions ..............................( )4 |
| | Article............................................( )6 | | Poster or Billboard ..............................( )5 |
| Television: | Commercial ....................................( )7 | | The soundtrack album ...........................( )6 |
| | Review...........................................( )8 | | Recommended from a friend or relative ......( )7 |
| | MTV or other Music Videos ...............( )9 | | Computer on-line service .......................( )8 |
| | Other cable ....................................( )0 | | Internet ...............................................( )9 |
| | Talk Show, other.............................( )x | | Other:..................................................( )0 |

2. Which of the following were important to you in deciding to come to this movie? (Mark as many as apply)

| Tony Curtis ...............( )1 | Billy Wilder, the producer....................( )4 | The visual effects.............( )7 |
|---|---|---|
| Marilyn Monroe ...............( )2 | The story............................................( )5 | The reviews ...................( )8 |
| Jack Lemmon ...............( )3 | The drama..........................................( )6 | Other_____( )9 |

3. What is your age? (Check one)    4.    What is the last grade of school you completed?

| Under 12 ...................( )1 | 30 to 34........................( )7 | Some high school or less.................( )1 |
|---|---|---|
| 12 to 14 ......................( )2 | 35 to 39........................( )8 | Completed high school.....................( )2 |
| 15 to 17 ......................( )3 | 40 to 44........................( )9 | Some college/currently in college... ( )3 |
| 18 to 20 ......................( )4 | 45 to 49........................( )0 | Completed 2 year college.................( )4 |
| 21 to 24 ......................( )5 | 50 to 59........................( )x | Completed 4 year college.................( )5 |
| 25 to 29 ......................( )6 | 60 & over......................( )y | Currently in/completed post-grad.....( )6 |

5. Are you…(Check One)    6.    What is your ethnic background? (Check One)    7. What is your marital status? (Check one)

| Male.....................( )1 | African-American ...........…...( )1  Caucasian .....................( )4 | Single.................................( )1 |
|---|---|---|
| Female .................( )2 | Asian ......................................( )2  Native American( )5Married | ( )2 |
| | Latino .....................................( )3 | Divorced/Separated ......................( )3 |
| | | Widowed ...................................( )4 |

8. Before today, how familiar were you with the storyline? Very familiar…( )1  Somewhat familiar…( )2 Not at all familiar…( )3

9. How many times before today have you seen the movie?

| None....................... ( )1 | Once...................... ( )2 | Twice........... ( )3 | Three or more times................( )4 |
|---|---|---|---|

## COMPLETE QUESTIONS 10 THROUGH 14 AFTER THE MOVIE

10. How would you rate the movie?    11. Would you recommend the movie?  12. How did the movie measure up to your expectations?    Excellent

| ( )1......................... | Definitely..........( )1 | Better than expected..........( )1 |
|---|---|---|
| Very Good .............. ( )2 | Probably............( )2 | About what I expected.......( )2 |
| Good...............( )3 | Probably Not.....( )3 | Not as good as expected.....( )3 |
| Fair.................( )4 | Definitely Not...( )4 | |
| Poor...............( )5 | | |

13. Which of the following words or phrases best describe the movie you just saw? (Mark as many as apply)

| Entertaining.............................. ( )1 | Worn-out theme ...........................( )1 | Well acted................................( )1 |
|---|---|---|
| Boring/dull.............................. ( )2 | Surprising.....................................( )2 | Not enough drama..................( )2 |
| Dramatic ............................. ( )3 | Too long .......................................( )3 | Has a good story....................( )3 |
| Interesting settings ................... ( )4 | Different/original ..........................( )4 | Unrealistic.............................( )4 |
| Too slow in parts ...................... ( )5 | Not my kind of movie...................( )5 | Educational............................( )5 |
| Offensive ............................... ( )6 | Controversial.................................( )6 | Nothing new/done before.......( )6 |
| Confusing.............................. ( )7 | Moved just right...........................( )7 | Thought provoking.................( )7 |
| Interesting characters .............. ( )8 | Humorous .....................................( )8 | Too predictable......................( )8 |
| Too silly/stupid....................... ( )9 | Good cast .....................................( )9 | Believable..............................( )9 |
| Action packed........................... ( )0 | In bad taste...................................( )0 | Depressing.............................( )0 |

14. Would you pay to see this movie again?  Yes.......................................( )1  No.......................................( )2

Figure 2.1.3

## ✐ 2.1 Exercises

### Basic Concepts

1.  What are the two fundamental problems of measurement?

2.  When measurements are used to help solve a problem, what desirable characteristics should the measurements possess?

3.  Name and briefly describe three measurement systems commonly used in business.

4.  When you encounter any type of data, what three questions should you ask to determine the quality of the measurements?

5.  What is the scientific method?

6.  What is a confounding variable?

7.  How does statistics interact with the steps in the scientific method?

8.  Name and briefly describe the two main branches of statistics.

9.  What is the decision-making method?

10. What is different between the scientific method and the decision-making method?

11. Are problems that can be solved by collecting data always the result of a system malfunction? Explain.

12. Give an example of how statistics can be used to improve a process.

13. What are fuzzy concepts? What are the measurement problems associated with fuzzy concepts?

14. Give an example of a tool that has been widely accepted as an instrument used to measure a fuzzy concept.

15. What are the two ways of obtaining data?

16. What are the dangers of making conclusions based on poorly collected data?

17. How do you treat the problem of a confounding variable?

18. Explain the difference between the control group and the experimental group in a controlled experiment.

19. What is an explanatory variable?

20. What is a response variable?

21. What is bias? How can it be controlled?

22. What is a completely randomized design? What are the advantages of using a completely randomized design?

23. What is a before and after study?

24. What is the placebo effect? Give an example.

25. What is a double blind study?

26. How do observational studies differ from controlled experiments?

27. What kinds of problems can be associated with an observational study?

28. Researchers use surveys for two main purposes. Name and give an example of each.

## Exercises

29. Specify whether the following variables are well-defined or not. Justify your answer.

    a. Height

    b. Weight

    c. Hot

    d. Temperature

    e. Beauty

30. A researcher has developed a test that reportedly measures intelligence. The test includes questions such as:

    *What is the lowest common denominator of the fractions $\frac{5}{32}$ and $\frac{6}{9}$?*

    *Who invented the digital computer?*

    Is it reasonable to measure intelligence with these questions? Discuss.

31. A hotel manager is interested in getting feedback from guests. Two variables of interest to the manager are cleanliness and aesthetics of the rooms. Discuss what problems you would encounter when measuring those variables.

32. Suppose you want to determine the proportion of college students in the state of Virginia that pays more than $500 per year on textbooks. Using the scientific method, how would you conduct the experiment?

33. The manager of an electronics company was interested in determining the reason for the increase in sales volume over the last three years. The manager randomly selected data on the advertising budget, number of salespeople, and average product costs. When examining the data, the manager found that her average product costs were fairly stable but the advertising budget steadily increased over the last two years along with the number of salespeople. Are there any confounding variables in this study? If so, what are they and why do you consider them confounding?

34. A company that produces bulbs for projectors wanted to conduct an experiment to determine the length of life of its bulbs. The company's leading competitor's bulbs have an average life of 1000 hours. The company sampled its bulbs and found that the average life of the bulbs was 1200 hours. Thus, the company has concluded and advertises that its bulbs last longer than the competition by at least 100 hours. Were the results of this experiment an example of descriptive or inferential statistics? Explain your answer.

35. The health and social problems associated with obesity can be a severe hindrance in attaining many of life's goals. Methods for treating obesity were compared in "One Year Behavioral Treatment of Obesity: Comparison of Moderate and Severe Caloric Restriction and the Effect of Weight Maintenance Therapy," in the *Journal of Consulting and Clinical Psychology*. In the study, a group of 25 women, each of whom was at least 25 kilograms (kg) overweight, were randomly split into two groups. The first group received behavior therapy and was placed on a 1200 calorie per day diet for a period of one year. The second group received behavior therapy and was placed on a 420 calorie per day diet for the first 16 weeks of the year. Then they returned to a 1200 calorie per day diet for the remainder of the year. At the end of a 26-week period, the average weight lost was 11.86 kg for the first group and 21.45 kg for the second group. But after 52 weeks, the average weight lost was 10.94 kg for the first group and 12.18 kg for the second group.

    a. Why is this study an example of a controlled experiment?

    b. What is the explanatory variable?

    c. What is the response variable?

     **d.** Is there a control group in the study? Explain.

     **e.** Suppose that the data were gathered from an observational study instead of from a controlled experiment. How would this affect the conclusions that might be made from the study?

**36.** An article appearing in the *New England Journal of Medicine* investigated whether the academic performance of asthmatic children being treated with the drug Theophylline was inferior to a non-asthmatic group. In one part of the study, 72 children were identified as being treated for asthma. For each child with asthma, a non-asthmatic sibling was also identified. (The use of sibling controls allows for control of family environment and certain genetic factors on academic achievement.) All 144 children were then given a test to measure academic achievement. There were no significant differences on the test between the two groups.

     **a.** Why is this study an example of a controlled experiment?

     **b.** What is the explanatory variable?

     **c.** What is the response variable?

     **d.** Is there a control group in the study? Explain.

     **e.** Suppose that the data were gathered from an observational study instead of from a controlled experiment. How would this affect the conclusions that might be made from the study?

**37.** A small clinical pilot study was conducted by a research team from Harvard Medical School and the School of Public Health. Fifteen individuals in the early stages of Multiple Sclerosis were fed bovine myelin, a substance containing two antigens thought to be the target of the immune system's attack in Multiple Sclerosis. Another fifteen were given a placebo. In the study, fewer members of the group fed bovine myelin had major attacks of the disease.
**Source:** Science, Vol. 259, No. 5099

     **a.** Which phase of the Scientific Method best describes this study?

     **b.** Is this an observational study or a controlled experiment?

     **c.** What is the response variable?

     **d.** What is the explanatory variable?

     **e.** Which group is the treatment group?

     **f.** Which group is the control group?

**38.** London scientists conducted a study to determine if chocolate can trigger migraines. Twelve migraine-prone subjects were given a peppermint-laced chocolate candy and eight migraine-prone subjects were given a peppermint-laced placebo made of carob, peppermint, and vegetable fat. Five subjects from the group given chocolate developed a migraine headache within one day. No one from the group given the placebo developed a migraine in the same time period.
**Source:** Self magazine

     **a.** Which phase of the Scientific Method best describes this study?

     **b.** Is this an observational study or a controlled experiment?

     **c.** What is the response variable?

     **d.** What is the explanatory variable?

     **e.** Which group is the treatment group?

     **f.** Which group is the control group?

**39.** Jacob normally plays basketball three days a week and has begun to develop patellar tendinitis, which is inflammation in the patellar tendon and results in nagging knee pain. In an effort to relieve his knee pain, Jacob decides to take a week away from playing basketball and rest his knee. However, after about four days, his friend offers him an analgesic rub and insists that his knee will feel better in two to three days. After using the analgesic rub for a couple of days, Jacob's knee begins to feel better. Did the analgesic rub work? Explain how confounding variables might have played a role on Jacob's knee getting better.

**40.** The Nurse's Health Study conducted on 87,245 women at Boston's Brigham and Women's Hospital revealed that women who eat a cup of beta carotene-rich food a day have 40 percent fewer strokes and 22 percent fewer heart attacks than those who consume a quarter of a cupful per day.

Source: Self magazine

   **a.** Which phase of the Scientific Method best describes this study?

   **b.** Is this an observational study or a controlled experiment?

   **c.** What is the response variable?

   **d.** What is the explanatory variable?

   **e.** Which group is the treatment group?

   **f.** Which group is the control group?

**41.** A religious group conducted a survey with two of the questions asking "Do you go to church?" and "Are you happy?" After conducting the survey, the group concluded that those who go to church are generally happier than those that do not go to church. Do you think going to church makes one happier? Describe how confounding variables could play a role with the conclusion drawn by the religious group.

**42.** In May 2011, Internet Explorer reversed its trend in the United States and gained usage share (the percentage of users using a particular Internet browser). In June of 2011, the trend reversal became global. Internet Explorer gained 0.57% in June across all operating systems with Internet Explorer 8.0 gaining 0.86% globally. The gains for Internet Explorer came primarily at the expense of Mozilla Firefox (−0.51%). Google Chrome's pace of usage share gains slowed to +0.2% for June. The gains for IE were the largest in Europe and Asia:

   Internet Explorer in Europe: +0.88%

   Internet Explorer in Asia: +0.81%

This increase may be the result of a marketing campaign. In early June, Microsoft launched their "Confidence" campaign aimed at showing the security features of Internet Explorer 8.

Source: netmarketshare.com

   **a.** Are the results stated above likely to have come from an observational study?

   **b.** How can Microsoft (and other companies) benefit from this information?

**43.** A survey was conducted by an investment firm asking participants the following questions: "Are you financially secure?" and "Do you independently make decisions about your investments?" After analyzing the data from the survey, the firm concluded that people who make investment decisions independently tend to be not as financially secure as those who make decisions with the help of an investment advisor. What confounding variables could have played a role in this conclusion?

# 2.2 Big Data and Business Analytics

## What is Big Data?

When it comes to business decision making, we normally have two options. One is to make decisions based on our intuition. Of course, using this approach is a function of one's experience and expertise associated with the decision that needs to be made. The other option is to make decisions based on data. There are advantages and disadvantages to both. However, many believe that in spite of one's expertise, the best way to make an informed decision is by using data. **Big Data** (sometimes referred to as data science) is used in industry to make informed decisions. The insurance industry has been analyzing Big Data for many years. Think about how insurance companies provide rates for automobile insurance. The premiums are based on risk. That is, the risk of a teenager getting in an accident is higher than that of an adult (or an experienced driver) and thus, the insurance premiums for a teenager are significantly more expensive than that of an adult or very experienced driver. The insurance industry also uses Big Data to determine premiums for life insurance, homeowner's insurance, and any other items that one wants to insure. Manufacturing firms use Big Data to determine sales levels; banks use Big Data when determining the risk associated with lending and borrowing money. As you can see, Big Data plays a part in our everyday lives—whether it is actively or passively.

What is Big Data? Big Data is characterized by any or all of three of the following characteristics: *volume*, *velocity*, and *variety*. Sometimes a fourth characteristic, called *veracity*, is included when describing Big Data. A data set possessing any or all of these characteristics will constitute Big Data. The first characteristic, volume, indicates that the amount of data is voluminous. Examples of voluminous data are GPS data from cellular phones, images, grocery store data from customer account cards, or web surfing activities. These represent lots of data. The second characteristic, velocity, indicates that the data come in so fast that there are not many methods to constantly update the data to match the new information. Examples of data with high velocity are social media streams (YouTube, Instagram, or Facebook). The third characteristic, variety, represents a diverse set of data. You have data of many forms—multimedia, images, audio, heart readings, etc.—data that do not fit in the regular rows and columns of a conventional data set. The fourth 'v', which is not always listed as a characteristic of Big Data, is veracity. Veracity refers to the overall quality of the data and to what extent it can be trusted.

When working with Big Data, one must remember the four v's and understand that it is unlikely that you will have data in a traditional spreadsheet model that fit nicely in labeled rows and columns. You will have three types of data with Big Data: structured, unstructured, and semi-structured. **Structured data** is the traditional type of data to which we have grown accustomed—highly organized, has labels, and fits in a spreadsheet such that each cell in the sheet is of an identifiable format. Unstructured data does not have a predictable organization. **Unstructured data** can be text, images, video, etc. The last type of Big Data is **semi-structured** which is a combination of both structured and unstructured data. One would usually see semi-structured data on Twitter—the number of followers and tweets are structured; the images that are shared and posted are unstructured.

**Definition**

**Big Data**

**Big Data** is any data that can be characterized by any or all of the following characteristics: *volume*, *velocity*, *variety*, and *veracity*.

**Definition**

**Structured Data**

The traditional type of data to which we have grown accustomed—highly organized, has labels, and fits in a spreadsheet such that each cell in the sheet is of an identifiable format is called **structured data**.

**Definition**

**Unstructured Data**

Data that does not have a predictable organization is called **unstructured data**.

**Example 2.2.1**

**Identifying Structured Data**

Suppose researchers wanted to study if there was a relationship between teens' social media habits and their friendships. To do this, researchers compiled a questionnaire to be completed by the teens in the United States of America (with parental consent for all teens under 18 years of age). Some of the questions contained in the survey asked for demographic information such as age, race, zip code, number of close friends, number of social media accounts, number of people that follow you on your social media sites, number of people that you follow on social media sites, and if they consider their friends/followers on social media sites to be close friends. A sample of the collected data is shown below. What type of Big Data does this represent?

## Flash Boys: Data Velocity

*Flash Boys* is a book written by Michael Lewis about high frequency stock trading. Part of the book describes the great lengths a firm went to reduce the time it takes to send a buy or sell order between New York and Chicago. The best available time in 2008 was 14.65 milliseconds (14.65 thousands of a second). But theoretically it should be possible to communicate between the two cities over a fiber line in 12 milliseconds. The book tells the story of what it took to build and market a "direct" fiber run between New York and Chicago at a cost of 300 million dollars. Note, that is 300 million dollars spent to improve data velocity by slightly more than 2 milliseconds.

### Table 2.2.1 – U.S. Teens Social Media Survey Results

| ID Number | Age | Race | Zip Code | No. Friends | No. Social Media Accounts | No. Following | No. Followers | Consider Friends |
|---|---|---|---|---|---|---|---|---|
| 65600355 | 18 | Black or African American | 88310 | 5 | 8 | 328 | 163 | 15 |
| 92110187 | 13 | White | 47722 | 9 | 9 | 141 | 159 | 19 |
| 64381638 | 15 | Asian | 51023 | 3 | 7 | 343 | 348 | 17 |
| 68647528 | 16 | Native Hawaiian or Pacific Islander | 90210 | 5 | 3 | 184 | 220 | 11 |
| 73692314 | 15 | Hispanic or Latino | 32003 | 6 | 10 | 307 | 172 | 27 |

### SOLUTION

Note that in the table above, each response by a teenager (represented by a randomly assigned ID Number) is identifiable. For example, for ID Number 65600355, we can see that the teen is black or African American, has 5 friends, 8 social media accounts, etc. This is the typical format of structured data.

## Example 2.2.2

### Identifying Unstructured Data

The image below contains a couple of reviews of a Nikon D850 FX camera which is sold on Amazon's website. What type of Big Data does this represent?

**Customer reviews**

★★★★★  4.8 out of 5

330 global ratings

| | |
|---|---|
| 5 star | 91% |
| 4 star | 5% |
| 3 star | 1% |
| 2 star | 0% |
| 1 star | 4% |

**Nikon D850 FX-format Digital SLR Camera Body**
by **Nikon**

Style: Body Only | Configuration: Base | **Change**

**Top positive review**

★★★★★ **A total game changer -- almost perfect**

I have owned a lot of digital cameras over the years - Nikon D50, D90, D3, D800, Df, Canon 5DMKIII, etc. Some of those cameras have disappointed -- the D90 with it's CMOS sensor seemed less sharp than the old school CCD in the D50, and even the D800 with its anti-alias filter seemed less sharp than I expected.

With the D850, I was impressed the second I took the first shot inside my dimly lit home after dark. Incredible detail, excellent noise, perfectly metered, impressively fast autofocus. 10/10.

**Top critical review**

★☆☆☆☆ **Beyond Disappointed**

Beyond disappointed.
This camera has given me problems from the start. I already serviced it once with Nikon; the problems are solved temporarily and then return. I am convinced I got a bad copy of the camera. Back focusing, noisy at ISO 125, motion blur on still subjects...I'm so unhappy with this camera. Amazon is telling me to reach out to Adorama. Adorama is telling me to talk to Nikon. No one is really helping. I'm so regretful of this purchase. If I knew it was going to be this way, I would have gotten aonther D750. It's a terrible thing to spend $3000+ and slowly watch it go down the drain.

### SOLUTION

You can see that there are a total of 330 global reviews currently. The reviews are helpful to Amazon customers interested in buying a camera, and also helpful to Amazon in that they can use the reviews to analyze the text data to understand the sentiment (positive, negative, or neutral) of its customers in the reviews. However, unlike the structured data in the previous example, the reviews are considered unstructured data because they do not have a specific format—other than text and perhaps pictures.

In addition to the unstructured data in Example 2.2.2, from the same reviews, Amazon can also collect demographic information about the reviewers such as age, race, zip code, marital status, and income or income range. Note that this data is structured. Thus, the semi-structured data from an Amazon product review could be a combination of the structured data containing the demographic information and the unstructured data containing the product review.

| Table 2.2.2 – Amazon's Reviewers Demographic Data | | | | | | | |
|---|---|---|---|---|---|---|---|
| Product ID | Age | Race | Zip Code | Marital Status | Income Range | No. of Stars | Product Review |
| B07524LHMT | 35 | White/ Caucasian | 33476 | Married | > $150,000 | 4 | See Image Below |

The picture below is also included with the customer's review, which is another component of unstructured data. Thus, the first seven columns of data are structured but the actual product review is unstructured, thus classifying this Amazon product review as semi-structured data.

**Definition**

**Semi-Structured Data**
Data that is a combination of both structured and unstructured data is called **semi-structured data**.

★★★★☆  **Good for the Brag Points**

I will say compared to my D810 the improved focal point system is noticeably better, but it can still be tricky to use during an active photo shoot. I still have not managed to use the touch screen or flip screen feature during a photo shoot. I was pleasantly surprised to see so many low ISO settings than the D810: 64, 80, 100, 125, 160, 200, 250, 320, and so on. About 3X more settings to choose from. Although, if you are shooting over ISO 200 that is a waste of this camera's potential. They claim this has a much better high ISO capability, but I would not agree there is a noticeable difference. These images still look like crap at anything over 400 ISO. You would be better off just using your smartphone.

The SnapBridge seems to work fine, but drains the battery life a lot faster, and not a terribly user-friendly app. It is too convoluted to just delete images that pile up and has very limited features. The scroll through images does not work well. It would be nice to have a much easier way to start and stop SnapBridge.

The wireless advantage for the SB 5000 speedlite may work, but this was quite misleading, since you then have to buy the special adapter and transmitter, which has to plug into the outside of the camera still, and was tricky to set up.

Twitter is an excellent example of a company collecting and analyzing Big Data. It has hundreds of millions of users who are posting text messages, pictures, and videos. Given the number of users, Twitter collects more than 1 terabyte per hour of user data—a statistic that represents the enormous size of the data, the speed at which the data changes, and the diversity of the data.

## What is Business Analytics & Its Role in Statistics?

Having discussed Big Data, the next logical question is, "How does one analyze Big Data?" One technical definition of **analytics** is that it is the science of examining raw data to draw conclusions about information contained in the data. Note that this sounds very similar to the definition of statistics. It sounds similar because many of the techniques in analytics are derived from classical statistical methodologies. Like statistics, analytics is used in many industries to allow companies and organizations to make better business decisions. Some other names for analytics are *business intelligence* and *data analytics*. Analytics is sometimes considered the analysis of Big Data with its purpose being to make sense of Big Data by converting it into valuable information.

**Definition**

**Analytics**
The science of examining raw data to draw conclusions about information contained in the data is called **analytics**.

Almost all companies and organizations, such as Google, eBay, PayPal, Microsoft, Apple, to name a few, use **business analytics** to help them make better business decisions. These companies obtain data from a variety of sources—mobile phones, social media networks, websites, online shopping and surfing, electronic communications, and GPS. These methods produce torrents of data just from normal, ordinary operations and use. With analytics,

**Definition**

**Business Analytics**
Analytics used by companies to make better decisions is called **business analytics**.

### Definition

**Predictive Analytics**

**Predictive analytics** uses past data to develop models that can help determine what future events are most likely to happen.

### Definition

**Prescriptive Analytics**

**Prescriptive analytics** is the development of models that help us answer the question, "What should we do moving forward?"

### Data Resources

We live in a data rich society. Anyone with access to a personal computer can access thousands of different databases throughout the internet. These databases are packed full of observational data. The website for this textbook, stat.hawkeslearning.com, provides numerous links to data resources. However, some of the largest, most credible, and most commonly used databases are:

- Amazon Web Services
- Centers for Disease Control and Prevention (CDC)
- Data.gov (Data regarding the U.S.)
- Federal Reserve Economic Data (FRED)
- Organization of Economic Cooperation and Development (OECD)
- The World Bank
- The World Factbook (CIA)
- UNdata (United Nations)
- United States Census Bureau
- World Health Organization

companies can gain a deeper understanding of their customers to better meet their needs. Companies also use analytics to develop models to drive insights into the past, present, and future.

There are primarily two types of analytics—predictive and prescriptive. We use both types of analytics to help us gain insight into the future. **Predictive analytics** answers the question of what could happen. **Prescriptive analytics** answers the question of what should happen.

Retail stores use predictive analytics to predict products that customers will buy, times that they will log on to specific sites, or even the amount of time that a customer may spend on a site. Being able to make these predictions allow retailers to better tend to customer needs as well as make the business more profitable.

After prediction, prescriptive analytics focuses on how to take advantage of future opportunities of the decision-making process. Prescriptive analytics is considered the future of business analytics. It provides an organization with adaptive, automated, and time-dependent courses of action to take advantage of business opportunities. For example, in finance, prescriptive analytics can be used to help investors select which investments to purchase. In sports, prescriptive analytics can help teams determine which player to draft or trade.

Analytics plays a major role in the past, present, and future of the decision-making processes of many organizations. Businesses of all kinds that use Big Data and analytics can improve their decision-making processes. Applications such as banking, health care, etc., need predictive and prescriptive analytics to improve their standards and quality to help their customers and themselves.

## ✎ 2.2 Exercises

### Basic Concepts

1. Where is Big Data used?

2. What are the three (sometimes four) characteristics of Big Data?

3. Give another example of where we can find semi-structured data.

4. What is business analytics?

5. How is business analytics used?

6. What is the difference between predictive and prescriptive analytics?

7. Give an example of a company using predictive analytics to make business decisions.

8. Give an example of a company using prescriptive analytics to help their organization moving forward.

### Exercises

9. GOES satellites (GOES-16 & GOES-17) provide continuous weather imagery and monitoring of meteorological and space environment data across North America. These satellites provide the kind of continuous monitoring necessary for intensive data analysis. They hover continuously over one position on the surface. Describe three characteristics of Big Data that would be produced by these satellites.

10. For the data in question 9, would the data collected be described as: structured, unstructured, or semi-structured? Explain your choice of answer.

11. The following sample of data about BMI (Body Mass Index) was obtained by the WHO (World Health Organization). What kind of data analytics can be done on these data?

| Mean BMI (kg/m²) [age-standardized estimate] 18+ years, 2016 | | | |
|---|---|---|---|
| Country | Both sexes | Male | Female |
| Afghanistan | 23.4 [22.0–24.8] | 22.6 [20.1–25.1] | 24.1 [23.0–25.3] |
| Albania | 26.7 [25.8–27.5] | 27.0 [25.8–28.2] | 26.3 [25.0–27.6] |
| Algeria | 25.5 [24.5–26.5] | 24.7 [23.4–26.1] | 26.4 [24.9–27.8] |
| Andorra | 26.7 [24.6–28.7] | 27.3 [24.8–29.8] | 26.1 [22.8–29.5] |
| Angola | 23.3 [21.2–25.6] | 22.3 [19.7–25.0] | 24.3 [20.9–27.7] |
| Antigua and Barbuda | 26.7 [24.6–28.8] | 25.7 [23.2–28.2] | 27.7 [24.4–31.0] |
| Argentina | 27.7 [26.8–28.6] | 27.8 [26.6–29.0] | 27.6 [26.3–28.8] |
| Armenia | 26.3 [25.8–26.9] | 25.6 [24.8–26.3] | 27.0 [26.1–27.8] |
| Australia | 27.1 [26.6–27.6] | 27.6 [26.9–28.2] | 26.7 [26.0–27.4] |
| Austria | 25.6 [24.3–26.8] | 26.5 [24.8–28.2] | 24.6 [22.6–26.5] |

**Source:** World Health Organization https://apps.who.int/gho/data/view.main.CTRY12461?lang=en

12. JetBlue Airlines collects data which includes passenger ID number, name, date of birth, country of birth, country of residence, frequent flyer number, departure airport, destination airport, airfare paid, and flight number. How can this information be used to make business decisions?

13. Mobile phone companies use the GPS feature to determine the location of users and to provide location-based services (LBS) such as information, entertainment, and security. Describe how a marketing company can use location-based services analytics to provide targeted ads about a nearby retail store to a user.

14. Describe how prescriptive analytics can be used in sports to determine which player to draft or trade.

15. Credit card companies are some of the biggest users of data analytics. How can location-based services (LBS) be used to prevent credit card fraud?

16. Many hospitals and health care providers are now utilizing electronic health records (EHR). This of course generates an immense amount of patient data; hence the need for data analytics. Describe how this data can be used to improve service and better patient care.

# 2.3 Data Classifications

Data or variables can be classified as **qualitative** or **quantitative**. If data or variables are quantitative, they can be further identified as **discrete** or **continuous**. Since the kind of data available affects the kind of analyses that can be performed, it is important to recognize data attributes.

## Qualitative and Quantitative Data

Qualitative data are measured on the nominal and ordinal scales (see levels of measurement on the next page). Some examples of qualitative data are shirt sizes (S, M, L, XL, XXL), occupation (doctor, lawyer, professor, etc.) and eye color (brown, blue, and green). As discussed earlier, regardless of how the measurements are coded, arithmetic operations cannot be performed with qualitative data.

Quantitative data are measured on the interval or ratio scale (see levels of measurement on the next page) and we can perform arithmetic operations with quantitative variables. Some examples of quantitative data are the number of students in class today, the number of dependents claimed on a tax return, the number of fans at a sporting event, and the time it takes to complete a particular task.

## Discrete and Continuous Data

Quantitative data or variables can be categorized as either discrete or continuous. To illustrate, suppose that observations are taken on two variables: the number of stocks and bonds in a mutual fund and the annual yield for the fund. In looking at the data, one difference is that the number of stocks and bonds is a whole number, whereas the annual yield is decimal-valued. Since it is impossible for a mutual fund to have 10.167 stocks and bonds, or any fractional number for that matter, there are gaps in the values that the variable "Number of Stocks" can assume.

| Table 2.3.1 – Mutual Funds | | | | | |
|---|---|---|---|---|---|
| **Number of Stocks** | 10 | 20 | 30 | 40 | 50 |
| **Annual Yield (%)** | 4.25 | 3.86 | 13.52 | 15.00 | 9.62 |

The stock data given in the first row of Table 2.3.1 are **discrete**. While the data representing the number of stocks can only assume integer values, it would be misleading to create the impression that a discrete variable can only assume integer values. Discrete data may assume decimal values. For example, a variable that takes on only the values 1, 1.5, 2, 2.5 is discrete.

The annual yield data given in Table 2.3.1 are **continuous**. It may appear that the annual yields are discrete, since values appear to only take on values up to the second decimal place. However, it is only the inadequacy of the measuring instrument that creates this illusion. The first annual yield is listed as 4.25%, but the actual annual yield of the first mutual fund may be 4.2531642%. If the measuring device can only detect differences in the hundredths place, then all measurements will be given to the hundredths place. Any digits beyond the hundredths place will be ignored. However, just because these digits are ignored doesn't mean they don't exist.

---

**Definition**

**Qualitative Data**

**Qualitative data** are measurements that can change in kind, but not in degree. Qualitative measurements often consist of labels or descriptions and do not have naturally occurring numerical values.

---

**Definition**

**Quantitative Data**

**Quantitative data** are measurements that change in magnitude from trial to trial such that some order or ranking can be applied. Quantitative variables can be measured using a naturally occurring numerical scale.

---

**Definition**

**Discrete Data**

Data in which the observations are restricted to a set of values (such as 1, 2, 3, 4) that possesses gaps are **discrete**.

---

**Definition**

**Continuous Data**

Data that can take on any value within some interval are **continuous**.

---

**Example 2.3.1**

**Classifying Data**

State whether each of the following variables is qualitative or quantitative. If the variable is quantitative, indicate whether it is discrete or continuous.

**a.** Political party affiliation

**b.** Amount of time spent on the computer

    **c.** Marital status

    **d.** Number of online orders made per month

**SOLUTION**

    **a.** Examples of political party affiliation are Republican, Democrat, Independent, etc. These are qualitative measurements because they identify the party by name only and we are unable to perform any arithmetic operations or establish distances between parties. Thus, political party affiliation is a qualitative variable.

    **b.** Examples of time spent on the computer could be any value from a few seconds to a large number of hours. Also, depending on the quality of the measurement, time could be recorded (or reported) in any number of significant digits. With that said, time spent on the computer is a quantitative, continuous random variable.

    **c.** For marital status, examples could be single, married, or divorced. These are qualitative measurements since one classification cannot be considered to be more than another and we also cannot perform any arithmetic operations on this variable.

    **d.** The number of online orders is countable, starting at 0. Thus, this is a quantitative, discrete random variable.

## Levels of Measurement

Like most things in life, data come in different qualities. Some measurements are purely numerical and are based on well-defined standards, such as pounds, inches, dollars, and percentages. On the other hand, some measurements are exceedingly "fuzzy" and the standard of measure is ill-defined, if defined at all. For example, consider the response to the following question.

    What is your opinion of the President's performance?

1. Extremely disappointing
2. Disappointing
3. Satisfactory
4. Good
5. Extraordinary

Someone's *good* response may be equal to someone else's *extraordinary*. There can be no guarantee of a common scale and thus the level of consistency between measurements is unreliable. This type of data is of much lower quality than measurements made on some standard scale, such as pounds or inches.

> **Definition**
>
> **Level of Measurement**
> The quality of data is referred to as the **level of measurement**.

The terms used to describe the quality of data are **nominal**, **ordinal**, **interval**, and **ratio**. When analyzing data you must be exceedingly conscious of the data's level of measurement because many statistical analyses can only be applied to data that possess a certain level of measurement.

Because of different levels of measurement, not all data are created equally. Unfortunately, once data are in numerical form, many believe that a number is a number. Everyone knows that you can add, subtract, multiply, and divide numbers. But for some measurements, adding, subtracting, multiplying, and dividing are simply meaningless.

Standard mathematical operations (such as addition and subtraction) are not defined for nominal and ordinal data, and, consequently, many forms of statistical analysis (descriptive and inferential) are not appropriate.

## Nominal Data

Dexterity (left-handed or right-handed) and hair color (blond, brunette, or redhead) are examples of nominal variables. If the dexterity variable is coded numerically, say Left-handed = 1, Right-handed = 0, then what do the numbers mean?

Is it meaningful to add, subtract, multiply, or divide these numbers? Let's try addition.

| 1 | + | 0 | = | 1 | |
|---|---|---|---|---|---|
| Left-handed | + | Right-handed | = | Left-handed | ? |

Is it meaningful to add left-handed and right-handed together and get left-handed? If it is, it is certainly a bizarre interpretation.

Let's try subtraction.

| 1 | − | 0 | = | 1 | |
|---|---|---|---|---|---|
| Left-handed | − | Right-handed | = | Left-handed | ? |

Is it meaningful to subtract right-handed from left-handed and get left-handed? If this kind of mathematics works, it doesn't say a lot for being right-handed!

Similarly, absurd conclusions can be reached for multiplication and division. For strictly nominal data, none of the arithmetic operators can be applied. Statistically speaking, only some graphical and very few numerical statistical procedures can be applied to nominal data. Keep in mind that the nominal level of measurement identifies the variable in name only.

## Ordinal Data

Consider a response to the fill in the blank question.

Frosty Pops taste ____.

1. Very Bad
2. Bad
3. Fair
4. Good
5. Very Good

The response to this question is often referred to as being measured on the Likert Scale, in which one rates the taste of Frosty Pops on a scale from 1 to 5. These responses are ordered on the basis of an individual's impression of the goodness of Frosty Pops. The response *bad* is perceived to be better than *very bad*, *good* better than *fair*, and so forth. If numerical codes are used to represent the responses, should the properties of addition, subtraction, multiplication, and division be applied to them? Let's try addition.

| 1 | + | 1 | + | 1 | + | 1 | + | 1 | = | 5 | |
|---|---|---|---|---|---|---|---|---|---|---|---|
| Very bad | + | Very bad | + | Very bad | + | Very bad | + | Very bad | = | Very good | ? |

This means that if you thought that a Frosty Pop tasted *very bad* (1) you should eat five of them and doing so would be equivalent to eating something that you thought was *very good* (5). Addition of ordinal values is not reasonable. And for that matter, neither are subtraction, division, and multiplication. So the only difference between ordinal and nominal data is that ordinal data possess order. Here again, like nominal data, only a very limited number of statistical analyses can be performed.

Note that ordinal data are also nominal, but they possess the additional property of ordinality (or ranking).

## Interval Data

One example of interval data is temperature (measured on the Fahrenheit scale).

| | | | | |
|---|---|---|---|---|
| 48 degrees | − | 45 degrees | = | 3 degrees |
| 72 degrees | − | 69 degrees | = | 3 degrees |

> **Definition**
>
> **Interval Data**
>
> If the data can be ordered and the arithmetic difference is meaningful, the data are **interval**.

In the case above, the difference between the temperatures is three degrees, and it is true that the difference in kinetic energy between 48 and 45 degrees is the same as the difference between 72 and 69 degrees.

Interval data also have another interesting property, an arbitrary zero value. You do not have to be a mathematician to appreciate the usual meaning of the zero concept, having zero dollars means that you do not have any money. However, a temperature of zero degrees on the Celsius or Fahrenheit scale does not mean there is no kinetic energy, and thus in the case of temperature, zero has been arbitrarily selected. In fact, the Celsius temperature scale places the value of zero at a temperature equivalent to 32 degrees Fahrenheit.

One implication of an arbitrary zero point is that the ratio of two variables has no meaning. For example, the kinetic energy associated with a temperature of four degrees on the Fahrenheit scale is not twice as great as the kinetic energy associated with a temperature of two degrees. The property that distinguishes interval data is the notion that equal intervals represent equal amounts. For example, the interval between four degrees and one degree represents the same difference in kinetic energy as the difference between 74 degrees and 71 degrees.

Interval data are numerical data that possess both the property of ordinality (ranking) and the interval property. However, interval data do not possess a meaningful origin (zero value).

## Ratio Data

> **Definition**
>
> **Ratio Data**
>
> **Ratio data** are similar to interval data, except that they have a meaningful zero point and the ratio of two data points is meaningful.

The operations of addition, subtraction, multiplication, and division are reasonable on ratio data. Many of the variables we commonly encounter are ratio variables: volumes, heights, weights, pressure. Being aware of the data's level of measurement is an extremely important part of any statistical analysis, since many statistical measures that are meaningful for interval and ratio data are not meaningful for ordinal and nominal measurements. Ratio data are the most desired and meaningful type of data for decision makers. We will explore analytical techniques for each type of data later in subsequent chapters.

Money is a good example of a ratio variable. Say that a friend has $40 and you have $20. Thus, $\frac{\$40}{\$20} = 2$. According to the ratio we just computed, your friend has twice as much money as you. Is this really true? Money is a ratio variable because ratios (quotients) are meaningful. If someone does have $40 and you have $20, they do have twice as much money as you.

---

Determine the level of measurement (nominal, ordinal, interval, or ratio) for each of the following variables.

a. The number of gigabytes of data that a person uses on their mobile data plan.

b. A student's response on a faculty evaluation: strongly disagree, disagree, somewhat agree, agree, strongly agree.

c. A patient's temperature (in degrees Fahrenheit) who has a fever.

d. A customer's response to a survey asking for their preferred brand of golf shoes: Footjoy, Nike, Adidas, Puma, or Ecco.

**Example 2.3.2**

**Determining the Level of Measurement of a Variable**

**SOLUTION**

**a.** The number of gigabytes used has a meaningful zero point and the ratio of two values for gigabyte usage is meaningful. That is, one who uses 16 gigabytes when compared to one who uses 4 gigabytes, a ratio of 4 can be calculated indicating that the customer using 16 gigabytes has used four times as much data as the customer using 4 gigabytes. Thus, the number of gigabytes used is measured on the ratio scale.

**b.** The student's response on the faculty evaluation is measured on the ordinal scale since we can have some order associated with the responses but cannot perform arithmetic operations.

**c.** As stated earlier in the text, temperature (measured in degrees Celsius or Fahrenheit) is measured on the interval scale. In this case, 0 degrees Fahrenheit does not mean the absence of temperature which prevents us from calculating meaningful ratios.

**d.** The customer's response will be only the name of the shoe brand. We cannot perform any arithmetic operations or ranking. Additionally, by the name only, one brand cannot be considered more (or better) than the other. Thus, shoe brand is measured on the nominal scale.

## ✍ 2.3 Exercises

### Basic Concepts

1. What are qualitative data? Give an example.

2. What are quantitative data? Give an example.

3. Which levels of measurement are associated with qualitative data? Which levels are associated with quantitative data?

4. If data are quantitative, they can be further classified into two categories. Name and briefly describe these categories.

5. What is the difference between discrete and continuous data?

6. What is a level of measurement?

7. What are the four levels of measurement? Give an example of each.

8. For which level(s) of measurement is arithmetic appropriate?

9. What is the primary difference between nominal and ordinal data?

10. What is an arbitrary zero value? Which level of measurement has this property?

11. What is the fundamental difference between interval and ratio data?

12. Decision makers usually prefer to consider data that possess which level of measurement?

### Exercises

13. Identify the following variables as discrete or continuous.

    **a.** The number of doctors who wash their hands between patient visits.

    **b.** The amount of liquid consumed by the average American each day.

    **c.** The weight of a newborn baby at a local hospital.

    **d.** The time it takes a person to react to a stimulus.

    **e.** The number of voters who favor a particular candidate.

**14.** Identify the following variables as discrete or continuous.

   **a.** The number of on-time flights at the Hartsfield International Airport in Atlanta.

   **b.** The height of skyscrapers in New York City.

   **c.** The price of General Electric's common stock.

   **d.** The temperature of U.S. cities.

   **e.** The number of alcoholics who are men.

**15.** The results of a study investigating the nutritional status of mid-nineteenth century Americans were reported in "The Height and Weight of West Point Cadets: Dietary Changes in Antebellum America," in the *Journal of Economic History*. The data are based upon physical examination lists for West Point applicants from 1843 to 1894. Some of the information obtained from each cadet were his height, weight, the state from which the cadet was appointed, the occupation of the father, the income of the parents, and the type of home residence (city, town, or rural) of the cadet.

   **a.** List the different variables measured on the cadets.

   **b.** Which variables are quantitative and which are qualitative?

   **c.** Give the levels of measurement for these variables.

   **d.** Why is some method of data summary necessary here?

**16.** The major television networks regularly conduct polls in order to ascertain the feelings of Americans on current political issues. In May of 1993, such a poll was conducted by ABC concerning United States involvement in Bosnia. The respondent's gender, political affiliation, and opinion (approve, disapprove, or no opinion) on how President Clinton was handling the situation in Bosnia represented some of the information supplied by the respondent on the survey. Each respondent was also asked to rate the job that the news media had done (excellent, good, not so good, poor) in covering the situation in Bosnia.

   **a.** List the different variables measured on the respondents.

   **b.** Which variables are quantitative and which are qualitative?

   **c.** Give the levels of measurement for these variables.

   **d.** What are some problems associated with collecting data in polls such as the one described in this exercise?

**17.** Under most states' auto lemon laws, dealers or car makers must replace defective autos that aren't successfully repaired after three attempts or that remain in the shop for 30 days. The table below shows data for Hawaii for the year 2010, weighing car makers' lemons against statewide market share. Assume the "lemon index" is the share of the complaints divided by the total market share for each manufacturer.

| Lemon Index: Hawaii, 2010 | | | |
|---|---|---|---|
| **Best** | **Lemon Index** | **Worst** | **Lemon Index** |
| Toyota (includes Lexus) | 0.212 | Chrysler (includes Dodge and Jeep) | 6.512 |
| Honda | 0.462 | Kia | 2.750 |
| Ford (includes Lincoln) | 0.868 | GM (includes Chevrolet, GMC, Buick) | 2.375 |
| Nissan (includes Infinity) | 1.056 | BMW | 2.129 |
| Mazda | 1.833 | Hyundai | 2.000 |

**Source:** Hawaii.gov

Answer the following questions for the variable "Lemon Index".

   **a.** Are the data quantitative or qualitative? Why?

   **b.** What is the highest level of measurement these data could have?

18. Determine the level of measurement (nominal, ordinal, interval, or ratio) for each of the following variables.

   a. The temperature (in degrees Fahrenheit) of patients with pneumonia.

   b. The age at which the average male marries.

   c. Client satisfaction survey responses: Poor, Average, Good, and Excellent.

   d. The region of the U.S. in which an individual lives: North, South, East, or West.

   e. The number of people with a Type A personality.

19. Determine the level of measurement (nominal, ordinal, interval, or ratio) for each of the following variables.

   a. The time it takes for a student to complete an exam.

   b. Majors of randomly selected students at a university.

   c. The category which best describes how frequently a person eats chocolate: Frequently, Occasionally, Seldom, Never.

   d. The number of pounds of snack food eaten by an individual in his or her lifetime.

20. Given the table below on browser usage, what is the highest level of measurement that these data could have? Justify your answer.

| Browser Usage Share (%) | | | | |
|---|---|---|---|---|
| Month | Microsoft Internet Explorer | Mozilla Firefox | Google Chrome | Apple Safari |
| July 2010 | 60.74 | 22.91 | 7.16 | 5.09 |
| August 2010 | 60.48 | 22.90 | 7.50 | 5.15 |
| September 2010 | 59.62 | 22.97 | 7.99 | 5.27 |
| October 2010 | 59.18 | 22.83 | 8.50 | 5.36 |
| November 2010 | 58.44 | 22.76 | 9.26 | 5.55 |
| December 2010 | 57.08 | 22.81 | 9.98 | 5.89 |
| January 2011 | 56.00 | 22.75 | 10.70 | 6.30 |
| February 2011 | 56.77 | 21.74 | 10.93 | 6.36 |
| March 2011 | 55.92 | 21.80 | 11.57 | 6.61 |
| April 2011 | 55.11 | 21.63 | 11.94 | 7.15 |
| May 2011 | 54.27 | 21.71 | 12.52 | 7.28 |
| June 2011 | 54.84 | 21.20 | 12.72 | 7.41 |

# 2.4 Time Series Data and Cross-Sectional Data

## Time Series Data

Recall from Chapter 1, the science of statistics is divided into two categories: descriptive statistics and inferential statistics. Fundamental to the concept of statistical inference is the notion of population—the total collection of measurements. **Time series data** originate as measurements usually taken from some process over equally spaced intervals of time.

Because measurements are taken over time, the concept of a population gets a little blurry. Suppose we want to examine the divorce rate in the United States from 2000 through 2018. What is the population we are studying?

Presumably the members of the population under study would be the residents of the United States. In 2000 there were about 281 million people in this country. Now there are over 332 million. Moreover, the population from which the divorce data are drawn is certainly not fixed. Why is this important? If a population doesn't contain a fixed set of members or subjects, how can inferences be made about it? The concept of population is not sufficiently broad to cope with time series measurements.

**Definition**

**Time Series Data**

**Time series data** are measurements taken from a process over equally spaced intervals of time.

| Table 2.4.1 – Divorce Rate in the United States 2000–2018 (Per 1000 People) | | | | | |
|------|--------------|------|--------------|------|--------------|
| Year | Divorce Rate | Year | Divorce Rate | Year | Divorce Rate |
| 2000 | 4.0 | 2007 | 3.6 | 2014 | 3.2 |
| 2001 | 4.0 | 2008 | 3.5 | 2015 | 3.1 |
| 2002 | 3.9 | 2009 | 3.5 | 2016 | 3.0 |
| 2003 | 3.8 | 2010 | 3.6 | 2017 | 2.9 |
| 2004 | 3.7 | 2011 | 3.6 | 2018 | 2.9 |
| 2005 | 3.6 | 2012 | 3.4 | | |
| 2006 | 3.7 | 2013 | 3.3 | | |

**Definition**

**Stationary Process**

In a **stationary process** the time series varies around some central value and has approximately the same variation over the series.

Time series data originate from processes. Processes can be divided into two categories: stationary and nonstationary. All time series that are interesting vary, and the nature of the variability determines how the process is characterized. In a **stationary process** the time series varies around some central value and has approximately the same variation over the series. In a **nonstationary process** the time series possesses a **trend**—the tendency for the series to either increase or decrease over time.

**Definition**

**Nonstationary Process**

In a **nonstationary process** the time series possesses a trend—the series either increases over time or decreases over time.

The gross domestic product (GDP) is the value of the goods and services produced by the nation's economy less the value of the goods and services used up in production. GDP is also equal to the sum of personal consumption expenditures, gross private domestic investment, net exports of goods and services, and gross investment. Real values are inflation–adjusted estimates—that is, estimates that exclude the effects of price changes. The data in Table 2.4.2 represent the percent change in GDP from 2000 to 2019. Determine if this time series is stationary or nonstationary.

**Example 2.4.1**

**Identifying a Time Series as Stationary or Nonstationary**

**⋮ Data**

This data set can be found on stat.hawkeslearning.com under **Discovering Business Statistics, Second Edition > Data Sets > Real Gross Domestic Product**.

| Table 2.4.2 – Real Gross Domestic Product, Percent Change from Preceding Period, Quarterly, Seasonally Adjusted Annual Rate | | | | | |
|--------|-------------------|--------|-------------------|--------|-------------------|
| Period | Percent Change | Period | Percent Change | Period | Percent Change |
| Jan-00 | 1.5 | Oct-06 | 3.5 | Jul-13 | 3.2 |
| Apr-00 | 7.5 | Jan-07 | 0.9 | Oct-13 | 3.2 |
| Jul-00 | 0.5 | Apr-07 | 2.3 | Jan-14 | −1.1 |
| Oct-00 | 2.5 | Jul-07 | 2.2 | Apr-14 | 5.5 |
| Jan-01 | −1.1 | Oct-07 | 2.5 | Jul-14 | 5.0 |
| Apr-01 | 2.4 | Jan-08 | −2.3 | Oct-14 | 2.3 |
| Jul-01 | −1.7 | Apr-08 | 2.1 | Jan-15 | 3.8 |
| Oct-01 | 1.1 | Jul-08 | −2.1 | Apr-15 | 2.7 |
| Jan-02 | 3.5 | Oct-08 | −8.4 | Jul-15 | 1.5 |
| Apr-02 | 2.4 | Jan-09 | −4.4 | Oct-15 | 0.6 |
| Jul-02 | 1.8 | Apr-09 | −0.6 | Jan-16 | 2.3 |

| Table 2.4.2 – Real Gross Domestic Product, Percent Change from Preceding Period, Quarterly, Seasonally Adjusted Annual Rate | | | | | |
|---|---|---|---|---|---|
| Period | Percent Change | Period | Percent Change | Period | Percent Change |
| Oct-02 | 0.6 | Jul-09 | 1.5 | Apr-16 | 1.3 |
| Jan-03 | 2.2 | Oct-09 | 4.5 | Jul-16 | 2.2 |
| Apr-03 | 3.5 | Jan-10 | 1.5 | Oct-16 | 2.5 |
| Jul-03 | 7.0 | Apr-10 | 3.7 | Jan-17 | 2.3 |
| Oct-03 | 4.7 | Jul-10 | 3.0 | Apr-17 | 1.7 |
| Jan-04 | 2.2 | Oct-10 | 2.0 | Jul-17 | 2.9 |
| Apr-04 | 3.1 | Jan-11 | −1.0 | Oct-17 | 3.9 |
| Jul-04 | 3.8 | Apr-11 | 2.9 | Jan-18 | 3.8 |
| Oct-04 | 4.1 | Jul-11 | −0.1 | Apr-18 | 2.7 |
| Jan-05 | 4.5 | Oct-11 | 4.7 | Jul-18 | 2.1 |
| Apr-05 | 1.9 | Jan-12 | 3.2 | Oct-18 | 1.3 |
| Jul-05 | 3.6 | Apr-12 | 1.7 | Jan-19 | 2.9 |
| Oct-05 | 2.6 | Jul-12 | 0.5 | Apr-19 | 1.5 |
| Jan-06 | 5.4 | Oct-12 | 0.5 | Jul-19 | 2.6 |
| Apr-06 | 0.9 | Jan-13 | 3.6 | Oct-19 | 2.4 |
| Jul-06 | 0.6 | Apr-13 | 0.5 | | |

**Source:** U.S. Bureau of Economic Analysis, Real Gross Domestic Product [A191RL1Q225SBEA], retrieved from FRED, Federal Reserve Bank of St. Louis; https://fred.stlouisfed.org/series/A191RL1Q225SBEA, September 6, 2020.

### SOLUTION

The GDP data from Table 2.4.2 are plotted in Figure 2.4.1. Notice that the time series plot seems to fluctuate up and down around some central value and the dispersion around the central value is reasonably constant throughout the timeframe. Thus, the time series appears to be stationary.

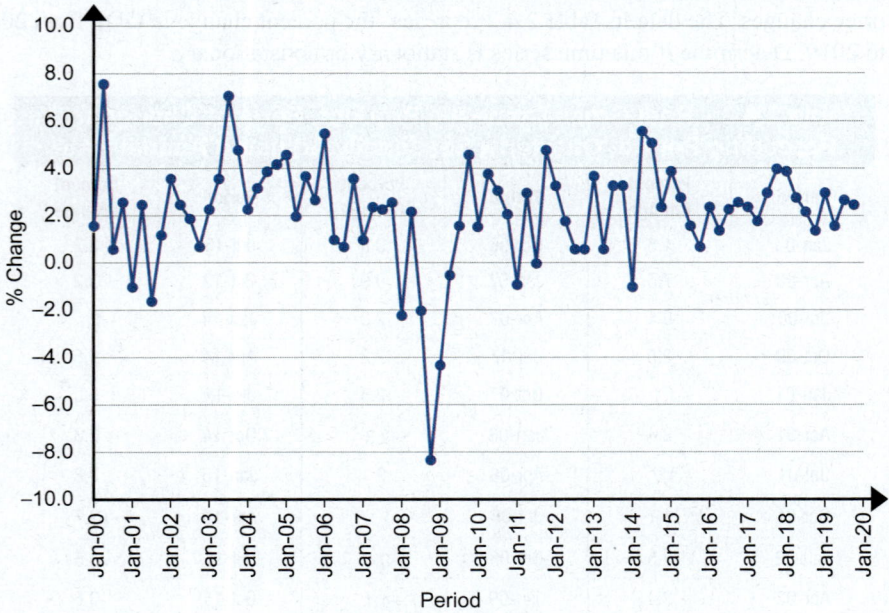

**Real Gross Domestic Product, Percent Change from Preceding Period, Quarterly, Seasonally Adjusted Annual Rate**

Figure 2.4.1

⅋ **Technology**

For instructions on how to create a line graph or time series plot in Excel, JMP, or other technologies, please visit stat.hawkeslearning.com and navigate to **Discovering Business Statistics, Second Edition > Technology Instructions > Graphs > Line Graphs**.

Determine if the time series corresponding to the divorce data given in Table 2.4.1 is stationary or nonstationary. Table 2.4.1 is repeated here for your convenience.

**Example 2.4.2**

**Identifying a Time Series as Stationary or Nonstationary**

| Table 2.4.1 – Divorce Rate in the United States 2000–2018 (Per 1000 People) | | | |
|---|---|---|---|
| Year | Divorce Rate | Year | Divorce Rate |
| 2000 | 4.0 | 2010 | 3.6 |
| 2001 | 4.0 | 2011 | 3.6 |
| 2002 | 3.9 | 2012 | 3.4 |
| 2003 | 3.8 | 2013 | 3.3 |
| 2004 | 3.7 | 2014 | 3.2 |
| 2005 | 3.6 | 2015 | 3.1 |
| 2006 | 3.7 | 2016 | 3.0 |
| 2007 | 3.6 | 2017 | 2.9 |
| 2008 | 3.5 | 2018 | 2.9 |
| 2009 | 3.5 | | |

**SOLUTION**

Figure 2.4.2 shows the divorce data from Table 2.4.1. There is strong evidence of a downward trend. Thus, the time series is nonstationary. (Note these data are not affected by population increases over the years since it is given in divorces per 1000 people.)

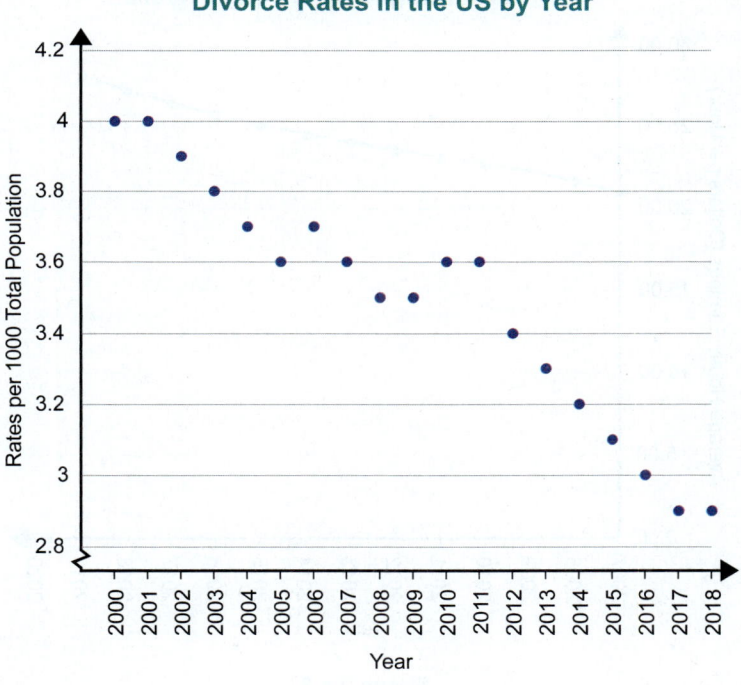

Figure 2.4.2

**Example 2.4.3**

**Identifying a Time Series as Stationary or Nonstationary**

Consider the following data regarding the average hourly earnings for all employees. Determine if the time series associated with these data is stationary or nonstationary.

| Table 2.4.3 – Average Hourly Earnings for All Employees, Total Private (US Dollars) | | | |
|---|---|---|---|
| Year | Average Hourly Earnings | Year | Average Hourly Earnings |
| 2007 | 20.91 | 2014 | 24.46 |
| 2008 | 21.56 | 2015 | 25.01 |
| 2009 | 22.17 | 2016 | 25.65 |
| 2010 | 22.58 | 2017 | 26.31 |
| 2011 | 23.03 | 2018 | 27.10 |
| 2012 | 23.47 | 2019 | 28.00 |
| 2013 | 23.96 | | |

**Source:** Federal Reserve Bank of St. Louis

### SOLUTION

As can be seen in Figure 2.4.3, the average hourly earnings for all employees has been on a steady increase since 2007. Given that the time series increases from year to year, this is obviously not a stationary series.

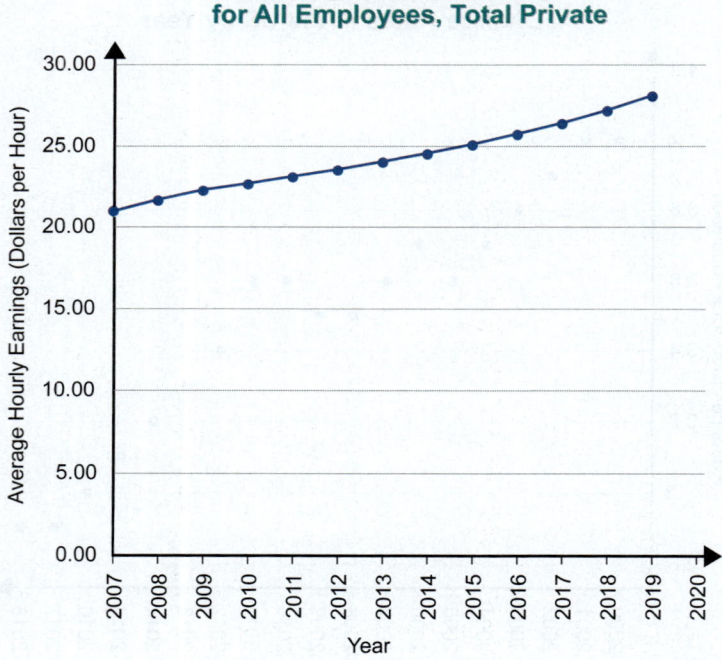

Figure 2.4.3

## Cross-Sectional Data

**Cross-sectional data** are measurements created at approximately the same period of time. For example, consider the life expectancy at birth (in 2020) for selected countries given in Table 2.4.4. These data represent cross-sectional measurements since the measurements were made in the same time period (2020). People in Andorra are expected to live on average until nearly 83 years old, approximately three years longer than the average for Americans. Developed countries such as Australia and the United States generally have higher life expectancies than developing countries such as Botswana and Kenya. But according to the World Health Organization, life expectancies in developing countries are on the rise due to medical interventions based on advanced technology and drugs. In fact, developing countries are expected to experience a 200 to 300 percent increase in their elderly populations in the next 35 years.

### Table 2.4.4 – Life Expectancy at Birth, 2020

| Country | Life Expectancy (Years) |
|---|---|
| Afghanistan | 52.8 |
| Andorra | 83.0 |
| Australia | 82.7 |
| Botswana | 64.8 |
| Egypt | 73.7 |
| Guatemala | 72.4 |
| Kenya | 69.0 |
| Sri Lanka | 77.5 |
| Sweden | 82.4 |
| United Kingdom | 81.1 |
| United States | 80.3 |

**Source:** The World Fact Book

**⠿ Data**

This data set is part of a larger data set of life expectancies for 227 countries. The entire data set of life expectancies can be found on stat.hawkeslearning.com under **Discovering Business Statistics, Second Edition > Data Sets > Life Expectancy 2020**.

Another example of cross-sectional data is greenhouse gas emissions of each state on a per capita basis. The air that surrounds the earth consists of a mixture of oxygen and nitrogen, interspersed with small amounts of carbon dioxide, methane, and other trace gases. The trace gases capture heat as the sun warms the earth and holds part of that heat in the atmosphere in what is known as the "greenhouse effect."

These gases prevent the sun's heat from simply hitting the ground and being rechanneled back into space. For most of the last several thousand years, the earth's greenhouse gases have been stable. The most abundant trace gas, carbon dioxide, was processed by plants and maritime organisms at approximately the same rate as it was given off by other organisms, until about the time of the Industrial Revolution. The natural world's ability to absorb carbon dioxide has been unable to keep up, and consequently the level of carbon dioxide in the atmosphere has been rising. A time series of the atmospheric carbon dioxide concentrations at Mauna Loa Observatory reveals a disturbing trend, as seen in Figure 2.4.4.

**Definition**

**Cross-Sectional Data**

**Cross-sectional data** are measurements created at approximately the same period of time.

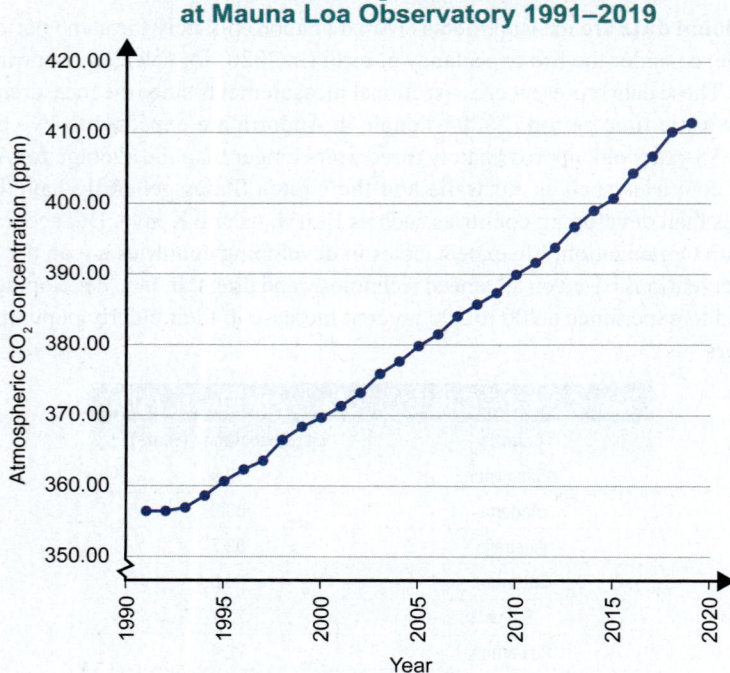

Atmospheric $CO_2$ Concentrations (ppm)
at Mauna Loa Observatory 1991–2019

**Source:** Carbon Dioxide Research Group, Scripps Institution of Oceanography,
University of California

**Figure 2.4.4**

The U.S. produces 15.7% of the world's carbon dioxide. It would be interesting to examine
the greenhouse emissions of each state on a per capita basis.

| Table 2.4.5 – Per Capita Carbon Dioxide Emissions by State in 2017 (Metric Tons) | | | | | |
|---|---|---|---|---|---|
| **State** | **Emissions** | **State** | **Emissions** | **State** | **Emissions** |
| Wyoming | 105.4 | Kansas | 20.0 | Virginia | 11.6 |
| North Dakota | 74.9 | Utah | 18.9 | Maine | 11.6 |
| West Virginia | 50.3 | Ohio | 17.6 | New Jersey | 11.4 |
| Louisiana | 48.8 | Wisconsin | 17.0 | North Carolina | 11.3 |
| Alaska | 46.3 | Pennsylvania | 16.9 | Florida | 10.9 |
| Montana | 28.9 | South Dakota | 16.7 | Idaho | 10.8 |
| Indiana | 26.6 | Minnesota | 15.9 | Washington | 10.6 |
| Kentucky | 25.7 | Illinois | 15.8 | New Hampshire | 10.0 |
| Texas | 25.1 | Colorado | 15.6 | Rhode Island | 9.5 |
| Nebraska | 25.0 | Michigan | 15.3 | Connecticut | 9.4 |
| Iowa | 24.4 | Tennessee | 14.6 | Vermont | 9.3 |
| Oklahoma | 23.7 | South Carolina | 13.8 | Oregon | 9.3 |
| New Mexico | 23.4 | Delaware | 12.9 | Massachusetts | 9.3 |
| Mississippi | 22.7 | Georgia | 12.7 | California | 9.2 |
| Alabama | 22.4 | Hawaii | 12.4 | Maryland | 8.6 |
| Arkansas | 21.4 | Nevada | 12.2 | New York | 8.1 |
| Missouri | 20.2 | Arizona | 12.2 | District of Columbia | 3.8 |

**Source:** U.S. Energy Information Administration

## ✏ 2.4 Exercises

### Basic Concepts

1. What are time series measurements?

2. What problems are associated with the concept of population when studying time series data?

3. What is a stationary process?

4. What is a nonstationary process?

5. What is a trend? If a time series has an 'upward trend' what does this mean?

6. What are cross-sectional data?

7. What is the difference between cross-sectional data and time series data?

### Exercises

8. Consider the following graph of long-term interest rates (10-year treasury notes) and inflation rates.

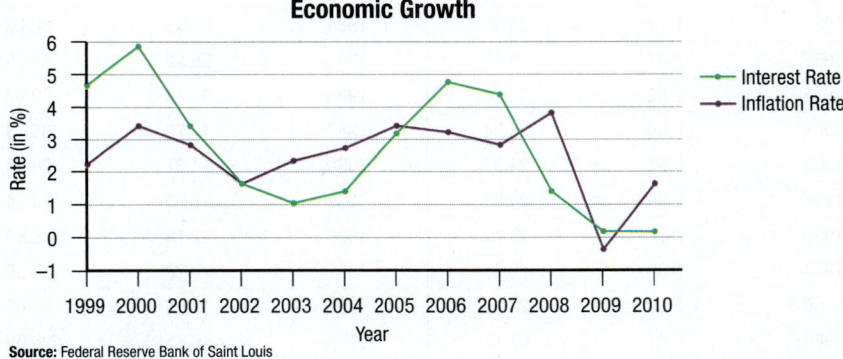

Source: Federal Reserve Bank of Saint Louis

   a. Are the interest rate data presented above time series data?

   b. Are the inflation rate data presented above time series data?

   c. For each of parts a. and b. if the data is time series data, does the series appear to be stationary or nonstationary?

9. Consider the following graph of total exports.

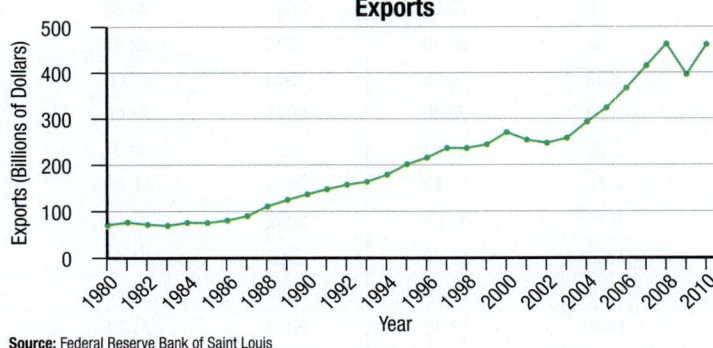

Source: Federal Reserve Bank of Saint Louis

   a. Are these data time series data?

   b. If the data are time series data, does the series appear to be stationary or nonstationary?

10. Using a newspaper, journal, or website as your source, give an example of time series data. Be sure to reference your source and give a brief description of the data.

11. The following table shows the annual average crude oil price from 1946 through 2011. Prices are adjusted for inflation to April 2011 prices using the Consumer Price Index (CPI-U) as presented by the Bureau of Labor Statistics. Inflation adjusted prices were at an all-time high in 1980, reaching $102.26 dollars per barrel. Crude oil prices reached an all-time low in 1998 (lower than the price in 1946!) when the price per barrel dipped to $16.44. Using the data in the table, discuss if the data set contains time series or cross-sectional data. Also, discuss the data and make some inferences. That is, can you explain some of the fluctuations in the oil prices?

| Annual Average Domestic Crude Oil Prices ($ per Barrel) | | | | | |
|---|---|---|---|---|---|
| Year | Nominal | Inflation Adjusted (April 2011) | Year | Nominal | Inflation Adjusted (April 2011) |
| 1946 | 1.63 | 18.49 | 1979 | 25.10 | 77.05 |
| 1947 | 2.16 | 21.73 | 1980 | 37.42 | 102.26 |
| 1948 | 2.77 | 25.92 | 1981 | 35.75 | 88.55 |
| 1949 | 2.77 | 26.17 | 1982 | 31.83 | 74.24 |
| 1950 | 2.77 | 25.90 | 1983 | 29.08 | 65.69 |
| 1951 | 2.77 | 24.00 | 1984 | 28.75 | 62.26 |
| 1952 | 2.77 | 23.47 | 1985 | 26.92 | 56.28 |
| 1953 | 2.92 | 24.50 | 1986 | 14.44 | 29.62 |
| 1954 | 2.99 | 25.04 | 1987 | 17.75 | 35.13 |
| 1955 | 2.93 | 24.57 | 1988 | 14.87 | 28.32 |
| 1956 | 2.94 | 24.35 | 1989 | 18.33 | 33.24 |
| 1957 | 3.14 | 25.12 | 1990 | 23.19 | 39.80 |
| 1958 | 3.00 | 23.38 | 1991 | 20.20 | 33.36 |
| 1959 | 3.00 | 23.15 | 1992 | 19.25 | 30.85 |
| 1960 | 2.91 | 22.15 | 1993 | 16.75 | 26.09 |
| 1961 | 2.85 | 21.44 | 1994 | 15.66 | 23.76 |
| 1962 | 2.85 | 21.19 | 1995 | 16.75 | 24.73 |
| 1963 | 2.91 | 21.39 | 1996 | 20.46 | 29.32 |
| 1964 | 3.00 | 21.75 | 1997 | 18.64 | 26.12 |
| 1965 | 3.01 | 21.47 | 1998 | 11.91 | 16.44 |
| 1966 | 3.10 | 21.48 | 1999 | 16.56 | 22.30 |
| 1967 | 3.12 | 21.04 | 2000 | 27.39 | 35.76 |
| 1968 | 3.18 | 20.53 | 2001 | 23.00 | 29.23 |
| 1969 | 3.32 | 20.36 | 2002 | 22.81 | 28.50 |
| 1970 | 3.39 | 19.65 | 2003 | 27.69 | 33.86 |
| 1971 | 3.60 | 20.00 | 2004 | 37.66 | 44.81 |
| 1972 | 3.60 | 21.44 | 2005 | 50.04 | 57.57 |
| 1973 | 4.75 | 23.87 | 2006 | 58.30 | 65.03 |
| 1974 | 9.35 | 42.58 | 2007 | 64.20 | 69.51 |
| 1975 | 12.21 | 51.00 | 2008 | 91.48 | 95.25 |
| 1976 | 13.10 | 51.78 | 2009 | 53.48 | 55.96 |
| 1977 | 14.40 | 53.41 | 2010 | 71.21 | 73.44 |
| 1978 | 14.95 | 51.58 | 2011 (Partial) | 86.84 | – |

**Source:** www.inflationdata.com

12. Do you think the pay of executives working for digital companies increases/decreases as the company's stock price increases/decreases? Examine the following table.

| CEO Compensation and Stock Performance | | | | | | |
|---|---|---|---|---|---|---|
| Exec | Salary/ Bonus ($) | Stock/ Options ($) | Other Non-Equity Compensation ($) | Total 2007 Compensation ($) | Change from 2006 Compensation (%) | 2007 Stock Performance (%) |
| Tom Rogers (Tivo) | 800,000 | 6,200,000 | 495,075 | 7,495,075 | +102 | +32 |
| Mel Karmazin (Sirius) | 5,250,000 | – | 18,743 | 5,268,743 | +23 | −23 |
| Paul Sagan (Akamai) | 403,651 | 3,554,264 | 497,362 | 4,455,277 | −40 | −48 |
| Reed Hastings (Netflix) | 850,000 | 1,568,307 | 270 | 2,418,577 | +5 | −6 |
| Rob Glaser (RealNetworks) | 1,169,384 | 643,400 | 354,200 | 2,166,984 | −26 | −45 |
| Bobby Kotick (Activision Blizzard) | 899,560 | 1,188,467 | – | 2,088,027 | +6 | +49 |
| Magid M. Abraham (comScore) | 421,952 | 1,125,000 | – | 1,546,952 | +185 | −16 |
| Barry Diller (IAC) | 500,000 | – | 927,429 | 1,427,429 | +270 | +21 |
| John S. Riccitiello (Electronic Arts) | 750,000 | – | 625,350 | 1,375,350 | −37 | −38 |
| Steve Ballmer (Microsoft) | 1,340,833 | – | 10,001 | 1,350,834 | N/A | 0 |
| Wayne T. Gattinella (WebMD) | 830,000 | – | 9214 | 839,214 | +6 | +10 |

**Source:** paidContent.org

What type of data is in the Salary/Bonus column? What do you think about executive salaries as a function of the company's stock performance? Justify your responses.

## T   Discovering Technology

**Note:** If you are not familiar with Excel, Appendix B is a tutorial section that will familiarize you with the controls in Excel.

### Using Excel

Line Chart

For this exercise, use the information from Example 2.4.1.

1. Enter the time period data into Column A – label as "Period".

2. Enter the percentage change in real gross domestic product into Column B – label as "Percentage Change".

3. Highlight the data in Columns A and B. On the **Insert** tab, select **Insert Line or Area Chart** and then **Line with Markers**.

4. Click the title and replace it with "Real Gross Domestic Product, Percent Change from Preceding Period, Quarterly, Seasonally Adjusted Annual Rate".

5. To format the *y*-axis, right-click the axis numbers and select **Format Axis**. Under **Horizontal axis crosses**, click the **Axis value** button and enter −10 so the horizontal axis crosses below all the data.

6. Click the chart. On the **Chart Design** tab, select **Add Chart Element**, and select **Axis Titles > Primary Vertical** and type "% Change".

7. Click the chart. On the **Chart Design** tab, select **Add Chart Element**, and select **Axis Titles > Primary Horizontal** and type "Period".

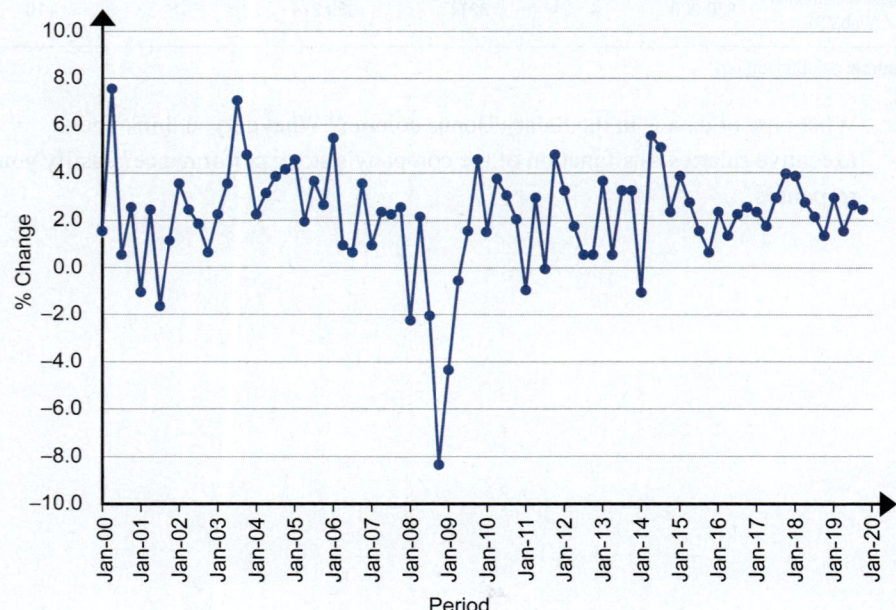

## Using JMP

## Line Graphs

For this exercise, use the data in Table 2.4.1 on Divorce Rate in the United States.

1.  Enter the data for the year into the first column of the JMP worksheet. Double click the label **Column 1** and input "**Year**" as the **Column Name**. Click **OK**.

2.  Enter the data for the divorce rate into the second column of the JMP worksheet. Double click the label **Column 2** and input "**Divorce Rate**" as the **Column Name**. Click **OK**.

3.  Click on **Graph** in the top row of the JMP spreadsheet and then select **Graph Builder**.

4.  Under **Variables**, click and drag the variable **Year** to the area at the bottom labeled **X**.

5.  Click and drag the variable **Divorce Rate** to the area on the left labeled **Y**.

6.  To change the *x*-axis scale to show each year, right-click on the *x*-axis of the graph and select **Axis Settings**. Choose **1** as the **Increment** and **0** as the **# Minor Ticks**. Click **OK**.

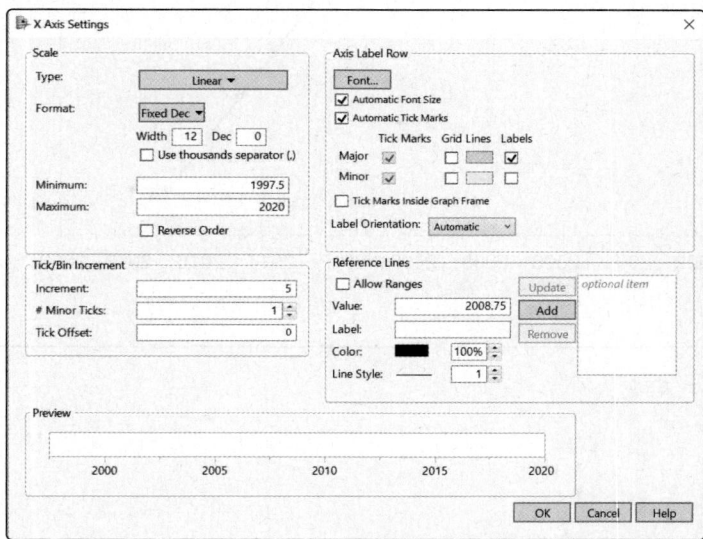

7. Since **Graph Builder** automatically adds a **Smoother** curve, to instead connect the points and remove the smoother, right-click in the graph area and select **Smoother**, **Change to**, and then **Line**.

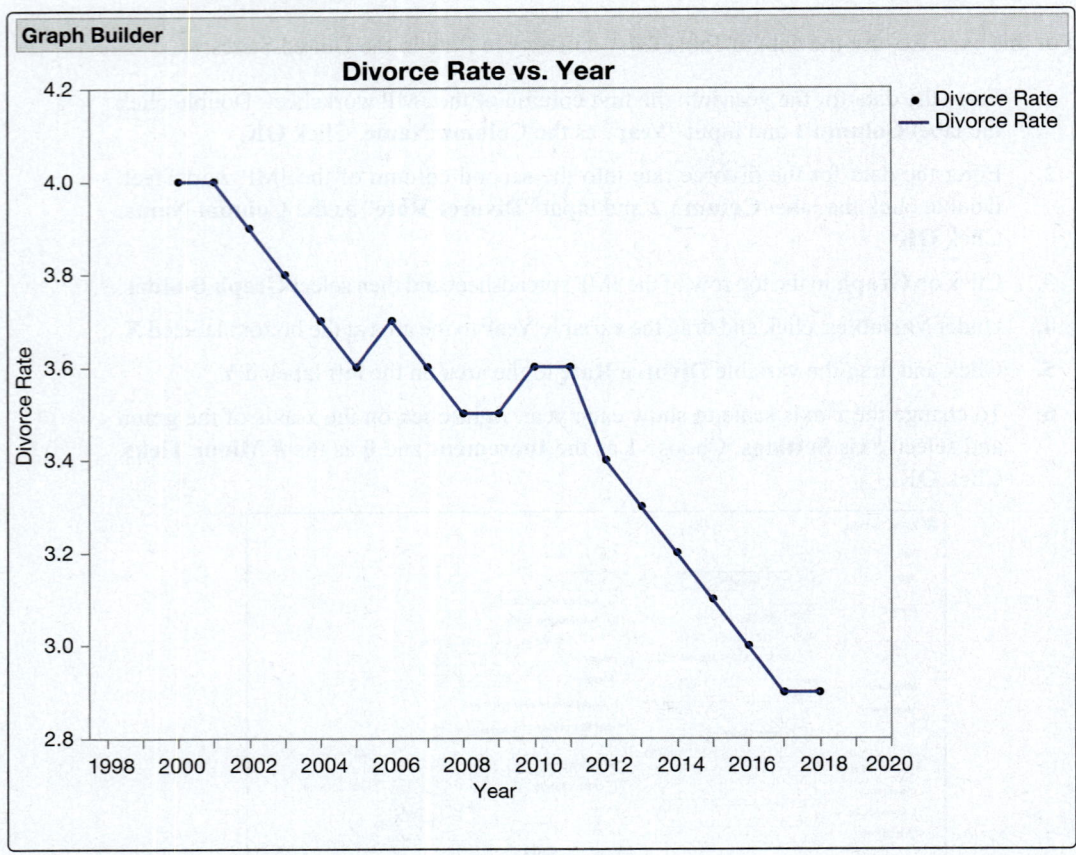

## Using Minitab

Time Series Plot

For this exercise, we will use the data from Table 2.4.1.

1. Enter the years into Column C1.

2. Enter the Divorce Rate into Column C2.

3. Under **Stat**, choose **Time Series**, then **Time Series Plots**

4. Then choose **Simple** and click **OK**

5. Enter **C2** in the box for Series.

6. Click **Time/Scale**, choose **Stamp** and enter **C1** for the Stamp Columns. Click **OK**

7. Click **OK** in the Series page to display the plot.

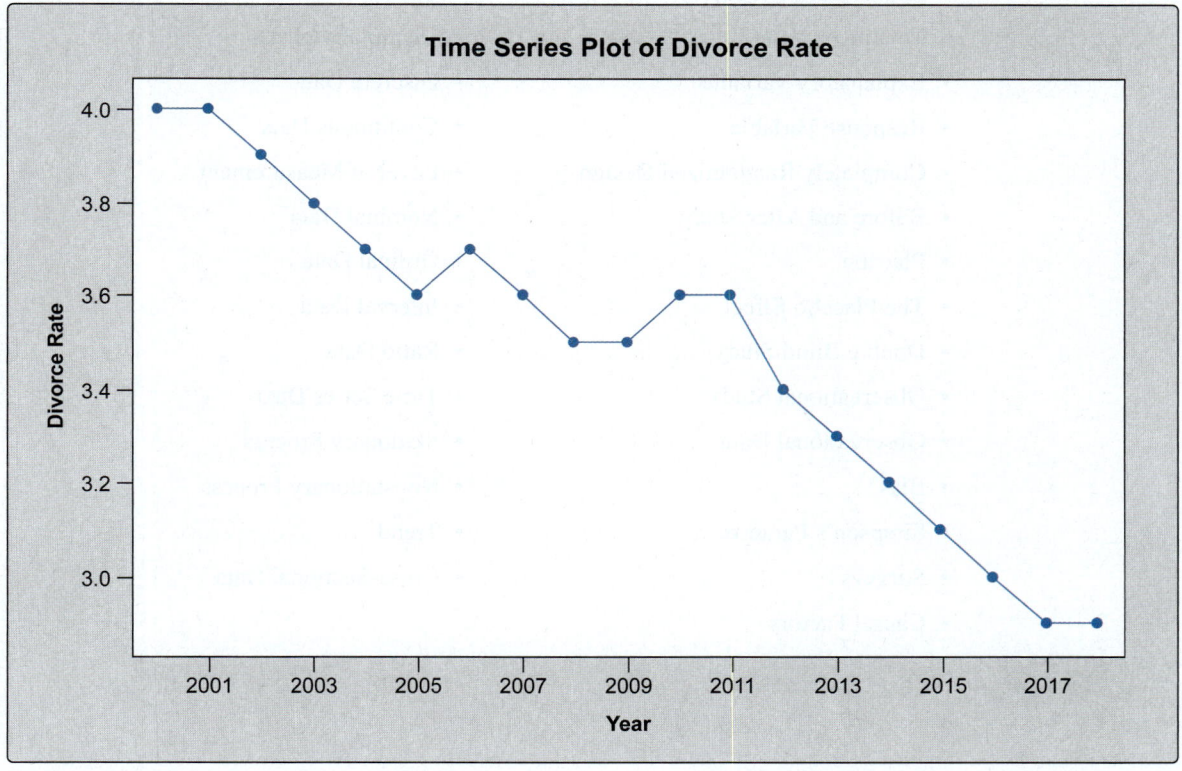

# R    Chapter 2 Review

## Key Terms and Ideas

- The Scientific Method
- Confounding Variable
- Descriptive Statistics
- Inferential Statistics
- The Decision-Making Method
- Fuzzy Concepts
- Controlled Experiment
- Control Group
- Experimental Group
- Treatment
- Explanatory Variable
- Response Variable
- Completely Randomized Design
- Before and After Study
- Placebo
- The Placebo Effect
- Double Blind Study
- Observational Study
- Observational Data
- Bias
- Simpson's Paradox
- Surveys
- Causal Factors
- Big Data
- Structured Data
- Unstructured Data
- Semi-Structured Data
- Analytics
- Business Analytics
- Predictive Analytics
- Prescriptive Analytics
- Qualitative Data
- Quantitative Data
- Discrete Data
- Continuous Data
- Level of Measurement
- Nominal Data
- Ordinal Data
- Interval Data
- Ratio Data
- Time Series Data
- Stationary Process
- Nonstationary Process
- Trend
- Cross-Sectional Data

# AE  Additional Exercises

1. Suppose you were the administrator of a public school system. What kinds of variables would you measure and how would you collect the measurements on the following subjects:

   a. Student learning

   b. School discipline

   c. Teacher preparation

   d. Absenteeism (pupil and teacher)

   e. Cafeteria food quality

2. The head of the Veterans Administration has been receiving complaints from a Vietnam Veterans organization concerning disability checks. The organization claims that checks are continually late. The checks are to arrive no later than the tenth of each month.

   a. What variables would you measure to explore this problem?

   b. How would you collect measurements on these variables?

3. A family member has unexpectedly bequeathed you a sizable sum of money.

   a. What criteria might you wish to evaluate in deciding how to invest the money?

   b. What data might be useful in your considerations?

4. Flying Eagle Airlines advertises that it surpasses all other airlines in flights that arrive on time. A competitor states that it has a better on-time record than any other airline. Can they both be correct? Explain.

5. Two local grocery stores both claim to have the lowest prices in town. Develop a measurement that you believe could be used as a criterion to determine which store actually has the lowest prices.

6. At the end of 2001, the United States had 32.9 million people living in poverty according to the Census Bureau (www.census.gov). This was an increase of 1.3 million from the previous year. Poverty was defined by the Census Bureau as having a cash income less than $14,255 a year. The Census Bureau does not include in their income measurement any part of $167 billion spent on Medicaid, a federal program by which medical care is provided to the poor. The Census Bureau only includes $34.9 billion out of the $205 billion spent annually on public welfare. Forty percent of those classified as impoverished own their own homes. How do you think poverty should be defined?

7. The quality movement has compelled American businesses to address the problem of measuring customer satisfaction. How would you measure customer satisfaction if you owned a car dealership?

8. Identify the following variables as discrete or continuous.

   a. Average test score on a test ranging from 0 to 100

   b. Number of boot errors on a computer

   c. Investment ratios for earnings per share

   d. Energy usage in a production process

9. Determine the level of measurement for each of the following variables.

   a. Golf score in relation to par

   b. SAT score

   c. Rating from 1 to 5 of quality of service in a restaurant

   d. Make and model of a vehicle

   e. The number of students with a business major

10. According to a Danish researcher, if you drop your average daily activity level by taking elevators instead of stairs, by parking your car in the closest space, or by never walking to run errands, you increase your risk of diabetes, heart disease, and premature death. The researcher studied two groups of healthy men (eight in the first group with an average age of 27 and an average body mass index (BMI) of 22.9, which is well within the normal range; and ten in the second group with an average age of 23.8 years and a BMI of 22.1). In addition to age and BMI, researchers also collected information such as number of steps per day (each group of men was fitted with pedometers), height, weight, and race. With the first group of men, the researchers asked that they reduce their daily activity (steps) by taking cars on short trips and elevators instead of stairs. The insulin levels were also measured for each group and the researchers found that with the reduced activity, insulin levels rose by nearly 60 percent after two weeks of inactivity, thus increasing the risk of diabetes and heart disease. However, the good news is that by increasing activity over a two-week period of time, one can begin to reduce his or her risk of diabetes and heart disease.
Source: U.S. News and World Report

   a. List the different variables measured in this study.

   b. Which variables are quantitative and which are qualitative?

   c. Of the variables that are quantitative, are they discrete or continuous?

   d. Give the levels of measurement for these variables.

   e. Why is some method of data summary necessary here?

11. Consider the world production of crude oil given in millions of barrels per day.

| World Production of Crude Oil | | | |
|---|---|---|---|
| Year | Total World Production (Millions of Barrels per Day) | Year | Total World Production (Millions of Barrels per Day) |
| 1980 | 63.987 | 1995 | 70.274 |
| 1981 | 60.602 | 1996 | 71.919 |
| 1982 | 58.098 | 1997 | 74.160 |
| 1983 | 57.934 | 1998 | 75.656 |
| 1984 | 59.568 | 1999 | 74.853 |
| 1985 | 59.172 | 2000 | 77.768 |
| 1986 | 61.407 | 2001 | 77.686 |
| 1987 | 62.086 | 2002 | 76.994 |
| 1988 | 64.380 | 2003 | 79.598 |
| 1989 | 65.508 | 2004 | 83.105 |
| 1990 | 66.426 | 2005 | 84.595 |
| 1991 | 66.399 | 2006 | 84.661 |
| 1992 | 66.564 | 2007 | 84.543 |
| 1993 | 67.091 | 2008 | 85.507 |
| 1994 | 68.590 | 2009 | 84.389 |

Source: Energy Information Administration

    **a.** What is the level of measurement of the data?

    **b.** Are the data time series or cross-sectional? If the data are time series, plot the data. Does the series appear to be stationary or nonstationary? Explain your answer.

**12.** Consider the graph of the number of respondents (in percentages) who think things in the U.S. are now on the wrong track versus those that think the economy is going in the right direction. The data were collected using a survey asking the question, *In general, are you satisfied or dissatisfied with the way things are going in the United States at this time?*

**Source:** Gallup Poll

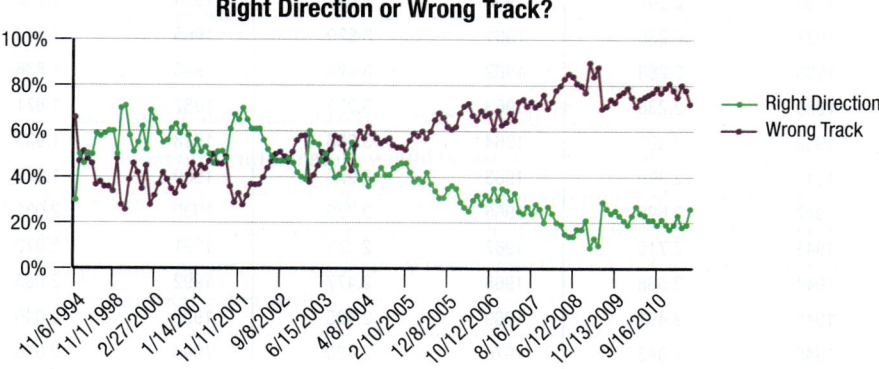

**Right Direction or Wrong Track?**

    **a.** Are the opinions on the outlook of the economy presented in time series or cross-sectional data? Justify your answer.

    **b.** If the data are time series data, does the series appear to be stationary or nonstationary? Explain your answer.

**13.** Can you think of a process that would yield measurements that did not have any variability? Would studying such a process be very interesting?

**Discovery Project**

## Bachelor's Degrees Conferred by Race/Ethnicity

Review the following graph using data from IPEDS (Integrated Postsecondary Education Data) collected from Fall 2000 through Fall 2016 and answer the questions below.

| Bachelor's Degrees Conferred by Postsecondary Institutions, by Race/Ethnicity of Student Selected Years: 2000 through 2016 | | | | | |
|---|---|---|---|---|---|
| Year | White | Black | Hispanic | Asian/Pacific Islander | American Indian/Alaskan Native |
| 2000 | 929,102 | 108,018 | 75,063 | 77,909 | 8,717 |
| 2001 | 927,357 | 111,307 | 77,745 | 78,902 | 9,049 |
| 2002 | 958,597 | 116,623 | 82,966 | 83,093 | 9,165 |
| 2003 | 994,616 | 124,253 | 89,029 | 87,964 | 9,875 |
| 2004 | 1,026,114 | 131,241 | 94,644 | 92,073 | 10,638 |
| 2005 | 1,049,141 | 136,122 | 101,124 | 97,209 | 10,307 |
| 2006 | 1,075,561 | 142,420 | 107,588 | 102,376 | 10,940 |
| 2007 | 1,099,850 | 146,653 | 114,936 | 105,297 | 11,455 |
| 2008 | 1,122,675 | 152,457 | 123,048 | 109,058 | 11,509 |
| 2009 | 1,144,628 | 156,603 | 129,473 | 112,581 | 12,221 |
| 2010 | 1,167,322 | 164,789 | 140,426 | 117,391 | 12,405 |
| 2011 | 1,182,690 | 172,731 | 154,450 | 121,118 | 11,935 |
| 2012 | 1,212,417 | 185,916 | 169,736 | 126,177 | 11,498 |
| 2013 | 1,221,908 | 191,233 | 186,677 | 130,129 | 11,432 |
| 2014 | 1,218,998 | 191,437 | 202,425 | 131,662 | 10,784 |
| 2015 | 1,210,071 | 192,829 | 218,098 | 133,916 | 10,202 |
| 2016 | 1,197,399 | 194,473 | 235,014 | 138,270 | 9,737 |

**Source:** U.S. Department of Education, National Center for Education Statistics, Higher Education General Information Survey (HEGIS), "Degrees and Other Formal Awards Conferred" surveys, 1976-77 and 1980-81; Integrated Postsecondary Education Data System (IPEDS), "Completions Survey" (IPEDS-C:90-99); and IPEDS Fall 2000 through Fall 2016, Completions component. (This table was prepared August 2017).

⋮ **Data**

This data set can be found on stat.hawkeslearning.com under **Discovering Business Statistics, Second Edition > Data Sets > Bachelor's Degrees Conferred by Race and Ethnicity**.

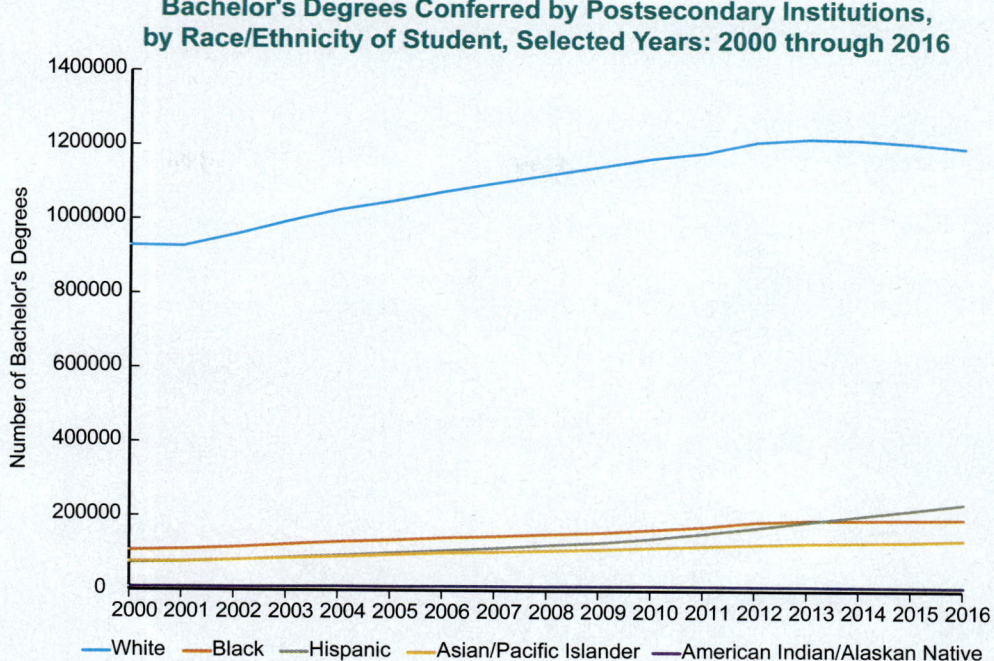

**Bachelor's Degrees Conferred by Postsecondary Institutions, by Race/Ethnicity of Student, Selected Years: 2000 through 2016**

1. Are the data displayed in the graph above discrete or continuous?

2. What is the level of measurement of the data?

3. Are the data above time series or cross-sectional data?

4. Examine the data for each race/ethnicity group. Do the data represent a stationary or nonstationary process? Explain your reasoning for each.

5. Do any of the race/ethnicity groups exhibit a decreasing trend? Try to think of some reasons why this is true.

6. Do any of the race/ethnicity groups show a strictly increasing trend over the entire time period from 2000 to 2016?

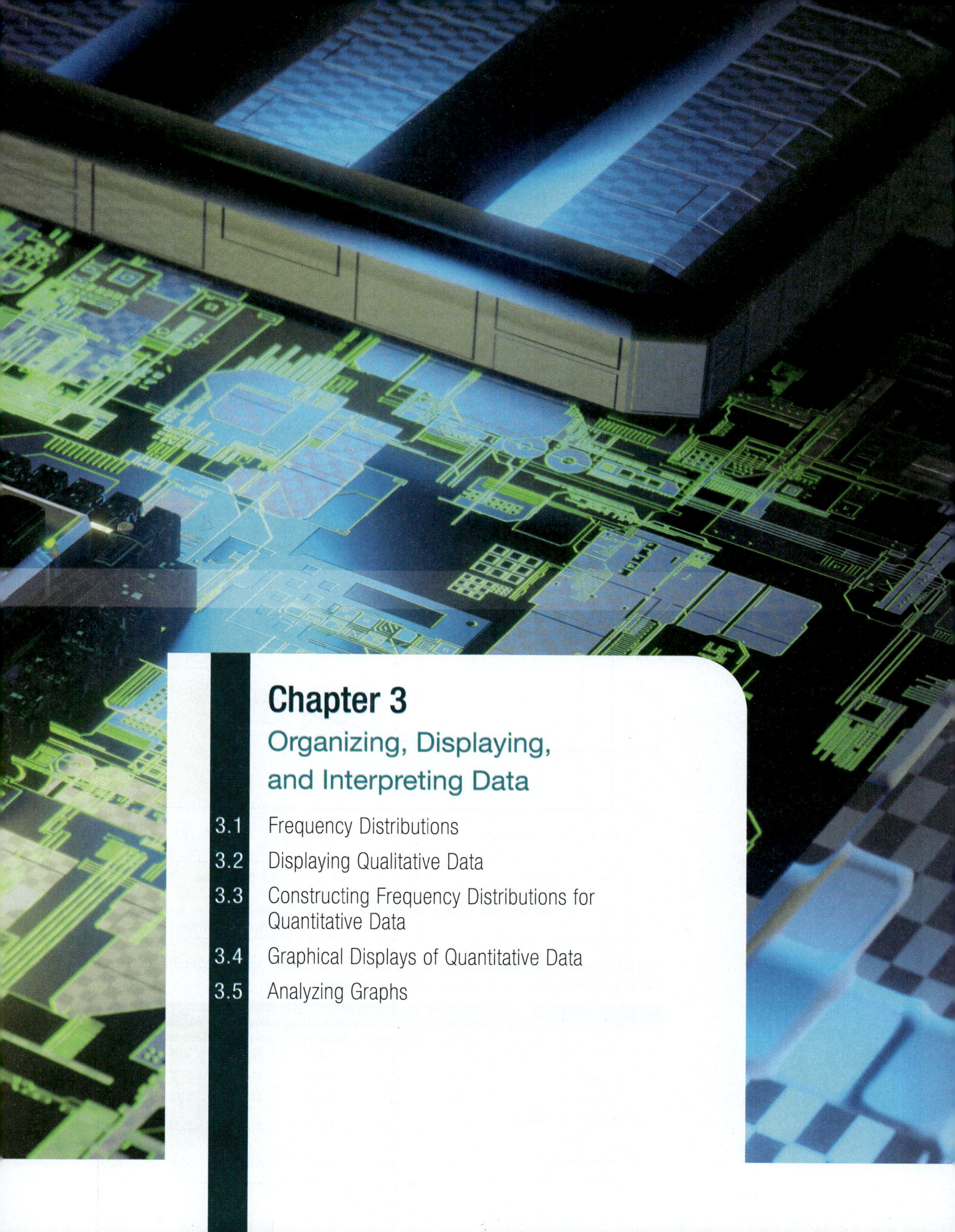

# Chapter 3

## Organizing, Displaying, and Interpreting Data

# Discovering the Real World

Apple Inc., a company traded on the United States New York Stock Exchange, is one of the world's most valuable organizations, surpassing a valuation of $2 trillion in August 2020. The company designs, manufactures, and markets smartphones, personal computers, tablets, wearables, and accessories, and sells a variety of related services (such as digital content stores and streaming, fee-based services that extend coverage of its products, cloud service, and licensing). Apple's customers are primarily in the consumer, small and mid-sized business, education, enterprise, and government markets. The company employs a variety of indirect distribution channels, such as third-party cellular network carriers, wholesalers, retailers, and resellers.

In sharing the company's performance with its investors, Apple Inc. will normally provide a prospectus that summarizes the previous year's income and expenses as well as compare performances year-over-year and with other major stock indexes.

The graph below shows a comparison of cumulative total shareholder return, calculated on a dividend-reinvested basis, against the S&P 500 Index, the S&P Information Technology Index, and the Dow Jones U.S. Technology Supersector Index for five years ending September 28, 2019.

## COMPARISON OF 5-YEAR CUMULATIVE TOTAL RETURN
### Among Apple Inc., the S&P 500 Index, the S&P Information Technology Index, and the Dow Jones U.S. Technology Supersector Index

The graph assumes $100 was invested in each of the company's common stock as of the market close on September 26, 2014.

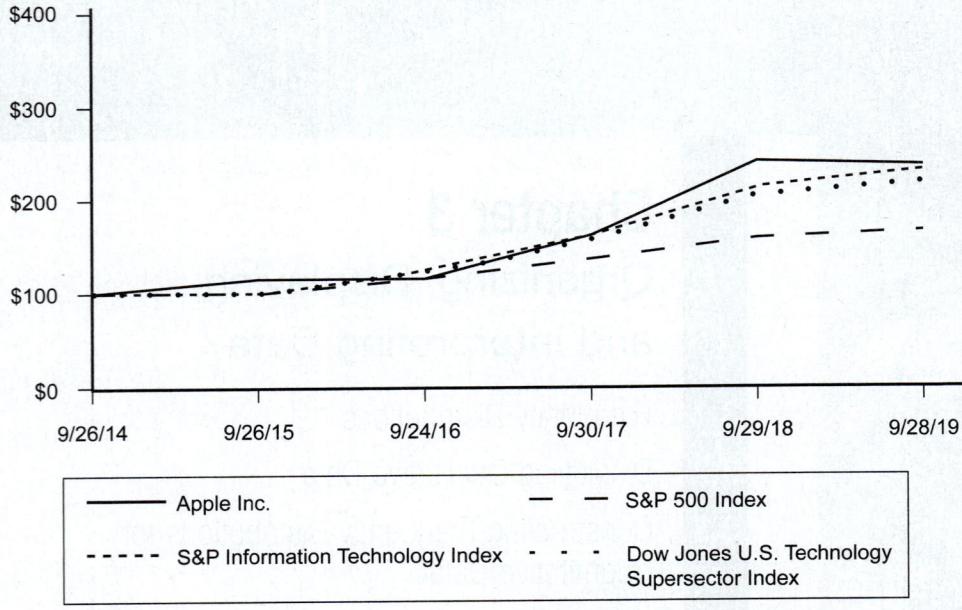

As can be seen in the graph above and the table below, Apple Inc.'s 5-year cumulative return has outperformed the three major technology indexes.

| Comparison of 5-Year Cumulative Total Return | | | | | | |
|---|---|---|---|---|---|---|
| | September 2014 | September 2015 | September 2016 | September 2017 | September 2018 | September 2019 |
| Apple Inc. | $100 | $116 | $116 | $162 | $240 | $237 |
| S&P 500 Index | $100 | $99 | $115 | $136 | $160 | $167 |
| S&P Information Technology Index | $100 | $102 | $125 | $162 | $213 | $231 |
| Dow Jones U.S. Technology Supersector Index | $100 | $100 | $122 | $156 | $205 | $218 |

The side-by-side bar graphs below depict Apple's sales by category. Beginning in the first quarter of 2019, Apple classified the amortization of the deferred value of Maps, Siri, and free iCloud services, which are bundled in the sales price of iPhone, Mac, iPad, and certain other products, into Services net sales. Historically, the company classified the amortization of these amounts in Products net sales consistent with its management reporting framework. As a result, Products and Services net sales for 2018 and 2017 were reclassified to conform to the 2019 presentation.

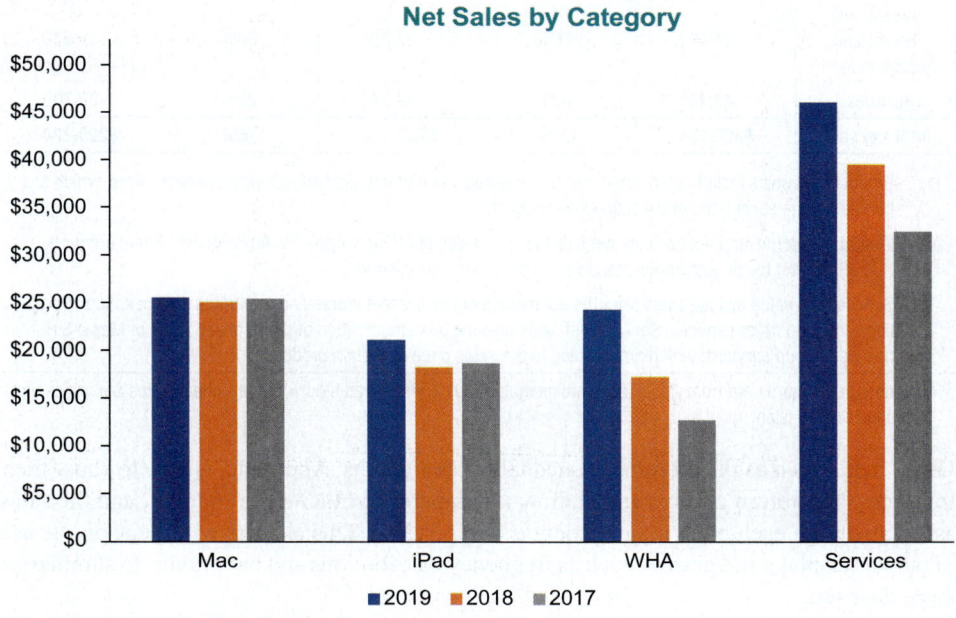

Net Sales by Category

Note: WHA = Wearables, Home, and Accessories

By viewing the graphs, it can be seen that iPhone net sales decreased during 2019 compared to 2018. However, Mac, iPad, WHA, and Services all increased during 2019 compared to 2018.

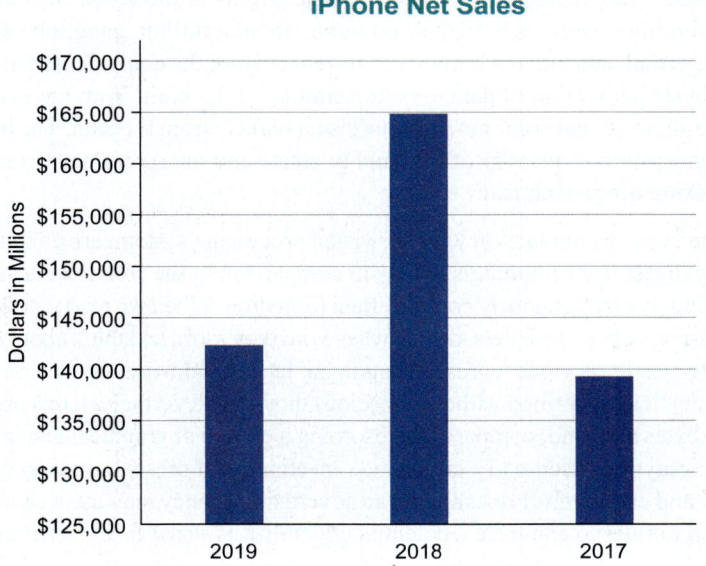

iPhone Net Sales

The table below contains the data used to create the graphs above.

| Net Sales by Category (Dollars in Millions) | | | | | |
|---|---|---|---|---|---|
| Category | 2019 | Change | 2018 | Change | 2017 |
| iPhone[1] | $142,381 | (14)% | $164,888 | 18% | $139,337 |
| Mac[1] | 25,740 | 2% | 25,198 | (1)% | 25,569 |
| iPad[1] | 21,280 | 16% | 18,380 | (2)% | 18,802 |
| Wearables, Home, and Accessories [1][2] | 24,482 | 41% | 17,381 | 36% | 12,826 |
| Services[3] | 46,291 | 16% | 39,748 | 22% | 32,700 |
| Total net sales | $260,174 | (2)% | $265,595 | 16% | $229,234 |

(1)    Products net sales include amortization of the deferred value of unspecified software upgrade rights, which are bundled in the sales price of the respective product.

(2)    Wearables, Home, and Accessories net sales include sales of AirPods, Apple TV, Apple Watch, Beats products, HomePod, iPod touch, and Apple-branded and third-party accessories.

(3)    Services net sales include sales from the company's digital content stores and streaming services, AppleCare, licensing, and other services. Services net sales also include amortization of the deferred value of Maps, Siri, and free iCloud services, which are bundled in the sales price of certain products.

**Source:** These figures are from Apple Inc.'s prospectus of 2019 which was filed with the United States Securities and Exchange Commission.

Using items such as the aforementioned tables and graphs, Apple Inc. is able to show their investors their return on investment, how it compares to other stock indexes, and the sales associated with each of its major products and services. This chapter will focus on the use of tabular displays and graphs, such as frequency distributions and bar graphs, to summarize large data sets.

## Introduction

Statistics is about understanding data. Graphical images are universally regarded as a powerful form of communicating data because the eyes and brain process visual data with amazing speed. The processing of a visual image begins in the retina of the eye where a few hundred million neurons pass their output to about a million ganglion cells, which in turn pass the visual data into the brain. This system enables the equivalent of approximately 90 million bytes per second of data to be transmitted to the brain from the eyes. The brain creates three-dimensional color models (our visual reality) from the data. The brain's ability to process an enormous quantity of data and to create and interpret images in their totality is an astonishing processing feat.

Although the eyes, in conjunction with our visual processing system, are a fabulous sensory system, they digest text or numbers slowly in comparison to the comprehension of a visual image. Reading is a tremendously complex transformation. When we read words or numbers, the brain must recognize the letters, decide what word they form, and think about the definition and how it relates to previous words. Recognizing letters and words is such an integral part of our lives that it is performed without conscious thought. Nevertheless, this process is slow (perhaps 50 bytes a second) compared to absorbing a picture or graphical display. Therefore, graphical displays are used daily in business meetings and presentations to communicate data quickly and effectively. For example, an advertising agency may use a chart showing an upward trend in sales to convince a potential client that its services are beneficial.

In financial markets, technical analysts (sometimes called technicians) use graphs, tables, and other tools to identify price patterns, trends, and relationships in an effort to forecast price movements over certain periods of time to improve investment decisions.

# 3.1 Frequency Distributions

Statistics exists because of variation. The statistician's job is to comprehend variation by looking for structure. Frequency distributions are one method of examining a data set's structure. To examine structural characteristics, ask questions such as, *Where are most of the observations located? Do the data cluster around one central point or are there several points that data seem to cluster around? Do the data seem to be uniformly spread out over some interval or bunched in some range?* These questions all relate to the concept of "distribution."

The process of refining information is interesting. The analyst begins with raw data, then organizes that data by counting the number of observations in each classification. In Table 3.1.1, the raw data consist of population counts in each state for the years 2010 and 2019. By comparing the populations in 2019 with the populations in 2010, a percentage population growth can be computed for each state over the 10-year period.

When the growth rate data is classified in the frequency distribution (see Table 3.1.2), the actual magnitudes of the data values disappear. Losing information may not seem to be a desirable result, but without some lumping together, it is difficult to comprehend large amounts of data.

> **Definition**
>
> Frequency Distribution
>
> A **frequency distribution** summarizes data into classes and provides in tabular form a list of the classes along with the number of observations in each class.

| Table 3.1.1 – Population Counts of Individual States in 2010 and 2019 (in Thousands) | | | | |
|---|---|---|---|---|
| State | 2010 Population | 2019 Population | Growth | Percent Growth |
| Alabama | 4780 | 4903 | 123 | 2.57 |
| Alaska | 710 | 732 | 21 | 2.96 |
| Arizona | 6392 | 7279 | 887 | 13.88 |
| Arkansas | 2916 | 3018 | 102 | 3.50 |
| | | … | | |
| Puerto Rico | 3726 | 3194 | − 532 | − 14.28 |

> **Data**
>
> The entire data set of population counts can be found at stat.hawkeslearning.com by navigating to **Discovering Business Statistics, Second Edition > Data Sets > State Population Counts**.

| Table 3.1.2 – Frequency Distribution of State 10-Year Population Growth Rates | |
|---|---|
| Percentage Population Growth 2010–2019 | Frequency |
| Less than or equal to 0 | 5 |
| 0.01 to 5 | 23 |
| 5.01 to 10 | 12 |
| 10.01 to 15 | 9 |
| 15.01 to 20 | 3 |

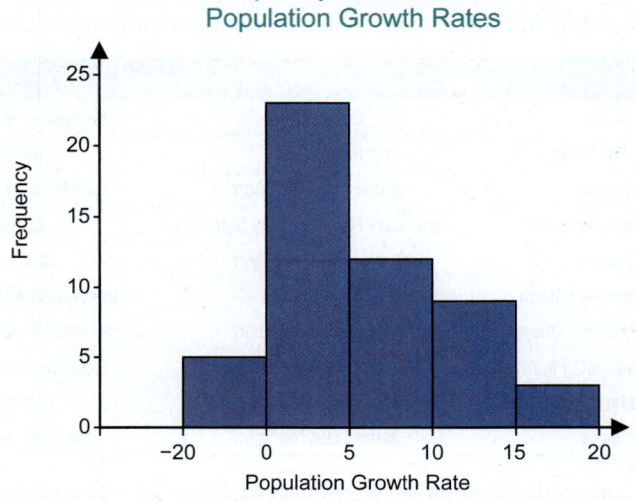

Frequency Distribution of Population Growth Rates

Figure 3.1.1

With the frequency distribution table, we are able to see the broader structure of the data. It is now easy to see that most state population 10-year growth rates are between 0 and 5 percent. Further, growth rates above 20% are uncommon. Without the organization that a frequency distribution provides, these conclusions would be more difficult to establish. If there were 10,000 data values instead of 52, it would be much more difficult to make similar conclusions.

There are only two steps in the construction of a frequency distribution.

**Step 1:**   Choose the classifications.

**Step 2:**   Count the number in each class.

For simple data, such as the results from tossing a coin, the choice of classifications is easy. Heads is one category and tails the other. However, for continuous data, such as weights, heights, and volumes, the choice of classification scheme becomes less obvious, since there are an enormous number of possibilities. There are two requirements that should be met when setting up the categories for classification: the categories must be both mutually exclusive and exhaustive. Essentially, this means categories should not overlap and should cover all possible values.

Since choosing the classification depends on whether the data are qualitative (nominal or ordinal) or quantitative (interval or ratio), the discussion of frequency distributions will be presented on the basis of these data types. This section will present how to construct a frequency distribution for qualitative data, and Section 3.3 will introduce constructing frequency distributions for quantitative data.

## Constructing Frequency Distributions for Qualitative Data

To construct a frequency distribution for qualitative data, choose the categories to classify the data. In many instances, the problem at hand will suggest the classification scheme. For instance, in the coin-tossing example, there are only two classes: heads and tails. If we are classifying a company's size based on its market capitalization, we could have three classes: small cap, mid cap, and large cap. For qualitative data, it would be unusual if a reasonable set of categories is *not* relatively obvious.

After the categories have been chosen, count the items belonging to each class in order to construct the frequency distribution.

| Example 3.1.1 |
|---|

**Creating a Frequency Distribution of Survey Responses**

Apple Inc. introduced its third generation iPad in March of 2012. After months of anticipation, owners were thrilled with their new iPads. Two of the main features that separated the 3rd generation iPad from its predecessors were an improved camera and a higher resolution display. In spite of the excitement of the launch, there were still things that owners did not like about the new device. The following table contains the responses from a survey of 30 new iPad owners when asked what they dislike about the iPad.

| Table 3.1.3 – Survey Responses | | |
|---|---|---|
| Cost of Device | Size/Weight | Excessive Heat from Device |
| Battery Life Too Short | Size/Weight | Cost of Device |
| Cost of Device | Battery Life Too Short | Excessive Heat from Device |
| Cost of Wireless Data Plan | Amount of Flash Memory Storage | Size/Weight |
| Cost of Device | Battery Life Too Short | Cost of Device |
| Amount of Flash Memory Storage | Cost of Device | Integration with Other Devices |
| Amount of Flash Memory Storage | Battery Life Too Short | Integration with Other Devices |
| Cost of Wireless Data Plan | Cost of Device | Excessive Heat from Device |
| Amount of Flash Memory Storage | Cost of Wireless Data Plan | Cost of Device |
| Cost of Device | Battery Life Too Short | Cost of Wireless Data Plan |

Create a frequency distribution using the data in Table 3.1.3 to summarize what iPad owners dislike about their devices.

**SOLUTION**

Examining the table of responses, one can see that there are seven responses (or categories) that were given by the 30 new iPad owners. Summarizing the responses in a table (by counting the number of times that each reason was given), we can create a frequency distribution.

| Table 3.1.4 – Frequency Distribution of Survey Responses | |
|---|---|
| **Response** | **Frequency** |
| Cost of Device | 9 |
| Battery Life Too Short | 5 |
| Cost of Wireless Data Plan | 4 |
| Amount of Flash Memory Storage | 4 |
| Size/Weight | 3 |
| Excessive Heat from Device | 3 |
| Integration with Other Devices | 2 |

The frequency distribution makes it easy to see the greatest concern for iPad owners. From the table, it is evident that in spite of the anticipation and the rush for many consumers to purchase the new iPad, the cost of the device was the number one reason that new owners disliked it. We can also see that iPad integration with other devices is not much of a concern for new customers. These conclusions would have been much more difficult to make if we considered the raw data alone without consolidating the responses into a frequency distribution.

**Example 3.1.2**

**A Model to Determine Virtual Security Practices**

Our increased reliance on digital information and our expansive use of the internet for a steadily rising number of tasks require that more emphasis be placed on digital information security. The importance of securing digital information is apparent but the success in persuading individual users to adopt and utilize tools to improve security has been arguably more difficult. A recent study conducted by Information Security Associates wanted to determine the extent to which students are aware of security measures available to protect personal information on their computers and the importance of such measures. One of the survey items was, "I am aware that there are measures that I can take to help protect my personal information on my personal computer." The frequency of each response category is shown in Table 3.1.5. The frequency distribution for the second item, "I am aware that I can reduce exposure to system compromise by restricting who uses my personal computer," is given in Table 3.1.6. The summary tables are much more informative than looking at more than a thousand observations for each question (note that the number of observations for each question is different because every student did not answer every question on the survey).

| Table 3.1.5 – Frequency Distribution of Responses | |
|---|---|
| **"I am aware that there are measures that I can take to help protect my personal information on my personal computer."** | |
| Strongly Agree | 419 |
| Slightly Agree | 327 |
| Neutral | 250 |
| Slightly Disagree | 124 |
| Strongly Disagree | 85 |

| Table 3.1.6 – Frequency Distribution of Responses | |
|---|---|
| "I am aware that I can reduce exposure to system compromise by restricting who uses my personal computer." | |
| Strongly Agree | 520 |
| Slightly Agree | 435 |
| Neutral | 310 |
| Slightly Disagree | 115 |
| Strongly Disagree | 90 |

**Definition**

**Relative Frequency Distribution**

A **relative frequency distribution** summarizes data into classes and provides in tabular form a list of the classes along with the proportion (or percentage) of observations in each class.

In Table 3.1.7 and Table 3.1.8, **relative frequency distributions** are calculated. In these tables, the frequencies are converted into percentages. These are defined as **relative frequencies**, i.e., the proportion relative to the total. They are valuable in assessing the data quickly in terms we use frequently. (See Section 3.3 for an additional discussion of relative frequency distributions.)

| Table 3.1.7 – Relative Frequency Distribution of Responses | |
|---|---|
| "I am aware that there are measures that I can take to help protect my personal information on my personal computer." | |
| Strongly Agree | 35% |
| Slightly Agree | 27% |
| Neutral | 21% |
| Slightly Disagree | 10% |
| Strongly Disagree | 7% |

| Table 3.1.8 – Relative Frequency Distribution of Responses | |
|---|---|
| "I am aware that I can reduce exposure to system compromise by restricting who uses my personal computer." | |
| Strongly Agree | 35% |
| Slightly Agree | 30% |
| Neutral | 21% |
| Slightly Disagree | 8% |
| Strongly Disagree | 6% |

As one can see in Tables 3.1.5 through 3.1.8, the majority of students are aware that there are measures that they can take to protect the information on their personal computers. Summarizing the qualitative data (via a frequency distribution table or relative frequency distribution table) allows the researcher to make conclusions about the data without having to view each observation.

# 📝 3.1 Exercises

## Basic Concepts

1. From a comprehension standpoint, what are the advantages of visual images over the written word?

2. Describe two situations in which graphical displays are used in business.

3. Describe the purpose of a frequency distribution.

4. What are the basic questions to ask when examining the structure of a data set?

5. What are the two steps to constructing a frequency distribution?

6. In the construction of a frequency distribution, what are the two requirements that the classification categories must meet?

## Exercises

**7.** In order to help him decide when and where to advertise, a local repairman decided to pull his invoices for the month of June and tally what types of machines he had worked on. There were forty-eight items repaired that month.

| | | |
|---|---|---|
| Office copier | Washing machine | Air conditioner |
| Air conditioner | Fan | Lawn mower |
| Lawn mower | Air conditioner | Fan |
| DVD Player | Fan | Air conditioner |
| Air conditioner | Lawn mower | Washing machine |
| Lawn mower | Air conditioner | Stereo |
| Exercise bike | DVD Player | Air conditioner |
| Air conditioner | Lawn mower | Lawn mower |
| Lawn mower | Air conditioner | Fan |
| Radio | Washing machine | Air conditioner |
| Air conditioner | Radio | Stereo |
| Fan | Air conditioner | Lawn mower |
| Washing machine | Lawn mower | Air conditioner |
| Air conditioner | Fan | Fan |
| Lawn mower | Air conditioner | DVD player |
| Washing machine | Washing machine | Air conditioner |

**a.** What level of measurement do the data possess?

**b.** Are the data qualitative or quantitative?

**c.** Construct a frequency distribution for the data. Any machine types worked on three or fewer times are classified as miscellaneous.

**8.** Parkinsonism is an affliction of the aged and is frequently caused by Parkinson's disease, Alzheimer's disease, or other illnesses. The results from a recent study on Parkinsonism were reported in "Prevalence of Parkinsonian Signs and Associated Mortality in a Community Population of Older People," *New England Journal of Medicine*. A sample of 467 people, all 65 years of age or older, was selected from East Boston, Massachusetts. Each person was clinically evaluated and various signs of Parkinsonism, if any, were noted. The following table is a frequency distribution for some of the signs of Parkinsonism.

| Signs of Parkinsonism | |
|---|---|
| **Sign** | **Frequency** |
| Reduced arm swing | 210 |
| Prolonged turning | 153 |
| Right leg rigidity | 141 |
| Left leg rigidity | 154 |
| Slow finger taps | 197 |
| Shuffling gait | 83 |

**a.** What level of measurement do the data possess?

**b.** What percent of the sample suffered from left leg rigidity? Round your answer to two decimal places.

**c.** Add up the frequencies. Why does the sum of the frequencies exceed the total sample size of 467?

**d.** Suppose 30 people suffer from both left leg rigidity and right leg rigidity. How many people in the sample suffer from rigidity in at least one of their legs?

9.  A small commuter airline in the West keeps records of complaints received from its customers. Complaints for March and July are listed in the following table.

| Customer Complaints | | |
|---|---|---|
| Type of Complaint | March | July |
| Tickets cost too much | 11 | 15 |
| Stewardess did not provide blankets | 8 | 3 |
| Schedules not convenient | 12 | 17 |
| Plane often late | 17 | 16 |
| Seats too stiff | 3 | 3 |
| Airplane too hot | 6 | 20 |
| Airplane too cold | 8 | 5 |
| Poor reservation system | 5 | 5 |
| Plane interior looks shabby | 5 | 6 |

a.  Classify the items by the following categories: comfort, price, service, and schedule, and develop a qualitative frequency distribution.

b.  Classify the items by the following categories: plane, personnel, building/equipment, and other, and develop a qualitative frequency distribution.

c.  Would another person necessarily assign the same items to the same categories as you have? Discuss the implications of this when reviewing data collected and distributed by someone else for open answer questions.

d.  Do the categories chosen in parts **a.** and **b.** meet the requirement that categories be mutually exclusive and exhaustive? Discuss.

# 3.2 Displaying Qualitative Data

Graphical analysis is a trade-off. We lose sight of the individual observations (the raw data). In return, we are able to see a representation of the totality of observations. The trade is almost always beneficial since a well-designed graph gives our visual processing system the kind of image it processes best, a picture.

Because a set of data can be graphically represented in many different ways, selecting and creating graphical displays requires a certain amount of artistic judgment. Fortunately, the development of graphics software has made the creation of sophisticated graphs quite easy.

Several types of graphs and tabular displays will be discussed in this chapter. Bar charts, stacked bar charts, 3-D bar charts, and pie charts are effective, visually appealing methods of graphically displaying qualitative data. An examination of publications such as *Time*, *USA Today*, *The Wall Street Journal*, *Scientific American*, or *Forbes* provides convincing evidence of the frequent and beneficial usage of these graphical display techniques.

## Definition

### Bar Chart

The **bar chart** is a simple graphical display in which the length of each bar corresponds to the number of observations in a category.

## Bar Charts

Bar charts are often used to illustrate a frequency distribution for qualitative data.

Bar charts are valuable as presentation tools and are especially effective at reinforcing differences in magnitudes, since they permit the visual comparison of data by displaying the magnitude of each category by a vertical or horizontal bar. Figure 3.2.1 is a bar chart constructed from majors of the students in a business statistics course.

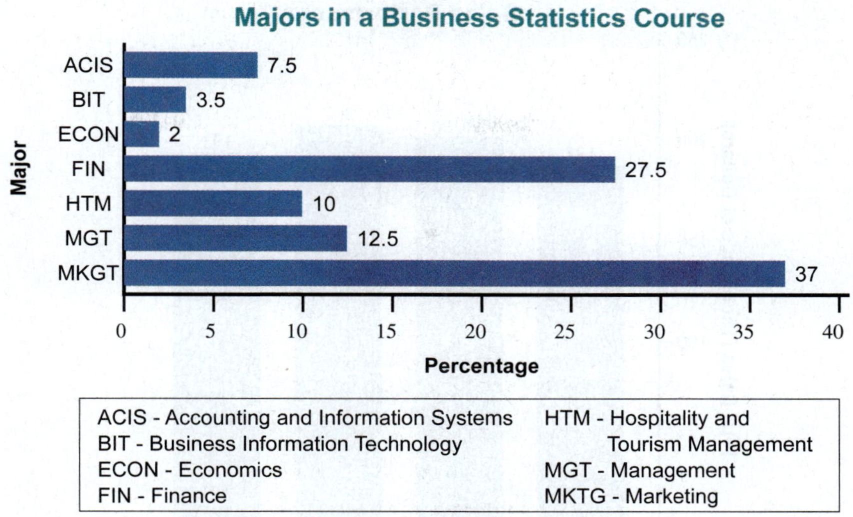
Majors in a Business Statistics Course
ACIS
BIT
ECON
FIN
HTM
MGT
MKGT
7.5
3.5
2
27.5
10
12.5
37
Major
0
5
10
15
20
25
30
35
40
Percentage
ACIS - Accounting and Information Systems
BIT - Business Information Technology
ECON - Economics
FIN - Finance
HTM - Hospitality and Tourism Management
MGT - Management
MKTG - Marketing

Figure 3.2.1

### Technology

For instructions on how to create a bar chart in Excel, JMP, or other technologies, please visit stat.hawkeslearning.com and navigate to **Discovering Business Statistics, Second Edition > Technology Instructions > Graphs > Bar Charts**.

## The Aesthetics of Bar Chart Construction

Bar chart construction requires numerous layout decisions such as size, use of color, and labeling locations. These decisions are frequently made on a trial-and-error basis. However, certain conventions have been developed that improve the quality and effectiveness of charts. They are presented below as suggestions, not rules. Actually, several of the points are general in nature and would serve as useful guidelines in the construction of any graph.

- Bar charts can be constructed horizontally or vertically. Customarily, horizontal orientation is used for categories that are descriptively labeled, and vertical (or columnar) orientation is used for categories that are numerical. It is important to remember that this idea is only a suggestion. If you believe that a vertical bar chart is more appealing, use it.
- If the categories have some associated order, they maintain that order in the bar chart. Otherwise, the categories may be listed alphabetically, in either ascending or descending order, or in some other pattern related to the nature of the data.
- Miscellaneous or "other" categories should be listed at the bottom of the chart (if oriented horizontally) or at the far right (if oriented vertically).
- The difference in bar length is the principal visual feature in comparing differences in category amounts. Consequently, scales for the axes should be chosen that will most effectively allow for the desired comparisons. **Unless there is a good reason, the axis used to measure the bars should start at zero. Otherwise, the axis can be stretched to exaggerate differences in the bar lengths.** For example, suppose the data in the following table were plotted.

**Table 3.2.1 – Sales Performance**

| Salesperson | Total Sales (Thousands of Dollars) |
|---|---|
| Susan | 187 |
| William | 201 |
| Beth | 207 |
| Rob | 193 |

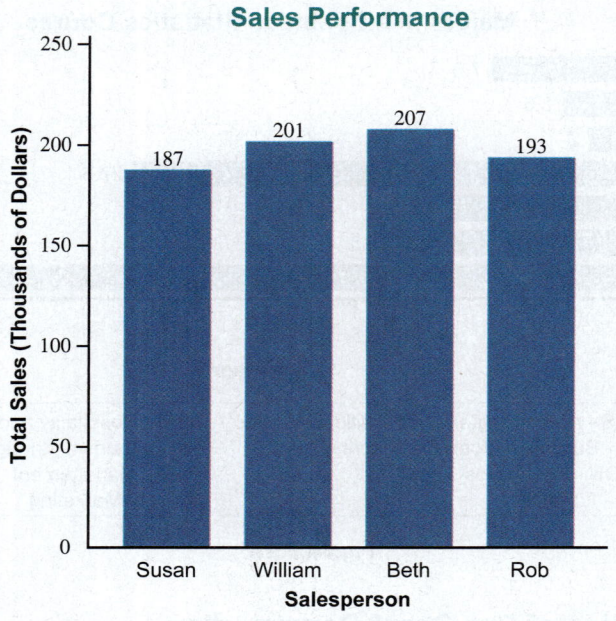

Figure 3.2.2

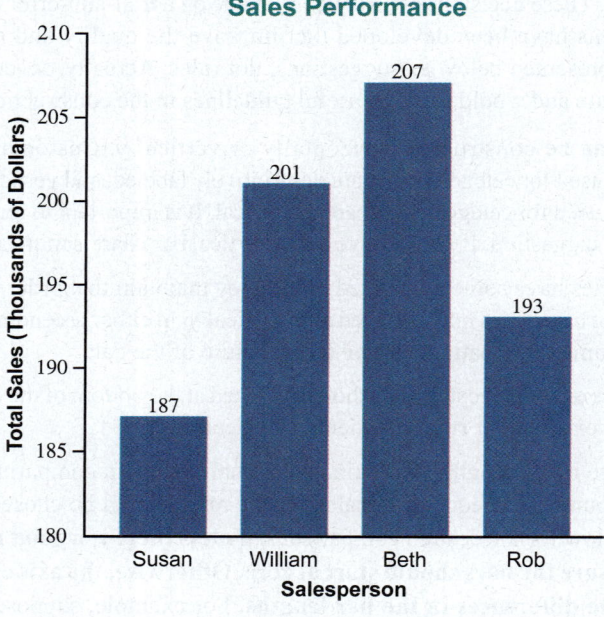

Figure 3.2.3

Figures 3.2.2 and 3.2.3 are plots of the same data (Table 3.2.1). What a difference axis selection can make on perception! In Figure 3.2.3 the *y*-axis begins at 180 instead of zero. If you want to emphasize similarity, use Figure 3.2.2 to do the job. If you want to emphasize differences, use Figure 3.2.3. However, it is difficult to imagine any legitimate reason for using Figure 3.2.3 to represent the data. Axis stretching is often employed to mislead. **When you see an axis that does not start at zero, you should be a bit skeptical as to the conclusions the author intends for you to make.**

- Bar widths should be chosen that are visually pleasing and should not be allowed to vary within a particular chart.

- Appropriate shading, crosshatching, and coloring of the bars can help in presenting data. Many spreadsheet programs incorporate sophisticated graphing programs which make changes in shading, color, and crosshatching patterns extremely easy.

- The spacing between bars can dramatically affect the perception of the graph. Spacing should be set at approximately one-half the width of a bar. This, however, is not a rigid rule. Artistic judgment is needed.

- Gridlines extended into the body of the chart are often useful and may be included if deemed helpful. Study Figure 3.2.4 and Figure 3.2.5. Both bar charts illustrate the same data. Notice that the readability of the graph in Figure 3.2.5 is improved by adding gridlines.

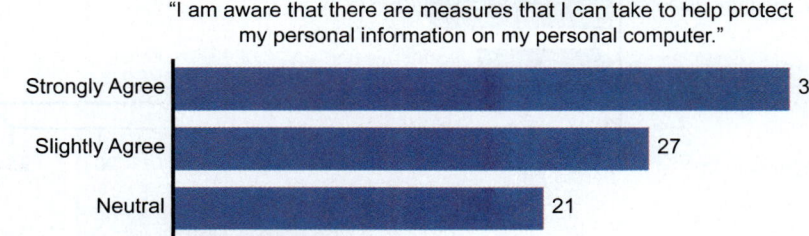
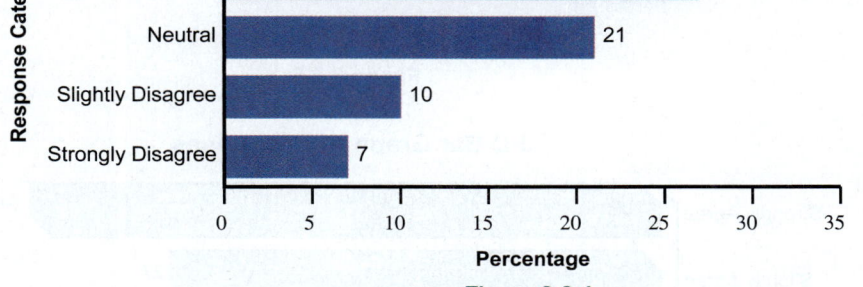

**Figure 3.2.4**

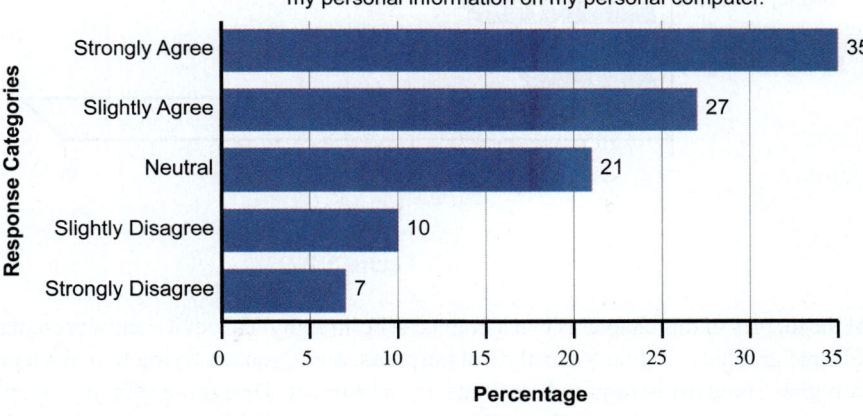

**Figure 3.2.5**

- Labels should be provided for each bar (category) and for each axis.
- Notes on sources of data or other footnotes should be given below the chart.

Bar charts used to be flat and simple. Now, computer graphics packages offer three-dimensional, stacked, side-by-side, colored, and other variations of bars. These capabilities offer the user the chance to create spectacular, multidimensional, eye-catching graphics.

Rotating the axes on a bar chart can create a very different graphical perspective. Figures 3.2.6 and 3.2.7 are identical graphs with a slight difference in perspective. (Note also the change in perspective by making the graphs 3-D. These are the same data that are graphed in Figures 3.2.4 and 3.2.5.) If you are using a large number of graphics in a document, even 3-D graphs can be burdensome. It would be more visually interesting to change the perspective of the graphic from time to time. Making the changes from 2-D to 3-D and performing the rotation are simple operations in most software packages.

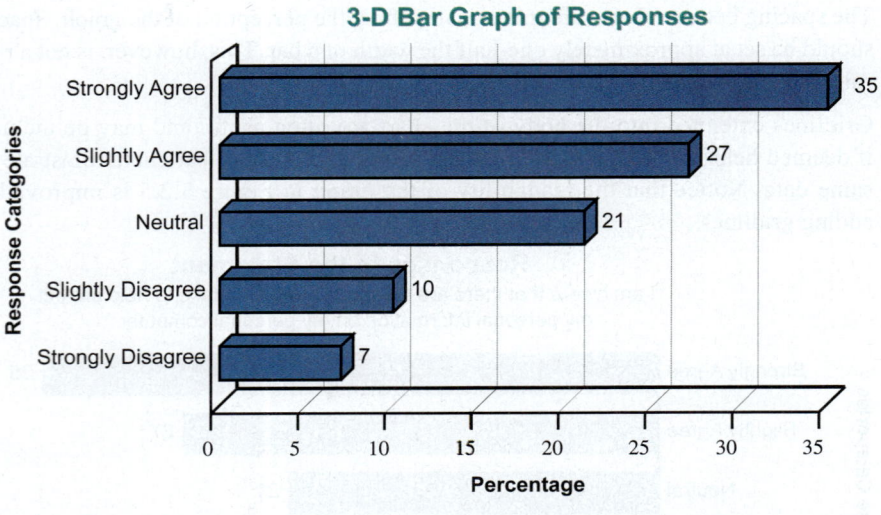

**Figure 3.2.6**

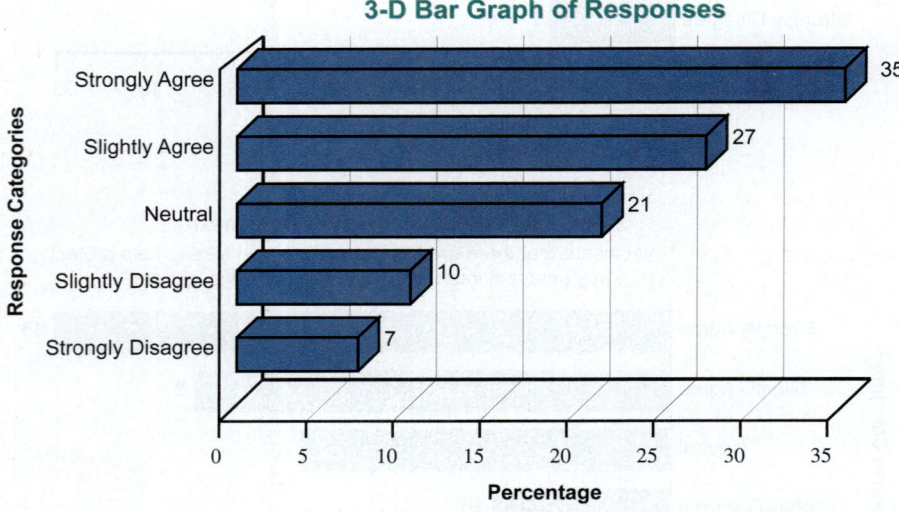

**Figure 3.2.7**

One of the themes of this chapter is that a graph can be an analytical device and a presentation tool. Simple graphics are fine for analytical purposes, but if you are trying to make a point, then a higher standard is required to create visual impact. Designing effective graphics is not just about making things look better; graphics can help the reader comprehend information. Using graphics well can emphasize meaning and organize content. It pays to know your audience's likes and dislikes in viewing graphics so that the maximum impact can be achieved.

## Stacked Bar Charts

Stacked bar charts are an interesting variation on the standard bar chart. The number of medals (gold, silver, bronze) won during the 2016 Summer Olympics for selected countries is given in Figure 3.2.8.

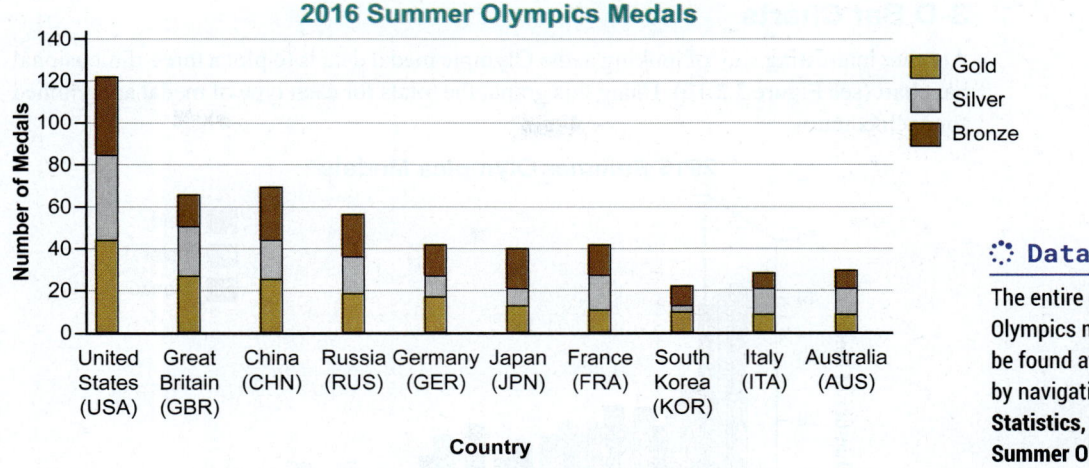

Figure 3.2.8

Using the stacked bar chart enables the reader to compare the total number of medals for each country as well as observe the number of each type of medal won.

Without the stacked bar chart, the reader would have to view either three different charts tallying medal counts for the 2016 Summer Olympics or a much "busier" chart plotting each medal on the horizontal axis above each country.

Figure 3.2.9 is another example of a stacked bar chart. The chart displays the number of grandchildren by age group living with their grandparents based on which parent is in the household. From this chart we can see that the majority of grandchildren living with their grandparents live in the same household with their mother only. We can also see that it is fairly rare that grandchildren live in the same household with their grandparents and their father only. Stacked bar charts are useful when there are three components to the data (in this case the percentage of children, the age group, and the parents in the household).

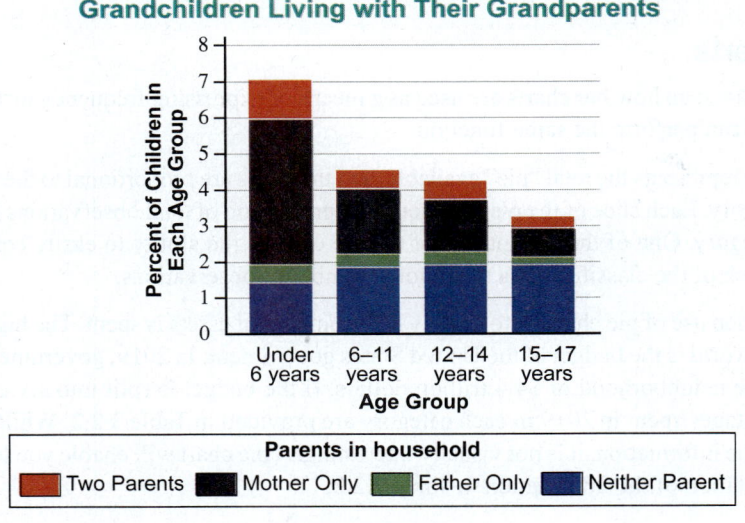

Figure 3.2.9

:•: **Data**

The entire data set of 2016 summer Olympics medals won by country can be found at stat.hawkeslearning.com by navigating to **Discovering Business Statistics, Second Edition > Data Sets > Summer Olympic Medals 2016**.

## 3-D Bar Charts

Another interesting way of looking at the Olympic medal data is to plot a three-dimensional bar chart (see Figure 3.2.10). Using this graph, the totals for each type of medal are graphed for each country.

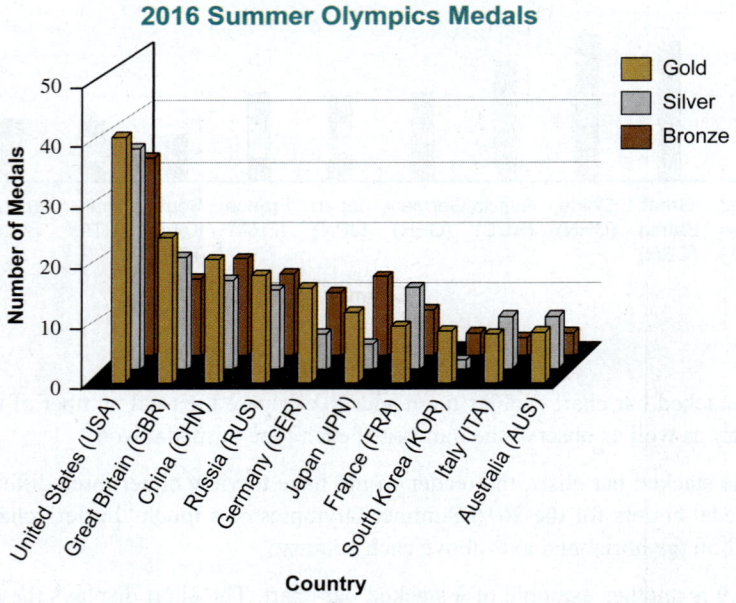

**Figure 3.2.10**

When constructing a 3-D graph there are a number of perspective issues that affect the visual quality of the graph. For example, in Figure 3.2.10, we are barely able to see the number of bronze medals won by Great Britain. All the major software packages permit you to spin and tilt the graph until you find a perspective you like, and that allows you to see all of the categories.

## Pie Charts

We have just seen how bar charts are used as a means of expressing frequency distributions. Pie charts can perform the same function.

The circle represents the total "pie" available, and the slices are proportional to the amount in each category. Each slice of the pie represents the proportion of total observations belonging to the category. One of the advantages of the pie chart is the ability to easily compare the total in each of the classifications to the total number of observations.

One common use of pie charts is to display how some set of assets is spent. The biggest asset pie in the world is the budget of the United States government. In 2019, government outlays were in the neighborhood of $4.4 trillion dollars. If the budget is split into six categories, the percentages spent in 2019 in each category are provided in Table 3.2.2. While the table provides the information, it is not visually interesting. A pie chart will enable you to improve the presentation of the information in Table 3.2.2.

Oftentimes, the percentages in tables such as Table 3.2.2 will either sum to more than 100% or be slightly less than 100% due to rounding. Such is the case for this table.

**⚭ Technology**

For instructions on how to create a pie chart in Excel, JMP, or other technologies, please visit stat.hawkeslearning.com and navigate to **Discovering Business Statistics, Second Edition > Technology Instructions > Graphs > Pie Chart**.

| Table 3.2.2 – Percentage Spent by the Federal Government in 2019 | |
|---|---|
| **Category** | **Percentage Spent** |
| Social Security | 23% |
| Safety Net Programs | 8% |
| Defense and International Security Assistance | 16% |

| Table 3.2.2 – Percentage Spent by the Federal Government in 2019 (cont.) ||
| Category | Percentage Spent |
|---|---|
| Other Entitlements | 19% |
| Medicare and Medicaid | 25% |
| Interest on Debt | 8% |

The pie chart in Figure 3.2.11 tells an interesting story about how our tax dollars are spent. In a glance at the pie chart, your eyes are drawn to the biggest slice of the pie, the 25% spent on Medicare and Medicaid. If you would like to look at current information on how government monies are spent, go to www.whitehouse.gov/omb/budget.

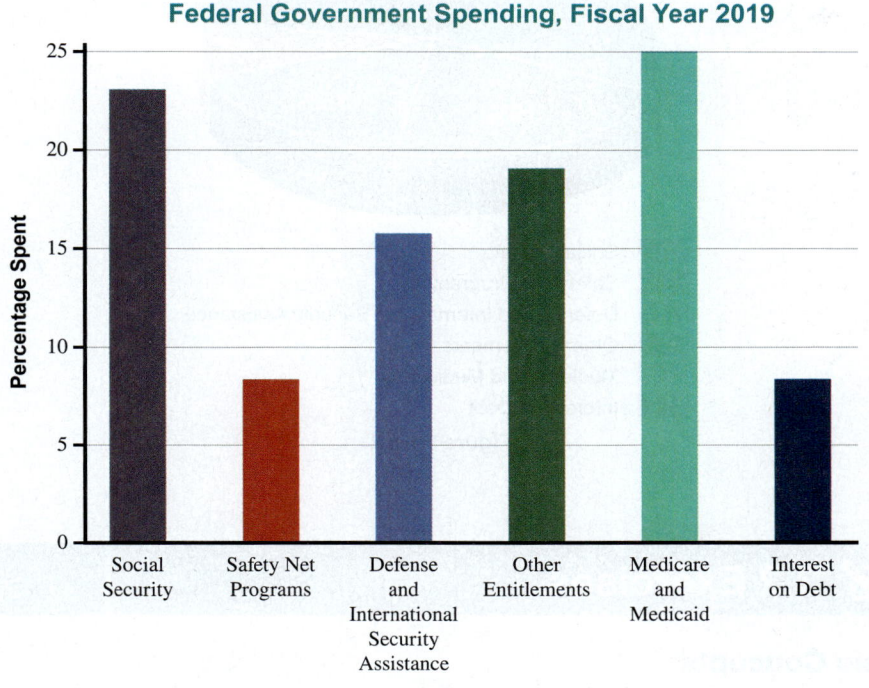

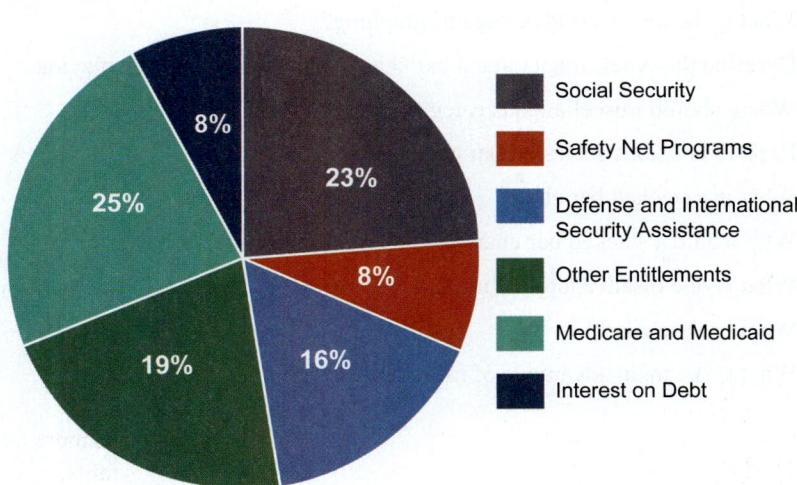

Figure 3.2.11

A three-dimensional version of the pie chart (see Figure 3.2.12) adds some spice to any presentation. If you are working with a computer application that can perform graphics, changing the image from a two-dimensional pie chart to a three-dimensional chart takes only a few mouse clicks. A 3-D pie chart of federal government spending is given in Figure 3.2.12.

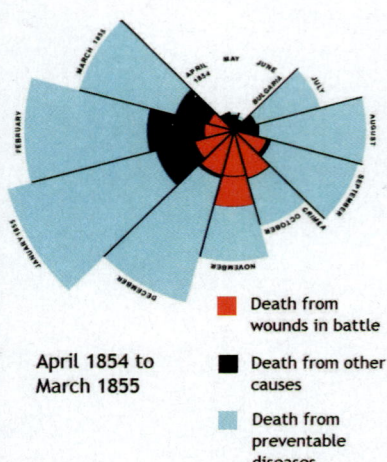

Death from wounds in battle

Death from other causes

Death from preventable diseases

April 1854 to March 1855

## A Passion for Compassion

In the 19th century, statistics was not widely seen as an applicable skill. That is, until Florence Nightingale came onto the scene. When she arrived at the front line of the Crimean War, she was appalled by the situation. The mortality rate was too high, and the hospitals were in complete disarray. She immediately set about organizing what little records were kept, and started to gather a lot of new data. Upon analyzing this new data, she discovered that the majority of deaths that were occurring in British military hospitals were due to preventable diseases. Using this new information, Nightingale was able to present a case to Parliament for improving the sanitary practices in British hospitals. She utilized data analysis and visualization to literally save thousands of lives, and in the process, her "rose diagram", also known as a "coxcomb chart", became an iconic data visualization.

In order to determine from the pie chart how much was spent in a particular category, multiply the total amount by the proportion given for that category in the pie chart. For example, to find the dollar amount of government funds spent on Medicare and Medicaid, multiply the total amount of government expenditures ($4.4 trillion) by the proportion spent on Medicare and Medicaid. This means that $4.4(0.25) = $1.1 trillion of all government expenditures goes to Medicare and Medicaid.

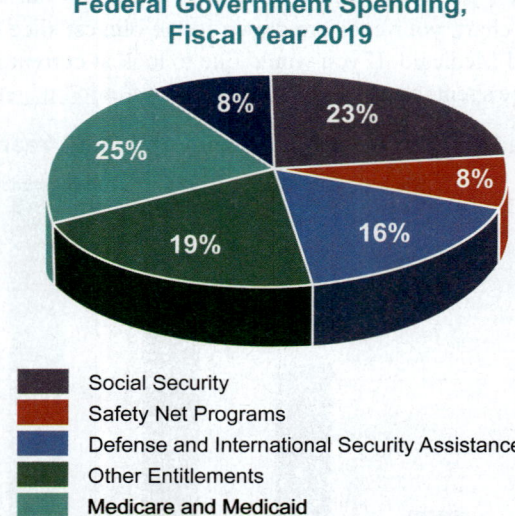

**Federal Government Spending, Fiscal Year 2019**

- Social Security
- Safety Net Programs
- Defense and International Security Assistance
- Other Entitlements
- Medicare and Medicaid
- Interest on Debt

Figure 3.2.12

## 3.2 Exercises

### Basic Concepts

1.  What are some benefits of graphing?

2.  What is the major disadvantage of graphing?

3.  Describe the types of data that a bar chart would be useful in displaying.

4.  Where should miscellaneous categories be displayed in a bar chart?

5.  Explain how axis scales on bar charts can be misleading.

6.  What is a stacked bar chart?

7.  Why would a stacked bar chart be preferred over a normal bar chart?

8.  What is one disadvantage of using a 3-D chart?

9.  What is a pie chart?

10. What is the main advantage of using a pie chart?

# Exercises

11. A consumer magazine uses bar charts to compare four popular brands of automobiles. This particular bar chart represents a comparison of the miles per gallon (mpg) for the four brands.

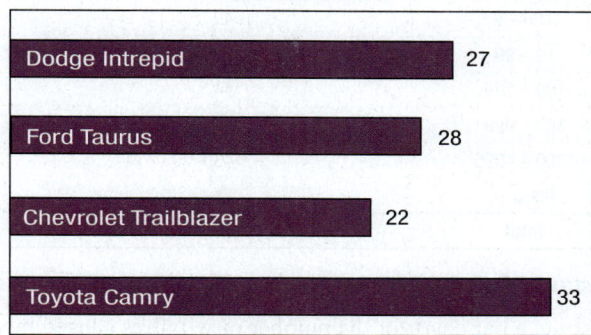

    a. What is wrong with this picture?

    b. Evaluate the bar chart using the guidelines suggested in the section on the aesthetics of bar chart construction.

12. The following bar chart presents the median income of U.S. employees by education level and gender. Evaluate the bar chart using the guidelines suggested in the section on the aesthetics of bar chart construction.

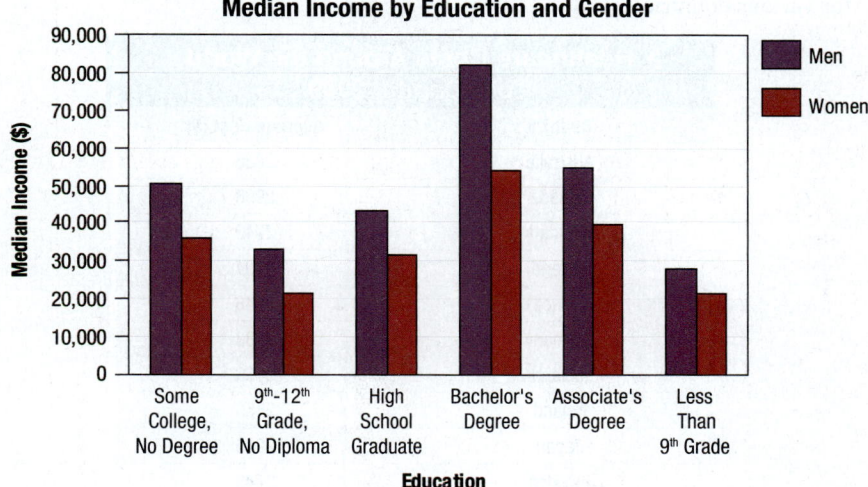

13. Consider the following data regarding the number of wildfires in the U.S. categorized by the size class in acres and the cause of the fire.

| Number of Wildfires in the U.S. | | |
|---|---|---|
| Size Class (Acres) | Lightning-Caused | Person-Caused |
| 0.25 or less | 4637 | 2367 |
| 0.26 – 9 | 1940 | 1904 |
| 10 – 99 | 219 | 571 |
| 100 – 299 | 44 | 103 |
| 300 – 999 | 26 | 52 |
| 1000 – 4999 | 43 | 17 |
| 5000 + | 21 | 9 |
| Total | 6930 | 5023 |

a. Construct a bar chart for the number of wildfires caused by lightning.

b. Construct a bar chart for the number of wildfires caused by people.

c. Construct a stacked bar chart for the number of wildfires caused by lightning and the number of wildfires caused by people.

d. Construct a pie chart for the number of wildfires caused by lightning.

e. Construct a pie chart for the number of wildfires caused by people.

f. What did you learn from the charts created in parts a. through e.?

14. Consider the following data regarding the average spending on healthcare per person for various countries.

| Healthcare Costs Around the World per Capita, 2009 | |
|---|---|
| Country | Average Cost ($) |
| Australia | 2886 |
| Canada | 2998 |
| Denmark | 2743 |
| Finland | 2104 |
| France | 3048 |
| Germany | 2983 |
| Iceland | 3159 |
| Ireland | 2455 |
| Japan | 2249 |
| Sweden | 2745 |
| Switzerland | 3847 |
| United Kingdom | 2317 |
| United States | 5711 |

Source: www.creditloan.com

a. Construct a bar chart for the average healthcare cost per person for the various countries.

b. What did you learn from the chart?

With the frequency distribution, we are able to see the broader structure of the data. It is now easy to see that the overwhelming majority of revenues are between $0 and $122 million. Further, revenues of $164 million and above are uncommon. These conclusions would be considerably more difficult to establish without the organization that the frequency distribution provides.

---

The purpose of the frequency distribution is to condense the set of data into a meaningful summary form. There are only two steps in the construction of a frequency distribution.

**Step 1:**  Choose the classifications.

**Step 2:**  Count the number in each class.

## Selecting the Number of Classes

Choosing the number of classes is arbitrary and should depend on the amount of data available. In general, the more observations one has in a data set, the more classes or intervals that can be used in the frequency table or histogram. Consequently, only very general guidelines exist. Generally, fewer than 4 classes would be too much compression of the data and greater than 20 classes provides too little summary information. To start, a good rule of thumb for the number of classes to create is to round $\sqrt{n}$ or $\sqrt[3]{2n}$ to the nearest whole number.

## Determining Class Width

After deciding on the desired number of classes, the next step is to specify the width of each class. Using classes that are of equal widths makes the frequency distribution or histogram easier to interpret. (We will discuss constructing histograms in a later lesson.) There is really no perfect formula for class width that will work for every data set. However, a good starting point for determining class width is to divide the difference between the largest and the smallest observations by the number of classes.

$$\text{Class Width} = \frac{\text{Largest Value} - \text{Smallest Value}}{\text{Number of Classes}}$$

Suppose we wanted to create a frequency distribution from the revenue data in Example 3.3.1. If there are to be 10 classes, determine a class width.

$$\text{Class Width} = \frac{\text{Largest Value} - \text{Smallest Value}}{\text{Number of Classes}} = \frac{408 - 1}{10} = \frac{407}{10} = 40.7$$

Class endpoints with fractional values will make the graph slightly harder to understand. If possible, try to keep the width to an integer value (i.e., round up to the next largest integer). If the calculated class width is 40.7, you might try a class width of 41. An interval width of 41 is used for the revenue data in Example 3.3.1.

Generally, class widths should be equal. The lower class limit is the smallest number that can belong to a particular class and the upper class limit is the largest number that can belong to a class. Using the smallest observation, or a smaller number, as the lower limit of the first class is a good place to start. However, you will need to use some judgment. You should choose the first lower limit of the first class, add the class width to it to find the lower limit of the second class, and continue until you have the desired number of lower class limits. The upper limit of each class is then determined such that the classes do not overlap. Once you have created the class intervals, if there are any data values that fall outside the class limits, you must adjust either the class width or your choice of the lower class limit of the first interval.

In the example, we have selected the starting point as 0 because our smallest value is 1, and obviously we want the first interval to contain the smallest observation. Adding the class width to the lower class limit creates the next interval. If this pattern is followed, the result will be intervals that will not overlap and will capture all the data (i.e., the classes will be mutually exclusive and exhaustive).

The frequency distribution given in Table 3.3.2 is just one way of organizing the data. There are three other distributions that can be calculated from the frequency distribution: the relative frequency, cumulative frequency, and cumulative relative frequency distributions. Each gives a slightly different perspective on the data.

## Relative Frequency Distribution

Relative frequency represents the proportion of the total observations in a given class. The **relative frequency distribution** enables the reader to view the number in each category in relation to the total number of observations. Relative frequency is a standardizing technique. Converting the frequency in each class to a proportion in each class enables us to compare data sets with different numbers of observations.

> **Formula**
>
> ### Relative Frequency
>
> The **relative frequency** of any class is the number of observations in the class divided by the total number of observations.
>
> $$\text{Relative Frequency} = \frac{\text{Number in Class}}{\text{Total Number of Observations}}$$

A relative frequency distribution of the revenue data from Example 3.3.1 is given in Table 3.3.3. Notice that the relative frequencies are obtained by dividing the frequencies in Table 3.3.2 by the total number of observations, which is 100.

### Table 3.3.3 – Relative Frequency Distribution of Revenue Data

| Revenue (Millions of Dollars) | Relative Frequency |
|---|---|
| 0 to 40 | $\frac{50}{100} = 0.50$ |
| 41 to 81 | $\frac{30}{100} = 0.30$ |
| 82 to 122 | $\frac{14}{100} = 0.14$ |
| 123 to 163 | $\frac{3}{100} = 0.03$ |
| 164 to 204 | $\frac{1}{100} = 0.01$ |
| 205 to 245 | $\frac{0}{100} = 0.00$ |
| 246 to 286 | $\frac{1}{100} = 0.01$ |
| 287 to 327 | $\frac{0}{100} = 0.00$ |
| 328 to 368 | $\frac{0}{100} = 0.00$ |
| 369 to 409 | $\frac{1}{100} = 0.01$ |

## Cumulative Frequency Distribution

The cumulative frequency distribution gives the reader an opportunity to look at any category and determine immediately the number of observations that belong to a particular category and all categories below it.

| Table 3.3.4 – Cumulative Frequency Distribution of Revenue Data | | |
|---|---|---|
| Revenue (Millions of Dollars) | Frequency | Cumulative Frequency |
| 0 to 40 | 50 | 50 |
| 41 to 81 | 30 | 80 |
| 82 to 122 | 14 | 94 |
| 123 to 163 | 3 | 97 |
| 164 to 204 | 1 | 98 |
| 205 to 245 | 0 | 98 |
| 246 to 286 | 1 | 99 |
| 287 to 327 | 0 | 99 |
| 328 to 368 | 0 | 99 |
| 369 to 409 | 1 | 100 |

**Definition**

**Cumulative Frequency**

The **cumulative frequency** is the sum of the frequency of a particular class and all preceding classes.

In this example, the reader can easily see in Table 3.3.4 that 97 out of 100 revenues are less than or equal to $163 million.

## Cumulative Relative Frequency

To obtain the cumulative relative frequency, add the relative frequencies of all preceding classes to the relative frequency of the current class.

| Table 3.3.5 – Cumulative Relative Frequency Distribution of Revenue Data | | | |
|---|---|---|---|
| Revenue (Millions of Dollars) | Frequency | Relative Frequency | Cumulative Relative Frequency |
| 0 to 40 | 50 | 0.50 | 0.50 |
| 41 to 81 | 30 | 0.30 | 0.80 |
| 82 to 122 | 14 | 0.14 | 0.94 |
| 123 to 163 | 3 | 0.03 | 0.97 |
| 164 to 204 | 1 | 0.01 | 0.98 |
| 205 to 245 | 0 | 0.00 | 0.98 |
| 246 to 286 | 1 | 0.01 | 0.99 |
| 287 to 327 | 0 | 0.00 | 0.99 |
| 328 to 368 | 0 | 0.00 | 0.99 |
| 369 to 409 | 1 | 0.01 | 1.00 |

**Definition**

**Cumulative Relative Frequency**

The **cumulative relative frequency** is the proportion of observations in a particular class and all preceding classes.

From the cumulative relative frequency in Table 3.3.5, it is easy to see that 97% of the revenues are less than or equal to $163 million.

## 🖋 3.3 Exercises

### Basic Concepts

1. What are the fundamental decisions in constructing frequency distributions for quantitative data?

2. Describe the general guidelines for selecting the number of classes for a quantitative frequency distribution.

3. What is a good starting point for determining the class width?

4.  What is a relative frequency distribution? How do you calculate relative frequencies from raw frequencies?

5.  What is a cumulative frequency distribution?

6.  What is a cumulative relative frequency distribution?

## Exercises

7.  A business magazine was conducting a study into the amount of travel required for mid-level managers across the U.S. Seventy-five managers were surveyed for the number of days they spent traveling each year.

| Mid-Level Manager Travel | |
|---|---|
| Days Traveling | Frequency |
| 0 – 6 | 15 |
| 7 – 13 | 21 |
| 14 – 20 | 27 |
| 21 – 27 | 9 |
| 28 – 34 | 2 |
| 35 and above | 1 |

a.  Construct a relative frequency distribution.

b.  Construct a cumulative frequency distribution.

8.  The closing prices (in pence) for selected stocks trading on the London Stock Exchange were as follows. Construct a frequency distribution for the stock prices.

| Closing Prices | |
|---|---|
| Stock | Closing Price (Pence) |
| Allied Lyons | 439 |
| Babcock | 208 |
| Barclays Bank | 543 |
| Bass Ltd | 992 |
| British GE | 238 |
| Cadbury Sch | 257 |
| Guinness | 379 |
| Hanson Trust | 169 |
| Lucas Indus | 655 |
| Reed Int'l | 467 |
| STC | 318 |
| Tate & Lyle | 833 |
| Thorm EMI | 741 |
| Utd. Biscuit | 326 |

9.  Every year, the average temperatures of 100 selected U.S. cities are published by the National Oceanic and Atmospheric Administration. The average temperature (°F) for the month of October for 15 randomly selected cities from the list of 100 are listed in the following table.

| Average Temperatures (°F) | | | | |
|---|---|---|---|---|
| 68.5 | 50.9 | 67.5 | 57.5 | 56.0 |
| 47.1 | 50.1 | 65.8 | 51.5 | 49.5 |
| 75.2 | 56.0 | 62.3 | 53.0 | 46.1 |

   a.  Construct a frequency distribution for the average temperatures for the month of October.

   b.  Construct a relative frequency distribution for the average temperatures for the month of October.

   c.  Construct a cumulative frequency distribution for the average temperatures for the month of October.

10. Consider the assets (in billions of dollars) of the 10 largest life insurance companies listed in the following table.

| Assets (Billions of Dollars) | | | | |
|---|---|---|---|---|
| 148.4 | 110.8 | 55.6 | 52.4 | 50.4 |
| 42.7 | 41.7 | 36.2 | 35.7 | 35.7 |

   a.  Construct a frequency distribution for the assets (in billions of dollars) of the 10 largest life insurance companies.

   b.  Construct a relative frequency distribution for the assets (in billions of dollars) of the 10 largest life insurance companies.

   c.  Construct a cumulative frequency distribution for the assets (in billions of dollars) of the 10 largest life insurance companies.

# 3.4 Graphical Displays of Quantitative Data

Several types of graphs and tabular displays will be discussed in this section such as histograms, line graphs, stem-and-leaf displays, and dot plots.

## Histograms

A histogram is a common graphical method that reveals the distribution of the data. Histograms are often constructed based on frequency distributions of quantitative data. Histograms look similar to bar graphs but are used to analyze quantitative data rather than qualitative data.

Each of the classes in the frequency distribution is represented by a vertical bar whose height is proportional to the frequency of the interval. The horizontal boundaries of each vertical bar correspond to the class boundaries. Once the frequency distribution has been calculated, all the information necessary for plotting a histogram is available. In Figure 3.4.1, the histogram is created from the frequency distribution of the revenue data in Table 3.3.2.

**Definition**

**Histogram**

A **histogram** is a bar graph of a frequency or relative frequency distribution in which the height of each bar corresponds to the frequency or relative frequency of each class.

## ⟳ Technology

For instructions on how to create a histogram in Excel, JMP, or other technologies, please visit stat.hawkeslearning.com and navigate to **Discovering Business Statistics, Second Edition > Technology Instructions > Graphs > Histograms**.

**Histogram of Revenue Data**

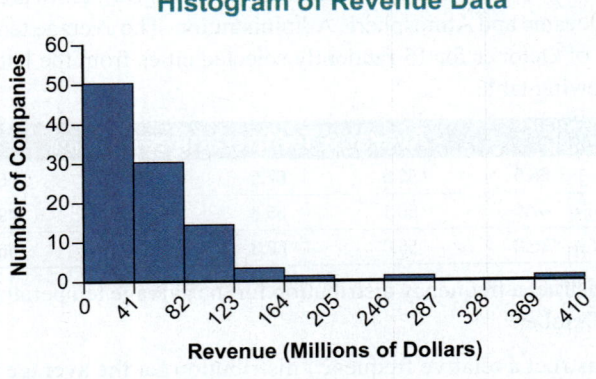

Figure 3.4.1

### Definition

**Symmetric Distribution**

A **symmetric distribution** is one in which, if a vertical line were drawn down the middle of the distribution, the two sides would mirror each other.

### Definition

**Skewed Distribution**

A **skewed distribution** is one in which, if a vertical line were drawn down the middle of the distribution, it would have a long tail to the right or to the left.

Histograms are one of the more frequently used statistical tools. A histogram is not only easy to interpret; it also reveals a great deal about the structure of the data. By examining the histogram, one can determine the shape of the distribution of the data. That is, the data could be symmetric or skewed (left skewed or right skewed). A **symmetric distribution**— for example, a bell-shaped distribution—is one in which if a line were drawn down the middle of the distribution, the two sides would mirror each other. A **skewed distribution** is represented by a group of observations that is not equal on both sides (sometimes called asymmetric). One can also get an idea about the modality of a distribution by examining the histogram. The mode of a set of observations is the value that occurs with the greatest frequency. (We will discuss the mode in further detail in Section 4.1.)

You can quickly see from Figure 3.4.1 that most of the revenues are in the first and second categories (0 to 40 and 41 to 81). The distribution of the data appears to be skewed to the right.

You will see many histograms throughout this course. When we look at a histogram, what features are important?

1.  Is the distribution symmetric or skewed (one tail is longer than the other)?

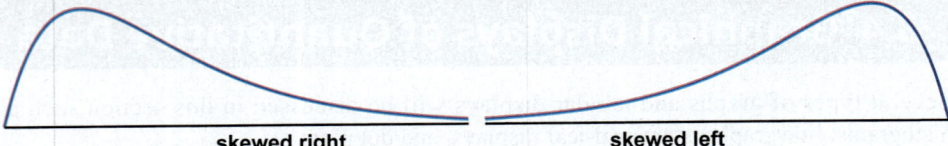

2.  Is it bell-shaped?

3.  Does the distribution have several peaks or modes?

4.  Where is the center of the distribution?

5.  Are there outliers (data values that are very different from the others)?

There are many statistical software programs that can be used to construct histograms. The Discovering Technology section at the end of this chapter explains how to create a histogram from raw data using JMP, Microsoft Excel, and Minitab. In spite of the popularity of using computer software packages to summarize data, there may be times when you need to manually create a histogram. Although there isn't a unified method for constructing a histogram, some steps are provided in the following box that can help you create one.

## Procedure

### Constructing a Histogram

1. Determine the number of classes (or intervals) for the histogram. There isn't a rule of thumb for the number of classes to use, and most of the time the number of classes is determined by trial and error. However, a good starting point could be to use $\sqrt{n}$ or $\sqrt[3]{2n}$.

2. Determine the largest and the smallest observations. Depending on the size of the data set, it may be a good idea to sort the observations from smallest to largest. This will also make it easier later when trying to determine the interval in which an observation falls.

3. Calculate the difference between the smallest and largest observations by subtracting the smallest observation from the largest observation. (Note that this measure is called the range, which will be discussed further in Chapter 4.)

4. Calculate the class width, using the following formula.

$$\text{Class Width} = \frac{\text{Largest Value} - \text{Smallest Value}}{\text{Number of Classes}}$$

5. Calculate the limits of the class intervals. Choose a starting value for the histogram such that the smallest observation will fall into your first class interval given the class width calculated in the previous step. Each subsequent interval can be found by adding the class width to the lower and upper class limits of the first class. The class intervals should be such that the first interval contains the smallest observation and the last interval contains the largest observation.

6. Tally each observation to determine in which interval it should fall. This will give you the class frequency (denoted by $f_i$), which is the number of observations in each interval.

7. All of the information for a histogram can be displayed in a frequency distribution table or a relative frequency distribution table, as can be seen in the following example.

8. Construct the histogram with the lower class limits on the horizontal axis and the frequency (or relative frequency) on the vertical axis. The heights of the bars will represent the class frequencies (or relative frequencies).

Suppose you are given a random sample of 28 applicants who took an examination designed to measure their aptitude for a job in sales. The following scores (measured in percentages) were obtained. Note that the observations are already sorted from smallest to largest.

**Example 3.4.1**

**Creating a Relative Frequency Distribution and Histogram of Aptitude Scores**

| Table 3.4.1 – Aptitude Scores (%) | | | | | | |
|---|---|---|---|---|---|---|
| 38 | 49 | 53 | 56 | 58 | 58 | 60 |
| 62 | 66 | 67 | 69 | 69 | 71 | 74 |
| 75 | 76 | 77 | 77 | 77 | 78 | 78 |
| 81 | 82 | 83 | 84 | 87 | 88 | 88 |

Construct a relative frequency distribution table for the test scores using 6 class intervals, and construct a histogram.

**SOLUTION**

To solve this problem, we will work through the 8-step procedure for constructing a histogram.

1. First, we need to determine the number of classes for the histogram. We are given that we want to construct a histogram using 6 intervals, so there will be 6 classes.

2. The second step is to identify the largest and the smallest observations in the data set. This is relatively straightforward in this case since the data are ordered from smallest to largest. From the data in Table 3.4.1, the smallest test score is 38 and the largest score is 88.

3. Next, we need to calculate the difference between the smallest and the largest observations. Subtracting the smallest score from the largest score, we have the following.

$$\text{Largest Score} - \text{Smallest Score} = 88 - 38 = 50$$

4. Now we need to calculate the class width. Given that the difference between the largest and smallest scores is 50, we have the following.

$$\text{Class Width} = \frac{\text{Largest Score} - \text{Smallest Score}}{\text{Number of Classes}} = \frac{50}{6} \approx 8.33$$

For simplicity, if the class width is not an integer we round up to the next largest integer. Thus, we will use a class width of 9 when constructing the histogram.

5. The next step is to construct the class intervals. In general, we can use the smallest data value, a smaller value, or 0 as the lower limit of the first class. Again, judgment is required in setting up the first interval so that the first interval includes the smallest observation. The lower limit of the first class interval will not be 0 in this case, since the smallest score, 38, is much larger than the class width. Therefore, since the smallest observation is 38, we will arbitrarily use 36 as the lower limit of the first class interval. So, the first interval is 36–44, the second is 45–53, and so on. Thus, the class limits are determined to be the following.

| Table 3.4.2 – Class Limits |
| --- |
| 36–44 |
| 45–53 |
| 54–62 |
| 63–71 |
| 72–80 |
| 81–89 |

6. Tally each observation to determine the interval in which each observation falls. Remember that the observation falls in the interval if it is greater than or equal to the lower class limit and less than or equal to the upper class limit. This gives you the class frequencies, as seen in Table 3.4.3.

| Table 3.4.3 – Class Frequencies | |
| --- | --- |
| **Class** | **Frequency** |
| 36–44 | 1 |
| 45–53 | 2 |
| 54–62 | 5 |
| 63–71 | 5 |
| 72–80 | 8 |
| 81–89 | 7 |

7. The relative frequency distribution table (Table 3.4.4) provides a summary of the data and allows the histogram to be constructed.

| Table 3.4.4 – Relative Frequency Distribution | | |
|---|---|---|
| Class Limits | Frequency ($f_i$) | Relative Frequency |
| 36–44 | 1 | 0.0357 |
| 45–53 | 2 | 0.0714 |
| 54–62 | 5 | 0.1786 |
| 63–71 | 5 | 0.1786 |
| 72–80 | 8 | 0.2857 |
| 81–89 | 7 | 0.2500 |
| **Total** | **28** | **1.0000** |

Keep in mind that the sum of the frequency column should be $n$, the total number of observations. Also, the sum of the relative frequency column should be equal to 1. If the sum of the relative frequency column is different from 1, but fairly close, it is likely due to rounding.

8. Finally, construct the histogram.

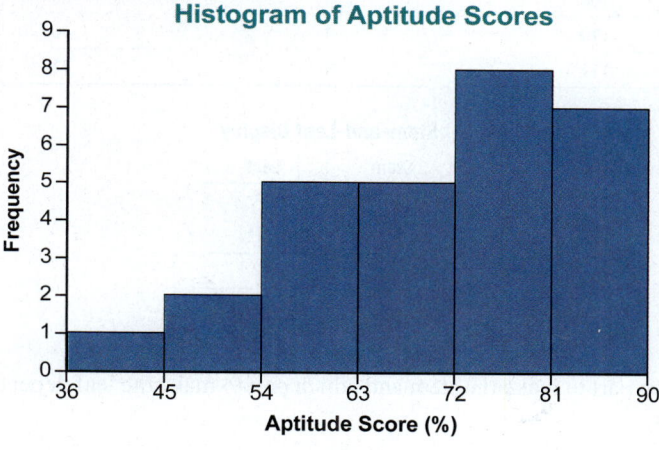

Figure 3.4.2

## Stem-and-Leaf Displays

The **stem-and-leaf display** is a hybrid graphical method. The display is similar to a histogram, but the data remain visible to the user. Like all graphical displays, the stem-and-leaf display is useful for both ordering and detecting patterns in the data. It is one of the few graphical methods in which the raw data are not lost in the construction of the graph. As the name implies, there is a "stem" to which "leaves" will be attached in some pattern.

Consider the following data: 97, 99, 108, 110, and 111. If we are interested in the variation of the last digit, the stems and leaves are as shown in Table 3.4.5 and displayed in Figure 3.4.3. The leaves in this case are the *ones* digits and the stems are the *tens* digits. All of the data values that have common stems are grouped together, and their leaves branch out from the common stem.

| Table 3.4.5 – Data, Stems, and Leaves | | |
|---|---|---|
| Data | Stem | Leaf |
| 97 | 09 | 7 |
| 99 | 09 | 9 |
| 108 | 10 | 8 |
| 110 | 11 | 0 |
| 111 | 11 | 1 |

⌘ **Technology**

For instructions on how to create a stem-and-leaf display in JMP, Minitab, or other technologies, please visit stat.hawkeslearning.com and navigate to **Discovering Business Statistics, Second Edition > Technology Instructions > Graphs > Stem-and-Leaf Plot**.

**Stem-and-Leaf Display**

| Stem | Leaf |
|---|---|
| 09 | 7 9 |
| 10 | 8 |
| 11 | 0 1 |
| Key:    09 | 7 = 97 |

Figure 3.4.3

If we are interested in the variation of the last two digits, the stems and leaves are shown in Table 3.4.6 and displayed in Figure 3.4.4. The leaves in this case are the *last two* digits and the stems are the *hundreds* digits. Again, all of the data values that have common stems are grouped together, and their leaves branch out from the common stem.

| Table 3.4.6 – Data, Stems, and Leaves | | |
|---|---|---|
| Data | Stem | Leaf |
| 97 | 0 | 97 |
| 99 | 0 | 99 |
| 108 | 1 | 08 |
| 110 | 1 | 10 |
| 111 | 1 | 11 |

**Stem-and-Leaf Display**

| Stem | Leaf |
|---|---|
| 0 | 97 99 |
| 1 | 08 10 11 |
| Key:    0 | 97 = 97 |

Figure 3.4.4

Deciding which part to make the stem and which part to make the leaf depends on the focus of the analysis.

**Example 3.4.2**

**Creating a Stem-and-Leaf Display of MSFT Closing Stock Prices**

The closing stock prices for Microsoft Corporation from December 2000 through December 2019 are given in Table 3.4.7. Construct a stem-and-leaf display of the closing stock prices of Microsoft (rounded to the nearest dollar).

| Table 3.4.7 – Closing Stock Prices of Microsoft Corporation (MSFT) December 2000 through December 2019 | | | |
|---|---|---|---|
| Period | Closing Price (Dollars) | Period | Closing Price (Dollars) |
| December 2000 | 22 | December 2010 | 28 |
| December 2001 | 33 | December 2011 | 26 |
| December 2002 | 26 | December 2012 | 27 |
| December 2003 | 27 | December 2013 | 37 |
| December 2004 | 27 | December 2014 | 46 |
| December 2005 | 26 | December 2015 | 55 |
| December 2006 | 30 | December 2016 | 62 |
| December 2007 | 36 | December 2017 | 86 |
| December 2008 | 19 | December 2018 | 102 |
| December 2009 | 30 | December 2019 | 158 |

## SOLUTION

Using a stem of the *tens* unit will break the data into fifteen classes, which seems reasonable. The stem-and-leaf display is given in Figure 3.4.5.

**Stem-and-Leaf Display (MSFT)**

| Stem | Leaf |
|---|---|
| 1 | 9 |
| 2 | 2 6 6 6 7 7 7 8 |
| 3 | 0 0 3 6 7 |
| 4 | 6 |
| 5 | 5 |
| 6 | 2 |
| 7 | |
| 8 | 6 |
| 9 | |
| 10 | 2 |
| 11 | |
| 12 | |
| 13 | |
| 14 | |
| 15 | 8 |

Key:   15 | 8 = $158

**Figure 3.4.5**

From the stem-and-leaf display, we would guess that the average closing price for MSFT is somewhere around $30. When analyzing a stem-and-leaf display, you can guess the average of the data set by choosing the value with an equal number of leaves greater than it and less than it. To be more precise, you can find the median of the data fairly easily using a stem-and-leaf display.

## Ordered Arrays

An **ordered array** is a listing of all the data in either increasing or decreasing magnitude. Data listed in increasing order are said to be listed in **rank order**. If listed in decreasing order, they are listed in **reverse rank order**. Listing the data in an ordered way can be very helpful. It allows you to scan the data quickly for the largest and smallest values, for large gaps in the data, and for concentrations or clusters of values.

**Definition**
**Ordered Array**
An **ordered array** is a listing of a data set in either increasing or decreasing order of magnitude.

**Definition**
**Rank Order**
Data listed in increasing order are said to be listed in **rank order**.

**Definition**
**Reverse Rank Order**
Data listed in decreasing order are said to be listed in **reverse rank order**.

**Example 3.4.3**
**Creating An Ordered Array of Employee Ages**

The personnel records for a clothing department store located in the local mall are examined, and the current ages for all employees are noted. There are 25 employees, and their ages are listed in the following table. It is desired that their ages be placed in rank order.

**Table 3.4.8 – Ages (Raw)**

| 32 | 21 | 24 | 19 | 61 | 18 | 18 | 16 | 16 | 35 | 39 | 17 | 22 |
| 21 | 60 | 18 | 53 | 18 | 57 | 63 | 28 | 20 | 29 | 35 | 45 |

## SOLUTION

**Table 3.4.9 – Ages (Ordered)**

| 16 | 16 | 17 | 18 | 18 | 18 | 18 | 19 | 20 | 21 | 21 | 22 | 24 |
| 28 | 29 | 32 | 35 | 35 | 39 | 45 | 53 | 57 | 60 | 61 | 63 |

## ☁ Technology

In Microsoft Excel, the Sort tool allows the user to sort any number of data values in ascending or descending order. To learn how to do this with Excel or other technologies, please visit stat.hawkeslearning.com and navigate to **Discovering Business Statistics, Second Edition > Technology Instructions > Data Manipulation > Sorting.**

### Definition

**Dot Plot**

A graph where each data value is plotted as a dot above a horizontal axis is called a **dot plot**.

It is always a good idea to get a look at the ordered array of the data early in your analysis. Examining the ranked data produces a good intuitive sense for the data. Looking at the ordered array, it is evident that over half of the employees are younger than 25 and only three employees are within 5 years of retirement. We can also easily see that the youngest employee is 16 years old and the oldest employee is 63 years old. Ordering the data makes it possible to analyze the data quickly and easily.

Ordered arrays are easy to create. Virtually all statistics, spreadsheet, and database programs enable the user to quickly sort the data in ascending or descending order. For example, in Microsoft Excel, the Sort tool is found under the Data tab, and allows the user to sort any number of data values in ascending or descending order. If a spreadsheet or database program is not available, a stem-and-leaf display can be helpful in sorting the data.

## Dot Plots

A **dot plot** is a graph where each data value is plotted as a point (or a dot) above a horizontal axis. If there are multiple entries of the same data value, they are plotted one above another.

### Example 3.4.4

**Creating a Dot Plot of MSFT Closing Stock Prices**

Construct a dot plot of the closing stock price on December 1st for Microsoft Corporation from data presented earlier in the section. For convenience, the data are repeated below.

| Table 3.4.7 – Closing Stock Prices of Microsoft Corporation (MSFT) December 2000 through December 2019 | | | |
|---|---|---|---|
| **Period** | **Closing Price (Dollars)** | **Period** | **Closing Price (Dollars)** |
| December 2000 | 22 | December 2010 | 28 |
| December 2001 | 33 | December 2011 | 26 |
| December 2002 | 26 | December 2012 | 27 |
| December 2003 | 27 | December 2013 | 37 |
| December 2004 | 27 | December 2014 | 46 |
| December 2005 | 26 | December 2015 | 55 |
| December 2006 | 30 | December 2016 | 62 |
| December 2007 | 36 | December 2017 | 86 |
| December 2008 | 19 | December 2018 | 102 |
| December 2009 | 30 | December 2019 | 158 |

#### SOLUTION

We plot each data value on the horizontal axis. For values where there are multiple entries, such as $26, we stack the points on top of one another.

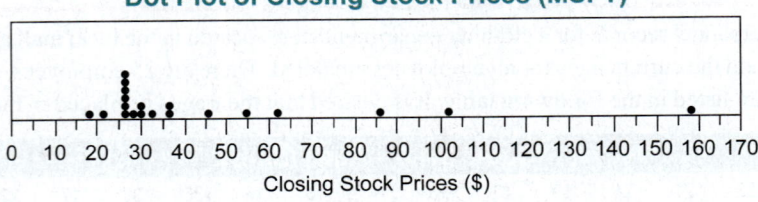

**Dot Plot of Closing Stock Prices (MSFT)**

Closing Stock Prices ($)

Figure 3.4.6

As we can see from the dot plot, about half of the prices are less than $30 and half are greater than $30. The values that occur most often are $26 and $27. There are also several values that are farther away from the main cluster of data values as this distribution is skewed (to the right). Dot plots are useful when you are interested in where the data are clustered and which values occur most often.

## Time Series Plots

Recall our discussion of time series data in a previous chapter. When looking at time series data, we are interested in whether the process is stationary or nonstationary. To determine whether a time series is stationary or nonstationary, we use a **time series plot**. A time series plot graphs data using time as the horizontal axis. Most often, a time series plot comes in the form of a **line graph**, which connects consecutive points over time with a line. You have already seen some line graphs as examples of time series plots in Chapter 2. A good example of time series data is the census measurements taken every decade since 1790. The population of the United States has been growing steadily since the first census (as shown in Table 3.4.11). Interestingly, the change in population between 2000 and 2010 was 27.3 million, which is greater than the entire population of the United States in 1850.

For a quick look at the current population of the U.S. and the world, go to www.census.gov.

| Table 3.4.11 – U.S. Census Data | | | |
|---|---|---|---|
| **Year** | **Population (Millions)** | **Year** | **Population (Millions)** |
| 1790 | 3.9 | 1910 | 92.2 |
| 1800 | 5.3 | 1920 | 106.0 |
| 1810 | 7.2 | 1930 | 123.2 |
| 1820 | 9.6 | 1940 | 132.2 |
| 1830 | 12.9 | 1950 | 151.3 |
| 1840 | 17.1 | 1960 | 179.3 |
| 1850 | 23.2 | 1970 | 203.3 |
| 1860 | 31.4 | 1980 | 226.5 |
| 1870 | 38.6 | 1990 | 248.7 |
| 1880 | 50.2 | 2000 | 281.4 |
| 1890 | 63.0 | 2010 | 308.7 |
| 1900 | 76.2 | | |

**Source:** U.S. Census Bureau; 2010

In a line graph, time is always labeled on the horizontal axis, with the variable being measured labeled on the vertical axis. Points are then plotted for each time period, and a line is drawn that connects each consecutive point. In Figure 3.4.7, each of the points is plotted with the year on the horizontal axis and the corresponding population (in millions) on the vertical axis. In Figure 3.4.8, a line connects consecutive points to give a line graph. From the line graph, we can easily see that the series is nonstationary, and that there is an upward trend in the data.

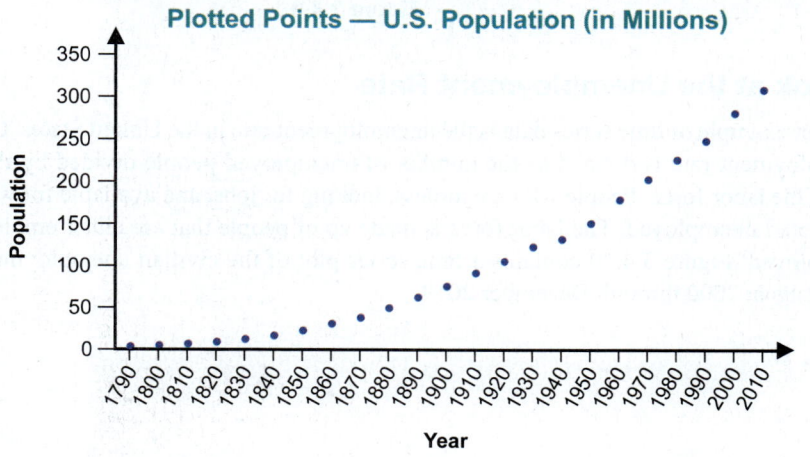

Figure 3.4.7

### Technology

For instructions on how to create a line graph in Excel, JMP, or other technologies, please visit stat.hawkeslearning.com and navigate to **Discovering Business Statistics, Second Edition > Technology Instructions > Graphs > Line Graphs**.

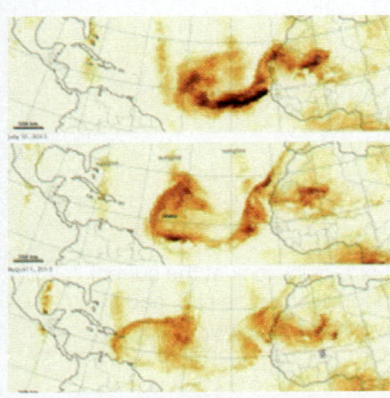

## How the Sahara Feeds the Amazon Jungle

The Sahara is one of the most desolate stretches of land on the planet, yet it breathes life into one of our most fertile jungles. It was a story that few people even knew existed until a group of researchers lead by Dr. Hongbin Yu, a research scientist at NASA's Goddard Space Flight Center, looked at the data. In 2015, the research group published a study that used satellite imagery to quantify the annual amount of phosphorous contained in Saharan dust that was swept into the atmosphere by winds, and transported over the Atlantic to the Amazon Basin. They created three-dimensional simulation models that allowed them to visualize the plumes of dust traveling over the ocean, and then they measured the trace amounts of phosphorous that remained in the dust from when water used to cover the Sahara. Phosphorous is vital in the growth of vegetation, and it is in low supply in tropical regions, making the link between the desert and the jungle vital to the stability of the global ecosystem. This is just one example of how our planet creates ecological balance, and it showcases how much we have left to learn about the place we call home. This discovery is only the beginning of a much larger revolution, and through the increased collection and analysis of data, we may one day be able to tell the true story about the ways of our world.

**Source:** NASA's Goddard Space Flight Center

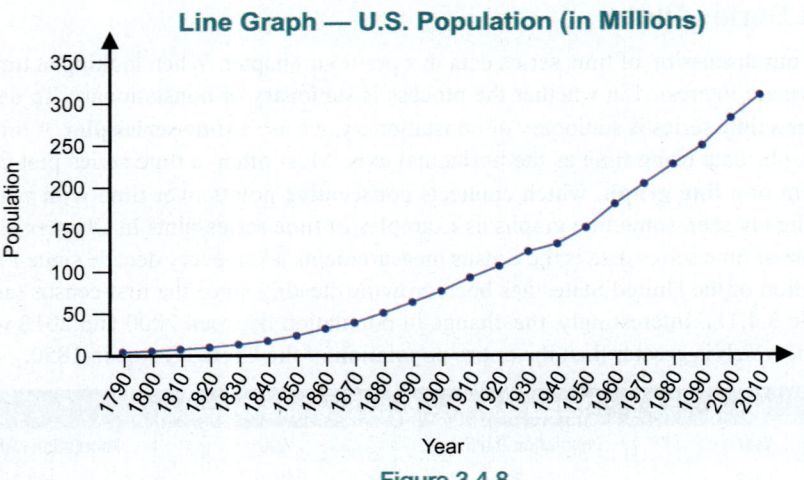

Figure 3.4.8

Line graphs can be analyzed with or without each point being visible. Another type of plot, called a **ribbon plot**, is another version of a line graph (see Figure 3.4.9). Although it is difficult to determine the population size in any one year, this graph makes an interesting visual statement about the population in the United States. Again, it is clear that the population has exhibited an upward trend since 1790.

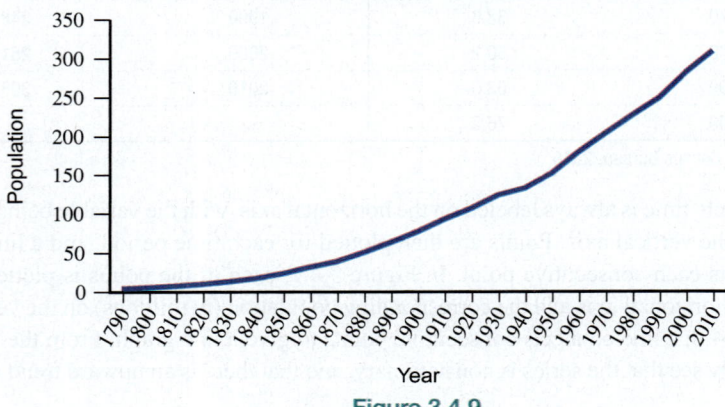

Figure 3.4.9

## A Look at the Unemployment Rate

Another example of time series data is the unemployment rate in the United States. Civilian unemployment rate is defined as the number of unemployed people divided by the total size of the labor force. People who are jobless, looking for jobs, and available for work are considered unemployed. The labor force is made up of people that are either employed or unemployed. Figure 3.4.10 contains a time series plot of the civilian unemployment rate from August 2000 through December 2019.

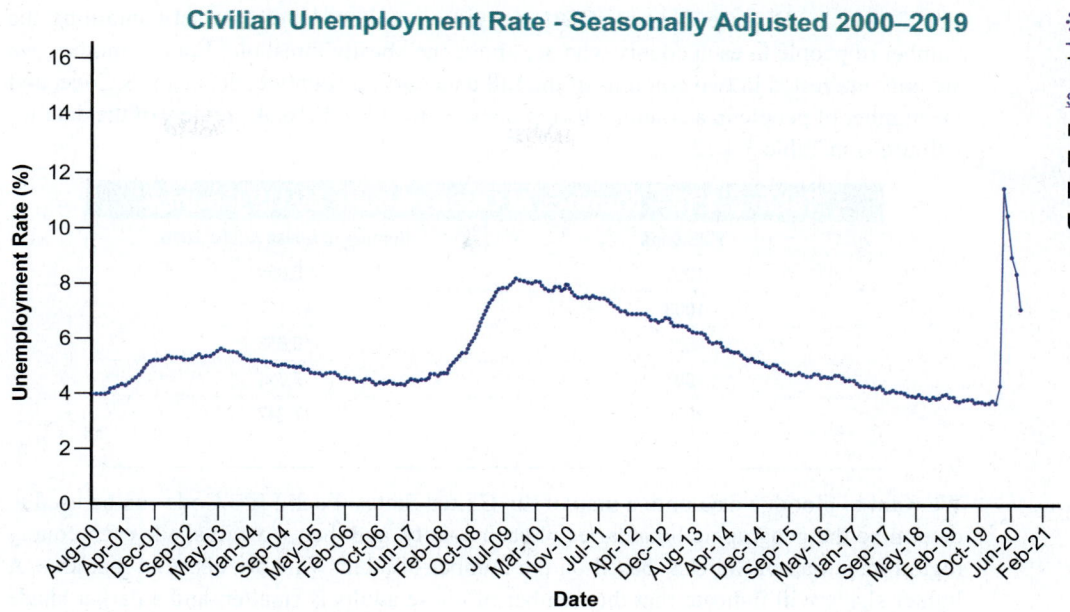

**Civilian Unemployment Rate - Seasonally Adjusted 2000–2019**

**Source:** U.S. Bureau of Labor Statistics

**Figure 3.4.10**

⋮⋮ **Data**

This data can be found at stat.hawkeslearning.com by navigating to **Discovering Business Statistics, Second Edition > Data Sets > Civilian Unemployement Rates**.

As can be seen in the figure, over the last 20 years, there was quite a bit of fluctuation in the unemployment rate. There was a steady increase in the unemployment rate from August 2000 through July 2003. From July 2003 through May 2007, there was a steady decline in civilian unemployment. As can be seen in the figure, there was a steady increase in civilian unemployment from June 2007 through November 2009. This steady increase was due to the subprime mortgage crisis which led to the collapse of the housing bubble in the United States. Falling housing-related assets contributed to a global financial crisis. The crisis led to a steady increase in civilian unemployment which ultimately rose to more than 10% and the failure of many of the United States' largest financial institutions in spite of the government bailout. The government bailed out many financial institutions with an unprecedented $700 billion bank bailout and $787 billion fiscal stimulus package. In November 2009, the unemployment rate began to drop. The figure only shows overall unemployment rates. If one wanted to further explore the unemployment rates, data could be collected (and displayed) by worker groups, gender, and race, for example.

## Geospatial Graphs

For centuries, humans have used maps to portray information about specific places or regions. Today, it is very common to see data that is associated with some type of geographic location such as zip codes, county codes, states, countries, or even specific longitudes and latitudes. This association allows us to layer the data values on top of a map of a certain area, or satellite image, ultimately providing us with a greater understanding of how different regions compare to one another based on a variety of different variables.

**Choropleth maps** are one type of geospatial graph that has become very popular, although their use dates back to the early 19th century. Choropleth maps utilize shades of color to measure a variable across several pre-defined areas, and those areas can be anywhere on a map as long as they are pre-defined and have data associated with them. Choropleth maps are similar to heat maps; however, heat maps use the given data to define regions whereas choropleth maps come with predefined regions with which data is already associated. An example of a heat map would be weather radar, where the shaded region is dynamically defined by the amount and type of precipitation in an area rather than by state or county lines.

Suppose we are interested in visualizing obesity in different regions of the United States.

✎ **NOTE**

You frequently hear data visualization specialists refer to encoding data into a graphic. For example, in a histogram the data is encoded into different categories. If you are creating a choropleth map, the data is encoded spatially by placing the data in a region (usually county, state, or country). The magnitude of the data is usually represented by a color shade. So, the data is encoded in two dimensions.

**Michael Jackson**

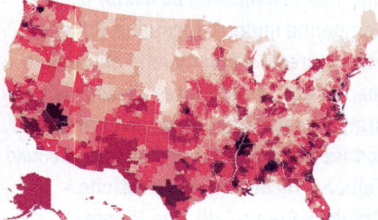

**Statistics Can Be Cool!**

If you enjoy music, you will likely enjoy looking at the geospatial popularity of various contemporary musical artists or groups. There is an article on the New York Times website entitled "What Music Do Americans Love the Most? 50 Detailed Fan Maps" that you are highly encouraged to examine.

The Center for Disease Control (CDC) provides a county level data set containing the number of people in each county who are above the obesity threshold. For our inquiry, we are only interested in two columns of the full data set: the county code, or FIPS Code, and the number of people in a county who are obese in the year 2016. A preview of the data we will use is in Table 3.4.12.

| Table 3.4.12 – Number of Obese Adults by US County ||
|---|---|
| FIPS Code | Number of Obese Adults 2016 |
| 1001 | 15,884 |
| 1003 | 47,117 |
| 1005 | 10,653 |
| 1007 | 7,611 |
| 1009 | 17,347 |
| … | … |

We want to plot this data onto a map of the United States. Each FIPS Code can be used to determine the geographic boundaries of each county, and then we will shade each county region on a map of the U.S. based on the number of adults who are obese in that area. A lighter shade will indicate that the number of obese adults is smaller, and a darker shade will indicate that the number is higher.

### County Obesity Population 2016

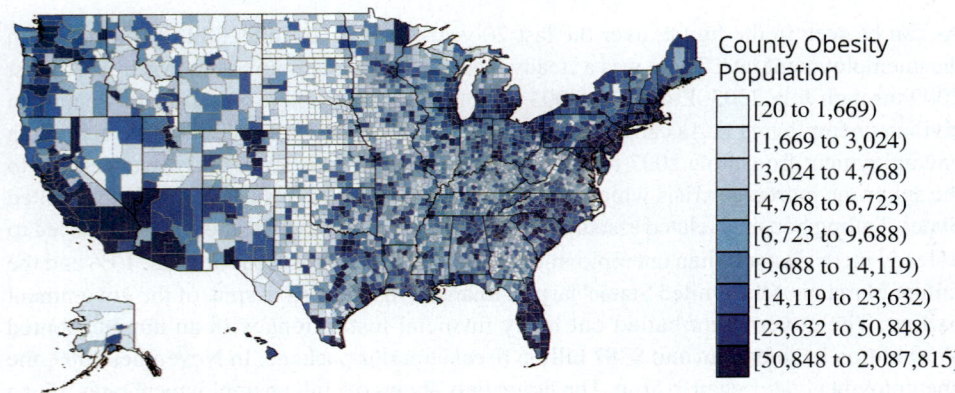

County Obesity
Population

[20 to 1,669)
[1,669 to 3,024)
[3,024 to 4,768)
[4,768 to 6,723)
[6,723 to 9,688)
[9,688 to 14,119)
[14,119 to 23,632)
[23,632 to 50,848)
[50,848 to 2,087,815]

Figure 3.4.11

Figure 3.4.11 suggests that the majority of obese adults in the United States reside in the Southwest and Northeast, and that the Midwest and Central Plains areas don't have very many obese adults at all. By looking at the legend to the right, you can see that there is an extremely wide range from the lowest class to the highest. This is because some counties contain millions of residents, while others contain as few as a couple hundred. So, for many of the rural counties in the center of the country, even if every single person in the county was obese, they still would not compare with, say, Los Angeles County which has a total population of nearly 10 million.

It is inaccurate to rank the counties simply by the total number of obese adults, and doing so will lead to faulty or incomplete conclusions. In order to get a better understanding of how the counties truly compare to one another, we need to *normalize* the values being graphed. Since our variable of interest in Figure 3.4.12, number of obese adults, is a subset of the total population in each county, we can use the total population in each county to find the percentage in each county that is obese. This can be done by dividing the number of people in a county who are obese by the total county population, however, the US County data set provides the percentages for us. A preview of the updated data set, with the new variable of interest, is shown in Table 3.4.13.

| Table 3.4.13 – Percentage of Obese Adults by US County | |
| --- | --- |
| FIPS Code | Percentage of Obese Adults 2016 |
| 1001 | 30.5 |
| 1003 | 26.6 |
| 1005 | 37.3 |
| 1007 | 34.3 |
| 1009 | 30.4 |
| … | … |

Using our new data, we generate the following map.

### County Obesity Percentages 2016

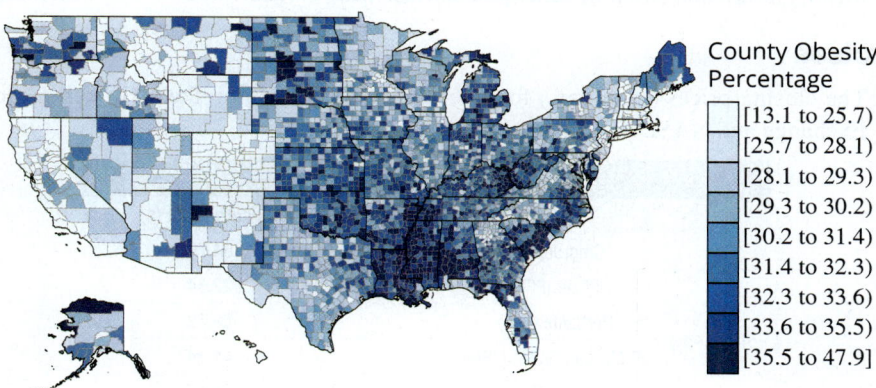

County Obesity Percentage

[13.1 to 25.7)
[25.7 to 28.1)
[28.1 to 29.3)
[29.3 to 30.2)
[30.2 to 31.4)
[31.4 to 32.3)
[32.3 to 33.6)
[33.6 to 35.5)
[35.5 to 47.9]

Figure 3.4.12

As you can see, Figure 3.4.12 is substantially different from Figure 3.4.11, and comes along with an entirely new set of conclusions. If we had used Figure 3.4.11 to come to a conclusion, we would have assumed that the Southwest and Northeast regions are the areas in the country that struggle with obesity the most. However, once we normalized our obesity variable by making it a ratio of the total county population, Figure 3.4.12 seems to suggest that it is actually the Southeast and the Midwest that struggle with obesity the most. The data shows that these regions have a higher percent of adults who are obese relative to their total population than the Northeast and Southwest do. This is a prime example of why it is necessary to normalize data when making comparisons.

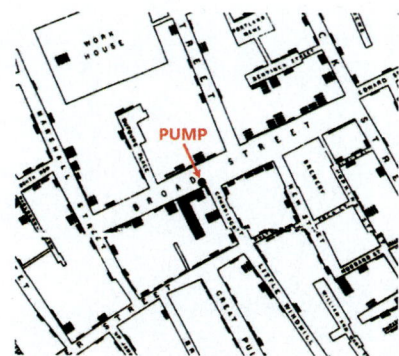

PUMP

### Absence of Evidence is Not Evidence of Absence

During the London cholera outbreaks of the mid-1800s, thousands of people died within a relatively short period. At the time, the prevailing theory regarding how cholera was spread was called the miasma theory. It stated that the disease was spread through "bad air" that emanated from rotting organic matter. However, Dr. John Snow suspected that unsanitary water from the River Thames was the true culprit. Unfortunately, germ theory had not been developed yet, so Dr. Snow didn't fully understand how the alternative transmission method worked. In 1854, Dr. Snow utilized sampling and data visualization to illustrate that most of the cholera outbreaks happening at the time were occurring in houses that were close to the water pump on Broad Street. Still, the skeptics endured. However, even though his examination of the water was absent of evidence for harmful microbes, that does not mean that the microbes themselves were absent. Over a decade later, Louis Pasteur would officially propose germ theory, vindicating the work of Dr. Snow.

## 📝 3.4 Exercises

### Basic Concepts

1. What is the main characteristic of data that a histogram reveals?

2. Describe the type of data that could be usefully described with a histogram.

3. True or false: A frequency distribution contains all of the information needed to construct a histogram.

4. List the important features to look for when studying a histogram.

5. Explain why the stem-and-leaf display is sometimes called a "hybrid graphical method."

6. Identify the advantages of a stem-and-leaf display.

7. Consider the following data value: 39. What would be the stem and leaf for this value if we identified the stem as the tens digit? What would be the steam and leaf if we identified the stem as the hundreds digit?

8. When constructing a stem-and-leaf display, how do you determine which part to make the stem and which part to make the leaf?

9. What is an ordered array?

10. What are some advantages of the ordered array?

11. What is a dot plot?

12. What are some advantages of using a dot plot?

13. How can the most frequently occurring value be identified by studying a dot plot?

14. Why is it important to plot time series data?

15. The time variable is always graphed on which axis?

16. Identify a variation on a time series plot that can make the data more visually interesting.

## Exercises

17. The closing prices (in dollars) for selected stocks trading on the New York Stock Exchange and NASDAQ Exchange were as follows.

| Closing Prices | |
| --- | --- |
| Stock | Closing Price ($) |
| Citigroup (C) | 34.70 |
| Pfizer (PFE) | 22.34 |
| Herbalife (HLF) | 69.72 |
| JP Morgan Chase (JPM) | 44.34 |
| Intel (INTC) | 28.07 |
| WalMart (WMT) | 60.67 |
| Microsoft (MSFT) | 31.52 |
| PepsiCo (PEP) | 66.17 |
| General Motors (GM) | 24.81 |
| Verizon Communications (VZ) | 37.66 |
| Southwest Airlines (LUV) | 8.31 |
| Sprint Nextel (S) | 2.76 |
| Yahoo! Inc (YHOO) | 15.07 |
| International Business Machines (IBM) | 205.47 |

a. Construct a frequency distribution for the closing prices.

b. Construct a histogram for the closing prices.

18. A sample of 80 laborers is selected from a large city and their annual salaries are determined. The following histogram summarizes the data.

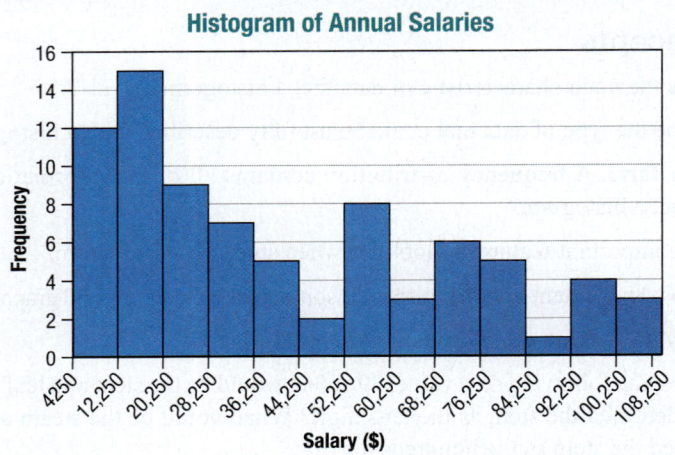

Histogram of Annual Salaries

a. What is the level of measurement of the variable?

b. How many of the laborers earn at least $36,250?

c. What percent of the laborers earn at most $28,249?

d. What percent of the laborers earn at least $84,250?

19. A nutritionist is interested in knowing the percent of calories from fat which Americans consume on a daily basis. To study this, the nutritionist randomly selects 25 Americans and evaluates the percent of calories from fat consumed in a typical day. The results of the study are as follows.

| Calories from Fat per Day (%) | | | | |
|---|---|---|---|---|
| 34 | 18 | 33 | 25 | 30 |
| 42 | 40 | 33 | 39 | 40 |
| 45 | 35 | 45 | 25 | 27 |
| 23 | 32 | 33 | 47 | 23 |
| 27 | 32 | 30 | 28 | 36 |

a. Construct a frequency distribution for the percent of calories from fat.

b. Construct a relative frequency distribution for the percent of calories from fat.

c. Construct a histogram of the relative frequency distribution.

d. Comment on any information about the percent of calories from fat consumed by the participants in the study which you were able to ascertain by examining the distributions and the histogram.

20. Consider the assets (in billions of dollars) of the 10 largest commercial banks listed in the following table.

| Assets (Billions of Dollars) | | | | |
|---|---|---|---|---|
| 216.9 | 138.9 | 115.5 | 110.3 | 103.5 |
| 98.2 | 76.4 | 64.0 | 53.5 | 49.0 |

a. Construct a frequency distribution for the assets (in billions of dollars) of the 10 largest commercial banks.

b. Construct a relative frequency distribution for the assets (in billions of dollars) of the 10 largest commercial banks.

c. Construct a histogram of the relative frequency distribution.

d. Comment on any information about the assets (in billions of dollars) of the 10 largest commercial banks which you were able to ascertain by examining the distributions and the histogram.

**21.** Fifty hospitals in a western state were polled as to their basic daily charges for a semi-private room. The results are listed in the following table, rounded to the nearest dollar.

| Daily Charges for Semi-Private Rooms (Dollars) | | | | | | | | | |
|---|---|---|---|---|---|---|---|---|---|
| 125 | 135 | 148 | 156 | 248 | 215 | 156 | 148 | 135 | 149 |
| 178 | 156 | 135 | 125 | 214 | 256 | 258 | 265 | 156 | 148 |
| 123 | 147 | 189 | 199 | 189 | 248 | 215 | 259 | 158 | 235 |
| 268 | 269 | 158 | 198 | 147 | 258 | 269 | 239 | 288 | 199 |
| 179 | 179 | 189 | 169 | 258 | 178 | 257 | 249 | 259 | 259 |

**a.** What level of measurement do the data possess?

**b.** Construct a stem-and-leaf display for the data using the tens digits as the stems.

**c.** Comment on the shape of the distribution.

**22.** The data in the following table are the toxic emissions (in thousands of tons) for 10 states in the United States.

| Toxic Emissions (Thousands of Tons) | | | | | | | | | |
|---|---|---|---|---|---|---|---|---|---|
| 206 | 147 | 441 | 128 | 127 | 133 | 422 | 152 | 114 | 134 |

**Source:** Toxics in the Community, U.S. Environmental Protection Agency

**a.** Construct a stem-and-leaf display for the data using the hundreds digits as the stems.

**b.** Comment on any information about the toxic emissions (in thousands of tons) of the 10 states that you were able to ascertain by examining the stem-and-leaf display.

**23.** Consider the following highway miles per gallon for 19 selected models of mini-compact, sub-compact, and compact cars.

| Miles per Gallon | | | | | | | | | |
|---|---|---|---|---|---|---|---|---|---|
| 26 | 46 | 36 | 31 | 28 | 28 | 27 | 38 | 42 | 36 |
| 37 | 33 | 23 | 29 | 37 | 34 | 29 | 40 | 28 | |

**a.** Construct a stem-and-leaf display for the data.

**b.** Comment on any information about the highway mpg of the selected models which you were able to ascertain by examining the stem-and-leaf display.

**24.** An instructor is interested in comparing exam scores for fraternity and non-fraternity males in her class. Meaningful comparisons between two sets of data can be made using a side-by-side stem-and-leaf display. To illustrate this, note the following display summarizing the scores.

| Leaf (Non-Fraternity) | Stem | Leaf (Fraternity) |
|---|---|---|
| | 0 | 9 |
| 2 | 1 | 4 0 8 |
| | 2 | 5 7 9 4 5 5 1 |
| 3 9 | 3 | 2 6 6 9 7 7 3 2 1 6 0 |
| | 4 | 2 7 5 |
| 5 6 4 8 9 9 0 2 | 5 | 4 7 6 7 |
| 4 4 7 8 1 0 3 2 2 6 8 9 | 6 | 6 8 9 9 5 |
| 5 4 7 8 4 3 8 8 9 1 | 7 | 3 4 2 7 8 6 7 4 3 |
| 2 9 7 4 | 8 | 4 5 3 8 9 9 6 4 2 1 1 4 5 |
| 4 2 | 9 | 4 3 5 1 6 7 7 0 3 |

Key:    Non-Fraternity 2 | 9 = 92

Fraternity 9 | 4 = 94

   **a.** What level of measurement do the data possess?

   **b.** Based upon the stem-and-leaf display, compare the two groups. Think of the several ways in which this can be done.

   **c.** Suppose that 60% is considered a passing score on the exam. What percent of the fraternity students passed the exam? Non-fraternity students?

   **d.** If someone scores 90 or higher on the exam, they will be exempt from taking the next exam. What percent of the fraternity students will be exempt from taking the next exam? Non-fraternity students?

**25.** Microsoft's consumer PC sales growth for the last 16 quarters are listed in the following table. Examine the data (sales growth, in percentages) and answer the following questions.

| PC Sales Growth | | | |
|---|---|---|---|
| Quarter | Sales Growth (%) | Quarter | Sales Growth (%) |
| 1 | 20 | 9 | 20 |
| 2 | 24 | 10 | 19 |
| 3 | 22 | 11 | 33 |
| 4 | 19 | 12 | 37 |
| 5 | 23 | 13 | 24 |
| 6 | 27 | 14 | 10 |
| 7 | 16 | 15 | 0 |
| 8 | 10 | 16 | −4 |

**Source:** Citi Investment Research and Analysis IDC, Company Reports, May 2011

   **a.** Construct an ordered array of the data in rank order.

   **b.** What conclusions can you make based on the ordered array?

**26.** *Fortune* magazine publishes a list of the top 100 best companies to work for. For the top 10 companies on this list, the average annual employee salaries are given in the following table (in thousands of dollars).

| Average Salaries (Thousands of Dollars) | | | | | | | | | |
|---|---|---|---|---|---|---|---|---|---|
| 121 | 122 | 136 | 74 | 118 | 101 | 114 | 61 | 95 | 132 |

   **a.** Construct a stem-and-leaf display for the data using the tens digits as the stems.

   **b.** Comment on any information about the average annual salaries (in thousands of dollars) of the top 10 companies which you were able to ascertain by examining the stem-and-leaf display.

   **c.** Construct an ordered array of the average annual salaries in rank order.

   **d.** Does the ordered array provide any additional insight into the nature of the data?

**27.** Construct a dot plot for the following set of data.

| | | | | |
|---|---|---|---|---|
| 23 | 19 | 15 | 20 | 17 |
| 16 | 18 | 14 | 23 | 22 |
| 19 | 23 | 19 | 16 | 25 |
| 17 | 20 | 21 | 23 | 24 |

**28.** Listed in the following table is the number of passing attempts per game by Super Bowl champion Aaron Rodgers in the 2010 NFL season. Construct a dot plot for the data set.

| Passing Attempts by Aaron Rodgers | | | | |
|---|---|---|---|---|
| 31 | 29 | 45 | 17 | 46 |
| 33 | 34 | 34 | 34 | 31 |
| 35 | 30 | 11 | 37 | 28 |

**Source:** ESPN

**29.** The following table contains the average monthly energy consumption (in kilowatt-hours) by household for nine South Atlantic states in 2007. Construct a dot plot for the data set.

| Monthly Energy Consumption (Kilowatt-Hours) | |
|---|---|
| 773 | 1143 |
| 960 | 1210 |
| 1163 | 1207 |
| 1171 | 1138 |
| 1086 | |

**Source:** U.S. Energy Information Administration

**30.** The following line graph displays the total IRA and Keogh accounts (in billions of dollars) in the U.S., charted from June 1990 to June 2011.

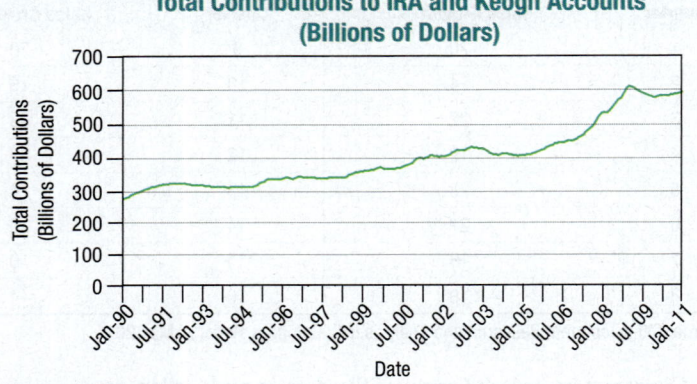

**Source:** www.economagic.com

   **a.** What conclusions can you make regarding the total contributed to the accounts?

   **b.** Are the data time series data?

   **c.** If the data are time series data, is the series stationary or nonstationary?

**31.** The following chart contains LIBOR (which stands for London Interbank Offered Rate) data for January 2011 through May 2011. LIBOR is the average interest rate that banks in London charge when lending funds to other banks. The line graphs in the figure represent 1-month, 3-month, 6-month, and 12-month interest rates.

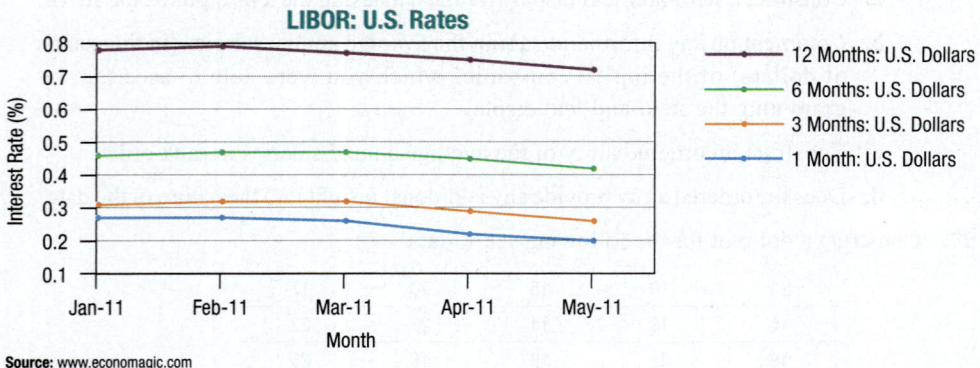

**Source:** www.economagic.com

   **a.** Examine the chart and discuss the data. What conclusions can you make?

   **b.** If the data are time series data, is it a stationary or nonstationary time series? Explain your reasoning.

**32.** The Gallup Poll frequently obtains responses to the question, *At the present time, do you think religion as a whole is increasing its influence on American life or losing its influence?* The percentage of the respondents who answered "increasing" is given below for various polls.

| Survey Responses | | | | | | | | | |
|---|---|---|---|---|---|---|---|---|---|
| Year | 2001 | 1995 | 1992 | 1991 | 1990 | 1988 | 1986 | 1984 | 1982 | 1980 |
| Percent | 71 | 38 | 27 | 27 | 33 | 36 | 48 | 42 | 41 | 35 |
| Year | 1978 | 1977 | 1975 | 1974 | 1970 | 1969 | 1968 | 1965 | 1962 | 1957 |
| Percent | 37 | 37 | 39 | 31 | 14 | 14 | 19 | 33 | 45 | 69 |

    **a.** What level of measurement do the responses to the question possess?

    **b.** Construct a time series plot for the data.

    **c.** What conclusions can you make from the plot?

**33.** The following table gives the number of immigrants (in thousands) and the average annual immigration rate per 1000 people in the U.S. population for the decade ending in the year given.

| Annual Immigration (per 1000 People) | | | | | | | | | | |
|---|---|---|---|---|---|---|---|---|---|---|
| Year | 1900 | 1910 | 1920 | 1930 | 1940 | 1950 | 1960 | 1970 | 1980 | 1990 | 2000 |
| Number | 3688 | 8795 | 5736 | 4107 | 528 | 1035 | 2515 | 3322 | 4493 | 7338 | 9095 |
| Rate | 5.3 | 10.4 | 5.7 | 3.5 | 0.4 | 0.7 | 1.5 | 1.7 | 2.1 | 2.9 | 3.2 |

    **a.** What levels of measurement do the three variables in this exercise possess?

    **b.** Construct a time series plot of the number of immigrants per decade.

    **c.** Find the percent change in the number of immigrants from the decade ending in 1900 to the decade ending in 2000.

    **d.** Find the percent change in the average annual immigration rate per 1000 people in the U.S. population from the decade ending in 1900 to the decade ending in 2000. Compare your answer to that which you obtained in part **c.** Can you explain why these answers are different?

# 3.5 Analyzing Graphs

Graphs that help us visualize data can either be enlightening, in the sense that they gives us insight and understanding of a set of data, or misleading, either intentionally or unintentionally. When you see graphs in the media, you need to be cautious to ensure the data has been accurately represented by the graph. This section will help you analyze graphs for accuracy and appropriate presentation of the given information. Here are a few key ideas to consider when interpreting information displayed in graphical form.

## Graph Labeling

Every graph should be properly labeled with an appropriate title that tells you what type of information is being displayed. Also, if the graph has a horizontal and vertical axis, these should be labeled and should include the unit of measurement when necessary for the understanding of the data. For example, in Figure 3.5.1 shown below, the title does not provide enough information about the data. Why were those countries chosen? Do they have relatively high or low prison populations compared to the rest of the world? Furthermore, we do not know whether this information is relevant to modern times. Is this data for a specific year? The countries are labeled along the horizontal axis, but note that the vertical axis is just labeled *Population*. We have no idea what the values along the vertical axis represent. Is the prisoner population in units of thousands, millions, or billions? In fact, this chart shows the countries with the top ten highest prisoner populations for the year 2016. The unit for the vertical axis should be thousands, which means that the United States had a prison population of approximately 2217 thousand, or 2.217 million, in the year 2016. Without these seemingly small pieces of information, the graph is not very informative. It is also good practice to use the largest possible unit for the scale of an axis, which in this case is correctly chosen to be thousands.

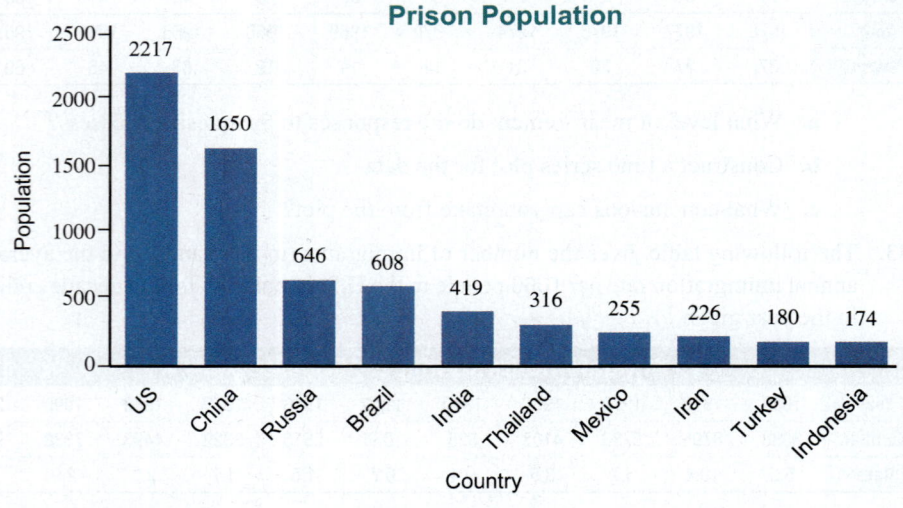

Figure 3.5.1

## Sources

When examining graphs in the media it is very important to consider the source of the information, i.e., who is telling the story. What good is a graph if the underlying data is not credible? The data on the US population from 1790 to 2010 from Section 3.4 came from census data collected by the U.S. Census Bureau, which is a highly reputable source of information. Generally, sources such as government entities and scientific journals are fairly reliable. When looking at the credibility of your data, keep in mind the old quote, "Large skepticism leads to large understanding. Small skepticism leads to small understanding. No skepticism leads to no understanding."

## Appropriateness of a Graph

Throughout this chapter we have looked at a wide variety of graphs. What we want to consider now is the *appropriateness* of a graph. In other words, we want to be able to determine whether the type of graph being used is best suited for the data being displayed. For example, let's contrast the different uses of line graphs and bar graphs. In Section 3.4, we looked at the US population since 1790 and chose to display that data using a line graph. We could have just as easily used the bar graph shown in Figure 3.5.2.

### Statistics Can Be Cool!

There is a YouTube video entitled "The best stats you've ever seen" with millions of views. The video is presented by Hans Rosling who was a global health expert. Mr. Rosling created some amazing data visualization about the state of the world, especially the "developing world".

If you want to see a statistical presentation that is truly inspiring, you will enjoy this video!

### ✎ NOTE

Our companion website, stat.hawkeslearning.com, has a link to a video called "The Art of Data Visualization" that was produced by PBS. While no particular data sets are featured in the video, it displays interesting data driven art and information. The video is only about seven minutes long and is worth taking a quick look.

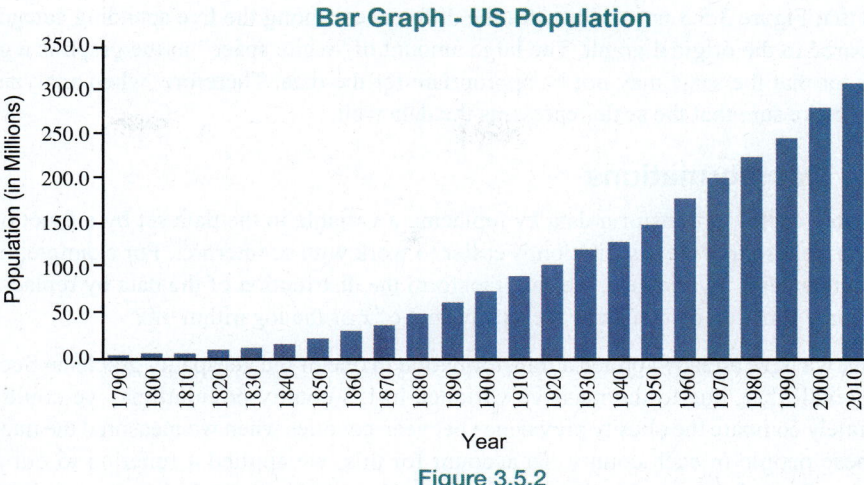

**Figure 3.5.2**

The bar graph also shows the increasing trend of the US population over time, but the trend is better seen using a line graph. The lengths and widths of the bars in the bar graph are somewhat distracting, and obscure the change in population from one time period to the next.

Bar graphs and pie charts are often interchangeable when displaying qualitative data, but if you want to accurately display the categories as parts of a whole, the pie chart will give the best representation, as it allows you to visually compare the slices of the "pie" very quickly. However, the pie chart becomes less effective as the number of categories increases.

## Scaling of Graphs

Another important feature to keep in mind when analyzing graphs is whether a graph is scaled appropriately. If you stretch or shrink the scale on either axis, the shape of the graph can change dramatically, and thus affect the interpretation of the graph. For example, suppose that we change the scale for the bar graph in Figure 3.2.11 on federal government spending to range from 0% to 80% instead of 0% to 25%.

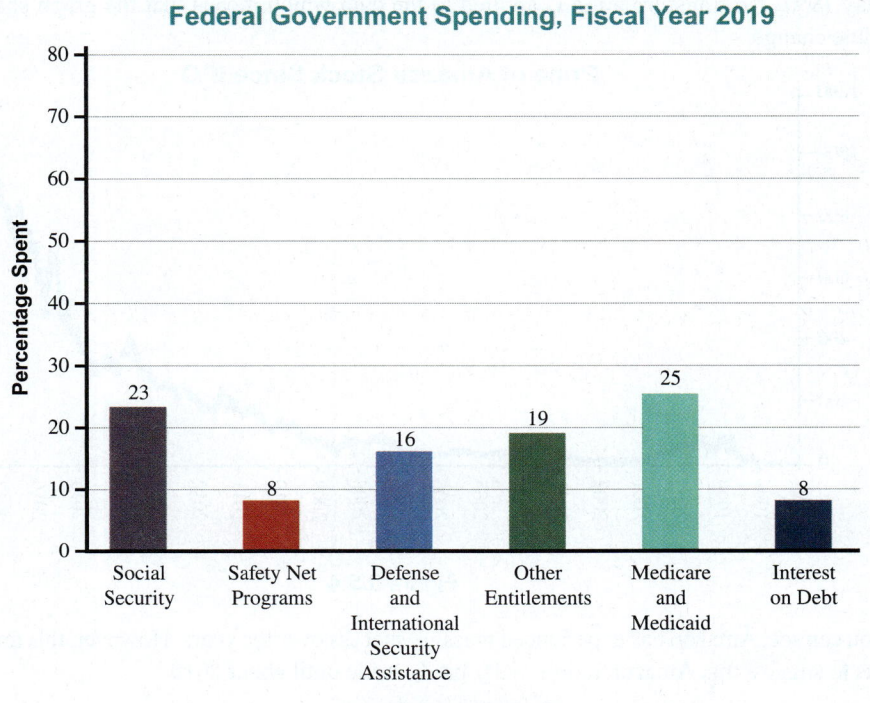

**Figure 3.5.3**

Note that Figure 3.5.3 now minimizes the differences among the five spending categories compared to the original graph. The large amount of "white space" in the graph is a good indicator that the scale may not be appropriate for the data. Therefore, when analyzing a graph make sure that the scale represents the data well.

## Data Transformations

It is often useful to transform data by replacing a variable in the data set by a function of that variable so that the distribution is easier to work with or interpret. For example, if the data set contains a variable $x$, we can transform the distribution of the data by replacing $x$ with some function of $x$, such as the square root of $x$ or the logarithm of $x$.

In fact, we have already applied a transformation to one of the geospatial graphs in Section 3.4. Recall that, due to the massive variance in US county populations, we could not accurately compare the obesity prevalence between counties when we measured the number of obese people in each county. To account for this, we applied a function to our data that transformed the variable of interest, the *number* of obese people in a county, into the *percentage* of obese people in a county. The function divided the number of obese people in a county by the total county population, multiplied the quotient by 100, and resulted in the percentage of the county population that was obese. The alternate perspective of the data then allowed us to easily compare counties since the variable of interest was *normalized* by total county population.

In this section, we will focus on another transformation, the log transformation, which is one of the most common transformations in statistics. The log transformation is often used to help "unclutter" data points for visualization purposes. When the data is tightly grouped together, it can be hard to visualize and make inferences about individual data points. Log transformations numerically "stretch" the portion of the axis closest to zero, and "compress" the portion of the axis farthest from zero. This allows us to better visualize each individual point while also maintaining the underlying relationship between the variables being graphed. The log transformation also allows us to visualize relative change as opposed to absolute change.

Figure 3.5.4 shows the price per share of Amazon stock since its initial public offering (IPO) in May 1997. No transformation is applied to the data which means that the graph shows absolute change.

**Figure 3.5.4**

As you can see, Amazon has experienced massive success over the years. However, this graph seems to suggest that Amazon didn't really hit its stride until about 2010.

In terms of data visualization, logarithms can be used to determine relative change between observations. As you can see in Figure 3.5.4, Amazon's stock price starts to become very

✎ **NOTE**

There is a very good TED Talk video on YouTube entitled "The Beauty of Data Visualization" by David McCandless. In the video, McCandless presents interesting data and a number of "normalization" techniques that are creative.

⸭ **Data**

The full Amazon stock price data set can be found on stat.hawkeslearning.com under **Discovering Business Statistics, Second Edition > Data Sets > Amazon Stock Price.**

substantial starting in 2010. In fact, the difference between the stock price in May 2015 and the price in May 2016 is a larger number than the maximum stock price was in any year prior to 2011. When we compare every data point to the original base, the stock price in May 1997, it becomes very hard to compare relative performances from year to year; a good amount of growth in the late 1990s and early 2000s would not be considered very good after 2010. However, transforming our price variable by applying a log function will allow us to analyze Amazon's yearly performance relative to the stock price for each year. After applying the log function to the y-axis of the previous graph, we generate Figure 3.5.5 which tells the story in a slightly different way.

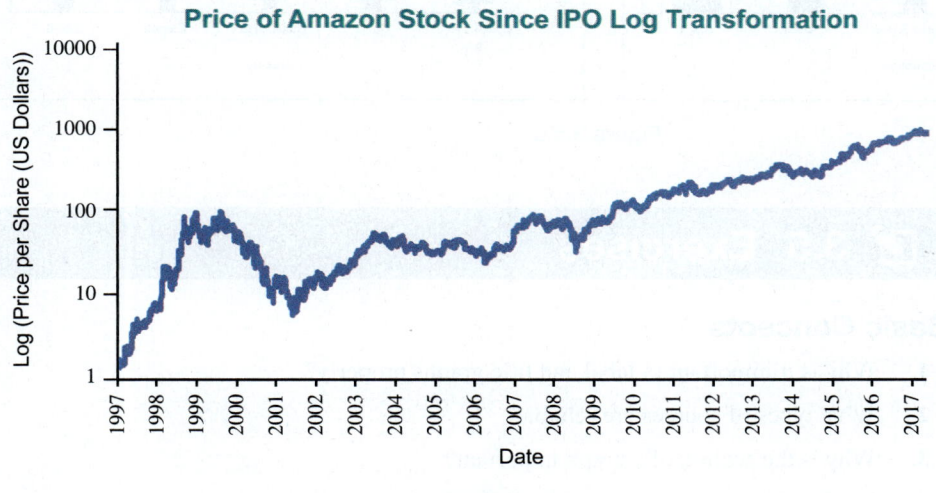

**Figure 3.5.5**

Notice that the y-axis of Figure 3.5.5 now portrays the price variable in factors of 10 as opposed to a continuous unit scale. It turns out that Amazon experienced its most rapid period of relative growth in the years immediately following its IPO. However, notice that the stock price began to sharply decline in 1999. This was when the dot-com bubble "burst". When the bubble burst, thousands of internet companies went out of business. Fortunately, Amazon's business model was built for long-term growth, and it allowed the company to survive the downward economic trend. In 2001, Amazon turned its first profit and restored the confidence of investors. Since then, Figure 3.5.5 reveals that Amazon continues to experience steady year over year growth on the log scale.

## Misleading Graphs

An issue related to scaling that is often used to mislead readers is to start the vertical scale at some value other than zero. We saw this earlier with the sales performance data in Section 3.2. By starting the scale on the vertical axis at $180,000 and using increments of $5,000, the sales performance differences were exaggerated (see Figures 3.2.2 and 3.2.3).

One type of graph commonly used in the media is a pictograph, because it is visually appealing and simple to understand. A **pictograph** is basically a bar graph that uses pictures of objects in place of the bars. For example, the graph on the left in Figure 3.5.6 shows a potentially misleading pictograph of the top five countries ranked in order by the amount of forest area they contain. The bars in this bar chart are represented by trees. Note how as the amount of forest area increases, the trees expand with regard to width and height, thus giving the illusion that the amount of increase is much larger than it is in reality. This is a common problem with pictographs found in the media. Often when the size of the object is increased or decreased, the change is not simply one-dimensional. So be very cautious when looking at data displayed with a pictograph to make sure the graph isn't misleading by representing increases or decreases along one dimension (height of the bar) using an object that is changing in area or volume. The graph on the right represents the correct way to scale a pictograph by only changing the heights of the trees.

**Definition**

**Pictograph**

A **pictograph** is a bar graph that uses images or symbols that are relevant to the data being graphed instead of bars.

**Incorrect Pictograph**                           **Correct Pictograph**

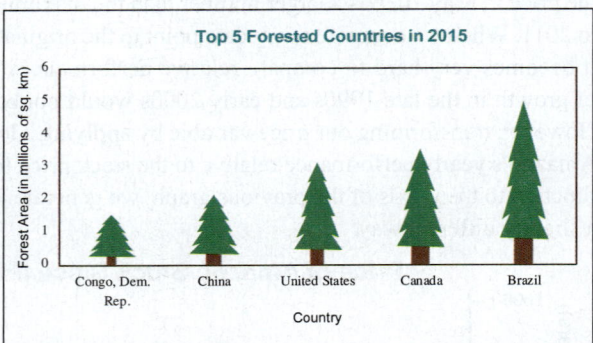

**Figure 3.5.6**

## ✎ 3.5 Exercises

### Basic Concepts

1. Why is it important to label and title graphs properly?

2. What types of sources are reliable?

3. Why is the scaling of a graph important?

4. Why are data transformations useful?

### Exercises

5. Do you see any issues with the scales used on the axes of the graph depicting banana prices per pound in July? Why or why not?

   **Source:** https://data.bls.gov/cgi-bin/surveymost

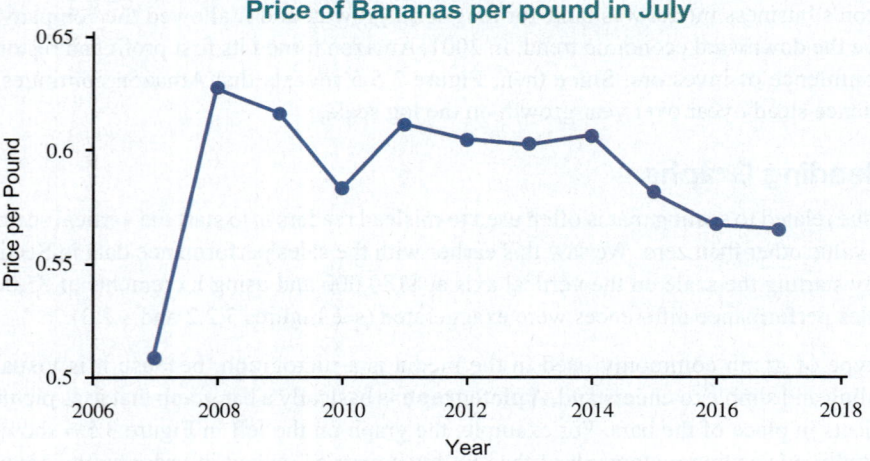

6. Using the San Francisco Salaries 2014 data set from the web resource, create a histogram for the variable TotalPayBenefits and answer the following:

   a. Does the distribution of the data in the histogram look bell-shaped, skewed right, or skewed left?

   b. Construct a new histogram for the variable LogTotalPayBenefits, which is a log transformation of the variable TotalPayBenefits.

   c. Does the distribution of the data in the log transformed histogram look bell-shaped, skewed right, or skewed left?

 **Data**

stat.hawkeslearning.com
**Discovering Business Statistics, Second Edition > Data Sets > San Francisco Salaries 2014**

7.  The US median home price increased from $219,600 in November 2010 to $318,700 in November 2017, as shown in the following pictograph.
    **Source:** U.S. Census Bureau

    a. What was the percentage increase in US median home price between November 2010 and November 2017?

    b. Is the pictograph shown an accurate depiction of this increase? Why or why not?

    c. How could you improve the pictograph so that it accurately represents the information?

### US Median Home Price

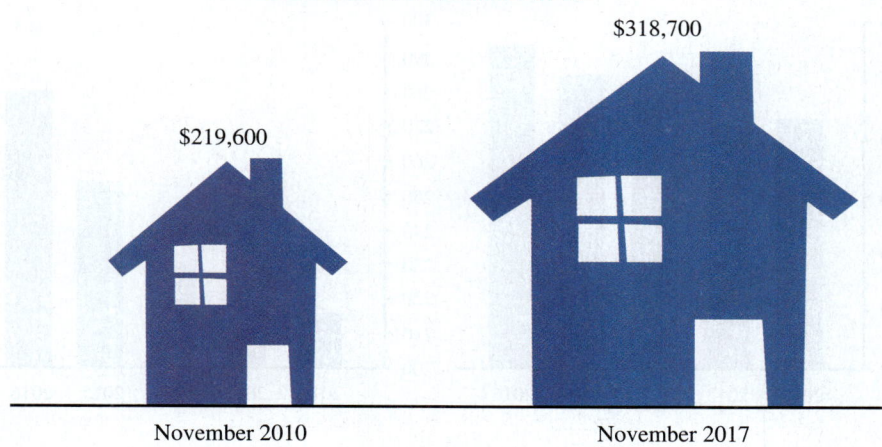

$318,700

$219,600

November 2010                    November 2017

8.  The following histogram uses the heart rate data from Example 3.3.1 but has different classes than were used in the example. What errors can you find in the histogram?

### Histogram of Student Heart Rate Data

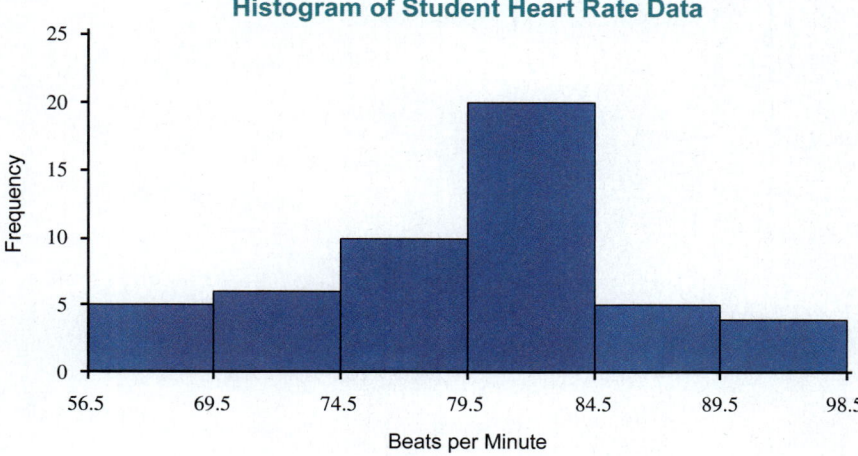

9.  The number of robberies in North Charleston, SC is depicted in two different graphs below. Use these graphs to answer the following.
    **Source:** <u>northcharleston.org</u>

    a.  Which graph do you feel better represents the data? Why?

    b.  If you lived in North Charleston, how concerned would each of these graphs make you feel? Explain.

    c.  Approximately how many times taller is the 2016 bar compared to the 2013 bar in Graph B? How many times more robberies were there actually in 2016 compared to 2013?

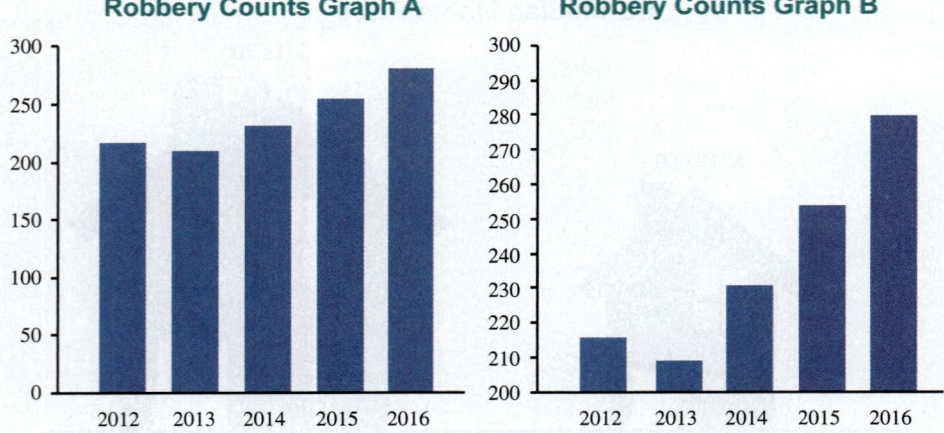

## T Discovering Technology

### Using Excel

Histogram

A histogram is a graphical image of a frequency distribution.

Use the data in Example 3.1.1 on company revenues for this exercise.

1.  Enter the data from Table 3.3.1 into Column A of a new worksheet.

2.  In Column B, you will enter the upper class limits (Excel calls these "Bins"). For the frequency distribution in Table 3.3.2, these are 41, 82, 123, 164, 205, 246, 287, 328, 369, and 410. However, you can experiment with different class widths if you wish.

3.  Under the **Data** tab, click on **Data Analysis**, and choose **Histogram** (See Getting Started with Excel).

4.  The Histogram dialog box will appear. Set the Input Range to **$A$1:$A$100**. You can click on the box in the address window to drag a selection rather than typing in cell addresses if you like.

5.  Set the Bin Range to **$B$1:$B$10**.

6.  In the Output Range, choose **$D$1**.

7.  Check the box next to **Chart Output** and click **OK**.

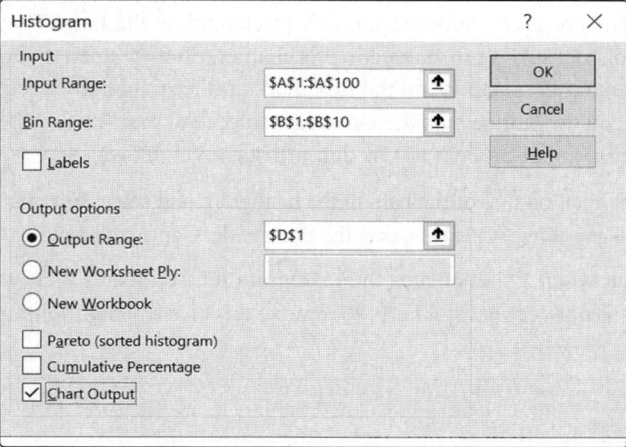

8.  Click **OK**. You should now have a new worksheet appear that looks like the following.

| | A | B | C | D | E | F | G | H | I | J | K | L | M |
|---|---|---|---|---|---|---|---|---|---|---|---|---|---|
| 1 | 116 | 41 | | *Bin* | *Frequency* | | | | | | | | |
| 2 | 408 | 82 | | 41 | 52 | | | | | | | | |
| 3 | 31 | 123 | | 82 | 28 | | | | | | | | |
| 4 | 26 | 164 | | 123 | 14 | | | | | | | | |
| 5 | 25 | 205 | | 164 | 4 | | | | | | | | |
| 6 | 29 | 246 | | 205 | 0 | | | | | | | | |
| 7 | 77 | 287 | | 246 | 0 | | | | | | | | |
| 8 | 99 | 328 | | 287 | 1 | | | | | | | | |
| 9 | 118 | 369 | | 328 | 0 | | | | | | | | |
| 10 | 35 | 410 | | 369 | 0 | | | | | | | | |
| 11 | 72 | | | 410 | 1 | | | | | | | | |
| 12 | 66 | | | More | 0 | | | | | | | | |
| 13 | 45 | | | | | | | | | | | | |
| 14 | 36 | | | | | | | | | | | | |
| 15 | 71 | | | | | | | | | | | | |
| 16 | 37 | | | | | | | | | | | | |
| 17 | 32 | | | | | | | | | | | | |
| 18 | 49 | | | | | | | | | | | | |
| 19 | 53 | | | | | | | | | | | | |
| 20 | 100 | | | | | | | | | | | | |
| 21 | 25 | | | | | | | | | | | | |
| 22 | 105 | | | | | | | | | | | | |

9.  You will notice that Excel creates a histogram with gaps between the bars, and the upper limits are given between the tick marks below the bars. This looks different from the histograms we discussed in this chapter. First we need to reformat the *x*-axis so that the numbers appear on the tick marks rather than between them. To do this, right click on the numbers on the *x*-axis and select **Format Axis**. Under Axis Position, select the radio button next to **On tick marks**. Click **Close**.

10.  Next, right click on one of the bars in the histogram and select **Format Data Series**. To remove the gap between the bars, drag the slider under Gap Width down to **0%**. Click **Close**.

11.  Notice that when Excel creates the histogram it also creates a class called "More." This class is not very helpful here, so remove this class by highlighting cells **D12:E12** and pressing **Delete**.

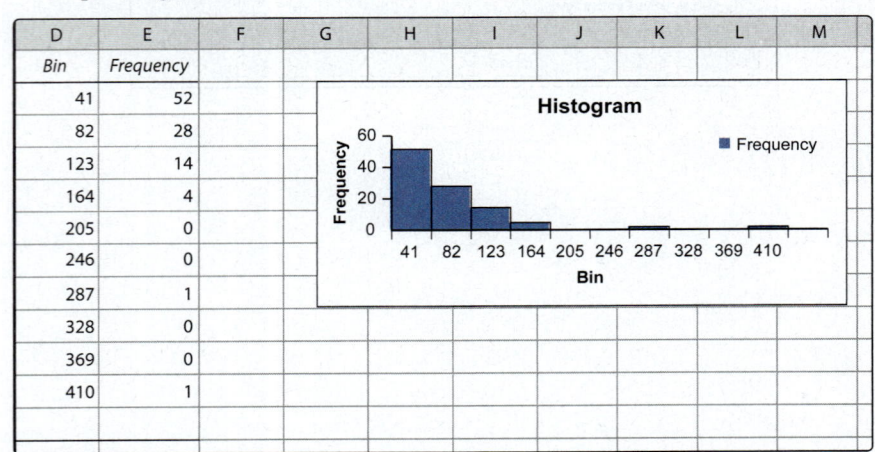

12. The histogram also has a legend, which is not very helpful, so remove it by clicking legend in the chart and pressing **Delete**. Now the histogram looks more similar to the histograms found in the chapter.

## Line Graph

Use the data in Table 3.4.11 for this exercise.

1. Enter the label **Year** in cell A1 and the label **Population (Millions)** in cell B1. Enter the data from Table 3.4.11 into columns A and B. Enter the years in Column A and the populations in Column B.

2. Highlight columns A and B, and under the **Insert** tab, select **Recommended Charts**.

3. Select **Line** and then **OK**.

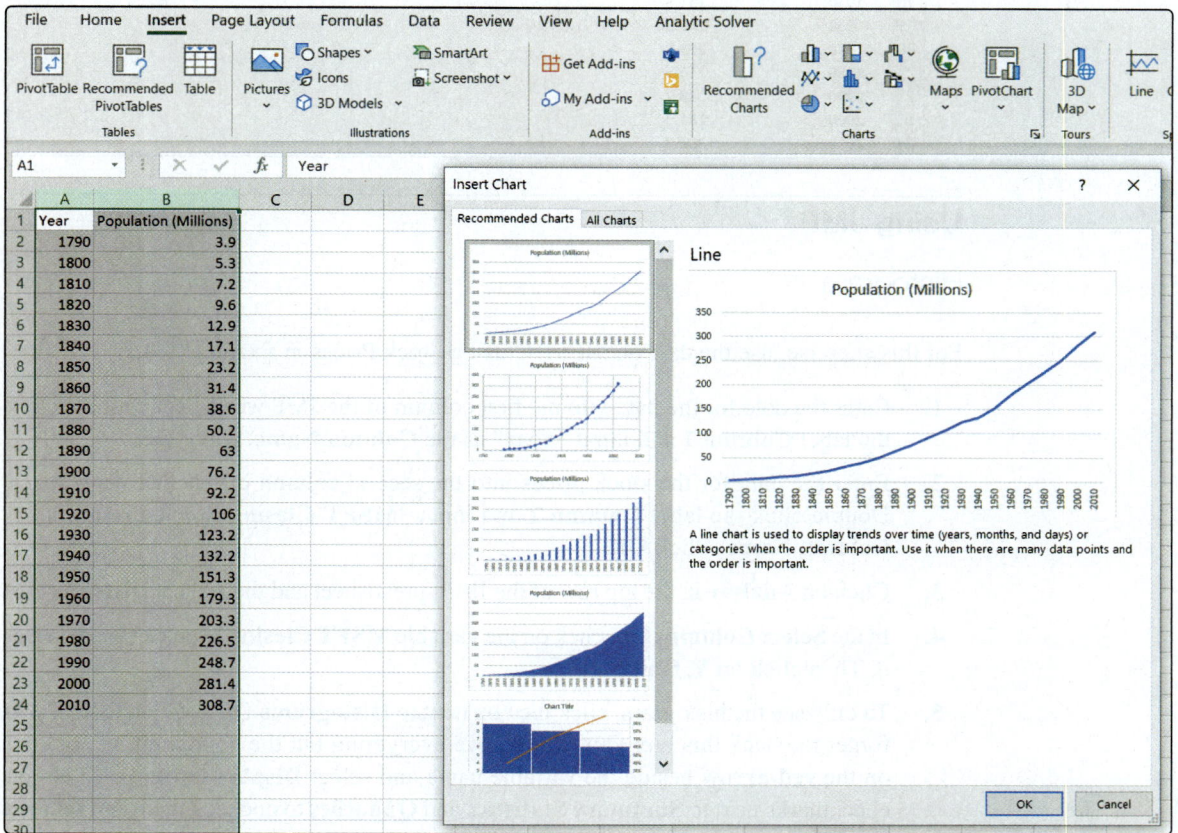

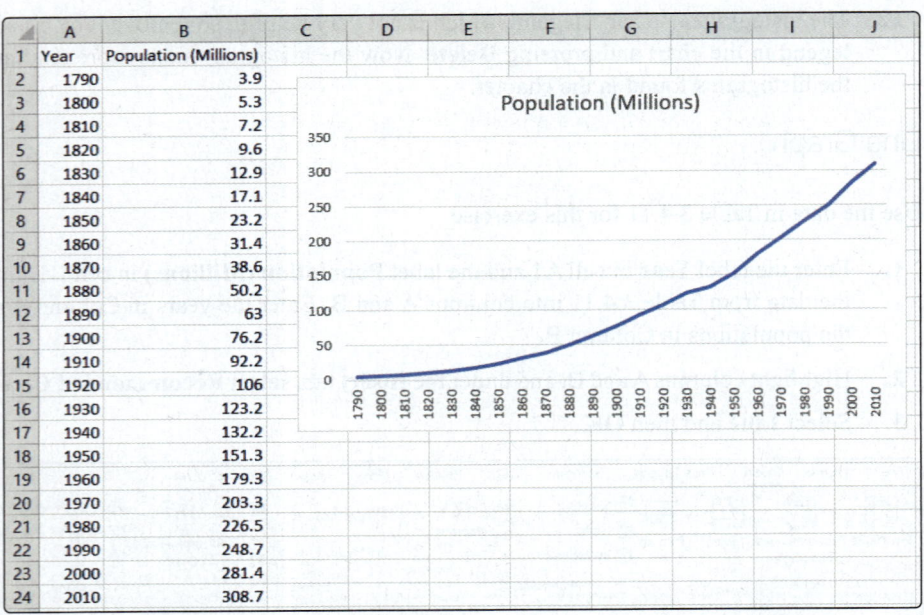

| | A | B |
|---|---|---|
| 1 | Year | Population (Millions) |
| 2 | 1790 | 3.9 |
| 3 | 1800 | 5.3 |
| 4 | 1810 | 7.2 |
| 5 | 1820 | 9.6 |
| 6 | 1830 | 12.9 |
| 7 | 1840 | 17.1 |
| 8 | 1850 | 23.2 |
| 9 | 1860 | 31.4 |
| 10 | 1870 | 38.6 |
| 11 | 1880 | 50.2 |
| 12 | 1890 | 63 |
| 13 | 1900 | 76.2 |
| 14 | 1910 | 92.2 |
| 15 | 1920 | 106 |
| 16 | 1930 | 123.2 |
| 17 | 1940 | 132.2 |
| 18 | 1950 | 151.3 |
| 19 | 1960 | 179.3 |
| 20 | 1970 | 203.3 |
| 21 | 1980 | 226.5 |
| 22 | 1990 | 248.7 |
| 23 | 2000 | 281.4 |
| 24 | 2010 | 308.7 |

## Using JMP

### Histogram

For this exercise, use the data on MFST Closing Stock Prices in Example 3.4.2.

1.  Enter the data for the date into the first column of the JMP worksheet. Double click the label **Column 1** and input **"Year"** as the **Column Name**. Click **OK**.

2.  Enter the data for the stock prices into the second column of the JMP worksheet. Double click the label **Column 2** and input "**MSFT Closing Stock Price**" as the **Column Name**. Click **OK**.

3.  Click on **Analyze** in the top row of the JMP spreadsheet and then select **Distribution**.

4.  In the **Select Columns** box click on the variable **MSFT Closing Stock Price** to select it. Then click on **Y, Columns**.

5.  To only see the histogram, click the box next to **Histograms Only**. Click **OK**. If you forget to check this box you can remove everything but the histogram by clicking on the **red arrow** beside the variable name and select **Display Options**. Click the checkmarks next to **Summary Statistics** and **Quantiles** to uncheck them and remove them from the output.

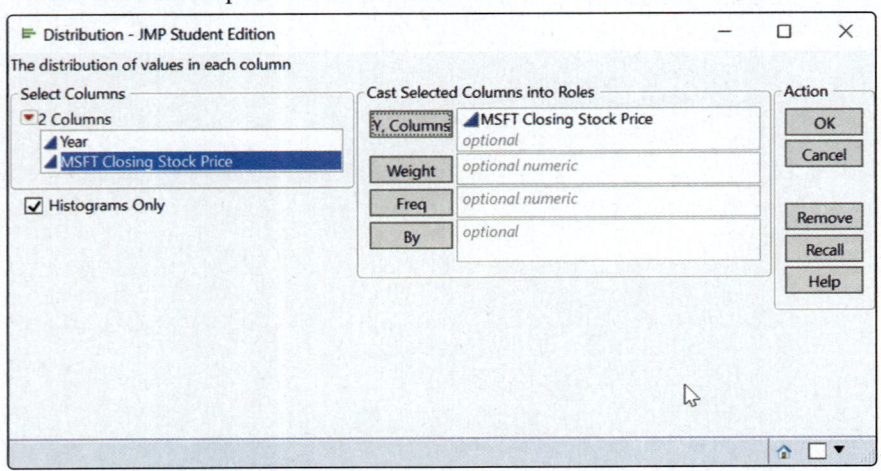

6. To adjust the class width, click the **red arrow** beside the variable name and select **Histogram Options** and then **Set Bin Width**. Enter your desired class width for **New Bin Width** and click **OK**. For this data try a class width of 25 and compare it to a class width of 30 to see how different the histograms look.

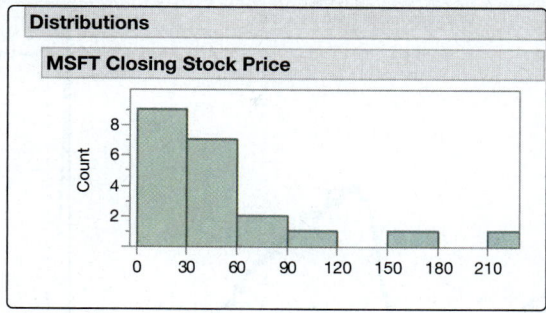

## Line Graphs

For this exercise, use the Civilian Unemployment Rate data used to construct Figure 3.4.10. This data can be found on stat.hawkeslearning.com under Data Sets for Discovering Business Statistics, Second Edition.

1. Enter the data for the date into the first column of the JMP worksheet. Double click the label **Column 1** and input "**Date**" as the **Column Name**. Click **OK**.

2. Enter the data for the unemployment rate into the second column of the JMP worksheet. Double click the label **Column 2** and input "**Unemployment Rate**" as the **Column Name**. Click **OK**.

3. Click on **Graph** in the top row of the JMP spreadsheet and then select **Graph Builder**.

4. Under **Variables**, click and drag the variable **Date** to the area at the bottom labeled **X**.

5. Click and drag the variable **Unemployment Rate** to the area on the left labeled **Y**.

6. To change the x-axis scale, right-click on the x-axis of the graph and select **Axis Settings**. Choose 1 as the **Increment** and 0 as the **# Minor Ticks**. Since there is so much data you can angle the values on the x-axis to make them fit better by choosing **Angled** from the dropdown menu under **Label Orientation**. Click **OK**.

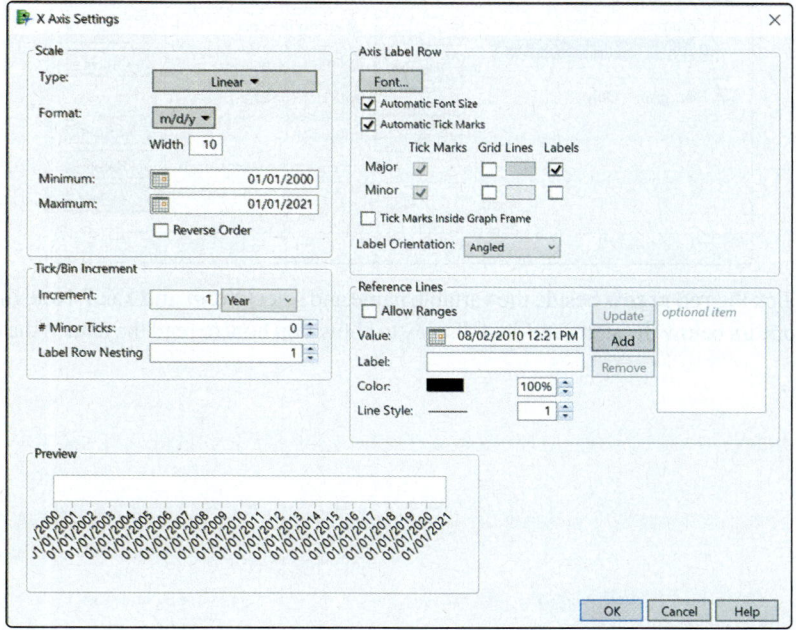

7.  Since **Graph Builder** automatically adds a **Smoother** curve, to instead connect the points and remove the smoother, right-click in the graph area and select **Smoother**, **Change to**, and then **Line**.

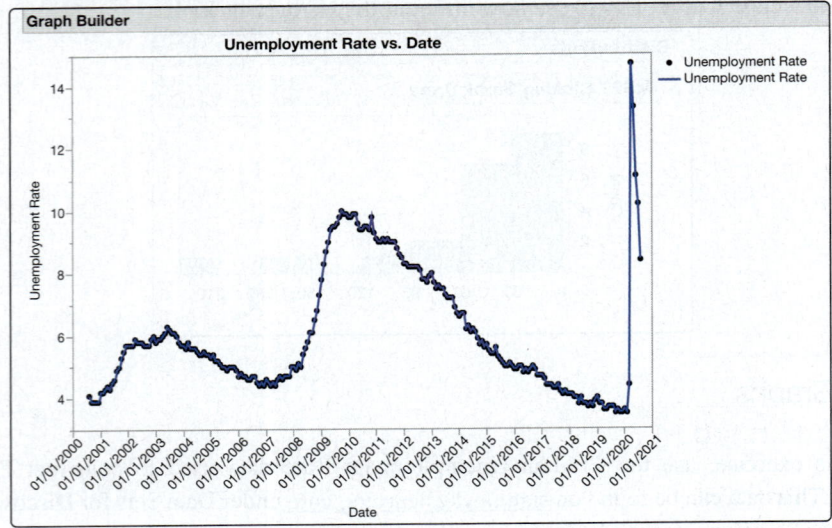

## Stem-and-Leaf Display

For this exercise, use the data on Aptitude Scores in Example 3.4.1.

1.  Enter the data into the first column of the JMP worksheet. Double click the label **Column 1** and input "**Aptitude Score**" as the **Column Name**. Click **OK**.

2.  Click on **Analyze** in the top row of the JMP spreadsheet and then select **Distribution**.

3.  In the **Select Columns** box click on the variable **Aptitude Score** to select it. Then click on **Y, Columns**.

4.  Click the box next to **Histograms Only**. Click **OK**.

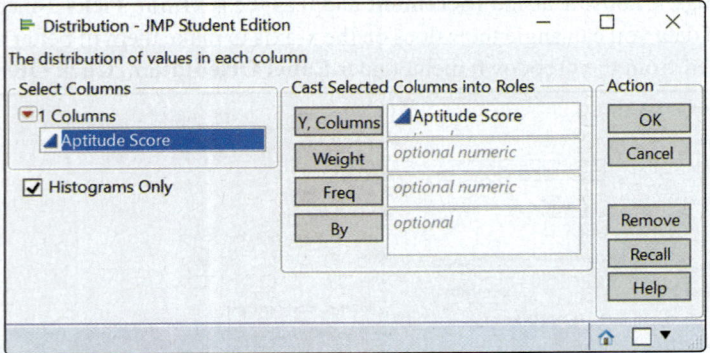

5.  Click the **red arrow** beside the variable name and select **Stem and Leaf**. Note that a key appears below the stem-and-leaf display to show you how to read the data in the display.

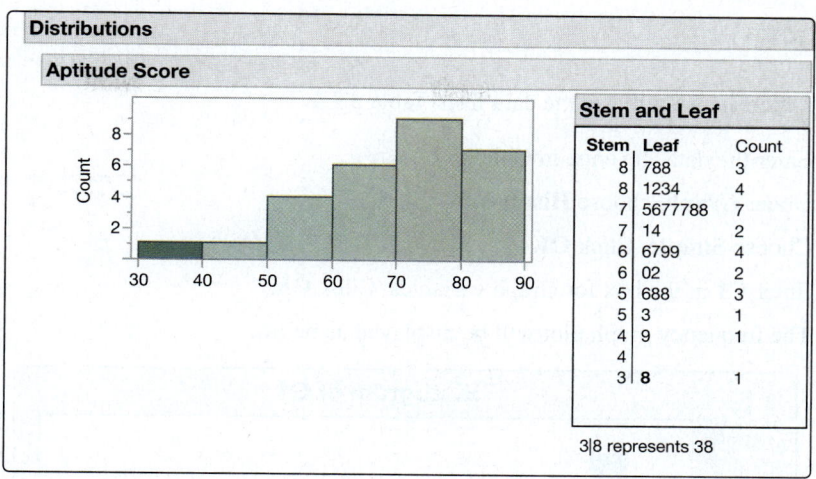

**Distributions**

**Aptitude Score**

**Stem and Leaf**

| Stem | Leaf | Count |
|---|---|---|
| 8 | 788 | 3 |
| 8 | 1234 | 4 |
| 7 | 5677788 | 7 |
| 7 | 14 | 2 |
| 6 | 6799 | 4 |
| 6 | 02 | 2 |
| 5 | 688 | 3 |
| 5 | 3 | 1 |
| 4 | 9 | 1 |
| 4 | | |
| 3 | **8** | 1 |

3|8 represents 38

## Using Minitab

Stem-and-Leaf Display

For this exercise we will use following data: 97, 99, 108, 110, and 111 (as in Table 3.4.5).

1. Enter the data in Column C1.

2. Under **Graph**, choose **Stem-and-Leaf**.

3. Enter **C1** in the box for Graph variables. Press **OK**.

4. The graph plot will be displayed as below with a specific key for the leaf unit in the plot.

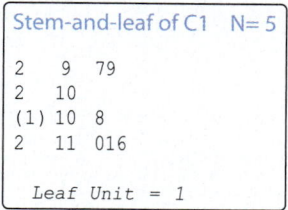

```
Stem-and-leaf of C1    N= 5

2     9   79
2    10
(1)  10   8
2    11   016

  Leaf Unit = 1
```

## Histogram

For this exercise we will use the data from Table 3.3.1.

1.  Enter the data Revenue in Column C1.

2.  Under **Graph**, choose **Histogram**.

3.  Choose **Simple**. Click **OK**.

4.  Enter **C1** in the box for Graph variables. Click **OK**.

5.  The frequency graph plot will be displayed as below.

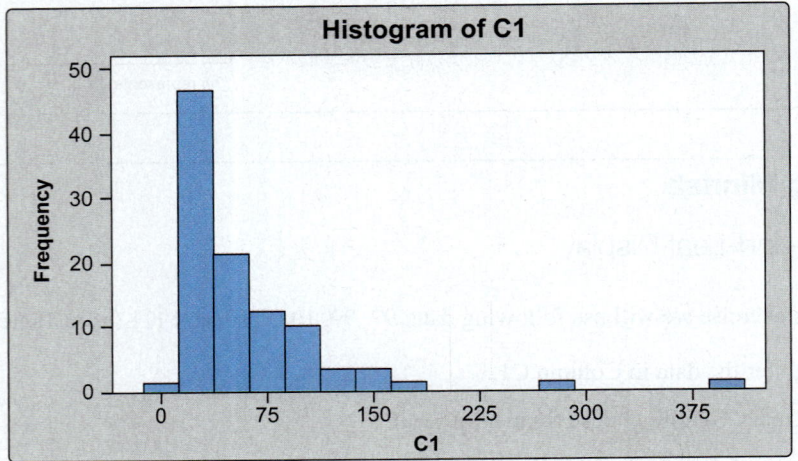

# R  Chapter 3 Review

## Key Terms and Ideas

- Frequency Distribution
- Bar Chart
- 3-D Bar Chart
- Stacked Bar Chart
- Pie Chart
- Relative Frequency
- Cumulative Frequency
- Cumulative Relative Frequency
- Histogram
- Class Limits
- Class Width
- Symmetric Distribution

- Skewed Distribution
- Stem-and-Leaf Display
- Ordered Array
- Rank Order
- Reverse Rank Order
- Dot Plot
- Time Series Plot
- Line Graph
- Ribbon Plot
- Choropleth Map
- Pictograph

## Key Formulas

|  | Section |
|---|---|
| **Number of Classes** | |
| Round $\sqrt{n}$ or $\sqrt[3]{2n}$ to the nearest whole number. | 3.4 |
| **Class Width** | |
| $$\frac{\text{Largest Value} - \text{Smallest Value}}{\text{Number of Classes}}$$ | 3.4 |
| **Relative Frequency** | |
| $$\frac{\text{Number in Class}}{\text{Total Number of Observations}}$$ | 3.4 |

**5.** *Billboard* magazine, in cooperation with Arbitron, produces a national radio format rating. The following data were gathered from radio listeners 12 and older.

| Radio Formats | Mon – Fri 6 AM – 10 AM | Mon – Fri 10 AM – 3 PM | Mon – Fri 3 PM – 7 PM | Mon – Fri 7 PM – 12 AM | Mon – Sun 12 AM – 6 AM | Mon – Sun 6 AM – 12 AM |
|---|---|---|---|---|---|---|
| Adult Contemporary | 17.2% | 19.7% | 17.7% | 15.0% | 16.2% | 20.0% |
| News/Talk | 17.9 | 13.1 | 12.5 | 14.3 | 5.3 | 15.6 |
| Country | 13.0 | 13.2 | 13.2 | 10.3 | 11.7 | 14.3 |
| Album Rock | 10.0 | 10.4 | 10.9 | 9.8 | 18.7 | 10.2 |
| Top 40 | 8.9 | 9.7 | 10.9 | 12.9 | 14.3 | 4.7 |
| Urban | 7.5 | 7.6 | 8.9 | 14.1 | 11.8 | 7.1 |
| Oldies | 6.0 | 6.8 | 6.9 | 6.5 | 4.3 | 10.2 |
| Classic Rock | 4.7 | 3.6 | 3.7 | 3.9 | 6.1 | 2.9 |
| Spanish | 4.5 | 4.2 | 3.7 | 2.2 | 4.9 | 4.2 |
| Adult Standards | 3.4 | 4.2 | 3.7 | 2.7 | 0.3 | 2.8 |
| Religious | 2.1 | 1.7 | 1.8 | 1.8 | 1.3 | 2.5 |
| Classical | 1.4 | 1.7 | 1.7 | 1.9 | 0.5 | 2.3 |
| Easy Listening | 0.9 | 1.1 | 0.9 | 0.8 | 0.2 | 1.2 |
| Modern Rock | 1.0 | 1.1 | 1.3 | 1.6 | 2.4 | 0.4 |
| Adult Alternative | 1.5 | 1.9 | 2.2 | 2.2 | 2.0 | 1.6 |

a. What kinds of graphs would be appropriate for displaying the data? Explain your choices.

b. Graph a column of the data. Briefly analyze your graph.

c. Create a graph that would be useful in visually comparing two columns of the data. Briefly analyze your graph.

**6.** The Caribbean has been a favorite vacation spot for affluent North Americans and Europeans, especially during the winter months. The following table lists the number of tourists during the first six months of the year for a number of Caribbean destinations.

| Number of Tourists | U.S. | Canada | Europe |
|---|---|---|---|
| Antigua & Barbuda | 53,811 | 10,709 | 18,591 |
| Aruba | 94,028 | 1320 | 4681 |
| Barbados | 105,236 | 51,830 | 34,562 |
| Bermuda | 250,390 | 21,241 | 11,715 |
| Bonaire | 12,210 | 352 | 2266 |
| Cayman Islands | 81,180 | 3791 | 3025 |
| Curacao | 15,186 | 572 | 6543 |
| Guadeloupe | 15,596 | 10,654 | 25,409 |
| Trinidad & Tobago | 29,110 | 12,470 | 11,820 |

a. Create a stacked bar graph that shows where tourists from the U.S., Canada, and Europe travel in the Caribbean.

b. Create three separate bar charts, one for American tourists, one for Canadian tourists, and one for European tourists, that show the number of people traveling to each Caribbean destination.

**7.** The following table contains a list of the top 20 global corporations, ranked by the amount spent on research and development in 2009.

| Amount Spent on Research and Development (R&D) in 2009 (Millions of Dollars) | | | | |
|---|---|---|---|---|
| Rank | Company | R&D Spending | Spending as a Percentage of Sales | Headquarters Location | Industry |
| 1 | Roche Holding | 9120 | 20.1 | Europe | Healthcare |
| 2 | Microsoft | 9010 | 15.4 | N. America | Software and Internet |
| 3 | Nokia | 8240 | 14.4 | Europe | Computing and Electronics |
| 4 | Toyota | 7822 | 3.8 | Japan | Auto |
| 5 | Pfizer | 7739 | 15.5 | N. America | Healthcare |
| 6 | Novartis | 7469 | 16.9 | Europe | Healthcare |
| 7 | Johnson & Johnson | 6986 | 11.3 | N. America | Healthcare |
| 8 | Sanofi-Aventis | 6391 | 15.6 | Europe | Healthcare |
| 9 | GlaxoSmithKline | 6187 | 13.9 | Europe | Healthcare |
| 10 | Samsung | 6002 | 5.5 | S. Korea | Computing and Electronics |
| 11 | General Motors | 6000 | 5.7 | N. America | Auto |
| 12 | IBM | 5820 | 6.1 | N. America | Computing and Electronics |
| 13 | Intel | 5653 | 16.1 | N. America | Computing and Electronics |
| 14 | Merck | 5613 | 20.5 | N. America | Healthcare |
| 15 | Volkswagen | 5359 | 3.7 | Europe | Auto |
| 16 | Siemens | 5285 | 5.1 | Europe | Industrials |
| 17 | Cisco Systems | 5208 | 14.4 | N. America | Computing and Electronics |
| 18 | Panasonic | 5143 | 6.4 | Japan | Computing and Electronics |
| 19 | Honda | 4996 | 5.4 | Japan | Auto |
| 20 | Ford | 4900 | 4.1 | N. America | Auto |

**Source:** Booz & Company

    **a.** For comparative purposes, which of the two columns reporting R&D spending is more useful, and why?

    **b.** What types of graphs would be useful in presenting these data? Explain your answers.

    **c.** Develop a histogram for the spending as a percentage of sales.

    **d.** Use computer software to develop pie charts for the headquarters location and industry categories of the top 20 global R&D spenders.

8. In New York, a group of women challenged the state's ban on topless sunbathing. The legal issue was whether the ban was discriminatory. During the controversy, the Gallup poll conducted a survey asking the following question: *Do you think women should be permitted to sunbathe topless on public beaches, if they choose to, or do you think topless sunbathing on public beaches should be banned?*

| Responses to Survey Question | | | | |
|---|---|---|---|---|
| | Permitted | Banned | No Opinion | Number of Interviews |
| National | 33% | 63% | 4% | 1001 |
| **Gender** | | | | |
| Male | 50 | 45 | 5 | 500 |
| Female | 18 | 79 | 3 | 501 |
| **Age** | | | | |
| 18 – 29 | 47 | 51 | 2 | 219 |
| 30 – 49 | 39 | 58 | 3 | 411 |
| 50 – 64 | 18 | 76 | 6 | 206 |
| 65 + | 18 | 77 | 5 | 357 |
| **Region** | | | | |
| East | 39 | 59 | 2 | 247 |
| Midwest | 34 | 62 | 4 | 254 |
| South | 25 | 71 | 4 | 301 |
| West | 38 | 57 | 5 | 199 |
| **Community** | | | | |
| Urban | 42 | 55 | 3 | 345 |
| Suburban | 35 | 62 | 3 | 351 |
| Rural | 23 | 72 | 5 | 298 |
| **Race** | | | | |
| White | 33 | 64 | 3 | 871 |
| Non-white | 39 | 57 | 4 | 121 |
| **Education** | | | | |
| College Grads | 46 | 48 | 6 | 288 |
| Some College | 35 | 62 | 3 | 233 |
| No College | 28 | 69 | 3 | 475 |
| **Sex/Education** | | | | |
| Male/College | 56 | 40 | 4 | 238 |
| Male/ No College | 45 | 49 | 6 | 238 |
| Female/ College | 26 | 70 | 4 | 264 |
| Female/ No College | 13 | 85 | 2 | 237 |

a. Suggest two different types of graphs that might be useful in graphing the data.

b. Create two different graphs using the data.

c. Write a short paragraph describing the data.

9. The nation's political identification (Republican, Democrat, or Independent) changes over time. The data in the following table represent Harris poll results on political identification from 1977 to 2008.

| Year | Republican | Democrat | Independent | Year | Republican | Democrat | Independent |
|------|-----------|----------|-------------|------|-----------|----------|-------------|
| \multicolumn{8}{c}{**Nation's Political Identification** (Percentage of the Population) **1977 – 2008**} | | | | | | | |
| 1977 | 21 | 48 | 25 | 1993 | 29 | 38 | 27 |
| 1978 | 22 | 43 | 30 | 1994 | 32 | 37 | 26 |
| 1979 | 22 | 41 | 31 | 1995 | 31 | 36 | 28 |
| 1980 | 24 | 41 | 29 | 1996 | 30 | 38 | 26 |
| 1981 | 28 | 39 | 28 | 1997 | 29 | 37 | 26 |
| 1982 | 26 | 40 | 28 | 1998 | 28 | 37 | 27 |
| 1983 | 26 | 41 | 27 | 1999 | 29 | 36 | 26 |
| 1984 | 27 | 40 | 24 | 2000 | 29 | 37 | 23 |
| 1985 | 30 | 39 | 26 | 2001 | 31 | 36 | 22 |
| 1986 | 30 | 39 | 25 | 2002 | 31 | 34 | 24 |
| 1987 | 29 | 38 | 28 | 2003 | 28 | 33 | 24 |
| 1988 | 31 | 39 | 25 | 2004 | 31 | 34 | 24 |
| 1989 | 33 | 40 | 23 | 2005 | 30 | 36 | 22 |
| 1990 | 33 | 38 | 25 | 2006 | 27 | 36 | 24 |
| 1991 | 32 | 37 | 26 | 2007 | 26 | 35 | 23 |
| 1992 | 30 | 36 | 29 | 2008 | 26 | 36 | 31 |

**Source:** Harris Interactive

a. What types of graphs would be useful in visualizing these data? Explain your answer.

b. Construct two different types of graphs from the data.

c. Examine the data and write a short paragraph on your conclusions.

10. Monaco is noted for having one of the highest population densities in the world, approximately 16,923 persons per square kilometer. Usually, dense urban areas have relatively high crime rates. This is not the case in Monaco. The following table gives crime data per 100,000 population for the year 2000 in Monaco as well as in other urban areas.

| Crime Data per 100,000 Population | | | | | |
|---|---|---|---|---|---|
| | Monaco | London | Chicago | New York | San Francisco |
| Homicide | 1.0 | 4.7 | 23.0 | 43.4 | 8.6 |
| Forcible Rape | 7.9 | 34.7 | – | 19.1 | 26.7 |
| Robbery | – | 625 | 635.6 | 352.5 | 409.9 |
| Aggravated Assault | – | 847.4 | 880.4 | 473.7 | 327.2 |
| Burglary | – | 938.9 | 895.4 | 394.6 | 764.8 |
| Larceny/Theft | 333.0 | 2675.3 | 3361.9 | 1674.2 | 803.0 |

**Source:** U.S. Department of Justice, CIA, BBC

a. What types of graphs would be useful in visualizing this data? Explain your answer.

b. Construct two different types of graphs from the data.

c. Examine the data and write a short paragraph on your conclusions.

# P   Discovery Project

Your manager asked you to look at your company performance verses other similar companies in your field. You feel that your company has had a successful year. To show this you decide to compare your company's performance with other successful companies. Choose a company that is traded on the United States New York Stock Exchange. This stock will then be considered "your company" for this project. The stock that your company will be compared against is the S&P 500 Information Technology Index or S&P 500 for short.

1. **Download Data**

   a. Download, copy and paste or transcribe the stock data for your company and S&P 500 for the prior year. For example, if it is 2021 then download 1/1/2020 to 12/31/2020 stock performance. Hint: use the internet and search how to download stock data.

      - When downloading your year of data, use the frequency of one week or one data point per week for approximately 52 data points.

      - Only use the Close of market (Close for short) data for this project.

2. **Plotting Time Domain Data**

   a. Create separate plots for your company's stock and the S&P 500 for the year. Remember to label the figure and don't forget to include units such as dollars for the y axis (two graphs).

   b. Create a time series plot including both companies' stocks (one graph).

3. **Hypothetical Investment**

   a. The stock prices of your company and the S&P 500 may vary greatly. To avoid this problem, you decide to make a hypothetical investment of $1,000 at the first of the year. Before doing that, normalize your data by dividing each value by the largest value in your dataset. All your values should be less than or equal to 1. Repeat Part 2 steps A and B plots using the hypothetical investment data (three graphs).

   b. Compare and contrast the plots from Parts 2 and 3.

4. **Report Findings to the Manager**

   After looking at the performance of your company and the S&P 500 write a 1-2 paragraph summary of your findings such as what was learned from the different plots, or how they stocks performed during different time frames. Include other suggestions to further benchmark performance.

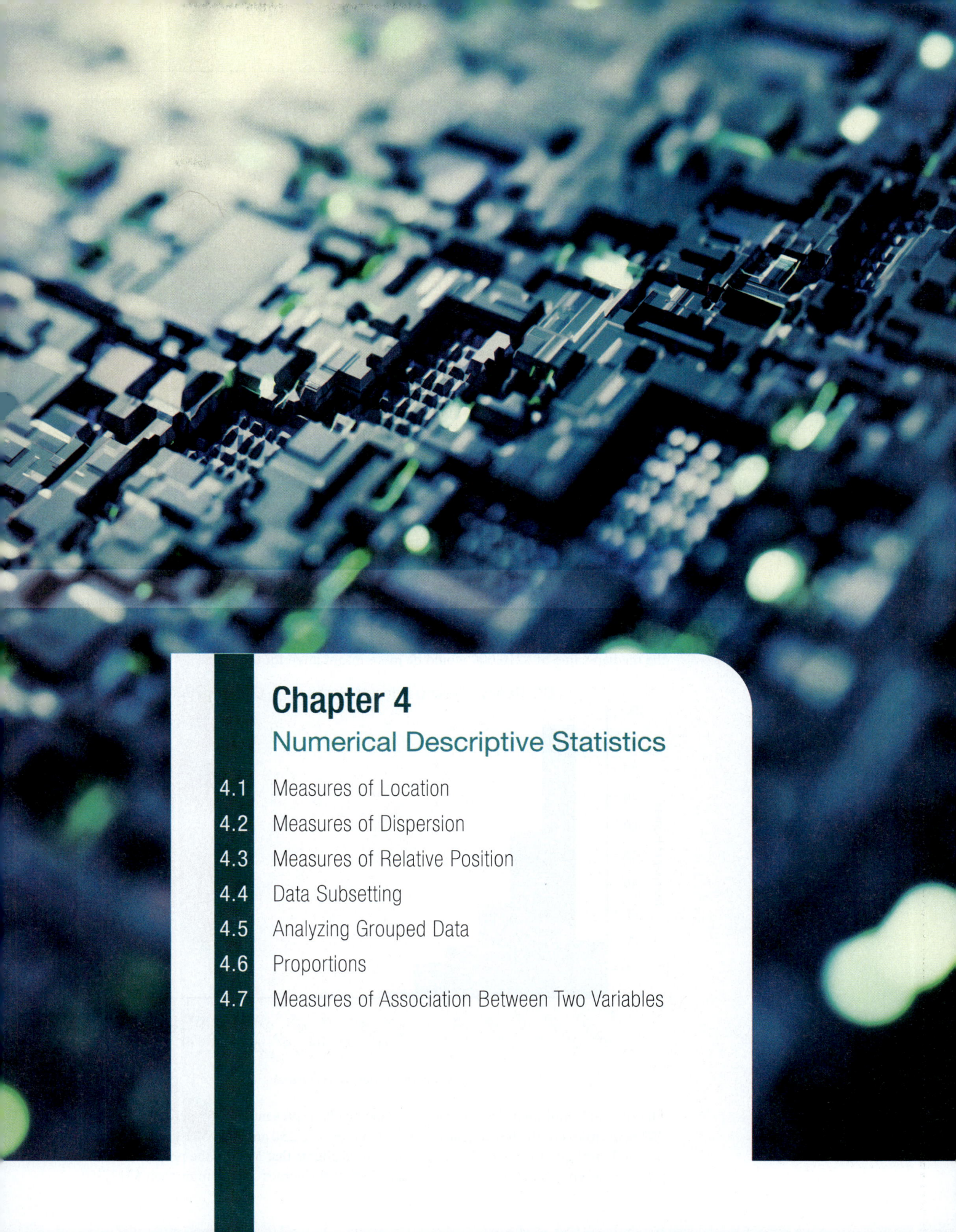

# Chapter 4
## Numerical Descriptive Statistics

## Discovering the Real World

### The Real Estate Market

JDC Realty is a real estate agency in southwest Virginia. JDC Realty is analyzing the data of home sales, commissions, etc., for the last quarter of 2020. Having discussed graphical techniques in Chapter 3, we now want to focus on numerical descriptive statistics in this chapter to get more details about the selling prices of the properties. Using the data from the fourth quarter sales of 2020, JDC Realty would like to provide some information for its clients to answer questions such as:

1. What is the average cost that properties are selling for in southwest Virginia?
2. What is the range of selling prices of properties in southwest Virginia?
3. What price can a client expect to pay in the upper 25% of properties in southwest Virginia?

JDC Realty, which is made up of approximately 10 real estate agents, sold 240 properties in the last quarter of 2020, ranging in price from $8,000 to $1,500,000. Examining the histogram (see the figure below), which is right skewed, it can be seen that a lot of properties sold for less than $200,000, but we can also see that there was one property that sold for a price between $1.5 and $1.6 million.

The questions above are all good questions and the information is not always readily available by examining a histogram. The average selling price of properties in southwest Virginia is $250,643.33. However, one home sold for $1.5 million, which is considered an **outlier**. Thus, the selling price of $1.5 million has inflated the average selling price of the properties. In a situation where outliers are present, reporting the median selling price ($210,000) is more representative of the center of the data than the mean selling price. If JDC Realty wanted to inform clients of a typical price they can expect to pay for a property in southwest Virginia, the median value of $210,000 would be more informative for the client.

 **Data**

This data set can be found on stat.hawkeslearning.com under **Discovering Business Statistics, Second Edition > Data Sets > JDC Realty Property Sales Prices**.

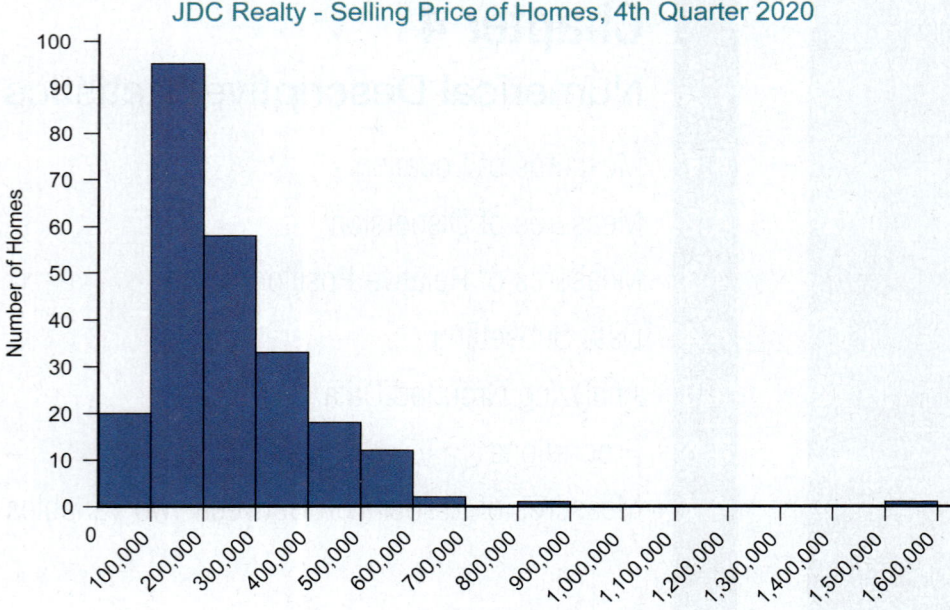

The first and third quartiles, $Q_1$ and $Q_3$, respectively, represent the 25th percentile and the 75th percentile. For the fourth quarter of 2020, $Q_1 = \$152,250$ and $Q_3 = \$313,750$. Using the first and third quartile, JDC Realty can inform its clients that 50% of the properties sell at prices between $152,250 and $313,750, and 25% of the homes sell for more than $313,750.

All of the aforementioned information cannot be deduced from the histogram; thus the need for numerical statistics.

In this chapter we will also discuss relationships and correlations. As JDC Realty examines the data on property sales as it relates to commissions, it is evident from the figure below that as the selling price of the property increases, the agent can expect to make more commission. For example, the agent who sold the property for $1.5 million received a commission of $45,000 while the agent who sold a property for $200,000 received a commission of $6,000.

In addition to discussing the center of the data and the relationship between measurements (such as sales price and commissions), we will also discuss the variability (or spread) of the measurements, the mode, correlation, quartiles, and percentiles, to name a few.

## Introduction

Frequency distributions, bar charts, pie charts, and histograms (all of which were discussed in Chapter 3) can be informative visual tools for examining the big picture when analyzing data. But there is a lack of exactness in the language that we use to describe these graphs. Suppose we say that one data set is more compact than another. This only leads to the question, *How much more compact is it?* Graphical analysis is ill-equipped to answer that question precisely.

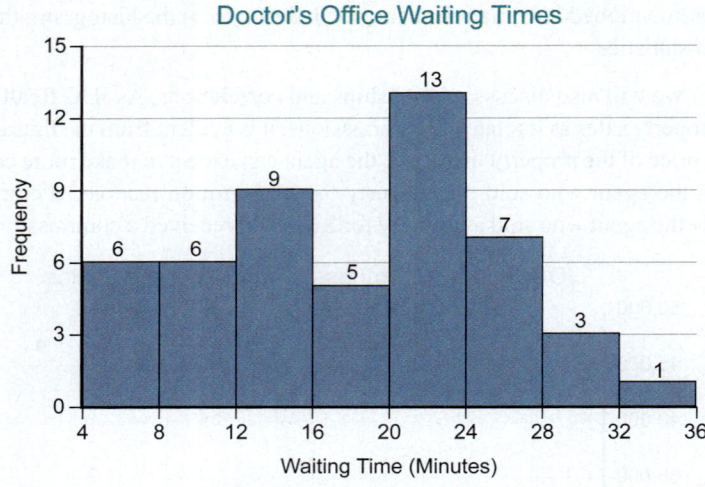

**Figure 4.1.1**

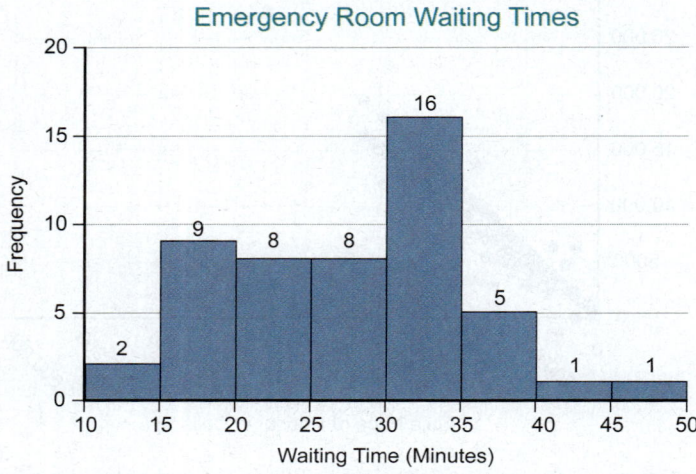

**Figure 4.1.2**

Figure 4.1.1 and Figure 4.1.2 are histograms of waiting times at a doctor's office and emergency room, respectively. While the two histograms are somewhat similar, there is a clear difference in the underlying data. It's clear that the range of waiting times for the doctor's office is smaller than that of the emergency room. Given the ubiquitous choices that patients have for their healthcare, it's important for clinics and hospitals to reduce these patient waiting times to improve patient satisfaction, thus, improving patient reviews and potentially increasing revenue. To precisely describe the differences, we need summary measures to characterize specific data attributes.

- Location: Where is the center of the data?

- Dispersion: Are the data widely scattered or tightly grouped around the central value?

- Shape: Are the data spread symmetrically about the central value? Are the data unbalanced or skewed (e.g., are the values much larger than the mean, but not much smaller)?

- Are there outliers (values that are drastically different from the mean) in the data?

- Do the data tend to cluster in several groups?

Different types of measurements will be developed for each of the attributes listed. For example, the mean (average) and the median measure the **location** or **central tendency** of a data set. Both of these statistical measures are trying to tell us something about the location of the middle of the data, but they use different ideas for defining that notion of middle.

**Definition**

**Measures of Location or Central Tendency**

Statistical measures that tell us something about the location of the middle of a set of data are called **measures of location** or **central tendency**.

From the morning paper to the evening news, general concepts are translated into specific statistical measures. Some examples are listed below.

- The proportion of United States residents earning below the poverty level was 11.8% in 2018.
  **Source:** U.S. Census Bureau

- The median price of new homes sold in 2018 was $326,400.
  **Source:** U.S. Census Bureau

- The average hourly manufacturing wage in the U.S. in 2018 was $27.04.
  **Source:** U.S. Bureau of Labor Statistics

Such measures are examples of **numerical descriptive statistics**.

There is a distinction between measures that are applied to populations (**parameters**) and measures that are applied to samples (**statistics**).

In most instances a data analyst will not know what the population parameters are, since the cost and feasibility of obtaining all the population data are usually prohibitive. **Inferential statistics** involves making conclusions regarding a population using data from a sample.

There are many formulas in this chapter that you will have to spend some time examining to appreciate. In most cases, the concepts which motivate these formulas are simple. Yet if these concepts are either ignored or forgotten, statistics becomes a meaningless assortment of symbols instead of a useful problem-solving and decision-making tool.

The data in Table 4.1.1 are costs (rounded to the nearest dollar) of fill-ups at a local gasoline station for 400 transactions. Looking at the 400 observations without using a graphical representation would be confusing. (It seems 400 measurements would contain a great deal of information, yet the sheer volume of the data obscures comprehension.) It is the old problem of not being able to see the forest for the trees.

**Definition**

**Numerical Descriptive Statistics**

**Numerical descriptive statistics** are numerical summaries of data

**Definition**

**Parameters**

Measures that apply to population data are called **parameters**.

**Definition**

**Statistics**

Measures that apply to sample data are called **statistics**.

**⠇ Data**

This data set can be found on stat.hawkeslearning.com under **Discovering Business Statistics, Second Edition > Data Sets > Costs of Fill-Ups.**

| Table 4.1.1 – Costs of Fill-Ups ($) | | | | | | | | | | | | | | | | |
|---|---|---|---|---|---|---|---|---|---|---|---|---|---|---|---|---|
| 22 | 39 | 25 | 35 | 43 | 36 | 52 | 44 | 37 | 49 | 23 | 21 | 41 | 45 | 43 | 50 | 53 |
| 32 | 32 | 35 | 44 | 33 | 51 | 44 | 28 | 29 | 43 | 51 | 56 | 51 | 35 | 67 | 49 | 38 |
| 37 | 45 | 44 | 28 | 51 | 38 | 47 | 34 | 52 | 50 | 43 | 35 | 11 | 48 | 37 | 27 | 32 |
| 33 | 45 | 28 | 54 | 38 | 40 | 37 | 42 | 38 | 44 | 24 | 37 | 36 | 41 | 30 | 31 | 44 |
| 30 | 40 | 36 | 38 | 32 | 38 | 28 | 23 | 51 | 28 | 51 | 46 | 32 | 45 | 55 | 20 | 48 |
| 43 | 43 | 25 | 24 | 47 | 36 | 22 | 24 | 42 | 55 | 36 | 24 | 48 | 41 | 31 | 31 | 54 |
| 39 | 45 | 36 | 36 | 38 | 32 | 43 | 32 | 34 | 54 | 25 | 19 | 28 | 14 | 41 | 50 | 43 |
| 45 | 58 | 49 | 23 | 55 | 43 | 29 | 11 | 34 | 45 | 42 | 25 | 43 | 53 | 28 | 32 | 47 |
| 48 | 36 | 22 | 34 | 18 | 27 | 37 | 50 | 39 | 30 | 25 | 37 | 34 | 44 | 40 | 42 | 37 |
| 57 | 48 | 36 | 64 | 62 | 33 | 33 | 58 | 31 | 47 | 43 | 38 | 19 | 40 | 72 | 40 | 37 |
| 36 | 38 | 33 | 41 | 26 | 43 | 28 | 42 | 43 | 39 | 41 | 34 | 45 | 45 | 33 | 53 | 46 |
| 50 | 32 | 47 | 22 | 42 | 35 | 21 | 56 | 46 | 21 | 50 | 35 | 43 | 50 | 26 | 41 | 30 |
| 50 | 23 | 52 | 46 | 31 | 41 | 35 | 33 | 30 | 47 | 45 | 48 | 38 | 41 | 30 | 31 | 37 |
| 54 | 43 | 43 | 44 | 51 | 49 | 31 | 56 | 27 | 49 | 36 | 45 | 42 | 34 | 31 | 39 | 41 |
| 34 | 51 | 38 | 34 | 28 | 36 | 43 | 35 | 29 | 47 | 31 | 43 | 34 | 39 | 38 | 52 | 32 |
| 31 | 32 | 37 | 22 | 44 | 34 | 19 | 42 | 47 | 25 | 36 | 53 | 46 | 35 | 36 | 42 | 54 |
| 29 | 48 | 32 | 30 | 25 | 61 | 30 | 21 | 33 | 42 | 60 | 28 | 52 | 35 | 34 | 46 | 55 |
| 41 | 24 | 42 | 39 | 25 | 45 | 36 | 43 | 49 | 41 | 40 | 34 | 44 | 47 | 40 | 44 | 37 |
| 31 | 46 | 55 | 27 | 57 | 32 | 56 | 45 | 66 | 30 | 36 | 47 | 50 | 38 | 53 | 44 | 35 |
| 30 | 33 | 37 | 33 | 41 | 59 | 33 | 24 | 29 | 42 | 35 | 49 | 51 | 39 | 46 | 26 | 46 |
| 44 | 58 | 53 | 47 | 50 | 46 | 30 | 36 | 47 | 51 | 45 | 43 | 20 | 43 | 38 | 33 | 32 |
| 17 | 54 | 59 | 54 | 41 | 39 | 60 | 40 | 36 | 32 | 46 | 43 | 39 | 42 | 49 | 37 | 31 |
| 49 | 25 | 42 | 38 | 55 | 28 | 49 | 45 | 43 | 50 | 35 | 24 | 47 | 42 | 55 | 48 | 54 |
| 52 | 43 | 40 | 29 | 37 | 43 | 27 | 45 | 36 | | | | | | | | |

Looking at a graph of the data (see Figure 4.1.3) is always a good first step. However, we need to learn more. To do this, we use statistical tools designed to reveal the data's fundamental characteristics. These statistical tools answer important questions such as, *Where is the center of the data?* and *How dispersed are the data?*

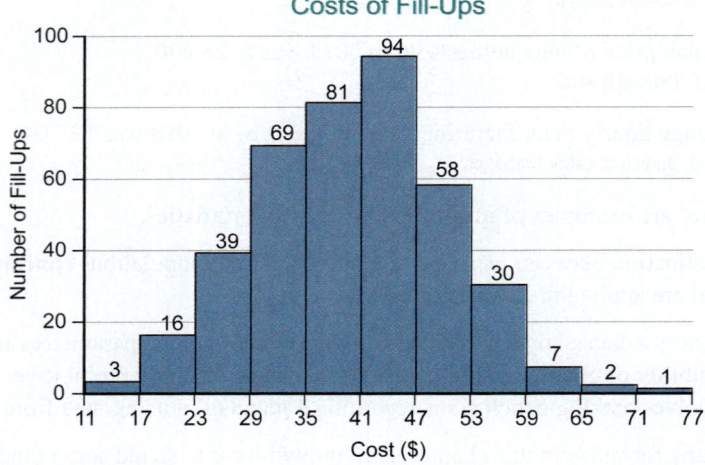

**Figure 4.1.3**

# 4.1 Measures of Location

Statistically speaking, the idea of location is similar to knowing the whereabouts of a person. If we think of a data set as a group of data values that cluster around some central value, then this central value provides a focal point for the data set—a location of sorts. Unfortunately, the notion of *central value* is a vague concept, which is as much defined by the way it is measured as by the notion itself. There are several statistical measures that can be used to define the notion of center: the arithmetic mean, weighted mean, trimmed mean, median, and mode.

## Arithmetic Mean

The arithmetic mean is one of the more commonly used statistical measures. It appears every day in newspapers, business publications, and frequently in conversation. For example, when your instructor returns an assessment, after viewing your grade, one of the first questions asked is, *What is the average?* The word average is often associated with the mean.

**Formula**

**Arithmetic Mean**

Suppose there are $n$ observations in a data set, consisting of the observations $x_1, x_2, \ldots, x_n$; then the arithmetic mean is defined to be

$$\frac{1}{n}\left(x_1 + x_2 + \cdots + x_n\right).$$

If we use some common mathematical notation (summation notation, represented by $\sum$), the formula can be simplified to

$$\frac{\sum x_i}{n},$$

where $x_i$ is the $i$th data value in the data set and $\sum$ (pronounced *sigma*) is a mathematical notation for adding values. There are two symbols that are associated with the expression given above:

$$\mu = \frac{1}{N}\left(x_1 + x_2 + \cdots + x_N\right)$$ the **population mean**, and

$$\bar{x} = \frac{1}{n}\left(x_1 + x_2 + \cdots + x_n\right)$$ the **sample mean**.

Here $N$ refers to the size of the population and $n$ refers to the size of the sample. Otherwise, the calculations are made in precisely the same way. The Greek letter $\mu$, representing the population mean, is pronounced *mu* and the symbol $\bar{x}$, representing the sample mean, is pronounced *x-bar*.

**Example 4.1.1**

**Calculating a Sample Mean of Wait Times**

Suppose the manager at a local automotive shop was observing the time customers spent in their waiting room while their vehicles were being serviced. Using the random sample of four customers' wait times, calculate the sample mean of the wait times.

Wait times: 4, 10, 7, 15.

**SOLUTION**

Note that $x_1 = 4$, $x_2 = 10$, $x_3 = 7$, $x_4 = 15$, and $n = 4$.

$$\bar{x} = \frac{1}{4}\left(4 + 10 + 7 + 15\right) = 9$$

$$= \frac{\sum x_i}{n} = \frac{4 + 10 + 7 + 15}{4} = \frac{36}{4} = 9$$

**⌾ Technology**

For technology instructions to calculate sample statistics like the mean, visit stat.hawkeslearning.com and navigate to **Discovering Business Statistics, Second Edition > Technology Instructions > Descriptive Statistics > One Variable**.

The sample mean is 9. But why does adding up a group of numbers and dividing by the number of observations measure central tendency? As unlikely as it sounds, the answer is related to balancing a scale.

Let's calculate the deviations from the mean for the data in Example 4.1.1. Examining the deviations from the mean in Table 4.1.2, we can see the deviations on the left side ($-5$ and $-2$), and right side (1 and 6), are in balance. In fact, the mean is considered a point of centrality because the deviations from the mean on the positive side and the negative side are equal (See Figure 4.1.4). The sample mean can be interpreted as a center of gravity.

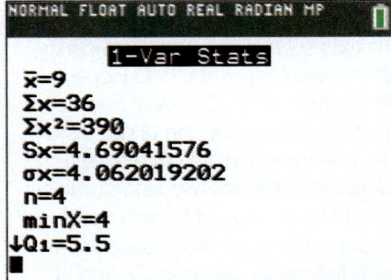

```
NORMAL FLOAT AUTO REAL RADIAN MP

           1-Var Stats
x̄=9
Σx=36
Σx²=390
Sx=4.69041576
σx=4.062019202
n=4
minX=4
↓Q₁=5.5
```

**Definition**

**Deviation**

Given some point $A$ and a data point $x$, then $x - A$ represents how far $x$ **deviates** from $A$. This difference is also called a **deviation**.

| Table 4.1.2 – Deviations from the Mean | |
|---|---|
| Data ($x_i$) | Deviations from the Mean ($x_i - 9$) |
| 4 | $-5$ |
| 10 | 1 |
| 7 | $-2$ |
| 15 | 6 |
| Total | $\sum (x_i - 9) = 0$ |

−5   −2                    1        6

Negative Deviations          Positive Deviations

$7 = |-5| + |-2|$         $\bar{x} = 9$         $(+6) + (+1) = 7$

**Figure 4.1.4**

On the other hand, if we calculate the deviations about any other value the deviations do not balance. For example, assume the central value is 8. The deviation from the alleged central value, 8, for each data value is calculated in Table 4.1.3 and shown in Figure 4.1.5.

## Will Rogers Phenomenon

Will Rogers was an American showman who unwittingly left a lasting impression on the mathematical community when he made a humorous remark regarding migration during the American Depression of the 1930s. He joked: "When the Okies left Oklahoma and moved to California, they raised the average intelligence level in both states." The idea played on the stereotype of the time that Californians were not particularly intelligent, and anyone migrating from any other state to California would raise the average intelligence of California. As it turns out, his joke is mathematically plausible. In fact, the phenomenon occurs so regularly in fields such as medical diagnostics that it was named the "Will Rogers Phenomenon" by Dr. Alvan Feinstein in 1985. The underlying principles of the phenomenon are quite simple. Suppose there are two sets of values; Set 1 has a mean of 50 and Set 2 has a mean of 30. Now, suppose we take five values between 35 and 40 out of Set 1; this will, by definition, raise the mean of Set 1 since the removed values were below the original mean. Now, if we insert those values removed from Set 1 into Set 2, it will, by definition, raise the mean of Set 2 since the inserted values are greater than the original mean of Set 2.

### Definition

**Outlier**

An **outlier** is an extremely large or extremely small data value relative to other data values and can have a dramatic impact on the mean.

### Definition

**Resistant Measures**

Statistical measures which are not affected by outliers are said to be **resistant**.

The positive deviations (2 and 7), are not counterbalanced by the negative deviations (−4 and −1). A desirable characteristic of a central value would be to have the positive and negative deviations equal to each other in absolute value.

| Table 4.1.3 – Deviations from Some Other Value | |
|---|---|
| Data ($x_i$) | Deviations from 8 ($x_i - 8$) |
| 4 | −4 |
| 10 | 2 |
| 7 | −1 |
| 15 | 7 |
| Total | $\sum (x_i - 8) = 4$ |

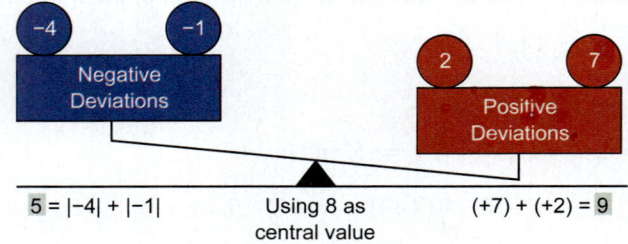

$5 = |-4| + |-1|$        Using 8 as central value        $(+7) + (+2) = 9$

**Figure 4.1.5**

Although the arithmetic mean is frequently used, there are times when it should not be employed. Since the mean requires that the data values be added, it should only be used for quantitative data. Furthermore, if one of the data values is extremely large or small relative to others, this could be considered an **outlier**. An outlier is a data value that can have a dramatic impact on the value of the mean.

The arithmetic mean is not a **resistant measure**.

## Weighted Mean

The weighted mean is similar to the arithmetic mean except it allows you to give different weights (or importance) to each data value. The weighted mean gives you the flexibility to assign weights when you find it inappropriate to treat each observation the same. The weights are usually positive numbers that sum to one, with the largest weight being applied to the observation with the greatest importance. The weights can be determined in a variety of ways, such as the number of employees, market value of a company, or some other objective or subjective method. There are occasions in which it is easier to assign the weights without worrying that they will sum to one. If you are concerned about your weights summing to one, you can make your weights sum to one by dividing each weight by the sum of all the weights.

### Formula

**Weighted Mean**

The weighted mean of a data set with values $x_1, x_2, x_3, \ldots, x_n$ is given by

$$\bar{x} = \frac{w_1 \cdot x_1 + w_2 \cdot x_2 + \cdots + w_n \cdot x_n}{w_1 + w_2 + \cdots + w_n} = \frac{\sum (w_i x_i)}{\sum w_i}$$

where $w_i$ is the weight of observation $x_i$.

The following table consists of the September 2020 unemployment rates and civilian labor force sizes for the Mid-Atlantic states.

| Table 4.1.4 – Unemployment Rates | | |
|---|---|---|
| State | Size of Civilian Labor Force (Thousands) | Unemployment Rate (%) |
| Delaware | 495.3 | 8.2 |
| Maryland | 3067.8 | 7.2 |
| New Jersey | 4394.4 | 6.7 |
| New York | 9134.0 | 9.7 |
| Pennsylvania | 6365.4 | 8.1 |
| Virginia | 4279.5 | 6.2 |
| District of Columbia | 396.8 | 8.7 |
| West Virginia | 773.7 | 8.6 |

**Source:** U.S. Department of Labor, Bureau of Labor Statistics

Using the weighted mean, calculate the average unemployment rate for the Mid-Atlantic states.

**SOLUTION**

The average unemployment rate is calculated as follows.

$$\bar{x} = \frac{\sum(w_i x_i)}{\sum w_i}$$

$$\sum(w_i x_i) = 493.5(8.2) + 3067.8(7.2) + 4394.4(6.7) + 9134.0(9.7) +$$
$$6365.4(8.1) + 4279.5(6.2) + 396.8(8.7) + 773.7(8.6) = 232375.76$$

$$\sum w_i = 495.3 + 3067.8 + 4394.4 + 9134.0 + 6365.4 + 4279.5 + 396.8 + 773.7 = 28906.9$$

$$\text{so, } \bar{x} = \frac{\sum(w_i x_i)}{\sum w_i} = \frac{232375.76}{28906.9} \approx 8.04\%$$

Thus, the average unemployment rate, calculated by the weighted mean, is 8.04%. It is appropriate to use the weighted mean to calculate the average unemployment rate since the size of the civilian labor force (the weight) is different for each state.

## Trimmed Mean

Since outliers can have an enormous effect on the value of the mean, the mean's usefulness as a typical measure of data is diminished if the data contain outliers.

### Finding the 10% Trimmed Mean

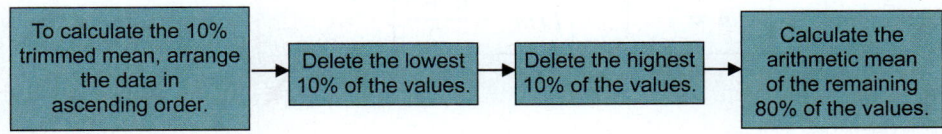

Before calculating the trimmed mean, the data are arranged in ascending order of magnitude. A 10% trimmed mean uses the middle 80% of the values. It is calculated by chopping off the top 10% and the bottom 10% of the data values, and finding the arithmetic mean of the remaining values. If the data set does not contain any outliers, the mean and the trimmed mean will be similar. Note that because the trimmed mean is not affected by outliers like the arithmetic mean is, the trimmed mean is considered a resistant measure.

---

**Example 4.1.2**

**Calculating a Weighted Mean of Unemployment Rates**

**☁ Technology**

For technology instructions to calculate a weighted mean, visit stat.hawkeslearning.com and navigate to **Discovering Business Statistics, Second Edition > Technology Instructions > Descriptive Statistics > Two Variables**.

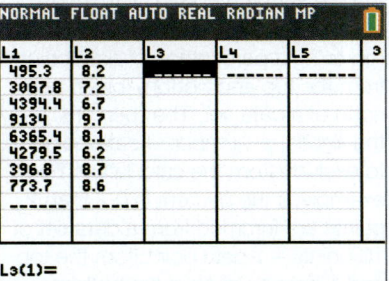

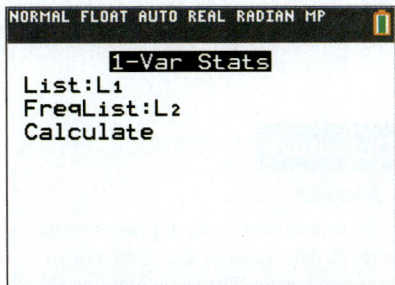

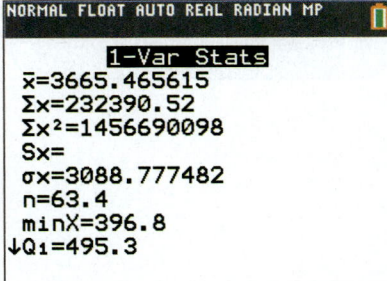

**✎ NOTE**

There is some discrepancy between the solution in the text and the calculator values due to rounding.

**Definition**

**Trimmed Mean**

The **trimmed mean** is a modification of the arithmetic mean which ignores an equal percentage of the highest and lowest data values in calculating the mean.

### Example 4.1.3

**Calculating a Trimmed Mean for Admission Applications**

Suppose that the following sample size of 10 represents the number of applications for admission from local high schools within 50 miles of State University.

$$15, 21, 25, 31, 35, 42, 48, 51, 54, 60$$

Find the 10% trimmed mean.

#### SOLUTION

Since there are 10 observations, removing the highest 10% and the lowest 10% means removing only one observation from each end of the data.

That is,

$$10\% \text{ of } 10 = 0.1 \cdot 10 = 1.$$

Note that the data are already sorted. If the mean is calculated without including the values of 15 and 60, the resulting measure is called the 10% trimmed mean.

$$\cancel{15}, 21, 25, 31, 35, 42, 48, 51, 54, \cancel{60}$$

$$10\% \text{ trimmed mean} = \frac{21 + 25 + 31 + 35 + 42 + 48 + 51 + 54}{8} = \frac{307}{8} = 38.375.$$

If there had been 100 observations, the largest 10% and the smallest 10% (a total of 20 data values) would have been removed before the mean was calculated.

## Median

The median of a set of data provides another measure of center. It is a simple idea. To find the median, place the data in ascending order and then find the observation that has an equal number of data values on either side. That is, half of the observations are less than the median and half of the observations are greater than the median. The median is the middle value.

**Finding the Median**

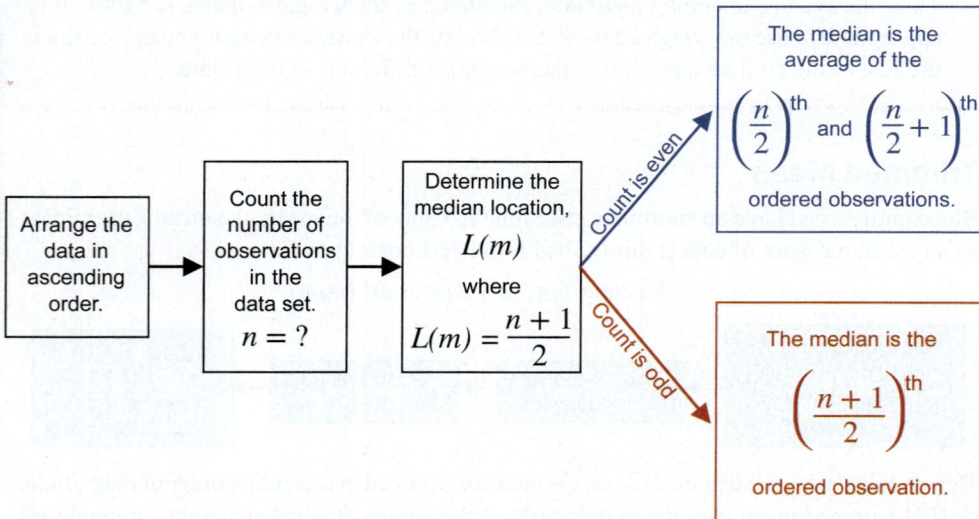

### Example 4.1.4

**Calculating the Median Number of Calls**

Suppose that the data below represent a random sample of the number of 911 calls to a police station over the past 30 days. Given the following eleven observations, find the median.

$$2, 4, 0, 3, 0, 1, 8, 5, 1, 5, 9$$

**SOLUTION**

First, the data set must be ordered.

$$0, 0, 1, 1, 2, 3, 4, 5, 5, 8, 9$$

The number of observations, $n = 11$.

Next, calculate the median location, $L(m) = \dfrac{n+1}{2} = \dfrac{11+1}{2} = 6$. Therefore, the median is the 6th ordered observation, which is 3.

$$\underbrace{0,0,1,1,2,}_{5 \text{ data values}} \boxed{3}, \underbrace{4,5,5,8,9}_{5 \text{ data values}}$$

☁ **Technology**

For technology instructions to calculate sample statistics like the median, visit stat.hawkeslearning.com and navigate to **Discovering Business Statistics, Second Edition > Technology Instructions > Descriptive Statistics > One Variable**.

```
NORMAL FLOAT AUTO REAL RADIAN MP
        1-Var Stats
↑Sx=3.077779601
 σx=2.934547708
 n=11
 minX=0
 Q₁=1
 Med=3
 Q₃=5
 maxX=9
```

Consider the following ten test scores on an aptitude test for a marketing position.

$$65, 98, 76, 83, 94, 79, 88, 72, 90, 85$$

Find the median.

**Example 4.1.5**

**Calculating the Median Test Score**

**SOLUTION**

The number of observations, $n = 10$.

The median location, $L(m) = \dfrac{10+1}{2} = \dfrac{11}{2} = 5.5$.

If there is an even number of observations, average the two center values in the ordered array. The median is the average of the $\dfrac{10}{2} = $ 5th and 6th ordered observations. Thus, we find the median as follows.

$$\underbrace{65,72,76,79,}_{4 \text{ data values}} \boxed{83}, \boxed{85}, \underbrace{88,90,94,98}_{4 \text{ data values}}$$

$$\dfrac{83+85}{2} = 84 \text{ (the median)}$$

The median possesses a rather obvious notion of centrality since it is defined as the central value in an ordered list. It is not affected by outliers and is thus a resistant measure. For example, if we replaced 98 with 200,000,000 in the data set from Example 4.1.5, the median would not change at all. The median does possess one limitation: it cannot be applied to nominal data. In order to calculate the median, the data must be placed in order. To accomplish this task meaningfully, the level of measurement must be at least ordinal.

Unless the data set is skewed or contains outliers, the median and the mean usually have similar values.

**a.** Consider the following data which represent a random sample of the number of product returns each month for the last 10 months from March to December.

$$16, 18, 20, 21, 23, 23, 24, 32, 36, 42$$
$$\text{mean} = 25.5$$

Find the median and 10% trimmed mean.

**Example 4.1.6**

**Calculating the Median and Trimmed Mean of Product Returns**

### A Knower of the Secret of the Dice

Oftentimes, it is easy for us to take simple mathematical concepts, such as the mean, for granted. However, imagine living in a time when these concepts were not so formally defined. The people with the ability to perform these seemingly basic functions were hailed as the magicians of their time.

An ancient Indian story recounts how King Rtuparna, known to be a frequent gambler, estimated the number of leaves and fruit on two large branches of a tree by using the leaves and fruit of a single twig as the average. He then multiplied the number of leaves and fruit on the single twig by the estimated number of twigs on each branch, and his estimation turned out to be very close to the actual value. When asked how he was able to do this, he replied, "Know that I am a knower of the secret of the dice, and therefore adept in the art of enumeration."

The invention and practice of estimation and measurement are arguably some of the most important milestones in human history, and it might come as a surprise that many of the concepts we use today stemmed from gambling scenarios. Technology, cartography, healthcare, and many other facets of our modern lives are almost entirely supported by our ability to estimate the parameters of a population by using descriptive and inferential statistics.

**SOLUTION**

First, find the median.

The number of observations, $n = 10$.

The median location, $L(m) = \dfrac{10+1}{2} = \dfrac{11}{2} = 5.5$.

If there is an even number of observations, average the two center values in the ordered array. The median is the average of the $\dfrac{10}{2} = 5^{\text{th}}$ and $6^{\text{th}}$ ordered observations.

$$16, 18, 20, 21, \boxed{23}, \boxed{23}, 24, 32, 36, 42$$
$$\underbrace{\qquad\qquad}_{\text{4 data values}} \qquad \underbrace{\qquad\qquad}_{\text{4 data values}}$$

$$\frac{23 + 23}{2} = 23 \ \text{(the median)}$$

Now find the 10% trimmed mean.

Since there are 10 observations, removing the highest 10% and the lowest 10% means removing only one observation from each end of the data.

$$10\% \text{ of } 10 = 0.1 \cdot 10 = 1$$

If the mean is calculated without including the values 16 and 42 in the data, the resultant measure is the 10% trimmed mean.

$$\cancel{16}, 18, 20, 21, 23, 23, 24, 32, 36, \cancel{42}$$

$$10\% \text{ trimmed mean} = \frac{18 + 20 + 21 + 23 + 23 + 24 + 32 + 36}{8} = 24.625$$

**b.** Consider the same data set, except the last data value is replaced with an outlier (490). Compare the mean, median, and 10% trimmed mean for this data set to the mean, median, and 10% trimmed mean of the original data set.

$$16, 18, 20, 21, 23, 23, 24, 32, 36, 490$$

$$\text{mean} = 70.3 \quad \text{median} = 23 \quad 10\% \text{ trimmed mean} = 24.625$$

**SOLUTION**

The median and 10% trimmed mean are not affected by the addition of the outlier, while the mean increases dramatically. This illustrates why the median and 10% trimmed mean are said to be resistant measures while the arithmetic mean is not.

## Mode

The **mode** is another measure of location. It is not used as frequently as the mean or the median, and its relation to these values is not so predictable. The mode is the only measure of location that can be used for nominal data. Of the three measures of location, the mode is used the least due to the limited information it provides. Sometimes sorting the data (in ascending or descending order) makes it easier to find the mode.

Find the mode of the following data set, which represents the number of paint chips found in a random sample of vehicles after production.

$$0, 1, 4, 3, 9, 8, 10, 0, 1, 3, 0$$

### SOLUTION

Since the value of 0 occurs more than any other value, it is the mode. In this instance, as a measure of location, the modal value is not a particularly appealing choice. However, the mode does possess one very favorable property—it is the only measure of location that can be applied to nominal data. Thus, for nominal measurements like color preferences, it would be perfectly reasonable to discuss the modal color.

> **Example 4.1.7**
>
> **Determining the Mode of Paint Defects**

> **Definition**
>
> **Mode**
> The **mode** of a data set is the most frequently occurring value.

Suppose we added one more value to the data set in Example 4.1.7. If this value were a 1, then both 0 and 1 would be repeated three times and there would be two modes. When this occurs, the data is said to be **bimodal**. Any time a data set has more than two modes it is said to be **multimodal**. If all observations in a data set occur with the same frequency, then there is **no mode** for that data set.

> **Definition**
>
> **Bimodal**
> A data set that has two modes is said to be **bimodal**.

## The Relationship Between the Mean, Median, and Mode

Oftentimes, the shape of the data determines how the mean, median, and mode are related. For a bell-shaped distribution, the mean, median, and mode are identical.

> **Definition**
>
> **Multimodal**
> A data set that has more than two modes is said to be **multimodal**.

### Bell-Shaped Distribution

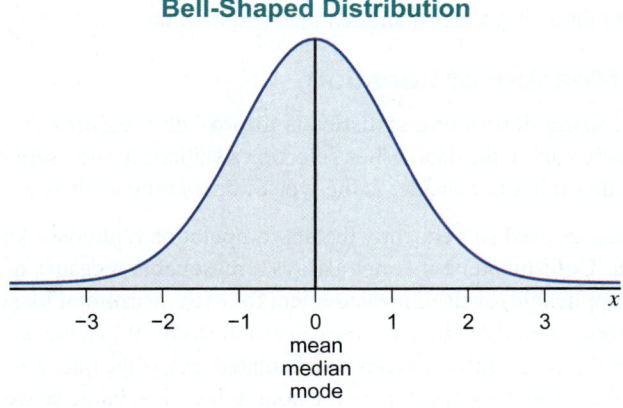

**Figure 4.1.6**

> **Definition**
>
> **No Mode**
> A data set is said to have **no mode** if all observations occur with the same frequency.

Certainly, not all data produce distributions that follow a bell-shaped curve. If the distribution of the data has a long tail on the right, it is said to be **skewed to the right**, or **positively skewed**. Conversely, if the distribution has a long tail on the left, it is said to be **skewed to the left**, or **negatively skewed**.

> **Definition**
>
> **Skewed to the Right or Positively Skewed**
> If the distribution of a data set has a long tail on the right, it is said to be **skewed to the right** or **positively skewed**.

### Positively Skewed Curve

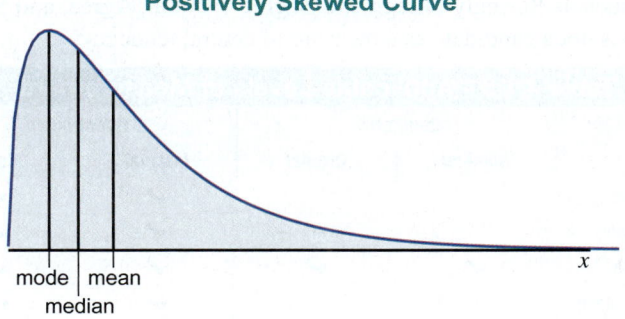

**Figure 4.1.7**

> **Definition**
>
> **Skewed to the Left or Negatively Skewed**
> If the distribution of a data set has a long tail on the left, it is said to be **skewed to the left** or **negatively skewed**.

### Measuring Figure Skating Performances

Almost every figure skating competition has some scoring controversy. The Winter Olympics of 2002 were no exception. French judge, Marie-Reine Le Gougne, said she was "pressured to vote a certain way" when she scored the Russian couple, Elena Berezhnaya and Anton Sikharulidze, over the Canadian pair, Jamie Sale and David Pelletier. In addition, very few people understood exactly how Sarah Hughes won the gold medal and how Michelle Kwan dropped to third after leading the event.

For almost a century figure skating has used a scoring method that is similar to the methodology of the trimmed mean in order to remove bias. Skaters are scored on a 0 to 6 scale. The highest and lowest scores are discarded (the data is trimmed) and the resulting score is computed.

The intent of trimming the data is to avoid bias caused by judges with nationalistic or political agendas.

Since the controversy, the International Skating Union has replaced this scoring method with a new system which, though it is different, still utilizes trimmed data to eliminate bias.

If the data are positively skewed, the median will be smaller than the mean.

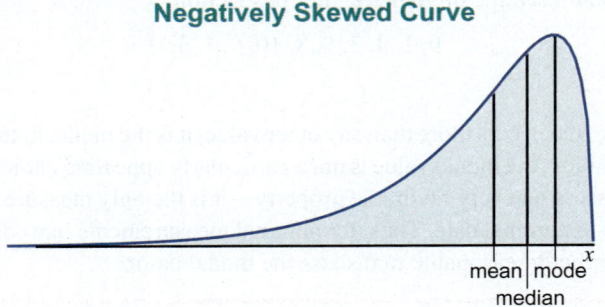

Figure 4.1.8

If the data are negatively skewed, the mean will be smaller than the median.

Where does the mode fall on these graphs? The area containing the greatest number of observations contains the mode. That area is represented by a large peak in the curve. In Figure 4.1.7 and Figure 4.1.8 the obvious peaks in the curves are the portions of the distributions that will contain the mode. The highest point on the curve will be the mode of the distribution. Notice that in a bell-shaped symmetrical distribution (Figure 4.1.6), the mode is equal to the mean and median. If the distribution is positively skewed (Figure 4.1.7), the mode is less than the mean and median. If the distribution is negatively skewed (Figure 4.1.8), the mode is greater than the mean and median.

## Selecting a Measure of Location

The objective of using descriptive statistics is to provide measures that convey useful summary information about the data. When selecting a statistic to represent the central value of a data set, the first thing to consider is the type of data being analyzed.

The arithmetic mean is used so frequently that its computation is almost a knee-jerk reaction to analyzing data. Unfortunately, it is not always a reasonable measure of location. Table 4.1.5 defines the applicable levels of measurement for each measure of location. Table 4.1.6 defines the sensitivity to outliers for each measure of location. When the data are qualitative (nominal or ordinal), the mean should not be calculated and, if the data are quantitative and contain outliers, the mean does not convey the notion of typical value as well as some other measures. The only time in which the mean should be used without any explanation is when the distribution of the data is symmetrical or nearly so. In that event, the mean and median should be about the same value.

While the median is not the best measure of central tendency for ordinal data, it can be used with some ordinal data. For example, variables that are measured on a Likert scale such as 1–5 can be summarized using the median. If one views variables measured on a Likert scale to be nominal, such as Strongly Disagree, Disagree, Neutral, Agree, and Strongly Agree, then the median is not a candidate as a measure of central tendency.

| Table 4.1.5 – Applicable Levels of Measurement | | | | |
|---|---|---|---|---|
| | Qualitative | | Quantitative | |
| | Nominal | Ordinal | Interval | Ratio |
| Mean | | | ✓ | ✓ |
| Median | | ✓ | ✓ | ✓ |
| Mode | ✓ | ✓ | ✓ | ✓ |
| Trimmed Mean | | | ✓ | ✓ |

## Table 4.1.6 – Sensitivity to Outliers

|  | Not Sensitive | Very Sensitive |
|---|---|---|
| Mean |  | ✔ |
| Median | ✔ |  |
| Mode | ✔ |  |
| Trimmed Mean | ✔ |  |

The median is also a good measure of central tendency. It is not sensitive to outliers and can be applied to data gathered from all levels of measurement except nominal.

If the level of measurement of the data is interval or ratio and there are no outliers, the mean is a reasonable choice. If the data set appears to have any unusual values, then the trimmed mean or the median would be more appropriate.

If the data's level of measurement is nominal or ordinal (the data are qualitative), appropriate measures of center are limited. If the data are ordinal, then the median is the best choice. If the data are nominal, there is only one choice, the mode. The mode is applicable to any level of data, although it is usually not very useful for quantitative data.

## Time Series Data and Measures of Location

We discussed two types of time series data in Chapter 2, stationary and nonstationary. Stationary time series wobbled around some central value, so calculating a central value is perfectly reasonable, and the methods we previously discussed are applicable. A nonstationary time series is another story. Nonstationary time series possess trend. That means there is no central value for the time series. Instead, the series trends in one direction or another. Computing a central value using the methods discussed earlier would be inappropriate for such data.

Table 4.1.7 shows the average U.S. gas price over a 20-year period. In this nonstationary time series, the central value of the process is trending upward as shown in Figure 4.1.9. One way to capture this movement is with a **moving average**.

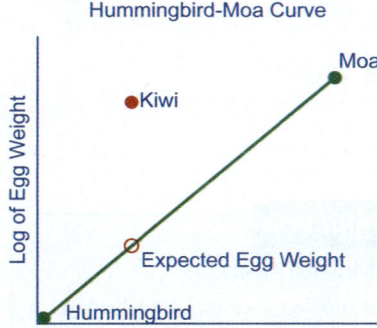

Hummingbird-Moa Curve

### Kiwi Eggs Are Outliers!

Both plant and animal kingdoms offer spectacularly odd and beautiful sights. A kiwi bird (one of the many interesting life forms from New Zealand) lays eggs that are close to 25% of its body weight and sometimes lays two or three such eggs at a time. For most species of birds, eggs usually correspond to about 5% of the bird's body weight.

If you draw a graph relating (log) egg weight to (log) body weight you get a so-called hummingbird-moa curve (moa is an extinct ostrich-like bird of the New Zealand area). In this curve, the kiwi bird is an outlier. Using the kiwi body weight (about 5 lb.) one expects an egg weight of about 55 to 100 grams while the actual weight of kiwi eggs is about 400 to 435 grams. This egg weight matches an expected body weight of about 40 lb. according to the hummingbird-moa curve. Why is this the case, and what accounts for such an anomaly?

## Table 4.1.7 – Average U.S. Gas Price 1996–2015 (Dollars per Gallon)

| Year | Average U.S. Gas Price | 2-Period Moving Average | 3-Period Moving Average |
|---|---|---|---|
| 1996 | 1.23 |  |  |
| 1997 | 1.23 | 1.23 |  |
| 1998 | 1.06 | 1.15 | 1.17 |
| 1999 | 1.17 | 1.11 | 1.15 |
| 2000 | 1.51 | 1.34 | 1.24 |
| 2001 | 1.46 | 1.49 | 1.38 |
| 2002 | 1.36 | 1.41 | 1.44 |
| 2003 | 1.59 | 1.47 | 1.47 |
| 2004 | 1.88 | 1.74 | 1.61 |
| 2005 | 2.30 | 2.09 | 1.92 |
| 2006 | 2.59 | 2.44 | 2.25 |
| 2007 | 2.80 | 2.70 | 2.56 |
| 2008 | 3.27 | 3.03 | 2.89 |
| 2009 | 2.35 | 2.81 | 2.81 |
| 2010 | 2.79 | 2.57 | 2.80 |
| 2011 | 3.53 | 3.16 | 2.89 |
| 2012 | 3.64 | 3.59 | 3.32 |
| 2013 | 3.53 | 3.59 | 3.57 |
| 2014 | 3.37 | 3.45 | 3.51 |
| 2015 | 2.45 | 2.91 | 3.11 |

**Source:** Bureau of Labor Statistics

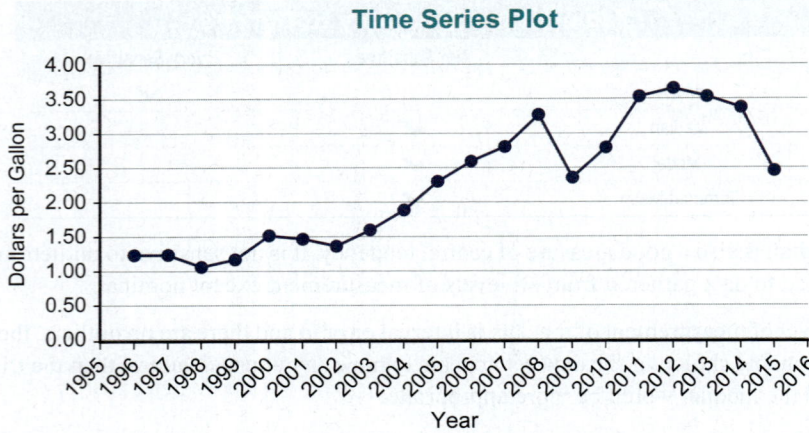

**Figure 4.1.9**

### Definition

**Moving Average**

A **moving average** is obtained by adding consecutive observations for a number of periods and dividing the result by the number of periods included in the average.

We will assume the moving average is to be used as a method of forecasting the next level of the time series. Suppose a two-period moving average is calculated for the gas price data and is used to specify the level of the series at a given point in time. The two-period moving average for 1997 averages the values of the time series in 1996 and 1997.

$$\frac{1.23 + 1.233}{2} = 1.23$$

Similarly, the two-period moving average for 1998 would be the average of the time series values in 1997 and 1998.

$$\frac{1.23 + 1.06}{2} = 1.145$$

Since data are not provided for 1994 or 1995, the three-period moving average for 1996 cannot be calculated. The three-period moving average associated with 1998 is the average of the time series values in 1996, 1997, and 1998.

$$\frac{1.23 + 1.23 + 1.06}{3} \approx 1.173$$

The three-period moving average for 1999 would be the average of the time series values in 1997, 1998, and 1999.

$$\frac{1.23 + 1.06 + 1.17}{3} \approx 1.153$$

The chart in Figure 4.1.10 displays the time series and the two- and three-period moving averages. Both of the averages follow the time series quite closely. However, notice that the two-period moving average (the blue line) follows the actual data values (the black line) more closely than the three-period moving average (the yellow line).

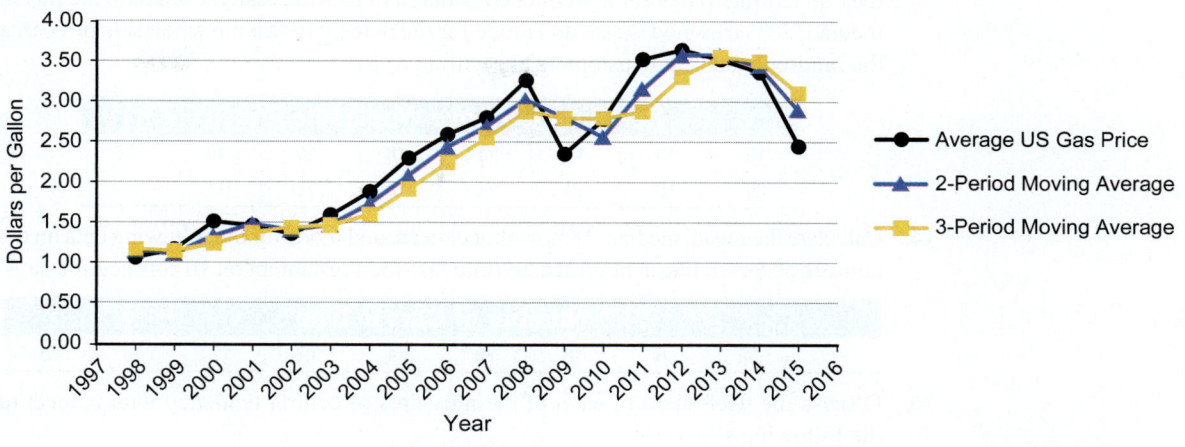

**Average U.S. Gas Price with Moving Averages**

Figure 4.1.10

---

## ✎ 4.1 Exercises

### Basic Concepts

1. Describe the difference between statistics and parameters.

2. Discuss three major attributes used in summarizing a data set.

3. What are numerical descriptive statistics and why are they important?

4. Identify and describe five measures of location. List the advantages and disadvantages of each.

5. List the types of data that are appropriate for each of the measures of location discussed in the previous question.

6. Why is the mean a measure of central tendency?

7. What is a resistant measure?

8. Describe a situation in which using the weighted mean as a measure of location would be appropriate.

9. What does it mean if we say that a data set is positively skewed? Negatively skewed?

10. Explain why the mean should not be calculated for a nonstationary time series.

11. What is a moving average? When is it useful?

### Exercises

12. The data in the table below represent the percentage growth of assets 20 years after the initial investment. Calculate the mean, median, 10% trimmed mean, and mode for percentage growth.

| Percentage Growth of Assets | | | | | | | | | |
|---|---|---|---|---|---|---|---|---|---|
| 90.25 | 93.83 | 91.41 | 92.27 | 90.89 | 99.12 | 92.88 | 97.74 | 96.28 | 95.33 |
| 91.16 | 94.30 | 95.51 | 92.27 | 97.63 | 95.94 | 90.95 | 94.76 | 92.27 | 92.88 |

13.  A survey was taken of customers asking what percentage above wholesale price would they be willing to pay for a product considered to be a necessity. Calculate the mean, median, 20% trimmed mean, and mode for the percentage above wholesale price that the randomly selected customers are willing to pay.

| Percentage Above Wholesale Price | | | | | | | |
|---|---|---|---|---|---|---|---|
| 19 | 14 | 11 | 11 | 18 | 20 | 10 | 15 |
| 20 | 10 | 19 | 11 | 18 | 18 | 11 | |

14.  Calculate the mean, median, 10% trimmed mean, and mode for the following data on the number of cars in line at noon at a favorite fast-food restaurant on 10 consecutive days.

| Number of Cars in Line | | | | | | | | | |
|---|---|---|---|---|---|---|---|---|---|
| 2 | 22 | 6 | 18 | 10 | 14 | 12 | 12 | 16 | 8 |

15.  Discuss the usefulness of each of the measures of central tendency with respect to the following situations.

   a.  A company is considering a move into a regional market for specialty soft drinks. In analyzing the size containers that his competitors are currently offering, would the company be more interested in the mean, median, or mode of their containers?

   b.  The creative director for an advertising agency is trying to target an ad campaign that will be shown in one city only. Would he be more interested in the mean or median family income in the city?

   c.  A young economist was assigned the task of comparing the interest rates of ninety-day certificates of deposit (CDs) in three major cities. Should she compare the mean, median, or modal interest for the banks in the three cities?

   d.  A telephone company is interested in knowing how customers rate their service: excellent, good, average, or poor. Would the company be more interested in studying the mean, median, or mode of the customer service ratings?

16.  Discuss the usefulness of each of the measures of central tendency with respect to the following situations.

   a.  A doctor is interested in analyzing the increase in systolic blood pressure caused by a certain antibiotic. Would the manufacturer be more interested in the mean, median, or mode of the ratings?

   b.  A car manufacturer is trying to decide in what colors it should offer its new sports coupe. In analyzing the preferred colors of other sports coupes, would the manufacturer be more interested in the mean, median, or mode of the colors?

   c.  A manufacturer of chocolate bars is interested in knowing how people rate its chocolate: the best, above average, average, below average, or the worst. Would the company be more interested in the mean, median, or mode of the ratings?

   d.  A realtor is interested in studying the prices of recent home sales in an area which has many diverse neighborhoods. Would the mean, median, or mode of the prices of recent home sales be the best measure of central tendency?

17. The following table contains the daily high temperatures for a southern city in July (measured in degrees Fahrenheit).

| High Temperatures in July (°F) | | | | | | | | | |
|---|---|---|---|---|---|---|---|---|---|
| 84 | 85 | 84 | 88 | 94 | 100 | 97 | 102 | 97 | 89 |
| 89 | 90 | 88 | 95 | 91 | 95 | 99 | 93 | 97 | 99 |
| 90 | 94 | 90 | 88 | 91 | 88 | 106 | 99 | 102 | 85 |

    **a.** Calculate the mean of the daily high temperatures.

    **b.** Calculate the median of the daily high temperatures.

    **c.** Calculate the mode of the daily high temperatures.

    **d.** Calculate the 10% trimmed mean of the daily high temperatures.

    **e.** Which measure of central tendency do you think best describes the center of the data set? Why?

18. A tour guide informs his group that the "average" temperature at their destination is 60 degrees Fahrenheit. Once they arrive, they discover that the daytime highs are about 120 degrees Fahrenheit and the nighttime lows are about 0 degrees Fahrenheit. Do you feel the tour guide accurately described the temperatures to the group? Discuss.

19. A worker is participating in a test on a new machine. Her daily production, measured in numbers of units, for the twenty-day test is listed in the following table. On days 4 and 5, the worker was ill and went home shortly after coming to work.

| Daily Production | | | | | | | | | |
|---|---|---|---|---|---|---|---|---|---|
| **Day** | 1 | 2 | 3 | 4 | 5 | 6 | 7 | 8 | 9 | 10 |
| **Units** | 100 | 104 | 117 | 20 | 20 | 111 | 105 | 106 | 115 | 101 |
| **Day** | 11 | 12 | 13 | 14 | 15 | 16 | 17 | 18 | 19 | 20 |
| **Units** | 101 | 102 | 115 | 116 | 113 | 103 | 104 | 119 | 118 | 108 |

    **a.** What level of measurement does the data possess?

    **b.** Compute the 10% trimmed mean and the 20% trimmed mean.

    **c.** Considering the worker's illness, which measure computed in part **b.** best describes the production capability of the machine? Discuss.

20. Consider the following per capita greenhouse emissions (in tons of carbon dioxide equivalent per capita) for 10 randomly selected states.

| Greenhouse Emissions per Capita (Tons) | | | | |
|---|---|---|---|---|
| 11.76 | 15.65 | 22.93 | 24.75 | 21.22 |
| 18.72 | 22.55 | 27.99 | 12.23 | 114.40 |

    **a.** What level of measurement do the data possess?

    **b.** Compute the 10% trimmed mean and the 20% trimmed mean.

    **c.** Considering the data, which measure computed in part **b.** best describes the per capita greenhouse emissions? Discuss.

21. Consider the following monthly sales for a small clothing store in a resort community.

| Clothing Store Sales | | | |
|---|---|---|---|
| Month | Sales ($) | Month | Sales ($) |
| January | 100,500 | July | 200,000 |
| February | 120,000 | August | 185,000 |
| March | 133,000 | September | 175,000 |
| April | 145,000 | October | 120,000 |
| May | 160,000 | November | 180,000 |
| June | 180,000 | December | 330,000 |

   a. Draw a line graph of the data.

   b. Calculate the two-period moving averages for the data.

   c. Calculate the three-period moving averages for the data.

   d. Add line graphs for the two-period moving averages and three-period moving averages to the graph which you constructed in part **a.**

   e. Which series of data (the original sales data, the two-period moving averages, or the three-period moving averages) do you think best represents sales for the year? Why?

22. Late in the summer of 1996, Tiger Woods became a professional golfer. This highly publicized event followed a sensational college career at Stanford University, where Tiger won three United States Amateur championships. Tiger was not a professional very long before he had his first win on the pro tour, the Las Vegas Invitational. He received a total of $297,000 for his accomplishment. Since becoming a professional, Tiger has won more than 82 times and has surmassed a net worth of more than $1 billion. The table below contains the prize money (in millions, US dollars) that Tiger has won on the golf course each year from 1996 through 2016.

| Career Earnings of Tiger Woods from 1996 to 2016 (in Million U.S. Dollars) | | | |
|---|---|---|---|
| Year | On Course | Year | On Course |
| 1996 | 0.89 | 2007 | 22.9 |
| 1997 | 2.38 | 2008 | 7.74 |
| 1998 | 2.93 | 2009 | 21.02 |
| 1999 | 7.68 | 2010 | 2.29 |
| 2000 | 11.03 | 2011 | 2.07 |
| 2001 | 7.77 | 2012 | 9.12 |
| 2002 | 8.29 | 2013 | 12.09 |
| 2003 | 6.7 | 2014 | 0.61 |
| 2004 | 6.37 | 2015 | 0.55 |
| 2005 | 11.99 | 2016 | 0.11 |
| 2006 | 11.94 | | |

   a. Find the mean.

   b. Find the median.

   c. Find the mode.

   d. Find the 10% trimmed mean and compare it to the mean and the median.

   e. Comment on the skewness of the distribution.

# 4.2 Measures of Dispersion

Suppose all people looked alike, all cars looked alike, everyone wore the same kind of clothes, and there was only one kind of hamburger (plain). Without diversity it would be a boring world and a world in which statistics would be of little value. Since much of statistics is devoted to describing, analyzing, and explaining variability, understanding how variability is measured is essential to understanding statistics.

The concept of **variability** (also referred to as **dispersion** or **spread**) is as vague as the concept of central tendency. And vague concepts lead to different measurement ideas. The same issues that are important in evaluating measures of location are meaningful in evaluating measures of dispersion.

Many of the good measures of dispersion use the concept of deviation from the mean. The distance that a value is from its mean is called a **deviation from the mean**. A data set and its deviations from the mean are calculated in Table 4.2.1.

> **Definition**
>
> Variability
> **Variability** is the amount an individual observation or set of observations fluctuates about the mean or center of a set of data.

### Table 4.2.1 – Calculating Deviations from the Mean

| Data Set: 3, 12, 20, 15, 0    Mean = 10 | |
| --- | --- |
| **Data Values** | **Deviations from the Mean (Data − Mean = Deviation)** |
| 3 | $3 - 10 = -7$ |
| 12 | $12 - 10 = 2$ |
| 20 | $20 - 10 = 10$ |
| 15 | $15 - 10 = 5$ |
| 0 | $0 - 10 = -10$ |

> **Definition**
>
> Deviation from the Mean
> The distance that a value is from the mean of the data set is called a **deviation from the mean**.

Because the mean is the point at which the sum of the positive deviations equals the sum of the absolute values of the negative deviations, the deviations will always sum to zero. Many of the variability measures average the deviations in some form.

## Range

The range is the simplest measure of dispersion. It does not provide much depth or understanding of the measure of spread and does not use the deviation concept.

> **Definition**
>
> Range
> The **range** is the difference between the largest and smallest data values.

---

The Pop Bottling Company wanted to know the monthly consumption of sodas by families living nearby. A random sample of nine homes were taken. The data below represent the number of sodas consumed in the last month by the families surveyed. Calculate the range of the following data set.

**Example 4.2.1**

**Calculating the Range of Sodas Consumed**

$$4, 6, 16, 9, 24, 8, 0, 12, 1$$

**SOLUTION**

The largest value equals 24 and the smallest value equals 0. Thus, the range is calculated as follows.

$$\text{Range} = 24 - 0 = 24$$

---

The problem with the range is that it is also affected by outliers, and it does not bring all the information in the data directly to bear on the problem of measuring variation. That is, the range only uses two values (the largest and smallest) to measure spread rather than all of the observations. The other measures of dispersion discussed in this lesson are generally more appropriate measures of spread.

## Mean Absolute Deviation

One of the ways of obtaining information about the spread of the data is to analyze the deviations from the mean. Instead of adding the raw deviations, suppose the absolute values of the deviations (which can be interpreted as distance from the mean) are summed and divided by the number of deviations. This new measure computes the average distance from the mean for the data set. This measure is called the **mean absolute deviation**. If data set A has a larger average deviation than B, then it is reasonable to believe that data set A has more variability than data set B.

### Formula

### Mean Absolute Deviation

The sample mean absolute deviation (MAD) is given by

$$MAD = \frac{\sum |x_i - \overline{x}|}{n}.$$

Example 4.2.2

**Calculating the Mean Absolute Deviation of Run Times**

Suppose six people participated in a 1000-meter run. Their times, measured in minutes, are given below.

$$4, 10, 9, 11, 9, 7$$

The mean time is approximately 8.3 minutes. Find the mean absolute deviation.

### SOLUTION

In Table 4.2.2 we do the basic calculations needed to compute the mean absolute deviation.

| Table 4.2.2 – Calculating Mean Absolute Deviation | | | |
|---|---|---|---|
| Time (Minutes) | Deviation $x_i - \overline{x}$ | Absolute Deviation $\|x_i - \overline{x}\|$ | % of Total Deviation |
| 4 | 4 − 8.3 | 4.3 | 37.72 |
| 10 | 10 − 8.3 | 1.7 | 14.91 |
| 9 | 9 − 8.3 | 0.7 | 6.14 |
| 11 | 11 − 8.3 | 2.7 | 23.68 |
| 9 | 9 − 8.3 | 0.7 | 6.14 |
| 7 | 7 − 8.3 | 1.3 | 11.40 |
| **Total** | | **11.4** | **100%** |

✎ **NOTE**

The percentages in Table 4.2.2 and 4.2.4 actually add up to 99.99 due to rounding.

$$\text{Mean Absolute Deviation} = \frac{11.4}{6} = 1.9 \text{ minutes}$$

Thus, on average, the values are 1.9 units from the mean. Note that the contribution to the sum of the deviations is proportional to the size of the deviation. That is, if one absolute deviation is twice as large as another, it contributes twice as much to the value of the statistic. For example, compare the data value 7, which is 1.3 units from the mean, to the data value 11, which is 2.7 units from the mean. The percentage contribution to the total deviation is 11.40% for 7 and 23.68% for 11, which is in proportion to their respective distances from the mean. A variability measure in which each data value contributes proportionally to its distance from the mean seems reasonable.

Suppose the value 200 is added to the data set given in Example 4.2.2.

The mean is drastically affected, increasing from 8.3 to 35.7. In Table 4.2.3 we redo the basic calculations for the mean absolute deviation. What effect, if any, does the value of 200 have on the MAD?

### SOLUTION

| Table 4.2.3 – Calculating Mean Absolute Deviation | | |
|---|---|---|
| Data | Deviation $x_i - \bar{x}$ | Absolute Deviation $\lvert x_i - \bar{x} \rvert$ |
| 4 | $4 - 35.7$ | 31.7 |
| 10 | $10 - 35.7$ | 25.7 |
| 9 | $9 - 35.7$ | 26.7 |
| 11 | $11 - 35.7$ | 24.7 |
| 9 | $9 - 35.7$ | 26.7 |
| 7 | $7 - 35.7$ | 28.7 |
| 200 | $200 - 35.7$ | 164.3 |
| | Total | 328.5 |

The mean absolute deviation changes dramatically, increasing to 46.9, which is calculated by the sum of the absolute deviations (328.5) divided by the sample size (7). Therefore, the mean absolute deviation is sensitive to outliers and is not a resistant measure. The mean absolute deviation is a very intuitive measure of variation.

## Variance and Standard Deviation

The **variance** and **standard deviation** are the most common measures of variability. Since the standard deviation is computed directly from the variance, our discussion will center on the variance. Like the MAD, the variance and standard deviation provide numerical measures of how the data vary around the mean. If the data are tightly packed around the mean, the variance and standard deviation will be relatively small. On the other hand, if the data are widely dispersed about the mean, the variance and standard deviation will be relatively large.

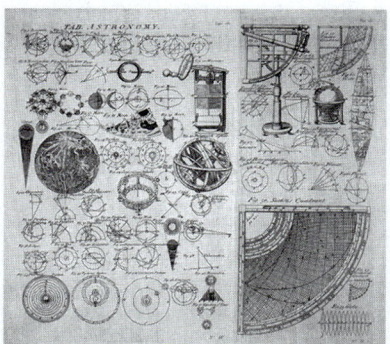

**The Introduction of Variance**

### Formula

#### Variance

The **variance** of a data set containing the complete set of population data is given by

$$\sigma^2 = \frac{\sum (x_i - \mu)^2}{N},$$

where $\mu$ is the population mean of the data set, $N$ is the size of the population, and $x_i$ is a particular value in the data set. $\sigma^2$ is pronounced *sigma squared*, and is called the **population variance**.

The **variance** of a data set containing sample data is given by

$$s^2 = \frac{\sum (x_i - \bar{x})^2}{n - 1}.$$

where $\bar{x}$ is the mean of the sample data, $n$ is the size of the sample, and $x_i$ is a particular value in the sample. $s^2$ is called the **sample variance**.

Carl Gauss (1777-1855) was a German mathematician who introduced the notion of variance. However, some believe that the work may have been prepared by Tycho Brahe while he was working on the problem of trying to estimate the position of a star using a series of measurements of its location. Both men lived highly interesting, yet quite contrasting, lives. Gauss was a child prodigy who rose out of poverty, while Brahe was a Dutch nobleman and astronomer. The Wikipedia articles outlining their lives and accomplishments are fascinating and well worth reading.

Both these definitions can be construed to be averages, although at first glance it may not be readily apparent. That is, for the population variance, we are adding up the sum of the squared deviations and dividing by the number of items that are added. Thus, the population variance is the average squared deviation from the mean. For the sample variance, we are dividing by $n - 1$ because it gives us an unbiased estimate of the population variance.

It is usually not necessary to compute a variance by manual methods, except to become familiar with the definition.

---

**Example 4.2.4**

**Calculating the Sample Variance of Run Times**

Given the following times in minutes of six persons running a 1000-meter course, compute the sample variance.

$$4, 10, 9, 11, 9, 7$$

**SOLUTION**

We previously computed the mean of this sample as 8.3. In Table 4.2.4 we do the basic calculations needed to compute the sample variance.

| Table 4.2.4 – Calculating the Sample Variance | | | |
|---|---|---|---|
| Data | Deviation $x_i - \overline{x}$ | Squared Deviation $(x_i - \overline{x})^2$ | % of Total Squared Deviation |
| 4 | $4 - 8.3 = -4.3$ | 18.49 | 59.00 |
| 10 | $10 - 8.3 = 1.7$ | 2.89 | 9.22 |
| 9 | $9 - 8.3 = 0.7$ | 0.49 | 1.56 |
| 11 | $11 - 8.3 = 2.7$ | 7.29 | 23.26 |
| 9 | $9 - 8.3 = 0.7$ | 0.49 | 1.56 |
| 7 | $7 - 8.3 = -1.3$ | 1.69 | 5.39 |
| **Total** | | **31.34** | **100%** |

$$s^2 = \frac{\sum (x_i - \overline{x})^2}{n-1} = \frac{31.34}{5} = 6.268 \text{ minutes squared}$$

Thus, the average squared deviation of the data is 6.268 minutes squared. The phrase "minutes squared" in the last sentence may seem a bit odd. No one carries out transactions in square minutes, or square dollars, or square tons, so it is difficult to interpret the significance of the measurement in this form. That is why the standard deviation exists. It converts the measure into the original units by taking the square root of the variance. The standard deviation for the previous data is then $\sqrt{6.368} \approx 2.503$ minutes.

---

But since there are two measures of variance, there will be two standard deviations, one for population data and one for sample data.

$$\sigma = \sqrt{\sigma^2} \quad \text{the \textbf{population standard deviation}}$$

$$s = \sqrt{s^2} \quad \text{the \textbf{sample standard deviation}}$$

It is important to remember these symbols ($\sigma$ and $s$) since the standard deviation is a fundamental statistical concept.

Describing the standard deviation in an intuitive way is not easy. It is not the average deviation from the mean (which always equals 0), although in most cases it will be reasonably close to the mean absolute deviation. The fact is that the standard deviation is the square root of the average squared deviation. It is not an intuitive concept! Certainly, a reasonable question at this juncture is, if the standard deviation is not very intuitive, why use it to measure deviation? Part of the answer lies in the fact that it is expressed in the same units as the data. The variance is an important theoretical measure of variability because it has "nice" mathematical properties (in contrast to the more intuitive MAD).

---

**⌁ Technology**

For technology instructions to calculate sample statistics like the variance, visit stat.hawkeslearning.com and navigate to **Discovering Business Statistics, Second Edition > Technology Instructions > Descriptive Statistics > One Variable.**

| Column1 | |
|---|---|
| Mean | 8.333333333 |
| Standard Error | 1.021980648 |
| Median | 9 |
| Mode | 9 |
| Standard Deviation | 2.503331114 |
| Sample Variance | 6.266666667 |
| Kurtosis | 1.137392485 |
| Skewness | -1.138907999 |
| Range | 7 |
| Minimum | 4 |
| Maximum | 11 |
| Sum | 50 |
| Count | 6 |

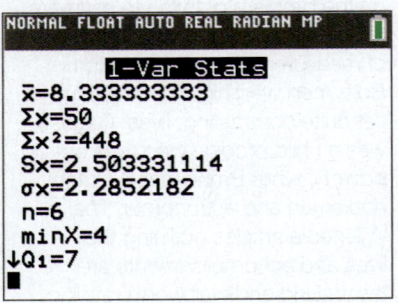

The standard deviation can be used to measure how far data values are from their mean. You will often see the majority of data values within one standard deviation of the mean. Further, relatively few data values will be more than two standard deviations from the mean.

The variance and standard deviation both suffer from the same problem as the mean: they are very sensitive to outliers. Suppose the value 200 were added to the data in Example 4.2.4. The sample variance would increase from 6.268 to 5253.238. The new sample variance (which includes the outlier 200) then is 838 times as large as the original variance. The standard deviation increases from 2.50 to 72.50. The presence of the outlier tarnishes the interpretation of the standard deviation as a measure of variability.

Another interesting property of the variance is that values further from the mean contribute a disproportionate amount to the value of the statistic. In Example 4.2.4, one data value, 4, which is 4.3 units from the mean, contributes 59% of the variation in the data (see the column labeled "% of Total Squared Deviation" in Table 4.2.4). Compare this to the data value 7, which is 1.3 units from the mean, yet only contributes 5.39% of the total variation. The reason that 4 contributes so heavily to the total variation is because the deviations are squared. By squaring the deviations, values farther from the mean have disproportionate effects on the sum of the squared deviations.

While there are a number of descriptive tools available for summarizing variability, the variance and standard deviation are the most frequently used statistics.

## Dispersion and Time Series Data

If the time series is stationary, then the methods discussed in this section can be used to measure dispersion. However, if the series is nonstationary, the dispersion measures we have discussed are not applicable. Measuring variation in a nonstationary time series is beyond the scope of this text.

## Using the Standard Deviation

Although the standard deviation is not an intuitive concept, knowing the mean and standard deviation of a data set provides a great deal of information about the data. If the histogram of the measurements is bell-shaped, the **empirical rule** describes the variability of a set of measurements. **Chebyshev's Theorem** is a more general rule describing the variability of *any* set of data regardless of the shape of its distribution.

## Empirical Rule

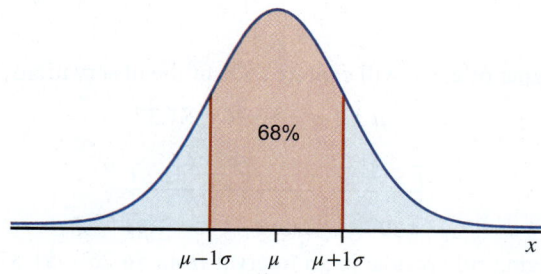

**Figure 4.2.1 – One Sigma**

**One sigma rule:** If the distribution of the data is bell-shaped, about 68% of the data should lie within one standard deviation of the mean.

A deviation of more than one sigma from the mean is to be expected about once in every three observations.

**Definition**
Standard Deviation
The **standard deviation** is the square root of the variance.

**Definition**
Empirical Rule
The **empirical rule** describes the variability of a set of measurements for a bell-shaped distribution.

**Definition**
Chebyshev's Theorem
**Chebyshev's Theorem** is a general rule that describes the variability of any set of data regardless of the shape of its distribution.

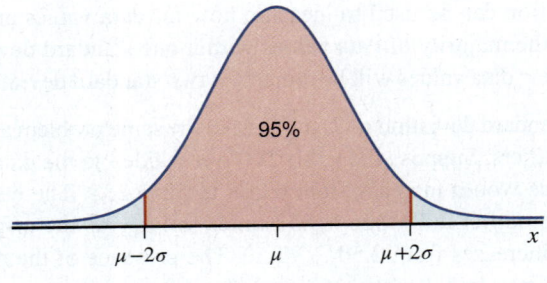

**Figure 4.2.2 – Two Sigma**

**Two sigma rule:** If the distribution of data is bell-shaped, about 95% of the data should lie within two standard deviations of the mean.

A deviation of more than two sigma from the mean is to be expected about once in every twenty observations.

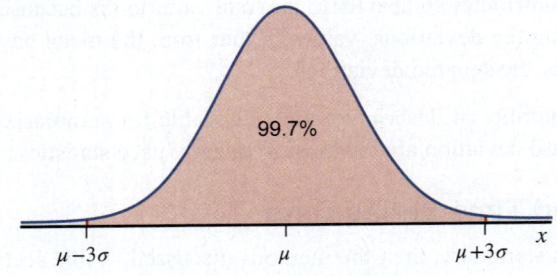

**Figure 4.2.3 – Three Sigma**

**Three sigma rule:** If the distribution of the data is bell-shaped, about 99.7% of the observations should lie within three standard deviations of the mean.

A deviation of more than three sigma from the mean is to be expected about once in every 333 observations, slightly more than 0.3% of the time.

**Example 4.2.5**

**Applying the Empirical Rule to Stock Earnings**

Suppose a group of high technology stocks has an average earnings per share of $6.26, with a standard deviation of $1.37. If the data possess a bell-shaped distribution, which interval contains 68% of the earnings? Which interval contains about 95% of the earnings?

**SOLUTION**

Using the one sigma rule, we will capture 68% of the observations.

$$\mu \pm 1 \cdot \sigma = \$6.26 \pm \$1.37$$

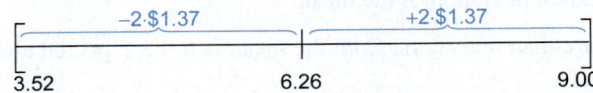

Using the one sigma rule results in an interval from $6.26 − $1.37 to $6.26 + $1.37. Doing the arithmetic produces an interval from $4.89 to $7.63.

To capture 95% of the earnings, use the two sigma rule, $6.26 ± 2·$1.37. Doing the arithmetic results in an interval from $3.52 to $9.00.

Note that to increase the percentage of data captured from 68% to 95% requires an interval that is twice as large.

## Chebyshev's Theorem

It is important to remember that the empirical rule applies only to bell-shaped distributions. For *any* distribution, regardless of shape, Chebyshev's Theorem may be used, although its results are much more approximate.

> **Theorem**
>
> Chebyshev's Theorem
>
> The proportion of any data set lying within $k$ standard deviations of the mean is at least
> $$1 - \frac{1}{k^2}, \text{ for } k > 1.$$

For example, if $k = 2$ at least $1 - \frac{1}{2^2} = \frac{3}{4}$ (or 75%) of the data values lie within 2 standard deviations of the mean, for any data set. Similarly, if $k = 3$ at least $1 - \frac{1}{3^2} = \frac{8}{9}$ (or approximately 88.9% ) of the data values lie within 3 standard deviations of the mean, for any data set. Also note that $k$ does not have to be an integer value. If $k = 1.5$, at least $1 - \frac{1}{1.5^2} = \frac{5}{9}$ (or approximately 55.6%) of the data values will lie within 1.5 standard deviations of the mean, for any data set.

The distribution histogram of tuition and fees of US colleges and universities in 2019–2020 is shown in Figure 4.2.4. The mean of the data is $7498, while the standard deviation is $3639. What can we conclude from Chebyshev's Theorem using $k = 2$?

**Example 4.2.6**

**Applying Chebyshev's Theorem to College Tuition and Fees**

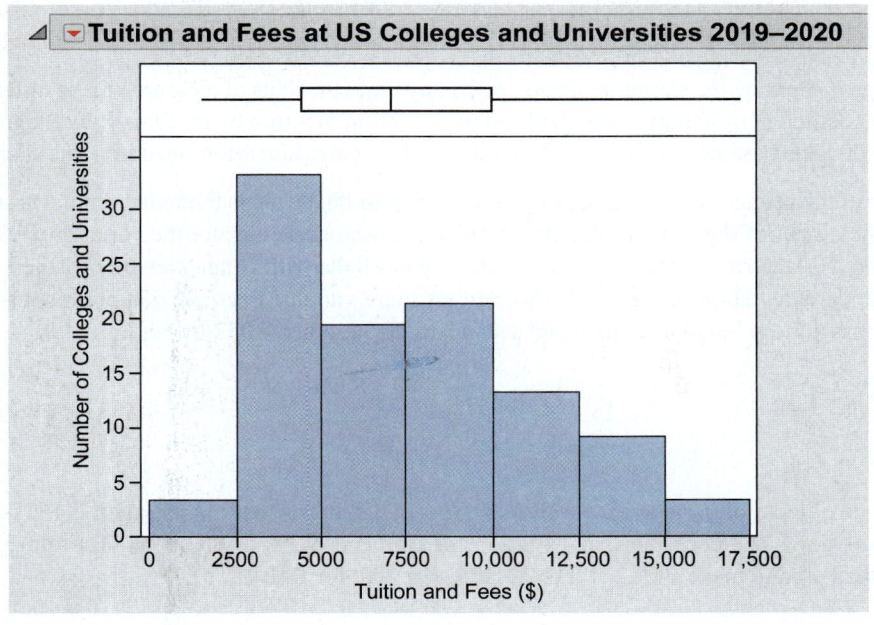

Figure 4.2.4

⸭ **Data**

This data set can be found on stat.hawkeslearning.com under **Discovering Business Statistics, Second Edition > Data Sets > Tuition and Fees 2019–2020**.

### SOLUTION

Because we are interested in $k = 2$, we will look at the values two standard deviations above and below the mean.

Two standard deviations above the mean is

$$\mu + 2\sigma = 7498 + 2(3639) = \$14{,}776$$

and two standard deviations below the mean is

$$\mu - 2\sigma = 7498 - 2(3639) = \$220.$$

Therefore, by Chebyshev's Theorem, we can say that at least 75% of the tuition and fees of colleges and universities in the United States is between \$220 and \$14,776 for 2019–2020.

## The Coefficient of Variation

Sometimes a data analyst wants to compare the variation of two or more data sets. The **coefficient of variation** is a unit-free statistical measure that enables the comparison of the variation in two or more data sets.

### Who Is King of the Hill?

In 1961, Wilt Chamberlain was the National Basketball Association (NBA) rebounding leader with 27 rebounds per game. In 1992, the colorful Dennis Rodman won the same honor with 18.7 rebounds per game. Common sense suggests that professional basketball in the 1990s is played at a much higher level than in the 1960s. So why has the rebounds per game for the rebounding leader fallen? Is it another case of "less is more"?

Researchers investigating this interesting puzzle considered two other variables: the number of rebounding opportunities (this had gone down since the field goal percentage has increased historically) and the average number of minutes played per game, which has also fallen.

Thus, when we adjust the actual rebounds obtained by the rebounding leaders to the number of minutes played and the total number of rebounding opportunities, we see a completely different picture. The adjusted rebound numbers for Chamberlain and Rodman are 35.42 and 51.06 respectively.

> ### Formula
>
> #### Coefficient of Variation
>
> The coefficient of variation, another statistical measure, compares the variation in data sets.
>
> For population data, the coefficient of variation is defined as
>
> $$CV = \left( \frac{\sigma}{\mu} \cdot 100 \right)\%,$$
>
> and for sample data,
>
> $$CV = \left( \frac{s}{\bar{x}} \cdot 100 \right)\%.$$

When comparing the variation of data sets, many times the units of measure will be different. The coefficient of variation standardizes the variation measure by dividing it by the mean. The division has one interesting side effect: the unit of measure is removed from the statistic.

One of the primary focuses of quality control in manufacturing is the reduction in variation of the output of the process. A bolt manufacturer wants to compare the variability of two bolt manufacturing processes. One process creates bolts with a mean length of 2.5 cm and a standard deviation of 0.2 cm. Is this process more variable than one that produces a bolt that has a mean length of 1 inch and a standard deviation of 0.052 inches?

$$CV_{\text{Bolt 1}} = \frac{0.2}{2.5} \cdot 100 = 8.0\%$$

$$CV_{\text{Bolt 2}} = \frac{0.052}{1} \cdot 100 = 5.2\%$$

The coefficient of variation for Bolt 1 is 8.0% . This means that the variation is 8% of the mean value. The coefficient of variation for Bolt 2 is 5.2% of the mean. Therefore, the process used to make Bolt 2 is less variable than that for Bolt 1.

## ✎ 4.2 Exercises

### Basic Concepts

1. Describe three measures of variation. Discuss the strengths and weaknesses of each.

2. What does the standard deviation measure?

3. Why are the variance and standard deviation more commonly used as measures of variability than the MAD?

4.   Explain how the variance can be construed as an average.

5.   True or false: The variance and standard deviation are resistant measures.

6.   When is it appropriate to calculate the variance of a time series?

7.   What is the empirical rule? When is it appropriate to use the empirical rule?

8.   What is Chebyshev's Theorem?

9.   Discuss the purpose of the coefficient of variation.

10.  How is the coefficient of variation calculated?

11.  Why is the coefficient of variation important?

## Exercises

12.  Find the missing age in the following set of four student ages.

| Student Ages | | |
|---|---|---|
| **Student** | **Age** | **Deviation from the Mean** |
| A | 19 | −4 |
| B | 20 | −3 |
| C | ? | +1 |
| D | 29 | +6 |

13.  Find the missing weight in the following data set.

| Weights | | |
|---|---|---|
| **Person** | **Weight** | **Deviation from the Mean** |
| A | 144 | −20 |
| B | 156 | −8 |
| C | ? | +1 |
| D | 176 | +12 |

14.  Consider the following time until failure for 10 randomly selected car batteries (measured in years).

| Years Until Failure for Car Batteries |
|---|
| 5   3   4   6   2   5   7   10   8   4 |

   a.  Calculate the sample variance of the time until failure.

   b.  Calculate the sample standard deviation of the time until failure.

   c.  Calculate the range of the time until failure.

   d.  What are some of the factors which might contribute to the variation in the observations?

15.  Consider the following distances jumped (in feet) by 8 randomly selected long jumpers.

| Jump Distances (Feet) |
|---|
| 21   15   12   18   10   14   17   11 |

   a.  Calculate the sample variance of the distances jumped.

   b.  Calculate the sample standard deviation of the distances jumped.

   c.  Calculate the range of the distances jumped.

   d.  What are some of the factors which might contribute to the variation in the observations?

**16.** The interest rates on 30-year mortgages offered by seven randomly selected banks in a large metropolitan area are recorded in the following table.

| Interest Rates (%) | | | | | | |
|---|---|---|---|---|---|---|
| 7.5 | 8.0 | 7.0 | 7.25 | 8.5 | 8.25 | 7.75 |

   **a.** Calculate the sample variance of the interest rates.

   **b.** Calculate the sample standard deviation of the interest rates.

   **c.** Calculate the range of the interest rates.

   **d.** What are some of the factors which might contribute to the variation in the observations?

**17.** A researcher has hypothesized that female college students are more disciplined than male college students. The researcher believes that a reasonable measure of discipline is performance on a statistics test in terms of both absolute scores and consistency of scores. Seven male statistics students and seven female statistics students are randomly selected and their scores on a statistics test are observed.

| Test Scores | | | | | | | |
|---|---|---|---|---|---|---|---|
| **Males** | 65 | 100 | 75 | 45 | 85 | 73 | 95 |
| **Females** | 75 | 80 | 95 | 85 | 82 | 72 | 49 |

   **a.** Calculate the average test score for male students and female students separately.

   **b.** Calculate the variance of the test scores for male students and female students separately.

   **c.** Calculate the standard deviation of the test scores for male students and female students separately.

   **d.** Do you think that the data tend to support the hypothesis that female college students are more disciplined than male college students based on the researcher's measurement?

   **e.** What do you think about this particular measurement of discipline?

**18.** Consider the following market values of two portfolios of stocks at five randomly selected times during a year.

| Market Values ($) | | | | | |
|---|---|---|---|---|---|
| **Portfolio A** | 150,000 | 155,000 | 145,000 | 160,000 | 140,000 |
| **Portfolio B** | 130,000 | 175,000 | 100,000 | 150,000 | 195,000 |

   **a.** What statistical criteria might you use to select the better portfolio? Justify your answer.

   **b.** Calculate the statistics you proposed in part **a.**

   **c.** Which portfolio has the least amount of risk? Why?

**19.** Add 20 to each of the following data values.

| | | | | | |
|---|---|---|---|---|---|
| 81 | 99 | 97 | 81 | 85 | 86 |
| 99 | 93 | 96 | 83 | 82 | 91 |

   **a.** Compute the mean and standard deviation for both the original data and adjusted data.

   **b.** Compare the mean and standard deviation of the adjusted data to the mean and standard deviation of the original data.

   **c.** Describe the effect on the mean and standard deviation of adding a constant to a data set.

**20.** Adjust the following data values by subtracting 20 from each data value.

| 745 | 789 | 712 | 764 | 736 |
|-----|-----|-----|-----|-----|
| 758 | 722 | 773 | 751 | 741 |

    **a.** Calculate the mean and variance for the original and adjusted data.

    **b.** Compare the mean and variance of the adjusted data to the mean and the variance of the original data.

    **c.** Describe the effect of subtracting a constant value from each member of a data set on the mean and variance of the data.

**21.** The average score on a pre-employment test is 26 with a standard deviation of 7. Using Chebyshev's Theorem, state the range in which at least 88.89% of the data will reside.

**22.** The daily average number of phone calls to a call center is 972 with a standard deviation of 127. Using Chebyshev's Theorem, state the range in which at least 75% of the data will reside.

**23.** There is an annual chowder eating contest in a small New England town. The average amount of chowder eaten at the contest was 32 ounces with a variance of 64 ounces. Given that one hundred people participated in the contest, find:

    **a.** The approximate number of people who ate between 24 and 40 ounces of chowder.

    **b.** The approximate number of people who ate between 16 and 48 ounces of chowder.

    **c.** What assumptions did you make about the amount of chowder eaten by each contestant in answering parts **a.** and **b.**?

**24.** The manager of a local diner has calculated his average daily sales to be $4500 with a standard deviation of $750.

    **a.** In what range can the manager expect his daily sales to be 68% of the time?

    **b.** In what range can the manager expect his daily sales to be 95% of the time?

    **c.** In what range can the manager expect his daily sales to be 99.7% of the time?

    **d.** What assumption did you make about daily sales when answering parts **a.**, **b.**, and **c.**?

**25.** A management consulting firm is evaluating the salary structure for a large insurance company. The goal of the study is to develop salary ranges for each of the possible job grades within the company. The company and the firm have agreed that a reasonable salary range for each job grade can be determined by finding the salary range in which 95% of the current salaries for that job grade fall. The average salary and the standard deviation of the salaries are listed in the following table for three of the job grades.

| Salaries ($) | | | |
|---|---|---|---|
| **Job Grade** | **25** | **33** | **40** |
| $\bar{x}$ | 22,000 | 35,000 | 45,000 |
| $s$ | 1500 | 2000 | 5000 |

    **a.** Determine the appropriate salary ranges for the three job grades.

    **b.** What assumption did you make about the salaries in each of the job grades in answering part **a.**?

26. A consumer interest group is interested in comparing two brands of vitamin C. One brand of vitamin C advertises that its tablets contain 500 mg of vitamin C. The other brand advertises that its tablets contain 250 mg of vitamin C. Tablets for each brand are randomly selected and the milligrams of vitamin C for each tablet are measured with the following results.

| Vitamin C Content (mg) | | |
|---|---|---|
| | Brand A (500 mg) | Brand B (250 mg) |
| $\bar{x}$ | 500 | 250 |
| $s$ | 10 | 7 |

a. Calculate the coefficient of variation for Brand A.

b. Calculate the coefficient of variation for Brand B.

c. Which brand more consistently produces tablets as advertised? Explain.

27. A manufacturer of bolts has two different machines. One machine is used to produce $\frac{1}{4}$-inch bolts; the other machine is used to produce $\frac{1}{2}$-inch bolts. It is very important that the machines consistently produce bolts of the correct diameters, or the bolts will not fit on the corresponding nuts. In order to compare the two machines, management randomly selects bolts produced from each machine and computes the average diameter of the bolts and the standard deviation of the bolts. The results of the study are shown in the following table.

| Bolt Diameter | | |
|---|---|---|
| | Machine X $\left(\frac{1}{4}"\right)$ | Machine Y $\left(\frac{1}{2}"\right)$ |
| $\bar{x}$ | 0.25" | 0.50" |
| $s$ | 0.03" | 0.05" |

a. Calculate the coefficient of variation for Machine X.

b. Calculate the coefficient of variation for Machine Y.

c. Which machine more consistently produces bolts of the correct diameter? Explain.

# 4.3 Measures of Relative Position

Suppose you want to know where an observation stands in relation to other values in a data set. For example, on many standardized tests such as the SAT, GMAT, and ACT, the test scores themselves are rather meaningless unless they are associated with some measure that tells you how well you did relative to others taking the same test. There are two principal methods of communicating relative position: **percentiles** and **z-scores**. Both of these methods are data transformations which change the scale of the data in some way.

**Definition**

$P^{th}$ Percentile

Given a set of data $x_1, x_2, \ldots, x_n$, the $P^{th}$ **percentile** is a value, say $x$, such that approximately $P$ percent of the data is less than or equal to $x$ and approximately $(100 - P)$ percent of the data is greater than or equal to $x$.

## Percentiles

The most commonly used measure of relative position is the percentile. In fact, we have already discussed the 50th percentile; it is the median. For example, in data sets that do not contain significant quantities of identical data, the 30th percentile is a value such that about 30 percent of the values are below it, and about 70 percent are above it.

## Finding the $P^{th}$ Percentile

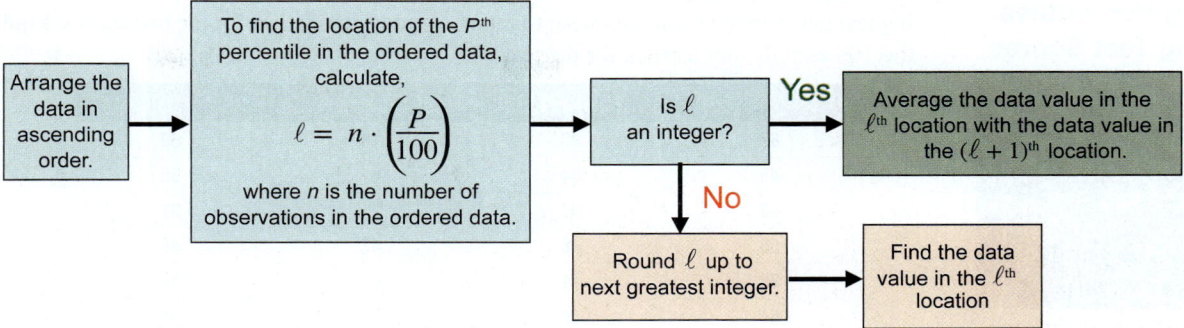

To determine the $P^{th}$ percentile, perform the following steps.

### Procedure

#### Finding the $P^{th}$ Percentile

1. Form an ordered array by placing the data in order from smallest to largest.
2. To find the location of the $P^{th}$ percentile in the ordered array, let

$$\ell = n \cdot \left(\frac{P}{100}\right)$$

   where $n$ is the number of observations in the ordered data.

3. If $\ell$ is not an integer, then round $\ell$ up to the next greatest integer. For example, if $\ell = 7.1$, then round $\ell$ up to 8 and find the data value in the $\ell^{th}$ location. If $\ell$ is an integer value, then average the data value in the $\ell^{th}$ location with the data value in the $(\ell + 1)^{th}$ location.

It is important to remember that when you find the value of $\ell$, this result is not the percentile. It is the *location* of the percentile in the ordered array. Thus, if the result of calculating (and rounding up) $\ell$ is 15, then the desired percentile would be the fifteenth value in the ordered list.

---

The grocery store manager wanted to know if the new bar code scanner had improved in its ability to scan products without error. The data below represent the number of scanning errors found in a random sample taken from the scanner. Find the 50th percentile for the following data set.

**Example 4.3.1**

**Determining Percentiles of Scanning Errors**

$$3, 5, 0, 1, 9, 2, 7$$

#### SOLUTION

The number of observations, $n = 7$.

The percentile, $P = 50$.

The location of the percentile, $\ell = 7 \cdot \left(\frac{50}{100}\right) = 3.5$ .

Since the location of the percentile is not an integer, the value is rounded up to 4.

Thus, the fourth observation in the ordered array is the 50th percentile.

$$0, 1, 2, \boxed{3}, 5, 7, 9$$

fourth observation

Therefore, the median value (which is the 50th percentile) is 3.

## Example 4.3.2

### Determining Percentiles of Screening Test Scores

### Interpreting Percentiles

When students take the SAT Reading Test, they receive a copy of their scores as well as the percentile they fall into. This percentile can sometimes be confusing. If a student receives a score of 620 on the Reading Test, they might fall into the 84th percentile. This means that they received a higher score than 84 percent of the students. The same score on the Math Test might place the student into the 80th percentile. Receiving a maximum score of 800 on the Reading Test or the Math Test will put the student in the 99th percentile. This means that less than 1 percent of the students taking the SAT had the same score.

### The Zen of Statistics

Douglas Hofstadter in his book *Gödel, Escher, Bach* describes Zen as an attitude in which words and truth are incompatible, or at least that no words can capture truth. If we think of the collected data as truth, statistics is a language whose "words" are pictures and numerical measures which seek to describe that "truth." Despite our best efforts, the statistical language suffers from the same inadequacies as our own language. We ignore the totality of the data in order to summarize it. There is a trade off—the loss of the "truth" for a better understanding.

Suppose the 40 members of your company are given a screening test for a new position. These scores are reported in Table 4.3.1. To inform potential employees of their screening test performance you may wish to report various percentiles for the test scores. Find the 10th and 88th percentiles for the test.

| Table 4.3.1 – Test Scores | | | |
|---|---|---|---|
| 67 | 45 | 18 | 82 |
| 45 | 54 | 61 | 55 |
| 63 | 47 | 21 | 31 |
| 58 | 46 | 43 | 49 |
| 35 | 71 | 69 | 56 |
| 54 | 80 | 73 | 77 |
| 27 | 70 | 41 | 29 |
| 66 | 32 | 44 | 33 |
| 21 | 64 | 52 | 81 |
| 48 | 55 | 57 | 62 |

| Table 4.3.2 – Ordered Test Scores | | | |
|---|---|---|---|
| 18 | 43 | 54 | 66 |
| 21 | 44 | 55 | 67 |
| 21 | 45 | 55 | 69 |
| 27 | 45 | 56 | 70 |
| 29 | 46 | 57 | 71 |
| 31 | 47 | 58 | 73 |
| 32 | 48 | 61 | 77 |
| 33 | 49 | 62 | 80 |
| 34 | 52 | 63 | 81 |
| 41 | 54 | 64 | 82 |

### SOLUTION

In order to calculate the percentiles, the data must be placed in an ordered array (Table 4.3.2). To compute the 10th percentile, its position in the ordered array must be determined.

The number of observations, $n = 40$.

The percentile, $P = 10$.

The location of the percentile, $\ell = 40 \cdot \left( \frac{10}{100} \right) = 4$. Since $\ell$ is an integer, the 4th and 5th observations in the array must be averaged. Since the fourth data value is 27 and the fifth data value is 29, then the 10th percentile is calculated as follows.

$$10^{th} \text{ percentile} = \frac{27+29}{2} = 28$$

To determine the 88th percentile, first calculate its location in the ordered array.

$$\ell = 40 \cdot \left( \frac{88}{100} \right) = 35.2$$

Since the location is not an integer, its value is rounded up to 36. The 36th observation in the ordered array will correspond to the 88th percentile. The 36th value is 73 in Table 4.3.2, so 73 is the 88th percentile.

A slightly different problem connected with percentiles involves taking a raw score and determining its corresponding percentile. Raw scores are usually not very meaningful. If someone scores a 56 on the screening test in the previous example, is that substantially less or about the same as someone who scored a 67 on the same test? To compare these two scores, find the percentile of each.

> **Formula**
>
> Percentile
>
> The **percentile** of some data value $x$ is given by:
>
> $$\text{percentile of } x = \frac{\text{number of data values less than or equal to } x}{\text{total number of data values}} \cdot 100$$

Note that when finding the percentile of a specific value, if there are multiple occurrences of that value in the data, they all need to be counted in the numerator in order to calculate the percentile. To determine the percentile for a score of 56, the number of data values less than or equal to 56 must be counted. Since there are 24 data values less than or equal to 56, the resulting percentile would be

$$\text{percentile of a score of } 56 = \frac{24}{40} \cdot 100 = 60.$$

Hence a score of 56 on the screening test corresponds to the $60^{\text{th}}$ percentile. Thus, approximately 60 percent of the scores are less than or equal to 56. Next, compute the percentile for a score of 67.

$$\text{percentile of a score of } 67 = \frac{32}{40} \cdot 100 = 80$$

A score of 67 on the screening test corresponds to the $80^{\text{th}}$ percentile. The score was better than or equal to approximately 80 percent of all other scores on the test. By computing percentiles, we have changed the data's scaling. We see the data from a new perspective. Using percentiles it is clear that a score of 67 is significantly better than a score of 56. The 11-point difference in raw score is translated into a 20 percent differential on the percentile scale.

## Quartiles

The $25^{\text{th}}$, $50^{\text{th}}$, and $75^{\text{th}}$ percentiles are known as **quartiles** and are denoted as $Q_1$, $Q_2$, and $Q_3$. They serve as markers that divide the data into four equal parts. $Q_1$ separates the lowest 25 percent, $Q_2$ represents the median ($50^{\text{th}}$ percentile), and $Q_3$ marks the beginning of the top 25 percent of the data.

> **Definition**
>
> Quartiles
>
> **Quartiles** are the $25^{\text{th}}$, $50^{\text{th}}$, and $75^{\text{th}}$ percentiles. They divide the data into four equal parts and are denoted as $Q_1$, $Q_2$, and $Q_3$.

Since quartiles are nothing more than percentiles ($25^{\text{th}}$, $50^{\text{th}}$, and $75^{\text{th}}$), the same methods used to calculate percentiles will also produce quartiles. For the screening test data in the previous example, the location of the $25^{\text{th}}$ percentile would be

$$\ell = 40 \cdot \left(\frac{25}{100}\right) = 10.$$

Since the location is an integer, we average the $10^{\text{th}}$ and $11^{\text{th}}$ observations in the ordered data to find the $25^{\text{th}}$ percentile.

$$Q_1 = 25^{\text{th}} \text{ percentile} = \frac{41 + 43}{2} = 42$$

Therefore, we would expect 25 percent of the data to be less than or equal to 42.

The location of the $50^{\text{th}}$ percentile is given by

$$\ell = 40 \cdot \left(\frac{50}{100}\right) = 20.$$

Since the location is an integer, we must average the $20^{\text{th}}$ and $21^{\text{st}}$ observations in order to calculate the percentile.

$$Q_2 = 50^{\text{th}} \text{ percentile} = \frac{54 + 54}{2} = 54,$$

which means approximately half the data is at or below 54.

The location of the 75th percentile is given by

$$\ell = 40 \cdot \left( \frac{75}{100} \right) = 30.$$

Since the location is an integer, average the 30th and 31st observations in the ordered array.

$$Q_3 = 75^{\text{th}} \text{ percentile} = \frac{64 + 66}{2} = 65$$

This means that approximately 75 percent of the data is less than or equal to 65. The quartiles are useful descriptions of data. They provide a good idea of how the data vary. The **interquartile range** is a measure of dispersion that is calculated using the first and third quartiles.

### Formula

#### Interquartile Range

The interquartile range is a measure of dispersion which describes the range of the middle fifty percent of the data.

$$\text{IQR} = Q_3 - Q_1$$

For the screening test data, the interquartile range is $65 - 42 = 23$, indicating that the middle 50 percent of the data spans a 23-unit range.

## Box Plots — Graphing with Quartiles

A very important use of quartiles is in the construction of box plots. As the name implies, box plots are graphical summaries of the data which, when constructed, have a box-like shape. They provide an alternative method to the histogram for displaying data. A **box plot** is a graphical summary of the central tendency, the spread, the skewness, and the potential existence of outliers in the data. Figure 4.3.1 displays a box plot of the screening test data from Example 4.3.2.

**Definition**

Box Plots

A **box plot** is a graphical summary of the data that has a box-like shape and displays the central tendency, the spread, the skewness, and the potential existence of outliers in the data.

### Box Plot of Screening Test Scores

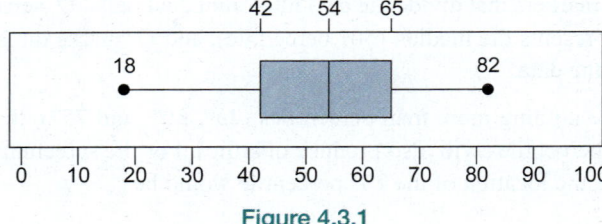

**Figure 4.3.1**

The box plot is constructed from five summary measures: the largest data value, the smallest data value, the 25th percentile, the 75th percentile, and the median.

The lower boundary of the box is the 25th percentile, which is 42 for the screening test data. The upper boundary of the box is the 75th percentile, which is 65 for the screening test data. The median is marked with a line through the box. The median of the test scores data is 54. Notice that the box itself represents the middle 50% of the data, and the length of the box is the interquartile range.

In Figure 4.3.1, a line is drawn from the 25th percentile to the smallest test score of 18, and another line is drawn from the 75th percentile to the largest score of 82. These lines are often referred to as **whiskers** and the box plot is often referred to as the **box and whisker plot**. The box plot for the screening test scores shows that the test score data are slightly skewed to the left. Why? Because the whisker extending from $Q_1$ appears to be longer than the whisker that extends from $Q_3$.

Although the box plot can be used to display data for a single data set, the histogram is probably more useful for this purpose. The real power of the box plot is the ease with which it allows the comparison of several data sets. Consider the number of wins per season for the New York Yankees, Los Angeles Dodgers, Atlanta Braves, and Chicago Cubs. The four data sets are displayed by box plots in Figure 4.3.2. It is easy to see from the box plots that the center of the Yankees' number of wins is slightly higher than that for the Dodgers, which has a higher center than the center of the Braves, and finally the Cubs. Also it appears that the spread of the data, or the variation within the observed values, is not the same for all four data sets. This type of comparison will be used in later chapters to help confirm assumptions which must be made about the data in order to perform statistical inference.

**:: Data**

This data set can be found on stat.hawkeslearning.com under **Discovering Business Statistics, Second Edition > Data Sets > Number of Franchise Wins 1961–2019**.

**Box Plots of the Number of Franchise Wins per Season 1961–2019**

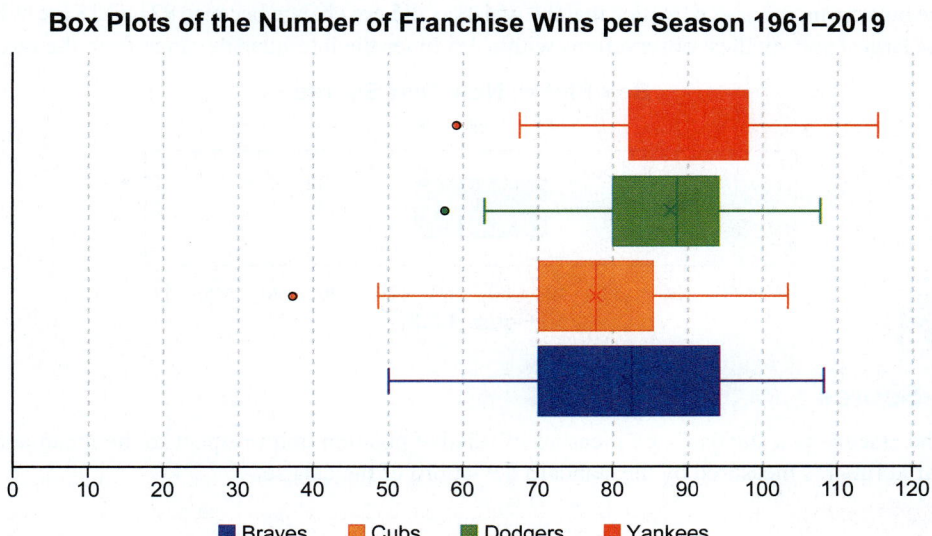

Braves    Cubs    Dodgers    Yankees

Figure 4.3.2

**⌕ Technology**

Microsoft Excel is capable of creating a multitude of different charts. For technology instructions on how to calculate a box plot using Excel or other technologies, visit stat.hawkeslearning.com and navigate to **Discovering Business Statistics, Second Edition > Technology Instructions > Graphs > Box Plot**.

**Definition**

Outlier

A data value is considered an **outlier** if it is 1.5 times the interquartile range above the 75th percentile or 1.5 times the interquartile range below the 25th percentile.

## Detecting Outliers

The concept of an **outlier** is arbitrary. What you consider an outlier and what someone else considers an outlier may not be the same thing. However, one definition of an outlier which has gained some acceptance is developed in the context of a box plot.

If there is an outlier in the data set, the whiskers are drawn to the largest or smallest data value which is within 1.5 times the interquartile range from the box, and the outliers are marked with an open circle. For example, suppose test scores of 110 and 2 were added to the screening test data.

| Table 4.3.3 – Ordered New Test Scores | | | |
|---|---|---|---|
| 2 | 43 | 55 | 69 |
| 18 | 44 | 55 | 70 |
| 21 | 45 | 56 | 71 |
| 21 | 45 | 57 | 73 |
| 27 | 46 | 58 | 77 |
| 29 | 47 | 61 | 80 |
| 31 | 48 | 62 | 81 |
| 32 | 49 | 63 | 82 |
| 33 | 52 | 64 | 110 |
| 34 | 54 | 66 | |
| 41 | 54 | 67 | |

**Baseball Players, Better or Worse?**

In the history of Major League Baseball (MLB), the last time someone had a batting average above .400 was Ted Williams in 1941. Before that, Ty Cobb had hit .420 in 1911. Finally, in 1980 George Brett came close but hit only .390. From this it appears that with time hitting for high averages is becoming more difficult. Is this then evidence of progress or lack of progress? The late evolutionary biologist Stephen Jay Gould, himself an avid baseball fan, had an interesting statistical angle to this riddle. He computed and observed:

*continued on next page...*

*continued...*

The yearly batting averages have remained more or less stable (around .267) over the history of the league. The annual standard deviations have declined steadily over the same period. Finally, the yearly batting averages are all normally distributed. Gould then computed the z-scores for all three players and observed they were all well above 4.0. His conclusion: Hitting has improved since the standard deviations of the averages are getting smaller. The best hitters are still at the fence of the normal distribution and are the best of their times. Therefore, as a result of the decline in standard deviation, the apparent decrease in batting average (disappearance of .400 hitters) is actually a sign of improvement! Less can actually be more in some situations!

For the new screening test data, the 25th percentile is 41 and the 75th percentile is 66, meaning that the interquartile range is $66 - 41 = 25$. A value is considered an outlier if the data value is:

- larger than the 75th percentile + 1.5 times the interquartile range.

$$66 + 1.5 \cdot 25 = 103.5$$

- smaller than the 25th percentile − 1.5 times the interquartile range.

$$41 - 1.5 \cdot 25 = 3.5$$

Since 110 is larger than 103.5, it is considered an outlier. Since 2 is smaller than 3.5, it is also considered an outlier. Figure 4.3.3 shows the box plot of the screening test data with the outliers incorporated. Notice that the whiskers did not change because 82 and 18 are still the largest and smallest observations within 1.5 times the interquartile range from the box.

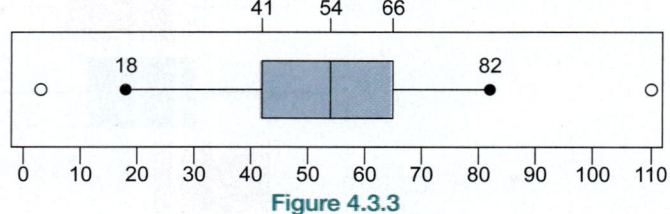

**Box Plot of New Test Scores**

Figure 4.3.3

## z-Scores

The **z-score** is a standardized measure of relative position, with respect to the mean and variability (as measured by the standard deviation) of the data set.

> **Formula**
>
> ### z-Score
>
> The z-score transforms a data value into the number of standard deviations that value is from the mean.
>
> $$z = \frac{x - \mu}{\sigma}$$

Describing a data value by its number of standard deviations from the mean is a fundamental concept in statistics that is found throughout this book. It is used as a standardization technique to describe properties of data sets and to compare the relative values of data from different data sets.

### Example 4.3.3

**Calculating a z-Score for Test Scores**

Suppose you scored an 86 on your marketing test and a 94 on your management test. The mean and standard deviations of the two tests are given below.

| Table 4.3.4 – Test Scores | | |
|---|---|---|
| Course | Mean | Standard Deviation |
| Marketing | 74 | 10 |
| Management | 82 | 11 |

What are the z-scores for your two tests? On which of the tests did you perform relatively better?

**SOLUTION**

The $z$-score for the marketing test is $z = \dfrac{86 - 74}{10} = 1.20$

The $z$-score for the management test is $z = \dfrac{94 - 82}{11} \approx 1.09$

On the marketing test you scored 1.20 standard deviations above the mean, compared to only 1.09 standard deviations above the mean for the management test. Even though the raw score on the management test is larger than the raw score on the marketing test, relative to the means of the data sets, the performance on the marketing test was slightly better. Once again, changing the scale of the data has beneficial effects. It enables the comparison of two measurements that are drawn from different populations.

---

If a $z$-score is negative, the data value is less than the mean. Conversely, if the $z$-score is positive, the data value is greater than the mean. The $z$-score is also a unit-free measure. That is, regardless of the original units of measurement (whether the data are measured in centimeters, meters, or kilometers), an observation's $z$-score will be the same.

## 📝 4.3 Exercises

### Basic Concepts

1. What are two methods for describing relative position?

2. If a data value is calculated to be the $72^{nd}$ percentile, what does this mean?

3. Describe how to find the percentile of a particular data value.

4. What are quartiles? Are they equivalent to percentiles? If so, how?

5. What is the interquartile range? What does it measure?

6. What are the advantages of using a box plot to display a data set?

7. What are the key calculations needed in order to construct a box plot?

8. What is an outlier? How can outliers be identified?

9. What is a $z$-score? Why is it useful?

### Exercises

10. The following test scores were recorded for an economics final examination.

| Test Scores | | | | | | | | | | | | | | |
|---|---|---|---|---|---|---|---|---|---|---|---|---|---|---|
| 60 | 81 | 100 | 44 | 90 | 56 | 71 | 42 | 64 | 100 | 69 | 80 | 90 | 87 | 94 |
| 41 | 78 | 100 | 50 | 96 | 77 | 61 | 38 | 41 | 68 | 50 | 69 | 85 | 47 | 86 |

   a. Calculate the $20^{th}$ percentile.

   b. Calculate the $95^{th}$ percentile.

   c. Interpret the meaning of each of these percentiles.

   d. Determine the percentile rank for the student who scored 56.

   e. Determine the percentile rank for the student who scored 80.

**11.** Copiers Etc. collects data on the number of copiers sold each day by each salesperson. The number of copiers sold for each salesperson for a small office on a randomly selected day is listed below.

| Numbers of Copiers Sold | | | | | | | | | | | |
|---|---|---|---|---|---|---|---|---|---|---|---|
| 1 | 5 | 2 | 3 | 7 | 6 | 1 | 0 | 0 | 3 | 4 | 5 |

    **a.** Calculate the $25^{th}$ percentile.

    **b.** Calculate the $90^{th}$ percentile.

    **c.** Interpret the meaning of each of these percentiles.

    **d.** Determine the percentile rank for the salesperson who sold 5 copiers.

    **e.** Determine the percentile rank for the salesperson who sold 1 copier.

**12.** Subjects in a marketing study were shown a film and at the end of the film were given a test to measure their recall. The scores are listed in the following table.

| Test Scores | | | | | | | | | | | | | | |
|---|---|---|---|---|---|---|---|---|---|---|---|---|---|---|
| 97 | 31 | 61 | 49 | 61 | 85 | 35 | 57 | 31 | 26 | 27 | 40 | 86 | 78 | 28 |
| 61 | 87 | 62 | 92 | 58 | 38 | 95 | 81 | 68 | 64 | 72 | 45 | 57 | 84 | 100 |

    **a.** Calculate $Q_1$, the first quartile.

    **b.** Calculate $Q_2$, the second quartile.

    **c.** Calculate $Q_3$, the third quartile.

    **d.** Explain the meaning of these quartiles in the context of the marketing study.

    **e.** Calculate the interquartile range.

    **f.** Construct a box plot for the test scores. Are there any outliers?

    **g.** Compute the $z$-score for a test score of 81.

    **h.** Compute the $z$-score for a test score of 62.

    **i.** Explain what the $z$-scores in parts **g.** and **h.** are measuring.

**13.** A baseball recruiter is interested in 20 perspective players. He goes to several games and determines the batting average for each player. The batting averages are displayed in the following table.

| Batting Averages | | | | | | | | | |
|---|---|---|---|---|---|---|---|---|---|
| .330 | .260 | .180 | .150 | .200 | .400 | .020 | .190 | .290 | .200 |
| .170 | .150 | .250 | .270 | .320 | .280 | .270 | .220 | .270 | .300 |

    **a.** Calculate $Q_1$, the first quartile.

    **b.** Calculate $Q_2$, the second quartile.

    **c.** Calculate $Q_3$, the third quartile.

    **d.** Explain the meaning of these quartiles in the context of the batting averages.

    **e.** Calculate the interquartile range.

    **f.** Construct a box plot for the batting averages. Are there any outliers? (Guess which player is the pitcher.)

    **g.** Compute the $z$-score for a batting average of .020.

    **h.** Compute the $z$-score for a batting average of .330.

    **i.** Explain what the $z$-scores in parts **g.** and **h.** are measuring.

    **j.** Determine the percentile rank for a player who had a batting average of .270.

    **k.** Determine the percentile rank for a player who had a batting average of .150.

14. Consider a set of data in which the sample mean is 64 and the sample standard deviation is 21. For the following specific values, calculate the $z$-score and interpret the results.

   a.  $x = 80$              b.  $x = 64$              c.  $x = 40$

15. A statistics student scored a 75 on the first exam of the semester and an 82 on the second exam of the semester. The average score and standard deviation of scores for the two exams are given in the following table. On which exam did the student perform relatively better?

| Test Scores | | |
|---|---|---|
| | First Exam | Second Exam |
| $\mu$ | 74 | 85 |
| $\sigma$ | 10 | 7 |

16. A hospital measures babies' heights when they are born in both inches and centimeters. Eight baby girls are randomly selected and the following heights are recorded in both inches and centimeters.

| Newborn Heights | | | | | | | | |
|---|---|---|---|---|---|---|---|---|
| Baby | 1 | 2 | 3 | 4 | 5 | 6 | 7 | 8 |
| Inches | 17.75 | 18.50 | 19.25 | 19.75 | 20.25 | 20.50 | 20.50 | 20.75 |
| Centimeters | 45.09 | 46.99 | 48.90 | 50.17 | 51.44 | 52.07 | 52.07 | 52.71 |

   a.  Calculate the mean height in inches and centimeters for the baby girls.

   b.  Calculate the standard deviation of the heights of baby girls in both inches and centimeters.

   c.  Calculate the $z$-score for the height of Baby Girl 3 measured in inches.

   d.  For Baby Girl 3, calculate the $z$-score for the height measured in centimeters.

   e.  Consider the $z$-scores calculated in parts c. and d. Are the $z$-scores as you expected them to be? Explain.

# 4.4 Data Subsetting

Data subsetting is used to provide more clarity and structure to the data. Referring to the histogram in Example 4.2.6, we know that the data consists of tuition and fees of two-year and four-year institutions. The histogram below (Figure 4.4.1) depicts the same data in Example 4.2.6 but has more intervals to give us a better spread of the data. Due to the large number of observations between $3,750 and $5,000, the histogram appears to be right-skewed. Were the histogram bell-shaped, it would be easier to identify the center of the data. Thus, it is sometimes prudent to separate the tuition data (i.e., data subsetting) into two groups—tuition for two-year institutions and tuition for four-year institutions.

## Where is the Central Value of the Tuition Data?

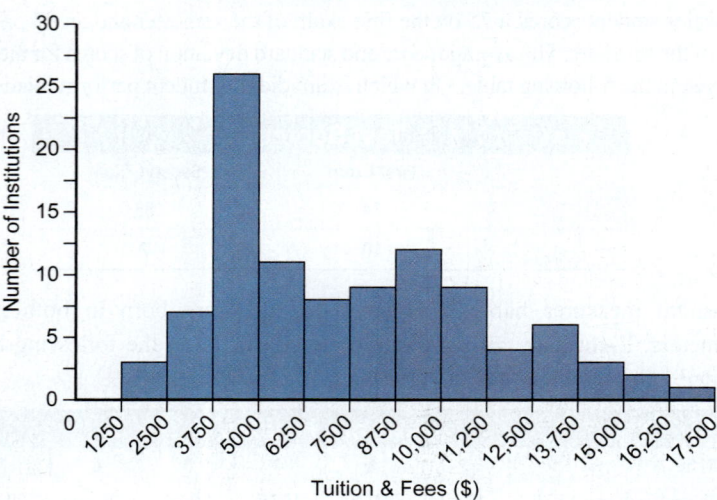

**Average Tuition and Fees at 2- and 4-Year
US Colleges and Universities**

Figure 4.4.1

There are several measurements that can be used to describe the location or central value of the tuition and fees data. The mode is not considered because the data are quantitative and because there are a large number of observations. Moreover, the mode is not a very informative measure of central tendency. Since the median of the data ($7,098) is a bit smaller than the mean ($7,498) and smaller than the trimmed mean ($7,206), the choice will make a difference.

Which one of the three candidate measurements should be considered as the central value of the data?

The choice is not easy. You are justified in selecting any of the three statistical measures (median, mean, or trimmed mean). However, because of the difference between the median and the other two measures, reporting two measures (the mean and median) would not be a bad idea. If only one measure is reported, identify the measure as the mean or the median rather than the average. The term "average" is a bit too ambiguous in current usage.

## Describing Dispersion of the Tuition Data

Several measurements which assess variation have been discussed, including variance, standard deviation, mean absolute deviation, and percentiles. Since the variance and standard deviation are different forms of the same measurement, usually only the standard deviation is reported. For the tuition data, a reasonable description of dispersion for the 101 values would be given by the following statistics.

| Table 4.4.1 – Tuition Data | |
|---|---|
| **Measure of Location** | **Value** |
| Mean | $7498 |
| Median | $7098 |
| 10% Trimmed Mean | $7206 |

Standard Deviation: $3,639

Minimum: $1,428

First Quartile ($Q_1$): $4,471

Median ($Q_2$): $7,098

Third Quartile ($Q_3$): $9,984

Maximum: $17,186

| Table 4.4.2 – Tuition Data Percentiles | |
|---|---|
| **Percentile** | **Value** |
| 100th (Max) | $17,186 |
| 90th | $13,244 |
| 75th ($Q_3$) | $9,984 |
| 50th ($Q_2$) | $7,098 |
| 25th ($Q_1$) | $4,471 |
| 10th | $3,600 |
| 0 (Min) | $1,428 |

The histogram in Figure 4.4.1 reveals an interesting characteristic of the data. There appears to be a large number of colleges and universities with tuition and fees less than $5,000. This seems a bit unusual given the costs of colleges and universities. That is, there seems to be a bundle of data between $0 and $5,000. What might be the reason for this? We could be looking at the difference between two-year and four-year institutions. This supposition could be checked rather easily if the data are divided (subsetted) into two-year and four-year schools.

The original data contained the average annual in-state tuition and fees by state for two-year and four-year institutions. The structure of a batch of data is often exposed by using another variable to break the data into smaller groups. This is called **subsetting**. A natural structuring of the tuition data is a grouping of two-year and four-year institutions.

**Definition**

**Subsetting**

**Subsetting** occurs when a variable is used to break up the data into smaller groups to show more structure in the data.

**Two-year Institutions (n = 49)**

Mean Tuition: $4,488

Median Tuition: $4,495

Trimmed Mean Tuition: $4,456

Standard Deviation: $1,286

$Q_1 = \$3,779$

$Q_3 = \$5,245$

Minimum: $1,428

Maximum: $8,205

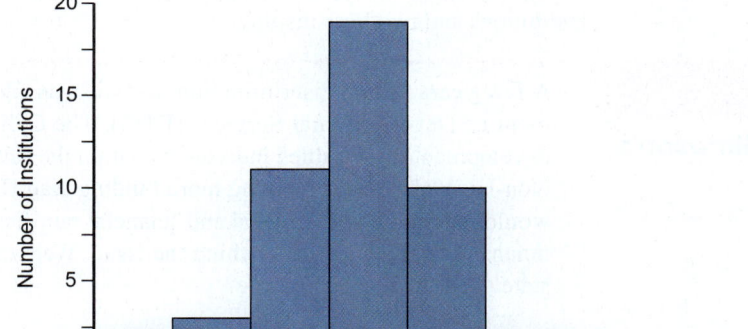

**Average Tuition and Fees at Two-Year Institutions 2019–2020**

Figure 4.4.2

**Four-year Institutions ($n = 52$)**

Mean Tuition: $10,335

Median Tuition: $9,847

Trimmed Mean Tuition: $10,214

Standard Deviation: $2,737

$Q_1 = \$8{,}511$

$Q_3 = \$11{,}881$

Minimum: $4,299

Maximum: $17,186

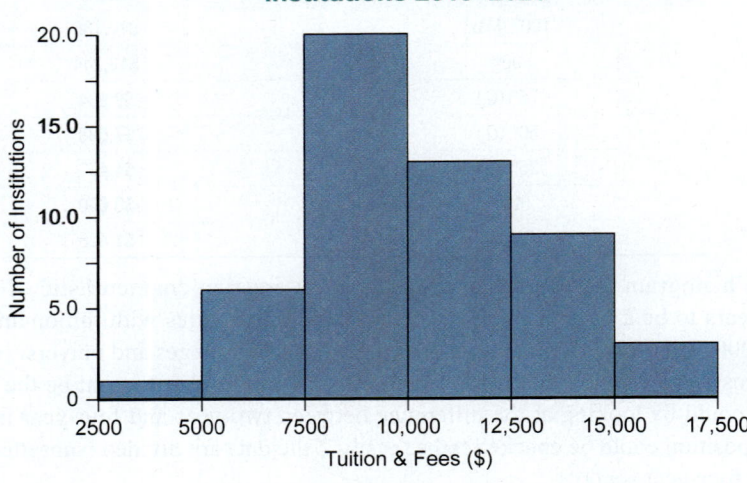

Figure 4.4.3

The data subsetting suggests that the data cluster between $0 and $5,000 revealed in the histogram in Figure 4.4.1 is the result of merging data from different kinds of educational institutions. Once the data are subsetted, the cluster at the lower end of the data disappears. For both two-year and four-year institutions, the majority of the values tend to cluster in the middle. The data tell quite an interesting story. From a tuition and fees point of view, two-year institutions are substantially less expensive than four-year institutions, on average. There is roughly a $6,000 average annual tuition and fees difference between four-year institutions and two-year institutions.

---

**Example 4.4.1**

**An Example of Simpson's Paradox**

A few years ago, a discrimination lawsuit was filed against the California Department of Developmental Services (DDS). The California DDS provides funding for developmentally-disabled individuals within the state. The lawsuit claimed that White Non-Hispanics were receiving more funding than Hispanics. If this was true, the DDS would face some serious legal and financial ramifications, so they hired several statisticians to analyze the data behind the issue. Were these claims of discrimination by a state department actually true?

**SOLUTION**

First, we will create box plots to visualize how the lawsuit arose. We will look at the relationship between ethnicity and expenditure by plotting expenditure grouped by the different ethnicities.

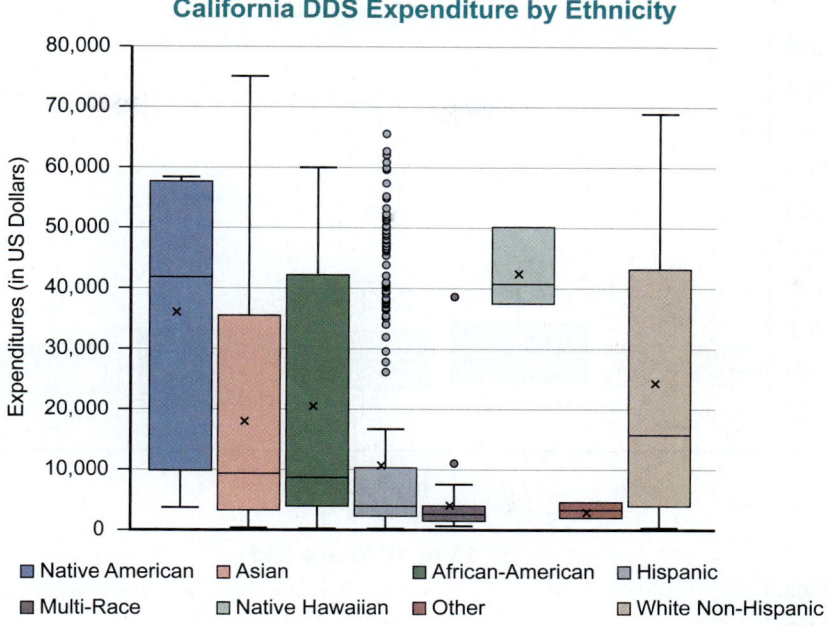

**Figure 4.4.4**

**⁘ Data**

The full California DDS data set can be found on stat.hawkeslearning.com under **Discovering Business Statistics, Second Edition > Data Sets > California DDS Expenditures**.

If the data is examined on a purely ethnic basis, then the figure above suggests that there is discrimination towards multiple different ethnic groups. But, now we need to think about any variables that could be confounding this picture.

The statisticians discovered that age played a massive role in the expenditure per person, because more costs are associated with caring for older developmentally-disabled individuals. To account for the variation in expenditure between different ages, the statisticians decided to subset the whole data set into six different age groups, and then examine the expenditure by ethnicity within each age group to see if the discrimination claim still held true. The box plots in the following figures illustrate their findings.

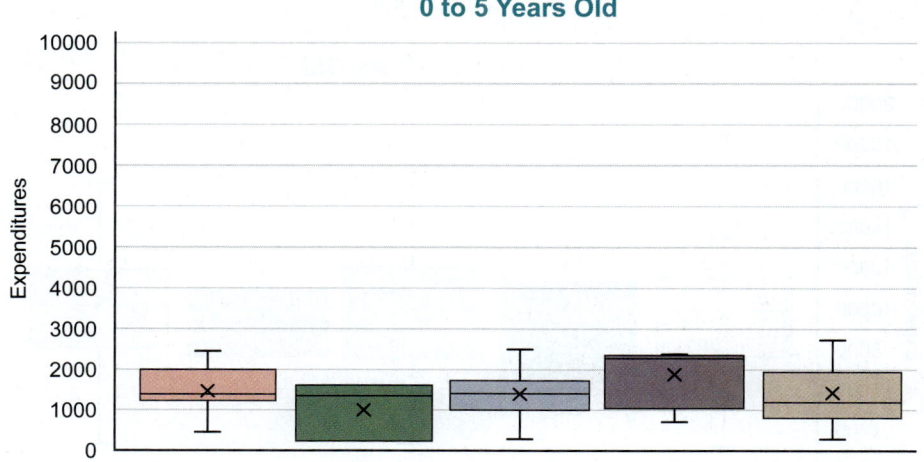

**Figure 4.4.5**

**⌳ Technology**

To learn how to perform calculations on a filtered data set, please visit stat.hawkeslearning.com and navigate to **Discovering Business Statistics, Second Edition > Technology Instructions > Data Manipulation > Subset Calculations**.

## 6 to 12 Years Old

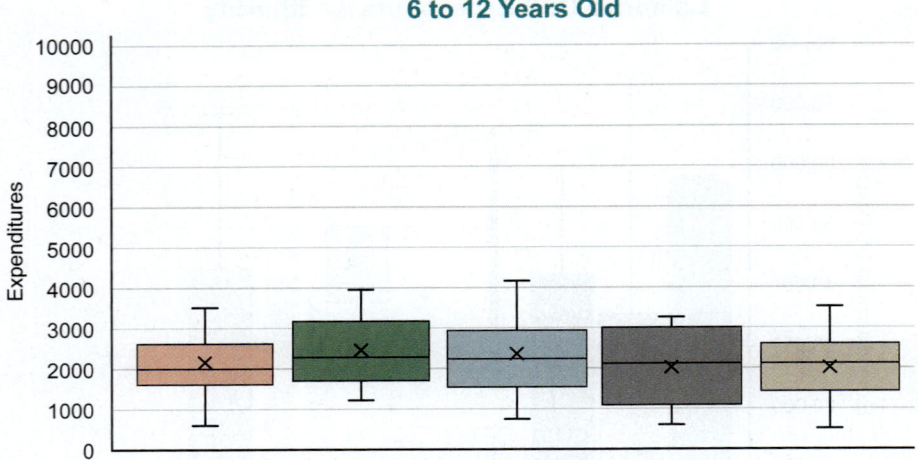

Figure 4.4.6

## 13 to 17 Years Old

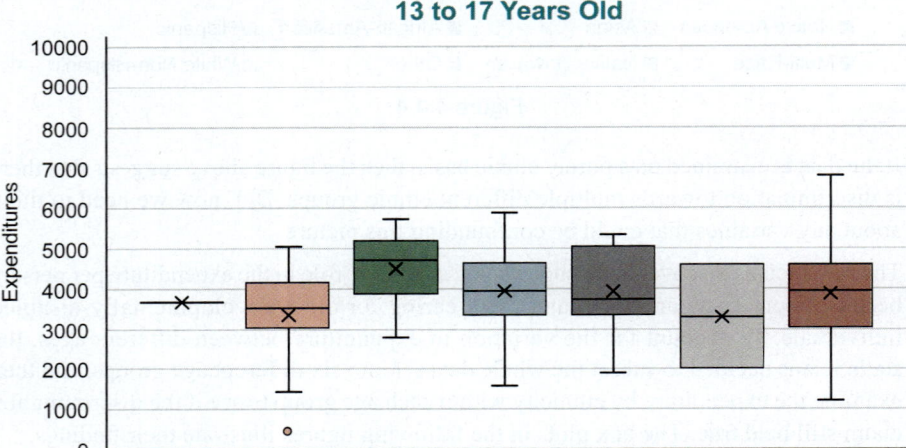

Figure 4.4.7

## 18 to 21 Years Old

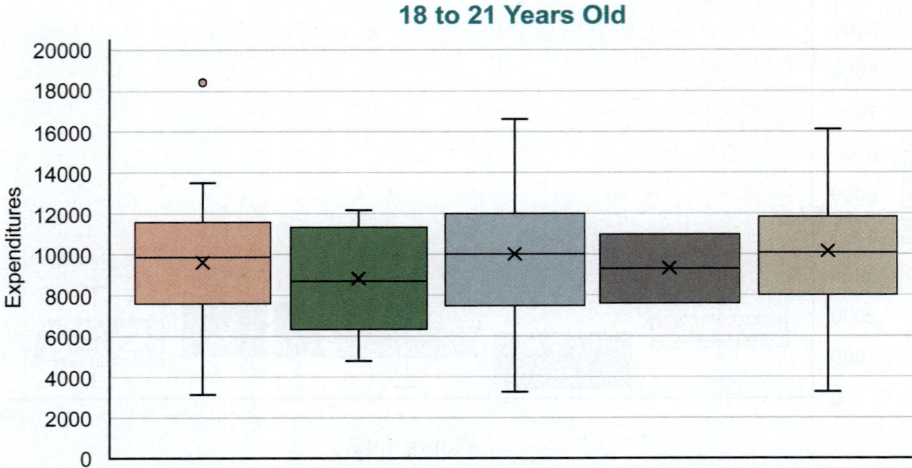

Figure 4.4.8

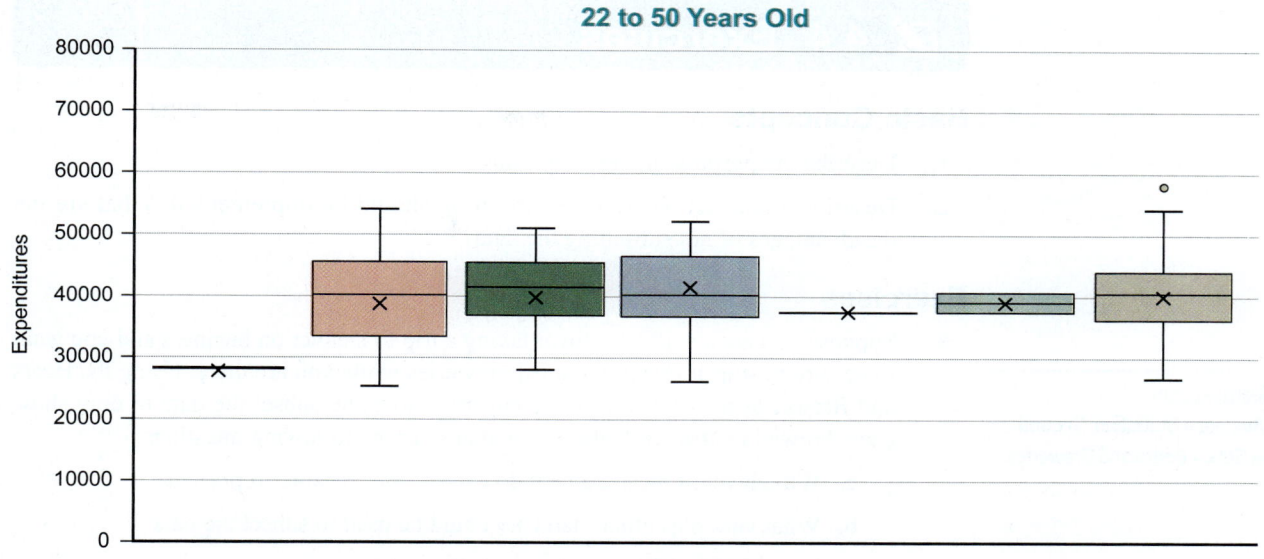

Figure 4.4.9

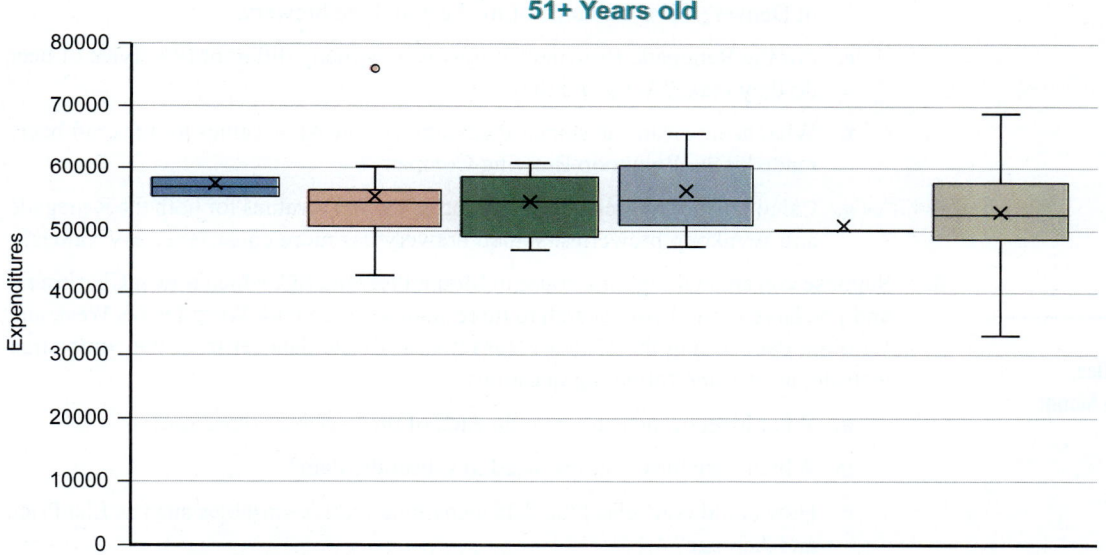

Figure 4.4.10

To the surprise of many involved, the box plots were roughly the same for each ethnic group once age was considered. There was no evidence of discrimination once the confounding variable "age" was accounted for.

This is a classic example of Simpson's paradox, which states that the perceived association between two variables (ethnicity and expenditures per person in this case) can be drastically affected by a third variable. When the data is partitioned by the third variable (age in our case) the "alleged" relationship may not exist or be just the opposite. We encountered this in a previous chapter when we discussed the gender discrimination lawsuit regarding admissions at the University of California, Berkeley. When analyzing data, it is good to keep in mind Oscar Wilde's cautionary quote, "the truth is rarely pure, and never simple". We must always be very cautious when attributing causation.

✎ **NOTE**

The scale for each age group is different to better display the data. Within each age group, the box plots are roughly the same for each ethnic group.

### ✏️ 4.4 Exercises

## Basic Concepts

1. Describe the purpose of data subsetting.

2. Describe a data set where data subsetting should be implemented. What are the disadvantages of not subsetting the data?

## Exercises

**Data**

stat.hawkeslearning.com

**Discovering Business Statistics, Second Edition > Data Sets > Beers and Breweries**

3. Suppose you are a craft beer lover taking a trip to Denver on business and you want to be sure to stop at one of the local breweries while you are there. Using the Beers and Breweries data set from the companion website, subset the data to only show beers brewed in Denver, Colorado, and answer the following questions.

   a. What level of measurement do each of the variables represent?

   b. What variables other than City could be used to subset the data?

   c. How many craft breweries are in Denver?

   d. Which craft beer has the highest Alcohol by Volume (ABV) of the beers brewed in Denver? Give the name of the beer and the brewery.

   e. For the Renegade Brewing Company, how many different IPA styles of beer do they make? What are they?

   f. What is the mean and standard deviation of the ABV values for the craft beers made by the Wynkoop Brewing Company?

   g. Calculate the coefficient of variation of the ABV values for both the Renegade and Wynkoop breweries. Which brewery has more consistent ABV values?

**Data**

stat.hawkeslearning.com

**Discovering Business Statistics, Second Edition > Data Sets > Mount Pleasant Real Estate Data**

4. Suppose you are looking for a house in Mount Pleasant, SC, which is near Charleston, and you have limited your search to three subdivisions: Park West, Dunes West, and Carolina Park. Using the Mount Pleasant Real Estate data set from the companion website, answer the following questions.

   a. What level of measurement do each of the variables represent?

   b. Which variables could be used to subset the data?

   c. How could you subset the data using quantitative variables such as List Price and Acreage?

   d. How many different house styles are represented in these three subdivisions? What are the styles?

   e. How many of the houses are newly built (2015–2017)? Which subdivision has the most new homes?

   f. What is the average price of new homes (2015–2017) in Carolina Park? Round your answer to the nearest whole dollar.

   g. For all new homes (2015–2017) in the three subdivisions, what is the minimum and maximum priced homes and in which subdivision are they?

   h. What is the price per square foot of the two homes in part **g.**?

   i. What variables do you think may contribute to the high price of the house with the maximum price?

5.  You are asked to evaluate whether there are issues with how funds are distributed to individuals with developmental disabilities in California. There is a concern that expenditures may not be allocated equitably across various demographic groups. Using the California DDS Expenditures data set, answer the following questions:

    a.  What are the mean, mode, and median for the variable Expenditures of the entire data set?

    b.  What are the variance, standard deviation, and range for the variable Expenditures of the entire data set?

    c.  What are the 1st and 3rd quartile for the variable Expenditures of the entire data set?

    d.  Which variables could be used to subset the data?

    e.  Find the average Expenditure for each Age Group

    f.  Which Age Group has the highest average Expenditure? Do you notice any trends by Age Group? What might account for differences that exist?

    g.  Which age group has the highest level of dispersion in Expenditure as measured by the standard deviation and coefficient of variation? Do you notice any trends by Age Group? What might account for differences that exist?

    h.  Find the average Expenditure for each Ethnicity.

    i.  What proportion of Expenditures is allocated to each Ethnicity?

    j.  Briefly discuss your findings based on the analysis in the previous sections.

**Data**

stat.hawkeslearning.com
**Discovering Business Statistics, Second Edition > Data Sets > California DDS Expenditures**

6.  You have been hired by a large company to investigate their employees' satisfaction level. There is some concern that there is high turnover with experienced employees. Using the Employee Satisfaction data set, answer the following questions:

    a.  What are the mean, mode, and median for the variable Satisfaction Level?

    b.  What are the variance, standard deviation, and range for the variable Satisfaction Level?

    c.  What are the 1st and 3rd quartile for the variable Satisfaction Level?

    d.  Which variables could be used to subset the data?

    e.  What is the correlation coefficient between employee satisfaction and the employee's last evaluation score? What does this correlation tell you about the relationship?

    f.  Find the average Satisfaction Level for each Department?

    g.  Which salary grouping (low, medium, high) has the highest level of dispersion for Satisfaction Level as measured by standard deviation and coefficient of variation?

    h.  Find the average Satisfaction Level of each year of experience. Are there differences in Satisfaction Level based on years spent at the company?

    i.  Briefly discuss your findings based on the analysis in the previous sections.

**Data**

stat.hawkeslearning.com
**Discovering Business Statistics, Second Edition > Data Sets > Employee Satisfaction**

**Data**

stat.hawkeslearning.com
**Discovering Business Statistics, Second Edition > Data Sets > San Francisco Salaries 2014**

7.  You have been applying for jobs in San Francisco. You want to research to understand what salary level you can expect to be offered. Using the San Francisco Salaries 2014 data set, answer the following questions:

    **a.** What are the mean, mode, and median for the variable Total Pay and Benefits?

    **b.** What are the variance, standard deviation, and range for the variable Total Pay and Benefits?

    **c.** What are the 1st and 3rd quartile for the variable Total Pay and Benefits?

    **d.** Which variables could be used to subset the data?

    **e.** Construct a frequency distribution for Base Pay? Include the relative frequency of each class.

    **f.** From part e. which pay group has the highest relative frequency? What trends do you notice? What might account for differences that exist?

    **g.** Determine the percentage of jobs that have overtime.

    **h.** Which group (overtime or no overtime) has the highest level of dispersion for total pay as measured by standard deviation and coefficient of variation?

    **i.** Briefly discuss your findings based on the analysis in the previous sections.

# 4.5 Analyzing Grouped Data

All of the statistical measurements we have discussed so far presume that raw data measurements are readily available. However, there may be instances in which only a frequency distribution of the data is available. When data are presented in that form, they are called **grouped data**. It is important to be able to compute measures such as the mean and variance for this type of data. Note that because the raw data observations are not available, the measures will be approximate.

## Finding the Mean and Variance of Grouped Data

The strategy for finding the mean of grouped data involves finding the midpoint of each of the classes in the frequency distribution and then weighting each of these midpoints by the number of observations in the class

**Formula**

### Mean of Grouped Data

The population **mean of grouped data** is given by

$$\mu = \frac{\Sigma(f_i M_i)}{N}$$

where

$f_i$ = the number of observations in the $i^{\text{th}}$ class,

$N$ = the total number of observations in all classes, $N = \Sigma f_i$, and

$M_i$ = the midpoint of the $i^{\text{th}}$ class.

The sample mean of grouped data is given by

$$\bar{x} = \frac{\Sigma(f_i M_i)}{n}.$$

where $n$ is the number of observations in the sample.

We can also estimate the variance of grouped data using the following formulas.

**Formula**

### Variance of Grouped Data

The population **variance of grouped data** is given by the following expression.

$$\sigma^2 = \frac{\Sigma(M_i - \mu)^2 f_i}{N}$$

The corresponding formula for the sample variance is as follows.

$$s^2 = \frac{\Sigma(M_i - \bar{x})^2 f_i}{n-1}$$

where $n$ equals the total number of observations in the sample.

There are also computational formulas that can be used to calculate the variance of grouped data. These formulas may make it easier to calculate the variance by hand. We will use these formulas in Example 4.5.1 to calculate the variance for a grouped data set.

## Formula

### Computational Formulas for the Variance of Grouped Data

The computational formulas for the population and sample variances of grouped data are as follows.

$$\sigma^2 = \frac{\sum\left(f_i M_i^2\right) - \dfrac{\left(\sum\left(f_i M_i\right)\right)^2}{N}}{N} = \frac{\sum\left(f_i M_i^2\right)}{N} - \left(\frac{\sum\left(f_i M_i\right)}{N}\right)^2$$

$$s^2 = \frac{\sum\left(f_i M_i^2\right) - \dfrac{\left(\sum\left(f_i M_i\right)\right)^2}{n}}{n-1}$$

**Example 4.5.1**

**Calculating the Mean and Variance of Grouped Data**

Table 4.5.1 presents, in grouped form, the amount of cash on hand for 45 technology companies in the business software and services industry. Compute the mean and variance for these data.

### Table 4.5.1 – Cash on Hand for Technology Companies

| Cash on Hand (Millions of Dollars) | Frequency |
|---|---|
| 0–10 | 10 |
| 10–20 | 7 |
| 20–30 | 7 |
| 30–40 | 7 |
| 40–50 | 1 |
| 50–60 | 4 |
| 60–70 | 2 |
| 70–80 | 2 |
| 80–90 | 2 |
| 90–100 | 3 |

### SOLUTION

To compute the mean and variance, the midpoints of each interval must be calculated. The class midpoint is determined as follows.

$$\text{midpoint} = \frac{\text{lower class boundary} + \text{upper class boundary}}{2}$$

The midpoints as well as the other required calculations are presented in Table 4.5.2.

### Table 4.5.2 – Midpoints and Other Required Calculations

| Cash on Hand (Millions of Dollars) | Midpoint $M_i$ | Frequency $f_i$ | $f_i \cdot M_i$ | $M_i^2$ | $f_i \cdot M_i^2$ |
|---|---|---|---|---|---|
| 0–10 | 5 | 10 | 50 | 25 | 250 |
| 10–20 | 15 | 7 | 105 | 225 | 1575 |
| 20–30 | 25 | 7 | 175 | 625 | 4375 |
| 30–40 | 35 | 7 | 245 | 1225 | 8575 |
| 40–50 | 45 | 1 | 45 | 2025 | 2025 |
| 50–60 | 55 | 4 | 220 | 3025 | 12,100 |
| 60–70 | 65 | 2 | 130 | 4225 | 8450 |
| 70–80 | 75 | 2 | 150 | 5625 | 11,250 |
| 80–90 | 85 | 2 | 170 | 7225 | 14,450 |
| 90–100 | 95 | 3 | 285 | 9025 | 27,075 |
| **Totals** | | 45 | 1575 | 33,250 | 90,125 |

Assuming the data in Table 4.5.1 are population data, the mean cash on hand for the 45 companies is calculated as follows.

$$\mu = \frac{\Sigma(f_i M_i)}{N} = \frac{1575}{45} = \$35 \text{ million}$$

The variance of the grouped data is calculated as follows.

$$\sigma^2 = \frac{\Sigma(f_i M_i^2) - \frac{(\Sigma(f_i M_i))^2}{N}}{N}$$

$$= \frac{90125 - \frac{1575^2}{45}}{45} \approx 777.7778$$

If the data are sample data, then the variance is:

$$s^2 = \frac{\Sigma(f_i M_i^2) - \frac{(\Sigma(f_i M_i))^2}{n}}{n-1}$$

$$= \frac{90125 - \frac{1575^2}{45}}{44} \approx 795.4545.$$

It is important to remember that the calculations of the mean and variance are approximate. That is, if the raw data are available, the actual mean and variance would differ from the measures calculated using the grouped data.

### Technology

For technology instructions to calculate the sample statistics for grouped data, like the mean and standard deviation, visit stat.hawkeslearning.com and navigate to **Discovering Business Statistics, Second Edition > Technology Instructions > Descriptive Statistics > Two Variables.**

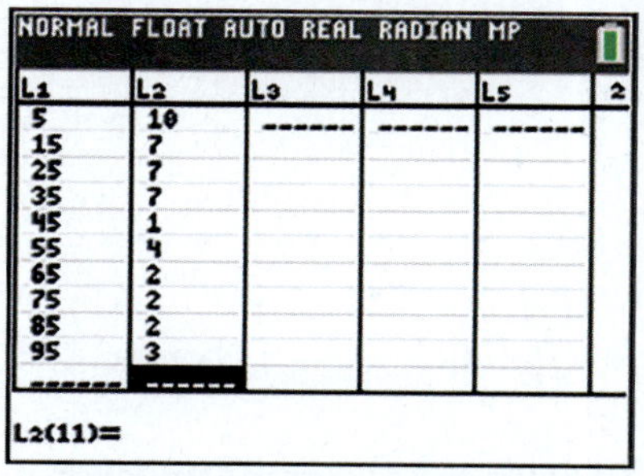


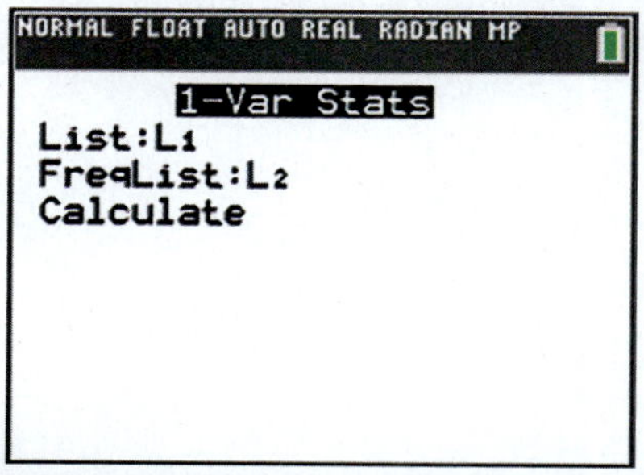


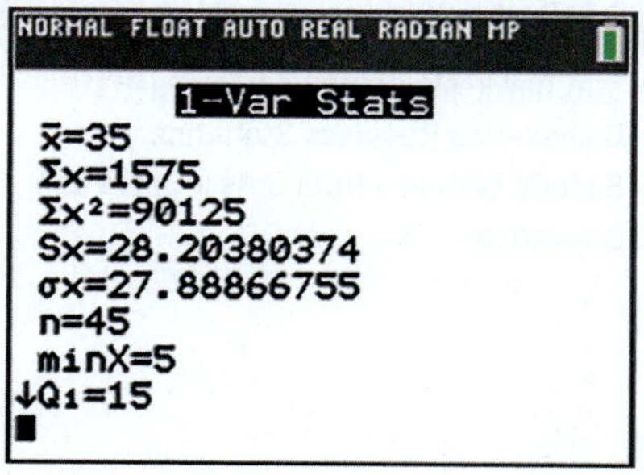


## 4.5 Exercises

### Basic Concepts

1. When analyzing grouped data, are the measurements exact? Why or why not?
2. What calculations are required in order to analyze grouped data?

### Exercises

3. A client of a commercial rose grower has been keeping records on the shelf-life of a rose. The client sent the frequency distribution to the grower. Calculate the mean and variance for the shelf-life given the following frequency distribution.

**Rose Shelf-Life**

| Days of Shelf-Life | Frequency |
|---|---|
| 1 – 6 | 2 |
| 7 – 12 | 3 |
| 13 – 18 | 9 |
| 19 – 24 | 6 |
| 25 – 30 | 3 |
| 31 – 36 | 1 |

4. An article in *Business Week* discussed the large spread between the federal funds rate and the average credit card rate. The table below is a frequency distribution of the credit card rate charged by the top 100 issuers. Note that at the time these figures were published, the average federal funds rate was well below 5%.

| Credit Card Rates | |
|---|---|
| Credit Card Rate | Frequency |
| 19% – 24% | 36 |
| 18% – 18.9% | 8 |
| 17% – 17.9% | 15 |
| 16% – 16.9% | 12 |
| 15% – 15.9% | 29 |

a. Calculate the average credit card rate charged by the top 100 issuers based on the frequency distribution.

b. Calculate the variance of the credit card rate charged by the top 100 issuers based on the frequency distribution.

c. Calculate the standard deviation of the credit card rate charged by the top 100 issuers based on the frequency distribution.

5. A frequency distribution for the Beers and Breweries data set from the companion website is shown below. Use the frequency distribution to perform the following.

| ABV Frequencies | |
|---|---|
| ABV | Frequency |
| 0.0010–0.017 | 1 |
| 0.0175–0.033 | 6 |
| 0.0335–0.049 | 402 |
| 0.0495–0.065 | 1228 |
| 0.0655–0.081 | 565 |
| 0.0815–0.097 | 146 |
| 0.0975–0.113 | 45 |
| 0.1135–0.129 | 3 |

a. Calculate the average ABV of all beers based on the frequency distribution. Round your answer to three decimal places.

b. Calculate the variance of the ABVs of the different beers based on the frequency distribution. Round your answer to four decimal places.

c. Calculate the standard deviation of the ABVs of the different beers based on the frequency distribution. Round your answer to three decimal places.

# 4.6 Proportions

The **proportion** is one of the more common summary measures.

To calculate a proportion, simply count the number in the group that possess the characteristic and divide the count by the total number in the group. Let

$x$ = number of observations that possess the characteristic,

$N$ = number of observations in the population, and

$n$ = number of observations in the sample, then

$p = \dfrac{x}{N} = $ the population proportion, and

$\hat{p} = \dfrac{x}{n} = $ the sample proportion.

The symbol $\hat{p}$ is pronounced *p-hat*.

---

Suppose the store manager wants to know the percentage of customers who buy something on days when she has most of the inventory prices reduced by as much as 50%. Of a random sample of 48 customers on a particular sale-day, only 4 of them did not purchase at least one item. What proportion of the customers purchased at least one item?

**Example 4.6.1**

**Calculating the Proportion of Customers**

#### SOLUTION

There are 48 pieces of data in the group which represents customers. Think of the data as composed of 0s and 1s. Any customer who purchased an item will be a 1, and any customer who did not purchase an item will be a 0. In our data set, there will be forty-four 1s and four 0s.

$$\text{Assuming } x_i = \begin{cases} 1 \text{ if customer purchased an item} \\ 0 \text{ if customer did not purchase an item} \end{cases}$$

$$\text{then } \sum x_i = 1 + 1 + \cdots + 1 + 0 + 0 + 0 + 0 = 44.$$

In the notation we used earlier, $x$ equals the number of observations that possess the characteristic. Therefore,

$$x = 44 \text{ and } p = \frac{x}{N} = \frac{44}{48} \approx 0.9167.$$

Thus, approximately $0.9167(100) = 91.67\%$ of customers purchased at least one item.

Note that we are using $\hat{p}$, the sample proportion in this case because we are considering the sample of 48 customers that came into the store. If we were using the data that represented all of the customers that came into the store on sale-day, the proportion would be the population proportion $p$ given by $\dfrac{x}{N}$, where $N$ represents the size of the population.

---

Suppose a Major League Baseball player has kept records on each plate appearance. According to his records, he batted 216 times this season. Of these 216 plate appearances, he walked 24 times, got on base by a fielding error 7 times, and reached base on a hit 64 times. Compute the player's batting average, which is a proportion.

**Example 4.6.2**

**Calculating a Batting Average**

#### SOLUTION

The batting average is the proportion of times the player reached base on a hit, excluding walks and errors. In this case the number in the group of at-bats we will consider is

$$N = \text{Plate appearances} - \text{Walks} - \text{Bases by fielding errors}$$
$$= 216 - 24 - 7$$
$$= 185 \text{ at-bats.}$$

The proportion of times he got a hit (excluding walks and errors) is

$$p = \frac{64}{185} \approx 0.3459.$$

Hence, his batting average is .346.

---

This chapter has been devoted to summarizing data. Yet, with the exception of the mode, none of the summary methods discussed should be applied to nominal data. Using proportions is one of the few summary methods available for analyzing qualitative data.

## ✎ 4.6 Exercises

### Basic Concepts

1. What is a proportion?

2. What is the difference in notation between a population proportion and a sample proportion?

3. Other than the mode, proportions are one of the few summary methods available to analyze what type of data?

### Exercises

4. A survey of shoppers at a local mall was taken to study the shopping habits of consumers. The mall contains a variety of specialty stores, especially stores that specialize in electronics and gadgets. One question in particular asked, "Do you enjoy shopping for electronics?" Of the 300 men surveyed, 175 answered "Yes." Of the 200 women surveyed, 55 answered "Yes."

   a. What proportion of men enjoy shopping for electronics?

   b. What proportion of women enjoy shopping for electronics?

   c. What is the overall proportion of consumers at the Tech Mall that enjoy shopping for electronics?

5. A survey released in April 2011 conducted by the consulting firm Booz & Company regarding the automobile industry found that U.S. automotive executives were skeptical about the industry's economic recovery. Suppose that 118 original equipment manufacturers (OEM) and 82 supplier executives participated in the study. When asked whether the overall state of the industry is fairly similar to, or somewhat better than, its low point in January 2009, 59 OEMs and 28 supplier executives answered that the current state of the industry was "about the same," and 55 OEMs and 52 supplier executives answered that the current state of the industry was "somewhat better."

   **Source:** Booz & Company; 2011

   a. Calculate the sample proportion of original equipment manufacturers that believe the state of the automobile industry is "about the same" as in January 2009.

   b. Calculate the sample proportion of supplier executives that believe the state of the automobile industry is "about the same" as in January 2009.

   c. Calculate the sample proportion of original equipment manufacturers that believe the state of the automobile industry is "somewhat better" than in January 2009.

   d. Calculate the sample proportion of supplier executives that believe the state of the automobile industry is "somewhat better" than in January 2009.

   e. Do these responses seem to support Booz & Company's conclusion that automotive executives are skeptical about the industry's economic recovery? Discuss.

6. An experiment was conducted to study how investors selected mutual funds. Two groups of investors were selected. Group 1 consisted of investors that used online brokerages and Group 2 was made up of investors that used full-service brokerages. Of the 150 investors in Group 1, 120 indicated that they selected mutual funds on their own, while 30 stated that they selected mutual funds using recommendations of the brokerage, family, and friends. Of the 200 investors in Group 2, 25 indicated that they selected their funds on their own, while 175 indicated that they selected the mutual funds using recommendations of the brokerage, family, and friends.

   a. What proportion of Group 1 investors selected mutual funds on their own?

   b. What proportion of investors in Group 2 selected mutual funds on the recommendation of others?

   c. What does this tell you about investors that use online brokerages versus those using full-service brokerages?

7. According to a study administered by the National Bureau of Economic Research, half of Americans would struggle to come up with $2000 in the event of a financial emergency. The majority of the 1900 Americans surveyed said they would rely on more than one method to come up with emergency funds if required. In the survey, 532 people said that they "certainly" would not be able to cope with an unexpected $2000 bill if they had to come up with the money in 30 days, and 418 people said they "probably" would not be able to cope.

   Source: CNNMoney.com; 2011

   a. What percentage of Americans "certainly" would not be able to produce $2000 in the event of an emergency according to the study?

   b. What percentage of Americans would "probably" not be able to pay a $2000 bill in 30 days if required?

   c. What does this say about the savings habits of Americans?

8. What college football conference has the right to brag about putting players in the NFL? A random sample of 100 current NFL players was surveyed to determine the conference in which they played college football. The following table displays the results of the survey.

| NFL Players and College Football Conferences ||
| Conference | Number of Players |
| --- | --- |
| SEC | 20 |
| Big 12 | 16 |
| Big 10 | 12 |
| Pac 10 | 6 |
| ACC | 6 |
| MAC | 2 |
| Big East | 2 |
| Other | 36 |

   a. What proportion of players are from the SEC?

   b. What proportion of players are from a conference other than the first 6 listed?

   c. Is it true that a player in the SEC has a better chance of being drafted in the NFL than a player from any other conference? Explain.

9. According to a survey administered by the market research group ChangeWave in February 2011, approximately 27% of respondents reported that they plan on buying a tablet device in the future. This result was 2 percentage points higher than in a similar survey administered in November of 2010. In the February study, 3091 customers were surveyed on tablet demand and future buying trends. Suppose that in the November study, 721 people said they planned on buying a tablet device in the future.

**Source:** InvestorPlace; 2011

   a. In the February study, how many people said that they planned on buying a tablet device in the future?

   b. In the November study, how many total customers were surveyed?

10. It is no secret that Wall Street firms compete aggressively to lure their clients. Having more high-end clients translates into fees and revenues that turn into profits. A survey of 150 high-end clients asked what lured them to their respective Wall Street firm. The following table shows the results.

| High-End Client Response | |
|---|---|
| **Perk Received** | **Client Response** |
| Pay a Kick-Back | 25 |
| Lucrative Golf Outings | 12 |
| Lavish Dinners | 8 |
| Free Private Jet Use | 33 |
| Prime Seats at Sports Events | 20 |
| Other | 22 |
| No Perk Received | 30 |

   a. Which type of perk appears to be most successful in luring clients?

   b. What proportion of clients were lured to a Wall Street firm by the perk identified in part **a.**?

   c. What proportion of clients did not receive a perk at all?

   d. Given that these perks aren't inexpensive, what conclusion can you make about providing perks to clients? Explain.

# 4.7 Measures of Association Between Two Variables

Oftentimes a manager or decision maker is interested in the relationship between two variables. Such relationships could be the amount of sales and advertising expenditures, education level and income, number of contacts made and sales, sales and earnings, number of real estate foreclosures and home prices, and so on. In earlier sections, we discussed methods used to study each of these measurements individually. Now we want to study the measurements of two different variables at the same time to see if there is a relationship between them. The purpose of studying the relationship between two variables is three-fold: to describe and understand the relationship, to forecast and predict a new observation, and if one is working with a process, not understanding the relationship could be detrimental to making necessary adjustments. There are statistical tools that can aid in the discovery of relationships.

Thinking about relationships is something that everyone does. Why, for example, does an admissions counselor want to know the relationship between SAT scores and college performance? The admissions counselor's task is to select students who will be successful at that college. If college performance is related to SAT scores and the relationship can be specified explicitly, then SAT scores can be used to *predict* performance. Consequently, the relationship would be helpful in selecting students for admission. On the other hand, if SAT scores are not useful in predicting college performance, then they should not be considered very important in making admissions decisions. Discovering whether SAT scores are related to college performance could improve the admissions process. Accurate predictions of a variable often suggest methods for process improvement.

Most students view grades as important and work to make good grades for self-satisfaction and to enhance career opportunities. Because grades are perceived as important, students often wonder about the relationship between the amount of time spent studying and the resulting exam grade. If study time is related to the grade received, a student could use the relationship to predict a grade based on study time. In addition, if the student is able to predict the grade this way, the student could adjust study time to obtain whatever grade is desired. Being able to predict a low grade could lead to corrective action and an improvement in grade point average. The ability to predict often leads to the ability to control.

## Bivariate Data

In earlier sections, all of the statistical summary measurements, like the mean, variance, and proportions, were concerned with describing **univariate** data (measurements of one variable). To understand the relationship between two variables, data on both variables need to be collected. This type of data is called **bivariate** data. With bivariate data, two observations are recorded from some entity.

> **Definition**
>
> Univariate data
>
> **Univariate data** are measurements collected on one variable.

Table 4.7.1 contains 60 bivariate data points of monthly rent and vacancy rates for various locations in the southeastern part of the United States.

> **Definition**
>
> Bivariate data
>
> **Bivariate data** are measurements collected on two variables.

| Table 4.7.1 – Monthly Rents and Vacancy Rates for Various Locations | | |
|---|---|---|
| Location | Monthly Rent ($) | Vacancy Rent (%) |
| Accomack | 639.00 | 9.0 |
| Airport | 752.00 | 17.3 |
| Albemarle | 717.00 | 11.7 |
| Anson | 589.00 | 9.4 |
| Appomattox | 605.00 | 14.5 |
| ... | | |
| York | 672.00 | 7.2 |

## Looking for Patterns in the Data

Detecting a relationship between two variables often begins with a graph. In the case of bivariate data, a **scatterplot** (or **scatter diagram**) is the traditional graphical method used to display the relationship between two variables. In a **scatterplot**, measurements are plotted in pairs with one variable plotted on each axis. When examining a scatterplot we are trying to draw conclusions concerning the overall pattern of the data. Does the pattern roughly follow a straight line? Is the pattern upward sloping or downward sloping? Are the data values tightly clustered in the pattern or widely dispersed? Are there significant deviations from the pattern?

> **Definition**
>
> Scatterplot
>
> A **scatterplot** is a graphical method used to display bivariate data by plotting the paired data on a set of coordinate axes.

A number of different scatterplots are shown in the following figures. In the first two scatterplots (Figures 4.7.1 and 4.7.2) the data are strongly related and, in fact, fall on a straight line. In Figure 4.7.1 the slope of the relationship is positive, that is, as the $x$-variable increases the $y$-variable also increases.

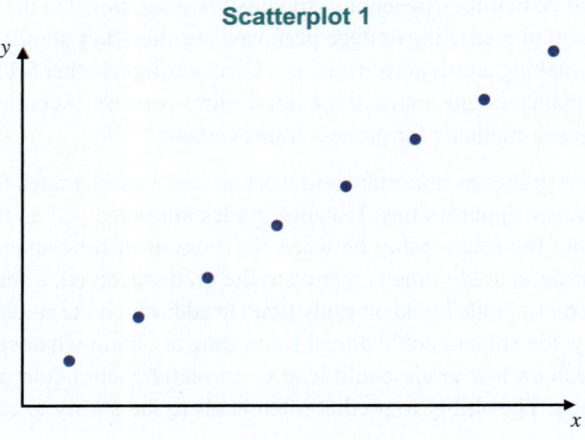

Figure 4.7.1

In Figure 4.7.2 the relationship is negative; as the $x$-variable increases, the $y$-variable decreases. This is also called an **inverse relationship**.

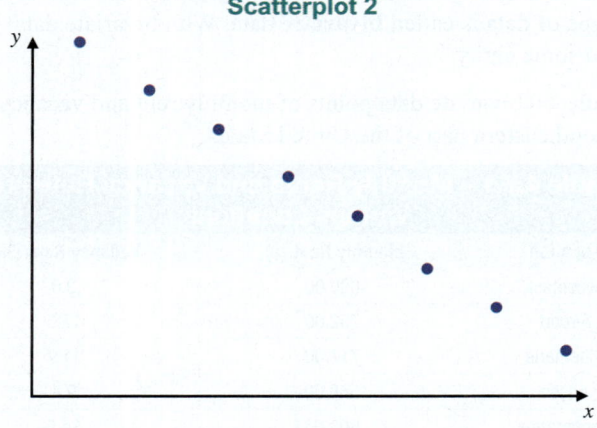

Figure 4.7.2

Figures 4.7.3 and 4.7.4 show less obvious relationships between the data. Figure 4.7.3 reveals a very imprecise relationship between $x$ and $y$, although as $x$ increases, $y$ tends to increase. The relationship between $x$ and $y$ is much more apparent in Figure 4.7.4 than in Figure 4.7.3.

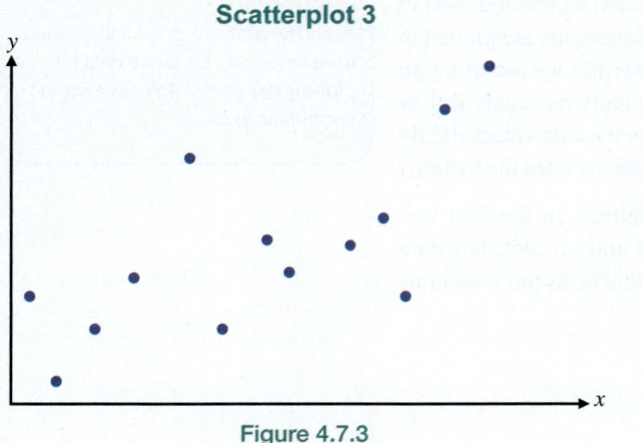

Figure 4.7.3

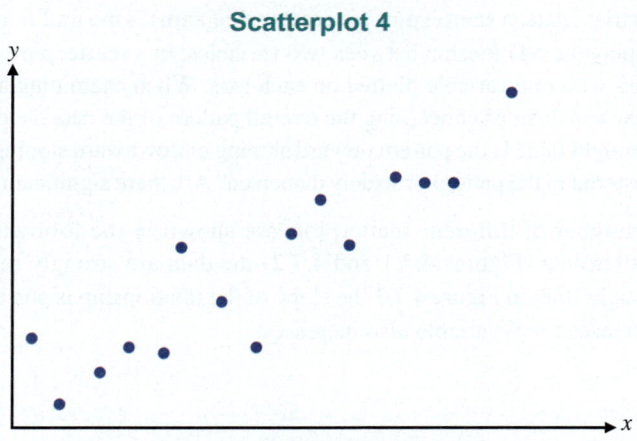

Figure 4.7.4

Figure 4.7.5 reveals a downward sloping relationship between $x$ and $y$. That is, as $x$ increases, $y$ tends to decrease. The relationship is not as exact as the relationship in Figure 4.7.2. In Figure 4.7.6, there is no apparent relationship between $x$ and $y$. That is, there is no tendency for $y$ to increase or decrease as $x$ increases.

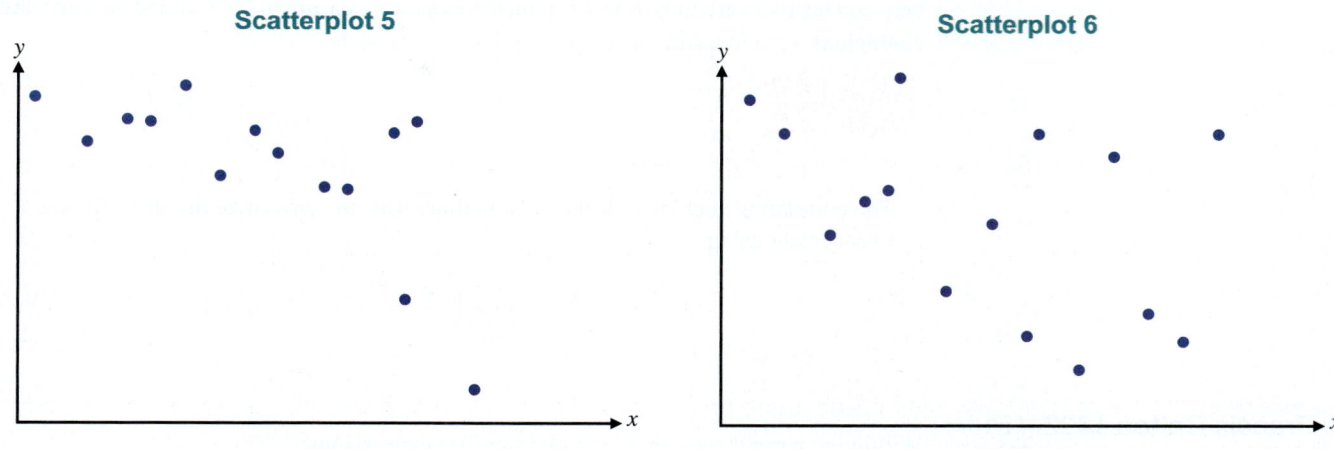

Figure 4.7.5                                          Figure 4.7.6

Let's explore the data in Table 4.7.1 with a scatterplot (see Figure 4.7.7). Examining the scatterplot reveals a slightly upward sloping relationship. That is, as monthly rent increases, the vacancy rate also increases.

Figure 4.7.7

While the relationship between monthly rent and vacancy rate is apparent, the relationship is not very strong. Interestingly, the law of supply and demand for rental space suggests that higher rents are a result of an increase in demand, contributing to higher prices. Thus, one would expect to see lower vacancy rates as rents rise, which could result in the development of more apartment units. However, that doesn't appear to be the case with these data. The fact that there appears to be a positive relationship between monthly rent and vacancy rate does not prove that high rent causes high vacancy rates even though such a conclusion may seem reasonable. One of the temptations that must be avoided is equating association with causation. There are many potential confounding variables in these observational data. However, we can quantify the strength of this relationship using the correlation coefficient, which is a measure of the linear association between two variables.

## Measuring the Degree of Linear Relationship: The Correlation Coefficient

A scatter diagram is a useful exploratory tool for detecting relationships between two variables. Eventually, however, a researcher will want to know the strength of the relationship between the two variables. Karl Pearson developed a measure in 1896 called the **correlation coefficient**, $r$, to measure the degree of linear relationship.

### Formula

#### Correlation Coefficient

The correlation coefficient is an index number used to summarize the strength of a linear relationship.

$$r = \frac{1}{n-1}\left\{ \sum_{i=1}^{n} \left( \frac{x_i - \bar{x}}{s_x} \right) \left( \frac{y_i - \bar{y}}{s_y} \right) \right\} \quad -1 \leq r \leq 1$$

### Francis Galton 1822–1911

Galton was the ninth child of a wealthy English family in Birmingham, England. He did extensive traveling early in his life and became interested in heritability of human traits. He was also greatly influenced by his first cousin, Charles Darwin. He believed that physical characteristics such as weight, height, and intelligence, as well as some personality traits were inherited. He collected data on these traits from mothers, fathers, and their children and found some interesting results. Tall fathers, for example, tended to have shorter children than themselves. Short fathers tended to have children that were taller than themselves. He also found this same property in seeds of wheat. He named this phenomena regression toward the mean. This is how the term regression entered statistics.

Galton was partially responsible for the development of the correlation coefficient. He needed some method of assessing the strength of the physical relationships between parents and their children.

He enlisted a young English statistician named Karl Pearson to work on a measure that could be used to determine association. It was Pearson that developed the correlation coefficient for this purpose. It is unclear exactly what role Galton played in the development of the correlation coefficient, however in the social science literature Galton is recognized as a co-developer of the statistic.

Within the parentheses, there are two familiar expressions:

$$\frac{y_i - \bar{y}}{s_y},$$

which is a $z$-score that shows how far $y$ deviates from its mean measured in standard deviation units ($s_y$ is the standard deviation of $y$), and

$$\frac{x_i - \bar{x}}{s_x},$$

which is a $z$-score that shows how far $x$ deviates from its mean measured in standard deviation units ($s_x$ is the standard deviation of $x$).

Summing the products of these **deviation measures** for each data pair determines the sign of the correlation coefficient.

**Note:** The formula presented for the correlation coefficient is difficult to calculate by hand. Often, you will see the correlation coefficient presented as the following equivalent computational formula. However, there are many statistical software programs that can calculate the correlation coefficient, so it is not often you will need to do it by hand.

### Formula

#### Computational Formula for the Correlation Coefficient

The computational formula for the correlation coefficient is as follows.

$$r = \frac{n\sum x_i y_i - \left(\sum x_i\right)\left(\sum y_i\right)}{\sqrt{n\sum x_i^2 - \left(\sum x_i\right)^2}\sqrt{n\sum y_i^2 - \left(\sum y_i\right)^2}}$$

## Positive Relationships

When $r$ is positive, there is a tendency for $y$ to increase as $x$ increases. If both of the deviations are positive, then each of the observations is above its mean. If both are negative, then each is below its mean. In a positive linear relationship, when one of the variables is above its mean, the other variable tends to be above its mean. Similarly, if one variable is below its mean, the other tends to be below its mean. Such is the case in Figure 4.7.8.

**Positive Relationship Between *x* and *y***

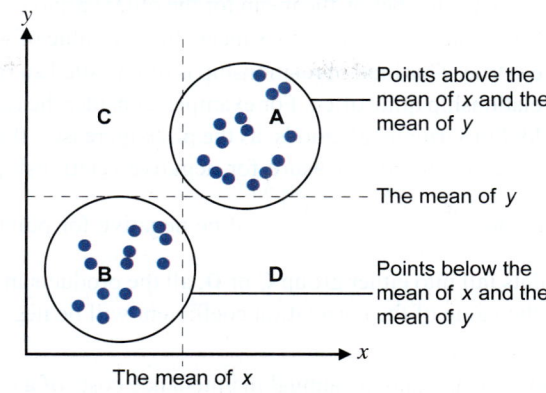

Figure 4.7.8

The points in the group labeled **A** are points whose *x*-values are greater than the mean of *x* and whose *y*-values are greater than the mean of *y*. Since for each point in group **A**, the deviations $x_i - \bar{x}$ and $y_i - \bar{y}$ will be positive numbers, the expression $\left(\dfrac{x_i - \bar{x}}{s_x}\right)\left(\dfrac{y_i - \bar{y}}{s_y}\right)$ will be the product of two positive numbers, which will be *positive*.

The points in the group labeled **B** are points whose *x*-values are below the mean of *x* and whose *y*-values are below the mean of *y*. Since $x_i - \bar{x}$ and $y_i - \bar{y}$ will both be negative numbers for members of this group, the expression $\left(\dfrac{x_i - \bar{x}}{s_x}\right)\left(\dfrac{y_i - \bar{y}}{s_y}\right)$ will be the product of two negative numbers, which will be *positive*.

Since all the points fall into either group **A** or **B**, all the products in the summation are positive. Thus, the correlation coefficient (*r*) will have a positive value for an upward sloping (positive) relationship. Now, let's look at a downward sloping relationship.

## Negative Relationships

**Negative Relationship Between *x* and *y***

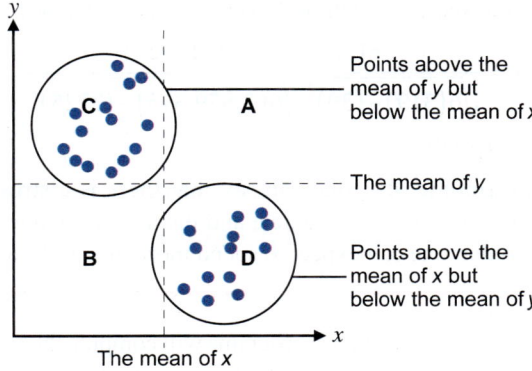

Figure 4.7.9

| Table 4.7.4 – Summary Statistics | | |
|---|---|---|
| **Property** | **Value** | **Accuracy** |
| Mean of $x$ | 9 | exact |
| Sample variance of $x$ | 11 | exact |
| Mean of $y$ | 7.50 | to 2 decimal places |
| Sample variance of $y$ | 4.125 | $+/-0.003$ |
| Correlation between $x$ and $y$ | 0.816 | to 3 decimal places |
| Linear regression line | $y = 3.000 + 0.500x$ | to 3 decimal places |

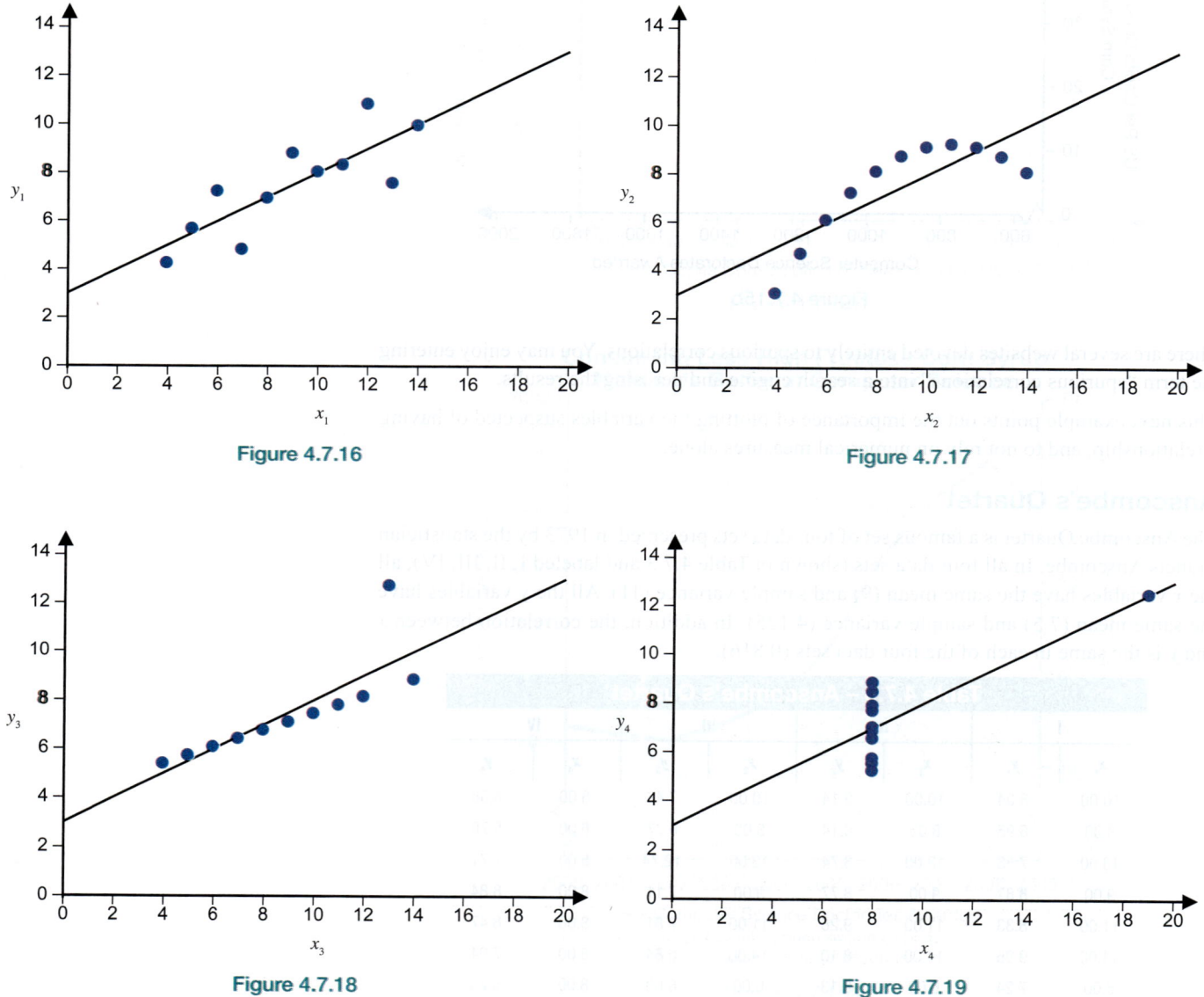

**Figure 4.7.16**

**Figure 4.7.17**

**Figure 4.7.18**

**Figure 4.7.19**

The Anscombe Quartet brings out an important data analytic principle. No matter what kind of numeric data is being analyzed, it is critically important to plot it before making any conclusions. For single variables, an analyst should at least be looking at histograms and box plots. When we begin to look for relationships in bivariate or multivariate data, creating scatterplots to display the relationship between the variables is more important than calculating summary measures of linear relationship, i.e., the correlation coefficient.

# 4.7  Exercises

## Basic Concepts

1. Give an example of a business situation in which knowledge of a relationship between two variables is desired.

2. If a relationship can be uncovered, what are the potential benefits?

3. What are bivariate data? How is bivariate data different from univariate data?

4. What graphical tool is often used in the discovery of relationships?

5. What are four common questions you should ask when studying a graphical representation of bivariate data?

6. If bivariate data exhibit an inverse relationship, what does that mean?

7. How do you construct exact relationships between two variables?

8. In what range is the value of $r$ when bivariate data exhibit a positive relationship? A negative relationship?

9. If the value of $r$ is small, does this always mean that no relationship exists? Explain.

10. What is confounding? Why is confounding a problem?

## Exercises

11. Consider the following scatterplots and answer the following questions regarding the overall pattern of the data for each of the graphs.

    • Does the pattern roughly follow a straight line?

    • Is the pattern upward sloping or downward sloping?

    • Are the data values tightly clustered in the pattern or widely dispersed?

    • Are there significant deviations from the pattern?

    a.
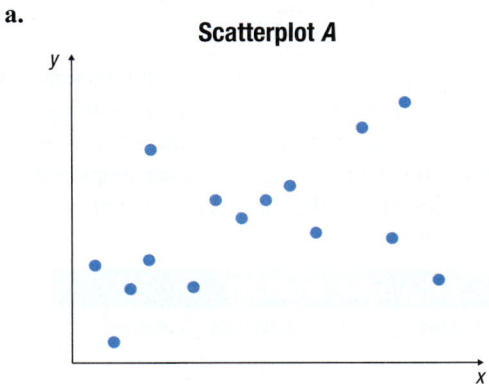

    b.
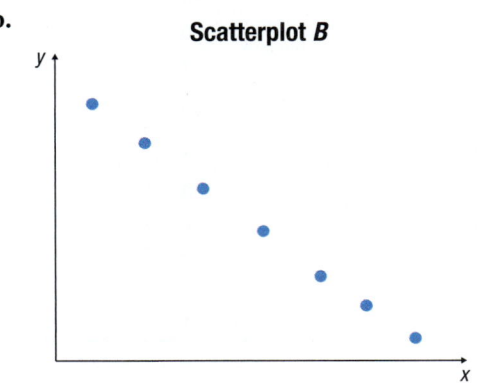

12. Consider the following scatterplots and answer the following questions regarding the overall pattern of the data for each of the graphs.

• Does the pattern roughly follow a straight line?

• Is the pattern upward sloping or downward sloping?

• Are the data values tightly clustered in the pattern or widely dispersed?

• Are there significant deviations from the pattern?

a.
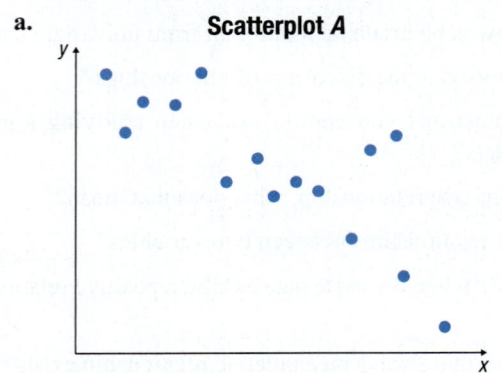
Scatterplot A

b.
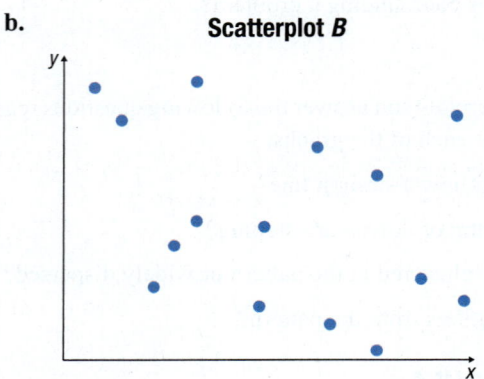
Scatterplot B

13. A manufacturing company which produces laminate for countertops is interested in studying the relationship between the number of hours of training an employee receives and the number of defects per countertop produced. Ten employees are randomly selected. The number of hours of training which each employee has received is recorded and the number of defects on the most recent countertop produced is determined. The results are as follows.

| Employee Training | |
|---|---|
| Hours of Training | Defects per Countertop |
| 1 | 1 |
| 4 | 4 |
| 7 | 0 |
| 3 | 3 |
| 2 | 5 |
| 2 | 4 |
| 5 | 3 |
| 5 | 2 |
| 1 | 5 |
| 6 | 1 |

    **a.** Analyze the data collected for the study by answering the following questions.

       **i.** Do the variables selected for measurement seem appropriate for answering the question that the manufacturing company is interested in?

       **ii.** What biases or errors might be present in the data?

       **iii.** How are the data collected – through observation or controlled experiment?

    **b.** Plot the data points on a scatterplot.

    **c.** Based on the scatterplot in part **b.**, answer the following questions regarding the overall pattern of the data.

       **i.** Does the pattern roughly follow a straight line?

       **ii.** Is the pattern upward sloping or downward sloping? Are the data values tightly clustered in the pattern or widely dispersed?

       **iii.** Are there significant deviations from the pattern?

**14.** Illustrate, using a scatterplot, a data set that would have a correlation coefficient of 1.

**15.** Illustrate, using a scatterplot, a data set that would have a correlation coefficient of $-1$.

**16.** Describe the relationships indicated by the correlation coefficients as tightly clustered in a positive linear fashion, tightly clustered in a negative linear fashion, loosely clustered in a positive linear fashion, loosely clustered in a negative linear fashion, or no linear relationship.

    **a.** $r = 0.9$         **c.** $r = -0.9$         **e.** $r = 0$

    **b.** $r = 0.5$         **d.** $r = -0.5$

**17.** Describe the relationships indicated by the correlation coefficients as tightly clustered in a positive linear fashion, tightly clustered in a negative linear fashion, loosely clustered in a positive linear fashion, loosely clustered in a negative linear fashion, or no linear relationship.

    **a.** $r = 0.8$         **c.** $r = -0.8$         **e.** $r = 0.1$

    **b.** $r = 0.4$         **d.** $r = -0.4$

**18.** A sample of 10 female swimmers, all 17 years old, is selected from a local swim league. Each swimmer's best time (in seconds) in the 50-yard freestyle and in the 100-yard individual medley are obtained. The 100-yard individual medley consists of swimming 25 yards with each of the four major strokes. The data are given in the following table.

| Best Times | | | | | | | | | |
|---|---|---|---|---|---|---|---|---|---|
| **Freestyle** | 27.4 | 27.0 | 26.8 | 30.7 | 28.5 | 28.6 | 29.6 | 30.8 | 31.5 | 29.8 |
| **Medley** | 66.3 | 66.4 | 66.7 | 78.7 | 69.4 | 72.0 | 73.5 | 81.1 | 78.6 | 73.5 |

    **a.** Construct a scatterplot of the data.

    **b.** Does there appear to be a negative or positive relationship between the variables?

    **c.** Compute the correlation coefficient.

19. A personnel director is interested in studying the relationship (if any) between age and salary. Sixteen employees are randomly selected and their ages and salaries are recorded.

| Ages and Salaries | | | |
|---|---|---|---|
| Age | Salary ($) | Age | Salary ($) |
| 25 | 22,000 | 49 | 39,000 |
| 55 | 45,000 | 37 | 45,000 |
| 27 | 43,000 | 62 | 60,000 |
| 30 | 30,000 | 40 | 35,000 |
| 22 | 24,000 | 35 | 34,000 |
| 33 | 53,000 | 29 | 30,000 |
| 19 | 18,000 | 58 | 73,000 |
| 45 | 38,000 | 52 | 42,000 |

a. Plot the data points on a scatterplot.

b. Determine the correlation coefficient.

c. Describe the relationship indicated by the correlation coefficient and the scatterplot.

20. The following variables have high positive linear correlations. Is it reasonable to conclude that an increase in one variable causes an increase in the other variable? Explain what could be causing this apparent relationship.

a. Height and vocabulary

b. Absenteeism from school and sale of cough syrup

c. Sale of turkey and sale of toys

21. The following variables have high positive linear correlations. Is it reasonable to conclude that an increase in one variable causes an increase in the other variable? Explain what could be causing this apparent relationship.

a. Sale of air conditioners and sale of tomatoes

b. Sale of greeting cards and sale of chocolates

c. The number of wrecks on a local highway and absenteeism from work

# T Discovering Technology

## Using the TI-84 Plus Calculator

Sample Mean and Standard Deviation

Use the data from Example 4.2.2 (4, 10, 9, 11, 9, 7) for this exercise.

1. Press **STAT**, and select **Edit**. Press **ENTER**. Then input the data into list **L1**.

2. Press **STAT**, and then press the right arrow key so that **CALC** is highlighted. Choose **1-Var Stats** and press **ENTER**. Press **2ND** and **1** to choose the data in list L1 and press **ENTER**.

3. Observe the output screen for the sample mean and sample standard deviation.

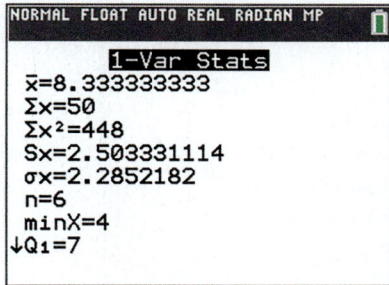

```
NORMAL FLOAT AUTO REAL RADIAN MP

      1-Var Stats
 x̄=8.333333333
 Σx=50
 Σx²=448
 Sx=2.503331114
 σx=2.2852182
 n=6
 minX=4
↓Q₁=7
```

## Correlation Coefficient

For this exercise, use the data from Example 4.7.1.

1. Press **STAT** and select **Edit**.

2. Enter the $x$ data values (ages) in **L1** and the $y$ data values (costs) in **L2**.

3. Access the catalog by pressing **2ND** then **0**.

4. Press the down arrow and find **DiagnosticOn**. Press **ENTER**, then **ENTER** again.

5. Press **STAT**, and then press the right arrow key so that **CALC** is highlighted.

6. Highlight **LinReg (ax + b)**. Press **ENTER**.

7. Press **ENTER** again. The calculator will display values for $a$ and $b$ in the equation $y = ax + b$ for the data entered in the lists.

8. The last value listed will be the correlation coefficient for the two sets of data, $r$.

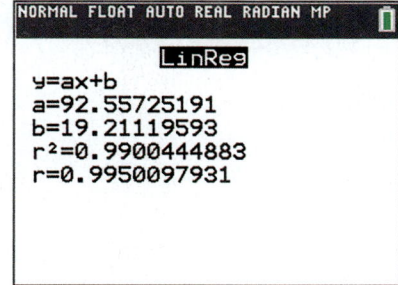

```
NORMAL FLOAT AUTO REAL RADIAN MP

         LinReg
 y=ax+b
 a=92.55725191
 b=19.21119593
 r²=0.9900444883
 r=0.9950097931
```

## Using Excel

### Sample Mean and Standard Deviation

Use the data from Example 4.2.2 (4, 10, 9, 11, 9, 7) for this exercise.

1.  Enter the data into Column A.

2.  Under the **Data** tab, choose **Data Analysis**, then **Descriptive Statistics**. Select **OK**.

3.  Enter the Input Range **$A$1:$A$6**, select the radio button next to **Output Range** and enter **A8**. Check the box next to **Summary Statistics**. Select **OK**.

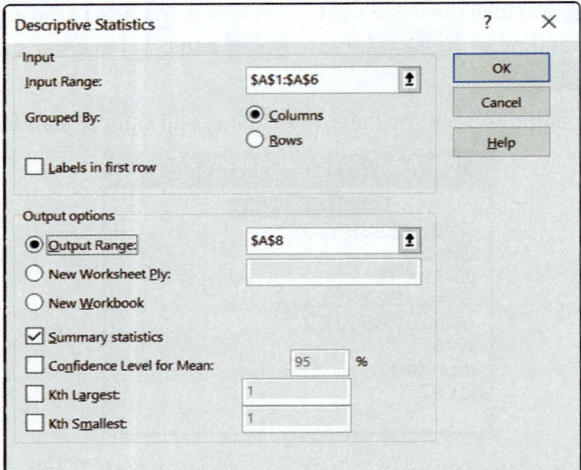

4.  Expand the width of column A by clicking and **dragging** the line between column A and column B to ensure the text in column A is visible and observe the output screen for summary statistics. The sample mean appears in cell B10 and the sample standard deviation appears in cell B14.

| | A | B |
|---|---|---|
| 1 | 4 | |
| 2 | 10 | |
| 3 | 9 | |
| 4 | 11 | |
| 5 | 9 | |
| 6 | 7 | |
| 7 | | |
| 8 | *Column 1* | |
| 9 | | |
| 10 | Mean | 8.333333 |
| 11 | Standard Error | 1.021981 |
| 12 | Median | 9 |
| 13 | Mode | 9 |
| 14 | Standard Deviation | 2.503331 |
| 15 | Sample Variance | 6.266667 |
| 16 | Kurtosis | 1.137392 |
| 17 | Skewness | −1.13891 |
| 18 | Range | 7 |
| 19 | Minimum | 4 |
| 20 | Maximum | 11 |
| 21 | Sum | 50 |
| 22 | Count | 6 |

## Percentiles and Quartiles

Use the data from Example 4.3.2 for this exercise.

Microsoft Excel has a function (PERCENTILE) to calculate the $k^{th}$ percentile of a data set and a function (QUARTILE) to calculate the quartiles of a data set. The method that Excel uses to calculate percentiles and quartiles is different from that which was shown in this chapter. The method is beyond the scope of this text, but it is important to realize that the percentiles and quartiles calculated by Excel may be different from those calculated by hand. In addition, Excel 2010 has two percentile functions, PERCENTILE.EXC and PERCENTILE.INC and two quartile functions, QUARTILE.EXC and QUARTILE.INC. The .EXC functions return the $k^{th}$ percentile in a range where $k$ is between 0 and 1, exclusive. The .INC functions return the $k^{th}$ percentile in a range where $k$ is between 0 and 1, inclusive. The regular PERCENTILE function is the one found in Excel 2007 and earlier, and calculates the percentiles in the same way that the PERCENTILE.INC function does. Just as we do in this chapter, notice that Excel equates the first quartile with the $25^{th}$ percentile, the second quartile with the $50^{th}$ percentile, and the third quartile with the $75^{th}$ percentile.

1. In cell A1 type the label **Data**.
2. Enter the data values in Column A, beginning in cell A2.
3. In cell B1 type the label **Percentile**.
4. In Column B, enter the following percentiles, beginning in cell B2. **10**, **25**, **40**, **50**, **60**, **75**, **90**.
5. In cell D1 type the label **Quartile**.
6. In Column D, enter the following quartiles, beginning in cell D2. **1**, **2**, **3**, **4**.

7.  The percentile function is **PERCENTILE.INC(array, k)** where *array* corresponds to the data set you are interested in and *k* is the $k^{th}$ percentile, expressed as a number between 0 and 1.

8.  Now, calculate the $10^{th}$, $25^{th}$, $40^{th}$, $50^{th}$, $60^{th}$, $75^{th}$, and $90^{th}$ percentiles. Starting with the $10^{th}$ percentile, enter the following formula into cell **C2**.

**=PERCENTILE.INC($A$2:$A$41, B2/100)**

As calculated in Excel, the $10^{th}$ percentile is 28.8. To calculate the other percentiles, **drag** the formula in cell C2 down to cells C3 to C8.

|   | A | B | C | D | E |
|---|---|---|---|---|---|
| 1 | Data | Percentile | | Quartile | |
| 2 | 67 | 10 | 28.8 | 1 | |
| 3 | 45 | 25 | 42.5 | 2 | |
| 4 | 63 | 40 | 47.6 | 3 | |
| 5 | 58 | 50 | 54 | 4 | |
| 6 | 35 | 60 | 56.4 | | |
| 7 | 54 | 75 | 64.5 | | |
| 8 | 27 | 90 | 73.4 | | |
| 9 | 66 | | | | |
| 10 | 21 | | | | |

9.  The quartile function is **QUARTILE.INC(array, quart)** where *array* corresponds to the data set you are interested in and *quart* is the quartile (1 through 4).

10. Now, calculate the $1^{st}$, $2^{nd}$, $3^{rd}$, and $4^{th}$ quartiles. Starting with the $1^{st}$ quartile, enter the following formula into cell E2.

**=QUARTILE.INC($A$2:$A$41, D2)**

As calculated in Excel, the $1^{st}$ percentile is 42.5. To calculate the other percentiles, **drag** the formula in cell E2 down to cells E3 to E5.

11. As calculated by Excel, the first quartile is 42.5. Notice this value is the same as the $25^{th}$ percentile. Calculate the second, third, and fourth quartiles using the appropriate function in cells E3 through E5, respectively. Notice that the second and third quartiles are equal to the $50^{th}$ and $75^{th}$ percentile, respectively, and that the fourth quartile is the maximum value in the data set (corresponding to the $100^{th}$ percentile).

|   | A | B | C | D | E |
|---|---|---|---|---|---|
| 1 | Data | Percentile | | Quartile | |
| 2 | 67 | 10 | 28.8 | 1 | 42.5 |
| 3 | 45 | 25 | 42.5 | 2 | 54 |
| 4 | 63 | 40 | 47.6 | 3 | 64.5 |
| 5 | 58 | 50 | 54 | 4 | 82 |
| 6 | 35 | 60 | 56.4 | | |
| 7 | 54 | 75 | 64.5 | | |
| 8 | 27 | 90 | 73.4 | | |
| 9 | 66 | | | | |
| 10 | 21 | | | | |

## Correlation Coefficient

For this exercise, use the data from Example 4.7.1.

1. Enter the label **Age** in cell A1, and the *x* data values in cells A2 through A7.

2. Enter the label **Annual Maintenance Cost (Dollars)** in cell B2, and the *y* data values in cells B2 through B7.

3. Under the **Data** tab select **Data Analysis**.

4. Select **Correlation** and press **OK**.

5. Enter **$A$2:$B$7** as the Input Range. Make sure that the radio button next to **Columns** is selected.

6. Select the radio button next to **Output Range** and enter **$E$1**. Press **OK**.

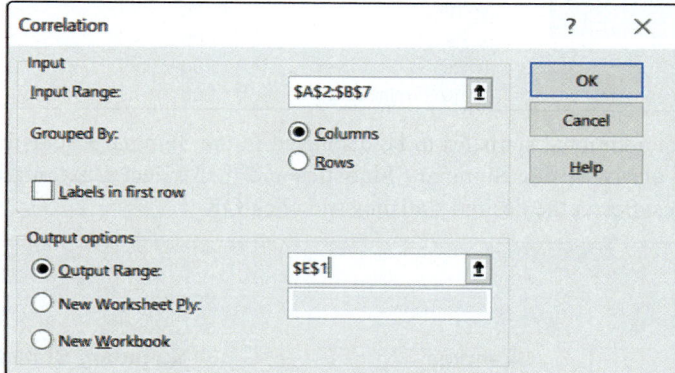

7. In the summary output, the correlation coefficient, *r*, is displayed (0.99501).

| | A | B | C | D | E | F | G |
|---|---|---|---|---|---|---|---|
| 1 | Age | Annual Maintenance Cost (Dollars) | | | | Column 1 | Column 2 |
| 2 | 2 | 225 | | | Column 1 | 1 | |
| 3 | 4 | 400 | | | Column 2 | 0.99501 | 1 |
| 4 | 5 | 475 | | | | | |
| 5 | 7 | 650 | | | | | |
| 6 | 9 | 800 | | | | | |
| 7 | 12 | 1175 | | | | | |

## Using JMP

### Summary Statistics and Quartiles

For this exercise, use the data on Average US Gas Prices from Table 4.1.7.

1. Enter the data for the gas prices into the first column of the JMP worksheet. Double click the label **Column 1** and input "Gas Price" as the **Column Name**. Click **OK**.

2. Click on **Analyze** in the top row of the JMP spreadsheet and then select **Distribution**.

3. In the **Select Columns** box click on the variable **Gas Price** to select it. Then click on **Y, Columns**. Click **OK**.

4. You will get a histogram of the data, boxplot, quantiles, and sample summary statistics that include the mean and standard deviation.

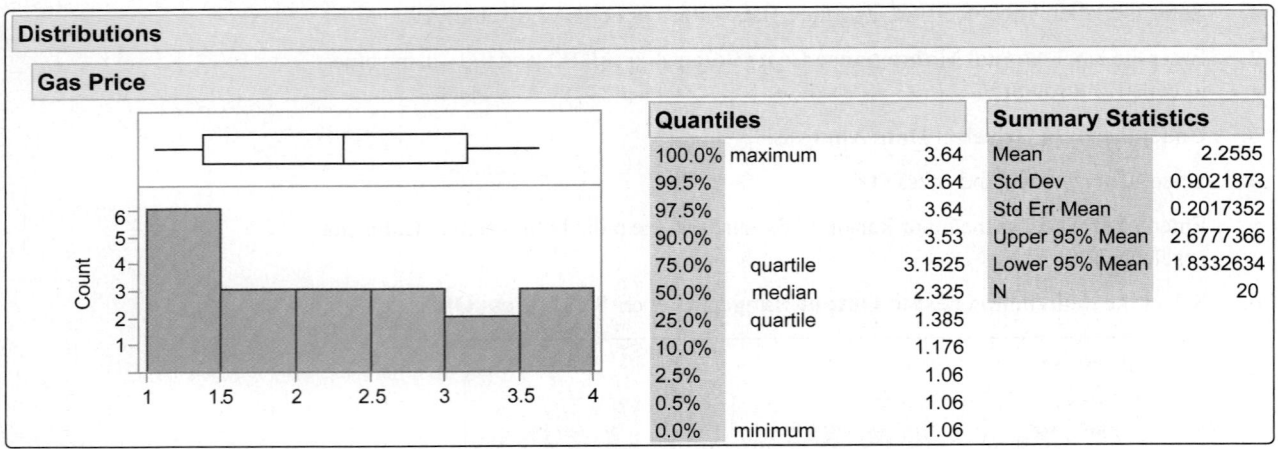

5. To select additional statistics to be displayed in the Summary Statistics box, click the **red arrow** beside Summary Statistics and then select **Customize Summary Statistics**. Select the desired statistics and click **OK**

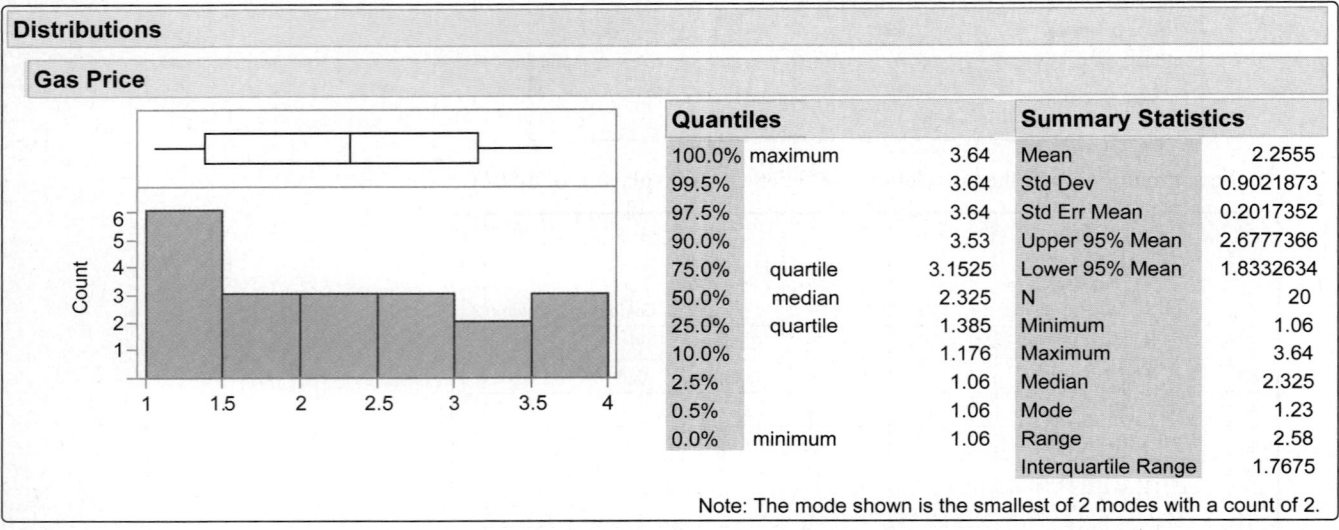

Note: The mode shown is the smallest of 2 modes with a count of 2.

Note that the method used in JMP to calculate percentiles and quartiles is different from the method shown in this chapter. Your results calculating the quartiles manually using the method shown in the book will likely give you different results than shown in the Quantiles list.

## Correlation Coefficient

For this exercise, use the data for Monthly Rent ($) and Vacancy Rate (%) from Table 4.7.1.

1. Enter the data for Monthly Rent into the first column of the JMP worksheet. Double click the label **Column 1** and input "**Monthly Rent**" as the **Column Name**. Click **OK**.

2. Enter the data for Vacancy Rate into the second column of the JMP worksheet. Double click the label **Column 2** and input "**Vacancy Rate**" as the **Column Name**. Click **OK**.

3. Click on **Analyze** in the top row of the JMP spreadsheet and then select **Multivariate**.

4. In the **Select Columns** box click on the variable **Monthly Rent** to select it. Then click on **Y, Columns**.

5. In the **Select Columns** box click on the variable **Vacancy Rate** to select it. Then click on **Y, Columns**. Click **OK**.

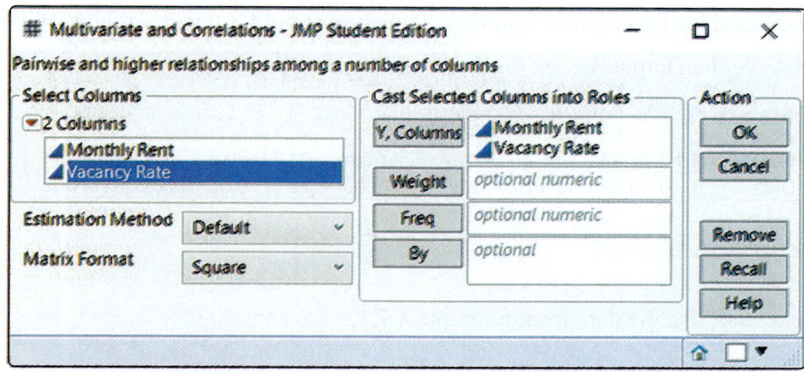

6. You will get a correlation matrix as well as a scatterplot matrix. The correlation coefficient between Monthly Rent and Vacancy Rate is 0.4786.

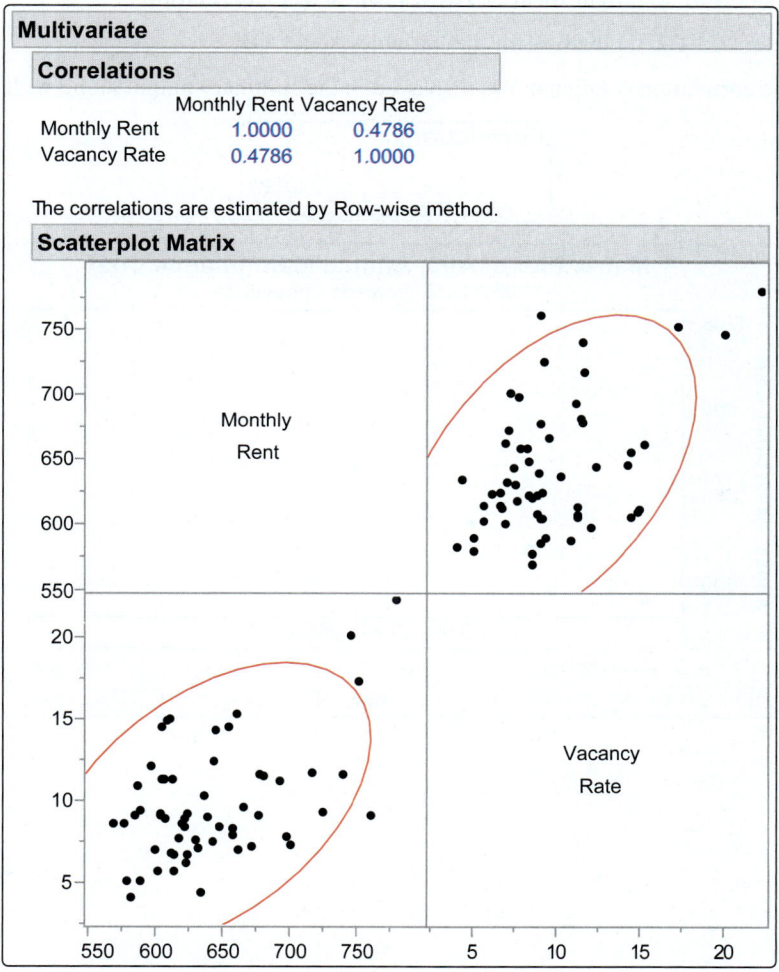

---

## Using Minitab

### Sample Mean and Standard Deviation

Use the data from Example 4.2.2 (4, 10, 9, 11, 9, 7) for this exercise.

1. Enter the data into Column C1.

2. Under **Stat**, choose **Basic Statistics**, then **Display Descriptive Statistics**.

3.  In the Display Descriptive Statistics dialog box, input **C1** under Variables. Click **OK**.

4.  Observe the Output Screen for the summary statistics.

**Statistics**

| Variable | N | N* | Mean | SE Mean | StDev | Minimum | Q1 | Median | Q3 | Maximum |
|---|---|---|---|---|---|---|---|---|---|---|
| Data | 6 | 0 | 8.33 | 1.02 | 2.50 | 4.00 | 6.25 | 9.00 | 10.25 | 11.00 |

## Correlation Coefficient

For this exercise, use the data from Example 4.7.1.

1.  Enter the $x$ data values (ages) in Column C1 and label it "Age".

2.  Enter the $y$ data values (costs) in Column C2 and label it "Annual Maintenance Cost".

3.  Under **Stat**, highlight **Basic Statistics**, and select **Correlation**.

4.  Enter "**C1 C2**" in the box for the Variables. Click **OK**.

5.  The correlation coefficient $r$ is displayed in the summary output along with a plot.

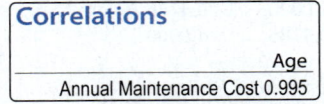

**Correlations**

|  | Age |
|---|---|
| Annual Maintenance Cost | 0.995 |

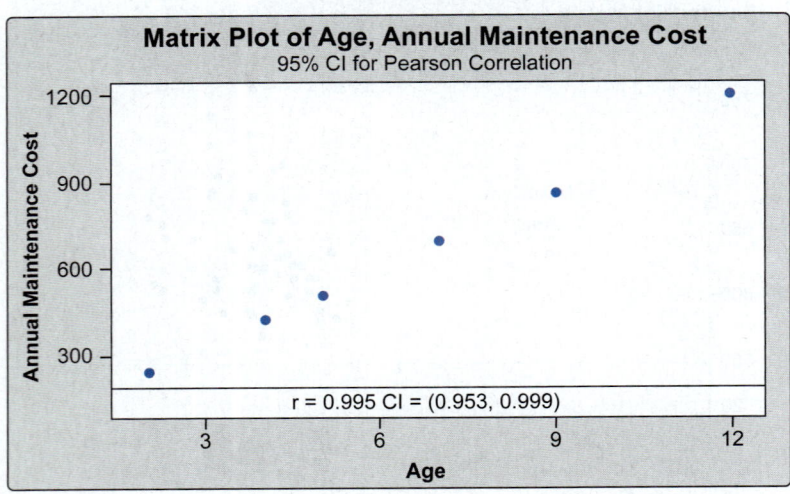

# R Chapter 4 Review

## Key Terms and Ideas

- Numerical Descriptive Statistics
- Parameters
- Statistics
- Inferential Statistics
- Measures of Location (Central Tendency)
- Arithmetic Mean
- Sample Mean
- Population Mean
- Deviation
- Weighted Mean
- Trimmed Mean
- Median
- Resistant Measure
- Mode
- Bimodal
- Multimodal
- Positively Skewed (Skewed to the Left)
- Negatively Skewed (Skewed to the Right)
- Moving Average
- Variability (Dispersion or Spread)
- Deviation from the Mean
- Range
- Mean Absolute Deviation
- Population Variance
- Sample Variance
- Population Standard Deviation
- Sample Standard Deviation
- Empirical Rule
- Chebyshev's Theorem
- $P^{th}$ Percentile
- Percentile
- Quartile
- Interquartile Range
- Outlier
- $z$-Score
- Shape
- Subsetting
- Coefficient of Variation
- Mean of Grouped Data
- Variance of Grouped Data
- Population Proportion
- Sample Proportion
- Univariate Data
- Bivariate Data
- Scatterplot (Scatter Diagram)
- Linear Relationship
- Positive Relationship
- Negative Relationship (Inverse Relationship)
- Correlation Coefficient
- Quadratic Relationship
- Common Response
- Confounding

## Key Formulas

|  | Section |
|---|---|

**Arithmetic Mean**

$$\frac{1}{n}\left(x_1 + x_2 + \cdots + x_N\right)$$

4.1

**Population Mean**

$$\mu = \frac{1}{N}\left(x_1 + x_2 + \cdots + x_N\right) = \frac{\sum x_i}{N}$$

4.1

**Sample Mean**

$$\bar{x} = \frac{1}{n}\left(x_1 + x_2 + \cdots + x_n\right) = \frac{\sum x_i}{n}$$

4.1

**Weighted Mean**

$$\bar{x} = \frac{w_1 x_1 + w_2 x_2 + \cdots + w_n x_n}{w_1 + w_2 + \cdots + w_n} = \frac{\sum\left(w_i x_i\right)}{\sum w_i}$$

4.1

**Median Location**

$$L(m) = \frac{n+1}{2}$$

4.1

**Range**

largest value − smallest value

4.2

**Mean Absolute Deviation**

$$\text{MAD} = \frac{\sum\left|x_i - \bar{x}\right|}{n}$$

4.2

**Population Variance**

$$\sigma^2 = \frac{\sum\left(x_i - \mu\right)^2}{N}$$

4.2

**Sample Variance**

$$s^2 = \frac{\sum\left(x_i - \bar{x}\right)^2}{n-1}$$

4.2

## Key Formulas (cont.)

| | Section |
|---|---|
| **Population Standard Deviation** | |
| $$\sigma = \sqrt{\sigma^2}$$ | 4.2 |
| **Sample Standard Deviation** | |
| $$s = \sqrt{s^2}$$ | 4.2 |
| **Empirical Rule** | |
| **One Sigma Rule:** $\mu \pm 1\sigma$ contains about 68% of the data<br>**Two Sigma Rule:** $\mu \pm 2\sigma$ contains about 95% of the data<br>**Three Sigma Rule:** $\mu \pm 3\sigma$ contains about 99.7% of the data | 4.2 |
| **Chebyshev's Theorem** | |
| The proportion of any data set lying within $k$ standard deviations of the mean is at least $1 - \dfrac{1}{k^2}$ for $k > 1$ | 4.2 |
| **Coefficient of Variation** | |
| Population data:  $CV = \left( \dfrac{\sigma}{\mu} \cdot 100 \right)\%$<br><br>Sample data:  $CV = \left( \dfrac{s}{\overline{x}} \cdot 100 \right)\%$ | 4.2 |
| **Location of the $P^{th}$ Percentile** | |
| $$\ell = n \cdot \left( \dfrac{P}{100} \right)$$ | 4.3 |
| **Percentile** | |
| $\text{percentile of } x = \dfrac{\text{number of data values less than or equal to } x}{\text{total number of data values}} \cdot 100$ | 4.3 |
| **Location of $Q_1$** | |
| $$\ell = n \cdot \left( \dfrac{25}{100} \right)$$ | 4.3 |
| **Location of $Q_2$** | 4.3 |
| $$\ell = n \cdot \left( \dfrac{50}{100} \right)$$ | |

## Key Formulas (cont.)

|  | Section |
|---|---|

**Location of $Q_3$**

$$\ell = n \cdot \left( \frac{75}{100} \right)$$

4.3

**Interquartile Range**

$$\text{IQR} = Q_3 - Q_1$$

4.3

**Outlier**

Data value greater than $Q_3 + 1.5 \cdot$ Interquartile Range

or

Data value less than $Q_1 - 1.5 \cdot$ Interquartile Range

4.3

**z-Score**

$$z = \frac{x - \mu}{\sigma}$$

4.3

**Mean of Grouped Data**

Population data: $\mu = \dfrac{\Sigma(f_i M_i)}{N}$

Sample data: $\bar{x} = \dfrac{\Sigma(f_i M_i)}{n}$

4.5

**Variance of Grouped Data**

Population data: $\sigma^2 = \dfrac{\Sigma(M_i - \mu)^2 f_i}{N}$

Sample data: $s^2 = \dfrac{\Sigma(M_i - \bar{x})^2 f_i}{n-1}$

4.5

**Computational Formulas for the Variance of Grouped Data**

Population Data:

$$\sigma^2 = \frac{\Sigma(f_i M_i^2) - \dfrac{\left(\Sigma(f_i M_i)\right)^2}{N}}{N} = \frac{\Sigma(f_i M_i^2)}{N} - \left(\frac{\Sigma(f_i M_i)}{N}\right)^2$$

4.5

Sample Data:

$$s^2 = \frac{\Sigma(f_i M_i^2) - \dfrac{\left(\Sigma(f_i M_i)\right)^2}{n}}{n-1}$$

## Key Formulas (cont.)

| | Section |
|---|---|
| **Class Midpoint** | |
| $$M_i = \dfrac{\text{lower class boundary} + \text{upper class boundary}}{2}$$ | 4.5 |
| **Population Proportion** | |
| $$p = \dfrac{x}{N}$$ | 4.6 |
| **Sample Proportion** | |
| $$\hat{p} = \dfrac{x}{n}$$ | 4.6 |
| **Correlation Coefficient** | |
| $$r = \dfrac{1}{n-1}\left\{\sum_{i=1}^{n}\left(\dfrac{x_i - \bar{x}}{s_x}\right)\left(\dfrac{y_i - \bar{y}}{s_y}\right)\right\}$$ | 4.7 |
| **Computational Formula for the Correlation Coefficient** | |
| $$r = \dfrac{n\sum x_i y_i - (\sum x_i)(\sum y_i)}{\sqrt{n\sum x_i^2 - (\sum x_i)^2}\,\sqrt{n\sum y_i^2 - (\sum y_i)^2}}$$ | 4.7 |

## AE    Additional Exercises

1. A carpenter is attempting to repair a porch and needs twenty boards which are eight feet long. The salesman at the hardware store says he has twenty boards that "average" eight feet long. When the carpenter checks what he has bought, there are ten boards at six feet and ten boards at ten feet. Do you feel the salesman accurately represented the lengths? Discuss.

2. The maximum heart rates achieved while performing a particular aerobic exercise routine are measured (in beats per minute) for 9 randomly selected individuals.

| Maximum Heart Rates (Beats per Minute) | | | | | | | | |
|---|---|---|---|---|---|---|---|---|
| 145 | 155 | 130 | 185 | 170 | 165 | 150 | 160 | 125 |

   a. Calculate the sample variance of the maximum heart rate achieved.

   b. Calculate the sample standard deviation of the maximum heart rate achieved.

   c. Calculate the range of the maximum heart rate achieved.

   d. What are some of the factors which might contribute to the variation in the observations?

3. A sample of teenagers was asked how many times they went to the movies in the past 3 months. The frequency distribution table summarizes the results.

| Teenager Movie Visits | |
|---|---|
| Number of Visits | Frequency |
| 0 | 13 |
| 1 | 18 |
| 2 | 11 |
| 3 | 7 |
| 4 | 4 |
| 5 | 3 |
| 6 | 0 |
| 7 | 3 |
| 8 | 3 |
| 9 | 0 |
| 10 | 2 |

   a. What proportion of the sample visited the movies at least 3 times in the previous 3 months?

   b. Find the mean and standard deviation of the number of visits using the formulas for grouped data.

   c. Compute the interval one standard deviation about the mean.

   d. Find the percent of data falling in the interval one standard deviation about the mean.

   e. Is the percent of the data falling in the interval one standard deviation about the mean close to what the Empirical Rule predicts? What is the reason for the discrepancy, if any?

4. A high school math teacher summarized the 35 math SAT scores for the students in her calculus class. The mean for the class was 521 and the median was 535. The range of the scores was 235 and the highest score in the entire class was 675. Approximately 40% of the class scored higher than 562. State whether each of the following is true or false.

    a. The 45th percentile exceeds 540.

    b. The lowest score in the class was 440.

    c. The z-score for a score of 510 is a negative number.

    d. The third quartile exceeds 562.

    e. The percentile rank of 562 is 40.

5. Consider the following number of defective circuit boards produced by two different machines on seven randomly selected days.

| Defective Circuit Boards | | | | | | | |
|---|---|---|---|---|---|---|---|
| Machine A | 2 | 3 | 7 | 4 | 5 | 1 | 0 |
| Machine B | 2 | 3 | 4 | 3 | 4 | 2 | 4 |

    a. Calculate the average number of defective circuit boards produced by each machine.

    b. Calculate the variance of the number of defective circuit boards produced by each machine.

    c. Calculate the standard deviation of the number of defective circuit boards produced by each machine.

    d. Which machine do you think is better? Why?

6. A basketball coach has one remaining scholarship to offer and has narrowed his choice to two players. Listed in the following table are the points scored per game over the last season for each player.

| Points Scored | | |
|---|---|---|
| Game Number | Braudrick | Douglas |
| 1 | 27 | 35 |
| 2 | 34 | 21 |
| 3 | 29 | 50 |
| 4 | 25 | 28 |
| 5 | 28 | missed |
| 6 | 35 | 32 |
| 7 | 31 | 29 |
| 8 | 33 | missed |
| 9 | 33 | 23 |
| 10 | 25 | 35 |
| 11 | 28 | 31 |
| 12 | 32 | 36 |
| Total | 360 | 320 |

    a. What level of measurement does the data possess?

    b. What statistical criteria might you use to select the better player? Justify your answer.

    c. Calculate the statistics you proposed in b.

    d. Which player is more consistent? Why?

7.  Consider the literacy data given in the following table.

| Literacy Rates | | | |
|---|---|---|---|
| Country | Literacy Rate (%) | Country | Literacy Rate (%) |
| Australia | 99.0 | Luxembourg | 99.0 |
| Bolivia | 90.7 | Mexico | 92.8 |
| Canada | 99.0 | Netherlands | 99.0 |
| Denmark | 99.0 | Peru | 89.6 |
| France | 99.0 | Saudi Arabia | 85.0 |
| India | 74.0 | United States of America | 99.0 |
| Kenya | 73.0 | Zimbabwe | 91.2 |

**Source**: United Nations Development Programme Report, 2009

a.  What is the mean literacy rate for these selected countries?

b.  What is the standard deviation of these literacy rates?

c.  How many countries in this group would we expect to have literacy rates between one standard deviation below the mean and one standard deviation above the mean?

d.  How many countries in this group actually have literacy rates between one standard deviation below the mean and one standard deviation above the mean?

e.  What assumption did you make in answering part **c.** above?

8.  A manufacturer considers her production process to be "in control" if the proportion of defective items is less than 3%. She randomly selects 200 items and determines that 9 of the items are defective.

a.  Calculate the sample proportion of defective items.

b.  Based on the sample, do you think it is reasonable for the manufacturer to conclude that the production process is "out of control"? Why or why not?

9.  A pharmacist is interested in studying the relationship between the amount of a particular drug in the bloodstream (in mg) and reaction time (in seconds) of subjects taking the drug. Ten subjects are randomly selected and administered various doses of the drug. The reaction times (in seconds) are measured 15 minutes after the drug is administered with the following results.

| Reaction Times | |
|---|---|
| Amount of Drug (mg) | Reaction Time (Seconds) |
| 1 | 0.5 |
| 2 | 0.7 |
| 3 | 0.6 |
| 4 | 0.7 |
| 5 | 0.8 |
| 6 | 0.8 |
| 7 | 0.9 |
| 8 | 0.6 |
| 9 | 0.9 |
| 10 | 1.0 |

a.  Analyze the data collected for the study by answering the following questions:

i.  Do the variables selected for measurement seem appropriate for answering the question the pharmacist is interested in?

ii.  What biases or errors might be present in the data?

iii.  What level of measurement (nominal, ordinal, interval, ratio) does the data possess?

    **b.** Plot the data points on a scatterplot.

    **c.** Based on the scatterplot in part **b.**, answer the following questions regarding the overall pattern of the data.

        **i.** Does the pattern roughly follow a straight line?

        **ii.** Is the pattern upward sloping or downward sloping? Are the data values tightly clustered in the pattern or widely dispersed?

        **iii.** Are there significant deviations from the pattern?

**10.** Sometimes the following descriptions are assigned to the correlation coefficient.

$$r = 0 \quad \text{no linear relationship}$$
$$-0.5 < r < 0 \quad \text{weak negative linear relationship}$$
$$0 < r < 0.5 \quad \text{weak positive linear relationship}$$
$$-0.8 < r \leq -0.5 \quad \text{moderate negative linear relationship}$$
$$0.5 \leq r < 0.8 \quad \text{moderate positive linear relationship}$$
$$-1.0 < r \leq -0.8 \quad \text{strong negative linear relationship}$$
$$0.8 \leq r < 1.0 \quad \text{strong positive linear relationship}$$
$$r = 1 \quad \text{exact positive linear relationship}$$
$$r = -1 \quad \text{exact negative linear relationship}$$

Describe the relationships indicated by the correlation coefficients below using the descriptions defined above.

    **a.** $r = 0.9$         **c.** $r = -0.9$         **e.** $r = 0$

    **b.** $r = 0.5$         **d.** $r = -0.5$

**11.** Describe the relationships indicated by the correlation coefficients below using the descriptions defined in problem 10 above.

    **a.** $r = 0.8$         **c.** $r = -0.8$         **e.** $r = 0.1$

    **b.** $r = 0.4$         **d.** $r = -0.4$

**12.** Consider the following data.

| x | 1 | 2 | 3 | 4 | 5 | 6 | 7 |
|---|---|---|---|---|---|---|---|
| y | 1 | 4 | 9 | 16 | 25 | 36 | 49 |

    **a.** Plot the data points on a scatterplot.

    **b.** Determine the correlation coefficient.

    **c.** Describe the relationship between $x$ and $y$.

**13.** Consider the following data.

| x | 1 | 2 | 3 | 4 | 5 | 6 | 7 |
|---|---|---|---|---|---|---|---|
| y | 1.00 | 1.41 | 1.73 | 2.00 | 2.24 | 2.45 | 2.65 |

    **a.** Plot the data points on a scatterplot.

    **b.** Determine the correlation coefficient.

    **c.** Describe the relationship between $x$ and $y$.

## P    Discovery Project

## Describing Real Estate Data

Answer the following questions regarding the realty data gathered on the selling prices and commissions of properties sold by JDC Realty in Southwest Virginia.

Use the JDC Realty Property Sales Prices data set. This data set contains information about 240 properties sold between January 2020 and November 2020.

Note that the gross commission paid is equal to the commission percentage multiplied by the sales price. Also note that the amount paid to the agent, the amount paid for referrals, agent liability costs, and the net office amount add up to the gross commission paid.

1.  Create a histogram of the property sales prices. Do the data appear symmetric or skewed? Are there any outliers?

2.  Calculate the following measures of location for sales prices and compare them: mean, median, 10% trimmed mean. Which of these measures do you think gives us the best estimate of the center of the data and why?

3.  Calculate the variance and standard deviation of sales prices.

4.  Use Chebyshev's Theorem to find the range of values in which at least 75% of the sales price data will reside.

5.  Calculate the five summary measures needed to construct a boxplot of sales prices and create the boxplot.

6.  Calculate the interquartile range for sales prices and use this value to identify any potential outliers.

7.  Create a scatterplot of Pay Agents vs. Sales Price.

8.  Calculate the correlation coefficient between Pay Agents and Sales Price. Describe the relationship indicated by the scatterplot and the correlation coefficient. Can you think of any factors that might affect the value of the correlation coefficient?

9.  Subset the data to only include the observations that have nonzero values for Pay Agents. Create a new scatterplot and calculate the correlation coefficient for the subsetted data. Describe your results.

# Chapter 5
## Probability, Randomness, and Uncertainty

# Discovering the Real World

## The Credit Card Industry

The majority of students on college campuses are approached by vendors representing credit card companies soliciting students to apply for credit cards. The solicitors preach to the students the importance of establishing a credit history before graduating. Having this credit history will make it easier for the students to obtain financing for cars, homes, and even apartment leases upon graduation. However, based on the survey by Sallie Mae and Ipsos, young adults are confident and optimistic in money management skills. Below is a list of facts allowing you to see how spending behaviors, saving habits, and financial literacy skills differ among college students, college graduates, and young adults who didn't finish their degree (non-completers).

- The #1 reason all young adults give for opening a credit card is to build and establish credit (58% of college students, 74% of college graduates, and 77% of non-completers).
  **Source:** Sallie Mae and Ipsos, "Majoring in Money 2019"

- In 2016, college students reported having and using an average of three credit cards. In 2019 they had, on average, five credit cards.
  **Source:** Sallie Mae and Ipsos, "Majoring in Money 2019"

- In 2019 college students had an average balance of $1,183 on their credit cards, which is an increase of 31% over the average credit card balance in 2016.
  **Source:** Sallie Mae and Ipsos, "Majoring in Money 2019"

- The majority of young adults have credit cards, including 57% of college students, 83% of college graduates, and 61% of non-completers.
  **Source:** Sallie Mae and Ipsos, "Majoring in Money 2019"

- According to data released November 1, 2019, there were 374 million open credit card accounts in the U.S. as of the middle of 2019.
  **Source:** www.creditcards.com

- For general purpose credit cards, the average credit card balance was $6,194 at the end of 2019.
  **Source:** www.experian.com

- The average American has 4 credit cards according to the 2019 Experian Consumer Credit Review.
  **Source:** www.experian.com

- The average FICO Score in the U.S. hit a record high of 703 in 2019.
  **Source:** www.experian.com

- The average credit card interest rate is 16.04% on a credit card with a balance on it as of November 2020.
  **Source:** www.creditcards.com

- Household debt balances through March 2020 totaled $14.3 trillion.
  **Source:** Federal Reserve Bank of New Your (www.newyorkfed.org)

- The top 10 U.S. credit card issuers held an 81.41% percent market share of $972.73 billion in general purpose card outstandings in 2019. That list includes Citi, JPMorgan (Chase), Capital One, American Express, and Discover.
  **Source:** Nilson Report, February 2020

- The U.S. credit card default rate was as high as 11.8% in June 2020.
  **Source:** Fitch Ratings, June 2020

- The U.S. credit card 60-day delinquency rate was as high as 4.2% in June 2020.
  **Source:** Fitch Ratings, June 2020

Like any other industry, banks take profit maximization as their ultimate goal. Given the statistics just presented, it is clear that over the past several years, competition in the U.S. credit card industry has been fierce because there is a great deal of money to be made. However, maximizing profit doesn't come without risk. This competition has led many banks, including large, full-service institutions, to lose customers to issuers that are aggressively expanding their card portfolios. Some of the more visible winners of this competition are Discover, Citi, American Express, JPMorgan (Chase), and Capital One. The tremendous volume of card loans generated has created an equally immense need for inexpensive, reliable funding. To be competitive and successful in this industry, credit card issuers must have sophisticated risk-based and risk-controlling technology models to determine not only who is issued a credit card but also the rate it can offer to a particular customer based on that customer's risk profile.

The traditional approach of assessing a customer's risk is by examining their credit score. By comparing the applicant's/borrower's credit history to all other borrowers' information, banks check whether the applicant's credit history is the same as those who regularly default and even declare bankruptcy to determine their credit worthiness. There are also other factors that are examined such as income, lengths of continuous employment at current (and past) jobs, number of credit cards, and whether one owns or rents their home. Rather than simply seeking to minimize the loss of an issued credit card, examining these factors allows the banks to determine an acceptable level of risk. Thus, when issuing a card to a customer, banks attempt to find the best balance or combination of risks and revenue. The capacity and the technology of controlling risks are the key to a bank's profitability.

Once the accounts are acquired (i.e., a customer has been approved for a credit card), banks begin managing the customer's account, which includes authorization of repeated credit purchases, credit line management, fraud detection, sales promotion, cross-selling, and collections. Risk management closely monitors the delinquency and utilization patterns along with charge-offs, bankruptcies, and fraud occurrences. Based on the aforementioned items, authorization and credit line policies are modified to reduce loss or improve profitability. Behavioral score cards, vintage performance dashboards, and reports are developed to support the decision process. Marketing analyzes activation and sales patterns and prompts the customers to buy more through rewards and other offers. Cross-selling of financial products like personal loans and mortgages to existing customers is another area marketing analytics focuses on. The collections analytics team identifies the likely defaulters using behavioral score cards and, based on the risk profiles, devises strategies for collections. They use the past experiences to improve the collection efficiency and work closely with the operations team to implement data driven strategies. The operations analytics team, typically aligned with customer service operations, forecasts the call volumes and analyzes call logs to improve the call routing and support processes. Collection efficiency of the analysts in terms of number of calls and amount collected are some of the metrics monitored by this department.

In summary, card issuers use a variety of tools and utilities in consumer lending to perform an array of analytics activities ranging from simple reporting to complex statistical modeling. Assessing customers' risk is the foundation during the whole process of risk control including risk identification, risk estimation, and risk assessment. Only by assessing customers' risk and issuing cards to the applicants can the bank card business begin. Accurately assessing customers' risk is an important basis for credit. It can play a role of early warning. Based on it, banks can develop appropriate risk control policies to improve their revenues and profit levels.

## Introduction

We all have to cope with the uncertainty. Uncertainty is that uneasy feeling we get when our gas tank is nearly empty a few miles from the closest gas station. You remember in the past that your car continued to run when the gauge was even lower, but you are not quite sure it will this time. In everyday usage, the word *probably* describes an event or circumstance the speaker believes will occur. At the same time the speaker reserves the possibility that it may not occur. In this sense, *probably* reflects a strong but nonspecific degree of belief.

> **Definition**
>
> **Probability**
>
> **Probability** is the likelihood of occurrence of an event.

**Probability** is used to quantify uncertainty. If a person says he believes there is a 0.95 probability that his car will make it to a gas station before it runs out of gas, he has made a precise statement which, no doubt, reflects his past experiences and indicates a strong belief in his chances of finding a gas station. This statement is vastly different from the statement that there is a 0.40 probability the car will make it to a gas station, which casts considerably more doubt on his prospects. The probability statement provides more precise information than phrases like *maybe I'll make it, I might make it,* or *I should make it.* Therein lies its value.

Uncertainty doesn't necessarily have to be associated with events in the future. There can be plenty of uncertainty about the past. What was the Native American population of the continental United States in the year 1492? Nobody knows, so ascribing a probability to the statement *the population of Native Americans in 1492 was between 8 and 15 million* attaches someone's degree of belief to the statement.

> **Definition**
>
> **Randomness**
>
> **Randomness** is said to occur when the outcome of an event cannot be predicted with certainty.

The word **randomness** suggests a certain haphazardness or unpredictability. Randomness and uncertainty are both vague concepts that deal with variation. And even though these words are not synonyms, their discussion leads to the concept of probability.

A simple example of randomness involves a coin toss. Unless the person tossing the coin possesses magical powers, the outcome of the toss (heads or tails) is uncertain. Since the coin tossing experiment is unpredictable, the outcome is said to exhibit randomness. There are many different kinds of randomness. Some are easily described, like the toss of a coin, and others are extraordinarily complex, like molecular motion or changes in stock prices.

Even though individual flips of a coin are unpredictable, if we flip the coin a large number of times, a pattern will emerge. For most coins, roughly half of the flips will be heads and half will be tails. This long-run regularity of a random event is described with probability. Our discussions of randomness will be limited to phenomena that in the short run are not exactly predictable but do exhibit long-run regularity.

# 5.1 Introduction to Probability

Games of chance, such as tossing a coin, provide a way to demonstrate some of the fundamental laws (rules) of probability. Statistically speaking, the playing of the game is called an **experiment**.

If you were to toss a single coin there are only 2 outcomes {Head, Tail} in the experiment. The sample space of an experiment contains every potential outcome that could occur in one trial of the experiment. For a coin tossing experiment, the sample space would be $S =$ {Head, Tail}. The sample space may be called the **outcome set**, since it contains all possible outcomes of an experiment. The result of a random experiment must be an outcome.

Examples of random experiments, events, and outcomes follow. These experiments will be used throughout the chapter to illustrate the laws of probability.

> **Definition**
>
> **Random Experiment**
>
> A **random experiment** is defined as any activity or phenomenon that meets the following conditions.
>
> 1. There is one distinct outcome for each trial of the experiment.
> 2. The outcome of the experiment is uncertain.
> 3. The set of all distinct outcomes of the experiment can be specified and is called the **sample space**, denoted by $S$.

**Experiment Number 1:** Toss a coin and observe the outcome. Have we met the three conditions of a random experiment?

1. There will only be one outcome for each trial of the experiment since it is not possible to observe both a head and a tail on the same toss.

2. The outcome is unknown before the toss.

3.  The sample space can be specified and contains two outcomes, $S$ = {Head, Tail}.

Because each of the three conditions is satisfied, for all practical purposes, this experiment meets the conditions of a random experiment.

It is possible for a coin to land on its edge and thus neither be heads nor tails. Any theory, however, involves a certain degree of idealization, which means, in this case, that landing on an edge will not be considered a possible outcome.

> **Definition**
>
> Outcome
> An **outcome** is any member of the sample space.

**Experiment Number 2:** Toss a coin three times and observe the number of heads. Have we met the three conditions of a random experiment?

1.  There will be only one outcome since, for example, it is not possible to have **exactly** one and **exactly** two heads on the same trial.

2.  The outcome will be unknown before tossing the coin three times.

3.  The sample space can be specified and is composed of eight outcomes,
    $S$ = {TTT, TTH, THT, THH, HTT, HTH, HHT, HHH}.

This experiment meets the conditions of a random experiment.

An **event** could be obtaining more than one head, which involves the following set of outcomes: {THH, HTH, HHT, HHH}.

> **Definition**
>
> Event
> An **event** is a set of outcomes.

**Experiment Number 3:** Select a student from a class of size 100 and observe his or her grade point average (GPA). Have we met the conditions of a random experiment?

1.  Although we can only select one student at a time, we are measuring GPA, which is a continuous random variable between zero and four.

2.  The outcome will be unknown before selecting the student.

3.  The sample space cannot be specified with certainty.

Given points **1.** and **3.**, this experiment does not meet the conditions of a random experiment.

**Experiment Number 4:** Assume we have a deck of playing cards consisting of 13 hearts, 13 clubs, 13 spades, and 13 diamonds. Draw a card from a well-shuffled deck and observe the suit of the card. Have we met the three conditions of a random experiment?

1.  There will be only one outcome.

2.  The suit will be unknown since the card will be drawn at random.

3.  The sample space consists of the set of outcomes,
    $S$ = {heart, club, spade, diamond}.

This experiment meets the conditions of a random experiment.

If the random experiment involves drawing a card and observing a spade or a club, then the **event** would be given by the set of outcomes {spade, club}.

**Experiment Number 5:** Inspect a transistor to determine if it meets quality control standards. Have we met the three conditions of a random experiment?

1.  There will only be one outcome.

2.  The outcome of the experiment will be unknown if the transistor is selected from a manufacturing process that occasionally produces defective parts.

3.  The sample space consists of the set of outcomes,
    $S$ = {meets standards, does not meet standards}.

This experiment meets the conditions of a random experiment.

## What Is Probability?

If someone is asked, *What is probability?*, they are likely to respond correctly with *It is the chance of something happening.* A more formal definition of probability is the likelihood of the occurrence of a particular event. However, there are many ideas of what probability is and how it should be calculated. There is **subjective probability**, which is one's personal belief about the occurrence of a specific event. With subjective probability, there is no specific calculation, and the probability simply reflects a person's opinion and/or past experiences. On the other hand, **objective probability** is the likelihood of the occurrence of a particular event that is based on recorded outcomes. Objective probabilities are more accurate than subjective probabilities. In this section we will discuss two types of objective probability, relative frequency and the classical approach, as well as subjective probability.

> **Definition**
>
> **Subjective Probability**
>
> One's personal belief about the occurrence of a specific event is referred to as **subjective probability**.

> **Definition**
>
> **Objective Probability**
>
> The likelihood of the occurrence of a particular event based on recorded outcomes is called **objective probability**.

## Relative Frequency

Someone who wanted to determine the probability of getting a head on the toss of a coin could toss a coin a large number of times and observe the number of times that a head appeared. The probability could be computed as the number of times a head was observed divided by the number of times the coin was flipped. This is the relative frequency interpretation of probability.

> **Formula**
>
> **Relative Frequency**
>
> If an experiment is performed $n$ times, under identical conditions, and the event $A$ happens $k$ times, the **relative frequency** of $A$ is given by the following expression.
>
> $$\text{Relative Frequency of } A = \frac{k}{n}$$

Let's flip a coin 42 times and observe the relative frequency of a head during those tosses.

### Relative Frequency of Heads for 42 Flips

| Coin | | | | | | | |
|---|---|---|---|---|---|---|---|
| Flip Number | 1 | 2 | 3 | 4 | 5 | 6 | 7 |
| Relative Frequency | 1.00 | 0.5 | 0.6667 | 0.75 | 0.6 | 0.5 | 0.4286 |
| Coin | | | | | | | |
| Flip Number | 8 | 9 | 10 | 11 | 12 | 13 | 14 |
| Relative Frequency | 0.375 | 0.4444 | 0.4 | 0.3636 | 0.4167 | 0.3846 | 0.3571 |
| Coin | | | | | | | |
| Flip Number | 15 | 16 | 17 | 18 | 19 | 20 | 21 |
| Relative Frequency | 0.3333 | 0.375 | 0.4118 | 0.3889 | 0.3684 | 0.35 | 0.3810 |

| **Relative Frequency of Heads for 42 Flips** (cont.) | | | | | | |
|---|---|---|---|---|---|---|
| Coin | | | | | | |
| Flip Number | 22 | 23 | 24 | 25 | 26 | 27 | 28 |
| Relative Frequency | 0.3636 | 0.3478 | 0.375 | 0.36 | 0.3462 | 0.3333 | 0.3214 |
| Coin | | | | | | |
| Flip Number | 29 | 30 | 31 | 32 | 33 | 34 | 35 |
| Relative Frequency | 0.3448 | 0.3333 | 0.3548 | 0.375 | 0.3636 | 0.3529 | 0.3429 |
| Coin | | | | | | |
| Flip Number | 36 | 37 | 38 | 39 | 40 | 41 | 42 |
| Relative Frequency | 0.3611 | 0.3514 | 0.3421 | 0.3590 | 0.35 | 0.3415 | 0.3571 |

## Origins of Probability

Galileo performed one of the earliest probability analyses. Galileo was approached by an Italian nobleman who had observed that three dice were more likely to obtain a sum of 10 than a sum of 9 when thrown. Galileo became interested in the uncertainties of throwing dice and wrote a short work outlining his findings. Galileo's work set forth some of the theory of probability. The next step in the birth of probability came from the French.

A French nobleman, Chevalier de Mere, won quite a few francs by getting unwitting souls to bet him that he would not roll at least one six in a sequence of four tosses of a die. He then lost his profits by wagering that he would get at least one double 6 in a sequence of 24 tosses of two dice. Chevalier de Mere asked two of the leading mathematicians of the time, Pierre de Fermat (1607–1665) and Blaise Pascal (1623–1662), to consider his problem. The exchange of letters between Fermat and Pascal led to some of the first analysis of random phenomena and development of the first principles of probability theory.

In Figure 5.1.1, you can see that the proportion of heads is very unstable during the first 15 flips. The proportion of heads begins to stabilize at around flip 20, although there is still some fluctuation in its value. Looking at the first 42 flips makes you wonder whether this is a "fair" coin, since heads is occurring only $0.3571(100) = 35.71\%$ of the time. In Figure 5.1.2, which starts at about 200 flips, the proportion of heads becomes very stable. By flip number 296, there are 141 heads and 155 tails which equate to (approximately) a 0.4764 probability (or relative frequency) of heads. Although this is slightly less than the expected 0.5 for a "fair" coin, such a proportion is reasonable considering the randomness of the coin toss.

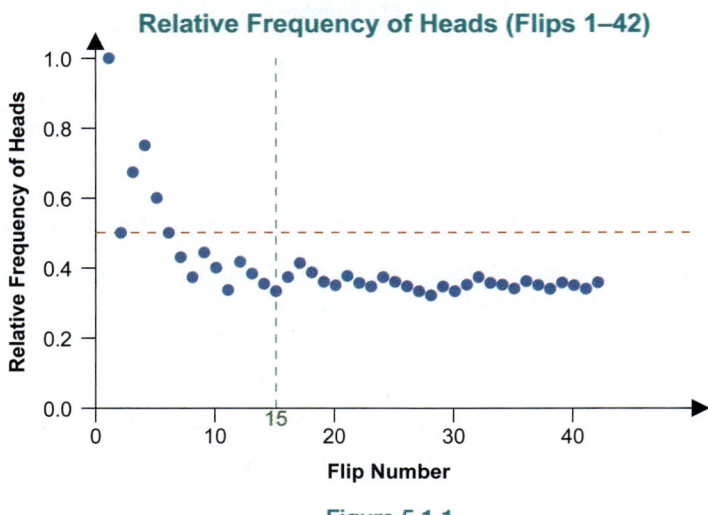

Figure 5.1.1

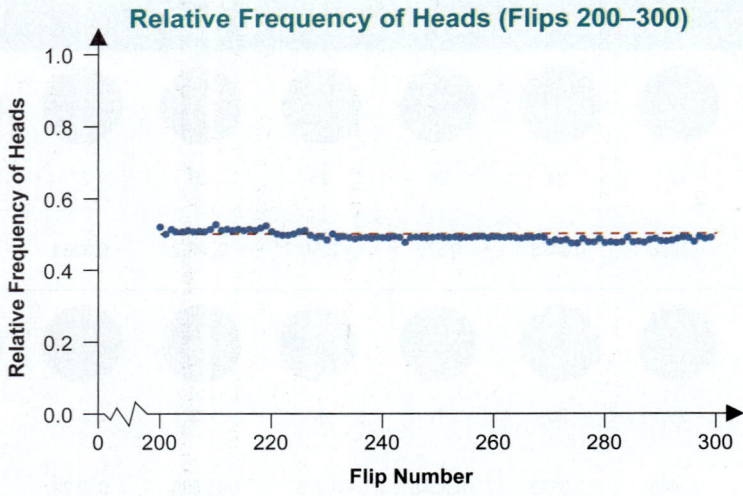

Figure 5.1.2

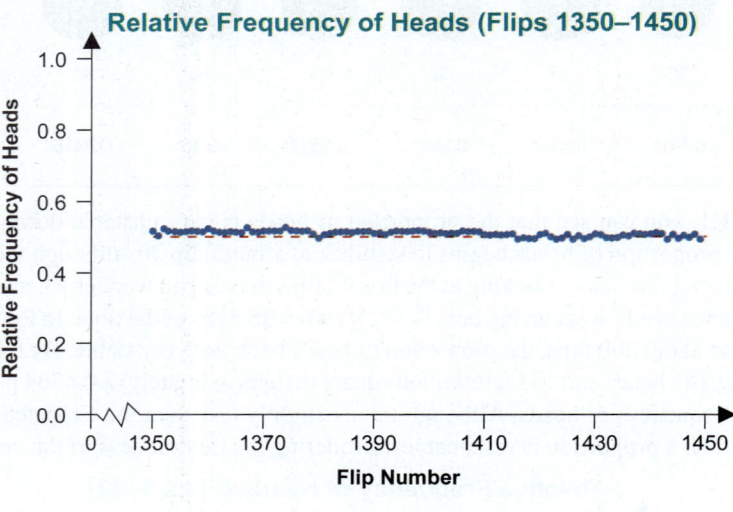

Figure 5.1.3

As can be seen in Figure 5.1.2 and Figure 5.1.3, the proportion of heads converges on some point which is close to 0.5. Figure 5.1.3 starts at 1350 flips. As you can see, the proportion of heads remains very stable. At flip number 1450 there are 718 heads and 732 tails which (approximately) equate to a 0.4952 probability (or relative frequency) of heads.

What is the relative frequency of a head on our coin? Our best available guess is 0.4952, since it is the observed relative frequency of heads using all 1450 tosses.

$$\text{Relative Frequency of } A = \frac{k}{n} = \frac{718}{1450} \approx 0.4952$$

How good is this guess for the relative frequency of heads? Since the coin has been tossed a large number of times and the observed frequency is very stable, the guess should be very good.

Will the observed relative frequency ever reach 0.5? No mathematical or physical law requires the observed relative frequency to ever reach some predetermined level. But if the probability of observing a head was really 0.5, the observed relative frequency should closely approach this value after a large number of flips.

## A Summary

**The experiment:** Toss a coin and observe which side of the coin appears on top.

**Duration of the experiment:** Toss the coin 1450 times. $n = 1450$

**Observe the event "getting a head":** The event was observed 718 times. $k = 718$

**Relative frequency of the event "getting a head":** $\dfrac{k}{n} = \dfrac{718}{1450} \approx 0.4952$

The relative frequency of a head seems to converge to the expected relative frequency of 0.5. This kind of convergence is sometimes called **statistical regularity**. Although the outcomes of the experiment may vary, in the long run the relative frequency of an outcome tends to some value, its probability.

### Problems with the Relative Frequency Approach

The problem with the relative frequency approach to defining probability is that probability only exists for events that can be repeated under the same conditions. Coin, dice, and card experiments can easily be repeated. However, because of the strict requirements of identical and repeatable experiments, many events in which it would be desirable to have relative frequency probabilities do not satisfy the requirement of repetition. Whether the next launch of the rocket will be successful, or whether you will make an A in your statistics course, are examples of experiments that are not repeatable under the exact same conditions. Thus, they are not appropriate for the application of the relative frequency idea. This perspective greatly limits the application of the relative frequency interpretation of probability. Despite its limitations, the relative frequency approach is a widely held interpretation of probability.

### Definition

**Statistical Regularity**

The belief that random events will exhibit regularity (i.e., when repeated enough times, the relative frequency of the outcomes of the experiment will approximate the true probability of the outcomes) is referred to as **statistical regularity**.

---

### Example 5.1.1

**Drawing a Card in Monopoly**

Suppose in the standard game of Monopoly, you have 16 Community Chest cards. Ten of these cards pay (or have the player collect) money if chosen, four have the player pay money if chosen, and two are associated with jail (either a "Go to Jail" or a "Get Out of Jail Free" card). Suppose a player lands on the Community Chest square when playing Monopoly and has to draw a card from the well-shuffled deck of 16 cards. Have we met the conditions of a random experiment?

**SOLUTION**

1. There will be only one outcome.

2. The type of card will be unknown since the card will be drawn at random.

3. The sample space consists of the 16 cards associated with the deck of Community Chest cards.

This experiment meets the conditions of a random experiment.

---

### Example 5.1.2

**Drawing Numbers Without Replacement**

Suppose we perform an experiment drawing three numbers, one number at a time, without replacement, from an urn containing 64 numbers. Have we met the conditions of a random experiment?

**SOLUTION**

1. For each draw from the urn, there will be only one outcome.

2. The number (or numbers) that will be drawn will be unknown and drawn at random.

3. The initial sample space consists of 64 numbers. After the first number is drawn (and

not placed back into the deck), the sample space changes; and will change again after the second number is drawn.

Thus, this experiment is not a random experiment. The reason it is not a random experiment is because it cannot be repeated under the same conditions. That is, as each number is selected, it is not returned to the urn prior to the subsequent selection, thus decreasing the amount of numbers remaining in the urn. For this reason, the experiment is different for each number drawn and cannot be considered a random experiment.

> **⚠ CAUTION**
>
> When using the classical approach, always ensure that each outcome in the sample space is equally likely.

## Classical Approach

The second objective approach commonly used in probability is the classical approach. **Classical probability** can be measured as a simple proportion: the number of outcomes that compose the event divided by the number of outcomes in the sample space, when it can be assumed that all of the outcomes are equally likely.

> **Formula**
>
> ### Classical Probability
>
> Using the **classical approach** to probability, the probability of an event $A$, denoted $P(A)$, is given by
>
> $$P(A) = \frac{\text{number of outcomes in } A}{\text{total number of outcomes in the sample space}}$$

### Example 5.1.3

**Determining Classical Probability with Coins**

In experiment number 2, a coin was tossed three times and the number of heads was observed. The sample space consists of 8 outcomes {TTT, TTH, THT, THH, HTT, HTH, HHT, HHH}. Let $A$ be the event of getting at least one head. What is $P(A)$?

**SOLUTION**

Since the event $A$ consists of 7 outcomes, {TTH, THT, THH, HTT, HTH, HHT, HHH}, and there are 8 equally likely outcomes in the sample space,

$$P(A) = \frac{7}{8} = 0.875 .$$

### Example 5.1.4

**Determining Classical Probability with Cards**

In experiment number 4, let $A$ be the event of drawing a heart. What is $P(A)$?

**SOLUTION**

Since there are 13 outcomes in $A$ (i.e., 13 hearts in a deck of cards) and 52 outcomes in the sample space (i.e., 52 cards in the deck) the probability of event $A$ is as follows.

$$P(A) = \frac{13}{52} = \frac{1}{4} = 0.25$$

If the sample space is composed of equally likely outcomes, then once the set of outcomes is determined, computing a probability is simply a matter of counting the members in each set and dividing by the total number of outcomes.

It is very important to remember that the classical approach rests on the assumption of equally likely outcomes. If the assumption is not reasonable, some other method of determining the probability must be used.

**Example 5.1.5**

**Formulating the Sample Space**

In experiment number 2, a coin was tossed three times and the number of heads was observed. The outcomes in the sample space were as follows.

$S = \{TTT, TTH, THT, THH, HTT, HTH, HHT, HHH\}$

Could the sample space in this experiment have been formulated differently? Could, for example, the outcomes be defined as the number of heads in three tosses, $S = \{0, 1, 2, 3\}$?

### SOLUTION

Using the classical definition of probability requires that the outcomes be equally likely. Since the sample space $S = \{0, 1, 2, 3\}$ does not contain equally likely events, the classical definition of probability is not applicable.

Another way to determine the number of outcomes of the previous example is to use a **tree diagram**. When an experiment is conducted in stages, as is the case of Example 5.1.5, a tree diagram can be used to organize the outcomes in a systematic manner. The tree begins with the possible outcomes of the first stage and then branches extend from each of these outcomes for the possible outcomes of the second stage, and continues for each stage of the experiment. The sample space is found by following each branch of the tree to identify all possible outcomes of the experiment. For Example 5.1.5, the tree diagram of the experiment and its sample space are provided in the following figure.

**Tree Diagram for Tossing a Coin 3 Times**

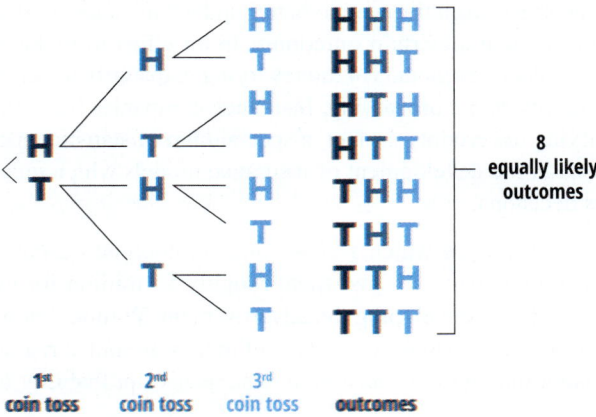

Figure 5.1.4

### Quantum Reality

From 1925 to 1927, Niels Bohr and Werner Heisenberg worked together to understand the ramifications of the new theory of quantum mechanics and revolutionized how we think about the universe. They proposed that the nature of reality is inherently probabilistic, an idea that is now referred to as the Copenhagen interpretation of quantum mechanics. One consequence of this interpretation is that an electron has a probability to be anywhere in the universe until it is measured and "forced" to be in one location. At the most fundamental level, the universe can only be understood in terms of probabilistic outcomes. This uncertainty prompted one of Einstein's most famous quotes, "I, at any rate, am convinced that He does not throw dice." Today, other interpretations of how the quantum world works have been entertained, but the probabilistic nature of the subatomic world remains a constant fixture.

## Subjective Approach

The subjective viewpoint regards the probability of an event as a measure of the degree of belief that the event has occurred or will occur. Someone's degree of belief in some event will depend on his or her life experiences. Different life experiences produce different degrees of belief. Hence, the subjective approach must allow for differences in the degree of belief among reasonable people examining the same evidence. One of the significant advantages of this view is the ability to discuss the probability of events that cannot be repeated. Thus, a subjectivist would be willing to assign the probability of making an A in your statistics course. Someone who adopts the subjective view could use the frequency interpretation to influence the determination of a subjective probability. For example, suppose that a coin had been tossed 20,000 times and had come up heads 63% of the time. It would certainly be reasonable for a subjectivist to use this information in the formulation of a statement of probability about the outcome of the next toss.

**Lloyd's of London**

This very modern looking building is the home of the world's second largest commercial insurer and the sixth largest reinsurance group, Lloyd's of London. At Lloyd's, like all other insurers, risk is measured in probabilities, which are usually subjective. Lloyd's differs from other insurers in the kinds of policies they write. Lloyd's has written policies on nuclear reactors, space shuttle cargo, oil tankers, art treasures, kidnap and ransom, as well as the legs of ballerinas and football players.

Insurance has a very important place in commerce, and without it, many business activities would not be possible. If a shipping company could not insure its ships, raising the money to buy them would be virtually impossible. Insurance is big business. Lloyd's annual marine insurance premiums amount to more than $30 billion a year, and that represents little more than one-third of their aggregate income. In addition to sizable revenues, Lloyd's employs about 70,000 people. Lloyd's is a market, rather than an entity. It houses underwriters who evaluate insurance risk for the syndicates they represent. A syndicate is a group of individuals, called Names, who individually assume a small amount of risk in return for a commensurate portion of the premium. To become a Name you must have a net worth in excess of $550,000 (excluding the value of your home) and apply to Lloyd's committee for approval. For large policies, like an ocean cargo vessel, even a syndicate does not usually underwrite the entire policy; more often groups of syndicates each take a small percentage—thus further diluting each individual's risk.

## Criticism of the Subjective View

If science is defined as finding out what is probably true, there should be a probability criterion on which all reasonable persons could agree. But if probability is subjective, how can it be used as a universally accepted criterion? Two reasonable persons might examine the same data and reach different conclusions about their degree of belief about some proposition.

## Probability, Statistics, and Business

Most of the time, when working with samples, statisticians try to deduce from the samples the population parameters (means, proportions, variances, etc.) of certain variables. This process of making judgments about population parameters is called **statistical inference**. Because samples are random, there is no guarantee that the sample will be representative of the population. If the sample is not representative, then using the sample mean as an estimate (inference) of the population mean would not be very wise. Probability is used to assess the quality of our inference. All statistical conclusions must be endowed with a degree of uncertainty. Because probability is used to assess the reliability of sample inferences, it is the foundation of all inferential statistics.

The probability concept also has many direct applications in business. When a manager wonders whether dropping a bid price by 5% will increase the probability of winning the bid, he or she is thinking about chance. Probability is also used as a criterion in designing and evaluating product reliability, evaluating insurance, inventory management, project management, and in the study of queuing theory (a probabilistic analysis of waiting lines).

Probability theory emerged from the need to better understand a game of chance. Business decisions, like games, have uncertain outcomes. In an effort to make better decisions, businesses spend considerable amounts of money trying to quantify uncertainty. This means trying to turn uncertainty into a probability. Insurance companies have historically done a good job of quantifying uncertainty. In fact, a special kind of statistician called an actuary has emerged to assist in the development of insurance models which quantify uncertainty and aid in business decisions.

For example, the next time you watch a 30-second commercial during the Super Bowl, consider the fact that a company has just spent roughly $3 million for the airtime plus a substantial amount of money developing the advertisement. Without knowing the effect of the advertisement in advance, extensive amounts of money are put at risk with an uncertain outcome. The manager making the decision uses subjective probability to assess the risk and reward.

## 📝 5.1 Exercises

### Basic Concepts

1. Describe randomness.

2. What is probability?

3. What are the conditions of a random experiment?

4. Consider the random experiment of flipping a fair coin twice. What is the sample space for this experiment?

5. What is an event?

6. Consider the random experiment of rolling a fair die once. Give an example of an event for this experiment and list the outcomes associated with that particular event.

7. What are the two main branches of probability?

8. What are the two approaches to objective probability?

9. What are some of the problems associated with the relative frequency approach?

10. True or false: According to the mathematical law of probability, the observed relative frequency of heads when flipping a coin will eventually reach 0.5 since the probability of heads is 0.5.

11. What is statistical regularity?

12. Describe the classical approach to probability.

13. Using the classical approach, describe how you would determine the probability of event $A$.

14. What is the subjective approach to probability? Discuss the problems of applying the subjective interpretation.

15. What is statistical inference?

16. Discuss the relationship between probability and statistics.

17. Give three applications of probability in business.

18. Describe the importance of probability in the insurance industry.

19. What type of probability does the manager of a company use when purchasing a commercial spot during the Super Bowl? Explain why.

## Exercises

20. Consider the following random experiment. A potato chip manufacturer is interested in determining if the brand of potato chip which it manufactures is preferred over three of its major competitors. Several customers are randomly selected and asked which brand of potato chip they prefer: Brand A, Brand B, Brand C, or Brand D.

    a. Determine the sample space for the experiment described.

    b. If the manufacturer makes Brand A, list the outcomes in the event $M =$ {customer does not prefer the manufacturer's brand}.

21. Consider the following random experiment. A doctor is interested in determining whether or not his patients think that he listens attentively to what they are saying. He randomly selects several patients and administers an anonymous survey that asks which of the following categories best describes his attentiveness: Very Attentive, Somewhat Attentive, Not Attentive.

    a. Determine the sample space for the above experiment.

    b. Determine all possible outcomes for the event $A =$ {the doctor is not described as very attentive}.

22. A gambler has made a weighted die. In order to decide which of the six sides is most likely to turn up, he tosses the die 33 times and notes the number of dots on the upper-most surface. The results of the experiment are shown in the following table.

| Rolls of a Weighted Die | | | | | | | | | | |
|---|---|---|---|---|---|---|---|---|---|---|
| 1 | 2 | 1 | 3 | 1 | 4 | 1 | 5 | 6 | 3 | 1 |
| 3 | 1 | 5 | 1 | 2 | 1 | 3 | 1 | 2 | 1 | 2 |
| 2 | 1 | 3 | 5 | 1 | 2 | 1 | 2 | 1 | 4 | 6 |

    a. Using the relative frequency approach, what is the probability of observing each side?

    b. Which side do you think the gambler will bet on when the die is tossed?

23. Assume there are two red, two yellow, and two blue buttons in a hat. A button is drawn out of the hat, the color is noted, and the button is returned. This is repeated fifty times. The results are listed in the following table.

| Button Drawing | | | | |
|---|---|---|---|---|
| Yellow | Yellow | Red | Yellow | Red |
| Red | Red | Blue | Red | Blue |
| Blue | Red | Red | Yellow | Red |
| Red | Blue | Yellow | Red | Yellow |
| Yellow | Blue | Red | Blue | Red |
| Red | Red | Red | Red | Yellow |
| Blue | Yellow | Yellow | Blue | Red |
| Yellow | Red | Red | Red | Yellow |
| Red | Yellow | Yellow | Yellow | Red |
| Red | Red | Blue | Red | Blue |

Using the relative frequency approach, what is the probability of drawing each color?

24. Twenty-five insurance agents are randomly selected and asked if they own a handgun. Twenty-two of those surveyed said that they do own a handgun. If an insurance agent is randomly selected, estimate the probability that the agent will own a handgun.

25. Thirty elementary school teachers are randomly selected and asked if they favor standardized testing of elementary school children. Twenty of those surveyed said that they did favor standardized testing of elementary school children. If an elementary school teacher is randomly selected, estimate the probability that the teacher will favor standardized testing for elementary school children.

26. Fifty chief executive officers (CEOs) of publicly traded companies are randomly selected and their salaries are determined. Forty-five of the CEOs selected have salaries in excess of $500,000. If a CEO from one of the selected publicly traded companies is randomly selected, find the probability that the CEO will have a salary in excess of $500,000.

27. Forty emergency calls to which a local police department responded were randomly selected. Of the forty emergency calls fifteen were categorized as domestic arguments. Estimate the probability that the next emergency call to which the local police department responds will be a domestic argument.

28. For the following situations, decide which probability interpretation is most reasonable to use: relative frequency, subjective, or classical.

    a. Whether or not you will have a wreck on your next trip to the mall.

    b. Whether or not a car coming off the Ford assembly line will have a defect.

    c. The probability that you will graduate from college in four calendar years.

    d. Whether a person will be in an automobile accident during the next year.

    e. The probability that you will be dealt a full house from a well-shuffled deck of cards.

29. For the following situations, decide which probability interpretation is most reasonable to use: relative frequency, subjective, or classical.

    a. Suppose you have purchased a lottery ticket. Describe your chances of winning the lottery.

    b. The probability you will enjoy a vacation trip to Mexico.

    c. The probability your company's sales will exceed seven million dollars this year.

**d.** One hundred people receive keys to a new car in a radio contest. Only one key actually fits the car. The probability that key number 25 will open the car door.

**e.** The probability that you will get a ticket if you drive 70 mph on the interstate between work and home this coming Tuesday.

**f.** The probability that the S&P 500 will increase or decrease by at least 25 points in one day.

**30.** A couple plans to have two children.

    **a.** List all possible outcomes for the sexes of the two children.

    **b.** Find the probability that the couple will have 2 boys.

    **c.** Find the probability that the couple will have at least 1 girl.

**31.** Consider a student who is taking a multiple choice examination where there are five possible answers for each question. Since the student has not studied or attended any of the classes, the student decides to randomly guess at each question.

    **a.** Find the probability that the student will answer the first question correctly.

    **b.** Find the probability that the student will answer the first question incorrectly.

**32.** A game show contestant has to choose one of three doors to win a prize. Behind one door the prize is a trip to Hawaii; behind another door, the prize is a color TV; behind the final door, the prize is a bag of potatoes. If a contestant randomly selects a door,

    **a.** Find the probability that the contestant will win a trip to Hawaii.

    **b.** Find the probability that the contestant will not win a trip to Hawaii.

# 5.2 Laws of Probability

Interpreting probability using the classical approach is a good way of thinking about the basic probability principles. In this section we will discuss certain laws that probabilities must obey, regardless of how probability is defined.

## Probability Law 1

A probability of zero means the event cannot happen.

For example, the probability of observing three heads in two tosses of a coin is zero.

## Probability Law 2

A probability of one means the event must happen.

For example, if we toss a coin, the probability of getting either a head or tail is one.

## Probability Law 3

All probabilities must be between zero and one, inclusively. That is, $0 \leq P(A) \leq 1$.

### Black Swan Events

Black Swan events are unexpected extreme events. The term stems from 16th century London: at this time, all known swans in the Euro-centric world were white. Subsequently, upon colonization of Western Australia, black swans were unexpectedly discovered. The term was popularized in Nassim Nicholas Taleb's book *The Black Swan: The Impact of the Highly Improbable* (2007). While the discovery of black swans did not adversely impact society, the black swan term today carries the connotations that the event is damaging, unexpected, and in hindsight, quite explainable. Two events occurring since 2000 that arguably qualify for black swan status are 9/11 and the disappearance of the MH-370 aircraft.

Statisticians quantify how rare events are via return periods. For example, if a 50-year earthquake at a fixed location has Richter magnitude 7.0, then the probability that a Richter magnitude 7.0 or greater earthquake occurs at the location over one year is roughly 1 / 50. Statisticians have a sub-discipline called extreme value theory that contains justifiable methods to estimate return periods (see Coles, 2001; *An Introduction to Statistical Modeling of Extreme Values*). This said, the field is often controversial and data void. Imagine trying to estimate a 200 year earthquake from only 50 years of data — a 200-year earthquake event is probably not contained in the data record!

While extreme value statisticians seldom refer to black swan events, the term is common in financial and insurance settings today. There, it often simply serves as a reminder that unexpected rare events do happen and are difficult to quantify.

*Courtesy of Robert Lund*

The closer the probability is to 1, the more likely the event will occur. The closer the probability is to 0, the less likely the event will occur.

> ### Probability Law 4
>
> The sum of the probabilities of all outcomes in a sample space must equal one. That is, if $P(A_i)$ is the probability of outcome $A_i$, and there are $n$ such outcomes, then
>
> $$P(A_1) + P(A_2) + \cdots + P(A_n) = 1.$$

There are a number of rules concerning the relationships between events that are useful in determining probabilities.

Suppose that the marketing director of *Sports Illustrated* believed that anyone who possessed an income greater than $50,000 and/or subscribed to more than one other sports magazine could potentially be a good prospect for a direct mail marketing campaign.

Let the events

$A$ = {annual income is greater than $50,000}

and

$B$ = {subscribes to more than one other sports magazine}.

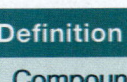

**Definition**

**Compound Event**

A **compound event** is an event that is defined by combining two or more events.

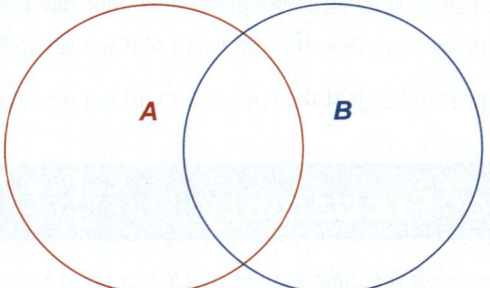

Figure 5.2.1

There are several different types of **compound events**. To illustrate the concepts, consider the two events $A$ and $B$ in Figure 5.2.1. The set of outcomes in which either or both of these events occurs is called the **union** of the two sets.

**Definition**

**Union**

The **union** of the events $A$ and $B$ is the set of outcomes that are included in $A$ or $B$ or both, and is denoted $A \cup B$.

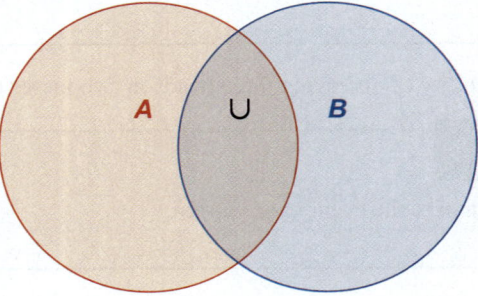

Figure 5.2.2

Notice in Figure 5.2.2 that the union includes all outcomes in both $A$ and/or $B$.

Suppose the marketing director was interested in persons who possessed an annual income greater than $50,000 and subscribed to more than one other sports magazine; that set would be called the **intersection** of $A$ and $B$.

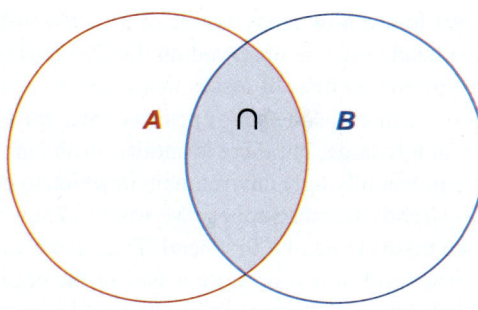

Figure 5.2.3

**Definition**

**Intersection**

The **intersection** of the events $A$ and $B$ is the set of all outcomes that are included in both $A$ and $B$. Symbolically, the intersection of $A$ and $B$ is denoted $A \cap B$ and is read "$A$ intersect $B$."

Notice in Figure 5.2.3 that the intersection includes only those outcomes in both $A$ and $B$.

Two other useful concepts are the notions of the **complement** of an event and events which are **mutually exclusive**.

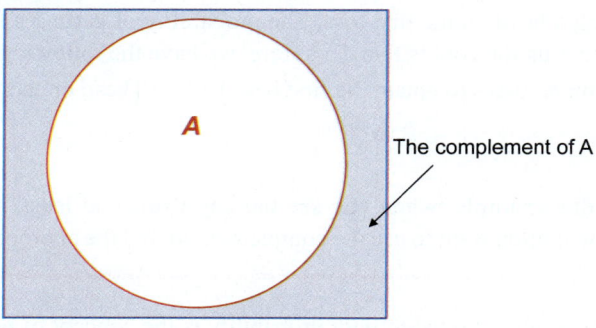

The complement of A

Figure 5.2.4

**Definition**

**Complement**

The **complement** of an event $A$ is the set of all outcomes in the sample space that are not in $A$.

The complement of the set $A$ is written as $A^c$. Notice in Figure 5.2.4 that the complement of event $A$ includes all outcomes which are not in $A$. For the event $A = \{$annual income is greater than \$50,000$\}$, the complement of $A$ would be

$$A^c = \{\text{annual income is less than or equal to } \$50{,}000\}.$$

Also note that $A^c \cup A = S$.

**Probability Law 5**

The probability of $A^c$ is given by $P(A^c) = 1 - P(A)$.

Sometimes it is much easier to calculate the probability of the complement of an event than the probability of the actual event.

Consider the event $A = \{$annual income is greater than \$50,000$\}$. Suppose the probability of $A$ was 0.08. Determine the probability of observing someone whose income was less than or equal to \$50,000.

**Example 5.2.1**

**Finding the Probability of the Complement**

**SOLUTION**

$P(\text{annual income is less than or equal to } \$50{,}000) = P(A^c)$
$= 1 - P(A)$
$= 1 - 0.08$
$= 0.92$

| Example 5.2.2 | Consider an experiment to see how many tosses of a coin will be required to obtain the first head. The first head could be observed on the first toss or the second toss but there is no upper limit on the number of tosses that could be required. Therefore, the sample space for this experiment is the set of positive integers $\{1, 2, 3, …\}$. Not only is the sample space infinitely large, but there is another problem: the outcomes are not equally likely. This is a potentially ugly environment in which to compute a probability. But let's make matters slightly worse. Suppose we want to know the probability that it will require at least two tosses to get the first head. That is, we want to know the probability of getting the first head in two or more tosses of the coin. This means that we must compute the probability of 2, the probability of 3, and so on up to infinity and add them up in some way. The problem is rather insidious if approached directly. |
|---|---|
| **Finding a Probability Using the Complement** | |

**SOLUTION**

The problem becomes rather trivial by determining the complement of the event and computing its probability. The complement of obtaining the first head in two or more tosses is getting a head on the first toss. The probability of getting a head on the first toss is 0.5, assuming the coin is fair. Therefore, we have the following.

$$P(\text{two or more tosses to obtain the first head}) = 1 - P(\text{head on the first toss})$$
$$= 1 - 0.5$$
$$= 0.5$$

As shown in this example, when you see the key words "at least" in a probability problem, you will often want to use the complement to find the appropriate probability.

Another topic that is often discussed with probability is the concept of **odds**. Oftentimes, one will hear odds spoken about at horse races, casinos, racetracks, and lottery games. For example, with regard to lottery games, you might see the following: "The odds of winning the Mega Millions Lottery game are 1 in nearly 303 million." Interestingly, there are two types of odds in gambling situations: *odds in favor of* and *odds against*.

Both odds are calculated as shown below.

> **Formula**
>
> **Odds**
>
> The **odds in favor of** the occurrence of an event $A$ is given by $\dfrac{P(A)}{P(\text{not } A)} = \dfrac{P(A)}{P(A^c)}$.
>
> The **odds against** the occurrence of an event $A$ is given by $\dfrac{P(\text{not } A)}{P(A)} = \dfrac{P(A^c)}{P(A)}$.

Recall that $A^c$ represents the complement of $A$ or *not A*. That is, $P(A^c) = 1 - P(A)$. For example, suppose $P(A) = 0.25$. Then $P(A^c) = 1 - 0.25 = 0.75$. Thus, the odds in favor of $A$ would be $\dfrac{P(A)}{P(A^c)} = \dfrac{0.25}{0.75} = \dfrac{1}{3}$ which is typically written as "1 to 3" or "1 : 3."

Similarly, if you are given the odds in favor of an event $A$ as $a : b$, then the probability of event $A$ can be calculated as $\dfrac{a}{a+b}$. Thus, if one has 1 : 5 odds of winning a game at a carnival then the probability of winning the game is $\dfrac{1}{1+5} = 16 = 0.1667$ or approximately a 17% chance of winning the game.

Conversely, if your odds are 5 : 1 of winning the particular game at the carnival, your likelihood of winning would be $\dfrac{5}{5+1} = \dfrac{5}{6} = 0.8333$, implying that you have an 83% chance of winning the game.

Having won the 2020 NBA Championship, the Los Angeles Lakers are favored to win the 2021 championship with 2 : 1 odds due to many off-season player acquisitions of several all-star players.

**a.** What is the probability that the Lakers will win the 2021 NBA Championship based on those odds?

**b.** Suppose that the consensus of many other NBA analysts is that there is a 60% chance that the Lakers will win the 2021 NBA Championship. What are the odds of the Lakers winning the championship?

### SOLUTION

**a.** The probability that the Lakers will win the championship is given by $\dfrac{a}{a+b} = \dfrac{2}{1+2} = 0.67$. Thus, there is a 67% chance that the Lakers will win the 2021 NBA Championship.

**b.** Given the probability that the Lakers will win the championship is 60%, we can calculate the odds by $\dfrac{P(\text{win})}{1-P(\text{win})} = \dfrac{0.60}{1-0.60} = 1.5$.

Therefore, the odds of the Lakers winning the 2021 NBA Championship is 1.5 to 1.

> **Example 5.2.3**
>
> **Determining the Odds of Winning the NBA Championship**

> **Definition**
>
> **Mutually Exclusive**
> Two events are **mutually exclusive** if they have no outcomes in common.

Another idea that is helpful in determining probabilities is the notion of **mutual exclusivity**. Mutual exclusivity is also called **disjointedness**. Figure 5.2.5 represents two disjoint events.

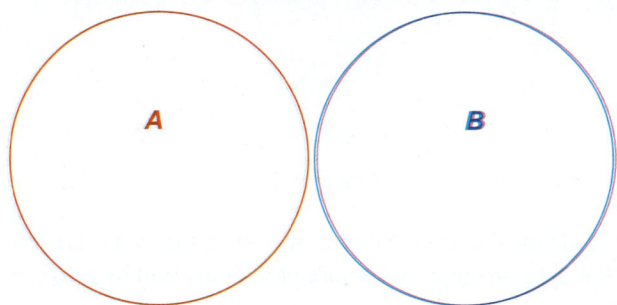

Figure 5.2.5

**Mutually Exclusive**

Two events are mutually exclusive if they cannot occur at the same time.

For example, if you were to select one card from a standard deck, the two outcomes:

- The card is a Jack
- The card is a Seven

are mutually exclusive, because you cannot select a card that is both a Jack *and* a Seven.

However, the two outcomes

- The card is a Jack
- The card is a Club

are not mutually exclusive, because you can select a card that is both a Jack *and* a Club.

> **Probability Law 6**
>
> **Union of Mutually Exclusive Events**
> If the events $A$ and $B$ are mutually exclusive, then $P(A \cup B) = P(A) + P(B)$.

## Probability Law 7

### Intersection of Mutually Exclusive Events

If the events $A$ and $B$ are mutually exclusive, then $P(A \cap B) = 0.$

---

**Example 5.2.4**

**Calculating the Probability of the Union of Mutually Exclusive Events**

Suppose $P(A) = 0.27$ and $P(B) = 0.19$. If $A$ and $B$ are mutually exclusive, what is the probability of $A \cup B$?

**SOLUTION**

Since these are mutually exclusive events,

$$P(A \cup B) = P(A) + P(B) = 0.27 + 0.19 = 0.46.$$

---

There is a more generalized rule that eliminates the assumption of mutual exclusivity between the sets.

## Probability Law 8

### The Addition Rule

For any two events $A$ and $B$, $P(A \cup B) = P(A) + P(B) - P(A \cap B).$

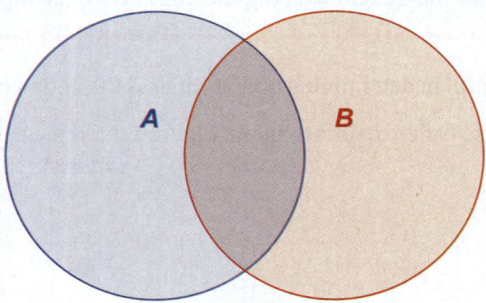

**Figure 5.2.6**

The addition rule is identical to Probability Law 6 when the events are mutually exclusive, since $A$ intersect $B$ will be an empty set whose probability will be zero (i.e., $P(A \cap B) = 0$).

---

**Example 5.2.5**

**Calculating the Probability of the Union of Two Events**

Suppose that the marketing manager mentioned earlier believed that the probability that someone earns more than \$50,000 is 0.2 and the probability that someone will subscribe to more than one sports magazine is 0.3. If the probability of finding someone in both categories is 0.08, what is the probability of finding someone who is earning over \$50,000 or subscribes to more than one sports magazine, or both?

**SOLUTION**

The problem involves the union of two events. Let us define the events as follows.

$A = \{$the event that someone earns more than \$50,000$\}$

$B = \{$the event that someone will subscribe to more than one sports magazine$\}$

From the problem statement, we know $P(A) = 0.2$, $P(B) = 0.3$, and $P(A$ and $B) = 0.08$.

We want to know the probability of event $A$ or event $B$ or both. Thus, the desired probability is the union of the two events given by

$$P(A \cup B) = P(A) + P(B) - P(A \cap B) = 0.2 + 0.3 - 0.08 = 0.42.$$

Therefore, the probability of finding someone who is earning over $50,000 or subscribes to more than one sports magazine, or both, is 0.42.

## ✍ 5.2 Exercises

### Basic Concepts

1. What laws must probability obey, regardless of the methodology used to derive the probabilities?

2. Suppose you are taking a test next week. Interpret each of the following statements.

   a. $P(\text{receiving an A on the test}) = 0$

   b. $P(\text{receiving an A on the test}) = 1$

   c. $P(\text{receiving an A on the test}) = 0.3$

3. What is a compound event?

4. Define the following set operations: union, intersection, and complement.

5. If you know the probability of two events, what else must you know in order to calculate the probability of *one event or the other*?

6. If events A and B are mutually exclusive, what is $P(A \cap B)$?

### Exercises

7. Determine if the following values could be probabilities. If the value cannot be a probability, explain why.

   a. 0

   b. $\dfrac{36}{25}$

   c. $\dfrac{7}{8}$

   d. $-0.4$

   e. 0.23

8. Determine if the following values could be probabilities. If the value cannot be a probability, explain why.

   a. 1

   b. $\dfrac{15}{16}$

   c. $\dfrac{4}{3}$

   d. 0.99

   e. $-0.05$

9. Interpret the following probabilities with respect to the occurrence of some event.

   a. $P(\text{event}) = 0$

   b. $P(\text{event}) = 1.0$

   c. $P(\text{event}) = 0.45$

   d. $P(\text{event}) = 65\%$

   e. $P(\text{event}) = -1.0$

10. Find the following probabilities.

   a. The probability of an event that must happen.

   b. The probability of an event that cannot happen.

   c. The probability of having a boy or a girl in a single birth.

   d. The probability of rolling a two and a five in a single toss of a die.

**11.** The annual premium amounts charged by life insurance companies to their clients are set very carefully. If the amount is too high, the client will take his or her business to another company. If it is too low, the insurance company may not make enough profit to stay in business. In order to properly determine a premium, the company often relies on life tables. These tables allow one to compute the probabilities of death at various ages. They are constructed only after collecting and reviewing extensive data on age at death from a large group of people. A life table is normally constructed assuming that 100,000 people are alive at age 0. This number is simply a reference value used to make comparisons throughout the table. Other numbers could be used. The table then gives the number of people of the original 100,000 that are alive at the beginning of various years of life. In order for the insurance company to optimally set premiums, a separate table should be constructed for the different genders and races. The following abbreviated life table is valid only for females.

| Life Table | | | | | | | | |
|---|---|---|---|---|---|---|---|---|
| Year | 0 | 1 | 5 | 10 | 15 | 20 | 25 | 30 | 35 |
| Number Alive | 100,000 | 99,090 | 98,912 | 98,815 | 98,716 | 98,477 | 98,204 | 97,897 | 97,500 |
| Year | 40 | 45 | 50 | 55 | 60 | 65 | 70 | 75 | 80 |
| Number Alive | 96,958 | 96,097 | 94,766 | 92,623 | 89,449 | 84,565 | 77,772 | 68,200 | 55,535 |

**a.** What is the probability that a newborn female lives until the age of 40?

**b.** What is the probability that a newborn female dies before she reaches the age of 50?

**12.** A health care provider classifies its customers by their housing situation and whether they have health insurance coverage. The market research department has gathered data from a random sample of 759 customers.

| Health Care Consumers | | |
|---|---|---|
| | Housing Situation | |
| Have Health Insurance Coverage | Rent | Own |
| Yes | 196 | 298 |
| No | 92 | 173 |

**a.** What is the probability that a customer rents their home?

**b.** What is the probability that a customer owns their home?

**c.** What is the probability that a customer has health insurance coverage and rents their home?

**d.** What is the probability that a customer owns their home and does not have health insurance coverage?

**e.** What is the probability that a customer has health insurance coverage and rents their home or does not have health insurance coverage and owns their home?

**f.** What is the probability that a customer does not have health insurance coverage?

**g.** What approach to probability did you use to calculate your answers?

**h.** Are the events {rents their home} and {owns their home} mutually exclusive? Explain.

**13.** A large life insurance company is interested in studying the insurance policies held by married couples. In particular, the insurance company is interested in the amount of insurance held by the husbands and the wives. The insurance company collects data for all of its 1000 policies where both the husband and the wife are insured. The results are summarized in the following table.

| Life Insurance Coverage | | Amount of Life Insurance on Husband ($) | | | |
|---|---|---|---|---|---|
| | | 0 – 50,000 | 50,000 – 100,000 | 100,000 – 150,000 | More than 150,000 |
| Amount of Life Insurance on Wife ($) | 0 – 50,000 | 400 | 200 | 50 | 50 |
| | 50,000 – 100,000 | 50 | 50 | 30 | 30 |
| | 100,000 – 150,000 | 20 | 10 | 25 | 25 |
| | More than 150,000 | 20 | 10 | 15 | 15 |

   **a.** For a randomly selected policy, what is the probability that the husband will have between $50,000 and $100,000 of insurance?

   **b.** For a randomly selected policy, what is the probability that the wife will have between $100,000 and $150,000 of insurance?

   **c.** For a randomly selected policy, what is the probability that the wife will have more than $150,000 of insurance or the husband will have more than $150,000 of insurance?

   **d.** For a randomly selected policy, what is the probability that the wife will have between $0 and $50,000 of insurance and the husband will have between $0 and $50,000 of insurance?

   **e.** For a randomly selected policy, what is the probability that the wife will not have between $0 and $50,000 of insurance?

   **f.** For a randomly selected policy, what is the probability that the husband will have more than $50,000 of insurance?

   **g.** What approach to probability did you use to calculate your answers?

   **h.** Are the events {the wife has more than $150,000 in insurance} and {the husband has between $50,000 and $100,000 of insurance} mutually exclusive? Explain.

# 5.3 Conditional Probability

Researchers often want to examine a limited portion of the sample space. For example, consider the question of whether cigarette smoking harms those that are indirectly exposed to the smoke. Suppose that 3 percent of women who do not smoke die of cancer. However, if a nonsmoking woman is married to a smoking husband (not to be confused with a husband who is on fire), the probability of dying of cancer is 0.08. This probability is a **conditional probability**, because the sample space is being limited by some condition—in this case, limited to only wives of smoking husbands. In this instance, the dramatic effect of a smoking husband on cancer rates is readily evident.

$$P(\text{a nonsmoking woman dies of cancer})$$

$$\neq$$

$$P(\text{a nonsmoking woman dies of cancer given that her husband smokes})$$

Similarly, the results from a market survey indicate that 39 percent of the customers surveyed believe a product is of high quality. However, if the analysis is limited to only women, 54

**Definition**

**Conditional Probability**
The probability that one event will occur given that some other event has occurred is a **conditional probability**.

percent of women surveyed believe the product is of high quality. Based on the survey, it appears that women have a much higher regard for the company's product than men. The difference in attitude would probably be something that could affect how the company spends its marketing dollars.

To compute a conditional probability, apply the following rule.

### Probability Law 9

**Conditional Probability**

The conditional probability of $A$, given that $B$ has occurred, is $P(A \mid B) = \dfrac{P(A \cap B)}{P(B)}$.

The notation $P(A \mid B)$ is read as *the probability of A given the occurrence of B*. The vertical bar within a probability statement will always mean *given*.

Note that since we are finding the conditional probability of $A$, given that $B$ has occurred, $P(B) \neq 0$.

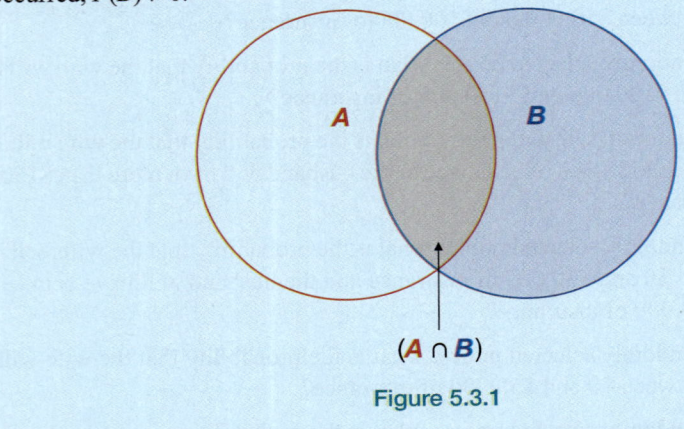

$(A \cap B)$

**Figure 5.3.1**

The events $A$ and $B$ can be reversed in the preceding rule to compute $P(B \mid A)$. It would be very unlikely that $P(A \mid B)$ equals $P(B \mid A)$.

### Example 5.3.1

**Calculating a Conditional Probability**

Suppose a marketing research firm has surveyed a panel of consumers to test a new product and produced the following cross tabulation indicating the number of panelists that liked the product, the number that did not like the product, and the number that were undecided.

| Table 5.3.1 – Market Research Survey | | | | |
|---|---|---|---|---|
| **Age** | **Like** | **Not Like** | **Undecided** | **Total** |
| 18–34 | 213 | 197 | 103 | 513 |
| 35–50 | 193 | 184 | 67 | 444 |
| Over 50 | 144 | 219 | 83 | 446 |
| Total | 550 | 600 | 253 | 1403 |

If an individual is between 35 and 50 years old, what is the probability he or she will like the product?

#### SOLUTION

Let the events

$A$ = {like the product}, and

$B$ = {age between 35 and 50}.

Then the desired probability can be formulated as

$$P(A \mid B) = \frac{P(A \cap B)}{P(B)} \, .$$

The $P(A \cap B)$ is called a joint probability since it is the probability of the occurrence of more than one event. To compute $P(A \cap B)$ use the empirical approach.

$$P(A \cap B) = \frac{193}{1403} \approx 0.1376$$

Similarly, $P(B)$ can be computed as

$$P(B) = \frac{444}{1403} \approx 0.3165 \, .$$

Consequently, $P(A \mid B)$ is

$$P(A \mid B) \approx \frac{0.1376}{0.3165} \approx 0.4348 \, .$$

Note that this answer could have also been obtained by simply dividing 193 by 444.

## Let's Make a Deal

A long time ago, back in the 70s, there was a television show called "Let's Make a Deal" starring Monty Hall as the host. This show produced an interesting problem in probability which someone submitted to Marilyn vos Savant which she answered in her column in Parade magazine. Incidentally, Ms. Savant is in the Guinness Book of World Records as having the highest recorded IQ (228). Here's the problem that was posed to Ms. Savant.

"Suppose you're on a game show, and you're given a choice of three doors: Behind one door is a car: behind the others, goats. You pick a door, say number 1, and the host, who knows what's behind the other doors, opens another door, say number 3, which has a goat. He then says to you, 'Do you want to pick door number 2?' Is it to your advantage to take the switch?"

Marilyn vos Savant answered the question in her column saying that it was to your advantage to switch. This set off a firestorm of mail telling Ms. Savant that she was incorrect. Much of this mail came from people with Ph.D.s behind their names. The New York Times printed a front page article in 1991 discussing the problem.

What do you think? To find the answer to this problem type "The Monty Hall problem" in your search engine and go to some of the web sites and try some of the simulations.

# 📝 5.3 Exercises

## Basic Concepts

1. Define conditional probability.

2. How do you calculate $P(A \mid B)$?

## Exercises

3. The following table was given in Section 5.2, Exercise 12.

| Health Care Consumers | | |
|---|---|---|
| | Housing Situation | |
| Have Health Insurance Coverage | Rent | Own |
| Yes | 196 | 298 |
| No | 92 | 173 |

   a. Given that the customer rents their home, what is the probability that the customer does not have health insurance?

   b. Given that the customer does not have health insurance, what is the probability that the customer rents their home?

   c. Given that the customer owns their home, what is the probability that the customer has health insurance?

   d. Given that the customer has health insurance, what is the probability that the customer owns their home?

4. The following table was given in Section 5.2, Exercise 13.

| Life Insurance Coverage | | | | |
|---|---|---|---|---|
| | | Amount of Life Insurance on Husband ($) | | |
| Amount of Life Insurance on Wife ($) | | 0 – 50,000 | 50,000 – 100,000 | 100,000 – 150,000 | More than 150,000 |
| | 0 – 50,000 | 400 | 200 | 50 | 50 |
| | 50,000 – 100,000 | 50 | 50 | 30 | 30 |
| | 100,000 – 150,000 | 20 | 10 | 25 | 25 |
| | More than 150,000 | 20 | 10 | 15 | 15 |

    **a.** Given the wife has between $100,000 and $150,000 of insurance, what is the probability that the husband has more than $150,000 of insurance?

    **b.** Given the wife has between $0 and $50,000 of insurance, what is the probability that the husband has between $0 and $150,000 of insurance?

    **c.** Given that the husband has between $0 and $50,000 of insurance, what is the probability that the wife will have more than $150,000 of insurance?

    **d.** Given that the husband has more than $150,000 of insurance, what is the probability that the wife will have more than $150,000 of insurance?

**5.** A computer software company receives hundreds of support calls each day. There are several common installation problems, call them A, B, C, and D. Several of these problems result in the same symptom, *lock up* after initiation. Suppose that the probability of a caller reporting the symptom *lock up* is 0.7 and the probability of a caller having problem A and a *lock up* is 0.6.

    **a.** Given that the caller reports a lock up, what is the probability that the cause is problem A?

    **b.** What is the probability that the cause of the malfunction is not problem A given that the caller is experiencing a lock up?

**6.** A television advertising representative has determined the following probabilities based on past experience. The probability that an individual will watch an ad during the Super Bowl is 0.10. Given that the individual watches the ad, the probability that the individual will buy the product is 0.005. It is also known that the probability that an individual would buy the product is 0.02. Given that an individual buys the product, find the probability that the individual watched the television ad during the Super Bowl.

**7.** Medical researchers have determined that there is a 2% chance that an individual will have a gene which gives him a predisposition for heart disease. Given that an individual has the gene, the probability that heart disease will develop is 25%. It is also known that the probability that an individual has heart disease is 12%.

    **a.** Find the probability that an individual will have the gene and develop heart disease.

    **b.** Given that a person has heart disease, what is the probability that they have the gene?

# 5.4 Independence

An extremely important concept in statistical analysis is **independence**. It describes a special kind of relationship between two events. Two events are said to be independent if knowledge of one event does not provide information of the other event's occurrence. In other words, the occurrence of one event does not affect the occurrence of another event if the events are independent.

**Example 5.4.1**

**Determining the Independence of Events**

Experiment: roll a fair die two times. Consider the two events

    $A = \{$rolling a six on the first roll of a fair die$\}$ and

    $B = \{$rolling a four on the second roll of a fair die$\}$.

Are these two events independent?

**SOLUTION**

Since knowledge of the outcome of the first roll does not help one make an inference of the outcome of the second roll, events $A$ and $B$ are independent.

Symbolically, the idea of independence is expressed in the following definition.

> ### Formula
>
> #### Independent Events
> Two events, $A$ and $B$, are **independent** if and only if $P(A \mid B) = P(A)$ and $P(B \mid A) = P(B)$.

In many cases, regarding the independence of two events, intuition and common sense will lead you to the correct determination. However, there are situations in which independence can only be discovered by formal application of the definition.

What does it mean if two events are not independent? The obvious response is to say that they are **dependent**, a term that is just as much a part of statistical vocabulary as independent. If events are dependent, they are related; the nature of the relationship and whether the relationship can be used for predictive purposes are problems often examined by statisticians.

During the course of a business negotiation, both negotiators may exhibit numerous types of idiosyncratic behavior. If they have jewelry, they might manipulate it. If they smoke cigarettes, they might play with their lighters or packs of cigarettes. Are their mannerisms independent of the importance of the issue they are negotiating? Does the negotiator tend to smoke a cigarette or play with jewelry when he or she has a strong position? Good negotiators will pick up dependencies and use the information to their advantage. However, the concept of association implied by dependence must not be confused with the idea of causation. It may be that one of the events does indeed cause the other, but the fact that they are not independent (dependent) is not evidence of causation.

Now let's consider the probability of two events both happening. For example, what is the probability of choosing a queen and then a king from a deck of cards? The key word in these problems is the word "and".

Before we can calculate these types of probabilities, we need to introduce a couple new terms. The phrase **with replacement** refers to placing objects back into consideration, such as choosing a card from a deck and then returning it to the deck for the next choice. In the same manner, the phrase **without replacement** means your first choice is not put back in for consideration, such as drawing a card and then drawing a second card from those left over. These two phrases affect the number of possible outcomes in the sample space.

Recall that two events are **independent** if one event happening does not influence the probability of the other event happening. For example, if after drawing a card you replace the card drawn and shuffle the deck, then the probability of the next card drawn is not affected by what was picked first; therefore, choosing the two cards from a deck with replacement are independent events.

When two events are independent, the probability that both events occur can easily be calculated by multiplying their respective probabilities together. Written mathematically, the formula referred to as the Multiplication Rule for Independent Events is as shown in the box below.

> ### Probability Law 10
>
> #### Multiplication Rule for Independent Events
> If two events, $A$ and $B$, are independent, then
> $$P(A \cap B) = P(A)P(B).$$
> If $n$ events, $A_1, A_2, \ldots, A_n$, are independent, then
> $$P(A_1 \cap A_2 \cap \cdots \cap A_n) = P(A_1)P(A_2) \cdots P(A_n).$$

> ### Definition
>
> #### Independent Events
> Two events are said to be **independent** if the occurrence of one event *does not* affect the occurrence of the other event.

> ### Definition
>
> #### Dependent Events
> Two events are said to be **dependent** if the occurrence of one event *does* affect the occurrence of the other event.

### Basketball and Dependence

If you were playing basketball and made a large number of consecutive shots, then there might be a temptation to boast of your skills. Alternatively, someone might suggest that you were on a lucky streak. Two psychologists examined the "lucky streak" phenomenon by analyzing the sequence of made and missed shots by professional basketball players. The selected players made roughly 50 percent of their shots. Their analysis found that there was no evidence of the hot hand—that is, there was no evidence of a dependent relationship between the consecutive shots. They did not find more long streaks of made baskets than would be expected to occur by chance.

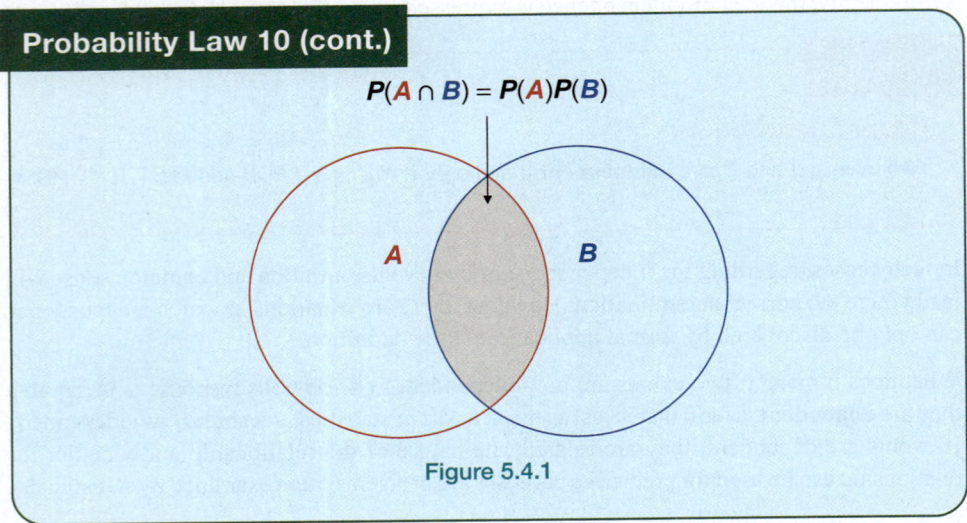

**Probability Law 10 (cont.)**

$$P(A \cap B) = P(A)P(B)$$

Figure 5.4.1

The multiplication rule for independent events (Probability Law 10) is sometimes called the **product rule**. The rule simply states that the probability of the joint occurrence of independent events is the product of their probabilities.

**Example 5.4.2**

**Finding the Probability of Independent Events**

A coin is flipped, a die is rolled, and a card is drawn from a deck of 52 cards. Find the probability of getting tails on the coin, a five on the die, and a Jack of clubs from the deck of cards.

**SOLUTION**

Since the three events (flipping a coin, rolling a die, and selecting a card) are independent, we can use the product rule.

We know the following.

$$P(\text{tails on coin}) = \frac{1}{2},$$

$$P(\text{five on die}) = \frac{1}{6}, \text{ and}$$

$$P(\text{Jack of clubs}) = \frac{1}{52}.$$

Therefore, we can calculate the probability as follows using the product rule.

$$P(\text{tails on coin} \cap \text{five on die} \cap \text{Jack of clubs}) = P(\text{tails on coin})P(\text{five on die})P(\text{Jack of clubs})$$

$$= \left(\frac{1}{2}\right)\left(\frac{1}{6}\right)\left(\frac{1}{52}\right)$$

$$= \frac{1}{624}$$

$$= 0.0016$$

So the probability of getting tails on the coin, rolling a five on the die, and then selecting the Jack of clubs from the deck of cards is approximately 0.0016.

**Example 5.4.3**

**Using the Probability of Independent Events in a Court Case**

This is an actual case that stirred up quite a controversy.

**People v. Collins (1968)**

On June 18, 1964, at about 11:30 AM, Mrs. Juanita Brooks was assaulted and robbed while walking through an alley in the San Pedro area of Los Angeles. Mrs. Brooks described her assailant as a young woman with a blonde pony tail. At about the same time John Bass was watering his lawn and witnessed the assault. He described the assailant as a Caucasian woman with dark-blonde hair. As she ran from the alley she jumped into a yellow automobile driven by a black man with a mustache and a beard.

Several days later the police arrested two individuals based on the descriptions provided by the assailant and the witness. The two suspects were eventually charged with the crime. During the trial the prosecution called a professor of mathematics to testify. The prosecutor set forth the following probabilities for the characteristics of the assailants:

| Table 5.4.1 – Assailant Characteristics Data ||
| Characteristic | Probability |
| --- | --- |
| Yellow automobile | 0.10 |
| Man with mustache | 0.25 |
| Girl with ponytail | 0.10 |
| Girl with blonde hair | 0.33 |
| Black man with beard | 0.10 |
| Interracial couple in a car | 0.001 |

How did the prosecution use these probabilities to argue its case?

**Struck by Lightning**

Many people have a greater chance of meeting someone who survived a lightning strike than someone who won the lottery. There are an estimated 1800 new thunderstorms being created around the world every minute and the odds against someone being struck by lightning are 606,944 to 1. That is unless you are Roy C. Sullivan. Mr. Sullivan was a former U.S. park ranger who was struck by lightning 7 times in less than 36 years.

**SOLUTION**

If the events are assumed to be independent, then the product rule can be used to calculate the likelihood of observing their joint occurrence.

$$P\left(\begin{array}{l}\text{Yellow automobile} \cap \text{Man with}\\ \text{mustache} \cap \text{Girl with ponytail} \cap \text{Girl}\\ \text{with blonde hair} \cap \text{Black man with}\\ \text{beard} \cap \text{Interracial couple in car}\end{array}\right) = (0.10)(0.25)(0.33)(0.10)(0.001)$$

$$= 0.0000000825$$

Based on the product rule, the mathematician testified that there was about a 1 in 12 million chance that a couple selected at random would possess these characteristics. The prosecution added that the probability was the chance that "any other couple possessed the distinctive characteristics of the defendants." The jury convicted the defendants. On appeal, the Supreme Court of California reversed the decision, based on two main points. First there was no proof offered that the probabilities used in the probability calculation were correct and there was no evidence that the events were independent. Second, the prosecution's evidence pertaining to a randomly selected couple was not pertinent to the problem of the existence of any other couple possessing the same characteristics.

**Example 5.4.4**

**Calculating the Probability of Defective Work**

In a production process, a product is assembled by using four independent parts ($A$, $B$, $C$, and $D$). In order for the product to operate properly, each part must be free of defects. The probability that each part is defect-free is given by $P(A) = 0.9$, $P(B) = 0.7$, $P(C) = 0.8$, and $P(D) = 0.9$.

**a.** What is the probability that all four parts have defects?

**b.** What is the probability that the product does not work?

**SOLUTION**

**a.** Since there are four parts with the probability of each part working (i.e., being defect-free) given to be $P(A) = 0.9$, $P(B) = 0.7$, $P(C) = 0.8$, and $P(D) = 0.9$, for all parts to have defects, we need the complement for each of the parts. That is,

$$P(\text{all four parts have defects}) = P(A^c \cap B^c \cap C^c \cap D^c).$$

## Winning the Lottery Twice!

In 1986 a woman won the New Jersey state lottery twice and in 1988 a man in Pennsylvania also won the State lottery twice for a total of 5.4 and 6.8 million dollars respectively. The New York Times reported the odds to be 1 in 17 trillion. How is this to be explained?

When millions of people play a lottery daily it can be shown that the odds of winning twice is about 1 in 30 in a 4-month period and the odds are even better in a 7 year period. Thus what appeared to be an almost theoretical impossibility, turns out to be quite a probable event. What one can say is that even when the probability of an event is very small, if there are millions of possibilities, then the rare event rarely remains rare. This is what has been called "The Law of Real Large Numbers" by Diaconis and Mosteller.

**Source:** Diaconis, P. and Mosteller, F. (1989). "Methods for Studying Coincidences," Journal of the American Statistical Association, 84, 853-861.

Since each part operates independently of the others, the probability that all four parts have defects is the product of the probabilities of each part's complement.

$$P(\text{all four parts have defects}) = P(A^c) \cdot P(B^c) \cdot P(C^c) \cdot P(D^c)$$
$$= (1-0.9)(1-0.7)(1-0.8)(1-0.9) = 0.0006$$

Thus, the probability that all four parts have defects is 0.0006, or 0.06% .

**b.** The probability that the product does not work is the probability that at least one of the parts does not work (since each part must be defect-free for the product to work).

$$P(\text{product does not work}) = P(\text{at least one part does not work})$$
$$= 1 - P(\text{all parts work})$$
$$= 1 - P(A \cap B \cap C \cap D)$$
$$= 1 - (0.9)(0.7)(0.8)(0.9)$$
$$= 1 - 0.4536 = 0.5464$$

Thus, given the above probabilities, there is nearly a 55% chance that the product will not work.

---

When two or more events are not independent, the occurrence of one event influences the occurrence of the other event. You may recall when we were drawing numbers from an urn without replacement in Example 5.1.2, that once the first number was drawn and not replaced, the outcome from drawing the second number was affected by the occurrence of the first number, and so on and so forth. The events are not independent; they are dependent. Thus, to calculate the probability of the occurrence of dependent events, we still multiply the probabilities, but we must consider the outcome of the first event when calculating the probability of occurrence of the second event. This concept brings us to the Multiplication Rule for Dependent Events.

### Probability Law 11

**Multiplication Rule for Dependent Events**

If two events, $A$ and $B$, are dependent, then

$$P(A \cap B) = P(A) \cdot P(B \mid A) = P(B) \cdot P(A \mid B).$$

### Example 5.4.5

**Calculating the Probability of Dependent Events**

Suppose that it is known that 30% of college students have an American Express charge card. It is also known that if a college student already has an American Express charge card, he/she also has a Citi credit card with a probability of 80%. What is the probability that a college student will have both an American Express charge card and a Citi credit card?

**SOLUTION**

Let $A$ = {college student has an American Express charge card}.

Let $C$ = {college student has a Citi credit card}.

We know that $P(A) = 0.30$ and we know that $P(C \mid A) = 0.80$.

Using the Multiplication Rule for Dependent Events, we want to find $P(A \cap C) = P(A) \cdot P(C \mid A) = (0.3)(0.8) = 0.24$.

Thus, we know that 24% of the college students hold both an American Express charge card and a Citi credit card.

## 5.4 Exercises

### Basic Concepts

1. Explain the difference between dependent and independent events.

2. Are mutually exclusive events dependent or independent? Explain your answer.

3. If events $A$ and $B$ are independent, what is $P(A|B)$ equal to?

4. What is the product rule?

5. In the case *People v. Collins* an appeals court overturned the conviction. What flaws did the appeals court detect in the case against the accused assailants?

### Exercises

6. The following table was given in Section 5.2, Exercise 12.

| Health Care Consumers | | |
|---|---|---|
| | Housing Situation | |
| Have Health Insurance Coverage | Rent | Own |
| Yes | 196 | 298 |
| No | 92 | 173 |

Are the events {customer rents their home} and {customer owns their home} independent? Explain.

7. The following table was given in Section 5.2, Exercise 13.

| Life Insurance Coverage | | | | | |
|---|---|---|---|---|---|
| | | Amount of Life Insurance on Husband ($) | | | |
| | | 0 – 50,000 | 50,000 – 100,000 | 100,000 – 150,000 | More than 150,000 |
| Amount of Life Insurance on Wife ($) | 0 – 50,000 | 400 | 200 | 50 | 50 |
| | 50,000 – 100,000 | 50 | 50 | 30 | 30 |
| | 100,000 – 150,000 | 20 | 10 | 25 | 25 |
| | More than 150,000 | 20 | 10 | 15 | 15 |

Are the events {the husband has more than $150,000 in insurance} and {the wife has more than $50,000 in insurance} independent? Explain.

8. Suppose you were flipping a coin. What is the probability that you would observe a head:

   a. on two consecutive flips?
   c. on four consecutive flips?
   b. on three consecutive flips?
   d. on 100 consecutive flips?

9. Suppose an atomic reactor has two independent cooling systems. The probability that Cooling System A will fail is 0.01 and the probability that Cooling System B will fail is 0.01. What is the probability that both systems will fail simultaneously?

10. Mandy is 30, and the probability that she will survive until age 65 is 0.90. Ashley is 45, and the probability that she will survive until age 65 is 0.95.

    a. Find the probability that both Mandy and Ashley will survive until age 65.

    b. Find the probability that only Mandy will survive until age 65.

    c. Find the probability that neither Mandy nor Ashley will survive until age 65.

    d. What assumption about the lives of Mandy and Ashley did you make in answering the above questions?

11. An insurance company is considering insuring two large oil tankers against spills. The limit of the liability on the coverage is $10,000,000. The company believes that the probability of an oil spill requiring the maximum liability coverage during the policy period is 0.001 per tanker.

   a. What is the probability that neither tanker would have a spill requiring the maximum liability coverage during the policy period?

   b. What is the probability that only one tanker would have a spill requiring the maximum liability coverage during the policy period?

   c. What is the probability that both tankers would have spills requiring the maximum liability coverage during the policy period?

12. Coin flipping can be used to model other real-life phenomena and aid in certain probability calculations. An example of this would be to compute the probability that the World Series ends in some specified number of games. The World Series is a best of seven game series played at the end of the regular baseball season between the champion of the American League and the champion of the National League. The first team to win four games is declared the champion of baseball for that year. If we assume the probability of either team winning a game is approximately 0.5 and the games are independent events, the probability that the series ends in either 4, 5, 6, or 7 games can be computed.

   a. What is the probability that the series ends in exactly 4 games? Write the sample space consisting of 16 equally likely simple events similar to the sample space resulting from tossing a coin four times.

   b. What is the probability that the series ends in exactly 5 games?

   c. Assume the probability that the series ends in exactly 6 games is $\frac{5}{16}$. Use this information together with your answers to the first two parts of this problem to compute the probability that the series ends in exactly 7 games.

13. Drug usage in the workplace costs employers incredible amounts of money each year. Drug testing potential employees has become so prevalent that drug users are finding it extremely hard to find jobs. Drug tests, however, are not completely reliable. The most common test used to detect drugs is approximately 98% accurate. To decrease the likelihood of making an error, all potential employees are screened through two tests, which are independent, and each has about 98% accuracy.

   a. If a person were drug free, what is the probability he or she would fail both tests?

   b. If a person were a drug user, what is the probability he or she would pass both tests?

# 5.5 Bayes' Theorem

We have completed the discussions about conditional probability and independent events in Section 5.3 and Section 5.4. **Bayes' theorem** (also referred to as **Bayes' rule** or **Bayes' law**) is somewhat of an extension of conditional probability in which we calculate probabilities based on new information. Please note and understand that the additional information is obtained for a subsequent event, and the new information is used to revise the initial probability. Recall the following formulas for conditional probability.

$$P(B \mid A) = \frac{P(A \cap B)}{P(A)} \quad \text{provided that } P(A) > 0$$

$$P(A \mid B) = \frac{P(A \cap B)}{P(B)} \quad \text{provided that } P(B) > 0$$

Bayes' theorem is developed from extending the definition of conditional probability. If we rearrange the formula for the probability of $A$ given $B$ and solve for $A$ intersect $B$, we have the following.

$$P(A \cap B) = P(A \mid B)P(B)$$

Substituting this expression for $A$ intersect $B$ in the formula for the conditional probability of $B$ given $A$, we have the following.

$$P(B \mid A) = \frac{P(A \cap B)}{P(A)} = \frac{P(A \mid B)P(B)}{P(A)}$$

Now recall the multiplication rule for independent events. That is, we know that if $A$ and $B$ are independent, then $P(A \cap B) = P(A)P(B)$. If this rule holds for two events, A and B, then we can generalize the multiplication rule as follows:

$$P(A) = P(A \cap B_1) + P(A \cap B_2) + \cdots + P(A \cap B_k)$$

where $B_1$, $B_2$, …, $B_k$ are $k$ mutually exclusive and collectively exhaustive events.

Using conditional probabilities,

$$P(A \cap B_i) = P(A \mid B_i)P(B_i).$$

Using the substitution, we can generalize the multiplication rule as follows.

$$P(A) = P(A \mid B_1)P(B_1) + P(A \mid B_2)P(B_2) + \cdots + P(A \mid B_k)P(B_k)$$

Substituting this formula for $P(A)$ into the formula for the conditional probability of $B$ given $A$, we can derive Bayes' theorem.

## Theorem

### Bayes' Theorem

$$P(B_i \mid A) = \frac{P(A \mid B_i)P(B_i)}{P(A \mid B_1)P(B_1) + P(A \mid B_2)P(B_2) + \cdots + P(A \mid B_k)P(B_k)}$$

where $B_i$ is the $i^{th}$ event of $k$ mutually exclusive and collectively exhaustive events.

The following example will illustrate how Bayes' theorem can be applied.

**Example 5.5.1**

**Using Bayes' Theorem to Calculate Probabilities**

Of the travelers arriving at a small airport, 60% fly on major airlines, 30% fly on privately owned planes, and the remainder fly on commercially owned planes not belonging to a major airline. Of those traveling on the major airlines, 50% are traveling for business reasons, whereas 60% of those arriving on private planes and 90% of those arriving on the other commercially owned planes are traveling for business reasons. Suppose that we randomly select one person arriving at the airport. What is the probability that the person

a. is traveling on business?

b. is traveling for business on a privately owned plane?

c. arrived on a privately owned plane given that the person is traveling for business reasons?

**SOLUTION**

First of all, we should define the events associated with this problem. Once we have defined the events, we can then note the probabilities that are given to us.

Let

$M$ = Major Airline

$P$ = Private Plane

$C$ = Commercial Airline

$B$ = Travel for Business Reasons.

In the context of Bayes' theorem, $k = 3$ for this example, since major airline, private plane, and commercial airline are the only ways in which travelers can arrive at the airport, and a particular passenger cannot arrive on more than one type of airplane (thus $M, P,$ and $C$ are mutually exclusive and exhaustive events). When we apply the theorem, $P(B_1)$ will be $P(M)$, $P(B_2)$ will be $P(P)$, and $P(B_3)$ will be $P(C)$. We know the following.

$$P(M) = 0.6$$
$$P(P) = 0.3$$
$$P(C) = 0.1$$
$$P(B \mid M) = 0.5$$
$$P(B \mid P) = 0.6$$
$$P(B \mid C) = 0.9$$

a. The first question asks what the probability is that a randomly selected person is traveling on business. We know that a person is traveling on business if they are traveling on business via any of the flight methods. Thus we have

$$P(B) = P(B \cap M) + P(B \cap P) + P(B \cap C)$$
$$= P(B \mid M)P(M) + P(B \mid P)P(P) + P(B \mid C)P(C)$$
$$= (0.5)(0.6) + (0.6)(0.3) + (0.9)(0.1)$$
$$= 0.57.$$

Thus, there is a 57% chance that the randomly selected traveler will be traveling for business.

b. To determine if the traveler is traveling for business on a privately owned plane, we want to find

$$P(B \cap P) = P(B \mid P)P(P) = (0.6)(0.3) = 0.18.$$

So, there is an 18% chance that the traveler will be traveling for business on a privately owned plane.

c. This part of the problem wants us to determine the probability that the person arrived on a privately owned plane, given that he or she is traveling for business reasons. We can write this probability statement as $P(P \mid B)$.

By the definition of conditional probability, we can write

$$P(P \mid B) = \frac{P(P \cap B)}{P(B)} = \frac{P(B \mid P)P(P)}{P(B)} .$$

From the items calculated in parts **a.** and **b.**, we know

$$P(P \mid B) = \frac{P(P \cap B)}{P(B)} = \frac{P(B \mid P)P(P)}{P(B)}$$

$$= \frac{P(B \mid P)P(P)}{P(B \mid M)P(M) + P(B \mid P)P(P) + P(B \mid C)P(C)}$$

$$= \frac{(0.6)(0.3)}{0.57}$$

$$\approx 0.3158.$$

Thus, we know that if the passenger is traveling on business, there is about a 32% chance that he or she will be traveling by private plane.

Even though it was fairly subtle (given that we performed the calculations in parts **a.** and **b.**), please note the use of Bayes' theorem in the previous calculation.

# 📝 5.5 Exercises

## Basic Concepts

1. Briefly explain the relationship between conditional probability and Bayes' theorem.

2. Other than conditional probability, which other rule which you previously studied is used in the derivation of Bayes' theorem?

3. What is Bayes' theorem?

4. How is Bayes' theorem used to "revise" a probability based on additional information?

## Exercises

5. The issue of Corporate Tax Reform has been cause for much debate in the United States, especially in the House Ways and Means Committee as well as the Senate Finance Committee. Among those in the legislature, 45% are Republicans and 55% are Democrats. It is reported that 30% of the Republicans and 70% of the Democrats favor some type of Corporate Tax Reform to prevent American companies from operating in foreign countries. Suppose a member of Congress is randomly selected and they are found to favor some type of corporate tax reform. What is the probability that this person is a Democrat?

6. Adults (18 years and older) and kids (under 13 years of age) are observed to react differently to sad, emotional movies. It has been observed that 70% of the kids say they cry at some point during those types of movies, whereas only 40% of the adults admit to crying during those types of movies. A group of 40 people, of whom 25 are kids, was shown a sad, emotional movie and the subjects were asked if they cried. A response picked at random from the 40 indicated that they cried. What is the probability that it was an adult?

7. As items come to the end of a production line, an inspector chooses which items are to go through a complete inspection. Eight percent of all items produced are defective. Sixty percent of all defective items go through a complete inspection, and 20% of all good items go through a complete inspection. Given that an item is completely inspected, what is the probability that it is defective?

## T    Discovering Technology

### Using the TI-84 Plus Calculator

Factorials

Calculate 10! using a TI-84 Plus calculator.

1.    To calculate the value of 10! using a TI-84 Plus calculator, first enter [1] [0] in the calculator.

2.    Press **MATH**, scroll to **PRB**, and select option **4:!**.

3.    Press **ENTER** twice. The value of 10! (3628800) is displayed.

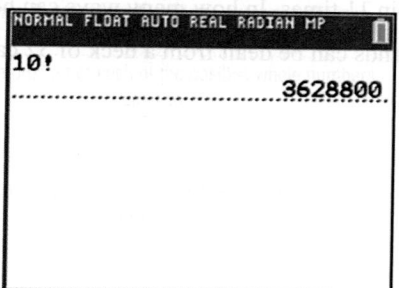

Combinations

Use the information from Example 5.6.4 for this exercise.

1.    In order to calculate the number of winning combinations for the Mega Millions lottery, we have to determine the number of ways of selecting 5 numbers from 70. This is a combination of 70 objects, taken 5 at a time, or $_{70}C_5$. To find the number of combinations using the calculator, first enter [7] [0], press **MATH**, scroll to **PRB**, and choose option **nCr**.

2.    Press **ENTER**. Enter the number [5] and press **ENTER** again.

3.    The number of combinations is given as 12103014. We then need to multiply this by 25, which is the number of ways you can select the Mega Ball. Multiply the answer of 12103014 by 46 to get 302575350, which is the total number of winning combinations. Compare these results to those in Example 5.6.4.

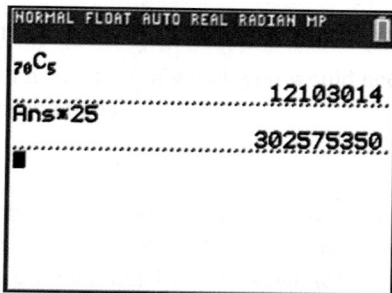

## Permutations

Use the information from Example 5.6.6 for this exercise.

1. Seven bids were placed for a commercial construction job, and we want to know in how many ways the bids can be selected in first second and third place. Since order is important, this is equivalent to a permutation of 7 objects, taken 3 at a time, or $_7P_3$. To calculate the number of permutations using the calculator, first enter ⬚7⬚, press ▉MATH▉, scroll to **PRB**, and choose option **2:nPr**.

2. Press ▉ENTER▉. Enter the number ⬚3⬚ and press ▉ENTER▉ again. The number of permutations is then displayed. Compare this to our solution to Example 5.6.6.

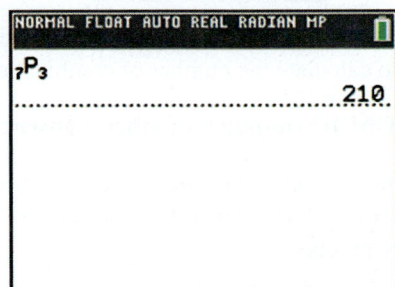

## Using Excel

### Probability and Independence

A coin is flipped, a die is rolled, and a card is drawn from a deck. Find the probability of getting heads on the coin, a 2 on the die, and drawing a spade from the deck of cards.

1. Fill in the data in a worksheet as shown here.

| | A | B | C | D |
|---|---|---|---|---|
| | | Coin | Die | Card |
| 1 | | | | |
| 2 | Chance of Success | 1 | 1 | 13 |
| 3 | Possible Outcomes | 2 | 6 | 52 |

2. In cell B4 divide B2 by B3 by entering the following formula.

$$=B2/B3$$

3. Copy the formula in cell B4 to C4 and D4.

4. Now we have the probability of success for each event listed in Row 4. In cell E4, multiply cells B4 through D4 using the following formula.

$$=B4*C4*D4$$

5. This gives us the result that the probability of getting heads on the coin, a 2 on the die and drawing a spade from the deck of cards is 0.020833 or about 2.1%.

| | A | B | C | D | E |
|---|---|---|---|---|---|
| | | Coin | Die | Card | |
| 1 | | | | | |
| 2 | Chance of Success | 1 | 1 | 13 | |
| 3 | Possible Outcomes | 2 | 6 | 52 | |
| 4 | | 0.5 | 0.166667 | 0.25 | 0.020833 |

## Combinations

For this exercise, use the information from Example 5.6.4.

1. The first step in this example was to determine the number of ways of selecting 5 numbers from 70. This is equivalent to determining the number of combinations of 70 unique objects taken 5 at a time. In an Excel spreadsheet, enter *n* and *k* in cells A1 and B1, respectively, and **70** and **5** in cells A2 and B2, respectively.

| | A | B |
|---|---|---|
| 1 | n | k |
| 2 | 70 | 5 |

2. Label cell C1 *nCk* to represent the number of combinations of *n* objects taken *k* at a time. The formula to calculate the number of combinations in Microsoft Excel is

**COMBIN(number, number_chosen)**

where *number* is the number of items and *number_chosen* is the number of items in each combination. In cell C2, enter the formula to calculate the number of combinations of lottery numbers.

**=COMBIN(A2, B2)**

3. Observe the number of combinations as calculated by Microsoft Excel in cell C2. Compare this to our solution in part **a.** of Example 5.6.4.

| | A | B | C |
|---|---|---|---|
| 1 | n | k | nCk |
| 2 | 70 | 5 | 12103014 |

## Permutations

For this exercise, use the information from Example 5.6.7.

1. Since order is important and we are interested in finding the number of ways the bids can be selected in first, second, and third place, we need to calculate the number of permutations of 7 unique objects taken 3 at a time. In an Excel spreadsheet, enter *n* and *k* in cells A1 and B1, respectively, and **7** and **3** in cells A2 and B2, respectively.

| | A | B |
|---|---|---|
| 1 | n | k |
| 2 | 7 | 3 |

2. Label cell C1 *nPk* to represent the number of permutations of *n* objects taken *k* at a time. The formula to calculate the number of permutations in Microsoft Excel is

**PERMUT(number, number_chosen)**

where *number* is the number of objects and *number_chosen* is the number of objects in each permutation. In cell C2, enter the formula to calculate the number of permutations of bid orders.

**=PERMUT(A2, B2)**

3.  Observe the number of permutations as calculated by Microsoft Excel. Compare this to our solution in Example 5.6.7.

| | A | B | C |
|---|---|---|---|
| 1 | n | k | nPk |
| 2 | 7 | 3 | 210 |

## Using JMP

### Combinations

To install the JMP add-in for calculating permutations and combinations go to the following URL and download the Combinations and Permutations Calculator: http://t.ly/pZAJ.

For this exercise, use the information from Example 5.6.4.

1.  The total number of ways to select 5 different numbers from 75 numbers is $_{75}C_5$. To calculate $_{75}C_5$ using JMP click on **Add-Ins** in the top row and select the **Combinations and Permutations Calculator** Add-In.

2.  Under **Calculator** input **75** for **N** and **5** for **K**. Under **Calculation** select **Combinations**.

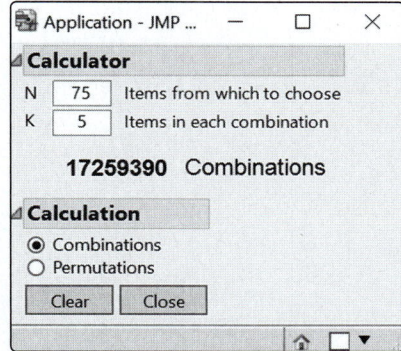

3.  The number 17259390 appears in the center of the window. Click **Close**.

4.  To find the total number of winning combinations we multiply $15 \cdot {_{75}C_5}$ to get 258,890,850.

### Permutations

To install the JMP add-in for calculating permutations and combinations go to the following URL and download the Combinations and Permutations Calculator: http://t.ly/pZAJ.

For this exercise, use the information from Example 5.6.6.

1.  To find the number of different three-digit codes possible on a five-button lock you need to calculate $_5P_3$. To calculate $_5P_3$ using JMP click on **Add-Ins** in the top row and select the **Combinations and Permutations Calculator** Add-In.

2.  Under **Calculator** input **5** for **N** and **3** for **K**. Under **Calculation** select **Permutations**.

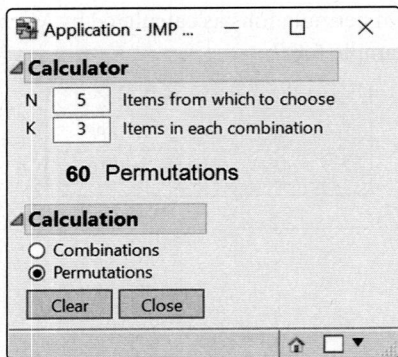

3. The number 60 appears in the center of the window. Click **Close**.

4. There are 60 possible three-digit codes.

# R Chapter 5 Review

## Key Terms and Ideas

- Probability
- Randomness
- Random Experiment
- Sample Space
- Outcome
- Event
- Subjective Probability
- Objective Probability
- Relative Frequency
- Statistical Regularity
- Classical Probability
- Tree Diagram
- Statistical Inference
- Probability Law 1
  (Probability of 0)
- Probability Law 2
  (Probability of 1)
- Probability Law 3
  $(0 \leq P(A) \leq 1)$
- Probability Law 4
  $(P(A_1) + P(A_2) + \cdots + P(A_n) = 1)$
- Compound Event
- Union
- Intersection
- Complement
- Probability Law 5
  $(P(A^c) = 1 - P(A))$

- Odds in Favor of
- Odds Against
- Mutual Exclusivity
- Probability Law 6
  (Union of Mutually Exclusive Events)
- Probability Law 7
  (Intersection of Mutually Exclusive Events)
- Probability Law 8
  (The Addition Rule)
- Conditional Probability
- Probability Law 9
  (Conditional Probability)
- Independent Events
- Dependent Events
- Probability Law 10
  (Multiplication Rule for Independent Events)
- Probability Law 11
  (Multiplication Rule for Dependent Events)
- Bayes' Theorem
- Fundamental Counting Principle
- Factorial
- Combination
- Permutation
- Distinguishable Permutation

## Key Formulas

Section

### Relative Frequency

If an experiment is performed $n$ times and event $A$ happens $k$ times, then

$$\text{Relative Frequency of } A = \frac{k}{n}.$$

5.1

## Key Formulas (cont.)

|  | Section |
|---|---|
| **Classical Probability** | 5.1 |

$$P(A) = \frac{\text{number of outcomes in } A}{\text{total number of outcomes in the sample space}}$$

| **Probability Law 3** | 5.2 |
|---|---|

$$0 \le P(A) \le 1$$

| **Probability Law 4** | 5.2 |
|---|---|

$$P(A_1) + P(A_2) + \cdots + P(A_n) = 1$$

| **Probability Law 5** | 5.2 |
|---|---|

$$P(A^c) = 1 - P(A)$$

| **Odds in Favor of** | 5.2 |
|---|---|

$$\frac{P(A)}{P(\text{not A})} = \frac{P(A)}{P(A^c)}$$

| **Odds Against** | 5.2 |
|---|---|

$$\frac{P(\text{not A})}{P(A)} = \frac{P(A^c)}{P(A)}$$

| **Union of Mutually Exclusive Events** | 5.2 |
|---|---|

$$P(A \cup B) = P(A) + P(B)$$

| **Intersection of Mutually Exclusive Events** | 5.2 |
|---|---|

$$P(A \cap B) = 0$$

| **The Addition Rule** | 5.2 |
|---|---|

$$P(A \cup B) = P(A) + P(B) - P(A \cap B)$$

| **Conditional Probability** | 5.3 |
|---|---|

$$P(A \mid B) = \frac{P(A \cap B)}{P(B)}$$

## Key Formulas (cont.)

| | Section |
|---|---|

**Independent Events** — 5.4

Two events are independent if and only if

$$P(A \mid B) = P(A) \text{ and } P(B \mid A) = P(B).$$

---

**Multiplication Rule for Independent Events** — 5.4

$$P(A \cap B) = P(A)P(B)$$

---

**Multiplication Rule for Dependent Events** — 5.4

$$P(A \cap B) = P(A) \cdot P(B \mid A) = P(B) \cdot P(A \mid B)$$

---

**Bayes' Theorem** — 5.5

$$P(B_i \mid A) = \frac{P(A \mid B_i)P(B_i)}{P(A \mid B_1)P(B_1) + P(A \mid B_2)P(B_2) + \cdots + P(A \mid B_k)P(B_k)}$$

where $B_i$ is the $i^{\text{th}}$ event out of $k$ mutually exclusive and collectively exhaustive events.

---

**Fundamental Counting Principle** — 5.6

If $E_1$ is an event with $n_1$ possible outcomes and $E_2$ is an event with $n_2$ possible outcomes, the number of ways the events can occur in sequence is $n_1 \cdot n_2$.

---

**n Factorial** — 5.6

$$n! = n(n-1)(n-2)\cdots(3)(2)(1)$$

---

**Combination** — 5.6

$$_nC_k = \frac{n!}{(n-k)!k!}$$

---

**Permutation** — 5.6

$$_nP_k = \frac{n!}{(n-k)!}$$

---

**Number of Distinguishable Permutations** — 5.6

$$\frac{n!}{(n_1!)(n_2!)(n_3!)\cdots(n_k!)}$$

## AE    Additional Exercises

1. A couple plans to have three children.

    a. List all possible outcomes for the sexes of the three children.

    b. Find the probability that the couple will have three girls.

    c. Find the probability that the couple will have at least one boy.

2. 671 registered voters were surveyed and asked their political affiliation and whether or not they favor a national healthcare policy. The results of the survey are displayed in the table below.

| Survey Results | | | |
|---|---|---|---|
| Position on National Healthcare | Democrat | Independent | Republican |
| Favor | 161 | 40 | 130 |
| Do Not Favor | 110 | 40 | 190 |

If one of the surveyed voters is randomly selected, answer the following questions.

    a. What is the probability that the voter will be a Republican?

    b. What is the probability that the voter will not favor a national healthcare policy?

    c. What is the probability that the voter will be a Democrat or an Independent?

    d. What is the probability that the voter will be a Democrat and favor a national healthcare policy?

    e. Given that the voter is a Republican, what is the probability that the voter will favor a national healthcare policy?

    f. If the voter does not favor a national healthcare policy, what is the probability that the voter is an Independent?

    g. Are the events {voter is a Democrat} and {voter favors national healthcare policy} independent? Explain.

3. A roulette wheel has 38 outcomes labeled 1 through 36 plus 0 and 00. The wheels are supposed to be designed so that each outcome is equally likely. The numbers 0 and 00 are often referred to as house numbers because the only way that a player can win when these outcomes are observed is by directly betting on the numbers. A great deal of the money wagered on a roulette wheel is wagered on odd or even numbers, or columns or rows of numbers. The numbers 0 and 00 are not in any row or column, nor are they odd or even.

    a. What is the probability of observing an even number (0 and 00 are neither odd nor even)?

    b. What is the probability of observing a number between 1 and 12, inclusive?

    c. What is the probability of observing 0 or 00?

    d. What is the probability of observing a 4?

    e. What is the probability of not observing 7, 13, or 21?

4. A survey of customers in a particular retail store showed that 10% were dissatisfied with the customer service. Half of the customers who were dissatisfied dealt with Bill, the senior customer service representative. If Bill responds to 40% of all customer service inquiries in the retail store, find the following probabilities.

   a. The probability that a customer will be unhappy, given that the representative was Bill.

   b. The probability that the service representative was not Bill, given that the customer complained.

5. A package of documents needs to be sent to a given destination, and it is important that it arrive within one day. To maximize the chances of on-time delivery, three copies of the documents are sent via three different delivery services. Service A is known to have a 90% on-time delivery record, Service B has an 88% on-time delivery record, and Service C has a 91% on-time delivery record. Assuming that the delivery services and their records are independent, what is the probability that at least one copy of the documents will arrive at its destination on time?

6. A boxcar contains six complex electronic systems. Two of the six are to be randomly selected for thorough testing and then classified as defective or not defective. If two of the six systems are actually defective:

   a. find the probability that at least one of the two systems tested will be defective.

   b. find the probability that both are defective.

7. *Odds in favor of* and *odds against* are often used to express chances of occurrences. For example, if the odds are 5 to 2 that it will rain tomorrow then we would be wise to carry an umbrella with us. How exactly are odds related to probabilities? If the probability of event $A$ occurring is $p$, then the odds in favor of $A$ occurring are $a$ to $b$ such that $\dfrac{a}{b} = \dfrac{p}{(1-p)}$. The odds against $A$ occurring are $b$ to $a$.

   a. What are the odds of rolling a six when a single die is thrown?

   b. What are the odds against getting a head when a coin is tossed?

   c. What are the odds against getting 3 consecutive heads when a coin is tossed 3 times?

   d. Suppose the odds in favor of your favorite athletic team winning this weekend are 8 to 3. What is the probability that they will win?

8. Consider a well-shuffled deck of cards with 13 hearts, 13 spades, 13 clubs, and 13 diamonds.

   a. Find the probability that the first card dealt is a heart.

   b. Find the probability that the first card dealt is a spade.

   c. Find the probability that the first card dealt is not a spade.

   d. If you know that the first card dealt will not be a spade, find the probability that it will be a heart.

   e. Suppose you saw the bottom card, and it was the queen of hearts. What is the probability that the first card dealt will be a heart?

9. A box contains eighteen large marbles and ten small marbles. Each marble is either green or white. Twelve of the large marbles are green and four of the small marbles are white. If a marble is randomly selected from the box, what is the probability that it is white or large?

10. User passwords for a certain computer network consist of four letters followed by two numbers. How many different passwords are possible?

11. Hydraulic assemblies for landing coming from an aircraft rework facility are each inspected for defects. Historical records indicate that 8% have defects in shafts only, 6% have defects in bushings only, and 2% have defects in both shafts and bushings. One of the hydraulic assemblies is selected randomly. What is the probability that:

   a. the assembly has a bushing defect?

   b. the assembly has a shaft or bushing defect?

   c. the assembly has exactly one of the two types of defects?

   d. the assembly has neither type of defect?

# P  Discovery Project

The cost of higher education at both public and private institutions is increasing and adversely affecting college students. Utilizing an institutional dataset called *Financial Survey of Students 2006* compiled at one of the largest public universities in the southwestern United States, this study examined financial satisfaction among an undergraduate student population. Understanding the financial satisfaction of college students can help improve the efforts of university administrators and educators. The data used come from an online survey that was conducted at a large public university in the southwestern United States. The data were created and administered by the Institutional Research and Informational Management (IRIM) department, in conjunction with the Office of Financial Aid and the Office of the Provost, to examine college students' financial characteristics. All enrolled students (22,851 students) were invited to participate via a mass email, and a link to the online survey was included after a disclaimer and description of the research project. As a participation incentive, the email included a drawing for two $500 scholarships. A total of 1,976 usable responses were received, yielding a response rate of roughly 8.7%. Of those, 1,935 were college students. The sample was further limited to those who responded to the financial satisfaction question *"How satisfied are you with your financial situation right now?"* The possible responses were satisfied, neutral or dissatisfied. After removing neutral responses from the sample, the final sample was 1,498 responses.

| Characteristics | n | % | % Satisfied | % Dissatisfied |
|---|---|---|---|---|
| **Student Loan** | | | | |
| No Student Loan | 575 | 38.4 | 49.2 | 50.8 |
| Student Loan | 923 | 61.6 | 28.9 | 71.1 |
| **Student Credit Card Debt** | | | | |
| No credit card debt | 812 | 54.2 | 42.5 | 57.5 |
| Credit card debt | 686 | 45.8 | 29.6 | 70.4 |
| **Student Credit Card Amount** | | | | |
| 0 | 812 | 54.2 | 42.5 | 57.5 |
| 1-500 | 209 | 14.0 | 36.4 | 63.6 |
| 501-1000 | 121 | 8.1 | 31.4 | 68.6 |
| 1001-2000 | 122 | 8.1 | 30.3 | 69.7 |
| >2001 | 231 | 15.4 | 25.4 | 76.6 |
| Missing | 3 | 0.2 | 0 | 100 |

**Source:** "The relationship of student loan and card debt on financial satisfaction of college students," Solis, O. and Ferguson, R., *College Student Journal*, Volume 51, Number 3, Fall 2017, pp. 329-336(8).

Using the tables above, answer the following questions (give percentages accurate to one decimal place):

1.  What percentage of college students who had student loans were dissatisfied with their financial situation?

2.  What percentage of college students who had credit card debt were dissatisfied with their financial situation?

3.  What percentage of college students who had credit card debt above $1000 were dissatisfied with their financial situation? (First calculate how many such students there are and round it to the nearest whole number, then find the percentage.)

4.  What percentage of college students are satisfied with their financial situation? (First calculate how many such students there are and round it to the nearest whole number, then find the percentage.)

5.  What percentage of college students are dissatisfied with their financial situation? (First calculate how many such students there are and round it to the nearest whole number, then find the percentage.)

6.  If a student is selected at random from this particular college, what is the likelihood that the student was included in the final sample?

7.  Given what you know about the data and its collection, what are some limitations of this study?

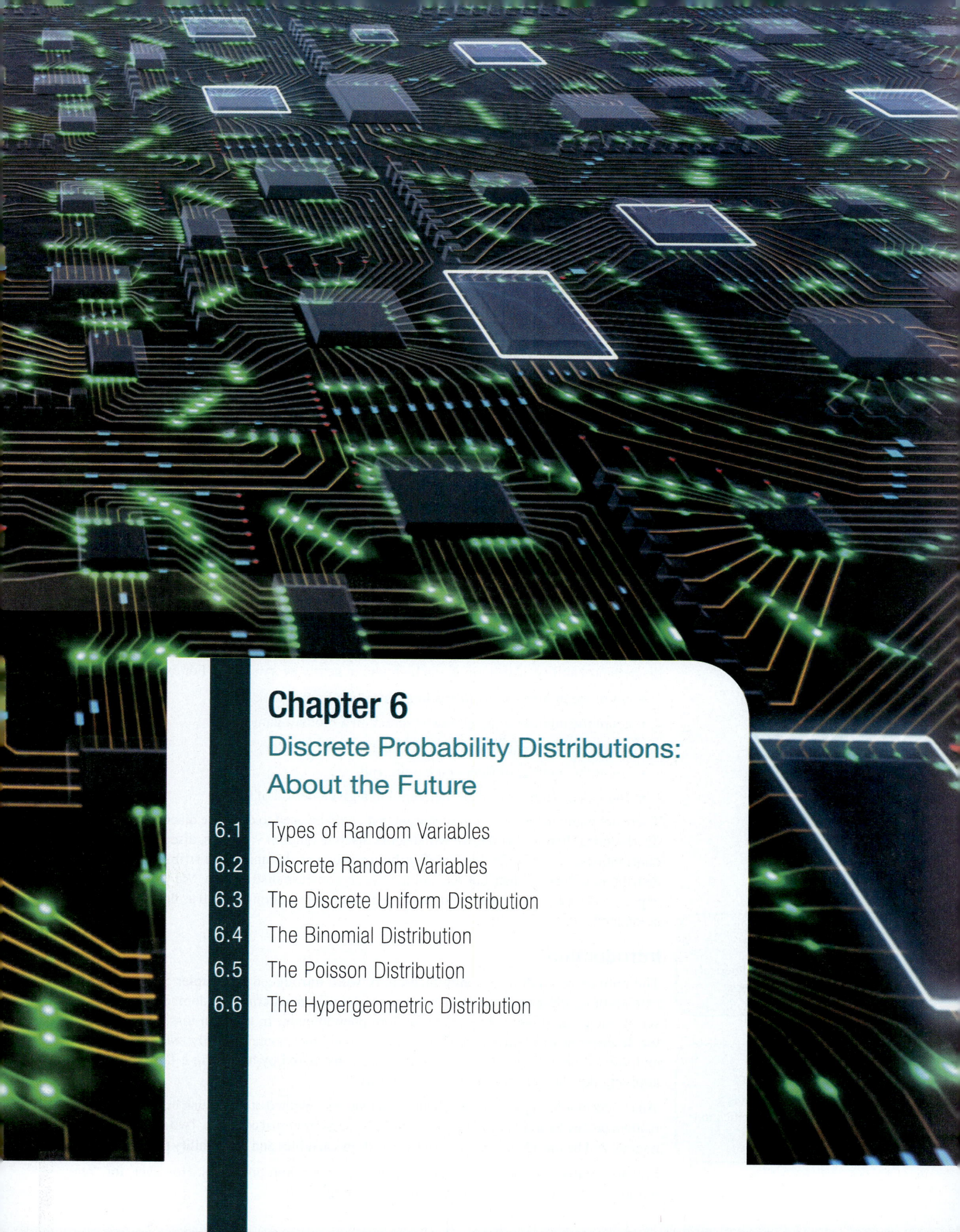

# Chapter 6
## Discrete Probability Distributions: About the Future

# Discovering the Real World

## What is your chance of getting COVID-19 (vaccinated versus unvaccinated)?

If you followed the news about the COVID-19 vaccines, there was a big announcement in November 2020 from the company Moderna. Moderna was one of the pharmaceutical companies that got a head start on developing a vaccine for the Novel Coronavirus (COVID-19). Moderna created the vaccine and then engaged in a large, randomized trial to check for effectiveness and safety of the vaccine.

Moderna's test involved 30,000 participants – half of the participants received the vaccine and half were given a placebo. Of the 15,000 vaccinated participants, it was revealed that five caught COVID-19. Of the 15,000 participants who got the placebo, 95 caught COVID-19.

The good news: The likelihood of contracting COVID-19 after being vaccinated is 5/15000 which is 0.0003. This implies that there is a 3 out of 10,000 chance that a participant will contract COVID-19 if vaccinated. Also, if a person is not vaccinated, there is a 95/15000 chance of contracting COVID-19 which is 0.0063. This implies that if a person is not vaccinated, they have a 63 out of 10000 chance of contracting COVID-19. This is a clear indication that the vaccine is effective.

This was great news from Moderna! Shortly thereafter, Pfizer also released very similar results with their vaccine trials.

The above scenario is an example of using discrete probability in the real world. A discrete random variable is one that is countable or one that has a countable number of outcomes. For example, we were able to count the number of participants that contracted COVID-19 if they were (or were not) vaccinated. A discrete probability distribution is one in which we can assign a probability for each possible outcome of a random variable. Like the example above, discrete probability distributions can be found in nearly every setting. For example, we can

- count the number of accidents in a given timeframe
- count the number of touchdowns scored in a football game
- count the number of hurricanes during hurricane season
- count the number of riders on a train, and
- count the number of pizzas delivered in a given period of time.

There are many more excellent scenarios that could be used to describe discrete probability distributions. In this chapter, we will discuss discrete random variables, discrete probability distributions, and specific discrete distributions such as the uniform distribution, binomial distribution, Poisson distribution, and the hypergeometric distribution. We will discuss the expected value (i.e., the mean) and variance of these distributions as well as find probabilities associated with each of them.

## Introduction

The notions of randomness and uncertainty were introduced in Chapter 5. This chapter extends those ideas by developing concepts to describe a pattern of randomness for an entire set of outcomes produced by some random phenomenon. In the coin tossing experiment the description was rather easy. The totality of outcomes contained only two values, *heads* and *tails*. Probabilities for these events were constructed by assuming a fair coin and by applying the classical definition of probability.

Analyzing more complex random phenomena requires a method of organizing information about random processes and a vocabulary to describe the organizational concepts. Two such descriptive notions will be introduced in this chapter: **random variables** and **probability distributions**.

Probability distributions are the best descriptors of random processes. However, for "real world" random processes they are often difficult to obtain.

**Definition**

**Random Variable**

A **random variable** is a numerical outcome of a random process.

**Definition**

**Probability Distribution**

A **probability distribution** is a model that describes a specific kind of random process.

# 6.1 Types of Random Variables

Quantitative random variables are classified as discrete or continuous. This categorization refers to the types of values that outcomes of the random variable can assume. **Discrete random variables** are analyzed in this chapter, and the next chapter discusses **continuous random variables**.

In defining random variables, there is a naming convention. Capital letters, such as $X$, will be used to refer to the random variable, while small letters, such as $x$ will refer to specific values of the random variable. Often the specific values will be subscripted, $x_1, x_2, \ldots, x_n$.

## Discrete Random Variables

With a discrete random variable, you can count the number of outcomes that the random variable might possess.

In fact, the values that many discrete random variables assume are the counting numbers from 0 to $N$, where $N$ depends upon the nature of the variable.

> **Definition**
>
> Discrete Random Variable
>
> A **discrete random variable** is a random variable which has a countable number of possible outcomes.

> **Procedure**
>
> Discrete Random Variables
>
> When describing a discrete random variable, you should do the following.
>
> 1. State the variable.
> 2. List all of the possible values of the variable.
> 3. Determine the probabilities of these values.

**Random Phenomenon:** Toss a six-sided die and observe the outcome of the toss.

1. *Identify the random variable*: $X$ = outcome of the toss of a six-sided die.

2. *Range of values*: Integers between 1 and 6, inclusive.

In this instance, $x_1 = 1$, $x_2 = 2$, $\ldots$, $x_6 = 6$.

**Example 6.1.1**

**Identifying Random Variables**

| Table 6.1.1 – Tossing a Die | |
|:---:|:---:|
| Value of $X$ | Probability |
| 1 | $\frac{1}{6}$ |
| 2 | $\frac{1}{6}$ |
| 3 | $\frac{1}{6}$ |
| 4 | $\frac{1}{6}$ |
| 5 | $\frac{1}{6}$ |
| 6 | $\frac{1}{6}$ |

3. *Probability distribution*: The outcomes of the toss of a six-sided die and their probabilities are given in Table 6.1.1. The probabilities are deduced using the classical method and the assumption of a fair die.

Not all discrete random variables have probability distributions that are easy to determine. In Examples 6.1.2 and 6.1.3, neither random variable has an easily determined probability distribution.

---

## Example 6.1.2

### Identifying Random Variables

**Random Phenomenon:** Suppose we observe the number of defective integrated circuits received in a batch of 1000.

1. *Identify the random variable*: $X$ = the number of defective integrated circuits in a batch of 1000.

2. *Range of values*: Integers between 0 and 1000, where $N = 1000$. If symbols were chosen to represent the values they could be given as $x_1 = 0$, $x_2 = 1$, ..., $x_{N+1} = 1000$. That is, there could be 0 defective circuits or they could all (1000) be defective.

3. *Probability distribution*: Unknown.

---

## Example 6.1.3

### Identifying Random Variables

**Random Phenomenon:** A stock market analyst is interested in the number of stocks on the New York Stock Exchange (NYSE) that increased in price on the previous day. She realizes that the number that increases is a random variable. She will have to develop a description of the randomness in order to study the stocks.

1. *Identify the random variable*: $X$ = number of stocks that increased in price.

2. *Range of values*: Integers between 0 and the number of stocks trading on the NYSE on the previous day.

3. *Probability distribution*: Unknown, but could be estimated using a relative frequency distribution from Section 5.2 in conjunction with historical data of the NYSE.

---

### Definition

**Continuous Random Variable**

A **continuous random variable** is a random variable whose measurements can assume any one of a countless number of values in an interval.

## Continuous Random Variables

Heights, weights, volumes, and time measurements, for example, are usually measured on a continuous scale. These measurements can take on any value in some interval.

---

## Example 6.1.4

### Identifying Random Variables

✏ **NOTE**

For continuous random variables, we specify probabilities with probability density functions.

**Random Phenomenon:** Sylars Watch Manufacturer has a process that indicates that all watches will be assembled between 20 to 45 minutes. Obviously, the time to assemble a watch will be a function of the style and other accoutrements that are selected by the customer. Since time is measured on a continuous scale and the variability of watch assembly is not predictable due to different workers, the overall time to assemble a watch is considered to be a continuous random variable.

1. *Identify the random variable*: $X$ = time to assemble a watch.

2. *Range of values*: Between 0 and $\infty$. Note that $X$ is measured on a continuous scale.

3. *Probability distribution*: Unknown.

# ✍ 6.1 Exercises

## Basic Concepts

1. What is a random variable?

2. What is a probability distribution?

3. Do all random variables have easily determined probability distributions? Explain.

4. What are the two types of random variables discussed in the chapter? What distinguishes the two types?

## Exercises

5. Classify the following as either a discrete random variable or a continuous random variable.

   **a.** The number of pages in a standard math textbook.

   **b.** The amount of electricity used daily in a home.

   **c.** The number of customers entering a restaurant in one day.

   **d.** The time spent daily on the phone after supper by a teenager.

   **e.** Campers at a state park over Labor Day weekend.

6. Classify the following as either a discrete random variable or a continuous random variable.

   a. The speed of a train.

   b. The possible scores on the SAT reasoning test.

   c. The number of pizzas delivered on a college campus each day.

   d. The daily takeoffs at Chicago's O'Hare Airport.

   e. The high temperatures in Maine and Florida tomorrow.

7. Classify the following as either a discrete random variable or a continuous random variable.

   a. The number of emergency phone calls received per day by a local fire department.

   b. The speed of pitches of major league baseball pitchers.

   c. The weight of a lobster caught in Maine.

   d. The number of defective circuits on a computer chip.

   e. The time it takes for a 5-year battery to die.

8. Classify the following as either a discrete random variable or a continuous random variable.

   a. The total points scored per football game for a local high school team.

   b. The daily price of a stock.

   c. The interest rate charged by local banks for 30-year mortgages.

   d. The number of times a backup of the computer network is performed in a month.

   e. The amount of sugar imported by the U.S. in a day.

# 6.2 Discrete Random Variables

**Definition**

**Discrete Probability Distribution**

A **discrete probability distribution** consists of all possible values of the discrete random variable along with their associated probabilities.

So far, the random variable concept is so general that it is not very useful by itself. What would make it useful is to determine what numerical values the random variable could assume and to assess the probability of each of these values. This information defines a probability distribution for a discrete random variable.

Discrete probability distributions always have three characteristics.

## Properties of Discrete Probability Distributions

1. The sum of all the probabilities must equal 1. That is,
$$P(X = x_1) + P(X = x_2) + \cdots + P(X = x_n) = 1.$$

2. The probability of any value must be between 0 and 1, inclusively. That is,
$$0 \leq P(X = x_i) \leq 1.$$

3. The probabilities are additive. That is,
$$P(X = x_i \text{ or } X = x_j) = P(X = x_i) + P(X = x_j).$$

The association of the possible values with their respective probabilities can be expressed in three different forms: in a table, in a graph, and in an equation.

**Example 6.2.1**

**Determining the Probability Distribution of Three Coin Tosses**

Consider the random phenomenon of tossing a fair coin three times and counting the number of heads. Since the coin is assumed to be fair, $P(H) = P(T) = 0.5$. What is the probability distribution for the number of heads observed in three tosses of a coin?

**SOLUTION**

The random variable is $X$ = the number of heads in three tosses of a coin.

| Table 6.2.1 – Tossing a Coin | | |
|:---:|:---:|:---:|
| **Value of $X$** | **$P(X = x)$** | **Simple Events** |
| 0 | $\frac{1}{8}$ | TTT |
| 1 | $\frac{3}{8}$ | HTT THT TTH |
| 2 | $\frac{3}{8}$ | HHT HTH THH |
| 3 | $\frac{1}{8}$ | HHH |
| **Total** | $\sum P(X = x_i) = 1.0$ | |

The probabilities given in Table 6.2.1 can be deduced using the classical approach to probability.

**Example 6.2.2**

**Determining the Probability Distribution of Daily Sales**

K.J. Johnson is a computer salesperson. During the last year he has kept records of his computer sales for the last 200 days. He recognizes that his daily sales constitute a random process and he wishes to determine the probability distribution for daily sales. Then, from the probability distribution, he would like to know what the probability is that he will sell **a.** at least 2 computers each day and **b.** at most two computers each day.

## SOLUTION

The random variable is $X$ = the number of computers sold each day.

| Table 6.2.2 – Frequency Distribution | |
|---|---|
| Sales | Frequency |
| 0 | 40 |
| 1 | 20 |
| 2 | 60 |
| 3 | 40 |
| 4 | 40 |

The probabilities for this random variable are computed in Table 6.2.3 based upon 200 days of sales data obtained from Mr. Johnson's records using the relative frequency approach.

| Table 6.2.3 – Probability Distribution | |
|---|---|
| Sales $X$ | Probability $P(X = x)$ |
| 0 | $\frac{40}{200} = 0.2$ |
| 1 | $\frac{20}{200} = 0.1$ |
| 2 | $\frac{60}{200} = 0.3$ |
| 3 | $\frac{40}{200} = 0.2$ |
| 4 | $\frac{40}{200} = 0.2$ |
| Total | $\sum P(X = x_i) = 1.0$ |

**Fat the Butch**

John Scarne, in his book *Scarne's Complete Guide to Gambling*, tells a story about a New York City gambler named Fat the Butch. It seems Fat the Butch lost $49,000 in virtually the same game that led Chevalier de Mere to his famous consultation with Pascal and Fermat. In this particular game another well known gambler offered Fat the chance to bet $1000 that he would not roll one double six in 21 tosses of the dice. After 12 hours of dice rolling, Fat lost $49,000 and decided to quit. Later, Scarne discussed the fact that Fat needed 24.6 rolls to break even and that he had a significant expected loss on each play. According to Scarne, Fat shrugged his shoulders and said, "Scarne, in gambling you got to pay and learn, but $49,000 was a lot of dough to pay just to learn that."

**a.** The probability that Mr. Johnson will sell at least 2 computers each day is calculated as follows.

$$P(X \geq 2) = P(X = 2) + P(X = 3) + P(X = 4) = 0.3 + 0.2 + 0.2 = 0.7$$

**b.** To find the probability that Mr. Johnson will sell at most 2 computers each day, the calculation is as follows.

$$P(X \leq 2) = P(X = 0) + P(X = 1) + P(X = 2) = 0.2 + 0.1 + 0.3 = 0.6$$

---

Phone operators of a TV shopping network usually receive about three calls per minute from interested shoppers. Suppose we are interested in the next three telephone calls and the number of callers that place an order. The probability distribution was created based on historical data. Given the probability distribution, what is the probability that at least one caller will purchase something each minute?

**Example 6.2.3**

**Determining the Probability Distribution of Callers Placing an Order**

$X$ = the number of callers that place an order

| Table 6.2.4 – Caller Data | |
|---|---|
| $X$ | $P(X = x)$ |
| 0 | 0.2 |
| 1 | 0.5 |
| 2 | 0.1 |
| 3 | 0.2 |

**SOLUTION**

The probability that at least one caller will purchase something is:

$$P(X \geq 1) = P(X = 1) + P(X = 2) + P(X = 3) = 0.5 + 0.1 + 0.2 = 0.8$$

or

$$P(X \geq 1) = 1 - P(X = 0) = 1 - 0.2 = 0.8.$$

Thus, the probability that at least one caller will purchase an item is 80% .

## Where Do Probability Distributions Come From?

In the previous examples, the probability distributions were given. But in the real world there are very few instances in which the probability distribution is conveniently available. Substantial effort is usually required to obtain the probability distribution. Probabilities are determined using the same techniques described in Chapter 5: classical, relative frequency, or subjective.

Sometimes, however, you get lucky. The random variable you wish to analyze either conforms to or can be approximated by an experiment that has a known probability distribution. Four well known discrete distributions will be discussed in this chapter: uniform, binomial, Poisson, and hypergeometric.

Each of the discrete distributions possesses a **probability distribution function**. These functions assign probabilities to each value of the random variable.

> **Definition**
>
> **Probability Distribution Function**
>
> A **probability distribution function** assigns a probability to each value of a random variable.

### Example 6.2.4

**Evaluating a Probability Distribution Given as a Function**

The following function is a discrete probability distribution function.

$$P(X = x) = \begin{cases} \dfrac{x^2}{30}, & \text{if } x = 1,2,3,4 \\ 0 & \text{otherwise} \end{cases}$$

Summarize the probability distribution for this function.

**SOLUTION**

To determine the probability for a value, use the value as the argument to the function. For example, to determine the probability that $X = 3$, calculate the following.

$$P(X = 3) = \frac{3^2}{30} = \frac{9}{30}$$

The probability that $X = 4$ can be computed in the same way.

$$P(X = 4) = \frac{4^2}{30} = \frac{16}{30}$$

The resulting probability distribution is summarized in Table 6.2.5.

| Table 6.2.5 – Probability Distribution | |
|:---:|:---:|
| **X** | **P(X = x)** |
| 1 | $\dfrac{1}{30}$ |
| 2 | $\dfrac{4}{30}$ |
| 3 | $\dfrac{9}{30}$ |
| 4 | $\dfrac{16}{30}$ |
| **Total** | $\sum P(X = x_i) = \dfrac{30}{30} = 1.0$ |

Note that the distribution possesses the essential properties of all probability distributions; that is, the probabilities sum to one, and all the probabilities are between 0 and 1.

## Expected Value and Variance

The notion of **expected value** is one of the most important concepts in the analysis of random phenomena. Expected value is important because it is a summary statistic for a probability distribution that can be used as a criterion for comparing alternative decisions in the presence of uncertainty. Conceptually, expected value is closely allied with the notion of mean or average.

The expected value of a random variable should be very close to the average value of a large number of observations from the random process, and the larger the number of observations collected, the more likely the average of the observations will be close to the expected value. It should not be interpreted as the value of the random variable we *expect* to see. In fact, for discrete random variables the expected value is rarely one of the possible outcomes of the random variable.

### Formula

**Expected Value**

The **expected value** of a discrete random variable $X$ is the mean of the random variable $X$. It is denoted $E(X)$ and is given by computing the expression

$$\mu = E(X) = \sum[x_i p(x_i)]$$

where $p(x_i) = P(X = x_i)$.

### ⌘ Technology

Notice that an expected value is just a weighted mean where the weights sum to **1**. For instructions on calculating a weighted mean using technology go to stat.hawkeslearning.com and navigate to **Discovering Business Statistics, Second Edition > Technology Instructions > Descriptive Statistics > Two Variable**.

Essentially the expected value is a weighted average, in which each possible value of the random variable is weighted by its probability.

| Table 6.2.6 – Calculating K.J. Johnson's Expected Number of Sales per Day | | |
|---|---|---|
| x | p(x) | xp(x) |
| 0 | 0.2 | 0 |
| 1 | 0.1 | 0.1 |
| 2 | 0.3 | 0.6 |
| 3 | 0.2 | 0.6 |
| 4 | 0.2 | 0.8 |
| **Total** | | **E(X) = 2.1** |

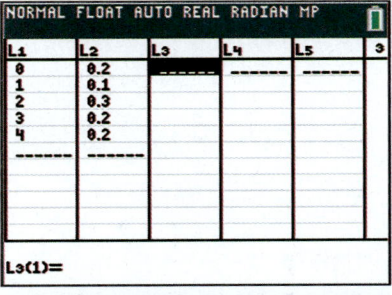

$$E(X) = \sum[x_i p(x_i)] = 0(0.2) + 1(0.1) + 2(0.3) + 3(0.2) + 4(0.2) = 2.1$$

The expected value of the probability distribution given in Example 6.2.2 is computed in Table 6.2.6. In the long run, K.J. Johnson will average 2.1 sales per day, which is the expected value of the probability distribution of K.J. Johnson's daily sales. That is, the expected value of a distribution can be considered the long run average of that distribution.

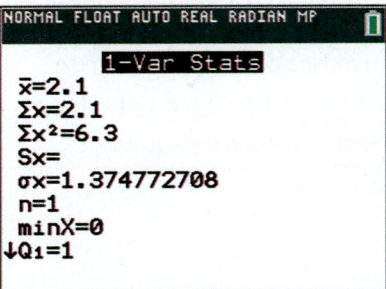

## Using Expected Values to Compare Alternatives

Expected value analysis can also be used to make comparisons when choosing between uncertain alternatives.

Suppose you are confronted with two investment alternatives that possess uncertain outcomes described by the probability distributions given in Table 6.2.7.

**Example 6.2.5**

**Comparing Alternatives Using Expected Values**

**SOLUTION**

| Table 6.2.7 – Investment Alternatives | | | |
|---|---|---|---|
| **Option A** | | **Option B** | |
| **Profit (Dollars)** | **Probability** | **Profit (Dollars)** | **Probability** |
| −2000 | 0.2 | −3000 | 0.2 |
| 0 | 0.1 | −1000 | 0.1 |
| 1000 | 0.3 | 2000 | 0.2 |
| 2000 | 0.3 | 3000 | 0.3 |
| 4000 | 0.1 | 4000 | 0.2 |

$$E(X_A) = \Sigma\left[x_i \cdot p(x_i)\right]$$
$$= (-2000)0.2 + (0)0.1 + (1000)0.3$$
$$+ (2000)0.3 + (4000)0.1$$
$$= \$900$$

$$E(X_B) = \Sigma\left[x_i \cdot p(x_i)\right]$$
$$= (-3000)0.2 + (-1000)0.1$$
$$+ (2000)0.2 + (3000)0.3 + (4000)0.2$$
$$= \$1400$$

Because of the randomness of the profit variable, it is difficult to evaluate the investments by merely eyeballing the two distributions. However, by calculating the expected values of the two alternatives the information in each distribution is condensed to a single point. This point characterizes the center of the distribution and facilitates comparison. The expected values of Options A and B are $900 and $1400, respectively. Thus, in the long run Option B would be $500 more profitable. The phrase *in the long run* is a significant qualifier. It means that under repeated investments you would receive an average profit of $1400 from Option B. But on any one investment in Option B, you may lose as much as $3000 or make as much as $4000.

## Variance of a Discrete Random Variable

In Example 6.2.5, while Option B has a greater expected value, it may differ in risk from Option A. Option B offers a greater chance of making a substantial gain, but it also has the potential for significant loss. The expected value of a distribution measures only one dimension of the random variable, namely its central value. To gauge the variability of a random variable we need another measure similar to the variance measure previously constructed, but one which accounts for the difference in probabilities of the variable.

**NOTE**

Sometimes it is easier to calculate the variance of a discrete random variable using the computational formula
$$\sigma^2 = V(X) = \Sigma\left[x_i^2 \cdot p(x_i)\right] - \mu^2.$$
Both equations are equivalent.

**Formula**

**Variance**

The **variance** of a discrete random variable $X$ is given by the following formula.

$$\sigma^2 = V(X) = \Sigma[(x_i - \mu)^2 \cdot p(x_i)]$$

Once again, the variance can be considered an average. In this case it is the weighted average of the squared deviations about the mean. This is very similar to the computations for the sample variance ($s^2$) and population variance ($\sigma^2$) given by $s^2 = \dfrac{\Sigma(x_i - \bar{x})^2}{n-1}$ and $\sigma^2 = \dfrac{\Sigma(x_i - \mu)^2}{N}$, respectively. The larger the variance, the more variability in the outcomes. To manually compute the variance of the random variable, it's often a good idea to construct a table.

The calculation of the variances of the random variables described in Example 6.2.5 are given in Tables 6.2.8 and 6.2.9.

**Example 6.2.6**

**Calculating the Variance of a Random Variable**

**SOLUTION**

| Table 6.2.8 – Variance of Option A | | |
|---|---|---|
| **Option A** | | |
| **Profit (Dollars)** | **Probability** | $(x - \mu)^2 \cdot p(x)$ |
| −2000 | 0.2 | $(-2000 - 900)^2 \cdot 0.2 = 1{,}682{,}000$ |
| 0 | 0.1 | $(0 - 900)^2 \cdot 0.1 = 81{,}000$ |
| 1000 | 0.3 | $(1000 - 9000)^2 \cdot 0.3 = 3000$ |
| 2000 | 0.3 | $(2000 - 900)^2 \cdot 0.3 = 363{,}000$ |
| 4000 | 0.1 | $(4000 - 900)^2 \cdot 0.1 = 961{,}000$ |
| **Total** | | $\sigma^2 = V(X) = 3{,}090{,}000$ |

The standard deviation is computed by taking the square root of the variance. In this instance, the standard deviation of Option A is given by

$$\sigma = \sqrt{V(X)} = \sqrt{3{,}090{,}000} \approx \$1757.84.$$

| Table 6.2.9 – Variance of Option B | | |
|---|---|---|
| **Option B** | | |
| **Profit (Dollars)** | **Probability** | $(x - \mu)^2 \cdot p(x)$ |
| −3000 | 0.2 | $(-3000 - 1400)^2 \cdot 0.2 = 3{,}872{,}000$ |
| −1000 | 0.1 | $(-1000 - 1400)^2 \cdot 0.1 = 576{,}000$ |
| 2000 | 0.2 | $(2000 - 1400)^2 \cdot 0.2 = 72{,}000$ |
| 3000 | 0.3 | $(3000 - 1400)^2 \cdot 0.3 = 768{,}000$ |
| 4000 | 0.2 | $(4000 - 1400)^2 \cdot 0.1 = 1{,}352{,}000$ |
| **Total** | | $\sigma^2 = V(X) = 6{,}640{,}000$ |

The standard deviation of Option B is given by

$$\sigma = \sqrt{6{,}640{,}000} \approx \$2576.82.$$

A larger standard deviation reflects greater variability in profits and increased risk. When risk is considered, the decision becomes more difficult. The option with the largest expected value (Option B) is also the option with the greatest risk. Hence, the decision maker must subjectively evaluate the trade-off between greater expected return and increased risk.

∽ **Technology**

Calculation of the variance of a discrete random variable can be done using technology. For instructions go to stat. hawkeslearning.com and navigate to **Discovering Business Statistics, Second Edition >Technology Instructions > Descriptive Statistics > Two Variable.**

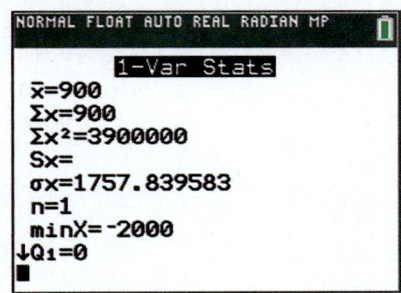

```
NORMAL FLOAT AUTO REAL RADIAN MP
            1-Var Stats
 x̄=900
 Σx=900
 Σx²=3900000
 Sx=
 σx=1757.839583
 n=1
 minX=-2000
↓Q₁=0
■
```

EJN Trucking Company is trying to decide which of two trucks to purchase. Each truck costs $125,000 but comes with different features and amenities. EJN's financial advisor estimates that the returns from each truck for the next five years will follow the probability distribution in the following table.

**Example 6.2.7**

**Determining Which Truck to Purchase**

| Table 6.2.10 – Return on Truck Purchase | | |
|---|---|---|
| **Return (Dollars)** | **Truck 1** | **Truck 2** |
| −5000 | 0.02 | 0.15 |
| 0 | 0.03 | 0.10 |
| 5000 | 0.20 | 0.10 |
| 10,000 | 0.50 | 0.10 |

### Table 6.2.10 – Return on Truck Purchase (cont.)

| Return (Dollars) | Truck 1 | Truck 2 |
|---|---|---|
| 15,000 | 0.20 | 0.30 |
| 20,000 | 0.03 | 0.20 |
| 25,000 | 0.02 | 0.05 |

Calculate the expected return and standard deviation for each truck and recommend which truck to purchase.

### Table 6.2.11 – Calculations

| Return (Dollars) | Truck 1 | Truck 2 | Truck 1 $xp(x)$ | Truck 2 $xp(x)$ | Truck 1 $(x-\mu)^2 \cdot p(x)$ | Truck 2 $(x-\mu)^2 \cdot p(x)$ |
|---|---|---|---|---|---|---|
| −5000 | 0.02 | 0.15 | −100 | −750 | 4,500,000 | 36,037,500 |
| 0 | 0.03 | 0.10 | 0 | 0 | 3,000,000 | 11,025,000 |
| 5000 | 0.20 | 0.10 | 1000 | 500 | 5,000,000 | 3,025,000 |
| 10,000 | 0.50 | 0.10 | 5000 | 1000 | 0 | 25,000 |
| 15,000 | 0.20 | 0.30 | 3000 | 4500 | 5,000,000 | 6,075,000 |
| 20,000 | 0.03 | 0.20 | 600 | 4000 | 3,000,000 | 18,050,000 |
| 25,000 | 0.02 | 0.05 | 500 | 1250 | 4,500,000 | 10,512,500 |
| | | $E(X) =$ | $10,000 | $10,500 | | |
| | | | $V(X) = \sigma^2 =$ | | 25,000,000 | 84,750,000 |
| | | | $\sqrt{V(X)} = \sigma =$ | | $5000 | $9206 |

The expected value and standard deviation of purchasing Truck 1 are $10,000 and $5000, respectively. Similarly, the expected value and standard deviation of purchasing Truck 2 are $10,500 and $9206. Given that the risk (standard deviation) of Truck 2 is nearly twice that of Truck 1, it appears that selecting Truck 1 is the best decision, even though the expected return is slightly less.

## ✍ 6.2  Exercises

### Basic Concepts

1.  Discrete probability distributions always have three characteristics. What are they?

2.  What is the value of describing a random variable with a probability distribution?

3.  What are three different ways to express possible values of a random variable along with their associated probabilities?

4.  How is a probability distribution created?

5.  Identify four discrete probability distributions.

6.  What is a probability distribution function?

7.  Why is the notion of expected value important in the analysis of random phenomena?

8.  True or false: the expected value of a random variable is usually one of the possible outcomes of the random variable.

9.  Suppose the expected value of a random variable was known to be 6.3. Interpret the meaning of the expected value.

10. Give an example of a situation in which expected value would be useful to compare alternatives.

11. How is the variance (or standard deviation) of a random variable related to risk?

## Exercises

12. Determine whether or not the following distribution is a probability distribution. If the distribution is not a probability distribution, give the characteristic which is not satisfied by the distribution.

| x | P(X = x) |
|---|---|
| 1 | $\frac{1}{3}$ |
| 2 | $\frac{2}{3}$ |
| 3 | $\frac{1}{3}$ |

13. Determine whether or not the following distribution is a probability distribution. If the distribution is not a probability distribution, give the characteristic which is not satisfied by the distribution.

| x | P(X = x) |
|---|---|
| −2 | 0.25 |
| 2 | 0.50 |
| 3 | 0.25 |

14. Determine whether or not the following distribution is a probability distribution. If the distribution is not a probability distribution, give the characteristic which is not satisfied by the distribution.

| x | P(X = x) |
|---|---|
| 2 | 0.30 |
| 3 | −0.50 |
| 4 | 0.50 |
| 5 | 0.70 |

15. Determine whether or not the following distribution is a probability distribution. If the distribution is not a probability distribution, give the characteristic which is not satisfied by the distribution.

| x | P(X = x) |
|---|---|
| 5 | 0.46 |
| 10 | 0.25 |
| 15 | 0.25 |

16. Determine whether or not the following distribution is a probability distribution. If the distribution is not a probability distribution, give the characteristic which is not satisfied by the distribution.

| x | P(X = x) |
|---|---|
| −10 | 0.18 |
| −5 | 0.39 |
| 3 | 0.08 |
| 8 | 0.35 |

17. Determine whether or not the following distribution is a probability distribution. If the distribution is not a probability distribution, give the characteristic which is not satisfied by the distribution.

| $x$ | $P(X = x)$ |
|-----|-----------|
| 100 | −0.10 |
| 200 | 0.50 |
| 300 | 0.50 |

18. Determine whether or not the following distribution is a probability distribution. If the distribution is not a probability distribution, give the characteristic which is not satisfied by the distribution.

$$P(X = x) = \frac{x}{16}, \text{ for } x = 1, 2, 3, 4, 5$$

19. Determine whether or not the following distribution is a probability distribution. If the distribution is not a probability distribution, give the characteristic which is not satisfied by the distribution.

$$P(X = x) = \frac{x^2}{30}, \text{ for } x = 1, 2, 3, 4$$

20. Find the expected value, the variance, and the standard deviation for a random variable with the following probability distribution.

| $x$ | -5 | -2 | 0 | 2 | 5 |
|------|------|------|------|------|------|
| $p(x)$ | 0.06 | 0.15 | 0.58 | 0.18 | 0.03 |

21. Find the expected value, the variance, and the standard deviation for a random variable with the following probability distribution.

| $x$ | 400 | 420 | 440 | 460 | 480 | 500 |
|------|------|------|------|------|------|------|
| $p(x)$ | 0 | 0.1 | 0.1 | 0.2 | 0.2 | 0.4 |

22. A regional hospital is considering the purchase of a helicopter to transport critical patients. The relative frequency of $X$, the number of times the helicopter is used to transport critical patients each month, is derived for a similarly sized hospital and is given in the following probability distribution.

| Number of Helicopter Transports | | | | | | | |
|------|------|------|------|------|------|------|------|
| $x$ | 0 | 1 | 2 | 3 | 4 | 5 | 6 |
| $p(x)$ | 0.15 | 0.20 | 0.34 | 0.19 | 0.06 | 0.05 | 0.01 |

a. Find the average number of times the helicopter is used to transport critical patients each month.

b. Find the variance of the number of times the helicopter is used to transport critical patients.

c. Find the standard deviation of the number of times the helicopter is used to transport critical patients.

d. Find the probability that the helicopter will not be used at all during a month to transport critical patients.

e. Find the probability that the helicopter will be used at least once to transport critical patients.

f. Find the probability that the helicopter will be used at most twice to transport critical patients.

g. Find the probability that the helicopter will be used more than three times to transport critical patients.

**23.** Based on past experience, an architect has determined a probability distribution for $X$, the number of times a drawing must be examined by a client before it is accepted.

| Number of Times Examined | | | | | |
|---|---|---|---|---|---|
| **x** | 1 | 2 | 3 | 4 | 5 |
| **p(x)** | 0.1 | 0.2 | 0.3 | 0.2 | 0.2 |

**a.** Find the average number of times a drawing must be examined by a client before it is accepted.

**b.** Find the variance of the number of times a drawing must be examined by a client before it is accepted.

**c.** Find the standard deviation of the number of times a drawing must be examined by a client before it is accepted.

**d.** What is the probability that a drawing must be examined five times before being accepted by the client?

**e.** Find the probability that the drawing must be examined at least twice before being accepted by the client.

**f.** Find the probability that a drawing must be examined at most three times before being accepted by the client.

**g.** Find the probability that a drawing must be examined less than twice before being accepted by the client.

**24.** The manager of a retail clothing store has determined the following probability distribution for $X$, the number of customers who will enter the store on Saturday.

| Customers on Saturday | | | | | |
|---|---|---|---|---|---|
| **x** | 10 | 20 | 30 | 40 | 50 | 60 |
| **p(x)** | 0.10 | 0.20 | 0.30 | 0.20 | 0.10 | 0.10 |

**a.** Find the expected number of customers who will enter the store on Saturday.

**b.** Find the standard deviation of the number of customers who will enter the store on Saturday.

**c.** Find the variance of the number of customers who will enter the store on Saturday.

**d.** Find the probability that more than 30 customers will enter the store on Saturday.

**e.** Find the probability that at most 20 customers will enter the store on Saturday.

**f.** Find the probability that at least 40 customers will enter the store on Saturday.

**g.** What is the probability that exactly 10 customers will enter the store on Saturday?

**25.** An entrepreneur is considering investing in a new venture. If the venture is successful, he will make $50,000. However, if the venture is not successful, he will lose his investment of $10,000. Based on past experience, he believes that there is a 40% chance that the venture will be successful.

**a.** Use the information in the problem to determine the probability distribution of the amount of money to be made (or lost) on the venture.

**b.** Determine the expected amount of money to be made on the venture.

**c.** Determine the standard deviation of the amount of money to be made on the venture.

**26.** An investor is considering two alternative investment options with the following payoff distributions.

| | Option 1 | | | Option 2 | | |
|---|---|---|---|---|---|---|
| Payoff | −$100,000 | $30,000 | $100,000 | −$20,000 | $0 | $20,000 |
| P(Payoff) | $\frac{1}{3}$ | $\frac{1}{3}$ | $\frac{1}{3}$ | 0.25 | 0.50 | 0.25 |

**a.** Calculate the expected payoff for each of the investment options.

**b.** Calculate the standard deviation of the payoff for each of the investment options.

**c.** Which investment option would you choose? Explain.

**27.** A cereal manufacturer has two new brands of cereal which it would like to produce. Because resources are limited, the cereal manufacturer can only afford to produce one of the new brands. A marketing study produced the following probability distributions for the amount of sales for each of the new brands of cereal.

| Cereal A | | Cereal B | |
|---|---|---|---|
| Sales | P(Sales) | Sales | P(Sales) |
| $150,000 | 0.2 | $10,000 | 0.40 |
| $200,000 | 0.3 | $300,000 | 0.40 |
| $300,000 | 0.3 | $600,000 | 0.10 |
| $400,000 | 0.2 | $1,000,000 | 0.10 |

**a.** What are the expected sales of each of the new brands of cereal?

**b.** What is the standard deviation of the sales for each of the brands of cereal?

**c.** If both of the brands of cereal cost the same amount to produce, which brand of cereal do you think the cereal manufacturer should produce? Explain.

# 6.3 The Discrete Uniform Distribution

The **discrete uniform distribution** is one of the simplest probability distributions. Each value of the random variable is assigned an identical probability. There are many situations in which the discrete uniform distribution arises. Some common examples are rolling a fair die or flipping a fair coin.

**Formula**

**Discrete Uniform Probability Distribution Function**

Mathematically, the **discrete uniform probability distribution function** is given by

$$P(X = x) = \frac{1}{n}$$

where $n$ = the number of values that the random variable may assume.

**Example 6.3.1**

**Determining the Probability Distribution of Throwing a Die**

What is the probability distribution for the outcome of the throw of a single six-sided die?

**SOLUTION**

If the die is fair, then each of the outcomes is equally likely, and thus we have a discrete uniform distribution in which all probabilities equal $\frac{1}{6}$. The probability distribution is given in Table 6.3.1.

| Table 6.3.1 – Throwing a Die | |
|:---:|:---:|
| x | P(X = x) |
| 1 | $\frac{1}{6}$ |
| 2 | $\frac{1}{6}$ |
| 3 | $\frac{1}{6}$ |
| 4 | $\frac{1}{6}$ |
| 5 | $\frac{1}{6}$ |
| 6 | $\frac{1}{6}$ |

**Example 6.3.2**

In many college and professional sports, athletes are randomly selected to undergo drug testing. Cameron is on a college basketball team that has 12 players which will randomly select one player to undergo drug testing. What is the chance that Cameron will be chosen to be tested?

**SOLUTION**

Since the players are chosen at random, the chance of any player being selected is $\frac{1}{12}$.

Thus, Cameron's chance of being selected to undergo drug testing is $\frac{1}{12}$.

**Example 6.3.3**

**Determining the Probability Distribution of Delivery Time**

Suppose a purchasing agent has just received a pricing and delivery schedule from a new vendor. The delivery schedule was quoted as 1 to 4 weeks.

a. Construct the probability distribution for the time until delivery.

b. Calculate the expected value and standard deviation of delivery time.

**SOLUTION**

a. The probabilities of the random variable $X$ = the number of weeks until delivery are given in Table 6.3.2.

| Table 6.3.2 – Delivery Distribution | |
|:---:|:---:|
| x | P(X = x) |
| 1 | $\frac{1}{4}$ |
| 2 | $\frac{1}{4}$ |
| 3 | $\frac{1}{4}$ |
| 4 | $\frac{1}{4}$ |

Without any prior information, the agent believes any time frame is as likely as any other. Hence, the number of weeks until delivery will be assumed to have a discrete uniform distribution. Over time the purchasing agent will undoubtedly revise the distribution as more information is gathered about the company's delivery schedule.

**b.** The expected value is calculated as follows.

$$E(X) = \Sigma\left[x_i P(x_i)\right]$$
$$= (1)\frac{1}{4} + (2)\frac{1}{4} + (3)\frac{1}{4} + (4)\frac{1}{4}$$
$$= 0.25 + 0.50 + 0.75 + 1$$
$$= 2.5 \text{ weeks}$$

The variance is calculated as follows.

$$\sigma^2 = V(X) = \Sigma\left[(x_i - \mu)^2 P(x_i)\right]$$
$$= (1-2.5)^2\frac{1}{4} + (2-2.5)^2\frac{1}{4} + (3-2.5)^2\frac{1}{4} + (4-2.5)^2\frac{1}{4}$$
$$= (2.25)\frac{1}{4} + (0.25)\frac{1}{4} + (0.25)\frac{1}{4} + (2.25)\frac{1}{4}$$
$$= 1.25$$

Therefore, the standard deviation is $\sqrt{1.25} \approx 1.12$ weeks.

Example 6.3.3 illustrates an important principle in the application of the discrete uniform distribution. That is, when there is little or no information concerning the outcome of a random variable, the discrete uniform distribution may be a reasonable initial alternative.

## 6.3 Exercises

### Basic Concepts

1. What is the most significant property of the uniform distribution?
2. What is the discrete uniform probability distribution function?
3. Explain why the uniform distribution is often used when there is little or no information concerning the outcome of a random variable.

### Exercises

4. In the casino game of roulette, a wheel is spun and a ball is set in motion, ultimately coming to rest in one of the 38 slots on the wheel. Any slot is as likely as any other to capture the ball. Of the 38 slots, 18 are red, 18 are black, and 2 are green. Suppose the entry fee to play a single game is $1 and the participant bets on red. If the ball comes to rest in one of the red slots, he wins $1 in addition to getting back the original $1 entry fee. If the ball does not end up in a red slot, the $1 entry fee is lost. Let $X$ denote the monetary gain when betting $1 on red, in a single game of roulette. Gain is defined as the amount won minus the fee to play.

   a. What are the possible values of $X$?
   b. Is $X$ a discrete or continuous random variable? Explain.
   c. Construct the probability distribution of $X$.
   d. Find the expected value of $X$ and interpret this number.
   e. Do you feel that in any casino games you would have a positive expected gain? Why?

5.  An experiment consists of tossing two coins and a die simultaneously.

    a.  List the 24 equally likely outcomes.

    b.  Define the random variable $X$ as the sum of the number of heads on the two coins and the number of dots on the die. What are the possible values of $X$?

    c.  Construct the probability distribution of $X$ in the form of a table.

    d.  Find the expected value of $X$.

6.  A classmate walks into class and states that he has an extra ticket to a rock concert on Friday night. He asks everyone in the class to put their name on a piece of paper and put it in a basket. He plans to draw from the basket to choose the person who will attend the concert with him. If there are 16 people in class that night, what is your chance of being chosen to attend the concert?

7.  Sharlene has just put a down payment on a lot in a small subdivision. There are 10 lots in the subdivision and all are approximately 0.25 acres in size. Five builders have been contracted by the subdivision manager to each build two homes in order to finish the subdivision in 6 months. Sharlene's uncle is one of the builders contracted by the subdivision manager. What is the probability that Sharlene's uncle will be the builder that builds her house?

8.  You order some clothing online and get an estimated delivery date of June 6–June 11. You know you will be out of town June 8th and 9th and are a little concerned about the package arriving when you are away. Assuming the delivery date follows a discrete uniform distribution, what is the likelihood your package will be delivered while you are out of town?

9.  An experiment consists of tossing a coin and rolling a six-sided die simultaneously.

    a.  List the sample space for the experiment.

    b.  What is the probability of getting a head on the coin and the number 3 on the die?

    c.  What is the probability of getting a tail on the coin and at least a 4 on the die?

10. Given the following discrete uniform probability distribution, find the expected value and standard deviation of the random variable.

| $x$ | 0 | 1 | 2 | 3 | 4 |
|---|---|---|---|---|---|
| $P(X = x)$ | $\frac{1}{5}$ | $\frac{1}{5}$ | $\frac{1}{5}$ | $\frac{1}{5}$ | $\frac{1}{5}$ |

# 6.4 The Binomial Distribution

The binomial distribution arises from experiments with repeated two-outcome trials, where only one of the outcomes is counted. Experiments of this kind are rather common in the business world. In market research, a survey respondent (a trial) either will or will not recognize a company's brand. The number that recognize the brand is a count that may be modeled as a **binomial random variable**. When a customer (a trial) enters a bank for service, he or she may have to wait. If we are counting customers who have to wait, then the count may conform to the binomial model. Experiments are required to meet several conditions in order to qualify as a binomial experiment.

## Procedure

### Binomial Experiment

A **binomial experiment** is a random experiment which satisfies all of the following conditions.

1. There are only two outcomes in each trial of the experiment. (One of the outcomes is usually referred to as a *success*, and the other as a *failure*.)
2. The experiment consists of $n$ identical trials as described in Condition 1.
3. The probability of success on any one trial is denoted by $p$ and does not change from trial to trial. (Note that the probability of a failure is $(1 - p)$ and also does not change from trial to trial.)
4. The trials are independent.
5. The binomial random variable is the count of the number of successes in $n$ trials.

A binomial random variable is formed by counting the number of successes in $n$ trials of an experiment with two outcomes. One of the simplest of binomial random variables is produced by tossing a coin.

---

**Example 6.4.1**

**Identifying a Binomial Random Variable**

Toss a coin 4 times and record the number of heads. Is the number of heads in 4 tosses a binomial random variable?

**SOLUTION**

1. There are only two outcomes, heads or tails.
2. The experiment will consist of 4 tosses of a coin. (Hence, $n = 4$.)
3. The probability of getting a head (success) is $\dfrac{1}{2}$ and does not change from trial to trial. (Hence, $p = \dfrac{1}{2}$.)
4. The outcome of one toss will not affect other tosses.
5. The variable of interest is the count of the number of heads in 4 tosses.

All the conditions of a binomial experiment are met, so the number of heads in 4 tosses of a coin is a binomial random variable.

The probability distribution for this experiment is given in the following table.

| Table 6.4.1 – Tossing a Coin | | |
|---|:---:|:---:|
| **Events** | **Number of Heads** | **Probability** |
| TTTT | 0 | $\dfrac{1}{16}$ |
| HTTT<br>THTT<br>TTHT<br>TTTH | 1 | $\dfrac{4}{16}$ |
| HHTT<br>HTHT<br>HTTH<br>THHT<br>THTH<br>TTHH | 2 | $\dfrac{6}{16}$ |
| THHH<br>HTHH<br>HHTH<br>HHHT | 3 | $\dfrac{4}{16}$ |
| HHHH | 4 | $\dfrac{1}{16}$ |

The probability distribution can be derived using the classical method described in the previous chapter. To derive the binomial distribution by listing the simple events is unnecessarily tedious (see Table 6.4.1). Instead of tossing the coin 4 times, suppose the coin is tossed 10 times in the experiment. The number of simple events for an experiment with 10 tosses would be 1024, and an experiment with 20 tosses would require listing a staggering 1,048,576 simple events. Fortunately, there is a far simpler method of obtaining the probability distribution. The binomial probability distribution function provides a relatively simple method of calculating binomial probabilities.

---

### Formula

#### Binomial Probability Distribution Function

The **binomial probability distribution function** is

$$P(X = x) = {}_nC_x \, p^x \left(1-p\right)^{n-x}$$

where ${}_nC_x$ represents the number of possible combinations of $n$ objects taken $x$ at a time (without replacement) and is given by

$${}_nC_x = \frac{n!}{x!(n-x)!} \text{ where } n! = n(n-1)(n-2)\cdots(2)(1) \text{ and } 0! = 1;$$

$n$ = the number of trials,

$p$ = the probability of a success, and

$x$ = the number of successes in $n$ trials.

---

To calculate a binomial probability, the parameters of the distribution ($n$ and $p$) as well as the value of the random variable must be specified. For example, to determine the probability of 3 heads in 4 tosses of a coin, you would substitute the values $X = 3$, $n = 4$, and $p = \frac{1}{2}$ into the binomial probability distribution function as follows.

$$P(X = 3) = {}_4C_3 \left(\frac{1}{2}\right)^3 \left(1-\frac{1}{2}\right)^{4-3}$$

Since ${}_4C_3 = \dfrac{4!}{3!(4-3)!} = \dfrac{(4)(3!)}{(3!)(1!)} = 4$, then

$$P(X = 3) = 4\left(\frac{1}{2}\right)^3 \left(\frac{1}{2}\right) = \frac{4}{16} = \frac{1}{4} = 0.25.$$

The probability that 2 heads would be tossed would be computed in a similar manner.

$$P(X = 2) = {}_4C_2 \left(\frac{1}{2}\right)^2 \left(1-\frac{1}{2}\right)^{4-2}$$

$$= \frac{4!}{(2!)(2!)}\left(\frac{1}{2}\right)^2 \left(\frac{1}{2}\right)^2 = \frac{6}{16} = \frac{3}{8} = 0.375$$

The complete distribution can be computed by substituting the remaining values of the random variable into the probability distribution function.

### ⚭ Technology

To find a binomial probability using technology go to stat.hawkeslearning.com and navigate to **Discovering Business Statistics, Second Edition > Technology Instructions > Binomial Distribution > Binomial Probability (pdf)**.

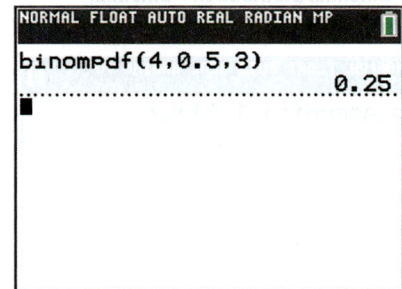

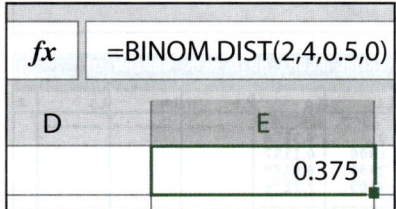

| Table 6.4.2 – Tossing a Coin | |
|---|---|
| **Number of Heads** | **Probability** |
| 0 | $\frac{1}{16}$ |
| 1 | $\frac{4}{16}$ |

| Example 6.4.4 | Compute the expected value and the variance of the number of customers that will approve the change in return policy in Example 6.4.3. |
|---|---|

**Calculating the Expected Value and Variance of a Binomial Random Variable**

**SOLUTION**

Since the random variable is binomial, we can use the shortcuts $E(X) = np$ and $V(X) = np(1 - p)$. Since $n = 10$ and $p = 0.7$, the expected value is given by the following expression.

$$E(X) = np = 10(0.7) = 7$$

The variance is

$$\sigma^2 = V(X) = np(1 - p) = 10(0.7)(0.3) = 2.1,$$

which implies that the standard deviation is $\sqrt{2.1} \approx 1.4491$.

Thus, if 10 randomly selected customers are polled, we would expect 7 of the 10 to approve the change in return policy, and the standard deviation would be 1.4491 customers.

## ✏️ 6.4  Exercises

### Basic Concepts

1. Describe the characteristics of a binomial experiment.

2. What are the parameters of a binomial probability model?

3. Give an example of a binomial experiment in a business context.

4. What is the binomial probability distribution function?

5. Describe the shape of a binomial distribution. Does the shape change? What influences the shape of the distribution?

6. How do you calculate the expected value of a binomial random variable? The variance? The standard deviation?

### Exercises

7. Calculate $_nC_x$ for each of the following combinations of $x$ and $n$.

   a. $n = 5, x = 4$        c. $n = 15, x = 1$

   b. $n = 10, x = 8$       d. $n = 20, x = 0$

8. Calculate $_nC_x$ for each of the following combinations of $x$ and $n$.

   a. $n = 4, x = 2$        c. $n = 18, x = 15$

   b. $n = 12, x = 8$       d. $n = 23, x = 20$

9. The random variable $X$ is a binomial random variable with n = 9 and p = 0.1.

   a. Find the expected value of $X$.

   b. Find the standard deviation of $X$.

   c. Find the probability that $X$ equals 2. (Use the formula for $P(X = x)$.)

   d. Find the probability that $X$ is at most 3.

   e. Find the probability that $X$ is at least 2.

   f. Find the probability that $X$ is less than 5.

**10.** The random variable $X$ is a binomial random variable with $n = 12$ and $p = 0.8$.

    **a.** Find the expected value of $X$.

    **b.** Find the standard deviation of $X$.

    **c.** Find the probability that $X$ equals 7. (Use the formula for $P(X = x)$.)

    **d.** Find the probability that $X$ is at most 4.

    **e.** Find the probability that $X$ is at least 1.

    **f.** Find the probability that $X$ is more than 10.

**11.** A real estate agent has ten properties that she shows. She feels that there is a ten percent chance of selling any one property during a week. The chance of selling any one property is independent of selling another property.

    **a.** What probability model would be appropriate for describing the number of properties sold each week?

    **b.** Compute the expected number of properties to be sold in a week.

    **c.** Compute the standard deviation of the number of properties sold each week.

    **d.** Compute the probability of selling one property in one week.

    **e.** Compute the probability of selling five properties in one week.

    **f.** Compute the probability of selling at least three properties in one week.

**12.** A small commuter airline is concerned about reservation no-shows and, correspondingly, how much they should overbook flights to compensate. Assume their commuter planes will hold 15 people. Industry research indicates that 20% of the people making a reservation will not show up for a flight. Whether or not one person takes the flight is considered to be independent of other persons holding reservations.

    **a.** What probability model would be appropriate for the number of passengers that actually take the flight?

    **b.** If the airlines decide to book 18 people for each flight, how often will there be at least one person who will not get a seat?

    **c.** If they book 17 people, how often will there be at least one person who will not get a seat?

    **d.** If they book 16 people, how often will there be at least one person who will not get a seat?

    **e.** If they book 18 people for each flight, how often will there be one or more empty seats?

    **f.** If they book 17 people, how often will there be one or more empty seats?

    **g.** If they book 16 people, how often will there be one or more empty seats?

    **h.** Based on the results from parts **b.** to **g.** above, which booking policy do you prefer? Explain your answer.

**13.** Seven plants are operated by a garment manufacturer. They feel there is a ten percent chance for a strike at any one plant and the risk of a strike at one plant is independent of the risk of a strike at another plant. Let $X =$ number of plants of the garment manufacturer that strike.

    **a.** Determine the probability distribution for $X$.

    **b.** Interpret the results for $P(X = 0)$, $P(X = 4)$, and $P(X = 7)$.

    **c.** Compute the expected value of $X$.

    **d.** Compute the standard deviation for $X$. Is this value large in relation to the expected value? In what units is the standard deviation expressed?

14. A company that makes traffic signal lights buys switches from a supplier. Out of each shipment of 1000 switches, the company will take a random sample of 10 switches. Let $X$ equal the number of defective switches in the sample.

    a. The company has a policy of rejecting a lot if they find any defective switches in the sample. What is the probability that the shipment will be accepted if, in fact, 2% of the switches are actually defective?

    b. What is the probability that the shipment will be accepted if the percent of defective switches is actually 5%?

    c. The company decides to change their policy and will accept the lot if they find no more than one defective switch. Repeat parts **a.** and **b.** for this new policy.

15. Parents have always wondered about the sex of a child before it is born. Suppose that the probability of having a male child was 0.5, and that the sex of one child is independent of the sex of other children.

    a. Determine the probability of having exactly two girls out of four children.

    b. What is the probability of having four boys out of four children?

16. A certain aspirin is advertised as being preferred by 4 out of 5 doctors. If the advertisement is assumed to be true, answer the following questions.

    a. What is the probability that at least half of ten doctors chosen at random will prefer this brand of aspirin?

    b. What is the probability that 9 out of 10 of the doctors will prefer this brand?

17. In manufacturing integrated circuits, the yield of the manufacturing process is the percentage of good chips produced by the process. The probability that an integrated circuit manufactured by the Ace Electronics Company will be defective is $p = 0.05$. If a random sample of 15 circuits is selected for testing, answer the following questions.

    a. What is the probability that no more than one integrated circuit will be defective in the sample?

    b. What is the expected number of defective integrated circuits in the sample?

18. The Alvin Secretarial Service procures temporary office personnel for major corporations. They have found that 90% of their invoices are paid within 10 working days. If a random sample of 12 invoices is checked, answer the following questions.

    a. What is the probability that all of the invoices will be paid within 10 working days?

    b. What is the probability that six or more of the invoices will be paid within 10 working days?

19. An experiment consists of rolling a pair of dice 10 times. On each roll the sum of the dots on the two dice is noted.

    a. Find the probability that on any roll of the two dice the sum of the dots is either 7 or 11.

    b. Find the probability that in the 10 rolls of the pair of dice, a 7 or 11 occurs 5 times.

    c. Find the probability that in the 10 rolls of the pair of dice, a 7 or 11 does not occur at all.

    d. Find the mean and variance of the number of times we see a 7 or 11 in the 10 rolls of the dice.

**20.** *Would you say you eat to live or live to eat?* was asked to each person in a sample of 1001 adults in a Gallup Poll taken in April 1996. Seventy-four percent of the respondents answered eat to live, 23% answered live to eat, and 3% had no opinion. Assuming these percentages are accurate, find the probability, in 12 randomly chosen adults, that the number who would answer "eat to live" is:

**a.** exactly 7.

**b.** no more than 10.

**c.** at most 11.

**d.** at least 3.

# 6.5 The Poisson Distribution

The binomial random variable requires a fixed number of repetitions of the experiment, where the outcomes are either successes or failures. The **Poisson distribution** is similar to the binomial in that the random variable represents a count of the total number of successes. The major difference between the two distributions is that the Poisson does not have a fixed number of trials. Instead, the Poisson uses a fixed interval of time or space in which the number of successes are recorded. Thus, there is no theoretical upper limit on the number of successes, although large numbers of successes are not very likely. The word *success* in the Poisson context can sometimes take on rather unpleasant connotations. For example, the randomness exhibited by the number of airplane crashes, oil tanker spills, and car accidents in some fixed period of time seem to conform to the randomness described by a Poisson random variable.

In business environments many variables seem to follow a pattern of randomness similar to that described by the Poisson distribution. One of the Poisson's principal areas of use in business is the analysis of waiting lines. Other random phenomena, such as airplane arrivals at an airport, trucks arriving at a loading dock, users logging on to a computer system, or the number of defects in a given surface area, can be modeled with a Poisson distribution. These variables are often of interest in determining personnel requirements, inventories, and quality control.

**Kicked by Horses**

A real world example of the Poisson distribution involves the distribution of Prussian cavalry deaths from getting kicked by horses, in the period 1875–1894. The Prussian military kept meticulous records on horse-kick deaths in each of its army corps, and the data are neatly summarized in a 1963 book called Lady Luck, by the late Warren Weaver. There were a total of 196 kicking deaths—these being the successes. The trials were each army corps's observations on the number of kicking deaths sustained during the year. With 14 army corps and data for 20 years, there were 280 trials. The Poisson formula predicts, for example, that there will be 34.1 instances of having exactly two deaths in a year. In fact, there were 32 such cases. Pretty good, eh?

## Procedure

### Poisson Random Variable

In order to qualify as a **Poisson random variable** an experiment must meet two conditions.

1. Successes occur one at a time. (That is, two or more successes cannot occur at exactly the same point in time or exactly at the same point in space.)

2. The occurrence of a success in any interval is independent of the occurrence of a success in any other interval.

If these two conditions are met, it can be proven that the random variable for the number of successes follows the Poisson probability distribution function.

> **Formula**
>
> ### Poisson Probability Distribution Function
>
> The **Poisson probability distribution function** is given by:
>
> $$P(X = x) = \frac{e^{-\lambda}\lambda^x}{x!}, \text{ for } x = 0, 1, 2, \ldots$$
>
> where $e = 2.71828\ldots$, and
>
> $\lambda$ = the mean number of successes.

## ∽ Technology

To find a Poisson probability distribution using technology go to stat.hawkeslearning.com and navigate to **Discovering Business Statistics, Second Edition > Technology Instructions > Poisson Distribution > Poisson Probability Distribution.**

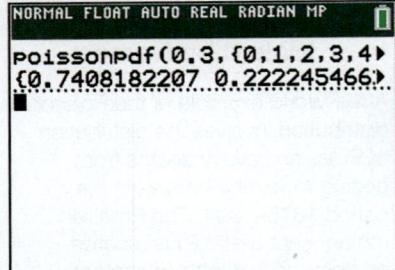

The Poisson distribution has only one parameter, $\lambda$, pronounced *lambda*. One peculiar feature of the distribution is that the variance of the distribution is equal to the mean (lambda). That is, $\mu = \sigma^2 = \lambda$.

Tables for the Poisson distribution are found in Appendix A, Table F. The tables give $P(X \leq x)$ for particular values of $x$ and $\lambda$. Poisson probabilities can also be computed using technology such as a TI-84 Plus calculator, Microsoft Excel, or Minitab. The Discovering Technology section at the end of this chapter shows how to calculate Poisson probabilities using available technology.

The shape of the distribution varies dramatically with the parameter $\lambda$. If $\lambda$ is small, say 0.3, then the corresponding distribution is found in Table 6.5.1 and displayed in Figure 6.5.1.

| Table 6.5.1 $\lambda = 0.3$ | |
|---|---|
| **x** | **P(X = x)** |
| 0 | 0.7408 |
| 1 | 0.2222 |
| 2 | 0.0333 |
| 3 | 0.0033 |
| 4 | 0.0003 |
| 5 | 0.000 |

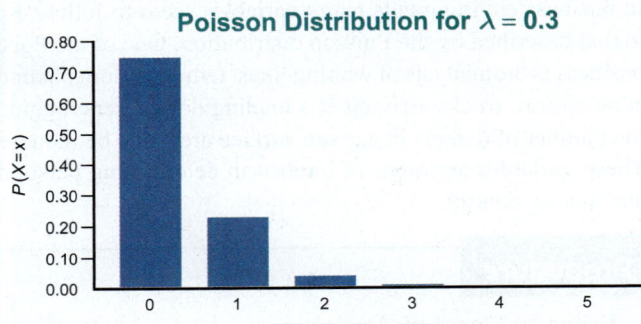

**Poisson Distribution for $\lambda = 0.3$**

**Figure 6.5.1**

As $\lambda$ increases to 3 (see Table 6.5.2), the distribution in Figure 6.5.2 exhibits a mound shape with skewness.

| Table 6.5.2 $\lambda = 3$ | |
|---|---|
| **x** | **P(X = x)** |
| 0 | 0.0498 |
| 1 | 0.1494 |
| 2 | 0.2240 |
| 3 | 0.2240 |
| 4 | 0.1680 |
| 5 | 0.1008 |
| 6 | 0.0504 |
| 7 | 0.0216 |
| 8 | 0.0081 |
| 9 | 0.0027 |
| 10 | 0.0008 |
| 11 | 0.0002 |
| 12 | 0.0001 |

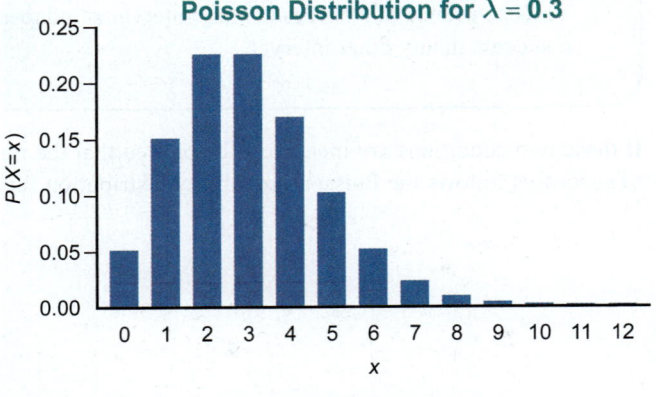

**Poisson Distribution for $\lambda = 0.3$**

**Figure 6.5.2**

As $\lambda$ becomes even larger, say $\lambda = 12$, the distribution begins to closely resemble a bell-shaped distribution as shown in Figure 6.5.3.

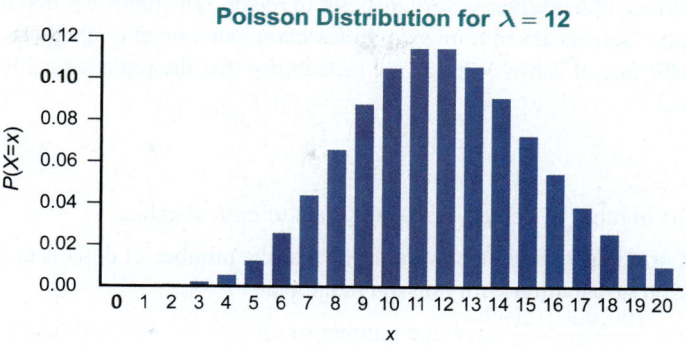

**Poisson Distribution for $\lambda = 12$**

Figure 6.5.3

## Poisson Random Variables for Time

Most Poisson applications relate to the number of occurrences of some event in a specific duration of time.

Suppose a bank has one automatic teller machine. Customers arrive at the machine at a rate of 20 per hour and according to a Poisson pattern.

**a.** What is the probability that no one will arrive in a 15-minute interval?

**b.** What is the probability that in a 15-minute period at least 3 persons will use the automated teller machine?

**SOLUTION**

**a.** Let $X$ = number of arrivals in a 15-minute period.

This problem contains one of the standard techniques used in working with Poisson random variables, that is, translating the arrival rate to correspond to the desired time interval. In this problem, the rate is given at 20 per hour which corresponds to a rate of 5 per $\frac{1}{4}$ hour (15 minutes). Thus, $\lambda = 5$ and the desired probability is

$$P(X = 0) = \frac{e^{-5}5^0}{0!}, \text{ or}$$

$P(X = 0) \approx 0.0067$. (**Note:** 0! is defined to be 1.)

**b.** To find the probability that in a 15-minute period at least 3 persons will use the automated teller machine, we are interested in $P(X \geq 3)$. This probability statement can be rewritten in terms of a cumulative probability so that we can use Appendix A, Table F. That is,

$$P(X \geq 3) = 1 - P(X \leq 2) = 1 - 0.1247 = 0.8753.$$

Note that $P(X \leq 2)$ can be read directly from Appendix A, Table F. Locate $\lambda$ across the top of Table F and then locate the appropriate value of $x$ in the leftmost column (in this case, 2) and then find the value in the table that is the intersection between $\lambda = 5$ and 2 (which is 0.1247).

### Example 6.5.1

**Calculating Probabilities Using the Poisson Distribution**

**Technology**

To find the Poisson probability in part **a.** using technology go to stat.hawkeslearning.com and navigate to **Discovering Business Statistics, Second Edition > Technology Instructions > Poisson Distribution > Poisson Probability (pdf)**.

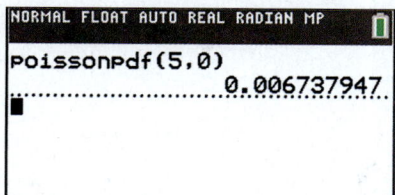

**Technology**

To find the Poisson probability in part **b.** using technology go to stat.hawkeslearning.com and navigate to **Discovering Business Statistics, Second Edition > Technology Instructions > Poisson Distribution > Poisson Probability (cdf)**.

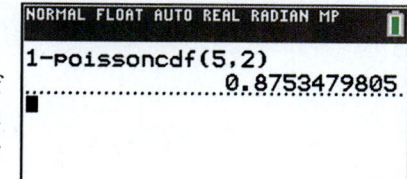

## Poisson Random Variables for Length or Space

Instead of counting the number of successes in a time interval, there are a number of applications of the Poisson distribution that measure the number of successes in some area or length. The average number of successes in the area or length will define the parameter of the Poisson random variable.

| Example 6.5.2 | The telephone company is considering purchasing optical cable from Optica, Inc. The company wishes to replace approximately 100,000 feet of conventional cable with optical fiber. Since optical fiber is very difficult to repair, it is important that the number of optical cable defects are minimized. Optica claims that on average there is one defect per 200,000 feet of cable. What is the probability that the replaced cable will contain no defects? |
|---|---|
| **Calculating a Probability Using the Poisson Distribution** | |

**SOLUTION**

Let $\lambda$ = the number of defects in 100,000 feet of optical cable.

Based on previous experience, we assume that the number of defects is approximated by a Poisson distribution with Poisson parameter

$$\lambda = \frac{100,000}{200,000} = \frac{1}{2} \text{ (average number of defects per 100,000 ft of cable).}$$

Using Table F in Appendix A or technology,

$$P(X = 0) = 0.6065.$$

# 6.5  Exercises

## Basic Concepts

1. How is the Poisson distribution similar to the binomial distribution?

2. What are the two conditions that an experiment must meet in order to be considered a Poisson random variable?

3. What are some uses of the Poisson probability model in business?

4. What is the Poisson probability distribution function?

5. What is the parameter of the Poisson probability model?

6. What is the expected value of a Poisson random variable? The variance? The standard deviation?

## Exercises

7. Suppose that, on average, 5 students enrolled in a small liberal arts college have their automobiles stolen during the semester. What is the probability that exactly 2 students will have their automobiles stolen during the current semester?

8. The number of calls received by an office on Monday morning between 8:00 AM and 9:00 AM has a Poisson distribution with $\lambda$ equal to 4.0.

    a. Determine the probability of getting no calls between eight and nine in the morning.

    b. Calculate the probability of getting exactly five calls between eight and nine in the morning.

    c. What will be the expected number of calls received by the office during this time period? What is the variance?

    d. Graph the probability distribution of the number of calls using values from Appendix A, Table F.

9. The director of a local hospital is studying the occurrence of medication errors. Medication errors are deemed to occur when a patient is given the wrong amount of medication or the wrong medication is given to a patient. Based on past experience, the director believes that medication errors follow a Poisson process with an average rate of 2 per week. (For the following problems, assume that 1 month = 4 weeks.)

   a. What is the probability that there are no medication errors in one week?

   b. What is the probability that there are no medication errors in one month?

   c. Find the average number of medication errors in one week.

   d. Find the average number of medication errors in one month.

   e. Find the standard deviation of the number of medication errors in one month.

   f. How likely is it that at least 4 medication errors will be observed in one month?

10. The number of weaving errors in a twenty foot by ten foot roll of carpet has a Poisson distribution with $\lambda = 0.1$.

   a. Using Appendix A, Table F, construct the probability distribution for the carpet.

   b. What is the probability of observing fewer than 2 errors in the carpet?

   c. What is the probability of observing more than 5 errors in the carpet?

11. A bank is evaluating their staffing policy to assure they have sufficient staff for their drive up window during the lunch hour. If the number of people who arrive at the window in a 15-minute period has a Poisson distribution with $\lambda = 5$, answer the following questions.

   a. How many people are expected to arrive during the lunch hour?

   b. What is the probability that no one will show up during the lunch hour of 12:00 PM to 1:00 PM?

   c. What is the probability that more than 6 people will show up in any 15-minute period?

12. An aluminum foil manufacturer wants to improve the quality of his product and is trying to develop a probability model for the flaws that occur in a sheet of foil. Assume that $X$, the number of flaws per square foot, has a Poisson distribution. If flaws occur randomly at an average of one flaw per 50 square feet, what is the probability that a box containing a 200 square foot roll will contain one flaw? More than one flaw?

13. A manufacturing company is concerned about the high rate of accidents that occurred on the production line last week. There were 6 accidents in the last week and this may require a report to be sent to the government agency for safety. Calculate the probability of 6 accidents occurring in a week when the average number of accidents per week has been 3.5. Assume that the number of accidents per week follows a Poisson distribution.

# 6.6 The Hypergeometric Distribution

The binomial and the hypergeometric random variables are very similar. Both random variables have only two outcomes in each trial of the experiment. They both count the number of successes in $n$ trials of an experiment. The hypergeometric distribution differs from the binomial distribution in the lack of independence between trials, which also implies that the probability of success will vary between trials. In addition, hypergeometric distributions have finite populations in which the total number of successes and failures are known.

Because the binomial and hypergeometric distributions are closely related, a small change in an experiment can switch the distribution of the random variable. A binomial experiment, such as counting the number of red cards drawn in 8 draws from a deck with replacement, can easily be modified to a hypergeometric by not replacing the cards. Since there are 26 red cards (successes) and 26 black cards (failures), the probability of drawing a red card on the first draw is $\frac{26}{52}$ or $\frac{1}{2}$. If a red card is drawn on the first draw and not replaced, the probability of drawing a red card on the next draw is slightly less $\left(\frac{25}{51}\right)$ since there is one less red card in the deck. If the next card drawn is also red, then the probability of a red card on the third draw will be diminished to $\frac{24}{50}$. The probability distribution function of the hypergeometric distribution is given below.

Generally, the hypergeometric distribution can be used when one is sampling $n$ items from a population of size $N$ without replacement and it is known that there are $k$ successes in the population (thus, $N - k$ failures in the population). Using the hypergeometric distribution, you can find the probability of $x$ successes in the sample of size $n$.

---

**Formula**

**Hypergeometric Probability Distribution Function**

The **hypergeometric probability distribution function** is given by:

$$P(X = x) = \frac{{}_k C_x \; {}_{N-k} C_{n-x}}{{}_N C_n},$$

where

$k$ = the total number of successes possible in the population,

$N$ = the size of the total population,

$n$ = the size of the sample drawn,

$x$ = the number of successes in the sample of size $n$, and

maximum of $(0, n + k - N) \leq x \leq$ minimum of $(k, n)$.

---

There cannot be more successes than there are potential successes in the population, nor can there be more successes than the total size of the sample. Thus, the maximum value of $X$ is the smaller of $k$ and $n$.

---

**Example 6.6.1**

**Determining the Probability Distribution of Stocks in a Mutual Fund**

In a volatile stock market, suppose a mutual fund contains 15 stocks. Four of the stocks in the fund have positive gains, while 11 of the stocks are losing money. If two of the stocks from the mutual fund are chosen at random (without replacement), what is the probability distribution for the number of stocks in the sample that will have a positive gain?

**SOLUTION**

The random variable under consideration is given as

$X$ = the number of stocks in the mutual fund that have positive gains.

The parameters of the distribution are

$k = 4$ (a success in this case is a money-making stock)

$N = 15$, and

$n = 2$.

The maximum value of $X$ in this case is 2. Using the hypergeometric distribution function, we have the following.

$$P(X = 0) = \frac{{}_4C_0\ {}_{15-4}C_{2-0}}{{}_{15}C_2} \approx 0.5238$$

$$P(X = 1) = \frac{{}_4C_1\ {}_{15-4}C_{2-1}}{{}_{15}C_2} \approx 0.4190$$

$$P(X = 2) = \frac{{}_4C_2\ {}_{15-4}C_{2-2}}{{}_{15}C_2} \approx 0.0571$$

The probability distribution is summarized in the following table.

| Table 6.6.1 – Probability Distribution | |
|---|---|
| *x* | *P(X = x)* |
| 0 | 0.5238 |
| 1 | 0.4190 |
| 2 | 0.0571 |

**∞ Technology**

To find hypergeometric probabilities using technology, go to stat.hawkeslearning.com and navigate to **Discovering Business Statistics, Second Edition > Technology Instructions > Hypergeometric Distribution.**

| *fx* | =HYPGEOM.DIST(2,2,4,15,FALSE) | |
|---|---|---|
| D | E | F |
| | 0.057143 | |

## Expected Value and Variance

It is apparent that the hypergeometric distribution can be rather tedious to calculate, based on the calculations in Example 6.6.1. Since there are no tables for the distribution, determining an expected value using the definition ($\sum[(x_i p(x_i)]$) would consume a considerable amount of time. Fortunately, there is a simpler method.

### Formula

#### Expected Value

The **expected value** of a hypergeometric random variable can be obtained using the following expression.

$$\mu = E(X) = n\left(\frac{k}{N}\right)$$

### Formula

#### Variance

The **variance** of a hypergeometric random variable is given by the following expression.

$$\sigma^2 = V(X) = n\left(\frac{k}{N}\right)\left(1 - \frac{k}{N}\right)\frac{(N-n)}{(N-1)}$$

**Example 6.6.2**

**Calculating the Expected Value and Variance of a Hypergeometric Random Variable**

Compute the expected value and variance for the random variable defined in Example 6.6.1.

**SOLUTION**

$$E(X) = 2\left(\frac{4}{15}\right) \approx 0.5333$$

$$\sigma^2 = V(X) = 2\left(\frac{4}{15}\right)\left(1 - \frac{4}{15}\right)\frac{(15-2)}{(15-1)} \approx 0.3632$$

Thus, if the experiment were repeated many times, the average number of stocks in the mutual fund that had positive gains would be 0.5333. This leads us to believe that this particular fund is not very promising.

## ✏ 6.6 Exercises

### Basic Concepts

1. How does the hypergeometric model differ from the binomial model?

2. What is the hypergeometric probability distribution function?

3. What are the parameters of the hypergeometric model?

4. How do you calculate the expected value of a hypergeometric random variable? The variance?

### Exercises

5. Suppose a batch of 50 light bulbs contains 3 light bulbs that are defective. Let $X =$ the number of defective light bulbs in a random sample of 10 light bulbs (where the sample is taken without replacement).

    a. What probability model would be appropriate for describing the number of defective light bulbs in the sample?

    b. Find the expected number of defective bulbs.

    c. Find the standard deviation of the number of defective bulbs.

    d. Find the probability that at least 1 of the bulbs sampled will be defective.

    e. Find the probability that at most 2 of the bulbs sampled will be defective.

    f. Find the probability that more than 3 of the bulbs sampled will be defective.

6. A small electronics firm has 60 employees. Ten of the employees are older than 55. An attorney is investigating a client's claim regarding age discrimination. The attorney randomly selects 15 employees without replacement and records the number of employees over age 55.

    a. What probability model would be appropriate for describing the number of employees over age 55 in a sample of 15 selected without replacement?

    b. Find the average number of employees over age 55 in the sample.

    c. Find the standard deviation of the number of employees over age 55 in the sample.

    d. Find the probability that at least 2 of the employees selected will be over age 55.

    e. Find the probability that less than 2 of the employees selected will be over age 55.

    f. Find the probability that at most 4 of the employees will be over age 55.

7.  A bank has to repossess 100 homes. Fifty of the repossessed homes have market values that are less than the outstanding balance of the mortgage. An auditor randomly selects 10 of the repossessed homes (without replacement) and records the number of homes that have market values less than the outstanding balance of the mortgage.

    a.  Find the expected number of homes the auditor will find with market values less than the outstanding balance of the mortgage.

    b.  Find the standard deviation of the number of homes the auditor will find with market values less than the outstanding balance of the mortgage.

    c.  What is the probability that all of the audited homes will have outstanding balances in excess of the mortgage?

    d.  What is the probability that none of the audited homes will have outstanding balances in excess of the mortgage?

8.  A small liberal arts college in the Northeast has 200 freshmen. Eighty of the freshmen are female. Suppose thirty freshmen are randomly selected (without replacement).

    a.  Find the expected number of females in the sample.

    b.  Find the standard deviation of the number of females in the sample.

    c.  Find the probability that none of the selected students will be female.

    d.  Find the probability that all of the selected students will be female.

## T    Discovering Technology

### Using the TI-84 Plus Calculator

## Binomial Distribution

Roll a single six-sided die 4 times and record the number of threes observed. What is the probability that you roll exactly 3 threes in the 4 rolls of the die? What is the probability that you roll less than 3 threes in the 4 rolls of the die?

1. Let $X$ be the number of threes in 4 rolls of the die. Then $X$ has a binomial probability distribution with $n = 4$ and $p = \frac{1}{6}$. To find the probability of rolling exactly 3 threes in 4 rolls of the die, we need to determine $P(X = 3)$. To do this, press **2ND**, and **VARS** to access the **DISTR** menu.

2. Since we are first interested in the probability of $X$ being equal to a particular value, namely 3, select option **binompdf**. Press **ENTER**.

3. The syntax for entering the values into the calculator is binompdf($n, p, x$). Thus, to find the probability of rolling exactly 3 threes in 4 rolls of the die, enter **binompdf(4,1/6,3)** and press **ENTER**.

4. The probability is displayed. Thus, $P(X = 3) \approx 0.0154$.

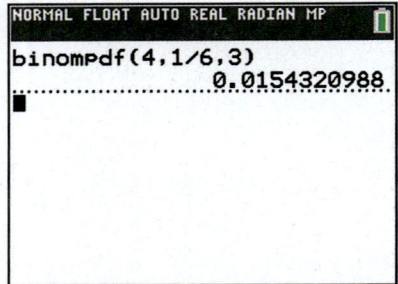

5. To find the probability of rolling less than 3 threes in 4 rolls of the die, we need to find $P(X \leq 2)$. To do this, again press **2ND**, and **VARS** to access the **DISTR** menu.

6. This time, we need the cumulative probability. Thus, select option **binomcdf**. Press **ENTER**.

7. The syntax here is the same, binomcdf($n, p, x$), except the cdf function returns the cumulative binomial probability up to and including $x$. Thus, to find the probability of rolling less than 3 threes in 4 rolls of the die, enter **binomcdf(4,1/6,2)** and press **ENTER**.

8. The probability is displayed. Thus, $P(X \leq 2) \approx 0.9838$.

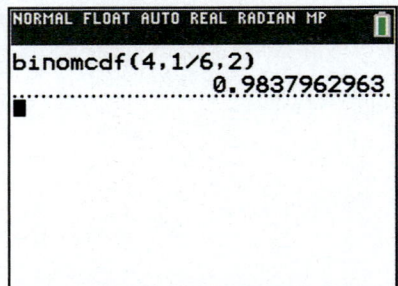

## Poisson Distribution

For this exercise, use the information from Example 6.5.1.

1. Let $X$ = the number of arrivals in a 15-minute period, and we know that $X$ has a Poisson distribution with $\lambda = 5$. We want to find the probability that there are no arrivals in a 15-minute period, i.e., $P(X = 0)$. To do this, press 2ND, and VARS to access the **DISTR** menu.

2. Since we are first interested in the probability of $X$ being equal to a particular value, namely 0, select option **poissonpdf**. The syntax is poissonpdf($\lambda$, $x$). Thus, to find the probability that there are no arrivals in a 15-minute interval, enter **poissonpdf(5,0)**. Press ENTER.

3. The probability is displayed. Thus, $P(X = 0) \approx 0.0067$. Compare this result with that in Example 6.5.1.

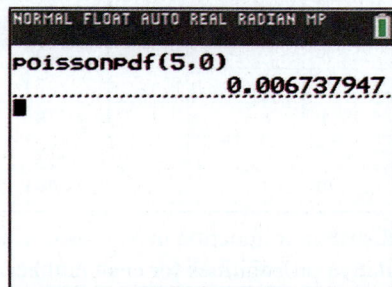

4. Next we want to find the probability that at least three people will use the machine in a fifteen minute interval, i.e., $P(X \geq 3)$. This is equivalent to finding $1 - P(X \leq 2)$. To find this probability, enter 1 and —, and then press 2ND, and VARS to access the **DISTR** menu.

5. This time, we need the cumulative probability. Thus, select option **poissoncdf**. Press ENTER.

6. The syntax is the same, namely poissonpdf($\lambda$, $x$), except the cdf function returns the cumulative poisson probability up to and including $x$. Enter **poissoncdf(5,2)**. Press ENTER.

7. The probability is displayed. Thus, $P(X \geq 3) \approx 0.8753$. Note that the answer obtained on the calculator is slightly different from that obtained in Example 6.5.1. This is due to rounding error when doing the calculations by hand using values from the table in Appendix A. Answers obtained using technology will be more accurate than those obtained using the tables in the Appendix.

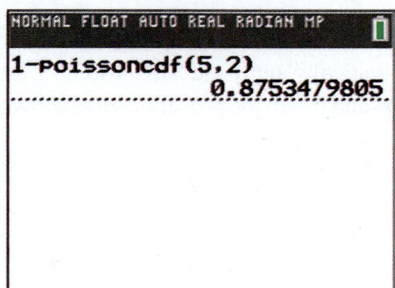

## Using Excel

### Binomial Distribution

For this exercise, use the data from Example 6.4.3.

$$n = 10, p = 0.7$$

1. Set up the worksheet as shown below.

| | A | B | C | D | E |
|---|---|---|---|---|---|
| 1 | x | P(X = x) | | x | P(X <= x) |
| 2 | 0 | 5.9049E-06 | | 0 | |
| 3 | 1 | | | 1 | |
| 4 | 2 | | | 2 | |
| 5 | 3 | | | 3 | |
| 6 | 4 | | | 4 | |
| 7 | 5 | | | 5 | |
| 8 | 6 | | | 6 | |
| 9 | 7 | | | 7 | |
| 10 | 8 | | | 8 | |
| 11 | 9 | | | 9 | |
| 12 | 10 | | | 10 | |

2. Using the binomial distribution function in Microsoft Excel, you can calculate both individual and cumulative probabilities for each number of successes. The function in Microsoft Excel that calculates binomial probabilities is

   **BINOM.DIST(number_s, trials, probability_s, cumulative)**

   where **number_s** = the number of successes,

   **trials** = the number of trials,

   **probability_s** = the probability of success, and

   **cumulative** = a logical value which is either TRUE (Excel returns the cumulative probability) or FALSE (Excel returns the probability there are exactly **number_s** successes).

3. Begin with the individual probabilities. In cell B2, enter the formula to return the probability that there are 0 successes in 10 trials as follows.

   **=BINOM.DIST(A2,10,0.7,FALSE)**

   Press **Enter**.

4. Click cell B2 and click and drag the little square on the bottom-right corner down to cell B12. Excel will automatically apply the BINOM.DIST function with the first input as the x value in each row and return the probabilities as shown in the following worksheet.

| ◢ | A | B | C | D | E |
|---|---|---|---|---|---|
| 1 | x | P(X = x) | | x | P(X <= x) |
| 2 | 0 | 5.9049E-06 | | 0 | |
| 3 | 1 | 0.00013778 | | 1 | |
| 4 | 2 | 0.0014467 | | 2 | |
| 5 | 3 | 0.00900169 | | 3 | |
| 6 | 4 | 0.03675691 | | 4 | |
| 7 | 5 | 0.10291935 | | 5 | |
| 8 | 6 | 0.20012095 | | 6 | |
| 9 | 7 | 0.26682793 | | 7 | |
| 10 | 8 | 0.23347444 | | 8 | |
| 11 | 9 | 0.12106082 | | 9 | |
| 12 | 10 | 0.02824752 | | 10 | |
| 13 | | | | | |

5. Next, calculate the cumulative probabilities. In cell E2, enter the formula to return the cumulative probability that there are no more than 0 successes in 10 trials.

**=BINOM.DIST(D2,10,0.7,TRUE)**

Press **Enter**.

6. Click cell E2 and click and drag the little square on the bottom-right corner down to cell E12. Excel will automatically apply the BINOM.DIST function with the first input as the x value in each row and return the cumulative probabilities as shown in the worksheet below. Compare these to the values we found in Example 6.4.3.

| ◢ | A | B | C | D | E |
|---|---|---|---|---|---|
| 1 | x | P(X = x) | | x | |
| 2 | 0 | 5.9049E-06 | | 0 | 5.9049E-06 |
| 3 | 1 | 0.00013778 | | 1 | 0.00014369 |
| 4 | 2 | 0.0014467 | | 2 | 0.00159039 |
| 5 | 3 | 0.00900169 | | 3 | 0.01059208 |
| 6 | 4 | 0.03675691 | | 4 | 0.04734899 |
| 7 | 5 | 0.10291935 | | 5 | 0.15026833 |
| 8 | 6 | 0.20012095 | | 6 | 0.35038928 |
| 9 | 7 | 0.26682793 | | 7 | 0.61721721 |
| 10 | 8 | 0.23347444 | | 8 | 0.85069165 |
| 11 | 9 | 0.12106082 | | 9 | 0.97075248 |
| 12 | 10 | 0.02824752 | | 10 | 1 |

## Poisson Distribution

Use Microsoft Excel to create the Poisson probability distribution in Table 6.5.1 ($\lambda = 0.3$).

1. Set up the worksheet as shown below.

| ◢ | A | B |
|---|---|---|
| 1 | x | P(X = x) |
| 2 | 0 | |
| 3 | 1 | |
| 4 | 2 | |
| 5 | 3 | |
| 6 | 4 | |
| 7 | 5 | |

2. The Excel function for the Poisson distribution is:

**POISSON.DIST(x, mean, cumulative)**

where **x** = the number of successes in the sample,

**mean** = the mean number of successes, and

**cumulative** = a logical value which is either TRUE (Excel returns the cumulative probability) or FALSE (Excel returns the probability there are **x** successes).

3. In cell B2, enter the formula to find the probability of 0 successes when the mean number of successes is 0.3.

**=POISSON.DIST(0,0.3,FALSE)**

Press **Enter**. The corresponding worksheet is shown below.

| | A | B |
|---|---|---|
| 1 | x | P(X = x) |
| 2 | 0 | 0.740818 |
| 3 | 1 | |
| 4 | 2 | |
| 5 | 3 | |
| 6 | 4 | |
| 7 | 5 | |

4. Click cell B2 and click and drag the little square on the bottom-right corner down to cell B7. Excel will automatically apply the POISSON.DIST function with the first input as the $x$ value in each row and return the probabilities as shown in the worksheet below as well as in Table 6.5.1.

| | A | B |
|---|---|---|
| 1 | x | P(X = x) |
| 2 | 0 | 0.740818 |
| 3 | 1 | 0.222245 |
| 4 | 2 | 0.033337 |
| 5 | 3 | 0.003334 |
| 6 | 4 | 0.00025 |
| 7 | 5 | 1.5E-05 |

## Hypergeometric Distribution

$$k = 2, n = 16, N = 30$$

1. Set up the worksheet as below.

| | A | B |
|---|---|---|
| 1 | x | P(X = x) |
| 2 | 0 | |
| 3 | 1 | |
| 4 | 2 | |

2. The Excel function for the hypergeometric distribution is:

**HYPGEOM.DIST** (**sample_s**, **number_sample**, **population_s**, **number_pop**, **cumulative**) where **sample_s** = the number of successes in the sample,

**number_sample** = the size of the sample,

**population_s** = the number of successes in the population,

**number_pop** = the population size, and

**cumulative** = a logical value which is either TRUE (Excel returns the cumulative probability) or FALSE (Excel returns the probability there are **sample_s** successes).

3. In cell B2, enter the formula as follows.

**=HYPGEOM.DIST(0,16,2,30,FALSE)**

Press **Enter**. The corresponding worksheet is shown below.

| | A | B |
|---|---|---|
| 1 | x | P(X = x) |
| 2 | 0 | 0.209195 |
| 3 | 1 | |
| 4 | 2 | |

4. Click in cell B2 and click and drag the little square on the bottom-right corner down to cell B4. Excel will automatically apply the HYPGEOM.DIST function with the first input as the $x$ value in each row and return the probabilities as shown in the worksheet below.

| | A | B |
|---|---|---|
| 1 | x | P(X = x) |
| 2 | 0 | 0.209195 |
| 3 | 1 | 0.514943 |
| 4 | 2 | 0.275862 |

## Using JMP

### Binomial Distribution

For this exercise, calculate the binomial probability in Example 6.4.3.

The probability to be calculated is that at least three of the ten customers will approve the return policy. For this example, $n = 10$ and $p = 0.7$. The desired probability can be calculated by finding the probability that 2 or less customers approve the return policy and subtracting this probability from 1. To find $1 - P(X \leq 2)$ with $n = 10$ and $p = 0.7$ using JMP follow these steps.

1. With a JMP **Data Table** open, click in the first cell of **Column 1**. Click on **Cols** in the top row and then select **New Columns**.

2. In the popup window that opens, click on **Column Properties** at the bottom of the window. Select **Formula** from the dropdown menu.

3. Click in the formula box in the center of the screen and enter **1**. Then click on the **minus sign** at the top of the screen. Now click on the arrow next to **Discrete Probability** and select **Binomial Distribution**. The **Binomial Distribution** formula will appear after the minus sign. The letter $p$ has a blue box around it.

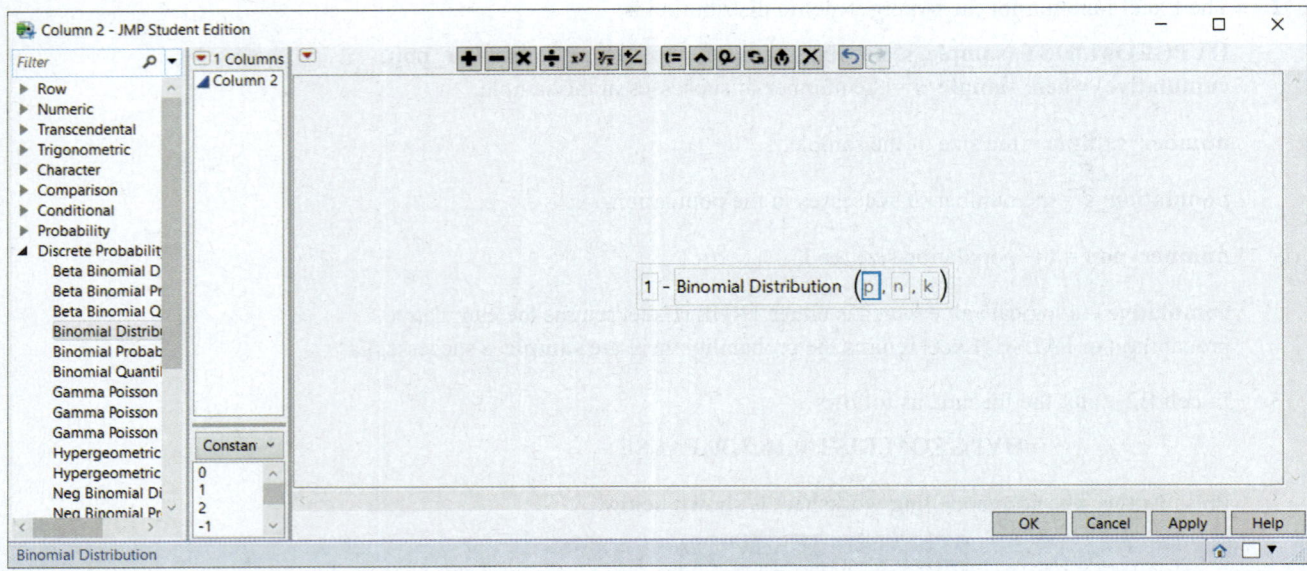

4.   Input **0.7** for **p** and press **Enter**.

5.   Click on **n** in the formula, input **10**, and press **Enter**.

6.   Click on **k** in the formula, input **2**, and press **Enter**.

7.   Click **Apply** and then **OK** in the first window. The binomial formula with the values displayed for *p*, *n*, and *k* will appear at the bottom. Click **Apply** and then **OK**.

8.   The value of 0.9984096136 should appear in row 1 of **Column 2**. If the value does not appear then double-click in the first row to the left of **Column 2**.

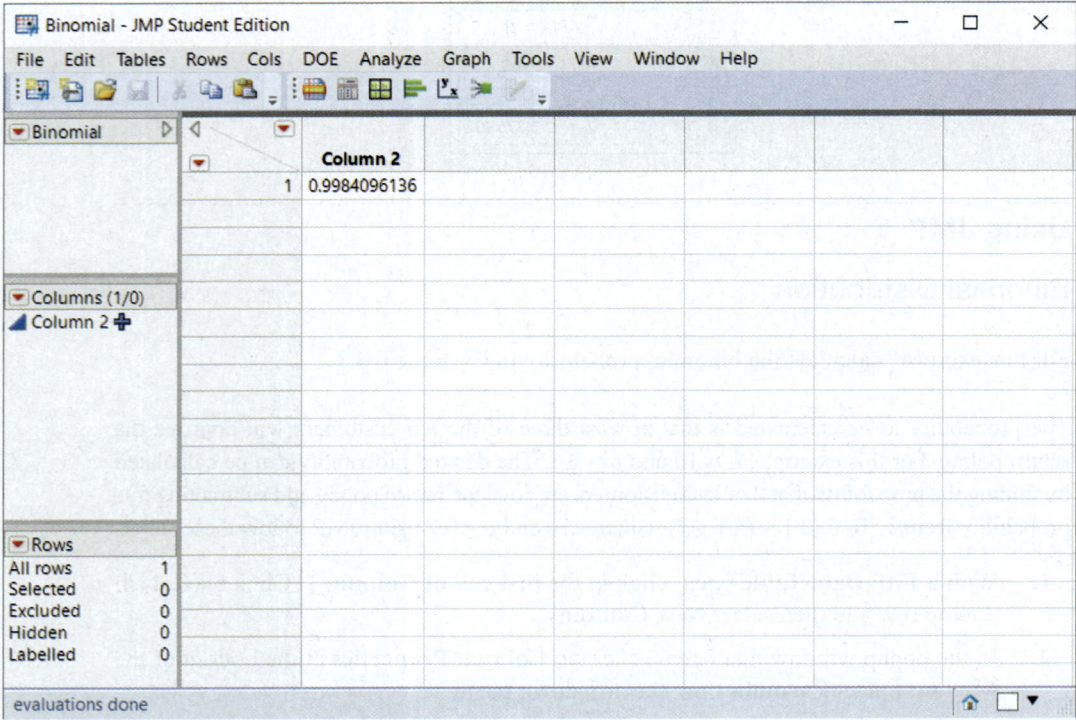

## Hypergeometric Distribution

For this exercise, calculate the hypergeometric probabilities in Example 6.6.1.

The problem asks you to calculate the probability distribution of drawing 2 stocks (without replacement) with positive gain from a total of 15 stocks in which 4 have a positive gain. For this example, $N = 15$, $k = 4$, and $n = 2$. The desired probability distribution for $X = 0$, 1, and 2 can be calculated in JMP as follows.

1.   With a JMP **Data Table** open, click in the first cell of **Column 1**. Click on **Cols** in the top row and then select **New Columns**.

2.   In the popup window that opens, click on **Column Properties** at the bottom of the window. Select **Formula** from the dropdown menu.

3.   Click on the arrow next to **Discrete Probability** and select **Hypergeometric Probability**. The **Hypergeometric Probability** formula will appear in the center of the screen. The letter $N$ has a blue box around it.

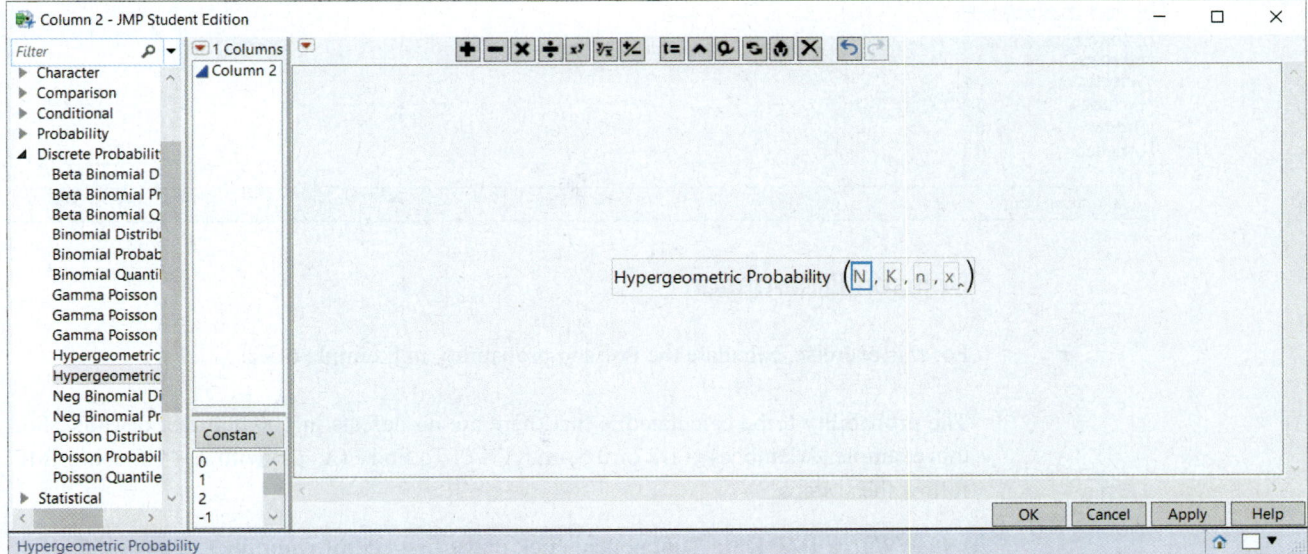

4.   Input 15 for $N$ and press **Enter**.

5.   Click on $K$ in the formula, input **4**, and press **Enter**.

6.   Click on $n$ in the formula, input **2**, and press **Enter**.

7.   Click on $x$ in the formula, input **0**, and press **Enter**.

8.   Click **Apply** and then **OK** in the first window. The hypergeometric formula with the values displayed for $N$, $K$, $n$, and $x$ will appear at the bottom. Click **Apply** and then **OK**.

9.   The value of 0.5238095238 should appear in row 1 of **Column 2**. If the value does not appear then double-click in the first row to the left of **Column 2**.

10.  Continue adding new columns to the data table and use the **Hypergeometric Probability Formula** with $N = 15$, $K = 4$, and $n = 2$ to calculate the probability for $x = 1$ and $x = 2$.

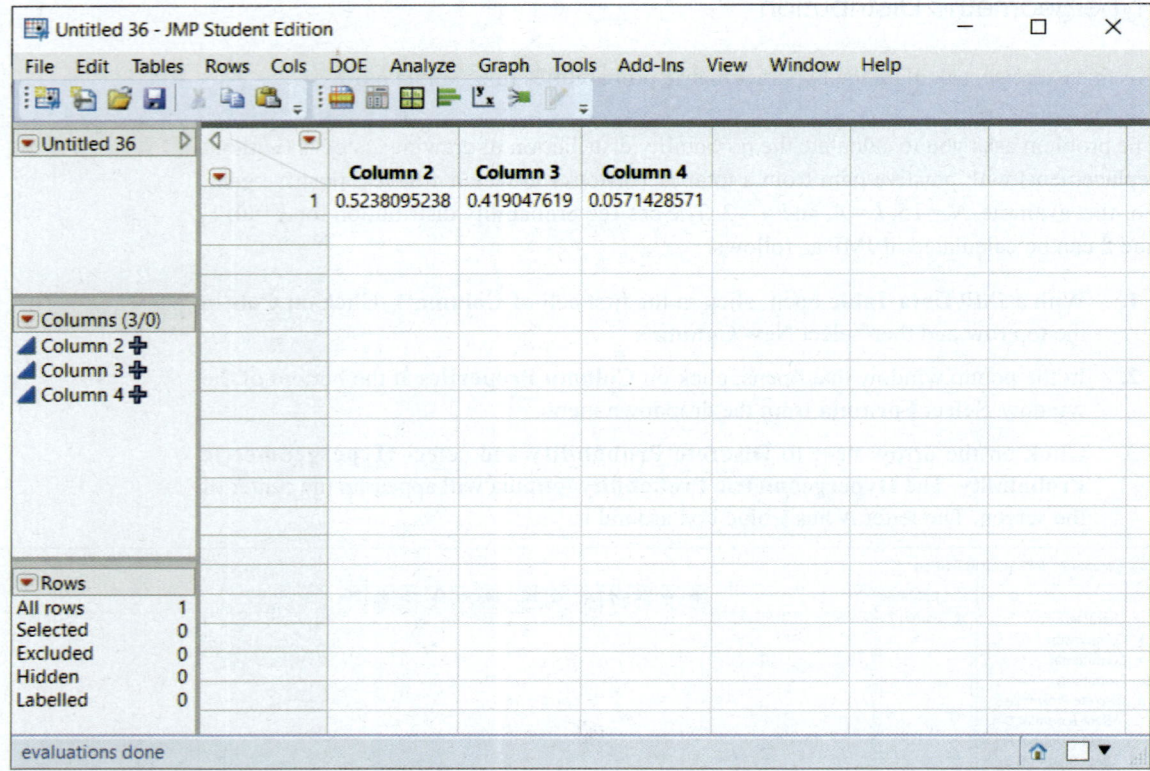

## Poisson Distribution

For this exercise, calculate the Poisson probability in Example 6.5.2.

The probability to be calculated is that there are no defects in 100,000 feet of cable. For this example, $\lambda$(lambda) = 1/2 or 0.5, and $X = 0$. To find $P(X = 0)$ with $\lambda = 0.5$ using JMP follow these steps.

1. With a JMP **Data Table** open, click in the first cell of **Column 1**. Click on **Cols** in the top row and then select **New Columns**.

2. In the popup window that opens, click on **Column Properties** at the bottom of the window. Select **Formula** from the dropdown menu.

3. Click on the arrow next to **Discrete Probability** and select **Poisson Probability**. The **Poisson Probability** formula will appear with a blue box around **lambda**.

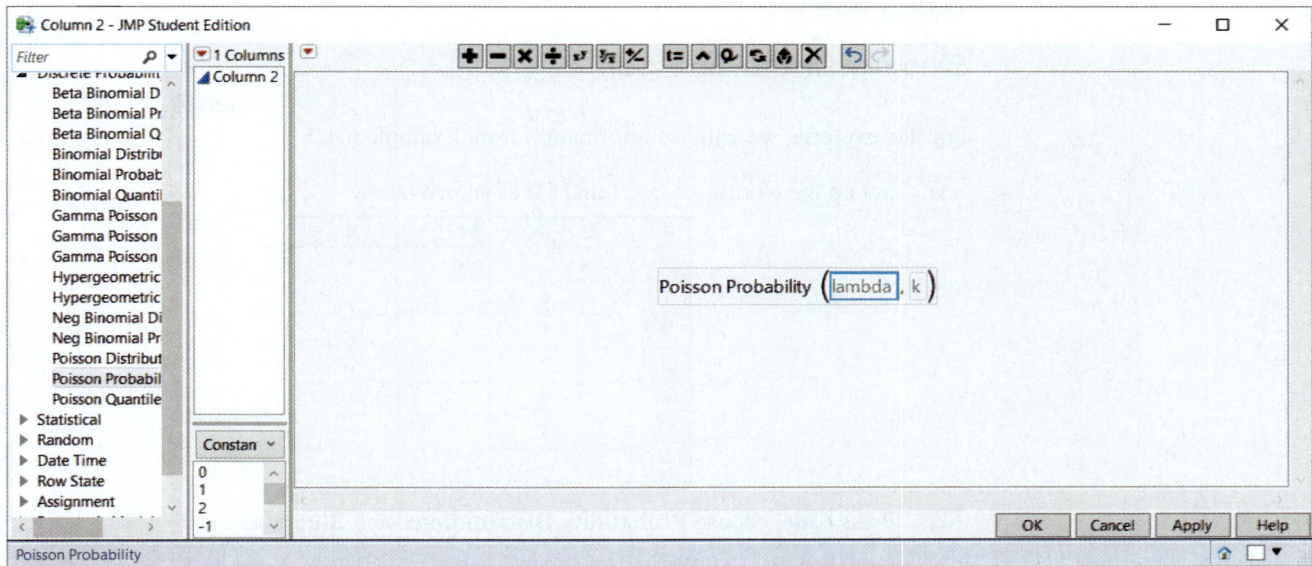

4.  Input **0.5** for **lambda** and press **Enter**.

5.  Click on **k** in the formula, input **0**, and press **Enter**.

6.  Click **Apply** and then **OK** in the first window. The Poisson probability formula with the values displayed for lambda and *k* will appear at the bottom. Click **Apply** and then **OK**.

7.  The value of 0.6065306597 should appear in row 1 of **Column 2**. If the value does not appear then double-click in the first row to the left of **Column 2**.

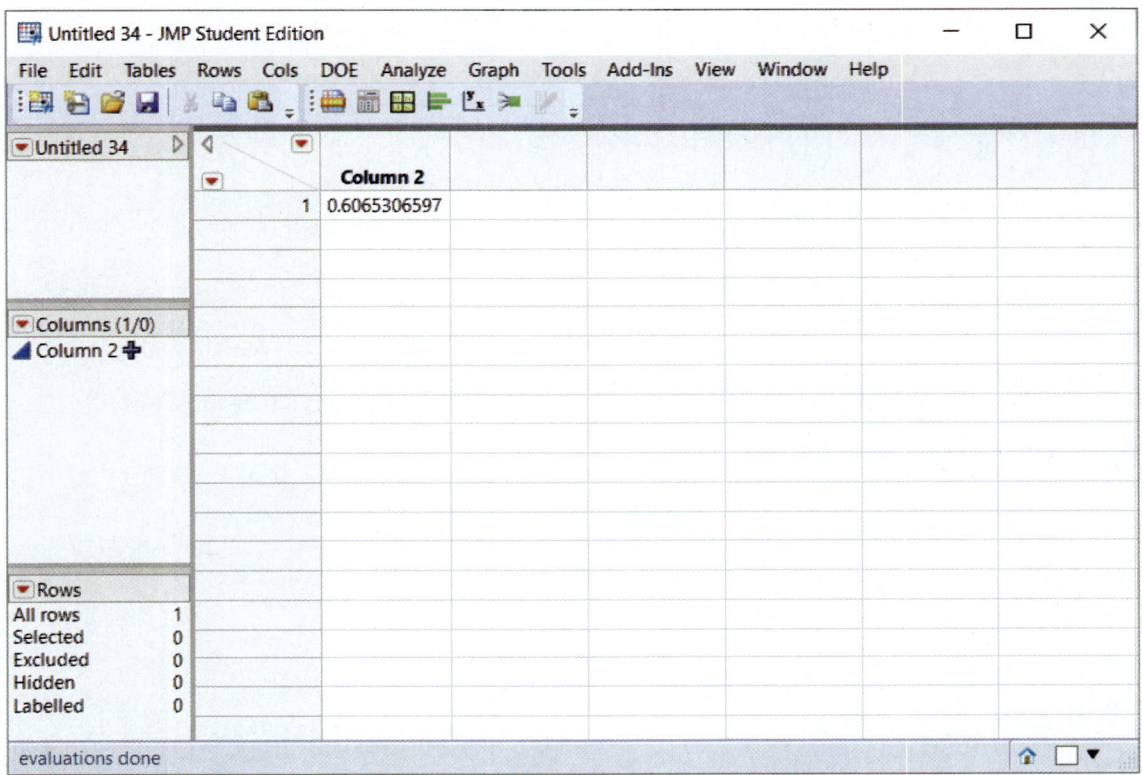

## Using Minitab

## Binomial Distribution (pdf)

For this exercise, we will use information from Example 6.4.2.

1. Set up the worksheet in C1 and C2 as shown below.

| ↓ | C1 | C2 | C3 |
|---|----|----|----|
|   | x | p(x) | |
| 1 | 0 | | |
| 2 | 1 | | |
| 3 | 2 | | |
| 4 | 3 | | |
| 5 | 4 | | |
| 6 | | | |

2. Press **Calc**, choose **Probability Distributions**, and **Binomial**.

3. Designate **Probability** and enter the Number of Trials as **4**. Input the Event probability as **0.16667**.

4. Enter **C1** for Input Column and **C2** for Optional Storage. Press **OK**

5. The output will be as follows (showing Binomial pdf).

| C1 | C2 | C3 |
|----|----|----|
| x | p(x) | |
| 0 | 0.482245 | |
| 1 | 0.385206 | |
| 2 | 0.115744 | |
| 3 | 0.015433 | |
| 4 | 0.000772 | |
| | | |

## Poisson Distribution (pdf)

For this exercise, we will use information from Example 6.5.1.

1. Set up the worksheet in C1 and C2 as shown below.

| ↓ | C1 | C2 | C3 |
|---|----|----|----|
|   | x  | p(x) |  |
| 1 | 0  |  |  |
| 2 | 1  |  |  |
| 3 | 2  |  |  |
| 4 | 3  |  |  |
| 5 | 4  |  |  |
| 6 |    |  |  |

2. Under **Calc**, choose **Probability Distributions**, and **Poisson**.

3. Designate **Probability** and enter the Mean as **5**.

4. Enter **C1** for the Input Column and **C2** for Optional Storage. Click **OK.**

5. The output will be as follows (showing Poisson pdf).

| ↓ | C1 | C2 | C3 |
|---|----|----|----|
|   | x  | p(x) |  |
| 1 | 0  | 0.006738 |  |
| 2 | 1  | 0.033690 |  |
| 3 | 2  | 0.084224 |  |
| 4 | 3  | 0.140374 |  |
| 5 | 4  | 0.175467 |  |
| 6 |    |  |  |

## R    Chapter 6 Review

### Key Terms and Ideas

- Random Variable
- Probability Distribution
- Discrete Random Variable
- Continuous Random Variable
- Discrete Probability Distribution
- Probability Distribution Function
- Expected Value of a Discrete Random Variable
- Variance of a Discrete Random Variable
- Probability Distribution Function
- Discrete Uniform Distribution
- Discrete Uniform Distribution Function
- Binomial Distribution
- Binomial Experiment
- Binomial Random Variable
- Binomial Probability Distribution Function
- Binomial Tables
- Expected Value of a Binomial Random Variable
- Variance of a Binomial Random Variable
- Standard Deviation of a Binomial Random Variable
- Poisson Distribution
- Poisson Random Variable
- Lambda
- Poisson Probability Distribution Function
- Poisson Tables
- Hypergeometric Distribution
- Hypergeometric Probability Distribution Function

### Key Formulas

|  | Section |
|---|---|

**Expected Value of a Discrete Random Variable X**

$$\mu = E(X) = \sum \left[ x_i p(x_i) \right]$$

6.2

where $p(x) = P(X = x)$

**Variance of a Discrete Random Variable X**

$$\sigma^2 = V(X) = \sum \left[ (x_i - \mu)^2 \, p(x_i) \right]$$

6.2

**Computational Formula for Variance of a Discrete Random Variable X**

$$\sigma^2 = V(X) = \sum \left[ x_i^2 p(x_i) \right] - \mu^2$$

6.2

**Discrete Uniform Probability Distribution Function**

$$P(X = x) = \frac{1}{n}$$

6.3

where $n$ = the number of values that the random variable may assume.

## Key Formulas (cont.)

|  | Section |
|---|---|

**Binomial Probability Distribution Function**

$$P(X = x) = {}_nC_x p^x (1-p)^{n-x}$$

where ${}_nC_x = \dfrac{n!}{x!(n-x)!}$, $n$ = the number of trials, and $p$ = the probability of a success.

Section 6.4

**Expected Value of a Binomial Random Variable**

$$\mu = E(X) = np$$

Section 6.4

**Variance and Standard Deviation of a Binomial Random Variable**

$$\sigma^2 = V(X) = np(1-p)$$

$$\sigma = \sqrt{V(X)} = \sqrt{np(1-p)}$$

Section 6.4

**Poisson Probability Distribution Function**

$$P(X = x) = \frac{e^{-\lambda}\lambda^x}{x!}$$

where $e = 2.71828\ldots$ and $\lambda =$ the mean number of successes.

Section 6.5

**Expected Value of a Poisson Random Variable**

$$\mu = E(X) = \lambda$$

Section 6.5

**Variance and Standard Deviation of a Poisson Random Variable**

$$\sigma^2 = V(X) = \lambda$$

$$\sigma = \sqrt{V(X)} = \sqrt{\lambda}$$

Section 6.5

**Hypergeometric Probability Distribution Function**

$$P(X = x) = \frac{{}_kC_x \, {}_{N-k}C_{n-x}}{{}_NC_n}$$

where $k$ = the total number of successes possible, $N$ = the size of the total population, and $n$ = the size of the sample drawn.

Section 6.6

**Expected Value of a Hypergeometric Random Variable**

$$\mu = E(X) = n\left(\frac{k}{N}\right)$$

Section 6.6

**Variance of a Hypergeometric Random Variable**

$$\sigma^2 = V(X) = n\left(\frac{k}{N}\right)\left(1 - \frac{k}{N}\right)\frac{(N-n)}{(N-1)}$$

Section 6.6

# AE    Additional Exercises

1.  A statistics professor has determined the following probability distribution for $X$, the grade which a student will earn in a business statistics class.

| Grade Distribution | | |
|---|---|---|
| Grade | x | $P(X = x)$ |
| A | 4.0 | 0.15 |
| B | 3.0 | 0.35 |
| C | 2.0 | 0.25 |
| D | 1.0 | 0.15 |
| F | 0.0 | 0.10 |

   a.  What is the average grade that a student will earn in a business statistics class?

   b.  Find the variance of the grades that students will earn in a business statistics class.

   c.  Find the standard deviation of the grades which students will earn in a business statistics class.

   d.  What is the probability that a student will earn a grade of 4.0?

   e.  Find the probability that a student will earn a grade of at least 2.0.

   f.  Find the probability that a student will earn a grade of at most 1.0.

   g.  Find the probability that a student will earn a grade of more than 3.0.

2.  The U.S. Department of Labor has issued a new set of guidelines governing certain work practices for employees. It estimates that only 20% of all firms will be subject to the new guidelines. To validate the estimate of the number of firms that will be affected by the new guidelines, the department randomly selects a sample of twenty firms for a study. Assuming their initial estimate of 20% is correct, answer the following questions.

   a.  What is the probability that 1 or fewer of the sampled firms will be subject to the new rules?

   b.  What is the probability that between 15 and 25 percent of the sampled firms will be subject to the rules?

   c.  One of the directors in the department remarked he thought that ten firms out of the sample would be subject to the rules. If the initial estimate is correct, what is the chance of this occurring?

3.  Historically, the probability that a library book will be returned in one week is $p = 0.50$. The head librarian for the University Staff Hospital library is monitoring a random sample of 10 books to determine if the historical proportion of the books returned within one week, 0.50, has changed. Assuming the historical return rate is still the same, answer the following questions.

   a.  What is the probability that between four and six books will be returned in one week?

   b.  What is the chance that eight or more books will be returned in one week?

   c.  What is the probability that only one book will be returned in one week?

4. The number of fatalities resulting from automobile accidents for a 10-mile stretch of an interstate highway averages 1 per 100,000 automobiles. During a particular holiday weekend, 500,000 automobiles traveled over the 10-mile segment. Using a Poisson distribution, find the probability of each of the following.

   a. No fatalities

   b. 3 fatalities

   c. At least one fatality

5. Compute the mean and variance for the following random variables.

   a. The number of sixes obtained in 10 rolls of a single die.

   b. The number of hearts in a 13 card bridge hand. (Draw 13 cards from a standard deck without replacement.)

   c. The number of free throws made by a professional basketball player in his next 10 attempts. (Assume the player makes 88% of his free throws in the long run.)

   d. The number of cracked eggs selected when randomly selecting 5 eggs from a 12-egg carton containing 2 cracked eggs.

   e. The number of dots on the upper face when a single die is thrown.

6. A manufacturer of digital cameras knows that a shipment of 30 cameras sent to a large discount store contains eight defective cameras. The manufacturer also knows that the store will choose two of the cameras at random, test them, and accept the shipment if neither one is defective.

   a. Find the probability that at least one is defective.

   b. What is the probability that the shipment is accepted?

7. In a certain shipment of sixteen radios, four are defective. Eight of the radios are selected at random without replacement. What is the probability that at least one of the eight radios is defective?

8. According to the American Hotel and Lodging Association (AH&LA), women accounted for 31% of business travelers in the year 2009. Suppose that to attract these women business travelers, the AH&LA found that 80% of hotels offer hair dryers in the bathrooms. Consider a random and independent sample of 15 hotels.

   **Source:** American Hotel & Lodging Association

   a. Based on the information given, how many of the 15 hotels are expected to offer hair dryers in the bathrooms?

   b. Find the probability that all of the hotels in the sample offer hair dryers in the bathrooms.

   c. Find the probability that more than 5 but less than 9 of the hotels in the sample offer hair dryers in the bathrooms.

9. A carnival has a game of chance: a fair coin is tossed. If it lands heads, you win $1, and if it lands tails, you lose $0.50. How much should a ticket cost to play this game if the carnival wants to break even?

10. You are working on a multiple choice test which consists of 15 problems. Each of the problems has five answers, only one of which is correct. If you are totally unprepared for the test and are guessing, what is the probability that your first correct answer is within the first fifteen problems?

11. An automobile manufacturer is always trying to improve the quality of its vehicles. Assume that the number of defects per vehicle follows a Poisson distribution. If these defects occur randomly at an average rate of five per vehicle, what is the probability that a randomly selected vehicle will have at least one defect?

12. When proofreading a statistics textbook, one can expect to find a number of errors, whether they are typographical, symbolic, or even incorrect mathematical calculations. On average, a statistics textbook will contain 30 errors. What is the probability that when proofreading a text, one finds at least three errors? Assume that the number of errors found follows a Poisson distribution.

13. While on a shopping spree, you randomly select five portable music players from an electronics store that sells 20 portable music players. Of these 20 music players, 12 will last beyond the 1-year limited warranty and will not need to be replaced or repaired. What is the probability that at least three of the five portable music players selected will not last beyond the limited warranty period without needing to be replaced or repaired?

14. A jeweler was given a collection of twelve diamonds, of which three were synthetic (fake). If the jeweler selected two of these diamonds at random (without replacement), what is the probability that neither jewel is found to be synthetic?

15. L-Mart Inspections is a building inspection company. There were ten new commercial construction buildings completed in the last month and the sites are now available for inspection. L-Mart plans to inspect some of the new constructions for code violations and believes that half of the buildings will have violations.

    a. What probability model would be appropriate for describing the number of buildings in the sample that have code violations? Explain your answer.

    b. If L-Mart randomly selects four buildings to inspect, what is the probability that three of the buildings will have violations?

# P Discovery Project

## Take Me Out to the Ball Game!

Use the Moneyball data set which contains selected statistics for Major League Baseball teams from 1962–2012.

**∴ Data**

The data can be found by visiting stat.hawkeslearning.com and navigating to **Discovering Business Statistics, Second Edition > Data Sets > MoneyBall**.

1. Select the variable *Number of wins*, *W*, and compare the distribution of *W* for the American League (AL) with that of the National League (NL). Use side-by-side boxplots as described in Chapter 4.

2. Identify the outliers in both leagues (i.e., the teams that have a total number of wins far from the rest of the teams in their league).

3. Compare the distribution of the *Number of wins, W,* for NYM and TEX using a side-by-side boxplot and by investigating the numerical summaries of each. (Compare the shapes, means, medians, and the variability).

4. Discuss why the discrepancy in variability between the performance of NYM and the performance of TEX didn't cause a similar discrepancy in their respective leagues.

5. Based on historical data, the probability that in a given year the NYM will make the playoffs is $p = 7/47 = 0.149$. Let $X$ be the discrete random variable that gives the total number of Playoffs made by NYM in the last 20 years, i.e., from 1993 to 2012.

    a. Assume that the outcomes for the NYM in these years are unknown for us. Also assume that the outcome in any of the years is independent of the outcome in any other year. Under these assumptions, what would be the distribution of $X$? Why?

    b. What is the probability that the total number of playoffs made by NYM during this 20-year period is exactly three?

    c. What is the probability that the total number of playoffs made by NYM is at most 3?

    d. What is the probability that the total number of playoffs made by NYM is at most 18?

    e. What is the probability that the total number of playoffs made by NYM is at least 15?

    f. What is the expected number of playoffs that NYM will make in this 20-year period?

    g. Find the variance of the number of playoffs that NYM is expected to make in this 20–year period?

    h. Can we use the Poisson distribution with $\lambda = 2.98$ to model the number of playoffs that NYM will make? Why?

    **Source:** https://www.baseball-reference.com/

## Discovering the Real World

### How are Automobile Tire Warranties Determined?

In businesses today, we can use the normal distribution in many aspects. We can use the normal distribution to help us determine probabilities associated with costs, profits, revenue, salaries, warranties, and so on.

Many automobile tire manufacturers advertise that their tires come with warranties that guarantee tread wear for a certain number of miles. For example, a set of Michelin tires could be purchased with a 50,000-mile warranty. Similarly, a set of tires from another manufacturer, say Goodyear, comes with a 75,000-mile warranty. With the warranty, if at least one of the tires wears out prior to the stated mileage of the warranty, the manufacturer will replace them free of charge. Thus, it is important for the manufacturer to determine a reliable number of miles for their warranty or they will replace a lot of tires, which would decrease the profits and potentially tarnish their brand.

So, how do tire manufacturers determine a reliable warranty to include with their tires? To determine the warranty of the tires, a manufacturer will test drive a relatively large sample of the tires to determine the mileage each tire was driven before needing to be replaced. Using the data from this sample, the manufacturer can determine its distribution and then calculate the likelihood that the tires will need to be replaced at a given number of miles. Using this calculated mileage, the manufacturer can then reliably state a warranty so that they are not replacing a large number of tires for free. All of these calculations can be done using the normal distribution as follows.

1.  Select a random sample of 30 sets of four tires to be tested. To test the tires, they will be installed on a vehicle and the vehicle will be driven until at least three of the tires need to be replaced.

2.  Determine/record the mileage at which at least three of the tires need to be replaced.

3.  Using the replacement mileage of each sample (remember, a sample is a set of four tires), ensure that the replacement mileage follows a normal distribution. This can be determined graphically via a histogram or even a stem-and-leaf plot.

4.  Now, under the premise that the data follow a normal distribution, the tire manufacturer can determine a mileage for which 99% of the tires were driven before needing to be replaced. Using this mileage, the manufacturer can then set a reasonable warranty that will not only prevent them from having to replace a large number of tires for free but also (hopefully) satisfy the customer. Of course, it is feasible that the manufacturer would like to put a lower mileage warranty on the tires so that they would determine the mileage for which at least 90% of the tires were driven before needing replacement.

We will discuss this more in the Discovery Project at the end of this chapter.

In this chapter, we will learn about continuous random variables and specific distributions associated with continuous random variables, such as the uniform and normal distributions. We will also discuss how to calculate probabilities associated with each of the distributions. Lastly, we will discuss the use of the normal distribution to approximate probabilities associated with discrete probability distributions such as the binomial and Poisson.

### Introduction

In the previous chapter random variables were primarily counts of some phenomenon called a *success*. These counts could only take on discrete values, usually starting at zero. In this chapter the focus will be on variables that can take on any value in some continuum, i.e., a range of numbers on the real number line. The range of adult heights, for example, lies on a continuum between about 26 and 100 inches.

Observations measured in a continuum can be very close together. For example, two heights could be

$$68.234175245987149399 \text{ inches and}$$

$$68.234175245987149398 \text{ inches.}$$

As a practical matter, it is hard to imagine a situation that would require the knowledge of a person's height to 18 decimal places. Variables like heights and weights are **continuous random variables** even though heights are usually given to the nearest inch or centimeter and weights to the nearest pound or kilogram. Variables measured in these units give the appearance of being discrete, yet they are continuous.

One of the striking differences between discrete and continuous variables concerns the way probability is defined. In a probability distribution for a discrete random variable, each possible outcome of the random variable is assigned its own probability. However, for continuous random variables, there are infinitely many outcomes, and each has *no* probability. Outcomes of a continuous random variable do not have probability assigned to any one point, because there are simply too many points. If we attempted to assign each value even an infinitesimal probability, the sum of all probabilities would exceed one. Thus, for continuous random variables, probability is only assigned to intervals.

**Definition**

**Continuous Random Variable**

A **continuous random variable** is a random variable whose measurements can assume any one of a countless number of values in an interval.

# 7.1 The Uniform Distribution

There are two types of uniform distributions, discrete and continuous. You already studied the discrete uniform distribution in Chapter 6. Recall that for the discrete uniform distribution, we assign the same probability to each possible value of the random variable. Conceptually, both distributions distribute probability evenly across a sample space. For the continuous uniform distribution, the probability is spread out over some range from $a$ to $b$, as shown in Figure 7.1.1.

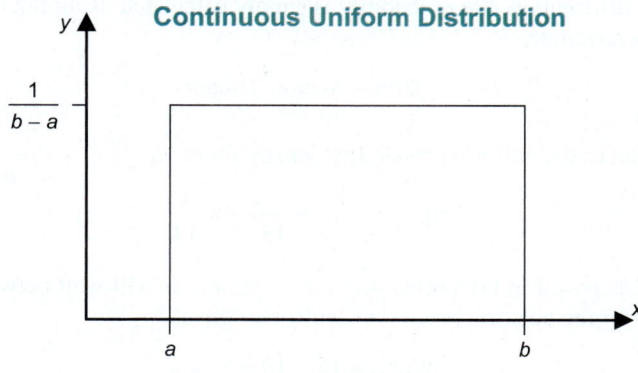

**Continuous Uniform Distribution**

**Figure 7.1.1**

Continuous random variables do not have probability distribution functions. Instead, they have **probability density functions**, which are denoted by $f(x)$. The probability density function is used to calculate probabilities for continuous random variables. The probability density function for the uniform random variable, its expected value (mean), and its standard deviation are given in the following formula. The parameters of the probability density function are the minimum and maximum values of the uniform random variable, and are referred to as $a$ and $b$, respectively.

**Formula**

## Uniform Probability Density Function

The **uniform probability density function** is as follows.

$$f(x) = \begin{cases} \dfrac{1}{b-a} & \text{for } a \leq x \leq b \\ 0 & \text{otherwise.} \end{cases}$$

The mean and standard deviation are given by the following expressions.

$$\mu = \frac{a+b}{2} \quad \text{and} \quad \sigma = \frac{b-a}{\sqrt{12}}$$

When the uniform probability density function is graphed, it produces a rectangle or square. The probability of observing a random variable in some interval is expressed as the area under the density function associated with the interval. Since the density function for the uniform distribution produces a rectangle, calculating the probability of an interval is as simple as calculating the area of a rectangle, which does not require complicated geometry.

**Example 7.1.1**

**Calculating Probability, Expected Value, and Standard Deviation of a Continuous Uniform Probability Density Function**

The Blacksburg Bus Company is interested in the wait time of its passengers between 8:00 am and 8:00 pm. Suppose that the wait time follows a uniform distribution with the minimum wait time being one minute and the maximum wait time 15 minutes.

a.  What is the probability that a passenger will wait between 10 and 15 minutes?

b.  What is the expected wait time between 8:00 am and 8:00 pm?

c.  What is the standard deviation of the wait time between 8:00 am and 8:00 pm?

**SOLUTION**

a.  Let $X$ = the number of minutes a passenger waits for the bus. The probability of a random variable is given by the area under the probability density function. For the uniform distribution, the probability is simply calculated using the formula for the area of a rectangle.

$$\text{Area} = \text{Width} \cdot \text{Height}$$

According to the uniform probability density function $f(x) = \dfrac{1}{b-a}$. Therefore,

$$\text{Height} = \frac{1}{b-a} = \frac{1}{15-1} = \frac{1}{14}.$$

We are interested in the probability that a passenger will wait between 10 and 15 minutes for the bus, so

$$\text{Width} = 15 - 10 = 5, \text{ and}$$

$$\text{Area} = \text{Width} \cdot \text{Height} = 5\left(\frac{1}{14}\right) = \frac{5}{14}.$$

Since the probability of observing a uniform random variable is the area under the density function associated with the interval, we calculate the probability of a passenger waiting between 10 and 15 minutes as follows (see Figure 7.1.2).

$$P(10 \leq X \leq 15) = \frac{5}{14}$$

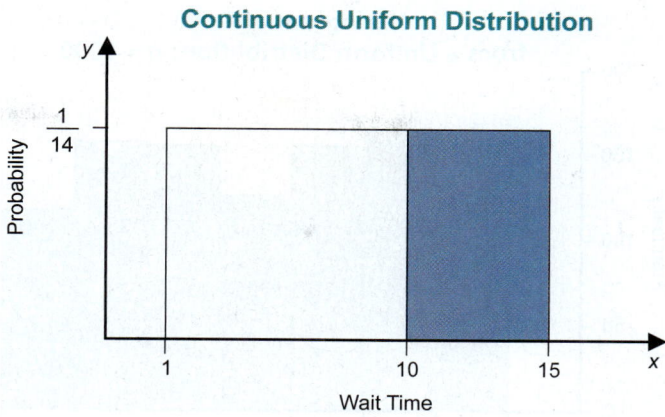

**Continuous Uniform Distribution**

Figure 7.1.2

**b.** The expected value of a continuous uniform random variable is given by

$$E(X) = \mu = \frac{a+b}{2} = \frac{(1+15)}{2} = 8.$$

Therefore, the expected wait time for a passenger is 8 minutes.

**c.** The standard deviation of a continuous uniform random variable is given by

$$\sigma = \frac{b-a}{\sqrt{12}} = \frac{15-1}{\sqrt{12}} \approx 4.0415.$$

Therefore, the standard deviation for the wait time of a passenger for a bus operated by the Blacksburg Bus Company is 4.0415 minutes.

## What Do Data from a Uniform Random Variable Look Like?

While the density function for the uniform distribution has a flat top, a histogram of data from a uniformly distributed random process will not be perfectly flat. Suppose that we generated 100 observations from a random process that was uniformly distributed between 2 and 15. Clearly, the frequency of each category is not identical (see Figure 7.1.3).

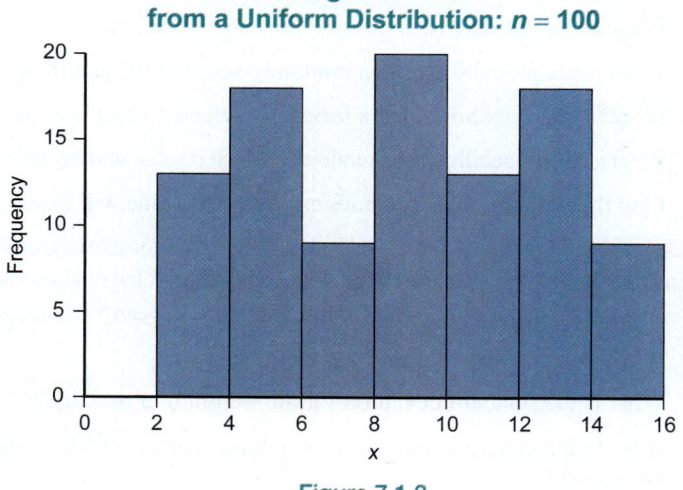

**Histogram of Data from a Uniform Distribution: $n = 100$**

Figure 7.1.3

However, if we were to generate 1000 observations, the distribution would begin to level (see Figure 7.1.4).

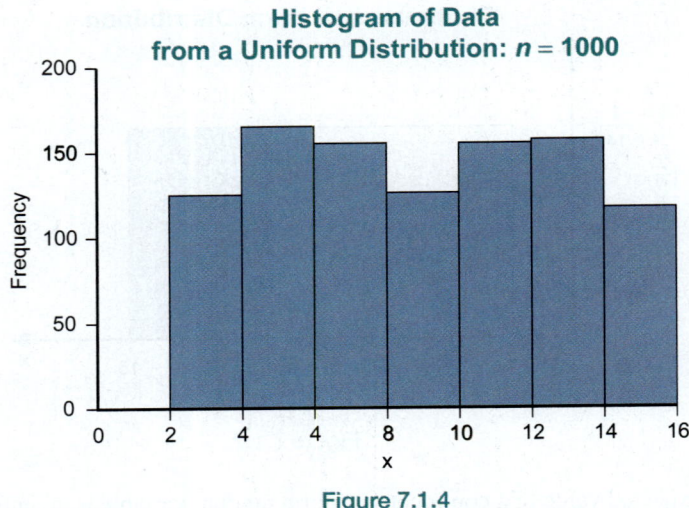

Figure 7.1.4

## ✎ 7.1 Exercises

### Basic Concepts

1. Probability is defined differently for discrete and continuous random variables. Describe this difference.

2. What is a probability density function?

3. How is the continuous uniform distribution different from the discrete uniform distribution?

4. What is the uniform probability density function?

5. Describe the shape of the density function for a uniform distribution.

### Exercises

6. Suppose a continuous random variable is uniformly distributed between 10 and 70.

    a. What is the mean of the distribution?

    b. What is the standard deviation of the distribution?

    c. What is the probability that a randomly selected value will be above 45?

    d. What is the probability that a randomly selected value will be less than 30?

    e. What is the probability that a randomly selected value will be between 25 and 50?

    f. Find the probability that a randomly selected value will exactly equal 35.

7. Polar Bear Frozen Foods manufactures frozen French fries for sale to grocery store chains. The final package weight is thought to be a uniformly distributed random variable. Assume $X$, the weight of French fries has a uniform distribution between 57 ounces and 63 ounces.

    a. What is the mean weight for a package?

    b. What is the standard deviation for the weight of a package?

    c. What is the probability that a store will receive a package weighing less than 59 ounces?

    d. What is the probability that a package will contain between 60 and 63 ounces?

    e. What is the probability that a package will contain more than 62 ounces?

    f. Find the probability that a package will contain exactly 60 ounces.

8. The annual increase in height of cedar trees is believed to be distributed uniformly between six and eleven inches.

   **a.** Draw a picture of the distribution of growth in height of cedar trees.

   **b.** What is the mean growth per year?

   **c.** What is the standard deviation of the growth per year?

   **d.** What is the probability that a randomly selected cedar tree will grow between 9 and 10 inches in a given year?

   **e.** Find the probability that a randomly selected cedar tree will grow less than 8 inches in a given year.

   **f.** Find the probability that a randomly selected cedar tree will grow more than 9 inches in a given year.

   **g.** Find the probability that a randomly selected cedar tree will grow exactly 7 inches in a given year.

9. A particular employee arrives to work sometime between 8:00 am and 8:30 am. Based on past experience the company has determined that the employee is equally likely to arrive at any time between 8:00 am and 8:30 am.

   **a.** On average, what time does the employee arrive?

   **b.** What is the standard deviation of the time at which the employee arrives?

   **c.** If a call comes in for the employee at 8:10 am, find the probability that the employee will be there to take the call.

   **d.** Find the probability that the employee will arrive between 8:20 am and 8:25 am.

   **e.** Find the probability that the employee will arrive after 8:15 am.

   **f.** Find the probability that the employee will arrive at exactly 8:10 am.

## 7.2 The Normal Distribution

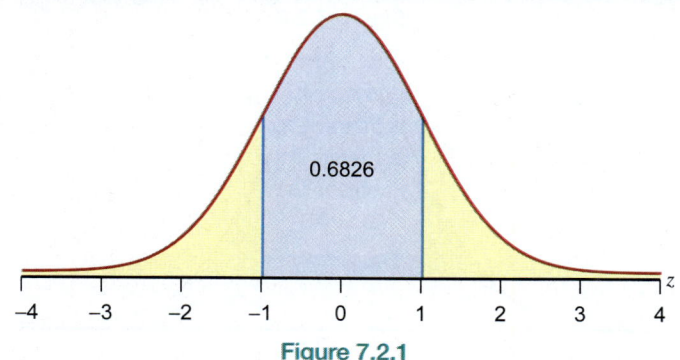

Figure 7.2.1

The **normal distribution**, originally called the Gaussian distribution, was named after Karl Gauss who published a work in 1833 describing the mathematical definition of the distribution. Gauss developed this distribution to describe the error in predicting the orbits of planets.

Normal distributions are all bell-shaped, but the bells come in various shapes and sizes. Since all normal distributions are symmetric, the mean, median, and mode are all equal.

Although normally distributed random variables can range in value from negative infinity to positive infinity, values that are a great distance from the mean rarely occur. You may recall that when we discussed box plots, these values were called outliers.

### The Origins of the Normal Distribution: Abraham de Moivre 1667–1754

De Moivre was born in France but lived most of his life in England. In a paper in 1733 de Moivre published the equation that describes the normal curve. He allegedly was doing calculations using the binomial distribution for gamblers and was looking for a shortcut in very arduous calculations. He discovered the normal distribution as the limit of the binomial distribution. De Moivre was a highly respected mathematician and friend of Issac Newton.

De Moivre's discovery received little attention until Laplace began writing on probability in the 1770s. There are two other mathematicians who discovered the equation of the normal curve, Adrain in 1808 and Gauss in 1809. Even though de Moivre published the equation for the normal distribution more than 75 years earlier than Gauss, the normal curve was called the Gaussian distribution for many years. Even now, you will hear the normal curve referred to as the Gaussian distribution.

> ### Properties
>
> #### Properties of the Normal Distribution
>
> 1. The normal distribution is symmetric. That is, the curve's shape to the left of the mean is the mirror image of the curve's shape to the right of the mean.
> 2. The highest point on the normal curve is located at the mean, which is also the location of the median and the mode of the distribution.
> 3. The area under the curve of the normal distribution equals 1.
> 4. Due to symmetry, the area to the right of the mean equals the area to the left of the mean, and each of these areas equals 0.5.
> 5. The shape of the normal distribution is defined by its two parameters, the mean ($\mu$) and the standard deviation ($\sigma$).

In Figure 7.2.1 the shaded area represents the probability of being within $\pm 1\sigma$ of the mean. Regardless of the values of the mean and standard deviation of the normal distribution, the area under the curve and the probability of being within one standard deviation ($\pm 1\sigma$) of the mean equals 0.6826.

Figure 7.2.2 illustrates the area under the curve within two standard deviations of the mean. The probability of being within $\pm 2\sigma$ of the mean equals 0.9544 for every normal distribution.

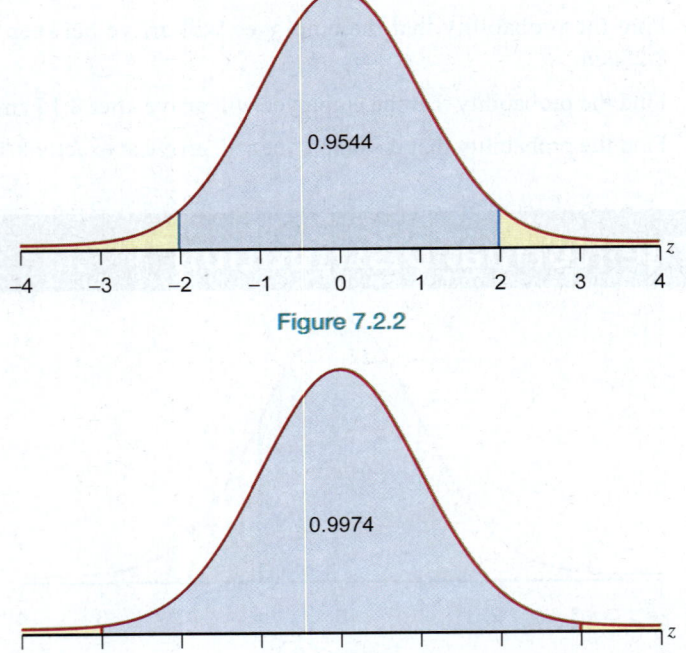

**Figure 7.2.2**

### Astronomy and the Normal Distribution

Although de Moivre derived the equation of the normal distribution as a consequence of problems involving the binomial distribution, Gauss and Laplace were inspired by their desire to predict the locations of planets, stars, meteors, and comets. Their predictions of the location of celestial bodies produced errors and they needed to describe those errors. In the 19th century the normal distribution was referred to as the astronomers' error law. Much of the early scientific use of the normal distribution was in the analysis of errors.

**Figure 7.2.3**

Figure 7.2.3 illustrates the area under the curve within three standard deviations of the mean. As you can see, virtually all of the area under the curve is within three standard deviations of the mean. The probability of being within $\pm 3\sigma$ of the mean equals 0.9974.

Since the empirical rule given in Chapter 4 is based on the normal distribution, these results are identical to the empirical rule frequencies.

Although a normal distribution has a bell shape, a bell shape does not imply a normal distribution. A number of non-normal distributions with bell shapes will be introduced in later chapters. In fact the word *normal* can be somewhat misleading. The name suggests that the distribution is a fact of nature. It is not. However, many variables seem to possess a shape that resembles the normal distribution.

The normal distribution is the preeminent distribution used in the statistical theory we will examine. Many statistical inference procedures, either directly or indirectly, have been developed based on the normal distribution. These procedures usually assume that the population from which a random sample is drawn is normally distributed.

Like other theoretical distributions, a normal distribution is completely defined by its probability density function.

### Formula

#### Normal Probability Density Function

The **normal probability density function** is given by:

$$f(x) = \frac{1}{\sigma\sqrt{2\pi}} e^{-\frac{(x-\mu)^2}{2\sigma^2}}$$

At first glance, this function does not look very *normal*! The distribution has two parameters, $\mu$ and $\sigma$, which are the mean and standard deviation, respectively. The mean defines the location, and the standard deviation determines the dispersion. Figure 7.2.4 illustrates three normal distributions with identical standard deviations. The only difference in the distributions is the central location, the mean.

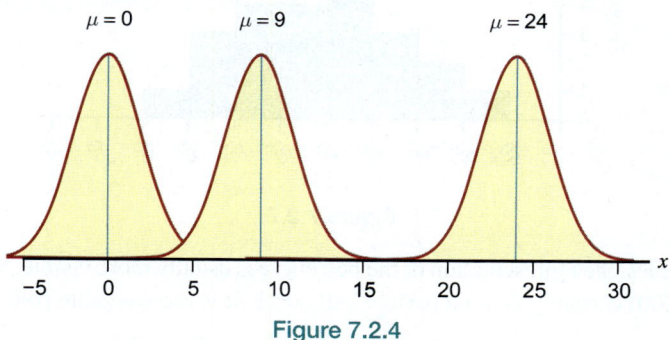

Figure 7.2.4

In Figure 7.2.5, there are two distributions with identical means, but with different standard deviations. Changing the standard deviation parameter can have rather significant effects on the shape of the distribution.

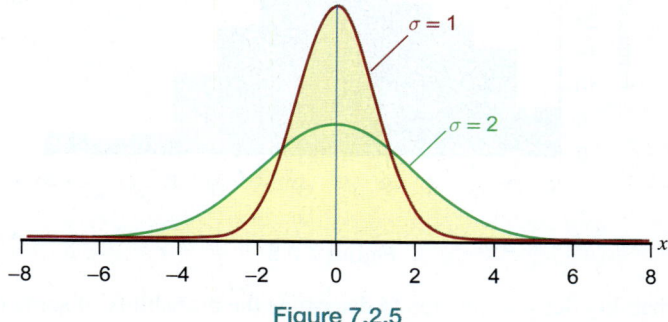

Figure 7.2.5

**Bringing the Normal Distribution Out of the Closet: Adolphe Quetelet 1796–1874**

Quetelet was born in Belgium and received his doctorate from the University of Ghent in 1819. In 1824 he went to Paris for three months to study astronomy and was exposed to the theory of probability by Laplace and Fourier.

Quetelet, however, became deeply interested in social science and believed the astronomers' error law and other laws of physics could be applied to phenomena in the social world. In 1844 Quetelet announced that the astronomers' error law (the normal distribution) could be applied to human features such as height, weight, and girth. He believed that the models that astronomers were using could eventually be used in a new science of "social mechanics." Quetelet was instrumental in popularizing the normal distribution in disciplines other than astronomy.

## Looking at Data from Normal Distributions

The normal distribution is a theoretical construct with a bell shape. Therefore, it would not be unreasonable to expect data drawn from a normal population to exhibit the bell-shaped characteristic. In two small samples taken from a normal population, shown in Figures 7.2.6 and 7.2.7, the data do show a faint resemblance of a bell shape. But as you can see, the shapes of the histograms developed from these small samples are somewhat unpredictable, even though the bell-shaped pattern is to some extent apparent.

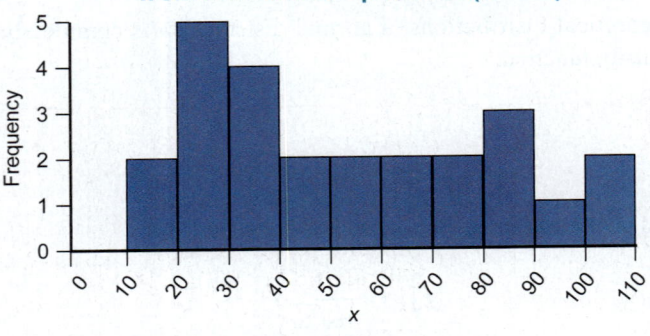

Figure 7.2.6

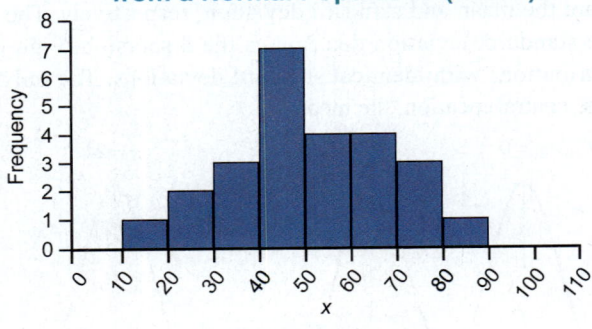

Figure 7.2.7

For large samples, the representation of the bell curve is usually more visible. While the large sample ($n = 200$) certainly is not a perfect bell curve, it is recognizable (see Figure 7.2.8).

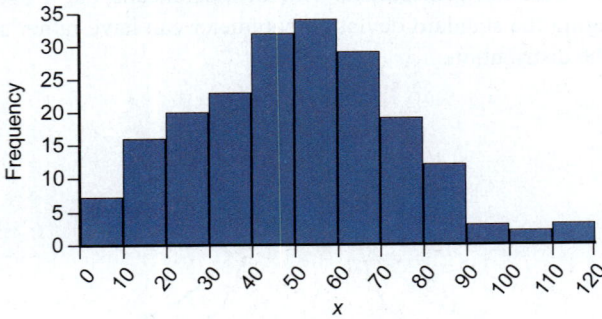

Figure 7.2.8

Using the probability density function to determine the probability of some interval would be complicated. Fortunately, there is an easier way. A special normal distribution, called the standard normal, can be used to determine probabilities for any normal random variable. The standard normal distribution will be discussed in Section 7.3.

## ✎ 7.2 Exercises

### Basic Concepts

1. How was the normal distribution developed?
2. Are the normal and uniform distributions probability models?
3. List the properties of the normal distribution.
4. What is the shape of the normal distribution?
5. What are the parameters of the normal distribution?
6. If the variance of a normal distribution is constant, what affect will changes in the mean have on the distribution?
7. If the mean of a normal distribution is constant, what effect will changes in the standard deviation have on the distribution?

### Exercises

8. Sketch a normal curve and mark each of the following on the $x$-axis.
      **a.** $\mu$           **b.** $\mu + \sigma$           **c.** $\mu - \sigma$
9. Sketch a normal curve and use labels to illustrate the empirical rule.
10. Sketch three normal curves on a single axis that have the same standard deviation but different means.
11. Sketch three normal curves on a single axis that have the same mean, but different standard deviations.

## 7.3 Assessing Normality Graphically

The normal distribution is the most important continuous probability distribution. The distribution follows a bell-shaped curve, centered about its mean, and usually, outliers do not have a large impact on the value of the mean. Numerous statistical methods used to analyze data make assumptions about normality including $t$-tests, regression analysis, and analysis of variance, which is the reason we first test the normality of the data before performing any analysis. If the data follow a normal distribution, then we can use parametric procedures to analyze the data; if not, then we will use nonparametric methods to analyze the data.

As stated above, an assessment of normality is the prerequisite for many statistical tests. Typically, a visual check is sufficient to assess normality. In this section, we will discuss graphical techniques that can be used to assess the normality of the data. Graphically assessing normality has the advantage of using the good judgment and expertise of the analyst. Of course, the more expertise the analyst has, the less likely it is that there will be an incorrect interpretation of the data.

In this section, we will discuss the histogram, box plot, and normal probability plot which are graphical techniques used to assess the normality of data. There are many statistical software programs that can be used to create the histogram, box plot, and normal probability plot such as JMP, Excel, and Minitab. We will use JMP to graphically assess normality in the example that follows.

### Histogram

As discussed in Chapter 3, a histogram is a graphical method that shows the distribution of a continuous random variable. Even though histograms resemble bar graphs, we analyze them quite differently given that we are using quantitative data. When analyzing the histogram, we can visually see the structure of the data—the shape (symmetric or skewed), the modality of the data, and the existence of outliers.

## Technology

For instructions on creating a histogram using technology, please visit stat.hawkeslearning.com and navigate to **Discovering Business Statistics, Second Edition > Technology Instructions > Graphs > Histogram.**

To assess normality using a histogram, if the histogram is approximately bell-shaped and symmetric about the mean, we can assume that the data are normally distributed. If the histogram is not bell-shaped, we cannot use classical statistical procedures for analysis that rely on the data to be normally distributed.

## Box Plot

The box plot is another way to assess the normality of a data set. Recall the box plot from Chapter 4 which provided a graphical summary of the central tendency, the spread, the skewness, and the potential existence of outliers. The box plot (see Figure 7.3.1) shows the median as a vertical line inside the box, the interquartile range (IQR), which is the range between the first and third quartile, and the whiskers (lines extending from the left and right of the box) which represent the minimum and maximum values of the data.

**General Layout of a Box Plot**

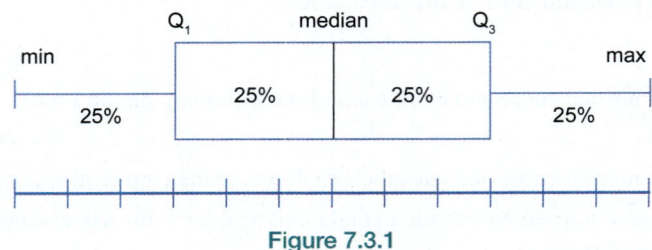

**Figure 7.3.1**

## Technology

For instructions on creating a box plot using technology, please visit stat.hawkeslearning.com and navigate to **Discovering Business Statistics, Second Edition > Technology Instructions > Graphs > Box Plot.**

When an observation is more than 1.5 times the IQR from either end of the box (i.e., $Q_1 - 1.5 \cdot \text{IQR}$ or $Q_3 + 1.5 \cdot \text{IQR}$) it is called an outlier. If an observation is more than three times the IQR outside the box, it is called an extreme outlier.

A box plot that is symmetric, with the median line at approximately the center of the box and with symmetric whiskers, tends to indicate that the data may have come from a normal distribution. If many outliers are present in the data, either the outliers should be removed or the data should not be treated as normally distributed.

## Normal Probability Plot

The normal probability plot is a plot of the $z$-scores (i.e., the normal scores) against the actual data. When analyzing the normal probability plot to assess normality, if the data follow a diagonal line this implies that you have normally distributed data. If the data are skewed (right or left) and do not follow a linear pattern then it's likely that you do not have normally distributed data. With a normal probability plot, it's relatively easy to see the individual observations that don't fit a normal distribution.

## Technology

For instructions on creating a normal probability plot using technology, please visit stat.hawkeslearning.com and navigate to **Discovering Business Statistics, Second Edition > Technology Instructions > Graphs > Normal Probability Plot.**

Figure 7.3.2 shows a general normal probability plot. The line represents the theoretical line for normally distributed data. The dots represent the real, empirical data that we are checking for normality. If all of the dots roughly follow the line, we can be confident that the data follow a normal distribution. However, if the data deviates significantly from the line, the assumption of normality may not hold.

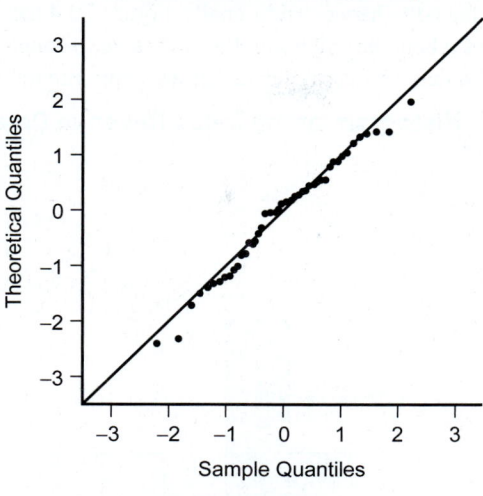

**Generic Normal Probability Plot**

Figure 7.3.2

The steps below describe the details of how one can manually create a normal probability plot.

> **Procedure**
>
> **Creating a Normal Probability Plot by Hand**
>
> 1. Arrange the data values in ascending order.
> 2. Calculate $f_i = \dfrac{(i - 0.5)}{n}$, where $i$ is the position of the data value in the ordered list and $n$ is the number of observations in the data set.
> 3. Find the $z$-score for each value of $f_i$.
> 4. Plot the data values on the horizontal axis and the corresponding $z$-score on the vertical axis.

The normal probability plot is rarely drawn by hand because the exact normal $z$-scores used for the plot can't be looked up in a table. Thus, we use software packages such as JMP, Excel, or Minitab to create normal probability plots.

---

**Example 7.3.1**

**Assessing Normality of Sales Revenue for CWK Consulting**

Cindy, the President and CEO of CWK Consulting, a midsize accounting firm, conducted a brief survey of her salespeople to analyze their annual sales revenue for 2019. She took a random sample of the sales revenue of 25 employees and the summary statistics are provided in Table 7.3.1. Determine if the sales revenue data follow a normal distribution using a histogram, box plot, and normal probability plot.

**SOLUTION**

The summary statistics of the sales revenue data can be found in Table 7.3.1. Note that the mean is $51,728.36 and the standard deviation is $1,061.47.

**Data**

The data can be found at stat.hawkeslearning.com by navigating to **Discovering Business Statistics, Second Edition > Data Sets > CWK Sales Revenue Data**.

| Table 7.3.1: Summary Statistics of the Sales Revenue Data | |
|---|---|
| Mean | 51,728.36 |
| Std Dev | 5307.3679 |
| Std Err Mean | 1061.4736 |
| Upper 95% Mean | 53919.134 |
| Lower 95% Mean | 49537.586 |
| N | 25 |

Using JMP, the histogram of the sales revenue data is shown in Figure 7.3.3. The histogram appears to be somewhat bell-shaped. Additionally, Figure 7.3.4 has a smooth curve drawn on it which resembles a bell-shaped curve. It's fairly safe to conclude by looking at both Figures 7.3.3 and 7.3.4 that the sales revenue data are approximately normally distributed.

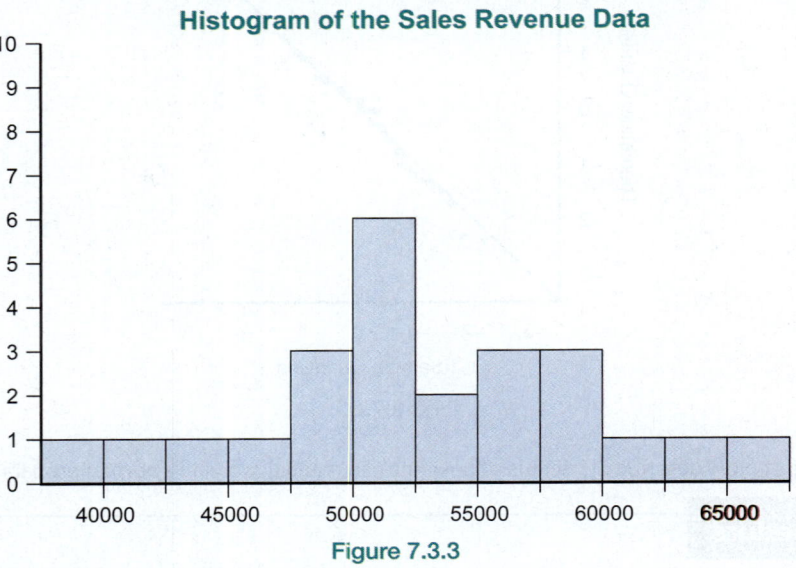

Figure 7.3.3

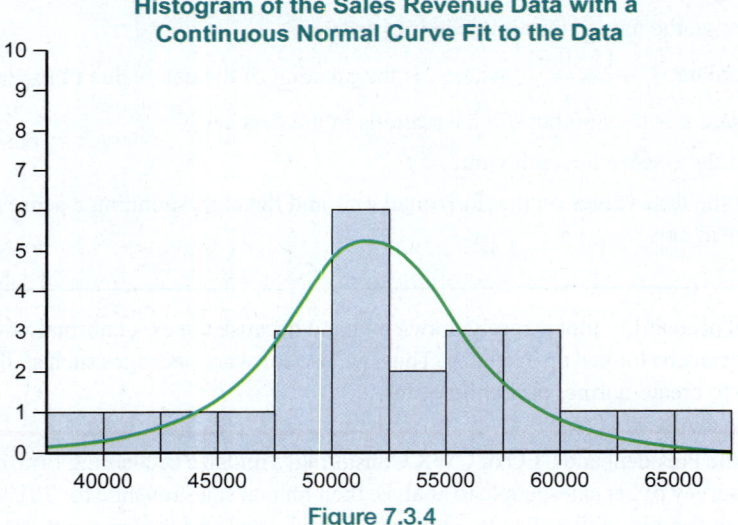

Figure 7.3.4

A box plot of the sales revenue data is presented in Figure 7.3.5. You can see that $Q_1$ and $Q_3$ are approximately equidistant from the median of the data. The whiskers also appear to be equidistant from $Q_1$ and $Q_3$, respectively. The box plot is yet another graphical display that supports that the data follow a normal distribution.

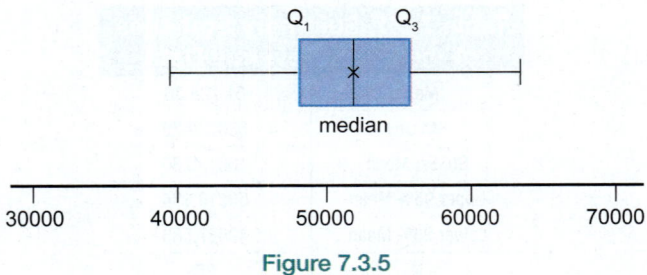

Figure 7.3.5

Lastly, the normal probability plot is displayed in Figure 7.3.6. The solid line represents the theoretical normal data and we can see that the actual data do not stray too far from the line. Again, the normal probability plot supports that the data follow a normal distribution.

**Normal Probability Plot of the Sales Revenue Data**

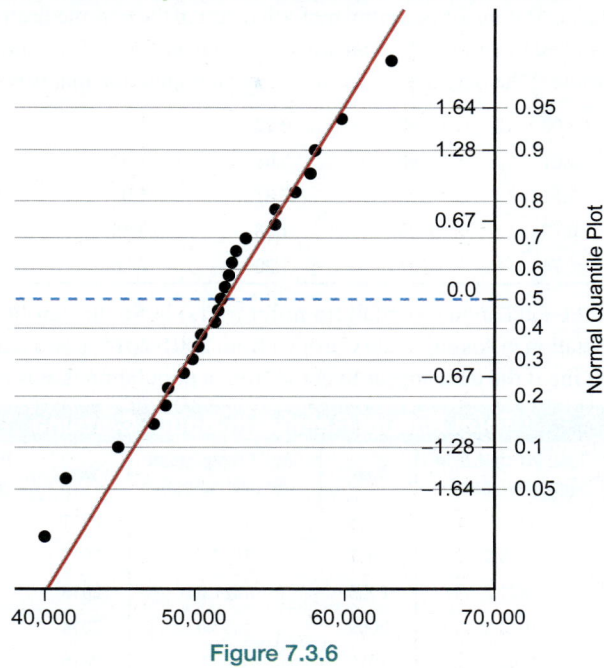

Figure 7.3.6

Given that all three graphical displays used in the example tend to support that the data follow a normal distribution, it is safe to continue with any analysis based on the assumption that the data are normally distributed.

More formal and precise tests are available which calculate the probability that a sample is collected from a normal population. To name a few, these tests are the Anderson-Darling Test, Kolmogorov-Smirnov Test, and Shapiro-Wilk W Test. These tests have the advantage of allowing the analyst to make an objective judgment of normality. However, the primary disadvantage is that these tests are sensitive to small sample sizes.

We have discussed using graphical techniques to determine if the data are collected from a normal distribution. What if the data do not follow a normal distribution? You can still analyze the data but first you must perform some type of transformation on the data or use nonparametric procedures. Keep in mind that if you use nonparametric procedures, they have less power than the classical parametric procedures. Lastly, parametric tests are robust to violations of normality when you have large sample sizes.

## ✏️ 7.3 Exercises

### Basic Concepts

1. List three ways to graphically assess the normality of a data set.

2. Describe the general procedure for creating a normal probability plot.

3. How should a normal probability plot look to indicate normality?

### Exercises

4. Construct a histogram using the "BA" (batting average) column of the Moneyball data set. Can we assume batting averages have a normal distribution?

 **Data**

The Moneyball data set can be found by visiting stat.hawkeslearning.com and navigating to **Discovering Business Statistics, Second Edition > Data Sets > Moneyball**.

5. Create a normal probability plot of the housefly wing lengths data. What do you observe?

6. A pharmaceutical company wants to test whether a new cold medication will perform better than an existing medication. Laboratory technicians observe a sample of 25 patients and record the number of hours it takes for each patient to feel symptom relief after taking the medicine. Before the company performs a test of the new medication against the current one, they need to know if the data are normally distributed. Use a normal probability plot to determine if the data appear to come from a population that is normally distributed.

| | | | | |
|---|---|---|---|---|
| 3.00 | 1.50 | 0.20 | 1.62 | 1.06 |
| 3.01 | 2.45 | 0.66 | 1.94 | 0.21 |
| 1.51 | 3.08 | 5.37 | 6.96 | 1.32 |
| 0.79 | 7.20 | 1.36 | 4.45 | 3.29 |
| 1.74 | 3.87 | 1.90 | 3.50 | 3.09 |

7. Data on the total annual rainfall (in millimeters) in South Carolina was gathered by a weather station in Aiken, South Carolina from 2001-2015. Use a normal probability plot to determine if the data appear to come from a population that is normally distributed.

| Total Annual Rainfall in South Carolina | | | | | |
|---|---|---|---|---|---|
| Year | Total Precipitation (in millimeters) | Year | Total Precipitation (in millimeters) | Year | Total Precipitation (in millimeters) |
| 2001 | 895.7 | 2006 | 1031.3 | 2011 | 991.8 |
| 2002 | 1106.9 | 2007 | 1002.7 | 2012 | 1089.5 |
| 2003 | 1681.3 | 2008 | 1321.6 | 2013 | 1584.0 |
| 2004 | 1003.6 | 2009 | 1434.0 | 2014 | 1070.2 |
| 2005 | 1166.1 | 2010 | 946.2 | 2015 | 1537.4 |

Source: The United States Historical Climatology Network.

8. A professor is interested in examining the distribution of the grades his students received on the midterm exam. There are 18 students in the class, and no time limit was given for the exam. Use a normal probability plot to determine if the students' grades are normally distributed.

| | | | | | |
|---|---|---|---|---|---|
| 80.8 | 81.7 | 81.7 | 81.7 | 81.7 | 82.5 |
| 83.3 | 83.3 | 84.2 | 84.2 | 85 | 86.7 |
| 86.7 | 87.5 | 87.5 | 90.3 | 90.4 | 90.8 |

Source: https://openmv.net/info/unlimited-time-test

9. A group of students and professors are studying conifers in the Pacific Northwest United States. They take a sample of 25 Douglas Fir trees and record several metrics, including the circumference of the trunks (in meters). Use a normal probability plot to determine if the trunk circumference values are normally distributed.

| | | | | |
|---|---|---|---|---|
| 4.97 | 0.45 | 0.40 | 0.15 | 2.84 |
| 6.65 | 0.62 | 0.39 | 0.86 | 1.24 |
| 4.93 | 0.64 | 0.62 | 2.22 | 2.23 |
| 0.29 | 0.18 | 0.27 | 1.97 | 2.45 |
| 0.19 | 0.55 | 0.41 | 2.85 | 9.09 |

Source: Biometrics of Douglas firs. http://seattlecentral.edu/qelp/sets/076/076.html. White River Valley, Washington, 20-Apr-01. Students: Ingrid McNeely and Dylan Morgan, Seattle Central Community College.

**10.** A group of friends decide to run a marathon together. There are 16 runners in the group, and they are all in relatively good shape. Use a normal probability plot to determine if their marathon times are normally distributed.

| | | | |
|---|---|---|---|
| 4:07:58 | 4:18:34 | 4:21:15 | 4:24:23 |
| 4:08:07 | 4:18:40 | 4:22:17 | 4:25:12 |
| 4:16:28 | 4:19:39 | 4:23:52 | 4:25:14 |
| 4:17:30 | 4:19:45 | 4:23:55 | 4:26:34 |

**11.** The underwriters for a new type of auto insurance policy gathered monthly mileage data from 18 city drivers. What conclusions can be drawn about the distribution of the data?

    **a.** Using the mileage data provided below, construct a box plot to assess the normality of the data set.

    **b.** Using the mileage data provided below, construct a histogram with eight classes to assess the normality of the data set.

| | | | | | |
|---|---|---|---|---|---|
| 385 | 410 | 416.5 | 421 | 433.5 | 451.5 |
| 408.5 | 411 | 416.5 | 425.2 | 437.5 | 452 |
| 408.5 | 412.5 | 421 | 433.5 | 437.5 | 460 |

**12.** Billings Marketing is asked to develop a recruiting campaign for ABC University. Using the age data from recent college applications shown below, construct a box plot and a histogram with eight classes. Describe the normality of the data set. The results will aid in the development of a marketing strategy.

| | | | | | |
|---|---|---|---|---|---|
| 18 | 19 | 20 | 26 | 22 | 20 |
| 18 | 19 | 20 | 27 | 22 | 19 |
| 18 | 28 | 20 | 19 | 23 | 20 |
| 19 | 19 | 21 | 29 | 19 | 30 |
| 24 | 19 | 21 | 18 | 24 | 19 |
| 19 | 19 | 21 | 19 | 25 | 18 |

**13.** Using the data from Exercise 8, construct a box plot and a histogram with 8 columns. The results seem to suggest that the data are not normally distributed. A normal probability plot indicated that the data was normally distributed. Why do these graphical methods for accessing normality seem to contradict one another?

| | | | | | |
|---|---|---|---|---|---|
| 80.8 | 81.7 | 81.7 | 81.7 | 81.7 | 82.5 |
| 83.3 | 83.3 | 84.2 | 84.2 | 85 | 86.7 |
| 86.7 | 87.5 | 87.5 | 90.3 | 90.4 | 90.8 |

# 7.4 The Standard Normal Distribution

Given that the normal distribution is a function of two continuous parameters $\mu$ and $\sigma$, there are an infinite number of combinations for $\mu$ and $\sigma$, and thus, an infinite number of normal distributions. The **standard normal distribution** in Figure 7.4.1 is a special version of the normal distribution.

**Definition**

Standard Normal Distribution

The **standard normal distribution** is a normal distribution with a mean of zero and a standard deviation of one.

$$\mu = 0 \text{ and } \sigma = 1$$

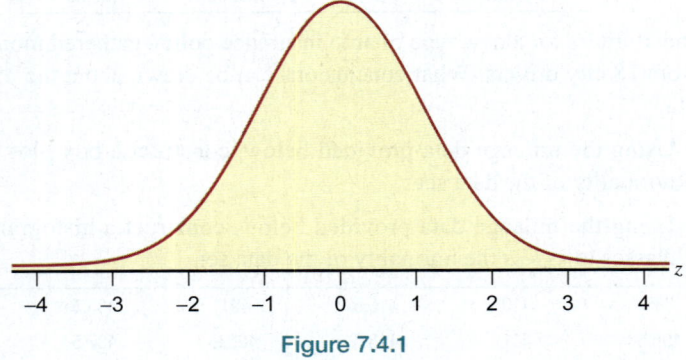

**Figure 7.4.1**

## Technology

For instructions on computing normal probabilities using technology, please visit stat.hawkeslearning.com and navigate to **Discovering Business Statistics, Second Edition > Technology Instructions > Normal Distribution > Normal Probability (cdf)**.

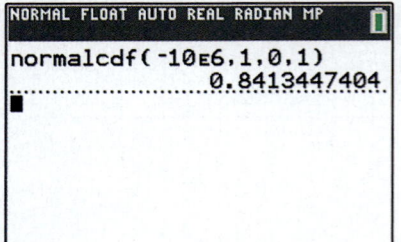

The standard normal distribution, also called the $z$-distribution, provides a basis for computing probabilities for all normal distributions. The technique used to convert any normal random variable into a standard normal random variable is called "standardizing" the random variable and was discussed earlier in Chapter 4.

Appendix A, Tables A, B, and C contain probability calculations for various areas under the standard normal curve. Specifically, Appendix A, Tables A and B provide the probability that a standard normal random variable will be less than a specified value. For example, to compute the probability that a standard normal random variable will be less than 1 (see Figure 7.4.2), look up the value 1.00 in Table B. The table value of 0.8413 is the area under the curve between negative infinity and 1, which is also the probability that the random variable will assume a value in that interval.

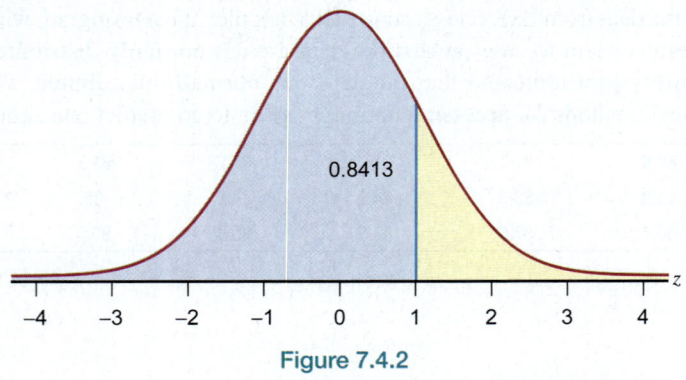

**Figure 7.4.2**

✏ **NOTE**

We will primarily use Tables A and B in the following examples to determine areas and probabilities under the standard normal curve, but note that Table C (areas between 0 and $z$) could also be used.

The $z$-distribution will be used throughout the remainder of this text. Thus, it is important to comprehend the use of the $z$-tables in the determination of probabilities for a variety of problems. Because of the manner in which the tables are constructed, some types of problems will require manipulation of the table values.

Compute the probability that a standard normal random variable is less than 1.27.

**SOLUTION**

Drawing a picture, even when the problem is rather simple, is a good idea. Remember that the probability is represented by the area under the standard normal curve.

**Example 7.4.1**

**Calculating a Probability Using the Standard Normal Distribution**

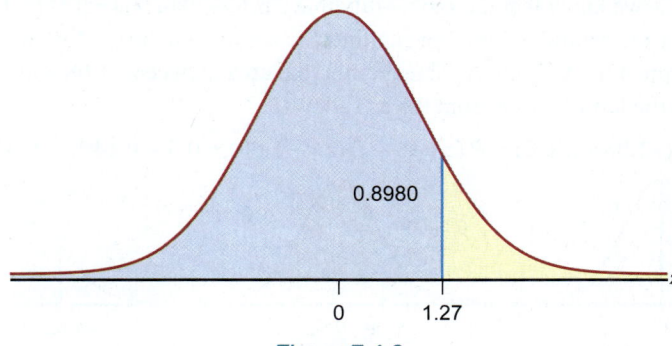

0.8980

0        1.27

**Figure 7.4.3**

Determining the area under the standard normal curve to the left of a particular value requires little effort since the tables are constructed to give the cumulative probabilities. That is, the table gives probabilities that the random variable $z$ is less than (or less than or equal to) some value (i.e., $P(z < z_0)$) where $z_0$ is the number of standard deviations above or below the mean). In this case, the construction of Table B exactly matches the kind of interval we are examining. Thus, merely looking up the value corresponding to 1.27 in the table is sufficient to obtain the probability.

⚭ **Technology**

Standard normal probabilities can be found using the NORM.S.DIST function in Excel. For instructions, please visit stat.hawkeslearning.com and navigate to **Discovering Business Statistics, Second Edition > Technology Instructions > Normal Distribution > Normal Probability (cdf)**.

| z | 0.00 | 0.01 | ... | 0.06 | 0.07 |
|-----|--------|--------|-----|--------|--------|
| 0.0 | 0.5000 | 0.5040 | | 0.5239 | 0.5279 |
| 0.1 | 0.5398 | 0.5438 | | 0.5636 | 0.5675 |
| ... | | | | | |
| 1.1 | 0.8643 | 0.8665 | | 0.8770 | 0.8790 |
| 1.2 | 0.8849 | 0.8869 | | 0.8962 | 0.8980 |
| 1.3 | 0.9032 | 0.9049 | | 0.9131 | 0.9147 |

| $fx$ | =NORM.S.DIST(1.27, TRUE) | | |
|------|------|------|------|
| | D | E | F | G |
| | | 0.897958 | | |

$$P(z < 1.27) = 0.8980$$

Determine the probability that a standard normal random variable is between −1.08 and 0.

**SOLUTION**

First, draw a picture.

**Example 7.4.2**

**Calculating a Probability Using the Standard Normal Distribution**

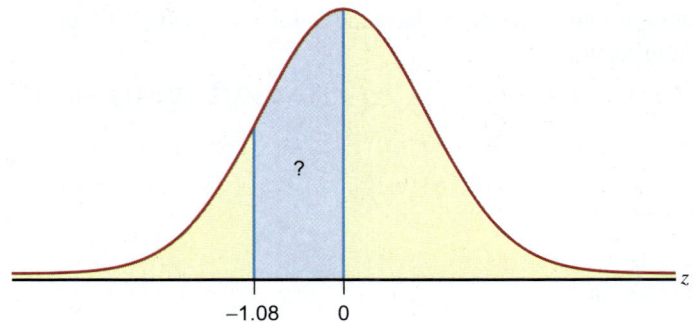

?

−1.08      0

**Figure 7.4.4**

## Technology

This probability can also be found using a TI-83/84 Plus calculator. For instructions, please visit stat.hawkeslearning.com and navigate to **Discovering Business Statistics, Second Edition > Technology Instructions >Technology Instructions > Normal Distribution > Normal Probability (cdf)**.

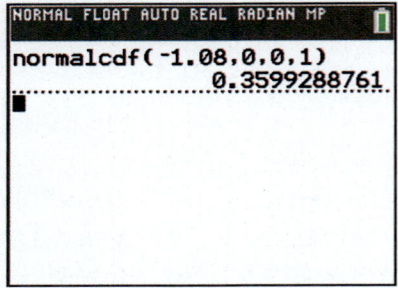

In this case, we cannot simply look up the value in the table to obtain the probability of interest. Understanding that the table gives us cumulative probabilities, if we find the probability that $z$ is less than 0 and then subtract the probability that $z$ is less than $-1.08$, we will get the probability that $z$ is between $-1.08$ and 0.

Because the standard normal distribution is symmetric and the total area under the curve is equal to 1, we know that the probability that $z$ is less than 0 is equal to 0.5 (since 0 is the mean or the central value). We can find $P(z < -1.08)$ to be 0.1401 from looking up $-1.08$ in Appendix A, Table A. Thus, to find the area between $-1.08$ and 0 we subtract the area to the left of $-1.08$ from 0.5 as follows.

$$P(-1.08 < z < 0) = P(z < 0) - P(z < -1.08) = 0.5 - 0.1401 = 0.3599$$

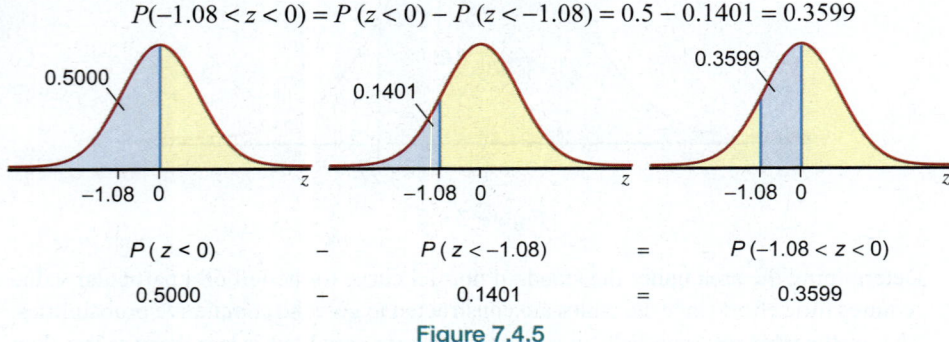

| $P(z < 0)$ | $-$ | $P(z < -1.08)$ | $=$ | $P(-1.08 < z < 0)$ |
|---|---|---|---|---|
| 0.5000 | $-$ | 0.1401 | $=$ | 0.3599 |

**Figure 7.4.5**

---

| Example 7.4.3 |
|---|

**Calculating a Probability Using the Standard Normal Distribution**

What is the probability that a standard normal random variable will be between 1 and 2?

**SOLUTION**

Again, first draw a picture.

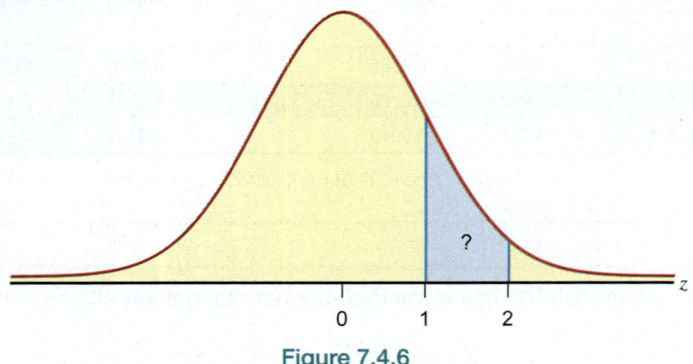

**Figure 7.4.6**

## Technology

For instructions on computing normal probabilities using technology, please visit stat.hawkeslearning.com and navigate to **Discovering Business Statistics, Second Edition >Technology Instructions > Normal Distribution > Normal Probability (cdf)**.

| $fx$ | =NORM.S.DIST(2, TRUE)-NORM.S.DIST(1, TRUE) | | | | |
|---|---|---|---|---|---|
| | D | E | F | G | H |
| | | 0.135905 | | | |

Because Table B gives us cumulative probabilities, we can find the probability that $z$ is less than 2 and then subtract the probability that $z$ is less than 1, yielding the probability that $z$ is between 1 and 2. Thus, we have the following. Figure 7.4.7 illustrates the areas involved in this calculation.

$$P(1 < z < 2) = P(z < 2) - P(z < 1) = 0.9772 - 0.8413 = 0.1359$$

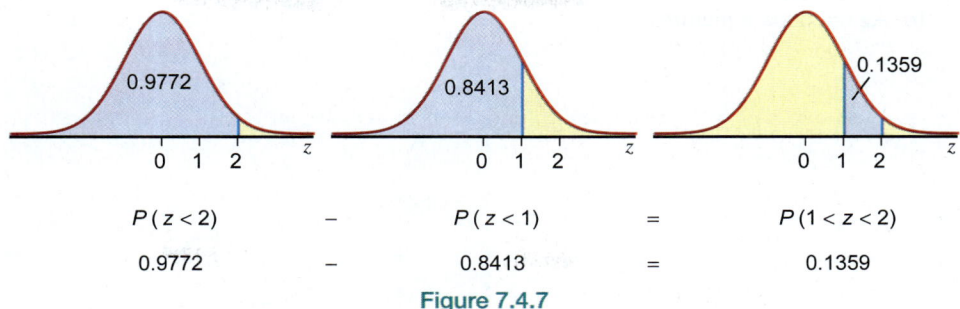

**Figure 7.4.7**

Again, drawing the picture proves to be invaluable in this example.

---

Given that $z$ is a standard normal random variable, find the value of $z$ for each situation.

**a.** The area to the left of $z$ is 0.9147.

**b.** The area between 0 and $z$ is 0.3665.

**c.** The area to the left of $z$ is 0.1469.

**d.** The area to the right of $z$ is 0.7967.

**Example 7.4.4**

**Determining the z-Value that corresponds to an Area Under the Standard Normal Distribution**

**SOLUTION**

**a.** First, draw a picture.

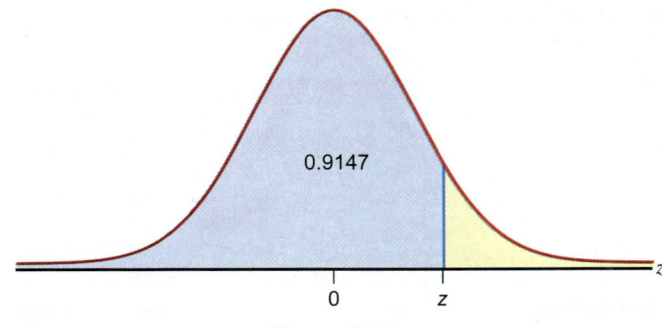

**Figure 7.4.8**

Note that this problem is slightly different from the previous one. In Example 7.4.3, you were asked to find a probability, given that you know the value of $z$. In this example, you are given a probability and asked to find the corresponding value of $z$. Recall that Appendix A, Table B gives you the cumulative probability of the area less than some value of $z$.

In order to find the value of $z$, look in the body of Appendix A, Table B and find the probability value 0.9147. Once you've found the value (the probability), determine the corresponding value of $z$. In this case, the value of $z$ is 1.37. So,

$$P(z < 1.37) = 0.9147$$

and the value of $z$ is 1.37 with the area to the left of it being 0.9147.

### ∞ Technology

To find the value of $z$ given a particular area with a TI-83/84 Plus calculator, use the invNorm function within the DISTR menu. For instructions, please visit stat.hawkeslearning.com and navigate to **Discovering Business Statistics, Second Edition > Technology Instructions > Normal Distribution > Inverse Normal**.

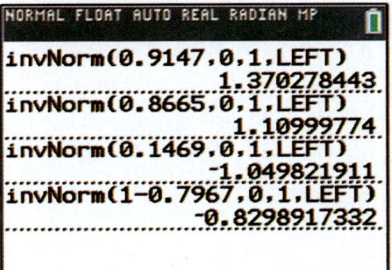

**b.** Again, draw a picture.

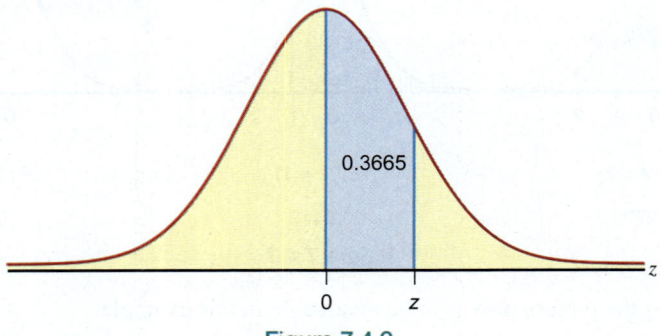

**Figure 7.4.9**

Recall that the table gives us cumulative probabilities. Since the area between 0 and *z* is 0.3665, if we write this as a cumulative probability, we need to add the area to the left of 0. Thus, the area to the left of *z* is 0.8665. This is the probability (less than some value of *z*) that is provided in Table B. Therefore, find 0.8665 in the body of Table B and locate the corresponding value of *z*. In this case,

$$P(z < 1.11) = 0.8665.$$

So, the value of *z* with the area 0.3665 between 0 and *z* is 1.11.

**c.** Just as in parts **a.** and **b.**, a picture can be helpful.

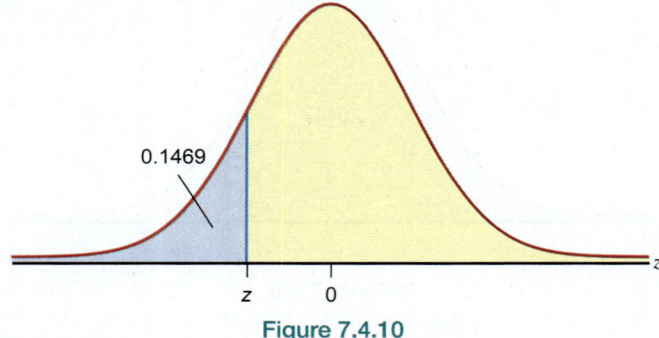

**Figure 7.4.10**

Please note that the value of *z* is to the left of 0. Thus, the value of *z* is going to be negative. Note that the area to the left of *z* represents the cumulative probability (in Figure 7.4.10). So, to find the value of *z*, we only need to find 0.1469 in the body of Appendix A, Table A. The value of *z* with the area 0.1469 to the left of it is −1.05. That is, $P(z < -1.05) = 0.1469$.

**d.** Once again, a picture can be very helpful.

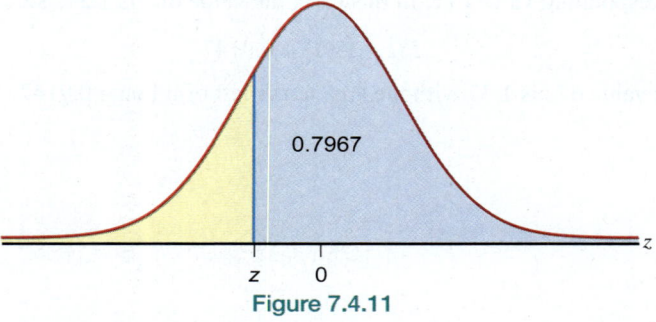

**Figure 7.4.11**

Note that from the picture, we have the area to the right of *z*. However, we know that the total area under the curve is 1. Thus, if the area to the right of *z* is 0.7967, then the area to the left of *z* is $1 - 0.7967 = 0.2033$. From the picture, it is clear that if we find 0.2033 in the body of Appendix A, Table A, the corresponding value of $z$ is the value we are interested in. This value of $z$ is −0.83. Therefore, the value of $z$ with the area 0.7967 to the right is −0.83.

## Formula

### Standardizing a Normal Random Variable

The following formula can transform any normal random variable into a **standard normal random variable**, $z$.

$$z = \frac{x - \mu}{\sigma}$$

where $x$ is a normal random variable with mean $\mu$ and standard deviation $\sigma$.

If we look at the individual pieces, exactly how the transformation works is not very mysterious. First, the numerator, $x - \mu$, centers the $z$-distribution around zero. By subtracting the mean of the random variable from each data value, the mean of the resulting random variable will be zero. A short example illustrates this point. Suppose that a population contained the following data values shown in Table 7.4.1.

| Table 7.4.1 | |
|---|---|
| Data Set A | Data Set A − 6 |
| 1 | $1 - 6 = -5$ |
| 5 | $5 - 6 = -1$ |
| 6 | $6 - 6 = 0$ |
| 12 | $12 - 6 = 6$ |
| Mean = 6 | Mean = 0 |

Data set A has a mean of 6. If 6 is subtracted from each of the data values, the resulting deviations are shown in the second column of the table. The deviations have a mean of zero. Essentially the location of the data set has been shifted to zero. The interrelationship of the data points to one another has not changed. Try this experiment on larger sets of data to convince yourself that subtracting the mean from each value of a data set will produce a data set that always has a mean of zero.

The standard deviation of data set A is approximately 3.9370. Let's standardize each data value in data set A (see Table 7.4.2). The resulting $z$-values indicate how far the data values in Table 7.4.1 are from the mean, measured in standard deviation units. The first $z$-value in Table 7.4.2 indicates that 1 is −1.27 standard deviation units from the mean. The mean and standard deviation of the transformed values in Table 7.4.2 are zero and one, respectively. You can verify that the mean and standard deviation of the $z$-scores are actually zero and one. Compute the population standard deviation rather than the sample standard deviation.

| Table 7.4.2 – Standardizing Data Set A | | |
|---|---|---|
| Data | Formula | Value of z |
| 1 | $\frac{1-6}{3.9370}$ | −1.27 |
| 5 | $\frac{5-6}{3.9370}$ | −0.25 |
| 6 | $\frac{6-6}{3.9370}$ | 0 |
| 12 | $\frac{12-6}{3.9370}$ | 1.52 |

Now that we have seen how to convert any normal random variable into a standard normal random variable, let's apply it.

---

<table>
<tr>
<td>

**Example 7.4.5**

**Calculating the Probability of Payroll Processing Time Using the Standard Normal Distribution**

</td>
<td>

Suppose that the times it takes for an accountant to process payroll are normally distributed with a mean of 10 minutes and a standard deviation of 20 minutes. Find the probability that the payroll processing time on a randomly selected payroll date will take between 10 minutes and 40 minutes.

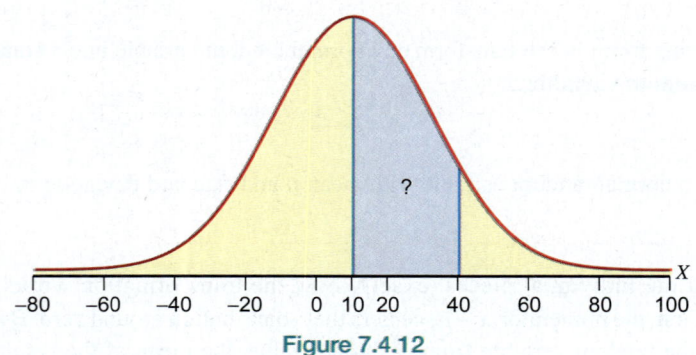

Figure 7.4.12

</td>
</tr>
</table>

## 🔗 Technology

To draw the area under a normal curve using the TI-83/84 calculator, please see the instructions on stat.hawkeslearning.com under **Discovering Business Statistics, Second Edition > Technology Instructions > Normal Distribution > Normal Probability Graph.**

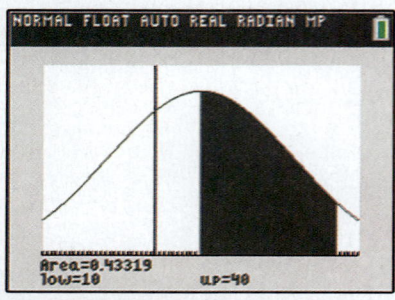

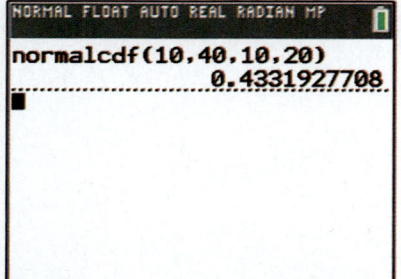

### SOLUTION

Suppose $X$, the payroll processing time, is a normally distributed random variable with a mean of 10 minutes and a standard deviation of 20 minutes.

Standardizing the random variable $X$ yields the following.

$$P(10 < X < 40) = P\left(\frac{10-10}{20} < \frac{x-\mu}{\sigma} < \frac{40-10}{20}\right)$$
$$= P(0 < z < 1.5)$$

Note that for each argument in the above probability statement, we subtracted the mean and divided by the standard deviation.

Once the problem has been converted to a problem involving $z$ (see Figure 7.4.13), the appropriate probability can be determined from the standard normal table in Appendix A, Table B.

We find the probability that $z$ is less than 1.5 and subtract 0.5 (the probability that $z$ is less than 0) as follows.

$$P(0 < z < 1.5) = P(z < 1.5) - P(z < 0) = 0.9332 - 0.5 = 0.4332$$

Thus, the probability that the payroll processing time on a randomly selected payroll date will take between 10 and 40 minutes is 0.4332.

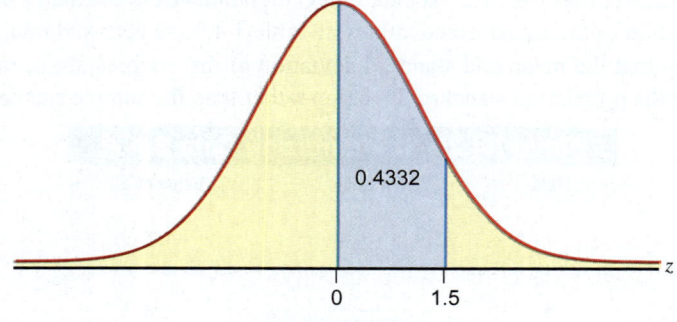

Figure 7.4.13

Continuing Example 7.4.5, find the probability that the payroll on a randomly selected payroll date will take more than 30 minutes to process by the accountant.

**Example 7.4.6**

**Calculating the Probability of Payroll Processing Time Using the Standard Normal Distribution**

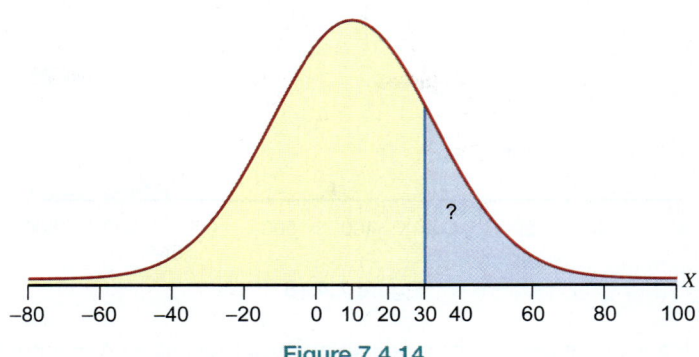

**Figure 7.4.14**

**SOLUTION**

Suppose $X$, the payroll processing time, is a random variable with a mean of 10 minutes and a standard deviation of 20 minutes. Standardizing the random variable, we have the following.

$$P(X > 30) = P\left(z > \frac{30 - 10}{20}\right)$$
$$= P(z > 1)$$

To find the probability that the accountant will take more than 30 minutes to process the payroll, we need to find the probability that the accountant will take less than 30 minutes, and subtract this from 1. Using Appendix A, Table B, we have the following.

$$P(z > 1) = 1 - P(z < 1)$$
$$= 1 - 0.8413$$
$$= 0.1587$$

**⌘ Technology**

For instructions on how to compute this probability using technology, please visit stat.hawkeslearning.com and navigate to **Discovering Business Statistics, Second Edition > Technology Instructions > Normal Distribution > Normal Probability(cdf)**.

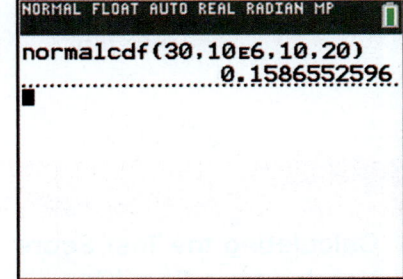

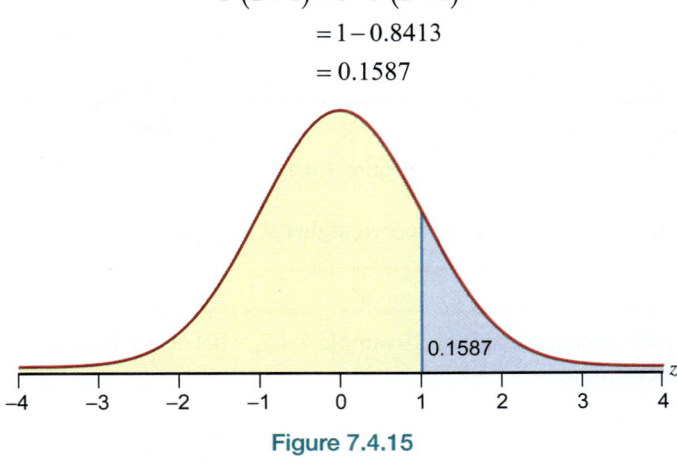

**Figure 7.4.15**

Note that the $X$-value of 30 transformed into the $z$-value of 1. In other words, 30 is one standard deviation away from the mean.

Suppose that a national testing service gives a test in which the results are normally distributed with a mean of 400 and a standard deviation of 100. If you score a 644 on the test, what percentage of the students taking the test exceeded your score?

**Example 7.4.7**

**Calculating the Probability of Exceeding a Test Score Using the Standard Normal Distribution**

**SOLUTION**

Let $X$ = a student's score on the test.

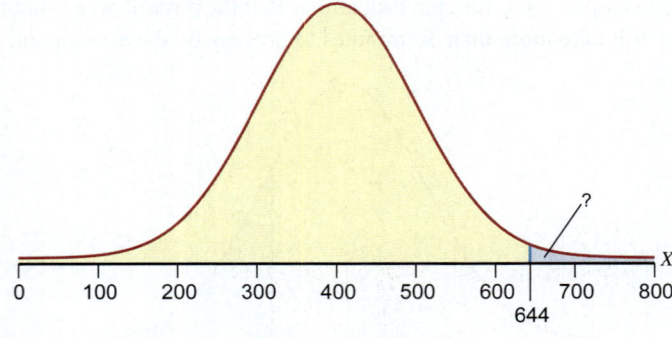

**Figure 7.4.16**

**Technology**

For instructions on how to compute this probability using technology, please visit stat.hawkeslearning.com and navigate to Discovering Business Statistics, Second Edition > Technology Instructions > Normal Distribution > Normal Probability(cdf).

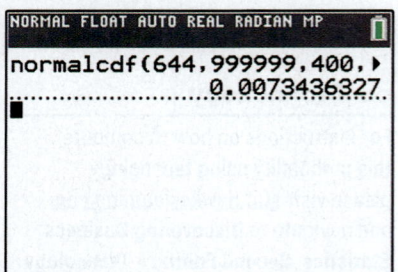

The first step is to standardize the random variable. Then, using Appendix A, Table B, find the appropriate probability.

$$P(X > 644) = P\left(z > \frac{644 - 400}{100}\right)$$
$$= P(z > 2.44)$$
$$= 1 - P(z < 2.44)$$
$$= 1 - 0.9927$$
$$= 0.0073$$

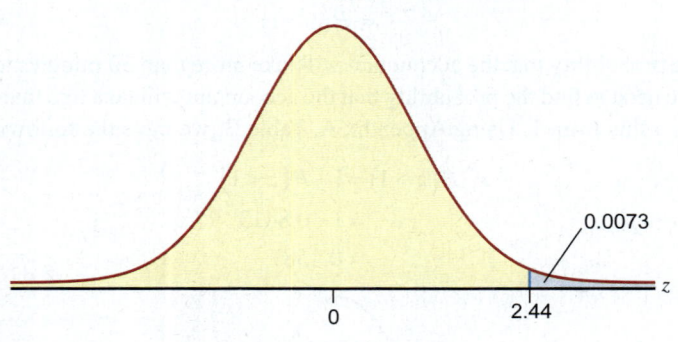

**Figure 7.4.17**

Thus, only 0.73% of the students scored higher than your score of 644.

## Example 7.4.8

### Calculating the Test Score Needed for a Specific Percentile

Using the information provided in Example 7.4.7, what score must a student get to be in the 90th percentile?

**SOLUTION**

Recall that $X$ = a student's score on the test follows a normal distribution with a mean of 400 and a standard deviation of 100. As with the examples before, a picture can be very helpful.

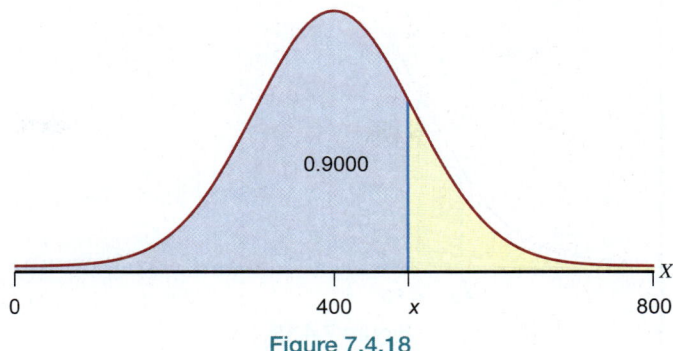

Figure 7.4.18

From the picture above, it can be seen that we want to find the value of $X$ (the student's score) that represents the 90th percentile (i.e., the student scored the same or better than 90% of other students taking the test). We would write $P(X \leq x) = 0.90$.

If we standardize the probability statement, we get the following.

$$P\left(z \leq \frac{x - \mu}{\sigma}\right) = 0.90$$

Next, substitute the values for the mean and standard deviation to get the following.

$$P\left(z \leq \frac{x - 400}{100}\right) = 0.90$$

Note that we can rewrite the above probability statement as

$$P(z \leq z_0) = 0.90 \text{ where } z_0 = \frac{x - 400}{100}.$$

Therefore, we need to find the value of $z$ with the area 0.90 to the left of it. To do this, we look in the body of the standard normal table (Table B in Appendix A) for the value 0.9000. It's not often that we can find the exact probability in the body of the table. When this is the case, we find the closest probability along with the corresponding value of $z$. In this case, the value of $z = 1.28$.

Thus,

$$1.28 = \frac{x - 400}{100}.$$

Solving for $x$, we get $x = 528$.

Therefore, a student who scores 528 on the test will be in the 90th percentile.

### Technology

To find the value of $z$ given a percentile with a TI-83/84 Plus calculator, use the invNorm function within the DISTR menu. For instructions, please visit stat.hawkeslearning.com and navigate to **Discovering Business Statistics, Second Edition > Technology Instructions > Normal Distribution > Inverse Normal.**

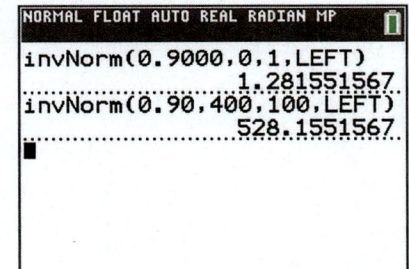

---

Suppose that for 132 space shuttle missions, the flight duration follows a normal distribution with a mean of 234 hours and a standard deviation of 94 hours. If you just read in the newspaper that a space shuttle will be launched tomorrow, what is the probability that the duration of the flight will be between 100 and 150 hours?

**Example 7.4.9**

**Calculating the Probability of Shuttle Flight Duration Using the Standard Normal Distribution**

#### SOLUTION

Let $X$ = the duration of a space shuttle flight. We are interested in the probability that $X$ is between 100 and 150 hours. Writing this probability statement and then standardizing the random variable $X$, we have the following.

$$P(100 < X < 150) = P\left(\frac{100 - 234}{94} < z < \frac{150 - 234}{94}\right)$$
$$= P(-1.43 < z < -0.89)$$

## Technology

For instructions on how to compute this probability using technology, please visit stat.hawkeslearning.com and navigate to **Discovering Business Statistics, Second Edition > Technology Instructions > Normal Distribution > Normal Probability (cdf)**.

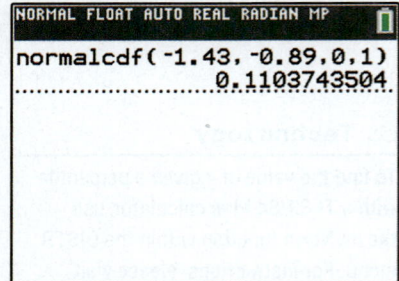

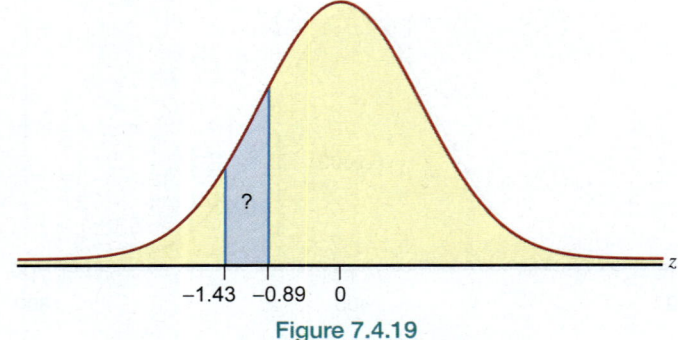

**Figure 7.4.19**

To find the probability that $z$ is between $-1.43$ and $-0.89$, we will need to find the probability that $z$ is less than $-0.89$ and subtract the probability that $z$ is less than $-1.43$. Using Table A in Appendix A, we have the following.

$$P(-1.43 < z < -0.89) = P(z < -0.89) - P(z < -1.43) = 0.1867 - 0.0764 = 0.1103$$

Thus, there is approximately an 11% chance that the flight will last between 100 and 150 hours.

## 📝 7.4 Exercises

### Basic Concepts

1.  What is the standard normal distribution? What are the parameters of the distribution?

2.  Why is the standard normal distribution important?

3.  Describe the connection between the z-transformation and the standard normal random variable.

### Exercises

4.  What proportion of the area under the standard normal curve falls between the following z-values?

    a.  0 and 0.67

    b.  0 and 1.645

    c.  0 and 1.96

    d.  0 and 2.575

5.  What proportion of the area under the standard normal curve falls between the following z-values?

    a.  −0.67 and 0

    b.  −1.645 and 0

    c.  −1.96 and 0

    d.  −2.575 and 0

6.  What proportion of the area under the standard normal curve falls between the following z-values?

    a.  −0.85 and 0.85

    b.  −0.55 and 0.55

    c.  −1.56 and 1.98

    d.  −2.23 and 2.96

7.  What proportion of the area under the standard normal curve falls between the following z-values?

    a.  −0.97 and 0.97

    b.  −0.54 and 1.82

    c.  −1.95 and 2.28

    d.  −2.89 and 1.59

8.  Using the standard normal tables in Appendix A, determine the following probabilities. Sketch the associated areas.

    **a.** $z \leq 0$          **c.** $z \leq -1$          **e.** $z \geq -1$

    **b.** $z \geq 0$          **d.** $z \leq 1$          **f.** $z \geq 1$

9.  Using the standard normal tables in Appendix A, determine the following probabilities. Sketch the associated areas.

    **a.** $z \leq -0.44$          **d.** $z \leq -0.67$

    **b.** $z \geq 0.44$          **e.** $z \geq 0.67$

    **c.** $-0.44 \leq z \leq 0.44$          **f.** $-0.67 \leq z \leq 0.67$

10. Using the standard normal tables in Appendix A, determine the following probabilities. Sketch the associated areas.

    **a.** $z \leq -1.28$          **d.** $z \leq -1.96$

    **b.** $z \geq 1.28$          **e.** $z \geq 1.96$

    **c.** $-1.28 \leq z \leq 1.28$          **f.** $-1.96 \leq z \leq 1.96$

11. Using the standard normal tables in Appendix A, determine the following probabilities. Sketch the associated areas.

    **a.** $P(0 \leq z \leq 0.79)$          **c.** $P(z \geq 1.89)$

    **b.** $P(-1.57 \leq z \leq 2.33)$          **d.** $P(z \leq -2.77)$

12. Using the standard normal tables in Appendix A, determine the following probabilities. Sketch the associated areas.

    **a.** $P(0 \leq z \leq 1.24)$          **c.** $P(z \geq 3.22)$

    **b.** $P(-2.64 \leq z \leq 3.32)$          **d.** $P(z \leq -3.39)$

13. Find the value of $z$ such that 0.05 of the area under the curve lies to the right of $z$.

14. Find the value of $z$ such that 0.01 of the area under the curve lies to the right of $z$.

15. Find the value of $z$ such that 0.10 of the area under the curve lies to the right of $z$.

16. Find the value of $z$ such that 0.05 of the area under the curve lies to the left of $z$.

17. Find the value of $z$ such that 0.01 of the area under the curve lies to the left of $z$.

18. Find the value of $z$ such that 0.10 of the area under the curve lies to the left of $z$.

19. Find the value of $z$ such that 0.7458 of the area under the curve lies between $-z$ and $z$.

20. Find the value of $z$ such that 0.9505 of the area under the curve lies between $-z$ and $z$.

21. Find the value of $z$ such that 0.90 of the area under the curve lies between $-z$ and $z$.

22. The random variable $X$ has a normal distribution with a mean of 30 and a standard deviation of 5.

    **a.** Find the probability that $X$ is between 25 and 35.

    **b.** Find the probability that $X$ is greater than 40.

    **c.** Find the probability that $X$ is less than 20.

23. The random variable $X$ has a normal distribution with a mean of 200 and a standard deviation of 25.

    **a.** Find the probability that $X$ is between 160 and 220.

    **b.** Find the probability that $X$ is greater than 240.

    **c.** Find the probability that $X$ is less than 150.

24. The Arc Electronic Company had an income of $200,000 last year. Suppose the mean income of firms in the industry for the year is $1,000,000 with a standard deviation of $500,000. If incomes for the industry are normally distributed, what proportion of the firms in the industry earned less than Arc?

25. A certain component for the newly developed electronic diesel engine is considered to be defective if its diameter is less than 8.0 mm or greater than 10.5 mm. The distribution of the diameters of these parts is known to be normal with a mean of 9.0 mm and a standard deviation of 1.5 mm. If a component is randomly selected, what is the probability that it will be defective?

26. A television manufacturer is studying television remote control unit usage. One of the criteria they are measuring is the distance at which people attempt to activate the television set with the remote unit. They have discovered that activation distances are normally distributed with an average activation distance of six feet with a standard deviation of three feet. If a remote unit's maximum range is ten feet, what fraction of the time will users attempt to operate the remote outside of the operating limit?

27. According to the Bureau of Labor Statistics, the mean weekly earnings for people working in a sales related profession in 2010 was $631. Assume that the weekly earnings are approximately normally distributed with a standard deviation of $90.

    **Source:** Bureau of Labor Statistics

    a. What are the mean weekly earnings for people working in a sales related profession in 2010?

    b. If a salesperson was randomly selected, find the probability that his or her weekly earnings exceed $700.

    c. If a salesperson was randomly selected, find the probability that his or her weekly earnings are at most $525.

    d. If a salesperson was randomly selected, find the probability that his or her weekly earnings are between $400 and $615.

    e. Do you feel that it is reasonable to assume that the weekly earnings have a normal distribution? Why or why not?

28. The repair time for air conditioning units is believed to have a normal distribution with a mean of 38 minutes.

    a. What is the standard deviation of repair time if 40% of the units are repaired between 33 and 43 minutes?

    b. Using the value of the standard deviation that you calculated in **a.**, what is the probability that a repair will be longer than an hour?

    c. Using the value of the standard deviation that you calculated in **a.**, what is the probability that the repair time for an air conditioning unit will be less than 25 minutes?

29. VGA monitors manufactured by TSI Electronics have life spans which have a normal distribution with an average life span of 15,000 hours and a standard deviation of 2000 hours. If a VGA monitor is selected at random, find the following probabilities.

    a. The probability that the life span of the monitor will be less than 12,000 hours.

    b. The probability that the life span of the monitor will be more than 18,000 hours.

    c. The probability that the life span of the monitor will be between 13,000 hours and 17,000 hours.

30. A beer distributor believes the amount of beer in a 12-ounce can of beer has a normal distribution with a mean of 12 ounces and a standard deviation of 1 ounce. If a 12-ounce beer can is randomly selected, find the following probabilities.

    **a.** The probability that the 12-ounce can of beer will actually contain less than 11 ounces of beer.

    **b.** The probability that the 12-ounce can of beer will actually contain more than 12.5 ounces of beer.

    **c.** The probability that the 12-ounce can of beer will actually contain between 10.5 and 11.5 ounces of beer.

31. A statistics teacher believes that the final exam grades for her business statistics class have a normal distribution with a mean of 82 and a standard deviation of 8.

    **a.** Find the score which separates the top 10% of the scores from the lowest 90% of the scores.

    **b.** The teacher plans to give all students who score in the top 10% of scores an A. Will a student who scored a 90 on the exam receive an A? Explain.

    **c.** Find the score which separates the lowest 20% of the scores from the highest 80% of the scores.

    **d.** The teacher plans to give all students who score in the lowest 10% of scores an F. Will a student who scored a 65 on the exam receive an F? Explain.

32. An investor believes that the yields of his mutual funds have a normal distribution with an average yield of 10% and a standard deviation of 2%. The investor would like to identify the stocks which yield the highest 5% to keep in his portfolio.

    **a.** Calculate the yield which separates the highest 5% of yields from the lowest 95% of yields.

    **b.** If a stock yielded 14% would it be kept? Explain.

    **c.** If a stock yielded 13% would it be kept? Explain.

33. In order for you to become a member of Mensa, a worldwide organization with approximately 100,000 members, your IQ score must be in the top 2%. The word *mensa* is Latin for "table," and was chosen to denote a group or round table of people with equal ability. In 1996, Mensa, which was founded by two British barristers, celebrated its 50th birthday. American Mensa Ltd., which was founded in 1960 has almost 50,000 members. Marilyn vos Savant, who is reputed to have the highest recorded IQ, is a member. Assuming that IQ scores have an approximately normal distribution with a mean and standard deviation of 100 and 15, respectively, answer the following questions.

    **a.** What IQ must one have in order to become a member of Mensa?

    **b.** What percent of all Americans have an IQ of at least 145?

    **c.** What percent of all members of Mensa have an IQ of at least 145?

    **d.** If Mensa decided to become more exclusive, and accepted only the top 1% instead of the top 2% as members, what IQ would one need in order to become a member of Mensa?

# 7.5 Approximations to Other Distributions

To approximate other distributions, the normal distribution can be very useful. Although it is a continuous distribution, it is used to approximate discrete distributions, specifically the binomial and the Poisson.

## The Binomial Distribution

Calculating binomial probabilities can be quite time consuming if $n$ is large. For example, suppose that you intend to sample 2000 subjects for a marketing research survey. If 50 percent of the population believes your product is superior to the competition's, what is the probability of obtaining 600 or fewer subjects who believe your company's product is superior?

$$P(X \le 600) = P(X = 0) + P(X = 1) + P(X = 2) + \cdots + P(X = 599) + P(X = 600)$$

Determining the appropriate probability using the binomial distribution would require the calculation of 601 individual probabilities, many of which would have extremely large combinations such as the following.

$$_{2000}C_{400}\, 0.5^{400}\left(1 - 0.5\right)^{1600}$$

Computing this and the other 600 similar calculations would be a formidable task. The normal distribution is useful in approximating binomial probabilities. The larger the binomial parameter, $n$, the more accurate the approximation. Determining the probability described above using the normal approximation is trivial in comparison to calculating the exact probability using the binomial.

Recall that the normal distribution is a function of two parameters, the mean and the standard deviation. Thus, if the normal distribution is used to approximate the binomial distribution, it seems reasonable that the mean and standard deviation of the normal should be the same as the mean and standard deviation of the binomial that is being approximated. Specifically, let

$$\mu = E(X) = np, \text{ and}$$

$$\sigma = \sqrt{V(X)} = \sqrt{np(1-p)}.$$

To approximate a binomial with $n = 20$ and $p = 0.5$ would require a normal distribution with

$$\mu = (20)(0.5) = 10$$

$$\sigma = \sqrt{(20)(0.5)(1-0.5)} = \sqrt{5} \approx 2.2361.$$

In this example, the shapes of the distributions are quite similar and consequently the approximation will be good.

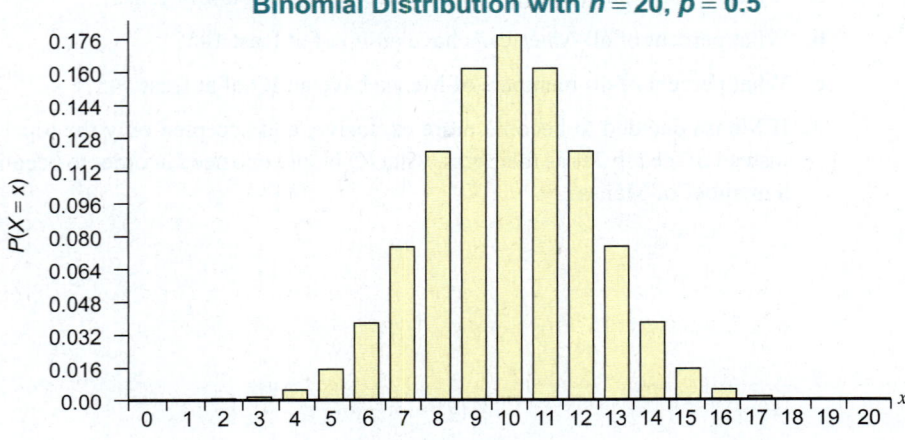

**Binomial Distribution with $n = 20$, $p = 0.5$**

**Figure 7.5.1**

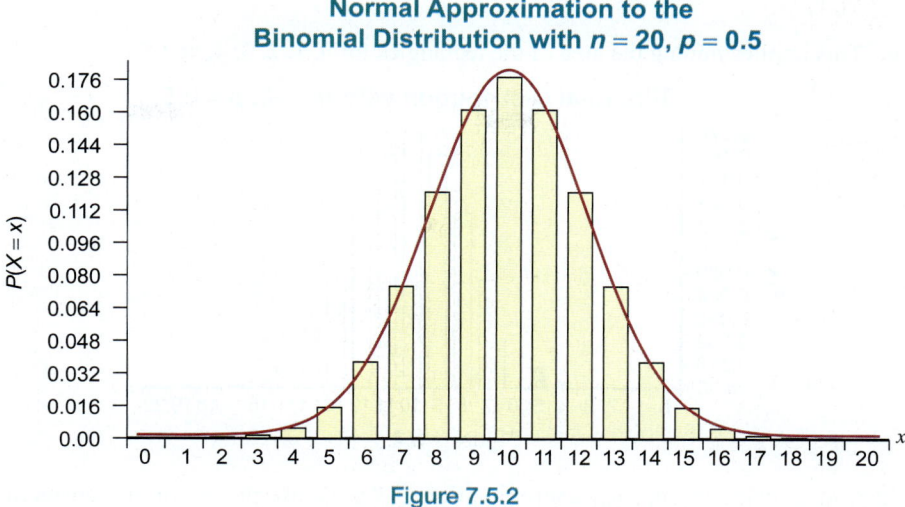

Figure 7.5.2

So, when should the normal distribution be used to approximate the binomial distribution? Generally, the approximation is reasonable when the mean of the binomial, $np$, is greater than or equal to 5 and $n(1 - p)$ is greater than or equal to 5. The approximation becomes quite good when $np$ is greater than or equal to 10 and $n(1 - p)$ is greater than or equal to 10.

The normal approximation to the binomial can be improved by using **continuity correction**.

Suppose that you wished to determine the probability that a binomial random variable ($n = 20$ and $p = 0.5$) is equal to 5. Recall that for a continuous random variable $X$, the probability that $X$ is equal to some specific value is equal to zero since there is no area under the curve for a single point, say $X = 5$. Therefore, to approximate the probability using the normal would be equivalent to approximating the area of the shaded region given in Figure 7.5.3. To approximate the area of the region using the normal would require finding the area under the curve between 4.5 and 5.5.

> **Definition**
>
> ### Continuity Correction
>
> **Continuity correction** is used when a discrete distribution is approximated using a continuous distribution. To apply continuity correction, subtract or add 0.5 (depending on the question at hand) to a selected value in order to find the desired probability.

**Normal Approximation to the Binomial, $n = 20$, $p = 0.5$**

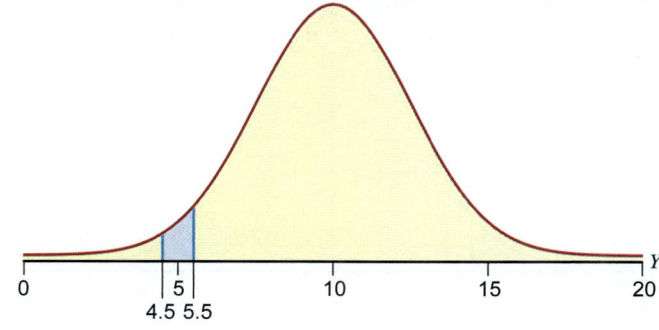

Figure 7.5.3

The continuity correction should be used whenever the normal distribution is used to approximate the binomial distribution. The following examples will illustrate how to approximate the binomial distribution using the normal distribution with continuity correction.

---

**a.** Assuming $n = 20$ and $p = 0.5$, use the normal distribution to approximate the probability that a binomial random variable is 5 or less.

**b.** Find the probability that the same random variable from part **a.** is greater than 4.

**Example 7.5.1**

**Using a Normal Distribution to Approximate a Binomial Probability**

**SOLUTION**

**a.** This implies finding the area of the rectangles for 0, 1, 2, 3, 4, and 5.

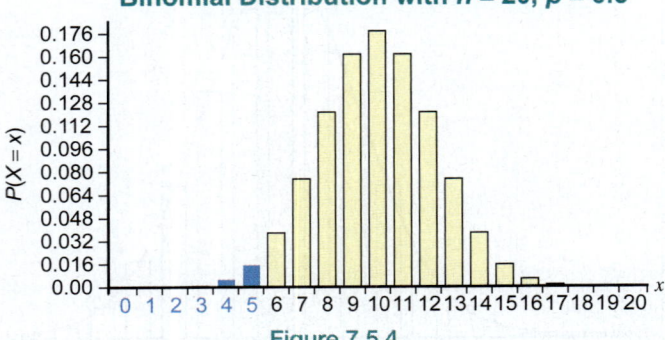

Figure 7.5.4

🌥 **Technology**

For instructions on how to compute binomial probabilities using technology, please visit stat.hawkeslearning.com and navigate to **Discovering Business Statistics, Second Edition > Technology Instructions > Binomial Distribution > Binomial Probability(cdf)**.

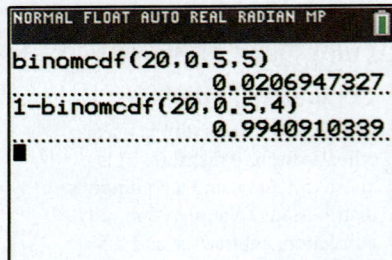

✏ **NOTE**

There is some discrepancy between the solution in the text and the calculator values due to rounding.

Instead of using the normal approximation $P(Y \le 5)$, use the continuity correction $P(Y \le 5.5)$ in order to accumulate all of the probabilities under the normal curve that correspond to the region associated with the point 5.

To use the normal approximation the mean and standard deviation of the binomial must be calculated.

$$\mu = E(X) = np = (20)(0.5) = 10$$

$$\sigma = \sqrt{np(1-p)} = \sqrt{(20)(0.5)(1-0.5)} = \sqrt{5} \approx 2.2361$$

Using the normal distribution, $Y$, with a mean of 10 and a standard deviation of 2.2361 to approximate the binomial using continuity correction,

$$P(Y \le 5.5) = P\left(z \le \frac{5.5 - 10}{2.2361}\right)$$

$$\approx P(z \le -2.01)$$

$$= 0.0222.$$

**Normal Approximation to the Binomial, $n = 20$, $p = 0.5$**

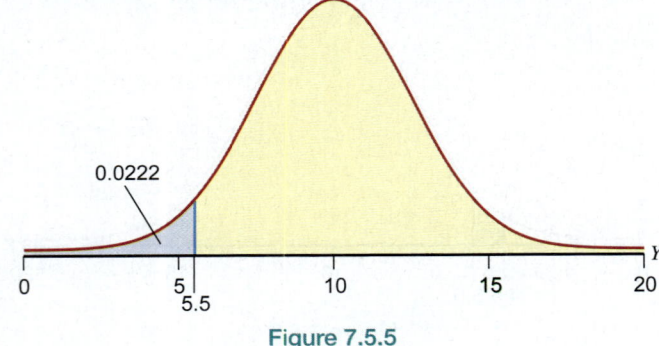

Figure 7.5.5

Thus, the probability that the random variable is 5 or less is 0.0222.

**b.**

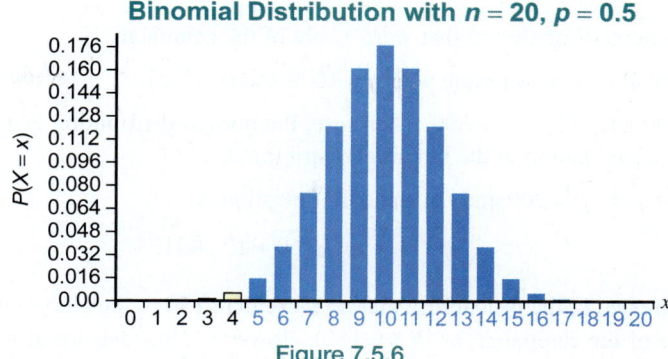

Figure 7.5.6

We are interested in the probability that a binomial random variable, $X$, is greater than 4. Since this is a discrete distribution, the probability that $X$ is greater than 4 is equal to the probability that $X$ is greater than or equal to 5. Thus, when using the normal approximation, we need to apply continuity correction and consider the probability that the normal random variable is greater than or equal to 4.5.

Using the normal distribution, $Y$, with a mean of 10 and a standard deviation of 2.2361 to approximate the binomial using continuity correction,

$$P(Y \geq 4.5) = P\left(z \geq \frac{4.5 - 10}{2.2361}\right)$$
$$\approx P(z \geq -2.46)$$
$$= 1 - P(z < -2.46)$$
$$= 1 - 0.0069$$
$$= 0.9931.$$

**Normal Approximation to the Binomial, $n = 20$, $p = 0.5$**

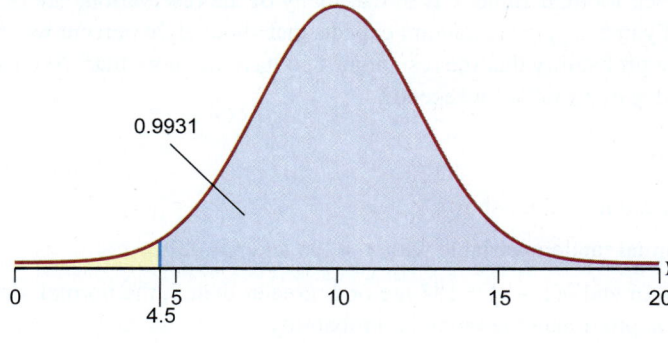

Figure 7.5.7

Thus, the probability that the random variable is greater than 4 is 0.9931.

An advertising agency hired on behalf of Tech's development office conducted an ad campaign aimed at making alumni aware of their new capital campaign. Upon completion of the new campaign, the agency claimed that 20% of alumni in the state of Virginia were aware of the new campaign. To validate the claim of the agency, the development office surveyed 1000 alumni in the state and found that 150 were aware of the campaign. Assuming that the ad agency's claim is true, what is the probability that no more than 150 of the alumni in the random sample were aware of the new campaign?

**Example 7.5.2**

**Using a Normal Distribution to Approximate a Binomial Probability about Campaign Awareness**

**SOLUTION**

Let $X$ = the number of alumni that were aware of the campaign.

$X$ is a binomial random variable with $n = 1000$ and $p = 0.20$.

So, $np = 200$ and $n(1 - p) = 800$. Therefore, the normal distribution is appropriate to use as an approximation to the binomial distribution.

The mean is $\mu = np = 200$ and the standard deviation is

$$\sigma = \sqrt{np(1-p)} = \sqrt{160} \approx 12.6491.$$

We are interested in the probability that no more than 150 of the alumni in the sample were aware of the campaign, or $P(X \le 150)$. However, since we are using the normal distribution to approximate the binomial, continuity correction must be applied.

Let $Y$ be a normally distributed random variable with a mean of 200 and a standard deviation of 12.6491. Applying continuity correction, we are interested in the following probability.

$$P(Y \le 150.5) = P\left(z \le \frac{150.5 - 200}{12.6491}\right) \approx P(z \le -3.91) \approx 0.$$

Thus, if the marketing agency's claim is true, the probability that 150 or fewer alumni are aware of the campaign is practically zero. This would lead the development office to believe that the agency's claim is false.

> ✏ **NOTE**
>
> The $z$-value $-3.91$ is not listed in the tables given in Appendix A. However, using technology such as a calculator or computer software, it can be calculated that the actual probability is approximately 0.000046.

The smaller the sample size, the more the binomial distribution deviates from the normal distribution. For this reason, continuity correction is especially useful for small sample sizes. Example 7.5.3 illustrates the difference that continuity correction makes when using the normal distribution to approximate the binomial.

## Example 7.5.3

**Using a Normal Distribution to Approximate a Binomial Probability about Restaurant No-Shows**

A popular restaurant near Tech's campus accepts 200 reservations on Saturdays, the day of a Tech football game. Given that many of the reservations are made weeks in advance of game day, the restaurant expects that about eight percent will be no-shows. What is the probability that the restaurant will have no more than 20 no-shows on the next Saturday of a football weekend?

**SOLUTION**

Let $X$ = the number of no-shows.

$X$ is a binomial random variable with $n = 200$ and $p = 0.08$.

Since $np = 16$ and $n(1 - p) = 184$ are both greater than 5, the normal distribution can be used to approximate the binomial probability.

For the binomial, $\mu = np = 16$ and

$$\sigma = \sqrt{np(1-p)} = \sqrt{(200)(0.08)(1-0.08)} = \sqrt{14.72} \approx 3.8367.$$

Using the normal distribution, $Y$, with a mean of 16 and a standard deviation of 3.8367, to approximate the binomial without continuity correction results in

$$P(Y \le 20) = P\left(z \le \frac{20 - 16}{3.8367}\right) \approx P(z \le 1.04) = 0.8508.$$

Using continuity correction,

$$P(Y \le 20.5) = P\left(z \le \frac{20.5 - 16}{3.8367}\right) \approx P(z \le 1.17) = 0.8790.$$

Thus, using the normal approximation and continuity correction, the probability that the restaurant will have no more than 20 no-shows is 0.8790. Notice that the continuity

> ∞ **Technology**
>
> For instructions on how to compute the exact binomial probabilitiy using technology, please visit stat.hawkeslearning.com and navigate to **Discovering Business Statistics, Second Edition > Technology Instructions > Binomial Distribution > Binomial Probability(cdf)**.
>
> ```
> NORMAL FLOAT AUTO REAL RADIAN MP
>
> binomcdf(200,0.08,20)
>                    0.877543355
> ```

correction has a significant impact on the accuracy of the approximation. Using the binomial distribution, the exact probability is 0.8775.

## The Poisson Distribution

Approximating the Poisson distribution with the normal distribution is similar to approximating the binomial distribution. To use this approximation, the mean and standard deviation of the normal should be set to the mean and standard deviation of the Poisson. Since the mean and variance of the Poisson are both $\lambda$, the appropriate mean, variance, and standard deviation for the normal would be as follows.

$$\mu = \lambda, \sigma^2 = \lambda, \sigma = \sqrt{\lambda}$$

Just as with the binomial distribution, when using the normal distribution to approximate the Poisson distribution, continuity correction should be applied for the best possible approximation. This is because we are using a continuous distribution to approximate a discrete distribution. The approximation becomes increasingly accurate as the mean becomes larger. However, it is not very accurate when the mean of the Poisson distribution is less than 5.

The manager of a sporting goods store wants to determine the best method of staffing employees without being wasteful. The problem that is often encountered is having too many employees and too few customers, and vice versa. The manager realized that during the initial kick-off of certain sports seasons such as football, baseball, and basketball, the store attracts, on average, 30 customers per hour. On some days, the number is higher, on others, the number is lower. The manager would like to determine the probability of at least 40 customers arriving in a given hour. Use the normal approximation to the Poisson distribution to find the probability.

**Example 7.5.4**

**Using a Normal Distribution to Approximate a Poisson Probability about Customer Arrivals**

### SOLUTION

Let $X =$ the number of customers that enter the store in a given hour.

Notice that since we are counting the number of customers in a specific time interval, $X$ has a Poisson distribution with

$$\mu = 30 \text{ and } \sigma = \sqrt{30} \approx 5.4772.$$

**Normal Approximation to the Poisson Distribution with $\lambda = 30$**

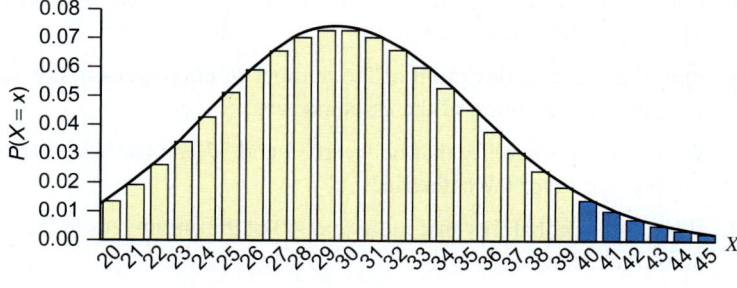

**Figure 7.5.8**

If $Y$ represents the normal random variable with a mean of 30 and a standard deviation of 5.4772, it should be a good approximation to the Poisson. We are interested in the probability that at least 40 customers arrive in a given hour, or $P(X \geq 40)$. Because we are using the normal distribution to approximate the Poisson distribution, we must use continuity correction. Thus, we are interested in the following probability.

$$P(X \geq 40) \approx P(Y \geq 39.5) = P\left(z \geq \frac{39.5 - 30}{5.4772}\right) \approx P(z \geq 1.73) = 1 - P(z < 1.73) = 1 - 0.9582 = 0.0418$$

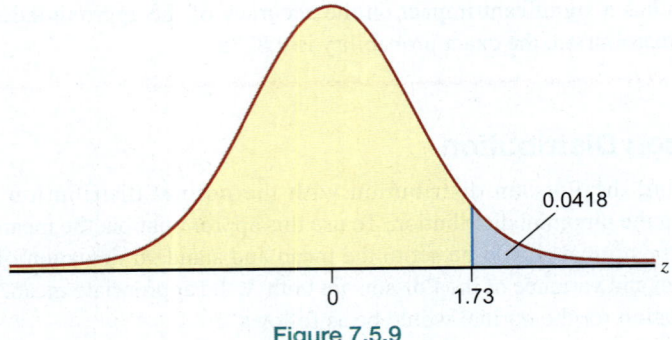

Figure 7.5.9

Thus, using the normal approximation to the Poisson, the probability that at least 40 customers arrive in a given hour is 0.0418.

## 📝 7.5 Exercises

### Basic Concepts

1. Why would you want to use the normal distribution to approximate a binomial distribution or a Poisson distribution?

2. What are the parameters of a normal distribution used to approximate a binomial distribution?

3. What are the parameters of a normal distribution used to approximate a Poisson distribution?

4. What is continuity correction? How does it improve the normal approximation to the binomial or Poisson?

### Exercises

5. Management at a small engineering company is considering the addition of a company cafeteria area. A random sample of 50 persons out of the total number of persons employed by the firm will be surveyed to see if they are in favor of the addition. Assume that the true percentage of persons that favor the addition is 90%.

   a. Find the expected number of employees in the sample who will favor the addition of the cafeteria area.

   b. Find the standard deviation of the number of employees in the sample who will favor the addition of the cafeteria area.

   c. What is the probability that between 35 and 37 employees (inclusive) in the sample will favor the cafeteria?

   d. What is the probability that more than 40 of the employees in the sample will favor the cafeteria?

   e. What is the probability that at most 38 of the employees in the sample will favor the cafeteria?

6. The accounting department of a large corporation checks the addition of expense reports submitted by executives before paying them. Historically, they have found that 15% of the reports contain addition errors. An auditor randomly selects 60 expense reports and audits them for addition errors.

   a. Find the expected number of reports in the sample that will have addition errors.

    **b.** Find the standard deviation of the number of reports sampled that will have addition errors.

    **c.** Find the probability that fewer than 10 of the sampled expense reports will have addition errors.

    **d.** Find the probability that at least 30 of the sampled expense reports will have addition errors.

    **e.** Find the probability that between 5 and 15 (inclusive) of the sampled expense reports will have addition errors.

**7.** A local electronics store purchased a market research study which suggests that 60 percent of all homes have DVD recorders/players. A sample of 200 homes is selected to confirm the study's findings. If the marketing study is correct, answer the following questions.

    **a.** Find the expected number of homes sampled which will have DVD recorders/players.

    **b.** Find the standard deviation of the number of homes in the sample which will have video recorders/players.

    **c.** What is the probability that at most 80 of the sampled homes will have DVD recorders/players?

    **d.** What is the probability that between 100 and 120 (inclusive) homes sampled will have DVD recorders/players?

    **e.** What is the probability that at least 130 of the sampled homes will have DVD recorders/players?

**8.** Suppose a virus is believed to infect two percent of the population. If a sample of 3000 randomly selected subjects are tested, answer the following questions.

    **a.** Find the expected number of subjects sampled that will be infected.

    **b.** Find the standard deviation of the number of subjects sampled that will be infected.

    **c.** What is the probability that fewer than 30 of the subjects in the sample will be infected?

    **d.** What is the probability that between 40 and 80 (inclusive) of the subjects in the sample will be infected?

    **e.** Find the probability that at least 70 of the subjects in the sample will be infected.

**9.** A company manufacturing metal sheets believes that the number of defects on a 10' by 10' sheet of metal follows a Poisson distribution with an average defect rate of 5 per sheet.

    **a.** Find the standard deviation of the number of defects per sheet.

    **b.** Using the Poisson table in Appendix A, Table F, find the probability of observing at least 10 defects per sheet.

    **c.** Using the normal approximation to the Poisson, find the probability of observing at least 10 defects per sheet.

    **d.** How do the answers in parts **b.** and **c.** compare?

**10.** Service calls arriving at an electric company follow a Poisson distribution with an average arrival rate of 60 per hour.

    **a.** Find the average number of service calls in a 30-minute period.

    **b.** Find the standard deviation of the number of service calls in a 30-minute period.

c. Using the normal approximation to the Poisson, find the probability that the electric company receives at least 40 service calls in a 30-minute period.

d. Using the normal approximation to the Poisson, find the probability that the electric company receives at most 20 service calls in a 30-minute period.

e. Using the normal approximation to the Poisson, find the probability that the electric company receives between 25 and 50 (inclusive) service calls in a 30-minute period.

11. Patients arriving at the emergency room of a local hospital follow a Poisson distribution with an average arrival rate of 15 per half hour.

a. Find the average number of patients that arrive at the emergency room in one hour.

b. Find the standard deviation of the number of patients that arrive at the emergency room in one hour.

c. Find the probability that at least 15 patients will arrive at the emergency room in one hour.

d. Find the probability that between 30 and 50 patients (inclusive) will arrive at the emergency room in one hour.

e. Find the probability that at most 35 patients will arrive at the emergency room in one hour

# T Discovering Technology

## Using the TI-84 Plus Calculator

### Normal Distribution: Finding the Area between Two Values

Use the information from Example 7.4.5 for this exercise. Suppose that the times it takes for an accountant to process payroll is normally distributed with a mean of 10 minutes and a standard deviation of 20 minutes. Find the probability that the payroll processing time on a randomly selected payroll date will take between 10 minutes and 40 minutes.

1.  To use a TI-84 Plus calculator to find this probability, you will use the normalcdf function. The syntax for the normalcdf function on the calculator is as follows.

    **normalcdf(lower bound, upper bound, mean, standard deviation)**

    Press 2ND and then VARS to access the **DISTR** menu.

2.  Select option **normalcdf**. Press ENTER.

3.  We are interested in the probability that the payroll will take between 10 and 40 minutes to process, assuming the times are normally distributed with a mean of 10 minutes and a standard deviation of 20 minutes. Thus, the function you want to enter is **normalcdf(10,40,10,20)**. Press ENTER and observe the results.

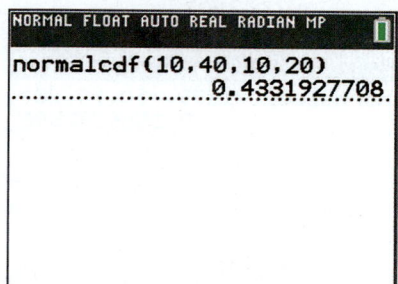

### Normal Distribution: Finding the Area to the Right of Some Value

Use the information from Example 7.4.7 for this exercise. Suppose that a national testing service gives a test in which the results are normally distributed with a mean of 400 and a standard deviation of 100. If you score a 644 on the test, what percentage of the students taking the test exceeded your score?

1.  Press 2ND and then VARS to access the **DISTR** menu. Select option **normalcdf**.

2.  We are interested in the probability of scoring greater than 644 on a test where the results are normally distributed with a mean of 400 and a standard deviation of 100. To enter this on the calculator, we enter a very large positive number as the upper bound in the normalcdf function. Enter **normalcdf(644,1E99,400,100)** into the calculator. Here, 1E99 represents infinity, and is entered by pressing 1 , 2ND, , , 9 , 9 .

3.  Press ENTER and observe the results.

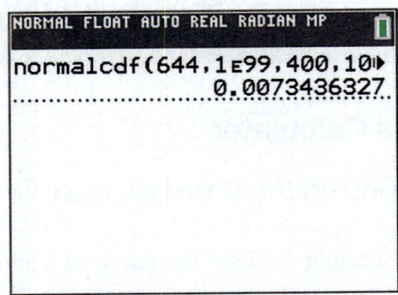

## Normal Distribution: Finding the Area to the Left of Some Value

Find the probability that a normal random variable with a mean of 10 and a standard deviation of 20 will lie below −10.

1. Press **2ND** and then **VARS** to access the **DISTR** menu. Select option **normalcdf**.

2. We are interested in the probability that a normal random variable with a mean of 10 and a standard deviation of 20 will lie below −10. To enter this on the calculator, we enter a very large negative number as the lower bound in the normalcdf function to represent negative infinity. Enter **normalcdf(−1E99,−10,10,20)** into the calculator. Here, −1E99 represents negative infinity and is entered by pressing **(−)**, **1**, **2ND**, **,**, **9**, **9**.

3. Press **ENTER** and observe the results.

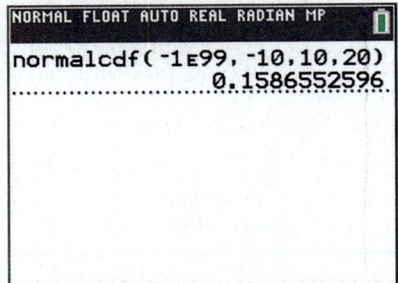

## Normal Distribution: Finding the Value of z Given an Area

Use the information from Example 7.4.4 for this exercise.

1. To find the value of z given a particular area, use the **invNorm** function within the **DISTR** menu. The syntax for the invNorm function is as follows.

    **invNorm(area to the left of z)**

    Part **a.** of Example 7.4.4 asks to find the value of z if the area to the left of z is 0.9147. Thus, to find this probability, press **2ND** and then **VARS** to access the **DISTR** menu and select option **invNorm**.

2. Since we are given the area to the left of z, simply enter **invNorm(0.9147)** into the calculator. Press **ENTER** and observe the results. The value of z is approximately 1.37.

3. Part **b.** of Example 7.4.4 asks to find the value of $z$ if the area between 0 and $z$ is 0.3665. Since the function in the calculator uses the area to the left of $z$, we need to add 0.5 to 0.3665 to get the area from negative infinity to $z$. Thus, press 2ND and then VARS to access the **DISTR** menu and select option **invNorm**. Enter **invNorm(0.8665)** into the calculator and press ENTER to observe the results. The value of $z$ is approximately 1.11.

4. Part **d.** asks for the value of $z$ if the area to the right of $z$ is 0.7967. Since the invNorm function uses the area to the left of $z$, we will need to enter $1-0.7967$ as the area. Enter **invNorm(1−0.7967)** and press ENTER to observe the results. The value of $z$ is approximately −0.83.

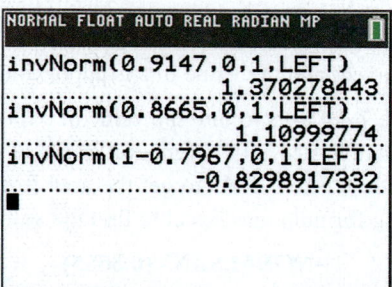

## Using Excel

### Normal Distribution

Consider a normally distributed random variable, $X$, with a mean of 15 and a standard deviation of 5. Use Microsoft Excel to find **a.** the area to the left of 8.4, **b.** the area between 9 and 11, and **c.** the area to the right of 17.

1. The function in Microsoft Excel that gives the area under a normal curve to the left of some value is as follows.

    **NORM.DIST(*x*, mean, standard_dev, cumulative)**

    where $x$ is the value of the random variable, **mean** is the mean of the random variable, **standard_dev** is the standard deviation of the random variable, and **cumulative** = TRUE if we are interested in the area to the left of $x$ and FALSE if we are interested in the value of the probability density function at the particular value of $x$ (this is usually not very useful). To find the area to the left of 8.4, enter the following formula into Excel.

    **=NORM.DIST(8.4,15,5,TRUE)**

    The result is the area to the left of 8.4, which is approximately 0.0934.

2. To find the area between 9 and 11, we will need to find the area to the left of 11 and subtract the area to the left of 9, since the formula in Excel only calculates areas to the left of $x$. To find the area between 9 and 11, enter the following formula into Excel.

    **=NORM.DIST(11,15,5,TRUE)−NORM.DIST(9,15,5,TRUE)**

    The result is the area between 9 and 11, which is approximately 0.0968.

3. To find the area to the right of 17, we will need to use Excel to find the area to the left of 17, and then subtract this area from 1. To find the area to the right of 17, enter the following formula into Excel.

    **=1−NORM.DIST(17,15,5,TRUE)**

    The result is the area to the right of 17, which is approximately 0.3446.

## Normal Distribution: Finding the Value of z Given an Area

Use the information from Example 7.4.4 for this exercise.

1.   The function in Microsoft Excel that returns the value of $z$ given the area to the left of $z$ for a standard normal random variable is as follows.

**=NORM.S.INV(probability)**

where **probability** is the area to the left of $z$. Part **a.** of Example 7.4.4 asks for the value of $z$ if the area to the left of $z$ is 0.9147. To find the value of $z$ in this case, enter the following formula in Microsoft Excel.

**=NORM.S.INV(0.9147)**

The result is displayed, and the value of $z$ is approximately 1.37.

2.   Part **b.** of Example 7.4.4 asks to find the value of $z$ if the area between 0 and $z$ is 0.3665. Since the NORM.S.INV function in Excel uses the area to the left of $z$, we need to add 0.5 to 0.3665 to get the area from negative infinity to $z$. Enter the following formula into Excel to find the value of $z$.

**=NORM.S.INV(0.8665)**

The result is displayed, and the value of $z$ is approximately 1.11.

3.   Part **d.** asks for the value of $z$ if the area to the right of $z$ is 0.7967. Since the NORM.S.INV function uses the area to the left of $z$, we will need to enter 1−0.7967 as the area. Enter the following formula into Excel to find the value of $z$.

**=NORM.S.INV(1−0.7967)**

The result is displayed, and the value of $z$ is approximately −0.83.

**Note:** For any normally distributed random variable, the excel function =NORM.INV(probability, mean, standard_dev) returns the value with the corresponding probability (area) to the left.

## Using JMP

## Normal Distribution: Finding the Area Between Two Values

Use the information from Example 7.4.5 for this exercise. Suppose that the times it takes for an accountant to process payroll is normally distributed with a mean of 10 minutes and a standard deviation of 20 minutes. Find the probability that the payroll processing time on a randomly selected payroll date will take between 10 minutes and 40 minutes.

1.   With a JMP **Data Table** open, right-click on **Column 1** and select **Formula** to access the Formula Editor.

2.   From the function list on the left, select **Probability > Normal Distribution**. The normal distribution formula appears with a blue box around the variable $x$.

3.   Click the **caret button** (^) on the keypad at the top of the screen twice to add the fields for the mean and standard deviation to the formula.

4.   In the fields provided in the formula enter 40 for $x$, 10 for the mean, and 20 for the standard deviation.

5.   Click the **minus sign** on the keypad at the top of the screen and then select **Normal Distribution** again from the function list on the left.

6.  Click the **caret button** (^) on the keypad at the top of the screen twice to add the fields for the mean and standard deviation to the formula.

7.  In the fields provided in the formula enter 10 for $x$, 10 for the mean, and 20 for the standard deviation. The formula will appear as follows on the screen:

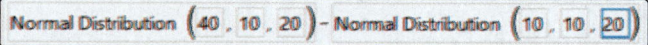

8.  Click **OK**. JMP will populate the first cell of the data table with the area between 10 and 40 for a normal distribution with a mean of 10 and standard deviation of 20. If the value does not appear then double-click in the first row to the left of **Column 1**.

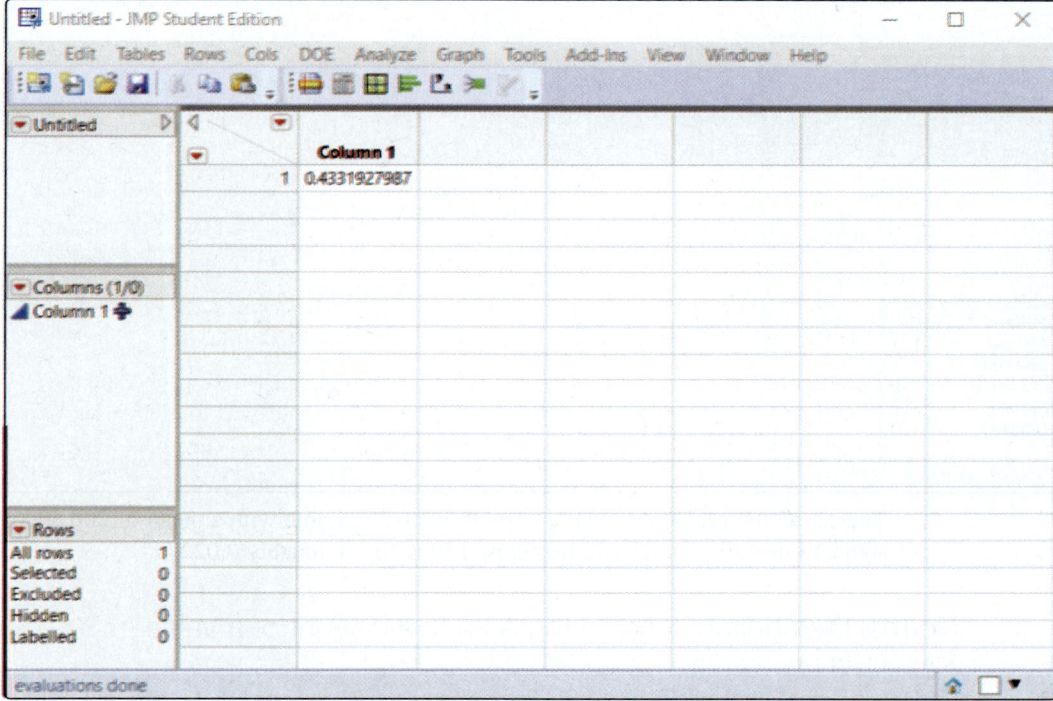

9.  Therefore, the probability that a randomly selected payroll date will take between 10 minutes and 40 minutes to process is approximately 0.4332.

## Normal Distribution: Finding the Area to the Left of a Value

Find the probability that a normal random variable with a mean of 200 and a standard deviation of 25 will lie below 150.

1.  With a JMP **Data Table** open, right-click on **Column 1** and select **Formula** to access the Formula Editor.

2.  From the function list on the left, select **Probability > Normal Distribution**. The normal distribution formula appears with a blue box around the variable $x$.

3.  Click the **caret button** (^) on the keypad at the top of the screen twice to add the fields for the mean and standard deviation to the formula.

4.  In the fields provided in the formula enter 150 for $x$, 200 for the mean, and 25 for the standard deviation.

5.  The formula will appear as follows on the screen:

Normal Distribution ( 150 , 200 , 25 )

6.  Click **OK**. JMP will populate the first cell of the data table with the area to the left of 150 for a normal distribution with a mean of 200 and standard deviation of 25. If the value does not appear then double-click in the first row to the left of **Column 1**.

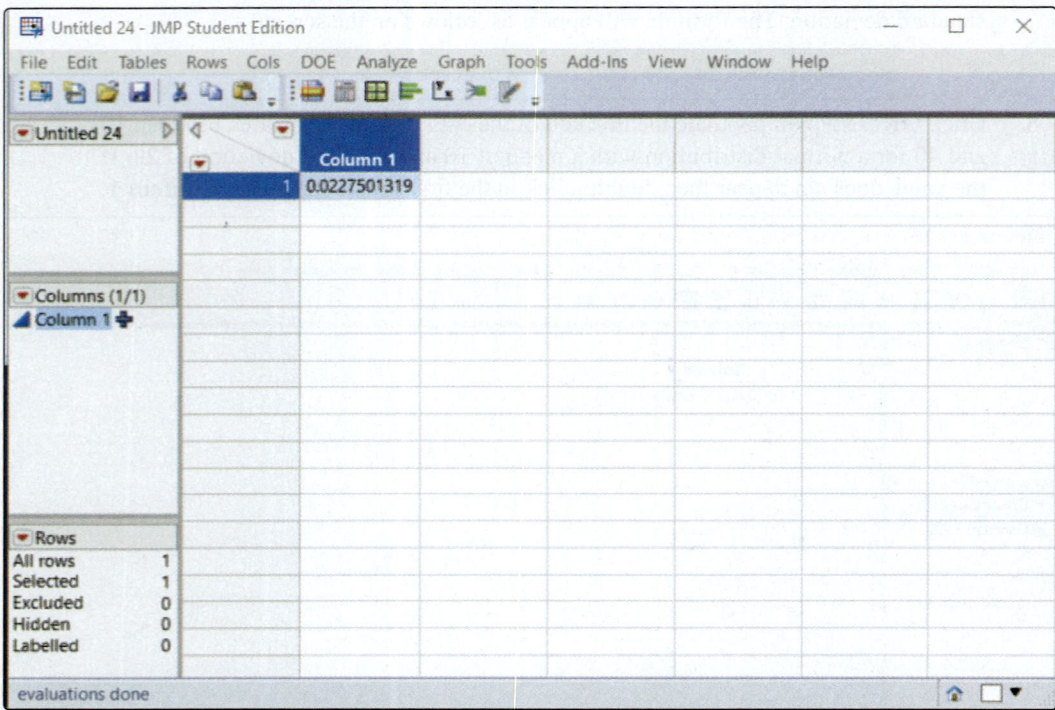

7.  Therefore, the probability that a normal random variable with a mean of 200 and a standard deviation of 25 will lie below 150 is approximately 0.02275.

## Normal Distribution: Finding the Area to the Right of a Value

Use the information from Example 7.4.7 for this exercise. Suppose that a national testing service gives a test in which the results are normally distributed with a mean of 400 and a standard deviation of 100. If you score a 644 on the test, what percentage of the students taking the test exceeded your score?

1.  With a JMP **Data Table** open, right-click on **Column 1** and select **Formula** to access the Formula Editor.

2.  Enter 1 in the formula box in the center of the screen. Now click the **minus sign** on the keypad at the top of the screen.

3.  From the function list on the left, select **Probability > Normal Distribution**. The normal distribution formula appears with a blue box around the variable *x*.

4.  Click the **caret button** (^) on the keypad at the top of the screen twice to add the fields for the mean and standard deviation to the formula.

5.  In the fields provided in the formula enter 644 for *x*, 400 for the mean, and 100 for the standard deviation.

6.  The formula will appear as follows on the screen:

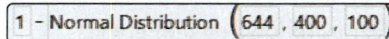

7.  Click **OK**. JMP will populate the first cell of the data table with the area to the right of 644 for a normal distribution with a mean of 400 and standard deviation of 100. If the value does not appear then double-click in the first row to the left of **Column 1**.

11. A cell phone manufacturer has developed a new type of battery for its phones. Extensive testing indicates that the population battery life (in days) obtained by all batteries of this new type is normally distributed with a mean of 700 days and a standard deviation of 100 days. The manufacturer wishes to offer a guarantee providing a discount on batteries if the original battery purchased does not exceed the days stated in the guarantee. What should the guaranteed battery life be (in days) if the manufacturer desires that no more than 5% of the batteries will fail to meet the guaranteed number of days?

12. The manager of a retail store wants to determine the best method of staffing employees without being wasteful. The problem that is often encountered is having too many employees and too few customers, and vice versa. The manager realized that during the holiday season they attract, on average, 90 customers per hour. On some days, the number is higher, on others, the number is lower. The manager would like to determine the probability of at least 2 customers arriving in a given minute during holiday season. Use the normal approximation to find the probability.

13. A machine used to regulate the amount of dye dispensed for mixing shades of paint can be set so that it discharges an average of $\mu$ milliliters of dye per can of paint. The amount of dye discharged is known to have a normal distribution with a variance equal to 0.0160. If more than 6 milliliters of dye are discharged when making a particular shade of blue paint, the shade is unacceptable. Determine the setting of $\mu$ so that no more than 1% of the cans of paint will be unacceptable.

14. The length of time required to complete a college achievement test is found to be normally distributed with a mean of 75 minutes and a standard deviation of 15 minutes. When should the test be terminated if we wish to allow sufficient time for 95% of the students to complete the test?

15. A manufacturing plant utilizes 3000 electric light bulbs that have a length of life that is normally distributed with a mean of 500 hours and a standard deviation of 50 hours. To minimize the number of bulbs that burn out during operation hours, all the bulbs are replaced after a given period of operation. How often should the bulbs be replaced if we want not more than 2% of the bulbs to burn out between replacement periods?

16. Howe's Finance Corporation provides financing for customers at an automotive dealership. The average loan amount is $24,000 with a standard deviation of $8000. Assuming that the loan amount is normally distributed, what is the probability that a randomly selected consumer buying a car will want to finance at least $20,000?

17. Suppose that the income of families in a large community follows a normal distribution. Two families are randomly selected and their incomes are $55,000 and $85,000, respectively. The two incomes correspond to $z$-scores of −0.5 and 2.0 respectively. Calculate the mean and standard deviation of the income of families in the neighborhood.

18. Suppose that the 30th percentile of a normal distribution is equal to 756 and that the 90th percentile of this normal distribution is 996. Find the mean and standard deviation of the normal distribution.

## P   **Discovery Project**

### How are tire mileage warranties calculated?

At the beginning of this chapter, we discussed the methodology of how automobile tire manufacturers determined the warranties associated with a particular tire that they made. The data set named Tire Manufacturer Warranty contains a sample of 30 tire mileages from each of 12 tire manufacturers. Using these data, please answer the following questions:

1.  Determine the mean and standard deviation of the mileages for each of the tire manufacturers.

2.  Do the mileages for each of the tire manufacturers follow a normal distribution? Justify your answer using graphical techniques.

3.  In the event that the mileages for a tire manufacturer do not follow a normal distribution, will that prevent us from calculating probabilities associated with the average mileages? Justify your answer.

4.  What is the distribution for each of the sample means of the mileages for each of the tire manufacturers.

5.  Determine the warranty mileage for each manufacturer if they want no more than 1% of the tires to need replacement.

6.  Answer question 5. if they want no more than 10% of the tires to need replacement.

**a.** Were the responses to this survey obtained using voluntary sampling techniques? Explain your answer.

**b.** What types of biases may be present in the responses?

**c.** Is 13% a reasonable estimate of the proportion of all Americans who eat chocolate frequently? Explain.

**14.** A magazine reported the results of a survey in which readers were asked to send in their responses to several questions regarding anger. Consider the reported results to the question, *How long do you usually stay angry?*

| Survey Responses | |
|---|---|
| **Category** | **% of Responses** |
| A few hours or less | 48 |
| A day | 12 |
| Several days | 9 |
| A month | 1 |
| I hold a grudge indefinitely | 22 |
| It depends on the situation | 8 |

**a.** Were the responses to this survey obtained using voluntary sampling techniques? Explain your answer.

**b.** What types of biases may be present in the responses?

**c.** Is 22% a reasonable estimate of the proportion of all Americans who hold a grudge indefinitely? Explain.

**15.** Students in a marketing class have been asked to conduct a survey to determine whether or not there is a demand for an insurance program at a local college. The students decide to randomly select students from the local college and mail them a questionnaire regarding the insurance program. Of the 150 surveys that were mailed, 50 students responded to the following survey item: *Pick the category which best describes your interest in an insurance program.*

| Survey Responses | |
|---|---|
| **Category** | **% of Responses** |
| Very Interested | 50 |
| Somewhat Interested | 15 |
| Interested | 10 |
| Not Very Interested | 5 |
| Not At All Interested | 20 |

**a.** What types of biases may be present in the responses?

**b.** Is 50% a reasonable estimate of the proportion of all students who would be very interested in an insurance program at the local college? Explain.

**c.** Is 50% a reasonable estimate of the proportion of all business majors who would be very interested in an insurance program at the local college? Explain.

**d.** What strategies do you think the marketing students could have used to get a less biased response to their survey?

**e.** Suppose the program was created and only a few people registered. How could the survey question have been reworded to better predict actual enrollment?

16. Television news programs often conduct opinion surveys by announcing some question on the air and advising viewers to call different numbers for a *yes* or *no* response. National television programs do the same thing except they use 900 numbers and the respondent must pay for the call. Suppose that a national news program asks its viewers to phone in a response to the following: *Women should be permitted to assume combat roles in the military*. The results of the particular survey were 34% *yes* and 66% *no*. Is it reasonable to believe that the results of the survey reflect the attitudes of the nation on this issue? What biases exist in this sampling method?

17. A local politician wants to know what the residents of his community think about an increase in the local property tax to pay for improvements to the highway. He decides to conduct a survey.

   a. What is the population of interest to the politician?

   b. Can you think of any good sources for a sampling frame?

   c. What are the shortcomings (if any) of the sources you picked for the sampling frame?

# 8.2 The Distribution of the Sample Mean and the Central Limit Theorem

Sample means vary because sample data vary from sample to sample. As an illustration, suppose that an automobile manufacturer wished to determine the average miles per gallon (mpg) of a specific vehicle model that it manufactures. Since determining the mpg of each vehicle is very time consuming, the manufacturer has decided to select two vehicles from a batch of six. Suppose that the actual mpg of the six vehicles are given in Table 8.2.1.

| Table 8.2.1 – Miles per Gallon | |
|---|---|
| **Car** | **MPG** |
| A | 25 |
| B | 27 |
| C | 40 |
| D | 29 |
| E | 28 |
| F | 30 |
| **Mean** | **29.8** |
| **Variance** | **23.14** |
| **Standard Deviation** | **4.81** |

It is important to realize that we are assuming the above set of data constitutes a population. The mean mpg rating of the population in Table 8.2.1 is approximately 29.8 and the population standard deviation is approximately 4.81.

$$\mu \approx 29.8$$
$$\sigma \approx 4.81$$

Both of these measures are considered population parameters. The mpg ratings given in Table 8.2.1 are not known by the manufacturer when the shipment arrives. The manufacturer's job is to estimate the population mean using a sample estimate, in this case using the sample mean from a sample of size two.

How many different samples of size two can be drawn? Assuming no replacement, there would be 15 possible samples of size two if order does not matter. A list of all possible samples and the resulting sample means is given in Table 8.2.2.

| Sample Number | Car 1 | Car 2 | First Observation | Second Observation | Mean, $\bar{x}$ |
|:---:|:---:|:---:|:---:|:---:|:---:|
| 1 | A | B | 25 | 27 | 26.0 |
| 2 | A | C | 25 | 40 | 32.5 |
| 3 | A | D | 25 | 29 | 27.0 |
| 4 | A | E | 25 | 28 | 26.5 |
| 5 | A | F | 25 | 30 | 27.5 |
| 6 | B | C | 27 | 40 | 33.5 |
| 7 | B | D | 27 | 29 | 28.0 |
| 8 | B | E | 27 | 28 | 27.5 |
| 9 | B | F | 27 | 30 | 28.5 |
| 10 | C | D | 40 | 29 | 34.5 |
| 11 | C | E | 40 | 28 | 34.0 |
| 12 | C | F | 40 | 30 | 35.0 |
| 13 | D | E | 29 | 28 | 28.5 |
| 14 | D | F | 29 | 30 | 29.5 |
| 15 | E | F | 28 | 30 | 29.0 |

Table 8.2.2 – MPG Sample Measurements ($n = 2$)

The sample means in Table 8.2.2 vary and when something varies there are at least three questions to ask:

1.  What is the central value of the variable?
2.  What is the variability of the variable?
3.  Is there a pattern (distribution) to the variability?

## What is the Central Value of $\bar{x}$?

Intuitively, you would expect the sample mean to be larger than $\mu$ some of the time and smaller than $\mu$ some of the time. For large samples, the distribution of $\bar{x}$ will be relatively symmetrical, and consequently, $\bar{x}$ should be larger than $\mu$ about 50% of the time and smaller than $\mu$ about 50% of the time. But for small samples, the distribution of the sample mean may not be symmetrical. This is the case for the sample means in Table 8.2.2. Ten of the 15 means are below 29.8, which is the population mean for the population in Table 8.2.1. Generally, however, the sample means should be near the population mean, or symbolically, $\bar{x}$ should be near $\mu$. It can be shown theoretically that the mean of the $\bar{x}$'s equals $\mu$. In the example, the mean of the sample means is approximately 29.8, which equals the population mean. This is not a coincidence.

Estimators are similar to marksmen. When you shoot, you want to hit what you are shooting at.

When you estimate, you want to get as close as possible to the population characteristic you are estimating. But, bullets do not always land exactly where the marksman aims. If the gun sights are properly adjusted, then the shots will be dispersed around the middle of the target area.

> **Definition**
>
> **Unbiased**
> If the average value of an estimator equals the population parameter being estimated, the estimator is said to be **unbiased**.

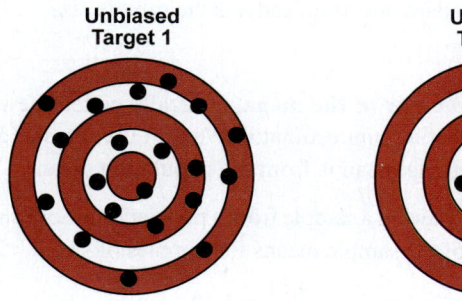

**Unbiased Target 1**     **Unbiased Target 2**

Figure 8.2.1

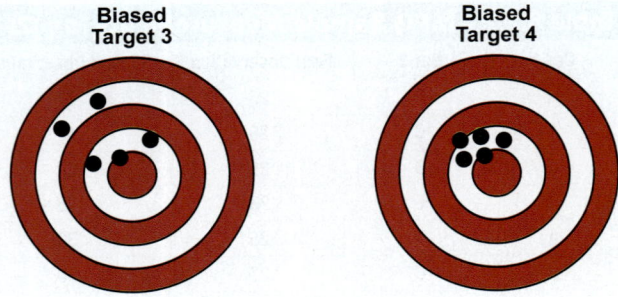

Figure 8.2.2

An estimator which produces estimates centered around the true value is said to be unbiased.

Unbiasedness is desirable property for an estimator to possess. Just as the ideal target rifle is one that would hit in exactly the same place every shot, the ideal estimator is one that is unbiased. Since the mean of $\bar{x}$ is always equal to $\mu$, $\bar{x}$ is an unbiased estimator of $\mu$.

### Properties

#### Unbiased Estimators

1. The sample mean, $\bar{x}$, is an unbiased estimator of $\mu$.
2. The sample proportion, $\hat{p}$, is an unbiased estimator $p$.
3. The sample variance, $s^2$, is an unbiased estimator of $\sigma^2$.

If an estimator is unbiased, its variability determines its reliability. If an unbiased estimator is extremely variable, then the individual estimates it produces may not be as close to the parameter being estimated as estimates produced by a biased estimator with little variability. This is why it is important to not only consider the central value, but also the variability of a random variable, in this case the sample mean.

## What is the Variability of $\bar{x}$?

The variability of an estimator reveals a great deal about the quality of that estimator. In order to assess how well the sample mean estimates the population mean, the standard deviation of the sample means must be determined.

### Formula

#### Standard Deviation of the Sample Mean: Infinite Population

It can be shown that for a population of infinite size, the standard deviation of $\bar{x}$, denoted as $\sigma_{\bar{x}}$, is

$$\sigma_{\bar{x}} = \frac{\sigma}{\sqrt{n}},$$

where $\sigma$ is the population standard deviation and $n$ is the sample size.

We refer to $\sigma_{\bar{x}}$ as the **standard error of the mean**, generally called the **standard error**, which is the standard deviation of a point estimator. We will use the standard error of the mean to indicate how far the sample mean is from the population mean.

Suppose, for example, you were drawing a sample from a population whose standard deviation is 4.81. The standard deviation of the sample means for samples of size $n = 2$ would be

$$\sigma_{\bar{x}} = \frac{\sigma}{\sqrt{n}} = \frac{4.81}{\sqrt{2}} \approx 3.40.$$

If the population is finite, as in Table 8.2.1, then the finite population correction factor $\left(\sqrt{\dfrac{N-n}{N-1}}\right)$ must be applied to the calculation of the standard deviation of the sample mean.

**Formula**

### Standard Deviation of the Sample Mean: Finite Population

For a finite population the standard deviation of $\bar{x}$ is

$$\sigma_{\bar{x}} = \sqrt{\frac{N-n}{N-1}} \cdot \frac{\sigma}{\sqrt{n}},$$

where $N$ = the size of the population and $n$ = the size of the sample.

Correcting for the finite population represented in Table 8.2.1, the standard deviation of the sample means for samples of size 2 would be

$$\sigma_{\bar{x}} = \sqrt{\frac{N-n}{N-1}} \cdot \frac{\sigma}{\sqrt{n}} = \sqrt{\frac{6-2}{6-1}} \cdot \frac{4.81}{\sqrt{2}} \approx 3.04.$$

Examining the formula for $\sigma_{\bar{x}}$, we notice that as the size of the sample, $n$, increases, the variability of the sample mean decreases. The possibility of changing the variability of an estimator means the accuracy of the estimator can be manipulated. For example, suppose that instead of using a sample of size two, the manufacturer decides to use a sample of size three. Six items chosen three at a time could produce 20 samples of size three. The samples and means of each sample are shown in Table 8.2.3. The standard deviation of the sample means for $n = 3$ in Table 8.2.3 is approximately 2.15, using the finite population correction factor.

For both samples ($n = 2$ and $n = 3$), the mean of the sample means is approximately 29.8, which equals the population mean. But the standard deviation of the sample means for samples of size three is 2.15, which is smaller than the standard deviation for samples of size two.

For $n = 2$, $\sigma_{\bar{x}} \approx 3.04$    For $n = 3$, $\sigma_{\bar{x}} \approx 2.15$

| Table 8.2.3 – MPG Sample Measurements ($n = 3$) | | | | | | |
|---|---|---|---|---|---|---|
| Sample Number | Car 1 | Car 2 | Car 3 | First Observation | Second Observation | Third Observation | Mean, $\bar{x}$ |
| 1 | A | B | C | 25 | 27 | 40 | 30.67 |
| 2 | A | B | D | 25 | 27 | 29 | 27.00 |
| 3 | A | B | E | 25 | 27 | 28 | 26.67 |
| 4 | A | B | F | 25 | 27 | 30 | 27.33 |
| 5 | A | C | D | 25 | 40 | 29 | 31.33 |
| 6 | A | C | E | 25 | 40 | 28 | 31.00 |
| 7 | A | C | F | 25 | 40 | 30 | 31.67 |
| 8 | A | D | E | 25 | 29 | 28 | 27.33 |
| 9 | A | D | F | 25 | 29 | 30 | 28.00 |
| 10 | A | E | F | 25 | 28 | 30 | 27.67 |
| 11 | B | C | D | 27 | 40 | 29 | 32.00 |
| 12 | B | C | E | 27 | 40 | 28 | 31.67 |
| 13 | B | C | F | 27 | 40 | 30 | 32.33 |
| 14 | B | D | E | 27 | 29 | 28 | 28.00 |
| 15 | B | D | F | 27 | 29 | 30 | 28.67 |
| 16 | B | E | F | 27 | 28 | 30 | 28.33 |
| 17 | C | D | E | 40 | 29 | 28 | 32.33 |
| 18 | C | D | F | 40 | 29 | 30 | 33.00 |
| 19 | C | E | F | 40 | 28 | 30 | 32.67 |
| 20 | D | E | F | 29 | 28 | 30 | 29.00 |

Since both estimators are unbiased and are centered around the population mean, the standard deviation of the estimator is a measure of how close the estimator (in this case the sample mean) is to the population mean. In other words, the lower $\sigma_{\bar{x}}$ is, the better the estimator is.

As in many problems, there is an accuracy versus dollars trade-off. Greater accuracy can be achieved by taking a larger sample, but larger samples normally cost more to collect and examine.

Suppose a population had a mean of 43,660 and a standard deviation of 2500. The distributions in Figure 8.2.3 are the distributions of the sample mean for samples of size 25, 100, and 200, respectively, from this population. Based on the graphs in Figure 8.2.3, it seems that the estimator shown for $n = 200$ would be preferred. Because it has less variability, the estimates of $\mu$ ($\bar{x}$'s) from samples of $n = 200$ should be closer (on average) to $\mu$.

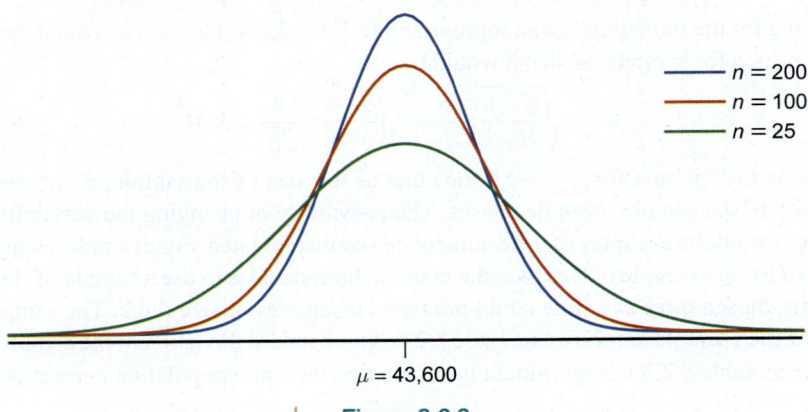

Sampling Distribution of the Sample Mean, $\sigma = 2500$

- $n = 200$
- $n = 100$
- $n = 25$

$\mu = 43{,}600$

**Figure 8.2.3**

---

**Properties**

## Characteristics of the Sample Mean

If an estimate of the population mean is required, the sample mean possesses two desirable characteristics.

1. The mean of the sample means is the population mean. Another way of expressing this concept is to say that the expected value of $\bar{x}$ is equal to the population mean. Symbolically, this can be expressed as

$$E(\bar{x}) = \mu \quad \text{(unbiasedness)}.$$

2. If the sample size is increased, the standard error of the sample mean decreases. This implies that the quality of the estimator tends to improve as the sample size increases.

$$\sigma_{\bar{x}} = \frac{\sigma}{\sqrt{n}}$$

---

The second characteristic has an important consequence for estimating a population parameter. By choosing a sufficiently large sample size, an estimate can be obtained with some specified level of accuracy. Being able to predetermine the accuracy of an estimate is an important topic in statistics and will be discussed in the next chapter.

### Is There a Familiar Pattern to the Variability?

The **Central Limit Theorem** is a very important theorem that summarizes the distribution of the sample mean. The distribution of the sample mean becomes closer to a normal distribution as the sample size becomes larger, regardless of the distribution of the population from which the sample is drawn.

The most important feature of the Central Limit Theorem is that it can be applied to any population. Because the theorem does not have any distributional assumptions, it is widely applicable and is one of the cornerstones of statistical inference. Many of the statistical techniques discussed in subsequent chapters will have their theoretical basis in this theorem.

### Theorem

#### Central Limit Theorem

If a sufficiently large random sample (i.e., $n \geq 30$) is drawn from a population with mean $\mu$ and standard deviation $\sigma$, the distribution of the sample mean will have the following characteristics.

1. An approximately normal distribution regardless of the distribution of the underlying population.

2. $\mu_{\bar{x}} = E(\bar{x}) = \mu$      The mean of the sample means equals the population mean.

3. $\sigma_{\bar{x}} = \dfrac{\sigma}{\sqrt{n}}$      The standard deviation of the sample means equals the standard deviation of the population divided by the square root of the sample size.

The only restrictive feature of the theorem is that the sample size must be sufficiently large for the theorem to be applicable. Even if the distribution of the population deviates substantially from the normal distribution, a sample size of at least 30 will be sufficiently large to produce a sampling distribution for $\bar{x}$ that is approximately normal. Additionally, points 2. and 3. are true even for small samples when it can be shown that the samples are drawn from a population that is normally distributed. If the population is known to be normally distributed, then the sampling distribution of $\bar{x}$ will be normally distributed for any sample size.

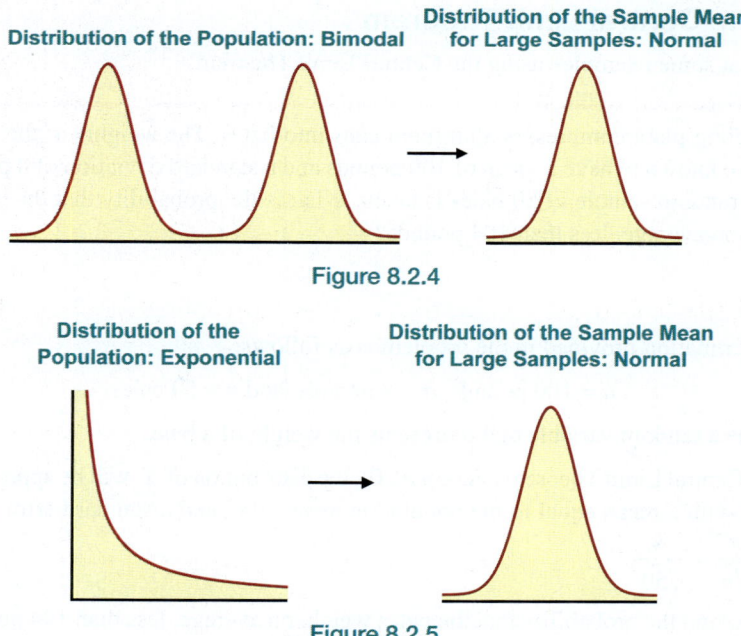

**Distribution of the Population: Bimodal**  **Distribution of the Sample Mean for Large Samples: Normal**

Figure 8.2.4

**Distribution of the Population: Exponential**  **Distribution of the Sample Mean for Large Samples: Normal**

Figure 8.2.5

## The History of the Central Limit Theorem

Pierre-Simon Laplace is credited with the initial statement of the Central Limit Theorem in 1776. He developed the theorem while working on the probability distribution of the sum of meteor inclination angles.

Although the theorem is stated with respect to the sample mean, it is a more general theorem regarding the sum of random variables. If you add up a sufficiently large number of random variables the sum will be normally distributed.

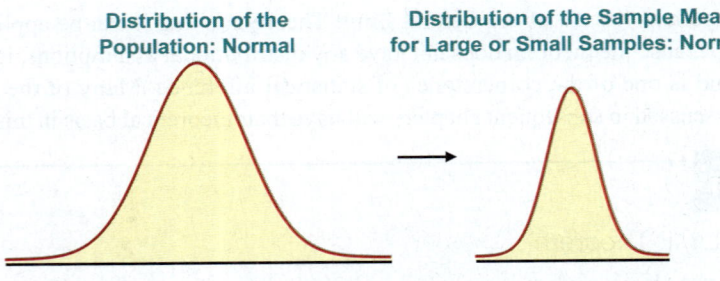

**Figure 8.2.6**

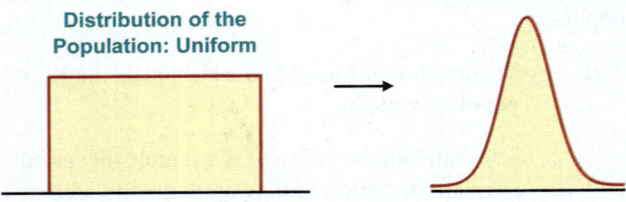

**Figure 8.2.7**

If we are dealing with samples that are sufficiently large, once we have determined that the Central Limit Theorem applies and that the sampling distribution of the sample means is approximately normally distributed, we can standardize any sample mean to find probabilities that we are interested in. Thus, we can adapt the z-score formula as follows.

$$z = \frac{\bar{x} - \mu_{\bar{x}}}{\sigma_{\bar{x}}} = \frac{\bar{x} - \mu}{\frac{\sigma}{\sqrt{n}}}$$

## Using the Central Limit Theorem

Let's look at some examples using the Central Limit Theorem.

### Example 8.2.1

**Calculating a Probability Using the Central Limit Theorem**

A recycling plant compresses aluminum cans into bales. The weights of the resulting bales are known to have a mean of 100 pounds and a standard deviation of 8 pounds. A simple random sample of 50 bales is taken. What is the probability that the bales will weigh, on average, less than 104 pounds?

#### SOLUTION

The information provided in the problem is as follows:

$$\mu = 100 \text{ pounds, } \sigma = 8 \text{ pounds, and } n = 50 \text{ bales.}$$

Let $X$ be a random variable that represents the weight of a bale.

By the Central Limit Theorem (since $n \geq 30$), the distribution of $\bar{x}$ will be approximately normal with a mean equal to the population mean, 100, and a standard error given by

$$\sigma_{\bar{x}} = \frac{\sigma}{\sqrt{n}} = \frac{8}{\sqrt{50}}.$$

Thus, to find the probability that the bales weigh, on average, less than 104 pounds, we are interested in the following probability.

$$P(\bar{x} < 104) = ?$$

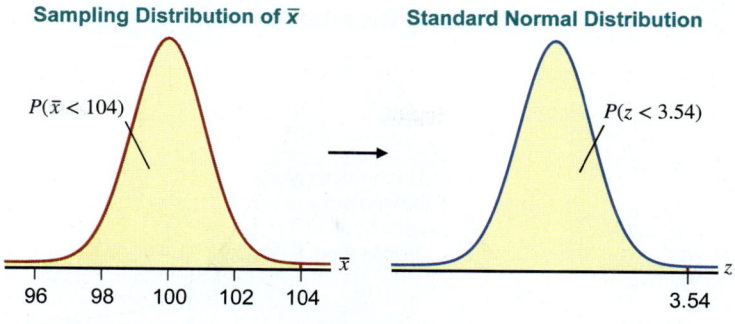

**Figure 8.2.8**

Since $\bar{x}$ is a normal random variable, the probability that $\bar{x}$ is less than 104 pounds is determined using a $z$-transformation.

$$P\left(\bar{x} < 104\right) = P\left(z < \frac{\bar{x} - \mu}{\sigma / \sqrt{n}}\right)$$

$$= P\left(z < \frac{104 - 100}{8 / \sqrt{50}}\right)$$

$$= P\left(z < 3.54\right)$$

$$\approx 1$$

Since Table B's largest value is 3.49, yielding a probability of 0.9998, we say that the probability is approximately 1.

---

The concept of using a statistic to estimate a population parameter introduces the concept of the **error of estimation** (also referred to as **sampling error** or **estimation error**).

The following example will further illustrate the error of estimation and how it is used.

> **Definition**
>
> Error of Estimation
>
> In general, the **error of estimation** is the difference between a sample statistic and the parameter that it is estimating.

---

Suppose that a population has an unknown mean and a standard deviation of 5. A sample of size 100 is drawn from the population. If the sample mean is used as an estimate of the population mean, what is the probability that the sample mean will be within one unit of the true mean?

> **Example 8.2.2**
>
> **Using the Error of Estimation to Calculate a Probability**

**SOLUTION**

Since the error of estimation is defined to be the difference between the sample statistic and the parameter that it is estimating, for this problem the error of estimation is one. In the problem at hand, the error is given by $\mu - \bar{x}$ since we will be using $\bar{x}$ to estimate $\mu$.

However, we are just as interested in negative errors as positive errors. Thus, to satisfy the condition that an error of less than one unit has been made, we are interested in the following probability.

$$P\left(\mu - 1 < \bar{x} < \mu + 1\right)$$

To find the probability that the sample mean is within one unit of the population mean, we must standardize the equation above so that we can use the standard normal distribution. Standardizing, we obtain the following.

$$P\left(\frac{(\mu - 1) - \mu}{5 / \sqrt{100}} < z < \frac{(\mu + 1) - \mu}{5 / \sqrt{100}}\right) = P\left(-2 < z < 2\right)$$

$$P\left(-2 < z < 2\right) = P\left(z < 2\right) - P\left(z < -2\right) = 0.9772 - 0.0228 = 0.9544$$

**Sampling Distribution of $\bar{x}$**

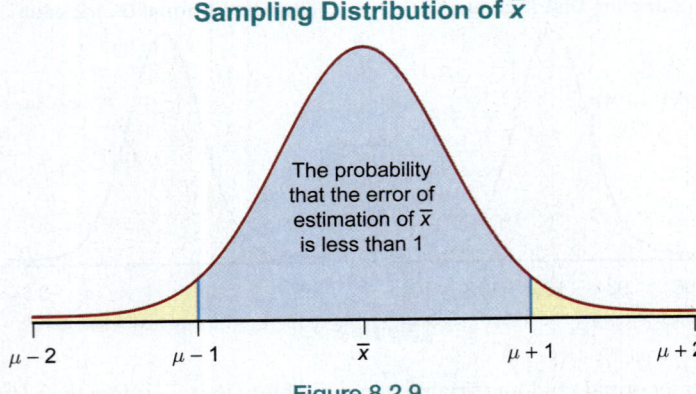

The probability that the error of estimation of $\bar{x}$ is less than 1

$\mu - 2$    $\mu - 1$    $\bar{x}$    $\mu + 1$    $\mu + 2$

Figure 8.2.9

**Standard Normal Distribution**

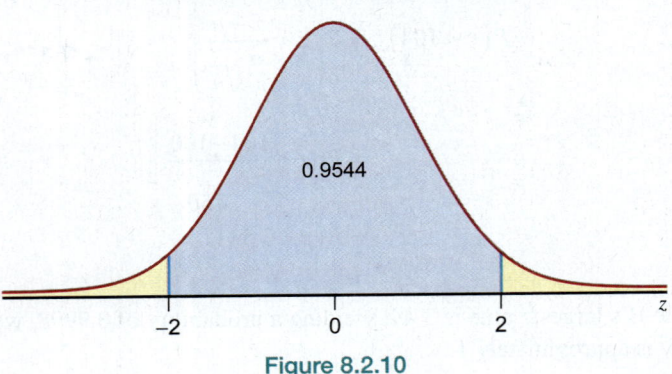

0.9544

$-2$    $0$    $2$    $z$

Figure 8.2.10

Thus, if a sample size of 100 is drawn from the population given, the probability that the sample mean will be within one unit of the population mean is 0.9544.

In this problem, we were able to determine the probability of making an error of less than one unit without knowing the mean of the population. This probability (0.9544) tells us a lot about the quality of the estimate $\bar{x}$. The estimate of $\mu$ obtained from a sample of size 100 is more than just an estimate now, it is an estimate with a "level of confidence" in its quality.

---

**Example 8.2.3**

**Calculating the Standard Deviation Given a Probability**

The lifetime of a certain type of micro transistor is normally distributed with a mean of 156 hours. It is known that 1.5 percent of the transistors have a lifetime greater than 167 hours.

a. What is the standard deviation of the distribution of the lifetimes of the transistors?

b. Using the standard deviation calculated in part **a.**, what is the probability that the average lifetime of a sample of 25 transistors is at least 155 hours?

**SOLUTION**

a. Let $X$ = the lifetime of a micro transistor.

$X$ is normally distributed with a mean of 156 hours. (Note that even though $n < 30$ we can still use the methods presented because we are told the underlying population is normal.) We also know that 1.5% of the transistors have a lifetime greater than 167 hours. This can be written as

$$P(X > 167) = 0.015. \quad (1)$$

Since is $\sigma$ is unknown, we can find $\sigma$ using (1) such that

$$P\left(z > \frac{167-156}{\sigma}\right) = 0.015.$$

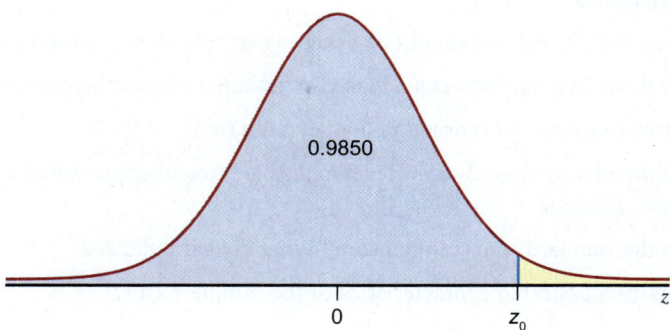

0.9850

$z_0$

0

**Figure 8.2.11**

$P(z > z_0) = 0.0150$ and $P(z < z_0) = 0.9850$ where $z_0 = \dfrac{167-156}{\sigma}$

Finding the value 0.9850 in Table B, we see that $z_0 = 2.17$. Thus, $\dfrac{167-156}{\sigma} = 2.17$, and solving for $\sigma$, we get $\sigma \approx 5.0691$.

**b.**  $P(\bar{x} \geq 155) = ?$

Since $X$ is normally distributed, then $\bar{x}$ is also normally distributed. Thus,

$$P(\bar{x} \geq 155) = P\left(z \geq \frac{155-156}{5.0691/\sqrt{25}}\right)$$
$$= P(z \geq -0.99)$$
$$= 1 - P(z < -0.99)$$
$$= 1 - 0.1611$$
$$= 0.8389$$

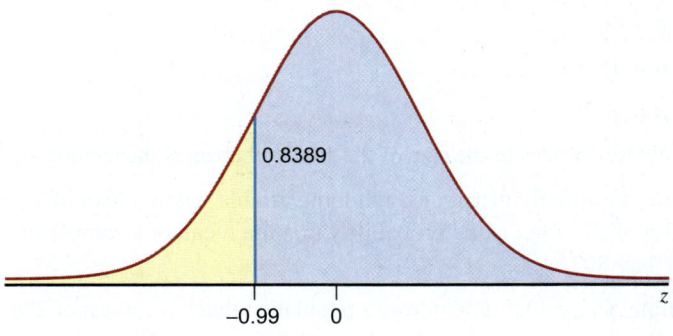

0.8389

$-0.99$   0

**Figure 8.2.12**

Thus, the probability that the average lifetime of a sample of 25 transistors is at least 155 hours is 0.8389.

## ✒ 8.2 Exercises

### Basic Concepts

1. What key three questions should be asked when considering a random variable?

2. Explain the difference between a biased estimator and an unbiased estimator.

3. Give three examples of estimators that are unbiased.

4. Is an unbiased estimator always closer to the parameter being estimated than a biased estimator? Explain.

5. What is the standard error of the mean? What does it indicate?

6. What are two desirable characteristics of the sample mean?

7. Explain the Central Limit Theorem.

8. What effect does increasing the sample size have on the accuracy of an estimate?

9. What is the error of estimation?

### Exercises

10. Suppose the random variable $X$ has a mean of 20 and a standard deviation of 5. Calculate the mean and the standard deviation of the sample mean for each of the following sample sizes (assume the population is infinite).

    a. $n = 35$

    b. $n = 50$

    c. $n = 75$

    d. What happens to the size of the standard deviation of the sample mean as the sample size increases?

11. Suppose the random variable $X$ has a mean of 50 and a standard deviation of 10. Calculate the mean and standard error for each of the following sample sizes (assume the population is infinite).

    a. $n = 40$

    b. $n = 55$

    c. $n = 100$

    d. What happens to the size of the standard error as the sample size increases?

12. If there is a normally distributed random variable with a mean of 75 and a standard deviation of 22, what is the probability that the mean of a sample of size 19 will be greater than 80?

13. If a sample of size 40 is drawn from a population that has a mean of 276 and a variance of 81, what is the probability that the mean of the sample will be less than 273?

14. Suppose there is a normally distributed population with a mean of 250 and a standard deviation of 50. If $\bar{x}$ is the average of a sample of 36, find the following probabilities.

    a. $P(\bar{x} \leq 240)$

    c. $P(246 \leq \bar{x} \leq 260)$

    b. $P(\bar{x} \geq 255)$

    d. $P(234 \leq \bar{x} \leq 245)$

15. Suppose there is a normally distributed population with a mean of 100 and a standard deviation of 10. If $\bar{x}$ is the average of a sample of 50, find the following probabilities.

    a. $P(\bar{x} \leq 110)$

    c. $P(95 \leq \bar{x} \leq 115)$

    b. $P(\bar{x} \geq 90)$

    d. $P(85 \leq \bar{x} \leq 98)$

**16.** A company fills bags with fertilizer for retail sale. The weights of the bags of fertilizer have a normal distribution with a mean weight of 15 lb and standard deviation of 1.70 lb.

    **a.** What is the probability that a randomly selected bag of fertilizer will weigh between 14 and 16 pounds?

    **b.** If 35 bags of fertilizer are randomly selected, find the probability that the average weight of the 35 bags will be between 14 and 16 pounds.

**17.** A travel agency conducted a survey of the prices charged by ocean cruise ship lines and determined they were approximately normally distributed with a mean of $110 per day and a standard deviation of $20 per day.

    **a.** If an ocean cruise ship line is chosen at random, find the probability that it will charge less than $99 per day.

    **b.** What is the probability that the average charge for a randomly selected sample of 35 ocean cruise ship lines will be less than $99 per day?

**18.** The turkeys found in a particular county have an average weight of 15.6 pounds with a standard deviation of 4.00 pounds. Forty-five turkeys are randomly selected for a county fair.

    **a.** Find the probability that the average weight of the turkeys will be less than 14.5 pounds.

    **b.** What is the probability that the average weight of the turkeys will be more than 17 pounds?

    **c.** Find the probability that the average weight of the turkeys will be between 13 and 18 pounds.

**19.** The average score for a water safety instructor (WSI) exam is 75 with a standard deviation of 12. Fifty scores for the WSI exam are randomly selected.

    **a.** Find the probability that the average of the fifty scores is at least 80.

    **b.** Find the probability that the average of the fifty scores is at most 70.

    **c.** Find the probability that the average of the fifty scores is between 72 and 78.

**20.** A college food service buys frozen fish in boxes labeled 10 pounds. The true average weight of the boxes is 8 pounds with a standard deviation of 2 pounds. The food service director suspects that the boxes do not contain as much fish as advertised. He decides to inspect 40 boxes from the next shipment. If the average weight is less than 10 pounds he will reject the entire shipment. Find the probability that the food service director will not reject the shipment.

**21.** The AQI, or the Air Quality Index, is an index used to determine the ozone level in a city. Depending upon the AQI reading, it may not be safe to jog or even to go outside. Readings in the 0–50 range mean that the air quality conditions are considered "good," 51–100 are "moderate," 101–150 means "unhealthy for sensitive groups," 151–200 means "unhealthy," 201–300 means "very unhealthy," and 301–500 means "hazardous." Suppose that an industrial region has an average AQI reading of 102 with a standard deviation of 40. Find the probability that for a random sample of 50 days, the average AQI reading is:

**Source:** airnow.gov

    **a.** at least 105.

    **b.** at most 90.

    **c.** between 100 and 115.

# 8.3 The Distribution of the Sample Proportion

There are many instances in which the variable of interest is a proportion. A manufacturer might be interested in knowing what fraction of his manufacturing components are defective. A market researcher might be interested in knowing what proportion of persons on a mailing list will buy the company's product. A college might be concerned with the fraction of freshmen that may be struggling academically after the first year. Population proportions must be estimated just like population means.

The symbols used to represent the population and sample proportions are

$p$, the **population proportion**, and

$\hat{p}$, the **sample proportion**.

**✎ NOTE**

Note: $\hat{p}$ is pronounced *p*-hat.

## Determining the Sample Proportion, $\hat{p}$

Suppose you are trying to determine the fraction of manufacturing components that are defective. If you select 120 components at random and 38 components are defective, then

$$\hat{p} = \text{the proportion in the sample that are defective} = \frac{38}{120} \approx 0.3167 .$$

In general, when calculating a proportion, the number in the sample that possess the characteristic of interest goes in the numerator, and the size of the sample is placed in the denominator.

---

**Formula**

### Sample Proportion

The **sample proportion** is given by

$$\hat{p} = \frac{x}{n} ,$$

where $x$ is the number of observations in the sample possessing the characteristic of interest and $n$ is the total number of observations in the sample.

---

Just as the sample mean was a good estimate of the population mean, the sample proportion is a good estimate of the population proportion. Also, recall that the sample mean varied depending on the sample selected. The sample proportion varies in the same manner.

Since $\hat{p}$ varies, three familiar questions must be examined.

1. What is the central value?
2. What is the variability?
3. Is there a pattern to the variability?

## What is the Central Value of $\hat{p}$?

The expected value of the sample proportion, $\hat{p}$, is the population proportion, $p$. Symbolically, this is expressed as

$$E\left(\hat{p}\right) = p .$$

Since the expected value of the estimator $\hat{p}$ is equal to $p$, $\hat{p}$ is an unbiased estimator of $p$.

## What is the Variability of $\hat{p}$?

> **Formula**
>
> ### Standard Deviation of the Sample Proportion
> The standard deviation of $\hat{p}$ is given by
> $$\sigma_{\hat{p}} = \sqrt{\frac{p(1-p)}{n}},$$
> where $p$ is the population proportion and $n$ is the sample size.

Note that $\sigma_{\hat{p}}$ is affected by the values of $p$ and $n$.

The standard deviation of $\hat{p}$ decreases as $n$ becomes larger. Also, the numerator reaches a maximum when $p = 0.5$ and declines as you move away from that figure. So for a fixed value of $n$, $\hat{p}$ has its greatest standard deviation when the population proportion equals 0.5. If the population proportion is unknown (which is usually the case), $p$ can be estimated by $\hat{p}$, and the standard deviation of the sample proportion is estimated as

$$\sigma_{\hat{p}} \approx \sqrt{\frac{\hat{p}(1-\hat{p})}{n}}.$$

## Is There a Familiar Pattern to the Variability of $\hat{p}$?

The sampling distribution of $\hat{p}$ approaches normality as $n$ becomes sufficiently large. The sample size is generally considered "sufficiently large" if $np \geq 5$ and $n(1-p) \geq 5$.

**Sampling Distribution of $\hat{p}$**

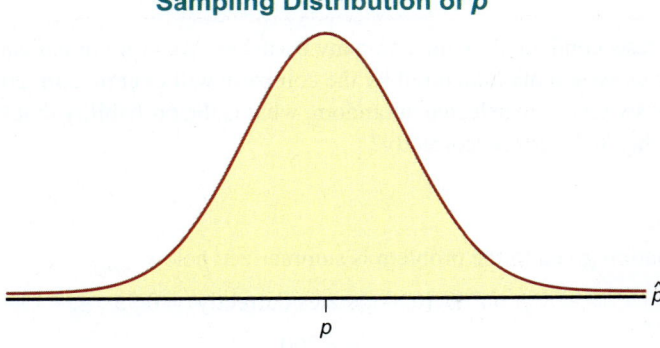

Figure 8.3.1

The sampling distribution of $\hat{p}$ is summarized below, for both finite and infinite populations.

> ## Properties
>
> ### Sampling Distribution of the Sample Proportion
>
> If the population is infinite and the sample is sufficiently large ($np \geq 5$ and $n(1-p) \geq 5$), the distribution of $\hat{p}$ has the following characteristics.
>
> 1. An approximately normal distribution.
>
> 2. $\mu_{\hat{p}} = E(\hat{p}) = p.$   The mean of the sample proportions equals the population proportion.
>
> 3. $\sigma_{\hat{p}} = \sqrt{\dfrac{p(1-p)}{n}} \approx \sqrt{\dfrac{\hat{p}(1-\hat{p})}{n}}.$
>
> If the population is finite and the sample is sufficiently large, the distribution of $\hat{p}$ has the following characteristics.
>
> 1. An approximately normal distribution.
>
> 2. $\mu_{\hat{p}} = E(\hat{p}) = p.$
>
> 3. $\sigma_{\hat{p}} = \sqrt{\dfrac{N-n}{N-1}} \cdot \sqrt{\dfrac{p(1-p)}{n}} \approx \sqrt{\dfrac{N-n}{N-1}} \cdot \sqrt{\dfrac{\hat{p}(1-\hat{p})}{n}},$
>
>    where $N$ is the size of the population.

Since $\hat{p}$ is a good estimator of $p$, one of the natural questions to ask is, can limits be established for the error of estimation? Since the sampling distribution of $\hat{p}$ is known, probabilities for various errors of estimation can be determined.

## Example 8.3.1

**Calculating a Probability for the Proportion of Switches**

A series of tests conducted by the company Switches, We Got'em indicates that a particular type of switch manufactured by the company will operate correctly 95% of the time. If 200 switches are selected at random, what is the probability that less than 90% of the switches will operate correctly?

### SOLUTION

The information given in the problem is summarized below.

$$p = P(\text{switch operates correctly}) = 0.95$$

$$n = 200$$

We want to determine the probability that less than 90% of the switches in the sample will operate correctly. This translates to $P(\hat{p} < 0.90)$. Since $np = 200(0.95) = 190 \geq 5$ and $n(1-p) = 200(0.05) = 10 \geq 5$, the sample size is sufficiently large, and we can assume that the sample proportion is approximately normally distributed with $\mu_{\hat{p}} = p = 0.95$ and

$$\sigma_{\hat{p}} = \sqrt{\frac{p(1-p)}{n}} = \sqrt{\frac{0.95(1-0.95)}{200}}.$$

Therefore, we can standardize the distribution of the sample proportion as follows.

$$P(\hat{p} < 0.90) = P\left(z < \frac{\hat{p} - p}{\sqrt{\dfrac{(p)(1-p)}{n}}}\right) = P\left(z < \frac{0.90 - 0.95}{\sqrt{\dfrac{(0.95)(1-0.95)}{200}}}\right)$$

$$= P(z < -3.24)$$

$$= 0.0006$$

**z-Distribution**

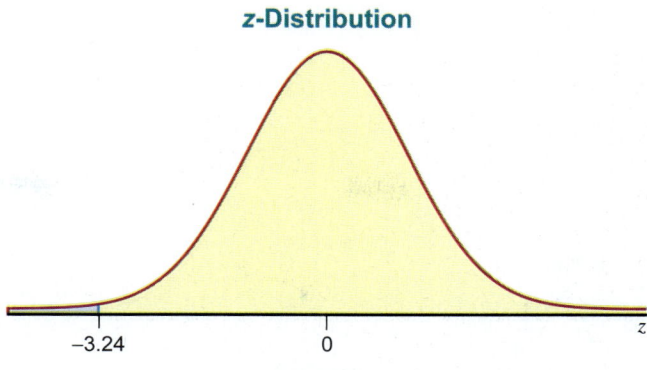

Figure 8.3.2

Thus, the probability that $\hat{p} < 0.90$ is 0.0006.

With the information we have developed thus far, we can begin to draw conclusions (i.e., make inferences about the population proportion). If it is true that this particular type of switch works 95% of the time, then it is unlikely that we would observe a sample proportion of less than 90% in a sample of 200 switches.

---

Suppose a sample of 400 people is used to perform a taste test. If the true proportion of the population that prefers Pepsi is 0.5, what is the probability that less than 44% of the persons in the sample will prefer Pepsi?

### SOLUTION

Assume the population from which the sample is drawn is extremely large and the finite population correction factor is not applicable. The distribution of $\hat{p}$ would then be normal with $\mu_{\hat{p}} = E\left(\hat{p}\right) = p = 0.5$ and

$$\sigma_{\hat{p}} = \sqrt{\frac{(0.5)(1-0.5)}{400}} = 0.025.$$

The probability that the sample proportion is less than 0.44 is given by

$$P\left(\hat{p} < 0.44\right) = P\left(z < \frac{0.44 - 0.5}{0.025}\right)$$
$$= P(z < -2.40)$$
$$= 0.0082.$$

**Sampling Distribution of $\hat{p}$**

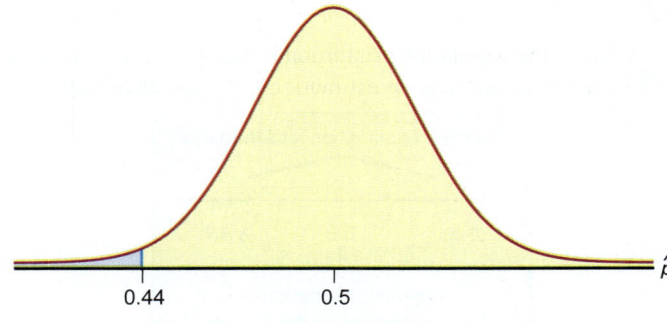

Figure 8.3.3

**Example 8.3.2**

**Calculating a Probability for the Proportion who Prefer Pepsi**

### ∞ Technology

For instructions on calculating this probability using technology, please visit stat.hawkeslearning.com and navigate to **Discovering Business Statistics, Second Edition > Technology Instructions > Normal Distributions > Normal Probability (cdf)**.

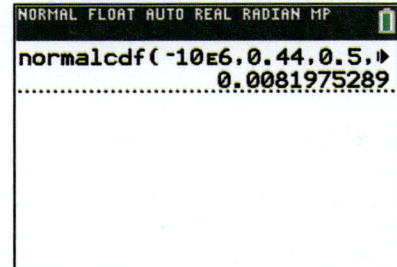

**z-Distribution**

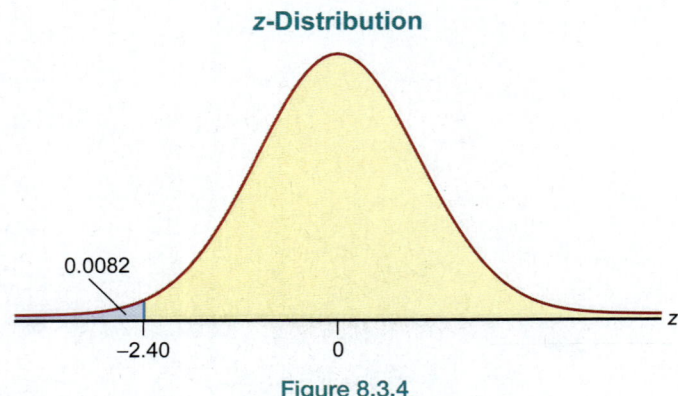

0.0082

−2.40       0       z

Figure 8.3.4

If the true proportion of people in the population who prefer Pepsi is 0.5, it is extremely unlikely (0.0082 is less than 1 in 100) to observe a sample proportion as low as 0.44. Suppose that you had to make a decision as to whether cola drinkers were indifferent between Pepsi and Coke. If they were indifferent, the proportion who prefer Pepsi should be around 0.5. If you used a sample of 400 people and observed a sample proportion of 0.438 that preferred Pepsi, which of the conclusions would you believe?

**Conclusion A:** Cola drinkers are indifferent between Pepsi and Coke. Stated another way, the proportion of persons that favor Pepsi is about 0.5.

**Conclusion B:** Cola drinkers are not indifferent between Pepsi and Coke. In other words, people prefer one brand of cola over the other.

The likelihood of observing a sample with $\hat{p}$ less than 0.44 is very rare (0.0082) given that the true proportion who prefer Pepsi is $P = 0.5$. This leads us to doubt Conclusion A, and select Conclusion B. The decision-making problem given above is really a statistical inference problem. Although the procedure for analyzing inference problems will be presented in subsequent chapters, the problem illustrates the connection between probability and inference. To reach a decision (make the inference) we used the fact that if the true proportion is really 0.5, a proportion below 0.44 is highly improbable for a sample of 400.

---

**Example 8.3.3**

**Calculating a Probability Concerning the Error of Estimation**

Suppose a sample of 150 is used to estimate the proportion of U.S. citizens over 18 that favors expanding government regulation of major financial institutions. If the proportion that favors expanding government regulation is really 0.6, what is the probability that the error of estimation will be less than five percentage points?

**SOLUTION**

Since the true value of the population proportion is 0.6, the value of $\hat{p}$ must fall between 0.55 and 0.65 in order for the error of estimation to be less than 0.05.

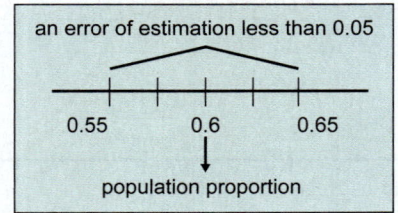

an error of estimation less than 0.05

0.55       0.6       0.65

population proportion

Figure 8.3.5

In order to determine the probability that $\hat{p}$ will fall in this interval, its distribution must be determined. Since the distribution of $\hat{p}$ is approximately normal for large samples

($np \geq 5$ and $n(1-p) \geq 5$), the distribution of $\hat{p}$ will be approximately normal with

$$\mu_{\hat{p}} = p = 0.6$$

$$\sigma_{\hat{p}} = \sqrt{\frac{p(1-p)}{n}} = \sqrt{\frac{0.6(1-0.6)}{150}} = 0.04.$$

To find the probability that $\hat{p}$ is within 0.05 of the true proportion, we must find

$$P\left(p - 0.05 < \hat{p} < p + 0.05\right) = P\left(0.55 < \hat{p} < 0.65\right).$$

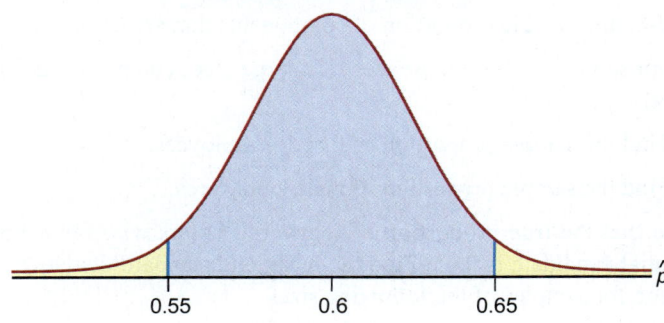

**Sampling Distribution of $\hat{p}$**

Figure 8.3.6

Using the $z$-transformation,

$$= P\left(\frac{0.55 - 0.6}{0.04} < z < \frac{0.65 - 0.6}{0.04}\right)$$
$$= P\left(-1.25 < z < 1.25\right)$$
$$= P\left(z < 1.25\right) - P\left(z < -1.25\right)$$
$$= 0.8944 - 0.1056$$
$$= 0.7888.$$

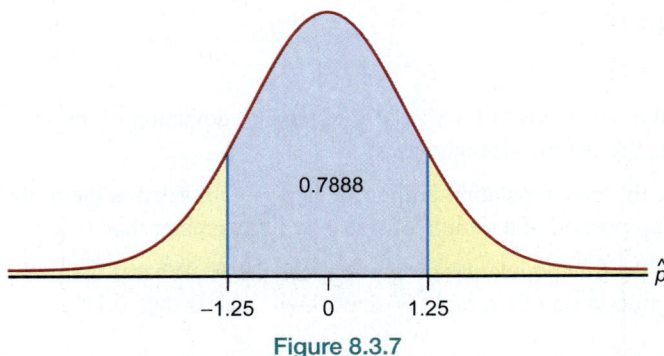

Figure 8.3.7

### ∞ Technology

For instructions on calculating this probability using technology, please visit stat.hawkeslearning.com and navigate to **Discovering Business Statistics, Second Edition > Technology Instructions > Normal Distributions > Normal Probability (cdf)**.

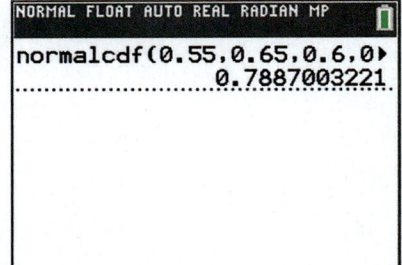

Therefore, for a sample of 150 U.S. citizens over 18 years of age, it is likely (0.7888) that the error of estimation will be less than five percentage points.

## ✎ 8.3 Exercises

### Basic Concepts

1. What does the symbol $\hat{p}$ represent?

2. What is the connection between $\hat{p}$ and $p$?

3. Is $\hat{p}$ an unbiased estimator? If so, of what?

4. What are the conditions that make the sample size $n$ "sufficiently large" for a sample proportion?

5. Describe the sampling distribution of $\hat{p}$ if $n$ is sufficiently large.

## Exercises

6. A random sample of 40 electronic components has 5 defective components.

   a. Find the sample proportion of components that are defective.

   b. Find the sample proportion of components that are not defective.

7. A random sample of 100 employees of a large steel company has 30 females and 70 males.

   a. Find the sample proportion of female employees.

   b. Find the sample proportion of male employees.

8. Suppose that the true proportion of registered voters who favor the Republican presidential candidate is 0.45. Find the mean and standard deviation of the sample proportion for samples of the following sizes.

   a. $n = 30$

   b. $n = 45$

   c. $n = 65$

   d. What happens to the size of the standard deviation of the sample proportion as the sample size increases?

9. Suppose that the true proportion of Americans over 25 years old that have a 4-year college degree is 0.35. Find the mean and the standard deviation of the sample proportion for samples of the following sizes.

   a. $n = 38$

   b. $n = 52$

   c. $n = 75$

   d. What happens to the size of the standard deviation of the sample proportion as the sample size increases?

10. Suppose the true population proportion is $p = 0.50$. What is the probability that the sample proportion of a sample of size 20 will be greater than 0.60?

11. Suppose the true population proportion is $p = 0.30$. What is the probability that the sample proportion of a sample of size 30 will be less than 0.20?

12. Suppose that the true proportion of Americans who save at least 10% of their income is 0.15. If $\hat{p}$ is the sample proportion of Americans surveyed who save at least 10% of their income from a sample of size 38, find the following probabilities.

   a. $P\left(\hat{p} > 0.25\right)$

   b. $P\left(\hat{p} < 0.09\right)$

   c. $P\left(0.10 < \hat{p} < 0.20\right)$

   d. $P\left(0.18 < \hat{p} < 0.25\right)$

13. Suppose that the true proportion of airline pilots between the ages of 35 and 45 is 0.60. If $\hat{p}$ is the sample proportion of airline pilots between the ages of 35 and 45 from a sample of size 100, find the following probabilities.

   a. $P\left(\hat{p} > 0.55\right)$

   b. $P\left(\hat{p} < 0.45\right)$

   c. $P\left(0.50 < \hat{p} < 0.60\right)$

   d. $P\left(0.60 < \hat{p} < 0.75\right)$

14. The director of a radio station in a large metropolitan area believes that the proportion of young professionals (his target market) in the area who prefer rock 'n' roll music has increased from 25% to 35%. The director randomly decides to select 50 young professionals and ask them if they prefer rock 'n' roll to any other type of music. If the sample proportion is greater than 0.35, he will switch to a new format emphasizing rock 'n' roll.

 a. If the true proportion of young professionals who prefer rock 'n' roll has not changed, find the probability that the radio director will switch to the new format.

 b. If the true proportion of young professionals who prefer rock 'n' roll has changed as the director suspects, find the probability that the radio director will switch to the new format.

15. The property manager of a large office building would like to make the building smoke free; however, he does not want to upset too many of his customers. He decides to randomly select 50 of the workers in the building and ask them whether or not they smoke. If the sample proportion of workers who smoke is less than 0.30, the property manager will make the building smoke free.

 a. Find the probability that the property manager will make the building smoke free when the true proportion of smokers is 0.5.

 b. Find the probability that the property manager will not make the building smoke free when the true proportion of smokers is 0.2.

16. Eighty percent of the flights arriving in Atlanta for a large U.S. airline are on time. If the FAA randomly selects 50 of the airline's flights, find the probability that:

 a. at least 85% of the sampled flights will be on time.

 b. at most 70% of the sampled flights will be on time.

 c. between 75% and 85% of the sampled flights will be on time.

17. Approximately 7% of the nation's public school children in grades 2 through 5 took medication in 2003 for attention deficit hyperactivity disorder (ADHD), a developmental disorder characterized by impulsiveness or difficulty concentrating or sitting still. The main treatment prescribed for ADHD is Ritalin, a relatively safe drug with few side effects. Assume that a suburban elementary school had an enrollment of 286 students in 2003.

 a. Find the probability that at least 4% of the school children took medication for ADHD.

 b. Find the probability that between 5% and 8% of the school children took medication for ADHD.

# 8.4 Sampling Methods

Random sampling is an effective means of obtaining a sample that is representative of the population. As we discussed previously, acquiring an exact sampling frame for the population under consideration is a requirement for simple random sampling, a requirement which can be time-consuming and expensive. In addition, it is sometimes not even possible to list all the members of a population. There are other sampling strategies that are designed to reduce the cost of sampling or add control to the sampling procedure. These techniques can be categorized as probability samples or non-probability samples.

**Probability samples** enable an analyst to determine the probable errors that an estimator might generate. Essentially, they allow the analyst a known degree of confidence in his or her estimation. All of statistical inference relies on probability sampling. **Non-probability samples**

**Definition**

Probability Sample
A **probability sample** is a sample used to estimate a population parameter that has known errors, allowing a statement about the reliability of the estimate to be made.

> **Definition**
>
> **Non-Probability Sample**
>
> A **non-probability sample** is a convenient sample used to estimate a population parameter in which no statement about the estimate's reliability can be made.

> **Definition**
>
> **Judgment Sample**
>
> A **judgment sample** is a sample collected by an expert in a specific field of study. The quality of the sample depends on the competence of the expert.

> **Definition**
>
> **Convenience Sample**
>
> A **convenience sample** is a non-probability sample that can be easily collected but may not be representative of the population.

are convenient means of obtaining sample data. If data from a non-robability sample are used to estimate a population parameter, there is no statistical theory that helps define the potential error of the estimate, and hence no statement about the estimate's reliability can be made.

## Non-Probability Samples

Non-probability samples come in several forms. A **judgment sample** is a sample in which sample values are selected by an expert in the field. One of the common uses of judgment samples is in auditing. When auditing a company's accounts receivable, an auditor may use a judgment sample to verify the accounts outstanding. The quality of the sample data is related to the competence of the expert. If the expert is good at what he or she does, this type of sampling can produce reasonable representations of the population and correspondingly good estimates of the population parameters.

Another type of non-probability sample is the **convenience sample**. As the name implies, a convenience sample is nothing more than a convenient group of observations. For example, the students in your statistics class would be a convenience sample of students at your college. Although convenience samples could be representative of the population, they tend to possess more bias than other forms of sampling. Consider your statistics class. It is likely that the class is dominated by some particular group of majors and has a disproportionate number of sophomores and juniors. Despite their shortcomings, convenience samples are convenient, and certainly a convenient sample is often better than no sample at all. There are two disadvantages to using a convenience sample. First, it's difficult to determine how representative the sample is of the population, or if it is representative at all. Secondly, even though the convenience sample may not be representative of the population, statistical methods are often applied. Since it's not a probability sample, the results will lack validity and are questionable.

**Convenience Sampling**

**Figure 8.4.1**

The worst forms of non-probability samples are voluntary or self-selected samples. Those samples were discussed in Section 8.1.

## Systematic Sampling

> **Definition**
>
> **Systematic Sampling**
>
> **Systematic sampling** involves including every $k^{th}$ member of the population in the sample.

One type of sampling technique, the **systematic sample**, does not belong to the probability or non-probability sample categories.

For example, suppose a sample of 1000 names is to be selected from a mailing list that contains 80,000 names. If every $80^{th}$ person in the mailing list is selected for inclusion in the sample, the result will be a sample of 1000 names. If the names are in random order, then the systematic sample will produce a random sample. Unfortunately, it is difficult to be sure that the list does not possess some pattern that would bias the sample. Still systematic samples are regularly used in sampling mailing lists. They are also used in controlling production quality. Production-oriented quality control plans frequently call for the inspection (sample) of every $k^{th}$ item on an assembly line. Conceptually, inspection of every $k^{th}$ item is no different from sampling every $k^{th}$ item from a mailing list. A sample using this technique is often called an "almost random sample."

**Systematic Sampling**

Every 4th bottle

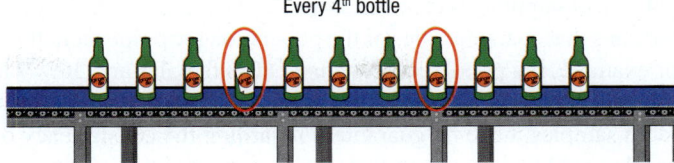

Figure 8.4.2

Systematic samples are generally good samples. But if there is some pattern in the sampling frame that corresponds to the sampling pattern, an unrepresentative sample can result. For example, suppose you wished to sample a list containing total daily sales. If you sampled every 7th item, the result would be a sample that contained sales from only one day of the week.

## Cluster Sampling

Although simple random sampling may be feasible, the traveling cost required to obtain the sample information may be prohibitive. Suppose McDonald's wishes to perform quality control inspection on its franchises. If a random sample of 500 is drawn from the 31,000 McDonald's restaurants scattered throughout the world, the quality inspection team could be in for quite a bit of traveling. Instead of selecting the stores individually, suppose we divide the world into 200 regions. Then we could select 20 of these regions and inspect every store in those regions. This technique is called **cluster sampling**. Using this technique, an inspector would be required to travel much less and a substantial reduction in cost would result.

Clusters are not always geographic. Suppose you were interested in soliciting student opinion regarding bookstore prices. If everyone in the college had a 9:00 AM class, one reasonable method of clustering would be to use each class as a cluster. Randomly select a set of clusters (classes) and interview each member of the selected cluster. The selection of clusters depends greatly on the population and variables of interest.

**Cluster Sampling**

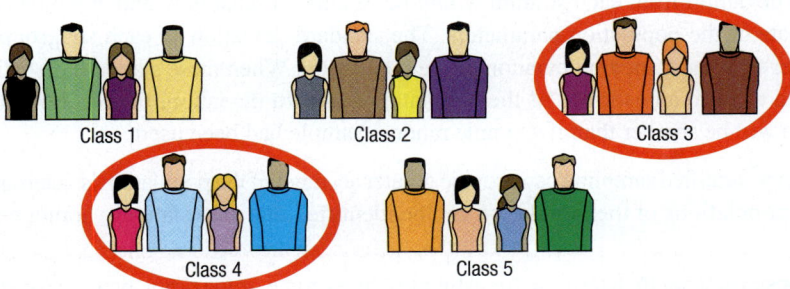

Figure 8.4.3

Cluster sampling can be as effective as simple random sampling if the clusters are as heterogeneous as the population. Unfortunately, clusters are almost never as diverse as the population. If the clusters are geographic, people in the same geographic area tend to be less diverse than the population. If the clusters are not heterogeneous, cluster sampling is less efficient than simple random sampling for a given sample size. Generally, smaller cluster sizes will result in more representative samples.

In simple random sampling, constructing the sampling frame is frequently cumbersome. Cluster sampling simplifies the task, since the initial frame is composed only of clusters. Only those clusters selected as part of the sample must be completely enumerated and sampled.

Cluster sampling is a good alternative to simple random sampling when the population's geographic area is spread out over a large area or when the sampling frame for a simple random sample is difficult to construct.

**Definition**

**Cluster Sampling**

**Cluster sampling** involves dividing the population into clusters, and randomly selecting a sample of clusters to represent the population.

## Stratified Sampling

The fundamental goal of sampling is to obtain a sample that is representative of the population. *Representative* means that characteristics of the population are proportionally represented in the sample. For example, if a population contained 60% females and 30% with blood type A, then it would be desirable for the sample to have 60% females and 30% with blood type A. Simple random samples make no guarantees regarding the constituency of the sample. If a random sample could be taken such that it was assured that population characteristics were properly represented, the resulting samples would tend to be more representative of the population. This is precisely the objective of **stratified sampling**.

**Stratified Sampling**

**Figure 8.4.4**

Suppose you wished to estimate the average profit of companies on the New York Stock Exchange (NYSE). Given that all companies are not the same size, it would be best to stratify the companies by some characteristic such as market capitalization or number of employees. For example, if companies were classified in intervals based on market cap, stratification would mean that a fixed percentage of the companies would come from each interval (stratum), thus, giving us a representative sample of the population of companies on the NYSE. The mean and standard deviation from each stratum would be calculated separately and combined to form an estimate of the population parameters. The standard deviation of each subgroup should be smaller than the standard deviation of the population. When these standard deviations are combined to form an estimate of the population standard deviation, the resulting standard deviation will be smaller than if a simple random sample had been used.

In summary, stratified sampling can provide greater accuracy if the population is heterogeneous, and sub-populations of the population can be identified that are relatively homogeneous.

---

**Example 8.4.1**

**Sampling College Students to Determine Grade Point Average**

Suppose we want to determine (or estimate) the average grade point average for students at Tech University by year (i.e., freshmen, sophomores, juniors, and seniors).

**a.** What is the population of interest?

**b.** What variable will be measured?

**c.** Discuss the best method of sampling for this scenario.

**SOLUTION**

**a.** The population of interest is all students at Tech University.

**b.** The grade point average for students that are included in the sample.

**c.** Stratified random sampling may be the best approach given that we have "built-in" strata (the class year of the student). Therefore, one would stratify the data by class year (freshmen, sophomores, juniors, and seniors). Once you have your strata, obtain an estimate of the size of each stratum and sample so that you have a good representation of the population. That is, if you have a total of 15,000 students and 5,000 are freshmen, you would want $\frac{1}{3}$ of your sample to be from freshman students.

---

**Example 8.4.2**

**Sampling a Large City to Determine Average Household Income**

Suppose we are interested in determining the average household income for a large city.

a. What is the population of interest?

b. What variable will be measured?

c. Discuss how we would use each of the following sampling methods.

   i. Simple random sampling

   ii. Cluster sampling

   iii. Stratified sampling

**SOLUTION**

a. The population of interest is all households in the city.

b. The variable to be measured is household income.

c. While there may be a right way to do sampling, time and costs are often a major consideration. For this scenario, let's discuss how we could perform the sampling techniques mentioned above.

   i. We can use simple random sampling, but would need a sampling frame (i.e., a list of all houses in the city) and then randomly select from the frame. This could be time-consuming and costly. Also, you may not get a fair representation of the population if you are going from door-to-door to collect your data.

   ii. To use cluster sampling for this situation, we could divide the city into blocks or neighborhoods and then randomly select a number of clusters (i.e., blocks) and collect data from each selected cluster.

   iii. To use stratified sampling, we would also need a sampling frame to separate into strata and then sample from each stratum.

---

**Example 8.4.3**

**Determining if a Sample is Random**

City officials sample the opinions of homeowners in a community about the possibility of raising taxes to improve the quality of local schools. A directory of all homes in the city is used; a computer generates random numbers to identify the addresses to be sampled. An interviewer visits each home between the hours of 3 pm and 5 pm on weekdays. If no one is home, the address is eliminated from the sample and replaced by another randomly chosen address. Does this process approximate random sampling?

**SOLUTION**

This is NOT random sampling. The 3 pm to 5 pm timeframe is indicative that this is not random sampling because we have not made it such that each homeowner in the community has the same likelihood of being selected. For example, if a homeowner works from 9 am to 5 pm, they will not have a chance to participate in this study.

## ✐ 8.4  Exercises

### Basic Concepts

1. What are the advantages and disadvantages of non-probability samples?

2. What is a judgment sample? Give an example not in the text of when a judgment sample would be appropriate.

3. What is a convenience sample? Are these samples usually representative of the population?

4. What are the worst forms of non-probability samples?

5. Explain the idea of systematic sampling. What are the advantages and disadvantages of this sampling procedure?

6. Explain the idea of cluster sampling. What are the advantages and disadvantages of this sampling procedure?

7. Explain the idea of stratified sampling. What are the advantages and disadvantages of this sampling procedure?

### Exercises

8. An employee-owned company has 6000 female employees and 2000 male employees. The human resources department decides to develop a survey on several different benefit plans, including child care and retirement benefits, that they may offer employees in the future. The results of the survey are to be presented to the board of directors for consideration. Because the human resources department wants to be sure of equal representation of the sexes, it has decided to randomly select 500 females and 500 males. What kind of sampling method is the human resources department using? If the sample is used to make inferences regarding the desirability of various benefits packages for all employees, discuss any deficiencies in the sampling procedure.

9. Explain why a systematic sample is not a random sample.

10. Suppose you were instructed to draw a simple random sample from a metropolitan area in order to gain information on the citizens' view on a proposed amendment to the state constitution. To create a simple random sample, you must create a sampling frame. You have decided to use the telephone directory as your frame for the metropolitan area.

    a. Identify the population under consideration.

    b. What kinds of people will be omitted from your frame?

    c. What kinds of biases will be introduced in your sample as a consequence of the omission you described in part b.? Can you think of ways of compensating for the bias?

11. A social researcher in Florida wants to determine the average number of children per family in the state.

    a. What is the population of interest?

    b. What variable will be measured?

    c. What level of measurement is the variable of interest?

    d. Discuss the steps that would be necessary for each of the following sampling methods.

        i. Simple random sampling

        ii. Cluster sampling

        iii. Stratified sampling

    e. What sampling method do you believe would be the most cost-effective? Justify your answer.

12. A stock analyst wants to estimate the average yearly earnings of stocks on the New York Stock Exchange.

    a. What is the population of interest?

    b. Discuss the steps necessary to apply each of the following sampling methods.

        i. Simple random sampling

        ii. Cluster sampling

        iii. Stratified sampling

13. A news reporter in Orlando, Florida wants to conduct a survey to determine how local residents feel about the institution of a state income tax. Since there will be a lot of people from which to choose, he goes to Disney World and randomly selects individuals entering the complex. He asks the selected people whether or not they favor a state income tax in Florida. The responses to the survey are as follows.

| Survey Responses | |
| --- | --- |
| Category | % of Responses |
| Favor a Florida State Income Tax | 50 |
| Do Not Favor a Florida State Income Tax | 50 |

    a. What sampling technique was used for this survey?

    b. What biases may be present in the responses?

    c. Is 50% a reasonable point estimate of the proportion of Orlando residents who favor the state income tax? Explain.

# AE    Additional Exercises

1. A national news network is interested in the opinion which Americans have regarding a national healthcare policy. During the evening news, they display a 900 telephone number and ask their viewers to call in and respond to the question: *Do you favor a national healthcare policy in the U.S.?*

| Survey Responses | |
|---|---|
| Category | % of Responses |
| Yes | 45 |
| No | 45 |
| Do not have enough information to decide | 10 |

   a. What sampling technique was used for this survey?

   b. What biases may be present in the responses?

   c. Is 45% a reasonable estimate of the proportion of all Americans who favor a national healthcare policy? Explain.

2. An entrepreneur wants to open a new Indian restaurant in a resort community. To determine if there is a market for the new restaurant, the entrepreneur decides to conduct a survey.

   a. What is the population of interest to the entrepreneur?

   b. Can you think of any good sources for a sampling frame?

   c. What are the shortcomings (if any) of the sources you picked for the sampling frame?

3. A manufacturer is developing a new type of paint. Test panels were exposed to various corrosive conditions to measure the protective ability of the paint. Based on the results of the test, the manufacturer has concluded that the mean life before corrosive failure for the new paint is 168 hours with a standard deviation of 30 hours. If the manufacturer's conclusions are correct, find the probability that the paint on a sample of 60 test panels will have a mean life before corrosive failure of less than 150 hours.

4. Seventy-five percent of the students graduating from high school in a small Iowa farm town attend college. The town's Chamber of Commerce randomly selects 30 recent graduates and inquires whether or not they will attend college.

   a. Find the probability that at least 80% of the surveyed students will be attending college.

   b. Find the probability that at most 70% of the surveyed students will be attending college.

   c. Find the probability that between 65% and 85% of the surveyed students will be attending college.

   d. Why might the data gathered from this sample misestimate the proportion of students who will actually be attending college?

5. A biology professor is interested in the proportion of students at his college who are pre-med majors. In his next class, he asks for the students who are pre-med majors to raise their hands. Fifty percent of the students raised their hands.

   a. What type of sampling technique was used for this survey?

   b. What types of biases may be present in the responses?

   c. Is 50% a reasonable point estimate of the proportion of students at the college who are pre-med majors? Explain.

6. A report released by the U.S. Census Bureau in November of 2001 stated that fewer families fit the traditional family makeup in 2000 as compared with 1970. The report stated that in 1970, 5.6 million families were headed by women with no husband present, and 16.6% of the households consisted of people living alone. In 2000, these figures increased to 16.2 million and 26.7% respectively. The report also stated that in 1970, 40% of the households consisted of married couples with no children while in the year 2000 this decreased to 25%. Suppose that a random sample of 150 households is chosen in 2000, and the percent of these that consist of people living alone is determined.

   a. Find the probability that at least 30% of those households sampled consist of people living alone.

   b. Find the probability that at most 28% of those sampled households consist of people living alone.

   c. Find the probability that between 26% and 36% of those households consist of people living alone.

7. It is known that the percentage return for a group of stocks in the technology sector is normally distributed with a mean of 15 percent and a standard deviation of 22 percent. Suppose you selected a random sample of 10 stocks from this sector.

   a. What are the mean and standard deviation of $\bar{x}$?

   b. Find the interval containing 68.26% of all possible sample mean returns.

8. A restaurant wants to determine the average time to prepare meals for its customers. To aid in this process, the restaurant randomly selects the meal preparation time of 150 of its customers and finds that the average preparation time is 18 minutes with a standard deviation of eight minutes. Describe the distribution of the sample mean of preparation time for its customers.

9. With such a large number of people using text messages as a means of communication, a company is interested in determining the number of work hours lost due to text messaging. Based on a survey of 30 randomly selected employees (anonymously, of course), the company has determined that the average amount of time spent texting over a one-month period is 180 minutes with a standard deviation of 60 minutes.

   a. What is the probability that the average amount of time spent using text messages is more than 210 minutes in this one-month period?

   b. Thinking that it's practically impossible for her employees to spend, on average, three hours a month texting while at work, the manager conducts another survey. She randomly samples 45 employees and finds that the average amount of time spent texting while at work over a one-month period is less than 180 minutes. Is it reasonable to conclude that the average amount of time spent using text messaging has decreased since the initial survey? Justify your answer.

   c. How might the data gathered from this sample not accurately depict the loss of productivity from text messaging?

10. Suppose that a random sample of size 64 was selected and the researcher found that the mean was 30 and the standard deviation was 4.

   a. What is the probability that the sample mean is more than 31.25?

   b. What assumptions were made in part a.?

11. A town is considering building a high school football stadium approximately one-half mile from a well-established housing development. The residents of the development opposed the stadium construction due to the noise coming from the stadium during games. In presenting their argument, the residents indicated that any noise more than 103 decibels would be unacceptable. Using a sample of 35 games previously played in the old arena, the town found that the average decibels were 100 with a standard deviation of 8 decibels.

   a. What is the probability that a randomly selected game will generate noise in excess of 103 decibels at the stadium?

   b. What is the probability that a randomly selected game will generate a noise level of exactly 103 decibels?

   c. Suppose a compromise was made that required the noise level to be lower than 103 decibels 95% of the time. Will the mean level of the noise have to be lowered to comply with the new regulation? If so, by how much? Assume that the standard deviation remains at 8 decibels.

12. The town manager believes that 60% of the residents will approve the construction of the proposed high school football stadium. A random sample of 100 residents will be used to estimate the proportion of residents that will approve the construction.

   a. Assuming that the town manager is correct and that $p = 0.6$, describe the sampling distribution of $\hat{p}$.

   b. What is the probability that between 50% and 70% of the residents will approve the stadium construction?

13. It is believed that 90% of all adults and 85% of all kids between the ages of 12 and 17 have cellular phones. Suppose a sample of 500 adults and 400 kids was taken.

   a. Describe the sampling distribution of the proportion of adults that have cellular phones. Assume that the stated probabilities above are true.

   b. Describe the sampling distribution of the proportion of kids that have cellular phones. Assume that the stated probabilities above are true.

   c. What is the probability that the sample proportion of adults having cell phones will be within 2% of the true proportion?

   d. What is the probability that the sample proportion of kids having cell phones will be within 4% of the true proportion?

14. A survey of college students was conducted to learn about their attitudes toward alcohol abuse on college campuses. Sixty-two percent of student respondents indicated that they believe there was a high rate of alcohol abuse on college campuses. Suppose that a sample of 250 college students was taken. What is the probability that more than seventy percent believed that there was a high rate of alcohol abuse on college campuses?

15. A credit card issuer believes that 75% of college students between the ages of 18 and 22 have more than $5000 of credit card debt. The credit card issuer conducted a survey of 500 college students between the ages of 18 and 22.

   a. What is the probability that at least 70% of college students between the ages of 18 and 22 have credit card debt in excess of $5000?

   b. Assuming that the credit card issuer is correct, what is the probability that the proportion will be within three percent of the population proportion?

   c. What is the probability that the proportion will not be within three percent of the population proportion?

**16.** A marketing firm conducts a survey by mail with a 20% response rate. If the firm mailed 1000 surveys for a new study, what is the probability that at least 220 individuals will respond?

**17.** Suppose that it has been reported by a group of researchers that the average number of hours of TV viewing per household per week in the United States is 50.4 hours. Suppose the standard deviation is 11.8 hours, and a random sample of 42 U.S. households is taken.

    **a.** What is the probability that the sample average is more than 35 hours? If the sample average is actually more than 35 hours, what would it mean in terms of the figures presented by the researchers?

    **b.** Suppose the population standard deviation is unknown. If 71% of all sample means are greater than 49 hours and the population mean is still 50.4 hours, what is the value of the population standard deviation? Use a sample size of 42.

# P  Discovery Project

## Data

For an example data set, please visit stat.hawkeslearning.com and navigate to **Discovering Business Statistics, Second Edition > Data Sets > EV Company Financials**.

## Industry Statistics

Pick an industry (software companies, computer manufacturers, oil, etc.) and research the financial information about companies in that industry. The internet, your library, and the prospectus from each of the companies may be useful resources to help you with your research. Also, select a statistic to track or follow about the companies such as sales, profit, revenue, etc. Using the information that you gather about the companies, answer the following questions.

1.  Identify the population of interest.

2.  Select a random sample of 10 companies in the industry that you've selected and record at least three variables of interest for each of the companies.

3.  Compute the average and standard deviation of each variable of interest.

4.  Compute the standard error of each variable of interest.

5.  Discuss what you have learned about the statistics you studied and the companies in your population.

6.  Identify the sources used for this project.

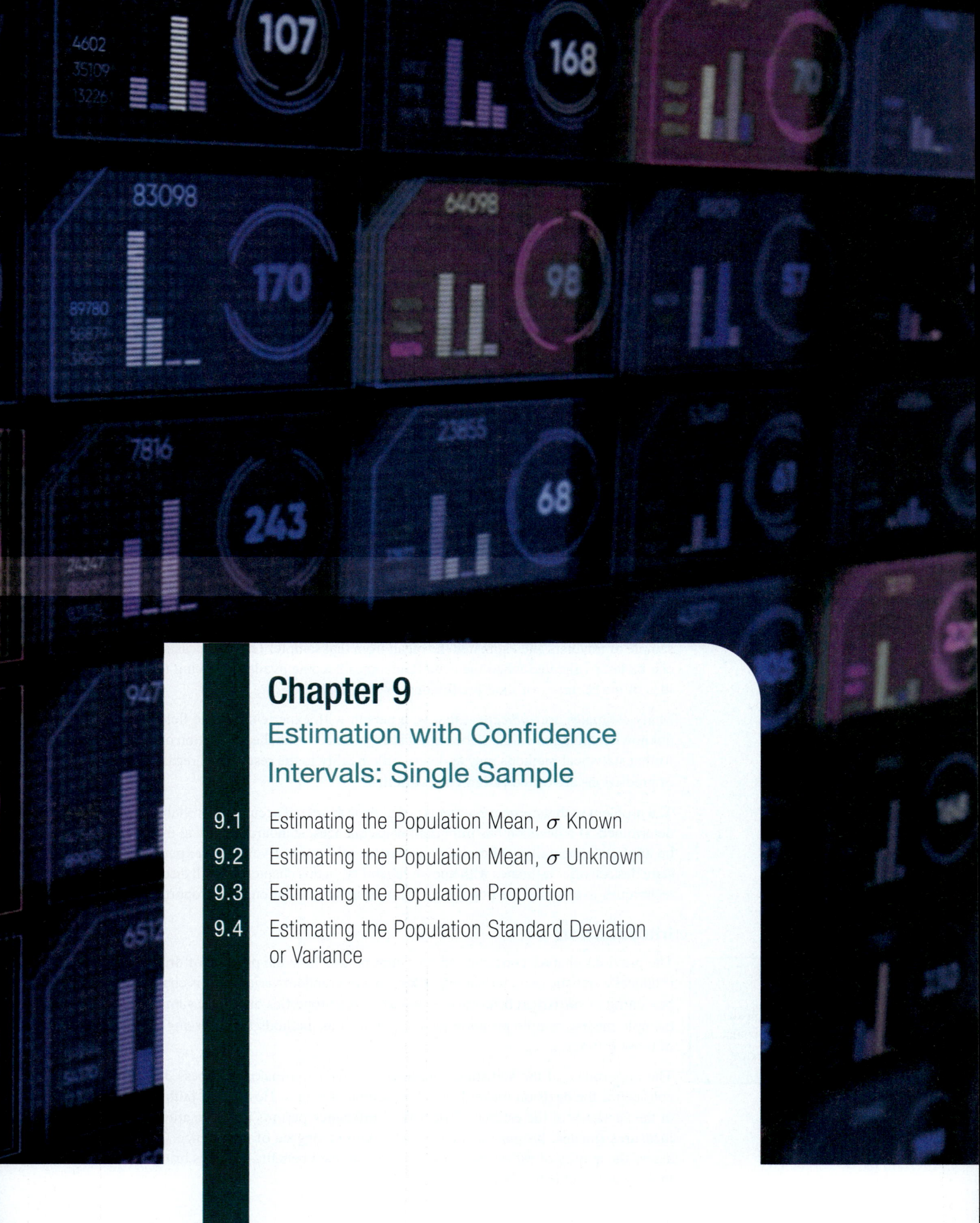

# Chapter 9

## Estimation with Confidence Intervals: Single Sample

# Discovering the Real World

## The Travel Industry

The U.S. Travel Association is a national, nonprofit organization that represents all components of the travel industry. Each year, the U.S. Travel Association makes forecasts on travel. They forecast the times when people will travel the most, the lengths of the vacations, the method of transportation, and many other factors that interest travelers and the travel industry. The travel industry depends on travelers to support its businesses, and travelers rely on the travel industry to provide them with services such as transportation, lodging, food and beverage, and entertainment. Some businesses rely on forecasts from the U.S. Travel Association to provide insight on travel-related expenditures, employment expectations within the industry, and earnings of various travel-related occupations. Of course, when making the predictions, these forecasts are evaluated based on their validity and reliability.

In 2019, Americans spent a total of $994 billion on domestic travel, of which $724 billion was for leisure and $270 billion was for business travel. Americans also took a total of 2.4 billion trips in 2019, of which 2.32 billion of the trips were domestic and 79.4 million were international arrivals. Of the domestic trips, 1.85 billion were for leisure, 464 million were business-related, 2.3 billion traveled by automobile, and 189 million trips were taken via airplane. Using the travel dollars mentioned above, Americans spent an average of $3,029 on domestic travel, an average of $2,206 traveling for leisure, and an average of $823 for business travel.

**Source:** www.ustravel.org

The estimates provided by the American Travel Association are viewed as the "best possible" information available. That is, they don't know the true population mean amount that Americans spend on domestic, leisure, or business travel. However, they took a random sample of travelers and estimated the mean from that sample. These forecasts (or estimates) can be later validated when the actual outcomes become available, giving the analysts an idea of the accuracy of their predictions.

Many estimates are *subjective*; that is, a person with experience in the field estimates an unknown population value. If the expert is experienced with the population or process, then formal statistical methods may add very little quality to the resulting forecast. These types of predictions are called **judgment estimates**.

The problem with judgment estimates is that their degree of accuracy or reliability cannot be determined. If a decision has important consequences, standard statistical methods should be applied, especially if there is no expert to rely upon. Even if an expert is available, statistics can offer estimates with known reliability. In this chapter, we will discuss estimation techniques to estimate the population mean, standard deviation, and proportion.

## Introduction

The previous chapter concentrated on sampling and how the practice of drawing samples produces a statistic (e.g., the sample mean) that is a random variable. This chapter marks the beginning of **statistical inference**. It discusses the properties of sample summary statistics (sample means, sample proportions, etc.) as well as methods for assessing the reliability of those estimates.

The importance of the statistical measures to a decision-making process depends on the confidence the decision maker has in the estimated values. How much faith can be placed in the accuracy of the estimate? **Statistical inference** permits the estimation of statistical measures (means, proportions, etc.) with a known degree of confidence. This ability to assess the quality of estimates is one of the significant benefits statistics brings to decision making and problem solving.

**Definition**

Statistical Inference
Using properly drawn sample data to draw conclusions about a population is called **statistical inference**.

Applications of statistics are usually concerned with learning about populations or processes. Populations are described using summary measures called parameters, such as the mean, the standard deviation, and the proportion. Processes are also described with summary measures and with various models. Unfortunately, determining the exact mean and standard deviation of a population is seldom an easy task and often is not feasible. Statistical methods rely upon samples of the population to obtain information about the population's parameters. Whether the sample data will produce reliable information about a process or population depends on the methods used to collect the data.

All forms of statistical inference are tarnished with a degree of uncertainty. If samples are selected randomly, the uncertainty of the inference is measurable. Drawing random samples allows us to measure the level of confidence associated with an estimate. If samples are not selected randomly, there will be no known relationship between the sample estimates and the population parameters they are supposed to estimate. Therefore, if a statistical technique requires a sample, assume the sample will be a random sample from a population or process.

This chapter is devoted to the one-sample problem. That is, a sample consisting of $n$ measurements, $x_1, x_2, \ldots, x_n$, of some population will be analyzed with the objective of making inferences about the population. Inference will be introduced with the one-sample problem, not because it is the most common, but because the procedures and basic principles of inference are easier to understand.

# 9.1 Estimating the Population Mean, $\sigma$ Known

**Definition**

**Estimator and Estimate**

An **estimator** is a strategy or rule that is used to estimate a population parameter.

If the strategy or rule is applied to a specific set of data, the result is an **estimate**.

What is meant by the terms **estimator** and **estimate**?

The sample mean is an *estimator* of the population mean. A specific sample mean, $\bar{x}$, such as 103.4, is an *estimate* of the population mean ($\mu$).

In this chapter, we will study two different kinds of estimators, point and interval. A **point estimator** uses a single point (or value) to estimate a population parameter. For example, $\bar{x}$ is a *point estimator* of a population mean ($\bar{x} = 12.7$ is a *point estimate* of a population mean).

**Definition**

**Point Estimator**

A **point estimator** is an estimator that uses a single point (or value) to estimate a population parameter.

| Table 9.1.1 – Point Estimators | | |
|---|---|---|
| Point Estimator | Parameter Being Estimated | Point Estimate |
| $\bar{x}$ | $\mu$ | $\bar{x} = 12.7$ |
| $\hat{p}$ | $p$ | $\hat{p} = 0.37$ |
| $s$ | $\sigma$ | $s = 6.4$ |

An **interval estimate** defines an upper and lower boundary for an interval that will hopefully contain the population parameter. Oftentimes, the interval estimate for a parameter is a function of the point estimate of that parameter.

The value of drawing random samples resides in the ability to assess the reliability of sample inference. Reliability is expressed in probability. That is why so much of the text has been devoted to probabilistic ideas.

**Definition**

**Interval Estimate**

An **interval estimate** is an interval with an upper and lower boundary that hopefully contains the population parameter of interest.

## Point Estimation of the Population Mean

Like other statistical inference methods, estimation begins with the collection of data. Two important questions come to mind.

1.  How should the data be used to estimate the population mean?
2.  How can you tell a good estimator from a bad one?

Good estimators conform to the rules of horseshoes: the closer, the better. If the objective is to estimate a population mean, closeness is measured in terms of the distance the estimate is from the actual population mean.

One of the more puzzling questions is, *how can you judge the accuracy of your estimate without knowing the true value of the population parameter?* It's like shooting an arrow at a bull's-eye without being able to see it. If you can't see the bull's-eye, how do you know how close you are?

> ## Definition
>
> ### Mean Squared Error
>
> An estimator's average squared distance from the true parameter is referred to as its **mean squared error** (MSE). The mean squared error for the sample mean is given by
>
> $$MSE\left(\overline{x}\right) = E\left(\overline{x} - \mu\right)^2.$$

A perfect estimator would have a mean squared error of zero, but there is no such thing as a perfect estimator. Since statistical estimators depend on data that is randomly drawn, estimates are random variables and will seldom be equal to the true population characteristic. The goal is to find an estimator whose average squared error is the smallest. Unfortunately, there are a number of different estimators, and without restricting the kinds of estimators that will be considered, very little progress can be made. One desirable restriction is **unbiasedness**. As was discussed in Chapter 8, to be an **unbiased estimator**, the expected value of the estimator must be equal to the parameter that is being estimated. For example, $\overline{x}$ is an unbiased estimator of the population mean since

$$E\left(\overline{x}\right) = \mu.$$

Unfortunately, there are a number of estimators that are unbiased estimators of the population mean, including the sample mean, sample median, or any single sample value. *Among unbiased estimators of the population mean, the sample mean has the smallest mean squared error.* There is no other unbiased estimator that can consistently do a better job of estimating the population mean.

The fact that the sample mean is a good estimator of the population mean should not be surprising, since you would expect most sample statistics to be reasonably good estimators of their population counterparts. One of the exceptions is the sample range, which is a poor estimate of the population range. *In short, the best estimate is the one that is unbiased with the smallest mean squared error.*

## Interval Estimation of the Population Mean

Rarely will a point estimate of the population mean result in a value that exactly matches the population mean, $\mu$. In fact, the probability that a point estimate for the population mean exactly equals the population mean is zero for data drawn from continuous distributions. Yet, if an estimate is used for decision making, it is desirable that there be some indication of its potential error. One of the significant limitations of simply reporting a point estimate is the lack of information concerning the estimator's accuracy.

Interval estimates, however, are constructed to provide additional information about the precision of the estimate. An interval estimator is made by developing an upper and lower boundary for an interval that will hopefully contain the population parameter. It would be easy to construct an interval estimator that would always contain the population parameter: for example, the interval from negative infinity to positive infinity. But this particular estimator would not contain any useful information about the location of the population parameter. In interval estimation, the narrower the width of the interval for a given level of confidence, the better the estimator.

The foundations of a good interval estimator for the population mean can be found in Chapter 8, which defined how $\overline{x}$ should vary. Recall that if the sample size is reasonably large ($n \geq 30$) the Central Limit Theorem ensures that $\overline{x}$ has an approximately normal distribution with mean $\mu$ and standard deviation $\dfrac{\sigma}{\sqrt{n}}$.

**Sampling Distribution of $\bar{x}$**

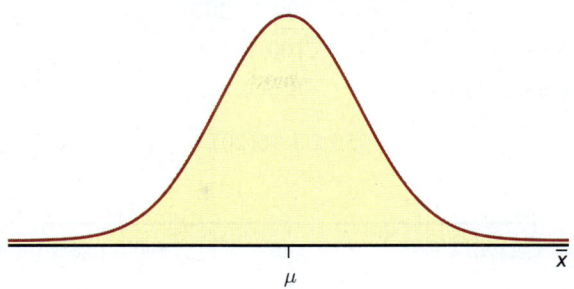

**Figure 9.1.1**

The sampling distribution can be used to develop an interval estimator. For a standard normal random variable, we know that

$$P(-1.96 < z < 1.96) = 0.95.$$

Since $\bar{x}$ can be transformed into a standard normal random variable by using the z-transformation, $z = \dfrac{\bar{x} - \mu}{\sigma_{\bar{x}}}$, then by substitution,

$$P\left(-1.96 < \frac{\bar{x} - \mu}{\sigma_{\bar{x}}} < 1.96\right) = 0.95,$$

and with some algebraic manipulation we obtain

$$P\left(\bar{x} - 1.96\sigma_{\bar{x}} < \mu < \bar{x} + 1.96\sigma_{\bar{x}}\right) = 0.95.$$

If the sample size is greater than or equal to 30, then by the Central Limit Theorem there is a 0.95 probability that the sample mean will be within 1.96 standard deviations of the population mean *before a particular sample is selected*.

**Sampling Distribution of $\bar{x}$**

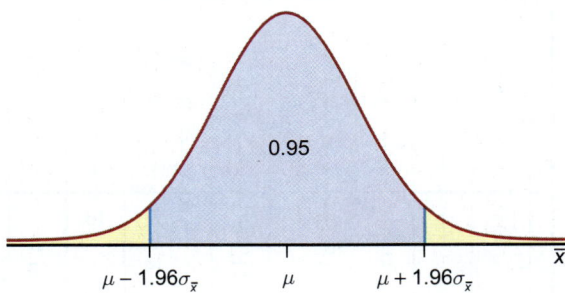

0.95

$\mu - 1.96\sigma_{\bar{x}}$      $\mu$      $\mu + 1.96\sigma_{\bar{x}}$

**Figure 9.1.2**

Since we discussed the sampling distribution of the sample mean in Chapter 8, the idea of defining the probability that the sample mean should fall within some specified distance of the population mean is not a new one. The expression does suggest a specific form for the interval since it provides an interval and the associated probability that the population mean will fall within the interval:

$$\bar{x} \pm 1.96\sigma_{\bar{x}}.$$

However, the provision *before a particular sample is selected* modifies the interpretation of the interval for a specific sample. After the sample is selected, the sample mean is no longer a random variable. Suppose a sample has been drawn from a population with a standard deviation of 200, and the following characteristics have been observed:

$$\sigma = 200, \quad \text{(given)}$$
$$n = 100, \quad \text{(chosen by researcher)}$$
$$\bar{x} = 150, \quad \text{(obtained from the sample)}$$

**The Father of Confidence Intervals: Jerzy Neyman 1894–1981**

Jerzy Neyman grew up in Poland. However, a significant part of Poland was under Russian control during his youth and Neyman received his training in mathematics in Russia.

In her book *Neyman—From Life*, Constance Reid attributes the development of the confidence interval to Jerzy Neyman.

"During the years 1934–38 Neyman made four fundamental contributions to the science of Statistics. Each of them would have been sufficient to establish an international reputation, both for their immediate effect and for the impetus which the new ideas and methods had on the thinking of young and old alike. He put forward the theory of confidence intervals, the importance of which in statistical theory and analysis of data cannot be overemphasized. His contribution to the theory of contagious distributions is still of great utility in the interpretation of biological data. His paper on sampling stratified populations paved the way for a statistical theory which, among other things, gave us the Gallup poll. [His] work, and that of Fisher, each with a different model for randomized experiments, led to the whole new field of experimentation so much used in agriculture, biology, medicine, and physical sciences."

Remember,

$$\sigma_{\bar{x}} = \frac{\sigma}{\sqrt{n}} = \frac{200}{\sqrt{100}} = \frac{200}{10} = 20.$$

The resulting interval would be

$$150 \pm 1.96(20).$$

That is,

$$\overset{(\phantom{xx}\underset{150}{\mid}\phantom{xx})}{\underset{150 - 1.96(20) \approx 111 \qquad\qquad 150 \qquad\qquad 150 + 1.96(20) \approx 189}{}}$$

Is the population mean, $\mu$, inside this interval? If not, what fraction of the time will $\mu$ be inside the interval? Even though the interval is calculated using a technique that captures the population mean 95% of the time, it would not be appropriate from a relative frequency point of view, to state that

$$P(111 < \mu < 189) = 0.95 \qquad \text{Wrong Interpretation}$$

since the population mean is an unknown but constant quantity. Either $\mu$ will always be inside the interval or will always be outside the interval. If this is true, then what information do we have about the interval? We say that this interval was constructed with a **confidence level** of 95% (i.e., $100(1-\alpha)\%$) or a **confidence coefficient** of 0.95 (i.e., $1 - \alpha$). Hence, the term **confidence interval** is used to describe the method of construction rather than a particular interval.

In summary, a 95% confidence interval can be interpreted to mean that if all possible samples of a given size are taken from a population, 95% of the samples would produce intervals that captured the true population mean and 5% would not. In Figure 9.1.3 we have plotted, as line segments, the confidence intervals from 20 random samples. All but one of the intervals, number 5, capture the mean, $\mu$.

> **Definition**
>
> Confidence Interval
>
> A **confidence interval** is an interval estimate for a population parameter that is associated with a certain confidence level (or confidence coefficient).

**Confidence Intervals from 20 Random Samples**

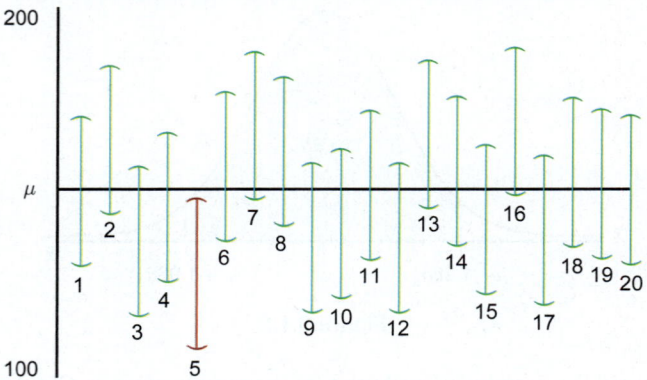

**Figure 9.1.3**

So far we have examined only the 95% confidence interval, yet the idea is a general one and can be extended to any specified degree of confidence.

In a practical sense, the selection of the degree of confidence depends upon the importance of the decision for which the confidence interval will be utilized. If we are launching a space vehicle, we would want to be very certain that the vehicle would have sufficient fuel to return safely. For other decisions, we might be willing to accept an 80% confidence of correctly estimating the population mean, especially if the cost of gathering additional data is large.

The expression $\bar{x} \pm z_{\alpha/2} \dfrac{\sigma}{\sqrt{n}}$ creates the "generalized" confidence interval.

**Formula**

**Confidence Interval for the Population Mean, $\sigma$ Known**

A $100(1 - \alpha)\%$ **confidence interval** for the population mean when $\sigma$ is known is given by

$$\bar{x} \pm z_{\alpha/2} \frac{\sigma}{\sqrt{n}}$$

if either of the following conditions are true:

1.  $n \geq 30$ or

2.  the data are collected from a normal distribution.

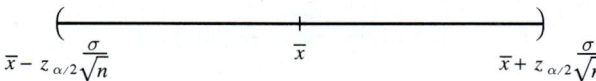

$$\bar{x} - z_{\alpha/2}\frac{\sigma}{\sqrt{n}} \qquad \bar{x} \qquad \bar{x} + z_{\alpha/2}\frac{\sigma}{\sqrt{n}}$$

The term $z_{\alpha/2}$ represents the $z$-value required to obtain an area of $\dfrac{\alpha}{2}$ in the right tail under the standard normal curve. The $z$-values for obtaining various $1 - \alpha$ areas centered under the standard normal curve are given in Table 9.1.2 and graphed in Figure 9.1.4.

| Table 9.1.2 – Critical Values of z | |
|---|---|
| **Confidence $(1 - \alpha)$** | **$z_{\alpha/2}$** |
| 0.80 | 1.28 |
| 0.90 | 1.645 |
| 0.95 | 1.96 |
| 0.99 | 2.575 |

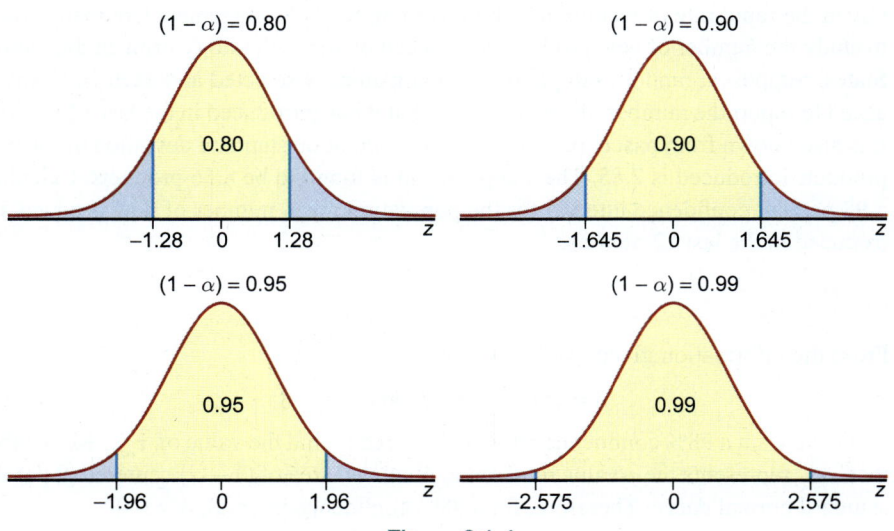

Figure 9.1.4

Alex was interested in learning the range that he should expect to pay for a lawnmower. He collected a sample of 100 lawnmower prices from various retailers within a 100-mile radius and found that the average cost was $425. Construct 80%, 90%, 95%, and 99% confidence intervals for the true average cost of a lawnmower if the population standard deviation is 900.

$$n = 100$$
$$\bar{x} = 425$$

**Example 9.1.1**

**Constructing a Confidence Interval for the Population Mean with $\sigma$ Known**

## ∽ Technology

Confidence intervals can be calculated using technology. For instructions, please visit stat.hawkeslearning.com and navigate to **Discovering Business Statistics, Second Edition > Technology Instructions > Confidence Intervals > z-Interval**.

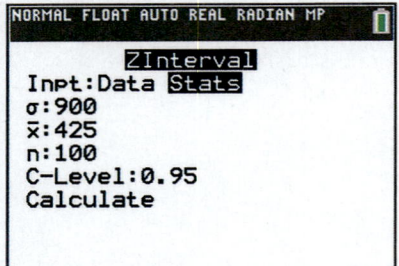

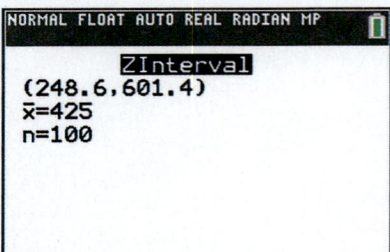

**SOLUTION**

**80% Confidence Interval:**

$$425 \pm 1.28 \cdot \frac{900}{\sqrt{100}} \text{ or } 310 \text{ to } 540$$

310      425      540

**90% Confidence Interval:**

$$425 \pm 1.645 \cdot \frac{900}{\sqrt{100}} \text{ or } 277 \text{ to } 573$$

277      425      573

**95% Confidence Interval:**

$$425 \pm 1.96 \cdot \frac{900}{\sqrt{100}} \text{ or } 249 \text{ to } 601$$

249      425      601

**99% Confidence Interval:**

$$425 \pm 2.575 \cdot \frac{900}{\sqrt{100}} \text{ or } 193 \text{ to } 657$$

193      425      657

The intervals in Example 9.1.1 illustrate that to achieve more confidence we must pay a price. For a given sample size, the only way to achieve greater confidence is to widen the interval. However, the resulting information provides a less precise location of the population mean.

### Example 9.1.2

**Calculating a Confidence Interval for the Number of New Technology Products**

Given the rapid rate of technological innovation, SWN Management Company wants to study the number of new products introduced by top technology firms in the United States. Suppose a random sample of 150 companies is selected and each company is asked to report the number of new products that it has introduced in the last 12 months. It is also known from past experience that the population standard deviation of the new products introduced is 7.85. The sample mean is found to be 8.56 products. Calculate a 98 percent confidence interval for the population mean number of new products introduced in the last 12 months.

**SOLUTION**

From the information given, we know that

$$n = 150, \ \overline{x} = 8.56, \text{ and } \sigma = 7.85.$$

Since we want a 98% confidence interval, we need to find the value of $z_{\alpha/2}$. Remember that $z_{\alpha/2}$ represents the z-value required to obtain an area of $(1 - \alpha)$ centered under the standard normal curve. Therefore, for a 98% confidence interval, $\alpha = 0.02$.

Similar to finding the values in Table 9.1.2 we can use Table A to find the corresponding value of z. That is, if we are to use Table A, we look up $0.01 \left( \text{i.e., } \frac{\alpha}{2} \right)$ in the body of the standard normal table.

Since the exact value of 0.01 is not in the table, we use the value that is closest. In this case, we can see that 0.0099 corresponds to a z-value of $-2.33$. Since the standard normal distribution is symmetrical, $z_{\alpha/2}$ capturing an area of 0.01 in the upper tail of the distribution will be 2.33.

$$\text{For } \alpha = 0.02, \ z_{\alpha/2} = z_{0.01} = 2.33.$$

A 98% confidence interval is then calculated as follows.

$$\bar{x} \pm z_{\alpha/2} \frac{\sigma}{\sqrt{n}}$$

$$8.56 \pm 2.33 \frac{7.85}{\sqrt{150}}$$

$$7.07 \text{ to } 10.05$$

```
 (————————————+————————————————————)
7.07         8.56                 10.05
```

Thus, we are 98% confident that the true mean number of new products introduced in the last 12 months will be contained in the above interval.

**⚭ Technology**

For instructions on calculating this confidence interval using technology, please visit stat.hawkeslearning.com and navigate to **Discovering Business Statistics, Second Edition > Technology Instructions > Confidence Intervals > z-Interval**.

So far, the confidence interval has been discussed as a way of placing bounds on the location of a parameter with a specific degree of confidence. But we can also think about the confidence interval as a means of describing the quality of a point estimate. Let's look at the expression for the confidence interval for the population mean.

**Confidence Interval for $\mu$**

$$\underbrace{\bar{x}}_{\text{point estimate}} \pm \underbrace{z_{\alpha/2} \frac{\sigma}{\sqrt{n}}}_{\substack{\text{margin of error with a} \\ \text{specific level of} \\ \text{confidence}}}$$

Another interpretation of the confidence interval is given below the expression of the confidence interval for $\mu$. The part of the expression that is added and subtracted to the point estimate, $z_{\alpha/2} \frac{\sigma}{\sqrt{n}}$, can be thought of as the **margin of error** (also known as the **maximum error of estimation**) using the point estimate $\bar{x}$ with a specified level of confidence. For example, the 95% confidence interval in Example 9.1.1 was given as

$$425 \pm 1.96 \cdot \frac{900}{\sqrt{100}}$$

$$425 \pm 176.4.$$

We could say that we are 95% confident that the point estimate of $\mu$, $\bar{x} = 425$, has a margin of error of 176.4 or an error of estimation no larger than 176.4. Being able to assess the error of an estimate is one of the most useful applications of statistical methods.

**Definition**

**Margin of Error**

The **margin of error**, or **maximum error of estimation** (often denoted as $E$), is the largest possible distance from the point estimate that a confidence interval will cover.

# ✍ 9.1 Exercises

## Basic Concepts

1. What is statistical inference?

2. What is an estimator?

3. What is a judgment estimate? What are some drawbacks of judgment estimates?

4. Explain, in your own words, the difference between the terms *estimator* and *estimate*.

5. What is the difference between a point estimate and an interval estimate?

6. Give three examples of point estimators. Identify the parameters being estimated by these estimators.

7. Describe the primary advantages of *random* sampling procedures.

8. What are two important questions to consider when estimating a population mean?

9. What is mean squared error?

10. What is an unbiased estimator? Give an example.

11. Why is the sample mean considered the best point estimate of the population mean?

12. Are all estimators unbiased? Explain.

13. Generally, we expect most sample statistics to be good estimators of their population counterparts. Which statistic is the exception to this idea?

14. What are two characteristics of the best available estimate for a parameter?

15. What is an interval estimator?

16. What is the distinction between probability and confidence?

17. What is the role of the $z$-value in the confidence interval expression?

18. Describe in words the ideas behind the construction of a confidence interval.

19. Consider the following statement: *If the sample size is greater than or equal to 30, then by the Central Limit Theorem there is a 0.95 probability that the sample mean will be within 1.96 standard deviations of the population mean before a particular sample is selected.* Explain why the phrase "before a particular sample is selected" is important here.

20. Explain what is wrong with the following expression: $P(111 < \mu < 189) = 0.95$.

21. Define the following terms: confidence level, confidence coefficient, confidence interval.

22. What are the conditions required in order to construct a $100(1-\alpha)\%$ confidence interval using the expression $\bar{x} \pm z_{\alpha/2}\dfrac{\sigma}{\sqrt{n}}$?

23. Describe the effect on the width of a confidence interval as each of the following increases:

    a. $n$            b. $1 - \alpha$            c. $\alpha$            d. $\bar{x}$

24. What expression indicates the margin of error? Is this the same as the maximum error of estimation?

## Exercises

25. Find $z_{\alpha/2}$ for the following levels of $\alpha$.

    a. $\alpha = 0.05$            b. $\alpha = 0.01$            c. $\alpha = 0.10$

26. Find $z_{\alpha/2}$ for the following levels of $\alpha$.

    a. $\alpha = 0.04$            b. $\alpha = 0.02$            c. $\alpha = 0.08$

27. Find $z_{\alpha/2}$ for the following confidence levels.

    a. 98%            b. 94%            c. 92%

28. Find $z_{\alpha/2}$ for the following confidence levels.

    a. 96%            b. 88%            c. 85%

29. Consider a normally distributed population with a standard deviation of 64. If a random sample of size 90 from the population produces a sample mean of 250, construct a 95% confidence interval for the true mean of the population.

30. Construct a 90% confidence interval for the true mean of a normal population if a random sample of size 40 from the population yields a sample mean of 75 and the population has a standard deviation of 5.

31.   A psychologist is studying learning in rats. The psychologist wants to determine the average time required for rats to learn to traverse a maze. She randomly selects 40 rats and records the time it takes for the rats to traverse the maze in minutes. The sample average time required for the rats to traverse the maze is 5 minutes with a population standard deviation of 1 minute. Estimate the average time required for rats to learn to traverse the maze with a 90% confidence interval.

32.   A paint manufacturer is developing a new type of paint. Thirty panels were exposed to various corrosive conditions to measure the protective ability of the paint. The mean life for the samples was 168 hours before corrosive failure. The life of paint samples is assumed to be normally distributed with a population standard deviation of 30 hours. Find the 95% confidence interval for the mean life of the paint.

33.   The chief purchaser for the State Education Commission is reviewing test data for a metal link chain which will be used on children's swing sets in elementary school playgrounds. The average breaking strength for a sample of 50 pieces of chain is 5000 pounds. Based on past experience, the breaking strength of metal chains is known to be normally distributed with a standard deviation of 100 pounds. Estimate the actual mean breaking strength of the metal link chain with 99% confidence.

34.   Tomatoes are grown in Florida for shipment to other parts of the country by the Anderson Produce Company. A random sample of 40 boxes is selected at one warehouse for weighing. The average weight for the sample is 33.5 pounds per box with a population standard deviation of 2.1 pounds. Find a 90% confidence interval for the true average weight of the boxes of tomatoes.

35.   Thirty-five strands of piano wire were selected at random from a recent shipment by the quality control department at Elkins Piano Company. The strands of piano wire were tested to failure in tests of tensile strength. The mean tensile strength of the sample was 30,000 pounds per square inch (psi) with a population standard deviation of 1950 psi. Find the 98% confidence interval for the true mean tensile strength.

36.   When preparing a standardized test to be given to all the sixth graders, the Standard Test Company gave a version of the test to a random sample of 45 sixth graders and timed how long it took them to finish the test. The average time required to finish the test for the sample was 2 hours and 15 minutes with a population standard deviation of 30 minutes. Estimate the true average time required to finish the test with 95% confidence.

37.   According to the 2009 College Senior Survey administered by the Higher Education Research Institute at UCLA, 56.4% of college seniors spend 10 hours or less studying or doing homework in a typical week. Suppose a random sample of 50 college seniors was selected from all the college seniors in the Southeast region to determine the homework habits of college seniors in the Southeast region of the United States. Each student in the sample is asked approximately how many hours per week he or she spends studying or doing homework. If the mean is 9.6 hours and the population standard deviation is 3.1 hours, construct a 99% confidence interval for the mean number of hours a week that a college senior in the region spends studying or doing homework per week.

      **Source:** Cooperative Institutional Research Program at the Higher Education Research Institute at UCLA

# 9.2 Estimating the Population Mean, $\sigma$ Unknown

In the previous section, we assumed that the population standard deviation was known. In practice this assumption is not very realistic, since the standard deviation describes variability about the mean. If the population standard deviation is known, the mean is usually also known, and there is no need to create an interval estimate for it. Why estimate something we already know?

If $\sigma$ is not known and the distribution of the population is assumed to be normal, then the derivation of the confidence interval must be changed slightly. Provided that the population from which the sample is drawn is normally distributed, the distribution of the quantity

$$t = \frac{\bar{x} - \mu}{\dfrac{s}{\sqrt{n}}} \text{ where } s \text{ is the standard deviation of the sample,}$$

has a **Student's *t*-distribution**. The *t*-distribution was discovered by W.S. Gosset in 1908. Gosset published the result under the pen name of Student while working as an employee of the Guinness Brewery in Dublin.

The *t*-distribution is very much like the normal distribution (see Figure 9.2.1). It is a symmetrical, bell-shaped distribution with slightly thicker tails than a normal distribution. The *t*-distribution has one parameter, **degrees of freedom**. The shape of the *t*-distribution approaches the normal distribution as the sample size *n*, and thus the degrees of freedom, become larger.

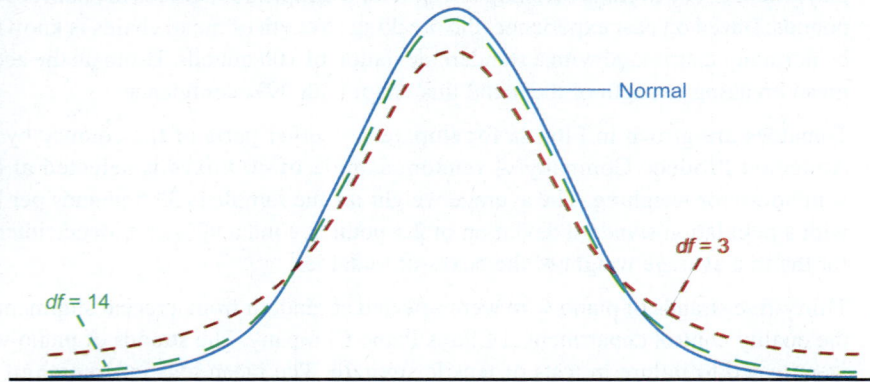

**Figure 9.2.1**

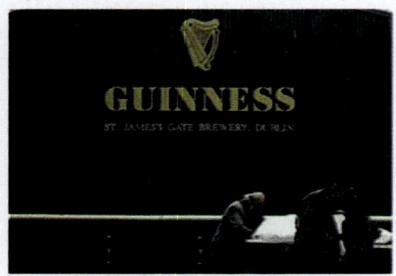

### William Sealy Gosset: The Student

Upon graduating from New College, Oxford with a strong understanding of mathematics, W.S. Gosset began working at the Guinness brewery in Dublin, Ireland. While working at Guinness, Gosset applied his statistical knowledge to find the best yielding varieties of barley, and in 1908, he developed the *t*-distribution. Few other statisticians at the time saw the merit in developing small-sample methods since most of their work required large data sets; however, Gosset was convinced of the importance of his work. Unfortunately, Guinness had prohibited its employees from publishing papers to protect trade secrets, and thus did not originally allow Gosset to publish his findings. After convincing the brewery that his statistical methods would be of no use to competing breweries, Guinness allowed Gosset to publish his conclusions on the *t*-distribution, but only under the pseudonym, "Student", to avoid issues with other staff members. To this day, Gosset's most noteworthy achievement is known simply as the "Student's *t*-distribution."

> ### Formula
>
> #### Degrees of Freedom
>
> The **degrees of freedom** for any *t*-distribution are computed in the following manner.
> $$df = \text{number of sample observations} - 1 = n - 1$$

> ### Formula
>
> #### Confidence Interval for the Population Mean, $\sigma$ Unknown
>
> If $\sigma$ is unknown, but the distribution of the population is assumed to be normal, a $100(1 - \sigma)\%$ confidence interval for the population mean is given by
>
> $$\bar{x} \pm t_{\alpha/2,\, df} \frac{s}{\sqrt{n}},$$
>
> where $t_{\alpha/2,\, df}$ is the critical value for a *t*-distribution with $n - 1$ degrees of freedom which captures an area of $\dfrac{\alpha}{2}$ in the right tail of the distribution. Note that this interval is only valid if $\bar{x}$ is normally distributed.

The form of the confidence interval is identical to the previous confidence interval for the mean of a population except that the *z* has been replaced by a *t* and $\sigma$ has been replaced by *s*. The interpretation of the confidence interval is identical to the previous interval.

A luxury car company wanted to determine a range for the age of people that purchase their vehicles. Over the course of several weeks, the company sampled seven customers who bought vehicles and asked their age. Given the following ages below, assume that they were collected from a normal population with unknown mean and variance, construct a 95% confidence interval for the population mean.

25, 19, 37, 29, 40, 28, 31

### SOLUTION

The sample mean and standard deviation of the customers ages are $\bar{x} \approx 29.8571$ and $s \approx 7.0811$, respectively. The degrees of freedom associated with the problem is

$$df = n - 1 = 7 - 1 = 6.$$

The $t$-value corresponding to 6 degrees of freedom and 95% confidence ($\alpha = 0.05$) is given in Table D of Appendix A as $t_{0.025,6} = 2.447$. We also know that the general form of the confidence interval for the population mean when $\sigma$ is unknown and the data are collected from a normal distribution is

$$\bar{x} \pm t_{\alpha/2,df} \frac{s}{\sqrt{n}}$$

$$29.8571 \pm 2.447 \left( \frac{7.0811}{\sqrt{7}} \right)$$

$$29.8571 \pm 6.5492.$$

Thus, we are 95% confident that the interval

23.3079 to 36.4063

will contain the population mean.

An alternate interpretation would be that we are 95% confident that the point estimate, 29.8571, has a maximum error of estimation of 6.5492.

**Example 9.2.1**

**Constructing a Confidence Interval for the Mean with $\sigma$ Unknown**

**∞ Technology**

For instructions on calculating this confidence interval using technology, please visit stat.hawkeslearning.com and navigate to **Discovering Business Statistics, Second Edition > Technology Instructions> Confidence Intervals > *t*-Interval**.

---

A manufacturing company is interested in the amount of time it takes to complete a certain stage of the production process. The project manager randomly samples 10 products as they come from the production line and notes the time of completion. The average completion time is 23.45 minutes with a sample standard deviation of 4.32 minutes. Based on this sample, construct a 95% confidence interval for the average completion time for that stage in the production process. Assume that the population distribution of the completion times is approximately normal.

### SOLUTION

Since the company wants to calculate a 95% confidence interval for the average completion time, $\mu$, we know that $\alpha = 0.05$. We also have a sample size of $n = 10$ and thus, $n - 1 = 10 - 1 = 9$ degrees of freedom. Therefore, we use $t_{\alpha/2,df} = t_{0.025,9} = 2.262$ from Table D in the Appendix.

We also know that the general form of the confidence interval for the population mean when $\sigma$ is unknown and it is assumed that the population distribution of completion times is approximately normal, is given by

$$\bar{x} \pm t_{\alpha/2,df} \frac{s}{\sqrt{n}}$$

$$23.45 \pm 2.262 \frac{4.32}{\sqrt{10}}$$

$$23.45 \pm 3.0901$$

$$20.36 \text{ to } 26.54.$$

Thus, we are 95% confident that the true average completion time of that stage of the process is between 20.36 minutes and 26.54 minutes.

**Example 9.2.2**

**Constructing a Confidence Interval for Completion Time**

**∞ Technology**

The margin of error can be found using Excel's **CONFIDENCE.T** function. For instructions on calculating the confidence interval using Excel, please visit stat.hawkeslearning.com and navigate to **Discovering Business Statistics, Second Edition > Technology Instructions > Confidence Intervals > *t*-Interval**.

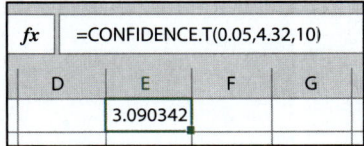

| $fx$ | =CONFIDENCE.T(0.05,4.32,10) | | |
|---|---|---|---|
| D | E | F | G |
| | 3.090342 | | |

## Interval Estimation of the Population Mean: A Summary

In Sections 9.1 and 9.2 we have outlined two procedures for determining an interval estimate for the population mean. In Section 9.1 we used the interval $\bar{x} \pm z_{\alpha/2} \dfrac{\sigma}{\sqrt{n}}$ for normally distributed populations in which the population standard deviation, $\sigma$, is known. In Section 9.2 we introduced the $t$-distribution with $n-1$ degrees of freedom and used the interval $\bar{x} \pm t_{\alpha/2,df} \dfrac{s}{\sqrt{n}}$ to find an interval estimate for the population mean for a normally distributed population where $\sigma$ is unknown. The following flowchart is useful when deciding how to construct an interval estimate for the population mean.

### Finding a Confidence Interval for the Population Mean

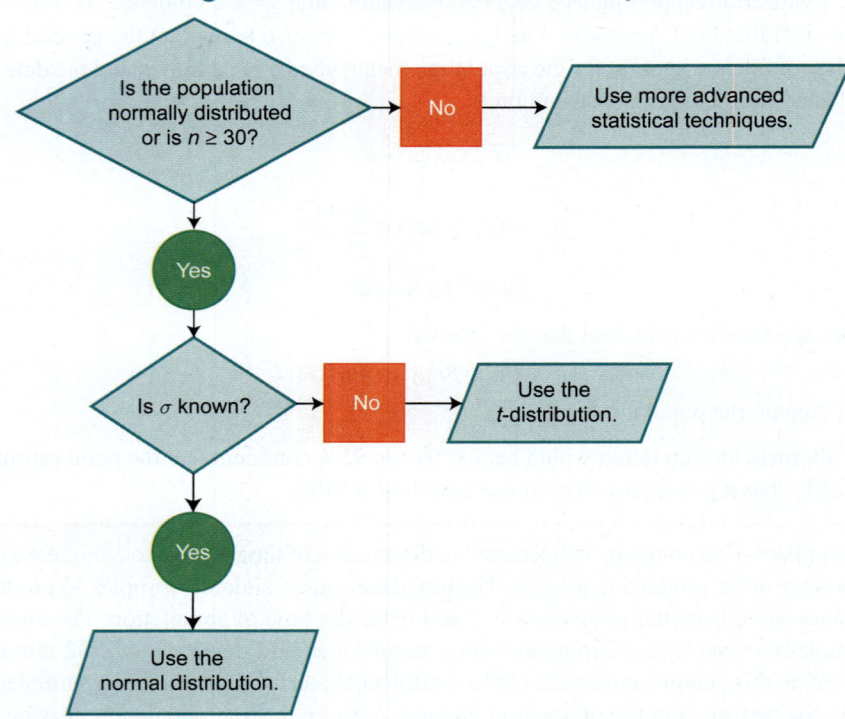

Figure 9.2.2

## Precision and Sample Size: Means

The more accurate an estimate, the greater its potential value in decision making. The only way to accurately determine an unknown population parameter is to perform a census, though this is usually impractical because of cost or time considerations. Realistically, the best interval estimate a decision maker could hope for would be an interval with a small width possessing a large amount of confidence. The width of the confidence interval defines the precision with which the population mean is estimated, the smaller the interval, the greater the precision. If the width of a confidence interval could be controlled, we could achieve estimates with a level of accuracy that is appropriate for the decision at hand.

## Determining the Sample Size: $\sigma$ Known

There are three components that affect the width of the confidence interval for the population mean:

| | |
|---|---|
| $z_{\alpha/2}$ | Represents the distance the confidence interval boundary is from the sample mean $\bar{x}$ in standard deviation units. The distance is related to the specific level of confidence. |
| $\sigma$ | Represents the population standard deviation. |
| $n$ | Represents the sample size. |

As discussed in Section 9.1, the level of confidence will affect confidence interval width. That is, the higher the level of confidence, the wider the confidence interval. The population standard deviation, $\sigma$, is a constant, and does not change. The sample size, however, is selected by the decision maker. The larger the sample, the smaller the width of the resulting confidence interval for some given level of confidence. Since the sample size can be increased, which reduces the width of the confidence interval, how large should the sample be? Taking too large a sample wastes money, while taking too small a sample produces an estimator that does not possess sufficient reliability.

The sample size should be selected in relation to the size of the margin of error the decision maker is willing to accept. This can be achieved by setting the error equal to one-half the confidence interval width.

$$E = \text{margin of error} = z_{\alpha/2}\frac{\sigma}{\sqrt{n}}$$

The preceding equation can be solved for the sample size, $n$.

$$E = z_{\alpha/2}\frac{\sigma}{\sqrt{n}}$$

$$\sqrt{n} = z_{\alpha/2}\frac{\sigma}{E}$$

$$n = \left(\frac{z_{\alpha/2}\sigma}{E}\right)^2$$

By selecting a level of confidence and the maximum error, the relationship can be used to determine the sample size necessary to estimate the population mean with the desired accuracy. *In order to assure the desired level of confidence, always round the value obtained for the sample size up to the next integer.*

**Example 9.2.3**

**Calculating the Sample Size Needed for 95% Confidence of the Mean**

Consider a population having a standard deviation of 15. We want to estimate the mean of the population. How large of a sample is needed to construct a 95% confidence interval for the mean of this population if the margin of error is equal to 1.5?

**SOLUTION**

$$n = \left(\frac{z_{\alpha/2}\sigma}{E}\right)^2 = \left(\frac{1.96 \cdot 15}{1.5}\right)^2 = 384.16$$

Rounding up, we have to have a sample size of 385 to ensure that we get at least a 95% confidence interval with a margin of error equal to 1.5.

**Example 9.2.4**

**Calculating the Sample Size Needed for 90% Confidence of the Mean Amount of Cleaning Fluid**

Suppose that a quality control manager at Argon Chemical Company wishes to measure the average amount of cleaning fluid the company is placing in their 12-ounce bottles. The manager is concerned that if the bottles are overfilled, then there could be a chance of an explosion. At the same time, if they are underfilled customers would be unhappy and begin to buy the competitor's product. From previous samples, they believe that the standard deviation is 0.3 ounce. How large a sample must be taken in order to be 90% confident of estimating the mean amount of cleaning fluid in a 12-ounce bottle to within 0.05 ounce?

**SOLUTION**

$$n = \left(\frac{z_{\alpha/2}\sigma}{E}\right)^2 = \left(\frac{1.645 \cdot 0.3}{0.05}\right)^2 = 97.4169$$

To be assured of finding the desired level of confidence, always round up. Thus, we are 90% confident that a sample of $n = 98$ observations would produce an estimate of the mean amount of cleaning fluid in a 12-ounce bottle to within 0.05 ounce. Being able to know the accuracy of your estimate is one of the significant benefits of inferential statistics. In this case, if 98 bottles are measured, we will be 90% confident that the resulting sample mean is within five one-hundredths of an ounce of the true mean. That's close.

## Determining the Sample Size: $\sigma$ Unknown

In the previous discussion of determining the sample size necessary to estimate a population mean with a desired accuracy, $\sigma$ was assumed to be known. This assumption is usually unreasonable in most problem-solving environments.

The most obvious method for obtaining an estimate of $\sigma$ is to take a small sample and use the sample standard deviation as an estimate of the population standard deviation. Replacing $\sigma$ with $s$ in the sample size determination relationship will provide an initial estimate of the required sample size. Another alternative is to use the value of the sample standard deviation obtained in a previous study, sometimes called a **pilot study**. Keep in mind that in order to construct a confidence interval for a population mean when the population standard deviation is unknown, the distribution of the population is assumed to be normal.

### Example 9.2.5

**Calculating the Sample Size Needed for 99% Confidence of the Mean Time Required to Replace a Jet Engine**

An airline's maintenance manager desires to estimate the average time (in hours) required to replace a jet engine in a Boeing 767. How large a sample would be necessary if the manager wishes to be 99% confident of estimating the population mean to within one-quarter of an hour ($E = 0.25$)? Assume a preliminary sample of size $n = 30$ has a mean replacement time of 16.7 hours with a standard deviation of 4.3 hours.

#### SOLUTION

Using the results from the initial sample,

$$n = \left(\frac{z_{\alpha/2}s}{E}\right)^2 = \left(\frac{2.575 \cdot 4.3}{0.25}\right)^2 = 1961.6041$$

$n = 1962$ (Always round up to assure required confidence.)

Notice that while the sample data values are being collected they can be used to improve the estimate of the population standard deviation. For example, suppose the sample standard deviation after sampling the first 1000 observations was 4.1. Using this estimate of $s$ instead of 4.3 results in a sample size of 1784 compared to the original specification of 1962. The notion of modifying the sample size estimate as additional data are observed can be applied at regular intervals during the sampling process until the estimate of the standard deviation stabilizes.

### Six Degrees of Separation: A Law of Small Worlds

What is the number of people a randomly chosen person in Omaha, Nebraska needs to contact before she can find a connection with a randomly chosen housewife in New England? How many intervening people do you think separates you from the President of the United States? Unsuspecting readers might guess very large numbers but the actual numbers are quite small. The answer to both of these questions may very well be less than 6! Psychologists have done ingenious experiments and have actually calculated this degree of separation, on average, to be six. What is amazing about this degree of separation is that it is equally true for the President of the United States and a sweet vendor in Bangladesh.

continued on next page...

## 9.2 Exercises

### Basic Concepts

1. Why is the assumption that the population standard deviation is known when estimating the population mean not very realistic?

2. What effect does knowing the standard deviation of the population have on the construction of the confidence interval?

3. What is the Student's *t*-distribution?

4. What are the conditions in which the *t*-distribution is used in interval estimation of the population mean?

5. What is the parameter of the *t*-distribution? How is it calculated?

6. What is the value of having a confidence interval with a small width?

7. Can a confidence interval be constructed with a width of your choice? Explain.

8. What are the three components that affect the width of the confidence interval for the population mean? Describe how changes in these three components affect the width of the confidence interval.

9. What is the margin of error? What is the connection between the expression for the margin of error and the equation to determine the sample size?

10. What is the rounding rule regarding the determination of the sample size?

11. What is the difference between the method of determining the sample size when σ is known versus when σ is unknown?

12. What is a pilot study?

13. Note that $E$ and $s$ can be viewed as measures of variation. Compare and contrast the meanings of $E$ and $s$ in layman's terms.

*continued...*

The degree of separation for the number of clicks that you will need to make to get to a website that interests you, as well as the analysis of terrorist networks, turn out to be of similar nature. The new science of networks can shed useful light and help to derive general laws applicable to many of these types of questions.

## Exercises

14. Find the *t*-value such that 0.025 of the area under the curve is to the right of the *t*-value. Assume the degrees of freedom equal 13.

15. Find the *t*-value such that 0.01 of the area under the curve is to the right of the *t*-value. Assume the degrees of freedom equal 21.

16. Find $t_{\alpha/2, df}$ for the following combinations of $\alpha$ and $n$.

    a. $\alpha = 0.05, n = 15$

    b. $\alpha = 0.01, n = 20$

    c. $\alpha = 0.10, n = 8$

17. Find $t_{\alpha/2, df}$ for the following combinations of $\alpha$ and $n$.

    a. $\alpha = 0.05, n = 12$

    b. $\alpha = 0.01, n = 18$

    c. $\alpha = 0.10, n = 22$

18. A random sample, consisting of the values listed below, was taken from a normally distributed population. Assuming the standard deviation of the population is unknown, construct a 99% confidence interval for the population mean.

| | | | |
|---|---|---|---|
| 27.4 | 26.5 | 25.7 | 31.4 |
| 28.2 | 21.9 | 16.3 | 22.7 |
| 18.8 | 34.4 | 29.2 | 20.5 |

19. Construct an 80% confidence interval for the mean of a normal population assuming that the values listed below comprise a random sample taken from the population. The population standard deviation is unknown.

| | | | | |
|---|---|---|---|---|
| 83.9 | 87.4 | 65.2 | 86.0 | 73.1 |
| 80.3 | 92.7 | 87.5 | 69.3 | 77.5 |
| 91.9 | 71.1 | 79.1 | 72.4 | 88.2 |

20. An FDA representative randomly selects 8 packages of ground chuck from a grocery store and measures the fat content (as a percent) of each package. The resulting measurements are given below.

| Fat Contents | | | |
|---|---|---|---|
| 13% | 12% | 14% | 17% |
| 15% | 16% | 18% | 15% |

   a. Calculate the sample mean and the sample standard deviation of the fat contents.

   b. Construct a 90% confidence interval for the true mean fat content of all the packages of ground beef.

   c. What assumption did you make about the fat content in constructing your interval?

21. A hospital would like to determine the mean length of stay for its patients having abdominal surgery. A sample of 15 patients revealed a sample mean of 6.4 days and a sample standard deviation of 1.4 days.

   a. Find a 95% confidence interval for the mean length of stay for patients with abdominal surgery.

   b. Interpret this interval and state any assumptions that were made in the construction of the interval.

22. An independent group of food service personnel conducted a survey on tipping practices in a large metropolitan area. They collected information on the percentage of the bill left as a tip for 25 randomly selected bills. The average tip was 12.3% of the bill with a standard deviation of 2.7%.

   a. Construct an interval to estimate the true average tip (as a percent of the bill) with 99% confidence.

   b. Interpret the interval, and state any assumptions that were made in the construction of the interval.

23. A travel agent is interested in the average price of a hotel room during the summer in a resort community. The agent randomly selects 15 hotels from the community and determines the price of a regular room with a king size bed. The average price of the room for the sample was $115 with a standard deviation of $30.

   a. Construct an interval to estimate the true average price of a regular room with a king size bed in the resort community with 90% confidence.

   b. Interpret the interval, and state any assumptions that were made in the construction of the interval.

24. In 2010 the median home price in all regions of the United States was $221,800. It is commonly thought that better schools are found in wealthier areas. In *Forbes* magazine's list of the "Best Schools for your Real Estate Buck," the top 10 cities in America were identified where your housing dollar will go the furthest in getting your children a great education. 17,589 towns and cities were analyzed using results from the most recent National Assessment for Educational Progress data, and the top 10 school districts were identified. The list counteracted the idea that more money equals better schools, as Falmouth, Maine topped the list beating out high-dollar school districts like Manhattan Beach, California. The top 10 cities are given below, along with the median home price for each city.

| Best Schools for Your Real Estate Buck | | |
|---|---|---|
| Education Rank | City | Median Home Price ($) |
| 1 | Falmouth, Maine | 351,550 |
| 2 | Mercer Island, Washington | 708,740 |
| 3 | Pella, Iowa | 148,200 |

| Best Schools for Your Real Estate Buck (cont.) | | |
|---|---|---|
| **Education Rank** | **City** | **Median Home Price ($)** |
| 4 | Barrington, Rhode Island | 296,010 |
| 5 | Bedford, New Hampshire | 293,730 |
| 6 | Manhattan Beach, California | 1,278,980 |
| 7 | Moraga, California | 722,010 |
| 8 | Parkland, Florida | 426,390 |
| 9 | St. Johns, Florida | 181,700 |
| 10 | Southlake, Texas | 476,880 |

**Source:** Forbes magazine

a. Construct a 90% confidence interval for the median home price of cities on the top 10 list.

b. Is the average median price for these cities higher than the median price for the U.S. as a whole?

c. What population assumption needs to be made here?

d. How would your solutions to **a.** and **b.** change if these were mean rather than median values?

25. A technician working for the Chase-National Food Additive Company would like to estimate the preserving ability of a new additive. This additive will be used for Auntie's brand preserves. Based on past tests, it is believed that the time to spoilage for this additive has a standard deviation of 6 days. To be 90% confident of the true mean time to spoilage, what sample size will be needed to estimate the mean time to spoilage with an accuracy of one day?

26. A computer software company would like to estimate how long it will take a beginner to become proficient at creating a graph using their new spreadsheet package. Past experience has indicated that the time required for a beginner to become proficient with a particular function of the new software product has an approximately normal distribution with a standard deviation of 15 minutes. Find the sample size necessary to estimate the true average time required for a beginner to become proficient at creating a graph with the new spreadsheet package to within 5 minutes with 95% confidence.

27. A hot-dog vendor is evaluating a downtown location by counting the number of people who walk past the prospective location on a particular day during lunch time (i.e. 11:00 AM to 2:00 PM). A preliminary study has indicated a standard deviation of about 30 people per lunch period. How many lunch periods will be needed to estimate the average number of people who walk past the prospective location during the lunch period to within 9 people with 90% confidence?

# 9.3 Estimating the Population Proportion

An attribute is a characteristic that members of a population either possess or do not possess. Attributes are almost always measured as the **proportion** of the population that possesses the characteristic.

Many decisions require a measure of a population attribute. Television and radio stations base their advertising charges on ratings reflecting the *percentage of television viewers who are watching a particular program*. A political analyst wants to know the *fraction of voters who favor a particular candidate*. A social researcher needs the *fraction of teachers who believe group learning is a beneficial instructional method*. An insurance company is interested in estimating the *fraction of their policies that will result in claims*. A quality control engineer requires the *percentage defective in a lot of goods*. A marketing researcher demands the *fraction of persons on a mailing list that will purchase the product as a result of a direct mail marketing campaign.* The items in italics are measures of attributes of some population. Researchers estimate the proportion of population members possessing those characteristics.

Estimating the proportion of the population that possesses an attribute is straightforward. A random sample is selected, and the sample proportion is computed as follows:

$x$ = number in the sample that possesses the attribute,

$n$ = sample size, and

$$\hat{p} = \frac{x}{n}$$

The symbol above the $p$ indicates an estimate of the quantity specified. Since $\hat{p}$ is computed from a random sample, $\hat{p}$ is a random variable whose value depends on which random sample is selected.

## Example 9.3.1

**Determining a Point Estimate for the Proportion of Defective Transistors**

Estimate the fraction of defective transistors in a lot containing 100,000 transistors. Suppose a sample of size 800 is drawn from the lot, and 5 transistors were found to be defective.

### SOLUTION

$x$ = number in the sample that possesses the attribute = 5

$n$ = sample size = 800

Then,

$$\hat{p} = \frac{5}{800} \approx 0.0063,$$

which is an estimate of the proportion of defective transistors in the lot of 100,000.

A natural question to ask is, *How good is the estimate of the fraction of defective transistors?* The answer to this question naturally arises in the discussion of interval estimation for proportions.

## Interval Estimation of the Population Proportion

The concept of confidence intervals, used to apprise a decision maker of the reliability of estimates of a population mean, can also be applied to estimating proportions. In order to develop the confidence interval for a population proportion, the sampling distribution of the point estimate must be developed. (See Section 8.3 for review.)

The random variable, $\hat{p}$, has a binomial distribution which is approximated with a normal random variable.

Thus, the sample proportion, $\hat{p}$, is normally distributed with mean, $p$, and standard deviation,

$$\sigma_{\hat{p}} = \sqrt{\frac{p(1-p)}{n}}.$$

If the true population proportion is unknown, the standard deviation of the sample proportion, $\hat{p}$, is denoted symbolically as $\sigma_{\hat{p}}$ and is given by

$$\sigma_{\hat{p}} = \sqrt{\frac{p(1-p)}{n}} \approx \sqrt{\frac{\hat{p}(1-\hat{p})}{n}},$$

where $\hat{p}$ is used as an estimate of $p$.

As before,

$$P(-1.96 < z < 1.96) = 0.95.$$

Substituting

$$z = \frac{\hat{p} - p}{\sigma_{\hat{p}}}$$

results in

$$P\left(-1.96 < \frac{\hat{p} - p}{\sigma_{\hat{p}}} < 1.96\right) = 0.95.$$

Manipulating the inequalities results in

$$P\left(\hat{p} - 1.96\sigma_{\hat{p}} < p < \hat{p} + 1.96\sigma_{\hat{p}}\right) = 0.95,$$

which suggests that the interval

$$\hat{p} \pm 1.96\sigma_{\hat{p}}$$

would be a good choice for a 95% confidence interval for the population proportion. As before, the probability that the interval will contain the true population proportion is 0.95 *before a specific sample is drawn*. After a specific sample is drawn, the only available information about the interval is that the technique which generated it will bound the true proportion 95% of the time.

## Formula

### Confidence Interval for the Population Proportion

If the sample size is sufficiently large, i.e., $np \geq 5$ and $n(1 - p \geq 5)$, the $100(1 - \alpha)\%$ confidence interval for the population proportion is given by the expression

$$\hat{p} \pm z_{\alpha/2}\sigma_{\hat{p}},$$

where $z_{\alpha/2}$ is the value of $z$ which captures an area of $\frac{\alpha}{2}$ in the right tail of the standard normal distribution, and $\sigma_{\hat{p}}$ is the standard deviation of $\hat{p}$.

---

**Example 9.3.2**

**Calculating a 95% Confidence Interval for the Proportion of Radio Listeners**

Suppose a sample of 410 randomly selected radio listeners revealed that 48 listened to WJLN.

$$\hat{p} = \frac{48}{410} \approx 0.1171$$

This is a point estimate of the proportion that listen to WJLN.

To obtain an interval estimate, the amount of confidence to be placed in the interval must be specified. Suppose we desire 95% confidence.

### SOLUTION

$$z_{\alpha/2} = z_{0.05/2} = z_{0.025} = 1.96, \text{ and } \sigma_{\hat{p}} \approx \sqrt{\frac{\hat{p}(1 - \hat{p})}{n}} = \sqrt{\frac{0.1171(1 - 0.1171)}{410}} \approx 0.0159.$$

Note that the sample proportion $\hat{p}$ is used in place of $p$ in the computation of $\sigma_{\hat{p}}$. For any realistic problem, this will always be the case. Fortunately, unless $\hat{p}$ and $p$ are far apart, the value of $\sigma_{\hat{p}}$ will not be greatly affected.

Computing the confidence interval $\hat{p} \pm z_{\alpha/2}\sigma_{\hat{p}}$ results in

$$0.1171 \pm 1.96(0.0159)$$
$$0.1171 \pm 0.0312$$
$$0.0859 \text{ to } 0.1482.$$

```
 (--------------------+--------------------)
0.0859              0.1171               0.1482
```

## Technology

For instructions on how to calculate a confidence interval for a proportion using technology, please visit stat.hawkeslearning.com and navigate to **Discovering Business Statistics, Second Edition > Technology Instructions > Confidence Intervals > Proportion**.

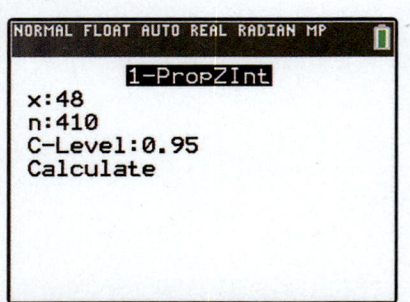

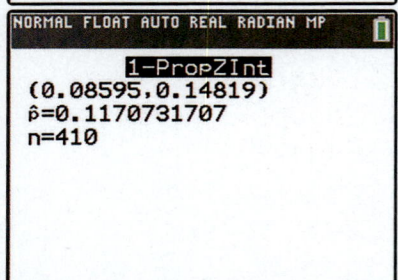

We are 95% confident in the procedure that created this interval. Another interpretation would be that we are 95% confident that the point estimate, 0.1171, has a maximum error of estimation of 0.0312. A maximum error of only 0.0312 with 95% confidence suggests a rather high level of accuracy in the estimation of the proportion.

## Precision and Sample Size: Proportions

Just as for the population mean, a specific level of accuracy in estimating a population proportion is desirable. Suppose, for example, that a direct-mail marketer would like to estimate the fraction of a mailing list that will purchase the company's product. To be profitable, a purchase response of at least 0.008 is required. Because the proportion to be estimated is of such a small magnitude, a high degree of precision in estimating the proportion is necessary. How large a sample would be required if the population proportion (the actual proportion of persons on the mailing list that will buy the product) is to be estimated with an accuracy of 0.002? We are saying that we want our maximum error to be less than two one-thousandths. That would seem to be a highly precise estimate. But, the quantity we are trying to estimate (the proportion of people on the list that will buy the product) could easily be near 0.008. The maximum error is about 25% as large as the value we are trying to estimate. When we estimate extremely small quantities, highly precise estimates are necessary.

The technique for deriving the sample size parallels the discussion of precision and sample size for the sample mean (Section 9.2). Setting one-half the entire width of the confidence interval equal to the maximum allowable error yields

$$E = \text{margin of error} = z_{\alpha/2}\sigma_{\hat{p}} = z_{\alpha/2}\sqrt{\frac{p(1-p)}{n}}.$$

Solving for $n$ yields

$$n = \frac{z_{\alpha/2}^2\, p(1-p)}{E^2}.$$

Generally, the population proportion is unknown and is estimated from a pilot study. In this case the sample size necessary to estimate the population proportion to within a particular error with a certain level of confidence is given by

$$n \approx \frac{z_{\alpha/2}^2\, \hat{p}(1-\hat{p})}{E^2}$$

where $\hat{p}$ is the estimate of the population proportion obtained from the pilot study.

If an estimate of the population proportion is not available, then the population proportion is set equal to 0.5. The value 0.5 maximizes the quantity $p(1-p)$ and thus provides the most conservative estimate of the sample size possible. Hence, if no estimate of the population proportion is available, the sample size necessary to estimate the population proportion to within a particular error with a certain level of confidence is given by

$$n = \frac{z_{\alpha/2}^2\,(0.5)(1-0.5)}{E^2} = \frac{z_{\alpha/2}^2\,(0.25)}{E^2}.$$

By selecting a level of confidence and an error, a sample size can be determined that will likely (at the level of confidence) produce an estimate with at least the desired accuracy. Remember to always round the sample size to the next largest integer to assure the desired level of accuracy.

---

**Example 9.3.3**

**Calculating the Sample Size Needed to Estimate the Proportion of Buyers**

How large a sample would be required to estimate the proportion of buyers on a mailing list that will buy the product with an accuracy of 0.002 with a 95% degree of confidence if the true proportion is approximately 0.008?

**SOLUTION**

$$p \approx 0.008,$$
$$z_{\alpha/2} = 1.96 \text{ for 95\% confidence, and}$$
$$E = 0.002$$

Using the sample size determination expression yields

$$n = \frac{z_{\alpha/2}^2 p(1-p)}{E^2}$$
$$= \frac{1.96^2 (0.008)(1-0.008)}{0.002^2}$$
$$= 7621.7344 \approx 7622 \quad \text{(always round up).}$$

Thus, to be 95% confident that the proportion is estimated with an error of at most 0.002 requires a sample size of at least 7622.

---

Using Example 9.3.2, suppose that the radio station WJLN desires to estimate the proportion of the market they hold with a maximum error of 0.01 and a confidence coefficient of 0.95. How large a sample would be required to estimate the fraction of listeners to within the desired level of accuracy? Since we don't know the true population proportion, let's assume the previous point estimate of 0.1171 is the true proportion.

**Example 9.3.4**

**Calculating the Sample Size Needed to Estimate the Proportion of the Market Held**

**SOLUTION**

$$p \approx 0.1171,$$
$$z_{\alpha/2} = 1.96 \text{ for 95\% confidence, and}$$
$$E = 0.01$$

Using the sample size determination expression yields

$$n = \frac{z_{\alpha/2}^2 p(1-p)}{E^2}$$
$$= \frac{1.96^2 (0.1171)(1-0.1171)}{0.01^2}$$
$$= 3971.7377 \approx 3972 \quad \text{(always round up).}$$

Thus, to be 95% confident that the proportion of listeners is estimated with an error of at most 0.01 would require a sample size of at least 3972.

---

Suppose we did not have a previous estimate of the population proportion in Example 9.3.4. In this case we would estimate $p$ with 0.5. The sample size necessary to estimate the true proportion of listeners to within 1% with 95% confidence is given by

$$n = \frac{z_{\alpha/2}^2 (0.5)(1-0.5)}{E^2} = \frac{1.96^2 (0.5)(1-0.5)}{0.01^2} = 9604.$$

Notice that the required sample size is significantly larger when an estimate of the population proportion is not available.

## ✐ 9.3 Exercises

### Basic Concepts

1. What is a proportion? What type of information does it give us about the population?

2. How is the sample proportion found?

3. Describe, in layman's terms, how a confidence interval is constructed for a population proportion.

4. It seems that estimating proportions produces estimates which are much more precise than those for means. Explain why this is the case.

5. The population proportion is often unknown. How is this issue dealt with when determining sample size?

6. What is the guideline to follow when there is no estimate available for the population proportion? Why is this done?

7. How do the resulting required sample sizes differ when there is an estimate available versus when there is no estimate available for the population proportion?

### Exercises

8. Acid rain accumulations in lakes and streams in the northeastern part of the United States are a major environmental concern. A researcher wants to know what fraction of lakes contain hazardous pollution levels. He randomly selects 200 lakes and determines that 45 of the selected lakes have an unsafe concentration of acid rain pollution.

   a. Calculate the best point estimate of the population proportion of lakes that have unsafe concentrations of acid rain pollution.

   b. Determine a 95% confidence interval for the population proportion.

   c. If a local politician states that only 20% of the lakes are contaminated, does the study provide overwhelming evidence at the 95% level to contradict his views?

9. *The Richland Gazette*, a local newspaper, conducted a poll of 1000 randomly selected readers to determine their views concerning the city's handling of snow removal. The paper found that 650 people in the sample felt the city did a good job.

   a. Compute the best point estimate for the percentage of readers who believe the city is doing a good job of snow removal.

   b. Construct a 90% confidence interval for this percentage.

10. The clinical testing of drugs involves many factors. For example, patients that have been given placebos, which are harmless compounds that have no effect on the patient, often will still report that they feel better. Assume that in a study of 500 random subjects conducted by the Poppins Sucre Drug Company, the percentage of patients reporting improvement when given a placebo was 37%.

    a. What would be a 95% confidence interval for the true proportion of patients who exhibit the placebo effect? Interpret this interval in terms of the problem.

    b. What would the 99% confidence interval be?

    c. To gain the additional 4% of confidence how much wider did the interval become?

11. The Peacock Cable Television Company thinks that 40% of their customers have more outlets wired than they are paying for. A random sample of 400 houses reveals that 110 of the houses have excessive outlets.

    a. Construct a 99% confidence interval for the true proportion of houses having too many outlets.

    **b.** Do you feel the company is accurate in its belief about the proportion of customers who have more outlets wired than they are paying for? Justify your answer.

12. Running continues to be a very popular sport in America. At a major race, like the Peachtree Road Race in Atlanta, there may be over 10,000 people entered to run. The race promoters for a road race in the Pacific Northwest took a random sample of 750 runners out of the 5000 runners entered to estimate the number of runners who will need hotel accommodations. Five hundred runners indicated they would need hotel accommodations.

    **a.** Construct a 90% confidence interval for the true proportion of runners who will need hotel accommodations.

    **b.** Is the confidence interval obtained sufficiently narrow to be of help in planning the number of hotel rooms which will be necessary to accommodate the runners? Justify your answer.

13. In the fourth quarter of 2010 the home ownership rate was 66.5%. This rate is 2.7 percentage points lower than the 2004 peak of 69.2%, and the lowest rate since 1998. Home ownership fell at an alarming pace in the fourth quarter of the year, despite the fact that home prices fell, affordability was much improved, and inventories of new and existing homes were running quite high. Suppose that a random sample of 120 households was selected from an area in the Midwest that is particularly economically depressed. Suppose that 57 of the households sampled were owned by the residents of the homes.
**Source:** U.S. Census Bureau

    **a.** Construct a 95% confidence interval for the proportion of households in the area sampled that are owned by the residents of the homes.

    **b.** Is there evidence at the 95% level that the proportion of the households in the area sampled that are owned by residents is less than the national rate?

14. In the *Gallup Poll Monthly*, it was reported that 31% of the people surveyed in a recent poll claimed that vegetables were their least favorite food. Surprisingly, only 14% responded with liver, and 10% of those surveyed did not submit a response because they claimed that they liked everything. The poll was based upon a sample of 1001 people. Assuming that a random sample was chosen, construct a 90% confidence interval for the percentage of all Americans who say that vegetables are their least favorite food.

15. The Federal Trade Commission (FTC) conducted a study investigating the accuracy of bar-code scanners. It was concluded that these computer scanners, used mainly at grocery, department store, and drugstore checkout counters, ring up the wrong price about 5% of the time. In most instances, however, the error was in favor of the shopper, according to the FTC. Suppose that your local grocery store conducts a study to determine the accuracy of its scanners. Assume 13 shoppers are randomly chosen and their bills, as indicated by the scanner, are checked against the correct bill computed by conventional means. Suppose that of the 200 items scanned, 21 of the items were charged incorrectly by the scanner.

    **a.** Construct a 95% confidence interval for the proportion of items that were rung up incorrectly by the scanner.

    **b.** Does it appear that the local grocery store has a larger error rate than 5%?

16. The Big Green Poster Company wants to estimate the fraction of poster sites controlled by their competition, Bird's Billboard Service. What sample size would be necessary to estimate this fraction to within 3% with 95% confidence? (They think Bird's controls about 33 percent of the boards.)

17. Researchers working in a remote area of Africa feel that 40% of families in the area are without adequate drinking water either through contamination or unavailability. What sample size will be necessary to estimate the percentage without adequate water to within 5% with 99% confidence?

18. Companies that provide environmental cleanup for hazardous waste and toxic chemicals are growing rapidly. W.R. Gross is thinking about entering this field with a subsidiary called Saf-t-Soil. They wish to estimate the true proportion of U.S. corporations that produce hazardous waste as a by-product of their manufacturing process to within 10% with 80% confidence. What sample size will be needed?

19. The public relations manager for a political candidate would like to determine if the registered voters in the candidate's district agree with the politician's view on a particular issue. Find the sample size necessary for the public relations manager to estimate the true proportion to within 5% with 85% confidence.

# 9.4 Estimating the Population Standard Deviation or Variance

Recall that the sample variance is

$$s^2 = \frac{\sum (x_i - \bar{x})^2}{n-1}$$

and it serves as the point estimate of the population variance, $\sigma^2$. As with the other tests developed for the population mean and the population proportion, we first need to develop a sampling distribution for

$$\frac{(n-1)s^2}{\sigma^2}$$

that will allow us to calculate a confidence interval for the population variance.

---

**Formula**

### $\chi^2$ Test Statistic

If we have a random sample of size $n$ taken from a normal population, then the sampling distribution of the test statistic is given by

$$\chi^2 = \frac{(n-1)s^2}{\sigma^2}$$

which has a **chi-square distribution** with $n-1$ degrees of freedom.

---

The chi-square distribution is a positively skewed (or skewed to the right) distribution. Like the $t$-distribution, the shape of the distribution is a function of its degrees of freedom. See Figure 9.4.1 which illustrates the chi-square distributions with 4 and 10 degrees of freedom.

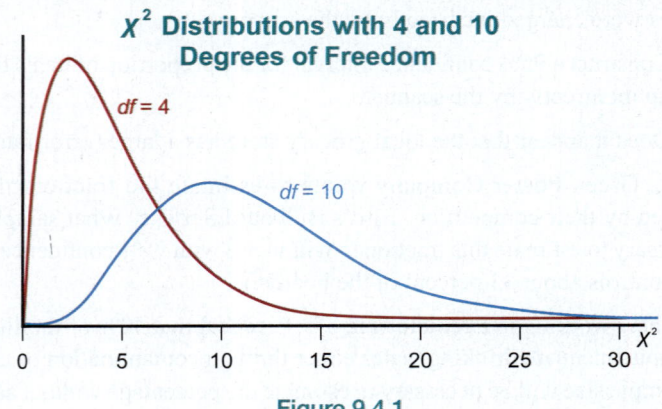

**Figure 9.4.1**

To use the chi-square distribution, we need a chi-square value, denoted by $\chi_\alpha^2$ (the Greek letter chi, pronounced *Ki*). We'll later call this our critical value for the chi-square distribution. As shown in Figure 9.4.2, $\chi_\alpha^2$ is the point on the horizontal axis under the curve with an area of $\alpha$ to the right of it. The value of $\chi_\alpha^2$ depends on the right-hand tail area, $\alpha$, and the number of degrees of freedom of the chi-square distribution. The values are tabulated in Table G in Appendix A. Looking at the chi-square table, the rows correspond to the appropriate number of degrees of freedom (the first column listed down the left side of the table), while the columns represent the right-hand tail area.

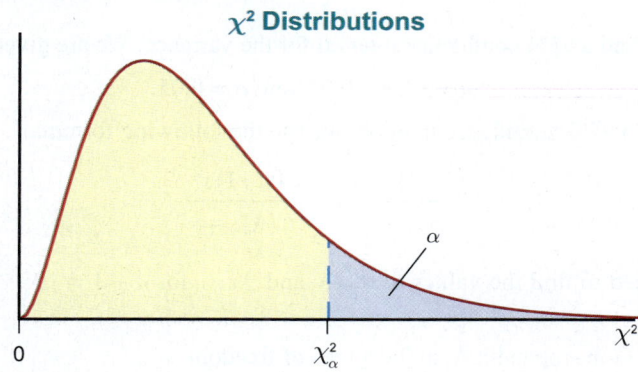

$\chi^2$ **Distributions**

**Figure 9.4.2**

Using Table G in Appendix A, suppose we want to find the chi-square value that gives us a right-hand tail area of 0.05 with 5 degrees of freedom. To do this, we would look down the leftmost column for 5 degrees of freedom and then the column labeled $\chi_{0.05}^2$. Doing so, we find that $\chi_{0.05}^2$ is 11.070.

| df | | $\chi_{0.050}^2$ | | | |
|---|---|---|---|---|---|
| 1 | 2.706 | 3.841 | 5.024 | 6.635 | 7.879 |
| 2 | 4.605 | 5.991 | 7.378 | 9.210 | 10.597 |
| 3 | 6.251 | 7.815 | 9.348 | 11.345 | 12.838 |
| 4 | 7.779 | 9.488 | 11.143 | 13.277 | 14.860 |
| 5 | 9.236 | 11.070 | 12.833 | 15.086 | 16.750 |
| 6 | 10.645 | 12.592 | 14.449 | 16.812 | 18.548 |

**⬡ Technology**

For instructions on finding the chi-square critical value using technology go to stat. hawkeslearning.com and navigate to **Discovering Business Statistics, Second Edition > Technology Instructions > Chi-Square Distribution > Critical Value**.

Now that we've established the sampling distribution associated with the sample variance, we can make inferences about the population variance. Suppose we have a random sample of size $n$ taken from a normal population and that $s^2$ is the estimate of the population variance, $\sigma^2$.

---

**Formula**

**$100(1-\alpha)\%$ Confidence Interval for $\sigma^2$**

A $100(1-\sigma)\%$ confidence interval for $\sigma^2$ is given by

$$\frac{(n-1)s^2}{\chi_{\alpha/2}^2} < \sigma^2 < \frac{(n-1)s^2}{\chi_{1-\alpha/2}^2}$$

where $\chi_{\alpha/2}^2$ and $\chi_{1-\alpha/2}^2$ are values under the curve of the chi-square distribution with $n-1$ degrees of freedom.

**✏ NOTE**

Recall that although $s^2$ is an unbiased point estimate for $\sigma^2$, $s$ is a biased estimator for $\sigma$.

## Example 9.4.1

**Calculating a Confidence Interval for the Population Standard Deviation**

The quality control supervisor of a bottling plant is concerned about the variance of fill per bottle. Regulatory agencies specify that the standard deviation of the amount of fill should be less than 0.1 ounce. To determine whether the process is meeting this specification, the supervisor randomly selects ten bottles, weighs the contents of each, and finds that the sample standard deviation of these measurements is 0.04. Assume that the data are collected from a normal population and compute a 95% confidence interval for the standard deviation of ounces of fill for the bottling plant.

### SOLUTION

We want to find a 95% confidence interval for the variance. We are given that

$$n = 10, s = 0.04, \text{ and } \alpha = 0.05.$$

To calculate a 95% confidence interval, we use the following formula.

$$\frac{(n-1)s^2}{\chi^2_{\alpha/2}} < \sigma^2 < \frac{(n-1)s^2}{\chi^2_{1-\alpha/2}}$$

Thus, we need to find the values of $\chi^2_{0.025}$ and $\chi^2_{0.975}$ for $n - 1 = 10 - 1 = 9$ degrees of freedom.

Using Table G in Appendix A, at 9 degrees of freedom,

$$\chi^2_{0.025} = 19.023$$
$$\chi^2_{0.975} = 2.700.$$

Substituting the values in the formula above, we have

$$\frac{(10-1)(0.04)^2}{19.023} < \sigma^2 < \frac{(10-1)(0.04)^2}{2.700}$$
$$0.000757 < \sigma^2 < 0.00533$$

So, a 95% confidence interval for the variance of fill of the bottles is between 0.000757 and 0.00533 ounce. However, the problem mentions the tolerance for the standard deviation of fill. So, to ensure that we make our interpretation in terms of the problem, to find a 95% confidence interval for the standard deviation, we take the square root of the confidence interval for the variance, yielding

$$0.0275 < \sigma < 0.0730.$$

The 95% confidence interval for the standard deviation of fill for the bottles is between 0.0275 and 0.0730 ounce, indicating that the process is meeting the specifications of being less than 0.1 ounce.

### ⚭ Technology

The confidence interval for the population variance can be obtained using Minitab. For detailed instructions, visit stat.hawkeslearning.com and navigate to **Discovering Business Statistics, Second Edition > Technology Instructions > Confidence Intervals > Variance**.

## 🖉 9.4 Exercises

### Basic Concepts

1. What is the sampling distribution for $\frac{(n-1)s^2}{\sigma^2}$?

2. What assumption must hold to use the chi-square distribution to make inferences about the population variance?

3. True or false: the chi-square distribution is skewed to the right.

4. Give an example where we would want to calculate a confidence interval for $\sigma^2$

## Exercises

5. A bolt manufacturer is very concerned about the consistency with which his machines produce bolts that are ¾ inch in diameter. When the manufacturing process is working normally the standard deviation of the bolt diameter is 0.05 inch. A random sample of 30 bolts has an average diameter of 0.25 inch with a standard deviation of 0.07 inch.

    a. Construct a 95% confidence interval for the standard deviation of the bolt diameter. Interpret the interval.

    b. What assumption did you make about the diameters of the bolts in constructing the confidence interval in part **a.**?

6. A drug that is used for treating cancer has potentially dangerous side effects if it is taken in doses that are larger than the required dosage for the treatment. The pharmaceutical company that manufactures the drug must be certain that the standard deviation of the drug content in the tablet is not more than 0.1 mg. Twenty-five tablets are randomly selected and the amount of drug in each tablet is measured. The sample has a mean of 20 mg and a variance of 0.015 mg.

    a. Construct a 99% confidence interval for the variance of the amount of drug in each tablet. Interpret the interval.

    b. What assumption did you make about the amounts of drug contained in the tablets in constructing the confidence interval in part **a.**?

7. A conservative investor would like to invest some money in a bond fund. The investor is concerned about the safety of her principal (the original money invested). Colonial Funds claims to have a bond fund which has maintained a consistent share price of $7. They claim that this share price has not varied by more than $0.25 on average since its inception. To test this claim, the investor randomly selects 25 days during the last year and determines the share price for the bond fund. The average share price of the sample is $7 with a standard deviation of $0.35.

    a. Construct a 90% confidence interval for the standard deviation of the share price of the bond fund. Interpret the interval.

    b. What assumption did you make about the share prices of the bond fund in constructing the confidence interval in part **a.**?

8. A manufacturer of automobile batteries is concerned about the life of the batteries that are produced. The manufacturer is comfortable with the average life of the batteries but more concerned about the standard deviation. Research has shown that the average life of the automobile batteries is 60 months. However, the manufacturer would like the standard deviation of the life of the automobile batteries to be relatively small, say, approximately six months. To determine a reliable range of the standard deviation of the batteries currently being produced, the manufacturer took a random sample of 15 batteries and found that the average life was 58 months with a standard deviation of seven months.

    a. Construct a 98% confidence interval for the standard deviation of the life of their automobile batteries. Interpret this interval.

    b. What assumptions did you make about the life of a battery being produced by the manufacturer?

9.   Almost all smart devices (phones, tablets, and computers) are made with touch screens. A concern of many consumers is the shelf life of the "touch" component of the screens. A consumer advocacy group wanted to inform its members of a range that they can expect their touch screens to last. The group took a sample of 29 screens and measured the life of the "touch" function of the screens. That is, they used digital devices to simulate billions of touches to determine the life of the screens. Of the 29 screens sampled, the average "touch" life was 90 months with a standard deviation of six months. Construct an 80% confidence interval for the standard deviation of the life of the touch screens. Interpret this interval.

10.   Photographers are always concerned about the number of shutter actuations that they will get from their cameras before they need to be serviced or the shutter needs to be replaced. To get an idea of the variability associated with the number of actuations, a photographer took a random sample of 20 cameras and found that the average number of actuations before failure was 200,000 with a standard deviation of 50,000.

   a.   Construct a 95% confidence interval for the standard deviation of the shutter actuations. Interpret the interval.

   b.   What assumptions did you make about the number of shutter actuations for the cameras?

## T  Discovering Technology

### Using the TI-84 Plus Calculator

#### Confidence Intervals for the Population Mean, $\sigma$ Known

Construct the 90% confidence interval for the population mean of a normal population if the population standard deviation is 900, the sample mean is 425, and the sample size is 100.

1. Choose **STAT**, then select **TESTS** and choose option **ZInterval**.

2. Press **ENTER** and choose **Stats**. Input **900** for $\sigma$, **425** for $\bar{x}$, **100** for $n$, and **0.90** for the C-Level. Highlight **Calculate**, and press **ENTER**.

3. The confidence interval, (276.96, 573.04) is listed on the output screen.

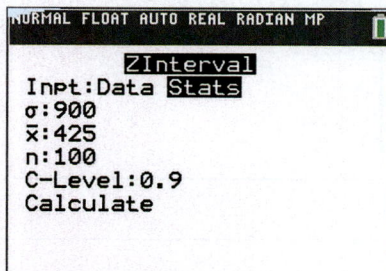
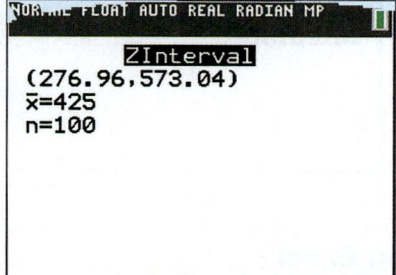

#### Confidence Intervals for the Population Mean, $\sigma$ Unknown

Construct the 95% confidence interval for the population mean of a normal population if the sample standard deviation is 6.5, the sample mean is 125, and the sample size is 15.

1. Choose **STAT**, then select **TESTS** and choose option **TInterval**.

2. Press **ENTER** and choose **Stats**. Input **125** for $\bar{x}$, **6.5** for $Sx$, **15** for $n$, and **0.95** for the C-Level. Highlight **Calculate**, and press **ENTER**.

3. The confidence interval, (121.4, 128.6) is listed on the output screen.

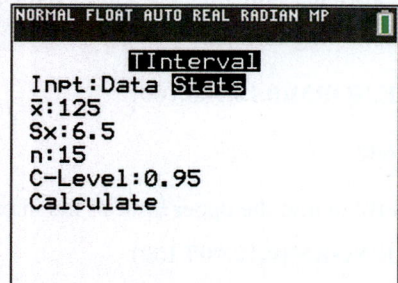
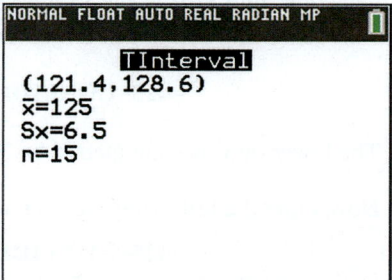

## Confidence Intervals for the Population Proportion

Suppose a sample of 410 randomly selected radio listeners revealed that 48 listened to WJLN. Construct a 95% confidence interval for the population proportion of radio listeners that listen to WJLN.

1. Choose **STAT** , then select **TESTS** and choose option **1-PropZInt**.

2. Enter **48** for $x$ (the number in the sample that listened to WJLN), **410** for $n$ (the sample size), and **0.95** for C-Level. Highlight **Calculate**, and press **ENTER** .

3. The confidence interval, (0.08595, 0.14819), is listed on the output screen, along with the sample proportion (0.1171), and the sample size (410).

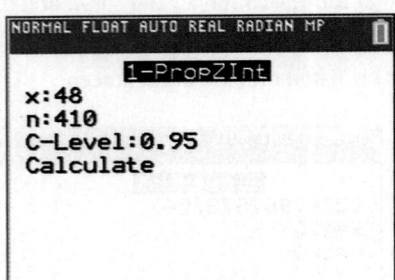
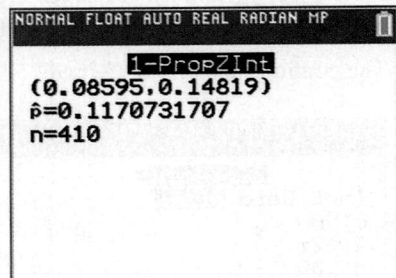

---

## Using Excel

## Confidence Intervals for the Population Mean, $\sigma$ Known

Construct a 90% confidence interval for the population mean of a normal population if the population standard deviation is 900, the sample mean is 425, and the sample size is 100.

1. Label cells A1 and B1 with **Lower Limit** and **Upper Limit**, respectively.

2. The function that returns the margin of error for a confidence interval in Microsoft Excel is

**CONFIDENCE.NORM(alpha, standard_dev, size)**

where **alpha** is the level of significance, **standard_dev** is the population standard deviation, and **size** is the sample size. First we will find the lower limit of the confidence interval. In cell B1, enter the following formula.

**=425−CONFIDENCE.NORM(0.10,900,100)**

The lower limit is computed to be 276.9632.

3. Now, enter the following formula in cell B2 to find the upper limit of the interval.

**=425+CONFIDENCE.NORM(0.10,900,100)**

The upper limit is computed to be 573.0368. Thus, we are 90% confident that the true population mean is between 276.9632 and 573.0368.

|   | A | B |
|---|---|---|
| 1 | Lower Limit | 276.9632 |
| 2 | Upper Limit | 573.0368 |

## Confidence Intervals for the Population Mean, $\sigma$ Unknown

Construct the 95% confidence interval for the population mean of a normal population if the sample standard deviation is 6.5, the sample mean is 125, and the sample size is 15.

1. Label cells A1 and B1 with **Lower Limit** and **Upper Limit**, respectively.

2. The function that returns the margin of error for a confidence interval in Microsoft Excel when sigma is unknown, and the data are drawn from a normally distributed population is

$$\textbf{CONFIDENCE.T(alpha, standard\_dev, size)}$$

where **alpha** is the level of significance, **standard_dev** is the sample standard deviation, and **size** is the sample size. First we will find the lower limit of the confidence interval. In cell B1, enter the following formula.

$$\textbf{=125--CONFIDENCE.T(0.05,6.5,15)}$$

The lower limit is computed to be 121.4004.

3. Now, enter the following formula in cell B2 to find the upper limit of the interval.

$$\textbf{=125+CONFIDENCE.T(0.05,6.5,15)}$$

The upper limit is computed to be 128.5996. Thus, we are 95% confident that the true population mean is between 121.4004 and 128.5996.

|   | A | B |
|---|---|---|
| 1 | Lower Limit | 121.4004 |
| 2 | Upper Limit | 128.5996 |

## Using JMP

## Confidence Intervals for the Population Mean, $\sigma$ Known

Construct the 90% confidence interval for the population mean of a normal population if the population standard deviation is 900, the sample mean is 425, and the sample size is 100.

1. With a JMP **Data Table** open, click on the first cell of **Column 1** and input the sample mean of 425.

2. Right click at the top of the next column and click on **New Columns**. Click **OK** in the next pop-up window to add **Column 2**.

3. Click on the first cell of **Column 2** and input the sample size of 100.

4. Select **Analyze** in the top row of the JMP spreadsheet and then select **Distribution**.

5. From the **Select Columns** box, click on **Column 1**. Then click on **Y, Columns**.

6. From the **Select Columns** box, click on **Column 2**. Then click on **Freq**. Click **OK**.

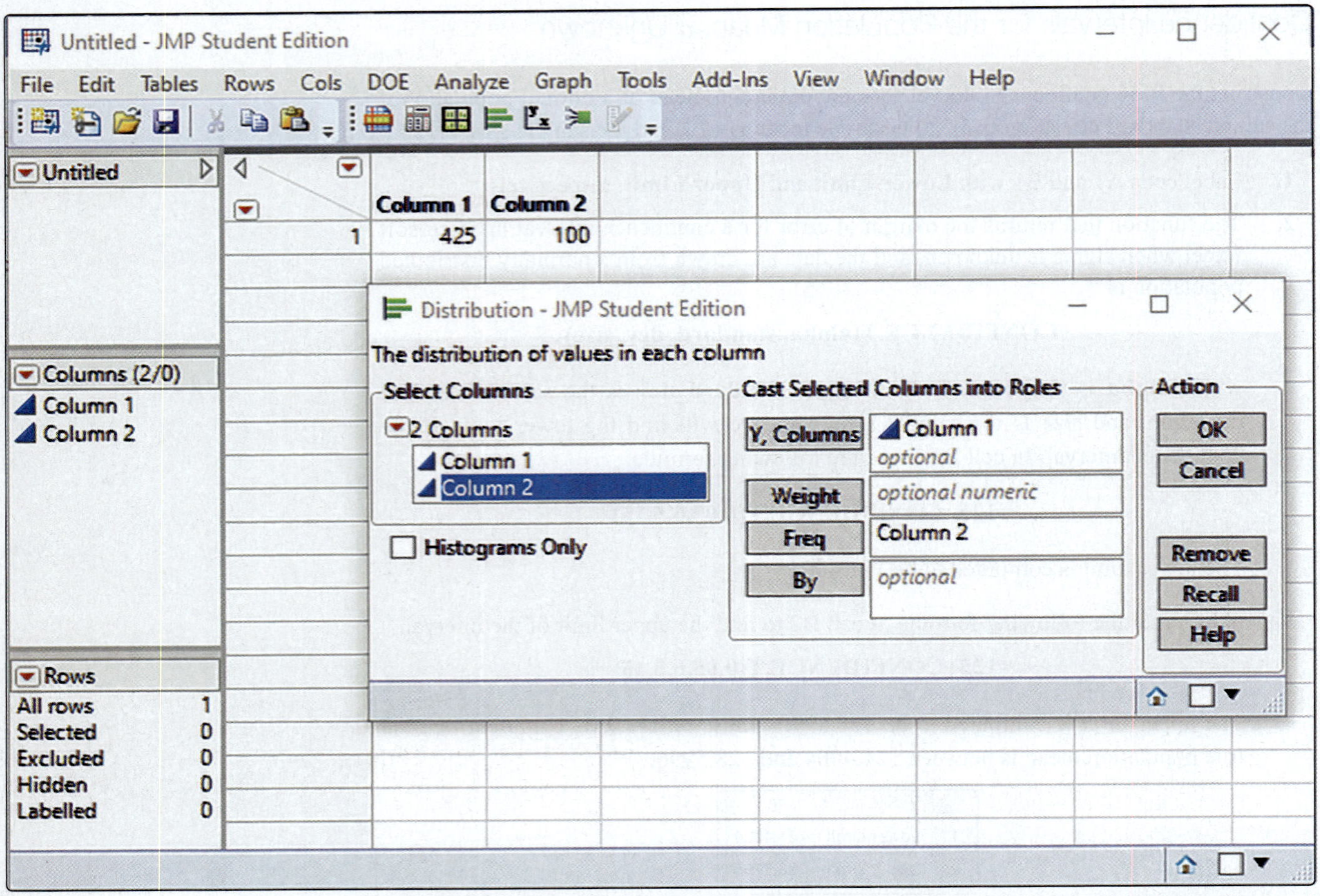

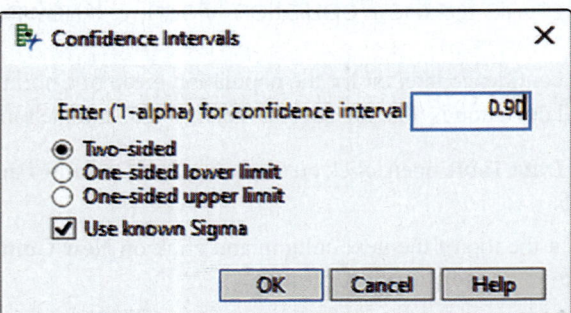

7. Click on the **red down arrow** next to **Column 1** and choose **Confidence Interval** from the list. Choose **Other** from the menu that appears to the right.

8. Input your confidence level as 0.90 and check the box labeled **Use known Sigma**. Click **OK**.

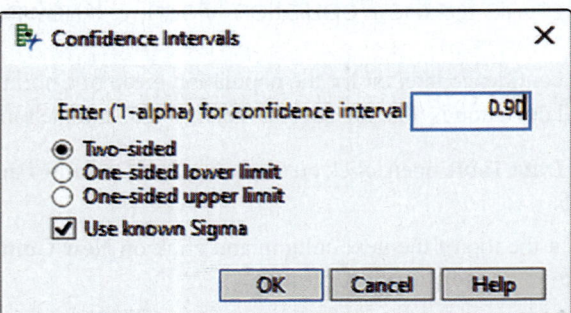

9. In the box that opens up input the known standard deviation of 900 and click **OK**.

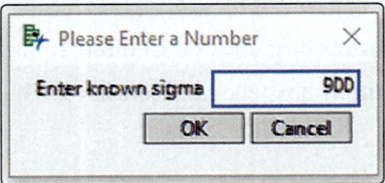

10. The interval is given in the top right corner of the output under **Confidence Intervals**. Therefore, the 90% confidence interval for the population mean is approximately (276.96, 573.04).

## Key Formulas (cont.)

Section

### Margin of Error for the Population Proportion (Maximum Error of Estimation)

$$E = z_{\alpha/2}\sigma_{\hat{p}} = z_{\alpha/2}\sqrt{\frac{p(1-p)}{n}}$$

9.3

where $z_{\alpha/2}$ is the critical value for a $z$-distribution which captures an area of $\alpha/2$ in the right tail of the distribution, $p$ is the population proportion, and $n$ is the sample size.

### Determining the Sample Size for the Population Proportion

$$n = \frac{z_{\alpha/2}^2 p(1-p)}{E^2} \approx \frac{z_{\alpha/2}^2 \hat{p}(1-\hat{p})}{E^2}$$

9.3

where $z_{\alpha/2}$ is the critical value for a $z$-distribution which captures an area of $\alpha/2$ in the right tail of the distribution, $p$ is the population proportion, and $\hat{p}$ is the sample proportion.

### Determining the Sample Size for the Population Proportion: No Estimate $\left(\hat{p}\right)$ Available

9.3

$$n = \frac{z_{\alpha/2}^2 (0.5)(1-0.5)}{E^2} \approx \frac{z_{\alpha/2}^2 (0.25)}{E^2}$$

### $\chi^2$ Test Statistic

If we have a random sample of size $n$ taken from a normal population, then the sampling distribution of the test statistic is given by

9.4

$$\chi^2 = \frac{(n-1)s^2}{\sigma^2}$$

which has a chi-square distribution with $n-1$ degrees of freedom.

### 100$(1-\alpha)$% Confidence Interval for $\sigma^2$

A 100$(1-\sigma)$% confidence interval for $\sigma^2$ is given by

$$\frac{(n-1)s^2}{\chi_{\alpha/2}^2} < \sigma^2 < \frac{(n-1)s^2}{\chi_{1-\alpha/2}^2}$$

9.4

where $\chi_{\alpha/2}^2$ and $\chi_{1-\alpha/2}^2$ are values under the curve of the chi-square distribution with $n-1$ degrees of freedom.

# AE   Additional Exercises

1.  The owner of Sloppy Jack's bar is thinking about installing some video game machines. To estimate the profitability of the machines, he measures the number of times a competitor's machines are played over a randomly selected sample of days. The preliminary sample showed that the standard deviation of the number of times the machines are played is 10 times per day. Find the sample size (in days) necessary to estimate the average number of times the machines will be played in a day to within 5 plays with 99% confidence.

2.  A random sample of fifteen eleven-year-old boys is selected in order to estimate the mean height for boys belonging to that age group. The resulting measurements in inches are given in the table below.

| Heights (Inches) | | | | |
|---|---|---|---|---|
| 55 | 58 | 52 | 58 | 54 |
| 57 | 56 | 54 | 58 | 56 |
| 52 | 59 | 55 | 61 | 57 |

   a.  Calculate the sample mean and the sample standard deviation of the heights.

   b.  Construct a 95% confidence interval for the mean height of all eleven-year-old boys.

   c.  What assumption did you make about the heights in constructing your interval?

3.  R. Cramden, chief development officer for Fontana Area Transport bus company, is concerned about the declining use of the bus system. He wishes to estimate the percentage of Fontana residents who consider safety a significant factor in their decision about whether or not to ride a bus. This will be a preliminary study so he is willing to develop an estimate with an error of 10% at a confidence level of 90%.

   a.  What sample size will be needed?

   b.  If 150 residents in a random sample of 500 Fontana residents say that they consider safety a significant factor in their decision about whether or not to ride a bus, estimate the true proportion of Fontana residents who think safety is a significant factor in their decision about whether or not to ride a bus with 95% confidence.

4.  According to a 2001 study conducted by the American Stock Exchange, 87% of 500 young Americans surveyed said that they can't count on Social Security as a source of income when they retire. Construct a 90% confidence interval for the proportion of young Americans who feel they can't count on Social Security as a source of income when they retire.

5.  The State Bureau of Standards must inspect gasoline station pumps on a regular basis to be sure they are operating properly. A recent survey of a randomly selected group of 61 pumps produced a sample mean of 9.75 gallons dispensed for a pump reading ten gallons. If the sample had a standard deviation of 1.12 gallons, find the 80% confidence interval for the mean amount of gas dispensed when a gas pump reads ten gallons.

6.  In a population of non-unionized employees, 55% are sympathetic toward unionization. The American Federation of Labor has drawn a random sample of 250 persons selected from this population to investigate union interest. Construct a 90% confidence interval for the proportion of the sample that will be sympathetic toward unionization.

7. Suppose a study designed to collect data on smokers and nonsmokers uses a preliminary estimate of the proportion that smoke of 22%. How large a sample should be taken to estimate the proportion of smokers in the population with a margin of error of 0.02 with 88% confidence?

8. As part of an annual review of its accounts, a discount brokerage firm selects a random sample of 15 customers. Their accounts are reviewed for a total account valuation, which showed a mean of $32,000 with a sample standard deviation of $8200.

   a. What is a 99% confidence interval for the mean account valuation of the population of customers? Interpret the interval in terms of the problem.

   b. What assumption about the account distribution is necessary to solve this problem?

9. Direct Music has 250 retail outlets throughout the United States. The firm is evaluating a potential location for a new outlet, based in part, on the mean annual income of the individuals in the marketing area of the new location. A sample of size 36 was taken; the sample mean income is $31,100. The population standard deviation is estimated to be $4500. Construct a confidence interval using a confidence coefficient of 0.95.

10. Suppose we want to determine the sample size required to give us a 95% confidence interval that estimates, to within $500, the average salary of a Virginia Tech employee. Also, suppose that from a previous experiment, we know that $s = \$6300$. What is the minimum sample size required?

11. A reporter for a student newspaper is writing an article on the cost of off-campus housing. A sample of 16 efficiency apartments within a half-mile of campus resulted in a sample mean of $650 per month and a sample standard deviation of $55. Construct a 95% confidence interval estimate of the mean rent per month for the population of efficiency apartments within a half-mile of campus. We will assume that this population is normally distributed.

12. A stock market analyst wants to estimate the average return on a certain stock. A random sample of 15 days yields an average return of 10.37% and a standard deviation of 3.5%. Give a 90% confidence interval for the true average return on the stock. Assume that the stock returns are normally distributed.

13. Voting, Inc. specializes in voter polls and surveys designed to keep political office seekers informed of their position in a race. Using telephone surveys, interviewers ask registered voters who they would vote for if the election were held that day. In a current election campaign, Voting, Inc. has found that 220 registered voters, out of 500 contacted, favor a particular candidate. Find a 95% confidence interval estimate for the proportion of the population of registered voters that favor the candidate.

14. Before beginning a pension program for its workers, a corporation wishes to estimate the proportion of its workers who have been employed at the company for at least 20 years. A random sample of 138 employees yielded 16 who have been with the corporation for at least 20 years. Construct a 92.2% confidence interval for the proportion of employees who have worked for this corporation for at least 20 years. Interpret your results.

## P  Discovery Project

### Home Sweet Home: Using Confidence Intervals to Analyze and Compare Home Prices

⠿ **Data**

The data can be found by visiting stat.hawkeslearning.com and navigating to **Discovering Business Statistics, Second Edition > Data Sets > Mount Pleasant Real Estate Data**.

One of the biggest purchases we make in our lives is a home. As we buy a home we ask ourselves many questions such as:

*How much should I spend for a home?*

*How many bathrooms are there?*

*What is the cost per square foot?*

Suppose you are looking for a house near Charleston in Mount Pleasant, SC, and you have narrowed your search to three subdivisions: Carolina Park, Dunes West, and Park West.

1.   Download the Mount Pleasant Real Estate data set.

2.   Import the data into Minitab, Excel or other statistical software.

3.   For the variable *List Price,* calculate the sample mean, the sample standard deviation, and the sample size for the three different subdivisions. Put the calculations in a table and round to the nearest dollar for the sample standard deviation and the mean.

4.   Based on the data set and the information we have, which confidence interval should we use here, a *z* or a *t* interval? Why?

5.   Find the critical value for a 95% confidence level for each subdivision for the variable *List Price.*

6.   Construct an interval to estimate the true average *List Price* for each subdivision with 95% confidence. Based on these confidence intervals, is it possible that Carolina Park and Dunes West have the same average *List Price*. Discuss.

7.   Do you think a *List Price* of $520,000 is a reasonable value for the Carolina Park subdivision?

8.   Do you think a *List Price* of $670,000 is a reasonable value for the Dunes West subdivision?

9.   Do you think a *List Price* of $568,000 is a reasonable value for both the Carolina Park and Park West subdivisions?

10

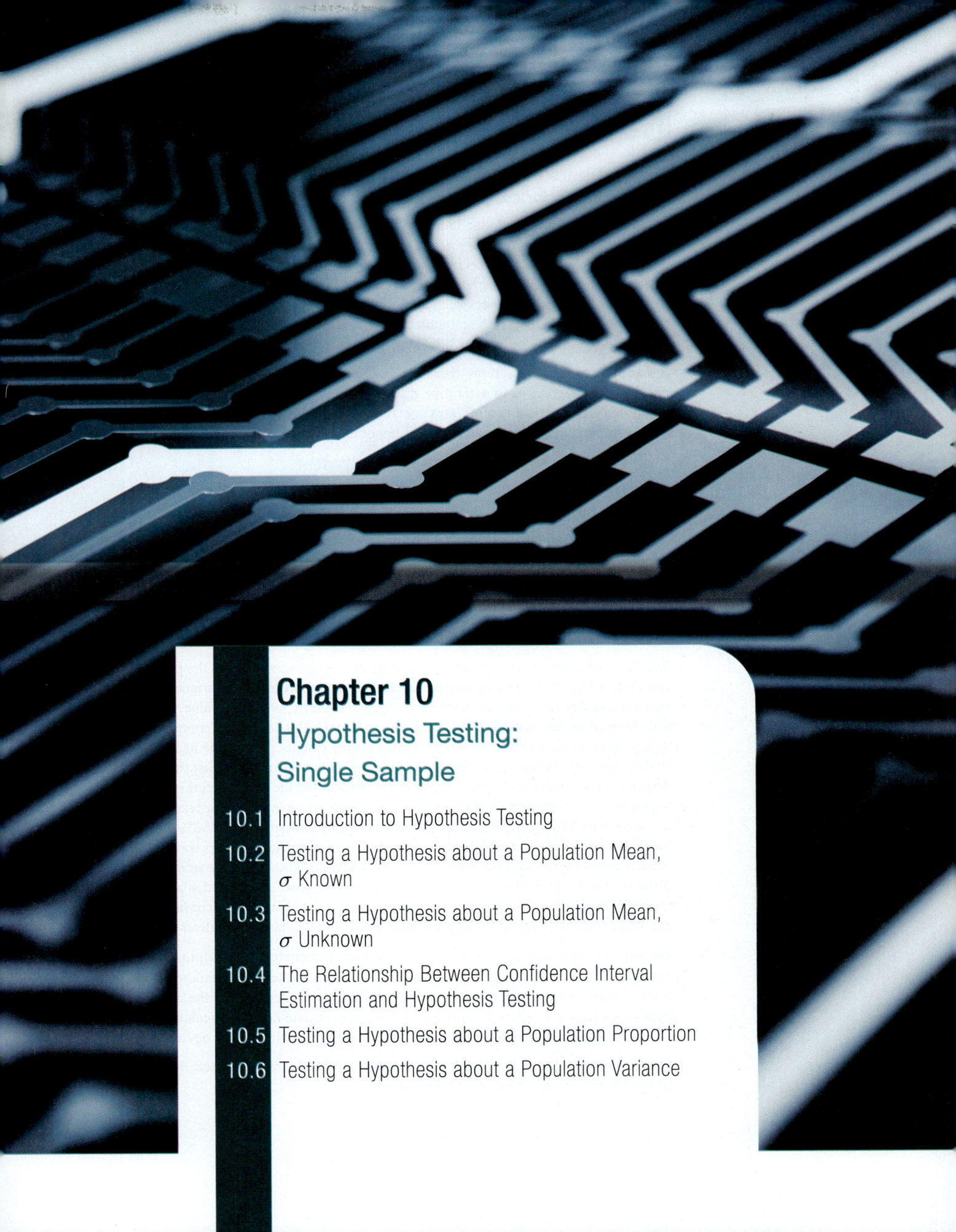

# Chapter 10
## Hypothesis Testing:
## Single Sample

# Discovering the Real World

## Do Members of the Sports Industry Support the Use of Wearable Technology?

Sports is a multibillion-dollar industry worldwide with athletes at all levels turning to technology to give them an advantage over their competition. Prior to the availability of using technology to track workouts, athletes and coaches would perform laborious tasks to determine how long an athlete ran, how much time was spent in the gym, or how much food was eaten. Smart devices such as Fitbit and Apple Watch have made it much easier to track speeds and distances that athletes run, nutrition, and other metrics needed to help athletes avoid fatigue or injury while helping to maximize performance.

A startup, which manufactures smart devices, collected data from athletes, trainers, team physicians, and coaches to determine if the implementation of smart devices or wearable technology has assisted in athletes' performances. While the survey contained many questions, the three primary questions of importance for this chapter are:

1.  Do you use smart devices to train?

2.  Do you feel that smart devices have helped you improve your performance?

3.  Would you recommend the use of smart devices to train?

A sample of 500 athletes, trainers, team physicians, and coaches was collected. The athletes referenced in the survey ranged in age from 18 through 23 years old (from high school through college-aged). It was indicated that a total of 245 survey respondents used smart devices to train. Of the 245 survey respondents that used the smart devices to train, 130 believed that their performances were improved. Lastly, of the 500 survey respondents, 249 indicated that they would recommend the use of smart devices as part of a training regimen.

The focus of this chapter is hypothesis testing of a single sample from the population. We will discuss hypothesis testing on population means, proportions, and variances. That is, we will test whether a population parameter is equal to some hypothesized value. In the scenario above, smart device manufacturers such as Apple and Fitbit have an enormous opportunity to capitalize on the use of their smart devices if they can get them in the hands of athletes. Before doing so, however, these manufacturers should test to ensure that there is a "real" demand for such devices. Thus, it would be prudent for the manufacturers to perform a hypothesis test to determine if more than 50% of athletes would use (or recommend) their smart devices. If the hypothesis test shows that significantly more than 50% of the athletes would use and/or recommend the use of smart devices, the manufacturers should then target athletes as another market for their devices (thus, adding another income stream). However, if the results of the hypothesis test indicate that there is no evidence that athletes would overwhelmingly use their device, then the manufacturers should not incur large marketing and manufacturing expenditures to get their devices in the hands of athletes.

We will discuss hypothesis testing of population parameters, developing the null and alternative hypotheses, significance (or lack thereof) of the hypothesis tests, $P$-values, and the interpretation of the results of hypothesis tests. Understanding the meaning of the hypothesis tests will allow businesses to make appropriate actions—such as whether the aforementioned manufacturers should target athletes to use their smart devices.

This is just a sample of the data for the wearable tech that was generated. These data will be used in the project at the end of the chapter.

| Do you use smart devices to train? | |
| --- | --- |
| **Role** | **Yes** |
| Athlete | 58 |
| Coach | 62 |
| Team Physician | 60 |
| Trainer | 65 |

| Smart device improved performance? | |
| --- | --- |
| **Role** | **Yes** |
| Athlete | 29 |
| Coach | 34 |
| Team Physician | 28 |
| Trainer | 39 |

| Recommend use of smart device? | |
| --- | --- |
| **Role** | **Yes** |
| Athlete | 52 |
| Coach | 66 |
| Team Physician | 59 |
| Trainer | 72 |

## Introduction

Given the global pandemic caused by COVID-19, one would think that the travel industry would face a significant decrease in number of travelers, amount of money spent on vacations (or the number of vacations taken), or the average number of daily passengers on an airline. As part of the domino effect, one would also expect the number of jobs available in the travel industry to witness a significant decrease. If we wanted to determine if the previous assumptions were valid, we would be interested in making inferences about the average number of travelers, the average amount of money spent on vacations in 2020 versus the amount of money spent on vacations in 2019, or the average number of daily airline passengers in 2020 when compared to 2019. Similarly, we would also want to make inferences about the number of jobs available in the travel industry in 2020 versus 2019. For any of these scenarios, we are interested in testing how the value of a parameter relates to (i.e., whether it is less than, equal to, or greater than) some specific numerical value. This type of inference is called **hypothesis testing**, which is the subject of this chapter.

Like the above examples, we perform some type of hypothesis test on a daily basis. Whether it's trying to determine if the shower temperature is reasonable (acceptable, too hot, or too cold) or if we want to confirm if the day's weather forecast is accurate (will it be too hot to wear a sweater, adequate, or if you'll be underdressed), this type of subjective decision making is always a part of our daily lives. We rely on personally accumulated data to help us make informed decisions. What distinguishes statistical hypothesis testing from the everyday variety is the use of statistical measures in the statement of the hypotheses, the collection of the sample data, and the use of the sample data in a well-defined decision-making process.

# 10.1 Introduction to Hypothesis Testing

It should come as no surprise that the first step in all statistical hypothesis testing is a statement of hypothesis, regardless of the nature of the problem. Further, if the hypothesis test is conducted using statistical methods, the hypotheses must be stated in terms of statistical measures such as the population mean, population proportion, or population variance.

---

**Example 10.1.1**

**Determining the Null and Alternative Hypotheses for a Right-Tailed Test of the Population Mean**

Tech Travel, which is a travel agency managed by C. Dubya, is interested in determining how much money people are spending on travel and entertainment. From previous data, the agency knows that in 2018, U.S. households spent, on average, a total of $1949 on transportation, food, lodging, and pleasure trips. However, the agency believes that people will spend more in 2019 on travel than they did the previous year. The agency's question can be investigated with two hypotheses.

- Travel expenses have not increased since 2018.

- Travel expenses have increased since 2018.

**SOLUTION**

To apply statistical methodology to answer the question, the preceding hypotheses must be translated into a problem statement concerning a statistical measure (e.g., mean, proportion, or variance). The statistical measure will be used in the definition of a criterion to decide the issue. There is a great deal of variability in the amount of money one spends on vacations, depending on the location, whether they are flying or driving, etc. Comparing the mean amount spent per family with that which was spent previously is a reasonable method of evaluating the differences. The average amount spent in 2018 is known to be $1949. However, the average amount spent in 2019 is currently unknown, and that uncertainty is what makes the problem difficult. If $\mu$ is defined as

$$\mu = \text{average amount spent on travel expenses in 2019,}$$

then the two claims can be written in the following manner.

$H_0: \mu = \$1949$     Money spent on 2019 travel expenses is equal to that spent in 2018.

$H_a: \mu > \$1949$     More money is spent on travel expenses in 2019 than in 2018.

$H_0$ is called the **null hypothesis**, and contends that the mean amount spent on travel expenses in 2019 is equal to that spent in 2018. This would contradict the agency's assertion. $H_a$ is called the **alternative hypothesis** (sometimes denoted by $H_1$) and declares that the mean amount spent in 2019 is more than that spent in 2018. Once the problem statement is formulated in terms of the population parameter (in this case, the parameter is $\mu$), sample data can be developed to help determine which hypothesis is more reasonable.

---

**Definition**

**Null Hypothesis**

The **null hypothesis**, denoted by $H_0$, represents the status quo and will not be rejected unless supported by the data.

**Definition**

**Alternative Hypothesis**

The **alternative hypothesis**, denoted by $H_a$ or $H_1$, contradicts the null hypothesis.

**Definition**

**One-Sided Alternative**

The **one-sided alternative** hypothesis is one in which the researcher is interested in whether the parameter of interest is significantly more than the hypothesized value, or when the researcher is interested in whether the parameter of interest is significantly less than the hypothesized value.

**Definition**

**Two-Sided Alternative**

The **two-sided alternative** hypothesis is one in which the researcher is interested in whether the parameter of interest is significantly more or less than the hypothesized value.

---

This example illustrates another important component of a hypothesis test, namely, deciding whether $H_a$ should be one-sided or two-sided. In Example 10.1.1, $H_a$ involves a **one-sided alternative**, since the agency is only interested in whether travelers spent *more* money. If the agency was interested in whether travelers spent *less* money in 2019, then $H_a$ would also have been one-sided, but the inequality would have been in the other direction ($H_a: \mu < \$1949$). A **two-sided alternative** ($H_a: \mu \neq \$1949$) would indicate that the agency is concerned with expenditures that are *above or below* the average in 2018. One-sided alternatives require a **one-tailed hypothesis test** and two-sided alternatives require a **two-tailed hypothesis test**.

The hypothesis testing procedure is a method for choosing between two competing hypotheses. Although the procedure will change for different kinds of hypotheses, there are common elements among all tests.

## Elements of Hypothesis Testing: Common Elements and Comments about Hypothesis Testing

- The null hypothesis is presumed to be true unless sample data produces overwhelming evidence to the contrary. That is, the test statistic is calculated under the assumption that the null hypothesis is true.

- Sample data are used to calculate a test statistic. The form of the test statistic will change depending upon the statistical measure used in the hypothesis statement, $\mu$, $p$, or $\sigma^2$, as well as with the assumptions about knowledge of the population. The **test statistic** is a component of the criteria used to evaluate the hypothesis. If the test statistic falls into a **rejection region**, the null hypothesis, $H_0$, will be rejected in favor of the alternative hypothesis, $H_a$.

- The test statistic is usually designed so that if the null hypothesis is true, the value of the test statistic will "probably" be close to zero. (The notion of *close* is arbitrary. Developing a formal test of a hypothesis will draw upon the theory of sampling distributions and the language of probability to more precisely define the meaning of close.)

- The conclusion of the hypothesis test results in a decision to either **reject** or **fail to reject** the null hypothesis. Note that we do not accept the null hypothesis; we only conclude that there is insufficient evidence to support the alternative.

- There is no way to be absolutely certain your decision is correct, no matter which hypothesis is selected.

**Definition**

### One-Tailed Hypothesis Test

A **one-tailed hypothesis test** is one in which the rejection region (or significance level) is in only one tail of the distribution.

**Definition**

### Two-Tailed Hypothesis Test

A **two-tailed hypothesis test** is one in which the rejection region (or significance level) is divided into two parts and contained in both tails of the distribution.

**Definition**

### Test Statistic

The **test statistic** is a value calculated from the sample data that is used to evaluate the null hypothesis.

**Definition**

### Rejection Region

The **rejection region** is an area containing the set of values that the test statistic can have that will lead us to reject the null hypothesis. If the test statistic falls within the rejection region, then the null hypothesis will be rejected in favor of the alternative hypothesis.

## Roles of the Null and Alternative Hypotheses

In the previous example, we treated the null and alternative hypotheses equally. In the practice of statistical inference, this is not the case. In a statistical test of a hypothesis, the null hypothesis, $H_0$, is a statement that is presumed to be true unless there is overwhelming evidence in favor of the alternative. In other words, the null hypothesis is given the benefit of the doubt.

A familiar example of the disparate treatment of $H_0$ and $H_a$ is found in our judicial system.

$H_0$: Defendant is not guilty.　　(null hypothesis)

$H_a$: Defendant is guilty.　　　　(alternative hypothesis)

A jury must believe that the evidence (data) demonstrates guilt "beyond a reasonable doubt" in order to convict a defendant ($H_a$). However, the defendant only has to demonstrate there is insufficient evidence of guilt in order to be acquitted ($H_0$). In other words, the null hypothesis ($H_0$: Defendant is not guilty) is presumed to be true unless there is overwhelming evidence to the contrary. The defendant is innocent until proven guilty.

In Example 10.1.1 the hypotheses were formulated as follows.

| Correct | | Incorrect |
|---|---|---|
| $H_0$: $\mu = \$1949$ | instead of | ~~$H_0$: $\mu > \$1949$~~ |
| $H_a$: $\mu > \$1949$ | | ~~$H_a$: $\mu = \$1949$~~ |

## The *Incorrect* Formulation: $H_0$: $\mu > \$1949$

The statement $H_0$: $\mu > \$1949$ says that the average travel expenditures, $\mu$, are greater than they were in 2018. The essence of the problem at hand is whether the statement $\mu > \$1949$ is true or untrue. By placing this statement in $H_0$, the statement is presumed to be true unless there is overwhelming evidence to the contrary. This does not make sense! Why would the agency try to prove that travel expenditures are greater in 2019 starting out by assuming that expenditures in 2019 are greater?

## The *Correct* Formulation: $H_0$: $\mu$ = \$1949

The agency must demonstrate overwhelming evidence that travel expenditures are greater in 2019 than in 2018 in order to be correct. To achieve this goal, the statement $\mu > \$1949$ must be placed in the alternative hypothesis. In the correct version, the null hypothesis, $H_0$: $\mu = \$1949$, states that we believe that travel expenditures are equal to the amount spent in 2018. Because the null hypothesis is presumed to be true, this hypothesis is not rejected unless there is overwhelming evidence to the contrary. If $H_0$: $\mu = \$1949$ is rejected in favor of $H_a$: $\mu > \$1949$, then there is overwhelming evidence that households spent more on travel in 2019 than in 2018.

To summarize, the agency wants to know if the travel expenditures are greater in 2019 than in 2018. If the agency is correct, the null hypothesis will be rejected in favor of the alternative. Rejecting the null in favor of the alternative means the sample data overwhelmingly support the idea that expenditures in 2019 are greater than those in 2018. Failing to reject the null hypothesis would mean that there is insufficient evidence to warrant concluding that people spent more on travel in 2019 than they did in 2018.

This type of formulation is typical of a test of a hypothesis in a research environment (testing a new theory or idea against an established one). In many research hypothesis tests, the investigator hopes to reject the null hypothesis in order to demonstrate the significance of a new idea. It is often the case that if there is a **hypothesized value** for a population parameter, the null hypothesis states that the population parameter is equal to the hypothesized value.

> **Definition**
>
> **Hypothesized Value**
>
> The **hypothesized value** is the value of the parameter that is believed to be true in the null hypothesis.

When formulating hypotheses, remember that the equality statement will always be contained in the null hypothesis. That is, the null hypothesis will contain = , whereas the alternative hypothesis will contain < , > , or ≠ . This is because the only way to support an equality statement in the alternative hypothesis would be to perform a census, which defeats the purpose of sampling.

### Example 10.1.2
**Determining the Null and Alternative Hypotheses for a Two-Tailed Test of the Population Mean**

Suppose a potato chip manufacturer is concerned that the bagging equipment is not functioning properly when filling 10-ounce bags. He wants to test a hypothesis that will help determine if there is a problem with the bagging equipment. What are the correct hypotheses?

$$H_0: \mu = 10 \text{ oz} \qquad H_0: \mu \neq 10 \text{ oz}$$
$$\text{or}$$
$$H_a: \mu \neq 10 \text{ oz} \qquad H_a: \mu = 10 \text{ oz}$$

**SOLUTION**

Since the hypothesized value in this problem is 10 ounces, the null hypothesis is $H_0$: $\mu = 10$ oz. The bagging equipment ordinarily functions properly; thus, the manufacturer requires overwhelming evidence that the machine is overfilling or underfilling the bags before shutting down the equipment.

When the alternative hypothesis allows values above and below the hypothesized value, it is a two-sided alternative. In this case, the manufacturer hopes that he *fails to reject* the null hypothesis, since that would suggest that his equipment is putting the right amount in the bags. This type of hypothesis is quite common in quality control and other kinds of system monitoring.

### Example 10.1.3
**Determining the Null and Alternative Hypotheses for a Right-Tailed Test of a Proportion**

Analysts at JBN Technologies want to assess the use of Apple and Windows computers in their line of work. With the rise in popularity of Apple and the Mac OS operating system, analysts believe that businesses are more likely to allow their employees to deploy Apple computers as their enterprise desktops. How should JBN Technologies analysts formulate an appropriate hypothesis to determine if more businesses will deploy Apple computers as their enterprise desktops?

**SOLUTION**

The correct formulation of the hypotheses is as follows, where $p$ is the fraction of enterprises that would support the use of Apple computers as their desktops.

$H_0: p = 0.5$ — The number of businesses using Apple and Windows computers as their enterprise desktops is the same.

$H_a: p > 0.5$ — A larger percentage of businesses deploy Apple computers than Windows computers as their enterprise desktops.

The problem's formulation requires the use of a different statistical measure, $p$, the population proportion. The crucial value of the proportion is 0.5, since in order to believe that more businesses will support Apple computers, this proportion has to be more than 50%. The null hypothesis contends that there is not sufficient evidence that more businesses will support the use of Apple computers. If the null hypothesis is rejected in favor of the alternative, $H_a: p > 0.5$, then there is overwhelming evidence that businesses are prepared to adopt Apple computers.

The alternative is one-sided, since the intent of the study is to learn if Apple computers are more popular in the workplace. (A two-sided alternative would imply interest in finding out if more companies or fewer companies are willing to accept Apple computers in the workplace, a rather useless idea.) The analysts hope the conclusion of the test will be to *reject* the null hypothesis, since that would support the analysts' claim.

### Formulating Hypothesis Testing Problems

To be successful at formulating hypothesis testing problems you must be able to do the following.

- Determine the appropriate statistical measure to test the desired hypothesis (population mean, proportion, or variance).
- Determine the appropriate value to use in the null hypothesis. (This may be stated in the problem as a hypothesized value or may need to be deduced from the information at hand.)
- Decide whether the alternative should be one-sided or two-sided.

The point of a hypothesis test is to select one of the competing hypotheses ($H_0$ or $H_a$) as the *correct* decision. Once you have reached a conclusion, there will remain a possibility that the inference (decision) is incorrect. The idea of performing the correct statistical test, exactly following the procedure, and still possibly making an incorrect decision may be bothersome. Uncertainty, however, is a part of every statistical inference.

### How Uncertain Are Our Conclusions?

Errors can happen in two ways. In statistical terminology, we call these **Type I errors** and **Type II errors**.

In statistical tests of hypotheses, we can only control Type I errors. This fact often influences how we construct hypotheses.

Obviously, no one will be perfect when testing hypotheses—we will make mistakes. Thus, we have the Type I and Type II errors to help us define the type of mistakes that we make when testing hypotheses. The probability of a Type I error is denoted by the Greek symbol $\alpha$. Similarly, the probability of a Type II error is denoted by the Greek symbol $\beta$. When testing hypotheses, the table below illustrates the four situations.

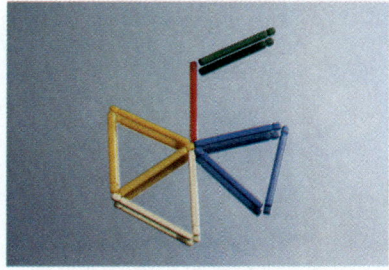

### A Proof by Contradiction?

Imagine a circumstance in which there are two conflicting propositions of which only one can be true. If you could prove that one of these is not true, then implicitly, the other is the truthful proposition. This is called a proof by contradiction and there are many famous examples in mathematics. This method of proof dates back to Aristotle.

Some important concepts that have been classically proven by contradiction include:

- Irrationality of the square root of 2,
- The length of the hypothenuse of a triangle is less than the sum of the lengths of the two remaining sides,
- No smallest rational number greater than 0.

Hypothesis testing is similar to a proof by contradiction in that the null is assumed to be true unless there is overwhelming evidence to the contrary. Frequently, the aim of the hypothesis test is to disprove the null hypothesis, hence the connection to the proof by contradiction.

### Definition

**Type I Error**

A **Type I error** occurs by rejecting $H_0$ when in fact $H_0$ is the correct choice.

### Definition

**Type II Error**

A **Type II error** occurs by failing to reject $H_0$ when in fact $H_a$ is the correct choice.

### The Trial of the Pyx

A very early instance of the hypothesis testing approach occurred in England in the twelfth century and is mysteriously named the Trial of the Pyx. There was a lot of concern over the weight of the gold and silver coins distributed by the London Mint. The king would provide gold and silver to the Mint where the metals were melted down and turned into coins. To ensure the coins were meeting the proper standards, every few months a trial would occur in front of a panel of judges, where recently minted coins selected seemingly at random were weighed—placed inside a box called a Pyx—to make sure they were within a fixed tolerance.

The null hypothesis in this test is that the weight of the coins equals the target weight. The alternative hypothesis would be that the weight of the coins is not equal to the target weight.

> ### Definition
> #### Level of the Test or Significance Level of the Test
> The probability of a Type I error is denoted by $\alpha$ (the Greek letter alpha), and is referred to as the **level of the test**, or **significance level of the test**.

> ### Definition
> #### Probability of a Type II Error
> The **probability of a Type II error** is denoted by $\beta$ (the Greek letter beta). Unfortunately, the value of $\beta$ depends on the actual value of the population parameter and thus cannot be determined unless the actual value of the population parameter is known, which is rarely the case.

### True State of Nature

|  | $H_0$ is True | $H_0$ is False |
|---|---|---|
| Fail to Reject $H_0$ | Correct Decision | Type II Error $\beta$ |
| Reject $H_0$ | Type I Error $\alpha$ | Correct Decision |

(left margin label: **Decision**)

The more serious type of error depends on the situation. Suppose you are a manufacturer that makes bolts that are to be three inches in length. If the bolts are too long or too short, they are deemed unsatisfactory to the customer. The null hypothesis is that the average length of the bolts is three inches. With this scenario, a Type I error occurs when we conclude that the bolts are not three inches in length when they actually are three inches. Committing a Type I error in this situation is costly to the manufacturer because believing that the bolts are not three inches in length will require them to stop production to unnecessarily search for a problem that does not exist. A Type II error occurs when the manufacturer concludes that the bolts are three inches in length when, in fact, they are not three inches in length. In this situation, committing a Type II error would have the manufacturer shipping bolts to consumers that are not being made to specifications while believing that they are of the required length. This type of error is also costly to the manufacturer by sending faulty bolts to the customer.

Another scenario could be associated with a police department utilizing resources to monitor traffic. Suppose on a specific stretch of road, the speed limit is 65 mph. However, the Chief of Police believes that drivers tend to speed more than 10 mph, on average, above the speed limit. In this scenario, a Type I error would be that the Chief of Police concludes that the average speed limit of drivers on this stretch of road is at least 10 mph over the speed limit when, in fact, traffic is averaging 65 mph which is the speed limit. The result of this error is that the Chief is allocating resources (police cars monitoring traffic) unnecessarily to find speeders. Committing a Type II error would be such that the Chief concludes that the average speed is 65 mph when, in fact, the average speed is significantly more than 65 mph on this stretch of road. The result of committing a Type II error is that the Chief is not allocating resources where they should be (specifically, on this stretch of road) and thus, more accidents are occurring due to vehicles traveling at high rates of speed.

Thus, it's not always easy to determine which error is worse. Determining which error is worse will always depend on the situation, the cost, or the point of view of the researcher. It is also up to the researcher to select the value of $\alpha$ prior to the start of the hypothesis test.

In examining the relationship between $\alpha$ and $\beta$, a natural question which might arise is:

*Why not make both $\alpha$ and $\beta$ as small as possible?*

Unfortunately, the probability of making a Type I error, $\alpha$, and the probability of making a Type II error, $\beta$, are inversely related. Thus, the smaller we make the probability of making a Type I error, the more likely we are to make a Type II error. In general, choose the largest level of $\alpha$ which is tolerable to avoid unnecessarily increasing the probability of making a Type II error.

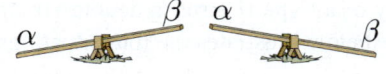

## A Procedure for Testing a Hypothesis

A procedure for testing a hypothesis is given below. In examining this procedure, you will notice we have already discussed the first two steps. The next four steps assist us in defining the **decision rule**, which is a criterion used to determine whether the null or the alternative hypothesis will be chosen. As you look over these steps, do not be overly concerned if you do not understand everything. You will learn by following the examples.

### Steps in the Test of a Hypothesis

**Step 1:** Determine the null hypothesis. In this process, select the appropriate statistical measure, such as the population mean, proportion, or variance.

**Step 2:** Determine the alternative hypothesis and whether it should be one-sided or two-sided.

**Step 3:** Select the appropriate test statistic based on the information at hand and the assumptions you are willing to make.

**Step 4:** Determine the critical value of the test statistic. Two factors must be considered.

1. The type of alternative hypothesis: two-sided, one-sided left, one-sided right. If the alternative hypothesis is two-sided, the hypothesis test will be **two-tailed**. If the alternative hypothesis is one-sided left, the hypothesis test will be a **left-tailed** or **lower-tailed** test. If the alternative hypothesis is one-sided right, the hypothesis test will be a **right-tailed** or **upper-tailed** test.

2. The specification of $\alpha$, the significance level of the test.

**Step 5:** Collect the sample data and compute the value of the test statistic.

**Step 6:** Make the decision and state the conclusion in terms of the original question.

- If the value of the test statistic is in the rejection region, reject the null hypothesis in favor of the alternative.

- If the value of the test statistic is not in the rejection region, fail to reject the null hypothesis.

### NOTE

Note at **Step 4** there are two options; you can find the critical value of the test statistic or the *P*-value of the test statistic. Both methods will always produce equivalent results; meaning, the decision regarding the hypothesis test will always be the same with both methods. We will often cover both methods in an example to illustrate this. Even though we may show a critical value and a *P*-value, only one of these is required to make the decision to reject or fail to reject the null hypothesis. You or your instructor may have a preference of one method over another.

### Type I and Type II Errors in the Trial of the Pyx

If the coins did in fact weigh less than they were intended to, the currency would become debased, and the Mint would be making a profit because they would be pocketing some of the metals they should be turning into coins. If the coins weighed more than they were supposed to, someone could collect these overweight coins, and sell them back to the Mint for a profit. Either way, the king is not happy that someone besides him is able to profit. And in those days, if the king is not happy, there is a high likelihood of important body parts being involuntarily cut off. So, if the coins are found to be off from the standard value, it could mean serious consequences for the head of the Mint.

The hypotheses are set up such that a Type I error implies that the coins were believed to be off from the standard value, when in fact they were meeting the standard. A Type II error would mean that the panel has believed the coins to be matching the weight standard, when in fact they are overweight or underweight. This is the preferred formulation for the head of the Mint, as the Type I error is the one he would like to control so that he does not lose his extremities for no reason! A Type II error would serve the Mint well, as it means the coins were in error, but it went undetected.

## 10.1 Exercises

### Basic Concepts

1. What is a hypothesis?

2. What is the first step in the test of a hypothesis?

3. Describe the common elements present in all hypothesis tests.

4. Summarize the difference between the null and alternative hypotheses.

5. Define and give an example of a one-sided alternative. How does this differ from a two-sided alternative?

6. What is the connection between one and two-sided alternatives and one and two-tailed tests?

7. Is there a way to be absolutely certain your decision is correct when performing a hypothesis test? Explain.

8. What are the three important things you must be able to do in order to be successful at formulating hypothesis testing problems?

9. Describe a Type I error.

10. Describe a Type II error.

11. Explain how Type I and Type II errors influence the construction of a hypothesis.

12. Can both Type I and Type II errors be controlled in the hypothesis testing procedure? Explain.

13. What is the level of the test?

14. Why is a Type II error difficult to express numerically?

## Exercises

15. The town mayor believes that more than 47% of the town residents favor annexation of a new community. How should she formulate the hypotheses to test her claim?

16. A chocolate chip manufacturer would like to know if its bag filling machine works correctly at the 450 gram setting. Assume the population is normally distributed. How should the manufacturer formulate the hypotheses to test if the bags are being overfilled?

17. A hospital director believes that 29% of the lab reports contain errors and feels an audit is required. A sample of 300 reports found 99 errors. Is there sufficient evidence at the 0.02 level to refute the hospital director's claim? State the null and alternate hypotheses for this test.

18. An engineer has designed a valve that will regulate water pressure on an automobile engine. The valve was tested on 140 engines and the mean pressure was 7.7 lbs/square inch. Assume the variance is known to be 0.64. If the valve was designed to produce a mean pressure of 7.9 lbs/square inch, is there sufficient evidence at the 0.10 level that the valve performs below the specifications? State the null and alternative hypotheses.

19. Using traditional methods it takes 10.9 hours to receive a basic flying license. A new license training method using Computer Aided Instruction (CAI) has been proposed. Set up the hypotheses to test the claim at the 0.05 level that the new technique performs differently than the traditional method. State the null and alternative hypotheses.

20. Our environment is very sensitive to the amount of ozone in the upper atmosphere. The level of ozone normally found is 7.6 parts/million (ppm). A researcher believes that the current ozone level is higher than the normal level. Set up the hypotheses to test the researcher's claim.

21. An automobile manufacturer claims that their van has a 56.8 miles/gallon (MPG) rating. An independent testing firm has been contracted to test the MPG for this van. After testing 99 vans they found a mean MPG of 56.4 with a standard deviation of 1.2 MPG. Is there sufficient evidence at the 0.025 level that the vans underperform the manufacturer's MPG rating? State the null and alternative hypotheses for this test.

22. A restaurant owner believes that tardiness has become a problem with her staff. In past years around 5% of her employees showed up late for their shift. She believes that the current rate is much higher. How should she formulate the hypotheses to test her belief?

23. For the following situations, develop the appropriate $H_0$ and $H_a$ and state what the consequences would be for Type I and Type II errors.

    a. The Standard Tire Company has introduced a new tire in Europe that will be guaranteed to last at least 30,000 kilometers. Standard Tire has hired an independent agency to determine if there is overwhelming evidence that their tires will last through the warranty period.

    b. Mrs. Russell, head product tester for Hathaway Tool Corporation, is testing a newly designed series of bar hooks. The hooks have been designed to give way if they get too hot. The previous design gave way at 240 degrees. Develop a test to determine if the newly designed hooks give way at a higher temperature than the previous design.

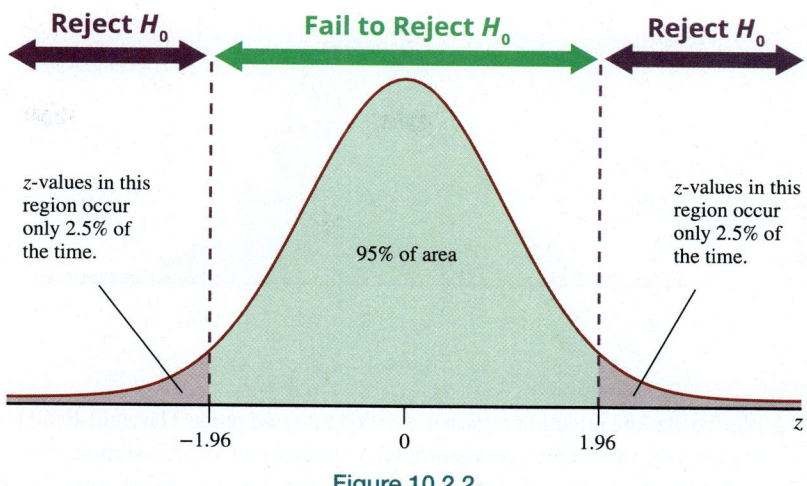

**Figure 10.2.2**

z-values in this region occur only 2.5% of the time.

It is important to recognize that this strategy attempts to define the notion of close (for $\alpha = 0.05$) as being within 1.96 standard deviation units. Using this definition of closeness, if $\bar{x}$ is within 1.96 standard deviations of the hypothesized value, it is close enough to the hypothesized value in the null hypothesis to have occurred from ordinary sampling variability; otherwise it is too far to have happened by chance, and we reject the null hypothesis. In our example, 1.96 is a **critical value** of $z$.

**A Decision Strategy**

Suppose we claim that any value of the $z$-test statistic less than 1.96 in absolute value represents ordinary sampling variability of $\bar{x}$. Any value of $|z|$ greater than or equal to 1.96 is attributed to a false null hypothesis.

**Definition**

**Critical Value**

The **critical value** is the value to which the test statistic is compared to determine whether to reject or fail to reject the null hypothesis. Traditionally, if the absolute value of the test statistic is greater than the critical value, we reject the null hypothesis.

**Step 4:** Determine the critical value of the test statistic. Two factors must be considered.

1. The type of alternative hypothesis: two-sided, one-sided left, one-sided right. If the alternative hypothesis is two-sided, the hypothesis test will be **two-tailed**. If the alternative hypothesis is one-sided left, the hypothesis test will be a **left-tailed** or **lower-tailed** test. If the alternative hypothesis is one-sided right, the hypothesis test will be a **right-tailed** or **upper-tailed** test.

2. The specification of $\alpha$, the significance level of the test.

In determining the critical value of the test statistic, we must take into account whether the alternative hypothesis is one-sided or two-sided, the level of the test, and the distribution of the test statistic.

**Continuing Example 10.2.1**

**Performing a Hypothesis Test for the Mean Cost of Textbooks**

| Table 10.2.1 - Critical Values of the $z$-Test Statistic for Two-Sided Alternatives | | |
|---|---|---|
| Level of the Test | Definition of Ordinary Variability | $z_{\alpha/2}$ |
| 0.20 | 80% interval around hypothesized mean | 1.28 |
| 0.10 | 90% interval around hypothesized mean | 1.645 |
| 0.05 | 95% interval around hypothesized mean | 1.96 |
| 0.01 | 99% interval around hypothesized mean | 2.575 |

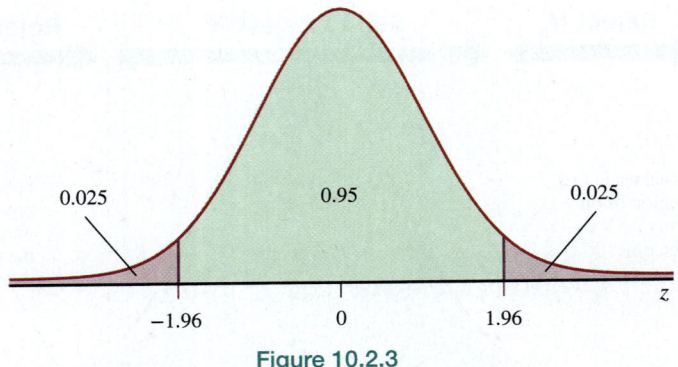

**Figure 10.2.3**

## Technology

For instructions on how to conduct a hypothesis test using the *z*-test statistic, please visit stat.hawkeslearning.com and navigate to **Discovering Business Statistics, Second Edition > Technology Instructions > Hypothesis Testing > z-Test.**

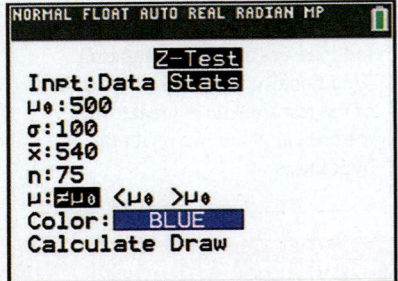

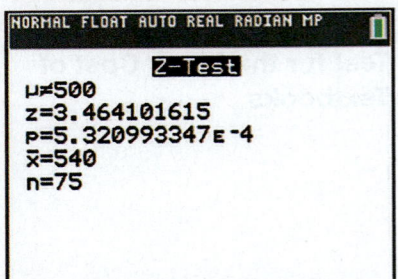

Notice that the rejection region is divided into two parts. The right-hand rejection region indicates above average amounts of money spent on textbooks per semester and, conversely, the left-hand rejection region indicates below average amounts of money spent on textbooks per semester. Also observe that the probability associated with the level of the test is divided equally between the two rejection regions, each region receiving 0.025. If the probabilities in these two regions are added together, $0.025 + 0.025 = 0.05$, the sum equals the level of the test ($\alpha$).

The level of the test can be thought of as a tolerance for rareness. The level of the test defines a rejection region which can be thought of as an intolerance zone. If the test statistic falls in this "intolerance" region, then the data have produced a sample mean that has in turn produced a test statistic that is too rare to have occurred by chance if the null hypothesis were true.

Essentially, we lose faith in the null hypothesis; the data have cast too much doubt. Consequently, the null hypothesis is rejected. The "fail to reject" zone can be interpreted as a zone of ordinary sampling variation given the null is true. For two-tailed tests with the level of the test set to $\alpha = 0.05$, the investigator is stating that any value of the test statistic which falls in a 95% interval around the hypothesized mean represents ordinary sampling variability. For the *z*-test statistic, this corresponds to a critical value of 1.96 standard deviation units (see Table 10.2.1). If the test statistic falls into the ordinary sampling variation zone, the null hypothesis will not be rejected.

**Rejection Region for $\alpha$ = 0.05**

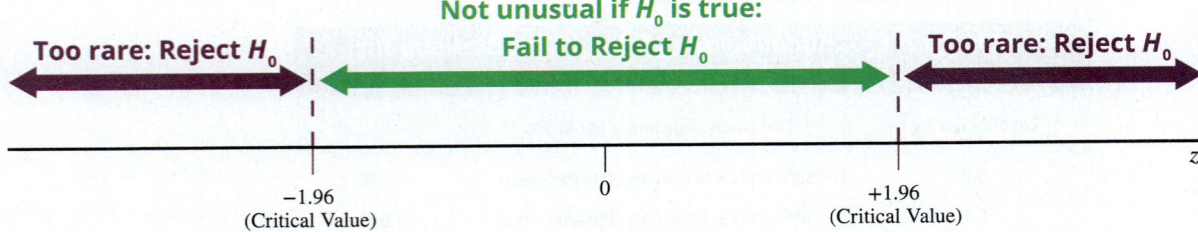

**Step 5:** Collect the sample data and compute the value of the test statistic.

A random sample of 75 students revealed a mean of $540 per semester spent on textbooks at the local university. As discussed earlier, the *z*-statistic is given by

$$z = \frac{\bar{x} - \mu_0}{\frac{\sigma}{\sqrt{n}}} = \frac{540 - 500}{\frac{100}{\sqrt{75}}} \approx 3.46 .$$

The sample mean is 3.46 standard deviations from the hypothesized value. Is the sample mean too far away from the value of the mean specified in the null hypothesis for us to believe that the null is true?

**Step 6:** Make the decision and state the conclusion in terms of the original question.

The critical values of the test statistic are ±1.96. If the null hypothesis were true, observing a value of $z$ equal to or larger in absolute value than 1.96 would occur only 5% of the time.

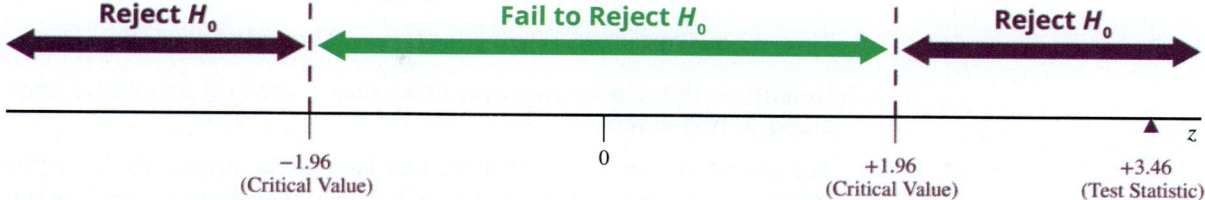

The test statistic $z = 3.46$ implies $\bar{x}$ is 3.46 standard deviation units from the mean which is substantially more than 1.96 standard deviations from the hypothesized value. The decision must be to reject $H_0$. The sample mean is too far from the hypothesized value for us to believe the difference is caused by ordinary sampling variation. Essentially, $\bar{x}$ exceeds the tolerance for rareness that we have imposed by setting $\alpha = 0.05$.

*Conclusion and Interpretation:* There is significant evidence at the 0.05 level that the students at the local university do not spend, on average, $500 per semester on textbooks. It would be tempting to conclude that the students at the local university spend more since the sample mean is greater than the national average. However, we did not test a hypothesis for spending more; we tested whether the students at the local university were spending *more or less* than the national average, and the conclusion must be consistent with the hypothesis tested.

### Stating a Conclusion

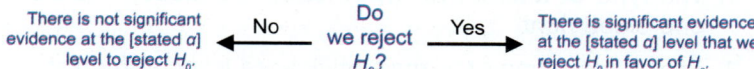

One temptation might be to reformulate the hypothesis and test to see if the students are spending more on textbooks with the same data. This would be practically unethical. If you already know the $z$-value for a given set of data, you should not formulate a hypothesis and use the data to "support" that hypothesis.

Part of every hypothesis test is arbitrary. The level of the test, $\alpha$, is arbitrarily defined by the researcher. As you have seen (e.g., Table 10.2.1) in the last example, the level of the test affects the decision rule. Since the level of the test is arbitrarily set, the decision rule inherits the arbitrariness.

> **Example 10.2.2**
>
> **Performing a Hypothesis Test for the Mean Life of a PMP**

The manufacturer of a portable music player (PMP) has shown that the average life of the product is 72 months with a standard deviation of 12 months. The manufacturer is considering using a new parts supplier for the PMPs and wants to test that the new hard drives will increase the life of the PMP. Before manufacturing the PMPs on a large scale, the manufacturer sampled 200 PMPs and found the average life to be 78 months. Test the hypothesis using $\alpha = 0.01$ that the new hard drives will increase the life of the PMPs. Assume that the standard deviation of the new PMPs is the same as the standard deviation of the older model.

### SOLUTION

**Step 1:** Determine the null hypothesis. In this process, select the appropriate statistical measure, such as the population mean, proportion, or variance.

The hypothesis is fairly straightforward. The null hypothesis is that the new hard drives do not change the average life of the PMPs, which is 72 months.

Once again, the population mean is the parameter of interest. If the average life of the new PMP is greater than the average life of the older model, all other things being equal, the new PMP will be better.

Let

$$\mu = \text{mean life of the new PMP.}$$

Thus, the null hypothesis is written as $H_0$: $\mu = 72$ months.

**Step 2:** Determine the alternative hypothesis and whether it should be one-sided or two-sided.

The alternative hypothesis is that the new hard drives increase the life of the PMPs. The goal is to determine if there is sufficient evidence to conclude that the new PMPs will last longer, on average, than the older model. The alternative hypothesis will be one-sided, and this will be a one-tailed test.

Thus, the alternative hypothesis is written as $H_a$: $\mu > 72$ months.

**Step 3:** Select the appropriate test statistic based on the information at hand and the assumptions you are willing to make.

The problem states that the standard deviation of the new model is equivalent to the standard deviation of the older model, which is known. Therefore, the standard deviation of the new PMP's life is assumed to be known. Since the sample is relatively large, the Central Limit Theorem asserts that the distribution of $\bar{x}$ should be approximately normal. Because the standard deviation is known and the sample size is greater than or equal to 30, the $z$-test statistic can be used.

**Step 4:** Determine the critical value of the test statistic.

Two factors must be considered.

1. The type of alternative hypothesis: two-sided, one-sided left, one-sided right.

2. The specification of $\alpha$, the significance level of the test.

Since the alternative hypothesis is one-sided, we want to know if there is evidence that $\mu$ (the mean life of the new PMP model) is greater than the hypothesized value, $\mu_0$ (72 months, which is the standard life of the existing PMP model). If $\bar{x}$ (the mean life of the 200 PMP sample) is much larger than $\mu_0$, that would suggest that $H_a$ is a more reasonable choice than $H_0$. Of course, through ordinary sampling variation, $\bar{x}$ could be larger than $\mu_0$. How much greater than $\mu_0$ does $\bar{x}$ have to be in order for us to believe that the mean life of the new PMP is greater than 72 months ($H_a$)? To locate the critical value of $z$ for a one-sided "greater than" alternative, find the value of $z$ that cuts off $\alpha$ worth of probability in the right-hand tail of the distribution. The critical values for "greater than" alternative hypotheses are given in Table 10.2.2 for typical values of $\alpha$.

| Table 10.2.2 - Critical Values of the $z$-Test Statistic for One-Sided (Greater Than) Alternatives | | |
|:---:|:---:|:---:|
| **Level of the Test** | **Definition of Ordinary Variability** | $z_\alpha$ |
| 0.20 | Lower 80% of the distribution | 0.84 |
| 0.10 | Lower 90% of the distribution | 1.28 |
| 0.05 | Lower 95% of the distribution | 1.645 |
| 0.01 | Lower 99% of the distribution | 2.33 |

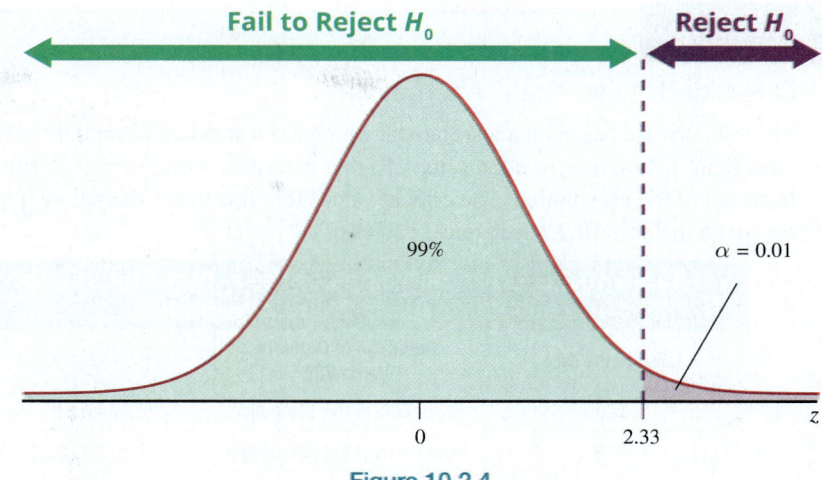

**Figure 10.2.4**

The $z$-test statistic has a standard normal distribution; so critical values are determined from the probability distribution of the standard normal distribution. The alternative hypothesis is one-sided ($H_a$: $\mu > 72$) and $\alpha = 0.01$. From Table 10.2.2, we find that the appropriate critical value for the test is 2.33. This critical value means that if $\bar{x}$ is 2.33 standard deviations larger than $\mu_0$, then $H_0$ should be rejected in favor of $H_a$. However, if $H_0$ is true, ordinary variation would cause $\bar{x}$ to be greater than or equal to 2.33 about 1% of the time. Thus, 1% of the time $H_0$ will be rejected when it is true, a Type I error.

**Step 5:** Collect the sample data and compute the value of the test statistic.

Suppose a random sample of 200 PMPs revealed a mean life of 78 months. The resulting $z$-test statistic is

$$z = \frac{\bar{x} - \mu_0}{\dfrac{\sigma}{\sqrt{n}}} = \frac{78 - 72}{\dfrac{12}{\sqrt{200}}} \approx 7.07$$

indicating that the sample mean life of the new PMP model is more than 7 standard deviations larger than the life of the current PMP model. It is quite unlikely that ordinary sampling variation could have caused the sample mean to be more than 7 standard deviations from the hypothesized value.

**Step 6:** Make the decision and state the conclusion in terms of the original question.

Since the test statistic is larger than the critical value, the decision is to reject $H_0$ in favor of $H_a$.

*Conclusion and Interpretation:* There is sufficient evidence at the $\alpha = 0.01$ level to conclude that the life of the new PMP model is superior to the life of the older model.

In the previous two examples, we have considered a two-sided alternative and a one-sided "greater than" alternative. When considering a one-sided "less than" alternative, the procedure is very similar to that of a one-sided "greater than" alternative. The null hypothesis will be rejected if the calculated value of the test statistic, $z$, is less than or equal to the critical value, $-z_\alpha$, for the specified level of significance.

## Procedure

### One-sided "Less Than" Alternatives

For tests that are based on a test statistic which has a standard normal distribution and a "less than" alternative, find the value of $z$ that cuts off $\alpha$ worth of probability in the left-hand tail of the distribution. The critical values for "less than" alternative hypotheses are given in Table 10.2.3 for typical values of $\alpha$.

| Table 10.2.3 - Critical Values of the $z$-Test Statistic for One-Sided (Less Than) Alternatives | | |
|---|---|---|
| **Level of the Test** | **Definition of Ordinary Variability** | $-z_\alpha$ |
| 0.20 | Upper 80% of the distribution | −0.84 |
| 0.10 | Upper 90% of the distribution | −1.28 |
| 0.05 | Upper 95% of the distribution | −1.645 |
| 0.01 | Upper 99% of the distribution | −2.33 |

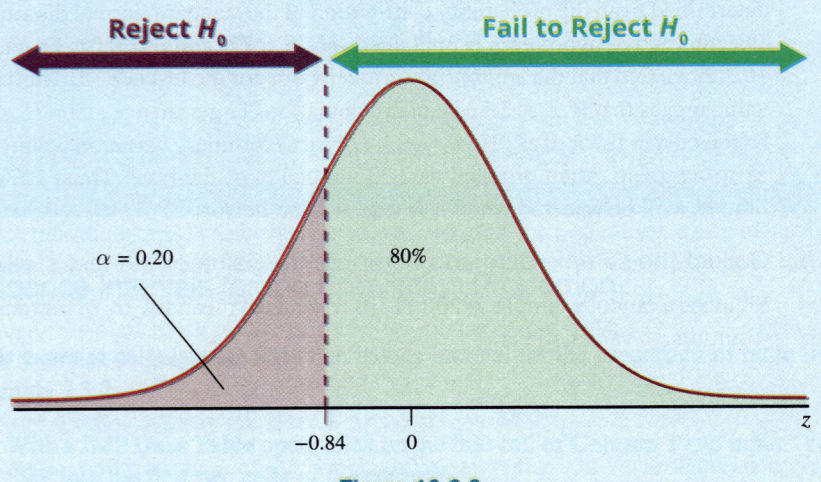

**Figure 10.2.6**

The figure above shows the rejection region for a test with a one-sided "less than" alternative hypothesis, and a significance level of 0.20. For this test, the null hypothesis will be rejected if the calculated value of the test statistic is less than or equal to −0.84.

## P-Values

We have been examining the classical approach to hypothesis testing. This approach results in a conclusion to reject or fail to reject the null hypothesis. What is missing from this type of conclusion is the degree to which the data confirm the inference. For example, suppose we are testing a hypothesis with the following rejection region.

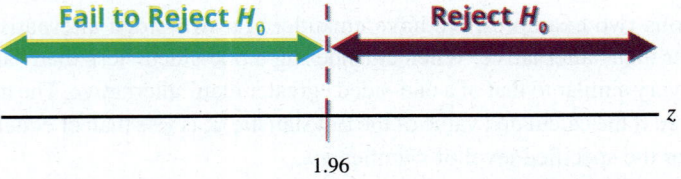

If the test statistic were $z = 2.03$, it would fall into the rejection region (but just barely), and the null hypothesis would be rejected.

Compare this to a test statistic of $z = 3.09$, which falls far into the rejection region and results in the same conclusion, rejecting the null hypothesis. Clearly, the sample data that produced a test statistic of 3.09 provide much stronger evidence against the null hypothesis than the sample that produced a test statistic of 2.03. Expressing the difference in the strength of our conclusion is accomplished using **P-values**.

For a left-tailed (or lower-tailed) test, the P-value is given by $P(z \le z_0)$, where $z_0$ is the observed value of the test statistic. For a right-tailed (or upper-tailed) test, the P-value is given by $P(z \ge z_0)$, where $z_0$ is the observed value of the test statistic. For two-tailed tests, the P-value is given by $2P(z \ge |z_0|)$ , where $z_0$ is the observed value of the test statistic. P-values can be determined using the tables in Appendix A, or technology such as a TI-84 Plus calculator, Microsoft Excel, JMP, or Minitab. The Discovering Technology section at the end of this chapter shows how to calculate P-values for $z$ and $t$ test statistics.

The following graph displays the application of the definition to the test statistics $z = 2.03$ and $z = 3.09$.

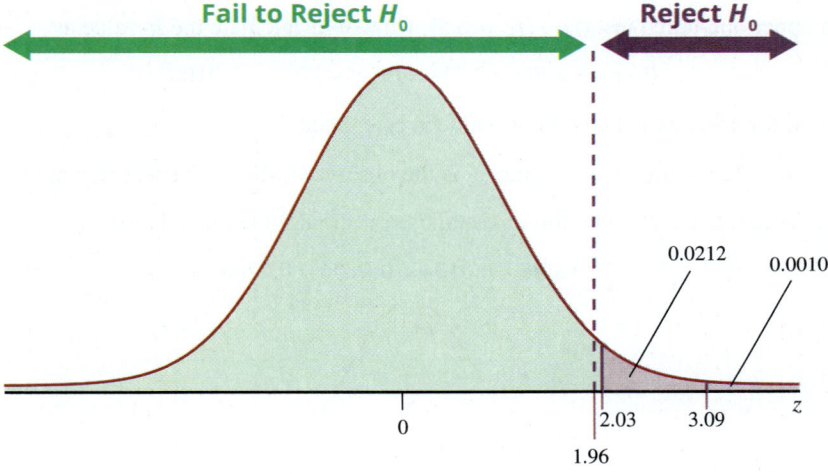

Figure 10.2.7

If the null hypothesis is true, the probability of observing a test statistic greater than or equal to 2.03 is 0.0212, which is the P-value. One interpretation of the P-value is that if 10,000 researchers were to draw similar size samples, assuming the null hypothesis is true, ordinary sampling variation would cause roughly 212 researchers to obtain samples that would produce test statistics as large or larger than 2.03.

Compare this to the test statistic 3.09. If the null hypothesis is true, the chance of sampling variation causing a test statistic greater than or equal to 3.09 is 0.0010, which is the P-value. If 10,000 researchers drew similar size samples, ordinary sampling variation would cause only about 10 of the researchers to observe test statistics as large or larger than 3.09. The further the test statistic penetrates the rejection region, the more confidence the researcher can place in his or her conclusions.

Many statistical computer programs provide P-values in their output. It is a fairly easy task to utilize these values to perform a classical hypothesis test.

## Performing a Hypothesis Test Using P-Values

- If the computed P-value is less than or equal to $\alpha$, reject the null hypothesis in favor of the alternative.

- If the computed P-value is greater than $\alpha$, fail to reject the null hypothesis.

### Definition

**P-value**

A **P-value** is the probability of observing a value of a test statistic as extreme or more extreme than the one observed, assuming the null hypothesis is true.

### An English Taste Test

Muriel Bristol was an English woman who, like many English women, often enjoyed an afternoon cup of tea. So developed were her tea taste buds, that she insisted she could tell the difference between a cup of tea which had the tea poured before the milk, from a cup which had the milk poured before the tea. Muriel Bristol, a psychologist, also happened to be a coworker of statistician Ronald Fisher at the Rothamsted Experimental Station. Hearing her rather bold boasts of tasting finesse, Fisher devised an experiment. Eight cups of tea were prepared, four in which the tea was poured first and four in which the milk was first added. The cups were presented to Mrs. Bristol in a random order. She had the opportunity to taste each cup and then identify four which were prepared in the same fashion. The pressure was high. Using a significance level of $\alpha = 0.05$, her ability would only be acknowledged if she did not make a mistake. (The probability of missing none given she randomly guessed is P-value = 170 ≈ 0.014, but the probability of missing only one given she randomly guessed is P-value = 17/70 ≈ 0.243.) But Muriel knew her tea. She correctly categorized all eight cups.

The rationale for this approach is that if the *P*-value is less than or equal to $\alpha$, we have observed a test statistic that is more unusual than the level of the test, which defines how rare an event we must observe in order to reject the null hypothesis. Conversely, if the *P*-value is larger than $\alpha$, the test statistic is not sufficiently rare (it could have been caused by ordinary sampling variation) to reject the null hypothesis.

## *P*-Values for a Two-Sided Hypothesis Test

So far, we have only discussed calculating a *P*-value for a one-tailed hypothesis test. To compute the *P*-value for a two-tailed test, simply double the tail probability of the test statistic. For example, suppose that we are testing the following hypotheses.

$$H_0: \mu = 5 \qquad \text{(null hypothesis)}$$

$$H_a: \mu \neq 5 \qquad \text{(alternative hypothesis)}$$

Further, suppose the computed test statistic is $z = 2.31$.

For an upper one-tailed test (i.e., $H_a: \mu > 5$), we would calculate the *P*-value as

$$P\text{-value} = P(z \geq 2.31) = P(z \leq -2.31) = 0.0104.$$

However, for a two-tailed test, the *P*-value is calculated as

$$P\text{-value} = 2P(z \geq |z_0|), \text{ where } z_0 \text{ is the observed value of the test statistic.}$$

Thus, to compute the *P*-value for a two-tailed test, we double the tail area.

$$P\text{-value} = 0.0104 + 0.0104 = 0.0208$$

### ∞ Technology

*P*-values for a *z*-test can be found using the standard normal tables or from the output from a hypothesis test using the *z*-test statistic as described previously. Also recall from Chapter 7 that a probability given a *z*-score can be obtained from technology using the cumulative normal distribution. For instructions on finding the *P*-value for a given *z*-test statistic, go to stat.hawkeslearning.com and navigate to **Discovering Business Statistics, Second Edition > Technology Instructions > Normal Distributions > Normal Probability (cdf).**

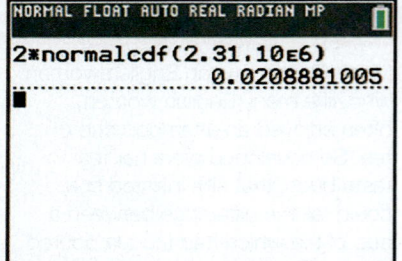

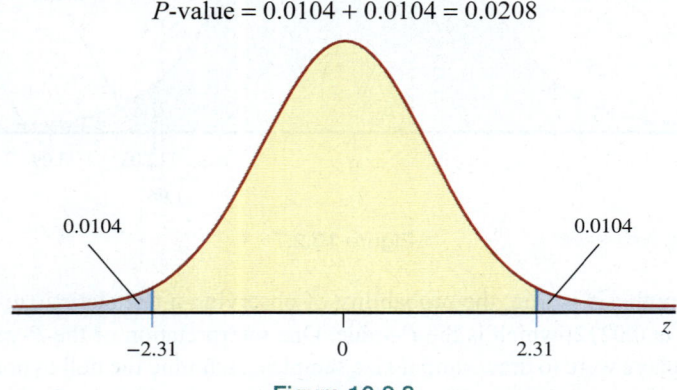

**Figure 10.2.8**

The *P*-value of 0.0208 is the likelihood of observing a value of the test statistic greater than or equal to 2.31 or less than or equal to −2.31 given the null hypothesis is true.

### Steps in the Test of a Hypothesis Using the *P*-Value Approach

**Step 1:** Determine the null hypothesis. In this process, select the appropriate statistical measure, such as the population mean, proportion, or variance.

**Step 2:** Determine the alternative hypothesis and whether it should be one-sided or two-sided.

**Step 3:** Select the appropriate test statistic based on the information at hand and the assumptions you are willing to make.

**Step 4:** Collect the sample data and compute the value of the test statistic.

## Steps in the Test of a Hypothesis Using the *P*-Value Approach (cont.)

**Step 5:** Calculate the *P*-value using the test statistic. For the sake of this setup, suppose we are performing a hypothesis test on the mean when $\sigma$ is known. Thus, the observed value of the test statistic will be

$$z_0 = \frac{\bar{x} - \mu_0}{\frac{\sigma}{\sqrt{n}}}.$$

The *P*-value will be found as follows:

a.  If the alternative hypothesis is $H_a\colon \mu < \mu_0$, then the *P*-value is calculated as $P(z \le z_0)$.

b.  If the alternative hypothesis is $H_a\colon \mu > \mu_0$, then the *P*-value is calculated as $P(z \ge z_0)$.

c.  If the alternative hypothesis is $H_a\colon \mu \ne \mu_0$, then the *P*-value is calculated as $2P(z \ge |z_0|)$.

Note that in this example, we are performing a test on the population mean with $\sigma$ known. If the parameter that is being tested changes so that the test statistic changes, then we would use the appropriate test statistic in the above probability statements.

**Step 6:** Make the decision and state the conclusion in terms of the original question. The decision-making process is as follows.

- If the *P*-value is less than or equal to $\alpha$, reject the null hypothesis in favor of the alternative hypothesis.

- If the *P*-value is greater than $\alpha$, fail to reject the null hypothesis.

### Different *P*-Values for Different Folks

In particle physics, the standard for "discovery" is a *P*-value less than 0.0000003. That is the probability which corresponds to observing a value that is at least 5 standard deviations from the mean for a one-tailed test. Particle physicists consider a *P*-value less than 0.003, which is the probability of observing a value at least 2.75 standard deviations from the mean, "evidence of a particle"—an encouraging result, but not "discovery".

The common significance levels we have used in this book are $\alpha = 0.05$ and $\alpha = 0.01$ which correspond to a value at least 1.645 and 2.33 standard deviations from the mean, respectively. This goes to show you that there is not any one significance level that everyone agrees on. Difference disciplines have different comfort levels with the idea of significance.

## ✐ 10.2 Exercises

### Basic Concepts

1.  What is the rationale for the *z*-statistic?

2.  What are the three key questions to be asked in the hypothesis testing procedure in order to determine which test statistic is appropriate?

3.  Describe the distribution of the *z*-test statistic.

4.  What are critical values? How do critical values influence the decision rule in the hypothesis testing procedure?

### Exercises

5.  Determine the critical value(s) of the test statistic for each of the following tests for the population mean when the population standard deviation is known.

    a.  Left-tailed test, $\alpha = 0.01$        c.  Two-tailed test, $\alpha = 0.05$

    b.  Right-tailed test, $\alpha = 0.10$

6.  Determine the critical value(s) of the test statistic for each of the following tests for the population mean when the population standard deviation is known.

    a.  Left-tailed test, $\alpha = 0.05$        c.  Two-tailed test, $\alpha = 0.08$

    b.  Right-tailed test, $\alpha = 0.02$

7. A random sample of 1000 observations produces a sample mean of 53.5 with a population standard deviation of 5.3. Test the hypothesis that the mean is not equal to 55 at $\alpha = 0.05$.

8. A random sample of 200 observations indicate a sample mean of 4117 with a population standard deviation of 300. Test the hypothesis that the mean is greater than 4100 at $\alpha = 0.01$.

9. The head of the Veterans Administration has been receiving complaints from a Vietnam veterans' organization concerning disability checks. The organization claims that checks are continually late. The checks are supposed to arrive no later than the tenth of each month. The administrator randomly selects 100 disabled veterans and measures the arrival time in relation to the tenth of the month for each check. If the check arrives early, it receives a negative value. For example, if the check arrives on the eighth of the month, it is measured as −2. If the check arrives on the twelfth of the month, it is measured as +2.

   a. What statistical measure should you use in your statement of hypothesis?

   b. Formulate hypotheses to test the veterans' organization's claim.

   c. Suppose in the sample of 100 disabled veterans receiving checks, the average number of days late was 1.2 with a population standard deviation of 1.4. Calculate the test statistic for your hypothesis.

   d. If the test is conducted at the 0.05 level, construct the decision rule for the test statistic.

   e. Is there overwhelming evidence at the 0.05 level that the checks arrive late?

   f. If you are the head of the Veterans Administration, what is your conclusion?

10. Hurricane Andrew swept through southern Florida causing billions of dollars of damage. Because of the severity of the storm and the type of residential construction used in this semitropical area, there was some concern that the average claim size would be greater than the historical average hurricane claim of $24,000. Several insurance companies collaborated in a data gathering experiment. They randomly selected 84 homes and sent adjusters to settle the claims. In the sample of 84 homes, the average claim was $27,500 with a population standard deviation of $2400.

   a. What is the population being studied?

   b. What statistical measure should you use in your hypothesis?

   c. State your hypotheses.

   d. Test the hypothesis at the 0.01 level.

   e. Is there overwhelming evidence (at the 0.01 level) that home damage is greater than the historical average? Write your conclusion in the context of the original problem.

11. A retail computer store is considering offering a two-year service warranty, instead of its current one-year plan. In order to do this, they must determine the average service costs for their systems in the second year of operation. A committee of technicians, sales, and management staff believe that the average repair cost in the second year should be approximately $50. Seventy-five customers who purchased machines between two and four years earlier are randomly selected. Tracking the service needs of these customers reveals an average service cost of $38 with a population standard deviation of $10.

   a. What is the population being studied?

   b. What variable is being measured in this problem?

   c. What level of measurement does the variable possess?

   d. Test the committee's claim at the 0.10 level.

   e. What concerns might you have about the data that were collected?

12. In preparation for upcoming wage negotiations with the union, the managers for the Bevel Hardware Company want to establish the time required to assemble a kitchen cabinet. A first line supervisor believes that the job should take 45 minutes on average to complete. A random sample of 125 cabinets has an average assembly time of 47 minutes with a population standard deviation of 10 minutes.

    a. Is there overwhelming evidence to contradict the first line supervisor's belief at a 0.05 significance level? Make your conclusion using the $P$-value approach.

    b. What is the lowest average assembly time that would allow the union to conclude that the supervisor is incorrect?

13. The Better Business Bureau has received several complaints that a flour company is underfilling its five pound bags of flour. The Bureau randomly selects 750 bags of flour and determines the weight of each bag. The sample average weight of the bags is 4.80 pounds with a population standard deviation of 0.15 pounds.

    a. Is there overwhelming evidence at the 0.01 level that the bags are underfilled?

    b. What is the lowest average bag weight that would allow the Bureau to conclude that the bags are underfilled?

14. A horticulturist working for a large plant nursery is conducting experiments on the growth rate of a new shrub. Based on previous research, the horticulturist feels the average daily growth rate of the new shrub is 1 cm per day. A random sample of 45 shrubs has an average growth of 0.90 cm per day with a population standard deviation of 0.30 cm. Will a test of hypothesis at the 0.05 significance level support the claim that the growth rate is less than 1 cm per day?

15. Del Valley Foods requires that corn supplied for canning must weigh more than 5 ounces per ear. South Valley Farms claims that the corn they supply meets the required specifications. 200 ears of corn are selected at random from a delivery. The sample has a mean of 5.01 ounces and a population standard deviation of 0.30 ounce. Will a test of hypothesis at $\alpha = 0.10$ support South Valley Farms' claim?

16. Government regulations restrict the amount of pollutants that can be released to the atmosphere through industrial smokestacks. To demonstrate that their smokestacks are releasing pollutants below the mandated limit of 5 parts per billion pollutants, REM Industries collects a random sample of 300 readings. The mean pollutant level for the sample is 4.85 parts per billion with a population standard deviation of 0.30 parts per billion. Do the data support the claim that the average pollutants produced by REM Industries are below the mandated level at a 0.01 significance level?

17. The director of the IRS has been flooded with complaints that people must wait more than 45 minutes before seeing an IRS representative. To determine the validity of these complaints, the IRS randomly selects 400 people entering IRS offices across the country and records the times that they must wait before seeing an IRS representative. The average waiting time for the sample is 55 minutes with a population standard deviation of 15 minutes.

    a. What is the population being studied?

    b. Are the complaints substantiated by the data at $\alpha = 0.10$?

18. The manufacturer of Brand X floor polish is developing a new polish that it hopes will dry faster than the competition's polish. The competition's polish is advertised to have an average drying time of 10 minutes. A random sample of 1000 Brand X polishes has an average drying time of 9.3 minutes with a population standard deviation of 0.5 minute. Based on the data, can the manufacturer conclude that the drying time for Brand X is faster than the competition's brand at a 0.05 significance level?

19. For each of the following combinations of the *P*-value and $\alpha$, decide whether you would reject or fail to reject the null hypothesis.

     **a.** *P*-value = 0.0935, $\alpha = 0.10$      **c.** *P*-value = 0.0545, $\alpha = 0.01$

     **b.** *P*-value = 0.0311, $\alpha = 0.05$      **d.** *P*-value = 0.0489, $\alpha = 0.05$

20. Consider the following hypothesis tests for the population mean. Compute the *P*-value for each test and decide whether you would reject or fail to reject the null hypothesis at $\alpha = 0.05$.

     **a.** $H_0: \mu = 15, H_a: \mu > 15, z = 1.58$

     **b.** $H_0: \mu = 1.9, H_a: \mu < 1.9, z = -2.25$

     **c.** $H_0: \mu = 100, H_a: \mu \neq 100, z = 1.90$

21. Consider the following hypothesis tests for the population mean. Compute the *P*-value for each test and decide whether you would reject or fail to reject the null hypothesis at $\alpha = 0.01$.

     **a.** $H_0: \mu = 10, H_a: \mu > 10, z = 2.00$

     **b.** $H_0: \mu = 82, H_a: \mu < 82, z = -2.45$

     **c.** $H_0: \mu = 100, H_a: \mu \neq 100, z = 2.70$

# 10.3 Testing a Hypothesis about a Population Mean, $\sigma$ Unknown

The hypothesis testing strategy from the previous section assumes that the population standard deviation, $\sigma$, is known. However, in most instances, the population standard deviation is just as *unknown* as the population mean. Despite the added uncertainty, the general approach to testing a hypothesis is the same provided that it is reasonable to *assume* that the population from which you are sampling is normal. Some of the technical details concerning the distribution of the test statistic change, since the sample standard deviation, *s*, will be used in place of the population standard deviation, $\sigma$. This modification will cause a change in the distribution of the test statistic.

## Formula

### *t*-Test Statistic

If the standard deviation of the population is unknown, but the distribution of the population is assumed to be normal, then the test statistic is given by

$$t = \frac{\bar{x} - \mu_0}{\frac{s}{\sqrt{n}}}$$

where
*n* is the sample size,
$\bar{x}$ is the sample mean,
$\mu_0$ is the hypothesized value of the population mean, and
*s* is the sample standard deviation.

It should be noted that this formula is only valid if $\bar{x}$ is normally distributed. The test statistic has a *t*-distribution with $n - 1$ degrees of freedom.

If we assume the population we are sampling from is normally distributed, the test statistic has a $t$-distribution. Consequently, the test statistic will be called a $t$-test statistic. The fundamental notion embodied in the test statistic has not changed. It measures how far the sample mean is from the hypothesized mean in standard deviation units.

In the previous chapter, we discussed the $t$-distribution in conjunction with estimating the population mean. **Step 3** in our hypothesis testing procedure must be modified to reflect the change in distribution of the test statistic.

---

Ally Hirsch, a website developer, has indicated to potential clients that for the sites she has developed, visitors spend an average of 45 minutes per day on the sites. One of her potential clients conducted a survey of 35 visitors to several of the sites created by the website developer and found that the average time spent on the sites was 35 minutes with a standard deviation of 7 minutes. Determine if there is sufficient evidence to conclude that the average time spent on her sites is different from what she has indicated. Conduct the test at the 0.05 significance level.

**Example 10.3.1**

**Performing a Hypothesis Test for the Mean Time Spent On Websites**

**SOLUTION**

**Step 1:** Determine the null hypothesis. In this process, select the appropriate statistical measure, such as the population mean, proportion, or variance.

In this case, the population consists of the time spent per visitor on all of the sites created by the website developer. Since the test is specified as having an average amount of time spent on the sites of 45 minutes, the parameter of interest will be the population mean. Let

$\mu$ = mean time spent on sites created by the website developer.

Thus, the null hypothesis should be written as $H_0$: $\mu = 45$ minutes which states that the average time spent on the sites is 45 minutes.

**Step 2:** Determine the alternative hypothesis and whether it should be one-sided or two-sided.

From the information given, we must assume that the client is interested in learning if the time spent on the sites is either significantly longer than 45 minutes or significantly less than 45 minutes. Thus, the alternative hypothesis will be two-sided, and this will be a two-tailed test. The alternative hypothesis should be written as $H_a$: $\mu \neq 45$ minutes which states that the average time spent on the sites is not 45 minutes.

**Step 3:** Select the appropriate test statistic based on the information at hand and the assumptions you are willing to make.

Since the standard deviation of the population is not known, the $t$-test statistic is utilized. However, to know that the test statistic possesses a $t$-distribution, we must be willing to make the assumption that the population possesses a reasonably normal distribution.

**Step 4:** Determine the critical value of the test statistic. Two factors must be considered.

1. The type of alternative hypothesis: two-sided, one-sided left, one-sided right.

2. The specification of $\alpha$, the significance level of the test.

The level of the test is specified in the problem as 0.05.

## Technology

For instructions on how to find the critical value for a *t*-distribution, please visit stat.hawkeslearning.com and navigate to **Discovering Business Statistics, Second Edition > Technology Instructions > Hypothesis Testing > Inverse *t*.**

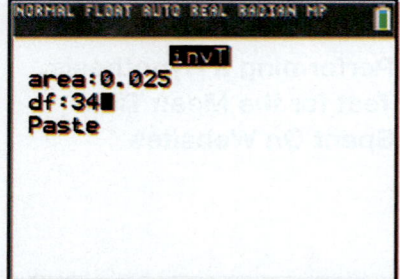

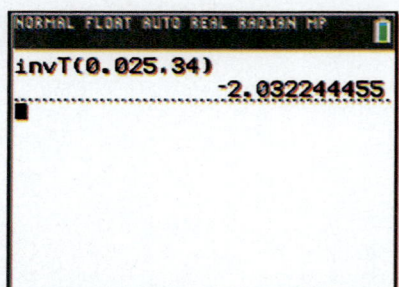

The *t*-distribution has a parameter called degrees of freedom, which must be defined in order to utilize the distribution. Since 35 visitors were sampled, the degrees of freedom are

$$df = n - 1 = 35 - 1 = 34.$$

**t-Distribution, df = 34**

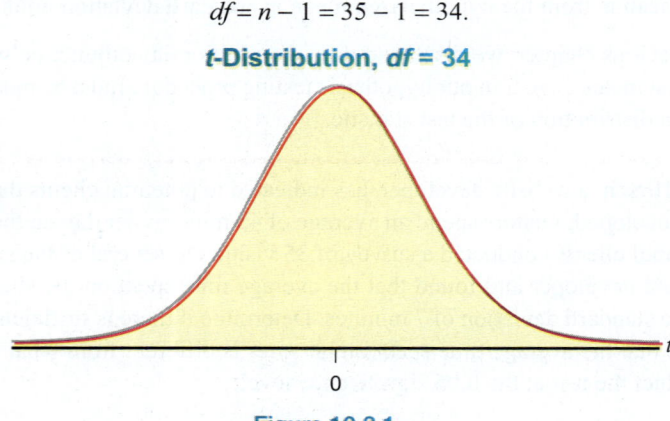

**Figure 10.3.1**

Since the alternative hypothesis ($H_a$) is two-sided, two tails of the distribution must be cut off as rejection regions. Each tail will receive half of the allotted level of the test. The value of *t* which begins the right-hand side rejection region is $t_{0.025,34}$, which corresponds to a critical value of 2.032. Since the *t*-distribution is symmetrical, the left-hand side rejection region begins at −2.032.

**t-Distribution, df = 34**

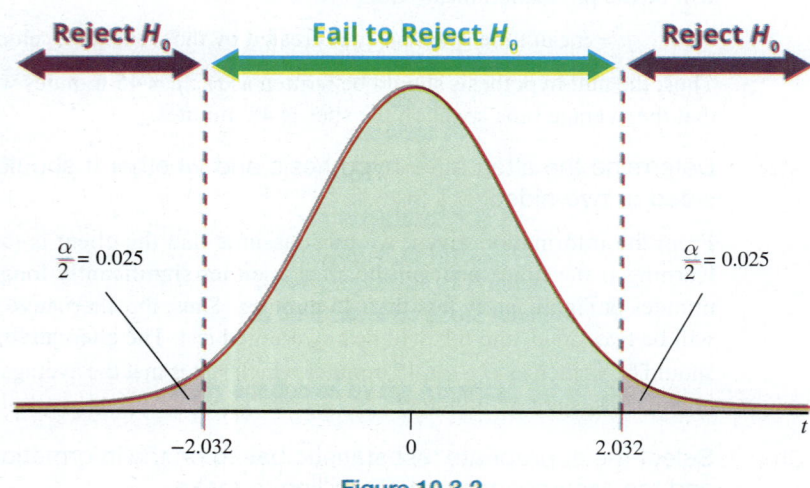

**Figure 10.3.2**

What the figure says is that it is unlikely (only occurs 5% of the time) that ordinary sampling variability will cause $\bar{x}$ to differ as much as 2.032 standard deviation units from the hypothesized mean of 45.

**Step 5:** Collect the sample data and compute the value of the test statistic.

Since 35 visitors were sampled and the mean time spent on the sites was 35 minutes with a standard deviation of 7 minutes, the computed value of the test statistic is

$$t = \frac{\bar{x} - \mu_0}{\frac{s}{\sqrt{n}}} = \frac{35 - 45}{\frac{7}{\sqrt{35}}} \approx -8.452$$

The observed sample mean is more than 8 standard deviations less than the hypothesized mean. The value of the test statistic signifies a striking difference between $\bar{x}$ and the hypothesized value, $\mu_0$. It is very unlikely that the cause of such a difference would be ordinary sampling variation.

**Step 6:** Make the decision and state the conclusion in terms of the original question.

Since the test statistic falls in the rejection region, we must reject $H_0$ in favor of $H_a$.

**⌘ Technology**

For instructions on how to conduct a hypothesis test using the *t*-test statistic, please visit stat.hawkeslearning.com and navigate to **Discovering Business Statistics, Second Edition > Technology Instructions > Hypothesis Testing > *t*-Test.**

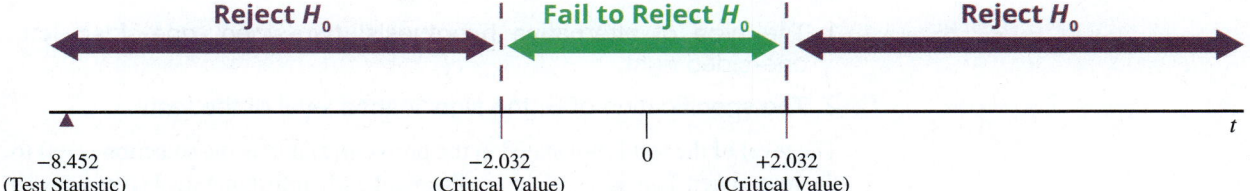

| | | |
|---|---|---|
| −8.452 (Test Statistic) | −2.032 (Critical Value)      0 | +2.032 (Critical Value) |

*Conclusion and Interpretation:* There is sufficient evidence at the $\alpha = 0.05$ level to conclude that the average time spent on the developer's sites is different from 45 minutes. Since the sample mean is much less than the indicated 45 minutes, it is tempting to conclude that the time spent on the sites is less than 45 minutes. However, we did not test the one-sided hypothesis. If the potential client would like to demonstrate that visitors spend significantly less time on the developer's sites, then they must start over by testing a one-sided hypothesis and obtaining new sample data.

---

The Alexander Bolt Company produces half-inch A-class stainless steel bolts. The specified quality standard is that they have a mean tensile strength of more than 4000 pounds per square inch (psi). These bolts are primarily used in the manufacturing of farm implements. The company is very concerned about quality and wants to be sure that its A-class product does not fall below the standard. A sample of 25 bolts is to be randomly selected from stock and tested for tensile strength. Design a test of hypothesis to determine if there is overwhelming evidence that the bolts meet the specified quality standard.

**Example 10.3.2**

**Performing a Hypothesis Test for the Mean Tensile Strength**

**SOLUTION**

**Step 1:** Determine the null hypothesis. In this process, select the appropriate statistical measure, such as the population mean, proportion, or variance.

Although there is nothing in the problem that specifically defines the population parameter of interest, there is mention of a "standard value." By comparing the mean breaking strength of the population to the hypothesized value, 4000 psi, a conclusion about the quality can be drawn.

Let

$\mu$ = mean tensile strength of the half-inch A-class stainless steel bolts.

The null hypothesis is that the half-inch A-class stainless steel bolts have a mean tensile strength of 4000 psi which can be written as $H_0$: $\mu = 4000$ psi.

**Step 2:** Determine the alternative hypothesis and whether it should be one-sided or two-sided.

If the population mean is significantly above the hypothesized value, then there would be evidence that the desired quality is being produced. Thus, the alternative hypothesis will be one-sided, and this will be a one-tailed test. The alternative hypothesis can be written as $H_a$: $\mu > 4000$ psi.

> **Reminder**
>
> The null hypothesis ($H_0$) always contains an equality. Thus, the null must contain an = .

**Step 3:** Select the appropriate test statistic based on the information at hand and the assumptions you are willing to make.

The population standard deviation is unknown and will be estimated with the sample standard deviation. Assuming the distribution of the tensile strengths is reasonably normal, the *t*-test statistic is appropriate.

**Step 4:** Determine the critical value of the test statistic. Two factors must be considered.

1. The type of alternative hypothesis: two-sided, one-sided left, one-sided right.

2. The specification of $\alpha$, the significance level of the test.

The level of the test is not stated in the problem, and thus the selection is left to the researcher. This is a typical predicament and, unfortunately, leaves us with what appears to be an arbitrary decision regarding the level of the test. What is a reasonable value for the level of the test?

In this problem, making a Type I error (rejecting the null hypothesis when it is true) means that management believes that the bolt production system is not flawed, when it is. If we set $\alpha$ too high, management will too frequently believe the quality level is being met when it is not, and poor quality bolts will not be readily detected. After all is said and done, the decision remains arbitrary. A Type I error could be calamitous, so a small level of $\alpha$ would be desirable. Let's use $\alpha = 0.01$.

Since the alternative hypothesis implies that we are interested in discovering if the mean is above the hypothesized value, the rejection region will only be on the right-hand side of the sampling distribution. Because the level of the test has been chosen to be 0.01, the interval encompassing 0.01 of the area on the right-hand side of the curve will define the rejection region. Remember that the test statistic has a *t*-distribution.

To obtain the critical value, first determine the degrees of freedom.

$$df = n - 1 = 25 - 1 = 24$$

Using Table D in Appendix A, $t_{0.01,24} = 2.492$. If the observed *t*-value is greater than or equal to 2.492 then the observed mean is presumed to be too far from the hypothesized value to have occurred from ordinary sample variation. If $t \geq 2.492$, the presumption will be that the mean tensile strength is greater than 4000 psi.

**t-Distribution, df = 24**

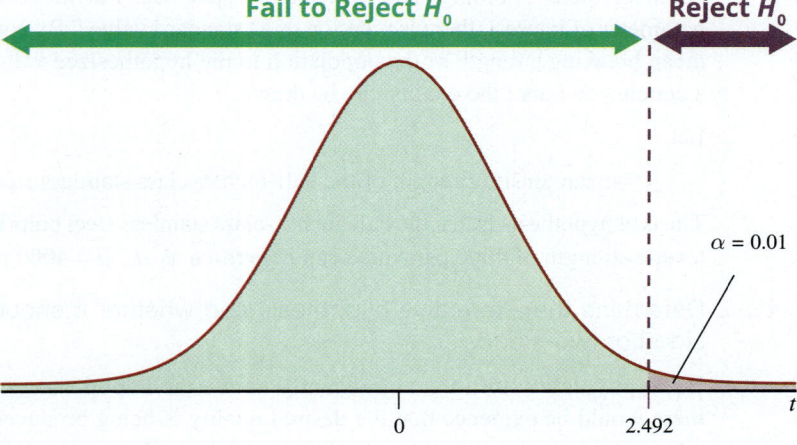

**Figure 10.3.3**

**Step 5:** Collect the sample data and compute the value of the test statistic.

Suppose that after randomly selecting and testing 25 bolts, the average tensile strength is 4014 psi, with a standard deviation of 20 psi. Symbolically this is expressed as

$$\bar{x} = 4014, \; s = 20, \text{ and } n = 25.$$

The resulting test statistic is

$$t = \frac{\bar{x} - \mu_0}{\frac{s}{\sqrt{n}}} = \frac{4014 - 4000}{\frac{20}{\sqrt{25}}} = 3.5.$$

The sample mean is 3.5 standard deviations more than the hypothesized value of the mean. Is this difference caused by ordinary sampling variation, or is this evidence of a false null hypothesis? Keep in mind that rejecting the null hypothesis ($H_0$) will be a desirable outcome since the null hypothesis states that the bolts are below quality standards.

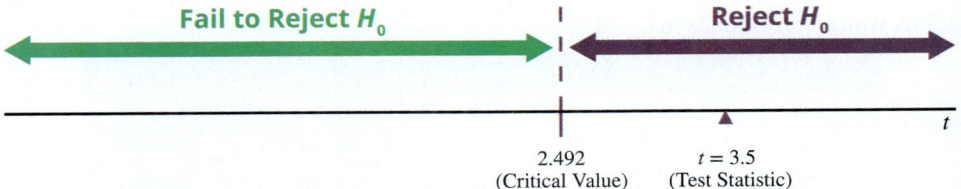

Fail to Reject $H_0$        Reject $H_0$

2.492 (Critical Value)        $t = 3.5$ (Test Statistic)

**Step 6:** Make the decision and state the conclusion in terms of the original question.

Since the test statistic falls in the rejection region, it indicates that the sample mean is too far from the hypothesized value to believe it is due to ordinary sampling variation. (In other words, $\bar{x} = 4014$ is a rare observation and has exceeded our tolerance for rareness.) The null hypothesis must be rejected in favor of the alternative hypothesis.

*Conclusion and Interpretation:* There is sufficient evidence to conclude that the bolt tensile strength is greater than 4000 psi, and there is at most a 1% chance that this conclusion is incorrect.

This conclusion is stated in a rather absolute manner. However, there is always a degree of subjectivity in any statistical conclusion. In this instance, we arbitrarily selected the value of $\alpha$, and we assumed that the distribution of the population was normal.

### Technology

For instructions on how to conduct a hypothesis test using the $t$-test statistic, please visit stat.hawkeslearning.com and navigate to **Discovering Business Statistics, Second Edition > Technology Instructions > Hypothesis Testing > $t$-Test.**

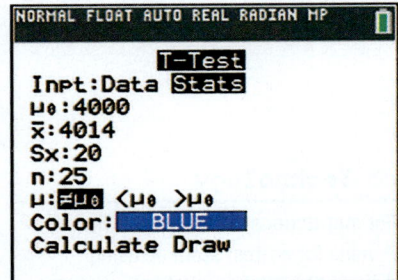

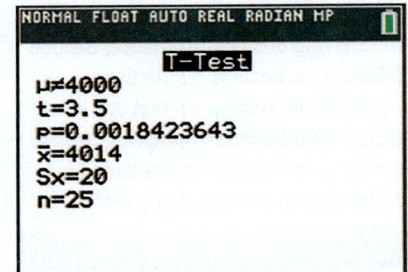

## P-Values for t-Test Statistics

Because there are numerous $t$-distributions, one for each different degree of freedom, the $t$-tables are constructed differently from the $z$-tables. The $t$-table only provides $t$-values for frequently used tail probabilities. Note in the $t$-table (Appendix A, Table D) that we only have six tail areas. Because of this limitation, in most instances the exact value of the $t$-test statistic will not be in the table. When this circumstance arises, find the closest $t$-values with the appropriate degrees of freedom that surround the test statistic. For example, suppose the value of the test statistic was

$$t = 2.40 \text{ with 17 degrees of freedom.}$$

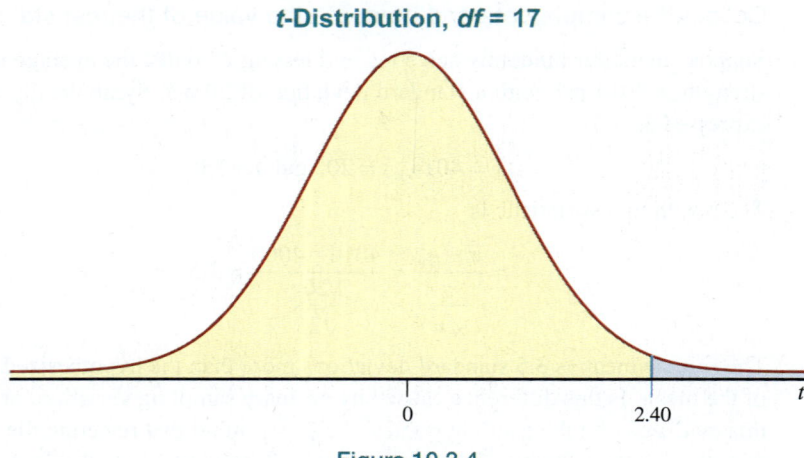

**Figure 10.3.4**

```
NORMAL FLOAT AUTO REAL RADIAN MP

tcdf(2.40,1E99,17)
                0.0140632427
■
```

"No scientific worker has a fixed level of significance at which from year to year, and in all circumstances, he rejects hypotheses; he rather gives his mind to each particular case in the light of his evidence and his ideas."

—*Ronald Fisher*

The value of the test statistic falls between 2.110, which corresponds to $\alpha = 0.025$, and 2.567, which corresponds to $\alpha = 0.01$.

| df | Area in One Tail | | | |
|---|---|---|---|---|
|  | 0.100 | 0.050 | 0.025 | 0.010 |
| 1 | 3.078 | 6.314 | 12.706 | 31.821 |
| 2 | 1.886 | 2.920 | 4.303 | 6.965 |
| 3 | 1.638 | 2.353 | 3.182 | 4.541 |
| … | | | | |
| 17 | 1.333 | 1.740 | 2.110 | 2.567 |
| 18 | 1.330 | 1.734 | 2.101 | 2.552 |
| … | | | | |

Thus, we report a bound for the *P*-value. In this case, the *P*-value is $0.01 < P\text{-value} < 0.025$. That is, the *P*-value falls between 0.01 and 0.025. If the level of the test, $\alpha$, is greater than 0.025, then our test is significant at the 0.025 level. Thus, the *P*-value would be reported as being significant at the 0.025 level, but not significant at the 0.01 level. Exact *P*-values for *t*-test statistics can be found using technology such as a TI-84 Plus calculator, Microsoft Excel, JMP, and Minitab. Methods for finding exact *P*-values can be found in the Discovering Technology section at the end of this chapter.

Because of the arbitrariness of assigning the level of the test, researchers will often report the significance of their findings as *P*-values. Determining the significance of the results is left to the researcher. There are few universal standards when it comes to evaluating the significance of a *P*-value. However, most researchers would view a *P*-value of less than 0.01 as significant and a *P*-value of greater than 0.10 as insignificant. Interpreting *P*-values between 0.01 and 0.10 will depend on the circumstances. However, once the *P*-value gets above 0.05, the conclusion starts to become unconvincing.

## Practical Significance versus Statistical Significance

A hypothesis is rarely exactly true. As the sample size becomes larger, the likelihood of rejecting the null becomes greater. As a result, sometimes even though the results of a hypothesis test are statistically significant, they may not be practically significant.

---

**Example 10.3.3**

**Determining Practical vs. Statistical Significance**

Suppose that Fife Police Department was concerned about speeds in the town school zone, especially during the beginning and end of the school day. The posted speed limit is 25 mph and the deputy officer, Benny, believes that the speed is being exceeded. To test his belief, Benny randomly samples (via radar) 2000 vehicles in the school zone over a 30-day period and finds that the average speed is 25.01 mph with a standard deviation of 0.10 mph.

## SOLUTION

To test Benny's claim, the hypothesis test is carried out with the following hypotheses.

$$H_0: \mu = 25 \text{ mph}$$

$$H_a: \mu > 25 \text{ mph}$$

The resulting test statistic is

$$z = \frac{\bar{x} - \mu_0}{\frac{s}{\sqrt{n}}} = \frac{25.01 - 25}{\frac{0.10}{\sqrt{2000}}} = 4.47.$$

The $P$-value for this test is 0.000004 and suggests that the test statistic is extremely rare, if the null hypothesis is true. Considering that a test statistic this large would result from ordinary sampling variation in only about 4 in a million samples, we should reject the null hypothesis ($H_0: \mu = 25$ mph) in favor of the alternative. With Benny being a "strictly by the book" officer, he would conclude that the speeds are significantly higher than 25 mph in the school zone and the department should allocate extra resources in that area to reduce speeds. However, from a practical perspective, there isn't much difference (other than random variation) between 25 mph and 25.01 mph, and it isn't likely to be detected by radar. So, despite the "statistical significance" of the test, the practical significance is negligible.

It is important to keep the practical significance of the hypothesis test in mind when making conclusions. This is one of the reasons why in the 6-step procedure for hypothesis testing, **Step 6** is to state the conclusion in terms of the original problem. This step may help shed light on the practical significance of the hypothesis test.

# 📝 10.3 Exercises

## Basic Concepts

1. Suppose a null hypothesis were rejected at $\alpha = 0.05$. Would it be rejected at 0.10? Explain.

2. Suppose a null hypothesis were rejected at $\alpha = 0.05$. Would it be rejected at 0.01? Explain.

3. What is a $P$-value?

4. Discuss how $P$-values are used in the test of a hypothesis.

5. Describe the difference between statistical significance and practical significance.

6. Give an example of a situation in which results could be statistically significant but not practically significant.

## Exercises

7. Determine the critical value(s) of the test statistic for each of the following tests for the population mean where the population standard deviation is unknown and the assumption of normality is satisfied.

   a. Left-tailed test, $\alpha = 0.01$, $n = 15$

   b. Right-tailed test, $\alpha = 0.10$, $n = 20$

   c. Two-tailed test, $\alpha = 0.05$, $n = 8$

8. Determine the critical value(s) of the test statistic for each of the following tests for the population mean where the population standard deviation is unknown and the assumption of normality is satisfied.

 a. Left-tailed test, $\alpha = 0.005$, $n = 12$

 b. Right-tailed test, $\alpha = 0.025$, $n = 5$

 c. Two-tailed test, $\alpha = 0.10$, $n = 25$

9. Consider the following random sample of size six from a normal population. Based on the sample, perform a hypothesis test to test the claim that the mean of the population is not equal to 10 at $\alpha = 0.05$.

| 10 | 15 | 12 | 9 | 11 | 10 |

10. Consider the following random sample of size eight from a normal population. Based on the sample, perform a hypothesis test to test the claim that the mean of the population is greater than 100 at $\alpha = 0.05$. Calculate the $P$-value for this hypothesis test.

| 100 | 150 | 120 | 90 | 95 | 110 | 100 | 80 |

11. Consider the following random sample of size seven from a normal population. Based on the sample, perform a hypothesis test to test the claim that the mean of the population is less than 0.5 at $\alpha = 0.10$. Calculate the $P$-value for this hypothesis test.

| 0.3 | 0.5 | 0.4 | 0.6 | 0.5 | 0.4 | 0.4 |

12. NarStor, a computer disk drive manufacturer, claims that the average time to failure for its hard drives is 14,400 hours. You work for a consumer group that has decided to examine this claim. Technicians ran 16 drives continuously for three years. Recently the last drive failed. The time to failure (in hours) are given below.

| Time Until Failure (Hours) | | | | | | | |
|------|------|------|------|------|------|------|------|
| 330 | 620 | 1870 | 2410 | 4620 | 6396 | 7822 | 8102 |
| 8309 | 12,882 | 14,419 | 16,092 | 18,384 | 20,916 | 23,812 | 25,814 |

 a. What is the population being studied?

 b. What is the variable being measured?

 c. What level of measurement does the variable possess?

 d. Conduct a hypothesis test to determine whether there is overwhelming evidence that the average time to failure is less than the manufacturer's claim. Use $\alpha = 0.01$.

 e. What assumption did you make in performing the test in part d.?

13. The admitting office at Sisters of Mercy Hospital wants to be able to inform patients of the average level of expenses they can expect per day. Historically, the average has been approximately $1240. The office would like to know if there is evidence of an increase in the average daily billing. Twenty randomly selected patients have an average daily charge of $1491 with a standard deviation of $342.

 a. What is the population being studied?

 b. Conduct a hypothesis test to determine whether there is evidence that average daily charges have increased at $\alpha = 0.10$.

 c. What assumption did you make in performing the test in part b.?

14. A supplier has agreed to provide the manager of a large hospital with light bulbs that he claims will last more than 1000 hours. Twenty-five bulbs are randomly selected and tested by the hospital's maintenance department. The sample has an average life of 1099 hours with a standard deviation of 99 hours.

    a. Perform a hypothesis test to determine whether the data support the supplier's claim at $\alpha = 0.05$.

    b. What assumption did you make in performing the test in part **a.**?

    c. What is the $P$-value for the hypothesis test performed in part **a.**?

15. The managers of a large department store wish to test reactions of shoppers to a new in-store video screen which will broadcast continuous information about the store and the items currently on sale. In past promotions, the video production company has indicated that the average shopper watched for five minutes. The managers randomly select 17 shoppers and determine how long they watch the video. The average time is 4.5 minutes with a standard deviation of 2.5 minutes.

    a. Perform a hypothesis test to determine whether there is overwhelming evidence to indicate that the shoppers will watch for less than five minutes. Use $\alpha = 0.01$.

    b. What assumption did you make in performing the test in part **a.**?

16. A group of local businessmen is thinking about developing land into a shopping mall. To evaluate the desirability of the location, they count the number of shoppers who visit the neighboring shopping center each day. A random sample of 25 days reveals a daily average of 107 shoppers with a standard deviation of 23 shoppers. The businessmen will develop the land if the average number of shoppers per day is more than 100.

    a. Based on the sample data, should the businessmen develop the land? Perform a hypothesis test and use $\alpha = 0.10$.

    b. What assumption did you make in performing the test in part **a.**?

17. The Dodge Reports are used by many companies in the construction field to estimate the time required to complete various jobs. The company has received several complaints that the time required to install 130 square feet of bathroom tile is greater than the eight hours reported in the current manual. A researcher for Dodge randomly selects 10 construction workers and determines the time required to install 130 square feet of bath tile. The average time required to install the tile for the sample is 8.5 hours with a standard deviation of 1 hour.

    a. Use a hypothesis test to determine whether the customers' complaints are substantiated by the data. Use $\alpha = 0.05$.

    b. What assumption did you make in performing the test in part **a.**?

18. Officials in charge of televising an international chess competition in South America want to determine if the average time per move for the top players has remained under five minutes over the last two years. Video tapes of matches which have been played over the two-year period are reviewed and a random sample of 50 moves are timed. The sample mean is 3.5 minutes with a standard deviation of 1.5 minutes.

    a. What is the population under study?

    b. Can the officials conclude at $\alpha = 0.05$ that the time per move is still under five minutes?

19. Buckshot Heaven is developing a new shotgun shell that they hope will have a significantly tighter pellet pattern than their competition. Twenty-five shells are tested at fifty yards. The average pellet pattern of the sample was 8.7 inches in diameter with a standard deviation of 2.0 inches. Their competitor advertises that the average pellet pattern of their shells is nine inches.

    a. Does the test completed by Buckshot Heaven support the claim that their shell pattern is tighter than the competition at a level of significance of 0.10?

    b. What assumption did you make in performing the test in part **a.**?

20. For each of the following combinations of the $P$-value and $\alpha$, decide whether you would reject or fail to reject the null hypothesis.

    a. $P$-value $= 0.0839$, $\alpha = 0.05$          c. $P$-value $= 0.0444$, $\alpha = 0.10$

    b. $P$-value $= 0.0174$, $\alpha = 0.02$          d. $P$-value $= 0.0374$, $\alpha = 0.01$

21. Consider the following hypothesis tests for the population mean. Compute the $P$-value for each test and decide whether you would reject or fail to reject the null hypothesis at $\alpha = 0.01$. See the Discovering Technology section at the end of this chapter for instructions on finding exact $P$-values for $t$-statistics.

    a. $H_0$: $\mu = 25$, $H_a$: $\mu > 25$, $t = 2.7$, $n = 15$

    b. $H_0$: $\mu = 0.85$, $H_a$: $\mu < 0.85$, $t = -2.5$, $n = 7$

    c. $H_0$: $\mu = 1000$, $H_a$: $\mu \neq 1000$, $t = 2.0$, $n = 15$

22. Consider the following small sample hypothesis tests for the population mean. Compute the $P$-value for each test and decide whether you would reject or fail to reject the null hypothesis at $\alpha = 0.05$. See the Discovering Technology section at the end of this chapter for instructions on finding exact $P$-values for $t$-statistics.

    a. $H_0$: $\mu = 120$, $H_a$: $\mu > 120$, $t = 1.5$, $n = 20$

    b. $H_0$: $\mu = 0.2$, $H_a$: $\mu < 0.2$, $t = -2.75$, $n = 18$

    c. $H_0$: $\mu = 50$, $H_a$: $\mu \neq 50$, $t = 2.4$, $n = 5$

23. A.C. Bone has developed a duck hunting boot which it claims can remain immersed for more than 12 hours without leaking. Five hundred pairs of the boots are tested and the time until first leakage is measured. The average time until first leakage for the sample is 12.25 hours with a standard deviation of 3.0 hours.

    a. Find the $P$-value to test the claim that the average time until first leakage for the hunting boot is more than 12 hours.

    b. Does this sample support A.C. Bone's claim at $\alpha = 0.10$?

24. In preparation for upcoming wage negotiations with the union, the managers for the Bevel Hardware Company want to establish the time required to assemble a kitchen cabinet. A first line supervisor believes that the job should take 45 minutes on average to complete. A random sample of 125 cabinets has an average assembly time of 47 minutes with a standard deviation of 10 minutes. Is there overwhelming evidence to contradict the first line supervisor's belief at a 0.05 significance level?

Discuss the statistical and practical significance for this problem.

25. A horticulturist working for a large plant nursery is conducting experiments on the growth rate of a new shrub. Based on previous research, the horticulturist feels the average daily growth rate of the new shrub is 1 cm per day. A random sample of 45 shrubs has an average growth of 0.90 cm per day with a standard deviation of 0.30 cm. Will a test of hypothesis at the 0.05 significance level support the claim that the growth rate is less than 1 cm per day?

    Discuss the statistical and practical significance for this problem.

26. The director of the IRS has been flooded with complaints that people must wait more than 45 minutes before seeing an IRS representative. To determine the validity of these complaints, the IRS randomly selects 400 people entering IRS offices across the country and records the times which they must wait before seeing an IRS representative. The average waiting time for the sample is 55 minutes with a standard deviation of 15 minutes. Are the complaints substantiated by the data at $\alpha = 0.10$?

    Discuss the statistical and practical significance for this problem.

27. The managers of a large department store wish to test reactions of shoppers to a new in-store video screen which will broadcast continuous information about the store and the items currently on sale. In past promotions, the video production company has indicated that the average shopper watched for five minutes. The managers randomly select 17 shoppers and determine how long they watch the video. The average time is 4.5 minutes with a standard deviation of 2.5 minutes. Perform a hypothesis test to determine whether there is overwhelming evidence to indicate that the shoppers watch for less than five minutes. Use $\alpha = 0.01$.

    Discuss the statistical and practical significance for this problem.

# 10.4 The Relationship Between Confidence Interval Estimation and Hypothesis Testing

Previously, we discussed interval estimation for the population mean and the population proportion. We know that when estimating the population mean, a $100(1 - \alpha)\%$ confidence interval for $\mu$ is given by

$$\bar{x} \pm z_{\alpha/2} \frac{\sigma}{\sqrt{n}} \, .$$

In this chapter, we have shown that the two-sided hypothesis test about the population mean $\mu$ is

$$H_0: \mu = \mu_0$$

$$H_a: \mu \neq \mu_0$$

where $\mu_0$ is some specific value of the population mean.

From the previous chapter, we know that $100(1 - \alpha)\%$ of the confidence intervals will contain $\mu$. Therefore, if we reject the null hypothesis when the confidence interval does not contain the value of $\mu_0$, we will reject the null hypothesis when it is actually true with probability of $\alpha$. You may recall that $\alpha$ represents the probability of committing a Type I error (i.e., we reject the null hypothesis when the null hypothesis is true). So, when we construct a $100(1 - \alpha)\%$ confidence interval and reject the null hypothesis when the interval does not contain $\mu_0$, this is equivalent to performing a two-tailed hypothesis test using $\alpha$ as the level of the test.

## Using a Confidence Interval to Test a Hypothesis

**Step 1:** If the hypothesis test is of the form

$$H_0: \mu = \mu_0$$
$$H_a: \mu \neq \mu_0.$$

**Step 2:** Calculate the $100(1-\alpha)\%$ confidence interval.

If the population standard deviation, $\sigma$, is known, then the confidence interval for the population mean, $\mu$, is given by

$$\bar{x} \pm z_{\alpha/2} \frac{\sigma}{\sqrt{n}}.$$

If the population standard deviation, $\sigma$, is unknown, then the confidence interval for the population mean, $\mu$, is given by

$$\bar{x} \pm t_{\alpha/2} \frac{s}{\sqrt{n}}.$$

**Step 3:** If $\mu_0$ falls within the interval, then we fail to reject the null hypothesis; however, if $\mu_0$ falls outside the calculated interval, then reject the null hypothesis in favor of the alternative.

### ∞ Technology

For instructions on how to calculate a confidence interval using the z-test statistic, please visit stat.hawkeslearning.com and navigate to **Discovering Business Statistics, Second Edition > Technology Instructions > Confidence Intervals > z-Interval.**

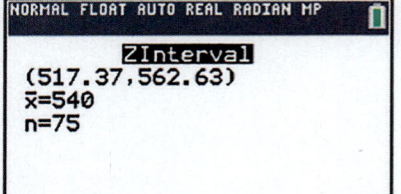

Let's look at Example 10.2.1. Recall that the hypotheses were

$$H_0: \mu = 500$$
$$H_a: \mu \neq 500.$$

The test statistic for this example was $z = 3.46$ and the null hypothesis was rejected at the 0.05 level. Note that the value $z_{\alpha/2} = z_{0.025} = 1.96$. Calculating a 95% confidence interval for $\mu$, we get

$$\bar{x} \pm z_{\alpha/2} \frac{\sigma}{\sqrt{n}}$$

$$540 \pm 1.96 \frac{100}{\sqrt{75}}$$

$$517.37 \text{ to } 562.63.$$

Because the hypothesized value ($\mu = 500$) does not fall in this interval, we reject the null hypothesis, which is consistent with the earlier conclusion. Although we presented this demonstration using the confidence interval for the population mean, the same relationship exists for other population parameters.

### Example 10.4.1

**Performing a Hypothesis Test Using a Confidence Interval on Average Income**

In a disagreement with his roommate, Sir believed that the average salary between New Yorkers was no different than that of Virginians (which is believed to be $74,231). Performing an internet search, he learned that from a sample of 3,351 New Yorkers the average salary was $73,707 with a standard deviation of $2,583. Calculate a 99% confidence interval to determine if there is a difference in average salaries between New Yorkers and Virginians.

#### SOLUTION

This is a situation where Sir believes the population average salary of Virginians is $74,231 and he would like to compare that with the average salary of New Yorkers to see if it is different. If Sir were performing a hypothesis test, his null and alternative hypotheses would be written as

$$H_0: \mu = \$74{,}231$$
$$H_a: \mu \neq \$74{,}231.$$

At the 1% level, a 99% confidence interval for the mean New Yorker salary is obtained by

$$\bar{x} \pm t_{df,\alpha/2} \frac{s}{\sqrt{n}}.$$

We know that the sample size is 3,351, the sample mean, $\bar{x} = \$73{,}707$, and the sample standard deviation is \$2,583. We also know that $\alpha = 0.01$. Thus, $t_{3350,0.005}$ = 2.577. Using these data, we find that the 99% confidence interval is given by $73{,}707 \pm (2.577)(2{,}583) / \sqrt{3{,}351}$. Performing the calculations, you will get a confidence interval of (\$73,592, \$73,822). Note that the hypothesized mean of \$74,231 does not fall in the confidence interval above. Thus, we reject the null hypothesis and conclude that the average salary of residents of New York is significantly different than the average salary of residents of Virginia.

## 10.4 Exercises

### Basic Concepts

1.  How can a confidence interval be used to test a hypothesis?

### Exercises

2.  AAA Controls makes a switch that is advertised to activate a warning light if the power supplied to a machine reaches 100 volts. A random sample of 250 switches is tested and the mean voltage at which the warning light occurs is 98 volts with a sample standard deviation of 3 volts. Using the confidence interval approach, test the hypothesis that the mean voltage activation is different from AAA Controls' claim at the 0.05 level.

3.  Researchers studying the effects of diet on growth would like to know if a vegetarian diet affects the height of a child. The researchers randomly selected 12 vegetarian children that were six years old. The average height of the children is 42.5 inches with a standard deviation of 3.8 inches. The average height for all six-year-old children is 45.75 inches.

    a.  Using confidence intervals, test to determine whether there is overwhelming evidence at $\alpha = 0.05$ that six-year-old vegetarian children are not the same height as other six-year-old children.

    b.  What assumption did you make in performing the test?

4.  High-power experimental engines are being developed by the Stevens Motor Company for use in its new sports coupe. The engineers have calculated the maximum horsepower for the engine to be 600 HP. Sixteen engines are randomly selected for horsepower testing. The sample has an average maximum HP of 620 with a standard deviation of 50 HP.

    a.  Use the confidence interval approach to determine whether the data suggest that the average maximum HP for the experimental engine is significantly different than the maximum horsepower calculated by the engineers. Use a significance level of $\alpha = 0.01$.

    b.  What assumption did you make in performing the test?

5. The nutrition label for Oriental Spice Sauce states that one package of sauce has 1190 milligrams of sodium. To determine if the label is accurate, the FDA randomly selects two hundred packages of Oriental Spice Sauce and determines the sodium content. The sample has an average of 1167.34 milligrams of sodium per package with a sample standard deviation of 252.94 milligrams.

   a. Calculate a 99% confidence interval for the mean sodium content in Oriental Spice Sauce.

   b. Using the confidence interval approach, is there evidence that the sodium content is different than the nutrition label states?

6. Officials in charge of televising an international chess competition in South America want to determine if the average time per move for the top players has remained at five minutes over the last two years. Video tapes of matches which have been played over the two-year period are reviewed and a random sample of 50 moves are timed. The sample mean is 3.5 minutes. Assume the population standard deviation is 1.5 minutes. Using the confidence interval approach, test the hypothesis that the average time per move is different from 5 minutes at a 0.01 significance level.

7. In example 9.2.2 of Chapter 9, we found the 95% confidence interval for mean $\mu$, the population average completion time for the stage in the production process, in minutes, to be (20.36, 26.54).

$$n = 10, \ \bar{x} = 23.45, \ s = 4.32$$

   Conduct a hypothesis test, using a 5% level of significance, to see if the population average completion time for the stage in the production process, in minutes, differs from the following values.

   a. The null and the alternate hypotheses are $H_0$: $\mu = 18.27$ versus $H_1$: $\mu \neq 18.27$.

   b. The null and the alternate hypotheses are $H_0$: $\mu = 24.96$ versus $H_1$: $\mu \neq 24.96$.

   c. The null and the alternate hypotheses are $H_0$: $\mu = 29.53$ versus $H_1$: $\mu \neq 29.53$.

8. The owner of an upscale restaurant in Atlanta, Georgia wanted to study the dining characteristics of her customers. She found that in a random sample of 290 customers, 60 purchased dessert. Find a 98% confidence interval for the proportion of customers who purchased dessert. Use this confidence interval to test if the proportion of customers who purchase dessert differs from 25%. Use a 2% level of significance.

9. The chief purchaser for the State Education Commission is reviewing test data for a metal link chain which will be used on children's swing sets in elementary school playgrounds. The average breaking strength for a sample of 50 pieces of chain is 5000 pounds. Based on past experience, the breaking strength of metal chains is known to be normally distributed with a standard deviation of 100 pounds. Estimate the actual mean breaking strength of the metal link chain with 99% confidence. Use this confidence interval to test if the mean breaking strength of the metal link chain is different from 5020 pounds. Use a 1% level of significance.

**10.** An FDA representative randomly selects 8 packages of ground chuck from a grocery store and measures the fat content (as a percent) of each package. The rating measurements are given below.

| Fat Contents | | | | | | | |
|---|---|---|---|---|---|---|---|
| 13% | 12% | 14% | 17% | 15% | 16% | 18% | 15% |

    **a.** Assuming that the population distribution of the fat content is approximately normal, construct a 90% confidence interval for the true mean fat content of all the packages of ground beef.

    **b.** Use the confidence interval in part (a) to test if the true mean fat content of all the packages of ground beef differs from 17.24%. Use a 10% level of significance.

**11.** A hospital would like to determine the mean length of stay for its patients having abdominal surgery. A sample of 15 patients revealed a sample mean of 6.4 days and a sample standard deviation of 1.4 days. Assume that the lengths of stay are approximately normally distributed.

    **a.** Construct a 95% confidence interval for the mean length of stay for patients with abdominal surgery.

    **b.** Use the confidence interval in part (a) to test if the mean length of stay for patients having abdominal surgery differs from 5.4 days. Use a 5% level of significance.

# 10.5 Testing a Hypothesis about a Population Proportion

The topic that we will develop in this section will be a hypothesis testing approach for categorical values (nominal data). The inferences that we will make with these data will concern one population. We will use the information in the sample proportion ($\hat{p}$) to test hypotheses about the population proportion, $p$.

Testing hypotheses about a population proportion could involve a variety of problems.

- What fraction of a student's grades will be A's?
- What fraction of graduating seniors obtain jobs with starting salaries in excess of $38,000?
- What fraction of products that a company produces are defective?
- What fraction of the voters favor the incumbent in the next election?
- What fraction of the customers who purchase a Ford Focus are extremely satisfied?
- What fraction of the time will a baseball player get a hit?
- What fraction of the time will a drug be successful in treating a specific disorder?

## Developing the Test

Testing a hypothesis concerning a population proportion is nearly identical to testing a hypothesis about a population mean. The major changes in the procedure include the use of the population proportion ($p$) in the formulation of the hypotheses rather than the population mean ($\mu$), and the calculation of the test statistic. Let's try an example.

**Example 10.5.1**

**Performing a Hypothesis Test for the Proportion of Stock Increases**

Brad is a novice at trading stocks and feels that if the Dow Jones Industrial Average (DJIA) is up on a given day, then it's likely that his portfolio will also be up, regardless of the stocks in his portfolio. In a random sample of 520 stocks on the New York Stock Exchange (NYSE), 350 increased on days that the DJIA increased. Brad would like to know if there is sufficient evidence to conclude that more than 60% of stocks in the NYSE increase on days that the DJIA increases.

### SOLUTION

**Step 1:** Determine the null hypothesis. In this process, select the appropriate statistical measure, such as the population mean, proportion, or variance.

Since the problem concerns the percentage of stocks on the NYSE that increase on days that the DJIA increases, the appropriate statistical measure will be a proportion.

Let $p$ = percentage of the stocks on the NYSE that increase on days when the DJIA increases.

The **null hypothesis** is that 60% of the stocks on the NYSE increase on days when the DJIA increases. The null hypothesis can be written as $H_0$: $p = 0.60$.

Note that $p = 60\%$ is equivalent to $p = 0.60$ so that the null hypothesis can be written as $H_0$: $p = 0.60$.

**Step 2:** Determine the alternative hypothesis and whether it should be one-sided or two-sided.

Since the problem is concerned with finding evidence that the percentage is greater than 60%, the formation of the hypothesis will be one-sided. The alternative hypothesis should be written as $H_a$: $p > 0.60$, which represents that more than 60% of the stocks increase on days when the DJIA increases.

**Step 3:** Select the appropriate test statistic based on the information at hand and the assumptions you are willing to make.

If $np_0 \geq 5$ and $n(1 - p_0) \geq 5$, the appropriate test statistic is given by

$$z = \frac{\hat{p} - p_0}{\sqrt{\dfrac{p_0(1 - p_0)}{n}}}.$$

The test statistic measures how far the sample proportion, $\hat{p}$, is from the hypothesized value of the population proportion, $p_0$, measured in standard deviation units. This is exactly the same concept used in testing hypotheses about a population mean.

**Step 4:** Determine the critical value of the test statistic. Two factors must be considered.

1. The type of alternative hypothesis: two-sided, one-sided left, one-sided right.

2. The specification of $\alpha$, the significance level of the test.

Since the level of the test is not specified in the problem, let's use $\alpha = 0.05$.

The test statistic, $z$, has a normal distribution with a mean of zero and a standard deviation of one. So, designating the decision rule is exactly like what has been done in previous problems. The test is a one-tailed test, and we decided to set the level of the test at 0.05.

The decision rule will be to reject $H_0$: $p = 0.60$ if the value of $z$ is greater than or equal to 1.645. In essence, we are saying that ordinary sampling variation might account for a $\hat{p}$ up to 1.645 standard deviations larger than the hypothesized

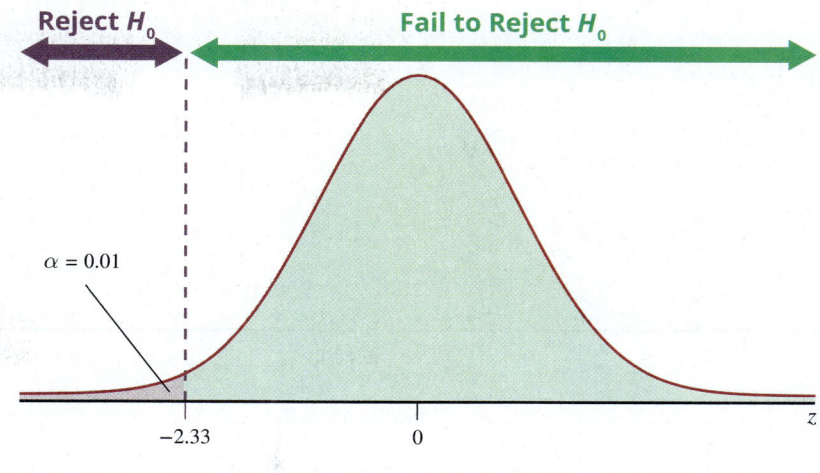

**Figure 10.5.2**

From Table 10.2.3, we know that the critical value of the test statistic is −2.33 (because it's a lower one-sided alternative hypothesis). The rejection region is the shaded area in the above graph. The decision will be to reject the null hypothesis if the value of the test statistic is less than or equal to −2.33.

**Step 5:** Collect the sample data and compute the value of the test statistic.

In a random sample of 60 online shoppers, 15 indicated that they did not have a good experience with the sites and thus, did not purchase anything. The value of $\hat{p}$ is given by

$$\hat{p} = \frac{15}{60} = 0.25, \text{ and}$$

$$z = \frac{\hat{p} - p_0}{\sqrt{\dfrac{p_0(1-p_0)}{n}}} = \frac{0.25 - 0.39}{\sqrt{\dfrac{0.39(1-0.39)}{60}}} \approx -2.22 \, .$$

The $z$-value indicates that the sample proportion is 2.22 standard deviation units below the hypothesized proportion. Could this have happened by ordinary sampling variation, or is this overwhelming evidence that the null hypothesis is not true, and the alternative should be selected?

**Step 6:** Make the decision and state the conclusion in terms of the original question.

Note that the value of $z$ does not fall in the rejection region. Thus, we fail to reject $H_0$: $p = 0.39$. There is not sufficient evidence (at the 0.01 level) to reject the null hypothesis. Another way of thinking about this is that the sample proportion was not sufficiently rare to reject $H_0$: $p = 0.39$.

*Conclusion and Interpretation:* At the 0.01 level of significance, there is not overwhelming evidence to conclude that shoppers' experiences during online shopping deterred them from making a purchase.

## Calculating a *P*-Value for a Proportion

We can use exactly the same idea in determining a *P*-value for a test about a population proportion as we did in calculating *P*-values for the test statistic of a hypothesis concerning a population mean. Let's calculate the *P*-value using the test statistic we computed in **Step 5** of the previous example. If the null is true, how rare is the sample proportion that produced a $z$-value of −2.22?

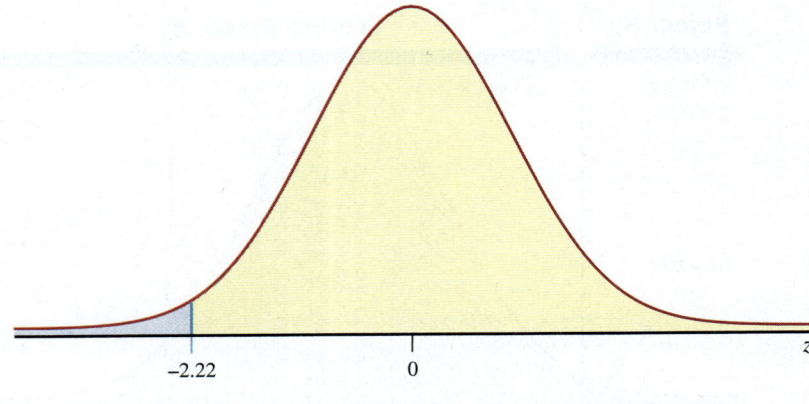

Figure 10.5.3

## Technology

P-values for a z-test can be found using the standard normal tables or from the output from a hypothesis test using the z-test statistic. For instructions on how to conduct a hypothesis test for a proportion, please visit stat.hawkeslearning.com and navigate to **Discovering Business Statistics, Second Edition > Technology Instructions > Hypothesis Testing > One Proportion z-Test.**

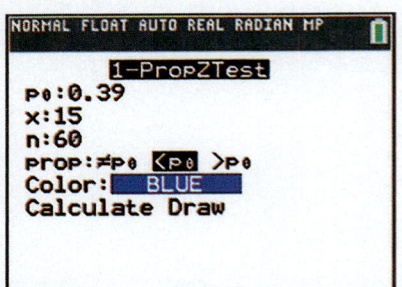

Assuming the null is true, $\widehat{p}$ has a normal distribution that is centered around 0.39. If the null is really true, a z-value of $-2.22$ is uncommon. The probability of observing a value as small or smaller than $-2.22$ is the P-value. Using Table A in the Appendix,

$$P\text{-value} = P(z \leq -2.22) = 0.0132.$$

| Table 10.5.1 – If P-Value = 0.0132 | |
|---|---|
| **Level of the Test** | **Reject or Fail to Reject $H_0$** |
| 0.10 | Reject |
| 0.05 | Reject |
| 0.01 | Fail to Reject |
| 0.005 | Fail to Reject |

If the null is true, the P-value measures the rareness of the test statistic under ordinary sampling variation. In other words, how often would we see a test statistic as small or smaller than the test statistic we have observed. Presuming the null is true, ordinary sampling variation produces a test statistic less than or equal to $-2.22$ about 1 time out of every 100. Should $H_0$ be rejected? Have we observed a test statistic that is too "rare" for $H_0$ to be true? The level of the test defines an unacceptable level of rareness for the test statistic. In the previous example $\alpha = 0.01$. Setting $\alpha = 0.01$ implies we are only willing to make a Type I error (reject the null hypothesis when it is true) once in every 100 trials of the experiment. Since the P-value of our test statistic, 0.0132, is greater than 0.01, the null hypothesis was not rejected.

If the level of the test had been 0.05, then to reject $H_0$ in favor of $H_a$ requires a test statistic whose rareness under ordinary sampling variation is less than or equal to 0.05. For $\alpha = 0.05$, the null hypothesis is rejected in favor of the alternative, since the test statistic has a P-value (0.0132) less than $\alpha$. In general, if the P-value is less than or equal to the level of the test, $\alpha$, then $H_0$ is rejected in favor of $H_a$. If the level of the test is less than the P-value, then $H_0$ is not rejected.

Note that if $H_a$ had required a two-tailed test we would double the single tail area. Thus, for a test statistic of $-2.22$ and a two-tailed $H_a$ the resulting P-value would be $2(0.0132) = 0.0264$.

# 10.5 Exercises

## Basic Concepts

1. How does testing a hypothesis about a proportion differ from testing a hypothesis about a mean?

2. What is the appropriate test statistic to be used in hypothesis testing of a population proportion?

3. What conditions must be met in order to perform a hypothesis test about a population proportion?

4. How are $P$-values determined for a proportion?

## Exercises

5. Determine the critical value(s) of the test statistic for each of the following large sample tests for the population proportion.

   a. Left-tailed test, $\alpha = 0.05$

   b. Right-tailed test, $\alpha = 0.01$

   c. Two-tailed test, $\alpha = 0.10$

6. Determine the critical value(s) of the test statistic for each of the following large sample tests for the population proportion.

   a. Left-tailed test, $\alpha = 0.07$

   b. Right-tailed test, $\alpha = 0.04$

   c. Two-tailed test, $\alpha = 0.09$

7. A commercial airline is concerned about the increase in usage of carry-on luggage. For years, the percentage of passengers with one or more pieces of carry-on luggage has been stable at approximately 38%. The airline recently selected 300 passengers at random and determined that 148 possessed carry-on luggage. Is there overwhelming evidence of an increase in carry-on luggage at a significance level of 0.01?

8. Ordinarily, when a company recruits a technical staff member, about 25% of the applicants are qualified. However, based on the information in 120 recently received resumes, 18 appear to be technically qualified.

   a. Is there overwhelming evidence that the percentage of qualified applicants is less than 25%? Test at the 0.05 level.

   b. What concerns might you have about the data in this problem?

9. The National Center for Drug Abuse is conducting a study to determine if heroin usage among teenagers has changed. Historically, about 1.3 percent of teenagers between the ages of 15 and 19 have used heroin one or more times. In a recent survey of 1824 teenagers, 37 indicated they had used heroin one or more times.

   a. Is there overwhelming evidence of a change in heroin usage among teenagers? Test at the 0.05 level.

   b. What concerns might you have about the data in this problem?

10. Paper International, Inc. has a large staff of salespeople nationwide. Top officials of the company believe that 75% of their salespeople have met their monthly sales goals by the end of the third week of each month. To investigate this, they randomly select 250 salespeople and examine their sales records at the end of the third week of the current month. One-hundred seventy-five of the 250 salespeople surveyed had already met their monthly sales goals.

   a. Does this sample support the belief of the top officials at the company at $\alpha = 0.10$?

   b. What concerns might you have about the manner in which the data were collected?

**11.** Ships arriving in U.S. ports are inspected by customs officials for contaminated cargo. Assume, for a certain port, that 20% of the ships arriving in the previous year contained cargo that was contaminated. A random selection of 50 ships in the current year included five that had contaminated cargo.

    **a.** Do the data suggest that the proportion of ships arriving in the port with contaminated cargoes has decreased in the current year at $\alpha = 0.01$?

    **b.** Do you have any concerns about the sample size? Explain.

**12.** Grain elevators store hundreds of thousands of bushels of grain each year that are waiting to be processed. It is critical to control the amount of moisture in the grain so that it does not spoil. A large storage facility is deemed to be "in control" if 1% of the grain elevators have a moisture content of 10%. One-hundred fifty grain elevators are randomly selected and the moisture content is measured. Two of the grain elevators sampled have a moisture content in excess of 10%.

    **a.** Is there sufficient evidence for the manager to conclude that the storage facility has a moisture content significantly greater than 1%? Use $\alpha = 0.05$.

    **b.** Do you have any concerns about the sample size? Explain.

**13.** Electronic circuit boards are randomly selected each day to determine if any of the boards are defective. A random sample of 100 boards from one day's production has four boards that are defective.

    **a.** Based on the data, is there overwhelming evidence that more than 5% of the circuit boards are defective? Test at the $\alpha = 0.10$ level.

    **b.** Do you have any concerns about the sample size? Explain.

**14.** Loch Ness Fish Farm breeds fish for commercial sale. The fish are kept in breeder tanks until more than 70% of the fish are five inches long at which time they are transferred to outdoor ponds. To determine if it is the appropriate time to transfer the fish, 50 fish are randomly selected and measured. If 33 of the fish are found to be over five inches long, does the sample data suggest that it is the appropriate time to transfer the fish at $\alpha = 0.05$?

**15.** Digger and Digger, a precious metals mining company, is considering the development of a new mining area. They have a lease on an area which they believe contains gypsum. The area will be profitable to mine if more than 15% of the rocks contain more than trace amounts of the mineral. Eighty rocks are randomly selected and the amount of gypsum is measured. Thirteen rocks in the sample are observed to have more than trace amounts of the mineral. Based on the sample data, should Digger and Digger conclude that the area will be profitable to mine? Use $\alpha = 0.01$.

**16.** A socially conscious corporation wants to relocate their headquarters to another part of town. One concern expressed by workers is that their commuting distance will increase. The corporation has decided that if more than 50% of the employees will have to drive farther to the proposed new location, they will cancel the move. In a random sample of 398 employees, 201 indicated that their commuting distance to the new office will be longer. Based on the sample data, should the corporation cancel the move? Use a significance level of 0.01.

**17.** A production process will normally produce defective parts 0.2% of the time. In a random sample of 1400 parts, three defectives are observed.

    **a.** Is this overwhelming evidence at the 0.05 level to indicate that the defective rate of the process has increased?

    **b.** Compute the $P$-value for the test statistic.

    **c.** Based on the $P$-value, would the decision change at $\alpha = 0.01$?

**18.** Bombay Charlie's, a fast food Indian restaurant, is thinking about adding a certain spice to their chicken curry dish to attract more customers. The restaurant manager has decided to add the spice if more than 80% of his customers prefer the taste of the chicken curry with the spice added. Sixty-five customers are randomly selected to participate in a blind taste test. Fifty-four of these customers prefer the chicken curry with the added spice.

    **a.** Find the $P$-value for the hypothesis test that the manager will perform to decide if more than 80% of the customers prefer the taste of the chicken curry with the added spice.

    **b.** Do the data suggest that more than 80% of the customers prefer the curry with the new spice at $\alpha = 0.05$?

**19.** The news program for KOPE, the local television station, claims to have 40% of the market. A random sample of 500 viewers conducted by an independent testing agency found 192 who claim to watch the KOPE news program on a regular basis.

    **a.** Find the $P$-value for testing the hypothesis that the news program for KOPE does not have more than 40% of the market as it claims.

    **b.** Is there sufficient evidence to reject the hypothesis that KOPE does not have at least 40% of the market at a significance level of 0.05?

**20.** The length of time that a storm window will last before beginning to leak is of interest to a window manufacturer who wishes to guarantee his windows. He believes that more than 50% of the windows will last at least four years. To research this, 931 windows, which were installed at least four years ago, are randomly selected and checked for leakage. Five hundred of the windows are found to still be leak-free.

    **a.** Find the $P$-value for testing the hypothesis that more than 50% of the windows will be leak-free in four years.

    **b.** Does the sample support the hypothesis that more than 50% of the windows will be leak-free in four years at $\alpha = 0.05$?

**21.** In order to discourage soldiers from smoking, the Pentagon raised the price of cigarettes by $4 a carton in October of 1996. This increased the average price of a carton of brand-name cigarettes to $17.50, an increase of about 30%. Prior to the price increase, about 32% of military personnel smoked, as opposed to 25% of all adult Americans. Suppose that following the price increase, a random sample of military personnel is selected to determine smoking habits. With $\alpha = 0.05$, can we conclude that the price increase was effective in decreasing the percentage of smokers if 50 of the 200 military personnel sampled smoke?

**22.** Wearing bright or fluorescent orange colored clothing clearly reduces the risk of being shot or killed by hunting. According to an October 1996 article appearing in *The Augusta Chronicle* (Georgia), about two-thirds of the hunters shot in Georgia and South Carolina during the preceding five years were not wearing bright clothes. Of the 52 that were killed, only 19 wore orange. Suppose that a random sample of 100 hunters in Georgia are surveyed and it is determined that of the 100, 62 routinely wear fluorescent orange colored clothing while hunting. With $\alpha = 0.10$, can it be concluded that over half the hunters in Georgia routinely wear fluorescent orange colored clothing while hunting?

23. According to the Federal Communications Commission, about 49% of the households in the United States had cable television in 1985. Suppose that a sample of 200 households is selected in 2003 and it is determined that 125 of them have cable television.

   a. With $\alpha = 0.05$, can it be concluded that a higher proportion of households in 2003 have cable television as compared with 1985?

   b. In the sample of 200, what is the fewest number of people who have cable television that would allow the conclusion that a higher proportion of households in 2003 have cable television as compared to 1985?

24. Selling autographed sports memorabilia has become a multimillion dollar industry in the United States. But just how does the purchaser of an autographed football jersey or an autographed baseball know that the autograph is indeed authentic? Unfortunately, the sports memorabilia market is teeming with con artists who prey upon the trusting nature of sports fans. In 1996, the FBI said that 70% of all autographed sports memorabilia is fraudulent. Assume that 50 pieces of autographed sports memorabilia are sampled at a large memorabilia show and that 40 of them are determined to be fraudulent.

   a. With $\alpha = 0.05$, can it be concluded that the proportion of fraudulent autographed sports memorabilia at the show differs from the FBI claim?

   b. What is the greatest number of fraudulent items in the sample of 50 that would not allow a conclusion that sports memorabilia at the show differs from the FBI claim?

# 10.6  Testing a Hypothesis about a Population Variance

In this section we want to adapt the hypothesis testing procedure to test a hypothesis concerning a population variance. Before we can perform the test about the population variance, we need to review a few topics that were discussed earlier in the text.

Recall that the sample variance is

$$s^2 = \frac{\Sigma\left(x_i - \bar{x}\right)^2}{n-1}$$

and it serves as the point estimate of the population variance, $\sigma^2$. In Section 9.4 we determined that the sampling distribution of

$$\frac{(n-1)s^2}{\sigma^2}$$

is a chi-square distribution with $n-1$ degrees of freedom.

### Friedrich Robert Helmert

Friedrich Robert Helmert was born in Germany in 1843. His interests were in geodesy, which is a discipline concerned with measuring the earth on a global scale. He studied engineering science at the Polytechnische Schule and while still a student had the opportunity to work on some important geodesy projects with one of his teachers, August Nagel. He later studied mathematics and astronomy to earn his doctorate. Geodesy led him into statistics, first writing a book on least squares. In 1876 he discovered the chi-square as the distribution of the sample variance for a normal distribution. His work was in German and was not translated to English, so later in 1900 English statisticians rediscovered the chi-square distribution (Karl Pearson) and its application to the sample variance (William Gosset, Ronald Fisher).

> **Formula**
>
> ### $\chi^2$-Test Statistic
>
> If we have a random sample of size $n$ taken from a normal population, then the sampling distribution of the test statistic is given by
>
> $$\chi^2 = \frac{(n-1)s^2}{\sigma^2}$$
>
> which has a chi-square distribution with $n-1$ degrees of freedom.

To adapt the hypothesis testing procedure to test a hypothesis concerning a population variance, let's look at the following example.

A drug manufacturer believes that the manufacturing process is in control if the standard deviation of the dosage in each tablet is at most 0.10 milligram. The quality control manager is willing to shut down the manufacturing process if there is overwhelming evidence that the process has excessive variation. A sample of 30 tablets is evaluated and the sample standard deviation is found to be 0.14 milligram. Assuming that the data are collected from a normal population, should the quality control manager shut down the manufacturing process? Use $\alpha = 0.01$.

**Example 10.6.1**

**Performing a Hypothesis Test for a Population Variance**

### SOLUTION

**Step 1:** Determine the null hypothesis. In this process, select the appropriate statistical measure, such as the population mean, proportion, or variance.

The null hypothesis is that the manufacturing process does not have excessive variation. Since the issue in this example is variation and specifically discusses the standard deviation, the hypothesis can be stated in terms of the variance. The null hypothesis is written as $H_0$: $\sigma^2 = 0.01$.

**Step 2:** Determine the alternative hypothesis and whether it should be one-sided or two-sided.

Most hypothesis tests concerning a variance will be one-sided. Small variation is desirable, while too much variation is undesirable. Because we are interested in determining if there is evidence that the process has excessive variation, the test will be one-tailed. The alternative hypothesis is written as $H_a$: $\sigma^2 > 0.01$.

**Step 3:** Select the appropriate test statistic based on the information at hand and the assumptions you are willing to make.

The test statistic will change rather dramatically. In the previous test statistics, the goal was to measure how far the sample statistic, $\bar{x}$ or $\hat{p}$, was from the hypothesized value. The test statistic for a population variance is given by

$$\chi^2 = \frac{(n-1)s^2}{\sigma_0^2} \text{ with } n-1 \text{ degrees of freedom.}$$

Let's look at the pieces of this statistic. The term $\sigma_0^2$ refers to the hypothesized value of the variance. In this sample, $\sigma_0 = 0.10$, which implies $\sigma_0^2 = 0.01$. The actual variance of the population is unknown. But the sample variance, $s^2$, should be reasonably close to the unknown population variance, $\sigma^2$. If the null is true, then $\sigma_0^2 = 0.01$ and the ratio $\frac{s^2}{\sigma_0^2}$ should be near 1, since $s^2$ should be close to $\sigma_0^2$. Assuming the null is true, multiplying this ratio by $n-1$ should produce a result near $n-1$. If the $\chi^2$ expression is a great deal larger than $n-1$, then $s^2$ will be a great deal larger than $\sigma_0^2$. Such an event would cast doubt on the validity of the null hypothesis.

**Step 4:** Determine the critical value of the test statistic. Two factors must be considered.

1. The type of alternative hypothesis: two-sided, one-sided left, one-sided right.

2. The specification of $\alpha$, the significance level of the test.

The role of the critical value in this test is no different from other hypothesis tests discussed earlier. It defines a range of values for the test statistic, the rejection region, that will be too rare to have likely occurred from ordinary sampling variability. From a probabilistic standpoint, the level of the test defines the size of the rejection region. Should the value of the test statistic fall in this region, the null hypothesis will be rejected. Determining the critical value will require knowledge of the sample size. The company studied a sample of 30 tablets. Since the level of the test is 0.01 and there are $df = 30 - 1 = 29$ degrees of freedom, the critical value is 49.588 (see Table G). The test statistic will exceed this value because of ordinary variation only 1% of the time.

| Area to the Right of the Critical Value of $\chi^2$ | | | | | |
|---|---|---|---|---|---|
| $df$ | ... | $\chi_{0.025}$ | $\chi_{0.01}$ | $\chi_{0.005}$ |
| 1 | | 5.024 | 6.635 | 7.879 |
| 2 | | 7.378 | 9.210 | 10.597 |
| 3 | | 9.348 | 11.345 | 12.838 |
| | ... | | | |
| 29 | | 45.722 | 49.588 | 52.336 |
| 30 | | 46.979 | 50.892 | 53.672 |
| | ... | | | |

**Technology**

To find the critical value for a chi-square distribution using technology, please visit stat.hawkeslearning.com and navigate to **Discovering Business Statistics, Second Edition > Technology Instructions > Chi-Square Distribution > Critical Value**.

**NOTE**

For large degrees of freedom, the chi-square distribution looks very similar to a normal distribution. Notice, however, that the right tail is a bit thicker than the left.

$\chi^2$-Distribution, $df = 29$

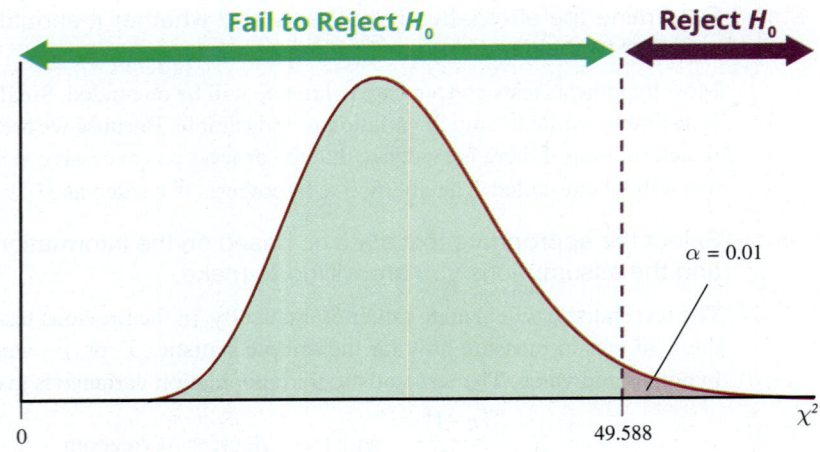

Figure 10.6.3

**Step 5:** Collect the sample data and compute the value of the test statistic.

A sample of 30 tablets is evaluated, and the sample standard deviation is found to be 0.14 milligram.

$$\chi^2 = \frac{(n-1)s^2}{\sigma_0^2} = \frac{(30-1)(0.14)^2}{(0.10)^2} = 56.84$$

**Step 6:** Make the decision and state the conclusion in terms of the original question.

Since the test statistic, $\chi^2 = 56.84$, exceeds the critical value, 49.588, we will conclude that the test statistic is too rare to have been caused by ordinary sampling variation. The null hypothesis is rejected in favor of the alternative hypothesis.

**Technology**

To find the *P*-value using technology, please visit stat.hawkeslearning.com and navigate to **Discovering Business Statistics, Second Edition > Technology Instructions > Chi-Square Distribution > Right Tailed Probability (cdf)**.

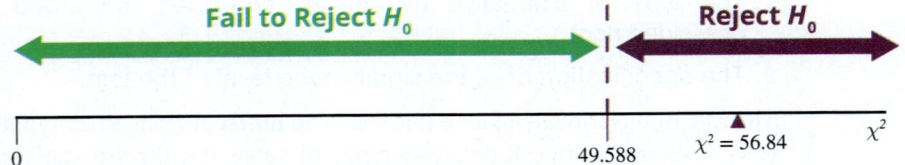

*Conclusion and Interpretation:* There is overwhelming evidence that the process variation exceeds the desired level. The quality control manager will likely shut down the manufacturing process.

## *P*-values for $\chi^2$ Test Statistics

Similar to the *t*-Test in Section 10.2, we have numerous chi-square distributions, one for each degree of freedom. The chi-square table is constructed such that it provides us with chi-square values for frequently used tail probabilities. The chi-square table in Appendix A, Table G, has only ten tail areas. Thus, in most instances, we will not be able to determine the exact *P*-value. Instead, we will find the closest chi-square values with the appropriate degrees of freedom surrounding the test statistic. For example, in Example 10.6.1, the value of the test statistic was

$$\chi^2 = 56.84$$

which has 29 degrees of freedom. Since the *P*-value is a function of the alternative hypothesis, we write

$$P\text{-value} = P(\chi^2 > 56.84)$$

As stated earlier, we are unable to find the exact probability for the *P*-value so we will need to put bounds on the *P*-value. The excerpt from the chi-square table above Figure 10.6.3 highlights the chi-square values with 29 degrees of freedom. Note that the chi-square table provides the value of the chi-square with the area of $\alpha$ to the right. Also, note that the values of $\alpha$ across the top row decrease from left to right and the chi-square values in the body of the table increase from left to right. At 29 degrees of freedom, we see that the value of the test statistic falls to the right (i.e., it is greater than) of 52.336 at 29 degrees of freedom corresponding to $\alpha = 0.005$. In this case, we report that the *P*-value is less than 0.005. In Example 10.6.1, we conducted the test using $\alpha = 0.01$. Therefore, since the *P*-value is less than 0.005, it is obviously less than 0.01 which would lead us to reject the null hypothesis and conclude that the variance is significantly more than 0.01 mg.

The exact *P*-value can be found using technology. In Example 10.6.1, using technology we obtain a *P*-value of 0.0015, which is less than $\alpha = 0.01$. Again, this leads us to reject the null hypothesis.

## 🖉 10.6 Exercises

### Basic Concepts

1. How does testing a hypothesis about a variance differ from testing a hypothesis about a mean?

2. What is the symbol for a critical value for the chi-square distribution? Describe the meaning of this critical value.

### Exercises

3. Determine the critical value(s) of the test statistic for each of the following tests for a population variance where the assumption of normality is satisfied.

    a. Right-tailed test, $\alpha = 0.01, n = 20$

    b. Right-tailed test, $\alpha = 0.05, n = 24$

    c. Right-tailed test, $\alpha = 0.005, n = 5$

4. Determine the critical value(s) of the test statistic for each of the following tests for a population variance where the assumption of normality is satisfied.

    a. Right-tailed test, $\alpha = 0.025, n = 18$

    b. Right-tailed test, $\alpha = 0.10, n = 24$

    c. Right-tailed test, $\alpha = 0.05, n = 41$

5. A bolt manufacturer is very concerned about the consistency with which his machines produce bolts that are $\frac{3}{4}$ inch in diameter. When the manufacturing process is working normally the standard deviation of the bolt diameter is 0.05 inch. A random sample of 30 bolts has an average diameter of 0.25 inch with a standard deviation of 0.07 inch.

   a. Can the manufacturer conclude that the standard deviation of bolt diameters is greater than 0.05 inches at $\alpha = 0.05$?

   b. What assumption did you make about the diameter of the bolts in performing the test in part **a.**?

6. A drug that is used for treating cancer has potentially dangerous side effects if it is taken in doses that are larger than the required dosage for the treatment. The pharmaceutical company that manufactures the drug must be certain that the standard deviation of the drug content in the tablet is not more than 0.1 mg. Twenty-five tablets are randomly selected and the amount of drug in each tablet is measured. The sample has a mean of 20 mg and a variance of 0.015 mg.

   a. Do the data suggest at $\alpha = 0.01$ that the standard deviation of drug content in the tablets is greater than 0.1 mg?

   b. What assumption did you make about the amount of drug contained in the tablets in performing the test in part **a.**?

7. A conservative investor would like to invest some money in a bond fund. The investor is concerned about the safety of her principal (the original money invested). Colonial Funds claims to have a bond fund which has maintained a consistent share price of $7. They claim that this share price has not varied by more than $0.25 on average since its inception. To test this claim, the investor randomly selects 25 days during the last year and determines the share price for the bond fund. The average share price of the sample is $7 with a standard deviation of $0.35.

   a. Can the investor conclude that the standard deviation of share price of the bond fund is greater than 0.25? Test at the 0.01 level.

   b. What assumption did you make about the share price of the bond fund in your test in part **a.**?

# T  Discovering Technology

## Using the TI-84 Plus Calculator

Hypothesis Test for a Population Mean: $\sigma$ Known

$$z = \frac{\bar{x} - \mu_0}{\frac{\sigma}{\sqrt{n}}}$$

$H_0: \mu = 16{,}000$   $n = 1000$   $\bar{x} = 16{,}500$
$H_a: \mu > 16{,}000$   $\sigma = 2500$

1. Choose **STAT**, then select **TESTS** and **Z-Test**.

2. Press **ENTER**, and choose **Stats**. Input **16000** for $\mu_0$, **2500** for $\sigma$, **16500** for $\bar{x}$, **1000** for $n$, and select $> \mu_0$. Press **Calculate**.

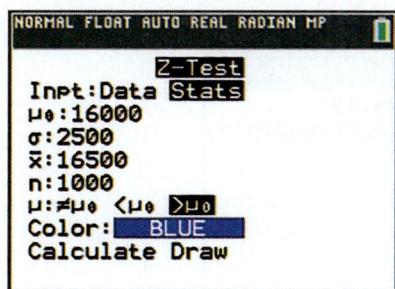
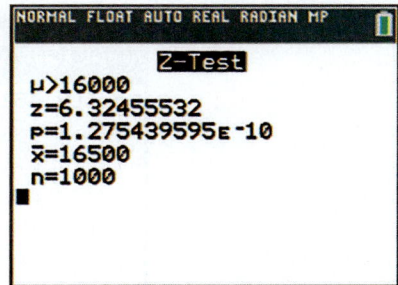

3. The results of the hypothesis test are displayed. The test statistic is $z \approx 6.32$, and the $P$-value is extremely small. Thus, there is sufficient evidence to reject the null hypothesis. The sample mean and sample size are also repeated for convenience.

Hypothesis Test for a Population Mean: $\sigma$ Unknown

$$t = \frac{\bar{x} - \mu_0}{\frac{s}{\sqrt{n}}}$$

$H_0: \mu = 35$   $n = 20$   $\bar{x} = 27$
$H_a: \mu > 35$   $s = 4$

1. Choose **STAT**, then select **TESTS** and **T-Test**.

2. Press **ENTER**, and choose **Stats**. Input **35** for $\mu_0$, **27** for $\bar{x}$, **4** for $s$ (designated on the calculator as Sx), **20** for $n$, and select $\neq \mu_0$. Press **Calculate**.

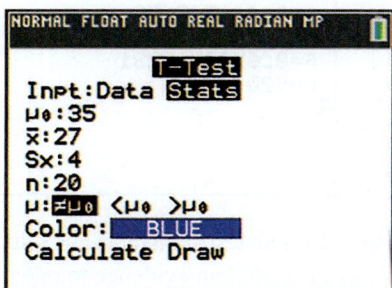
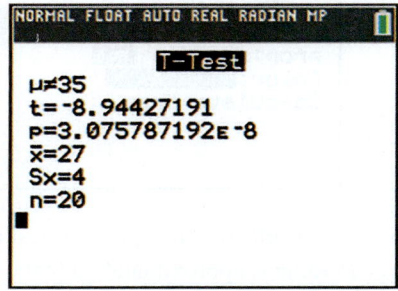

3.    The results of the hypothesis test are displayed. The test statistic is $t \approx -8.944$, and the $P$-value is extremely small. Thus, there is sufficient evidence to reject the null hypothesis. The sample mean, sample standard deviation, and sample size are also repeated for convenience.

## $P$-Values for $t$-Test Statistics

Calculate the $P$-value for a test statistic of $t = 2.451$ with 17 degrees of freedom. Assume that the hypothesis test was an upper-tailed test.

1.    Access the **DISTR** menu by pressing 2ND and then VARS .

2.    Select option **tcdf**. The syntax to find the area under the $t$-distribution between two $t$-values is **tcdf(lower bound, upper bound, degrees of freedom)**. To find the $P$-value for a test statistic of $t = 2.451$ with 17 degrees of freedom, assuming we are performing a one-sided test, enter **tcdf(2.451,1E99,17)**. Note that 1E99 represents a large positive number, and is entered into the calculator by pressing 1 , 2ND , , , 9 , 9 .

3.    The $P$-value is given as approximately 0.0127. This is the area to the right of $t = 2.451$ with 17 degrees of freedom. Note that for a two-tailed test, the value displayed would need to be multiplied by 2 to get the $P$-value in both tails.

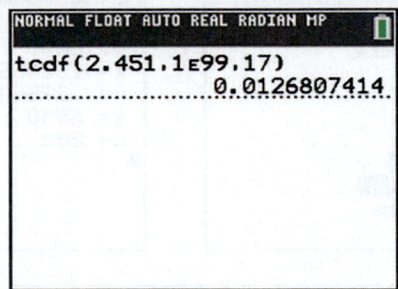

## Hypothesis Test for a Population Proportion

$$z = \frac{\hat{p} - p_0}{\sqrt{\dfrac{p_0(1 - p_0)}{n}}}$$

$$H_0: p = 0.60 \qquad\qquad n = 520$$
$$H_a: p > 0.60 \qquad\qquad x = 350$$

1.    Choose STAT , then select **TESTS** and **1-PropZTest**.

2.    Input **0.60** for $p_0$, **350** for $x$, **520** for $n$, and select **>$p_0$**. Press **Calculate**.

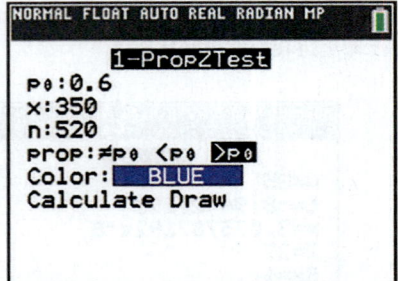 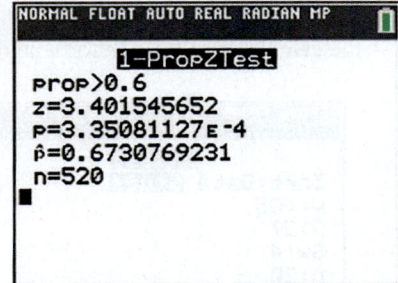

3.    The results of the hypothesis test are displayed. The test statistic is $z \approx 3.40$, and the $P$-value is approximately 0.0003. Thus, there is sufficient evidence to reject the null hypothesis. The sample proportion and sample size are also displayed.

## Using Excel

Hypothesis Test for a Population Mean: $\sigma$ Known

Finding the $z$-test statistic can be one of the more laborious steps in hypothesis testing. Use Excel to compute the $z$-statistic for Example 10.2.1.

1. Create a worksheet with these labels.

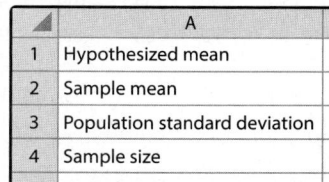

| | A |
|---|---|
| 1 | Hypothesized mean |
| 2 | Sample mean |
| 3 | Population standard deviation |
| 4 | Sample size |

2. Referring to the example, enter **500** for the hypothesized mean in cell **B1**. Enter **540** for the sample mean in cell **B2**. Enter **100** for the population standard deviation in cell **B3**. Enter **75** for the sample size in cell **B4**.

3. Given these values, we have enough data to compute the $z$-test statistic. In cell **A5**, write $z$-**test statistic** and in cell **B5** compute it with the formula:

$$=(B2-B1)/(B3/(SQRT(B4)))$$

4. Press **Enter**. The resulting $z$-test statistic is approximately 3.46. Compare this to our calculations in Example 10.2.1

| | A | B |
|---|---|---|
| 1 | Hypothesized mean | 500 |
| 2 | Sample mean | 540 |
| 3 | Population standard deviation | 100 |
| 4 | Sample size | 75 |
| 5 | z-test statistic | 3.464102 |

You can use this worksheet to compute $z$ for other hypothesis testing questions by changing the values for the hypothesized mean, sample mean, population standard deviation ($\sigma$), and sample size

### P-Values for t-Test Statistics

Calculate the $P$-value if the observed value of the test statistic is $t = 2.378$ with 15 degrees of freedom. Assume the hypothesis test is a one-sided, upper-tailed test.

1. The function in Microsoft Excel that returns the probability in the lower (or left) tail of a $t$-distribution is

$$=T.DIST(x, deg\_freedom, cumulative)$$

where $x$ is the $t$-statistic, *deg_freedom* is the degrees of freedom for the $t$-distribution, and *cumulative* is a logical value that if TRUE returns the cumulative distribution function, and if FALSE returns the probability density function. When finding the $P$-value, cumulative will always be TRUE. To find the $P$-value for a one-sided, upper-tailed test when the observed value of the test statistic is $t = 2.378$ with 15 degrees of freedom, enter the following formula into Excel.

$$=1-T.DIST(2.378,15,TRUE)$$

2. The $P$-value is displayed as approximately 0.0156. Remember that the function returns the area to the left of the test statistic. This is why we had to subtract the function from 1 in order to find the area in the upper tail.

## Using JMP

### Hypothesis Test for a Population Mean: $\sigma$ Known

Consider the following sample data.

| 95 | 110 | 60 | 75 | 85 | 90 | 110 | 90 | 115 | 100 | 90 | 100 |
|----|-----|----|----|----|----|-----|----|-----|-----|----|-----|
| 90 | 50 | 60 | 110 | 90 | 90 | 55 | 100 | 40 | 115 | 50 | 100 |
| 70 | 70 | 80 | 95 | 90 | 70 | 95 | 80 | 105 | 90 | 55 | 65 |

$$H_0: \mu = 80 \qquad\qquad \alpha = 0.05$$
$$H_a: \mu > 80 \qquad\qquad \sigma = 20.15$$

1.  With a JMP **Data Table** open, click on the first cell of **Column 1** and input the data.

2.  Select **Analyze** in the top row of the JMP spreadsheet and then select **Distribution**.

3.  From the **Select Columns** box, click on **Column 1**. Then click on **Y, Columns**. Click **OK**.

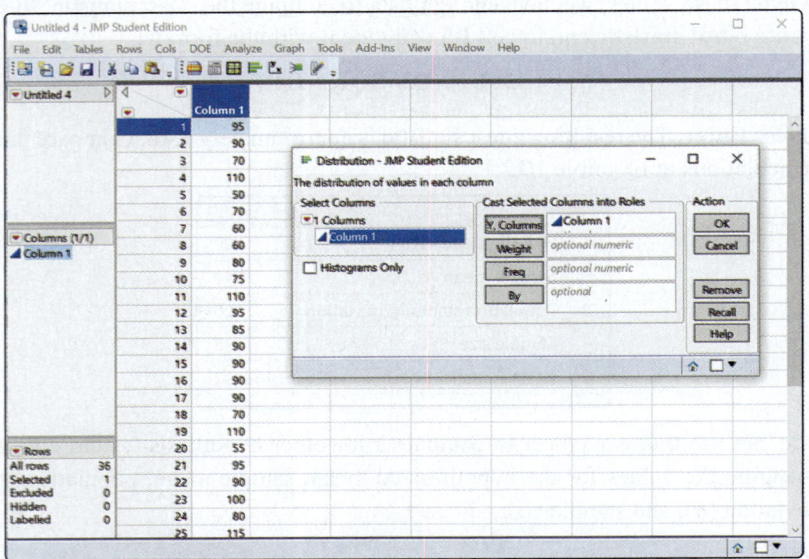

4.  Click on the **red down arrow** next to **Column 1** and choose **Test Mean** from the list.

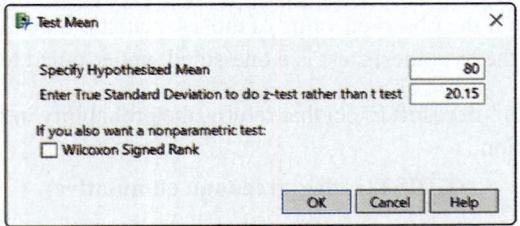

5.  Enter **80** for the Hypothesized Mean and enter **20.15** for the True Standard Deviation. Click **OK**.

6.  Observe the output screen for the results. The z-test statistic for this sample data is 1.2821 and the *P*-value is 0.0999.

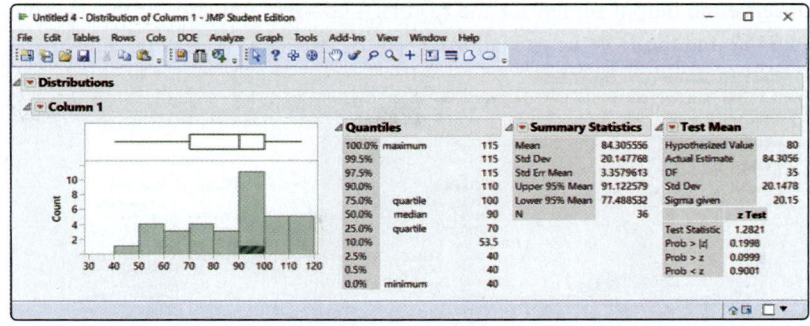

## Hypothesis Test for a Population Mean: $\sigma$ Unknown

Consider the following sample data.

| | | | | | | | | | |
|---|---|---|---|---|---|---|---|---|---|
| 24.9 | 26.8 | 27.2 | 34.1 | 28.9 | 25.9 | 25.1 | 23.9 | 23.0 | 30.9 |
| 31.9 | 35.1 | 27.4 | 25.9 | 19.1 | 23.7 | 22.0 | 25.9 | 29.0 | 29.3 |

$$H_0: \mu = 80 \qquad \alpha = 0.05$$
$$H_a: \mu > 80 \qquad \sigma = 20.15$$

1. With a JMP **Data Table** open, click on the first cell of **Column 1** and input the data.

2. Select **Analyze** in the top row of the JMP spreadsheet and then select **Distribution**.

3. From the **Select Columns** box, click on **Column 1**. Then click on **Y, Columns**. Click **OK**.

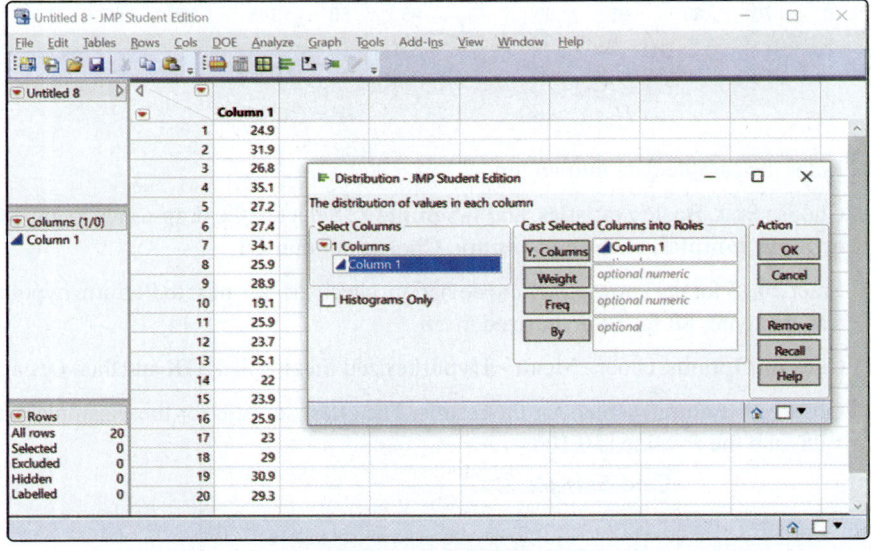

4. Click on the **red down arrow** next to **Column 1** and choose **Test Mean** from the list.

5. Enter **35** for the Hypothesized Mean and leave the input box for the True Standard Deviation blank. Click **OK**.

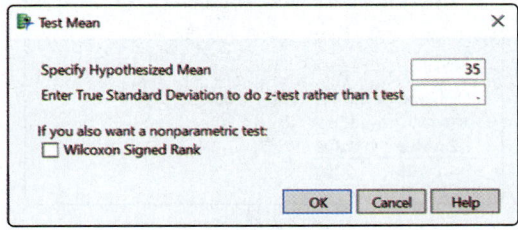

6. Observe the output screen for the results. The *t*-test statistic for this sample data is −8.9452 and the *P*-value is <0.0001.

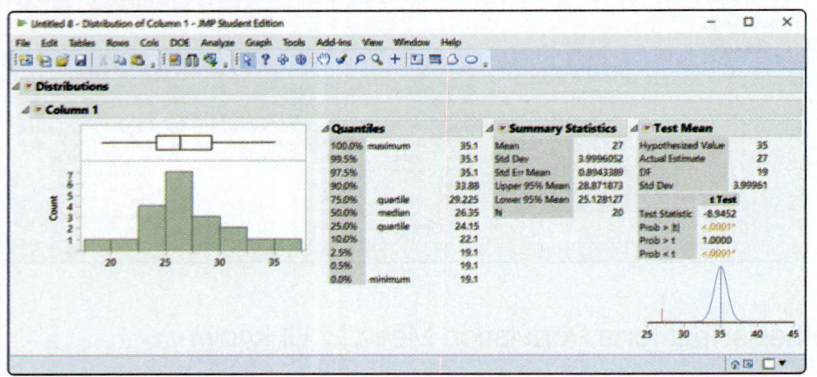

## Using Minitab

### Hypothesis Test for a Population Mean: $\sigma$ Known

Minitab will compute the *z*-test statistic for a set of sample data.

Consider the following sample data.

| 95 | 110 | 60 | 75 | 85 | 90 | 110 | 90 | 115 | 100 | 90 | 100 |
|----|-----|----|----|----|----|-----|----|-----|-----|----|-----|
| 90 | 50  | 60 | 110 | 90 | 90 | 55  | 100 | 40 | 115 | 50 | 100 |
| 70 | 70  | 80 | 95 | 90 | 70 | 95  | 80 | 105 | 90 | 55 | 65  |

$$H_0: \mu = 80 \qquad\qquad \alpha = 0.05$$
$$H_a: \mu > 80 \qquad\qquad \sigma = 20.15$$

1. Enter the sample data into column C1.

2. Choose **Stat**, **Basic Statistics**, and **1-Sample Z**. Select in the drop-down menu – **One or more samples, each in a column**. Choose column **C1**.

3. Enter **20.15** for the known standard deviation, check the box next to Perform hypothesis test, and enter **80** for Hypothesized mean.

4. Click on **Options**, choose **Mean > Hypothesized mean**. Click **OK** and then **OK** again.

5. Observe the output screen for the results. The *z*-test statistic for these sample data is 1.28, and the *P*-value is 0.100.

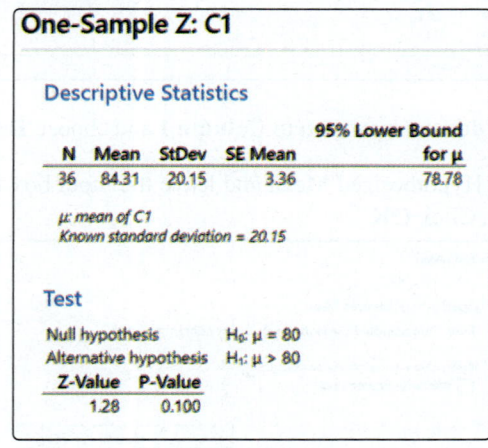

## Hypothesis Test for a Population Mean: $\sigma$ Unknown

Consider the following sample data.

| 24.9 | 26.8 | 27.2 | 34.1 | 28.9 | 25.9 | 25.1 | 23.9 | 23.0 | 30.9 |
|------|------|------|------|------|------|------|------|------|------|
| 31.9 | 35.1 | 27.4 | 25.9 | 19.1 | 23.7 | 22.0 | 25.9 | 29.0 | 29.3 |

$$H_0: \mu = 35 \qquad\qquad \alpha = 0.05$$
$$H_a: \mu \neq 35$$

1. Enter the sample data into column C1.

2. Choose **Stat**, **Basic Statistics**, and **1-Sample t**. Select in the drop-down menu – **One or more samples, each in a column**. Choose column **C1**.

3. Check the box next to Perform hypothesis test, and enter **35** for Hypothesized mean. .

4. Click on Options, choose **Mean $\neq$ Hypothesized mean**. Click **OK** and then **OK** again.

5. Observe the output screen for the results. The $t$-test statistic for these sample data is 8.95, and the $P$-value is 0.000.

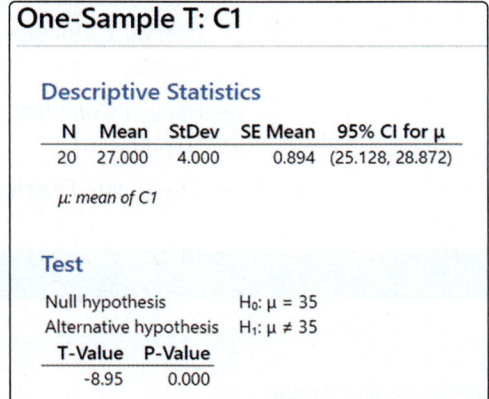

**One-Sample T: C1**

**Descriptive Statistics**

| N | Mean | StDev | SE Mean | 95% CI for $\mu$ |
|----|--------|-------|---------|--------------------|
| 20 | 27.000 | 4.000 | 0.894 | (25.128, 28.872) |

$\mu$: mean of C1

**Test**

Null hypothesis      $H_0: \mu = 35$
Alternative hypothesis    $H_1: \mu \neq 35$

| T-Value | P-Value |
|---------|---------|
| -8.95 | 0.000 |

**Note:** Performing a hypothesis test for a population proportion using Minitab is very similar to the steps taken to perform a test about a population mean. The tool is found under **Stat**, **Basic Statistics**, and **1 Proportion**. Minitab can either perform the test using raw data values, or summarized data. This is also true for the tests about population means. Summarized data may be used in place of the raw data.

## R   Chapter 10 Review

## Key Terms and Ideas

- Hypothesis Testing
- Null Hypothesis ($H_0$)
- Alternative Hypothesis ($H_a$)
- One-Sided Alternative
- Two-Sided Alternative
- One-Tailed Test
- Two-Tailed Test
- Test Statistic
- Rejection Region
- Reject $H_0$
- Fail To Reject $H_0$
- Standard Value
- Type I Error
- Type II Error

- Level of the Test (Significance Level of the Test)
- Decision Rule
- Left-Tailed (Lower-Tailed) Test
- Right-Tailed (Upper-Tailed) Test
- $z$-Test Statistic
- Critical Value
- $t$-Test Statistic
- $P$-Value
- Practical vs. Statistical Significance
- The Relationship between Confidence Interval Estimation and Hypothesis Testing
- Using a Confidence Interval to Test a Hypothesis
- Chi-Square Distribution

## Key Formulas

|  | Section |
|---|---|

**$z$-Test Statistic for a Population Mean**

$$z = \frac{\bar{x} - \mu_0}{\frac{\sigma}{\sqrt{n}}}$$

10.2

**$t$-Test Statistic**

$$t = \frac{\bar{x} - \mu_0}{\frac{s}{\sqrt{n}}}$$

10.3

**$z$-Test Statistic for a Population Proportion**

If $np_0 \geq 5$ and $n(1 - p_0) \geq 5$, the appropriate test statistic is given by

$$z = \frac{\hat{p} - p_0}{\sqrt{\frac{p_0(1 - p_0)}{n}}}$$

10.5

**Test Statistic for a Population Variance**

$$\chi^2 = \frac{(n-1)s^2}{\sigma_0^2} \text{ with } n-1 \text{ degrees of freedom}$$

10.6

# AE    Additional Exercises

1.  A tire company has found that the mean time required for a mechanic to replace a set of four tires is 18 minutes. After instituting a new installation procedure, the company unfortunately believes that the expected time required to replace the set of four tires remains unchanged. A test of the company's belief will be performed.

    a.  What are the null and alternative hypotheses for the test of the company's belief?

    b.  Describe, in terms of the problem, how a Type I error could occur.

    c.  Describe, in terms of the problem, how a Type II error could occur.

2.  Tech Transit wishes to test whether the mean number of passenger miles on a particular route exceeds 66,000 passenger miles, the number of passenger miles the company needs on that route to cover all allocated costs. A random sample of 25 trips on the route yields a mean of 70,250 miles and a standard deviation of 9000 miles. It is desired to control the significance level at 1%.

    a.  State the appropriate hypotheses for this problem.

    b.  Describe, in terms of the problem, how a Type I error could occur.

    c.  Describe, in terms of the problem, how a Type II error could occur.

3.  A pain reliever currently being used in a hospital is known to bring relief to patients in a mean time of 3.5 minutes. To compare a new pain reliever with the one currently being used, the new drug is administered to a random sample of 50 patients. The mean time to relief for the sample of patients is 2.8 minutes and the population standard deviation is 1.14 minutes. Do the data provide sufficient evidence to conclude that the new drug was effective in reducing the mean time until a patient receives relief from pain? Test using $\alpha = 0.10$.

4.  Tech uses thousands of fluorescent light bulbs each year. The brand of bulb it currently uses has a mean life of 900 hours. A manufacturer claims that its new brand of bulbs, which cost the same as the brand the university currently uses, has a mean life of more than 900 hours. The university has decided to purchase the new brand if, when tested, the test evidence supports the manufacturer's claim at the 0.05 significance level. Suppose 64 bulbs were tested and they were found to have an average life of 920 hours and the population standard deviation is 80 hours. Will the university purchase the new brand of fluorescent bulbs?

5.  The daily wages in a particular industry are normally distributed with a mean of $13.20 and a population standard deviation of $2.50. If a company in this industry employing 40 workers pays these workers, on average, $12.20, can this company be accused of paying inferior wages? Use the $P$-value approach with a significance level of 1%.

6.  A coin-operated soft drink machine was designed to discharge, on average, 12 ounces of beverage per cup. In a test of the machine, ten cupfuls of beverage were drawn from the machine and measured. The mean and standard deviation of the ten measurements were 12.1 ounces and 0.12 ounce, respectively. Do these data present sufficient evidence to indicate that the mean discharge differs from 12 ounces? Test using $\alpha = 0.10$.

7. Techside Real Estate, Inc. is a research firm that tracks the cost of apartment rentals in Southwest Virginia. In mid-2002, the regional average apartment rental rate was $895 per month. Assume that, based on the historical quarterly surveys, it is reasonable to assume that the population standard deviation is $225. In a current study of apartment rental rates, a sample of 180 apartments in the region provided the apartment rental rates. Do the sample data enable Techside Real Estate, Inc. to conclude that the population mean apartment rental rate now exceeds the level reported in 2002? The sample mean is $915 and the sample standard deviation is $227.50. Make your decision based on $\alpha = 0.10$.

8. Suppose that the national average price for used cars is $10,192. A manager of a local used car dealership reviewed a sample of 25 recent used car sales at the dealership in an attempt to determine whether the population mean price for the used cars at this particular dealership differed from the national mean. The prices for the sample of 25 cars are given in the data with a mean of $9750 and standard deviation of $1400. Test using $\alpha = 0.05$ whether a difference exists in the mean price for used cars at the dealership.

9. Suppose you are responsible for auditing invoices. Historically, about 0.003 of the invoices possessed material errors. During the last audit cycle a number of suggestions were made and implemented, and you hope that the next audit will provide evidence of improvement in the error rate. An audit of 6000 recent invoices reveals 12 material errors.

    a. Do the data suggest an improvement in the invoice error rate at $\alpha = 0.05$?

    b. Compute the $P$-value of the test statistic.

    c. Based on the $P$-value, would the decision change at $\alpha = 0.10$?

10. After completing Chemistry 101, Tommy Walker decides to conduct an experiment on his favorite brand of whiskey to determine if the proof rating on the bottle is accurate. He selects eight small eighty-proof bottles from different stores around town and measures the percent of alcohol in each bottle. (**Note:** 80-proof alcohol contains 40% alcohol.)

    The resulting measurements are as follows.

| Percent of Alcohol per Bottle | | | | | | | |
|---|---|---|---|---|---|---|---|
| 38% | 40% | 42% | 41% | 39% | 38% | 40% | 38% |

    a. What is the population being studied?

    b. What is the variable being measured?

    c. What level of measurement does the data possess?

    d. Can Tommy conclude that the actual proof of whiskey is not equal to 80 at $\alpha = 0.05$?

    e. What assumption did Tommy make in performing the test in part **d.**?

11. You have decided to become a professional gambler specializing in roulette. If the roulette wheel is fair (each number has a $\frac{1}{38}$ chance) then you will lose in the long run. However, you plan to locate wheels that are not balanced properly. An unbalanced wheel will produce some numbers more often than expected. You believe that you have found such a wheel and have started keeping track of the number 29. After 420 spins of the wheel, the number 29 has been observed 14 times. Is this overwhelming evidence at the $\alpha = 0.05$ level that you should start betting heavily on the number 29?

12. A commercial airline is concerned over the increase in weight of a carry-on luggage. In the past, the airline has estimated that the average piece of carry-on luggage will weigh 12 pounds. A random selection of 148 pieces of carry-on luggage has an average weight of 14.2 pounds and the population standard deviation is 3.4 pounds. Do you think that the airline's concern is justified? Use $\alpha = 0.01$.

13. Consider the following large sample hypothesis tests for the population mean. Compute the $P$-value for each test and decide whether you would reject or fail to reject the null hypothesis at $\alpha = 0.01$.

   a. $H_0: \mu = 15$, $H_a: \mu > 15$, $z = 2.50$

   b. $H_0: \mu = 80$, $H_a: \mu < 80$, $z = -1.95$

   c. $H_0: \mu = 1200$, $H_a: \mu \neq 1200$, $z = 3.70$

14. Deli Delivery delivers sandwiches to neighboring office buildings during lunch time in New York City. The deli claims that the sandwiches will be delivered within 20 minutes from receiving the order. Given the hectic schedules of their customers, consistent delivery time is a must. The owner has decided that the standard deviation of delivery times should be at most 4 minutes. To determine how consistently the sandwiches are being delivered, the manager randomly selects 27 orders and measures the time from receiving the order to delivery of the sandwich. The average time to delivery of the sample was 20 minutes with a standard deviation of 4.5 minutes.

   a. Will the manager conclude at $\alpha = 0.10$ that the delivery times vary more than the owner desires?

   b. What assumption did you make about the delivery times in performing the test in part **a.**?

15. Consider the following small sample hypothesis tests for the population mean. Compute the $P$-value for each of the tests and decide whether you would reject or fail to reject the null hypothesis at $\alpha = 0.05$.

   a. $H_0: \mu = 12$, $H_a: \mu > 12$, $t = 1.75$, $n = 25$

   b. $H_0: \mu = 0.12$, $H_a: \mu < 0.12$, $t = -2.95$, $n = 16$

   c. $H_0: \mu = 55$, $H_a: \mu \neq 55$, $t = 2.35$, $n = 8$

16. In each of the following experimental situations, give the appropriate null and alternative hypotheses to be tested. Define all terms that appear in these hypotheses.

   a. A random sample of 100 customers in a bank are selected and their times to be served are noted. The bank has recently retrained its tellers to be more efficient with the hope of decreasing its average time in servicing its customers, which has been 4 minutes in the past.

   b. A local driver training school claims that at least 75% of its pupils pass the driving test on their first attempt. A sample of 60 students from the school are selected, and their performances on the driving test are noted. Based upon the data collected, we would like to refute the claim of the school.

   c. A spokesperson for a popular diet claims that the average weight lost for someone on the diet will be at least 15 pounds over a two-month period. The amount of weight lost for each person in a sample of 10 people on the diet is determined in order to try to refute the claim of the diet spokesperson.

   d. A tire company tests 68 of its new premium tires to determine if the average lifespan of the tire is more than the average lifespan of its major competitor's best tire. The average lifespan of the competitor's tire is 63,000 miles.

   e. An elementary statistics student conducts an experiment in order to show that a coin from a magic kit is biased. The student flips the coin 500 times.

17. It is essential in the manufacture of machinery to utilize parts that conform to specifications. In the past, diameters of the ball bearings produced by a certain manufacturer had a variance of 0.00156. To cut costs, the manufacturer instituted a less expensive production method. The variance of the diameters of 101 randomly sampled bearings produced by the new process was 0.0021. Do the data provide sufficient evidence to indicate that the diameters of ball bearings produced by the new process are more variable than those produced by the old process? Test using $\alpha = 0.10$.

18. A national news magazine is interested in the proportion of counties in which the cost of living has decreased in the past 24 months. The news magazine believes that the true proportion is less than 30%. In a random sample of 100 counties, 20 counties had cost of living decreases.

    a. Test the news magazine's claim at $\alpha = 0.08$.

    b. Find the P-value for this test.

19. An increasing number of businesses are offering child-care benefits for their workers. However, one union claims that more than 90% of firms in the manufacturing sector still do not offer any child-care benefits to their workers. A random sample of 350 manufacturing firms is selected, and only 28 of them offer child-care benefits.

    a. Does this sample result support the claim of the union? Test using $\alpha = 0.10$.

    b. Calculate the P-value associated with this test.

20. During the holiday season, law enforcement officials estimated that 500 people would be killed and 25,000 injured on the nation's roads. They claimed that more than 50% of the accidents would be caused by drunk driving. A sample of 120 accidents showed that 67 were caused by drunk driving. Use these data to test their claim with $\alpha = 0.05$.

21. The quality control supervisor of a cannery is concerned about the variance of fill per can. Regulatory agencies specify that the standard deviation of the amount of fill should be less than 0.1 ounce. To determine whether the process is meeting this specification, the supervisor randomly selects ten cans, weighs the contents of each, and finds that the sample standard deviation of these measurements is 0.04.

    Do these data provide sufficient evidence to indicate that the variability is as small as desired? Test using $\alpha = 0.05$.

## P  Discovery Project

### Wearables in the Sports Industry?

Apolo, a startup which manufactures smart devices, recorded 500 responses from athletes, trainers, coaches, and team physicians regarding their interest in wearing Apolo smart devices and their beliefs on whether wearables improve their performance. These responses are recorded in the Wearables in the Sports Industry data set. As discussed at the beginning of this chapter, the use of wearables is a multibillion-dollar industry worldwide for companies such as Apple, Fitbit, and other organizations that produce wearable technology to track fitness and nutrition.

The individuals who participated in the sample were athletes or those who work with athletes. All participants were 18 years or older. The variables measured were Age and Role (coach, team physician, athlete, or trainer), and three questions were asked:

1.  Do you use Apolo smart devices to train?
2.  Do you feel that Apolo smart devices have helped you improve your performance?
3.  Would you recommend the use of Apolo smart devices to train?

The responses to these questions were recorded as 0 (No) or 1 (Yes). If the answer to question 1 was yes, then a fourth question was asked:

4.  How many hours per week do you use Apolo smart devices to train?

If the participant did not use Apolo smart devices, then 0 was recorded for the hours.

Using the aforementioned data, please do the following:

1.  Summarize the data by role indicating the number of coaches, team physicians, athletes, and trainers that chose to participate in the survey.
2.  Summarize the data according to the role of the participants and the first three questions asked of them. That is, create a table indicating the number of people in each role who answered "Yes" to each of the first three questions.
3.  Apolo believes that they will be profitable if more than 40% of survey respondents use smart devices. Test that Apolo will be profitable using a significance level of 5%.
4.  Some managers at Apolo believe that the responses of the athletes are the only ones that matter. Find the $P$-value for the hypothesis test that at least 40% of the athletes would use Apolo smart devices. Do the data suggest that more than 40% of the athletes will use Apolo smart devices at a 5% level of significance?
5.  For Apolo customers (i.e., the survey participants who answered "Yes" to "Do you use Apolo smart devices to train?"), calculate the sample mean and sample standard deviation for the number of hours that they use an Apolo smart device per week.
6.  Apolo believes that in order to retain customers, the customers need to be using smart devices more than 6 hours per week. Test the hypothesis that, on average, Apolo customers use their smart device more than 6 hours per week. Use a significance level of 1%.

### Data

The data set can be found on stat.hawkeslearning.com under **Discovering Business Statistics, Second Edition > Data Sets > Wearables in the Sports Industry**.

7. Apolo would like to know a probable range of values for the mean hours customers use a smart device per week. Construct a 99% confidence interval for the mean hours Apolo customers use a smart device per week. What can Apolo conclude about the mean?

| Sample of the Data Collected | | | | | |
|---|---|---|---|---|---|
| Gender | Age | Role | Do you use Apolo smart devices to train? | Apolo smart device improved performance? | Recommend use of Apolo smart device? |
| Male | 40 | Coach | 1 | 0 | 1 |
| Male | 35 | Team Physician | 1 | 1 | 1 |
| Female | 48 | Team Physician | 0 | 0 | 0 |
| Male | 60 | Team Physician | 0 | 0 | 1 |
| Female | 18 | Athlete | 1 | 0 | 1 |
| Male | 47 | Team Physician | 0 | 0 | 1 |
| Female | 39 | Trainer | 0 | 0 | 0 |
| Female | 20 | Athlete | 1 | 0 | 1 |

11

# Chapter 11

## Inferences about Two Samples

## Discovering the Real World

### Smartphone Screen Time

In an article in The Drum Network (www.thedrum.com), it was stated that nearly four out of ten people now use screen time tracking as a way to monitor their own or children's phone usage. However, in spite of using these apps, there does not appear to be much concern about how long they are spending online. According to findings, the average screen time in the UK is three hours and 23 minutes per day which amounts to more than 50 days per year (actually, it's 51.45 days). Seeing these exorbitant screen times, it was interesting to see how screen time tracking varied by age group. The table below contains the percentage of respondents who track their screen time usage by age group.

The link below is the source used for the data on the right.

http://hawkes.biz/screentimestats

| Screen Time Tracking | | | | | |
|---|---|---|---|---|---|
| Age Group (in Years) | 16-24 | 25-34 | 35-44 | 45-54 | 55+ |
| Percentage | 59.37 | 53.75 | 41.88 | 25.38 | 18.45 |

In the scenario above, we may want to know if there is a significant difference between any two of the age groups. For example, do the survey participants in the 45–54-year-old age group track their screen time more than the participants in the 55 and over age group? While we can see that the percentage of 45–54-year-olds who track their screen time is more than the percentage of those 55 years old and older, we would need to conduct a formal hypothesis test to determine if these two percentages (i.e., proportions) are significantly different. This type of hypothesis test would be carried out using a two-sample test of proportions by collecting samples from two independent populations (those survey participants in each of the respective age groups).

In Chapter 10, we examined how to test if the means and proportions were equal to some prespecified value from a single population. In this chapter, we will examine comparing means, proportions, and variances between two independent populations. Additionally, we will discuss a paired differences test in which the subjects in the sample are considered homogeneous but separated into two groups or we have two measurements (say, before and after) on the same subject.

### Introduction

Basic concepts of hypothesis testing were introduced in Chapter 10. In this chapter we will continue to follow the same basic hypothesis testing procedure developed in the previous chapter. The inferential methods that were previously developed concerned one population and the value of its mean, proportion, or variance. We use the information in the sample statistics $\bar{x}$, $\hat{p}$, and $s^2$ to test hypotheses about the population parameters.

There are many instances in which we wish to compare two population means or two population proportions. For example, in Chapter 2 we discussed controlled experiments in which we tried to develop data that would be used to efficiently test hypotheses about two or more proportions. In this chapter, we will begin to explore methods for testing population parameters of this type.

In particular, we will develop methods for comparing two population means, and try to answer questions such as:

Is $\mu_1 > \mu_2$?             Is $\mu_1 < \mu_2$?             Is $\mu_1 \neq \mu_2$?

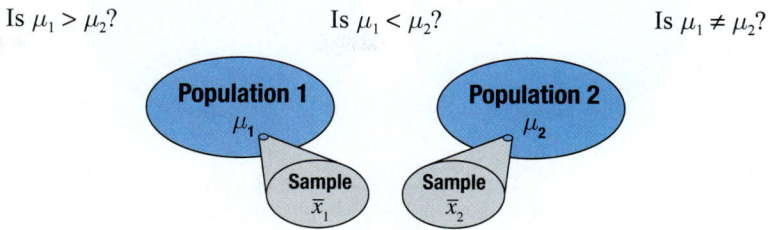

Similarly, we will develop methods for comparing two population proportions, and try to answer questions such as:

Is $p_1 > p_2$?             Is $p_1 < p_2$?             Is $p_1 \neq p_2$?

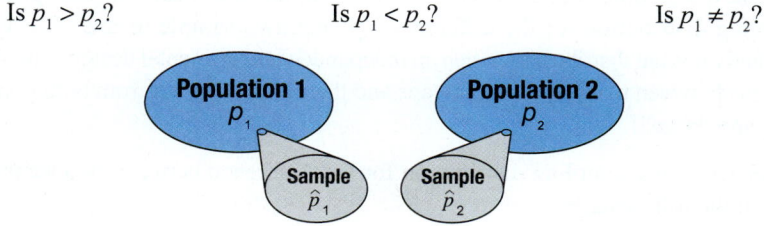

# 11.1 Comparing Two Population Means, $\sigma_1$ and $\sigma_2$ Known

We developed procedures for testing a hypothesis about a single population mean in Section 10.2. In this section, a procedure is developed for comparing two population means when independent samples are drawn from populations with known population standard deviations. In Section 11.2 we will develop a procedure that is used when the samples are independent samples drawn from populations where the population standard deviations are unknown.

There are many situations where our interest is in comparing two population means or the average response of experimental units to two different treatments. For example, a marketing professor may be interested in students' grades when using PowerPoint slides as a delivery method during lecture as opposed to when using the blackboard. A fleet manager may be interested in comparing the average gas mileage for two different makes of cars. An economist may be interested in the difference between costs of living in industrialized and nonindustrialized countries.

In each of the above situations, an experiment can be designed for sampling from two separate populations (two different makes of cars or groups of students), or the experiment can be designed for randomly assigning experimental units to two different treatments (students receiving lectures virtually vs. in-person).

When observations are randomly selected from two independent populations or experimental units are randomly assigned to two different treatments where the variation between experimental units is small, the sampling design is called an **independent experimental design** or a **completely randomized design**.

> **Definition**
>
> Independent Experimental Design
>
> When experimental units are randomly assigned to two different treatments where the variation between experimental units is small, the sampling design is called an **independent experimental design**.

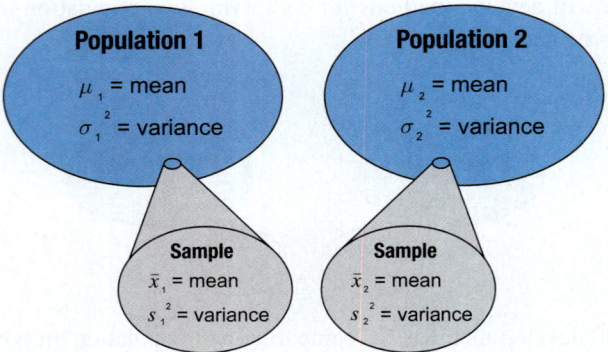

We will use the sample means $\bar{x}_1$ and $\bar{x}_2$ to compare the means of two populations. The sampling distribution for the difference between two sample means, $\bar{x}_1 - \bar{x}_2$, has an approximately normal distribution (when an independent experimental design is used to make comparisons between two population means and the samples drawn from both populations, $n_1$ and $n_2$, are "large").

The properties of the sampling distribution for the difference between two sample means are given in the following box.

## Properties of the Sampling Distribution of $\bar{x}_1 - \bar{x}_2$

1. If $n_1 \geq 30$ and $n_2 \geq 30$, the sampling distribution of $\bar{x}_1 - \bar{x}_2$ has an approximately normal distribution.

2. $\mu_{\bar{x}_1 - \bar{x}_2} = \mu_1 - \mu_2$

3. $\sigma_{\bar{x}_1 - \bar{x}_2} = \sqrt{\dfrac{\sigma_1^2}{n_1} + \dfrac{\sigma_2^2}{n_2}}$ if the two samples are independent.

**Note:** Regardless of the sample size, if a sample of size $n_1$ and a sample of size $n_2$ are collected from two independent normal distributions, respectively, then $\bar{x}_1 - \bar{x}_2$ follows a normal distribution.

The sampling distribution of $\bar{x}_1 - \bar{x}_2$ will be used in the development of the confidence interval and test statistic for comparing two population means.

## Interval Estimation of $\mu_1 - \mu_2$

Given the sampling distribution of $\bar{x}_1 - \bar{x}_2$, we can develop the confidence interval estimate of the difference between two population means.

### Formula

#### $100(1 - \alpha)\%$ Confidence Interval for $\mu_1 - \mu_2$

The $100(1 - \alpha)\%$ confidence interval estimate for the difference in the population means of two independent populations is given by

$$(\bar{x}_1 - \bar{x}_2) \pm z_{\alpha/2} \sqrt{\dfrac{\sigma_1^2}{n_1} + \dfrac{\sigma_2^2}{n_2}}$$

if the population variances, $\sigma_1^2$ and $\sigma_2^2$, are known and $z_{\alpha/2}$ is the critical value of the standard normal distribution with an area of $\alpha/2$ in the upper tail.

## Hypothesis Testing about $\mu_1 - \mu_2$

When comparing the means of two independent populations, you take a random sample of size $n_1$ from the first population and a random sample of size $n_2$ from the second population.

When both $\sigma_1$ and $\sigma_2$ are known, the following test statistic is valid.

---

**Formula**

### Test Statistic for a Hypothesis Test about $\mu_1 - \mu_2$

If $\sigma_1^2$ and $\sigma_2^2$ are known, then the value of the test statistic for a hypothesis test about $\mu_1 - \mu_2$ is given by

$$z = \frac{(\bar{x}_1 - \bar{x}_2) - (\mu_1 - \mu_2)}{\sqrt{\dfrac{\sigma_1^2}{n_1} + \dfrac{\sigma_2^2}{n_2}}}$$

where $\bar{x}_1$ is the sample mean for Population 1, $n_1$ is the sample size for Population 1, $\sigma_1$ is the standard deviation of Population 1, $\bar{x}_2$ is the sample mean for Population 2, $n_2$ is the sample size for Population 2, and $\sigma_2$ is the standard deviation of Population 2.

Note that $(\mu_1 - \mu_2)$ represents the hypothesized difference between the two population means and that the $z$-statistic follows the standard normal distribution.

---

Often, the goal of a hypothesis test for the difference between two population means is to determine if the population means are different. In this case, the hypothesized difference between the population means will be 0. When the hypothesized difference between the population means is zero, the formula for the test statistic is simplified to

$$z = \frac{\bar{x}_1 - \bar{x}_2}{\sqrt{\dfrac{\sigma_1^2}{n_1} + \dfrac{\sigma_2^2}{n_2}}}.$$

Constructing a confidence interval and performing a hypothesis test for two population means will be developed in Example 11.1.1.

---

A telecommunications analyst is interested in knowing if there is a significant difference in the average quality of service (QOS) between cable television subscribers and satellite television subscribers. She randomly selects 50 cable subscribers and 50 satellite subscribers for the study from normally distributed populations. She gives each subscriber a survey and asks them to complete it. She tallies the results of the survey and calculates a mean QOS for each service. The QOS is ranked on a scale of 1–10, with the higher value implying that subscribers are more satisfied with their service. The results of the study are shown in Table 11.1.1.

**Example 11.1.1**

**Calculating a Confidence Interval and Performing a Hypothesis Test for the Difference in Means**

### Table 11.1.1 — QOS Scores for Subscribers

|           | $n$ | $\bar{x}$ | $\sigma$ |
|-----------|-----|-----------|----------|
| Cable     | 50  | 8.8       | 3        |
| Satellite | 50  | 9.5       | 2        |

a. Calculate a 95% confidence interval for the mean difference in average QOS between cable and satellite subscribers.

b. Is there persuasive evidence for the analyst to conclude at $\alpha = 0.05$ that there is a difference in mean QOS between cable and satellite subscribers?

## ∞ Technology

For technology instructions to calculate a confidence interval using the normal ($z$) distribution, visit stat.hawkeslearning.com and navigate to **Discovering Business Statistics, Second Edition > Technology Instructions > Confidence Intervals > Two Sample $z$-Interval.**

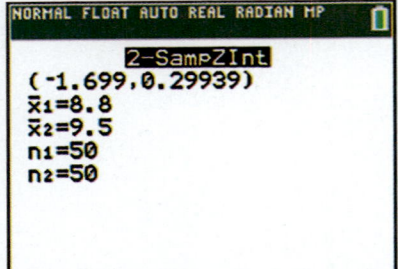

### Hypothesis Testing Is Not Loved by All Statisticians

The view is expressed in an article by Marks R. Nester entitled "A Myopic View and History of Hypothesis Testing".

"I contend that the general acceptance of statistical hypothesis testing is one of the most unfortunate aspects of 20th century applied science. Tests for the identity of population distributions, for equality of treatment means, for presence of interactions, for the nullity of a correlation coefficient, and so on, have been responsible for much bad science, much lazy science, and much silly science. A good scientist can manage with, and will not be misled by, parameter estimates and their associated standard errors or confidence limits. A theory dealing with the statistical behavior of populations should be supported by rational argument as well as data. In such cases, accurate statistical evaluation of the data is hindered by null hypothesis testing. The scientist must always give due thought to the statistical analysis, but must never let statistical analysis be a substitute for thinking!"

## SOLUTION

**a.** From the information given in the problem, we know the following.

$$n_1 = 50, \; n_2 = 50, \; \sigma_1 = 3, \; \sigma_2 = 2, \; \overline{x}_1 = 8.8, \text{ and } \overline{x}_2 = 9.5$$

We know that the population standard deviations are known and the data are collected from two normally distributed populations.

Since we want a 95% confidence interval, we need to find the value of $z_{\alpha/2}$. Remember that $z_{\alpha/2}$ represents the $z$-value required to obtain an area of $1 - \alpha$ centered under the standard normal curve. Therefore, for a 95% confidence interval, $\alpha = 0.05$.

For $\alpha = 0.05$, $z_{\alpha/2} = z_{0.025} = 1.96$.

A 95% confidence interval for the difference in the average QOS between cable and satellite providers is given by

$$\left( \overline{x}_1 - \overline{x}_2 \right) \pm z_{\alpha/2} \sqrt{ \frac{\sigma_1^2}{n_1} + \frac{\sigma_2^2}{n_2} }.$$

This gives:

$$\left( 8.8 - 9.5 \right) \pm 1.96 \sqrt{ \frac{3^2}{50} + \frac{2^2}{50} }$$

$$-0.7 \pm 1.96 \sqrt{ \frac{13}{50} }$$

$$-1.7 \text{ to } 0.3.$$

Thus, we are 95% confident that the true mean difference in QOS between cable and satellite subscribers is between $-1.7$ and $0.3$. Another way to interpret this interval in terms of the problem is that the average QOS for satellite subscribers is between 1.7 rating points higher and 0.3 rating points lower than cable subscribers. Since the confidence interval includes zero, this indicates that the difference between the average QOS for cable and satellite television subscribers is not statistically significant at the 0.05 level.

**b. Step 1:** Determine the null hypothesis. In this process, select the appropriate statistical measure, such as the population means, proportions, or variances.

The premise of the problem is that there is no difference in average QOS between cable and satellite subscribers. Since the analyst is interested in comparing the average QOS between cable and satellite subscribers, the appropriate statistical measures are the following.

$$\mu_1 = \text{the true mean QOS for cable subscribers}$$

$$\mu_2 = \text{the true mean QOS for satellite subscribers}$$

Thus, the null hypothesis is that there is no difference in average QOS between cable and satellite subscribers which should be written as

$$H_0 \colon \mu_1 - \mu_2 = 0.$$

**Step 2:** Determine the alternative hypothesis and whether it should be one-sided or two-sided.

The analyst is interested in whether or not there is a difference in the QOS between cable and satellite subscribers. Thus, the alternative hypothesis is two-sided, and the test is two-tailed and should be written as

$$H_a \colon \mu_1 - \mu_2 \neq 0.$$

**Step 3:** Select the appropriate test statistic based on the information at hand and the assumptions you are willing to make.

The difference in sample means, $\bar{x}_1 - \bar{x}_2$, is an unbiased point estimate of the difference in population means, $\mu_1 - \mu_2$. The sampling distribution for the point estimate, $\bar{x}_1 - \bar{x}_2$, provides essential information for determining the test statistic. Since the data are collected from two normally distributed populations and the variances for each population are known, the sampling distribution $\bar{x}_1 - \bar{x}_2$ follows a normal distribution with mean, $\mu_{\bar{x}_1 - \bar{x}_2} = \mu_1 - \mu_2$, and standard deviation,

$$\sigma_{\bar{x}_1 - \bar{x}_2} = \sqrt{\frac{\sigma_1^2}{n_1} + \frac{\sigma_2^2}{n_2}}.$$

Thus, the test statistic is a standard normal random variable which is calculated by using the familiar $z$-transformation. The test statistic is given by

$$z = \frac{\left(\bar{x}_1 - \bar{x}_2\right) - \mu_{\bar{x}_1 - \bar{x}_2}}{\sigma_{\bar{x}_1 - \bar{x}_2}} = \frac{\left(\bar{x}_1 - \bar{x}_2\right) - \overbrace{\left(\mu_1 - \mu_2\right)}^{\substack{\text{Hypothesized}\\ \text{Difference}\\ \text{in Means}}}}{\sqrt{\dfrac{\sigma_1^2}{n_1} + \dfrac{\sigma_2^2}{n_2}}}.$$

If the null hypothesis is true, $z$ has an approximately normal distribution. If the observed value of $\bar{x}_1 - \bar{x}_2$ is significantly larger or smaller than $\mu_1 - \mu_2$, the hypothesized value of the difference, this will produce a large or small value of the test statistic, causing us to doubt whether or not the null hypothesis is in fact true. How large is *large*? The critical value of the test statistic specified in **Step 4** will implicitly define the notion of "large".

**Step 4:** Determine the critical value of the test statistic.

The significance level of the test is specified in the problem as $\alpha = 0.05$.

The role of the critical value(s) in this test is exactly the same as for all of the hypothesis tests discussed earlier. It defines a range of values for the test statistic, the rejection region, that will be so rare that it is unlikely that the test statistic occurred from ordinary sampling variability, assuming the null is true. The level of the test defines the size of the rejection region. Should the computed value of the test statistic fall within the rejection region, the null hypothesis will be rejected.

If the null hypothesis is true, the test statistic has an approximately normal distribution. Thus the critical value is determined in the same way as for the other tests of hypothesis in which the test statistic had an approximately normal distribution. The rejection region for a two-tailed test with level of significance, $\alpha = 0.05$, is displayed in Figure 11.1.1.

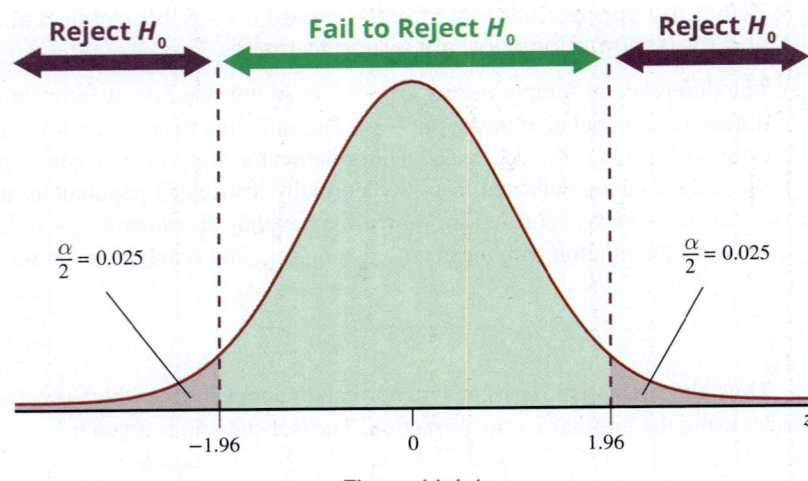

**Figure 11.1.1**

## ☁ Technology

For technology instructions to do a two-sample hypothesis test of the means using the normal (*z*) distribution, visit stat.hawkeslearning.com and navigate to **Discovering Business Statistics, Second Edition > Technology Instructions > Hypothesis Testing > Two Sample *z*-Test**.

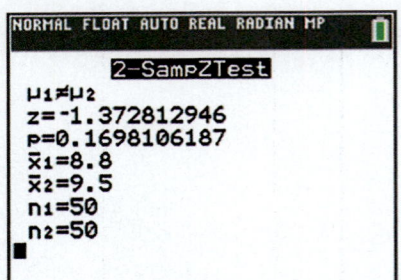

The null hypothesis will be rejected if the computed value of the test statistic is greater than or equal to 1.96 or less than or equal to −1.96. In other words, we will reject the null hypothesis if the observed difference in the average QOS scores is at least 1.96 standard deviations above or below the hypothesized value of 0 (no difference in QOS between cable and satellite).

**Step 5:** Collect the sample data and compute the value of the test statistic.

Based on the data in Table 11.1.1, the computed value of the test statistic is given by

$$z = \frac{(8.8 - 9.5) - 0}{\sqrt{\dfrac{3^2}{50} + \dfrac{2^2}{50}}} \approx -1.37 \cdot$$

**Step 6:** Make the decision and state the conclusion in terms of the original question.

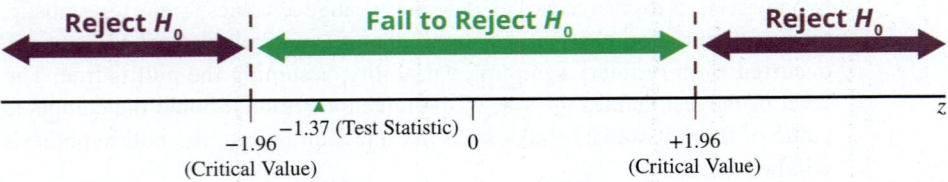

**Figure 11.1.2**

As shown in Figure 11.1.2, the calculated value of the test statistic does not fall in the rejection region because −1.37 falls between the critical values −1.96 and 1.96. There is insufficient evidence to conclude that the difference between the observed value and the hypothesized value is due to anything other than ordinary sampling variation. Thus, we fail to reject the null hypothesis at $\alpha = 0.05$.

Using the *P*-value approach, we want to find

$$2P(Z > |-1.37|) = 2P(Z > 1.37) = 2(0.0853) = 0.1706.$$

You may recall that the decision rule when using the *P*-value approach is that we reject the null hypothesis if the *P*-value is $< \alpha$. Thus, since the *P*-value is 0.1706, which is greater than 0.05, we fail to reject the null hypothesis.

*Conclusion and Interpretation:* There is not sufficient evidence at the 0.05 level to conclude that the average QOS is significantly different between cable and satellite television subscribers.

# 📝 11.1 Exercises

## Basic Concepts

1. What questions are we interested in answering when comparing two population means?

2. What is an independent experimental design?

3. Which sampling distribution do we use in the formulation of the test statistic when comparing two population means with population variances known? What are the properties of this distribution?

4. Does the determination of the critical value(s) for two-sample hypothesis tests differ from one-sample hypothesis tests?

5. What conditions are necessary to perform a test for the difference between two population means?

## Exercises

6. Determine the critical value(s) of the test statistic for each of the following tests for the comparison of two population means where the population standard deviations are known.

   a. Left-tailed test, $\alpha = 0.05$

   b. Right-tailed test, $\alpha = 0.10$

   c. Two-tailed test, $\alpha = 0.01$

7. Determine the critical value(s) of the test statistic for each of the following tests for the comparison of two population means where the population standard deviations are known.

   a. Left-tailed test, $\alpha = 0.04$

   b. Right-tailed test, $\alpha = 0.08$

   c. Two-tailed test, $\alpha = 0.02$

8. A luxury car dealer is considering two possible locations for a new auto mall. The rent on the south side of town is cheaper. However, the dealer believes that the average household income is significantly higher on the north side of town. The dealer has decided that he will locate the new auto mall on the north side of town if the results of a study that he commissioned show that the average household income is significantly higher on the north side of town. The results of the study are as follows.

| Income (Thousands of Dollars) | | | |
|---|---|---|---|
| | $n$ | $\bar{x}$ | $\sigma$ |
| North Side | 35 | 50 | 10 |
| South Side | 40 | 43 | 5 |

   a. Calculate a 90% confidence interval for the difference in average income between the north and south sides of town. Interpret the interval.

   b. Based on the study, will the auto dealer decide to locate the new auto mall on the north side of town? Use $\alpha = 0.05$.

9. An internal auditor for Tiger Enterprises has been asked to determine if there is a difference in the average amount charged for daily expenses by two top salesmen, Mr. Ellis and Mr. Ford. The auditor randomly selects 45 days and determines the daily expenses for each of the salesmen.

| Expenses (Dollars) | | | |
| --- | --- | --- | --- |
| | $n$ | $\bar{x}$ | $\sigma$ |
| Mr. Ellis | 45 | $55 | $8 |
| Mr. Ford | 45 | $60 | $3 |

a. Calculate a 95% confidence interval for the difference in the average amounts charged for daily expenses between Mr. Ellis and Mr. Ford. Interpret the interval.

b. Based on the survey, can the auditor conclude that there is a difference in the average amounts charged for daily expenses by the two top salesmen? Use $\alpha = 0.05$.

c. Explain how the 95% confidence interval in part **a.** would lead you to make the same decision that was made in part **b.**

10. The military has two different programs for training aircraft personnel. A government regulatory agency has been commissioned to evaluate any differences that may exist between the two programs. The agency administers standardized tests to randomly selected groups of students from the two programs. The results of the tests for the students in each of the programs are as follows

| Military Training Programs | | | |
| --- | --- | --- | --- |
| | $n$ | $\bar{x}$ | $\sigma$ |
| Program A | 50 | 85 | 10 |
| Program B | 55 | 87 | 9 |

a. Calculate a 99% confidence interval for the difference between the average scores of the two military programs. Interpret the interval.

b. Can the agency conclude that there is a difference in the average test scores of students in the two programs? Use $\alpha = 0.01$.

11. Tom Sealack, a supply clerk with the Navy, has been asked to determine if a new battery that has been offered to the Navy (at a reduced price) has a shorter average life than the battery they are currently using. He randomly selects batteries of each type and allows them to run continuously so that he can measure the time until failure for each battery. The results of the test are as follows.

| Battery Life (Hours) | | | |
| --- | --- | --- | --- |
| | $n$ | $\bar{x}$ | $\sigma$ |
| New Battery | 35 | 700 | 30 |
| Old Battery | 35 | 710 | 35 |

a. Do the data suggest at $\alpha = 0.10$ that the time until failure for the new battery is significantly less than the time until failure for the old battery?

b. Calculate the $P$-value for the test in **a.**

c. Based on the $P$-value, would the decision change at $\alpha = 0.05$?

**12.** The City Bank believes that checking account balances are significantly larger for customers who are aged 40 to 49 than those who are aged 30 to 39. To investigate this belief, they randomly select customers from each age group and determine the average daily account balance for each customer for the current month. The results of the study are as follows.

| Checking Account Balances | | | |
|---|---|---|---|
| Age Group | $n$ | $\bar{x}$ | $\sigma$ |
| 30 – 39 | 200 | $2500 | $550 |
| 40 – 49 | 150 | $3500 | $950 |

  **a.** Do the data suggest at $\alpha = 0.05$ that the average daily account balances are significantly higher for the 40 to 49 age group than the 30 to 39 age group?

  **b.** Calculate the $P$-value for the test in **a.**

  **c.** Based on the $P$-value, would the decision change at $\alpha = 0.10$?

# 11.2 Comparing Two Population Means, $\sigma_1$ and $\sigma_2$ Unknown

It is still possible to make comparisons between two population means if the population standard deviations are unknown.

There are several assumptions that must be met which are outlined below.

## Assumptions

**Assumptions for Inferences about $\mu_1 - \mu_2$ when the Population Standard Deviations are Unknown**

1. An independent experimental design is used.

2. Both populations of interest are approximately normal.

3. Both of the populations have approximately equal (but unknown) variances, $\sigma_1^2 = \sigma_2^2 = \sigma^2$.

In order to determine if both populations of interest are approximately normal, it is helpful to draw histograms of the sample observations from each population. If these histograms appear to be approximately normal, then it is reasonable to infer this assumption is satisfied. With limited data, it is sometimes difficult to determine if the sample data are from a normal population. In these situations, you may have to assume normality and recognize that your inferences are predicated on the validity of the assumption. Figure 11.2.1 shows three histograms of sample data drawn from normal populations.

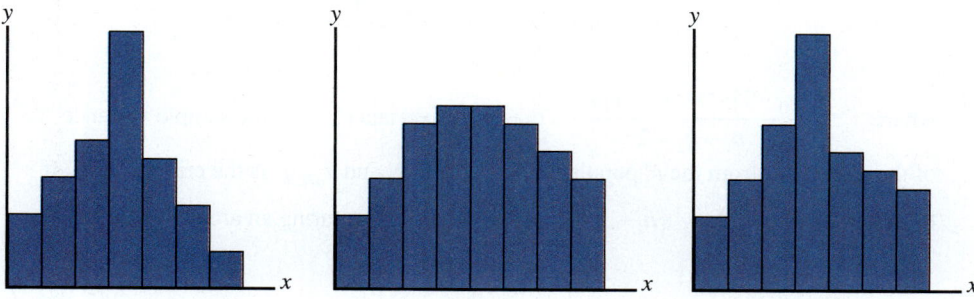

**Figure 11.2.1**

To examine the equal variance assumption, we can compare the variances of the samples drawn from the populations. In addition to comparing the sample variances we can draw box plots of the data from each of the distributions and compare the spread of the box plots. If the box plots look approximately the same, it is reasonable to assume that the variances of the two distributions are approximately equal. Figure 11.2.2 displays two box plots of sample data from populations with approximately equal variances, but with slightly different means.

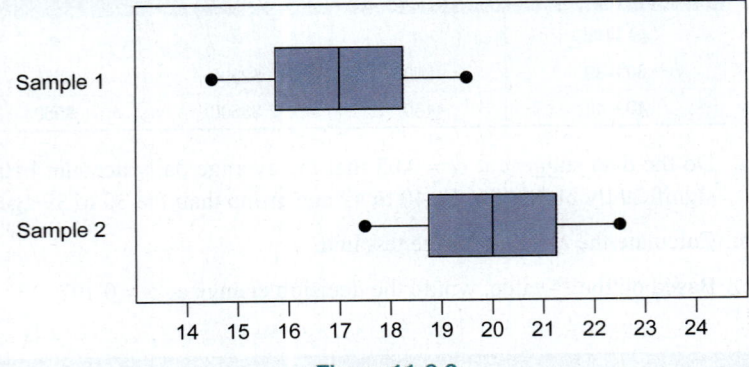

**Figure 11.2.2**

## Interval Estimation of $\mu_1 - \mu_2$, $\sigma_1$ and $\sigma_2$ Unknown

We can use an interval estimate to determine the difference in two independent population means. In Section 11.1 we used the following interval estimate for the case when samples were drawn from normal populations and $\sigma_1$ and $\sigma_2$ were known.

$$\left(\bar{x}_1 - \bar{x}_2\right) \pm z_{\alpha/2} \sqrt{\frac{\sigma_1^2}{n_1} + \frac{\sigma_2^2}{n_2}}$$

Assuming that the standard deviations, $\sigma_1$ and $\sigma_2$, are unknown, we will use the sample standard deviations, $s_1$ and $s_2$ to estimate the population standard deviations. However, since we are assuming that the variances are equal, we can estimate the standard error by

$$\sqrt{s_p^2 \left(\frac{1}{n_1} + \frac{1}{n_2}\right)}$$

where $s_p^2 = \dfrac{(n_1 - 1)s_1^2 + (n_2 - 1)s_2^2}{n_1 + n_2 - 2}$, and is called the **pooled variance**. In addition, we will assume that the data follow a normal distribution and replace $z_{\alpha/2}$ with $t_{\alpha/2, df}$.

> ### Definition
>
> **Pooled Variance**
>
> Since the variances are assumed to be equal, the **pooled variance**, $s_p^2$, is an estimate of the common variance with each variance being weighted by a function of its sample size. That is, the weight on $s_1^2$ is $(n_1 - 1)$ and the weight on $s_2^2$ is $(n_2 - 1)$.

> ### Formula
>
> #### $100(1 - \alpha)\%$ Confidence Interval for $\mu_1 - \mu_2$ Assuming Equal Variances
>
> Assuming equal variances, the $100(1 - \alpha)\%$ interval estimate for the difference between two population means if $\sigma_1$ and $\sigma_2$ are unknown is given by
>
> $$\left(\bar{x}_1 - \bar{x}_2\right) \pm t_{\alpha/2, df} \sqrt{s_p^2 \left(\frac{1}{n_1} + \frac{1}{n_2}\right)}$$
>
> where $s_p^2 = \dfrac{(n_1 - 1)s_1^2 + (n_2 - 1)s_2^2}{n_1 + n_2 - 2}$ is the pooled variance, $s_i^2$ is the sample variance
>
> of the data taken from the $i^{\text{th}}$ population ($i = 1$ and $2$), and $t_{\alpha/2, df}$ is the critical value of
>
> the $t$-distribution with $n_1 + n_2 - 2$ degrees of freedom, capturing an area of $\alpha/2$ in the upper tail.

## Hypothesis Test about $\mu_1 - \mu_2$, $\sigma_1$ and $\sigma_2$ Unknown and Variances Assumed Equal

Providing the assumptions previously outlined have been met, the test procedure shown below and developed in Example 11.2.1 may be used for making comparisons between two population means when the variances are assumed to be equal.

### Formula

**Inferences about $\mu_1 - \mu_2$ Assuming Equal Variances**

**Test Statistic:**

$$t = \frac{(\bar{x}_1 - \bar{x}_2) - (\mu_1 - \mu_2)}{\sqrt{s_p^2\left(\dfrac{1}{n_1} + \dfrac{1}{n_2}\right)}}$$

where $s_p^2 = \dfrac{(n_1 - 1)s_1^2 + (n_2 - 1)s_2^2}{n_1 + n_2 - 2}$ is the pooled variance, $s_i^2$ is the sample variance of the data taken from the $i^{th}$ population ($i = 1$ and $2$), and $t_{\alpha/2, df}$ is the critical value of the $t$-distribution with $n_1 + n_2 - 2$ degrees of freedom.

For a consumer product, the mean dollar sales per retail outlet last year in a sample of 15 stores was $3425 with a standard deviation of $200. For a second product, the mean dollar sales per outlet in a sample of 16 stores was $3250 with a standard deviation of $175. The sales amounts per outlet are assumed to be approximately normally distributed for both products. Also, assume the two products have equal population variances.

**Example 11.2.1**

**Calculating a Confidence Interval and Performing a Hypothesis Test for Two Population Means with Equal Variances**

| Table 11.2.1 – Retail Sales (Dollars) | | | |
|---|---|---|---|
| | $n$ | $\bar{x}$ | $s$ |
| Product 1 | 15 | $3425 | 200 |
| Product 2 | 16 | $3250 | 175 |

**a.** Calculate a 95% confidence interval for the difference in average dollar sales between the two products.

**b.** Test to see if Product 1 has a higher mean dollar sales record than Product 2. Use $\alpha = 0.01$.

### SOLUTION

**a.** From the information given in the problem, we know that

$$n_1 = 15, \; n_2 = 16, \; s_1 = 200, \; s_2 = 175, \; \bar{x}_1 = \$3425, \text{ and } \bar{x}_2 = \$3250.$$

With $\sigma_1$ and $\sigma_2$ unknown and assuming that the data follow a normal distribution with equal population variances, we estimate the population standard deviations with the sample standard deviations. Since we want a 95% confidence interval, we need to find the value of $t_{\alpha/2, df}$. Note that $df = n_1 + n_2 - 2 = 15 + 16 - 2 = 29$. Thus, $t_{0.025, 29} = 2.045$.

A 95% confidence interval for the difference in average dollar sales between the two products is given by

$$(\bar{x}_1 - \bar{x}_2) \pm t_{\alpha/2, df}\sqrt{s_p^2\left(\frac{1}{n_1} + \frac{1}{n_2}\right)} \quad \text{where } s_p^2 = \frac{(n_1 - 1)s_1^2 + (n_2 - 1)s_2^2}{n_1 + n_2 - 2}.$$

## ∞ Technology

For technology instructions to calculate a confidence interval for the difference between two population means using the *t*-distribution, visit stat.hawkeslearning.com and navigate to **Discovering Business Statistics, Second Edition > Technology Instructions > Confidence Intervals > Two Sample *t*-Interval**.

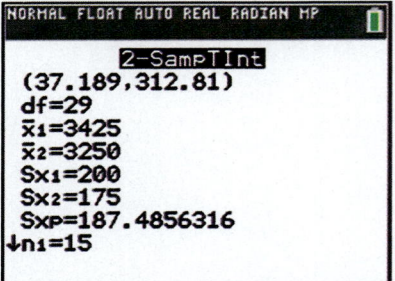

First we need to calculate $s_p^2$.

$$s_p^2 = \frac{(15-1)200^2 + (16-1)175^2}{15+16-2} \approx 35150.8621$$

The confidence interval is then calculated as follows.

$$(3425 - 3250) \pm 2.045\sqrt{35150.8621\left(\frac{1}{15} + \frac{1}{16}\right)}$$

$$175 \pm 2.045\sqrt{35150.8621\left(\frac{1}{15} + \frac{1}{16}\right)}$$

$$37.20 \text{ to } 312.80$$

Thus, we are 95% confident that the true mean difference in dollar sales between Product 1 and Product 2 is between \$37.20 and \$312.80. That is, on average, Product 1 generates between \$37.20 and \$312.80 more dollar sales than Product 2.

**b. Step 1:** Determine the null hypothesis. In this process, select the appropriate statistical measure, such as the population means, proportions, or variances.

Since the retailer is interested in comparing the average retail sales between the two products, the appropriate statistical measures are:

$$\mu_1 = \text{the true mean retail sales of Product 1}$$

$$\mu_2 = \text{the true mean retail sales of Product 2}$$

The null hypothesis is that the average sales for Product 1 is equal to the average sales for Product 2 which should be written as

$$H_0: \mu_1 - \mu_2 = 0.$$

**Step 2:** Determine the alternative hypothesis and whether it should be one-sided or two-sided.

The retailer is interested in whether the average sales of Product 1 is higher than the average sales of Product 2. Thus, the alternative hypothesis is one-sided and the test is one-tailed.

One way to state this in terms of the statistical measures is $\mu_1 > \mu_2$. But in order to perform the hypothesis test, this statement must be rewritten in a form for which we have a point estimate, or $\mu_1 - \mu_2 > 0$. The resulting alternative hypothesis is written as

$$H_a: \mu_1 - \mu_2 > 0.$$

**Step 3:** Select the appropriate test statistic based on the information at hand and the assumptions you are willing to make.

## ✎ NOTE

$\mu_1 - \mu_2$ in the *t*-test statistic is the hypothesized difference in means. In most problems this difference is hypothesized to be 0.

To develop the appropriate test statistic, a random variable whose value will be used to make the decision to reject or fail to reject $H_0$ must be found. The sampling distribution $\bar{x}_1 - \bar{x}_2$ provides essential information in assessing the rareness of the test statistic. If the assumptions previously outlined are satisfied, the sampling distribution of $\bar{x}_1 - \bar{x}_2$ has a *t*-distribution. The *t*-test statistic is given by

$$t = \frac{(\bar{x}_1 - \bar{x}_2) - (\mu_1 - \mu_2)}{\sqrt{s_p^2\left(\frac{1}{n_1} + \frac{1}{n_2}\right)}},$$

where $s_p^2 = \dfrac{(n_1 - 1)s_1^2 + (n_2 - 1)s_2^2}{n_1 + n_2 - 2}$, and is the pooled variance.

In this hypothesis test, the population variances are assumed to be equal. Thus $s_p^2$, which is the weighted average of the two sample variances, is used to approximate the unknown population variance.

If the null hypothesis is true, $t$ has a $t$-distribution with $n_1 + n_2 - 2$ degrees of freedom. If the observed value of $\bar{x}_1 - \bar{x}_2$ is significantly larger than the hypothesized value of the difference $H_0: \mu_1 - \mu_2 = 0$, this will produce a large value of the test statistic, causing us to doubt whether the null hypothesis is in fact true. How large is *large*? The critical value of the test statistic specified in **Step 4** will implicitly define the notion of "large."

**Step 4:** Determine the critical value of the test statistic.

The significance level of the test is specified as $\alpha = 0.01$. The role of the critical value in this test is exactly the same as for all of the hypothesis tests discussed earlier. It defines a range of values for the test statistic, the rejection region, that will be so rare that it is unlikely that it occurred from ordinary sampling variability if the null hypothesis is true. The level of the test defines the size of the rejection region. Should the computed value of the test statistic fall in the rejection region, we will presume that the value of the test statistic is too large to have occurred from ordinary sampling variation, and the null hypothesis will be rejected.

If the null hypothesis is true, the test statistic has a $t$-distribution. Thus, the critical value is determined in the same way as for the other tests of hypothesis where the test statistic had a $t$-distribution, except that the degrees of freedom are $n_1 + n_2 - 2$, or $15 + 16 - 2 = 29$. The rejection region corresponding to the alternative hypothesis, $H_a: \mu_1 - \mu_2 > 0$, with $\alpha = 0.01$ for a $t$-test statistic with 29 degrees of freedom is given in Figure 11.2.3.

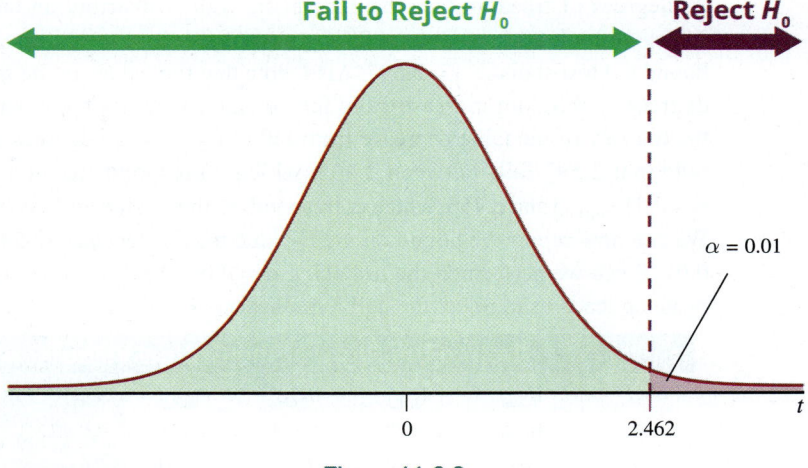

*t*-Distribution, *df* = 29

**Figure 11.2.3**

We will reject the null hypothesis if the computed value of the test statistic is greater than or equal to 2.462.

## Technology

For technology instructions to do a two-sample hypothesis test of the means using the *t*-distribution, please visit stat.hawkeslearning.com and navigate to **Discovering Business Statistics, Second Edition > Technology Instructions > Hypothesis Testing > Two Sample *t*-Test.**

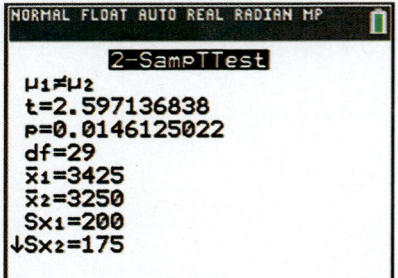

**Step 5:** Collect the sample data and compute the value of the test statistic.

Based on the data in Table 11.2.1, the computed value of the test statistic is given by

$$t = \frac{(3425 - 3250) - 0}{\sqrt{(35150.8621)\left(\frac{1}{15} + \frac{1}{16}\right)}} \approx 2.597$$

where $s_p^2 = \dfrac{(14)(200)^2 + (15)(175)^2}{15 + 16 - 2} \approx 35150.8621$.

**Step 6:** Make the decision and state the conclusion in terms of the original question.

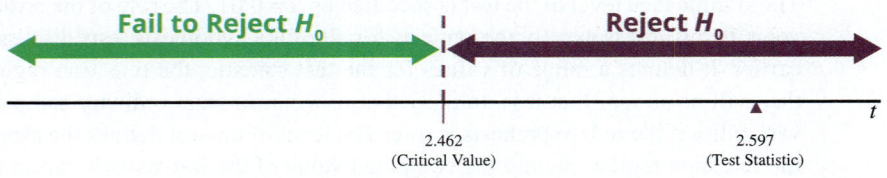

**Figure 11.2.4**

As displayed in Figure 11.2.4, the value of the test statistic does fall in the rejection region. In fact, a test statistic of 2.597 says that the observed difference between the means is about 2.6 standard deviations higher than the hypothesized difference of zero. It is highly unlikely that the difference between the observed value and the hypothesized value is due to ordinary sampling variation. Thus, we reject the null hypothesis at $\alpha = 0.01$.

Using the *P*-value approach, we have *P*-value $= P(t > 2.597)$. For this problem, however, we cannot find the exact *P*-value when using the *t*-table. Instead, we need to put bounds on the *P*-value. Note that for this problem, we have 29 degrees of freedom (see the excerpt from the *t*-distribution table below). Thus, we look at the *t*-table at 29 degrees of freedom to find the values that bound the test statistic, $t = 2.597$. Also, note that the values in the table for any degrees of freedom increase from left to right, whereas the $\alpha$ values across the top row of the table decrease from left to right. At 29 degrees of freedom, note that 2.597 falls between 2.462, which corresponds to the *t*-value with $\alpha = 0.010 (t_{0.010})$ and 2.756, which corresponds to the *t*-value with $\alpha = 0.005 (t_{0.005})$. We can now report the bound on the *P*-value for this test as $0.005 < P\text{-value} < 0.01$. Since we performed the test using $\alpha = 0.01$, the *P*-value is still less than $\alpha$ which leads us to reject the null hypothesis.

| df | $t_{0.200}$ | $t_{0.100}$ | $t_{0.050}$ | $t_{0.025}$ | $t_{0.010}$ | $t_{0.005}$ |
|---|---|---|---|---|---|---|
| \multicolumn{7}{c}{**Area to the Right of the Critical Value**} |
| 1 | 1.376 | 3.078 | 6.314 | 12.706 | 31.821 | 63.657 |
| 2 | 1.061 | 1.886 | 2.920 | 4.303 | 6.965 | 9.925 |
| | | | ... | | | |
| 27 | 0.855 | 1.314 | 1.703 | 2.052 | 2.473 | 2.771 |
| 28 | 0.855 | 1.313 | 1.701 | 2.048 | 2.467 | 2.763 |
| 29 | 0.854 | 1.311 | 1.699 | 2.045 | 2.462 | 2.756 |
| 30 | 0.854 | 1.310 | 1.697 | 2.042 | 2.457 | 2.750 |
| | | | ... | | | |

*Conclusion and Interpretation:* There is sufficient evidence at the 0.01 level to conclude that the average sales for Product 1 is significantly higher than the average sales for Product 2.

## *t*-Test about $\mu_1 - \mu_2$ Assuming Unequal Variances

Recall that one of the assumptions when performing the two-sample *t*-test is that the population variances are unknown but equal (that is, $\sigma_1^2 = \sigma_2^2 = \sigma^2$). If you cannot make the assumption that the variances are equal, then it is not appropriate to pool the two sample variances into one common variance, $s_p^2$. Instead, you must use a separate variance test developed by Satterthwaite (Satterthwaite, F.E., "An Approximate Distribution for Estimates of Variance Components," *Biometrics Bulletin*, 2 [1946]: 110–114). Satterthwaite's test procedure uses a series of computations that involve using the two separate sample variances to calculate the degrees of freedom for the test statistic. The computations for the $100(1 - \alpha)\%$ confidence interval and test statistic are as follows.

---

### Formula

#### Inferences about $\mu_1 - \mu_2$ Assuming Unequal Variances

**Confidence Interval:**

$$\left( \overline{x}_1 - \overline{x}_2 \right) \pm t_{\alpha/2, df} \sqrt{\frac{s_1^2}{n_1} + \frac{s_2^2}{n_2}}$$

**Test Statistic:**

$$t = \frac{\left( \overline{x}_1 - \overline{x}_2 \right) - \left( \mu_1 - \mu_2 \right)}{\sqrt{\dfrac{s_1^2}{n_1} + \dfrac{s_2^2}{n_2}}}.$$

which follows a *t*-distribution with degrees of freedom equal to:

$$df = \frac{\left( \dfrac{s_1^2}{n_1} + \dfrac{s_2^2}{n_2} \right)^2}{\dfrac{1}{n_1 - 1}\left( \dfrac{s_1^2}{n_1} \right)^2 + \dfrac{1}{n_2 - 1}\left( \dfrac{s_2^2}{n_2} \right)^2}.$$

The calculation for *df* should be rounded down to the nearest integer. Remember, these formulas should only be used if $\sigma_1$ and $\sigma_2$ are unknown, the data follow a normal distribution, and the population variances are assumed to be unequal.

---

**Example 11.2.2**

**Performing a Hypothesis Test for Two Population Means with Unequal Variances**

Consumers have long been interested in the difference in gas mileage between hybrid model vehicles and traditional gasoline engine models. The data below represent miles per gallon from two models of the Porsche Cayenne—one hybrid model and one gasoline model. The miles per gallon (MPG) readings are from ten vehicles that were deployed at rental agencies for everyday, normal use.

| Porsche Cayenne Hybrid Model (MPG) | Porsche Cayenne Gas Model (MPG) |
|---|---|
| 25 | 21 |
| 32 | 24 |
| 25 | 24 |
| 25 | 24 |
| 27 | 23 |
| 30 | 19 |
| 23 | 25 |
| 24 | 23 |
| 32 | 23 |
| 34 | 22 |

We want to test that the average MPG for the gasoline vehicle is the same as the hybrid vehicle. The summary statistics for this data are in the following table.

|  | Hybrid | Gasoline |
|---|---|---|
| $n$ | 10 | 10 |
| Mean MPG | 27.70 | 22.80 |
| Variance | 15.57 | 3.07 |

By examining the above statistics, assume that the data are collected from a normal distribution and that the variances of each type of vehicle are not equal. Test using a significance level of 5%.

### SOLUTION

**Step 1:** Determine the null hypothesis. In this process, select the appropriate statistical measure, such as the population means, proportions, or variances.

We want to compare the average MPG for each of the vehicle models. Thus, the statistical measure is the population mean MPG for the Porsche Cayenne Hybrid model ($\mu_1$) and the population mean MPG of the Porsche Cayenne Gas model ($\mu_2$). The null hypothesis is that there is no difference in average MPG between the Porsche Cayenne Hybrid and the Porsche Cayenne Gas models which should be written as

$$H_0: \mu_1 - \mu_2 = 0.$$

**Step 2:** Determine the alternative hypothesis and whether it should be one-sided or two-sided.

The alternative hypothesis contradicts the null hypothesis. There is nothing to indicate that we are interested in whether one model gets better gas mileage than the other. Thus, our research hypothesis is that there is a difference in average MPG between the Porsche Cayenne Hybrid and the Porsche Cayenne Gas models. The alternative hypothesis is two-sided and should be written as

$$H_a: \mu_1 - \mu_2 \neq 0.$$

**Step 3:** Select the appropriate test statistic based on the information at hand and the assumptions you are willing to make.

Assuming that the data from each of the populations follow a normal distribution and that the population variances are not equal, the test statistic will be

$$t = \frac{(\bar{x}_1 - \bar{x}_2) - (\mu_1 - \mu_2)}{\sqrt{\frac{s_1^2}{n_1} + \frac{s_2^2}{n_2}}},$$

which follows a $t$-distribution with the degrees of freedom equal to

$$df = \frac{\left(\frac{s_1^2}{n_1} + \frac{s_2^2}{n_2}\right)^2}{\frac{1}{n_1 - 1}\left(\frac{s_1^2}{n_1}\right)^2 + \frac{1}{n_2 - 1}\left(\frac{s_2^2}{n_2}\right)^2}.$$

Note that it is unlikely that the degrees of freedom will be an integer, thus, we will round $df$ down to the nearest integer.

**Step 4:** Determine the critical value of the test statistic.

Since we have a two-tailed alternative hypothesis and a significance level of 5% (i.e., $\alpha = 0.05$), the critical value will be $t_{\alpha/2, df}$ where the degrees of freedom is calculated as follows.

$$df = \frac{\left(\dfrac{15.57}{10} + \dfrac{3.07}{10}\right)^2}{\dfrac{1}{10-1}\left(\dfrac{15.57}{10}\right)^2 + \dfrac{1}{10-1}\left(\dfrac{3.07}{10}\right)^2} \approx 12.42 \, .$$

We will use 12 degrees of freedom for this $t$-test which yields a critical value of $t_{0.05/2,12} = t_{0.025,12} = 2.179$. We will reject the null hypothesis if the test statistic is greater than or equal to 2.179 or if the test statistic is less than or equal to $-2.179$.

**Step 5:** Collect the sample data and compute the value of the test statistic.

Using the values in the table above, the test statistic is

$$t = \frac{(27.7 - 22.8) - 0}{\sqrt{\dfrac{15.57}{10} + \dfrac{3.07}{10}}} \approx 3.59 \, .$$

**Step 6:** Make the decision and state the conclusion in terms of the original question.

The critical values of the test statistic are $\pm 2.179$. Thus, since the test statistic, $t$, is greater than 2.179 we reject the null hypothesis in favor of the alternative.

Using the Technology Instructions for conducting the Two-Sample $t$-Test, we get $P$-value = 0.0035. Since we are performing the test using a significance level of 0.05, the $P$-value approach still leads us to reject the null hypothesis in favor of the alternative.

*Conclusion and Interpretation:* This indicates that there is evidence to conclude that the average miles per gallon between the Porsche Cayenne Hybrid model and the Porsche Cayenne Gas model are significantly different.

Note that we did not test whether the average miles per gallon for one model was more or less than the other, thus, the conclusion remains consistent with the stated hypotheses.

> ## ∞ Technology
>
> For technology instructions to do a two-sample hypothesis test of the means using the $t$-distribution, visit stat.hawkeslearning.com and navigate to **Discovering Business Statistics, Second Edition > Technology Instructions > Hypothesis Testing > Two-Sample $t$-Test.**

# ✏ 11.2 Exercises

## Basic Concepts

1. Why might large samples not be available when attempting to make inferences about two population means?

2. What assumptions are necessary to perform a test for the difference between two population means when the population variances are unknown?

3. What is the test statistic for an hypothesis test about two population means when the population variances are unknown? How does this statistic differ from the test statistic used in Section 11.1?

4. What is a pooled variance? Why is it used?

## Exercises

5.  Determine the critical value(s) of the test statistic for each of the following tests for the comparison of two population means where the assumptions of normality and equal variance have been satisfied.

    a.  Left-tailed test, $\alpha = 0.05$, $n_1 = 10$, $n_2 = 15$

    b.  Right-tailed test, $\alpha = 0.10$, $n_1 = 8$, $n_2 = 12$

    c.  Two-tailed test, $\alpha = 0.01$, $n_1 = 5$, $n_2 = 7$

6.  Determine the critical value(s) of the test statistic for each of the following tests for the comparison of two population means where the assumptions of normality and equal variance have been satisfied.

    a.  Left-tailed test, $\alpha = 0.025$, $n_1 = 13$, $n_2 = 25$

    b.  Right-tailed test, $\alpha = 0.005$, $n_1 = 7$, $n_2 = 18$

    c.  Two-tailed test, $\alpha = 0.10$, $n_1 = 15$, $n_2 = 15$

7.  *Popular Science* (Vol. 242, No. 3) reported the results of a comparison of several popular minivans. One of the features that they compared was the time required to accelerate from 0 to 60 miles per hour in seconds. The Dodge Grand Caravan ES was able to accelerate from 0 to 60 mph in 11.3 seconds, on average. The Volkswagen Eurovan took 16.5 seconds on average to accelerate from 0 to 60 mph. Suppose that 15 minivans of each type were tested and that the sample standard deviation of the times required to accelerate from 0 to 60 for each minivan was 4 seconds. Assume that the population variances are approximately equal.

    a.  Calculate a 95% confidence interval for the difference in average acceleration time between the two types of minivans. Interpret the interval.

    b.  Do the data suggest that there is a significant difference in the time required to accelerate from 0 to 60 between the two types of minivans at $\alpha = 0.05$?

    c.  What assumptions did you make about the time required to accelerate from 0 to 60 mph in calculating the confidence interval in part **a.** and for performing the test in part **b.**?

8.  A cereal manufacturer has advertised that its product, Fiber Oat Flakes, has a lower fat content than its competitor, Bran Flakes Plus. Because of complaints from the manufacturers of Bran Flakes Plus, the FDA has decided to test the claim that Fiber Oat Flakes has a lower average fat content than Bran Flakes Plus. Several boxes of each cereal are selected and the fat content per serving is measured. The results of the study are as follows. Assume that the population variances are approximately equal.

| Fat Content (Grams) | | | |
| --- | --- | --- | --- |
| | $n$ | $\bar{x}$ | $s$ |
| Fiber Oat Flakes | 16 | 5 | 1 |
| Bran Flakes Plus | 15 | 6 | 2 |

    a.  Calculate a 90% confidence interval for the difference in average fat content between Fiber Oat Flakes and Bran Flakes Plus. Interpret the interval.

    b.  Does the study performed by the FDA substantiate the claim made by the manufacturer of Fiber Oat Flakes at $\alpha = 0.10$?

    c.  What assumptions must be made in order to calculate the confidence interval in part **a.** and perform the hypothesis test in part **b.**?

9.   A large construction company would like to expand its operations into a new geographic area. The company has narrowed the choice of locations down to two cities. A major consideration in deciding between the two cities will be the average hourly wage they must pay for general laborers. The company randomly selects laborers from each city and determines their hourly wage with the following results. Assume that the population variances are approximately equal.

| Hourly Wages (Dollars) | | | |
|---|---|---|---|
| | $n$ | $\bar{x}$ | $s$ |
| City A | 20 | $7 | $3 |
| City B | 20 | $8 | $2 |

   a.  Calculate a 99% confidence interval for the difference in average hourly wage between City A and City B. Interpret the interval.

   b.  Do the data indicate that there is a significant difference in hourly wages at $\alpha = 0.05$?

   c.  Calculate the $P$-value for the test performed in part **b.**

   d.  What assumptions must be made in order to calculate the confidence interval in part **a.** and perform the hypothesis test in part **b.**?

10.  A Hollywood studio believes that a movie that is considered a drama will draw a larger crowd on average than a movie that is a comedy. To test this theory, the studio randomly selects several movies that are classified as dramas and several movies that are classified as comedies and determines the box office revenue for each movie. The results of the survey are as follows. Assume that the population variances are approximately equal.

| Box Office Revenues (Millions of Dollars) | | | |
|---|---|---|---|
| | $n$ | $\bar{x}$ | $s$ |
| Drama | 15 | 180 | 50 |
| Comedy | 13 | 150 | 30 |

   a.  Calculate a 95% confidence interval for the difference in average revenue at the box office for drama and comedy movies. Interpret the interval.

   b.  Do the data substantiate the studio's belief that dramas will draw a larger crowd on average than comedies at $\alpha = 0.01$?

   c.  Calculate the $P$-value for the test you conducted in part **b.**

   d.  What assumptions must be made in order to calculate the confidence interval in part **a.** and to perform the hypothesis test in part **b.**?

11.  *Consumer Magazine* is reviewing the top of the line amplifiers produced by two major stereo manufacturers. One of the most important qualities of the amplifiers is the maximum power output. Brand A has redone their internal design and claims to have a higher maximum power level than Brand B. To test this claim, *Consumer Magazine* randomly selects amplifiers from each brand and determines the maximum power output. The results of the test are as follows. Assume that the population variances are approximately equal.

| Amplifier Power Output (Watts) | | | |
|---|---|---|---|
| | $n$ | $\bar{x}$ | $s$ |
| Brand A | 12 | 800 | 25 |
| Brand B | 10 | 780 | 25 |

   a.  What assumptions must be made in order to perform the hypothesis test?

   b.  Do the data substantiate the claim that the Brand A amplifier has a higher average maximum power output than Brand B at $\alpha = 0.05$?

12. The State Environmental Board wants to compare pollution levels in two of its major cities. Sunshine City thrives on the tourist industry and Service City thrives on the service industry. The environmental board randomly selects several areas within the cities and measures the pollution levels in parts per million with the following results. Assume that the population variances are approximately equal.

| Pollution Levels (ppm) | | | |
|---|---|---|---|
| | $n$ | $\bar{x}$ | $s$ |
| Sunshine City | 15 | 8.5 | 0.57 |
| Service City | 10 | 7.9 | 0.50 |

a. What assumptions must be made in order to perform a hypothesis test for the difference between these two population means?

b. Will the State Environmental Board conclude at $\alpha = 0.01$ that Service City has a lower pollution level on average than Sunshine City?

c. Repeat part b., assuming that the population variances are not equal.

d. Compare the results of part b. and part c.

13. In 2009 U.S. charitable giving fell 3.6 percent to $303.75 billion for the year. Total charitable contributions from American individuals, corporations, and foundations fell to $303.75 billion from $315.08 billion for 2008. The largest share of contributions went to religious organizations, representing 33 percent of total giving. The next largest shares went to educational organizations, receiving an estimated 13 percent of the total, and foundations, which received 10 percent of the total. Suppose a sample of 6 employees is randomly chosen from a large corporation and their charitable contributions in 2008 and 2009 are determined. The following table gives these amounts (in dollars). Assume that the population variances are approximately equal.

| Charitable Contributions ($) | |
|---|---|
| Giving in 2008 | Giving in 2009 |
| 232 | 215 |
| 150 | 125 |
| 50 | 50 |
| 400 | 350 |
| 325 | 210 |
| 175 | 150 |

**Source:** Giving USA Foundation, the Center on Philanthropy at Indiana University.

a. Can we conclude with $\alpha = 0.01$ that the average contribution to charity has decreased in this corporation from 2008 to 2009?

b. Give the assumptions for your test.

c. Repeat part a., assuming that the population variances are not equal.

d. Compare the results of part a. and part c.

# 11.3 Paired Difference Test

Suppose we are interested in comparing the durability of the soles of two brands of tennis shoes, Spikes and Kickers. One approach to making this comparison is the independent experimental design discussed in Section 11.1. Using this design one may randomly select 10 people to wear the Spikes brand of shoes for six months and then randomly select 10 other people to wear the Kickers brand of shoes for six months. After the six-month period, the average wear for the 10 pairs of Spikes shoes is measured and compared to the average wear for the 10 pairs of Kickers tennis shoes.

While this is a perfectly reasonable approach, it does have shortcomings. For example, what if one of the people selected to wear the Spikes brand of shoes is a cross-country runner? Certainly the wear on the runner's pair of shoes will be much greater than the wear on the shoes for someone who just wears the shoes in the evenings after work. Certainly many factors could have a large effect on the observed wear of the tennis shoes.

What can we do to help reduce the effect of factors that cloud the issue of tennis shoe durability? One approach is to have each person wear a Spikes brand of shoe on one foot and a Kickers brand of shoe on the other foot, where the feet are randomly selected. After six months, the wear on the Spikes shoe would be compared to the wear on the Kickers shoe for each person. By designing the experiment in this fashion, the external effects of weight, amount of use, and so forth have been significantly reduced when comparing the two brands of tennis shoes. This type of design is an example of a **paired difference experimental design**.

In a paired design, experimental units are paired such that units within a pair are much more alike than in the population as a whole. If we have been successful at significantly reducing the variation among the sample observations by pairing, we will create very potent data to make decisions.

The paired difference design is frequently used in experiments where we are interested in the average response of an experimental unit before and after some treatment. For example, a physician might be interested in comparing a patient's heart rate before and after treatment with some drug. A state legislator may be interested in comparing a person's reaction time before and after drinking one ounce of 100-proof alcohol. An instructor may be interested in comparing a student's math skills before and after being taught a particular course. A company may be interested in a customer's response to a product before and after a particular marketing campaign.

Generally, the sample size for a paired difference experimental design is small, and we must make several assumptions for the test to be valid.

> ### Definition
> 
> **Paired Difference Experimental Design**
> 
> A **paired difference experimental design**, sometimes called matched-pairs samples, is when the samples are paired (before or after measurements on the same subject) or somehow matched (i.e., you divide a set of homogeneous subjects into two groups).

> ### Assumptions
> 
> **Assumptions for the Paired Difference Experimental Design**
> 1. Experimental units can be paired such that they are more alike within the pair than within the population as a whole.
> 2. The differences have an approximately normal distribution.

To test the mean difference between two related population means, we treat the difference scores, denoted by $d_i$, as values from a single sample. The general data setup is shown in Table 11.3.1.

### Table 11.3.1 — General Setup for the Difference Between Paired Samples

| Observation | Sample 1 | Sample 2 | Difference |
|---|---|---|---|
| 1 | $x_{11}$ | $x_{21}$ | $d_1 = x_{11} - x_{21}$ |
| 2 | $x_{12}$ | $x_{22}$ | $d_2 = x_{12} - x_{22}$ |
| 3 | $x_{13}$ | $x_{23}$ | $d_3 = x_{13} - x_{23}$ |
| ... | | | |
| $i$ | $x_{1i}$ | $x_{2i}$ | $d_i = x_{1i} - x_{2i}$ |
| ... | | | |
| $n$ | $x_{1n}$ | $x_{2n}$ | $d_n = x_{1n} - x_{2n}$ |

Note that $d_i$ is the difference between the $i^{th}$ observations in each sample.

We can use an interval estimate to determine the difference in two related population means. Given the differences, we can use the sample standard deviation of the difference to estimate the population standard deviation of the difference. We can construct a $100(1 - \alpha)\%$ confidence interval for the mean difference $\mu_d$ using the following formula.

---

**Formula**

**$100(1 - \alpha)\%$ Confidence Interval for $\mu_d$**

A $100(1 - \alpha)\%$ confidence interval for the mean difference is given by

$$\overline{x}_d \pm t_{\alpha/2,df} \frac{s_d}{\sqrt{n_d}}$$

where $\overline{x}_d$ is the sample mean of the differences, $n_d$ is the number of differences,

$$\overline{x}_d = \frac{\sum d_i}{n_d}, \ s_d = \sqrt{\frac{\sum \left(d_i - \overline{x}_d\right)^2}{n_d - 1}},$$ and $t_{\alpha/2,df}$ is the critical value of the $t$-distribution

with an area of $\alpha/2$ in the upper tail with $n_d - 1$ degrees of freedom ($df$).

---

**Formula**

**Inferences about $\mu_d$**

**Test Statistic:**

$$t = \frac{\overline{x}_d - \mu_d}{s_d / \sqrt{n_d}},$$

where $\overline{x}_d$ is the sample mean of the differences, $n_d$ is the number of differences,

$$\overline{x}_d = \frac{\sum d_i}{n_d}, \ s_d = \sqrt{\frac{\sum \left(d_i - \overline{x}_d\right)^2}{n_d - 1}},$$ and $t_{\alpha/2,df}$ is the critical value of the $t$-distribution

with an area of $\alpha/2$ in the upper tail with $n_d - 1$ degrees of freedom ($df$).

---

The methodology for the paired difference test of hypothesis and the calculation of the confidence interval for the paired difference will be presented in Example 11.3.1.

---

**Example 11.3.1**

**Calculating a Confidence Interval and Performing a Hypothesis Test for the Mean Difference**

Bull & Bones Brewhaus is a microbrewery that has two restaurants located within 15 miles of each other. The owner of the microbrewery wants to compare the average daily food sales of the two restaurants. To do so, the owner randomly selects 10 days over a five-month period (college football season) and records the daily food sales. The results are given in Table 11.3.2.

a. Calculate a 95% confidence interval for the mean difference in restaurant sales.

b. The owner wants to know if there is evidence of a difference between the average daily food sales of the two restaurants. Test using $\alpha = 0.01$.

| Table 11.3.2 — Daily Food Sales for Two Restaurants ($) | | | |
|---|---|---|---|
| Day | Restaurant 1 | Restaurant 2 | Difference = Restaurant 1 − Restaurant 2 |
| 1 | 5828 | 7894 | −2066 |
| 2 | 9836 | 11,573 | −1737 |
| 3 | 3984 | 5319 | −1335 |
| 4 | 5845 | 6389 | −544 |
| 5 | 5210 | 6055 | −845 |
| 6 | 9668 | 10,631 | −963 |
| 7 | 6768 | 7866 | −1098 |
| 8 | 6726 | 7976 | −1250 |
| 9 | 4399 | 5652 | −1253 |
| 10 | 6692 | 8083 | −1391 |

### SOLUTION

**a.** From the information given, we know that $n = 10$ and $\alpha = 0.05$. We now need to calculate the mean and standard deviation of the differences, $\bar{x}_d$ and $s_d$, respectively.

$$\bar{x}_d = \frac{\sum d_i}{n_d} = \frac{(-2066)+(-1737)+\cdots+(-1253)+(-1391)}{10} = -1248.20$$

$$s_d = \sqrt{\frac{\sum(d_i - \bar{x}_d)^2}{n_d - 1}}$$

$$= \sqrt{\frac{\left(-2066-(-1248.20)\right)^2 + \cdots + \left(-1391-(-1248.20)\right)^2}{10-1}}$$

$$\approx 434.3631$$

Since we want a 95% confidence interval, we need to find the value of $t_{\alpha/2,df}$. Note that $df = n_d - 1 = 9$. Thus, $t_{\alpha/2,df} = t_{0.025,9} = 2.262$.

A 95% confidence interval for the average difference in sales record between the two restaurants is calculated as follows.

$$\bar{x}_d \pm t_{\alpha/2,df} \frac{s_d}{\sqrt{n_d}}$$

$$-1248.20 \pm 2.262\left(\frac{434.3631}{\sqrt{10}}\right)$$

$$-1558.90 \text{ to } -937.50$$

We are 95% confident that the true mean difference in sales between Restaurant 1 and Restaurant 2 is between −$1558.90 and −$937.50. That is, on average, Restaurant 2 averages between $937.50 and $1558.90 more in sales.

**b.** Before performing the hypothesis test, we need to ensure that the assumption that the differences have a normal distribution is reasonable for the sales data. A histogram of differences is provided in Figure 11.3.1. Based on this histogram, it appears that the differences are approximately normal and it is safe to proceed with the paired difference test.

### ⊗ Technology

For technology instructions to calculate a confidence interval for the differences using the *t*-distribution, visit stat.hawkeslearning.com and navigate to **Discovering Business Statistics, Second Edition > Technology Instructions > Confidence Intervals > *t*-Interval**.

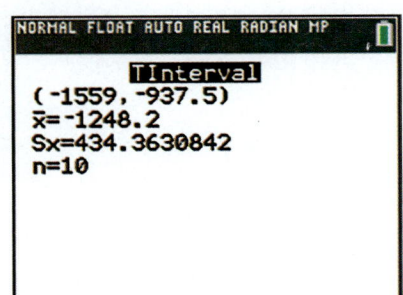

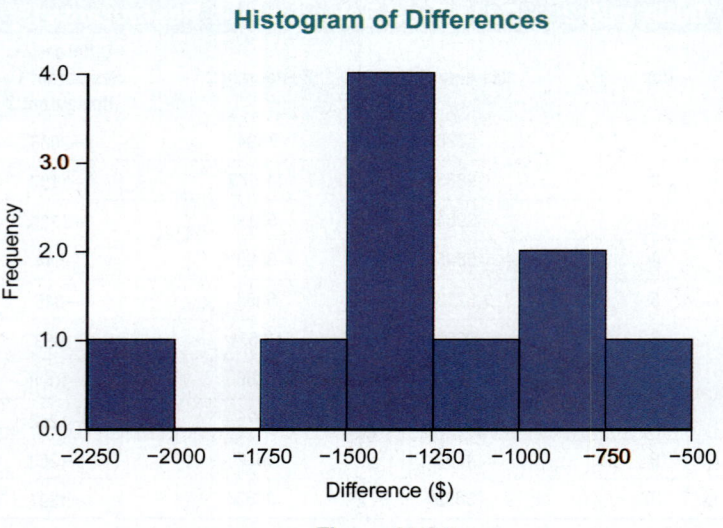

**Figure 11.3.1**

**Step 1:** Determine the null hypothesis. In this process, select the appropriate statistical measure, such as the population mean, proportion, or variance.

In a paired difference experimental design, the population parameter of interest is the population mean of the differences. Thus, the appropriate statistical measure is given by

$\mu_d$ = the average of the differences in daily sales between the two restaurants.

The null hypothesis is that there is no difference in the average sales between the two restaurants on a given day, which should be written as

$$H_0: \mu_d = 0.$$

**Step 2:** Determine the alternative hypothesis and whether it should be one-sided or two-sided.

Since the owner is interested in whether or not the average daily sales are different between the two restaurants, the alternative hypothesis will be two-sided and this will be a two-tailed test.

The alternative hypothesis is that there is a difference in average daily sales between the two restaurants and should be written as

$$H_a: \mu_d \neq 0.$$

**Step 3:** Select the appropriate test statistic based on the information at hand and the assumptions you are willing to make.

To develop the appropriate test statistic, a random variable whose value will be used to make the decision to reject or fail to reject $H_0$ must be developed. It is interesting to notice that by evaluating the differences of the paired data, we have effectively reduced the two-sample problem into a small, one-sample $t$-test for a population mean. The point estimate for the population mean of the differences, $\mu_d$, is $\overline{x}_d$, the sample mean of the differences. Hence, the sampling distribution for $\overline{x}_d$ provides essential information for testing the hypothesis. If the differences are normally distributed and the null hypothesis is assumed to be true, the sampling distribution of $\overline{x}_d$ has a $t$-distribution with $n_d - 1$ degrees of freedom. Note that $n_d$ represents the number of differences. The $t$-test statistic is given by

$$t = \frac{\bar{x}_d - \mu_d}{s_d/\sqrt{n_d}},$$

where $s_d$ is the sample standard deviation of the differences.

If the observed value of $\bar{x}_d$ is significantly smaller or larger than $\mu_d$, this will produce a large negative (or positive) value of the test statistic, causing us to question whether the null hypothesis is in fact true. How large is *large*? This is answered by the critical value of the test statistic specified in **Step 4**.

**Step 4:** Determine the critical value of the test statistic.

The significance level of the test is specified in the problem to be $\alpha = 0.01$.

The role of the critical value in this test is exactly the same as for all of the hypothesis tests discussed earlier. It defines a range of values for the test statistic, the rejection region, that will be so rare that it is unlikely that it occurred from ordinary sampling variability. The level of the test defines the size of the rejection region. Should the computed value of the test statistic fall in the rejection region, its value will be presumed to be too rare to have occurred because of ordinary sampling variation, and the null hypothesis will be rejected.

If the null hypothesis is true, the test statistic has a *t*-distribution with $n_d - 1$ degrees of freedom. Thus, the critical value is determined in the same way as for the other tests of hypothesis where the test statistic had a *t*-distribution, except that the degrees of freedom are $n_d - 1 = 10 - 1 = 9$. The rejection region for the alternative hypothesis, $H_a: \mu_d \neq 0$, with $\alpha = 0.01$ and 9 degrees of freedom is displayed in Figure 11.3.2. Because we have a two-sided hypothesis, we reject $H_0$ if the test statistic is less than or equal to $-3.250$ or if it is greater than or equal to 3.250.

*t*-Distribution, *df* = 9

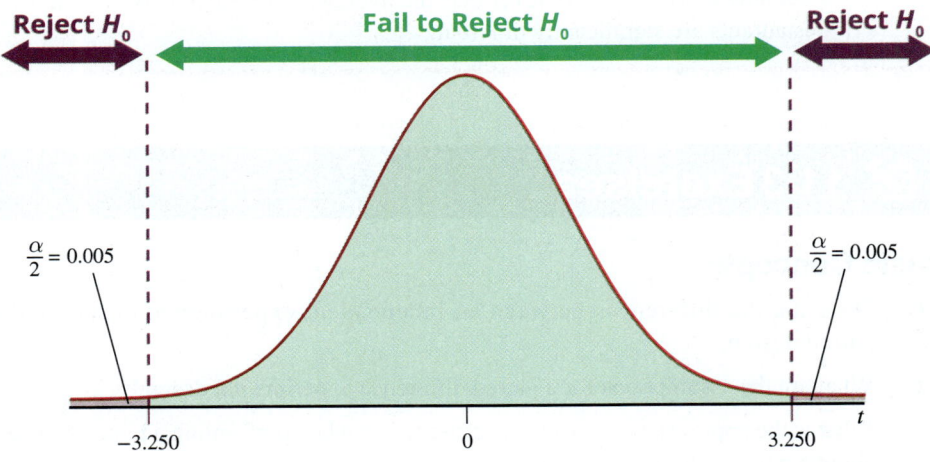

**Figure 11.3.2**

**Step 5:** Collect the sample data and compute the value of the test statistic.

Based on the data in Table 11.3.2, $\bar{x}_d = -1248.20$, and the computed value of the test statistic is as follows.

$$t = \frac{\bar{x}_d - \mu_d}{s_d/\sqrt{n_d}} = \frac{-1248.20 - 0}{434.3631/\sqrt{10}} \approx -9.087$$

**🔗 Technology**

For technology instructions on performing a hypothesis test for dependent samples using a *t*-test, please visit stat.hawkeslearning.com and navigate to **Discovering Business Statistics, Second Edition > Technology Instructions > Hypothesis Testing > *t*-Test**.

**Step 6:** Make the decision and state the conclusion in terms of the original question.

As shown in Figure 11.3.3, the value of the test statistic falls in the rejection region to the left. The test statistic indicates that the observed average daily sales are more than 9 standard deviations below the hypothesized value of 0. It is highly unlikely that the difference between the observed value and the hypothesized value is due to ordinary sampling variation. Thus, the null hypothesis is rejected at $\alpha = 0.01$.

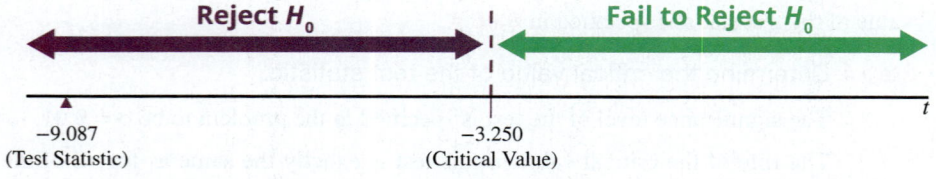

**Reject $H_0$**           **Fail to Reject $H_0$**

$-9.087$                $-3.250$
(Test Statistic)         (Critical Value)

**Figure 11.3.3**

## Technology

For technology instructions on finding the *P*-value for a given *t*-statistic, please visit stat.hawkeslearning. com and navigate to **Discovering Business Statistics, Second Edition > Technology Instructions > *t*-Distribution > *t*-Probability(cdf)**.

To find the *P*-value for this test, we use the same methodology as in Example 11.2.1. The *P*-value is given by

$$P\text{-value} = 2P(t < |-9.087|) = 2P(t > 9.087).$$

We cannot use the *t*-table to find $P(t > 9.087)$; we can only put bounds on the probability. Thus, at 9 degrees of freedom, the largest value in the respective row is 3.250. The best information that we can gain from the *t*-table is that $P(t > 9.087) < 0.005$. Since we need to multiply the probability by 2, we have *P*-value $< 0.01$.

Since we are performing the test at $\alpha = 0.01$, we reject the null hypothesis because the *P*-value is less than $\alpha$—the same decision we made using the rejection region approach.

*Conclusion and Interpretation:* There is sufficient evidence for the owner to conclude at the $\alpha = 0.01$ level that the average daily sales between the two restaurants are significantly different.

## 11.3 Exercises

### Basic Concepts

1. Describe the differences between an independent experimental design and a paired design.

2. What are the assumptions for a paired difference experimental design?

3. What is the appropriate statistical measure to use when performing a hypothesis test about a paired difference experiment?

4. How does the hypothesis testing procedure for a paired difference experiment differ from that of a two-sample *t*-test?

5. What is the test statistic used in a paired difference hypothesis test?

## Exercises

6. Determine the critical value(s) of the test statistic for each of the following paired difference tests (assume the differences have an approximately normal distribution).

   a. Left-tailed test, $\alpha = 0.01$, $n_d = 15$

   b. Right-tailed test, $\alpha = 0.10$, $n_d = 20$

   c. Two-tailed test, $\alpha = 0.05$, $n_d = 8$

7. Determine the critical value(s) of the test statistic for each of the following paired difference tests (assume the differences have an approximately normal distribution).

   a. Left-tailed test, $\alpha = 0.005$, $n_d = 12$

   b. Right-tailed test, $\alpha = 0.025$, $n_d = 5$

   c. Two-tailed test, $\alpha = 0.10$, $n_d = 25$

8. Given that most textbooks can now be purchased online, one wonders if students can save money by comparison shopping for textbooks at online retailers and at their local bookstores. To investigate, students at Tech University randomly sampled 25 textbooks on the shelves of their local bookstores. The students then found the "best" available price for the same textbooks via online retailers. The prices for the textbooks are listed in the following table.

| Textbook Prices | | | | | |
|---|---|---|---|---|---|
| | Price ($) | | | Price ($) | |
| Textbook | Bookstore | Online Retailer | Textbook | Bookstore | Online Retailer |
| 1 | 70 | 60 | 14 | 85 | 75 |
| 2 | 38 | 36 | 15 | 100 | 85 |
| 3 | 88 | 89 | 16 | 68 | 62 |
| 4 | 165 | 149 | 17 | 67 | 69 |
| 5 | 80 | 136 | 18 | 140 | 142 |
| 6 | 103 | 95 | 19 | 49 | 40 |
| 7 | 42 | 50 | 20 | 149 | 127 |
| 8 | 98 | 111 | 21 | 126 | 130 |
| 9 | 89 | 65 | 22 | 92 | 93 |
| 10 | 97 | 86 | 23 | 144 | 129 |
| 11 | 140 | 130 | 24 | 98 | 84 |
| 12 | 40 | 30 | 25 | 40 | 52 |
| 13 | 175 | 150 | | | |

∴ **Data**

This data set can be found on stat.hawkeslearning.com under **Discovering Business Statistics, Second Edition > Data Sets > Textbook Prices**.

   a. Is a paired design appropriate for the above study? Explain.

   b. What assumption must be made in order to perform the test of hypothesis?

   c. Do the data appear to satisfy the assumption described in part **b.**? Why or why not?

   d. Based on the data, is it less expensive for the students to purchase textbooks from the online retailers than from local bookstores? Use $\alpha = 0.01$.

   e. Calculate a 99% confidence interval for the mean difference in cost between the bookstores and the online retailers. Interpret the interval.

9.  The management for a large grocery store chain would like to determine if a new cash register will enable cashiers to process a larger number of items on average than the cash register they are currently using. Seven cashiers are randomly selected, and the number of grocery items they can process in three minutes is measured for both the old cash register and the new cash register. The results of the test are as follows.

| Number of Grocery Items Processed in Three Minutes | | | | | | | |
|---|---|---|---|---|---|---|---|
| Cashier | 1 | 2 | 3 | 4 | 5 | 6 | 7 |
| Old Cash Register | 60 | 70 | 55 | 75 | 62 | 52 | 58 |
| New Cash Register | 65 | 71 | 55 | 75 | 65 | 57 | 57 |

a.  Is a paired design appropriate for the above experiment? Explain.

b.  What assumption must be made in order to perform the test of hypothesis?

c.  Do the data appear to satisfy the assumption described in part **b.**? Why or why not?

d.  Calculate a 95% confidence interval for the mean difference between the number of items processed using the old cash register and the new cash register. Interpret this interval.

e.  Can the management conclude that the new cash register will allow cashiers to process a significantly larger number of items on average than the old cash register at $\alpha = 0.05$?

10. An auto dealer is marketing two different models of a high-end sedan. Since customers are particularly interested in the safety features of the sedans, the dealer would like to determine if there is a difference in the braking distance (the number of feet required to go from 60 mph to 0 mph) of the two sedans. Six drivers are randomly selected and asked to participate in a test to measure the braking distance for both models. Each driver is asked to drive both models and brake once they have reached exactly 60 mph. The distance required to come to a complete halt is then measured in feet. The results of the test are as follows.

| Braking Distance of High-End Sedans (Feet) | | | | | | |
|---|---|---|---|---|---|---|
| Driver | 1 | 2 | 3 | 4 | 5 | 6 |
| Model A | 150 | 145 | 160 | 155 | 152 | 153 |
| Model B | 152 | 146 | 160 | 157 | 154 | 155 |

a.  Is a paired design appropriate for the above experiment? Explain.

b.  What assumption must be made in order to perform the test of hypothesis?

c.  Do the data appear to satisfy the assumption described in part **b.**? Why or why not?

d.  Calculate a 90% confidence interval for the average difference between braking distances for Model A and Model B. Interpret the interval.

e.  Can the auto dealer conclude that there is a significant difference in the braking distances of the two models of high-end sedans? Use $\alpha = 0.10$.

# 11.4 Comparing Two Population Proportions

Techniques are developed in this section for comparing two population proportions. A methodology for comparing two population proportions is particularly useful because proportions are among the few measures that can be used for summarizing categorical data. For a more extensive treatment of comparisons for categorical data, see Chapter 16.

There are many situations where comparing two population proportions may be of interest. For example, a sociologist may be interested in comparing the proportion of females who believe it is okay to cry in public to the proportion of males who think it is okay to cry in public. A marketing manager may be interested in comparing the proportion of customers who favor Product A to the proportion of customers who favor Product B.

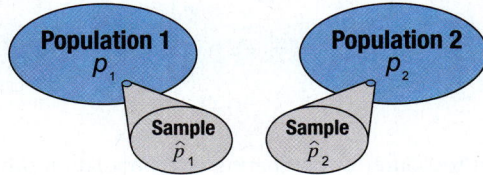

In order to perform a comparison of two population proportions, the assumptions outlined below must be met.

## Assumptions

### Assumptions for Comparing Two Population Proportions

1. An independent experimental design is used.
2. The samples are large enough such that $n_1 \hat{p}_1 \geq 5$, $n_1\left(1-\hat{p}_1\right) \geq 5$, $n_2 \hat{p}_2 \geq 5$, and $n_2\left(1-\hat{p}_2\right) \geq 5$ where $\hat{p}_1$ and $n_1$ are the sample proportion and sample size, respectively, from the first population and $\hat{p}_2$ and $n_2$ are the sample proportion and sample size, respectively, from the second population.

## Interval Estimation of $p_1 - p_2$

### Formula

#### 100(1 − α)% Confidence Interval for $p_1 - p_2$

We can construct a $100(1-\alpha)\%$ confidence interval estimate for the difference between two population proportions using the following:

$$\left(\hat{p}_1 - \hat{p}_2\right) \pm z_{\alpha/2} \sqrt{\frac{\hat{p}_1\left(1-\hat{p}_1\right)}{n_1} + \frac{\hat{p}_2\left(1-\hat{p}_2\right)}{n_2}}$$

where $\hat{p}_1$ and $n_1$ are the sample proportion and sample size, respectively, from the first population, $\hat{p}_2$ and $n_2$ are the sample proportion and sample size, respectively, from the second population, $n_1 \hat{p}_1 \geq 5$, $n_1\left(1-\hat{p}_1\right) \geq 5$, $n_2 \hat{p}_2 \geq 5$, $n_2\left(1-\hat{p}_2\right) \geq 5$, and $z_{\alpha/2}$ is the critical value for the $z$-distribution that captures an area of $\alpha/2$ in the upper tail.

## Hypothesis Testing About $p_1 - p_2$

### Formula

**Inferences about $p_1 - p_2$**

**Test Statistic:**

$$z = \frac{\left(\hat{p}_1 - \hat{p}_2\right) - \left(p_1 - p_2\right)}{\sqrt{\overline{p}\left(1 - \overline{p}\right)\left(\dfrac{1}{n_1} + \dfrac{1}{n_2}\right)}},$$

where $\hat{p}_1$ and $n_1$ are the sample proportion and sample size, respectively, from the first population, $\hat{p}_2$ and $n_2$ are the sample proportion and sample size, respectively, from the second population, $n_1\hat{p}_1 \geq 5$, $n_1\left(1 - \hat{p}_1\right) \geq 5$, $n_2\hat{p}_2 \geq 5$, $n_2\left(1 - \hat{p}_2\right) \geq 5$, $z_{\alpha/2}$ is the critical value for the $z$-distribution, and $\overline{p} = \dfrac{x_1 + x_2}{n_1 + n_2}$ is the weighted average of the two sample proportion estimates, $\hat{p}_1$ and $\hat{p}_2$.

The hypothesis testing procedure for comparing two population proportions is developed in Example 11.4.1.

### Example 11.4.1

**Calculating a Confidence Interval and Performing a Hypothesis Test for the Difference in Two Proportions**

A cell phone executive has recently been bombarded with complaints from his customers about defective phones. He has two plants that produce the cell phones, and he is not sure where the defective phones are coming from. In the past, the plants have had good control over the number of defective phones produced. Because of the recent flurry of complaints, he thinks that one of the plants may have lost control over its production process. To test this theory, he randomly selects 200 phones from each of the plants and counts the number of defective phones. The results of the survey are displayed in Table 11.4.1.

| Table 11.4.1 - Cell Phone Survey Data | | |
|---|---|---|
| | **Number Sampled** | **Number of Defectives** |
| **Plant A** | 200 | 8 |
| **Plant B** | 200 | 10 |

**a.** Are the sample sizes large enough to assume that the sample proportions are approximately normally distributed? Why is this necessary?

**b.** Calculate a 95% confidence interval for the difference between the proportions of defective phones from Plant A and Plant B.

**c.** Is there sufficient evidence for the cell phone executive to conclude that there is a difference in the proportion of defective cell phones produced by the two plants at $\alpha = 0.10$?

**SOLUTION**

**a.** Let

$\hat{p}_1$ = the sample proportion of defective phones produced by Plant A,

$\hat{p}_2$ = the sample proportion of defective phones produced by Plant B,

$x_1$ = the sample number of defective phones produced in Plant A,

$x_2$ = the sample number of defective phones produced in Plant B,

$n_1$ = the number of phones sampled from Plant A, and

$n_2$ = the number of phones sampled from Plant B.

Before we can determine a confidence interval or perform a hypothesis test for the difference between the population proportions, we need to be sure that the sample sizes are large enough. You may recall that when dealing with single samples of proportions, the sample size, $n$ must be large enough such that the sampling distribution of the sample proportion will be approximately normal. To determine whether the sample sizes are large enough, we need to show that $n_1 \hat{p}_1$, $n_1\left(1-\hat{p}_1\right)$, $n_2 \hat{p}_2$, and $n_2\left(1-\hat{p}_2\right)$ are all greater than or equal to 5. If both samples are large enough, then we know that the sampling distribution of the difference between the sample proportions is approximately normal, and we can proceed with constructing the confidence interval and performing the hypothesis test. First, we need to calculate the sample proportions for the defective phones produced in Plant A and Plant B.

$$\hat{p}_1 = \frac{x_1}{n_1} = \frac{8}{200} = 0.04$$

$$\hat{p}_2 = \frac{x_2}{n_2} = \frac{10}{200} = 0.05$$

Now, we can use the sample proportions along with the sample sizes to verify that the samples are large enough such that the sampling distribution of the difference between the sample proportions is approximately normal.

$$n_1 \hat{p}_1 = 200\left(0.04\right) = 8$$

$$n_1\left(1-\hat{p}_1\right) = 200\left(1-0.04\right) = 192$$

$$n_2 \hat{p}_2 = 200\left(0.05\right) = 10$$

$$n_2\left(1-\hat{p}_2\right) = 200\left(1-0.05\right) = 190$$

Since $n_1 \hat{p}_1$, $n_1\left(1-\hat{p}_1\right)$, $n_2 \hat{p}_2$, and $n_2\left(1-\hat{p}_2\right)$ are all greater than or equal to 5, we can conclude that $\hat{p}_1 - \hat{p}_2$ has an approximately normal distribution. Now that we have verified the conditions required to proceed with the calculation of a confidence interval, we can find the 95% confidence interval for the difference between the population proportions.

b. We know that $n_1 = n_2 = 200$, $x_1 = 8$, and $x_2 = 10$. With the information given, we can calculate the sample proportions.

$$\hat{p}_1 = \frac{x_1}{n_1} = \frac{8}{200} = 0.04$$

$$\hat{p}_2 = \frac{x_2}{n_2} = \frac{10}{200} = 0.05$$

For a 95% confidence interval, $\alpha = 0.05$ and $z_{\alpha/2} = z_{0.025} = 1.96$.
Therefore, the 95% confidence interval is calculated as follows.

$$\left(\hat{p}_1 - \hat{p}_2\right) \pm z_{\alpha/2}\sqrt{\frac{\hat{p}_1\left(1-\hat{p}_1\right)}{n_1} + \frac{\hat{p}_2\left(1-\hat{p}_2\right)}{n_2}}$$

$$\left(0.04 - 0.05\right) \pm 1.96\sqrt{\frac{0.04\left(1-0.04\right)}{200} + \frac{0.05\left(1-0.05\right)}{200}}$$

$$-0.01 \pm 1.96\sqrt{\frac{0.04\left(1-0.04\right)}{200} + \frac{0.05\left(1-0.05\right)}{200}}$$

$$-0.0506 \text{ to } 0.0306$$

Thus, we are 95% confident that the true difference in the proportion of defectives between Plant A and Plant B is between −0.0506 and 0.0306.

### ∞ Technology

For technology instructions to calculate a confidence interval for the difference in two proportions, visit stat.hawkeslearning.com and navigate to **Discovering Business Statistics, Second Edition > Technology Instructions > Confidence Intervals > Two Sample Proportions z-Interval.**

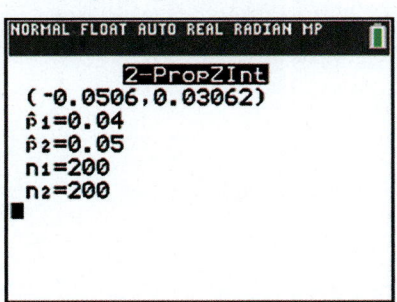

**c. Step 1:** Determine the null hypothesis. In this process, select the appropriate statistical measure, such as the population means, proportions, or variances.

Since the executive is interested in comparing the proportion of defective phones produced at Plant A to the proportion of defective phones produced at Plant B, the appropriate statistical measures are as follows.

$p_1$ = the true proportion of defective phones produced at Plant A

$p_2$ = the true proportion of defective phones produced at Plant B

Thus, the null hypothesis is that there is no difference in the proportion of defective cell phones produced at the two plants and should be written as

$$H_0: p_1 - p_2 = 0.$$

**Step 2:** Determine the alternative hypothesis and whether it should be one-sided or two-sided.

The executive's interest is in whether or not there is a difference in the proportion of defective phones produced by the two plants. Thus, the alternative hypothesis is two-sided and this is a two-tailed test which should be written as

$$H_a: p_1 - p_2 \neq 0.$$

**Step 3:** Select the appropriate test statistic based on the information at hand and the assumptions you are willing to make.

To develop the appropriate test statistic, a random variable whose value will be used to help make the decision to reject or fail to reject $H_0$ must be found. The point estimate of $p_1 - p_2$ is $\hat{p}_1 - \hat{p}_2$. The sampling distribution of $\hat{p}_1 - \hat{p}_2$ will be used in determining the critical values of the test statistic. If the assumptions previously outlined are met, and we assume the null hypothesis is true, the sampling distribution of $\hat{p}_1 - \hat{p}_2$ has an approximately normal distribution with mean 0 and standard deviation

$$\sigma_{\hat{p}_1 - \hat{p}_2} = \sqrt{\bar{p}(1-\bar{p})\left(\frac{1}{n_1} + \frac{1}{n_2}\right)} \, ,$$

where $\bar{p} = \dfrac{x_1 + x_2}{n_1 + n_2}$.

Note that $\bar{p}$ is the weighted average of the two sample proportion estimates, $\hat{p}_1$ and $\hat{p}_2$. In hypothesis tests comparing two population proportions, we will always assume that the hypothesized difference between the two proportions is zero.

If the null hypothesis is assumed to be true, then $p_1 - p_2 = 0$, which implies that $p_1 = p_2$. Thus $\hat{p}_1$ and $\hat{p}_2$ are estimating the same quantity. Therefore, $\hat{p}_1$ and $\hat{p}_2$ are pooled to derive a better estimate of the population proportion. The test statistic is a standard normal random variable given by

$$z = \frac{\left(\hat{p}_1 - \hat{p}_2\right) - \left(p_1 - p_2\right)}{\sqrt{\bar{p}(1-\bar{p})\left(\dfrac{1}{n_1} + \dfrac{1}{n_2}\right)}} \, .$$

If the null hypothesis is true, $z$ has an approximately normal distribution. If the observed value of $\hat{p}_1 - \hat{p}_2$ is significantly larger or smaller than 0, this will produce a large or small value of the test statistic, causing us to question whether the null hypothesis is true. How large is *large*? This is answered by the critical value of the test statistic specified in **Step 4**.

**Step 4:** Determine the critical value of the test statistic.

The significance level of the test is specified in the problem to be $\alpha = 0.10$.

The role of the critical value in this test is exactly the same as for all of the hypothesis tests discussed earlier. It defines a range of values for the test statistic, the rejection region, that will be so rare that it is unlikely that it occurred from ordinary sampling variability, assuming $H_0$ is true. The level of the test defines the size of the rejection region. Should the computed value of the test statistic fall in the rejection region, the null hypothesis will be rejected.

If the null hypothesis is true, the test statistic has a standard normal distribution. Thus the critical value is determined in the same way as for other tests of hypothesis where the test statistic had a standard normal distribution. The rejection region for a two-tailed test with $\alpha = 0.10$ is displayed in Figure 11.4.1. We will reject the null hypothesis if the computed value of the test statistic is larger than 1.645 or smaller than $-1.645$.

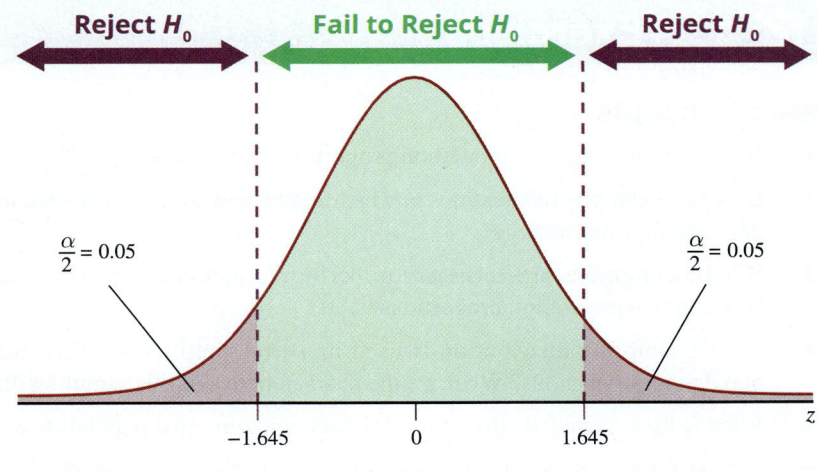

Figure 11.4.1

**Step 5:** Collect the sample data and compute the value of the test statistic.

Based on the data in Table 11.4.1, the computed value of the test statistic is given by

$$z = \frac{\frac{8}{200} - \frac{10}{200} - 0}{\sqrt{0.045(1-0.045)\left(\frac{1}{200} + \frac{1}{200}\right)}} \approx -0.48$$

where $\overline{p} = \frac{8+10}{200+200} = 0.045$.

**Step 6:** Make the decision and state the conclusion in terms of the original question.

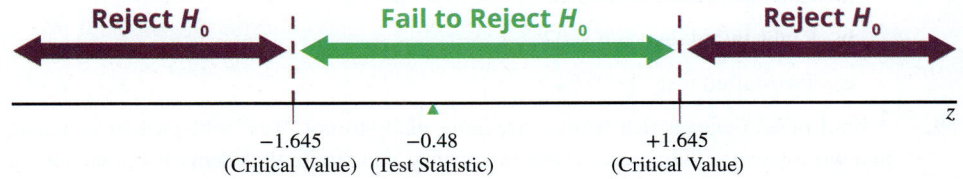

Figure 11.4.2

### ⛅ Technology

For technology instructions to do a hypothesis test for the difference between two proportions, visit stat.hawkeslearning.com and navigate to **Discovering Business Statistics, Second Edition > Technology Instructions > Hypothesis Testing > Two Proportion z-Test.**

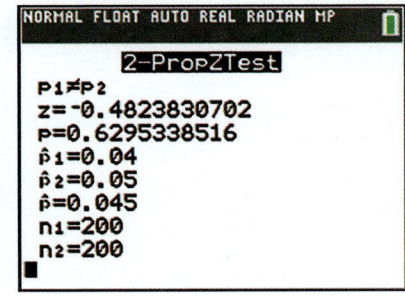

As displayed in Figure 11.4.2, the value of the test statistic does not fall in the rejection region because $-1.645 < -0.48 < 1.645$. Thus, the difference between the observed value and the hypothesized value is likely due to ordinary sampling variation. We fail to reject the null hypothesis at $\alpha = 0.10$.

Using the *P*-value approach, we want to find $2P(Z < -0.48) = 2(0.3156) = 0.6312$. You may recall that the decision rule when using the *P*-value approach is that we reject the null hypothesis if the *P*-value is less than $\alpha$. Thus, since the *P*-value is 0.6312 which is greater than 0.10, we fail to reject the null hypothesis.

*Conclusion and Interpretation:* There is insufficient evidence at $\alpha = 0.10$ for the cell phone executive to conclude that the proportion of defective phones produced differs between the two plants.

# ✍ 11.4 Exercises

## Basic Concepts

1. Why is comparing two population proportions particularly useful?

2. Give two examples of situations in which someone would be interested in comparing population proportions.

3. What assumptions are necessary to perform a hypothesis test for the difference between two population proportions?

4. Which sampling distribution is used in a two-sample test of hypothesis about population proportions? What are the characteristics of this sampling distribution?

5. What is the test statistic that is used when comparing two population proportions?

6. True or false: in order to use the specified test statistic, the hypothesized difference in the null hypothesis between the two population proportions must be zero.

## Exercises

7. Determine the critical value(s) of the test statistic for each of the following large sample tests for the comparison of two population proportions.

   a. Left-tailed test, $\alpha = 0.01$

   b. Right-tailed test, $\alpha = 0.05$

   c. Two-tailed test, $\alpha = 0.10$

8. Determine the critical value(s) of the test statistic for each of the following large sample tests for the comparison of two population proportions.

   a. Left-tailed test, $\alpha = 0.025$

   b. Right-tailed test, $\alpha = 0.02$

   c. Two-tailed test, $\alpha = 0.04$

9. A fund-raiser believes that women are more likely to say "Yes" when asked to donate to a worthy cause than men. To test this theory, she randomly selects 100 men and 95 women and asks for donations to the same cause. The results of the survey are as follows.

| Fund-Raiser Survey | | |
|---|---|---|
| | Number Surveyed | # of "Yes" Responses |
| Men | 100 | 6 |
| Women | 95 | 9 |

**a.** Are the sample sizes large enough such that a hypothesis test for the difference between two population proportions may be performed? If so, do the data substantiate the fund-raiser's theory at $\alpha = 0.10$?

**b.** Calculate the $P$-value for the test and interpret its meaning.

**c.** Calculate a 95% confidence interval for the difference in the proportion of men and women who would most likely donate to a worthy cause. Interpret the interval.

**10.** A poll is conducted to determine if U.S. citizens think that there should be a national health care system in the U.S. 69% of the 300 women surveyed and 63% of the 250 men surveyed think that there should be a national health care system in the U.S. Are the sample sizes large enough such that a hypothesis test for the difference between two population proportions may be performed? If so, is there sufficient evidence to conclude at $\alpha = 0.05$ that men and women feel differently about this issue?

**11.** Major television networks have never seemed to have issues showing commercials for beer and other alcoholic beverages. Even though adult viewers tend to enjoy the commercials, most adults seem to think that the commercials target teenagers and young adults (those under 21 years old). To study this belief, the networks conducted a joint poll of viewers and asked them if they felt that beer and other alcoholic beverage commercials targeted teenagers and young adults. The results of the survey are as follows.

| Network Advertising Survey | | |
|---|---|---|
| Age Group | Number Surveyed | Number of "Yes" Responses |
| 30 or Younger | 1000 | 450 |
| Older than 30 | 1000 | 655 |

**a.** Are the sample sizes large enough such that inferences about the difference between two population proportions can be made? If so, calculate a 99% confidence interval for the difference in the proportions of those older than 30 and those 30 or younger that believe alcoholic beverage commercials targeted teenagers and young adults. Interpret the interval.

**b.** Based on the data, can the networks conclude that the percentage of viewers who believe beer and alcoholic beverage commercials target teenagers and young adults is significantly higher in the over 30 age group than in the 30 or younger age group at $\alpha = 0.01$?

**12.** A manufacturer is comparing shipments of machine parts from two suppliers. The parts from Supplier A are less expensive; however, the manufacturer is concerned that the parts may be of a lower quality than those from Supplier B. The manufacturer has decided that he will purchase his supplies from Supplier A unless he can show that the proportion of defective parts is significantly higher for Supplier A than for Supplier B. He randomly selects parts from each supplier and inspects them for defects. The results are as follows. Determine whether the sample sizes are large enough such that inferences about the difference between the population proportions can be made. If so, which supplier will the manufacturer choose at $\alpha = 0.05$? Explain.

| Number of Defective Parts | | |
|---|---|---|
| | Number Surveyed | Number of Defective Parts |
| Supplier A | 400 | 8 |
| Supplier B | 300 | 5 |

# 11.5 Comparing Two Population Variances

Having introduced and performed tests comparing means and proportions, it's a natural transition to discuss comparing variances. This is especially useful given that when we compare two population means, we make the assumption that the variances are either equal or unequal and then perform the test accordingly. An interesting question is, *When do we even care if the variances (or variability) of two populations are equal or not?* This question often arises in manufacturing or production industries. For example, one may be interested in whether the variability in the number of products from a production line differs significantly between the morning shift and the night shift. Specifically, an automobile tire manufacturer may be interested in the variability of the number of tires produced at one plant versus another. As another example, a financial advisor may measure the risk of a portfolio by examining the variance of the portfolio. Thus, if an advisor wants to compare the risk between two portfolios, the advisor can compare the variances—the one with the largest variance would be deemed the one that exhibits the greatest amount of risk.

The procedure in this section introduces a test that indicates if there are any violations of the homogeneous variance assumption. When testing the equality of two population variances, the natural inclination is to write $\sigma_1^2 = \sigma_2^2$, which is equivalent to looking at the ratio of the two variances given by $\dfrac{\sigma_1^2}{\sigma_2^2} = 1$. Logically, if the ratio of the variances is not significantly different from one, the assumption of homogeneity of variances holds; otherwise, it's reasonable to conclude that the variances are not equal. Lastly, when carrying out the test, we will use the ratio of the sample variances, $\dfrac{s_1^2}{s_2^2}$, as an estimate of the ratio of the population variances, $\dfrac{\sigma_1^2}{\sigma_2^2}$, assuming that the data are collected from two independent normal distributions. Before formalizing the test for equality of variances, we need to discuss the sampling distribution of the ratio of sample variances.

## The Sampling Distribution of $\dfrac{s_1^2}{s_2^2}$

Just as we did with the sampling distributions of other statistics that estimated their respective population parameters, we want to discuss the sampling distribution of $\dfrac{s_1^2}{s_2^2}$. We discussed in Chapter 9 that the sample variance, $s^2$, follows a chi-square distribution. Thus, $\dfrac{s_1^2}{s_2^2}$ is the ratio of two chi-square distributed statistics, which follows an $F$-distribution. Like the chi-square distribution, the $F$-distribution is a positively skewed distribution with values that are always greater than or equal to zero. $F$-distributions are associated with test statistics that are quotients. The $F$-distribution is a function of two degrees of freedom—the numerator degrees of freedom and the denominator degrees of freedom. The common practice is to refer to the $F$-distribution as $F_{df_{num}, df_{den}}$ where $df_{num}$ represents the numerator degrees of freedom and $df_{den}$ represents the denominator degrees of freedom. The $df_{num}$ is equal to $n_1-1$, the denominator of the variance term in the numerator ($s_1^2$) and $df_{den}$ is equal to $n_2-1$, the denominator of the variance term in the denominator ($s_2^2$).

The shape of the $F$-distribution is a function of its degrees of freedom. In Figure 11.5.1, the $F$-distributions for $df_{sum} = 6$, $df_{den} = 40$, and $df_{num} = 10$, $df_{den} = 4$ are given.

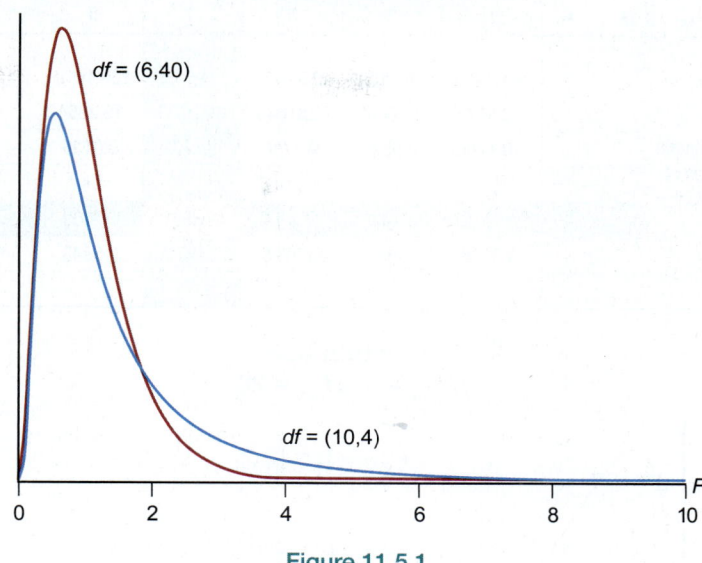

**Figure 11.5.1**

Tables have been compiled (see Appendix A) that show critical values of the $F$-distribution at common levels of significance. The notation $F_\alpha$ will indicate the value of $F$ such that $\alpha$ of the area under the curve lies to the right of this value.

Find a value on the $F$-distribution with $df_{num} = 4$ and $df_{den} = 20$ such that 0.05 of the area lies to the right of this value. (See Figure 11.5.2.)

**Example 11.5.1**

**Finding an $F$-Critical Value**

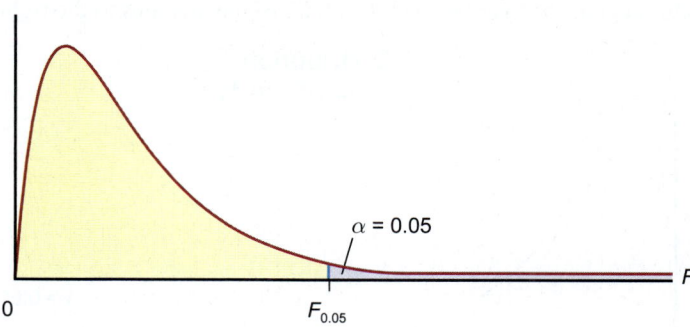

**Figure 11.5.2**

### SOLUTION

Use the $F$-table corresponding to $\alpha = 0.05$. The leftmost column of the table corresponds to the degrees of freedom in the denominator, while the top row of the table corresponds to the degrees of freedom in the numerator. Looking up the critical value in the 0.05 table corresponding to $df_{num} = 4$ and $df_{den} = 20$ produces an $F$-value of 2.8661.

| $\alpha = 0.05$ | | Numerator Degrees of Freedom | | | | | |
|---|---|---|---|---|---|---|---|
| | | 1 | 2 | 3 | 4 | 5 | ... |
| Denominator Degrees of Freedom | 1 | 161.4476 | 199.5000 | 215.7073 | 224.5832 | 230.1619 | |
| | 2 | 18.5128 | 19.0000 | 19.1643 | 19.2468 | 19.2964 | |
| | 3 | 10.1280 | 9.5521 | 9.2766 | 9.1172 | 9.0135 | |
| | | | | ... | | | |
| | 20 | 4.3512 | 3.4928 | 3.0984 | 2.8661 | 2.7109 | |
| | 21 | 4.3248 | 3.4668 | 3.0725 | 2.8401 | 2.6848 | |
| | | | | ... | | | |

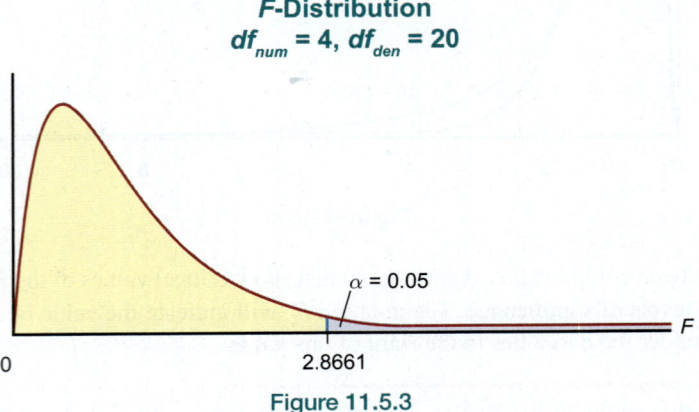

**F-Distribution**
$df_{num} = 4$, $df_{den} = 20$

$\alpha = 0.05$

0        2.8661        F

Figure 11.5.3

---

## Example 11.5.2

### Finding an *F*-Critical Value

Find a value on the *F*-distribution with 4 numerator degrees of freedom and 10 denominator degrees of freedom such that 0.01 of the area lies to the right.

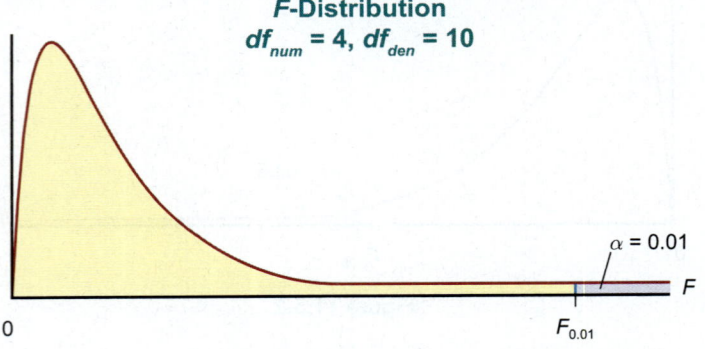

**F-Distribution**
$df_{num} = 4$, $df_{den} = 10$

$\alpha = 0.01$

0        $F_{0.01}$        F

Figure 11.5.4

### SOLUTION

Using the *F*-table for $\alpha = 0.01$ corresponding to $df_{num} = 4$ and $df_{den} = 10$ results in an *F*-value of 5.9943.

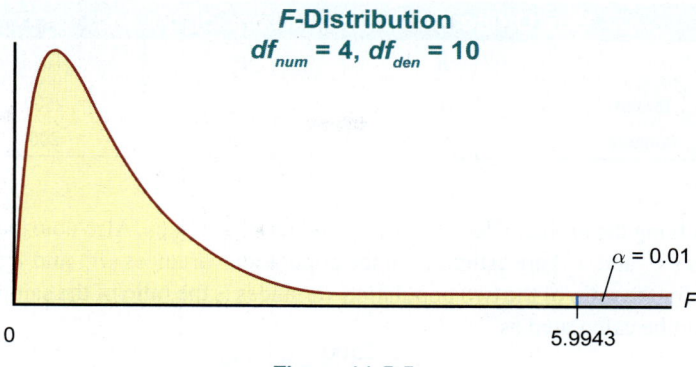

**Figure 11.5.5**

Now that we know the sampling distribution of $\frac{s_1^2}{s_2^2}$, we can make inferences about the ratio of two population variances, as well as construct confidence intervals and perform hypothesis testing.

## Interval Estimation of the Ratio of Two Population Variances

Like the formula for the confidence interval for the population variance, $\sigma^2$, the formula for the ratio of two population variances, $\frac{\sigma_1^2}{\sigma_2^2}$, is not symmetric. For example, unlike the confidence intervals for the means and proportions, we do not subtract the margin of error from and add the margin of error to the point estimate to obtain the confidence limits.

### Formula

#### Confidence Interval for $\frac{\sigma_1^2}{\sigma_2^2}$

A $100(1-\alpha)\%$ confidence interval for $\frac{\sigma_1^2}{\sigma_2^2}$, the ratio of two population variances, is given by

$$\left(\left(\frac{s_1^2}{s_2^2}\right)\frac{1}{F_{\alpha/2,df_{num},df_{den}}}, \left(\frac{s_1^2}{s_2^2}\right)\frac{1}{F_{1-\alpha/2,df_{num},df_{den}}}\right)$$

where $s_1^2$ and $s_2^2$ are the two sample variances, and $df_{num} = n_1 - 1$ (the denominator of $s_1^2$) and $df_{den} = n_2 - 1$ (the denominator of $s_2^2$).

These $F$-values can be found in the $F$-distribution table in Appendix A Table H. Also note that the data are collected from two independent, normally distributed populations.

**Example 11.5.3**

**Constructing a Confidence Interval for the Ratio of Variation in Box Office Revenues**

A Hollywood studio believes that a movie that is considered a drama will draw a larger crowd on average than a movie that is a comedy. Thus, the studio expects that the spread in revenue will be quite large between dramas and comedies. To test this theory the studio randomly selects several movies that are classified as dramas and several movies that are classified as comedies and determines the box office revenue for each movie. The results of the survey follow. Construct a 95% confidence interval for the ratio of population variances. Assume that the samples are collected from two independent normal populations.

| Box Office Revenues (Millions of Dollars) | | | |
|---|---|---|---|
| | $n$ | $\bar{x}$ | $s^2$ |
| Drama | 20 | 180 | 2500 |
| Comedy | 15 | 150 | 900 |

## SOLUTION

Before solving the problem, let $s_1^2 = s_{Drama}^2$ and let $s_2^2 = s_{Comedy}^2$. Also note that the sample variances ($s_1^2$ and $s_2^2$) are estimates of the population variances ($\sigma_1^2$ and $\sigma_2^2$). The point estimate for the ratio of the two population variances is the ratio of the sample variances and would be calculated as

$$\frac{s_1^2}{s_2^2} = \frac{2500}{900} = 2.7778.$$

While it appears that the variability of the revenue for dramas is more than twice that of the revenue for comedies, let's examine the confidence interval to see if that gives us a bit more information. The following formula is used to construct a 95% confidence interval on the ratio of variances for the box office revenues.

$$\left( \left( \frac{s_1^2}{s_2^2} \right) \frac{1}{F_{\alpha/2, df_{num}, df_{den}}}, \left( \frac{s_1^2}{s_2^2} \right) \frac{1}{F_{1-\alpha/2, df_{num}, df_{den}}} \right)$$

We will let drama be Population 1 and comedy be Population 2, so that $n_1 = 20$, $s_1^2 = 2500$, $n_2 = 15$, and $s_2^2 = 900$. Given the different sample sizes, we can find $df_{num} = n_1 - 1 = 20 - 1 = 19$, and $df_{den} = n_2 - 1 = 15 - 1 = 14$. Since $\alpha = 0.05$, we need to find the $F$-table values that correspond to $F_{0.025,19,14} = 2.8607$ and $F_{0.975,19,14} = 0.3778$. Using these two $F$-values, the confidence interval is as follows:

$$\left( \left( \frac{2500}{900} \right) \frac{1}{2.8607}, \left( \frac{2500}{900} \right) \frac{1}{0.3778} \right) = (0.9710, 7.3525)$$

Thus, we are 95% confident that the ratio of the two population variances is between 0.9710 and 7.3525. Note that the interval (0.9710, 7.3525) includes 1. This means that the ratio of the two variances is not significantly different from 1. Therefore, there is insufficient evidence of a difference between the population variances.

## Formula

### Inferences about $\frac{\sigma_1^2}{\sigma_2^2}$

**Test Statistic:**

$$F = \frac{s_1^2}{s_2^2}$$

where $s_1^2$ and $s_2^2$ are the sample variances from the two samples of sizes $n_1$ and $n_2$, respectively, collected from two independent normally distributed populations. The degrees of freedom for the $F$-test are $df_{num} = n_1 - 1$ and $df_{den} = n_2 - 1$.

## Hypothesis Testing of the Ratio of Two Population Variances

As we did in the previous section, we will let $s_1^2$ be the sample variance of $n_1$ independent observations taken from a normally distributed population with variance of $\sigma_1^2$. Similarly, let $s_2^2$ be the sample variance of $n_2$ independent observations taken from a normally distributed population with variance of $\sigma_2^2$. Lastly, assume that the populations are independent.

When comparing two population variances, we want to test the null hypothesis that

$$H_0: \sigma_1^2 = \sigma_2^2.$$

This null hypothesis can be rewritten in the form of

$$H_0: \frac{\sigma_1^2}{\sigma_2^2} = 1.$$

We will test the null hypothesis against one of the following three alternative hypotheses.

1.  $H_a: \dfrac{\sigma_1^2}{\sigma_2^2} < 1$, left-sided test

2.  $H_a: \dfrac{\sigma_1^2}{\sigma_2^2} > 1$, right-sided test

3.  $H_a: \dfrac{\sigma_1^2}{\sigma_2^2} \neq 1$, two-sided test

The point estimate for $\dfrac{\sigma_1^2}{\sigma_2^2}$ is $\dfrac{s_1^2}{s_2^2}$, which follows an $F$-distribution. Thus, the value of the test statistic for the hypothesis test for the ratio of two population variances is given by

$$F = \frac{s_1^2}{s_2^2}$$

where $n_1$ and $n_2$ samples are collected from two independent normally distributed populations. The degrees of freedom for the $F$-test are $df_{num} = n_1 - 1$ and $df_{den} = n_2 - 1$.

The direction of the alternative hypothesis determines the rejection region. For the alternative hypotheses listed above, the corresponding rejection regions are as follows.

1.  Reject $H_0$ if $F \leq F_{1-\alpha, df_{num}, df_{den}}$.
2.  Reject $H_0$ if $F \geq F_{\alpha, df_{num}, df_{den}}$.
3.  Reject $H_0$ if $F \leq F_{1-\alpha/2, df_{num}, df_{den}}$ or if $F \geq F_{\alpha/2, df_{num}, df_{den}}$.

The $F$-values can be found in the $F$-distribution table in Appendix A Table H.

---

Using the Hollywood studio data in Example 11.5.3, determine whether the variances of the two revenue streams (drama and comedy) differ using a 5% level of significance.

**Example 11.5.4**

**Performing a Hypothesis Test of Variances for Box Office Revenues**

**SOLUTION**

We will follow the same six steps as we did in the previous sections of this chapter (as well as in Chapter 10) to work through this example.

**Step 1:** Determine the null hypothesis. In this process, select the appropriate statistical measure, such as the population mean, proportion, or variance.

The population consists of all revenue generated from movies that were dramas and comedies. The test specifies that we want to compare the population variance in revenues generated from movies that were dramas to the population variance of revenues generated from movies that were comedies. The parameters of interest in this case are $\sigma_1^2$ and $\sigma_2^2$. The null hypothesis should be written as $H_0: \sigma_1^2 = \sigma_2^2$ which is equivalent to $H_0: \dfrac{\sigma_1^2}{\sigma_2^2} = 1$. Please note that we will let $s_1^2 = s_{Drama}^2$ (the variance associated with revenue from dramas) and let $s_2^2 = s_{Comedy}^2$ (the variance associated with revenue from comedies).

**Step 2:** Determine the alternative hypothesis and whether it should be one-sided or two-sided.

From the information provided, we are interested in whether the variances are different, thus, the alternative hypothesis should be two-sided and written as

$$H_a: \frac{\sigma_1^2}{\sigma_2^2} \neq 1 .$$

**Step 3:** Select the appropriate test statistic based on the information at hand and the assumptions you are willing to make.

Understanding that the data are collected from two independent normally distributed observations and that we are testing the ratio of two variances, the test statistic is given by $F = \dfrac{s_1^2}{s_2^2} .$

**Step 4:** Determine the critical value of the test statistic.

We know that we are testing at the 5% level of significance (i.e., $\alpha = 0.05$). Additionally, we know that we have a two-sided test based on the alternative hypothesis. Since the alternative hypothesis is two-sided, two tails of the $F$-distribution must be determined as rejection regions. That is, we want to determine if

$F \leq F_{1-\alpha/2, df_{num}, df_{den}}$ or if $F \geq F_{\alpha/2, df_{num}, df_{den}}$ . These critical values are

$$F_{1-\alpha/2, df_{num}, df_{den}} = F_{0.975, 19, 14} = 0.3778$$

and

$$F_{\alpha/2, df_{num}, df_{den}} = F_{0.025, 19, 14} = 2.8607 .$$

The rejection region is that we will reject the null hypothesis if the $F$-test statistic is less than or equal to 0.3778, or if the $F$-test statistic is greater than or equal to 2.8607.

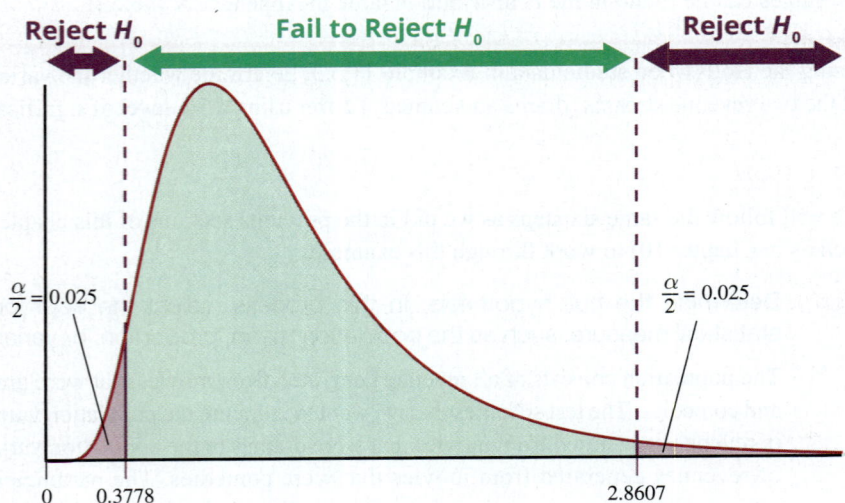

*F*-Distribution
$df_{num} = 19$, $df_{den} = 14$

**Figure 11.5.6**

**⌀ Technology**

To find the critical value for an *F*-distribution using technology, please visit stat.hawkeslearning.com and navigate to **Discovering Business Statistics, Second Edition > Technology Instructions > F-Distribution > Critical Value.**

**Step 5:** Collect sample data and compute the value of the test statistic.

The test statistic is given by

$$F = \frac{s_1^2}{s_2^2} = \frac{2500}{900} = 2.7778.$$

This statistic indicates that the variance of revenue for dramas is close to three times that of the variance of the revenue of comedies. Does that indicate that the ratio of the variances is significantly different at the 5% level? We will discuss in **Step 6**.

**Technology**

To find the *P*-value for an *F*-distribution using technology, please visit stat.hawkeslearning.com and navigate to **Discovering Business Statistics, Second Edition > Technology Instructions > F-Distribution > F-Probability (cdf)**.

**Step 6:** Make the decision and state the conclusion in terms of the original question.

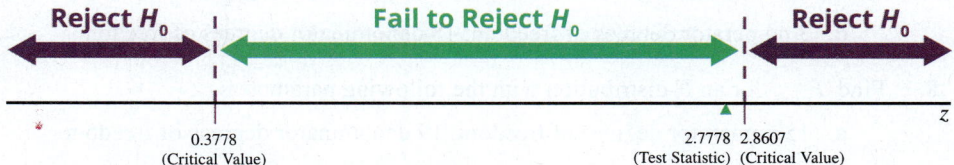

**Figure 11.5.7**

Since the test statistic does not fall in the rejection region, we fail to reject the null hypothesis.

Using the *P*-value approach, we want to find $2P(F > 2.7778)$ since this is a two-tailed test. This results in a *P*-value of 0.0564. Since the *P*-value is greater than the significance level of 0.05, we fail to reject the null hypothesis.

*Conclusion and Interpretation*: We conclude that there is insufficient evidence to indicate that the variances in revenue between dramas and comedies are significantly different.

Please note that the methods presented in this section work very poorly when the normality assumption is violated. It is very important to validate the assumption of normality before developing confidence intervals or performing hypothesis tests on the ratio of the variances.

# 11.5 Exercises

## Basic Concepts

1. Give two examples of situations in which someone would be interested in comparing population variances (or standard deviations).

2. What assumptions are necessary to perform a hypothesis test for two population variances?

3. What is the test statistic that is used when comparing two population variances?

4. What are the parameters of the distribution of the test statistic in the previous question?

## Exercises

5. Find a point on the *F*-distribution with 7 numerator degrees of freedom and 22 denominator degrees of freedom such that the following area lies to the right of this value.

   **a.** $\alpha = 0.100$     **c.** $\alpha = 0.025$

   **b.** $\alpha = 0.050$     **d.** $\alpha = 0.010$

6. Find a point on the $F$-distribution with 30 numerator degrees of freedom and 8 denominator degrees of freedom such that the following area lies to the right of this value.

   a. $\alpha = 0.100$          c. $\alpha = 0.025$

   b. $\alpha = 0.050$          d. $\alpha = 0.010$

7. Find $F_{0.025}$ for an $F$-distribution with the following parameters.

   a. 1 numerator degree of freedom, 25 denominator degrees of freedom

   b. 6 numerator degrees of freedom, 11 denominator degrees of freedom

   c. 8 numerator degrees of freedom, 40 denominator degrees of freedom

   d. 3 numerator degrees of freedom, 18 denominator degrees of freedom

8. Find $F_{0.010}$ for an $F$-distribution with the following parameters.

   a. 15 numerator degrees of freedom, 19 denominator degrees of freedom

   b. 10 numerator degrees of freedom, 29 denominator degrees of freedom

   c. 60 numerator degrees of freedom, 24 denominator degrees of freedom

   d. 12 numerator degrees of freedom, 21 denominator degrees of freedom

9. State the null and alternative hypotheses for each scenario.

   a. A professor believes that the variance of SAT scores of honor students is less than that of all students who take the SAT. Let $\sigma_1^2$ represent the population variance for honor students.

   b. A quality control inspector believes that the variance in the diameters of soda cans produced by Machine 1 is greater than the variance in the diameters of soda cans produced by Machine 2. Let $\sigma_1^2$ represent the population variance for Machine 1.

10. Calculate the test statistic for a hypothesis test for two population variances using the given information. Assume that both population distributions are approximately normal.

    $$n_1 = 4, \quad s_1^2 = 0.961, \quad n_2 = 6, \quad s_2^2 = 0.899$$

11. State the critical value(s) of the test statistic, and determine the rejection region for the hypothesis test for the two population variances using the given information. Then give the appropriate conclusion for the hypothesis test. Assume that both population distributions are approximately normal.

    a. $n_1 = 14, \quad s_1^2 = 3.152, \quad n_2 = 11, \quad s_2^2 = 9.300, \quad H_a: \sigma_1^2 < \sigma_2^2, \quad \alpha = 0.05$

    b. $n_1 = 12, \quad s_1^2 = 1893, \quad n_2 = 26, \quad s_2^2 = 1066, \quad H_a: \sigma_1^2 > \sigma_2^2, \quad \alpha = 0.01$

    c. $n_1 = 20, \quad s_1^2 = 27.08, \quad n_2 = 29, \quad s_2^2 = 11.77, \quad H_a: \sigma_1^2 \neq \sigma_2^2, \quad \alpha = 0.05$

    For exercises 12-16, complete the following steps. Assume that both population distributions are approximately normal in each scenario.

    a. State the null and alternative hypotheses.

    b. Determine which distribution to use for the test statistic and state the level of significance.

    c. Calculate the test statistic.

    d. Draw a conclusion and interpret the decision.

**12.** A golf pro believes that the variances of his driving distances are different for different brands of golf balls. In particular, he believes that his driving distances, measured in yards, have a smaller variance when he uses Titleist golf balls than when he uses a generic store brand. He hits 10 Titleist golf balls and records a sample variance of 201.65. He hits 10 generic golf balls and records a sample variance of 364.57. Test the golf pro's claim using a 0.05 level of significance. Assume the samples are from populations that are approximately normally distributed. Does the evidence support the golf pro's claim?

**13.** A quality control inspector believes that the variance in the diameters of soda cans, measured in millimeters, is greater for soda cans produced by Machine A than for soda cans produced by Machine B. The sample variance of a random sample of 15 soda cans from Machine A is 2.788. The sample variance for a random sample of 17 soda cans from Machine B is 1.982. Test the inspector's claim using a 0.10 level of significance. Assume the samples are from populations that are approximately normally distributed. Does the evidence support the inspector's claim?

**14.** A medical researcher believes that the variance of total cholesterol levels in men is greater than the variance of total cholesterol levels in women. The sample variance for a random sample of 8 men's cholesterol levels, measured in mg/dL, is 277. The sample variance for a random sample of 7 women is 89. Test the researcher's claim using a 0.10 level of significance. Assume the samples are from populations that are approximately normally distributed. Does the evidence support the researcher's belief?

**15.** A basketball coach believes that the variance of the heights of adult male basketball players is different from the variance of heights for the general population of men. The sample variance of heights, measured in inches, for a random sample of 12 basketball players is 24.76. The sample variance for a random sample of 13 other men is 25.87. Test the coach's claim using a 0.01 level of significance. Assume the samples are from populations that are approximately normally distributed. Does the evidence support the coach's claim?

**16.** One study claims that the variance in the resting heart rates of smokers is different than the variance in the resting heart rates of nonsmokers. A medical student decides to test this claim. The sample variance of resting heart rates, measured in beats per minute, for a random sample of 5 smokers is 545.1. The sample variance for a random sample of 5 nonsmokers is 103.7. Test the study's claim using a 0.01 level of significance. Assume the samples are from populations that are approximately normally distributed. Does the evidence support the study's claim?

> ## T    Discovering Technology

### Using the TI-84 Plus Calculator

Comparing Two Population Means, $\sigma_1$ and $\sigma_2$ Known

Use the information from Example 11.1.1 for this exercise.

$$n_1 = 50, \; n_2 = 50, \; s_1 = 3, \; s_2 = 2, \; \bar{x}_1 = 8.8, \; \text{and} \; \bar{x}_2 = 9.5$$

1. Part **a.** of Example 11.1.1 asks for the 95% confidence interval for the difference in the population mean QOS for cable and satellite subscribers. To calculate the interval using a TI-84 Plus calculator, press **STAT**, scroll to **TESTS**, and select option **2-SampZInt**.

2. Select **Stats**, and enter **3** for $\sigma_1$, **2** for $\sigma_2$, **8.8** for $\bar{x}_1$, **50** for $n_1$, **9.5** for $\bar{x}_2$, **50** for $n_2$, and **0.95** for C-Level. Press **Calculate**. (Note that since both sample sizes are greater than or equal to 30 we can use $s_1$ and $s_2$ as approximations for $\sigma_1$ and $\sigma_2$.)

3. The 95% confidence interval is displayed as (−1.699, 0.29939). Compare this to the confidence interval we calculated in Example 11.1.1.

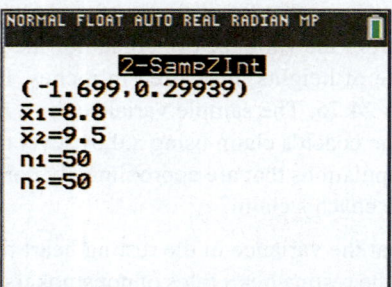

4. Part **b.** of Example 11.1.1 asks if there is sufficient evidence for the analyst to conclude at $\alpha = 0.05$ that there is a difference in average QOS between cable and satellite subscribers. To perform this test using a TI-84 Plus calculator, press **STAT**, scroll to **TESTS**, and select option **2-SampZTest**.

5. Select **Stats**, and enter **3** for $\sigma_1$, **2** for $\sigma_2$, **8.8** for $\bar{x}_1$, **50** for $n_1$, **9.5** for $\bar{x}_2$, **50** for $n_2$, and $\neq \mu_2$ as the alternative hypothesis. Press **Calculate**. (Note that since both sample sizes are greater than or equal to 30 we can use $s_1$ and $s_2$ as approximations for $\sigma_1$ and $\sigma_2$.)

6. The results of the hypothesis test are displayed. The $z$-statistic is calculated to be −1.37 and the $P$-value is given as 0.1698. Thus, there is not sufficient evidence at the 0.05 level to conclude that the mean QOS is significantly different between cable and satellite subscribers.

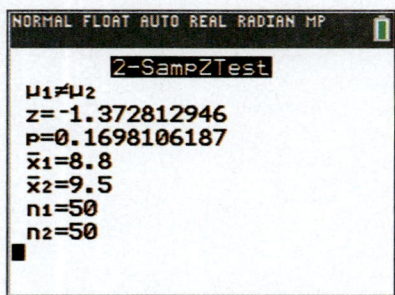

## Comparing Two Population Means, $\sigma_1$ and $\sigma_2$ Unknown

Use the information from Example 11.2.1 for this exercise.

$$n_1 = 15, \; n_2 = 16, \; s_1 = 200, \; s_2 = 175, \; \bar{x}_1 = 3425, \text{ and } \bar{x}_2 = 3250$$

1. Part **a.** of Example 11.2.1 asks for the 95% confidence interval for the difference in the population mean dollar sales between the two products. To calculate the interval using a TI-84 Plus calculator, press **STAT**, scroll to **TESTS**, and select option **2-SampTInt**.

2. Select **Stats**, and enter **3425** for $\bar{x}_1$, **200** for $s_1$, **15** for $n_1$, **3250** for $\bar{x}_2$, **175** for $s_2$, **16** for $n_2$, **0.95** for C-Level, and **Yes** for Pooled. Press **Calculate**.

3. The 95% confidence interval is displayed as (37.189, 312.81). Compare this to the confidence interval we calculated in Example 11.2.1. The slight difference in the values of the interval endpoints is due to rounding error that results from rounding the pooled variance in the hand calculation.

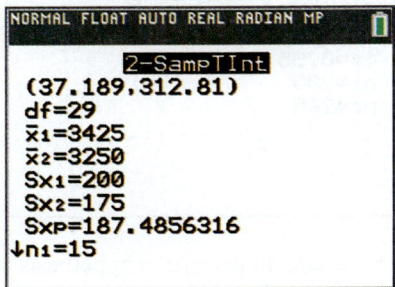

4. Part **b.** of Example 11.2.1 asks to perform a hypothesis test to determine if Product 1 has higher mean dollar sales than Product 2 using $\alpha = 0.01$. To perform this test using a TI-84 Plus calculator, press **STAT**, scroll to **TESTS**, and select option **2-SampTTest**.

5. Select **Stats**, and enter **3425** for $\bar{x}_1$, **200** for $s_1$, **15** for $n_1$, **3250** for $\bar{x}_2$, **175** for $s_2$, **16** for $n_2$, $> \mu_2$ for the alternative hypothesis, and **Yes** for Pooled. Press **Calculate**.

6. The results of the hypothesis test are displayed. The $t$-statistic is calculated to be 2.597 and the $P$-value is given as 0.0073. Thus, there is sufficient evidence at the 0.01 level to conclude that the mean dollar sales for Product 1 are significantly higher than the mean dollar sales for Product 2.

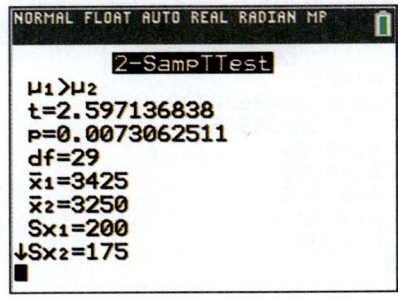

## Comparing Two Population Proportions

Use the information from Example 11.4.1 for this exercise.

$$x_1 = 8, \ n_1 = 200, \ x_2 = 10, \ n_2 = 200$$

1.  Part **b.** of Example 11.4.1 asks for the 95% confidence interval for the difference between the population proportions of defective phones from Plant A and Plant B. To calculate the interval using a TI-84 Plus calculator, press **STAT**, scroll to **TESTS**, and select option **2-PropZInt**.

2.  Enter **8** for $x_1$, **200** for $n_1$, **10** for $x_2$, **200** for $n_2$, and **0.95** for C-Level. Press **Calculate**.

3.  The 95% confidence interval is displayed as (−0.0506, 0.03062). Compare this to the confidence interval we calculated in Example 11.4.1.

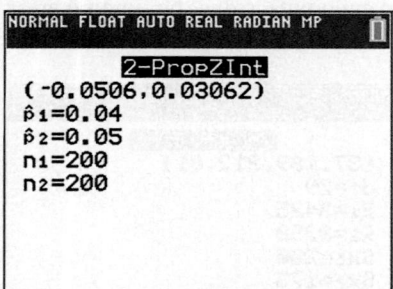

4.  Part **c.** of Example 11.4.1 asks to perform a hypothesis test to determine if there is sufficient evidence for the cell phone executive to conclude that there is a difference in the proportion of defective phones produced by the two plants at $\alpha = 0.10$. To perform this test using a TI-84 Plus calculator, press **STAT**, scroll to **TESTS**, and select option **2-PropZTest**.

5.  Enter **8** for $x_1$, **200** for $n_1$, **10** for $x_2$, **200** for $n_2$, and $\neq p_2$ for the alternative hypothesis. Press **Calculate**.

6.  The results of the hypothesis test are displayed. The $z$-statistic is calculated to be −0.48 and the $P$-value is given as 0.6295. Thus, there is not sufficient evidence at the 0.10 level for the cell phone executive to conclude that the proportion of defective phones produced differs between the two plants.

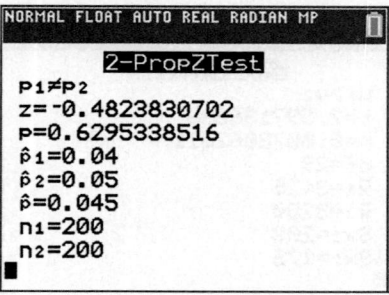

## Using Excel

When using these data analysis tools in Excel, the raw data must be available and you will not be able to plug in summary statistics. You will be asked to select a range of data to compute the answer. The data analysis tools that are available in Excel are displayed in the following table.

| Excel Data Analysis Tools for this Chapter | | |
|---|---|---|
| Section | Topic | Excel Data Analysis Tool |
| 11.1 | Comparing Two Population Means, $\sigma_1$ and $\sigma_2$ Known | z-Test: Two Sample for Means |
| 11.2 | Comparing Two Population Means, $\sigma_1$ and $\sigma_2$ Unknown and $\sigma_1 = \sigma_2$ | t-Test: Two-Sample Assuming Equal Variances |
| 11.2 | Comparing Two Population Means, $\sigma_1$ and $\sigma_2$ Unknown and $\sigma_1 \neq \sigma_2$ | t-Test: Two-Sample Assuming Unequal Variances |
| 11.3 | Paired Difference | t-Test: Paired Two Sample for Means |

### Paired Difference

For this exercise, use the data from Example 11.3.1.

1. Enter the daily food sales for each restaurant found in Table 11.3.2 into Column A and Column B in an Excel worksheet.

| | A | B |
|---|---|---|
| 1 | Restaurant 1 | Restaurant 2 |
| 2 | 5828 | 7894 |
| 3 | 9836 | 11573 |
| 4 | 3984 | 5319 |
| 5 | 5845 | 6389 |
| 6 | 5210 | 6055 |
| 7 | 9668 | 10631 |
| 8 | 6768 | 7866 |
| 9 | 6726 | 7976 |
| 10 | 4399 | 5652 |
| 11 | 6692 | 8083 |

2. Select **Data Analysis** from the Data tab and select **t-Test: Paired Two Sample for Means**.

3. Input the range **A1:A11** as the Variable 1 Range or click and drag in the worksheet to select the sales for Restaurant 1 as the first variable. Input the range **B1:B11** as the Variable 2 Range or click and drag in the worksheet to select the sales for Restaurant 2 as the second variable.

4. Check the checkbox for **Labels** since we have the restaurant numbers listed in row 1.

5. Enter **0** for the Hypothesized Mean Difference.

6. Specify the significance level of the test by entering **0.01** for Alpha.

7. Select **Output Range** and enter **D1** into the box.

8. Press **OK**. The results of the test are displayed beside the data. You can widen columns **D** to **F** to display the data more effectively.

   Notice that the test statistic (t Stat) and the critical values for one-tailed (t Critical one-tail) and two-tailed (t Critical two-tail) tests are displayed. Compare the test statistic to that which we obtained in Example 11.3.1.

| ◢ | A | B | C | D | E | F |
|---|---|---|---|---|---|---|
| 1 | Restaurant 1 | Restaurant 2 | | t-Test: Paired Two Sample for Means | | |
| 2 | 5828 | 7894 | | | | |
| 3 | 9836 | 11573 | | | Restaurant 1 | Restaurant 2 |
| 4 | 3984 | 5319 | | Mean | 6495.6 | 7743.8 |
| 5 | 5845 | 6389 | | Variance | 3845984.044 | 4238663.733 |
| 6 | 5210 | 6055 | | Observations | 10 | 10 |
| 7 | 9668 | 10631 | | Pearson Correlation | 0.977817104 | |
| 8 | 6768 | 7866 | | Hypothesized Mean Difference | 0 | |
| 9 | 6726 | 7976 | | df | 9 | |
| 10 | 4399 | 5652 | | t Stat | -9.087224765 | |
| 11 | 6692 | 8083 | | P(T<=t) one-tail | 3.94487E-06 | |
| 12 | | | | t Critical one-tail | 2.821437921 | |
| 13 | | | | P(T<=t) two-tail | 7.88974E-06 | |
| 14 | | | | t Critical two-tail | 3.249835541 | |

## Using JMP

### Paired Difference

For this exercise, use the data from Example 11.3.1.

1. With a JMP data table open, enter the daily food sales for each restaurant in Table 11.3.2 into Column 1 and Column 2.

| | Column 1 | Column 2 |
|---|---|---|
| 1 | 5828 | 7894 |
| 2 | 9836 | 11573 |
| 3 | 3984 | 5319 |
| 4 | 5845 | 6389 |
| 5 | 5210 | 6055 |
| 6 | 9668 | 10631 |
| 7 | 6768 | 7866 |
| 8 | 6726 | 7976 |
| 9 | 4399 | 5652 |
| 10 | 6692 | 8083 |

2. Select **Analyze** in the top row of the JMP spreadsheet and then select **Matched Pairs**.

3. From the **Select Columns** box, click on **Column 1**, then click on **Y, Paired Response**. Click on **Column 2**, then click on **Y, Paired Response**. Click **OK**.

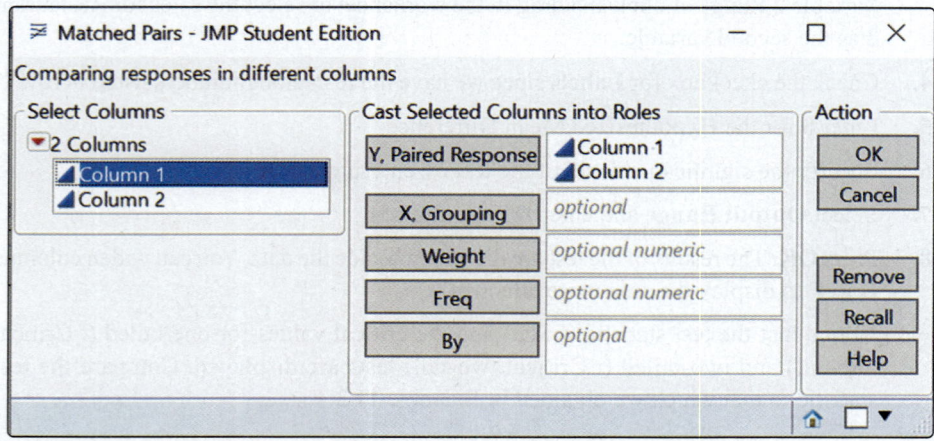

4. Observe the output screen for the results. By default, JMP produces a 95% confidence interval for the mean difference. Notice that the test statistic (t-Ratio) and the *P*-values are displayed. Compare the test statistic to that which we obtained in Example 11.3.1.

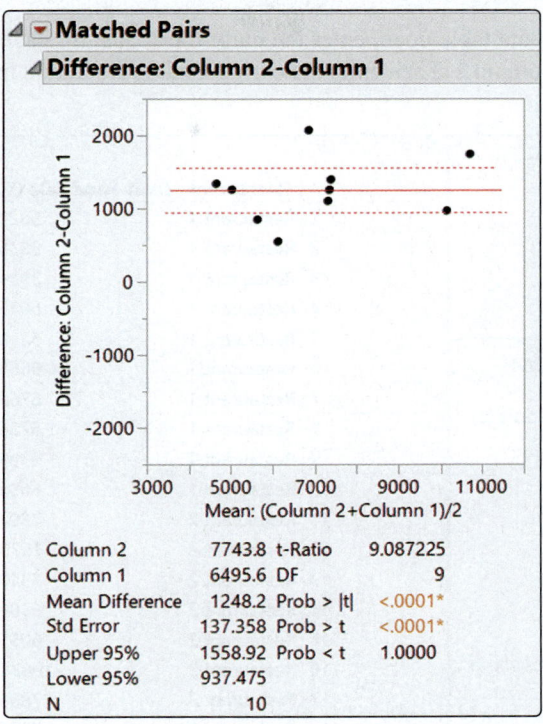

**Matched Pairs**

**Difference: Column 2-Column 1**

| | | | |
|---|---|---|---|
| Column 2 | 7743.8 | t-Ratio | 9.087225 |
| Column 1 | 6495.6 | DF | 9 |
| Mean Difference | 1248.2 | Prob > |t| | <.0001* |
| Std Error | 137.358 | Prob > t | <.0001* |
| Upper 95% | 1558.92 | Prob < t | 1.0000 |
| Lower 95% | 937.475 | | |
| N | 10 | | |

5. To adjust the confidence level, click on the **red arrow** next to Matched Pairs, click on **Set $\alpha$ Level** and select **0.01**. The results of the test are displayed.

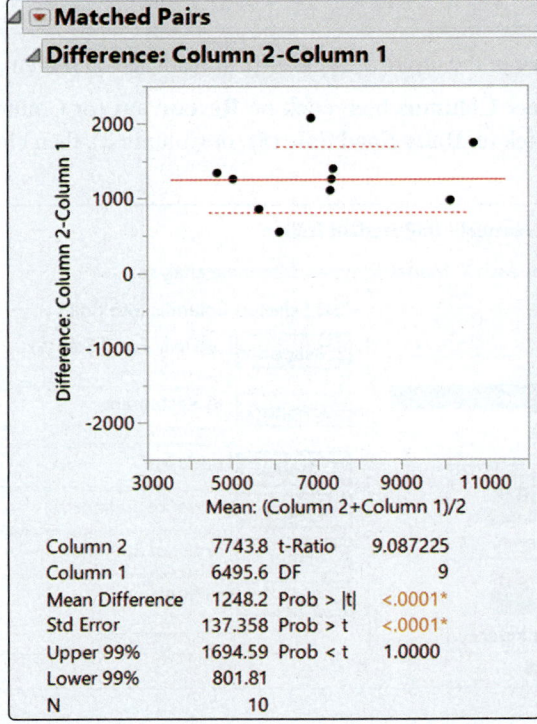

**Matched Pairs**

**Difference: Column 2-Column 1**

| | | | |
|---|---|---|---|
| Column 2 | 7743.8 | t-Ratio | 9.087225 |
| Column 1 | 6495.6 | DF | 9 |
| Mean Difference | 1248.2 | Prob > |t| | <.0001* |
| Std Error | 137.358 | Prob > t | <.0001* |
| Upper 99% | 1694.59 | Prob < t | 1.0000 |
| Lower 99% | 801.81 | | |
| N | 10 | | |

Two Sample *t*-Test

For this exercise, use the data from Example 11.3.1.

1.  With a JMP data table open, enter the daily food sales for each restaurant in Table 11.3.2 into Column 2. Then enter the corresponding restaurant in Column 1 for each daily food sale.

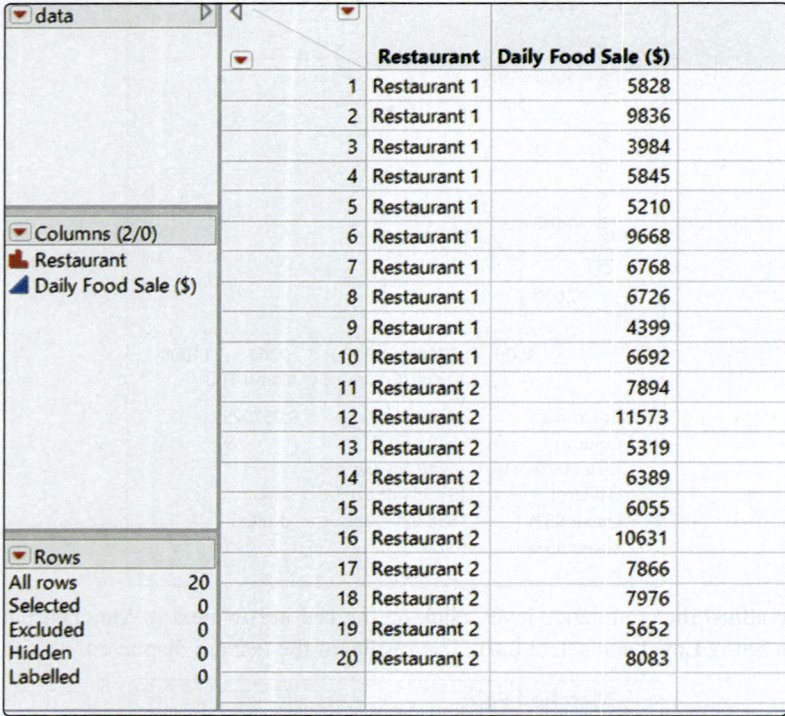

| | Restaurant | Daily Food Sale ($) |
|---|---|---|
| 1 | Restaurant 1 | 5828 |
| 2 | Restaurant 1 | 9836 |
| 3 | Restaurant 1 | 3984 |
| 4 | Restaurant 1 | 5845 |
| 5 | Restaurant 1 | 5210 |
| 6 | Restaurant 1 | 9668 |
| 7 | Restaurant 1 | 6768 |
| 8 | Restaurant 1 | 6726 |
| 9 | Restaurant 1 | 4399 |
| 10 | Restaurant 1 | 6692 |
| 11 | Restaurant 2 | 7894 |
| 12 | Restaurant 2 | 11573 |
| 13 | Restaurant 2 | 5319 |
| 14 | Restaurant 2 | 6389 |
| 15 | Restaurant 2 | 6055 |
| 16 | Restaurant 2 | 10631 |
| 17 | Restaurant 2 | 7866 |
| 18 | Restaurant 2 | 7976 |
| 19 | Restaurant 2 | 5652 |
| 20 | Restaurant 2 | 8083 |

2.  Select **Analyze** in the top row of the JMP spreadsheet and then select **Fit Y by X**.

3.  From the **Select Columns** box, click on **Restaurant** (or Column 1), then click on **X, Factor**. Click on **Daily Food Sale ($)** (or Column 2), then click on **Y, Response**. Click **OK**.

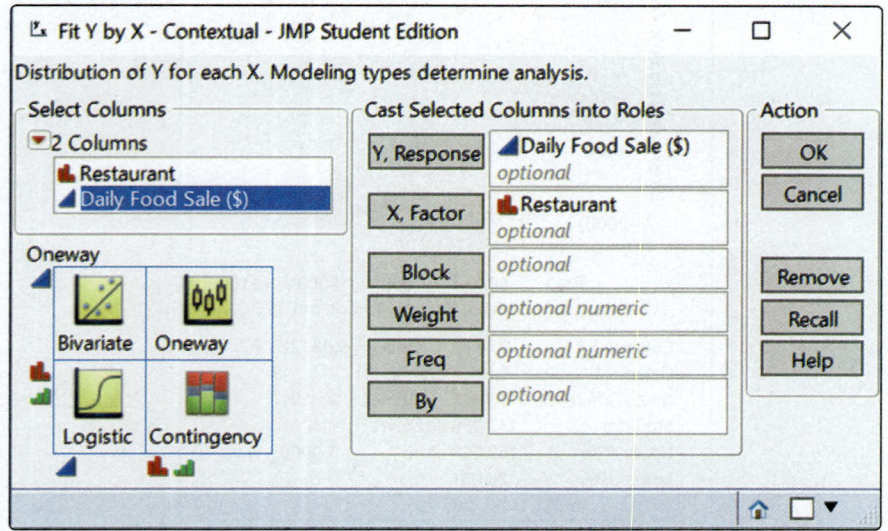

4.  Observe the output screen for the results. From the Oneway Analysis output, click on the **red arrow** next to Oneway Analysis of Daily Food Sale ($) By Restaurant and click on **Means/Anova/Pooled t**. The results of the test are displayed. Notice that the test statistic (t Ratio) and the *P*-values are displayed.

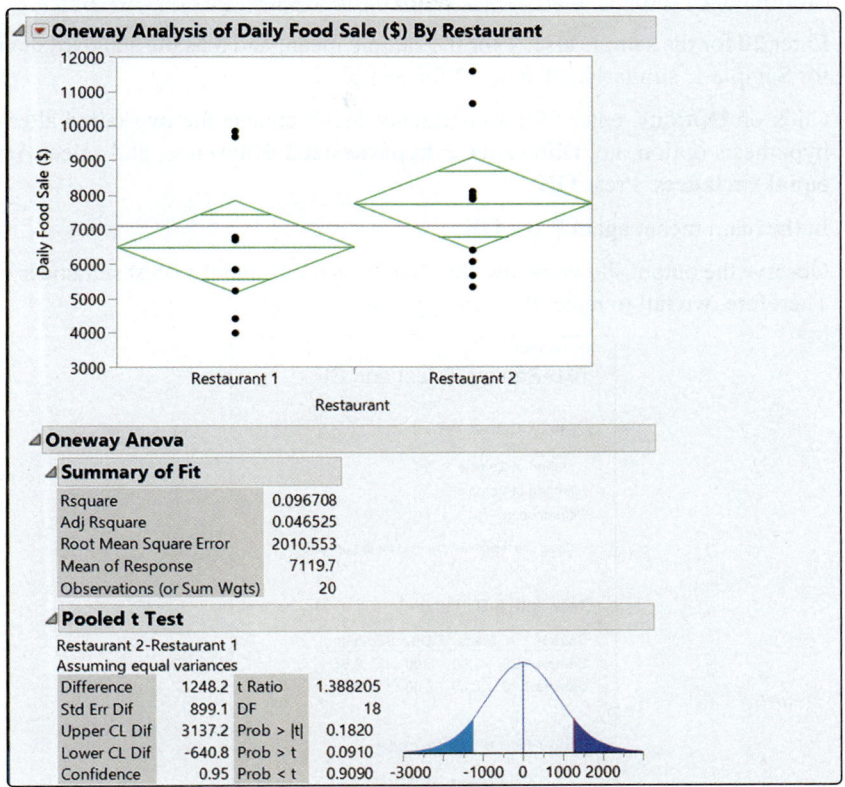

## Using Minitab

Using Minitab you can make inferences about the difference between two population means or proportions using either summarized data or raw data. The Minitab functions that are available to perform hypothesis tests and construct confidence intervals for two populations are given in the following table.

| Minitab Functions for this Chapter | | |
|---|---|---|
| Section | Topic | Minitab Function |
| 11.1 – 11.2 | Comparing Two Population Means | 2-Sample *t* |
| 11.3 | Paired Difference | Paired *t* |
| 11.4 | Comparing Two Population Proportions | 2 Proportions |

## Comparing Two Population Means

For this we will use the example data of hourly wages in two cities City A and City B. Assume we are given the summarized data as follows:

| | *n* | $\bar{x}$ | *s* |
|---|---|---|---|
| City A | 20 | $7 | $3 |
| City B | 20 | $8 | $2 |

Note: If we are given the raw data instead of summarized data, then enter the data in columns C1 and C2 and choose the option **Each sample is in a column** in Step 1.

Assuming the population variances are approximately equal, we would like to test if there is difference in hourly wages between the two cities at $\alpha = 0.05$

1. Choose **Stat**, **Basic Statistics** and **2-Sample t**. In the drop-down menu select **Summarized data**.

2. Enter **20** for the sample size, **7** for the sample mean, and **3** as the standard deviation for Sample 1; similarly, **20**, **8** and **3** for Sample 2.

3. Click on **Options**, enter **95** for confidence level, choose the two-tailed alternative hypothesis option, i.e, **Difference ≠ hypothesized difference**, and select **Assume equal variances**. Press **OK**.

4. In the main menu, again Press **OK**.

5. Observe the output shown below, the *P*-value is 0.222 and the *t*-test statistic is −1.24. Therefore, we fail to reject the null hypothesis.

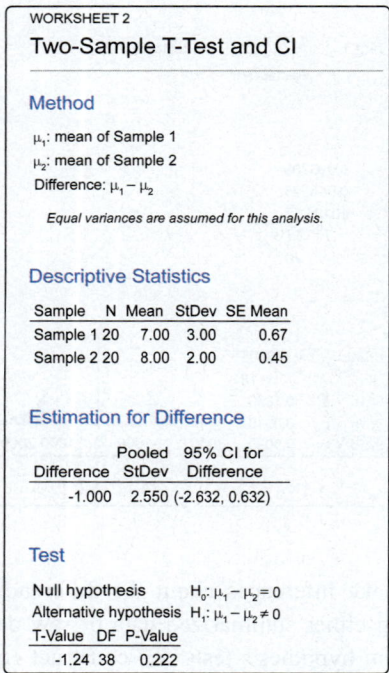

WORKSHEET 2

**Two-Sample T-Test and CI**

**Method**

$\mu_1$: mean of Sample 1
$\mu_2$: mean of Sample 2
Difference: $\mu_1 - \mu_2$

*Equal variances are assumed for this analysis.*

**Descriptive Statistics**

| Sample | N | Mean | StDev | SE Mean |
|--------|---|------|-------|---------|
| Sample 1 | 20 | 7.00 | 3.00 | 0.67 |
| Sample 2 | 20 | 8.00 | 2.00 | 0.45 |

**Estimation for Difference**

| Difference | Pooled StDev | 95% CI for Difference |
|------------|--------------|------------------------|
| -1.000 | 2.550 | (-2.632, 0.632) |

**Test**

Null hypothesis        $H_0: \mu_1 - \mu_2 = 0$
Alternative hypothesis  $H_1: \mu_1 - \mu_2 \neq 0$

| T-Value | DF | P-Value |
|---------|-----|---------|
| -1.24 | 38 | 0.222 |

## Paired Difference

For this exercise we will use the example data of food sales at two restaurants from Example 11.3.1.

| Day | Restaurant 1 | Restaurant 2 |
|-----|--------------|--------------|
| 1 | 5828 | 7894 |
| 2 | 9836 | 11573 |
| 3 | 3984 | 5319 |
| 4 | 5845 | 6389 |
| 5 | 5210 | 6055 |
| 6 | 9668 | 10631 |
| 7 | 6768 | 7866 |
| 8 | 6726 | 7976 |
| 9 | 4399 | 5652 |
| 10 | 6692 | 8083 |

The owner wants to know if there is a difference in sales between the two restaurants at $\alpha = 0.01$.

1. Enter the sample sales data into column **C1** and **C2**.

2. Choose **Stat**, **Basic Statistics** and **Paired t**. In the drop-down menu select **Each sample is in a column**. Choose column **C1** for Sample 1 and **C2** for Sample 2.

3. Click on **Options**, enter **99** for confidence level, and choose **Difference ≠ Hypothesized difference**. Press **OK**.

4. In the main menu, again Press **OK**.

5. Observe the output shown below, the *P*-value is 0.000 and the *t*-test statistic is −9.09. Therefore, we reject the null hypothesis.

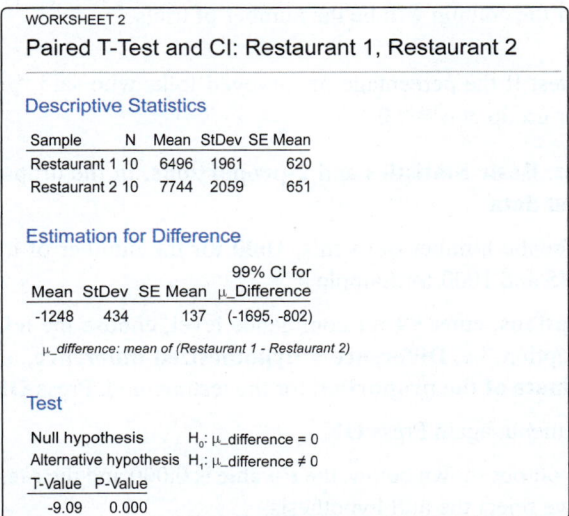

Note: Let's say you already have the summarized data of the sample. For example, for the above sample, the summary is as follows: sample mean difference $(\overline{x}_d) = -1248.20$, sample standard deviation $(s_d) = 434.36$, sample size $= 10$. Then do the following:

1. Choose **Stat**, **Basic Statistics** and **Paired t**. In the drop-down menu select **Summarized data**.

2. Enter **10** for the sample size, **−1248.20** for the sample mean, and **434.36** as the standard deviation.

3. Click on **Options**, enter **99** for the confidence level, and choose **Difference ≠ Hypothesized difference**. Press **OK**.

4. In the main menu, again Press **OK**.

5. Observe the output shown below, the *P*-value is 0.000 and the *t*-test statistic is −9.09. Therefore, we reject the null hypothesis.

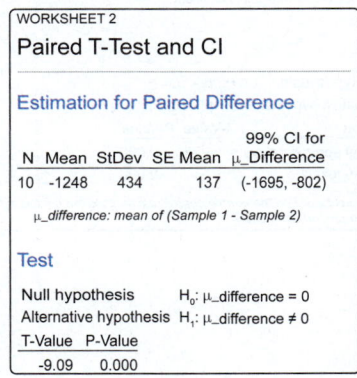

## Comparing Two Population Proportions

For this we will use the example data from the network advertising survey for two age groups. Assume we are given the summarized data as follows:

| Age Group | Number Surveyed | Number of "Yes" Responses |
|---|---|---|
| 30 or Younger | 1000 | 450 |
| Older than 30 | 1000 | 655 |

Note: If we are given the raw data instead of summarized data then enter them in columns C1 and C2 and choose the option **Each sample is in a column** in Step 1. The columns will be a basic "Yes" or "No" indicating the success of the binomial distribution for each sample. The size of the column will be the number of trials.

We would like to test if the percentage of surveyed folks who said "yes" is significantly higher for the older group at $\alpha = 0.05$

1.  Choose **Stat**, **Basic Statistics** and **2-proportions**. In the drop-down menu select **Summarized data**.

2.  Enter **450** for the number of events, **1000** for the number of trials for Sample 1; similarly, **655** and 1000 for Sample 2.

3.  Click on **Options**, enter **99** for confidence level, choose the left tailed alternative hypothesis option, i.e, **Difference < hypothesized difference**, and choose **Use the Pooled estimate of the proportion** for the test method. Press **OK**.

4.  In the main menu, again Press **OK**.

5.  Observe the output shown below, the *P*-value is 0.000 and the *t*-test statistic is −9.22. Therefore, we reject the null hypothesis.

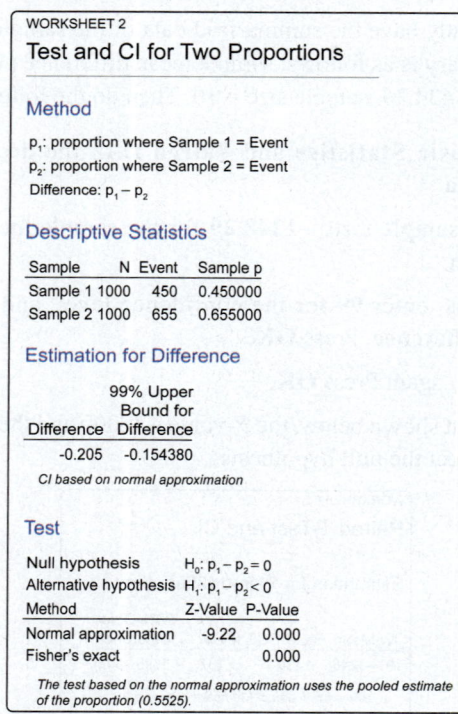

# R | Chapter 11 Review

## Key Terms and Ideas

- Independent Experimental Design
- Completely Randomized Design
- Sampling Distribution of $\bar{x}_1 - \bar{x}_2$
- Interval Estimation of $\mu_1 - \mu_2$, $\sigma_1$ and $\sigma_2$ Known
- Hypothesized Difference
- Hypothesis Test about $\mu_1 - \mu_2$, $\sigma_1$ and $\sigma_2$ Known
- Assumptions for Inferences about $\mu_1 - \mu_2$
- Interval Estimation of $\mu_1 - \mu_2$, $\sigma_1$ and $\sigma_2$ Unknown
- Small Sample Hypothesis Test about $\mu_1 - \mu_2$, $\sigma_1$ and $\sigma_2$ Unknown
- Pooled Variance
- $t$-Test about $\mu_1 - \mu_2$ Assuming Unequal Variances
- Paired Difference Experimental Design
- Assumptions for the Paired Difference Experimental Design
- Assumptions for Comparing Two Population Proportions
- Interval Estimation of $p_1 - p_2$,
- Hypothesis Test about $p_1 - p_2$,
- Interval Estimation of $\sigma_1^2 / \sigma_2^2$
- Hypothesis Test about $\sigma_1^2 / \sigma_2^2$

## Key Formulas

| | Section |
|---|---|

**Sampling Distribution of $\bar{x}_1 - \bar{x}_2$**

$$\mu_{\bar{x}_1 - \bar{x}_2} = \mu_1 - \mu_2$$

$$\sigma_{\bar{x}_1 - \bar{x}_2} = \sqrt{\frac{\sigma_1^2}{n_1} + \frac{\sigma_2^2}{n_2}}$$

11.1

**$100(1 - \alpha)\%$ Confidence Interval for $\mu_1 - \mu_2$**

$$\left(\bar{x}_1 - \bar{x}_2\right) \pm z_{\alpha/2} \sqrt{\frac{\sigma_1^2}{n_1} + \frac{\sigma_2^2}{n_2}}$$

11.1

where $\sigma_1$ and $\sigma_2$ are known and the samples follow a normal distribution.

**Test Statistic for a Hypothesis Test about $\mu_1 - \mu_2$**

$$z = \frac{\left(\bar{x}_1 - \bar{x}_2\right) - \mu_{\bar{x}_1 - \bar{x}_2}}{\sigma_{\bar{x}_1 - \bar{x}_2}} = \frac{\left(\bar{x}_1 - \bar{x}_2\right) - \left(\mu_1 - \mu_2\right)}{\sqrt{\frac{\sigma_1^2}{n_1} + \frac{\sigma_2^2}{n_2}}}$$

11.1

where $\sigma_1$ and $\sigma_2$ are known and the samples follow a normal distribution.

## Key Formulas (cont.)

Section

$100(1 - \alpha)\%$ Confidence Interval for $\mu_1 - \mu_2$, $\sigma_1$ and $\sigma_2$ Unknown $(\sigma_1 = \sigma_2)$

$$\left(\bar{x}_1 - \bar{x}_2\right) \pm t_{\alpha/2,df}\sqrt{s_p^2\left(\frac{1}{n_1} + \frac{1}{n_2}\right)}$$

11.2

where $s_p^2 = \dfrac{\left(n_1 - 1\right)s_1^2 + \left(n_2 - 1\right)s_2^2}{n_1 + n_2 - 2}$, and $t_{\alpha/2,df}$ is the critical value of the

$t$-distribution with $n_1 + n_2 - 2$ degrees of freedom capturing an area of $\alpha/2$ in the upper tail.

Test Statistic for a Hypothesis Test about $\mu_1 - \mu_2$, $\sigma_1$ and $\sigma_2$ Unknown $(\sigma_1 = \sigma_2)$

$$t = \frac{\left(\bar{x}_1 - \bar{x}_2\right) - \left(\mu_1 - \mu_2\right)}{\sqrt{s_p^2\left(\frac{1}{n_1} + \frac{1}{n_2}\right)}}$$

11.2

$100(1 - \alpha)\%$ Confidence Interval for $\mu_1 - \mu_2$, $\sigma_1$ and $\sigma_2$ Unknown $(\sigma_1 \neq \sigma_2)$

$$\left(\bar{x}_1 - \bar{x}_2\right) \pm t_{\alpha/2,df}\sqrt{\frac{s_1^2}{n_1} + \frac{s_2^2}{n_2}},$$

where $t_{\alpha/2,df}$ is the critical value capturing an area of $\alpha/2$ in the upper tail of $t$-distribution with degrees of freedom equal to:

11.2

$$df = \frac{\left(\dfrac{s_1^2}{n_1} + \dfrac{s_2^2}{n_2}\right)^2}{\dfrac{1}{n_1 - 1}\left(\dfrac{s_1^2}{n_1}\right)^2 + \dfrac{1}{n_2 - 1}\left(\dfrac{s_2^2}{n_2}\right)^2}.$$

Test Statistic for a Hypothesis Test for $\mu_1 - \mu_2$, $\sigma_1$ and $\sigma_2$ Unknown $(\sigma_1 \neq \sigma_2)$

$$t = \frac{\left(\bar{x}_1 - \bar{x}_2\right) - \left(\mu_1 - \mu_2\right)}{\sqrt{\frac{s_1^2}{n_1} + \frac{s_2^2}{n_2}}},$$

which follows a $t$-distribution with degrees of freedom equal to:

11.2

$$df = \frac{\left(\dfrac{s_1^2}{n_1} + \dfrac{s_2^2}{n_2}\right)^2}{\dfrac{1}{n_1 - 1}\left(\dfrac{s_1^2}{n_1}\right)^2 + \dfrac{1}{n_2 - 1}\left(\dfrac{s_2^2}{n_2}\right)^2}.$$

## Key Formulas (cont.)

**100(1 − α)% Confidence Interval for $\mu_d$**

$$\bar{x}_d \pm t_{\alpha/2,df}\frac{s_d}{\sqrt{n_d}},$$

11.3

where $\bar{x}_d$ is the sample mean of the differences, $s_d$ is the sample standard deviation of the differences, $n_d$ is the number of differences, and $t_{\alpha/2,df}$ is the critical value of the $t$-distribution with an area of $\alpha/2$ in the upper tail with $n_d - 1$ degrees of freedom ($df$).

**Test Statistic for a Paired Difference Hypothesis Test**

$$t = \frac{\bar{x}_d - \mu_d}{s_d\Big/\sqrt{n_d}},$$

11.3

where $\bar{x}_d$ is the mean of the sample differences, $\mu_d$ is the hypothesized mean of the differences, $s_d$ is the standard deviation of the sample differences, and $n_d$ is the number of differences.

**100(1 − α)% Confidence Interval for $p_1 - p_2$**

$$\left(\hat{p}_1 - \hat{p}_2\right) \pm z_{\alpha/2}\sqrt{\frac{\hat{p}_1\left(1-\hat{p}_1\right)}{n_1} + \frac{\hat{p}_2\left(1-\hat{p}_2\right)}{n_2}}$$

11.4

where $n_1\hat{p}_1 \geq 5$, $n_1\left(1-\hat{p}_1\right) \geq 5$, $n_2\hat{p}_2 \geq 5$, and $n_2\left(1-\hat{p}_2\right) \geq 5$ and $z_{\alpha/2}$ is the critical value for the $z$-distribution that captures an area of $\alpha/2$ in the upper tail.

**Sampling Distribution of $\hat{p}_1 - \hat{p}_2$**

$$\mu_{\hat{p}_1-\hat{p}_2} = p_1 - p_2$$

$$\sigma_{\hat{p}_1-\hat{p}_2} = \sqrt{\bar{p}\left(1-\bar{p}\right)\left(\frac{1}{n_1}+\frac{1}{n_2}\right)}$$

11.4

where $\bar{p} = \dfrac{x_1 + x_2}{n_1 + n_2}$

**Test Statistic for a Large Sample Hypothesis Test about $p_1 - p_2$**

$$z = \frac{\left(\hat{p}_1 - \hat{p}_2\right) - \left(p_1 - p_2\right)}{\sqrt{\bar{p}\left(1-\bar{p}\right)\left(\frac{1}{n_1}+\frac{1}{n_2}\right)}}$$

11.4

## Key Formulas (cont.)

Section

**Confidence Interval for** $\dfrac{\sigma_1^2}{\sigma_2^2}$

A $100(1 - \alpha)\%$ confidence interval for $\dfrac{\sigma_1^2}{\sigma_2^2}$, the ratio of two population variances, is given by

11.5

$$\left( \left( \frac{s_1^2}{s_2^2} \right) \frac{1}{F_{\alpha/2, df_{num}, df_{den}}}, \left( \frac{s_1^2}{s_2^2} \right) \frac{1}{F_{1-\alpha/2, df_{num}, df_{den}}} \right)$$

where $s_1^2$ and $s_2^2$ are the two sample variances, and $df_{num} = n_1 - 1$ (the denominator of $s_1^2$) and $df_{den} = n_2 - 1$ (the denominator of $s_2^2$).

**Inferences about** $\dfrac{\sigma_1^2}{\sigma_2^2}$

**Test Statistic:**

$$F = \frac{s_1^2}{s_2^2}$$

11.5

where $s_1^2$ and $s_2^2$ are the sample variances from the two samples of sizes $n_1$ and $n_2$, respectively, collected from two independent normally distributed populations. The degrees of freedom for the $F$-test are $df_{num} = n_1 - 1$ and $df_{den} = n_2 - 1$.

# AE | Additional Exercises

1. Black Bark, a Colorado based company, makes wood burning stoves. They are interested in comparing two designs to determine which design will produce a stove with a greater average burning time. Several prototypes of each design are tested and the time required to burn 15 pounds of wood was measured (the burning time is measured in hours). The results of the test are as follows.

| Burning Time for Stoves (Hours) | | | |
|---|---|---|---|
| | $n$ | $\bar{x}$ | $\sigma$ |
| Stove A | 32 | 9.35 | 0.50 |
| Stove B | 35 | 9.75 | 0.75 |

Is there sufficient evidence at $\alpha = 0.05$ for Black Bark to conclude that the mean burning time for Stove B is greater than for Stove A?

2. In each of the following experimental situations give the appropriate null and alternative hypotheses to be tested. Define all terms that appear in these hypotheses.

   a. Independent random samples of 50 male nurses and 50 female nurses are selected from the hospitals in a Southern state. Each nurse is asked whether he or she is satisfied with the working conditions in the hospital. It is of interest to see if there is a difference between male nurses and female nurses on satisfaction with working conditions.

   b. A group of 45 high school seniors take the SAT reasoning test both before and after a 3-month training course, which is designed to improve SAT scores. We wish to determine if the training course is effective.

   c. Starting salaries are determined for 40 female and 40 male electrical engineers. It is of interest to determine if female electrical engineers tend to have higher starting salaries than their male counterparts.

   d. Random and independent samples of younger (age $\leq 30$) and older (age $> 30$) automobile drivers are chosen and asked whether they have had a speeding ticket in the past 12 months. It is intended to show that younger drivers are more likely than older drivers to have had a speeding ticket in the past 12 months.

   e. Do women have a shorter reaction time than men when exposed to a certain stimulus? Random and independent samples of 10 men and 10 women are included in an experiment that measures reaction time to the stimulus.

3. A nutritionist is interested in determining the decrease in cholesterol level which a person can achieve by following a particular diet that is low in fat and high in fiber. Seven subjects are randomly selected to try the diet for six months, and their cholesterol levels are measured both before and after the diet. The results of the study are as follows.

| Cholesterol Levels | | | | | | | |
|---|---|---|---|---|---|---|---|
| Subject | 1 | 2 | 3 | 4 | 5 | 6 | 7 |
| Before Diet | 155 | 170 | 145 | 200 | 162 | 180 | 160 |
| After Diet | 152 | 168 | 148 | 195 | 162 | 178 | 157 |

   a. Is a paired design appropriate for the above experiment? Explain.

   b. What assumption must be made in order to perform the test of hypothesis?

   c. Do the data appear to satisfy the assumption described in part b.? Why or why not?

   d. Can the nutritionist conclude that there is a significant decrease in average cholesterol level when the diet is used? Use $\alpha = 0.01$.

4. The design group for a monofilament cord manufacturer is testing two possible compositions of the cord for tensile strength. Composition A is more difficult to manufacture than Composition B, so the design group has decided that it will recommend Composition A only if the mean tensile strength for Composition A is shown to be significantly greater than the mean tensile strength for Composition B. Several monofilament cords of each sample are tested and the tensile strengths are measured in pounds per square inch. Assume that the population variances are approximately equal.

| Tensile Strength (Pounds per Square Inch) | | | |
|---|---|---|---|
| | $n$ | $\bar{x}$ | $s$ |
| Composition A | 20 | 52,907 | 2575 |
| Composition B | 20 | 50,219 | 1210 |

a. What assumptions must be made in order to perform the hypothesis test?

b. Will the design group recommend Composition A or Composition B for the monofilament cord at $\alpha = 0.10$?

5. Consider Example 11.3.1. If you were to perform a two-sample $t$-test, you would find that you would fail to reject the null hypothesis and conclude that there is no difference in average daily sales between the two restaurants. Of course, it would be difficult to believe such a test given that if we examine the data in Table 11.3.2, we see that each of the daily sales figures from Restaurant 2 is more than that from Restaurant 1. From this observation, it is clear that the average daily sales between the restaurants are different. Why, then, would the $t$-test be unable to detect this difference? The answer: **an independent samples $t$-test is not a valid procedure to use with paired data**. The $t$-test is inappropriate because the assumption of independent samples is invalid since the dependence between restaurants is a function of the days. Perform the two-sample $t$-test to verify that the independent samples test would lead the owner to fail to reject the null hypothesis and conclude that there is not a significant difference between average daily sales. The data from Table 11.3.2 are replicated below for your convenience. Use $\alpha = 0.01$.

| Daily Food Sales for Two Restaurants ($) | | | |
|---|---|---|---|
| Day | Restaurant 1 | Restaurant 2 | Difference = Restaurant 1 − Restaurant 2 |
| 1 | 5828 | 7894 | −2066 |
| 2 | 9836 | 11,573 | −1737 |
| 3 | 3984 | 5319 | −1335 |
| 4 | 5845 | 6389 | −544 |
| 5 | 5210 | 6055 | −845 |
| 6 | 9668 | 10,631 | −963 |
| 7 | 6768 | 7866 | −1098 |
| 8 | 6726 | 7976 | −1250 |
| 9 | 4399 | 5652 | −1253 |
| 10 | 6692 | 8083 | −1391 |

6. Two independent random samples have been selected, 100 from Population 1 and 150 from Population 2. The sample mean from the first population is 1025 with a population standard deviation of 10. For the second sample, the mean is 1039 with a population standard deviation of 12.

a. Test that there is no difference between the groups using $\alpha = 0.01$.

b. Construct a 95% confidence interval for the true mean difference between Population 1 and Population 2.

7.  The manufacturer of Brand 1 cigarettes claims that his cigarettes are no more harmful to health than Brand 2 (filtered) cigarettes. Assuming harmfulness is to be associated with nicotine content, the FDA took random samples of 125 cigarettes from Brand 1 and 180 cigarettes from Brand 2. The average nicotine content in the sample of Brand 1 was 24.6 mg with a population standard deviation of 1.4 mg; the average nicotine content in the sample of Brand 2 was 24.3 mg with a population standard deviation of 1.1 mg.

    a.  Is there evidence to refute the manufacturer's claim at $\alpha = 0.05$?

    b.  Construct an 85% confidence interval for the true mean difference in nicotine content between Brand 1 and Brand 2.

8.  A team of organizational behavior managers investigated the effects of an orientation program on "first day of work" anxiety levels of new employees. 72 new employees were randomly assigned to receive or not to receive a two-day company orientation program prior to their first day at work. Two hours after beginning work, each employee was given a test to measure his or her level of anxiety. The mean score was 1002 for the 37 receiving orientation and 1018 for the 35 who did not receive the orientation. Scores of employees who attended similar orientation programs in the past have had a standard deviation of 142. Scores of employees that did not attend the orientation program had this same standard deviation.

    a.  Test to see if there is evidence of a difference in the mean test scores between those who participate in an orientation program and those who do not. Use a level of significance equal to 0.05.

    b.  Calculate the observed significance level (*P*-value) of this test.

9.  A property manager of thousands of apartments wants to test the difference in the mean net annual income between two types of leasing arrangements. Arrangement A is to charge a lower rent but to require the tenants to make repairs. Arrangement B is to charge a higher rent and to state that the landlord will make the repairs. A sample of 25 apartments using Arrangement A had a mean net annual income of $1532.50 with a standard deviation of $400. A sample of 22 apartments using Arrangement B had a mean net annual income of $1489.20 with a standard deviation of $100. Test at $\alpha = 0.025$ that Arrangement B will have a lower mean net annual income by **at least $10** than Arrangement A. Assume that the incomes are normally distributed and that the population variances are equal.

10. For a consumer product, the mean dollar sales per retail outlet last year in a sample of 25 stores were $3425 with a standard deviation of $400. For a second product, the mean dollar sales per outlet in a sample of 16 stores were $3250 with a standard deviation of $175. The sales amounts per outlet are assumed to be approximately normally distributed for both products. Test to see if the first product has a better mean dollar sales record than the second product. Use the *P*-value approach and base your decision on a significance level of 0.01. Assume that the population variances are not equal.

11. A random sample of 10 filled sports drink bottles is taken in one bottling plant, and the mean weight of the bottles is found to be 22 ounces with a variance of 0.09 ounces squared. At another plant, 10 randomly selected bottles have a mean weight of 21 ounces with a variance of 0.04 ounces squared. Assuming the weights in both populations are normally distributed and the population variances are equal, test whether there is a difference between the average weights of the bottles being filled at the two plants. Use $\alpha = 0.05$.

**16.** A company believes that the variance in revenue from products produced in two facilities, measured in millions of dollars, is greater for Facility A than for Facility B. The sample standard deviation of a random sample of 19 products from Facility A is 1.5984 million of dollars. The sample standard deviation for a random sample of 18 products from Facility B is 1.0426 million of dollars. Assume that both population distributions are approximately normal and test the company's claim using a 0.10 level of significance. Does the evidence support the company's claim? Let the products produced in Facility A be Population 1 and let the products produced in Facility B be Population 2.

**17.** Shirley is analyzing her family's budget regarding how much they spend when eating out. She believes that the variance in expenditures when eating out is less when she uses cash as compared to when she uses her credit card. The following data represent a random sample of her family's cash and credit card purchases when eating out last month. Assume that both population distributions are approximately normal and test Shirley's claim using a 0.05 level of significance. Let Population 1 be the cash purchases and let Population 2 be the credit card purchases.

| Cash | $24.24 | $26.96 | $22.48 | $26.45 | $26.74 | $23.99 | $25.70 | $26.73 | $25.12 | $24.23 |
|---|---|---|---|---|---|---|---|---|---|---|
| Credit Card | $20.46 | $25.02 | $26.36 | $23.95 | $25.84 | $24.96 | $20.82 | $23.41 | $24.70 | $23.58 |

# P   Discovery Project

## Understanding Credit Scores

There are many factors that determine one's eligibility to obtain credit. Several factors are considered such as education level and credit scores. An individual's credit score is a good predictor of one's ability to manage their finances and pay their debts responsibly. See Credit Score: Definition, Factors, & Improving It (investopedia.com) for more information on understanding credit scores. Credit scores are used to determine your eligibility for credit cards, car loans, and even for some types of insurance. Often when an individual has a lower credit score, they will have higher interest rates and lower borrowing capacity for loans and credit cards.

Suppose you are working in the marketing department for a large credit card company. Your company is launching a new credit card and will be mailing information to prospective customers. You are examining information from your current customers and are interested in understanding the differences in credit scores among groups of customers. You are looking at factors such as marital status, how many credit cards are used by the customer, and whether they rent or own their home.

Using the Credit Card Data file, answer the following questions to better understand your current customer base. The data set includes credit scores and data on nine (9) predictor (159 data points).

### ⁚⁚ Data

This data set can be found on stat.hawkeslearning.com under **Discovering Business Statistics, Second Edition > Data Sets > Credit Card Data**.

1. Download the data file and open it in Microsoft Excel.

2. Determine the mean, mode, median, maximum, minimum, standard deviation, and the coefficient of variation of the following variables: age, total credit limit, total balance, credit score, annual household income, and number of children and briefly discuss the results. (Hint: these values can be quickly calculated using the Data Analysis Add-in: Descriptive Statistics in Excel).

3. Fully summarize the qualitative variables (i.e., What percent of the sample has a college degree) and briefly discuss your findings. (Hint: These values can be quickly determined using the Data Analysis Add-in: Histogram in Excel).

4. Determine if there is a difference in credit scores for those that are single (marital status = 0) versus those that are married (marital status = 1) using the appropriate hypothesis test. Use a significance level of 0.05.

5. Is there a higher proportion of customers that own (housing = 1) their home as opposed to renting (housing = 0)? Conduct the appropriate hypothesis test using a significance level of 0.10.

6. Determine if there is a difference greater than $2,000 in the total balance on all credit cards between those that have children versus those that do not have any children using the appropriate hypothesis test. Use a significance level of 0.01.

7. Is there a difference in the proportion of customers that have some college (education level = 2) or a college degree (education level = 3) versus those that have a high school diploma (education level = 1)? Conduct the appropriate hypothesis test using a significance level of 0.05.

8. Determine if customers under 40 have a fewer number of credit cards issued versus those customers 40 or older using the appropriate hypothesis test. Use the 0.10 significance level.

9. Is there a difference in household income based on being married (marital status = 1) or separated/divorced (marital status = 2)? Conduct the appropriate hypothesis test using a significance level of 0.05.

10. Determine whether the variances of the total credit limit differ by housing status (own = 1/rent = 0). Conduct the appropriate hypothesis test using the 0.01 significance level.

11. Briefly summarize your findings from this data set to your manager.

# Chapter 12
## Analysis of Variance (ANOVA)

# Discovering the Real World

## Smartphone Screen Time and ANOVA

In Chapter 11, we introduced two-sample testing from two independent populations in which we compared two population parameters. For example, in Chapter 11, we took two samples from two independent populations and wanted to compare the population means with the following hypotheses.

$$H_0: \mu_1 - \mu_2 = 0$$

$$H_a: \mu_1 - \mu_2 \neq 0$$

Extending the tests from Chapter 11, we are now at a point where we would like to test the equality of three or more means when the samples are taken from three or more independent populations. Thus, rather than making pairwise comparisons of average screen time between two age groups as we did in Chapter 11, using analysis of variance (ANOVA), we can now compare the average screen time among all age groups at one time.

| Screen Time Tracking | | | | | |
|---|---|---|---|---|---|
| Age Group (in Years) | 16–24 | 25–34 | 35–44 | 45–54 | 55+ |
| Average Screen Time (in Minutes) | 59.37 | 53.75 | 41.88 | 25.38 | 18.45 |

The new set of hypotheses could be written as follows.

$$H_0: \mu_1 = \mu_2 = \mu_3 = \mu_4 = \mu_5$$

$H_a$: At least one of the means is different.

where $\mu_i$ represents the average screen time of each of the age groups in the table above containing respondents who track their screen time usage.

The results of the ANOVA table will indicate whether we should reject the null hypothesis (i.e., all of the means are not the same and at least one of the means is significantly different from the other means) or fail to reject the null hypothesis (i.e., all means are the same).

Additionally, in this chapter, once we have rejected the null hypothesis and concluded that at least one of the means is different from the others, we will discuss multiple comparison procedures that will help us determine which of the means are different.

## Introduction

Chapter 11 described methods for comparing two population means. In that chapter we also began to explore some experimental design issues. For example, if variation among sample observations could be significantly reduced by pairing experimental units, the paired difference test was used. But we paid a price for pairing experimental units by giving up degrees of freedom. If variation could not be significantly reduced by pairing, then experimental units should be randomly selected from each of the populations of interest. Often, we will be interested in comparing more than two population means. In the sections that follow, we will extend the concepts presented in Chapter 11 to situations where there are more than two populations of interest.

There are many situations where someone might be interested in comparing several population means. A cellular phone manufacturer may wish to compare the average talk times for several different models of cellular phone. A marketing director may want to compare the average numbers of tickets sold for various promotional campaigns. A stock analyst may be interested in the average yields of various types of equities (small, medium, and large cap). A manufacturer may wish to compare the average outputs of operators during various work shifts. An accounting firm may want to compare auditor proficiency resulting from three different training methods. Comparisons of several population means can be made using the *F*-test, which will be discussed in Section 12.3.

Additionally, there are situations where we are interested in the average response of a variable that depends on two factors. For example, a manager might be interested in the number of defects in a plant with respect to different machines used by different operators. This situation can be analyzed using a **two-way ANOVA**, which will be discussed in sections 12.5 and 12.6.

# 12.1 Introduction to Analysis of Variance (ANOVA)

In comparing population means, a natural question arises: *is there a significant difference between the means?* A preliminary answer to this question can be obtained by drawing a random sample from each of the populations of interest and computing the sample means. The larger the differences between the sample means, the more likely it would seem that there is a difference among the population means. How large is large enough to conclude that a difference is significant? Since a sample has been drawn from each of the populations, we must decide whether or not the observed differences between the sample means are simply due to random variation among the sample observations or to a real difference in the population means. Often, we wish to perform a hypothesis test to determine if a difference among population means can be attributed to some **treatment.**

**Definition**

**Experimental Units**
Individuals or objects on which the experiment is performed are called **experimental units**.

Suppose a marketing firm wants to know if there are differences between the sales of a certain brand of vehicles resulting from three different advertising strategies. The firm experimented with the marketing campaigns for a 12-month period and noted the sales for the three strategies. The marketing firm selected dealerships that have been using the strategies and computed the summary measures for sales that result in the box plots in Figure 12.1.1. The firm finds that the sample mean of the sales for Strategy 1 is approximately $8.1 million, the sample mean of the sales for Strategy 2 is approximately $6.4 million, and the sample mean of the sales for Strategy 3 is approximately $7.7 million. Is there a significant difference between the sales resulting from the three advertising strategies? Note that in this example the sales are the experimental units and the advertising strategies are the treatments.

**Definition**

**Treatment**
An experimental condition applied to the units is called a **treatment**.

| Table 12.1.1 – Sales by Strategy (Millions of Dollars) | | |
|:---:|:---:|:---:|
| **Strategy 1** | **Strategy 2** | **Strategy 3** |
| 3 | 2 | 4 |
| 6 | 5 | 2 |
| 7 | 5 | 5 |
| 4 | 3 | 6 |
| 6 | 7 | 6 |
| 7 | 8 | 7 |
| 10 | 6 | 9 |
| 6 | 4 | 8 |
| 15 | 10 | 14 |
| 8 | 6 | 8 |
| 9 | 9 | 7 |
| 16 | 12 | 16 |

**Box Plots – Sales by Strategy**

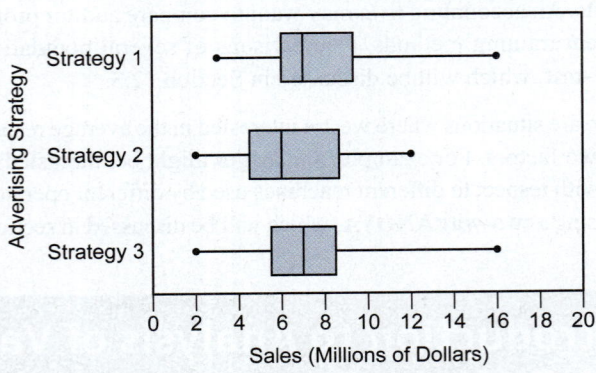

Figure 12.1.1

As another example, suppose that an investor is interested in comparing the average portfolio rates of return achieved by two different local investment brokers, Mr. Morgan and Mr. Stanley. To help him make this comparison, the investor randomly selects returns for each of these investment brokers over a 25-year period. From these samples he calculates the summary measures, which result in the box plots shown in Figure 12.1.2. The sample median portfolio rate of return for Mr. Morgan is 10.9%, which is somewhat higher than the sample median portfolio rate of return for Mr. Stanley, which is 10.1%. Should the investor conclude that Mr. Morgan's average portfolio rate of return is significantly higher than Mr. Stanley's? Note that in this example, the portfolio rates of return are the experimental units and the investment broker is the treatment.

**Box Plots – Rates of Return**

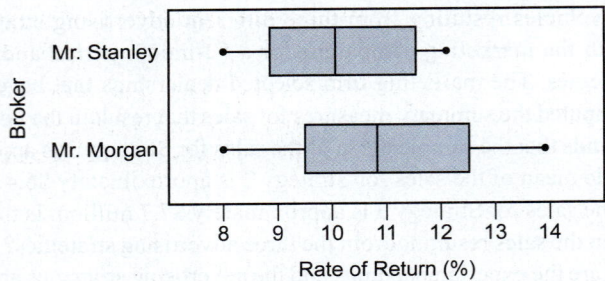

Figure 12.1.2

The investor might be tempted to simply compare the sample median rates of return for Mr. Morgan and Mr. Stanley before making a decision. However, this would not be a completely informed decision. Examining the box plots will give him a much better understanding of the situation. The box plots show that the majority of the observed rates of return for Mr. Morgan and Mr. Stanley are similar. The center of the box plot of observed rates of return for Mr. Morgan is slightly higher than for Mr. Stanley, but it is very possible that the difference can be attributed to sampling variation. In other words, the difference in the observed sample means is small when compared to the variation in the observed rates of return for the two samples.

Although box plots can provide a sense of whether or not there is a difference among the population means, they cannot help us evaluate how likely it is that the observed difference is due to ordinary sampling variation. The following sections will develop measures that will help us assess how likely it is that the observed differences among the sample means are due to ordinary sampling variation.

## Analysis of Variance (ANOVA)

**Analysis of variance (ANOVA)** deals with differences between or among population means. You may recall that you were restricted to testing the difference between two population means when using the $t$-test in Chapter 11. Unlike the $t$-test, analysis of variance does not restrict the number of means. Instead of asking whether two means differ, we ask if three, four, five, etc., means differ. ANOVA allows us to deal with two or more independent variables simultaneously, asking not only about the effects of each variable separately, but also about how two or more variables interact. In studying ANOVA, we will analyze the differences in means by breaking apart the total variation. The total variation will be separated into two pieces: **sum of squares for treatments, SST** (the variation attributed to the treatments) and **sum of squares for error, SSE** (the variation not explained by the treatments).

The variation among all the sample observations, without regard to which treatment or population they are from, can be summarized by the **sample variance**, which is given by the following formula.

### Formula

**Sample Variance**

$$s^2 = \frac{\sum\limits_{j=1}^{k}\sum\limits_{i=1}^{n_j}\left(x_{ij} - \overline{\overline{x}}\right)^2}{n_T - 1}$$

where

$\overline{\overline{x}}$ is the sample mean of all of the observations (the grand mean),

$n_j$ is the number of observations in the $j^{\text{th}}$ treatment,

$k$ is the number of treatments, and

$n_T$ is the total number of observations in all samples.

The numerator of $s^2$ is called the **total sum of squares**, or **TSS**, since it describes the total variation among all of the sample observations. The denominator of $s^2$ gives the degrees of freedom associated with TSS, $n_T - 1$.

The total variation in the sample measurements can be partially attributed to the treatments from which they came and partially attributed to random variation inherent in sampling. Thus, the total sum of squares can be divided into two components. The first component measures the variation that can be attributed to the treatments and is called the **sum of squares for treatments**, or **SST**.

### Formula

**Sum of Squares for Treatments**

The mathematical expression for the sum of squares for treatments is given by

$$\text{SST} = \sum_{j=1}^{k} n_j \left(\overline{x}_j - \overline{\overline{x}}\right)^2$$

where

$n_j$ is the number of observations in the $j^{\text{th}}$ treatment,

$\overline{x}_j$ is the sample mean of the observations in the $j^{\text{th}}$ treatment, and

$\overline{\overline{x}}$ is the grand mean.

SST is a summary measure of how much each treatment or population mean differs from the grand mean, $\bar{\bar{x}}$, which is the mean of all of the sample observations. The larger the differences between the treatment means $\left(\bar{x}_j\right)$ and the grand mean $\left(\bar{\bar{x}}\right)$, the more likely the variation in sample observations is due to the treatments rather than to ordinary sampling variation. Thus, SST measures the variation between the treatments. If using statistical software to perform ANOVA, the variation between treatments may also be referred to as "Among variation" or "Between variation." In Microsoft Excel, the source of variation associated with treatments is called "Between Groups" variation. In Minitab, the output yields an ANOVA table that refers to treatment variation as "Factor" variation.

The degrees of freedom associated with SST is $k-1$, where $k$ is the number of treatments. The **mean square for treatments**, **MST**, is the sum of squares for treatments divided by its degrees of freedom.

$$MST = \frac{SST}{k-1}$$

The MST represents the average weighted squared deviation of the sample treatment means from the grand mean. The MST is basically the variance of the weighted sample means.

The second component measures the random variation attributable to sampling and is called the **sum of squares for error**, or **SSE**.

---

### Formula

#### Sum of Squares for Error

The mathematical expression for the sum of squares for error is given by the following formula.

$$SSE = \underbrace{\sum_{i=1}^{n_1}\left(x_{i1}-\bar{x}_1\right)^2}_{\substack{\text{Note: This is the} \\ \text{numerator of} \\ \text{the sample variance} \\ \text{for Sample 1.}}} + \sum_{i=1}^{n_2}\left(x_{i2}-\bar{x}_2\right)^2 + \cdots + \sum_{i=1}^{n_k}\left(x_{ik}-\bar{x}_k\right)^2$$

---

The SSE is a summary measure of how much of the total variation in the sample data is not explained by SST. The expression is a direct calculation of SSE which shows that the SSE summarizes how much the sample observations within each treatment vary from the treatment mean. In other words, SSE is a measure of the variation within the treatments. In practice, it is much easier to calculate SSE as the total variation in the sample, TSS, minus the variation explained by the treatments, SST. The variation associated with the error term is oftentimes called "Within Groups" variation or simply "Error" variation.

The degrees of freedom associated with SSE are $n_T - k$. The **mean square for error**, **MSE**, is the SSE divided by its degrees of freedom. The MSE represents the total variance (or common variance) of the model. That is, the estimate of $\sigma^2$ is $s^2$, which is given by the MSE of the ANOVA model.

$$MSE = \frac{SSE}{n_T - k}$$

Thus, we have the following results.

## Formula

### Sum of Squares and Degrees of Freedom

**Sum of Squares**

$$TSS = SST + SSE$$

$$\sum_{j=1}^{k}\sum_{i=1}^{n_j}\left(x_{ij}-\overline{\overline{x}}\right)^2 = \sum_{j=1}^{k}n_j\left(\overline{x}_j-\overline{\overline{x}}\right)^2 + \left\{\sum_{i=1}^{n_1}\left(x_{i1}-\overline{x}_1\right)^2 + \sum_{i=1}^{n_2}\left(x_{i2}-\overline{x}_2\right)^2 + \cdots + \sum_{i=1}^{n_k}\left(x_{ik}-\overline{x}_k\right)^2\right\}$$

**Degrees of Freedom**

$$Total = Treatment + Error$$

$$n_T - 1 = (k-1) + (n_T - k)$$

As you might expect, these measures of variability are the fundamental pieces which will be used to develop a hypothesis test for determining whether or not there is a significant difference among the population means. This is precisely the reason that the test which we will use to make this decision is often called an **analysis of variance**, or **ANOVA**. The expressions are complicated. Fortunately, most statistical analysis programs will compute these quantities.

In the example about the marketing firm that wanted to determine if there was a difference in the sales resulting from the three advertising strategies, there were two overriding factors which contributed to the analysis. First, the firm considered if there were significant differences between the average sales for the three strategies, which is summarized by SST. The second factor considered was whether or not these differences were larger than the differences among the sample observations for each of the strategies, which is summarized by SSE. These sources of variation (SST and SSE) will be the basis for answering the question of whether or not we can conclude that there is a significant difference among population means.

$$\text{Is } \mu_1 = \mu_2 = \mu_3 ?$$

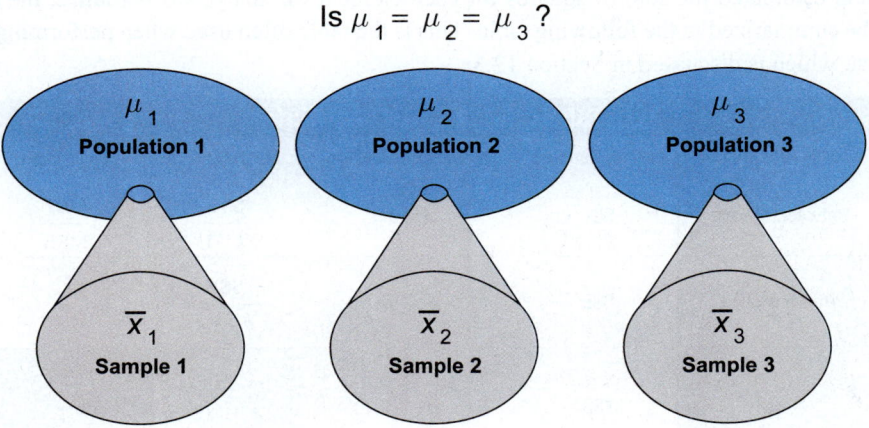

The null hypothesis will state that the population means are equal.

## Formula

### ANOVA Formulas

Grand Mean:

$$\overline{\overline{x}} = \frac{\sum_{j=1}^{k}\sum_{i=1}^{n_j} x_{ij}}{n_T}$$

Sum of Squares for Treatments:

$$SST = \sum_{j=1}^{k} n_j \left(\overline{x}_j - \overline{\overline{x}}\right)^2$$

Mean Square for Treatments:

$$MST = \frac{SST}{k-1}$$

Sum of Squares for Error:

$$SSE = \sum_{i=1}^{n_1}\left(x_{i1} - \overline{x}_1\right)^2 + \sum_{i=1}^{n_2}\left(x_{i2} - \overline{x}_2\right)^2 + \cdots + \sum_{i=1}^{n_k}\left(x_{ik} - \overline{x}_k\right)^2$$

Mean Square for Error:

$$MSE = \frac{SSE}{n_T - k}$$

Total Sum of Squares:

$$TSS = SST + SSE = \sum_{j=1}^{k}\sum_{i=1}^{n_j}\left(x_{ij} - \overline{\overline{x}}\right)^2$$

Having calculated the sum of squares for each element for analysis of variance, the data can be summarized in the following table. This is the table often used when performing the $F$-test, which is discussed in Section 12.3.

| Table 12.1.2 – ANOVA Table | | | | |
|---|---|---|---|---|
| Source of Variation | Sum of Squares | Degrees of Freedom | Mean Square | $F$-Statistic |
| Between Groups | SST | $k-1$ | $\frac{SST}{k-1}$ | $\frac{MST}{MSE}$ |
| Within Groups | SSE | $n_T - k$ | $\frac{SSE}{n_T - k}$ | |
| Total | TSS | $n_T - 1$ | Notes: TSS = SST + SSE; Reject $H_0$ if $F \geq F_\alpha$ | |

## Example 12.1.1

**Performing an ANOVA on Media Screen Time Usage by Tweens, Teens, and Adults**

Data were collected to study the use of media by three age groups – Tweens (8–12 years old), Teens (13–18 years old), and Adults (over 18 years old). A sample of 50 participants in each age group was asked how frequently they engaged in activities such as time spent on cell phones, watching online videos, watching television, and playing mobile games. The data are in the following table.

| 8–12 Years Old | 13–18 Years Old | Over 18 Years Old | 8–12 Years Old | 13–18 Years Old | Over 18 Years Old | 8–12 Years Old | 13–18 Years Old | Over 18 Years Old |
|---|---|---|---|---|---|---|---|---|
| 120 | 100 | 86 | 103 | 97 | 90 | 106 | 118 | 118 |
| 133 | 161 | 106 | 139 | 120 | 89 | 165 | 113 | 95 |
| 146 | 120 | 91 | 125 | 119 | 106 | 137 | 139 | 93 |
| 89 | 117 | 92 | 124 | 152 | 61 | 128 | 113 | 99 |
| 161 | 176 | 84 | 156 | 143 | 86 | 149 | 155 | 116 |
| 117 | 115 | 137 | 113 | 93 | 112 | 144 | 138 | 75 |
| 156 | 127 | 99 | 120 | 133 | 97 | 103 | 96 | 100 |
| 116 | 148 | 92 | 152 | 147 | 103 | 88 | 154 | 113 |
| 120 | 181 | 50 | 148 | 120 | 112 | 106 | 151 | 127 |
| 112 | 158 | 96 | 148 | 153 | 72 | 117 | 155 | 80 |
| 142 | 147 | 103 | 171 | 129 | 73 | 145 | 172 | 104 |
| 137 | 97 | 121 | 133 | 157 | 111 | 109 | 144 | 120 |
| 97 | 114 | 118 | 106 | 173 | 159 | 112 | 121 | 82 |
| 129 | 146 | 145 | 151 | 176 | 78 | 121 | 123 | 87 |
| 118 | 157 | 107 | 124 | 190 | 75 | 116 | 145 | 95 |
| 137 | 138 | 114 | 87 | 114 | 122 | 160 | 154 | 102 |
| 123 | 88 | 103 | 141 | 177 | 110 | | | |

Table 12.1.3 – Screen Time (Minutes/Day) by Age Group

A summary table showing the sample size, the average time spent on devices, and the standard deviation of the data by age group is given below.

### Table 12.1.4 - Summary of Screen Time by Age Group

| | 8–12 Years Old | 13–18 Years Old | Over 18 Years Old |
|---|---|---|---|
| $n$ | 50 | 50 | 50 |
| Mean | 128 | 137.48 | 100.12 |
| Standard Deviation | 20.90 | 25.57 | 20.32 |

**:: Data**

This data set can be found on stat.hawkeslearning.com under **Discovering Business Statistics, Second Edition > Data Sets > Screen Time by Age Group**.

Test to determine if there is a significant difference between average time spent on devices between age groups.

### SOLUTION

Let

$\mu_1$ = population mean time spent on devices for participants 8–12 years old

$\mu_2$ = population mean time spent on devices for participants 13–18 years old

$\mu_3$ = population mean time spent on devices for participants over 18 years old.

Thus, the null and alternative hypotheses for this scenario should be written as

$H_0$: $\mu_1 = \mu_2 = \mu_3$

$H_a$: At least one $\mu_i$ is different, for $i = 1, 2, 3$.

Assuming that the variances of time spent on devices are equal, the figure below shows the JMP output for a one-way ANOVA comparing means.

**One way Anova**

Summary of Fit

| | |
|---|---|
| RSquare | 0.338598 |
| Adj Rsquare | 0.3296 |
| Root Mean Square Error | 22.38679 |
| Mean of Response | 121.8667 |
| Observations (or Sum Wgts) | 150 |

Analysis of Variance

| Source | DF | Sum of Squares | Mean Square | F-Ratio | Prob > F |
|---|---|---|---|---|---|
| Age Group | 2 | 37715.57 | 18857.8 | 37.6276 | < 0001* |
| Error | 147 | 73671.76 | 501.2 | | |
| C. Total | 149 | 111387.33 | | | |

Means for One way Anova

| Level | Number | Mean | Std Error | Lower 95% | Upper 95% |
|---|---|---|---|---|---|
| 8-12 Years Old | 50 | 128.000 | 3.1660 | 121.74 | 134.26 |
| 13-18 Years Old | 50 | 137.480 | 3.1660 | 131.22 | 143.74 |
| Over 18 Years Old | 50 | 100.120 | 3.1660 | 93.86 | 106.38 |

Std Error uses a pooled estimate of error variance

**Figure 12.1.3**

### Technology

For instructions on performing an ANOVA test visit stat.hawkeslearning.com and navigate to **Discovering Business Statistics, Second Edition > Technology Instructions > ANOVA > One-Way**.

Note that the $P$-value is less than 0.0001 which indicates that we would reject the null hypothesis and conclude that the average viewing time between age groups is significantly different. However, from the results of the one-way ANOVA, we only know that the mean viewing time between age groups is different. We will discuss in Section 12.4 some popular multiple comparison procedures that will let us know specifically which of the group means is significantly different.

## 12.1 Exercises

### Basic Concepts

1. What is the price that is paid when pairing experimental units in a paired difference experiment?

2. Give two examples of business situations in which a manager would be interested in comparing several population means.

3. Give an example of a business situation in which a manager would be interested in the average response of a variable that depends on more than one factor.

4. What are experimental units?

5. What is a treatment?

6. Why does simply comparing the sample means for multiple populations not suffice when determining if there is a significant difference in the population means?

7. Explain how box plots can be useful in analyzing data when comparing population means.

8. How is the total variation in the dependent variable broken down in analysis of variance?

9. What does the total sum of squares describe? What are its degrees of freedom?

10. What is the mathematical expression for the sum of squares for treatments?

11. What is the grand mean? How is it calculated?

12. What is the mean square for treatments?

13. What is the relationship between TSS, SST, and SSE? Explain why this relationship makes sense.

## Exercises

14. Consider the following box plots for data collected to compare the average fat contents (in grams) per serving (2 tablespoons) of three popular brands of peanut butter.

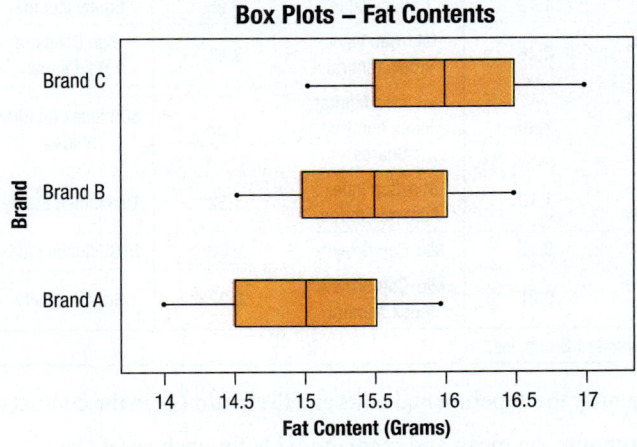

**Box Plots – Fat Contents**

a. Based on the box plots, do you think that there may be a significant difference in the average fat contents per serving of Brand A and Brand B? Explain.

b. Based on the box plots, do you think that there may be a significant difference in the average fat contents per serving of Brand B and Brand C? Explain.

c. Based on the box plots, do you think that there may be a significant difference in the average fat content per serving of Brand A and Brand C? Explain.

15. Consider the following box plots for data collected to compare the average scores achieved on a standardized aptitude test by freshmen, sophomores, juniors, and seniors at a large university.

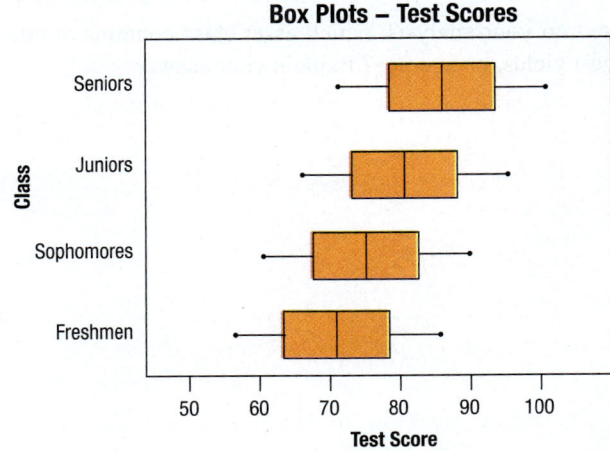

**Box Plots – Test Scores**

a. Based on the box plots, do you think that there may be a significant difference in the average scores achieved by freshmen and sophomores on the standardized test? Explain.

b. Based on the box plots, do you think that there may be a significant difference in the average scores achieved by freshmen and juniors on the standardized test? Explain.

c. Based on the box plots, do you think that there may be a significant difference in the average scores achieved by juniors and seniors on the standardized test? Explain.

16. Consider the following table containing yields for mutual funds in different asset classes (small, mid, and large cap).

| Fund Yields by Asset Class | | | | | |
|---|---|---|---|---|---|
| Small Cap | | Mid Cap | | Large Cap | |
| Fund | Yield (%) | Fund | Yield (%) | Fund | Yield (%) |
| Explorer Value | 1.13 | Capital Value | 0.96 | Equity Income | 3.24 |
| Small-Cap Value Index Admiral | 2.46 | Mid-Cap Value Index Admiral | 2.47 | High Dividend Yield Index | 3.50 |
| Small-Cap Index Admiral Shares | 1.49 | Extended Market Index Admiral Shares | 1.22 | 500 Index Admiral Shares | 2.35 |
| Strategic Small-Cap Equity | 1.10 | Mid-Cap Index Admiral Shares | 1.52 | Diversified Equity | 1.23 |
| Explorer | 0.17 | Mid-Cap Growth | 0.10 | FTSE Social Index | 1.42 |
| Small-Cap Growth Index Admiral | 0.21 | Mid-Cap Growth Index Admiral | 0.32 | Growth Equity | 0.60 |

**Source:** The Vanguard Group, Inc.

a. Identify the experimental units and the treatment in the context of this problem.

b. Compute the mean and median yields for each asset class.

c. Compute the values of the minimum, maximum, first, and third quartiles for each asset class.

d. Construct side-by-side box plots for the three asset classes.

e. Based on the box plots, do you think that there may be a significant difference in the average yields of small-cap and mid-cap funds? Explain.

f. Based on the box plots, do you think that there may be a significant difference in the average yields of mid-cap and large-cap funds? Explain.

g. Based on the box plots, do you think that there may be a significant difference in the average yields of small-cap and large-cap funds? Explain.

h. Based on your analysis, which asset class contains mutual funds with the largest yields, on average? Explain your answer.

**17.** Consider the following table containing daily production data from a particular week for three different employee shifts.

| Items Produced | | | |
|---|---|---|---|
| | First Shift (7 AM–3 PM) | Second Shift (3 PM–11 PM) | Third Shift (11 PM–7 AM) |
| Monday | 140 | 168 | 77 |
| Tuesday | 181 | 224 | 123 |
| Wednesday | 127 | 162 | 77 |
| Thursday | 172 | 182 | 101 |
| Friday | 161 | 219 | 147 |
| Saturday | 152 | 171 | 145 |
| Sunday | 173 | 217 | 111 |

a. Identify the experimental units and the treatment in the context of this problem.

b. Compute the mean and median numbers of items produced for each shift.

c. Compute the values of the minimum, maximum, first, and third quartiles for each shift.

d. Construct side-by-side box plots for the three shifts.

e. Based on the box plots, do you think that there may be a significant difference in the average numbers of items produced during the first and second shifts? Explain.

f. Based on the box plots, do you think that there may be a significant difference in the average numbers of items produced during the second and third shifts? Explain.

g. Based on the box plots, do you think that there may be a significant difference in the average numbers of items produced during the first and third shifts? Explain.

h. Based on your analysis, which shift would you say is the most productive, on average? Explain your answer.

**18.** The sales by strategy data given in Table 12.1.1 yield the following statistics.

| Sales by Strategy (Millions of Dollars) | | |
|---|---|---|
| Strategy 1 | Strategy 2 | Strategy 3 |
| 3 | 2 | 4 |
| 6 | 5 | 2 |
| 7 | 5 | 5 |
| 4 | 3 | 6 |
| 6 | 7 | 6 |
| 7 | 8 | 7 |
| 10 | 6 | 9 |
| 6 | 4 | 8 |
| 15 | 10 | 14 |
| 8 | 6 | 8 |
| 9 | 9 | 7 |
| 16 | 12 | 16 |

$$\text{SST} \approx 18.0556$$
$$\text{SSE} = 438.5$$

a. What are the degrees of freedom associated with the total sum of squares?

b. What are the degrees of freedom associated with the sum of squares for treatments?

c. Find the mean square for treatments, MST.

d. Find the mean square for error, MSE.

**19.** The fund yield data given in Exercise 16 give the following summary statistics.

| Fund Yields by Asset Class | | | | | |
|---|---|---|---|---|---|
| **Small Cap** | | **Mid Cap** | | **Large Cap** | |
| **Fund** | **Yield (%)** | **Fund** | **Yield (%)** | **Fund** | **Yield (%)** |
| Explorer Value | 1.13 | Capital Value | 0.96 | Equity Income | 3.24 |
| Small-Cap Value Index Admiral | 2.46 | Mid-Cap Value Index Admiral | 2.47 | High Dividend Yield Index | 3.50 |
| Small-Cap Index Admiral Shares | 1.49 | Extended Market Index Admiral Shares | 1.22 | 500 Index Admiral Shares | 2.35 |
| Strategic Small-Cap Equity | 1.10 | Mid-Cap Index Admiral Shares | 1.52 | Diversified Equity | 1.23 |
| Explorer | 0.17 | Mid-Cap Growth | 0.10 | FTSE Social Index | 1.42 |
| Small-Cap Growth Index Admiral | 0.21 | Mid-Cap Growth Index Admiral | 0.32 | Growth Equity | 0.60 |

**Source:** The Vanguard Group, Inc.

$$MST \approx 1.8464$$
$$MSE \approx 0.9423$$

**a.** Interpret the value of MST.

**b.** What are the degrees of freedom associated with the sum of squares for treatments?

**c.** Find the sum of squares for treatments.

**d.** What are the degrees of freedom for the sum of squares for error?

**e.** Find the sum of squares for error.

**20.** Consider the production data given in Exercise 17.

| Items Produced | | | |
|---|---|---|---|
| | **First Shift (7 AM–3 PM)** | **Second Shift (3 PM–11 PM)** | **Third Shift (11 PM–7 AM)** |
| **Monday** | 140 | 168 | 77 |
| **Tuesday** | 181 | 224 | 123 |
| **Wednesday** | 127 | 162 | 77 |
| **Thursday** | 172 | 182 | 101 |
| **Friday** | 161 | 219 | 147 |
| **Saturday** | 152 | 171 | 145 |
| **Sunday** | 173 | 217 | 111 |

**a.** What is the value of the grand mean, $\bar{\bar{x}}$?

**b.** What is the value of $n_j$?

**c.** What is the value of $k$?

**d.** What is the value of $n_T$?

**e.** For these data, identify the degrees of freedom associated with the total sum of squares, the degrees of freedom associated with the sum of squares for treatments, and the degrees of freedom associated with the sum of squares for error. Verify that the relationship between the degrees of freedom $\left(\text{Total} = \text{Treatment} + \text{Error}\right)$ holds.

# 12.2 Assumptions in an ANOVA Test

Before proceeding with a description of the hypothesis testing procedure for determining whether or not there is a significant difference among several population means, it is important to point out the assumptions upon which the test is based. If the data do not appear to conform to these assumptions, the hypothesis testing procedures described in the subsequent sections will not be valid inferential techniques.

The first assumption is that the distributions of all $k$ populations of interest are approximately normal. The best way to determine whether or not this assumption is satisfied is to construct a histogram of the sample data for each of the $k$ populations of interest.

If the histograms appear to be approximately normal, then it is reasonable to proceed. Figure 12.2.1 gives examples of histograms of sample data drawn from normal populations.

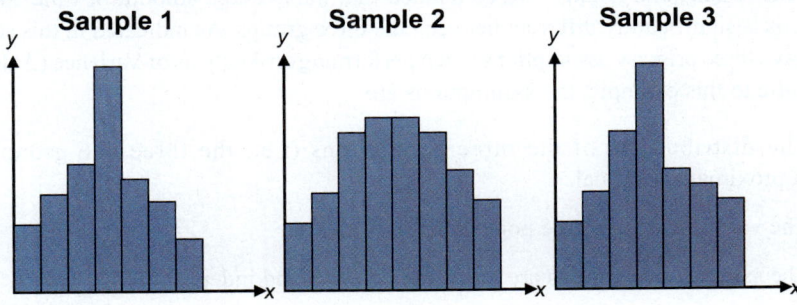

**Figure 12.2.1**

The second assumption is that the variances of the $k$ populations of interest are equal. We can determine if this assumption is reasonable by drawing box plots of the $k$ samples of data and comparing the spread of the data for each sample. If the spread or variability of the data is approximately the same for each of the $k$ samples, it is reasonable to proceed. For example, consider the box plots for three samples from three populations of interest in Figure 12.2.2. Although the spread of the box plot is not exactly the same for each of the samples, it is similar enough that it is likely that the observed difference in spread is due to sampling variation, and it is safe to proceed. There is a hypothesis test which can be used to determine if there is a significant difference among the variances, which will be discussed in Example 12.2.1.

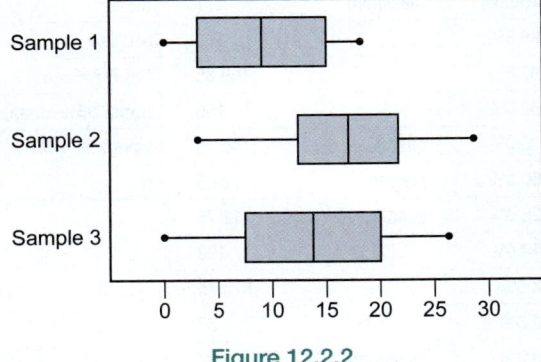

**Figure 12.2.2**

**✎ NOTE**

There is a simple "rule of thumb" that you can use to check the variance assumption. If the largest standard deviation is no more than twice the smallest, then the presumption is that the assumption holds.

The third assumption that must be met is that each of the $k$ samples must be selected independently from each other and in a random fashion from each of the respective populations.

> ### Procedure
>
> ### ANOVA Assumptions
>
> In order for an analysis of variance to be valid, the following three assumptions must hold.
>
> 1. The distributions of all $k$ populations of interest are approximately normal.
> 2. The variances of all $k$ populations are equal.
> 3. The sample observations are randomly selected and independent.

### Example 12.2.1

### Validating Assumptions for ANOVA

Let's return to Example 12.1.1 where we compared the media use screen time between Tweens, Teens, and Adults. We concluded that the average amount of time spent on screens is significantly different between the three groups. As indicated in this section, we have three primary assumptions when performing an Analysis of Variance (ANOVA). Specific to this example, the assumptions are

1. The distributions of the three populations (i.e., the three age groups) are approximately normal.

2. The variances of all three populations are equal.

3. The sample observations are randomly selected and independent.

#### SOLUTION

Let's discuss and validate each of these assumptions. Given that the three age groups do not overlap and that it's stated that the samples are randomly selected, the assumption of independence between the three groups holds.

The assumptions of normality and homogeneity of variances are not as easy to show. The assumption of normality can be shown using descriptive statistics. See the histogram and box plot in the JMP output below for the screen time for each of the age groups.

**Distributions Age Group = 8–12 Years Old**
**Screen Time**

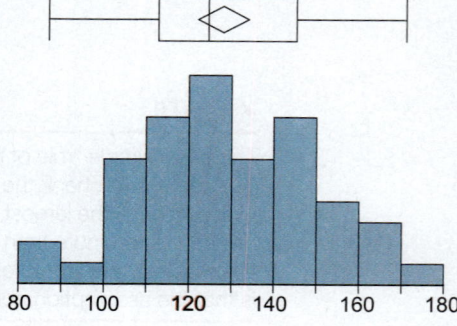

| | Quantiles | |
|---|---|---|
| 100.0% | maximum | 171 |
| 99.5% | | 171 |
| 97.5% | | 169.35 |
| 90.0% | | 156 |
| 75.0% | quartile | 145.25 |
| 50.0% | median | 124.5 |
| 25.0% | quartile | 112.75 |
| 10.0% | | 103 |
| 2.5% | | 87.275 |
| 0.5% | | 87 |
| 0.0% | minimum | 87 |

| Summary Statistics | |
|---|---|
| Mean | 128 |
| Std Dev | 20.895268 |
| Std Err Mean | 2.9550372 |
| Upper 95% Mean | 133.93837 |
| Lower 95% Mean | 122.06163 |
| N | 50 |

### ⚙ Technology

To create a histogram using technology visit stat.hawkeslearning.com and navigate to **Discovering Business Statistics, Second Edition > Technology Instructions > Graphs > Histogram**.

**Distributions Age Group = 13–18 Years Old**
**Screen Time**

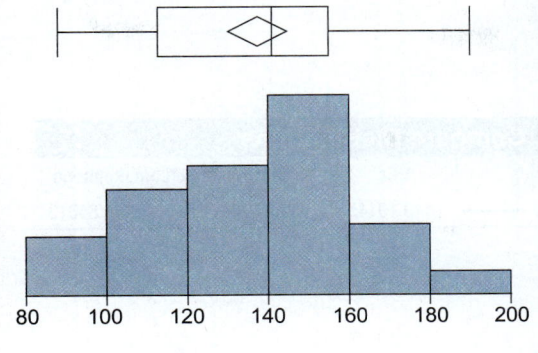

| Quantiles | | |
|---|---|---|
| 100.0% | maximum | 190 |
| 99.5% | | 190 |
| 97.5% | | 187.525 |
| 90.0% | | 175.7 |
| 75.0% | quartile | 155 |
| 50.0% | median | 141 |
| 25.0% | quartile | 117.75 |
| 10.0% | | 97.3 |
| 2.5% | | 89.375 |
| 0.5% | | 88 |
| 0.0% | minimum | 88 |

| Summary Statistics | |
|---|---|
| Mean | 137.48 |
| Std Dev | 25.570423 |
| Std Err Mean | 3.6162039 |
| Upper 95% Mean | 144.74703 |
| Lower 95% Mean | 130.21297 |
| N | 50 |

**Distributions Age Group = Over 18 Years Old**
**Screen Time**

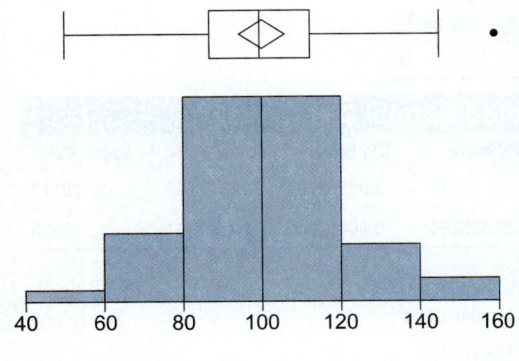

| Quantiles | | |
|---|---|---|
| 100.0% | maximum | 159 |
| 99.5% | | 159 |
| 97.5% | | 155.15 |
| 90.0% | | 121.9 |
| 75.0% | quartile | 112.25 |
| 50.0% | median | 99.5 |
| 25.0% | quartile | 86.75 |
| 10.0% | | 75 |
| 2.5% | | 53.025 |
| 0.5% | | 50 |
| 0.0% | minimum | 50 |

| Summary Statistics | |
|---|---|
| Mean | 100.12 |
| Std Dev | 20.323546 |
| Std Err Mean | 2.8741835 |
| Upper 95% Mean | 105.89589 |
| Lower 95% Mean | 94.344112 |
| N | 50 |

**Figure 12.2.3**

The histogram for each of the age groups appears to follow a bell-shaped curve (i.e., a normal distribution). Additionally, if you observe the continuous fit to the data in the JMP output below, you will see that each of the continuous fits appears to follow a normal distribution for each age group. Lastly, JMP offers a Goodness-of-Fit Test which is a formal test of normality for each age group. The null hypothesis of the test is

$H_0$: The data are from a normal distribution.

Thus, small $P$-values listed under **Prob < W** (associated with the Shapiro-Wilk test) and under **Simulated $p$-Value** (associated with the Anderson-Darling test) indicate that the data are less likely to have come from a normal distribution. As can be seen in the output, the $P$-values associated with the Goodness-of-Fit Test for each age group yield a $P$-value greater than 0.05, which implies that we would fail to reject the null hypothesis at a 0.05 significance level and conclude that the data follow a normal distribution.

**Distributions Age Group = 8–12 Years Old**
**Screen Time**

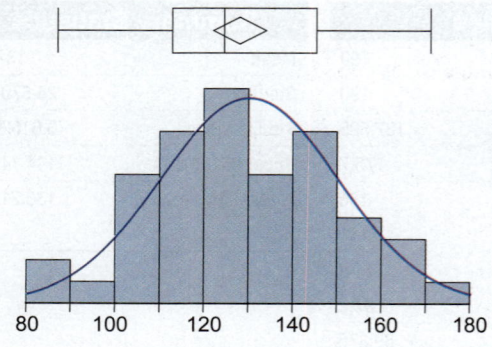

| Compare Distributions | | | | | |
|---|---|---|---|---|---|
| Show | Distribution | | AICc | BIC | -2*LogLikelihood |
| ☑ | Normal | —— | 449.10145 | 452.67017 | 444.84613 |

| Quantiles | | |
|---|---|---|
| 100.0% | maximum | 171 |
| 99.5% | | 171 |
| 97.5% | | 169.35 |
| 90.0% | | 156 |
| 75.0% | quartile | 145.25 |
| 50.0% | median | 124.5 |
| 25.0% | quartile | 112.75 |
| 10.0% | | 103 |
| 2.5% | | 87.275 |
| 0.5% | | 87 |
| 0.0% | minimum | 87 |

| Summary Statistics | |
|---|---|
| Mean | 128 |
| Std Dev | 20.895268 |
| Std Err Mean | 2.9550372 |
| Upper 95% Mean | 133.93837 |
| Lower 95% Mean | 122.06163 |
| N | 50 |

| Fitted Normal Distribution | | | | | |
|---|---|---|---|---|---|
| Parameter | | Estimate | Std Error | Lower 95% | Upper 95% |
| Location | $\mu$ | 128 | 2.9550372 | 122.20823 | 133.79177 |
| Dispersion | $\sigma$ | 20.895268 | 0.4641284 | 20.005111 | 21.713568 |

| Measures | |
|---|---|
| -2*LogLikelihood | 444.84613 |
| AICc | 449.10145 |
| BIC | 452.67017 |

| Goodness-of-Fit Test | | |
|---|---|---|
| | W | Prob<W |
| Shapiro-Wilk | 0.9815568 | 0.6192 |

| | A2 | Simulated p-Value |
|---|---|---|
| Anderson-Darling | 0.2769295 | 0.6692 |

### 🔗 Technology

For instructions on performing a test for normality using technology visit stat.hawkeslearning.com and navigate to **Discovering Business Statistics, Second Edition > Technology Instructions > Normal Distribution > Test for Normality**.

**Distributions Age Group = 13–18 Years Old**
 **Screen Time**

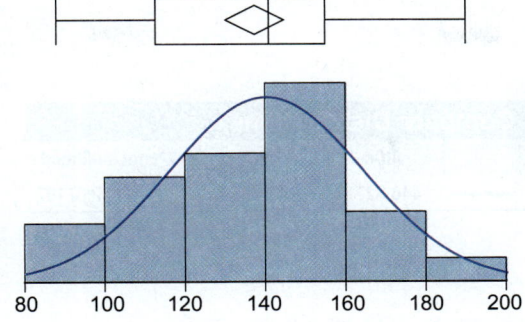

| Compare Distributions | | | | | |
|---|---|---|---|---|---|
| **Show** | **Distribution** | | **AICc** | **BIC** | **-2*LogLikelihood** |
| ☑ | Normal | —— | 469.29281 | 472.86153 | 465.03749 |

| Quantiles | | |
|---|---|---|
| 100.0% | maximum | 190 |
| 99.5% | | 190 |
| 97.5% | | 187.525 |
| 90.0% | | 175.7 |
| 75.0% | quartile | 155 |
| 50.0% | median | 141 |
| 25.0% | quartile | 117.75 |
| 10.0% | | 97.3 |
| 2.5% | | 89.375 |
| 0.5% | | 88 |
| 0.0% | minimum | 88 |

| Summary Statistics | |
|---|---|
| **Mean** | 137.48 |
| **Std Dev** | 25.570423 |
| **Std Err Mean** | 3.6162039 |
| **Upper 95% Mean** | 144.74703 |
| **Lower 95% Mean** | 130.21297 |
| **N** | 50 |

| Fitted Normal Distribution | | | | | |
|---|---|---|---|---|---|
| **Parameter** | | **Estimate** | **Std Error** | **Lower 95%** | **Upper 95%** |
| **Location** | $\mu$ | 137.48 | 3.6162039 | 130.39237 | 144.56763 |
| **Dispersion** | $\sigma$ | 25.570423 | 0.5134322 | 24.583658 | 26.571811 |

| Measures | |
|---|---|
| -2*LogLikelihood | 465.03749 |
| AICc | 469.29281 |
| BIC | 472.86153 |

| Goodness-of-Fit Test | | |
|---|---|---|
| | **W** | **Prob<W** |
| **Shapiro-Wilk** | 0.9716919 | 0.2708 |
| | **A2** | **Simulated p-Value** |
| **Anderson-Darling** | 0.4978111 | 0.1920 |

**Distributions Age Group = Over 18 Years Old**
**Screen Time**

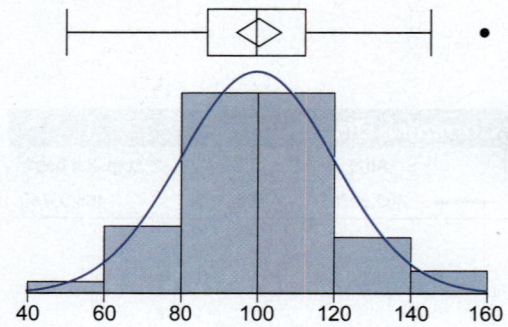

| Compare Distributions | | | | | |
|---|---|---|---|---|---|
| Show | Distribution | | AICc | BIC | -2*LogLikelihood |
| ☑ | Normal | —— | 446.32719 | 449.89591 | 442.07187 |

| Quantiles | | |
|---|---|---|
| 100.0% | maximum | 159 |
| 99.5% | | 159 |
| 97.5% | | 155.15 |
| 90.0% | | 121.9 |
| 75.0% | quartile | 112.25 |
| 50.0% | median | 99.5 |
| 25.0% | quartile | 86.75 |
| 10.0% | | 75 |
| 2.5% | | 53.025 |
| 0.5% | | 50 |
| 0.0% | minimum | 50 |

| Summary Statistics | |
|---|---|
| Mean | 100.12 |
| Std Dev | 20.323546 |
| Std Err Mean | 2.8741835 |
| Upper 95% Mean | 105.89589 |
| Lower 95% Mean | 94.344112 |
| N | 50 |

| Fitted Normal Distribution | | | | | |
|---|---|---|---|---|---|
| Parameter | | Estimate | Std Error | Lower 95% | Upper 95% |
| Location | $\mu$ | 100.12 | 2.8741835 | 94.486704 | 105.7533 |
| Dispersion | $\sigma$ | 20.323546 | 0.4577347 | 19.445916 | 21.119456 |

| Measures | |
|---|---|
| -2*LogLikelihood | 442.07187 |
| AICc | 446.32719 |
| BIC | 449.89591 |

| Goodness-of-Fit Test | | |
|---|---|---|
| | W | Prob<W |
| Shapiro-Wilk | 0.9858148 | 0.8060 |

| | A2 | Simulated p-Value |
|---|---|---|
| Anderson-Darling | 0.2258447 | 0.8188 |

Figure 12.2.4

The last assumption that needs to be validated is that of homogeneous variances. That is, we performed the ANOVA under the assumption that

$$H_0: \sigma_1^2 = \sigma_2^2 = \sigma_3^2$$

against the alternative hypothesis that at least one of the variances is different, which can be written as

$$H_a: \text{At least one } \sigma_i^2 \text{ is different.}$$

To test the equality of the variances, we will use the procedure discussed in Section 11.5 when we compared two variances. In this particular situation, since we have three group variances (in general, we can have three or more groups when performing ANOVA), we want to compare the largest variance to the smallest variance. That is, we will test using a null hypothesis of

$$H_0: \frac{\sigma^2_{Max}}{\sigma^2_{Min}} = 1$$

against an alternative of

$$H_a: \frac{\sigma^2_{Max}}{\sigma^2_{Min}} \neq 1.$$

where $\sigma^2_{Max}$ represents the largest variance of the three age groups and $\sigma^2_{Min}$ represents the smallest variance of the three age groups. The rationale is that if there is not a significant difference between the largest and smallest variances, then there won't be a significant difference among all group variances.

Of course, if there is a significant difference between the largest and smallest variances, then our assumption is violated, and we may need to do one of the following: transform the data by taking the natural log or square root of the responses; use nonparametric procedures; or use some alternative statistics such as Welch's or Brown-Forsythe procedures which use an alternative $F$-statistic to determine if you have statistical significance.

Understanding that the data are collected from three independent normally distributed populations and that we are testing the ratio of two variances, the test statistic is given by

$$F = \frac{s^2_{Max}}{s^2_{Min}}.$$

Suppose we are testing at the 5% level of significance (i.e., $\alpha = 0.05$). Additionally, we know that we have a two-sided test based on the alternative hypothesis. Since the alternative hypothesis is two-sided, two tails of the $F$-distribution must be determined as rejection regions. That is, we want to determine if $F \leq F_{1-\alpha/2, df_{num}, df_{den}}$ or if $F \leq F_{\alpha/2, df_{num}, df_{den}}$. These critical values are

$$F_{1-\alpha/2, df_{num}, df_{den}} = F_{0.975, 49, 49} = 0.5675$$
$$F_{\alpha/2, df_{num}, df_{den}} = F_{0.025, 49, 49} = 1.7622$$

The rejection region is that we will reject the null hypothesis if the $F$-test statistic is less than or equal to 0.5675 or if the $F$-test statistic is greater than or equal to 1.7622.

As can be seen in the JMP output, the standard deviation (and thus, the variance) is largest for the group of teens (13–18 years old) and the standard deviation is smallest for the adults (more than 18 years old).

The test statistic is given by

$$F = \frac{s^2_{Max}}{s^2_{Min}} = \frac{(25.5704)^2}{(20.3235)^2} \approx 1.5830.$$

Since the test statistic does not fall in the rejection region, we fail to reject the null hypothesis and conclude that there is no evidence to indicate that the variances of screen time among tweens, teens, and adults are significantly different.

## ⚭ Technology

Using the Excel function, F.INV.RT, we can find the critical value $F_{0.975, 49, 49}$ by typing the following into a cell of the spreadsheet "=F.INV.RT(0.975, 49, 49)" which equals 0.5675.

For instructions on finding $F$ critical values using technology visit stat.hawkeslearning.com and navigate to **Discovering Business Statistics, Second Edition > Technology Instructions > F-Distribution > Critical Value.**

# ✏ 12.2 Exercises

## Basic Concepts

1. Why is it important to validate the assumptions upon which a hypothesis test is based?

2. What is the first assumption on which ANOVA is based?

3. How can we test to see if the data reasonably satisfy the first assumption?

4. What is the second assumption on which ANOVA is based?

5. How can we determine if the second assumption is reasonable for the data we are interested in?

6. What is a simple "rule of thumb" that may be used to check the second assumption?

7. What is the third assumption that must be met before performing ANOVA?

## Exercises

8. For each of the following histograms of sample data, decide whether or not you think it is reasonable to assume that the data were drawn from a population that has an approximately normal distribution.

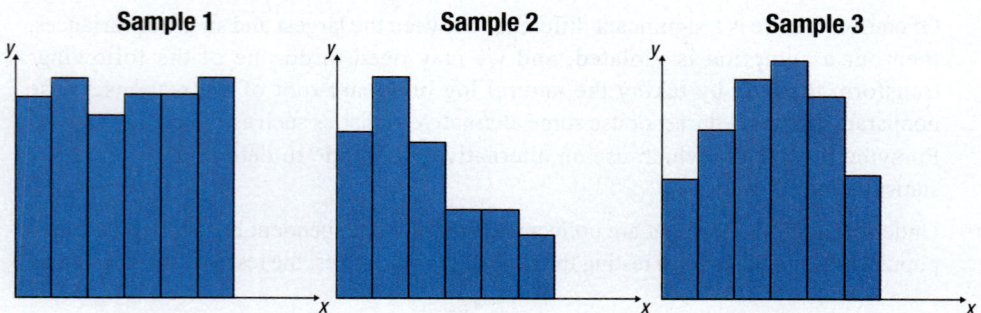

9. For each of the following histograms of sample data, decide whether or not you think it is reasonable to assume that the data were drawn from a population that has an approximately normal distribution.

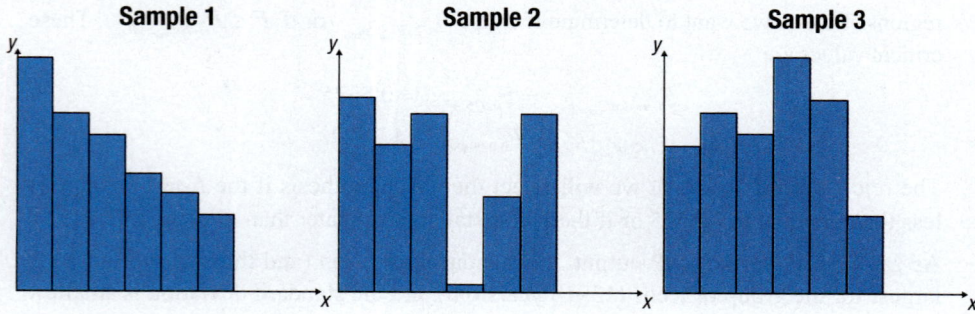

10. Consider the following box plots.

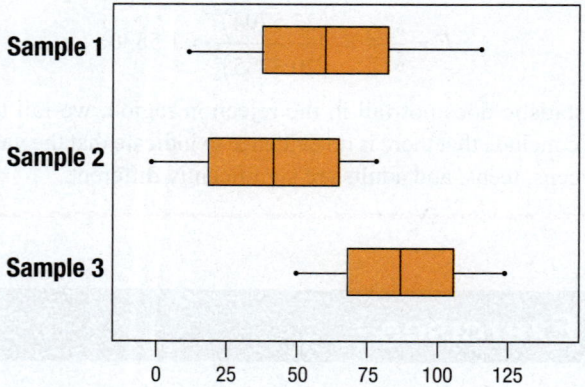

Do you think it is reasonable to assume that the three populations represented by the sample data in these box plots have equal variances? Explain.

**11.** Consider the following box plots.

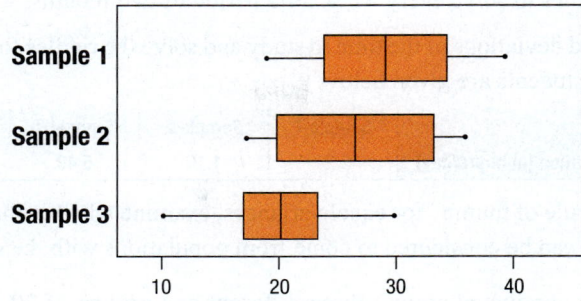

Do you think it is reasonable to assume that the three populations represented by the sample data in these box plots have equal variances? Explain.

**12.** Consider the following data on the diameter measurements (in inches) of soft drink bottles of three different brands.

| Pepsi | Coca-Cola | Dr. Pepper |
|-------|-----------|------------|
| 1.17 | 0.14 | 2.17 |
| 1.20 | 0.07 | 2.20 |
| 1.15 | 0.11 | 2.16 |
| 1.21 | 0.08 | 2.20 |
| 1.07 | 0.21 | 2.31 |
| 1.18 | 0.22 | 2.09 |
| 1.12 | 0.15 | 2.24 |
| 1.09 | 0.08 | 2.07 |
| 1.30 | 0.12 | 2.05 |
| 1.11 | 0.28 | 2.18 |

Test the assumption of homogeneity required to conduct the ANOVA test, that is, the three samples come from populations with equal variances at a 0.05 significance level.

**13.** Consider the given data on the foot lengths (in cms) of adult males obtained from three different states of the U.S.

| Iowa | Hawaii | California |
|------|--------|-----------|
| 177.23 | 172.24 | 170.28 |
| 175.79 | 172.20 | 177.46 |
| 176.76 | 176.46 | 174.67 |
| 180.52 | 180.23 | 186.57 |
| 185.28 | 171.68 | 189.29 |
| 179.62 | 177.98 | 178.63 |
| 188.47 | 179.68 | 179.91 |
| 179.63 | 177.00 | 184.72 |
| 175.24 | 179.25 | 171.52 |
| 180.81 | 181.97 | 179.08 |
| 178.50 | 182.45 | 185.49 |
| 190.01 | 175.97 | 180.24 |
| 188.36 | 176.84 | 177.64 |
| 182.49 | 169.96 | 176.85 |
| 180.73 | 167.56 | 178.72 |

Do the data satisfy the assumption that the samples come from normally distributed populations? (Use the Shapiro-Wilk test or the Anderson-Darling test.)

14. Four different samples of 10 students each from the same age group are given four types of riddles to solve every week for a period of two months.

    The standard deviations of the time to study and solve the riddles for each of the four samples of students are given below.

    |  | Sample 1 | Sample 2 | Sample 3 | Sample 4 |
    |---|---|---|---|---|
    | Standard deviation (in hours/day) | 2.24 | 1.10 | 5.49 | 2.72 |

    Using the "rule of thumb" for equal variances, examine whether the four samples of study hours can be considered to come from populations with the same variance.

15. A survey is conducted among three different age groups of 20 people each at a shopping center located near the researcher's residence. The three samples are from the following age groups: Teen (15-19 years), Young Adult (20-30 years), and Adult (31 years and above). The researcher asks the participants about the amount of money they spend monthly on shopping.

    A one-way ANOVA is conducted to test the difference between the mean amounts spent by the respondents for the three different age groups. Consider the assumptions necessary to perform an ANOVA and identify which assumption, if any, is violated in this scenario.

16. An assumption of the ANOVA test is that the k samples considered should come from populations with the same variance. If a boxplot is created for each sample on the same scale, what characteristic should be examined to conclude that the populations have the same variance?

17. Three samples of ten college students were randomly selected from the Japanese club, the Computer Science club, and the soccer team. Each sample was surveyed on their political views of the president. Can the samples in this scenario be considered independent? Explain your answer.

# 12.3 The *F*-Distribution and the *F*-Test

In Section 12.1, we introduced most of the formulas that are used to analyze data from more than two samples in order to determine whether or not there is a significant difference among population means. We developed MST, a measure for summarizing the variability among the sample means, and MSE, a measure for summarizing the variability within the samples themselves. We determined that if the variability among the sample means is much larger than the variability within the sample observations, we will doubt the hypothesis that the population means are the same. Alternatively, if the variability among the sample means is small when compared to the variability within the sample observations, it is not likely that the population means are significantly different.

Consider the ratio of the MST (mean square for treatments), the summary measure of the variability among the sample means, to the MSE (mean square for error), the summary measure of the variability within the samples.

$$\frac{\text{MST}}{\text{MSE}}$$

The *F*-distribution, named after the English statistician Sir Ronald Fisher, is a continuous distribution. It will be used in this and subsequent chapters to analyze variation in test statistics formed as ratios of two random variables. The *F*-distribution is not symmetrical; rather, it is skewed to the right. Like the *t*-distribution, its parameters are degrees of freedom. *F*-distributions are associated with test statistics that are quotients. What distinguishes the *F*-distribution is that it has a pair of values for its degrees of freedom. The number of degrees of freedom associated with the numerator is $df_{num}$ and the number of degrees of freedom associated with the denominator is $df_{den}$.

The shape of the *F*-distribution is a function of its degrees of freedom. Below, the *F*-distributions for $df_{num} = 6$, $df_{den} = 40$ and $df_{num} = 10$, $df_{den} = 4$ are given.

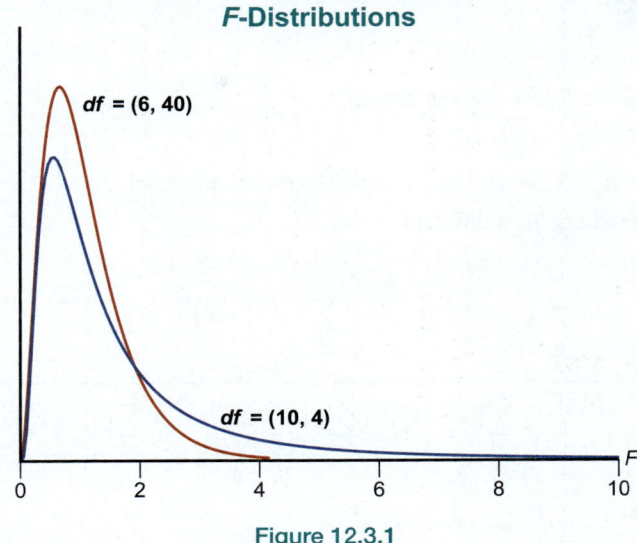

**Figure 12.3.1**

Like the *t*-distribution, tables have been compiled (see Appendix A) that show critical values of the *F*-distribution at common levels of significance. The notation $F_\alpha$ will indicate the value of *F* such that $\alpha$ of the area under the curve lies to the right of this value.

For example, to find the critical value at $\alpha = 0.05$ for an *F*-distribution with 3 numerator degrees of freedom and 7 denominator degrees of freedom, consult the table in Appendix A, Table H, with $\alpha = 0.05$. The leftmost column corresponds to the denominator degrees of freedom, and the top row corresponds to the numerator degrees of freedom.

| | | **Numerator Degrees of Freedom** | | | |
|---|---|---|---|---|---|
| | | **1** | **2** | **3** | **4** |
| | **1** | 161.4476 | 199.5000 | 215.7073 | 224.5832 |
| | **2** | 18.5128 | 19.0000 | 19.1643 | 19.2468 |
| | **3** | 10.1280 | 9.5521 | 9.2766 | 9.1172 |
| **Denominator Degrees of Freedom** | **4** | 7.7086 | 6.9443 | 6.5914 | 6.3882 |
| | **5** | 6.6079 | 5.7861 | 5.4095 | 5.1922 |
| | **6** | 5.9874 | 5.1433 | 4.7571 | 4.5337 |
| | **7** | 5.5914 | 4.7374 | 4.3468 | 4.1203 |
| | **8** | 5.3177 | 4.4590 | 4.0662 | 3.8379 |

From the table, we see that $F_{0.05}$ for 3 numerator degrees of freedom and 7 denominator degrees of freedom is 4.3468. Thus, for this *F*-distribution, an area of 0.05 lies to the right of 4.3468.

If the assumptions discussed in Section 12.2 are met and we assume that the population means are equal, $\dfrac{MST}{MSE}$ will have an *F*-distribution with $k - 1$ numerator degrees of freedom (associated with MST) and $n_T - k$ denominator degrees of freedom (associated with MSE). If the variability among the sample means is close to the variability within the sample observations, *F* will be close to 1. However, as the variability among the sample means increases relative to the variability within the sample observations, the value of *F* will become large, causing us to doubt the assumption that the population means are equal.

Thus, $F = \dfrac{MST}{MSE}$ is a natural test statistic to use in determining whether or not a difference exists among the population means. We will reject the null hypothesis that the population means are equal for large values of the $F$-test statistic. This is why the analysis of variance (or ANOVA) test for differences among population means is often referred to as the **$F$-test**.

## Procedure

### $F$-Test

A summary of the $F$-test is given below.

**Hypotheses:**

$H_0: \mu_1 = \mu_2 = \cdots = \mu_k$ The $k$ population means are equal.

$H_a:$ At least one $\mu_i$ is different.

**Test Statistic:**

$$F = \frac{\text{MST}}{\text{MSE}} = \frac{\dfrac{\displaystyle\sum_{j=1}^{k} n_j \left( \bar{x}_j - \bar{\bar{x}} \right)^2}{k-1}}{\dfrac{\displaystyle\sum_{i=1}^{n_1} \left( x_{i1} - \bar{x}_1 \right)^2 + \sum_{i=1}^{n_2} \left( x_{i2} - \bar{x}_2 \right)^2 + \cdots + \sum_{i=1}^{n_k} \left( x_{ik} - \bar{x}_k \right)^2}{n_T - k}}$$

**Assumptions:**

The $k$ populations of interest are normally distributed with equal variances, and independent, random samples are drawn from each population.

**Rejection Region:**

$H_0$ will be rejected for large values of $F = \dfrac{\text{MST}}{\text{MSE}}$. In particular, we will reject $H_0$ if

$F \geq F_\alpha$ with $(k-1)$ numerator degrees of freedom and $(n_T - k)$ denominator degrees of freedom.

Unfortunately, the formulas for MST and MSE given in Section 12.1 are difficult to use for actual calculations. So, we present computational formulas for these summary measures.

## Formula

### Computational Formulas for MST and MSE

Let $k$ be the total number of treatments and $n_T = n_1 + n_2 + \cdots + n_k$ be the total number of observations. Then the computational formulas are as follows.

$$\text{MST} = \frac{\left[ \dfrac{\left( \displaystyle\sum_{i=1}^{n_1} x_{i1} \right)^2}{n_1} + \dfrac{\left( \displaystyle\sum_{i=1}^{n_2} x_{i2} \right)^2}{n_2} + \cdots + \dfrac{\left( \displaystyle\sum_{i=1}^{n_k} x_{ik} \right)^2}{n_k} \right] - \dfrac{\left( \displaystyle\sum_{j=1}^{k} \sum_{i=1}^{n_j} x_{ij} \right)^2}{n_T}}{k-1}$$

$$\text{MSE} = \frac{\displaystyle\sum_{j=1}^{k} \sum_{i=1}^{n_j} x_{ij}^2 - \left[ \dfrac{\left( \displaystyle\sum_{i=1}^{n_1} x_{i1} \right)^2}{n_1} + \dfrac{\left( \displaystyle\sum_{i=1}^{n_2} x_{i2} \right)^2}{n_2} + \cdots + \dfrac{\left( \displaystyle\sum_{i=1}^{n_k} x_{ik} \right)^2}{n_k} \right]}{n_T - k}$$

Hopefully, you will never have to use these formulas in practice, for there are many statistical packages available for this purpose.

You have just been promoted to sales manager of a company that manufactures robots used to assemble automobiles. Although your sales force is given a suggested price at which to sell the robots, they have considerable leeway in negotiating the final price. Past sales records indicate that sometimes there is a large difference in the selling prices that different sales reps are able to negotiate. You are interested in knowing if this difference is significant, possibly because of a more effective negotiating strategy or exceptional interpersonal skills, or whether this observed difference in selling prices is just due to random variation. You decide to randomly select four sales over the last year for each of your three sales representatives and observe the actual selling prices of the robot. Table 12.3.1 shows the amounts at which the robots sold in thousands of dollars.

**Example 12.3.1**

**Comparing Selling Prices by Salespersons Using ANOVA**

| Table 12.3.1 – Selling Prices (Thousands of Dollars) | | |
|---|---|---|
| **Salesperson 1** | **Salesperson 2** | **Salesperson 3** |
| 10 | 11 | 11 |
| 14 | 16 | 13 |
| 13 | 14 | 12 |
| 12 | 15 | 15 |
| **Total**        49 | 56 | 51 |

Based on the results of your survey, can you conclude that there is a significant difference among the average selling prices that the three sales reps have been able to negotiate? Use $\alpha = 0.05$.

**SOLUTION**

**Step 1:** Determine the null hypothesis. In this process, select the appropriate statistical measure, such as the population mean, proportion, or variance.

The hypotheses are fairly straightforward since you are interested in comparing the average selling prices for the three sales reps, the population parameters of interest are the true mean selling prices for the three sales reps.

$\mu_1$ = true mean selling price for sales person 1

$\mu_2$ = true mean selling price for sales person 2

$\mu_3$ = true mean selling price for sales person 3

The null hypothesis is written as $H_0$: $\mu_1 = \mu_2 = \mu_3$ which implies that the mean selling prices are the same for all three sales persons.

**Step 2:** Determine the alternative hypothesis and whether it should be one-sided or two-sided.

The alternative hypothesis states that there is a difference among the average selling prices for the three sales persons. Based on the way the test statistic is constructed, we will reject the null hypothesis for large values of the test statistic, meaning that the variability among the sample means is much larger than the variability within the sample observations. The *F*-test is always a one-tailed test. The alternative hypothesis should be written as

$H_a$: At least one $\mu_i$ is different.

**Step 3:** Select the appropriate test statistic based on the information at hand and the assumptions you are willing to make.

There are three key questions we must ask.

1. Are the selling prices for each sales person normally distributed?

2. Do the selling prices for each sales person have approximately equal variances?

3. Were the sample selling prices for each sales person collected in an independent and random fashion?

Based on prior studies, you have reason to believe that the selling prices of the robots for each of the sales persons have an approximately normal distribution and that the variances of the three distributions are approximately equal. Also, you used random sampling to collect your data. Thus, the appropriate test statistic is the $F$-statistic.

$$F = \frac{MST}{MSE}$$

**Step 4:** Determine the critical value of the test statistic.

The level of the test is specified in the problem as $\alpha = 0.05$.

The level of the test is $\alpha = 0.05$, and there are $(k - 1) = (3 - 1) = 2$ numerator degrees of freedom and $(n_T - k) = (12 - 3) = 9$ denominator degrees of freedom. Thus, the critical value is $F_{0.05} = 4.2565$. We will reject $H_0$ if the computed value of the test statistic is greater than or equal to 4.2565. Figure 12.3.2 displays the rejection region.

### F-Distribution
### $df_{num} = 2$, $df_{den} = 9$

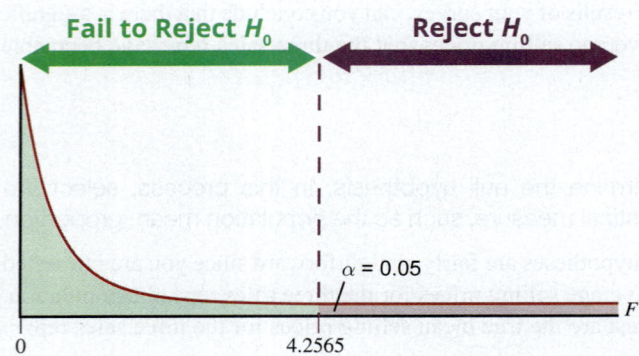

Figure 12.3.2

**Step 5:** Collect sample data and compute the value of the test statistic.

Using the computational formulas presented previously, we have the following.

$$MST = \frac{\left[ \frac{\left( \sum\limits_{i=1}^{n_1} x_{i1} \right)^2}{n_1} + \frac{\left( \sum\limits_{i=1}^{n_2} x_{i2} \right)^2}{n_2} + \cdots + \frac{\left( \sum\limits_{i=1}^{n_k} x_{ik} \right)^2}{n_k} \right] - \frac{\left( \sum\limits_{j=1}^{k} \sum\limits_{i=1}^{nj} x_{ij} \right)^2}{n_T}}{k - 1}$$

$$= \frac{\left[ \frac{(49)^2}{4} + \frac{(56)^2}{4} + \frac{(51)^2}{4} \right] - \frac{(156)^2}{12}}{3 - 1}$$

$$= 3.25$$

$$MSE = \frac{\sum_{j=1}^{k}\sum_{i=1}^{n_j} x_{ij}^2 - \left[ \frac{\left(\sum_{i=1}^{n_1} x_{i1}\right)^2}{n_1} + \frac{\left(\sum_{i=1}^{n_2} x_{i2}\right)^2}{n_2} + \cdots + \frac{\left(\sum_{i=1}^{n_k} x_{ik}\right)^2}{n_k} \right]}{n_T - k}$$

$$= \frac{(10)^2 + (14)^2 + \cdots + (15)^2 - \left[ \frac{(49)^2}{4} + \frac{(56)^2}{4} + \frac{(51)^2}{4} \right]}{12 - 3}$$

$$= 3.5$$

The resulting calculated value of the test statistic is as follows.

$$F = \frac{MST}{MSE} = \frac{3.25}{3.5} \approx 0.9286$$

**Step 6:** Make the decision and state the conclusion in terms of the original question.

Since the resulting value of the test statistic, 0.9286, is less than the critical value of 4.2565, we fail to reject the null hypothesis (see Figure 12.3.3).

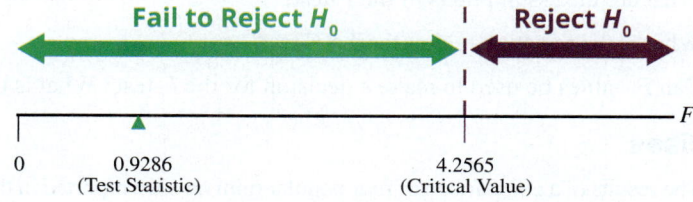

Figure 12.3.3

*Conclusion and Interpretation*: There is not sufficient evidence at $\alpha = 0.05$ to reject the null hypothesis. Thus, we cannot conclude that there is a difference among the average selling prices for the three sales persons.

It is often more convenient to use statistical software (rather than the computational formulas) to perform an analysis of variance such as the one in this example. The summary output from Microsoft Excel is given in Figure 12.3.4. (See the Discovering Technology section at the end of this chapter for directions on performing ANOVA using Microsoft Excel, JMP, and Minitab.)

ANOVA

| Source of Variation | SS | df | MS | F | P-value | F crit |
|---|---|---|---|---|---|---|
| Between Groups | 6.5 | 2 | 3.25 | 0.928571429 | 0.429903823 | 4.256494729 |
| Within Groups | 31.5 | 9 | 3.5 | | | |
| Total | 38 | 11 | | | | |

Figure 12.3.4

Note that the *F*-statistic is approximately 0.9286 and the *P*-value associated with the test is approximately 0.4299. Since $0.4299 > 0.05$, there is not sufficient evidence to conclude that there is a difference between the selling prices negotiated by the three salespersons.

Note that if the null hypothesis that all population (or group) means are equal is rejected, the analysis of variance procedure does not specifically tell you *which* population means are not equal. There are procedures however, called **multiple comparison procedures**, that allow us to assess differences among specified means. Some of the more popular multiple comparison procedures are Fisher's Least Significant Difference method and Tukey's Honest Significant Difference test, which are procedures specifically for pairwise comparisons when the sample sizes of the treatments are equal. These procedures will be discussed in the next section.

🔗 **Technology**

For instructions on performing an ANOVA test using technology visit stat.hawkeslearning.com and navigate to **Discovering Business Statistics, Second Edition > Technology Instructions > ANOVA > One-Way**.

## 📝 12.3 Exercises

### Basic Concepts

1. If you found that MST is much larger than MSE, would you tend to think that the population means were similar or different? Explain how this ratio brings you to this conclusion.

2. What kind of distribution does the ratio $\dfrac{\text{MST}}{\text{MSE}}$ have?

3. What are the degrees of freedom associated with $\dfrac{\text{MST}}{\text{MSE}}$?

4. If the variability among the sample means is very similar to the variability among the sample observations, what value will $F$ be close to? Explain why.

5. Is the null hypothesis generally rejected for large or small values of the $F$-statistic? Explain why this is the case.

6. What are the null and alternative hypotheses for the $F$-test?

7. What are the assumptions of the $F$-test?

8. What is the rejection region for the $F$-test?

9. Can $P$-values be used to make a decision for the $F$-test? What is the decision rule?

### Exercises

10. The results of a comparison of four popular minivans are reported in the following table. One of the features the researchers compared was the distance (in feet) required for the minivan to come to a complete stop when traveling at a speed of 60 miles per hour (braking distance). Suppose the braking distances were measured for five minivans of each type with the following results.

| Braking Distances (Feet) | | | |
|---|---|---|---|
| Minivan A | Minivan B | Minivan C | Minivan D |
| 150 | 153 | 155 | 167 |
| 152 | 150 | 150 | 164 |
| 151 | 156 | 157 | 169 |
| 149 | 151 | 158 | 162 |
| 153 | 155 | 155 | 173 |

   a. Can the researchers conclude at $\alpha = 0.10$ that there is a difference among average braking distances for the four minivan models?

   b. What assumptions did the researchers make in performing the test procedure in part a.? Do the data appear to satisfy these assumptions? Explain.

11. A steel company is considering the relocation of one of its manufacturing plants. The company's executives have selected four areas that they believe are suitable locations. However, they want to determine if the average wages are significantly different in any of the locations, since this could have a major impact on the cost of production. A survey of hourly wages of similar workers in each of the four areas is performed with the following results.

| Hourly Wages ($) | | | |
|---|---|---|---|
| Area 1 | Area 2 | Area 3 | Area 4 |
| 10 | 15 | 13 | 20 |
| 12 | 16 | 14 | 16 |
| 11 | 18 | 15 | 18 |
| 13 | 17 | 15 | 17 |
| 10 | 14 | 12 | 16 |

**a.** Do the data indicate a significant difference among the average hourly wages in the four areas at $\alpha = 0.05$?

**b.** What assumptions were made in performing the test in part **a.**? Do the data appear to satisfy these assumptions? Explain.

12. A director of training at a large temporary services company has learned of three different methods for teaching a person to type. He is interested in determining if there is a difference in the average typing speeds for employees who are taught to type using each of the three methods. He randomly selects 15 new employees and then randomly assigns five employees to learn to type by each of the training methods. At the end of the course, he measures the number of correct words per minute for each employee. The results are as follows.

| Typing Speeds (Correct Words per Minute) | | |
|---|---|---|
| Method 1 | Method 2 | Method 3 |
| 45 | 50 | 60 |
| 50 | 55 | 63 |
| 40 | 49 | 55 |
| 43 | 52 | 52 |
| 47 | 53 | 58 |

**a.** Can the director of training conclude that there is a difference among the average typing speeds of the employees for the three methods at $\alpha = 0.10$?

**b.** What assumptions did the director of training make in performing the test in part **a.**? Do the data appear to satisfy these assumptions? Explain.

13. A physical trainer has four workouts that he recommends for his clients. The workouts have been designed so that the average maximum heart rate achieved is the same for each workout. To test this design he randomly selects 12 people and randomly assigns three of them to use each of the workouts. During each workout, he measures the maximum heart rate in beats per minute with the following results.

| Maximum Heart Rates (Beats per Minute) | | | |
|---|---|---|---|
| Workout #1 | Workout #2 | Workout #3 | Workout #4 |
| 180 | 160 | 175 | 185 |
| 185 | 170 | 180 | 190 |
| 170 | 175 | 170 | 180 |

**a.** Can the physical trainer conclude at $\alpha = 0.05$ that there is a difference among the average maximum heart rates which are achieved during the four workouts?

**b.** What assumptions did the physical trainer make in performing the test procedure in part **a.**? Do the data appear to satisfy these assumptions? Explain.

**14.** The results of a survey comparing the costs of staying one night in a full-service hotel (including food, beverages, and telephone calls, but not taxes or gratuities) for several major cities are given in the following table.

| Hotel Costs per Night ($) | | | | |
|---|---|---|---|---|
| New York | Los Angeles | Atlanta | Houston | Phoenix |
| 300 | 240 | 190 | 195 | 238 |
| 320 | 250 | 198 | 190 | 240 |
| 325 | 230 | 185 | 200 | 236 |
| 350 | 245 | 195 | 192 | 248 |
| 275 | 235 | 182 | 198 | 228 |

a.  Do the data suggest that there is a significant difference among the average costs of one night in a full-service hotel for the five major cities at $\alpha = 0.05$?

b.  What assumptions were made in performing the test procedure in part **a.**? Do the data appear to satisfy these assumptions? Explain.

c.  Based on the analysis you performed in part **b.**, which cities, if any, do you think have significantly different average costs for a one-night stay in a full-service hotel? Explain.

**15.** Consider the following information regarding the dividends paid per share by companies in the banking, transportation, and energy industries.

| Dividends per Share ($) | | |
|---|---|---|
| Banking | Transportation | Energy |
| 1.52 | 1.00 | 2.08 |
| 3.12 | 1.20 | 2.68 |
| 1.32 | 0.20 | 0.70 |
| 0.60 | 0.40 | 2.00 |
| 1.20 | 1.09 | 1.91 |
| 1.00 | 0.61 | 1.60 |
| 1.19 | 0.35 | 1.28 |

a.  Do the data provide sufficient evidence to conclude that there is a significant difference among the average dividends paid per share for the three different industries? Use $\alpha = 0.10$.

b.  What assumptions were made in performing the test procedure in part **a.**? Do the data appear to satisfy these assumptions? Explain.

c.  Based on the analysis you performed in part **b.**, which industries, if any, do you think pay significantly different average dividends per share? Explain.

# 12.4 Multiple Comparison Procedures

In the previous sections, we used one-way ANOVA to test whether differences existed between population means. In the earlier examples, when we rejected the null hypothesis that all of the population means were equal, we were *only* testing if differences existed. However, the results of the one-way ANOVA test do not indicate which population means are different. To determine which population means are different, we need to perform more tests to determine if there are statistically significant differences between two population means such as $\mu_1 - \mu_2$, for example. Multiple comparison procedures present several options to the analyst when comparing means after finding significance when performing a one-way ANOVA. The ones we will consider are Fisher's Least Significance Difference (LSD) Method, Tukey's Honest Significant Difference (HSD), and performing *t*-tests to make pairwise comparisons between means. We will discuss each of these methods using the data from Example 12.1.1.

In Example 12.1.1, we compared the average amount of time spent on devices among three age groups (Tweens, Teens, and Adults). The table below presents some summary statistics from the data collected showing the sample size, the sample mean time spent on devices, and the sample standard deviation of the data by age group.

| Table 12.4.1 – Screen Time Summary Statistics by Age Group | | | |
|---|---|---|---|
| | Tweens 8–12 Years Old | Teens 13–18 Years Old | Adults Over 18 Years Old |
| *n* | 50 | 50 | 50 |
| Mean | 128 | 137.48 | 100.12 |
| Standard Deviation | 20.90 | 25.57 | 20.32 |

Test to determine if there is a significant difference between average time spent on devices among the three age groups.

Let

$\mu_1$ = population mean time spent on devices for participants 8–12 years old,

$\mu_2$ = population mean time spent on devices for participants 13–18 years old, and

$\mu_3$ = population mean time spent on devices for participants over 18 years old.

Thus, the null and alternative hypotheses for this scenario should be written as

$H_0$: $\mu_1 = \mu_2 = \mu_3$

$H_a$: At least one $\mu_i$ is different for $i$ = 1, 2, 3.

The results from the one-way ANOVA yielded a *P*-value less than 0.0001 (see the JMP output below), which indicated that we would reject the null hypothesis and conclude that at least one of the means is different.

**Technology**

For instructions on performing an ANOVA test visit stat.hawkeslearning.com and navigate to **Discovering Business Statistics, Second Edition > Technology Instructions > ANOVA > One-Way**.

**One way Anova**

Summary of Fit

| | |
|---|---|
| Rsquare | 0.338598 |
| Adj Rsquare | 0.3296 |
| Root Mean Square Error | 22.38679 |
| Mean of Response | 121.8667 |
| Observations (or Sum Wgts) | 150 |

Analysis of Variance

| Source | DF | Sum of Squares | Mean Square | F-Ratio | Prob > F |
|---|---|---|---|---|---|
| Age Group | 2 | 37715.57 | 18857.8 | 37.6276 | < 0001* |
| Error | 147 | 73671.76 | 501.2 | | |
| C. Total | 149 | 111387.33 | | | |

Means for One way Anova

| Level | Number | Mean | Std Error | Lower 95% | Upper 95% |
|---|---|---|---|---|---|
| 8-12 Years Old | 50 | 128.000 | 3.1660 | 121.74 | 134.26 |
| 13-18 Years Old | 50 | 137.480 | 3.1660 | 131.22 | 143.74 |
| Over 18 Years Old | 50 | 100.120 | 3.1660 | 93.86 | 106.38 |

Std Error uses a pooled estimate of error variance

Figure 12.4.1

We can now answer the natural question, *Which of these means is different?* One method that was introduced in Chapter 11 would be the two-sample *t*-test. Using a series of two-sample *t*-tests, we would have the following hypotheses.

1. $H_0: \mu_1 - \mu_2 = 0$

    $H_a: \mu_1 - \mu_2 \neq 0$

2. $H_0: \mu_1 - \mu_3 = 0$

    $H_a: \mu_1 - \mu_3 \neq 0$

3. $H_0: \mu_2 - \mu_3 = 0$

    $H_a: \mu_2 - \mu_3 \neq 0$

For each set of hypotheses above, the test statistic is given by

$$t = \frac{\bar{x}_i - \bar{x}_j}{s_p\sqrt{\dfrac{1}{n_i} + \dfrac{1}{n_j}}}$$

where $\bar{x}_i$ is the mean of the $i^{th}$ group, $s_p$ is the pooled variance, and $n_i$ is the sample size of the $i^{th}$ group. The decision would be to reject the null hypothesis if the test statistic is greater than $t_{\alpha/2, n_i + n_j - 2}$ or if the test statistic is less than $-t_{\alpha/2, n_i + n_j - 2}$. For this data set, the decision is to reject the null hypothesis if the test statistic is less than $-1.9845$ or if the test statistic is greater than $1.9845$.

The test statistics for each set of hypotheses numbered above along with their respective $P$-values are presented in the table below.

| Table 12.4.2 – Hypothesis Testing Summary | | |
|---|---|---|
| **Hypotheses** | **Test Statistic** | **P-Value** |
| 1. Tweens and Teens | −2.030 | 0.0451 |
| 2. Tweens and Adults | 6.763 | < 0.0001 |
| 3. Teens and Adults | 8.088 | < 0.0001 |

At the 5% significance level (i.e., $\alpha = 0.05$), we will reject each of the hypotheses and conclude that the average screen time for tweens is significantly different from the average screen time of teens; the average screen time for tweens is significantly different from the average screen time of adults; and the average screen time of teens is significantly different from the average screen time of adults.

While the two-sample $t$-test may be appropriate when comparing means for a small number of groups, the probability of a Type I Error increases as the number of groups increases. Using this current example comparing screen time between the three age groups, we are making three pairwise comparisons. You may recall that $\alpha$ is the probability of committing a Type I Error; thus, the probability of not committing a Type I Error is $1 - \alpha$. For this example, let $\alpha = 0.05$. The probability of committing at least one Type I Error is given by

$$P(at\ least\ one\ Type\ I\ Error) = 1 - P(no\ Type\ I\ Error\ is\ committed)$$

$$= 1 - (1 - \alpha)^3$$

$$= 1 - (0.95)^3$$

$$= 0.1426$$

Even though we made our conclusions using $\alpha = 0.05$ (which is the $P$(Type I Error)), by performing three individual $t$-tests to compare the means, the likelihood of a Type I Error has increased to 0.1426. The solution to comparing the means when performing ANOVA is to use a multiple comparison procedure so that the likelihood of a Type I Error remains fixed at the $\alpha$ level of significance.

# Fisher's Least Significant Difference (LSD) Method

The first multiple comparison procedure that we will discuss is Fisher's Least Significant Difference (LSD) method. Fisher's LSD is a least significant difference method that allows you to compare the difference between two means to a value similar to the margin of error. Suppose we want to compare the population means of three groups, say Tweens, Teens, and Adults which are the three groups in Example 12.1.1.

## Formula

### Fisher's Least Significant Difference Method

We consider populations $i$ and $j$ to be significantly different if

$$\left| \bar{x}_i - \bar{x}_j \right| \ge t_{\alpha/2, n_T - k} \sqrt{MSE\left( \frac{1}{n_i} + \frac{1}{n_j} \right)}$$

where

$\bar{x}_i$ is the sample mean of the observations in the $i^{th}$ treatment,

$\bar{x}_j$ is the sample mean of the observations in the $j^{th}$ treatment,

$n_i$ is the number of observations in the $i^{th}$ treatment,

$n_j$ is the number of observations in the $j^{th}$ treatment,

$k$ is the number of treatments, and

$n_T$ is the total number of observations in all samples.

The MSE is the mean squared error which can be found in the ANOVA table. Please note that the significance level ($\alpha$) applies to each individual comparison and the right-hand side of the expression above is equivalent to the margin of error for a confidence interval.

Comparing the Tweens and Teens means using the data in Example 12.1.1, we calculate the left-hand side of the argument above which represents the absolute difference in sample means by

$$\left| \bar{x}_1 - \bar{x}_2 \right| = \left| 128 - 137.48 \right| = 9.48.$$

We then calculate the right-hand side of the equation above using $\alpha = 0.05$ by

$$t_{\alpha/2, df_{Error}} \sqrt{MSE\left( \frac{1}{n_i} + \frac{1}{n_j} \right)} = t_{0.025, 147} \sqrt{MSE\left( \frac{1}{n_1} + \frac{1}{n_2} \right)}$$

$$= 1.976 \sqrt{501.2 \left( \frac{1}{50} + \frac{1}{50} \right)} \approx 8.8475$$

Thus, since 9.48 is more than 8.8475, we conclude that the average screen time between tweens and teens is significantly different.

Similarly, when comparing Tweens to Adults, we calculate an absolute value of their differences to be

$$\left| \bar{x}_1 - \bar{x}_3 \right| = \left| 128 - 100.12 \right| = 27.88$$

and when comparing Teens to Adults, we get

$$\left| \bar{x}_2 - \bar{x}_3 \right| = \left| 137.48 - 100.12 \right| = 37.36$$

Since both of these absolute differences are larger than 8.8475, we conclude that the average screen time between tweens and adults is significantly different, and that the average screen time between teens and adults is significantly different.

Another approach to using Fisher's LSD Method is to calculate a $100(1 - \alpha)\%$ confidence interval for the difference between two population means $\mu_i - \mu_j$.

> **Formula**
>
> ### Fisher's LSD Method, Confidence Interval Approach
>
> The $100(1-\alpha)\%$ confidence interval for the difference between two population means $\mu_i - \mu_j$ given by
>
> $$\left(\overline{x}_i - \overline{x}_j\right) \pm t_{\alpha/2, n_T - k} \sqrt{\text{MSE}\left(\frac{1}{n_i} + \frac{1}{n_j}\right)}$$
>
> where
>
> $\overline{x}_i$ is the sample mean of the observations in the $i^{\text{th}}$ treatment,
>
> $\overline{x}_j$ is the sample mean of the observations in the $j^{\text{th}}$ treatment,
>
> $n_i$ is the number of observations in the $i^{\text{th}}$ treatment,
>
> $n_j$ is the number of observations in the $j^{\text{th}}$ treatment,
>
> $k$ is the number of treatments, and
>
> $n_T$ is the total number of observations in all samples.

The MSE is the mean squared error from the ANOVA table. Since we are testing the hypotheses

$$H_0: \mu_i - \mu_j = 0$$
$$H_a: \mu_i - \mu_j \neq 0$$

∞ **Technology**

For instructions on performing Fisher's LSD method visit stat.hawkeslearning.com and navigate to **Discovering Business Statistics, Second Edition > Technology Instructions > ANOVA > Fisher's LSD**.

then we fail to reject the null hypothesis if the interval contains the value zero. If the interval does not contain zero then we reject the null hypothesis and conclude that the two population means are significantly different.

Similar to the example above, let's compare Tweens and Teens. We have

$$\left(\overline{x}_1 - \overline{x}_2\right) \pm t_{\alpha/2, n_T - k} \sqrt{\text{MSE}\left(\frac{1}{n_i} + \frac{1}{n_j}\right)}$$

$$\left(128 - 137.48\right) \pm t_{0.025, 147} \sqrt{\text{MSE}\left(\frac{1}{n_1} + \frac{1}{n_2}\right)}$$

$$\left(128 - 137.48\right) \pm 1.976 \sqrt{501.2\left(\frac{1}{50} + \frac{1}{50}\right)}$$

$$-9.48 \pm 8.8475$$
$$\left(-18.3275, \ -0.6325\right)$$

Since the interval above does not contain zero, we reject the null hypothesis and conclude that the average screen time between tweens and teens is significantly different.

When comparing Tweens to Adults, we calculate the 95% confidence interval for the mean differences as

$$\left(\overline{x}_1 - \overline{x}_2\right) \pm t_{\alpha/2, n_T - k} \sqrt{\text{MSE}\left(\frac{1}{n_i} + \frac{1}{n_j}\right)}$$

$$\left(128 - 100.12\right) \pm 1.976 \sqrt{501.2\left(\frac{1}{50} + \frac{1}{50}\right)}$$

$$27.88 \pm 8.8475$$
$$\left(19.0325, 36.7275\right)$$

and when comparing Teens to Adults, we get

$$\left(\bar{x}_2 - \bar{x}_3\right) \pm t_{\alpha/2, n_T - k} \sqrt{MSE\left(\frac{1}{n_2} + \frac{1}{n_3}\right)}$$

$$\left(137.48 - 100.12\right) \pm 1.976 \sqrt{501.2\left(\frac{1}{50} + \frac{1}{50}\right)}$$

$$37.36 \pm 8.8475$$

$$\left(28.5125,\ 46.2075\right)$$

Since neither of the last two intervals contain zero, we conclude that the average screen time between tweens and adults is significantly different, and that the average screen time between teens and adults is also significantly different.

## Tukey's Honest Significant Difference (HSD) Method

Like Fisher's LSD, Tukey's HSD test is used to determine group differences when the ANOVA gives a significant result, indicating that at least one group mean differs from the other group means. Tukey's HSD test is a pairwise comparison that is used to compute the honestly significant difference between two means using a statistical distribution called the **q-distribution** (also known as the **studentized range distribution**). The $q$-distribution is similar to the $t$-distribution but has more variability (i.e., thicker) in the tails, yielding a larger studentized range value, $q_\alpha$, than $t_\alpha$. To ensure that all tests are compared using the same significance level, Tukey's HSD keeps the likelihood of making at least one Type I Error constant (i.e., all tests are carried out using the significance level). Critical values of the $q$-distribution table can be found in Table N of Appendix A.

Please note that the hypotheses for all three pairwise comparisons are given by:

1. $H_0: \mu_1 - \mu_2 = 0$
   $H_a: \mu_1 - \mu_2 \neq 0$

2. $H_0: \mu_1 - \mu_3 = 0$
   $H_a: \mu_1 - \mu_3 \neq 0$

3. $H_0: \mu_2 - \mu_3 = 0$
   $H_a: \mu_2 - \mu_3 \neq 0$

The three hypotheses above are equivalent to $H_0: \mu_i - \mu_j = 0$ vs $H_a: \mu_i - \mu_j \neq 0$ for $i \neq j$ with $i, j = 1, 2, 3$.

### Formula

#### Tukey's Honest Significant Difference Method

Two means, $\mu_i$ and $\mu_j$, are significantly different if

$$\left|\bar{x}_i - \bar{x}_j\right| \geq q_{\alpha, k, n_T - k} \sqrt{\frac{MSE}{2}\left(\frac{1}{n_i} + \frac{1}{n_j}\right)}$$

where

$\bar{x}_i$ is the sample mean of the observations in the $i^{th}$ treatment,

$\bar{x}_j$ is the sample mean of the observations in the $j^{th}$ treatment,

$q_{\alpha, k, n_T - k}$ is the studentized range value,

$n_i$ is the number of observations in the $i^{th}$ treatment,

$n_j$ is the number of observations in the $j^{th}$ treatment,

$k$ is the number of treatments, and

$n_T$ is the total number of observations in all samples.

> **Formula**
>
> ### Tukey's HSD Method, Confidence Interval Approach
>
> The $100(1 - \alpha)\%$ confidence interval for the difference between two population means $\mu_i - \mu_j$ is given by
>
> $$\left(\overline{x}_i - \overline{x}_j\right) \pm q_{\alpha,k,n_T-k}\sqrt{\frac{MSE}{n}} \text{ for \textbf{balanced data} } (n = n_i = n_j)$$
>
> or
>
> $$\left(\overline{x}_i - \overline{x}_j\right) \pm q_{\alpha,k,n_T-k}\sqrt{\frac{MSE}{2}\left(\frac{1}{n_i} + \frac{1}{n_j}\right)} \text{ for \textbf{unbalanced data} } (n_i \neq n_j)$$
>
> where
>
> $\overline{x}_i$ is the sample mean of the observations in the $i^{th}$ treatment,
>
> $\overline{x}_j$ is the sample mean of the observations in the $j^{th}$ treatment,
>
> $q_{\alpha,k,n_T-k}$ is the studentized range value,
>
> $n_i$ is the number of observations in the $i^{th}$ treatment,
>
> $n_j$ is the number of observations in the $j^{th}$ treatment,
>
> $k$ is the number of treatments, and
>
> $n_T$ is the total number of observations in all samples.

To illustrate the use of Tukey's HSD, we will use the data in Example 12.1.1 in which we performed a one-way ANOVA to compare average screen time between tweens, teens, and adults. The table below contains the summary statistics of the data which is a duplicate of the table at the beginning of this section.

| Table 12.4.1 – Screen Time Summary Statistics by Age Group | | | |
|---|---|---|---|
| | **Tweens**<br>**8–12 Years Old** | **Teens**<br>**13–18 Years Old** | **Adults**<br>**Over 18 Years Old** |
| $n$ | 50 | 50 | 50 |
| **Mean** | 128 | 137.48 | 100.12 |
| **Standard Deviation** | 20.90 | 25.57 | 20.32 |

### ⚙ Technology

For instructions on performing Tukey's HSD method visit stat.hawkeslearning.com and navigate to **Discovering Business Statistics, Second Edition > Technology Instructions > ANOVA > Tukey's HSD**.

### ⚙ Technology

To find critical values for the $q$-distribution visit stat.hawkeslearning.com and navigate to **Discovering Business Statistics, Second Edition > Technology Instructions > q-Distribution > Critical Value**.

### ✎ NOTE

The critical value used for these calculations is the value from Table N with $k = 3$ and error $df = \infty$.

Using the confidence interval approach for Tukey's HSD for the first hypothesis above, a 95% confidence interval for the mean differences between Tweens and Teens is given by

$$\left(\overline{x}_i - \overline{x}_j\right) \pm q_{\alpha,k,n_T-k}\sqrt{\frac{MSE}{n}}$$

$$\left(\overline{x}_1 - \overline{x}_2\right) \pm q_{0.05,3,147}\sqrt{\frac{MSE}{n}}$$

$$\left(128 - 137.48\right) \pm 3.314\sqrt{\frac{501.2}{50}}$$

$$-9.48 \pm 10.4924$$

$$\left(-19.9724, 1.0124\right)$$

Since the interval above includes zero, we fail to reject the null hypothesis and conclude that there is not enough evidence to indicate that the amount of screen time for tweens is different from that of teens.

Note that the MSE is the mean squared error which comes from the ANOVA.

Similarly, comparing the average screen time for Tweens and Adults using Tukey's HSD, a 95% confidence interval is given by

$$(128-100.12)\pm3.314\sqrt{\frac{501.2}{50}}$$
$$27.88\pm10.4924$$
$$(17.3876, 38.3724)$$

Since the interval does not include zero, we reject the null hypothesis and conclude that the average screen time between tweens and adults is significantly different.

Lastly, when comparing Teens and Adults using Tukey's HSD, a 95% confidence interval yields the following.

$$(137.48-100.12)\pm3.314\sqrt{\frac{501.2}{50}}$$
$$37.36\pm10.4924$$
$$(26.8676, 47.8524)$$

Again, the interval does not contain zero, thus we reject the null hypothesis and conclude that the average screen time between teens and adults is significantly different.

If we examine the "Connecting Letters Report" in the JMP output in the figure below, we can see that tweens and teens have the same letter (A) and adults have the letter B. As stated in the output, the levels that are not connected by the same letter are significantly different. Thus, the average screen times between tweens and teens are not significantly different, but the average screen time between tweens and adults are significantly different. Also, the average screen times between teens and adults are significantly different. This was supported in the calculations presented above.

**Comparisons for all pairs using Tukey-Kramer HSD**

Confidence Quantile

| q* | Alpha |
|---|---|
| 2.36773 | 0.05 |

HSD Threshold Matrix

Abs(Dif)-HSD

| | 13-18 Years Old | 8-12 Years Old | Over 18 Years Old |
|---|---|---|---|
| 13-18 Years Old | -10.601 | -1.121 | 26.759 |
| 8-12 Years Old | -1.121 | -10.601 | 17.279 |
| Over 18 Years Old | 26.759 | 17.279 | -10.601 |

Positive values show pairs of means that are significantly different.

Connecting Letters Report

| Level | | Mean |
|---|---|---|
| 13-18 Years Old | A | 137.48000 |
| 8-12 Years Old | A | 128.00000 |
| Over 18 Years Old | B | 100.12000 |

Levels not connected by same letter are significantly different.

Ordered Differences Report

| Level | -Level | Difference | Std Err Dif | Lower CL | Upper CL | p-Value |
|---|---|---|---|---|---|---|
| 13-18 Years Old | Over 18 Years Old | 37.36000 | 4.477358 | 26.7588 | 47.96118 | <.0001* |
| 8-12 Years Old | Over 18 Years Old | 27.88000 | 4.477358 | 17.2788 | 38.48118 | <.0001* |
| Over 18 Years Old | 8-12 Years Old | 9.48000 | 4.477358 | -1.1212 | 20.08118 | 0.0898 |

**Figure 12.4.2**

It is interesting to note that the two multiple comparison procedures (Fisher's LSD and Tukey's HSD) in the aforementioned example do not produce the same results. This is not uncommon. Specifically, as the number of comparisons increases, the probability of committing a Type I Error increases when using Fisher's LSD. However, regardless of the number of comparisons being made, the probability of committing a Type I Error remains the same (i.e., $\alpha$) when using Tukey's HSD.

✎ **NOTE**

Note that the confidence intervals in the Ordered Differences Report vary slightly from our previous calculations due to the exact $q$-distribution critical value being used instead of the value from the table with error $df = \infty$.

## ✏ 12.4 Exercises

### Basic Concepts

1. What is the purpose of multiple comparison procedures?

2. When should multiple comparison procedures be used?

3. What are the hypotheses tested if there are four population means in the ANOVA?

4. Define the concepts of balanced and unbalanced data when conducting a test to compare the pairwise sample means for a given set of samples.

### Exercises

5. How many individual pairwise comparisons would need to be made if there are four population means in the ANOVA? What would be the probability of at least one Type I error if performing individual pairwise comparisons at a 0.01 significance level?

6. A two-sample $t$-test is conducted to test the pairwise differences in the mean number of candies consumed per family (average size of four family members) per day. The families belong to four different states. The following output is obtained (differences are computed in the given order of the states).

| | Null hypothesis | Difference in Means | $t$-Test Statistic | P-value |
|---|---|---|---|---|
| Alabama and Los Angeles | $(H_0: \mu_A - \mu_{LA} = 0)$ | 5.6378 | 2.3479 | 0.046835 |
| New York and Los Angeles | $(H_0: \mu_{NY} - \mu_{LA} = 0)$ | −12.5798 | 7.8741 | 0.000049 |
| Alabama and Texas | $(H_0: \mu_A - \mu_T = 0)$ | 41.2156 | 12.3721 | 0.000002 |
| New York and Texas | $(H_0: \mu_{NY} - \mu_T = 0)$ | 0.4132 | 1.1553 | 0.281305 |
| Texas and Los Angeles | $(H_0: \mu_T - \mu_{LA} = 0)$ | −24.8714 | 9.2496 | 0.000015 |
| Alabama and New York | $(H_0: \mu_A - \mu_{NY} = 0)$ | 32.6741 | 11.7420 | 0.000003 |

Assuming a significance level of $\alpha = 0.05$, answer the following questions.

   a. Is there evidence to conclude that, on average, families in Alabama consume more candies per day than families in New York?

   b. Which state appears to have the highest candy consumption per family per day according to the output.

7. Fisher's Least Significant Difference method examines the pairwise difference in the mean values of four treatment groups at a 0.05 level of significance. Determine the critical value if the total number of observations in all the samples is 30.

8. The number of paint defects found in a sample of 50 cars produced by three different car manufacturers (labeled A, B and C) are studied. The analysis of variance was significant at the 0.05 level indicating a difference in the average number of paint defects among the car manufacturers. Determine which car manufacturers are different using Fisher's Least Significant Difference method. Assume that the value calculated for Fisher's LSD is 4.4763, which is the same for each pair.

The following table shows the sample mean number of paint defects for each of the manufacturers.

| Manufacturer | Mean Number of Paint Defects |
|---|---|
| A | 7 |
| B | 12 |
| C | 9 |

9. The mean effect of three treatments on fasting blood sugar levels for three samples of 10 patients are shown below.

| Treatments | Mean Fasting Blood Glucose Levels (mg/dL) |
|---|---|
| A | 87.5 |
| B | 86.5 |
| C | 78.2 |

The ANOVA output for this experiment using R is as follows.

| | df | Sum of Squares | Mean Square | F-value | Pr(>F) |
|---|---|---|---|---|---|
| Treatment | 2 | 521.3 | 260.6 | 2.549 | 0.0968 |
| Residuals | 27 | 2760.6 | 102.2 | | |

Assuming the level of significance is $\alpha = 0.10$, compare the pairwise differences in the mean blood glucose level for the three treatments using Fisher's Least Significant Difference method.

10. List one advantage of Tukey's HSD method over the two-sample $t$-test when the pairwise differences between the sample means are to be examined.

11. Compute the studentized range value for conducting Tukey's HSD test when the level of significance is equal to 0.05, the number of treatments is equal to 4, and the sample size of each of the four samples is equal to 16.

12. The cholesterol level of a total of 45 subjects is measured. The subjects were randomly divided into three groups and given different doses of medication (0 mg, 5 mg. 10 mg).

The one-way ANOVA table for testing if there is a significant difference in the mean cholesterol level for the different doses of medication is shown below.

| | df | Sum of Squares | Mean Square | F-value | Pr(>F) |
|---|---|---|---|---|---|
| Dosage | 2 | 53402 | 26701 | 3.57566 | 0.036813 |
| Residuals | 42 | 313632 | 7467.42857 | | |

Is it wise to conduct a Tukey's HSD test to compare the difference in the mean cholesterol level at the following levels of significance?

a. 1%

b. 5%

13. Consider the test scores of a group of 15 students divided into three samples based on the type of curriculum studied. The following output is obtained after conducting a one-way ANOVA test.

| | df | Sum of Squares | Mean Square | F-value | Pr(>F) |
|---|---|---|---|---|---|
| Curriculum | 2 | 1301.7 | 650.9 | 28.18 | 2.93 E-05 |
| Residuals | 12 | 277.2 | 23.1 | | |

The mean scores for the three samples are tabulated below.

| Type of Curriculum | Mean Test Score |
|---|---|
| A | 87.2 |
| B | 76.6 |
| C | 64.4 |

Determine if the mean tests scores are different for the following curriculum types using the confidence interval approach for Tukey's HSD with a 0.05 level of significance.

a. Sample A and Sample B

b. Sample B and Sample C

# 12.5 Two-Way ANOVA: The Randomized Block Design

> **Definition**
>
> **Completely Randomized Design**
>
> A **completely randomized design** is a design that randomly assigns treatments to each of the study participants so that each participant has an equal chance of being assigned to each treatment.

An accounting firm wants to study auditor proficiency using three training methods for auditing: in-home training, on-site training, and off-site training. One approach in designing this study is to randomly select a number of auditors, say 30, and then randomly assign 10 auditors to each training method. At the end of the training period, the firm could then assign them to a case and measure their proficiency. We could then compare the average proficiency for each of the training methods using the $F$-test described in Section 12.3. This type of design is called a **completely randomized design**. Although this may be a perfectly reasonable approach, it does have some shortcomings. For example, what if all of the people selected for the off-site training method had more than 20 years of experience, whereas all of the people selected for in-home training were with the company less than two years? Certainly, a person's experience will have an effect on his or her proficiency to evaluate a case, regardless of the training method.

What can be done to reduce the effect of this factor that affects proficiency, but is not directly related to the training methods? One approach would be to divide our study participants into several experience bands called **blocks:** those people with one year of experience, those people with two years of experience, and so on. Then randomly select three people from each experience level (block) and assign each a different training method. Once the data have been collected, compare the average proficiency of each of the training methods for individuals in each of the experience bands (blocks).

This design is an extension of the paired difference design for comparing two means that we discussed in Chapter 11. Although the effect that current proficiency has on the proficiency after training has not been eliminated, it has probably been greatly reduced by this blocking. As with the paired difference design, we give up degrees of freedom when we create blocks. Thus, we must be sure that the reduction in variation among sample observations is enough to compensate for the loss of degrees of freedom. This type of design is an example of a **randomized block design.**

> **Definition**
>
> **Randomized Block Design**
>
> A **randomized block design** is a design that uses blocks such that the experimental units within the blocks are as alike as possible. The experimental units within each block are then randomly assigned to the treatments of interest.

In a randomized block design, we use blocks such that the experimental units within the blocks are as much alike as possible. Then we randomly assign the experimental units within each block to the $k$ populations or treatments of interest. Finally, we compare the response of the experimental units to each of the treatments of interest within each of the blocks. In this way, we eliminate possible variation due to some of the extraneous factors that are unrelated to the treatments. Table 12.5.1 gives the data collected by the accounting firm for the randomized block design. Note that the data in the body of the table are proficiency scores (out of 100).

| Block | Treatment | | |
|---|---|---|---|
| Years of Experience | In-Home Training | On-Site Training | Off-Site Training |
| 1 | 91 | 72 | 79 |
| 2 | 95 | 75 | 77 |
| 3 | 63 | 63 | 84 |
| 4 | 84 | 89 | 77 |
| 5 | 72 | 92 | 73 |
| 6 | 85 | 94 | 68 |
| 7 | 72 | 89 | 76 |
| 8 | 79 | 92 | 92 |
| 9 | 87 | 71 | 91 |
| 10 | 76 | 70 | 68 |

Table 12.5.1 – Randomized Block Design

Our interest is still in comparing the population (or treatment) means. For the accounting firm, the treatments of interest are the training methods. Thus, the null and alternative hypotheses are the same as they were in the $F$-test.

$H_0$: $\mu_1 = \mu_2 = \mu_3$ The average proficiency resulting from each of the training methods is the same.

$H_a$: At least one $\mu_i$ is different.

The test statistic appears to be exactly the same as that used for the $F$-test.

$$F = \frac{MST}{MSE}$$

However, the denominator of the $F$-statistic is calculated differently because it takes into account the fact that we were able to eliminate some of the variation among our sample responses by blocking the experimental units. The total sum of squares (TSS) and the degrees of freedom that we discussed in Section 12.1 can be broken down further as follows. Note that in the equations below **SST** is the **sum of squares for treatments**, **SSBL** is the **sum of squares for blocks**, and **SSE** is the **sum of squares for error**.

**Sum of Squares:** $\quad TSS = SST + SSBL + SSE$

**Degrees of Freedom:** $\quad n-1 = (k-1) + (b-1) + (b-1)(k-1)$

where $n$ is the total number of observations,

$k$ is the number of treatments, and

$b$ is the number of blocks.

If any of the observed variation in the sample observations has been reduced by blocking, then the SSE is reduced. But it is important to note that the degrees of freedom that are used by the SSBL are taken from the SSE; thus, the reduction in variation achieved by blocking must offset this loss in degrees of freedom. If we are successful at significantly reducing variation by blocking and there is a difference among the sample means, we will be more likely to detect it. The test procedure for the randomized block design is outlined in the following box.

## Procedure

### Two-Way ANOVA: Randomized Block Design

**Hypotheses:**

$H_0$: $\mu_1 = \mu_2 = \mu = \mu_k$ The $k$ population means are equal.

$H_a$: At least one $\mu_i$ is different.

**Test Statistic:**

$$F = \frac{MST}{MSE} = \frac{\dfrac{SST}{k-1}}{\dfrac{SSE}{(k-1)(b-1)}}$$

where $SSE = TSS - SST - SSBL$.

**Assumptions:**

The differences in observed responses to treatments for blocked units are normally distributed with equal variances.

**Rejection Region:**

$H_0$ will be rejected for large values of $F = \dfrac{MST}{MSE}$. In particular, we will reject $H_0$ if $F \geq F_\alpha$ with $(k-1)$ numerator degrees of freedom and $(k-1)(b-1)$ denominator degrees of freedom.

The calculation formulas for the $F$-test statistic in the randomized block design are beyond the scope of this text and, therefore, will not be presented in this chapter. We will assume that the reader has access to a statistical package that will produce the results of the $F$-test for a randomized block design. For reference, the general form of an ANOVA table for a randomized block design is given in Table 12.5.2.

| Table 12.5.2 – ANOVA Summary Table for a Randomized Block Design | | | | |
|---|---|---|---|---|
| Source of Variation | SS | df | MS | F |
| Block | SSBL | $b-1$ | MSBL | $\dfrac{\text{MSBL}}{\text{MSE}}$ |
| Treatment | SST | $k-1$ | MST | $\dfrac{\text{MST}}{\text{MSE}}$ |
| Error | SSE | $(b-1)(k-1)$ | MSE | |
| Total | TSS | $n-1$ | | |

The results of the accounting firm's test are given in Figure 12.5.1.

**Anova: Two-Factor Without Replication**

| SUMMARY | Count | Sum | Average | Variance |
|---|---|---|---|---|
| 1 | 3 | 242 | 80.66667 | 92.33333 |
| 2 | 3 | 247 | 82.33333 | 121.3333 |
| 3 | 3 | 210 | 70 | 147 |
| 4 | 3 | 250 | 83.33333 | 36.33333 |
| 5 | 3 | 237 | 79 | 127 |
| 6 | 3 | 247 | 82.33333 | 174.3333 |
| 7 | 3 | 237 | 79 | 79 |
| 8 | 3 | 263 | 87.66667 | 56.33333 |
| 9 | 3 | 249 | 83 | 112 |
| 10 | 3 | 214 | 71.33333 | 17.33333 |
| In-Home | 10 | 804 | 80.4 | 96.48889 |
| On-Site | 10 | 807 | 80.7 | 133.3444 |
| Off-Site | 10 | 785 | 78.5 | 70.05556 |

| ANOVA | | | | | | |
|---|---|---|---|---|---|---|
| Source of Variation | SS | df | MS | F | P-value | F crit |
| Rows | 801.4667 | 9 | 89.05185 | 0.844746 | 0.586668 | 2.456281 |
| Columns | 28.46667 | 2 | 14.23333 | 0.135017 | 0.874577 | 3.554557 |
| Error | 1897.533 | 18 | 105.4185 | | | |
| Total | 2727.467 | 29 | | | | |

**Figure 12.5.1**

The $F$-statistic has numerator degrees of freedom equal to 2 and denominator degrees of freedom equal to 18. Thus, for $\alpha = 0.05$, the $F$-critical value is approximately 3.5546. The calculated value of the test statistic is given by

$$F = \frac{\text{MST}}{\text{MSE}} \approx \frac{14.2333}{105.4185} \approx 0.1350.$$

Identify these values in Figure 12.5.1. MST and MSE are given in the *MS* column, the $F$-statistic is given in the *F* column, and the critical value is given in the *F crit* column.

## Technology

For instructions on performing an ANOVA test with two factors visit stat.hawkeslearning.com and navigate to **Discovering Business Statistics, Second Edition > Technology Instructions > ANOVA > Two-Way**.

## NOTE

For the computer output shown in Figure 12.5.1, "Rows" corresponds to the blocks, since in Table 12.5.1 each row represented a block. "Columns" corresponds to the treatments, since the treatments were organized in columns. When performing an ANOVA in Excel or any other statistical software, it is important to be aware of whether the blocks and treatments are organized in rows or columns to accurately interpret the results.

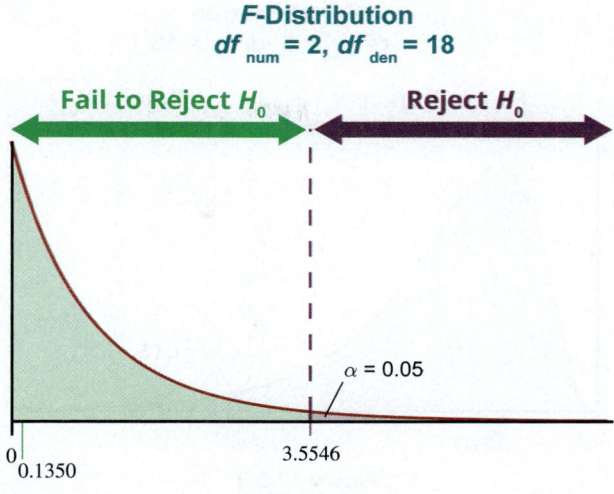

**Figure 12.5.2**

As shown in Figure 12.5.2, the value of our test statistic, 0.1350, is less than the $F$-critical value of 3.5546. Therefore, the accounting firm will fail to reject the null hypothesis. Also note that the $P$-value is given in the output as approximately 0.8746. Since the $P$-value is greater than the level of significance, 0.05, the accounting firm will fail to reject the null hypothesis.

There is not sufficient evidence to conclude that there is a significant difference in average proficiency among the three training programs. Because the accounting firm failed to reject the null hypothesis, they may be concerned that they were not able to significantly reduce the variation by blocking.

To determine if the blocking was successful at reducing variation among the sample observations, we use the test statistic $F = \dfrac{\text{MSBL}}{\text{MSE}}$. Under the null hypothesis that the block means are the same (we were not successful in reducing variation by blocking because the block means were not significantly different), the $F$-test statistic has an $F$-distribution with $(b-1)$ numerator degrees of freedom and $(k-1)(b-1)$ denominator degrees of freedom.

At $\alpha = 0.05$, the accounting firm will reject the null hypothesis that block means are all equal (blocking was unsuccessful) if the calculated value of the test statistic is larger than the $F$-critical value with 9 numerator degrees of freedom and 18 denominator degrees of freedom, which is approximately 2.4563. The calculated value of the test statistic is given by

$$F = \frac{\text{MSBL}}{\text{MSE}} \approx \frac{89.0519}{105.4185} \approx 0.8447 .$$

Again, MSBL and MSE are given in the $MS$ column, the $F$-statistic is given in the $F$ column, and the critical value is given in the $F$ $crit$ column in Figure 12.5.1.

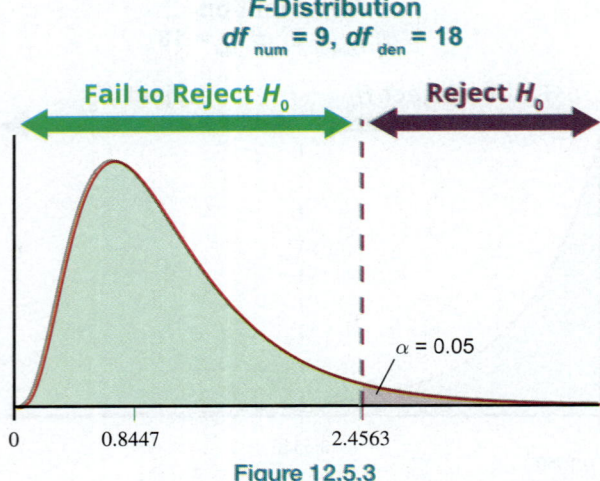

**Figure 12.5.3**

Figure 12.5.3 displays the rejection region and the calculated value of the test statistic. Since 0.8447 is less than 2.4563, we fail to reject the null hypothesis that the block means are all equal. This decision can also be made using the reported $P$-value, 0.5867, which is greater than the level of significance, 0.05. The firm was unable to reduce a significant amount of variation among sample observations by blocking.

The hypothesis test just presented is one example of a **two-way analysis of variance (two-way ANOVA)**. This means there are two independent factors considered in the analysis, namely the blocks (the different experience levels) and the treatments (training methods), and treatment was applied once within each block. Note that the researcher's primary interest was to determine if there was a significant difference between training methods. However, the researcher wanted to reduce variation in the model by adding the blocking factor (experience). The researcher was not interested in studying the effect of experience on audit proficiency. Section 12.6 will discuss situations in which the researcher is interested in the effects of both factors.

> **Definition**
>
> **Two-Way Analysis of Variance**
>
> A **two-way analysis of variance** is an experiment or study in which the means of two independent factors are compared.

**Example 12.5.1**

**Performing an ANOVA Test with Blocking on Media Type**

Suppose that in addition to studying screen time by age group (Tweens, Teens, and Adults), data were also collected indicating the particular media type on which the participants spend their time. For this experiment, five participants were randomly assigned to each age group-media type combination. There were seven media types (watching tv, gaming, listening to music, reading, browsing websites, using social media, and video chatting) used. However, we are more interested in just the screen time and consider the media type as a nuisance parameter. That is, the media type may add unnecessary variation to the model. One approach to reducing the effect of media type on the screen time is to use blocking. Our interest is still in comparing the average screen time between age groups which will serve as our treatments. Based on the data in Table 12.5.3, can you conclude that there is a significant difference in average screen time among the age groups? Use a significance level of 0.05.

**⠿ Data**

This data set can be found on stat.hawkeslearning.com by navigating to **Discovering Business Statistics, Second Edition > Data Sets > Screen Time by Age Group and Media Type**.

| Table 12.5.3 – Screen Time by Age Group and Media Type (Minutes per Day) | | | |
|---|---|---|---|
| **Media Type** | **Age Group** | | |
| | **8–12 Years Old** | **13–18 Years Old** | **Over 18 Years Old** |
| **Watching TV** | 135 | 99 | 94 |
| | 112 | 100 | 84 |
| | 138 | 129 | 109 |
| | 142 | 127 | 92 |
| | 105 | 121 | 97 |

| Table 12.5.3 – Screen Time by Age Group and Media Type (Minutes per Day) (cont.) | | | |
|---|---|---|---|
| **Media Type** | **Age Group** | | |
| | **8–12 Years Old** | **13–18 Years Old** | **Over 18 Years Old** |
| Gaming | 98 | 149 | 63 |
| | 136 | 155 | 90 |
| | 100 | 136 | 126 |
| | 156 | 129 | 88 |
| | 141 | 127 | 122 |
| Listening to Music | 150 | 155 | 107 |
| | 116 | 84 | 101 |
| | 136 | 195 | 81 |
| | 165 | 113 | 105 |
| | 141 | 101 | 115 |
| Reading | 107 | 164 | 98 |
| | 124 | 162 | 127 |
| | 138 | 121 | 126 |
| | 111 | 147 | 112 |
| | 106 | 209 | 97 |
| Browsing Websites | 120 | 118 | 79 |
| | 130 | 143 | 98 |
| | 110 | 143 | 102 |
| | 146 | 165 | 122 |
| | 145 | 148 | 106 |
| Using Social Media | 155 | 135 | 120 |
| | 144 | 131 | 99 |
| | 165 | 125 | 100 |
| | 137 | 125 | 66 |
| | 142 | 144 | 103 |
| Video Chatting | 83 | 126 | 132 |
| | 165 | 134 | 101 |
| | 145 | 138 | 110 |
| | 151 | 106 | 90 |
| | 109 | 159 | 113 |

### SOLUTION

The null and alternative hypotheses will be the same as in Example 12.2.1, given by

$H_0$: $\mu_1 = \mu_2 = \mu_3$

$H_a$: At least one $\mu_i$ is different.

The test statistic appears to be exactly the same as that used for the $F$-test which is

$$F = \frac{\text{MST}}{\text{MSE}}$$

However, the denominator of the $F$-statistic is calculated differently because it takes into account the fact that we will eliminate some of the variation among our samples by blocking the experimental units. The total sum of squares (TSS) and the degrees of freedom will be partitioned as SST (sum of squares for treatments), SSBL (sum of squares for blocks), and SSE (sum of squares for error). The partitioning of the sum of squares and degrees of freedom are shown below.

$$\textbf{Sum of Squares: } \text{TSS} = \text{SST} + \text{SSBL} + \text{SSE}$$
$$\textbf{Degrees of Freedom: } n-1 = (k-1) + (b-1) + (b-1)(k-1)$$

where $n = kb$ = total number of observations, $k$ = number of treatments, and $b$ = number of blocks. However, in situations where there is replication (i.e., we have more than one observation for each treatment-block combination), represented by $r$, the degrees of freedom are partitioned as follows:

**Degrees of Freedom with Replication**: $rkb - 1 = (k-1) + (b-1) + (n-k-b+1)$

where $n = rkb$.

We used JMP to obtain the output in the table below.

**Technology**

For instructions on performing an ANOVA test with two factors visit stat.hawkeslearning.com and navigate to **Discovering Business Statistics, Second Edition > Technology Instructions > ANOVA > Two-Way**.

Analysis of Variance

| Source | DF | Sum of Squares | Mean Square | F-Ratio | Prob > F |
|---|---|---|---|---|---|
| Model | 8 | 26511.600 | 3313.95 | 7.2277 | < 0001* |
| Error | 96 | 44016.457 | 458.50 | | |
| C. Total | 104 | 70528.057 | | | |

Effects Tests

| Source | Nparm | DF | Sum of Squares | F-Ratio | Prob > F |
|---|---|---|---|---|---|
| Media Type | 6 | 6 | 2745.257 | 0.9979 | 0.4313 |
| Age Group | 2 | 2 | 23766.343 | 25.9172 | < 0001* |

**Figure 12.5.4**

Note that in the Analysis of Variance table, the P-value (indicated by *Prob > F*) is less than 0.0001 which indicates that the overall model is significant. More importantly, the "Effects Tests" table shows that Media Type, which is the blocking factor, is not significant (P-value = 0.4313) but the Age Group is significant (P-value < 0.0001). Thus, we reject the null hypothesis and conclude that there is a significant difference in the average screen time spent among the age groups.

Now that we know that we have rejected the null hypothesis and know that there is a difference in screen time among the age groups, the natural question is, *Which of the three age groups is different*? This is where we use multiple comparison procedures. For this particular example, the JMP output below uses Tukey's HSD to compare the average screen time of the three age groups. JMP produces a connected letters report to compare means. Levels (or age groups in our example) that are not connected by the same letter are significantly different. Thus, we see that the 8–12 Years Old and the 13–18 Years Old age groups have the letter A, indicating that they are not significantly different. However, the Over 18 Years Old age group has the letter B, indicating that this age group is significantly different from the other two age groups.

**LSMeans Differences Tukey HSD**

$\alpha = 0.050$  Q = 2.38063

|  | | LSMean[j] | |
|---|---|---|---|
| Mean[i]-Mean[j]<br>Std Err Dif<br>Lower CL Dif<br>Upper CL Dif | 8-12 Years Old | 13-18 Years Old | Over 18 Years Old |
| 8-12 Years Old | 0 | -4.5429 | 29.4 |
|  | 0 | 5.11862 | 5.11862 |
|  | 0 | -16.728 | 17.2145 |
|  | 0 | 7.64267 | 41.5855 |
| 13-18 Years Old | 4.54286 | 0 | 33.9429 |
|  | 5.11862 | 0 | 5.11862 |
|  | -7.6427 | 0 | 21.7573 |
|  | 16.7284 | 0 | 46.1284 |
| Over 18 Years Old | -29.4 | -33.943 | 0 |
|  | 5.11862 | 5.11862 | 0 |
|  | -41.586 | -46.128 | 0 |
|  | -17.214 | -21.757 | 0 |

(row label on left axis: LSMean[i])

| Level | | Least Sq Mean |
|---|---|---|
| 13-18 Years Old | A | 136.08571 |
| 8-12 Years Old | A | 131.54286 |
| Over 18 Years Old | B | 102.14286 |

Levels not connected by same letter are significantly different.

**Figure 12.5.5**

# 📝 12.5 Exercises

## Basic Concepts

1. What is a completely randomized design? Give an example.

2. Identify a shortcoming that could arise from using a completely randomized design for the example you gave in Exercise 1.

3. What are blocks? What is their purpose?

4. What is a randomized block design? How is it different from a completely randomized design?

5. What are the null and alternative hypotheses when comparing means using a randomized block design?

6. What is the test statistic for the hypothesis test described in Exercise 5?

7. What is the breakdown of the sum of squares for a randomized block design? Does this breakdown make sense? Explain.

8. How are the corresponding degrees of freedom for TSS, SST, SSBL, and SSE related? Verify that this relationship is true.

9. If blocking is successful, how does the value of SSE change?

10. What are the assumptions when performing a two-way ANOVA for a randomized block design?

11. What is the rejection region for the test when performing a two-way ANOVA for a randomized block design?

12. What are the degrees of freedom associated with the test statistic for the two-way ANOVA described in Exercise 11?

## Exercises

13. A car dealer is interested in comparing the average gas mileages of four different car models. The dealer believes that the average gas mileage of a particular car will vary depending on the person who is driving the car due to different driving styles. Because of this, he decides to use a randomized block design. He randomly selects six drivers and asks them to drive each of the cars. He then determines the average gas mileage for each car and each driver. The results of the study are as follows.

| Gas Mileage (MPG) | | | | |
|---|---|---|---|---|
| | Car A | Car B | Car C | Car D |
| Driver 1 | 33 | 29 | 27 | 37 |
| Driver 2 | 36 | 32 | 30 | 40 |
| Driver 3 | 34 | 30 | 28 | 38 |
| Driver 4 | 31 | 27 | 25 | 35 |
| Driver 5 | 33 | 29 | 27 | 37 |
| Driver 6 | 35 | 33 | 31 | 41 |

a. Do you think a randomized block design is appropriate for the car dealer's study? Explain.

b. The results of the two-way ANOVA for the dealer's survey of the average gas mileages of the different car models are given in the following table.

ANOVA

| Source of Variation | SS | df | MS |
|---|---|---|---|
| Rows | 84.8333 | 5 | 16.9667 |
| Columns | 348.5000 | 3 | 116.1667 |
| Error | 2.5000 | 15 | 0.1667 |
| | | | |
| Total | 435.8333 | 23 | |

Can the dealer conclude that there is a significant difference in average gas mileages of the four car models? Use $\alpha = 0.05$.

c. Was the dealer able to significantly reduce variation among the observed gas mileages by blocking? Use $\alpha = 0.05$.

14. A banana grower has three fertilizers from which to choose. He would like to determine which fertilizer produces banana trees with the largest yield (measured in pounds of bananas produced). The banana grower has noticed that there is a difference in the average yields of the banana trees depending on which side of the farm they are planted (South Side, North Side, West Side, or East Side). Because of the variation in yields among the areas on the farm, the farmer has decided to randomly select three trees within each area and then randomly assign the fertilizers to the trees. After harvesting the bananas, he calculates the yields of the trees within each of the areas. The results are as follows.

| Banana Yields (Pounds) | | | |
|---|---|---|---|
| | Fertilizer A | Fertilizer B | Fertilizer C |
| South Side | 53 | 51 | 58 |
| North Side | 48 | 47 | 53 |
| West Side | 50 | 48 | 56 |
| East Side | 50 | 47 | 54 |

**a.** Do you think a randomized block design is appropriate for the banana grower's study? Explain.

**b.** The results of the two-way ANOVA for the banana grower's study are given in the following table.

ANOVA

| Source of Variation | SS | df | MS |
|---|---|---|---|
| Rows | 36.2500 | 3 | 12.0833 |
| Columns | 104.0000 | 2 | 52.0000 |
| Error | 2.0000 | 6 | 0.3333 |
| Total | 142.2500 | 11 | |

Can the banana grower conclude that there is a significant difference among the average yields of the banana trees for the three fertilizers? Use $\alpha = 0.10$.

**c.** Was the banana grower able to significantly reduce variation among the observed yields by blocking? Use $\alpha = 0.10$.

**15.** The FAA is interested in knowing if there is a difference in the average numbers of on-time arrivals for four of the major airlines. The FAA believes that the number of on-time arrivals varies by airport. To control for this variation, they randomly select 100 flights for each of the major airlines at each of four randomly selected airports and record the number of on-time flights. The results of the study are as follows.

| On-Time Flights | | | |
|---|---|---|---|
| | Airline A | Airline B | Airline C | Airline D |
| Airport A | 87 | 82 | 79 | 81 |
| Airport B | 88 | 84 | 81 | 82 |
| Airport C | 89 | 84 | 83 | 82 |
| Airport D | 90 | 86 | 85 | 83 |

**a.** Do you think a randomized block design is appropriate for the FAA's study? Explain.

**b.** The results of the two-way ANOVA for the FAA's study are given in the following table.

ANOVA

| Source of Variation | SS | df | MS |
|---|---|---|---|
| Rows | 29.2500 | 3 | 9.7500 |
| Columns | 112.7500 | 3 | 37.5833 |
| Error | 5.7500 | 9 | 0.6389 |
| Total | 147.7500 | 15 | |

Can the FAA conclude that there is a significant difference among the average number of on-time arrivals for the four major airlines? Use $\alpha = 0.01$.

**c.** Was the FAA able to significantly reduce variation among the observed number of on-time arrivals by blocking? Use $\alpha = 0.01$.

16. A psychologist is interested in determining if there is a difference in the average numbers of suicides for several age groups. The psychologist believes that there may be some variation in the numbers of suicides depending on the region of the country (Northeast, Northwest, Southeast, or Southwest). The psychologist randomly selects 100,000 deaths from each region of the country for each of the age groups of interest and determines the number of suicides. The results of the study are as follows.

| Suicides | | | | | | | |
|---|---|---|---|---|---|---|---|
| | Age 15–24 | Age 25–34 | Age 35–44 | Age 45–54 | Age 55–64 | Age 65–74 | Age 75–84 |
| Northeast | 15 | 17 | 16 | 17 | 55 | 22 | 27 |
| Northwest | 13 | 16 | 16 | 16 | 49 | 19 | 26 |
| Southeast | 12 | 14 | 15 | 15 | 47 | 17 | 24 |
| Southwest | 13 | 15 | 15 | 16 | 53 | 20 | 25 |

a. Do you think a randomized block design is appropriate for the psychologist's study? Explain.

b. The results of the two-way ANOVA for the psychologist's study are given in the following table.

ANOVA

| Source of Variation | SS | df | MS |
|---|---|---|---|
| Rows | 44.9643 | 3 | 14.9881 |
| Columns | 4223.3571 | 6 | 703.8929 |
| Error | 25.7857 | 18 | 1.4325 |
| | | | |
| Total | 4294.1071 | 27 | |

Can the psychologist conclude that there is a significant difference among the average number of suicides for the different age groups? Use $\alpha = 0.10$.

c. Was the psychologist able to significantly reduce variation among the observed number of suicides by blocking? Use $\alpha = 0.05$.

17. In an experiment designed to compare automated blood pressure devices with those of the standard cuff method, each man in a sample of six patients has his systolic blood pressure determined by three different automated devices and by the standard cuff method. The data are given in the following table.

| Blood Pressure (mmHg) | | | |
|---|---|---|---|
| | Device 1 | Device 2 | Device 3 | Standard Cuff |
| Patient 1 | 126 | 128 | 132 | 131 |
| Patient 2 | 134 | 138 | 137 | 140 |
| Patient 3 | 145 | 144 | 150 | 152 |
| Patient 4 | 129 | 134 | 132 | 136 |
| Patient 5 | 154 | 160 | 162 | 160 |
| Patient 6 | 144 | 144 | 148 | 145 |

a. Why was a randomized block design used in this experiment?

b. From the data, SST and SSE were computed to be 106.4583 and 53.2917, respectively. With $\alpha = 0.05$, can we conclude that the four different methods of determining systolic blood pressure have different mean readings?

c. SSBL was computed to be 2412.8750. With $\alpha = 0.05$, can we conclude that using people as blocks significantly reduced variation in this study?

# 12.6 Two-Way ANOVA: The Factorial Design

The techniques that we discussed for a randomized block design can be extended to the situation where there are two factors of interest. A director of personnel might be interested in relating average salary to two factors: age and experience. A supervisor at a manufacturing plant might be interested in relating the average number of defective products to two factors: operator and machine. A doctor might be interested in relating the increase in average patient heart rate to two factors: medication and age.

In order for the director of personnel to evaluate the relationship between average salary, age, and experience, he chooses the following experimental design. He selects four different age groups and three different experience levels and observes two salaries for each of the possible combinations of age and experience. The resulting data are displayed in Table 12.6.1.

| Table 12.6.1 – Salaries (Thousands of Dollars) | | | | |
|---|---|---|---|---|
| | **Age** | | | |
| **Years of Experience** | **25–34** | **35–44** | **45–54** | **55–64** |
| 0–4 | 22 | 25 | 34 | 37 |
| | 27 | 35 | 36 | 43 |
| 5–9 | 34 | 35 | 42 | 49 |
| | 36 | 45 | 48 | 51 |
| 10–14 | 39 | 40 | 53 | 51 |
| | 41 | 50 | 57 | 59 |

Again, this type of design involves a **two-way analysis of variance** because there are two classifications. It is called a **complete factorial experiment** since there is at least one observation for every possible combination of age and years of experience. Factorial experiments provide valuable information by enabling the interaction between the two variables to be estimated.

Interaction between the two variables means that the average salary is affected by the combination of age and experience. An example of two variables that interact is shown in Figure 12.6.1. This is called a **profile plot** (or **interaction plot**) which plots the means by factors. If there is no interaction at all between two variables, the lines in the profile plot will be perfectly parallel. An example of two variables that do not interact is shown in Figure 12.6.2. A similar graph for the salary data is shown in Figure 12.6.3. Based on this graph, there appears to be slight interaction between age and experience when age is between 45 and 65 years and experience is between 10 and 14 years.

> **Definition**
>
> **Complete Factorial Experiment**
>
> A **complete factorial experiment** is an experiment involving at least two factors with at least one observation for every possible combination of the factor levels.

## Profile Plot - Variables That Interact

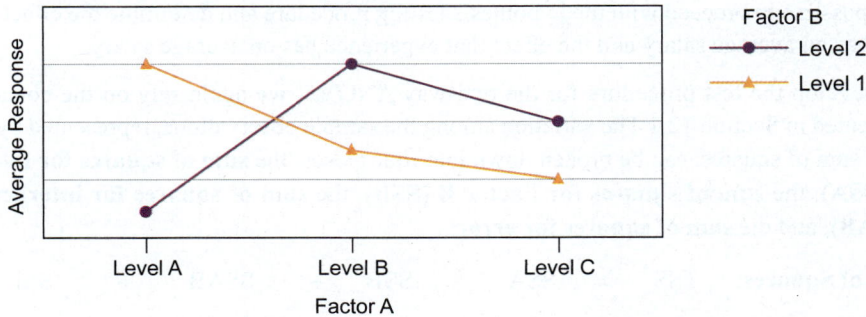

Figure 12.6.1

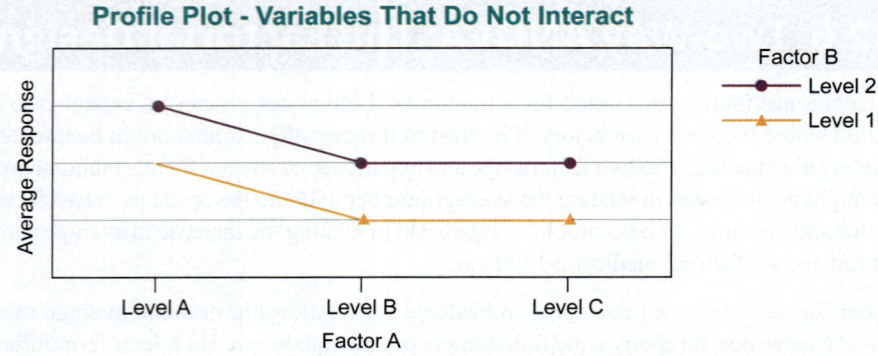

**Figure 12.6.2**

There are three effects on average salary that will interest the personnel director. The first is the effect that the interaction between age and experience has on average salary, called the **main effect for interaction**. The second is the effect that experience has on average salary, called the **main effect for experience (Factor A)**. The third is the effect that age has on average salary, called the **main effect for age (Factor B)**.

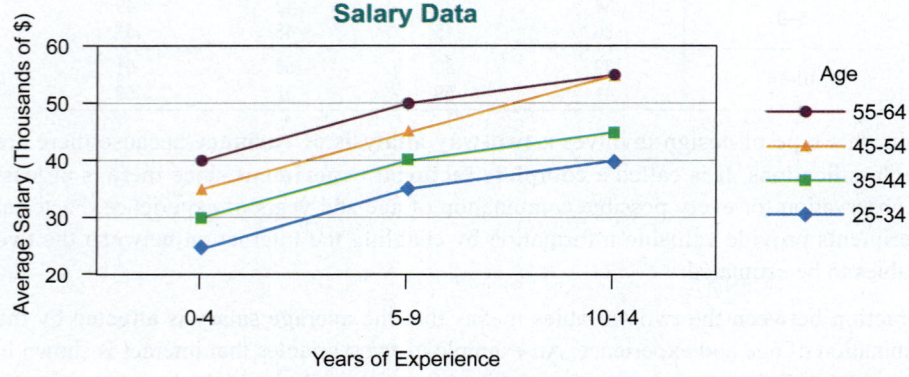

**Figure 12.6.3**

The test procedure is somewhat different than those we have discussed previously because of the potential presence of interaction between the two variables. Of primary importance is determining whether or not there is *any* interaction between the two variables. If there is interaction, we will not be able to separate out the effects that age and experience have on average salary, and the hypothesis testing procedure is halted. If there is not interaction, then it is possible to proceed with the hypothesis testing procedure and determine the effect that age has on average salary and the effect that experience has on average salary.

To develop the test procedure for the two-way ANOVA, we again rely on the concepts presented in Section 12.1 The variation among the sample observations, represented by the total sum of squares, can be broken down into four pieces: the **sum of squares for Factor A (SSA)**, the **sum of squares for Factor B (SSB)**, the **sum of squares for interaction (SSAB)**, and the **sum of squares for error**.

**Sum of Squares:**    $\text{TSS} = \text{SSA} + \text{SSB} + \text{SSAB} + \text{SSE}$

**Degrees of Freedom:**    $n-1 = a-1 + b-1 + (a-1)(b-1) + ab(r-1)$

where

$n$ is the total number of observations,

$a$ is the number of levels of Factor A,

*b* is the number of levels of Factor B, and

*r* is the number of observations in the combinations of levels of Factor A and Factor B (i.e., the number of observations in each cell).

The test statistics for a two-way ANOVA for a factorial design are derived by dividing the sums of squares by the appropriate degrees of freedom to produce mean squares, and then dividing each of the respective mean squares by the mean square for error. The test procedures are outlined in the following boxes.

## Procedure

### Test for Interaction Between Factors

**Hypotheses:**

$H_0$: There is no interaction between Factor A and Factor B.

$H_a$: There is interaction between Factor A and Factor B.

**Test Statistic:**

$$F = \frac{\dfrac{SSAB}{(a-1)(b-1)}}{\dfrac{SSE}{ab(r-1)}} = \frac{MSAB}{MSE}$$

**Rejection Region:**

Reject the null hypothesis if $F \geq F_a$ with $(a-1)(b-1)$ numerator degrees of freedom and $ab(r-1)$ denominator degrees of freedom.

**Note:** If the null hypothesis is rejected, then interaction exists. Do not proceed with the main effects tests for Factor A and Factor B if this is the case.

## Procedure

### Test for Main Effects for Factor A

**Hypotheses:**

$H_0$: Factor A has no effect on average response.

$H_a$: Factor A has an effect on average response.

**Test Statistic:**

$$F = \frac{\dfrac{SSA}{(a-1)}}{\dfrac{SSE}{ab(r-1)}} = \frac{MSA}{MSE}$$

**Rejection Region:**

Reject the null hypothesis if $F \geq F_a$ with $(a-1)$ numerator degrees of freedom and $ab(r-1)$ denominator degrees of freedom.

**Procedure**

**Test for Main Effects for Factor B**

**Hypotheses:**

$H_0$: Factor B has no effect on average response.

$H_a$: Factor B has an effect on average response.

**Test Statistic:**

$$F = \frac{\dfrac{SSB}{(b-1)}}{\dfrac{SSE}{ab(r-1)}} = \frac{MSB}{MSE}$$

**Rejection Region:**

Reject the null hypothesis if $F \geq F_a$ with $(b-1)$ numerator degrees of freedom and $ab(r-1)$ denominator degrees of freedom.

The calculations of the $F$-statistics for a two-way analysis of variance are beyond the scope of this text. We will assume that the reader has access to a statistical software package that will produce the results of a two-way ANOVA. For reference, the general form of the ANOVA table for a factorial experiment is given in Table 12.6.2.

### Table 12.6.2 – ANOVA Summary Table for a Factorial Experiment

| Source of Variation | SS | df | MS | F |
|---|---|---|---|---|
| Factor A | SSA | $a-1$ | MSA | $\dfrac{MSA}{MSE}$ |
| Factor B | SSB | $b-1$ | MSB | $\dfrac{MSB}{MSE}$ |
| Interaction | SSAB | $(a-1)(b-1)$ | MSAB | $\dfrac{MSAB}{MSE}$ |
| Error | SSE | $ab(r-1)$ | MSE | |
| Total | TSS | $n-1$ | | |

**Technology**

For instructions on performing an ANOVA test with two factors visit stat.hawkeslearning.com and navigate to **Discovering Business Statistics, Second Edition > Technology Instructions > ANOVA > Two-Way**.

The results of the two-way ANOVA for the personnel director's salary data are given in Figure 12.6.4.

ANOVA

| Source of Variation | SS | df | MS | F | P-value | F crit |
|---|---|---|---|---|---|---|
| Sample | 1092.583 | 2 | 546.2917 | 26.59432 | 3.89E-05 | 3.885294 |
| Columns | 828.4583 | 3 | 276.1528 | 13.44354 | 0.000382 | 3.490295 |
| Interaction | 24.41667 | 6 | 4.069444 | 0.198107 | 0.971006 | 2.99612 |
| Within | 246.5 | 12 | 20.54167 | | | |
| Total | 2191.958 | 23 | | | | |

**Figure 12.6.4**

**Note:** In the ANOVA table given in Figure 12.6.4, "Sample" corresponds to Experience (Factor A) since the experience levels were organized by row. "Columns" corresponds to Age (Factor B) since the age groups are organized by column. "Interaction" corresponds to the interaction between experience and age, and "Within" corresponds to the variation within the sample observations, or the error. When using Excel or other statistical software programs to perform a two-way ANOVA, it is important to pay close attention to how the data are organized so that you can identify Factor A and Factor B and accurately interpret the results.

The personnel director must first decide if there is interaction between age and experience. The appropriate test statistic for testing for interaction, which is the mean square of the interaction term divided by the mean square for error, is given by

$$F = \frac{\text{MSAB}}{\text{MSE}}.$$

Under the null hypothesis that there is no interaction, the $F$-test statistic has an $F$-distribution with 6 numerator degrees of freedom and 12 denominator degrees of freedom. At $\alpha = 0.05$, the null hypothesis will be rejected if the calculated value of the test statistic is greater than or equal to 2.9961.

### ∾ Technology

For instructions on finding *F* critical values using technology visit stat.hawkeslearning.com and navigate to **Discovering Business Statistics, Second Edition > Technology Instructions > *F*-Distribution > Critical Value**.

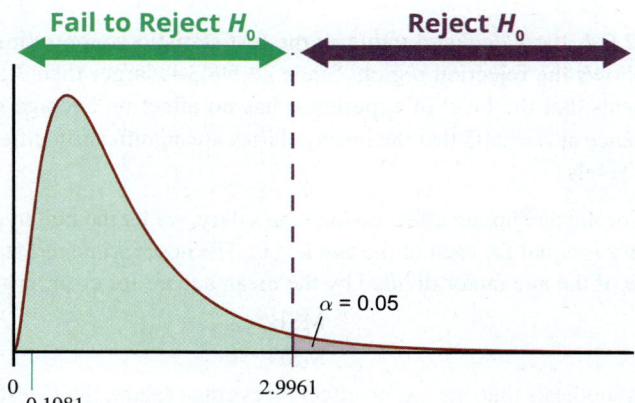

**F-Distribution**
$df_{num} = 6$, $df_{den} = 12$

**Fail to Reject $H_0$**          **Reject $H_0$**

$\alpha = 0.05$

0          2.9961
0.1981

**Figure 12.6.5**

From Figure 12.6.4, the calculated value of the test statistic is approximately 0.1981. Figure 12.6.5 shows the rejection region and the calculated value of the test statistic. Since 0.1981 is less than 2.9961, we fail to reject the null hypothesis that there is no interaction between experience and age. This is what the personnel director expected based on Figure 12.6.3. Because there is not evidence of significant interaction between age and experience, it is safe to use the $F$-test statistic to test whether or not each of the factors, age and experience, has an effect on average salary.

To test whether or not experience has an effect on average salary, we let the null hypothesis be that the average salary is equal for each of the experience levels. The appropriate test statistic, which is the mean square of the experience factor divided by the mean square for error, is given by

$$F = \frac{\text{MSA}}{\text{MSE}}.$$

Under the null hypothesis that level of experience has no effect on average salary, the $F$-test statistic has an $F$-distribution with 2 numerator degrees of freedom and 12 denominator degrees of freedom. At $\alpha = 0.05$, the null hypothesis will be rejected if the calculated value of the test statistic is greater than or equal to 3.8853.

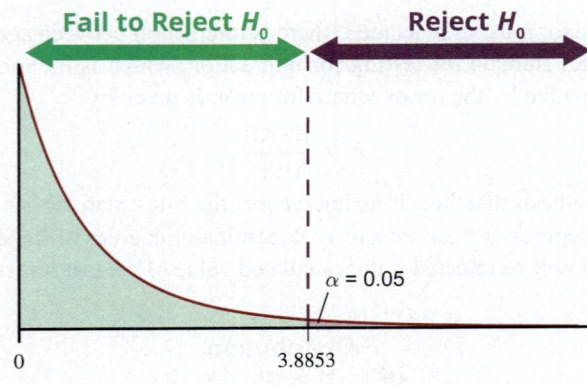

**Figure 12.6.6**

From Figure 12.6.4, the calculated value of the test statistic is approximately 26.5943. Figure 12.6.6 shows the rejection region. Since 26.5943 is larger than 3.8853, we reject the null hypothesis that the level of experience has no effect on average salary. There is persuasive evidence at $\alpha = 0.05$ that the mean salaries are significantly different for at least two experience levels.

To test whether or not age has an effect on average salary, we let the null hypothesis be that the average salary is equal for each of the age levels. The appropriate test statistic, which is the mean square of the age factor divided by the mean square for error, is given by

$$F = \frac{\text{MSB}}{\text{MSE}}.$$

Under the null hypothesis that age has no effect on average salary, the $F$-test statistic has an $F$-distribution with 3 numerator degrees of freedom and 12 denominator degrees of freedom. At $\alpha = 0.05$, the null hypothesis will be rejected if the calculated value of the test statistic is greater than or equal to 3.4903.

From Figure 12.6.4, the calculated value of the test statistic is approximately 13.4435. Figure 12.6.7 shows the rejection region. Since 13.4435 is larger than 3.4903, we reject the null hypothesis that age has no effect on average salary. There is persuasive evidence at $a = 0.05$ that there is a significant difference between the mean salaries for at least two age levels.

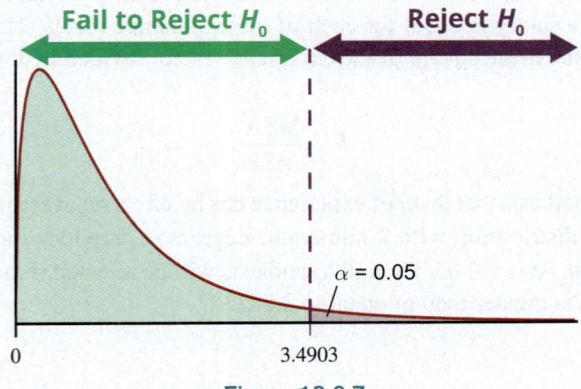

**Figure 12.6.7**

When using a two-way ANOVA, it is important to remember that the assumptions of normality, equal variances, and random sampling must be satisfied in order for the test to produce meaningful results.

## ∽ Technology

For instructions on finding $F$ critical values using technology visit stat. hawkeslearning.com and navigate to **Discovering Business Statistics, Second Edition > Technology Instructions > F-Distribution > Critical Value**.

## ✎ NOTE

For each of these tests, we could also use the $P$-values to determine whether to reject or fail to reject the null hypothesis. $P$-values for Factor A, Factor B, and the interaction (AB) are given in the $P$-value column of the Excel output. We can compare these $P$-values to the significance levels rather than comparing the test statistics to the critical values, if we wish. That is, for any given test, if the $P$-value is less than or equal to $\alpha$, we would reject the null hypothesis.

The analysis in Example 12.5.1 (using blocking) is equivalent to a two-way ANOVA without interaction. The added variability of the Media Type on screen time reduced the variability of screen time on Age Group. In this example, we will use the same data as Example 12.5.1 but we will perform a two-way ANOVA with interaction. That is, we are now interested if there is a relationship (i.e., interaction) between the Age Group and Media Type factors.

**Example 12.6.1**

**Performing a Two-Way ANOVA with Interaction**

### SOLUTION

It's important to see the means that are being compared and the associated hypotheses. The table below should prove helpful. Note that $\mu_{ij}$, the cells within the table, represent the population mean of the $i^{th}$ row and the $j^{th}$ column. If replication exists ($r > 1$; in this example, we have 5 replications so $r = 5$), the estimate of $\mu_{ij}$ is the sample mean of the observations in the $i^{th}$ row and $j^{th}$ column, which is given by

$$\overline{x}_{ij} = \frac{\sum x_{ij1} + x_{ij2} + x_{ij3} + x_{ij4} + x_{ij5}}{5}$$

where $x_{ijk}$ is the $k^{th}$ observation in the $i^{th}$ row and $j^{th}$ column.

Similarly, the estimate of the row means $\mu_{i.}$ and column means $\mu_{.j}$ can be obtained by taking the average of the observations in each row and column, respectively.

| | Table 12.6.3 Two-way Table Showing Means | | | |
|---|---|---|---|---|
| | **Age Group ($j$)** | | | |
| **Media Type ($i$)** | **(1) 8–12 Years Old** | **(2) 13–18 Years Old** | **(3) Over 18 Years Old** | **Row Mean** |
| (1) Watching TV | $\mu_{11}$ | $\mu_{12}$ | $\mu_{13}$ | $\mu_{1.}$ |
| (2) Gaming | $\mu_{21}$ | $\mu_{22}$ | $\mu_{23}$ | $\mu_{2.}$ |
| (3) Listening to Music | $\mu_{31}$ | $\mu_{32}$ | $\mu_{33}$ | $\mu_{3.}$ |
| (4) Reading | $\mu_{41}$ | $\mu_{42}$ | $\mu_{43}$ | $\mu_{4.}$ |
| (5) Browsing Websites | $\mu_{51}$ | $\mu_{52}$ | $\mu_{53}$ | $\mu_{5.}$ |
| (6) Using Social Media | $\mu_{61}$ | $\mu_{62}$ | $\mu_{63}$ | $\mu_{6.}$ |
| (7) Video Chatting | $\mu_{71}$ | $\mu_{72}$ | $\mu_{73}$ | $\mu_{7.}$ |
| Column Mean | $\mu_{.1}$ | $\mu_{.2}$ | $\mu_{.3}$ | $\mu_{..}$ |

We have three hypotheses that need to be tested: a test for interaction and a test for each of the main effects. The null and alternative hypotheses for the test for interaction are given by:

$H_0$: The interaction of Media Type and Age Group is zero.

$H_a$: The interaction of Media Type and Age Group is not zero.

The test statistic for the interaction effect is given by

$$F = \frac{MS_{Interaction}}{MSE}$$

which follows an $F$-distribution with $df_{num} = (7 - 1)(3 - 1) = 12$ and $df_{den} = ab(r - 1) = 84$. Note that

$$MS_{Interaction} = \frac{SS_{Interaction}}{(a-1)(b-1)}$$

where $a$ is the number of levels of Media Type, $b$ is the number of levels of Age Group, and $r$ is the number of replications.

The critical value, using a significance level of $\alpha = 0.05$, for this test statistic is $F_{0.05,12,84} = 1.8693$. Thus, we reject the null hypothesis that there is an interaction effect if the $F$-statistic is greater than 1.8693.

### ⚭ Technology

For instructions on finding $F$ critical values using technology visit stat.hawkeslearning.com and navigate to **Discovering Business Statistics, Second Edition > Technology Instructions > F-Distribution > Critical Value.**

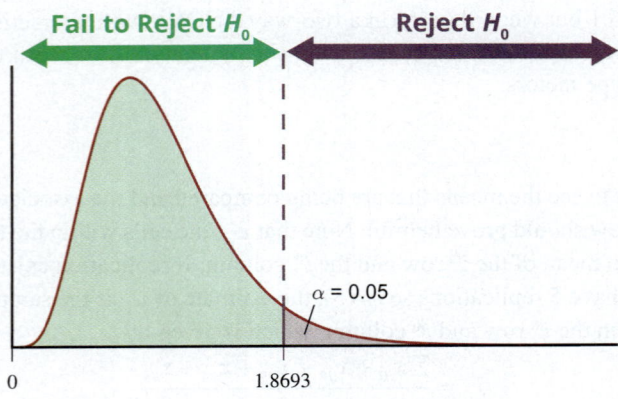

**F-Distribution**
$df_{num} = 12, df_{den} = 84$

**Fail to Reject $H_0$**          **Reject $H_0$**

$\alpha = 0.05$

0          1.8693

**Figure 12.6.8**

**Definition**

**Masking**

**Masking** occurs when a significant interaction term in an ANOVA hides (or masks) the variation in the main effects.

If the interaction term is insignificant, we then perform the test for main effects. That is, we test to see if there are significant differences in average screen time by Age Group and test to see if there are significant differences in average screen time by Media Type. If the interaction term is significant, then concern over the main effects is lost because any significance in the main effects could be *masked* by the significant interaction term. **Masking** is when the significant interaction term hides (or masks) the variation of the main effects or that there is little value in proceeding with testing the main effects.

The hypotheses for testing the Media Type effect are given by:

$H_0: \mu_{1.} = \mu_{2.} = \mu_{3.} = \mu_{4.} = \mu_{5.} = \mu_{6.} = \mu_{7.}$

$H_1$: At least one $\mu_{i.}$ is different.

The test statistic is

$$F = \frac{MS_{Media\ Type}}{MSE}$$

which follows an $F$-distribution with $df_{num} = a - 1 = 7 - 1 = 6$ and $df_{den} = ab(r-1) = 84$.

$$MS_{Media\ Type} = \frac{SS_{Media\ Type}}{a-1}$$

**Technology**

For instructions on finding $F$ critical values using technology visit stat.hawkeslearning.com and navigate to **Discovering Business Statistics, Second Edition > Technology Instructions > F-Distribution > Critical Value**.

where a is the number of levels of Media Type.

The critical value, using a significance level of $\alpha = 0.05$, for this test statistic is $F_{0.05, 6, 84}$ = 2.2086. Thus, we reject the null hypothesis that there is no media type effect if the $F$-statistic is greater than 2.2086.

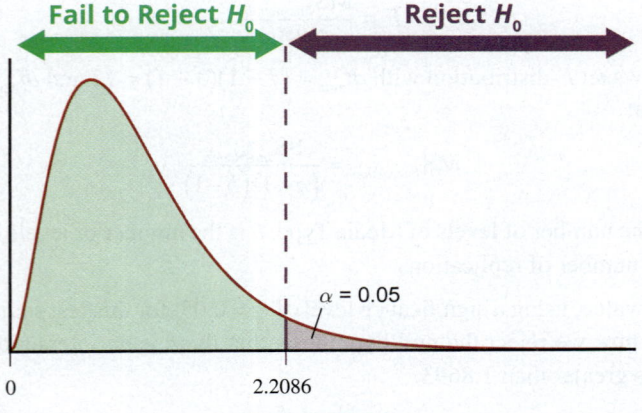

**F-Distribution**
$df_{num} = 6, df_{den} = 84$

**Fail to Reject $H_0$**          **Reject $H_0$**

$\alpha = 0.05$

0          2.2086

**Figure 12.6.9**

Similarly, the hypotheses for testing the Age Group effect are given by:

$H_0$: $\mu_{.1} = \mu_{.2} = \mu_{.3}$

$H_1$: At least one $\mu_{.j}$ is different.

The test statistic is

$$F = \frac{\text{MS}_{Media\ Type}}{\text{MSE}}$$

which follows an $F$-distribution with $df_{num} = b - 1 = 3 - 1 = 2$ and $df_{den} = ab(r-1) = 84$. The critical value, using a significance level of $\alpha = 0.05$, for this test statistic is $F_{0.05, 2, 84}$ = 3.1052. Thus, we reject the null hypothesis that there is an age group effect if the $F$-statistic is greater than 3.1052.

## ⟳ Technology

For instructions on finding F critical values using technology visit stat.hawkeslearning.com and navigate to **Discovering Business Statistics, Second Edition > Technology Instructions > F-Distribution > Critical Value**.

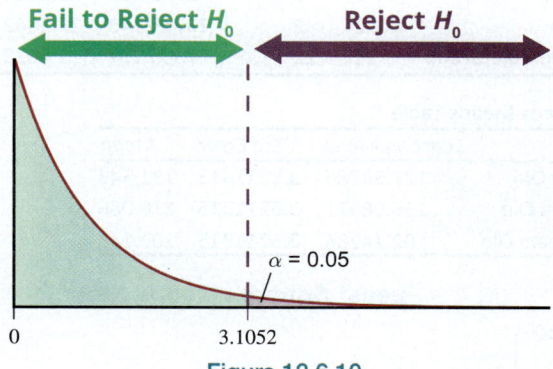

**F-Distribution**
**$df_{num} = 2$, $df_{den} = 84$**

Fail to Reject $H_0$    Reject $H_0$

$\alpha = 0.05$

0          3.1052

**Figure 12.6.10**

Using the JMP output below, the table titled "Effects Tests" is the two-way ANOVA in which we are interested. The test for interaction (Media Type*Age Group) has a test statistic, indicated by *F-Ratio*, of 1.4528 with an associated *P*-value = 0.1590. At the 5% level of significance, we fail to reject the null hypothesis and conclude that the interaction between Media Type and Age Group is zero. This conclusion can also be drawn by comparing the test statistic to the critical value of 1.8693. That is, since the test statistic is less than 1.8693, we would still fail to reject the null hypothesis.

Now that the interaction term is insignificant, we can focus our attention on the main effects. Referring to the "Effects Tests" table, we see that the test statistic for Media Type is 1.0544 with an associated *P*-value of 0.3966 indicating that we would fail to reject the null hypothesis and conclude that there is no evidence to indicate that the average screen time between Media Types is different. We would also draw this conclusion when comparing the test statistic $F = 1.0544$ to the critical value of 2.2086.

Lastly, when testing for an Age Group effect, we have a test statistic of $F = 27.3842$ with an associated *P*-value of less than 0.0001. When compared with the critical value of 3.1052, we reject the null hypothesis and conclude there is a significant difference in average screen time by Age Group. Additionally, the *P*-value would also lead us to reject the null hypothesis. See the end of Example 12.5.1 in which we performed Tukey's HSD multiple comparison procedure to indicate which of the three age groups were significantly different.

Summary of Fit

| | |
|---|---|
| RSquare | 0.483167 |
| RSquare Adj | 0.360112 |
| Root Mean Square Error | 20.8313 |
| Mean of Response | 123.2571 |
| Observations (or Sum Wgts) | 105 |

Analysis of Variance

| Source | DF | Sum of Squares | Mean Square | F-Ratio | Prob > F |
|---|---|---|---|---|---|
| Model | 20 | 34076.857 | 1703.84 | 3.9264 | < .0001* |
| Error | 84 | 36451.200 | 433.94 | | |
| C. Total | 104 | 70528.057 | | | |

Effects Tests

| Source | Nparm | DF | Sum of Squares | F-Ratio | Prob > F |
|---|---|---|---|---|---|
| Media Type | 6 | 6 | 2745.257 | 1.0544 | 0.3966 |
| Age Group | 2 | 2 | 23766.343 | 27.3842 | < .0001* |
| Media Type*Age Group | 12 | 12 | 7565.257 | 1.4528 | 0.1590 |

Least Squares Means Table

| Level | Least Sq Mean | Std Error | Mean |
|---|---|---|---|
| 8-12 Years Old | 131.54286 | 3.5211315 | 131.543 |
| 13-18 Years Old | 136.08571 | 3.5211315 | 136.086 |
| Over 18 Years Old | 102.14286 | 3.5211315 | 102.143 |

**Least Squares Means Plot**

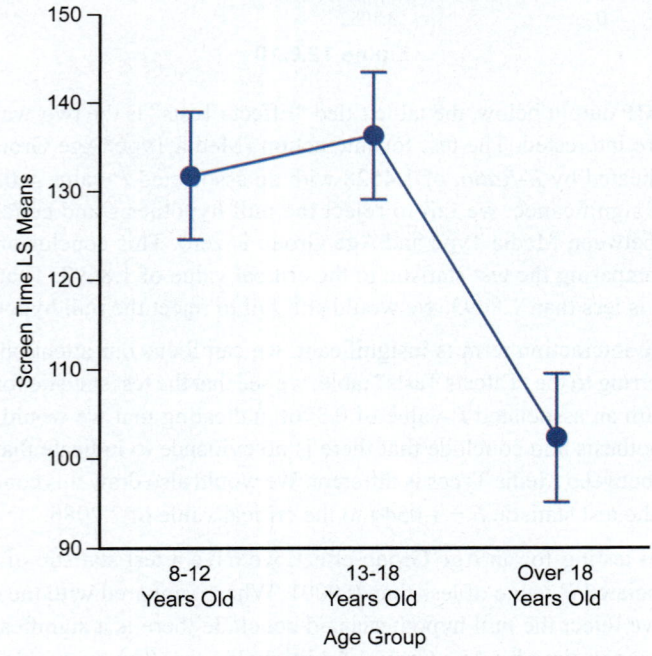

**Figure 12.6.11**

## 12.6 Exercises

### Basic Concepts

1. What is the difference between a randomized block design and a factorial design?

2. What is a complete factorial experiment?

3. What is a profile plot? What kind of information does this plot give us?

4. Why is it so important to determine if there is interaction between the two variables of interest in a factorial design?

5. Is it possible to perform a two-way analysis of variance if interaction exists between the two variables of interest? Explain why or why not.

6. Identify the four components that make up the total sum of squares in a complete factorial model. Also, give the acronym associated with each component.

7. Give the degrees of freedom associated with each component of the total sum of squares.

8. What is the test statistic for a test of interaction between factors? What are the degrees of freedom associated with this test statistic?

9. If there is enough evidence to reject the null hypothesis in a test for interaction, may we proceed with the main effects tests? Explain.

10. What is the test statistic for the main effects test for Factor A? What are the degrees of freedom associated with this test statistic?

11. What is the test statistic for the main effects test for Factor B? What are the degrees of freedom associated with this test statistic?

12. What are the rejection rules for the main effects tests? Can $P$-values be used as rejection criteria?

### Exercises

13. The following table contains the results of a survey of daily rental rates of a mid-size car for three major rental car companies at three airport locations on three different days during the year.

| Daily Rental Rates of Mid-Size Cars ($) | | | |
|---|---|---|---|
| | **New York** | **Chicago** | **Miami** |
| **Hertz** | 93.99 | 54.99 | 71.99 |
| | 90.99 | 63.99 | 87.99 |
| | 96.99 | 57.99 | 68.99 |
| **Avis** | 58.86 | 81.99 | 61.99 |
| | 52.10 | 85.99 | 70.99 |
| | 68.98 | 71.99 | 66.99 |
| **National** | 56.00 | 64.99 | 66.00 |
| | 63.00 | 67.00 | 58.99 |
| | 52.00 | 52.99 | 71.99 |

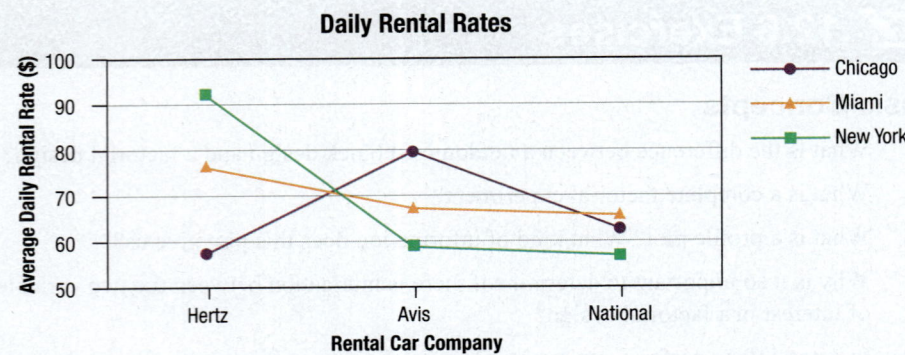

a. Consider the graph of the average daily rental rates for each of the major car rental companies by airport location. Does there appear to be any interaction between the variables airport location and major car rental company?

b. The results of the two-way ANOVA for the study are given in the following table.

ANOVA

| Source of Variation | SS | df | MS |
|---|---|---|---|
| Sample | 1011.7730 | 2 | 505.8865 |
| Columns | 58.7126 | 2 | 29.3563 |
| Interaction | 2514.3099 | 4 | 628.5775 |
| Within | 819.1289 | 18 | 45.5072 |
| Total | 4403.9244 | 26 | |

Perform a hypothesis test to determine if there is any interaction between the variables major rental car company and airport location at $\alpha = 0.05$. Does this agree with your observation in part **a.**?

c. If there is no interaction found in part **b.**, is there sufficient evidence to conclude that there is a significant difference among the average daily rental rates for mid-size cars for the three rental car companies at the 0.05 level?

**14.** A doctor is interested in determining the increase in average heart rate caused by a medication used for treating high blood pressure. The doctor believes that the increase in heart rate will be related to two factors: the age of a person and the weight of a person. To test this theory, the doctor randomly selects two patients in each of the age and weight categories listed in the following table and determines the increase in heart rate (in beats per minute) of each patient 15 minutes after administering the drug. The results of the study are as follows.

| Increase in Heart Rate (Beats per Minute) | | | |
|---|---|---|---|
| | **25–39 Years** | **40–54 Years** | **55–69 Years** |
| **100–149 Pounds** | 2 | 7 | 11 |
| | 2 | 6 | 7 |
| **150–199 Pounds** | 7 | 11 | 16 |
| | 7 | 9 | 12 |
| **200–249 Pounds** | 10 | 13 | 18 |
| | 8 | 11 | 14 |

**a.** Consider the following graph of the average increase in heart rate for each of the weight and age categories. Does there appear to be any interaction between the age and weight variables? Explain.

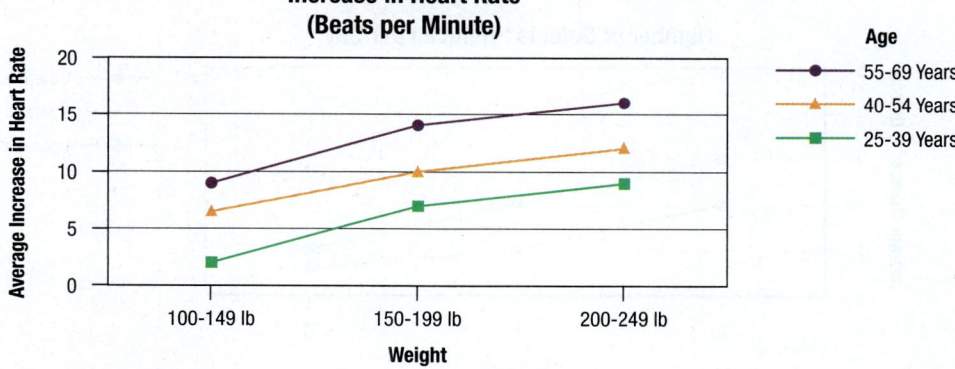

**b.** The results of the two-way ANOVA for the study are given in the following table.

| ANOVA | | | |
|---|---|---|---|
| *Source of Variation* | *SS* | *df* | *MS* |
| Sample | 133.0000 | 2 | 66.5000 |
| Columns | 147.0000 | 2 | 73.5000 |
| Interaction | 2.0000 | 4 | 0.5000 |
| Within | 30.5000 | 9 | 3.3889 |
| | | | |
| Total | 312.5000 | 17 | |

Perform a hypothesis test to determine if there is any interaction between the variables age and weight at $\alpha = 0.01$. Does this agree with your observation in part **a.**?

**c.** Is there sufficient evidence to conclude that there is a significant difference among the average increases in heart rate for the different weight categories? Use $\alpha = 0.01$.

**d.** Is there sufficient evidence to conclude that there is a significant difference among the average increases in heart rate for the different age groups? Use $\alpha = 0.01$.

**15.** A supervisor of a manufacturing plant is interested in relating the average number of defects produced per day to two factors: the operator working the machine and the machine itself. The supervisor randomly assigns each operator to use each machine for three days and records the number of defects produced per day. The results of the study are as follows.

| Number of Defects Produced per Day | | | |
|---|---|---|---|
| | **Operator A** | **Operator B** | **Operator C** |
| **Machine A** | 3 | 7 | 3 |
| | 3 | 5 | 2 |
| | 3 | 3 | 1 |
| **Machine B** | 2 | 6 | 2 |
| | 2 | 4 | 1 |
| | 2 | 2 | 0 |
| **Machine C** | 1 | 5 | 1 |
| | 1 | 3 | 0 |
| | 1 | 2 | 1 |

**a.** Consider the following graph of the average number of defects produced per day for each of the operators by machine. Does there appear to be any interaction between the variables operator and machine?

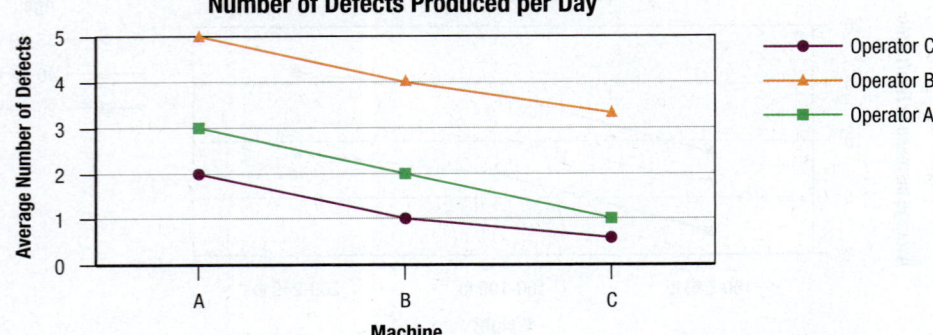

**b.** The results of the two-way ANOVA for the supervisor's survey of the number of defects produced per day are given in the following table.

ANOVA

| Source of Variation | SS | df | MS |
|---|---|---|---|
| Sample | 12.6667 | 2 | 6.3333 |
| Columns | 40.2222 | 2 | 20.1111 |
| Interaction | 0.4444 | 4 | 0.1111 |
| Within | 25.3333 | 18 | 1.4074 |
| | | | |
| Total | 78.6667 | 26 | |

Perform a hypothesis test to determine if there is any interaction between the machine and operator variables. Use $\alpha = 0.10$. Does this agree with your observation in part **a.**?

**c.** Is there sufficient evidence to conclude that there is a significant difference among the average number of defects produced per day for the different machines? Use $\alpha = 0.10$.

**d.** Is there sufficient evidence to conclude that there is a significant difference among the average number of defects produced per day for the different operators? Use $\alpha = 0.10$.

16. A dairy farmer thinks that the average weight gain of his cows depends on two factors: the type of grain that they are fed and the type of grass that they are fed. The dairy farmer has four different types of grain from which to choose and three different types of grass from which to choose. He would like to determine if there is a particular combination of grain and grass that would lead to the greatest weight gain on average for his cows. He randomly selects three one-year-old cows and assigns them to each of the possible combinations of grain and grass. After one year he records the weight gain for each cow (in pounds) with the following results.

| Cow Weight Gain (Pounds) | | | |
|---|---|---|---|
| | Grass A | Grass B | Grass C |
| Grain A | 175 | 225 | 250 |
| | 160 | 215 | 240 |
| | 185 | 230 | 260 |
| Grain B | 190 | 245 | 275 |
| | 185 | 240 | 260 |
| | 195 | 255 | 285 |
| Grain C | 210 | 255 | 300 |
| | 200 | 245 | 310 |
| | 220 | 265 | 295 |
| Grain D | 225 | 275 | 350 |
| | 235 | 270 | 360 |
| | 220 | 280 | 345 |

a. Consider the following graph of the average weight gain of the cows for each of the possible combinations of grass and grain. Does there appear to be any interaction between the grass and grain variables?

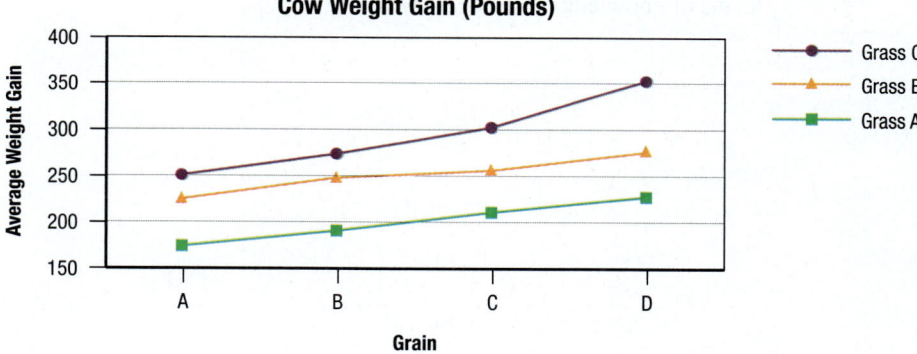

b. The results of the two-way ANOVA for the farmer's study are given in the following table.

ANOVA

| Source of Variation | SS | df | MS |
|---|---|---|---|
| Sample | 23097.2222 | 3 | 7699.0741 |
| Columns | 53272.2222 | 2 | 26636.1111 |
| Interaction | 3127.7778 | 6 | 521.2963 |
| Within | 1916.6667 | 24 | 79.8611 |
| | | | |
| Total | 81413.8889 | 35 | |

Perform a hypothesis test to determine if there is any interaction between the variables grass and grain at $\alpha = 0.05$. Does this agree with your observation in part a.?

   c. If there is no interaction found in part **b.**, is there sufficient evidence to conclude that there is a significant difference in the average weight gains among the cows for the four different types of grain? Use $\alpha = 0.05$.

   d. Is there sufficient evidence to conclude that there is a significant difference in the average weight gains among the cows for the three different types of grass? Use $\alpha = 0.05$.

17. The partially completed analysis of variance table given below is taken from the article, "Power and Status, Exchange, Attribution, and Expectation States (Small Group Research)." The experimenters investigated the effects of power and knowledge on one's emotional reaction in a study involving 52 students selected from a large private university. Each of the factors was run at two levels, with 13 subjects at each of the four different factor combinations.

ANOVA

| Source of Variation | SS | df | MS | F |
|---|---|---|---|---|
| Power | 1.2700 | 1 | | |
| Knowledge | 0.2500 | 1 | | |
| Interaction | | 1 | | |
| Error | 4.1400 | 48 | | |
| Total | 5.6700 | 51 | | |

   a. Complete the ANOVA table.

   b. Can we conclude, with $\alpha = 0.10$, that there is interaction between power and knowledge?

   c. With $\alpha = 0.05$, can we conclude that there is a significant difference in the two levels of power?

   d. With $\alpha = 0.05$, can we conclude that there is a significant difference in the two levels of knowledge?

**Discovering Technology**

## Using the TI-84 Plus Calculator

### One-Way ANOVA

Use the information from Example 12.3.3 for this exercise.

1.   Press **STAT**, select **Edit**, and press **ENTER**. Then input the four sales for the three sales reps into lists L1, L2, and L3.

2.   Press **STAT** and select **TESTS**, then choose option **ANOVA** and press **ENTER**.

3.   Enter **ANOVA(L1, L2, L3)**. (Note that lists can be entered by pressing **2ND** followed by the number of the list. For example, **2ND** and **1** gives L1.) Press **ENTER**.

4.   Observe the output screen for the one-way analysis of variance. Compare the results to those which we found in Example 12.3.3.

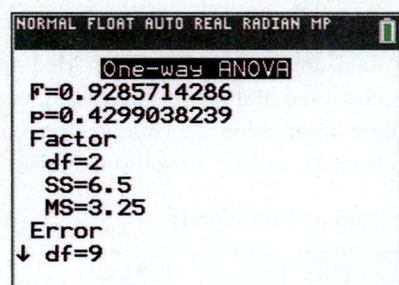 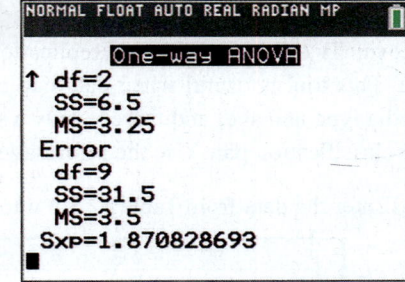

## Using Excel

### One-Way ANOVA

Use the information from Example 12.3.3 for this exercise.

1.   Enter the sales for each of the three sales reps into columns **A**, **B**, and **C**. Cells **A1**, **B1**, and **C1** should contain the column titles **Salesperson 1**, **Salesperson 2**, and **Salesperson 3**, respectively.

|   | A | B | C |
|---|---|---|---|
| 1 | Salesperson 1 | Salesperson 2 | Salesperson 3 |
| 2 | 10 | 11 | 11 |
| 3 | 14 | 16 | 13 |
| 4 | 13 | 14 | 12 |
| 5 | 12 | 15 | 15 |

2.   Under the **Data** tab, choose **Data Analysis**, and **Anova: Single Factor**.

3.   Enter the sales data **A1:C5** for the Input Range.

4.   Be sure that the radio button next to **Columns** is selected, and that Alpha is **0.05**. Check the box **Labels in First Row** since we are also including the column titles in the input range.

5.   Click **Output Range** and input **E1**.

6.   Click **OK**. You can widen columns **E** to **K** to display the data more effectively

7.   Observe the output screen for the one-way analysis of variance. Compare these results to the ones presented in Example 12.3.3.

| | A | B | C | D | E | F | G | H | I | J | K |
|---|---|---|---|---|---|---|---|---|---|---|---|
| 1 | Salesperson 1 | Salesperson 2 | Salesperson 3 | | Anova: Single Factor | | | | | | |
| 2 | 10 | 11 | 11 | | | | | | | | |
| 3 | 14 | 16 | 13 | | SUMMARY | | | | | | |
| 4 | 13 | 14 | 12 | | *Groups* | *Count* | *Sum* | *Average* | *Variance* | | |
| 5 | 12 | 15 | 15 | | Salesperson 1 | 4 | 49 | 12.25 | 2.916666667 | | |
| 6 | | | | | Salesperson 2 | 4 | 56 | 14 | 4.666666667 | | |
| 7 | | | | | Salesperson 3 | 4 | 51 | 12.75 | 2.916666667 | | |
| 8 | | | | | | | | | | | |
| 9 | | | | | | | | | | | |
| 10 | | | | | ANOVA | | | | | | |
| 11 | | | | | *Source of Variation* | *SS* | *df* | *MS* | *F* | *P-value* | *F crit* |
| 12 | | | | | Between Groups | 6.5 | 2 | 3.25 | 0.928571429 | 0.429903823 | 4.256494729 |
| 13 | | | | | Within Groups | 31.5 | 9 | 3.5 | | | |
| 14 | | | | | | | | | | | |
| 15 | | | | | Total | 38 | 11 | | | | |

## Two-Way ANOVA: Randomized Block Design

The Anova: Two-Factor Without Replication data analysis tool will be used for this exercise. This tool is useful when data can be classified along two different dimensions (i.e., media type and age) and there is only a single observation (i.e., average screen time) for each classification pair. Use the information from Table 12.5.3 for this exercise.

1.    1. Enter the data from Table 12.5.3 into columns **A** through **D**.

| | A | B | C | D |
|---|---|---|---|---|
| 1 | Media Type | 8-12 Years Old | 13-18 Years Old | Over 18 Years Old |
| 2 | Watching TV | 140 | 148 | 79 |
| 3 | | 116 | 140 | 90 |
| 4 | | 147 | 123 | 78 |
| 5 | | 131 | 132 | 71 |
| 6 | | 100 | 156 | 57 |
| 7 | Gaming | 103 | 117 | 116 |
| 8 | | 141 | 146 | 79 |
| 9 | | 132 | 124 | 80 |
| 10 | | 124 | 138 | 86 |
| 11 | | 106 | 122 | 100 |
| 12 | Listening to Music | 161 | 123 | 87 |
| 13 | | 153 | 176 | 140 |
| 14 | | 126 | 152 | 85 |
| 15 | | 76 | 115 | 86 |
| 16 | | 107 | 131 | 98 |
| 17 | Reading | 137 | 93 | 101 |
| 18 | | 101 | 160 | 90 |
| 19 | | 167 | 131 | 121 |
| 20 | | 165 | 102 | 91 |
| 21 | | 124 | 135 | 115 |
| 22 | Browsing Websites | 98 | 115 | 90 |
| 23 | | 113 | 116 | 103 |
| 24 | | 131 | 125 | 95 |
| 25 | | 122 | 95 | 105 |
| 26 | | 117 | 147 | 82 |

| 27 | Using Social Media | 131 | 122 | 109 |
| 28 | | 135 | 97 | 100 |
| 29 | | 110 | 106 | 81 |
| 30 | | 149 | 147 | 112 |
| 31 | | 133 | 162 | 125 |
| 32 | Video Chatting | 115 | 125 | 103 |
| 33 | | 91 | 111 | 125 |
| 34 | | 132 | 141 | 145 |
| 35 | | 117 | 113 | 97 |
| 36 | | 174 | 139 | 124 |

2. Here, we have the raw data, but we need to convert it to sample means for each media type and age range to use the randomized block experimental design. In cell **F1**, type **Average Screen Time Data**, select cells **F1** to **I1** and choose **Merge & Center** from the **Home** tab, copy and paste the age range titles into cells **G2** to **I2** and type the media types into cells **F3** to **F9**.

| | F | G | H | I |
|---|---|---|---|---|
| 1 | | Average Screen Time Data | | |
| 2 | | 8-12 Years Old | 13-18 Years Old | Over 18 Years Old |
| 3 | Watching TV | 126.8 | 139.8 | 75 |
| 4 | Gaming | 121.2 | 129.4 | 92.2 |
| 5 | Listening to Music | 124.6 | 139.4 | 99.2 |
| 6 | Reading | 138.8 | 124.2 | 103.6 |
| 7 | Browsing Websites | 116.2 | 119.6 | 95 |
| 8 | Using Social Media | 131.6 | 126.8 | 105.4 |
| 9 | Video Chatting | 125.8 | 125.8 | 118.8 |

3. In cell **G3**, average the 8–12 Year Old sample screen times from Watching TV with the formula

$$=AVERAGE(B2:B6)$$

In cell **G4**, average the 8–12 Year Old sample screen times from Gaming with the formula

$$=AVERAGE(B7:B11)$$

In cell **G5**, average the 8–12 Year Old sample screen times from Listening to Music with the formula

$$=AVERAGE(B12:B16)$$

In cell **G6**, average the 8–12 Year Old sample screen times from Reading with the formula

$$=AVERAGE(B17:B21)$$

In cell **G7**, average the 8–12 Year Old sample screen times from Browsing Websites with the formula

$$=AVERAGE(B22:B26)$$

In cell **G8**, average the 8–12 Year Old sample screen times from Using Social Media with the formula

$$=AVERAGE(B27:B31)$$

In cell **G9**, average the 8–12 Year Old sample screen times from Video Chatting with the formula

$$=AVERAGE(B32:B36)$$

4. **Select** the averages in cells **G3** to **G9**. **Click and drag** the small rectangle at the bottom right of your selection to the right by two cells to compute the averages for the other two age groups.

5. Under the **Data** tab, choose **Data Analysis**, and **Anova: Two-Factor Without Replication**.

6. Enter the averaged data table **F2:I9** as the Input Range.

7. Enter 0.05 for Alpha and click the **checkbox next to Labels** since we are including the row and column titles in the input range. Enter **F11** for the Output Range. Click **OK**.

8. Observe the output screen for the two-way analysis of variance. Notice that the ANOVA table is given following the summary statistics. "Rows" corresponds to the blocks (media type) and "Columns" corresponds to the treatments (age range). Compare these results to those that we discussed in Section 12.5.

| | F | G | H | I | J | K | L |
|---|---|---|---|---|---|---|---|
| 11 | Anova: Two-Factor Without Replication | | | | | | |
| 12 | | | | | | | |
| 13 | *SUMMARY* | *Count* | *Sum* | *Average* | *Variance* | | |
| 14 | Watching TV | 3 | 341.6 | 113.8666667 | 1175.213 | | |
| 15 | Gaming | 3 | 342.8 | 114.2666667 | 382.0133 | | |
| 16 | Listening to Music | 3 | 363.2 | 121.0666667 | 413.3733 | | |
| 17 | Reading | 3 | 366.6 | 122.2 | 312.76 | | |
| 18 | Browsing Websites | 3 | 330.8 | 110.2666667 | 177.6933 | | |
| 19 | Using Social Media | 3 | 363.8 | 121.2666667 | 194.5733 | | |
| 20 | Video Chatting | 3 | 370.4 | 123.4666667 | 16.33333 | | |
| 21 | | | | | | | |
| 22 | Watching TV | 7 | 885 | 126.4285714 | 52.60571 | | |
| 23 | Gaming | 7 | 905 | 129.2857143 | 58.4781 | | |
| 24 | Listening to Music | 7 | 689.2 | 98.45714286 | 181.7295 | | |
| 25 | | | | | | | |
| 26 | | | | | | | |
| 27 | ANOVA | | | | | | |
| 28 | *Source of Variation* | *SS* | *df* | *MS* | *F* | *P-value* | *F crit* |
| 29 | Rows | 475.2114286 | 6 | 79.20190476 | 0.741551 | 0.626939 | 2.99612 |
| 30 | Columns | 4062.251429 | 2 | 2031.125714 | 19.01701 | 0.00019 | 3.885294 |
| 31 | Error | 1281.668571 | 12 | 106.8057143 | | | |
| 32 | | | | | | | |
| 33 | Total | 5819.131429 | 20 | | | | |

## Two-Way ANOVA: Factorial Design

The Anova: Two-Factor With Replication data analysis tool will be used for this exercise. This tool is useful when data can be classified along two different dimensions (i.e. age and experience) and there is more than one observation for each combination of classifications. Use the information from Table 12.6.1 for this exercise.

1. Enter the age ranges, **25−34**, **35−44**, **45−54**, and **55−64** in cells **B1** through **E1**.

2. Select column **A** and, on the **Home** tab in the **Number** section, choose **Text** from the formatting dropdown menu to ensure Excel does not automatically convert the years of experience to dates after we input them.

3. Enter the years of experience, **0−4**, **5−9**, and **10−14** in cells **A2**, **A4**, and **A6**.

4. Input the sample data into cells **B2:E7** as given in Table 12.6.1.

| ▲ | A | B | C | D | E |
|---|---|---|---|---|---|
| 1 | | 25-34 | 35-44 | 45-54 | 55-64 |
| 2 | 0-4 | 22 | 25 | 34 | 37 |
| 3 | | 27 | 35 | 36 | 43 |
| 4 | 5-9 | 34 | 35 | 42 | 49 |
| 5 | | 36 | 45 | 48 | 51 |
| 6 | 10-14 | 39 | 40 | 53 | 51 |
| 7 | | 41 | 50 | 57 | 59 |

5. Under the **Data** tab, choose **Data Analysis**, and **Anova: Two-Factor With Replication**.

6. Enter the range **A1:E7** as the Input Range.

7. Enter **2** for Rows per sample, enter **0.05** for Alpha, enter **G1** for Output Range, and click **OK**.

8. Observe the output screen for the two-way analysis of variance. Notice that the ANOVA table is given following the summary statistics that are presented for each group. "Sample" corresponds to years of experience (Factor A) and "Columns" corresponds to age (Factor B). Compare these results to those that we discussed in Section 12.6.

| ▲ | G | H | I | J | K | L | M |
|---|---|---|---|---|---|---|---|
| 1 | Anova: Two-Factor Without Replication | | | | | | |
| 2 | | | | | | | |
| 3 | SUMMARY | 25-34 | 35-44 | 45-54 | 55-64 | Total | |
| 4 | *0-4* | | | | | | |
| 5 | Count | 2 | 2 | 2 | 2 | 8 | |
| 6 | Sum | 49 | 60 | 70 | 80 | 259 | |
| 7 | Average | 24.5 | 30 | 35 | 40 | 32.375 | |
| 8 | Variance | 12.5 | 50 | 2 | 18 | 49.6964 | |
| 9 | | | | | | | |
| 10 | *5-9* | | | | | | |
| 11 | Count | 2 | 2 | 2 | 2 | 8 | |
| 12 | Sum | 70 | 80 | 90 | 100 | 340 | |
| 13 | Average | 35 | 40 | 45 | 50 | 42.5 | |
| 14 | Variance | 2 | 50 | 18 | 2 | 46 | |
| 15 | | | | | | | |
| 16 | *10-14* | | | | | | |
| 17 | Count | 2 | 2 | 2 | 2 | 8 | |
| 18 | Sum | 80 | 90 | 110 | 110 | 390 | |
| 19 | Average | 40 | 45 | 55 | 55 | 48.75 | |
| 20 | Variance | 2 | 50 | 8 | 32 | 61.3571 | |
| 21 | | | | | | | |
| 22 | *Total* | | | | | | |
| 23 | Count | 6 | 6 | 6 | 6 | | |
| 24 | Sum | 199 | 230 | 270 | 290 | | |
| 25 | Average | 33.1667 | 38.3333 | 45 | 48.3333 | | |
| 26 | Variance | 53.3667 | 76.6667 | 85.6 | 57.0667 | | |
| 27 | | | | | | | |

| | | | | | | |
|---|---|---|---|---|---|---|
| 28 | | | | | | |
| 29 | ANOVA | | | | | |
| 30 | *Source of Variation* | *SS* | *df* | *MS* | *F* | *P-value* | *F crit* |
| 31 | Sample | 1092.58 | 2 | 546.292 | 26.5943 | 3.9E-05 | 3.88529 |
| 32 | Columns | 828.458 | 3 | 276.153 | 13.4435 | 0.00038 | 3.49029 |
| 33 | Interaction | 24.4167 | 6 | 4.06944 | 0.19811 | 0.97101 | 2.99612 |
| 34 | Within | 246.5 | 12 | 20.5417 | | | |
| 35 | | | | | | | |
| 36 | Total | 2191.96 | 23 | | | | |

## F-Critical Value

Using a significance level of 0.025, numerator degrees of freedom of 40, and denominator degrees of freedom of 49, find the *F*-critical value with 0.025 area to the right.

1. In any cell in Microsoft Excel, type **=F.INV.RT(probability, deg_freedom1, deg_freedom2)** and press Tab to select the function.

2. For the first value, enter the significance level 0.025 followed by a comma.

3. For the second value, enter the numerator degrees of freedom of 40 followed by a comma.

4. For the third value, enter the denominator degrees of freedom of 49.

5. Close out the parenthesis and press **Enter**.

6. The value of 1.802689 should appear in the cell chosen, which corresponds to the *F*-critical value with a probability of 0.025 in the right tail.

## Using JMP

### Test for Normality

For this exercise, use the data from Example 12.2.1.

1. With a JMP data table open, enter the Screen Time by Age Group data into Column 1 and Column 2.

| | Age Group | Screen Time |
|---|---|---|
| 1 | 8 - 12 Years Old | 120 |
| 2 | 8 - 12 Years Old | 133 |
| 3 | 8 - 12 Years Old | 146 |
| 4 | 8 - 12 Years Old | 89 |
| 5 | 8 - 12 Years Old | 161 |
| 6 | 8 - 12 Years Old | 117 |
| 7 | 8 - 12 Years Old | 156 |
| 8 | 8 - 12 Years Old | 116 |
| 9 | 8 - 12 Years Old | 120 |
| 10 | 8 - 12 Years Old | 112 |
| 11 | 8 - 12 Years Old | 142 |
| 12 | 8 - 12 Years Old | 137 |
| 13 | 8 - 12 Years Old | 97 |
| 14 | 8 - 12 Years Old | 129 |
| 15 | 8 - 12 Years Old | 118 |
| 16 | 8 - 12 Years Old | 137 |
| 17 | 8 - 12 Years Old | 123 |
| 18 | 8 - 12 Years Old | 103 |
| 19 | 8 - 12 Years Old | 139 |

2.  Right click on the Age Group title (or Column 1) and select Column Info. Set the Data Type to Character and the Modeling Type to Ordinal. Click OK. Right click on the Screen Time title (or Column 2) and select Column Info. Set the Data Type to Numeric and the Modeling Type to Continuous. Click OK.

3.  Select Analyze in the top row of the JMP output table, and then select Distribution. From the Select Columns box, click on Screen Time, then click on Y, Response . Click on Age Group, then click on By. Click OK.

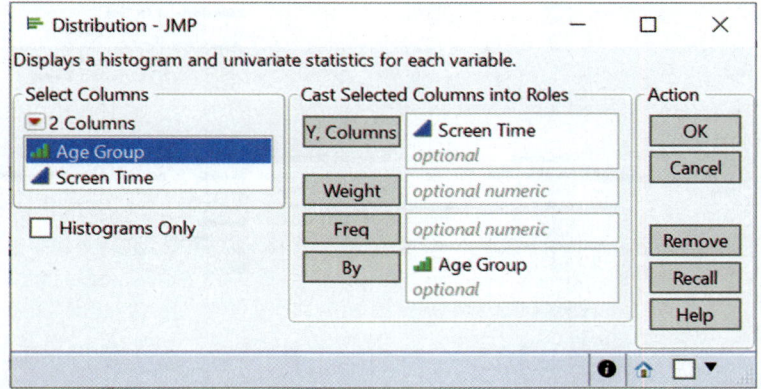

4.  From the output, click on the red triangle next to Screen Time for each of the Age Groups. Click on Continuous Fit and select Fit Normal. Then, click on the red triangle next to Fitted Normal Distribution and select Goodness of Fit. Repeat this for each of the Age Groups. Note: If the first outputs show vertically, click on the red triangle next to Screen Time > Display Options > Horizontal Layouts.

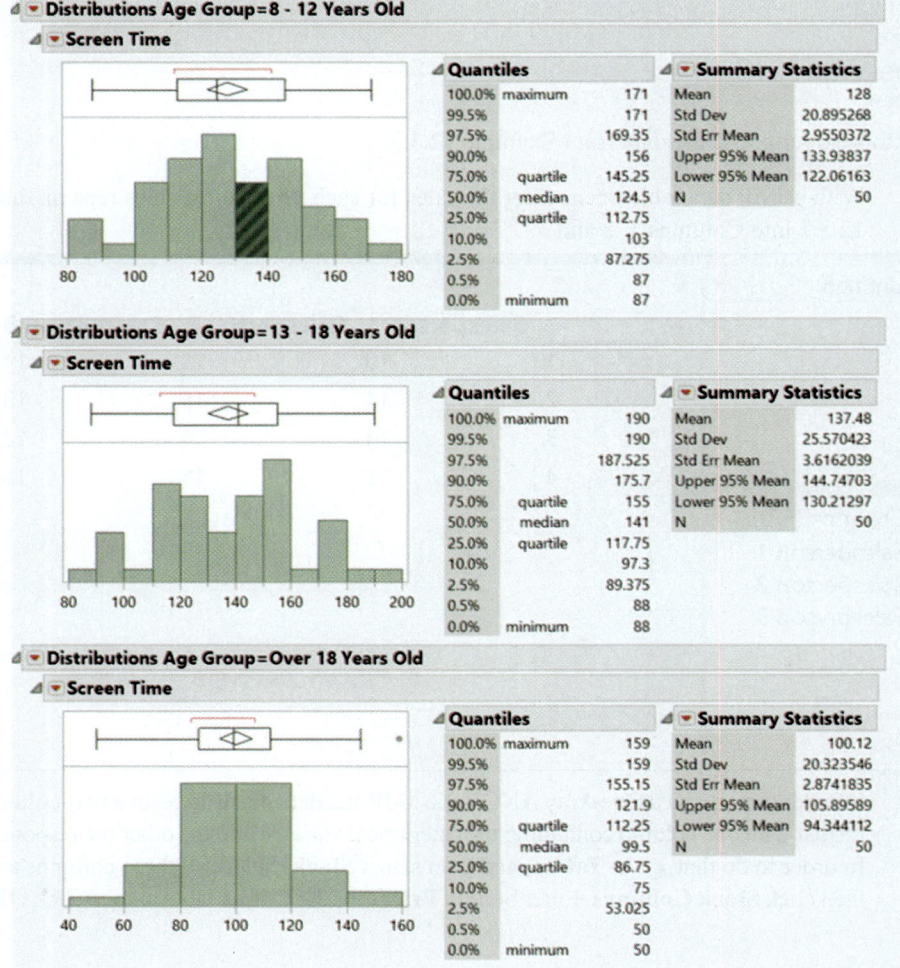

5.    Observe the outputs for each Age Group and compare them to the results obtained in Example 12.2.1

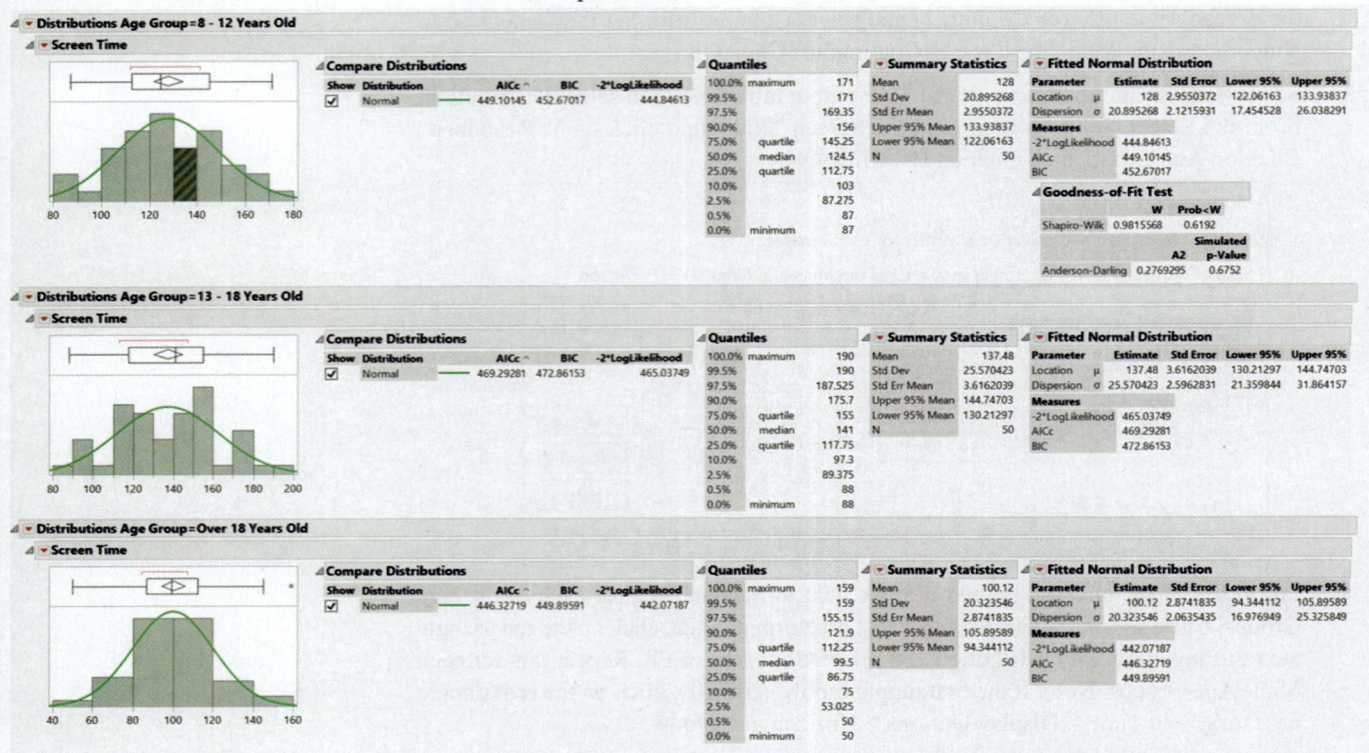

## One-Way ANOVA

For this exercise, use the data from Example 12.3.3.

1.    With a JMP data table open, enter the sales for each of the three sales reps in Table 12.3.1 into Columns 1, 2 and 3.

| | Salesperson 1 | Salesperson 2 | Salesperson 3 |
|---|---|---|---|
| 1 | 10 | 11 | 11 |
| 2 | 14 | 16 | 13 |
| 3 | 13 | 14 | 12 |
| 4 | 12 | 15 | 15 |

Columns (3/0)
Salesperson 1
Salesperson 2
Salesperson 3

2.    In order to perform a One-Way ANOVA in JMP, the data needs to be in a two-column format with one column containing the categorical variable and the other the response. In order to do that, go to **Tables**, and then select **Stack**. Select all three columns and then click **Stack Columns**. Enter **Selling Prices** for the Output table name. Click **OK**.

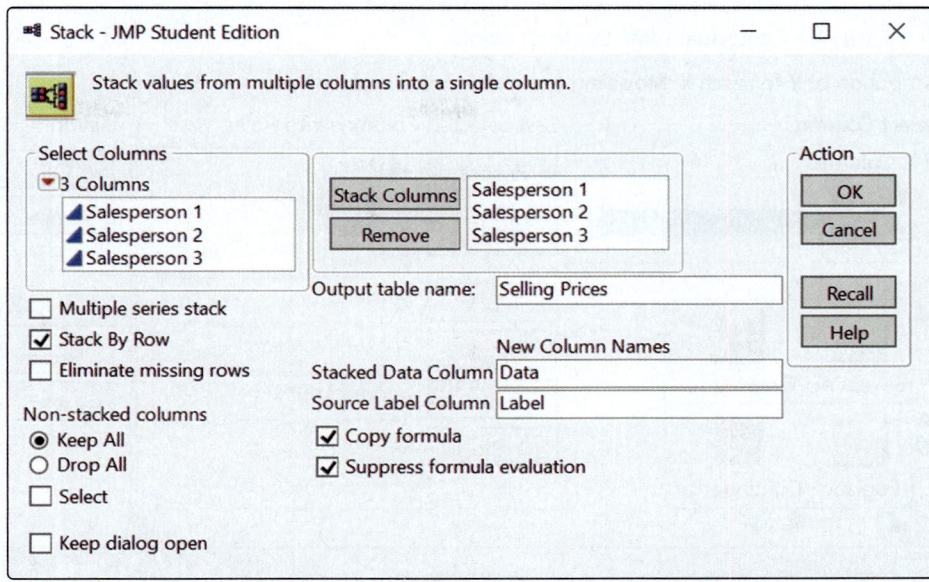

| | Label | Data | |
|---|---|---|---|
| 1 | Salesperson 1 | 10 | |
| 2 | Salesperson 2 | 11 | |
| 3 | Salesperson 3 | 11 | |
| 4 | Salesperson 1 | 14 | |
| 5 | Salesperson 2 | 16 | |
| 6 | Salesperson 3 | 13 | |
| 7 | Salesperson 1 | 13 | |
| 8 | Salesperson 2 | 14 | |
| 9 | Salesperson 3 | 12 | |
| 10 | Salesperson 1 | 12 | |
| 11 | Salesperson 2 | 15 | |
| 12 | Salesperson 3 | 15 | |

3. Select **Analyze** in the top row of the JMP output table Selling Prices, and then select **Fit Y by X**.

4. From the **Select Columns** box, click on **Label**, then click on **X, Factor**. Click on **Data**, then click on **Y, Response**. Click **OK**.

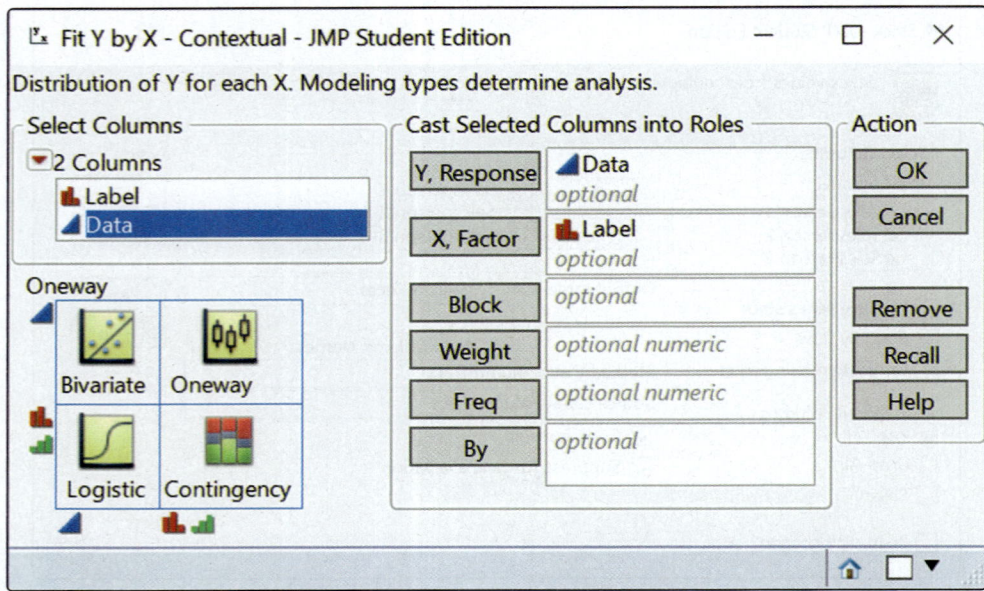

5.  From the Oneway Analysis output, click on the **red triangle** next to Oneway Analysis of Data By Label and click on **Means/Anova**. The results of the test are displayed below. Compare these results to the ones presented in Example 12.3.3.

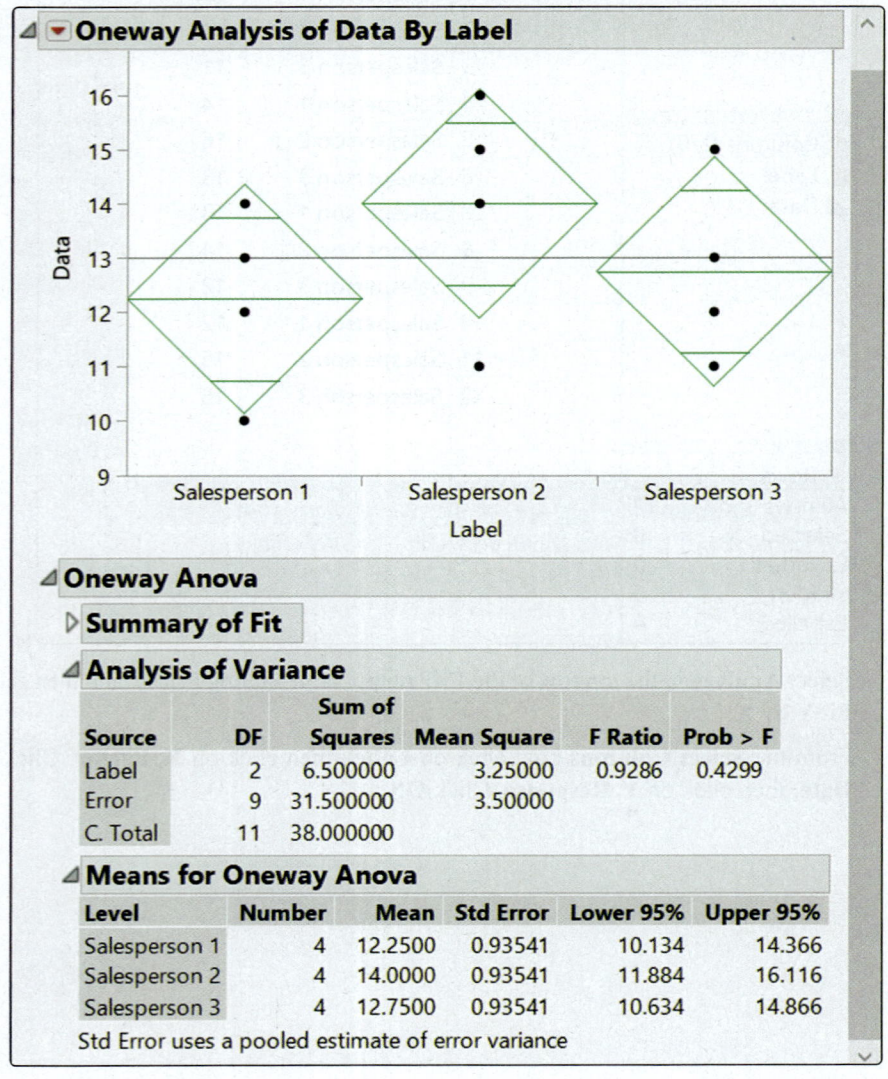

### Oneway Anova

▷ **Summary of Fit**

### Analysis of Variance

| Source | DF | Sum of Squares | Mean Square | F Ratio | Prob > F |
|--------|-----|----------------|-------------|---------|----------|
| Label | 2 | 6.500000 | 3.25000 | 0.9286 | 0.4299 |
| Error | 9 | 31.500000 | 3.50000 | | |
| C. Total | 11 | 38.000000 | | | |

### Means for Oneway Anova

| Level | Number | Mean | Std Error | Lower 95% | Upper 95% |
|-------|--------|---------|-----------|-----------|-----------|
| Salesperson 1 | 4 | 12.2500 | 0.93541 | 10.134 | 14.366 |
| Salesperson 2 | 4 | 14.0000 | 0.93541 | 11.884 | 16.116 |
| Salesperson 3 | 4 | 12.7500 | 0.93541 | 10.634 | 14.866 |

Std Error uses a pooled estimate of error variance

## Tukey's HSD

For this exercise, use the data from Example 12.1.3.

1.  With a JMP data table open, enter the Age Group data in Column 1, and the Screen Time data in Column 2.

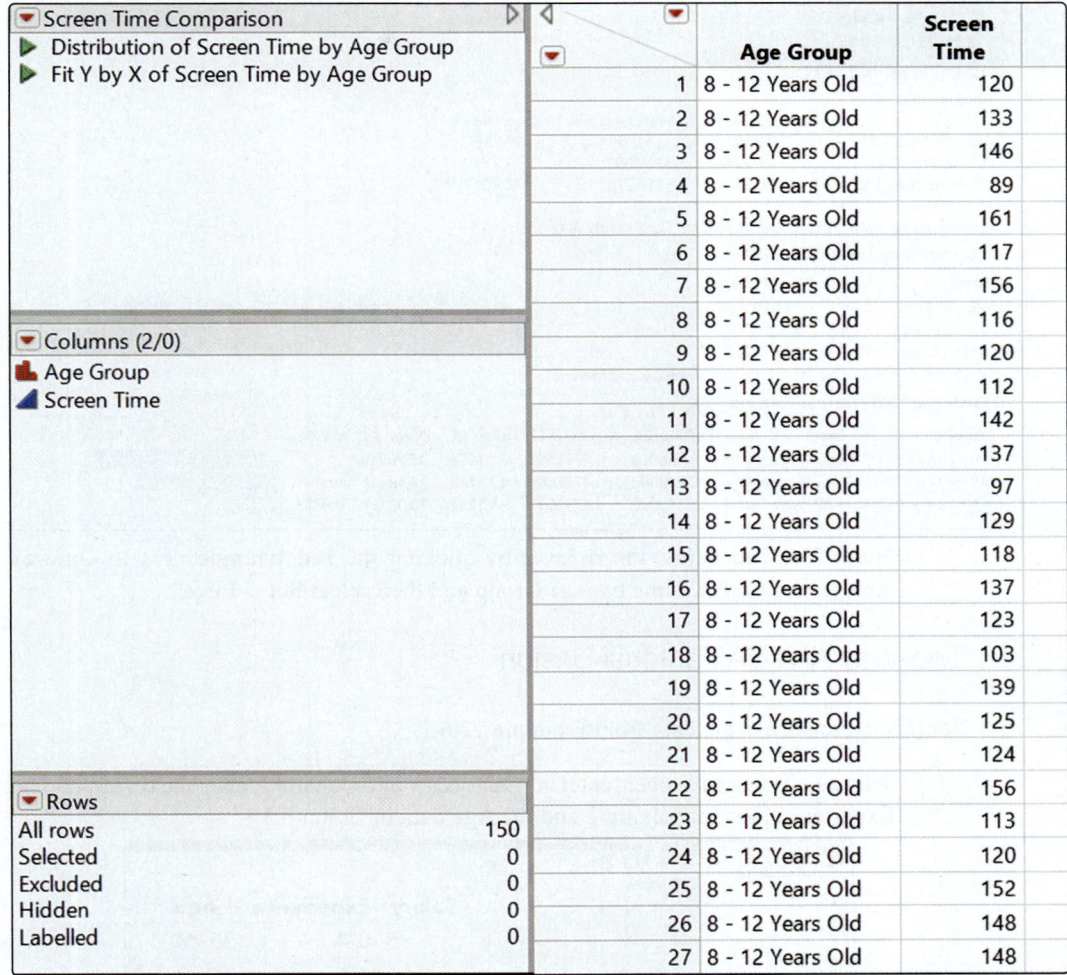

| | Age Group | Screen Time |
|---|---|---|
| 1 | 8 - 12 Years Old | 120 |
| 2 | 8 - 12 Years Old | 133 |
| 3 | 8 - 12 Years Old | 146 |
| 4 | 8 - 12 Years Old | 89 |
| 5 | 8 - 12 Years Old | 161 |
| 6 | 8 - 12 Years Old | 117 |
| 7 | 8 - 12 Years Old | 156 |
| 8 | 8 - 12 Years Old | 116 |
| 9 | 8 - 12 Years Old | 120 |
| 10 | 8 - 12 Years Old | 112 |
| 11 | 8 - 12 Years Old | 142 |
| 12 | 8 - 12 Years Old | 137 |
| 13 | 8 - 12 Years Old | 97 |
| 14 | 8 - 12 Years Old | 129 |
| 15 | 8 - 12 Years Old | 118 |
| 16 | 8 - 12 Years Old | 137 |
| 17 | 8 - 12 Years Old | 123 |
| 18 | 8 - 12 Years Old | 103 |
| 19 | 8 - 12 Years Old | 139 |
| 20 | 8 - 12 Years Old | 125 |
| 21 | 8 - 12 Years Old | 124 |
| 22 | 8 - 12 Years Old | 156 |
| 23 | 8 - 12 Years Old | 113 |
| 24 | 8 - 12 Years Old | 120 |
| 25 | 8 - 12 Years Old | 152 |
| 26 | 8 - 12 Years Old | 148 |
| 27 | 8 - 12 Years Old | 148 |

2.  Select **Analyze** in the top row of the JMP spreadsheet and then select **Fit Y by X**.

3.  From the **Select Columns** box, click on **Age Group** (or Column 1), then click on **X, Factor**. Click on **Screen Time** (or Column 2), then click on **Y, Response**. Click **OK**.

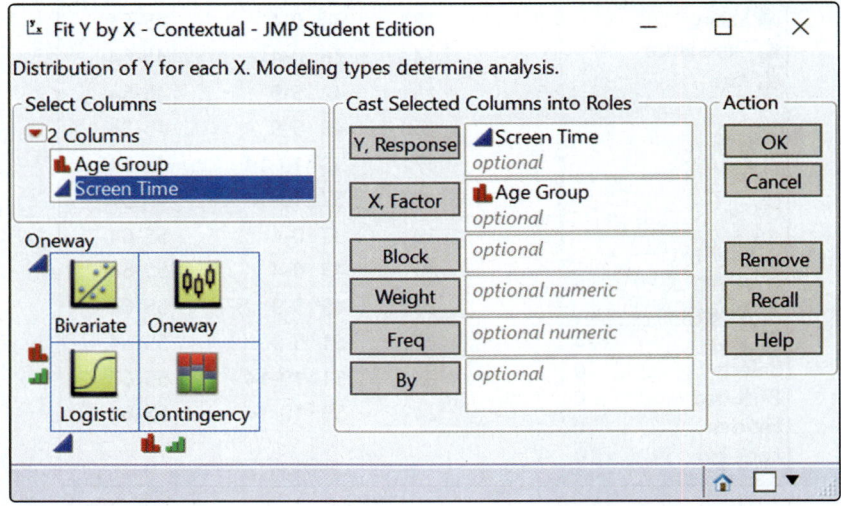

4. From the Oneway Analysis output, click on the **red triangle** next to Oneway Analysis of Screen Time By Age Group, then click on **Compare Means** and select **All Pairs, Tukey HSD**. The results of the test are displayed. Compare these results to the ones presented in Section 12.4.

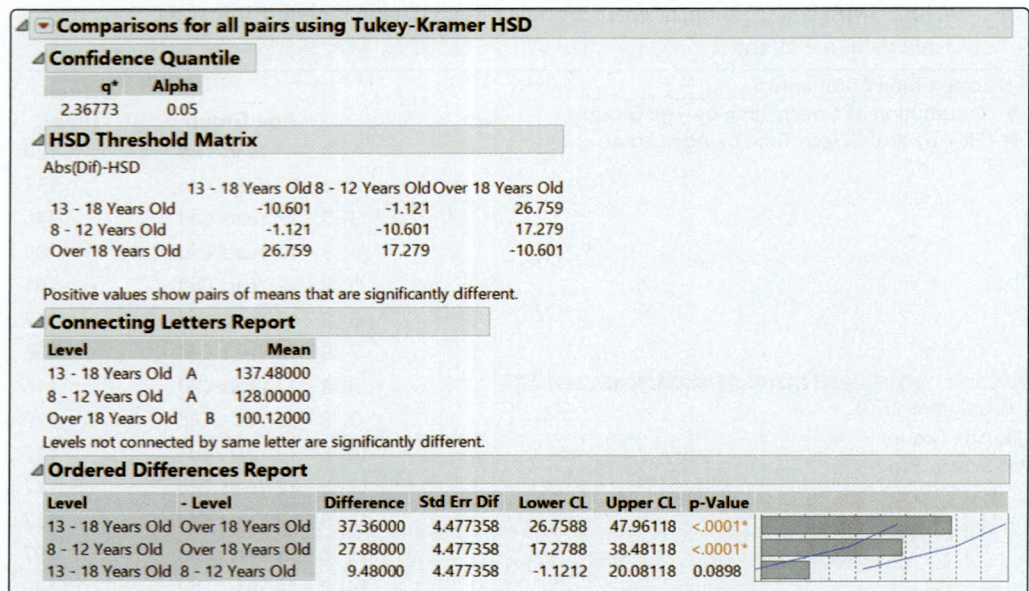

**▽ Comparisons for all pairs using Tukey-Kramer HSD**

**⊿ Confidence Quantile**

| q* | Alpha |
|---|---|
| 2.36773 | 0.05 |

**⊿ HSD Threshold Matrix**

Abs(Dif)-HSD

| | 13 - 18 Years Old | 8 - 12 Years Old | Over 18 Years Old |
|---|---|---|---|
| 13 - 18 Years Old | -10.601 | -1.121 | 26.759 |
| 8 - 12 Years Old | -1.121 | -10.601 | 17.279 |
| Over 18 Years Old | 26.759 | 17.279 | -10.601 |

Positive values show pairs of means that are significantly different.

**⊿ Connecting Letters Report**

| Level | | Mean |
|---|---|---|
| 13 - 18 Years Old | A | 137.48000 |
| 8 - 12 Years Old | A | 128.00000 |
| Over 18 Years Old | B | 100.12000 |

Levels not connected by same letter are significantly different.

**⊿ Ordered Differences Report**

| Level | - Level | Difference | Std Err Dif | Lower CL | Upper CL | p-Value | |
|---|---|---|---|---|---|---|---|
| 13 - 18 Years Old | Over 18 Years Old | 37.36000 | 4.477358 | 26.7588 | 47.96118 | <.0001* | |
| 8 - 12 Years Old | Over 18 Years Old | 27.88000 | 4.477358 | 17.2788 | 38.48118 | <.0001* | |
| 13 - 18 Years Old | 8 - 12 Years Old | 9.48000 | 4.477358 | -1.1212 | 20.08118 | 0.0898 | |

**Note:** You can adjust the $\alpha$ level by clicking the **red triangle** next to Oneway Analysis of Screen Time by Age Group and then select **Set $\alpha$ Level**.

## Two-Way ANOVA Factorial Design

For this exercise, use the data from Example 12.6.1.

1. With a JMP data table open, enter the Salary data into Column 1, enter the corresponding Experience data in Column 2 and the Age data in Column 3.

| ▽ Untitled 3 | | | Salary | Experience | Age |
|---|---|---|---|---|---|
| | | 8 | 35 | 0-4 | 35-44 |
| | | 9 | 35 | 5-9 | 35-44 |
| | | 10 | 45 | 5-9 | 35-44 |
| | | 11 | 40 | 10-14 | 35-44 |
| | | 12 | 50 | 10-14 | 35-44 |
| ▽ Columns (3/0) | | 13 | 34 | 0-4 | 45-54 |
| ⊿ Salary | | 14 | 36 | 0-4 | 45-54 |
| ▮ Experience | | 15 | 42 | 5-9 | 45-54 |
| ▮ Age | | 16 | 48 | 5-9 | 45-54 |
| | | 17 | 53 | 10-14 | 45-54 |
| | | 18 | 57 | 10-14 | 45-54 |
| | | 19 | 37 | 0-4 | 55-64 |
| | | 20 | 43 | 0-4 | 55-64 |
| ▽ Rows | | 21 | 49 | 5-9 | 55-64 |
| All rows | 24 | 22 | 51 | 5-9 | 55-64 |
| Selected | 0 | 23 | 51 | 10-14 | 55-64 |
| Excluded | 0 | 24 | 59 | 10-14 | 55-64 |
| Hidden | 0 | | | | |
| Labelled | 0 | | | | |

2. Click on **Analyze** and then **Fit Model**. From the **Select Columns** box, click on **Salary**, and then click **Y**. Then, select both **Experience** and **Age** from the **Select Columns** box, and click on **Macros** and then **Full Factorial**. Click **Run**.

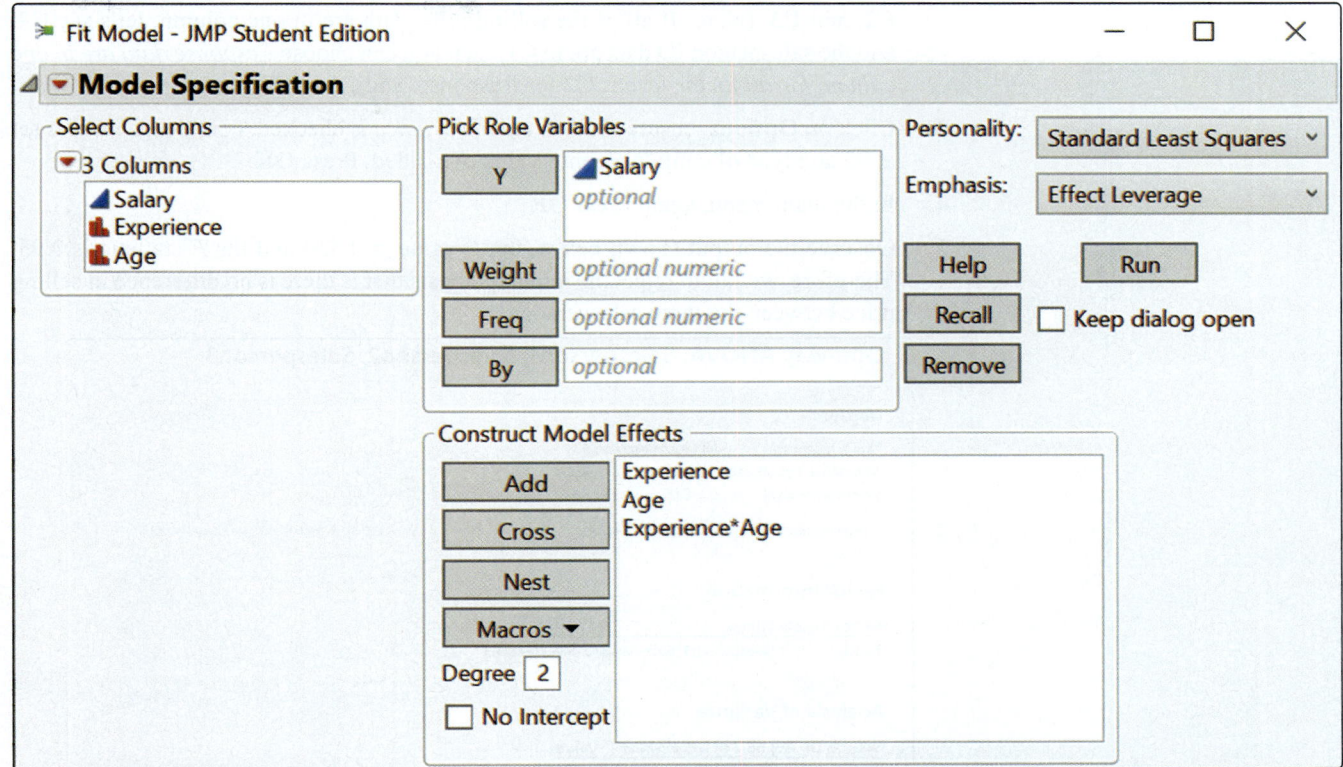

3. Notice that Analysis of Variance is in the bottom left of the output screen. Compare these results to those discussed in Section 12.6.

### Analysis of Variance

| Source | DF | Sum of Squares | Mean Square | F Ratio |
|---|---|---|---|---|
| Model | 11 | 1945.4583 | 176.860 | 8.6098 |
| Error | 12 | 246.5000 | 20.542 | Prob > F |
| C. Total | 23 | 2191.9583 | | 0.0004* |

### ▷ Parameter Estimates

### Effect Tests

| Source | Nparm | DF | Sum of Squares | F Ratio | Prob > F |
|---|---|---|---|---|---|
| Experience | 2 | 2 | 1092.5833 | 26.5943 | <.0001* |
| Age | 3 | 3 | 828.4583 | 13.4435 | 0.0004* |
| Experience*Age | 6 | 6 | 24.4167 | 0.1981 | 0.9710 |

## Using Minitab

### One-Way ANOVA

Use the information from Example 12.3.3 for this exercise.

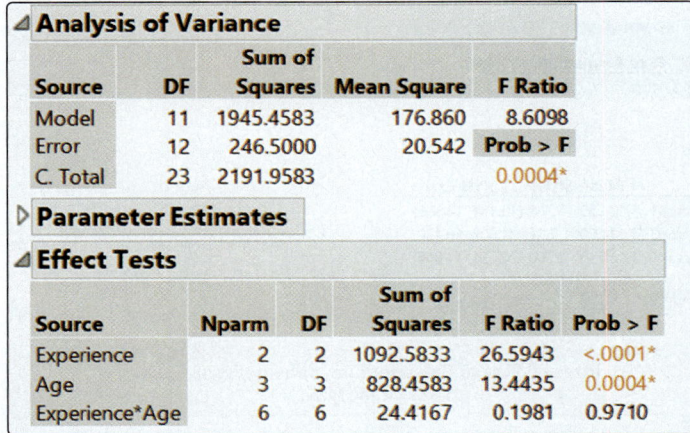

| Salesperson 1 | Salesperson 2 | Salesperson 3 |
|---|---|---|
| 10 | 11 | 11 |
| 14 | 16 | 13 |
| 13 | 14 | 12 |
| 12 | 15 | 15 |

1. Enter the sample data into columns **C1**, **C2** and **C3**.

2. Choose **Stat**, **ANOVA** and **One-Way ANOVA**. Select in the drop-down menu: **Response data are in a separate column for each factor level**. Choose column **C1**, **C2**, and **C3**. (Note: If all of the selling price data are in one column, let's say C2, and the salesperson ID data are in C1, then you can choose *Response data are in one column for all factor levels*, C2 for Response, and C1 for Factor)

3. Click on **Options**, ensure *Assume equal variance* is checked, confidence level is set at 95 and type of confidence interval is two-sided. Press **OK**.

4. In the main menu, again Press **OK**.

5. Observe the output shown below, the *P*-value is 0.430 and the *F*-statistic is 0.93. Therefore, we fail to reject the null hypothesis that is there is no difference in selling price between the three salespersons.

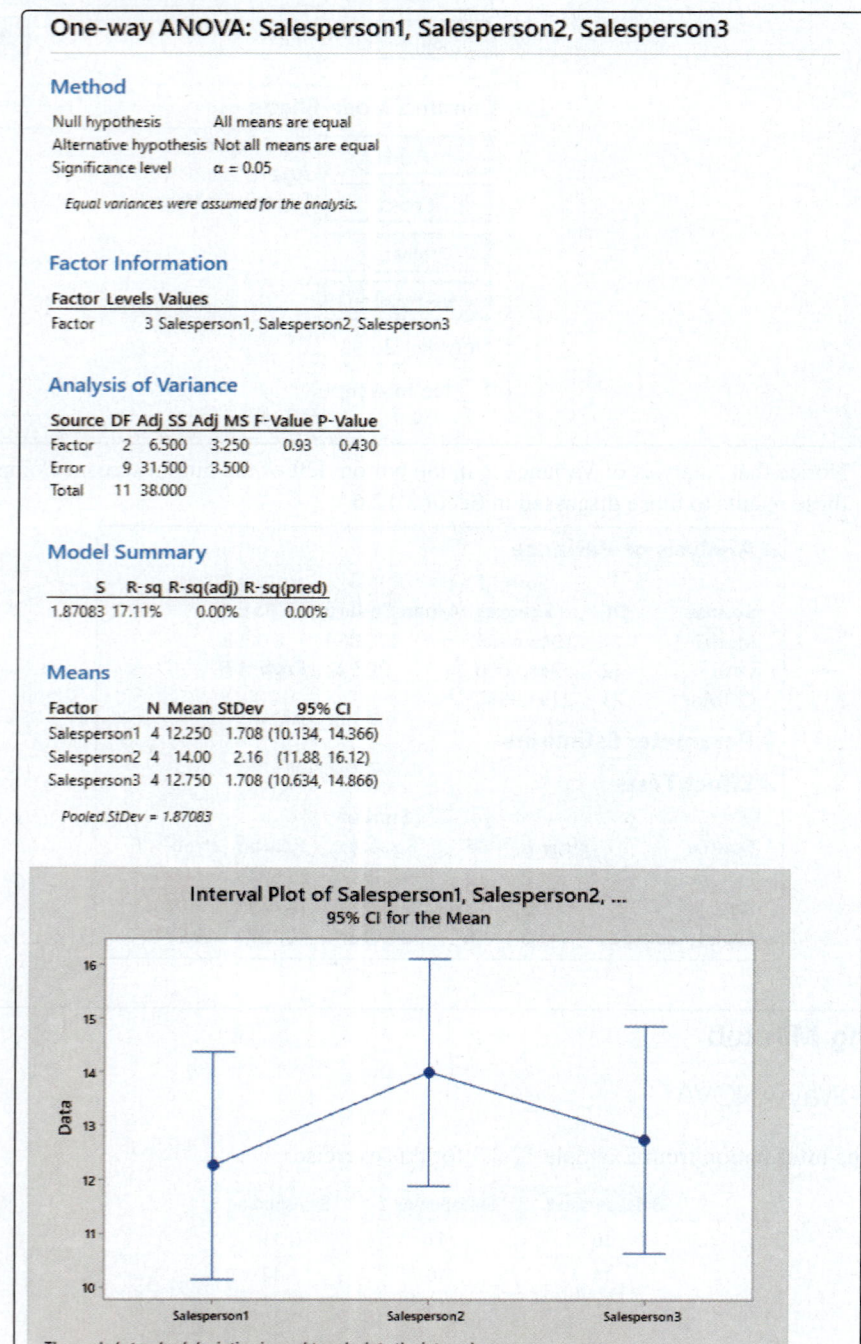

### One-way ANOVA: Salesperson1, Salesperson2, Salesperson3

#### Method

| | |
|---|---|
| Null hypothesis | All means are equal |
| Alternative hypothesis | Not all means are equal |
| Significance level | $\alpha = 0.05$ |

*Equal variances were assumed for the analysis.*

#### Factor Information

| Factor | Levels | Values |
|---|---|---|
| Factor | 3 | Salesperson1, Salesperson2, Salesperson3 |

#### Analysis of Variance

| Source | DF | Adj SS | Adj MS | F-Value | P-Value |
|---|---|---|---|---|---|
| Factor | 2 | 6.500 | 3.250 | 0.93 | 0.430 |
| Error | 9 | 31.500 | 3.500 | | |
| Total | 11 | 38.000 | | | |

#### Model Summary

| S | R-sq | R-sq(adj) | R-sq(pred) |
|---|---|---|---|
| 1.87083 | 17.11% | 0.00% | 0.00% |

#### Means

| Factor | N | Mean | StDev | 95% CI |
|---|---|---|---|---|
| Salesperson1 | 4 | 12.250 | 1.708 | (10.134, 14.366) |
| Salesperson2 | 4 | 14.00 | 2.16 | (11.88, 16.12) |
| Salesperson3 | 4 | 12.750 | 1.708 | (10.634, 14.866) |

*Pooled StDev = 1.87083*

Interval Plot of Salesperson1, Salesperson2, ...
95% CI for the Mean

*The pooled standard deviation is used to calculate the intervals.*

## Two-Way ANOVA

For this we will use the salary data by age group and years of experience (see below). We are interested in knowing if there is a relationship between average salary, age and experience at $\alpha = 0.05$.

| | Salary | Years of Experience | Age |
|---|---|---|---|
| 1 | 22 | 0-4 | 25-34 |
| 2 | 27 | 0-4 | 25-34 |
| 3 | 34 | 5-9 | 25-34 |
| 4 | 36 | 5-9 | 25-34 |
| 5 | 39 | 10-14 | 25-34 |
| 6 | 41 | 10-14 | 25-34 |
| 7 | 25 | 0-4 | 35-44 |
| 8 | 35 | 0-4 | 35-44 |
| 9 | 35 | 5-9 | 35-44 |
| 10 | 45 | 5-9 | 35-44 |
| 11 | 40 | 10-14 | 35-44 |
| 12 | 50 | 10-14 | 35-44 |
| 13 | 34 | 0-4 | 45-54 |
| 14 | 36 | 0-4 | 45-54 |
| 15 | 42 | 5-9 | 45-54 |
| 16 | 48 | 5-9 | 45-54 |
| 17 | 53 | 10-14 | 45-54 |
| 18 | 57 | 10-14 | 45-54 |
| 19 | 37 | 0-4 | 55-64 |
| 20 | 43 | 0-4 | 55-64 |
| 21 | 49 | 5-9 | 55-64 |
| 22 | 51 | 5-9 | 55-64 |
| 23 | 51 | 10-14 | 55-64 |
| 24 | 59 | 10-14 | 55-64 |

1. Enter the data into columns **C1 (Salary)**, **C2 (Years of Experience)** and **C3 (Age)**.

2. Choose **Stat**, **ANOVA**, **General Linear Model** and **Fit General Linear Model**. Enter **C1** for Response, **C2** and **C3** for Factors. This would include only the main effects. If we need interaction, click on **Model**. Select the **two factors** listed in the factors and covariates box, click on **Add**. Make sure it has **2** in the drop-down box to ensure two-way interaction. Press **OK**.

3. In the main menu, again Press **OK**.

4. Observe the output shown below, the adjusted $R$-sq. is 78.45% and both main effects of Age and Years of Experience are significant.

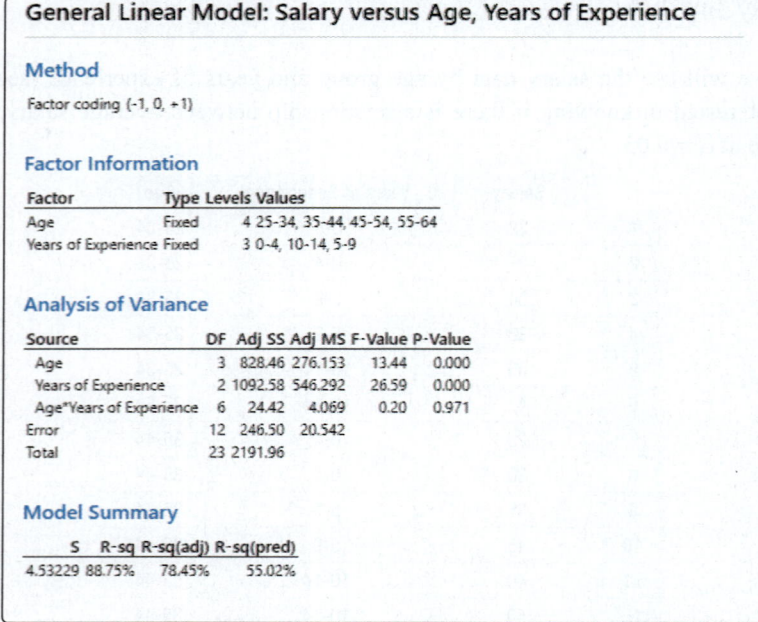

**General Linear Model: Salary versus Age, Years of Experience**

**Method**

Factor coding (-1, 0, +1)

**Factor Information**

| Factor | Type | Levels | Values |
|---|---|---|---|
| Age | Fixed | 4 | 25-34, 35-44, 45-54, 55-64 |
| Years of Experience | Fixed | 3 | 0-4, 10-14, 5-9 |

**Analysis of Variance**

| Source | DF | Adj SS | Adj MS | F-Value | P-Value |
|---|---|---|---|---|---|
| Age | 3 | 828.46 | 276.153 | 13.44 | 0.000 |
| Years of Experience | 2 | 1092.58 | 546.292 | 26.59 | 0.000 |
| Age*Years of Experience | 6 | 24.42 | 4.069 | 0.20 | 0.971 |
| Error | 12 | 246.50 | 20.542 | | |
| Total | 23 | 2191.96 | | | |

**Model Summary**

| S | R-sq | R-sq(adj) | R-sq(pred) |
|---|---|---|---|
| 4.53229 | 88.75% | 78.45% | 55.02% |

## Using RStudio

### *F*-Critical Value

Using a significance level of 0.05, numerator degrees of freedom of 40 and denominator degrees of freedom of 49, find the *F*-critical value.

1. In the console, type "qf("

2. In the first value, enter the significance level "p=0.025" followed by a comma.

3. In the second value, enter the numerator degrees of freedom "df1=40" followed by a comma.

4. In the third value, enter the denominator degrees of freedom "df2=49" followed by a comma.

5. Finally, enter "lower.tail=FALSE" and close the parentheses.

6. It should look like this: qf(p=0.025, df1=40, df2=49, lower.tail=FALSE). Note: You could also do qf(0.025, 40, 49, lower.tail = F).

7. Press **Enter** and the value of 1.802689 should appear in the cell chosen, which corresponds to the *F*-critical value.

# R Chapter 12 Review

## Key Terms and Ideas

- Experimental Units
- Treatment
- Analysis of Variance (ANOVA)
- Sum of Squares for Treatments (SST)
- Sum of Squares for Error (SSE)
- Sample Variance
- Grand Mean
- Total Sum of Squares (TSS)
- Mean Square for Treatments (MST)
- Mean Square for Error (MSE)
- Goodness-of-Fit Test for Normality
- $F$-Statistic
- $F$-Test
- $F$-Distribution
- Numerator Degrees of Freedom
- Denominator Degrees of Freedom
- Multiple Comparison Procedures
- Fisher's Least Significant Difference (LSD)
- Tukey's Honest Significant Difference (HSD)
- Completely Randomized Design

- Blocks
- Randomized Block Design
- Sum of Squares for Blocks (SSBL)
- Mean Square for Blocks (MSBL)
- Two-Way Analysis of Variance (Two-Way ANOVA)
- Complete Factorial Experiment
- Profile Plot (or Interaction Plot)
- Main Effect for Interaction
- Main Effect for Factor A
- Main Effect for Factor B
- Sum of Squares for Factor A (SSA)
- Sum of Squares for Factor B (SSB)
- Sum of Squares for Interaction (SSAB)
- Test for Interaction between Factors
- Mean Square for Interaction (MSAB)
- Masking
- Test for Main Effects for Factor A
- Mean Square for Factor A (MSA)
- Test for Main Effects for Factor B
- Mean Square for Factor B (MSB)

## Key Formulas

| | Section |
|---|---|
| **Grand Mean** $$\overline{\overline{x}} = \frac{\sum_{j=1}^{k}\sum_{i=1}^{n_j} x_{ij}}{n_T}$$ | 12.1 |
| **Sample Variance** $$s^2 = \frac{\sum_{j=1}^{k}\sum_{i=1}^{n_j}\left(x_{ij} - \overline{\overline{x}}\right)^2}{n_T - 1}$$ | 12.1 |

## Key Formulas (cont.)

| | Section |
|---|---|

**Sum of Squares for Treatments**

$$\text{SST} = \sum_{j=1}^{k} n_j \left( \bar{x}_j - \bar{\bar{x}} \right)^2$$

12.1

**Mean Square for Treatments**

$$\text{MST} = \frac{\text{SST}}{k-1}$$

12.1

**Sum of Squares for Error**

$$\text{SSE} = \sum_{i=1}^{n_1} \left( x_{i1} - \bar{x}_1 \right)^2 + \sum_{i=1}^{n_2} \left( x_{i2} - \bar{x}_2 \right)^2 + \cdots + \sum_{i=1}^{n_k} \left( x_{ik} - \bar{x}_k \right)^2$$

12.1

**Total Sum of Squares**

$$\text{TSS} = \text{SST} + \text{SSE} = \sum_{j=1}^{k} \sum_{i=1}^{n_j} \left( x_{ij} - \bar{\bar{x}} \right)^2$$

12.1

**Mean Square for Error**

$$\text{MSE} = \frac{\text{SSE}}{n_T - k}$$

12.1

***F*-Statistic**

$$F = \frac{\text{MST}}{\text{MSE}}$$

12.3

***F*-Statistic for One-Way ANOVA**

$$F = \frac{\text{MST}}{\text{MSE}} = \frac{\dfrac{\sum_{j=1}^{k} n_j \left( \bar{x}_j - \bar{\bar{x}} \right)^2}{k-1}}{\dfrac{\sum_{i=1}^{n_1} \left( x_{i1} - \bar{x}_1 \right)^2 + \sum_{i=1}^{n_2} \left( x_{i2} - \bar{x}_2 \right)^2 + \cdots + \sum_{i=1}^{n_k} \left( x_{ik} - \bar{x}_k \right)^2}{n_T - k}}$$

12.3

## Key Formulas (cont.)

| | Section |
|---|---|

### Computational Formula for MST

$$\text{MST} = \frac{\left[\dfrac{\left(\sum\limits_{i=1}^{n_1} x_{i1}\right)^2}{n_1} + \dfrac{\left(\sum\limits_{i=1}^{n_2} x_{i2}\right)^2}{n_2} + \cdots + \dfrac{\left(\sum\limits_{i=1}^{n_k} x_{ik}\right)^2}{n_k}\right] - \dfrac{\left(\sum\limits_{j=1}^{k}\sum\limits_{i=1}^{n_j} x_{ij}\right)^2}{n_T}}{k-1}$$

12.3

### Computational Formula for MSE

$$\text{MSE} = \frac{\sum\limits_{j=1}^{k}\sum\limits_{i=1}^{n_j} x_{ij}^2 - \left[\dfrac{\left(\sum\limits_{i=1}^{n_1} x_{i1}\right)^2}{n_1} + \dfrac{\left(\sum\limits_{i=1}^{n_2} x_{i2}\right)^2}{n_2} + \cdots + \dfrac{\left(\sum\limits_{i=1}^{n_k} x_{ik}\right)^2}{n_k}\right]}{n_T - k}$$

12.3

### Fisher's Least Significant Difference Method

$$\left|\overline{x}_i - \overline{x}_j\right| \geq t_{\alpha/2,\, n_T - k}\sqrt{\text{MSE}\left(\frac{1}{n_i} + \frac{1}{n_j}\right)}$$

12.4

### Fisher's LSD Method, Confidence Interval Approach

$$\left(\overline{x}_i - \overline{x}_j\right) \pm t_{\alpha/2,\, n_T - k}\sqrt{\text{MSE}\left(\frac{1}{n_i} + \frac{1}{n_j}\right)}$$

12.4

### Tukey's Honest Significant Difference Method

$$\left|\overline{x}_i - \overline{x}_j\right| \geq q_{\alpha, k,\, n_T - k}\sqrt{\frac{\text{MSE}}{2}\left(\frac{1}{n_i} + \frac{1}{n_j}\right)}$$

12.4

### Tukey's HSD Method, Confidence Interval Approach

$$\left(\overline{x}_i - \overline{x}_j\right) \pm q_{\alpha, k,\, n_T - k}\sqrt{\frac{\text{MSE}}{n}} \quad \text{for \textbf{balanced data} } (n = n_i = n_j)$$

or

$$\left(\overline{x}_i - \overline{x}_j\right) \pm q_{\alpha, k,\, n_T - k}\sqrt{\frac{\text{MSE}}{2}\left(\frac{1}{n_i} + \frac{1}{n_j}\right)} \quad \text{for \textbf{unbalanced data} } (n_i \neq n_j)$$

12.4

## Key Formulas (cont.)

| | Section |
|---|---|

**F-Statistic for Two-Way ANOVA (Randomized Block Design)**

$$F = \frac{\text{MST}}{\text{MSE}} = \frac{\dfrac{\text{SST}}{k-1}}{\dfrac{\text{SSE}}{(k-1)(b-1)}}$$

12.5

where $\text{SSE} = \text{TSS} - \text{SST} - \text{SSBL}$

**Test Statistic for Interaction between Factors**

$$F = \frac{\dfrac{\text{SSAB}}{(a-1)(b-1)}}{\dfrac{\text{SSE}}{ab(r-1)}} = \frac{\text{MSAB}}{\text{MSE}}$$

12.6

**Test Statistic for Main Effects for Factor A**

$$F = \frac{\dfrac{\text{SSA}}{(a-1)}}{\dfrac{\text{SSE}}{ab(r-1)}} = \frac{\text{MSA}}{\text{MSE}}$$

12.6

**Test Statistic for Main Effects for Factor B**

$$F = \frac{\dfrac{\text{SSB}}{(b-1)}}{\dfrac{\text{SSE}}{ab(r-1)}} = \frac{\text{MSB}}{\text{MSE}}$$

12.6

# AE  Additional Exercises

1.  An experimenter often uses a randomized block design to reduce variation by comparing the treatments in homogeneous groups of experimental units called blocks. In many cases, differences in treatments are more likely to be detected with such a design than if the blocking factor were ignored. In each of the following situations, give an example of how one would run a randomized block design to make the comparison in each case. Be aware that there may be more than one correct answer in each example.

    a.  Three different methods of teaching science are to be analyzed by comparing final exam scores from classes taught by the different methods. Assume that the same final exam is given to each class.

    b.  Five hypertensive treatments are to be compared based on their ability to reduce systolic blood pressure. It is felt that the performance of the drugs is affected by the weight of the participant in the study.

    c.  It is desired to compare three different ethnic groups on their knowledge of American history. Each person in the study will be given a 50-question multiple choice test to determine their overall knowledge of the subject.

2.  A pharmacist is interested in studying the rate at which three different sinus headache drugs are absorbed into the bloodstream. She randomly selects 12 people, and then randomly assigns four people to try each drug. She administers the drug to each participant and measures the time it takes for the drug to be absorbed into the patient's bloodstream (in minutes). The results of the study are as follows.

| Drug Absorption Time (Minutes) | | |
|---|---|---|
| Drug 1 | Drug 2 | Drug 3 |
| 5 | 10 | 6 |
| 4 | 11 | 7 |
| 6 | 9 | 5 |
| 3 | 8 | 5 |

    a.  Can the pharmacist conclude at $\alpha = 0.01$ that there is a significant difference among the average times required for absorption into the bloodstream for the three drugs?

    b.  What assumptions did the pharmacist make in performing the test procedure in part a.? Do the data appear to satisfy these assumptions? Explain.

    c.  Describe an alternate design that the pharmacist could have used for the above analysis. What are the advantages and disadvantages of this design?

3.  An FDA representative is interested in knowing if there is a difference in the average fat contents of three different brands of margarine. The representative randomly selects six samples of each of the brands of margarine and measures the average fat contents per serving. The results of the study are displayed in the following table.

| Fat Content per Serving (Grams) | | |
|---|---|---|
| Margarine #1 | Margarine #2 | Margarine #3 |
| 6 | 5 | 9 |
| 7 | 6 | 8 |
| 6 | 5 | 7 |
| 8 | 4 | 8 |
| 6 | 6 | 9 |
| 8 | 5 | 7 |

**a.** Do the data indicate a difference among average fat contents per serving for the three brands of margarine at $\alpha = 0.01$?

**b.** What assumptions were made for the test in part **a.**? Do the data appear to satisfy these assumptions? Explain.

**c.** Why wouldn't a randomized block design be appropriate for this experiment?

**4.** Psychological reactance may be viewed as the motivational state resulting when someone's freedom is threatened or eliminated. A study relating psychological reactance to one's age was reported in "Psychological Reactance: Effects of Age and Gender" in the *Journal of Social Psychology*. In order to determine the degree of psychological reactance, participants were asked to fill out a questionnaire which was then scored. The higher the score, the more acute the degree of psychological reactance. The means, standard deviations, and group sizes (for different age groups) are given in the following table.

| Psychological Reactance | | | |
|---|---|---|---|
| Age Group | Mean | Standard Deviation | Group Size |
| 18–24 | 3.36 | 0.60 | 1011 |
| 24–29 | 3.28 | 0.65 | 321 |
| 30–40 | 3.16 | 0.64 | 385 |

Although the summary statistics were given in the article, the actual data values upon which the statistics were based were not listed. This is standard procedure in many scientific journals.

**a.** Compute the sums of squares and their degrees of freedom for treatments and error based upon the statistics given in the table.

**b.** Compute MST and MSE.

**c.** With $\alpha = 0.01$, can we conclude that there is a significant difference among the degrees of psychological reactance for the different age groups?

**d.** What assumptions are necessary for performing the test in part **c.**? Can they be checked in this instance?

**5.** Interviews of fans following an Australian Football League game were summarized in the article "On Being a Sore Loser: How Fans React to Their Team's Failure" in the *Australian Journal of Psychology*. The study divided the fans interviewed into losers (those who supported the losing team), winners (those who supported the winning team), and non-partisans (those who were indifferent to the outcome of the game). Each fan was asked several questions, all dealing with the fan's perceptions of the game. The purpose of the study was to see if the groups differed on their responses to any of the questions. One question asked the fans to rate the umpire's performance on a five-point scale from *very bad* (1) to *very good* (5). The mean responses and group sizes associated with this question are given in the following table.

| Umpire Performance | | |
|---|---|---|
| Group | Mean | Group Size |
| Losers | 2.8 | 49 |
| Winners | 3.7 | 35 |
| Non-Partisans | 3.5 | 57 |

**a.** Compute the grand mean and the sum of squares for treatments (SST).

**b.** Compute the $F$-statistic for testing for equality of the group means. The sum of squares for error (SSE) was given in the article as 39.3.

**c.** With $\alpha = 0.01$, can we conclude that there is a significant difference among the groups in the perception of the umpire?

**d.** What assumptions are necessary for performing the test in part **c.**? Can they be checked in this instance?

**6.** Consider the following data regarding median starting and mid-career salaries for graduates from schools in different regions of the United States.

| Starting and Mid-Career (Salaries by Region) | | |
|---|---|---|
| **School** | **Starting Median Salary ($)** | **Mid-Career Median Salary ($)** |
| **Midwest** | | |
| Notre Dame | 52,900 | 107,000 |
| Carleton College | 42,800 | 98,300 |
| Illinois Institute of Technology | 52,000 | 96,000 |
| Denison University | 40,600 | 94,000 |
| University of Chicago | 46,900 | 92,700 |
| Northwestern University | 49,900 | 88,300 |
| Washington University in St. Louis | 51,200 | 87,700 |
| **Northeast** | | |
| Princeton | 56,900 | 130,000 |
| Harvard | 54,100 | 116,000 |
| Massachusetts Institute of Technology | 69,700 | 115,000 |
| Dartmouth | 51,600 | 114,000 |
| Bucknell University | 52,600 | 108,000 |
| Manhattan College | 53,900 | 107,000 |
| Williams College | 51,800 | 105,000 |
| **South** | | |
| Duke University | 54,400 | 113,000 |
| Vanderbilt University | 51,300 | 100,000 |
| Washington and Lee University | 48,600 | 99,800 |
| Wake Forest University | 46,000 | 98,800 |
| Rice University | 51,100 | 97,400 |
| Georgetown University | 50,300 | 96,900 |
| College of William and Mary | 45,000 | 96,500 |
| **West** | | |
| University of Colorado – Boulder | 45,900 | 90,400 |
| University of Washington | 46,700 | 88,400 |
| Gonzaga University | 44,200 | 87,700 |
| Brigham Young University | 47,400 | 86,800 |
| University of Arizona | 45,400 | 81,600 |
| University of Oregon | 39,700 | 79,200 |
| Santa Clara University | 52,900 | 105,000 |

**Source:** Payscale.com

a. Using $\alpha = 0.05$, is there a significant difference among average starting salaries for the four different regions?

b. What is the value of $F$ for this test?

c. Using $\alpha = 0.05$, is there a significant difference among average mid-career salaries for the four different regions?

d. State the assumptions made for the two hypothesis tests in parts **a.** and **c.**

e. Do you have any concerns about these data? Explain.

7. Consider the following data regarding the median starting and mid-career salaries by type of major.

### Starting and Mid-Career Salaries by Major

| Major | Starting Median Pay ($) | Mid-Career Median Pay ($) | Major | Starting Median Pay ($) | Mid-Career Median Pay ($) |
|---|---|---|---|---|---|
| **Engineering** | | | **Math and Science** | | |
| Petroleum Engineering | 97,900 | 155,000 | Applied Mathematics | 52,600 | 98,600 |
| Chemical Engineering | 64,500 | 109,000 | Computer Science | 56,600 | 97,900 |
| Electrical Engineering | 61,300 | 103,000 | Statistics | 49,000 | 93,800 |
| Aerospace Engineering | 60,700 | 102,000 | Mathematics | 47,000 | 89,900 |
| Computer Engineering | 61,800 | 101,000 | Physics | 49,800 | 101,000 |
| Nuclear Engineering | 65,100 | 97,800 | Biochemistry | 41,700 | 84,700 |
| Biomedical Engineering | 53,800 | 97,800 | Food Science | 43,300 | 83,700 |
| Mechanical Engineering | 58,400 | 94,500 | Geology | 45,300 | 83,300 |
| Industrial Engineering | 57,400 | 93,100 | Molecular Biology | 40,500 | 81,200 |
| Civil Engineering | 53,100 | 90,200 | Chemistry | 42,000 | 80,900 |
| Environmental Engineering | 51,700 | 88,600 | **Other** | | |
| **Business** | | | Economics | 47,300 | 94,700 |
| Finance | 46,500 | 87,300 | Film Production | 41,600 | 80,700 |
| Supply Chain Management | 50,200 | 84,700 | Political Science | 39,900 | 80,100 |
| International Business | 41,600 | 83,700 | International Relations | 40,500 | 79,400 |
| Accounting | 44,700 | 75,700 | Philosophy | 39,800 | 75,600 |
| Advertising | 37,700 | 74,700 | History | 37,800 | 69,000 |
| Marketing | 38,200 | 73,500 | Communications | 38,000 | 66,900 |
| Business | 41,000 | 70,500 | Journalism | 36,100 | 66,400 |
| Public Relations | 35,500 | 65,700 | Spanish | 36,400 | 58,400 |

**Source:** Payscale.com

a. State the null and alternative hypotheses to test if there is a significant difference in starting salary among the four types of majors.

b. Use $\alpha = 0.01$ to test for a significant difference among average starting salaries for the different types of majors.

c. State the null and alternative hypotheses to test if there is a significant difference among average mid-career salaries for the four types of majors.

d. Use $\alpha = 0.01$ to test for a significant difference among average mid-career salaries for the different types of majors.

8. Consider the following partially completed ANOVA table for a $3 \times 4$ factorial experiment with two replications.

ANOVA

| Source of Variation | SS | df | MS | F |
|---|---|---|---|---|
| Factor A | 0.800 | 2 | | |
| Factor B | 5.300 | 3 | | |
| Interaction | 9.600 | | | |
| Within | | | | |
| Total | 17.000 | 26 | | |

a. Complete the ANOVA table.

b. At the 0.05 level, is there evidence of significant interaction between A and B? Justify your answer.

c. At the 0.05 level, is there evidence of a significant Factor A effect? Justify your answer.

d. At the 0.05 level, is there evidence of a significant Factor B effect? Justify your answer.

e. Does the result of the test for interaction suggest further investigation? Justify your answer.

9. In an experiment to determine the best method by which to assess college students, a group of students were exposed to one of three types of tests. The three methods were: all multiple choice questions, all free-response questions, and mixed questions (a mixture of multiple choice and free-response questions). The scores were recorded for each test taken. Fifteen students were used in the study and were grouped by class level (freshman, sophomore, junior, senior, and graduate). The following table contains the results of the experiment.

| | Testing Methods | | |
|---|---|---|---|
| Class Level | Multiple Choice | Free-Response | Mixed |
| Freshman | 78 | 84 | 90 |
| Sophomore | 82 | 90 | 95 |
| Junior | 90 | 94 | 98 |
| Senior | 88 | 96 | 100 |
| Graduate | 95 | 98 | 99 |

a. Graphically plot the test scores by class level and testing method. Discuss the graph.

b. Perform an analysis of the data using the class-level blocks. Are blocking effects significant at the 0.05 level of significance? Explain.

c. Is the experiment useful having been analyzed as a completely randomized block design? Explain.

10. A randomized block design yielded the following ANOVA table.

| ANOVA | | | | |
|---|---|---|---|---|
| Source of Variation | SS | df | MS | F |
| Treatment | 500.000 | 5 | 100 | 7.502 |
| Block | 230.000 | 3 | 76.67 | 5.752 |
| Error | 120.000 | 9 | 13.33 | |
| Total | 850.000 | 17 | | |

a. How many blocks are used in the experiment?

b. How many treatments are used in the experiment?

c. How many observations are used in the experiment?

d. What are the null and alternative hypotheses to test if there is a difference among the treatment means?

e. What test statistic should be used to conduct the test in part d.?

f. What is the rejection region for the test in parts d. and e.?

g. Carry out the test and state your conclusion based on a significance level of 0.05.

11. JAS & Associates, a commercial developer, usually gets three cost estimates for many of the jobs for their building projects. Even though one contractor normally works on each potential job, it is in the best interest of the company to get additional estimates and compare them for consistency, no matter who gets the job. To check the consistency of the estimates, several projects are selected and three contractors are asked to submit estimates. The estimates (in thousands of dollars) for the 10 jobs are given in the following table.

| Contractor Cost Estimates (Thousands of Dollars) | | | |
|---|---|---|---|
| Job | Contractor A | Contractor B | Contractor C |
| 1 | 27 | 26 | 28 |
| 2 | 20 | 18 | 22 |
| 3 | 14 | 13 | 17 |
| 4 | 18 | 21 | 20 |
| 5 | 23 | 20 | 22 |
| 6 | 19 | 17 | 19 |
| 7 | 12 | 14 | 15 |
| 8 | 10 | 12 | 13 |
| 9 | 16 | 20 | 19 |
| 10 | 40 | 42 | 47 |

a. Perform the appropriate analysis on the data given and generate the ANOVA table for the analysis.

b. Do the data provide sufficient evidence to conclude that there is a difference among the cost estimates supplied by the contractors? Use a significance level of 0.05 to make your decision.

c. What is the $P$-value for the test performed in part a.? Interpret this value.

12. The following table is a $3 \times 3$ factorial design with three observations for each factor level.

| Factorial Design Data | | | |
|---|---|---|---|
| | | Factor B | |
| Factor A | 1 | 2 | 3 |
| 1 | 30 | 47 | 36 |
| | 30 | 42 | 37 |
| | 30 | 42 | 38 |
| 2 | 12 | 27 | 35 |
| | 14 | 24 | 31 |
| | 15 | 22 | 33 |
| 3 | 10 | 34 | 24 |
| | 13 | 31 | 20 |
| | 12 | 31 | 22 |

a. Plot the treatment means using Factor A as the $x$-axis and Factor B as plotting symbols. Do the means appear to be different? Does interaction between factors A and B appear to be present? Justify your answers.

b. Perform the analysis using a software package, generating the ANOVA table.

c. Test for significant interaction using a 0.05 level of significance. Discuss your findings.

d. Test for A and B effects using a 0.05 level of significance. Discuss your findings.

13. Tech SportsPlex (TSP) is conducting a study to determine the effectiveness of three types of marketing/advertising methods: e-coupons, newspaper ads, and price discounts. Three counties (believed to be of equal size and close driving distance to TSP) were selected for the marketing campaign. Each strategy was used for a three-month period. It is known that the sales would be seasonal (i.e. TSP's management expects less activity during the summer months). The revenue data (in thousands of dollars) from the study are given in the following table.

| TSP Revenues by Marketing Strategy (Thousands of Dollars) | | | |
|---|---|---|---|
| Quarter | e-Coupons | Newspaper Ads | Price Discounts |
| 1 | 48 | 42 | 37 |
| 2 | 25 | 18 | 21 |
| 3 | 20 | 15 | 18 |
| 4 | 40 | 30 | 24 |

a. Specify the null and alternative hypotheses to determine if there is a significant difference among average revenues for the three advertising strategies.

b. Generate the ANOVA table to test the hypotheses in a.

c. Conduct the test in a. using a significance level of 0.05.

d. Was the variation among the observed revenues significantly reduced by blocking? Explain using $\alpha = 0.05$.

## P   Discovery Project

Economists use several metrics to understand the state of the economy and what might lie ahead. Some of those metrics specifically focus on consumer spending. One such measure is consumer consumption. This value gives economists an idea of how much consumers are spending and is tracked monthly, in order to understand how consumer spending impacts the overall economy.

One way that consumers choose to pay for goods and services is through using credit cards. Obtaining a credit card is generally simple and most consumers can qualify for some type of card. The terms of the credit issued can vary though based on variables such as household income, credit score, and educational status.

∴ **Data**

This data set can be found at stat.hawkeslearning.com by navigating to **Discovering Business Statistics, Second Edition > Data Sets > Credit Card Data.**

Using the data set Credit Card Data, answer the following questions to better understand how differences in certain factors can impact spending and credit limits. The data set has nine variables (159 data points).

1. Download the data file and open it in Microsoft Excel.

2. Determine the mean, mode, median, maximum, minimum, standard deviation, and the coefficient of variation of the following variables: age, total credit limit, total balance, credit score, annual household income, and number of children and briefly discuss the results. (**Hint**: These values can be quickly calculated using the Data Analysis Add-in: Descriptive Statistics in Excel.)

3. Fully summarize the qualitative variables (i.e., what percent of the sample has a college degree?) and briefly discuss your findings. (**Hint**: These values can be quickly determined using the Data Analysis Add-in: Histogram in Excel.)

4. Determine if there is a significant difference in credit card balances based on educational status using the appropriate hypothesis test. Use a significance level of 0.05. If significant differences are found, determine which groups are different from each other. (Note: It will be necessary to rearrange the dataset to conduct the hypothesis test.)

5. Choose one of the other qualitative variables (i.e., housing status) and determine if there is a significant difference in credit card limits based on that variable. Use the significance level of 0.01. If significant differences are found, determine which groups are different from each. (Note: It will be necessary to rearrange the dataset to conduct the hypothesis test.)

6. Conduct the appropriate hypothesis test to see if there is a difference in credit card limits by both marital status and housing status. Use the significance level of 0.01. (Note: It will be necessary to rearrange the dataset to conduct the hypothesis test.)

7. Partition and sort the data for credit score based on the marital status and number of children in the household. Randomly select 10 data points where there are no kids in the household, then do the same for those households with children, also include the marital status. This data selection will be used in the next problem.

8. Conduct the appropriate hypothesis test to see if there is a difference in credit scores by both marital status and number of children in the household. Use the significance level of 0.05. Is there any interaction present between the variables? (Note: You will use the data subset you selected in the previous problem. Be sure the data is formatted appropriately for the selected hypothesis test.)

9. Briefly summarize your findings from these analyses and discuss the limitations present in the data set.

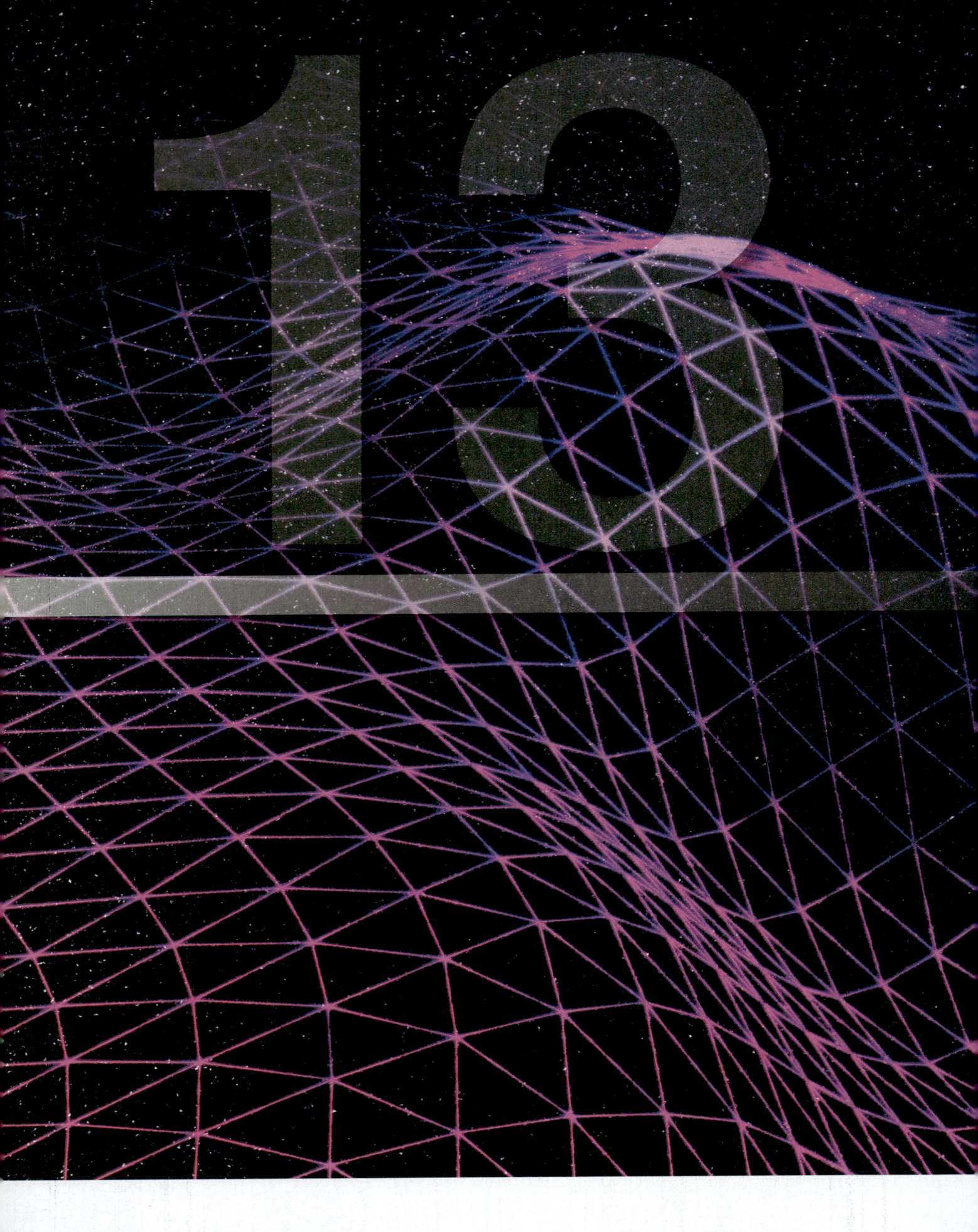

# Chapter 13
## Regression, Inference, and Model Building

## Discovering the Real World

### What is the Relationship Between Oil Prices and Gas Prices?

You may have noticed that the gas prices at the pump tend to fluctuate daily. You may wonder what impacts the gas prices. You may also wonder about the relationship between crude oil prices and the price of gas at the pump. Crude oil prices tend to create huge changes in prices at the pump. In addition to the price of crude oil, the fluctuation at the pump is a function of several things such as refinery and distribution costs, corporate profits, state and federal taxes, as well as location. In addition to the aforementioned items that are factored into the price of gas at the pump, the type of crude oil also makes a difference in the price we pay at the pump. In the United States, West Texas Intermediate (WTI) is the crude oil that is used as the benchmark in oil pricing. The international (primarily in Europe and Africa) benchmark for oil pricing is Brent Blend. WTI is higher quality because it is lightweight and has low sulfur content. The Brent Blend is a combination of crude oil from a number of oil fields in the North Sea which is heavier than WTI but still good for making gasoline.

The figure below will give you an indication of what you pay for per gallon of regular grade gas.

**What we pay for in a gallon of:**

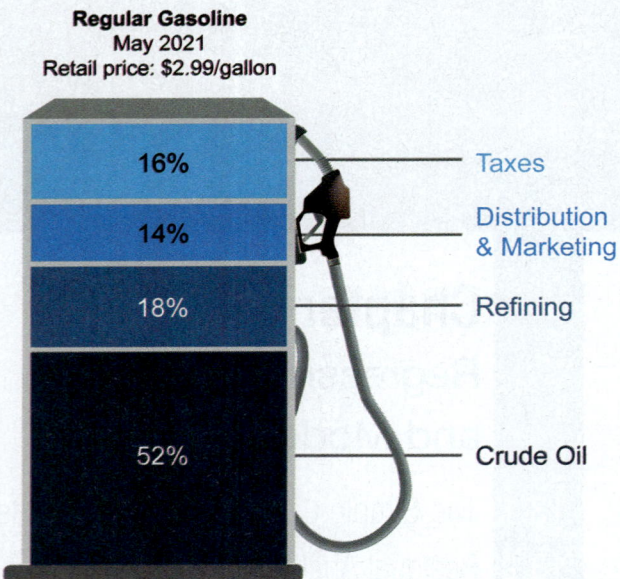

In May of 2021, the average price per barrel of Brent Crude oil was $68.53 (or $1.63 per gallon since there are 42 gallons in a barrel). As noted in the figure, the majority of the price is a function of the price of crude oil. Please note that the percentages in the figure are national averages.

The two figures below are scatterplots of the price of regular gas per gallon against the price of WTI crude and the Brent Blend crude. You will notice that there is a linear (i.e., straight line) relationship between the price of Brent Blend and the price of regular gas; there is also a linear relationship between the price of WTI and the price of regular gas. Viewing the scatterplots, you can also see that the intercept of Brent Blend is higher, but the scatterplot of the Brent Blend has a slightly flatter slope. The flatter slope may be an indication that the Brent Blend may not have as much impact on the price at the pump as the Brent Blend price increases in relation to the increase in price per gallon of WTI.

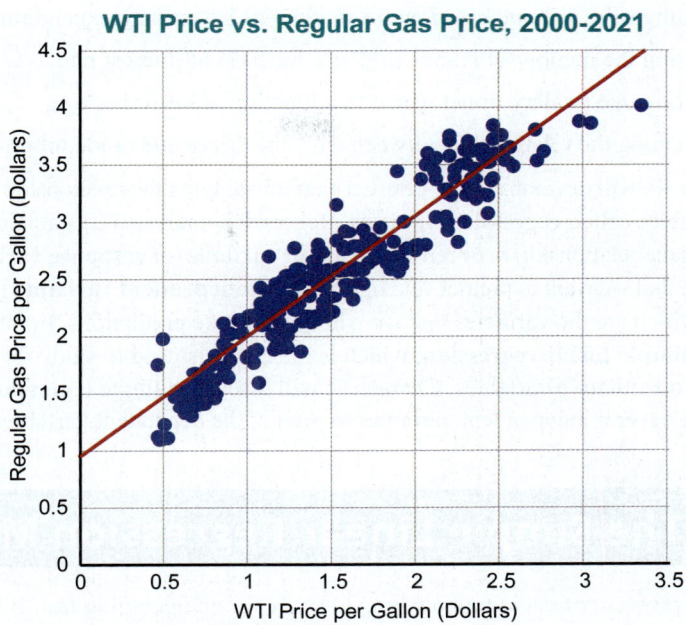

WTI Price vs. Regular Gas Price, 2000-2021

Brent Blend Price vs. Regular Gas Price, 2000-2021

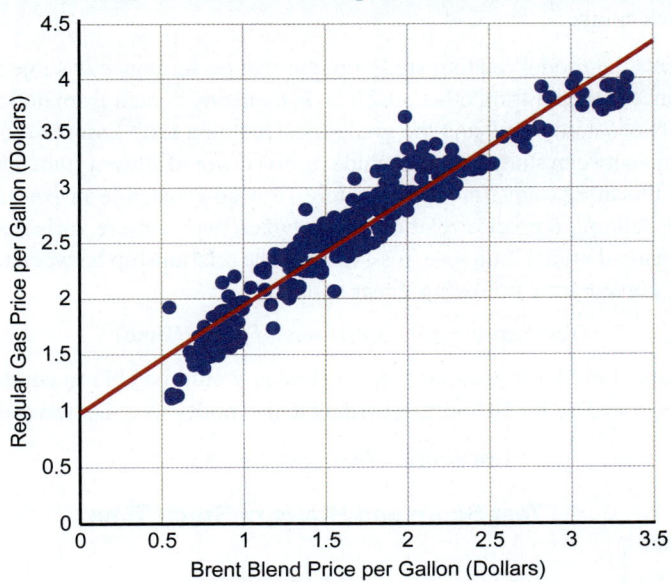

Given the brief introduction you have to gas prices, you are now prepared to answer the question about the relationship between crude oil prices and gas prices. In this chapter, we will study simple linear regression which we can use to model the relationship between crude oil (WTI or Brent Blend) prices and gas prices. We can develop a regression model, fit the model to the data (i.e., find the best line to fit the data as in the figures above), validate the assumptions, test if the model is appropriate to make predictions, and predict the price at the pump for a given price of crude oil.

## Introduction

In Chapter 4, we discussed bivariate data and how to look for patterns in data to help us establish a relationship between two variables. In particular, we discussed establishing a linear relationship between two variables. Businesses are interested in developing relationships between two variables for many reasons such as the following.

- Predicting sales volume based on the amount of advertising expenditures
- Predicting the number of homes sold as a function of interest rate
- Predicting one's salary or net worth as a function of his or her age
- Determining the relationship between gasoline prices and crude oil prices

In this chapter we will develop a more detailed plan for studying the relationship between two or more variables called **regression analysis**. Regression analysis is a statistical technique that describes the relationship between a **dependent variable** (or **response variable**), which is the variable that we want to predict, and one or more **independent variables** (or **predictor variables**), which are the variables that we will use to make predictions. In this chapter we will discuss **simple linear regression**, which is the analysis used to study the relationship between two quantitative variables. Chapter 14 will discuss multiple regression models, in which we use several independent variables to predict the dependent variable.

# 13.1 The Simple Linear Regression Model

Consider the problem of deciding how long to study for an upcoming test. If we knew the exact relationship between time spent studying and the grade received, it could be useful in allocating study time. But the exact relationship between these variables is unknown and different for each course.

For most students the model relating study time to test performance is subjective, relying on past experience and data from other students. Is there any benefit from defining an exact relationship between study time and the grade received on a test? Essentially, knowing an exact relationship between study time and grade received would allow a student to choose his or her grade by allocating study time appropriately. How do you define an exact relationship? One method of defining a precise relationship between two or more variables is with the use of a mathematical model. Suppose, for example, the relationship between test score and study time was given by the following linear equation.

$$\text{Test Score} = 45 + 3.8(\text{Hours of Study Time})$$

If this mathematical model is accurate, then a student would be able to control his or her destiny. If a person studied for 10 hours, according to the model his or her test score would be:

$$\text{Test Score} = 45 + 3.8(10) = 83.$$

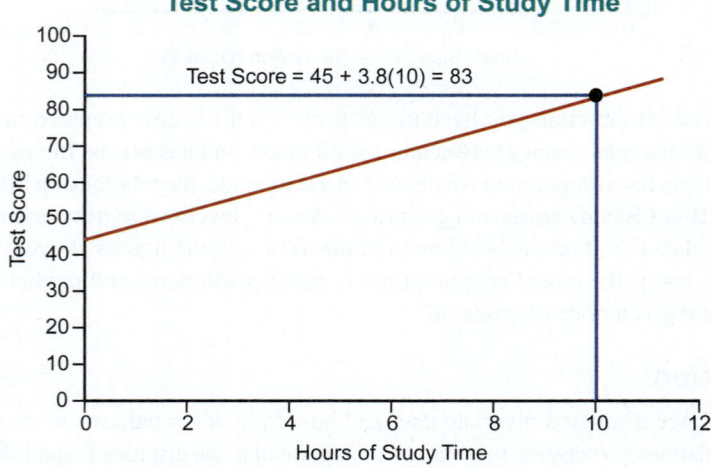

**Test Score and Hours of Study Time**

Figure 13.1.1

To get a higher test score, then study for 12 hours.

$$\text{Test Score} = 45 + 3.8(12) = 90.6$$

Since the relationship is known in advance, students would be able to control their grade by choosing their study time appropriately. In fact, if the exact relationship were known in advance, taking the test would be necessary only as an intellectual exercise, since the grade would already be determined by the time studied and the model. Admittedly, there is no model that can precisely predict a test score solely on the basis of time studied; since there are many other variables that affect test scores. But suppose a model were available which, although imperfect, fairly reliably predicted test scores based on hours studied.

$$\text{Test Score} = 45 + 3.8(\text{Hours of Study Time}) + \text{Error}$$

The new model introduces the error term. Now, if someone studies 10 hours, the model would predict

$$\text{Test Score} = 45 + 3.8(10) + \text{Error} = 83 + \text{Error}.$$

The predicted test score would still be 83, but there is an unknown random error associated with the prediction. If the error is reasonably small (say, at most 5 points), then the prediction will still be useful for planning purposes. But if the error is too large, then it will be difficult to rely on the model's predictions. If a model admits the possibility of an error, then gauging the expected magnitude of the error is essential in determining the model's usefulness. Estimating the mean and variance of the errors will be an important part of determining model utility. A model with a mean error of zero and small variation in the error terms would be desirable and should yield useful predictions. The model we have been using is simple. If two variables appear to be related in a straight-line manner, we can use a **simple linear regression model** to describe their relationship.

The model-building process begins with a desire to find a relationship between two or more variables. Since there are a great number of possible models that can be selected, choosing the type of model to represent the relationship is a complex problem. A straight line is the simplest relationship between two variables. This straight-line relationship is modeled by the simple linear regression model given by the following linear equation.

## Formula

### Simple Linear Regression Model

The **simple linear regression model** is given by the linear equation

$$y_i = \beta_0 + \beta_1 x_i + \varepsilon_i$$

where

$\beta_0$ is the $y$-intercept for the population data,

$\beta_1$ is the slope coefficient for the population data,

$x_i$ is the value of the independent (or predictor) variable for observation $i$,

$\varepsilon_i$ is the random error in $y$ for observation $i$, and

$y_i$ is the value of the dependent (or response) variable for observation $i$.

$\beta_0$ is pronounced *beta sub-zero* (or *beta naught*) and $\beta_1$ is pronounced *beta sub-one*. $y_i = \beta_0 + \beta_1 x_i$ is the portion of the model that represents a straight line. The slope of the line, denoted by $\beta_1$, is the expected (or average) change in $y$ for each unit change in $x$. That is, it represents the average amount that $y$ changes for every one unit that $x$ changes. The $y$-intercept, denoted by $\beta_0$, represents the average value of $y$ when $x$ is equal to zero. The last term in the model, $\varepsilon_i$, represents the random error in $y$ for each observation. That is, $\varepsilon_i$ represents the vertical distance that the *actual* value of $y_i$ is above or below the *average* value of $y_i$ on the regression line.

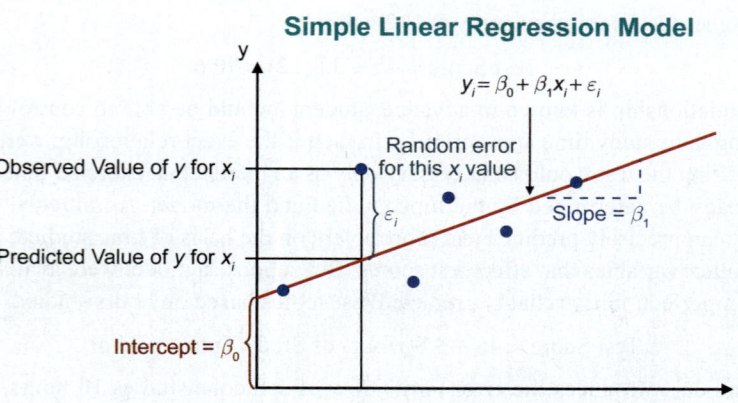

**Figure 13.1.2**

The slope and the intercept, $\beta_0$ and $\beta_1$, respectively, are the **parameters** of the simple linear regression model. The parameters completely define the equation of the line. Developing a model to fit real-world measurements is not trivial. Nature doesn't cooperate by requiring all relationships to be straight lines. Seldom, in fact, do pairs of measurements fall on perfectly straight lines. If a linear relationship exists, the data will have some general tendency to move together, or in opposite directions, as in Figure 13.1.3. In the following sections we will look at ways of measuring the degree of a linear relationship between two variables, as well as estimating the parameters (slope and intercept) of a line for a specific set of data.

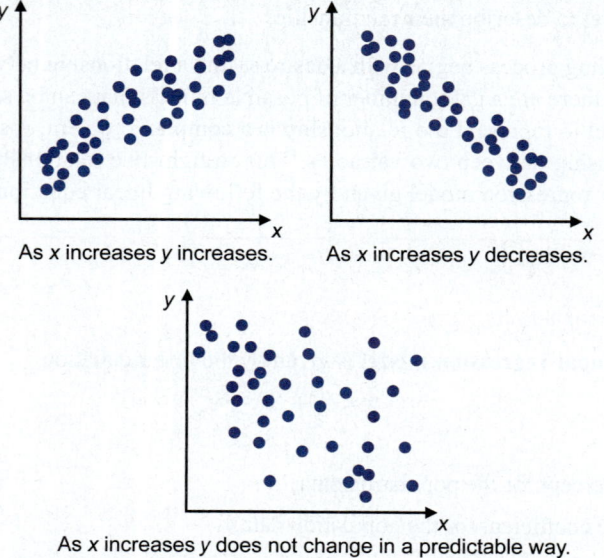

**Figure 13.1.3**

More often than not, bivariate data for an entire population are not available. Therefore, we must estimate the simple linear regression model using sample data. Using the data collected from the sample, we can estimate the coefficients of the **simple linear regression equation** and use this model for estimation and prediction.

> **Formula**
>
> ### Estimated Simple Linear Regression Equation
>
> The **estimated simple linear regression equation** is
>
> $$\hat{y}_i = b_0 + b_1 x_i \,,$$
>
> where $b_0$ and $b_1$ are estimates of their population counterparts. Specifically,
>
> $b_0$ is an estimate of $\beta_0$, and $b_1$ is an estimate of $\beta_1$.
>
> $\hat{y}_i$ is the predicted value of $y$ for a given value of $x_i$, and is pronounced *y-hat*. The symbol $y_i$ is reserved for the observed value of $y$.

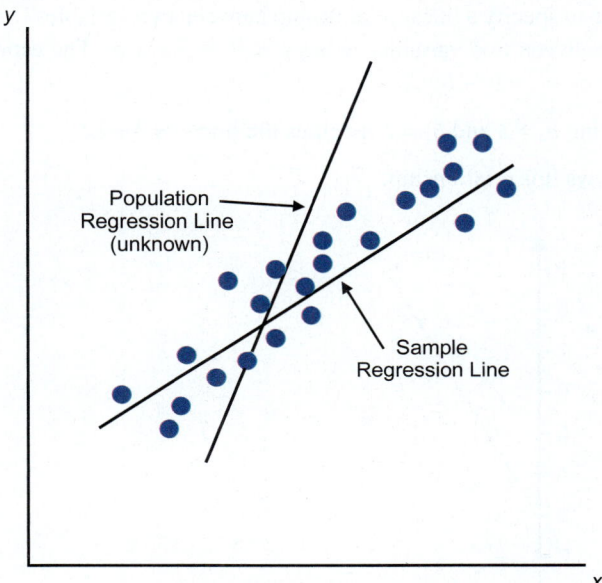

**Figure 13.1.4**

Essentially the regression model is used as a descriptive tool to summarize the relationship between the two variables. By providing estimates of $\beta_0$ and $\beta_1$, the simple linear relationship between $x$ and $y$ is specified. After data are collected, the **method of least squares** is the technique used to estimate $\beta_0$ and $\beta_1$. This method will be discussed later in the section.

Suppose we want to estimate the relationship between weight and height throughout the world for persons over the age of 18. Assuming the relationship is fundamentally linear, how is the relationship estimated? The number of persons on the earth over age 18 is somewhere in the neighborhood of 5.5 billion. Collecting 5.5 billion heights and weights could be a formidable task. In its current form, the problem is too large. Suppose the problem is reduced by only considering the population of the United States, where there are roughly 250 million persons over age 18. Even though this reduces the size from 5.5 billion to 250 million, it is still too large.

 To make progress on this problem requires the use of sampling and inference. It is quite conceivable that a small random sample of bivariate measurements (height and weight) can be drawn from persons over 18 in the United States. Since all 250 million measurements will never be available, the best that we will be able to do is to infer that the relationship constructed from the sample will be representative of the 250 million population members.

If all 250 million bivariate measures were available, then the exact values of $\beta_0$ and $\beta_1$ could be determined. Since the entire population will never be at our disposal, we are confronted with a familiar sampling problem. Even if a random sample from the population is drawn, there can be no guarantee the sample will be exactly representative of the population. Since there is no guarantee that the sample will be representative, when a linear relationship is estimated from sample data we need to address the accuracy of the sample estimates ($b_0$ and $b_1$). Two familiar inferential techniques, confidence intervals and hypothesis testing, will be used to analyze the model's estimated coefficients and predicted values.

## Defining a Linear Relationship

In Chapter 4, the correlation coefficient was used to measure the degree of linear relationship between two variables. However, it does not describe the exact linear association between $x$ and $y$. That is the role of regression analysis. By determining a specific relationship between $x$ and $y$, we may be able to use $x$ to help predict $y$.

What does it mean to specify a linear relationship between two variables? Suppose we model the relationship between two variables using $y_i = \beta_0 + \beta_1 x_i + \varepsilon_i$. The estimated regression equation is given by $\hat{y}_i = b_0 + b_1 x_i$,

For example, letting $b_0 = 3$ and $b_1 = 2$ specifies the line $\hat{y} = 3 + 2x$.

Figure 13.1.5 shows this relationship.

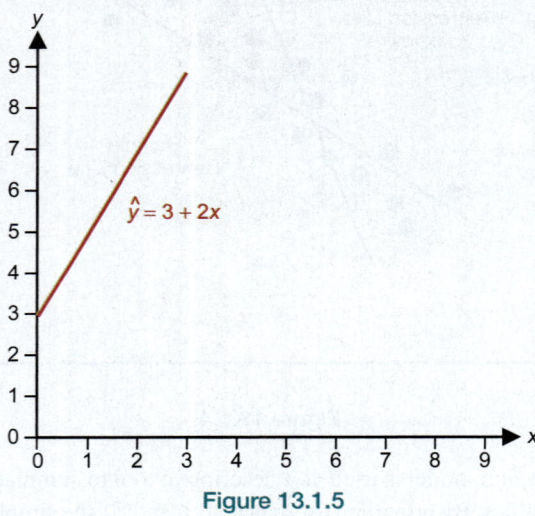

**Figure 13.1.5**

Letting $b_0 = 8$ and $b_1 = -2$ specifies the line $\hat{y} = 8 - 2x$.

Figure 13.1.6 shows this relationship.

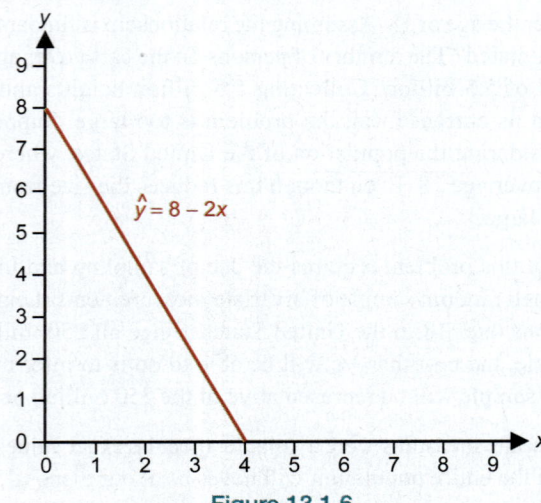

**Figure 13.1.6**

Consider the data plotted in Figure 13.1.7.

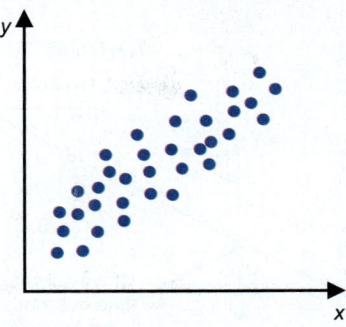

Figure 13.1.7

The data in Figure 13.1.7 seem to be related. Specifying the relationship between $x$ and $y$ with a linear model means finding a line that best fits the data in some way. The problem is that there are many lines that could be interpreted as fitting the data.

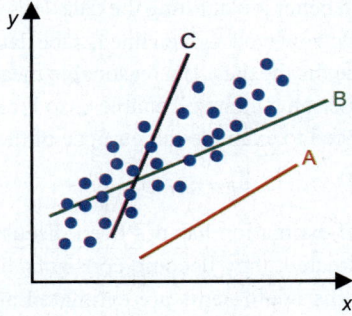

Figure 13.1.8

Which one of the lines in Figure 13.1.8 do you think best fits the data? Clearly, Line $A$ doesn't do a very good job. The points don't cluster around the line at all. Although Lines $B$ and $C$ are very different lines, they both seem to go through the data and it would be difficult to choose between them. To find the best line, we must develop a method of summarizing how close each line is to the data. The closeness measure can then be used as a criterion to choose between various lines.

## How Do We Measure How Close a Line Is to the Data?

One possible method of choosing the best fitting line is to use the line to predict the $y$-value for each observation. The superior line is the one that does the best job of predicting the observed $y$-values. Let's look at a small data set.

| x | y |
|---|---|
| 2 | 3 |
| 4 | 2 |
| 5 | 6 |
| 8 | 5 |
| 9 | 8 |

We plotted these data in Figure 13.1.9 and then tried to draw a line through the points. There are infinitely many lines that can be fitted to the data. However, there is no straight line that passes through all of the data points. One line that seems to fit the data reasonably well is:

$$y = 1 + 0.7x.$$

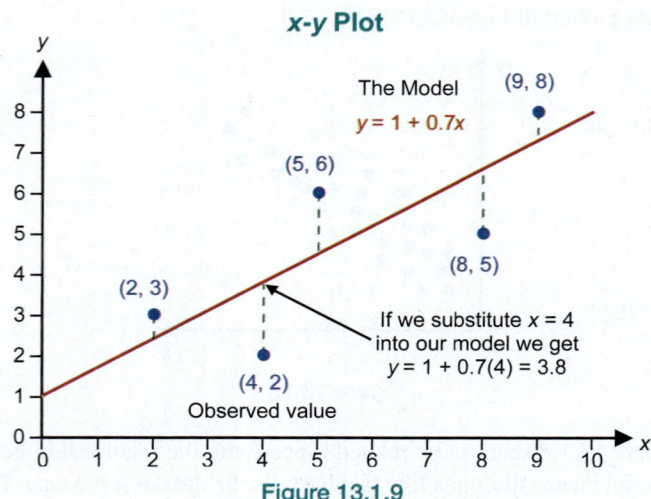

**Figure 13.1.9**

Although other lines might do a better job of fitting the data, let's see how well this line predicts the $y$-values. In order to evaluate how well a given line fits the data we must construct a method of measuring how well any line fits the data. If a reasonable measurement is developed, it can serve as a decision-making criterion which will enable us to look for the *best* line. To develop the appropriate criterion, we need to examine the purpose of the regression line

$$\hat{y}_i = b_0 + b_1 x_i \,.$$

Although the exact method of estimation has not been discussed, once the coefficients of the model ($b_0$ and $b_1$) are estimated, they become constants. The value of $x$ is selected by the user of the model. Once the coefficients are estimated and the value of $x$ is chosen, the corresponding value of $y$ can be predicted. For instance, if $x = 2$, then using the model $\hat{y}_i = 1 + 0.7x_i$, the predicted $y$-value is

$$\hat{y} = 1 + 0.7(2) = 2.4 \,.$$

The purpose of the model is to predict $y$ for some given value of $x$. According to the model, $y$ should be 2.4 when $x = 2$. But the model is wrong. In the first historical observation, the observed data for $x = 2$ is $y = 3$, not 2.4 as the model predicts. The difference between the observed value of $y$ and the predicted value of $y$ is called the **error**, **estimated error**, or **residual** ($e_i$). Remember, the predicted value of $y$, the dependent variable, is referred to as $\hat{y}$. The symbol $y$ is reserved for the observed value of $y$. Therefore, the error for each observation is given by the following formula.

---

**Formula**

**Residual**

A **residual** is calculated by

$$\text{Residual} = e_i = \text{Observed } y - \text{Predicted } y = y_i - \hat{y}_i \,.$$

---

For the first observation, the error is given by

$$e_i = y_1 - \hat{y}_1 = 3 - 2.4 = 0.6 \,.$$

The errors reflect how far each observation is from the line. Examining the errors suggests how well the line fits the data.

If we take each of the observed values of $x$ in the data and use the regression line to predict the corresponding value of $y$, we can compare the model's predicted value of $y$ for each $x$ to the actual value of $y$ for each $x$. These calculations are presented in Table 13.1.1.

| | | Predicted $y$ | Error | Squared Error |
|---|---|---|---|---|
| Observed $x$ | Observed $y$ | $\hat{y}_i = 1 + 0.7x_i$ | $y_i - \hat{y}_i$ | $\left(y_i - \hat{y}_i\right)^2$ |
| 2 | 3 | $2.4 = 1 + 0.72$ | $3 - 2.4 = 0.6$ | 0.36 |
| 4 | 2 | $3.8 = 1 + 0.74$ | $2 - 3.8 = -1.8$ | 3.24 |
| 5 | 6 | $4.5 = 1 + 0.75$ | $6 - 4.5 = 1.5$ | 2.25 |
| 8 | 5 | $6.6 = 1 + 0.78$ | $5 - 6.6 = -1.6$ | 2.56 |
| 9 | 8 | $7.3 = 1 + 0.79$ | $8 - 7.3 = 0.7$ | 0.49 |

**Table 13.1.1 – Observed versus Predicted Values**

$$\Sigma\left(y_i - \hat{y}_i\right) = -0.6 \qquad \Sigma\left(y_i - \hat{y}_i\right)^2 = 8.90$$

One possible criterion that could be used to compare different lines that might fit the data would be to sum all the errors and choose the line with the smallest sum. But as you can see in the previous table some of the larger positive errors were canceled out by equally large negative errors. It is possible to develop a line that has a small sum of errors, but the errors themselves could be quite large. Thus, the sum of the errors is not the criterion we have been looking for, but it is close. By squaring the error terms and then adding the squared errors, there are no negative errors to cancel out the positive ones.

Let's examine the measurement of the **sum of squared errors (SSE)**.

## Formula

### Sum of Squared Errors (SSE)

The **sum of squared errors (SSE)** is given by

$$\text{SSE} = \Sigma \text{error}_i^2 = \Sigma\left(y_i - \hat{y}_i\right)^2 = \Sigma\left(y_i - \left(b_0 + b_1 x_i\right)\right)^2 .$$

SSE can be used as a criterion for selecting the best fitting line through a set of points. If SSE is zero, then the model fits the data exactly and the observed data must lie in a straight line. If Line A's SSE is larger than Line B's, then Line B fits the data better than Line A. For the model we have been studying, the value of SSE is 8.90. Without studying other possible lines, there is no way to judge whether this is a large or small value for SSE. However, if there were a method that would find the *best* line (the line with the smallest sum of squared errors), then our problem would be solved. This line of best fit is called the **least squares line** since it has the smallest SSE.

## Finding the Least Squares Line

To define a line you must specify its slope and y-intercept. Fortunately, the problem of finding the least squares line was solved by Karl Gauss several hundred years ago. The line that best fits the data is the least squares line that is obtained by minimizing the sum of squared errors (SSE). That is, we want to find the values of $b_0$ and $b_1$ that minimize

$$\Sigma\left(y_i - \hat{y}_i\right)^2 .$$

Because $\hat{y}_i = b_0 + b_1 x_i$, we want to minimize

$$\Sigma\left(y_i - \left(b_0 + b_1 x_i\right)\right)^2 .$$

In the above equation, we have two unknowns, $b_0$ and $b_1$. The **method of least squares** determines the values of $b_0$ and $b_1$ that minimize the sum of squared errors around the prediction line. Using calculus, the computational formulas for $b_0$ and $b_1$ are given as follows.

**Adrien-Marie Legendre 1752-1833**

In 1806, Legendre was investigating the orbits of comets and published a book on the subject. In the appendix he gave the method of least squares curve fitting. In 1809 Karl Gauss also published his version of the least squares method. Although acknowledging Legendre's work, Gauss claimed priority on the discovery. This greatly hurt Legendre who fought for many years to have his discovery recognized as his contribution to statistics.

## Definition

**Least Squares Line**

The **least squares line** is the line of best fit to a set of data that results in a line with the smallest sum of squared errors (SSE).

## Formula

### Slope and *y*-Intercept of the Least Squares Line

The equation for finding the slope is given by

$$b_1 = \frac{SS_{xy}}{SS_{xx}}$$

where

$$SS_{xy} = \sum (x_i - \bar{x})(y_i - \bar{y}) = \sum x_i y_i - \frac{(\sum x_i)(\sum y_i)}{n}$$

and

$$SS_{xx} = \sum (x_i - \bar{x})^2 = \sum x_i^2 - \frac{(\sum x_i)^2}{n}.$$

The slope can also be calculated using

$$b_1 = \frac{n\sum x_i y_i - \sum x_i \sum y_i}{n\sum x_i^2 - (\sum x_i)^2}.$$

The estimate of the intercept is given by

$$b_0 = \bar{y} - b_1\bar{x} = \frac{1}{n}\left(\sum y_i - b_1 \sum x_i\right).$$

The $x_i$ and $y_i$ referred to in the expressions are the observed data values of $x$ and $y$, respectively.

You may never have to use these formulas, since many calculators and computer programs compute the coefficients of the least squares line. However, should manual calculation become necessary, the slope coefficient $b_1$ must be calculated prior to calculating $b_0$.

The Discovering Technology section at the end of this chapter shows how to estimate simple linear regression equations using a TI-84 Plus calculator, Microsoft Excel, JMP, and Minitab. Using statistical software to estimate linear regression equations will give the most accurate results and is highly recommended.

## Example 13.1.1

**Estimating the Slope and Intercept**

Table 13.1.2 contains data on the number of items produced during 10 randomly selected weeks in a production factory and the total production costs. It seems obvious that in a production operation there is a relationship between the number of items produced (the independent variable, $x$) and the total production cost (the dependent variable, $y$). Using the data in Table 13.1.2 and the summary statistics below, answer the following questions.

a.  Calculate the slope and intercept for the simple linear regression equation to predict total production cost.

b.  Create a scatter plot of the data and include the estimated regression line.

c.  Using the estimated regression model, calculate the predicted cost for weeks 1 through 10, and the associated errors and squared errors for weeks 1 through 10.

d.  Interpret the estimate of the intercept and the estimate of the slope.

| Table 13.1.2 – Weekly Production | | |
|---|---|---|
| Week | Items Produced | Cost ($) |
| 1 | 22 | 3500 |
| 2 | 30 | 3800 |
| 3 | 36 | 4500 |
| 4 | 41 | 4200 |
| 5 | 27 | 3700 |
| 6 | 45 | 4600 |
| 7 | 30 | 3600 |
| 8 | 37 | 4550 |
| 9 | 32 | 3990 |
| 10 | 31 | 3675 |

Some summary statistics are given below.

$$\sum x_i = 331$$
$$\sum y_i = 40{,}115$$
$$\sum x_i y_i = 1{,}350{,}055$$
$$\sum x_i^2 = 11{,}369$$

## SOLUTION

**a.** Using the least squares equations and the summary calculations given above, we can estimate the following coefficients:

$$b_1 = \frac{SS_{xy}}{SS_{xx}}$$

$$= \frac{n\sum x_i y_i - \sum x_i \sum y_i}{n\sum x_i^2 - \left(\sum x_i\right)^2}$$

$$= \frac{10(1{,}350{,}055) - 331(40{,}115)}{10(11{,}369) - (331)^2}$$

$$= \frac{222{,}485}{4129} \approx 53.8835$$

and

$$b_0 = \bar{y} - b_1 \bar{x}$$

$$= \frac{1}{n}\left(\sum y_i - b_1 \sum x_i\right)$$

$$= \frac{1}{10}\left(40{,}115 - 53.8835(331)\right)$$

$$\approx 2227.9562.$$

Therefore, the model to estimate the total production cost is as follows.

$$\text{Estimated Cost} = \$2227.96 + \$53.88\,(\text{Age})$$

Dependent Variable            Independent Variable

### ∽ Technology

For instructions on how to estimate the least squares regression line using technology, visit stat.hawkeslearning.com and navigate to **Discovering Business Statistics, Second Edition > Technology Instructions > Regression > Simple Linear Regression**.

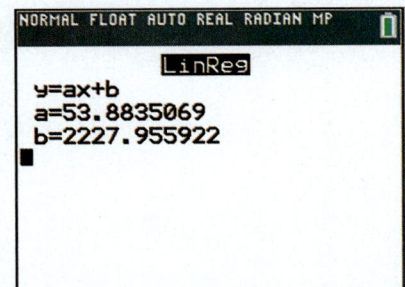

```
NORMAL FLOAT AUTO REAL RADIAN MP
                 LinReg
 y=ax+b
 a=53.8835069
 b=2227.955922
■
```

**b.** Figure 13.1.10 shows a graph of this model using the estimated coefficients.

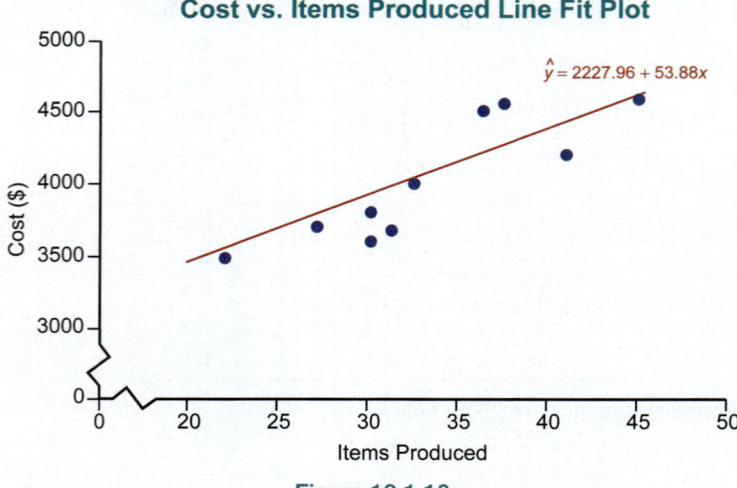

**Figure 13.1.10**

This model has a smaller SSE for these data than any other line.

**c.** In Table 13.1.3 we examine the errors produced by the least squares model. Note that the sum of the errors equals zero.

**NOTE**

Data values for predicted cost, error, and squared error in Table 13.1.3 are given based on the residual output given in Microsoft Excel. Values may differ slightly if computed by hand.

| | | | Table 13.1.3 — Errors Produced by the Least Squares Model | | |
|---|---|---|---|---|---|
| Week | Items Produced $x$ | Cost($) $y$ | Predicted Cost ($) $\hat{y}_i = 2227.96 + 53.88 x_i$ | Error $y_i - \hat{y}_i$ | Squared Error $\left(y_i - \hat{y}_i\right)^2$ |
| 1 | 22 | 3500 | 3413.39 | 86.61 | 7500.76 |
| 2 | 30 | 3800 | 3844.46 | −44.46 | 1976.79 |
| 3 | 36 | 4500 | 4167.76 | 332.24 | 110,381.98 |
| 4 | 41 | 4200 | 4437.18 | −237.18 | 56,254.21 |
| 5 | 27 | 3700 | 3682.81 | 17.19 | 295.48 |
| 6 | 45 | 4600 | 4652.71 | −52.71 | 2778.74 |
| 7 | 30 | 3600 | 3844.46 | −244.46 | 59,761.24 |
| 8 | 37 | 4550 | 4221.65 | 328.35 | 107,816.56 |
| 9 | 32 | 3990 | 3952.23 | 37.77 | 1426.71 |
| 10 | 31 | 3675 | 3898.34 | −223.34 | 49,882.83 |

$$\sum y_i - \hat{y}_i \approx 0 \qquad \sum \left(y_i - \hat{y}_i\right)^2 \approx 398,075.30$$

As you can see, in the building of the linear model, there is potential for a great deal of calculation. For most problems you will use some type of statistical analysis package or spreadsheet to compute the least squares coefficients as well as related statistics and diagnostic measures. These packages will compute a wealth of summary and diagnostic measures concerning the estimated model.

**d.** Using the data and the estimated regression model, if the number of items produced is 0, then the predicted cost of production would be $2227.96, the value of $b_0$.

$$\text{Estimated Cost} = \$2227.96 + \$53.88(0) = \$2227.96$$

The value of $b_0$ is always interpreted as the average value of the dependent variable, in this case, the total production cost, when the independent variable is set equal to zero. Since $b_1$ is the estimated slope of the line, it is interpreted as the average change in the dependent variable (total production cost) for a one-unit change in the independent variable (items produced). In our example, the independent variable is expressed in units.

Therefore, for every additional unit, the total production cost is expected to increase by $53.88. If this interpretation is correct, the model's predicted cost for 27 items should be $53.88 more than that for 26 items.

If 27 items are produced, then $x = 27$, and

$$\text{Estimated Cost} = \$2227.96 + \$53.88(27) = \$3682.72.$$

If 26 items are produced, then $x = 26$, and

$$\text{Estimated Cost} = \$2227.96 + \$53.88(26) = \$3628.84.$$

According to the model, the difference in cost between producing 27 items and 26 items is exactly equal to the slope.

| Table 13.1.4 – Difference in Production Cost | |
| --- | --- |
| Predicted cost for producing 27 items | $3682.72 |
| Predicted cost for producing 26 items | $3628.84 |
| Difference | $53.88 |

**Definition**

**Interpretation of the Regression Coefficients**

The **intercept coefficient**, $b_0$, is the average value of the dependent variable, $y$, when the independent variable, $x$, is equal to zero.

The **slope coefficient**, $b_1$, is the average change in the dependent variable, $y$, for a one-unit change in the independent variable, $x$.

## The Importance of Errors

The usefulness of the estimated model depends on the magnitude of the prediction errors you expect the model to produce. The production model is

$$\text{Cost} = \beta_0 + \beta_1(\text{Items Produced}) + \varepsilon_i.$$

Yet we ignored the error component previously when we predicted the production cost for different numbers of items. For instance, when we predicted the price of producing 26 items, we found

$$\text{Estimated Cost} = \$2227.96 + \$53.88(26) = \$3628.84.$$

Since the model is not going to be a perfect predictor, we should incorporate the possibility of error in the model. Thus,

$$\text{Estimated Cost} = \$2227.96 + \$53.88(26) + e_i = \$3628.84 + e_i$$

would have been a more precise statement. It is important to assess the magnitude of the error when the model is used for predictive purposes. If the errors are too large, then it will not be advantageous to use the model for prediction.

## How Do We Assess the Magnitude of the Errors?

You may have wondered why we have been trying to predict observations that we already know. For example, an observation in the production data was items produced equals 27 and cost equals $3700. If we already know the cost is $3700, why try to predict it? Using the model to predict the observed outcomes of the dependent variable, $y$, produces error data that can be used to evaluate the errors the model produces. Knowledge of the errors will provide understanding of the predictive quality of the model.

## How Do We Summarize the Errors a Model Produces?

For the production model, the mean error is zero. This is true for all least squares models and hence mean error won't provide any useful information. What about the average variation in the error terms? Large variation in the errors would indicate that a model's prediction is not very reliable. On the other hand, small variation in the errors would indicate the model is capable of producing more trustworthy predictions.

Computing the variation of the error data is not much different from computing the variation of any data set. Recall that the formula for sample variance is

$$s^2 = \frac{\sum(x_i - \bar{x})^2}{n-1}.$$

In regression, we are discussing errors. For simple linear regression the degrees of freedom for the errors are $(n - 2)$, where the "2" represents the number of parameters estimated in the model (i.e., $\beta_0$ and $\beta_1$ were estimated). Making this slight adjustment for degrees of freedom produces the definition for the **variance of the error terms**, also known as the **mean square error**.

---

### Formula

#### Mean Square Error

The variance of the error terms is also known as the **mean square error** and is given by:

$$s_e^2 = \frac{\sum\left(y_i - \hat{y}_i\right)^2}{n-2} = \frac{\text{SSE}}{n-2}.$$

---

The numerator is divided by $n - 2$ (the degrees of freedom of the error data) instead of $n - 1$ to account for an additional constraint imposed by least squares estimation. In the production data the mean square error is

$$s_e^2 = \frac{398075.2967}{10-2} \approx 49759.41$$

and the **standard error** (standard deviation of the error terms) is given by

$$s_e = \sqrt{49759.41} \approx 223.0682 \,.$$

SUMMARY OUTPUT

| Regression Statistics | |
|---|---|
| Multiple R | 0.866441198 |
| R Square | 0.75072035 |
| Adjusted R Square | 0.719560393 |
| Standard Error | 223.0681781 |
| Observations | 10 |

ANOVA

| | df | SS | MS | F |
|---|---|---|---|---|
| Regression | 1 | 1198827.203 | 1198827 | 24.09247 |
| Residual | 8 | 398075.2967 | 49759.41 | |
| Total | 9 | 1596902.5 | | |

| | Coefficients | Standard Error | t Stat | P-value |
|---|---|---|---|---|
| Intercept | 2227.955922 | 370.1487713 | 6.019082 | 0.000317 |
| Items Produced | 53.8835069 | 10.97779658 | 4.908408 | 0.001181 |

**Figure 13.1.11**

**Technology**

For instructions on how to obtain this regression output from Excel or other technologies, visit stat.hawkeslearning.com and navigate to **Discovering Business Statistics, Second Edition > Technology Instructions > Regression > Simple Linear Regression.**

Figure 13.1.11 shows the summary output from Microsoft Excel for the production data. (See the Discovering Technology section at the end of this chapter for information on how to generate a summary output using Excel.) Notice that the mean square error is given in the ANOVA table in the column labeled *MS* and the row labeled *Residual*. The standard error of the model is given in the *Regression Statistics* table and is labeled *Standard Error*.

Should you be satisfied with a model whose standard error is $223.07? It would certainly be more desirable if the model's standard deviation of error was only $100. If you know a great deal about the production costs (you have an intuitive model), then perhaps the current model would not be of value. However, if you are uninformed, the current model provides a basic understanding of the relationship between items produced and production cost.

What constitutes small and large variation of the errors is dependent on the nature of the problem. Using a model to predict the diameter of a valve to be used in a heart-lung machine is certainly going to produce a different idea of *small* than estimating production costs. It would be ideal to develop a measure that would summarize the degree of fit on some standardized scale. Such a measure is introduced in Section 13.3.

## Assumptions of the Simple Linear Model

A number of assumptions are necessary to make inferences about the linear model. An error term was incorporated in the model because virtually no real set of bivariate data is exactly linear. Incorporating the error term in the population regression line produces the simple linear regression model:

$$y_i = \beta_0 + \beta_1 x_i + \varepsilon_i.$$

The error term, $\varepsilon_i$, represents the variation in $y$ not accounted for by the linear regression model. In order to perform inference on the model, some assumptions about the nature of the error term are required.

**Gauss 1777-1855**

In statistics Karl Friedrich Gauss is best known for invoking the normal error distribution in conjunction with solving the least squares regression problem. Although modern statistical historians find much to criticize in Gauss' work, it was his publication of the normal error curve in relation to the regression problem that led Pierre Simon Laplace to see the connection between the Central Limit Theorem and linear estimation. For several generations of statistics, linear estimation based on the normal error distribution was a fundamental tool of data analysis.

### Assumptions

#### Assumptions about the Error Term in the Linear Model

1. The average response at each value of the independent variable is a linear function. That is, there is a linear relationship between $x$ and $y$.

2. The errors, $\varepsilon_i$, are assumed to be independent of each other.

3. The errors, $\varepsilon_i$, at each value of $x_i$ are normally distributed.

4. The errors, $\varepsilon_i$, at each value of $x_i$ have equal variances, $\sigma_\varepsilon^2$.

Oftentimes, you might see these assumptions referred to using the acronym L.I.N.E. for

L = Linearity

I = Independent

N = Normally distributed

E = Equal variances

We write the notation as $\varepsilon \sim N\left(0, \sigma_\varepsilon^2\right)$ which means that the error term is normally distributed with a mean of zero and a variance of $\sigma_\varepsilon^2$. The residual ($e_i$) is an estimate of the individual error term ($\varepsilon_i$). The residual for observation $i$ is the difference between the observed value of $y$ and the predicted value given by

$$e_i = y_i - \hat{y}_i$$

where $y_i$ is the $i^{\text{th}}$ observation and $\hat{y}_i$ is the estimated value of $y_i$ for a given value of $x_i$.

The model's parameters are $\beta_0$, $\beta_1$, and $\sigma_\varepsilon^2$. The estimation of these quantities was discussed earlier.

In particular,

$$b_1 = \frac{n \sum x_i y_i - \sum x_i \sum y_i}{n \sum x_i^2 - \left(\sum x_i\right)^2},$$

$$b_0 = \bar{y} - b_1 \bar{x} = \frac{1}{n}\left(\sum y_i - b_1 \sum x_i\right),$$

$$s_e^2 = \frac{\sum\left(y_i - \hat{y}_i\right)^2}{n-2} = \frac{\text{SSE}}{n-2}.$$

In addition to the formal assumptions previously stated, a linear model should only be used to fit data that appear to be reasonably linear. Because of the wide availability of computer programs that calculate least squares estimates, you will not need to manually calculate estimates very often.

## 📝 13.1 Exercises

### Basic Concepts

1.  What is regression analysis?

2.  Give two examples of why businesses might be interested in studying the relationship between two variables.

3.  What is the difference between a dependent and an independent variable?

4.  What is a simple linear regression model? Give the equation that describes a simple linear regression model and define all terms in the equation.

5.  What is the estimated simple linear regression equation and how is it used?

6.  What is $\hat{y}$? How does this differ from $y$?

7.  What is the technique used to estimate the simple linear regression coefficients?

8.  What is the relationship between scatterplots and simple linear regression?

9.  Why is it often difficult to accurately describe real world situations using a simple linear regression equation?

10. What is the correlation coefficient? Why is the correlation coefficient insufficient when describing an exact linear relationship between $x$ and $y$?

11. What is the residual of a model?

12. What is the sum of squared errors and what does it measure?

13. Explain why the best line is referred to as the least squares line.

14. What measure should be minimized in order to find the least squares line?

15. What is the equation for finding the slope of the least squares line?

16. What is the equation for finding the intercept of the least squares line?

17. When finding the least squares line manually, which must be calculated first: the slope or the $y$-intercept?

18. Interpret the intercept coefficient, $b_0$.

19. Interpret the slope coefficient, $b_1$.

20. Why is the magnitude of the prediction errors important when estimating a regression model?

21. What is the mean error for a least squares model?

22. Describe what the magnitude of the variation in the error terms tells us about the reliability of the regression model.

23. What is mean square error?

24. How many degrees of freedom are associated with the error term in a simple linear regression model?

25. What is the square root of the mean square error known as?

26. Describe where the summary statistics for the standard error and mean square error are found in a standard regression summary output in Microsoft Excel.

27. Is there a universal rule on how large is *large* with regard to standard error in a model?

28. What is estimated by the mean square error and what is estimated by the standard error?

29. Why is there an error term incorporated in the simple linear model?

30. What does the error term represent?

31. List the four assumptions about the error term in the simple linear model.

32. List the parameters of the simple linear regression model, and identify their estimates.

## Exercises

33. Consider the following simple linear regression model. Write the estimated simple linear regression equation that corresponds to this model.

$$y_i = \beta_0 + \beta_1 x_i + \varepsilon_i$$

34. Consider the following estimated simple linear regression equation.

$$\hat{y}_i = b_0 + b_1 x_i$$

   a. What population parameter does $b_0$ estimate?

   b. What population parameter does $b_1$ estimate?

   c. Is error incorporated into the estimated model? Explain.

35. Suppose that a company wishes to predict sales volume based on the amount of advertising expenditures. The sales manager thinks that sales volume and advertising expenditures are modeled according to the following linear equation. Both sales volume and advertising expenditures are in thousands of dollars.

   Estimated Sales Volume $= 49.25 + 0.51\big($Advertising Expenditures$\big)$

   a. What is the dependent variable in this model? Explain.

   b. What is the independent variable in this model? Explain.

   c. What is the estimated sales volume for this company when the marketing department spends $40,000 on advertising?

   d. If the company had a target sales volume of $100,000, how much should the sales manager allocate for advertising in the budget?

   e. What is the sales manager forgetting to account for when using this linear equation to determine sales volume? What kinds of problems could this cause for the company?

36. Suppose the following estimated regression equation was determined to predict salary based on years of experience.

   Estimated Salary $= 25689.10 + 2148.35\big($Years of Experience$\big)$

   a. What is the dependent variable?

   b. What is the independent variable?

   c. What is the value that estimates $\beta_0$ in this particular equation?

   d. What is the value that estimates $\beta_1$ in this particular equation?

   e. What is the estimated salary for an employee with 15 years of experience?

37. Plot the following lines.

   a. $y = 2 + 3x$

   b. $y = 4 + 8x$

   c. $y = 9 - 2x$

   d. $y = x$

**38.** Plot the following lines.

    **a.** $y = 100 + 50x$

    **b.** $y = 0.5 + 0.7x$

    **c.** $y = 20 - 5x$

**39.** Consider the following estimated regression equation.

$$\hat{y}_i = 10x_i - 5$$

    **a.** Complete the following table.

| Predicted Values | |
|---|---|
| *x* | $\hat{y}$ |
| 2 | |
| 5 | |
| 7 | |
| 9 | |
| 10 | |

    **b.** Do these two variables appear to have a positive or negative relationship?

    **c.** For these two variables, what sign would you expect the correlation coefficient to have? Explain.

**40.** Consider the following data.

| Observed Values | |
|---|---|
| *x* | *y* |
| 0 | 2 |
| 1 | 4 |
| 5 | 9 |
| 6 | 7 |
| 8 | 8 |

    **a.** Draw a scatterplot of the data.

    **b.** Draw a line which you believe fits the data.

    **c.** Suppose that $\hat{y}_i = 3 + 0.8x_i$ is a line that fits the data reasonably well. Complete the following table.

| Observed and Predicted Values | | | | |
|---|---|---|---|---|
| Observed *x* | Observed *y* | Predicted *y* | Error | Squared Error |
| 0 | 2 | | | |
| 1 | 4 | | | |
| 5 | 9 | | | |
| 6 | 7 | | | |
| 8 | 8 | | | |

    **d.** What is the sum of squared errors for these data?

**41.** Consider the following data regarding home sale prices and square footage.

| Housing Prices and Square Footage | |
| --- | --- |
| Selling Price (Thousands of Dollars) | Square Footage |
| 199.9 | 1065 |
| 228.0 | 1254 |
| 235.0 | 1300 |
| 285.0 | 1577 |
| 239.0 | 1600 |
| 293.0 | 1750 |
| 285.0 | 1800 |
| 365.0 | 1870 |
| 295.0 | 1935 |
| 290.0 | 1948 |
| 385.0 | 2254 |
| 505.0 | 2600 |
| 425.0 | 2800 |
| 415.0 | 3000 |

**a.** Suppose we want to predict selling price based on square footage. Write the estimated regression equation in terms of selling price and square footage. (Assume the parameters of this model have not been estimated.)

**b.** Create a scatterplot of the data and draw a line of best fit.

**c.** Suppose we determine that an equation that fits the data reasonably well is

$$\text{Estimated Selling Price} = 52.35 + 0.14(\text{Square Footage}).$$

Complete the following table.

| Housing Prices and Square Footage | | | | |
| --- | --- | --- | --- | --- |
| Observed Selling Price (Thousands of Dollars) | Observed Square Footage | Predicted Selling Price (Thousands of Dollars) | Error | Squared Error |
| 199.9 | 1065 | | | |
| 228.0 | 1254 | | | |
| 235.0 | 1300 | | | |
| 285.0 | 1577 | | | |
| 239.0 | 1600 | | | |
| 293.0 | 1750 | | | |
| 285.0 | 1800 | | | |
| 365.0 | 1870 | | | |
| 295.0 | 1935 | | | |
| 290.0 | 1948 | | | |
| 385.0 | 2254 | | | |
| 505.0 | 2600 | | | |
| 425.0 | 2800 | | | |
| 415.0 | 3000 | | | |

**d.** Compute the sum of squared errors for these data.

**42.** Consider the following data.

| x | 1 | 2 | 3 | 4 | 5 |
|---|---|---|---|---|---|
| y | 1 | 3 | 4 | 4 | 6 |

    **a.** Plot the data points on a scatterplot.

    **b.** Determine the least squares line. Use $x$ as the independent variable.

    **c.** Plot the least squares line on the scatterplot.

    **d.** Use the model to compute the error for each data point.

**43.** Consider the following data.

| x | -2 | -1 | 0 | 3 | 5 |
|---|----|----|---|---|---|
| y | 1 | 3 | 5 | 4 | 8 |

    **a.** Plot the data points on a scatterplot.

    **b.** Determine the least squares line. Use $x$ as the independent variable.

    **c.** Plot the least squares line on the scatterplot.

    **d.** Use the model to compute the error for each data point.

**44.** Comparing the least squares lines in Exercises 42 and 43, which line fits the data better? Explain your answer.

**45.** Suppose a linear regression analysis produced the following equation relating an individual's salary to the current value of his or her home.

$$\text{Estimated Current Value of Home} = 12331 + 3.14(\text{Annual Salary})$$

    **a.** Which of the variables in the model is the dependent variable?

    **b.** Which of the variables in the model is the independent variable?

    **c.** What would be the predicted current value of home for someone earning a salary of $32,000?

    **d.** If a person earned $5000 additional income, how much of an increase in home value would be predicted?

    **e.** In terms of the problem, interpret the estimate of the slope in the model.

    **f.** In terms of the problem, interpret the estimate of the intercept in the model.

    **g.** Do you believe annual salary is a causal factor in explaining the price of someone's home? Explain.

**46.** Suppose a linear regression analysis produced the following equation relating a basketball player's total points scored to the number of minutes played in a season.

$$\text{Estimated Points Scored} = -97.2 + 0.645(\text{Minutes Played})$$

    **a.** Which of the variables in the model is the dependent variable?

    **b.** Which of the variables in the model is the independent variable?

    **c.** What would be the predicted value of total points scored for a basketball player who plays 500 minutes in a season?

    **d.** If a basketball player played an additional 100 minutes, how much of an increase in total points scored would be predicted?

    **e.** In the model, which of the coefficients is the slope?

    **f.** In the model, which of the coefficients is the intercept?

    **g.** Do you believe the number of minutes played is a causal factor in explaining the total points scored? Explain.

**47.** Suppose you were studying the educational level of husbands and wives (measured in number of years of education). You have randomly selected 10 couples and have obtained the data in the following table.

| Education Level | | | | | | | | | | |
|---|---|---|---|---|---|---|---|---|---|---|
| **Husband** | 12 | 16 | 16 | 18 | 20 | 17 | 23 | 14 | 12 | 16 |
| **Wife** | 14 | 16 | 14 | 16 | 16 | 18 | 18 | 12 | 16 | 20 |

**a.** Suppose you wanted to predict the husband's years of education based on the wife's. Use the data to estimate the appropriate model.

**b.** Use the model in part **b.** to predict the husband's educational level if married to a woman with 16 years of education.

**c.** Suppose you wanted to predict the years of education for the wife based on the husband's years of education. Use the data to create the appropriate model. Did you get the same model as in part **b.**?

**d.** Use the model created in part **d.** to predict the wife's educational level if married to a husband with 16 years of education.

**e.** Do you believe there is a causal relationship between the two variables? If so, which direction is the causality? Does the husband's education cause the wife to have more or less education, or vice versa?

**48.** Consider the following summary output.

SUMMARY OUTPUT

| *Regression Statistics* | |
|---|---|
| Multiple R | 0.911653228 |
| R Square | 0.831111609 |
| Adjusted R Square | 0.79733393 |
| Standard Error | 0.253142413 |
| Observations | 7 |

ANOVA

| | df | SS | MS | F |
|---|---|---|---|---|
| Regression | 1 | 1.576737452 | 1.576737 | 24.60535 |
| Residual | 5 | 0.320405405 | 0.064081 | |
| Total | 6 | 1.897142857 | | |

| | Coefficients | Standard Error | t Stat | P-value |
|---|---|---|---|---|
| Intercept | 4.021621622 | 0.181401491 | 22.16973 | 3.47E-06 |
| X Variable 1 | -0.22297297 | 0.044950802 | -4.96038 | 0.004247 |

**a.** What is the mean square error for these data?

**b.** What is the standard error of the model?

**49.** Consider the following data.

| Observed Values | |
|---|---|
| *x* | *y* |
| 15 | 110 |
| 18 | 135 |
| 25 | 150 |
| 24 | 149 |
| 26 | 158 |
| 40 | 169 |

**a.** Suppose that, using statistical software, we determine that $b_0 = 93.2922$ and $b_1 = 2.1030$. Complete the following table.

| Observed versus Predicted Values | | | | |
|---|---|---|---|---|
| Observed *x* | Observed *y* | Predicted *y* | Error | Squared Error |
| 15 | 110 | | | |
| 18 | 135 | | | |
| 25 | 150 | | | |
| 24 | 149 | | | |
| 26 | 158 | | | |
| 40 | 169 | | | |

**b.** Compute the sum of squared errors.

**c.** Compute the mean square error.

**d.** Compute the standard error of the model.

**e.** Do you believe these estimates of $b_0$ and $b_1$ provide a reliable estimated regression equation for these data? Explain.

**50.** Consider the following data regarding students' college GPAs and high school GPAs.

| GPAs | |
|---|---|
| College GPA | High School GPA |
| 2.80 | 3.42 |
| 3.54 | 3.56 |
| 2.88 | 3.13 |
| 2.15 | 3.27 |
| 2.22 | 3.38 |
| 3.31 | 4.13 |
| 2.13 | 3.95 |
| 2.39 | 3.81 |
| 3.01 | 4.33 |
| 2.68 | 2.85 |

**a.** Suppose we want to predict college GPA based on high school GPA. Write the estimated regression equation in terms of college GPA and high school GPA. (Assume the parameters of the model have not been estimated.)

**b.** Suppose we determine, using statistical software, that the estimated regression equation is

Estimated College GPA $= 1.88 + 0.2319($High School GPA$)$.

Complete the following table.

| GPAs | | | | |
|---|---|---|---|---|
| Observed College GPA | Observed High School GPA | Predicted College GPA | Error | Squared Error |
| 2.80 | 3.42 | | | |
| 3.54 | 3.56 | | | |
| 2.88 | 3.13 | | | |
| 2.15 | 3.27 | | | |
| 2.22 | 3.38 | | | |
| 3.31 | 4.13 | | | |
| 2.13 | 3.95 | | | |
| 2.39 | 3.81 | | | |
| 3.01 | 4.33 | | | |
| 2.68 | 2.85 | | | |

    **c.** Compute the sum of squared errors for the model.

    **d.** Compute the standard error of the model.

**51.** The regression equation that relates the delivery time with number of pizzas and distance is given by $\widehat{\text{Delivery Time}} = 1.79 + 1.95(\text{Number of Pizzas}) + 1.57(\text{Distance})$.

    **a.** Estimate the delivery time to deliver 5 pizzas at a distance of 2 miles.

    **b.** The observed data shows that the time taken to deliver 5 pizzas at a distance of 2 miles is 16 minutes. Find the residual.

    **c.** Interpret the meaning of the residual in the context of the problem.

# 13.2 Residual Analysis

When performing regression analysis, residual analysis is a useful technique to help us determine if the model we are using is appropriate. By studying the estimated errors (i.e., the residuals), we can check the underlying assumptions of the regression model. Before one can adequately make predictions with the estimated regression model, the analyst should ensure that the assumptions of the model are valid. Residual analysis is the method used to validate those assumptions.

All estimates, intervals, and hypothesis tests in regression analysis are based on assuming that the model is correct. If the model is not correct (i.e., at least one of the assumptions is not valid), the formulas and methods will also be incorrect.

Validating the assumptions of the simple linear regression model revolve around the error term ($\varepsilon$). You may recall that the assumptions of the simple linear regression model are:

1. The average response at each value of the independent variable is a linear function. That is, there is a linear relationship between $x$ and $y$.

2. The errors, $\varepsilon_i$, are assumed to be independent of each other.

3. The errors, $\varepsilon_i$, at each value of $x_i$ are normally distributed.

4. The errors, $\varepsilon_i$, at each value of $x_i$ have equal variances, $\sigma_\varepsilon^2$.

One method of validating the assumptions is by performing a graphical analysis of the residuals. The most frequently used graph is that of the residuals ($e_i$) vs. the fitted values ($\hat{y}_i$). It is a scatter plot with the residuals on the vertical axis and the fitted values on the horizontal axis. This plot is used to detect non-linearity, unequal variances, and outliers.

In Figure 13.2.1, (a) is a scatterplot of the raw data, $y$ vs. $x$, with a simple linear regression line fit through the data. Note that the scatterplot is somewhat curvilinear. When we plot the residuals against the fitted values (see (b) in Figure 13.2.1) for the same data, the scatterplot seems to also follow a curvilinear pattern indicating that the data are not linear, thus violating the assumption of linearity.

### Plot of y vs. x and Plot of Residuals vs. Fitted Values - Not Linear

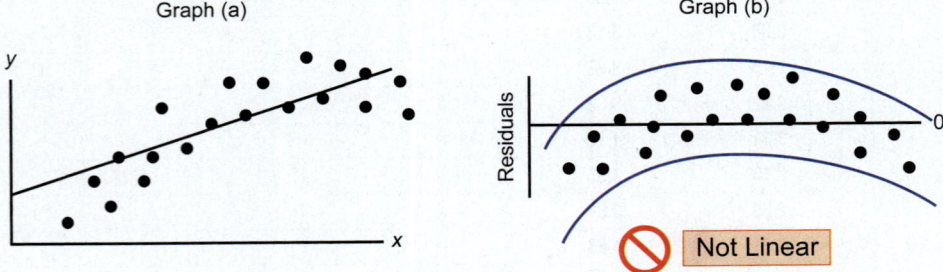

Figure 13.2.1

In Figure 13.2.2, (a) is a scatterplot of $y$ vs. $x$ with a straight line fit through the data. The raw data seem to follow a linear pattern. In plot (b), which is a scatterplot of the residuals vs. the fitted values, the data are randomly scattered above and below a horizontal line ($y = 0$) and do not follow a pattern. This indicates that the data are linear, therefore, the assumption of linearity is upheld. In general, a well-behaved residual plot will have data points that are randomly scattered around the zero line if the data follow a linear pattern.

### Plot of y vs. x and Plot of Residuals vs. Fitted Values - Linear

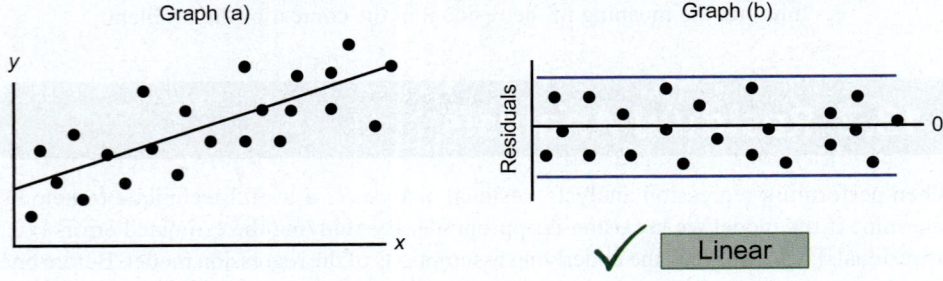

Figure 13.2.2

We can also look at the residual plots to determine if the errors are independent. Figure 13.2.3 is a plot of the residuals vs. $x$ and shows a random scatter of the observations about the horizontal line $y = 0$. Given that there is no pattern associated with the residual plot, it is an indication that the errors are independent of each other, upholding the independence assumption.

### Residuals vs. x Plot - Independent

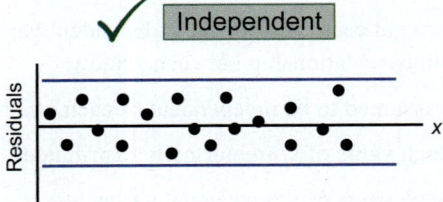

Figure 13.2.3

Figure 13.2.4, however, shows two residual plots (residuals vs. $x$) that indicate the existence of a pattern. Figure 13.2.4(a) shows that there is a linear trend between the residuals and the independent variable, $x$. If a pattern exists, this is an indication that there is a relationship between the independent variable and the residual which violates the assumption of independence.

Similarly, Figure 13.2.4(b) also shows a **sigmoidal relationship** between the residuals and the independent variable—another example of a violation of the assumption of independence.

### Residuals vs. *x* Plot - Not Independent

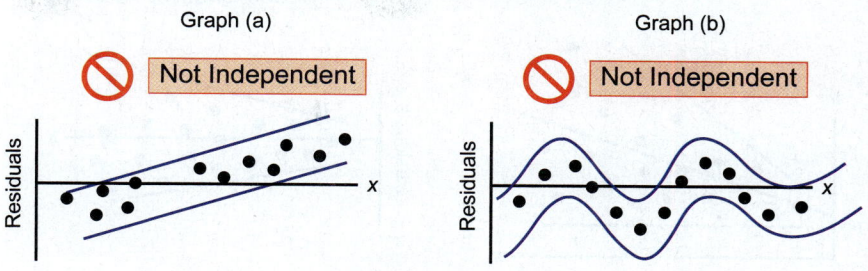

Figure 13.2.4

The third assumption listed above is that for each value of the independent variable, $x_i$, the errors are normally distributed. There are several techniques that we can employ to verify the assumption of normality. We could do the following:

- Examine the stem-and-leaf display of the residuals

- Examine the boxplot of the residuals

- Examine the histogram of the residuals

- Construct a **normal probability plot** of the residuals

- Perform a formal test in JMP called the Shapiro-Wilk W Test

One of the most common graphical techniques, and the one that we will employ in this section, is that of the normal probability plot (or normal quantile plot). A normal probability plot is a plot of the theoretical percentiles of the normal distribution versus the observed sample percentiles. While this plot can be created manually (see Section 7.3), it's best (and simplest) to have the plot created using a software package . When examining a normal probability plot, it should be approximately linear. Given that we are interested in the normality of the residuals, we create a plot of the residuals. See Figure 13.2.5 which is an example of a normal probability plot. The residuals vs. percentiles nearly fall on a straight line and thus the assumption of normality is upheld.

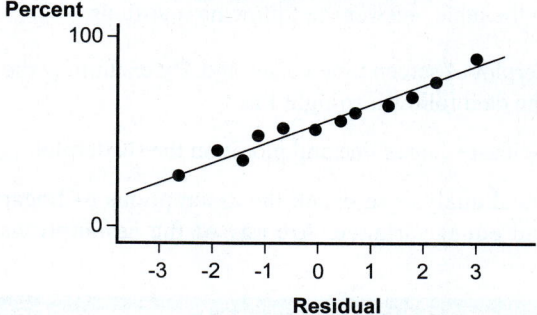

Figure 13.2.5

The last assumption listed above is the equality of variance. Figure 13.2.6 contains a plot of the raw data (*y* vs. *x*) as well as a plot of the residuals vs. *x*. The assumption is that for each value of $x_i$, the variance of the of the error term is constant. The plot in Figure 13.2.6 shows that as the independent variable *x* increases, the spread of the responses (*y*) also increases. This is an indication that the variance is not constant across all values of *x*, thus violating the equal variance assumption.

**Plots Showing Non-Constant Variance**

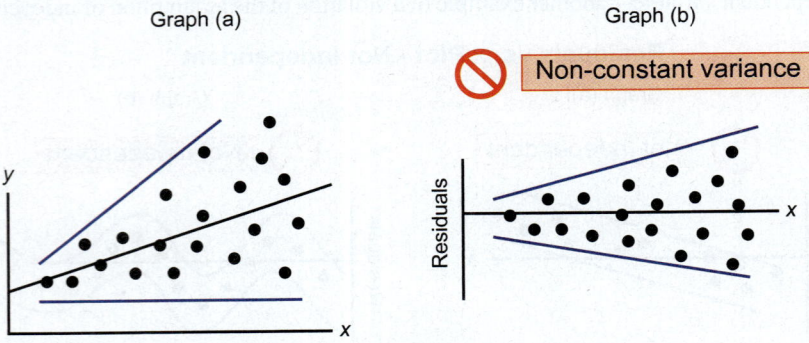

Figure 13.2.6

Figure 13.2.7 contains a plot of the raw data ($y$ vs. $x$) and a plot of the residuals vs. $x$. Note that in both plots, as $x$ increases, the spread is constant which indicates that the variance is constant. Thus, the assumption of equal variance across all values of $x$ is not violated.

**Plots Showing Constant Variance**

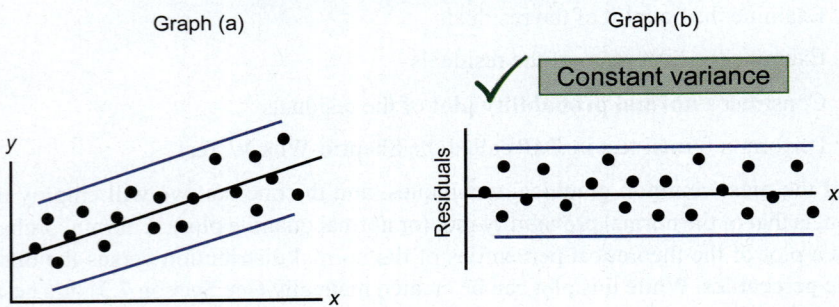

Figure 13.2.7

| Example 13.2.1 | Please recall the examples in Chapter 12 regarding media use by tweens, teens, and adults. The data in the table below shows the age and the amount of time spent on cell phones, watching online videos, watching television, and playing mobile video games. Using the data in the table, answer the following questions. |

**Simple Linear Regression of Age by Screen Time**

**a.** Create a scatterplot of screen time versus age. By examining the scatterplot, do you believe that the data follow a straight line?

**b.** Determine the least squares line and plot it on the scatterplot.

**c.** Perform residual analysis to check the assumptions of linearity, independence, normality, and equal variance. Are any of the assumptions violated? Justify your answer.

⁛ **Data**

For the full data set visit stat.hawkeslearning.com and navigate to **Discovering Business Statistics, Second Edition > Data Sets > Screen Time Comparison Data (Regression).**

| Table 13.2.1 - Screen Time Comparison Data | | |
| --- | --- | --- |
| Age Group | Age | Screen Time (in Minutes) |
| 8–12 Years Old | 10 | 135 |
| 8–12 Years Old | 11 | 112 |
| 8–12 Years Old | 10 | 138 |
| 8–12 Years Old | 8 | 142 |
| 8–12 Years Old | 10 | 105 |
| ... | | |
| Over 18 Years Old | 59 | 113 |

## SOLUTION

**a.** Figure 13.2.8 shows the scatterplot of the data along with the least squares line. The scatterplot appears to be linear in that as the age of the respondent increases, the amount of time spent on mobile devices decreases. Thus, the scatterplot shows that fitting a straight line to the data is appropriate.

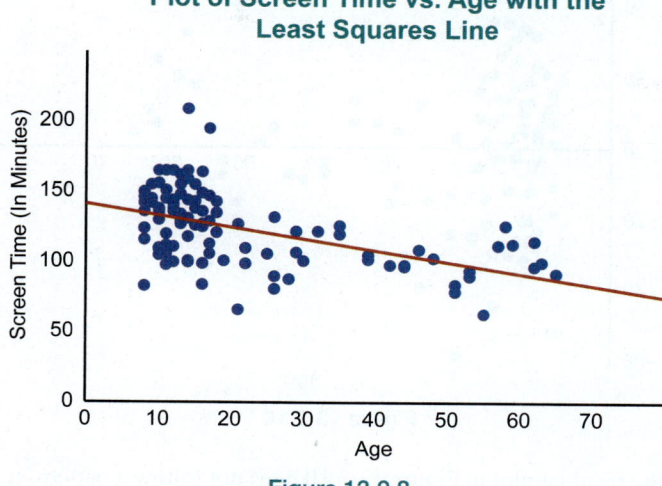

Figure 13.2.8

**b.** Figure 13.2.9 is the JMP output containing the estimates of the slope and intercept. The fitted regression line is

$$\widehat{Screen\ Time} = 141.39 - 0.82\,Age.$$

### Least Squares Estimates of the Slope and Intercept

Parameter Estimates

| Term | Estimate | Std Error | t Ratio | Prob > |t| |
|------|----------|-----------|---------|-----------|
| Intercept | 141.38892 | 3.811268 | 37.10 | < .0001* |
| Age | -0.824529 | 0.141103 | -5.84 | < .0001* |

Figure 13.2.9

Additionally, the *P*-value associated with the hypotheses

$$H_0: \beta_1 = 0$$
$$H_a: \beta_1 \neq 0$$

is less than 0.0001 which implies that we reject the null hypothesis and conclude that the slope is significantly different from zero. Rejecting this hypothesis implies that fitting a straight line through the data is appropriate.

**c.** To test the assumptions of the regression analysis, we must check for linearity, independence of the errors, normality of the errors, and equal variance. First of all, the scatterplot in part a. shows that the data follow a linear pattern which validates the linearity assumption. Also, if we examine the residual plot of Age vs. Screen Time Residuals in Figure 13.2.10, it shows a random scatter which is an indication that the straight-line fit to the data is appropriate, upholding the linearity assumption.

### ⌘ Technology

For instructions on how to estimate (and test for significance) the slope and intercept of the least squares regression line using technology, visit stat.hawkeslearning.com and navigate to **Discovering Business Statistics, Second Edition > Technology Instructions > Regression > Simple Linear Regression**.

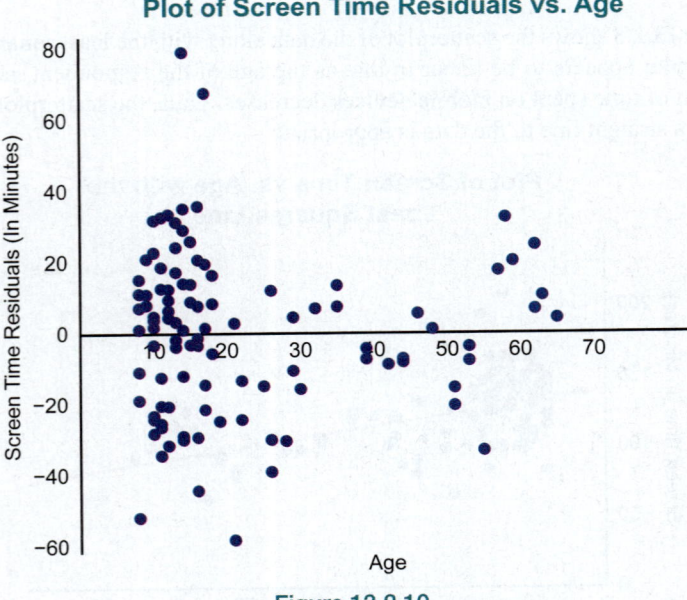

**Figure 13.2.10**

Given that the residual plot in Figure 13.2.10 does not follow a pattern, it also upholds the assumption that the data are independent (i.e., the errors at each value of age are independent).

The normal probability plot of the residuals is given in Figure 13.2.11. Note that almost all of the data fall on the diagonal line, which is an indication that the errors are normally distributed.

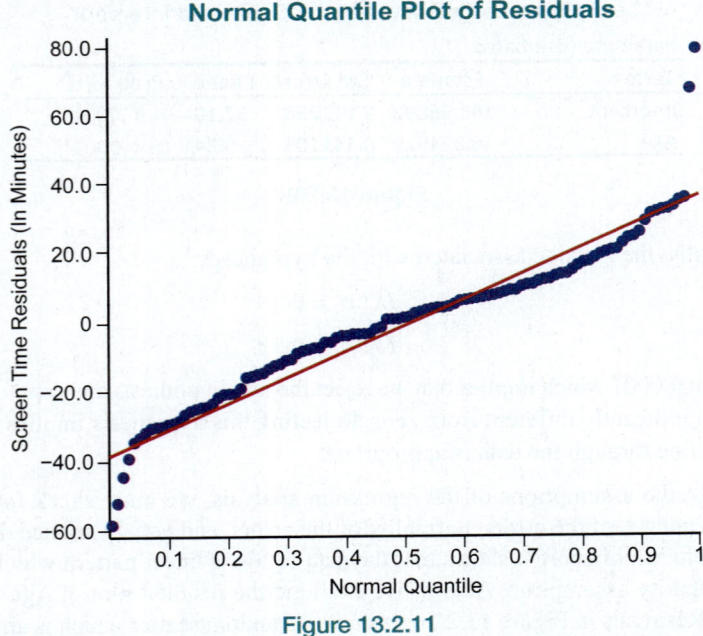

**Figure 13.2.11**

Lastly, we need to examine the equal variance assumption. Examining the residual plot (Figure 13.2.10), there appears to be random scatter above and below the zero-line. The nonexistence of a pattern between the independent variable (Age) and the residuals leads us to conclude that the variance is constant across the values of the independent variable.

It is important to validate the assumptions when performing a regression analysis. Depending on which assumption is violated, the results may still be meaningful. Note that hypothesis tests, confidence intervals, and prediction intervals are sensitive to departures from independence and departures from equal variance. Hypothesis tests and confidence intervals for the slope and intercept are robust against departures from normality. Lastly, prediction intervals are very sensitive to departures from normality.

## ✎ 13.2 Exercises

### Basic Concepts

1. What are the assumptions of the simple linear model that need to be validated when doing a residual analysis?

2. What should a well-behaved residual plot look like?

3. List three ways to determine if the errors are normally distributed in a regression analysis.

4. How should a normal probability plot look to indicate normality?

### Exercises

5. A scatterplot of $y$ versus $x$ for a dataset is given below. Which regression assumption is violated in this plot?

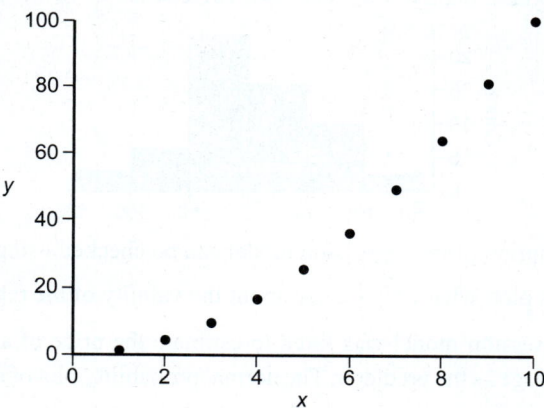

6. Based on the above plot, would you recommend fitting the regression model to predict the response given the predictor? Please explain.

7. A linear regression model was fitted to estimate the salary of an employee based on his/her experience. The plot of residuals of the regression model against the experience is given below. Which regression assumption is violated in this plot?

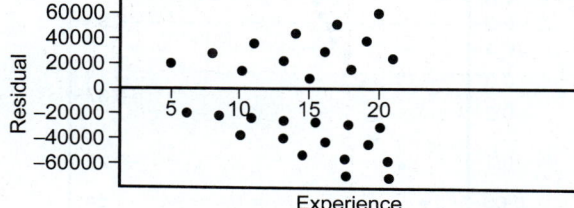

8. A regression model was fitted to predict the price of a used car using the mileage as the predictor. The plot of residuals of the regression model against the mileage is given below. State the regression assumption, if any, violated in this plot.

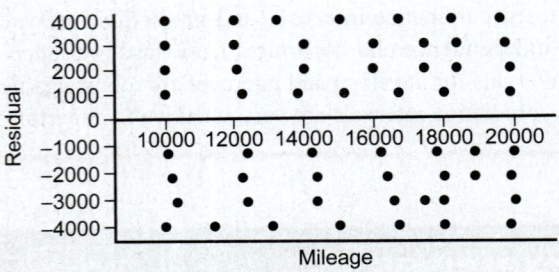

9.  Observe the residuals vs. the fitted plot for the regression model of the price of a car against the age of the car. Is this model appropriate for predicting the price of the car using the age of the car? Explain.

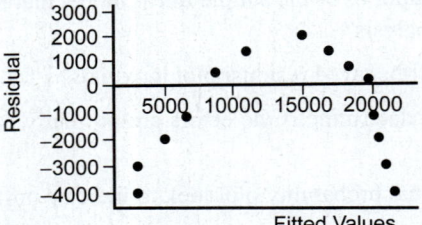

A simple regression model was fitted to estimate the credit score of customers based on their income. The histogram of residuals of the regression model is shown below. Use the histogram to answer the next two exercises.

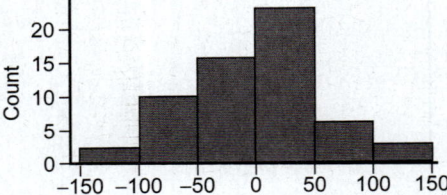

10.  Which assumption of the regression model can be checked using this plot?

11.  Based on this plot, what can you say about the validity of the regression model?

A simple regression model was fitted to estimate the price of a used Honda Civic using the mileage as the predictor. The normal probability plot of regression residuals is shown below. Use this to answer the next two exercises.

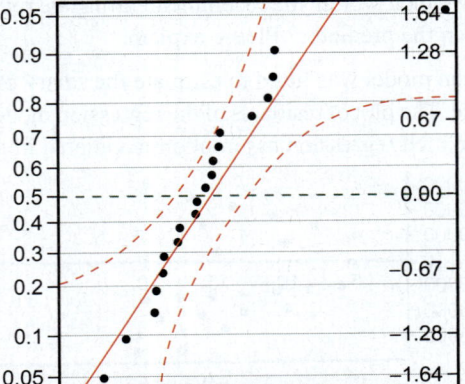

12.  Which assumption of the regression model can be checked using this plot?

13.  Based on this plot, what can you say about the validity of the regression model?

14. Download the Pizza Delivery Data, which describes the relationship between Delivery Time (Minutes), the Number of Pizzas delivered, and the Distance (Miles). Use the data to answer the following questions.

    a. Create a scatterplot of Delivery Time vs. Number of Pizzas. By examining the scatterplot, do you believe that the data follow a linear pattern?

    b. Perform a residual analysis to check the assumptions of linearity, independence, normality, and equal variance. Are any of the assumptions violated? Justify your answer.

15. Download the Marathon Time Data, which has the finishing Marathon Times of 44 runners along with the total number of kilometers they run in training the 4 weeks prior to the race. Use the data to answer the following questions.

    a. Create a scatterplot of Marathon Time vs. Km Run in 4 Weeks Prior. By examining the scatterplot, do you believe that the data follow a linear pattern?

    b. Perform a residual analysis to check the assumptions of linearity, independence, normality, and equal variance. Are any of the assumptions violated? Justify your answer.

**Data**

The data can be found by visiting stat.hawkeslearning.com and navigating to **Discovering Business Statistics, Second Edition > Data Sets > Pizza Delivery Data**.

**Data**

The data can be found by visiting stat.hawkeslearning.com and navigating to **Discovering Business Statistics, Second Edition > Data Sets > Marathon Time**.

# 13.3 Evaluating the Fit of the Linear Regression Model

The goal in constructing most linear models is to use the independent variable, $x$, to explain or predict the dependent variable, $y$. The question we want to consider is, how much of the variation in $y$ can be explained with the model? Before determining how much variation the model explains, it will be necessary to evaluate how much variability exists in the $y$-variable. This quantity is called the **total sum of squares (TSS)** and represents the total variation in the dependent variable, $y$.

**Formula**

**Total Sum of Squares (TSS)**

The total variation in $y$ is given by the **total sum of squares (TSS)**.

$$TSS = \Sigma(y_i - \bar{y})^2$$

If you think TSS looks a great deal like the numerator of the formula for the sample variance, you are right. TSS is the sum of the squared deviations about the mean of the dependent variable, $y$. If TSS were divided by $n - 1$ it would be the sample variance of $y$.

## What is an Error?

An error $(y_i - \hat{y}_i)$ represents the model's inability to predict the variation in the dependent variable, $y$. If $y$ didn't vary, for example if all $y$'s were 6, its value would be easy to predict and the model's errors would all be zero. Adding all of the squared errors accumulates the total of all *unexplained* variation.

$$SSE = \Sigma(y_i - \hat{y}_i)^2$$

The variation in $y$ can be divided into two categories, unexplained and explained. Total variation must equal unexplained variation plus explained variation.

$$\text{TSS} = \text{Unexplained Variation} + \text{Explained Variation}$$

or

$$\text{TSS} = \text{SSE} + \text{Explained Variation}$$

Denoting explained variation as **SSR (sum of squares of regression)** produces

$$\text{TSS} = \text{SSE} + \text{SSR} \quad (\text{TSS is the total unexplained and explained variation in } y).$$

Solving this equation for SSR results in

$$\text{SSR} = \text{TSS} - \text{SSE} \quad (\text{the explained variation, SSR, is equal to the total variation minus the unexplained variation}).$$

---

### Formula

#### Sum of Squares of Regression (SSR)

The explained variation in $y$ is given by the **sum of squares of regression**, which is equal to the total variation minus the unexplained variation.

$$\text{SSR} = \text{TSS} - \text{SSE}$$

---

**Measures of Variation**

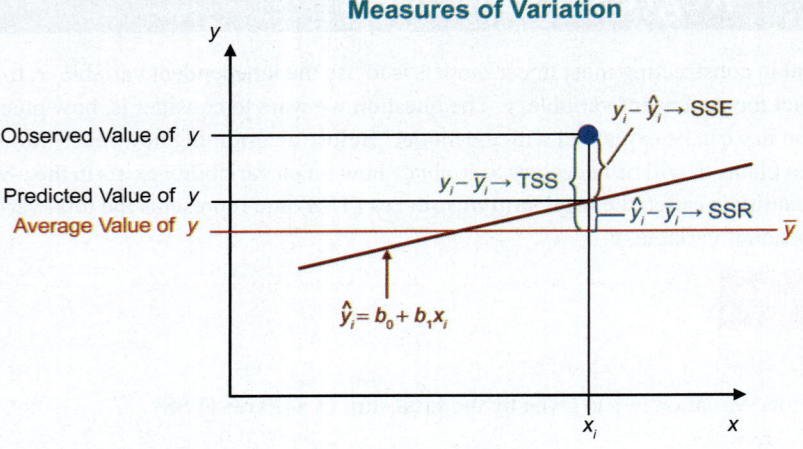

Figure 13.3.1

## Interpreting SSR

It would be delightful if the model would explain all of the variability in $y$, but unless all of the observed data points fall in a straight line, this will not happen. In virtually all models, there will be errors, i.e., unexplained variation. The difference between the total variation in $y$ and the unexplained variation must be the variation that is explained by the regression model. That's why explained variation is called the sum of squares of regression, SSR.

In the production example,

$$\text{TSS} = 1,596,902.5 \text{ and } \text{SSE} = 398,075.2967.$$

Therefore,

$$\text{SSR} = \text{TSS} - \text{SSE} = 1,596,902.5 - 398,075.2967 = 1,198,827.2033.$$

These values are found in the ANOVA table of the summary output in the *SS* (sum of squares) column.

SUMMARY OUTPUT

| Regression Statistics | |
|---|---|
| Multiple R | 0.866441198 |
| R Square | 0.75072035 |
| Adjusted R Square | 0.719560393 |
| Standard Error | 223.0681781 |
| Observations | 10 |

ANOVA

| | df | SS | MS | F |
|---|---|---|---|---|
| Regression | 1 | 1198827.203 | 1198827 | 24.09247 |
| Residual | 8 | 398075.2967 | 49759.41 | |
| Total | 9 | 1596902.5 | | |

| | Coefficients | Standard Error | t Stat | P-value |
|---|---|---|---|---|
| Intercept | 2227.955922 | 370.1487713 | 6.019082 | 0.000317 |
| Items Produced | 53.8835069 | 10.97779658 | 4.908408 | 0.001181 |

**Figure 13.3.2**

Of the roughly 1.6 million units of total variation in $y$, the model explains about 1.2 million. The proportion of the variation explained by the model is called the **coefficient of determination** and is denoted as $R^2$.

> **Formula**
>
> ### Coefficient of Determination
>
> The **coefficient of determination**, $R^2$, is given by
>
> $$R^2 = \frac{\text{SSR}}{\text{TSS}} = 1 - \frac{\text{SSE}}{\text{TSS}}.$$
>
> The coefficient of determination is a value between 0 and 1, inclusive. That is, $0 \le R^2 \le 1$.

The $R^2$ measurement summarizes the degree of fit on a standardized scale. The largest value $R^2$ can attain is 1, which will occur when the model explains all of the variation in $y$ and consequently SSR = TSS. The smallest value of $R^2$ is 0, which occurs when the model does not explain any of the variation in $y$ and, consequently, SSR = 0. Thus, the $R^2$ value is the proportion of the variation in $y$ explained by the model.

For the production data,

$$R^2 = \frac{1,198,827.2033}{1,596,902.5} \approx 0.7507.$$

In other words, the estimated model explains about 75 percent of the variation in costs. That's pretty good. One of the interesting features of the $R^2$ statistic is the ability to compare the fits of two models. If one model explains 75 percent of the data and another explains 82 percent, then the second model is preferred, all other things being equal.

---

**Example 13.3.1**

**Explaining Variation in a Linear Model**

An entrance exam given at a private college has been used for years as a predictor of academic success. If these test scores are predictors of academic success, they should be positively related to the grade point average upon graduation. Thirty graduates of the college were sampled and their grade point averages (GPA) upon graduation and entrance exam test scores reported upon admission are recorded in the following table.

**Data**

The Test Scores and Graduating GPA data set can be found on stat.hawkeslearning.com by navigating to **Discovering Business Statistics, Second Edition > Data Sets > Test Scores and Graduating GPA**.

| | | | Table 13.3.1 – Test Scores and Graduating GPA | | | | |
|---|---|---|---|---|---|---|---|
| Student | Verbal | Math | Total | College GPA | Predicted GPA | Error | Error Squared |
| 1 | 680 | 554 | 1234 | 3.42 | 2.8647 | 0.5553 | 0.30835809 |
| 2 | 486 | 562 | 1048 | 2.37 | 2.4741 | −0.1041 | 0.01083681 |
| 3 | 500 | 564 | 1064 | 2.52 | 2.5077 | 0.0123 | 0.00015129 |
| 4 | 501 | 564 | 1065 | 2.25 | 2.5098 | −0.2598 | 0.06749604 |
| 5 | 503 | 583 | 1086 | 2.9 | 2.5539 | 0.3461 | 0.11978521 |
| | | | | ... | | | |
| 30 | 549 | 564 | 1113 | 2.34 | 2.6106 | −0.2706 | 0.07322436 |

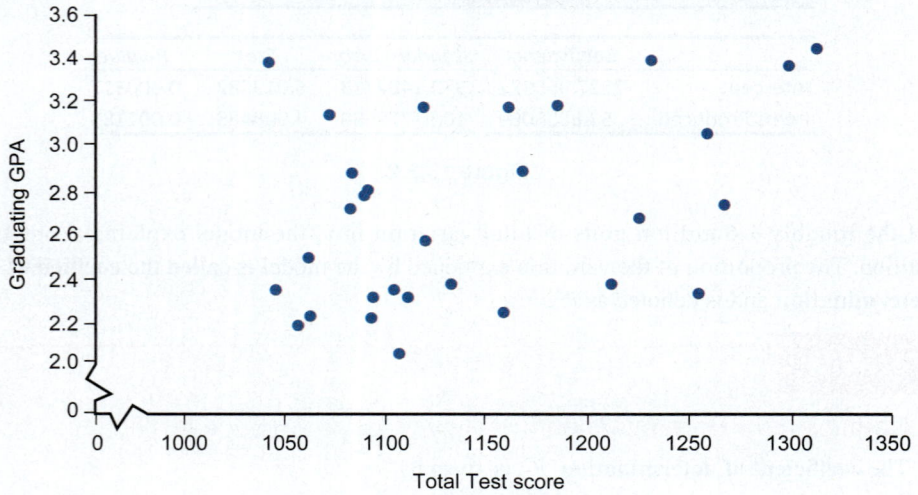

**Scatterplot of Graduating GPA and Total Test Score**

Figure 13.3.3

The scatterplot in Figure 13.3.3 suggests that as test scores on the entrance exam increase the GPA tends to increase, although there is a substantial amount of variability in the relationship. The upward sloping pattern of the data suggests a linear model could be constructed. However, a great deal of variation in the model's errors should be expected. What percent of the variation in final grade point average can be explained by the model relating total test score to graduating GPA?

### SOLUTION

Using the least squares method, the estimated model is given by

$$\text{Estimate Graduating GPA} = 0.2733 + 0.0021(\text{Total Test Score}).$$

SUMMARY OUTPUT

| Regression Statistics | |
| --- | --- |
| Multiple R | 0.3966 |
| R Square | 0.1597 |
| Adjusted R Square | 0.1297 |
| Standard Error | 0.4008 |
| Observations | 30 |

ANOVA

| | df | SS | MS | F |
| --- | --- | --- | --- | --- |
| Regression | 1 | 0.8548 | 0.8548 | 5.3218 |
| Residual | 28 | 4.4976 | 0.1606 | |
| Total | 29 | 5.3524 | | |

| | Coefficients | Standard Error | t Stat | P-value |
| --- | --- | --- | --- | --- |
| Intercept | 0.2733 | 1.0677 | 0.2560 | 0.7998 |
| Total Test Score | 0.0021 | 0.0009 | 2.3069 | 0.0287 |

**Figure 13.3.4**

When we studied the production model, a list of predicted values and errors for each observed value was given. Instead of providing a list of the errors for each of the observed values in the GPA model, let's summarize the errors from the model. In particular,

$$s_e^2 \approx 0.1606 \quad \text{and} \quad s_e \approx 0.4008.$$

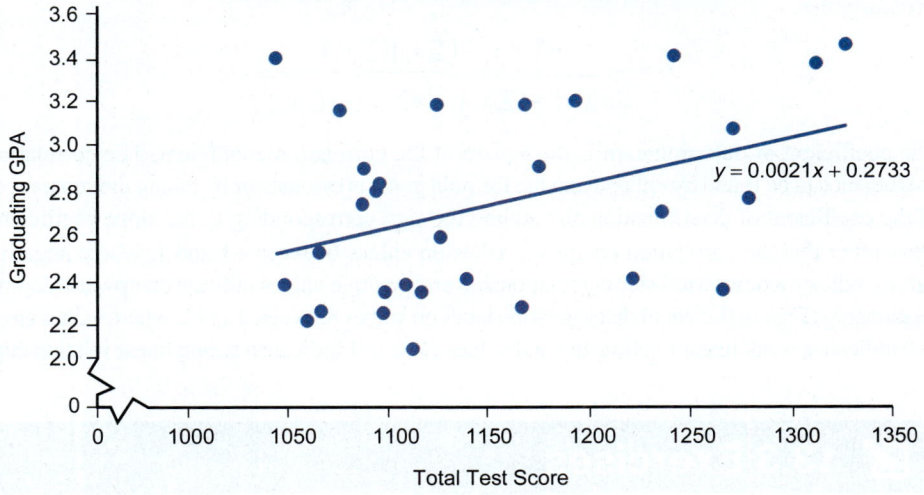

**Figure 13.3.5**

One of the differences in the production model and the GPA model is the manner in which the models seem to fit the data. In the production model, the data seemed to fit closely around the line, while in the GPA model the data are loosely clustered about the line. While *tight* and *loose* are interesting portrayals of the relative fit of the models to the data, it would be desirable to have a numerical measure to describe fit. $R^2$ is such a measure.

$$R^2 = \frac{\text{SSR}}{\text{TSS}} \approx \frac{0.8548}{5.3524} \approx 0.1597$$

Thus, approximately 16% of the variation in graduating GPA is explained by the linear model.

Because $R^2$ is a unit-free measure, it can be used to compare the fit of two models. The GPA model only explains approximately 16% of the variation in the dependent variable. Compared to the production model, which had an $R^2$ of approximately 0.7507, this model seems dramatically inferior. Using $R^2$ as a criterion, the production model seems to have a substantially better fit (0.7507 versus 0.1597) than the GPA model. The real question is whether you can predict more accurately with the model than other available alternatives. If so, models with relatively low coefficients of determination (such as the GPA model) are useful. For example, if you could develop a model to predict stock prices, minute-by-minute, achieving an $R^2$ value of only 0.20, you could be a very wealthy person.

$R^2$ can also be found using the following computational formula.

**Technology**

For instructions on how to find the coefficient of determination using technology, visit stat.hawkeslearning.com and navigate to **Discovering Business Statistics, Second Edition > Technology Instructions > Regression > Coefficient of Determination**.

> **Formula**
>
> Coefficient of Determination
>
> The **coefficient of determination**, $R^2$, can be calculated using the equation
>
> $$R^2 = \left( \frac{n \sum x_i y_i - \sum x_i \sum y_i}{\sqrt{\left( n \sum x_i^2 - \left( \sum x_i \right)^2 \right) \left( n \sum y_i^2 - \left( \sum y_i \right)^2 \right)}} \right)^2.$$

**Technology**

For instructions on how to find the correlation coefficient using technology, visit stat.hawkeslearning.com and navigate to **Discovering Business Statistics, Second Edition > Technology Instructions > Regression > Correlation Coefficient**.

Normally you will not have to use this formula since calculators and computer programs can calculate the coefficient of determination. Recall the computational formula for the correlation coefficient, discussed in Section 4.7, that measures the degree of linear relationship between two variables.

$$r = \frac{n \sum x_i y_i - \left( \sum x_i \right) \left( \sum y_i \right)}{\sqrt{n \sum x_i^2 - \left( \sum x_i \right)^2} \sqrt{n \sum y_i^2 - \left( \sum y_i \right)^2}}$$

The coefficient of determination is the square of the correlation coefficient. The correlation coefficient can be found by either using the formula given previously or by taking the square root of the coefficient of determination and adding the sign corresponding to the slope coefficient. Remember that the correlation coefficient takes on values between −1 and 1, where negative values indicate a downward sloping relationship and positive values indicate an upward sloping relationship. The coefficient of determination takes on values between 0 and 1, where values close to 0 indicate a weak linear relationship and values close to 1 indicate a strong linear relationship.

## ✏ 13.3 Exercises

### Basic Concepts

1.  What is the total sum of squares?

2.  How are the total sum of squares and the sample variance related?

3.  Define error in terms of a regression model.

4.  What part of the simple linear regression model captures the unexplained variation?

5.  Describe the total sum of squares in terms of explained and unexplained variation.

6.  What is the sum of squares of regression?

7.  Express SSR in terms of the total sum of squares and the sum of squared errors. Interpret this in terms of model variation.

8.  Why will there be errors in virtually all regression models?

9. What is the coefficient of determination? What kinds of values can the coefficient of determination take?

10. Suppose that regression analysis is performed and the resulting model has an $R^2$ value of 0.856. Interpret this value.

11. How is the coefficient of determination related to the correlation coefficient?

## Exercises

12. A direct mail marketing company has been experimenting with the effect of price on sales. Five different direct mail prices have been sent to different sets of customers. They have carefully tracked the customers from each group and have recorded the proportion from each price category that purchased the product. The results are given in the following table.

| Direct Mail | |
|---|---|
| Proportion That Purchased Product | Price of Product ($) |
| 0.032 | 29.95 |
| 0.028 | 34.95 |
| 0.026 | 39.95 |
| 0.015 | 44.95 |
| 0.009 | 49.95 |

   a. What level of measurement do the two variables in the table possess?

   b. Specify the model that the marketing manager would be interested in estimating.

   c. Which of the variables is the dependent variable in the model?

   d. Which of the variables is the independent variable in the model?

   e. Draw a scatterplot of the data.

   f. Use the data in the table to estimate the model.

   g. Predict the proportion that will buy the product if the price is $35.00.

   h. Compute the mean error for the model you estimated in part f.

   i. Determine the mean square error.

   j. What is the coefficient of determination? Interpret this value in terms of the problem.

   k. Consider exercise 12 parts f and j. Use the information in these two parts to compute the correlation coefficient between the Proportion that Purchased Product and the Price of Product.

13. An economist is studying the relationship between income and savings. He has randomly selected seven subjects and obtained income and savings data from them. He wishes to use a simple linear regression model to predict savings based on annual income.

| Income and Savings | |
|---|---|
| Income (Thousands of Dollars) | Savings (Thousands of Dollars) |
| 28 | 0.2 |
| 25 | 0 |
| 34 | 0.8 |
| 43 | 1.2 |
| 48 | 3.1 |
| 39 | 2.1 |
| 74 | 8.3 |

   a. What level of measurement do the two variables in the table possess?

**b.** Which of the variables is the dependent variable in the model?

**c.** Which of the variables is the independent variable in the model?

**d.** Draw a scatterplot of the data. Does the scatterplot suggest that a linear model is appropriate? Explain.

**e.** Use the data to estimate the appropriate model.

**f.** Predict the savings for someone who earns fifty thousand dollars annually.

**g.** Interpret the meaning of the slope coefficient in the problem.

**h.** What fraction of the variation in savings is explained by income?

14. The Road Warrior Trucking Company has kept careful records on ten hauls. The traffic manager has recorded the haul weight of each truck and its miles per gallon during ten runs with the intent of building a regression model. He wants to predict the miles per gallon for a haul based on the haul weight. The haul weights and miles per gallon information is given in the following table. Haul weights are given in thousands of pounds.

| Trucking | |
|---|---|
| Miles per Gallon | Haul Weight (Thousands of Pounds) |
| 4.6 | 36 |
| 4.8 | 33 |
| 5.1 | 31 |
| 4.0 | 42 |
| 4.7 | 33 |
| 5.2 | 30 |
| 4.5 | 37 |
| 4.6 | 37 |
| 4.2 | 40 |
| 4.5 | 36 |

**a.** What is the dependent variable in the model?

**b.** What is the independent variable in the model?

**c.** Construct a scatterplot of the data. Based on the scatterplot, does a linear model seem appropriate?

**d.** Write the model in terms of miles per gallon and haul weight. (Assume the parameters of the model have not been estimated.)

**e.** Use the data provided and estimate the coefficients of the linear model.f. Interpret the coefficient of the independent variable.

**g.** Use the model to predict the miles per gallon for a truck hauling 38,000 pounds.

**h.** Do you believe there is a causal relationship between haul weight and the miles per gallon? If so, which direction is the causality? Do greater haul weights cause reduced mileage, or vice versa? Does the regression analysis prove the causality?

**15.** An agricultural research station is trying to determine the relationship between the yield of sunflower seeds and the amount of fertilizer applied. To determine the relationship, three different fields were planted. In each field four different plots were defined. In each plot a different amount of fertilizer was used. The plot assignments for the fertilizer application were randomly selected in each field.

| Agricultural Research | |
|---|---|
| Pounds of Fertilizer (per Acre) | Pounds of Sunflower Seeds (per Acre) |
| 200 | 420 |
| 200 | 445 |
| 200 | 405 |
| 400 | 580 |
| 400 | 540 |
| 400 | 550 |
| 600 | 580 |
| 600 | 600 |
| 600 | 610 |
| 800 | 630 |
| 800 | 620 |
| 800 | 626 |

**a.** Are the data developed through a controlled experiment or are the data observational?

**b.** Draw a scatterplot of the data.

**c.** If a linear model is developed, which of the variables will be the dependent variable? Why?

**d.** Use the method of least squares to estimate the appropriate model.

**e.** Interpret the meaning of the slope coefficient in the model.

**f.** What fraction of the variation in pounds of sunflower seeds per acre can be explained by the amount of fertilizer used?

**g.** Predict the sunflower seed yield per acre if 500 pounds of fertilizer are applied.

**16.** Since 2009, the average term for a new-car loan was nearly 64 months. This leaves the buyer vulnerable to owing more on the car than it is worth. When applying for an automobile loan, it is oftentimes recommended to sign up for the shortest term you can afford. It is believed that along with one's credit rating, the length of the loan will help the buyer get a favorable interest rate. The following table contains interest rates and lengths of loans for 20 randomly selected auto purchases. Using the data in the table, answer the following questions.

| Lengths of Loans and Interest Rates | |
|---|---|
| Months Financed | Interest Rate (%) |
| 12 | 4.00 |
| 24 | 4.40 |
| 36 | 5.24 |
| 12 | 3.43 |
| 24 | 4.40 |
| 36 | 5.79 |
| 36 | 5.98 |
| 48 | 6.58 |
| 36 | 5.31 |
| 36 | 5.91 |

| Lengths of Loans and Interest Rates (cont.) | |
|---|---|
| Months Financed | Interest Rate (%) |
| 48 | 6.51 |
| 48 | 6.68 |
| 60 | 7.13 |
| 60 | 7.48 |
| 72 | 8.31 |
| 60 | 7.85 |
| 72 | 8.07 |
| 72 | 8.48 |
| 48 | 6.12 |
| 72 | 8.07 |

a. Using statistical software, estimate the coefficients of the least squares regression equation.

b. Interpret the meaning of the slope and the intercept in part **a**.

c. Predict the interest rate for a person interested in a four-year auto loan.

d. Should you use the model to predict interest rates for an eight-year loan? Justify your answer.

e. Determine the coefficient of determination and explain its meaning in terms of the problem.

f. Calculate the correlation coefficient for this model. What does it mean?

g. What interest rate would one expect to get if they were planning to apply for a five-year auto loan?

17. A sample data shows that the correlation coefficient between the number of pizzas and the delivery time is 0.64. If you would fit a regression model for the data to predict the delivery time given the number of pizzas, what percentage of the variation in the delivery time would be explained by the regression model?

# 13.4  Fitting a Linear Time Trend

In Chapter 4 we discussed the notion that the mean is not a reasonable descriptor for nonstationary time series data. Recall that nonstationary time series do not meander around some central value. Instead, the data tend to get larger or smaller over time. How can you describe time series data that possess a trend? For some time series, a **linear time trend** is a useful model. A linear time trend is nothing but a line that is used to model the changes in some phenomenon measured over time. In a linear time trend model, the independent variable is always a time index. The following example will illustrate the estimation of a **linear time trend model**.

**Example 13.4.1**

**Modeling Data with a Linear Time Trend**

 **Data**

This data set can be found on stat.hawkeslearning.com under **Discovering Business Statistics, Second Edition > Data Sets > Tuition Consumer Price Index**.

Many analysts believe that college tuition prices may soon be in the same situation as housing prices were when the housing bubble burst (causing home prices to drop significantly). Table 13.4.1 contains data for the Tuition Consumer Price Index (TCPI) from 1978 to 2020. Use a linear time trend to model the data.

| Table 13.4.1 – Tuition Consumer Price Index, 1978-2020 | |
|---|---|
| Year | TCPI |
| 1978 | 59.9 |
| 1979 | 64.7 |

| Table 13.4.1 – Tuition Consumer Price Index, 1978-2020 (cont.) | |
| --- | --- |
| 1980 | 70.8 |
| 1981 | 79.6 |
| 1982 | 90.3 |
| ... | |
| 2020 | 877.3 |

**SOLUTION**

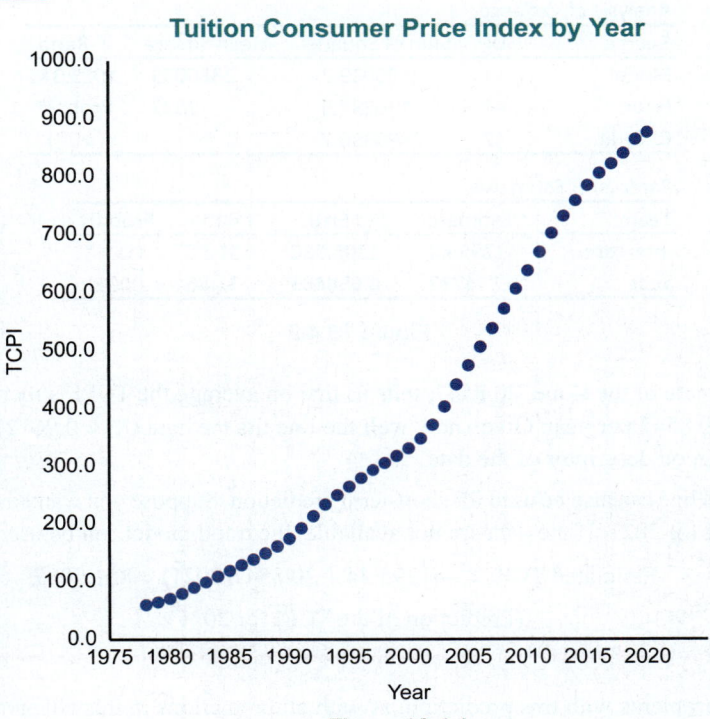

**Figure 13.4.1**

A graph of the data reveals an upward trend in the tuition consumer price index. The data appear to be a nonstationary time series with an upward trend. To describe the data, we will model the trend by fitting a line through the data with the notion of capturing how fast (on average) the series is changing over time. Estimating the slope of the line will provide the average rate of change per year in the TCPI. The line is fitted using least squares estimates in exactly the same way as other regression models have been constructed. The independent variable in a linear trend model is always time. In this case, the dependent variable is TCPI.

The estimated least squares equation is

$$\text{Estimated TCPI} = -41295.44 + 20.8547(\text{Year})$$

The JMP output for the problem is given in Figure 13.4.2.

Linear Fit

TCPI = −41295.44 + 20.854732*Year

Summary of Fit

| | |
|---|---|
| RSquare | 0.961187 |
| RSquare Adj | 0.96024 |
| Root Mean Square Error | 53.25903 |
| Mean of Response | 393.1699 |
| Observations (or Sum Wgts) | 43 |

Analysis of Variance

| Source | DF | Sum of Squares | Mean Square | F-Ratio |
|---|---|---|---|---|
| Model | 1 | 2880039.2 | 2880039 | 1015.341 |
| Error | 41 | 116297.5 | 2837 | Prob > F |
| C. Total | 42 | 2996336.7 | | < .0001* |

Parameter Estimates

| Team | Estimate | Std Error | t Ratio | Prob>|t| |
|---|---|---|---|---|
| Intercept | −41295.44 | 1308.338 | -31.56 | < .0001* |
| Year | 20.854732 | 0.654483 | 31.86 | < .0001* |

**Figure 13.4.2**

The estimate of the slope, 20.8547, tells us that on average the TCPI is increasing at a rate of 20.8547 per year. Given how well the line fits the data ($R^2 \approx 0.9612$), the trend line is a good descriptor of the data.

The trend line can also be used for short-term prediction. Suppose you wanted to estimate the TCPI for 2021. If the data are not available, the trend model can be used.

$$\text{Estimated TCPI} = -41295.44 + 20.8547(2021) = 851.9087$$

(Prediction of the TCPI for 2021)

One of the problems with this prediction, as with all predictions in this chapter, is that the accuracy of the prediction is unknown. It might be very close to the true value or it could be very inaccurate. If there were some knowledge about the accuracy of the prediction it would be more useful. In later sections, we will return to this topic and study inferential methods.

## ✏️ 13.4 Exercises

### Basic Concepts

1. Why is the mean not a reasonable descriptor for nonstationary time series data?

2. What is a linear time trend?

3. What is the independent variable in a linear trend model?

4. Is there a difference between the way the best fit line is determined for time series data and the way it is determined for other types of data?

5. Identify a problem with predictions that are made using a time trend model.

### Exercises

6. Consider the following table containing the Consumer Price Index (CPI) for all urban consumers in the United States from 1990 to 2010. The index is based on 1982–84 prices.

| Consumer Price Index | | | |
|---|---|---|---|
| Year | Consumer Price Index (CPI) | Year | Consumer Price Index (CPI) |
| 1990 | 130.7 | 2001 | 177.1 |
| 1991 | 136.2 | 2002 | 179.9 |
| 1992 | 140.3 | 2003 | 184.0 |
| 1993 | 144.5 | 2004 | 188.9 |
| 1994 | 148.2 | 2005 | 195.3 |
| 1995 | 152.4 | 2006 | 201.6 |
| 1996 | 156.9 | 2007 | 207.34 |
| 1997 | 160.5 | 2008 | 215.30 |
| 1998 | 163.0 | 2009 | 214.54 |
| 1999 | 166.6 | 2010 | 218.06 |
| 2000 | 172.2 | | |

**Source:** Bureau of Labor Statistics

**Data**

This data set can be found on stat.hawkeslearning.com under **Discovering Business Statistics, Second Edition > Data Sets > Consumer Price Index**.

   a. Looking at the data in the table, do you believe the trend line will slope upward or downward?

   b. Suppose we are interested in constructing a linear trend model for these data. Identify the independent and dependent variables for this model.

   c. Write the general equation for the time trend model in terms of year and CPI.

   d. Using statistical software, the following least squares model was determined.

$$\text{Estimated CPI} = -8647.4245 + 4.4107(\text{Year})$$

   Use this model to predict the price level in 2015.

   e. Can we determine the accuracy of this prediction? Explain.

**7.** Consider the following monthly sales data for an up-and-coming technology company.

| Sales Data | |
|---|---|
| Month | Sales (Thousands of Dollars) |
| 1 | 321 |
| 2 | 542 |
| 3 | 540 |
| 4 | 581 |
| 5 | 641 |
| 6 | 700 |
| 7 | 698 |
| 8 | 710 |
| 9 | 799 |
| 10 | 821 |
| 11 | 833 |
| 12 | 850 |

   a. Identify the independent and dependent variables for the linear time trend model.

   b. Using statistical software, the following summary output was produced.

SUMMARY OUTPUT

| Regression Statistics | |
|---|---|
| Multiple R | 0.949341195 |
| R Square | 0.901248704 |
| Adjusted R Square | 0.891373575 |
| Standard Error | 51.20789475 |
| Observations | 12 |

ANOVA

| | df | SS | MS | F |
|---|---|---|---|---|
| Regression | 1 | 239318.1818 | 239318.1818 | 91.26449427 |
| Residual | 10 | 26222.48485 | 2622.248485 | |
| Total | 11 | 265540.6667 | | |

| | Coefficients | Standard Error | t Stat | P-value |
|---|---|---|---|---|
| Intercept | 403.7575758 | 31.51628057 | 12.81107949 | 1.57569E-07 |
| Month | 40.90909091 | 4.282219283 | 9.553245222 | 2.41268E-06 |

Write the estimated regression equation.

c.  What is the mean square error for this model? The standard error?

d.  Using this model, predict the company's sales for the 13$^{th}$ month.

e.  What percent of the variation in sales is explained by the linear time trend model? Does this model seem to accurately fit the data?

8.  Consider the following table containing unemployment rates for North Carolina and South Carolina in 2000 through 2010.

### Unemployment Rates 2000–2010

| | Unemployment Rate (%) | |
|---|---|---|
| Year | North Carolina | South Carolina |
| 2000 | 3.7 | 3.6 |
| 2001 | 5.6 | 5.2 |
| 2002 | 6.6 | 6.0 |
| 2003 | 6.5 | 6.7 |
| 2004 | 5.5 | 6.8 |
| 2005 | 5.3 | 6.8 |
| 2006 | 4.8 | 6.4 |
| 2007 | 4.7 | 5.6 |
| 2008 | 6.2 | 6.8 |
| 2009 | 10.8 | 11.3 |
| 2010 | 10.6 | 11.2 |

Source: Bureau of Labor Statistics

a.  Using statistical software, estimate the following linear time trend model:

$$\text{N.C. Unemployment Rate} = \beta_0 + \beta_1(\text{Year}) + \varepsilon_i.$$

Write the estimated regression equation using the least squares estimates for $\beta_0$ and $\beta_1$.

b.  Using statistical software, estimate the following linear time trend model:

$$\text{S.C. Unemployment Rate} = \beta_0 + \beta_1(\text{Year}) + \varepsilon_i.$$

Write the estimated regression equation using the least squares estimates for $\beta_0$ and $\beta_1$.

c.  Use the equations in parts **a.** and **b.** to estimate the unemployment rates for North and South Carolina in the year 2013.

  d.  What is the coefficient of determination for the regression model in part **a.**?

  e.  What is the coefficient of determination for the regression model in part **b.**?

  f.  Do you think that these regression models are reliable in predicting future unemployment rates? Of the two models, which seems to fit the data better?

# 13.5 Inference Concerning the Slope

Since $\beta_1$ specifies the rate of change between $x$ and $y$, in most linear models the parameter of interest is $\beta_1$. Two inferential techniques are useful in evaluating the estimate of $\beta_1$. Confidence intervals, similar in structure to those used for means and proportions, will be developed. In addition, a hypothesis testing procedure will be presented to test whether $\beta_1$ is equal to some particular value.

## The Confidence Interval for $\beta_1$

Developing a confidence interval for $\beta_1$ requires thinking about the estimate $b_1$ as a random variable. Each random sample from the population will produce different data and hence different estimates of $b_0$ and $b_1$. The confidence interval will serve two purposes: to place bounds on the location of $\beta_1$ and to provide information about the quality of the point estimate, $b_1$. The form of the confidence interval is familiar.

$$\begin{pmatrix} \text{Sample} \\ \text{estimate of} \\ \text{parameter} \end{pmatrix} \pm \begin{pmatrix} \text{A certain number of standard} \\ \text{deviation units depending on} \\ \text{the desired confidence} \end{pmatrix} \cdot \begin{pmatrix} \text{The standard} \\ \text{deviation of the} \\ \text{sample estimate} \end{pmatrix}$$

The sample estimate of $\beta_1$ is $b_1$. The variance of $b_1$ is given by

$$\sigma_{b_1}^2 = \frac{\sigma_\varepsilon^2}{\sum (x_i - \bar{x})^2}$$

but like all population measurements, $\sigma_{b_1}^2$ usually has to be estimated from the data. Notice that the denominator of the expression above is equal to the variance of $x$, multiplied by the sample size, $n$. This indicates that the variance of $b_1$ is reduced if the variance of the error terms decreases, the sample size increases, or the variance of $x$ increases.

The sample estimate of the variance of $b_1$ is given by

$$s_{b_1}^2 = \frac{s_e^2}{\sum (x_i - \bar{x})^2} \,.$$

The only difference in the computation of $\sigma_{b_1}^2$ and $s_{b_1}^2$ is the replacement of the population variance of the error terms, $\sigma_\varepsilon^2$, with the corresponding sample statistic, $s_e^2$. The standard deviation (standard error) of the sample estimate $b_1$ is

$$s_{b_1} = \sqrt{\frac{s_e^2}{\sum (x_i - \bar{x})^2}} \,.$$

---

**Formula**

### 100(1−α)% Confidence Interval for $\beta_1$

The $100(1 - \alpha)\%$ confidence interval for $\beta_1$ is given by

$$b_1 \pm t_{\alpha/2, df} s_{b_1},$$

where $t_{\alpha/2, df}$ is the critical value for a $t$-distribution with $n - 2$ degrees of freedom.

✎ **NOTE**

The degrees of freedom are reduced for each parameter estimated in the linear model. Since we estimated two parameters, $\beta_0$ and $\beta_1$, we have $n - 2$ degrees of freedom.

## 🔗 Technology

The $t$-value corresponding to a particular area under the $t$-distribution curve can be found using the **invT** function on the TI-84 Plus calculator. For instructions, please visit stat.hawkeslearning.com and navigate to **Discovering Business Statistics, Second Edition > Technology Instructions > $t$-Distribution > Inverse $t$**.

```
NORMAL FLOAT AUTO REAL RADIAN MP

invT(0.975,18)
                         2.100922015
```

The expression $b_1 \pm t_{\alpha/2,df} s_{b_1}$ creates the following interval.

$$\langle \overset{\longleftarrow}{\underset{b_1 - t_{\alpha/2,df}\, s_{b_1}}{}} \quad \overset{|}{\underset{b_1}{}} \quad \overset{\longrightarrow}{\underset{b_1 + t_{\alpha/2,df}\, s_{b_1}}{}} \rangle$$

The expression $t_{\alpha/2,df}$ relates the width of the interval to the amount of confidence required.

Recall that the $t$-distribution is very similar to the $z$-distribution. To use the $t$-distribution requires that its degrees of freedom be specified. In this case the degrees of freedom are $df = n - 2$.

If a 95% confidence interval is desired, then $1 - \alpha = 0.95$, which implies $\alpha = 0.05$ and $\alpha/2 = 0.05/2 = 0.025$. Suppose the sample size is $n = 20$. Using the tables for the $t$-distribution in Appendix A, $t_{\alpha/2,df}$ would be

$$t_{\alpha/2,df} = t_{0.025,18} = 2.101.$$

| df | $t_{0.100}$ | $t_{0.050}$ | $t_{0.025}$ | $t_{0.010}$ |
|---|---|---|---|---|
| 1 | 3.078 | 6.314 | 12.706 | 31.821 |
| 2 | 1.886 | 2.920 | 4.303 | 6.965 |
| | | ... | | |
| 17 | 1.333 | 1.740 | 2.110 | 2.567 |
| 18 | 1.330 | 1.734 | 2.101 | 2.552 |
| | | ... | | |

Hence, to be 95% confident in capturing the true value of $\beta_1$ in the confidence interval will require placing the interval endpoints 2.101 standard deviation units from the point estimate, $b_1$.

## Example 13.5.1

### Estimating a Confidence Interval for the Slope

| Table 13.5.1 – Weekly Production | | |
|---|---|---|
| **Week** | **Items Produced** | **Cost ($)** |
| 1 | 22 | 3500 |
| 2 | 30 | 3800 |
| 3 | 36 | 4500 |
| 4 | 41 | 4200 |
| 5 | 27 | 3700 |
| 6 | 45 | 4600 |
| 7 | 30 | 3600 |
| 8 | 37 | 4550 |
| 9 | 32 | 3990 |
| 10 | 31 | 3675 |

In Section 13.1, a model relating the number of items produced to total cost was constructed.

$$\text{Cost} = \beta_0 + \beta_1(\text{Items Produced}) + \varepsilon_i$$

If the relationship is to be applicable for the entire production process, then a substantial amount of data will be required, more than we could hope to collect. If the data given in Table 13.5.1 are considered a random sample of weekly production, then a relationship can be constructed from the sample data. Specifically, the estimated least squares regression line relating items produced to total cost is

$$\text{Estimated Cost} = \$2227.96 + \$53.88(\text{Items Produced}),$$

where

$b_0 = \$2227.96$ (the sample estimate of $\beta_0$, the $y$-intercept), and

$b_1 = \$53.88$ (the sample estimate of $\beta_1$, the slope).

**Note:** Both estimates were determined using Microsoft Excel and rounded to the nearest hundredth.

To draw conclusions about the relationship between items produced and total cost for the entire population, statistical inference must be applied. We will begin by estimating a 95% confidence interval for $\beta_1$.

## SOLUTION

Three pieces of information are required to calculate a confidence interval for $\beta_1$.

1. A sample estimate of $\beta_1$

2. A $t$-value corresponding to the level of confidence and the degrees of freedom associated with the data used to estimate the model

3. A sample estimate of the standard deviation of $b_1$, $s_{b_1}$.

Estimating the least squares line from the sample data produces $b_1 = 53.88$.

Since 95% confidence is required, $\alpha = 0.05$, and $\dfrac{\alpha}{2} = \dfrac{0.05}{2} = 0.025$. Degrees of freedom will be

$$df = n - 2 = 10 - 2 = 8.$$

Thus, $t_{0.025,8} = 2.306$.

The remaining piece of missing information is the standard deviation of $b_1$,

$$s_{b_1} = \sqrt{\frac{s_e^2}{\sum (x_i - \bar{x})^2}}.$$

### Table 13.5.2 — Squared Deviation

| Items Produced | Squared Deviation $(x_i - \bar{x})^2$ |
|---|---|
| 22 | $(22 - 33.1)^2 = 123.21$ |
| 30 | $(30 - 33.1)^2 = 9.61$ |
| 36 | $(36 - 33.1)^2 = 8.41$ |
| 41 | $(41 - 33.1)^2 = 62.41$ |
| 27 | $(27 - 33.1)^2 = 37.21$ |
| 45 | $(45 - 33.1)^2 = 141.61$ |
| 30 | $(30 - 33.1)^2 = 9.61$ |
| 37 | $(37 - 33.1)^2 = 15.21$ |
| 32 | $(32 - 33.1)^2 = 1.21$ |
| 31 | $(31 - 33.1)^2 = 4.41$ |
| **Total** | $\sum (x_i - \bar{x})^2 = 412.90$ |

Determining $s_{b_1}$ will require the knowledge of two quantities: the mean square error, $s_e^2$, and the sum of the squared deviations of the independent variable, $\sum (x_i - \bar{x})^2$.

The mean square error was computed in Section 13.1 as follows.

$$s_e^2 = \frac{\sum (y_i - \hat{y}_i)^2}{n - 2} = \frac{398075.2967}{10 - 2} \approx 49759.41$$

The sum of the squared deviations of the independent variable (items produced), $\sum (x_i - \bar{x})^2$, is given in Table 13.5.2. Thus, the standard deviation of $b_1$ is calculated as follows.

$$s_{b_1} = \sqrt{\frac{s_e^2}{\sum (x_i - \bar{x})^2}} = \sqrt{\frac{49759.41}{412.90}} \approx 10.9778$$

### ⌥ Technology

For instructions on obtaining regression output using Microsoft Excel or other technologies, visit stat.hawkeslearning.com and navigate to **Discovering Business Statistics, Second Edition > Technology Instructions > Regression > Simple Linear Regression**.

SUMMARY OUTPUT

| Regression Statistics | |
| --- | --- |
| Multiple R | 0.866441198 |
| R Square | 0.75072035 |
| Adjusted R Square | 0.719560393 |
| Standard Error | 223.0681781 |
| Observations | 10 |

ANOVA

| | df | SS | MS | F | Significance F |
| --- | --- | --- | --- | --- | --- |
| Regression | 1 | 1198827.203 | 1198827 | 24.09247121 | 0.00118116 |
| Residual | 8 | 398075.2967 | 49759.41 | | |
| Total | 9 | 1596902.5 | | | |

| | Coefficients | Standard Error | t Stat | P-value | Lower 95% | Upper 95% |
| --- | --- | --- | --- | --- | --- | --- |
| Intercept | 2227.955922 | 370.1487713 | 6.019082 | 0.000316596 | 1374.391324 | 3081.520519 |
| Items Produced | 53.8835069 | 10.97779658 | 4.908408 | 0.00118116 | 28.5686626 | 79.1983512 |

**Figure 13.5.1**

The manual calculation of $s_{b_1}$ is tedious. Virtually every statistical analysis program that performs regression analysis calculates $s_{b_1}$. The summary output from Microsoft Excel is given in Figure 13.5.1. Most software packages will automatically include a confidence interval for $\beta_1$ or it will include the pieces required to compute a confidence interval. Microsoft Excel automatically displays the 95% confidence interval for $\beta_1$ and is capable of displaying an interval for any level of confidence you choose.

### 95% Confidence Interval for $\beta_1$

$$b_1 \pm t_{\alpha/2, df} s_{b_1}$$
$$53.88 \pm 2.306(10.9778)$$
$$53.88 \pm 25.3148$$
$$28.57 \text{ to } 79.19$$

```
(                          |                          )
28.57                    53.88                      79.19
```

Putting the pieces together results in an interval which spans from $28.57 to $79.19. We are 95% confident that this interval contains the true value of $\beta_1$. There are two possible interpretations of this interval.

1.  We are 95% confident that the true increase in total cost from producing one additional item is between $28.57 and $79.19.

2.  We are 95% confident the maximum error of the point estimate ($b_1 = 53.88$) in estimating the unknown $\beta_1$ (the true increase in total cost from producing one additional item) is at most $25.31.

**Notes:** The "confidence" we are discussing is in the procedure, not in the specific interval 28.57 to 79.19. Since the hand calculation of the confidence interval for $\beta_1$ used values that were rounded, the interval varies slightly from what is reported by Microsoft Excel. Using unrounded values in Excel, the confidence interval for $\beta_1$ is given as 28.57 to 79.20, rounded to two decimal places (see Figure 13.5.1).

Construct a 99% confidence interval for $\beta_1$ using the model described in Example 13.5.1.

**Example 13.5.2**

**Constructing a Confidence Interval for the Slope**

### SOLUTION

The difference between a 95% confidence interval and a 99% confidence interval is expressed in the value of $t_{\alpha/2,df}$. For 99% confidence, $1 - \alpha = 0.99$, which implies that $\alpha = 0.01$.

Hence, $\dfrac{\alpha}{2} = \dfrac{0.01}{2} = 0.005$.

The degrees of freedom will remain $df = 10 - 2 = 8$.

The appropriate $t$-value is $t_{\alpha/2,df} = t_{0.005,8} = 3.355$.

**99% Confidence Interval for $\beta_1$**

$$b_1 \pm t_{\alpha/2,df} s_{b_1}$$
$$53.88 \pm 3.355(10.9778)$$
$$53.88 \pm 36.8305$$
$$17.05 \text{ to } 90.71$$

17.05     53.88     90.71

**Note:** The confidence interval in this example was calculated using rounded values from the summary output. Microsoft Excel calculates the confidence interval using unrounded values as 17.05 to 90.72.

**Technology**

For instructions on obtaining regression confidence intervals for the slope and intercept using Microsoft Excel or other technologies, visit stat.hawkeslearning.com and navigate to **Discovering Business Statistics, Second Edition > Technology Instructions > Regression > Simple Linear Regression**.

Requiring more confidence results in a wider interval. The trade-off is less precision (larger interval) for more confidence.

It is important to remember that the assumptions of the linear model given in Section 13.1 must hold in order to retain the specified degree of confidence in the interval.

In Example 13.3.1 we examined a model relating entrance exam test scores to graduating GPA.

| Table 13.3.1 – Test Scores and Graduating GPA | | | |
|---|---|---|---|
| Student | Verbal | Math | Total | Graduating GPA |
| 1 | 680 | 554 | 1234 | 3.42 |
| 2 | 486 | 562 | 1048 | 2.37 |
| 3 | 500 | 564 | 1064 | 2.52 |
| 4 | 501 | 564 | 1065 | 2.25 |
| 5 | 503 | 583 | 1086 | 2.9 |
| ... | | | | |
| 30 | 549 | 564 | 1113 | 2.34 |

**Data**

The full data set is available on stat.hawkeslearning.com under **Discovering Business Statistics, Second Edition > Data Sets > Test Scores and Graduating GPA**.

Assuming the goal is to estimate a model that could be applied to all students that have graduated from college during the last five years, the population under consideration would be quite large, too large to obtain all the measurements. If we assume the data are from a sample, then we can estimate a model using the sample data and make inferences about the population.

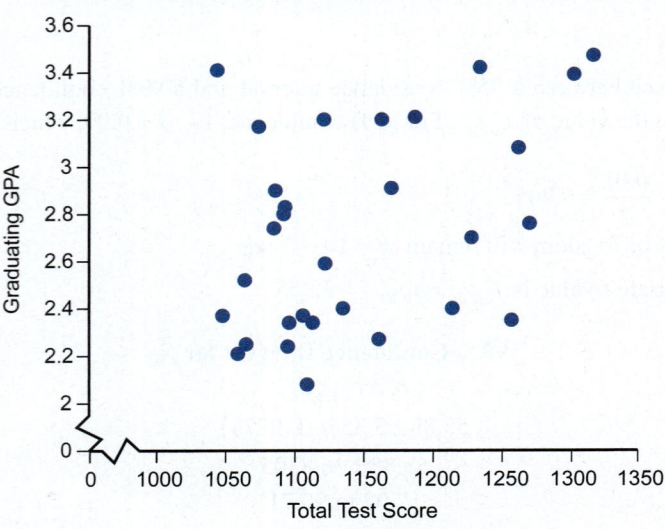

**Scatterplot of Graduating GPA and Total Test Score**

Figure 13.5.2

The population regression model to be estimated is

$$\text{Graduating GPA} = \beta_0 + \beta_1 \cdot (\text{Total Test Score}).$$

We will begin by making an inference concerning $\beta_1$. Construct a 95% confidence interval for $\beta_1$.

### SOLUTION

The paired data on test score and graduating GPA are taken from a random sample of students. The scatterplot does not show an overwhelming linear pattern, but a slight linear trend is apparent.

Using Excel's statistical analysis package to estimate the linear regression model, the output is given in Figure 13.5.3.

**ANOVA**

|  | df | SS | MS | F | Significance F |
|---|---|---|---|---|---|
| Regression | 1 | 0.854832198 | 0.854832198 | 5.321848064 | 0.028670474 |
| Residual | 28 | 4.497554469 | 0.160626945 | | |
| Total | 29 | 5.352386667 | | | |

|  | Coefficients | Standard Error | t Stat | P-value | Lower 95% | Upper 95% |
|---|---|---|---|---|---|---|
| Intercept | 0.273342604 | 1.067710391 | 0.256008189 | 0.799816021 | -1.913762987 | 2.460448195 |
| Total Test Score | 0.002144823 | 0.000929737 | 2.306913103 | 0.028670474 | 0.000240343 | 0.004049304 |

Figure 13.5.3

Using the computer output will make the job of calculating a confidence interval easy. Two of the three pieces of information required to calculate a confidence interval by hand are given in the output.

✏ **NOTE**

In the summary output from Excel, the confidence interval is already given (see the highlighted values above), but some technologies may require you to pull the pieces from the output to calculate the confidence interval.

- The sample estimate of $\beta_1$ is given in the *Coefficients* column associated with the variable Total Test Score. That is, $b_1$ is estimated to be 0.0021.

- The standard deviation of $b_1$ is given in the *Standard Error* column associated with the variable Total Test Score. That is, $s_{b_1}$ is estimated to be 0.00093.

The missing piece of information is the value of $t_{\alpha/2,df}$. Determining $t_{\alpha/2,df}$ requires the calculation of $\dfrac{\alpha}{2}$ and the degrees of freedom. Since the level of confidence is specified to be 95%, $\alpha = 0.05$ and $\dfrac{\alpha}{2} = \dfrac{0.05}{2} = 0.025$. The degrees of freedom are

$$df = n - 2 = 30 - 2 = 28.$$

Consequently, using technology

$$t_{\alpha/2,df} = t_{0.025,28} = 2.0484.$$

**∞ Technology**

The *t*-value corresponding to a particular area under the *t*-distribution curve can be found using the **invT** function on the TI-84 Plus calculator. For instructions please visit visit stat.hawkeslearning.com and navigate to **Discovering Business Statistics, Second Edition > t-Distribution > Inverse t**.

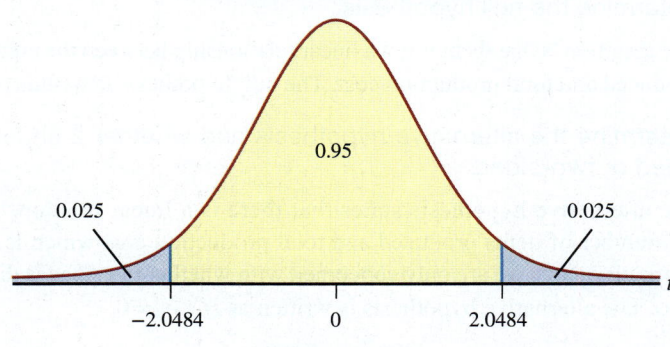

**t-Distribution, df = 28**

Figure 13.5.4

Now, let's assemble all the pieces. The $100(1 - \alpha)\%$ confidence interval for $\beta_1$ is

$$b_1 \pm t_{\alpha/2,df} \cdot s_{b_1}$$
$$0.0021 \pm 2.0484 \left(0.00093\right)$$
$$0.0021 \pm 0.001905$$

$$\vdash\!\!\!\!\!\overset{\textstyle 0.0002 \qquad\qquad 0.0021 \qquad\qquad 0.0040}{\rule{8cm}{0.4pt}}\!\!\!\!\!\dashv$$

We are 95% confident that the interval contains the true value of $\beta_1$. There are two possible interpretations of this interval.

1. We are 95% confident that the true increase in GPA for a one-point increase in total test score is between 0.0002 and 0.0040.

2. We are 95% confident the maximum error of the point estimate ($b_1 = 0.0021$) in estimating the unknown $\beta_1$ (the true increase in GPA for a one-point increase in total test score) is at most 0.001905.

## Testing a Hypothesis Concerning $\beta_1$

In constructing a regression model we must ask the question: *Does a linear relationship exist between y and x?* In answering this question, remember the linear model connects $x$ to $y$ through the slope parameter, $\beta_1$:

$$y_i = \beta_0 + \beta_1 x_i + \varepsilon_i.$$

If $\beta_1 = 0$, then there is no linear relationship between $x$ and $y$ since the term $\beta_1 x_i = 0$. Regardless of the value of $x$, the model becomes

$$y_i = \beta_0 + \varepsilon_i.$$

This says that $y$ is equal to a constant, $\beta_0$, plus a random error. Most of the time when developing a linear model used for predictive purposes, discovering that $\beta_1 = 0$ is bad news. Essentially this says that $x$ is not a useful predictor of $y$. Since $\beta_1$ is a model parameter and cannot be known unless all the bivariate population measurements are obtained, the sample estimate, $b_1$, will be used to make an inference concerning $\beta_1$.

If the assumptions of the linear model have been met sufficiently, statistical inference methods can be used to aid in answering the question, *is $b_1$ close enough to 0 to believe that $\beta_1 = 0$?* We will follow the hypothesis testing procedure used in Chapter 10.

---

**Example 13.5.4**

**Conducting a Hypothesis Test of the Slope**

Using the data in Example 13.5.1, determine if there is overwhelming evidence at the $\alpha = 0.05$ significance level of a relationship between the number of items produced and the total production cost.

**SOLUTION**

**Step 1:** Determine the null hypothesis.

The assertion is that there is not a linear relationship between the number of items produced and total production cost. The null hypothesis is written as $H_0: \beta_1 = 0$.

**Step 2:** Determine the alternative hypothesis and whether it should be one-sided or two-sided.

The alternative hypothesis states that there is a linear relationship between the number of items produced and total production cost which is a two-sided alternative since we are only concerned with whether the slope is different from zero. The alternative hypothesis is written as $H_a: \beta_1 \neq 0$.

**Step 3:** Select the appropriate test statistic based on the information at hand and the assumptions you are willing to make.

---

**Formula**

**Test Statistic for Testing the Hypothesis $\beta_1 \neq 0$**

The test statistic for testing the hypothesis $\beta_1 \neq 0$ is given by

$$t = \frac{b_1 - 0}{s_{b_1}} = \frac{b_1}{s_{b_1}}.$$

The test statistic follows a *t*-distribution with $n - 2$ degrees of freedom.

---

The test statistic is similar in nature to the other test statistics developed in Chapter 10. It measures how far $b_1$ is from the hypothesized value of $\beta_1$, which is 0. This distance is measured in standard deviation units. If $t$ is close to 0, then $b_1$ is close to 0 and $H_0: \beta_1 = 0$ is the more reasonable conclusion. However, if $t$ is far from zero, then $b_1$ is far from its hypothesized value and $H_a: \beta_1 \neq 0$ would seem more reasonable. This criterion is defined by the critical value of the test statistic.

**Step 4:** Determine the critical value of the test statistic.

The test is two-tailed and the significance level of the test is specified to be 0.05, which implies

$$\alpha = 0.05 \text{ and } \frac{\alpha}{2} = \frac{0.05}{2} = 0.025.$$

The test statistic has a *t*-distribution with

$$df = n - 2 = 10 - 2 = 8.$$

The critical value corresponds to $t_{0.025,8} = 2.306$.

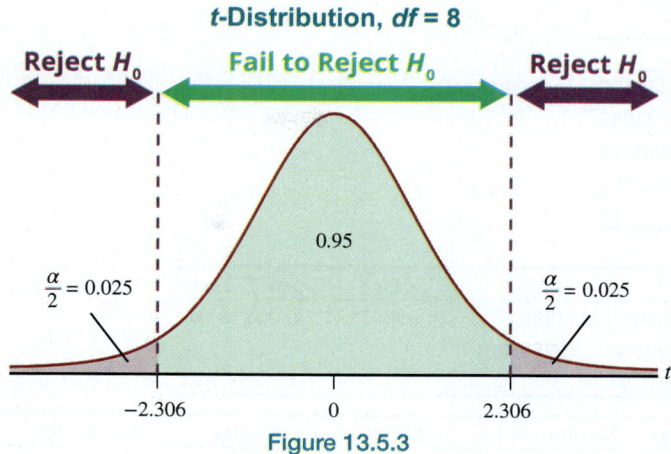

**Figure 13.5.3**

**Step 5:** Collect the sample data and compute the value of the test statistic.

| Table 13.5.3 – Regression Results | | | |
|---|---|---|---|
| **Predictor** | **Coefficient** | **Standard Deviation of Coefficient** | **t-value** |
| Intercept | 2227.96 | 370.1488 | 6.019 |
| Items Produced | 53.88 | 10.9778 | 4.908 |

$$t = \frac{b_1 - 0}{s_{b_1}} = \frac{b_1}{s_{b_1}} \approx \frac{53.88}{10.9778} \approx 4.908$$

The estimated value of $b_1$ is almost five standard deviations above zero. This is very persuasive evidence that $\beta_1 \neq 0$.

**Step 6:** Make the decision and state the conclusion in terms of the original question.

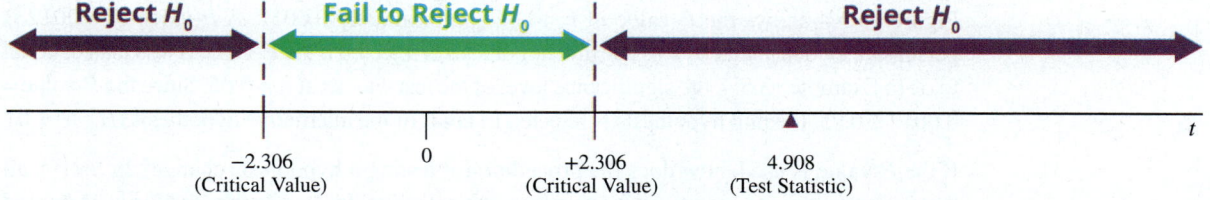

Since the value of the test statistic falls into the rejection region, reject the null hypothesis in favor of the alternative.

*Conclusion and Interpretation*: There is overwhelming evidence at the 0.05 significance level that $\beta_1 \neq 0$ so we reject the null hypothesis in favor of the alternative. This implies that it is reasonable to believe (at the 0.05 level) that there is a linear relationship between the number of items produced and total cost. In fact, there appears to be a positive linear relationship between items produced and production cost. However, our hypothesis test did not address the issue of a *positive* relationship, so we cannot make this conclusion.

Thus far, the focus has been on inference about $\beta_1$. What about $\beta_0$? Since $\beta_0$ is merely a constant term, in most problems its value is not of great concern. However, if a confidence interval or test of hypothesis is needed, the methods used would be virtually identical to those presented for analyzing $\beta_1$.

## Using *P*-Values to Test a Hypothesis Concerning $\beta_1$

The simplest method of testing any hypothesis is the *P*-value method. Since most computer outputs for regression analysis contain *P*-values, there are no calculations to make. Choosing $\alpha$ is all that is required to perform the test.

**Technology**

For instructions on obtaining regression output using technology, visit stat.hawkeslearning.com and navigate to **Discovering Business Statistics, Second Edition > Technology Instructions > Regression > Simple Linear Regression.**

**Technology**

For instructions on obtaining regression output using technology, visit stat.hawkeslearning.com and navigate to **Discovering Business Statistics, Second Edition > Technology Instructions > Regression > Simple Linear Regression.**

SUMMARY OUTPUT

| Regression Statistics | |
|---|---|
| Multiple R | 0.866441198 |
| R Square | 0.75072035 |
| Adjusted R Square | 0.719560393 |
| Standard Error | 223.0681781 |
| Observations | 10 |

ANOVA

| | df | SS | MS | F | Significance F |
|---|---|---|---|---|---|
| Regression | 1 | 1198827.203 | 1198827 | 24.09247121 | 0.00118116 |
| Residual | 8 | 398075.2967 | 49759.41 | | |
| Total | 9 | 1596902.5 | | | |

| | Coefficients | Standard Error | t Stat | P-value | Lower 95% | Upper 95% |
|---|---|---|---|---|---|---|
| Intercept | 2227.955922 | 370.1487713 | 6.019082 | 0.000316596 | 1374.391324 | 3081.520519 |
| Items Produced | 53.8835069 | 10.97779658 | 4.908408 | 0.00118116 | 28.5686626 | 79.1983512 |

**Figure 13.5.4**

The *P*-value measures the probability that the test statistic is as large as it is (in magnitude) under the assumption that the null hypothesis is true. Specifically, a *P*-value is the probability of observing a test statistic as large or larger (in absolute value) than what has been observed, given that the null hypothesis is true. In Example 13.5.4, the value of the test statistic was 4.908. The probability of observing a test statistic this large (in absolute value) or larger, given that the true value of the slope is zero, is very small. Fortunately, virtually all statistical analysis programs that perform regression analysis calculate *P*-values for the two-tailed test of hypothesis

$$H_0: \beta_1 = 0$$

$$H_a: \beta_1 \neq 0.$$

Figure 13.5.4 shows the *P*-value of $b_1$ to be approximately 0.0012. A *P*-value of 0.0012 is persuasive evidence that $\beta_1 \neq 0$. The null hypothesis is rejected if the *P*-value is less than or equal to $\alpha$. In Example 13.5.4, the significance level of the test was set at $\alpha = 0.05$. Since the *P*-value = $0.0012 \leq 0.05$, the null hypothesis is rejected in favor of the alternative hypothesis ($H_a: \beta_1 \neq 0$).

If the *P*-value is used, how does the procedure for testing a hypothesis change? In **Step 4** all that is necessary is to specify $\alpha$. It serves as a critical value. In **Step 6**, the *P*-value is compared to $\alpha$. Everything else remains the same, provided the *P*-value has been calculated for you.

If a data analyst feels that the assumptions of the simple linear model have been met and decides to make an inference about the model, the *P*-value of $b_1$ will be one of the first pieces of the computer output that will be examined.

## ✎ 13.5 Exercises

### Basic Concepts

1. Give an example of a practical application of the confidence interval for $\beta_1$.

2. Identify two purposes that confidence intervals for the estimated regression coefficients serve.

3. What is the sampling distribution for $b_1$? Give the mean and standard deviation.

4. What is the expression for determining the $100(1-\alpha)\%$ confidence interval for $\beta_1$?

5. Suppose a 95% confidence interval for $\beta_1$ is found to be $(15.11, 20.11)$. Give two interpretations of this interval.

6. If there is no linear relationship between two variables, what is the value of $\beta_1$? Explain.

7. What is the test statistic for testing the hypothesis that $\beta_1 \neq 0$? Describe how this test statistic is similar to other test statistics used in hypothesis testing.

8. What are the degrees of freedom associated with the simple linear regression model?

9. Can we make inferences about $\beta_0$? Explain why we are more interested in inferences about $\beta_1$.

10. Describe why the *P*-value corresponding to $b_1$, which is displayed by many regression summary outputs, is one of the first values examined by data analysts.

## Exercises

11. Consider the summary output for the monthly sales data given in Exercise 7 in Section 13.4.

SUMMARY OUTPUT

| Regression Statistics | |
|---|---|
| Multiple R | 0.949341195 |
| R Square | 0.901248704 |
| Adjusted R Square | 0.891373575 |
| Standard Error | 51.20789475 |
| Observations | 12 |

ANOVA

| | df | SS | MS | F |
|---|---|---|---|---|
| Regression | 1 | 239318.1818 | 239318.1818 | 91.26449427 |
| Residual | 10 | 26222.48485 | 2622.248485 | |
| Total | 11 | 265540.6667 | | |

| | Coefficients | Standard Error | t Stat | P-value |
|---|---|---|---|---|
| Intercept | 403.7575758 | 31.51628057 | 12.81107949 | 1.57569E-07 |
| Age | 40.9090901 | 4.282219283 | 9.55324522 | 2.41268E-06 |

a. Compute a 90% confidence interval for $\beta_1$.

b. Interpret this interval.

12. Consider the data in the following table regarding the age of a particular model of car and the asking price for that car.

| Car Data | |
|---|---|
| Age (Years) | Asking Price ($) |
| 1 | 11,875 |
| 1 | 10,995 |
| 2 | 9995 |
| 2 | 8500 |
| 3 | 8995 |
| 4 | 6995 |
| 5 | 4450 |
| 5 | 5500 |
| 6 | 4400 |
| 6 | 4800 |

a. Using statistical software, determine the sample estimate of $\beta_1$.

**b.** What is the standard error of $b_1$?

**c.** Find a 99% confidence interval for $\beta_1$.

**d.** Interpret the confidence interval found in part **c**.

**13.** An economist is studying the relationship between income and IRA contributions. He has randomly selected eight subjects and obtained annual income and IRA contribution data from them. He wishes to predict the amount of money contributed to an IRA based on annual income.

| Income and IRA Contributions | |
|---|---|
| Annual Income (Thousands of Dollars) | IRA Contribution (Thousands of Dollars) |
| 28 | 0.3 |
| 25 | 0 |
| 34 | 1.0 |
| 43 | 1.3 |
| 48 | 3.3 |
| 39 | 2.2 |
| 74 | 8.5 |

**a.** Draw a scatterplot of the data. Describe the relationship that you observe between income and IRA contribution.

**b.** Estimate the parameters of the following model using statistical software.

$$\text{IRA Contribution} = \beta_0 + \beta_1 (\text{Income}) + \varepsilon_i$$

**c.** Calculate and interpret a 95% confidence interval for $\beta_1$.

**d.** What assumptions are being made in the construction of the confidence interval for $\beta_1$?

**e.** Use the confidence interval you obtained to test the hypothesis that the IRA contribution increases with the increase in income of the subject.

**14.** Consider the following summary output, which was generated from a sample of 8 employees relating age to annual salary.

SUMMARY OUTPUT

*Regression Statistics*

| | |
|---|---|
| Multiple R | 0.732431223 |
| R Square | 0.536455496 |
| Adjusted R Square | 0.459198079 |
| Standard Error | 15.60374155 |
| Observations | 8 |

ANOVA

| | df | SS | MS | F |
|---|---|---|---|---|
| Regression | 1 | 1690.639497 | 1690.639 | 6.943741 |
| Residual | 6 | 1460.860503 | 243.4768 | |
| Total | 7 | 3151.5 | | |

| | Coefficients | Standard Error | t Stat | P-value |
|---|---|---|---|---|
| Intercept | -2.132440745 | 20.99597109 | -0.10156 | 0.922412 |
| Age | 1.564320608 | 0.593648001 | 2.635098 | 0.038794 |

**a.** What is the estimated regression equation?

**b.** Is there evidence of a linear relationship between age and salary at the 0.05 significance level?

**c.** Does the decision in part **b**. change at the 0.01 significance level? Explain.

**d.** What percentage of the variation in annual salary is explained by the model?

**15.** The college placement office is developing a model to relate grade point average (GPA) to starting salary for liberal arts majors. Ten recent graduates have been randomly selected, and their graduating GPAs and starting salaries were recorded.

| GPA and Starting Salary | |
|---|---|
| **GPA** | **Starting Salary (Thousands of Dollars)** |
| 2.2 | 35.1 |
| 3.5 | 45.2 |
| 2.1 | 36.3 |
| 2.8 | 39.3 |
| 3.2 | 41.4 |
| 2.5 | 37.6 |
| 2.4 | 34.8 |
| 2.9 | 25.7 |
| 3.1 | 40.1 |
| 3.7 | 39.5 |

**a.** Plot the data. Describe the relationship you observe between GPA and starting salary.

**b.** Using statistical software, estimate the parameters of the model
$$\text{Starting Salary} = \beta_0 + \beta_1(\text{GPA}) + \varepsilon_i.$$

**c.** Is there evidence of a linear relationship between GPA and starting salary? Test at the 0.05 significance level.

**d.** Predict the starting salary for a student with a GPA of 2.5.

**e.** Interpret the coefficient of GPA in the model.

**f.** What fraction of the variation in starting salaries is explained by GPA?

**g.** To perform statistical inference on the model, what assumptions are being made?

**16.** An experienced census official feels that she can accurately estimate the number of inhabitants of a city block by simply noting the size of the block and the types of buildings (single family homes, apartments, businesses, etc.) that are found on the block. This procedure, if accurate, would be much quicker and cheaper than visiting each residence and taking a survey of its inhabitants. The data below are estimates of block populations provided by the official for 10 blocks in a large city. Also given are the actual numbers of inhabitants for the same 10 blocks. These were found at a later point in time by conventional methods.

| Inhabitants of City Blocks | | | | | | | | | | |
|---|---|---|---|---|---|---|---|---|---|---|
| **Estimate** | 115 | 234 | 215 | 97 | 78 | 134 | 78 | 129 | 170 | 67 |
| **Actual** | 100 | 225 | 190 | 99 | 92 | 125 | 75 | 130 | 155 | 82 |

**a.** Draw a scatterplot of points of the actual number against the estimated number of inhabitants. Does the relationship appear to be linear?

**b.** Estimate the slope and intercept of the following regression equation using statistical software.
$$\text{Actual Inhabitants} = \beta_0 + \beta_1(\text{Estimated Inhabitants}) + \varepsilon_i$$

**c.** Is there evidence of a linear relationship between the actual number and the estimated number? Test at the $\alpha = 0.01$ significance level.

**d.** Interpret the regression coefficient for the estimated number of inhabitants.

**e.** Construct a 95% confidence interval for the slope of the regression equation. Interpret the interval.

**f.** Compute and interpret the $R^2$ value.

**g.** Predict the actual number of inhabitants on a block when the estimated number is 150. Round your answer to the nearest whole number.

**17.** A statistics professor would like to build a model relating student scores on the first test to the scores on the second test. The test scores from a random sample of 21 students who have previously taken the course are given in the table.

| | | | Test Scores | | | |
|---|---|---|---|---|---|---|
| **Student** | **First Test Grade** | **Second Test Grade** | **Student** | **First Test Grade** | **Second Test Grade** |
| 1 | 69 | 73 | 12 | 54 | 67 |
| 2 | 66 | 56 | 13 | 57 | 65 |
| 3 | 69 | 65 | 14 | 85 | 67 |
| 4 | 75 | 51 | 15 | 75 | 67 |
| 5 | 57 | 59 | 16 | 79 | 77 |
| 6 | 75 | 76 | 17 | 44 | 51 |
| 7 | 75 | 76 | 18 | 82 | 84 |
| 8 | 82 | 76 | 19 | 57 | 81 |
| 9 | 91 | 82 | 20 | 75 | 90 |
| 10 | 66 | 73 | 21 | 69 | 73 |
| 11 | 88 | 67 | | | |

**a.** Using statistical software, estimate the parameters of the model

$$\text{Second Test Grade} = \beta_0 + \beta_1 \left( \text{First Test Grade} \right) + \varepsilon_i.$$

**b.** What fraction of the variation in the grades on the second test is explained by the grades on the first test?

**c.** Is there a linear relationship between the first test grades and the second test grades? Test at the 0.05 significance level.

**d.** Suppose you're enrolled in the professor's course this semester. If you scored a 75 on the first test, use the model to predict your second test score. Round your answer to the nearest whole number.

# 13.6 Inference Concerning the Model's Prediction

Many regression models are developed for predictive purposes. For example, if you built the model relating the number of items produced to total cost, it was probably because you want to use it to predict total cost. While it is important to evaluate $b_1$, the estimate of the slope, the real concern of the model builder is the accuracy of the model's predictions. In the case of the production model, how accurate are the costs the model predicts? If the assumptions of the linear model (detailed in Section 13.1) have been met, then it is possible to make inferences as to the quality of a model's predictions.

## The Regression Line as the Mean Value of *y* Given *x*

Examining the production data in Table 13.5.1 reveals two weeks in which 30 items were produced. For a given value of items produced (say 30) the costs of producing 30 items were $3800 and $3600. For anyone who has observed a production process, price variation is not unexpected. If you use the model

$$\text{Estimated Cost} = \$2227.96 + \$53.88(\text{Items Produced})$$

for predictive purposes, then the predicted cost of producing 30 items will be

$$\text{Estimated Cost} = \$2227.96 + \$53.88(30) = \$3844.36.$$

Using this model, producing 30 items will have a predicted cost of $3844.36. Since the actual cost of producing 30 items varies, how do you interpret the predicted cost of $3844.36, which happens to be above the two observed values? The model's predicted value when 30 items are produced is considered to be the average cost of producing 30 items. In other words, it is the mean value of $y$ (cost) when $x$ (items produced) equals 30. But wait a minute! The costs of producing 30 items were $3800 and $3600, and the average of these numbers is not $3844.36. What kind of average is this? What we are essentially saying is that if we are willing to acknowledge that the relationship between $x$ and $y$ is linear, then all data (not just the data for $x = 30$) will be useful in establishing the estimated relationship. Once the relationship is established it will be used to estimate the mean value of $y$ for any given $x$. In fact, if items produced = 35, the predicted cost would be

$$\text{Estimated Cost} = \$2227.96 + \$53.88(35) = \$4113.76,$$

which would be interpreted to be the average cost of producing 35 items.

The entire regression line can be considered a collection of means of $y$ for different values of $x$. Unfortunately, the true linear relationship is unknown since all the data in the population will not be available. Since only the estimated model is available, only estimated values of the mean value of $y$ for some given $x$ will be available. How good are these estimates? How good is the estimate of a mean value of $y = \$4113.76$ when $x = 35$? To answer this question, we will once again rely on the notion of a confidence interval.

## Confidence Intervals for the Mean Value of $y$ Given $x$

### Formula

**$100(1-\alpha)\%$ Confidence Interval for the Mean Value of $y$ Given $x$**

If $x_p$ is a value of $x$ for which we wish to know the mean value of $y$, then the $100(1 - \alpha)\%$ confidence interval is given by

$$\hat{y}_p \pm t_{\alpha/2, df} \, s_e \sqrt{\frac{1}{n} + \frac{\left(x_p - \bar{x}\right)^2}{\sum \left(x_i - \bar{x}\right)^2}} \text{ , where}$$

| | |
|---|---|
| $\hat{y}_p$ | is the predicted value of $y$ when $x = x_p$, i.e., $\hat{y}_p = b_0 + b_1 x_p$, |
| $t_{\alpha/2, df}$ | is the $t$-value associated with $100(1 - \alpha)\%$ confidence (the same $t$ used in constructing confidence intervals for $\beta_1$), |
| $s_e$ | is the standard deviation of the error terms, and |
| $\dfrac{\left(x_p - \bar{x}\right)^2}{\sum \left(x_i - \bar{x}\right)^2}$ | measures how far $x_p$ is from $\bar{x}$ in relation to the total variation of $x$. The further $x_p$ is from $\bar{x}$, the larger this ratio will be, resulting in a wider confidence interval. |

For the production model, calculate the 95% confidence interval for the average cost of producing 35 items.

**Example 13.6.1**

**Calculating a Confidence Interval for the Average Cost**

**SOLUTION**

Four pieces of information are required to calculate the 95% confidence interval for the mean value of $y$ given $x = x_p$.

1.  Use the estimated regression line to calculate $\hat{y}_p$ for the value given for $x = x_p$.

2.  Find a $t$-value corresponding to the level of confidence and the degrees of freedom associated with the data used to estimate the model.

3.  Determine the standard deviation of the error terms.

4.  Compute the term $\dfrac{\left(x_p - \bar{x}\right)^2}{\Sigma\left(x_i - \bar{x}\right)^2}$.

In Example 13.5.1, Estimated Cost = \$2227.96 + \$53.88(Items Produced), $x_p = 35$, and the predicted value of $y$ for this given $x$ is

$$\hat{y}_p = b_0 + b_1 x_p = \$2227.96 + \$53.88(35) = \$4113.76 \, .$$

For a 95% confidence interval, $\alpha = 0.05$, and the degrees of freedom are $n - 2 = 10 - 2 = 8$. Thus, $t_{\alpha/2, df} = t_{0.025, 8} = 2.306$.

The standard deviation of the error terms is found in the output in Figure 13.5.1. In particular, $s_e \approx 223.0682$. Note that $s_e$ can also be computed by taking the square root of the mean square error, $s_e^2 \approx 49759.41$, which was computed in Section 13.5.

The last piece of information is the quotient

$$\frac{\left(x_p - \bar{x}\right)^2}{\Sigma\left(x_i - \bar{x}\right)^2} = \frac{(35 - 33.1)^2}{412.90} \approx 0.0087 \, .$$

Table 13.6.1 contains the calculation for the sum of the squared deviations, $\Sigma\left(x_i - \bar{x}\right)^2$.

| Table 13.6.1 – Squared Deviation | |
| --- | --- |
| Items Produced | Squared Deviation $\left(x_i - \bar{x}\right)^2$ |
| 22 | $(22 - 33.1)^2 = 123.21$ |
| 30 | $(30 - 33.1)^2 = 9.61$ |
| 36 | $(36 - 33.1)^2 = 8.41$ |
| 41 | $(41 - 33.1)^2 = 62.41$ |
| 27 | $(27 - 33.1)^2 = 37.21$ |
| 45 | $45 - 33.1)^2 = 141.61$ |
| 30 | $(30 - 33.1)^2 = 9.61$ |
| 37 | $(37 - 33.1)^2 = 15.21$ |
| 32 | $(32 - 33.1)^2 = 1.21$ |
| 31 | $(31 - 33.1)^2 = 4.41$ |
| Total | $\Sigma\left(x_i - \bar{x}\right)^2 = 412.90$ |

Assembling the pieces for the 95% confidence interval we have the following.

**95% Confidence Interval for the Mean Value of $y$ Given $x = 35$**

$$\hat{y}_p \pm t_{\alpha/2, df} \, s_e \sqrt{\frac{1}{n} + \frac{\left(x_p - \bar{x}\right)^2}{\Sigma\left(x_i - \bar{x}\right)^2}}$$

$$4113.76 \pm 2.306 \left( 223.0682 \sqrt{\frac{1}{10} + 0.0087} \right)$$

$$4113.76 \pm 169.5945$$

$$3944.17 \text{ to } 4283.35$$

```
 |---------------------+---------------------|
3944.17            4113.76              4283.35
```

The confidence interval can be interpreted in two ways.

1. We are 95% confident that the average cost of producing 35 items is between $3944.17 and $4283.35. (Note that like all confidence intervals, the confidence is in the method not in a particular interval.)

2. We are 95% confident that the maximum error of estimation for the average cost of producing 35 items is $169.59.

Figure 13.6.1 shows the fitted line plot of the production data along with the 95% confidence interval. Notice that as you approach the edges of the data, the confidence interval gets wider. This illustrates why it is important to only make predictions within the scope of the sample data.

### ◇ Technology

In order to graph the regression line, along with the confidence interval bands, using Minitab, visit stat.hawkeslearning.com and navigate to **Discovering Business Statistics, Second Edition > Technology Instructions > Regression > Linear Regression Fitted Line Plot with Confidence Interval**.

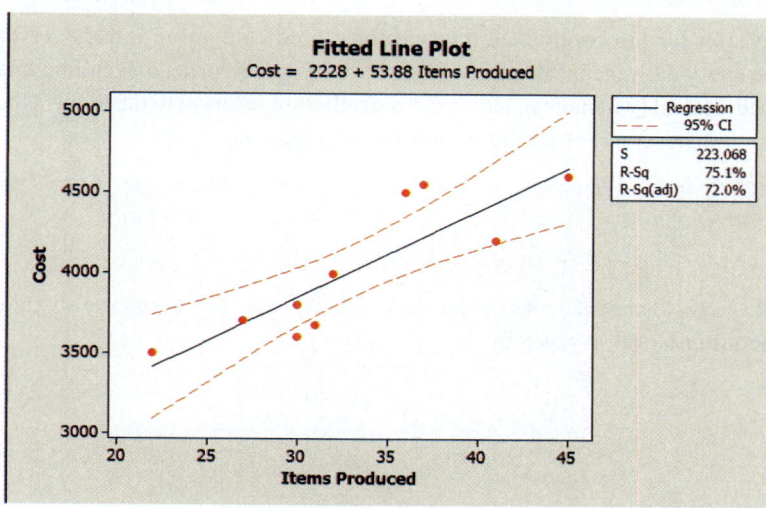

**Figure 13.6.1**

Many statistical analysis packages compute a confidence interval for the mean value of $y$ for a given value of $x$. Figure 13.6.2 is an example of a Minitab output. The output uses the estimated production model when items produced = 35. The value labeled "Fit" is the predicted value of $y$ when $x = 35$. (Note that values differ slightly to the ones previously presented due to rounding.) The value labeled "SE Fit" is the standard error of the fitted value when $x = 35$. The "95% CI" is the 95% confidence interval for the mean value of $y$ when $x = 35$. The "95% PI" will be discussed next.

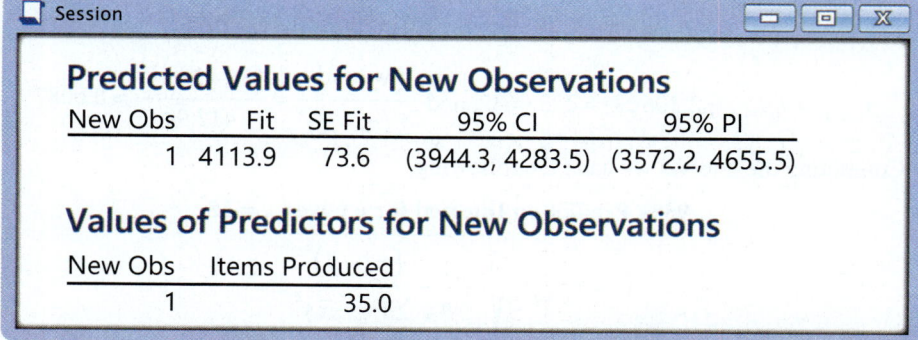

**Figure 13.6.2**

## Confidence Intervals for the Predicted Value of *y* Given *x*

We previously discussed confidence intervals for the mean value of $y$ given $x$, rather than individual outcomes. Suppose you wish to produce 35 items, and you wish to predict the total cost of production and compute a 95% confidence interval for the total cost. Using the model,

$$\text{Estimated Cost} = \$2227.96 + \$53.88(35) = \$4113.76$$

for producing 35 items. The actual cost of producing 35 items in a particular week may be above or below average, the model has no way of knowing. Consequently, the actual cost will likely be different from the average value, and that difference will be an error in the model's prediction. That error needs to be accounted for in the confidence interval for the predicted value of $y$ given $x$.

The expression for the confidence interval for a predicted value is very similar to the confidence interval for the mean value of $y$ given $x$. However, instead of calling this interval a confidence interval (which it is), let's call it a **prediction interval** to distinguish the interval from the confidence interval for the mean value of $y$ given $x$.

---

### Formula

#### $100(1-\alpha)\%$ Prediction Interval for the Value of *y* Given *x*

A $100(1 - \alpha)\%$ confidence interval for the predicted value of $y$ given $x$, also known as a **prediction interval**, is given by

$$\hat{y}_p \pm t_{\alpha/2,df}\, s_e \sqrt{1 + \frac{1}{n} + \frac{\left(x_p - \bar{x}\right)^2}{\sum\left(x_i - \bar{x}\right)^2}}\,.$$

---

Since the only difference between the prediction interval and the confidence interval for the mean value of $y$ given $x$ is the "1" inside the square root, the same information is required to compute the interval. However, since many statistical analysis packages will compute the prediction interval, you may not have to perform the computations very often.

---

**Example 13.6.2**

**Calculating a Prediction Interval for the Total Cost**

Suppose you, as the production manager, wish to produce 35 items this week. Compute the 95% prediction interval for the total cost of production.

#### SOLUTION

The predicted cost of producing 35 items is given by

$$\text{Estimated Cost} = \$2227.96 + \$53.88(35) = \$4113.76,$$

$$t_{\alpha/2,df} = t_{0.025,8} = 2.306 \,, \; s_e = 223.0682, \text{ and } \frac{\left(x_p - \bar{x}\right)^2}{\sum\left(x_i - \bar{x}\right)^2} = \frac{(35 - 33.1)^2}{412.90} \approx 0.0087\,.$$

Computing the interval we have the following.

**95% Prediction Interval for *y* Given *x* = 35**

$$\hat{y}_p \pm t_{\alpha/2,df}\, s_e \sqrt{1 + \frac{1}{n} + \frac{\left(x_p - \bar{x}\right)^2}{\sum\left(x_i - \bar{x}\right)^2}}$$

$$4113.76 \pm 2.306 \left( 223.0682 \sqrt{1 + \frac{1}{10} + 0.0087} \right)$$

$$4113.76 \pm 541.6316$$

$$3572.13 \text{ to } 4655.39$$

We are 95% confident that the actual cost of production will be within $541.63 of the average price for producing 35 items ($4113.76). Specifically, we are 95% confident that the production cost will be between $3572.13 and $4655.39.

Figure 13.6.3 shows a fitted line plot of the production data along with the 95% prediction interval. Notice the difference between the plots of the 95% confidence interval (in Figure 13.6.1) and the 95% prediction interval.

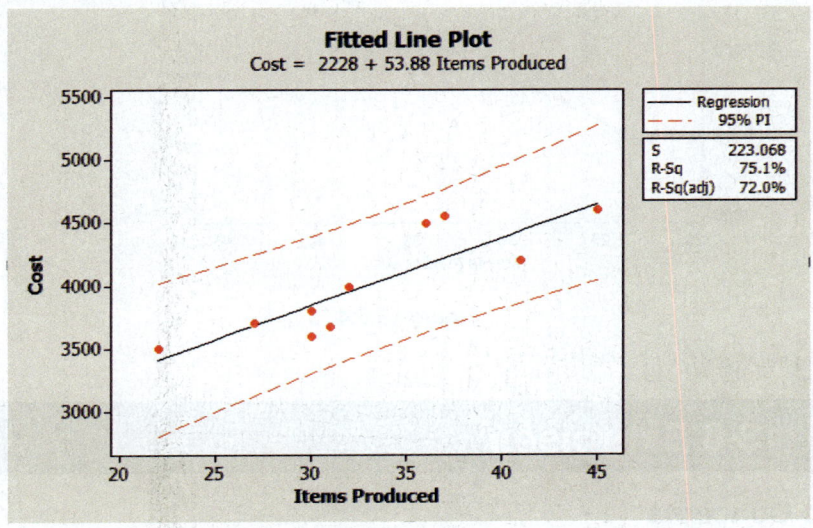

Figure 13.6.3

**Technology**

In order to graph the regression line, along with the prediction interval bands, using Minitab, visit stat.hawkeslearning.com and navigate to **Discovering Business Statistics, Second Edition > Technology Instructions > Regression > Linear Regression Fitted Line Plot with Prediction Interval**.

The graphs below compare the confidence intervals for the mean value of $y$ given $x = 35$ and the predicted value of $y$ given $x = 35$.

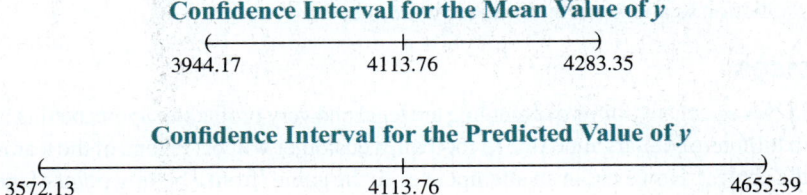

The prediction interval is more than three times wider than the confidence interval for the average value. A high price has been paid in order to account for individual variability.

Using the model for prediction outside the range of $x$ values used to create the model can be dangerous. The nature of the relationship may not be linear outside of the range of $x$ used to define the model. In the production example, the range of $x$-values spans from 22 to 45 items produced. Using the model to predict the cost of producing 100 items would no doubt have a sizable error. Inferential methods are not valid outside the range of $x$. Figure 13.6.4 illustrates the uncertainty associated with the regression line outside the range of the data.

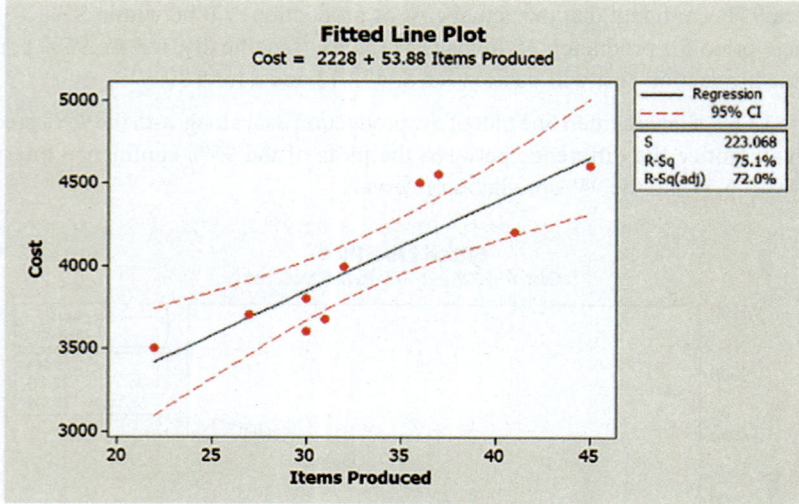

**Figure 13.6.4**

# 📝 13.6 Exercises

## Basic Concepts

1. What is the main concern for the model builder when performing regression analysis?

2. Distinguish between the mean value of $y$ given $x$ and the predicted value of $y$ given $x$.

3. Distinguish between a confidence interval and a prediction interval. Which interval is wider? Explain why.

4. Is there a particular range of $x$-values for which using the regression model for prediction is appropriate? Explain your answer.

## Exercises

5. In Nevada, many forms of gambling are legal and very profitable. Sports betting amounts to billions of dollars annually. In football, a customer will bet on one of the teams to win the contest. However, in an attempt to even the game (from a betting point of view) one of the teams is selected as the favorite. The favorite's score in the game is reduced by an amount called the line. For example, if the Cowboys are favored over the Falcons by four points, then four points are subtracted from the Cowboys' score to determine the outcome of the game for betting purposes. Thus, if the Cowboys defeat the Falcons 32 to 30, in so far as settling any bets, the Cowboys score would be reduced by the spread and the Cowboys would be the loser $32 - 4 = 28$ to 30. Where does the betting line come from? The line is created by a betting market. If too many people are betting on the Cowboys before the game starts, the bookmaker will try to make the game more attractive to potential Falcon bettors by increasing the spread say from four points to five points. On the other hand, if too many people are betting on the Falcons, the spread will diminish from four to perhaps three points. How accurate is the betting spread at predicting the actual spread, which is the actual difference in points between the favorite and the underdog? In the example of the Cowboys and the Falcons, the actual spread was +2 (32 – 30). To examine this question, we want to build the following model:

$$\text{Actual Point Spread} = \beta_0 + \beta_1\left(\text{Betting Spread}\right) + \varepsilon_i.$$

If the betting spread is a good predictor of the actual spread, it should be able to account for a substantial portion of the variation in the actual spreads. The following table contains betting and actual spreads from 15 randomly selected football games.

| Betting vs. Actual Spreads | | | | | | | | | | | | | | |
|---|---|---|---|---|---|---|---|---|---|---|---|---|---|---|
| **Betting** | 4 | 1 | 3 | 2 | 1 | 2 | 5 | 5 | 3 | 4 | 2 | 3 | 5 | 7 | 6 |
| **Actual** | 12 | −2 | 6 | 7 | 3 | 1 | 14 | 3 | −7 | 5 | 14 | 9 | 2 | 21 | 8 |

a. Draw a scatterplot of the data. Describe the relationship you observe between actual point spread and the betting spread.

b. Estimate the parameters of the model using statistical software.

c. Is there evidence at the 0.05 significance level of a linear relationship between the betting spread and the actual spread?

d. What fraction of the variation in the actual point spread is explained by the betting spread?

e. Interpret the coefficient of the betting spread in the model $(\beta_1)$

f. Construct and interpret a 95% confidence interval for $\beta_1$.

g. If the betting spread is five, what is the predicted actual spread?

h. Construct and interpret a 95% prediction interval for a betting spread of five.

i. Construct a 95% confidence interval for the average value of the actual spread when the betting spread is five.

6. Net income is the level of actual profit that a company reports for the year. Net sales is the total sales less adjustment for returns. What is the relationship between net income and net sales for large corporations? Suppose a random sample of 27 large corporations has been selected, and the net income and net sales have been recorded. A regression analysis has been performed to estimate the model, and the output is given.

$$\text{Net Income} = \beta_0 + \beta_1(\text{Net Sales}) + \varepsilon_i$$

## Regression Analysis: Income versus Sales

The regression equation is
Income = 84 + 18.4 Sales

| Predictor | Coef | SE Coef | T | P |
|---|---|---|---|---|
| Constant | 83.6 | 118.1 | 0.71 | 0.486 |
| Sales | 18.434 | 4.446 | 4.15 | 0.000 |

S = 372.478       R-Sq = 40.7%       R-Sq(adj) = 38.4%

Analysis of Variance

| Source | DF | SS | MS | F | P |
|---|---|---|---|---|---|
| Regression | 1 | 2384660 | 2384660 | 17.19 | 0.000 |
| Residual Error | 25 | 3468497 | 138740 | | |
| Total | 26 | 5853157 | | | |

Predicted Values for New Observations

| New Obs | Fit | SE Fit | 95% CI | 95% PI |
|---|---|---|---|---|
| 1 | 1005.3 | 147.1 | (702.4, 1308.2) | (180.5, 1830.0) |

Values of Predictors for New Observations

| New Obs | Sales |
|---|---|
| 1 | 50.0 |

a. Find and interpret the standard deviation of the error terms in the output.

b. Interpret the slope coefficient. (The data used to estimate the model was in millions of dollars.)

**c.** What fraction of the variation in net income is explained by net sales?

**d.** Is there evidence of a linear relationship between net income and net sales? Test at the 0.05 significance level.

**e.** Construct and interpret a 95% confidence interval for $\beta_1$, the slope of the line.

**f.** The output also contains a predicted value for net income when sales are $50,000,000. Find the predicted value of net income when sales are $50,000,000. (Note that in the original data all observations were measured in millions of dollars. Thus a predicted value of 10,000,000 would be displayed in the output as 10.)

**g.** Find and interpret the 95% confidence interval for the average value of net income given that sales are $50,000,000.

**h.** Suppose your firm generated $50,000,000 in sales. What would be the 95% prediction interval for your firm's net income?

**i.** Use the model to predict net income for a company with $60,000,000 in sales. (Note that you must compute this manually.)

**7.** The personnel director of a large hospital is interested in determining the relationship (if any) between an employee's age and the number of sick days the employee takes per year. The director randomly selects eight employees and records their age and the number of sick days which they took in the previous year.

### Sick Days and Age

| Employee | 1 | 2 | 3 | 4 | 5 | 6 | 7 | 8 |
|---|---|---|---|---|---|---|---|---|
| Age | 30 | 50 | 40 | 55 | 30 | 28 | 60 | 25 |
| Sick Days | 7 | 4 | 3 | 2 | 9 | 10 | 0 | 8 |

A regression analysis has been performed to estimate the model and the output is given.

$$\text{Sick Days} = \beta_0 + \beta_1 (\text{Age}) + \varepsilon_i$$

**Regression Analysis: Sick Days versus Age**

The regression equation is
Sick Days = 15.2 - 0.247 Age

| Predictor | Coef | SE Coef | T | P |
|---|---|---|---|---|
| Constant | 15.186 | 1.713 | 8.86 | 0.000 |
| Age | -0.24681 | 0.04105 | -6.01 | 0.001 |

S = 1.47652    R-Sq = 85.8%    R-Sq(adj) = 83.4%

**Analysis of Variance**

| Source | DF | SS | MS | F | P |
|---|---|---|---|---|---|
| Regression | 1 | 78.794 | 78.794 | 36.14 | 0.001 |
| Residual Error | 6 | 13.081 | 2.180 | | |
| Total | 7 | 91.875 | | | |

**Predicted Values for New Observations**

| New Obs | Fit | SE Fit | 95% CI | 95% PI |
|---|---|---|---|---|
| 1 | 6.547 | 0.557 | (5.184, 7.911) | (2.686, 10.409) |

**Values of Predictors for New Observations**

| New Obs | Age |
|---|---|
| 1 | 35.0 |

a. Draw a scatterplot of the data. Describe the relationship you observe between the number of sick days and age.

b. Find and interpret the standard deviation of the error terms in the output.

c. Interpret the slope coefficient.

d. What fraction of the variation in the number of sick days an employee takes per year is explained by age?

e. Is there evidence of a linear relationship between the number of sick days an employee takes per year and age? Test at the significance 0.05 level.

f. Construct and interpret a 95% confidence interval for $\beta_1$, the slope of the line.

g. Find the predicted value of the number of sick days an employee will take per year if the employee is 35 years old.

h. Find and interpret the 95% confidence interval for the average number of sick days an employee will take per year, given the employee is 35.

i. Suppose a new employee is 35. Find a 95% prediction interval for the number of sick days this employee will take this year.

j. Use the model to predict the number of sick days per year for an employee who is 45 years old. Round to the nearest whole number.

8. A manufacturing company that produces laminate for countertops is interested in studying the relationship between the number of hours of training that an employee receives and the number of defects per countertop produced. Ten employees are randomly selected. The number of hours of training each employee has received is recorded and the number of defects on the most recent countertop produced is determined. The results are as follows.

| Training Hours and Countertop Defects | |
|:---:|:---:|
| Hours of Training | Defects per Countertop |
| 1 | 1 |
| 4 | 4 |
| 7 | 0 |
| 3 | 3 |
| 2 | 5 |
| 2 | 4 |
| 5 | 3 |
| 5 | 2 |
| 1 | 5 |
| 6 | 1 |

A regression analysis has been performed to estimate the model, and the following output is produced.

$$\text{Defects per Countertop} = \beta_0 + \beta_1 \left( \text{Hours of Training} \right) + \varepsilon_i$$

---

**Regression Analysis: Defects per Countertop versus Hours of Training**

The regression equation is
Defects per Countertop = 4.65 - 0.515 Hours of Training

| Predictor | Coef | SE Coef | T | P |
|---|---|---|---|---|
| Constant | 4.6535 | 0.9426 | 4.94 | 0.001 |
| Hours of Training | -0.5149 | 0.2286 | -2.25 | 0.054 |

S = 1.45306        R-Sq = 38.8%              R-Sq(adj) = 31.2%

**Analysis of Variance**

| Source | DF | SS | MS | F | P |
|---|---|---|---|---|---|
| Regression | 1 | 10.709 | 10.709 | 5.07 | 0.054 |
| Residual Error | 8 | 16.891 | 2.111 | | |
| Total | 9 | 27.600 | | | |

**Predicted Values for New Observations**

| New Obs | Fit | SE Fit | 95% CI | 95% PI |
|---|---|---|---|---|
| 1 | 2.594 | 0.469 | (1.514, 3.674) | (-0.927, 6.115) |

**Values of Predictors for New Observations**

| New Obs | Hours of Training |
|---|---|
| 1 | 4.00 |

---

a. Draw a scatterplot of the data. Describe the relationship you observe between the number of defects per countertop and hours of training. Are there any unusual observations?

b. Find and interpret the standard deviation of the error terms in the output.

c. Interpret the slope coefficient.

d. What fraction of the variation in the number of defects per countertop is explained by the hours of training? What other factors might affect the number of defects?

e. Is there evidence of a linear relationship between the number of hours of training and the number of defects per countertop? Test at the 0.05 significance level and the 0.10 significance level.

f. Construct and interpret a 95% confidence interval for $\beta_1$, the slope coefficient.

g. Find the predicted value of the number of defects per countertop for an employee who has had 4 hours of training.

h. Find and interpret the 95% confidence interval for the average number of defects per countertop for employees who have had 4 hours of training.

i. Suppose a new employee has had 4 hours of training. What would be the 95% prediction interval for the number of defects per countertop?

j. Use the model to predict the number of defects per countertop for an employee who has had 7 hours of training. Round your answer to the nearest whole number.

9.   Use the following data regarding the age of a particular model of car and the asking price for that car. Construct a confidence interval for the slope to test if there is a significant relation between the age of the car and its price. Use 1% level of significance.

| Car Data | |
|---|---|
| Age (Years) | Asking Price ($) |
| 1 | 11,875 |
| 1 | 10,995 |
| 2 | 9995 |
| 2 | 8500 |
| 3 | 8995 |
| 4 | 6995 |
| 5 | 4450 |
| 5 | 5500 |
| 6 | 4400 |
| 6 | 4800 |

## T   Discovering Technology

### Using the TI-84 Plus Calculator

Regression Analysis

For this exercise, use the production data from Table 13.1.2.

1. Before performing regression analysis using the TI-84 Plus calculator, we need to turn on diagnostics so that the coefficient of determination is displayed in the output. Press **2ND** and **0** to access the catalog.

2. Scroll down until you reach the option that reads **DiagnosticOn**. With DiagnosticOn selected, press **ENTER**. Press **ENTER** again. You are now ready to perform regression analysis.

3. Press **STAT**, select **EDIT**, **Edit**, and **ENTER**. Then input the *x* and *y* data into lists **L1** and **L2**.

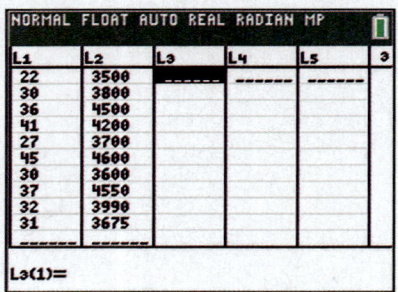

4. Press **STAT** and select **CALC**, then choose option **LinReg(ax+b)** and press **ENTER**.

5. Press **2ND** and **1** to select L1 as the independent variable. Press the **,** button and then **2ND** and **2** to select L2 as the dependent variable. Press **ENTER**.

6. Observe the output screen for the regression analysis. The estimate for $\beta_1$ is given as 53.88 and the estimate for $\beta_0$ is given as 2227.96 (rounded to two decimal places). The coefficient of determination is also given as 0.7507. Compare these results to those we calculated for the production data.

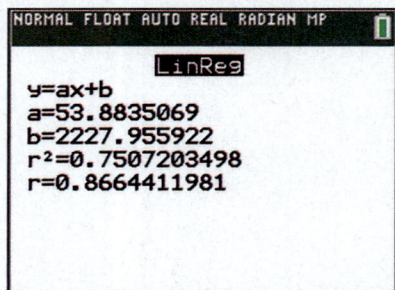

## Using Excel

Regression Analysis

For this exercise, use the production data from Table 13.1.2.

1.  Enter the *x* (items produced) and *y* (cost) data into columns A and B as shown below.

| | A | B |
|---|---|---|
| 1 | Items Produced (x) | Cost (y) |
| 2 | 22 | 3500 |
| 3 | 30 | 3800 |
| 4 | 36 | 4500 |
| 5 | 41 | 4200 |
| 6 | 27 | 3700 |
| 7 | 45 | 4600 |
| 8 | 30 | 3600 |
| 9 | 37 | 4550 |
| 10 | 32 | 3990 |
| 11 | 31 | 3675 |

2.  Under the **Data** tab, choose **Data Analysis**. In the Data Analysis dialog box, choose **Regression**.

3.  Select the data in Column B (**$B$2:$B$11**) for Input Y Range and the data in Column A (**$A$2:$A$11**) for Input X Range.

4.  Click the checkbox beside **Labels** so the titles of the columns will be included in the regression output in Excel.

5.  Excel automatically produces a 95% confidence interval for the estimate of $\beta_0$ and $\beta_1$. If you require another level of confidence, check the box next to **Confidence Level** and enter the desired confidence level in the box. For this exercise, enter **99** in the Confidence Level input box.

6.  Select the radio button next to **Output Range** and enter **$D$1**. Select the box for **Normal Probability Plots**.

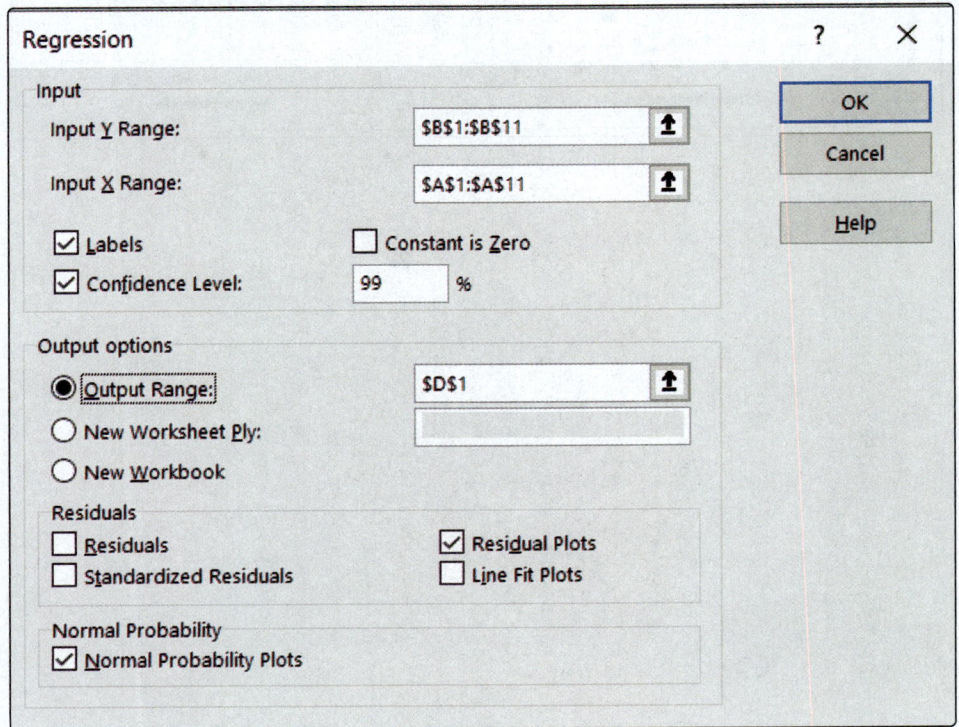

7. Press OK. Highlight columns D to L and double-click the vertical bar separating column D from column E to expand the columns. Observe the summary output. The Regression Statistics table contains the coefficient of determination, standard error of the model, and the number of observations. The ANOVA table contains the degrees of freedom, sums of squares, and mean squares values for regression and error (residual). Finally, the least squares estimates of the coefficients are given along with their standard errors, $t$-statistics, and $P$-values. Notice that the 95% and 99% confidence intervals for $\beta_1$ are given in the *Lower 95%*, *Upper 95%*, *Lower 99.0%*, and *Upper 99.0%* columns of the output.

SUMMARY OUTPUT

| Regression Statistics | |
| --- | --- |
| Multiple R | 0.866441198 |
| R Square | 0.75072035 |
| Adjusted R Square | 0.719560393 |
| Standard Error | 223.0681781 |
| Observations | 10 |

ANOVA

| | df | SS | MS | F | Significance F |
| --- | --- | --- | --- | --- | --- |
| Regression | 1 | 1198827.203 | 1198827 | 24.09247121 | 0.00118116 |
| Residual | 8 | 398075.2967 | 49759.41 | | |
| Total | 9 | 1596902.5 | | | |

| | Coefficients | Standard Error | t Stat | P-value | Lower 95% | Upper 95% | Lower 99.0% | Upper 99.0% |
| --- | --- | --- | --- | --- | --- | --- | --- | --- |
| Intercept | 2227.955922 | 370.1487713 | 6.019082 | 0.000316596 | 1374.391324 | 3081.520519 | 985.9634238 | 3469.948419 |
| Items Produced | 53.8835069 | 10.97779658 | 4.908408 | 0.00118116 | 28.5686626 | 79.1983512 | 17.04874735 | 90.71826646 |

8. Notice the residual plot and normal probability plot confirm the assumptions of the linear regression model are satisfied, as we see below. The residual plot shows the residuals have a mean of 0 and have a consistent standard deviation for low or high values of $x$. The normal probability plot is approximately linear, confirming the residuals are normally distributed.

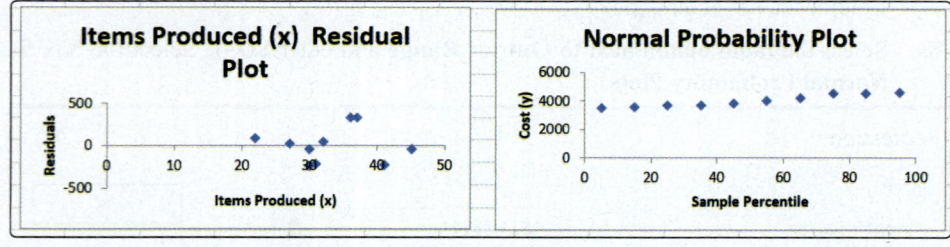

## Using JMP

Prediction Interval

For this exercise, use the data from Table 13.1.2.

1. With a JMP data table open, enter the Items Produced data in Column 1, and the Cost data in Column 2. Enter the value of **Items Produced** that you want to predict the cost for at the end of your Items Produced column. Leave the cost blank.

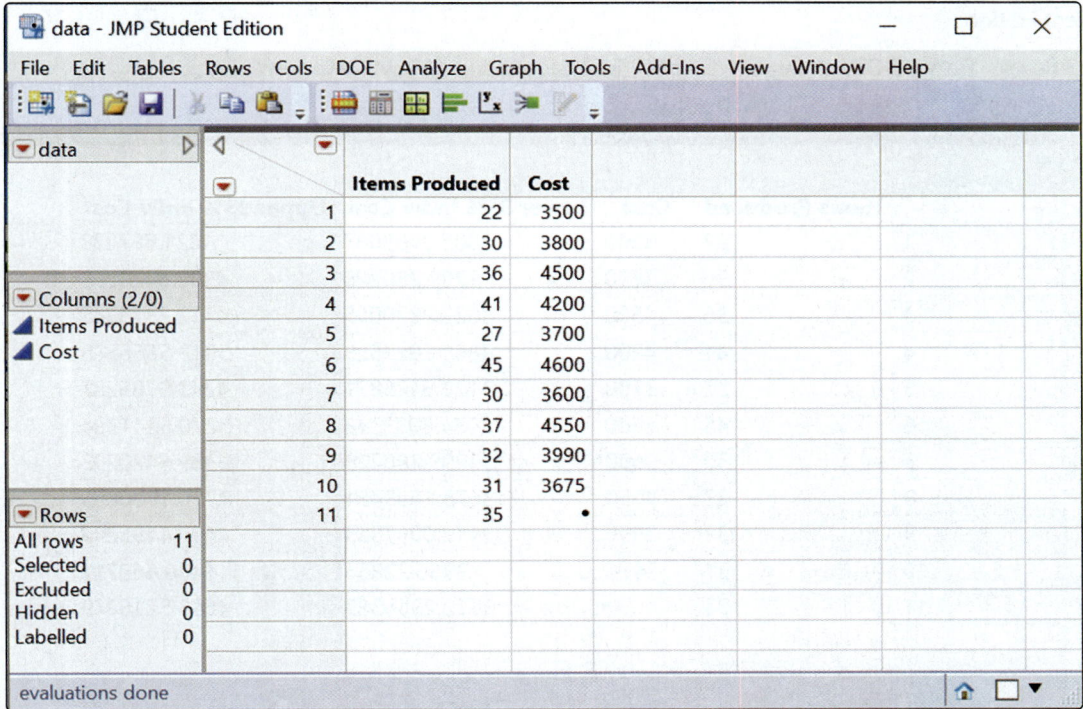

2. Select **Analyze** in the top row of the JMP spreadsheet and then select **Fit Y by X**.

3. From the **Select Columns** box, click on **Items Produced** (or Column 1), then click on **X, Factor**. Click on **Cost** (or Column 2), then click on **Y, Response**. Click **OK**.

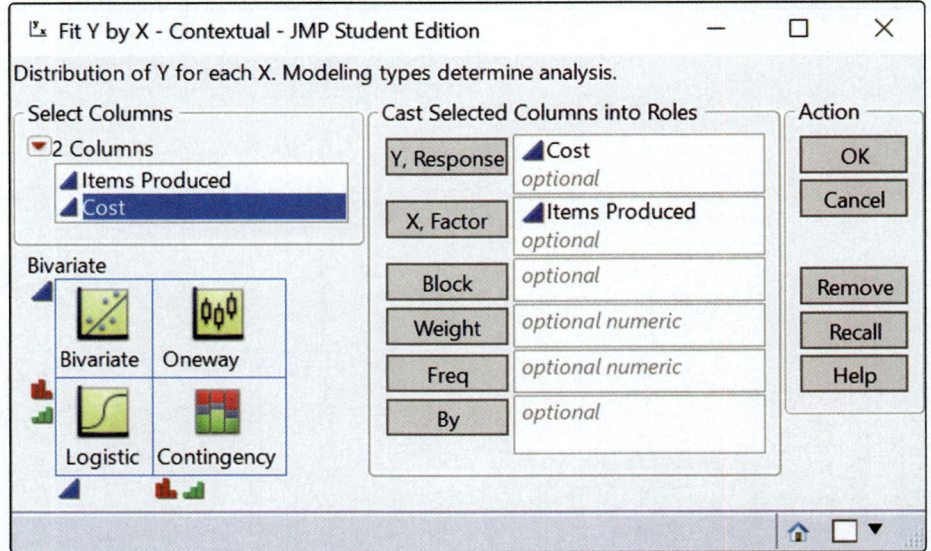

4. Observe the output screen for the results. To fit a regression line click on **the red triangle** next to Bivariate Fit of Cost By Items Produced and select **Fit Line**. By default, JMP will ignore your extra Items Produced value and produce the regression equation (under Linear Fit), Summary of Fit, the ANOVA table, and the Parameter Estimates. Now click on the **red triangle** next to Linear Fit and select **Indiv Confidence Limit Formula**. Look for the results back in your JMP data table.

**Note:** By default, JMP produces a 95% confidence interval. You can adjust the $\alpha$ level by clicking the **red triangle** next to Linear Fit and then select **Set $\alpha$ Level**.

---

**data - JMP Student Edition** — □ ×

File   Edit   Tables   Rows   Cols   DOE   Analyze   Graph   Tools   Add-Ins   View   Window   Help

▼data

Columns (4/0)
- Items Produced
- Cost
- Lower 9...div Cost
- Upper ...div Cost

Rows

| All rows | 11 |
| Selected | 0 |
| Excluded | 0 |
| Hidden | 0 |
| Labelled | 0 |

| | Items Produced | Cost | Lower 95% Indiv Cost | Upper 95% Indiv Cost |
|---|---|---|---|---|
| 1 | 22 | 3500 | 2805.0989087 | 4021.687238 |
| 2 | 30 | 3800 | 3299.280228 | 4389.6420292 |
| 3 | 36 | 4500 | 3623.2870055 | 4712.2373345 |
| 4 | 41 | 4200 | 3861.8027523 | 5012.5566568 |
| 5 | 27 | 3700 | 3121.6426871 | 4243.9785287 |
| 6 | 45 | 4600 | 4034.8033248 | 5270.6241394 |
| 7 | 30 | 3600 | 3299.280228 | 4389.6420292 |
| 8 | 37 | 4550 | 3673.1833484 | 4770.1080055 |
| 9 | 32 | 3990 | 3412.0067553 | 4492.4495295 |
| 10 | 31 | 3675 | 3356.22856 | 4440.460711 |
| 11 | 35 | • | 3572.2356383 | 4655.5216879 |

## Regression Analysis

For this exercise, use the data from Table 13.1.2.

1. With a JMP data table open, enter the Items Produced data in Column 1, and the Cost data in Column 2.

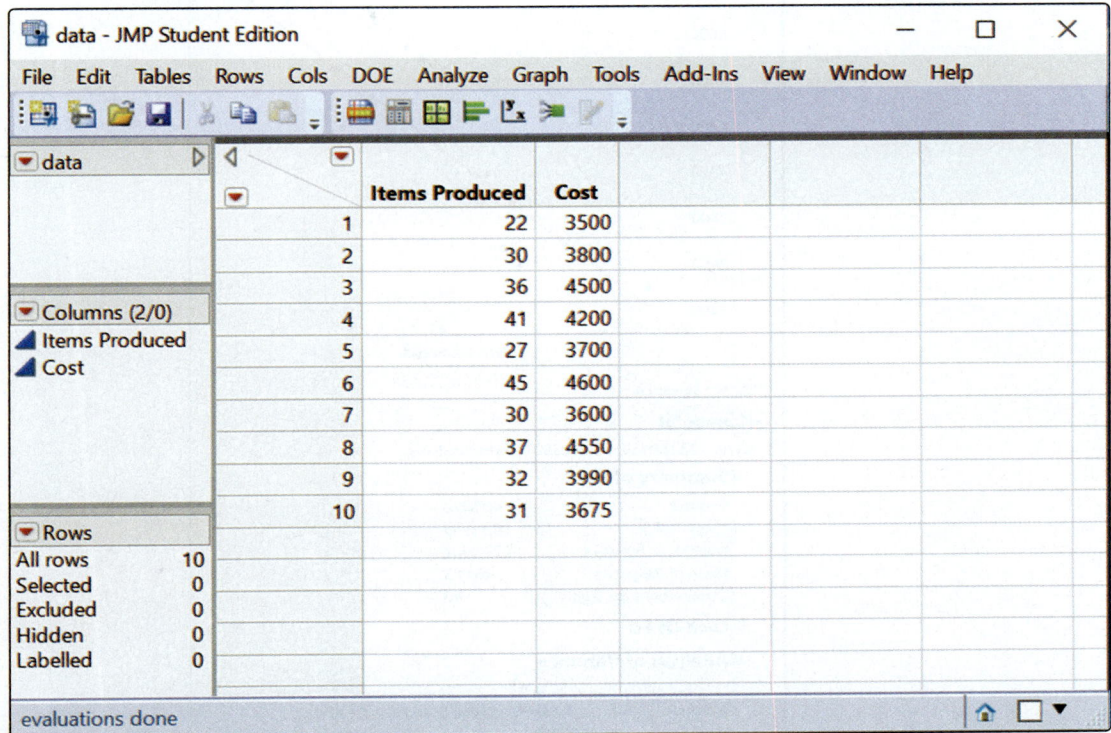

2. Select **Analyze** in the top row of the JMP spreadsheet and then select **Fit Y by X**.

3. From the **Select Columns** box, click on **Items Produced** (or Column 1), then click on **X, Factor**. Click on **Cost** (or Column 2), then click on **Y, Response**. Click **OK**.

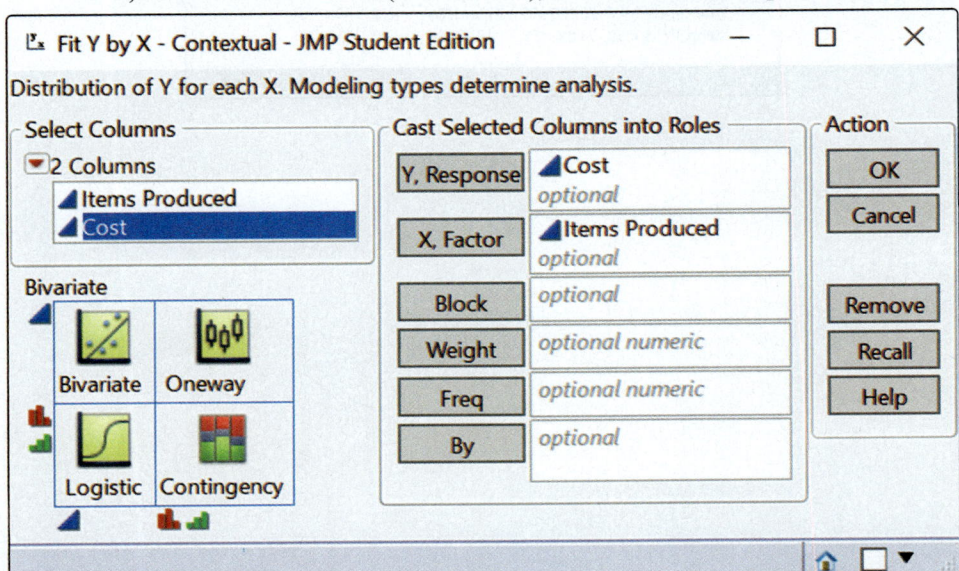

4. Observe the output screen for the results. To fit a regression line to the scatterplot, click on **the red triangle** next to Bivariate Fit of Cost By Items Produced and select **Fit Line**. By default, JMP will produce the regression equation (under Linear Fit), Summary of Fit, the ANOVA table, and the Parameter Estimates.

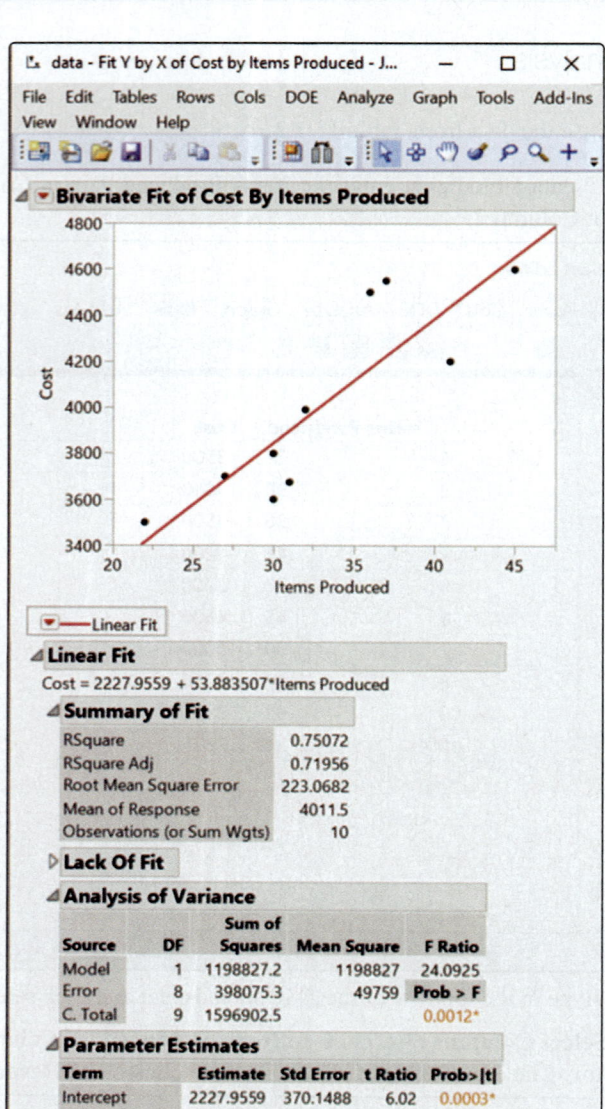

## Using Minitab

Regression Analysis

For this exercise, use the production data from Example 13.1.2.

1. Enter the data from Table 13.1.2 into the Worksheet. The heading for column C1 is **Items produced** and the heading for column C2 is **Cost**. Enter data into columns C1 and C2.

| | A | B |
|---|---|---|
| | Items Produced (x) | Cost (y) |
| 1 | 22 | 3500 |
| 2 | 30 | 3800 |
| 3 | 36 | 4500 |
| 4 | 41 | 4200 |
| 5 | 27 | 3700 |
| 6 | 45 | 4600 |
| 7 | 30 | 3600 |
| 8 | 37 | 4550 |
| 9 | 32 | 3990 |
| 10 | 31 | 3675 |

2. Choose **Stat**, **Regression**, and **Fitted Line Plot**. A dialog box will appear. Enter **C2** (Items Produced) for Response (Y) and enter **C1** (Cost) for Predictor (X). Press OK.

3. Observe the least squares coefficients, $R^2$ value, and standard error, as well as the regression plot (Cost vs Items Produced). Notice that the coefficient of determination is given as a percentage.

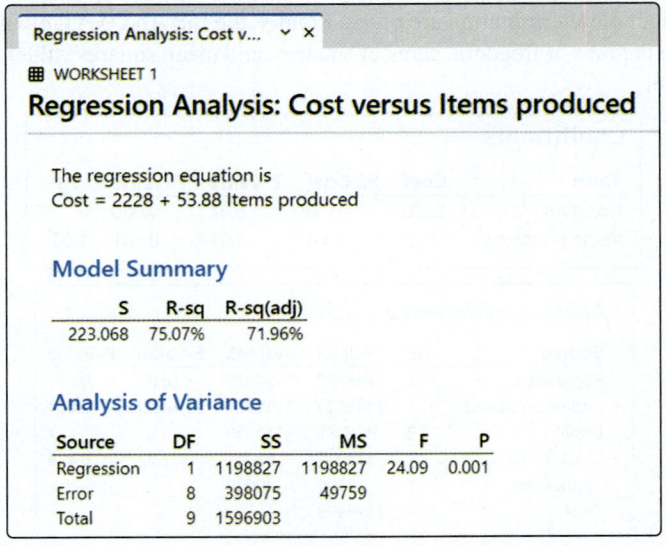

Regression Analysis: Cost v... ∨ ✕

⊞ WORKSHEET 1

### Regression Analysis: Cost versus Items produced

The regression equation is
Cost = 2228 + 53.88 Items produced

#### Model Summary

| S | R-sq | R-sq(adj) |
|---|---|---|
| 223.068 | 75.07% | 71.96% |

#### Analysis of Variance

| Source | DF | SS | MS | F | P |
|---|---|---|---|---|---|
| Regression | 1 | 1198827 | 1198827 | 24.09 | 0.001 |
| Error | 8 | 398075 | 49759 | | |
| Total | 9 | 1596903 | | | |

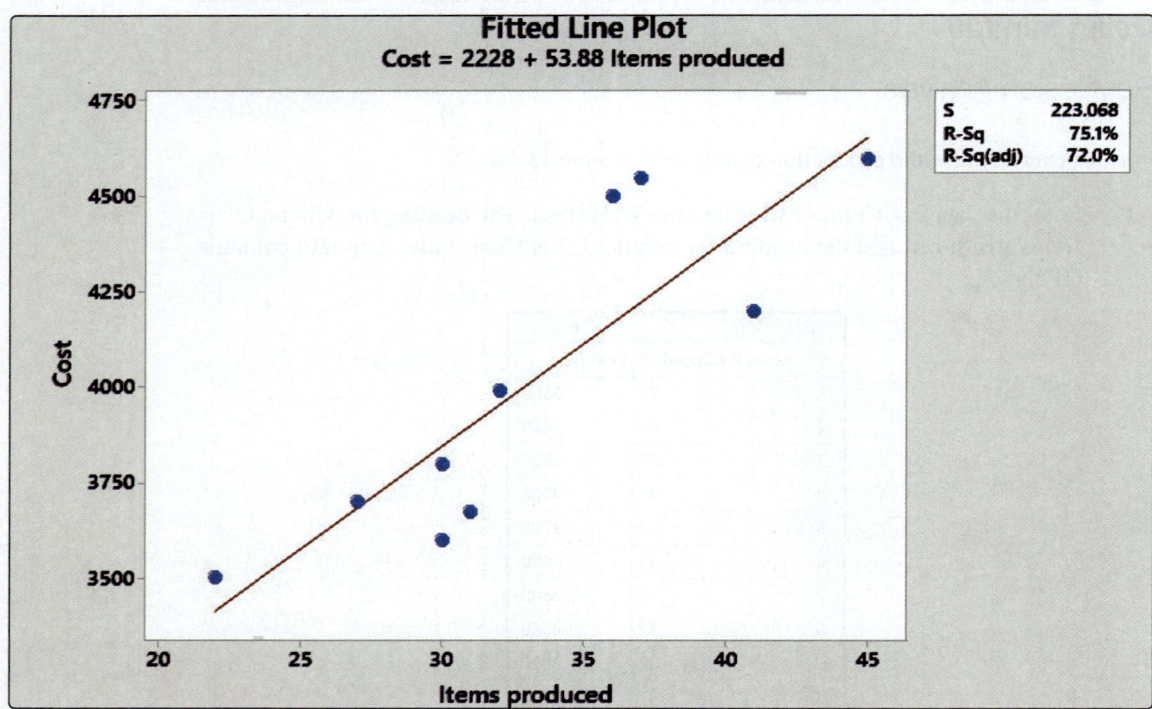

4.  For additional output Choose **Stat**, **Regression**, **Regression** and **Fit Regression Model.** Enter **C2** (Cost) for Response and **C1** (Items Produced) for Continuous Predictors and press **OK**.

5.  Observe the regression analysis output (Cost vs Items Produced). In addition to the previous outputs, we get the estimated coefficients along with their standard errors, t-statistics, and $P$-values. Then, the standard error for the regression model and the coefficient of determination are given. Finally, the full ANOVA table is given, which lists the degrees of freedom, sums of squares, and mean squares values for regression and error.

### Coefficients

| Term | Coef | SE Coef | T-Value | P-Value | VIF |
|---|---|---|---|---|---|
| Constant | 2228 | 370 | 6.02 | 0.000 | |
| Items produced | 53.9 | 11.0 | 4.91 | 0.001 | 1.00 |

### Analysis of Variance

| Source | DF | Adj SS | Adj MS | F-Value | P-Value |
|---|---|---|---|---|---|
| Regression | 1 | 1198827 | 1198827 | 24.09 | 0.001 |
| Items produced | 1 | 1198827 | 1198827 | 24.09 | 0.001 |
| Error | 8 | 398075 | 49759 | | |
| Lack-of-Fit | 7 | 378075 | 54011 | 2.70 | 0.438 |
| Pure Error | 1 | 20000 | 20000 | | |
| Total | 9 | 1596903 | | | |

## Confidence and Prediction Intervals

For this exercise, use the production data from Table 13.1.2.

1.  Enter the data from Table 13.1.2 into the Worksheet. The heading for column C1 is **Items produced** and the heading for column C2 is **Cost**. Enter data into columns C1 and C2.

| ◢ | A | B |
|---|---|---|
| | Items Produced (x) | Cost (y) |
| 1 | 22 | 3500 |
| 2 | 30 | 3800 |
| 3 | 36 | 4500 |
| 4 | 41 | 4200 |
| 5 | 27 | 3700 |
| 6 | 45 | 4600 |
| 7 | 30 | 3600 |
| 8 | 37 | 4550 |
| 9 | 32 | 3990 |
| 10 | 31 | 3675 |

2.  Choose **Stat**, **Regression**, **Regression** and **Fit Regression Model**. (Note the Predict option is gray until steps 4 and 5 are completed)

3.  The Regression dialog box will appear. Enter **C2** (Cost) in the Responses box (Y) and enter **C1** (Items Produced) in the Continuous predictors box (X).

4.  In the Regression dialog box, choose Options and set the Confidence Level to **99**. Click **OK**. Click **OK**. This will produce Regression Analysis results which must be set so that predictions can be made.

5.  Choose again **Stat**, **Regression**, **Regression** and then **Predict**. The Predict dialog box will appear provided steps 4 and 5 are completed. Enter **35** in the column under Items Produced. (This is the value used to predict its cost)

6.  Choose Options and Set the Confidence level at **99**. Press **OK** to exit Option box. Press **OK** to exit the Predict box.

7.  Observe the predicted value, along with its confidence and prediction interval.

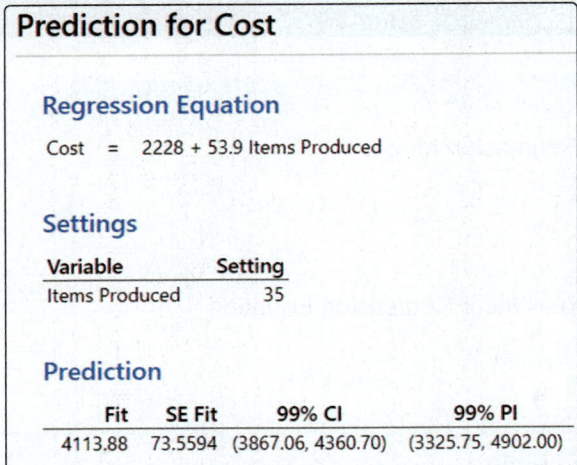

## Prediction for Cost

### Regression Equation

Cost  =  2228 + 53.9 Items Produced

### Settings

| Variable | Setting |
|---|---|
| Items Produced | 35 |

### Prediction

| Fit | SE Fit | 99% CI | 99% PI |
|---|---|---|---|
| 4113.88 | 73.5594 | (3867.06, 4360.70) | (3325.75, 4902.00) |

## R    Chapter 13 Review

### Key Terms and Ideas

- Regression Analysis
- Dependent Variable (Response Variable)
- Independent Variable (Predictor Variable)
- Simple Linear Regression
- Simple Linear Regression Model
- Parameters
- Error Term
- Estimated Simple Linear Regression Equation
- Method of Least Squares
- Correlation Coefficient
- Residual
- Sum of Squared Errors
- Least Squares Line
- Variance of the Error Terms

- Mean Square Error
- Standard Error
- L.I.N.E.
- Sigmoidal Relationship
- Normal Probability Plot
- Total Sum of Squares
- Sum of Squares of Regression
- Coefficient of Determination
- Linear Time Trend
- Assumptions about the Error Term
- Confidence Interval for $\beta_1$
- Testing a Hypothesis Concerning $\beta_1$
- Confidence Interval for the Mean Value of $y$ Given $x$
- Prediction Interval for the Value of $y$ Given $x$

### Key Formulas

|  | Section |
|---|---|

**Simple Linear Regression Model**

$$y_i = \beta_0 + \beta_1 x_i + \varepsilon_i$$

13.1

**Estimated Simple Linear Regression Equation**

$$\hat{y}_i = b_0 + b_1 x_i$$

13.1

**Sum of Squared Errors**

$$\text{SSE} = \sum \left( y_i - \hat{y}_i \right)^2 = \sum \left( y_i - \left( b_0 + b_1 x_i \right) \right)^2$$

13.1

## Key Formulas (cont.)

|  | Section |
|---|---|

### Slope of the Least Squares Line

$$b_1 = \frac{SS_{xy}}{SS_{xx}}$$

where

$$SS_{xy} = \sum(x_i - \bar{x})(y_i - \bar{y}) = \sum x_i y_i - \frac{\left(\sum x_i\right)\left(\sum y_i\right)}{n}$$

and

13.1

$$SS_{xx} = \sum(x_i - \bar{x})^2 = \sum x_i^2 - \frac{\left(\sum x_i\right)^2}{n}$$

or

$$b_1 = \frac{n\sum x_i y_i - \sum x_i \sum y_i}{n\sum x_i^2 - \left(\sum x_i\right)^2}.$$

### y-Intercept of the Least Squares Line

13.1

$$b_0 = \bar{y} - b_1\bar{x} = \frac{1}{n}\left(\sum y_i - b_1\sum x_i\right)$$

### Mean Square Error

13.1

$$s_e^2 = \frac{\sum(y_i - \hat{y}_i)^2}{n-2} = \frac{SSE}{n-2}$$

### Standard Error

13.1

$$s_e = \sqrt{\frac{\sum(y_i - \hat{y}_i)^2}{n-2}} = \sqrt{\frac{SSE}{n-2}}$$

### Residual

13.2

$$\text{Residual} = e_i = \text{Observed } y - \text{Predicted } y = y_i - \hat{y}_i$$

### Total Sum of Squares

$$TSS = \sum(y_i - \bar{y})^2$$

13.3

$$TSS = SSE + SSR$$

## Key Formulas (cont.)

|  | Section |
|---|---|

**Sum of Squares of Regression**

$$SSR = TSS - SSE$$

13.3

**Coefficient of Determination**

$$R^2 = \frac{SSR}{TSS} = 1 - \frac{SSE}{TSS}$$

or

$$R^2 = \left( \frac{n\sum x_i y_i - \sum x_i \sum y_i}{\sqrt{\left(n\sum x_i^2 - \left(\sum x_i\right)^2\right)\left(n\sum y_i^2 - \left(\sum y_i\right)^2\right)}} \right)^2$$

13.3

13.3

**Correlation Coefficient**

$$r = \frac{n\sum x_i y_i - \sum x_i \sum y_i}{\sqrt{\left(n\sum x_i^2 - \left(\sum x_i\right)^2\right)\left(n\sum y_i^2 - \left(\sum y_i\right)^2\right)}}$$

13.3

**Sample Estimate of the Variance of $b_1$**

$$s_{b_1}^2 = \frac{s_e^2}{\sum (x_i - \bar{x})^2}$$

13.5

**Sample Estimate of the Standard Deviation (Standard Error) of $b_1$**

$$s_{b_1} = \sqrt{\frac{s_e^2}{\sum (x_i - \bar{x})^2}}$$

13.5

**$100(1-\alpha)\%$ Confidence Interval for $\beta_1$**

$$b_1 \pm t_{\alpha/2, df}\, s_{b_1}$$

13.5

**Test Statistic for Testing the Hypothesis $\beta_1 \neq 0$**

$$t = \frac{b_1 - 0}{s_{b_1}} = \frac{b_1}{s_{b_1}}$$

13.5

## Key Formulas (cont.)

| | Section |
|---|---|

$100(1-\alpha)\%$ Confidence Interval for the Mean Value of $y$ Given $x$

13.6

$$\hat{y}_p \pm t_{\alpha/2,df} s_e \sqrt{\frac{1}{n} + \frac{\left(x_p - \bar{x}\right)^2}{\sum\left(x_i - \bar{x}\right)^2}}$$

$100(1-\alpha)\%$ Prediction Interval for the Value of $y$ Given $x$

13.6

$$\hat{y}_p \pm t_{\alpha/2,df} s_e \sqrt{1 + \frac{1}{n} + \frac{\left(x_p - \bar{x}\right)^2}{\sum\left(x_i - \bar{x}\right)^2}}$$

# AE    Additional Exercises

1.    A pharmacist is interested in studying the relationship between the amount of a particular drug in the bloodstream (in mg) and reaction time (in seconds) of subjects taking the drug. Ten subjects are randomly selected and administered various doses of the drug. The reaction times (in seconds) are measured 15 minutes after the drug is administered with the following results.

| Reaction Times | | | |
|---|---|---|---|
| Amount of Drug (mg) | Reaction Time (Seconds) | Amount of Drug (mg) | Reaction Time (Seconds) |
| 1 | 0.5 | 6 | 0.8 |
| 2 | 0.7 | 7 | 0.9 |
| 3 | 0.6 | 8 | 0.6 |
| 4 | 0.7 | 9 | 0.9 |
| 5 | 0.8 | 10 | 1.0 |

A regression analysis has been performed to estimate the model, and the following output was produced.

$$\text{Reaction Time} = \beta_0 + \beta_1 \left(\text{Amount of Drug}\right) + \varepsilon_i$$

**Regression Analysis: Reaction Time (Seconds) versus Amount of Drug (mg)**

The regression equation is
Reaction Time (Seconds) = 0.533 + 0.0394 Amount of Drug (mg)

| Predictor | Coef | SE Coef | T | P |
|---|---|---|---|---|
| Constant | 0.53333 | 0.07521 | 7.09 | 0.000 |
| Amount of Drug (mg) | 0.03939 | 0.01212 | 3.25 | 0.012 |

S = 0.110096    R-Sq = 56.9%    R-Sq(adj) = 51.5%

**Analysis of Variance**

| Source | DF | SS | MS | F | P |
|---|---|---|---|---|---|
| Regression | 1 | 0.12803 | 0.12803 | 10.56 | 0.012 |
| Residual Error | 8 | 0.09697 | 0.01212 | | |
| Total | 9 | 0.22500 | | | |

**Predicted Values for New Observations**

| New Obs | Fit | SE Fit | 95% CI | 95% PI |
|---|---|---|---|---|
| 1 | 0.6909 | 0.0393 | (0.6003, 0.7815) | (0.4214, 0.9605) |

**Values of Predictors for New Observations**

| New Obs | Amount of Drug (mg) |
|---|---|
| 1 | 4.00 |

   a.    Draw a scatterplot of the data. Describe the relationship you observe between the reaction time and the amount of drug in the bloodstream. Are there any unusual observations?

   b.    Find and interpret the standard deviation of the error terms in the output.

   c.    Interpret the slope coefficient.

   d.    What fraction of the variation in reaction time is explained by the amount of drug in the bloodstream? What other factors might affect reaction time?

    e. Is there evidence of a linear relationship between the amount of drug in the bloodstream and reaction time? Test at the 0.05 significance level and the 0.01 significance level.

    f. Construct and interpret a 95% confidence interval for $\beta_1$, the slope of the line.

    g. Find the predicted value of the reaction time of an individual who has 4 mg of the drug in the bloodstream.

    h. Find and interpret a 95% confidence interval for the average reaction time of all individuals who have 4 mg of the drug in their bloodstreams.

    i. Suppose a particular individual has 4 mg of the drug in the bloodstream. What would be the 95% prediction interval for the reaction time?

**2.** A sample of 11 lonely hearts advertisements, all placed by males, was selected from the local newspaper. In each of the selected ads, the males gave their heights, along with other physical characteristics and preferences. Some of the males obviously felt that being taller than average might result in more responses to the ad. Suppose that $y$, the number of responses to the ad over the next 30 days, was determined for each male. The following table contains the data.

| Height and Response | | | | | | | | | | | |
|---|---|---|---|---|---|---|---|---|---|---|---|
| Height (Inches) | 70 | 62 | 67 | 75 | 78 | 69 | 70 | 64 | 66 | 69 | 75 |
| $y$ | 14 | 7 | 10 | 18 | 17 | 12 | 15 | 9 | 12 | 14 | 17 |

    a. Draw a scatterplot of the data. Does the relationship appear to be linear?

    b. Estimate the slope and intercept of the regression equation using statistical software.

    c. Is there evidence of a linear relationship between the number of responses and height? Test at the $\alpha = 0.01$ significance level.

    d. Interpret the regression coefficient corresponding to height.

    e. Construct a 95% confidence interval for the slope.

    f. Compute $R^2$ and interpret this value.

    g. Estimate the number of responses for a male 6 feet tall. Round your answer to the nearest whole number.

    h. Construct and interpret a 95% prediction interval for the number of responses for a male who is 6 feet tall.

    i. Construct and interpret a 95% confidence interval for the average number of responses for a male who is 6 feet tall.

3.    It is believed that when one is in the process of buying a home, the interest rate that is given on the loan is a function of his or her credit score. The Fair Isaac Corporation (FICO) is a major producer of credit scores. They have collected data from major lenders about buyers' history of borrowing and paying back credit. The following table contains 20 randomly selected loan applicants along with their FICO scores and the interest rate that they were given when financing their homes. With the data given, answer the following questions.

### Credit Scores and Interest Rates

| Observation | FICO Score | Interest Rate (%) |
|---|---|---|
| 1 | 756 | 6.32 |
| 2 | 679 | 7.85 |
| 3 | 527 | 10.20 |
| 4 | 839 | 5.52 |
| 5 | 677 | 7.30 |
| 6 | 686 | 7.37 |
| 7 | 512 | 9.67 |
| 8 | 590 | 8.40 |
| 9 | 765 | 5.82 |
| 10 | 502 | 10.01 |
| 11 | 819 | 5.86 |
| 12 | 630 | 8.51 |
| 13 | 704 | 6.83 |
| 14 | 679 | 7.72 |
| 15 | 663 | 7.68 |
| 16 | 542 | 9.53 |
| 17 | 575 | 6.86 |
| 18 | 508 | 9.65 |
| 19 | 689 | 7.75 |
| 20 | 750 | 6.89 |

a.  Draw a scatterplot of the data. Does there appear to be a linear relationship between FICO score and interest rate?

b.  Estimate the simple linear regression equation using statistical software.

c.  What is the estimate of the mean square error? Interpret this value.

d.  Test at the 5% significance level if a linear relationship exists between FICO scores and interest rates.

e.  Interpret the regression coefficient corresponding to FICO score.

f.  Construct a 95% confidence interval for the slope. Interpret the interval.

g.  Compute the coefficient of determination. Interpret this value.

h.  Calculate the correlation coefficient. Interpret this value.

i.  What is the average interest rate for a credit score of 725?

j.  Construct a 90% confidence interval for the average interest rate for people who have FICO scores of 725. Interpret this interval.

k.  Construct a 90% prediction interval for the interest rate for a person with a FICO score of 725. Interpret this interval.

4.   It appears that many cellular phone service providers are making huge profits from customers using their messaging services such as text and multimedia messaging services (MMS). To that end, the cellular phone companies are using their marketing campaigns to target kids rather than adults. The belief is that kids tend to utilize their messaging services much more than adults. In fact, it is the belief that the younger one is, the more texts and MMS sent via his or her cell phone. Using the data given which reports the number of monthly messages sent by age, formulate a simple linear regression model to answer the following questions.

| Age and Message Use | | | |
|---|---|---|---|
| Age | Number of Messages | Age | Number of Messages |
| 78 | 7 | 37 | 1541 |
| 36 | 1607 | 69 | 6 |
| 11 | 3037 | 69 | 25 |
| 69 | 26 | 55 | 517 |
| 56 | 491 | 39 | 1439 |
| 74 | 0 | 20 | 2505 |
| 22 | 2373 | 14 | 2845 |
| 74 | 5 | 10 | 3048 |
| 10 | 3059 | 80 | 0 |
| 26 | 2155 | 59 | 295 |
| 18 | 2619 | 40 | 1374 |
| 68 | 17 | 67 | 35 |
| 10 | 3067 | | |

a.   Draw a scatterplot of the data. Does there appear to be a linear relationship between age and the number of messages that one sends?

b.   What is the estimated simple linear regression equation?

c.   What is the estimate of the coefficient of determination? Interpret this value.

d.   Test at the 5% significance level if a linear relationship exists between age and the number of messages sent via a cellular phone.

e.   Interpret the regression coefficient corresponding to age.

f.   Construct a 95% confidence interval for the slope. Interpret this interval.

g.   Calculate the correlation coefficient. Interpret this value.

h.   What is the average number of messages sent by a 15-year-old? Round your answer to the nearest whole number.

i.   Construct a 95% confidence interval for the average number of messages sent by a 15-year-old. Interpret this interval.

j.   Suppose Jacob's parents are contemplating giving him a cell phone but with a limited messaging plan at 500 per month. Eager to get the cell phone, Jacob, at 15 years old, promises that he won't send more than 500 messages per month and he'll also limit the number of friends that will have his phone number. In spite of Jacob's honesty and loyalty, should his parents believe that he won't send more than 500 messages per month? Explain your answer.

5.  For the last 10 years, the Virginia Department of Mines, Minerals, and Energy (VDMME) has been promoting Energy Star, a resource for energy-efficient products and solutions. VDMME wants all energy consumers to take responsibility and exercise leadership by practicing conservation and efficiency on a daily basis. The average annual energy usage for a 1800 square foot home is 18,000 kilowatt hours. VDMME believes that this number can be significantly reduced if consumers started using Energy Star appliances. Answer the following questions based on data of 25 randomly selected homes with Energy Star appliances built within the last five years.

| Home Size and Energy Usage | | | |
|---|---|---|---|
| Home Size (Square Feet) | Annual Energy Usage (kWh) | Home Size (Square Feet) | Annual Energy Usage (kWh) |
| 2895 | 15,200 | 2180 | 13,227 |
| 3650 | 17,333 | 4492 | 19,492 |
| 2927 | 15,050 | 6450 | 25,353 |
| 6289 | 24,763 | 1583 | 11,075 |
| 7252 | 27,098 | 4170 | 18,557 |
| 4147 | 18,291 | 4189 | 18,636 |
| 6505 | 25,028 | 3920 | 18,210 |
| 1413 | 11,099 | 6833 | 26,075 |
| 2279 | 13,110 | 4469 | 19,232 |
| 3251 | 15,844 | 6141 | 24,225 |
| 2992 | 14,904 | 5084 | 21,530 |
| 6912 | 26,329 | 6746 | 26,333 |
| 2503 | 13,765 | | |

a.  Draw a scatterplot of the data. Does there appear to be a linear relationship between home size and the amount of annual kWh used?

b.  What is the estimated simple linear regression equation?

c.  What is the estimate of the coefficient of determination? Interpret this value.

d.  Test at the 5% significance level if a linear relationship exists between home size and the annual amount of kWh used.

e.  Interpret the regression coefficient corresponding to home size.

f.  Construct a 99% confidence interval for the slope. Interpret the interval.

g.  Calculate the correlation coefficient. Interpret this value.

h.  Suppose the James family constructed a 3200 square foot home using all Energy Star appliances. How many kilowatt hours should they expect to use in their first year in the home?

i.  Construct a 95% confidence interval for the average number of kWh that will be used by the James family. Interpret this interval.

**6.** With grade inflation being a major problem in many U.S. high schools, college admissions offices are beginning to look at other performance measures when evaluating student applications. It is believed that many students with high grade point averages in high school will not necessarily score high on the SAT. Using the data of 30 randomly selected students that took the SAT, answer the following questions to determine if there is a linear relationship between high school GPA and SAT score.

| High School GPA and SAT Score | | | |
|---|---|---|---|
| High School GPA | SAT Score | High School GPA | SAT Score |
| 3.21 | 1448 | 4.95 | 1960 |
| 2.23 | 1435 | 4.69 | 1717 |
| 2.89 | 1411 | 2.49 | 1365 |
| 1.84 | 1291 | 2.45 | 1561 |
| 3.34 | 1462 | 2.57 | 1474 |
| 2.42 | 1357 | 1.28 | 1328 |
| 2.75 | 1396 | 1.94 | 1302 |
| 2.35 | 1549 | 4.75 | 1622 |
| 4.80 | 1829 | 1.91 | 1499 |
| 1.98 | 1508 | 4.25 | 1566 |
| 2.92 | 1514 | 1.15 | 1413 |
| 4.18 | 1658 | 2.17 | 1428 |
| 4.50 | 1694 | 4.73 | 1720 |
| 4.42 | 1686 | 4.39 | 1783 |
| 4.78 | 1840 | 2.92 | 1614 |

**a.** Draw a scatterplot of the data. Does there appear to be a linear relationship between high school GPA and SAT score?

**b.** What is the estimated simple linear regression equation?

**c.** What is the coefficient of determination? Interpret this value.

**d.** Test at the 5% significance level if a linear relationship exists between high school GPA and SAT score.

**e.** Interpret the regression coefficient corresponding to high school GPA.

**f.** Construct a 95% confidence interval for the slope. Interpret the interval.

**g.** Calculate the correlation coefficient. Interpret this value.

**h.** What SAT score would you expect for students with a GPA of 3.5? Round your answer to the nearest whole number.

## P    Discovery Project

### Gas vs. Oil Prices

Managing expenses is often a critical job duty for Chief Financial Officers (CFOs). Sometimes costs are fairly easy to forecast. Items such as office supplies and staffing costs are generally more manageable costs to estimate for budgeting purposes. Other expenses such as raw materials and utility costs can be harder to forecast because the prices can fluctuate widely during a given time period.

Gasoline is another such expense. Gasoline prices are determined in part by the price per barrel of crude oil. There are two predominate types of crude oil used in the industry to determine the price of gas we see at the pump: West Texas Intermediate (WTI) and Brent Blend. WTI is the benchmark used in the United States and Brent Blend is mostly used in Europe and Africa. An interesting side note is that about 20 gallons of gasoline are made from one barrel of crude oil (Source: Frequently Asked Questions (FAQs) – U.S. Energy Information Administration (EIA) ).

Industries that rely heavily on gasoline to provide their product or service can find it difficult to estimate the gasoline expense from month to month and sometimes even from day to day. Some examples of industries that have fluctuating gas price impacts are transportation, landscaping, and logistics. Being able to reliably forecast the gasoline expense is a critical component of estimating company profits.

Imagine you are tasked with determining reliable estimates of the price per gallon of gasoline to be used in forecasting company profits. Using the data file called Gas Prices vs. Oil Prices, answer the following questions to better understand the relationship between the two types of crude oil and their impact on gasoline prices. The data set includes monthly gasoline and crude oil prices between January 2000 and March 2021 (255 data points).

1.  Download the data file and open it in Microsoft Excel.

2.  Determine the mean, mode, median, maximum, minimum, range, standard deviation, and the coefficient of variation of the price per gallon (to 2 decimal places) and briefly discuss the results. (Hint: These values can be quickly calculated using the Data Analysis Add-in: Descriptive Statistics in Excel).

3.  Create scatterplots of each of the crude oils against the price per gallon.

4.  Fit a simple linear regression (SLR) model for Brent Blend vs. Reg Gas Price per Gallon and WTI vs. Reg Gas Price per Gallon.

5.  Obtain the residuals for each regression analysis and plot them.

6.  Validate the assumptions — of linearity, independence, normality, constant variance.

7.  Test the slopes for each crude oil model to see if they are statistically significant.

8.  Obtain confidence intervals for the slopes using a 5% significance level.

9.  Make some predictions. Pick a price per gallon (or barrel) for Brent Blend and WTI and predict the price of a regular gallon of gas.

10. Which crude oil would you recommend as the best predictor of regular gasoline prices for your company to use to forecast expenses?

### :: Data

This data set can be found at stat.hawkeslearning.com by navigating to **Discovering Business Statistics, Second Edition > Data Sets > Gas Prices vs Oil Prices.**

### ∞ Technology

For instructions on performing a regression analysis using technology, visit stat.hawkeslearning.com and navigate to **Discovering Business Statistics, Second Edition > Technology Instructions > Data Sets > Regression > Simple Linear Regression.**

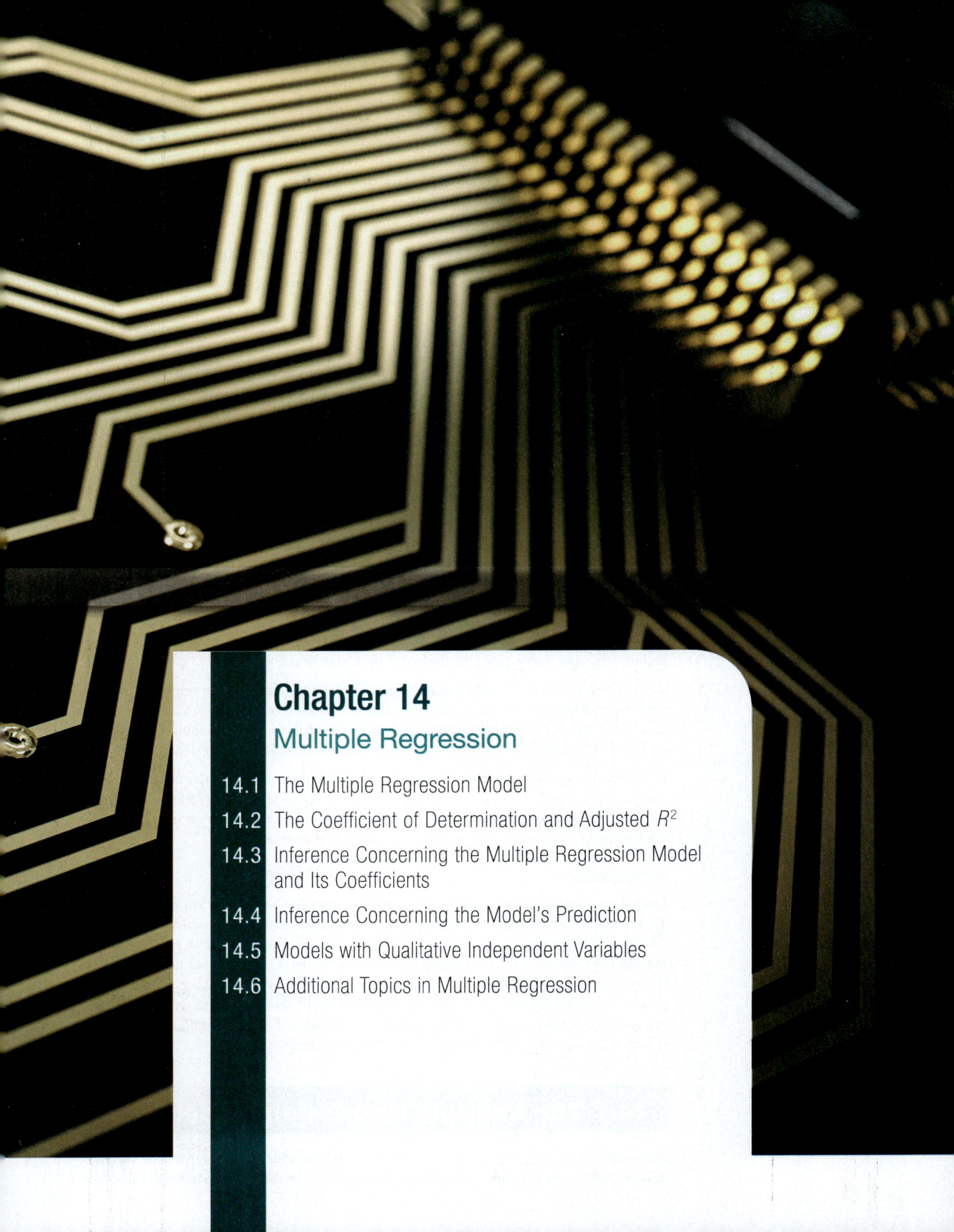

# Chapter 14

## Multiple Regression

# Discovering the Real World

## How are Credit Card Limits Determined?

Rarely does one give much thought of how a credit card limit is determined when applying for a credit card. Some questions worth pondering are:

- What is the minimum credit card limit for a college student who is unemployed?
- What is the minimum credit card limit for an individual who is unemployed with a high school diploma but without a college education?
- What should be the minimum credit card limit for a person who is employed, has a high salary, and owns their home?

These are just three questions of many that can be pondered when discussing spending limits on credit cards. Credit card companies use statistics to help determine these spending limits. When a consumer applies for a credit card, the card issuers require lots of information (independent variables) such as age, gender, ethnicity, income, marital status, number of years at current place of employment, number of years of education, housing situation (own or rent), student (yes or no), and perhaps a number of other variables to help them determine a spending limit (dependent variable) for the applicant.

There are also other factors that will be analyzed when determining a spending limit. For example, given the independent variables listed above, the card issuer can also predict the likelihood of default (i.e., that a cardholder will exhaust the available credit and not pay the credit card bill). The likelihood of default will not only determine if a credit card is issued to an applicant, but it will also determine the spending limit. If an applicant has a high likelihood of default, rather than not issuing a card, the issuer may give the applicant a relatively low spending limit. Similarly, if the applicant has a very low likelihood of default and a high income, the card issuer may give the applicant a high spending limit.

Regression analysis, particularly multiple regression analysis, is a technique employed by credit card issuers when determining credit limits and the likelihood of default on a credit card. The card issuer will use the aforementioned independent variables to build a multiple regression model to make predictions to allow them to answer some of the questions listed above.

In this chapter, we will discuss multiple regression techniques and statistics used to evaluate the quality of fit of the multiple regression model. Once the model is deemed adequate, we will discuss making predictions, confidence intervals, and prediction intervals of the dependent variable at specific values of the independent variables.

## Introduction

Oftentimes in business, we need to examine the relationship between two or more variables. This chapter is an extension of the topic of simple linear regression discussed in Chapter 13. We will begin this section with a discussion of the multiple regression model and then discuss the statistics used to evaluate the fit of the model. Once we have determined that the multiple regression model fit is adequate, we will then discuss making inferences and predictions using the multiple regression model. We will also discuss performing multiple regression analysis with qualitative independent variables. Finally, we end this chapter by discussing some more advanced topics in multiple regression.

# 14.1 The Multiple Regression Model

Previously, we constructed linear regression models in which one independent variable, $x$, is related to one dependent variable, $y$. For example, in the production model from the previous chapter, only one independent variable (items produced) is used to explain total cost. But there are many other variables, the number of labor hours required or energy costs, for example, which may also be useful in explaining total cost. If a model is to accurately represent real world phenomena, then the model must potentially be able to accommodate multiple independent variables. **Multiple regression** is an extension of simple regression techniques, allowing more than one independent variable in the regression equation.

## Formula

### Multiple Regression Model

The **multiple regression model** is given by

$$y_i = \beta_0 + \beta_1 x_{1i} + \beta_2 x_{2i} + \cdots + \beta_k x_{ki} + \varepsilon_i$$

where $\beta_0, \beta_1, \beta_2, \ldots, \beta_k$ are the model's parameters. (They are unknown constants that will require estimation.)

$x_{1i}, x_{2i}, \ldots, x_{ki}$ are independent variables which are measured without error.

$\varepsilon_i$ is a random error which is normally distributed with a mean of zero and a standard deviation $\sigma_\varepsilon$. (The errors are independent of each other.)

The multiple regression model contains additional parameters that must be estimated. Recall from Chapter 13 that the **method of least squares** is used to find the estimated regression line that minimizes the sum of squared errors, $\left( \text{SSE} = \Sigma \left( y_i - \hat{y}_i \right)^2 \right)$. Even though we are now discussing multiple regression, least squares will remain the method of estimation. Because of the complexity of the calculations, formulas for the least squares coefficients will not be presented. Instead, estimates of the model's parameters will be obtained from one of the many statistical analysis programs that perform multiple regression analysis. Although the format of the output varies among the programs, the program outputs mostly contain the same fundamental information. The Discovering Technology section at the end of this chapter illustrates how to perform and interpret multiple regression equations using Microsoft Excel, JMP, and Minitab. Throughout the examples in this chapter, we will be using outputs from these statistical programs to construct and interpret multiple regression models.

Just as with the simple linear regression model, it is not often that an entire population is available. Therefore, we must estimate the parameters of the multiple regression model using sample data. Using data from the sample, we can estimate the coefficients of the **estimated multiple regression equation**.

## Formula

### Estimated Multiple Regression Equation

The **estimated multiple regression equation** is

$$\hat{y}_i = b_0 + b_1 x_{1i} + b_2 x_{2i} + \cdots + b_k x_{ki}$$

where $b_0, b_1, b_2, \ldots, b_k$ are estimates of their population counterparts. Specifically, $b_0$ is an estimate of $\beta_0$, $b_1$ is an estimate of $\beta_1$, $b_2$ is an estimate of $\beta_2$, etc.

$\hat{y}_i$ is the predicted value of $y$ for given values of $x_1, x_2, \ldots, x_k$, and is pronounced *y-hat*. The symbol $y_i$ is reserved for the observed value of $y$.

Model building is a process. The greatest challenge in building a multiple regression model is in determining the appropriate independent variables needed to explain the dependent variable. In practical applications, this usually requires a great deal of experimentation with the model. Analysts examine the effects of adding and removing independent variables in order to determine which model is the best predictor of the dependent variable.

## Example 14.1.1

**Modeling Pizza Delivery Time**

A pizza delivery manager is analyzing the delivery routes in her system. She is interested in predicting the amount of time required for the driver to deliver the pizzas on a specific route. The driver has to make several stops on the route because the manager doesn't want to send several vehicles/drivers to cover the same area/route. Fit a multiple linear regression model using the data in Table 14.1.1 to predict delivery time.

| Table 14.1.1 – Pizza Delivery Time Data | | | |
|---|---|---|---|
| Observation | Delivery Time (Minutes) | Number of Pizzas | Distance (Miles) |
| 1 | 16.68 | 7 | 5.60 |
| 2 | 11.50 | 3 | 2.20 |
| 3 | 12.03 | 3 | 3.40 |
| 4 | 14.88 | 8 | 0.80 |
| 5 | 13.75 | 6 | 1.50 |
| 6 | 18.11 | 7 | 3.30 |
| 7 | 8.00 | 2 | 1.10 |
| 8 | 17.83 | 7 | 2.10 |
| 9 | 79.24 | 30 | 14.60 |
| 10 | 21.50 | 5 | 6.05 |
| 11 | 40.33 | 16 | 6.88 |
| 12 | 21.00 | 10 | 2.15 |
| 13 | 13.50 | 4 | 2.55 |
| 14 | 19.75 | 6 | 4.62 |
| 15 | 24.00 | 9 | 4.48 |
| 16 | 29.00 | 10 | 7.76 |
| 17 | 15.35 | 6 | 2.00 |
| 18 | 19.00 | 7 | 1.32 |
| 19 | 9.50 | 3 | 0.36 |
| 20 | 35.10 | 17 | 7.70 |
| 21 | 17.90 | 10 | 1.40 |
| 22 | 52.32 | 26 | 8.10 |
| 23 | 18.75 | 9 | 4.50 |
| 24 | 19.83 | 8 | 6.35 |
| 25 | 10.75 | 4 | 1.50 |

**⋮ Data**

This data set can be found at stat.hawkeslearning.com by navigating to **Discovering Business Statistics, Second Edition > Data Sets > Pizza Delivery Time**.

### SOLUTION

The linear multiple regression model is as follows.

$$\text{Delivery Time} = \beta_0 + \beta_1(\text{Number of Pizzas}) + \beta_2(\text{Distance}) + \varepsilon_i$$

To have a useful model, the parameters $\beta_0$, $\beta_1$, and $\beta_2$ must be estimated. Estimating these parameters requires the collection of historical data on pizza delivery times. If inferences concerning the model's predicted values or parameters are desired, then random sampling methods must be used during data collection. Assume that the data in Table 14.1.1 have been collected using random sampling techniques. The summary output from JMP is given in Figure 14.1.1.

Summary of Fit

| | |
|---|---|
| RSquare | 0.964022 |
| R Square Adj | 0.960751 |
| Root Mean Square Error | 3.075694 |
| Mean of Response | 22.384 |
| Observations (or Sum Wgts) | 25 |

Analysis of Variance

| Source | DF | Sum of Squares | Mean Square | F-Ratio |
|---|---|---|---|---|
| Model | 2 | 5576.4249 | 2788.21 | 294.7403 |
| Error | 22 | 208.1177 | 9.46 | Prob > F |
| C. Total | 24 | 5784.5426 | | < .0001* |

Parameter Estimates

| Term | Estimate | Std Error | t Ratio | Prob > \|t\| |
|---|---|---|---|---|
| Intercept | 1.7929033 | 1.048779 | 1.71 | 0.1014 |
| Number of Pizzas | 1.589101 | 0.156346 | 10.16 | < .0001* |
| Distance (Miles) | 1.5677081 | 0.327535 | 4.79 | < .0001* |

Figure 14.1.1

The estimated model parameters are $b_0 = 1.7929$, $b_1 = 1.5891$, and $b_2 = 1.5677$. The estimated multiple regression model is as follows.

Estimated Delivery Time = 1.7929 + 1.5891(Number of Pizzas) + 1.5677(Distance)

**∽ Technology**

For instructions on obtaining this output for multiple regression using technology, please visit stat.hawkeslearning.com and navigate to **Discovering Business Statistics, Second Edition > Technology Instructions > Regression > Multiple Regression**.

There are numerous questions that could be asked about the multiple regression model in Example 14.1.1.

- Can the model explain a substantial portion of the variation in delivery times? If not, it will not be very useful.

- Do the signs and magnitudes of the estimated coefficients appear to be reasonable? The answer to this question is a function of the data. For the pizza delivery example, the estimate of the coefficient on number of pizzas is 1.5891. This means that for a fixed distance, we would expect the average delivery time to increase by approximately 1.59 minutes for each additional pizza that is ordered. Does this make sense? Certainly, one would expect the average delivery time to increase as the number of pizzas increases. Therefore, the sign (positive, in this case) and magnitude are reasonable.

- Are the estimates of the coefficients reliable or do the estimates have substantial sampling variation?

- Are both independent variables necessary? Do any of the variables not contribute to the explanation of delivery times? This question will be addressed using the familiar hypothesis testing procedure.

- Are there other independent variables that could be included that would enhance the model's ability to accurately predict delivery times?

- Can the model make a useful prediction of delivery time? How much confidence can be placed in the prediction?

We will address these questions as we continue to discuss multiple regression throughout this chapter.

### Interpreting the Coefficients of the Multiple Regression Model

In interpreting the coefficients of the model, we ask the question, *Do the signs and magnitudes of the estimated coefficients appear to be reasonable?*

In the simple linear regression model, the estimated coefficient, $b_1$, is the slope of the line. It is interpreted to be the average change in the dependent variable associated with a one-unit change in the independent variable. This interpretation remains basically valid for the multiple regression model as well. For the pizza delivery model, the coefficient $b_1$ is the estimated change in delivery time for a one-unit increase in the number of pizzas, given that distance traveled is constant. Is it reasonable to believe that each additional pizza would add approximately 1.6 minutes to the delivery time? While the delivery time varies, 1.6 minutes seems sensible.

The coefficient $b_2$ is the estimated change in delivery time for a one-unit increase in distance (measured in miles), given a specific number of pizzas (i.e., the number of pizzas to be delivered is constant). That is, for each additional mile added to the delivery, we should expect the average delivery time to increase by approximately 1.57 minutes. All other conditions being equal, do we find that the signs of the coefficients are reasonable? If we add more pizzas and distance traveled to the delivery route, it seems reasonable to expect the delivery time to increase. Thus, the positive signs on the coefficients seem to make sense. Were the signs of the coefficients unexpected, a reasonable question would be, *Is the estimate accurate?* Are there other factors that have not been considered that could reasonably change the signs of the coefficients? We will consider this question later in Section 14.6.

## 📝 14.1 Exercises

### Basic Concepts

1. Explain why a simple linear regression model might not always suffice when attempting to establish a relationship between variables in a business environment.

2. What is the general multiple regression model?

3. What are the assumptions about the error term in a multiple regression model? Are these different from the assumptions required for the simple linear model?

4. What method is used to find the estimated regression equation? Is this method different from the one used to find the simple linear regression equation?

5. What is the greatest challenge in building a multiple regression model?

6. What are some questions that should be asked once a multiple regression model is estimated? Give at least four.

7. In the simple linear regression model, what is the interpretation of $b_1$? Does this interpretation change in the multiple regression model?

8. When interpreting the coefficient of an independent variable in a multiple regression model, what assumption are we making regarding the other independent variables?

9. What two aspects of the model coefficients are usually analyzed first when studying a multiple regression model?

### Exercises

10. Consider the following computer output of a multiple regression analysis relating annual salary to years of education and years of work experience.

SUMMARY OUTPUT

| Regression Statistics | |
|---|---|
| Multiple R | 0.566946595 |
| R Square | 0.321428441 |
| Adjusted R Square | 0.29192533 |
| Standard Error | 10909.996 |
| Observations | 49 |

ANOVA

| | df | SS | MS | F | Significance F |
|---|---|---|---|---|---|
| Regression | 2 | 2593556200 | 1296778100 | 10.89473033 | 0.000133875 |
| Residual | 46 | 5475288584 | 119028012.7 | | |
| Total | 48 | 8068844784 | | | |

| | Coefficients | Standard Error | t Stat | P-value | Lower 95% | Upper 95% |
|---|---|---|---|---|---|---|
| Intercept | 11214.19915 | 5625.172956 | 1.993574106 | 0.052147881 | -108.6867382 | 22537.08504 |
| Education (Years) | 2854.891271 | 689.6666061 | 4.139523715 | 0.000146836 | 1466.664395 | 4243.118147 |
| Experience (Years) | 839.6360369 | 261.7094444 | 3.208275646 | 0.002433357 | 312.842248 | 1366.429826 |

    **a.** Identify the estimated values of the coefficients $b_0$, $b_1$, and $b_2$.

    **b.** Write the estimated multiple regression equation.

    **c.** Can you think of other independent variables that may be useful in predicting annual salary?

**11.** The manager of a publishing company would like to conduct cost analysis on the most recent books the company has published. He would like to estimate a multiple regression model to relate the cost of printing (per book) to the number of pages in the book and the number of copies printed. A computer output of the multiple regression model for the manager's data is given in the following table.

SUMMARY OUTPUT

| Regression Statistics | |
|---|---|
| Multiple R | 0.987606014 |
| R Square | 0.975365639 |
| Adjusted R Square | 0.972467479 |
| Standard Error | 0.445885396 |
| Observations | 20 |

ANOVA

| | df | SS | MS | F | Significance F |
|---|---|---|---|---|---|
| Regression | 2 | 133.8201656 | 66.91008281 | 336.5464936 | 2.12863E-14 |
| Residual | 17 | 3.379834375 | 0.198813787 | | |
| Total | 19 | 137.2 | | | |

| | Coefficients | Standard Error | t Stat | P-value | Lower 95% | Upper 95% |
|---|---|---|---|---|---|---|
| Intercept | 6.134155476 | 3.993435752 | 1.536059638 | 0.142925974 | -2.291257484 | 14.55956844 |
| Number of Pages | 0.010801 | 0.004147682 | 2.604105041 | 0.018522101 | 0.002050156 | 0.019551845 |
| Number of Copies | -0.009954478 | 0.005271436 | -1.888380579 | 0.07616193 | -0.021076236 | 0.00116728 |

    **a.** Identify the estimated regression coefficients.

    **b.** Write the estimated multiple regression equation.

    **c.** Do the magnitudes and signs of the coefficients seem reasonable? Explain.

    **d.** What other variables do you think could be useful in explaining printing cost per book?

12. A nutritionist wishes to study body weight based on height, age, average calories consumed per day, and the average number of minutes spent exercising per day.

   a. Write the multiple regression model the nutritionist is interested in in terms of weight, height, age, calories, and exercise. Assume the coefficients have not yet been estimated.

   b. Identify the independent variables in the multiple regression model.

   c. Predict the sign of the coefficient for each of the independent variables in the model. Explain your answers.

   d. Can you think of any other variables that might be useful for the nutritionist to take into account before performing the regression analysis?

13. Suppose the CEO of an electronics company wants to study the effects of various business practices on annual revenue.

   a. Make a list of independent variables the CEO might be interested in studying.

   b. Suppose the CEO has narrowed his list of factors down, and decided he wants to mainly study the effects of research and development expenditures, advertising expenditures, and the average annual salary paid to employees. Write the multiple regression model in terms of the dependent and independent variables, assuming the coefficients have not yet been estimated.

   c. Make a guess of the sign of the coefficient of research and development expenditures. Explain your prediction.

   d. Why should the CEO be cautious when using this model for revenue estimation and prediction?

14. Consider the following estimated multiple regression equation relating the number of study hours and GPA to a student's ACT score.

   $$\text{Estimated ACT Score} = 8.35 + 1.53(\text{Study Hours}) + 0.30(\text{GPA})$$

   a. Identify the values of $b_0$, $b_1$, and $b_2$.

   b. Interpret the value of $b_0$ in terms of the problem.

   c. Interpret the value of $b_1$ in terms of the problem.

   d. Interpret the value of $b_2$ in terms of the problem.

15. Consider the following estimated regression model relating annual salary to years of education and work experience, which was presented in Exercise 10.

   $$\text{Estimated Salary} = 11214.20 + 2854.89(\text{Education}) + 839.64(\text{Experience})$$

   a. Consider the coefficient for the education variable. Do the sign and magnitude of the coefficient seem to make sense? Explain.

   b. Consider the coefficient for the experience variable. Do the sign and magnitude of the coefficient seem to make sense? Explain.

   c. Interpret the regression coefficient for years of experience.

   d. Suppose an employee with 8 years of education (note that education years are the number of years after $8^{\text{th}}$ grade) has been with the company for 5 years. According to this model, what is his estimated annual salary?

   e. How would you expect his salary to change if he stays at the company for another year?

   f. Suppose two employees at the company have been working there for five years. One has a bachelor's degree (8 years of education) and one has a master's degree (10 years of education). Which employee would you expect to earn a higher salary? How much more money would you expect him to make?

**16.** Suppose the owner of a car dealership wishes to study how certain factors affect the number of new cars sold. Specifically, he wishes to construct a multiple regression model relating car price, average income per capita in the surrounding area, and the interest rate to the quantity of new cars sold. After compiling historical data, he obtains the following summary output for the multiple regression model. In the original data, car price and average income per capita were in thousands of dollars, and interest rates were reported as percentages.

SUMMARY OUTPUT

| Regression Statistics | |
| --- | --- |
| Multiple R | 0.581495041 |
| R Square | 0.338136482 |
| Adjusted R Square | 0.305043307 |
| Standard Error | 227.5372802 |
| Observations | 64 |

ANOVA

| | df | SS | MS | F | Significance F |
| --- | --- | --- | --- | --- | --- |
| Regression | 3 | 2361116.922 | 787038.9742 | 10.21771025 | 1.57438E-05 |
| Residual | 60 | 4621616.515 | 77026.94192 | | |
| Total | 63 | 6982733.438 | | | |

| | Coefficients | Standard Error | t Stat | P-value | Lower 95% | Upper 95% |
| --- | --- | --- | --- | --- | --- | --- |
| Intercept | -308.6097834 | 971.3332832 | -0.317717707 | 0.751802188 | -2251.565634 | 1634.346068 |
| Price | -18.29845156 | 6.196928308 | -2.952826085 | 0.004489897 | -30.69415375 | -5.902749358 |
| Income | 458.1641011 | 137.8510094 | 3.323618036 | 0.001518198 | 182.4210273 | 733.907175 |
| Interest Rate | -26.29956832 | 11.93107528 | -2.204291541 | 0.031350669 | -50.16527221 | -2.43386442 |

a. Write the estimated multiple regression equation.

b. Consider the coefficient of the price variable. Do you think the magnitude and sign of this coefficient make sense? Explain your answer.

c. Consider the coefficient of the income variable. Do you think the magnitude and sign of this coefficient make sense? Explain your answer.

d. Consider the coefficient of the interest rate variable. Do you think the magnitude and sign of this coefficient make sense? Explain your answer.

e. How many additional cars would the dealership owner expect to sell if per capita income for the area increased by $1000?

f. How would the dealership owner expect the quantity of cars sold to change if the interest rate increased by 2 percentage points?

# 14.2 The Coefficient of Determination and Adjusted $R^2$

Just as for simple linear regression, we will discuss methods that can be used to evaluate the overall effectiveness of multiple regression models. For the pizza delivery model in the previous section, one of the questions to ask is, how do we determine whether the model explains a substantial portion of the variation in the delivery times? The overall effectiveness and usefulness of multiple regression models can be addressed using the coefficient of determination ($R^2$) and the adjusted $R^2$ ($R_a^2$) statistics.

## Coefficient of Determination ($R^2$)

Recall our discussion about the coefficient of determination ($R^2$) in the previous chapter. In Section 13.3, we defined the $R^2$ statistic as the statistic that directly measures the degree to which the model explains the dependent variable. In multiple regression, the **coefficient of determination** (sometimes called the **multiple coefficient of determination**), still denoted by $R^2$, represents the proportion of variation in the response variable, $y$, that is explained by the set of independent variables, $x_1, x_2, ..., x_k$. $R^2$ is defined in the same way as for the simple linear regression model.

## ∞ Technology

The values for $R^2$ and adjusted $R^2$ can be obtained from the multiple regression output produced by Excel, JMP, and Minitab. For instructions on obtaining this output for multiple regression using technology, please visit
stat.hawkeslearning.com and navigate to **Discovering Business Statistics, Second Edition > Technology Instructions > Regression > Multiple Regression**.

> ### Formula
>
> #### Coefficient of Determination
>
> The coefficient of determination, $R^2$, is given by
>
> $$R^2 = \frac{\text{SSR}}{\text{TSS}} = 1 - \frac{\text{SSE}}{\text{TSS}}$$
>
> where SSR = the sum of squares of regression, TSS = the total sum of squares, and SSE = the sum of squared errors. $R^2$ represents the proportion of variation in the dependent variable explained by the set of independent variables in a multiple regression model.

Just as for simple linear regression,

$$0 \le R^2 \le 1.$$

If all of the slopes are zero ($b_i = 0$ for $i = 1, 2, ..., k$) then $R^2$ is also zero, indicating that there is no relationship between the $x_i$'s and the response variable, $y$. Similarly, if $y_i = \hat{y}_i$ for all $i$, the value of the coefficient of determination is one ($R^2 = 1$). Note that since we are fitting a multiple regression model, we are not fitting a line but a plane or surface.

The output in Figure 14.1.1 reveals $R^2 \approx 0.9640$. Thus, in the pizza delivery model, the two independent variables (number of pizzas and distance) can explain approximately 96.4% of the variation in delivery times. Accounting for such a large amount of variation in the dependent variable would seem to demonstrate substantial explanatory power.

It should be noted, however, that a large value of $R^2$ does not necessarily imply that the fitted model is a useful one. For instance, observations may have been taken at only a few levels of the independent variable(s). Despite a large $R^2$ in this case, the fitted model may not be useful because most predictions would require extrapolations outside the region of the observations. Again, even though $R^2$ is large, the mean square error (MSE) may still be too large for inferences to be useful when high precision is required. Thus, one should examine more than one statistic when evaluating the adequacy of a regression model.

## Adjusted $R^2$ ($R_a^2$)

Adding more independent variables to a regression model will always increase the $R^2$ value. $R^2$ will never decrease as variables are added because the SSE can never become smaller with the addition of independent variables and the TSS is always the same for a given set of responses. Since $R^2$ can be made larger by including a large number of independent variables, it is sometimes suggested that a modified measure be used that adjusts for the number of independent variables in the model. The **adjusted coefficient of determination** (denoted by $R_a^2$) adjusts $R^2$ by dividing each sum of squares by its associated degrees of freedom. Thus, ($R_a^2$) is given by the following formula.

## Formula

### Adjusted $R^2$

The adjusted $R^2$ statistic takes into account the number of independent variables in the model by dividing each sum of squares by its associated degrees of freedom.

$$R_a^2 = 1 - \left(\frac{n-1}{n-k-1}\right)\frac{SSE}{TSS}$$

where $n$ is the number of observations and $k$ is the number of independent variables in the model.

**Example 14.2.1**

**Assessing the Fit of the Pizza Delivery Time Model using $R^2$ and Adjusted $R^2$**

a. Using the data in Table 14.1.1, fit a simple linear regression model using just Number of Pizzas as the independent variable and Delivery Time as the dependent variable. What is the value of $R^2$ and what does it mean?

b. Fit a multiple regression model with both Number of Pizzas and Distance. What are the values of $R^2$ and $R_a^2$ and what do they mean?

c. Did the addition of Distance improve the model fit? Explain your answer.

#### SOLUTION

a. Fitting a simple linear regression model to pizza delivery time using only number of pizzas as the independent variable yields the output provided in Figure 14.2.1.

**Simple Linear Regression Output Modeling Delivery Time by Number of Pizzas**

Summary of Fit

| | |
|---|---|
| RSquare | 0.926556 |
| R Square Adj | 0.923363 |
| Root Mean Square Error | 4.297823 |
| Mean of Response | 22.384 |
| Observations (or Sum Wgts) | 25 |

Analysis of Variance

| Source | DF | Sum of Squares | Mean Square | F-Ratio |
|---|---|---|---|---|
| Model | 1 | 5359.7031 | 5359.70 | 290.1641 |
| Error | 23 | 424.8395 | 18.47 | Prob > F |
| C. Total | 24 | 5784.5426 | | < .0001* |

Parameter Estimates

| Term | Estimate | Std Error | t Ratio | Prob > \|t\| |
|---|---|---|---|---|
| Intercept | 2.8170193 | 1.43469 | 1.96 | 0.0618 |
| Number of Pizzas | 2.1936077 | 0.128777 | 17.03 | < .0001* |

**Figure 14.2.1**

Given the output in Figure 14.2.1, we see that the $R^2 \approx 0.9266$ and $R_a^2 \approx 0.9233$. Note that the adjusted $R^2$ value does not have much meaning in simple linear regression. The value of $R^2$ implies that 92.66% of the variation in delivery time can be explained by the number of pizzas that the driver has to deliver.

b. When we add the distance independent variable to the model, the JMP output is given in Figure 14.1.1 in Section 14.1. $R^2 \approx 0.9640$ and $R_a^2 \approx 0.9608$. With the addition of the independent variable, distance, to the model, we have $R^2 \approx 0.9640$ and $R_a^2 \approx 0.9608$. Similar to what was stated above, the value of $R^2$ implies that 96.40% of the variability in delivery time can be explained by the number of pizzas being delivered and the

**Testing a Model's Predictive Ability**

If you use all your data to create a model then the "real world" predictive ability of the model is not known since the model was fitted to your data. How would your model perform on data that it was not fitted to? If another set of data is not readily available for this purpose, then researchers often use a hold-out sample for evaluating predictive performance. A sample of observations are withheld from the model estimation process and used for assessing the predictive performance of the model. When multiple models are being considered for the same purpose, the hold-out sample can provide a means of deciding which one is the best predictor.

distance that is driven. The interpretation of $R_a^2$ is the same—96.08% of the variation in delivery time is explained by the two independent variables in the model.

c.  With both number of pizzas and distance in the model, the value of $R^2$ increased by 0.0374, indicating that adding the variable distance to the model helped explain more variability in delivery times. The value of $R_a^2$ increased by slightly more (0.0375). Using both variables in the model explained nearly 4% more variability in delivery time.

As the number of independent variables increases, the difference between the $R^2$ and adjusted $R^2$ values also increases. $R_a^2$ is commonly used as a method of comparison between multiple regression models when one is attempting to find the model that best fits the data. Unlike the $R^2$ value, the adjusted coefficient of determination may actually become smaller when another independent variable is added to the model. Thus, the adjusted $R^2$ value is most useful when comparing multiple regression models with different numbers of independent variables.

# ✍ 14.2 Exercises

## Basic Concepts

1.  What is the purpose of the $R^2$ and adjusted $R^2$ statistics?

2.  What values can the coefficient of determination take?

3.  If a particular regression model explains 68% of variation in the dependent variable, what is the value of $R^2$?

4.  If the coefficient of determination has a value of zero, is it possible for a regression coefficient to have a value other than zero? Explain why.

5.  If the coefficient of determination has a value of one, what is the relationship between the sum of squares of regression and the total sum of squares? Explain why.

6.  Does a large value of $R^2$ always indicate that the fitted model is useful? Explain.

7.  Explain the difference between $R^2$ and adjusted $R^2$.

8.  Explain why the adjusted $R^2$ statistic is sometimes a better measure to use to evaluate the fit of a regression model.

9.  Will there ever be a situation in which the adjusted $R^2$ statistic is greater than $R^2$ statistic? Explain your answer.

## Exercises

10. Consider the following ANOVA table for a multiple regression model relating housing prices (in thousands of dollars) to the number of bedrooms in the house and the size of the lot on which the house was built (in square feet). There were 88 total observations.

$$\text{Estimated Price} = 63.26 + 57.31(\text{Bedrooms}) + 0.0029(\text{Lot Size})$$

ANOVA

|  | df | SS | MS | F | Significance F |
|---|---|---|---|---|---|
| Regression | 2 | 309148.8902 | 154574.4451 | 21.58486376 | 2.62677E-08 |
| Residual | 85 | 608705.618 | 7161.242565 | | |
| Total | 87 | 917854.5083 | | | |

a.  Identify the values of SSR, SSE, and TSS from the table.

b.  What is the coefficient of determination for this model? Interpret this value in terms of the problem.

**c.** What is $R_a^2$? Interpret this value.

**d.** Compare the $R^2$ and $R_a^2$ values. Which value should be used to evaluate the fit of the multiple regression model? Explain why.

**11.** Suppose an additional variable, Square Feet, was added to the housing price model from Exercise 10. The summary output is given below.

SUMMARY OUTPUT

| Regression Statistics | |
|---|---|
| Multiple R | 0.819976968 |
| R Square | 0.672362228 |
| Adjusted R Square | 0.660660879 |
| Standard Error | 59.83347988 |
| Observations | 88 |

ANOVA

| | df | SS | MS | F | Significance F |
|---|---|---|---|---|---|
| Regression | 3 | 617130.7018 | 205710.2339 | 57.46023188 | 2.69597E-20 |
| Residual | 84 | 300723.8065 | 3580.045315 | | |
| Total | 87 | 917854.5083 | | | |

| | Coefficients | Standard Error | t Stat | P-value | Lower 95% | Upper 95% |
|---|---|---|---|---|---|---|
| Intercept | -21.7703086 | 29.47504196 | -0.738601446 | 0.462207782 | -80.38466199 | 36.84404478 |
| Bedrooms | 13.85252186 | 9.010145446 | 1.537435988 | 0.127945059 | -4.065140472 | 31.7701842 |
| Lot Size | 0.002067707 | 0.000642126 | 3.220095719 | 0.001822929 | 0.000790769 | 0.003344644 |
| Square Feet | 0.122778185 | 0.013237407 | 9.275092996 | 1.65802E-14 | 0.096454149 | 0.149102222 |

**a.** What is $R_a^2$ for this model?

**b.** How does the adjusted $R^2$ value for this model compare to the adjusted $R^2$ value for the model in Exercise 10?

**c.** Do you think adding the additional independent variable, Square Feet, improved the model? Explain your answer.

**12.** The owner of a new pizzeria in town wants to study the relationship between weekly revenues and advertising expenditures. Both measures were recorded in thousands of dollars. The computer output for the simple linear regression model is given below.

SUMMARY OUTPUT

| Regression Statistics | |
|---|---|
| Multiple R | 0.858179902 |
| R Square | 0.736472743 |
| Adjusted R Square | 0.692551534 |
| Standard Error | 1.058296197 |
| Observations | 8 |

ANOVA

| | df | SS | MS | F | Significance F |
|---|---|---|---|---|---|
| Regression | 1 | 18.78005496 | 18.78005496 | 16.76804334 | 0.006394067 |
| Residual | 6 | 6.719945042 | 1.11999084 | | |
| Total | 7 | 25.5 | | | |

| | Coefficients | Standard Error | t Stat | P-value | Lower 95% | Upper 95% |
|---|---|---|---|---|---|---|
| Intercept | 74.69887795 | 7.104358625 | 10.51451396 | 4.34789E-05 | 57.31513863 | 92.08261726 |
| Advertising Expenditures | 1.854820243 | 0.452960815 | 4.094880138 | 0.006394067 | 0.746465058 | 2.963175428 |

**a.** Write the estimated regression equation.

**b.** What is the coefficient of determination for this model? Interpret this value.

**c.** What is the value of the adjusted $R^2$ statistic? Is this statistic useful for the pizzeria owner as he studies this model? Explain.

    **d.** Do you believe this model is useful in explaining revenues based on advertising expenditures? Explain your answer.

    **e.** How could the restaurant owner improve this model? Are there other independent variables that he should consider including?

**13.** The owner of the pizzeria discussed in Exercise 12 wishes to build on the model relating revenues to advertising expenditures by breaking the advertising expenditures into three categories: television advertising, newspaper advertising, and direct mail advertising.

    **a.** Write the new regression model in terms of television, newspaper, and mail expenditures. Assume the coefficients have not yet been estimated.

    **b.** Consider the following summary output for the new model. Write the estimated multiple regression equation.

SUMMARY OUTPUT

| Regression Statistics | |
| --- | --- |
| Multiple R | 0.967040091 |
| R Square | 0.935166537 |
| Adjusted R Square | 0.88654144 |
| Standard Error | 0.64289449 |
| Observations | 8 |

ANOVA

| | df | SS | MS | F | Significance F |
| --- | --- | --- | --- | --- | --- |
| Regression | 3 | 23.8467467 | 7.948915566 | 19.23217829 | 0.007708883 |
| Residual | 4 | 1.653253302 | 0.413313326 | | |
| Total | 7 | 25.5 | | | |

| | Coefficients | Standard Error | t Stat | P-value | Lower 95% | Upper 95% |
| --- | --- | --- | --- | --- | --- | --- |
| Intercept | 73.93199827 | 4.523870838 | 16.34264127 | 8.20538E-05 | 61.37171922 | 86.49227731 |
| Television | 2.383047934 | 0.318133378 | 7.490719616 | 0.001698799 | 1.499768074 | 3.266327793 |
| Newspaper | 1.454439994 | 0.355820285 | 4.087569076 | 0.015004989 | 0.466524505 | 2.442355483 |
| Mail | 1.815990841 | 0.276487962 | 6.568064755 | 0.002780349 | 1.048337191 | 2.58364449 |

    **c.** Interpret the coefficient for television advertising expenditures. Remember that revenues and expenditures are in thousands of dollars.

    **d.** What is the adjusted coefficient of determination? Interpret this value.

    **e.** How does the coefficient of determination of this model compare to the coefficient of determination for the simple linear regression model in Exercise 12? Does this appear to be a more useful model? Explain.

    **f.** What is the value of the $R^2$ statistic for this model? Should we use the $R^2$ value or the adjusted $R^2$ value when evaluating the usefulness of this model? Explain why.

# 14.3 Inference Concerning the Multiple Regression Model and Its Coefficients

When fitting a multiple regression model, we must ask the question, *is the overall model worthwhile?*

If the model has no redeeming merit, then the set of independent variables will not be related to the dependent variable. Consider the multiple regression model

$$y_i = \beta_0 + \beta_1 x_{1i} + \beta_2 x_{2i} + \cdots + \beta_k x_{ki} + \varepsilon_i.$$

Suppose all of the $\beta_1 = \beta_2 = \cdots = \beta_k = 0$, then the regression equation becomes

$$y_i = \beta_0 + 0x_{1i} + 0x_{2i} + \cdots + 0x_{ki} + \varepsilon_i.$$

Since multiplying any number by 0 results in 0, all of the independent variables disappear from the model, and we are left with

$$y_i = \beta_0 + \varepsilon_i,$$

which says that $y$ is a constant, $\beta_0$, plus a random error, $\varepsilon_i$. So, if $\beta_1 = \beta_2 = \cdots = \beta_k = 0$, then the model is not useful. We will develop a methodology for testing the hypotheses

$$H_0: \beta_1 = \beta_2 = \cdots = \beta_k = 0$$

$$H_a: \text{At least one } \beta_i \neq 0.$$

## Steps in the Test of Hypothesis

**Step 1:** Determine the null hypothesis.

For multiple regression, the null hypothesis is stated to indicate that the overall model is not useful in explaining variation in the dependent variable. The null hypothesis is written as

$$H_0: \beta_1 = \beta_2 = \cdots = \beta_k = 0.$$

**Step 2:** Determine the alternative hypothesis.

The alternative hypothesis is stated to indicate that the overall model is useful in explaining variation in the dependent variable. That is, at least one of the coefficients is different from zero, and thus, the model can be useful in explaining the variation in $y$. The alternative hypothesis is written as

$$H_a: \text{At least one } \beta_i \neq 0.$$

If some of the model's independent variables are useful predictors of $y$, then some of the coefficients, $\beta_i$, of these variables will have nonzero values.

**Step 3:** Select the appropriate test statistic based on the information at hand and the assumptions you are willing to make.

If none of the independent variables are useful predictors of $y$, then the model will not explain much (if any) of the variation in the dependent variable. One way of testing whether the overall model is useful is to examine whether the model explains a sufficient portion of the variation in the dependent variable. We have already studied $R^2$, which measures the fraction of variation explained by the model. However, the sampling distributions of $R^2$ and $R_a^2$ for the multiple regression model are too complex. Instead, we will use the **F-statistic.**

> **Formula**
>
> *F*-Statistic
>
> The **F-statistic** is the test statistic used to test the hypothesis $H_a$: At least one $\beta_i \neq 0$.
>
> $$F = \dfrac{\dfrac{\text{Sum of Squares Regression}}{k}}{\dfrac{\text{Sum of Squared Errors}}{n-(k+1)}}$$
>
> $$= \dfrac{\dfrac{\text{SSR}}{k}}{\dfrac{\text{SSE}}{n-(k+1)}} = \dfrac{\text{Mean Square Regression}}{\text{Mean Square Error}}$$
>
> The *F*-statistic is a ratio which compares the variation explained by the model (SSR) to the unexplained variation (SSE). The **sum of squares of regression** (SSR) and **the sum of squared errors** (SSE) were discussed in Section 13.3. The symbol *k* represents the number of independent variables in the model and *n* represents the number of observations.

**Definition**

**Mean Square Regression (MSR)**

**MSR** is the sum of squares associated with the regression term in the model divided by the degrees of freedom associated with the regression term. The degrees of freedom for the regression term is *k* which represents the number of independent variables in the model.

A small value of *F* means that **mean square regression** (MSR) is small in relation to **mean square error** (MSE). This would mean that the SSR, which is the amount of variation explained by the model, is small in relation to SSE (which represents the variation not explained by the model). A small value of *F* would indicate that the model is of little value in explaining the variation in the dependent variable, *y*. On the other hand, a large *F*-value indicates that the variation explained by the model is large in relation to the unexplained variation. The fundamental question is how large must the *F*-value be in order to believe the model has some explanatory power. This question is complicated by sampling variation. Even if there is no relationship between the independent variables and the dependent variable, sampling variation will produce a model that explains some portion of the variation in *y*.

### Calculating Degrees of Freedom in Multiple Regression Models

**Definition**

**Mean Square Error (MSE)**

**MSE** is the sum of squares associated with the error term in the model divided by the degrees of freedom associated with the error term. The degrees of freedom for the error term is $n - (k + 1)$.

The *F*-statistic has an *F*-distribution with *k* numerator degrees of freedom and $n - (k + 1)$ denominator degrees of freedom, where *k* is the number of independent variables in the model. We will reject the null hypothesis if the *F*-statistic is larger than the *F*-value corresponding to a one-tailed test for some prescribed significance level, $\alpha$.

**Step 4:** Determine the critical value of the test statistic.

For $df_{\text{num}} = 4$, $df_{\text{den}} = 10$ and $\alpha = 0.01$, the *F*-value from the *F*-distribution table is 5.9943. If the null hypothesis is true ($\beta_1 = \beta_2 = \cdots = \beta_k = 0$), and there is no relationship between the independent variables and the dependent variable, sampling variation would cause *F* to be as large or larger than 5.9943 only 1% of the time. If an *F*-value larger than 5.9943 is observed, then there is a choice to make. Is the null hypothesis really true, and we have observed a rare phenomenon (occurs less than 1% of the time due to normal sampling variation)? Or is the null hypothesis false? Suppose the following decision rule is used.

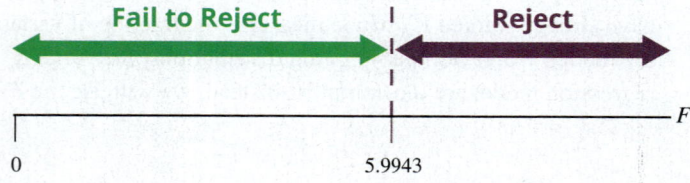

The decision rule defines "rareness" for the $F$-statistic assuming the null hypothesis is true. Essentially what the rule says is that $F$-values as large or larger than 5.9943 would be too rare for us to believe that the null is reasonable. When such a value occurs, the null hypothesis will be rejected in favor of the alternative.

**Step 5:** Collect the sample data and compute the value of the test statistic.

Fortunately, the $F$-statistic is computed by virtually every statistical analysis program. It is found in the analysis of variance (ANOVA) table under the heading "F Ratio" in the JMP output, for example.

**Step 6:** Make the decision state the conclusion in terms of the original question.

If the value of the $F$-statistic falls in the rejection region, then reject the null hypothesis. If not, then we will fail to reject the null hypothesis in favor of the alternative.

Anyone trying to build a model for predictive purposes hopes the null hypothesis ($H_0: \beta_1 = \beta_2 = \cdots = \beta_k = 0$) is rejected, since rejecting the null implies the model can explain some of the variation in the dependent variable.

Additionally, the $P$-value associated with this test can be found in the ANOVA table under the heading "Prob > F." Using the $P$-value approach, we reject the null hypothesis if the $P$-value is $\leq \alpha$; otherwise, we fail to reject the null hypothesis.

---

**Example 14.3.1**

**Determining the Significance of a Multiple Regression Model**

For the pizza delivery model from Section 14.1.1, determine if there is sufficient evidence at the 0.05 significance level that the overall model is useful in explaining the variation in delivery times.

**SOLUTION**

**Step 1:** Determine the null hypothesis.

The null hypothesis states that using number of pizzas and distance in the model is not useful in explaining the variation in delivery times. The null hypothesis is written as

$$H_0: \beta_1 = \beta_2 = 0.$$

**Step 2:** Determine the alternative hypothesis.

The alternative hypothesis states that the overall model with number of pizzas and distance is useful in explaining the variation in delivery times. The alternative hypothesis is written as

$$H_a: \text{At least one } \beta_i \neq 0.$$

If some of the model's independent variables are useful predictors of $y$, then the coefficients ($\beta_i$) of these variables will have nonzero values.

**Step 3:** Select the appropriate test statistic.

We will use the $F$-statistic as the test statistic for this hypothesis test.

**Step 4:** Determine the critical value of the test statistic.

Since there are two independent variables (number of pizzas and distance) in the model, the degrees of freedom are $df_{num} = 2$ and $df_{den} = 25 - (2 + 1) = 22$. The critical value for $F$ at the 0.05 level with 2 numerator degrees of freedom and 22 denominator degrees of freedom is 3.4434.

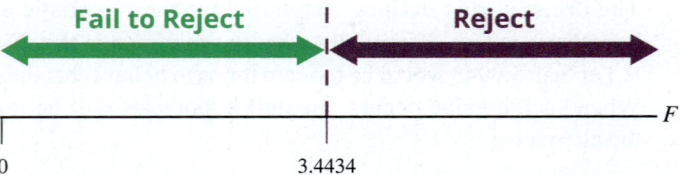

An $F$-value larger than 3.4434 will indicate a value of $F$ that is too rare to have occurred by chance if the null were true, and thus we will reject the null hypothesis.

**Step 5:** Compute the test statistic.

The ANOVA table for the pizza delivery model is given in Figure 14.3.1. The value of $F$ is approximately 294.7403.

SUMMARY OUTPUT

| Regression Statistics | |
|---|---|
| Multiple R | 0.981846095 |
| R Square | 0.964021754 |
| Adjusted R Square | 0.960751004 |
| Standard Error | 3.075694294 |
| Observations | 25 |

ANOVA

| | df | SS | MS | F | Significance F |
|---|---|---|---|---|---|
| Regression | 2 | 5576.424901 | 2788.212 | 294.7403 | 1.30749E-16 |
| Residual | 22 | 208.1176986 | 9.459895 | | |
| Total | 24 | 5784.5426 | | | |

**Figure 14.3.1**

**Step 6:** Make the decision and state the conclusion in terms of the original question.

Since the value of $F$, 294.7403, falls in the rejection region, the null hypothesis will be rejected in favor of the alternative.

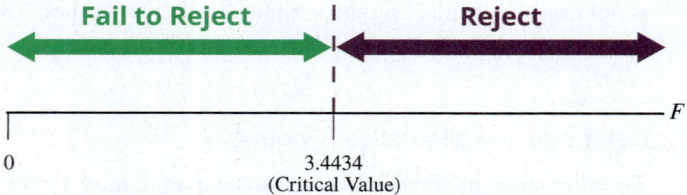

The $P$-value associated with this test is extremely small and therefore less than $\alpha = 0.05$. Thus, we also reject the null hypothesis in favor of the alternative.

*Conclusion and Interpretation*: The null hypothesis is rejected in favor of the alternative, which states that at least one of the independent variables is useful in explaining the variation in delivery times.

## Testing Hypotheses Concerning Individual $\beta_i$

In Section 13.5, we discussed the implications of testing whether the slope coefficient $\beta_1 = 0$. In the case of the simple linear model, if $\beta_1 = 0$, then the independent variable is not related to the dependent variable. However, in multiple regression there is more than one independent variable. In multiple regression models, deciding whether a particular independent variable is a useful predictor of the dependent variable is an important part of the model-building process. For example, is the distance variable useful in predicting delivery time? The rationale for testing whether an independent variable is useful in predicting the dependent variable is the same as that used in testing whether $\beta_1 = 0$ for the simple linear model.

## ⬡ Technology

For instructions on obtaining this output for multiple regression using technology, please visit stat.hawkeslearning.com and navigate to **Discovering Business Statistics, Second Edition > Technology Instructions > Regression > Multiple Regression**.

**Formula**

**Test Statistic for Testing the Hypothesis** $\beta_i \neq 0$

In a test of hypothesis about an individual coefficient, $\beta_i$, the test statistic is computed as

$$t = \frac{b_i - 0}{s_{b_i}} = \frac{b_i}{s_{b_i}},$$

where $b_i$ is the estimated coefficient and $s_{b_i}$ is the standard deviation (standard error) of the estimated coefficient. The test statistic follows a $t$-distribution with $n - (k + 1)$ degrees of freedom.

The technique will be illustrated with an example.

For the pizza delivery model,

$$\text{Delivery Time} = \beta_0 + \beta_1(\text{Number of Pizzas}) + \beta_2(\text{Distance}) + \varepsilon_i.$$

Is the distance variable a useful predictor of delivery time? Use $\alpha = 0.05$ as the level of significance.

**Example 14.3.2**

**Determining the Significance of Coefficients in a Multiple Regression Model**

**SOLUTION**

**Step 1:** Determine the null hypothesis.

The null hypothesis is that distance is not a useful predictor of delivery time. The null hypothesis should be written as

$$H_0: \beta_2 = 0.$$

**Step 2:** Determine the alternative hypothesis and whether it should be one-sided or two-sided.

The alternative hypothesis, which contradicts the null hypothesis, is that distance is a useful predictor of delivery time. For the pizza delivery model, if distance is **not** a useful predictor of delivery time, then its coefficient in the model ($\beta_2$) will equal 0. The sample estimate of $\beta_2$, namely $b_2$, will be used to evaluate the reasonableness of the null hypothesis. The alternative hypothesis should be written as

$$H_a: \beta_2 \neq 0.$$

The alternative hypothesis is two-sided since the relationship between distance and delivery time can be positive or negative. Thus, we will conduct a two-tailed test.

**Step 3:** Select the appropriate test statistic based on the information at hand and the assumptions you are willing to make.

The test statistic is $t = \frac{b_2 - 0}{s_{b_2}} = \frac{b_2}{s_{b_2}}$. The test statistic is identical in form to the test statistic used in Section 13.5 for the simple linear model. If the value of the test statistic is near zero, then there is evidence that the distance variable is not a significant predictor of delivery time.

**Step 4:** Determine the critical value of the test statistic.

Since the test is two-tailed and the level of significance is specified to be $\alpha = 0.05$, $\frac{\alpha}{2} = \frac{0.05}{2} = 0.025$. The test statistic has a $t$-distribution with $df = 25 - (2 + 1) = 22$. The critical values correspond to $t_{0.025,22} = \pm 2.074$.

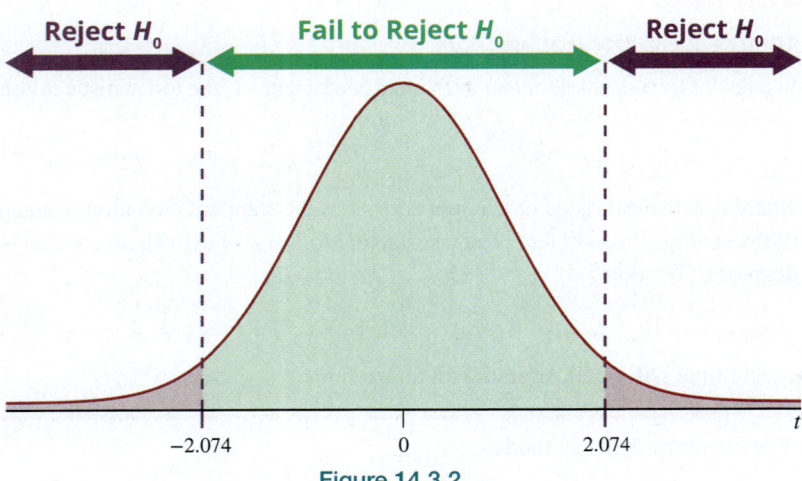

Figure 14.3.2

**Step 5:** Compute the test statistic

The summary output for the pizza delivery model is given in Figure 14.3.3.

| Summary of Fit | |
| --- | --- |
| RSquare | 0.964022 |
| R Square Adj | 0.960751 |
| Root Mean Square Error | 3.075694 |
| Mean of Response | 22.384 |
| Observations (or Sum Wgts) | 25 |

| Analysis of Variance | | | | |
| --- | --- | --- | --- | --- |
| Source | DF | Sum of Squares | Mean Square | F-Ratio |
| Model | 2 | 5576.4249 | 2788.21 | 294.7403 |
| Error | 22 | 208.1177 | 9.46 | Prob > F |
| C. Total | 24 | 5784.5426 | | < .0001* |

| Parameter Estimates | | | | |
| --- | --- | --- | --- | --- |
| Term | Estimate | Std Error | t Ratio | Prob > |t| |
| Intercept | 1.7929033 | 1.048779 | 1.71 | 0.1014 |
| Number of Pizzas | 1.589101 | 0.156346 | 10.16 | < .0001* |
| Distance (Miles) | 1.5677081 | 0.327535 | 4.79 | < .0001* |

Figure 14.3.3

Using the values of $b_2$ and the standard error of $b_2$ from the summary output,

$$t = \frac{b_2}{s_{b_2}} \approx \frac{1.56771}{0.32753} \approx 4.786.$$

The estimated value of $b_2$ is approximately 4.786 standard deviations from zero. Is this persuasive evidence that $\beta_2 \neq 0$?

✏ **NOTE**

The *t*-statistic is also given in the t Ratio column in the JMP summary output.

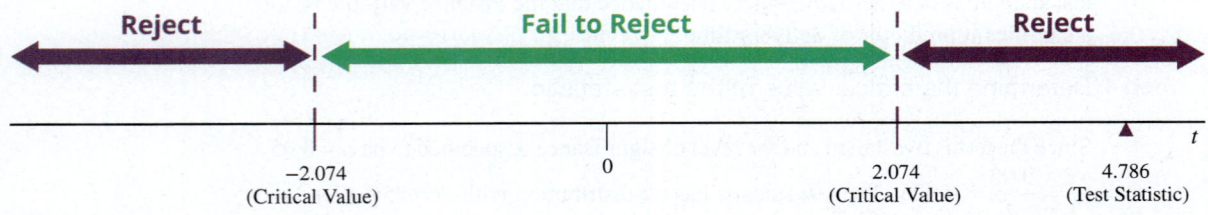

**Step 6:** Make the decision and state the conclusion in terms of the original question.

Since the value of the test statistic falls into the rejection region, we reject the null hypothesis, $H_0$: $\beta_2 = 0$, at the 0.05 level.

The $P$-value is less than 0.0001 (which is less than the significance level of $\alpha = 0.05$) and leads us to also reject the null hypothesis.

*Conclusion and Interpretation*: Since we rejected the null hypothesis, $H_0$: $\beta_2 = 0$, the distance variable is a significant predictor of delivery time, given that the other variable currently in the model (number of pizzas) is constant.

We can apply the exact same $t$-test to the other variable in the model. The estimated coefficient of the number of pizzas variable, $b_1$, is also significant ($t = 10.164$) with a $P$-value less than 0.0001, which is less than the significance level of $\alpha = 0.05$. This suggests that the two-variable model is a reasonable model to use to predict delivery times.

## Confidence Intervals for Individual Coefficients

Confidence intervals for individual $\beta_i$ in a multiple regression model are almost identical to the confidence interval for $\beta_1$ in the simple regression model introduced in Section 13.5.

> **Formula**
>
> ### 100$(1-\alpha)$% Confidence Interval for an Individual Coefficient, $\beta_i$
>
> The 100$(1 - \alpha)$% confidence interval for each coefficient, $\beta_i$, in a multiple regression model is given by
>
> $$b_i \pm t_{\alpha/2, df}\, s_{b_i}\, .$$

The computation of the confidence interval requires three pieces of information:

1.  An estimate of the coefficient $\beta_i$, namely $b_i$.

2.  The standard deviation of the estimate $s_{b_i}$. Both $b_i$ and $s_{b_i}$ are reported in the summary output.

3.  The value of $t_{\alpha/2, df}$, which is the number of standard deviations the endpoint of the interval is from the point estimate. (The degrees of freedom for this $t$ is $n - (k + 1)$, which is the same as the degrees of freedom associated with the error terms (residuals) in the ANOVA table.)

---

Compute and interpret the 95% confidence interval for $\beta_1$ in the pizza delivery model.

**Example 14.3.3**

**Calculating a Confidence Interval for a Coefficient in the Regression Model**

**SOLUTION**

The $\beta_1$ model coefficient relates to the number of pizzas variable in the pizza delivery model. The JMP output in Figure 14.3.4 provides the information necessary to develop a confidence interval for $\beta_1$.

Parameter Estimates

| Term | Estimate | Std Error | t Ratio | Prob > \|t\| | Lower 95% | Upper 95% |
|------|----------|-----------|---------|------------|-----------|-----------|
| Intercept | 1.7929033 | 1.048779 | 1.71 | 0.1014 | -0.382131 | 3.9679381 |
| Number of Pizzas | 1.589101 | 0.156346 | 10.16 | < 0001* | 1.2648599 | 1.9133421 |
| Distance (Miles) | 1.5677081 | 0.327535 | 4.79 | < 0001* | 0.8884431 | 2.2469731 |

**Figure 14.3.4**

$b_1$ is given in the output to be approximately 1.5891.

$s_{b_i}$ is given in the output to be approximately 0.1563.

In order to compute the $t$-value, the degrees of freedom must be determined.

$$df = n - (k + 1) = 25 - (2 + 1) = 22$$

For a 95% confidence interval, $t_{\alpha/2,df}$ will be $t_{0.025,22} = 2.074$. The resulting confidence interval will be

$$1.5891 \pm 2.074(0.1563)$$
$$1.5891 \pm 0.3242$$
$$1.2649 \text{ to } 1.9133$$

We are 95% confident that the true value of $\beta_1$, the increase in the delivery time for each additional pizza (given that distance is held constant), will be between 1.2649 and 1.9133 minutes.

Notice that JMP automatically generates confidence intervals for each individual coefficient. The endpoints for the 95% confidence intervals are given in the Lower 95% and Upper 95% columns of the output. Compare the upper and lower limits given by JMP to the ones just calculated by hand. JMP has the ability to calculate confidence intervals for individual coefficients for any level of significance.

# ✍ 14.3 Exercises

## Basic Concepts

1. If the overall multiple regression model is not useful, what does this tell us about the coefficients of the independent variables?

2. What is the hypothesis being tested when we test to determine if the overall multiple regression model is useful?

3. When testing the overall model, describe the null and alternative hypotheses in plain English.

4. Why is the $R^2$ value not used in the test statistic for a hypothesis test to determine if a multiple regression model is significant?

5. What is the test statistic used in a hypothesis test to determine if an overall model is significant? What is the distribution of this test statistic?

6. Give two equivalent formulas for the test statistic in a hypothesis test about an overall regression model.

7. Explain the significance of the ratio of the mean square regression to the mean square error.

8. True or false: Even if there is no relationship between any of the independent variables and the dependent variable, sampling variation will explain some portion of the variation in the dependent variable.

9. How are the degrees of freedom calculated for a multiple regression model?

10. When testing the overall model for significance, do you perform a one or two-tailed test?

11. What is the rejection rule in tests of hypothesis for model significance?

12. What is the expression for a confidence interval for an individual coefficient, $\beta_i$?

13. Outline the three pieces of information needed to compute a confidence interval for an individual coefficient.

14. What is the test statistic used to test a hypothesis about an individual coefficient in a multiple regression model? How many degrees of freedom are associated with this test statistic?

**15.** If we fail to reject the null hypothesis in a hypothesis test about an individual coefficient, should this variable remain in the regression model? Explain.

## Exercises

**16.** An article appearing in the *Journal of Wildlife Management* summarized the percent fat of 75 arctic foxes. According to the authors, "Storage of fat to provide energy during regular periods of food shortage, or to insulate against low ambient temperatures, is essential for survival in severe arctic homeotherms." Computing percent fat was a laborious process. In one analysis, the author regressed $y$ = percent fat on rump fat thickness (RFT), which was measured in millimeters and was much easier to determine than percent fat. It was noted that a plot of percent fat versus RFT was indeed linear, and the resulting regression equation was $y = 7.40 + 1.36(\text{RFT})$. $R^2$ and $s_e^2$ were determined to be 0.88 and 3.70, respectively.

ANOVA

|  | df | SS | MS | F |
|---|---|---|---|---|
| Regression | $k$ | SSR | SSR/$k$ | MSR/MSE |
| Residual | $n-(k+1)$ | SSE | SSE/$(n-(k+1))$ | |
| Total | $n-1$ | SST | | |

Use the table shown to help answer the following questions.

**a.** Compute the sum of squares of regression and the sum of squared errors. Note that $R^2$ is SSR/TSS and $s_e^2$ is the same as MSE.

**b.** Give the degrees of freedom for the regression and for the error (residual).

**c.** Compute MSR and MSE.

**d.** Compute the $F$ ratio for testing the significance of the regression line. With $\alpha = 0.05$, can we conclude that the relationship between percent fat and RFT is significant?

**e.** Compute a point estimate for percent fat if RFT = 10 millimeters.

**f.** For a 2-millimeter increase in RFT, what would be the expected change in percent fat?

**17.** Consider the model from Exercise 10 in Section 14.1 relating annual salary to years of work experience and years of education.

SUMMARY OUTPUT

| Regression Statistics | |
|---|---|
| Multiple R | 0.566946595 |
| R Square | 0.321428441 |
| Adjusted R Square | 0.29192533 |
| Standard Error | 10909.996 |
| Observations | 49 |

ANOVA

|  | df | SS | MS | F | Significance F |
|---|---|---|---|---|---|
| Regression | 2 | 2593556200 | 1296778100 | 10.89473033 | 0.000133875 |
| Residual | 46 | 5475288584 | 119028012.7 | | |
| Total | 48 | 8068844784 | | | |

|  | Coefficients | Standard Error | t Stat | P-value | Lower 95% | Upper 95% |
|---|---|---|---|---|---|---|
| Intercept | 11214.19915 | 5625.172956 | 1.993574106 | 0.052147881 | -108.6867382 | 22537.08504 |
| Education (Years) | 2854.891271 | 689.6666061 | 4.139523715 | 0.000146836 | 1466.664395 | 4243.118147 |
| Experience (Years) | 839.6360369 | 261.7094444 | 3.208275646 | 0.002433357 | 312.842248 | 1366.429826 |

**a.** Formulate the hypotheses for testing the multiple regression model for overall significance.

**b.** Find the value of the test statistic for a hypothesis test about the overall model.

c. Is there evidence at the 5% level of significance that the overall model is useful in predicting annual salary?

d. Consider the coefficient for years of education. Find a 95% confidence interval for the value of $\beta_1$. Interpret this interval.

e. Formulate the hypotheses for testing the significance of the coefficient $\beta_1$.

f. Is there sufficient evidence at the 0.05 level that years of education is useful in predicting annual salary?

18. Consider the printing cost model discussed in Exercise 11 of Section 14.1.

SUMMARY OUTPUT

| Regression Statistics | |
|---|---|
| Multiple R | 0.987606014 |
| R Square | 0.975365639 |
| Adjusted R Square | 0.972467479 |
| Standard Error | 0.445885396 |
| Observations | 20 |

ANOVA

| | df | SS | MS | F | Significance F |
|---|---|---|---|---|---|
| Regression | 2 | 133.8204656 | 66.91008281 | 336.5464936 | 2.12863E-14 |
| Residual | 17 | 3.379834375 | 0.198813787 | | |
| Total | 19 | 137.2 | | | |

| | Coefficients | Standard Error | t Stat | P-value | Lower 95% | Upper 95% |
|---|---|---|---|---|---|---|
| Intercept | 6.134155476 | 3.993435752 | 1.536059638 | 0.142925974 | -2.291257484 | 14.55956844 |
| Number of Pages | 0.010801 | 0.004147682 | 2.604105041 | 0.018522101 | 0.002050156 | 0.019551845 |
| Number of Copies | -0.009954478 | 0.005271436 | -1.888380579 | 0.07616193 | -0.021076236 | 0.00116728 |

a. What percentage of the variation in printing price is explained by the two independent variables number of pages and number of copies?

b. Is the overall model significant at the 1% level?

c. Consider the estimated regression coefficient for the number of pages. Construct a 99% confidence interval for $\beta_1$. Interpret this interval.

d. Is the number of pages variable useful in predicting printing cost at the 5% level? Would the decision change at the 1% level?

e. Construct a 95% confidence interval for $\beta_2$. Interpret this interval.

f. Is the number of copies useful in explaining the variation in printing cost at the 5% level of significance? Do you think the publisher should consider removing this variable from the model? Explain your answer.

19. The following table contains data from selected cities regarding rental rates of two-bedroom apartments, city populations, and median incomes. Monthly rent is given in dollars, population is given in thousands of people, and median income is given in thousands of dollars. Suppose we wish to build a multiple regression model to predict the cost of rent based on population and median income.

| Monthly Rent, Population, and Median Income in Selected Cities | | | |
|---|---|---|---|
| City | Monthly Rent ($) | 2010 Population (Thousands) | 2010 Median Income (Thousands of Dollars) |
| Denver, CO | 868 | 600.158 | 45.438 |
| Birmingham, AL | 711 | 212.237 | 31.704 |
| San Diego, CA | 1414 | 1307.402 | 61.962 |
| Gainesville, FL | 741 | 124.354 | 28.653 |
| Winston-Salem, NC | 707 | 229.617 | 41.979 |
| Memphis, TN | 819 | 646.889 | 36.535 |

| Monthly Rent, Population, and Median Income in Selected Cities (cont.) | | | |
|---|---|---|---|
| City | Monthly Rent ($) | 2010 Population (Thousands) | 2010 Median Income (Thousands of Dollars) |
| Austin, TX | 966 | 790.390 | 50.236 |
| Seattle, WA | 1219 | 608.660 | 58.990 |
| Richmond, VA | 735 | 204.214 | 37.735 |
| Charleston, SC | 812 | 120.083 | 47.799 |
| College Park, MD | 1407 | 30.413 | 66.900 |
| Savannah, GA | 789 | 136.286 | 33.778 |
| Minneapolis, MN | 988 | 382.578 | 45.625 |
| Detroit, MI | 805 | 713.777 | 29.447 |
| Baton Rouge, LA | 827 | 229.493 | 35.436 |

Source: U.S. Census Bureau

a. Write the multiple regression model in terms of rent, population, and income. Assume the regression coefficients have not yet been estimated.

b. Predict the signs of the coefficients $\beta_1$ and $\beta_2$. Explain your answers.

c. Using statistical software, estimate the multiple regression equation. Identify the values of $b_0$, $b_1$, and $b_2$ and write the estimated multiple regression equation. Interpret the estimated coefficients.

d. At the 1% level of significance, is the overall model useful in predicting monthly rent? Identify the test statistic for this test.

e. Find a 95% confidence interval for $\beta_2$. Interpret this interval.

f. Determine if each independent variable is related to the dependent variable at the 0.05 level of significance.

g. Should we consider removing any independent variables from this regression model? If yes, identify the variable(s) that should be removed and explain why.

20. Using the information from Exercise 19, estimate the simple linear regression equation relating monthly rent to median income only.

a. Write the estimated simple regression equation.

b. Is the simple linear regression model significant at $\alpha = 0.01$?

c. Is median income related to the monthly rental rate at $\alpha = 0.01$? Identify the test statistic used in this hypothesis test.

d. What percent of the variation in monthly rent is explained by median income? Compare this to the percent of variation in monthly rent explained by both population and median income in Exercise 19.

e. Which model do you think is a better model to use to predict monthly rental rates? Explain your answer.

# 14.4 Inference Concerning the Model's Prediction

Many regression models are developed solely to predict the dependent variable. To use the multiple regression model for prediction, insert the values of the independent variables in the model and calculate the predicted value. Recall the estimated multiple regression equation for our pizza delivery model:

Delivery Time = 1.7929 + 1.5891(Number of Pizzas) + 1.5677(Distance).

For 5 pizzas being delivered 6 miles away, the model would predict the delivery time to be

Estimated Delivery Time = 1.7929 + 1.5891(5) + 1.5677(6) = 19.1446 minutes.

This is the point estimate. How good is this estimate? The answer to this question depends on what you are trying to predict. Are you trying to predict the average delivery time for delivering 5 pizzas 6 miles away, or are you trying to predict the delivery time for a particular order?

### Confidence Interval for the Mean Value of *y* Given *x*

In Section 13.6, we discussed a confidence interval for the mean value of $y$ given $x = x_p$ for the simple regression model. In our multiple regression model, the point estimate, 19.1446 minutes, is the average value of $y$ given $x_1 = 5$ and $x_2 = 6$. In other words, the delivery time of 19.1446 minutes is the estimated average delivery time for *all* deliveries of 5 pizzas to locations 6 miles away. Since we do not have all the deliveries in the sample, the predicted average delivery time of 19.1446 is only an estimate of the true mean value. How good is the estimate? For multiple regression, the expression for the confidence interval of the mean value of $y$ given $x$ is beyond the scope of this text.

Fortunately, statistical analysis programs such as Minitab will compute a confidence interval for the mean value of $y$ given $x$ for the multiple regression model. (See Figure 14.4.1.)

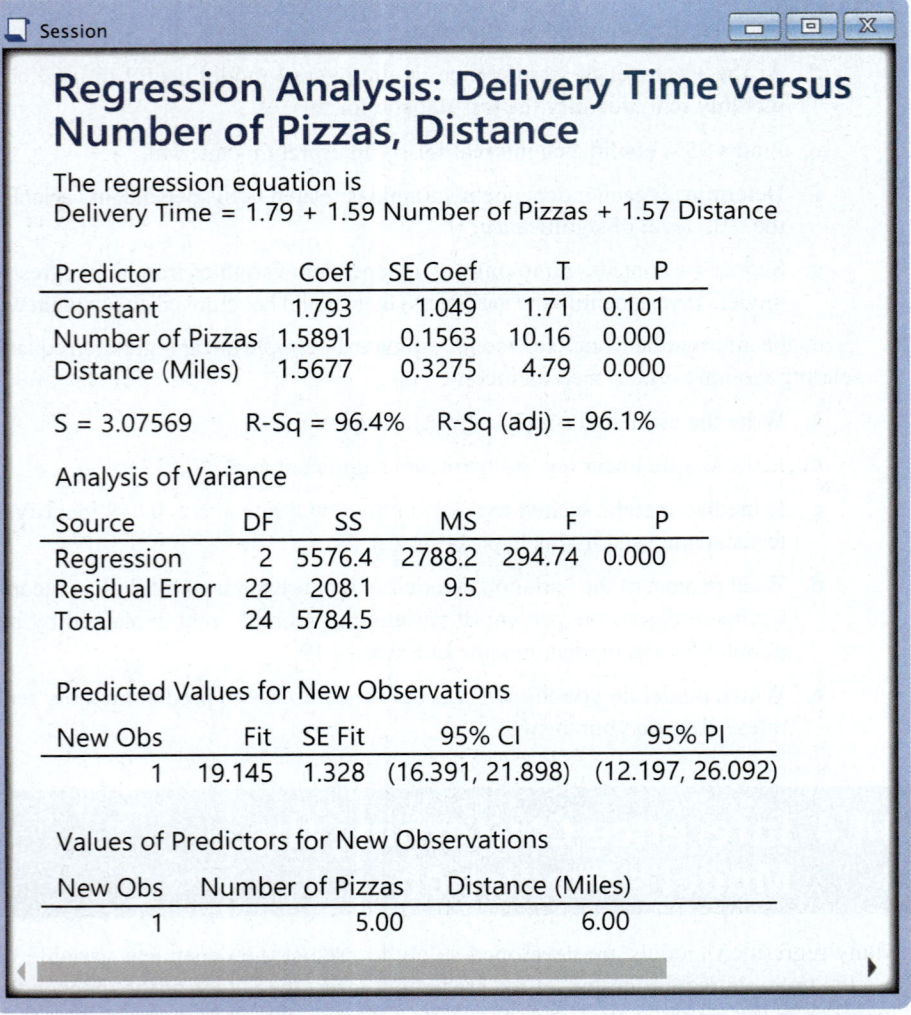

Figure 14.4.1

According to the Minitab output given in Figure 14.4.1, the 95% confidence interval for the mean delivery time for 5 pizzas being delivered 6 miles away is 16.391 minutes to 21.898 minutes.

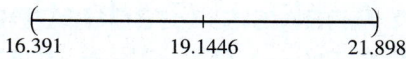

## Confidence Interval for the Predicted Value of *y* Given *x*

A caller asks to speak to the manager of the pizza restaurant. The caller wants to know how long it would take to deliver five pizzas to his specific location, which is six miles away. As the manager, you would like to guarantee how long it will take to make this delivery of 5 pizzas to this customer. You are not especially interested in the *average* delivery time for such a delivery. Instead, it would be preferable to create a confidence interval for the time it is going to take for this particular order to be delivered. Once again, for multiple regression, the expression for the prediction interval is beyond the scope of this text. Fortunately, statistical analysis programs such as Minitab will also produce a prediction interval. Using the output shown in Figure 14.4.1, the 95% prediction interval for the delivery of 5 pizzas to a location 6 miles away is 12.197 minutes to 26.092 minutes. As we observed in Section 13.6, to account for individual variation, the prediction interval for *y* given *x* is substantially wider than the confidence interval for the mean value of *y* given *x*.

**Confidence Interval for Average Delivery Time**

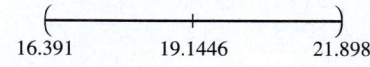

**Prediction Interval for Individual Delivery Time**

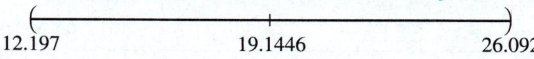

Can the model make a useful prediction of the delivery time? Although the model has an $R^2$ of 0.9640, the prediction interval is fairly wide. This indicates that not a great deal of confidence can be placed in the estimated value of 19.1446 minutes as a delivery time for 5 pizzas traveling 6 miles.

# 📝 14.4 Exercises

## Basic Concepts

1. What is a point estimate for a multiple regression model?

2. Explain how a point estimate is interpreted as an "average" value.

3. Distinguish between a confidence interval and a prediction interval for a multiple regression model.

4. What is the price that is paid when making predictions regarding individual values?

5. Suppose an estimated multiple regression model, $\hat{y}_i = b_0 + b_1 x_{1i} + b_2 x_{2i}$, produces a 95% confidence interval of (3.292, 7.072) and a 95% prediction interval of (0.364, 10.000) when $x_1 = 6$ and $x_2 = 6$. Interpret both of these intervals.

## Exercises

6. Consider the multiple regression model predicting graduating GPA using both the SAT critical reading score and the SAT math score. Computer output of the model

$$\text{GPA} = \beta_0 + \beta_1 \left(\text{SAT Reading}\right) + \beta_2 \left(\text{SAT Math}\right) + \varepsilon_i$$

is given.

<head>
</head>

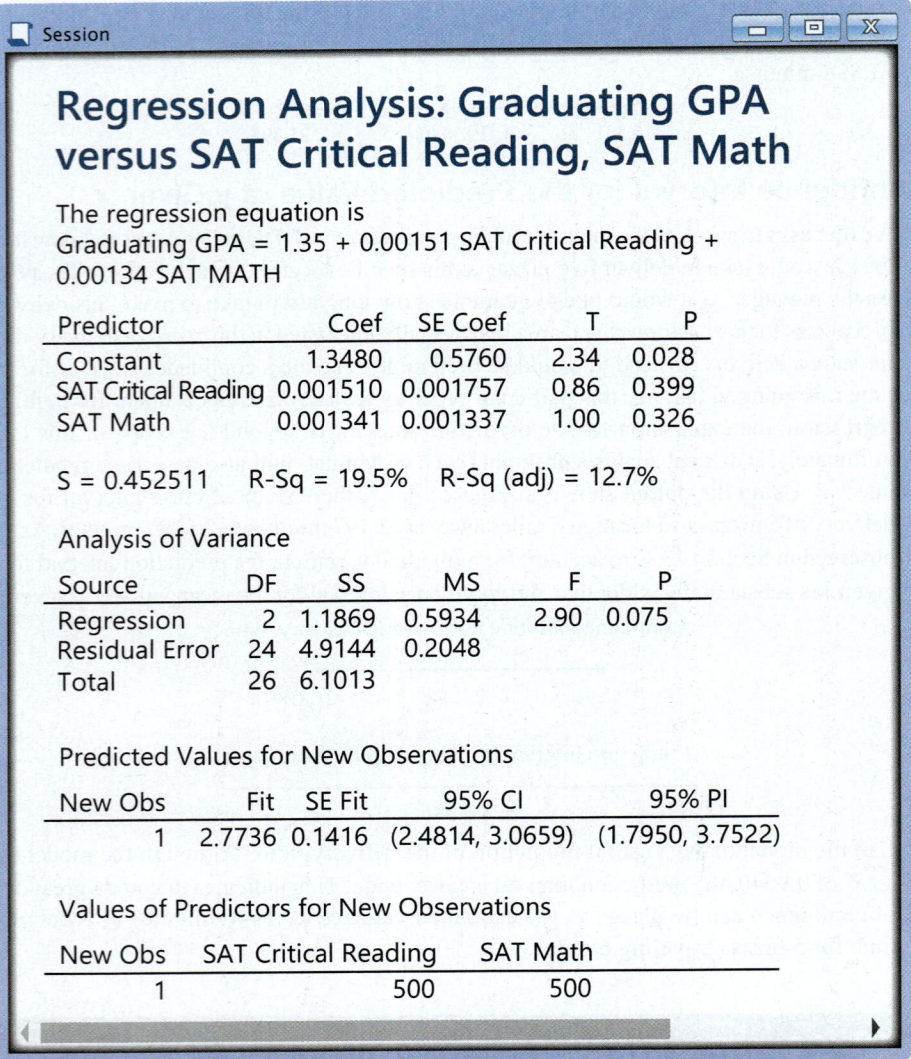

**Regression Analysis: Graduating GPA versus SAT Critical Reading, SAT Math**

The regression equation is
Graduating GPA = 1.35 + 0.00151 SAT Critical Reading + 0.00134 SAT MATH

| Predictor | Coef | SE Coef | T | P |
|---|---|---|---|---|
| Constant | 1.3480 | 0.5760 | 2.34 | 0.028 |
| SAT Critical Reading | 0.001510 | 0.001757 | 0.86 | 0.399 |
| SAT Math | 0.001341 | 0.001337 | 1.00 | 0.326 |

S = 0.452511    R-Sq = 19.5%    R-Sq (adj) = 12.7%

Analysis of Variance

| Source | DF | SS | MS | F | P |
|---|---|---|---|---|---|
| Regression | 2 | 1.1869 | 0.5934 | 2.90 | 0.075 |
| Residual Error | 24 | 4.9144 | 0.2048 | | |
| Total | 26 | 6.1013 | | | |

Predicted Values for New Observations

| New Obs | Fit | SE Fit | 95% CI | 95% PI |
|---|---|---|---|---|
| 1 | 2.7736 | 0.1416 | (2.4814, 3.0659) | (1.7950, 3.7522) |

Values of Predictors for New Observations

| New Obs | SAT Critical Reading | SAT Math |
|---|---|---|
| 1 | 500 | 500 |

a. Find the standard deviation of the error terms in the output.

b. Interpret the coefficient of SAT Critical Reading. What would it mean if the coefficient was negative?

c. Determine if the overall model is useful in explaining GPA. Test at the 0.05 level.

d. What fraction of the variation in GPA is explained by the model?

e. Determine if the SAT Critical Reading variable is a useful predictor of GPA. Test at the 0.05 level.

f. The output includes a predicted GPA for someone scoring 500 on both the SAT Critical Reading and SAT Math portions. Find the predicted value in the output.

g. What is the average GPA for an individual who scored 500 on both the SAT Critical Reading and SAT Math sections? Find the 95% confidence interval for this average. Interpret this interval.

h. Suppose your nephew scored 500 on both the critical reading and math sections. What would be the model's prediction for his graduating GPA? Find the 95% prediction interval for your nephew in the output. Interpret this interval.

i. Why is the prediction interval so much wider than the confidence interval in part g.?

j. Summarize the strengths and weaknesses of the estimated model.

12. Consider the following sales data regarding weekly sales, the number of sales reps, and whether or not the sales were made in the first, second, third, or fourth quarter of the year. For each column containing an indicator variable, the variable is equal to 1 if that particular week was in that particular quarter, and equal to zero otherwise. For example, if the weekly data were recorded in January, the 1st quarter indicator variable would be equal to 1 and the indicator variables for the 2nd, 3rd, and 4th quarters would be equal to zero. The first quarter comprises January through March, the second quarter April through June, the third quarter July through September, and the fourth quarter October through December.

**∴ Data**

This data set can be found at stat.hawkeslearning.com by navigating to **Discovering Business Statistics, Second Edition > Data Sets > Weekly Sales by Quarter**.

| Weekly Sales by Quarter | | | | | |
|---|---|---|---|---|---|
| Weekly Sales ($) | Number of Sales Reps | 1st Quarter | 2nd Quarter | 3rd Quarter | 4th Quarter |
| 4272.90 | 3 | 1 | 0 | 0 | 0 |
| 5069.70 | 9 | 1 | 0 | 0 | 0 |
| 6067.70 | 11 | 1 | 0 | 0 | 0 |
| 6680.55 | 17 | 1 | 0 | 0 | 0 |
| 9725.05 | 20 | 1 | 0 | 0 | 0 |
| 4107.10 | 3 | 0 | 1 | 0 | 0 |
| 7520.25 | 9 | 0 | 1 | 0 | 0 |
| 12,135.00 | 11 | 0 | 1 | 0 | 0 |
| 13,016.55 | 17 | 0 | 1 | 0 | 0 |
| 13,673.90 | 20 | 0 | 1 | 0 | 0 |
| 3272.05 | 3 | 0 | 0 | 1 | 0 |
| 5074.40 | 9 | 0 | 0 | 1 | 0 |
| 7505.45 | 11 | 0 | 0 | 1 | 0 |
| 8272.75 | 17 | 0 | 0 | 1 | 0 |
| 10,020.40 | 20 | 0 | 0 | 1 | 0 |
| 4925.75 | 3 | 0 | 0 | 0 | 1 |
| 10,018.10 | 9 | 0 | 0 | 0 | 1 |
| 12,505.85 | 11 | 0 | 0 | 0 | 1 |
| 15,329.05 | 17 | 0 | 0 | 0 | 1 |
| 19,477.20 | 20 | 0 | 0 | 0 | 1 |

a. How many indicator variables should be included in the multiple regression model relating weekly sales to the number of sales reps and the quarter of the year? Explain why.

b. What sign would you expect the coefficient for the sales reps variable to have? Explain your reasoning.

c. Using statistical software, estimate the following multiple regression model.

$$\text{Sales} = \beta_0 + \beta_1\left(\text{Reps}\right) + \beta_2\left(\text{Quarter 1}\right) + \beta_3\left(\text{Quarter 2}\right) + \beta_4\left(\text{Quarter 3}\right) + \varepsilon$$

Write the estimated multiple regression equation.

d. Interpret the coefficient of the indicator variable representing the first quarter.

e. Is there sufficient evidence that sales in the second quarter tend to be different from the sales in the fourth quarter? Use $\alpha = 0.05$.

f. What concerns should we have when predicting weekly sales using this model?

# 14.6 Additional Topics in Multiple Regression

When fitting multiple regression models, there are some potential problems that one should be aware of and that one must consider. One of the problems that must be considered is that of **multicollinearity**. Multicollinearity is when some of the independent variables ($x_1$, $x_2$, ..., $x_k$) are similar to each other (i.e., highly correlated). When this is the case, the individual regression coefficients are not properly estimated because they are so similar to each other. That is, since you have two or more variables that are related, perhaps these variables are measuring the same information, and thus you only need one of these variables in the model to explain the dependent variable. Thus, omitting highly correlated variables will help minimize the risk of multicollinearity. Multicollinearity can be detected by studying a correlation matrix for the variables in question, which is easily generated using technology such as Microsoft Excel or Minitab. If the correlation coefficient between two of the independent variables is close to 1 or −1, it is likely that multicollinearity could be obscuring the regression results.

---

### Procedure

#### Potential Problems in Multiple Regression

- **Multicollinearity**: Multicollinearity occurs when two or more independent variables are similar to each other or highly correlated.

- **Parameter Estimability**: Parameter estimability is the inability to estimate the parameters of the regression model because the data are concentrated in one area.

- **Variable Selection**: Variable selection is the process by which variables are added (or removed) from the regression model to meet a specified criteria with the goal of obtaining the best (relatively speaking) multiple regression model.

- **Stepwise Regression**: Stepwise regression is a step-by-step iterative procedure that is used to build multiple regression model by adding one independent variable at a time until the final model is obtained. This procedure involves adding (or removing) independent variables after each iteration.

- **Extrapolation**: Extrapolation is the process by which the analyst uses the multiple regression model to make predictions outside the range of values of the independent variables.

- **Correlated Errors**: Correlated errors occur in regression when the responses (the dependent variables) are correlated (i.e., there is a relationship between the responses in the regression model).

---

Another concern in multiple regression is **parameter estimability**. Parameter estimability is the inability to estimate the parameters of the regression model because the data are concentrated in one area. To ensure that the parameters are estimable, the data must include at least one more level of the independent variable than the highest order of the independent variable included in the model. This means that the sample size, $n$, must exceed $k + 1$ to ensure that the degrees of freedom are not equal to zero.

Another concern or pitfall of multiple regression involves **variable selection**. The problem of variable selection arises when you have a large number of independent variables and need to decide which ones to include in the model. As we have seen in earlier sections, the $R^2$ value increases as more variables are added to the model. However, the quality of the model may degrade as more variables are added to the model because information is being wasted when estimating unnecessary parameters. Therefore, it would be best to include only important variables in the model or variables that are clearly necessary. A solution to avoid over-fitting a model is to utilize a model-building procedure such as **stepwise regression**. Stepwise regression involves selecting independent variables using an automated procedure, which is beyond the scope of this text.

Some other issues with fitting multiple regression models are **extrapolation** (which was discussed in Chapter 13) and **correlated errors**. Extrapolation can be a concern when the regression model is used to predict values outside the range of the data used to estimate the model. Be sure to only use the model within an appropriate range of $x$-values. The problem with correlated errors arises when measurements of the dependent variable are correlated. That is, since the observations (the responses) of the regression model are assumed to be independent, it is problematic if there is a relationship between responses. One will often see this type of dependency with time series data. Current measurements are often dependent on measurements in the previous time period.

## ✎ 14.6 Exercises

### Basic Concepts

1. Define multicollinearity. How can you detect if multicollinearity exists in a regression model?

2. Why is multicollinearity a concern when performing regression analysis?

3. How can you attempt to correct the problem of multicollinearity?

4. What is parameter estimability?

5. How can you alleviate concerns about parameter estimability?

6. Why is variable selection difficult when building a multiple regression model?

7. Does the $R^2$ value always increase as additional variables are added? Does this mean that adding additional variables always produces a more useful model? Explain.

8. What is extrapolation? Why is this a concern?

9. With what type of data do you often encounter issues with correlated errors?

### Exercises

10. In Exercise 8 of Section 14.4, we modeled the relationship between total points and rushing yards, passing yards, and first downs.

    a. Using the correlation matrix below, discuss whether collinearity might play a role in estimating total points using rushing yards, passing yards, and first downs in the model.

| Correlation Matrix | | | |
|---|---|---|---|
| | **Rushing Yards** | **Passing Yards** | **First Downs** |
| **Rushing Yards** | 1.0000 | −0.1943 | 0.3789 |
| **Passing Yards** | | 1.0000 | 0.5744 |
| **First Downs** | | | 1.0000 |

    b. How would you determine if there is a relationship (and if so, the strength of such relationship) between the independent variables in the model?

    c. Given the multiple regression model that was fit in Exercise 8, what would the total points be if a team had 30 rushing yards, 100 passing yards, and 5 first downs?

    d. Should you have any concerns about the estimate in part **c.**? Explain your answer.

11. In Exercise 10 of Section 14.5, salary was modeled as a function of age, experience, and gender.

    a. Discuss how collinearity might play a role in estimating salary using age, experience, and gender in the model.

    b. How would you determine if there is a relationship (and if so, the strength of such relationship) between the independent variables in the model?

    c. Given the multiple regression model that was fit in the problem, what would the expected salary be for a 60-year-old male employee with 25 years of experience?

    d. Should you have any concerns about the estimate in part c.? Explain your answer.

12. In Exercise 11 of Section 14.5, we attempted to predict the number of crimes on a college/university campus based on the number of police, the enrollment at the university, and if it was a private institution.

    a. Examine the correlation matrix below and discuss whether collinearity might play a role in estimating the number of crimes based on the number of police, enrollment, and if the institution is private.

| Correlation Matrix | | | |
|---|---|---|---|
| | Police | Enrollment | Private |
| Police | 1.0000 | 0.8042 | −0.4942 |
| Enrollment | | 1.0000 | −0.8795 |
| Private | | | 1.0000 |

    b. How would you determine if there is a relationship (and if so, the strength of such relationship) between the independent variables in the model?

    c. Given the multiple regression model that was fit in the problem, what would the expected number of crimes be for a private university with a police force of 100 officers and an enrollment of 50,000?

    d. Should you have any concerns about the estimate in part c.? Explain your answer.

13. Suppose you fit a multiple regression model of the form

$$y_i = \beta_0 + \beta_1 x_{1i} + \beta_2 x_{2i} + \beta_3 x_{3i} + \beta_4 x_{4i} + \varepsilon_i$$

The correlation matrix for the pairs of independent variables is given in the following table. Discuss if you detect multicollinearity between any of the variables.

| Correlation Matrix | | | |
|---|---|---|---|
| | $x_1$ | $x_2$ | $x_3$ | $x_4$ |
| $x_1$ | 1.00 | 0.18 | 0.86 | 0.45 |
| $x_2$ | | 1.00 | 0.35 | 0.22 |
| $x_3$ | | | 1.00 | 0.50 |
| $x_4$ | | | | 1.00 |

# Discovering Technology

## Using Excel

Regression

1. Use the data in Table 14.1.1 to perform multiple regression analysis.

|  | A | B | C | D |
|---|---|---|---|---|
| 1 | | Table 14.1.1 – Pizza Delivery Time Data | | |
| 2 | Observation | Delivery Time (Minutes) | Number of Pizzas | Distance (Miles) |
| 3 | 1 | 16.68 | 7 | 5.6 |
| 4 | 2 | 11.5 | 3 | 2.2 |
| 5 | 3 | 12.03 | 3 | 3.4 |
| 6 | 4 | 14.88 | 8 | 0.8 |
| 7 | | ... | | |
| 27 | 25 | 10.75 | 4 | 1.5 |

**Data**

This data set can be found at stat.hawkeslearning.com by navigating to **Discovering Business Statistics, Second Edition > Data Sets > Pizza Delivery Time**.

2. Under the **Data** tab, choose **Data Analysis**, and **Regression**.

3. Select the delivery time for each pizza order in Column B (**$B$2:$B$27**) for Input Y Range and the number of pizzas in each order as well as the distance from the pizza restaurant to each customer's locations in Columns C and D (**$C$2:$D$27**) for Input X Range.

4. Click the checkbox beside **Labels** so the titles of the columns will be included in the regression output in Excel.

5. Select the radio button next to **Output Range** and enter **$F$1**.

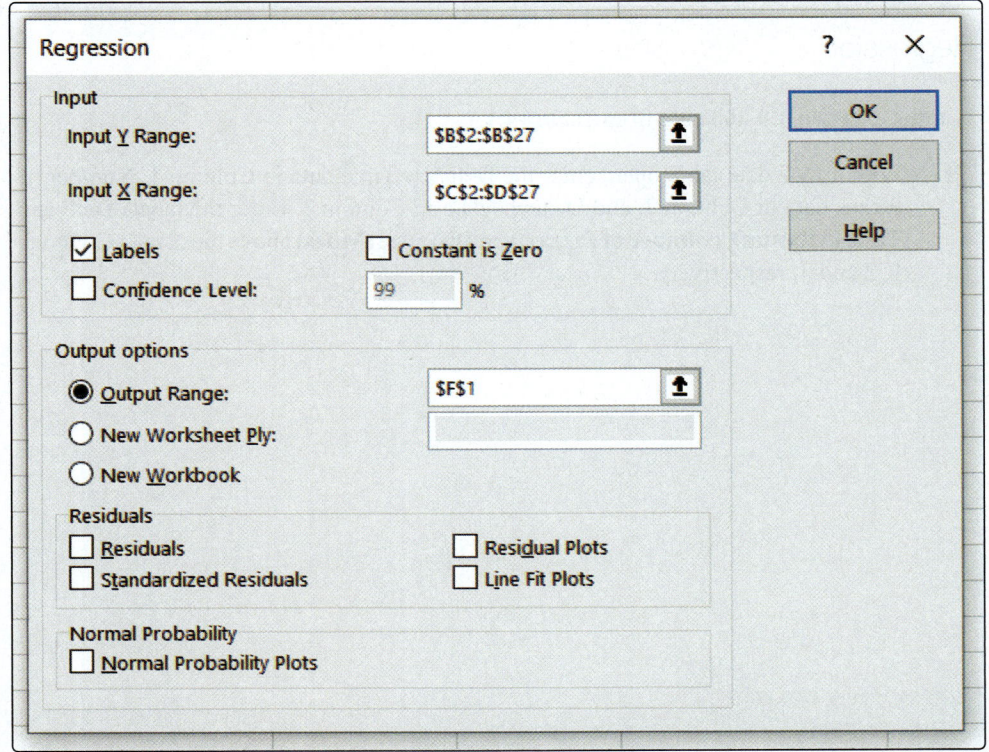

6.   Press **OK**. Highlight columns **F** to **L** and double-click the vertical bar separating column **D** from column **E** to expand the columns. Observe the summary output. The Regression Statistics table contains the coefficient of determination, standard error of the model, and the number of observations. The ANOVA table contains the degrees of freedom, sums of squares, and mean squares values for regression and error (residual). Finally, the least squares estimates of the coefficients are given along with their standard errors, $t$-statistics, and $P$-values. Notice that the 95% confidence intervals for $\beta_0$ and $\beta_1$ are given in the *Lower 95%* and *Upper 95%* columns of the output.

SUMMARY OUTPUT

| *Regression Statistics* | |
|---|---|
| Multiple R | 0.981846095 |
| R Square | 0.964021754 |
| Adjusted R Square | 0.960751004 |
| Standard Error | 3.075694294 |
| Observations | 25 |

ANOVA

| | df | SS | MS | F | Significance F |
|---|---|---|---|---|---|
| Regression | 2 | 5576.424901 | 2788.212451 | 294.7403048 | 1.30749E-16 |
| Residual | 22 | 208.1176986 | 9.45989539 | | |
| Total | 24 | 5784.5426 | | | |

| | Coefficients | Standard Error | t Stat | P-value | Lower 95% | Upper 95% |
|---|---|---|---|---|---|---|
| Intercept | 1.792903307 | 1.048779122 | 1.709514681 | 0.101425505 | -0.382131469 | 3.967938082 |
| Number of Pizzas | 1.589101026 | 0.156345691 | 10.16402191 | 8.97113E-10 | 1.264859909 | 1.913342144 |
| Distance | 1.567708058 | 0.327534513 | 4.786390423 | 8.84845E-05 | 0.888443052 | 2.246973064 |

## Using JMP

Regression

For this exercise, use the data from Table 14.1.1.

1.   With a JMP data table open, enter the Delivery Time data in Column 1, Number of Pizzas data in Column 2, and Distance data in Column 3. Enter the labels **Delivery Time (Minutes)**, **Number of Pizzas**, and **Distance (Miles)** above the data in Columns 1, 2 and 3, respectively.

| | Delivery Time (Minutes) | Number of Pizzas | Distance (Miles) |
|---|---|---|---|
| 1 | 16.68 | 7 | 5.6 |
| 2 | 11.5 | 3 | 2.2 |
| 3 | 12.03 | 3 | 3.4 |
| 4 | 14.88 | 8 | 0.8 |
| 5 | 13.75 | 6 | 1.5 |
| 6 | 18.11 | 7 | 3.3 |
| 7 | 8 | 2 | 1.1 |
| 8 | 17.83 | 7 | 2.1 |
| 9 | 79.24 | 30 | 14.6 |
| 10 | 21.5 | 5 | 6.05 |
| 11 | 40.33 | 16 | 6.88 |
| 12 | 21 | 10 | 2.15 |
| 13 | 13.5 | 4 | 2.55 |
| 14 | 19.75 | 6 | 4.62 |
| 15 | 24 | 9 | 4.48 |
| 16 | 29 | 10 | 7.76 |
| 17 | 15.35 | 6 | 2 |
| 18 | 19 | 7 | 1.32 |
| 19 | 9.5 | 3 | 0.36 |
| 20 | 35.1 | 17 | 7.7 |
| 21 | 17.9 | 10 | 1.4 |
| 22 | 52.32 | 26 | 8.1 |
| 23 | 18.75 | 9 | 4.5 |
| 24 | 19.83 | 8 | 6.35 |
| 25 | 10.75 | 4 | 1.5 |

**data**

**Columns (3/0)**
- Delivery ... (Minutes)
- Number of Pizzas
- Distance (Miles)

**Rows**

| | |
|---|---|
| All rows | 25 |
| Selected | 0 |
| Excluded | 0 |
| Hidden | 0 |
| Labelled | 0 |

2. Select **Analyze** in the top row of the JMP spreadsheet and then select **Fit Model**.

3. From the **Select Columns** box, click on **Delivery Time (Minutes)** (or Column 1), then click on **Y** under the heading Pick Role Variables. Click on **Number of Pizzas** (or Column 2), then click on **Add** under the heading Construct Model Effects. Click on **Distance (Miles)** (or Column 3), then click on **Add** again. Click **Run**.

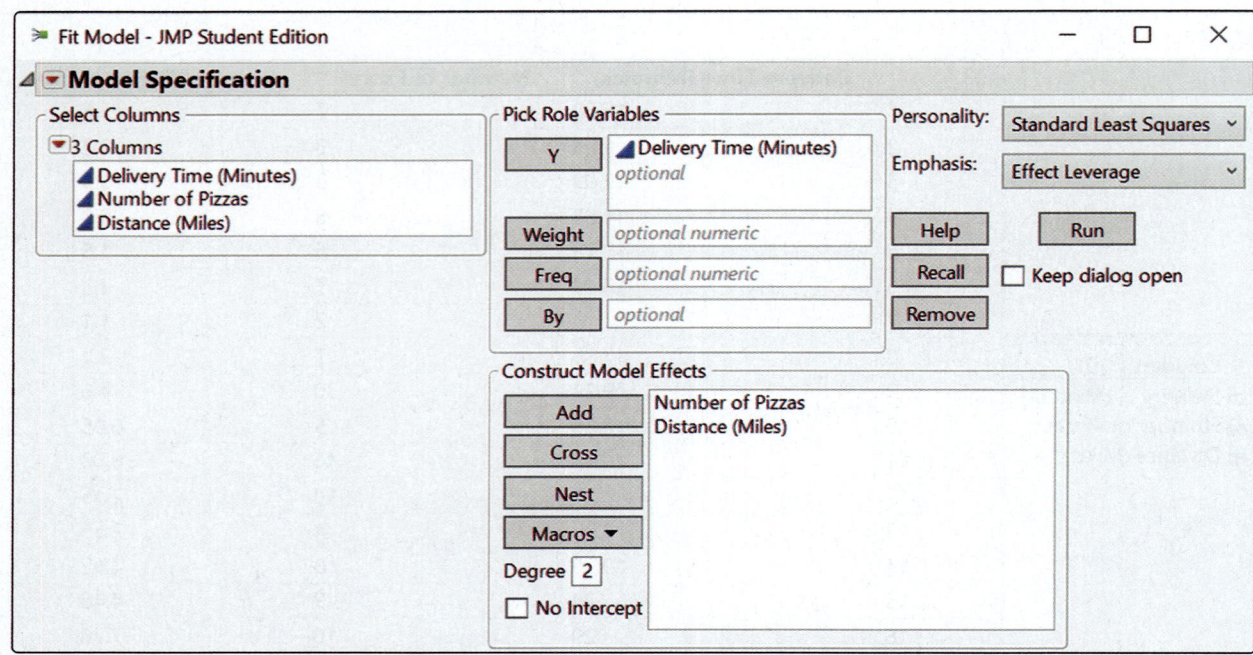

4. Observe the output screen for the results.. By default, JMP will produce the regression equation (under Linear Fit), Summary of Fit, the ANOVA table, and the Parameter Estimates.

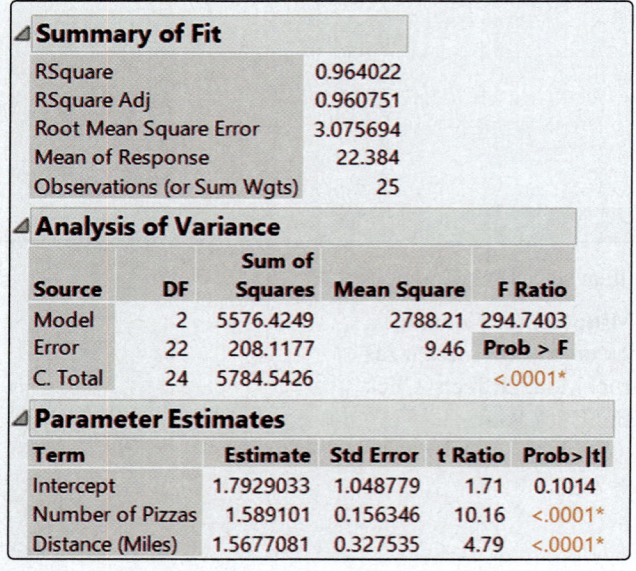

### Summary of Fit

| | |
|---|---|
| RSquare | 0.964022 |
| RSquare Adj | 0.960751 |
| Root Mean Square Error | 3.075694 |
| Mean of Response | 22.384 |
| Observations (or Sum Wgts) | 25 |

### Analysis of Variance

| Source | DF | Sum of Squares | Mean Square | F Ratio |
|---|---|---|---|---|
| Model | 2 | 5576.4249 | 2788.21 | 294.7403 |
| Error | 22 | 208.1177 | 9.46 | Prob > F |
| C. Total | 24 | 5784.5426 | | <.0001* |

### Parameter Estimates

| Term | Estimate | Std Error | t Ratio | Prob>|t| |
|---|---|---|---|---|
| Intercept | 1.7929033 | 1.048779 | 1.71 | 0.1014 |
| Number of Pizzas | 1.589101 | 0.156346 | 10.16 | <.0001* |
| Distance (Miles) | 1.5677081 | 0.327535 | 4.79 | <.0001* |

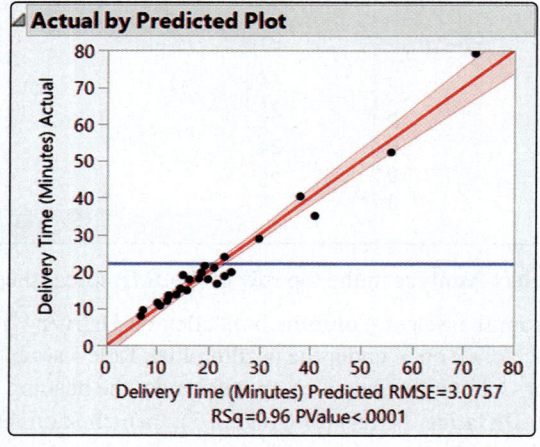

Confidence Intervals, Predicted Values and Mean Confidence Interval

For this exercise, use the data from Table 14.1.1.

1. Enter the labels **Delivery Time (Minutes)**, **Number of Pizzas**, and **Distance (Miles)** in Columns 1, 2 and 3, respectively. With a JMP data table open, enter the data for Delivery Time in Column 1, the data for Number of Pizzas in Column 2, and the data for Distance in Column 3. Enter the values for **Number of Pizzas** and **Distance** that you want to predict the Delivery Time for in the last row of the data table in the appropriate columns. Leave the **Delivery Time** column blank.

| | Delivery Time (Minutes) | Number of Pizzas | Distance (Miles) |
|---|---|---|---|
| 1 | 16.68 | 7 | 5.6 |
| 2 | 11.5 | 3 | 2.2 |
| 3 | 12.03 | 3 | 3.4 |
| 4 | 14.88 | 8 | 0.8 |
| 5 | 13.75 | 6 | 1.5 |
| 6 | 18.11 | 7 | 3.3 |
| 7 | 8 | 2 | 1.1 |
| 8 | 17.83 | 7 | 2.1 |
| 9 | 79.24 | 30 | 14.6 |
| 10 | 21.5 | 5 | 6.05 |
| 11 | 40.33 | 16 | 6.88 |
| 12 | 21 | 10 | 2.15 |
| 13 | 13.5 | 4 | 2.55 |
| 14 | 19.75 | 6 | 4.62 |
| 15 | 24 | 9 | 4.48 |
| 16 | 29 | 10 | 7.76 |
| 17 | 15.35 | 6 | 2 |
| 18 | 19 | 7 | 1.32 |
| 19 | 9.5 | 3 | 0.36 |
| 20 | 35.1 | 17 | 7.7 |
| 21 | 17.9 | 10 | 1.4 |
| 22 | 52.32 | 26 | 8.1 |
| 23 | 18.75 | 9 | 4.5 |
| 24 | 19.83 | 8 | 6.35 |
| 25 | 10.75 | 4 | 1.5 |
| 26 | • | 5 | 6 |

data

Columns (3/0)
- Delivery ... (Minutes)
- Number of Pizzas
- Distance (Miles)

Rows
| All rows | 26 |
| Selected | 0 |
| Excluded | 0 |
| Hidden | 0 |
| Labelled | 0 |

2. Select **Analyze** in the top row of the JMP spreadsheet and then select **Fit Model**.

3. From the **Select Columns** box, click on **Delivery Time (Minutes)** (or Column 1), then click on **Y** under the heading Pick Role Variables. Click on **Number of Pizzas** (or Column 2), then click on **Add** under the heading Construct Model Effects. Click on **Distance (Miles)** (or Column 3), then click on **Add** again. Click **Run**.

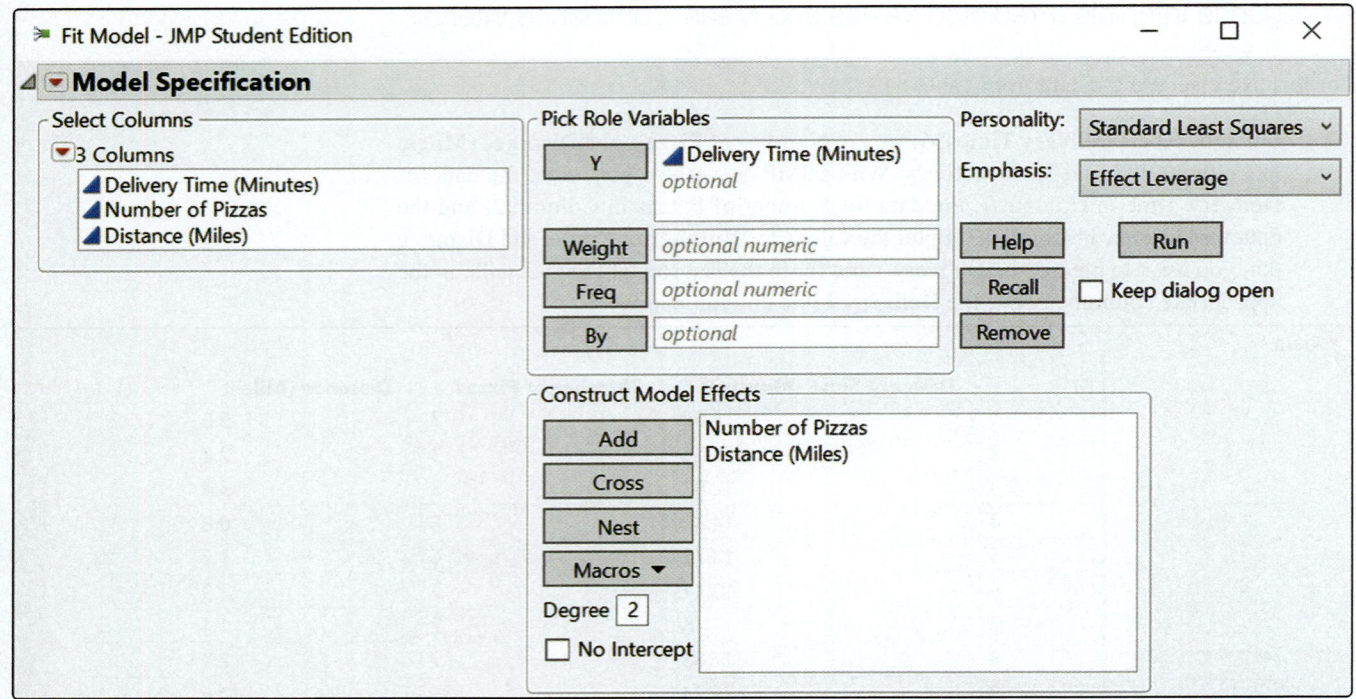

4.  Observe the output screen for the results. By default, JMP will produce the Summary of Fit, the ANOVA table, and the Parameter Estimates. Click on the red triangle next to **Response Delivery Time (Minute)**, click on **Save Columns** and then:

- Click on **Indiv Confidence Interval** to get a 95% Confidence Interval for each estimate.

- Click on **Predicted Values** to get the predicted Delivery Times from the model.

- Click on **Mean Confidence Interval** to get confidence intervals for the mean Delivery Time.

| | Delivery Time (Minutes) | Number of Pizzas | Distance (Miles) | Lower 95% Indiv Delivery Time (Minutes) | Upper 95% Indiv Delivery Time (Minutes) | Predicted Delivery Time (Minutes) | Lower 95% Mean Delivery Time (Minutes) | Upper 95% Mean Delivery Time (Minutes) |
|---|---|---|---|---|---|---|---|---|
| 1 | 16.68 | 7 | 5.6 | 15.004033733 | 28.387517497 | 21.695775615 | 19.672689593 | 23.718861637 |
| 2 | 11.5 | 3 | 2.2 | 3.4012647437 | 16.617063482 | 10.009164113 | 8.2835332981 | 11.734794928 |
| 3 | 12.03 | 3 | 3.4 | 5.2000802277 | 18.580747337 | 11.890413783 | 9.8719909548 | 13.90883661 |
| 4 | 14.88 | 8 | 0.8 | 8.9534262641 | 22.56632966 | 15.759877962 | 13.384745891 | 18.135010033 |
| 5 | 13.75 | 6 | 1.5 | 7.0746623055 | 20.283480796 | 13.679071551 | 11.96685395 | 15.391289152 |
| 6 | 18.11 | 7 | 3.3 | 11.574686738 | 24.605407424 | 18.090047081 | 16.762120865 | 19.417973298 |
| 7 | 8 | 2 | 1.1 | 0.0539151447 | 13.337253301 | 6.6955842228 | 4.8448449707 | 8.5463234749 |
| 8 | 17.83 | 7 | 2.1 | 9.6382252288 | 22.779369594 | 16.208797412 | 14.632127996 | 17.785466827 |
| 9 | 79.24 | 30 | 14.6 | 64.534193711 | 80.174749767 | 72.354471739 | 67.83006945 | 76.878874028 |
| 10 | 21.5 | 5 | 6.05 | 12.26409293 | 26.181991447 | 19.223042189 | 16.440877129 | 22.005207248 |
| 11 | 40.33 | 16 | 6.88 | 31.360065907 | 44.648636422 | 38.004351165 | 36.144245195 | 39.864457135 |
| 12 | 21 | 10 | 2.15 | 14.351890944 | 27.757080842 | 21.054485893 | 18.995785651 | 23.113186135 |
| 13 | 13.5 | 4 | 2.55 | 5.5700226808 | 18.723903238 | 12.146962959 | 10.543962316 | 13.749963603 |
| 14 | 19.75 | 6 | 4.62 | 11.945638739 | 25.195002644 | 18.570320692 | 16.78150024 | 20.359141144 |
| 15 | 24 | 9 | 4.48 | 16.608703862 | 29.627585422 | 23.118144642 | 21.81957354 | 24.416715744 |
| 16 | 29 | 10 | 7.76 | 22.976744419 | 36.721911778 | 29.849328099 | 27.290837515 | 32.407818683 |
| 17 | 15.35 | 6 | 2 | 7.9011601678 | 21.024690991 | 14.46292558 | 12.923369489 | 16.00248167 |
| 18 | 19 | 7 | 1.32 | 8.3260579194 | 21.645912333 | 14.985985126 | 13.070757193 | 16.90121306 |
| 19 | 9.5 | 3 | 0.36 | 0.448928363 | 13.800234209 | 7.1245812859 | 5.1553658117 | 9.0937967602 |
| 20 | 35.1 | 17 | 7.7 | 34.18492507 | 47.573020527 | 40.878972798 | 38.848272763 | 42.909672834 |
| 21 | 17.9 | 10 | 1.4 | 13.036398569 | 26.72101113 | 19.878704849 | 17.402695383 | 22.354714316 |
| 22 | 52.32 | 26 | 8.1 | 48.32206445 | 63.293866064 | 55.807965257 | 51.88977964 | 59.726150874 |
| 23 | 18.75 | 9 | 4.5 | 16.63953869 | 29.659458917 | 23.149498803 | 21.848326903 | 24.450670704 |
| 24 | 19.83 | 8 | 6.35 | 17.715828445 | 31.205486925 | 24.460657685 | 22.268355444 | 26.652959925 |
| 25 | 10.75 | 4 | 1.5 | 3.911370567 | 17.09036843 | 10.500869498 | 8.8470969017 | 12.154642095 |
| 26 | • | 5 | 6 | 12.197133492 | 26.09218008 | 19.144656786 | 16.391195772 | 21.8981178 |

data

Columns (8/0)
Delivery ... (Minutes)
Number of Pizzas
Distance (Miles)
Lower 95...(Minutes)
Upper 95...(Minutes)
Predicted...(Minutes)
Lower 95...(Minutes)
Upper 95...(Minutes)

Rows
All rows    26
Selected    0
Excluded    0
Hidden    0
Labelled    0

## Using Minitab

### Regression

Use the data in Table 14.1.1 to perform multiple regression analysis.

1. Enter the data of Table 14.1.1 into the Minitab worksheet. Enter **Delivery Time** for column C1, **Number of Pizzas** for column C2, and **Distance** for column C3.

2. Choose **Stat, Regression, Regression** and **Fit Regression Model**.

3. Enter **C1** (Delivery Time) in the Response box, and **C2 C3** in the Continuous predictors box. Press **OK**.

4. Observe the output for the regression analysis. The estimated regression equation is written first. Then, the estimated values of the coefficients are reported along with their corresponding standard errors, $t$-statistics, and $P$-values. Then the standard error of the model is given, S, along with the coefficient of determination, R-Sq, and the adjusted coefficient of determination, R-Sq(adj). Finally, the ANOVA table is given with the degrees of freedom for regression and error, along with SSR, SSE, TSS, MSR, MSE, and the $F$-statistic. The P-Value column contains the $P$-value corresponding to the $F$-statistic. This is the $P$-value we are interested in when testing the overall model for significance.

### Regression Analysis: Delivery Time versus Number of Pizzas, Distance

**Regression Equation**

Delivery Time = 1.79 + 1.589 Number of Pizzas + 1.568 Distance

**Coefficients**

| Term | Coef | SE Coef | T-Value | P-Value | VIF |
|---|---|---|---|---|---|
| Constant | 1.79 | 1.05 | 1.71 | 0.101 | |
| Number of Pizzas | 1.589 | 0.156 | 10.16 | 0.000 | 2.88 |
| Distance | 1.568 | 0.328 | 4.79 | 0.000 | 2.88 |

**Model Summary**

| S | R-sq | R-sq(adj) | R-sq(pred) |
|---|---|---|---|
| 3.07569 | 96.40% | 96.08% | 92.98% |

**Analysis of Variance**

| Source | DF | Adj SS | Adj MS | F-Value | P-Value |
|---|---|---|---|---|---|
| Regression | 2 | 5576.4 | 2788.21 | 294.74 | 0.000 |
| Number of Pizzas | 1 | 977.3 | 977.28 | 103.31 | 0.000 |
| Distance | 1 | 216.7 | 216.72 | 22.91 | 0.000 |
| Error | 22 | 208.1 | 9.46 | | |
| Total | 24 | 5784.5 | | | |

**Fits and Diagnostics for Unusual Observations**

| Obs | Delivery Time | Fit | Resid | Std Resid | | |
|---|---|---|---|---|---|---|
| 9 | 79.24 | 72.35 | 6.89 | 3.18 | R | X |
| 22 | 52.32 | 55.81 | -3.49 | -1.44 | | X |

R Large residual
X Unusual X

## Confidence and Prediction Intervals

For this exercise, use the production data from Table 14.1.1.

1. Enter the data of Table 14.1.1 into the Minitab worksheet. Enter **Delivery Time** for column C1, **Number of Pizzas** for column C2, and **Distance** for column C3.

2. Choose **Stat, Regression, Regression and Fit Regression Model**. (Note the Predict option is gray until steps 4 and 5 are completed)

3. Enter **C1** (Delivery Time) in the Response box, and **C2 C3** in the Continuous predictors box. Press **OK**.

4. In the Regression dialog box, choose Options and set the Confidence Level to **99**. Click **OK**. Click **OK**. This will produce Regression Analysis results which must be set so that predictions can be made.

5. Choose again **Stat, Regression, Regression** and then **Predict**. The Predict dialog box will appear provided steps 4 and 5 are completed.

6. In the Predict dialog box, enter **5** for number of Pizzas and, **6** for Distance. This tells Minitab to compute a point estimate and confidence and prediction intervals for delivering 5 pizzas to a location 6 miles away. Choose Options and Set the Confidence level at **99**. Press **OK** to exit the Option box. Press **OK** to exit the Predict box.

7. Observe the predicted value, along with its confidence and prediction interval.

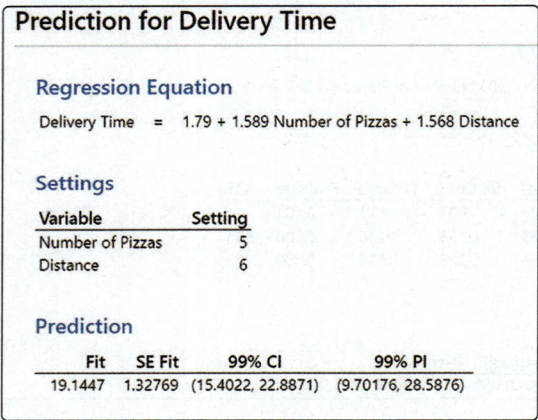

**Prediction for Delivery Time**

**Regression Equation**

Delivery Time  =   1.79 + 1.589 Number of Pizzas + 1.568 Distance

**Settings**

| Variable | Setting |
|---|---|
| Number of Pizzas | 5 |
| Distance | 6 |

**Prediction**

| Fit | SE Fit | 99% CI | 99% PI |
|---|---|---|---|
| 19.1447 | 1.32769 | (15.4022, 22.8871) | (9.70176, 28.5876) |

# R Chapter 14 Review

## Key Terms and Ideas

- Multiple Regression
- Multiple Regression Model
- Method of Least Squares
- Estimated Multiple Regression Equation
- Coefficient of Determination (Multiple Coefficient of Determination)
- Adjusted $R^2$
- $F$-Distribution
- Numerator Degrees of Freedom
- Denominator Degrees of Freedom
- $F$-Statistic
- Sum of Squares of Regression
- Sum of Squared Errors
- Total Sum of Squares
- Mean Square Regression
- Mean Square Error
- Calculating Degrees of Freedom in Multiple Regression Models
- Hypothesis Tests Concerning Individual Coefficients
- Test Statistic for Testing the Hypothesis $\beta_i \neq 0$
- Confidence Intervals for Individual Coefficients
- Confidence Interval for the Mean Value of $y$ Given $x$
- Confidence Interval for the Predicted Value of $y$ Given $x$
- Indicator (Dummy) Variable
- Base Level Variable
- Interaction Terms
- Polynomial (Nonlinear) Regression Models
- Multicollinearity
- Parameter Estimability
- Variable Selection
- Stepwise Regression
- Extrapolation
- Correlated Errors

## Key Formulas

| | Section |
|---|---|

**Multiple Regression Model**

$$y_i = \beta_0 + \beta_1 x_{1i} + \beta_2 x_{2i} + \ldots + \beta_k x_{ki} + \varepsilon_i$$

14.1

where $\beta_0$, $\beta_1$, $\beta_2$, ..., $\beta_k$ are the model's parameters, $x_{1i}$, $x_{2i}$, ..., $x_{ki}$ are the independent variables, and $\varepsilon_i$ is a random error.

**Estimated Multiple Regression Equation**

$$\hat{y}_i = b_0 + b_1 x_{1i} + b_2 x_{2i} + \cdots + b_k x_k$$

14.1

where $b_0$, $b_1$, $b_2$, ..., $b_k$ are estimates of their population counterparts.

## Key Formulas (cont.)

|  | Section |
|---|---|

**Sum of Squared Errors**

$$\text{SSE} = \sum \left(y_i - \hat{y}_i\right)^2$$

14.1

**Coefficient of Determination**

$$R^2 = \frac{\text{SSR}}{\text{TSS}} = 1 - \frac{\text{SSE}}{\text{TSS}}$$

14.2

**Adjusted $R^2$**

$$R_a^2 = 1 - \left(\frac{n-1}{n-k-1}\right)\frac{\text{SSE}}{\text{TSS}}$$

14.2

where $n$ is the number of observations and $k$ is the number of independent variables in the model.

**$F$-Statistic**

$$F = \frac{\dfrac{\text{Sum of Squares of Regression}}{k}}{\dfrac{\text{Sum of Squared Errors}}{n-(k+1)}}$$

14.5

$$= \frac{\dfrac{\text{SSR}}{k}}{\dfrac{\text{SSE}}{n-(k+1)}} = \frac{\text{Mean Square Regression}}{\text{Mean Square Error}}$$

**Test Statistic for Testing the Hypothesis $\beta_i \neq 0$**

$$t = \frac{b_i - 0}{s_{b_i}} = \frac{b_i}{s_{b_i}}$$

14.5

where $b_i$ is the estimated coefficient and $s_{b_i}$ is the standard deviation (standard error) of the estimated coefficient.

**$100(1-\alpha)\%$ Confidence Interval for an Individual Coefficient, $\beta_i$**

$$b_i \pm t_{\alpha/2, df} s_{b_i}$$

14.5

# AE    Additional Exercises

1.   Drew is undecided about whether to go back to school and get his master's degree. He is trying to perform a cost-benefit analysis to determine whether the cost of attending the school of his choice will be outweighed by the increase in salary he will receive after he attains his degree. He does research and compiles data on annual salaries in the industry he currently works in (he has been working for 10 years), along with the years of experience for each employee and whether or not the employee has a master's degree. Earning his master's degree will require him to take out approximately $20,000 worth of student loans. He has decided that if the multiple regression model shows, with 95% confidence, that earning a master's degree is significant in predicting annual salary, and the estimated increase in salary is at least $10,000, he will enroll in a degree program.

**⁝ Data**

This data set can be found at stat.hawkeslearning.com by navigating to **Discovering Business Statistics, Second Edition > Data Sets > Industry Salaries**.

| Industry Salaries | | |
|---|---|---|
| **Salary ($)** | **Years of Experience** | **Master's Degree** |
| 37,620 | 22 | No |
| 67,080 | 27 | Yes |
| 31,280 | 15 | No |
| 21,500 | 2 | No |
| 75,120 | 28 | Yes |
| 59,820 | 25 | Yes |
| 40,180 | 15 | Yes |
| 81,360 | 32 | Yes |
| 35,080 | 19 | No |
| 36,080 | 12 | Yes |
| 36,680 | 22 | No |
| 29,200 | 11 | Yes |
| 33,040 | 18 | No |
| 30,060 | 14 | No |
| 53,300 | 21 | Yes |
| 22,820 | 7 | No |
| 72,900 | 31 | Yes |
| 55,920 | 22 | Yes |
| 19,280 | 0 | No |
| 26,000 | 7 | No |

a.   Create an indicator variable, degree, that is equal to 1 if the employee has a master's degree and equal to 0 if the employee does not have a master's degree.

b.   Using statistical software, estimate the following multiple regression model.

$$\text{Salary} = \beta_0 + \beta_1 \left(\text{Experience}\right) + \beta_2 \left(\text{Degree}\right) + \varepsilon_i$$

Write the estimated multiple regression equation.

c.   According to the model, how much does salary increase on average with each additional year of experience?

d.   According to this model, will Drew decide to enroll in a master's program? Explain your answer.

e.   Why should Drew be cautious when using this model to make his decision?

2. A chain of sports clubs wishes to use regression analysis to help determine which features should be included in their new location. They believe that median income in the area is a significant factor in determining the number of people who join a neighborhood sports club. The CEO of the chain gathered data from existing sports clubs regarding the number of members each club had, the median income in the area in which they were located, and whether or not the clubs had a pool, racquetball courts, or group fitness classes. If management can determine with 90% confidence that a pool, racquetball courts, or group fitness classes produces significantly more memberships than sports clubs without those features, they will include them in the new location.

| Sports Club Membership | | | | |
|---|---|---|---|---|
| Number of Members | Median Income ($) | Pool? | Racquetball Courts? | Fitness Classes? |
| 1258 | 32,223 | No | No | No |
| 1479 | 34,975 | No | No | No |
| 1480 | 43,187 | No | Yes | No |
| 1701 | 44,337 | No | No | No |
| 2014 | 52,167 | No | No | Yes |
| 2271 | 57,521 | No | No | Yes |
| 2615 | 58,347 | No | Yes | No |
| 2632 | 60,960 | Yes | No | No |
| 2737 | 62,201 | Yes | No | Yes |
| 2810 | 67,993 | No | No | Yes |
| 3563 | 68,770 | No | No | Yes |
| 3765 | 81,289 | Yes | Yes | Yes |
| 3792 | 83,902 | No | No | Yes |
| 4069 | 84,594 | Yes | No | Yes |
| 4393 | 86,855 | Yes | Yes | Yes |
| 4787 | 88,381 | Yes | Yes | Yes |

a. What sign do you expect the coefficient of median income to have? Explain why.

b. Create three dummy variables, pool, courts, and classes, that are equal to 1 if the observation contains this feature and equal to 0 if the observation does not contain this feature.

c. Use statistical software to estimate the following regression models. In each case, write the estimated regression equation and state whether the coefficient of the independent variable is significant at the 0.10 level.

   i.   $\text{Members} = \beta_0 + \beta_1(\text{Pool}) + \varepsilon_i$

   ii.  $\text{Members} = \beta_0 + \beta_1(\text{Courts}) + \varepsilon_i$

   iii. $\text{Members} = \beta_0 + \beta_1(\text{Classes}) + \varepsilon_i$

d. Estimate the following multiple regression model.

   $\text{Members} = \beta_0 + \beta_1(\text{Income}) + \beta_2(\text{Pool}) + \beta_3(\text{Courts}) + \beta_4(\text{Classes}) + \varepsilon$

   Write the estimated regression equation.

e. Are any of the coefficients of the indicator variables significant at the 0.10 level?

f. Explain why it is important to include the income variable in the regression model.

g. After studying these regression results, how would you suggest the management of the sports club chain go about building their new location? Should they use any of the regression models you have estimated? Explain why or why not.

3. The amount of a certain additive injected into a chemical process has a direct effect on the yield. The following table contains data on the amount of additive and yield.

| Amount of Additive and Yield | | | | | | | | | | |
|---|---|---|---|---|---|---|---|---|---|---|
| Additive | 12.0 | 6.7 | 5.6 | 13.2 | 8.9 | 7.8 | 12.9 | 16.4 | 4.5 | 9.6 | 5.8 |
| Yield | 96 | 50 | 42 | 82 | 76 | 70 | 89 | 94 | 15 | 75 | 32 |

a. Assuming that yield is the dependent variable, plot yield against additive. Does the relationship appear to be linear?

b. Using statistical software, estimate the simple linear regression model. Identify $R^2$ and $s_e^2$.

c. In instances such as this where linearity does not hold, polynomial regression can be used to provide a better fit to the data. Polynomial regression is a special case of multiple regression where new predictor variables are formed by raising other predictor variables to integral powers. In this exercise, a new predictor will be formed by squaring the values of additive (Add_sq). Yield will then be fitted to the predictors Additive and Add_sq. The prediction equation based upon the polynomial regression is Estimated Yield $= -67.53 + 23.04$ (Additive) $- 0.82$ (Add_sq). $R^2$ and $s_e^2$ are 0.95 and 47.53, respectively. Predict the yield when Additive $= 16$. Make this prediction using both the linear and polynomial fits. Compare your results.

d. Compare the linear and polynomial fits to the data by the values for $R^2$ and $s_e^2$.

e. Which model do you believe is best to use for estimation and prediction? Explain your answer.

4. Suppose that an association of real estate professionals has reported home sales for 2011 in a data set titled Home Sales. The table contains the current sales by region and the inventory for existing-home sales (single-family and condos/co-ops). An excerpt of the full table is given below.

| Home Sales | | | | | | | | 
|---|---|---|---|---|---|---|---|
| Sale Price | Region | Home Type | Inventory | Sale Price | Region | Home Type | Inventory |
| $237,000 | NE | Condo/Co-op | 185,000 | $239,600 | NE | Single-Family | 550,000 |
| $225,400 | NE | Condo/Co-op | 188,000 | $242,400 | NE | Single-Family | 550,000 |
| $235,200 | NE | Condo/Co-op | 205,000 | $244,600 | NE | Single-Family | 560,000 |
| $144,900 | MW | Condo/Co-op | 80,000 | $138,800 | MW | Single-Family | 850,000 |
| $145,000 | MW | Condo/Co-op | 79,000 | $138,900 | MW | Single-Family | 840,000 |
| $139,200 | MW | Condo/Co-op | 82,000 | $138,600 | MW | Single-Family | 900,000 |
| $110,400 | S | Condo/Co-op | 194,000 | $153,400 | S | Single-Family | 1,520,000 |
| $108,100 | S | Condo/Co-op | 176,000 | $153,100 | S | Single-Family | 1,520,000 |
| $112,100 | S | Condo/Co-op | 200,000 | $150,800 | S | Single-Family | 1,570,000 |
| $154,600 | W | Condo/Co-op | 90,000 | $223,100 | W | Single-Family | 940,000 |
| $152,900 | W | Condo/Co-op | 91,000 | $216,300 | W | Single-Family | 940,000 |
| $146,800 | W | Condo/Co-op | 74,000 | $216,900 | W | Single-Family | 1,050,000 |

:: **Data**

The full table can be found at stat.hawkeslearning.com by navigating to **Discovering Business Statistics, Second Edition > Data Sets > Home Sales**.

a. Suggest a regression model that would allow you to predict sale price as a function of inventory, region, and whether you have a condo/co-op or a single-family home.

b. Estimate the model that you suggested in part **a.**

c. What is the estimated equation for predicting sale price by region?

d. What is the estimated equation for predicting sale price by type of home?

**e.** Is your model estimated in part **b.** statistically useful for predicting sale price at a 1% significance level? Explain your answer.

∴ **Data**

This data set can be found at stat.hawkeslearning.com by navigating to **Discovering Business Statistics, Second Edition > Data Sets > Tablet Survey**.

**5.** Given the digital revolution in the United States and the trend towards textbooks and lecture materials being made available through electronic means, one wonders if having these technologies will help improve students' grades. To help answer this question, a survey was taken on a university campus inquiring if students used a tablet for their classes, which tablet was used, and their grade point average, household income, and the highest level of education attained by their parents. Using the data from the survey presented in the following table, answer the following questions.

| Key for Student Survey | | |
|---|---|---|
| **Tablet:** | **Household Income:** | **Parent's Highest Level of Education:** |
| 1 = Motorola Xoom | 1 = < $30,000 | 1 = Some High School |
| 2 = Samsung Galaxy | 2 = $30,000–$49,999 | 2 = High School Diploma |
| 3 = Apple iPad | 3 = $50,000–$74,999 | 3 = Some College |
| 4 = No Tablet | 4 = > $75,000 | 4 = College Graduate |

| Student Data | | | | | | | |
|---|---|---|---|---|---|---|---|
| **GPA** | **Tablet** | **Income** | **Education** | **GPA** | **Tablet** | **Income** | **Education** |
| 3.9622 | 4 | 3 | 1 | 3.0530 | 2 | 3 | 4 |
| 2.9555 | 1 | 2 | 3 | 3.8034 | 2 | 3 | 3 |
| 3.2058 | 4 | 2 | 2 | 2.3986 | 2 | 2 | 4 |
| 3.6487 | 1 | 4 | 2 | 3.1191 | 1 | 3 | 2 |
| 3.4459 | 1 | 4 | 4 | 2.9556 | 1 | 3 | 3 |
| 3.5222 | 4 | 1 | 1 | 3.3100 | 2 | 3 | 1 |
| 3.9964 | 2 | 1 | 3 | 3.5477 | 1 | 3 | 2 |
| 2.4374 | 1 | 3 | 4 | 3.7710 | 3 | 3 | 1 |
| 2.9262 | 4 | 3 | 3 | 3.2706 | 1 | 3 | 1 |
| 3.8684 | 2 | 2 | 3 | 3.8039 | 2 | 1 | 3 |
| 3.2102 | 1 | 3 | 3 | 3.5149 | 1 | 1 | 2 |
| 3.4394 | 2 | 3 | 4 | 2.0290 | 2 | 3 | 1 |
| 3.8169 | 4 | 1 | 3 | 3.2985 | 3 | 1 | 4 |
| 3.6107 | 2 | 2 | 1 | 2.2877 | 2 | 2 | 2 |
| 2.7475 | 3 | 2 | 3 | 2.4569 | 3 | 2 | 1 |

**a.** Suggest a multiple regression model to predict GPA from tablet use and which type of tablet is being used.

**b.** What is the estimated regression equation for the model proposed in part **a.**?

**c.** Is the model useful in predicting GPA from whether a student uses a tablet or not and which type of tablet is being used at the 0.05 level? Justify your answer.

**d.** Discuss whether there is a difference between tablet use and household income regarding how the variables affect one's GPA.

**e.** Discuss whether there is a difference in the extent to which each type of tablet affects one's GPA.

6. The following table contains a list of high-dividend exchange-traded funds (ETFs). Exchange-traded funds are investment funds traded on stock exchanges, much like stocks. ETFs are traditionally index funds, but ETFs can hold assets such as stocks, commodities, or bonds, and trade at approximately the same price as the net asset value of their underlying assets over the course of the trading day. ETFs may be attractive as investments because of their low costs, tax efficiency, and stock-like features.

**⠿ Data**

This data set can be found at stat.hawkeslearning.com by navigating to **Discovering Business Statistics, Second Edition > Data Sets > Exchange-Traded Funds**.

| ETF | Share Price ($) | Dividend Per Share ($) | Dividend Yield (%) | ETF | Share Price ($) | Dividend Per Share ($) | Dividend Yield (%) |
|---|---|---|---|---|---|---|---|
| 1 | 4.32 | 0.28 | 6.49 | 26 | 23.82 | 0.78 | 3.28 |
| 2 | 15.38 | 0.92 | 6.02 | 27 | 19.58 | 0.64 | 3.26 |
| 3 | 25.25 | 1.43 | 5.66 | 28 | 136.94 | 4.41 | 3.22 |
| 4 | 21.28 | 1.13 | 5.31 | 29 | 83.57 | 2.65 | 3.17 |
| 5 | 698.75 | 36.88 | 5.28 | 30 | 6.44 | 0.20 | 3.12 |
| 6 | 120.55 | 6.22 | 5.16 | 31 | 13.50 | 0.42 | 3.07 |
| 7 | 23.00 | 1.09 | 4.74 | 32 | 47.47 | 1.45 | 3.06 |
| 8 | 13.34 | 0.62 | 4.66 | 33 | 30.28 | 0.92 | 3.03 |
| 9 | 24.82 | 1.15 | 4.63 | 34 | 14.56 | 0.43 | 2.94 |
| 10 | 22.96 | 1.04 | 4.53 | 35 | 36.06 | 1.06 | 2.93 |
| 11 | 78.51 | 3.21 | 4.09 | 36 | 14.26 | 0.41 | 2.90 |
| 12 | 24.22 | 0.99 | 4.08 | 37 | 369.35 | 10.56 | 2.86 |
| 13 | 8.70 | 0.35 | 4.02 | 38 | 7.07 | 0.20 | 2.83 |
| 14 | 15.08 | 0.58 | 3.86 | 39 | 20.12 | 0.57 | 2.82 |
| 15 | 23.44 | 0.90 | 3.84 | 40 | 1.00 | 0.03 | 2.81 |
| 16 | 22.81 | 0.87 | 3.80 | 41 | 104.15 | 2.91 | 2.79 |
| 17 | 22.80 | 0.84 | 3.70 | 42 | 32.4 | 0.89 | 2.75 |
| 18 | 113.84 | 4.21 | 3.69 | 43 | 47.58 | 1.30 | 2.73 |
| 19 | 13.62 | 0.49 | 3.60 | 44 | 18.43 | 0.49 | 2.64 |
| 20 | 22.86 | 0.82 | 3.58 | 45 | 27.78 | 0.73 | 2.63 |
| 21 | 15.58 | 0.56 | 3.57 | 46 | 30.10 | 0.76 | 2.52 |
| 22 | 121.33 | 4.16 | 3.43 | 47 | 26.55 | 0.66 | 2.49 |
| 23 | 15.27 | 0.51 | 3.34 | 48 | 19.53 | 0.46 | 2.35 |
| 24 | 8.70 | 0.29 | 3.30 | 49 | 27.43 | 0.62 | 2.26 |
| 25 | 45.40 | 1.50 | 3.30 | 50 | 30.33 | 0.62 | 2.04 |

a. Using the data in the table, can dividend yield be predicted by share price and dividend per share? Is it a useful model? Justify your answers.

b. Which variable explains the greatest amount of variability in dividend yield? Explain your answer.

c. Can you detect any multicollinearity in the model containing share price and dividend per share? Explain your answer.

**Data**

This data set can be found at
stat.hawkeslearning.com by navigating to
**Discovering Business Statistics, Second
Edition > Data Sets > SNAP Benefits**.

7.  The Supplemental Nutrition Assistance Program (SNAP) provides monthly benefits that help eligible low-income households buy the food they need for good health. For most households, SNAP funds account for only a portion of their food budgets, so they must also use their own funds to buy enough food to last throughout the month. Eligible households can receive food assistance through regular SNAP or through the Louisiana Combined Application Project (LaCAP). Using the data in the table, answer the following questions to help predict monthly benefits to eligible households.

| SNAP Benefits | | | | | |
|---|---|---|---|---|---|
| Monthly Benefit ($) | Family Size | Gross Monthly Income ($) | Monthly Benefit ($) | Family Size | Gross Monthly Income ($) |
| 603.41 | 5 | 3753 | 556.42 | 1 | 3098 |
| 560.69 | 3 | 3778 | 569.05 | 8 | 3707 |
| 623.24 | 6 | 3609 | 365.80 | 8 | 2071 |
| 416.12 | 5 | 2262 | 489.08 | 5 | 3166 |
| 323.90 | 1 | 1966 | 495.86 | 4 | 3126 |
| 418.78 | 4 | 2736 | 642.77 | 4 | 3933 |
| 506.46 | 2 | 3274 | 364.81 | 8 | 1925 |
| 552.53 | 2 | 3480 | 619.30 | 6 | 3736 |
| 586.46 | 7 | 3741 | 238.71 | 1 | 1453 |
| 637.18 | 8 | 3684 | 378.94 | 4 | 2538 |
| 244.49 | 2 | 1476 | 302.58 | 1 | 1798 |
| 507.19 | 5 | 2835 | 231.74 | 8 | 1189 |
| 512.56 | 5 | 2873 | 428.67 | 6 | 2247 |
| 312.89 | 4 | 1618 | 286.99 | 5 | 1460 |
| 329.05 | 4 | 1565 | 268.81 | 1 | 1567 |
| 243.49 | 6 | 1582 | 329.81 | 6 | 1622 |
| 560.37 | 8 | 3380 | 627.25 | 3 | 3828 |
| 599.90 | 3 | 3922 | 421.52 | 6 | 2782 |
| 657.09 | 5 | 3845 | 656.38 | 2 | 3978 |
| 394.82 | 5 | 2233 | 400.64 | 3 | 2493 |

a.  Suggest a regression model that will assist SNAP administrators in providing a monthly benefit to eligible households.

b.  Fit the model that you suggested in part **a**. Is this model useful in predicting monthly benefits? Justify your answer.

c.  Are all independent variables in the model helpful in explaining the variation in monthly benefits? Explain your answer.

d.  Give a 95% confidence interval for average monthly benefits for a four-member household with a gross monthly income of $2500. Interpret this interval.

e.  Provide a 99% prediction interval for a four-member household with a gross monthly income of $2500. Interpret this interval.

f.  What is the difference between the intervals found in parts **d**. and **e**.?

# P Discovery Project

⋮⋮ **Data**

For the full data set visit
stat.hawkeslearning.com and navigate to
**Discovering Business Statistics, Second
Edition > Data Sets > Mount Pleasant
Real Estate Data**.

## Home Sweet Home: Using Multiple Regression to Analyze and Predict Home Prices

An important problem in real estate is determining how to price homes to be sold. There are so many factors—size, age, and style of the home; number of bedrooms and bathrooms; size of the lot; and so on—which makes setting a price a challenging task. In this project, we will try to help realtors in this task by determining how different characteristics of homes relate to home prices, identifying the key variables in pricing, and building multiple-variable regression models to predict prices based on property characteristics.

Our analysis will be based on the Mount Pleasant Real Estate Data. This data set includes information about 245 properties for sale in three communities in the suburban town of Mount Pleasant, South Carolina, in 2017.

### Phase 1: Data Preparation

1. Download the Mount Pleasant Real Estate Data from stat.hawkeslearning.com and open it with Microsoft Excel.

2. Determine the mean, mode, median, maximum, minimum, standard deviation, and the coefficient of variation of the following variables: price, number of bedrooms, number of bathrooms, number of stories, and square footage, and briefly discuss the results. (Hint: these values can be quickly calculated using the Data Analysis Add-in: Descriptive Statistics in Excel).

3. Fully summarize the qualitative variables (i.e., What percent of the sample has a pool?) and briefly discuss your findings.

4. To ensure the data contain comparable properties, eliminate duplexes and properties whose prices are outliers. What limitations does this impose on our analysis? How did you determine which prices were outliers?

   Consider the following variables associated with each property:

   $x_1$ = number of bedrooms    $x_5$ = square footage

   $x_2$ = number of bathrooms    $x_6$ = age (based on year built)

   $x_3$ = number of stories    $x_7$ = acreage

   $x_4$ = subdivision    $x_8$ = new owned

5. For the qualitative variables, adjust this data in a reasonable, quantitative way for use in a regression analysis.

6. Use the following correlation matrix and describe any issues with multicollinearity.

| | Bedrooms | Baths - Total | Baths - Full | Baths - Half | Stories | Subdivision | Square Footage | Age | Acreage | New Owned? | House Style | Covered Parking Spots | Fenced Yard | Screened Porch? | Golf Course? | Fireplace? |
|---|---|---|---|---|---|---|---|---|---|---|---|---|---|---|---|---|
| Bedrooms | 1 | | | | | | | | | | | | | | | |
| Baths - Total | 0.70 | 1.00 | | | | | | | | | | | | | | |
| Baths - Full | 0.67 | 0.95 | 1.00 | | | | | | | | | | | | | |
| Baths - Half | 0.06 | 0.13 | -0.18 | 1.00 | | | | | | | | | | | | |
| Stories | 0.43 | 0.50 | 0.42 | 0.24 | 1.00 | | | | | | | | | | | |
| Subdivision | 0.08 | -0.05 | -0.04 | -0.01 | 0.06 | 1.00 | | | | | | | | | | |
| Square Footage | 0.71 | 0.74 | 0.68 | 0.19 | 0.44 | 0.11 | 1.00 | | | | | | | | | |
| Age | 0.05 | -0.07 | -0.12 | 0.14 | -0.07 | 0.28 | 0.18 | 1.00 | | | | | | | | |
| Acreage | 0.12 | 0.19 | 0.16 | 0.11 | 0.00 | -0.03 | 0.35 | 0.33 | 1.00 | | | | | | | |
| New Owned? | -0.14 | 0.00 | 0.01 | -0.05 | 0.04 | -0.29 | -0.22 | -0.78 | -0.15 | 1.00 | | | | | | |
| House Style | -0.22 | -0.22 | -0.22 | 0.03 | -0.15 | 0.07 | -0.22 | -0.06 | -0.12 | 0.11 | 1.00 | | | | | |
| Covered Parking Spots | 0.28 | 0.35 | 0.31 | 0.13 | 0.16 | 0.03 | 0.47 | 0.14 | 0.24 | -0.15 | 0.07 | 1.00 | | | | |
| Fenced Yard | 0.04 | -0.12 | -0.13 | 0.04 | -0.11 | 0.01 | -0.03 | 0.29 | -0.10 | -0.41 | -0.01 | -0.04 | 1.00 | | | |
| Screened Porch? | 0.17 | 0.24 | 0.23 | 0.02 | 0.14 | -0.25 | 0.14 | -0.04 | 0.03 | -0.05 | -0.18 | -0.03 | 0.12 | 1.00 | | |
| Golf Course? | 0.30 | 0.26 | 0.27 | -0.02 | 0.19 | -0.14 | 0.37 | 0.34 | 0.48 | -0.19 | -0.08 | 0.23 | 0.12 | 0.13 | 1.00 | |
| Fireplace? | 0.18 | 0.18 | 0.20 | -0.08 | 0.06 | 0.04 | 0.22 | 0.14 | 0.09 | -0.13 | -0.17 | 0.05 | 0.11 | 0.30 | 0.22 | 1.00 |
| Number of Fireplaces | 0.16 | 0.21 | 0.22 | -0.03 | 0.04 | 0.11 | 0.35 | 0.24 | 0.14 | -0.20 | -0.13 | 0.15 | 0.13 | 0.28 | 0.17 | 0.80 |

**Phase 2: Constructing Predictive Models**

7.  Construct the multiple regression model with input variables $x_1$, $x_2$, $x_3$, and $x_4$.

8.  Examine the impact of adding additional variables to the model

    a.  Add $x_5$ to the model. Is the addition of $x_5$ to the model significant? How was the adjusted $R^2$ impacted? What is the $P$-value for $x_5$?

    b.  Add $x_6$ to the model. Is the addition of $x_6$ to the model significant? How was the adjusted $R^2$ impacted? What is the $P$-value for $x_6$?

    c.  Add $x_7$ to the model. Is the addition of $x_7$ to the model significant? How was the adjusted $R^2$ impacted? What is the $P$-value for $x_7$?

    d.  Add $x_8$ to the model. Is the addition of $x_8$ to the model significant? How was the adjusted $R^2$ impacted? What is the $P$-value for $x_8$?

9.  Perform a hypothesis test to determine if the model is useful for predicting home values at a significance level of $\alpha = 0.05$. State the $P$-value and interpret its meaning.

10. Are any variables not useful predictors of home price at a significance level of $\alpha = 0.05$? State the $P$-values of these variables. Intuitively, what does this mean with respect to pricing properties?

11. State the best model for the data and justify your answer.

**Phase 3: Applying and Interpreting the Model**

12. Suppose you own a 2000 square foot 2-story house in one of the communities in the data set with 3 bedrooms, 2.5 baths, a pool, and it is located on a golf course, but has no dock or fenced yard. What does the model predict the price of your house to be?

13. A common term in real estate is "comparables," or "comps" for short, which are properties that have similar characteristics. It is common for realtors to look up "comps" for a certain property to get an idea of how to price it. Locate the "comps" for your home in the data set. Create a box plot of the "comps" and estimate a price range for your house on this basis.

14. What advantages and disadvantages does this approach have to the multiple regression model above?

15

# Chapter 15

## Time Series Analysis and Forecasting

## Discovering the Real World

### The Financial Crisis of 2007–2008

Forecasting plays a major role in our everyday life. From choosing the clothes that we wear to making investment decisions, we are making forecasts as to what the weather will be or how the economic conditions might be.

Time series analysis is an essential part of forecasting and both have a wide range of applications. One can use time series models to forecast sales, operations, manufacturing, management decisions, and resource allocations in a variety of organizations and industries. In finance, time series analysis can be used to analyze accounts receivable, monitor travel and expense data, or examine cash flow. Stock market analysts use time series analysis to help them understand changes in stock prices, interest rates, or other market components over time. Healthcare organizations use time series analysis to help reduce operating costs by forecasting the number of beds that will be used as well as the number of nurses and physicians that will be needed at any given time.

With respect to finance, banking, forecasting, and the economy, one of the most common questions is, *Could the 2007-2008 banking meltdown have been predicted?* It depends on who you ask. We know that "hindsight is 20/20" so in looking back, economists can show policymakers the mistakes that were made in the past. More than a decade has passed since the risky mortgage lending, excessive borrowing, and soaring housing prices collided to begin the most severe financial crisis in American history. With that passage of time, much data have been collected to help analysts understand the countless mistakes that were made by the financial institutions. To predict this economic crisis, economists have collected data on the outstanding credit of borrowers, stock market values, home prices leading up to the crisis, and interest rates. Like time series data, all of the aforementioned variables were collected over time (i.e., time is the independent variable; outstanding credit, stock market values, home prices, and interest rates are the dependent variables). Using these data, analysts can identify trends, cycles, seasonality, and other charting characteristics to help predict the financial crisis.

In this chapter, we will discuss a variety of time series methods such as the moving average method, weighted moving average, exponential smoothing techniques, and seasonality. Once we make forecasts, we will then examine forecast accuracy techniques to help us determine the quality of any predictions. As you will see, the primary forecasting methods discussed will minimize the need for complex mathematics. The focus will be on the concepts associated with each of the methods in the hope that students will gain a deeper understanding of forecasting and time series methods.

### Introduction

**Definition**

Forecasting

**Forecasting** is a prediction of what will occur in the future.

**Forecasting** is a prediction of what will occur in the future. Some examples of forecasting include simple meteorology (*What should you wear today based on the weather? Is it going to rain?*) or stock market analytics (*Will a company's stock price increase or decrease?*). Prediction of growth, sales, demand, and interest rates are among the most common applications of forecasting. Even simple things like predicting the number of visitors to an amusement park is very crucial for staffing, planning for dining, and setting up transportation resources. Forecasting is one of the first types of analysis that a business makes, as this sets their agenda for the day, week, month or the year ahead. In general, forecasting methods will make predictions using past experiences. However, in quite a few instances, we make modifications based on information available about the future such as the weather, sporting events, etc.

## Methods Used in Forecasting

We will talk about two primary methods of forecasting in this chapter. We will begin the chapter with Time Series Analysis and end the chapter with Regression Analysis. **Time Series Analysis (TSA)** is a statistical technique that uses time series data for explaining the past or forecasting future events. "Time series" refers to data that is presented in a periodic fashion, which could be in days, weeks, months, quarters, or years. These predictions are also a function of time. With time series data there is no *causal* variable. In other words, we simply examine past behavior of a specific variable and attempt to predict its future behavior.

> **Definition**
>
> Time Series Analysis (TSA)
>
> **Time Series Analysis (TSA)** is a statistical technique that uses time series data for explaining the past or forecasting the future events.

# 15.1 Time Series Components

## Components of Time Series Analysis

Time series have several components. The first component is **timeframe (short, medium,** and **long)**—*how far can we predict into the future?* The short term would be 1–2 periods, a medium term would be 5–10 periods, and a long term is 12 or more periods. Interestingly, there is no line of demarcation in time series analysis. That is, depending on the size of your data, short term could be significantly more than 1–2 periods and your medium period could be more than 5–10 periods. The timeframe is a function of the number of observations; thus, there is no rule of thumb that states that our period ranges for each timeframe are exact. Remember, we represent the timeframe here as periods. If it is given weekly, then a short-term forecast represents 1 to 2 weeks ahead; if given yearly, then a short-term forecast represents 1 to 2 years ahead.

> **Definition**
>
> Timeframe
>
> **Timeframe** refers to how far we want to predict into the future. The timeframe will be a function of the number of observations and the structure of the data into periods.

Another component is **trend**. The trend is a gradual, long-term movement (up or down) of a variable. Trend is fairly easy to detect since we can observe whether the data values are going up (positive trend), or going down (negative trend), or whether the data values are neither going up nor down (stationary).

> **Definition**
>
> Trend
>
> **Trend** is a gradual, long-term movement (up or down) of a variable.

Another component of time series is **seasonal variation**. Seasonal variation is an up-and-down repetitive movement that occurs periodically within a trend. Oftentimes this trend is weather-related, but it could be daily, weekly, monthly, or whenever a trend is found. An example could be the demand for jet ski rentals. Jet ski rental demand will spike in the summer months but will gradually decrease as you approach the fall and winter months. Demand will likely begin to increase again during the spring months and will again reach its highest expected point of demand during the summer months.

> **Definition**
>
> Seasonal Variation
>
> **Seasonal variation** is an up-and-down repetitive movement that occurs within a trend and occurs periodically.

Similar to seasonal variation, the **cycle** (also referred to as cyclical variation) is another component of time series which refers to an up-and-down repetitive movement in a variable. A cycle repeats itself over a long period of time such as every few years (e.g., business cycles and inflation).

> **Definition**
>
> Cycle or Cyclical Variation
>
> **Cycle** or **cyclical variation** is an up-and-down repetitive movement in a variable that repeats itself over a long period of time.

Finally, **random variations** are erratic movements that are not predictable because they do not follow a pattern. In time series analysis we know we are measuring a variable, and we know that variable is subject to change from trial to trial. These are the random components of time series analysis that are difficult to predict, and in most cases, represent the random error in forecasting models.

> **Definition**
>
> Random Variations
>
> **Random variations** are erratic movements that are not predictable because they do not follow a pattern.

## Time Series Plot

Plotting the time series data should be the first step in any forecasting analysis. A simple plot will reveal various components of the time series. For example, Figure 15.1.1 is a plot which represents the new homes sold in the U.S. between 1963 and 2019. It is clear from the plot that there was an increasing trend until 2008, with some periodic ups and downs representing a business cycle (cyclical variation). After 2008, there was a steep drop (due to the subprime mortgage crisis) and over the last decade new home sales have been improving continuously. As history suggests, we should expect a cyclical drop in the near future. As is the case with home purchases over the long-term, there are regular seasonal variations in every year, and this will not be visible in a yearly plot. So, let's examine the monthly plot of home sales in the same period (see Figure 15.1.2). We observe that the home sales are lower during November and December, and higher in late-spring and summer. Please note that the time frequency is important to identify any patterns in the data.

**⋰ Data**

For the full data set, visit stat.hawkeslearning.com and navigate to **Discovering Business Statistics, Second Edition > Data Sets > New Home Sales**.

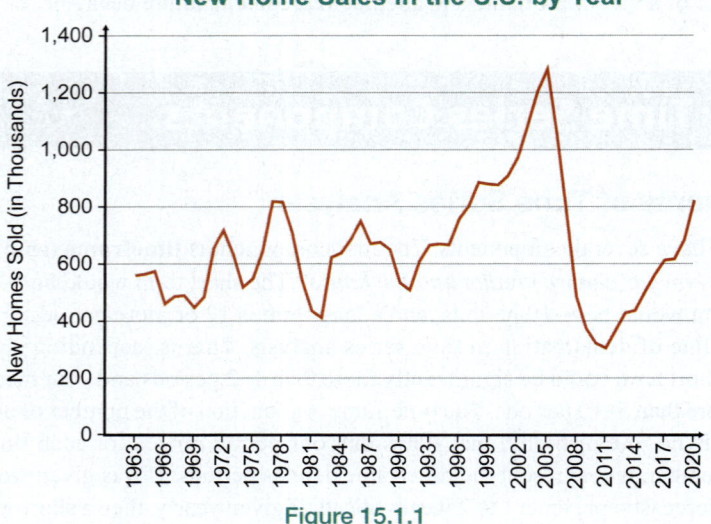

Figure 15.1.1

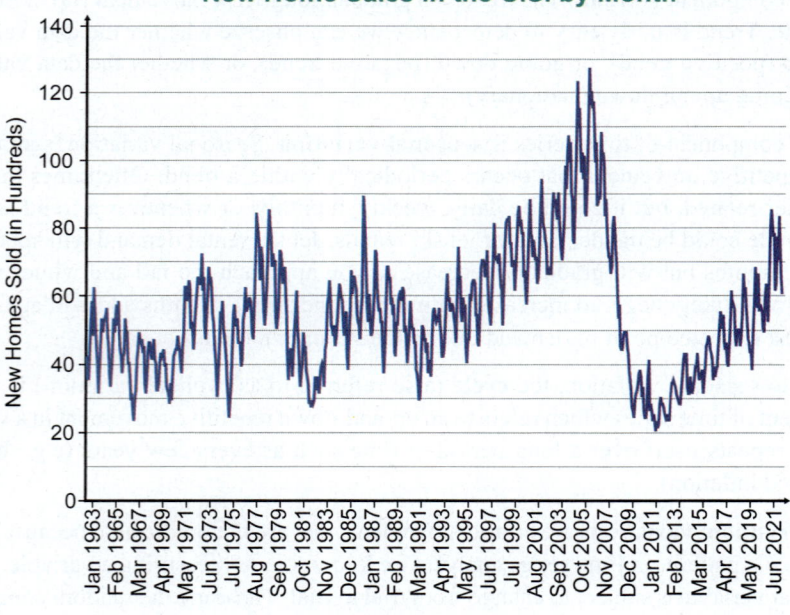

Figure 15.1.2

Another example uses data from a retail store in the southern U.S. Figure 15.1.3 is the monthly sales data of frozen pizza in a region with several stores. The actual data spans a two-year period. Figures 15.1.4 through Figure 15.1.6 display the data quarterly, weekly, and daily, respectively. We can see seasonality in the daily and weekly plots; i.e., sales are high during weekends and are low during midweek. Similarly, sales are high at the start of month, then sales decrease over time. In the monthly data we see a slight downward trend, but that becomes clearly visible in the quarterly plots. Thus, choosing a time frequency is crucial when visualizing the data.

Depending on the need of the company, the data can be aggregated and forecasted at various time frequencies. For example, central transportation resources could be planned at a weekly level, stocking (ordering) policies could be planned at a daily level, and hiring policies at a monthly level.

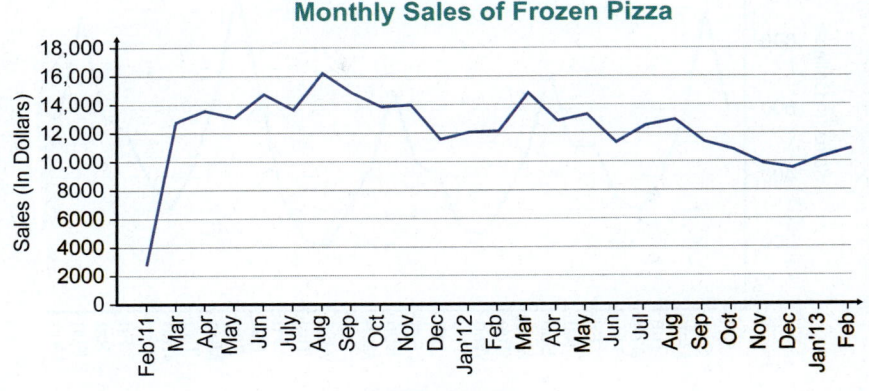

Figure 15.1.3

**⁙ Data**

For the full data set, visit stat.hawkeslearning.com and navigate to **Discovering Business Statistics, Second Edition > Data Sets > Frozen Pizza Sales**.

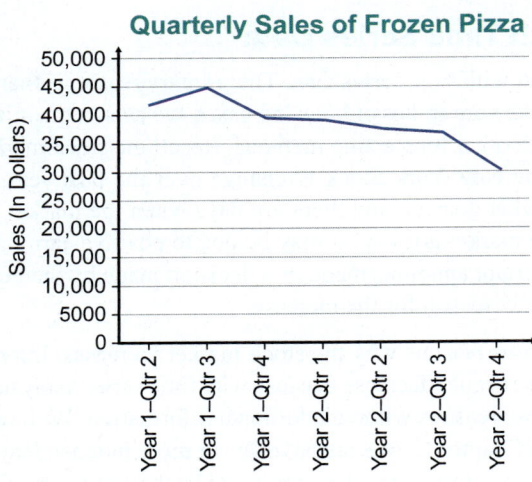

Figure 15.1.4

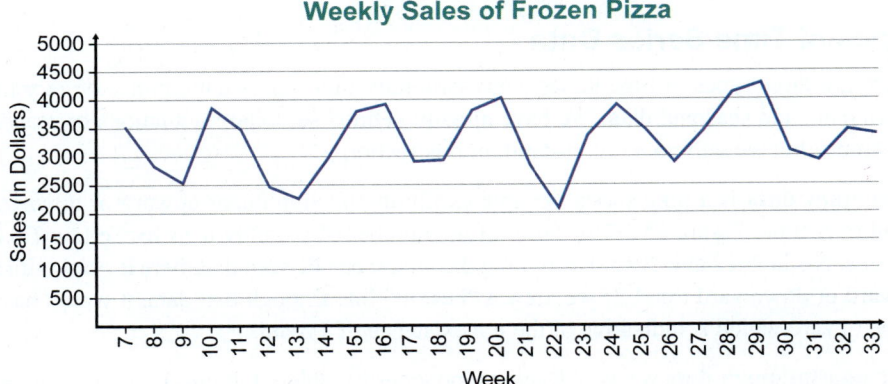

Figure 15.1.5

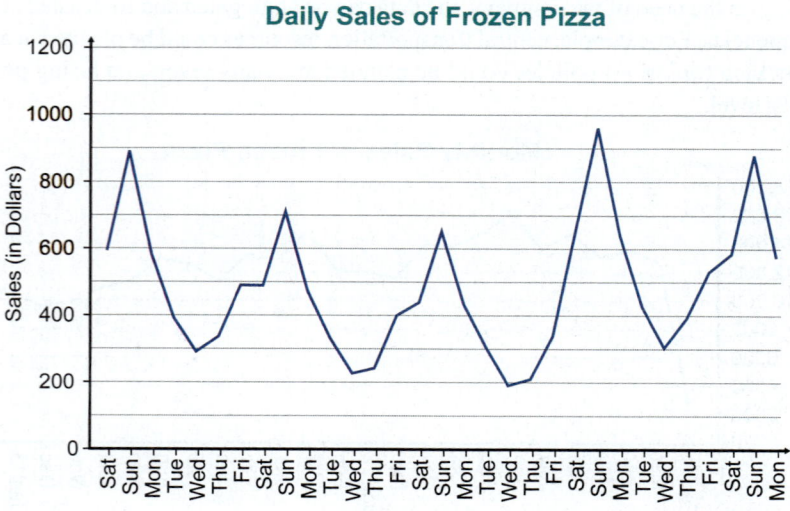

**Figure 15.1.6**

## Forecasting with Time Series Data

It is difficult to forecast with time series data. The primary reason is that there is no *causal* variable. Even when there are spikes or large increases, we cannot quantify it with a specific variable as we can with other forecasting methods. Recall the stock market example. If we look at the close of the New York Stock Exchange over the past year, we will see there are days when the market goes up and there are days when the market goes down. When there are days that the market goes up, it may be due to positive earnings by several large companies, or an important announcement, or a decision made by the Federal Reserve; it's hard to identify the exact reason for the increase.

There could be numerous reasons why the stock market increases. Interestingly, for those same reasons, the market could decrease because with time series analysis, we cannot easily quantify or qualify those reasons without additional information. With regression analysis, which we discussed in Chapter 13, we can have one or more independent variables that will help us make predictions or forecasts—i.e., one could be the price of stock, another could be related to the earnings report or competitive stock prices, etc. We can use these additional variables to make better predictions.

## Types of Time Series Data

There are three types of time series data: stationary time series data, nonstationary time series data, and seasonal data. We have already defined seasonal fluctuations or seasonal variations, but we will go more in-depth in this section.

**Stationary data** is a time series variable exhibiting no significant upward or downward trend over time. Figure 15.1.7 is a plot of the number of monthly trips by Yellow Taxi in Queens, NY in 2012 and 2013. Examining the plot, it can be seen that there is no significant upward or downward trend. If we drew a "best fit" line through that data, it would have a slope of zero which is indicative of the nonexistence of a trend.

With **nonstationary data** we would have a time series variable exhibiting a significant upward or downward trend over time. The plot in Figure 15.1.1 resembles nonstationary data. If we were to draw a "best fit" line through the data, it can be seen that as time increases our number of new homes sold increases until 2008, which is indicative of an upward trend over time.

**Seasonal data** is a time series variable exhibiting a repeating pattern at regular intervals over time. The time series plot in Figure 15.1.6 is an example of seasonality, given that sales are high during weekends and low during midweek days such as Wednesday and Thursday.

### Definition

#### Types of Time Series Data

**Stationary data** is the result of a time series variable exhibiting no significant upward or downward trend over time.

**Nonstationary data** is the result of a time series variable exhibiting a significant upward or downward trend over time.

**Seasonal data** is the result of a time series variable exhibiting a repeating pattern at regular intervals over time.

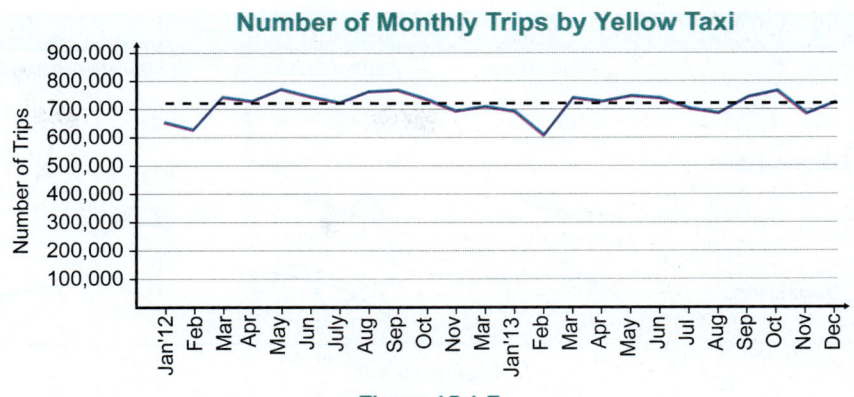

**Figure 15.1.7**

**⁘ Data**

For the full data set, visit
stat.hawkeslearning.com and navigate to
**Discovering Business Statistics, Second
Edition > Data Sets > Yellow Taxi Trips.**

## Time Series Methods

We will discuss four time series methods used to make forecasts: simple moving average, weighted moving average, exponential smoothing, and adjusted exponential smoothing. Even though one can use either of these methods to make forecasts, the best method is determined by tracking the error metrics which will be discussed in later sections. However, one should realize that the choice of appropriate method(s) can be made by just observing the patterns in the data. Gardner (1985) provided an excellent visual plot of different patterns in time series using Pegels' (1969) classification (see Figure 15.1.8). Through a preliminary examination of the time series plot, users can determine if the data have a linear, exponential, or damped trend, as well as whether the data are seasonal or nonseasonal – specifically if it has additive or multiplicative seasonality. Table 15.1.1 provides a brief description of the different trend and seasonal profiles seen in Figure 15.1.8.

| Table 15.1.1 – The Most Common Time Series Profiles | |
|---|---|
| **Pattern Profile** | **Description** |
| Constant Level | No upward or downward trend |
| Linear trend | Consistent increase in either upward or downward direction |
| Exponential trend | Growth rate increases to higher rates over time |
| Damped trend | Growth pattern gradually decays after a few periods to no increase |
| Nonseasonal | No seasonal variation |
| Additive seasonality | The seasonal variation does not grow or decline systemically over time |
| Multiplicative seasonality | The seasonal variation widens over time with the increase in level of data |

**Note:** These pattern profiles may also occur in combination.

| Time Series Profile Based on Pegels' and Gardner's Classification | | | |
|---|---|---|---|
| | **Nonseasonal** | **Additive seasonality** | **Multiplicative seasonality** |

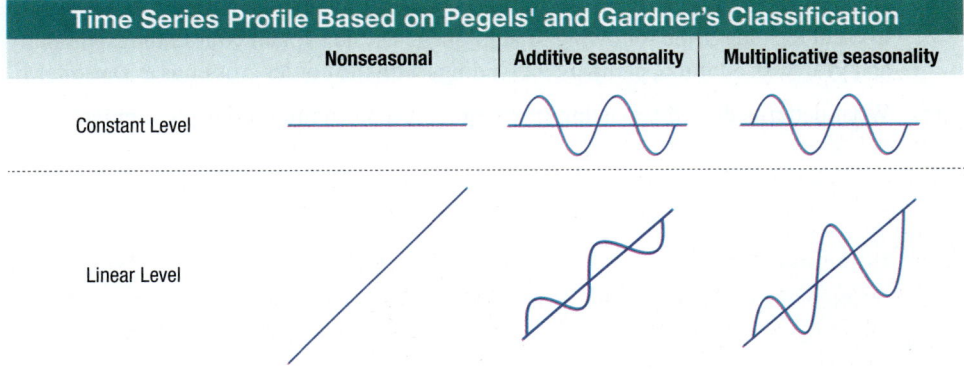

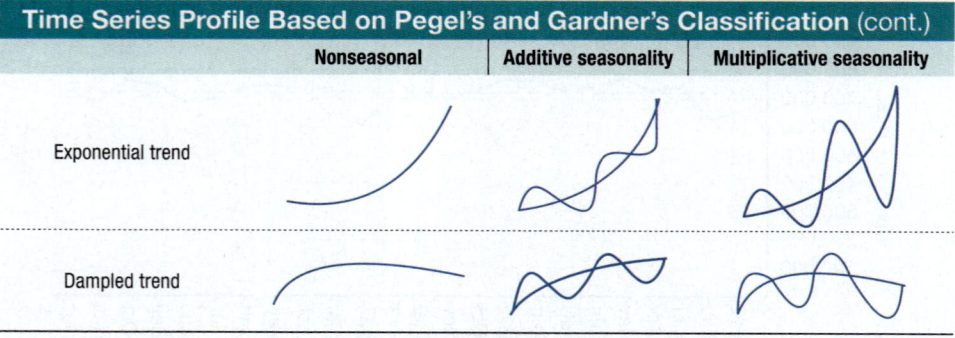

| Time Series Profile Based on Pegel's and Gardner's Classification (cont.) | | | |
|---|---|---|---|
| | **Nonseasonal** | **Additive seasonality** | **Multiplicative seasonality** |
| Exponential trend | | | |
| Dampled trend | | | |

**Figure 15.1.8**

## ✏ 15.1 Exercises

### Basic Concepts

1. What is the difference between seasonal variation and cyclical variation?

2. What is timeframe?

3. Give three examples of business variables that can be represented using a time series plot.

4. What is stationary data?

5. Give three examples of seasonal data in the business world.

6. Suppose a variable is exhibiting a significant upward trend over time. What type of time series data would this represent?

7. What are two ways to determine the best time series method to make a forecast?

### Exercises

 **Data**

The data set can be found by visiting stat.hawkeslearning.com and navigating to **Discovering Business Statistics, Second Edition > Data Sets > Border Crossings**.

8. Use the Border Crossings data set. Plot the truck crossings across the U.S.-Canada border at Detroit, MI and identify any time series patterns. Look at the data at the following time frequencies and explain your findings: monthly and yearly.

9. Use the Border Crossings data set. Plot the passenger vehicle crossings across the U.S.-Canada border at Detroit, MI and identify any time series patterns. Look at the data in the following time frequencies and explain your findings: monthly and yearly.

10. Use the Border Crossings data set. Plot the truck and passenger vehicle crossings across the U.S.-Mexico border at Laredo, TX and identify any time series patterns. Look at the data in the following time frequencies and explain your findings: monthly and yearly. In addition, compare the findings with the border crossings at Detroit, MI.

11. What patterns does the existing (not new) home sales time series plot depict?

**Existing Home Sales**

12. Use the Monthly Average Retail Gas Prices data set, which includes the average gas prices in the U.S. from April 1993 to July 2021.

    a. What patterns do you see in the data?

    b. Is monthly the right frequency to explore the data, or would you prefer quarterly or yearly? Explain your reasons.

13. Use the Mortgage Rates data set, which includes the yearly mortgage rate in the U.S. from 1971. Currently, there is a belief that the mortgage rate is at an all-time low; do you agree? What is the current trend showing?

**⋯ Data**

The data set can be found by visiting stat.hawkeslearning.com and navigating to **Discovering Business Statistics, Second Edition > Data Sets > Monthly Average Retail Gas Prices**.

**⋯ Data**

The data set can be found by visiting stat.hawkeslearning.com and navigating to **Discovering Business Statistics, Second Edition > Data Sets > Mortgage Rates**.

# 15.2 Moving Averages

## Simple Moving Average (SMA)

The first method we are going to talk about is the **Simple Moving Average (SMA)**. The simple moving average method uses several values (two or more) from the recent past to develop a forecast. It is a smoothing technique because we are taking two to three observations, or even more, and predicting one. Therefore, we are averaging these observations and smoothing out some of the variability.

When we are using the simple moving average, we actually compute the average from a chosen window of points and the resulting average is the forecast for that next period of time. The wider the window of points, the smoother the fit will be, because we are using more observations and turning them into one, thereby smoothing out the variability.

**Definition**

Simple Moving Average (SMA)

The **simple moving average (SMA)**, uses the average of several values (two or more) from the recent past to develop a forecast.

**Formula**

Simple Moving Average

$MA_n$ denotes the **moving average** over $n$ periods, and it is the sum of the most recent $n$ data values in the time series divided by the number of periods ($n$) that we use to calculate that moving average.

$$MA_n = \frac{\sum_{i=1}^{n} D_i}{n}$$

where $n$ = the number of periods used to compute the moving average and

$D_i$ = the actual data value of the time series in period $i$.

Let's look at an example.

**Example 15.2.1**

**Calculating a 3-Month Moving Average**

Suppose TixPixx is a company that sells concert tickets and would like to predict the sales for June using a 3-month moving average. The data for January through May can be found in the table below.

| Month | Tickets Purchased |
|---|---|
| January | 9 |
| February | 15 |
| March | 11 |
| April | 10 |
| May | 12 |
| June | * |

**SOLUTION**

Using the data for January through May, we want to predict the sales for June. To calculate a 3-month moving average for June, we use the data values for the previous three months of March, April, and May, which are 11, 10, and 12, respectively, to predict the tickets purchased for June. The computation is as follows:

$$\text{MA}_{\text{June}} = \frac{11+10+12}{3} = 11$$

The three-month moving average for the demand for June is 11.

It would be helpful to forecast the demand for the preceding months for benchmarking or if we would like to analyze the accuracy of the forecasts (see Section 15.4). It should be noted that we cannot compute a 3-month moving average for January, February, or March since we do not have three months of prior data. Thus, the first month that we can compute a forecast for demand is April. To compute the moving average for April, we need to use the data values for ticket purchases for January, February, and March, which are 9, 15, and 11, respectively, and divide by 3. The 3-month moving average for April is given by

$$\text{MA}_{\text{April}} = \frac{9+15+11}{3} = 11.67.$$

The 3-month moving average for May will be calculated using the data values from the previous three months, which will be 15, 11, and 10, divided by 3.

$$\text{MA}_{\text{May}} = \frac{15+11+10}{3} = 12.$$

What is the value of calculating the 3-month moving average for April and May? While we do not need them now, they will be helpful to determine the accuracy of this forecast when compared to other forecasting methods.

What would we do if we wanted to predict or forecast for several periods in the future? We would use the forecast for multiple periods and then update these forecasts as the data become available. For example, if we wanted a moving average for July, then we will use the actual data value for April plus the actual data value for May, and then we would use the forecast for June. For example, the 3-month moving average for July can be calculated as

$$\text{MA}_{\text{July}} = \frac{10+12+\mathbf{11}}{3} = 11.$$

Using the 3-month moving average, we predict that we will have 11 ticket purchases for the month of July. Note that the 11 in the numerator is the forecast for June. We would update this forecasted value for June when we have new data. Similarly, if we wanted to predict two months into the future, we would have to use the forecast for July to help us predict for August.

The advantages to using the simple moving average is that it is quick, it is easy, and it is simple. It does not take a lot of time to calculate these moving averages. One of the constraints in a simple moving average is that the weights are all equally distributed across every period. If we compute a 5-period moving average, then every period is weighted 1/5 or 20% . There are times, however, that we may want to weight recent periods more heavily than periods in the distant past. If this is the case, we may want to use a procedure called a weighted moving average.

## Weighted Moving Average (WMA)

With the weighted moving average procedure, each value in the chosen timeframe (number of periods) is multiplied by a weight between 0 and 1, with the sum of the weights equaling 1. The weighting determines the amount of emphasis given to the observations. In general, more weight is given to recent periods than those in the past.

> **Definition**
>
> **Weighted Moving Average**
>
> In a **weighted moving average**, each value in the time window (number of periods chosen) is multiplied by a weight between 0 and 1, with the sum of the weights equaling 1.

> **Formula**
>
> ### Weighted Moving Average
>
> The computational formula for **weighted moving average** involves summing the product of the weights and the corresponding time series data values as follows:
>
> $$\text{WMA}_n = \sum_{i=1}^{n} W_i D_i$$
>
> where $W_i$ = the weight for period $i$, $D_i$ = the data value for period $i$, $n$ = the number of periods chosen, and
>
> $$\sum W_i = 1.$$

> **Example 15.2.2**
>
> **Calculating a Weighted Moving Average for the TixPixx Data**

Using the TixPixx data, calculate a weighted moving average for June using the weights of 0.1, 0.3, and 0.6 on the most recent months.

#### SOLUTION

Note that the instructions imply that we want to put the 0.6 weight on the most recent month, 0.3 on the next most recent month, and 0.1 on the more distant month's data value. The sum of the weights is equal to 1 and it should be noted that this a three-period weighted moving average. Therefore, the forecast for June using a 3-month WMA is

$$\text{WMA}_{\text{June}} = 11(0.1) + 10(0.3) + 12(0.6) = 11.3.$$

Recall that the simple moving average forecast was 11.

## Choosing the Appropriate Weights and the Number of Periods

The choice of the number of periods ($n$) in the moving average and the appropriate weights for each of those periods of the WMA is determined by monitoring the forecast error, which we will discuss in Section 15.4. The $n$ that minimizes the forecast error, i.e., provides the most accurate forecast, is the "correct" number of periods for that particular time series. Similarly, we can vary weights in the WMA to minimize the error. There are two things to keep in mind when calculating weighted moving averages: the sum of the weights should equal 1 and the weights assigned to the recent periods should be greater than the weights assigned to the preceding periods.

## Shortcomings in Moving Average Methods

Using the moving average procedures has some disadvantages. The primary disadvantage of moving average methods is that they do *not react well to variations due to trend or seasonality* because they are considered smoothing methods (i.e., multiple data values are being used to calculate single forecasts). This is especially the case when the time series has a lot of variation due to trend, which always lags behind. We will demonstrate these disadvantages using two examples.

## Example 15.2.3

### Calculating Moving Averages (SMA and WMA) for Truck Crossings

**⁘ Data**

For the full data set, visit
stat.hawkeslearning.com and navigate to
**Discovering Business Statistics, Second
Edition > Data Sets > Border Crossings.**

The Department of Transportation wanted to predict the number of truck crossings at the Port of Laredo for 2019. Using the data in the table below which contain annual truck crossings from 2011 through 2018, predict the number of truck crossings in 2019 using a 3-period moving average.

| Year | Number of Truck Crossings | Year | Number of Truck Crossings |
|------|---------------------------|------|---------------------------|
| 2011 | 1,695,916 | 2015 | 2,015,773 |
| 2012 | 1,789,546 | 2016 | 2,083,964 |
| 2013 | 1,846,282 | 2017 | 2,182,984 |
| 2014 | 1,947,846 | 2018 | 2,313,967 |

### SOLUTION

For 2019, the SMA using 3 periods is $\frac{2,083,964 + 2,182,984 + 2,313,967}{3} \approx 2,193,638$.

However, we clearly know the number of crossings are increasing and have a positive trend, but the moving average is taking an average of the past three periods which are lower values—hence the forecast will always lag behind! Even if we chose WMA, the maximum weight we can give is about 100% or 1.0 to the most recent period, which in this case would be the number of crossings for 2018. Hence the best forecast would be 2,313,967—which is still not following a positive trend. It should be noted that when the weight is 1.0, the forecast is called the *naïve forecast*, where the forecast is nothing but the previous period's value. As the name suggests, it is simplistic in nature and is used for benchmarking forecasting methods.

## Example 15.2.4

### Calculating Moving Averages for Frozen Pizza Sales

Gigi's Grocery wants to track the number of frozen pizzas that she sells. The table below shows her quarterly pizza sales. Using the data provided in the table, use a 3-quarter moving average method to predict sales for the 5ᵗʰ quarter.

| Quarter | Frozen Pizza Sales |
|---------|--------------------|
| 1 | 38,828 |
| 2 | 37,302 |
| 3 | 36,642 |
| 4 | 30,022 |

### SOLUTION

If one examines the data in the table above, it's easy to see that the pizza sales follow a decreasing (or negative) trend. The three-period simple moving average forecast would be

$$\frac{37,302 + 36,642 + 30,022}{3} = 34,655.$$

Note that this forecast does not capture the decreasing trend. The same will be true for a weighted moving average method regardless of the weighting. In general, moving averages are not well suited for time series that depict trend or seasonality.

Another disadvantage of a moving average is the *need for data*. Given the TixPixx example earlier, suppose we wanted to predict the tickets purchased for the month of July. To do so, we have to use forecasted data. If we are going to predict for August without real data, we will still need to use forecasted data. Thus, it will be difficult to have reliable forecasts because we are using forecasts to predict forecasts. Thus, if we want to use more periods in the forecast, then we need to store that data and access it every time we forecast a particular time series.

Lastly, another disadvantage of a moving average is associated with the technique itself. In general, the forecast would be better if we incorporate more historical information. However as the number of periods increases in the moving average, the forecast will become smoother and will not capture any of the variation in the data. For example, look at the 3-period, 20-period and 50-period SMA forecast of the Laredo Truck crossings. The more periods we include in the forecast, the smoother the forecast is and does not capture the variations in the time series at all.

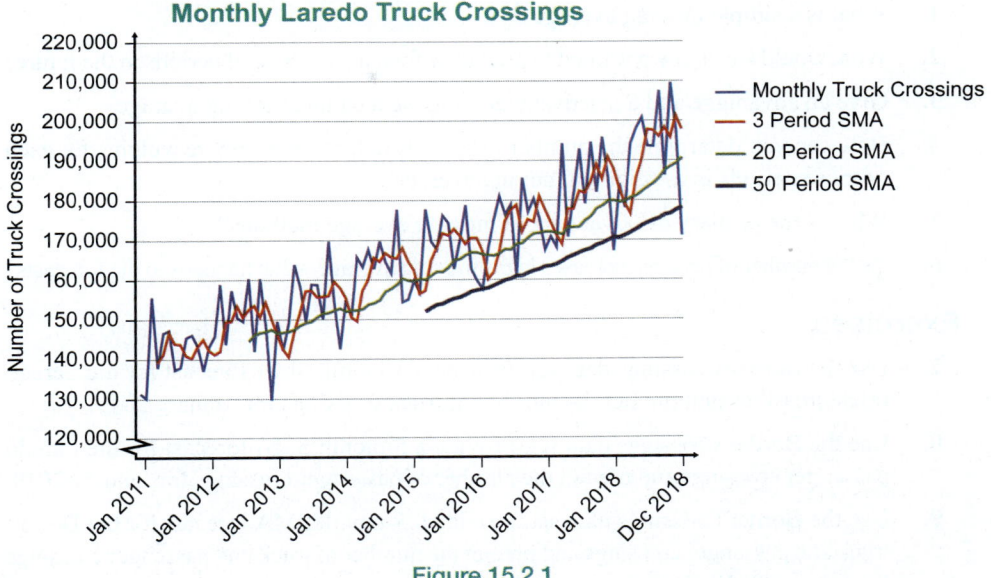

Figure 15.2.1

Regardless of all the disadvantages of the moving average techniques, it is still a popular method of forecasting and is used in several fields for simple analysis. For example, go to any financial website with some sort of analytical capability (see finance.yahoo.com, barchart.com, or any investment bank websites) and you will be able to plot several versions of moving average charts. They use the moving average plots to provide a first round of predictions, such as a trigger of sell or buy decisions, when a stock price touches a 200-day moving average.

Below is a sample chart of a stock price. It shows the daily price of MSFT, along with a 5-day SMA, 20-day SMA, 100-day SMA, and a 200-day SMA.

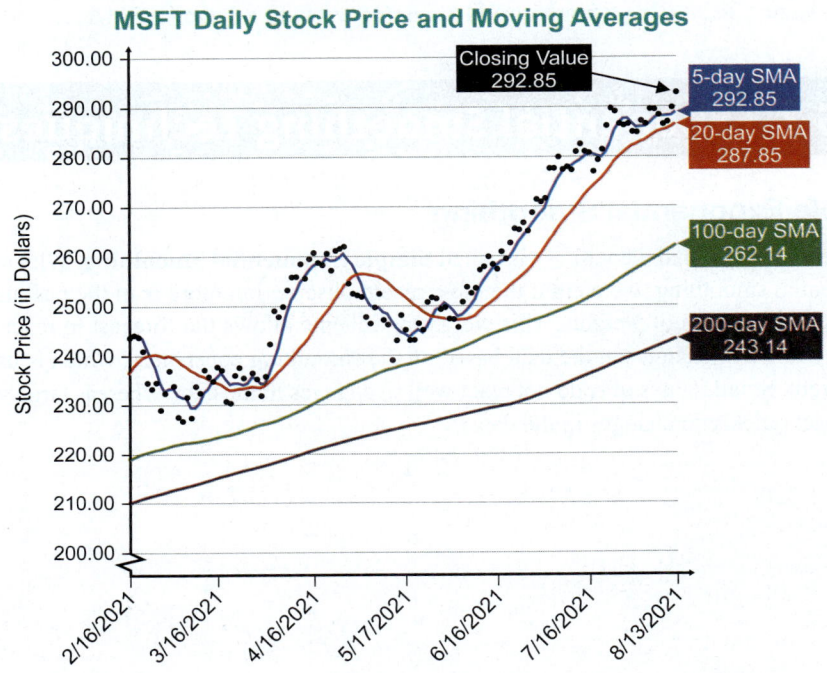

Figure 15.2.2

## ✏ 15.2 Exercises

### Basic Concepts

1. What is a simple moving average?

2. What would we do if we wanted to predict or forecast for several periods in the future?

3. Give an advantage and a disadvantage of using a simple moving average.

4. How can you determine the number of periods and the appropriate weights for each of those periods in a weighted moving average?

5. What is the primary disadvantage of moving average methods?

6. As the number of periods increases in the moving average, what happens to the forecasts?

### Exercises

 **Data**

The data set can be found by visiting stat.hawkeslearning.com and navigating to **Discovering Business Statistics, Second Edition > Data Sets > Border Crossings**.

7. Use the Border Crossings data set. Provide a 3-month SMA forecast for the Laredo truck crossings and predict the number of truck crossings for January 2019.

8. Use the Border Crossings data set. Provide a 5-month SMA forecast for the Laredo passenger crossings and predict the number of passenger crossings for January 2019.

9. Use the Border Crossings data set. Provide a 5-month SMA forecast for the Detroit truck and passenger crossings and predict the number of truck and passenger crossings for January 2019.

10. Use the Border Crossings data set. Provide a 3-month WMA and 5-month WMA forecast for the Laredo truck crossings and predict the number of truck crossings for January 2019. Note: Use the weights of 0.6, 0.3 and 0.1 for the 3-month WMA and 0.4, 0.3, 0.15, 0.1 and 0.05 for the 5-month WMA.

**Data**

The data set can be found by visiting stat.hawkeslearning.com and navigating to **Discovering Business Statistics, Second Edition > Data Sets > Mortgage Rates**.

11. Use the Mortgage Rates data set, which includes the yearly mortgage rate in the U.S. from 1971. Predict the U.S. mortgage rate for the year 2020 using a 4-year SMA and 4-year WMA. For WMA use the weights of 0.4, 0.3, 0.2 and 0.1, respectively.

**Data**

The data set can be found by visiting stat.hawkeslearning.com and navigating to **Discovering Business Statistics, Second Edition > Data Sets > Monthly Average Retail Gas Prices**.

12. Use the Monthly Average Retail Gas Prices data set, which includes the average gas prices in the U.S. from April 1993 to July 2021. Predict the retail gasoline price for August 2021 using a 5-month SMA, compare it with a 3-month SMA.

## 15.3  Exponential Smoothing Techniques

### Simple Exponential Smoothing

Another technique that we will use is called **Simple Exponential Smoothing**. With simple exponential smoothing, we weight the most recent observation more than the past using a convex combination of weights. This weighting scheme allows the forecast to react more strongly to quick changes in the data based on the smoothing constant $\alpha$, which is used as the weight. Small values of $\alpha$ do not react well to changes in the data, whereas large values of $\alpha$ react quickly to changes in the data.

**Definition**

Simple Exponential Smoothing
In **simple exponential smoothing**, we weight the most recent observation more than the past using a convex combination of weights.

> **Formula**
>
> ### Simple Exponential Smoothing
>
> The forecast for the next period is given by
> $$F_{t+1} = \alpha D_t + (1 - \alpha)F_t$$
> where
>
> $D_t$ = the actual data value for the present period,
>
> $F_t$ = the forecast for the present period, and
>
> $\alpha$ = the weight (smoothing constant).

Small values of $\alpha$ do not react well to changes in the data because very little weight is applied to the actual data value and more weight is given to the forecast. Larger values of $\alpha$ give more weight to the actual data value. Note that typical values of $\alpha$ are between 0.1 to 0.5, although $\alpha$ can be any value between 0 and 1, inclusive ($0 \le \alpha \le 1$).

The exponential smoothing technique is one of the basic fundamental blocks of many advanced forecasting methods used today. If we rewrite the formula, we have

$$\begin{aligned} F_{t+1} &= \alpha D_t + (1 - \alpha)F_t \\ &= F_t + \alpha(D_t - F_t) \\ &= \text{Previous Forecast} + \alpha(\text{Forecast Error in the previous period}). \end{aligned}$$

It can be easily seen that for every period, we correct for the potential mistake, or error, that was committed in the previous period. It is essentially a self-correcting method, i.e., the method "learns" from past mistakes. If the predicted forecast was below the actual data value, then the forecast increases in the next period. On the other hand, if we forecasted above the actual value, then the predicted forecast drops in the next period.

When implementing this formula for the forecast of period 2, $F_2$, we need a forecast for the first period, $F_1$. $F_2$ is written as

$$F_2 = \alpha D_1 + (1 - \alpha)F_1.$$

The initial forecast for the first period is not available to us. Thus, we make the forecast for the first period equal to the actual data value of the first period. That is, $F_1 = D_1$. Interestingly, it may seem strange to make a forecast using only data from the previous period (i.e., the forecast for period 4 needs only data from period 3, $D_3$ and $F_3$). However, it is more nuanced than that; in fact, all historical information is included, as shown below.

$$\begin{aligned} F_4 &= \alpha D_3 + (1 - \alpha)F_3 \\ &= \alpha D_3 + (1 - \alpha)\left(\alpha D_2 + (1 - \alpha)F_2\right) \\ &= \alpha D_3 + \alpha(1 - \alpha)D_2 + (1 - \alpha)^2 F_2 \\ &= \alpha D_3 + \alpha(1 - \alpha)D_2 + (1 - \alpha)^2\left(\alpha D_1 + (1 - \alpha)F_1\right) \\ &= \alpha D_3 + \alpha(1 - \alpha)D_2 + \alpha(1 - \alpha)^2 D_1 + (1 - \alpha)^3 F_1 \end{aligned}$$

The above expression shows that all the historical information in the forecast and the weights for the previous periods become smaller and smaller as we move back in time. The rate at which it reduces depends on the value of $\alpha$, which is constrained to be between 0 and 1, inclusive. It should be noted that $\alpha$ is known as the exponential smoothing constant due to the rate at which the weight decreases, as shown in Table 15.3.1 and Figure 15.3.1 for various $\alpha$ values. If $\alpha$ is high, the weight decreases rapidly (in other words, more weight is applied to recent periods) and if $\alpha$ is low, the weight decreases slowly (less weight is applied to recent periods).

| Table 15.3.1 – Weight and Time Period Relationship | | | | | |
|---|---|---|---|---|---|
| Time Period | Weight | $\alpha = 0.1$ | $\alpha = 0.3$ | $\alpha = 0.5$ | $\alpha = 0.9$ |
| $t$ | $\alpha$ | 0.1 | 0.3 | 0.5 | 0.9 |
| $t-1$ | $\alpha(1-\alpha)$ | 0.09 | 0.21 | 0.25 | 0.09 |
| $t-2$ | $\alpha(1-\alpha)^2$ | 0.081 | 0.147 | 0.125 | 0.009 |
| $t-3$ | $\alpha(1-\alpha)^3$ | 0.073 | 0.103 | 0.0625 | 0.0009 |
| | | ... | | | |
| $t-n$ | $\alpha(1-\alpha)^n$ | (will not be zero) | (will not be zero) | (will not be zero) | (will not be zero) |

**Weight and Period Relationship**

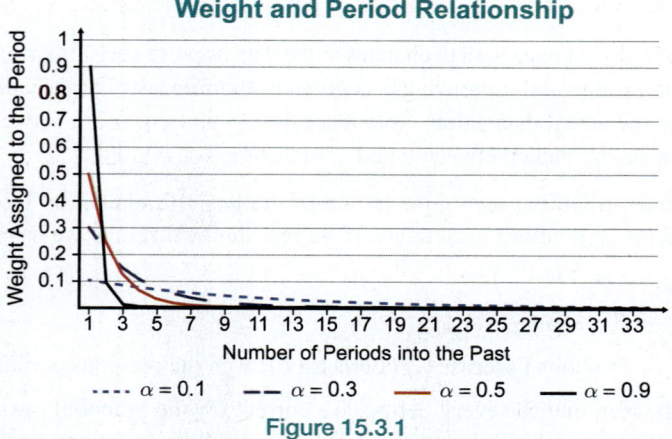

Figure 15.3.1

There is a variety of methods by which $\alpha$ can be determined. We can choose $\alpha$ to minimize a forecast error such as mean absolute deviation (MAD), mean absolute percentage error (MAPE), or the mean squared error (MSE). We will discuss these forecasting errors in Section 15.4. We can also determine $\alpha$ subjectively. The general rule of thumb is to keep the value of $\alpha$ between 0.2 and 0.5. Anything above $\alpha = 0.5$ would indicate the presence of trend and a different forecasting method is needed. Like the SMA and WMA, the simple exponential smoothing technique cannot predict trend. Note that when $\alpha = 1.0$, 100% weight is given to the last period's data, which is oftentimes called the naïve forecast.

---

**Example 15.3.1**

**Computing an Exponential Smoothing Forecast**

Using the TixPixx data from Example 15.2.1, compute an exponentially smoothed forecast for June using $\alpha = 0.3$.

**SOLUTION**

Recall that the formula for exponentially smoothed forecasting is given by $F_{t+1} = \alpha D_t + (1-\alpha)F_t$. To determine the forecast for June using the simple exponential smoothing method, we need to calculate the forecast for periods one through six, which would include June (period six). We assume that the forecast for the first period is the same as the actual time series data value. Therefore,

$$F_1 = D_1 = 9$$
$$F_2 = 0.3(9) + 0.7(9) = 9$$
$$F_3 = 0.3(15) + 0.7(9) = 10.8$$
$$F_4 = 0.3(11) + 0.7(10.8) = 10.86$$
$$F_5 = 0.3(10) + 0.7(10.86) \approx 10.60$$
$$F_6 = 0.3(12) + 0.7(10.6) = 11.02$$

Using previous forecasts, we eventually see that our forecast for the month of June ($F_6$) is 11.02.

The advantage of exponential smoothing techniques is that you don't need to keep or process a lot of data, you only need information from the previous period—the forecast and the actual value. The disadvantage of simple exponential smoothing lies in its inability to forecast trend or seasonality. For example, reviewing the simple exponential forecast of Laredo truck crossings—both monthly and yearly—it can be seen that the forecast lags behind the actual truck volume. The higher the $\alpha$, the closer the forecast is to the actual data. However, the simple exponential forecasting method cannot predict trend (lags behind) nor seasonality (lags behind and smoothens).

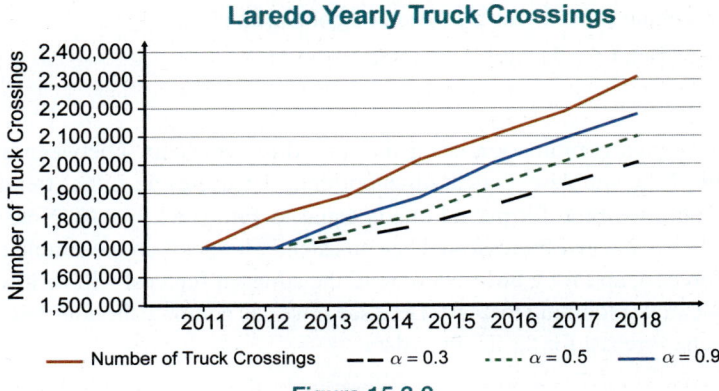

**Figure 15.3.2**

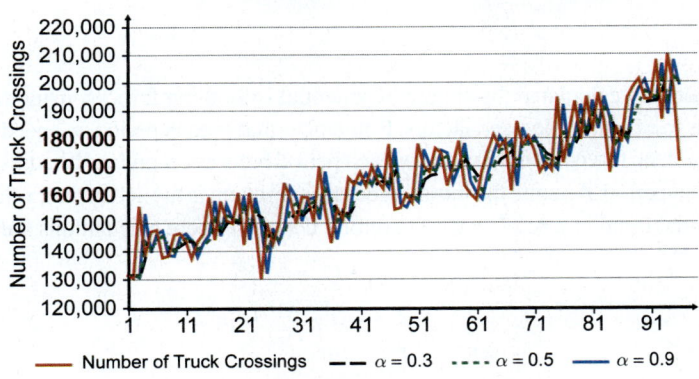

**Figure 15.3.3**

## Adjusted Exponential Smoothing

To predict trend, Holt developed a modified simple exponential smoothing technique, called the **Adjusted Exponential Smoothing** technique. It is popularly known as Holt's Method or exponential smoothing with trend adjustment. It is basically a simple exponential smoothing forecast that we adjust using a trend factor.

**Definition**

**Adjusted Exponential Smoothing**

**Adjusted exponential smoothing** is a simple exponential smoothing forecast that we adjust using a trend factor.

**Formula**

### Adjusted Exponential Smoothing

The adjusted forecast for the next period is given by

$$AF_{t+1} = F_{t+1} + T_{t+1}$$

where

$F_{t+1}$ = the forecast component of the period computed as a simple exponential smoothing forecast.

$$F_{t+1} = \alpha D_t + (1-\alpha) AF_t$$

$$= AF_t + \alpha (D_t - AF_t)$$

$$= \text{Previous Adjusted Forecast} + \alpha (\text{Forecast Error in the previous period}),$$

> **Formula** (cont.)
>
> ### Adjusted Exponential Smoothing
>
> $\alpha$ = the smoothing constant for the forecast component,
>
> $AF_t$ = the previous period's trend adjusted forecast,
>
> $T_{t+1}$ = the forecast for the trend factor,
>
> $T_{t+1} = T_t + \beta(F_{t+1} - AF_t)$ = Previous Trend + $\beta$(Trend Error),
>
> $\beta$ = the smoothing constant for the trend component, and
>
> $T_t$ = the trend factor for the previous period.

Trends are a convex combination using the smoothing constant $\beta$ between our forecast and our trend. Like simple exponential smoothing, it requires an initial trend value; we will assume that the trend for period one is equal to zero (i.e., $T_1 = 0$). Note that for an upward trend, the adjusted forecast will be consistently higher than the simple exponential smoothing forecast, and for a downward trend the adjusted forecast will be lower than the simple exponential smoothing forecast. Let's see if we can compute the adjusted exponential smoothing June forecast for the TixPixx data.

## Example 15.3.2

**Computing an Adjusted Exponential Smoothing Forecast**

Compute the adjusted exponential smoothing forecast for June using $\beta = 0.2$ and $\alpha = 0.3$ (the same $\alpha$ value as in Example 15.3.1).

### SOLUTION

We have already calculated the simple exponential smoothing forecast using $\alpha = 0.3$. The next step is to calculate the trend factors. Remember that in each period, we need to compute the trend and the forecast component for each period before computing the adjusted forecast ($AF$) for that period. Assume the initial forecast component is the actual data value ($AF_1 = 9$) and the initial trend is zero ($T_1 = 0$). Therefore, the subsequent calculations are as follows.

$$F_{t+1} = \alpha D_t + (1 - \alpha)AF_t$$
$$T_{t+1} = T_t + \beta(F_{t+1} - AF_t)$$
$$AF_{t+1} = F_{t+1} + T_{t+1}$$

Period 1:

$F_1 = D_1 = 9$

$T_1 = 0$

$AF_1 = 9 + 0 = 9$

Period 2:

$F_2 = 0.3(9) + 0.7(9) = 9$

$T_2 = 0 + 0.2(0) = 0$

$AF_2 = F_2 + T_2 = 9 + 0 = 9$

Period 3:

$F_3 = 0.3(15) + 0.7(9) = 10.8$

$T_3 = 0 + 0.2(10.8 - 9) = 0.36$

$AF_3 = 11.16$

The table below contains the remaining forecasts.

| Month | Actual Sales | Forecast Component | Trend Component | Adjusted Forecast |
|---|---|---|---|---|
| January | 9 | 9.00 | 0.00 | 9.00 |
| February | 15 | 9.00 | 0.00 | 9.00 |
| March | 11 | 10.80 | 0.36 | 11.16 |

| Month | Actual Sales | Forecast Component | Trend Component | Adjusted Forecast |
|-------|-------------|--------------------|-----------------|-------------------|
| April | 10 | 11.11 | 0.35 | 11.46 |
| May | 12 | 11.02 | 0.26 | 11.28 |
| June | - | 11.50 | 0.30 | 11.80 |

Therefore, the adjusted exponential smoothing forecast for June is 11.80.

Figure 15.3.4 contains the simple exponentially smoothed and the adjusted exponentially smoothed forecasts for the yearly Laredo truck crossings using $\alpha = 0.3$ and $\beta = 0.4$, respectively. Note that the forecast is very close to the actual data.

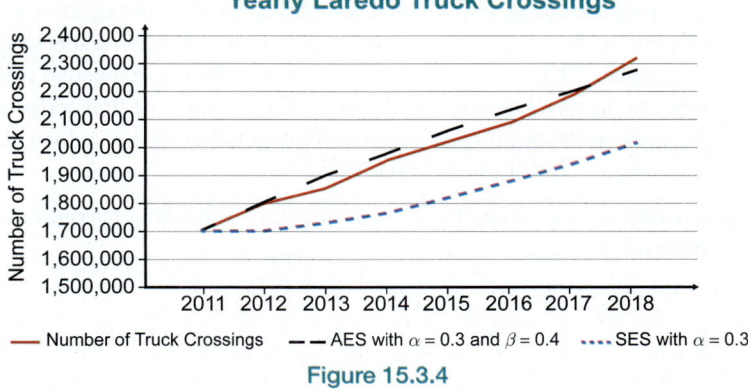

**Figure 15.3.4**

---

# 📝 15.3 Exercises

## Basic Concepts

1. What is a simple exponential smoothing?

2. How do different values of $\alpha$ react to changes in the data in simple exponential smoothing?

3. Describe the relationship between weight and period in simple exponential smoothing.

4. Give an advantage and a disadvantage of using simple exponential smoothing.

5. What do we adjust for in adjusted exponential smoothing?

6. For an upward trend, which forecast will be higher? The adjusted or the simple exponential smoothing? What about a downward trend?

## Exercises

7. Use the Border Crossings data set. Using $\alpha = 0.3$, calculate the simple exponential smoothing yearly forecast of Detroit truck crossings for 2011–2019. Assume the forecast for the year 2011 to be the actual truck crossing value of year 2011.

8. Use the Border Crossings data set. Using $\alpha = 0.3$ and $\beta = 0.4$, calculate the adjusted exponential smoothing yearly forecast of Laredo truck crossings for 2011–2019. Assume the forecast component for the year 2011 to be the actual truck crossing value of year 2011 minus the initial trend. Let the initial trend be 100,000 trucks. (Note: One way to compute the initial trend is to determine the slope of a linear regression line fit to the data. In this case, if you fit a line to the Laredo yearly truck crossing data, you will get a slope of 84,220. This could be used as the initial trend.)

## ⁘ Data

The data set can be found by visiting  and navigating to **Discovering Business Statistics, Second Edition > Data Sets > Border Crossings**.

9. Use the Mortgage Rates data set, which contains the yearly mortgage rate in the U.S. since 1971.

   a. Calculate the simple exponential smoothing forecast and the adjusted exponential smoothing forecast for the yearly mortgage rates for 2020. Assume $\alpha = 0.2$ and $\beta = 0.3$ for the respective methods.

   b. Create a time series plot with both forecasts. What conclusions can you draw from it?

10. Use the Monthly Average Retail Gas Prices data set, which includes the average gas prices in the U.S. from April 1993 to July 2021.

   a. Compare the simple exponential smoothing forecast and adjusted exponential smoothing forecast for the monthly gas price and their respective forecasts for August 2020. Assume $\alpha = 0.3$ and $\beta = 0$ for the respective methods and forecast the gasoline price for August 2021. Assume the first period forecasted gas price as the original gas price, and the initial trend is the slope of the linear regression line that fits the data.

   b. Create a time series plot with both forecasts. What conclusions can you draw from it?

# 15.4 Forecast Accuracy

## Accuracy of Forecasts

A forecast is only as good as it is accurate. However, one should realize before making predictions that the forecast is seldom perfect. Therefore, it is essential to know how good it is. To understand the quality of the forecast, we should examine the forecast error. In simple terms, the forecast error is the difference between the forecast and the actual data value. Small forecast errors mean we have a good (or accurate) forecast, while larger forecast errors imply that we do not have a good forecast. By changing the number of periods in a moving average or changing the $\alpha$ and $\beta$ in the exponential smoothing forecasts, we may be able to reduce the forecast errors.

Forecast errors can be broadly classified into two groups: one that measures error by looking at absolute deviation and the other that measures error in terms of bias. The absolute deviation measures include mean absolute deviation (MAD), mean absolute percent error (MAPE), weighted mean absolute percent error (WMAPE), and mean squared error (MSE). The absolute error quite broadly says how good the forecast is regardless of the sign of the error. A forecast is bad regardless of whether it is significantly above or significantly below the actual value. The bias measures include running sum of forecast error (RSFE) and tracking signal (TS). The bias error metrics reveal whether the forecast is consistently above or below the actual value.

**Definition**

**Forecast Error**

**Forecast error** is the difference between the forecast and the actual data value.

## Absolute Deviation Error Metrics

Notationally, the forecast error is the difference between actual and forecast, expressed as $(D_t - F_t)$, where $D_t$ is the actual value for period $t$ and $F_t$ is the forecast for period $t$. If the forecast error is positive, the forecast is less than the actual value; and if it is negative, the forecast is more than the actual value.

**Formula**

### Mean Absolute Deviation (MAD)

The mean absolute deviation is the sum of the absolute value of the differences between the actual data and the forecast, divided by $n$, the number of periods for which the error is computed.

$$\text{MAD} = \frac{\Sigma|D_t - F_t|}{n}$$

where

$D_t$ = the actual data for the $t^{\text{th}}$ period,

$F_t$ = the forecast for the $t^{\text{th}}$ period,

$n$ = the total number of periods for which we have computed the forecasts, and

$t$ = the time period number.

Consider the following time series data and forecast for the number of deliveries made by a driver over the last six months. Compute the MAD of the forecast.

**Example 15.4.1**

**Calculating the Mean Absolute Deviation of Forecasts**

| Time Period, $t$ | Actual Value, $D_t$ | Forecast, $F_t$ | Forecast Error, $FE_t$ | Absolute Error, $|FE_t|$ |
|---|---|---|---|---|
| 1 | 134 | 132 | 2 | 2 |
| 2 | 142 | 141 | 1 | 1 |
| 3 | 143 | 145 | −2 | 2 |
| 4 | 156 | 143 | 13 | 13 |
| 5 | 151 | 154 | −3 | 3 |
| 6 | 145 | 150 | −5 | 5 |
| | | | Total | 26 |

**SOLUTION**

$$\text{MAD} = \frac{\sum |D_t - F_t|}{n} = \frac{2+1+2+13+3+5}{6} = \frac{26}{6} \approx 4.333$$

If we were actually performing calculations using a 3-month simple moving average method, the first period we could forecast is period 4. In that case, we could compute the forecast for only periods 4, 5 and 6, in the above example. Thus, when MAD is computed, we would divide by 3, which represents the number of periods for which we have forecasts. Therefore, remember that $n$ in the denominator represents the number of periods for which we have forecasts.

In the above example the MAD is approximately 4.333. However, it is very difficult to know if this is a small or high error metric. It is relative to the actual value. For example, if the average value of the data was 10, then being off by 4.333 on an average equates to a 43% error. On the other hand, if the average is approximately 150 units, being off by 4.333 is approximately a 3% error. Without knowing the magnitude of the data, using the MAD will make it difficult to assess whether your forecast is good or not.

A variation of the MAD is the mean absolute percentage error (MAPE). The MAPE measures the absolute error as a percentage of the actual value. The MAPE eliminates the problem of interpreting the measure of accuracy relative to the magnitude of the actual and forecast values.

---

**Formula**

### Mean Absolute Percentage Error (MAPE)

In this error metric, we compute the percent by which we miss the actual data in every period. The formula is as follows.

$$\text{MAPE} = \frac{100}{n} \sum_{t=1}^{n} \frac{|D_t - F_t|}{D_t}$$

where

$D_t$ = the actual data for the $t^{th}$ period,

$F_t$ = the forecast for the $t^{th}$ period,

$n$ = the total number of periods for which we have computed the forecasts, and

$t$ = the time period number.

---

**Example 15.4.2**

**Calculating the MAPE of Forecasts**

For the previous time series data in Example 15.4.1, compute the MAPE.

| Time Period $t$ | Actual data $D_t$ | Forecast $F_t$ | Error $FE_t$ | Absolute Error $|FE_t|$ | Abs. Percent Error, $100\dfrac{|FE_t|}{D_t}$ |
|---|---|---|---|---|---|
| 1 | 134 | 132 | 2 | 2 | 1.493% |
| 2 | 142 | 141 | 1 | 1 | 0.704% |
| 3 | 143 | 145 | −2 | 2 | 1.399% |
| 4 | 156 | 143 | 13 | 13 | 8.333% |
| 5 | 151 | 154 | −3 | 3 | 1.987% |
| 6 | 145 | 150 | −5 | 5 | 3.448% |
|  |  |  |  | Total | 17.364% |

**SOLUTION**

$$\text{MAPE} = \frac{100}{n} \sum_{t=1}^{n} \frac{|D_t - F_t|}{D_t} = \frac{17.364}{6} = 2.894\%$$

The above error metric is useful as it gives a relative measure of accuracy. Anyone looking at this error metric would know if the time series forecast is good or not without looking at the actual data. However, this error metric is not popular with the primary drawback being that it is possible to have time series data with actual zeros. In that case, the absolute percent error would be indeterminate since we will be dividing by zero. It would also not be appropriate to omit that period's error metric since the forecast may be a non-zero value. Therefore, we need another metric that addresses this deficiency but still maintains the advantage of a scaled error metric.

## Mean Absolute Percentage Deviation (MAPD)

The mean absolute percentage deviation (MAPD) is sometimes preferred to the MAPE. We compute it as the sum of the absolute difference between the actual data and the forecast divided by the sum of the actual data; in other words, it is the MAD over the average actual value!

$$\text{MAPD} = \frac{\sum |D_t - F_t|}{\sum D_t} = \frac{\text{MAD}}{\text{Average Actual Value}}$$

For the above example the $\text{MAPD} = \frac{4.333}{145.16} = 2.98\%$.

The only way this error metric would be indeterminate is when the average is zero which is possible only when the actual value is zero for all periods. Therefore, MAPD is a better measure of forecast accuracy regardless of the distribution of the time series data. It is especially useful when the time series data is intermittent (i.e., has a lot of zeros as an actual value).

> **Definition**
>
> **Mean Absolute Percentage Deviation (MAPD)**
>
> The **mean absolute percentage deviation** is the sum of the absolute differences between the actual data and the forecasts divided by the sum of the actual data.

## Mean Squared Error (MSE)

The **Mean Squared Error (MSE)** is the average of the squared errors. It is the sum of the squared deviations between the forecast and the actual divided by $n$. Again, please note the $n$ represents the number of periods for which we have a forecast. This error metric is especially useful when we are trying to identify an outlier. In regression, we try to minimize this error. That is, we try to fit a straight line through the data series such that the errors are minimized.

> **Definition**
>
> **Mean Squared Error (MSE)**
>
> The **mean squared error** is the average of the squared errors.

> **Formula**
>
> **Mean Squared Error (MSE)**
>
> The MSE is computed as follows:
>
> $$\text{MSE} = \frac{\sum (D_t - F_t)^2}{n}$$
>
> where
>
> $D_t$ = the actual data for the $t^{\text{th}}$ period,
>
> $F_t$ = the forecast for the $t^{\text{th}}$ period,
>
> $n$ = the total number of time periods for which we have computed the forecasts, and
>
> $t$ = the time period number.

**Example 15.4.3**

**Calculating the Mean Squared Error of Forecasts**

Using the data in Example 15.4.1, compute the MSE.

| Time Period $t$ | Actual data $D_t$ | Forecast $F_t$ | Forecast Error $FE_t$ | Absolute Error $|FE_t|$ | Sq. Error $(FE_t)^2$ |
|---|---|---|---|---|---|
| 1 | 134 | 132 | 2 | 2 | 4 |
| 2 | 142 | 141 | 1 | 1 | 1 |
| 3 | 143 | 145 | −2 | 2 | 4 |
| 4 | 156 | 143 | 13 | 13 | 169 |
| 5 | 151 | 154 | −3 | 3 | 9 |
| 6 | 145 | 150 | −5 | 5 | 25 |
| | | | | Total | 212 |

**SOLUTION**

$$\text{MSE} = \frac{212}{6} \approx 35.333$$

Looking at the above table, we can clearly see that Period 4 is an outlier. If one wants to reduce the error associated with Period 4, one may try by changing the number of periods in a moving average method or by changing the $\alpha$ and $\beta$ of the exponential smoothing methods.

One point of caution is that lower mean squared errors may result from a technique that fits older values better than the most recent values. Keep in mind that sometimes it is better to compute the mean squared error using the most recent values rather than using the entire data set. That is, it may be better to pick a recent set of observations from the data set to compute the MSE. It is feasible to take a subset of the most recent observations and calculate the MSE to determine the quality of the forecasts if the user is more concerned about forecasting recent values rather than those values in the distant past.

Having a lower (or higher) error forecast, but consistently below (or above) the actual time series values, indicates that the forecast is biased. To identify bias, it is not enough to just look at one period's forecast but one must observe the direction of the errors over a period of time. There are two bias error metrics commonly used: cumulative error and tracking signal.

**Definition**

Cumulative Error ($E$)

The **cumulative error** ($E$) is the sum of the forecast error.

## Cumulative Error ($E$)

The **cumulative error** is the sum of the forecast errors. The formula involves calculating the forecast error for each period as the difference between the time series actual value and the forecasts, and adding them up. The computation formula is given as

$$E = \Sigma(D_t - F_t).$$

The resulting sign of the error indicates the direction of the bias. If the cumulative error is positive, then the forecast is underestimating the time series data. If the sign is negative, then the forecast is overestimating the time series data on average.

**Example 15.4.4**

**Calculating the Cumulative Error**

Using the data in Example 15.4.1, compute the cumulative error.

| Time Period, $t$ | Actual data, $D_t$ | Forecast, $F_t$ | Error, $FE_t$ |
|---|---|---|---|
| 1 | 134 | 132 | 2 |
| 2 | 142 | 141 | 1 |
| 3 | 143 | 145 | −2 |
| 4 | 156 | 143 | 13 |
| 5 | 151 | 154 | −3 |
| 6 | 145 | 150 | −5 |

## SOLUTION

$$E = 2 + 1 - 2 + 13 - 3 - 5 = 6$$

The cumulative error ($E$) for this dataset is 6, which indicates that the forecasts are consistently underestimating the actual data. However, a closer look at the forecast errors reveals that there are ups and downs and the positive value of $E$ is primarily a result of the forecast for Period 4 which underestimates the actual value by 13 units. Thus, examining just $E$ is somewhat of a disadvantage given that three of the forecasts are overestimations and three are underestimations. This drawback is overcome by the tracking signal error metric.

## Tracking Signal (*TS*)

As can be seen in Example 15.4.4, bias should never be confirmed using a single measure. Bias should be observed over time, which is the primary purpose of the **tracking signal** (*TS*). It is computed for each time period using the following formula.

$$TS = \frac{E}{\text{MAD}}$$

There are typically two ground rules to detect bias using the tracking signal, which are based on the control chart principles (see Chapter 18). The first rule is that any time the *TS* is above +4 or below −4, it is considered to be "out of control" and the forecast is biased. The second rule is to ensure that the *TS* has no trend, increasing upwards or decreasing downwards, for any considerable amount of time. A trend in the *TS* indicates that the forecasts are inching towards being biased and some corrective measures should be taken, such as using a different forecasting method.

Using the data in Example 15.4.1, compute the tracking signal (*TS*).

**Example 15.4.5**

**Calculating the Tracking Signal for Forecasts**

| Time Period, $t$ | Actual data, $D_t$ | Forecast, $F_t$ | Error, $FE_t$ | $E$ | MAD | $TS = \dfrac{E}{\text{MAD}}$ |
|---|---|---|---|---|---|---|
| 1 | 134 | 132 | 2 | 2 | 2.00 | 1.00 |
| 2 | 142 | 141 | 1 | 3 | 1.500 | 2.00 |
| 3 | 143 | 145 | −2 | 1 | 1.667 | 0.60 |
| 4 | 156 | 143 | 13 | 14 | 4.500 | 3.11 |
| 5 | 151 | 154 | −3 | 11 | 4.200 | 2.62 |
| 6 | 145 | 150 | −5 | 6 | 4.333 | 1.38 |

## SOLUTION

As you can see from the above table, the forecast is biased as indicated by $E$. But the *TS* is within +4 and −4 and there is no discernible trend; it goes up and down randomly.

There are several error metrics in forecasting analysis. They all let you make the same decisions—whether the forecast is good or not. Depending upon on our need, we could choose the appropriate error metric on which to focus. For example, MAPD is suggested for accuracy, *TS* is suggested for bias, and MSE is suggested for outliers.

# ✍ 15.4 Exercises

## Basic Concepts

1. What is forecast error?

2. If the forecast error is positive, what does it mean? What if it is negative?

3. Mean absolute deviation relies on which variable in order to assess if the forecast is good or not?

4.  Describe one key difference between mean absolute deviation and mean absolute percentage error.

5.  What is one draw back from the mean absolute percentage error that makes this error metric not so popular?

6.  What is the only way that mean absolute percentage deviation is indeterminate?

7.  When is the mean squared error especially useful?

8.  Having a lower (or higher) error forecast, but consistently below (or above) the time series actual value is an indicator of what?

9.  What does a positive/negative cumulative error indicate?

10. What is the drawback in cumulative error that is overcome by the tracking signal?

## Exercises

**:: Data**

The data set can be found by visiting stat.hawkeslearning.com and navigating to **Discovering Business Statistics, Second Edition > Data Sets > Border Crossings**.

11. Use the Border Crossings data set. For the 3-month SMA forecast of Laredo truck crossings, compute the MAPD and plot the $TS$. Is the forecast good?

12. Use the Border Crossings data set. For the simple exponential smoothing forecast of Detroit truck crossings, compute the MAPD and plot the $TS$. Is the forecast good? (Use $\alpha = 0.2$.)

13. Use the Border Crossings data set. For the adjusted exponential smoothing forecast of Laredo truck crossings, compute the MAPD and plot the $TS$. Is the forecast good? (Use $\alpha = 0.2$; $\beta = 0.4$.)

**:: Data**

The data set can be found by visiting stat.hawkeslearning.com and navigating to **Discovering Business Statistics, Second Edition > Data Sets > Monthly Average Retail Gas Prices**.

14. Use the Monthly Average Retail Gas Prices data set, which includes the average gas prices in the U.S. from April 1993 to July 2021. Perform a simple exponential smoothing forecast of retail gasoline price and compute the MSE. (Use $\alpha = 0.3$.)

15. Use the Monthly Average Retail Gas Prices data set, which includes the average gas prices in the U.S. from April 1993 to July 2021. Perform an adjusted exponential smoothing forecast of retail gasoline price and compute the MAPD. (Use $\alpha = 0.3$; $\beta = 0.4$.)

16. Use the Monthly Average Retail Gas Prices data set, which includes the average gas prices in the U.S. from April 1993 to July 2021. Calculate the best $\alpha$, $\beta$ combination that minimizes the MAPD of the forecast for retail gasoline price.

17. Which error metric(s) should we concentrate on for each forecasting objective?

| Objective | Error Metrics |
|---|---|
| Minimize Outliers | |
| Minimize Overall Forecast Errors | |
| Minimize Bias | |
| Minimize Overall Forecast Errors of intermittent items | |

# 15.5 Seasonality

Apart from trend, one of the most common time series patterns is one that is **seasonal**. A **seasonal pattern** is a repetitive up-and-down movement in a time series that can occur on a quarterly, monthly, weekly, or daily basis. Two common approaches to forecasting seasonality are the additive seasonal forecasting method and the multiplicative seasonal forecasting method.

> **Definition**
>
> Seasonal Pattern
>
> A **seasonal pattern** is a repetitive up-and-down movement in a time series that can occur on a quarterly, monthly, weekly, or daily basis.

## Additive Seasonal Forecasting

With the additive seasonal forecasting method, we assume the seasonal adjustments are constant and do not change over time. This method is accomplished by a regression technique, with each season being represented by a dummy variable. Suppose we have four quarters. In this case, we will use three dummy variables, $Q_1$, $Q_2$, and $Q_3$, which take the value 1, if it is that respective quarter, and 0, otherwise. When all three of the dummy variables are zero, then it represents the 4th quarter. Similarly, if we have daily seasonality, then we will have six dummy variables $(D_1, D_2, \ldots, D_6)$ and if we have monthly seasonality, we will have 11 dummy variables $(D_1, D_2, \ldots, D_{11})$. If trend is present, we will introduce another variable $t$, to represent the time period.

We will demonstrate this method using the quarterly truck volume of the Laredo truck crossings. Consider the 3-year truck crossings at the Laredo border (Note: For a good seasonality forecast, we need at least 3 to 5 seasons worth of data (i.e., 3 to 5 years' worth of data if it is monthly/quarterly data).

**Table 15.5.1 – Quarterly Truck Volume for the Laredo Truck Crossings**

|  | Quarter 1 | Quarter 2 | Quarter 3 | Quarter 4 |
|---|---|---|---|---|
| 2016 | 500,794 | 536,312 | 523,496 | 523,362 |
| 2017 | 533,245 | 545,223 | 555,682 | 548,834 |
| 2018 | 556,298 | 592,371 | 587,238 | 578,060 |

First we must rearrange the data so that we can introduce the dummy variables representing the quarter as shown in Table 15.5.2.

**Table 15.5.2 – Quarterly Truck Volume for the Laredo Truck Crossings with Dummy Variables**

| Year | Quarter | Truck Crossings | $Q_1$ | $Q_2$ | $Q_3$ |
|---|---|---|---|---|---|
| 2016 | 1 | 500,794 | 1 | 0 | 0 |
|  | 2 | 536,312 | 0 | 1 | 0 |
|  | 3 | 523,496 | 0 | 0 | 1 |
|  | 4 | 523,362 | 0 | 0 | 0 |
| 2017 | 1 | 533,245 | 1 | 0 | 0 |
|  | 2 | 545,223 | 0 | 1 | 0 |
|  | 3 | 555,682 | 0 | 0 | 1 |
|  | 4 | 548,834 | 0 | 0 | 0 |
| 2018 | 1 | 556,298 | 1 | 0 | 0 |
|  | 2 | 592,371 | 0 | 1 | 0 |
|  | 3 | 587,238 | 0 | 0 | 1 |
|  | 4 | 578,060 | 0 | 0 | 0 |

Next, we should fit a simple linear regression model to the data using truck crossings as the dependent variable and using the three dummy variables as the independent variables. Note that you should have an intercept in the simple linear regression model.

For the above data, the fitted simple linear regression model is given by

$$\text{Truck Crossings} = 550{,}085 - 19{,}973(Q_1) + 7883(Q_2) + 5387(Q_3).$$

The estimated number of Truck Crossings by quarter are:

$Q_1$: $550,085 - 19,973(1) + 7883(0) + 5387(0) = 530,112$

$Q_2$: $550,085 - 19,973(0) + 7883(1) + 5387(0) = 557,968$

$Q_3$: $550,085 - 19,973(0) + 7883(0) + 5387(1) = 555,472$

$Q_4$: $550,085 - 19,973(0) + 7883(0) + 5387(0) = 550,085$

The fitted regression model shows that the base forecast is for Quarter 4 (when $Q_1$, $Q_2$, and $Q_3$ are equal to zero). Quarter 1 is the lowest (valley) and the highest is Quarter 2 (peak), followed by Quarter 3 and Quarter 4. The slopes represent the respective seasonal adjustments.

As we know, the Laredo truck crossing data has both trend and seasonality. Therefore, we need to introduce the variable $t$ representing the time period.

| Table 15.5.3 – Quarterly Truck Volume for Laredo Truck Crossing using Dummy Variables | | | | | | |
|---|---|---|---|---|---|---|
| Year | Quarter | Truck Crossings | $Q_1$ | $Q_2$ | $Q_3$ | $t$ |
| 2016 | 1 | 500,794 | 1 | 0 | 0 | 1 |
| | 2 | 536,312 | 0 | 1 | 0 | 2 |
| | 3 | 523,496 | 0 | 0 | 1 | 3 |
| | 4 | 523,362 | 0 | 0 | 0 | 4 |
| 2017 | 1 | 533,245 | 1 | 0 | 0 | 5 |
| | 2 | 545,223 | 0 | 1 | 0 | 6 |
| | 3 | 555,682 | 0 | 0 | 1 | 7 |
| | 4 | 548,834 | 0 | 0 | 0 | 8 |
| 2018 | 1 | 556,298 | 1 | 0 | 0 | 9 |
| | 2 | 592,371 | 0 | 1 | 0 | 10 |
| | 3 | 587,238 | 0 | 0 | 1 | 11 |
| | 4 | 578,060 | 0 | 0 | 0 | 12 |

Fitting a simple linear regression model using the quarter, dummy variables, and time as the independent variables, yields the following additive seasonal model with trend.

$$\text{Truck Crossings} = 492,585 + 1590(Q_1) + 22,259(Q_2) + 12,574(Q_3) + 7188(t)$$

Note that there is a positive trend (7188) indicating that the number of truck crossings increases by approximately 7200 trucks every quarter. The disadvantage of an additive seasonal model is that the seasonal adjustment does not increase over time. That is, if there is a 1590 truck adjustment for Quarter 1 (see the estimated model above), it remains the same year-over-year—in 2016, 2017 and 2018. This shortcoming can be overcome by using the multiplicative seasonal model.

## Multiplicative Seasonal Forecasting

Definition

Seasonal Factor

A seasonal factor is a numerical value that is multiplied by the normal forecast to get a seasonally adjusted forecast.

The multiplicative seasonal forecasting method is one in which a seasonally adjusted forecast can be developed by multiplying the normal forecast by a seasonal factor. A **seasonal factor** is a numerical value that is multiplied by the normal forecast to get a seasonally adjusted forecast. The seasonal factor can be determined by dividing the actual data for each seasonal period by the sum of the total time series values.

$$S_i = \frac{D_i}{\sum D_i}$$

The resulting seasonal factors between 0 and 1 are, in effect, the portion of the total time series data assigned to each season. These seasonal factors are multiplied by the forecast value to yield seasonally adjusted forecasts for each period. The most common multiplicative seasonal forecasting method is *Holt–Winters Method* which takes the adjusted exponential forecast and multiplies it by the seasonal factor.

Let's look at a simple multiplicative seasonal model which is based on a simple linear regression using the Laredo crossings data. To perform a multiplicative seasonal forecast, we do the following:

1. Compute the seasonal factors, $S_i$, for each quarter.
2. Fit a linear regression model to the totals for each year.
3. Compute the trend forecast for 2019 using the simple linear regression model.
4. Multiply the forecast by the computed seasonal indices to obtain the final forecasts.

In a seasonal forecasting method, it is possible to forecast more than one period – we can forecast the entire season for the next year.

To demonstrate, we will use the same three years of data that were used for the previous Laredo truck crossing example.

|  | Quarter 1 | Quarter 2 | Quarter 3 | Quarter 4 | Total |
|---|---|---|---|---|---|
| 2016 | 500,794 | 536,312 | 523,496 | 523,362 | 2,083,964 |
| 2017 | 533,245 | 545,223 | 555,682 | 548,834 | 2,182,984 |
| 2018 | 556,298 | 592,371 | 587,238 | 578,060 | 2,313,967 |
| **Total** | **1,590,337** | **1,673,906** | **1,666,416** | **1,650,256** | **6,580,915** |

$$S_1 = \text{Seasonal factor of Quarter 1} = \frac{1,590,337}{\text{Overall Sum}} = \frac{1,590,337}{6,580,915} = 0.2417$$

$$S_2 = \frac{1,673,906}{6,580,915} = 0.2544$$

$$S_3 = \frac{1,666,416}{6,580,915} = 0.2532$$

$$S_4 = \frac{1,650,256}{6,580,915} = 0.2508$$

From the seasonal factors, we know that Quarter 2 is the peak (highest), while Quarter 1 is the valley (lowest).

We want to multiply the forecasted truck crossings for the next year by each of the seasonal factors to get the forecasted truck crossings for each quarter.

To do this, we need a forecast for 2019. We compute a linear trend line using the sum for each year to get a forecast for 2019 (4th year) of 2,423,641.

$$y = 1,963,635 + 115,002x = 1,963,635 + 115,002(4) = 2,423,643$$

The final forecasts for 2019 are obtained by multiplying by the seasonality factors.

Quarter 1 of 2019 = $2,423,643 \cdot S_1 = 2,423,643 \cdot 0.2417 = 585,795$

Quarter 2 of 2019 = $2,423,643 \cdot S_2 = 2,423,643 \cdot 0.2544 = 616,575$

Quarter 3 of 2019 = $2,423,643 \cdot S_3 = 2,423,643 \cdot 0.2532 = 613,666$

Quarter 4 of 2019 = $2,423,643 \cdot S_4 = 2,423,643 \cdot 0.2508 = 607,850$

Like the forecasts for 2016 through 2018, these quarterly forecasts for 2019 tend to follow a general upward trend. Note that Quarter 1 is still the valley and Quarter 2 is still the peak for 2019.

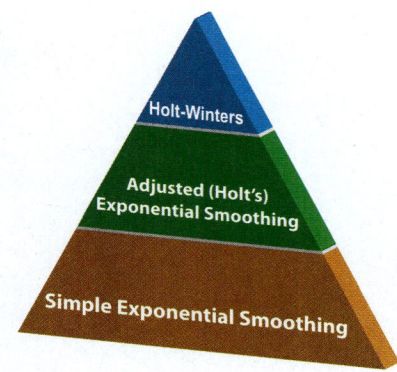

## The Holt–Winters Method

Charles C. Holt (1921–2010) was a professor at the Department of Management at the McCombs School of Business at the University of Texas at Austin.

During the late 1950sHolt modified the simple exponential smoothing model in order to account for a trend component. This is known today as Holt's exponential smoothing, adjusted exponential smoothing, or double exponential smoothing. However, with seasonal time series data, this technique didn't perform well. One of Holt's students, Peter Winters, then improved his teacher's model by introducing an additional parameter to account for the seasonality component.

This improved method with the seasonality component is known today as the Holt-Winters method, or triple exponential smoothing, because the three aspects of time series— value, trend, and seasonality—are represented by the three types of exponential smoothing. The model requires several parameters: one for each type of exponential smoothing ($\alpha$, $\beta$, $\gamma$), one for the length of a season, and one for the number of periods in a season.

## ✎ 15.5 Exercises

### Basic Concepts

1. What is seasonality?

2. Give a business scenario where we would see seasonality.

3. What assumption do we make with the additive seasonal forecasting method?

4. What is a disadvantage of the additive seasonal model? Which model can overcome this?

5. What is a seasonal factor? What does it represent?

6. What are the four steps to perform a multiplicative seasonal forecast?

### Exercises

**⋰ Data**

The data set can be found by visiting stat.hawkeslearning.com and navigating to **Discovering Business Statistics, Second Edition > Data Sets > Border Crossings**.

7. Use the Border Crossings data set. Calculate the monthly seasonality factor for the Laredo Truck Crossings data.

8. Use the Border Crossings data set. Calculate the monthly seasonality factor for the Detroit Truck Crossings data.

9. Use the Border Crossings data set. Compute the additive quarterly seasonal forecast model for the Detroit Truck Crossing data without trend.

10. Use the Border Crossings data set. Compute the additive quarterly seasonal forecast model for the Detroit Truck Crossing data with trend.

**⋰ Data**

The data set can be found by visiting stat.hawkeslearning.com and navigating to **Discovering Business Statistics, Second Edition > Data Sets > Monthly Average Retail Gas Prices**.

11. Use the Monthly Average Retail Gas Prices data set, which includes the average gas prices in the U.S. from April 1993 to July 2021. Using the data from January 1994 to December 2020, compute the monthly seasonality factor for the gas prices. Which month or months generally have the highest gas prices in the U.S.?

12. Use the Monthly Average Retail Gas Prices data set, which includes the average gas prices in the U.S. from April 1993 to July 2021. Use the data from January 2016 to December 2020 to compute the monthly forecast of gas prices for 2021 if the same trend and seasonality pattern continue.

# T  Discovering Technology

## Using Excel

### Simple Moving Average

Use the information in Example 15.2.1 for this exercise.

1.  Enter the values for **Month** and **Tickets Purchased** in columns **A** and **B**, respectively.

    |    | A        | B                |
    |----|----------|------------------|
    | 1  | Month    | Tickets Purchased |
    | 2  | January  | 9                |
    | 3  | February | 15               |
    | 4  | March    | 11               |
    | 5  | April    | 10               |
    | 6  | May      | 12               |

2.  To calculate the simple moving average for June, in cell **C7**, enter **=AVERAGE (B4:B6)** as shown below. Press **Enter**.

    |    | A        | B                | C               | D |
    |----|----------|------------------|-----------------|---|
    | 1  | Month    | Tickets Purchased |                 |   |
    | 2  | January  | 9                |                 |   |
    | 3  | February | 15               |                 |   |
    | 4  | March    | 11               |                 |   |
    | 5  | April    | 10               |                 |   |
    | 6  | May      | 12               |                 |   |
    | 7  | June     |                  | =AVERAGE(B4:B6) |   |

3.  The simple moving average for June is **11**. Compare this to the result obtained in Example 15.2.1.

    |    | A        | B                | C  |
    |----|----------|------------------|----|
    | 1  | Month    | Tickets Purchased |    |
    | 2  | January  | 9                |    |
    | 3  | February | 15               |    |
    | 4  | March    | 11               |    |
    | 5  | April    | 10               |    |
    | 6  | May      | 12               |    |
    | 7  | June     |                  | 11 |

### Weighted Moving Average

Use the information from Example 15.2.2 for this exercise.

1.  Enter the values for **Month** and **Tickets Purchased** in Columns **A** and **B**, respectively. Create a column for **Weight** in column **C** as shown below.

| | A | B | C |
|---|---|---|---|
| 1 | Month | Tickets Purchased | Weight |
| 2 | January | 9 | 0 |
| 3 | February | 15 | 0 |
| 4 | March | 11 | 0.1 |
| 5 | April | 10 | 0.3 |
| 6 | May | 12 | 0.6 |
| 7 | June | | |

2.    In cell **D7**, enter "**=SUMPRODUCT(B4:B6,C4:C6)**" and press **Enter**.

| | A | B | C | D | E | F |
|---|---|---|---|---|---|---|
| 1 | Month | Tickets Purchased | Weight | | | |
| 2 | January | 9 | 0 | | | |
| 3 | February | 15 | 0 | | | |
| 4 | March | 11 | 0.1 | | | |
| 5 | April | 10 | 0.3 | | | |
| 6 | May | 12 | 0.6 | | | |
| 7 | June | | | =SUMPRODUCT(B4:B6,C4:C6) | | |

3.    The desired weighted moving average for June for a period of 3 will appear as **11.3**. Compare this to the result obtained in Example 15.2.2.

| | A | B | C | D |
|---|---|---|---|---|
| 1 | Month | Tickets Purchased | Weight | |
| 2 | January | 9 | 0 | |
| 3 | February | 15 | 0 | |
| 4 | March | 11 | 0.1 | |
| 5 | April | 10 | 0.3 | |
| 6 | May | 12 | 0.6 | |
| 7 | June | | | 11.3 |

## Simple Exponential Smoothing

Use the information in Example 15.3.1 for this exercise.

1.    Enter the values for **Month** and **Tickets Purchased** in columns **A** and **B**, respectively.

| | A | B |
|---|---|---|
| 1 | Month | Tickets Purchased |
| 2 | January | 9 |
| 3 | February | 15 |
| 4 | March | 11 |
| 5 | April | 10 |
| 6 | May | 12 |
| 7 | June | |

2.    Create a column **C** called **Forecast** for each month with the initial forecast as **9** in cell **C2**, as shown below.

| | A | B | C |
|---|---|---|---|
| 1 | Month | Tickets Purchased | Forecast |
| 2 | January | 9 | 9 |
| 3 | February | 15 | |
| 4 | March | 11 | |
| 5 | April | 10 | |
| 6 | May | 12 | |
| 7 | June | | |

3. The forecast for the second period, February, will be computed using the formula $F_2 = \alpha*D_1 + (1-\alpha)*F_1$ for observed values $D_i$ and forecasted values $F_i$. Using $\alpha = 0.3$, write the formula in cell **C3** as **=0.3\*B2+(1-0.3)\*C2**. Press **Enter**.

| | A | B | C | D |
|---|---|---|---|---|
| 1 | Month | Tickets Purchased | Forecast | |
| 2 | January | 9 | 9 | |
| 3 | February | 15 | =0.3*B2+(1-0.3)*C2 | |
| 4 | March | 11 | | |
| 5 | April | 10 | | |
| 6 | May | 12 | | |
| 7 | June | | | |

4. The Forecast for February is **9**.

| | A | B | C |
|---|---|---|---|
| 1 | Month | Tickets Purchased | Forecast |
| 2 | January | 9 | 9 |
| 3 | February | 15 | 9 |
| 4 | March | 11 | |
| 5 | April | 10 | |
| 6 | May | 12 | |
| 7 | June | | |

5. Double-click on the bottom right corner of cell **C3**. The forecast values for all periods will be calculated automatically in Excel.

| | A | B | C |
|---|---|---|---|
| 1 | Month | Tickets Purchased | Forecast |
| 2 | January | 9 | 9 |
| 3 | February | 15 | 9 |
| 4 | March | 11 | 10.80 |
| 5 | April | 10 | 10.86 |
| 6 | May | 12 | 10.60 |
| 7 | June | | 11.02 |

6. The desired result for the forecast for June is **11.02**. Compare this to the result obtained in Example 15.3.1.

**Alternate method**:

Simple Exponential Smoothing in Excel using the Data Analysis Toolpack.

1. Enter the values for **Month** and **Tickets Purchased** in columns **A** and **B**, respectively.

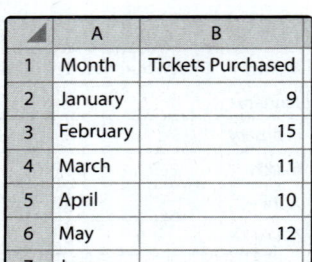

| | A | B |
|---|---|---|
| 1 | Month | Tickets Purchased |
| 2 | January | 9 |
| 3 | February | 15 |
| 4 | March | 11 |
| 5 | April | 10 |
| 6 | May | 12 |
| 7 | June | |

2. Exponential smoothing can be performed using the Data Analysis Toolpack in Excel. Go to the **Data** tab and select **Data Analysis**, as shown below.

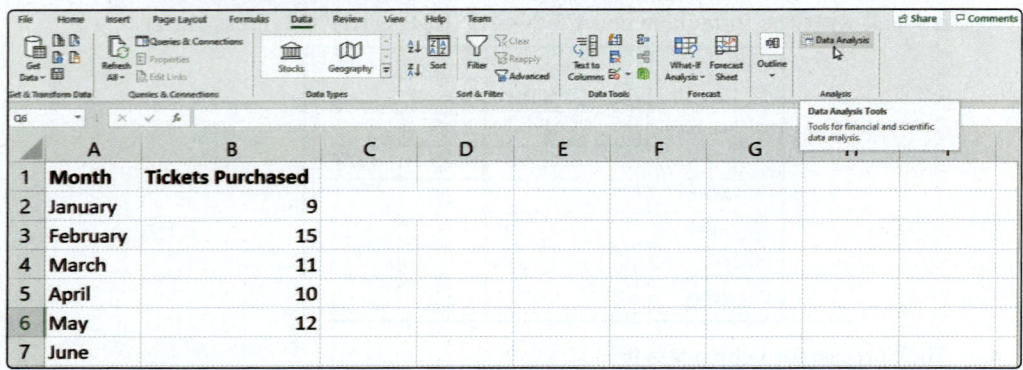

3. A pop-up opens. Select **Exponential Smoothing**. Click **OK**.

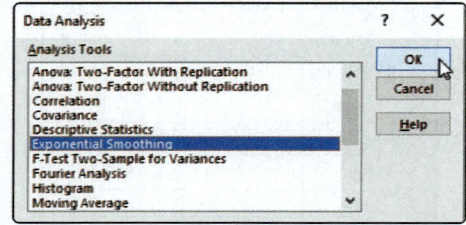

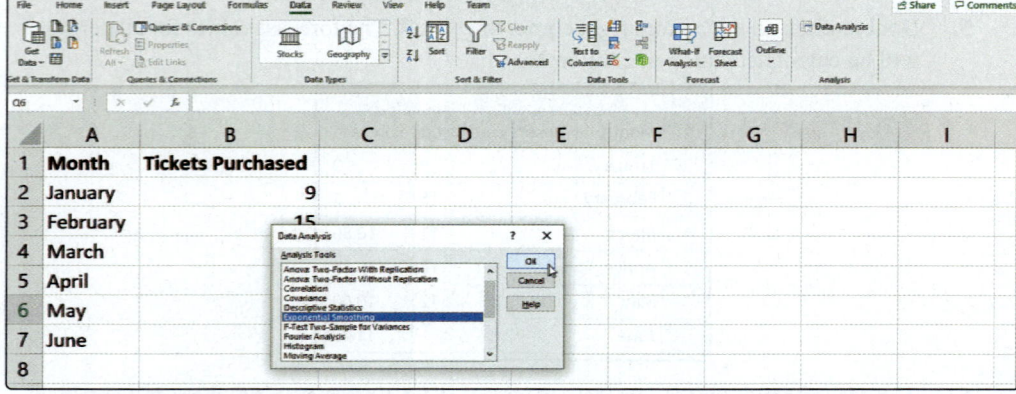

4. A pop-up opens. Provide the input range as **B2:B7**, insert the damping factor $1-\alpha = 0.7$, and provide the output range as **C2:C7**. Click **OK**.

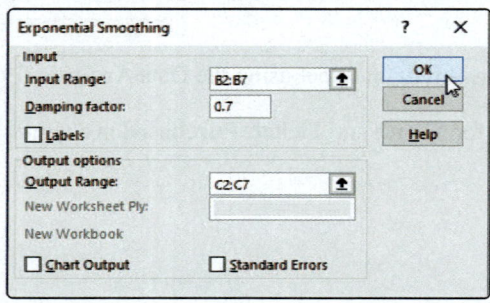

5.  Name column **C Forecast**. The first value shows as #N/A in Excel as there is no forecast available for the initial value. The desired result for the forecast for June appears in cell **E7** as **11.02**. Compare this to the result in Example 15.3.1.

| | A | B | C |
|---|---|---|---|
| 1 | Month | Tickets Purchased | Forecast |
| 2 | January | 9 | #N/A |
| 3 | February | 15 | 9 |
| 4 | March | 11 | 10.8 |
| 5 | April | 10 | 10.86 |
| 6 | May | 12 | 10.602 |
| 7 | June | | 11.0214 |

## Adjusted Exponential Smoothing

Use the information in Example 15.3.2 for this exercise.

1.  Enter the values for **Month** and **Tickets Purchased** in columns **A** and **B**, respectively.

| | A | B |
|---|---|---|
| 1 | Month | Tickets Purchased |
| 2 | January | 9 |
| 3 | February | 15 |
| 4 | March | 11 |
| 5 | April | 10 |
| 6 | May | 12 |
| 7 | June | |

2.  Create the columns named **Forecast**, **Trend**, and **Adjusted Forecast** in cells **C1**, **D1**, and **E1**. Enter the initial forecast in cell **C2** as **9**, trend in cell **D2** as **0**, and adjusted forecast in cell **E2** as **9**.

| | A | B | C | D | E |
|---|---|---|---|---|---|
| 1 | Month | Tickets Purchased | Forecast | Trend | Adjusted Forecast |
| 2 | January | 9 | 9 | 0 | 9 |
| 3 | February | 15 | | | |
| 4 | March | 11 | | | |
| 5 | April | 10 | | | |
| 6 | May | 12 | | | |
| 7 | June | | | | |

3.  Using the formula for Forecast, $F_2 = \alpha*(D_1) + (1-\alpha)*(AF_1)$, write a formula with $\alpha = 0.3$ in cell **C3** using cell value **B2** in place of $D_1$ and **E2** in place of $AF_1$, as shown below. Press **Enter**.

| | A | B | C | D | E |
|---|---|---|---|---|---|
| 1 | Month | Tickets Purchased | Forecast | Trend | Adjusted Forecast |
| 2 | January | 9 | 9 | 0 | 9 |
| 3 | February | 15 | =0.3*B2+(1-0.3)*E2 | | |
| 4 | March | 11 | | | |
| 5 | April | 10 | | | |
| 6 | May | 12 | | | |
| 7 | June | | | | |

4.  The Forecast for February is displayed as **9**.

| | A | B | C | D | E |
|---|---|---|---|---|---|
| 1 | Month | Tickets Purchased | Forecast | Trend | Adjusted Forecast |
| 2 | January | 9 | 9 | 0 | 9 |
| 3 | February | 15 | 9 | | |
| 4 | March | 11 | | | |
| 5 | April | 10 | | | |
| 6 | May | 12 | | | |
| 7 | June | | | | |

5.  Using the formula for Trend, $T_{t+1} = T_t + \beta*(F_{t+1} - AF_t)$, write a formula with $\beta = 0.2$ in cell **D3** using cell value **D2** in place of $T_t$ and **C3** and **E2** in place of $F_{t+1}$ and $AF_t$, respectively. Press **Enter**.

| | A | B | C | D | E |
|---|---|---|---|---|---|
| 1 | Month | Tickets Purchased | Forecast | Trend | Adjusted Forecast |
| 2 | January | 9 | 9 | 0 | 9 |
| 3 | February | 15 | 9 | =D2+0.2*(C3-E2) | |
| 4 | March | 11 | | | |
| 5 | April | 10 | | | |
| 6 | May | 12 | | | |
| 7 | June | | | | |

6.  The trend for February is **0**.

| | A | B | C | D | E |
|---|---|---|---|---|---|
| 1 | Month | Tickets Purchased | Forecast | Trend | Adjusted Forecast |
| 2 | January | 9 | 9 | 0 | 9 |
| 3 | February | 15 | 9 | 0 | |
| 4 | March | 11 | | | |
| 5 | April | 10 | | | |
| 6 | May | 12 | | | |
| 7 | June | | | | |

7.  In cell **E3**, enter "**=SUM(C3:D3)**" to find the Adjusted Forecast for February and press **Enter**.

| | A | B | C | D | E |
|---|---|---|---|---|---|
| 1 | Month | Tickets Purchased | Forecast | Trend | Adjusted Forecast |
| 2 | January | 9 | 9 | 0 | 9 |
| 3 | February | 15 | 9 | 0 | =SUM(C3:D3) |
| 4 | March | 11 | | | |
| 5 | April | 10 | | | |
| 6 | May | 12 | | | |
| 7 | June | | | | |

8.  The desired Adjusted Forecast for February is **9**.

| | A | B | C | D | E |
|---|---|---|---|---|---|
| 1 | Month | Tickets Purchased | Forecast | Trend | Adjusted Forecast |
| 2 | January | 9 | 9 | 0 | 9 |
| 3 | February | 15 | 9 | 0 | 9 |
| 4 | March | 11 | | | |
| 5 | April | 10 | | | |
| 6 | May | 12 | | | |
| 7 | June | | | | |

9.  We have performed the calculation for the first row, and it needs to be repeated for all other rows. Select the cells from **C3** to **E3** and double-click on the bottom-right corner of **E3**, which will autofill the values till **E7**, as shown below.

| | A | B | C | D | E |
|---|---|---|---|---|---|
| 1 | Month | Tickets Purchased | Forecast | Trend | Adjusted Forecast |
| 2 | January | 9 | 9 | 0 | 9 |
| 3 | February | 15 | 9.00 | 0.00 | 9.00 |
| 4 | March | 11 | 10.80 | 0.36 | 11.16 |
| 5 | April | 10 | 11.11 | 0.35 | 11.46 |
| 6 | May | 12 | 11.02 | 0.26 | 11.29 |
| 7 | June | | 11.50 | 0.31 | 11.81 |

10. The desired Adjusted Forecast for June is **11.81**. Compare this to the result obtained in Example 15.3.2.

## Mean Absolute Deviation

Use the information given in Example 15.4.1 for this exercise.

1.  Enter the values for **Period ($t$)**, **Actual Data ($D_i$)**, and **Forecast ($F_i$)** in Columns **A**, **B**, and **C**, respectively.

| | A | B | C |
|---|---|---|---|
| 1 | Period(t) | Actual Data(Di) | Forecast(Fi) |
| 2 | 1 | 134 | 132 |
| 3 | 2 | 142 | 141 |
| 4 | 3 | 143 | 145 |
| 5 | 4 | 156 | 143 |
| 6 | 5 | 151 | 154 |
| 7 | 6 | 145 | 150 |

2.  Calculate the **Absolute Error $|D_i - F_i|$** in column **D**. To do this, in cell **D2**, enter "**=ABS(B2-C2)**". Press **Enter**.

| | A | B | C | D |
|---|---|---|---|---|
| 1 | Period(t) | Actual Data(Di) | Forecast(Fi) | Absolute Error \|Di-Fi\| |
| 2 | 1 | 134 | 132 | =ABS(B2-C2) |
| 3 | 2 | 142 | 141 | |
| 4 | 3 | 143 | 145 | |
| 5 | 4 | 156 | 143 | |
| 6 | 5 | 151 | 154 | |
| 7 | 6 | 145 | 150 | |

3.  The value obtained is **2**. Double-click on the right corner of cell **D2** so that the formula gets auto-applied to the cells from **D3** to **D7**, as shown in the screenshot below.

| | A | B | C | D |
|---|---|---|---|---|
| 1 | Period(t) | Actual Data(Di) | Forecast(Fi) | Absolute Error \|Di-Fi\| |
| 2 | 1 | 134 | 132 | 2 |
| 3 | 2 | 142 | 141 | 1 |
| 4 | 3 | 143 | 145 | 2 |
| 5 | 4 | 156 | 143 | 13 |
| 6 | 5 | 151 | 154 | 3 |
| 7 | 6 | 145 | 150 | 5 |

4. In cell **D9**, compute the MAD by entering **=AVERAGE(D2:D7)**. Press **Enter**.

| | A | B | C | D |
|---|---|---|---|---|
| 1 | Period(t) | Actual Data(Di) | Forecast(Fi) | Absolute Error \|Di-Fi\| |
| 2 | 1 | 134 | 132 | 2 |
| 3 | 2 | 142 | 141 | 1 |
| 4 | 3 | 143 | 145 | 2 |
| 5 | 4 | 156 | 143 | 13 |
| 6 | 5 | 151 | 154 | 3 |
| 7 | 6 | 145 | 150 | 5 |
| 8 | | | | |
| 9 | | | | =AVERAGE(D2:D7) |

5. The value obtained for the MAD is **4.33**. Compare this to the result obtained in Example 15.4.1.

| | A | B | C | D |
|---|---|---|---|---|
| 1 | Period(t) | Actual Data(Di) | Forecast(Fi) | Absolute Error \|Di-Fi\| |
| 2 | 1 | 134 | 132 | 2 |
| 3 | 2 | 142 | 141 | 1 |
| 4 | 3 | 143 | 145 | 2 |
| 5 | 4 | 156 | 143 | 13 |
| 6 | 5 | 151 | 154 | 3 |
| 7 | 6 | 145 | 150 | 5 |
| 8 | | | | |
| 9 | | | | 4.33 |

## Mean Absolute Percentage Error

Use the information in Example 15.4.2 for this exercise.

1. Enter the values for **Period ($t$)**, **Actual Data ($D_i$)**, and **Forecast ($F_i$)** in Columns **A**, **B**, and **C**, respectively.

| | A | B | C |
|---|---|---|---|
| 1 | Period(t) | Actual Data(Di) | Forecast(Fi) |
| 2 | 1 | 134 | 132 |
| 3 | 2 | 142 | 141 |
| 4 | 3 | 143 | 145 |
| 5 | 4 | 156 | 143 |
| 6 | 5 | 151 | 154 |
| 7 | 6 | 145 | 150 |

2. Calculate the **Absolute Error, $|D_i\text{-}F_i|$**. In cell **D2**, enter "**=ABS (B2-C2)**". Press **Enter**.

| | A | B | C | D |
|---|---|---|---|---|
| 1 | Period(t) | Actual Data(Di) | Forecast(Fi) | Absolute Error \|Di-Fi\| |
| 2 | 1 | 134 | 132 | =ABS(B2-C2) |
| 3 | 2 | 142 | 141 | |
| 4 | 3 | 143 | 145 | |
| 5 | 4 | 156 | 143 | |
| 6 | 5 | 151 | 154 | |
| 7 | 6 | 145 | 150 | |

3. Double-click on the right corner of cell **D2** so that the formula gets auto-applied on the remaining cells (**D3-D7**), as shown below.

| | A | B | C | D |
|---|---|---|---|---|
| 1 | Period(t) | Actual Data(Di) | Forecast(Fi) | Absolute Error \|Di-Fi\| |
| 2 | 1 | 134 | 132 | 2 |
| 3 | 2 | 142 | 141 | 1 |
| 4 | 3 | 143 | 145 | 2 |
| 5 | 4 | 156 | 143 | 13 |
| 6 | 5 | 151 | 154 | 3 |
| 7 | 6 | 145 | 150 | 5 |

4. Calculate the **Absolute Percentage Error**, $100*|D_t - F_t|/D_t$. In cell **E2**, enter **=100*(D2/B2)**. Press **Enter**.

| | A | B | C | D | E |
|---|---|---|---|---|---|
| 1 | Period(t) | Actual Data(Di) | Forecast(Fi) | Absolute Error \|Di-Fi\| | Absolute Percentage Error |
| 2 | 1 | 134 | 132 | 2 | =100*(D2/B2) |
| 3 | 2 | 142 | 141 | 1 | |
| 4 | 3 | 143 | 145 | 2 | |
| 5 | 4 | 156 | 143 | 13 | |
| 6 | 5 | 151 | 154 | 3 | |
| 7 | 6 | 145 | 150 | 5 | |

5. Double-click on the right corner of the cell **E2** so that the formula gets auto-applied on the remaining cells (**E3-E7**) as shown below.

| | A | B | C | D | E |
|---|---|---|---|---|---|
| 1 | Period(t) | Actual Data(Di) | Forecast(Fi) | Absolute Error \|Di-Fi\| | Absolute Percentage Error |
| 2 | 1 | 134 | 132 | 2 | 1.492537313 |
| 3 | 2 | 142 | 141 | 1 | 0.704225352 |
| 4 | 3 | 143 | 145 | 2 | 1.398601399 |
| 5 | 4 | 156 | 143 | 13 | 8.333333333 |
| 6 | 5 | 151 | 154 | 3 | 1.986754967 |
| 7 | 6 | 145 | 150 | 5 | 3.448275862 |

6. MAPE is the average of all the values in the column Absolute Percentage Error. In cell **E9**, enter "**=AVERAGE (E2:E7)**". Press **Enter**.

| | A | B | C | D | E |
|---|---|---|---|---|---|
| 1 | Period(t) | Actual Data(Di) | Forecast(Fi) | Absolute Error \|Di-Fi\| | Absolute Percentage Error |
| 2 | 1 | 134 | 132 | 2 | 1.492537313 |
| 3 | 2 | 142 | 141 | 1 | 0.704225352 |
| 4 | 3 | 143 | 145 | 2 | 1.398601399 |
| 5 | 4 | 156 | 143 | 13 | 8.333333333 |
| 6 | 5 | 151 | 154 | 3 | 1.986754967 |
| 7 | 6 | 145 | 150 | 5 | 3.448275862 |
| 8 | | | | | |
| 9 | | | | | =AVERAGE(E2:E7) |

7. The desired MAPE is **2.8939**. Compare this to the results obtained in Example 15.4.2.

| | A | B | C | D | E |
|---|---|---|---|---|---|
| 1 | Period(t) | Actual Data(Di) | Forecast(Fi) | Absolute Error \|Di-Fi\| | Absolute Percentage Error |
| 2 | 1 | 134 | 132 | 2 | 1.492537313 |
| 3 | 2 | 142 | 141 | 1 | 0.704225352 |
| 4 | 3 | 143 | 145 | 2 | 1.398601399 |
| 5 | 4 | 156 | 143 | 13 | 8.333333333 |
| 6 | 5 | 151 | 154 | 3 | 1.986754967 |
| 7 | 6 | 145 | 150 | 5 | 3.448275862 |
| 8 | | | | | |
| 9 | | | | | 2.893954704 |

## Mean Squared Error

Use the information in Example 15.4.3 for this exercise.

1. Enter the values for **Period (*t*)**, **Actual Data (*D_i*)**, and **Forecast (*F_i*)** in Columns **A**, **B**, and **C**, respectively.

| | A | B | C |
|---|---|---|---|
| 1 | Period(t) | Actual Data(Di) | Forecast(Fi) |
| 2 | 1 | 134 | 132 |
| 3 | 2 | 142 | 141 |
| 4 | 3 | 143 | 145 |
| 5 | 4 | 156 | 143 |
| 6 | 5 | 151 | 154 |
| 7 | 6 | 145 | 150 |

2. Calculate the **Absolute Error, $|D_i\text{-}F_i|$**. To do this, in cell **D2**, enter "=ABS(B2-C2)". Press **Enter**.

| | A | B | C | D |
|---|---|---|---|---|
| 1 | Period(t) | Actual Data(Di) | Forecast(Fi) | Absolute Error \|Di-Fi\| |
| 2 | 1 | 134 | 132 | =ABS(B2-C2) |
| 3 | 2 | 142 | 141 | |
| 4 | 3 | 143 | 145 | |
| 5 | 4 | 156 | 143 | |
| 6 | 5 | 151 | 154 | |
| 7 | 6 | 145 | 150 | |

3. Double-click on the right corner of cell **D2** so that the formula gets auto-applied on the remaining cells (**D3-D7**), as shown below.

| | A | B | C | D |
|---|---|---|---|---|
| 1 | Period(t) | Actual Data(Di) | Forecast(Fi) | Absolute Error \|Di-Fi\| |
| 2 | 1 | 134 | 132 | 2 |
| 3 | 2 | 142 | 141 | 1 |
| 4 | 3 | 143 | 145 | 2 |
| 5 | 4 | 156 | 143 | 13 |
| 6 | 5 | 151 | 154 | 3 |
| 7 | 6 | 145 | 150 | 5 |

4. Now, calculate the Squared Error for each period by entering "**=D2^2**" in cell **E2**. Click **Enter**.

| | A | B | C | D | E |
|---|---|---|---|---|---|
| 1 | Period(t) | Actual Data(Di) | Forecast(Fi) | Absolute Error \|Di-Fi\| | Squared Error |
| 2 | 1 | 134 | 132 | 2 | =D2^2) |
| 3 | 2 | 142 | 141 | 1 | |
| 4 | 3 | 143 | 145 | 2 | |
| 5 | 4 | 156 | 143 | 13 | |
| 6 | 5 | 151 | 154 | 3 | |
| 7 | 6 | 145 | 150 | 5 | |

5. Double-click on the right corner of cell **E2** so that the formula gets auto-applied on the remaining cells (**E3-E7**) as shown below.

| | A | B | C | D | E |
|---|---|---|---|---|---|
| 1 | Period(t) | Actual Data(Di) | Forecast(Fi) | Absolute Error \|Di-Fi\| | Squared Error |
| 2 | 1 | 134 | 132 | 2 | 4 |
| 3 | 2 | 142 | 141 | 1 | 1 |
| 4 | 3 | 143 | 145 | 2 | 4 |
| 5 | 4 | 156 | 143 | 13 | 169 |
| 6 | 5 | 151 | 154 | 3 | 9 |
| 7 | 6 | 145 | 150 | 5 | 25 |

6. The **MSE** is the average of all the values in the column **Squared Error**. In cell **E9**, enter **=AVERAGE(E2:E7)**. Click **Enter**.

| | A | B | C | D | E |
|---|---|---|---|---|---|
| 1 | Period(t) | Actual Data(Di) | Forecast(Fi) | Absolute Error \|Di-Fi\| | Squared Error |
| 2 | 1 | 134 | 132 | 2 | 4 |
| 3 | 2 | 142 | 141 | 1 | 1 |
| 4 | 3 | 143 | 145 | 2 | 4 |
| 5 | 4 | 156 | 143 | 13 | 169 |
| 6 | 5 | 151 | 154 | 3 | 9 |
| 7 | 6 | 145 | 150 | 5 | 25 |
| 8 | | | | | |
| 9 | | | | | =AVERAGE(E2:E7) |

7. The desired result for the **MSE** is **35.3333**. Compare this with the result obtained in Example 15.4.3.

| | A | B | C | D | E |
|---|---|---|---|---|---|
| 1 | Period(t) | Actual Data(Di) | Forecast(Fi) | Absolute Error \|Di-Fi\| | Squared Error |
| 2 | 1 | 134 | 132 | 2 | 4 |
| 3 | 2 | 142 | 141 | 1 | 1 |
| 4 | 3 | 143 | 145 | 2 | 4 |
| 5 | 4 | 156 | 143 | 13 | 169 |
| 6 | 5 | 151 | 154 | 3 | 9 |
| 7 | 6 | 145 | 150 | 5 | 25 |
| 8 | | | | | |
| 9 | | | | | 35.3333 |

## Using JMP

### Simple Moving Average

Use the information in Example 15.2.1 for this exercise.

1.  Enter the values in the JMP data table as follows.

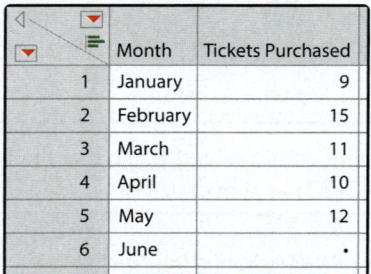

| | Month | Tickets Purchased |
|---|---|---|
| 1 | January | 9 |
| 2 | February | 15 |
| 3 | March | 11 |
| 4 | April | 10 |
| 5 | May | 12 |
| 6 | June | . |

2.  Right-Click on the **Tickets Purchased** Column, select **New Formula Column**, **Row**, and then **Moving Average**.

3.  A pop-up opens, as shown below. Under **Weighting**, select **Equal**. Under **Items Before** (i.e., Number of periods), select **Fixed** and enter **3**.For **Items After**, Select **None**. Click **OK**.

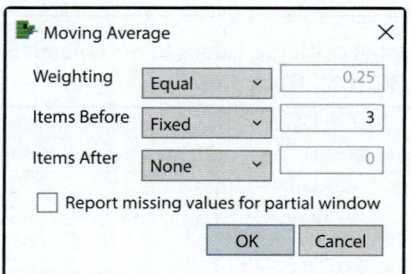

| Moving Average | | X |
|---|---|---|
| Weighting | Equal ∨ | 0.25 |
| Items Before | Fixed ∨ | 3 |
| Items After | None ∨ | 0 |
| ☐ Report missing values for partial window | | |
| | OK | Cancel |

4.  The desired Moving Average for June for a period of 3 will appear as 11. Compare this to the result obtained in Example 15.2.1.

| | Month | Tickets Purchased | Moving Average [...kets Purchased] |
|---|---|---|---|
| 1 | January | 9 | 9 |
| 2 | February | 15 | 12 |
| 3 | March | 11 | 11.666666667 |
| 4 | April | 10 | 11.25 |
| 5 | May | 12 | 12 |
| 6 | June | . | 11 |

### Simple Exponential Smoothing

Use the information in Example 15.3.1 for this exercise.

1.  Enter the values in the JMP data table as follows.

| | A | B |
|---|---|---|
| 1 | Month | Tickets Purchased |
| 2 | January | 9 |
| 3 | February | 15 |
| 4 | March | 11 |
| 5 | April | 10 |
| 6 | May | 12 |
| 7 | June | |

2. Right-click on the Tickets Purchased column, select **New Formula Column**, then **Row** and **Simple Exponential Smoothing**.

3. A pop-up opens, as shown below. Enter the smoothing weight ($\alpha$) as **0.3**. Click **OK**.

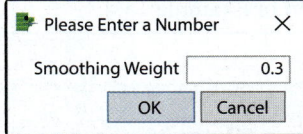

Please Enter a Number ✕

Smoothing Weight    0.3

OK    Cancel

4. Automatically, a new column with simple exponential smoothing values will be displayed, as shown below. Compare this to the results obtained in Example 15.3.1.

| | Month | Tickets Purchased | Simple Exponent [...kets Purchased] |
|---|---|---|---|
| 1 | January | 9 | • |
| 2 | February | 15 | 9 |
| 3 | March | 11 | 10.8 |
| 4 | April | 10 | 10.86 |
| 5 | May | 12 | 10.602 |
| 6 | June | • | 11.0214 |

## Mean Absolute Deviation

Use the information in Example 15.4.1 for this exercise.

1. In a JMP data table, enter the values for **Period (t)**, **Actual Data $D_t$**, **Forecast $F_t$**, but leave the values for **Forecast Error $FE_t$** and **Absolute Error** blank as shown below.

| | Period(t) | Actual Data Dt | Forecast Ft | Forecast Error FEt | Absolute Error |
|---|---|---|---|---|---|
| 1 | 1 | 134 | 132 | • | • |
| 2 | 2 | 142 | 141 | • | • |
| 3 | 3 | 143 | 145 | • | • |
| 4 | 4 | 156 | 143 | • | • |
| 5 | 5 | 151 | 154 | • | • |
| 6 | 6 | 145 | 150 | • | • |

2. As the objective is to calculate Mean Absolute Deviation (MAD), the first step is to calculate **Absolute Difference between Forecast & Actual Data** (i.e., $|D_t - F_t|$). To perform this, Right-Click on the **Forecast Error** Column and click on **Formula**. Enter the Formula: **Actual Data $D_t$ – Forecast $F_t$**. You can click on the Preview option to view results instantly. Click **OK**.

3. To calculate Absolute Error, Right-click on the **Absolute Error** Column and click on **Formula**. In the right-side pane, click on **Numeric** and then double-click on **Abs**. From the Columns list, select **Forecast Error**. You can click on the Preview option to view results instantly. Click **OK**. Your table should look like this now:

| Period(t) | Actual Data Dt | Forecast Ft | Forecast Error FEt | Absolute Error |
|---|---|---|---|---|
| 1 | 1 | 134 | 132 | 2 | 2 |
| 2 | 2 | 142 | 141 | 1 | 1 |
| 3 | 3 | 143 | 145 | -2 | 2 |
| 4 | 4 | 156 | 143 | 13 | 13 |
| 5 | 5 | 151 | 154 | -3 | 3 |
| 6 | 6 | 145 | 150 | -5 | 5 |

4. To calculate Mean Absolute Deviation, go to **Analyze**, **Distribution**, and enter **Absolute Error** in the **Y, Columns** box. Click **OK**.

5. The Mean value of 4.3333 corresponds to the Mean Absolute Deviation. Compare this to the results obtained in Example 15.4.1.

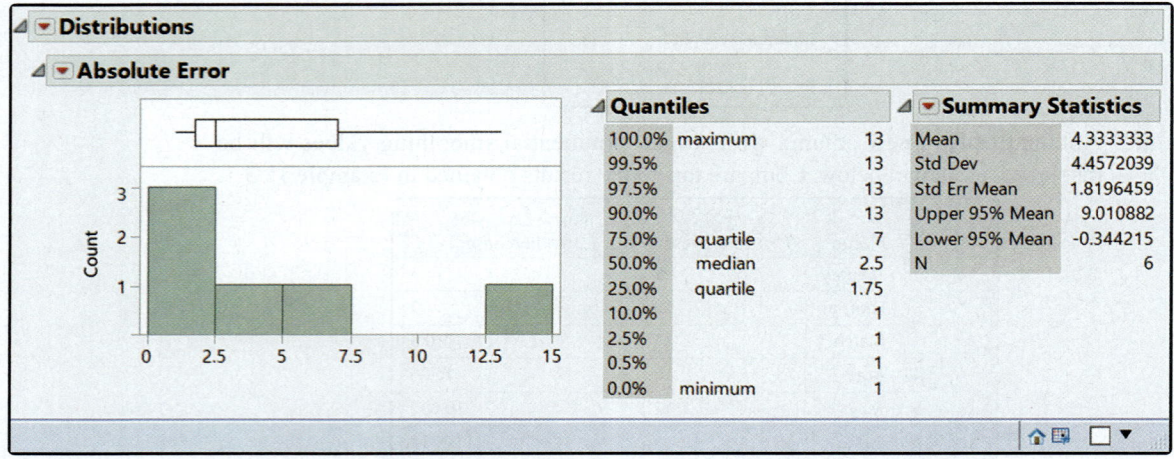

## Mean Absolute Percentage Error

Use the information in Example 15.4.3 for this exercise.

1. In a JMP data table, enter the values for **Period ($t$)**, **Actual Data $D_t$**, **Forecast $F_t$**, but leave the values for **Forecast Error $FE_t$** and **Absolute Error** blank as shown below.

| Period(t) | Actual Data Dt | Forecast Ft | Forecast Error FEt | Absolute Error |
|---|---|---|---|---|
| 1 | 1 | 134 | 132 | . | . |
| 2 | 2 | 142 | 141 | . | . |
| 3 | 3 | 143 | 145 | . | . |
| 4 | 4 | 156 | 143 | . | . |
| 5 | 5 | 151 | 154 | . | . |
| 6 | 6 | 145 | 150 | . | . |

2. As the objective is to calculate Mean Absolute Percent Error (MAPE), the first step is to calculate the Absolute Difference between Forecast & Actual Data (i.e., $|D_t - F_t|$). To perform this, Right-Click on the **Forecast Error** Column and click on **Formula**. Enter the Formula: **Actual Data $D_t$ – Forecast $F_t$**. You can click on the Preview option to view results instantly. Click **OK**.

3. To calculate Absolute Error, Right-click on the **Absolute Error** Column and click on **Formula**. In the right-side pane, click on **Numeric** and then double-click on **Abs**. From the Columns list, select **Forecast Error**. You can click on the Preview option to view results instantly. Click **OK**. Your table should look like this now:

| | Period(t) | Actual Data Dt | Forecast Ft | Forecast Error FEt | Absolute Error |
|---|---|---|---|---|---|
| 1 | 1 | 134 | 132 | 2 | 2 |
| 2 | 2 | 142 | 141 | 1 | 1 |
| 3 | 3 | 143 | 145 | -2 | 2 |
| 4 | 4 | 156 | 143 | 13 | 13 |
| 5 | 5 | 151 | 154 | -3 | 3 |
| 6 | 6 | 145 | 150 | -5 | 5 |

4. Create a Column with the name **Abs. Percentage Error**. To calculate the Absolute Percentage Error, right-click on the Column and then click on **Formula**. A Pop-up opens. Double-Click on **Absolute Error**, and then click on the **division symbol**. Double-Click on **Actual Data**. Click on the multiply symbol and enter **100**. Click **OK**.

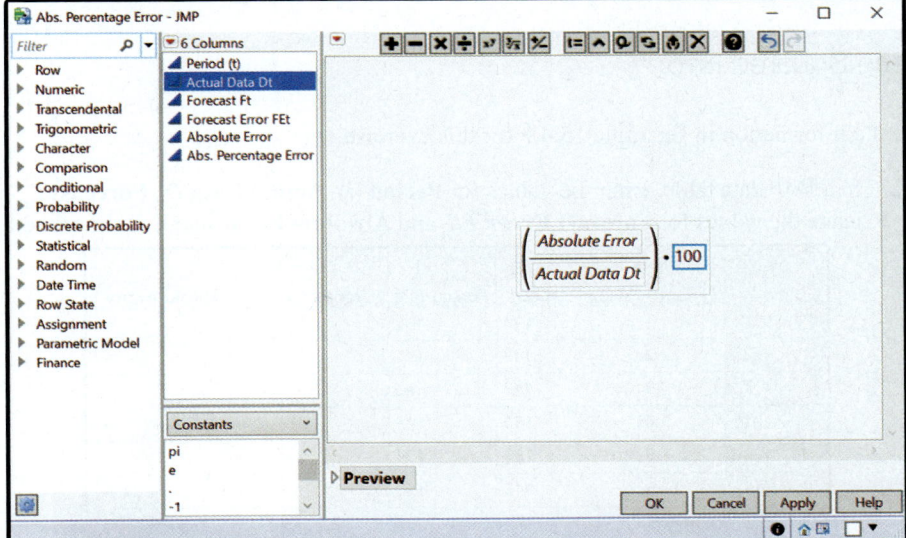

5. The results will look similar to the data table shown below.

| | Period(t) | Actual Data Dt | Forecast Ft | Forecast Error FEt | Absolute Error | Abs. Percentage Error |
|---|---|---|---|---|---|---|
| 1 | 1 | 134 | 132 | 2 | 2 | 1.4925373134 |
| 2 | 2 | 142 | 141 | 1 | 1 | 0.7042253521 |
| 3 | 3 | 143 | 145 | -2 | 2 | 1.3986013986 |
| 4 | 4 | 156 | 143 | 13 | 13 | 8.3333333333 |
| 5 | 5 | 151 | 154 | -3 | 3 | 1.9867549669 |
| 6 | 6 | 145 | 150 | -5 | 5 | 3.4482758621 |

6. MAPE is the Average of all values in the Column **Abs. Percentage Error**. To calculate MAPE, select **Analyze** and then **Distribution**. From the Select Columns, click on **Abs. Percentage Error** and then click on **Y, Columns**. Click **OK**.

7. The Mean value of 2.894 corresponds to the Mean Absolute Percentage Error. Compare this to the results obtained in Example 15.4.2.

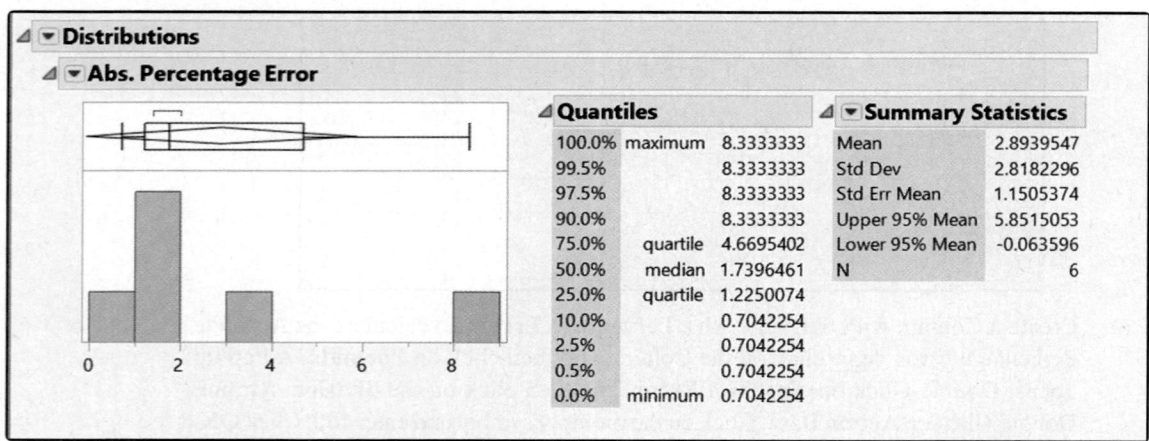

## Mean Squared Error

Use the information in Example 15.4.3 for this exercise.

1.  In a JMP data table, enter the values for **Period** (*t*), **Actual Data** $D_t$, **Forecast** $F_t$, but leave the values for **Forecast Error** $FE_t$ and **Absolute Error** blank as shown below.

| | Period(t) | Actual Data Dt | Forecast Ft | Forecast Error FEt | Absolute Error |
|---|---|---|---|---|---|
| 1 | 1 | 134 | 132 | . | . |
| 2 | 2 | 142 | 141 | . | . |
| 3 | 3 | 143 | 145 | . | . |
| 4 | 4 | 156 | 143 | . | . |
| 5 | 5 | 151 | 154 | . | . |
| 6 | 6 | 145 | 150 | . | . |

2.  As the objective is to calculate Mean Absolute Percent Error (MAPE), the first step is to calculate the Absolute Difference between Forecast & Actual Data (i.e., $|D_t-F_t|$). To perform this, Right-Click on the **Forecast Error** Column and click on **Formula**. Enter the Formula: **Actual Data** $D_t$ – **Forecast** $F_t$. You can click on the Preview option to view results instantly. Click **OK**.

3.  To calculate Absolute Error, Right-click on the **Absolute Error** Column and click on **Formula**. In the right-side pane, click on **Numeric** and then double-click on **Abs**. From the Columns list, select **Forecast Error**. You can click on the Preview option to view results instantly. Click **OK**. Your table should look like this now:

| | Period(t) | Actual Data Dt | Forecast Ft | Forecast Error FEt | Absolute Error |
|---|---|---|---|---|---|
| 1 | 1 | 134 | 132 | 2 | 2 |
| 2 | 2 | 142 | 141 | 1 | 1 |
| 3 | 3 | 143 | 145 | -2 | 2 |
| 4 | 4 | 156 | 143 | 13 | 13 |
| 5 | 5 | 151 | 154 | -3 | 3 |
| 6 | 6 | 145 | 150 | -5 | 5 |

4.  To calculate the Squared Error, create a column named **Sq. Error**. Right-click on the Column and then click on **Formula**.

5.  A pop up opens. Double-Click on **Absolute Error** and then click on the **Power** symbol. Enter the power value of **2**. Click **OK**.

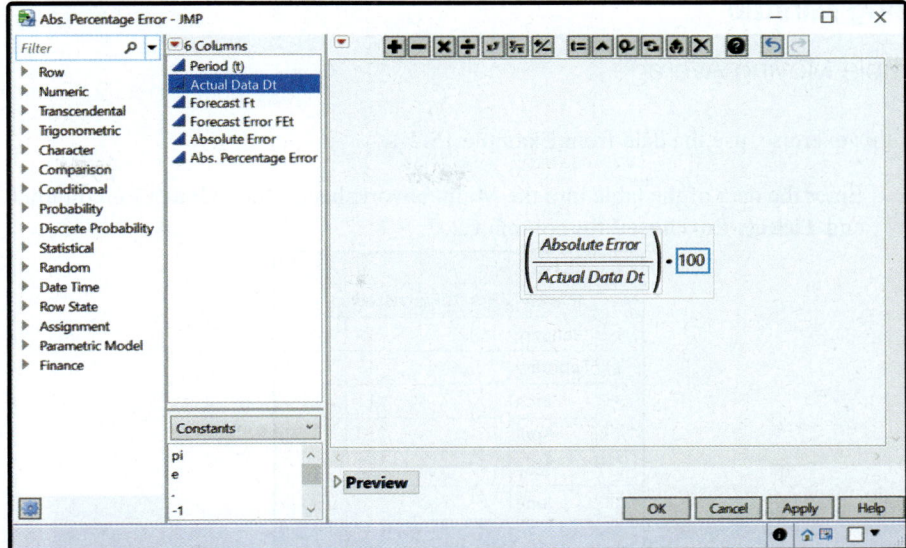

6. Now, the results will look similar to the Data table shown below.

| Period(t) | Actual Data Dt | Forecast Ft | Forecast Error FEt | Absolute Error | Sq. Error |
|---|---|---|---|---|---|
| 1 | 1 | 134 | 132 | 2 | 2 | 4 |
| 2 | 2 | 142 | 141 | 1 | 1 | 1 |
| 3 | 3 | 143 | 145 | -2 | 2 | 4 |
| 4 | 4 | 156 | 143 | 13 | 13 | 169 |
| 5 | 5 | 151 | 154 | -3 | 3 | 9 |
| 6 | 6 | 145 | 150 | -5 | 5 | 25 |

7. Mean Squared Error is the Average of all values in Column **Sq. Error**. Go to **Analyze** and then **Distribution**. From the Select Columns, click on **Sq. Error** and then click on **Y, Columns**. Click **OK**.

8. The Mean value of 35.333 corresponds to the Mean Squared Error. Compare this to the results obtained in Example 15.4.3.

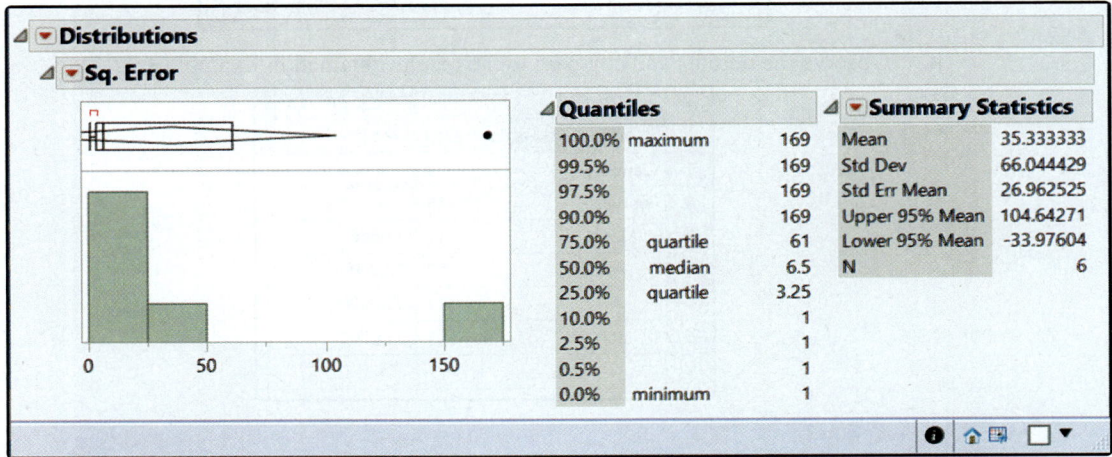

## Using Minitab

### Simple Moving Average

For this exercise, use the data from Example 15.2.1.

1. Enter the data of the table into the Minitab worksheet. Enter **Month** for column C1, and **Tickets Purchased** for column C2.

| ↓ | C1-D | C2 | C3 |
|---|------|-----|-----|
|   | Month | Tickets Purchased | |
| 1 | January | 9 | |
| 2 | February | 15 | |
| 3 | March | 11 | |
| 4 | April | 10 | |
| 5 | May | 12 | |
| 6 | June | * | |
| 7 | | | |

2. Choose **Stat**, **Time Series**, and **Moving Average**.

3. Enter **C2** (Tickets Purchased) in the **Variable** box, and **3** in the **MA length** box. Click on **Storage** and check the box for **Moving Averages**. Press **OK**. Press **OK** again.

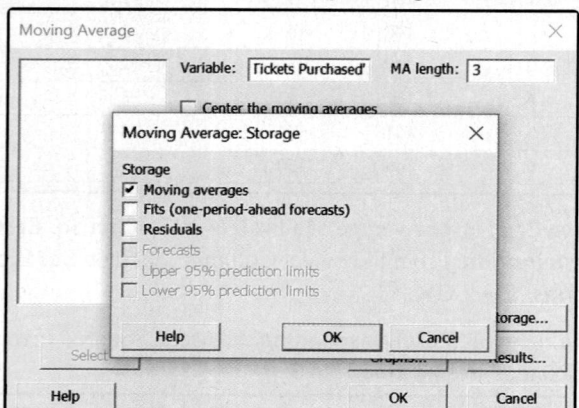

4. Observe the outputs and compare to the results obtained in Example 15.2.1.

| ↓ | C1-D | C2 | C3 | C4 |
|---|------|-----|------|-----|
|   | Month | Tickets Purchased | AVER1 | |
| 1 | January | 9 | * | |
| 2 | February | 15 | * | |
| 3 | March | 11 | 11.6667 | |
| 4 | April | 10 | 12.0000 | |
| 5 | May | 12 | 11.0000 | |
| 6 | June | * | 11.0000 | |
| 7 | | | | |

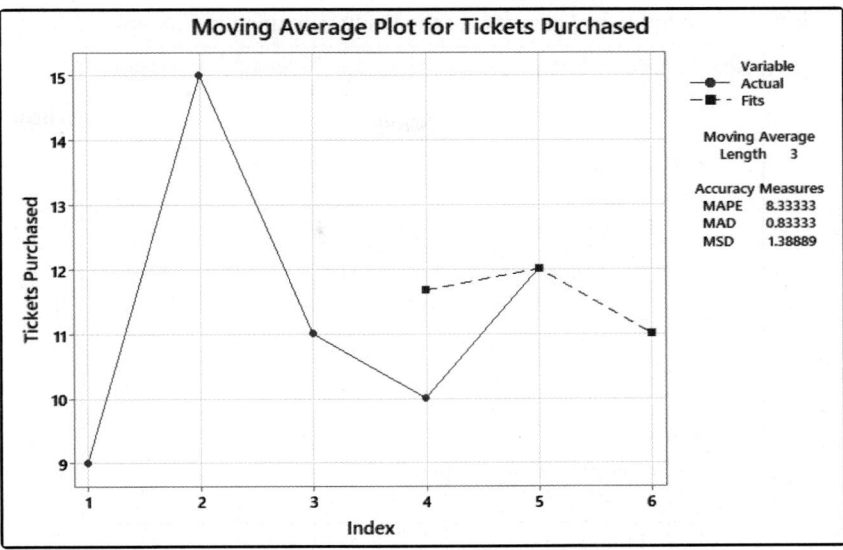

## Simple Exponential Smoothing

For this exercise, use the data from Example 15.3.1.

1. Enter the data of the table into the Minitab worksheet. Enter **Month** for column C1, and **Tickets Purchased** for column C2. Make sure the cell for June is blank (the * sign is not blank). To remove the * sign, right click on the cell and select **Clear Cells**.

| ↓ | C1-D | C2 | C3 |
|---|------|-----|----|
|   | Month | Tickets Purchased | |
| 1 | January | 9 | |
| 2 | February | 15 | |
| 3 | March | 11 | |
| 4 | April | 10 | |
| 5 | May | 12 | |
| 6 | June | | |
| 7 | | | |

2. Choose **Stat**, **Time Series**, and **Single Exp Smoothing**.

3. Enter **C2** (Tickets Purchased) in the **Variable** box. Under **Weight to Use in Smoothing**, select the radio button next to **Use** and enter **0.3**. Check the box for **Generate forecasts** and enter **1** for the **Number of forecasts**. Click on **Options** and enter **1** for K. Press **OK**. Click on **Storage** and check the box for **Forecasts**. Press **OK**. Click on **Results** and check the box for **Summary table and results able**. Press **OK**. Press **OK** again.

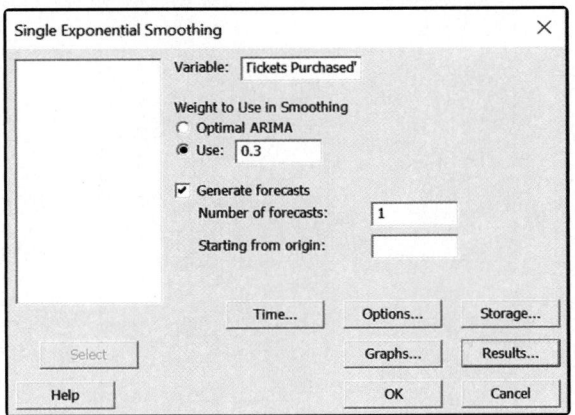

4. Observe the outputs and compare to the results obtained in Example 15.3.1.

**Single Exponential Smoothing for Tickets Purchased**

### Model Summary

| Time | Tickets Purchased | Smooth | Predict | Error |
|------|-------------------|---------|---------|--------|
| 1 | 9 | 9.0000 | 9.000 | 0.000 |
| 2 | 15 | 10.8000 | 9.000 | 6.000 |
| 3 | 11 | 10.8600 | 10.800 | 0.200 |
| 4 | 10 | 10.6020 | 10.860 | -0.860 |
| 5 | 12 | 11.0214 | 10.602 | 1.398 |

### Forecasts

| Period | Forecast | Lower | Upper |
|--------|----------|---------|---------|
| 6 | 11.0214 | 6.87706 | 15.1657 |

# R Chapter 15 Review

## Key Terms and Ideas

- Forecasting
- Time Series Analysis
- Timeframe
- Trend
- Seasonal Variation
- Cycle or Cyclical Variation
- Random Variations
- Stationary Data
- Nonstationary Data
- Seasonal Data
- Simple Moving Average (SMA)
- Weighted Moving Average (WMA)
- Simple Exponential Smoothing

- Adjusted Exponential Smoothing
- Forecast Error
- Mean Absolute Deviation (MAD)
- Mean Absolute Percentage Error (MAPE)
- Mean Absolute Percentage Deviation (MAPD)
- Mean Squared Error (MSE)
- Cumulative Error ($E$)
- Tracking Signal ($TS$)
- Seasonality
- Additive Seasonal Forecasting
- Multiplicative Seasonal Forecasting
- Holt's Winter Method

## Key Terms and Ideas

| | Section |
|---|---|

### Simple Moving Average

$$\text{MA}_n = \frac{\sum_{i=1}^{n} D_i}{n}$$

15.2

where $n$ = the number of periods used to compute the moving average and

$D_i$ = the actual data value of the time series in period $i$

### Weighted Moving Average

$$\text{WMA}_n = \sum_{i=1}^{n} W_i D_i$$

15.2

where $W_i$ = the weight for period $i$, $D_i$ = the data value for the $i^{\text{th}}$ period,
$n$ = the number of periods chosen, and
$$\sum W_i = 1.$$

### Simple Exponential Smoothing

$$F_{t+1} = \alpha D_t + (1 - \alpha)F_t$$

where

$D_t$ = the actual data value for the present period,

$F_t$ = the forecast for the present period, and

$\alpha$ = the weight (smoothing constant).

15.3

## Key Formulas (cont.)

Section

### Adjusted Exponential Smoothing

$$AF_{t+1} = F_{t+1} + T_{t+1}$$

where $F_{t+1}$ = the forecast component of the period computed as a simple exponential smoothed forecast

$$F_{t+1} = \alpha D_t + (1-\alpha) AF_t$$
$$= AF_t + \alpha (D_t - AF_t)$$
$$= \text{Previous Adjusted Forecast} + \alpha (\text{Forecast Error in the previous period}),$$

15.3

$\alpha$ = the smoothing constant

$AF_t$ = the previous period's trend adjusted forecast

$T_{t+1}$ = the forecast for the trend factor

$T_{t+1} = T_t + \beta(F_{t+1} - AF_t) = \text{Previous Trend} + \beta(\text{Trend Error})$

$\beta$ = the smoothing constant

$T_t$ = the trend factor for the previous period

### Mean Absolute Deviation (MAD)

$$\text{MAD} = \frac{\Sigma |D_t - F_t|}{n}$$

where $D_t$ = the actual data for the $t^{\text{th}}$ period,

$F_t$ = the forecast for the $t^{\text{th}}$ period,

$n$ = the total number of periods for which we have computed the forecasts, and

$t$ = the time period number.

15.4

### Mean Absolute Percentage Error (MAPE)

$$\text{MAPE} = \frac{100}{n} \sum_{t=1}^{n} \frac{|D_t - F_t|}{D_t}$$

where $D_t$ = the actual data for the $t^{\text{th}}$ period,

$F_t$ = the forecast for the $t^{\text{th}}$ period,

$n$ = the total number of periods for which we have computed the forecasts, and

$t$ = the time period number.

15.4

### Mean Squared Error (MSE)

$$\text{MSE} = \frac{\Sigma (D_t - F_t)^2}{n}$$

where $D_t$ = the actual data for the $t^{\text{th}}$ period,

$F_t$ = the forecast for the $t^{\text{th}}$ period,

$n$ = the total number of time periods for which we have computed the forecasts, and

$t$ = the time period number.

15.4

# AE  Additional Exercises

1. Compute the Mean Absolute Deviation for the 3-month SMA forecast using the TixPixx data.

2. Compute the Mean Absolute Percentage Deviation for the 3-month SMA and WMA forecasts using the TixPixx data.

3. Compute the Cumulative Error for the 3-month SMA and WMA forecasts using the TixPixx data.

4. Compute the Mean Squared Error for the 3-month SMA and WMA forecasts using the TixPixx data.

5. Triplett Farms is a company that raises turkeys, which it sells to a meat-processing company throughout the year. However, the peak season obviously occurs during the fourth quarter of the year, October to December. Triplett Farms has experienced a demand for turkeys for the past 3 years shown in the following table:

| Demand (1,000s) | | | | | |
|---|---|---|---|---|---|
| Year | Quarter 1 | Quarter 2 | Quarter 3 | Quarter 4 | Total |
| 2019 | 12.6 | 8.6 | 6.3 | 17.5 | 45.0 |
| 2020 | 14.1 | 10.3 | 7.5 | 18.2 | 50.1 |
| 2021 | 15.3 | 10.6 | 8.1 | 19.6 | 53.6 |
| Total | 42.0 | 29.5 | 21.9 | 55.3 | 148.7 |

Compute the seasonally adjusted forecasts for each quarter of 2022.

## P   Discovery Project

### The Global Financial Crisis of 2007 – 2008

The global financial crisis of 2007 – 2008 seemed to have been years in the making. By the summer of 2007, financial markets around the world were showing signs that a correction was overdue for several years due to companies (primarily financial institutions) taking advantage of "cheap credit." Several large banks were the first to collapse and investors were being warned that they might not be able to withdraw their money from stock market accounts, retirement funds, or even regular bank accounts. This was a stark reminder of the Great Depression between 1929 – 1932. Even with these warnings, investors did not anticipate the worst financial crisis in nearly 80 years was about to cripple the global financial system. The financial crisis cost many ordinary people their jobs, their life savings, their homes, and for some, all three were lost.

> **Data**
>
> The data set can be found by visiting stat.hawkeslearning.com and navigating to **Discovering Business Statistics, Second Edition > Data Sets > Percentage Change in Real Home Price Index since 1890**.

This global financial crisis will be written about for many, many years to come with many financial experts asking, *Should we have seen this coming*? To shed just a little light on the crisis and to help answer the question, please use the data titled "Percentage Change in Real Home Price Index since 1890" to answer to answer the following questions.

1. Plot the raw data (Percentage Change in Home Price Index (HPI) against Date) from January 1, 1970, through December 1, 2008. Do you see any patterns over the first five years; over the first 15 years; over the first 30 years?

2. Are these data stationary or nonstationary?

3. Do you see any patterns of variation in the data such as trends, cycles, or seasonality? If so, please identify the timeframe and whether these patterns might have been helpful in predicting the crisis (as a function of HPI).

4. Perform a 12-period moving average to predict HPI. Using these predictions, did you see any evidence that would help you predict the global financial crisis? Is the moving average method good for predicting HPI for this data? Please justify your answer.

5. Use the adjusted exponential smoothing procedure to predict the HPI for January 1, 2009. Using $\alpha = 0.3$ and $\beta = 0.7$. Using these predictions, did you see any evidence that would help you predict the global financial crisis? Is this forecasting method good for predicting HPI for this data? Please justify your answer.

6. Would the additive seasonal forecasting method be good to use for this data? Please justify your answer.

There are many more questions that could be asked. Please take the time to explore all of the data to find as many "stories" as possible. One can also examine productivity growth and the labor force growth, labor force participation rates, average household incomes, aggregate household debt, just to name a few. This is just one of many sets of data associated with the Global Financial Crisis of 2007 – 2008.

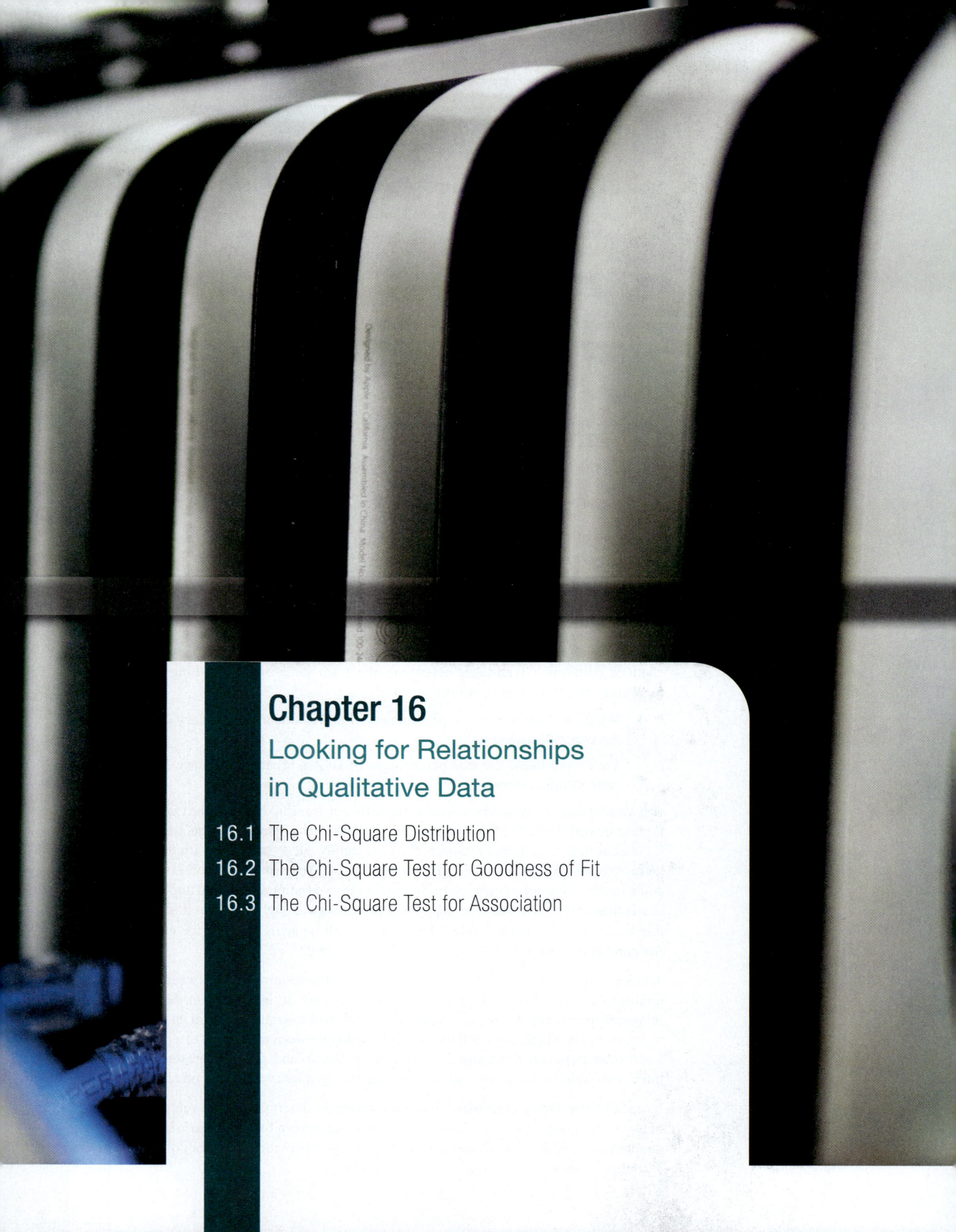

# Chapter 16

## Looking for Relationships in Qualitative Data

# Discovering the Real World

## Do Americans Show Support for Basic Research?

An online survey was conducted in the United States in January 2017 and again in January of 2021. The question presented to the survey participants was, *Do you agree or disagree with the following statement? Even if it brings no immediate benefits, basic scientific research that advances the frontiers of knowledge is necessary and should be supported by the federal government.*

The results of the nationwide poll are presented in the table below.

| Opinion Results on Support for Basic Research Survey | | | | | |
|---|---|---|---|---|---|
| | Strongly Agree | Somewhat Agree | Somewhat Disagree | Strongly Disagree | Don't Know |
| January 2017 | 24% | 39% | 16% | 5% | 16% |
| January 2021 | 46% | 39% | 8% | 2% | 5% |

For this survey, the variable of interest is a qualitative variable: opinion. Based on the survey, the ordinal variable (opinion), can take on five categorical values: Strongly Agree, Somewhat Agree, Somewhat Disagree, Strongly Disagree, and Don't Know.

In evaluating the survey question presented in the table above, we might be interested in knowing if the American opinion has changed on basic research from January 2017 to January 2021. There could be a number of reasons for this significant change of opinion, but most scientists would point to the fact that in 2021 the United States (and the rest of the world) was in the midst of the Coronavirus pandemic. In 2017, there weren't any major crises and thus, the "need" for basic research may not have been perceived to be very important. Of course, during the pandemic, there was a huge rush to develop a vaccine which may have changed opinions. Given these assertions, there are some important things to remember regarding the survey and its results.

- We do not know the true percentages of all Americans in January 2017 (we only have sample estimates).

- We do not know the true percentages of all Americans in January 2021 (we only have sample estimates).

Any conclusions we wish to draw concerning the relationship between the two opinion polls is complicated by the fact that we have sample estimates rather than population parameters. It could be that the difference in percentages is simply due to sampling variation. Population percentages for Americans could be unchanged and the sole cause of the observed differences could be that different subjects were chosen for the two samples. Thus, an important question is whether or not the difference in percentages is large enough such that it is unlikely to be due to sampling variation alone. This question will be analyzed with the **chi-square test for goodness of fit**, which is discussed in Section 16.2.

A slightly different but similar area of interest is the relationship between two types of qualitative data. For example, a brokerage firm may be interested in determining the relationship between the age of investors and their risk tolerance. A political candidate may be interested in whether or not there is a relationship between the gender of a particular voter and a voter's opinion about abortion. Here, the political candidate is interested in evaluating the relationship between two qualitative variables, gender and opinion on abortion.

In each of the above examples we are interested in determining if a relationship exists between two qualitative variables. The **chi-square test for association** between two qualitative variables, introduced in Section 16.3, will provide the methodology to test the existence of these types of qualitative relationships.

## Introduction

In Chapters 5 and 13 we discussed techniques for describing relationships between quantitative variables. However, up to this point we have not discussed any means of effectively summarizing relationships between qualitative variables. This chapter focuses on two methods of summarizing these qualitative relationships. The first method compares the actual proportion of observations appearing in a particular category with the proportion of observations which are expected to appear in that category. The second method determines whether or not two categorical variables are related. Both inferential methods rely on the chi-square distribution , which was discussed in Chapter 9.

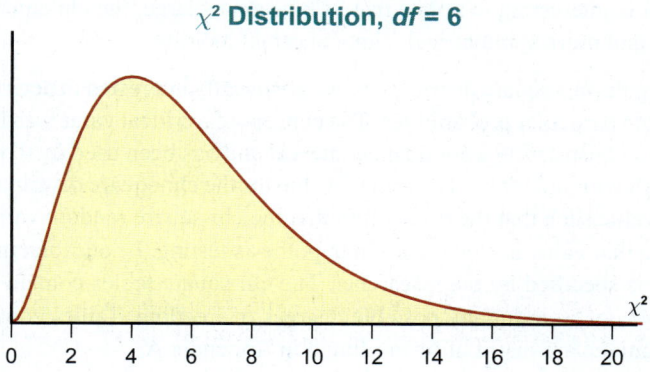

$\chi^2$ **Distribution, *df* = 6**

# 16.1 The Chi-Square Distribution

In Section 9.4 we discussed the chi-square distribution in the context of developing a confidence interval for the population variance. But the chi-square distribution has much broader applications. Before discussing these applications, let's review a few of the basics about the chi-square distribution.

---

**Formula**

### Chi-Square Statistic

If $n$ observations are randomly selected from a normal population with variance $\sigma^2$, and $s^2$ is computed for the sample, then the chi-square statistic

$$\chi^2 = \frac{(n-1)s^2}{\sigma^2}$$

has a **chi-square distribution** with $n - 1$ degrees of freedom.

---

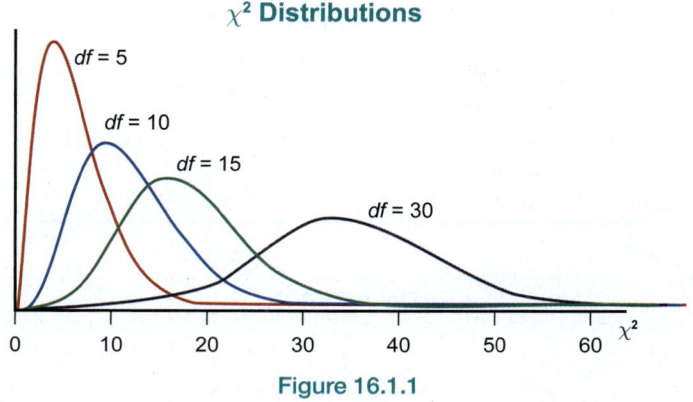

$\chi^2$ **Distributions**

**Figure 16.1.1**

A chi-square distribution is a continuous distribution. Unlike the normal distribution and the $t$-distribution, the chi-square distribution is not symmetric. In fact, for small values of $n$, it is very skewed, as you can see from Figure 16.1.1. Another property of the chi-square distribution which is different from the normal and $t$-distributions is that it only takes on values which are nonnegative. This seems reasonable given that the chi-square distribution is used to represent a ratio which is composed of variances ($s^2$ and $\sigma^2$) which must be nonnegative. In fact, $\sigma^2$ must be greater than zero for any population in which variation exists.

There are infinitely many chi-square distributions. Each chi-square distribution is uniquely defined by its degrees of freedom. Figure 16.1.1 shows chi-square distributions for various sample sizes. It is interesting to notice that as $n$ becomes large, the chi-square distribution becomes more and more symmetrical, almost normal looking.

When analyzing the chi-square distribution, we are usually interested in determining critical values rather than particular probabilities. The concept of a critical value was first introduced in the context of constructing a confidence interval and has been used in all of the chapters which have dealt with inference. The critical value for the chi-square distribution is denoted by $\chi_\alpha^2$. It is a value such that the probability that the chi-square random variable is greater than or equal to that value is equal to $\alpha$. In hypothesis testing, $\alpha$, our tolerance for making a Type I error, is specified by the researcher. The chi-square tables contain critical values for various levels of $\alpha$ and many possible degrees of freedom. Tables containing critical values for the chi-square distribution are found in Appendix A.

### Determining the Chi-Square Critical Values

Suppose a significance level of $\alpha = 0.05$ has been specified and our sample size is 20. The chi-square distribution has one parameter, degrees of freedom, which is equal to $n - 1$. If the null hypothesis is rejected for large values of the test statistic, we look in the table under the column labeled $\chi_{0.050}^2$ and find the critical value corresponding to $20 - 1$, or 19 degrees of freedom. The corresponding critical value is 30.144, as shown in Figure 16.1.2.

| df | ... | $\chi_{0.050}^2$ | $\chi_{0.025}^2$ | $\chi_{0.010}^2$ |
|---|---|---|---|---|
| 1 | | 3.841 | 5.024 | 6.635 |
| 2 | | 5.991 | 7.378 | 9.210 |
| 3 | | 7.815 | 9.348 | 11.345 |
| ... | | | | |
| 19 | | 30.144 | 32.852 | 36.191 |
| 20 | | 31.410 | 34.170 | 37.566 |
| ... | | | | |

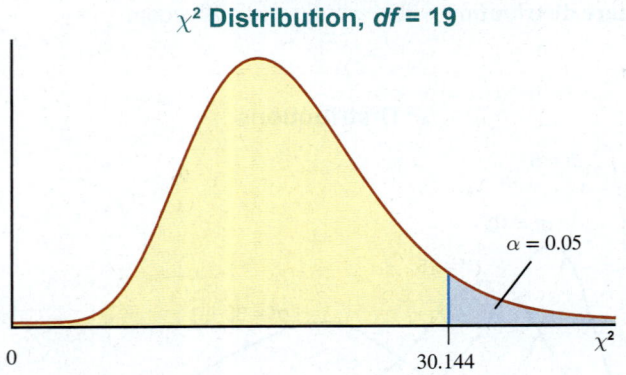

$\chi^2$ **Distribution, df = 19**

$\alpha = 0.05$

0          30.144          $\chi^2$

**Figure 16.1.2**

Suppose a significance level of $\alpha = 0.01$ has been specified and our sample size is 14. If we reject the null hypothesis for large values of the test statistic, we look in the table under the column labeled $\chi^2_{0.010}$ and find the critical value corresponding to $14 - 1$, or 13 degrees of freedom. The corresponding critical value is 27.688, as shown in Figure 16.1.3.

| df | ... | $\chi^2_{0.050}$ | $\chi^2_{0.025}$ | $\chi^2_{0.010}$ |
|---|---|---|---|---|
| 1 | | 3.841 | 5.024 | 6.635 |
| 2 | | 5.991 | 7.378 | 9.210 |
| 3 | | 7.815 | 9.348 | 11.345 |
| ... | | | | |
| 13 | | 22.362 | 24.736 | 27.688 |
| 14 | | 23.685 | 26.119 | 29.141 |
| ... | | | | |

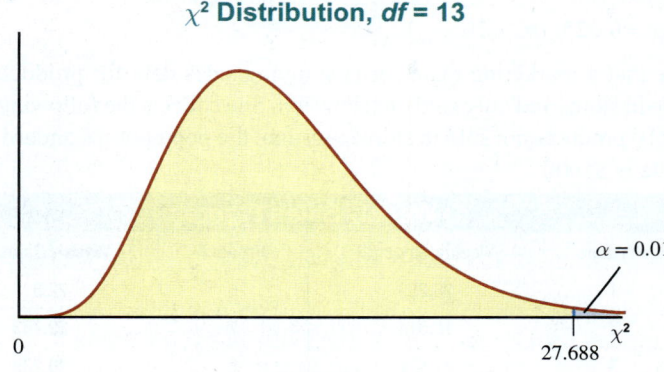

$\chi^2$ **Distribution, df = 13**

$\alpha = 0.01$

0    27.688    $\chi^2$

**Figure 16.1.3**

The analysis in the remainder of the chapter deals with comparing the *actual* number of observations falling into a particular category with the number of observations that is *expected* to fall in that category, based on our hypothesis. In certain circumstances, this type of formulation can be evaluated with a chi-square distribution.

# 🖉 16.1 Exercises

## Basic Concepts

1. Describe the shape of the chi-square distribution.

2. What is the sampling distribution of the sample variance?

3. What are the degrees of freedom associated with the chi-square distribution?

4. Can a chi-square statistic ever be negative? Explain why or why not.

5. Describe how the chi-square distribution changes in shape as $n$ becomes large.

6. Explain the meaning of $\chi^2_\alpha$.

7. Explain the procedure for determining chi-square critical values.

## Exercises

8. Find the chi-square critical value for each of the following.

   a. $\alpha = 0.01$, $df = 14$

   b. $\alpha = 0.01$, $df = 26$

   c. $\alpha = 0.05$, $df = 4$

   d. $\alpha = 0.05$, $df = 9$

   e. $\alpha = 0.005$, $df = 12$

9.  Find the chi-square critical value for each of the following.

    a.  $\alpha = 0.005$, $df = 21$          d.  $\alpha = 0.10$, $df = 90$

    b.  $\alpha = 0.025$, $df = 16$          e.  $\alpha = 0.10$, $df = 17$

    c.  $\alpha = 0.025$, $df = 1$

10. Find the chi-square critical value for each of the following.

    a.  $\alpha = 0.01$, $df = 10$           d.  $\alpha = 0.05$, $df = 11$

    b.  $\alpha = 0.01$, $df = 21$           e.  $\alpha = 0.005$, $df = 29$

    c.  $\alpha = 0.05$, $df = 6$

11. Find the chi-square critical value for each of the following.

    a.  $\alpha = 0.005$, $df = 40$          d.  $\alpha = 0.10$, $df = 24$

    b.  $\alpha = 0.025$, $df = 15$          e.  $\alpha = 0.10$, $df = 50$

    c.  $\alpha = 0.025$, $df = 2$

12. Suppose that a marketing manager is studying sales data for products that are not available in stores and only sold on television. She collects the following weekly sales data for 10 products not sold in stores. Assume the population standard deviation for these data is $5000.

| Weekly Sales Figures | | | |
|---|---|---|---|
| Product | Weekly Sales ($) | Product | Weekly Sales ($) |
| 1 | 26,259 | 6 | 22,511 |
| 2 | 18,514 | 7 | 29,753 |
| 3 | 21,579 | 8 | 20,235 |
| 4 | 18,739 | 9 | 16,258 |
| 5 | 27,821 | 10 | 15,990 |

    a.  Compute the sample standard deviation for these data. Round your answer to the nearest dollar.

    b.  Compute the value of $\chi^2$. Round your answer to three decimal places.

    c.  How many degrees of freedom are associated with this chi-square distribution?

    d.  What is the value of $\chi^2_{0.05}$ for these data?

13. Michael is studying 30-year fixed mortgage rates in Myrtle Beach, SC. He got quotes from 8 lenders, and the APR rates that were quoted to him are given in the following table.

| 30-Year Fixed Mortgage Rates | |
|---|---|
| Lender | APR (%) |
| EverBank | 3.918 |
| AimLoan | 3.925 |
| Great Western | 4.062 |
| Greenlight | 4.353 |
| Flagstar | 4.350 |
| AuroraBank | 4.040 |
| Quicken | 4.458 |
| Roundpoint | 4.125 |

    a.  Calculate the variance of the sample. Round your answer to six decimal places.

    b.  Assuming the population standard deviation for the rates is 0.1%, calculate the value of $\chi^2$.

    c.  Determine the value of $\chi^2_{0.025}$ for these data.

# 16.2 The Chi-Square Test for Goodness of Fit

An owner of a grocery store chain suspects that there has been a change in the shopping pattern of his customers. He has always believed that his stores are equally busy regardless of the day of the week, and he has geared his weekly advertised specials accordingly.

If the shopping pattern has changed and some days are in fact busier shopping days than others, he estimates that this could be costing him thousands of dollars in lost sales due to improperly scheduled advertisements. He decides to use statistical inference to test his theory. He randomly samples 105 customers from his customer database and asks them which day of the week they go grocery shopping. The results of the survey are listed in Table 16.2.1. Based on the data, what criteria can the grocer use to decide if some days of the week are preferred for grocery shopping over others?

**Karl Pearson (1857–1936)**

| Table 16.2.1 – Customer Shopping Data | | | | | | |
|---|---|---|---|---|---|---|
| Monday | Tuesday | Wednesday | Thursday | Friday | Saturday | Sunday |
| 10 | 15 | 14 | 16 | 11 | 20 | 19 |

In 1900, Karl Pearson developed the chi-square test for goodness of fit. The test statistic for the chi-square test for goodness of fit is a summary measure which compares the actual percentage of observations which fall into a particular category with the expected percentage for that category.

In order to use the chi-square test for goodness of fit for inference, our experiment must satisfy some basic conditions. First, we must be able to assume that the underlying distribution of the number of shoppers shopping on the various days of the week has a **multinomial probability distribution**.

The multinomial probability distribution is simply an extension of the familiar binomial probability distribution. The properties of the multinomial experiment are listed in the following box.

Pearson was educated at home until he was nine, and then he was sent to University College where he eventually earned a degree in mathematics. After studying in Germany he returned to University College as a teacher and lecturer. Karl Pearson is regarded as one of the founders of modern statistics. In addition to developing the correlation coefficient, he developed the chi-square test (which included the development of the chi-square sampling distribution) as a means for assessing relationships for categorical data.

## Properties

### Multinomial Experiment
1. The experiment consists of $n$ independent, identical trials.
2. There are $k$ possible outcomes for each trial.
3. The probabilities of the $k$ outcomes, $p_1, p_2, ..., p_k$, are constant from trial to trial.
4. The random variables of interest are the counts in each of the $k$ possible outcomes, $n_1, n_2, ..., n_k$.

## Definition

**Multinomial Probability Distribution**

The **multinomial probability distribution** is a type of probability distribution used to calculate the outcomes of experiments that have random variables with two or more outcomes.

The grocer's survey of the number of shoppers shopping on the various days of the week seems to satisfy the conditions of a multinomial experiment. Since the customers were randomly selected and each of them was asked the same question, there were 105 independent, identical trials. The true proportions of shoppers shopping on each day of the week are unknown and constant from trial to trial.

The proportion of shoppers shopping on each day of the week form a probability distribution whose probabilities sum to one.

Using the procedure outlined in Section 10.1 for testing a hypothesis, we must first define the hypotheses to be tested. Our grocer might formulate his hypotheses as follows.

**Step 1:** Determine the null hypothesis.

The null hypothesis is that the proportion of shoppers does not vary by day of the week. This hypothesis can be written as

$$H_0: p_1 = p_2 = p_3 = p_4 = p_5 = p_6 = p_7 = \frac{1}{7}.$$

**Step 2:** Determine the alternative hypothesis and whether it should be one-sided or two-sided.

In this step, the alternative hypothesis is that the proportion of shoppers varies by day of the week. The alternative hypothesis can be written as

$$H_a: \text{More than one proportion does not equal } \frac{1}{7}.$$

**Step 3:** Select the appropriate test statistic based on the information at hand and the assumptions you are willing to make.

In this situation, the proportion is the obvious measure of choice. However, there are several proportions of interest, namely, the proportion of shoppers shopping on each day of the week. Therefore, let

$$p_1 = \text{proportion of shoppers shopping on Monday,}$$
$$p_2 = \text{proportion of shoppers shopping on Tuesday,}$$
$$\vdots$$
$$p_7 = \text{proportion of shoppers shopping on Sunday.}$$

To determine the appropriate test statistic, let's consider the statement of the null hypothesis. If the null hypothesis is true, then

$$H_0: p_1 = p_2 = p_3 = p_4 = p_5 = p_6 = p_7 = \frac{1}{7},$$

and 15 customers $\left( \frac{1}{7} \text{ of the 105 customers} \right)$ would be expected to shop on each day of the week. If we let $n_1, n_2, ..., n_7$ represent the number of customers who shop on the seven days of the week, respectively, then this expectation would be formally stated as follows.

$$E(n_1) = n \cdot p_1 = 105 \cdot \frac{1}{7} = 15$$

Similarly, $E(n_2) = E(n_3) = \cdots = E(n_7) = 15$.

To test the hypothesis, we need to measure how close reality (the data given by the shoppers) is to what is stated in the null hypothesis. This measure will become a test statistic. The test statistic will be developed as part of a criterion which will be used to make a decision whether or not to reject the null hypothesis. Intuitively, the test statistic should compare the expected number of shoppers for each day of the week given that our null hypothesis is true ($E(n_i)$), to the actual number of shoppers reported in the grocer's survey ($n_i$).

A reasonable test statistic would consist of the differences between the number of shoppers the grocer expected to see each day and the number of shoppers the grocer actually observed each day. In attempting to summarize this information, it is tempting to simply add these differences. However, negative and positive differences will cancel each other out. Further, we have not yet adjusted these differences for any potential differences in scale among the days of the week. (**Note:** In our problem the null hypothesis implies there is no difference in scale. However, the methodology is designed to handle hypothesized differences in scale among the categories.) To compensate for these problems, square the differences so that positive and negative differences will not cancel each other out, and divide each squared difference by the appropriate $E(n_1)$ to adjust for any differences in scale. This will give us the following test statistic.

$$\chi^2 = \frac{\left[n_1 - E(n_1)\right]^2}{E(n_1)} + \frac{\left[n_2 - E(n_2)\right]^2}{E(n_2)} + \cdots + \frac{\left[n_7 - E(n_7)\right]^2}{E(n_7)}$$

If the null hypothesis is true and the data satisfy the conditions of a multinomial distribution, this test statistic has an approximate chi-square distribution with $7 - 1 = 6$ degrees of freedom.

If the differences between what is observed and what is expected are large enough, the chi-square test statistic quantity will become large, causing us to doubt the reasonableness of the null hypothesis.

**Step 4:** Determine the critical value of the test statistic.

Our grocer must choose a level of the test. Let's assume that he is willing to live with a 5% chance of making a Type I error. Namely, concluding that the proportion of shoppers varies by day of the week, when in fact it doesn't. Therefore, he chooses a significance level of $\alpha = 0.05$.

Note that we will reject the null hypothesis for large values of the test statistic (we have a one-tailed test). How large is *large*?

How large must the value of $\chi^2$ be in order to conclude the null hypothesis is not reasonable? We must have a criterion. The critical value for a chi-square distribution with a level of significance of 0.05, $\chi^2_{0.050}$, with 6 degrees of freedom is 12.592. Thus, if the null hypothesis is true, a test statistic as large or larger than 12.592 will occur only 5% of the time due to sampling variation. Hence, if we observe a value of the test statistic that is greater than or equal to 12.592, we will reject the null hypothesis. This rejection region is drawn in Figure 16.2.1.

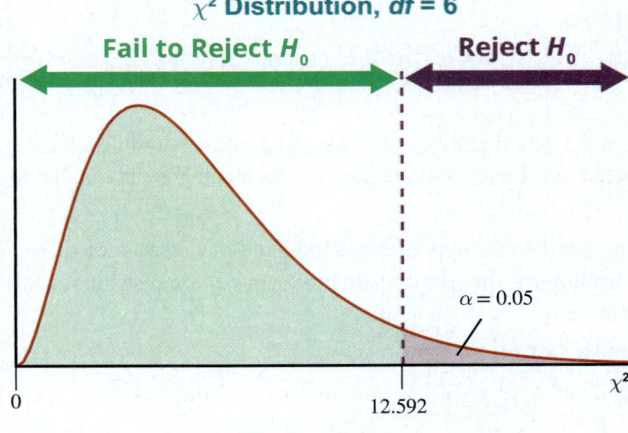

**$\chi^2$ Distribution, *df* = 6**

**Fail to Reject $H_0$**          **Reject $H_0$**

$\alpha = 0.05$

0          12.592          $\chi^2$

**Figure 16.2.1**

**Step 5:** Collect the sample data and compute the value of the test statistic.

The resulting value of the test statistic is as follows.

$$\chi^2 = \frac{\left[10-15\right]^2}{15} + \frac{\left[15-15\right]^2}{15} + \frac{\left[14-15\right]^2}{15} + \frac{\left[16-15\right]^2}{15} + \frac{\left[11-15\right]^2}{15} + \frac{\left[20-15\right]^2}{15} + \frac{\left[19-15\right]^2}{15}$$
$$= 5.6$$

**Step 6:** Make the decision and state the conclusion in terms of the original question.

Notice that the value of the test statistic does not fall in the rejection region since 5.6 is less than 12.592. Thus, we conclude that the data do not provide enough evidence to reject the null hypothesis.

## Technology

The chi-square test statistic and corresponding *P*-value can be found using technology. For instructions, please visit stat.hawkeslearning.com and navigate to **Discovering Business Statistics, Second Edition > Technology Instructions > Chi-Square Distribution > Test for Goodness of Fit**.

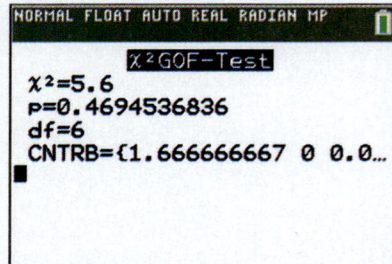

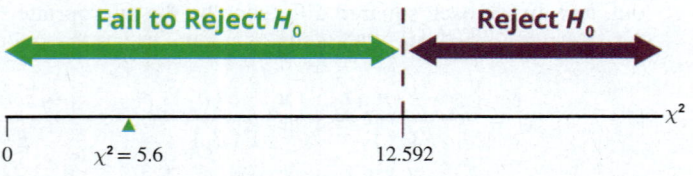

**Figure 16.2.2**

The exact *P*-value for the $\chi^2$ test statistic in this case can be found using technology. The *P*-value is approximately 0.4695, which is greater than the significance level of 0.05. Thus, we conclude that the data do not provide enough evidence to reject the null hypothesis, the same conclusion we arrived at using the rejection region.

*Conclusion and Interpretation:* Our conclusion is that the grocer's suspicion that some shopping days are busier than others was not substantiated. Based on the evidence gathered in his study he should not change his advertising practice.

The basic steps in performing a chi-square test for goodness of fit are outlined in the following procedure. It is important to note that in order to use this test we must assume that the sample size, *n*, is large enough that there will be at least five expected observations per category and that the underlying distribution is a multinomial probability distribution.

---

### Procedure

#### Chi-Square Test for Goodness of Fit

**Hypotheses:**

$H_0: p_1 = p_{1,0}, p_2 = p_{2,0}, \ldots, p_k = p_{k0}$

$H_a$: Any possible difference.

**Test Statistic:**

$$\chi^2 = \sum_{i=1}^{k} \frac{\left[ n_i - E(n_i) \right]^2}{E(n_i)}$$

where $n_i$ is the actual number of observations observed in each category, and $E(n_i)$ is the expected number of observations for each category given that the null hypothesis is true.

Assuming that the null hypothesis is true, and *n* is large enough so that $E(n_i)$ is at least 5 for each category, the test statistic has a chi-square distribution with $k - 1$ degrees of freedom.

**Rejection Region:**

Reject $H_0$ if $\chi^2 \geq \chi^2_{\alpha}$ with $k - 1$ degrees of freedom.

---

### Example 16.2.1

**Detecting a Change in a Qualitative Variable**

Let's return to the survey at the beginning of the chapter regarding support for basic research between surveys conducted in January of 2017 and again in January of 2021. A cursory look at Table 16.2.2 appears to suggest that the American public's perception of support for basic research has increased over time. This difference may simply be due to random variation in the samples or it could represent a real shift in the public sentiment. We will rely on statistical inference methods to determine if the differences are too large to be attributed to sampling variation alone. The chi-square test for goodness of fit can be used to answer the question of whether or not there had been any change in the American sentiment towards support for basic research.

| Table 16.2.2 – Opinion on Support for Basic Research | | |
|---|---|---|
| **Opinion** | **January 2017** | **January 2021** |
| **Strongly Agree** | 24% | 46% |
| **Somewhat Agree** | 39% | 39% |
| **Somewhat Disagree** | 16% | 8% |
| **Strongly Disagree** | 5% | 2% |
| **Don't know** | 16% | 5% |

The question: *Do you agree or disagree with the following statement? Even if it brings no immediate benefits, basic scientific research that advances the frontiers of knowledge is necessary and should be supported by the federal government.*

There were 1215 respondents in the January 2017 survey. Is there sufficient evidence to conclude at $\alpha = 0.10$ that American sentiment towards support for basic research changed from January 2017 to January 2021?

### SOLUTION

**Step 1:** Determine the null hypothesis.

The null hypothesis states that the American sentiment toward support for basic research did not change from January 2017 to January 2021.

The variable to be analyzed is opinion, which has five possible values: Strongly Agree, Somewhat Agree, Somewhat Disagree, Strongly Disagree, and Don't Know. Further, our interest is in the proportion (or percentage) of Americans that falls into each of the opinion categories.

$$p_1 = \text{Strongly Agree}$$
$$p_2 = \text{Somewhat Agree}$$
$$p_3 = \text{Somewhat Disagree}$$
$$p_4 = \text{Strongly Disagree}$$
$$p_5 = \text{Don't Know}$$

The null hypothesis can be written as

$$H_0: p_1 = 0.24, p_2 = 0.39, p_3 = 0.16, p_4 = 0.05, p_5 = 0.16.$$

**Step 2:** Determine the alternative hypothesis and whether it should be one-sided or two-sided.

The alternative hypothesis is that the American sentiment did change from January 2017 to January 2021. This can be written as

$H_a$: Any possible difference from the hypothesized proportions.

The chi-square test for goodness of fit is always a one-tailed test because of the way the test statistic is constructed. We will reject the null hypothesis in favor of the alternative hypothesis for large values of the test statistic.

**Step 3:** Select the appropriate test statistic based on the information at hand and the assumptions you are willing to make.

The number of Americans falling into a particular sentiment category satisfies the properties of a multinomial probability distribution, and the expected number of observations in each category is at least 5. Thus, we can use the chi-square test for goodness of fit. The chi-square test statistic is given by

$$\chi^2 = \frac{\left[n_1 - E(n_1)\right]^2}{E(n_1)} + \frac{\left[n_2 - E(n_2)\right]^2}{E(n_2)} + \frac{\left[n_3 - E(n_3)\right]^2}{E(n_3)} + \frac{\left[n_4 - E(n_4)\right]^2}{E(n_4)} + \frac{\left[n_5 - E(n_5)\right]^2}{E(n_5)}.$$

This test statistic has an approximate chi-square distribution with $5 - 1 = 4$ degrees of freedom, assuming the null hypothesis is true.

**Step 4:** Determine the critical value of the test statistic.

The significance level of the test is specified in the statement of the problem as $\alpha = 0.10$.

Since the level of the test is $\alpha = 0.10$ and we reject the null hypothesis for large values of the test statistic, the chi-square critical value is $\chi^2_{0.100}$ with 4 degrees of freedom, or 7.779.

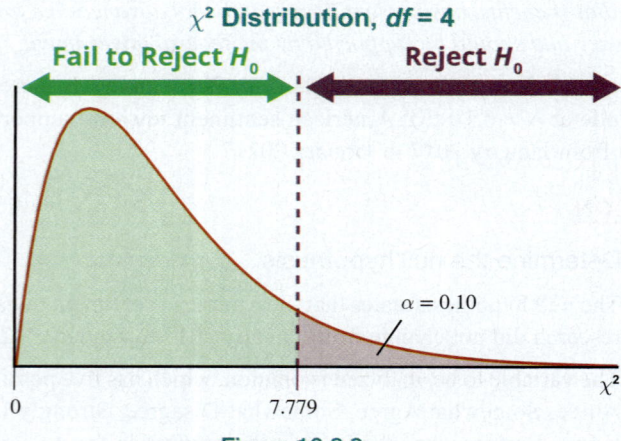

**$\chi^2$ Distribution, *df* = 4**

**Fail to Reject $H_0$**          **Reject $H_0$**

$\alpha = 0.10$

0          7.779          $\chi^2$

**Figure 16.2.3**

**Step 5:** Collect the sample data and compute the value of the test statistic.

To compute the test statistic, the expected number of observations in each category must be computed. Using the proportions from the January 2017 data, we would expect the number of respondents in each category in January 2021 to be as follows.

$$E(n_1) = 1215(0.24) \approx 292$$
$$E(n_2) = 1215(0.39) \approx 474$$
$$E(n_3) = 1215(0.16) \approx 194$$
$$E(n_4) = 1215(0.05) \approx 61$$
$$E(n_5) = 1215(0.16) \approx 194$$

The actual number of respondents in each category in the January 2021 sample is as follows.

$$n_1 = 1215(0.46) \approx 559$$
$$n_2 = 1215(0.39) \approx 474$$
$$n_3 = 1215(0.08) \approx 97$$
$$n_4 = 1215(0.02) \approx 24$$
$$n_5 = 1215(0.05) \approx 61$$

Our test statistic is calculated as

$$\chi^2 = \frac{(559-292)^2}{292} + \frac{(474-474)^2}{474} + \frac{(97-194)^2}{194} + \frac{(24-61)^2}{61} + \frac{(61-194)^2}{194}$$
$$\approx 406.2634.$$

**Step 6:** Make the decision and state the conclusion in terms of the original question.

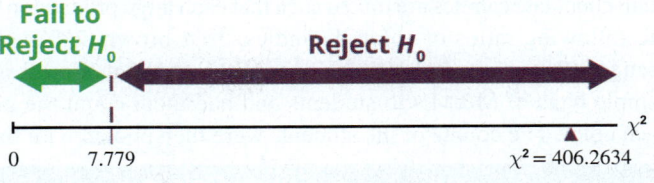

Figure 16.2.4

If the null hypothesis is true, the test statistic will be greater than or equal to the critical value of 7.779 only 10% of the time. Since $\chi^2 \approx 406.2634$ is larger than 7.779, we will reject $H_0$. Large values of the test statistic indicate that the proportion of Americans falling into a category changed from January 2017 to January 2021, and the change is too great to be due to ordinary sampling variation.

The *P*-value for a test statistic value of 406.2634 and degrees of freedom equal to 4 is approximately 0. Therefore, we reject the null hypothesis.

*Conclusion and Interpretation:* At the 10% level of significance, there is sufficient evidence to conclude that American sentiment towards support for basic research changed from January 2017 to January 2021. The difference in sentiment over the 4-year period is much too great to be attributed to ordinary sampling variation alone.

### ⌔ Technology

The *P*-value can be found in Excel using the CHISQ.DIST.RT function. For instructions, please visit stat.hawkeslearning.com and navigate to **Discovering Business Statistics, Second Edition > Technology Instructions > Chi-Square Distribution > Right-Tailed Probability (cdf)**.

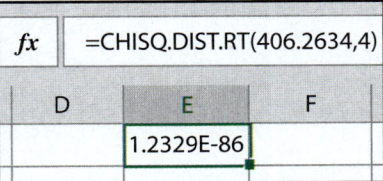

## ✎ 16.2 Exercises

### Basic Concepts

1. Describe what the test statistic for the chi-square test for goodness of fit measures.

2. What is a multinomial probability distribution? What more familiar probability distribution discussed previously in the text is a multinomial probability distribution related to?

3. List the four requirements for a multinomial experiment.

4. What are the null and alternative hypotheses for a chi-square test for goodness of fit?

5. What is the test statistic for a chi-square test for goodness of fit?

6. How many degrees of freedom does the test statistic for the chi-square test for goodness of fit have?

7. What assumptions are necessary for a chi-square test for goodness of fit?

8. How are the expected values determined in a chi-square test for goodness of fit?

### Exercises

9. A telephone company claims that the service calls they receive are equally distributed among the five working days of the week. A survey of 85 randomly selected service calls produced the following results.

| Service Calls | | | | |
|---|---|---|---|---|
| Monday | Tuesday | Wednesday | Thursday | Friday |
| **Number of Calls** 15 | 20 | 25 | 15 | 10 |

a. Is the company's claim refuted by the data at $\alpha = 0.05$?

b. What assumptions were made in the test for part **a.**?

10. Suppose a consumer affairs representative for Mars Incorporated claims that M&M's plain chocolate candies are mixed such that each large production batch has "precisely" the following ratios of colored candies: 30% brown, 20% yellow, 20% red, 10% orange, 10% green, and 10% blue. To test this claim, a professor distributed small sample bags of M&M's to students and had them count the number of candies of each color. The counts of the students were then pooled with the following results.

| Candy Colors | | | | | | | |
|---|---|---|---|---|---|---|---|
| | Brown | Yellow | Red | Orange | Green | Blue | Total |
| Number of Candies | 84 | 79 | 75 | 49 | 36 | 47 | 370 |

a. If the representative's claim is true, what would be the expected number of candies in each of the color categories for 370 candies?

b. Is the representative's claim refuted by the data at $\alpha = 0.01$?

c. What assumptions were made in performing the test for part b.?

11. A highway department executive claims that the number of fatal accidents which occur in her state does not vary from month to month. A survey of 170 fatal accidents produced the following results.

| Accidents | | | | | | | | | | | | |
|---|---|---|---|---|---|---|---|---|---|---|---|---|
| | Jan. | Feb. | Mar. | Apr. | May | Jun. | July | Aug. | Sept. | Oct. | Nov. | Dec. |
| Accidents | 18 | 16 | 7 | 5 | 8 | 12 | 15 | 18 | 15 | 11 | 20 | 25 |

a. Is the executive's claim refuted by the data at $\alpha = 0.01$?

b. What assumptions were made in the test for part a.?

12. The market research firm Nielson recently published market share figures for the operating systems in smartphones. The report stated the following results.

| Market Share for Smartphone Operating Systems | |
|---|---|
| Operating System | Market Share (%) |
| Android OS | 29 |
| iPhone OS | 27 |
| Blackberry OS | 27 |
| Microsoft Windows Mobile | 10 |
| HP Palm/WebOS | 4 |
| Symbian OS | 2 |
| Other | 1 |

**Source:** Nielson.com

Suppose that a marketing manager for a telecommunications company that uses one of the above operating systems doubts the Nielson findings. He collects his own data by surveying 400 people at a local mall. His findings are given in the following table.

| Survey Results | |
|---|---|
| Operating System | Number of People |
| Android OS | 125 |
| iPhone OS | 115 |
| Blackberry OS | 99 |
| Microsoft Windows Mobile | 56 |
| HP Palm/WebOS | 5 |
| Symbian OS | 0 |
| Other | 0 |

a. Compute the expected number of observations for each category for the survey conducted by the telecommunications marketing manager.

b. State the null and alternative hypotheses for the chi-square test for goodness of fit.

c. Using $\alpha = 0.05$, perform a goodness of fit test to determine if the survey conducted by the marketing manager is evidence that the market shares reported by Nielson have changed.

d. What assumptions were made in the test for part **c.**?

e. Do you have any concerns about the way in which the marketing manager's survey was conducted? Explain.

13. A psychologist conducted an attitude survey of 200 randomly selected individuals several years ago. The individuals were asked to pick the one category which most accurately described their attitudes. The results of the survey were as follows.

| 1st Attitude Survey | |
|---|---|
| **Attitude** | **Percent of Respondents** |
| Optimistic | 15% |
| Slightly Optimistic | 30% |
| Slightly Pessimistic | 30% |
| Pessimistic | 25% |

The psychologist believes that these attitudes have changed over time. To test this theory, he randomly selects 200 individuals and asks them the same questions. The results of the second survey are as follows.

| 2nd Attitude Survey | |
|---|---|
| **Attitude** | **Percent of Respondents** |
| Optimistic | 20% |
| Slightly Optimistic | 40% |
| Slightly Pessimistic | 30% |
| Pessimistic | 10% |

a. Can the psychologist conclude that the attitudes have changed over time at $\alpha = 0.01$?

b. What assumptions were made in the test for part **a.**?

# 16.3 The Chi-Square Test for Association

Sometimes, our interest extends beyond one variable to summarizing the relationship between two qualitative variables. For example, a radio executive might be interested in knowing if the age group in which an individual falls affects that person's preference for music. Here the qualitative variables of interest could be age group, which can take on the values Teens and Adults, and preference for music, which can take on the values Classical, Jazz, Easy Listening, Rock, Rap, and Country. Other examples of relationships between two qualitative variables which might be of interest are race and political preference, education level and job performance, income level and occupation, etc.

When we are interested in this type of relationship, we often make use of a **contingency table**. A contingency table organizes data on two characteristics simultaneously. Each cell in a contingency table contains a count (or a proportion) which represents the number of observations falling into that combination of categories. It is important to note that contingency tables are composed of two qualitative variables and each variable satisfies the properties of a multinomial distribution.

Table 16.3.1 shows the general form of a contingency table.

> **Definition**
>
> **Contingency Table**
>
> A **contingency table** organizes data on two characteristics simultaneously. Each cell in a contingency table contains a count (or a proportion) which represents the number of observations falling into that combination of categories.

| Table 16.3.1 – General Form of a Contingency Table | | | |
|---|---|---|---|
| | **Factor A** | | |
| | **Level A1** | **Level A2** | **Total** |
| **Factor B**  **Level B1** | $n_1$ | $n_2$ | $n_1 + n_2$ |
| **Level B2** | $n_3$ | $n_4$ | $n_3 + n_4$ |
| **Total** | $n_1 + n_3$ | $n_2 + n_4$ | $n = n_1 + n_2 + n_3 + n_4$ |

For example, the editor of a regional newspaper is concerned about dwindling sales. Due to her concerns, she's interested in the primary method by which people receive news. She surveyed 990 residents (in three different age groups) in the region and asked them how they get their news (newspaper, radio, television, or the internet). For those who responded, the contingency table shown in Table 16.3.2 resulted.

| Table 16.3.2 – Method of Receiving News | | | | | |
|---|---|---|---|---|---|
| **Age Group** | **Newspaper** | **Radio** | **Television** | **Internet** | **Total** |
| **18–24** | 30 | 20 | 60 | 200 | 310 |
| **25–40** | 100 | 75 | 75 | 150 | 400 |
| **Over 40** | 125 | 50 | 75 | 30 | 280 |
| **Total** | 255 | 145 | 210 | 380 | 990 |

One of the interesting features of this table is that the age variable, whose level of measurement would ordinarily be at the ratio level, has been divided into three categories. Age is now a qualitative variable measured at the ordinal level. What is the advantage of reducing the level of measurement of a variable? To use the test we are about to develop, both variables must be qualitative (i.e., nominal or ordinal levels of measurement). The other variable in the table is the method by which the person receives his or her news, which is also qualitative.

Notice that the column totals and the row totals both sum to the number of readers who responded to the survey, 990. This means that each respondent was allowed to choose only one category. This is necessary if the table is to satisfy the conditions of a multinomial experiment. If the survey is designed such that respondents are allowed to choose more than one category, the following analysis will not apply.

Suppose we are interested in knowing if there is some relationship between age group and method of receiving the news. Another way of thinking about the same problem is to ask if the primary method of receiving the news is dependent on the age group. Rephrasing the question in this manner is important because a formal definition of dependence was developed in Chapter 5. We will use this definition in the development of the hypothesis test. Further, stating that two variables are dependent implies that they are related.

Recall the multiplication rule for independent events from Section 5.4. It states that if two events $A$ and $B$ are independent, then $P(A \cap B) = P(A)P(B)$. If we consider each cell in our contingency table to be the intersection of two events, age group and primary method of receiving news, we can use this multiplication rule to help determine whether or not these two events are dependent. To see this, it is helpful to express our contingency table in terms of relative frequencies or proportions, as shown in Table 16.3.3.

| Table 16.3.3 – Method of Receiving News | | | | | |
|---|---|---|---|---|---|
| Age Group | Newspaper | Radio | Television | Internet | Total |
| 18–24 | $p_{1N} = \dfrac{30}{990}$ | $p_{1R} = \dfrac{20}{990}$ | $p_{1T} = \dfrac{60}{990}$ | $p_{1I} = \dfrac{200}{990}$ | $p_1 = \dfrac{310}{990}$ |
| 25–40 | $p_{2N} = \dfrac{100}{990}$ | $p_{2R} = \dfrac{75}{990}$ | $p_{2T} = \dfrac{75}{990}$ | $p_{2I} = \dfrac{150}{990}$ | $p_2 = \dfrac{400}{990}$ |
| Over 40 | $p_{3N} = \dfrac{125}{990}$ | $p_{3R} = \dfrac{50}{990}$ | $p_{3T} = \dfrac{75}{990}$ | $p_{3I} = \dfrac{30}{990}$ | $p_3 = \dfrac{280}{990}$ |
| Total | $p_N = \dfrac{255}{990}$ | $p_R = \dfrac{145}{990}$ | $p_T = \dfrac{210}{990}$ | $p_I = \dfrac{380}{990}$ | 1 |

The relative frequencies in Table 16.3.3 were determined in the following manner.

$$p_{1N} = P(18-24 \cap \text{Newspaper}) = \frac{30}{990} = \text{ the proportion of residents ages 18–24 that used the newspaper as their primary news source.}$$

$$p_N = P(\text{Newspaper}) = \frac{255}{990} = \text{ the proportion of residents that used the newspaper as their primary news source.}$$

$$p_1 = P(18-24) = \frac{310}{990} = \text{ the proportion of residents ages 18–24 in the sample.}$$

If age group and method of receiving the news are independent, then

$$p_{1N} = p_1 \cdot p_N$$
$$p_{1R} = p_1 \cdot p_R$$
$$\vdots$$
$$p_{3I} = p_3 \cdot p_I$$

Given that we are interested in determining whether or not the primary method of receiving the news is dependent on the age group, and we have developed some relationships that should hold true if the events are independent, it seems reasonable to formulate our hypotheses as follows.

**Null Hypothesis:** Age group and method of receiving the news are independent.

**Alternative Hypothesis:** Age group and method of receiving the news are dependent.

Assuming that the null hypothesis is true (age group and method of receiving the news are independent), then we would expect the number of respondents who received their news by newspapers and were in the 18–24 age group, $E(n_{1N})$, to be equal to the total number of people surveyed ($n$) times the true probability that a person is between 18–24 years old ($p_1$) times the true probability that a person receives his or her news via newspaper ($p_N$). Since the true probabilities of the age group 18–24 and that people receive their news from newspapers are unknown, the results of the survey can be used to estimate these probabilities.

This can be written symbolically as

$$E(n_{1N}) = (n)(p_1)(p_N) = (990)\left(\frac{310}{990}\right)\left(\frac{255}{990}\right) \approx 79.85 \, .$$

This is equivalent to multiplying the row total by the column total and dividing by the total number of observations. We will use the following formula to calculate the expected value for each cell (see Table 16.3.4).

$$E(n_{ij}) = \frac{(\text{row } i \text{ total})(\text{column } j \text{ total})}{\text{total number of observations}}$$

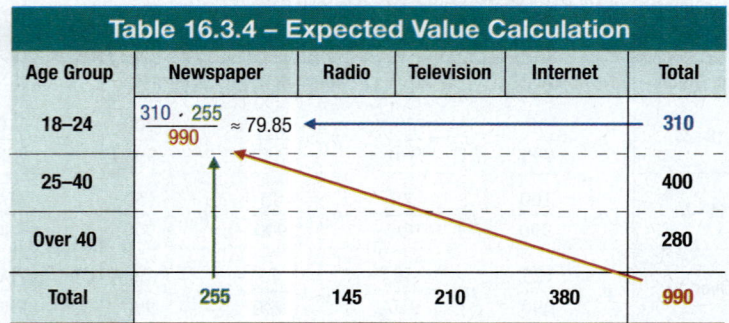

### Table 16.3.4 – Expected Value Calculation

| Age Group | Newspaper | Radio | Television | Internet | Total |
|---|---|---|---|---|---|
| 18–24 | $\frac{310 \cdot 255}{990} \approx 79.85$ | | | | 310 |
| 25–40 | | | | | 400 |
| Over 40 | | | | | 280 |
| Total | 255 | 145 | 210 | 380 | 990 |

Table 16.3.5 displays the expected values for each cell and Table 16.3.6 displays the observed values alongside the expected values.

### Table 16.3.5 – Expected Values

| Age Group | Newspaper | Radio | Television | Internet | Total |
|---|---|---|---|---|---|
| 18–24 | $\frac{310 \cdot 255}{990} \approx 79.85$ | $\frac{310 \cdot 145}{990} \approx 45.40$ | $\frac{310 \cdot 210}{990} \approx 65.76$ | $\frac{310 \cdot 380}{990} \approx 118.99$ | 310 |
| 25–40 | $\frac{400 \cdot 255}{990} \approx 103.03$ | $\frac{400 \cdot 145}{990} \approx 58.59$ | $\frac{400 \cdot 210}{990} \approx 84.85$ | $\frac{400 \cdot 380}{990} \approx 153.54$ | 400 |
| Over 40 | $\frac{280 \cdot 255}{990} \approx 72.12$ | $\frac{280 \cdot 145}{990} \approx 41.01$ | $\frac{280 \cdot 210}{990} \approx 59.39$ | $\frac{280 \cdot 380}{990} \approx 107.47$ | 280 |
| Total | 255 | 145 | 210 | 380 | 990 |

### Table 16.3.6 – Observed versus Expected Values

| Age Group | Newspaper Observed | Newspaper Expected | Radio Observed | Radio Expected | Television Observed | Television Expected | Internet Observed | Internet Expected |
|---|---|---|---|---|---|---|---|---|
| 18–24 | 30 | 79.85 | 20 | 45.40 | 60 | 65.76 | 200 | 118.99 |
| 25–40 | 100 | 103.03 | 75 | 58.59 | 75 | 84.85 | 150 | 153.54 |
| Over 40 | 125 | 72.12 | 50 | 41.01 | 75 | 59.39 | 30 | 107.47 |

Using the same reasoning employed in the chi-square test for goodness of fit, we can construct a chi-square statistic using the differences between the observed and expected values. Notationally, this statistic can be written as follows.

$$\chi^2 = \Sigma \frac{(\text{Observed} - \text{Expected})^2}{\text{Expected}}$$

Given Table 16.3.6, the chi-square statistic is calculated as follows.

$$\chi^2 = \frac{\left[n_{1N} - E(n_{1N})\right]^2}{E(n_{1N})} + \frac{\left[n_{2N} - E(n_{2N})\right]^2}{E(n_{2N})} + \cdots + \frac{\left[n_{3I} - E(n_{3I})\right]^2}{E(n_{3I})}$$

Big differences between the observed and expected values will cause the $\chi^2$ statistic to become large. If the statistic is too large to be due to ordinary sampling variation alone, this will cast doubt on the null hypothesis and cause us to reject the belief that age group and the method of receiving the news are independent. If the assumption of independence is true and the expected number of observations in each cell is at least five, then the sampling distribution of the test statistic has an approximate chi-square distribution with $(r-1)(c-1)$ degrees of freedom where $r$ is the number of rows and $c$ is the number of columns.

If we choose $\alpha = 0.01$, we will reject the null hypothesis if the chi-square test statistic is larger than or equal to the chi-square critical value of 16.812, which is $\chi^2_{0.01}$ with $(3-1)(4-1) = 6$ degrees of freedom.

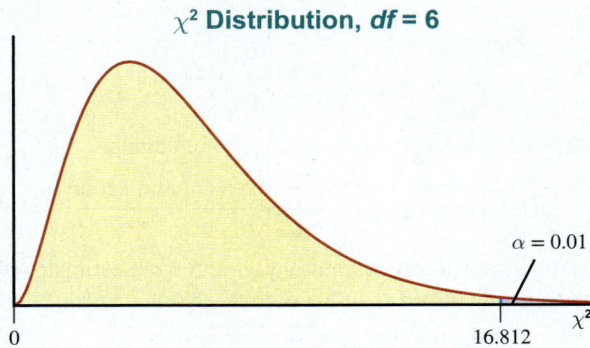

**Figure 16.3.1**

For our data, the chi-square test statistic is given by

$$\chi^2 = \frac{[30-79.85]^2}{79.85} + \frac{[100-103.03]^2}{103.03} + \cdots + \frac{[30-107.47]^2}{107.47} \approx 207.590$$

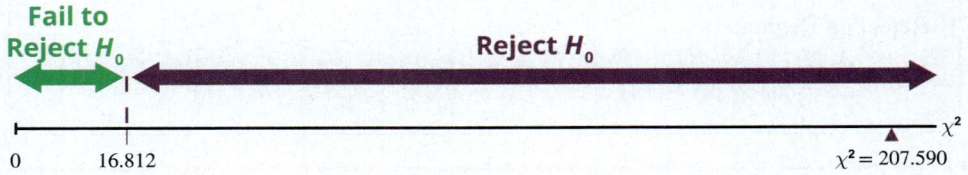

**Figure 16.3.2**

Since the value of the test statistic falls in the rejection region, we will reject the null hypothesis that age group and preferred method of getting the news are independent.

Using technology, the *P*-value for the test statistic of 207.590 with 6 degrees of freedom is extremely small, 4.59298E-42.

Therefore, there is sufficient evidence at the 0.01 level to conclude that age group and preferred method of receiving the news are dependent. In other words, there is a relationship between the two variables.

A summary of the chi-square test for association between two qualitative variables is given in the following box. It is important to note that $n$ should be large enough such that the expected cell count (number of observations) in each cell will be equal to five or more.

✎ **NOTE**

Using unrounded values, the test statistic is approximately 207.594.

☍ **Technology**

The *P*-value can be easily calculated from the test statistic using the $\chi^2$cdf function on a TI-84 Plus calculator. For instructions, please visit stat.hawkeslearning.com and navigate to **Discovering Business Statistics, Second Edition > Technology Instructions > Chi-Square Distribution > Right Tailed Probability (cdf)**.

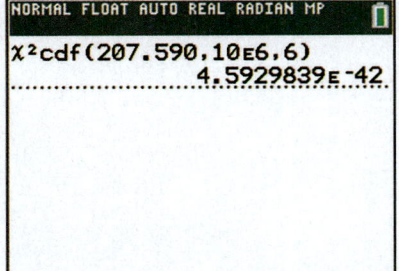

**Procedure**

**Chi-Square Test for Association Between Two Qualitative Variables**

**Hypotheses:**

$H_0$: The two qualitative variables are independent (not related).

$H_a$: The two qualitative variables are dependent (related).

**Test Statistic:**

$$\chi^2 = \sum_{i=1}^{r}\sum_{j=1}^{c} \frac{\left[n_{ij} - E\left(n_{ij}\right)\right]^2}{E\left(n_{ij}\right)}$$

where $r$ = the number of rows and $c$ = the number of columns.

$$E\left(n_{ij}\right) = np_i p_j = \frac{\left(\text{row } i \text{ total}\right)\left(\text{column } j \text{ total}\right)}{n}$$

where $n$ is the total number of observations and $p_i$ and $p_j$ are estimates of the true population proportions calculated as follows.

$$p_i = \frac{\text{the number of observations in row } i}{n}$$

$$p_j = \frac{\text{the number of observations in column } j}{n}$$

If the null hypothesis is true and $n$ is large enough so that the expected number of observations in each cell is at least 5, then the test statistic has an approximate chi-square distribution with $(r-1)(c-1)$ degrees of freedom.

**Rejection Region:**

Reject $H_0$ if $\chi^2 \geq \chi_\alpha^2$ with $(r-1)(c-1)$ degrees of freedom.

**Example 16.3.1**

**Detecting an Association Between Qualitative Variables**

Many grocery stores have self-checkout scanners so that customers can check out faster. Some customers are concerned that the scanners will be wrong and overcharge them. However, the manufacturer of the scanners indicates that the scanners will err in favor of the customer most of the time. In a random sample of customers checking themselves out of the grocery store as well as those that allow the employees to check them out, we have the following contingency table.

| Table 16.3.7 – Self-Checkout versus Employee Checkout at the Grocery Store | | | |
|---|---|---|---|
| **Price Charged** | **Self-Checkout** | **Employee Checkout** | **Total** |
| Undercharge | 20 | 10 | 30 |
| Overcharge | 15 | 30 | 45 |
| Correct Price | 200 | 225 | 425 |
| **Total** | **235** | **265** | **500** |

Based on the data, can we conclude that the price charged using scanners is independent of whether a customer checks themselves out or if a store employee checks them out?

**SOLUTION**

**Step 1:** Determine the null hypothesis.

The null hypothesis is that the price charged at checkout and method of checkout are independent (not related). The null hypothesis is written as

$H_0$: Price charged at checkout and method of checkout are independent (not related).

For this example, we are interested in the proportion of data in each of the Price Charged and Checkout Method categories.

**Step 2:** Determine the alternative hypothesis and whether it should be one-sided or two-sided.

The alternative hypothesis is that the price charged at checkout and method of checkout are dependent (related). The alternative hypothesis is written as

$H_a$: Price charged at checkout and method of checkout are dependent (related).

The chi-square test for association between two qualitative variables is always a one-tailed test because of the way we construct the test statistic. We will reject the null hypothesis that price charged at checkout is independent of checkout method for large values of the test statistic.

**Step 3:** Select the appropriate test statistic based on the information at hand and the assumptions that you are willing to make.

Since each piece of data cannot belong to more than one category and the expected cell count for each category is at least 5, we can use the chi-square test for association between two qualitative variables. The formula for the test statistic is given as follows.

$$\chi^2 = \frac{\left[ n_{US} - E\left(n_{US}\right) \right]^2}{E\left(n_{US}\right)} + \frac{\left[ n_{OS} - E\left(n_{OS}\right) \right]^2}{E\left(n_{OS}\right)} + \frac{\left[ n_{CS} - E\left(n_{CS}\right) \right]^2}{E\left(n_{CS}\right)}$$
$$+ \frac{\left[ n_{UE} - E\left(n_{UE}\right) \right]^2}{E\left(n_{UE}\right)} + \frac{\left[ n_{OE} - E\left(n_{OE}\right) \right]^2}{E\left(n_{OE}\right)} + \frac{\left[ n_{CE} - E\left(n_{CE}\right) \right]^2}{E\left(n_{CE}\right)}$$

This test statistic has an approximate chi-square distribution with $(3 - 1)(2 - 1) = 2$ degrees of freedom, assuming the null hypothesis is true.

We must specify the level of the test ourselves since we are not given one in the problem. Let's choose $\alpha = 0.10$.

**Step 4:** Determine the critical value of the test statistic.

Since the level of the test is $\alpha = 0.10$ and we will reject the null hypothesis for large values of the test statistic, the chi-square critical value is $\chi^2_{0.100}$ with $(3 - 1)(2 - 1) = 2$ degrees of freedom, or 4.605. Values of the test statistic greater than or equal to 4.605 would indicate that the dependence of the two classifications is not likely due to ordinary sampling variation alone.

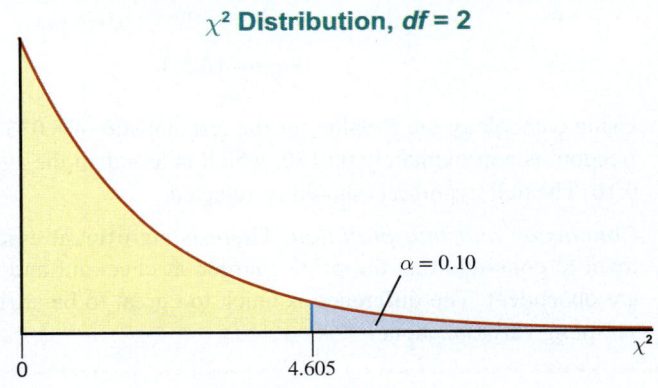

$\chi^2$ **Distribution, *df* = 2**

Figure 16.3.3

**Step 5:** Collect the sample data and compute the value of the test statistic.

The expected values are calculated in Table 16.3.8, and Table 16.3.9 gives the actual data values in each category versus the expected values in each category.

**Table 16.3.8 – Self-Checkout versus Employee Checkout at the Grocery Store: Expected Values**

| Price Charged | Self Checkout | Employee Checkout | Total |
|---|---|---|---|
| Undercharge | $\frac{30 \cdot 235}{500} = 14.10$ | $\frac{30 \cdot 265}{500} = 15.90$ | 30 |
| Overcharge | $\frac{45 \cdot 235}{500} = 21.15$ | $\frac{45 \cdot 265}{500} = 23.85$ | 45 |
| Correct Price | $\frac{425 \cdot 235}{500} = 199.75$ | $\frac{425 \cdot 265}{500} = 225.25$ | 425 |
| Total | 235 | 265 | 500 |

**Table 16.3.9 – Self-Checkout versus Employee Checkout at the Grocery Store: Observed and Expected Values**

| Price Charged | Self Checkout | | Employee Checkout | |
|---|---|---|---|---|
| | Observed | Expected | Observed | Expected |
| Undercharge | 20 | 14.10 | 10 | 15.90 |
| Overcharge | 15 | 21.15 | 30 | 23.85 |
| Correct Price | 200 | 199.75 | 225 | 225.25 |

Based on these results, the test statistic is calculated as follows.

$$\chi^2 = \frac{[20-14.10]^2}{14.10} + \frac{[15-21.15]^2}{21.15} + \frac{[200-199.75]^2}{199.75}$$
$$+ \frac{[10-15.90]^2}{15.90} + \frac{[30-23.85]^2}{23.85} + \frac{[225-225.25]^2}{225.25}$$
$$\approx 8.033$$

**Step 6:** Make the decision and state the conclusion in terms of the original question.

Since the $\chi^2$ test statistic is greater than the critical value and falls in the rejection region, the null hypothesis should be rejected.

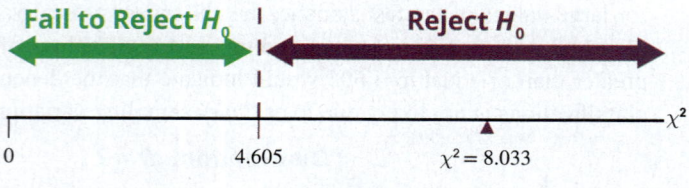

**Figure 16.3.4**

Using technology the *P*-value for the test statistic of 8.033 with 2 degrees of freedom is approximately 0.0180, which is less than the significance level of 0.10. The null hypothesis should be rejected.

*Conclusion and Interpretation*: There is significant evidence at the 0.10 level to conclude that the price charged at checkout and checkout method are dependent. The difference is much too great to be attributed to ordinary sampling variation alone.

## ⌘ Technology

The chi-square test statistic and corresponding *P*-value for can be obtained using technology. For instructions please visit stat.hawkeslearning.com and navigate to **Discovering Business Statistics, Second Edition > Technology Instructions > Chi-Square Distribution > Test for Association**.

| *fx* | =CHISQ.DIST.RT(8.033,2) | |
|---|---|---|
| D | E | F |
| | 0.018016 | |

# 📝 16.3 Exercises

## Basic Concepts

1. Give two examples of relationships between qualitative variables that would be of interest to a manager in a business setting.

2. Explain the difference between the chi-square test for goodness of fit and the chi-square test for association.

3. What is a contingency table?

4. Describe the information that each cell in a contingency table gives.

5. What properties must the two categories of the contingency table possess?

6. What level(s) of measurement may the categories of a contingency table have?

7. Consider the variable income. Describe how this variable could be transformed to be included in a contingency table. Is information lost during the transformation?

8. Explain why a test for association is not valid if single data points are allowed to belong to more than one category.

9. Restate the multiplication rule for independent events. Explain how this rule pertains to the chi-square test for association.

10. State the null and alternative hypotheses for a chi-square test for association between two qualitative variables.

11. What is the test statistic for the chi-square test for association?

12. How many degrees of freedom are associated with the test statistic given in Exercise 11?

## Exercises

13. A political analyst is interested in studying the relationship between age and political affiliation. The analyst randomly selects 200 people and determines their age and political affiliation. The number of responses in each of the categories is as follows.

| Age and Political Affiliation | | | |
|---|---|---|---|
| | **Political Affiliation** | | |
| **Age** | **Democrat** | **Republican** | **Independent** |
| 18–34 | 50 | 10 | 15 |
| 35–51 | 15 | 25 | 15 |
| 52–68 | 25 | 35 | 10 |

   a. Can the analyst conclude that age and political affiliation are dependent at $\alpha = 0.05$?

   b. What assumptions were made in the test for part **a.**?

14. A sociologist is interested in studying the relationship between education and crime. She randomly selects 150 people and asks their education level and whether or not they have ever been convicted of a felony. The following table displays the number of respondents in each category.

| Education and Crime | | |
|---|---|---|
| Have you ever been convicted of a felony? | | |
| Education Level | Response | |
| | Yes | No |
| Less Than 9 Years | 2 | 35 |
| 9 Years to 12 Years | 4 | 31 |
| 12 Years to 16 Years | 1 | 31 |
| 16+ Years | 4 | 42 |

a. Can the sociologist conclude that education level and crime are dependent at $\alpha = 0.10$?

b. What assumptions were made in the test for part **a.**?

15. A psychologist is preparing his thesis on child abuse. He thinks that there may be a relationship between various types of child abuse and the age of the child. To study this, he randomly selects the records of 197 abused children and determines both the age group in which each child falls (Tweens (ages 9-12) and Teens (ages 13-17)) and the documented type of child abuse. The results of the study are as follows.

| Child Abuse | | |
|---|---|---|
| Type of Abuse | Age Group | |
| | Tweens | Teens |
| Neglect | 50 | 50 |
| Physical | 20 | 30 |
| Sexual | 10 | 19 |
| Emotional | 10 | 8 |

a. Can the psychologist conclude that the type of child abuse and age group in which the child falls are dependent at $\alpha = 0.05$?

b. What assumptions were made in the test for part **a.**?

16. The National Fire Protection Association is interested in studying the relationship between the causes of fires and the region of the country in which the fires occur. They randomly select 500 fires and determine the region of the country in which the fire occurred and cause of the fire with the following results.

| Fires | | | | |
|---|---|---|---|---|
| Cause of Fire | Region | | | |
| | North | South | East | West |
| Smoking | 37 | 38 | 40 | 35 |
| Heating Equipment | 25 | 20 | 18 | 19 |
| Arson | 17 | 15 | 16 | 15 |
| Electrical | 12 | 13 | 12 | 13 |
| Children at Play | 10 | 11 | 12 | 11 |
| Other | 27 | 28 | 29 | 27 |

a. Can the association conclude that the cause of the fire and the region of the fire are dependent at $\alpha = 0.01$?

b. What assumptions were made in the test for part **a.**?

# T Discovering Technology

## Using the TI-84 Plus Calculator

### Chi-Square Test for Association

Use the information from Example 16.3.1 for this exercise.

1. Access the **MATRIX** menu by pressing 2ND and $[x^{-1}]$.

2. Choose **EDIT**, and [A]. Press ENTER.

3. Enter 3 × 2 as the matrix dimensions, and enter the data from Table 16.3.7 into this matrix. The total row and column are not needed.

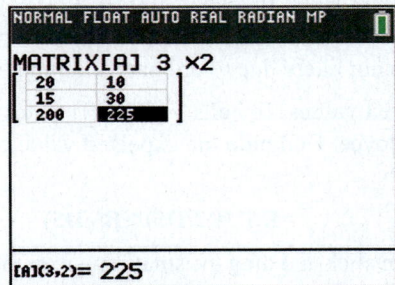

4. Press STAT, TESTS, and choose $\chi^2$ – **Test**.

5. Press ENTER. Choose **Calculate**. Press ENTER Observe the chi-square results. Compare these to the results which we determined in Example 16.3.1. Notice that the test statistic, $P$-value for the test, and the degrees of freedom are given in the calculator output. Since the $P$-value is less than the level of significance, we reject the null hypothesis and conclude that price charged at checkout and checkout method are dependent.

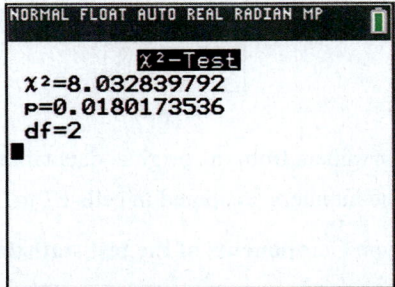

## Using Excel

### Chi-Square Test for Association

Use the information from Example 16.3.1 for this exercise.

1. Enter the data from Table 16.3.7 into a new worksheet in Excel.

|  | A | B | C | D |
|---|---|---|---|---|
| 1 | Price Charged | Self Checkout | Employee Checkout | Total |
| 2 | Undercharge | 20 | 10 | 30 |
| 3 | Overcharge | 15 | 30 | 45 |
| 4 | Correct Price | 200 | 225 | 425 |
| 5 | Total | 235 | 265 | 500 |

2.  The statements of the hypotheses are as follows.

$H_0$:  Price charged at checkout and method of checkout are independent (not related).

$H_a$:  Price charged at checkout and method of checkout are dependent (related).

3.  Specify the critical value with the function

**CHISQ.INV.RT(probability, deg_freedom)**

where **probability** is the significance level and **deg_freedom** is the number of degrees of freedom for the test.

The level of significance in Example 16.3.1 is $\alpha = 0.1$ and the number of degrees of freedom is $(3 - 1)(2 - 1) = 2$. Label cell A7 as **Critical Value** and find the critical value in cell B7 with the function

**=CHISQ.INV.RT(0.1, 2).**

This returns the critical value of the $\chi^2$ test as 4.605. Values of the test statistic greater than or equal to 4.605 would indicate that the apparent dependence of the two classifications is not likely due to ordinary sampling variation alone.

4.  Compute the expected values. In cells F1 and G1 enter the labels **Expected Self** and **Expected Employee**. Compute the expected value of undercharges using self checkout in cell F2.

**=D5*(D2/D5)*(B5/D5)**

Click cell F2 and then click and drag the small square at the lower-right corner of the cell down and to the right to cell G4 to compute the other expected values.

| | A | B | C | D | E | F | G |
|---|---|---|---|---|---|---|---|
| 1 | Price Charged | Self Checkout | Employee Checkout | Total | | Expected Self | Expected Employee |
| 2 | Undercharge | 20 | 10 | 30 | | 14.1 | 15.9 |
| 3 | Overcharge | 15 | 30 | 45 | | 21.15 | 23.85 |
| 4 | Correct Price | 200 | 225 | 425 | | 199.75 | 225.25 |
| 5 | Total | 235 | 265 | 500 | | | |

5.  Compute the components of the test statistic. Written mathematically as:

$$\frac{\left[n_{ij} - E\left(n_{ij}\right)\right]^2}{E\left(n_{ij}\right)}.$$

The $n_{ij}$ terms are the numbers from the original data table in cells B2 to C4 while the $E[n_{ij}]$ terms are the numbers computed in cells F2 to G4.

In cells A9 to A11, type **Components of the test statistic**. In cell B9, input

**=((B2-F2)^2)/F2**

Click cell B9 and then click and drag the small square at the lower-right corner of the cell down and to the right to cell C11 to compute these terms.

6.  Compute the test statistic. In cell A13, type **Test Statistic**. In cell B13, add the components computed in the previous step with the formula

**=SUM(B9:C11)**

This computes the test statistic to be 8.03284.

| | A | B | C | D | E | F | G |
|---|---|---|---|---|---|---|---|
| 1 | Price Charged | Self Checkout | Employee Checkout | Total | | Expected Self | Expected Employee |
| 2 | Undercharge | 20 | 10 | 30 | | 14.1 | 15.9 |
| 3 | Overcharge | 15 | 30 | 45 | | 21.15 | 23.85 |
| 4 | Correct Price | 200 | 225 | 425 | | 199.75 | 225.25 |
| 5 | Total | 235 | 265 | 500 | | | |
| 6 | | | | | | | |
| 7 | Critical Value | 4.60517 | | | | | |
| 8 | | | | | | | |
| 9 | Components | 2.468794 | 2.189308 | | | | |
| 10 | of the test | 1.788298 | 1.585849 | | | | |
| 11 | statistic | 0.000313 | 0.000277 | | | | |
| 12 | | | | | | | |
| 13 | Test Statistic | 8.03284 | | | | | |

7. Make a decision. The chi-square test statistic is approximately 8.033. Since the value of the test statistic is greater than the critical value and falls in the rejection region, the null hypothesis should be rejected.

## Using JMP

### Chi-Square Test for Association

Use the information from Example 16.3.1 for this exercise.

1. With a JMP **Data Table** open, enter the data into **Columns 1-3** and add the column labels as shown below.

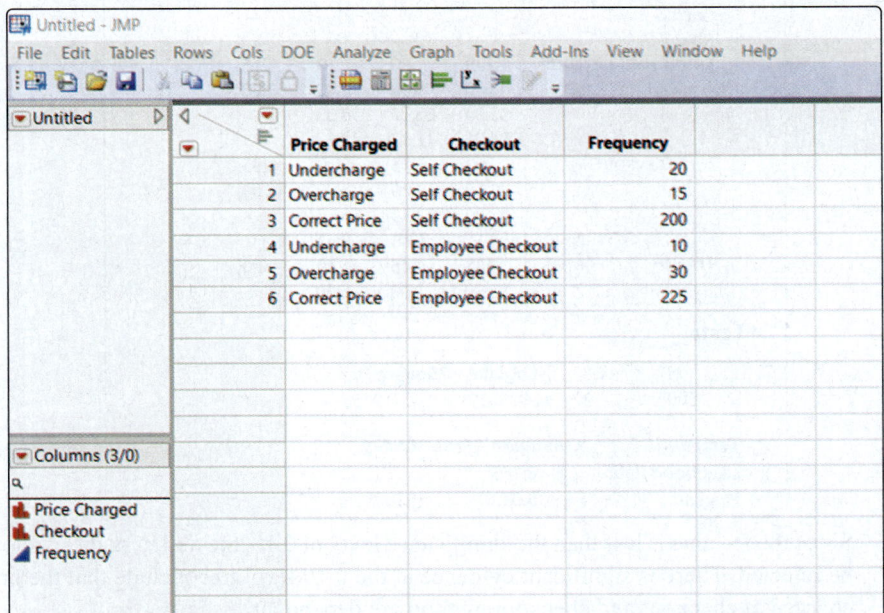

2. Select **Analyze** in the top row of the JMP spreadsheet and then select **Fit Y by X**. From the **Select Columns** box, click on **Price Charged** and then click on **Y, Response**.

3. From the **Select Columns** box, click on **Checkout** and then click on **X, Factor**. From the **Select Columns** box, click on **Frequency** and then click on **Freq**. Click **OK**.

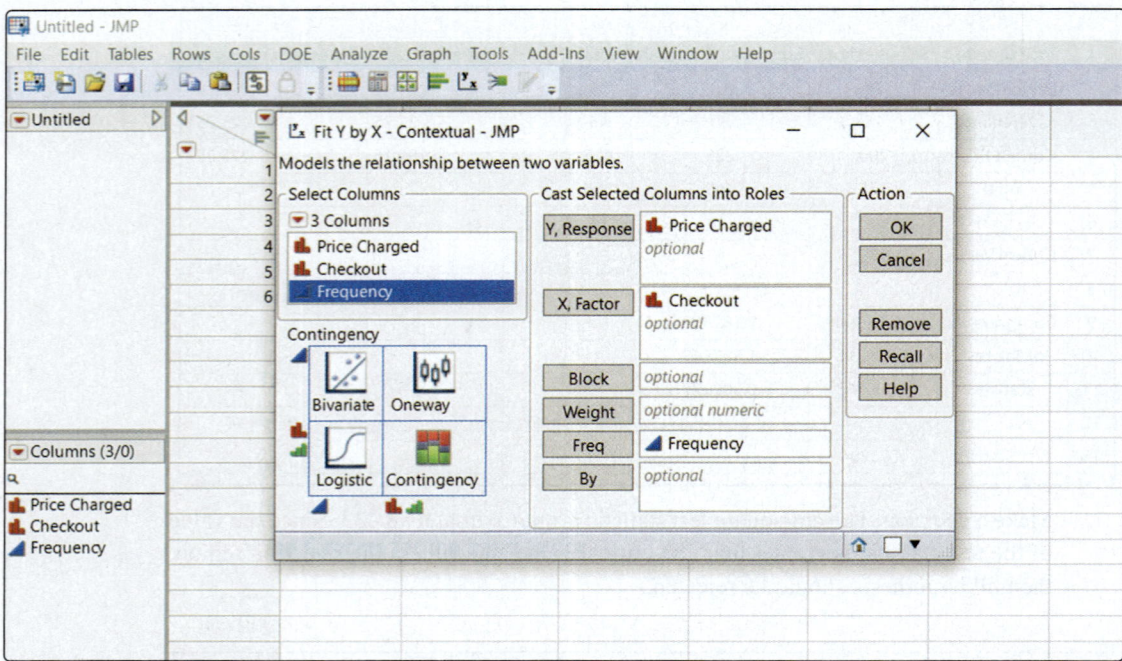

4.  At the bottom of the output, is the Pearson Test. The value of 8.033 is the value of the chi-square test statistic for testing whether the two variables are associated. The *P*-value for the test statistic is shown to the right as 0.0180.

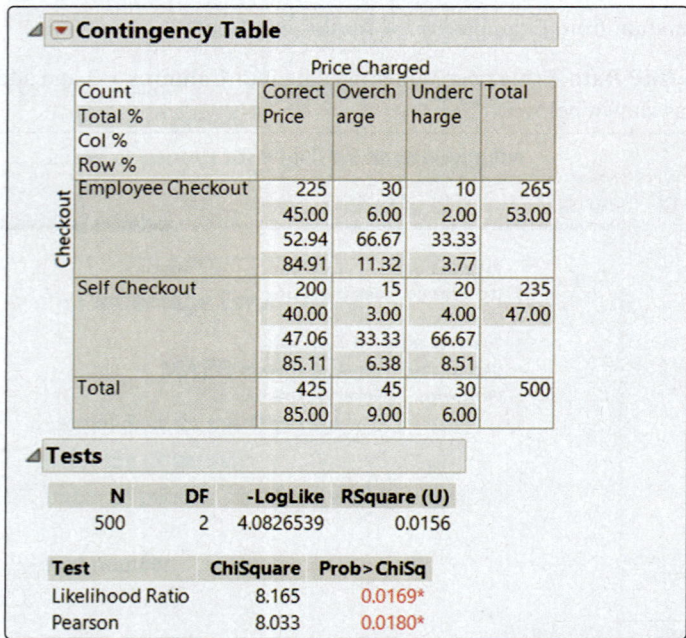

### Contingency Table

|  | | Price Charged | | | |
|---|---|---|---|---|---|
| Count<br>Total %<br>Col %<br>Row % | | Correct<br>Price | Overch<br>arge | Underc<br>harge | Total |
| **Checkout** | Employee Checkout | 225<br>45.00<br>52.94<br>84.91 | 30<br>6.00<br>66.67<br>11.32 | 10<br>2.00<br>33.33<br>3.77 | 265<br>53.00 |
| | Self Checkout | 200<br>40.00<br>47.06<br>85.11 | 15<br>3.00<br>33.33<br>6.38 | 20<br>4.00<br>66.67<br>8.51 | 235<br>47.00 |
| | Total | 425<br>85.00 | 45<br>9.00 | 30<br>6.00 | 500 |

### Tests

| N | DF | -LogLike | RSquare (U) |
|---|---|---|---|
| 500 | 2 | 4.0826539 | 0.0156 |

| Test | ChiSquare | Prob>ChiSq |
|---|---|---|
| Likelihood Ratio | 8.165 | 0.0169* |
| Pearson | 8.033 | 0.0180* |

5.  Since the *P*-value is less than the significance level of 0.10, the null hypothesis should be rejected. There is significant evidence at the 0.10 level to conclude that the price charged at checkout and checkout method are dependent.

## Chi-Square Test for Goodness of Fit

For this exercise, use the data from Table 16.2.1.

1.  With a JMP **Data Table** open, enter the data into **Column 1** and **Column 2** and add the column labels as shown below.

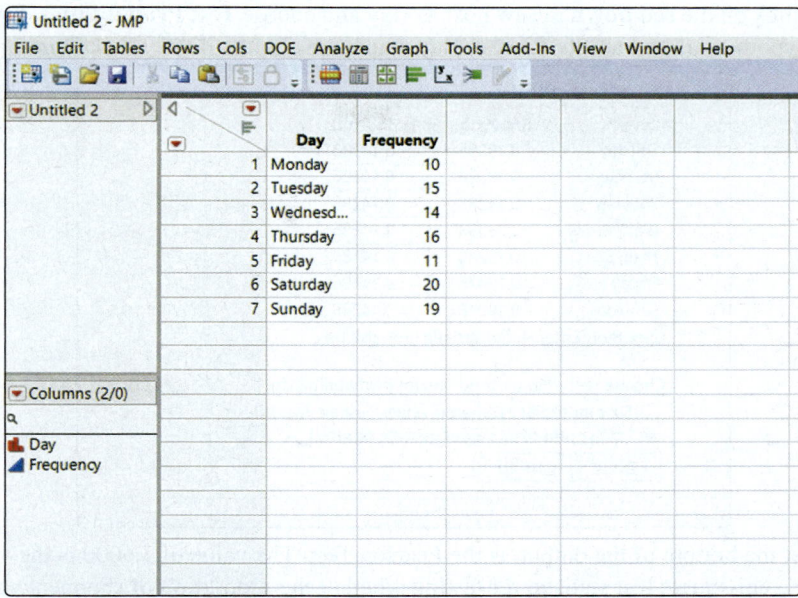

2.  Select **Analyze** in the top row of the JMP spreadsheet and then select **Distribution**. From the **Select Columns** box, click on **Day** and then click on **Y, Columns**.

3.  From the **Select Columns** box, click on **Frequency** and then click on **Freq**. Click **OK**.

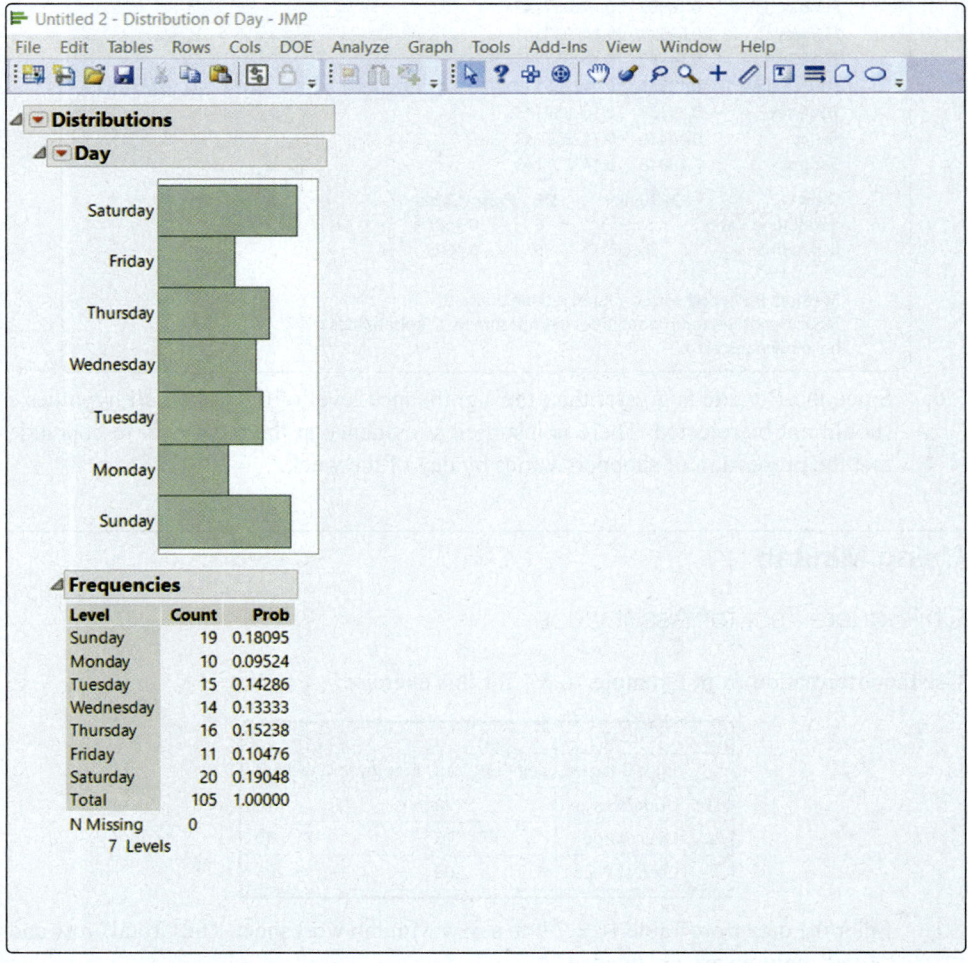

4.   Click on the **red down arrow** next to **Day** and choose **Test Probabilities**. Fill in the hypothetical probability for each day which is 1/7 or 0.14286. Click **Done**.

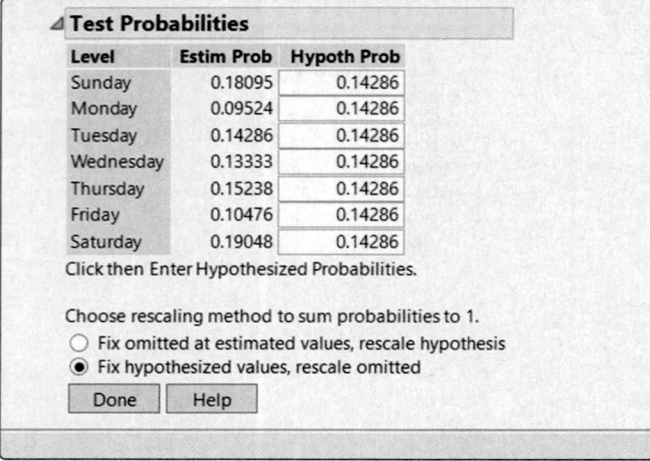

5.   At the bottom of the output, is the Pearson Test. The value of 5.6000 is the value of the chi-square test statistic for testing whether the proportion of shoppers varies by day of the week. The *P*-value for the test statistic is shown to the right as 0.4695.

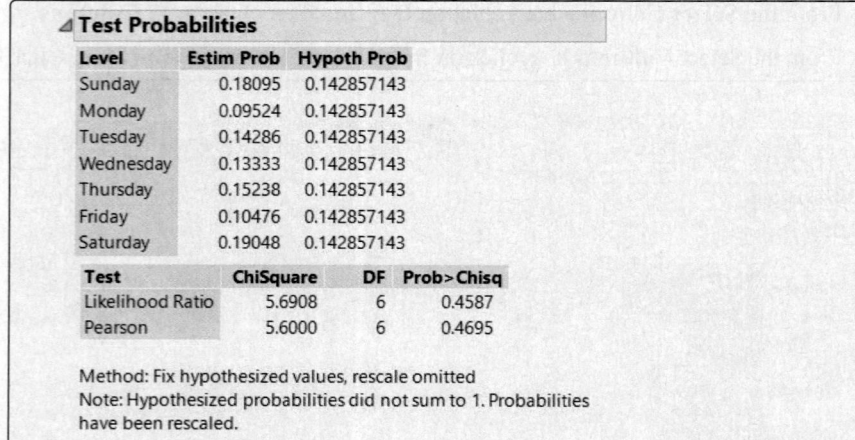

6.   Since the *P*-value is greater than the significance level of 0.05, the null hypothesis should not be rejected. There is insufficient evidence at the 0.05 level to conclude that the proportion of shoppers varies by day of the week.

## Using Minitab

### Chi-Square Test for Association

Use the information from Example 16.3.1 for this exercise.

| ↓ | C1-T | C2 | C3 |
|---|------|-----|-----|
|   | Price Charged | Self Checkout | Employee Checkout |
| 1 | Undercharge | 20 | 10 |
| 2 | Overcharge | 15 | 30 |
| 3 | Correct Price | 200 | 225 |

1.   Enter the data from Table 16.3.7 into a new Minitab worksheet. The "Total" row and "Total" column are not needed.

2. The statements of the hypothesis are as follows.

$H_0$: Price charged at checkout and method of checkout are independent (not related).

$H_a$: Price charged at checkout and method of checkout are dependent (related).

3. Choose **Stat**, **Tables**, and **Chi-Square Test for Association**.

4. Click on **summarized data in a two-way table** from the dropdown menu. Enter **C2** and **C3** as the columns containing the table.

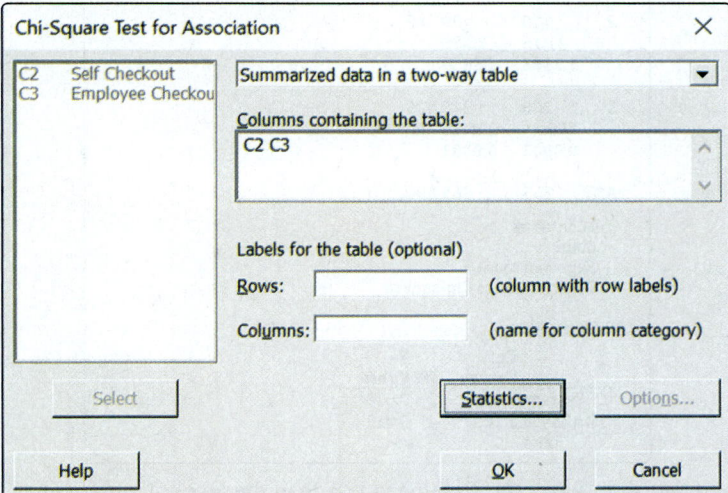

5. Click on Statistics. Select the radio button **Each cell's contribution to chi-square** and Press **OK**. Press **OK** again.

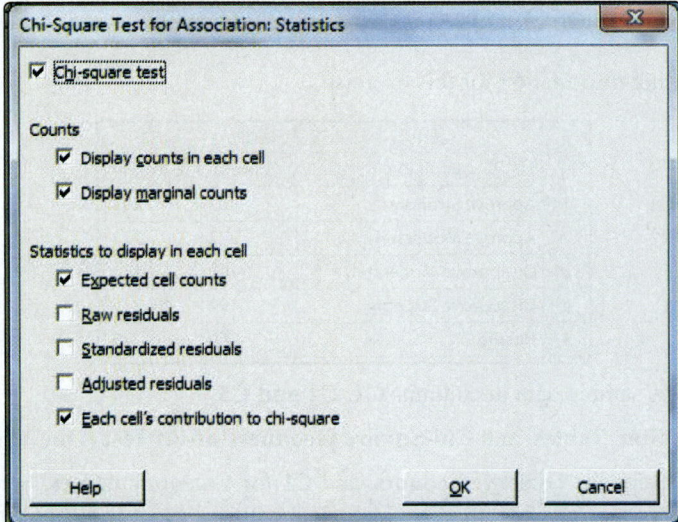

6. Observe the output shown below.

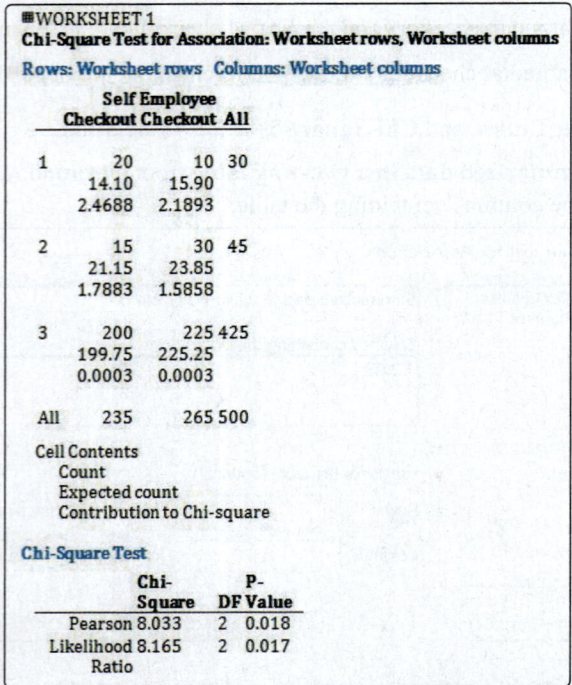

■ WORKSHEET 1
**Chi-Square Test for Association: Worksheet rows, Worksheet columns**

Rows: Worksheet rows   Columns: Worksheet columns

|   | Self Employee Checkout | Checkout | All |
|---|---|---|---|
| 1 | 20 | 10 | 30 |
|   | 14.10 | 15.90 | |
|   | 2.4688 | 2.1893 | |
| 2 | 15 | 30 | 45 |
|   | 21.15 | 23.85 | |
|   | 1.7883 | 1.5858 | |
| 3 | 200 | 225 | 425 |
|   | 199.75 | 225.25 | |
|   | 0.0003 | 0.0003 | |
| All | 235 | 265 | 500 |

Cell Contents
  Count
  Expected count
  Contribution to Chi-square

**Chi-Square Test**

| | Chi-Square | DF | P-Value |
|---|---|---|---|
| Pearson | 8.033 | 2 | 0.018 |
| Likelihood Ratio | 8.165 | 2 | 0.017 |

7. Make a decision. Since the $P$-value is less than the level of significance ($\alpha = 0.10$), the null hypothesis should be rejected.

## Chi-Square Test for Goodness of Fit

Use the following information for this exercise.

| ↓ | C1-T | C2 | C3 |
|---|---|---|---|
| | opinion | 2009 | 2010 |
| 1 | Approve Strongly | 403 | 292 |
| 2 | Approve Moderately | 272 | 222 |
| 3 | Disapprove Moderately | 101 | 131 |
| 4 | Disapprove Strongly | 192 | 343 |
| 5 | Unsure | 40 | 20 |

1. Enter this sample data in columns **C1, C2 and C3**.

2. Choose **Stat**, **Tables**, and **Chi-Square Goodness-of-Fit Test (One Variable)**.

3. Enter **C3** for the **Observed counts**, and **C1** for **Category names**. Select the radio button **Proportions specified by historical counts**, select **Input column**, and then enter **C2** as the historical counts.

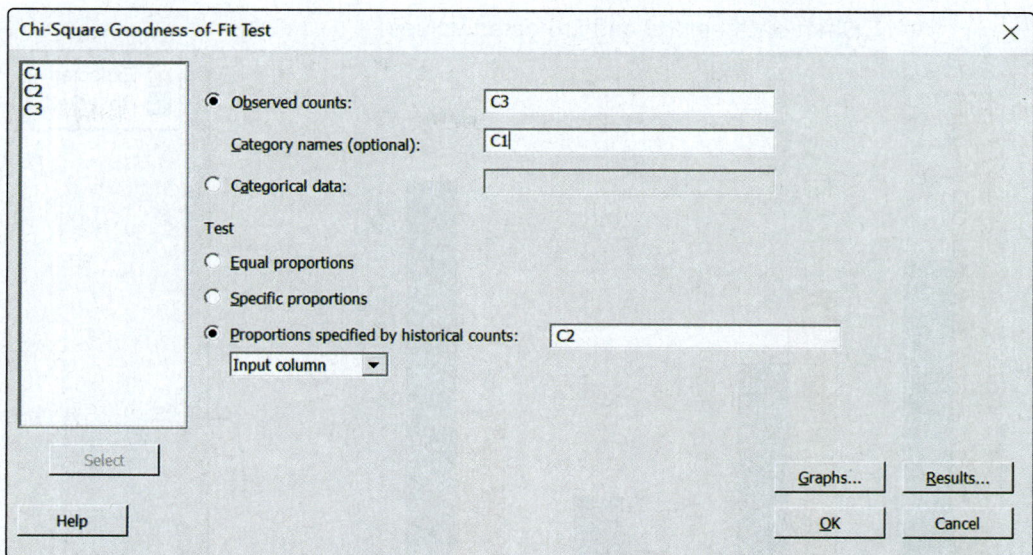

4. Press **OK**

5. Observe the output shown below. The calculated chi-square value is **177.430**.

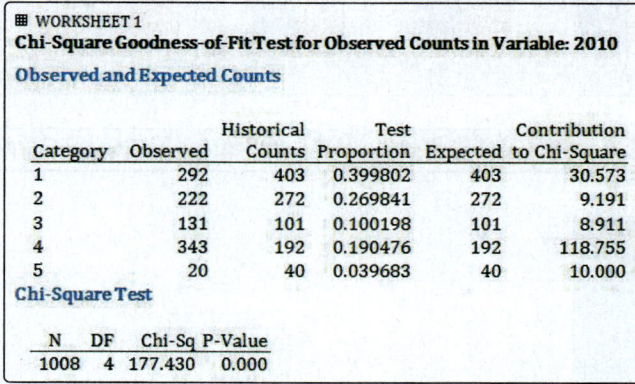

**WORKSHEET 1**
**Chi-Square Goodness-of-Fit Test for Observed Counts in Variable: 2010**
**Observed and Expected Counts**

| Category | Observed | Historical Counts | Test Proportion | Expected | Contribution to Chi-Square |
|---|---|---|---|---|---|
| 1 | 292 | 403 | 0.399802 | 403 | 30.573 |
| 2 | 222 | 272 | 0.269841 | 272 | 9.191 |
| 3 | 131 | 101 | 0.100198 | 101 | 8.911 |
| 4 | 343 | 192 | 0.190476 | 192 | 118.755 |
| 5 | 20 | 40 | 0.039683 | 40 | 10.000 |

**Chi-Square Test**

| N | DF | Chi-Sq | P-Value |
|---|---|---|---|
| 1008 | 4 | 177.430 | 0.000 |

6. Minitab gives two charts, the first one represents the observed and expected values for each category and the second chart shows the contribution to the chi-square value by category.

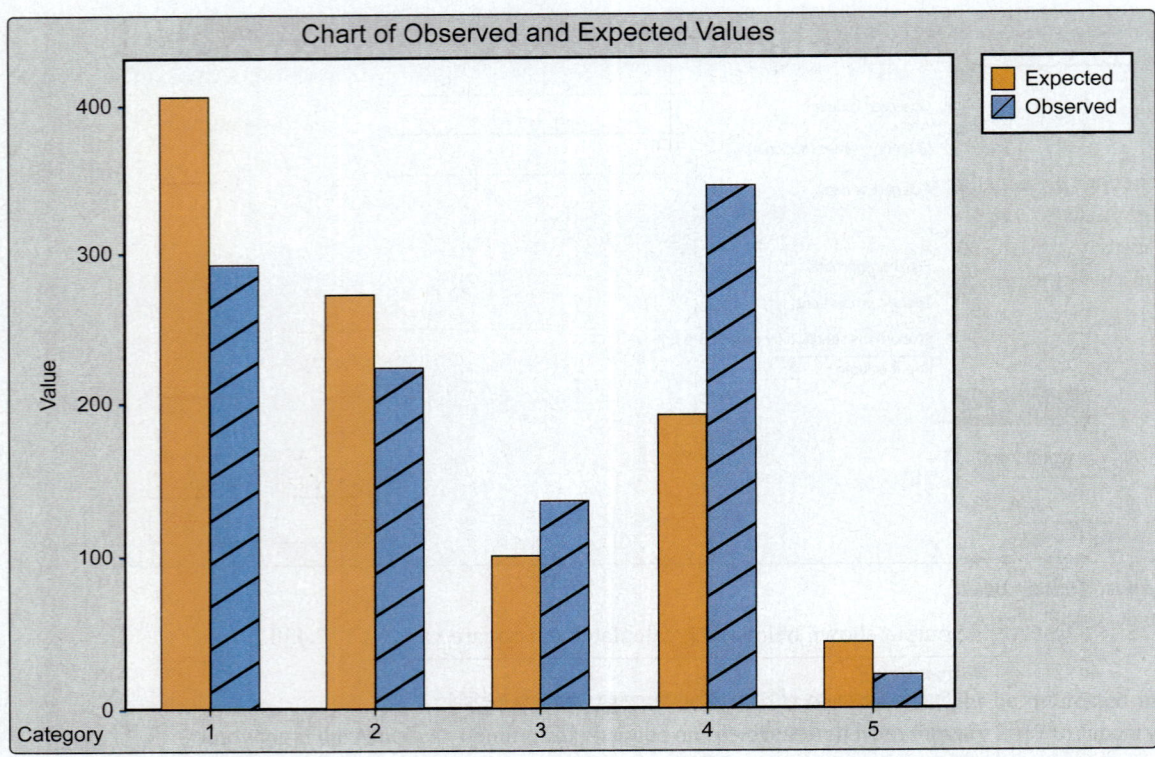

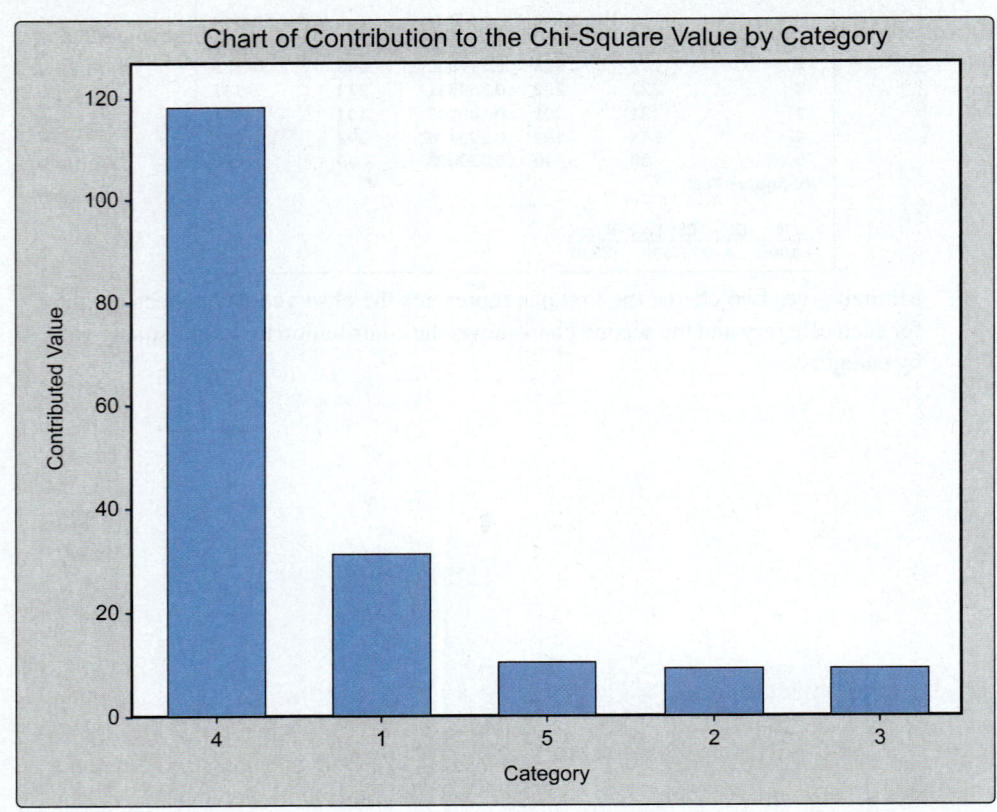

# R | Chapter 16 Review

## Key Terms and Ideas

- Chi-Square Test for Goodness of Fit
- Chi-Square Test for Association between Two Qualitative Variables
- Chi-Square Distribution
- Chi-Square Critical Values
- Multinomial Probability Distribution
- Multinomial Experiment
- Contingency Table
- Multiplication Rule for Independent Events

## Key Terms and Ideas

| | Section |
|---|---|

**Chi-Square Statistic**

$$\chi^2 = \frac{(n-1)s^2}{\sigma^2}$$

16.1

**Test Statistic for the Chi-Square Test for Goodness of Fit**

$$\chi^2 = \sum_{i=1}^{k} \frac{\left[n_i - E(n_i)\right]^2}{E(n_i)}$$

16.2

where $n_i$ is the actual number of observations for each category, and $E(n_i)$ is the expected number of observations for each category.

**Multiplication Rule for Independent Events**

$$P(A \cap B) = P(A)P(B)$$

16.3

**Test Statistic for the Chi-Square Test for Association between Two Qualitative Variables**

$$\chi^2 = \sum_{i=1}^{r} \sum_{j=1}^{c} \frac{\left[n_{ij} - E(n_{ij})\right]^2}{E(n_{ij})}$$

16.3

where $r$ = the number of rows, $c$ = the number of columns, and

$E(n_{ij}) = np_i p_j = \dfrac{(\text{row } i \text{ total})(\text{column } j \text{ total})}{n}$ where $n$ is the total number of

observations and $p_i$ and $p_j$ are estimates of the true population proportions.

## AE    Additional Exercises

1.  A sales manager for an insurance company believes that customers have the following preferences for life insurance products: 50% prefer Whole Life, 25% prefer Universal Life, and 25% prefer Life Annuities. A survey of 250 customers produced the following results.

| Insurance Preferences | |
| --- | --- |
| Product | Number |
| Whole Life | 60 |
| Universal Life | 100 |
| Life Annuities | 90 |

   a.  Is the sales manager's claim refuted by the data at $\alpha = 0.05$?

   b.  What assumptions were made in the test for part **a.**?

2.  Consider the following proportions of wives having affairs and husbands having affairs reported in *What the Odds Are* by Les Kranz.

| Affairs | | |
| --- | --- | --- |
| | Wives | Husbands |
| Have an Affair Before 2 Years of Marriage | 0.13 | 0.14 |
| Have an Affair Between 2 and 10 Years of Marriage | 0.20 | 0.25 |
| Have an Affair After 10 Years of Marriage | 0.20 | 0.33 |
| Have Not Had an Affair | 0.47 | 0.28 |

   a.  If the results were based on a survey of 200 wives and 200 husbands, do the data suggest that the proportion of wives in the various categories who have had affairs differs significantly from the proportion of husbands in the various categories who have had affairs at $\alpha = 0.10$?

   b.  What assumptions were made in the test for part **a.**?

3.  Do you think there are people somewhat like ourselves living on other planets in the universe? The responses to this question, which was asked in several different calendar years, were summarized in the *Gallup Poll Monthly*. For the year 1978, 51% answered "Yes," 33% "No," and the remainder had no opinion. Suppose that a sample of 100 people is chosen in 2011 in order to determine if opinions have changed concerning extraterrestrial life. Assume that the same question is asked and that 42 answer "Yes," 30 "No," and that the rest have no opinion. Assuming that the percentages given above accurately represent the attitudes of the people in 1978, can we conclude with $\alpha = 0.05$ that people's attitudes of toward extraterrestrial life have changed since 1978?

**4.** A traffic engineer feels that on a certain four-lane highway, the probability of being in the innermost lane is twice as great as any of the other lanes. Assume the other lanes have equal probabilities. A random sample of 200 motorists is chosen and the lanes in which they are traveling in are noted. The results (Lane 1 is the innermost lane) are given in the following table.

| Traffic Lanes | | | | |
|---|---|---|---|---|
| Lane | 1 | 2 | 3 | 4 |
| Frequency | 55 | 45 | 62 | 38 |

   **a.** Find the probabilities implied by the engineer's claim that a randomly chosen motorist will be in each of the four lanes.

   **b.** With $\alpha = 0.05$, can we refute the claim of the traffic engineer?

**5.** According to the Statistical Abstract of the United States, 22.6% of those 18 and over in the U.S. in 1992 were never married, 61.1% were married, 7.5% were widowed, and 8.8% were divorced. Suppose the following table summarizes the marital status of 90 randomly chosen adults in 2011.

| Marital Status | | | |
|---|---|---|---|
| | Never Married | Married | Widowed | Divorced |
| Frequency | 27 | 42 | 10 | 11 |

With $\alpha = 0.01$, can we conclude that in 2011 the U.S. is different than in 1992, with respect to marital status?

**6.** The National Restaurant Association is interested in determining if there is a relationship between the type of pizza pie Americans prefer and the region of the country in which they live. The association randomly selects 285 Americans and records the category of pizza pie which best describes their preference and the region of the country in which they live with the following results.

| Pizza Preference | | | | |
|---|---|---|---|---|
| | Region | | | |
| Type of Pizza Pie Preferred | North | South | East | West |
| Thin Crust | 40 | 30 | 35 | 45 |
| Thick Crust | 17 | 15 | 21 | 22 |
| Pan Pizza | 15 | 15 | 15 | 15 |

   **a.** Can the association conclude that the type of pizza pie Americans prefer and the region of the country in which they live are dependent at $\alpha = 0.10$?

   **b.** What assumptions were made in the test for part **a.**?

7. Do you think marriages between homosexuals should or should not be recognized by law as valid with the same rights as traditional marriages? The responses to this question and the region of the country where the respondent lived were summarized in the *Gallup Poll Monthly*. The following contingency table summarizes the 799 responses.

| Poll Results | | | |
|---|---|---|---|
| | | Response | |
| Region | Should | Should Not | No Opinion |
| East | 78 | 136 | 16 |
| Midwest | 63 | 183 | 16 |
| South | 74 | 23 | 16 |
| West | 50 | 138 | 6 |

With $\alpha = 0.01$, can we conclude that attitude about homosexual marriages is dependent on region?

8. Consider the following data regarding the percentage of mobile application usage and smartphone operating system. Suppose the data were based off of 100 responses.

| Mobile App Usage (%) | | | |
|---|---|---|---|
| Operating System | Multiple Times a Day | Once a Day | Several Times a Week |
| Apple iOS | 68 | 11 | 16 |
| Android OS | 60 | 12 | 21 |
| Palm OS | 48 | 14 | 25 |
| Blackberry OS | 45 | 13 | 29 |
| Microsoft Windows Mobile | 29 | 21 | 43 |
| Other | 47 | 13 | 21 |

a. Is there evidence that the frequency of mobile app usage is dependent on the smartphone operating system? Test at $\alpha = 0.05$.

b. What assumptions are made for the test in part **a.**? Are the assumptions reasonable for this problem? Why or why not?

# P Discovery Project

## Individual Stocks vs. Index-Matching Investments

In stock market investing, the traditional approach is to buy and sell stocks for individual companies, but this is sometimes risky as it is very difficult to predict when large changes to prices may occur. Large upswings or downswings in stock prices may happen for many reasons. Some positive examples leading to large stock price increases include the following.

- Retail sales on Amazon have frequently exceeded expectations.[1]

- Apple introduced its wildly popular iPhone, and continued to introduce new versions.[2]

- Netflix exploded in popularity; a growth of 7.41 million subscribers in the first quarter of 2018 (up from the expected 6.35 million) led to a stock increase of 60% in that quarter.[3]

Some negative events leading to plummeting stock prices include the following.

- Microsoft lost a federal antitrust lawsuit in 1999 and its stock price dropped 14% in a single day.[4]

- In 2015, the Environmental Protection Agency (EPA) discovered Volkswagen had intentionally programmed certain diesel engines to activate emission controls for certain nitrous oxides only during testing, which made It appear as though the vehicles released less than the legal limit of the polluting gases, but the vehicles actually released 40 times the legal limit! In the wake of the scandal, Volkswagen stock prices dropped from $162 to $105 in just three days.[5]

As we see, certain events may cause large fluctuations in stock prices, which may be good or bad for the investor, but in either case, it results in a volatile and sometimes risky investment.

Index-matching funds are portfolios (groups of multiple stocks) structured so that they match a market index, such as the Standard & Poor's 500 Index (S&P 500).[6] Gains or losses to the investment will be (proportionally) the same as that of the whole S&P 500. Such an index tends to be less volatile than investing in individual stocks, therefore index-matching funds are considered less risky investments. These funds are often used as parts of retirement funds or other long-term investments. But, how true is this assumption? Is an S&P 500 index-matching fund actually a safer investment? Our goal for this project is to test that assumption.

1.  Daily closing stock prices can be found on the website https://www.macrotrends.net/stocks/stock-screener.

 **Data**

For an example data set, please visit stat.hawkeslearning.com and navigate to **Discovering Business Statistics, Second Edition > Data Sets > Stock Comparison Data.**

The closing price data for three stocks (Amazon (AMZN), Starbucks (SBUX), and Coca-Cola (KO)), along with the S&P 500 index fund (SPY) from the beginning of 2000 to the end of 2017, can be found on our website.

2. We want to determine how the value of certain stocks and the S&P 500 compare over time, so notice that a new column called Price Change was created measuring the daily change in stock price for each stock, i.e.,

Day 2 Price – Day 1 Price

Day 3 Price – Day 2 Price

Day 4 Price – Day 3 Price

etc.

It does not make sense to simply compare stock prices of the different stocks because the prices are on very different scales. For example, if a $100 stock drops by $1, it is less consequential than if a $3 stock drops by $1. A reasonable adjustment is to consider a price change as a percentage of the previous day's price, so the example above would be a 1% drop compared to a 33.3% drop, which allows us to do a more reasonable comparison. A new column for each stock and the S&P 500 index has been created with these percentages and is labelled as *Return*.

3. Can you think of any descriptive statistics that would compare the volatility of the returns?

4. Create box plots of the returns for each stock and the S&P 500 index.

5. Do you notice any patterns in the box plots that suggest a difference in the stocks and the S&P 500 index? Explain their practical significance, if any.

6. Delete the outliers in the data for each stock and the S&P 500 index and create histograms for each on the same scale. [*Hint*. Excel's Data ToolPak add-in allows for easy histograms where we can specify the bins.]

7. Are any differences between the stocks and the S&P 500 index apparent from the histograms? Explain their practical significance, if any.

8. Use a test for goodness of fit to test whether the distributions of the non-outlying stock returns and index returns are different (at least over the range of the S&P 500 returns) at a significance level of $\alpha = 0.2$.

**References**

1 https://www.cnbc.com/2017/10/26/amazon-earnings-q3-2017.html

2 http://fortune.com/2016/09/09/apple-stock-iphone-launches/

3 https://www.thestreet.com/investing/stocks/netflix-shares-rise-after-beating-estimates-14557098

4 https://www.nytimes.com/2000/04/04/business/us-vs-microsoft-overview-us-judge-says-microsoft-violated-antitrust-laws-with.html

5 http://fortune.com/2015/09/23/volkswagen-stock-drop/

6 https://www.investopedia.com/terms/i/indexfund.asp

17

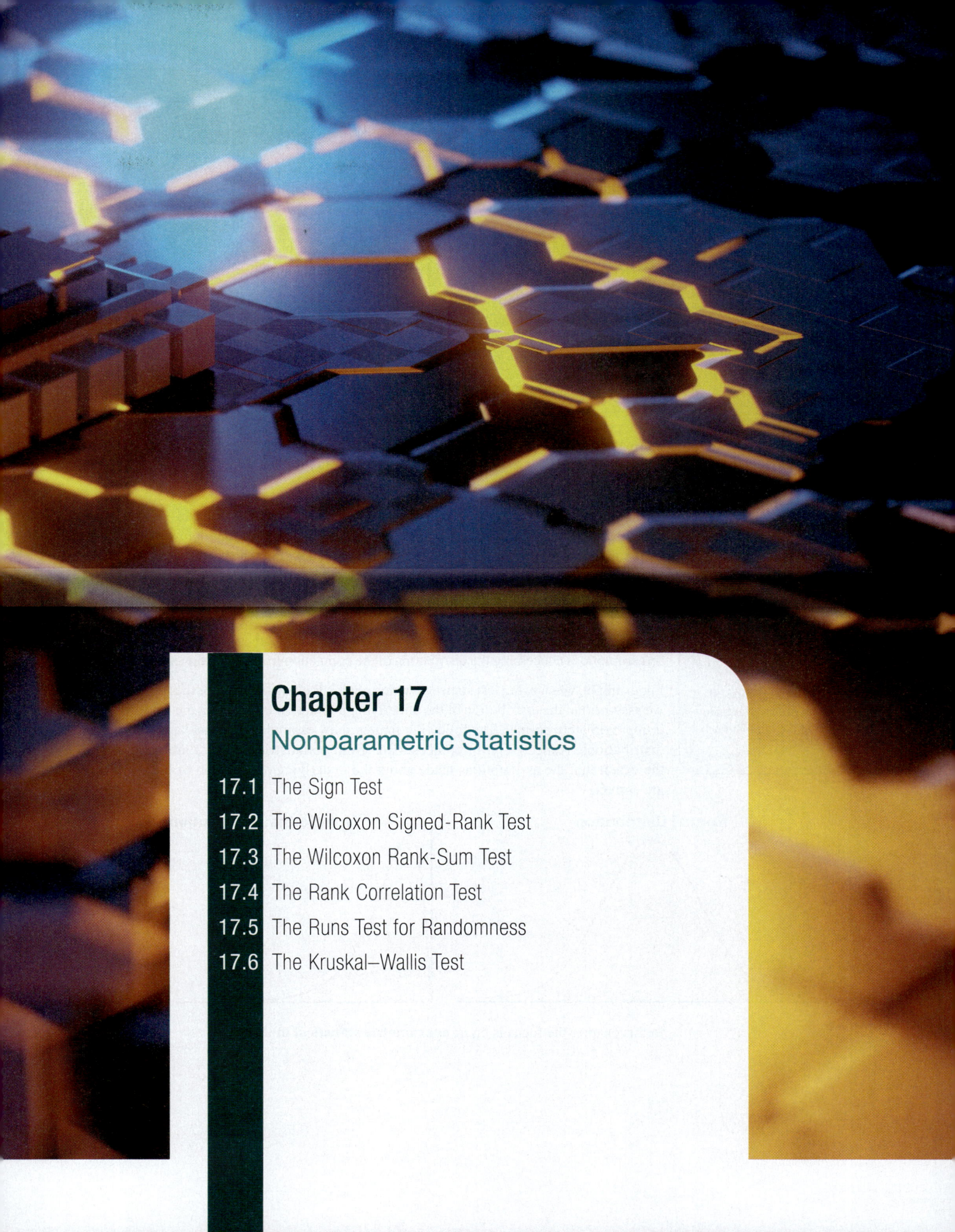

# Chapter 17

## Nonparametric Statistics

# Discovering the Real World

Throughout this textbook, almost all topics that were discussed have been based on knowing, or assuming, that the sample was collected from a known distribution, specifically, the normal distribution. Many classical statistical procedures are derived on the assumption that the sample follows a normal distribution. In these situations, the procedures used in the analysis are called **parametric procedures**. However, if the distribution of the sample is unknown, we need to use **nonparametric procedures** to perform the analysis.

> **Definition**
>
> ## Nonparametric Statistics
>
> **Nonparametric statistics** are methods that do not rely on an underlying distribution or assumptions about the distribution of the population from which a sample is taken.

In general, **nonparametric statistics** refers to a method that makes statistical inferences without regard to any underlying distribution or underlying assumptions. Often, nonparametric statistical methods use data that are ordinal because they depend on rankings instead of the raw, observed observations.

Suppose that a marketing research firm is interested in knowing whether an email marketing campaign or a brand awareness campaign is associated with how fast the company gains brand positioning. Randomly selecting the sample size, an experiment that measures the company's strategic goals to address market dynamics (which also determines brand positioning) cannot be assumed to follow a normal distribution. Thus, the analysis of brand positioning will be done using nonparametric procedures.

In this chapter we will discuss several nonparametric statistical procedures such as the sign test, the Wilcoxon signed-rank test, Wilcoxon rank-sum test, and a few others. These nonparametric statistical procedures are based on the rankings of the observations instead of making any underlying assumptions about the data.

## Introduction

> **Definition**
>
> ## Parametric Statistics
>
> **Parametric statistics** are methods that rely on assumptions concerning the distribution of the population from which a sample is taken.

One of the characteristics of **parametric statistics** is that hypothesis testing procedures rely on assumptions concerning the distribution of the population from which the sample is drawn.

For example, when using test statistics which rely on the $t$-distribution or the $F$-distribution, we assume that the distribution of the underlying population from which the data are drawn is approximately normal. The following figures show examples of normal and non-normal distributions, respectively. Inferences made from parametric statistics are valid only to the extent that the assumptions made about the underlying distribution of the population are correct.

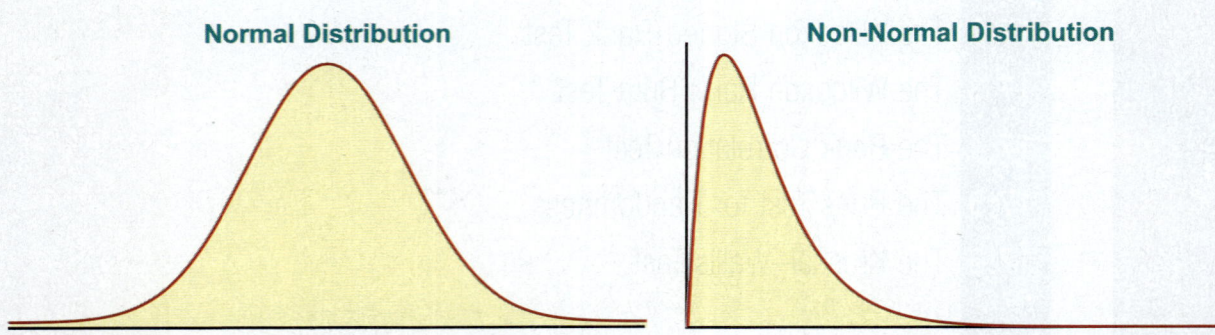

**Normal Distribution**                    **Non-Normal Distribution**

In this chapter the focus is on **nonparametric statistical methods**.

> ### Properties
>
> #### Nonparametric Statistical Methods
>
> There are three characteristics of nonparametric statistical methods.
>
> 1. They do not involve the estimation of specific population parameters.
> 2. These methods are tests of hypotheses using data measured on an ordinal scale (in most instances).
> 3. Assumptions regarding the underlying distribution of the data are not required.

The fact that a distributional assumption about the data is not necessary to make valid inferences is one of the significant advantages of nonparametric techniques.

However, there are some disadvantages of nonparametric statistics. Because nonparametric tests tend to waste information, they are not as powerful as their parametric counterparts. Some nonparametric tests reduce exact ratio level measurements to ordinal level measurements. For example, taking an income measurement of $48,200 (a ratio measurement) and converting it into an ordinal variable consisting of three categories (low, medium, high) results in a significant loss of information.

Several nonparametric techniques are discussed in this chapter. The first two, the **sign test** and the **Wilcoxon signed-rank test**, are used to conduct hypothesis tests involving paired data experiments. Paired data experiments are designed to detect the effect of a treatment on some population measurement. In a paired data experiment, a treatment is applied to a population or sample. There is a comparison of some population characteristic of interest before and after the treatment is applied. Suppose, for example, database entry personnel are given a course in speed-typing. If the course is successful, there should be a shift in the typing speed of the database personnel. A *before* and *after* measurement of typing speed would be desirable to measure the effect of the typing training. Since there are different typing speeds in the population of typists, how can we compare the average typing speed before and after the training? One way is to compare the population medians. If the distribution of typing speeds has shifted because of the treatment (training), then the median will also have shifted. The sign test is used for testing a hypothesis concerning two population medians or testing the hypothesis that a population median is equal to some predetermined value.

Other nonparametric tests introduced in this chapter include:

- The Wilcoxon rank-sum test, used for comparing two independent samples
- The Kruskal-Wallis test, a test similar to the ANOVA $F$-test
- The rank correlation test, a test similar to the parametric correlation test
- The runs test for randomness, a test to determine if a sequence exhibits randomness

## 17.1  The Sign Test

Sign tests are used in the comparison of paired data when it is not possible to assume that the differences have an approximately normal distribution. Suppose a company has two developmental gasoline products, say, Gasoline A and Gasoline B. The company wants to know if there is a significant difference in miles per gallon (mpg) between the two types of gasoline. Ten vehicles are randomly selected and driven using each of the gasolines in order to track the mpg. The data are as follows.

| Table 17.1.1 – Miles per Gallon | | |
|---|---|---|
| Car | MPG for Gasoline A | MPG for Gasoline B |
| 1 | 35 | 40 |
| 2 | 38 | 35 |
| ... | ... | ... |
| 10 | 41 | 38 |

Traditionally, to determine if there was a difference between the mpg between the two gasolines, the company would perform a paired *t*-test. However, since the company is not comfortable making assumptions about the distribution of the given data (i.e., assuming that the differences follow a normal distribution), the sign test is more appropriate to compare the differences in mpg between the two gasolines.

The methodology for the sign test is based on the assumption that the two populations have the same median. Suppose we were to subtract the paired sample values. If the population medians are identical, the resulting differences would be centered around zero. Further, we would expect the number of positive difference to be roughly equal to the number of negative differences. If the number of positive differences significantly outnumbers the number of negative differences or vice-versa, this will cause us to doubt that the two populations have the same median.

## Using the Sign Test on Paired Data

We will present the test procedure for the sign test by example. For the paired difference experimental design discussed in Section 11.3, there is an assumption that the differences have an approximately normal distribution. The sign test makes no such assumption.

### Example 17.1.1

**Performing the Sign Test on Paired Data**

Are the demands of college significantly less than the demands of high school? Students from Tech University were surveyed asking them the amount of time they spent watching TV per day in high school and the amount of time they spent watching TV in college. The data are listed in the following table. Can the researcher conclude at the 1% level of significance that the median TV-watching time (in minutes) is greater in college than in high school?

| Table 17.1.2 – TV-Watching Times (Minutes) | | | | |
|---|---|---|---|---|
| Student | High School | College | Difference | Sign |
| 1 | 30 | 120 | −90 | − |
| 2 | 60 | 60 | 0 | Ignore |
| 3 | 45 | 180 | −135 | − |
| 4 | 120 | 240 | −120 | − |
| 5 | 60 | 75 | −15 | − |
| 6 | 180 | 240 | −60 | − |
| 7 | 45 | 60 | −15 | − |
| 8 | 90 | 90 | 0 | Ignore |
| 9 | 60 | 90 | −30 | − |
| 10 | 90 | 120 | −30 | − |

#### SOLUTION

Normally, we would analyze these data with a paired difference test (see Section 11.3). However, if we doubt that the differences follow a normal distribution, we will use the sign test to determine if the populations are different.

The sign test is based on the idea that if two data sets have the same median, the number of differences with positive signs should approximately equal the number of differences with negative signs. We will conclude that the median number of minutes of TV-watching time increases in college if the number of negative signs is larger than the number of positive signs. How large is large? We will answer this question in **Step 4**.

**Step 1:** Determine the null hypothesis.

The null hypothesis is that the median TV-watching time in college is equal to the median TV-watching time in high school.

For the sign test, we would write the null hypothesis as

$H_0$: The number of negative signs is equal to the number of positive signs.

**Step 2:** Determine the alternative hypothesis.

The alternative hypothesis is that the median TV-watching time is greater in college than in high school.

For the sign test, we would write the alternative hypothesis as

$H_a$: The number of negative signs is greater than the number of positive signs.

Since the researcher is interested in whether the TV-watching time is longer in college than in high school, this is a one-tailed test. Thus, the alternative hypothesis will be one-sided.

**Step 3:** Select the appropriate test statistic based on the information at hand and the assumptions you are willing to take.

The procedure we are using is called the sign test because it relies on counting the number of positive differences and the number of negative differences. Only the sign of the difference matters, the magnitude is ignored. Differences of zero are also ignored.

Our discussion has focused on the signs of the differences. Suppose we let

$X$ = the number of times the less frequent sign occurs.

Assuming that the null hypothesis is true, the number of times the less frequent sign is observed in the paired difference experiment has a binomial distribution with

$$p = 0.5 \text{ and}$$

$$n = \text{the number of non-zero differences.}$$

If $X$ is too small, then either $H_0$ is true and a rare phenomenon has been observed, or $H_0$ is not true. To make the decision whether $X$ is too small, a decision rule must be developed.

If $n \leq 25$, where $n$ = the number of non-zero differences, the test statistic of the sign test is the variable $X$ described previously.

If $n > 25$, $X$ can be approximated by a normal distribution and the test statistic is given by

$$z = \frac{X + 0.5 - \left(\dfrac{n}{2}\right)}{\dfrac{\sqrt{n}}{2}}.$$

Assuming the null hypothesis is true, $z$ has an approximately standard normal distribution.

**Step 4:** Determine the critical value of the test statistic.

The researcher wants to determine if the median TV-watching time for college students is longer than that of high school students. The statistical measures being used are counts of the number of positive signs and negative signs. If the amount of TV-watching time is greater in college than in high school, then we would expect the negative differences to outnumber the positive differences. If the amounts of TV-watching time in high school and college are identical, then

**⟲ Technology**

For instructions on performing the sign test using technology, please visit stat.hawkeslearning.com and navigate to **Discovering Business Statistics, Second Edition > Technology Instructions > Nonparametrics > Sign Test.**

we would expect that any difference would be due to random variation, in which case we would have about the same number of positive and negative differences.

The level of the test is specified in the problem to be $\alpha = 0.01$.

The role of the critical value in this test is exactly the same as for all of the hypothesis tests discussed earlier. It defines a range of values for the test statistic, the rejection region, such that if the test statistic falls in this region it is unlikely to be due to ordinary sampling variation.

The level of the test defines the rareness criterion that will be used and implicitly determines the size of the rejection region. Should the computed value of the test statistic fall in the rejection region, its value will be presumed to be too rare to have occurred because of ordinary sampling variation, and the null hypothesis will be rejected.

If $n \leq 25$, the critical values are given in Appendix A, Table I for the sign test. Because the test statistic is defined to be the number of times the less frequent sign occurs, we will always reject the null hypothesis for small values of the test statistic. If the alternative hypothesis is *greater than* or *less than*, use the column in Table I labeled $\alpha$ *for a one-tailed test*. If the alternative is *not equal to*, use the column in Table I labeled $\alpha$ *for a two-tailed test*.

There are $n = 8$ nonzero differences in the TV-watching example, $\alpha = 0.01$, and the alternative hypothesis is *greater than*. From Table I in the Appendix, we find that the critical value is 0. The rejection region is drawn in Figure 17.1.1.

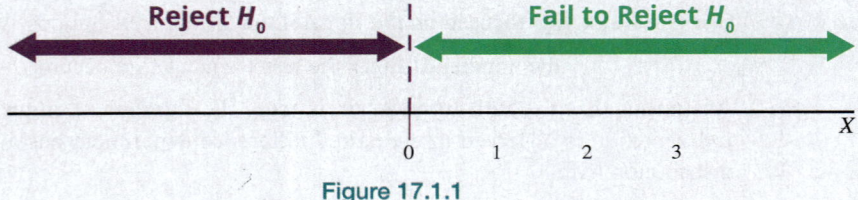

**Figure 17.1.1**

**NOTE**

Because of the manner in which the test statistic is defined, fail to reject the null hypothesis and do not proceed with the test procedure if the computed value of the test statistic does not tend to support the alternative hypothesis.

If $n > 25$, the critical values are determined in a similar manner to other hypothesis tests whose test statistics have standard normal distributions. However, because the test statistic is defined to be the number of times the less frequent sign occurs, we will always reject the null hypothesis for small values of the test statistic. Thus, the rejection region will always be established in the left tail, with the critical value defined by $\alpha$ for one-tailed tests, and $\dfrac{\alpha}{2}$ for two-tailed tests.

**Step 5:** Collect the sample data and compute the value of the test statistic.

Since $n \leq 25$, the value of the test statistic is given by

$X =$ the number of times the less frequent sign occurs.

The less frequent sign in this case is the positive sign. In fact, the positive sign does not occur at all in these data, meaning that the TV-watching time in college was always greater than in high school. Since there are eight negative signs (−) and zero positive signs (+), the value of $X$ is

$$X = 0.$$

Note that differences of zero are ignored.

Since the number of positive signs is zero, this tends to support the alternative hypothesis that the number of minutes watching TV in college is greater than in high school. It is safe to proceed with the test. Suppose there were not any negative signs, meaning that all of the TV-watching times in college were less than or equal to the times in high school. This evidence would not support the alternative hypothesis, and the test must be halted at this point with a conclusion that the null hypothesis cannot be rejected. This stage of a one-sided sign test will always require thinking to be sure that the sign with the minimum value is the sign that supports the alternative hypothesis.

**Step 6:** Make the decision and state the conclusion in terms of the original question.

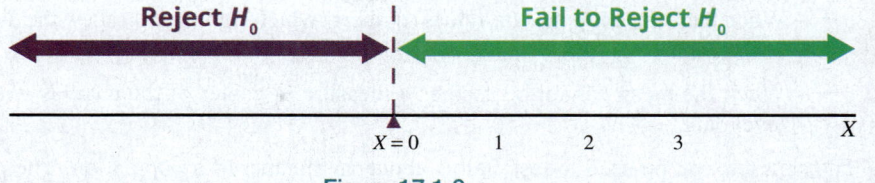

**Figure 17.1.2**

As shown in Figure 17.1.2, the value of the test statistic ($X = 0$) falls in the rejection region. Thus, we reject the null hypothesis at $\alpha = 0.01$. It is unlikely that the small number of positive differences could be attributed to ordinary sampling variation.

*Conclusion and Interpretation*: There is sufficient evidence for the researcher to conclude at $\alpha = 0.01$ that the median TV-watching time for college students is longer than the median TV-watching time when those students were in high school.

### Nonparametric Rejection Regions

Most of the rejection regions we will construct for nonparametric tests will look different from those we have previously constructed. Because the test statistics are counts or sums of rank data, the test statistic will usually be a discrete value, so the rejection "region" will not be an interval. Most of the test statistics in this chapter will have rejection regions with only a few points in them, sometimes only one point.

The test procedure for the sign test is outlined in the following box.

## Procedure

### Sign Test

**Hypotheses:**

$H_0$: The number of positive [negative] signs is equal to the number of negative [positive] signs.

$H_a$: The number of negative [positive] signs is greater than, less than, or not equal to the number of positive [negative] signs (depending on the claim which is being tested).

**Note:** No sign is given if the difference is zero and the difference is ignored.

**Test Statistic:**

If $n \le 25$, then

$X$ = the number of times the less frequent sign occurs.

If $n > 25$, then

$$z = \frac{X + 0.5 - \left(\dfrac{n}{2}\right)}{\dfrac{\sqrt{n}}{2}},$$

where $X$ is defined just as when $n \le 25$.

**Critical Value(s):**

If the data tend to support the alternative hypothesis and $n \le 25$:

Reject the null hypothesis if the test statistic is less than or equal to the critical value in Appendix A, Table I.

$n > 25$:

The critical values are based on the standard normal distribution with the critical values defined such that we reject the null hypothesis for small values of the test statistic.

**Assumptions:**

The pairs of data are randomly selected.

### Making the Decision in the Sign Test

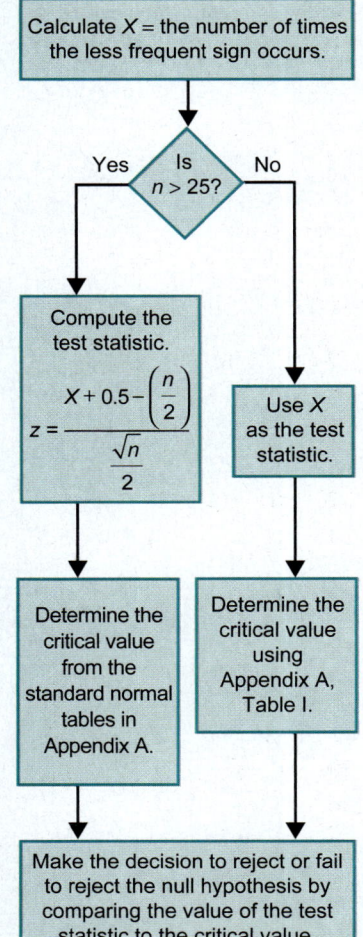

## Testing a Population Median with the Sign Test

The median is a more useful measure of the center for a population than the mean in certain situations such as the following.

- When there are extreme data values (outliers) which significantly skew the distribution of the data.

- When the mean cannot be used as a measure of center of the data, as with ordinal level data.

The sign test can be used to test claims about the median of a population. The procedure for using the sign test to evaluate claims about a single population median is presented in the following example.

---

**Example 17.1.2**

**Testing a Population Median with the Sign Test**

The CEO of Montgomery County Hospital (MCH) is interested in the length of patients' stays at her hospital. Most people tend to stay only a few days after being admitted to the hospital, but there are a few people who have hospital stays that are longer. She believes that the median length of stay at her hospital is less than the national median. In a recent National Health Statistics report, it was stated that the median hospital stay for hospitals in the United States is 4.7 days. Upon discharge, the lengths of stay for 50 randomly selected patients of MCH are noted. Of the randomly selected patients, the length of stay for 35 patients was less than 4.7 days. Can the CEO of Montgomery County Hospital conclude at the 10% significance level that the median length of stay at her hospital is shorter than the national median?

### SOLUTION

**Step 1:** Determine the null hypothesis.

The null hypothesis is that the median hospital stay (in days) at Montgomery County Hospital is equal to the median hospital stay for the United States. For the sign test, we would write the null hypothesis as

$H_0$: The number of positive signs is equal to the number of negative signs.

**Step 2** Determine the alternative hypothesis.

The alternative hypothesis is that the median hospital stay (in days) at Montgomery County Hospital is shorter than the median hospital stay for the United States. For the sign test, we would write the alternative hypothesis as

$H_a$: The number of negative signs outnumbers the number of positive signs.

Since the CEO's claim is that the median hospital stay is *shorter* than the median hospital stay for the United States, this test is a one-tailed test. Thus, the alternative hypothesis will be one-sided.

**Step 3:** Select the appropriate test statistic based on the information at hand and the assumptions you are willing to make.

Once again, we will look at differences, but in this case the differences will be created by subtracting the median from each data value. The sign test will rely on counting the number of positive signs (meaning that the length of hospital stay at MCH is longer than the median length of stay for the United States) and the number of negative signs (meaning that the length of stay at MCH is shorter than the median length of stay for the United States).

Since $n > 25$, the test statistic is given by

$$z = \frac{X + 0.5 - \left(\dfrac{n}{2}\right)}{\dfrac{\sqrt{n}}{2}}.$$

Assuming the null hypothesis is true, $z$ has an approximately standard normal distribution.

**Step 4:** Determine the critical value of the test statistic.

The level of the test is specified in the problem to be $\alpha = 0.10$.

More than half of the hospital stays were shorter than the United States median length of stay, which supports the alternative hypothesis. Thus, it is appropriate to proceed with the test procedure.

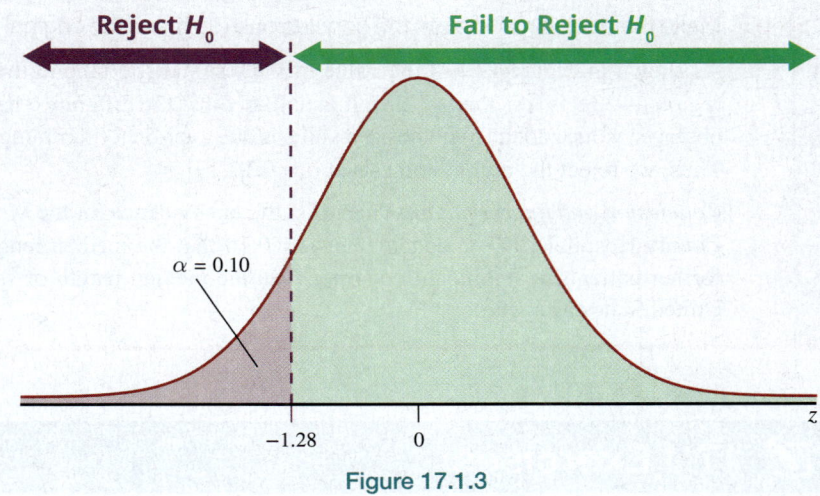

**Figure 17.1.3**

☍ **Technology**

For instructions on performing the sign test using technology, please visit stat.hawkeslearning.com and navigate to **Discovering Business Statistics, Second Edition > Technology Instructions > Nonparametrics > Sign Test.**

Since $n > 25$, the critical value is determined from the standard normal distribution. Because of the way the test statistic is defined, we will always reject the null hypothesis for small values of the test statistic. If the alternative hypothesis is *greater than* or *less than*, use an area of $\alpha$ in the left tail to determine the critical value. If the alternative is *not equal to*, use an area of $\dfrac{\alpha}{2}$ in the left tail to determine the critical value. In this example, the alternative is *less than* and $\alpha = 0.10$. The rejection region is shown in Figure 17.1.3. We will reject the null hypothesis if the calculated value of the test statistic is less than or equal to $-1.28$.

**Step 5:** Collect the sample data and compute the value of the test statistic.

Negative signs will occur if the length of stay for patients in MCH is shorter than the median length of stay for the United States. Thirty-five of the sampled patients had an actual length of stay shorter than the median length of stay for the United States. Thus, there will be 35 negative signs.

Positive signs will occur if the length of stay for patients in MCH is longer than the median length of stay for the United States. 15 of the sampled patients had an actual length of stay longer than the median length of stay for the United States. Thus, there will be 15 positive signs.

Since $X =$ the number of times the less frequent sign occurs, $X = 15$, and the calculated value of the test statistic is

$$z = \frac{15 + 0.5 - \left(\dfrac{50}{2}\right)}{\dfrac{\sqrt{50}}{2}} \approx -2.69 \, .$$

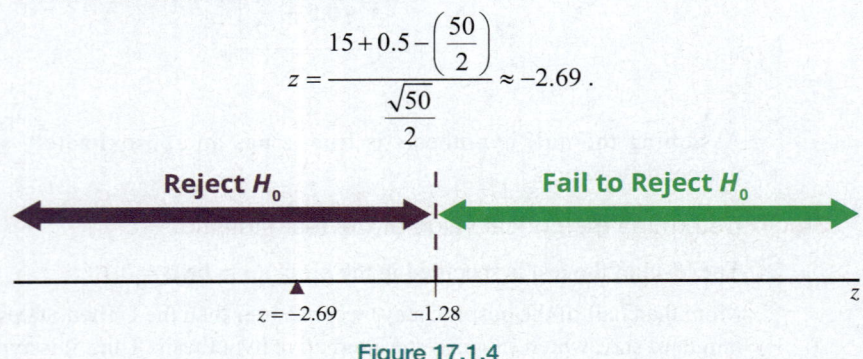

**Figure 17.1.4**

**Step 6:** Make the decision and state the conclusion in terms of the original question.

As shown in Figure 17.1.4, the value of the test statistic falls in the rejection region ( $-2.69$ is less than $-1.28$). It is unlikely that the difference between the observed value and the hypothesized value is due to ordinary sampling variation. Thus, we reject the null hypothesis at $\alpha = 0.10$.

*Conclusion and Interpretation*: There is sufficient evidence for the Montgomery County Hospital CEO to conclude at $\alpha = 0.10$ that the median length of stay for her patients is significantly shorter than the median length of stay for the United States as a whole.

## 📝 17.1 Exercises

### Basic Concepts

1.  What are parametric statistics?

2.  Identify and explain the main disadvantage of the sign test.

3.  Under what conditions are parametric statistical methods not appropriate for data analysis?

4.  Identify the three characteristics of nonparametric statistical methods.

5.  What are the disadvantages of nonparametric statistics?

6.  The sign test and the Wilcoxon signed-rank test are designed to conduct hypothesis tests involving which kind(s) of experiments? What is the corresponding parametric statistical technique used to analyze these types of experiments?

7.  What assumptions are made when conducting the sign test?

8.  Name the two ways that the sign test can be used to perform hypothesis tests.

9.  How do the rejection regions for nonparametric tests differ from those for parametric tests? Explain.

10. What is done with measurements that have a difference of zero in a paired difference experiment? Why is this the case?

11. What are the null and alternative hypotheses associated with the sign test?

12. What is the test statistic for the sign test for small samples? How small is a *small* sample?

13. What is the test statistic for the sign test for large samples? How large is a *large* sample?

14. Identify the critical values and rejection rules for both small and large samples with regard to the sign test.

## Exercises

15. Hurricane Hugo swept through the Lowcountry in South Carolina causing billions of dollars of damage. In the past, the median claim for homes damaged by hurricanes for an insurance company in the Lowcountry had been $25,000. The insurance company believes that the median claim will be significantly larger for homes damaged by Hugo than past hurricanes. In order to investigate this theory, the insurance company randomly selects 55 homes and sends adjusters to settle the claims. In the sample of 55 homes, 40 of the homes had a claim in excess of the historical median. Is there overwhelming evidence at $\alpha = 0.10$ that the median claim for home damage from Hurricane Hugo was greater than the historical median?

16. The manufacturer of Brand X floor polish is developing a new polish that they hope will dry faster than the competition's polish. The competition's polish is advertised to have an average (median) drying time of 10 minutes. In a random sample of 1000 polishes with the new polish, 700 of the polishes dried in less than 10 minutes. Based on the data, can the manufacturer conclude that the median drying time for Brand X is faster than the competition's brand at a 0.05 level of significance?

17. NarStor, a computer disk drive manufacturer, claims that the median time until failure for their hard drives is 14,400 hours. You work for a consumer group that has decided to examine this claim. Technicians ran 16 NarStor hard drives continuously for almost three years. Recently the last drive failed. The times to failure (in hours) are given in the following table.

| Time Until Hard Drive Failure (Hours) | | | | | | | |
|---|---|---|---|---|---|---|---|
| 330 | 620 | 1870 | 2410 | 4620 | 6396 | 7822 | 8102 |
| 8309 | 12,882 | 14,419 | 16,092 | 18,384 | 20,916 | 23,812 | 25,814 |

   a. Is there overwhelming evidence that the median time until failure is less than the manufacturer claims? Use $\alpha = 0.10$.

   b. What assumption did you make in performing the test in part **a.**?

18. A.C. Bone has developed a duck hunting boot which it claims can remain immersed for more than 12 hours without leaking. 15 of the boots are tested and the time until first leakage is measured. Nine of the boots last more than 12 hours without leaking.

   a. Do the data substantiate A.C. Bone's claim at $\alpha = 0.05$?

   b. What assumption did you make in performing the test in part **a.**?

19. Given that most textbooks can now be purchased online, one wonders if students can save money by comparison shopping for textbooks at online retailers and at their local bookstores. To investigate, students at Tech University randomly sampled 25 textbooks on the shelves of their local bookstores. The students then found the "best" available price for the same textbooks via online retailers. The prices for the textbooks are listed in the following table.

| | Textbook Prices | | | | | | | |
| --- | --- | --- | --- | --- | --- | --- | --- | --- |
| | Price ($) | | | Price ($) | | | Price ($) | |
| Textbook | Bookstore | Online Retailer | Textbook | Bookstore | Online Retailer | Textbook | Bookstore | Online Retailer |
| 1 | 70 | 60 | 10 | 97 | 86 | 19 | 49 | 40 |
| 2 | 38 | 36 | 11 | 140 | 130 | 20 | 149 | 127 |
| 3 | 88 | 89 | 12 | 40 | 30 | 21 | 126 | 130 |
| 4 | 165 | 149 | 13 | 175 | 150 | 22 | 92 | 93 |
| 5 | 80 | 136 | 14 | 85 | 75 | 23 | 144 | 129 |
| 6 | 103 | 95 | 15 | 100 | 85 | 24 | 98 | 84 |
| 7 | 42 | 50 | 16 | 68 | 62 | 25 | 40 | 52 |
| 8 | 98 | 111 | 17 | 67 | 69 | | | |
| 9 | 89 | 65 | 18 | 140 | 142 | | | |

Using the data in the table, and without making any distributional assumptions, is it less expensive for the students to purchase textbooks from the online retailers than the local bookstores? Use $\alpha = 0.01$.

20. The management for a large grocery store chain would like to determine if a new cash register will enable cashiers to process a larger number of items on average than the cash register which they are currently using. Seven cashiers are randomly selected, and the number of grocery items which they can process in three minutes is measured for both the old cash register and the new cash register. The results of the test are as follows.

| Number of Grocery Items Processed in Three Minutes | | | | | | | |
| --- | --- | --- | --- | --- | --- | --- | --- |
| Cashier | 1 | 2 | 3 | 4 | 5 | 6 | 7 |
| Old Register | 60 | 70 | 55 | 75 | 62 | 52 | 58 |
| New Register | 65 | 71 | 55 | 75 | 65 | 57 | 57 |

Without making any assumptions about the distribution, can management conclude that the new cash register will allow cashiers to process a significantly larger number of items on average than the old cash register at $\alpha = 0.05$.

21. An auto dealer is marketing two different models of a high-end sedan. Since customers are particularly interested in the safety features of the sedans, the dealer would like to determine if there is a difference in the braking distance (the number of feet required to go from 60 mph to 0 mph) of the two sedans. Six drivers are randomly selected and asked to participate in a test to measure the braking distance for both models. Each driver is asked to drive both models and brake once they have reached exactly 60 mph. The distance required to come to a complete halt is then measured in feet. The results of the test are as follows.

| Braking Distance of High-End Sedans (in Feet) | | | | | | |
| --- | --- | --- | --- | --- | --- | --- |
| Driver | 1 | 2 | 3 | 4 | 5 | 6 |
| Model A | 150 | 145 | 160 | 155 | 152 | 153 |
| Model B | 152 | 146 | 160 | 157 | 154 | 155 |

Without making assumptions about the distribution of the data, can the auto dealer conclude that there is a significant difference in the braking distance of the two models of high end sedans? Use $\alpha = 0.10$.

**22.** A nutritionist is interested in determining the decrease in cholesterol level which a person can achieve by following a particular diet which is low in fat and high in fiber. Seven subjects are randomly selected to try the diet for six months, and their cholesterol levels are measured both before and after the diet. The results of the study are as follows.

| Cholesterol Levels | | | | | | | |
|---|---|---|---|---|---|---|---|
| Subject | 1 | 2 | 3 | 4 | 5 | 6 | 7 |
| Before Diet | 155 | 170 | 145 | 200 | 162 | 180 | 160 |
| After Diet | 152 | 168 | 148 | 195 | 162 | 178 | 157 |

Can the nutritionist conclude that there is a significant decrease in average cholesterol level when the diet is used? We don't have any knowledge about the distribution of the data. Use $\alpha = 0.01$.

# 17.2  The Wilcoxon Signed-Rank Test

A disadvantage of the sign test is that it wastes information. The sign test merely counts the number of positive or negative signs in a paired difference experiment and ignores the magnitude of the differences. The **Wilcoxon signed-rank test** is a nonparametric technique which can also be used to evaluate a paired difference experiment. This test is designed to detect populations whose centers are shifted to the right or the left of each other. As with the sign test, no distributional assumption is required. However, the pairs of data must have been randomly selected, and it must be possible to rank the differences.

An advantage of the Wilcoxon signed-rank test is that it does not ignore the magnitudes of the differences. However, it does not take the magnitude directly into account. Instead, the ranks of the data are analyzed.

Ranking is nothing new. It simply requires putting the data in order from smallest to largest and attaching a rank to each data item. In general, the lowest value is assigned a rank of one and the highest value is assigned a rank of $n$, where $n$ is the number of nonzero differences. How do we handle ties? If there are two or more values with the same magnitude, these values will each be assigned the same rank, which is equal to the average of the ranks which would have been assigned to these values if they had slightly different consecutive values. The ranking procedure is explained more fully in the following example.

Rank the following stocks, traded on the New York Stock Exchange, from smallest price to largest price.

**Example 17.2.1**

**Ranking Quantitative Data**

| Table 17.2.1 – Stock Prices | |
|---|---|
| Stock | Price per Share ($) |
| Merck & Co., Inc. | 78.25 |
| AT&T, Inc. | 27.15 |
| SecureWorks Corp. | 28.04 |
| Micron Technology, Inc. | 73.21 |
| HP Inc. | 25.16 |
| AMC Entertainment Holdings, Inc. | 30.50 |
| CitiGroup, Inc. | 60.12 |
| Exxon Mobil Co. | 55.30 |
| AutoCanada Inc. | 30.50 |

## SOLUTION

The lowest priced stock is HP Inc. at $25.16. It is assigned a rank of one. Rankings of two and three go to AT&T, Inc. and SecureWorks Corp., respectively, because they have the next lowest prices. Fourth place is a tie between AMC Entertainment Holdings, Inc. and AutoCanada Inc. To assign their ranks, compute the average of the ranks of the positions for which they are tied. The rank for each is determined as the average of four and five. Thus,

$$\text{Rank} = \frac{4+5}{2} = 4.5.$$

### Table 17.2.2 – Ranked Stock Prices

| Stock | Price per Share ($) | Rank | |
|---|---|---|---|
| Merck & Co., Inc. | 78.25 | 9 | |
| AT&T, Inc. | 27.15 | 2 | |
| SecureWorks Corp. | 28.04 | 3 | |
| Micron Technology, Inc. | 73.21 | 8 | |
| HP Inc. | 25.16 | 1 | |
| AMC Entertainment Holdings, Inc. | 30.50 | 4.5 | Tied for 4th and 5th |
| CitiGroup, Inc. | 60.12 | 7 | |
| Exxon Mobil Co. | 55.30 | 6 | |
| AutoCanada Inc. | 30.50 | 4.5 | Tied for 4th and 5th |

There are no other ties. The ranking resumes with Exxon Mobile Co. being ranked sixth. The highest ranked stock is Merck & Co., Inc., which has a rank of nine.

The test procedure for the Wilcoxon signed-rank test will be developed in the following example.

### Example 17.2.2

**Performing the Wilcoxon Signed-Rank Test**

It is believed by many economists that the United States entered into a recession in December of 2007. In tough economic times, most families tend to cut back on dining outside the home in an effort to save money. Thus, one indication of a weak economy is the reduction in the number of times a family dines out. Randomly sampling nine customers at the local grocer, the patrons were asked the number of times that they dined out between January 2006 and December 2006 (prior to the recession) and from January 2008 to December 2008 (12 months after entering the recession). The results of the survey are shown in Table 17.2.3. Based on the data, can it be concluded that people tend to dine out less during tough economic times (i.e., during a recession) at $\alpha = 0.05$?

### Table 17.2.3 – Number of Times Dining Out

| Customer | 2006 (Y) | 2008 (X) | Difference 2006–2008 | Absolute Value of Difference | Rank of Absolute Value of Difference | Signed Rank |
|---|---|---|---|---|---|---|
| 1 | 52 | 47 | 5 | 5 | 7 | 7 |
| 2 | 68 | 68 | 0 | 0 | Ignore | Ignore |
| 3 | 72 | 66 | 6 | 6 | 8 | 8 |
| 4 | 64 | 67 | −3 | 3 | 4 | −4 |
| 5 | 70 | 66 | 4 | 4 | 5.5 | 5.5 |
| 6 | 34 | 36 | −2 | 2 | 2.5 | −2.5 |
| 7 | 64 | 60 | 4 | 4 | 5.5 | 5.5 |
| 8 | 66 | 64 | 2 | 2 | 2.5 | 2.5 |
| 9 | 34 | 33 | 1 | 1 | 1 | 1 |
| | | | | | Total + | 29.5 |
| | | | | | Total − | 6.5 |

**SOLUTION**

**Step 1:** Determine the null hypothesis.

The null hypothesis is that the number of times that people dined out in 2006 (pre-recession) is equal to the number of times that people dined out in 2008 (recession). The null hypothesis can be written as

$H_0$: The distribution of those dining out in 2008 after the U.S. economy entered into a recession is the same as the distribution of those dining out in 2006.

**Step 2:** Determine the alternative hypothesis.

The alternative hypothesis is that the number of times that people dined out in 2008 is less than the number of times that people dined out in 2006. The alternative hypothesis can be written as

$H_a$: The distribution of those dining out in 2008 after the U.S. economy entered into a recession is shifted to the left of the distribution of those dining out in 2006.

Since we are interested in knowing if the number of times people dined out has decreased, this test is a one-tailed test. Thus, the alternative hypothesis will be one-sided.

**Step 3:** Select the appropriate test statistic based on the information at hand and the assumptions you are willing to make.

The Wilcoxon signed-rank test compares the probability distributions for 2006 and 2008 (i.e., before recession and during recession). How this is accomplished will be discussed in **Step 6**.

If we let $n =$ the number of nonzero differences, the test statistic for the Wilcoxon signed-rank test is determined in the following manner.

1.  Compute the difference for each of the pairs.

2.  Rank the absolute values of the differences from lowest to highest (ignoring zero differences).

3.  Calculate $T_+ =$ the sum of the ranks associated with positive differences.

4.  Calculate $T_- =$ the sum of the ranks associated with negative differences.

5.  Determine the test statistic based on the following criteria.

The test statistic will vary depending on $n$ and the alternative hypothesis.

If $n \leq 25$:

*   If the alternative hypothesis is that Population $X$ is shifted to the right of Population $Y$, use

    $T_+ =$ the sum of the ranks associated with positive differences, where differences are defined as observations in Population $Y$ minus observations in Population $X$.

    The test statistic is defined in this manner because we expect the number of negative differences to outnumber the number of positive differences, and thus expect $T_+$ to be the rank sum with the smallest value.

*   If the alternative hypothesis is that Population $X$ is shifted to the left of Population $Y$, use

    $T_- =$ the sum of the ranks associated with negative differences, where differences are defined as observations in Population $Y$ minus observations in Population $X$.

### Technology

For instructions on performing the Wilcoxon signed-rank test using technology, please visit stat.hawkeslearning.com and navigate to **Discovering Business Statistics, Second Edition > Technology Instructions > Nonparametrics > Wilcoxon Signed-Rank Test**.

### NOTE

In the following discussion, Population $X$ will be the group in 2008 (during recession) and Population $Y$ will be the group in 2006 (before recession).

### NOTE

This is analogous to the *greater than* alternative hypothesis.

### NOTE

This is analogous to the *less than* alternative hypothesis.

The test statistic is defined in this manner because we expect the number of positive differences to outnumber the number of negative differences, and thus expect $T_-$ to be the rank sum with the smallest value.

- If the alternative hypothesis is that Population $X$ is shifted to the left or to the right of Population $Y$, use

$$T = \min(T_+, T_-) = \text{the rank sum with the smallest value.}$$

It is important to note that if the smaller sum of the ranks is not as expected, the test should be halted and the decision should be to fail to reject the null hypothesis since there is no evidence in favor of the alternative hypothesis.

If $n > 25$, the test statistic can be approximated by a standard normal distribution and is specified in the procedure for the signed-rank test following this example.

For the example at hand, $n \leq 25$ and the alternative hypothesis is that the probability distribution for dining out in 2008 (recession) is shifted to the left of dining out in 2006 (pre-recession). This implies that dining out in 2006 minus dining out in 2008 should generally be positive. If the rank sum with the smallest value is $T_+$, fail to reject the null hypothesis since there is no evidence that the alternative hypothesis is more reasonable. Otherwise, the test statistic is given by

$$T = T_- = \text{the sum of the ranks associated with negative differences.}$$

**Step 4:** Determine the critical value of the test statistic.

If $n \leq 25$ the critical values are determined from Appendix A, Table J for the Wilcoxon signed-rank test. The critical value is specified by the level of $\alpha$, $n$ = number of nonzero differences in the test, and the alternative hypothesis. If the alternative hypothesis is *greater than* or *less than*, use the column in Table J labeled $\alpha$ *for a one-tailed test*. If the alternative hypothesis is *not equal to*, use the column in Table J labeled $\alpha$ *for a two-tailed test*. Because of the way the test statistic is defined, the null hypothesis is always rejected if the test statistic that is calculated from the data, $T$, is less than or equal to the critical value in Table J, $T_c$.

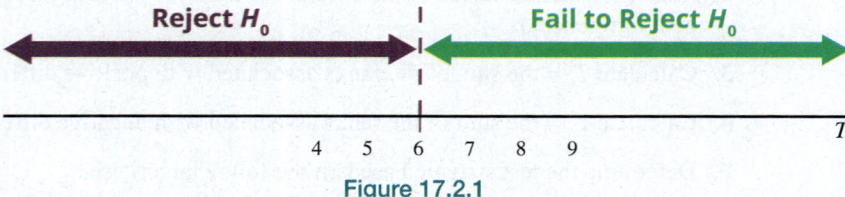

**Reject $H_0$**        **Fail to Reject $H_0$**

4   5   6   7   8   9     $T$

**Figure 17.2.1**

For the dining out example, $n = 8$ (recall there was one difference which was zero), $\alpha = 0.05$, and the test is one-tailed. Based on these specifications, the critical value determined from Table J is $T_c = 6$. The null hypothesis will be rejected if $T \leq 6$. This rejection region is displayed in Figure 17.2.1.

If $n > 25$, the critical values are determined in a similar manner to test statistics which have an approximate standard normal distribution under the null hypothesis. However, because of the way the test statistic is defined, we will always reject the null hypothesis for small values of the test statistic. Thus, the rejection region will always be established in the left tail, with the critical value defined by $\alpha$ for one-tailed tests, and $\dfrac{\alpha}{2}$ for two-tailed tests.

The level of the test is specified in the problem to be $\alpha = 0.05$.

**Step 5:** Collect the sample data and compute the value of the test statistic.

In order to calculate the test statistic, the signed ranks must be determined. To determine the signed ranks, first compute the difference for each pair of data values. Next, find the absolute value of each difference. Rank the absolute values of the differences using the ranking technique illustrated in Example 17.2.1 (ignore differences of zero). Once the absolute values of the differences have been ranked, reassign the ranks the sign which each associated difference had before the absolute value was computed. Finally, add all of the ranks with positive signs to determine $T_+$, and add all of the ranks with negative signs to determine $T_-$. This procedure is illustrated in Table 17.2.3.

Using the results from the table, the test statistic is given by

$T = T_- =$ the sum of the ranks associated with negative differences $= 6.5$.

**Step 6:** Make the decision and state the conclusion in terms of the original question.

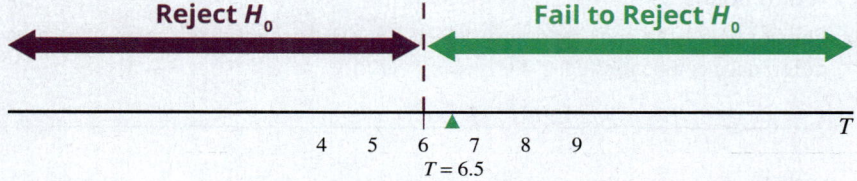

**Figure 17.2.2**

As shown in Figure 17.2.2, the value of the test statistic does not fall in the rejection region (that is, $T_- = 6.5$ is greater than $T_c = 6$). Thus, we fail to reject the null hypothesis at $\alpha = 0.05$. It is likely that the difference in the distributions of the two groups is due to ordinary sampling variation.

*Conclusion and Interpretation*: There is insufficient evidence to conclude at $\alpha = 0.05$ that the number of times people dined out was significantly higher before the recession than during the recession.

---

## Procedure

**Wilcoxon Signed-Rank Test**

**Hypotheses:**

$H_0$: The distributions of the two populations of interest are the same.

$H_a$: > One-Tailed: The distribution of Population $X$ is shifted to the *right* of the distribution of Population $Y$ (Difference $= Y - X$).

< One-Tailed: The distribution of Population $X$ is shifted to the *left* of the distribution of Population $Y$ (Difference $= Y - X$).

≠ Two-Tailed: The distribution of Population $X$ is shifted to the left or to the right of the distribution of Population $Y$.

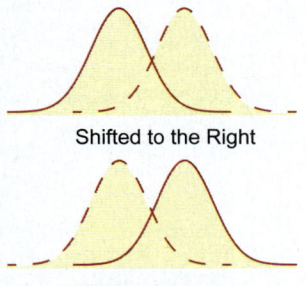

Shifted to the Right

Shifted to the Left

**Test Statistic:**

If $n \leq 25$:

If $H_a$ is > One-Tailed then

$T = T_+ =$ the sum of the ranks associated with the positive differences.

If $H_a$ is < One-Tailed then

$T = T_- =$ the sum of the ranks associated with the negative differences.

If $H_a$ is ≠ Two-Tailed, then $T = \text{Min}(T_+, T_-)$.

If $n > 25$ then $T = \text{Min}(T_+, T_-)$, and the test statistic is given by

$$z = \frac{T - \dfrac{n(n+1)}{4}}{\sqrt{\dfrac{n(n+1)(2n+1)}{24}}}.$$

> **Procedure** (cont.)
>
> ## Wilcoxon Signed-Rank Test
> **Critical Value(s):**
>
> If $n \leq 25$, reject $H_0$ if $T \leq T_c$, where $T_c$ is the critical value found in Appendix A, Table J.
>
> If $n > 25$:
>
>   One-Tailed Test: reject $H_0$ if $z \leq -z_\alpha$.
>
>   Two-Tailed Test: reject $H_0$ if $z \leq -z_{\alpha/2}$.
>
> **Assumptions:**
>
> Pairs of data have been randomly selected and are such that the absolute values of their differences can be ranked.

# 17.2 Exercises

## Basic Concepts

1. What assumptions are required for the Wilcoxon signed-rank test?

2. The Wilcoxon signed-rank test is primarily used to perform hypothesis tests about what type of experiment?

3. What are the advantages and disadvantages of the Wilcoxon signed-rank test?

4. Describe the procedure for assigning ranks to data in order to perform a Wilcoxon signed-rank test. What is to be done when two values are the same?

5. Describe how to calculate the rank sums for a paired difference experiment in order to perform a Wilcoxon signed-rank test.

6. If the sample size is less than or equal to 25, identify the three possible test statistics used for the Wilcoxon signed-rank test. How do you choose which statistic to use?

7. What are the null and alternative hypotheses associated with the Wilcoxon signed-rank test?

8. Explain why the population distributions are important when performing a Wilcoxon signed-rank test.

9. What is the test statistic for the Wilcoxon signed-rank test if the sample size is large? How large is *large* with regard to sample size?

10. Identify the critical values and rejection regions for both large and small samples with regard to the Wilcoxon signed-rank test.

## Exercises

11. Rank the following emerging markets mutual funds from lowest to highest price using the methodology presented for the signed-rank test.

| Emerging Markets Mutual Funds | | | |
|---|---|---|---|
| Mutual Fund | Price ($) | Mutual Fund | Price ($) |
| American Funds | 24.40 | DWS Investments | 15.57 |
| Columbia Management | 9.41 | UBS | 12.15 |
| Morgan Stanley | 23.74 | Prudential Investments | 9.23 |
| Fidelity Investments | 24.40 | Value Line Funds | 32.82 |
| John Hancock | 9.41 | The Vanguard Group | 34.72 |

12. Rank the following consumer price indexes (CPI) for selected groups of goods and services in September 2011 using the methodology presented for the signed-rank test. The data in the table represent the unadjusted percent change in price level from September 2010 to September 2011.

| Percent Change in CPI | |
|---|---|
| Expenditure Category | CPI (% Change 9/10 to 9/11) |
| Food | 4.7 |
| Alcoholic Beverages | 1.4 |
| Housing | 1.8 |
| Apparel | 3.5 |
| Public Transportation | 7.4 |
| Medical Care | 2.8 |
| Education | 4.4 |
| Tobacco and Smoking Products | 2.4 |
| Gasoline | 33.3 |
| New and Used Motor Vehicles | 3.6 |

Source: Bureau of Labor Statistics

13. A study conducted by the Orentreich Foundation found that women who practiced transcendental meditation (T.M.) for 20 minutes a day had high levels of DHEA-S, a hormone that may help prevent breast cancer and osteoporosis. Suppose eight women are randomly selected to participate in a study. The DHEA-S levels of the participants are measured prior to practicing transcendental meditation and then measured one year after practicing transcendental meditation for 20 minutes a day. The following table is a summary of the results of the study.

| Study Results | | |
|---|---|---|
| Study Participant | DHEA-S Level Before T.M. (mg) | DHEA-S Level After T.M. (mg) |
| A | 20 | 25 |
| B | 25 | 25 |
| C | 18 | 20 |
| D | 27 | 26 |
| E | 19 | 20 |
| F | 24 | 26 |
| G | 20 | 21 |
| H | 30 | 29 |

a. Using the sign test, do the data indicate that the DHEA-S level of women increases after practicing transcendental meditation for 20 minutes per day for one year at $\alpha = 0.05$?

b. What assumptions were necessary to perform the sign test?

c. Using the signed-rank test, do the data indicate that the DHEA-S level of women increases after practicing transcendental mediation for 20 minutes per day for one year at $\alpha = 0.05$?

d. What assumptions were necessary to perform the signed-rank test?

e. Which test do you think produces more accurate results? Why?

**14.** The management for a large grocery store chain would like to determine if a new cash register will enable cashiers to process a larger number of items on average than the cash register which they are currently using. Seven cashiers are randomly selected, and the number of grocery items which they can process in three minutes is measured for both the old cash register and the new cash register. The results of the test are as follows.

| Number of Grocery Items Processed in Three Minutes | | | | | | | |
|---|---|---|---|---|---|---|---|
| Cashier | 1 | 2 | 3 | 4 | 5 | 6 | 7 |
| Old Cash Register | 60 | 70 | 55 | 75 | 62 | 52 | 58 |
| New Cash Register | 65 | 71 | 55 | 75 | 65 | 57 | 57 |

**a.** What assumption must be made in order to perform the test of hypothesis using the paired difference $t$-test?

**b.** Using the signed-rank test, do the data provide conclusive evidence that the new cash register enables cashiers to process a significantly larger number of items than the old cash register at $\alpha = 0.05$?

**c.** What assumptions were made in performing the signed-rank test?

**d.** How do the results of the signed-rank test compare with the paired difference $t$-test performed in Section 11.3, Exercise 9?

**15.** An auto dealer is marketing two different models of a high-end sedan. Since customers are particularly interested in the safety features of the sedans, the dealer would like to determine if there is a difference in the braking distance (the number of feet required to go from 60 mph to 0 mph) of the two sedans. Six drivers are randomly selected and asked to participate in a test to measure the braking distance for both models. Each driver is asked to drive both models and brake once they have reached exactly 60 mph. The distance required to come to a complete halt is then measured in feet. The results of the test are as follows.

| Braking Distance of High-End Sedans (in Feet) | | | | | | |
|---|---|---|---|---|---|---|
| Driver | 1 | 2 | 3 | 4 | 5 | 6 |
| Model A | 150 | 145 | 160 | 155 | 152 | 153 |
| Model B | 152 | 146 | 160 | 157 | 154 | 155 |

**a.** What assumption must be made in order to perform a test of hypothesis using the paired difference $t$-test?

**b.** Using the signed-rank test, do the data provide conclusive evidence that there is a significant difference in the median braking distance of the two sedans at $\alpha = 0.10$?

**c.** What assumptions were made in performing the signed-rank test?

**d.** How do the results of the sign test performed in Section 17.1, Exercise 21 and the signed-rank test performed in part **b.** compare with the paired difference $t$-test performed in Section 11.3, Exercise 10?

# 17.3 The Wilcoxon Rank-Sum Test

We discussed nonparametric procedures for testing claims about a paired difference experiment in the previous two sections. In this section we will discuss a nonparametric procedure for hypothesis tests in which an independent experimental design is used to compare two population medians.

In Chapter 11 we discussed the small sample *t*-test for comparing two population means when independent random samples are drawn from two separate populations. In order to perform that test we need to make the important assumption that the distributions of both populations of interest are approximately normal.

The **Wilcoxon rank-sum test** is a nonparametric technique which can be used to compare two populations when we are either unwilling or unable to make the assumption of normality. It may also be used in the situation where the level of measurement is only ordinal, meaning we can only rank the data. Although the Wilcoxon rank-sum test does not require the assumption of normality, it does require that the two samples are drawn in a random and independent manner and that the data are such that they can be ranked from smallest to largest.

The test procedure for the Wilcoxon rank-sum test will be illustrated in the following example.

**Example 17.3.1**

**Performing the Wilcoxon Rank-Sum Test**

Audiophiles R Us (ARU) reviews high-end audio electronics as a service to customers and publishes their findings in *ARU Magazine*. One of the most popular issues is the one in which ARU reviews headphones. In its most recent magazine article about headphones, ARU reviewed two of the most popular high-end models of headphones on the market. Brand A headphones cost $449 and feature over the ear pads and crisp sound. Brand B headphones cost $300 and have on-ear pads and equally crisp sound. ARU has randomly selected 20 customers to evaluate the headphones (10 provided with Brand A headphones and 10 provided with Brand B headphones). The customers were told to provide a review of the headphones, giving a score between 0 and 100. The ratings are given in Table 17.3.1. Do the data provide sufficient evidence for ARU to conclude that there is a difference between Brand A and Brand B headphones at $\alpha = 0.05$?

| Table 17.3.1 – Headphone Ratings | | | |
|---|---|---|---|
| **Brand A** | **Rank** | **Brand B** | **Rank** |
| 93 | 15.5 | 99 | 19 |
| 100 | 20 | 84 | 8.5 |
| 75 | 3 | 76 | 4 |
| 92 | 14 | 93 | 15.5 |
| 55 | 1 | 87 | 12 |
| 98 | 18 | 95 | 17 |
| 85 | 10 | 77 | 5 |
| 84 | 8.5 | 79 | 6 |
| 82 | 7 | 61 | 2 |
| 90 | 13 | 86 | 11 |
| **Rank Sum** | 110 | | 100 |

**SOLUTION**

Histograms of the two ratings are provided in Figure 17.3.1 and Figure 17.3.2. The histograms indicate that the ratings for Brand A and Brand B headphones appear to follow a skewed distribution. Thus, the assumption of normality may not be reasonable for these data. If the assumption of normality is not met, the *t*-test for comparing two population means will not be valid. However, independent random samples were drawn for each set of customers and the data can be ranked. Thus, the assumptions for the Wilcoxon rank-sum test are satisfied and we can proceed with the test procedure.

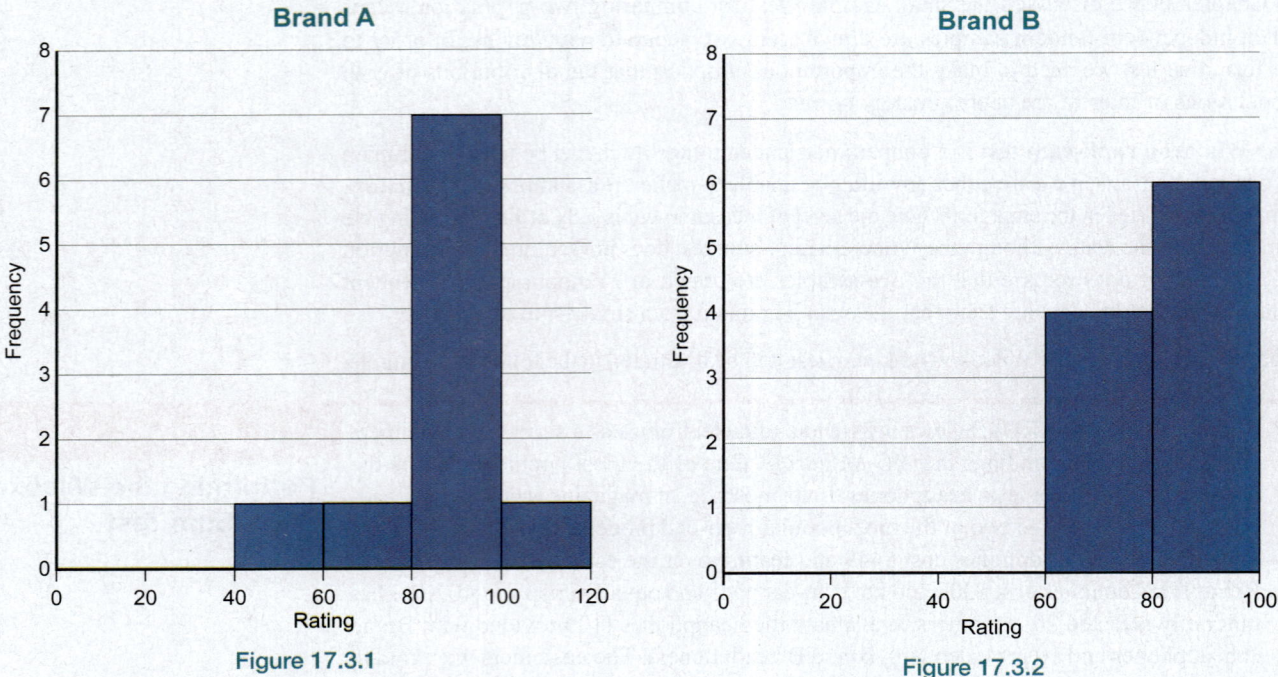

Figure 17.3.1

Figure 17.3.2

**Step 1:** Determine the null hypothesis.

The null hypothesis is that the ratings for Brand A and Brand B headphones are the same. The null hypothesis can be written as

$H_0$: The distributions of the ratings for Brand A headphones and Brand B headphones are the same.

**Step 2:** Determine the alternative hypothesis.

The alternative hypothesis is that the ratings for Brand A and Brand B headphones are different. The alternative hypothesis can be written as

$H_a$: The distribution of the ratings for Brand A headphones is shifted to the right or to the left of the distribution of the ratings for Brand B headphones.

Since ARU wants to determine if there is a difference between the headphones, this is a two-tailed test, and the hypothesis will be two-sided.

**Step 3:** Select the appropriate test statistic based on the information at hand and the assumptions you are willing to make.

The Wilcoxon rank-sum test uses the idea that if the null hypothesis is true, both samples come from the same population and you should be able to combine them. After the samples are combined, rank the data and compute the sum of the ranks for each sample. In our example, the Wilcoxon rank-sum test compares the sum of the ranks, the rank sum, for Brand A headphone ratings to the rank sum for Brand B headphone ratings.

If the null hypothesis is true, we would expect the rank sums to have approximately the same value. If the rank sums are very different, it will cause us to doubt that the observations come from the same population. Thus, the rank sum will be the measure which will be used for our test statistic. How large a difference in the rank sums is necessary before they are considered significantly different? This question is answered by the critical values specified in **Step 4**.

If $n \leq 10$, the test statistic is given by the rank sum associated with the population from which the smallest sample is taken.

$T$ = rank sum of the population with the smallest sample size

If the sample sizes are equal and the hypothesis is one-sided, use the rank sum associated with the population in the alternative hypothesis which is specified to be shifted to the right or shifted to the left. If the sample sizes are equal and the hypothesis is two-sided, either rank sum can be used.

If $n > 10$, the test statistic is given in the procedure following this example. The distribution of the test statistic can be approximated by a standard normal distribution.

**Step 4:** Determine the critical value of the test statistic.

If $n \leq 10$, the critical values are determined from Appendix A, Table K for the Wilcoxon rank-sum test. The critical value is specified by the level of $\alpha$, $n_1 =$ the size of the first sample, $n_2 =$ the size of the second sample, and whether the test is a one-tailed or two-tailed test.

- If the alternative hypothesis is one-sided and specifies that the distribution of Population $X$ is shifted to the right of the distribution of Population $Y$, the null hypothesis is rejected if $T \geq T_U$ from the Table K.

- If the alternative hypothesis is one-sided, and specifies that the distribution of Population $X$ is shifted to the left of the distribution of Population $Y$, the null hypothesis is rejected if $T \leq T_L$ from Table K.

- If the alternative hypothesis is two-sided and specifies that the distribution of Population $X$ is shifted to the left of the distribution of Population $Y$ or the distribution of Population $X$ is shifted to the right of the distribution of Population $Y$, the null hypothesis is rejected if $T \leq T_L$ or $T \geq T_U$ from Table K.

> ✎ **NOTE**
>
> The smaller sample size is associated with Population $X$.

The level of the test is specified in the problem to be $\alpha = 0.05$.

For the headphones example, $n_1 = 10$, $n_2 = 10$, $\alpha = 0.05$, and the alternative hypothesis is that there is a difference between the headphone ratings. Thus, this is a two-tailed test where the distribution of the ratings for Brand A headphones is hypothesized to be shifted to the right or to the left of the distribution of the ratings for Brand B headphones. Using Table K, we find that the null hypothesis should be rejected if $T \leq 79$ or $T \geq 131$. The rejection region is displayed in Figure 17.3.3.

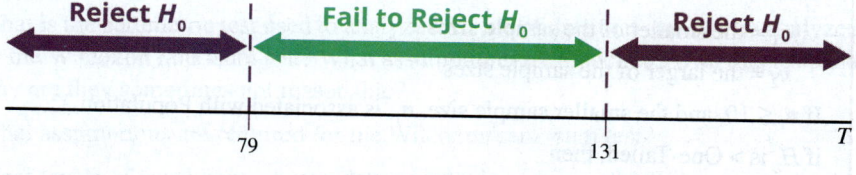

**Figure 17.3.3**

If $n > 10$, the critical values are determined in the usual manner for test statistics which have an approximate standard normal distribution under the null hypothesis. The box following this example describes the procedure for determining the rejection region in this case.

**Step 5:** Collect the sample data and compute the value of the test statistic.

For the headphones example, $n_1 = 10$ and $n_2 = 10$. Since the two sample sizes are the same, we will use the rank sum associated with the ratings for Brand A headphones. Since $n_1 = n_2$, we have the following.

$$T = \text{the rank sum for Brand A headphones} = 110$$

> ∞ **Technology**
>
> For instructions on performing the Wilcoxon rank-sum test using technology, please visit stat.hawkeslearning.com and navigate to **Discovering Business Statistics, Second Edition > Technology Instructions > Nonparametrics > Wilcoxon Rank-Sum Test**.

## Measure of Correlation

Let us denote the ranks assigned to data values $x_i$ and $y_i$ by $R(x_i)$ and $R(y_i)$ respectively. Then the **Spearman rank correlation coefficient**, also called **Spearman's rho**, is usually denoted by $r_s$, and is defined as follows.

> ### Formula
>
> #### Spearman Rank Correlation Coefficient
> The **Spearman rank correlation coefficient** is given by
>
> $$r_s = 1 - \frac{6 \sum_{i=1}^{n} d_i^2}{n(n^2 - 1)},$$
>
> where $d_i = R(x_i) - R(y_i)$.

This method is applicable when there are no ties in the rank data. The values of $r_s$ are always between $-1$ and $1$. The $r_s$ statistic behaves like the parametric correlation coefficient ($r$). If there is a positive relationship, $r_s$ will be positive but less than or equal to 1. If there is a negative relationship, $r_s$ will be negative but greater than or equal to $-1$.

### Example 17.4.1

**Calculating the Spearman Rank Correlation Coefficient**

A faculty member at State University is curious if there is a relationship between the number of hours that a student spends studying in their freshman math course versus the number of hours spent studying in their freshman English class. The data in Table 17.4.1 below is a random sample of four students in the faculty member's class. The variables $x$ and $y$ represent the number of hours spent studying for math and English, respectively.

| Table 17.4.1 – Time Spent Studying (hours) | |
|---|---|
| **Math (x)** | **English (y)** |
| 7 | 4 |
| 5 | 7 |
| 8 | 9 |
| 9 | 8 |

Calculate the Spearman rank correlation coefficient, $r_s$, for these data.

#### SOLUTION

Converting the data to ranks (from smallest to largest), we obtain

| Table 17.4.2 – Ranked Data | | | |
|---|---|---|---|
| **x** | **R (x)** | **y** | **R (y)** |
| 7 | 2 | 4 | 1 |
| 5 | 1 | 7 | 2 |
| 8 | 3 | 9 | 4 |
| 9 | 4 | 8 | 3 |

The value of $r_s$ is then given by

$$r_s = 1 - \frac{6\left((2-1)^2 + (1-2)^2 + (3-4)^2 + (4-3)^2\right)}{4(4^2-1)} = 1 - \frac{6(4)}{4(15)} = 0.60.$$

We want to test the null hypothesis that there is no correlation between the two variables against the alternative hypothesis that there is a correlation. More specifically, we test

$H_0: \rho_s = 0$   There is no correlation between the two variables.

$H_a: \rho_s \neq 0$   There is a correlation between the two variables.

where $\rho$ is the Greek letter rho. Note that $\rho_s$ represents Spearman's rho for the population and $r_s$ is the corresponding sample statistic.

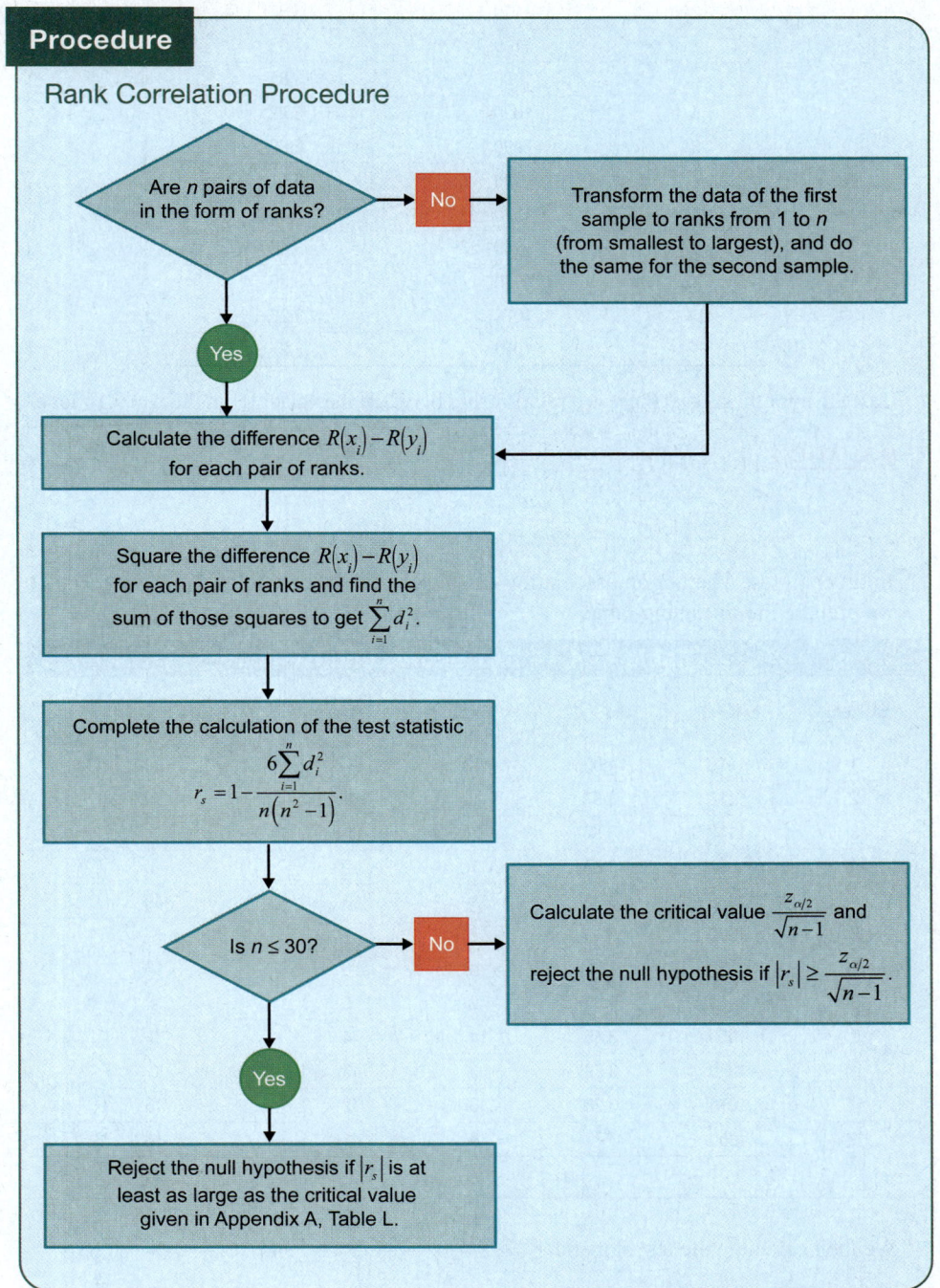

## Procedure

### Rank Correlation Procedure

Are $n$ pairs of data in the form of ranks?

No → Transform the data of the first sample to ranks from 1 to $n$ (from smallest to largest), and do the same for the second sample.

Yes

Calculate the difference $R(x_i) - R(y_i)$ for each pair of ranks.

Square the difference $R(x_i) - R(y_i)$ for each pair of ranks and find the sum of those squares to get $\sum_{i=1}^{n} d_i^2$.

Complete the calculation of the test statistic

$$r_s = 1 - \frac{6\sum_{i=1}^{n} d_i^2}{n(n^2 - 1)}.$$

Is $n \leq 30$?

No → Calculate the critical value $\dfrac{z_{\alpha/2}}{\sqrt{n-1}}$ and reject the null hypothesis if $|r_s| \geq \dfrac{z_{\alpha/2}}{\sqrt{n-1}}$.

Yes

Reject the null hypothesis if $|r_s|$ is at least as large as the critical value given in Appendix A, Table L.

✏ **NOTE**

The advantage of Spearman's rank correlation coefficient is that it can be used to test for a monotonic, nonlinear relationship, (i.e., $y$ increases as $x$ increases or $y$ decreases as $x$ increases, but not necessarily linearly).

**Example 17.4.2**

**Performing the Rank Correlation Test**

The academic performances of 12 college graduates are observed to examine the relationship between their SAT scores and their GPAs. The SAT scores (combined mathematics and critical reading sections which could yield a maximum score of 1600) and the GPAs are given in the following table.

| Student | SAT (x) | GPA (y) |
|---------|---------|---------|
| 1 | 1210 | 4.00 |
| 2 | 1110 | 3.97 |
| 3 | 1140 | 3.92 |
| 4 | 1080 | 3.85 |
| 5 | 1020 | 3.72 |
| 6 | 1090 | 3.60 |
| 7 | 1120 | 3.51 |
| 8 | 1030 | 3.49 |
| 9 | 1050 | 3.45 |
| 10 | 1040 | 3.39 |
| 11 | 1070 | 3.28 |
| 12 | 1060 | 3.27 |

Table 17.4.3 – SAT Scores and GPAs

Test the hypothesis that there is a relationship between the variables at the $\alpha = 0.10$ level.

**SOLUTION**

$$H_0: \rho_s = 0$$

$$H_a: \rho_s \neq 0$$

In order to test whether an association exists between these two variables at $\alpha = 0.10$ we prepare the following table.

| Student | SAT (x) | GPA (y) | R (x) | R (y) | $d_i^2 = \left[R\left(x_i\right) - R\left(y_i\right)\right]^2$ |
|---------|---------|---------|-------|-------|-----------------------------------------------------------------|
| 1 | 1210 | 4.00 | 12 | 12 | 0 |
| 2 | 1110 | 3.97 | 9 | 11 | 4 |
| 3 | 1140 | 3.92 | 11 | 10 | 1 |
| 4 | 1080 | 3.85 | 7 | 9 | 4 |
| 5 | 1020 | 3.72 | 1 | 8 | 49 |
| 6 | 1090 | 3.60 | 8 | 7 | 1 |
| 7 | 1120 | 3.51 | 10 | 6 | 16 |
| 8 | 1030 | 3.49 | 2 | 5 | 9 |
| 9 | 1050 | 3.45 | 4 | 4 | 0 |
| 10 | 1040 | 3.39 | 3 | 3 | 0 |
| 11 | 1070 | 3.28 | 6 | 2 | 16 |
| 12 | 1060 | 3.27 | 5 | 1 | 16 |
| | | | | Total | 116 |

Table 17.4.4 – SAT Scores and GPAs

We then calculate the test statistic $r_s = 1 - \dfrac{6(116)}{12(12^2 - 1)} \approx 0.5944$ .

Since our sample size is 12 ($n < 30$), the value of the statistic is compared with the critical value obtained using Appendix A, Table L.

| n | $\alpha = 0.10$ | $\alpha = 0.05$ | ... | $\alpha = 0.01$ |
|---|---|---|---|---|
| ... | | | | |
| 11 | 0.523 | 0.623 | | 0.818 |
| 12 | 0.497 | 0.591 | | 0.780 |
| 13 | 0.475 | 0.566 | | 0.745 |
| 14 | 0.457 | 0.545 | | 0.716 |
| 15 | 0.441 | 0.525 | | 0.689 |
| ... | | | | |

Note that Table L in Appendix A is constructed for a two-tailed test. So, we want to reject $H_0$ if $r_s \leq -0.497$ or $r_s \geq 0.497$. Since $0.5944 > 0.497$, we reject the null hypothesis at the 10% level of significance. Hence, there seems to be an association between SAT scores and GPAs.

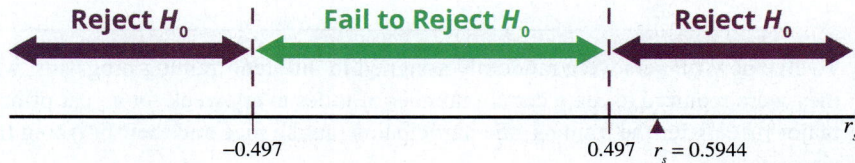

Figure 17.4.1

# 17.4 Exercises

## Basic Concepts

1. What is the correlation coefficient? How is this different from the Spearman rank correlation coefficient?

2. What is the formula for calculating Spearman's rho?

3. Can you calculate Spearman's rho if there are ties in the rank data?

4. Identify the difference in notation between Spearman's rho for population and sample data.

5. Explain the similarities in the behavior of the parametric correlation coefficient and Spearman's rho.

6. Identify one main advantage of the Spearman's rank correlation coefficient versus the parametric correlation coefficient.

7. Explain the procedure for ranking data when calculating Spearman's rho.

8. What are the null and alternative hypotheses for the rank correlation test?

9. Consider the value $r_s = 0.12$. Interpret this value in terms of the $x$ and $y$ variables used to calculate Spearman's rho.

## Exercises

10. Chris is a new cashier assigned to a cash register in a supermarket. Each day a sample of purchases at that register is examined and a percent of pricing errors is recorded along with the total number of customers who used that register. Do the following data indicate an association between Chris' performance and how busy his register was? Use $\alpha = 0.05$.

| % Pricing Errors and Total Customers | | | |
|---|---|---|---|
| Number of Customers | Errors (%) | Number of Customers | Errors (%) |
| 57 | 4.2 | 67 | 2.5 |
| 44 | 5.5 | 71 | 2.9 |
| 32 | 5.7 | 69 | 2.6 |
| 60 | 3.9 | 56 | 1.0 |
| 55 | 3.2 | 51 | 2.0 |
| 59 | 4.1 | 70 | 1.7 |
| 63 | 3.3 | | |

11. Twelve new runners were randomly assigned to different training programs, where they were required to run a certain number of miles every week for a year prior to a major race. After the training, the participants ran the race and their finishing times were recorded.

| Miles of Training and Race Times | | | |
|---|---|---|---|
| Miles Logged | Race Time (Minutes) | Miles Logged | Race Time (Minutes) |
| 35 | 198 | 30 | 189 |
| 25 | 165 | 29 | 240 |
| 45 | 155 | 42 | 224 |
| 60 | 148 | 24 | 201 |
| 70 | 135 | 19 | 246 |
| 21 | 243 | 55 | 166 |

a. With 95% confidence, is there evidence that the number of miles logged in a week during training affects the runner's race time?

b. Can the linear correlation coefficient, $r$, be calculated in order to fit a least squares regression line to the data in the table in an effort to predict the finish time of runners based on the number of miles logged during training? Why or why not?

12. The following data consist of college rankings of five universities by two different magazines. Is there a correlation between the rankings of the magazines? Use $\alpha = 0.10$.

| College Rankings by Magazines | | | | | |
|---|---|---|---|---|---|
| College | A | B | C | D | E |
| Magazine 1 | 1 | 4 | 2 | 3 | 5 |
| Magazine 2 | 4 | 3 | 1 | 5 | 2 |

**13.** An anthropologist records the heights (in inches) of ten fathers and their sons. Do the following data support (at the 5% level) that taller fathers tend to have taller sons?

| Heights of Fathers and Sons (Inches) | | | |
|:---:|:---:|:---:|:---:|
| Son's Height | Father's Height | Son's Height | Father's Height |
| 72 | 70 | 65 | 71 |
| 68 | 73 | 70 | 78 |
| 74 | 72 | 69 | 67 |
| 66 | 68 | 67 | 65 |
| 71 | 69 | 80 | 66 |

**14.** After a mother-daughter golf tournament, mothers and daughters were ranked among themselves. Do the following data show (at the 5% level) a correlation between the daughters' and mothers' golf skills?

| Golf Rankings | | | |
|:---:|:---:|:---:|:---:|
| Daughter's Ranking | Mother's Ranking | Daughter's Ranking | Mother's Ranking |
| 1 | 5 | 5 | 3 |
| 9 | 4 | 3 | 6 |
| 10 | 8 | 7 | 7 |
| 2 | 2 | 6 | 10 |
| 4 | 1 | 8 | 9 |

# 17.5  The Runs Test for Randomness

Randomness is an important concept in probability and statistics. In this section we are going to discuss a method for determining whether a sequence of observations exhibits randomness. To illustrate this concept, we will use the familiar coin-tossing experiment. One characteristic of the coin-tossing experiment is that in the long run there should be approximately equal numbers of heads and tails. In an ordered sequence, however, randomness implies more than compliance with this frequency criterion. For example, if the outcomes of 20 tosses of a coin were recorded as

$$H\,H\,H\,H\,H\,T\,T\,T\,T\,T\,T\,T\,T\,T\,T\,H\,H\,H\,H\,H,$$

we would suspect that the process was flawed. We would be equally surprised if the ordered outcomes were

$$H\,T\,H\,T\,H\,T\,H\,T\,H\,T\,H\,T\,H\,T\,H\,T\,H\,T\,H\,T,$$

but be reasonably happy with the sequence

$$H\,H\,T\,H\,T\,T\,T\,H\,T\,H\,H\,H\,T\,H\,T\,H\,H\,H\,H\,T\,T\,H.$$

A characteristic that reflects our reservations about the first two sequences is the number of **runs**, where a run is a subsequence of one or more heads (or tails).

In the first sequence, there are three runs: a run of 5 heads, then 10 tails, then 5 heads.

H H H H H T T T T T T T T T T H H H H H

Run          Run          Run

In the second sequence, there are 20 runs, each consisting of a single head or tail.

H T H T H T H T H T H T H T H T H T H T

R R R R R R R R R R R R R R R R R R R R

> **Definition**
>
> **Run**
>
> A **run** is a series of increasing values, a series of decreasing values, or a sequence of at least one symbol.

## Are There Streak Shooters?

Are there streak shooters in basketball? The answer to this question is remarkably consistent from respondents who answer it in the affirmative. Yes, there are streak shooters. But do the data support this strongly held belief?

Psychologists Tversky and Gilovich examined data from several professional NBA teams, especially data for free throwing. Free throwing was considered because it removed other dynamics of the game such as defense and various strategies.

Examining the data from the Boston Celtics and the Philadelphia 76'ers for an entire season, they failed to observe any streaks.

Successes and failures in free throws occurred randomly with each player. This study teaches that our concepts of randomness and non-randomness are very vague and often wrong.

**Source:** Tversky, A. and Gilovich, T. (1989). "The Cold Facts about the Hot Hand in Basketball."

Intuitively, we feel that these two sequences have respectively too few and too many runs for a truly random sequence. In the last sequence there are 13 runs.

$$\underset{R}{\text{H H}}\ \underset{R}{\text{T}}\ \underset{R}{\text{H}}\ \underset{R}{\text{T T T}}\ \underset{R}{\text{H}}\ \underset{R}{\text{T}}\ \underset{R}{\text{H H}}\ \underset{R}{\text{T}}\ \underset{R}{\text{H}}\ \underset{R}{\text{T}}\ \underset{R}{\text{H H H}}\ \underset{R}{\text{T T}}\ \underset{R}{\text{H}}$$

Given the number of times the coin was flipped, the number of runs does not seem excessively large or small. This intuitive notion of rejecting "randomness" in the sequence if there are too few or too many runs is the same notion we will use in the test of hypothesis.

In order to test the hypotheses

$H_0$: The sequence is random.

$H_a$: The sequence is not random.

we use a test based on the number of runs, $R$, in a sequence of $N$ ordered observations.

### Procedure

#### Runs Test for Randomness

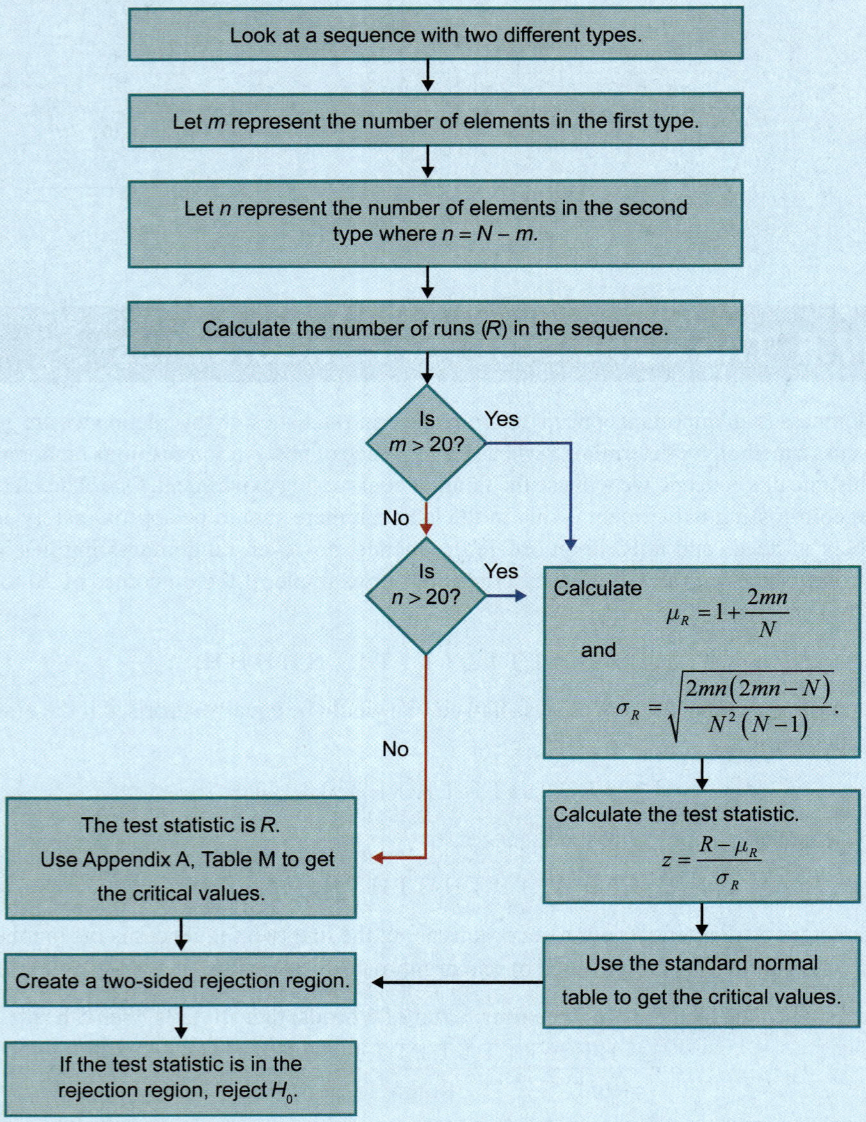

Look at a sequence with two different types.

↓

Let $m$ represent the number of elements in the first type.

↓

Let $n$ represent the number of elements in the second type where $n = N - m$.

↓

Calculate the number of runs ($R$) in the sequence.

↓

Is $m > 20$? — Yes →

No ↓

Is $n > 20$? — Yes → Calculate
$$\mu_R = 1 + \frac{2mn}{N}$$
and
$$\sigma_R = \sqrt{\frac{2mn(2mn - N)}{N^2(N-1)}}.$$

No ↓

The test statistic is $R$. Use Appendix A, Table M to get the critical values.

(from Calculate box) ↓

Calculate the test statistic.
$$z = \frac{R - \mu_R}{\sigma_R}$$

↓

Use the standard normal table to get the critical values.

→ Create a two-sided rejection region.

↓

If the test statistic is in the rejection region, reject $H_0$.

Let us apply the runs test for randomness to each of the sequences of heads and tails seen earlier in this section. In each test the hypotheses are as follows.

$H_0$: The sequence is random.

$H_a$: The sequence is not random.

**Example 17.5.1**

**Performing the Runs Test to Determine If a Sequence of Coin Flips Is Random**

### SOLUTION

**a.** H H H H H T T T T T T T T T T T H H H H H

$N = 20$ (the number of observations)

$m$ (the number of heads) $= n$ (the number of tails) $= 10$

$R = 3$ (the number of runs)

Using Appendix A, Table M, we see that the critical values are $R \leq 6$ or $R \geq 16$ at the 0.05 level of significance. Since $R = 3$, we reject the null hypothesis and conclude that there is strong evidence of non-randomness.

⌀ **Technology**

For instructions on performing the runs test for randomness using technology, please visit stat.hawkeslearning.com and navigate to **Discovering Business Statistics, Second Edition > Technology Instructions > Nonparametrics > Runs Test for Randomness**.

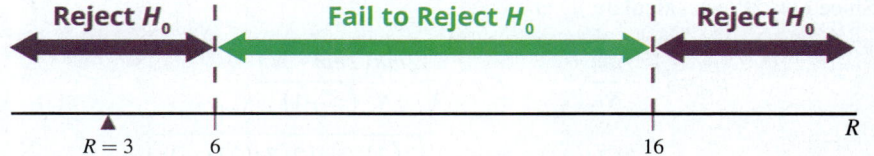

**Figure 17.5.1**

**b.** H T H T H T H T H T H T H T H T H T H T

$N = 20$ (the number of observations)

$m$ (the number of heads) $= n$ (the number of tails) $= 10$

$R = 20$ (the number of runs)

Using Appendix A, Table M, we see that the critical values are $R \leq 6$ or $R \geq 16$ at the 0.05 level of significance. Since $R = 20$, we reject the null hypothesis and conclude that there is strong evidence of non-randomness.

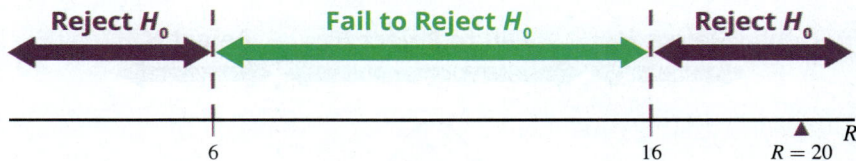

**Figure 17.5.2**

**c.** H H T H T T T H T H H H T H T H H H H T T H

$N = 20$ (the number of observations)

$m = 11$ (the number of heads)

$n = 9$ (the number of tails)

$R = 13$ (the number of runs)

Using Appendix A, Table M, we see that the critical values are $R \leq 6$ or $R \geq 16$ at the 0.05 level of significance. Since $R = 13$, we fail to reject the null hypothesis and conclude that there is not sufficient evidence of non-randomness.

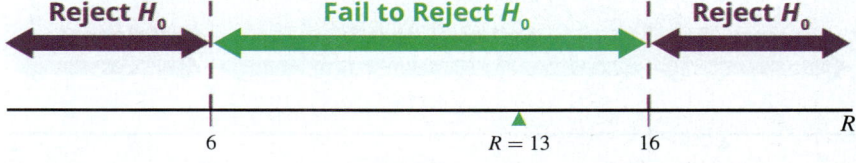

**Figure 17.5.3**

**Example 17.5.2**

**Performing the Runs Test to Detect a Pattern in College Applications**

The admissions office of a college records the in-state versus out-of-state status of 28 applicants in the order the applications arrive. The data set is given as follows.

$$\text{I O O I I I O O O O I I I I I I I I I I I O I I I I I I}$$

Is this sequence random? Test at the 0.05 significance level.

**SOLUTION**

To test the randomness of this sequence at the 5% level of significance, we obtain the following.

$N = 28$ (the number of observations)

$m = 21$ (the number of in-state applicants)

$n = 7$ (the number of out-of-state applicants)

$R = 7$ (the number of runs)

Since $m > 20$, we calculate $\mu_R$ and $\sigma_R$.

$$\mu_R = 1 + \frac{2mn}{N} \qquad \sigma_R = \sqrt{\frac{2mn(2mn-N)}{N^2(N-1)}}$$

$$= 1 + \frac{2(7)(21)}{28} \qquad = \sqrt{\frac{2(7)(21)(2(7)(21)-28)}{28^2(28-1)}}$$

$$= 11.5 \qquad \approx 1.92$$

Thus, we have the following.

$$z = \frac{R - \mu_R}{\sigma_R}$$

$$= \frac{7 - 11.5}{1.92}$$

$$\approx -2.34$$

⬱ **Technology**

For instructions on performing the runs test for randomness using technology, please visit stat.hawkeslearning.com and navigate to **Discovering Business Statistics, Second Edition > Technology Instructions > Nonparametrics > Runs Test for Randomness**.

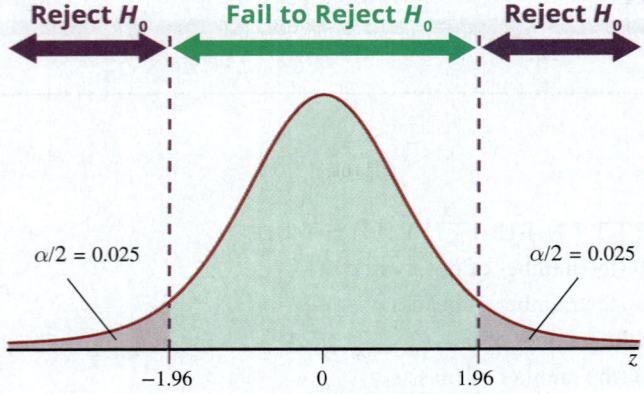

**Figure 17.5.4**

Since the test statistic is in the rejection region, we reject the null hypothesis that the status of the arriving student applications follows a random sequence.

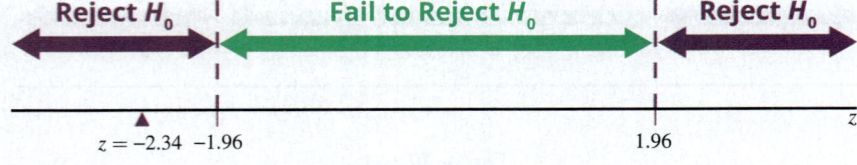

**Figure 17.5.5**

Are the following data random? Test at $\alpha = 0.05$.

$$16, 25, 52, 11, 38, 47, 12, 98, 4$$

**Example 17.5.3**

**Detecting Randomness of a Set of Numbers**

#### SOLUTION

How do you test randomness with a numerical set? Create a new data set comparing each value to the median value. To do this, substitute each value in the original data set with an A if it is above the median value, a B if it is below the median value, and eliminate any values that equal the median.

$H_0$: The data are random.

$H_a$: The data are not random.

Median $= 25$

| 16 | 25 | 52 | 11 | 38 | 47 | 12 | 98 | 4 |
|----|----|----|----|----|----|----|----|----|
| B  | ∅  | A  | B  | A  | A  | B  | A  | B |

$m = 4$ (the number of A's)

$n = 4$ (the number of B's)

$R = 7$ (the number of runs)

Using Appendix A, Table M the rejection region is $R \leq 1$ or $R \geq 9$ at the 0.05 level of significance. Since $R = 7$, we fail to reject $H_0$ and conclude that there is not sufficient evidence of non-randomness.

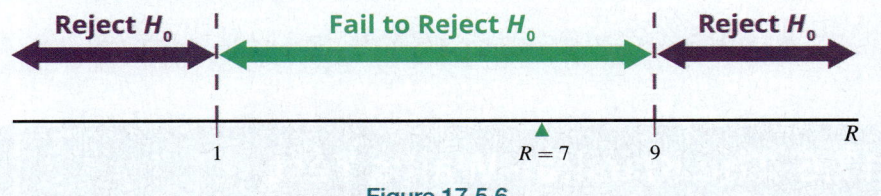

**Figure 17.5.6**

🔗 **Technology**

For instructions on performing the runs test for randomness using technology, please visit stat.hawkeslearning.com and navigate to **Discovering Business Statistics, Second Edition > Technology Instructions > Nonparametrics > Runs Test for Randomness**.

## ✎ 17.5 Exercises

### Basic Concepts

1. Describe in your own words what is being tested with the runs test.

2. Consider the following sequence of 10 coin tosses.

   H, H, T, T, H, H, H, T, T, H

   Without performing any kind of test, do you believe this sequence is random? Explain why or why not.

3. What are the null and alternative hypotheses associated with the runs test?

4. What parameters need to be calculated in order to perform a runs test?

5. What is the rejection rule for a small sample runs test? How small is a *small* sample?

6. What is the rejection rule for a large sample runs test? How large is a *large* sample?

7. If a numerical set of data is under consideration, which parameter are the data points compared to in order to perform the runs test?

## Exercises

8.  Suppose that in your city the number of deaths due to traffic accidents involving drunk driving from 1999 to 2011 were 75, 91, 54, 85, 79, 63, 12, 55, 63, 49, 89, 98, and 71. Use the runs test to examine non-randomness at the 0.05 level.

9.  A sociologist designs a study that involves a procedure of selecting families randomly from a phone book and then calling them to determine if they own or rent their residence. The results are recorded in the order of phone calls (O = Own, R = Rent).

    O O O R R O R O R R O R R R R O R R R O O R R R O R

    Does the sociologist have a random sequence of residential data at the 0.05 level?

10. A car tire manufacturer keeps track of the tires produced by one of the production lines. They observe the following sequence (D for defective items and N for non-defective items).

    D D D N N D N D N D D D

    Test the quality control manager's claim that there is no pattern in producing defective tires at the 0.05 level.

11. A marathon runner tries to run every day except when it is raining during the month of July. He observes the rainy (R) days and sunny (S) days to be able to predict the weather as follows.

    S S S R S S S R R R R S R S R R S S S R S R S R R S R S R S S

    Are the rainy days randomly scattered in the month of July at the 0.05 level?

# 17.6 The Kruskal–Wallis Test

In this section we present a procedure where $k$ random samples are obtained, one from each of the $k$ possibly different populations, and we are interested in testing whether all of the populations have identical distributions. Suitable hypotheses for this test would be as follows.

$H_0$: The populations from which the samples are drawn have identical distributions.

$H_a$: Not all populations have the same distribution.

The **Kruskal-Wallis test** is a method that can be used instead of the ANOVA $F$-test. The Kruskal-Wallis test does not need the assumption of normality of the populations. It does require that independent, random samples be drawn.

> **Definition**
>
> **Kruskal-Wallis Test**
>
> The **Kruskal-Wallis test** is a nonparametric procedure that can be used to determine if two or more distributions are different.

The Kruskal-Wallis test is similar to the Wilcoxon rank-sum test in that the test statistic will be based on the sums of the ranks of the groups being compared.

The data consist of $k$ random samples (not necessarily the same size) drawn from their respective populations. The data set may be arranged as follows.

| Group 1 | Group 2 | ... | Group $k$ |
|---------|---------|-----|-----------|
| $x_{1,1}$ | $x_{2,1}$ | ... | $x_{k,1}$ |
| $x_{1,2}$ | $x_{2,2}$ | ... | $x_{k,2}$ |
| ... | ... | ... | ... |
| $x_{1,n_1}$ | $x_{2,n_2}$ | ... | $x_{k,n_k}$ |

Let $N$ be the total number of observations, that is, $N = \sum_{i=1}^{k} n_i$.

To compute the Kruskal-Wallis test statistic we first combine all $N$ observations and order them from the smallest to largest. Next, we assign rank 1 to the smallest of all observations, rank 2 to the second smallest, and so on, until the largest observation is assigned the rank $N$. Let us denote the rank of $x_{i,j}$ by $r_{ij}$, and define

$$R_i = \sum_{j=i}^{n_i} r_{ij} \quad i = 1, 2, ..., k.$$

For example, $R_1$ is the sum of the ranks received by the Group 1 observations. The test statistic for the Kruskal-Wallis test is then given by the following formula.

$$H = \frac{12}{N(N+1)} \sum_{i=1}^{k} \frac{R_i^2}{n_i} - 3(N+1)$$

We reject the null hypothesis if $H \geq \chi_\alpha^2$, where $\chi_\alpha^2$ is obtained from the chi-square table (Appendix A, Table G) with $k-1$ degrees of freedom and a predetermined value of $\alpha$.

---

**Formula**

**The Kruskal-Wallis Test Statistic**

$$H = \frac{12}{N(N+1)} \sum_{i=1}^{k} \frac{R_i^2}{n_i} - 3(N+1)$$

where $R_i = \sum_{j=1}^{n_i} r_{ij}$ and $N = \sum_{i=1}^{k} n_i$.

---

**Example 17.6.1**

**Performing the Kruskal-Wallis Test to Determine If Stain Distributions Are Identical**

Ris Bocaj, the owner of Master Deck Sealer (MDS), specializes in staining and sealing wooden decks of residential homes. The company has developed a new stain, called MDS1, that not only stains but seals the decks from inclement weather for a period of 10 years. Even though there are other products on the market, MDS believes that its product is significantly better than the competition. To test his claim, Ris conducted accelerated-climate tests in a laboratory, exposing the wood (treated with his MDS1 and two other popular brands) to inclement weather such as rain, snow, high temperatures, and other conditions that would eventually result in the wood needing to be re-treated. The number of months before the stain began to wear and needed to be re-treated is recorded in the following table.

| Table 17.6.1 – Wood Stains: Months Before Re-Treatment | | |
|:---:|:---:|:---:|
| **MDS1** | **Competition Brand 1** | **Competition Brand 2** |
| 124 | 39 | 25 |
| 75 | 29 | 14 |
| 69 | 26 | 26 |
| 70 | 28 | 35 |
| 122 | 26 | 34 |
| 73 | 16 | 16 |
| | | 12 |

Use the Kruskal-Wallis test to determine whether there is a difference between the deck stains using $\alpha = 0.05$.

**SOLUTION**

The $N = 18$ data values are assigned overall ranks. In the following table, the ranks are shown in the column next to each observation.

| Table 17.6.2 – Wood Stain Rankings | | | | | |
|---|---|---|---|---|---|
| MDS1 | Rank | Competition Brand 1 | Rank | Competition Brand 2 | Rank |
| 124 | 18 | 39 | 12 | 25 | 4 |
| 75 | 16 | 29 | 9 | 14 | 2 |
| 69 | 13 | 26 | 6 | 26 | 6 |
| 70 | 14 | 28 | 8 | 35 | 11 |
| 122 | 17 | 26 | 6 | 34 | 10 |
| 73 | 15 | | | 16 | 3 |
| | | | | 12 | 1 |

The hypotheses of this test can be written as follows.

$H_0$: The three stains are equally durable.

$H_a$: At least one stain durability is different.

The sum of the ranks for each brand of stain are as follows.

MDS1:                              $18 + 16 + 13 + 14 + 17 + 15 = 93 = R_1$

Competition Brand 1:              $12 + 9 + 6 + 8 + 6 = 41 = R_2$

Competition Brand 2:      $4 + 2 + 6 + 11 + 10 + 3 + 1 = 37 = R_3$

Hence, we have the following test statistic.

$$H = \frac{12}{N(N+1)} \sum_{i=1}^{k} \frac{R_i^2}{n_i} - 3(N+1)$$

$$= \frac{12}{18(18+1)} \left( \frac{93^2}{6} + \frac{41^2}{5} + \frac{37^2}{7} \right) - 3(18+1)$$

$$\approx 12.238$$

There are $k - 1 = 3 - 1 = 2$ degrees of freedom. The critical value is given in Appendix A, Table G as $\chi_{0.05}^2 = 5.991$. Since $12.238 > 5.991$, we reject the null hypothesis and state that the durability of the stains does differ significantly.

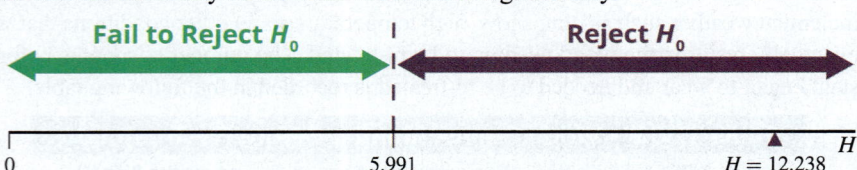

**Figure 17.6.1**

## Technology

For instructions on performing the Kruskal-Wallis test using technology, please visit stat.hawkeslearning.com and navigate to **Discovering Business Statistics, Second Edition > Technology Instructions > Nonparametrics > Kruskal-Wallis Test.**

---

**Example 17.6.2**

**Performing the Kruskal-Wallis Test to Detect Differences in Braking Distances**

A traffic safety engineer records the braking distances of a test vehicle at a fixed speed using three different brake pads. Braking distances (in feet) and their respective ranks (in parentheses) for brake pads of types A, B, and C are given in the following table. Is there sufficient evidence at the 0.05 level to conclude that the braking distances for the three pads differ?

| Table 17.6.3 – Braking Distances | | |
|---|---|---|
| Brake Pad A | Brake Pad B | Brake Pad C |
| 89 (1) | 97 (5) | 98 (6) |
| 99 (7) | 139 (15) | 116 (11) |
| 101 (8) | 119 (12) | 104 (10) |
| 94 (2) | 103 (9) | 96 (4) |
| 95 (3) | 127 (14) | 126 (13) |
| **Rank Sum**    21 | 55 | 44 |

**SOLUTION**

The null and alternative hypotheses for this test can be written as follows.

$H_0$: The braking distances for the three pads are the same.

$H_a$: At least one of the braking distances is different.

In Table 17.6.3 we are given the ranks of the observations. We then need to determine $R_1$, $R_2$, and $R_3$, corresponding to the sums of the ranks assigned to the observations for brake pads A, B, and C. The sum of the ranks for each brake pad is as follows.

$$R_1 = 21$$
$$R_2 = 55$$
$$R_3 = 44$$

Having the ranks, we can now calculate the test statistic, $H$.

$$H = \frac{12}{N(N+1)} \sum_{i=1}^{k} \frac{R_i^2}{n_i} - 3(N+1)$$
$$= \frac{12}{15(15+1)} \left( \frac{21^2}{5} + \frac{55^2}{5} + \frac{44^2}{5} \right) - 3(15+1)$$
$$= 6.02$$

Referring to Appendix A, Table G, we see that $\chi^2_{0.05} = 5.991$. Thus, we want to reject the null hypothesis if the test statistic, $H$, is greater than or equal to 5.991. Since the test statistic exceeds the critical value (6.02 > 5.991), we reject the null hypothesis and conclude that the braking distances are sufficiently different.

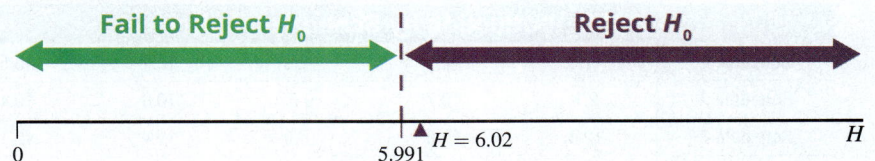

Figure 17.6.2

**🖙 Technology**

For instructions on performing the Kruskal-Wallis test using technology, please visit stat.hawkeslearning.com and navigate to **Discovering Business Statistics, Second Edition > Technology Instructions > Nonparametrics > Kruskal-Wallis Test**.

## ✍ 17.6 Exercises

## Basic Concepts

1. Which parametric test corresponds to the nonparametric Kruskal-Wallis test?

2. What are the null and alternative hypotheses associated with the Kruskal-Wallis test?

3. What are the assumptions associated with the Kruskal-Wallis test?

4. How is the Kruskal-Wallis test similar to the Wilcoxon rank-sum test?

5. What is the test statistic for the Kruskal-Wallis test? How is it calculated?

6. What is the rejection rule for the Kruskal-Wallis test?

7. How many populations can be compared using the Kruskal-Wallis test?

## Exercises

8.  An Internet service provider is considering four different servers for purchase. Potentially, the company would be purchasing hundreds of these servers, so it wants to make sure it is making the best decision. Initially, five of each type of server are borrowed, and each is randomly assigned to one of the 20 technicians (all technicians are similar in skill). Each server is then put through a series of tasks and rated using a standardized test. The higher the score on the test, the better the performance of the server. The data are as follows.

| Server Test Scores | | | |
|---|---|---|---|
| Server 1 | Server 2 | Server 3 | Server 4 |
| 48.5 | 56.4 | 52.1 | 64.3 |
| 46.5 | 68.2 | 56.3 | 68.3 |
| 52.4 | 68.5 | 48.3 | 72.2 |
| 54.1 | 64.2 | 52.2 | 70.6 |
| 58.9 | 60.1 | 54.8 | 56.5 |

Perform a Kruskal-Wallis test on these data using $\alpha = 0.10$. Are there differences between the servers?

9.  The following summary is obtained from an experiment where groups of cows were fed according to one of the four different feeding schedules, and their milk productions were recorded. The data given show the daily milk production in gallons for each cow. Test at $\alpha = 0.10$ to examine whether or not the milk production for all four schedules is the same.

| Milk Production by Schedule (Gallons) | | | | | |
|---|---|---|---|---|---|
| Schedule 1 | 11.5 | 12.7 | 12.9 | 10.1 | 10.5 |
| Schedule 2 | 9.1 | 10.7 | 9.5 | 10.9 | 10.4 |
| Schedule 3 | 12.4 | 11.9 | 10.0 | 11.4 | 12.1 |
| Schedule 4 | 12.8 | 12.6 | 11.7 | 11.3 | 10.9 |

10. The following data set contains the reading speed (in words per minute) of second grade students.

| Reading Speeds (wpm) | | |
|---|---|---|
| Public School | Private School | Home School |
| 54 | 66 | 65 |
| 67 | 55 | 64 |
| 63 | 62 | 60 |
| 105 | 69 | 72 |
| 61 | 71 | 68 |

Is there sufficient evidence at the 0.01 level of significance to conclude that the reading speeds vary by school type?

# T   Discovering Technology

## Using Excel

Testing a Population Median with the Sign Test

Use the information from Example 17.1.2 for this exercise.

1.  We will use the sign test to determine whether the median patient stay at MCH is less than the national median. The hypotheses for the sign test for the population median are:

    $H_0$: The median length of stay at MCH is the same as the national median.

    $H_a$: The median length of stay at MCH is shorter than the national median.

2.  Since the sample size $n = 50$ is greater than 25, the $z$-test statistic for the hypothesis test is computed as

    $$z = \frac{X + 0.5 - \dfrac{n}{2}}{\dfrac{\sqrt{n}}{2}}$$

    All patients stay either less time than the national median or at least the national median time. 35 patients in the sample stay less time than the national median and 15 patients in the sample stay at least the national median time. Here, $X = 15$ is the smaller of these two numbers and $n = 50$ is the sample size.

    Enter the labels **X**, **n**, and **z** into cells A1, A2, and A3, respectively. Enter **15** and **50** into cells B1 and B2, respectively.

3.  Compute the test statistic. Enter the following formula into cell B3:

    **=(B1 + 0.5 − B2/2)/(SQRT(B2)/2)**

    The result is −2.69 rounded to two decimal places. This is the value of the $z$-test statistic.

4.  Compute the $P$-value of the $z$-test statistic with the standard normal CDF. Type the label **p-value** into cell A5 and compute the $P$-value in cell B5 with the following formula:

    **=NORM.S.DIST(B3, 1)**

    The results are below:

| | A | B |
|---|---|---|
| 1 | X | 15 |
| 2 | n | 50 |
| 3 | z | -2.68701 |
| 4 | | |
| 5 | p-value | 0.003605 |

5.  Make a decision. The resulting $P$-value is 0.003605, which is less than the significance level $\alpha = 0.01$, so we reject the null hypothesis. Hence, we find evidence at the 10% significance level that the median length of patient stays at MCH is less than the national median length of stay.

## Spearman Rank Correlation Test

For this exercise, use the information from Example 17.4.2.

1. We will use the rank correlation test to determine whether SAT scores and GPAs are correlated. Specifically, the hypotheses are:

    $H_0$: There is no correlation between SAT scores and GPAs ($\rho = 0$).

    $H_a$: There is a correlation between SAT scores and GPAs ($\rho \neq 0$).

2. The Spearman's rho, $r_s$, is computed as

$$r_s = 1 - \frac{6\sum_{i=1}^{n} d_i^2}{n(n^2 - 1)}$$

where

$$d_i = R(x_i) - R(y_i)$$

and $R(x_i)$ and $R(y_i)$ are the ranks assigned to $x_i$ and $y_i$, respectively. The first step to computing the test statistic is to enter the sample data from Table 17.4.3 into Excel. Input the column titles **SAT (x)** and **GPA (y)** into cells A1 and B1, respectively. Enter the numerical data into columns A and B in Excel:

|    | A       | B       |
|----|---------|---------|
| 1  | SAT (x) | GPA (y) |
| 2  | 1210    | 4       |
| 3  | 1110    | 3.97    |
| 4  | 1140    | 3.92    |
| 5  | 1080    | 3.85    |
| 6  | 1020    | 3.72    |
| 7  | 1090    | 3.6     |
| 8  | 1120    | 3.51    |
| 9  | 1030    | 3.49    |
| 10 | 1050    | 3.45    |
| 11 | 1040    | 3.39    |
| 12 | 1070    | 3.28    |
| 13 | 1060    | 3.27    |

3. Next, we will compute the ranks sorting the data from largest to smallest. In cell C1, input the title **R(X)**. In cells C2 to C13, input the ranks of the SAT scores. In cell D1, input the title **R(Y)**. In cells D2 to C13, input the rank of the GPAs.

| | A | B | C | D |
|---|---|---|---|---|
| 1 | SAT (x) | GPA (y) | R(X) | R(Y) |
| 2 | 1210 | 4 | 12 | 12 |
| 3 | 1110 | 3.97 | 9 | 11 |
| 4 | 1140 | 3.92 | 11 | 10 |
| 5 | 1080 | 3.85 | 7 | 9 |
| 6 | 1020 | 3.72 | 1 | 8 |
| 7 | 1090 | 3.6 | 8 | 7 |
| 8 | 1120 | 3.51 | 10 | 6 |
| 9 | 1030 | 3.49 | 2 | 5 |
| 10 | 1050 | 3.45 | 4 | 4 |
| 11 | 1040 | 3.39 | 3 | 3 |
| 12 | 1070 | 3.28 | 6 | 2 |
| 13 | 1060 | 3.27 | 5 | 1 |

4. Then, compute the sum of $d_i^2$. In cell E1, input the title **[R(X) - R(Y)]^2**. In cell E2, input the formula

$$=(C2 - D2)^2$$

Click cell **E2**. Click the small square at the lower-right corner of the cell and drag it down to cell E13 to replicate the formula in cells E3 through E13.

5. Compute Spearman's rho. In cell D14, insert the text **rs =**. In cell E14, compute Spearman's rho with the following formula:

$$=1-((6*SUM(E2:E13))/(12*(12^2-1)))$$

| | A | B | C | D | E |
|---|---|---|---|---|---|
| 2 | 1210 | 4 | 12 | 12 | 0 |
| 3 | 1110 | 3.97 | 9 | 11 | 4 |
| 4 | 1140 | 3.92 | 11 | 10 | 1 |
| 5 | 1080 | 3.85 | 7 | 9 | 4 |
| 6 | 1020 | 3.72 | 1 | 8 | 49 |
| 7 | 1090 | 3.6 | 8 | 7 | 1 |
| 8 | 1120 | 3.51 | 10 | 6 | 16 |
| 9 | 1030 | 3.49 | 2 | 5 | 9 |
| 10 | 1050 | 3.45 | 4 | 4 | 0 |
| 11 | 1040 | 3.39 | 3 | 3 | 0 |
| 12 | 1070 | 3.28 | 6 | 2 | 16 |
| 13 | 1060 | 3.27 | 5 | 1 | 16 |
| 14 | | | | rs= | 0.594405594 |

6. Reference Table L in Appendix A to find the critical value. With N = 12 and $\alpha = 0.10$, the critical value comes out to be 0.497.

7. Make a decision. Since the test statistic is greater than the critical value, we reject the null hypothesis, indicating there is statistical evidence that SAT scores and GPAs are correlated.

## Using JMP

### Wilcoxon Signed-Rank Test

For this exercise, use the data from Example 17.2.2.

1. With a JMP data table open, enter the data from Table 17.2.3. Do not include the totals.

| | Customer | 2006(y) | 2008(x) | Difference 2006-2008 | Absolute Value of Difference | Rank of Absolute Value of Difference | Signed Rank |
|---|---|---|---|---|---|---|---|
| 1 | 1 | 52 | 47 | 5 | 5 | 7 | 7 |
| 2 | 2 | 68 | 68 | 0 | 0 | Ignore | Ignore |
| 3 | 3 | 72 | 66 | 6 | 6 | 8 | 8 |
| 4 | 4 | 64 | 67 | -3 | 3 | 4 | -4 |
| 5 | 5 | 70 | 66 | 4 | 4 | 5.5 | 5.5 |
| 6 | 6 | 34 | 36 | -2 | 2 | 2.5 | -2.5 |
| 7 | 7 | 64 | 60 | 4 | 4 | 5.5 | 5.5 |
| 8 | 8 | 66 | 34 | 2 | 2 | 2.5 | 2.5 |
| 9 | 9 | 34 | 33 | 1 | 1 | 1 | 1 |

2. Select **Analyze** in the top row of the JMP output table, then select **Specialized Modelling** and then **Matched pairs**. From the **Select Columns** box, select **2008(x)** and **2006(y)**. Then select **Y, Paired Response**. Note: 2008(x) must be selected first to perform 2006 – 2008. Click **OK**.

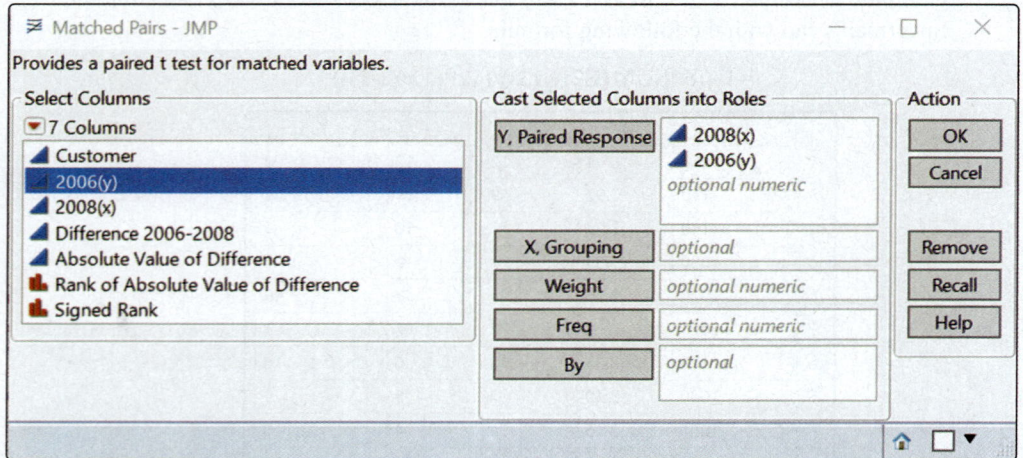

3. From the output, click on the red triangle next to **Matched Pairs** and select **Wilcoxon Signed Rank**. The output for the test is shown in the following screenshot.

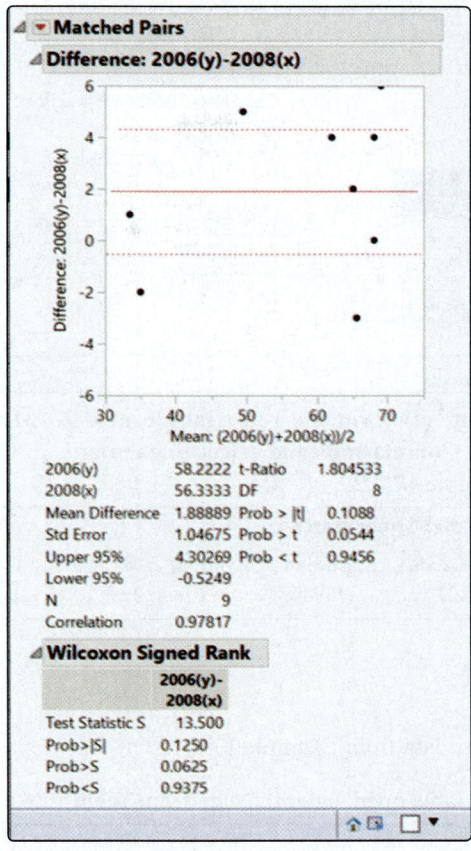

## Spearman Rank Correlation Test

For this exercise, use the data from Example 17.4.2.

1. With a JMP data table open, enter the data from Table 17.4.3 into **Columns 1-3**.

| | Student | SAT(x) | GPA(y) |
|---|---|---|---|
| 1 | 1 | 1210 | 4 |
| 2 | 2 | 1110 | 3.97 |
| 3 | 3 | 1140 | 3.92 |
| 4 | 4 | 1080 | 3.85 |
| 5 | 5 | 1020 | 3.72 |
| 6 | 6 | 1090 | 3.6 |
| 7 | 7 | 1120 | 3.51 |
| 8 | 8 | 1030 | 3.49 |
| 9 | 9 | 1050 | 3.45 |
| 10 | 10 | 1040 | 3.39 |
| 11 | 11 | 1070 | 3.28 |
| 12 | 12 | 1060 | 3.27 |

2. Select **Analyze** in the top row of the JMP output table, then select **Multivariate Methods** and then **Multivariate**. From the **Select Columns** box, click on **SAT(x)** and **GPA(y)**, then click on **Y, Response**. Click **OK**.

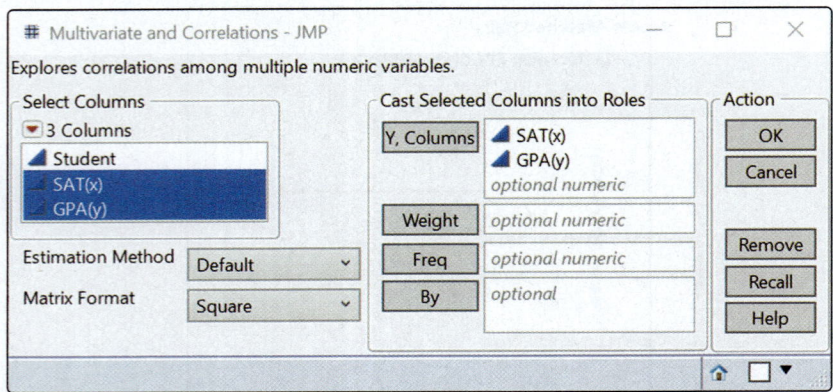

3.  From the output, click on the red triangle next to **Multivariate**. Click on **Nonparametric Correlations** and select **Spearman's ρ**. Observe the results and compare to Example 17.4.2.

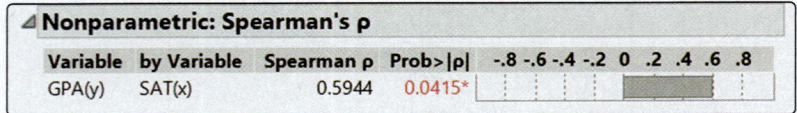

| Nonparametric: Spearman's ρ | | | | |
|---|---|---|---|---|
| Variable | by Variable | Spearman ρ | Prob>\|ρ\| | -.8 -.6 -.4 -.2 0 .2 .4 .6 .8 |
| GPA(y) | SAT(x) | 0.5944 | 0.0415* | |

## Kruskal-Wallis Test

For this exercise, use the data from Example 17.6.1.

1.  With a JMP data table open, enter the data from Table 17.6.1 into **Columns 1-3**.

| | MDS1 | Competition Brand 1 | Competition Brand 2 |
|---|---|---|---|
| 1 | 124 | 39 | 25 |
| 2 | 75 | 29 | 14 |
| 3 | 69 | 26 | 26 |
| 4 | 70 | 28 | 35 |
| 5 | 122 | 26 | 34 |
| 6 | 73 | • | 16 |
| 7 | • | • | 12 |

2.  Select **Table** in the top row, and then **Stack**. From the **Select Columns** box, click on **MDS1**, **Competition Brand 1**, and **Competition Brand 2**; then click on **Stack**. Make sure Stack By Row is selected. Click **OK**.

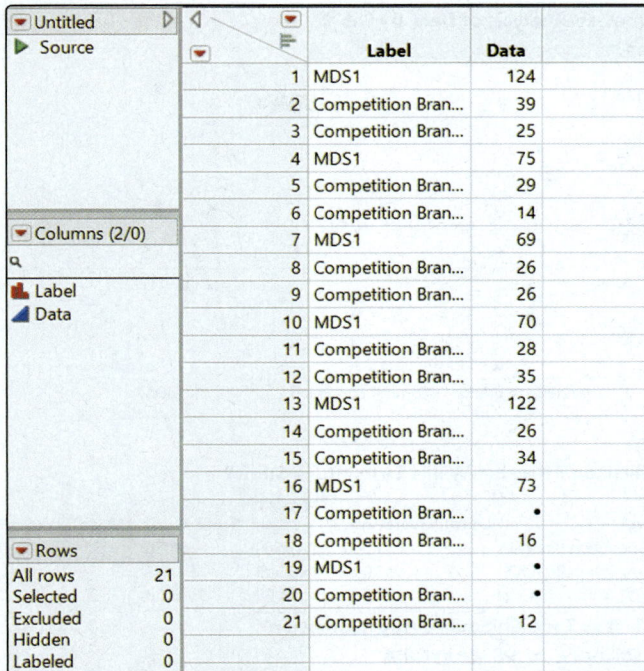

3.  Select **Analyze** in the top row of the JMP output table, then select **Fit Y by X**. From the **Select Columns** box, click on **Data**, then click on **Y, Response**. Click on **Label** and then click on **X, Factor**. Click **OK**.

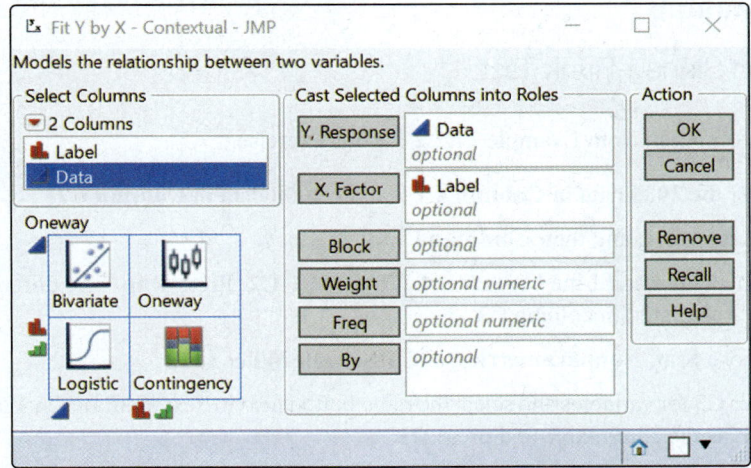

4.  From the output, click on the red triangle next to **Oneway Analysis of Data by Label**. Click on **Nonparametric** and select **Wilcoxon Test**. It performs the Wilcoxon rank-sum test if there are two groups and the Kruskal-Wallis test for more than two groups.

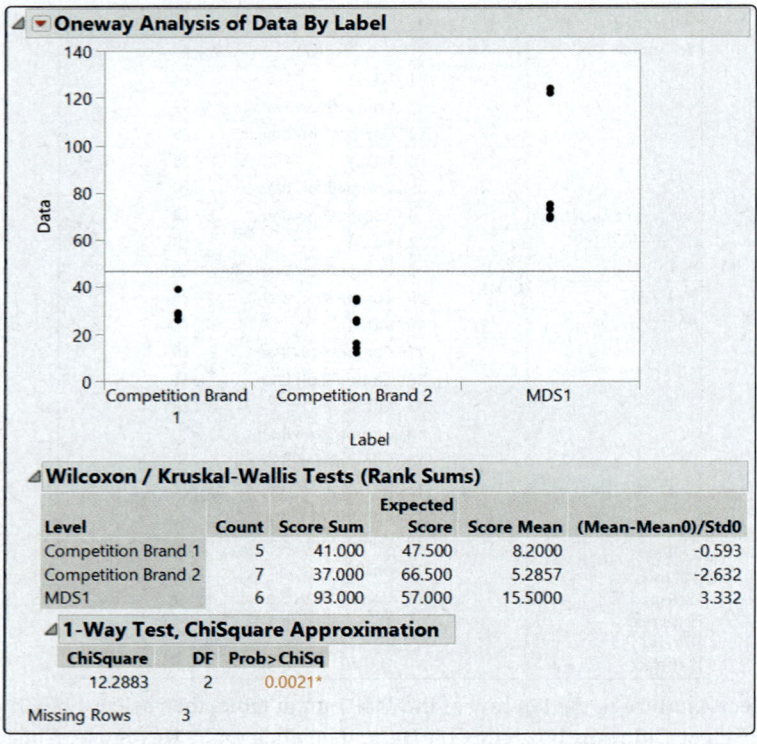

### Using Minitab

### Wilcoxon Signed-Rank Test

Use the information from Example 17.2.2 for this exercise.

1.   Enter the **2006** data in **Column C1** and the **2008** data in **Column C2**.

2.   Choose **View**, and then **Command Line/History**.

3.   In the Command Line box, enter **LET C3=C1-C2**. Press **Run**. The differences are now calculated in column C3.

4.   Choose **Stat**, **Nonparametrics**, and **1-Sample Wilcoxon**.

5.   Enter **C3** for variables and select the radio button next to **Test median**. Choose **greater than** as the Alternative and press **OK**.

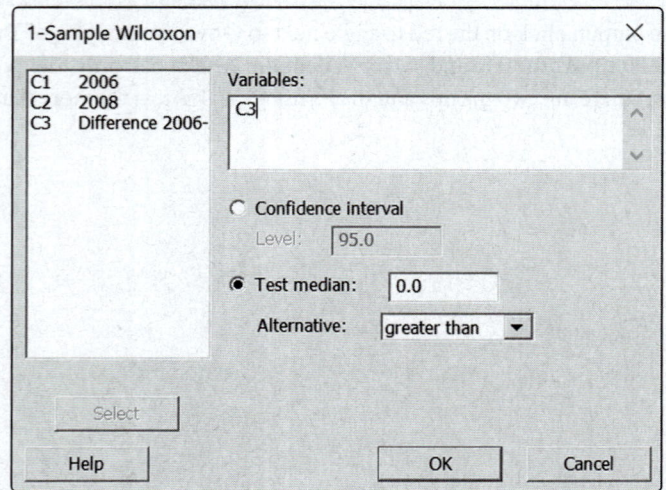

6. Observe the output screen for the Wilcoxon signed-rank test. The *P*-value is reported as 0.062 which is greater than the level of $\alpha$. Thus, we fail to reject the null hypothesis.

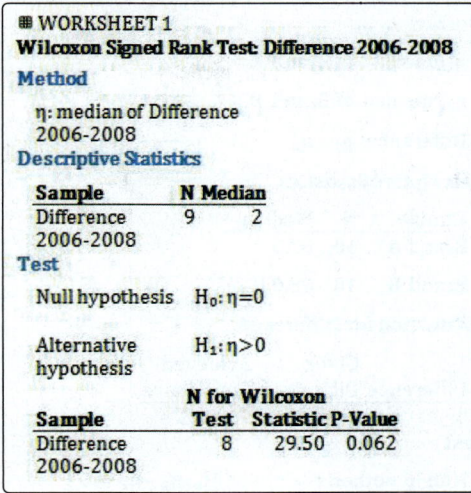

## Wilcoxon Rank-Sum Test

Use the information from Example 17.3.1 for this exercise.

1. Input the ratings for **Brand A** headphones in **Column C1** and the ratings for **Brand B** headphones in **Column C2**.

2. Choose **Stat**, **Nonparametrics**, and **Mann-Whitney**. Enter **C1** as the First Sample and **C2** as the Second Sample. Ensure that the confidence level is set to 95.0 and that the Alternative is **not equal**. Press **OK**.

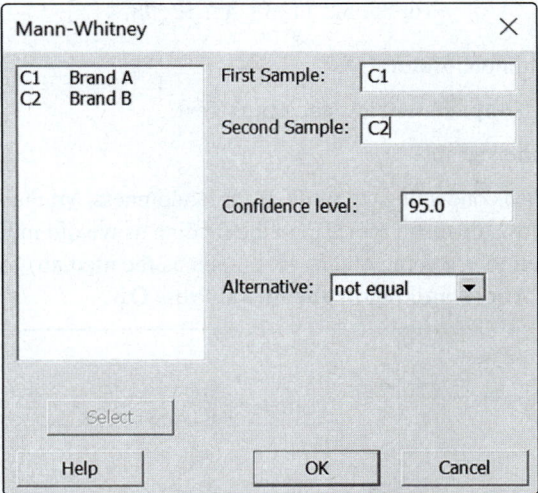

3. Observe the result of the Wilcoxon rank-sum test (also known as the Mann-Whitney test). Compare the value of the test statistic, W = 110.0, to that which we calculated in Example 17.3.1. Minitab reports the *P*-value of this test to be 0.734, causing us to fail to reject the null hypothesis.

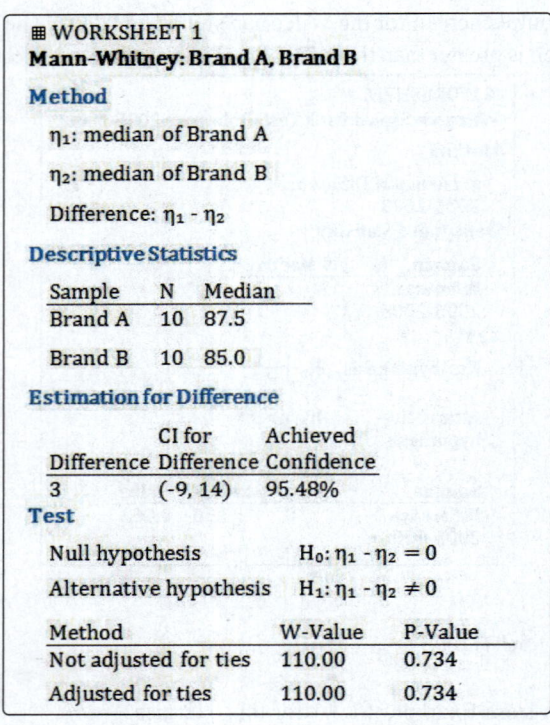

### WORKSHEET 1
**Mann-Whitney: Brand A, Brand B**

**Method**

$\eta_1$: median of Brand A

$\eta_2$: median of Brand B

Difference: $\eta_1 - \eta_2$

**Descriptive Statistics**

| Sample | N | Median |
|---|---|---|
| Brand A | 10 | 87.5 |
| Brand B | 10 | 85.0 |

**Estimation for Difference**

| Difference | CI for Difference | Achieved Confidence |
|---|---|---|
| 3 | (-9, 14) | 95.48% |

**Test**

| Null hypothesis | $H_0$: $\eta_1 - \eta_2 = 0$ |
|---|---|
| Alternative hypothesis | $H_1$: $\eta_1 - \eta_2 \neq 0$ |

| Method | W-Value | P-Value |
|---|---|---|
| Not adjusted for ties | 110.00 | 0.734 |
| Adjusted for ties | 110.00 | 0.734 |

## Runs Test for Randomness

Use the information from Example 17.5.3 for this exercise.

Are the following data random? Test at $\alpha = 0.05$.

$$16, 25, 52, 11, 38, 47, 12, 98, 4$$

1.  Enter the data into **Column C1**.

2.  Choose **Stat**, **Nonparametrics**, and **Runs Test**.

3.  Select **C1** as the variable.

4.  Notice that when conducting the runs test for randomness, Minitab counts observations above and below the mean rather than the median as we did in Example 17.5.3. You have the option to use some other value (such as the median) by choosing the radio button below **Above and below the mean**. Press **OK**.

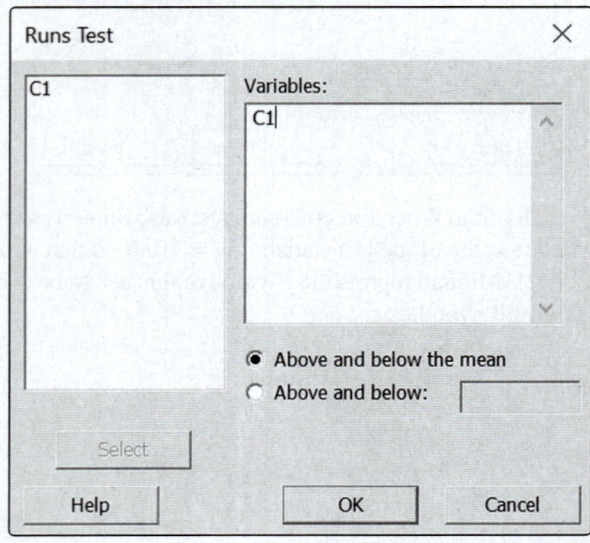

5. Observe the outcome of run test. The *P*-value is indicated as 0.261, thus we fail to reject the null hypothesis.

---

⊞ **WORKSHEET 1**
**Runs Test: Data**

**Descriptive Statistics**

| N | K | ≤ K | > K |
|---|---|---|---|
| | | **Number of Observations** | |
| 9 | 33.6667 | 5 | 4 |

K = sample mean

**Test**

| Null hypothesis | H$_0$: The order of the data is random |
|---|---|
| Alternative hypothesis | H$_1$: The order of the data is not random |

**Number of Runs**

| Observed | Expected | P-Value |
|---|---|---|
| 7 | 5.44 | 0.261 |

The p-value may not be accurate for samples with fewer than 11 observations above K or fewer than 11 below.

---

## Kruskal-Wallis Test

Use the information from Example 17.6.1 for this exercise.

1. Enter the data for the time until re-treatment is needed for all three brands of stains in **Column C1**. Enter the brand corresponding to each data value in **Column C2** as shown.

| C1 | C2-T |
|---|---|
| **Time** | **Brands** |
| 124 | MDS1 |
| 75 | MDS1 |
| 69 | MDS1 |
| 70 | MDS1 |
| 122 | MDS1 |
| 73 | MDS1 |
| 39 | BRAND1 |
| 29 | BRAND1 |
| 26 | BRAND1 |
| 28 | BRAND1 |
| 26 | BRAND1 |
| 25 | BRAND2 |
| 14 | BRAND2 |
| 26 | BRAND2 |
| 35 | BRAND2 |
| 34 | BRAND2 |
| 16 | BRAND2 |
| 12 | BRAND2 |

2.  Choose **Stat**, **Nonparametrics**, and **Kruskal-Wallis**. Enter **Time** for the Response and **Brands** for the Factor. Press **OK**.

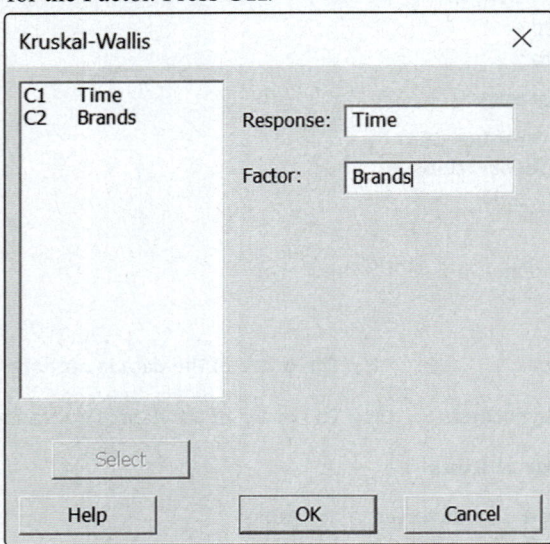

3.  Observe the results of the Kruskal-Wallis test. Compare the computed value of the test statistic to that which we found in Example 17.6.1. The *P*-value is given as 0.002, causing us to reject the null hypothesis at the 0.05 level of significance.

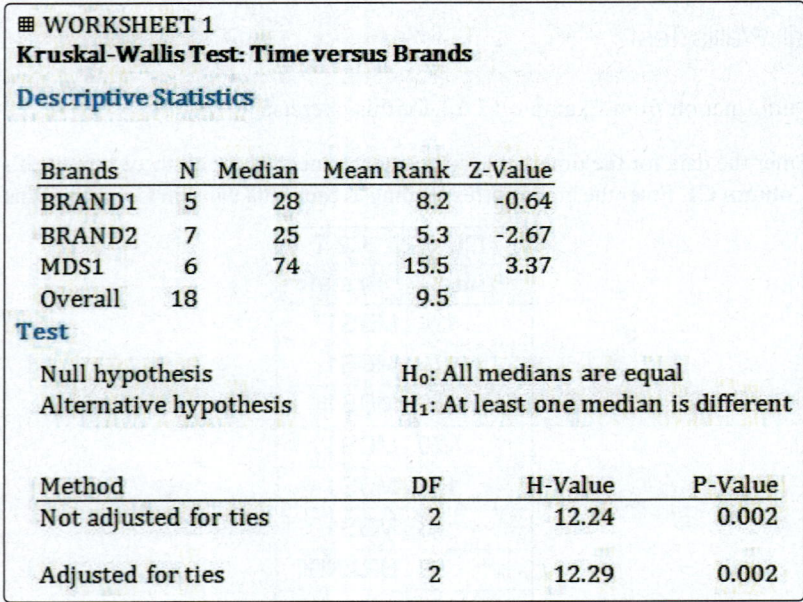

⊞ **WORKSHEET 1**
**Kruskal-Wallis Test: Time versus Brands**

**Descriptive Statistics**

| Brands | N | Median | Mean Rank | Z-Value |
|---|---|---|---|---|
| BRAND1 | 5 | 28 | 8.2 | -0.64 |
| BRAND2 | 7 | 25 | 5.3 | -2.67 |
| MDS1 | 6 | 74 | 15.5 | 3.37 |
| Overall | 18 | | 9.5 | |

**Test**

| Null hypothesis | $H_0$: All medians are equal |
|---|---|
| Alternative hypothesis | $H_1$: At least one median is different |

| Method | DF | H-Value | P-Value |
|---|---|---|---|
| Not adjusted for ties | 2 | 12.24 | 0.002 |
| Adjusted for ties | 2 | 12.29 | 0.002 |

# R Chapter 17 Review

## Key Terms and Ideas

- Parametric Statistics
- Nonparametric Statistical Methods
- Sign Test
- Hypothesis Test about a Population Median
- Wilcoxon Signed-Rank Test
- Rank Sum
- Wilcoxon Rank-Sum Test
- Rank Correlation Test

- Correlation Coefficient
- Spearman Rank Correlation Coefficient (Spearman's Rho)
- Runs Test for Randomness
- Runs
- Randomness
- Non-Randomness
- Kruskal-Wallis Test

## Key Formulas

| | Section |
|---|---|
| Test Statistic for the Sign Test, $n \leq 25$ <br><br> $X$ = the number of times the less frequent sign occurs | 17.1 |
| Test Statistic for the Sign Test, $n > 25$ <br><br> $$z = \frac{X + 0.5 - \left(\dfrac{n}{2}\right)}{\dfrac{\sqrt{n}}{2}}$$ | 17.1 |
| Test Statistic for the Wilcoxon Signed-Rank Test, $n \leq 25$ <br><br> If $H_a$ is > One-Tailed: <br> $T = T_+$ = the sum of the ranks associated with the positive differences. <br><br> If $H_a$ is < One-Tailed: <br> $T = T_-$ = the sum of the ranks associated with the negative differences. <br><br> If $H_a$ is ≠ Two-Tailed: <br> $T = \text{Min}\left(T_+, T_-\right)$. | 17.2 |
| Test Statistic for the Wilcoxon Signed-Rank Test, $n > 25$ <br><br> $$z = \frac{T - \dfrac{n(n+1)}{4}}{\sqrt{\dfrac{n(n+1)(2n+1)}{24}}}$$ <br><br> where $T = \text{Min}\left(T_+, T_-\right)$. | 17.2 |

## Key Formulas (cont.)

|  | Section |
|---|---|

Test Statistic for the Wilcoxon Rank-Sum Test, $n_1 \leq 10$

**If $H_a$ is > One-Tailed:**

$T = T_x$ = the rank sum of the sample with the fewest members. (If the sample sizes are the same, $T_x$ = the rank sum of the population hypothesized to be shifted to the right.)

**If $H_a$ is < One-Tailed:**

$T = T_x$ = the rank sum of the sample with the fewest members. (If the sample sizes are the same, $T_x$ = the rank sum of the population hypothesized to be shifted to the left.)

17.3

**If $H_a$ is ≠ Two-Tailed:**

$T = T_x$ = the rank sum of the sample with the fewest members. (If the sample sizes are the same, either rank sum can be used.)

---

Test Statistic for the Wilcoxon Rank-Sum Test, $n_1 > 10$

$$z = \frac{T - \dfrac{n_1(n_1 + n_2 + 1)}{2}}{\sqrt{\dfrac{n_1 n_2 (n_1 + n_2 + 1)}{12}}}$$

17.3

---

Spearman Rank Correlation Coefficient (Spearman's Rho)

$$r_s = 1 - \frac{6 \sum\limits_{i=1}^{n} d_i^2}{n(n^2 - 1)}$$

17.4

where $d_i = R(x_i) - R(y_i)$.

---

Test Statistic for the Runs Test for Randomness, $m > 20$ or $n > 20$

$$z = \frac{R - \mu_R}{\sigma_R}$$

17.5

where $\mu_R = 1 + \dfrac{2mn}{N}$ and $\sigma_R = \sqrt{\dfrac{2mn(2mn - N)}{N^2(N-1)}}$.

---

Test Statistic for the Kruskal-Wallis Test

$$H = \frac{12}{N(N+1)} \sum_{i=1}^{k} \frac{R_i^2}{n_i} - 3(N+1)$$

17.6

where $R_i = \sum\limits_{j=1}^{n_i} r_{ij}$ and $N = \sum\limits_{i=1}^{k} n_i$.

## AE  Additional Exercises

1. A new method for temporarily relieving the lung congestion of cystic fibrosis patients has been introduced. The traditional method of relieving the congestion involves a series of manual techniques where the chest and back area are pounded and massaged. The new method is a mechanical vest which has been designed to perform the manual techniques. A study is conducted to measure the effectiveness of the new vest. Five cystic fibrosis patients are randomly selected and the diameter of the blood vessels in their lungs is measured after using the traditional treatment and after using the vest treatment. The larger the diameter of the blood vessels within the lungs, the better the treatment. If the study provides conclusive evidence that the vest is more effective than the manual method in increasing the diameter of the blood vessels, the hospital will recommend the vest to its patients because the vest allows the patients to be much more independent. The results of the study are as follows.

| Diameter of Lung Blood Vessels (in mm) | | | | | |
|---|---|---|---|---|---|
| Subject | 1 | 2 | 3 | 4 | 5 |
| After Traditional Treatment | 0.5 | 0.4 | 0.7 | 0.6 | 0.2 |
| After Vest Treatment | 0.6 | 0.6 | 0.7 | 0.7 | 0.5 |

   a. What assumption must be made in order to perform the test of hypothesis using the paired difference $t$-test?

   b. Using the sign test, do the data provide conclusive evidence that the median diameter of blood vessels in the lungs is significantly larger after using the vest treatment than after using the traditional treatment at $\alpha = 0.01$?

   c. What assumptions were made in performing the sign test?

   d. Using the signed-rank test, do the data provide conclusive evidence that the median diameter of blood vessels in the lungs is significantly larger after using the vest treatment than after using the traditional treatment at $\alpha = 0.01$?

   e. What assumptions were made in performing the signed-rank test?

   f. Which test do you think produces more accurate results? Why?

   g. Perform a paired difference $t$-test. How do the results of the sign test and the signed-rank test compare with the results of the $t$-test?

2. As private companies prepare to go public, many analysts attempt to predict whether the stock will have a positive or negative return in the first day of trading. One particular analyst believes the return after the first day of trading is positive, on average. Another analyst wishes to test this analyst's claim using 15 recently offered public stocks.

| Initial Public Offerings | | |
|---|---|---|
| Company | Initial Offer Price ($) | Price After 1st Day of Trading ($) |
| Enduro Royalty Trust (NDRO) | 22.00 | 21.26 |
| ZELTIQ Aesthetics (ZLTQ) | 13.00 | 15.50 |
| Ubiquiti Networks (UBNT) | 15.00 | 17.50 |
| Tudou Holdings Limited (TUDO) | 29.00 | 25.56 |
| Carbonite (CARB) | 10.00 | 12.35 |
| SandRidge Permian Trust (PER) | 18.00 | 18.00 |
| American Capital Mortgage Investment (MTGE) | 20.00 | 18.41 |
| C&J Energy Services (CJES) | 29.00 | 30.50 |
| Chefs Warehouse Holdings (CHEF) | 15.00 | 17.50 |

| Initial Public Offerings | | |
|---|---|---|
| Company | Initial Offer Price ($) | Price After 1st Day of Trading ($) |
| Spirit Airlines (SAVE) | 12.00 | 11.55 |
| Pandora Media (P) | 16.00 | 17.42 |
| Wesco Aircraft Holdings (WAIR) | 15.00 | 14.92 |
| American Midstream Partners (AMID) | 21.00 | 20.95 |
| Dunkin Brands Group (DNKN) | 19.00 | 27.85 |
| Skullcandy (SKUL) | 20.00 | 20.00 |

**Source:** IPOScoop.com

a. Use the sign test to test the hypothesis that the return after the first day of trading is positive. Test at the 0.05 level.

b. Suppose that an analyst claimed that the median return on a stock in the first day of trading is +$1.00. Perform a test of hypothesis to determine if the median return is different than what the analyst claims. Use $\alpha = 0.05$.

c. Use the Wilcoxon signed-rank test to determine if the price after the first day of trading is generally greater than the initial offer price. Test at $\alpha = 0.05$.

d. Analyze the results of the sign test performed in part **a.** and the signed-rank test performed in part **c.** in terms of the problem. Which test do you think yields more accurate results?

e. What concerns do you have with the tests performed in this problem?

3. A weight loss center is trying to determine which of its diets results in higher client satisfaction. The center polled 20 clients (10 were on Diet A and 10 on Diet B) and had them rate their satisfaction in the diets from 1 to 100.

| Diet Ratings | |
|---|---|
| Diet A | Diet B |
| 84 | 94 |
| 77 | 81 |
| 89 | 95 |
| 98 | 93 |
| 97 | 97 |
| 100 | 99 |
| 75 | 82 |
| 85 | 92 |
| 96 | 95 |
| 78 | 89 |

a. Which nonparametric test do you think is most appropriate to test the claim that Diet B results in higher client satisfaction than Diet A? Explain why.

b. Write the null and alternative hypotheses for a rank-sum test to determine if Diet B results in higher client satisfaction than Diet A.

c. Using the rank-sum test, do the data provide sufficient evidence at $\alpha = 0.05$ that Diet B results in greater client satisfaction than Diet A?

d. What assumptions were made in the test performed in part **c.**?

4. Are students with higher GPAs more likely to get a higher paying job upon graduation? Consider the following data regarding student GPA and starting salary.

| GPA and Starting Salary | |
|---|---|
| GPA | Starting Salary ($) |
| 2.37 | 37,000 |
| 3.20 | 38,000 |
| 3.21 | 42,000 |
| 3.39 | 40,000 |
| 3.55 | 44,000 |
| 3.57 | 48,000 |
| 3.76 | 50,000 |
| 3.77 | 60,000 |
| 3.79 | 59,000 |
| 3.90 | 55,000 |

a. Determine the ranks for the *x*-variable, GPA.

b. Determine the ranks for the *y*-variable, Starting Salary.

c. Calculate the Spearman rank correlation coefficient.

d. Interpret the value of the coefficient. Is the relationship between these two variables positive or negative? Is this relationship what you expected? Explain.

e. Comment on the strength of the relationship between these two variables.

f. Is there evidence at $\alpha = 0.10$ that these two variables are related?

5. *Fortune* magazine releases a list of the world's most admired companies. In the survey they ask business people to vote for companies that they admire most from various industries. The table below lists the top 10 most admired companies of 2011 along with each company's Fortune 500 ranking. Note that the Fortune 500 ranks companies based on revenues.

| Top 10 Most Admired Companies and Profits | | |
|---|---|---|
| Company | Most Admired Ranking | Fortune 500 Ranking |
| Apple | 1 | 35 |
| Google | 2 | 92 |
| Berkshire Hathaway | 3 | 7 |
| Southwest Airlines | 4 | 205 |
| Procter & Gamble | 5 | 26 |
| Coca-Cola | 6 | 70 |
| Amazon.com | 7 | 78 |
| FedEx | 8 | 73 |
| Microsoft | 9 | 38 |
| McDonald's | 10 | 111 |

**Source:** CNN Money/Fortune Magazine

a. Compute the Spearman rank correlation coefficient. Interpret this value.

b. With 95% confidence, can we conclude that there is an association between company admiration and revenue?

**6.** The given table shows key dates of the Dow-Jones Industrial Average. Apply the runs test to check for randomness at the 0.05 level. (**Hint:** First find the median of the values, then label each value by A if it is above the median and B if it is below the median.)

| Significant Levels on the Dow (December 1974 to February 2009) | | |
|---|---|---|
| **Date** | **Dow Jones Industrial Average** | **Significance** |
| December 6, 1974 | 577 | The last Bear Market bottom |
| July 12, 1976 | 1011 | Highest point between January, 1973 and October, 1982 |
| August 12, 1982 | 776 | The start of the "Reagan Bull" |
| August 25, 1987 | 2722 | The 1987 high |
| October 19, 1987 | 1738 | The (508 point) crash of 1987 |
| February 2, 1994 | 3975 | The top of the post 1987 crash recovery |
| November 23, 1994 | 3674 | The start of the Clinton "super bull" |
| March 29, 1999 | 10,006 | The first Dow close above 10,000 |
| January 14, 2000 | 11,723 | The "Clinton bull" high |
| March 17, 2000 | 10,630 | The biggest one day gain (499 points) |
| March 20, 2001 | 9720 | Dow closes below previous year low for the first time since 1982 |
| September 11-14, 2001 | 9605 | Terrorist attack closed the Dow for four days |
| September 17, 2001 | 8920 | The biggest one day fall (685 points) |
| September 21, 2001 | 8235 | The Dow's second worst week ever (−14.26%) |
| December 31, 2001 | 10,021 | Dow up 21.7% from September 21 low but down 7.2% on the year |
| September 30, 2002 | 7591 | New 2002 low – all treasury yields (except 30-year bond) at 2002 lows |
| October 9, 2002 | 7286 | New 2002 low – Dow down 37.8% from the January, 2000 all-time high |
| October 31, 2002 | 8397 | Dow up 806 points (10.6%) for October – first positive month since March |
| November 6, 2002 | 8771 | Federal reserve cuts rates for the first time since December, 2001 |
| December 31, 2002 | 8341 | Dow down 16.8% for 2002 – first three consecutive year loss since 1939-41 |
| May 23, 2003 | 8601 | Senate passes bill raising the treasury debt limit |
| June 25, 2003 | 9011 | Federal reserve cuts rates by 0.25% |
| December 31, 2003 | 10,453 | Dow up 25.32% in 2003 |
| October 3, 2006 | 11,727 | Dow exceeds the previous all-time high in January, 2000 |
| October 9, 2007 | 14,164 | New all-time high on the Dow |
| July 2, 2008 | 11,215 | Dow closes more than 20% below the October, 2007 high |
| February 27, 2009 | 7062 | Dow closes more than 50% below the October, 2007 high |

**7.** The irrational number $\pi$ can be approximated by the rational number $\dfrac{22}{7}$. Test the randomness of odd and even digits in $\dfrac{22}{7}$ at the 0.05 level using the first nine digits.

**8.** A polling agency conducts exit interviews after an election. If R = Republican and D = Democrat, the first 20 voter responses in a random sample are as follows.

$$\text{D D D R D R R D D D D R D D R R R D R D}$$

Test for non-randomness using $\alpha = 0.05$.

9. Consider the following *U.S. News and World Report* college rankings for schools in the Big Ten, the Big 12, and the Atlantic Coast conferences.

| College Rankings by Conference | | |
|---|---|---|
| **Atlantic Coast** | **Big Ten** | **Big 12** |
| 29 | 55 | 101 |
| 10 | 62 | 45 |
| 101 | 42 | 58 |
| 71 | 28 | 143 |
| 68 | 45 | 90 |
| 31 | 71 | 94 |
| 55 | 45 | 101 |
| 25 | 12 | 75 |
| 38 | 68 | 101 |
| 101 | 71 | 132 |
| 36 | 75 | 160 |
| 25 | | 97 |

**Source:** *U.S. News and World Report*, 2011

With $\alpha = 0.10$, use the Kruskal-Wallis test to determine if there is a significant difference in *U.S. News and World Report* rankings among the three athletic conferences.

10. Consider the following scores reported by *Condé Nast Traveler* in their annual list of the top cities to visit around the world. The results were determined from more than 8 million votes cast in the Readers' Choice Awards survey.

| Readers' Choice City Rankings, 2011 | | | | | |
|---|---|---|---|---|---|
| **Asia** | | **Europe** | | **United States** | |
| Kyoto | 82.3 | Florence | 85.0 | Charleston, SC | 84.7 |
| Bangkok | 81.6 | Barcelona | 82.8 | San Francisco, CA | 83.7 |
| Hong Kong | 81.1 | Rome | 82.4 | Santa Fe, NM | 83.0 |
| Chiang Mai | 80.8 | Paris | 81.9 | Chicago, IL | 82.2 |
| Ubud | 80.0 | Bruges | 81.7 | Honolulu, HI | 80.9 |
| Singapore | 78.4 | Venice | 81.7 | New York, NY | 80.8 |
| Tokyo | 76.8 | Salzburg | 81.4 | Savannah, GA | 79.1 |
| Luang Prabang | 76.4 | Vienna | 81.0 | Carmel, CA | 78.5 |
| Thimphu | 75.1 | Prague | 79.7 | Seattle, WA | 78.4 |
| Shanghai | 74.9 | Siena | 79.7 | Boston, MA | 78.0 |

**Source:** *Condé Nast Traveler*, 2011

a. Using the Kruskal-Wallis test and $\alpha = 0.05$, test to determine if there is a difference in rankings between Asia, Europe, and the United States.

b. What assumptions were made for the test performed in part **a.**?

## P    Discovery Project

### Home Sweet Home: Using Nonparametric Tests to Compare Home Prices

**⋰ Data**

The data can be found by visiting stat.hawkeslearning.com and navigating to **Discovering Business Statistics, Second Edition > Data Sets > Mount Pleasant Real Estate**.

Use the Mount Pleasant Real Estate data which contains information about properties for sale in three subdivisons of Mount Pleasant, South Carolina in the year 2017.

1. Download the Mount Pleasant Real Estate data into a statistical software package like Excel or Minitab.

2. Classify the three variables *List Price*, *Square Footage*, and *Subdivision* as qualitative or quantatitive and provide the level of measurement (nominal, ordinal, interval, or ratio).

3. Which of the quantitative variable(s) should be considered as the dependent variable? Why?

4. Use statistical software to make a histogram for *List Price* and describe the distribution.

5. Can we use the *t*-test to see if the mean home price is significantly more than $500,000? Justify your answer.

6. Assuming that the underlying distribution is not normal, we have an opportunity to use nonparametric methods to analyze the data. Can we conclude that the median *List Price* in Mount Pleasant in 2017 is significantly more than half a million dollars? State your hypotheses and perform a sign test using $\alpha = 0.05$.

7. Create side-by-side boxplots of *List Price* for the three Mount Pleasant subdivisions: Carolina Park, Dunes West, and Park West. Describe the distributions of the three subdivisions and comment about their variability.

8. Use the Wilcoxon rank-sum test to see if the distribution of *List Price* in Park West in 2017 is to the left of that in Dunes West.

# Chapter 18

## Statistical Process Control

# Discovering the Real World

## Quality Control, Deming's 14 Points for Management, and Six Sigma Technology

Statistics and statistical methods have led to major discoveries and economic gains in agriculture, biology, business, engineering, and the sciences. Historically, the use of statistical techniques to improve quality began to evolve during World War II, due to the importance placed on war-time efficiency. Immediately after the war, manufacturers in the United States had a relatively easy time selling their goods throughout the world, regardless of quality, since there were very few international competitors. But by the 1960s, Japanese manufacturers began to challenge American companies. It was during this period that the Japanese gained a reputation for producing quality products. Also during this time, the modern field of **quality control**, known by names such as **total quality control**, **total quality management**, or simply **quality management**, started to emerge.

During the 1920s, Dr. Walter A. Shewhart of Bell Laboratories developed the concepts of **statistical quality control**. He introduced the concept of "controlling" the quality of a product as it was being manufactured rather than after it was manufactured. To control the quality, Shewhart developed charting techniques for controlling in-process manufacturing operations. Shewhart also introduced statistical sample inspection to estimate the quality of a product as it was being manufactured. This method replaced the older method of inspecting each item after it was manufactured.

World War II virtually destroyed Japanese production capabilities. After World War II, Dr. W. Edwards Deming and Joseph M. Juran convinced the management of several large Japanese companies that adoption of statistical methods would lead to increased quality, thereby building up manufacturing and improving the quality of living of the Japanese people. These manufacturers have done an excellent job of attaining high quality for their products, as evidenced by their cars, television sets, and many other electronic products. What made Japanese management different was its focus on process-oriented management. The strength of this system is that it acknowledges the ends (the results) but emphasizes the means (the process).

Deming's philosophy is an important framework for implementing quality and productivity improvement. His philosophy is summarized in his 14 Points for Management. **Deming's 14 Points** are as follows.

1. Create a constancy of purpose toward improvement of product and service, with the aim to become competitive and to stay in business, and to provide jobs.

2. Adopt the new philosophy. We are in a new economic age. Western management must awaken to the challenge, must learn their responsibilities, and take on leadership for change.

3. Cease dependence on inspection to achieve quality. Eliminate the need for inspection on a mass basis by building quality into the product in the first place.

4. End the practice of awarding business on the basis of price tag. Instead, minimize total cost. Move toward a single supplier for any one item, on a long-term relationship of loyalty and trust.

5. Improve constantly and forever the system of production and service, to improve quality and productivity, and thus constantly decrease costs.

6. Institute training on the job.

7. Institute leadership. The aim of supervision should be to help people and machines and gadgets to do a better job. Supervision of management is in need of an overhaul, as well as supervision of production workers.

---

**Definition**

### Statistical Quality Control

**Statistical quality control** was developed by Shewhart as a way to control the quality of a product as it is manufactured instead of after it has been manufactured.

8. Drive out fear, so that everyone may work effectively for the company.

9. Break down barriers between departments. People in research, design, sales, and production must work as a team, to foresee problems of production and in use that may be encountered with the product or service.

10. Eliminate slogans, exhortations, and targets for the work force asking for zero defects and new levels of productivity. Such exhortations only create adversarial relationships, as the bulk of the causes of low quality and low productivity belong to the system and thus lie beyond the power of the work force.

11. a. Eliminate work standards (quotas) on the factory floor. Substitute leadership.

    b. Eliminate management by objective. Eliminate management by numbers, or numerical goals. Substitute leadership.

12. a. Remove barriers that rob the hourly worker of his right to pride of workmanship. The responsibility of supervisors must be changed from sheer numbers to quality.

    b. Remove barriers that rob people in management and in engineering of their right to pride of workmanship. This means, *inter alia*, abolishment of the annual or merit rating and of management by objective.

13. Institute a vigorous program of education and self-improvement.

14. Put everybody in the company to work to accomplish the transformation. The transformation is everybody's job.

    **Source:** Deming, W. Edwards. *Out of the Crisis*. Cambridge, MA: MIT Press, 1982. 23-24. Print.

These 14 points apply anywhere, to small or large organizations, to the service or manufacturing industry.

The fathers of modern quality control are many and have come from various fields, but they have one common philosophy: confronting reality in manufacturing and service operations through the collection and analysis of data is the best way to improve quality.

In the 1980s, a program called **Six Sigma** was developed at Motorola after a Japanese firm took over the Motorola factory that manufactured television sets in the United States. The firm began to make drastic operational changes in response to Motorola having been consistently beaten in the television set market. With the new changes, the factory was soon producing TV sets with one-twentieth the number of defects they had produced under previous management.

> **Definition**
>
> **Six Sigma**
>
> **Six Sigma** is a program developed by Motorola in the 1980s that encourages companies to take a customer focus in order to improve business processes.

The improvement was a result of the implementation of Six Sigma methodology. Six Sigma methodology blends many of the key elements of past quality initiatives while adding its own special approach to business management. Six Sigma emphasizes the reduction in variation, a focus on doing the right things right, combining customer knowledge with the core process improvement efforts, and subsequent improvement in company sales and revenue growth. Basically, Six Sigma is driven by results and encourages companies to take a customer focus in order to improve their business processes.

Six Sigma professionals exist at every level and each has a different role to play. There are primarily four different roles in Six Sigma:

1. Yellow Belt

2. Green Belt

3. Black Belt

4. Master Black Belt

The diagram below shows the levels of each of the belts and their roles in conducting projects and implementing improvements. Practitioners can receive formal Six Sigma certifications from the American Society for Quality which recognize that an individual has demonstrated a proficiency within, and a comprehension of, a specific body of knowledge for each Belt.

**Belt Colors in Six Sigma**

 **Master Black Belt:** The highest level of Six Sigma expertise. Responsibilities involve implementation of statistical analysis, strategic and policy planning, and mentoring of Black Belts.

 **Black Belt:** A professional who has usually completed an examination and has been certified in Six Sigma methods. Responsibilities include implementation of Six Sigma methodology throughout all levels of the business, leading teams and projects, and providing training and mentoring to Green and Yellow Belts.

 **Green Belt:** A Six Sigma trained (often certified) professional who works on Six Sigma projects exclusively. In many organizations this is the entry level. Responsibilities include leading teams/projects and implementing Six Sigma methodology at the project level.

 **Yellow Belt:** This level is generally the lowest in the hierarchy, but holders are often closest to the project details and occupy a key role on the project team. They usually observe and ascertain that project details are completed in a timely manner.

The term "Lean" was introduced by John Krafcik of MIT while he was working with the International Motor Vehicle Program. During a study of Japanese companies, Krafcik observed that Toyota did "everything with half of everything": half the people, half the space, half the inventory, half the resources, etc., yet with very high quality. Thus, Krafcik called this method "Lean." Today, "Lean" is aimed at eliminating waste and streamlining business processes. Lean practitioners view waste as coming from process steps that do not add value and Six Sigma practitioners see waste as coming from variation in the process. Both approaches hold some truth; thus, the combination of the two methods has been successful. So much so, that they have been combined to be called Lean Six Sigma—eliminating waste while reducing variation in a process.

Companies using the Six Sigma methodology have seen an enhanced ability to provide value to their customers. Internally, the companies have a better understanding of their key business process and these processes have undergone process flow improvements. Improved process flow means reduced cycle times, reduction to elimination of defects, and increased capacity and productivity rates.

# 18.1  Basic Charts and Diagrams Used in Quality Control

The quality movement is not simply about using statistical tools to monitor quality. The main thrust of the quality movement is the improvement of processes. These improvements are often facilitated by the use of statistical techniques. In this section, we will discuss some of the basic charts and diagrams that are used to monitor processes.

## Flowcharts

A process cannot be improved unless everyone agrees on what the process is. One way of beginning to determine how a process should work is to use a flowchart to define how the process currently works.

One of the processes that some students are familiar with is the test preparation process.

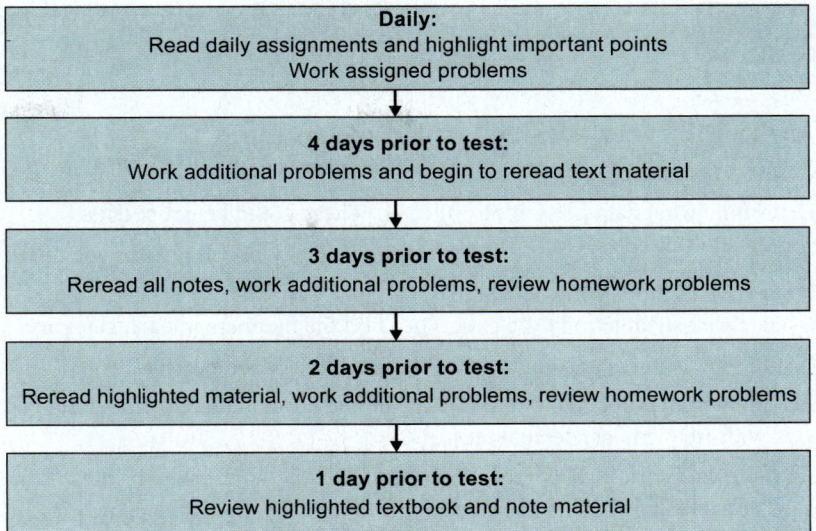

A student wanting to improve test-taking skills would define their process and improve upon it over time. There are two types of charts that aid in process improvement: Pareto charts and run charts.

## Pareto Charts

The **Pareto chart** (pronounced *pah-ray-toe*), named after Vilfredo Pareto, is a graphical tool for ranking the causes of problems from most significant to least significant. Pareto's research examined patterns of wealth and income in nineteenth-century England. He found that 20% of the population enjoyed 80% of the wealth, the top 10% of the population controlled 65% of the wealth, and the top 5% controlled 50% of the wealth. What really excited Pareto was that he saw this same pattern of imbalance repeated in different time periods and different countries.

> **Definition**
>
> **Pareto Chart**
>
> A **Pareto chart** is a bar graph with the categories ranked from largest to smallest.

What Pareto discovered with respect to wealth has also been seen in vastly different venues.

- 80% of the world's energy is consumed by 15% of the population.
- 20% of a country's population consumes 80% of the health resources.
- 20% of a company's customers account for 80% of a company's revenue.
- 80% of complaints come from 20% of customers.
- 80% of crimes are committed by 20% of criminals.

This 80/20 notion has been found to be very useful in analyzing process improvement. The notion has been deemed the **Pareto principle** (also known as the **80/20 rule** and the **law of the vital few**). Every process has problems to overcome. The causes and effects of these problems are not equally linked. If the Pareto principle holds, we would see that 20% of the causes produce 80% of the problems. Identifying the highly significant 20% of process problems should focus attention on these key areas and lead to substantial process improvement. While the split is not always 80/20, the Pareto chart is a visual method allowing us to identify which problems are most significant. The use of a Pareto chart also limits the tendency of people to focus on the most recent problems rather than on the most important problems.

> **Definition**
>
> **Pareto Principle**
>
> The **Pareto principle** is a general rule where approximately 20% of the causes produce 80% of the problems. It is used as a decision-making tool to focus attention on the key areas for process improvement.

## Procedure

### Pareto Chart

A Pareto chart is constructed using the following steps.

1. Create a list of problems that will be the subject of the chart.

2. Determine what data need to be collected. These could be count data, percentages, or costs.

3. Determine the timeframe to be used for collecting the data.

4. Use a check sheet to tally the data. Then add the numbers in each category.

5. List the items in decreasing order of the measure of comparison.

6. List the items (problems) being tracked on the horizontal axis. Label the vertical axis with frequencies, percentages, etc.

7. Analyze the chart(s). The largest bars represent the vital few problems. If there does not appear to be one or two major problems, re-check the categories to determine if another analysis is necessary.

Suppose you are a publishing company executive and you have just learned that your typesetting budget had been exceeded by 40% . By carefully studying typesetting budget overruns during the next six months, you observe the following causes.

**✎ NOTE**

The percentages given in Table 18.1.1 are approximate, as they are rounded to one decimal place.

| Table 18.1.1 – Typesetting Budget Overrun Causes | | |
|---|---|---|
| Causes of Problems | Frequency of Occurrence | Percentage (%) |
| Author late with corrections | 48 | 41.0 |
| Author late with original manuscript | 35 | 29.9 |
| Author made too many corrections | 12 | 10.3 |
| Book longer than planned | 9 | 7.7 |
| Proofreader late | 4 | 3.4 |
| Figures incorrectly done | 3 | 2.6 |
| Index compiler late | 3 | 2.6 |
| Permissions received late | 1 | 0.9 |
| Typesetter correction errors | 1 | 0.9 |
| Schedule changed by editor | 1 | 0.9 |

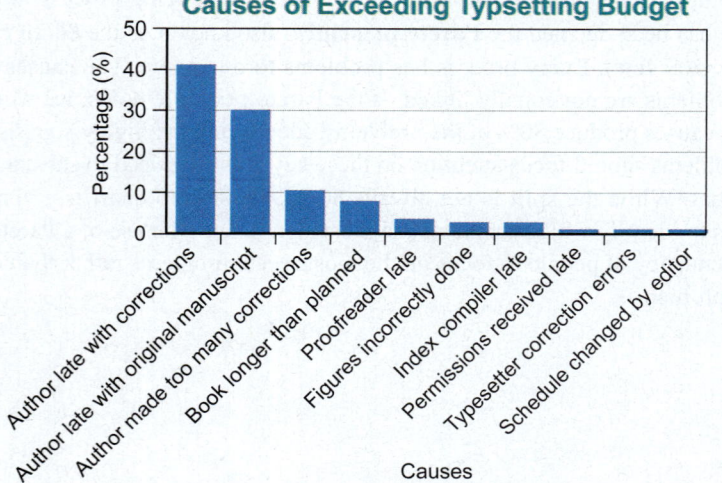

Figure 18.1.1

The Pareto chart in Figure 18.1.1 indicates that approximately 81.2% of the problems are associated with the author. The executives need to discuss issues with the author to determine if problems can be alleviated. The visual summary of these data is quite powerful and should provide a manager with an extremely sharp focus of what problems need to be attacked.

In addition to the main bar graph, the Pareto chart may also include an integrated line graph that shows the cumulative frequency (or cumulative relative frequency) of the observations at each bar on the bar graph. The line graph can be used to help determine which problems (the bars) belong to the *vital few* (the 20%) and the *trivial many* (the 80%).

## Run Charts

In Section 2.4 we discussed time series plots. In these plots, the variable of interest is graphed against a time period plotted on the horizontal axis. A **run chart** is a time series plot. The purpose of these charts in a quality control context is to monitor quality characteristics. Suppose you were manufacturing ball bearings.

You would be concerned with quality characteristics such as the following.

- The average diameter of the ball bearings

- The number of surface defects

- The average compression force measured in Newtons (crushing strength)

> **Definition**
>
> **Run Chart**
>
> A **run chart** is a form of time series plot that plots manufacturing or business process data in a time sequence.

Processes that are working acceptably produce measurable quality characteristics that are stable. When processes are not working properly, they are either becoming unstable or are unstable. To effectively manage a process (e.g., a production line manufacturing ball bearings), a manager would like to intervene when a process is becoming unstable. What would a process that is becoming unstable look like?

If you were manufacturing two-inch ball bearings, then a quality characteristic that you would be interested in is the diameter of the ball bearings being produced. Assume that each data point in the following diagrams represents the average diameter of a sample of ball bearings. Also, assume the samples of ball bearings are drawn every 15 minutes from the production line. The following run charts are graphs of the sample mean diameters for 40 samples of five ball bearings.

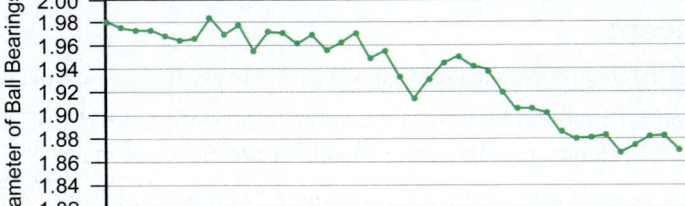

**Figure 18.1.2**

In Figure 18.1.2, the mean diameter of the ball bearings is trending downward. In Figure 18.1.3, the mean diameter of the ball bearings is trending upward. Both of these charts suggest that process output is unstable. These patterns indicate a gradual process shift that will eventually produce output that is unusable.

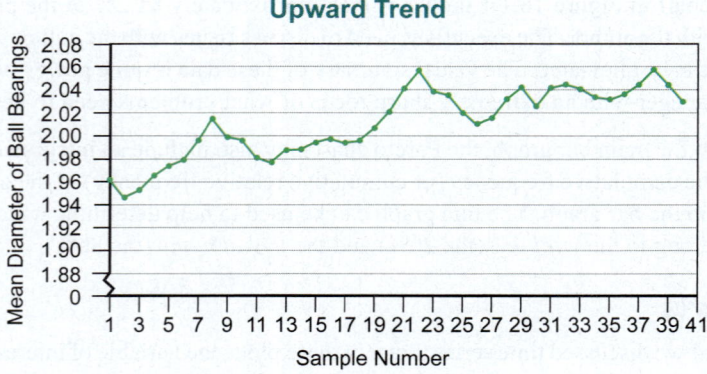

**Figure 18.1.3**

**Definition**

Cycle

A **cycle** is a systematic repeating pattern observed in a run chart.

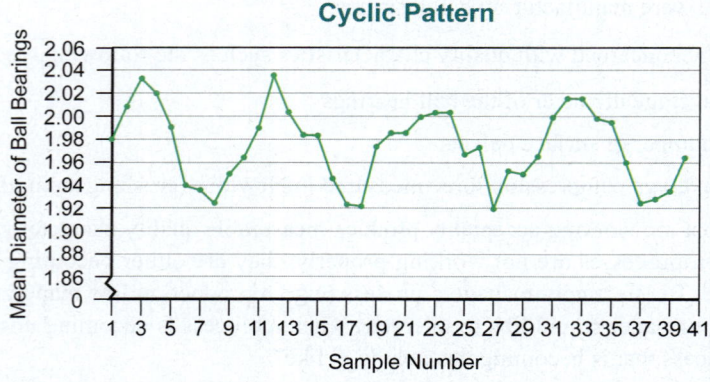

**Figure 18.1.4**

Notice how the data in the graph in Figure 18.1.4 move systematically up and down in a repeating pattern. This pattern (called a **cycle**) is also an indication of a process that is unstable.

## ✎ 18.1 Exercises

### Basic Concepts

1.  Describe the contributions made by Dr. Walter A. Shewhart in the field of quality control.

2.  What contributions did W. Edwards Deming make in the field of quality control? Of Deming's 14 Points, give five that you believe are most important.

3.  To which types of organizations do Deming's 14 Points apply?

4.  What is the common philosophy of the fathers of modern quality control?

5.  What is Six Sigma? Briefly describe the methodology behind Six Sigma.

6.  What is a flowchart? Why are flowcharts important in quality control?

7.  What is a Pareto chart? Explain how a Pareto chart can be used in the field of quality control.

8.  What is a run chart? Is a run chart the same as a time series plot? Explain.

9.  Explain how a run chart can be used in the field of quality control.

### Exercises

10. Create a flowchart for each of the following processes.

    a.  Getting ready for work in the morning

       **b.**  Getting married

       **c.**  Going on a week-long vacation in Bermuda

       **d.**  Going to a job interview

**11.**  Consider the following flowchart regarding incoming call routing for a software company.

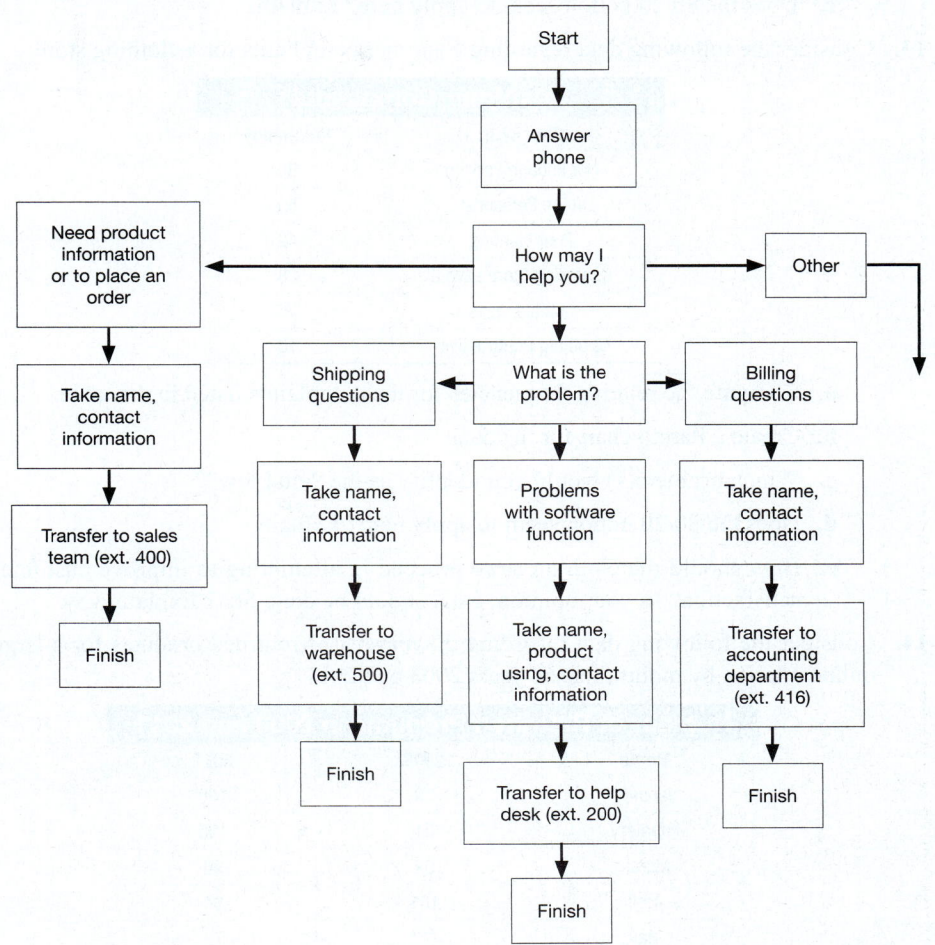

       **a.**  Explain why it is important for a new receptionist working at the company to understand this flowchart.

       **b.**  In what ways do you think this flowchart could be improved?

**12.**  Consider the following Pareto chart regarding reasons for a delay in processing credit card applications.

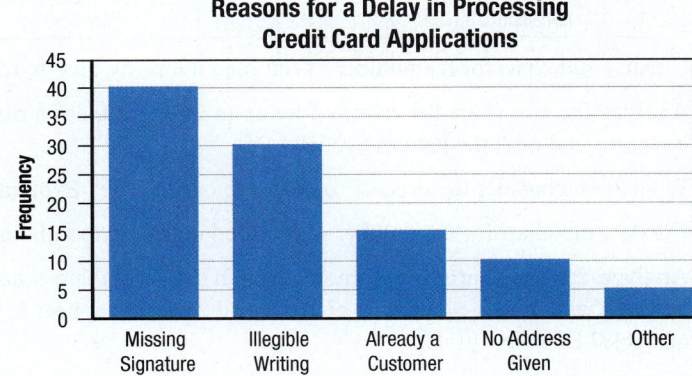

       **a.**  What percentage of delays is caused by a missing signature on the application?

**b.** What percentage of delays is caused by a missing signature or illegible writing on the application?

**c.** From the chart, what would you identify as the "vital few" problems?

**d.** How do you think the credit card company could attempt to correct these problems?

**e.** Does the 80/20 notion seem to apply here? Explain.

13. Consider the following data regarding customer complaints for a clothing store.

| Customer Complaints | |
|---|---|
| Complaint | Frequency |
| Not Enough Parking | 80 |
| Rude Personnel | 50 |
| Poor Lighting | 42 |
| Confusing Store Layout | 28 |
| Limited Sizes | 15 |
| Clothing Unattractive | 10 |

**a.** Compute the relative frequencies for the complaints listed in the table.

**b.** Create a Pareto chart for the data.

**c.** Which problem(s) would you identify as the "vital few"?

**d.** Does the 80/20 notion seem to apply here? Explain.

**e.** How should the clothing store proceed in attempting to improve customer satisfaction? In your opinion, what should be done first? Explain why.

14. Consider the following data regarding the number of returned products for a large online retailer, by month, for the years 2003 and 2011.

| Number of Returned Products in 2003 and 2011 | | |
|---|---|---|
| Month | 2003 | 2011 |
| January | 79 | 100 |
| February | 81 | 105 |
| March | 92 | 96 |
| April | 101 | 84 |
| May | 111 | 72 |
| June | 120 | 80 |
| July | 119 | 64 |
| August | 125 | 60 |
| September | 137 | 55 |
| October | 120 | 59 |
| November | 140 | 42 |
| December | 145 | 56 |

**a.** Create a run chart for the number of returned items, by month, in 2003.

**b.** Analyze the run chart for returned items in 2003. Is there a downward or upward trend or is the pattern cyclic?

**c.** Would you consider the process to be stable or unstable? Explain why.

**d.** Create a run chart for the number of returned items, by month, in 2011.

**e.** Analyze the run chart for returned items in 2011. Is there a downward or upward trend or is the pattern cyclic? Does the process appear to be stable or unstable? Explain.

**f.** Does it appear that the retailer has improved the process from 2003 to 2011? Explain.

**15.** Consider the following data regarding revenues, by quarter, for a popular local restaurant.

| Quarterly Revenues (Thousands of Dollars) | | |
|---|---|---|
| Year | Quarter | Revenue |
| 2008 | 1 | 270 |
| 2008 | 2 | 369 |
| 2008 | 3 | 468 |
| 2008 | 4 | 306 |
| 2009 | 1 | 285 |
| 2009 | 2 | 354 |
| 2009 | 3 | 525 |
| 2009 | 4 | 330 |
| 2010 | 1 | 261 |
| 2010 | 2 | 288 |
| 2010 | 3 | 375 |
| 2010 | 4 | 366 |
| 2011 | 1 | 303 |
| 2011 | 2 | 420 |
| 2011 | 3 | 471 |
| 2011 | 4 | 414 |

    **a.** Create a run chart for the quarterly revenues.

    **b.** Analyze the run chart. Is there a downward trend or an upward trend? Do revenues appear to be cyclical? Explain.

    **c.** Do you think restaurant sales depend on the time of year?

    **d.** Can you think of any reasons why there would be this type of trend for restaurant sales?

    **e.** Does this process appear to be unstable? If so, suggest ways that quality control could help the restaurant manager control the process.

# 18.2 Basic Concepts

The scientific method for attaining quality relies on two basic concepts.

1. No matter what the specifications are for a product, the process that produces the product will create output that has *variation*. (For example, suppose a manufacturer desires to produce ball bearings with a diameter of two inches. If a process is set up to produce ball bearings with a diameter of two inches, then each item, when actually measured, will show deviation from the *ideal* of two inches.)

2. Improving a process requires removing variation from it. (Though the ideal would be to remove all variation, this cannot be achieved. The goal then is to move towards the ideal. This notion is known as *continuous improvement*.)

With these two powerful ideas in mind, let's look at some important definitions.

> **Definition**
>
> ### Control Chart Terminology
> A **control chart** for a process consists of values plotted over time. This chart has an upper bound and a lower bound called the **upper control limit (UCL)** and the **lower control limit (LCL)**, respectively. The process is **out of control** when a measurement falls either above the UCL or below the LCL. The control chart also contains a **centerline** that represents the average value of the quality characteristic corresponding to the in-control state.

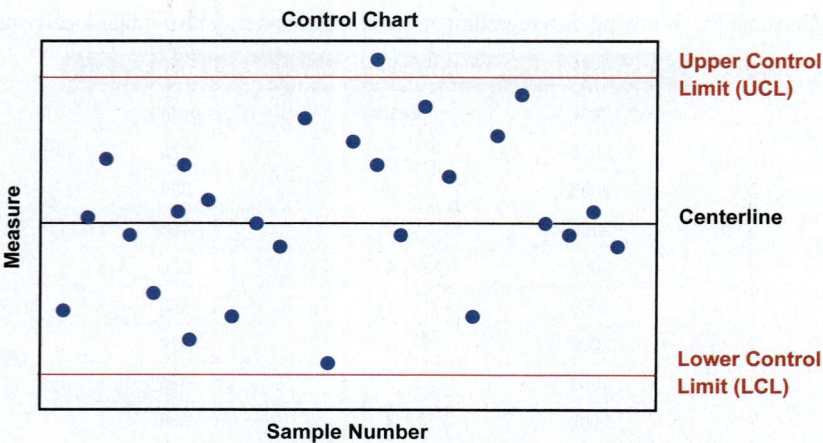

**Figure 18.2.1**

## Definition

### Types of Variation

**Normal process variation** is normal variation in a process in which the data falls within the control limits. **Assignable variation** is random variation that causes data to fall outside the control limits but can be reduced by determining the root cause of the variation.

## The Highway Control Chart

When you are driving a car and you stay in your lane, you could say that you are operating the car "in control." The white lines that define your lane are similar to the UCL and LCL. The car would be expected to move around within the lane, which would be normal process variation. Veering outside your lane might have assignable causes such as cell phone usage, children fighting in the back seat, or any number of other distractions.

The control chart with its limits tells us when to stop the process (if it is out of control) and when not to interrupt the process. When data points fall within the UCL and LCL, we think the variation is due to **normal process variation** (also called **common cause variation** or **chance variation**). But when a point or points fall outside the control limits, the cause is said to be **assignable variation** (or **special cause variation**). This type of variation is not random and can be eliminated (or reduced) by investigating the problem and determining the root cause(s). Reducing system variation is the surest path toward continuous improvement. For assignable causes, the system should be stopped and the cause(s) should be found and removed before the process is resumed.

It is important to emphasize that a control chart focuses on the process, not on the product. It does not ensure good quality, but instead allows management to check a quality characteristic of the process at regular intervals in order to determine if the statistical distribution of the characteristic has changed. If it has, then modifications may be needed to correct the process.

## ✐ 18.2 Exercises

### Basic Concepts

1.  Identify and describe the two concepts on which the scientific method for attaining quality is based on.

2.  What is a control chart? What is the basic purpose of control charts?

3.  Identify and define the three basic components of a control chart.

4.  What is normal process variation?

5.  What is assignable variation?

6.  Does a control chart give information about a process or a product? Explain.

7.  What does it mean to say that a process is in control?

8.  Does an in-control state guarantee quality output? Explain.

# Exercises

9. Consider the following control chart.

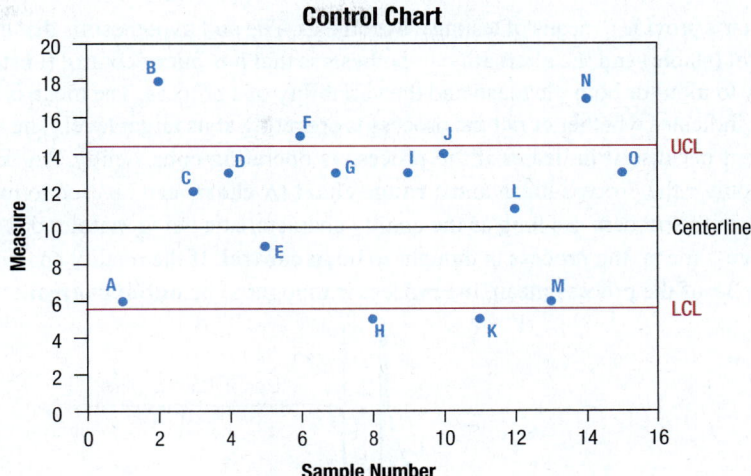

a. From the chart, estimate the values of the UCL, LCL, and centerline.

b. Interpret the estimated values of the UCL, LCL, and centerline.

c. Which points, if any, are out of control?

10. Consider the following values of the UCL, LCL, and centerline from a control chart.

13.56, 16.56, 10.56

a. Identify which value is the LCL, which is the UCL, and which is the centerline.

b. Plot the UCL, LCL, and centerline on a control chart.

c. Identify three points that would be considered in control and three points that would be considered out of control.

d. Plot these points on the chart that you made in part **b**.

11. Consider the following control chart.

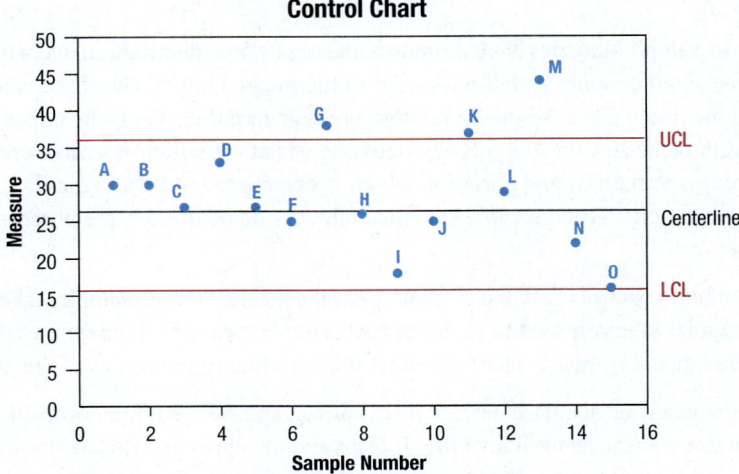

a. From the chart, estimate the values of the UCL, LCL, and centerline.

b. Interpret the estimated values of the UCL, LCL, and centerline.

c. Identify any points that can be attributed to assignable variation.

# 18.3 Monitoring with $\bar{x}$ and $R$ Charts

**Definition**

**$\bar{x}$ Chart and Range Chart ($R$ Chart)**

An $\bar{x}$ chart is a control chart used to monitor the process mean to make sure the process is operating at its target level.

A **range chart ($R$ chart)** is a control chart used to monitor the variability of a process where the variation is measured using the range of a set of observations.

Control charts provide a means of testing a hypothesis. The null hypothesis is that the process is in control (stable) and the alternative hypothesis is that it is out of control (unstable). It is customary to monitor both the mean and the variability of a process. The mean is important because it indicates whether or not the process is operating at its target level. The variability is important because it indicates if the process is operating consistently. An $\bar{x}$ chart is used to monitor the process mean and a **range chart ($R$ chart)** can be used to monitor the variability of the process. So long as the quality characteristic being measured is within $3\sigma$ of the process mean, the process is thought to be **in control**. If the quality characteristic is not within $3\sigma$ of the process mean, the process is thought to be **out of control**.

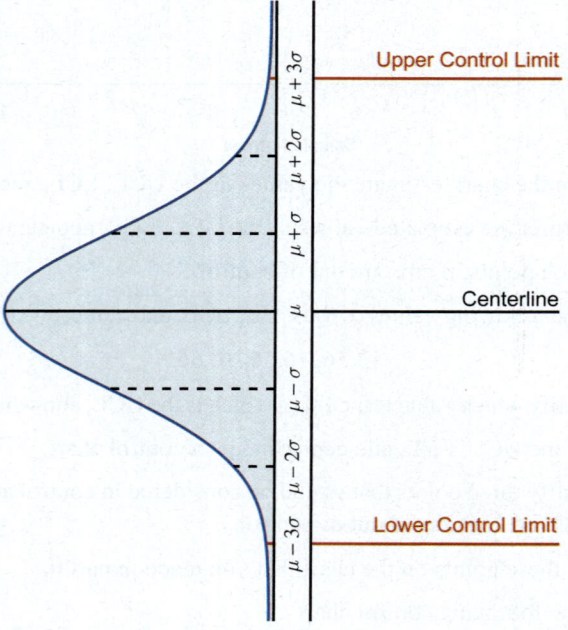

**Figure 18.3.1**

Assuming the sample statistics have an approximately normal distribution, we would expect 99.7% of the sample values to fall within $3\sigma$ of the mean. Only 0.3% of the values would fall outside the interval $\mu \pm 3\sigma$ due to random process variation. Thus, the upper and lower control limits represent the boundaries between variation which is considered random (normal process variation) and variation which is considered to be assignable (not caused by random variation). When sample statistics fall outside of $\mu \pm 3\sigma$, the process is said to be out of control.

The most common control chart, the $\bar{x}$ chart, plots the means of small samples taken from the process at regular intervals over time. Since each sample mean, $\bar{x}$, is an unbiased estimator of the process mean, $\mu$, the $\bar{x}$ chart monitors the variation in the center of the process.

If the process mean and standard deviation are known, and we assume a normal distribution, then we can use $\mu$ as the centerline of the $\bar{x}$ chart and the upper and lower control limits are given by $\text{UCL} = \mu + 3\sigma_{\bar{x}} = \mu + \dfrac{3\sigma}{\sqrt{n}}$ and $\text{LCL} = \mu - 3\sigma_{\bar{x}} = \mu - \dfrac{3\sigma}{\sqrt{n}}$, respectively. Constructing the control limits in this way ensures that about 99.7% of all values of $\bar{x}$ will lie within the control limits when the process is working properly.

**Formula**

**$3\sigma$ Control Limits for $\bar{x}$ When Standard Values of the Process Mean and Standard Deviation Are Known**

$$\text{Upper Control Limit (UCL)} = \mu + 3\sigma_{\bar{x}} = \mu + \frac{3\sigma}{\sqrt{n}}$$

$$\text{Lower Control Limit (LCL)} = \mu - 3\sigma_{\bar{x}} = \mu - \frac{3\sigma}{\sqrt{n}}$$

$$\text{Centerline} = \mu$$

where $\mu$ is the process mean, $\sigma$ is the process standard deviation, and $n$ is the number of observations in each sample.

Of course, it is rarely the case that the process mean and standard deviation are known. When $\mu$ and $\sigma$ are unknown, we estimate them based on samples taken when the process is believed to be in control. Taking at least 20 to 25 samples is recommended with a sample size of $n$. Once the data have been collected, we must calculate the two items that will be used to construct the control charts: $\bar{\bar{x}}$, the **grand mean** (think of this as the mean of the sample means), and the average range, $\bar{R}$ which is the average of the sample ranges (computed as the difference between the maximum and minimum data values in each sample). The grand mean can be used to estimate the population mean, and $\bar{R}$ can be used to estimate the population standard deviation when constructing an $\bar{x}$ chart when the process mean and standard deviation are unknown.

**Formula**

**$3\sigma$ Control Limits for $\bar{x}$ When Standard Values of the Process Mean and Standard Deviation Are Unknown**

$$UCL = \bar{\bar{x}} + A\bar{R}$$
$$LCL = \bar{\bar{x}} - A\bar{R}$$
$$\text{Centerline} = \bar{\bar{x}}$$

where

$\bar{R} =$ the mean of the sample ranges,

$\bar{\bar{x}} =$ the mean of the sample means, and

$A =$ a factor that relates $3\sigma$ to $\bar{R}$ and can be found using Table 18.3.1.

### Table 18.3.1 – $3\sigma$ Factors for Computing Control Chart Limits

| Sample Size | Mean Factor ($A$) | Lower Range ($D_3$) | Upper Range ($D_4$) |
|---|---|---|---|
| 2 | 1.880 | 0.000 | 3.267 |
| 3 | 1.023 | 0.000 | 2.574 |
| 4 | 0.729 | 0.000 | 2.282 |
| 5 | 0.577 | 0.000 | 2.114 |
| 6 | 0.483 | 0.000 | 2.004 |
| 7 | 0.419 | 0.076 | 1.924 |
| 8 | 0.373 | 0.136 | 1.864 |
| 9 | 0.337 | 0.184 | 1.816 |
| 10 | 0.308 | 0.223 | 1.777 |
| 15 | 0.223 | 0.348 | 1.652 |
| 20 | 0.180 | 0.414 | 1.586 |

### Table 18.3.1 – $3\sigma$ Factors for Computing Control Chart Limits (cont.)

| Sample Size | Mean Factor (A) | Lower Range ($D_3$) | Upper Range ($D_4$) |
|---|---|---|---|
| 25 | 0.153 | 0.459 | 1.541 |
| Over 25 | $0.75\left(\dfrac{1}{\sqrt{n}}\right)$ | $1.55 - 0.0015n$ | $0.45 + 0.001n$ |

In addition to the process mean, managers often are concerned with the variability in the samples. Even if the $\bar{x}$ chart indicates that the process is in control, the variation within the samples could be too much. To check this, managers often employ a range chart ($R$ chart). In a manner similar to the $\bar{x}$ chart, the upper and lower control limits for the $R$ chart are calculated using

$$\text{Centerline} = \bar{R} = \frac{\sum\limits_{i=1}^{m} R_i}{m}$$

$$\text{UCL}_R = \bar{R} + 3\sigma_R$$

$$\text{LCL}_R = \bar{R} - 3\sigma_R$$

where $\text{UCL}_R$ and $\text{LCL}_R$ are the upper and lower control limits of the $R$ chart, respectively. $R_i$ represents the range of the $i^{th}$ sample, $\sigma_R$ is the population standard deviation of the sample ranges, and $m$ is the number of samples. The population standard deviation of the sample ranges is often not available, so the upper and lower control limits for the $R$ chart are constructed using factors $D_3$ and $D_4$, which are also found in Table 18.3.1.

---

### Formula

**$3\sigma$ Control Chart for the Process Range**

$$\text{Upper Control Limit (UCL)} = \bar{R}D_4$$

$$\text{Lower Control Limit (LCL)} = \bar{R}D_3$$

$$\text{Centerline} = \bar{R}$$

where $\bar{R}$ is the mean of the sample ranges and $D_3$ and $D_4$ are factors which can be found in Table 18.3.1.

---

### Example 18.3.1

**Determining If a Process Is in Control Using an $\bar{x}$ Chart**

A manufacturing company uses a machine to punch out parts of a window hinge. Historically, the manufacturing process has produced hinges with a mean width of 0.45 inches with a process standard deviation of 0.022 inches. To monitor the production and to make sure the parts are acceptable for the next stage of home window assembly, a sample of three parts is taken each hour. The width of each part is then measured. As a result there are a total of eight samples with three observations in each sample. Table 18.3.2 contains the data for the widths of the hinges. Determine if the process mean is in control using an $\bar{x}$ chart. Indicate which (if any) samples are out of control.

### Table 18.3.2 – Sampling Data for Window Hinges (Inches)

| Sample Number | Observation 1 | Observation 2 | Observation 3 |
|---|---|---|---|
| 1 | 0.42 | 0.48 | 0.46 |
| 2 | 0.44 | 0.48 | 0.45 |
| 3 | 0.44 | 0.45 | 0.41 |
| 4 | 0.47 | 0.42 | 0.46 |
| 5 | 0.53 | 0.43 | 0.42 |
| 6 | 0.47 | 0.42 | 0.45 |
| 7 | 0.44 | 0.47 | 0.50 |
| 8 | 0.48 | 0.43 | 0.44 |

## SOLUTION

Control limits need to be established and plotted on the $\bar{x}$ chart. Typically, control limits for the $\bar{x}$ chart are three standard deviations above and below the process mean for the given quality characteristic.

In this problem, we have external standards to gauge our process. In particular, the process mean (0.45) and process standard deviation (0.022) are known. Thus, the control limits are determined as follows.

$$UCL = \mu + \frac{3\sigma}{\sqrt{n}}$$

$$= 0.45 + 3\left(\frac{0.022}{\sqrt{3}}\right)$$

$$\approx 0.4881$$

$$LCL = \mu - \frac{3\sigma}{\sqrt{n}}$$

$$= 0.45 - 3\left(\frac{0.022}{\sqrt{3}}\right)$$

$$\approx 0.4119$$

$$\text{Centerline} = \mu = 0.45$$

| Table 18.3.3 – Sampling Data for Window Hinges (Inches) | | | |
|---|---|---|---|
| Sample Number | Observation 1 | Observation 2 | Observation 3 | Sample Mean |
| 1 | 0.42 | 0.48 | 0.46 | 0.453 |
| 2 | 0.44 | 0.48 | 0.45 | 0.457 |
| 3 | 0.44 | 0.45 | 0.41 | 0.433 |
| 4 | 0.47 | 0.42 | 0.46 | 0.450 |
| 5 | 0.53 | 0.43 | 0.42 | 0.460 |
| 6 | 0.47 | 0.42 | 0.45 | 0.447 |
| 7 | 0.44 | 0.47 | 0.50 | 0.470 |
| 8 | 0.48 | 0.43 | 0.44 | 0.450 |

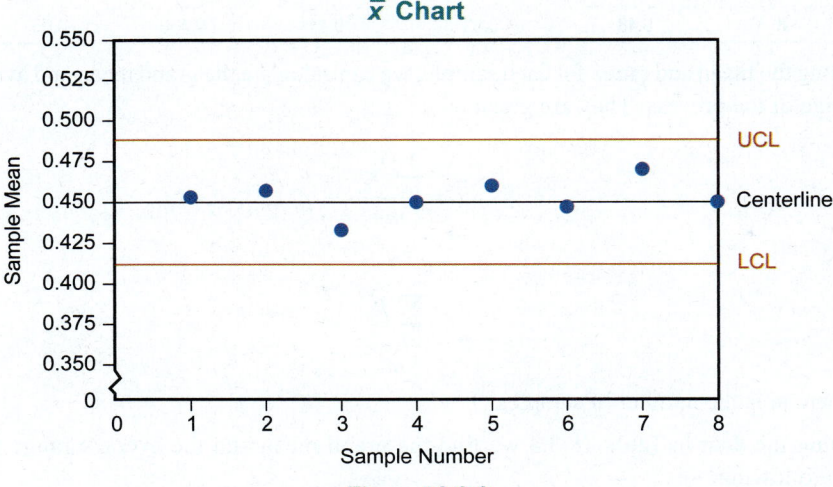

**Figure 18.3.2**

The mean for each sample is calculated and displayed in Table 18.3.3. The sample means are plotted in the Figure 18.3.2 along with the UCL, LCL, and centerline. Note that according to the $\bar{x}$ chart, since all of the sample means fall within the control limits, the process is in control.

In the previous example, it was assumed that the process mean and standard deviation were known. However, in most instances, this is not the case. In the next example, we will illustrate how to construct control charts when the process mean and variation are not known.

**Example 18.3.2**

**Constructing $\bar{x}$ and $R$ Charts for a Process**

Using the data in Example 18.3.1 and assuming that we do not know the mean or standard deviation of the process, construct the $\bar{x}$ chart and $R$ chart for this process.

**SOLUTION**

In this example, we do not know the process standard deviation. However, we are going to use another measure of variation, the sample range, to compute the UCL and LCL. Since we do not know the true process mean, we are going to estimate that mean with $\bar{\bar{x}}$, the mean of the sample means.

Before calculating the control limits for the $\bar{x}$ chart and $R$ chart, we need to perform some preliminary calculations. We need to calculate the mean and range for each sample. That is, the $i^{th}$ sample mean and range are calculated as follows.

$$\bar{x}_i = \frac{\sum\limits_{i=1}^{n} x_i}{n}$$

$R_i$ = Maximum Observation in $i^{th}$ Sample − Minimum Observation in $i^{th}$ Sample

Note that $n$ is the number of observations in each sample.

The mean and range for each sample are presented in Table 18.3.4.

| Table 18.3.4 – Sampling Data for Window Hinges (Inches) | | | | | |
|---|---|---|---|---|---|
| Sample | Observation 1 | Observation 2 | Observation 3 | Mean | Range |
| 1 | 0.42 | 0.48 | 0.46 | 0.453 | 0.06 |
| 2 | 0.44 | 0.48 | 0.45 | 0.457 | 0.04 |
| 3 | 0.44 | 0.45 | 0.41 | 0.433 | 0.04 |
| 4 | 0.47 | 0.42 | 0.46 | 0.450 | 0.05 |
| 5 | 0.53 | 0.43 | 0.42 | 0.460 | 0.11 |
| 6 | 0.47 | 0.42 | 0.45 | 0.447 | 0.05 |
| 7 | 0.44 | 0.47 | 0.50 | 0.470 | 0.06 |
| 8 | 0.48 | 0.43 | 0.44 | 0.450 | 0.05 |

Using the mean and range for each sample, we can calculate the grand mean and average range of the process. They are given by

$$\bar{\bar{x}} = \frac{\sum\limits_{i=1}^{m} \bar{x}_i}{m}$$

and

$$\bar{R} = \frac{\sum\limits_{i=1}^{m} R_i}{m}$$

where $m$ is the number of samples.

Using the data in Table 18.3.4 we find the grand mean and the average range to be the following.

$$\bar{\bar{x}} = 0.4525 \text{ (mean of the sample means)}$$
$$\bar{R} = 0.0575 \text{ (mean of the sample ranges)}$$

The 18.3.5 provides $3\sigma$ factors for computing control chart limits.

| Table 18.3.5 – $3\sigma$ Factors for Computing Control Chart Limits | | | |
|---|---|---|---|
| Sample Size | Mean Factor ($A$) | Lower Range ($D_3$) | Upper Range ($D_4$) |
| 2 | 1.880 | 0.000 | 3.267 |
| 3 | 1.023 | 0.000 | 2.574 |
| 4 | 0.729 | 0.000 | 2.282 |
| 5 | 0.577 | 0.000 | 2.114 |
| 6 | 0.483 | 0.000 | 2.004 |
| 7 | 0.419 | 0.076 | 1.924 |
| 8 | 0.373 | 0.136 | 1.864 |
| 9 | 0.337 | 0.184 | 1.816 |
| 10 | 0.308 | 0.223 | 1.777 |
| 15 | 0.223 | 0.348 | 1.652 |
| 20 | 0.180 | 0.414 | 1.586 |
| 25 | 0.153 | 0.459 | 1.541 |
| Over 25 | $0.75\left(\dfrac{1}{\sqrt{n}}\right)$ | $1.55 - 0.0015n$ | $0.45 + 0.001n$ |

For a sample size of 3, the mean factor, $A = 1.023$. Using a mean value of 0.4525 and an average range of 0.0575, the control limits for the $\bar{x}$ chart are determined as follows.

$$\text{UCL} = \bar{\bar{x}} + A\bar{R}$$
$$= 0.4525 + 1.023(0.0575)$$
$$\approx 0.5113$$
$$\text{LCL} = \bar{\bar{x}} - A\bar{R}$$
$$= 0.4525 - 1.023(0.0575)$$
$$\approx 0.3937$$
$$\text{Centerline} = \bar{\bar{x}} = 0.4525$$

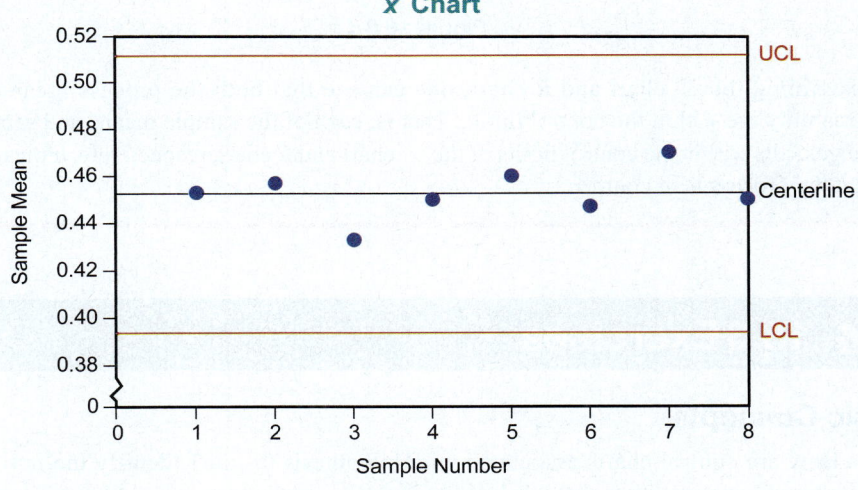

Figure 18.3.3

Computing the control limits for the $R$ chart, we have

$$\text{UCL} = \bar{R}D_4$$
$$= 0.0575(2.574)$$
$$\approx 0.1480$$
$$\text{LCL} = \bar{R}D_3$$
$$= 0.0575(0)$$
$$= 0$$
$$\text{Centerline} = \bar{R} = 0.0575.$$

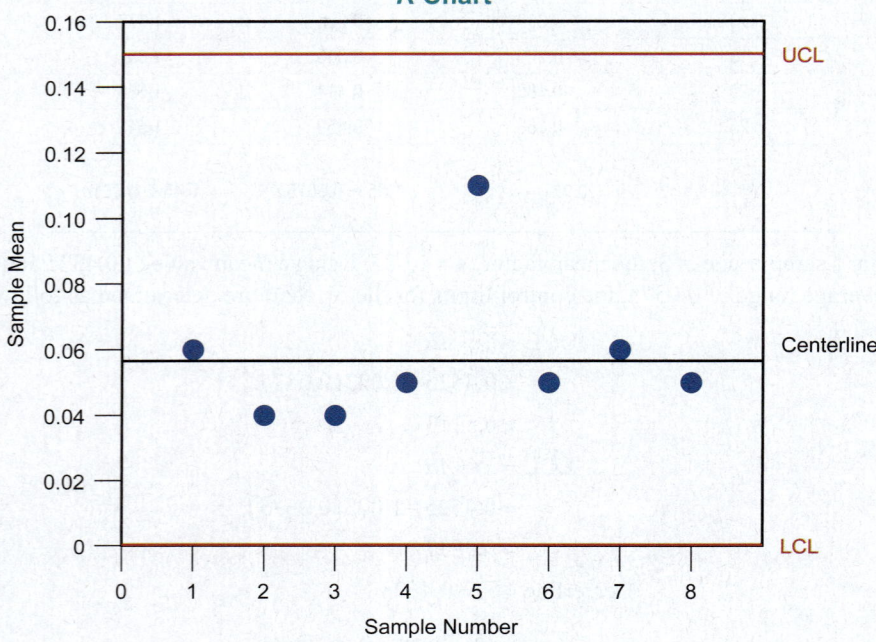

Figure 18.3.4

Examining the $\bar{x}$ chart and $R$ chart, one can see that both the process mean and variability are within the control limits. That is, each of the sample means and sample ranges falls within the control limits of the $\bar{x}$ chart and $R$ chart, respectively, indicating that the process is in control.

---

## ✎ 18.3 Exercises

### Basic Concepts

1.  How are control charts associated with hypothesis testing? Identify the null and alternative hypotheses that can be tested using control charts.

2.  What is an $\bar{x}$ chart?

3.  What is an $R$ chart?

4.  What does the Central Limit Theorem have to do with statistical quality control?

5.  What is a common interval used to measure in-control and out-of-control processes? What is the probability that random variation would cause a sample statistic to fall outside this interval?

6. How are the upper and lower control limits for an $\bar{x}$ chart calculated if the process mean and standard deviation are known?

7. How do the calculations for the control limits change for an $\bar{x}$ chart when the process mean and standard deviation are unknown?

8. When studying a control chart, how do you determine if the process is out of control?

9. Consider the process in Example 17.1. Give an example of something that might cause this process to be out of control.

10. Why do managers study $\bar{x}$ charts and $R$ charts together?

## Exercises

11. An automobile manufacturer requires the fuel injections to be adjusted so that its cars get an average of 30 miles per gallon with a standard deviation of 3 miles per gallon over a long period of time. Calculate the upper and lower control limits for an $\bar{x}$ chart if the quality control department starts sampling 64 cars each day during the month.

12. A pharmaceutical manufacturer requires the average active ingredient of allergy pills to be 0.03 grams, with a standard deviation of 0.002 grams. An FDA inspector inspects 10 batches of 100 pills and finds the following sample means.

| 0.032 | 0.028 | 0.031 | 0.032 | 0.026 | 0.027 | 0.030 | 0.033 | 0.034 | 0.026 |

   a. Determine which batches, if any, are out of control using an $\bar{x}$ chart.

   b. Does the process appear to be in statistical control? Explain.

13. A manufacturer of auto windows employs a constant quality control technique where the thickness of glass is checked every hour. A perfect piece of glass will have a thickness of 4 mm. From past experience, it is known that the standard deviation of thickness is 0.25 mm. The result of one shift's production is given in the following table.

| Glass Thickness (mm) | | | | | | | | | | | |
|---|---|---|---|---|---|---|---|---|---|---|---|
| Sample | Observations | | | | | Sample | Observations | | | | |
| 1 | 4 | 3 | 4 | 2 | 4 | 9 | 5 | 4 | 3 | 2 | 5 |
| 2 | 5 | 3 | 5 | 4 | 2 | 10 | 4 | 4 | 4 | 4 | 4 |
| 3 | 3 | 3 | 3 | 3 | 4 | 11 | 1 | 6 | 4 | 4 | 2 |
| 4 | 4 | 5 | 5 | 4 | 5 | 12 | 3 | 3 | 4 | 5 | 4 |
| 5 | 4 | 2 | 2 | 3 | 2 | 13 | 4 | 4 | 5 | 6 | 5 |
| 6 | 4 | 5 | 2 | 5 | 4 | 14 | 3 | 2 | 5 | 4 | 2 |
| 7 | 3 | 5 | 4 | 4 | 5 | 15 | 3 | 3 | 3 | 2 | 1 |
| 8 | 2 | 4 | 4 | 4 | 3 | 16 | 4 | 4 | 4 | 5 | 3 |

   a. Construct an $\bar{x}$ chart for these data.

   b. Construct an $R$ chart for these data.

   c. According to the $\bar{x}$ chart constructed in part a., which samples, if any, are out of control?

   d. According to the $R$ chart constructed in part b., which samples, if any, are out of control?

14. Princeton Manufacturing produces air conditioning units designed to maintain 45 degrees. Samples of 10 units are taken to monitor the process, and it is found that the units are maintaining 45 degrees as designed. The mean of the sample ranges is found to be 2 degrees.

   a. Find the UCL, LCL, and centerline for the $\bar{x}$ chart.

   b. Find the UCL, LCL, and centerline for the $R$ chart.

15. A trucking company tries to deliver its freight in 24 hours. Ten samples of 20 customers are taken with the following sample means.

| 22.6 | 24.5 | 24.1 | 23.8 | 25.3 | 25.0 | 23.8 | 23.6 | 23.0 | 25.2 |

The average range for these deliveries is 5.8 hours.

   a. Compute the $3\sigma$ control limits for the mean delivery time. Which samples, if any, are out of control?

   b. Compute the $3\sigma$ control limits for the process range.

16. Natural Life produces a variety of natural food products. The quality control department samples one cereal to ensure proper net weight. In the past when taking samples of 15 boxes, the average range was 0.45 ounces. Find the upper and lower control limits for an $R$ chart.

17. A paper products manufacturer makes 60-inch cores that are later cut into smaller lengths in the production of bathroom tissue. To monitor the production and to make sure the cores are acceptable for the cutting state, a sample of 25 cores is taken each hour of the day. Along with the core length, the range of the core length is recorded for each sample (as shown in the following table). Determine the upper and lower control limits for an $R$ chart and indicate which samples, if any, are out of control.

| Core Lengths | | | |
|---|---|---|---|
| Sample Number | Sample Range | Sample Number | Sample Range |
| 1 | 0.10 | 13 | 0.19 |
| 2 | 0.20 | 14 | 0.08 |
| 3 | 0.22 | 15 | 0.10 |
| 4 | 0.08 | 16 | 0.07 |
| 5 | 0.06 | 17 | 0.16 |
| 6 | 0.23 | 18 | 0.19 |
| 7 | 0.20 | 19 | 0.21 |
| 8 | 0.09 | 20 | 0.14 |
| 9 | 0.25 | 21 | 0.16 |
| 10 | 0.17 | 22 | 0.19 |
| 11 | 0.14 | 23 | 0.12 |
| 12 | 0.18 | 24 | 0.13 |

18. A manufacturer of small electric motors has a model that draws 1300 watts when working properly. To ensure conformity with this standard, samples of 20 motors are taken each hour during the day shift. Along with the average wattage, the range of wattage is recorded for each sample (as shown in the following table). Determine the upper and lower control limits for an $R$ chart and indicate which samples, if any, are out of control.

| Electric Motor Wattage | | | |
|---|---|---|---|
| Sample Number | Sample Range | Sample Number | Sample Range |
| 1 | 8.8 | 5 | 4.1 |
| 2 | 12.2 | 6 | 9.6 |
| 3 | 11.6 | 7 | 8.8 |
| 4 | 8.0 | 8 | 5.3 |

# 18.4 Monitoring with a *p* Chart

The $\bar{x}$ chart and the $R$ chart, which are control charts for variables, cannot be used when we are sampling attributes. Variables are some measurable characteristic of a product or service such as the length of a rod, diameter of a ball bearing, or ounces in a bottle of soda. Attributes, however, are characteristics associated with a product or service such as the number of leaking containers, number of scratches on a car's paint job, or number of errors on an invoice. **Attribute sampling** involves counting the defective units in a sample. With a ***p* chart**, the total number of defective units is divided by the total number of items sampled and multiplied by 100 to find the percent defective.

When there is a "standard" percent defective, or the population proportion is known, the control limits for a $3\sigma$ *p* chart are calculated as follows.

---

**Formula**

### $3\sigma$ Control Chart for *p* When the Process Proportion Is Known

$$\text{Upper Control Limit (UCL)} = p + 3\sigma_p = p + 3\sqrt{\frac{p(1-p)}{n}}$$

$$\text{Lower Control Limit (LCL)} = p - 3\sigma_p = p - 3\sqrt{\frac{p(1-p)}{n}}$$

$$\text{Centerline} = p$$

where $p$ = the standard process percent defective and $n$ = the sample size.

---

When the process proportion is unknown, the proportion is estimated using the sample data.

---

**Formula**

### $3\sigma$ Control Chart for *p* When the Process Proportion Is Unknown

$$\text{Upper Control Limit (UCL)} = \bar{p} + 3\sqrt{\frac{\bar{p}(1-\bar{p})}{n}}$$

$$\text{Lower Control Limit (LCL)} = \bar{p} - 3\sqrt{\frac{\bar{p}(1-\bar{p})}{n}}$$

$$\text{Centerline} = \bar{p}$$

where $\bar{p} = \dfrac{\text{Total number of defective units in all of the samples}}{\text{Total number of units sampled}}$ and

$n$ = the sample size.

---

**Example 18.4.1**

**Determining If a Process Is in Control Using a *p* Chart**

Silvia Garcia is monitoring the operation of a machine that makes radar components. Historically, she expects about 2% defectives and some chance variation. After studying the process, Silvia decides to construct a $3\sigma$ *p* chart. She takes 10 daily samples of 200 components. Her results are shown in Table 18.4.1. What does her *p* chart look like, and which samples (if any) are out of control?

| Table 18.4.1 – Sampling Data for Components | |
|---|---|
| Sample Number | Number Defective |
| 1 | 9 |
| 2 | 7 |
| 3 | 7 |
| 4 | 8 |
| 5 | 9 |
| 6 | 15 |
| 7 | 11 |
| 8 | 6 |
| 9 | 9 |
| 10 | 8 |

## SOLUTION

Since a "standard" percent defective has been specified in the problem, we will not have to estimate the process percent defective with the sample data. Silvia calculates the percent defective for each sample. For example, in Sample 1, the percent defective is $\frac{9}{200}(100) = 4.5\%$. The results are shown in Table 18.4.2. Using the data in Table 18.4.2 with the UCL and LCL, we construct the $p$ chart.

| Table 18.4.2 – Sampling Percent Defectives | |
|---|---|
| Sample Number | Percent Defective |
| 1 | 4.5 |
| 2 | 3.5 |
| 3 | 3.5 |
| 4 | 4.0 |
| 5 | 4.5 |
| 6 | 7.5 |
| 7 | 5.5 |
| 8 | 3.0 |
| 9 | 4.5 |
| 10 | 4.0 |

The $3\sigma$ control limits are set as follows.

$$\text{UCL} = p + 3\sqrt{\frac{p(1-p)}{n}}$$

$$= 0.02 + 3\sqrt{\frac{0.02(0.98)}{200}}$$

$$\approx 0.0497 = 4.97\%$$

$$\text{LCL} = p - 3\sqrt{\frac{p(1-p)}{n}}$$

$$= 0.02 - 3\sqrt{\frac{0.02(0.98)}{200}}$$

$$\approx -0.0097 \approx 0\%$$

✎ NOTE

Since the LCL was negative, we set LCL = 0, since proportions are never negative.

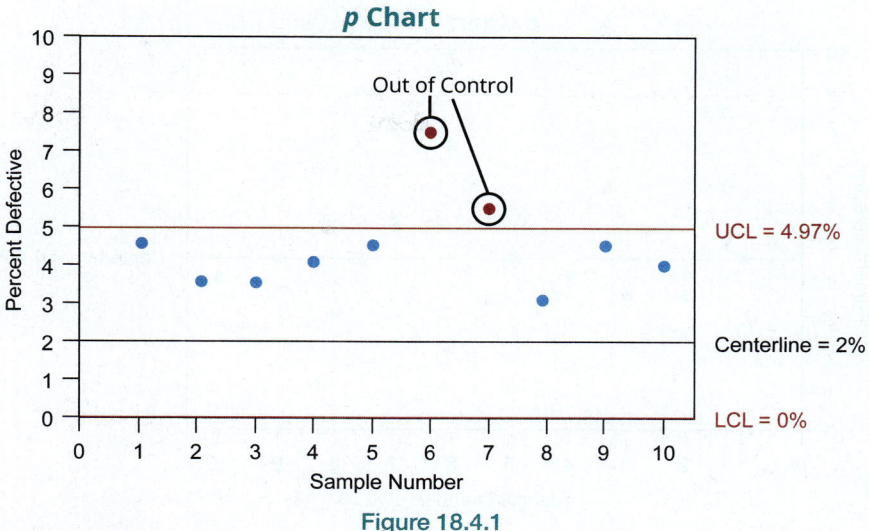

**Figure 18.4.1**

From the *p* chart, we can see that samples 6 and 7 are out of control.

---

**Example 18.4.2**

**Constructing a *p* Chart for a Process**

In Example 18.4.1, a standard process value was known. Use the sample data in Example 18.4.1 to create a $3\sigma$ control chart for *p* (*p* chart) assuming there is no standard process value.

**SOLUTION**

In many instances, the process proportion of interest is not known and has to be estimated from the data. In this example, there are 10 samples of size 200 and we wish to know the proportion of defective output the process is producing. To estimate the process proportion, we need to pool the data from all 10 samples.

$$\bar{p} = \frac{\text{Total number of defective units in all of the samples}}{\text{Total number of units sampled}}$$

$$= \frac{89}{2000}$$

$$= 0.0445$$

Using the estimate for *p* from the sample data, the control limits are calculated as follows.

$$\text{UCL} = \bar{p} + 3\sqrt{\frac{\bar{p}(1-\bar{p})}{n}}$$

$$= 0.0445 + 3\sqrt{\frac{0.0445(1-0.0445)}{200}}$$

$$\approx 0.0882 = 8.82\%$$

$$\text{LCL} = \bar{p} - 3\sqrt{\frac{\bar{p}(1-\bar{p})}{n}}$$

$$= 0.0445 - 3\sqrt{\frac{0.0445(1-0.0445)}{200}}$$

$$\approx 0.0008 = 0.08\%$$

Using the data in Table 18.4.2 and the UCL and LCL allows Silvia to draw the *p* chart and to plot each sample, as shown in Figure 18.4.2.

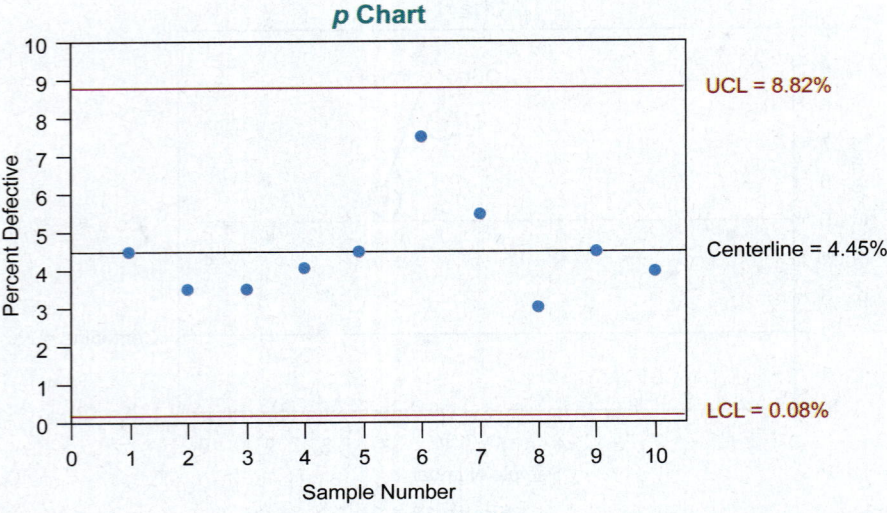

**Figure 18.4.2**

From the $p$ chart, we can see that for this data, when the standard percent defective is unknown, all of the samples fall within the control limits.

## ✏ 18.4 Exercises

### Basic Concepts

1. Explain the difference between control charts for attributes and control charts for variables.

2. What is a $p$ chart?

3. How are the upper and lower control limits calculated for a $p$ chart when the process proportion is known?

4. Is the allowable variation for a process involving a $p$ chart larger, smaller, or the same as the allowable variation for a process involving a mean chart or range chart? Explain.

5. Suppose the LCL for a process is computed to be $-0.07$. What value should be used for the LCL in the $p$ chart? Explain.

6. How does the procedure for constructing a $p$ chart change if the process proportion is not known?

7. What is $\bar{p}$? What other hypothesis testing procedure uses the concept of $\bar{p}$? Are these measures the same? Explain.

8. When examining a $p$ chart, how do you determine if samples are out of control?

### Exercises

9. In a paper products plant, 100 product samples are taken each hour and tested for being either acceptable or defective. In the past, 1% defective was considered normal. Find the upper and lower control limits for a $3\sigma$ $p$ chart.

10. To monitor the production of sheet metal screws by a particular machine in a large manufacturing company, a sample of 100 screws is examined each hour for three shifts of eight hours each. Each screw is inspected and designated as conforming or nonconforming according to specifications. Historically, the proportion of nonconforming screws has been 5%. Use the following results of one day's sampling to construct a $3\sigma$ $p$ chart. Which samples, if any, are out of control?

| Nonconforming Screws | | | | | |
|---|---|---|---|---|---|
| Sample Number | Number Defective | Sample Number | Number Defective | Sample Number | Number Defective |
| 1 | 4 | 9 | 10 | 17 | 9 |
| 2 | 7 | 10 | 5 | 18 | 11 |
| 3 | 9 | 11 | 5 | 19 | 14 |
| 4 | 10 | 12 | 4 | 20 | 5 |
| 5 | 8 | 13 | 12 | 21 | 6 |
| 6 | 6 | 14 | 6 | 22 | 12 |
| 7 | 5 | 15 | 7 | 23 | 15 |
| 8 | 1 | 16 | 13 | 24 | 5 |

11. A production process involves the manufacture of rubber gaskets for windows. When these gaskets are inspected, they are classified as conforming or nonconforming based on a number of different characteristics, such as thickness, consistency, overall size, and so on. To monitor the percentage of nonconforming gaskets being produced, a sample of 25 gaskets is inspected each hour. Management predetermines the acceptable fraction of nonconforming gaskets as 10%.

   a. Determine the UCL, LCL, and centerline.
   b. Use the following table to plot the samples on your control chart.

| Nonconforming Gaskets | | | |
|---|---|---|---|
| Sample Number | Percent Defective | Sample Number | Percent Defective |
| 1 | 16 | 13 | 12 |
| 2 | 16 | 14 | 8 |
| 3 | 16 | 15 | 8 |
| 4 | 12 | 16 | 12 |
| 5 | 8 | 17 | 8 |
| 6 | 8 | 18 | 12 |
| 7 | 4 | 19 | 12 |
| 8 | 0 | 20 | 4 |
| 9 | 8 | 21 | 8 |
| 10 | 4 | 22 | 12 |
| 11 | 4 | 23 | 16 |
| 12 | 4 | 24 | 4 |

   c. Are any samples out of control? If so, identify which ones.

12. The academic dean decides to sample 200 students each semester to study the drop rate at his institution. The numbers of drops for the last eight semesters are shown in the following table. Find the upper and lower control limits and construct a *p* chart. Indicate which semesters, if any, are out of control.

| Number of Drops | | | |
|---|---|---|---|
| Semester Number | Number of Drops | Semester Number | Number of Drops |
| 1 | 10 | 5 | 8 |
| 2 | 12 | 6 | 6 |
| 3 | 14 | 7 | 13 |
| 4 | 9 | 8 | 15 |

13. The Thompson Company makes voltage protectors at its Midland, Georgia plant. During each shift, 10 protectors are tested until failure, with some rated defective and others rated non-defective. The numbers of defective protectors for the last 20 shifts are given in the following table. Find the upper and lower control limits and construct a $p$ chart. Indicate which shifts, if any, are out of control.

| Defective Voltage Protectors | | | |
|---|---|---|---|
| Sample Number | Number Defective | Sample Number | Number Defective |
| 1 | 1 | 11 | 1 |
| 2 | 1 | 12 | 1 |
| 3 | 0 | 13 | 1 |
| 4 | 2 | 14 | 2 |
| 5 | 3 | 15 | 0 |
| 6 | 1 | 16 | 0 |
| 7 | 0 | 17 | 2 |
| 8 | 2 | 18 | 1 |
| 9 | 3 | 19 | 1 |
| 10 | 0 | 20 | 1 |

## T Discovering Technology

### Using Excel

Pareto Chart

This exercise will illustrate how to create a Pareto chart in Microsoft Excel with the integrated line graph of cumulative percentages. Note that the line graph portion of this chart is optional. Use the information from Table 18.1.1 for this exercise.

1. Enter the data from the first two columns of Table 18.1.1 into an Excel worksheet. **Causes of Problems** should be in Column A and **Frequency of Occurrence** should be in Column B.

2. Compute the percentages of each problem type from the full set of errors. In cell C1 enter the title Percentage (%). Type **Total** in cell A12. In cell C2 compute the percentage for the author being late with corrections by entering the following formula.

$$=B2/\$B\$12*100$$

Click cell C2. Click the small square at the lower-right corner of the cell and drag it down to cell C11 to replicate the formula for each row.

| | A | B | C |
|---|---|---|---|
| 1 | Causes of Problems | Frequency of Occurrence | Percentage (%) |
| 2 | Author late with corrections | 48 | 41.0 |
| 3 | Author late with original manuscript | 35 | 29.9 |
| 4 | Author made too many corrections | 12 | 10.3 |
| 5 | Book longer than planned | 9 | 7.7 |
| 6 | Proofreader late | 4 | 3.4 |
| 7 | Figures incorrectly done | 3 | 2.6 |
| 8 | Index compiler late | 3 | 2.6 |
| 9 | Permissions received late | 1 | 0.9 |
| 10 | Typesetter correction errors | 1 | 0.9 |
| 11 | Schedule changed by editor | 1 | 0.9 |
| 12 | Total | 117 | |

3. Compute the cumulative percentages. Type the title **Cumulative %** into cell D1. In cell D2, input the formula =C2. In cell D3, input the formula

$$=D2+C3$$

Click cell D3. Click the small square at the lower-right corner of the cell and drag it down to cell D11 to replicate the formula for each row.

| | A | B | C | D |
|---|---|---|---|---|
| 1 | Causes of Problems | Frequency of Occurrence | Percentage (%) | Cumulative (%) |
| 2 | Author late with corrections | 48 | 41.0 | 41.0 |
| 3 | Author late with original manuscript | 35 | 29.9 | 70.9 |
| 4 | Author made too many corrections | 12 | 10.3 | 81.2 |
| 5 | Book longer than planned | 9 | 7.7 | 88.9 |
| 6 | Proofreader late | 4 | 3.4 | 92.3 |
| 7 | Figures incorrectly done | 3 | 2.6 | 94.9 |
| 8 | Index compiler late | 3 | 2.6 | 97.4 |
| 9 | Permissions received late | 1 | 0.9 | 98.3 |
| 10 | Typesetter correction errors | 1 | 0.9 | 99.1 |
| 11 | Schedule changed by editor | 1 | 0.9 | 100.00 |
| 12 | Total | 117 | | |

4.   Highlight the data in columns A, C, and D. To do this, select cells A1:A11, hold the CTRL key, and select cells C1:D11 so that all three columns are selected.

5.   Insert a chart. Under the **Insert** tab, select the **Insert Combo Chart**, and then choose **Clustered Column – Line**.

6.   Customize the chart with labels. Click the chart title and replace it with **Pareto Chart**. Right-click the numbers on the vertical axis, choose **Format Axis…**, and input **100** for the maximum bound.

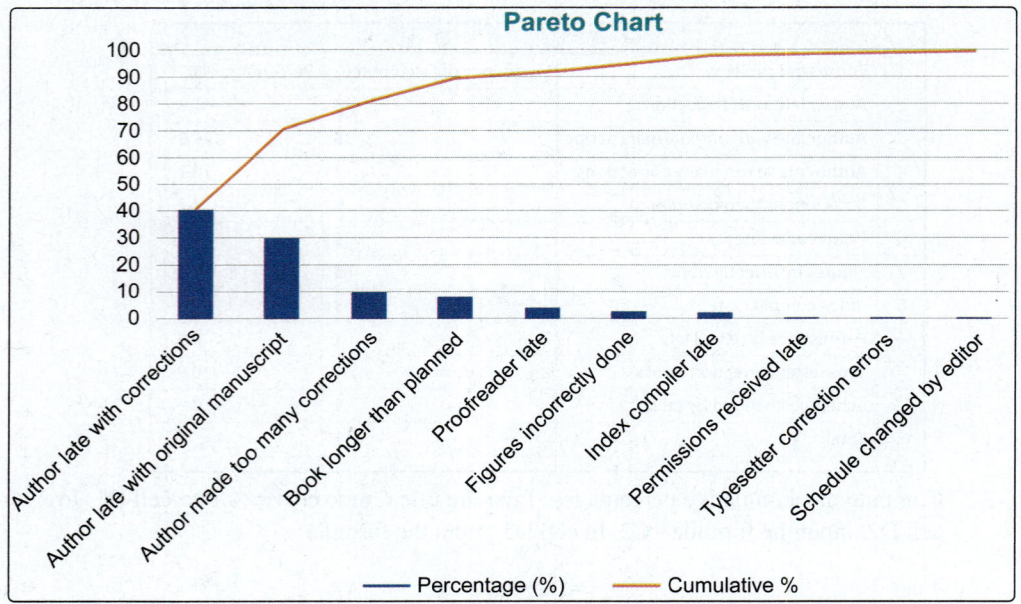

## R Chart

A steel manufacturer has decided to use an R chart to monitor the variability of their 52-pound steel rods. The shift manager randomly samples 6 rods at 14 successive time periods and checks the weight. The data below contain the results for each sample.

| ◢ | A | B | C | D | E | F | G | H |
|---|---|---|---|---|---|---|---|---|
| 1 | | | | | Observation | | | |
| 2 | Period | 1 | 2 | 3 | 4 | 5 | 6 | Sample Range |
| 3 | 1 | 51.99 | 51.99 | 51.97 | 52.01 | 52.03 | 52.05 | 0.08 |
| 4 | 2 | 52.04 | 52.00 | 51.98 | 52.04 | 51.99 | 52.05 | 0.07 |
| 5 | 3 | 52.02 | 51.96 | 52.01 | 52.00 | 52.01 | 52.01 | 0.06 |
| 6 | 4 | 52.01 | 52.03 | 51.95 | 52.04 | 51.99 | 52.03 | 0.09 |
| 7 | 5 | 51.98 | 52.02 | 51.95 | 51.97 | 52.04 | 51.97 | 0.09 |
| 8 | 6 | 52.03 | 52.00 | 52.01 | 52.05 | 52.04 | 52.04 | 0.05 |
| 9 | 7 | 51.99 | 52.02 | 51.95 | 51.96 | 52.04 | 52.00 | 0.09 |
| 10 | 8 | 51.95 | 52.05 | 51.98 | 52.04 | 52.01 | 51.98 | 0.10 |
| 11 | 9 | 51.96 | 52.03 | 51.97 | 52.03 | 52.04 | 52.01 | 0.08 |
| 12 | 10 | 51.98 | 52.00 | 51.97 | 51.99 | 52.02 | 51.98 | 0.05 |
| 13 | 11 | 51.99 | 52.00 | 51.99 | 51.97 | 52.01 | 52.05 | 0.08 |
| 14 | 12 | 51.95 | 52.05 | 51.99 | 51.95 | 52.01 | 52.04 | 0.10 |
| 15 | 13 | 52.02 | 51.97 | 51.96 | 51.95 | 52.00 | 52.03 | 0.08 |
| 16 | 14 | 51.96 | 52.02 | 51.96 | 52.03 | 52.00 | 52.05 | 0.09 |

Construct a worksheet like the one shown above.

1. Determine the centerline of the $R$ chart. To determine the centerline we need to average the values in the sample range column. In cell H17 enter the formula for $\bar{R}$.

**=AVERAGE(H3:H16)**

and press **Enter**. The result is 0.0793 rounded to four decimal places. This is the centerline of the control chart.

2. Determine the upper control limit. Use the factor table found in Table 18.2.1 to find the upper range $(D_4)$. According to the table, the value for a sample size of six is 2.004. In cell H18 enter the formula for the upper control limit:

**=H17*2.004**

and press **Enter**. The result is 0.1589 rounded to four decimal places. This is the upper control limit.

3. Determine the lower control limit. Use the factor table found in Table 18.2.1 to find the lower range $(D_3)$. According to the table, the value for a sample size of six is 0.000. In cell H19 enter the formula for the lower control limit:

**=H17*0**

and press **Enter**. The result is 0. This is the lower control limit.

| ▲ | A | B | C | D | E | F | G | H |
|---|---|---|---|---|---|---|---|---|
| 1 | | | | | Observation | | | |
| 2 | Period | 1 | 2 | 3 | 4 | 5 | 6 | Sample Range |
| 3 | 1 | 51.99 | 51.99 | 51.97 | 52.01 | 52.03 | 52.05 | 0.08 |
| 4 | 2 | 52.04 | 52.00 | 51.98 | 52.04 | 51.99 | 52.05 | 0.07 |
| 5 | 3 | 52.02 | 51.96 | 52.01 | 52.00 | 52.01 | 52.01 | 0.06 |
| 6 | 4 | 52.01 | 52.03 | 51.95 | 52.04 | 51.99 | 52.03 | 0.09 |
| 7 | 5 | 51.98 | 52.02 | 51.95 | 51.97 | 52.04 | 51.97 | 0.09 |
| 8 | 6 | 52.03 | 52.00 | 52.01 | 52.05 | 52.04 | 52.04 | 0.05 |
| 9 | 7 | 51.99 | 52.02 | 51.95 | 51.96 | 52.04 | 52.00 | 0.09 |
| 10 | 8 | 51.95 | 52.05 | 51.98 | 52.04 | 52.01 | 51.98 | 0.10 |
| 11 | 9 | 51.96 | 52.03 | 51.97 | 52.03 | 52.04 | 52.01 | 0.08 |
| 12 | 10 | 51.98 | 52.00 | 51.97 | 51.99 | 52.02 | 51.98 | 0.05 |
| 13 | 11 | 51.99 | 52.00 | 51.99 | 51.97 | 52.01 | 52.05 | 0.08 |
| 14 | 12 | 51.95 | 52.05 | 51.99 | 51.95 | 52.01 | 52.04 | 0.10 |
| 15 | 13 | 52.02 | 51.97 | 51.96 | 51.95 | 52.00 | 52.03 | 0.08 |
| 16 | 14 | 51.96 | 52.02 | 51.96 | 52.03 | 52.00 | 52.05 | 0.09 |
| 17 | | | | | | | R-bar | 0.079285714 |
| 18 | | | | | | | UCL | 0.158888571 |
| 19 | | | | | | | LCL | 0 |

4.    Create a scatterplot of the data. Highlight the data points in Column H (H3:H16). Under the **Insert** tab, choose **Scatter**, and **Scatter with only Markers**. A scatterplot is generated. Under the **Layout** tab, choose **Chart Title**, and **Above Chart**. Make the title of the chart **Control Chart**. Next, choose **Legend**, and **None** to turn off the legend.

5.    Select the $y$-axis of the control chart, right click, and select **Format Axis**. Using this dialog box, we will change the scale of the $y$-axis so that the upper and lower control limits can be displayed on the chart. In the "Axis Options" window, select the radio buttons next to "Fixed" for maximum value and enter **0.18**. Press **Close**.

6.    Next we need to draw the centerline and upper and lower control limits. Under the **Insert** tab, choose **Shapes**, and a line with no arrow. Draw a line in the chart area which intersects the $y$-axis at 0.1589 and extends to the right border of the graph. (Note: holding the shift key down as you draw the line ensures that the line is horizontal.) Once the line is drawn, under the **Format** tab, choose **Shape Outline**, **Weight**, and **1½ pt**. This is the upper control limit. With this line selected, press **Ctrl+C** to copy the line and then **Ctrl+V** to paste it. Now you have a copy of the line.

7.    Drag this copy of the line so that it intersects with 0.0793 on the $y$-axis. This is the centerline. Make another copy of the line and drag it so that it intersects the $y$-axis at 0. This is the lower control limit.

8.    Now we need to make labels for the lines. Under the **Insert** tab, choose **Shapes**, and **Text Box**. Draw a text box on the right-hand side of the graphing area just above the upper control limit. In the text box, label the line **UCL**. Make a copy of this text box by using **Ctrl+C** and paste it using **Ctrl+V**. Write **Centerline** in this text box and place it above the centerline of the control chart. Paste another copy of the text box using **Ctrl+V** and label the lower control limit **LCL**. For each of the text boxes, be sure that under the **Format** tab you choose **No Fill** for **Shape Fill** and **No Outline** for **Shape Outline**. You now have an $R$ chart. Notice that the process appears to be in control since all observations fall between the UCL and the LCL.

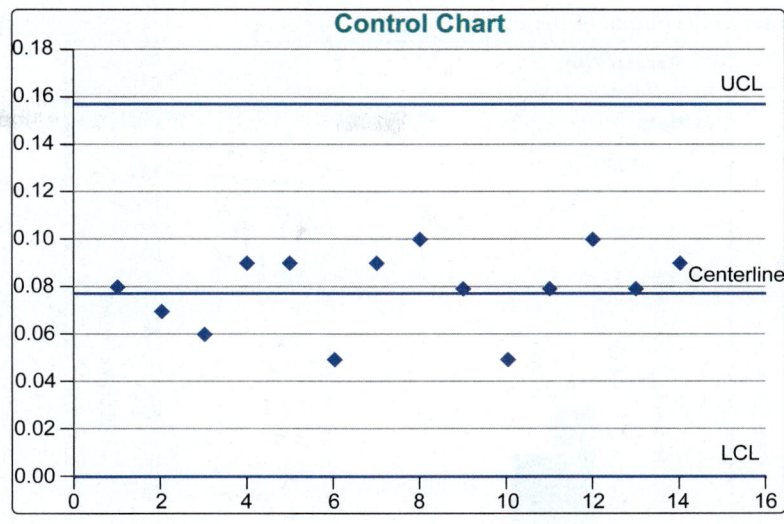

## Using JMP

## Pareto Chart

For this exercise, use the data from Table 18.1.1.

1. With a JMP data table open, enter the data from Table 18.1.1 into **Columns 1-3**.

| | Causes of Problems | Frequency of Occurrence | Percentage (%) |
|---|---|---|---|
| 1 | Author late with corrections | 48 | 41 |
| 2 | Author late with original manuscript | 35 | 29.9 |
| 3 | Author made too many corrections | 12 | 10.3 |
| 4 | Book longer than planned | 9 | 7.7 |
| 5 | Proofreader late | 4 | 3.4 |
| 6 | Figures incorrectly done | 3 | 2.6 |
| 7 | Index compiler late | 3 | 2.6 |
| 8 | Permissions received late | 1 | 0.9 |
| 9 | Typesetter correction errors | 1 | 0.9 |
| 10 | Schedule changed by editor | 1 | 0.9 |

2. Select **Analyze** in the top row of the JMP output table, then select **Quality and Process** and then **Pareto Plot**. From the **Select Columns** box, click on **Causes of Problems**, then click on **Y, Cause**. Click on **Frequency of Occurrence** and then click on **Freq**. Click **OK**.

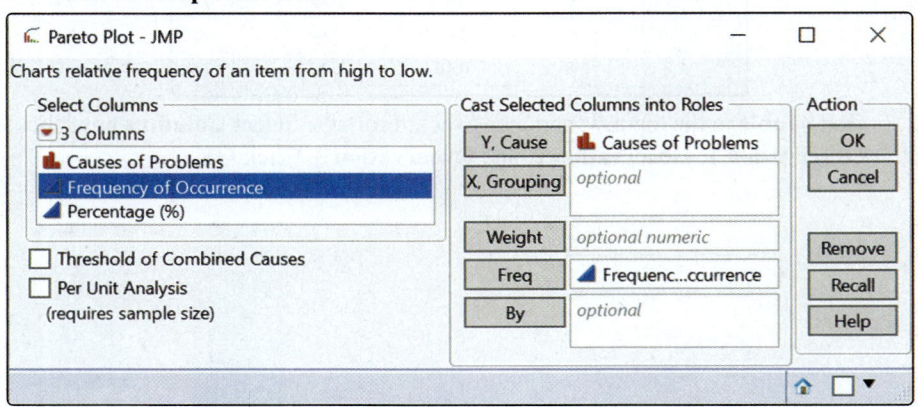

**3.** Observe the output of the Pareto chart.

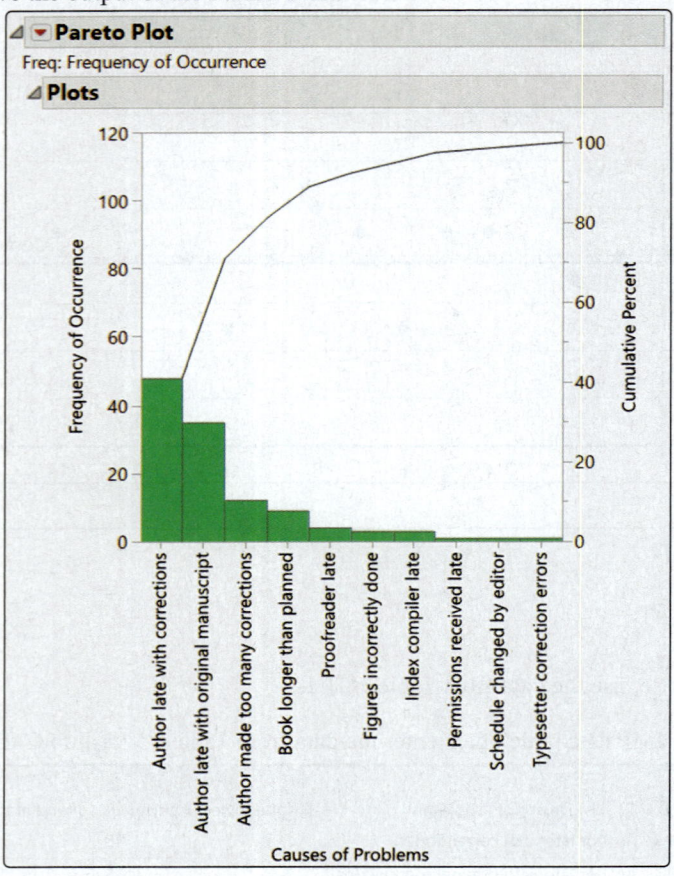

## x̄ Chart

For this exercise, use the data from Example 18.3.1.

**1.** With a JMP data table open, enter the data from Table 18.3.2 into **Columns 1-4**.

| | Sample | Observation 1 | Observation 2 | Observation 3 |
|---|---|---|---|---|
| 1 | 1 | 0.42 | 0.48 | 0.46 |
| 2 | 2 | 0.44 | 0.48 | 0.45 |
| 3 | 3 | 0.44 | 0.45 | 0.41 |
| 4 | 4 | 0.47 | 0.42 | 0.46 |
| 5 | 5 | 0.53 | 0.43 | 0.42 |
| 6 | 6 | 0.47 | 0.42 | 0.45 |
| 7 | 7 | 0.44 | 0.47 | 0.5 |
| 8 | 8 | 0.48 | 0.43 | 0.44 |

**2.** Go to **Table** in the top row, and then **Stack**. From the **Select Columns** box, click on **Observation 1**, **Observation 2**, and **Observation 3**. Click **OK**.

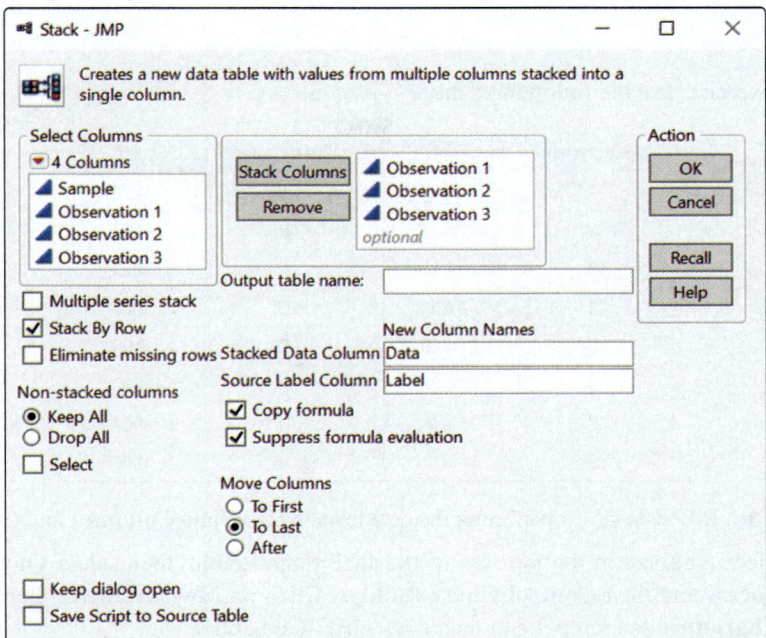

3. Select **Analyze** in the top row of the JMP output table, then select **Quality and Process**, then **Control Chart**, and then **XBar Control Chart**. From the **Select Columns** box, click on **Sample**, then click on **Subgroup**. Click on **Data** and then click on **Y**. Click **OK**.

4. Observe the output of the $\bar{x}$ chart.

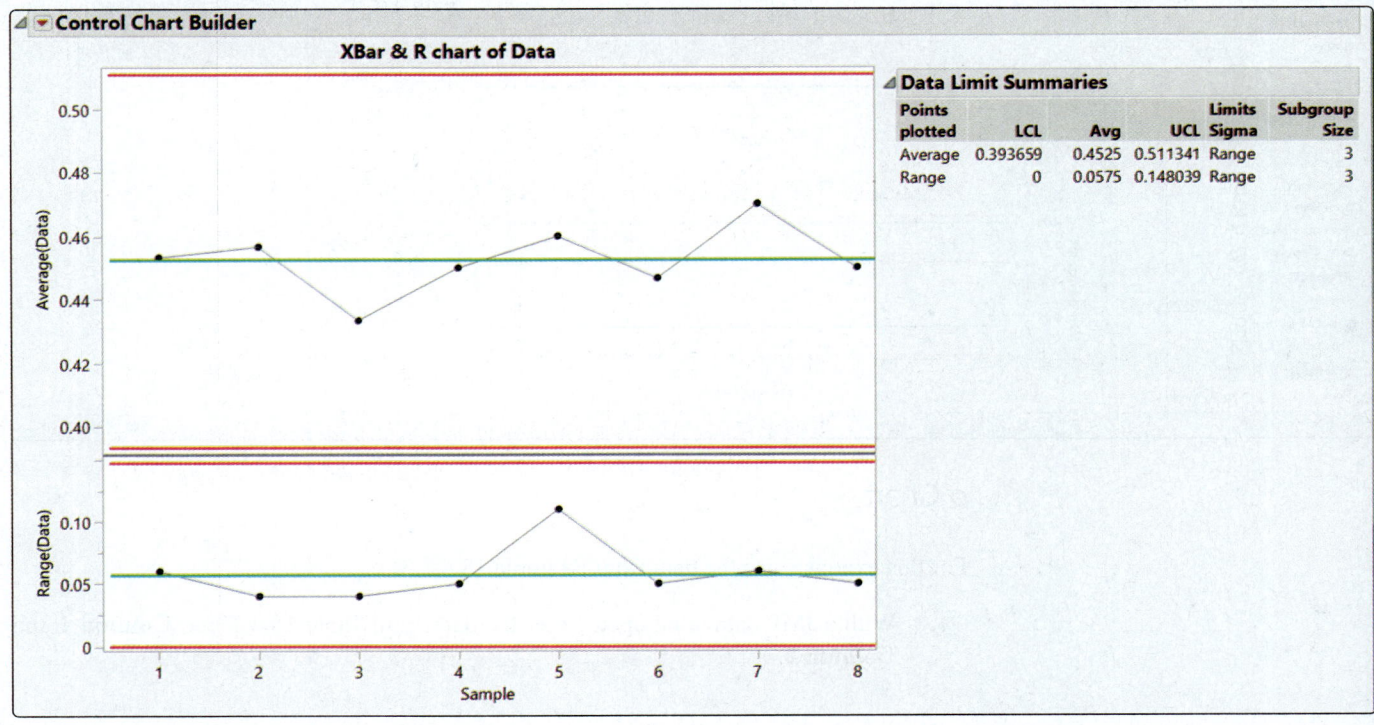

## R Chart

For this exercise, use the following data:

| Measurement | pH | Measurement | pH |
|---|---|---|---|
| 1 | 4.24 | 9 | 4.23 |
| 2 | 4.23 | 10 | 4.23 |
| 3 | 4.25 | 11 | 4.23 |
| 4 | 4.22 | 12 | 4.23 |
| 5 | 4.09 | 13 | 4.06 |
| 6 | 4.08 | 14 | 4.09 |
| 7 | 4.1 | 15 | 4.07 |
| 8 | 4.1 | 16 | 4.08 |

1. With a JMP data table open, enter the data from the table into **Column 1** and **Column 2**.

2. Select **Analyze** in the top row of the JMP output table, then select **Quality and Process** and then **Control Chart Builder**. Click on **Measurement**, then click on **Subgroup**. Click on **pH** and then click on **Y**. Click **OK**.

3. Observe the output of the *R* chart.

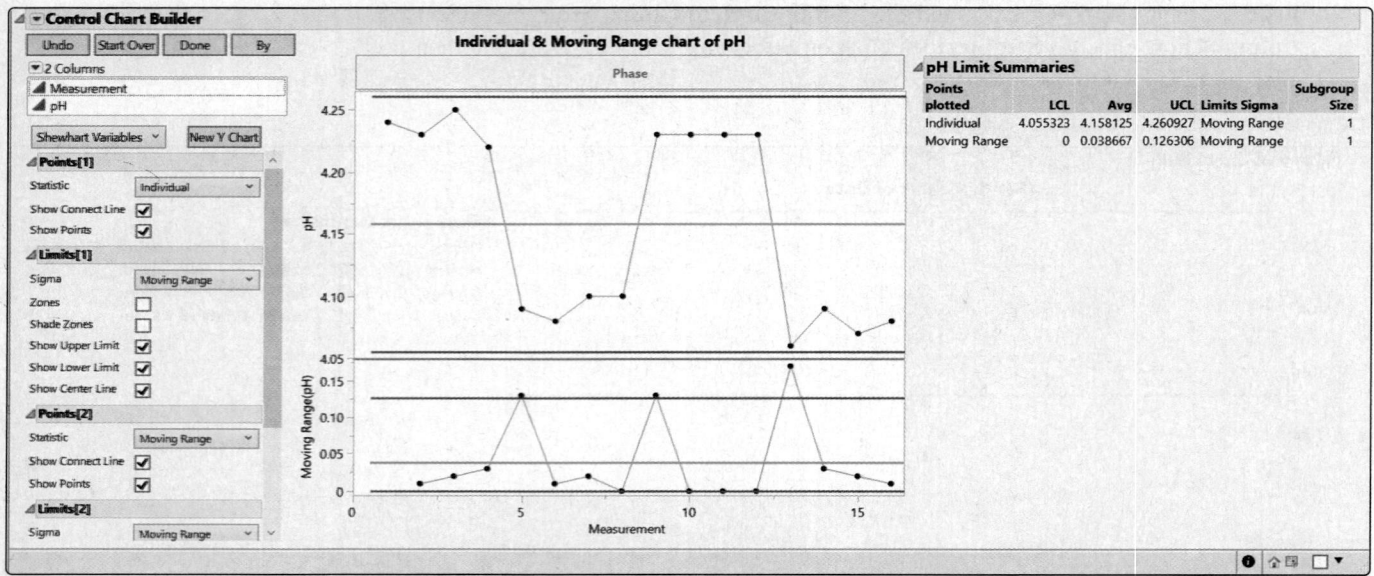

## p Chart

For this exercise, use the data from Example 18.4.1.

1. With a JMP data table open, enter the data from Table 18.4.1 into **Column 1** and **Column 2**.

| Sample Number | Number Defective |
|---|---|
| 1 | 9 |
| 2 | 7 |
| 3 | 7 |
| 4 | 8 |
| 5 | 9 |
| 6 | 15 |
| 7 | 11 |
| 8 | 6 |
| 9 | 9 |
| 10 | 8 |

2. Select **Analyze** in the top row of the JMP output table, then select **Quality and Process**, then **Control Chart** and then **P Control Chart**. From the **Select Columns** box, click on **Sample Number**, then click on **Subgroup**. Click on **Number Defective** and then click on **Y**. Click **OK**.

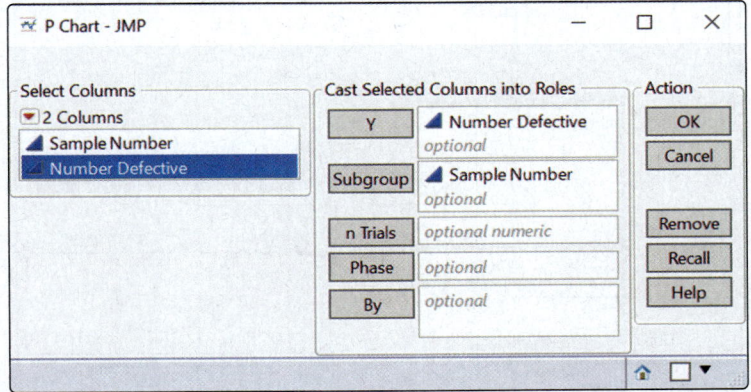

3. Observe the output of the *p* chart. On the bottom right, under **Warnings**, enter **200** for the **n Trials** and press **Enter**. Right click on the graph, click on **Limits**, then **Set Control Limits** and enter **0.02** for **Avg**.

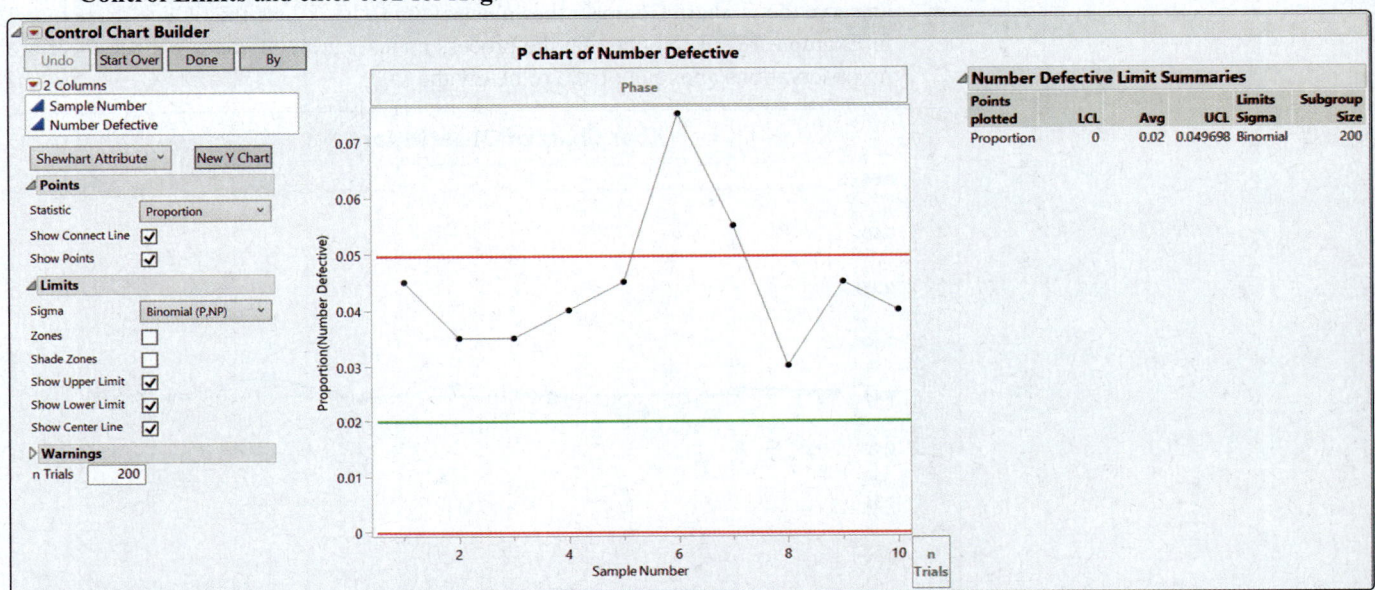

## Using Minitab

### $\bar{x}$ Chart

Use the information from Example 18.3.1 for this exercise.

1.  Enter the data from Table 18.3.2 into columns C1 through C4. Sample should be in column C1, Observation 1 should be in column C2, Observation 2 should be in column C3, Observation 3 should be in column C4.

2.  Choose **Stat**, **Control Charts**, **Variables Charts for Subgroups**, and **Xbar**. In the 'Xbar Chart' dialog box, be sure the option **Observation for a subgroup are in one row of columns** is selected from the drop down menu at the top. Note that you could also make the chart if all of the observations were in a single column. Select **C2-C4** as the columns to be charted.

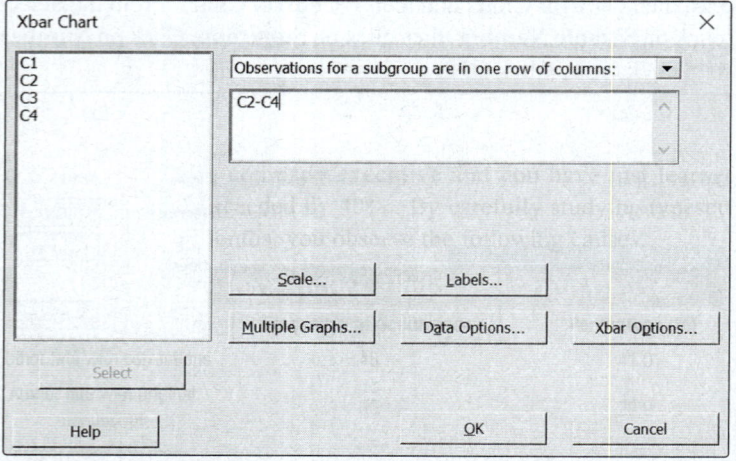

3.  Choose **Xbar Options**. Enter **0.45** for the "**Mean**" and **0.11** for the "**Standard deviation**". Press **OK**. Click on **Labels** and enter **Xbar Chart of Observations 1-3**. Press **OK**. Press **OK** again.

4.  Observe the $\bar{x}$ chart. Compare the values of the UCL, $\bar{\bar{x}}$, and the LCL to those found in Example 18.3.1. Notice that the process appears to be in control, since there are no observations above the UCL or below the LCL.

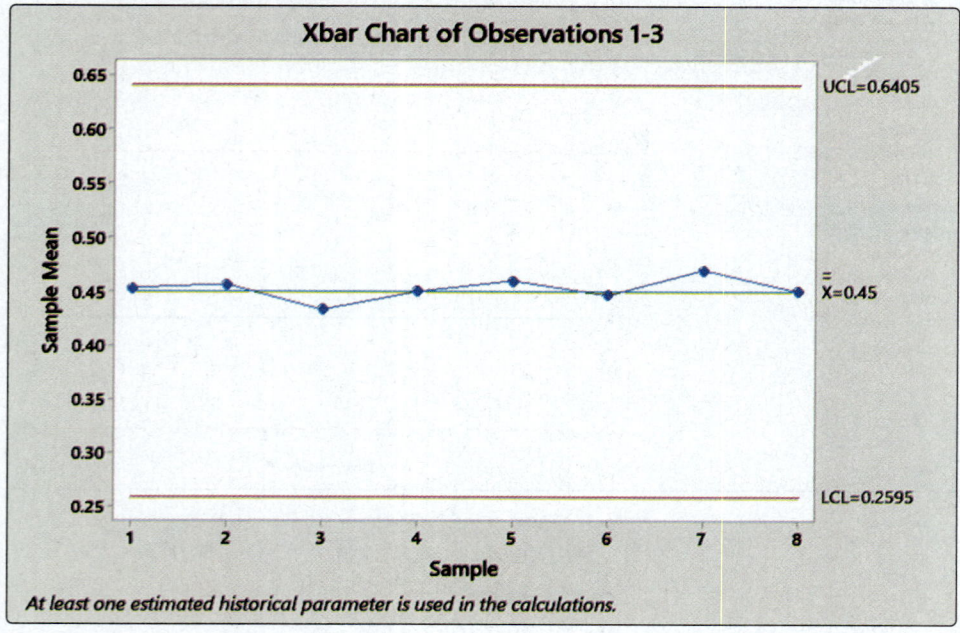

*At least one estimated historical parameter is used in the calculations.*

## *p* Chart

Use the information from Example 18.4.1 for this exercise.

1. Enter the data from the table into columns **C1** and **C2**. The **Sample Number** should be in column **C1** and the **Number Defective** should be in column **C2**.

2. Choose **Stat**, **Control Charts**, **Attributes Charts**, and **P**.

3. In the **P Chart** dialog box, select **C2** for Variable and enter **200** as the Subgroup size.

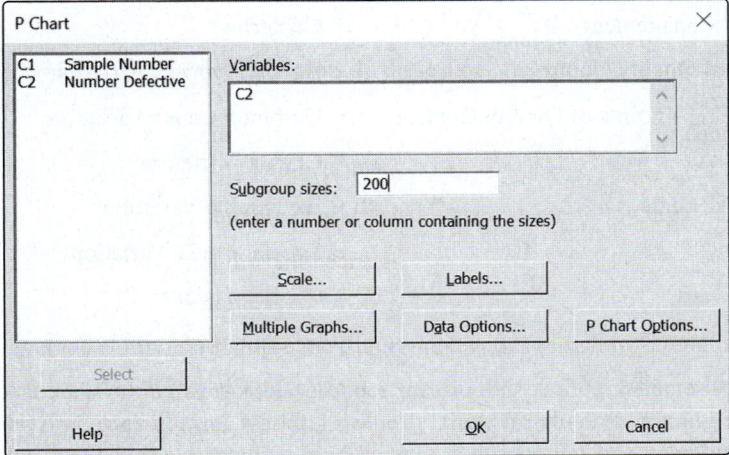

4. Click on the **P Chart Options** button. Enter **0.02** for the proportion since we are assuming the population proportion is known. Press **OK**. Click on **Labels** and enter **P Chart of Proportion Defective** for the Title. Press **OK**. Press **OK** again.

5. Observe the *p* chart. Compare the values for the UCL and the LCL to those which we found in Example 18.4.1. Notice that there are two points marked in red. These two samples are out of control.

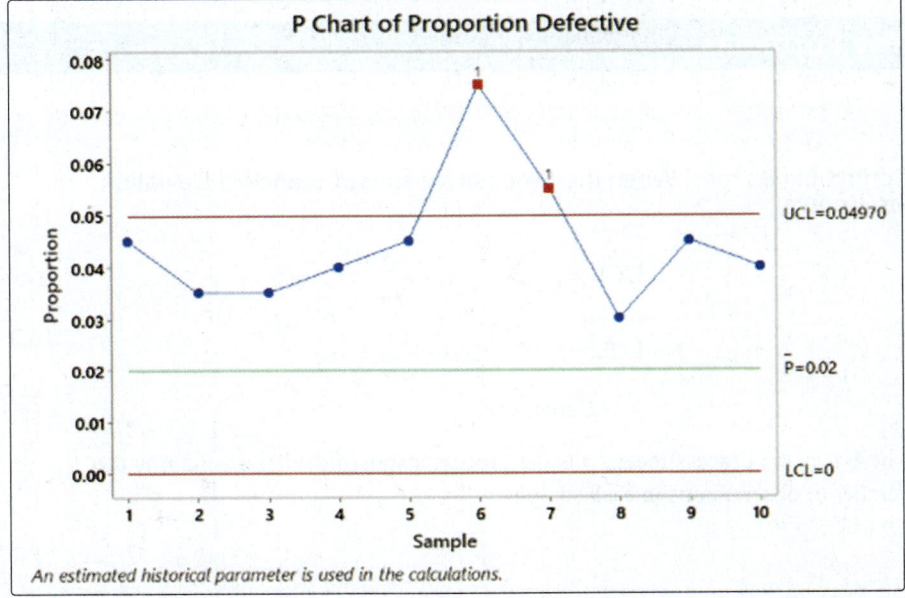

## R   Chapter 18 Review

### Key Terms and Ideas

- Quality Control
- Total Quality Control
- Total Quality Management
- Quality Management
- Statistical Quality Control
- Deming's 14 Points of Quality Control
- Six Sigma
- Lean Six Sigma
- Flowchart
- Pareto Chart
- Pareto Principle
- 80/20 Rule
- Vital Few
- Trivial Many
- Run Chart
- Cycle
- Variation
- Continuous Improvement

- Control Chart
- Upper Control Limit (UCL)
- Lower Control Limit (LCL)
- Centerline
- Normal Process Variation
- Common Cause Variation
- Chance Variation
- Assignable Variation
- Special Cause Variation
- Quality Theory
- $\bar{x}$ Chart
- Range Chart ($R$ Chart)
- In Control State
- Out of Control State
- Grand Mean $\left( \bar{\bar{x}} \right)$
- Attribute Sampling
- $p$ Chart

### Key Formulas

Section

Control Limits for $\bar{x}$ When the Process Mean and Standard Deviation Are Known

$$\text{UCL} = \mu + 3\sigma_{\bar{x}} = \mu + \frac{3\sigma}{\sqrt{n}}$$

$$\text{LCL} = \mu - 3\sigma_{\bar{x}} = \mu - \frac{3\sigma}{\sqrt{n}}$$

$$\text{Centerline} = \mu$$

18.3

where $\mu$ is the process mean, $\sigma$ is the process standard deviation, and $n$ is the number of observations in each sample.

## Key Formulas (cont.)

|  | Section |
|---|---|

### Control Limits for $\bar{x}$ When the Process Mean and Standard Deviation Are Unknown

$$UCL = \bar{\bar{x}} + A\bar{R}$$
$$LCL = \bar{\bar{x}} - A\bar{R}$$
$$\text{Centerline} = \bar{\bar{x}}$$

18.3

where

$\bar{R}$ = the mean of the sample ranges,

$\bar{\bar{x}}$ = the mean of the sample means, and

$A$ = a factor that relates $3\sigma$ to $\bar{R}$.

### Control Limits for the Process Range

$$\text{Upper Control Limit (UCL)} = \bar{R}D_4$$
$$\text{Lower Control Limit (LCL)} = \bar{R}D_3$$
$$\text{Centerline} = \bar{R}$$

18.3

where $\bar{R}$ is the mean of the sample ranges and $D_3$ and $D_4$ are relational factors.

### Control Limits for $p$ When the Process Proportion Is Known

$$UCL = p + 3\sigma_p = p + 3\sqrt{\frac{p(1-p)}{n}}$$

$$LCL = p - 3\sigma_p = p - 3\sqrt{\frac{p(1-p)}{n}}$$

18.4

$$\text{Centerline} = p$$

where $p$ is the standard process percent defective and $n$ is the sample size.

### Control Limits for $p$ When the Process Proportion Is Unknown

$$\text{Upper Control Limit (UCL)} = \bar{p} + 3\sqrt{\frac{\bar{p}(1-\bar{p})}{n}}$$

$$\text{Lower Control Limit (LCL)} = \bar{p} - 3\sqrt{\frac{\bar{p}(1-\bar{p})}{n}}$$

18.4

$$\text{Centerline} = \bar{p}$$

where $\bar{p} = \dfrac{\text{Total number of defective units in all of the samples}}{\text{Total number of units sampled}}$ and

$n$ = the sample size.

# AE    Additional Exercises

1.   Edwards Electrical Company produces voltage regulators designed to maintain 220 volts. Samples of 10 units are taken from production to monitor the process and the voltage regulators are found to maintain 220 volts as designed. The mean of the sample ranges is found to be 4 volts.

   **a.**  Determine the UCL, LCL, and centerline for an $\bar{x}$ chart.

   **b.**  Determine the UCL, LCL, and centerline for an $R$ chart.

2.   A steel products manufacturer makes 20-foot lengths of pipe that are later cut into smaller lengths in the production process. To monitor the production and to make sure the pipe is acceptable for the cutting state, a sample of 20 pipes is taken each hour of the day. Along with the pipe length, the range of pipe length is recorded for each sample. Determine the upper and lower control limits for an $R$ chart and indicate which samples, if any, are out of control.

| Pipe Length Ranges | | | | | |
|---|---|---|---|---|---|
| Sample Number | Sample Range | Sample Number | Sample Range | Sample Number | Sample Range |
| 1 | 0.03 | 9 | 0.08 | 17 | 0.05 |
| 2 | 0.07 | 10 | 0.06 | 18 | 0.06 |
| 3 | 0.07 | 11 | 0.05 | 19 | 0.07 |
| 4 | 0.03 | 12 | 0.06 | 20 | 0.05 |
| 5 | 0.02 | 13 | 0.06 | 21 | 0.05 |
| 6 | 0.08 | 14 | 0.03 | 22 | 0.06 |
| 7 | 0.07 | 15 | 0.03 | 23 | 0.04 |
| 8 | 0.03 | 16 | 0.02 | 24 | 0.04 |

3.   The league director decides to sample 200 matches each season to study forfeits in league games. The numbers of forfeits for the last eight seasons are shown in the following table. Find the upper and lower control limits and construct a $p$ chart. Indicate which seasons, if any, are out of control.

| Game Forfeits | | | |
|---|---|---|---|
| Season Number | Number Forfeited | Season Number | Number Forfeited |
| 1 | 8 | 5 | 7 |
| 2 | 10 | 6 | 5 |
| 3 | 9 | 7 | 12 |
| 4 | 8 | 8 | 16 |

4.  A tire manufacturer randomly samples 20 tires at the end of each shift to determine if the tires are defective. The numbers of defectives in 12 shifts are given in the following table. Construct an appropriate control chart to determine if the tire manufacturing process is in control. Identify any shifts that are out of control.

| Defective Tires | |
|---|---|
| Shift | Number of Defectives |
| 1 | 4 |
| 2 | 2 |
| 3 | 0 |
| 4 | 5 |
| 5 | 2 |
| 6 | 3 |
| 7 | 14 |
| 8 | 2 |
| 9 | 3 |
| 10 | 4 |
| 11 | 12 |
| 12 | 3 |

5.  The vice president of audit at a Fortune 500 firm customarily checks the financial statements for errors in 15 departments. The following table contains information about mistakes made on financial statements. Construct an appropriate control chart to determine if the VP should be concerned about the number of mistakes being made on the financial statements.

| Mistakes on Financial Statements | | | |
|---|---|---|---|
| Sample Size | Number of Mistakes | Sample Size | Number of Mistakes |
| 15 | 0 | 15 | 0 |
| 15 | 0 | 15 | 3 |
| 15 | 3 | 15 | 8 |
| 15 | 2 | 15 | 7 |
| 15 | 6 | 15 | 0 |
| 15 | 3 | 15 | 1 |
| 15 | 4 | 15 | 5 |
| 15 | 8 | 15 | 4 |
| 15 | 2 | | |

6.  15 samples of five items each were taken to monitor the amount of fill for a 12-ounce bottle of soda. As each bottle is selected from the line, it is measured. The data from the process is given in the following table. Create an $\bar{x}$ chart and an $R$ chart to determine if the process is in control.

### Fill Amounts for 12-Ounce Bottles

| Sample | Observations | | | | |
|---|---|---|---|---|---|
| | 1 | 2 | 3 | 4 | 5 |
| 1 | 13.11 | 11.17 | 11.35 | 13.71 | 13.00 |
| 2 | 13.19 | 13.50 | 13.63 | 12.36 | 11.98 |
| 3 | 12.83 | 12.12 | 12.65 | 12.22 | 11.43 |
| 4 | 11.62 | 13.72 | 13.73 | 11.36 | 13.93 |
| 5 | 13.65 | 12.44 | 13.15 | 12.42 | 11.21 |
| 6 | 12.60 | 11.07 | 11.17 | 12.88 | 12.33 |
| 7 | 11.07 | 13.22 | 11.41 | 13.21 | 13.14 |
| 8 | 11.56 | 12.16 | 12.47 | 13.41 | 13.13 |
| 9 | 13.23 | 12.05 | 11.25 | 12.55 | 13.00 |
| 10 | 12.30 | 12.18 | 12.00 | 13.59 | 12.18 |
| 11 | 14.00 | 12.33 | 11.52 | 13.29 | 13.51 |
| 12 | 12.05 | 11.52 | 12.80 | 12.21 | 12.67 |
| 13 | 13.29 | 11.91 | 11.11 | 11.32 | 13.85 |
| 14 | 11.21 | 11.89 | 13.07 | 11.22 | 13.09 |
| 15 | 13.37 | 13.25 | 11.48 | 13.46 | 11.26 |

7.  A company packages salt pellets for in-ground well water softeners in bags with a 40-pound label weight. During a typical day's operation of the filling process, 10 samples of five bag fills are selected and measured. Using the data in the following table, create an $\bar{x}$ and an $R$ chart to determine if the process is in control.

### Bag Weights (Pounds)

| Sample | Observations | | | | |
|---|---|---|---|---|---|
| | 1 | 2 | 3 | 4 | 5 |
| 1 | 41.41 | 40.63 | 38.83 | 40.57 | 39.94 |
| 2 | 40.31 | 40.20 | 41.28 | 40.13 | 42.53 |
| 3 | 41.64 | 41.02 | 39.49 | 41.64 | 39.14 |
| 4 | 40.34 | 39.23 | 41.96 | 42.26 | 40.97 |
| 5 | 40.77 | 42.73 | 41.83 | 42.59 | 40.43 |
| 6 | 42.54 | 42.87 | 40.04 | 40.00 | 40.84 |
| 7 | 39.84 | 42.65 | 42.86 | 41.75 | 39.68 |
| 8 | 42.25 | 40.96 | 39.11 | 41.66 | 39.69 |
| 9 | 42.94 | 38.52 | 41.50 | 39.13 | 40.57 |
| 10 | 39.42 | 39.98 | 38.90 | 41.17 | 41.03 |

## P Discovery Project

## Using Statistical Process Control to Improve Air Traffic Processes

1. Choose two cities of interest and research/report actual flight times over a particular weekend, Friday through Sunday. Data sets should include at least ten observations and should be organized in a table for later use.

2. Create a Pareto chart for your flight times.

3. Identify the Upper and Lower Control Limits for any of the three days and draw a graph to represent findings.

4. Calculate the mean and standard deviation of flight times for each day and draw an $\bar{x}$ chart.

5. Create a quality control workflow chart to present to air traffic clients on ways to improve tracking data. Ideas can include measurement improvements,

6. Think about Deming's points 7-9. How can you, as a leader, use statistical processes to improve the effectiveness of flight times?

7. Creatively organize all results and write a summary of your findings to be presented to a board for statistical improvement of air traffic processes.

   a. Include charts and label properly.

   b. Interpret your results from parts 1-4.

   c. Use parts 5-6 to write the narrative of your presentation.

   d. Identify different types of variation and reasons for such data points.

```
Sequence)

equence = nSequence;
ecvOffset = 0;
ext = NULL;

Sequence;
RecvOffset;
t* m_pNext;
Buffer[MAX_ELEMENT_SIZE];

nqueue()

* pQE;
riticalSection(&m_cs);
ail == NULL)
 m_pHead = m_pTail = m_pGarbage = new QElement(0);
(m_pTail->m_nRecvOffset >= MAX_ELEMENT_SIZE)
 m_pTail->m_pNext = new QElement(m_pTail->m_nSequence
il = pQE;
pTail->m_nSequence > m_nMaxQueueLength)
pHead = m_pHead->m_pNext;

= m_pTail;
riticalSection(&m_cs);
QE;

e(QElement*& pCurElement, int& nCurElementSendOffset)

riticalSection(&m_cs);
lement == NULL || pCurElement->m_nSequence <
nSequence)
bRealTime)
urElement = m_pTail, nCurElementSendOffset = m_pTail ?
nRecvOffset : 0;

Element = m_pHead, nCurElementSendOffset = 0;
```

# Appendix A

## Statistical Tables

# Appendix B

## Getting Started with Excel (Desktop)

# Appendix C

## Getting Started with Minitab

# Appendix D

## Getting Started with JMP

# Appendix A
## Statistical Tables

| A | Standard Normal Distribution |
|---|---|

Numerical entries represent the probability that a standard normal random variable is between $-\infty$ and $z$ where $z = \dfrac{x - \mu}{\sigma}$.

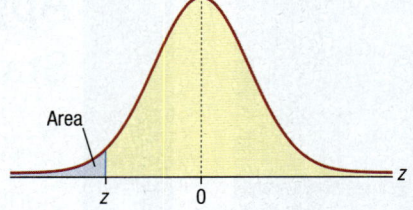

Area

| z | 0.09 | 0.08 | 0.07 | 0.06 | 0.05 | 0.04 | 0.03 | 0.02 | 0.01 | 0.00 |
|------|--------|--------|--------|--------|--------|--------|--------|--------|--------|--------|
| −3.4 | 0.0002 | 0.0003 | 0.0003 | 0.0003 | 0.0003 | 0.0003 | 0.0003 | 0.0003 | 0.0003 | 0.0003 |
| −3.3 | 0.0003 | 0.0004 | 0.0004 | 0.0004 | 0.0004 | 0.0004 | 0.0004 | 0.0005 | 0.0005 | 0.0005 |
| −3.2 | 0.0005 | 0.0005 | 0.0005 | 0.0006 | 0.0006 | 0.0006 | 0.0006 | 0.0006 | 0.0007 | 0.0007 |
| −3.1 | 0.0007 | 0.0007 | 0.0008 | 0.0008 | 0.0008 | 0.0008 | 0.0009 | 0.0009 | 0.0009 | 0.0010 |
| −3.0 | 0.0010 | 0.0010 | 0.0011 | 0.0011 | 0.0011 | 0.0012 | 0.0012 | 0.0013 | 0.0013 | 0.0013 |
| −2.9 | 0.0014 | 0.0014 | 0.0015 | 0.0015 | 0.0016 | 0.0016 | 0.0017 | 0.0018 | 0.0018 | 0.0019 |
| −2.8 | 0.0019 | 0.0020 | 0.0021 | 0.0021 | 0.0022 | 0.0023 | 0.0023 | 0.0024 | 0.0025 | 0.0026 |
| −2.7 | 0.0026 | 0.0027 | 0.0028 | 0.0029 | 0.0030 | 0.0031 | 0.0032 | 0.0033 | 0.0034 | 0.0035 |
| −2.6 | 0.0036 | 0.0037 | 0.0038 | 0.0039 | 0.0040 | 0.0041 | 0.0043 | 0.0044 | 0.0045 | 0.0047 |
| −2.5 | 0.0048 | 0.0049 | 0.0051 | 0.0052 | 0.0054 | 0.0055 | 0.0057 | 0.0059 | 0.0060 | 0.0062 |
| −2.4 | 0.0064 | 0.0066 | 0.0068 | 0.0069 | 0.0071 | 0.0073 | 0.0075 | 0.0078 | 0.0080 | 0.0082 |
| −2.3 | 0.0084 | 0.0087 | 0.0089 | 0.0091 | 0.0094 | 0.0096 | 0.0099 | 0.0102 | 0.0104 | 0.0107 |
| −2.2 | 0.0110 | 0.0113 | 0.0116 | 0.0119 | 0.0122 | 0.0125 | 0.0129 | 0.0132 | 0.0136 | 0.0139 |
| −2.1 | 0.0143 | 0.0146 | 0.0150 | 0.0154 | 0.0158 | 0.0162 | 0.0166 | 0.0170 | 0.0174 | 0.0179 |
| −2.0 | 0.0183 | 0.0188 | 0.0192 | 0.0197 | 0.0202 | 0.0207 | 0.0212 | 0.0217 | 0.0222 | 0.0228 |
| −1.9 | 0.0233 | 0.0239 | 0.0244 | 0.0250 | 0.0256 | 0.0262 | 0.0268 | 0.0274 | 0.0281 | 0.0287 |
| −1.8 | 0.0294 | 0.0301 | 0.0307 | 0.0314 | 0.0322 | 0.0329 | 0.0336 | 0.0344 | 0.0351 | 0.0359 |
| −1.7 | 0.0367 | 0.0375 | 0.0384 | 0.0392 | 0.0401 | 0.0409 | 0.0418 | 0.0427 | 0.0436 | 0.0446 |
| −1.6 | 0.0455 | 0.0465 | 0.0475 | 0.0485 | 0.0495 | 0.0505 | 0.0516 | 0.0526 | 0.0537 | 0.0548 |
| −1.5 | 0.0559 | 0.0571 | 0.0582 | 0.0594 | 0.0606 | 0.0618 | 0.0630 | 0.0643 | 0.0655 | 0.0668 |
| −1.4 | 0.0681 | 0.0694 | 0.0708 | 0.0721 | 0.0735 | 0.0749 | 0.0764 | 0.0778 | 0.0793 | 0.0808 |
| −1.3 | 0.0823 | 0.0838 | 0.0853 | 0.0869 | 0.0885 | 0.0901 | 0.0918 | 0.0934 | 0.0951 | 0.0968 |
| −1.2 | 0.0985 | 0.1003 | 0.1020 | 0.1038 | 0.1056 | 0.1075 | 0.1093 | 0.1112 | 0.1131 | 0.1151 |
| −1.1 | 0.1170 | 0.1190 | 0.1210 | 0.1230 | 0.1251 | 0.1271 | 0.1292 | 0.1314 | 0.1335 | 0.1357 |
| −1.0 | 0.1379 | 0.1401 | 0.1423 | 0.1446 | 0.1469 | 0.1492 | 0.1515 | 0.1539 | 0.1562 | 0.1587 |
| −0.9 | 0.1611 | 0.1635 | 0.1660 | 0.1685 | 0.1711 | 0.1736 | 0.1762 | 0.1788 | 0.1814 | 0.1841 |
| −0.8 | 0.1867 | 0.1894 | 0.1922 | 0.1949 | 0.1977 | 0.2005 | 0.2033 | 0.2061 | 0.2090 | 0.2119 |
| −0.7 | 0.2148 | 0.2177 | 0.2206 | 0.2236 | 0.2266 | 0.2296 | 0.2327 | 0.2358 | 0.2389 | 0.2420 |
| −0.6 | 0.2451 | 0.2483 | 0.2514 | 0.2546 | 0.2578 | 0.2611 | 0.2643 | 0.2676 | 0.2709 | 0.2743 |
| −0.5 | 0.2776 | 0.2810 | 0.2843 | 0.2877 | 0.2912 | 0.2946 | 0.2981 | 0.3015 | 0.3050 | 0.3085 |
| −0.4 | 0.3121 | 0.3156 | 0.3192 | 0.3228 | 0.3264 | 0.3300 | 0.3336 | 0.3372 | 0.3409 | 0.3446 |
| −0.3 | 0.3483 | 0.3520 | 0.3557 | 0.3594 | 0.3632 | 0.3669 | 0.3707 | 0.3745 | 0.3783 | 0.3821 |
| −0.2 | 0.3859 | 0.3897 | 0.3936 | 0.3974 | 0.4013 | 0.4052 | 0.4090 | 0.4129 | 0.4168 | 0.4207 |
| −0.1 | 0.4247 | 0.4286 | 0.4325 | 0.4364 | 0.4404 | 0.4443 | 0.4483 | 0.4522 | 0.4562 | 0.4602 |
| −0.0 | 0.4641 | 0.4681 | 0.4721 | 0.4761 | 0.4801 | 0.4840 | 0.4880 | 0.4920 | 0.4960 | 0.5000 |

# B | Standard Normal Distribution

Numerical entries represent the probability that a standard normal random variable is between $-\infty$ and $z$ where $z = \dfrac{x - \mu}{\sigma}$.

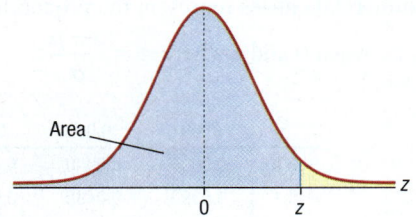

| z | 0.00 | 0.01 | 0.02 | 0.03 | 0.04 | 0.05 | 0.06 | 0.07 | 0.08 | 0.09 |
|---|---|---|---|---|---|---|---|---|---|---|
| 0.0 | 0.5000 | 0.5040 | 0.5080 | 0.5120 | 0.5160 | 0.5199 | 0.5239 | 0.5279 | 0.5319 | 0.5359 |
| 0.1 | 0.5398 | 0.5438 | 0.5478 | 0.5517 | 0.5557 | 0.5596 | 0.5636 | 0.5675 | 0.5714 | 0.5753 |
| 0.2 | 0.5793 | 0.5832 | 0.5871 | 0.5910 | 0.5948 | 0.5987 | 0.6026 | 0.6064 | 0.6103 | 0.6141 |
| 0.3 | 0.6179 | 0.6217 | 0.6255 | 0.6293 | 0.6331 | 0.6368 | 0.6406 | 0.6443 | 0.6480 | 0.6517 |
| 0.4 | 0.6554 | 0.6591 | 0.6628 | 0.6664 | 0.6700 | 0.6736 | 0.6772 | 0.6808 | 0.6844 | 0.6879 |
| 0.5 | 0.6915 | 0.6950 | 0.6985 | 0.7019 | 0.7054 | 0.7088 | 0.7123 | 0.7157 | 0.7190 | 0.7224 |
| 0.6 | 0.7257 | 0.7291 | 0.7324 | 0.7357 | 0.7389 | 0.7422 | 0.7454 | 0.7486 | 0.7517 | 0.7549 |
| 0.7 | 0.7580 | 0.7611 | 0.7642 | 0.7673 | 0.7704 | 0.7734 | 0.7764 | 0.7794 | 0.7823 | 0.7852 |
| 0.8 | 0.7881 | 0.7910 | 0.7939 | 0.7967 | 0.7995 | 0.8023 | 0.8051 | 0.8078 | 0.8106 | 0.8133 |
| 0.9 | 0.8159 | 0.8186 | 0.8212 | 0.8238 | 0.8264 | 0.8289 | 0.8315 | 0.8340 | 0.8365 | 0.8389 |
| 1.0 | 0.8413 | 0.8438 | 0.8461 | 0.8485 | 0.8508 | 0.8531 | 0.8554 | 0.8577 | 0.8599 | 0.8621 |
| 1.1 | 0.8643 | 0.8665 | 0.8686 | 0.8708 | 0.8729 | 0.8749 | 0.8770 | 0.8790 | 0.8810 | 0.8830 |
| 1.2 | 0.8849 | 0.8869 | 0.8888 | 0.8907 | 0.8925 | 0.8944 | 0.8962 | 0.8980 | 0.8997 | 0.9015 |
| 1.3 | 0.9032 | 0.9049 | 0.9066 | 0.9082 | 0.9099 | 0.9115 | 0.9131 | 0.9147 | 0.9162 | 0.9177 |
| 1.4 | 0.9192 | 0.9207 | 0.9222 | 0.9236 | 0.9251 | 0.9265 | 0.9279 | 0.9292 | 0.9306 | 0.9319 |
| 1.5 | 0.9332 | 0.9345 | 0.9357 | 0.9370 | 0.9382 | 0.9394 | 0.9406 | 0.9418 | 0.9429 | 0.9441 |
| 1.6 | 0.9452 | 0.9463 | 0.9474 | 0.9484 | 0.9495 | 0.9505 | 0.9515 | 0.9525 | 0.9535 | 0.9545 |
| 1.7 | 0.9554 | 0.9564 | 0.9573 | 0.9582 | 0.9591 | 0.9599 | 0.9608 | 0.9616 | 0.9625 | 0.9633 |
| 1.8 | 0.9641 | 0.9649 | 0.9656 | 0.9664 | 0.9671 | 0.9678 | 0.9686 | 0.9693 | 0.9699 | 0.9706 |
| 1.9 | 0.9713 | 0.9719 | 0.9726 | 0.9732 | 0.9738 | 0.9744 | 0.9750 | 0.9756 | 0.9761 | 0.9767 |
| 2.0 | 0.9772 | 0.9778 | 0.9783 | 0.9788 | 0.9793 | 0.9798 | 0.9803 | 0.9808 | 0.9812 | 0.9817 |
| 2.1 | 0.9821 | 0.9826 | 0.9830 | 0.9834 | 0.9838 | 0.9842 | 0.9846 | 0.9850 | 0.9854 | 0.9857 |
| 2.2 | 0.9861 | 0.9864 | 0.9868 | 0.9871 | 0.9875 | 0.9878 | 0.9881 | 0.9884 | 0.9887 | 0.9890 |
| 2.3 | 0.9893 | 0.9896 | 0.9898 | 0.9901 | 0.9904 | 0.9906 | 0.9909 | 0.9911 | 0.9913 | 0.9916 |
| 2.4 | 0.9918 | 0.9920 | 0.9922 | 0.9925 | 0.9927 | 0.9929 | 0.9931 | 0.9932 | 0.9934 | 0.9936 |
| 2.5 | 0.9938 | 0.9940 | 0.9941 | 0.9943 | 0.9945 | 0.9946 | 0.9948 | 0.9949 | 0.9951 | 0.9952 |
| 2.6 | 0.9953 | 0.9955 | 0.9956 | 0.9957 | 0.9959 | 0.9960 | 0.9961 | 0.9962 | 0.9963 | 0.9964 |
| 2.7 | 0.9965 | 0.9966 | 0.9967 | 0.9968 | 0.9969 | 0.9970 | 0.9971 | 0.9972 | 0.9973 | 0.9974 |
| 2.8 | 0.9974 | 0.9975 | 0.9976 | 0.9977 | 0.9977 | 0.9978 | 0.9979 | 0.9979 | 0.9980 | 0.9981 |
| 2.9 | 0.9981 | 0.9982 | 0.9982 | 0.9983 | 0.9984 | 0.9984 | 0.9985 | 0.9985 | 0.9986 | 0.9986 |
| 3.0 | 0.9987 | 0.9987 | 0.9987 | 0.9988 | 0.9988 | 0.9989 | 0.9989 | 0.9989 | 0.9990 | 0.9990 |
| 3.1 | 0.9990 | 0.9991 | 0.9991 | 0.9991 | 0.9992 | 0.9992 | 0.9992 | 0.9992 | 0.9993 | 0.9993 |
| 3.2 | 0.9993 | 0.9993 | 0.9994 | 0.9994 | 0.9994 | 0.9994 | 0.9994 | 0.9995 | 0.9995 | 0.9995 |
| 3.3 | 0.9995 | 0.9995 | 0.9995 | 0.9996 | 0.9996 | 0.9996 | 0.9996 | 0.9996 | 0.9996 | 0.9997 |
| 3.4 | 0.9997 | 0.9997 | 0.9997 | 0.9997 | 0.9997 | 0.9997 | 0.9997 | 0.9997 | 0.9997 | 0.9998 |

# C  Standard Normal Distribution

Numerical entries represent the probability that a standard normal random variable is between 0 and $z$ where $z = \dfrac{x - \mu}{\sigma}$.

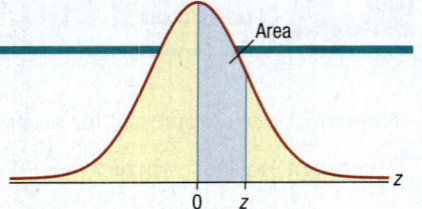

| z | 0.00 | 0.01 | 0.02 | 0.03 | 0.04 | 0.05 | 0.06 | 0.07 | 0.08 | 0.09 |
|---|---|---|---|---|---|---|---|---|---|---|
| 0.0 | 0.0000 | 0.0040 | 0.0080 | 0.0120 | 0.0160 | 0.0199 | 0.0239 | 0.0279 | 0.0319 | 0.0359 |
| 0.1 | 0.0398 | 0.0438 | 0.0478 | 0.0517 | 0.0557 | 0.0596 | 0.0636 | 0.0675 | 0.0714 | 0.0753 |
| 0.2 | 0.0793 | 0.0832 | 0.0871 | 0.0910 | 0.0948 | 0.0987 | 0.1026 | 0.1064 | 0.1103 | 0.1141 |
| 0.3 | 0.1179 | 0.1217 | 0.1255 | 0.1293 | 0.1331 | 0.1368 | 0.1406 | 0.1443 | 0.1480 | 0.1517 |
| 0.4 | 0.1554 | 0.1591 | 0.1628 | 0.1664 | 0.1700 | 0.1736 | 0.1772 | 0.1808 | 0.1844 | 0.1879 |
| 0.5 | 0.1915 | 0.1950 | 0.1985 | 0.2019 | 0.2054 | 0.2088 | 0.2123 | 0.2157 | 0.2190 | 0.2224 |
| 0.6 | 0.2257 | 0.2291 | 0.2324 | 0.2357 | 0.2389 | 0.2422 | 0.2454 | 0.2486 | 0.2517 | 0.2549 |
| 0.7 | 0.2580 | 0.2611 | 0.2642 | 0.2673 | 0.2704 | 0.2734 | 0.2764 | 0.2794 | 0.2823 | 0.2852 |
| 0.8 | 0.2881 | 0.2910 | 0.2939 | 0.2967 | 0.2995 | 0.3023 | 0.3051 | 0.3078 | 0.3106 | 0.3133 |
| 0.9 | 0.3159 | 0.3186 | 0.3212 | 0.3238 | 0.3264 | 0.3289 | 0.3315 | 0.3340 | 0.3365 | 0.3389 |
| 1.0 | 0.3413 | 0.3438 | 0.3461 | 0.3485 | 0.3508 | 0.3531 | 0.3554 | 0.3577 | 0.3599 | 0.3621 |
| 1.1 | 0.3643 | 0.3665 | 0.3686 | 0.3708 | 0.3729 | 0.3749 | 0.3770 | 0.3790 | 0.3810 | 0.3830 |
| 1.2 | 0.3849 | 0.3869 | 0.3888 | 0.3907 | 0.3925 | 0.3944 | 0.3962 | 0.3980 | 0.3997 | 0.4015 |
| 1.3 | 0.4032 | 0.4049 | 0.4066 | 0.4082 | 0.4099 | 0.4115 | 0.4131 | 0.4147 | 0.4162 | 0.4177 |
| 1.4 | 0.4192 | 0.4207 | 0.4222 | 0.4236 | 0.4251 | 0.4265 | 0.4279 | 0.4292 | 0.4306 | 0.4319 |
| 1.5 | 0.4332 | 0.4345 | 0.4357 | 0.4370 | 0.4382 | 0.4394 | 0.4406 | 0.4418 | 0.4429 | 0.4441 |
| 1.6 | 0.4452 | 0.4463 | 0.4474 | 0.4484 | 0.4495 | 0.4505 | 0.4515 | 0.4525 | 0.4535 | 0.4545 |
| 1.7 | 0.4554 | 0.4564 | 0.4573 | 0.4582 | 0.4591 | 0.4599 | 0.4608 | 0.4616 | 0.4625 | 0.4633 |
| 1.8 | 0.4641 | 0.4649 | 0.4656 | 0.4664 | 0.4671 | 0.4678 | 0.4686 | 0.4693 | 0.4699 | 0.4706 |
| 1.9 | 0.4713 | 0.4719 | 0.4726 | 0.4732 | 0.4738 | 0.4744 | 0.4750 | 0.4756 | 0.4761 | 0.4767 |
| 2.0 | 0.4772 | 0.4778 | 0.4783 | 0.4788 | 0.4793 | 0.4798 | 0.4803 | 0.4808 | 0.4812 | 0.4817 |
| 2.1 | 0.4821 | 0.4826 | 0.4830 | 0.4834 | 0.4838 | 0.4842 | 0.4846 | 0.4850 | 0.4854 | 0.4857 |
| 2.2 | 0.4861 | 0.4864 | 0.4868 | 0.4871 | 0.4875 | 0.4878 | 0.4881 | 0.4884 | 0.4887 | 0.4890 |
| 2.3 | 0.4893 | 0.4896 | 0.4898 | 0.4901 | 0.4904 | 0.4906 | 0.4909 | 0.4911 | 0.4913 | 0.4916 |
| 2.4 | 0.4918 | 0.4920 | 0.4922 | 0.4925 | 0.4927 | 0.4929 | 0.4931 | 0.4932 | 0.4934 | 0.4936 |
| 2.5 | 0.4938 | 0.4940 | 0.4941 | 0.4943 | 0.4945 | 0.4946 | 0.4948 | 0.4949 | 0.4951 | 0.4952 |
| 2.6 | 0.4953 | 0.4955 | 0.4956 | 0.4957 | 0.4959 | 0.4960 | 0.4961 | 0.4962 | 0.4963 | 0.4964 |
| 2.7 | 0.4965 | 0.4966 | 0.4967 | 0.4968 | 0.4969 | 0.4970 | 0.4971 | 0.4972 | 0.4973 | 0.4974 |
| 2.8 | 0.4974 | 0.4975 | 0.4976 | 0.4977 | 0.4977 | 0.4978 | 0.4979 | 0.4979 | 0.4980 | 0.4981 |
| 2.9 | 0.4981 | 0.4982 | 0.4982 | 0.4983 | 0.4984 | 0.4984 | 0.4985 | 0.4985 | 0.4986 | 0.4986 |
| 3.0 | 0.4987 | 0.4987 | 0.4987 | 0.4988 | 0.4988 | 0.4989 | 0.4989 | 0.4989 | 0.4990 | 0.4990 |
| 3.1 | 0.4990 | 0.4991 | 0.4991 | 0.4991 | 0.4992 | 0.4992 | 0.4992 | 0.4992 | 0.4993 | 0.4993 |
| 3.2 | 0.4993 | 0.4993 | 0.4994 | 0.4994 | 0.4994 | 0.4994 | 0.4994 | 0.4995 | 0.4995 | 0.4995 |
| 3.3 | 0.4995 | 0.4995 | 0.4995 | 0.4996 | 0.4996 | 0.4996 | 0.4996 | 0.4996 | 0.4996 | 0.4997 |
| 3.4 | 0.4997 | 0.4997 | 0.4997 | 0.4997 | 0.4997 | 0.4997 | 0.4997 | 0.4997 | 0.4997 | 0.4998 |

## Critical Values of z

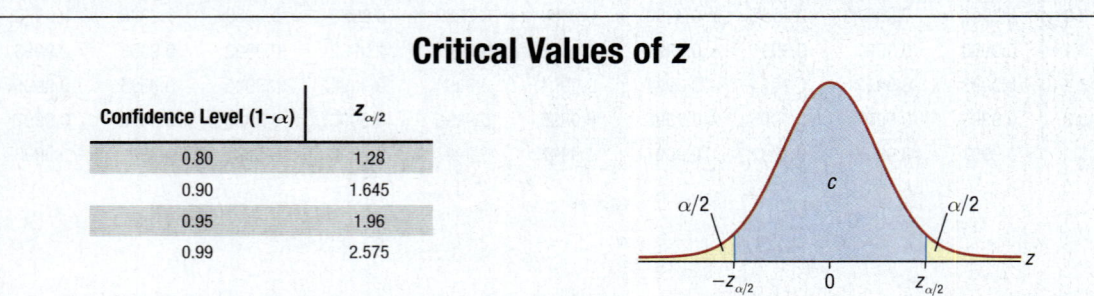

| Confidence Level $(1 - \alpha)$ | $z_{\alpha/2}$ |
|---|---|
| 0.80 | 1.28 |
| 0.90 | 1.645 |
| 0.95 | 1.96 |
| 0.99 | 2.575 |

# D  Critical Values of *t*

Numerical entries represent the value of *t* such that the area to the right of the *t* is equal to $\alpha$.

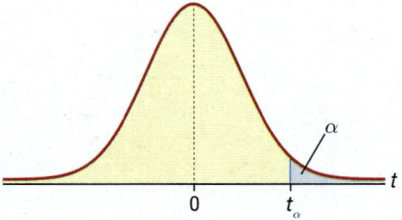

| Degrees of Freedom | Area to the Right of the Critical Value | | | | | |
|---|---|---|---|---|---|---|
| | $t_{0.200}$ | $t_{0.100}$ | $t_{0.050}$ | $t_{0.025}$ | $t_{0.010}$ | $t_{0.005}$ |
| 1 | 1.376 | 3.078 | 6.314 | 12.706 | 31.821 | 63.657 |
| 2 | 1.061 | 1.886 | 2.920 | 4.303 | 6.965 | 9.925 |
| 3 | 0.978 | 1.638 | 2.353 | 3.182 | 4.541 | 5.841 |
| 4 | 0.941 | 1.533 | 2.132 | 2.776 | 3.747 | 4.604 |
| 5 | 0.920 | 1.476 | 2.015 | 2.571 | 3.365 | 4.032 |
| 6 | 0.906 | 1.440 | 1.943 | 2.447 | 3.143 | 3.707 |
| 7 | 0.896 | 1.415 | 1.895 | 2.365 | 2.998 | 3.499 |
| 8 | 0.889 | 1.397 | 1.860 | 2.306 | 2.896 | 3.355 |
| 9 | 0.883 | 1.383 | 1.833 | 2.262 | 2.821 | 3.250 |
| 10 | 0.879 | 1.372 | 1.812 | 2.228 | 2.764 | 3.169 |
| 11 | 0.876 | 1.363 | 1.796 | 2.201 | 2.718 | 3.106 |
| 12 | 0.873 | 1.356 | 1.782 | 2.179 | 2.681 | 3.055 |
| 13 | 0.870 | 1.350 | 1.771 | 2.160 | 2.650 | 3.012 |
| 14 | 0.868 | 1.345 | 1.761 | 2.145 | 2.624 | 2.977 |
| 15 | 0.866 | 1.341 | 1.753 | 2.131 | 2.602 | 2.947 |
| 16 | 0.865 | 1.337 | 1.746 | 2.120 | 2.583 | 2.921 |
| 17 | 0.863 | 1.333 | 1.740 | 2.110 | 2.567 | 2.898 |
| 18 | 0.862 | 1.330 | 1.734 | 2.101 | 2.552 | 2.878 |
| 19 | 0.861 | 1.328 | 1.729 | 2.093 | 2.539 | 2.861 |
| 20 | 0.860 | 1.325 | 1.725 | 2.086 | 2.528 | 2.845 |
| 21 | 0.859 | 1.323 | 1.721 | 2.080 | 2.518 | 2.831 |
| 22 | 0.858 | 1.321 | 1.717 | 2.074 | 2.508 | 2.819 |
| 23 | 0.858 | 1.319 | 1.714 | 2.069 | 2.500 | 2.807 |
| 24 | 0.857 | 1.318 | 1.711 | 2.064 | 2.492 | 2.797 |
| 25 | 0.856 | 1.316 | 1.708 | 2.060 | 2.485 | 2.787 |
| 26 | 0.856 | 1.315 | 1.706 | 2.056 | 2.479 | 2.779 |
| 27 | 0.855 | 1.314 | 1.703 | 2.052 | 2.473 | 2.771 |
| 28 | 0.855 | 1.313 | 1.701 | 2.048 | 2.467 | 2.763 |
| 29 | 0.854 | 1.311 | 1.699 | 2.045 | 2.462 | 2.756 |
| 30 | 0.854 | 1.310 | 1.697 | 2.042 | 2.457 | 2.750 |
| 40 | 0.851 | 1.303 | 1.684 | 2.021 | 2.423 | 2.704 |
| 50 | 0.849 | 1.299 | 1.676 | 2.009 | 2.403 | 2.678 |
| 60 | 0.848 | 1.296 | 1.671 | 2.000 | 2.390 | 2.660 |
| 70 | 0.847 | 1.294 | 1.667 | 1.994 | 2.381 | 2.648 |
| 80 | 0.846 | 1.292 | 1.664 | 1.990 | 2.374 | 2.639 |
| 90 | 0.846 | 1.291 | 1.662 | 1.987 | 2.368 | 2.632 |
| 100 | 0.845 | 1.290 | 1.660 | 1.984 | 2.364 | 2.626 |
| 120 | 0.845 | 1.289 | 1.658 | 1.980 | 2.358 | 2.617 |
| $\infty$ | 0.842 | 1.282 | 1.645 | 1.96 | 2.326 | 2.576 |

# E   Cumulative Binomial Probabilities

Numerical entries represent $P(X \le x)$.

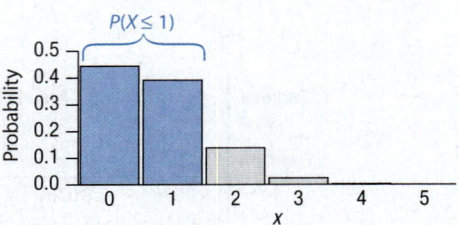

| n | x | 0.1 | 0.2 | 0.3 | 0.4 | 0.5 | 0.6 | 0.7 | 0.8 | 0.9 |
|---|---|-----|-----|-----|-----|-----|-----|-----|-----|-----|
| 1 | 0 | 0.9000 | 0.8000 | 0.7000 | 0.6000 | 0.5000 | 0.4000 | 0.3000 | 0.2000 | 0.1000 |
|   | 1 | 1.0000 | 1.0000 | 1.0000 | 1.0000 | 1.0000 | 1.0000 | 1.0000 | 1.0000 | 1.0000 |
| 2 | 0 | 0.8100 | 0.6400 | 0.4900 | 0.3600 | 0.2500 | 0.1600 | 0.0900 | 0.0400 | 0.0100 |
|   | 1 | 0.9900 | 0.9600 | 0.9100 | 0.8400 | 0.7500 | 0.6400 | 0.5100 | 0.3600 | 0.1900 |
|   | 2 | 1.0000 | 1.0000 | 1.0000 | 1.0000 | 1.0000 | 1.0000 | 1.0000 | 1.0000 | 1.0000 |
| 3 | 0 | 0.7290 | 0.5120 | 0.3430 | 0.2160 | 0.1250 | 0.0640 | 0.0270 | 0.0080 | 0.0010 |
|   | 1 | 0.9720 | 0.8960 | 0.7840 | 0.6480 | 0.5000 | 0.3520 | 0.2160 | 0.1040 | 0.0280 |
|   | 2 | 0.9990 | 0.9920 | 0.9730 | 0.9360 | 0.8750 | 0.7840 | 0.6570 | 0.4880 | 0.2710 |
|   | 3 | 1.0000 | 1.0000 | 1.0000 | 1.0000 | 1.0000 | 1.0000 | 1.0000 | 1.0000 | 1.0000 |
| 4 | 0 | 0.6561 | 0.4096 | 0.2401 | 0.1296 | 0.0625 | 0.0256 | 0.0081 | 0.0016 | 0.0001 |
|   | 1 | 0.9477 | 0.8192 | 0.6517 | 0.4752 | 0.3125 | 0.1792 | 0.0837 | 0.0272 | 0.0037 |
|   | 2 | 0.9963 | 0.9728 | 0.9163 | 0.8208 | 0.6875 | 0.5248 | 0.3483 | 0.1808 | 0.0523 |
|   | 3 | 0.9999 | 0.9984 | 0.9919 | 0.9744 | 0.9375 | 0.8704 | 0.7599 | 0.5904 | 0.3439 |
|   | 4 | 1.0000 | 1.0000 | 1.0000 | 1.0000 | 1.0000 | 1.0000 | 1.0000 | 1.0000 | 1.0000 |
| 5 | 0 | 0.5905 | 0.3277 | 0.1681 | 0.0778 | 0.0313 | 0.0102 | 0.0024 | 0.0003 | 0.0000 |
|   | 1 | 0.9185 | 0.7373 | 0.5282 | 0.3370 | 0.1875 | 0.0870 | 0.0308 | 0.0067 | 0.0005 |
|   | 2 | 0.9914 | 0.9421 | 0.8369 | 0.6826 | 0.5000 | 0.3174 | 0.1631 | 0.0579 | 0.0086 |
|   | 3 | 0.9995 | 0.9933 | 0.9692 | 0.9130 | 0.8125 | 0.6630 | 0.4718 | 0.2627 | 0.0815 |
|   | 4 | 1.0000 | 0.9997 | 0.9976 | 0.9898 | 0.9688 | 0.9222 | 0.8319 | 0.6723 | 0.4095 |
|   | 5 | 1.0000 | 1.0000 | 1.0000 | 1.0000 | 1.0000 | 1.0000 | 1.0000 | 1.0000 | 1.0000 |
| 6 | 0 | 0.5314 | 0.2621 | 0.1176 | 0.0467 | 0.0156 | 0.0041 | 0.0007 | 0.0001 | 0.0000 |
|   | 1 | 0.8857 | 0.6554 | 0.4202 | 0.2333 | 0.1094 | 0.0410 | 0.0109 | 0.0016 | 0.0001 |
|   | 2 | 0.9842 | 0.9011 | 0.7443 | 0.5443 | 0.3438 | 0.1792 | 0.0705 | 0.0170 | 0.0013 |
|   | 3 | 0.9987 | 0.9830 | 0.9295 | 0.8208 | 0.6563 | 0.4557 | 0.2557 | 0.0989 | 0.0159 |
|   | 4 | 0.9999 | 0.9984 | 0.9891 | 0.9590 | 0.8906 | 0.7667 | 0.5798 | 0.3446 | 0.1143 |
|   | 5 | 1.0000 | 0.9999 | 0.9993 | 0.9959 | 0.9844 | 0.9533 | 0.8824 | 0.7379 | 0.4686 |
|   | 6 | 1.0000 | 1.0000 | 1.0000 | 1.0000 | 1.0000 | 1.0000 | 1.0000 | 1.0000 | 1.0000 |
| 7 | 0 | 0.4783 | 0.2097 | 0.0824 | 0.0280 | 0.0078 | 0.0016 | 0.0002 | 0.0000 | 0.0000 |
|   | 1 | 0.8503 | 0.5767 | 0.3294 | 0.1586 | 0.0625 | 0.0188 | 0.0038 | 0.0004 | 0.0000 |
|   | 2 | 0.9743 | 0.8520 | 0.6471 | 0.4199 | 0.2266 | 0.0963 | 0.0288 | 0.0047 | 0.0002 |
|   | 3 | 0.9973 | 0.9667 | 0.8740 | 0.7102 | 0.5000 | 0.2898 | 0.1260 | 0.0333 | 0.0027 |
|   | 4 | 0.9998 | 0.9953 | 0.9712 | 0.9037 | 0.7734 | 0.5801 | 0.3529 | 0.1480 | 0.0257 |
|   | 5 | 1.0000 | 0.9996 | 0.9962 | 0.9812 | 0.9375 | 0.8414 | 0.6706 | 0.4233 | 0.1497 |
|   | 6 | 1.0000 | 1.0000 | 0.9998 | 0.9984 | 0.9922 | 0.9720 | 0.9176 | 0.7903 | 0.5217 |
|   | 7 | 1.0000 | 1.0000 | 1.0000 | 1.0000 | 1.0000 | 1.0000 | 1.0000 | 1.0000 | 1.0000 |

# E Cumulative Binomial Probabilities (cont.)

| n | x | 0.1 | 0.2 | 0.3 | 0.4 | p<br>0.5 | 0.6 | 0.7 | 0.8 | 0.9 |
|---|---|-----|-----|-----|-----|-----|-----|-----|-----|-----|
| 8 | 0 | 0.4305 | 0.1678 | 0.0576 | 0.0168 | 0.0039 | 0.0007 | 0.0001 | 0.0000 | 0.0000 |
|   | 1 | 0.8131 | 0.5033 | 0.2553 | 0.1064 | 0.0352 | 0.0085 | 0.0013 | 0.0001 | 0.0000 |
|   | 2 | 0.9619 | 0.7969 | 0.5518 | 0.3154 | 0.1445 | 0.0498 | 0.0113 | 0.0012 | 0.0000 |
|   | 3 | 0.9950 | 0.9437 | 0.8059 | 0.5941 | 0.3633 | 0.1737 | 0.0580 | 0.0104 | 0.0004 |
|   | 4 | 0.9996 | 0.9896 | 0.9420 | 0.8263 | 0.6367 | 0.4059 | 0.1941 | 0.0563 | 0.0050 |
|   | 5 | 1.0000 | 0.9988 | 0.9887 | 0.9502 | 0.8555 | 0.6846 | 0.4482 | 0.2031 | 0.0381 |
|   | 6 | 1.0000 | 0.9999 | 0.9987 | 0.9915 | 0.9648 | 0.8936 | 0.7447 | 0.4967 | 0.1869 |
|   | 7 | 1.0000 | 1.0000 | 0.9999 | 0.9993 | 0.9961 | 0.9832 | 0.9424 | 0.8322 | 0.5695 |
|   | 8 | 1.0000 | 1.0000 | 1.0000 | 1.0000 | 1.0000 | 1.0000 | 1.0000 | 1.0000 | 1.0000 |
| 9 | 0 | 0.3874 | 0.1342 | 0.0404 | 0.0101 | 0.0020 | 0.0003 | 0.0000 | 0.0000 | 0.0000 |
|   | 1 | 0.7748 | 0.4362 | 0.1960 | 0.0705 | 0.0195 | 0.0038 | 0.0004 | 0.0000 | 0.0000 |
|   | 2 | 0.9470 | 0.7382 | 0.4628 | 0.2318 | 0.0898 | 0.0250 | 0.0043 | 0.0003 | 0.0000 |
|   | 3 | 0.9917 | 0.9144 | 0.7297 | 0.4826 | 0.2539 | 0.0994 | 0.0253 | 0.0031 | 0.0001 |
|   | 4 | 0.9991 | 0.9804 | 0.9012 | 0.7334 | 0.5000 | 0.2666 | 0.0988 | 0.0196 | 0.0009 |
|   | 5 | 0.9999 | 0.9969 | 0.9747 | 0.9006 | 0.7461 | 0.5174 | 0.2703 | 0.0856 | 0.0083 |
|   | 6 | 1.0000 | 0.9997 | 0.9957 | 0.9750 | 0.9102 | 0.7682 | 0.5372 | 0.2618 | 0.0530 |
|   | 7 | 1.0000 | 1.0000 | 0.9996 | 0.9962 | 0.9805 | 0.9295 | 0.8040 | 0.5638 | 0.2252 |
|   | 8 | 1.0000 | 1.0000 | 1.0000 | 0.9997 | 0.9980 | 0.9899 | 0.9596 | 0.8658 | 0.6126 |
|   | 9 | 1.0000 | 1.0000 | 1.0000 | 1.0000 | 1.0000 | 1.0000 | 1.0000 | 1.0000 | 1.0000 |
| 10 | 0 | 0.3487 | 0.1074 | 0.0282 | 0.0060 | 0.0010 | 0.0001 | 0.0000 | 0.0000 | 0.0000 |
|   | 1 | 0.7361 | 0.3758 | 0.1493 | 0.0464 | 0.0107 | 0.0017 | 0.0001 | 0.0000 | 0.0000 |
|   | 2 | 0.9298 | 0.6778 | 0.3828 | 0.1673 | 0.0547 | 0.0123 | 0.0016 | 0.0001 | 0.0000 |
|   | 3 | 0.9872 | 0.8791 | 0.6496 | 0.3823 | 0.1719 | 0.0548 | 0.0106 | 0.0009 | 0.0000 |
|   | 4 | 0.9984 | 0.9672 | 0.8497 | 0.6331 | 0.3770 | 0.1662 | 0.0473 | 0.0064 | 0.0001 |
|   | 5 | 0.9999 | 0.9936 | 0.9527 | 0.8338 | 0.6230 | 0.3669 | 0.1503 | 0.0328 | 0.0016 |
|   | 6 | 1.0000 | 0.9991 | 0.9894 | 0.9452 | 0.8281 | 0.6177 | 0.3504 | 0.1209 | 0.0128 |
|   | 7 | 1.0000 | 0.9999 | 0.9984 | 0.9877 | 0.9453 | 0.8327 | 0.6172 | 0.3222 | 0.0702 |
|   | 8 | 1.0000 | 1.0000 | 0.9999 | 0.9983 | 0.9893 | 0.9536 | 0.8507 | 0.6242 | 0.2639 |
|   | 9 | 1.0000 | 1.0000 | 1.0000 | 0.9999 | 0.9990 | 0.9940 | 0.9718 | 0.8926 | 0.6513 |
|   | 10 | 1.0000 | 1.0000 | 1.0000 | 1.0000 | 1.0000 | 1.0000 | 1.0000 | 1.0000 | 1.0000 |
| 11 | 0 | 0.3138 | 0.0859 | 0.0198 | 0.0036 | 0.0005 | 0.0000 | 0.0000 | 0.0000 | 0.0000 |
|   | 1 | 0.6974 | 0.3221 | 0.1130 | 0.0302 | 0.0059 | 0.0007 | 0.0000 | 0.0000 | 0.0000 |
|   | 2 | 0.9104 | 0.6174 | 0.3127 | 0.1189 | 0.0327 | 0.0059 | 0.0006 | 0.0000 | 0.0000 |
|   | 3 | 0.9815 | 0.8389 | 0.5696 | 0.2963 | 0.1133 | 0.0293 | 0.0043 | 0.0002 | 0.0000 |
|   | 4 | 0.9972 | 0.9496 | 0.7897 | 0.5328 | 0.2744 | 0.0994 | 0.0216 | 0.0020 | 0.0000 |
|   | 5 | 0.9997 | 0.9883 | 0.9218 | 0.7535 | 0.5000 | 0.2465 | 0.0782 | 0.0117 | 0.0003 |
|   | 6 | 1.0000 | 0.9980 | 0.9784 | 0.9006 | 0.7256 | 0.4672 | 0.2103 | 0.0504 | 0.0028 |
|   | 7 | 1.0000 | 0.9998 | 0.9957 | 0.9707 | 0.8867 | 0.7037 | 0.4304 | 0.1611 | 0.0185 |
|   | 8 | 1.0000 | 1.0000 | 0.9994 | 0.9941 | 0.9673 | 0.8811 | 0.6873 | 0.3826 | 0.0896 |
|   | 9 | 1.0000 | 1.0000 | 1.0000 | 0.9993 | 0.9941 | 0.9698 | 0.8870 | 0.6779 | 0.3026 |
|   | 10 | 1.0000 | 1.0000 | 1.0000 | 1.0000 | 0.9995 | 0.9964 | 0.9802 | 0.9141 | 0.6862 |
|   | 11 | 1.0000 | 1.0000 | 1.0000 | 1.0000 | 1.0000 | 1.0000 | 1.0000 | 1.0000 | 1.0000 |

# E  Cumulative Binomial Probabilities (cont.)

| | | | | | | p | | | | |
|---|---|---|---|---|---|---|---|---|---|---|
| n | x | 0.1 | 0.2 | 0.3 | 0.4 | 0.5 | 0.6 | 0.7 | 0.8 | 0.9 |
| 12 | 0 | 0.2824 | 0.0687 | 0.0138 | 0.0022 | 0.0002 | 0.0000 | 0.0000 | 0.0000 | 0.0000 |
| | 1 | 0.6590 | 0.2749 | 0.0850 | 0.0196 | 0.0032 | 0.0003 | 0.0000 | 0.0000 | 0.0000 |
| | 2 | 0.8891 | 0.5583 | 0.2528 | 0.0834 | 0.0193 | 0.0028 | 0.0002 | 0.0000 | 0.0000 |
| | 3 | 0.9744 | 0.7946 | 0.4925 | 0.2253 | 0.0730 | 0.0153 | 0.0017 | 0.0001 | 0.0000 |
| | 4 | 0.9957 | 0.9274 | 0.7237 | 0.4382 | 0.1938 | 0.0573 | 0.0095 | 0.0006 | 0.0000 |
| | 5 | 0.9995 | 0.9806 | 0.8822 | 0.6652 | 0.3872 | 0.1582 | 0.0386 | 0.0039 | 0.0001 |
| | 6 | 0.9999 | 0.9961 | 0.9614 | 0.8418 | 0.6128 | 0.3348 | 0.1178 | 0.0194 | 0.0005 |
| | 7 | 1.0000 | 0.9994 | 0.9905 | 0.9427 | 0.8062 | 0.5618 | 0.2763 | 0.0726 | 0.0043 |
| | 8 | 1.0000 | 0.9999 | 0.9983 | 0.9847 | 0.9270 | 0.7747 | 0.5075 | 0.2054 | 0.0256 |
| | 9 | 1.0000 | 1.0000 | 0.9998 | 0.9972 | 0.9807 | 0.9166 | 0.7472 | 0.4417 | 0.1109 |
| | 10 | 1.0000 | 1.0000 | 1.0000 | 0.9997 | 0.9968 | 0.9804 | 0.9150 | 0.7251 | 0.3410 |
| | 11 | 1.0000 | 1.0000 | 1.0000 | 1.0000 | 0.9998 | 0.9978 | 0.9862 | 0.9313 | 0.7176 |
| | 12 | 1.0000 | 1.0000 | 1.0000 | 1.0000 | 1.0000 | 1.0000 | 1.0000 | 1.0000 | 1.0000 |
| 13 | 0 | 0.2542 | 0.0550 | 0.0097 | 0.0013 | 0.0001 | 0.0000 | 0.0000 | 0.0000 | 0.0000 |
| | 1 | 0.6213 | 0.2336 | 0.0637 | 0.0126 | 0.0017 | 0.0001 | 0.0000 | 0.0000 | 0.0000 |
| | 2 | 0.8661 | 0.5017 | 0.2025 | 0.0579 | 0.0112 | 0.0013 | 0.0001 | 0.0000 | 0.0000 |
| | 3 | 0.9658 | 0.7473 | 0.4206 | 0.1686 | 0.0461 | 0.0078 | 0.0007 | 0.0000 | 0.0000 |
| | 4 | 0.9935 | 0.9009 | 0.6543 | 0.3530 | 0.1334 | 0.0321 | 0.0040 | 0.0002 | 0.0000 |
| | 5 | 0.9991 | 0.9700 | 0.8346 | 0.5744 | 0.2905 | 0.0977 | 0.0182 | 0.0012 | 0.0000 |
| | 6 | 0.9999 | 0.9930 | 0.9376 | 0.7712 | 0.5000 | 0.2288 | 0.0624 | 0.0070 | 0.0001 |
| | 7 | 1.0000 | 0.9988 | 0.9818 | 0.9023 | 0.7095 | 0.4256 | 0.1654 | 0.0300 | 0.0009 |
| | 8 | 1.0000 | 0.9998 | 0.9960 | 0.9679 | 0.8666 | 0.6470 | 0.3457 | 0.0991 | 0.0065 |
| | 9 | 1.0000 | 1.0000 | 0.9993 | 0.9922 | 0.9539 | 0.8314 | 0.5794 | 0.2527 | 0.0342 |
| | 10 | 1.0000 | 1.0000 | 0.9999 | 0.9987 | 0.9888 | 0.9421 | 0.7975 | 0.4983 | 0.1339 |
| | 11 | 1.0000 | 1.0000 | 1.0000 | 0.9999 | 0.9983 | 0.9874 | 0.9363 | 0.7664 | 0.3787 |
| | 12 | 1.0000 | 1.0000 | 1.0000 | 1.0000 | 0.9999 | 0.9987 | 0.9903 | 0.9450 | 0.7458 |
| | 13 | 1.0000 | 1.0000 | 1.0000 | 1.0000 | 1.0000 | 1.0000 | 1.0000 | 1.0000 | 1.0000 |
| 14 | 0 | 0.2288 | 0.0440 | 0.0068 | 0.0008 | 0.0001 | 0.0000 | 0.0000 | 0.0000 | 0.0000 |
| | 1 | 0.5846 | 0.1979 | 0.0475 | 0.0081 | 0.0009 | 0.0001 | 0.0000 | 0.0000 | 0.0000 |
| | 2 | 0.8416 | 0.4481 | 0.1608 | 0.0398 | 0.0065 | 0.0006 | 0.0000 | 0.0000 | 0.0000 |
| | 3 | 0.9559 | 0.6982 | 0.3552 | 0.1243 | 0.0287 | 0.0039 | 0.0002 | 0.0000 | 0.0000 |
| | 4 | 0.9908 | 0.8702 | 0.5842 | 0.2793 | 0.0898 | 0.0175 | 0.0017 | 0.0000 | 0.0000 |
| | 5 | 0.9985 | 0.9561 | 0.7805 | 0.4859 | 0.2120 | 0.0583 | 0.0083 | 0.0004 | 0.0000 |
| | 6 | 0.9998 | 0.9884 | 0.9067 | 0.6925 | 0.3953 | 0.1501 | 0.0315 | 0.0024 | 0.0000 |
| | 7 | 1.0000 | 0.9976 | 0.9685 | 0.8499 | 0.6047 | 0.3075 | 0.0933 | 0.0116 | 0.0002 |
| | 8 | 1.0000 | 0.9996 | 0.9917 | 0.9417 | 0.7880 | 0.5141 | 0.2195 | 0.0439 | 0.0015 |
| | 9 | 1.0000 | 1.0000 | 0.9983 | 0.9825 | 0.9102 | 0.7207 | 0.4158 | 0.1298 | 0.0092 |
| | 10 | 1.0000 | 1.0000 | 0.9998 | 0.9961 | 0.9713 | 0.8757 | 0.6448 | 0.3018 | 0.0441 |
| | 11 | 1.0000 | 1.0000 | 1.0000 | 0.9994 | 0.9935 | 0.9602 | 0.8392 | 0.5519 | 0.1584 |
| | 12 | 1.0000 | 1.0000 | 1.0000 | 0.9999 | 0.9991 | 0.9919 | 0.9525 | 0.8021 | 0.4154 |
| | 13 | 1.0000 | 1.0000 | 1.0000 | 1.0000 | 0.9999 | 0.9992 | 0.9932 | 0.9560 | 0.7712 |
| | 14 | 1.0000 | 1.0000 | 1.0000 | 1.0000 | 1.0000 | 1.0000 | 1.0000 | 1.0000 | 1.0000 |

# E  Cumulative Binomial Probabilities (cont.)

| n | x | 0.1 | 0.2 | 0.3 | 0.4 | 0.5 | 0.6 | 0.7 | 0.8 | 0.9 |
|---|---|-----|-----|-----|-----|-----|-----|-----|-----|-----|
| 15 | 0 | 0.2059 | 0.0352 | 0.0047 | 0.0005 | 0.0000 | 0.0000 | 0.0000 | 0.0000 | 0.0000 |
|  | 1 | 0.5490 | 0.1671 | 0.0353 | 0.0052 | 0.0005 | 0.0000 | 0.0000 | 0.0000 | 0.0000 |
|  | 2 | 0.8159 | 0.3980 | 0.1268 | 0.0271 | 0.0037 | 0.0003 | 0.0000 | 0.0000 | 0.0000 |
|  | 3 | 0.9444 | 0.6482 | 0.2969 | 0.0905 | 0.0176 | 0.0019 | 0.0001 | 0.0000 | 0.0000 |
|  | 4 | 0.9873 | 0.8358 | 0.5155 | 0.2173 | 0.0592 | 0.0093 | 0.0007 | 0.0000 | 0.0000 |
|  | 5 | 0.9978 | 0.9389 | 0.7216 | 0.4032 | 0.1509 | 0.0338 | 0.0037 | 0.0001 | 0.0000 |
|  | 6 | 0.9997 | 0.9819 | 0.8689 | 0.6098 | 0.3036 | 0.0950 | 0.0152 | 0.0008 | 0.0000 |
|  | 7 | 1.0000 | 0.9958 | 0.9500 | 0.7869 | 0.5000 | 0.2131 | 0.0500 | 0.0042 | 0.0000 |
|  | 8 | 1.0000 | 0.9992 | 0.9848 | 0.9050 | 0.6964 | 0.3902 | 0.1311 | 0.0181 | 0.0003 |
|  | 9 | 1.0000 | 0.9999 | 0.9963 | 0.9662 | 0.8491 | 0.5968 | 0.2784 | 0.0611 | 0.0022 |
|  | 10 | 1.0000 | 1.0000 | 0.9993 | 0.9907 | 0.9408 | 0.7827 | 0.4845 | 0.1642 | 0.0127 |
|  | 11 | 1.0000 | 1.0000 | 0.9999 | 0.9981 | 0.9824 | 0.9095 | 0.7031 | 0.3518 | 0.0556 |
|  | 12 | 1.0000 | 1.0000 | 1.0000 | 0.9997 | 0.9963 | 0.9729 | 0.8732 | 0.6020 | 0.1841 |
|  | 13 | 1.0000 | 1.0000 | 1.0000 | 1.0000 | 0.9995 | 0.9948 | 0.9647 | 0.8329 | 0.4510 |
|  | 14 | 1.0000 | 1.0000 | 1.0000 | 1.0000 | 1.0000 | 0.9995 | 0.9953 | 0.9648 | 0.7941 |
|  | 15 | 1.0000 | 1.0000 | 1.0000 | 1.0000 | 1.0000 | 1.0000 | 1.0000 | 1.0000 | 1.0000 |
| 16 | 0 | 0.1853 | 0.0281 | 0.0033 | 0.0003 | 0.0000 | 0.0000 | 0.0000 | 0.0000 | 0.0000 |
|  | 1 | 0.5147 | 0.1407 | 0.0261 | 0.0033 | 0.0003 | 0.0000 | 0.0000 | 0.0000 | 0.0000 |
|  | 2 | 0.7892 | 0.3518 | 0.0994 | 0.0183 | 0.0021 | 0.0001 | 0.0000 | 0.0000 | 0.0000 |
|  | 3 | 0.9316 | 0.5981 | 0.2459 | 0.0651 | 0.0106 | 0.0009 | 0.0000 | 0.0000 | 0.0000 |
|  | 4 | 0.9830 | 0.7982 | 0.4499 | 0.1666 | 0.0384 | 0.0049 | 0.0003 | 0.0000 | 0.0000 |
|  | 5 | 0.9967 | 0.9183 | 0.6598 | 0.3288 | 0.1051 | 0.0191 | 0.0016 | 0.0000 | 0.0000 |
|  | 6 | 0.9995 | 0.9733 | 0.8247 | 0.5272 | 0.2272 | 0.0583 | 0.0071 | 0.0002 | 0.0000 |
|  | 7 | 0.9999 | 0.9930 | 0.9256 | 0.7161 | 0.4018 | 0.1423 | 0.0257 | 0.0015 | 0.0000 |
|  | 8 | 1.0000 | 0.9985 | 0.9743 | 0.8577 | 0.5982 | 0.2839 | 0.0744 | 0.0070 | 0.0001 |
|  | 9 | 1.0000 | 0.9998 | 0.9929 | 0.9417 | 0.7728 | 0.4728 | 0.1753 | 0.0267 | 0.0005 |
|  | 10 | 1.0000 | 1.0000 | 0.9984 | 0.9809 | 0.8949 | 0.6712 | 0.3402 | 0.0817 | 0.0033 |
|  | 11 | 1.0000 | 1.0000 | 0.9997 | 0.9951 | 0.9616 | 0.8334 | 0.5501 | 0.2018 | 0.0170 |
|  | 12 | 1.0000 | 1.0000 | 1.0000 | 0.9991 | 0.9894 | 0.9349 | 0.7541 | 0.4019 | 0.0684 |
|  | 13 | 1.0000 | 1.0000 | 1.0000 | 0.9999 | 0.9979 | 0.9817 | 0.9006 | 0.6482 | 0.2108 |
|  | 14 | 1.0000 | 1.0000 | 1.0000 | 1.0000 | 0.9997 | 0.9967 | 0.9739 | 0.8593 | 0.4853 |
|  | 15 | 1.0000 | 1.0000 | 1.0000 | 1.0000 | 1.0000 | 0.9997 | 0.9967 | 0.9719 | 0.8147 |
|  | 16 | 1.0000 | 1.0000 | 1.0000 | 1.0000 | 1.0000 | 1.0000 | 1.0000 | 1.0000 | 1.0000 |

## E  Cumulative Binomial Probabilities (cont.)

| n | x | p 0.1 | 0.2 | 0.3 | 0.4 | 0.5 | 0.6 | 0.7 | 0.8 | 0.9 |
|---|---|---|---|---|---|---|---|---|---|---|
| 17 | 0 | 0.1668 | 0.0225 | 0.0023 | 0.0002 | 0.0000 | 0.0000 | 0.0000 | 0.0000 | 0.0000 |
| | 1 | 0.4818 | 0.1182 | 0.0193 | 0.0021 | 0.0001 | 0.0000 | 0.0000 | 0.0000 | 0.0000 |
| | 2 | 0.7618 | 0.3096 | 0.0774 | 0.0123 | 0.0012 | 0.0001 | 0.0000 | 0.0000 | 0.0000 |
| | 3 | 0.9174 | 0.5489 | 0.2019 | 0.0464 | 0.0064 | 0.0005 | 0.0000 | 0.0000 | 0.0000 |
| | 4 | 0.9779 | 0.7582 | 0.3887 | 0.1260 | 0.0245 | 0.0025 | 0.0001 | 0.0000 | 0.0000 |
| | 5 | 0.9953 | 0.8943 | 0.5968 | 0.2639 | 0.0717 | 0.0106 | 0.0007 | 0.0000 | 0.0000 |
| | 6 | 0.9992 | 0.9623 | 0.7752 | 0.4478 | 0.1662 | 0.0348 | 0.0032 | 0.0001 | 0.0000 |
| | 7 | 0.9999 | 0.9891 | 0.8954 | 0.6405 | 0.3145 | 0.0919 | 0.0127 | 0.0005 | 0.0000 |
| | 8 | 1.0000 | 0.9974 | 0.9597 | 0.8011 | 0.5000 | 0.1989 | 0.0403 | 0.0026 | 0.0000 |
| | 9 | 1.0000 | 0.9995 | 0.9873 | 0.9081 | 0.6855 | 0.3595 | 0.1046 | 0.0109 | 0.0001 |
| | 10 | 1.0000 | 0.9999 | 0.9968 | 0.9652 | 0.8338 | 0.5522 | 0.2248 | 0.0377 | 0.0008 |
| | 11 | 1.0000 | 1.0000 | 0.9993 | 0.9894 | 0.9283 | 0.7361 | 0.4032 | 0.1057 | 0.0047 |
| | 12 | 1.0000 | 1.0000 | 0.9999 | 0.9975 | 0.9755 | 0.8740 | 0.6113 | 0.2418 | 0.0221 |
| | 13 | 1.0000 | 1.0000 | 1.0000 | 0.9995 | 0.9936 | 0.9536 | 0.7981 | 0.4511 | 0.0826 |
| | 14 | 1.0000 | 1.0000 | 1.0000 | 0.9999 | 0.9988 | 0.9877 | 0.9226 | 0.6904 | 0.2382 |
| | 15 | 1.0000 | 1.0000 | 1.0000 | 1.0000 | 0.9999 | 0.9979 | 0.9807 | 0.8818 | 0.5182 |
| | 16 | 1.0000 | 1.0000 | 1.0000 | 1.0000 | 1.0000 | 0.9998 | 0.9977 | 0.9775 | 0.8332 |
| | 17 | 1.0000 | 1.0000 | 1.0000 | 1.0000 | 1.0000 | 1.0000 | 1.0000 | 1.0000 | 1.0000 |
| 18 | 0 | 0.1501 | 0.0180 | 0.0016 | 0.0001 | 0.0000 | 0.0000 | 0.0000 | 0.0000 | 0.0000 |
| | 1 | 0.4503 | 0.0991 | 0.0142 | 0.0013 | 0.0001 | 0.0000 | 0.0000 | 0.0000 | 0.0000 |
| | 2 | 0.7338 | 0.2713 | 0.0600 | 0.0082 | 0.0007 | 0.0000 | 0.0000 | 0.0000 | 0.0000 |
| | 3 | 0.9018 | 0.5010 | 0.1646 | 0.0328 | 0.0038 | 0.0002 | 0.0000 | 0.0000 | 0.0000 |
| | 4 | 0.9718 | 0.7164 | 0.3327 | 0.0942 | 0.0154 | 0.0013 | 0.0000 | 0.0000 | 0.0000 |
| | 5 | 0.9936 | 0.8671 | 0.5344 | 0.2088 | 0.0481 | 0.0058 | 0.0003 | 0.0000 | 0.0000 |
| | 6 | 0.9988 | 0.9487 | 0.7217 | 0.3743 | 0.1189 | 0.0203 | 0.0014 | 0.0000 | 0.0000 |
| | 7 | 0.9998 | 0.9837 | 0.8593 | 0.5634 | 0.2403 | 0.0576 | 0.0061 | 0.0002 | 0.0000 |
| | 8 | 1.0000 | 0.9957 | 0.9404 | 0.7368 | 0.4073 | 0.1347 | 0.0210 | 0.0009 | 0.0000 |
| | 9 | 1.0000 | 0.9991 | 0.9790 | 0.8653 | 0.5927 | 0.2632 | 0.0596 | 0.0043 | 0.0000 |
| | 10 | 1.0000 | 0.9998 | 0.9939 | 0.9424 | 0.7597 | 0.4366 | 0.1407 | 0.0163 | 0.0002 |
| | 11 | 1.0000 | 1.0000 | 0.9986 | 0.9797 | 0.8811 | 0.6257 | 0.2783 | 0.0513 | 0.0012 |
| | 12 | 1.0000 | 1.0000 | 0.9997 | 0.9942 | 0.9519 | 0.7912 | 0.4656 | 0.1329 | 0.0064 |
| | 13 | 1.0000 | 1.0000 | 1.0000 | 0.9987 | 0.9846 | 0.9058 | 0.6673 | 0.2836 | 0.0282 |
| | 14 | 1.0000 | 1.0000 | 1.0000 | 0.9998 | 0.9962 | 0.9672 | 0.8354 | 0.4990 | 0.0982 |
| | 15 | 1.0000 | 1.0000 | 1.0000 | 1.0000 | 0.9993 | 0.9918 | 0.9400 | 0.7287 | 0.2662 |
| | 16 | 1.0000 | 1.0000 | 1.0000 | 1.0000 | 0.9999 | 0.9987 | 0.9858 | 0.9009 | 0.5497 |
| | 17 | 1.0000 | 1.0000 | 1.0000 | 1.0000 | 1.0000 | 0.9999 | 0.9984 | 0.9820 | 0.8499 |
| | 18 | 1.0000 | 1.0000 | 1.0000 | 1.0000 | 1.0000 | 1.0000 | 1.0000 | 1.0000 | 1.0000 |

# E Cumulative Binomial Probabilities (cont.)

| | | | | | | p | | | | |
|---|---|---|---|---|---|---|---|---|---|---|
| n | x | 0.1 | 0.2 | 0.3 | 0.4 | 0.5 | 0.6 | 0.7 | 0.8 | 0.9 |
| 19 | 0 | 0.1351 | 0.0144 | 0.0011 | 0.0001 | 0.0000 | 0.0000 | 0.0000 | 0.0000 | 0.0000 |
| | 1 | 0.4203 | 0.0829 | 0.0104 | 0.0008 | 0.0000 | 0.0000 | 0.0000 | 0.0000 | 0.0000 |
| | 2 | 0.7054 | 0.2369 | 0.0462 | 0.0055 | 0.0004 | 0.0000 | 0.0000 | 0.0000 | 0.0000 |
| | 3 | 0.8850 | 0.4551 | 0.1332 | 0.0230 | 0.0022 | 0.0001 | 0.0000 | 0.0000 | 0.0000 |
| | 4 | 0.9648 | 0.6733 | 0.2822 | 0.0696 | 0.0096 | 0.0006 | 0.0000 | 0.0000 | 0.0000 |
| | 5 | 0.9914 | 0.8369 | 0.4739 | 0.1629 | 0.0318 | 0.0031 | 0.0001 | 0.0000 | 0.0000 |
| | 6 | 0.9983 | 0.9324 | 0.6655 | 0.3081 | 0.0835 | 0.0116 | 0.0006 | 0.0000 | 0.0000 |
| | 7 | 0.9997 | 0.9767 | 0.8180 | 0.4878 | 0.1796 | 0.0352 | 0.0028 | 0.0000 | 0.0000 |
| | 8 | 1.0000 | 0.9933 | 0.9161 | 0.6675 | 0.3238 | 0.0885 | 0.0105 | 0.0003 | 0.0000 |
| | 9 | 1.0000 | 0.9984 | 0.9674 | 0.8139 | 0.5000 | 0.1861 | 0.0326 | 0.0016 | 0.0000 |
| | 10 | 1.0000 | 0.9997 | 0.9895 | 0.9115 | 0.6762 | 0.3325 | 0.0839 | 0.0067 | 0.0000 |
| | 11 | 1.0000 | 1.0000 | 0.9972 | 0.9648 | 0.8204 | 0.5122 | 0.1820 | 0.0233 | 0.0003 |
| | 12 | 1.0000 | 1.0000 | 0.9994 | 0.9884 | 0.9165 | 0.6919 | 0.3345 | 0.0676 | 0.0017 |
| | 13 | 1.0000 | 1.0000 | 0.9999 | 0.9969 | 0.9682 | 0.8371 | 0.5261 | 0.1631 | 0.0086 |
| | 14 | 1.0000 | 1.0000 | 1.0000 | 0.9994 | 0.9904 | 0.9304 | 0.7178 | 0.3267 | 0.0352 |
| | 15 | 1.0000 | 1.0000 | 1.0000 | 0.9999 | 0.9978 | 0.9770 | 0.8668 | 0.5449 | 0.1150 |
| | 16 | 1.0000 | 1.0000 | 1.0000 | 1.0000 | 0.9996 | 0.9945 | 0.9538 | 0.7631 | 0.2946 |
| | 17 | 1.0000 | 1.0000 | 1.0000 | 1.0000 | 1.0000 | 0.9992 | 0.9896 | 0.9171 | 0.5797 |
| | 18 | 1.0000 | 1.0000 | 1.0000 | 1.0000 | 1.0000 | 0.9999 | 0.9989 | 0.9856 | 0.8649 |
| | 19 | 1.0000 | 1.0000 | 1.0000 | 1.0000 | 1.0000 | 1.0000 | 1.0000 | 1.0000 | 1.0000 |
| 20 | 0 | 0.1216 | 0.0115 | 0.0008 | 0.0000 | 0.0000 | 0.0000 | 0.0000 | 0.0000 | 0.0000 |
| | 1 | 0.3917 | 0.0692 | 0.0076 | 0.0005 | 0.0000 | 0.0000 | 0.0000 | 0.0000 | 0.0000 |
| | 2 | 0.6769 | 0.2061 | 0.0355 | 0.0036 | 0.0002 | 0.0000 | 0.0000 | 0.0000 | 0.0000 |
| | 3 | 0.8670 | 0.4114 | 0.1071 | 0.0160 | 0.0013 | 0.0000 | 0.0000 | 0.0000 | 0.0000 |
| | 4 | 0.9568 | 0.6296 | 0.2375 | 0.0510 | 0.0059 | 0.0003 | 0.0000 | 0.0000 | 0.0000 |
| | 5 | 0.9887 | 0.8042 | 0.4164 | 0.1256 | 0.0207 | 0.0016 | 0.0000 | 0.0000 | 0.0000 |
| | 6 | 0.9976 | 0.9133 | 0.6080 | 0.2500 | 0.0577 | 0.0065 | 0.0003 | 0.0000 | 0.0000 |
| | 7 | 0.9996 | 0.9679 | 0.7723 | 0.4159 | 0.1316 | 0.0210 | 0.0013 | 0.0000 | 0.0000 |
| | 8 | 0.9999 | 0.9900 | 0.8867 | 0.5956 | 0.2517 | 0.0565 | 0.0051 | 0.0001 | 0.0000 |
| | 9 | 1.0000 | 0.9974 | 0.9520 | 0.7553 | 0.4119 | 0.1275 | 0.0171 | 0.0006 | 0.0000 |
| | 10 | 1.0000 | 0.9994 | 0.9829 | 0.8725 | 0.5881 | 0.2447 | 0.0480 | 0.0026 | 0.0000 |
| | 11 | 1.0000 | 0.9999 | 0.9949 | 0.9435 | 0.7483 | 0.4044 | 0.1133 | 0.0100 | 0.0001 |
| | 12 | 1.0000 | 1.0000 | 0.9987 | 0.9790 | 0.8684 | 0.5841 | 0.2277 | 0.0321 | 0.0004 |
| | 13 | 1.0000 | 1.0000 | 0.9997 | 0.9935 | 0.9423 | 0.7500 | 0.3920 | 0.0867 | 0.0024 |
| | 14 | 1.0000 | 1.0000 | 1.0000 | 0.9984 | 0.9793 | 0.8744 | 0.5836 | 0.1958 | 0.0113 |
| | 15 | 1.0000 | 1.0000 | 1.0000 | 0.9997 | 0.9941 | 0.9490 | 0.7625 | 0.3704 | 0.0432 |
| | 16 | 1.0000 | 1.0000 | 1.0000 | 1.0000 | 0.9987 | 0.9840 | 0.8929 | 0.5886 | 0.1330 |
| | 17 | 1.0000 | 1.0000 | 1.0000 | 1.0000 | 0.9998 | 0.9964 | 0.9645 | 0.7939 | 0.3231 |
| | 18 | 1.0000 | 1.0000 | 1.0000 | 1.0000 | 1.0000 | 0.9995 | 0.9924 | 0.9308 | 0.6083 |
| | 19 | 1.0000 | 1.0000 | 1.0000 | 1.0000 | 1.0000 | 1.0000 | 0.9992 | 0.9885 | 0.8784 |
| | 20 | 1.0000 | 1.0000 | 1.0000 | 1.0000 | 1.0000 | 1.0000 | 1.0000 | 1.0000 | 1.0000 |

# F  Cumulative Poisson Probabilities

Numerical entries represent $P(X \leq x)$.

$P(X \leq 2)$

| x | 0.02 | 0.03 | 0.04 | 0.05 | 0.06 | 0.07 | 0.08 | 0.09 | 0.10 | 0.20 | 0.30 | 0.40 |
|---|------|------|------|------|------|------|------|------|------|------|------|------|
| 0 | 0.9802 | 0.9704 | 0.9608 | 0.9512 | 0.9418 | 0.9324 | 0.9231 | 0.9139 | 0.9048 | 0.8187 | 0.7408 | 0.6703 |
| 1 | 0.9998 | 0.9996 | 0.9992 | 0.9988 | 0.9983 | 0.9977 | 0.9970 | 0.9962 | 0.9953 | 0.9825 | 0.9631 | 0.9384 |
| 2 | 1.0000 | 1.0000 | 1.0000 | 1.0000 | 1.0000 | 0.9999 | 0.9999 | 0.9999 | 0.9998 | 0.9989 | 0.9964 | 0.9921 |
| 3 | 1.0000 | 1.0000 | 1.0000 | 1.0000 | 1.0000 | 1.0000 | 1.0000 | 1.0000 | 1.0000 | 0.9999 | 0.9997 | 0.9992 |
| 4 | 1.0000 | 1.0000 | 1.0000 | 1.0000 | 1.0000 | 1.0000 | 1.0000 | 1.0000 | 1.0000 | 1.0000 | 1.0000 | 0.9999 |
| 5 | 1.0000 | 1.0000 | 1.0000 | 1.0000 | 1.0000 | 1.0000 | 1.0000 | 1.0000 | 1.0000 | 1.0000 | 1.0000 | 1.0000 |

| x | 0.50 | 0.60 | 0.70 | 0.80 | 0.90 | 1.00 | 1.10 | 1.20 | 1.30 | 1.40 | 1.50 | 1.60 |
|---|------|------|------|------|------|------|------|------|------|------|------|------|
| 0 | 0.6065 | 0.5488 | 0.4966 | 0.4493 | 0.4066 | 0.3679 | 0.3329 | 0.3012 | 0.2725 | 0.2466 | 0.2231 | 0.2019 |
| 1 | 0.9098 | 0.8781 | 0.8442 | 0.8088 | 0.7725 | 0.7358 | 0.6990 | 0.6626 | 0.6268 | 0.5918 | 0.5578 | 0.5249 |
| 2 | 0.9856 | 0.9769 | 0.9659 | 0.9526 | 0.9371 | 0.9197 | 0.9004 | 0.8795 | 0.8571 | 0.8335 | 0.8088 | 0.7834 |
| 3 | 0.9982 | 0.9966 | 0.9942 | 0.9909 | 0.9865 | 0.9810 | 0.9743 | 0.9662 | 0.9569 | 0.9463 | 0.9344 | 0.9212 |
| 4 | 0.9998 | 0.9996 | 0.9992 | 0.9986 | 0.9977 | 0.9963 | 0.9946 | 0.9923 | 0.9893 | 0.9857 | 0.9814 | 0.9763 |
| 5 | 1.0000 | 1.0000 | 0.9999 | 0.9998 | 0.9997 | 0.9994 | 0.9990 | 0.9985 | 0.9978 | 0.9968 | 0.9955 | 0.9940 |
| 6 | 1.0000 | 1.0000 | 1.0000 | 1.0000 | 1.0000 | 0.9999 | 0.9999 | 0.9997 | 0.9996 | 0.9994 | 0.9991 | 0.9987 |
| 7 | 1.0000 | 1.0000 | 1.0000 | 1.0000 | 1.0000 | 1.0000 | 1.0000 | 1.0000 | 0.9999 | 0.9999 | 0.9998 | 0.9997 |
| 8 | 1.0000 | 1.0000 | 1.0000 | 1.0000 | 1.0000 | 1.0000 | 1.0000 | 1.0000 | 1.0000 | 1.0000 | 1.0000 | 1.0000 |

| x | 1.70 | 1.80 | 1.90 | 2.00 | 2.10 | 2.20 | 2.30 | 2.40 | 2.50 | 2.60 | 2.70 | 2.80 |
|---|------|------|------|------|------|------|------|------|------|------|------|------|
| 0 | 0.1827 | 0.1653 | 0.1496 | 0.1353 | 0.1225 | 0.1108 | 0.1003 | 0.0907 | 0.0821 | 0.0743 | 0.0672 | 0.0608 |
| 1 | 0.4932 | 0.4628 | 0.4337 | 0.4060 | 0.3796 | 0.3546 | 0.3309 | 0.3084 | 0.2873 | 0.2674 | 0.2487 | 0.2311 |
| 2 | 0.7572 | 0.7306 | 0.7037 | 0.6767 | 0.6496 | 0.6227 | 0.5960 | 0.5697 | 0.5438 | 0.5184 | 0.4936 | 0.4695 |
| 3 | 0.9068 | 0.8913 | 0.8747 | 0.8571 | 0.8386 | 0.8194 | 0.7993 | 0.7787 | 0.7576 | 0.7360 | 0.7141 | 0.6919 |
| 4 | 0.9704 | 0.9636 | 0.9559 | 0.9473 | 0.9379 | 0.9275 | 0.9162 | 0.9041 | 0.8912 | 0.8774 | 0.8629 | 0.8477 |
| 5 | 0.9920 | 0.9896 | 0.9868 | 0.9834 | 0.9796 | 0.9751 | 0.9700 | 0.9643 | 0.9580 | 0.9510 | 0.9433 | 0.9349 |
| 6 | 0.9981 | 0.9974 | 0.9966 | 0.9955 | 0.9941 | 0.9925 | 0.9906 | 0.9884 | 0.9858 | 0.9828 | 0.9794 | 0.9756 |
| 7 | 0.9996 | 0.9994 | 0.9992 | 0.9989 | 0.9985 | 0.9980 | 0.9974 | 0.9967 | 0.9958 | 0.9947 | 0.9934 | 0.9919 |
| 8 | 0.9999 | 0.9999 | 0.9998 | 0.9998 | 0.9997 | 0.9995 | 0.9994 | 0.9991 | 0.9989 | 0.9985 | 0.9981 | 0.9976 |
| 9 | 1.0000 | 1.0000 | 1.0000 | 1.0000 | 0.9999 | 0.9999 | 0.9999 | 0.9998 | 0.9997 | 0.9996 | 0.9995 | 0.9993 |
| 10 | 1.0000 | 1.0000 | 1.0000 | 1.0000 | 1.0000 | 1.0000 | 1.0000 | 1.0000 | 0.9999 | 0.9999 | 0.9999 | 0.9998 |
| 11 | 1.0000 | 1.0000 | 1.0000 | 1.0000 | 1.0000 | 1.0000 | 1.0000 | 1.0000 | 1.0000 | 1.0000 | 1.0000 | 1.0000 |

| x | 2.90 | 3.00 | 3.10 | 3.20 | 3.30 | 3.40 | 3.50 | 3.60 | 3.70 | 3.80 | 3.90 | 4.00 |
|---|------|------|------|------|------|------|------|------|------|------|------|------|
| 0 | 0.0550 | 0.0498 | 0.0450 | 0.0408 | 0.0369 | 0.0334 | 0.0302 | 0.0273 | 0.0247 | 0.0224 | 0.0202 | 0.0183 |
| 1 | 0.2146 | 0.1991 | 0.1847 | 0.1712 | 0.1586 | 0.1468 | 0.1359 | 0.1257 | 0.1162 | 0.1074 | 0.0992 | 0.0916 |
| 2 | 0.4460 | 0.4232 | 0.4012 | 0.3799 | 0.3594 | 0.3397 | 0.3208 | 0.3027 | 0.2854 | 0.2689 | 0.2531 | 0.2381 |
| 3 | 0.6696 | 0.6472 | 0.6248 | 0.6025 | 0.5803 | 0.5584 | 0.5366 | 0.5152 | 0.4942 | 0.4735 | 0.4532 | 0.4335 |
| 4 | 0.8318 | 0.8153 | 0.7982 | 0.7806 | 0.7626 | 0.7442 | 0.7254 | 0.7064 | 0.6872 | 0.6678 | 0.6484 | 0.6288 |
| 5 | 0.9258 | 0.9161 | 0.9057 | 0.8946 | 0.8829 | 0.8705 | 0.8576 | 0.8441 | 0.8301 | 0.8156 | 0.8006 | 0.7851 |
| 6 | 0.9713 | 0.9665 | 0.9612 | 0.9554 | 0.9490 | 0.9421 | 0.9347 | 0.9267 | 0.9182 | 0.9091 | 0.8995 | 0.8893 |
| 7 | 0.9901 | 0.9881 | 0.9858 | 0.9832 | 0.9802 | 0.9769 | 0.9733 | 0.9692 | 0.9648 | 0.9599 | 0.9546 | 0.9489 |
| 8 | 0.9969 | 0.9962 | 0.9953 | 0.9943 | 0.9931 | 0.9917 | 0.9901 | 0.9883 | 0.9863 | 0.9840 | 0.9815 | 0.9786 |
| 9 | 0.9991 | 0.9989 | 0.9986 | 0.9982 | 0.9978 | 0.9973 | 0.9967 | 0.9960 | 0.9952 | 0.9942 | 0.9931 | 0.9919 |
| 10 | 0.9998 | 0.9997 | 0.9996 | 0.9995 | 0.9994 | 0.9992 | 0.9990 | 0.9987 | 0.9984 | 0.9981 | 0.9977 | 0.9972 |
| 11 | 0.9999 | 0.9999 | 0.9999 | 0.9999 | 0.9998 | 0.9998 | 0.9997 | 0.9996 | 0.9995 | 0.9994 | 0.9993 | 0.9991 |

# F | Cumulative Poisson Probabilities (cont.)

| x | 2.90 | 3.00 | 3.10 | 3.20 | 3.30 | 3.40 | 3.50 | 3.60 | 3.70 | 3.80 | 3.90 | 4.00 |
|---|---|---|---|---|---|---|---|---|---|---|---|---|
| 12 | 1.0000 | 1.0000 | 1.0000 | 1.0000 | 1.0000 | 0.9999 | 0.9999 | 0.9999 | 0.9999 | 0.9998 | 0.9998 | 0.9997 |
| 13 | 1.0000 | 1.0000 | 1.0000 | 1.0000 | 1.0000 | 1.0000 | 1.0000 | 1.0000 | 1.0000 | 1.0000 | 0.9999 | 0.9999 |
| 14 | 1.0000 | 1.0000 | 1.0000 | 1.0000 | 1.0000 | 1.0000 | 1.0000 | 1.0000 | 1.0000 | 1.0000 | 1.0000 | 1.0000 |

| x | 4.10 | 4.20 | 4.30 | 4.40 | 4.50 | 4.60 | 4.70 | 4.80 | 4.90 | 5.00 | 5.10 | 5.20 |
|---|---|---|---|---|---|---|---|---|---|---|---|---|
| 0 | 0.0166 | 0.0150 | 0.0136 | 0.0123 | 0.0111 | 0.0101 | 0.0091 | 0.0082 | 0.0074 | 0.0067 | 0.0061 | 0.0055 |
| 1 | 0.0845 | 0.0780 | 0.0719 | 0.0663 | 0.0611 | 0.0563 | 0.0518 | 0.0477 | 0.0439 | 0.0404 | 0.0372 | 0.0342 |
| 2 | 0.2238 | 0.2102 | 0.1974 | 0.1851 | 0.1736 | 0.1626 | 0.1523 | 0.1425 | 0.1333 | 0.1247 | 0.1165 | 0.1088 |
| 3 | 0.4142 | 0.3954 | 0.3772 | 0.3594 | 0.3423 | 0.3257 | 0.3097 | 0.2942 | 0.2793 | 0.2650 | 0.2513 | 0.2381 |
| 4 | 0.6093 | 0.5898 | 0.5704 | 0.5512 | 0.5321 | 0.5132 | 0.4946 | 0.4763 | 0.4582 | 0.4405 | 0.4231 | 0.4061 |
| 5 | 0.7693 | 0.7531 | 0.7367 | 0.7199 | 0.7029 | 0.6858 | 0.6684 | 0.6510 | 0.6335 | 0.6160 | 0.5984 | 0.5809 |
| 6 | 0.8786 | 0.8675 | 0.8558 | 0.8436 | 0.8311 | 0.8180 | 0.8046 | 0.7908 | 0.7767 | 0.7622 | 0.7474 | 0.7324 |
| 7 | 0.9427 | 0.9361 | 0.9290 | 0.9214 | 0.9134 | 0.9049 | 0.8960 | 0.8867 | 0.8769 | 0.8666 | 0.8560 | 0.8449 |
| 8 | 0.9755 | 0.9721 | 0.9683 | 0.9642 | 0.9597 | 0.9549 | 0.9497 | 0.9442 | 0.9382 | 0.9319 | 0.9252 | 0.9181 |
| 9 | 0.9905 | 0.9889 | 0.9871 | 0.9851 | 0.9829 | 0.9805 | 0.9778 | 0.9749 | 0.9717 | 0.9682 | 0.9644 | 0.9603 |
| 10 | 0.9966 | 0.9959 | 0.9952 | 0.9943 | 0.9933 | 0.9922 | 0.9910 | 0.9896 | 0.9880 | 0.9863 | 0.9844 | 0.9823 |
| 11 | 0.9989 | 0.9986 | 0.9983 | 0.9980 | 0.9976 | 0.9971 | 0.9966 | 0.9960 | 0.9953 | 0.9945 | 0.9937 | 0.9927 |
| 12 | 0.9997 | 0.9996 | 0.9995 | 0.9993 | 0.9992 | 0.9990 | 0.9988 | 0.9986 | 0.9983 | 0.9980 | 0.9976 | 0.9972 |
| 13 | 0.9999 | 0.9999 | 0.9998 | 0.9998 | 0.9997 | 0.9997 | 0.9996 | 0.9995 | 0.9994 | 0.9993 | 0.9992 | 0.9990 |
| 14 | 1.0000 | 1.0000 | 1.0000 | 0.9999 | 0.9999 | 0.9999 | 0.9999 | 0.9999 | 0.9998 | 0.9998 | 0.9997 | 0.9997 |
| 15 | 1.0000 | 1.0000 | 1.0000 | 1.0000 | 1.0000 | 1.0000 | 1.0000 | 1.0000 | 0.9999 | 0.9999 | 0.9999 | 0.9999 |
| 16 | 1.0000 | 1.0000 | 1.0000 | 1.0000 | 1.0000 | 1.0000 | 1.0000 | 1.0000 | 1.0000 | 1.0000 | 1.0000 | 1.0000 |

| x | 5.30 | 5.40 | 5.50 | 5.60 | 5.70 | 5.80 | 5.90 | 6.00 | 6.10 | 6.20 | 6.30 | 6.40 |
|---|---|---|---|---|---|---|---|---|---|---|---|---|
| 0 | 0.0050 | 0.0045 | 0.0041 | 0.0037 | 0.0033 | 0.0030 | 0.0027 | 0.0025 | 0.0022 | 0.0020 | 0.0018 | 0.0017 |
| 1 | 0.0314 | 0.0289 | 0.0266 | 0.0244 | 0.0224 | 0.0206 | 0.0189 | 0.0174 | 0.0159 | 0.0146 | 0.0134 | 0.0123 |
| 2 | 0.1016 | 0.0948 | 0.0884 | 0.0824 | 0.0768 | 0.0715 | 0.0666 | 0.0620 | 0.0577 | 0.0536 | 0.0498 | 0.0463 |
| 3 | 0.2254 | 0.2133 | 0.2017 | 0.1906 | 0.1800 | 0.1700 | 0.1604 | 0.1512 | 0.1425 | 0.1342 | 0.1264 | 0.1189 |
| 4 | 0.3895 | 0.3733 | 0.3575 | 0.3422 | 0.3272 | 0.3127 | 0.2987 | 0.2851 | 0.2719 | 0.2592 | 0.2469 | 0.2351 |
| 5 | 0.5635 | 0.5461 | 0.5289 | 0.5119 | 0.4950 | 0.4783 | 0.4619 | 0.4457 | 0.4298 | 0.4141 | 0.3988 | 0.3837 |
| 6 | 0.7171 | 0.7017 | 0.6860 | 0.6703 | 0.6544 | 0.6384 | 0.6224 | 0.6063 | 0.5902 | 0.5742 | 0.5582 | 0.5423 |
| 7 | 0.8335 | 0.8217 | 0.8095 | 0.7970 | 0.7841 | 0.7710 | 0.7576 | 0.7440 | 0.7301 | 0.7160 | 0.7017 | 0.6873 |
| 8 | 0.9106 | 0.9027 | 0.8944 | 0.8857 | 0.8766 | 0.8672 | 0.8574 | 0.8472 | 0.8367 | 0.8259 | 0.8148 | 0.8033 |
| 9 | 0.9559 | 0.9512 | 0.9462 | 0.9409 | 0.9352 | 0.9292 | 0.9228 | 0.9161 | 0.9090 | 0.9016 | 0.8939 | 0.8858 |
| 10 | 0.9800 | 0.9775 | 0.9747 | 0.9718 | 0.9686 | 0.9651 | 0.9614 | 0.9574 | 0.9531 | 0.9486 | 0.9437 | 0.9386 |
| 11 | 0.9916 | 0.9904 | 0.9890 | 0.9875 | 0.9859 | 0.9841 | 0.9821 | 0.9799 | 0.9776 | 0.9750 | 0.9723 | 0.9693 |
| 12 | 0.9967 | 0.9962 | 0.9955 | 0.9949 | 0.9941 | 0.9932 | 0.9922 | 0.9912 | 0.9900 | 0.9887 | 0.9873 | 0.9857 |
| 13 | 0.9988 | 0.9986 | 0.9983 | 0.9980 | 0.9977 | 0.9973 | 0.9969 | 0.9964 | 0.9958 | 0.9952 | 0.9945 | 0.9937 |
| 14 | 0.9996 | 0.9995 | 0.9994 | 0.9993 | 0.9991 | 0.9990 | 0.9988 | 0.9986 | 0.9984 | 0.9981 | 0.9978 | 0.9974 |
| 15 | 0.9999 | 0.9998 | 0.9998 | 0.9998 | 0.9997 | 0.9996 | 0.9996 | 0.9995 | 0.9994 | 0.9993 | 0.9992 | 0.9990 |
| 16 | 1.0000 | 0.9999 | 0.9999 | 0.9999 | 0.9999 | 0.9999 | 0.9999 | 0.9998 | 0.9998 | 0.9997 | 0.9997 | 0.9996 |
| 17 | 1.0000 | 1.0000 | 1.0000 | 1.0000 | 1.0000 | 1.0000 | 1.0000 | 0.9999 | 0.9999 | 0.9999 | 0.9999 | 0.9999 |
| 18 | 1.0000 | 1.0000 | 1.0000 | 1.0000 | 1.0000 | 1.0000 | 1.0000 | 1.0000 | 1.0000 | 1.0000 | 1.0000 | 1.0000 |

| x | 6.50 | 6.60 | 6.70 | 6.80 | 6.90 | 7.00 | 7.10 | 7.20 | 7.30 | 7.40 | 7.50 | 7.60 |
|---|---|---|---|---|---|---|---|---|---|---|---|---|
| 0 | 0.0015 | 0.0014 | 0.0012 | 0.0011 | 0.0010 | 0.0009 | 0.0008 | 0.0007 | 0.0007 | 0.0006 | 0.0006 | 0.0005 |
| 1 | 0.0113 | 0.0103 | 0.0095 | 0.0087 | 0.0080 | 0.0073 | 0.0067 | 0.0061 | 0.0056 | 0.0051 | 0.0047 | 0.0043 |
| 2 | 0.0430 | 0.0400 | 0.0371 | 0.0344 | 0.0320 | 0.0296 | 0.0275 | 0.0255 | 0.0236 | 0.0219 | 0.0203 | 0.0188 |
| 3 | 0.1118 | 0.1052 | 0.0988 | 0.0928 | 0.0871 | 0.0818 | 0.0767 | 0.0719 | 0.0674 | 0.0632 | 0.0591 | 0.0554 |

# F Cumulative Poisson Probabilities (cont.)

| | | | | | | λ | | | | | | |
|---|---|---|---|---|---|---|---|---|---|---|---|---|
| x | 6.50 | 6.60 | 6.70 | 6.80 | 6.90 | 7.00 | 7.10 | 7.20 | 7.30 | 7.40 | 7.50 | 7.60 |
| 4 | 0.2237 | 0.2127 | 0.2022 | 0.1920 | 0.1823 | 0.1730 | 0.1641 | 0.1555 | 0.1473 | 0.1395 | 0.1321 | 0.1249 |
| 5 | 0.3690 | 0.3547 | 0.3406 | 0.3270 | 0.3137 | 0.3007 | 0.2881 | 0.2759 | 0.2640 | 0.2526 | 0.2414 | 0.2307 |
| 6 | 1.0000 | 1.0000 | 1.0000 | 1.0000 | 1.0000 | 1.0000 | 1.0000 | 1.0000 | 1.0000 | 1.0000 | 1.0000 | 1.0000 |
| 7 | 0.6728 | 0.6581 | 0.6433 | 0.6285 | 0.6136 | 0.5987 | 0.5838 | 0.5689 | 0.5541 | 0.5393 | 0.5246 | 0.5100 |
| 8 | 0.7916 | 0.7796 | 0.7673 | 0.7548 | 0.7420 | 0.7291 | 0.7160 | 0.7027 | 0.6892 | 0.6757 | 0.6620 | 0.6482 |
| 9 | 0.8774 | 0.8686 | 0.8596 | 0.8502 | 0.8405 | 0.8305 | 0.8202 | 0.8096 | 0.7988 | 0.7877 | 0.7764 | 0.7649 |
| 10 | 0.9332 | 0.9274 | 0.9214 | 0.9151 | 0.9084 | 0.9015 | 0.8942 | 0.8867 | 0.8788 | 0.8707 | 0.8622 | 0.8535 |
| 11 | 0.9661 | 0.9627 | 0.9591 | 0.9552 | 0.9510 | 0.9467 | 0.9420 | 0.9371 | 0.9319 | 0.9265 | 0.9208 | 0.9148 |
| 12 | 0.9840 | 0.9821 | 0.9801 | 0.9779 | 0.9755 | 0.9730 | 0.9703 | 0.9673 | 0.9642 | 0.9609 | 0.9573 | 0.9536 |
| 13 | 0.9929 | 0.9920 | 0.9909 | 0.9898 | 0.9885 | 0.9872 | 0.9857 | 0.9841 | 0.9824 | 0.9805 | 0.9784 | 0.9762 |
| 14 | 0.9970 | 0.9966 | 0.9961 | 0.9956 | 0.9950 | 0.9943 | 0.9935 | 0.9927 | 0.9918 | 0.9908 | 0.9897 | 0.9886 |
| 15 | 0.9988 | 0.9986 | 0.9984 | 0.9982 | 0.9979 | 0.9976 | 0.9972 | 0.9969 | 0.9964 | 0.9959 | 0.9954 | 0.9948 |
| 16 | 0.9996 | 0.9995 | 0.9994 | 0.9993 | 0.9992 | 0.9990 | 0.9989 | 0.9987 | 0.9985 | 0.9983 | 0.9980 | 0.9978 |
| 17 | 0.9998 | 0.9998 | 0.9998 | 0.9997 | 0.9997 | 0.9996 | 0.9996 | 0.9995 | 0.9994 | 0.9993 | 0.9992 | 0.9991 |
| 18 | 0.9999 | 0.9999 | 0.9999 | 0.9999 | 0.9999 | 0.9999 | 0.9998 | 0.9998 | 0.9998 | 0.9997 | 0.9997 | 0.9996 |
| 19 | 1.0000 | 1.0000 | 1.0000 | 1.0000 | 1.0000 | 1.0000 | 0.9999 | 0.9999 | 0.9999 | 0.9999 | 0.9999 | 0.9999 |
| 20 | 1.0000 | 1.0000 | 1.0000 | 1.0000 | 1.0000 | 1.0000 | 1.0000 | 1.0000 | 1.0000 | 1.0000 | 1.0000 | 1.0000 |

| x | 7.70 | 7.80 | 7.90 | 8.00 | 8.10 | 8.20 | 8.30 | 8.40 | 8.50 | 8.60 | 8.70 | 8.80 |
|---|---|---|---|---|---|---|---|---|---|---|---|---|
| 0 | 0.0005 | 0.0004 | 0.0004 | 0.0003 | 0.0003 | 0.0003 | 0.0002 | 0.0002 | 0.0002 | 0.0002 | 0.0002 | 0.0002 |
| 1 | 0.0039 | 0.0036 | 0.0033 | 0.0030 | 0.0028 | 0.0025 | 0.0023 | 0.0021 | 0.0019 | 0.0018 | 0.0016 | 0.0015 |
| 2 | 0.0174 | 0.0161 | 0.0149 | 0.0138 | 0.0127 | 0.0118 | 0.0109 | 0.0100 | 0.0093 | 0.0086 | 0.0079 | 0.0073 |
| 3 | 0.0518 | 0.0485 | 0.0453 | 0.0424 | 0.0396 | 0.0370 | 0.0346 | 0.0323 | 0.0301 | 0.0281 | 0.0262 | 0.0244 |
| 4 | 0.1181 | 0.1117 | 0.1055 | 0.0996 | 0.0940 | 0.0887 | 0.0837 | 0.0789 | 0.0744 | 0.0701 | 0.0660 | 0.0621 |
| 5 | 0.2203 | 0.2103 | 0.2006 | 0.1912 | 0.1822 | 0.1736 | 0.1653 | 0.1573 | 0.1496 | 0.1422 | 0.1352 | 0.1284 |
| 6 | 0.3514 | 0.3384 | 0.3257 | 0.3134 | 0.3013 | 0.2896 | 0.2781 | 0.2670 | 0.2562 | 0.2457 | 0.2355 | 0.2256 |
| 7 | 0.4956 | 0.4812 | 0.4670 | 0.4530 | 0.4391 | 0.4254 | 0.4119 | 0.3987 | 0.3856 | 0.3728 | 0.3602 | 0.3478 |
| 8 | 0.6343 | 0.6204 | 0.6065 | 0.5925 | 0.5786 | 0.5647 | 0.5507 | 0.5369 | 0.5231 | 0.5094 | 0.4958 | 0.4823 |
| 9 | 0.7531 | 0.7411 | 0.7290 | 0.7166 | 0.7041 | 0.6915 | 0.6788 | 0.6659 | 0.6530 | 0.6400 | 0.6269 | 0.6137 |
| 10 | 0.8445 | 0.8352 | 0.8257 | 0.8159 | 0.8058 | 0.7955 | 0.7850 | 0.7743 | 0.7634 | 0.7522 | 0.7409 | 0.7294 |
| 11 | 0.9085 | 0.9020 | 0.8952 | 0.8881 | 0.8807 | 0.8731 | 0.8652 | 0.8571 | 0.8487 | 0.8400 | 0.8311 | 0.8220 |
| 12 | 0.9496 | 0.9454 | 0.9409 | 0.9362 | 0.9313 | 0.9261 | 0.9207 | 0.9150 | 0.9091 | 0.9029 | 0.8965 | 0.8898 |
| 13 | 0.9739 | 0.9714 | 0.9687 | 0.9658 | 0.9628 | 0.9595 | 0.9561 | 0.9524 | 0.9486 | 0.9445 | 0.9403 | 0.9358 |
| 14 | 0.9873 | 0.9859 | 0.9844 | 0.9827 | 0.9810 | 0.9791 | 0.9771 | 0.9749 | 0.9726 | 0.9701 | 0.9675 | 0.9647 |
| 15 | 0.9941 | 0.9934 | 0.9926 | 0.9918 | 0.9908 | 0.9898 | 0.9887 | 0.9875 | 0.9862 | 0.9848 | 0.9832 | 0.9816 |
| 16 | 0.9974 | 0.9971 | 0.9967 | 0.9963 | 0.9958 | 0.9953 | 0.9947 | 0.9941 | 0.9934 | 0.9926 | 0.9918 | 0.9909 |
| 17 | 0.9989 | 0.9988 | 0.9986 | 0.9984 | 0.9982 | 0.9979 | 0.9977 | 0.9973 | 0.9970 | 0.9966 | 0.9962 | 0.9957 |
| 18 | 0.9996 | 0.9995 | 0.9994 | 0.9993 | 0.9992 | 0.9991 | 0.9990 | 0.9989 | 0.9987 | 0.9985 | 0.9983 | 0.9981 |
| 19 | 0.9998 | 0.9998 | 0.9998 | 0.9997 | 0.9997 | 0.9997 | 0.9996 | 0.9996 | 0.9995 | 0.9995 | 0.9994 | 0.9992 |
| 20 | 0.9999 | 0.9999 | 0.9999 | 0.9999 | 0.9999 | 0.9999 | 0.9998 | 0.9998 | 0.9998 | 0.9998 | 0.9997 | 0.9997 |
| 21 | 1.0000 | 1.0000 | 1.0000 | 1.0000 | 1.0000 | 1.0000 | 0.9999 | 0.9999 | 0.9999 | 0.9999 | 0.9999 | 0.9999 |
| 22 | 1.0000 | 1.0000 | 1.0000 | 1.0000 | 1.0000 | 1.0000 | 1.0000 | 1.0000 | 1.0000 | 1.0000 | 1.0000 | 1.0000 |

| x | 8.90 | 9.00 | 9.10 | 9.20 | 9.30 | 9.40 | 9.50 | 9.60 | 9.70 | 9.80 | 9.90 | 10.00 |
|---|---|---|---|---|---|---|---|---|---|---|---|---|
| 0 | 0.0001 | 0.0001 | 0.0001 | 0.0001 | 0.0001 | 0.0001 | 0.0001 | 0.0001 | 0.0001 | 0.0001 | 0.0001 | 0.0000 |
| 1 | 0.0014 | 0.0012 | 0.0011 | 0.0010 | 0.0009 | 0.0009 | 0.0008 | 0.0007 | 0.0007 | 0.0006 | 0.0005 | 0.0005 |
| 2 | 0.0068 | 0.0062 | 0.0058 | 0.0053 | 0.0049 | 0.0045 | 0.0042 | 0.0038 | 0.0035 | 0.0033 | 0.0030 | 0.0028 |
| 3 | 0.0228 | 0.0212 | 0.0198 | 0.0184 | 0.0172 | 0.0160 | 0.0149 | 0.0138 | 0.0129 | 0.0120 | 0.0111 | 0.0103 |

# F  Cumulative Poisson Probabilities (cont.)

| | | | | | | | λ | | | | | |
|---|---|---|---|---|---|---|---|---|---|---|---|---|
| x | 8.90 | 9.00 | 9.10 | 9.20 | 9.30 | 9.40 | 9.50 | 9.60 | 9.70 | 9.80 | 9.90 | 10.00 |
| 4 | 0.0584 | 0.0550 | 0.0517 | 0.0486 | 0.0456 | 0.0429 | 0.0403 | 0.0378 | 0.0355 | 0.0333 | 0.0312 | 0.0293 |
| 5 | 0.1219 | 0.1157 | 0.1098 | 0.1041 | 0.0986 | 0.0935 | 0.0885 | 0.0838 | 0.0793 | 0.0750 | 0.0710 | 0.0671 |
| 6 | 0.2160 | 0.2068 | 0.1978 | 0.1892 | 0.1808 | 0.1727 | 0.1649 | 0.1574 | 0.1502 | 0.1433 | 0.1366 | 0.1301 |
| 7 | 0.3357 | 0.3239 | 0.3123 | 0.3010 | 0.2900 | 0.2792 | 0.2687 | 0.2584 | 0.2485 | 0.2388 | 0.2294 | 0.2202 |
| 8 | 0.4689 | 0.4557 | 0.4426 | 0.4296 | 0.4168 | 0.4042 | 0.3918 | 0.3796 | 0.3676 | 0.3558 | 0.3442 | 0.3328 |
| 9 | 0.6006 | 0.5874 | 0.5742 | 0.5611 | 0.5479 | 0.5349 | 0.5218 | 0.5089 | 0.4960 | 0.4832 | 0.4705 | 0.4579 |
| 10 | 0.7178 | 0.7060 | 0.6941 | 0.6820 | 0.6699 | 0.6576 | 0.6453 | 0.6329 | 0.6205 | 0.6080 | 0.5955 | 0.5830 |
| 11 | 0.8126 | 0.8030 | 0.7932 | 0.7832 | 0.7730 | 0.7626 | 0.7520 | 0.7412 | 0.7303 | 0.7193 | 0.7081 | 0.6968 |
| 12 | 0.8829 | 0.8758 | 0.8684 | 0.8607 | 0.8529 | 0.8448 | 0.8364 | 0.8279 | 0.8191 | 0.8101 | 0.8009 | 0.7916 |
| 13 | 0.9311 | 0.9261 | 0.9210 | 0.9156 | 0.9100 | 0.9042 | 0.8981 | 0.8919 | 0.8853 | 0.8786 | 0.8716 | 0.8645 |
| 14 | 0.9617 | 0.9585 | 0.9552 | 0.9517 | 0.9480 | 0.9441 | 0.9400 | 0.9357 | 0.9312 | 0.9265 | 0.9216 | 0.9165 |
| 15 | 0.9798 | 0.9780 | 0.9760 | 0.9738 | 0.9715 | 0.9691 | 0.9665 | 0.9638 | 0.9609 | 0.9579 | 0.9546 | 0.9513 |
| 16 | 0.9899 | 0.9889 | 0.9878 | 0.9865 | 0.9852 | 0.9838 | 0.9823 | 0.9806 | 0.9789 | 0.9770 | 0.9751 | 0.9730 |
| 17 | 0.9952 | 0.9947 | 0.9941 | 0.9934 | 0.9927 | 0.9919 | 0.9911 | 0.9902 | 0.9892 | 0.9881 | 0.9870 | 0.9857 |
| 18 | 0.9978 | 0.9976 | 0.9973 | 0.9969 | 0.9966 | 0.9962 | 0.9957 | 0.9952 | 0.9947 | 0.9941 | 0.9935 | 0.9928 |
| 19 | 0.9991 | 0.9989 | 0.9988 | 0.9986 | 0.9985 | 0.9983 | 0.9980 | 0.9978 | 0.9975 | 0.9972 | 0.9969 | 0.9965 |
| 20 | 0.9996 | 0.9996 | 0.9995 | 0.9994 | 0.9993 | 0.9992 | 0.9991 | 0.9990 | 0.9989 | 0.9987 | 0.9986 | 0.9984 |
| 21 | 0.9998 | 0.9998 | 0.9998 | 0.9998 | 0.9997 | 0.9997 | 0.9996 | 0.9996 | 0.9995 | 0.9995 | 0.9994 | 0.9993 |
| 22 | 0.9999 | 0.9999 | 0.9999 | 0.9999 | 0.9999 | 0.9999 | 0.9999 | 0.9998 | 0.9998 | 0.9998 | 0.9997 | 0.9997 |
| 23 | 1.0000 | 1.0000 | 1.0000 | 1.0000 | 1.0000 | 1.0000 | 0.9999 | 0.9999 | 0.9999 | 0.9999 | 0.9999 | 0.9999 |
| 24 | 1.0000 | 1.0000 | 1.0000 | 1.0000 | 1.0000 | 1.0000 | 1.0000 | 1.0000 | 1.0000 | 1.0000 | 1.0000 | 1.0000 |

| x | 10.10 | 10.20 | 10.30 | 10.40 | 10.50 | 10.60 | 10.70 | 10.80 | 10.90 | 11.00 | 11.10 | 11.20 |
|---|---|---|---|---|---|---|---|---|---|---|---|---|
| 0 | 0.0000 | 0.0000 | 0.0000 | 0.0000 | 0.0000 | 0.0000 | 0.0000 | 0.0000 | 0.0000 | 0.0000 | 0.0000 | 0.0000 |
| 1 | 0.0005 | 0.0004 | 0.0004 | 0.0003 | 0.0003 | 0.0003 | 0.0003 | 0.0002 | 0.0002 | 0.0002 | 0.0002 | 0.0002 |
| 2 | 0.0026 | 0.0023 | 0.0022 | 0.0020 | 0.0018 | 0.0017 | 0.0016 | 0.0014 | 0.0013 | 0.0012 | 0.0011 | 0.0010 |
| 3 | 0.0096 | 0.0089 | 0.0083 | 0.0077 | 0.0071 | 0.0066 | 0.0062 | 0.0057 | 0.0053 | 0.0049 | 0.0046 | 0.0042 |
| 4 | 0.0274 | 0.0257 | 0.0241 | 0.0225 | 0.0211 | 0.0197 | 0.0185 | 0.0173 | 0.0162 | 0.0151 | 0.0141 | 0.0132 |
| 5 | 0.0634 | 0.0599 | 0.0566 | 0.0534 | 0.0504 | 0.0475 | 0.0448 | 0.0423 | 0.0398 | 0.0375 | 0.0353 | 0.0333 |
| 6 | 0.1240 | 0.1180 | 0.1123 | 0.1069 | 0.1016 | 0.0966 | 0.0918 | 0.0872 | 0.0828 | 0.0786 | 0.0746 | 0.0708 |
| 7 | 0.2113 | 0.2027 | 0.1944 | 0.1863 | 0.1785 | 0.1710 | 0.1636 | 0.1566 | 0.1498 | 0.1432 | 0.1369 | 0.1307 |
| 8 | 0.3217 | 0.3108 | 0.3001 | 0.2896 | 0.2794 | 0.2694 | 0.2597 | 0.2502 | 0.2410 | 0.2320 | 0.2232 | 0.2147 |
| 9 | 0.4455 | 0.4332 | 0.4210 | 0.4090 | 0.3971 | 0.3854 | 0.3739 | 0.3626 | 0.3515 | 0.3405 | 0.3298 | 0.3192 |
| 10 | 0.5705 | 0.5580 | 0.5456 | 0.5331 | 0.5207 | 0.5084 | 0.4961 | 0.4840 | 0.4719 | 0.4599 | 0.4480 | 0.4362 |
| 11 | 0.6853 | 0.6738 | 0.6622 | 0.6505 | 0.6387 | 0.6269 | 0.6150 | 0.6031 | 0.5912 | 0.5793 | 0.5673 | 0.5554 |
| 12 | 0.7820 | 0.7722 | 0.7623 | 0.7522 | 0.7420 | 0.7316 | 0.7210 | 0.7104 | 0.6996 | 0.6887 | 0.6777 | 0.6666 |
| 13 | 0.8571 | 0.8494 | 0.8416 | 0.8336 | 0.8253 | 0.8169 | 0.8083 | 0.7995 | 0.7905 | 0.7813 | 0.7719 | 0.7624 |
| 14 | 0.9112 | 0.9057 | 0.9000 | 0.8940 | 0.8879 | 0.8815 | 0.8750 | 0.8682 | 0.8612 | 0.8540 | 0.8467 | 0.8391 |
| 15 | 0.9477 | 0.9440 | 0.9400 | 0.9359 | 0.9317 | 0.9272 | 0.9225 | 0.9177 | 0.9126 | 0.9074 | 0.9020 | 0.8963 |
| 16 | 0.9707 | 0.9684 | 0.9658 | 0.9632 | 0.9604 | 0.9574 | 0.9543 | 0.9511 | 0.9477 | 0.9441 | 0.9403 | 0.9364 |
| 17 | 0.9844 | 0.9830 | 0.9815 | 0.9799 | 0.9781 | 0.9763 | 0.9744 | 0.9723 | 0.9701 | 0.9678 | 0.9654 | 0.9628 |
| 18 | 0.9921 | 0.9913 | 0.9904 | 0.9895 | 0.9885 | 0.9874 | 0.9863 | 0.9850 | 0.9837 | 0.9823 | 0.9808 | 0.9792 |
| 19 | 0.9962 | 0.9957 | 0.9953 | 0.9948 | 0.9942 | 0.9936 | 0.9930 | 0.9923 | 0.9915 | 0.9907 | 0.9898 | 0.9889 |
| 20 | 0.9982 | 0.9980 | 0.9978 | 0.9975 | 0.9972 | 0.9969 | 0.9966 | 0.9962 | 0.9958 | 0.9953 | 0.9948 | 0.9943 |
| 21 | 0.9992 | 0.9991 | 0.9990 | 0.9989 | 0.9987 | 0.9986 | 0.9984 | 0.9982 | 0.9980 | 0.9977 | 0.9975 | 0.9972 |
| 22 | 0.9997 | 0.9996 | 0.9996 | 0.9995 | 0.9994 | 0.9994 | 0.9993 | 0.9992 | 0.9991 | 0.9990 | 0.9988 | 0.9987 |
| 23 | 0.9999 | 0.9998 | 0.9998 | 0.9998 | 0.9998 | 0.9997 | 0.9997 | 0.9996 | 0.9996 | 0.9995 | 0.9995 | 0.9994 |

## F  Cumulative Poisson Probabilities (cont.)

| x | 10.10 | 10.20 | 10.30 | 10.40 | 10.50 | 10.60 | 10.70 | 10.80 | 10.90 | 11.00 | 11.10 | 11.20 |
|---|---|---|---|---|---|---|---|---|---|---|---|---|
| 24 | 0.9999 | 0.9999 | 0.9999 | 0.9999 | 0.9999 | 0.9999 | 0.9999 | 0.9998 | 0.9998 | 0.9998 | 0.9998 | 0.9997 |
| 25 | 1.0000 | 1.0000 | 1.0000 | 1.0000 | 1.0000 | 1.0000 | 0.9999 | 0.9999 | 0.9999 | 0.9999 | 0.9999 | 0.9999 |
| 26 | 1.0000 | 1.0000 | 1.0000 | 1.0000 | 1.0000 | 1.0000 | 1.0000 | 1.0000 | 1.0000 | 1.0000 | 1.0000 | 1.0000 |

| x | 11.30 | 11.40 | 11.50 | 11.60 | 11.70 | 11.80 | 11.90 | 12.00 | 12.10 | 12.20 | 12.30 | 12.40 |
|---|---|---|---|---|---|---|---|---|---|---|---|---|
| 0 | 0.0000 | 0.0000 | 0.0000 | 0.0000 | 0.0000 | 0.0000 | 0.0000 | 0.0000 | 0.0000 | 0.0000 | 0.0000 | 0.0000 |
| 1 | 0.0002 | 0.0001 | 0.0001 | 0.0001 | 0.0001 | 0.0001 | 0.0001 | 0.0001 | 0.0001 | 0.0001 | 0.0001 | 0.0001 |
| 2 | 0.0009 | 0.0009 | 0.0008 | 0.0007 | 0.0007 | 0.0006 | 0.0006 | 0.0005 | 0.0005 | 0.0004 | 0.0004 | 0.0004 |
| 3 | 0.0039 | 0.0036 | 0.0034 | 0.0031 | 0.0029 | 0.0027 | 0.0025 | 0.0023 | 0.0021 | 0.0020 | 0.0018 | 0.0017 |
| 4 | 0.0123 | 0.0115 | 0.0107 | 0.0100 | 0.0094 | 0.0087 | 0.0081 | 0.0076 | 0.0071 | 0.0066 | 0.0062 | 0.0057 |
| 5 | 0.0313 | 0.0295 | 0.0277 | 0.0261 | 0.0245 | 0.0230 | 0.0217 | 0.0203 | 0.0191 | 0.0179 | 0.0168 | 0.0158 |
| 6 | 0.0671 | 0.0636 | 0.0603 | 0.0571 | 0.0541 | 0.0512 | 0.0484 | 0.0458 | 0.0433 | 0.0410 | 0.0387 | 0.0366 |
| 7 | 0.1249 | 0.1192 | 0.1137 | 0.1085 | 0.1035 | 0.0986 | 0.0940 | 0.0895 | 0.0852 | 0.0811 | 0.0772 | 0.0734 |
| 8 | 0.2064 | 0.1984 | 0.1906 | 0.1830 | 0.1757 | 0.1686 | 0.1617 | 0.1550 | 0.1486 | 0.1424 | 0.1363 | 0.1305 |
| 9 | 0.3089 | 0.2987 | 0.2888 | 0.2791 | 0.2696 | 0.2603 | 0.2512 | 0.2424 | 0.2338 | 0.2254 | 0.2172 | 0.2092 |
| 10 | 0.4246 | 0.4131 | 0.4017 | 0.3905 | 0.3794 | 0.3685 | 0.3578 | 0.3472 | 0.3368 | 0.3266 | 0.3166 | 0.3067 |
| 11 | 0.5435 | 0.5316 | 0.5198 | 0.5080 | 0.4963 | 0.4847 | 0.4731 | 0.4616 | 0.4502 | 0.4389 | 0.4278 | 0.4167 |
| 12 | 0.6555 | 0.6442 | 0.6329 | 0.6216 | 0.6102 | 0.5988 | 0.5874 | 0.5760 | 0.5645 | 0.5531 | 0.5417 | 0.5303 |
| 13 | 0.7528 | 0.7430 | 0.7330 | 0.7230 | 0.7128 | 0.7025 | 0.6920 | 0.6815 | 0.6709 | 0.6603 | 0.6495 | 0.6387 |
| 14 | 0.8313 | 0.8234 | 0.8153 | 0.8069 | 0.7985 | 0.7898 | 0.7810 | 0.7720 | 0.7629 | 0.7536 | 0.7442 | 0.7347 |
| 15 | 0.8905 | 0.8845 | 0.8783 | 0.8719 | 0.8653 | 0.8585 | 0.8516 | 0.8444 | 0.8371 | 0.8296 | 0.8219 | 0.8140 |
| 16 | 0.9323 | 0.9280 | 0.9236 | 0.9190 | 0.9142 | 0.9092 | 0.9040 | 0.8987 | 0.8932 | 0.8875 | 0.8816 | 0.8755 |
| 17 | 0.9601 | 0.9572 | 0.9542 | 0.9511 | 0.9478 | 0.9444 | 0.9408 | 0.9370 | 0.9331 | 0.9290 | 0.9248 | 0.9204 |
| 18 | 0.9775 | 0.9757 | 0.9738 | 0.9718 | 0.9697 | 0.9674 | 0.9651 | 0.9626 | 0.9600 | 0.9572 | 0.9543 | 0.9513 |
| 19 | 0.9879 | 0.9868 | 0.9857 | 0.9845 | 0.9832 | 0.9818 | 0.9803 | 0.9787 | 0.9771 | 0.9753 | 0.9734 | 0.9715 |
| 20 | 0.9938 | 0.9932 | 0.9925 | 0.9918 | 0.9910 | 0.9902 | 0.9893 | 0.9884 | 0.9874 | 0.9863 | 0.9852 | 0.9840 |
| 21 | 0.9969 | 0.9966 | 0.9962 | 0.9958 | 0.9954 | 0.9950 | 0.9945 | 0.9939 | 0.9934 | 0.9927 | 0.9921 | 0.9914 |
| 22 | 0.9985 | 0.9984 | 0.9982 | 0.9980 | 0.9978 | 0.9975 | 0.9972 | 0.9970 | 0.9966 | 0.9963 | 0.9959 | 0.9955 |
| 23 | 0.9993 | 0.9992 | 0.9992 | 0.9991 | 0.9989 | 0.9988 | 0.9988 | 0.9987 | 0.9985 | 0.9984 | 0.9980 | 0.9978 |
| 24 | 0.9997 | 0.9997 | 0.9996 | 0.9996 | 0.9995 | 0.9995 | 0.9994 | 0.9993 | 0.9992 | 0.9991 | 0.9990 | 0.9989 |
| 25 | 0.9999 | 0.9999 | 0.9998 | 0.9998 | 0.9998 | 0.9998 | 0.9997 | 0.9997 | 0.9997 | 0.9996 | 0.9996 | 0.9995 |
| 26 | 0.9999 | 0.9999 | 0.9999 | 0.9999 | 0.9999 | 0.9999 | 0.9999 | 0.9999 | 0.9998 | 0.9998 | 0.9998 | 0.9998 |
| 27 | 1.0000 | 1.0000 | 1.0000 | 1.0000 | 1.0000 | 1.0000 | 1.0000 | 0.9999 | 0.9999 | 0.9999 | 0.9999 | 0.9999 |
| 28 | 1.0000 | 1.0000 | 1.0000 | 1.0000 | 1.0000 | 1.0000 | 1.0000 | 1.0000 | 1.0000 | 1.0000 | 1.0000 | 1.0000 |

| x | 12.50 | 12.60 | 12.70 | 12.80 | 12.90 | 13.00 | 14.00 | 15.00 | 16.00 | 17.00 | 18.00 | 19.00 |
|---|---|---|---|---|---|---|---|---|---|---|---|---|
| 0 | 0.0000 | 0.0000 | 0.0000 | 0.0000 | 0.0000 | 0.0000 | 0.0000 | 0.0000 | 0.0000 | 0.0000 | 0.0000 | 0.0000 |
| 1 | 0.0001 | 0.0000 | 0.0000 | 0.0000 | 0.0000 | 0.0000 | 0.0000 | 0.0000 | 0.0000 | 0.0000 | 0.0000 | 0.0000 |
| 2 | 0.0003 | 0.0003 | 0.0003 | 0.0003 | 0.0002 | 0.0002 | 0.0001 | 0.0000 | 0.0000 | 0.0000 | 0.0000 | 0.0000 |
| 3 | 0.0016 | 0.0014 | 0.0013 | 0.0012 | 0.0011 | 0.0011 | 0.0005 | 0.0002 | 0.0001 | 0.0000 | 0.0000 | 0.0000 |
| 4 | 0.0053 | 0.0050 | 0.0046 | 0.0043 | 0.0040 | 0.0037 | 0.0018 | 0.0009 | 0.0004 | 0.0002 | 0.0001 | 0.0000 |
| 5 | 0.0148 | 0.0139 | 0.0130 | 0.0122 | 0.0115 | 0.0107 | 0.0055 | 0.0028 | 0.0014 | 0.0007 | 0.0003 | 0.0002 |
| 6 | 0.0346 | 0.0326 | 0.0308 | 0.0291 | 0.0274 | 0.0259 | 0.0142 | 0.0076 | 0.0040 | 0.0021 | 0.0010 | 0.0005 |
| 7 | 0.0698 | 0.0664 | 0.0631 | 0.0599 | 0.0569 | 0.0540 | 0.0316 | 0.0180 | 0.0100 | 0.0054 | 0.0029 | 0.0015 |
| 8 | 0.1249 | 0.1195 | 0.1143 | 0.1093 | 0.1044 | 0.0998 | 0.0621 | 0.0374 | 0.0220 | 0.0126 | 0.0071 | 0.0039 |
| 9 | 0.2014 | 0.1939 | 0.1866 | 0.1794 | 0.1725 | 0.1658 | 0.1094 | 0.0699 | 0.0433 | 0.0261 | 0.0154 | 0.0089 |
| 10 | 0.2971 | 0.2876 | 0.2783 | 0.2693 | 0.2604 | 0.2517 | 0.1757 | 0.1185 | 0.0774 | 0.0491 | 0.0304 | 0.0183 |
| 11 | 0.4058 | 0.3950 | 0.3843 | 0.3738 | 0.3634 | 0.3532 | 0.2600 | 0.1848 | 0.1270 | 0.0847 | 0.0549 | 0.0347 |

# F  Cumulative Poisson Probabilities (cont.)

$\lambda$

| x | 12.50 | 12.60 | 12.70 | 12.80 | 12.90 | 13.00 | 14.00 | 15.00 | 16.00 | 17.00 | 18.00 | 19.00 |
|---|---|---|---|---|---|---|---|---|---|---|---|---|
| 12 | 0.5190 | 0.5077 | 0.4964 | 0.4853 | 0.4741 | 0.4631 | 0.3585 | 0.2676 | 0.1931 | 0.1350 | 0.0917 | 0.0606 |
| 13 | 0.6278 | 0.6169 | 0.6060 | 0.5950 | 0.5840 | 0.5730 | 0.4644 | 0.3632 | 0.2745 | 0.2009 | 0.1426 | 0.0984 |
| 14 | 0.7250 | 0.7153 | 0.7054 | 0.6954 | 0.6853 | 0.6751 | 0.5704 | 0.4657 | 0.3675 | 0.2808 | 0.2081 | 0.1497 |
| 15 | 0.8060 | 0.7978 | 0.7895 | 0.7810 | 0.7724 | 0.7636 | 0.6694 | 0.5681 | 0.4667 | 0.3715 | 0.2867 | 0.2148 |
| 16 | 0.8693 | 0.8629 | 0.8563 | 0.8495 | 0.8426 | 0.8355 | 0.7559 | 0.6641 | 0.5660 | 0.4677 | 0.3751 | 0.2920 |
| 17 | 0.9158 | 0.9111 | 0.9062 | 0.9011 | 0.8959 | 0.8905 | 0.8272 | 0.7489 | 0.6593 | 0.5640 | 0.4686 | 0.3784 |
| 18 | 0.9481 | 0.9448 | 0.9414 | 0.9378 | 0.9341 | 0.9302 | 0.8826 | 0.8195 | 0.7423 | 0.6550 | 0.5622 | 0.4695 |
| 19 | 0.9694 | 0.9672 | 0.9649 | 0.9625 | 0.9600 | 0.9573 | 0.9235 | 0.8752 | 0.8122 | 0.7363 | 0.6509 | 0.5606 |
| 20 | 0.9827 | 0.9813 | 0.9799 | 0.9783 | 0.9767 | 0.9750 | 0.9521 | 0.9170 | 0.8682 | 0.8055 | 0.7307 | 0.6472 |
| 21 | 0.9906 | 0.9898 | 0.9889 | 0.9880 | 0.9870 | 0.9859 | 0.9712 | 0.9469 | 0.9108 | 0.8615 | 0.7991 | 0.7255 |
| 22 | 0.9951 | 0.9946 | 0.9941 | 0.9936 | 0.9930 | 0.9924 | 0.9833 | 0.9673 | 0.9418 | 0.9047 | 0.8551 | 0.7931 |
| 23 | 0.9975 | 0.9973 | 0.9970 | 0.9967 | 0.9964 | 0.9960 | 0.9907 | 0.9805 | 0.9633 | 0.9367 | 0.8989 | 0.8490 |
| 24 | 0.9988 | 0.9987 | 0.9985 | 0.9984 | 0.9982 | 0.9980 | 0.9950 | 0.9888 | 0.9777 | 0.9594 | 0.9317 | 0.8933 |
| 25 | 0.9994 | 0.9994 | 0.9993 | 0.9992 | 0.9991 | 0.9990 | 0.9974 | 0.9938 | 0.9869 | 0.9748 | 0.9554 | 0.9269 |
| 26 | 0.9997 | 0.9997 | 0.9997 | 0.9996 | 0.9996 | 0.9995 | 0.9987 | 0.9967 | 0.9925 | 0.9848 | 0.9718 | 0.9514 |
| 27 | 0.9999 | 0.9999 | 0.9999 | 0.9998 | 0.9998 | 0.9998 | 0.9994 | 0.9983 | 0.9959 | 0.9912 | 0.9827 | 0.9687 |
| 28 | 1.0000 | 0.9999 | 0.9999 | 0.9999 | 0.9999 | 0.9999 | 0.9997 | 0.9991 | 0.9978 | 0.9950 | 0.9897 | 0.9805 |
| 29 | 1.0000 | 1.0000 | 1.0000 | 1.0000 | 1.0000 | 1.0000 | 0.9999 | 0.9996 | 0.9989 | 0.9973 | 0.9941 | 0.9882 |
| 30 | 1.0000 | 1.0000 | 1.0000 | 1.0000 | 1.0000 | 1.0000 | 0.9999 | 0.9998 | 0.9994 | 0.9986 | 0.9967 | 0.9930 |
| 31 | 1.0000 | 1.0000 | 1.0000 | 1.0000 | 1.0000 | 1.0000 | 1.0000 | 0.9999 | 0.9997 | 0.9993 | 0.9982 | 0.9960 |
| 32 | 1.0000 | 1.0000 | 1.0000 | 1.0000 | 1.0000 | 1.0000 | 1.0000 | 1.0000 | 0.9999 | 0.9996 | 0.9990 | 0.9978 |
| 33 | 1.0000 | 1.0000 | 1.0000 | 1.0000 | 1.0000 | 1.0000 | 1.0000 | 1.0000 | 0.9999 | 0.9998 | 0.9995 | 0.9988 |
| 34 | 1.0000 | 1.0000 | 1.0000 | 1.0000 | 1.0000 | 1.0000 | 1.0000 | 1.0000 | 1.0000 | 0.9999 | 0.9998 | 0.9994 |
| 35 | 1.0000 | 1.0000 | 1.0000 | 1.0000 | 1.0000 | 1.0000 | 1.0000 | 1.0000 | 1.0000 | 1.0000 | 0.9999 | 0.9997 |
| 36 | 1.0000 | 1.0000 | 1.0000 | 1.0000 | 1.0000 | 1.0000 | 1.0000 | 1.0000 | 1.0000 | 1.0000 | 0.9999 | 0.9998 |
| 37 | 1.0000 | 1.0000 | 1.0000 | 1.0000 | 1.0000 | 1.0000 | 1.0000 | 1.0000 | 1.0000 | 1.0000 | 1.0000 | 0.9999 |
| 38 | 1.0000 | 1.0000 | 1.0000 | 1.0000 | 1.0000 | 1.0000 | 1.0000 | 1.0000 | 1.0000 | 1.0000 | 1.0000 | 1.0000 |

| x | 20.00 | 21.00 | 22.00 | 23.00 | 24.00 | 25.00 | 26.00 | 27.00 | 28.00 | 29.00 | 30.00 | 31.00 |
|---|---|---|---|---|---|---|---|---|---|---|---|---|
| 0 | 0.0000 | 0.0000 | 0.0000 | 0.0000 | 0.0000 | 0.0000 | 0.0000 | 0.0000 | 0.0000 | 0.0000 | 0.0000 | 0.0000 |
| 1 | 0.0000 | 0.0000 | 0.0000 | 0.0000 | 0.0000 | 0.0000 | 0.0000 | 0.0000 | 0.0000 | 0.0000 | 0.0000 | 0.0000 |
| 2 | 0.0000 | 0.0000 | 0.0000 | 0.0000 | 0.0000 | 0.0000 | 0.0000 | 0.0000 | 0.0000 | 0.0000 | 0.0000 | 0.0000 |
| 3 | 0.0000 | 0.0000 | 0.0000 | 0.0000 | 0.0000 | 0.0000 | 0.0000 | 0.0000 | 0.0000 | 0.0000 | 0.0000 | 0.0000 |
| 4 | 0.0000 | 0.0000 | 0.0000 | 0.0000 | 0.0000 | 0.0000 | 0.0000 | 0.0000 | 0.0000 | 0.0000 | 0.0000 | 0.0000 |
| 5 | 0.0001 | 0.0000 | 0.0000 | 0.0000 | 0.0000 | 0.0000 | 0.0000 | 0.0000 | 0.0000 | 0.0000 | 0.0000 | 0.0000 |
| 6 | 0.0003 | 0.0001 | 0.0001 | 0.0000 | 0.0000 | 0.0000 | 0.0000 | 0.0000 | 0.0000 | 0.0000 | 0.0000 | 0.0000 |
| 7 | 0.0008 | 0.0004 | 0.0002 | 0.0001 | 0.0000 | 0.0000 | 0.0000 | 0.0000 | 0.0000 | 0.0000 | 0.0000 | 0.0000 |
| 8 | 0.0021 | 0.0011 | 0.0006 | 0.0003 | 0.0002 | 0.0001 | 0.0000 | 0.0000 | 0.0000 | 0.0000 | 0.0000 | 0.0000 |
| 9 | 0.0050 | 0.0028 | 0.0015 | 0.0008 | 0.0004 | 0.0002 | 0.0001 | 0.0001 | 0.0000 | 0.0000 | 0.0000 | 0.0000 |
| 10 | 0.0108 | 0.0063 | 0.0035 | 0.0020 | 0.0011 | 0.0006 | 0.0003 | 0.0002 | 0.0001 | 0.0000 | 0.0000 | 0.0000 |
| 11 | 0.0214 | 0.0129 | 0.0076 | 0.0044 | 0.0025 | 0.0014 | 0.0008 | 0.0004 | 0.0002 | 0.0001 | 0.0001 | 0.0000 |
| 12 | 0.0390 | 0.0245 | 0.0151 | 0.0091 | 0.0054 | 0.0031 | 0.0018 | 0.0010 | 0.0006 | 0.0003 | 0.0002 | 0.0001 |
| 13 | 0.0661 | 0.0434 | 0.0278 | 0.0174 | 0.0107 | 0.0065 | 0.0038 | 0.0022 | 0.0013 | 0.0007 | 0.0004 | 0.0002 |
| 14 | 0.1049 | 0.0716 | 0.0477 | 0.0311 | 0.0198 | 0.0124 | 0.0076 | 0.0046 | 0.0027 | 0.0016 | 0.0009 | 0.0005 |
| 15 | 0.1565 | 0.1111 | 0.0769 | 0.0520 | 0.0344 | 0.0223 | 0.0142 | 0.0088 | 0.0054 | 0.0033 | 0.0019 | 0.0011 |
| 16 | 0.2211 | 0.1629 | 0.1170 | 0.0821 | 0.0563 | 0.0377 | 0.0248 | 0.0160 | 0.0101 | 0.0063 | 0.0039 | 0.0023 |
| 17 | 0.2970 | 0.2270 | 0.1690 | 0.1228 | 0.0871 | 0.0605 | 0.0411 | 0.0274 | 0.0179 | 0.0115 | 0.0073 | 0.0045 |

## F   Cumulative Poisson Probabilities (cont.)

| x | 20.00 | 21.00 | 22.00 | 23.00 | 24.00 | 25.00 | 26.00 | 27.00 | 28.00 | 29.00 | 30.00 | 31.00 |
|---|---|---|---|---|---|---|---|---|---|---|---|---|
| 18 | 0.3814 | 0.3017 | 0.2325 | 0.1748 | 0.1283 | 0.0920 | 0.0646 | 0.0445 | 0.0300 | 0.0199 | 0.0129 | 0.0083 |
| 19 | 0.4703 | 0.3843 | 0.3060 | 0.2377 | 0.1803 | 0.1336 | 0.0968 | 0.0687 | 0.0478 | 0.0326 | 0.0219 | 0.0144 |
| 20 | 0.5591 | 0.4710 | 0.3869 | 0.3101 | 0.2426 | 0.1855 | 0.1387 | 0.1015 | 0.0727 | 0.0511 | 0.0353 | 0.0239 |
| 21 | 0.6437 | 0.5577 | 0.4716 | 0.3894 | 0.3139 | 0.2473 | 0.1905 | 0.1436 | 0.1060 | 0.0767 | 0.0544 | 0.0379 |
| 22 | 0.7206 | 0.6405 | 0.5564 | 0.4723 | 0.3917 | 0.3175 | 0.2517 | 0.1952 | 0.1483 | 0.1104 | 0.0806 | 0.0577 |
| 23 | 0.7875 | 0.7160 | 0.6374 | 0.5551 | 0.4728 | 0.3939 | 0.3209 | 0.2559 | 0.1998 | 0.1529 | 0.1146 | 0.0844 |
| 24 | 0.8432 | 0.7822 | 0.7117 | 0.6346 | 0.5540 | 0.4734 | 0.3959 | 0.3242 | 0.2599 | 0.2042 | 0.1572 | 0.1188 |
| 25 | 0.8878 | 0.8377 | 0.7771 | 0.7077 | 0.6319 | 0.5529 | 0.4739 | 0.3979 | 0.3272 | 0.2637 | 0.2084 | 0.1615 |
| 26 | 0.9221 | 0.8826 | 0.8324 | 0.7723 | 0.7038 | 0.6294 | 0.5519 | 0.4744 | 0.3997 | 0.3301 | 0.2673 | 0.2124 |
| 27 | 0.9475 | 0.9175 | 0.8775 | 0.8274 | 0.7677 | 0.7002 | 0.6270 | 0.5509 | 0.4749 | 0.4014 | 0.3329 | 0.2708 |
| 28 | 0.9657 | 0.9436 | 0.9129 | 0.8726 | 0.8225 | 0.7634 | 0.6967 | 0.6247 | 0.5500 | 0.4753 | 0.4031 | 0.3355 |
| 29 | 0.9782 | 0.9626 | 0.9398 | 0.9085 | 0.8679 | 0.8179 | 0.7593 | 0.6935 | 0.6226 | 0.5492 | 0.4757 | 0.4047 |
| 30 | 0.9865 | 0.9758 | 0.9595 | 0.9360 | 0.9042 | 0.8633 | 0.8134 | 0.7553 | 0.6903 | 0.6206 | 0.5484 | 0.4761 |
| 31 | 0.9919 | 0.9848 | 0.9735 | 0.9564 | 0.9322 | 0.8999 | 0.8589 | 0.8092 | 0.7515 | 0.6874 | 0.6186 | 0.5476 |
| 32 | 0.9953 | 0.9907 | 0.9831 | 0.9711 | 0.9533 | 0.9285 | 0.8958 | 0.8546 | 0.8051 | 0.7479 | 0.6845 | 0.6168 |
| 33 | 0.9973 | 0.9945 | 0.9895 | 0.9813 | 0.9686 | 0.9502 | 0.9249 | 0.8918 | 0.8505 | 0.8011 | 0.7444 | 0.6818 |
| 34 | 0.9985 | 0.9968 | 0.9936 | 0.9882 | 0.9794 | 0.9662 | 0.9472 | 0.9213 | 0.8879 | 0.8465 | 0.7973 | 0.7411 |
| 35 | 0.9992 | 0.9982 | 0.9962 | 0.9927 | 0.9868 | 0.9775 | 0.9637 | 0.9441 | 0.9178 | 0.8841 | 0.8426 | 0.7936 |
| 36 | 0.9996 | 0.9990 | 0.9978 | 0.9956 | 0.9918 | 0.9854 | 0.9756 | 0.9612 | 0.9411 | 0.9144 | 0.8804 | 0.8389 |
| 37 | 0.9998 | 0.9995 | 0.9988 | 0.9974 | 0.9950 | 0.9908 | 0.9840 | 0.9737 | 0.9587 | 0.9381 | 0.9110 | 0.8768 |
| 38 | 0.9999 | 0.9997 | 0.9993 | 0.9985 | 0.9970 | 0.9943 | 0.9897 | 0.9825 | 0.9717 | 0.9562 | 0.9352 | 0.9077 |
| 39 | 0.9999 | 0.9999 | 0.9996 | 0.9992 | 0.9983 | 0.9966 | 0.9936 | 0.9887 | 0.9810 | 0.9697 | 0.9537 | 0.9322 |
| 40 | 1.0000 | 0.9999 | 0.9998 | 0.9996 | 0.9990 | 0.9980 | 0.9961 | 0.9928 | 0.9875 | 0.9795 | 0.9677 | 0.9513 |
| 41 | 1.0000 | 1.0000 | 0.9999 | 0.9998 | 0.9995 | 0.9988 | 0.9976 | 0.9955 | 0.9920 | 0.9864 | 0.9779 | 0.9657 |
| 42 | 1.0000 | 1.0000 | 1.0000 | 0.9999 | 0.9997 | 0.9993 | 0.9986 | 0.9973 | 0.9950 | 0.9911 | 0.9852 | 0.9763 |
| 43 | 1.0000 | 1.0000 | 1.0000 | 0.9999 | 0.9998 | 0.9996 | 0.9992 | 0.9984 | 0.9969 | 0.9944 | 0.9903 | 0.9840 |
| 44 | 1.0000 | 1.0000 | 1.0000 | 1.0000 | 0.9999 | 0.9998 | 0.9996 | 0.9991 | 0.9981 | 0.9965 | 0.9937 | 0.9894 |
| 45 | 1.0000 | 1.0000 | 1.0000 | 1.0000 | 1.0000 | 0.9999 | 0.9998 | 0.9995 | 0.9989 | 0.9978 | 0.9960 | 0.9931 |
| 46 | 1.0000 | 1.0000 | 1.0000 | 1.0000 | 1.0000 | 0.9999 | 0.9999 | 0.9997 | 0.9994 | 0.9987 | 0.9975 | 0.9956 |
| 47 | 1.0000 | 1.0000 | 1.0000 | 1.0000 | 1.0000 | 1.0000 | 0.9999 | 0.9998 | 0.9996 | 0.9992 | 0.9985 | 0.9972 |
| 48 | 1.0000 | 1.0000 | 1.0000 | 1.0000 | 1.0000 | 1.0000 | 1.0000 | 0.9999 | 0.9998 | 0.9996 | 0.9991 | 0.9983 |
| 49 | 1.0000 | 1.0000 | 1.0000 | 1.0000 | 1.0000 | 1.0000 | 1.0000 | 1.0000 | 0.9999 | 0.9998 | 0.9995 | 0.9990 |
| 50 | 1.0000 | 1.0000 | 1.0000 | 1.0000 | 1.0000 | 1.0000 | 1.0000 | 1.0000 | 0.9999 | 0.9999 | 0.9997 | 0.9994 |
| 51 | 1.0000 | 1.0000 | 1.0000 | 1.0000 | 1.0000 | 1.0000 | 1.0000 | 1.0000 | 1.0000 | 0.9999 | 0.9998 | 0.9996 |
| 52 | 1.0000 | 1.0000 | 1.0000 | 1.0000 | 1.0000 | 1.0000 | 1.0000 | 1.0000 | 1.0000 | 1.0000 | 0.9999 | 0.9998 |
| 53 | 1.0000 | 1.0000 | 1.0000 | 1.0000 | 1.0000 | 1.0000 | 1.0000 | 1.0000 | 1.0000 | 1.0000 | 0.9999 | 0.9999 |
| 54 | 1.0000 | 1.0000 | 1.0000 | 1.0000 | 1.0000 | 1.0000 | 1.0000 | 1.0000 | 1.0000 | 1.0000 | 1.0000 | 0.9999 |
| 55 | 1.0000 | 1.0000 | 1.0000 | 1.0000 | 1.0000 | 1.0000 | 1.0000 | 1.0000 | 1.0000 | 1.0000 | 1.0000 | 1.0000 |

The column headings are values of $\lambda$.

# G  Critical Values of $\chi^2$

Numerical entries represent the value of $\chi^2_\alpha$.

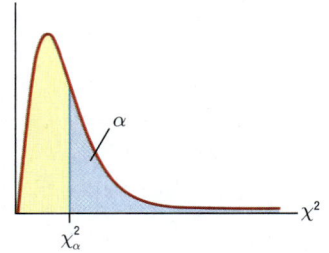

**Area to the Right of the Critical Value**

| df | $\chi^2_{0.995}$ | $\chi^2_{0.990}$ | $\chi^2_{0.975}$ | $\chi^2_{0.950}$ | $\chi^2_{0.900}$ | $\chi^2_{0.100}$ | $\chi^2_{0.050}$ | $\chi^2_{0.025}$ | $\chi^2_{0.010}$ | $\chi^2_{0.005}$ |
|---|---|---|---|---|---|---|---|---|---|---|
| 1 | 0.000 | 0.000 | 0.001 | 0.004 | 0.016 | 2.706 | 3.841 | 5.024 | 6.635 | 7.879 |
| 2 | 0.010 | 0.020 | 0.051 | 0.103 | 0.211 | 4.605 | 5.991 | 7.378 | 9.210 | 10.597 |
| 3 | 0.072 | 0.115 | 0.216 | 0.352 | 0.584 | 6.251 | 7.815 | 9.348 | 11.345 | 12.838 |
| 4 | 0.207 | 0.297 | 0.484 | 0.711 | 1.064 | 7.779 | 9.488 | 11.143 | 13.277 | 14.860 |
| 5 | 0.412 | 0.554 | 0.831 | 1.145 | 1.610 | 9.236 | 11.070 | 12.833 | 15.086 | 16.750 |
| 6 | 0.676 | 0.872 | 1.237 | 1.635 | 2.204 | 10.645 | 12.592 | 14.449 | 16.812 | 18.548 |
| 7 | 0.989 | 1.239 | 1.690 | 2.167 | 2.833 | 12.017 | 14.067 | 16.013 | 18.475 | 20.278 |
| 8 | 1.344 | 1.646 | 2.180 | 2.733 | 3.490 | 13.362 | 15.507 | 17.535 | 20.090 | 21.955 |
| 9 | 1.735 | 2.088 | 2.700 | 3.325 | 4.168 | 14.684 | 16.919 | 19.023 | 21.666 | 23.589 |
| 10 | 2.156 | 2.558 | 3.247 | 3.940 | 4.865 | 15.987 | 18.307 | 20.483 | 23.209 | 25.188 |
| 11 | 2.603 | 3.053 | 3.816 | 4.575 | 5.578 | 17.275 | 19.675 | 21.920 | 24.725 | 26.757 |
| 12 | 3.074 | 3.571 | 4.404 | 5.226 | 6.304 | 18.549 | 21.026 | 23.337 | 26.217 | 28.300 |
| 13 | 3.565 | 4.107 | 5.009 | 5.892 | 7.042 | 19.812 | 22.362 | 24.736 | 27.688 | 29.819 |
| 14 | 4.075 | 4.660 | 5.629 | 6.571 | 7.790 | 21.064 | 23.685 | 26.119 | 29.141 | 31.319 |
| 15 | 4.601 | 5.229 | 6.262 | 7.261 | 8.547 | 22.307 | 24.996 | 27.488 | 30.578 | 32.801 |
| 16 | 5.142 | 5.812 | 6.908 | 7.962 | 9.312 | 23.542 | 26.296 | 28.845 | 32.000 | 34.267 |
| 17 | 5.697 | 6.408 | 7.564 | 8.672 | 10.085 | 24.769 | 27.587 | 30.191 | 33.409 | 35.718 |
| 18 | 6.265 | 7.015 | 8.231 | 9.390 | 10.865 | 25.989 | 28.869 | 31.526 | 34.805 | 37.156 |
| 19 | 6.844 | 7.633 | 8.907 | 10.117 | 11.651 | 27.204 | 30.144 | 32.852 | 36.191 | 38.582 |
| 20 | 7.434 | 8.260 | 9.591 | 10.851 | 12.443 | 28.412 | 31.410 | 34.170 | 37.566 | 39.997 |
| 21 | 8.034 | 8.897 | 10.283 | 11.591 | 13.240 | 29.615 | 32.671 | 35.479 | 38.932 | 41.401 |
| 22 | 8.643 | 9.542 | 10.982 | 12.338 | 14.041 | 30.813 | 33.924 | 36.781 | 40.289 | 42.796 |
| 23 | 9.260 | 10.196 | 11.689 | 13.091 | 14.848 | 32.007 | 35.172 | 38.076 | 41.638 | 44.181 |
| 24 | 9.886 | 10.856 | 12.401 | 13.848 | 15.659 | 33.196 | 36.415 | 39.364 | 42.980 | 45.559 |
| 25 | 10.520 | 11.524 | 13.120 | 14.611 | 16.473 | 34.382 | 37.652 | 40.646 | 44.314 | 46.928 |
| 26 | 11.160 | 12.198 | 13.844 | 15.379 | 17.292 | 35.563 | 38.885 | 41.923 | 45.642 | 48.290 |
| 27 | 11.808 | 12.879 | 14.573 | 16.151 | 18.114 | 36.741 | 40.113 | 43.195 | 46.963 | 49.645 |
| 28 | 12.461 | 13.565 | 15.308 | 16.928 | 18.939 | 37.916 | 41.337 | 44.461 | 48.278 | 50.993 |
| 29 | 13.121 | 14.256 | 16.047 | 17.708 | 19.768 | 39.087 | 42.557 | 45.722 | 49.588 | 52.336 |
| 30 | 13.787 | 14.953 | 16.791 | 18.493 | 20.599 | 40.256 | 43.773 | 46.979 | 50.892 | 53.672 |
| 40 | 20.707 | 22.164 | 24.433 | 26.509 | 29.051 | 51.805 | 55.758 | 59.342 | 63.691 | 66.766 |
| 50 | 27.991 | 29.707 | 32.357 | 34.764 | 37.689 | 63.167 | 67.505 | 71.420 | 76.154 | 79.490 |
| 60 | 35.534 | 37.485 | 40.482 | 43.188 | 46.459 | 74.397 | 79.082 | 83.298 | 88.379 | 91.952 |
| 70 | 43.275 | 45.442 | 48.758 | 51.739 | 55.329 | 85.527 | 90.531 | 95.023 | 100.425 | 104.215 |
| 80 | 51.172 | 53.540 | 57.153 | 60.391 | 64.278 | 96.578 | 101.879 | 106.629 | 112.329 | 116.321 |
| 90 | 59.196 | 61.754 | 65.647 | 69.126 | 73.291 | 107.565 | 113.145 | 118.136 | 124.116 | 128.299 |
| 100 | 67.328 | 70.065 | 74.222 | 77.929 | 82.358 | 118.498 | 124.342 | 129.561 | 135.807 | 140.169 |

# H  Critical Values of the *F*-Distribution ($\alpha$ = 0.995)

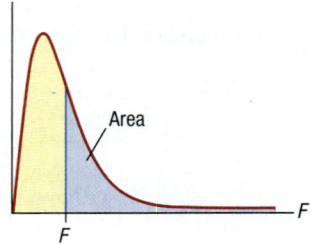

**Numerator Degrees of Freedom**

| | 1 | 2 | 3 | 4 | 5 | 6 | 7 | 8 | 9 |
|---|---|---|---|---|---|---|---|---|---|
| 1 | 0.0001 | 0.0050 | 0.0180 | 0.0319 | 0.0439 | 0.0537 | 0.0616 | 0.0681 | 0.0735 |
| 2 | 0.0001 | 0.0050 | 0.0201 | 0.0380 | 0.0546 | 0.0688 | 0.0806 | 0.0906 | 0.0989 |
| 3 | 0.0000 | 0.0050 | 0.0211 | 0.0412 | 0.0605 | 0.0774 | 0.0919 | 0.1042 | 0.1147 |
| 4 | 0.0000 | 0.0050 | 0.0216 | 0.0432 | 0.0643 | 0.0831 | 0.0995 | 0.1136 | 0.1257 |
| 5 | 0.0000 | 0.0050 | 0.0220 | 0.0445 | 0.0669 | 0.0872 | 0.1050 | 0.1205 | 0.1338 |
| 6 | 0.0000 | 0.0050 | 0.0223 | 0.0455 | 0.0689 | 0.0903 | 0.1092 | 0.1258 | 0.1402 |
| 7 | 0.0000 | 0.0050 | 0.0225 | 0.0462 | 0.0704 | 0.0927 | 0.1125 | 0.1300 | 0.1452 |
| 8 | 0.0000 | 0.0050 | 0.0227 | 0.0468 | 0.0716 | 0.0946 | 0.1152 | 0.1334 | 0.1494 |
| 9 | 0.0000 | 0.0050 | 0.0228 | 0.0473 | 0.0726 | 0.0962 | 0.1175 | 0.1363 | 0.1529 |
| 10 | 0.0000 | 0.0050 | 0.0229 | 0.0477 | 0.0734 | 0.0976 | 0.1193 | 0.1387 | 0.1558 |
| 11 | 0.0000 | 0.0050 | 0.0230 | 0.0480 | 0.0741 | 0.0987 | 0.1209 | 0.1408 | 0.1584 |
| 12 | 0.0000 | 0.0050 | 0.0230 | 0.0483 | 0.0747 | 0.0997 | 0.1223 | 0.1426 | 0.1606 |
| 13 | 0.0000 | 0.0050 | 0.0231 | 0.0485 | 0.0752 | 0.1005 | 0.1235 | 0.1441 | 0.1625 |
| 14 | 0.0000 | 0.0050 | 0.0232 | 0.0487 | 0.0757 | 0.1012 | 0.1246 | 0.1455 | 0.1642 |
| 15 | 0.0000 | 0.0050 | 0.0232 | 0.0489 | 0.0761 | 0.1019 | 0.1255 | 0.1468 | 0.1658 |
| 16 | 0.0000 | 0.0050 | 0.0233 | 0.0491 | 0.0764 | 0.1025 | 0.1263 | 0.1479 | 0.1671 |
| 17 | 0.0000 | 0.0050 | 0.0233 | 0.0492 | 0.0767 | 0.1030 | 0.1271 | 0.1489 | 0.1684 |
| 18 | 0.0000 | 0.0050 | 0.0233 | 0.0494 | 0.0770 | 0.1035 | 0.1278 | 0.1498 | 0.1695 |
| 19 | 0.0000 | 0.0050 | 0.0234 | 0.0495 | 0.0773 | 0.1039 | 0.1284 | 0.1506 | 0.1705 |
| 20 | 0.0000 | 0.0050 | 0.0234 | 0.0496 | 0.0775 | 0.1043 | 0.1290 | 0.1513 | 0.1715 |
| 21 | 0.0000 | 0.0050 | 0.0234 | 0.0497 | 0.0777 | 0.1046 | 0.1295 | 0.1520 | 0.1723 |
| 22 | 0.0000 | 0.0050 | 0.0234 | 0.0498 | 0.0779 | 0.1050 | 0.1300 | 0.1526 | 0.1731 |
| 23 | 0.0000 | 0.0050 | 0.0234 | 0.0499 | 0.0781 | 0.1053 | 0.1304 | 0.1532 | 0.1739 |
| 24 | 0.0000 | 0.0050 | 0.0235 | 0.0499 | 0.0782 | 0.1055 | 0.1308 | 0.1538 | 0.1745 |
| 25 | 0.0000 | 0.0050 | 0.0235 | 0.0500 | 0.0784 | 0.1058 | 0.1312 | 0.1543 | 0.1752 |
| 26 | 0.0000 | 0.0050 | 0.0235 | 0.0501 | 0.0785 | 0.1060 | 0.1315 | 0.1547 | 0.1758 |
| 27 | 0.0000 | 0.0050 | 0.0235 | 0.0501 | 0.0787 | 0.1063 | 0.1319 | 0.1552 | 0.1763 |
| 28 | 0.0000 | 0.0050 | 0.0235 | 0.0502 | 0.0788 | 0.1065 | 0.1322 | 0.1556 | 0.1768 |
| 29 | 0.0000 | 0.0050 | 0.0235 | 0.0502 | 0.0789 | 0.1067 | 0.1325 | 0.1560 | 0.1773 |
| 30 | 0.0000 | 0.0050 | 0.0235 | 0.0503 | 0.0790 | 0.1069 | 0.1327 | 0.1563 | 0.1778 |
| 40 | 0.0000 | 0.0050 | 0.0236 | 0.0506 | 0.0798 | 0.1082 | 0.1347 | 0.1590 | 0.1812 |
| 60 | 0.0000 | 0.0050 | 0.0237 | 0.0510 | 0.0806 | 0.1096 | 0.1368 | 0.1619 | 0.1848 |
| 120 | 0.0000 | 0.0050 | 0.0238 | 0.0514 | 0.0815 | 0.1111 | 0.1390 | 0.1649 | 0.1887 |
| ∞ | 0.0000 | 0.0050 | 0.0239 | 0.0517 | 0.0823 | 0.1126 | 0.1413 | 0.1681 | 0.1928 |

*Denominator Degrees of Freedom*

# H Critical Values of the *F*-Distribution ($\alpha = 0.995$) (cont.)

**Numerator Degrees of Freedom**

| | 10 | 11 | 12 | 13 | 14 | 15 | 16 | 17 | 18 |
|---|---|---|---|---|---|---|---|---|---|
| 1 | 0.0780 | 0.0818 | 0.0851 | 0.0879 | 0.0904 | 0.0926 | 0.0946 | 0.0963 | 0.0979 |
| 2 | 0.1061 | 0.1122 | 0.1175 | 0.1222 | 0.1262 | 0.1299 | 0.1331 | 0.1360 | 0.1386 |
| 3 | 0.1238 | 0.1316 | 0.1384 | 0.1444 | 0.1497 | 0.1544 | 0.1586 | 0.1625 | 0.1659 |
| 4 | 0.1362 | 0.1453 | 0.1533 | 0.1604 | 0.1667 | 0.1723 | 0.1774 | 0.1819 | 0.1861 |
| 5 | 0.1455 | 0.1557 | 0.1647 | 0.1727 | 0.1798 | 0.1861 | 0.1919 | 0.1971 | 0.2018 |
| 6 | 0.1528 | 0.1639 | 0.1737 | 0.1824 | 0.1902 | 0.1972 | 0.2035 | 0.2093 | 0.2145 |
| 7 | 0.1587 | 0.1705 | 0.1810 | 0.1904 | 0.1988 | 0.2063 | 0.2131 | 0.2193 | 0.2250 |
| 8 | 0.1635 | 0.1760 | 0.1871 | 0.1970 | 0.2059 | 0.2139 | 0.2212 | 0.2278 | 0.2339 |
| 9 | 0.1676 | 0.1806 | 0.1922 | 0.2026 | 0.2120 | 0.2204 | 0.2281 | 0.2351 | 0.2415 |
| 10 | 0.1710 | 0.1846 | 0.1966 | 0.2075 | 0.2172 | 0.2261 | 0.2341 | 0.2414 | 0.2481 |
| 11 | 0.1740 | 0.1880 | 0.2005 | 0.2117 | 0.2218 | 0.2310 | 0.2393 | 0.2469 | 0.2539 |
| 12 | 0.1766 | 0.1910 | 0.2038 | 0.2154 | 0.2258 | 0.2353 | 0.2439 | 0.2518 | 0.2591 |
| 13 | 0.1789 | 0.1936 | 0.2068 | 0.2187 | 0.2294 | 0.2392 | 0.2481 | 0.2562 | 0.2637 |
| 14 | 0.1810 | 0.1960 | 0.2094 | 0.2216 | 0.2326 | 0.2426 | 0.2517 | 0.2601 | 0.2678 |
| 15 | 0.1828 | 0.1981 | 0.2118 | 0.2242 | 0.2355 | 0.2457 | 0.2551 | 0.2636 | 0.2715 |
| 16 | 0.1844 | 0.2000 | 0.2139 | 0.2266 | 0.2381 | 0.2485 | 0.2581 | 0.2669 | 0.2749 |
| 17 | 0.1859 | 0.2017 | 0.2159 | 0.2287 | 0.2404 | 0.2511 | 0.2608 | 0.2698 | 0.2780 |
| 18 | 0.1873 | 0.2032 | 0.2177 | 0.2307 | 0.2426 | 0.2534 | 0.2634 | 0.2725 | 0.2809 |
| 19 | 0.1885 | 0.2047 | 0.2193 | 0.2325 | 0.2446 | 0.2556 | 0.2657 | 0.2749 | 0.2835 |
| 20 | 0.1896 | 0.2060 | 0.2208 | 0.2342 | 0.2464 | 0.2576 | 0.2678 | 0.2772 | 0.2859 |
| 21 | 0.1906 | 0.2072 | 0.2221 | 0.2357 | 0.2481 | 0.2594 | 0.2698 | 0.2793 | 0.2881 |
| 22 | 0.1916 | 0.2083 | 0.2234 | 0.2371 | 0.2496 | 0.2611 | 0.2716 | 0.2813 | 0.2902 |
| 23 | 0.1925 | 0.2093 | 0.2246 | 0.2384 | 0.2511 | 0.2627 | 0.2733 | 0.2831 | 0.2922 |
| 24 | 0.1933 | 0.2103 | 0.2257 | 0.2397 | 0.2524 | 0.2641 | 0.2749 | 0.2848 | 0.2940 |
| 25 | 0.1941 | 0.2112 | 0.2267 | 0.2408 | 0.2537 | 0.2655 | 0.2764 | 0.2864 | 0.2957 |
| 26 | 0.1948 | 0.2120 | 0.2276 | 0.2419 | 0.2549 | 0.2668 | 0.2778 | 0.2879 | 0.2973 |
| 27 | 0.1954 | 0.2128 | 0.2285 | 0.2429 | 0.2560 | 0.2680 | 0.2791 | 0.2893 | 0.2988 |
| 28 | 0.1961 | 0.2135 | 0.2294 | 0.2438 | 0.2570 | 0.2692 | 0.2803 | 0.2906 | 0.3002 |
| 29 | 0.1967 | 0.2142 | 0.2302 | 0.2447 | 0.2580 | 0.2702 | 0.2815 | 0.2919 | 0.3015 |
| 30 | 0.1972 | 0.2149 | 0.2309 | 0.2455 | 0.2589 | 0.2712 | 0.2826 | 0.2930 | 0.3028 |
| 40 | 0.2014 | 0.2197 | 0.2365 | 0.2519 | 0.2660 | 0.2789 | 0.2909 | 0.3020 | 0.3124 |
| 60 | 0.2058 | 0.2250 | 0.2425 | 0.2587 | 0.2736 | 0.2873 | 0.3001 | 0.3119 | 0.3230 |
| 120 | 0.2105 | 0.2306 | 0.2491 | 0.2661 | 0.2819 | 0.2965 | 0.3102 | 0.3229 | 0.3348 |
| ∞ | 0.2156 | 0.2367 | 0.2562 | 0.2742 | 0.2910 | 0.3067 | 0.3214 | 0.3351 | 0.3480 |

*Denominator Degrees of Freedom*

# H   Critical Values of the *F*-Distribution ($\alpha = 0.995$) (cont.)

**Numerator Degrees of Freedom**

| | 19 | 20 | 24 | 30 | 40 | 60 | 120 |
|---|---|---|---|---|---|---|---|
| 1 | 0.0993 | 0.1006 | 0.1047 | 0.1089 | 0.1133 | 0.1177 | 0.1223 |
| 2 | 0.1410 | 0.1431 | 0.1501 | 0.1574 | 0.1648 | 0.1726 | 0.1805 |
| 3 | 0.1690 | 0.1719 | 0.1812 | 0.1909 | 0.2010 | 0.2115 | 0.2224 |
| 4 | 0.1898 | 0.1933 | 0.2045 | 0.2163 | 0.2286 | 0.2416 | 0.2551 |
| 5 | 0.2061 | 0.2100 | 0.2229 | 0.2365 | 0.2509 | 0.2660 | 0.2818 |
| 6 | 0.2192 | 0.2236 | 0.2380 | 0.2532 | 0.2693 | 0.2864 | 0.3044 |
| 7 | 0.2302 | 0.2349 | 0.2506 | 0.2673 | 0.2850 | 0.3038 | 0.3239 |
| 8 | 0.2394 | 0.2445 | 0.2613 | 0.2793 | 0.2985 | 0.3190 | 0.3410 |
| 9 | 0.2474 | 0.2528 | 0.2706 | 0.2898 | 0.3104 | 0.3324 | 0.3561 |
| 10 | 0.2543 | 0.2599 | 0.2788 | 0.2990 | 0.3208 | 0.3443 | 0.3697 |
| 11 | 0.2603 | 0.2663 | 0.2860 | 0.3072 | 0.3302 | 0.3550 | 0.3819 |
| 12 | 0.2657 | 0.2719 | 0.2924 | 0.3146 | 0.3386 | 0.3647 | 0.3931 |
| 13 | 0.2706 | 0.2769 | 0.2982 | 0.3212 | 0.3463 | 0.3735 | 0.4033 |
| 14 | 0.2749 | 0.2815 | 0.3034 | 0.3272 | 0.3532 | 0.3816 | 0.4127 |
| 15 | 0.2788 | 0.2856 | 0.3081 | 0.3327 | 0.3596 | 0.3890 | 0.4215 |
| 16 | 0.2824 | 0.2893 | 0.3124 | 0.3377 | 0.3654 | 0.3959 | 0.4296 |
| 17 | 0.2856 | 0.2927 | 0.3164 | 0.3423 | 0.3708 | 0.4023 | 0.4371 |
| 18 | 0.2886 | 0.2958 | 0.3200 | 0.3466 | 0.3758 | 0.4082 | 0.4442 |
| 19 | 0.2914 | 0.2987 | 0.3234 | 0.3506 | 0.3805 | 0.4137 | 0.4508 |
| 20 | 0.2939 | 0.3014 | 0.3265 | 0.3542 | 0.3848 | 0.4189 | 0.4570 |
| 21 | 0.2963 | 0.3039 | 0.3295 | 0.3577 | 0.3889 | 0.4238 | 0.4629 |
| 22 | 0.2985 | 0.3062 | 0.3322 | 0.3609 | 0.3927 | 0.4283 | 0.4684 |
| 23 | 0.3006 | 0.3083 | 0.3347 | 0.3639 | 0.3963 | 0.4326 | 0.4737 |
| 24 | 0.3025 | 0.3104 | 0.3371 | 0.3667 | 0.3997 | 0.4367 | 0.4787 |
| 25 | 0.3043 | 0.3123 | 0.3393 | 0.3693 | 0.4029 | 0.4406 | 0.4834 |
| 26 | 0.3059 | 0.3140 | 0.3414 | 0.3718 | 0.4059 | 0.4442 | 0.4879 |
| 27 | 0.3075 | 0.3157 | 0.3434 | 0.3742 | 0.4087 | 0.4477 | 0.4922 |
| 28 | 0.3090 | 0.3173 | 0.3452 | 0.3764 | 0.4114 | 0.4510 | 0.4964 |
| 29 | 0.3104 | 0.3188 | 0.3470 | 0.3785 | 0.4140 | 0.4542 | 0.5003 |
| 30 | 0.3118 | 0.3202 | 0.3487 | 0.3805 | 0.4164 | 0.4572 | 0.5040 |
| 40 | 0.3220 | 0.3310 | 0.3616 | 0.3962 | 0.4356 | 0.4810 | 0.5345 |
| 60 | 0.3333 | 0.3429 | 0.3762 | 0.4141 | 0.4579 | 0.5096 | 0.5725 |
| 120 | 0.3459 | 0.3564 | 0.3927 | 0.4348 | 0.4846 | 0.5452 | 0.6229 |
| ∞ | 0.3602 | 0.3717 | 0.4119 | 0.4596 | 0.5177 | 0.5922 | 0.6988 |

**Denominator Degrees of Freedom**

# H   Critical Values of the *F*-Distribution ($\alpha = 0.990$)

**Numerator Degrees of Freedom**

| | 1 | 2 | 3 | 4 | 5 | 6 | 7 | 8 | 9 |
|---|---|---|---|---|---|---|---|---|---|
| 1 | 0.0002 | 0.0102 | 0.0293 | 0.0472 | 0.0615 | 0.0728 | 0.0817 | 0.0888 | 0.0947 |
| 2 | 0.0002 | 0.0101 | 0.0325 | 0.0556 | 0.0753 | 0.0915 | 0.1047 | 0.1156 | 0.1247 |
| 3 | 0.0002 | 0.0101 | 0.0339 | 0.0599 | 0.0829 | 0.1023 | 0.1183 | 0.1317 | 0.1430 |
| 4 | 0.0002 | 0.0101 | 0.0348 | 0.0626 | 0.0878 | 0.1093 | 0.1274 | 0.1427 | 0.1557 |
| 5 | 0.0002 | 0.0101 | 0.0354 | 0.0644 | 0.0912 | 0.1143 | 0.1340 | 0.1508 | 0.1651 |
| 6 | 0.0002 | 0.0101 | 0.0358 | 0.0658 | 0.0937 | 0.1181 | 0.1391 | 0.1570 | 0.1724 |
| 7 | 0.0002 | 0.0101 | 0.0361 | 0.0668 | 0.0956 | 0.1211 | 0.1430 | 0.1619 | 0.1782 |
| 8 | 0.0002 | 0.0101 | 0.0364 | 0.0676 | 0.0972 | 0.1234 | 0.1462 | 0.1659 | 0.1829 |
| 9 | 0.0002 | 0.0101 | 0.0366 | 0.0682 | 0.0984 | 0.1254 | 0.1488 | 0.1692 | 0.1869 |
| 10 | 0.0002 | 0.0101 | 0.0367 | 0.0687 | 0.0995 | 0.1270 | 0.1511 | 0.1720 | 0.1902 |
| 11 | 0.0002 | 0.0101 | 0.0369 | 0.0692 | 0.1004 | 0.1284 | 0.1529 | 0.1744 | 0.1931 |
| 12 | 0.0002 | 0.0101 | 0.0370 | 0.0696 | 0.1011 | 0.1296 | 0.1546 | 0.1765 | 0.1956 |
| 13 | 0.0002 | 0.0101 | 0.0371 | 0.0699 | 0.1018 | 0.1306 | 0.1560 | 0.1783 | 0.1978 |
| 14 | 0.0002 | 0.0101 | 0.0371 | 0.0702 | 0.1024 | 0.1315 | 0.1573 | 0.1799 | 0.1998 |
| 15 | 0.0002 | 0.0101 | 0.0372 | 0.0704 | 0.1029 | 0.1323 | 0.1584 | 0.1813 | 0.2015 |
| 16 | 0.0002 | 0.0101 | 0.0373 | 0.0707 | 0.1033 | 0.1330 | 0.1594 | 0.1826 | 0.2031 |
| 17 | 0.0002 | 0.0101 | 0.0373 | 0.0708 | 0.1037 | 0.1336 | 0.1603 | 0.1837 | 0.2045 |
| 18 | 0.0002 | 0.0101 | 0.0374 | 0.0710 | 0.1041 | 0.1342 | 0.1611 | 0.1848 | 0.2058 |
| 19 | 0.0002 | 0.0101 | 0.0374 | 0.0712 | 0.1044 | 0.1347 | 0.1618 | 0.1857 | 0.2069 |
| 20 | 0.0002 | 0.0101 | 0.0375 | 0.0713 | 0.1047 | 0.1352 | 0.1625 | 0.1866 | 0.2080 |
| 21 | 0.0002 | 0.0101 | 0.0375 | 0.0715 | 0.1050 | 0.1356 | 0.1631 | 0.1874 | 0.2090 |
| 22 | 0.0002 | 0.0101 | 0.0375 | 0.0716 | 0.1052 | 0.1360 | 0.1636 | 0.1881 | 0.2099 |
| 23 | 0.0002 | 0.0101 | 0.0376 | 0.0717 | 0.1054 | 0.1364 | 0.1641 | 0.1888 | 0.2107 |
| 24 | 0.0002 | 0.0101 | 0.0376 | 0.0718 | 0.1056 | 0.1367 | 0.1646 | 0.1894 | 0.2115 |
| 25 | 0.0002 | 0.0101 | 0.0376 | 0.0719 | 0.1058 | 0.1371 | 0.1651 | 0.1900 | 0.2122 |
| 26 | 0.0002 | 0.0101 | 0.0376 | 0.0720 | 0.1060 | 0.1374 | 0.1655 | 0.1905 | 0.2128 |
| 27 | 0.0002 | 0.0101 | 0.0377 | 0.0721 | 0.1062 | 0.1376 | 0.1659 | 0.1910 | 0.2135 |
| 28 | 0.0002 | 0.0101 | 0.0377 | 0.0721 | 0.1063 | 0.1379 | 0.1662 | 0.1915 | 0.2141 |
| 29 | 0.0002 | 0.0101 | 0.0377 | 0.0722 | 0.1065 | 0.1381 | 0.1666 | 0.1920 | 0.2146 |
| 30 | 0.0002 | 0.0101 | 0.0377 | 0.0723 | 0.1066 | 0.1383 | 0.1669 | 0.1924 | 0.2151 |
| 40 | 0.0002 | 0.0101 | 0.0379 | 0.0728 | 0.1076 | 0.1400 | 0.1692 | 0.1955 | 0.2190 |
| 60 | 0.0002 | 0.0101 | 0.0380 | 0.0732 | 0.1087 | 0.1417 | 0.1717 | 0.1987 | 0.2231 |
| 120 | 0.0002 | 0.0101 | 0.0381 | 0.0738 | 0.1097 | 0.1435 | 0.1743 | 0.2022 | 0.2274 |
| ∞ | 0.0002 | 0.0101 | 0.0383 | 0.0743 | 0.1109 | 0.1453 | 0.1770 | 0.2058 | 0.2320 |

*Denominator Degrees of Freedom*

## H  Critical Values of the *F*-Distribution ($\alpha = 0.990$) (cont.)

**Numerator Degrees of Freedom**

| | 10 | 11 | 12 | 13 | 14 | 15 | 16 | 17 | 18 |
|---|---|---|---|---|---|---|---|---|---|
| 1 | 0.0996 | 0.1037 | 0.1072 | 0.1102 | 0.1128 | 0.1152 | 0.1172 | 0.1191 | 0.1207 |
| 2 | 0.1323 | 0.1388 | 0.1444 | 0.1492 | 0.1535 | 0.1573 | 0.1606 | 0.1636 | 0.1663 |
| 3 | 0.1526 | 0.1609 | 0.1680 | 0.1742 | 0.1797 | 0.1846 | 0.1890 | 0.1929 | 0.1964 |
| 4 | 0.1668 | 0.1764 | 0.1848 | 0.1921 | 0.1986 | 0.2044 | 0.2095 | 0.2142 | 0.2184 |
| 5 | 0.1774 | 0.1881 | 0.1975 | 0.2057 | 0.2130 | 0.2195 | 0.2254 | 0.2306 | 0.2354 |
| 6 | 0.1857 | 0.1973 | 0.2074 | 0.2164 | 0.2244 | 0.2316 | 0.2380 | 0.2438 | 0.2491 |
| 7 | 0.1923 | 0.2047 | 0.2155 | 0.2252 | 0.2338 | 0.2415 | 0.2484 | 0.2547 | 0.2604 |
| 8 | 0.1978 | 0.2108 | 0.2223 | 0.2324 | 0.2415 | 0.2497 | 0.2571 | 0.2638 | 0.2699 |
| 9 | 0.2023 | 0.2159 | 0.2279 | 0.2386 | 0.2482 | 0.2568 | 0.2645 | 0.2716 | 0.2780 |
| 10 | 0.2062 | 0.2203 | 0.2328 | 0.2439 | 0.2538 | 0.2628 | 0.2709 | 0.2783 | 0.2850 |
| 11 | 0.2096 | 0.2241 | 0.2370 | 0.2485 | 0.2588 | 0.2681 | 0.2765 | 0.2842 | 0.2912 |
| 12 | 0.2125 | 0.2274 | 0.2407 | 0.2525 | 0.2631 | 0.2728 | 0.2815 | 0.2894 | 0.2967 |
| 13 | 0.2151 | 0.2303 | 0.2439 | 0.2561 | 0.2670 | 0.2769 | 0.2859 | 0.2941 | 0.3015 |
| 14 | 0.2174 | 0.2329 | 0.2468 | 0.2592 | 0.2704 | 0.2806 | 0.2898 | 0.2982 | 0.3059 |
| 15 | 0.2194 | 0.2352 | 0.2494 | 0.2621 | 0.2735 | 0.2839 | 0.2933 | 0.3020 | 0.3099 |
| 16 | 0.2212 | 0.2373 | 0.2517 | 0.2647 | 0.2763 | 0.2869 | 0.2966 | 0.3054 | 0.3134 |
| 17 | 0.2229 | 0.2392 | 0.2539 | 0.2670 | 0.2789 | 0.2897 | 0.2995 | 0.3085 | 0.3167 |
| 18 | 0.2244 | 0.2409 | 0.2558 | 0.2691 | 0.2812 | 0.2922 | 0.3022 | 0.3113 | 0.3197 |
| 19 | 0.2257 | 0.2425 | 0.2576 | 0.2711 | 0.2833 | 0.2945 | 0.3046 | 0.3139 | 0.3224 |
| 20 | 0.2270 | 0.2440 | 0.2592 | 0.2729 | 0.2853 | 0.2966 | 0.3069 | 0.3163 | 0.3250 |
| 21 | 0.2281 | 0.2453 | 0.2607 | 0.2745 | 0.2871 | 0.2985 | 0.3090 | 0.3185 | 0.3273 |
| 22 | 0.2292 | 0.2465 | 0.2620 | 0.2761 | 0.2888 | 0.3003 | 0.3109 | 0.3206 | 0.3295 |
| 23 | 0.2302 | 0.2476 | 0.2633 | 0.2775 | 0.2903 | 0.3020 | 0.3127 | 0.3225 | 0.3315 |
| 24 | 0.2311 | 0.2487 | 0.2645 | 0.2788 | 0.2918 | 0.3036 | 0.3144 | 0.3243 | 0.3334 |
| 25 | 0.2320 | 0.2497 | 0.2656 | 0.2800 | 0.2931 | 0.3050 | 0.3160 | 0.3260 | 0.3352 |
| 26 | 0.2328 | 0.2506 | 0.2667 | 0.2812 | 0.2944 | 0.3064 | 0.3174 | 0.3276 | 0.3369 |
| 27 | 0.2335 | 0.2515 | 0.2676 | 0.2823 | 0.2956 | 0.3077 | 0.3188 | 0.3290 | 0.3385 |
| 28 | 0.2342 | 0.2523 | 0.2685 | 0.2833 | 0.2967 | 0.3089 | 0.3201 | 0.3304 | 0.3399 |
| 29 | 0.2348 | 0.2530 | 0.2694 | 0.2842 | 0.2977 | 0.3101 | 0.3213 | 0.3317 | 0.3413 |
| 30 | 0.2355 | 0.2537 | 0.2702 | 0.2851 | 0.2987 | 0.3111 | 0.3225 | 0.3330 | 0.3426 |
| 40 | 0.2401 | 0.2591 | 0.2763 | 0.2919 | 0.3062 | 0.3193 | 0.3313 | 0.3424 | 0.3527 |
| 60 | 0.2450 | 0.2648 | 0.2828 | 0.2993 | 0.3143 | 0.3282 | 0.3409 | 0.3528 | 0.3637 |
| 120 | 0.2502 | 0.2710 | 0.2899 | 0.3072 | 0.3232 | 0.3379 | 0.3515 | 0.3642 | 0.3760 |
| ∞ | 0.2558 | 0.2776 | 0.2975 | 0.3159 | 0.3329 | 0.3486 | 0.3633 | 0.3769 | 0.3897 |

*Denominator Degrees of Freedom*

# H  Critical Values of the *F*-Distribution ($\alpha = 0.990$) (cont.)

**Numerator Degrees of Freedom**

| Denominator Degrees of Freedom | 19 | 20 | 24 | 30 | 40 | 60 | 120 |
|---|---|---|---|---|---|---|---|
| 1 | 0.1222 | 0.1235 | 0.1278 | 0.1322 | 0.1367 | 0.1413 | 0.1460 |
| 2 | 0.1688 | 0.1710 | 0.1781 | 0.1855 | 0.1931 | 0.2009 | 0.2089 |
| 3 | 0.1996 | 0.2025 | 0.2120 | 0.2217 | 0.2319 | 0.2424 | 0.2532 |
| 4 | 0.2222 | 0.2257 | 0.2371 | 0.2489 | 0.2612 | 0.2740 | 0.2874 |
| 5 | 0.2398 | 0.2437 | 0.2567 | 0.2703 | 0.2846 | 0.2995 | 0.3151 |
| 6 | 0.2539 | 0.2583 | 0.2727 | 0.2879 | 0.3039 | 0.3206 | 0.3383 |
| 7 | 0.2656 | 0.2704 | 0.2860 | 0.3026 | 0.3201 | 0.3386 | 0.3582 |
| 8 | 0.2754 | 0.2806 | 0.2974 | 0.3152 | 0.3341 | 0.3542 | 0.3755 |
| 9 | 0.2839 | 0.2893 | 0.3071 | 0.3261 | 0.3463 | 0.3679 | 0.3908 |
| 10 | 0.2912 | 0.2969 | 0.3156 | 0.3357 | 0.3571 | 0.3800 | 0.4045 |
| 11 | 0.2977 | 0.3036 | 0.3232 | 0.3442 | 0.3667 | 0.3908 | 0.4168 |
| 12 | 0.3033 | 0.3095 | 0.3299 | 0.3517 | 0.3753 | 0.4006 | 0.4280 |
| 13 | 0.3084 | 0.3148 | 0.3359 | 0.3585 | 0.3830 | 0.4095 | 0.4382 |
| 14 | 0.3130 | 0.3195 | 0.3413 | 0.3647 | 0.3901 | 0.4177 | 0.4476 |
| 15 | 0.3171 | 0.3238 | 0.3462 | 0.3703 | 0.3966 | 0.4251 | 0.4563 |
| 16 | 0.3209 | 0.3277 | 0.3506 | 0.3755 | 0.4025 | 0.4320 | 0.4643 |
| 17 | 0.3243 | 0.3313 | 0.3548 | 0.3802 | 0.4080 | 0.4384 | 0.4718 |
| 18 | 0.3274 | 0.3346 | 0.3585 | 0.3846 | 0.4131 | 0.4443 | 0.4788 |
| 19 | 0.3303 | 0.3376 | 0.3620 | 0.3886 | 0.4178 | 0.4498 | 0.4854 |
| 20 | 0.3330 | 0.3404 | 0.3652 | 0.3924 | 0.4221 | 0.4550 | 0.4915 |
| 21 | 0.3355 | 0.3430 | 0.3682 | 0.3959 | 0.4262 | 0.4598 | 0.4973 |
| 22 | 0.3378 | 0.3454 | 0.3710 | 0.3991 | 0.4301 | 0.4644 | 0.5028 |
| 23 | 0.3399 | 0.3476 | 0.3736 | 0.4022 | 0.4337 | 0.4687 | 0.5079 |
| 24 | 0.3419 | 0.3497 | 0.3761 | 0.4050 | 0.4371 | 0.4727 | 0.5128 |
| 25 | 0.3438 | 0.3517 | 0.3784 | 0.4077 | 0.4403 | 0.4766 | 0.5175 |
| 26 | 0.3455 | 0.3535 | 0.3805 | 0.4103 | 0.4433 | 0.4802 | 0.5219 |
| 27 | 0.3472 | 0.3553 | 0.3826 | 0.4127 | 0.4461 | 0.4836 | 0.5261 |
| 28 | 0.3487 | 0.3569 | 0.3845 | 0.4149 | 0.4488 | 0.4869 | 0.5301 |
| 29 | 0.3502 | 0.3584 | 0.3863 | 0.4171 | 0.4514 | 0.4900 | 0.5340 |
| 30 | 0.3516 | 0.3599 | 0.3880 | 0.4191 | 0.4538 | 0.4930 | 0.5376 |
| 40 | 0.3622 | 0.3711 | 0.4012 | 0.4349 | 0.4730 | 0.5165 | 0.5673 |
| 60 | 0.3739 | 0.3835 | 0.4161 | 0.4529 | 0.4952 | 0.5446 | 0.6040 |
| 120 | 0.3870 | 0.3973 | 0.4329 | 0.4738 | 0.5216 | 0.5793 | 0.6523 |
| ∞ | 0.4017 | 0.4130 | 0.4523 | 0.4984 | 0.5541 | 0.6247 | 0.7243 |

# H  Critical Values of the *F*-Distribution ($\alpha = 0.975$)

**Numerator Degrees of Freedom**

| | 1 | 2 | 3 | 4 | 5 | 6 | 7 | 8 | 9 |
|---|---|---|---|---|---|---|---|---|---|
| 1 | 0.0015 | 0.0260 | 0.0573 | 0.0818 | 0.0999 | 0.1135 | 0.1239 | 0.1321 | 0.1387 |
| 2 | 0.0013 | 0.0256 | 0.0623 | 0.0939 | 0.1186 | 0.1377 | 0.1529 | 0.1650 | 0.1750 |
| 3 | 0.0012 | 0.0255 | 0.0648 | 0.1002 | 0.1288 | 0.1515 | 0.1698 | 0.1846 | 0.1969 |
| 4 | 0.0011 | 0.0255 | 0.0662 | 0.1041 | 0.1354 | 0.1606 | 0.1811 | 0.1979 | 0.2120 |
| 5 | 0.0011 | 0.0254 | 0.0672 | 0.1068 | 0.1399 | 0.1670 | 0.1892 | 0.2076 | 0.2230 |
| 6 | 0.0011 | 0.0254 | 0.0679 | 0.1087 | 0.1433 | 0.1718 | 0.1954 | 0.2150 | 0.2315 |
| 7 | 0.0011 | 0.0254 | 0.0684 | 0.1102 | 0.1459 | 0.1756 | 0.2002 | 0.2208 | 0.2383 |
| 8 | 0.0010 | 0.0254 | 0.0688 | 0.1114 | 0.1480 | 0.1786 | 0.2041 | 0.2256 | 0.2438 |
| 9 | 0.0010 | 0.0254 | 0.0691 | 0.1123 | 0.1497 | 0.1810 | 0.2073 | 0.2295 | 0.2484 |
| 10 | 0.0010 | 0.0254 | 0.0694 | 0.1131 | 0.1511 | 0.1831 | 0.2100 | 0.2328 | 0.2523 |
| 11 | 0.0010 | 0.0254 | 0.0696 | 0.1137 | 0.1523 | 0.1849 | 0.2123 | 0.2357 | 0.2556 |
| 12 | 0.0010 | 0.0254 | 0.0698 | 0.1143 | 0.1533 | 0.1864 | 0.2143 | 0.2381 | 0.2585 |
| 13 | 0.0010 | 0.0254 | 0.0699 | 0.1147 | 0.1541 | 0.1877 | 0.2161 | 0.2403 | 0.2611 |
| 14 | 0.0010 | 0.0254 | 0.0700 | 0.1152 | 0.1549 | 0.1888 | 0.2176 | 0.2422 | 0.2633 |
| 15 | 0.0010 | 0.0254 | 0.0702 | 0.1155 | 0.1556 | 0.1898 | 0.2189 | 0.2438 | 0.2653 |
| 16 | 0.0010 | 0.0254 | 0.0703 | 0.1158 | 0.1562 | 0.1907 | 0.2201 | 0.2453 | 0.2671 |
| 17 | 0.0010 | 0.0254 | 0.0704 | 0.1161 | 0.1567 | 0.1915 | 0.2212 | 0.2467 | 0.2687 |
| 18 | 0.0010 | 0.0254 | 0.0704 | 0.1164 | 0.1572 | 0.1922 | 0.2222 | 0.2479 | 0.2702 |
| 19 | 0.0010 | 0.0254 | 0.0705 | 0.1166 | 0.1576 | 0.1929 | 0.2231 | 0.2490 | 0.2715 |
| 20 | 0.0010 | 0.0253 | 0.0706 | 0.1168 | 0.1580 | 0.1935 | 0.2239 | 0.2500 | 0.2727 |
| 21 | 0.0010 | 0.0253 | 0.0706 | 0.1170 | 0.1584 | 0.1940 | 0.2246 | 0.2510 | 0.2738 |
| 22 | 0.0010 | 0.0253 | 0.0707 | 0.1172 | 0.1587 | 0.1945 | 0.2253 | 0.2518 | 0.2749 |
| 23 | 0.0010 | 0.0253 | 0.0708 | 0.1173 | 0.1590 | 0.1950 | 0.2259 | 0.2526 | 0.2758 |
| 24 | 0.0010 | 0.0253 | 0.0708 | 0.1175 | 0.1593 | 0.1954 | 0.2265 | 0.2533 | 0.2767 |
| 25 | 0.0010 | 0.0253 | 0.0708 | 0.1176 | 0.1595 | 0.1958 | 0.2270 | 0.2540 | 0.2775 |
| 26 | 0.0010 | 0.0253 | 0.0709 | 0.1178 | 0.1598 | 0.1962 | 0.2275 | 0.2547 | 0.2783 |
| 27 | 0.0010 | 0.0253 | 0.0709 | 0.1179 | 0.1600 | 0.1965 | 0.2280 | 0.2552 | 0.2790 |
| 28 | 0.0010 | 0.0253 | 0.0710 | 0.1180 | 0.1602 | 0.1968 | 0.2284 | 0.2558 | 0.2797 |
| 29 | 0.0010 | 0.0253 | 0.0710 | 0.1181 | 0.1604 | 0.1971 | 0.2288 | 0.2563 | 0.2803 |
| 30 | 0.0010 | 0.0253 | 0.0710 | 0.1182 | 0.1606 | 0.1974 | 0.2292 | 0.2568 | 0.2809 |
| 40 | 0.0010 | 0.0253 | 0.0712 | 0.1189 | 0.1619 | 0.1995 | 0.2321 | 0.2604 | 0.2853 |
| 60 | 0.0010 | 0.0253 | 0.0715 | 0.1196 | 0.1633 | 0.2017 | 0.2351 | 0.2642 | 0.2899 |
| 120 | 0.0010 | 0.0253 | 0.0717 | 0.1203 | 0.1648 | 0.2039 | 0.2382 | 0.2682 | 0.2948 |
| ∞ | 0.0010 | 0.0253 | 0.0719 | 0.1211 | 0.1662 | 0.2062 | 0.2414 | 0.2725 | 0.3000 |

*Denominator Degrees of Freedom*

# H  Critical Values of the *F*-Distribution ($\alpha = 0.975$) (cont.)

**Numerator Degrees of Freedom**

| | 10 | 11 | 12 | 13 | 14 | 15 | 16 | 17 | 18 |
|---|---|---|---|---|---|---|---|---|---|
| 1 | 0.1442 | 0.1487 | 0.1526 | 0.1559 | 0.1588 | 0.1613 | 0.1635 | 0.1655 | 0.1673 |
| 2 | 0.1833 | 0.1903 | 0.1962 | 0.2014 | 0.2059 | 0.2099 | 0.2134 | 0.2165 | 0.2193 |
| 3 | 0.2072 | 0.2160 | 0.2235 | 0.2300 | 0.2358 | 0.2408 | 0.2453 | 0.2493 | 0.2529 |
| 4 | 0.2238 | 0.2339 | 0.2426 | 0.2503 | 0.2569 | 0.2629 | 0.2681 | 0.2729 | 0.2771 |
| 5 | 0.2361 | 0.2473 | 0.2570 | 0.2655 | 0.2730 | 0.2796 | 0.2855 | 0.2909 | 0.2957 |
| 6 | 0.2456 | 0.2577 | 0.2682 | 0.2774 | 0.2856 | 0.2929 | 0.2993 | 0.3052 | 0.3105 |
| 7 | 0.2532 | 0.2661 | 0.2773 | 0.2871 | 0.2959 | 0.3036 | 0.3106 | 0.3169 | 0.3226 |
| 8 | 0.2594 | 0.2729 | 0.2848 | 0.2952 | 0.3044 | 0.3126 | 0.3200 | 0.3267 | 0.3327 |
| 9 | 0.2646 | 0.2787 | 0.2910 | 0.3019 | 0.3116 | 0.3202 | 0.3280 | 0.3350 | 0.3414 |
| 10 | 0.2690 | 0.2836 | 0.2964 | 0.3077 | 0.3178 | 0.3268 | 0.3349 | 0.3422 | 0.3489 |
| 11 | 0.2729 | 0.2879 | 0.3011 | 0.3127 | 0.3231 | 0.3325 | 0.3409 | 0.3485 | 0.3554 |
| 12 | 0.2762 | 0.2916 | 0.3051 | 0.3171 | 0.3279 | 0.3375 | 0.3461 | 0.3540 | 0.3612 |
| 13 | 0.2791 | 0.2948 | 0.3087 | 0.3210 | 0.3320 | 0.3419 | 0.3508 | 0.3589 | 0.3663 |
| 14 | 0.2817 | 0.2977 | 0.3119 | 0.3245 | 0.3357 | 0.3458 | 0.3550 | 0.3633 | 0.3709 |
| 15 | 0.2840 | 0.3003 | 0.3147 | 0.3276 | 0.3391 | 0.3494 | 0.3587 | 0.3672 | 0.3750 |
| 16 | 0.2860 | 0.3026 | 0.3173 | 0.3304 | 0.3421 | 0.3526 | 0.3621 | 0.3708 | 0.3787 |
| 17 | 0.2879 | 0.3047 | 0.3196 | 0.3329 | 0.3448 | 0.3555 | 0.3652 | 0.3741 | 0.3821 |
| 18 | 0.2896 | 0.3066 | 0.3217 | 0.3352 | 0.3473 | 0.3582 | 0.3681 | 0.3770 | 0.3853 |
| 19 | 0.2911 | 0.3084 | 0.3237 | 0.3373 | 0.3496 | 0.3606 | 0.3706 | 0.3798 | 0.3881 |
| 20 | 0.2925 | 0.3100 | 0.3254 | 0.3393 | 0.3517 | 0.3629 | 0.3730 | 0.3823 | 0.3908 |
| 21 | 0.2938 | 0.3114 | 0.3271 | 0.3410 | 0.3536 | 0.3649 | 0.3752 | 0.3846 | 0.3932 |
| 22 | 0.2950 | 0.3128 | 0.3286 | 0.3427 | 0.3554 | 0.3668 | 0.3773 | 0.3868 | 0.3955 |
| 23 | 0.2961 | 0.3140 | 0.3300 | 0.3442 | 0.3570 | 0.3686 | 0.3792 | 0.3888 | 0.3976 |
| 24 | 0.2971 | 0.3152 | 0.3313 | 0.3456 | 0.3586 | 0.3703 | 0.3809 | 0.3907 | 0.3996 |
| 25 | 0.2981 | 0.3163 | 0.3325 | 0.3470 | 0.3600 | 0.3718 | 0.3826 | 0.3924 | 0.4014 |
| 26 | 0.2990 | 0.3173 | 0.3336 | 0.3482 | 0.3614 | 0.3733 | 0.3841 | 0.3940 | 0.4031 |
| 27 | 0.2998 | 0.3183 | 0.3347 | 0.3494 | 0.3626 | 0.3746 | 0.3856 | 0.3956 | 0.4048 |
| 28 | 0.3006 | 0.3191 | 0.3357 | 0.3505 | 0.3638 | 0.3759 | 0.3869 | 0.3970 | 0.4063 |
| 29 | 0.3013 | 0.3200 | 0.3366 | 0.3515 | 0.3649 | 0.3771 | 0.3882 | 0.3984 | 0.4077 |
| 30 | 0.3020 | 0.3208 | 0.3375 | 0.3525 | 0.3660 | 0.3783 | 0.3894 | 0.3997 | 0.4091 |
| 40 | 0.3072 | 0.3267 | 0.3441 | 0.3598 | 0.3739 | 0.3868 | 0.3986 | 0.4095 | 0.4194 |
| 60 | 0.3127 | 0.3329 | 0.3512 | 0.3676 | 0.3825 | 0.3962 | 0.4087 | 0.4201 | 0.4308 |
| 120 | 0.3185 | 0.3397 | 0.3588 | 0.3761 | 0.3919 | 0.4063 | 0.4196 | 0.4319 | 0.4433 |
| ∞ | 0.3247 | 0.3469 | 0.3670 | 0.3853 | 0.4021 | 0.4175 | 0.4317 | 0.4450 | 0.4573 |

Denominator Degrees of Freedom

# H  Critical Values of the *F*-Distribution ($\alpha = 0.975$) (cont.)

**Numerator Degrees of Freedom**

| | 19 | 20 | 24 | 30 | 40 | 60 | 120 |
|---|---|---|---|---|---|---|---|
| 1 | 0.1689 | 0.1703 | 0.1749 | 0.1796 | 0.1844 | 0.1892 | 0.1941 |
| 2 | 0.2219 | 0.2242 | 0.2315 | 0.2391 | 0.2469 | 0.2548 | 0.2628 |
| 3 | 0.2562 | 0.2592 | 0.2687 | 0.2786 | 0.2887 | 0.2992 | 0.3099 |
| 4 | 0.2810 | 0.2845 | 0.2959 | 0.3077 | 0.3199 | 0.3325 | 0.3455 |
| 5 | 0.3001 | 0.3040 | 0.3170 | 0.3304 | 0.3444 | 0.3589 | 0.3740 |
| 6 | 0.3153 | 0.3197 | 0.3339 | 0.3488 | 0.3644 | 0.3806 | 0.3976 |
| 7 | 0.3278 | 0.3325 | 0.3480 | 0.3642 | 0.3811 | 0.3989 | 0.4176 |
| 8 | 0.3383 | 0.3433 | 0.3598 | 0.3772 | 0.3954 | 0.4147 | 0.4349 |
| 9 | 0.3472 | 0.3525 | 0.3700 | 0.3884 | 0.4078 | 0.4284 | 0.4501 |
| 10 | 0.3550 | 0.3605 | 0.3788 | 0.3982 | 0.4187 | 0.4405 | 0.4636 |
| 11 | 0.3617 | 0.3675 | 0.3866 | 0.4069 | 0.4284 | 0.4513 | 0.4757 |
| 12 | 0.3677 | 0.3737 | 0.3935 | 0.4146 | 0.4370 | 0.4610 | 0.4867 |
| 13 | 0.3730 | 0.3792 | 0.3997 | 0.4215 | 0.4448 | 0.4698 | 0.4966 |
| 14 | 0.3778 | 0.3842 | 0.4052 | 0.4278 | 0.4519 | 0.4778 | 0.5057 |
| 15 | 0.3821 | 0.3886 | 0.4103 | 0.4334 | 0.4583 | 0.4851 | 0.5141 |
| 16 | 0.3860 | 0.3927 | 0.4148 | 0.4386 | 0.4642 | 0.4919 | 0.5219 |
| 17 | 0.3896 | 0.3964 | 0.4190 | 0.4434 | 0.4696 | 0.4981 | 0.5291 |
| 18 | 0.3928 | 0.3998 | 0.4229 | 0.4477 | 0.4747 | 0.5039 | 0.5358 |
| 19 | 0.3958 | 0.4029 | 0.4264 | 0.4518 | 0.4793 | 0.5093 | 0.5421 |
| 20 | 0.3986 | 0.4058 | 0.4297 | 0.4555 | 0.4836 | 0.5143 | 0.5480 |
| 21 | 0.4011 | 0.4084 | 0.4327 | 0.4590 | 0.4877 | 0.5190 | 0.5535 |
| 22 | 0.4035 | 0.4109 | 0.4356 | 0.4623 | 0.4914 | 0.5234 | 0.5587 |
| 23 | 0.4057 | 0.4132 | 0.4382 | 0.4653 | 0.4950 | 0.5275 | 0.5636 |
| 24 | 0.4078 | 0.4154 | 0.4407 | 0.4682 | 0.4983 | 0.5314 | 0.5683 |
| 25 | 0.4097 | 0.4174 | 0.4430 | 0.4709 | 0.5014 | 0.5351 | 0.5727 |
| 26 | 0.4115 | 0.4193 | 0.4452 | 0.4734 | 0.5044 | 0.5386 | 0.5769 |
| 27 | 0.4132 | 0.4210 | 0.4472 | 0.4758 | 0.5072 | 0.5419 | 0.5809 |
| 28 | 0.4148 | 0.4227 | 0.4491 | 0.4780 | 0.5098 | 0.5451 | 0.5847 |
| 29 | 0.4163 | 0.4243 | 0.4510 | 0.4802 | 0.5123 | 0.5481 | 0.5883 |
| 30 | 0.4178 | 0.4258 | 0.4527 | 0.4822 | 0.5147 | 0.5509 | 0.5917 |
| 40 | 0.4286 | 0.4372 | 0.4660 | 0.4978 | 0.5333 | 0.5734 | 0.6195 |
| 60 | 0.4406 | 0.4498 | 0.4808 | 0.5155 | 0.5547 | 0.6000 | 0.6536 |
| 120 | 0.4539 | 0.4638 | 0.4975 | 0.5358 | 0.5800 | 0.6325 | 0.6980 |
| ∞ | 0.4688 | 0.4795 | 0.5167 | 0.5597 | 0.6108 | 0.6747 | 0.7631 |

*Denominator Degrees of Freedom*

# H  Critical Values of the *F*-Distribution ($\alpha = 0.950$)

**Numerator Degrees of Freedom**

| | 1 | 2 | 3 | 4 | 5 | 6 | 7 | 8 | 9 |
|---|---|---|---|---|---|---|---|---|---|
| 1 | 0.0062 | 0.0540 | 0.0987 | 0.1297 | 0.1513 | 0.1670 | 0.1788 | 0.1881 | 0.1954 |
| 2 | 0.0050 | 0.0526 | 0.1047 | 0.1440 | 0.1728 | 0.1944 | 0.2111 | 0.2243 | 0.2349 |
| 3 | 0.0046 | 0.0522 | 0.1078 | 0.1517 | 0.1849 | 0.2102 | 0.2301 | 0.2459 | 0.2589 |
| 4 | 0.0045 | 0.0520 | 0.1097 | 0.1565 | 0.1926 | 0.2206 | 0.2427 | 0.2606 | 0.2752 |
| 5 | 0.0043 | 0.0518 | 0.1109 | 0.1598 | 0.1980 | 0.2279 | 0.2518 | 0.2712 | 0.2872 |
| 6 | 0.0043 | 0.0517 | 0.1118 | 0.1623 | 0.2020 | 0.2334 | 0.2587 | 0.2793 | 0.2964 |
| 7 | 0.0042 | 0.0517 | 0.1125 | 0.1641 | 0.2051 | 0.2377 | 0.2641 | 0.2857 | 0.3037 |
| 8 | 0.0042 | 0.0516 | 0.1131 | 0.1655 | 0.2075 | 0.2411 | 0.2684 | 0.2909 | 0.3096 |
| 9 | 0.0042 | 0.0516 | 0.1135 | 0.1667 | 0.2095 | 0.2440 | 0.2720 | 0.2951 | 0.3146 |
| 10 | 0.0041 | 0.0516 | 0.1138 | 0.1677 | 0.2112 | 0.2463 | 0.2750 | 0.2988 | 0.3187 |
| 11 | 0.0041 | 0.0515 | 0.1141 | 0.1685 | 0.2126 | 0.2483 | 0.2775 | 0.3018 | 0.3223 |
| 12 | 0.0041 | 0.0515 | 0.1144 | 0.1692 | 0.2138 | 0.2500 | 0.2797 | 0.3045 | 0.3254 |
| 13 | 0.0041 | 0.0515 | 0.1146 | 0.1697 | 0.2148 | 0.2515 | 0.2817 | 0.3068 | 0.3281 |
| 14 | 0.0041 | 0.0515 | 0.1147 | 0.1703 | 0.2157 | 0.2528 | 0.2833 | 0.3089 | 0.3305 |
| 15 | 0.0041 | 0.0515 | 0.1149 | 0.1707 | 0.2165 | 0.2539 | 0.2848 | 0.3107 | 0.3327 |
| 16 | 0.0041 | 0.0515 | 0.1150 | 0.1711 | 0.2172 | 0.2550 | 0.2862 | 0.3123 | 0.3346 |
| 17 | 0.0040 | 0.0514 | 0.1152 | 0.1715 | 0.2178 | 0.2559 | 0.2874 | 0.3138 | 0.3363 |
| 18 | 0.0040 | 0.0514 | 0.1153 | 0.1718 | 0.2184 | 0.2567 | 0.2884 | 0.3151 | 0.3378 |
| 19 | 0.0040 | 0.0514 | 0.1154 | 0.1721 | 0.2189 | 0.2574 | 0.2894 | 0.3163 | 0.3393 |
| 20 | 0.0040 | 0.0514 | 0.1155 | 0.1723 | 0.2194 | 0.2581 | 0.2903 | 0.3174 | 0.3405 |
| 21 | 0.0040 | 0.0514 | 0.1156 | 0.1726 | 0.2198 | 0.2587 | 0.2911 | 0.3184 | 0.3417 |
| 22 | 0.0040 | 0.0514 | 0.1156 | 0.1728 | 0.2202 | 0.2593 | 0.2919 | 0.3194 | 0.3428 |
| 23 | 0.0040 | 0.0514 | 0.1157 | 0.1730 | 0.2206 | 0.2598 | 0.2926 | 0.3202 | 0.3438 |
| 24 | 0.0040 | 0.0514 | 0.1158 | 0.1732 | 0.2209 | 0.2603 | 0.2932 | 0.3210 | 0.3448 |
| 25 | 0.0040 | 0.0514 | 0.1158 | 0.1733 | 0.2212 | 0.2608 | 0.2938 | 0.3217 | 0.3456 |
| 26 | 0.0040 | 0.0514 | 0.1159 | 0.1735 | 0.2215 | 0.2612 | 0.2944 | 0.3224 | 0.3465 |
| 27 | 0.0040 | 0.0514 | 0.1159 | 0.1737 | 0.2217 | 0.2616 | 0.2949 | 0.3231 | 0.3472 |
| 28 | 0.0040 | 0.0514 | 0.1160 | 0.1738 | 0.2220 | 0.2619 | 0.2954 | 0.3237 | 0.3479 |
| 29 | 0.0040 | 0.0514 | 0.1160 | 0.1739 | 0.2222 | 0.2623 | 0.2958 | 0.3242 | 0.3486 |
| 30 | 0.0040 | 0.0514 | 0.1161 | 0.1740 | 0.2224 | 0.2626 | 0.2962 | 0.3247 | 0.3492 |
| 40 | 0.0040 | 0.0514 | 0.1164 | 0.1749 | 0.2240 | 0.2650 | 0.2994 | 0.3286 | 0.3539 |
| 60 | 0.0040 | 0.0513 | 0.1167 | 0.1758 | 0.2257 | 0.2674 | 0.3026 | 0.3327 | 0.3588 |
| 120 | 0.0039 | 0.0513 | 0.1170 | 0.1767 | 0.2274 | 0.2699 | 0.3060 | 0.3370 | 0.3640 |
| ∞ | 0.0039 | 0.0513 | 0.1173 | 0.1777 | 0.2291 | 0.2726 | 0.3096 | 0.3416 | 0.3695 |

Denominator Degrees of Freedom

# H    Critical Values of the *F*-Distribution ($\alpha$ = 0.950) (cont.)

**Numerator Degrees of Freedom**

| | 10 | 11 | 12 | 13 | 14 | 15 | 16 | 17 | 18 |
|---|---|---|---|---|---|---|---|---|---|
| 1 | 0.2014 | 0.2064 | 0.2106 | 0.2143 | 0.2174 | 0.2201 | 0.2225 | 0.2247 | 0.2266 |
| 2 | 0.2437 | 0.2511 | 0.2574 | 0.2628 | 0.2675 | 0.2716 | 0.2752 | 0.2784 | 0.2813 |
| 3 | 0.2697 | 0.2788 | 0.2865 | 0.2932 | 0.2991 | 0.3042 | 0.3087 | 0.3128 | 0.3165 |
| 4 | 0.2875 | 0.2979 | 0.3068 | 0.3146 | 0.3213 | 0.3273 | 0.3326 | 0.3373 | 0.3416 |
| 5 | 0.3007 | 0.3121 | 0.3220 | 0.3305 | 0.3380 | 0.3447 | 0.3506 | 0.3559 | 0.3606 |
| 6 | 0.3108 | 0.3231 | 0.3338 | 0.3430 | 0.3512 | 0.3584 | 0.3648 | 0.3706 | 0.3758 |
| 7 | 0.3189 | 0.3320 | 0.3432 | 0.3531 | 0.3618 | 0.3695 | 0.3763 | 0.3825 | 0.3881 |
| 8 | 0.3256 | 0.3392 | 0.3511 | 0.3614 | 0.3706 | 0.3787 | 0.3859 | 0.3925 | 0.3984 |
| 9 | 0.3311 | 0.3453 | 0.3576 | 0.3684 | 0.3780 | 0.3865 | 0.3941 | 0.4009 | 0.4071 |
| 10 | 0.3358 | 0.3504 | 0.3632 | 0.3744 | 0.3843 | 0.3931 | 0.4010 | 0.4082 | 0.4146 |
| 11 | 0.3398 | 0.3549 | 0.3680 | 0.3796 | 0.3898 | 0.3989 | 0.4071 | 0.4145 | 0.4212 |
| 12 | 0.3433 | 0.3587 | 0.3722 | 0.3841 | 0.3946 | 0.4040 | 0.4124 | 0.4201 | 0.4270 |
| 13 | 0.3464 | 0.3621 | 0.3759 | 0.3881 | 0.3988 | 0.4085 | 0.4171 | 0.4250 | 0.4321 |
| 14 | 0.3491 | 0.3651 | 0.3792 | 0.3916 | 0.4026 | 0.4125 | 0.4214 | 0.4294 | 0.4367 |
| 15 | 0.3515 | 0.3678 | 0.3821 | 0.3948 | 0.4060 | 0.4161 | 0.4251 | 0.4333 | 0.4408 |
| 16 | 0.3537 | 0.3702 | 0.3848 | 0.3976 | 0.4091 | 0.4193 | 0.4285 | 0.4369 | 0.4445 |
| 17 | 0.3556 | 0.3724 | 0.3872 | 0.4002 | 0.4118 | 0.4222 | 0.4316 | 0.4402 | 0.4479 |
| 18 | 0.3574 | 0.3744 | 0.3893 | 0.4026 | 0.4144 | 0.4249 | 0.4345 | 0.4431 | 0.4510 |
| 19 | 0.3590 | 0.3762 | 0.3913 | 0.4047 | 0.4167 | 0.4274 | 0.4371 | 0.4459 | 0.4539 |
| 20 | 0.3605 | 0.3779 | 0.3931 | 0.4067 | 0.4188 | 0.4296 | 0.4395 | 0.4484 | 0.4565 |
| 21 | 0.3618 | 0.3794 | 0.3948 | 0.4085 | 0.4207 | 0.4317 | 0.4417 | 0.4507 | 0.4589 |
| 22 | 0.3631 | 0.3808 | 0.3964 | 0.4102 | 0.4225 | 0.4336 | 0.4437 | 0.4528 | 0.4612 |
| 23 | 0.3643 | 0.3821 | 0.3978 | 0.4117 | 0.4242 | 0.4354 | 0.4456 | 0.4548 | 0.4632 |
| 24 | 0.3653 | 0.3833 | 0.3991 | 0.4132 | 0.4258 | 0.4371 | 0.4473 | 0.4567 | 0.4652 |
| 25 | 0.3663 | 0.3844 | 0.4004 | 0.4145 | 0.4272 | 0.4386 | 0.4490 | 0.4584 | 0.4670 |
| 26 | 0.3673 | 0.3855 | 0.4015 | 0.4158 | 0.4286 | 0.4401 | 0.4505 | 0.4600 | 0.4687 |
| 27 | 0.3681 | 0.3864 | 0.4026 | 0.4170 | 0.4299 | 0.4415 | 0.4520 | 0.4616 | 0.4703 |
| 28 | 0.3689 | 0.3874 | 0.4036 | 0.4181 | 0.4311 | 0.4427 | 0.4533 | 0.4630 | 0.4718 |
| 29 | 0.3697 | 0.3882 | 0.4046 | 0.4191 | 0.4322 | 0.4439 | 0.4546 | 0.4643 | 0.4732 |
| 30 | 0.3704 | 0.3890 | 0.4055 | 0.4201 | 0.4332 | 0.4451 | 0.4558 | 0.4656 | 0.4746 |
| 40 | 0.3758 | 0.3951 | 0.4122 | 0.4275 | 0.4412 | 0.4537 | 0.4650 | 0.4753 | 0.4848 |
| 60 | 0.3815 | 0.4016 | 0.4194 | 0.4354 | 0.4499 | 0.4629 | 0.4749 | 0.4858 | 0.4959 |
| 120 | 0.3876 | 0.4085 | 0.4272 | 0.4440 | 0.4592 | 0.4730 | 0.4857 | 0.4973 | 0.5081 |
| ∞ | 0.3940 | 0.4159 | 0.4355 | 0.4532 | 0.4693 | 0.4841 | 0.4976 | 0.5101 | 0.5217 |

*Denominator Degrees of Freedom*

# H  Critical Values of the *F*-Distribution ($\alpha = 0.950$) (cont.)

**Numerator Degrees of Freedom**

| Denominator Degrees of Freedom | 19 | 20 | 24 | 30 | 40 | 60 | 120 |
|---|---|---|---|---|---|---|---|
| 1 | 0.2283 | 0.2298 | 0.2348 | 0.2398 | 0.2448 | 0.2499 | 0.2551 |
| 2 | 0.2839 | 0.2863 | 0.2939 | 0.3016 | 0.3094 | 0.3174 | 0.3255 |
| 3 | 0.3198 | 0.3227 | 0.3324 | 0.3422 | 0.3523 | 0.3626 | 0.3731 |
| 4 | 0.3454 | 0.3489 | 0.3602 | 0.3718 | 0.3837 | 0.3960 | 0.4086 |
| 5 | 0.3650 | 0.3689 | 0.3816 | 0.3947 | 0.4083 | 0.4222 | 0.4367 |
| 6 | 0.3805 | 0.3848 | 0.3987 | 0.4131 | 0.4281 | 0.4436 | 0.4598 |
| 7 | 0.3932 | 0.3978 | 0.4128 | 0.4284 | 0.4446 | 0.4616 | 0.4792 |
| 8 | 0.4038 | 0.4087 | 0.4246 | 0.4413 | 0.4587 | 0.4769 | 0.4959 |
| 9 | 0.4128 | 0.4179 | 0.4347 | 0.4523 | 0.4708 | 0.4902 | 0.5105 |
| 10 | 0.4205 | 0.4259 | 0.4435 | 0.4620 | 0.4814 | 0.5019 | 0.5234 |
| 11 | 0.4273 | 0.4329 | 0.4512 | 0.4705 | 0.4908 | 0.5122 | 0.5350 |
| 12 | 0.4333 | 0.4391 | 0.4580 | 0.4780 | 0.4991 | 0.5215 | 0.5453 |
| 13 | 0.4386 | 0.4445 | 0.4641 | 0.4847 | 0.5066 | 0.5299 | 0.5548 |
| 14 | 0.4433 | 0.4494 | 0.4695 | 0.4908 | 0.5134 | 0.5376 | 0.5634 |
| 15 | 0.4476 | 0.4539 | 0.4745 | 0.4963 | 0.5196 | 0.5445 | 0.5713 |
| 16 | 0.4515 | 0.4579 | 0.4789 | 0.5013 | 0.5253 | 0.5509 | 0.5785 |
| 17 | 0.4550 | 0.4615 | 0.4830 | 0.5059 | 0.5305 | 0.5568 | 0.5853 |
| 18 | 0.4582 | 0.4649 | 0.4868 | 0.5102 | 0.5353 | 0.5623 | 0.5916 |
| 19 | 0.4612 | 0.4679 | 0.4902 | 0.5141 | 0.5397 | 0.5674 | 0.5974 |
| 20 | 0.4639 | 0.4708 | 0.4934 | 0.5177 | 0.5438 | 0.5721 | 0.6029 |
| 21 | 0.4665 | 0.4734 | 0.4964 | 0.5210 | 0.5477 | 0.5765 | 0.6080 |
| 22 | 0.4688 | 0.4758 | 0.4991 | 0.5242 | 0.5512 | 0.5806 | 0.6129 |
| 23 | 0.4710 | 0.4781 | 0.5017 | 0.5271 | 0.5546 | 0.5845 | 0.6174 |
| 24 | 0.4730 | 0.4802 | 0.5041 | 0.5298 | 0.5577 | 0.5882 | 0.6217 |
| 25 | 0.4749 | 0.4822 | 0.5063 | 0.5324 | 0.5607 | 0.5916 | 0.6258 |
| 26 | 0.4767 | 0.4840 | 0.5084 | 0.5348 | 0.5635 | 0.5949 | 0.6297 |
| 27 | 0.4784 | 0.4858 | 0.5104 | 0.5371 | 0.5661 | 0.5980 | 0.6333 |
| 28 | 0.4799 | 0.4874 | 0.5123 | 0.5393 | 0.5686 | 0.6009 | 0.6368 |
| 29 | 0.4814 | 0.4890 | 0.5141 | 0.5413 | 0.5710 | 0.6037 | 0.6402 |
| 30 | 0.4828 | 0.4904 | 0.5157 | 0.5432 | 0.5733 | 0.6064 | 0.6434 |
| 40 | 0.4935 | 0.5016 | 0.5286 | 0.5581 | 0.5907 | 0.6272 | 0.6688 |
| 60 | 0.5052 | 0.5138 | 0.5428 | 0.5749 | 0.6108 | 0.6518 | 0.6998 |
| 120 | 0.5181 | 0.5273 | 0.5588 | 0.5940 | 0.6343 | 0.6815 | 0.7397 |
| ∞ | 0.5325 | 0.5425 | 0.5770 | 0.6164 | 0.6627 | 0.7198 | 0.7975 |

# H | Critical Values of the *F*-Distribution ($\alpha = 0.900$)

Numerator Degrees of Freedom

| | 1 | 2 | 3 | 4 | 5 | 6 | 7 | 8 | 9 |
|---|---|---|---|---|---|---|---|---|---|
| 1 | 0.0251 | 0.1173 | 0.1806 | 0.2200 | 0.2463 | 0.2648 | 0.2786 | 0.2892 | 0.2976 |
| 2 | 0.0202 | 0.1111 | 0.1831 | 0.2312 | 0.2646 | 0.2887 | 0.3070 | 0.3212 | 0.3326 |
| 3 | 0.0187 | 0.1091 | 0.1855 | 0.2386 | 0.2763 | 0.3041 | 0.3253 | 0.3420 | 0.3555 |
| 4 | 0.0179 | 0.1082 | 0.1872 | 0.2435 | 0.2841 | 0.3144 | 0.3378 | 0.3563 | 0.3714 |
| 5 | 0.0175 | 0.1076 | 0.1884 | 0.2469 | 0.2896 | 0.3218 | 0.3468 | 0.3668 | 0.3831 |
| 6 | 0.0172 | 0.1072 | 0.1892 | 0.2494 | 0.2937 | 0.3274 | 0.3537 | 0.3748 | 0.3920 |
| 7 | 0.0170 | 0.1070 | 0.1899 | 0.2513 | 0.2969 | 0.3317 | 0.3591 | 0.3811 | 0.3992 |
| 8 | 0.0168 | 0.1068 | 0.1904 | 0.2528 | 0.2995 | 0.3352 | 0.3634 | 0.3862 | 0.4050 |
| 9 | 0.0167 | 0.1066 | 0.1908 | 0.2541 | 0.3015 | 0.3381 | 0.3670 | 0.3904 | 0.4098 |
| 10 | 0.0166 | 0.1065 | 0.1912 | 0.2551 | 0.3033 | 0.3405 | 0.3700 | 0.3940 | 0.4139 |
| 11 | 0.0165 | 0.1064 | 0.1915 | 0.2560 | 0.3047 | 0.3425 | 0.3726 | 0.3971 | 0.4173 |
| 12 | 0.0165 | 0.1063 | 0.1917 | 0.2567 | 0.3060 | 0.3443 | 0.3748 | 0.3997 | 0.4204 |
| 13 | 0.0164 | 0.1062 | 0.1919 | 0.2573 | 0.3071 | 0.3458 | 0.3767 | 0.4020 | 0.4230 |
| 14 | 0.0164 | 0.1062 | 0.1921 | 0.2579 | 0.3080 | 0.3471 | 0.3784 | 0.4040 | 0.4253 |
| 15 | 0.0163 | 0.1061 | 0.1923 | 0.2584 | 0.3088 | 0.3483 | 0.3799 | 0.4058 | 0.4274 |
| 16 | 0.0163 | 0.1061 | 0.1924 | 0.2588 | 0.3096 | 0.3493 | 0.3812 | 0.4074 | 0.4293 |
| 17 | 0.0163 | 0.1060 | 0.1926 | 0.2592 | 0.3102 | 0.3503 | 0.3824 | 0.4089 | 0.4309 |
| 18 | 0.0162 | 0.1060 | 0.1927 | 0.2595 | 0.3108 | 0.3511 | 0.3835 | 0.4102 | 0.4325 |
| 19 | 0.0162 | 0.1059 | 0.1928 | 0.2598 | 0.3114 | 0.3519 | 0.3845 | 0.4114 | 0.4338 |
| 20 | 0.0162 | 0.1059 | 0.1929 | 0.2601 | 0.3119 | 0.3526 | 0.3854 | 0.4124 | 0.4351 |
| 21 | 0.0162 | 0.1059 | 0.1930 | 0.2604 | 0.3123 | 0.3532 | 0.3862 | 0.4134 | 0.4363 |
| 22 | 0.0162 | 0.1059 | 0.1930 | 0.2606 | 0.3127 | 0.3538 | 0.3870 | 0.4143 | 0.4373 |
| 23 | 0.0161 | 0.1058 | 0.1931 | 0.2608 | 0.3131 | 0.3543 | 0.3877 | 0.4152 | 0.4383 |
| 24 | 0.0161 | 0.1058 | 0.1932 | 0.2610 | 0.3134 | 0.3548 | 0.3883 | 0.4160 | 0.4392 |
| 25 | 0.0161 | 0.1058 | 0.1932 | 0.2612 | 0.3137 | 0.3553 | 0.3889 | 0.4167 | 0.4401 |
| 26 | 0.0161 | 0.1058 | 0.1933 | 0.2614 | 0.3140 | 0.3557 | 0.3894 | 0.4174 | 0.4408 |
| 27 | 0.0161 | 0.1058 | 0.1934 | 0.2615 | 0.3143 | 0.3561 | 0.3900 | 0.4180 | 0.4416 |
| 28 | 0.0161 | 0.1058 | 0.1934 | 0.2617 | 0.3146 | 0.3565 | 0.3904 | 0.4186 | 0.4423 |
| 29 | 0.0161 | 0.1057 | 0.1935 | 0.2618 | 0.3148 | 0.3568 | 0.3909 | 0.4191 | 0.4429 |
| 30 | 0.0161 | 0.1057 | 0.1935 | 0.2620 | 0.3151 | 0.3571 | 0.3913 | 0.4196 | 0.4435 |
| 40 | 0.0160 | 0.1056 | 0.1938 | 0.2629 | 0.3167 | 0.3596 | 0.3945 | 0.4235 | 0.4480 |
| 60 | 0.0159 | 0.1055 | 0.1941 | 0.2639 | 0.3184 | 0.3621 | 0.3977 | 0.4275 | 0.4528 |
| 120 | 0.0159 | 0.1055 | 0.1945 | 0.2649 | 0.3202 | 0.3647 | 0.4012 | 0.4317 | 0.4578 |
| ∞ | 0.0158 | 0.1054 | 0.1948 | 0.2659 | 0.3221 | 0.3674 | 0.4047 | 0.4362 | 0.4631 |

Denominator Degrees of Freedom

# H Critical Values of the *F*-Distribution ($\alpha = 0.900$) (cont.)

**Numerator Degrees of Freedom**

| | 10 | 11 | 12 | 13 | 14 | 15 | 16 | 17 | 18 |
|---|---|---|---|---|---|---|---|---|---|
| 1 | 0.3044 | 0.3101 | 0.3148 | 0.3189 | 0.3224 | 0.3254 | 0.3281 | 0.3304 | 0.3326 |
| 2 | 0.3419 | 0.3497 | 0.3563 | 0.3619 | 0.3668 | 0.3710 | 0.3748 | 0.3781 | 0.3811 |
| 3 | 0.3666 | 0.3759 | 0.3838 | 0.3906 | 0.3965 | 0.4016 | 0.4062 | 0.4103 | 0.4139 |
| 4 | 0.3838 | 0.3943 | 0.4032 | 0.4109 | 0.4176 | 0.4235 | 0.4287 | 0.4333 | 0.4375 |
| 5 | 0.3966 | 0.4080 | 0.4177 | 0.4261 | 0.4335 | 0.4399 | 0.4457 | 0.4508 | 0.4554 |
| 6 | 0.4064 | 0.4186 | 0.4290 | 0.4380 | 0.4459 | 0.4529 | 0.4591 | 0.4646 | 0.4696 |
| 7 | 0.4143 | 0.4271 | 0.4381 | 0.4476 | 0.4560 | 0.4634 | 0.4699 | 0.4758 | 0.4811 |
| 8 | 0.4207 | 0.4340 | 0.4455 | 0.4555 | 0.4643 | 0.4720 | 0.4789 | 0.4851 | 0.4907 |
| 9 | 0.4260 | 0.4399 | 0.4518 | 0.4621 | 0.4713 | 0.4793 | 0.4865 | 0.4930 | 0.4988 |
| 10 | 0.4306 | 0.4448 | 0.4571 | 0.4678 | 0.4772 | 0.4856 | 0.4931 | 0.4998 | 0.5058 |
| 11 | 0.4344 | 0.4490 | 0.4617 | 0.4727 | 0.4824 | 0.4910 | 0.4987 | 0.5056 | 0.5119 |
| 12 | 0.4378 | 0.4527 | 0.4657 | 0.4770 | 0.4869 | 0.4958 | 0.5037 | 0.5108 | 0.5172 |
| 13 | 0.4408 | 0.4560 | 0.4692 | 0.4807 | 0.4909 | 0.5000 | 0.5081 | 0.5154 | 0.5220 |
| 14 | 0.4434 | 0.4589 | 0.4723 | 0.4841 | 0.4945 | 0.5037 | 0.5120 | 0.5194 | 0.5262 |
| 15 | 0.4457 | 0.4614 | 0.4751 | 0.4871 | 0.4976 | 0.5070 | 0.5155 | 0.5231 | 0.5300 |
| 16 | 0.4478 | 0.4638 | 0.4776 | 0.4897 | 0.5005 | 0.5101 | 0.5187 | 0.5264 | 0.5334 |
| 17 | 0.4497 | 0.4658 | 0.4799 | 0.4922 | 0.5031 | 0.5128 | 0.5215 | 0.5294 | 0.5365 |
| 18 | 0.4514 | 0.4677 | 0.4819 | 0.4944 | 0.5054 | 0.5153 | 0.5241 | 0.5321 | 0.5394 |
| 19 | 0.4530 | 0.4694 | 0.4838 | 0.4964 | 0.5076 | 0.5176 | 0.5265 | 0.5346 | 0.5420 |
| 20 | 0.4544 | 0.4710 | 0.4855 | 0.4983 | 0.5096 | 0.5197 | 0.5287 | 0.5370 | 0.5444 |
| 21 | 0.4557 | 0.4725 | 0.4871 | 0.5000 | 0.5114 | 0.5216 | 0.5308 | 0.5391 | 0.5466 |
| 22 | 0.4569 | 0.4738 | 0.4886 | 0.5015 | 0.5131 | 0.5234 | 0.5327 | 0.5411 | 0.5487 |
| 23 | 0.4580 | 0.4750 | 0.4899 | 0.5030 | 0.5146 | 0.5250 | 0.5344 | 0.5429 | 0.5506 |
| 24 | 0.4590 | 0.4762 | 0.4912 | 0.5044 | 0.5161 | 0.5266 | 0.5360 | 0.5446 | 0.5524 |
| 25 | 0.4600 | 0.4773 | 0.4923 | 0.5056 | 0.5174 | 0.5280 | 0.5375 | 0.5462 | 0.5540 |
| 26 | 0.4609 | 0.4783 | 0.4934 | 0.5068 | 0.5187 | 0.5294 | 0.5390 | 0.5477 | 0.5556 |
| 27 | 0.4617 | 0.4792 | 0.4944 | 0.5079 | 0.5199 | 0.5306 | 0.5403 | 0.5491 | 0.5571 |
| 28 | 0.4625 | 0.4801 | 0.4954 | 0.5089 | 0.5210 | 0.5318 | 0.5415 | 0.5504 | 0.5584 |
| 29 | 0.4633 | 0.4809 | 0.4963 | 0.5099 | 0.5220 | 0.5329 | 0.5427 | 0.5516 | 0.5597 |
| 30 | 0.4639 | 0.4816 | 0.4971 | 0.5108 | 0.5230 | 0.5340 | 0.5438 | 0.5528 | 0.5610 |
| 40 | 0.4691 | 0.4874 | 0.5035 | 0.5177 | 0.5305 | 0.5419 | 0.5522 | 0.5616 | 0.5702 |
| 60 | 0.4746 | 0.4936 | 0.5103 | 0.5251 | 0.5384 | 0.5504 | 0.5613 | 0.5712 | 0.5803 |
| 120 | 0.4804 | 0.5001 | 0.5175 | 0.5331 | 0.5470 | 0.5597 | 0.5712 | 0.5817 | 0.5914 |
| ∞ | 0.4865 | 0.5071 | 0.5253 | 0.5417 | 0.5564 | 0.5698 | 0.5820 | 0.5932 | 0.6036 |

Denominator Degrees of Freedom

# H  Critical Values of the *F*-Distribution ($\alpha = 0.900$) (cont.)

**Numerator Degrees of Freedom**

| | 19 | 20 | 24 | 30 | 40 | 60 | 120 |
|---|---|---|---|---|---|---|---|
| 1 | 0.3345 | 0.3362 | 0.3416 | 0.3471 | 0.3527 | 0.3583 | 0.3639 |
| 2 | 0.3838 | 0.3862 | 0.3940 | 0.4018 | 0.4098 | 0.4178 | 0.4260 |
| 3 | 0.4172 | 0.4202 | 0.4297 | 0.4394 | 0.4492 | 0.4593 | 0.4695 |
| 4 | 0.4412 | 0.4447 | 0.4556 | 0.4668 | 0.4783 | 0.4900 | 0.5019 |
| 5 | 0.4596 | 0.4633 | 0.4755 | 0.4880 | 0.5008 | 0.5140 | 0.5275 |
| 6 | 0.4741 | 0.4782 | 0.4914 | 0.5050 | 0.5190 | 0.5334 | 0.5483 |
| 7 | 0.4859 | 0.4903 | 0.5044 | 0.5190 | 0.5340 | 0.5496 | 0.5658 |
| 8 | 0.4958 | 0.5004 | 0.5153 | 0.5308 | 0.5468 | 0.5634 | 0.5807 |
| 9 | 0.5041 | 0.5089 | 0.5246 | 0.5408 | 0.5578 | 0.5754 | 0.5937 |
| 10 | 0.5113 | 0.5163 | 0.5326 | 0.5496 | 0.5673 | 0.5858 | 0.6052 |
| 11 | 0.5176 | 0.5228 | 0.5397 | 0.5573 | 0.5757 | 0.5951 | 0.6154 |
| 12 | 0.5231 | 0.5284 | 0.5459 | 0.5641 | 0.5832 | 0.6033 | 0.6245 |
| 13 | 0.5280 | 0.5335 | 0.5514 | 0.5702 | 0.5900 | 0.6108 | 0.6328 |
| 14 | 0.5323 | 0.5380 | 0.5564 | 0.5757 | 0.5960 | 0.6175 | 0.6403 |
| 15 | 0.5363 | 0.5420 | 0.5608 | 0.5806 | 0.6015 | 0.6237 | 0.6472 |
| 16 | 0.5398 | 0.5457 | 0.5649 | 0.5851 | 0.6066 | 0.6293 | 0.6536 |
| 17 | 0.5431 | 0.5490 | 0.5686 | 0.5893 | 0.6112 | 0.6345 | 0.6595 |
| 18 | 0.5460 | 0.5521 | 0.5720 | 0.5930 | 0.6154 | 0.6393 | 0.6649 |
| 19 | 0.5487 | 0.5549 | 0.5751 | 0.5965 | 0.6194 | 0.6437 | 0.6700 |
| 20 | 0.5512 | 0.5575 | 0.5780 | 0.5998 | 0.6230 | 0.6479 | 0.6747 |
| 21 | 0.5535 | 0.5598 | 0.5807 | 0.6028 | 0.6264 | 0.6517 | 0.6792 |
| 22 | 0.5557 | 0.5621 | 0.5831 | 0.6056 | 0.6295 | 0.6554 | 0.6833 |
| 23 | 0.5576 | 0.5641 | 0.5854 | 0.6082 | 0.6325 | 0.6587 | 0.6873 |
| 24 | 0.5595 | 0.5660 | 0.5876 | 0.6106 | 0.6353 | 0.6619 | 0.6910 |
| 25 | 0.5612 | 0.5678 | 0.5896 | 0.6129 | 0.6379 | 0.6649 | 0.6945 |
| 26 | 0.5629 | 0.5695 | 0.5915 | 0.6150 | 0.6403 | 0.6678 | 0.6978 |
| 27 | 0.5644 | 0.5711 | 0.5933 | 0.6171 | 0.6427 | 0.6705 | 0.7010 |
| 28 | 0.5658 | 0.5726 | 0.5950 | 0.6190 | 0.6449 | 0.6730 | 0.7040 |
| 29 | 0.5672 | 0.5740 | 0.5966 | 0.6208 | 0.6469 | 0.6754 | 0.7068 |
| 30 | 0.5684 | 0.5753 | 0.5980 | 0.6225 | 0.6489 | 0.6777 | 0.7095 |
| 40 | 0.5781 | 0.5854 | 0.6095 | 0.6356 | 0.6642 | 0.6957 | 0.7312 |
| 60 | 0.5887 | 0.5964 | 0.6222 | 0.6504 | 0.6816 | 0.7167 | 0.7574 |
| 120 | 0.6003 | 0.6085 | 0.6364 | 0.6672 | 0.7019 | 0.7421 | 0.7908 |
| ∞ | 0.6132 | 0.6221 | 0.6524 | 0.6866 | 0.7263 | 0.7743 | 0.8385 |

Denominator Degrees of Freedom

# H  Critical Values of the *F*-Distribution ($\alpha$ = 0.100)

Numerical entries represent the value of $F_\alpha$.

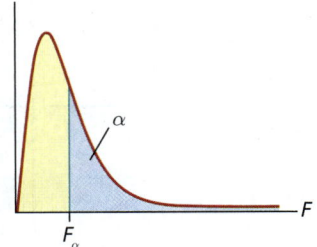

**Numerator Degrees of Freedom**

| | 1 | 2 | 3 | 4 | 5 | 6 | 7 | 8 | 9 |
|---|---|---|---|---|---|---|---|---|---|
| 1 | 39.8635 | 49.5000 | 53.5932 | 55.8330 | 57.2401 | 58.2044 | 58.9060 | 59.4390 | 59.8576 |
| 2 | 8.5263 | 9.0000 | 9.1618 | 9.2434 | 9.2926 | 9.3255 | 9.3491 | 9.3668 | 9.3805 |
| 3 | 5.5383 | 5.4624 | 5.3908 | 5.3426 | 5.3092 | 5.2847 | 5.2662 | 5.2517 | 5.2400 |
| 4 | 4.5448 | 4.3246 | 4.1909 | 4.1072 | 4.0506 | 4.0097 | 3.9790 | 3.9549 | 3.9357 |
| 5 | 4.0604 | 3.7797 | 3.6195 | 3.5202 | 3.4530 | 3.4045 | 3.3679 | 3.3393 | 3.3163 |
| 6 | 3.7759 | 3.4633 | 3.2888 | 3.1808 | 3.1075 | 3.0546 | 3.0145 | 2.9830 | 2.9577 |
| 7 | 3.5894 | 3.2574 | 3.0741 | 2.9605 | 2.8833 | 2.8274 | 2.7849 | 2.7516 | 2.7247 |
| 8 | 3.4579 | 3.1131 | 2.9238 | 2.8064 | 2.7264 | 2.6683 | 2.6241 | 2.5893 | 2.5612 |
| 9 | 3.3603 | 3.0065 | 2.8129 | 2.6927 | 2.6106 | 2.5509 | 2.5053 | 2.4694 | 2.4403 |
| 10 | 3.2850 | 2.9245 | 2.7277 | 2.6053 | 2.5216 | 2.4606 | 2.4140 | 2.3772 | 2.3473 |
| 11 | 3.2252 | 2.8595 | 2.6602 | 2.5362 | 2.4512 | 2.3891 | 2.3416 | 2.3040 | 2.2735 |
| 12 | 3.1765 | 2.8068 | 2.6055 | 2.4801 | 2.3940 | 2.3310 | 2.2828 | 2.2446 | 2.2135 |
| 13 | 3.1362 | 2.7632 | 2.5603 | 2.4337 | 2.3467 | 2.2830 | 2.2341 | 2.1953 | 2.1638 |
| 14 | 3.1022 | 2.7265 | 2.5222 | 2.3947 | 2.3069 | 2.2426 | 2.1931 | 2.1539 | 2.1220 |
| 15 | 3.0732 | 2.6952 | 2.4898 | 2.3614 | 2.2730 | 2.2081 | 2.1582 | 2.1185 | 2.0862 |
| 16 | 3.0481 | 2.6682 | 2.4618 | 2.3327 | 2.2438 | 2.1783 | 2.1280 | 2.0880 | 2.0553 |
| 17 | 3.0262 | 2.6446 | 2.4374 | 2.3077 | 2.2183 | 2.1524 | 2.1017 | 2.0613 | 2.0284 |
| 18 | 3.0070 | 2.6239 | 2.4160 | 2.2858 | 2.1958 | 2.1296 | 2.0785 | 2.0379 | 2.0047 |
| 19 | 2.9899 | 2.6056 | 2.3970 | 2.2663 | 2.1760 | 2.1094 | 2.0580 | 2.0171 | 1.9836 |
| 20 | 2.9747 | 2.5893 | 2.3801 | 2.2489 | 2.1582 | 2.0913 | 2.0397 | 1.9985 | 1.9649 |
| 21 | 2.9610 | 2.5746 | 2.3649 | 2.2333 | 2.1423 | 2.0751 | 2.0233 | 1.9819 | 1.9480 |
| 22 | 2.9486 | 2.5613 | 2.3512 | 2.2193 | 2.1279 | 2.0605 | 2.0084 | 1.9668 | 1.9327 |
| 23 | 2.9374 | 2.5493 | 2.3387 | 2.2065 | 2.1149 | 2.0472 | 1.9949 | 1.9531 | 1.9189 |
| 24 | 2.9271 | 2.5383 | 2.3274 | 2.1949 | 2.1030 | 2.0351 | 1.9826 | 1.9407 | 1.9063 |
| 25 | 2.9177 | 2.5283 | 2.3170 | 2.1842 | 2.0922 | 2.0241 | 1.9714 | 1.9292 | 1.8947 |
| 26 | 2.9091 | 2.5191 | 2.3075 | 2.1745 | 2.0822 | 2.0139 | 1.9610 | 1.9188 | 1.8841 |
| 27 | 2.9012 | 2.5106 | 2.2987 | 2.1655 | 2.0730 | 2.0045 | 1.9515 | 1.9091 | 1.8743 |
| 28 | 2.8938 | 2.5028 | 2.2906 | 2.1571 | 2.0645 | 1.9959 | 1.9427 | 1.9001 | 1.8652 |
| 29 | 2.8870 | 2.4955 | 2.2831 | 2.1494 | 2.0566 | 1.9878 | 1.9345 | 1.8918 | 1.8568 |
| 30 | 2.8807 | 2.4887 | 2.2761 | 2.1422 | 2.0492 | 1.9803 | 1.9269 | 1.8841 | 1.8490 |
| 40 | 2.8354 | 2.4404 | 2.2261 | 2.0909 | 1.9968 | 1.9269 | 1.8725 | 1.8289 | 1.7929 |
| 60 | 2.7911 | 2.3933 | 2.1774 | 2.0410 | 1.9457 | 1.8747 | 1.8194 | 1.7748 | 1.7380 |
| 120 | 2.7478 | 2.3473 | 2.1300 | 1.9923 | 1.8959 | 1.8238 | 1.7675 | 1.7220 | 1.6842 |
| ∞ | 2.7055 | 2.3026 | 2.0838 | 1.9449 | 1.8473 | 1.7741 | 1.7167 | 1.6702 | 1.6315 |

Denominator Degrees of Freedom

# H   Critical Values of the *F*-Distribution ($\alpha$ = 0.100) (cont.)

**Numerator Degrees of Freedom**

| | 10 | 11 | 12 | 13 | 14 | 15 | 16 | 17 | 18 |
|---|---|---|---|---|---|---|---|---|---|
| 1 | 60.1950 | 60.4727 | 60.7052 | 60.9028 | 61.0727 | 61.2203 | 61.3499 | 61.4644 | 61.5664 |
| 2 | 9.3916 | 9.4006 | 9.4081 | 9.4145 | 9.4200 | 9.4247 | 9.4289 | 9.4325 | 9.4358 |
| 3 | 5.2304 | 5.2224 | 5.2156 | 5.2098 | 5.2047 | 5.2003 | 5.1964 | 5.1929 | 5.1898 |
| 4 | 3.9199 | 3.9067 | 3.8955 | 3.8859 | 3.8776 | 3.8704 | 3.8639 | 3.8582 | 3.8531 |
| 5 | 3.2974 | 3.2816 | 3.2682 | 3.2567 | 3.2468 | 3.2380 | 3.2303 | 3.2234 | 3.2172 |
| 6 | 2.9369 | 2.9195 | 2.9047 | 2.8920 | 2.8809 | 2.8712 | 2.8626 | 2.8550 | 2.8481 |
| 7 | 2.7025 | 2.6839 | 2.6681 | 2.6545 | 2.6426 | 2.6322 | 2.6230 | 2.6148 | 2.6074 |
| 8 | 2.5380 | 2.5186 | 2.5020 | 2.4876 | 2.4752 | 2.4642 | 2.4545 | 2.4458 | 2.4380 |
| 9 | 2.4163 | 2.3961 | 2.3789 | 2.3640 | 2.3510 | 2.3396 | 2.3295 | 2.3205 | 2.3123 |
| 10 | 2.3226 | 2.3018 | 2.2841 | 2.2687 | 2.2553 | 2.2435 | 2.2330 | 2.2237 | 2.2153 |
| 11 | 2.2482 | 2.2269 | 2.2087 | 2.1930 | 2.1792 | 2.1671 | 2.1563 | 2.1467 | 2.1380 |
| 12 | 2.1878 | 2.1660 | 2.1474 | 2.1313 | 2.1173 | 2.1049 | 2.0938 | 2.0839 | 2.0750 |
| 13 | 2.1376 | 2.1155 | 2.0966 | 2.0802 | 2.0658 | 2.0532 | 2.0419 | 2.0318 | 2.0227 |
| 14 | 2.0954 | 2.0729 | 2.0537 | 2.0370 | 2.0224 | 2.0095 | 1.9981 | 1.9878 | 1.9785 |
| 15 | 2.0593 | 2.0366 | 2.0171 | 2.0001 | 1.9853 | 1.9722 | 1.9605 | 1.9501 | 1.9407 |
| 16 | 2.0281 | 2.0051 | 1.9854 | 1.9682 | 1.9532 | 1.9399 | 1.9281 | 1.9175 | 1.9079 |
| 17 | 2.0009 | 1.9777 | 1.9577 | 1.9404 | 1.9252 | 1.9117 | 1.8997 | 1.8889 | 1.8792 |
| 18 | 1.9770 | 1.9535 | 1.9333 | 1.9158 | 1.9004 | 1.8868 | 1.8747 | 1.8638 | 1.8539 |
| 19 | 1.9557 | 1.9321 | 1.9117 | 1.8940 | 1.8785 | 1.8647 | 1.8524 | 1.8414 | 1.8314 |
| 20 | 1.9367 | 1.9129 | 1.8924 | 1.8745 | 1.8588 | 1.8449 | 1.8325 | 1.8214 | 1.8113 |
| 21 | 1.9197 | 1.8956 | 1.8750 | 1.8570 | 1.8412 | 1.8271 | 1.8146 | 1.8034 | 1.7932 |
| 22 | 1.9043 | 1.8801 | 1.8593 | 1.8411 | 1.8252 | 1.8111 | 1.7984 | 1.7871 | 1.7768 |
| 23 | 1.8903 | 1.8659 | 1.8450 | 1.8267 | 1.8107 | 1.7964 | 1.7837 | 1.7723 | 1.7619 |
| 24 | 1.8775 | 1.8530 | 1.8319 | 1.8136 | 1.7974 | 1.7831 | 1.7703 | 1.7587 | 1.7483 |
| 25 | 1.8658 | 1.8412 | 1.8200 | 1.8015 | 1.7853 | 1.7708 | 1.7579 | 1.7463 | 1.7358 |
| 26 | 1.8550 | 1.8303 | 1.8090 | 1.7904 | 1.7741 | 1.7596 | 1.7466 | 1.7349 | 1.7243 |
| 27 | 1.8451 | 1.8203 | 1.7989 | 1.7802 | 1.7638 | 1.7492 | 1.7361 | 1.7243 | 1.7137 |
| 28 | 1.8359 | 1.8110 | 1.7895 | 1.7708 | 1.7542 | 1.7395 | 1.7264 | 1.7146 | 1.7039 |
| 29 | 1.8274 | 1.8024 | 1.7808 | 1.7620 | 1.7454 | 1.7306 | 1.7174 | 1.7055 | 1.6947 |
| 30 | 1.8195 | 1.7944 | 1.7727 | 1.7538 | 1.7371 | 1.7223 | 1.7090 | 1.6970 | 1.6862 |
| 40 | 1.7627 | 1.7369 | 1.7146 | 1.6950 | 1.6778 | 1.6624 | 1.6486 | 1.6362 | 1.6249 |
| 60 | 1.7070 | 1.6805 | 1.6574 | 1.6372 | 1.6193 | 1.6034 | 1.5890 | 1.5760 | 1.5642 |
| 120 | 1.6524 | 1.6250 | 1.6012 | 1.5803 | 1.5617 | 1.5450 | 1.5300 | 1.5164 | 1.5039 |
| ∞ | 1.5987 | 1.5705 | 1.5458 | 1.5240 | 1.5046 | 1.4871 | 1.4714 | 1.4570 | 1.4439 |

*Denominator Degrees of Freedom*

# H  Critical Values of the *F*-Distribution ($\alpha = 0.100$) (cont.)

**Numerator Degrees of Freedom**

| | 19 | 20 | 24 | 30 | 40 | 60 | 120 |
|---|---|---|---|---|---|---|---|
| 1 | 61.6579 | 61.7403 | 62.0020 | 62.2650 | 62.5291 | 62.7943 | 63.0606 |
| 2 | 9.4387 | 9.4413 | 9.4496 | 9.4579 | 9.4662 | 9.4746 | 9.4829 |
| 3 | 5.1870 | 5.1845 | 5.1764 | 5.1681 | 5.1597 | 5.1512 | 5.1425 |
| 4 | 3.8485 | 3.8443 | 3.8310 | 3.8174 | 3.8036 | 3.7896 | 3.7753 |
| 5 | 3.2117 | 3.2067 | 3.1905 | 3.1741 | 3.1573 | 3.1402 | 3.1228 |
| 6 | 2.8419 | 2.8363 | 2.8183 | 2.8000 | 2.7812 | 2.7620 | 2.7423 |
| 7 | 2.6008 | 2.5947 | 2.5753 | 2.5555 | 2.5351 | 2.5142 | 2.4928 |
| 8 | 2.4310 | 2.4246 | 2.4041 | 2.3830 | 2.3614 | 2.3391 | 2.3162 |
| 9 | 2.3050 | 2.2983 | 2.2768 | 2.2547 | 2.2320 | 2.2085 | 2.1843 |
| 10 | 2.2077 | 2.2007 | 2.1784 | 2.1554 | 2.1317 | 2.1072 | 2.0818 |
| 11 | 2.1302 | 2.1230 | 2.1000 | 2.0762 | 2.0516 | 2.0261 | 1.9997 |
| 12 | 2.0670 | 2.0597 | 2.0360 | 2.0115 | 1.9861 | 1.9597 | 1.9323 |
| 13 | 2.0145 | 2.0070 | 1.9827 | 1.9576 | 1.9315 | 1.9043 | 1.8759 |
| 14 | 1.9701 | 1.9625 | 1.9377 | 1.9119 | 1.8852 | 1.8572 | 1.8280 |
| 15 | 1.9321 | 1.9243 | 1.8990 | 1.8728 | 1.8454 | 1.8168 | 1.7867 |
| 16 | 1.8992 | 1.8913 | 1.8656 | 1.8388 | 1.8108 | 1.7816 | 1.7507 |
| 17 | 1.8704 | 1.8624 | 1.8362 | 1.8090 | 1.7805 | 1.7506 | 1.7191 |
| 18 | 1.8450 | 1.8368 | 1.8103 | 1.7827 | 1.7537 | 1.7232 | 1.6910 |
| 19 | 1.8224 | 1.8142 | 1.7873 | 1.7592 | 1.7298 | 1.6988 | 1.6659 |
| 20 | 1.8022 | 1.7938 | 1.7667 | 1.7382 | 1.7083 | 1.6768 | 1.6433 |
| 21 | 1.7840 | 1.7756 | 1.7481 | 1.7193 | 1.6890 | 1.6569 | 1.6228 |
| 22 | 1.7675 | 1.7590 | 1.7312 | 1.7021 | 1.6714 | 1.6389 | 1.6041 |
| 23 | 1.7525 | 1.7439 | 1.7159 | 1.6864 | 1.6554 | 1.6224 | 1.5871 |
| 24 | 1.7388 | 1.7302 | 1.7019 | 1.6721 | 1.6407 | 1.6073 | 1.5715 |
| 25 | 1.7263 | 1.7175 | 1.6890 | 1.6589 | 1.6272 | 1.5934 | 1.5570 |
| 26 | 1.7147 | 1.7059 | 1.6771 | 1.6468 | 1.6147 | 1.5805 | 1.5437 |
| 27 | 1.7040 | 1.6951 | 1.6662 | 1.6356 | 1.6032 | 1.5686 | 1.5313 |
| 28 | 1.6941 | 1.6852 | 1.6560 | 1.6252 | 1.5925 | 1.5575 | 1.5198 |
| 29 | 1.6849 | 1.6759 | 1.6465 | 1.6155 | 1.5825 | 1.5472 | 1.5090 |
| 30 | 1.6763 | 1.6673 | 1.6377 | 1.6065 | 1.5732 | 1.5376 | 1.4989 |
| 40 | 1.6146 | 1.6052 | 1.5741 | 1.5411 | 1.5056 | 1.4672 | 1.4248 |
| 60 | 1.5534 | 1.5435 | 1.5107 | 1.4755 | 1.4373 | 1.3952 | 1.3476 |
| 120 | 1.4926 | 1.4821 | 1.4472 | 1.4094 | 1.3676 | 1.3203 | 1.2646 |
| ∞ | 1.4318 | 1.4206 | 1.3832 | 1.3419 | 1.2951 | 1.2400 | 1.1686 |

*Denominator Degrees of Freedom*

# H  Critical Values of the *F*-Distribution ($\alpha = 0.050$)

**Numerator Degrees of Freedom**

| | 1 | 2 | 3 | 4 | 5 | 6 | 7 | 8 | 9 |
|---|---|---|---|---|---|---|---|---|---|
| 1 | 161.4476 | 199.5000 | 215.7073 | 224.5832 | 230.1619 | 233.9860 | 236.7684 | 238.8827 | 240.5433 |
| 2 | 18.5128 | 19.0000 | 19.1643 | 19.2468 | 19.2964 | 19.3295 | 19.3532 | 19.3710 | 19.3848 |
| 3 | 10.1280 | 9.5521 | 9.2766 | 9.1172 | 9.0135 | 8.9406 | 8.8867 | 8.8452 | 8.8123 |
| 4 | 7.7086 | 6.9443 | 6.5914 | 6.3882 | 6.2561 | 6.1631 | 6.0942 | 6.0410 | 5.9988 |
| 5 | 6.6079 | 5.7861 | 5.4095 | 5.1922 | 5.0503 | 4.9503 | 4.8759 | 4.8183 | 4.7725 |
| 6 | 5.9874 | 5.1433 | 4.7571 | 4.5337 | 4.3874 | 4.2839 | 4.2067 | 4.1468 | 4.0990 |
| 7 | 5.5914 | 4.7374 | 4.3468 | 4.1203 | 3.9715 | 3.8660 | 3.7870 | 3.7257 | 3.6767 |
| 8 | 5.3177 | 4.4590 | 4.0662 | 3.8379 | 3.6875 | 3.5806 | 3.5005 | 3.4381 | 3.3881 |
| 9 | 5.1174 | 4.2565 | 3.8625 | 3.6331 | 3.4817 | 3.3738 | 3.2927 | 3.2296 | 3.1789 |
| 10 | 4.9646 | 4.1028 | 3.7083 | 3.4780 | 3.3258 | 3.2172 | 3.1355 | 3.0717 | 3.0204 |
| 11 | 4.8443 | 3.9823 | 3.5874 | 3.3567 | 3.2039 | 3.0946 | 3.0123 | 2.9480 | 2.8962 |
| 12 | 4.7472 | 3.8853 | 3.4903 | 3.2592 | 3.1059 | 2.9961 | 2.9134 | 2.8486 | 2.7964 |
| 13 | 4.6672 | 3.8056 | 3.4105 | 3.1791 | 3.0254 | 2.9153 | 2.8321 | 2.7669 | 2.7144 |
| 14 | 4.6001 | 3.7389 | 3.3439 | 3.1122 | 2.9582 | 2.8477 | 2.7642 | 2.6987 | 2.6458 |
| 15 | 4.5431 | 3.6823 | 3.2874 | 3.0556 | 2.9013 | 2.7905 | 2.7066 | 2.6408 | 2.5876 |
| 16 | 4.4940 | 3.6337 | 3.2389 | 3.0069 | 2.8524 | 2.7413 | 2.6572 | 2.5911 | 2.5377 |
| 17 | 4.4513 | 3.5915 | 3.1968 | 2.9647 | 2.8100 | 2.6987 | 2.6143 | 2.5480 | 2.4943 |
| 18 | 4.4139 | 3.5546 | 3.1599 | 2.9277 | 2.7729 | 2.6613 | 2.5767 | 2.5102 | 2.4563 |
| 19 | 4.3807 | 3.5219 | 3.1274 | 2.8951 | 2.7401 | 2.6283 | 2.5435 | 2.4768 | 2.4227 |
| 20 | 4.3512 | 3.4928 | 3.0984 | 2.8661 | 2.7109 | 2.5990 | 2.5140 | 2.4471 | 2.3928 |
| 21 | 4.3248 | 3.4668 | 3.0725 | 2.8401 | 2.6848 | 2.5727 | 2.4876 | 2.4205 | 2.3660 |
| 22 | 4.3009 | 3.4434 | 3.0491 | 2.8167 | 2.6613 | 2.5491 | 2.4638 | 2.3965 | 2.3419 |
| 23 | 4.2793 | 3.4221 | 3.0280 | 2.7955 | 2.6400 | 2.5277 | 2.4422 | 2.3748 | 2.3201 |
| 24 | 4.2597 | 3.4028 | 3.0088 | 2.7763 | 2.6207 | 2.5082 | 2.4226 | 2.3551 | 2.3002 |
| 25 | 4.2417 | 3.3852 | 2.9912 | 2.7587 | 2.6030 | 2.4904 | 2.4047 | 2.3371 | 2.2821 |
| 26 | 4.2252 | 3.3690 | 2.9752 | 2.7426 | 2.5868 | 2.4741 | 2.3883 | 2.3205 | 2.2655 |
| 27 | 4.2100 | 3.3541 | 2.9604 | 2.7278 | 2.5719 | 2.4591 | 2.3732 | 2.3053 | 2.2501 |
| 28 | 4.1960 | 3.3404 | 2.9467 | 2.7141 | 2.5581 | 2.4453 | 2.3593 | 2.2913 | 2.2360 |
| 29 | 4.1830 | 3.3277 | 2.9340 | 2.7014 | 2.5454 | 2.4324 | 2.3463 | 2.2783 | 2.2229 |
| 30 | 4.1709 | 3.3158 | 2.9223 | 2.6896 | 2.5336 | 2.4205 | 2.3343 | 2.2662 | 2.2107 |
| 40 | 4.0847 | 3.2317 | 2.8387 | 2.6060 | 2.4495 | 2.3359 | 2.2490 | 2.1802 | 2.1240 |
| 60 | 4.0012 | 3.1504 | 2.7581 | 2.5252 | 2.3683 | 2.2541 | 2.1665 | 2.0970 | 2.0401 |
| 120 | 3.9201 | 3.0718 | 2.6802 | 2.4472 | 2.2899 | 2.1750 | 2.0868 | 2.0164 | 1.9588 |
| ∞ | 3.8415 | 2.9957 | 2.6049 | 2.3719 | 2.2141 | 2.0986 | 2.0096 | 1.9384 | 1.8799 |

Denominator Degrees of Freedom

# H  Critical Values of the F-Distribution ($\alpha$ = 0.050) (cont.)

**Numerator Degrees of Freedom**

| | 10 | 11 | 12 | 13 | 14 | 15 | 16 | 17 | 18 |
|---|---|---|---|---|---|---|---|---|---|
| 1 | 241.8817 | 242.9835 | 243.9060 | 244.6898 | 245.3640 | 245.9499 | 246.4639 | 246.9184 | 247.3232 |
| 2 | 19.3959 | 19.4050 | 19.4125 | 19.4189 | 19.4244 | 19.4291 | 19.4333 | 19.4370 | 19.4402 |
| 3 | 8.7855 | 8.7633 | 8.7446 | 8.7287 | 8.7149 | 8.7029 | 8.6923 | 8.6829 | 8.6745 |
| 4 | 5.9644 | 5.9358 | 5.9117 | 5.8911 | 5.8733 | 5.8578 | 5.8441 | 5.8320 | 5.8211 |
| 5 | 4.7351 | 4.7040 | 4.6777 | 4.6552 | 4.6358 | 4.6188 | 4.6038 | 4.5904 | 4.5785 |
| 6 | 4.0600 | 4.0274 | 3.9999 | 3.9764 | 3.9559 | 3.9381 | 3.9223 | 3.9083 | 3.8957 |
| 7 | 3.6365 | 3.6030 | 3.5747 | 3.5503 | 3.5292 | 3.5107 | 3.4944 | 3.4799 | 3.4669 |
| 8 | 3.3472 | 3.3130 | 3.2839 | 3.2590 | 3.2374 | 3.2184 | 3.2016 | 3.1867 | 3.1733 |
| 9 | 3.1373 | 3.1025 | 3.0729 | 3.0475 | 3.0255 | 3.0061 | 2.9890 | 2.9737 | 2.9600 |
| 10 | 2.9782 | 2.9430 | 2.9130 | 2.8872 | 2.8647 | 2.8450 | 2.8276 | 2.8120 | 2.7980 |
| 11 | 2.8536 | 2.8179 | 2.7876 | 2.7614 | 2.7386 | 2.7186 | 2.7009 | 2.6851 | 2.6709 |
| 12 | 2.7534 | 2.7173 | 2.6866 | 2.6602 | 2.6371 | 2.6169 | 2.5989 | 2.5828 | 2.5684 |
| 13 | 2.6710 | 2.6347 | 2.6037 | 2.5769 | 2.5536 | 2.5331 | 2.5149 | 2.4987 | 2.4841 |
| 14 | 2.6022 | 2.5655 | 2.5342 | 2.5073 | 2.4837 | 2.4630 | 2.4446 | 2.4282 | 2.4134 |
| 15 | 2.5437 | 2.5068 | 2.4753 | 2.4481 | 2.4244 | 2.4034 | 2.3849 | 2.3683 | 2.3533 |
| 16 | 2.4935 | 2.4564 | 2.4247 | 2.3973 | 2.3733 | 2.3522 | 2.3335 | 2.3167 | 2.3016 |
| 17 | 2.4499 | 2.4126 | 2.3807 | 2.3531 | 2.3290 | 2.3077 | 2.2888 | 2.2719 | 2.2567 |
| 18 | 2.4117 | 2.3742 | 2.3421 | 2.3143 | 2.2900 | 2.2686 | 2.2496 | 2.2325 | 2.2172 |
| 19 | 2.3779 | 2.3402 | 2.3080 | 2.2800 | 2.2556 | 2.2341 | 2.2149 | 2.1977 | 2.1823 |
| 20 | 2.3479 | 2.3100 | 2.2776 | 2.2495 | 2.2250 | 2.2033 | 2.1840 | 2.1667 | 2.1511 |
| 21 | 2.3210 | 2.2829 | 2.2504 | 2.2222 | 2.1975 | 2.1757 | 2.1563 | 2.1389 | 2.1232 |
| 22 | 2.2967 | 2.2585 | 2.2258 | 2.1975 | 2.1727 | 2.1508 | 2.1313 | 2.1138 | 2.0980 |
| 23 | 2.2747 | 2.2364 | 2.2036 | 2.1752 | 2.1502 | 2.1282 | 2.1086 | 2.0910 | 2.0751 |
| 24 | 2.2547 | 2.2163 | 2.1834 | 2.1548 | 2.1298 | 2.1077 | 2.0880 | 2.0703 | 2.0543 |
| 25 | 2.2365 | 2.1979 | 2.1649 | 2.1362 | 2.1111 | 2.0889 | 2.0691 | 2.0513 | 2.0353 |
| 26 | 2.2197 | 2.1811 | 2.1479 | 2.1192 | 2.0939 | 2.0716 | 2.0518 | 2.0339 | 2.0178 |
| 27 | 2.2043 | 2.1655 | 2.1323 | 2.1035 | 2.0781 | 2.0558 | 2.0358 | 2.0179 | 2.0017 |
| 28 | 2.1900 | 2.1512 | 2.1179 | 2.0889 | 2.0635 | 2.0411 | 2.0210 | 2.0030 | 1.9868 |
| 29 | 2.1768 | 2.1379 | 2.1045 | 2.0755 | 2.0500 | 2.0275 | 2.0073 | 1.9893 | 1.9730 |
| 30 | 2.1646 | 2.1256 | 2.0921 | 2.0630 | 2.0374 | 2.0148 | 1.9946 | 1.9765 | 1.9601 |
| 40 | 2.0772 | 2.0376 | 2.0035 | 1.9738 | 1.9476 | 1.9245 | 1.9037 | 1.8851 | 1.8682 |
| 60 | 1.9926 | 1.9522 | 1.9174 | 1.8870 | 1.8602 | 1.8364 | 1.8151 | 1.7959 | 1.7784 |
| 120 | 1.9105 | 1.8693 | 1.8337 | 1.8026 | 1.7750 | 1.7505 | 1.7285 | 1.7085 | 1.6904 |
| $\infty$ | 1.8307 | 1.7887 | 1.7522 | 1.7202 | 1.6918 | 1.6664 | 1.6435 | 1.6228 | 1.6039 |

Denominator Degrees of Freedom

## H    Critical Values of the *F*-Distribution ($\alpha$ = 0.050) (cont.)

**Numerator Degrees of Freedom**

| | 19 | 20 | 24 | 30 | 40 | 60 | 120 |
|---|---|---|---|---|---|---|---|
| 1 | 247.6861 | 248.0131 | 249.0518 | 250.0951 | 251.1432 | 252.1957 | 253.2529 |
| 2 | 19.4431 | 19.4458 | 19.4541 | 19.4624 | 19.4707 | 19.4791 | 19.4874 |
| 3 | 8.6670 | 8.6602 | 8.6385 | 8.6166 | 8.5944 | 8.5720 | 8.5494 |
| 4 | 5.8114 | 5.8025 | 5.7744 | 5.7459 | 5.7170 | 5.6877 | 5.6581 |
| 5 | 4.5678 | 4.5581 | 4.5272 | 4.4957 | 4.4638 | 4.4314 | 4.3985 |
| 6 | 3.8844 | 3.8742 | 3.8415 | 3.8082 | 3.7743 | 3.7398 | 3.7047 |
| 7 | 3.4551 | 3.4445 | 3.4105 | 3.3758 | 3.3404 | 3.3043 | 3.2674 |
| 8 | 3.1613 | 3.1503 | 3.1152 | 3.0794 | 3.0428 | 3.0053 | 2.9669 |
| 9 | 2.9477 | 2.9365 | 2.9005 | 2.8637 | 2.8259 | 2.7872 | 2.7475 |
| 10 | 2.7854 | 2.7740 | 2.7372 | 2.6996 | 2.6609 | 2.6211 | 2.5801 |
| 11 | 2.6581 | 2.6464 | 2.6090 | 2.5705 | 2.5309 | 2.4901 | 2.4480 |
| 12 | 2.5554 | 2.5436 | 2.5055 | 2.4663 | 2.4259 | 2.3842 | 2.3410 |
| 13 | 2.4709 | 2.4589 | 2.4202 | 2.3803 | 2.3392 | 2.2966 | 2.2524 |
| 14 | 2.4000 | 2.3879 | 2.3487 | 2.3082 | 2.2664 | 2.2229 | 2.1778 |
| 15 | 2.3398 | 2.3275 | 2.2878 | 2.2468 | 2.2043 | 2.1601 | 2.1141 |
| 16 | 2.2880 | 2.2756 | 2.2354 | 2.1938 | 2.1507 | 2.1058 | 2.0589 |
| 17 | 2.2429 | 2.2304 | 2.1898 | 2.1477 | 2.1040 | 2.0584 | 2.0107 |
| 18 | 2.2033 | 2.1906 | 2.1497 | 2.1071 | 2.0629 | 2.0166 | 1.9681 |
| 19 | 2.1683 | 2.1555 | 2.1141 | 2.0712 | 2.0264 | 1.9795 | 1.9302 |
| 20 | 2.1370 | 2.1242 | 2.0825 | 2.0391 | 1.9938 | 1.9464 | 1.8963 |
| 21 | 2.1090 | 2.0960 | 2.0540 | 2.0102 | 1.9645 | 1.9165 | 1.8657 |
| 22 | 2.0837 | 2.0707 | 2.0283 | 1.9842 | 1.9380 | 1.8894 | 1.8380 |
| 23 | 2.0608 | 2.0476 | 2.0050 | 1.9605 | 1.9139 | 1.8648 | 1.8128 |
| 24 | 2.0399 | 2.0267 | 1.9838 | 1.9390 | 1.8920 | 1.8424 | 1.7896 |
| 25 | 2.0207 | 2.0075 | 1.9643 | 1.9192 | 1.8718 | 1.8217 | 1.7684 |
| 26 | 2.0032 | 1.9898 | 1.9464 | 1.9010 | 1.8533 | 1.8027 | 1.7488 |
| 27 | 1.9870 | 1.9736 | 1.9299 | 1.8842 | 1.8361 | 1.7851 | 1.7306 |
| 28 | 1.9720 | 1.9586 | 1.9147 | 1.8687 | 1.8203 | 1.7689 | 1.7138 |
| 29 | 1.9581 | 1.9446 | 1.9005 | 1.8543 | 1.8055 | 1.7537 | 1.6981 |
| 30 | 1.9452 | 1.9317 | 1.8874 | 1.8409 | 1.7918 | 1.7396 | 1.6835 |
| 40 | 1.8529 | 1.8389 | 1.7929 | 1.7444 | 1.6928 | 1.6373 | 1.5766 |
| 60 | 1.7625 | 1.7480 | 1.7001 | 1.6491 | 1.5943 | 1.5343 | 1.4673 |
| 120 | 1.6739 | 1.6587 | 1.6084 | 1.5543 | 1.4952 | 1.4290 | 1.3519 |
| ∞ | 1.5865 | 1.5705 | 1.5173 | 1.4591 | 1.3940 | 1.3180 | 1.2214 |

*Denominator Degrees of Freedom*

# H Critical Values of the *F*-Distribution ($\alpha = 0.025$)

**Numerator Degrees of Freedom**

| | 1 | 2 | 3 | 4 | 5 | 6 | 7 | 8 | 9 |
|---|---|---|---|---|---|---|---|---|---|
| 1 | 647.7890 | 799.5000 | 864.1630 | 899.5833 | 921.8479 | 937.1111 | 948.2169 | 956.6562 | 963.2846 |
| 2 | 38.5063 | 39.0000 | 39.1655 | 39.2484 | 39.2982 | 39.3315 | 39.3552 | 39.3730 | 39.3869 |
| 3 | 17.4434 | 16.0441 | 15.4392 | 15.1010 | 14.8848 | 14.7347 | 14.6244 | 14.5399 | 14.4731 |
| 4 | 12.2179 | 10.6491 | 9.9792 | 9.6045 | 9.3645 | 9.1973 | 9.0741 | 8.9796 | 8.9047 |
| 5 | 10.0070 | 8.4336 | 7.7636 | 7.3879 | 7.1464 | 6.9777 | 6.8531 | 6.7572 | 6.6811 |
| 6 | 8.8131 | 7.2599 | 6.5988 | 6.2272 | 5.9876 | 5.8198 | 5.6955 | 5.5996 | 5.5234 |
| 7 | 8.0727 | 6.5415 | 5.8898 | 5.5226 | 5.2852 | 5.1186 | 4.9949 | 4.8993 | 4.8232 |
| 8 | 7.5709 | 6.0595 | 5.4160 | 5.0526 | 4.8173 | 4.6517 | 4.5286 | 4.4333 | 4.3572 |
| 9 | 7.2093 | 5.7147 | 5.0781 | 4.7181 | 4.4844 | 4.3197 | 4.1970 | 4.1020 | 4.0260 |
| 10 | 6.9367 | 5.4564 | 4.8256 | 4.4683 | 4.2361 | 4.0721 | 3.9498 | 3.8549 | 3.7790 |
| 11 | 6.7241 | 5.2559 | 4.6300 | 4.2751 | 4.0440 | 3.8807 | 3.7586 | 3.6638 | 3.5879 |
| 12 | 6.5538 | 5.0959 | 4.4742 | 4.1212 | 3.8911 | 3.7283 | 3.6065 | 3.5118 | 3.4358 |
| 13 | 6.4143 | 4.9653 | 4.3472 | 3.9959 | 3.7667 | 3.6043 | 3.4827 | 3.3880 | 3.3120 |
| 14 | 6.2979 | 4.8567 | 4.2417 | 3.8919 | 3.6634 | 3.5014 | 3.3799 | 3.2853 | 3.2093 |
| 15 | 6.1995 | 4.7650 | 4.1528 | 3.8043 | 3.5764 | 3.4147 | 3.2934 | 3.1987 | 3.1227 |
| 16 | 6.1151 | 4.6867 | 4.0768 | 3.7294 | 3.5021 | 3.3406 | 3.2194 | 3.1248 | 3.0488 |
| 17 | 6.0420 | 4.6189 | 4.0112 | 3.6648 | 3.4379 | 3.2767 | 3.1556 | 3.0610 | 2.9849 |
| 18 | 5.9781 | 4.5597 | 3.9539 | 3.6083 | 3.3820 | 3.2209 | 3.0999 | 3.0053 | 2.9291 |
| 19 | 5.9216 | 4.5075 | 3.9034 | 3.5587 | 3.3327 | 3.1718 | 3.0509 | 2.9563 | 2.8801 |
| 20 | 5.8715 | 4.4613 | 3.8587 | 3.5147 | 3.2891 | 3.1283 | 3.0074 | 2.9128 | 2.8365 |
| 21 | 5.8266 | 4.4199 | 3.8188 | 3.4754 | 3.2501 | 3.0895 | 2.9686 | 2.8740 | 2.7977 |
| 22 | 5.7863 | 4.3828 | 3.7829 | 3.4401 | 3.2151 | 3.0546 | 2.9338 | 2.8392 | 2.7628 |
| 23 | 5.7498 | 4.3492 | 3.7505 | 3.4083 | 3.1835 | 3.0232 | 2.9023 | 2.8077 | 2.7313 |
| 24 | 5.7166 | 4.3187 | 3.7211 | 3.3794 | 3.1548 | 2.9946 | 2.8738 | 2.7791 | 2.7027 |
| 25 | 5.6864 | 4.2909 | 3.6943 | 3.3530 | 3.1287 | 2.9685 | 2.8478 | 2.7531 | 2.6766 |
| 26 | 5.6586 | 4.2655 | 3.6697 | 3.3289 | 3.1048 | 2.9447 | 2.8240 | 2.7293 | 2.6528 |
| 27 | 5.6331 | 4.2421 | 3.6472 | 3.3067 | 3.0828 | 2.9228 | 2.8021 | 2.7074 | 2.6309 |
| 28 | 5.6096 | 4.2205 | 3.6264 | 3.2863 | 3.0626 | 2.9027 | 2.7820 | 2.6872 | 2.6106 |
| 29 | 5.5878 | 4.2006 | 3.6072 | 3.2674 | 3.0438 | 2.8840 | 2.7633 | 2.6686 | 2.5919 |
| 30 | 5.5675 | 4.1821 | 3.5894 | 3.2499 | 3.0265 | 2.8667 | 2.7460 | 2.6513 | 2.5746 |
| 40 | 5.4239 | 4.0510 | 3.4633 | 3.1261 | 2.9037 | 2.7444 | 2.6238 | 2.5289 | 2.4519 |
| 60 | 5.2856 | 3.9253 | 3.3425 | 3.0077 | 2.7863 | 2.6274 | 2.5068 | 2.4117 | 2.3344 |
| 120 | 5.1523 | 3.8046 | 3.2269 | 2.8943 | 2.6740 | 2.5154 | 2.3948 | 2.2994 | 2.2217 |
| ∞ | 5.0239 | 3.6889 | 3.1161 | 2.7858 | 2.5665 | 2.4082 | 2.2876 | 2.1918 | 2.1137 |

*Denominator Degrees of Freedom*

**H**  **Critical Values of the _F_-Distribution ($\alpha$ = 0.025)** (cont.)

**Numerator Degrees of Freedom**

| | 10 | 11 | 12 | 13 | 14 | 15 | 16 | 17 | 18 |
|---|---|---|---|---|---|---|---|---|---|
| 1 | 968.6274 | 973.0252 | 976.7079 | 979.8368 | 982.5278 | 984.8668 | 986.9187 | 988.7331 | 990.3490 |
| 2 | 39.3980 | 39.4071 | 39.4146 | 39.4210 | 39.4265 | 39.4313 | 39.4354 | 39.4391 | 39.4424 |
| 3 | 14.4189 | 14.3742 | 14.3366 | 14.3045 | 14.2768 | 14.2527 | 14.2315 | 14.2127 | 14.1960 |
| 4 | 8.8439 | 8.7935 | 8.7512 | 8.7150 | 8.6838 | 8.6565 | 8.6326 | 8.6113 | 8.5924 |
| 5 | 6.6192 | 6.5678 | 6.5245 | 6.4876 | 6.4556 | 6.4277 | 6.4032 | 6.3814 | 6.3619 |
| 6 | 5.4613 | 5.4098 | 5.3662 | 5.3290 | 5.2968 | 5.2687 | 5.2439 | 5.2218 | 5.2021 |
| 7 | 4.7611 | 4.7095 | 4.6658 | 4.6285 | 4.5961 | 4.5678 | 4.5428 | 4.5206 | 4.5008 |
| 8 | 4.2951 | 4.2434 | 4.1997 | 4.1622 | 4.1297 | 4.1012 | 4.0761 | 4.0538 | 4.0338 |
| 9 | 3.9639 | 3.9121 | 3.8682 | 3.8306 | 3.7980 | 3.7694 | 3.7441 | 3.7216 | 3.7015 |
| 10 | 3.7168 | 3.6649 | 3.6209 | 3.5832 | 3.5504 | 3.5217 | 3.4963 | 3.4737 | 3.4534 |
| 11 | 3.5257 | 3.4737 | 3.4296 | 3.3917 | 3.3588 | 3.3299 | 3.3044 | 3.2816 | 3.2612 |
| 12 | 3.3736 | 3.3215 | 3.2773 | 3.2393 | 3.2062 | 3.1772 | 3.1515 | 3.1286 | 3.1081 |
| 13 | 3.2497 | 3.1975 | 3.1532 | 3.1150 | 3.0819 | 3.0527 | 3.0269 | 3.0039 | 2.9832 |
| 14 | 3.1469 | 3.0946 | 3.0502 | 3.0119 | 2.9786 | 2.9493 | 2.9234 | 2.9003 | 2.8795 |
| 15 | 3.0602 | 3.0078 | 2.9633 | 2.9249 | 2.8915 | 2.8621 | 2.8360 | 2.8128 | 2.7919 |
| 16 | 2.9862 | 2.9337 | 2.8890 | 2.8506 | 2.8170 | 2.7875 | 2.7614 | 2.7380 | 2.7170 |
| 17 | 2.9222 | 2.8696 | 2.8249 | 2.7863 | 2.7526 | 2.7230 | 2.6968 | 2.6733 | 2.6522 |
| 18 | 2.8664 | 2.8137 | 2.7689 | 2.7302 | 2.6964 | 2.6667 | 2.6404 | 2.6168 | 2.5956 |
| 19 | 2.8172 | 2.7645 | 2.7196 | 2.6808 | 2.6469 | 2.6171 | 2.5907 | 2.5670 | 2.5457 |
| 20 | 2.7737 | 2.7209 | 2.6758 | 2.6369 | 2.6030 | 2.5731 | 2.5465 | 2.5228 | 2.5014 |
| 21 | 2.7348 | 2.6819 | 2.6368 | 2.5978 | 2.5638 | 2.5338 | 2.5071 | 2.4833 | 2.4618 |
| 22 | 2.6998 | 2.6469 | 2.6017 | 2.5626 | 2.5285 | 2.4984 | 2.4717 | 2.4478 | 2.4262 |
| 23 | 2.6682 | 2.6152 | 2.5699 | 2.5308 | 2.4966 | 2.4665 | 2.4396 | 2.4157 | 2.3940 |
| 24 | 2.6396 | 2.5865 | 2.5411 | 2.5019 | 2.4677 | 2.4374 | 2.4105 | 2.3865 | 2.3648 |
| 25 | 2.6135 | 2.5603 | 2.5149 | 2.4756 | 2.4413 | 2.4110 | 2.3840 | 2.3599 | 2.3381 |
| 26 | 2.5896 | 2.5363 | 2.4908 | 2.4515 | 2.4171 | 2.3867 | 2.3597 | 2.3355 | 2.3137 |
| 27 | 2.5676 | 2.5143 | 2.4688 | 2.4293 | 2.3949 | 2.3644 | 2.3373 | 2.3131 | 2.2912 |
| 28 | 2.5473 | 2.4940 | 2.4484 | 2.4089 | 2.3743 | 2.3438 | 2.3167 | 2.2924 | 2.2704 |
| 29 | 2.5286 | 2.4752 | 2.4295 | 2.3900 | 2.3554 | 2.3248 | 2.2976 | 2.2732 | 2.2512 |
| 30 | 2.5112 | 2.4577 | 2.4120 | 2.3724 | 2.3378 | 2.3072 | 2.2799 | 2.2554 | 2.2334 |
| 40 | 2.3882 | 2.3343 | 2.2882 | 2.2481 | 2.2130 | 2.1819 | 2.1542 | 2.1293 | 2.1068 |
| 60 | 2.2702 | 2.2159 | 2.1692 | 2.1286 | 2.0929 | 2.0613 | 2.0330 | 2.0076 | 1.9846 |
| 120 | 2.1570 | 2.1021 | 2.0548 | 2.0136 | 1.9773 | 1.9450 | 1.9161 | 1.8900 | 1.8663 |
| ∞ | 2.0483 | 1.9927 | 1.9447 | 1.9028 | 1.8657 | 1.8326 | 1.8028 | 1.7760 | 1.7515 |

Denominator Degrees of Freedom

# H Critical Values of the *F*-Distribution ($\alpha$ = 0.025) (cont.)

**Numerator Degrees of Freedom**

| | 19 | 20 | 24 | 30 | 40 | 60 | 120 |
|---|---|---|---|---|---|---|---|
| 1 | 991.7973 | 993.1028 | 997.2492 | 1001.4144 | 1005.5981 | 1009.8001 | 1014.0202 |
| 2 | 39.4453 | 39.4479 | 39.4562 | 39.4646 | 39.4729 | 39.4812 | 39.4896 |
| 3 | 14.1810 | 14.1674 | 14.1241 | 14.0805 | 14.0365 | 13.9921 | 13.9473 |
| 4 | 8.5753 | 8.5599 | 8.5109 | 8.4613 | 8.4111 | 8.3604 | 8.3092 |
| 5 | 6.3444 | 6.3286 | 6.2780 | 6.2269 | 6.1750 | 6.1225 | 6.0693 |
| 6 | 5.1844 | 5.1684 | 5.1172 | 5.0652 | 5.0125 | 4.9589 | 4.9044 |
| 7 | 4.4829 | 4.4667 | 4.4150 | 4.3624 | 4.3089 | 4.2544 | 4.1989 |
| 8 | 4.0158 | 3.9995 | 3.9472 | 3.8940 | 3.8398 | 3.7844 | 3.7279 |
| 9 | 3.6833 | 3.6669 | 3.6142 | 3.5604 | 3.5055 | 3.4493 | 3.3918 |
| 10 | 3.4351 | 3.4185 | 3.3654 | 3.3110 | 3.2554 | 3.1984 | 3.1399 |
| 11 | 3.2428 | 3.2261 | 3.1725 | 3.1176 | 3.0613 | 3.0035 | 2.9441 |
| 12 | 3.0896 | 3.0728 | 3.0187 | 2.9633 | 2.9063 | 2.8478 | 2.7874 |
| 13 | 2.9646 | 2.9477 | 2.8932 | 2.8372 | 2.7797 | 2.7204 | 2.6590 |
| 14 | 2.8607 | 2.8437 | 2.7888 | 2.7324 | 2.6742 | 2.6142 | 2.5519 |
| 15 | 2.7730 | 2.7559 | 2.7006 | 2.6437 | 2.5850 | 2.5242 | 2.4611 |
| 16 | 2.6980 | 2.6808 | 2.6252 | 2.5678 | 2.5085 | 2.4471 | 2.3831 |
| 17 | 2.6331 | 2.6158 | 2.5598 | 2.5020 | 2.4422 | 2.3801 | 2.3153 |
| 18 | 2.5764 | 2.5590 | 2.5027 | 2.4445 | 2.3842 | 2.3214 | 2.2558 |
| 19 | 2.5265 | 2.5089 | 2.4523 | 2.3937 | 2.3329 | 2.2696 | 2.2032 |
| 20 | 2.4821 | 2.4645 | 2.4076 | 2.3486 | 2.2873 | 2.2234 | 2.1562 |
| 21 | 2.4424 | 2.4247 | 2.3675 | 2.3082 | 2.2465 | 2.1819 | 2.1141 |
| 22 | 2.4067 | 2.3890 | 2.3315 | 2.2718 | 2.2097 | 2.1446 | 2.0760 |
| 23 | 2.3745 | 2.3567 | 2.2989 | 2.2389 | 2.1763 | 2.1107 | 2.0415 |
| 24 | 2.3452 | 2.3273 | 2.2693 | 2.2090 | 2.1460 | 2.0799 | 2.0099 |
| 25 | 2.3184 | 2.3005 | 2.2422 | 2.1816 | 2.1183 | 2.0516 | 1.9811 |
| 26 | 2.2939 | 2.2759 | 2.2174 | 2.1565 | 2.0928 | 2.0257 | 1.9545 |
| 27 | 2.2713 | 2.2533 | 2.1946 | 2.1334 | 2.0693 | 2.0018 | 1.9299 |
| 28 | 2.2505 | 2.2324 | 2.1735 | 2.1121 | 2.0477 | 1.9797 | 1.9072 |
| 29 | 2.2313 | 2.2131 | 2.1540 | 2.0923 | 2.0276 | 1.9591 | 1.8861 |
| 30 | 2.2134 | 2.1952 | 2.1359 | 2.0739 | 2.0089 | 1.9400 | 1.8664 |
| 40 | 2.0864 | 2.0677 | 2.0069 | 1.9429 | 1.8752 | 1.8028 | 1.7242 |
| 60 | 1.9636 | 1.9445 | 1.8817 | 1.8152 | 1.7440 | 1.6668 | 1.5810 |
| 120 | 1.8447 | 1.8249 | 1.7597 | 1.6899 | 1.6141 | 1.5299 | 1.4327 |
| ∞ | 1.7291 | 1.7085 | 1.6402 | 1.5660 | 1.4836 | 1.3883 | 1.2685 |

Denominator Degrees of Freedom

# H Critical Values of the *F*-Distribution ($\alpha = 0.010$)

**Numerator Degrees of Freedom**

| Denominator Degrees of Freedom | 1 | 2 | 3 | 4 | 5 | 6 | 7 | 8 | 9 |
|---|---|---|---|---|---|---|---|---|---|
| 1 | 4052.1807 | 4999.5000 | 5403.3520 | 5624.5833 | 5763.6496 | 5858.9861 | 5928.3557 | 5981.0703 | 6022.4732 |
| 2 | 98.5025 | 99.0000 | 99.1662 | 99.2494 | 99.2993 | 99.3326 | 99.3564 | 99.3742 | 99.3881 |
| 3 | 34.1162 | 30.8165 | 29.4567 | 28.7099 | 28.2371 | 27.9107 | 27.6717 | 27.4892 | 27.3452 |
| 4 | 21.1977 | 18.0000 | 16.6944 | 15.9770 | 15.5219 | 15.2069 | 14.9758 | 14.7989 | 14.6591 |
| 5 | 16.2582 | 13.2739 | 12.0600 | 11.3919 | 10.9670 | 10.6723 | 10.4555 | 10.2893 | 10.1578 |
| 6 | 13.7450 | 10.9248 | 9.7795 | 9.1483 | 8.7459 | 8.4661 | 8.2600 | 8.1017 | 7.9761 |
| 7 | 12.2464 | 9.5466 | 8.4513 | 7.8466 | 7.4604 | 7.1914 | 6.9928 | 6.8400 | 6.7188 |
| 8 | 11.2586 | 8.6491 | 7.5910 | 7.0061 | 6.6318 | 6.3707 | 6.1776 | 6.0289 | 5.9106 |
| 9 | 10.5614 | 8.0215 | 6.9919 | 6.4221 | 6.0569 | 5.8018 | 5.6129 | 5.4671 | 5.3511 |
| 10 | 10.0443 | 7.5594 | 6.5523 | 5.9943 | 5.6363 | 5.3858 | 5.2001 | 5.0567 | 4.9424 |
| 11 | 9.6460 | 7.2057 | 6.2167 | 5.6683 | 5.3160 | 5.0692 | 4.8861 | 4.7445 | 4.6315 |
| 12 | 9.3302 | 6.9266 | 5.9525 | 5.4120 | 5.0643 | 4.8206 | 4.6395 | 4.4994 | 4.3875 |
| 13 | 9.0738 | 6.7010 | 5.7394 | 5.2053 | 4.8616 | 4.6204 | 4.4410 | 4.3021 | 4.1911 |
| 14 | 8.8616 | 6.5149 | 5.5639 | 5.0354 | 4.6950 | 4.4558 | 4.2779 | 4.1399 | 4.0297 |
| 15 | 8.6831 | 6.3589 | 5.4170 | 4.8932 | 4.5556 | 4.3183 | 4.1415 | 4.0045 | 3.8948 |
| 16 | 8.5310 | 6.2262 | 5.2922 | 4.7726 | 4.4374 | 4.2016 | 4.0259 | 3.8896 | 3.7804 |
| 17 | 8.3997 | 6.1121 | 5.1850 | 4.6690 | 4.3359 | 4.1015 | 3.9267 | 3.7910 | 3.6822 |
| 18 | 8.2854 | 6.0129 | 5.0919 | 4.5790 | 4.2479 | 4.0146 | 3.8406 | 3.7054 | 3.5971 |
| 19 | 8.1849 | 5.9259 | 5.0103 | 4.5003 | 4.1708 | 3.9386 | 3.7653 | 3.6305 | 3.5225 |
| 20 | 8.0960 | 5.8489 | 4.9382 | 4.4307 | 4.1027 | 3.8714 | 3.6987 | 3.5644 | 3.4567 |
| 21 | 8.0166 | 5.7804 | 4.8740 | 4.3688 | 4.0421 | 3.8117 | 3.6396 | 3.5056 | 3.3981 |
| 22 | 7.9454 | 5.7190 | 4.8166 | 4.3134 | 3.9880 | 3.7583 | 3.5867 | 3.4530 | 3.3458 |
| 23 | 7.8811 | 5.6637 | 4.7649 | 4.2636 | 3.9392 | 3.7102 | 3.5390 | 3.4057 | 3.2986 |
| 24 | 7.8229 | 5.6136 | 4.7181 | 4.2184 | 3.8951 | 3.6667 | 3.4959 | 3.3629 | 3.2560 |
| 25 | 7.7698 | 5.5680 | 4.6755 | 4.1774 | 3.8550 | 3.6272 | 3.4568 | 3.3239 | 3.2172 |
| 26 | 7.7213 | 5.5263 | 4.6366 | 4.1400 | 3.8183 | 3.5911 | 3.4210 | 3.2884 | 3.1818 |
| 27 | 7.6767 | 5.4881 | 4.6009 | 4.1056 | 3.7848 | 3.5580 | 3.3882 | 3.2558 | 3.1494 |
| 28 | 7.6356 | 5.4529 | 4.5681 | 4.0740 | 3.7539 | 3.5276 | 3.3581 | 3.2259 | 3.1195 |
| 29 | 7.5977 | 5.4204 | 4.5378 | 4.0449 | 3.7254 | 3.4995 | 3.3303 | 3.1982 | 3.0920 |
| 30 | 7.5625 | 5.3903 | 4.5097 | 4.0179 | 3.6990 | 3.4735 | 3.3045 | 3.1726 | 3.0665 |
| 40 | 7.3141 | 5.1785 | 4.3126 | 3.8283 | 3.5138 | 3.2910 | 3.1238 | 2.9930 | 2.8876 |
| 60 | 7.0771 | 4.9774 | 4.1259 | 3.6490 | 3.3389 | 3.1187 | 2.9530 | 2.8233 | 2.7185 |
| 120 | 6.8509 | 4.7865 | 3.9491 | 3.4795 | 3.1735 | 2.9559 | 2.7918 | 2.6629 | 2.5586 |
| ∞ | 6.6349 | 4.6052 | 3.7816 | 3.3192 | 3.0173 | 2.8020 | 2.6393 | 2.5113 | 2.4074 |

# H  Critical Values of the *F*-Distribution ($\alpha = 0.010$) (cont.)

**Numerator Degrees of Freedom**

| | 10 | 11 | 12 | 13 | 14 | 15 | 16 | 17 | 18 |
|---|---|---|---|---|---|---|---|---|---|
| 1 | 6055.8467 | 6083.3168 | 6106.3207 | 6125.8647 | 6142.6740 | 6157.2846 | 6170.1012 | 6181.4348 | 6191.5287 |
| 2 | 99.3992 | 99.4083 | 99.4159 | 99.4223 | 99.4278 | 99.4325 | 99.4367 | 99.4404 | 99.4436 |
| 3 | 27.2287 | 27.1326 | 27.0518 | 26.9831 | 26.9238 | 26.8722 | 26.8269 | 26.7867 | 26.7509 |
| 4 | 14.5459 | 14.4523 | 14.3736 | 14.3065 | 14.2486 | 14.1982 | 14.1539 | 14.1146 | 14.0795 |
| 5 | 10.0510 | 9.9626 | 9.8883 | 9.8248 | 9.7700 | 9.7222 | 9.6802 | 9.6429 | 9.6096 |
| 6 | 7.8741 | 7.7896 | 7.7183 | 7.6575 | 7.6049 | 7.5590 | 7.5186 | 7.4827 | 7.4507 |
| 7 | 6.6201 | 6.5382 | 6.4691 | 6.4100 | 6.3590 | 6.3143 | 6.2750 | 6.2401 | 6.2089 |
| 8 | 5.8143 | 5.7343 | 5.6667 | 5.6089 | 5.5589 | 5.5151 | 5.4766 | 5.4423 | 5.4116 |
| 9 | 5.2565 | 5.1779 | 5.1114 | 5.0545 | 5.0052 | 4.9621 | 4.9240 | 4.8902 | 4.8599 |
| 10 | 4.8491 | 4.7715 | 4.7059 | 4.6496 | 4.6008 | 4.5581 | 4.5204 | 4.4869 | 4.4569 |
| 11 | 4.5393 | 4.4624 | 4.3974 | 4.3416 | 4.2932 | 4.2509 | 4.2134 | 4.1801 | 4.1503 |
| 12 | 4.2961 | 4.2198 | 4.1553 | 4.0999 | 4.0518 | 4.0096 | 3.9724 | 3.9392 | 3.9095 |
| 13 | 4.1003 | 4.0245 | 3.9603 | 3.9052 | 3.8573 | 3.8154 | 3.7783 | 3.7452 | 3.7156 |
| 14 | 3.9394 | 3.8640 | 3.8001 | 3.7452 | 3.6975 | 3.6557 | 3.6187 | 3.5857 | 3.5561 |
| 15 | 3.8049 | 3.7299 | 3.6662 | 3.6115 | 3.5639 | 3.5222 | 3.4852 | 3.4523 | 3.4228 |
| 16 | 3.6909 | 3.6162 | 3.5527 | 3.4981 | 3.4506 | 3.4089 | 3.3720 | 3.3391 | 3.3096 |
| 17 | 3.5931 | 3.5185 | 3.4552 | 3.4007 | 3.3533 | 3.3117 | 3.2748 | 3.2419 | 3.2124 |
| 18 | 3.5082 | 3.4338 | 3.3706 | 3.3162 | 3.2689 | 3.2273 | 3.1904 | 3.1575 | 3.1280 |
| 19 | 3.4338 | 3.3596 | 3.2965 | 3.2422 | 3.1949 | 3.1533 | 3.1165 | 3.0836 | 3.0541 |
| 20 | 3.3682 | 3.2941 | 3.2311 | 3.1769 | 3.1296 | 3.0880 | 3.0512 | 3.0183 | 2.9887 |
| 21 | 3.3098 | 3.2359 | 3.1730 | 3.1187 | 3.0715 | 3.0300 | 2.9931 | 2.9602 | 2.9306 |
| 22 | 3.2576 | 3.1837 | 3.1209 | 3.0667 | 3.0195 | 2.9779 | 2.9411 | 2.9082 | 2.8786 |
| 23 | 3.2106 | 3.1368 | 3.0740 | 3.0199 | 2.9727 | 2.9311 | 2.8943 | 2.8613 | 2.8317 |
| 24 | 3.1681 | 3.0944 | 3.0316 | 2.9775 | 2.9303 | 2.8887 | 2.8519 | 2.8189 | 2.7892 |
| 25 | 3.1294 | 3.0558 | 2.9931 | 2.9389 | 2.8917 | 2.8502 | 2.8133 | 2.7803 | 2.7506 |
| 26 | 3.0941 | 3.0205 | 2.9578 | 2.9038 | 2.8566 | 2.8150 | 2.7781 | 2.7451 | 2.7153 |
| 27 | 3.0618 | 2.9882 | 2.9256 | 2.8715 | 2.8243 | 2.7827 | 2.7458 | 2.7127 | 2.6830 |
| 28 | 3.0320 | 2.9585 | 2.8959 | 2.8418 | 2.7946 | 2.7530 | 2.7160 | 2.6830 | 2.6532 |
| 29 | 3.0045 | 2.9311 | 2.8685 | 2.8144 | 2.7672 | 2.7256 | 2.6886 | 2.6555 | 2.6257 |
| 30 | 2.9791 | 2.9057 | 2.8431 | 2.7890 | 2.7418 | 2.7002 | 2.6632 | 2.6301 | 2.6003 |
| 40 | 2.8005 | 2.7274 | 2.6648 | 2.6107 | 2.5634 | 2.5216 | 2.4844 | 2.4511 | 2.4210 |
| 60 | 2.6318 | 2.5587 | 2.4961 | 2.4419 | 2.3943 | 2.3523 | 2.3148 | 2.2811 | 2.2507 |
| 120 | 2.4721 | 2.3990 | 2.3363 | 2.2818 | 2.2339 | 2.1915 | 2.1536 | 2.1194 | 2.0885 |
| ∞ | 2.3209 | 2.2477 | 2.1848 | 2.1299 | 2.0815 | 2.0385 | 2.0000 | 1.9652 | 1.9336 |

**Denominator Degrees of Freedom**

# H   Critical Values of the *F*-Distribution ($\alpha = 0.010$) (cont.)

**Numerator Degrees of Freedom**

| | 19 | 20 | 24 | 30 | 40 | 60 | 120 |
|---|---|---|---|---|---|---|---|
| 1 | 6200.5756 | 6208.7302 | 6234.6309 | 6260.6486 | 6286.7821 | 6313.0301 | 6339.3913 |
| 2 | 99.4465 | 99.4492 | 99.4575 | 99.4658 | 99.4742 | 99.4825 | 99.4908 |
| 3 | 26.7188 | 26.6898 | 26.5975 | 26.5045 | 26.4108 | 26.3164 | 26.2211 |
| 4 | 14.0480 | 14.0196 | 13.9291 | 13.8377 | 13.7454 | 13.6522 | 13.5581 |
| 5 | 9.5797 | 9.5526 | 9.4665 | 9.3793 | 9.2912 | 9.2020 | 9.1118 |
| 6 | 7.4219 | 7.3958 | 7.3127 | 7.2285 | 7.1432 | 7.0567 | 6.9690 |
| 7 | 6.1808 | 6.1554 | 6.0743 | 5.9920 | 5.9084 | 5.8236 | 5.7373 |
| 8 | 5.3840 | 5.3591 | 5.2793 | 5.1981 | 5.1156 | 5.0316 | 4.9461 |
| 9 | 4.8327 | 4.8080 | 4.7290 | 4.6486 | 4.5666 | 4.4831 | 4.3978 |
| 10 | 4.4299 | 4.4054 | 4.3269 | 4.2469 | 4.1653 | 4.0819 | 3.9965 |
| 11 | 4.1234 | 4.0990 | 4.0209 | 3.9411 | 3.8596 | 3.7761 | 3.6904 |
| 12 | 3.8827 | 3.8584 | 3.7805 | 3.7008 | 3.6192 | 3.5355 | 3.4494 |
| 13 | 3.6888 | 3.6646 | 3.5868 | 3.5070 | 3.4253 | 3.3413 | 3.2548 |
| 14 | 3.5294 | 3.5052 | 3.4274 | 3.3476 | 3.2656 | 3.1813 | 3.0942 |
| 15 | 3.3961 | 3.3719 | 3.2940 | 3.2141 | 3.1319 | 3.0471 | 2.9595 |
| 16 | 3.2829 | 3.2587 | 3.1808 | 3.1007 | 3.0182 | 2.9330 | 2.8447 |
| 17 | 3.1857 | 3.1615 | 3.0835 | 3.0032 | 2.9205 | 2.8348 | 2.7459 |
| 18 | 3.1013 | 3.0771 | 2.9990 | 2.9185 | 2.8354 | 2.7493 | 2.6597 |
| 19 | 3.0274 | 3.0031 | 2.9249 | 2.8442 | 2.7608 | 2.6742 | 2.5839 |
| 20 | 2.9620 | 2.9377 | 2.8594 | 2.7785 | 2.6947 | 2.6077 | 2.5168 |
| 21 | 2.9039 | 2.8796 | 2.8010 | 2.7200 | 2.6359 | 2.5484 | 2.4568 |
| 22 | 2.8518 | 2.8274 | 2.7488 | 2.6675 | 2.5831 | 2.4951 | 2.4029 |
| 23 | 2.8049 | 2.7805 | 2.7017 | 2.6202 | 2.5355 | 2.4471 | 2.3542 |
| 24 | 2.7624 | 2.7380 | 2.6591 | 2.5773 | 2.4923 | 2.4035 | 2.3100 |
| 25 | 2.7238 | 2.6993 | 2.6203 | 2.5383 | 2.4530 | 2.3637 | 2.2696 |
| 26 | 2.6885 | 2.6640 | 2.5848 | 2.5026 | 2.4170 | 2.3273 | 2.2325 |
| 27 | 2.6561 | 2.6316 | 2.5522 | 2.4699 | 2.3840 | 2.2938 | 2.1985 |
| 28 | 2.6263 | 2.6017 | 2.5223 | 2.4397 | 2.3535 | 2.2629 | 2.1670 |
| 29 | 2.5987 | 2.5742 | 2.4946 | 2.4118 | 2.3253 | 2.2344 | 2.1379 |
| 30 | 2.5732 | 2.5487 | 2.4689 | 2.3860 | 2.2992 | 2.2079 | 2.1108 |
| 40 | 2.3937 | 2.3689 | 2.2880 | 2.2034 | 2.1142 | 2.0194 | 1.9172 |
| 60 | 2.2230 | 2.1978 | 2.1154 | 2.0285 | 1.9360 | 1.8363 | 1.7263 |
| 120 | 2.0604 | 2.0346 | 1.9500 | 1.8600 | 1.7628 | 1.6557 | 1.5330 |
| ∞ | 1.9048 | 1.8783 | 1.7908 | 1.6964 | 1.5923 | 1.4730 | 1.3246 |

**Denominator Degrees of Freedom**

# H Critical Values of the *F*-Distribution ($\alpha = 0.005$)

**Numerator Degrees of Freedom**

| Denominator Degrees of Freedom | 1 | 2 | 3 | 4 | 5 | 6 | 7 | 8 | 9 |
|---|---|---|---|---|---|---|---|---|---|
| 1 | 16210.7227 | 19999.5000 | 21614.7414 | 22499.5833 | 23055.7982 | 23437.1111 | 23714.5658 | 23925.4062 | 24091.0041 |
| 2 | 198.5013 | 199.0000 | 199.1664 | 199.2497 | 199.2996 | 199.3330 | 199.3568 | 199.3746 | 199.3885 |
| 3 | 55.5520 | 49.7993 | 47.4672 | 46.1946 | 45.3916 | 44.8385 | 44.4341 | 44.1256 | 43.8824 |
| 4 | 31.3328 | 26.2843 | 24.2591 | 23.1545 | 22.4564 | 21.9746 | 21.6217 | 21.3520 | 21.1391 |
| 5 | 22.7848 | 18.3138 | 16.5298 | 15.5561 | 14.9396 | 14.5133 | 14.2004 | 13.9610 | 13.7716 |
| 6 | 18.6350 | 14.5441 | 12.9166 | 12.0275 | 11.4637 | 11.0730 | 10.7859 | 10.5658 | 10.3915 |
| 7 | 16.2356 | 12.4040 | 10.8824 | 10.0505 | 9.5221 | 9.1553 | 8.8854 | 8.6781 | 8.5138 |
| 8 | 14.6882 | 11.0424 | 9.5965 | 8.8051 | 8.3018 | 7.9520 | 7.6941 | 7.4959 | 7.3386 |
| 9 | 13.6136 | 10.1067 | 8.7171 | 7.9559 | 7.4712 | 7.1339 | 6.8849 | 6.6933 | 6.5411 |
| 10 | 12.8265 | 9.4270 | 8.0807 | 7.3428 | 6.8724 | 6.5446 | 6.3025 | 6.1159 | 5.9676 |
| 11 | 12.2263 | 8.9122 | 7.6004 | 6.8809 | 6.4217 | 6.1016 | 5.8648 | 5.6821 | 5.5368 |
| 12 | 11.7542 | 8.5096 | 7.2258 | 6.5211 | 6.0711 | 5.7570 | 5.5245 | 5.3451 | 5.2021 |
| 13 | 11.3735 | 8.1865 | 6.9258 | 6.2335 | 5.7910 | 5.4819 | 5.2529 | 5.0761 | 4.9351 |
| 14 | 11.0603 | 7.9216 | 6.6804 | 5.9984 | 5.5623 | 5.2574 | 5.0313 | 4.8566 | 4.7173 |
| 15 | 10.7980 | 7.7008 | 6.4760 | 5.8029 | 5.3721 | 5.0708 | 4.8473 | 4.6744 | 4.5364 |
| 16 | 10.5755 | 7.5138 | 6.3034 | 5.6378 | 5.2117 | 4.9134 | 4.6920 | 4.5207 | 4.3838 |
| 17 | 10.3842 | 7.3536 | 6.1556 | 5.4967 | 5.0746 | 4.7789 | 4.5594 | 4.3894 | 4.2535 |
| 18 | 10.2181 | 7.2148 | 6.0278 | 5.3746 | 4.9560 | 4.6627 | 4.4448 | 4.2759 | 4.1410 |
| 19 | 10.0725 | 7.0935 | 5.9161 | 5.2681 | 4.8526 | 4.5614 | 4.3448 | 4.1770 | 4.0428 |
| 20 | 9.9439 | 6.9865 | 5.8177 | 5.1743 | 4.7616 | 4.4721 | 4.2569 | 4.0900 | 3.9564 |
| 21 | 9.8295 | 6.8914 | 5.7304 | 5.0911 | 4.6809 | 4.3931 | 4.1789 | 4.0128 | 3.8799 |
| 22 | 9.7271 | 6.8064 | 5.6524 | 5.0168 | 4.6088 | 4.3225 | 4.1094 | 3.9440 | 3.8116 |
| 23 | 9.6348 | 6.7300 | 5.5823 | 4.9500 | 4.5441 | 4.2591 | 4.0469 | 3.8822 | 3.7502 |
| 24 | 9.5513 | 6.6609 | 5.5190 | 4.8898 | 4.4857 | 4.2019 | 3.9905 | 3.8264 | 3.6949 |
| 25 | 9.4753 | 6.5982 | 5.4615 | 4.8351 | 4.4327 | 4.1500 | 3.9394 | 3.7758 | 3.6447 |
| 26 | 9.4059 | 6.5409 | 5.4091 | 4.7852 | 4.3844 | 4.1027 | 3.8928 | 3.7297 | 3.5989 |
| 27 | 9.3423 | 6.4885 | 5.3611 | 4.7396 | 4.3402 | 4.0594 | 3.8501 | 3.6875 | 3.5571 |
| 28 | 9.2838 | 6.4403 | 5.3170 | 4.6977 | 4.2996 | 4.0197 | 3.8110 | 3.6487 | 3.5186 |
| 29 | 9.2297 | 6.3958 | 5.2764 | 4.6591 | 4.2622 | 3.9831 | 3.7749 | 3.6131 | 3.4832 |
| 30 | 9.1797 | 6.3547 | 5.2388 | 4.6234 | 4.2276 | 3.9492 | 3.7416 | 3.5801 | 3.4505 |
| 40 | 8.8279 | 6.0664 | 4.9758 | 4.3738 | 3.9860 | 3.7129 | 3.5088 | 3.3498 | 3.2220 |
| 60 | 8.4946 | 5.7950 | 4.7290 | 4.1399 | 3.7599 | 3.4918 | 3.2911 | 3.1344 | 3.0083 |
| 120 | 8.1788 | 5.5393 | 4.4972 | 3.9207 | 3.5482 | 3.2849 | 3.0874 | 2.9330 | 2.8083 |
| ∞ | 7.8795 | 5.2983 | 4.2794 | 3.7151 | 3.3499 | 3.0913 | 2.8968 | 2.7444 | 2.6211 |

## H    Critical Values of the *F*-Distribution ($\alpha$ = 0.005) (cont.)

Numerator Degrees of Freedom

| | 10 | 11 | 12 | 13 | 14 | 15 | 16 | 17 | 18 |
|---|---|---|---|---|---|---|---|---|---|
| 1 | 24224.4868 | 24334.3581 | 24426.3662 | 24504.5356 | 24571.7673 | 24630.2051 | 24681.4673 | 24726.7982 | 24767.1704 |
| 2 | 199.3996 | 199.4087 | 199.4163 | 199.4227 | 199.4282 | 199.4329 | 199.4371 | 199.4408 | 199.4440 |
| 3 | 43.6858 | 43.5236 | 43.3874 | 43.2715 | 43.1716 | 43.0847 | 43.0083 | 42.9407 | 42.8804 |
| 4 | 20.9667 | 20.8243 | 20.7047 | 20.6027 | 20.5148 | 20.4383 | 20.3710 | 20.3113 | 20.2581 |
| 5 | 13.6182 | 13.4912 | 13.3845 | 13.2934 | 13.2148 | 13.1463 | 13.0861 | 13.0327 | 12.9850 |
| 6 | 10.2500 | 10.1329 | 10.0343 | 9.9501 | 9.8774 | 9.8140 | 9.7582 | 9.7086 | 9.6644 |
| 7 | 8.3803 | 8.2697 | 8.1764 | 8.0967 | 8.0279 | 7.9678 | 7.9148 | 7.8678 | 7.8258 |
| 8 | 7.2106 | 7.1045 | 7.0149 | 6.9384 | 6.8721 | 6.8143 | 6.7633 | 6.7180 | 6.6775 |
| 9 | 6.4172 | 6.3142 | 6.2274 | 6.1530 | 6.0887 | 6.0325 | 5.9829 | 5.9388 | 5.8994 |
| 10 | 5.8467 | 5.7462 | 5.6613 | 5.5887 | 5.5257 | 5.4707 | 5.4221 | 5.3789 | 5.3403 |
| 11 | 5.4183 | 5.3197 | 5.2363 | 5.1649 | 5.1031 | 5.0489 | 5.0011 | 4.9586 | 4.9205 |
| 12 | 5.0855 | 4.9884 | 4.9062 | 4.8358 | 4.7748 | 4.7213 | 4.6741 | 4.6321 | 4.5945 |
| 13 | 4.8199 | 4.7240 | 4.6429 | 4.5733 | 4.5129 | 4.4600 | 4.4132 | 4.3716 | 4.3344 |
| 14 | 4.6034 | 4.5085 | 4.4281 | 4.3591 | 4.2993 | 4.2468 | 4.2005 | 4.1592 | 4.1221 |
| 15 | 4.4235 | 4.3295 | 4.2497 | 4.1813 | 4.1219 | 4.0698 | 4.0237 | 3.9827 | 3.9459 |
| 16 | 4.2719 | 4.1785 | 4.0994 | 4.0314 | 3.9723 | 3.9205 | 3.8747 | 3.8338 | 3.7972 |
| 17 | 4.1424 | 4.0496 | 3.9709 | 3.9033 | 3.8445 | 3.7929 | 3.7473 | 3.7066 | 3.6701 |
| 18 | 4.0305 | 3.9382 | 3.8599 | 3.7926 | 3.7341 | 3.6827 | 3.6373 | 3.5967 | 3.5603 |
| 19 | 3.9329 | 3.8410 | 3.7631 | 3.6961 | 3.6378 | 3.5866 | 3.5412 | 3.5008 | 3.4645 |
| 20 | 3.8470 | 3.7555 | 3.6779 | 3.6111 | 3.5530 | 3.5020 | 3.4568 | 3.4164 | 3.3802 |
| 21 | 3.7709 | 3.6798 | 3.6024 | 3.5358 | 3.4779 | 3.4270 | 3.3818 | 3.3416 | 3.3054 |
| 22 | 3.7030 | 3.6122 | 3.5350 | 3.4686 | 3.4108 | 3.3600 | 3.3150 | 3.2748 | 3.2387 |
| 23 | 3.6420 | 3.5515 | 3.4745 | 3.4083 | 3.3506 | 3.2999 | 3.2549 | 3.2148 | 3.1787 |
| 24 | 3.5870 | 3.4967 | 3.4199 | 3.3538 | 3.2962 | 3.2456 | 3.2007 | 3.1606 | 3.1246 |
| 25 | 3.5370 | 3.4470 | 3.3704 | 3.3044 | 3.2469 | 3.1963 | 3.1515 | 3.1114 | 3.0754 |
| 26 | 3.4916 | 3.4017 | 3.3252 | 3.2594 | 3.2020 | 3.1515 | 3.1067 | 3.0666 | 3.0306 |
| 27 | 3.4499 | 3.3602 | 3.2839 | 3.2182 | 3.1608 | 3.1104 | 3.0656 | 3.0256 | 2.9896 |
| 28 | 3.4117 | 3.3222 | 3.2460 | 3.1803 | 3.1231 | 3.0727 | 3.0279 | 2.9879 | 2.9520 |
| 29 | 3.3765 | 3.2871 | 3.2110 | 3.1454 | 3.0882 | 3.0379 | 2.9932 | 2.9532 | 2.9173 |
| 30 | 3.3440 | 3.2547 | 3.1787 | 3.1132 | 3.0560 | 3.0057 | 2.9611 | 2.9211 | 2.8852 |
| 40 | 3.1167 | 3.0284 | 2.9531 | 2.8880 | 2.8312 | 2.7811 | 2.7365 | 2.6966 | 2.6607 |
| 60 | 2.9042 | 2.8166 | 2.7419 | 2.6771 | 2.6205 | 2.5705 | 2.5259 | 2.4859 | 2.4498 |
| 120 | 2.7052 | 2.6183 | 2.5439 | 2.4794 | 2.4228 | 2.3727 | 2.3280 | 2.2878 | 2.2514 |
| ∞ | 2.5188 | 2.4325 | 2.3583 | 2.2938 | 2.2371 | 2.1868 | 2.1417 | 2.1011 | 2.0643 |

Denominator Degrees of Freedom

# H  Critical Values of the *F*-Distribution ($\alpha = 0.005$) (cont.)

**Numerator Degrees of Freedom**

| | 19 | 20 | 24 | 30 | 40 | 60 | 120 |
|---|---|---|---|---|---|---|---|
| 1 | 24803.3549 | 24835.9709 | 24939.5653 | 25043.6277 | 25148.1532 | 25253.1369 | 25358.5735 |
| 2 | 199.4470 | 199.4496 | 199.4579 | 199.4663 | 199.4746 | 199.4829 | 199.4912 |
| 3 | 42.8263 | 42.7775 | 42.6222 | 42.4658 | 42.3082 | 42.1494 | 41.9895 |
| 4 | 20.2104 | 20.1673 | 20.0300 | 19.8915 | 19.7518 | 19.6107 | 19.4684 |
| 5 | 12.9422 | 12.9035 | 12.7802 | 12.6556 | 12.5297 | 12.4024 | 12.2737 |
| 6 | 9.6247 | 9.5888 | 9.4742 | 9.3582 | 9.2408 | 9.1219 | 9.0015 |
| 7 | 7.7881 | 7.7540 | 7.6450 | 7.5345 | 7.4224 | 7.3088 | 7.1933 |
| 8 | 6.6411 | 6.6082 | 6.5029 | 6.3961 | 6.2875 | 6.1772 | 6.0649 |
| 9 | 5.8639 | 5.8318 | 5.7292 | 5.6248 | 5.5186 | 5.4104 | 5.3001 |
| 10 | 5.3055 | 5.2740 | 5.1732 | 5.0706 | 4.9659 | 4.8592 | 4.7501 |
| 11 | 4.8863 | 4.8552 | 4.7557 | 4.6543 | 4.5508 | 4.4450 | 4.3367 |
| 12 | 4.5606 | 4.5299 | 4.4314 | 4.3309 | 4.2282 | 4.1229 | 4.0149 |
| 13 | 4.3008 | 4.2703 | 4.1726 | 4.0727 | 3.9704 | 3.8655 | 3.7577 |
| 14 | 4.0888 | 4.0585 | 3.9614 | 3.8619 | 3.7600 | 3.6552 | 3.5473 |
| 15 | 3.9127 | 3.8826 | 3.7859 | 3.6867 | 3.5850 | 3.4803 | 3.3722 |
| 16 | 3.7641 | 3.7342 | 3.6378 | 3.5389 | 3.4372 | 3.3324 | 3.2240 |
| 17 | 3.6372 | 3.6073 | 3.5112 | 3.4124 | 3.3108 | 3.2058 | 3.0971 |
| 18 | 3.5275 | 3.4977 | 3.4017 | 3.3030 | 3.2014 | 3.0962 | 2.9871 |
| 19 | 3.4318 | 3.4020 | 3.3062 | 3.2075 | 3.1058 | 3.0004 | 2.8908 |
| 20 | 3.3475 | 3.3178 | 3.2220 | 3.1234 | 3.0215 | 2.9159 | 2.8058 |
| 21 | 3.2728 | 3.2431 | 3.1474 | 3.0488 | 2.9467 | 2.8408 | 2.7302 |
| 22 | 3.2060 | 3.1764 | 3.0807 | 2.9821 | 2.8799 | 2.7736 | 2.6625 |
| 23 | 3.1461 | 3.1165 | 3.0208 | 2.9221 | 2.8197 | 2.7132 | 2.6015 |
| 24 | 3.0920 | 3.0624 | 2.9667 | 2.8679 | 2.7654 | 2.6585 | 2.5463 |
| 25 | 3.0429 | 3.0133 | 2.9176 | 2.8187 | 2.7160 | 2.6088 | 2.4961 |
| 26 | 2.9981 | 2.9685 | 2.8728 | 2.7738 | 2.6709 | 2.5633 | 2.4501 |
| 27 | 2.9571 | 2.9275 | 2.8318 | 2.7327 | 2.6296 | 2.5217 | 2.4079 |
| 28 | 2.9194 | 2.8899 | 2.7941 | 2.6949 | 2.5916 | 2.4834 | 2.3690 |
| 29 | 2.8847 | 2.8551 | 2.7594 | 2.6600 | 2.5565 | 2.4479 | 2.3331 |
| 30 | 2.8526 | 2.8230 | 2.7272 | 2.6278 | 2.5241 | 2.4151 | 2.2998 |
| 40 | 2.6281 | 2.5984 | 2.5020 | 2.4015 | 2.2958 | 2.1838 | 2.0636 |
| 60 | 2.4171 | 2.3872 | 2.2898 | 2.1874 | 2.0789 | 1.9622 | 1.8341 |
| 120 | 2.2183 | 2.1881 | 2.0890 | 1.9840 | 1.8709 | 1.7469 | 1.6055 |
| ∞ | 2.0307 | 1.9999 | 1.8983 | 1.7891 | 1.6692 | 1.5326 | 1.3638 |

Denominator Degrees of Freedom

# I  Critical Values for the Sign Test

**Note:** * denotes that it is not possible to have values in the critical region.

| α for a one-tailed test | 0.005 | 0.01 | 0.025 | 0.05 |
|---|---|---|---|---|
| α for a two-tailed test | 0.01 | 0.02 | 0.05 | 0.10 |
| *n* | | | | |
| 5 | * | * | * | 0 |
| 6 | * | * | 0 | 0 |
| 7 | * | 0 | 0 | 0 |
| 8 | 0 | 0 | 0 | 1 |
| 9 | 0 | 0 | 1 | 1 |
| 10 | 0 | 0 | 1 | 1 |
| 11 | 0 | 1 | 1 | 2 |
| 12 | 1 | 1 | 2 | 2 |
| 13 | 1 | 1 | 2 | 3 |
| 14 | 1 | 2 | 2 | 3 |
| 15 | 2 | 2 | 3 | 3 |
| 16 | 2 | 2 | 3 | 4 |
| 17 | 2 | 3 | 4 | 4 |
| 18 | 3 | 3 | 4 | 5 |
| 19 | 3 | 4 | 4 | 5 |
| 20 | 3 | 4 | 5 | 5 |
| 21 | 4 | 4 | 5 | 6 |
| 22 | 4 | 5 | 5 | 6 |
| 23 | 4 | 5 | 6 | 7 |
| 24 | 5 | 5 | 6 | 7 |
| 25 | 5 | 6 | 7 | 7 |

$$\text{For } n > 25, \ z = \frac{X + 0.5 - \dfrac{n}{2}}{\dfrac{\sqrt{n}}{2}}.$$

# J    Critical Values for the Wilcoxon Signed-Rank Test

**Note:** * denotes that it is not possible to have values in the critical region.

| $\alpha$ for a one-tailed test | 0.005 | 0.01 | 0.025 | 0.05 |
|---|---|---|---|---|
| $\alpha$ for a two-tailed test | 0.01 | 0.02 | 0.05 | 0.10 |
| **n** | | | | |
| 5 | * | * | * | 1 |
| 6 | * | * | 1 | 2 |
| 7 | * | 0 | 2 | 4 |
| 8 | 0 | 2 | 4 | 6 |
| 9 | 2 | 3 | 6 | 8 |
| 10 | 3 | 5 | 8 | 11 |
| 11 | 5 | 7 | 11 | 14 |
| 12 | 7 | 10 | 14 | 17 |
| 13 | 10 | 13 | 17 | 21 |
| 14 | 13 | 16 | 21 | 26 |
| 15 | 16 | 20 | 25 | 30 |
| 16 | 19 | 24 | 30 | 36 |
| 17 | 23 | 28 | 35 | 41 |
| 18 | 28 | 33 | 40 | 47 |
| 19 | 32 | 38 | 46 | 54 |
| 20 | 37 | 43 | 52 | 60 |
| 21 | 43 | 49 | 59 | 68 |
| 22 | 49 | 56 | 66 | 75 |
| 23 | 55 | 62 | 73 | 83 |
| 24 | 61 | 69 | 81 | 92 |
| 25 | 68 | 77 | 90 | 101 |
| 26 | 76 | 85 | 98 | 110 |
| 27 | 84 | 93 | 107 | 120 |
| 28 | 92 | 102 | 117 | 130 |
| 29 | 100 | 111 | 127 | 141 |
| 30 | 109 | 120 | 137 | 152 |

# K   Critical Values for the Wilcoxon Rank-Sum Test

### For one-tailed tests with $\alpha = 0.025$ or two-tailed tests with $\alpha = 0.05$

| $n_1$ | 3 | | 4 | | 5 | | 6 | | 7 | | 8 | | 9 | | 10 | |
|---|---|---|---|---|---|---|---|---|---|---|---|---|---|---|---|---|
| $n_2$ | $T_L$ | $T_U$ | $T_L$ | $T_U$ | $T_L$ | $T_U$ | $T_L$ | $T_U$ | $T_L$ | $T_U$ | $T_L$ | $T_U$ | $T_L$ | $T_U$ | $T_L$ | $T_U$ |
| 3 | 5 | 16 | 6 | 18 | 6 | 21 | 7 | 23 | 7 | 26 | 8 | 28 | 8 | 31 | 9 | 33 |
| 4 | 6 | 18 | 11 | 25 | 12 | 28 | 12 | 32 | 13 | 35 | 14 | 38 | 15 | 41 | 16 | 44 |
| 5 | 6 | 21 | 12 | 28 | 18 | 37 | 19 | 41 | 20 | 45 | 21 | 49 | 22 | 53 | 24 | 56 |
| 6 | 7 | 23 | 12 | 32 | 19 | 41 | 26 | 52 | 28 | 56 | 29 | 61 | 31 | 65 | 32 | 70 |
| 7 | 7 | 26 | 13 | 35 | 20 | 45 | 28 | 56 | 37 | 68 | 39 | 73 | 41 | 78 | 43 | 83 |
| 8 | 8 | 28 | 14 | 38 | 21 | 49 | 29 | 61 | 39 | 73 | 49 | 87 | 51 | 93 | 54 | 98 |
| 9 | 8 | 31 | 15 | 41 | 22 | 53 | 31 | 65 | 41 | 78 | 51 | 93 | 63 | 108 | 66 | 114 |
| 10 | 9 | 33 | 16 | 44 | 24 | 56 | 32 | 70 | 43 | 83 | 54 | 98 | 66 | 114 | 79 | 131 |

### For one-tailed tests with $\alpha = 0.05$ or two-tailed tests with $\alpha = 0.10$

| $n_1$ | 3 | | 4 | | 5 | | 6 | | 7 | | 8 | | 9 | | 10 | |
|---|---|---|---|---|---|---|---|---|---|---|---|---|---|---|---|---|
| $n_2$ | $T_L$ | $T_U$ | $T_L$ | $T_U$ | $T_L$ | $T_U$ | $T_L$ | $T_U$ | $T_L$ | $T_U$ | $T_L$ | $T_U$ | $T_L$ | $T_U$ | $T_L$ | $T_U$ |
| 3 | 6 | 15 | 7 | 17 | 7 | 20 | 8 | 22 | 9 | 24 | 9 | 27 | 10 | 29 | 11 | 31 |
| 4 | 7 | 17 | 12 | 24 | 13 | 27 | 14 | 30 | 15 | 33 | 16 | 36 | 17 | 39 | 18 | 42 |
| 5 | 7 | 20 | 13 | 27 | 19 | 36 | 20 | 40 | 22 | 43 | 24 | 46 | 25 | 50 | 26 | 54 |
| 6 | 8 | 22 | 14 | 30 | 20 | 40 | 28 | 50 | 30 | 54 | 32 | 58 | 33 | 63 | 35 | 67 |
| 7 | 9 | 24 | 15 | 33 | 22 | 43 | 30 | 54 | 39 | 66 | 41 | 71 | 43 | 76 | 46 | 80 |
| 8 | 9 | 27 | 16 | 36 | 24 | 46 | 32 | 58 | 41 | 71 | 52 | 84 | 54 | 90 | 57 | 95 |
| 9 | 10 | 29 | 17 | 39 | 25 | 50 | 33 | 63 | 43 | 76 | 54 | 90 | 66 | 105 | 69 | 111 |
| 10 | 11 | 31 | 18 | 42 | 26 | 54 | 35 | 67 | 46 | 80 | 57 | 95 | 69 | 111 | 83 | 127 |

# L  Critical Values of Spearman's Rank Correlation Coefficient, $r_s$

**Note:** * denotes that it is not possible to have values in the critical region.

| n | $\alpha = 0.10$ | $\alpha = 0.05$ | $\alpha = 0.02$ | $\alpha = 0.01$ |
|---|---|---|---|---|
| 5 | 0.900 | * | * | * |
| 6 | 0.829 | 0.886 | 0.943 | * |
| 7 | 0.714 | 0.786 | 0.893 | * |
| 8 | 0.643 | 0.738 | 0.833 | 0.881 |
| 9 | 0.600 | 0.683 | 0.783 | 0.833 |
| 10 | 0.564 | 0.648 | 0.745 | 0.794 |
| 11 | 0.523 | 0.623 | 0.736 | 0.818 |
| 12 | 0.497 | 0.591 | 0.703 | 0.780 |
| 13 | 0.475 | 0.566 | 0.673 | 0.745 |
| 14 | 0.457 | 0.545 | 0.646 | 0.716 |
| 15 | 0.441 | 0.525 | 0.623 | 0.689 |
| 16 | 0.425 | 0.507 | 0.601 | 0.666 |
| 17 | 0.412 | 0.490 | 0.582 | 0.645 |
| 18 | 0.399 | 0.476 | 0.564 | 0.625 |
| 19 | 0.388 | 0.462 | 0.549 | 0.608 |
| 20 | 0.377 | 0.450 | 0.534 | 0.591 |
| 21 | 0.368 | 0.438 | 0.521 | 0.576 |
| 22 | 0.359 | 0.428 | 0.508 | 0.562 |
| 23 | 0.351 | 0.418 | 0.496 | 0.549 |
| 24 | 0.343 | 0.409 | 0.485 | 0.537 |
| 25 | 0.336 | 0.400 | 0.475 | 0.526 |
| 26 | 0.329 | 0.392 | 0.465 | 0.515 |
| 27 | 0.323 | 0.385 | 0.456 | 0.505 |
| 28 | 0.317 | 0.377 | 0.448 | 0.496 |
| 29 | 0.311 | 0.370 | 0.440 | 0.487 |
| 30 | 0.305 | 0.364 | 0.432 | 0.478 |

# M   Critical Values for the Number of Runs ($\alpha = 0.05$)

**Value of $n$**

Value of $m$

| $m$ | 2 | 3 | 4 | 5 | 6 | 7 | 8 | 9 | 10 | 11 | 12 | 13 | 14 | 15 | 16 | 17 | 18 | 19 | 20 |
|---|---|---|---|---|---|---|---|---|---|---|---|---|---|---|---|---|---|---|---|
| 2 | 1 | 1 | 1 | 1 | 1 | 1 | 1 | 1 | 1 | 1 | 2 | 2 | 2 | 2 | 2 | 2 | 2 | 2 | 2 |
|   | 6 | 6 | 6 | 6 | 6 | 6 | 6 | 6 | 6 | 6 | 6 | 6 | 6 | 6 | 6 | 6 | 6 | 6 | 6 |
| 3 | 1 | 1 | 1 | 1 | 2 | 2 | 2 | 2 | 2 | 2 | 2 | 2 | 2 | 3 | 3 | 3 | 3 | 3 | 3 |
|   | 6 | 8 | 8 | 8 | 8 | 8 | 8 | 8 | 8 | 8 | 8 | 8 | 8 | 8 | 8 | 8 | 8 | 8 | 8 |
| 4 | 1 | 1 | 1 | 2 | 2 | 2 | 3 | 3 | 3 | 3 | 3 | 3 | 3 | 4 | 4 | 4 | 4 | 4 | 4 |
|   | 6 | 8 | 9 | 9 | 9 | 10 | 10 | 10 | 10 | 10 | 10 | 10 | 10 | 10 | 10 | 10 | 10 | 10 | 10 |
| 5 | 1 | 1 | 2 | 2 | 3 | 3 | 3 | 3 | 3 | 4 | 4 | 4 | 4 | 4 | 4 | 4 | 5 | 5 | 5 |
|   | 6 | 8 | 9 | 10 | 10 | 11 | 11 | 12 | 12 | 12 | 12 | 12 | 12 | 12 | 12 | 12 | 12 | 12 | 12 |
| 6 | 1 | 2 | 2 | 3 | 3 | 3 | 3 | 4 | 4 | 4 | 4 | 5 | 5 | 5 | 5 | 5 | 5 | 6 | 6 |
|   | 6 | 8 | 9 | 10 | 11 | 12 | 12 | 13 | 13 | 13 | 13 | 14 | 14 | 14 | 14 | 14 | 14 | 14 | 14 |
| 7 | 1 | 2 | 2 | 3 | 3 | 3 | 4 | 4 | 5 | 5 | 5 | 5 | 5 | 6 | 6 | 6 | 6 | 6 | 6 |
|   | 6 | 8 | 10 | 11 | 12 | 13 | 13 | 14 | 14 | 14 | 14 | 15 | 15 | 15 | 16 | 16 | 16 | 16 | 16 |
| 8 | 1 | 2 | 3 | 3 | 3 | 4 | 4 | 5 | 5 | 5 | 6 | 6 | 6 | 6 | 6 | 7 | 7 | 7 | 7 |
|   | 6 | 8 | 10 | 11 | 12 | 13 | 14 | 14 | 15 | 15 | 16 | 16 | 16 | 16 | 17 | 17 | 17 | 17 | 17 |
| 9 | 1 | 2 | 3 | 3 | 4 | 4 | 5 | 5 | 5 | 6 | 6 | 6 | 7 | 7 | 7 | 7 | 8 | 8 | 8 |
|   | 6 | 8 | 10 | 12 | 13 | 14 | 14 | 15 | 16 | 16 | 16 | 17 | 17 | 18 | 18 | 18 | 18 | 18 | 18 |
| 10 | 1 | 2 | 3 | 3 | 4 | 5 | 5 | 5 | 6 | 6 | 7 | 7 | 7 | 7 | 8 | 8 | 8 | 8 | 9 |
|   | 6 | 8 | 10 | 12 | 13 | 14 | 15 | 16 | 16 | 17 | 17 | 18 | 18 | 18 | 19 | 19 | 19 | 20 | 20 |
| 11 | 1 | 2 | 3 | 4 | 4 | 5 | 5 | 6 | 6 | 7 | 7 | 7 | 8 | 8 | 8 | 9 | 9 | 9 | 9 |
|   | 6 | 8 | 10 | 12 | 13 | 14 | 15 | 16 | 17 | 17 | 18 | 19 | 19 | 19 | 20 | 20 | 20 | 21 | 21 |
| 12 | 2 | 2 | 3 | 4 | 4 | 5 | 6 | 6 | 7 | 7 | 7 | 8 | 8 | 8 | 9 | 9 | 9 | 10 | 10 |
|   | 6 | 8 | 10 | 12 | 13 | 14 | 16 | 16 | 17 | 18 | 19 | 19 | 20 | 20 | 21 | 21 | 21 | 22 | 22 |
| 13 | 2 | 2 | 3 | 4 | 5 | 5 | 6 | 6 | 7 | 7 | 8 | 8 | 9 | 9 | 9 | 10 | 10 | 10 | 10 |
|   | 6 | 8 | 10 | 12 | 14 | 15 | 16 | 17 | 18 | 19 | 19 | 20 | 20 | 21 | 21 | 22 | 22 | 23 | 23 |
| 14 | 2 | 2 | 3 | 4 | 5 | 5 | 6 | 7 | 7 | 8 | 8 | 9 | 9 | 9 | 10 | 10 | 10 | 11 | 11 |
|   | 6 | 8 | 10 | 12 | 14 | 15 | 16 | 17 | 18 | 19 | 20 | 20 | 21 | 22 | 22 | 23 | 23 | 23 | 24 |
| 15 | 2 | 3 | 3 | 4 | 5 | 6 | 6 | 7 | 7 | 8 | 8 | 9 | 9 | 10 | 10 | 11 | 11 | 11 | 12 |
|   | 6 | 8 | 10 | 12 | 14 | 15 | 16 | 18 | 18 | 19 | 20 | 21 | 22 | 22 | 23 | 23 | 24 | 24 | 25 |
| 16 | 2 | 3 | 4 | 4 | 5 | 6 | 6 | 7 | 8 | 8 | 9 | 9 | 10 | 10 | 11 | 11 | 11 | 12 | 12 |
|   | 6 | 8 | 10 | 12 | 14 | 16 | 17 | 18 | 19 | 20 | 21 | 21 | 22 | 23 | 23 | 24 | 25 | 25 | 25 |
| 17 | 2 | 3 | 4 | 4 | 5 | 6 | 7 | 7 | 8 | 9 | 9 | 10 | 10 | 11 | 11 | 11 | 12 | 12 | 13 |
|   | 6 | 8 | 10 | 12 | 14 | 16 | 17 | 18 | 19 | 20 | 21 | 22 | 23 | 23 | 24 | 25 | 25 | 26 | 26 |
| 18 | 2 | 3 | 4 | 5 | 5 | 6 | 7 | 8 | 8 | 9 | 9 | 10 | 10 | 11 | 11 | 12 | 12 | 13 | 13 |
|   | 6 | 8 | 10 | 12 | 14 | 16 | 17 | 18 | 19 | 20 | 21 | 22 | 23 | 24 | 25 | 25 | 26 | 26 | 27 |
| 19 | 2 | 3 | 4 | 5 | 6 | 6 | 7 | 8 | 8 | 9 | 10 | 10 | 11 | 11 | 12 | 12 | 13 | 13 | 13 |
|   | 6 | 8 | 10 | 12 | 14 | 16 | 17 | 18 | 20 | 21 | 22 | 23 | 23 | 24 | 25 | 26 | 26 | 27 | 27 |
| 20 | 2 | 3 | 4 | 5 | 6 | 6 | 7 | 8 | 9 | 9 | 10 | 10 | 11 | 12 | 12 | 13 | 13 | 13 | 14 |
|   | 6 | 8 | 10 | 12 | 14 | 16 | 17 | 18 | 20 | 21 | 22 | 23 | 24 | 25 | 25 | 26 | 27 | 27 | 28 |

# N  Critical Values of the $q$-Distribution ($\alpha = 0.05$)

This table contains the critical values of the studentized range distribution where $k$ is the number of treatments and $df$ is the error degrees of freedom from the analysis of variance (total number of observations $- k$).

| $df$ | 2 | 3 | 4 | 5 | 6 | 7 | 8 | 9 | 10 | 11 | 12 | 13 | 14 | 15 | 16 | 17 | 18 | 19 | 20 |
|---|---|---|---|---|---|---|---|---|---|---|---|---|---|---|---|---|---|---|---|
| 1 | 17.97 | 26.98 | 32.82 | 37.08 | 40.41 | 43.12 | 45.40 | 47.36 | 49.07 | 50.59 | 51.96 | 53.20 | 54.33 | 55.36 | 56.32 | 57.22 | 58.04 | 58.83 | 59.56 |
| 2 | 6.085 | 8.331 | 9.798 | 10.88 | 11.74 | 12.44 | 13.03 | 13.54 | 13.99 | 14.39 | 14.75 | 15.08 | 15.38 | 15.65 | 15.91 | 16.14 | 16.37 | 16.57 | 16.77 |
| 3 | 4.501 | 5.910 | 6.825 | 7.502 | 8.037 | 8.478 | 8.853 | 9.177 | 9.462 | 9.717 | 9.946 | 10.15 | 10.35 | 10.53 | 10.69 | 10.84 | 10.98 | 11.11 | 11.24 |
| 4 | 3.927 | 5.040 | 5.757 | 6.287 | 6.707 | 7.053 | 7.347 | 7.602 | 7.826 | 8.027 | 8.208 | 8.373 | 8.525 | 8.664 | 8.794 | 8.914 | 9.028 | 9.134 | 9.233 |
| 5 | 3.635 | 4.602 | 5.218 | 5.673 | 6.033 | 6.330 | 6.582 | 6.802 | 6.995 | 7.168 | 7.324 | 7.466 | 7.596 | 7.717 | 7.828 | 7.932 | 8.030 | 8.122 | 8.208 |
| 6 | 3.461 | 4.339 | 4.896 | 5.305 | 5.628 | 5.895 | 6.122 | 6.319 | 6.493 | 6.649 | 6.789 | 6.917 | 7.034 | 7.143 | 7.244 | 7.338 | 7.426 | 7.508 | 7.587 |
| 7 | 3.344 | 4.165 | 4.681 | 5.060 | 5.359 | 5.606 | 5.815 | 5.998 | 6.158 | 6.302 | 6.431 | 6.550 | 6.658 | 6.759 | 6.852 | 6.939 | 7.020 | 7.097 | 7.170 |
| 8 | 3.261 | 4.041 | 4.529 | 4.886 | 5.167 | 5.399 | 5.597 | 5.767 | 5.918 | 6.054 | 6.175 | 6.287 | 6.389 | 6.483 | 6.571 | 6.653 | 6.729 | 6.802 | 6.870 |
| 9 | 3.199 | 3.949 | 4.415 | 4.756 | 5.024 | 5.244 | 5.432 | 5.595 | 5.739 | 5.867 | 5.983 | 6.089 | 6.186 | 6.276 | 6.359 | 6.437 | 6.510 | 6.579 | 6.644 |
| 10 | 3.151 | 3.877 | 4.327 | 4.654 | 4.912 | 5.124 | 5.305 | 5.461 | 5.599 | 5.722 | 5.833 | 5.935 | 6.028 | 6.114 | 6.194 | 6.269 | 6.339 | 6.405 | 6.467 |
| 11 | 3.113 | 3.820 | 4.256 | 4.574 | 4.823 | 5.028 | 5.202 | 5.353 | 5.487 | 5.605 | 5.713 | 5.811 | 5.901 | 5.984 | 6.062 | 6.134 | 6.202 | 6.265 | 6.326 |
| 12 | 3.082 | 3.773 | 4.199 | 4.508 | 4.751 | 4.950 | 5.119 | 5.265 | 5.395 | 5.511 | 5.615 | 5.710 | 5.798 | 5.878 | 5.953 | 6.023 | 6.089 | 6.151 | 6.209 |
| 13 | 3.055 | 3.735 | 4.151 | 4.453 | 4.690 | 4.885 | 5.049 | 5.192 | 5.318 | 5.431 | 5.533 | 5.625 | 5.711 | 5.789 | 5.862 | 5.931 | 5.995 | 6.055 | 6.112 |
| 14 | 3.033 | 3.702 | 4.111 | 4.407 | 4.639 | 4.829 | 4.990 | 5.131 | 5.254 | 5.364 | 5.463 | 5.554 | 5.637 | 5.714 | 5.786 | 5.852 | 5.915 | 5.974 | 6.029 |
| 15 | 3.014 | 3.674 | 4.076 | 4.367 | 4.595 | 4.782 | 4.940 | 5.077 | 5.198 | 5.306 | 5.404 | 5.493 | 5.574 | 5.649 | 5.720 | 5.785 | 5.846 | 5.904 | 5.958 |
| 16 | 2.998 | 3.649 | 4.046 | 4.333 | 4.557 | 4.741 | 4.897 | 5.031 | 5.150 | 5.256 | 5.352 | 5.439 | 5.520 | 5.593 | 5.662 | 5.727 | 5.786 | 5.843 | 5.897 |
| 17 | 2.984 | 3.628 | 4.020 | 4.303 | 4.524 | 4.705 | 4.858 | 4.991 | 5.108 | 5.212 | 5.307 | 5.392 | 5.471 | 5.544 | 5.612 | 5.675 | 5.734 | 5.790 | 5.842 |
| 18 | 2.971 | 3.609 | 3.997 | 4.277 | 4.495 | 4.673 | 4.824 | 4.956 | 5.071 | 5.174 | 5.267 | 5.352 | 5.429 | 5.501 | 5.568 | 5.630 | 5.688 | 5.743 | 5.794 |
| 19 | 2.960 | 3.593 | 3.977 | 4.253 | 4.469 | 4.645 | 4.794 | 4.924 | 5.038 | 5.140 | 5.231 | 5.315 | 5.391 | 5.462 | 5.528 | 5.589 | 5.647 | 5.701 | 5.752 |
| 20 | 2.950 | 3.578 | 3.958 | 4.232 | 4.445 | 4.620 | 4.768 | 4.896 | 5.008 | 5.108 | 5.199 | 5.282 | 5.357 | 5.427 | 5.493 | 5.553 | 5.610 | 5.663 | 5.714 |
| 24 | 2.919 | 3.532 | 3.901 | 4.166 | 4.373 | 4.541 | 4.684 | 4.807 | 4.915 | 5.012 | 5.099 | 5.179 | 5.251 | 5.319 | 5.381 | 5.439 | 5.494 | 5.545 | 5.594 |
| 30 | 2.888 | 3.486 | 3.845 | 4.102 | 4.302 | 4.464 | 4.602 | 4.720 | 4.824 | 4.917 | 5.001 | 5.077 | 5.147 | 5.211 | 5.271 | 5.327 | 5.379 | 5.429 | 5.475 |
| 40 | 2.858 | 3.442 | 3.791 | 4.039 | 4.232 | 4.389 | 4.521 | 4.635 | 4.735 | 4.824 | 4.904 | 4.977 | 5.044 | 5.106 | 5.163 | 5.216 | 5.266 | 5.313 | 5.358 |
| 60 | 2.829 | 3.399 | 3.737 | 3.977 | 4.163 | 4.314 | 4.441 | 4.550 | 4.646 | 4.732 | 4.808 | 4.878 | 4.942 | 5.001 | 5.056 | 5.107 | 5.154 | 5.199 | 5.241 |
| 120 | 2.800 | 3.356 | 3.685 | 3.917 | 4.096 | 4.241 | 4.363 | 4.468 | 4.560 | 4.641 | 4.714 | 4.781 | 4.842 | 4.898 | 4.950 | 4.998 | 5.044 | 5.086 | 5.126 |
| $\infty$ | 2.772 | 3.314 | 3.633 | 3.858 | 4.030 | 4.170 | 4.286 | 4.387 | 4.474 | 4.552 | 4.622 | 4.685 | 4.743 | 4.796 | 4.845 | 4.891 | 4.934 | 4.974 | 5.012 |

## N  Critical Values of the *q*-Distribution ($\alpha = 0.01$) (cont.)

*k*

| df | 2 | 3 | 4 | 5 | 6 | 7 | 8 | 9 | 10 | 11 | 12 | 13 | 14 | 15 | 16 | 17 | 18 | 19 | 20 |
|----|---|---|---|---|---|---|---|---|----|----|----|----|----|----|----|----|----|----|----|
| 1 | 90.03 | 135.0 | 164.3 | 185.6 | 202.2 | 215.8 | 227.2 | 237.0 | 245.6 | 253.2 | 260.0 | 266.2 | 271.8 | 277.0 | 281.8 | 286.3 | 290.4 | 294.3 | 298.0 |
| 2 | 14.04 | 19.02 | 22.29 | 24.72 | 26.63 | 28.20 | 29.53 | 30.68 | 31.69 | 32.59 | 33.40 | 34.13 | 34.81 | 35.43 | 36.00 | 36.53 | 37.03 | 37.50 | 37.95 |
| 3 | 8.261 | 10.62 | 12.17 | 13.33 | 14.24 | 15.00 | 15.64 | 16.20 | 16.69 | 17.13 | 17.53 | 17.89 | 18.22 | 18.52 | 18.81 | 19.07 | 19.32 | 19.55 | 19.77 |
| 4 | 6.512 | 8.120 | 9.173 | 9.958 | 10.58 | 11.10 | 11.55 | 11.93 | 12.27 | 12.57 | 12.84 | 13.09 | 13.32 | 13.53 | 13.73 | 13.91 | 14.08 | 14.24 | 14.40 |
| 5 | 5.702 | 6.976 | 7.804 | 8.421 | 8.913 | 9.321 | 9.669 | 9.972 | 10.24 | 10.48 | 10.70 | 10.89 | 11.08 | 11.24 | 11.40 | 11.55 | 11.68 | 11.81 | 11.93 |
| 6 | 5.243 | 6.331 | 7.033 | 7.556 | 7.973 | 8.318 | 8.613 | 8.869 | 9.097 | 9.301 | 9.485 | 9.653 | 9.808 | 9.951 | 10.08 | 10.21 | 10.32 | 10.43 | 10.54 |
| 7 | 4.949 | 5.919 | 6.543 | 7.005 | 7.373 | 7.679 | 7.939 | 8.166 | 8.368 | 8.548 | 8.711 | 8.860 | 8.997 | 9.124 | 9.242 | 9.353 | 9.456 | 9.554 | 9.646 |
| 8 | 4.746 | 5.635 | 6.204 | 6.625 | 6.960 | 7.237 | 7.474 | 7.681 | 7.863 | 8.027 | 8.176 | 8.312 | 8.436 | 8.552 | 8.659 | 8.760 | 8.854 | 8.943 | 9.027 |
| 9 | 4.596 | 5.428 | 5.957 | 6.348 | 6.658 | 6.915 | 7.134 | 7.325 | 7.495 | 7.647 | 7.784 | 7.910 | 8.025 | 8.132 | 8.232 | 8.325 | 8.412 | 8.495 | 8.573 |
| 10 | 4.482 | 5.270 | 5.769 | 6.136 | 6.428 | 6.669 | 6.875 | 7.055 | 7.213 | 7.326 | 7.485 | 7.603 | 7.712 | 7.812 | 7.906 | 7.993 | 8.076 | 8.153 | 8.226 |
| 11 | 4.392 | 5.146 | 5.621 | 5.970 | 6.247 | 6.476 | 6.672 | 6.842 | 6.992 | 7.128 | 7.250 | 7.362 | 7.465 | 7.560 | 7.649 | 7.732 | 7.809 | 7.883 | 7.952 |
| 12 | 4.320 | 5.046 | 5.502 | 5.836 | 6.101 | 6.321 | 6.507 | 6.670 | 6.814 | 6.943 | 7.060 | 7.167 | 7.265 | 7.356 | 7.441 | 7.520 | 7.594 | 7.665 | 7.731 |
| 13 | 4.260 | 4.964 | 5.404 | 5.727 | 5.981 | 6.192 | 6.372 | 6.528 | 6.667 | 6.791 | 6.903 | 7.006 | 7.101 | 7.188 | 7.269 | 7.345 | 7.417 | 7.485 | 7.548 |
| 14 | 4.210 | 4.895 | 5.322 | 5.634 | 5.881 | 6.085 | 6.258 | 6.409 | 6.543 | 6.664 | 6.772 | 6.871 | 6.962 | 7.047 | 7.126 | 7.199 | 7.268 | 7.333 | 7.395 |
| 15 | 4.168 | 4.836 | 5.252 | 5.556 | 5.796 | 5.994 | 6.162 | 6.309 | 6.439 | 6.555 | 6.660 | 6.757 | 6.845 | 6.927 | 7.003 | 7.074 | 7.142 | 7.204 | 7.264 |
| 16 | 4.131 | 4.786 | 5.192 | 5.489 | 5.722 | 5.915 | 6.079 | 6.222 | 6.349 | 6.462 | 6.564 | 6.658 | 6.744 | 6.823 | 6.898 | 6.967 | 7.032 | 7.093 | 7.152 |
| 17 | 4.099 | 4.742 | 5.140 | 5.430 | 5.659 | 5.847 | 6.007 | 6.147 | 6.270 | 6.381 | 6.480 | 6.572 | 6.656 | 6.732 | 6.806 | 6.873 | 6.937 | 6.997 | 7.053 |
| 18 | 4.071 | 4.703 | 5.094 | 5.379 | 5.603 | 5.788 | 5.944 | 6.081 | 6.201 | 6.310 | 6.407 | 6.497 | 6.579 | 6.655 | 6.725 | 6.792 | 6.854 | 6.912 | 6.968 |
| 19 | 4.046 | 4.670 | 5.054 | 5.334 | 5.554 | 5.735 | 5.889 | 6.022 | 6.141 | 6.247 | 6.342 | 6.430 | 6.510 | 6.585 | 6.654 | 6.719 | 6.780 | 6.837 | 6.891 |
| 20 | 4.024 | 4.639 | 5.018 | 5.294 | 5.510 | 5.688 | 5.839 | 5.970 | 6.087 | 6.191 | 6.285 | 6.371 | 6.450 | 6.523 | 6.591 | 6.654 | 6.714 | 6.771 | 6.823 |
| 24 | 3.956 | 4.546 | 4.907 | 5.168 | 5.374 | 5.542 | 5.685 | 5.809 | 5.919 | 6.017 | 6.106 | 6.186 | 6.261 | 6.330 | 6.394 | 6.453 | 6.510 | 6.563 | 6.612 |
| 30 | 3.889 | 4.455 | 4.799 | 5.048 | 5.242 | 5.401 | 5.536 | 5.653 | 5.756 | 5.849 | 5.932 | 6.008 | 6.078 | 6.143 | 6.203 | 6.259 | 6.311 | 6.361 | 6.407 |
| 40 | 3.825 | 4.367 | 4.696 | 4.931 | 5.114 | 5.256 | 5.392 | 5.502 | 5.599 | 5.686 | 5.764 | 5.835 | 5.900 | 5.961 | 6.017 | 6.069 | 6.119 | 6.165 | 6.209 |
| 60 | 3.762 | 4.282 | 4.595 | 4.818 | 4.991 | 5.133 | 5.253 | 5.356 | 5.447 | 5.528 | 5.601 | 5.667 | 5.728 | 5.785 | 5.837 | 5.886 | 5.931 | 5.974 | 6.015 |
| 120 | 3.702 | 4.200 | 4.497 | 4.709 | 4.872 | 5.005 | 5.118 | 5.214 | 5.299 | 5.375 | 5.443 | 5.505 | 5.562 | 5.614 | 5.662 | 5.708 | 5.750 | 5.790 | 5.827 |
| ∞ | 3.643 | 4.120 | 4.403 | 4.603 | 4.757 | 4.882 | 4.987 | 5.078 | 5.157 | 5.227 | 5.290 | 5.348 | 5.400 | 5.448 | 5.493 | 5.535 | 5.574 | 5.611 | 5.645 |

# N Critical Values of the *q*-Distribution ($\alpha$ = 0.001) (cont.)

| df | 2 | 3 | 4 | 5 | 6 | 7 | 8 | 9 | 10 | 11 | 12 | 13 | 14 | 15 | 16 | 17 | 18 | 19 | 20 |
|---|---|---|---|---|---|---|---|---|---|---|---|---|---|---|---|---|---|---|---|
| 1 | 900.3 | 1351. | 1643. | 1856. | 2022. | 2158. | 2272. | 2370. | 2455. | 2532. | 2600. | 2662. | 2718. | 2770. | 2818. | 2863. | 2904. | 2943. | 2980. |
| 2 | 44.69 | 60.42 | 70.77 | 78.43 | 84.49 | 89.46 | 93.67 | 97.30 | 100.5 | 103.3 | 105.9 | 108.2 | 110.4 | 112.3 | 114.2 | 115.9 | 117.4 | 118.9 | 120.3 |
| 3 | 18.28 | 23.32 | 26.65 | 29.13 | 31.11 | 32.74 | 34.12 | 35.33 | 36.39 | 37.34 | 38.20 | 38.98 | 39.69 | 40.35 | 40.97 | 41.54 | 42.07 | 42.58 | 43.05 |
| 4 | 12.18 | 14.99 | 16.84 | 18.23 | 19.34 | 20.26 | 21.04 | 21.73 | 22.33 | 22.87 | 23.36 | 23.81 | 24.21 | 24.59 | 24.94 | 25.27 | 25.58 | 25.87 | 26.14 |
| 5 | 9.714 | 11.67 | 12.96 | 13.93 | 14.71 | 15.35 | 15.90 | 16.38 | 16.81 | 17.18 | 17.53 | 17.85 | 18.13 | 18.41 | 18.66 | 18.89 | 19.10 | 19.31 | 19.51 |
| 6 | 8.427 | 9.960 | 10.97 | 11.72 | 12.32 | 12.83 | 13.26 | 13.63 | 13.97 | 14.27 | 14.54 | 14.79 | 15.01 | 15.22 | 15.42 | 15.60 | 15.78 | 15.94 | 16.09 |
| 7 | 7.648 | 8.930 | 9.768 | 10.40 | 10.90 | 11.32 | 11.68 | 11.99 | 12.27 | 12.52 | 12.74 | 12.95 | 13.14 | 13.32 | 13.48 | 13.64 | 13.78 | 13.92 | 14.04 |
| 8 | 7.130 | 8.250 | 8.978 | 9.522 | 9.958 | 10.32 | 10.64 | 10.91 | 11.15 | 11.36 | 11.56 | 11.74 | 11.91 | 12.06 | 12.21 | 12.34 | 12.47 | 12.59 | 12.70 |
| 9 | 6.762 | 7.768 | 8.419 | 8.906 | 9.295 | 9.619 | 9.897 | 10.14 | 10.36 | 10.55 | 10.73 | 10.89 | 11.03 | 11.18 | 11.30 | 11.42 | 11.54 | 11.64 | 11.75 |
| 10 | 6.487 | 7.411 | 8.006 | 8.450 | 8.804 | 9.099 | 9.352 | 9.573 | 9.769 | 9.946 | 10.11 | 10.25 | 10.39 | 10.52 | 10.64 | 10.75 | 10.85 | 10.95 | 11.03 |
| 11 | 6.275 | 7.136 | 7.687 | 8.098 | 8.426 | 8.699 | 8.933 | 9.138 | 9.319 | 9.482 | 9.630 | 9.766 | 9.892 | 10.01 | 10.12 | 10.22 | 10.31 | 10.41 | 10.49 |
| 12 | 6.106 | 6.917 | 7.436 | 7.821 | 8.127 | 8.383 | 8.601 | 8.793 | 8.962 | 9.115 | 9.254 | 9.381 | 9.498 | 9.606 | 9.707 | 9.802 | 9.891 | 9.975 | 10.06 |
| 13 | 5.970 | 6.740 | 7.231 | 7.595 | 7.885 | 8.126 | 8.333 | 8.513 | 8.673 | 8.817 | 8.948 | 9.068 | 9.178 | 9.281 | 9.376 | 9.466 | 9.550 | 9.629 | 9.704 |
| 14 | 5.856 | 6.594 | 7.062 | 7.409 | 7.685 | 7.915 | 8.110 | 8.282 | 8.434 | 8.571 | 8.696 | 8.809 | 8.914 | 9.012 | 9.103 | 9.188 | 9.267 | 9.343 | 9.414 |
| 15 | 5.760 | 6.470 | 6.920 | 7.252 | 7.517 | 7.736 | 7.925 | 8.088 | 8.234 | 8.365 | 8.483 | 8.592 | 8.693 | 8.786 | 8.872 | 8.954 | 9.030 | 9.102 | 9.170 |
| 16 | 5.678 | 6.365 | 6.799 | 7.119 | 7.374 | 7.585 | 7.766 | 7.923 | 8.063 | 8.189 | 8.303 | 8.407 | 8.504 | 8.593 | 8.676 | 8.755 | 8.828 | 8.897 | 8.963 |
| 17 | 5.608 | 6.275 | 6.695 | 7.005 | 7.250 | 7.454 | 7.629 | 7.781 | 7.916 | 8.037 | 8.148 | 8.248 | 8.342 | 8.427 | 8.508 | 8.583 | 8.654 | 8.720 | 8.784 |
| 18 | 5.546 | 6.196 | 6.604 | 6.905 | 7.143 | 7.341 | 7.510 | 7.657 | 7.788 | 7.906 | 8.012 | 8.110 | 8.199 | 8.283 | 8.361 | 8.434 | 8.502 | 8.567 | 8.628 |
| 19 | 5.492 | 6.127 | 6.525 | 6.817 | 7.049 | 7.242 | 7.405 | 7.549 | 7.676 | 7.790 | 7.893 | 7.988 | 8.075 | 8.156 | 8.232 | 8.303 | 8.369 | 8.432 | 8.491 |
| 20 | 5.444 | 6.065 | 6.454 | 6.740 | 6.966 | 7.154 | 7.313 | 7.453 | 7.577 | 7.688 | 7.788 | 7.880 | 7.966 | 8.044 | 8.118 | 8.186 | 8.251 | 8.312 | 8.370 |
| 24 | 5.297 | 5.877 | 6.238 | 6.503 | 6.712 | 6.884 | 7.031 | 7.159 | 7.272 | 7.374 | 7.467 | 7.551 | 7.629 | 7.701 | 7.768 | 7.831 | 7.890 | 7.946 | 7.999 |
| 30 | 5.156 | 5.698 | 6.033 | 6.278 | 6.470 | 6.628 | 6.763 | 6.880 | 6.984 | 7.077 | 7.162 | 7.239 | 7.310 | 7.375 | 7.437 | 7.494 | 7.548 | 7.599 | 7.647 |
| 40 | 5.022 | 5.528 | 5.838 | 6.063 | 6.240 | 6.386 | 6.509 | 6.616 | 6.711 | 6.796 | 6.872 | 6.942 | 7.007 | 7.067 | 7.122 | 7.174 | 7.223 | 7.269 | 7.312 |
| 60 | 4.894 | 5.365 | 5.653 | 5.860 | 6.022 | 6.155 | 6.268 | 6.366 | 6.451 | 6.528 | 6.598 | 6.661 | 6.720 | 6.774 | 6.824 | 6.871 | 6.914 | 6.956 | 6.995 |
| 120 | 4.771 | 5.211 | 5.476 | 5.667 | 5.815 | 5.937 | 6.039 | 6.128 | 6.206 | 6.276 | 6.339 | 6.396 | 6.448 | 6.496 | 6.542 | 6.583 | 6.623 | 6.660 | 6.695 |
| $\infty$ | 4.654 | 5.063 | 5.309 | 5.484 | 5.619 | 5.730 | 5.823 | 5.903 | 5.973 | 6.036 | 6.092 | 6.144 | 6.191 | 6.234 | 6.274 | 6.312 | 6.347 | 6.380 | 6.411 |

# Appendix B
## Getting Started with Microsoft Excel (Desktop)

## The Basics of Excel Office 365

Microsoft Excel is a spreadsheet program that allows users to track and analyze data. Spreadsheets such as those created with Microsoft Excel are widely used in the business world to perform various tasks such as accounting, budgeting, billing, reporting, planning, and tracking. The instructions provided in this Appendix are intended for non-mobile users.

When you open Excel, you will see various tabs along the top such as File, Home, Insert, Page Layout, Formulas, Data, etc. The numerous commands and options available when working within Excel can be found under these tabs. At the bottom, there is a tab labeled Sheet1. The icon to the right of the Sheet1 tab creates a new tab with a blank spreadsheet. Tabs can be renamed by double-clicking on them and entering new text. Figure B.1 shows what a new workbook looks like when Excel is first opened.

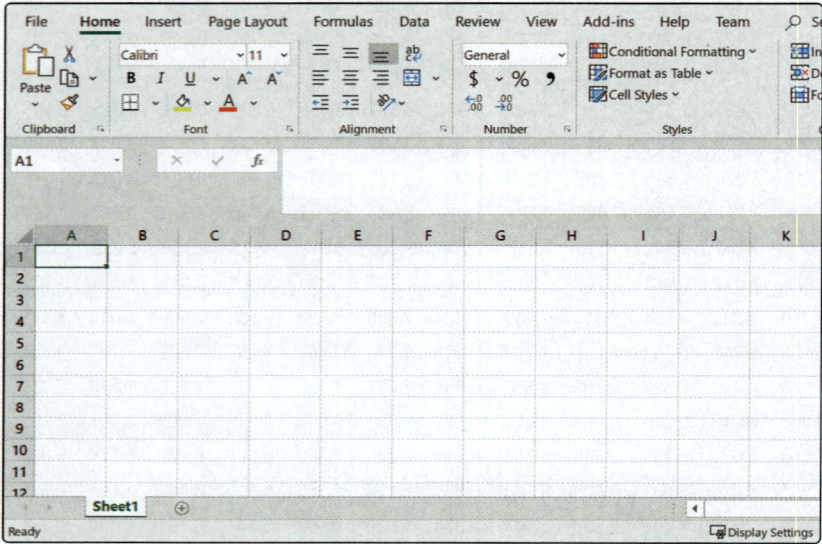

Figure B.1

## Cells

Cells in Microsoft Excel may contain numerical data and/or text. The cells' locations are described by their positions in terms of columns and rows. Columns are listed from left to right across the top of the worksheet and are labeled with letters. Rows are listed from top to bottom along the left side of the worksheet and are labeled with numbers. A column letter followed by a row number describes the active cell position. For example, **B2** would be referring to the cell found in the second row of Column B (see Figure B.2). A thick solid border will outline the active cell. To help you identify the cell address, the column and row headers are also highlighted. Another way to identify the active cell's address is to look at the window just above the header for Column A.

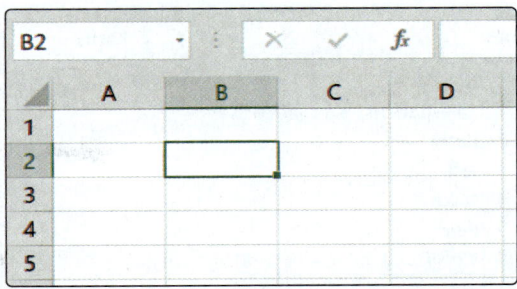

Figure B.2

The column width can be altered by placing your cursor between two column labels; when it turns into a vertical line with arrows pointing to the left and right, you can then click and drag to the left or right to make the column width larger or smaller. The same can be done with row heights. Alternatively, double-clicking between two column or row labels will autofit the row height or column width to the data it contains.

To change the active cell, move the mouse to the desired cell and click. The border will now be surrounding the new cell and the address will have changed in the active cell address window. The arrow keys can also control navigation of the active cell.

To the right of the cell address window is the formula window (labeled with $f_x$). This displays the contents of the active cell. The contents of the cell can be edited either within the formula window or the cell itself. Thus, as you enter data in the active cell, it is displayed within the cell and in the formula window at the top of the worksheet.

Suppose you wanted to sum the numbers 15, 10, and 3. To do this, enter the formula **=15+10+3** into cell **A1**. The equal sign, "=", is the command to start a formula in Excel. Notice that the formula is seen within in the cell as well as in the formula window (see Figure B.3a). When you press **Enter**, you will see that the active cell moves to cell **A2**, and the solution of 28 is now displayed in cell **A1** (see Figure B.3b).

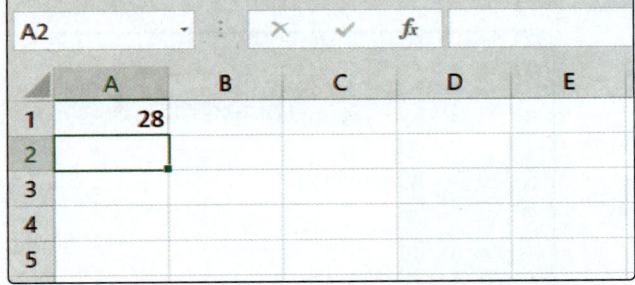

Figure B.3a                                                              Figure B.3b

When you finish entering a formula in a cell, press **Enter**. The active cell will now move one row down. The answer to the formula, rather than the formula itself, will be displayed in the cell. Cells or data can be added, subtracted, multiplied, divided, and manipulated in any combination through formulas.

## Filling Cells

Excel will try to anticipate a reoccurring word or phrase within a column. After you have entered some text in a column, if you type the first letter of a previous cell again, Excel will fill in the rest of the word or phrase. To accept the automatic completion of the word, just hit the **Enter** key or click in a new cell. If you do not want to use the automatic completion of the word, simply keep typing in the cell and the new word or phrase will appear. Under the Review tab, there is an option for Excel to spell check your spreadsheet.

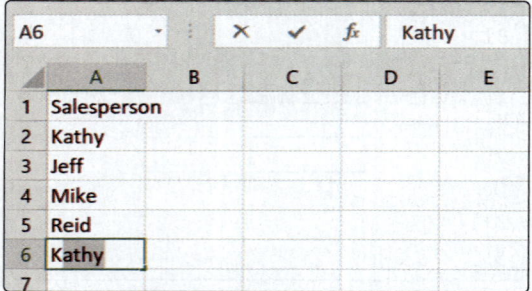

Figure B.4

Excel has a helpful tool when entering a series of data. Try entering 1 into cell **A1** and 2 into cell **A2**. Then highlight both of the cells. You will see a small square at the bottom right of the highlighted section.

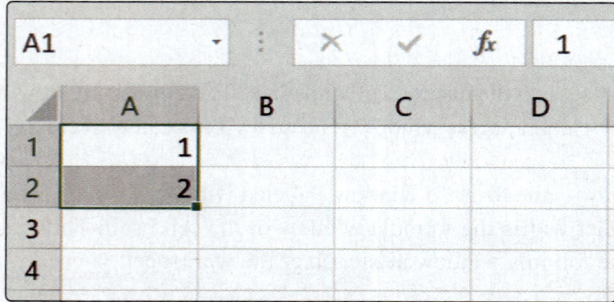

Figure B.5

If you move your cursor over this square, the cursor should change into a narrow plus sign. Now, click the left mouse button and drag the mouse down to include cells **A3** through **A10**. When you release the mouse button, the cells should now be filled with values from 1 to 10. We will refer to this as *filling cells*. Also notice that at the bottom of the screen, Excel gives some summary statistics about the highlighted cells. The average, count of the number of values, and sum are displayed for any highlighted cells.

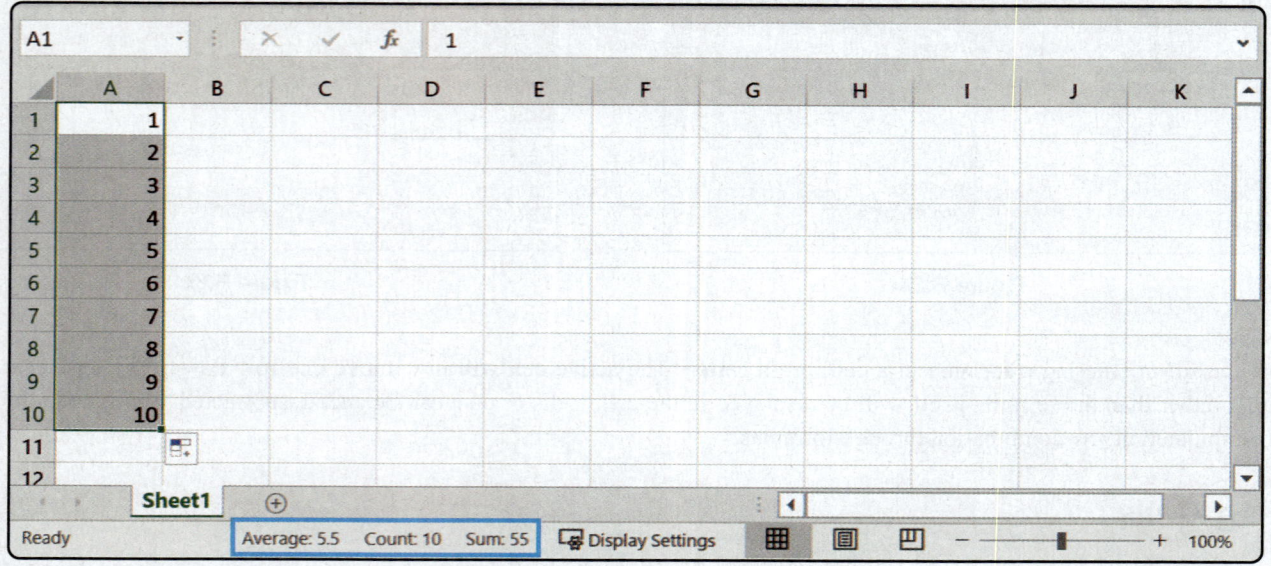

Figure B.6

The fill tool will work for other (nonsequential) values as well. Excel finds the relationship between two data points and replicates this when the tool is used. As an example, try this with multiples of 5. Enter 5 in cell **A1** and 10 in cell **A2**. Highlight cells **A1** and **A2** again and put your cursor over the square in the bottom right of the highlighted section. Click the left mouse button and drag down to cell **A10**. Release the button. Now the values should have changed to 5 through 50, in increments of 5.

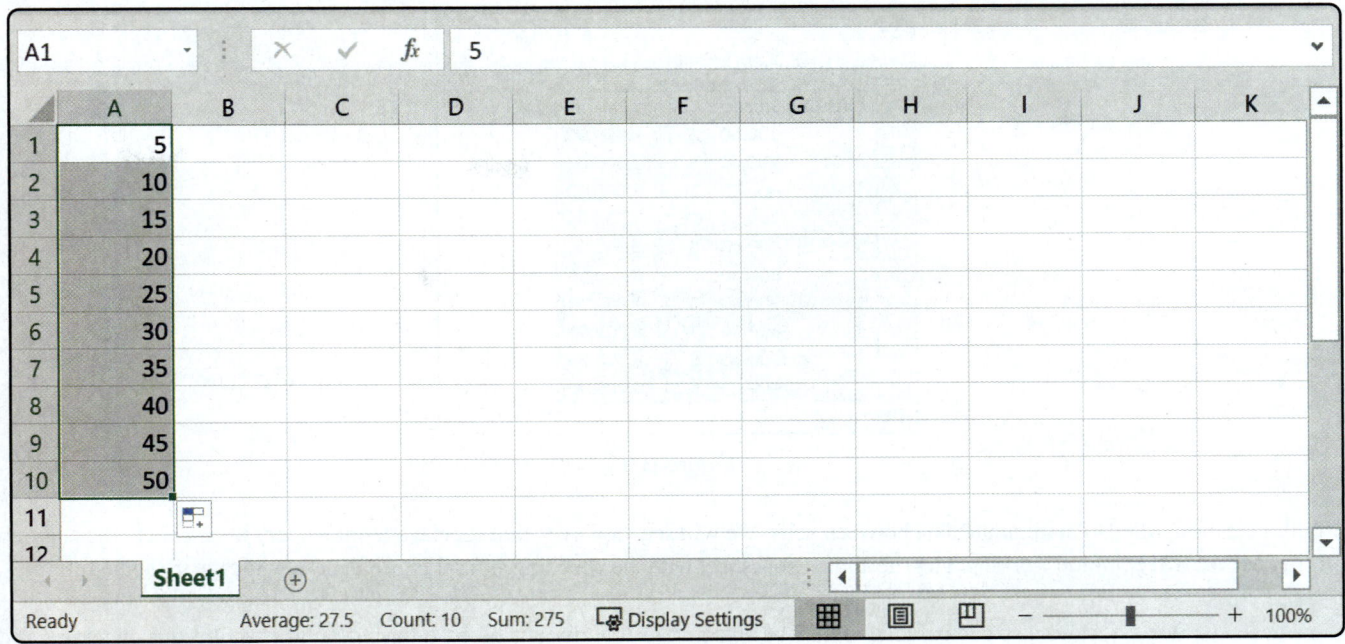

Figure B.7

You can also use this feature with sequential labels such as days of the week or months of the year. For example, if you type Monday into cell **A1** and use the fill tool to drag to cell **G1**, the intermediate cells will populate with all of the days of the week.

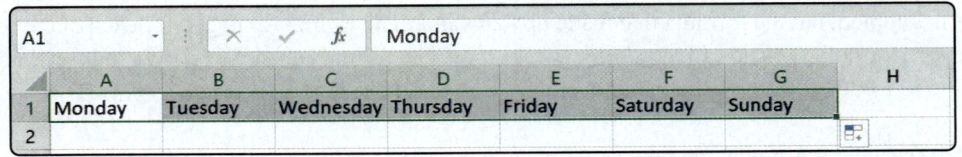

Figure B.8

Under the Home tab, you can change the appearance of text by altering the font, size, color, or alignment of the text within the cells. You also can use Microsoft Excel to sort a long list of items alphabetically or numerically or to find a particular item you are looking for. The sort and find features can be found under the Home tab as well.

Now that you have the basic skills to work with Excel, start a new worksheet and we will try some examples.

## Formulas and Addressing Using Excel

Enter the labels Checking Balance and Savings Balance in cells **A1** and **B1**, respectively. You will notice that the label you typed in cell **A1** is cut off. We need to resize the columns to allow for the entire label. Move your cursor to the line that separates the column labels for A and B. Your cursor will change to an arrow pointing off to the left and right. You can click and drag the cursor to the right to increase the column width, or simply double-click to have the width autofit the label. Repeat this for Column B.

Next, select the column labels A and B and you should see both columns highlighted. Under the Home tab there is an area labeled Number. The buttons in this area format any numbers you enter in the selected cells. With columns A and B highlighted, click on the $ button. Now all of the values put in these columns will be formatted as currency.

Next, we will use the fill tool to get some values to work with. Start by entering 100 in cell **A2** and 200 in cell **A3**. Now move to cell **B2**, enter 1000, and enter 2000 in cell **B3**. Highlight cells **A2** through **B3** by clicking in **A2** and dragging down and to the right to include **B3**. With those cells highlighted, put your cursor over the square in the bottom right of the highlighted section. When your mouse changes to a narrow plus sign, click and drag down to row 11. When you release the mouse button, the cells should fill with data and end with $10,000.00 in cell **B11**. You should have a screen that looks like the following.

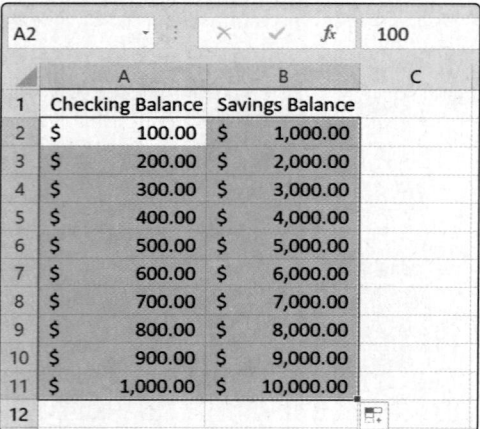

Figure B.9

Formulas can be applied to manipulate data between cells. We will see, however, that special attention needs to be paid to the copying of formulas. We will continue working with the above worksheet with the checking and savings balances while we try some formulas.

We want to see what your total bank balance would be if you had the amount listed in a particular row in your checking and savings accounts. In cell **C2**, type = to indicate you are entering a formula, and then click in cell **A2**, type +, and click in cell **B2**. Press **Enter**. You have just added cells **A2** and **B2** together, and the sum ($1,100.00) is displayed in cell **C2**. Under the Home tab, the Clipboard section contains buttons allowing you to cut, copy, and paste cells. With cell **C2** selected, either click on the **copy** button or use the keyboard shortcut **Ctrl+c** to copy the formula. Highlight cells **C3** through **C11**. Under the Home tab, press the **Paste** button or use the keyboard shortcut **Ctrl+v** to paste the formula into these cells. Notice that the value $1,100.00 that was displayed in **C2** was not copied, but the formula that made up that value. The formula changed for each of the rows and substituted the new row value in the formula. This is called *relative addressing*. The formula changes relative to its cell address. The last cell should show the value $11,000.00. If the value is not visible, the column needs to be resized.

Suppose we wanted to see what the annual interest would be on our savings account at the levels listed in Column B. Create a new label in cell **A13** called Interest Rate. Now in cell **B13**, input the interest rate 0.045. We will need to format the cell for percentages rather than currency. To do this, highlight cell **B13**, and under the Home tab in the Number area, select the **%** button next to the **$** button that we used to originally format the cells. The number will most likely be displayed as 5%. We can adjust the number of decimals displayed by clicking on the ⬅️.00 button, also located in the Number section under the Home tab. This button increases the number of decimal places displayed by one decimal place each time it is clicked. (Notice that there is a similar button next to the one mentioned that decreases the number of places displayed.) The value should now display as 4.5%.

Now we want to use this value to compute interest for each level of savings. Label cell **D1** Savings Interest. Resize the cell to fit the label. In cell **D2**, we will put our formula. Type = to start the formula, and then select cell **B13** to get the percentage rate. Type * to indicate multiplication, and select cell **B2**. Upon pressing **Enter**, the resulting value will be displayed ($45.00). Thus, the annual interest gained on $1,000.00 in savings at 4.5% APR is $45.00.

Since we have the formula, we can copy it to the rest of the column and get the interest income for each of the savings levels. With cell **D2** selected, click in the lower right-hand corner of the cell and drag down to **D3**. Is this the result you expected? Probably not. Remember the rules of relative addressing. As you copy the formula down one row, the formula values change by one row. So in cell **D3**, the formula is pulling from cell **B14** for the interest rate and cell **B3** for the principal amount. You will notice that cell **B14** is empty. Thus, Excel computes this formula as being equal to zero. However, want to use cell B13 for each of these formulas in Column D. So how do we lock in the address of cell **B13**? We use *absolute addressing*. The **$** symbol is the key to locking the position in the formula. If the $ is in front of the column indicator of a cell reference, it will lock the column ($B13), and if it is in front of the row indicator, it will lock the row (B$13). These can be used together to lock a cell reference to a specific cell. In cell **D2**, we need to change the formula to read **=$B$13*B2**. This will "lock" cell B13 into the formula when it is copied throughout Column D.

Go back and copy the new formula into cells **D3** through **D11** using the fill tool. The results should now look like Figure B.10.

| D11 | | | fx | =B$13*B11 | |
|---|---|---|---|---|---|
| | A | B | C | D | E |
| 1 | Checking Balance | Savings Balance | | Savings Interest | |
| 2 | $ 100.00 | $ 1,000.00 | $ 1,100.00 | $ 45.00 | |
| 3 | $ 200.00 | $ 2,000.00 | $ 2,200.00 | $ 90.00 | |
| 4 | $ 300.00 | $ 3,000.00 | $ 3,300.00 | $ 135.00 | |
| 5 | $ 400.00 | $ 4,000.00 | $ 4,400.00 | $ 180.00 | |
| 6 | $ 500.00 | $ 5,000.00 | $ 5,500.00 | $ 225.00 | |
| 7 | $ 600.00 | $ 6,000.00 | $ 6,600.00 | $ 270.00 | |
| 8 | $ 700.00 | $ 7,000.00 | $ 7,700.00 | $ 315.00 | |
| 9 | $ 800.00 | $ 8,000.00 | $ 8,800.00 | $ 360.00 | |
| 10 | $ 900.00 | $ 9,000.00 | $ 9,900.00 | $ 405.00 | |
| 11 | $ 1,000.00 | $ 10,000.00 | $11,000.00 | $ 450.00 | |
| 12 | | | | | |
| 13 | Interest Rate | 4.5% | | | |
| 14 | | | | | |

Figure B.10

Excel can compute a multitude of calculations on data values in the spreadsheet. If you press the $f_x$ button next to the formula window, you can explore all of the functions and formulas that Excel has to offer.

# Charts

Suppose you have the following data about ticket sales from your county fair. Adult tickets are $20 and child tickets are $12. Ticket sales for the week are listed in Figure B.11.

| C8 | | | fx | 7740 | |
|---|---|---|---|---|---|
| | A | B | C | D |
| 1 | | Adult | Child | | |
| 2 | Sunday | $ 10,120.00 | $ 5,760.00 | | |
| 3 | Monday | $ 9,040.00 | $ 3,600.00 | | |
| 4 | Tuesday | $ 8,380.00 | $ 3,360.00 | | |
| 5 | Wednesday | $ 7,620.00 | $ 3,132.00 | | |
| 6 | Thursday | $ 7,900.00 | $ 2,520.00 | | |
| 7 | Friday | $ 10,560.00 | $ 4,944.00 | | |
| 8 | Saturday | $ 13,260.00 | $ 7,740.00 | | |
| 9 | | | | | |

Figure B.11

From these data, you might see how the ticket sales change daily and wish to compare child and adult ticket sales. This information might be easier to understand if it were graphically displayed on a chart. Enter the data as seen above in a new worksheet and use a bar chart for this particular example.

To create a chart, we need to select the data to be graphed. We want to chart the ticket sales for both the adults and children. In a bar chart, the dollar amount will determine the height of the bar. The days of the week should also be included since they are the labels for the bars.

Highlight the data in Columns A, B, and C from rows 1 to 8. (Note that Excel can interpret the first row or column in a set of data as labels and not part of the data.) Under the Insert tab, there is a Charts group of icons.

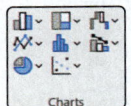

Figure B.12

Choose Column (in the upper left corner), and then select the top-left graph (the first one listed under the 2-D column heading). This is a clustered column graph. Excel creates a side-by-side bar graph based on the highlighted data. After the chart is created, a set of tabs labeled Chart Design and Format appear. Additionally, three icons appear to the right of the chart: Chart Elements, Chart Styles, and Chart Filters. These tabs and icons can be used to edit the chart that has been made.

Click on the chart title, type a new title of "Ticket Sales", and press **Enter**. Using the Chart Elements icon, you can edit the chart appearance by adding axis titles, data labels, or gridlines. Microsoft Excel makes it easy to create the chart you want.

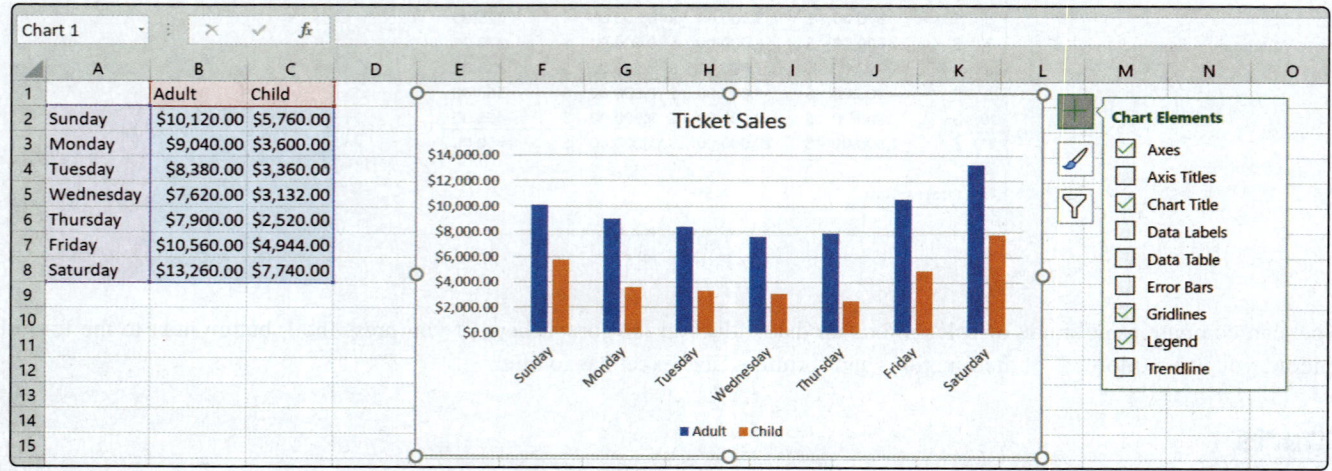

Figure B.13

Excel can create most types of charts, ranging from pie charts to line graphs to scatter plots. These types of charts are indicated by the menus seen when an icon in the Charts group is clicked on. Histograms can also be created using Excel, but require the **Data Analysis** tools, which will be discussed next.

## Installing Data Analysis Tools

To get the most out of Excel as a statistical tool, you will want to use the Analysis ToolPak. When Microsoft Excel is first installed, this feature is not included. The ToolPak is an application add-in. To install the Analysis ToolPak, under the File tab, select **Options**, and then **Add-ins** on the left-hand side. When **Add-ins** is selected, at the bottom of the dialog box there is a Manage drop-down menu that has Excel Add-ins selected. Press the **Go** button next to this drop-down menu. A list of available add-ins will appear. Check the boxes next to **Analysis ToolPak** and **Analysis ToolPak – VBA**. Press **OK**. The data analysis tools will be installed.

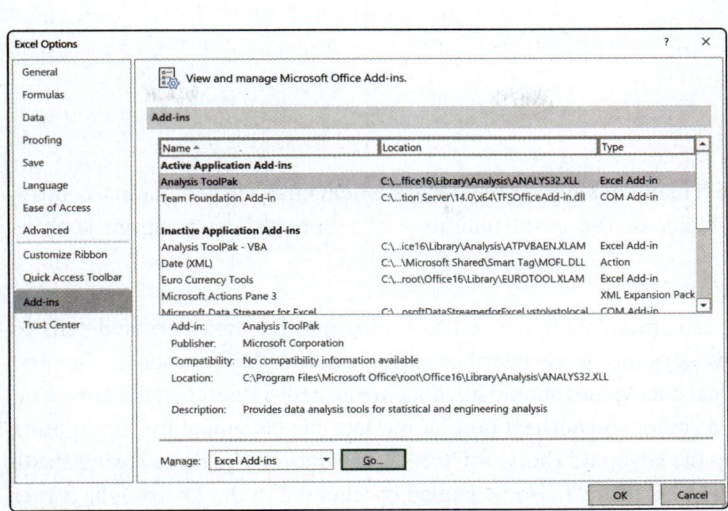

Figure B.14a

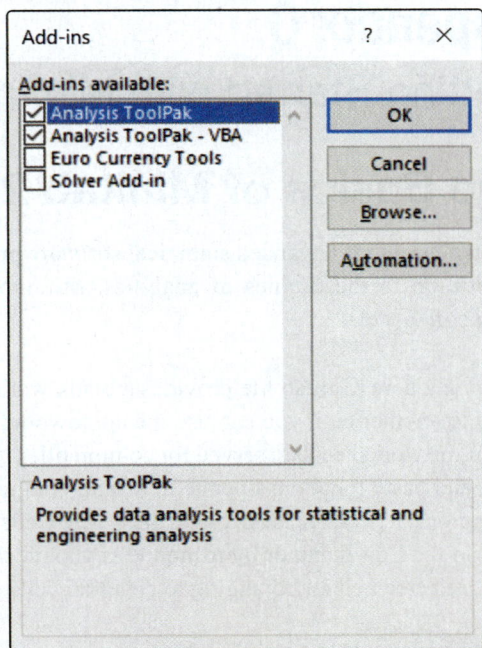

Figure B.14b

With these tools installed, you can perform many useful analyses in statistics. The tool is found under the Data tab once it is installed. The button is located in the Analysis group of icons and is labeled Data Analysis. When you click on the **Data Analysis** button under the Data tab, a dialog box will appear with the Excel analysis tools that are available (see Figure B.15). Using these tools, you can perform hypothesis tests, obtain basic descriptive statistics, create histograms, and run regression analysis. More detailed directions on using the Data Analysis tools to solve problems are found at stat.hawkeslearning.com by navigating to Discovering Business Statistics, Second Edition followed by Technology Instructions.

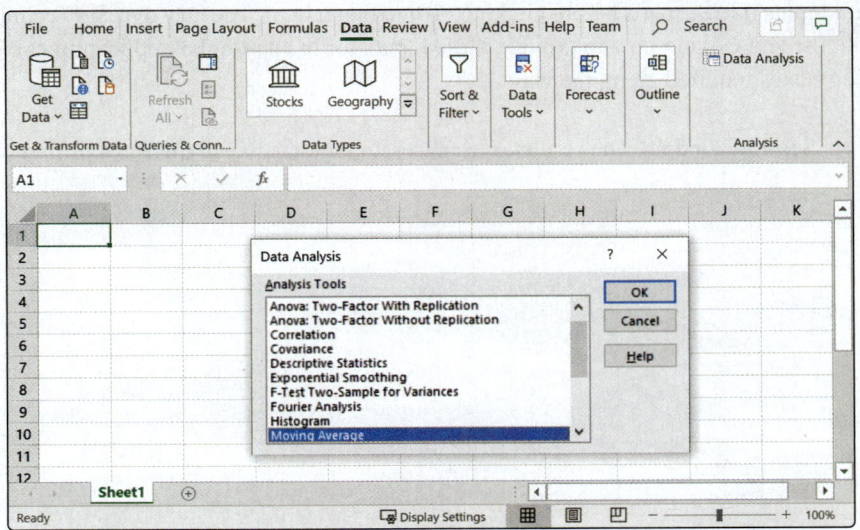

Figure B.15

# Appendix C
## Getting Started with Minitab

## The Basics of Minitab 21

Minitab is a more advanced statistical software program which many businesses use for statistical analysis and quality control. In addition to these types of analytics, Minitab performs many of the basic functions of a spreadsheet program such as Microsoft Excel.

Opening a new Minitab file provides a blank worksheet. You can enter data into a cell by typing into the dark-bordered cell. To move to another cell, you can use the up/down/left/right arrow keys on the keyboard or simply click with your mouse. The first row of the worksheet is reserved for column titles. Any numerical data values should go in the white cells rather than the grey. Cut, copy, and paste functions operate in Minitab in the same way as other spreadsheet programs. Data can be copied by highlighting the appropriate cell and selecting **Edit**, **Copy Cells**, or by using the keyboard shortcut **Ctrl+C**. The data can be pasted using **Paste Cells** in the Edit menu or by using the keyboard shortcut **Ctrl+V**. Cells can also be copied by clicking in the lower right corner of the selected cell and dragging to cover all cells for which you wish to have that value.

When you open Minitab you will notice a menu bar at the top of the screen. On that menu bar you will see menus such as **Edit**, **Data**, **Calc**, **Stat**, and **Graph**. The **Stat** menu contains many of the tools we utilize in this text. If you click on the **Stat** menu, you will see a list of concepts, some of which should be familiar such as **Basic Statistics**, **Regression**, and **ANOVA**. Clicking on any of these reveals other possibilities. Clicking on **Basic Statistics** leads to **Display Descriptive Statistics**, clicking on **Regression** allows you to make a **Fitted Line Plot**, and clicking on **ANOVA** allows you to perform a one or two-way analysis of variance.

If you choose **Calc** on the menu bar, you will see a list including options such as **Random Data** and **Probability Distributions**. Choosing **Random Data** will allow you to generate data according to some specified distribution. Clicking on Probability **Distributions** allows you to calculate probabilities for the binomial or **Poisson** distributions, among others.

Both Stat and Calc require inputs into the cells before clicking on that menu item. The Discovering Technology sections at the end of the chapters throughout the text explain how to provide these inputs for the particular application and give further explanation as to how to use the dialog boxes required along the way.

Above the worksheet is the Session window. The results of each session are displayed there. To print what is in the window, you can go to **File**, **Print**.

# Cells

To create a new worksheet, you can select **File**, **New**, and select **Worksheet**. This opens a new worksheet to use in the current project. Columns run left to right and are labeled with the letter C and a number. Rows run top to bottom and are labeled with numbers. To refer to a specific cell, we will use the format "Column name, Row name". For example, if we were referring to the third row in the second column, we would use "C2, 3" to reference the cell. Minitab provides a row that is not labeled for you to insert your column headings.

The cells in Minitab may contain data or text. If you are inputting text in the cells other than the cell for the column heading, the column label will add "-T" to the column name. In some cases, Minitab will recognize the type of data. For example, if we typed the days of the week in the first eight rows of Column C1, the column name would change to "C1-D" because the data is a date or time.

| ↓ | C1-D | C2 | C3 |
|---|---|---|---|
|  | Days of the Week |  |  |
| 1 | Monday |  |  |
| 2 | Tuesday |  |  |
| 3 | Wednesday |  |  |
| 4 | Thursday |  |  |
| 5 | Friday |  |  |
| 6 | Saturday |  |  |
| 7 | Sunday |  |  |
| 8 |  |  |  |

Figure C.1

# Filling Cells and Formulas

Now, let's input some data values into a new project to show some of the calculation features of Minitab. Choose **File**, **New**, and **Project**.

Minitab allows you to fill cells when entering a series of data. Enter **100** into cell **C1, 1** and **200** into cell **C1, 2**. Highlight both of the cells and locate the small square in the bottom right corner of the highlighted section. When you move your cursor over the square, a plus sign should appear. Click the left mouse button and drag the mouse down to include **C1, 3** through **C1, 10**. The cells should now be filled with values from 100 to 1000.

We can perform some basic functions on these values. From the menu bar, choose **Calc** and **Column Statistics**. This allows you to find the sum, the mean, the standard deviation, and many other summary measures corresponding to a particular column. Choose the **mean** and click inside the box next to **Input variable**. While your cursor is in this box, all the columns that contain data will be listed in the box to the left. You can click on the column in this box and press Select to specify **C1** as the "Input variable". Click **OK** and the mean of C1 will appear in the session window.

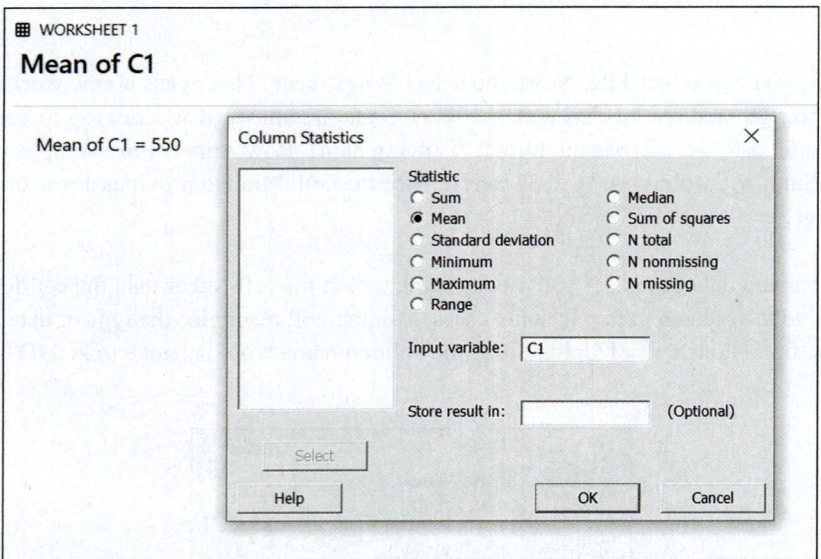

Figure C.2

To perform calculations on the data in the cells, select **Calc** and **Calculator** from the menu bar. In the "Store result in variable" box, enter **C2**. You will see a drop down menu of functions, as well as a list of functions on the right-hand side of the dialog box. In the list of functions, scroll down and select the function **Round**. Press **Select** and the text ROUND(number,decimals) will appear in the "Expression" box. For the parameter "number" enter **C1*7/3**, and for the parameter "decimals" enter **1**. This will multiply the data in Column C1 by $\frac{7}{3}$, round the result to one decimal place, and display the final value in Column C2.

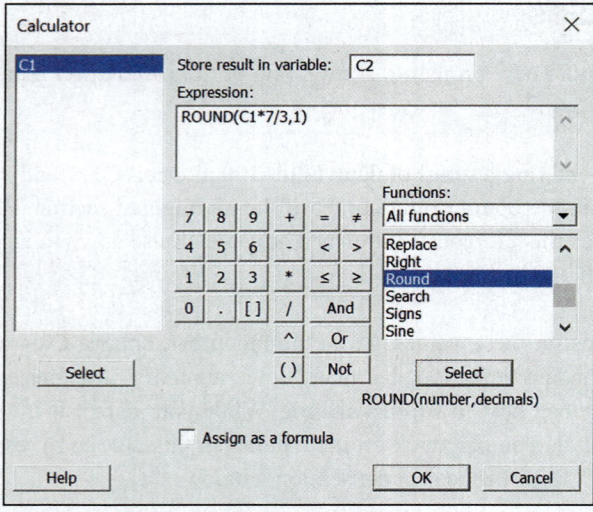

Figure C.3

Press **OK**, and you will see the values of this expression displayed in Column C2, rounded to one decimal place.

| ♦ | C1 | C2 | C3 |
|---|----|----|----|
| 1 | 100 | 233.3 | |
| 2 | 200 | 466.7 | |
| 3 | 300 | 700.0 | |
| 4 | 400 | 933.3 | |
| 5 | 500 | 1166.7 | |
| 6 | 600 | 1400.0 | |
| 7 | 700 | 1633.3 | |
| 8 | 800 | 1866.7 | |
| 9 | 900 | 2100.0 | |
| 10 | 1000 | 2333.3 | |
| 11 | | | |
| 12 | | | |

Figure C.4

# Graphs

Suppose you had the following data about a firm's sales per quarter for a particular year.

| ♦ | C1-T | C2 | C3 |
|---|------|-----|----|
| | Quarter | Sales | |
| 1 | First | 356210 | |
| 2 | Second | 349800 | |
| 3 | Third | 370355 | |
| 4 | Fourth | 402775 | |
| 5 | | | |
| 6 | | | |
| 7 | | | |
| 8 | | | |

Figure C.5

To graphically display this information, select **Graph** and **Time Series Plot**. Select the **Simple** time series plot and press **OK**. Put the cursor in the "Series" dialog box, and the available columns with data will appear in the box on the left. Since we only have one column with quantitative data in this case, only C2 appears. Select **C2** in the box and press **Select**. Press **OK**. A time series plot is created.

Figure C.6

## Data Analysis

One of the most important uses of Minitab is statistical analysis. Start a new Minitab project and enter the following data for the number of phone calls a company receives each hour of the workday.

| ♦ | C1 | C2 | C3 |
|---|---|---|---|
| | Phone Calls | | |
| 1 | 10 | | |
| 2 | 27 | | |
| 3 | 21 | | |
| 4 | 24 | | |
| 5 | 30 | | |
| 6 | 23 | | |
| 7 | 36 | | |
| 8 | 30 | | |
| 9 | 15 | | |
| 10 | | | |

Figure C.7

Select **Stat**, **Basic Statistics**, and **Display Descriptive Statistics**. In the "Variables" dialog box, enter **C1** and press **OK**. Basic descriptive statistics are displayed about the data in C1. This includes the number of data values, the mean, the standard deviation, the minimum value, the first quartile, the median, the third quartile, and the maximum value.

### Descriptive Statistics: Phone Calls

#### Statistics

| Variable | N | N* | Mean | SE Mean | StDev | Minimum | Q1 | Median | Q3 | Maximum |
|---|---|---|---|---|---|---|---|---|---|---|
| Phone Calls | 9 | 0 | 24.00 | 2.67 | 8.00 | 10.00 | 18.00 | 24.00 | 30.00 | 36.00 |

Figure C.8

There are many other useful statistical features of Minitab that you will encounter throughout the text. Refer to the Discovering Technology sections at the end of the chapters throughout the text for detailed instructions on using Minitab for statistical analysis.

# Appendix D
## Getting Started with JMP

## The Basics of JMP 16

When you open JMP, you will see the **Tip of the Day** and the **JMP Home** windows. The **Tip of the Day** provides helpful hints on using JMP. (Note: The **Tip of the Day** and the **JMP Home** windows appear by default when JMP opens. To change the default windows, go to **File > Preferences > General**.)

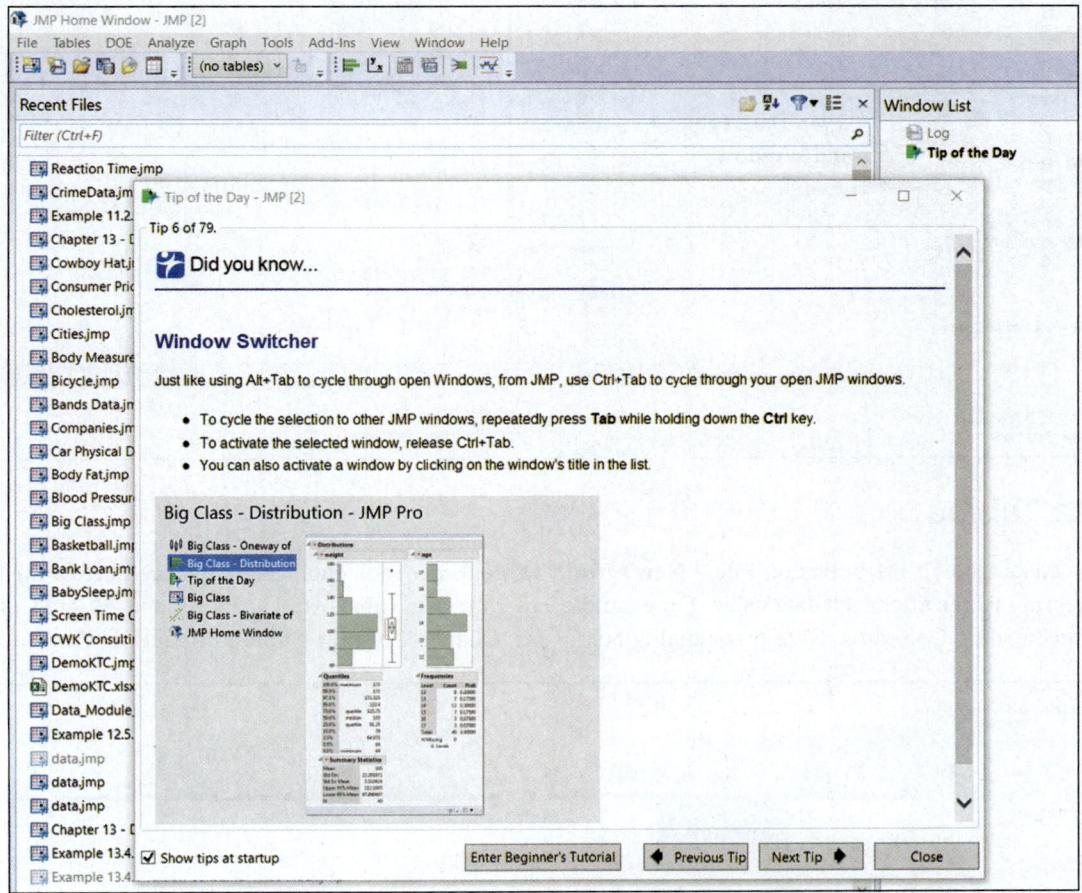

The **JMP Home** window displays menus and toolbars at the top, recently used files on the left, and all open data tables and windows on the right. The menus across the top are used to perform JMP functions. The last item in the menu bar is the **Help** menu where you can find such resources as JMP searchable documentation, tutorials, sample data sets, indexes of terms and functions, and a summary of new features. The toolbar located below the menu bar contains shortcuts and other helpful tools. (Note: To have menus and toolbars always display, go to **File > Preferences > Windows Specific** and change **Auto-hide menus and toolbars** from **Based on window size** to **Never**.)

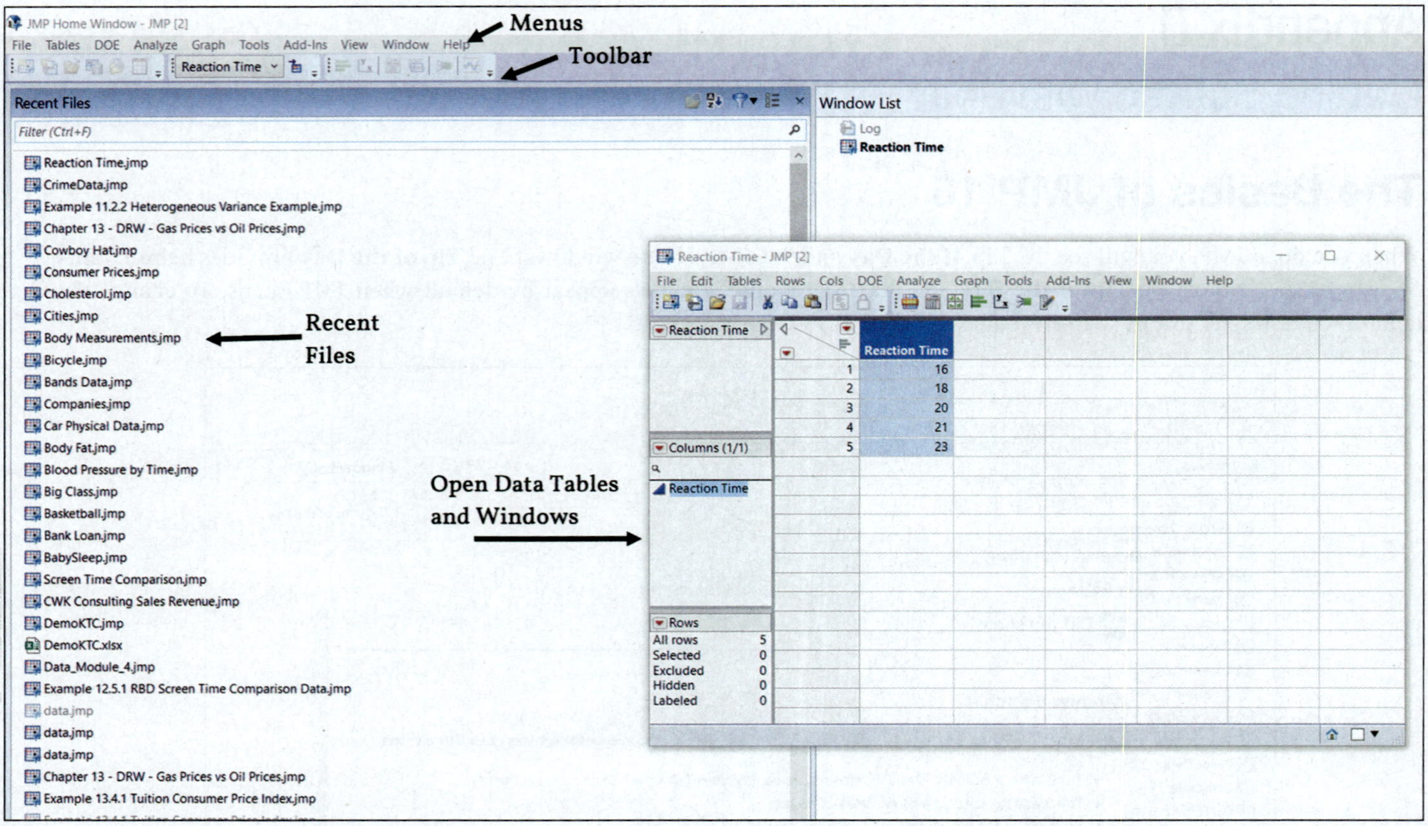

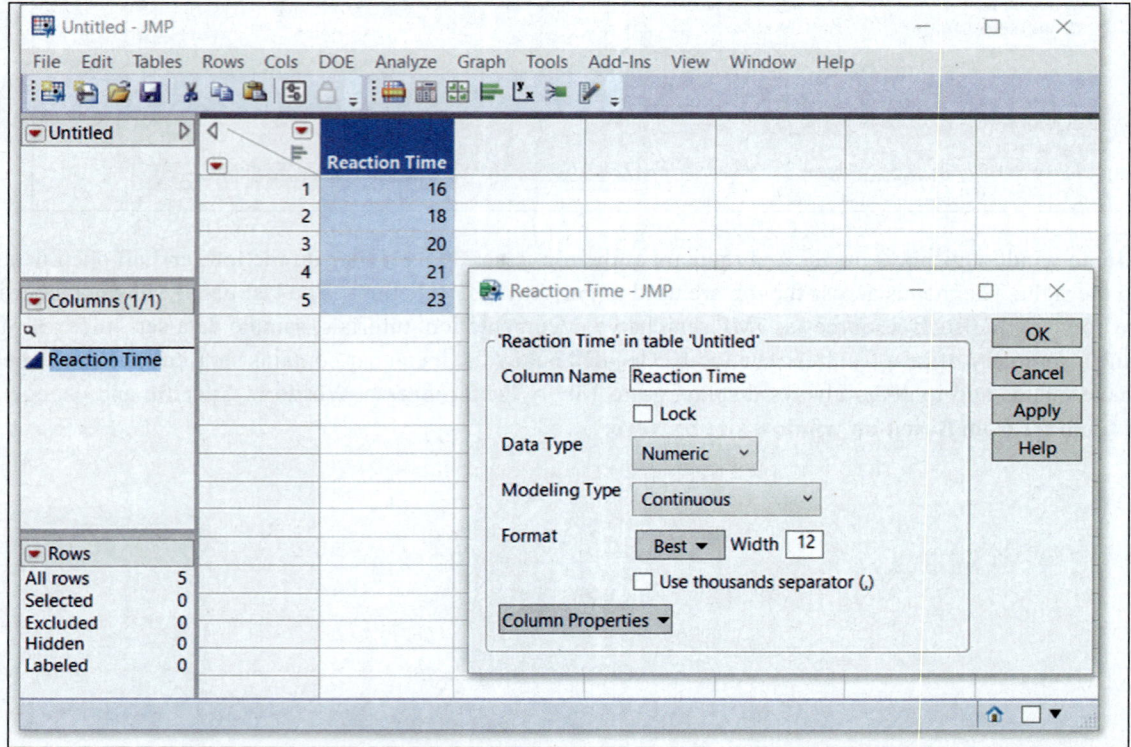

## JMP Data Tables

To create a new data table in JMP, click on **File > New > Data Table**. Enter each data value into a cell below the label heading **Column 1**, pressing **Enter** after each data value. For example, enter the data values 16, 18, 20, 21, and 23 into **Column 1**. Then double-click the heading Column 1. Note that a dialog box opens. Change **Column Name** to **Reaction Time**.

To open an existing JMP data table click on **File > Open**. Then navigate to the directory where your JMP, Excel, or other data files are located. Click on the file name and then click **Open**.

To open a file from the JMP Sample Data directory, go to **Help > Sample Data > Open the Sample Data Directory**. For example, opening the data file CrimeData you will see the following on your screen.

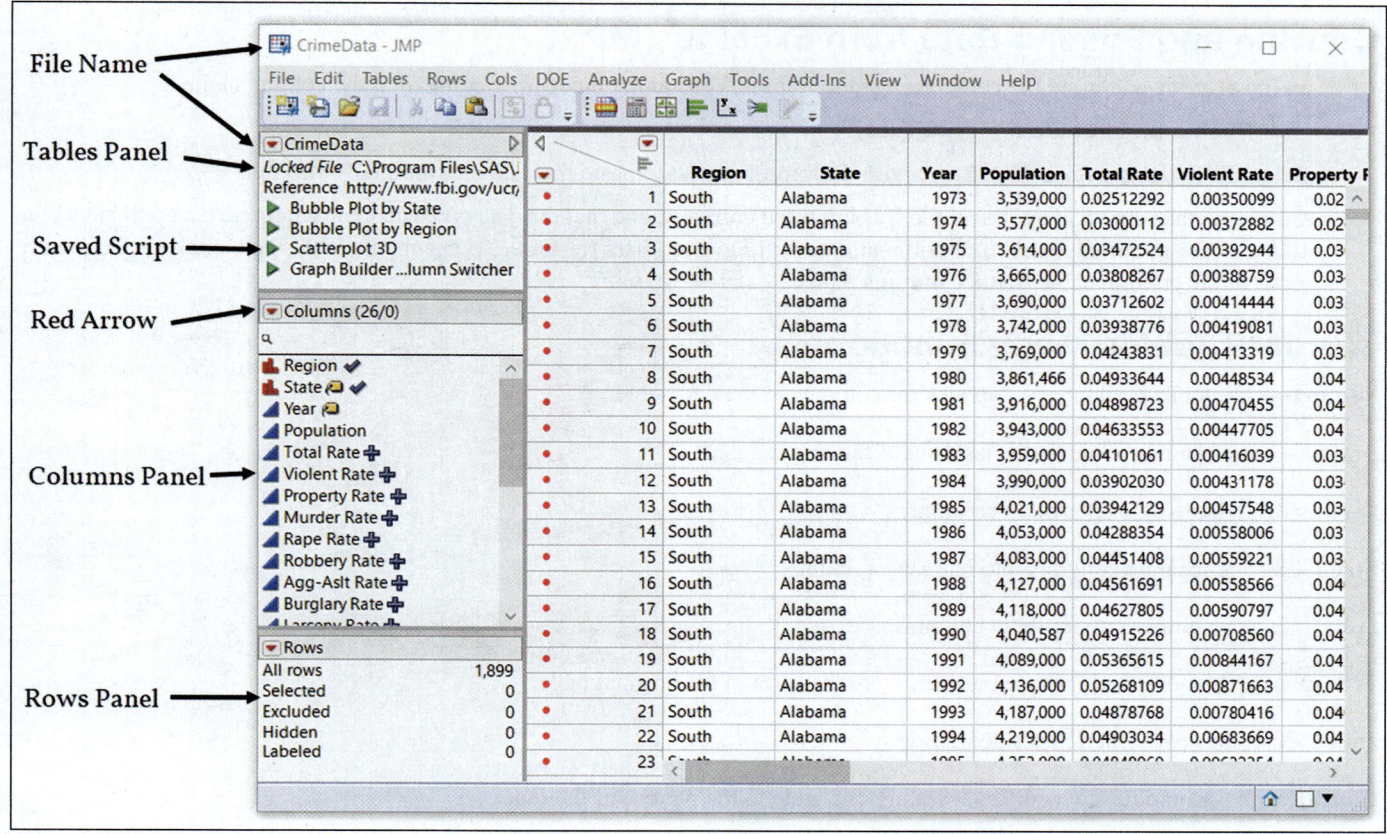

**Tables Panel**: The tables panel contains the data table name and a list of table properties and scripts. To display or open a property, double-click on it. Clicking on a green arrow "plays" a saved script and recreates the analysis.

**Columns Panel**: The columns panel contains the number of columns, the number of selected columns, column names, the type of data in the column (a red bar indicates nominal data, a green bar indicates ordinal data, and a blue triangle indicates continuous data), and column properties. A plus sign after a column name indicates that the column has a stored formula, and you can display the formula by clicking on the plus sign.

**Rows Panel**: The rows panel displays the number of rows (or observations), the number of selected rows, hidden rows (with a mask), excluded rows (with a don't sign), and labeled rows (with a tag). Note that hidden rows will not display on graphs and excluded rows will not be included in most future analyses.

Red arrows are used in JMP to gain access to other commands. **Gray arrows** are used to minimize the display area. Right-clicking in different areas of the data table or a graph provides a list of additional options.

## Importing Data from Excel to JMP

1.  **Select File > Open** and navigate to the directory where your Excel file is stored. JMP will display all the file types that JMP can import (unrecognized files will be grayed out). To see the complete list of file types JMP can read, click on the dropdown arrow to the right of **All JMP Files**.

2. Click **Open** or use the **Excel Wizard** (click the dropdown arrow to the right of **Open** and select **Use Excel Wizard**) to customize the import.

3. If you use the **Excel Wizard**, JMP will determine the file's structure and display the appropriate Text Import window. You can edit the information using the dialog box. Select **Import** at the bottom of the window.

4. If you have more than one worksheet in your Excel file, JMP will import each worksheet into a separate data table.

## Copying and Pasting Data from Excel to JMP

1. In Excel, copy and paste the cells you want to import. You can include 1 row containing the column name.

2. In JMP, click on **File > New > Data Table**.

3. Select **Edit > Paste** (or **Edit > Paste with Column Names**) to paste the data into the JMP data table.

4. Check to make sure that the Excel file was imported correctly, and that the data type is correct (numeric data will have blue triangles for continuous data and nominal data will have red bars for text). To change the data type, click on the icon in front of the column name in the **Columns** panel.

## The JMP Tables Menu Functions

**Summary** – Calculates summary statistics for columns in the data table.

**Subset** – Creates a new table that is a subset of the original table.

**Sort** – Sorts a table by one or more columns.

**Stack** – Stacks separate columns into one new column.

**Split** – Splits a column into multiple columns.

**Transpose** – Creates a table whose columns were the rows in the original table.

**Join** – Combines two data tables side by side.

**Concatenate** – Combines two tables by adding one table to the bottom of the other.

**Missing Data Pattern** – Allows you to explore patterns in missing data.

**Compare Data Tables** – Identifies any differences between two data tables.

## Calculating Summary Statistics for a Set of Data

1. Open or create a JMP data table and select **Tables > Summary**.

2. Select one or more variables from **Select Columns**. Then, click on **Statistics** and select a statistic.

For example, to find the mean and standard deviation for the Reaction Time data we used earlier, with the data table open click on **Tables** in the menu bar and then click on **Summary**. JMP automatically puts Reaction Time as the variable under Select Columns. Click the **Statistics** dropdown arrow on the right and click **Mean** and **Std Dev**. Click **OK**. The results are shown below.

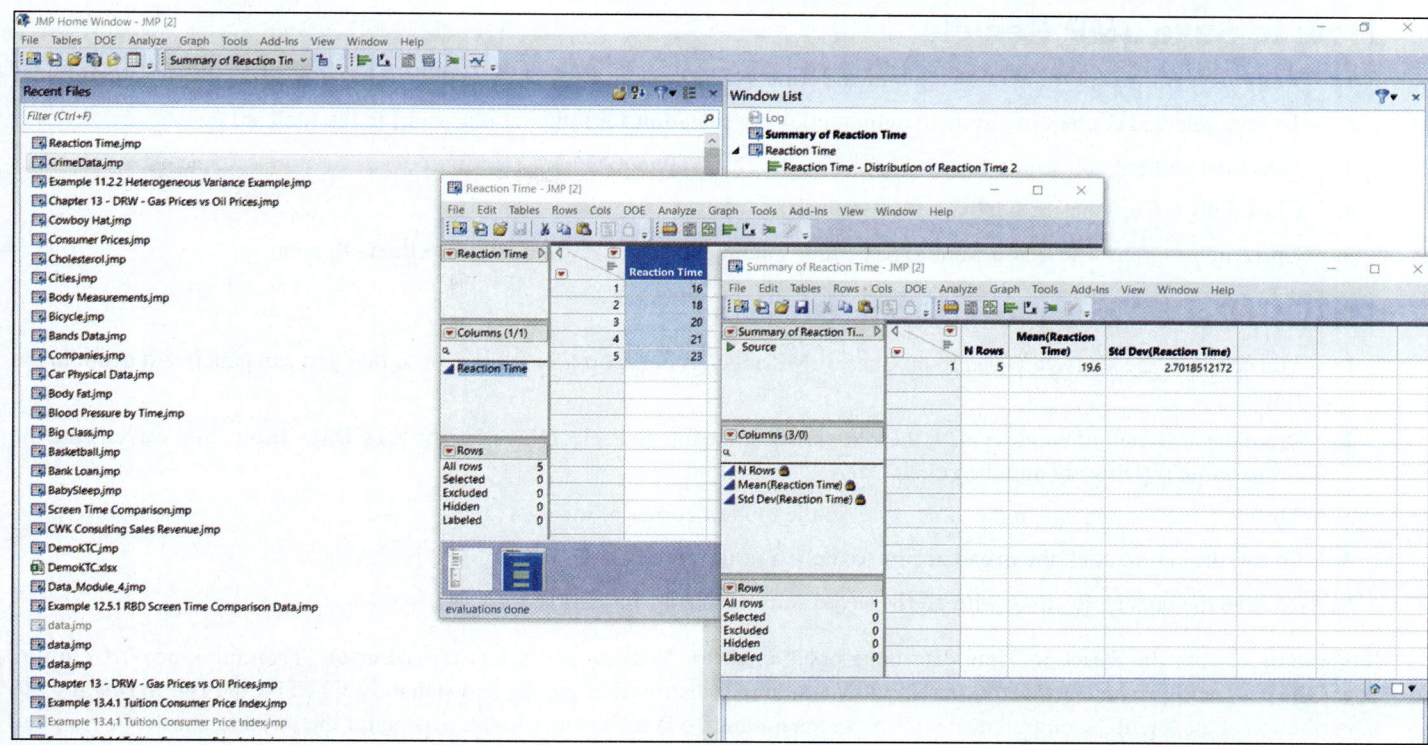

3.  To create a row for each category of a grouping variable, select the grouping variable from **Select Columns** and click **Group**.

4.  Click **OK** to create the summary table. The summary table will have a row for each category of the grouping variable. The column labeled **N Rows** shows the number of rows for each category of the Group variable that was in the original data table.

Let's add a categorical variable to the reaction time data. Add a second column called Test Group to the Reaction Time data table. The values are 'Control' for rows 1 and 2 and 'Treatment' for rows 3-5. Then click on **Tables** in the menu bar and then click on **Summary**. Note that both variables appear under Select Columns in the dialog box that opens. Click on **Reaction Time** in the Select Columns box and then the **Statistics dropdown arrow** on the right and click **Mean** and **Std Dev**. Click on **Test Group** in the Select Columns box and then click **Group**. Click **OK**. The results are shown below.

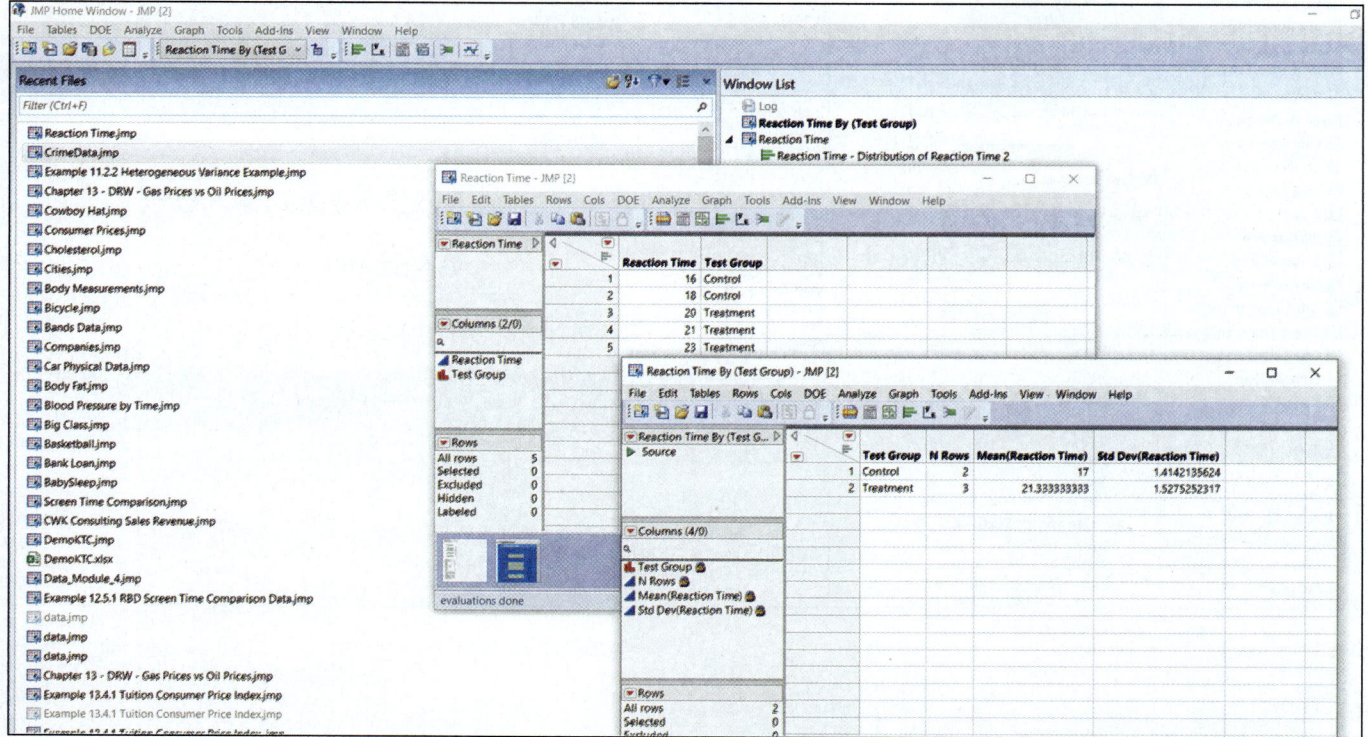

## How to Save JMP Results

1. To save an entire report in various graphical formats (PNG, SVG, EPS, HTML, TIFF, etc.) use **Edit > Save Selection As**.

2. To save selected content of any JMP output, click the **selection tool** (the fat plus sign) in the toolbar.

3. Select the content you would like to copy. (Use the shift key to extend a selection.)

4. Click **Edit > Copy** or use **Ctrl-C**.

5. Open the program where you want to paste the content and select **Paste** or **Paste > Paste Special**.

## How to Save Your Work Using Scripts

1. You can save the steps you used to produce a JMP analysis or report as a JSL script so that you can generate it again in the future.

2. From the JMP output window click the **red arrow** at the top and select **Save Script > To Data Table**. You can change the script name if you want and then click **OK**.

3. The saved script appears in the table panel in the top left corner of the data table.

4. To run the script, click the **green arrow** to the left of the script name.

5. To save the data table along with all the saved scripts, select **File > Save**.

For example, with the Reaction Time data table open, click on **Analyze** and select **Distribution**. Then click on **Y, Columns** and **OK**. This will produce a JMP output window containing distribution graphs and statistics. Click on the **red arrow** next to Distribution at the top of the output, then click **Save Script** and **To Data Table**. Choose a name for the script or use the one created by JMP in the dialog box that opens and click **OK**.

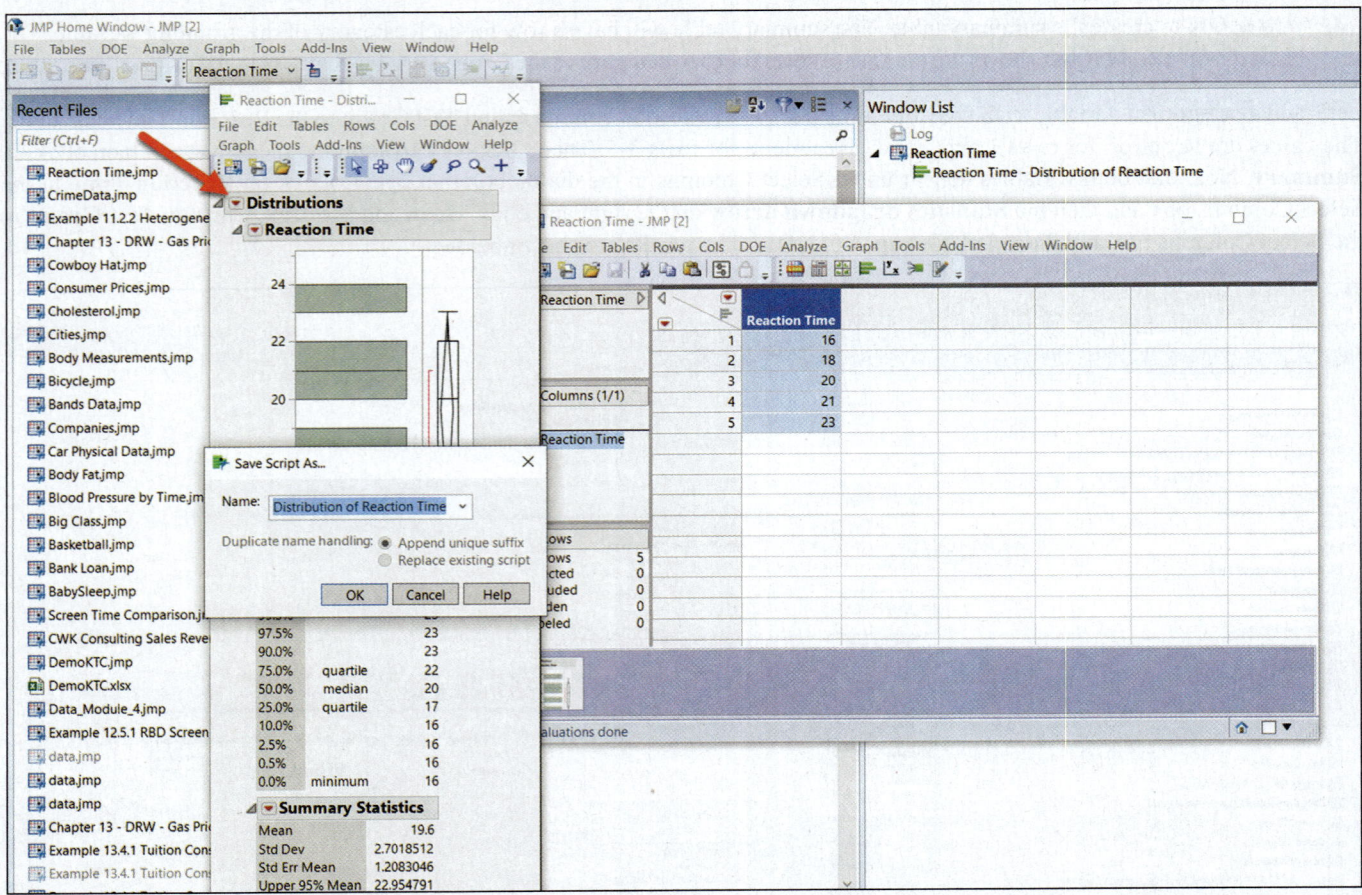

The saved script now appears in the table panel in the top left corner of the data table below the name with a green arrow to the left of it. To save the data table with all the saved scripts, select **File** and then **Save**.

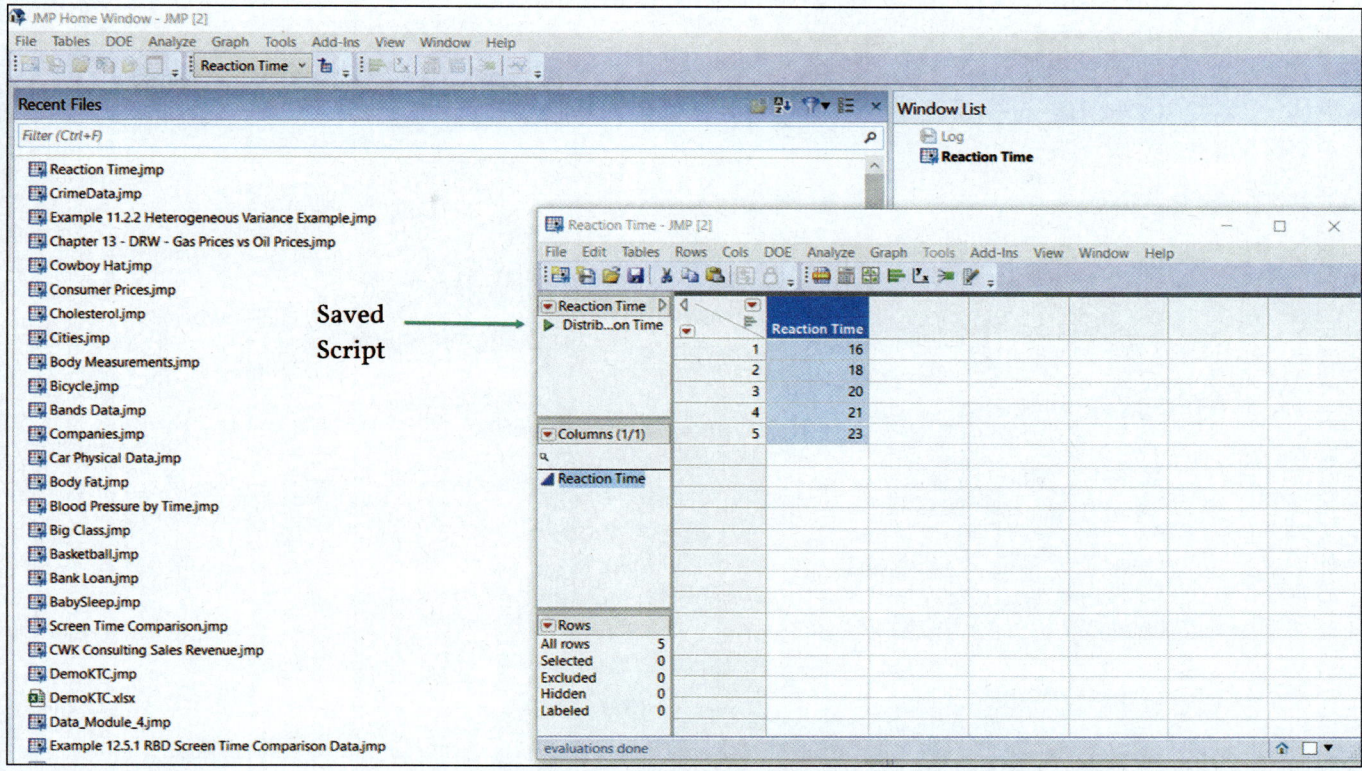

Saved
Script

# Answer Key

## Chapter 1

### Section 1.1

**15.** Recalling the definitions for population and sample, we know that the population is a particular group of interest, and a sample is a subset of the population from which data are collected. In this case, we are looking at the number of billable man-hours logged per week by employees at Deloitte. This is the group of interest, and therefore the statement describes a population.

**17.** Recalling the definitions for population and sample, we know that the population is a particular group of interest, and a sample is a subset of the population from which data are collected. In this case, we are looking at the final rankings of 5 candidates who applied for the open CFO position in your organization. This is a subset of the 22 candidates who applied, and therefore the statement describes a sample.

**19.** Recalling the definitions for population and sample, we know that the population is a particular group of interest, and a sample is a subset of the population from which data are collected. In this case, the particular group of interest is all the employees who work at a company in Silicon Valley. Therefore, the sample is the 35 employees who work at a company in Silicon Valley.

**21.** Recalling the definitions for population parameter and sample statistic, we know that a parameter is a numerical description of a particular population characteristic, and a statistic is the actual numerical description of a particular sample characteristic. Therefore, first consider whether the statement refers to a sample or a population. To be a population parameter, it must describe all members being studied, not just a portion of them.

In this case, the average number of hours is a population parameter because it is based on all of the students in your statistics class.

### Section 1.2

**5.** **a.** Answers will vary.
   **b.** Answers will vary.
   **c.** Answers will vary.

### Section 1.3

**3.** Recalling the definitions of descriptive and inferential statistics, we know that a descriptive statistic gathers, sorts, summarizes, and displays data while an inferential statistic involves using descriptive statistics to estimate population parameters.

In this case, the average price of a car at the new car dealership in town is a descriptive statistic because it describes all of the cars at the new dealership.

### Section 1.4

**3.** **a.** Answers will vary.
   **b.** Answers will vary.
   **c.** Answers will vary.
   **d.** Answers will vary.

### Chapter 1 Additional Exercises

**1.** **a.** Internet users
   **b.** The amount of time users view photos of products on websites
   **c.** Inferential

**3.** **a.** Elderly citizens
   **b.** Self-esteem after reading a news story
   **c.** 276 elderly citizens
   **d.** Answers will vary.
   **e.** Inferential

5.  a.  Mobile phone users
    b.  The percentage of people who use their mobile phones to access the Internet for different reasons
    c.  500 American adults 18 years of age and older
    d.  Answers will vary.
    e.  Inferential

7.  a.  Married couples
    b.  The percentage of married couples that met online
    c.  7000 adults married in the past 5 years
    d.  Answers will vary.

9.  a.  U.S. states with coastlines
    b.  Coastline length

# Chapter 2

## Section 2.1

29. a.  Well-defined
    b.  Well-defined
    c.  Not well-defined
    d.  Well-defined
    e.  Not well-defined

31. There is no well-defined scale to measure cleanliness or aesthetics. Answers will vary.

33. Answers will vary.

35. a.  By randomly assigning women to two groups and using one of the groups as a "control" group, the experiment should produce data that will reveal the impact of the different diets.
    b.  Difference in diet. (The first group received 1200 calorie per day diet for the entire period whereas the second group received 420 calorie per day diet for 16 weeks and then were shifted to 1200 calorie per day diet for the rest of the experimental period.)
    c.  Weight loss.
    d.  Yes, the women receiving the 1200 calorie per day diet represent a control group.
    e.  Observational studies are subject to self-selection bias. We would not necessarily know the cause of the weight reduction.

37. a.  Phase 1: Gather information about the phenomenon being studied.
    b.  Controlled experiment.
    c.  Number of major attacks of Multiple Sclerosis.
    d.  Bovine myelin.
    e.  Fifteen individuals in the early stages of MS fed bovine myelin.
    f.  Fifteen individuals in the early stages of MS given a placebo.

39. Jacob's knee could feel better simply because he took a week break from playing basketball. Answers will vary.

41. There are many factors that affect whether or not someone is happy. Additionally, since both questions require a yes or no reply there is no way to quantify happiness or going to church. Answers will vary.

43. Generally people that have more money will seek the help of a financial advisor since advisors are paid for their services. Answers will vary.

## Section 2.2

9.  Volume – the satellites would collect a huge volume of data since they are continuously monitoring one position.

    Velocity – the data are being collected every second.

    Veracity – the data will be of very good quality and can be trusted since it is automatically collected by a government source.

    Variety – images and numbers are part of the data.

11. Both predictive and prescriptive analytics can be done on these data. Predictive analytics can be used to develop a model to help determine any trends or patterns in the BMI by country and prescriptive analytics can be used to implement or prescribe nutritional programs.

13. Since most people tend to have their mobile phones with them at all times, the marketing company can use the location services in addition to what the user searched for to provide coupons or discounts for products at nearby retail stores. Based on purchasing history and other factors, the marketing company can offer very specific ads to individual users.

15. LBS can be used to prevent credit card fraud by matching the user location from the smartphone to a credit card transaction. Tying the smartphone's location to a credit card allows the credit card company to flag transactions made across several geographic locations over a short time or determine if the card is being used outside of its home zone.

## Section 2.3

**13. a.** Discrete

    **b.** Continuous

    **c.** Continuous

    **d.** Continuous

    **e.** Discrete

**15. a.** Cadet height, weight, state from which the cadet was appointed, father's occupation, parents' income, type of home residence.

    **b.** State from which the cadet was appointed, father's occupation, type of home residence are qualitative variables. Cadet height and weight, parents' income, are quantitative variables.

    **c.** State from which the cadet was appointed – Nominal; Father's occupation – Nominal; Type of home residence – Nominal; Cadet Weight – Ratio; Parent income – Ratio; Cadet Height – Ratio

    **d.** A lot of information is gathered about the cadets; to be able to derive some conclusions from the information, a data summary answering the relevant questions is required.

**17. a.** Quantitative

    **b.** Ratio

    **b.** Nominal

    **c.** Ordinal

**19. a.** Ratio

    **d.** Ratio

## Section 2.4

**9. a.** Yes

    **b.** Nonstationary

**11.** The table contains time series data. Answers will vary.

## Chapter 2 Additional Exercises

**1. a.** Standardized aptitude test, GPA, etc.

    **b.** Number of student suspensions, number of student detentions, etc.

    **c.** Survey students to rate teacher preparedness: Always prepared, Sometimes prepared, Never prepared; or measure the number of times teachers are unprepared on a spot check by the principal.

    **d.** Number of days missed, number of hours missed (for both teachers and students).

    **e.** Survey students, survey teachers: e.g. Rate cafeteria food as Excellent, Very Good, Good, Poor, Very Poor.

**3. a.** Types of investments available, rate of return, amount, etc.

    **b.** Rates of return on various types of investments: Returns for diversified portfolio of stocks: e.g. S&P 500, Dow Jones Index Returns for diversified portfolio of bonds.

**5.** One example: Make a list of standard grocery items for several randomly selected families. Have families buy groceries from one store and then buy exactly the same groceries from another store and then compare prices.

**7.** Survey of customer satisfaction: How would you rate the service you received when purchasing the car: Poor, Average, Good? How would you rate the service after you purchased the car: Poor, Average, Good? How would you rate the car's performance: Poor, Average, Good? How satisfied are you with your purchase: Not Satisfied, Satisfied, Very Satisfied?

**9. a.** Interval

    **b.** Interval

    **c.** Ordinal

    **d.** Nominal

    **e.** Ratio

**11. a.** The level of measurement for millions of barrels of oil per day is ratio.

    **b.** Time series, Nonstationary

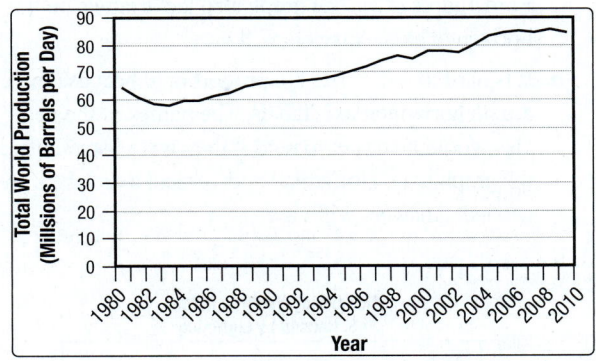

**13.** Answers will vary.

**15. a.** Nominal.

    **b.** Time Series.

    **c.** If respondents were owners or renters, answers will vary.

# Chapter 3

## Section 3.1

7. **a.** Nominal

   **b.** Qualitative

   **c.**

   | | |
   |---|---|
   | Air Conditioner | 16 |
   | Lawn Mower | 10 |
   | Fan | 7 |
   | Washing Machine | 6 |
   | Miscellaneous | 9 |

9. **a.** Answers will vary.

   | Type of Complaint | March | July |
   |---|---|---|
   | Comfort | 17 | 28 |
   | Price | 11 | 15 |
   | Service | 18 | 14 |
   | Schedule | 29 | 33 |

   **b.** Answers will vary.

   | Type of Complaint | March | July |
   |---|---|---|
   | Plane | 22 | 34 |
   | Personnel | 8 | 3 |
   | Building/ Equipment | 17 | 16 |
   | Other | 28 | 37 |

   **c.** No. Another person would not necessarily have assigned the various complaints to the same categories. Results may vary depending on who prepared the data.

   **d.** Yes. Given that you assign a complaint to only one category, the categories are mutually exclusive. Given that you assign each complaint to a category, the categories are exhaustive.

## Section 3.2

11. **a.** Bars are not proportional. The bar for the Dodge Intrepid (27) should be shorter than the bar for the Ford Taurus (28). The graph also lacks labels on the horizontal axis. Answers will vary.

    **b.** It is hard to tell if bars are proportional because there are no horizontal axis labels. The names of the vehicles would be easier to read if they were listed outside the graph area. Horizontal and vertical axis labels are needed. Answers will vary.

13. **a.**

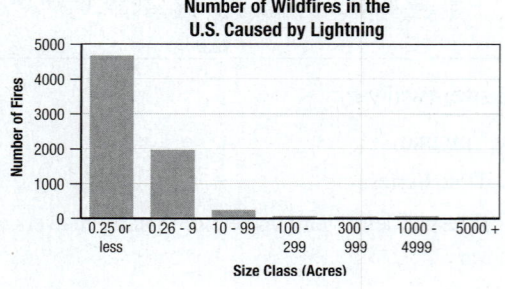

    **b.**

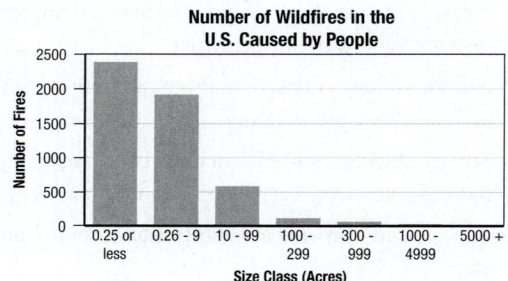

    **c.**

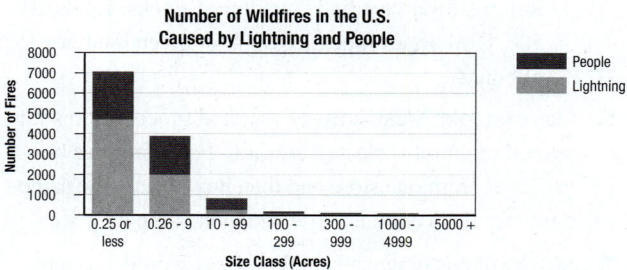

    **d.**

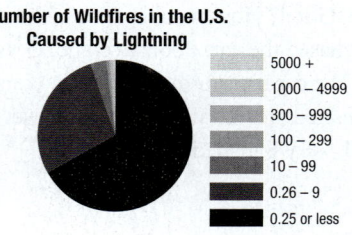

**e.**

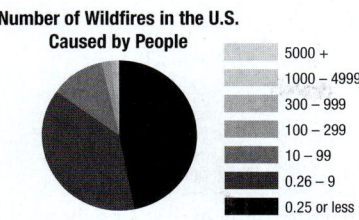

Number of Wildfires in the U.S. Caused by People

5000 +
1000 – 4999
300 – 999
100 – 299
10 – 99
0.26 – 9
0.25 or less

**f.** A majority of fires both lightning-caused and people-caused occur in the 0.25 or less acre size class. For the 0.25 or less acre size class, about twice as many fires are caused by lightning than by people. However, in the 0.26-9 acre size class, the number of wildfires caused by lightning and by people are almost the same. The number of wildfires caused by both lightning and people is much smaller for the size classes above 0.26-9 acres.

**15. a.**

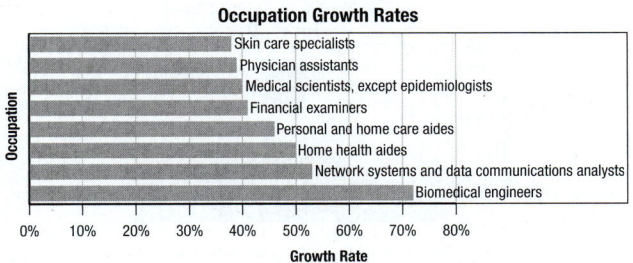

Occupation Growth Rates

**b.** Biomedical engineers have the highest projected growth rate. The rest of the growth rates appear to be between 35% and 55%, whereas the rate for biomedical engineers is above 70%.

## Section 3.3

Solutions given here are only examples of frequency distributions. Students' answers may be different if they choose different classes.

**7.**

| Days Traveling | Frequency | Relative Frequency | Cumulative Frequency |
|---|---|---|---|
| 0 – 6 | 15 | 0.20 | 15 |
| 7 – 13 | 21 | 0.28 | 36 |
| 14 – 20 | 27 | 0.26 | 63 |
| 21 – 27 | 9 | 0.12 | 72 |
| 28 – 34 | 2 | 0.03 | 74 |

| | | | |
|---|---|---|---|
| 35 and above | 1 | 0.01 | 75 |

**9.**

| Average Temp (°F) | Frequency | Relative Frequency | Cumulative Frequency |
|---|---|---|---|
| 40 – 49 | 3 | 0.20 | 3 |
| 50 – 59 | 7 | 0.467 | 10 |
| 60 – 69 | 4 | 0.267 | 14 |
| 70 – 79 | 1 | 0.067 | 15 |

## Section 3.4

**17. a.**

| Closing Price | Frequency |
|---|---|
| $0 – $44.00 | 10 |
| $45.00 – $89.00 | 3 |
| $90.00 – $134.00 | 0 |
| $135.00 – $179.00 | 0 |
| $180.00 – $224.00 | 1 |

**b.**

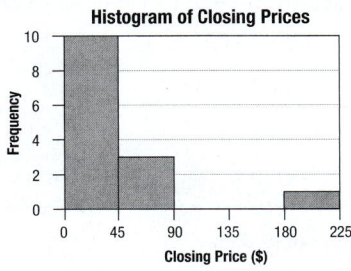

Histogram of Closing Prices

**19. a., b.**

| % of Calories From Fat | Frequency | Relative Frequency |
|---|---|---|
| 15% – 19% | 1 | 0.04 |
| 20% – 24% | 2 | 0.08 |
| 25% – 29% | 5 | 0.20 |
| 30% – 34% | 8 | 0.32 |
| 35% – 39% | 3 | 0.12 |
| 40% – 44% | 3 | 0.12 |
| 45% – 49% | 3 | 0.12 |

**c.**

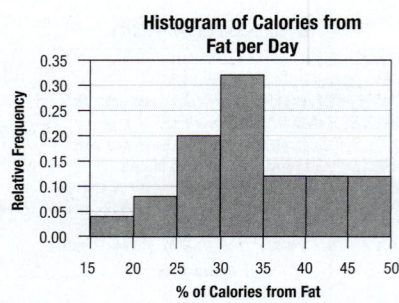

**Histogram of Calories from Fat per Day**

**d.** A majority of the subjects in the sample consumed between 30% and 34.9% of calories from fat per day. The sample percentages appear to have a bell-shaped distribution. Relatively few of the subjects consumed less than 20% of calories per day from fat. Answers will vary.

**21. a.** Ratio

**b.**

| Stem | Leaf |
|---|---|
| 12 | 3 5 5 |
| 13 | 5 5 5 |
| 14 | 7 7 8 8 8 9 |
| 15 | 6 6 6 6 8 8 |
| 16 | 9 |
| 17 | 8 8 9 9 |
| 18 | 9 9 9 |
| 19 | 8 9 9 |
| 20 | |
| 21 | 4 5 5 |
| 22 | |
| 23 | 5 9 |
| 24 | 8 8 9 |
| 25 | 6 7 8 8 8 9 9 9 |
| 26 | 5 8 9 9 |
| 27 | |
| 28 | 8 |

Key: 12 | 3 = $123

**c.** There appear to be two different clusters of daily rates for semi-private rooms. One of the clusters seems to center around $150 per day. The other cluster seems to center around $250 per day. One explanation of this could be the location of the surveyed hospitals, it could be that some of the hospitals surveyed were in large metropolitan areas and some were in smaller cities or suburban areas. Answers will vary.

## Section 3.5

**5.** Answers will vary. A possible answer is that the scale is not a good choice since a dramatic change is shown even though the change is only a few cents.

**23. a.**

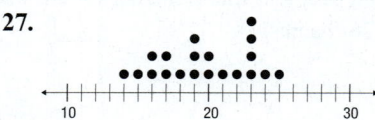

| Stem | Leaf |
|---|---|
| 2 | 3 6 7 8 8 8 9 9 9 |
| 3 | 1 3 4 6 6 7 7 8 |
| 4 | 0 2 6 |

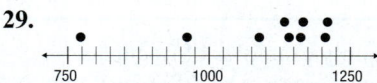

Key: 2 | 3 = 23 (Miles per Gallon)

**b.** The majority of the miles per gallon are between 25 and 38. Miles per gallon ratings above 40 are uncommon. Answers will vary.

**25. a.** −4, 0, 10, 10, 16, 19, 19, 20, 20, 22, 23, 24, 24, 27, 33, 37

**b.** Only 1 quarter showed a loss, and most quarters showed growth between 20% and 30%. Answers will vary.

**27.**

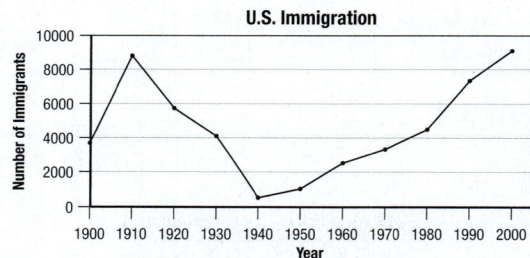

**29.**

**31. a.** All of the rates seem to change in the same way. The longer the period of time, the higher the interest rate charged. Answers will vary.

**b.** Stationary. Answers will vary.

**33. a.** Year – Ordinal; Number – Ratio; Rate – Ratio

**b.**

**U.S. Immigration**

**c.** 146.61%

**d.** −39.62%. Though the number of immigrants has increased drastically, the general population of the U.S. has increased also, so the rate of annual immigration (per 1000 people) has actually decreased from 1900 to 1990. Answers will vary.

**7. a.** Approximately 45%

**b.** No. The area of the graphic for November 2017 has an area more than double that of the November 2010 graphic, even though the value is only 45% more.

**c.** Answers will vary. Ensure that the width of the two graphics is consistent and only alter the height, thus ensuring that the area increase matches the percentage increase.

**9. a.** Answers will vary. Graph A is better because the vertical scale starts at 0.

**b.** Answers will vary. Graph B causes more concern as the increase in robberies appears to be much more dramatic.

**c.** The 2016 bar in Graph B is approximately 8 times taller than the 2013 bar. There were actually only about 1.3 times as many robberies in 2016 than in 2013.

## Chapter 3 Additional Exercises

**1. a.** Time sequence plot or line graph of both the median family income and percent change in median family income.

**b.** Answers will vary.

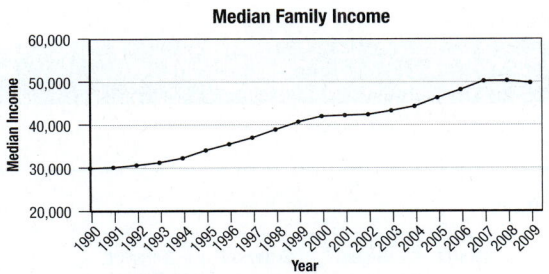

**c.** Median household income has an upward trend, but this could likely be due to inflation. 2009 is the only year in which median household income decreased from the previous year. Answers will vary.

**3. a.** A side-by-side bar chart showing the percentage of both men and women who ranked each company in their top 5. Answers will vary.

**b.**

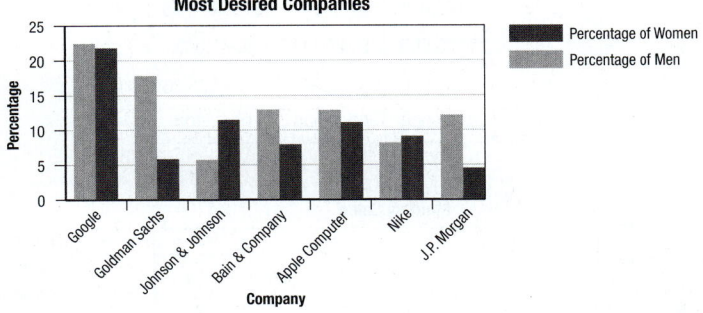

**c.** Google appears to be the most desirable company for both men and women. Goldman Sachs appears to be desirable for men, but not so much for women. Johnson & Johnson appears to be popular among women, but not so much among men. Answers will vary.

**5. a.** Bar charts and pie charts would both be appropriate for displaying the data. Answers will vary.

**b.** Answers will vary.

**c.** Answers will vary.

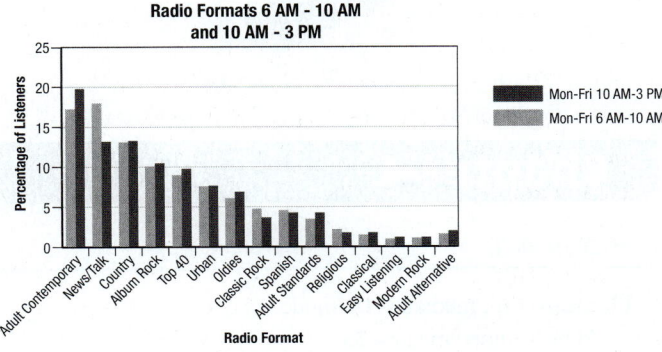

**7. a.** Spending as a percentage of sales is a more useful measure of research and development expenditures for comparative purposes because it standardizes research and development expenditures.

**b.** A bar chart for R&D expenditures, pie charts representing industries, and headquarters locations of the top 20 R&D spenders, a histogram to represent spending as a percentage of sales. Answers will vary.

**c.**

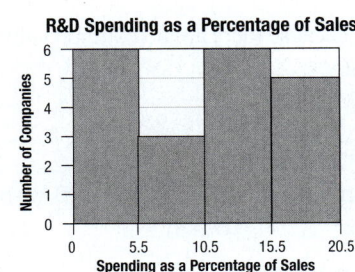

**d.**

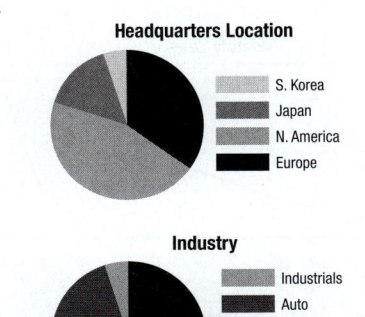

**9. a.** A line graph displaying the percentage of voters in each category and a pie chart showing political identification in a particular year might be helpful in visualizing the data. Answers will vary.

**b.**

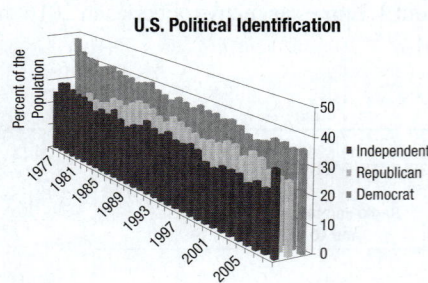

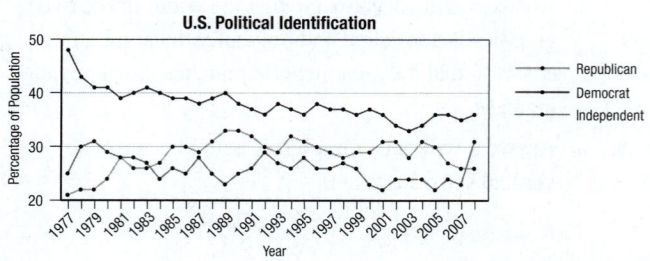

**c.** The time series appears to be stationary. Recently it appears that more people identify themselves as Democrats than Republicans or Independents. Answers will vary.

# Chapter 4

## Section 4.1

**13.** mean = 15, median = 15, mode = 11, 20% trimmed mean = 15

**15. a.** mode

**b.** median (some very high incomes may skew data)

**c.** mean

**d.** median or mode

**17. a.** 92.9667

**b.** 92

**c.** 88

**d.** 92.75

**e.** Mean, because there are no extreme values. Answers may vary.

**19. a.** Ratio

**b.** mean = 99.9

10 % trimmed mean = 107.5625, 20 % trimmed mean = 107.25

**c.** Answers will vary.

**21. a.**

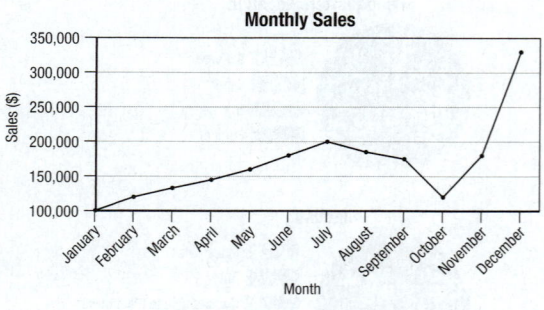

**b., c.**

| Month | Sales | 2-period | 3-period |
|-------|-------|----------|----------|
| Jan | $100,500 | | |
| Feb | $120,000 | $110,250 | |
| Mar | $133,000 | $126,500 | $117,833.33 |
| Apr | $145,000 | $139,000 | $132,666.66 |
| May | $160,000 | $152,500 | $146,000 |
| June | $180,000 | $170,000 | $161,666.66 |
| July | $200,000 | $190,000 | $180,000 |
| Aug | $185,000 | $192,500 | $188,333.33 |
| Sept | $175,000 | $180,000 | $186,666.66 |
| Oct | $120,000 | $147,500 | $160,000 |
| Nov | $180,000 | $150,000 | $158,333.33 |
| Dec | $330,000 | $255,000 | $210,000 |

**d.**

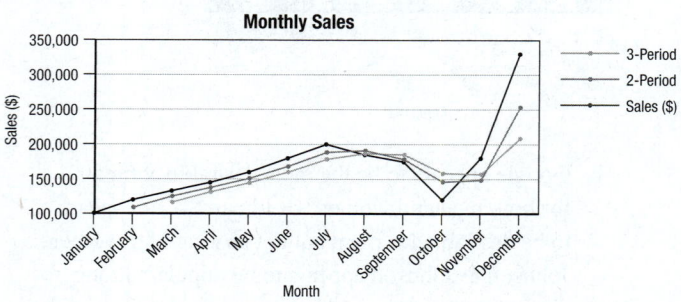

**e.** Answers will vary.

## Section 4.2

**13.** 180

**15. a.** 14.2143

    **b.** 3.7702 feet

    **c.** 11

    **d.** Athletic ability, sex, height, weight, etc. Answers will vary.

**17. a.** both averages = 76.8571

    **b.** male variance = 350.8095

        female variance = 205.8095

    **c.** male std. dev. = 18.7299

        female std. dev. = 14.3461

    **d.** Average scores are the same for males and females but the standard deviation of scores of females is lower, implying more consistent scores.

    **e.** Answers will vary.

**19. a.** mean of original data = 89.4167

        std. dev. of original data = 7.1916

        mean of adjusted data = 109.4167, std. dev. of adjusted data = 7.1916

    **b.** mean of adjusted data = mean of original data + 20; the standard deviations are the same.

    **c.** These results hold in general. If you add a constant to each data point, the mean of the adjusted data is equal to the mean of the original data + the constant. The standard deviation of the adjusted data is the same as the standard deviation of the original data (i.e., the way in which the data vary does not change if you simply add a constant to each data point).

**21.** 5 to 47

**23. a.** 68

    **b.** 95

    **c.** In order to use the empirical rule, we must assume that the distribution of the amounts of chowder eaten is approximately bell-shaped.

**25. a.** Job grade 25: \$19,000 to \$25,000

        Job grade 33: \$31,000 to \$39,000

        Job grade 40: \$35,000 to \$55,000

    **b.** In order to use the empirical rule, we must assume that the distribution of the salaries is approximately bell-shaped.

**27. a.** 12

    **b.** 10

    **c.** Since the coefficient of variation is smaller for Machine Y than Machine X, the standard deviation in the diameter of the bolts produced is smaller relative to the average diameter for Machine Y than for Machine X. Thus, Machine Y more consistently produces bolts of the correct diameter.

## Section 4.3

**11. a.** 1

    **b.** 6

    **c.** 25th percentile: approximately 25% of the salespeople sold one or fewer copiers in a day.

        90th percentile: approximately 90% of salespeople sold 6 or fewer copiers in a day.

    **d.** 83rd percentile

    **e.** 33rd percentile

**13. a.** .185

    **b.** .255

    **c.** .285

    **d.** 1st quartile: approximately 25% of the batting averages are at or below .185,

        2nd quartile: approximately 50% of the batting averages are at or below .255,

        3rd quartile: approximately 75% of the batting averages are at or below .285.

    **e.** 0.1

    **f.** 0.02 is an outlier. This likely represents the pitcher's batting average.

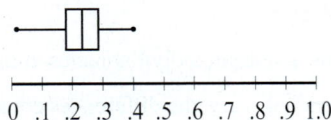

    0  .1  .2  .3  .4  .5  .6  .7  .8  .9  1.0

    **g.** −2.61

    **h.** 1.14

    **i.** The player with the .020 batting average has a batting average 2.61 standard deviations below the mean. The player with a .330 batting average has a batting average 1.14 standard deviations above the mean.

    **j.** 70th percentile

    **k.** 15th percentile

**15.** First exam: $z = 0.10$, Second exam: $z = -0.43$. On the first exam, the student's score was 0.1 standard deviations above the mean; but on the second exam, the student's score was 0.43 standard deviations below the mean. Thus, although the student achieved a higher absolute score on the second exam, the student performed relatively better on the first exam than on the second.

## Section 4.4

3. **a.** Nominal: Beer ID, Beer Name, Beer Style, Brewery ID, Brewery Name, City, and State; Ratio: ABV, IBU, and Ounces

   **b.** Style, Ounces, Brewery Name/ID, and State.

   **c.** 8

   **d.** Elevation Triple India Pale Ale, Renegade Brewing Company.

   **e.** 2, American IPA and American Double/Imperial IPA

   **f.** mean=0.059, st. deviation = 0.013

   **g.** CVRenegade=28.16%. CVWynkoop=21.30%; The Wynkoop Brewery has more consistent ABV values since it has a smaller coefficient of variation.

5. **a.** The mean is 18,065.79, the mode is 3,060, the median is 7,026.

   **b.** The variance is 381,922,239 the standard deviation is 19,542.83, and the range is 74,876.

   **c.** The 1st quartile is 2,898.75, the 3rd quartile is 37,712.75

   **d.** Age, age group, gender, ethnicity

   **e.**

   | Age Group | Average Expenditure |
   |-----------|---------------------|
   | 0 to 5    | 1415.28             |
   | 6 to 12   | 2226.86             |
   | 13 to 17  | 3922.61             |
   | 18 to 21  | 9888.54             |
   | 22 to 50  | 40209.28            |
   | 51+       | 53521.90            |

   **f.** The highest average Expenditure occurs in the 51+ age group.

   It seems that as an individual ages, more is spent.

   Answers will vary, the differences could be because as the person ages, they may no longer have family to care for them. Therefore, public institutions may have to assume the cost of their care. Also, as a person ages, perhaps there are more health issues to deal with that will increase the amount spent on that age group.

   **g.** The standard deviation and coefficient of variation for each age group are listed below.

   | Age Group | Standard Deviation | CV    |
   |-----------|--------------------|-------|
   | 0 to 5    | 612.6              | 43.3% |
   | 6 to 12   | 830.9              | 37.3% |
   | 13 to 17  | 1012.7             | 25.8% |

   | 18 to 21 | 2940.6 | 29.7% |
   |----------|--------|-------|
   | 22 to 50 | 6287.3 | 15.6% |
   | 51+      | 6283.8 | 11.7% |

   The highest level of dispersion is in the lowest age group.

   Answers will vary, the differences could be the result of greater levels of care needed at younger ages when trying to determine the appropriate level of service needed.

   **h.**

   | Ethnicity          | Expenditures |
   |--------------------|--------------|
   | American Indian    | 36438.25     |
   | Asian              | 18392.37     |
   | Black              | 20884.59     |
   | Hispanic           | 11065.57     |
   | Multi Race         | 4456.73      |
   | Native Hawaiian    | 42782.33     |
   | Other              | 3316.50      |
   | White not Hispanic | 24697.55     |

   **i.** Divide the average Expenditure for each Ethnicity by the total Expenditure to get the proportion for each group as shown below.

   | Ethnicity          | Proportion |
   |--------------------|------------|
   | American Indian    | 22.5%      |
   | Asian              | 11.4%      |
   | Black              | 12.9%      |
   | Hispanic           | 6.8%       |
   | Multi Race         | 2.8%       |
   | Native Hawaiian    | 26.4%      |
   | Other              | 2.0%       |
   | White not Hispanic | 15.2%      |

   **j.** Answers will vary. There seems to be some inequity by age and by ethnicity. The reasons for the inequity is unknown. Also, there is much more variability in the age group 6-12 in terms of spending.

7. **a.** The mean is 138,528.40, the mode is 93,896.18, the median is 127,850.10.

   **b.** The variance is 2,292,419,896, the standard deviation is 47,879.22, and the range is 447,679.40.

   **c.** The 1st quartile is 102,031.16, the 3rd quartile is 167,463.53

**d.** Base pay, total pay, whether there was any overtime, and benefits. The data could also be arranged by job title, although there are several job titles listed.

**e.** The relative frequency per category should be determined. The table could look like this:

| Class | Frequency | Relative Frequency |
|---|---|---|
| 0-25000 | 0 | 0.00% |
| 25001-50000 | 775 | 3.47% |
| 50001-75000 | 8054 | 36.06% |
| 75001-100000 | 6023 | 26.97% |
| 100001-125000 | 4244 | 19.00% |
| 125001-150000 | 2073 | 9.28% |
| 150001-175000 | 505 | 2.26% |
| 175001-200000 | 508 | 2.27% |
| 200001-225000 | 105 | 0.47% |
| 225001-250000 | 24 | 0.11% |
| 250001-275000 | 14 | 0.06% |
| 275001-300000 | 4 | 0.02% |
| 300001-350000 | 5 | 0.02% |

**f.** The highest group is the 75,000. Between 75,000 up to 125,000 makes up about 82% of the salaries for the data set. Only a small percentage is below 75,000 and above 175,000. The differences could be because of experience levels or the specific type of job. When the data were sorted by salaries, public service aides were among the lowest paid in the group. The higher paying jobs were managers and department heads.

## Section 4.5

**3.** Mean = 17.5 days, Variance = 55.3043 days$^2$

**5. a.** 0.061

## Section 4.6

**5. a.** 0.5

**b.** 0.3415

**c.** 0.4661

**d.** 0.6341

**e.** It appears that the supplier executives are optimistic, while the original equipment managers appear to be skeptical about the economic recovery. Answers will

## Section 4.7

**11. a.** Yes, the pattern roughly follows a straight line. The pattern is upward sloping; as $x$ increases, $y$ generally increases. The data values are widely dispersed. The fanning of the data could be considered a significant deviation from the pattern.

**g.** Approximately 60% of the jobs have OT pay listed. (No OT = 9010 rows)

**h.**

| | Mean | Standard Deviation | CV |
|---|---|---|---|
| No OT | 100711.41 | 39013.68 | 39% |
| OT | 105395.36 | 41734.74 | 40% |

Both have very similar dispersion about the mean. In addition, the averages are quite close.

**i.** Answers will vary. For example, the pay one could expect seems to be between about 75,000 and 125,000 based on the data set. Seventy-five percent (75%) of people are getting paid more than approximately 102,000 in total pay. However, this of course is contingent upon the position and other individual factors of the applicant such as experience and degree. Also, it seems that most of the positions do have OT associated with the job (60%). But that OT doesn't provide a huge difference in total pay for the individual. Finally, it must be noted that this data is from 2014 and is quite out of date at this point. So, this analysis can provide some perspective, but it is limited in its applicability to the job market today.

**b.** 0.0002

**c.** 0.014

vary.

**7. a.** 28%

**b.** 22%

**c.** Americans are not in the habit of saving money in case of a financial emergency. Around a quarter would not be able to cope at all. Answers will vary.

**9. a.** 835      **b.** 2884

**b.** Yes, the pattern roughly follows a straight line. The pattern is downward sloping; as $x$ increases, $y$ decreases. The data values are tightly clustered; in fact, they exactly fall in a straight line. There are no significant deviations from the pattern.

13. **a.** Yes, the data collected seem to be appropriate to study the relationship between training and countertop defects. An example of a bias could be that some employees could have more experience than others aside from the training. The data are collected using an observational study. Answers will vary.

**b.**

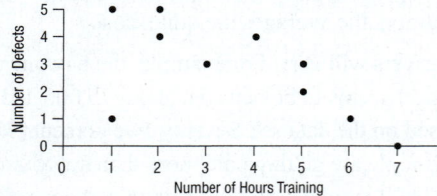

Effect of Training Time on Number of Defects

**c.** Yes, the pattern roughly follows a straight line. The pattern is downward sloping; as the number of hours of training increases the number of defects tends to decrease. The data values are tightly clustered. There is one significant deviation from the pattern: the employee with only one hour of training who had only one defect.

15. Answers may vary, but points should be plotted in a downward sloping, perfectly straight line.

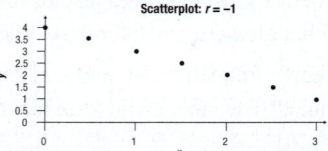

Scatterplot: $r = -1$

17. **a.** Tightly clustered in a positive linear fashion.

   **b.** Loosely clustered in a positive linear fashion.

   **c.** Tightly clustered in a negative linear fashion.

   **d.** Loosely clustered in a negative linear fashion.

   **e.** Loosely clustered in a positive linear fashion.

19. **a.**

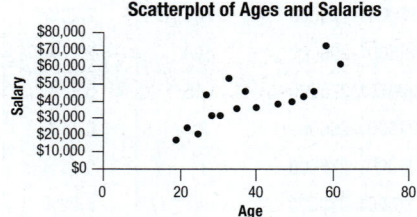

Scatterplot of Ages and Salaries

   **b.** $r = 0.7707$

   **c.** The correlation coefficient indicates a moderate positive linear relationship. This seems consistent with the scatterplot.

21. **a.** Summer

   **b.** Valentine's Day, Christmas

   **c.** Winter

## Chapter 4 Additional Exercises

1. No. Because you did not want any variation in the length of the boards. Answers will vary.

3. **a.** 0.3438

   **b.** Mean = 2.4531, Standard Deviation = 2.5754

   **c.** 0 to 5.0285

   **d.** 87.5%

   **e.** The empirical rule predicts that 68.26% of the data falls within one standard deviation of the mean. The percent of the data in this problem falling within one standard deviation is 87.5%, which is not very close to the empirical rule. Answers may vary.

5. **a.** Machine A = 3.1429, Machine B = 3.1429

   **b.** Machine A = 5.8095, Machine B = 0.8095

   **c.** Machine A = 2.4103, Machine B = 0.8997

   **d.** Machine B is probably a better machine because the average number of defects produced by the 2 machines is the same but Machine B is much more consistent in the number of defective circuit boards it produces. Answers may vary.

7. **a.** 92.0929

   **b.** 9.1383

   **c.** 9

   **d.** 12

   **e.** The literacy rates are normally distributed.

9. **a.** Yes, the variables measured seem appropriate to study the relationship between reaction time and the amount of drug in the bloodstream. Biases could include differences in reaction time without the drug. The data are ratio data. Answers will vary.

   **b.**

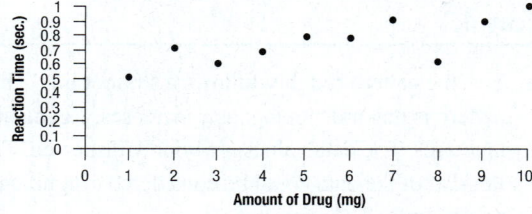

Scatterplot of Drug Amounts and Reaction Times

c. Yes, the pattern roughly follows a straight line. The pattern is upward sloping; as the amount of drug increases, the reaction time tends to increase. The data values are tightly clustered. There is one significant deviation from the pattern: the person who was administered 8 mg of the drug had a reaction time of only 0.6 sec. Answers will vary.

11. **a.** Strong positive linear relationship

**b.** Weak positive linear relationship

**c.** Strong negative linear relationship.

**d.** Weak negative linear relationship.

**e.** Weak positive linear relationship.

13. **a.**

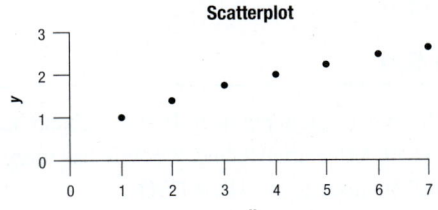

**b.** $r = 0.9916$

**c.** There is a strong positive relationship between $x$ and $y$. However, the scatterplot does not appear to be linear. It appears that $y$ is the square root of $x$. Answers will vary.

# Chapter 5

## Section 5.1

21. **a.** $S =$ {Very Attentive, Somewhat Attentive, Not Attentive}

**b.** $A =$ {Somewhat Attentive, Not Attentive}

23. $P(\text{Yellow}) = 0.3$, $P(\text{Red}) = 0.5$, $P(\text{Blue}) = 0.2$

25. 0.6667

27. 0.375

**d.** relative frequency

**e.** classical

29. **a.** classical

**b.** subjective

**c.** subjective

**d.** classical

**e.** subjective

**f.** relative frequency

31. **a.** $\dfrac{1}{5} = 0.2$

**b.** $\dfrac{4}{5} = 0.8$

## Section 5.2

7. **a.** Yes

**b.** No, probabilities cannot be greater than 1.

**c.** Yes

**d.** No, probabilities cannot be less than 0.

**e.** Yes

9. **a.** The event cannot occur.

**b.** The event is certain to occur.

**c.** Relative frequency interpretation: If an experiment is performed 100 times, the event will occur, on average, 45 times.

**d.** Relative frequency interpretation: If an experiment is performed 100 times, the event will occur, on average, 65 times.

**e.** Not a valid probability because it is negative.

11. **a.** 0.9696           **b.** 0.0523

13. **a.** 0.27

**b.** 0.08

**c.** 0.165

**d.** 0.4

**e.** 0.3

**f.** 0.51

**g.** Relative frequency

**h.** No, the wife could have more than $150,000 in insurance and the husband could have between $50,000 and $100,000 of insurance. Answers may vary.

## Section 5.3

3. **a.** 0.3194

**b.** 0.3472

**c.** 0.6327

**d.** 0.6032

5.  **a.** 0.8571

   **b.** 0.1429

## Section 5.4

7.  No, the events are dependent. If $A$ = husband has more than \$150,000 insurance and $B$ = wife has more than \$50,000 insurance, $P(A|B) \neq P(A)$.

9.  0.0001

11. **a.** 0.9980

## Section 5.5

5.  $P(\text{Dem}|\text{Favor}) = 0.7404$

7.  $P(\text{Def}|\text{Insp}) = 0.2069$

## Section 5.6

5.  120

7.  **a.** 1

   **b.** 6

   **c.** 120

## Chapter 5 Additional Exercises

1.  **a.** $S$ = {MMM, MMF, MFM, MFF, FMM, FMF, FFM, FFF}

   **b.** $\dfrac{1}{8} = 0.125$

   **c.** $\dfrac{7}{8} = 0.875$

3.  **a.** $\dfrac{18}{38} = \dfrac{9}{19} \approx 0.4737$     **b.** $\dfrac{12}{38} = \dfrac{6}{19} \approx 0.3158$

7.  **a.** 0.005

   **b.** 0.0417

   **b.** 0.0020

   **c.** 0.000001

13. **a.** 0.0004

   **b.** 0.0004

9.  $P(\text{Woman's name}|\text{Man chosen}) = 0.7347$

   **d.** 5040

9.  665,280

11. 50,400

13. 55

   **c.** $\dfrac{2}{38} = \dfrac{1}{19} \approx 0.0526$

   **d.** $\dfrac{1}{38} \approx 0.0263$

   **e.** $\dfrac{35}{38} \approx 0.9211$

5.  0.9989

7.  **a.** 1 to 5

   **b.** 1 to 1

   **c.** 7 to 1

   **d.** $\dfrac{8}{11} \approx 0.7273$

9.  $\dfrac{11}{14} \approx 0.7857$

11. **a.** 8%

   **b.** 16%

   **c.** 14%

   **d.** 84%

# Chapter 6

## Section 6.1

5.  **a.** Discrete

   **b.** Continuous

   **c.** Discrete

   **d.** Continuous

   **e.** Discrete

7.  **a.** Discrete

   **b.** Continuous

   **c.** Continuous

   **d.** Discrete

   **e.** Continuous

## Section 6.2

13. Yes

15. No. The sum of the probabilities is less than 1.

17. No. Probabilities cannot be negative.

19. Yes

| x | P(X = x) |
|---|---|
| 1 | $\frac{1}{30}$ |
| 2 | $\frac{4}{30}$ |
| 3 | $\frac{9}{30}$ |
| 4 | $\frac{16}{30}$ |

**21.**

| x | p(x) | xp(x) | $(x - \mu)^2 p(x)$ |
|---|---|---|---|
| 400 | 0.0 | 0 | 0.0 |
| 420 | 0.1 | 42 | 291.6 |
| 440 | 0.1 | 44 | 115.6 |
| 460 | 0.2 | 92 | 39.2 |
| 480 | 0.2 | 96 | 7.2 |
| 500 | 0.4 | 200 | 270.4 |
| Total | 1.0 | 474 | 724.0 |

$E(X) = 474$

$\sigma^2 = 724$

$\sigma = 26.9072$

**23.**

| x | p(x) | xp(x) | $(x - \mu)^2 p(x)$ |
|---|---|---|---|
| 1 | 0.1 | 0.1 | 0.484 |
| 2 | 0.2 | 0.4 | 0.288 |
| 3 | 0.3 | 0.9 | 0.012 |

## Section 6.3

**5. a.** HH1, HH2, HH3, HH4, HH5, HH6, HT1, HT2, HT3, HT4, HT5, HT6, TH1, TH2, TH3, TH4, TH5, TH6, TT1, TT2, TT3, TT4, TT5, TT6

**b.** $X$ = Sum of the number of heads on the two coins and number of dots on the die.

$X = \{1, 2, 3, 4, 5, 6, 7, 8\}$

**c.**

| x | P(X = x) |
|---|---|
| 1 | $\frac{1}{24}$ |
| 2 | $\frac{3}{24}$ |
| 3 | $\frac{4}{24}$ |
| 4 | $\frac{4}{24}$ |

## Section 6.4

**7. a.** 5

**b.** 45

| 4 | 0.2 | 0.8 | 0.128 |
|---|---|---|---|
| 5 | 0.2 | 1.0 | 0.648 |
| Total | 1.0 | 3.2 | 1.560 |

**a.** $E(X) = 3.2$

**b.** $\sigma^2 = 1.56$

**c.** $\sigma = 1.2490$

**d.** $P(X = 5) = 0.2$

**e.** $P(X \geq 2) = 0.9$

**f.** $P(X \leq 3) = 0.6$

**g.** $P(X < 2) = 0.1$

**25. a.**

| x | p(x) | xp(x) | $(x - \mu)^2 p(x)$ |
|---|---|---|---|
| $50,000 | 0.4 | 20,000 | 518,400,000 |
| −$10,000 | 0.6 | −6000 | 345,600,000 |
| Total | 1.0 | 14,000 | 864,000,000 |

**b.** $14,000

**c.** $29,393.88

**27. a.** Cereal A = $200,000; Cereal B = $276,000.

**b.** Cereal A = $188,414.40; Cereal B = $309,619.10

**c.** Cereal B has a greater value for expected sales, but also a much greater standard deviation. The difference in the expected sales is much smaller than the difference in the standard deviation, so Cereal A is probably the best choice. Answers will vary.

| 5 | $\frac{4}{24}$ |
|---|---|
| 6 | $\frac{4}{24}$ |
| 7 | $\frac{3}{24}$ |
| 8 | $\frac{1}{24}$ |

**d.** $E(X) = 4.5$

**7.** $\frac{2}{10} = 0.2$

**9. a.** $\{H1, H2, H3, H4, H5, H6, T1, T2, T3, T4, T5, T6\}$

**b.** $\frac{1}{12} \approx 0.0833$

**c.** $\frac{3}{12} = 0.25$

**c.** 15

**d.** 1

9.  **a.** $E(X) = 0.9$

    **b.** $\sigma = 0.9$

    **c.** $P(X = 2) = 0.1722$

    **d.** $P(X \le 3) = 0.9917$

    **e.** $P(X \ge 2) = 0.2252$

    **f.** $P(X < 5) = 0.9991$

11. **a.** Binomial distribution with $n = 10$ and $p = 0.10$

    **b.** $E(X) = 1$

    **c.** $\sigma = 0.9487$

    **d.** $P(X = 1) = 0.3874$

    **e.** $P(X = 5) = 0.0015$

    **f.** $P(X \ge 3) = 0.0702$

13. **a.** Binomial distribution with $n = 7$ and $p = 0.1$

    **b.** $P(X = 0) = 0.4783$.
    There is a 47.83% chance that none of the plants will strike.
    $P(X = 4) = 0.0026$. There is a 0.26% chance that

exactly 4 of the plants will strike.

$P(X = 7) = 0$.

There is a negligible chance that all 7 plants will strike.

    **c.** $E(X) = 0.7$

    **d.** $\sigma = 0.794$. The standard deviation is larger than the expected value. The standard deviation is expressed as the number of plants that strike. Answers may vary.

15. **a.** $P(X = 2) = 0.375$

    **b.** $P(X = 4) = 0.0625$

17. **a.** $P(X \le 1) = 0.8290$

    **b.** $E(X) = 0.75$

19. **a.** $\dfrac{2}{9}$ or $0.2222$

    **b.** $P(X = 5) = 0.0389$

    **c.** $P(X = 0) = 0.0810$

    **d.** $E(X) = 2.2222$,

    $\sigma^2 = 1.7284$

## Section 6.5

7.  $P(X = 2) = 0.0842$

9.  **a.** $P(X = 0) = 0.1353$

    **b.** $P(X = 0) = 0.0003$

    **c.** $\mu = \lambda = 2$

    **d.** $\mu = \lambda = 8$

    **e.** $\sigma = 2.8284$

    **f.** $P(X \ge 4) = 0.9576$

11. **a.** $\lambda = 20$ (If 5 people arrive on average in 15 minutes, then 20 will arrive on average in 60 minutes.)

    **b.** $P(X = 0) = 0$

    **c.** $P(X > 6) = 0.2378$

13. $P(X = 6) = 0.0771$

## Section 6.6

5.  **a.** $X$ has a hypergeometric distribution with $N = 50$, $k = 3$ and $n = 10$.

    **b.** $E(X) = 0.6$

    **c.** $\sigma = 0.6785$

    **d.** $P(X \ge 1) = 0.4959$

    **e.** $P(X \le 2) = 0.9939$

    **f.** $P(X > 3) = 0$

7.  **a.** $E(X) = 5$

    **b.** $\sigma = 1.5076$

    **c.** $P(X = 10) = 0.0006$

    **d.** $P(X = 0) = 0.0006$

## Chapter 6 Additional Exercises

1.  **a.** $E(X) = 2.3$

    **b.** $\sigma^2 = 1.41$

    **c.** $\sigma = 1.1874$

    **d.** $P(X = 4.0) = 0.15$

    **e.** $P(X \ge 2.0) = 0.75$

    **f.** $P(X \le 1.0) = 0.25$

    **g.** $P(X > 3.0) = 0.15$

3.  **a.** $P(4 \le X \le 6) = 0.6563$

    **b.** $P(X \ge 8) = 0.0547$

    **c.** $P(X = 1) = 0.0098$

5.  **a.** $E(X) = 1.6667$, $\sigma^2 = 1.3889$

    **b.** $E(X) = 3.25$, $\sigma^2 = 2.4375$

    **c.** $E(X) = 8.8$, $\sigma^2 = 1.056$

    **d.** $E(X) = 0.8333$, $\sigma^2 = 0.4419$

    **e.** $E(X) = 3.5$, $\sigma^2 = 2.9167$

7.  $P(X \ge 1) = 0.9615$

9.  $\$0.25$

11. $P(X \ge 1) = 0.9933$

13. $P(X \geq 3) = 0.2962$

b. $P(X = 3) = 0.25$

15. a. Binomial distribution with $n$ = the number of buildings inspected and $p = 0.5$. The binomial is used rather than the hypergeometric because it is not known how many buildings in the population of new buildings have violations. Answers may vary.

# Chapter 7

## Section 7.1

7. a. $\mu = 60$
   b. $\sigma = 1.7321$
   c. 0.3333
   d. 0.5
   e. 0.1667
   f. 0

9. a. $\mu$ = 8:15 am
   b. $\sigma = 0.1443$
   c. 0.3333
   d. 0.1667
   e. 0.5
   f. 0

## Section 7.2

9.

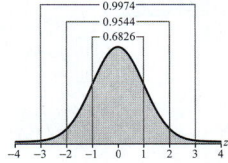

11.

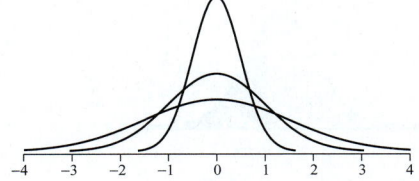

## Section 7.3

5. The data seem to fit a line very closely. We would assume from the normal probability plot that the population is normally distributed.

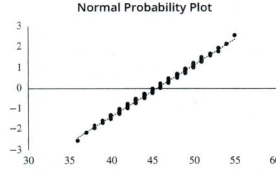

7. By examining the normal probability plot, we notice a substantial deviation from a linear pattern. The data do not appear to be normally distributed.

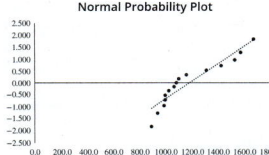

9. By examining the normal probability plot, we notice a substantial deviation from a linear pattern. The data do not appear to be normally distributed.

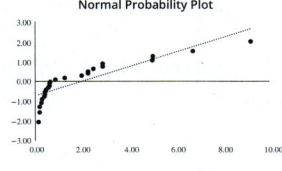

11. The data in this problem are not from a normal distribution. The histogram allows one to make a decision more easily.

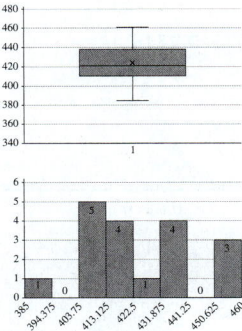

13. When the data set is small the box plot and the histogram are not definitive in assessing normality. Care needs to be taken when dealing with statistics not to jump to conclusions. An interesting read is *How to Lie With Statistics* by Darrell Huff.

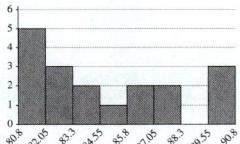

## Section 7.4

**5. a.** 0.2486

**b.** 0.4500

**c.** 0.4750

**d.** 0.4950

**7. a.** 0.6680

**b.** 0.6710

**c.** 0.9631

**d.** 0.9422

**9. a.** $P(z \le -0.44) = 0.3300$

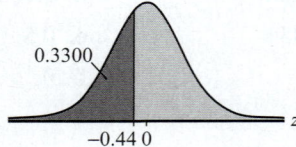

**b.** $P(z \ge 0.44) = 0.3300$

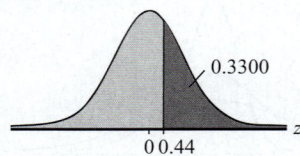

**c.** $P(-0.44 \le z \le 0.44) = 0.3400$

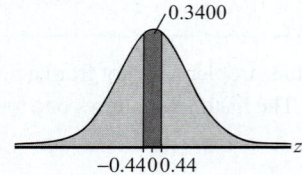

**d.** $P(z \le -0.67) = 0.2514$

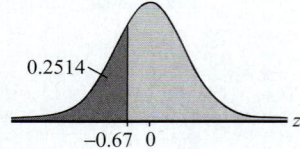

**e.** $P(z \ge 0.67) = 0.2514$

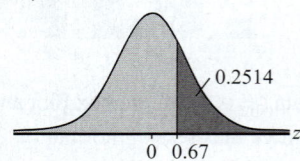

**f.** $P(-0.67 \le z \le 0.67) = 0.4972$

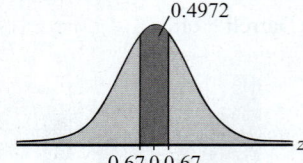

**11. a.** $P(0 \le z \le 0.79) = 0.2852$

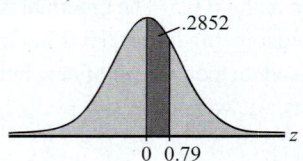

**b.** $P(-1.57 \le z \le 2.33) = 0.9319$

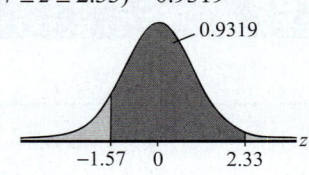

**c.** $P(z \ge 1.89) = 0.0294$

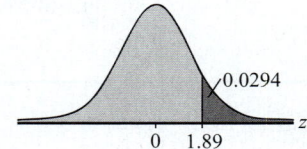

**d.** $P(z \le -2.77) = 0.0028$

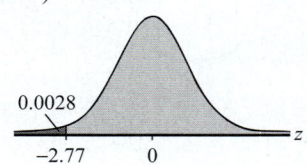

**13.** 1.645

**15.** 1.28

**17.** −2.33

**19.** 1.14

**21.** 1.645

**23. a.** 0.7333

**b.** 0.0548

**c.** 0.0228

**25.** 0.4101

**27. a.** $631

**b.** 0.2206

**c.** 0.1190

**d.** 0.4235

**e.** Answers will vary.

**29. a.** 0.0668

**b.** 0.0668

**c.** 0.6826

**31. a.** 92.24

**b.** No. The score must be at least a 92.24 to be in the top 10% of scores.

**c.** 75.28

**d.** The student who scored a 65 would receive an F because the score is less than 71.76 and thus is in the lowest 10% of the scores.

**33. a.** At least 131

## Section 7.5

**5. a.** 45

**b.** 2.1213

**c.** 0

**d.** 0.9830

**e.** 0.0011

**7. a.** 120

**b.** 6.9282

**c.** 0

**d.** 0.5264

**e.** 0.0853

**9. a.** 2.2361

**b.** 0.13%

**c.** 6.5%

**d.** At least 135

**b.** 0.0318

**c.** 0.0222

**d.** 0.0318 is the more accurate probability. When using the normal approximation, the probability is underestimated. Answers will vary.

**11. a.** 30

**b.** 5.4772

**c.** 0.9977

**d.** 0.5359

**e.** 0.8413

## Chapter 7 Additional Exercises

**1. a.** 20

**b.** 8.6603

**c.** 0.1667

**d.** 57.73%

**e.** According to the empirical rule 68% of the results will fall within one standard deviation of the mean. The discrepancy is due to the fact that the empirical rule applies to distributions that are bell-shaped. The uniform distribution is not bell-shaped.

**3. a.** $P(0 \leq z \leq 1.00) = 0.3413$

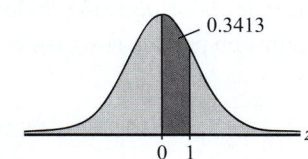

**b.** $P(-2.50 \leq z \leq 3.01) = 0.9925$

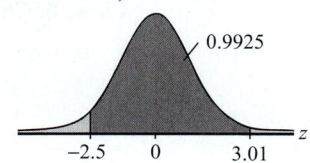

**c.** $P(z \geq 3.25) = 0.0006$

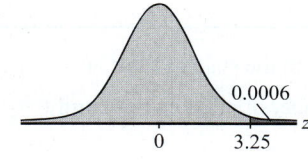

**d.** $P(z \leq -2.50) = 0.0062$

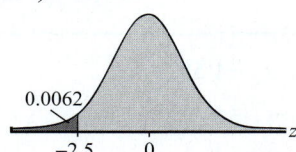

**5.** 1.28

**7. a.** 5

**b.** 2.2349

**c.** Using the normal approximation: 0.9778; Using tables: 0.9933

**d.** Using the normal approximation: 0.9418; Using tables: 0.9596

**e.** Yes, $np = 5$, which is almost too small for the approximation to the binomial to produce accurate results. Answers may vary.

**9.** $\mu \approx 507, \sigma \approx 114$

**11.** 535 days

**13.** $\mu \approx 5.71$ ml

**15.** The bulbs should be replaced after approximately 397 hours.

**17.** $\mu = \$61,000, \sigma = \$12,000$

# Chapter 8

## Section 8.1

**13. a.** Yes, readers voluntarily sent in their responses.

**b.** People with strong opinions would be more likely to reply and the categories are not specific so people could have different ideas of what "Frequently" means, for example. Answers will vary.

**c.** No. All Americans do not read the magazine, and even if they did, it is likely that only those readers with strong opinions would have responded to the survey.

**15. a.** Only students with very strong opinions may have responded, and the response categories are not well-defined. Answers will vary.

**b.** No. Only 150 surveys were mailed, and it is possible that only students with strong opinions responded. Answers will vary.

**c.** No. Nothing about business majors is mentioned so we don't know how many, if any, business majors responded to this survey. Answers will vary.

**d.** Instead of using a voluntary survey they could have surveyed all students in a particular class, for example. Answers will vary.

**e.** A question that asked whether or not the participant would enroll in the program if it were created, for example. Answers will vary.

**17. a.** Residents of the community

**b.** Post office, DMV, or the IRS, for example. Answers will vary.

**c.** Not everyone has a car so some would be excluded from the DMV frame. People move often so the frame obtained from the post office could be inaccurate. The IRS may be able to provide a list of tax payers for that community. This would probably be the best option since the politician is concerned about opinions on a tax increase. Answers will vary.

## Section 8.2

**11. a.** $\mu_{\bar{x}} = 50$, $\sigma_{\bar{x}} = 1.5811$

**b.** $\mu_{\bar{x}} = 50$, $\sigma_{\bar{x}} = 1.3484$

**c.** $\mu_{\bar{x}} = 50$, $\sigma_{\bar{x}} = 1$

**13.** 0.0174

**15. a.** 1      **b.** 1

**c.** 1

**d.** 0.0793

**17. a.** 0.2912

**b.** 0.0006

**19. a.** 0.0016

**b.** 0.0016

**c.** 0.9232

**21. a.** 0.2981

**b.** 0.0170

**c.** 0.6261

## Section 8.3

**7. a.** 0.3

**b.** 0.7

**9. a.** $\mu_{\hat{p}} = 0.35$,

$\sigma_{\hat{p}} = 0.0774$

**b.** $\mu_{\hat{p}} = 0.35$,

$\sigma_{\hat{p}} = 0.0661$

**c.** $\mu_{\hat{p}} = 0.35$,

$\sigma_{\hat{p}} = 0.0551$

**d.** The standard deviation decreases, reflecting the additional information provided by a larger sample size.

**11.** 0.1151

**13. a.** 0.8461

**b.** 0.0011

**c.** 0.4793

**d.** 0.4989

**15. a.** 0.0023

**b.** 0.0384

**17. a.** 0.9767

**b.** 0.6536

## Section 8.4

**9.** A systematic sample is not a random sample because it has potential for bias if there is some inherent data pattern.

**11. a.** Families in the state of Florida

**b.** The number of children per family

**c.** Ratio

**d.** Simple random sampling: develop sampling frame, choose participants randomly from sampling frame; Cluster sampling: create clusters, randomly select clusters, survey all members of the chosen clusters; Stratified sampling: create strata, randomly select participants from each stratum such that the population characteristics are adequately represented. Answers will vary.

**e.** Cluster sampling, because travel costs would likely be minimized. Answers will vary.

**13. a.** Convenience sampling

**b.** All of the people surveyed may not be residents of Orlando, Florida because there are a large number of tourists visiting the area.

**c.** No, the sample was not representative of the population of interest. Answers will vary.

## Chapter 8 Additional Exercises

**1. a.** Voluntary sampling

**b.** All Americans do not watch the news program. It is likely that only those with strong opinions responded.

**c.** No. Answers will vary.

**3.** 0

**5. a.** Convenience sample

**b.** Pre-med majors may be over-represented since there are more pre-med majors in a biology class than in most other classes. Answers will vary.

**c.** No, because the sample is biased.

**7. a.** $\mu_{\bar{x}} = 15\%$,
$\sigma_{\bar{x}} = 6.957\%$

**b.** 8.043% to 21.957%

**9. a.** 0.0031

**b.** Not necessarily. The samples may not have been representative of the population and employees may have not been honest about the amount of time they spend texting at work. Answers will vary.

**c.** Survey respondents might not be completely truthful when answering the survey since it is a representation of job performance. Answers will vary.

**11. a.** 0.0132

**b.** 0

**c.** No. Noise in excess of 103 decibels only occurs 1.32% of the time.

**13. a.** $\mu_{\hat{p}} = 0.90$, $\sigma_{\hat{p}} = 0.0134$

**b.** $\mu_{\hat{p}} = 0.85$, $\sigma_{\hat{p}} = 0.0179$

**c.** 0.8638

**d.** 0.9750

**15. a.** 0.9951

**b.** 0.8788

**c.** 0.1212

**17. a.** $P(\bar{x} > 35) \approx 1$, indicating that the researchers are likely correct in claiming that the U.S. average is greater than 35 hours (i.e., 50.4 hours).

**b.** $\sigma \approx 16.5$

# Chapter 9

## Section 9.1

**25. a.** 1.96

**b.** 2.575

**c.** 1.645

**29.** (237, 263)

**31.** (4.7, 5.3)

**27. a.** 2.33

**b.** 1.88

**c.** 1.75

**33.** (4964, 5036).

**35.** (29,232, 30,768)

**37.** (8.5, 10.7)

## Section 9.2

**15.** 2.518

**17. a.** 2.201

**b.** 2.898

**c.** 1.721

**19.** (77.3579, 83.3887)

**21. a.** (5.6, 7.2)

**b.** We are 95% confident that the true average length of stay for the hospital's abdominal surgery patients is between 5.6 days and 7.2 days. We are assuming that the lengths of stay are approximately normally distributed.

**23. a.** (101.36, 128.64)

    **b.** We are 90% confident that the true average price of a regular room with a king size bed in the resort community is between \$101.36 and \$128.64. We are

assuming the prices are normally distributed.

**25.** $n = 98$

**27.** $n = 31$

## Section 9.3

**9. a.** 65%

    **b.** (62.52, 67.48)

**11. a.** (0.2176, 0.3324)

    **b.** No, 0.40 falls outside the confidence interval.

**13. a.** (0.3856, 0.5644)

    **b.** Yes, 0.665 falls above the interval.

**15. a.** (0.0625, 0.1475)

    **b.** Yes, 0.05 falls below the interval.

**17.** $n = 637$

**19.** $n = 208$

## Section 9.4

**5. a.** (0.0557, 0.0941) We are 95% confident that the standard deviation of the bolt diameters is between 0.0557 inch and 0.0941 inch.

    **b.** The diameters of the bolts have an approximately normal distribution.

**7. a.** (0.2841, 0.4608) We are 90% confident that the standard deviation of the share prices of the bond fund is between \$0.28 and \$0.46.

    **b.** The share prices of the bond fund have an approximately normal distribution.

**9. a.** (5.1561, 7.2954) We are 80% confident that the standard deviation of the life of the touch screens is between 5.16 months and 7.30 months.

    **b.** The life of the touch screens as measured by the consumer advocacy group has an approximately normal distribution.

## Chapter 9 Additional Exercises

**1.** $n = 27$

**3. a.** $n = 68$

    **b.** (0.2598, 0.3402) We are 95% confident that the true proportion of Fontana residents who think safety is a significant factor in their decision about whether or not to ride a bus is between 0.2598 and 0.3402.

**5.** (9.56, 9.94)

**7.** $n = 1038$

**9.** (29,630, 32,570)

**11.** (621, 679)

**13.** (0.3965, 0.4835)

# Chapter 10

## Section 10.1

**15.** $H_0: p = 0.47$, $H_a: p > 0.47$

**17.** $H_0: p = 0.29$, $H_a: p \neq 0.29$

**19.** $H_0: \mu = 10.9$, $H_a: \mu \neq 10.9$

**21.** $H_0: \mu = 56.8$, $H_a: \mu < 56.8$

**23. a.** $H_0: \mu = 30,000$, $H_a: \mu > 30,000$

    Type I error: They will risk needing to replace tires.
    Type II error: The company will research ways to

make their tires last longer, even though that may be unnecessary.

    **b.** $H_0: \mu = 240$, $H_a: \mu > 240$

    Type I error: Mrs. Russell doesn't research ways to improve the bar hooks even though she may need to.
    Type II error: Mrs. Russell will do research to improve the bar hooks, even though that may be unnecessary.

## Section 10.2

**5. a.** $z = -2.33$

    **b.** $z = 1.28$

    **c.** $z = 1.96$ and $-1.96$

**7.** $H_0: \mu = 55$, $H_a: \mu \neq 55$, Critical values $= -1.96, 1.96$, $z = -8.95$, Reject $H_0$.

9. **a.** $\mu$ = the average number of days which the checks are late (or early).

   **b.** $H_0: \mu = 0, H_a: \mu > 0$,

   **c.** $z = 8.57$

   **d.** Reject the null hypothesis if the calculated value of $z$ is greater than or equal to 1.645.

   **e.** Yes

   **f.** There is sufficient evidence that the veterans organization's complaints are warranted and the checks arrive later than the 10th of the month, on average.

11. **a.** All computer systems sold by the retail computer store.

   **b.** Service costs of the systems in the second year of operation.

   **c.** Ratio

   **d.** $H_0: \mu = 50, H_a: \mu \neq 50$
   Critical values $= -1.645, 1.645, z = -10.39$, Reject $H_0$.

## Section 10.3

7. **a.** $t = -2.624$

   **b.** $t = 1.328$

   **c.** $t = 2.365$ and $-2.365$

9. $H_0: \mu = 10, H_a: \mu \neq 10$
   Critical values $= -2.571, 2.571, t = 1.337$, Fail to reject $H_0$.

11. $H_0: \mu = 0.5, H_a: \mu < 0.5$
   Critical values $= -1.440, t = -1.542$,
   $0.05 < P\text{-value} < 0.10$ (tables), $P\text{-value} = 0.0863$ (exact), Reject $H_0$.

13. **a.** The patients of the Sisters of Mercy Hospital.

   **b.** Yes. $H_0: \mu = \$1240, H_a: \mu > \$1240$
   Critical values $= 1.328, t = 3.282$, Reject $H_0$.

   **c.** The daily charges for patients of Sisters of Mercy Hospital have an approximately normal distribution.

15. **a.** No. $H_0: \mu = 5, H_a: \mu < 5$
   Critical values $= -2.583, t = -0.825$,
   Fail to reject $H_0$.

   **b.** The times customers spend watching the in-store video have an approximately normal distribution.

17. **a.** No. $H_0: \mu = 8, H_a: \mu > 8$
   Critical values $= 1.833, t = 1.581$, Fail to reject $H_0$.

   **b.** The times required to install 130 square feet of bathroom tile have an approximately normal distribution.

**e.** Answers will vary.

13. **a.** Yes. $H_0: \mu = 5, H_a: \mu < 5$, Critical value $= -2.33$, $z = -36.51$, Reject $H_0$.

   **b.** 4.99 lb

15. No. $H_0: \mu = 5, H_a: \mu > 5$
   Critical value $= 1.28, z = 0.47$, Fail to reject $H_0$.

17. **a.** IRS customers

   **b.** Yes. $H_0: \mu = 45, H_a: \mu > 45$
   Critical value $= 1.28, z = 13.33$, Reject $H_0$.

19. **a.** Reject $H_0$

   **b.** Reject $H_0$

   **c.** Fail to reject $H_0$

   **d.** Reject $H_0$

21. **a.** $P\text{-value} = 0.0228$, Fail to reject $H_0$.

   **b.** $P\text{-value} = 0.0071$, Reject $H_0$.

   **c.** $P\text{-value} = 0.0070$, Reject $H_0$.

19. **a.** No. $H_0: \mu = 9, H_a: \mu < 9$
   Critical values $= -1.318, t = -0.75$, Fail to reject $H_0$.

   **b.** The diameters of the pellet patterns have an approximately normal distribution.

21. **a.** $P\text{-value} = 0.0086$, Reject $H_0$.

   **b.** $P\text{-value} = 0.0233$, Fail to reject $H_0$.

   **c.** $P\text{-value} = 0.0652$, Fail to reject $H_0$.

23. **a.** 0.0314

   **b.** Yes, there is sufficient evidence to support the claim that the boots can remain immersed for more than 12 hours without leaking.

25. The test is statistically significant since the null hypothesis that the average daily growth of the shrub is equal to 1 cm per day was rejected. However, it is unlikely that a difference of 0.10 cm growth will be noticeable to the untrained eye. Answers will vary.

27. The test is not statistically significant because the conclusion was to fail to reject the null hypothesis that the average time customers watch the video is equal to 5 minutes The test is practically significant because it lets the store know that customers are spending time watching the new in-store video. Answers will vary.

## Section 10.4

**3. a.** (40.09, 44.91), Reject $H_0$

**b.** Heights of children are approximately normally distributed.

**5. a.** (1120.82, 1213.86)

**b.** No

**7. a.** Since the hypothesized value of 18.27 falls outside the given confidence interval (20.36, 26.54), we reject the null hypothesis at a 5% level of significance. There is sufficient evidence to conclude that the population average completion time is different from 18.27 minutes.

**b.** Since the hypothesized value of 24.96 falls within the given confidence interval (20.36, 26.54), we fail to reject the null hypothesis at a 5% level of significance. There is not sufficient evidence to conclude that the population average completion time is different from 24.96 minutes.

**c.** Since the hypothesized value of 29.53 falls outside the given confidence interval (20.36, 26.54), we reject the null hypothesis at a 5% level of significance. There is not sufficient evidence to conclude that the population average completion time is different from 29.53 minutes.

**9.** Since the hypothesized value of 5020 falls within the confidence interval (4964.6, 5036), we fail to reject the null hypothesis at a 1% level of significance. There is not sufficient evidence to conclude that the true mean breaking strength of the metal link chain is different from 5020 pounds.

**11. a.** (5.6, 7.2)

**b.** Since the hypothesized value of 5.4 falls outside the confidence interval (5.6, 7.2), we reject the null hypothesis at a 5% level of significance. There is sufficient evidence to conclude that the true mean length of stay for patients having abdominal surgery is different from 5.4 days.

## Section 10.5

**5. a.** $z = -1.645$

**b.** $z = 2.33$

**c.** $z = 1.645$ and $-1.645$

**7.** Yes. $H_0: p = 0.38, H_a: p > 0.38$

Critical values $= 2.33, z = 4.04$, Reject $H_0$.

**9. a.** Yes. $H_0: p = 0.013, H_a: p \neq 0.013$

Critical values $= -1.96, 1.96, z = 2.75$, Reject $H_0$.

**b.** Teenagers may not be honest when answering a survey like this. Answers will vary.

**11. a.** No. $H_0: p = 0.20, H_a: p < 0.20$

Critical values $= -2.33, z = -1.77$, Fail to reject $H_0$.

**b.** No, $np_0 \geq 5$ and $n(1 - p_0) \geq 5$.

**13. a.** No. $H_0: p = 0.05, H_a: p > 0.05$

Critical value $= 1.28, z = -0.46$, Fail to reject $H_0$.

**b.** Yes, $np_0 \geq 5$ and $n(1 - p_0) \geq 5$, but $np_0 = 5$, which could be a concern. Answers may vary.

**15.** No. $H_0: p = 0.15, H_a: p > 0.15$

Critical value $= 2.33, z = 0.31$, Fail to reject $H_0$.

**17. a.** No. $H_0: p = 0.002, H_a: p > 0.002$

Critical value $= 1.645, z = 0.12$, Fail to reject H0

**b.** 0.4522

**c.** No

**19. a.** $H_0: p = 0.40, H_a: p < 0.40$

$z = -0.73$, P-value $= 0.2327$

**b.** No

**21.** Yes. $H_0: p = 0.32, H_a: p < 0.32$

Critical value $= -1.645, z = -2.12$, Reject $H_0$.

**23. a.** Yes. $H_0: p = 0.49, H_a: p > 0.49$

Critical value $= 1.645, z = 3.82$, Reject $H_0$.

**b.** 110 people

## Section 10.6

**3. a.** $df = 19, \chi^2 = 36.191$

**b.** $df = 23, \chi^2 = 35.172$

**c.** $df = 4, \chi^2 = 14.860$

**5. a.** Yes. $H_0: \sigma^2 = 0.0025, H_a: \sigma^2 > 0.0025$

Critical value $= 42.557, \chi^2 = 56.84$,
P-value $= 0.0015$, Reject $H_0$.

**b.** The diameters of the bolts have an approximately normal distribution.

**7. a.** Yes. $H_0: \sigma^2 = 0.0625, H_a: \sigma^2 > 0.0625$

Critical value $= 42.980, \chi^2 = 47.04$,
P-value $= 0.0033$, Reject $H_0$.

**b.** The share prices of the bond fund have an approximately normal distribution.

## Chapter 10 Additional Exercises

1.  **a.** $H_0: \mu = 18, H_a: \mu \neq 18$

    **b.** The company believes that the average time to replace a set of 4 tires has changed when in fact the average time is unchanged.

    **c.** The company believes that the average time to replace a set of 4 tires remains unchanged when in fact the average time has changed.

3.  Yes. $H_0: \mu = 3.5, H_a: \mu < 3.5$

    Critical value $= -1.28, z = -4.34$, Reject $H_0$.

5.  Yes. $H_0: \mu = 13.20, H_a: \mu < 13.20$

    $z = -2.53$, P-value $= 0.0057$, Reject $H_0$.

7.  No. $H_0: \mu = 895, H_a: \mu > 895$

    Critical value $= 1.28, z = 1.19$, Fail to reject $H_0$.

9.  **a.** No. $H_0: p = 0.003, H_a: p < 0.003$

    Critical value $= -1.645, z = -1.42$, Fail to reject $H_0$.

    **b.** 0.0778

    **c.** Yes, $H_0$ would be rejected.

11. No. $H_0: p = 0.2632, H_a: p > 0.2632$

    Critical value $= 1.645, z = 0.90$, Fail to reject $H_0$.

13. **a.** P-value $= 0.0062$, Reject $H_0$.

    **b.** P-value $= 0.0256$, Fail to reject $H_0$.

    **c.** P-value $= 0.0002$, Reject $H_0$.

    **b.** Delivery times are approximately normally distributed.

15. **a.** P-value $= 0.0464$, Reject $H_0$.

    **b.** P-value $= 0.0050$, Reject $H_0$.

    **c.** P-value $= 0.0510$, Fail to reject $H_0$.

17. Yes. $H_0 : \sigma^2 = 0.00156, H_a : \sigma^2 > 0.00156$

    Critical value $= 118.498, \chi^2 = 134.615$, Reject $H_0$.

19. **a.** No. $H_0: p = 0.90, H_a: p > 0.90$

    Critical value $= 1.28, z = 1.25$, Fail to reject $H_0$.

    **b.** 0.1056

21. Yes. $H_0: \sigma^2 = 0.01, H_a: \sigma^2 < 0.01$

    Critical value $= 3.325, \chi^2 = 1.44$, Reject $H_0$.

## Chapter 11

### Section 11.1

7.  **a.** $z = -1.75$

    **b.** $z = 1.41$

    **c.** $z = 2.33$ and $-2.33$

9.  **a.** $(-7.50, -2.50)$ We are 95% confident that Mr. Ellis' expenses are between \$2.50 and \$7.50 less than Mr. Ford's.

    **b.** Yes. $H_0: \mu_1 - \mu_2 = 0, H_a: \mu_1 - \mu_2 \neq 0$,

    Critical values $= -1.96, 1.96, z = -3.93$, Reject $H_0$.

    **c.** The confidence interval only contains negative values indicating that with 95% confidence the expenses for Mr. Ellis will always be less than those of Mr. Ford.

11. **a.** Yes. $H_0: \mu_1 - \mu_2 = 0, H_a: \mu_1 - \mu_2 < 0$,

    Critical value $= -1.28, z = -1.283$, Reject $H_0$.

    **b.** P-value $= 0.0997$

    **c.** Yes, we would fail to reject $H_0$ at $\alpha = 0.05$.

### Section 11.2

5.  **a.** $df = 23, t = -1.714$

    **b.** $df = 18, t = 1.330$

    **c.** $df = 10, t = 3.169$ and $-3.169$

7.  **a.** $(-8.19, -2.21)$ We are 95% confident that the Dodge Grand Caravan ES takes between 8.19 and 2.21 fewer seconds to accelerate from 0 to 60 mph.

    **b.** Yes. $H_0: \mu_1 - \mu_2 = 0, H_a: \mu_1 - \mu_2 \neq 0$,

    Critical values $= -2.048, 2.048, t = -3.560$,

    Reject $H_0$.

    **c.** An independent experimental design is used, both populations are approximately normal, population variances are equal.

9.  **a.** $(-3.19, 1.19)$ We are 99% confident that the hourly wage in City A is between \$3.19 lower and \$1.19 higher than in City B.

    **b.** No. $H_0: \mu_1 - \mu_2 = 0, H_a: \mu_1 - \mu_2 \neq 0$,

    Critical values $= -2.024, 2.024, t = -1.240$,

    Fail to reject $H_0$.

    **c.** P-value $= 0.2226$

    **d.** An independent experimental design is used, both populations are approximately normal, population variances are equal.

**11. a.** An independent experimental design is used, both populations are approximately normal, population variances are equal.

   **b.** Yes. $H_0: \mu_1 - \mu_2 = 0, H_a: \mu_1 - \mu_2 > 0$,

     Critical value $= 1.725$, $t = 1.868$, Reject $H_0$.

                 **13.**     **a.**     No. $H_0: \mu_1 - \mu_2 = 0, H_a: \mu_1 - \mu_2 > 0$,

   Critical value $= 2.764$, $t = 0.585$, Fail to reject $H_0$.

**b.** An independent experimental design is used, both populations are approximately normal, population variances are equal.

   **c.** $H_0: \mu_1 - \mu_2 = 0, H_a: \mu_1 - \mu_2 > 0$,

     Critical value $= 2.764$, $t = 0.585$, Fail to reject $H_0$.

   **d.** The results of the hypothesis test did not change. Answers will vary.

## Section 11.3

**7. a.** $df = 11$, $t = -3.106$

   **b.** $df = 4$, $t = 2.776$

   **c.** $df = 24$, $t = 1.711$ and $-1.711$

**9. a.** Yes. The same cashier is using the old register and the new register so the samples can be paired. Answers will vary.

   **b.** The differences have an approximately normal distribution.

   **c.** Answers will vary.

   **d.** $(-4.15, 0.43)$ With 95% confidence, cashiers using the old cash register process between 4.15 fewer and 0.43 more items than using the new cash register.

   **e.** Yes. $H_0: \mu_d = 0, H_a: \mu_d < 0$,

     Critical value $= -1.943$, $t = -1.983$, Reject $H_0$.

## Section 11.4

**7. a.** $z = -2.33$

   **b.** $z = 1.645$

   **c.** $z = 1.645$ and $-1.645$

**9. a.** Yes, the sample sizes are sufficiently large.

     $H_0: p_1 - p_2 = 0, H_a: p_1 - p_2 < 0$,

     Critical value $= -1.28$, $z = -0.91$, Fail to reject $H_0$.

     There is not sufficient evidence to support the fund-raiser's theory.

   **b.** $P$-value $= 0.1814$. This is the probability of making a Type I error. Answers will vary.

   **c.** $(-0.1098, 0.0403)$ We are 95% confident that the proportion of men who answered "Yes" when asked to donate to a worthy cause is between 0.1098 less than and 0.0403 greater than the proportion of women who answered "Yes."

**11. a.** Yes, the sample sizes are sufficiently large. $(-0.261, -0.149)$ We are 99% confident that the proportion of people age 30 or younger that believe alcoholic beverage commercials are targeted at teenagers is between 0.149 and 0.261 lower than the proportion of people in the older than 30 age group.

   **b.** Yes. $H_0: p_1 - p_2 = 0, H_a: p_1 - p_2 < 0$,

     Critical value $= -2.33$, $z = -9.22$, Reject $H_0$.

## Section 11.5

**7. a.** 5.6864

   **b.** 3.8807

   **c.** 2.5289

   **d.** 3.9539

**9. a.** $H_0: \sigma_1^2 = \sigma_2^2; H_a: \sigma_1^2 < \sigma_2^2$

   **b.** $H_0: \sigma_1^2 = \sigma_2^2; H_a: \sigma_1^2 > \sigma_2^2$

**11. a.** $F_{0.950} = 0.3744$; reject $H_0$ if $F \le 0.3744$; reject $H_0$.

   **b.** $F_{0.010} = 3.0558$; reject $H_0$ if $F \ge 3.0558$; fail to reject $H_0$.

   **c.** $F_{0.975} = 0.4148$, $F_{0.025} = 2.2505$; reject $H_0$ if $F \le 0.4148$ or $F \ge 2.2505$; reject $H_0$.

**13.** $H_0: \sigma_1^2 = \sigma_2^2; H_a: \sigma_1^2 > \sigma_2^2$

   **b.** $F$-distribution; $\alpha = 0.10$

   **c.** $F \approx 1.4067$

   **d.** $F_{0.100} = 1.9532$; reject $H_0$ if $F \ge 1.9532$. $P$-value $\approx 0.2543$; fail to reject $H_0$. At the 0.10 level of significance, there is not sufficient evidence to support the inspector's claim that the variance in the diameters of soda cans is greater for soda cans produced by Machine A than for soda cans produced by Machine B.

**15. a.** $H_0: \sigma_1^2 = \sigma_2^2; H_a: \sigma_1^2 \ne \sigma_2^2$

   **b.** $F$-distribution; $\alpha = 0.01$

   **c.** $F \approx 0.9571$

**d.** $F_{0.995} = 0.1910$, $F_{0.005} = 4.9884$; reject $H_0$ if $F \leq 0.1910$ or $F \geq 4.9884$. $P$-value $\approx 0.9490$; fail to reject $H_0$. At the 0.01 level of significance, there is not sufficient evidence to support the coach's claim that the variance in heights of adult male basketball players is different than that of the general population of men.

## Chapter 11 Additional Exercises

**1.** Yes. $H_0: \mu_1 - \mu_2 = 0$, $H_a: \mu_1 - \mu_2 < 0$,
Critical value $= -1.645$, $z = -2.59$, Reject $H_0$.

**3. a.** Yes. The cholesterol levels are measured for the same person before and after the diet so the samples can be paired. Answers will vary.

**b.** The differences have an approximately normal distribution.

**c.** Answers will vary.

**d.** No. $H_0: \mu_d = 0$, $H_a: \mu_d > 0$,
Critical value $= 3.143$, $t = 1.769$, Fail to reject $H_0$.

**5.** $H_0: \mu_1 - \mu_2 = 0$, $H_a: \mu_1 - \mu_2 \neq 0$,
Critical values $= -2.878$, $2.878$, $t = -1.388$, Fail to reject $H_0$.

**7. a.** No. $H_0: \mu_1 - \mu_2 = 0$, $H_a: \mu_1 - \mu_2 > 0$,
Critical value $= 1.645$, $z = 2.00$, Reject $H_0$.

**b.** $(0.08, 0.52)$

**9.** $H_0: \mu_1 - \mu_2 = 10$, $H_a: \mu_1 - \mu_2 > 10$,
Critical value $= 2.014$, $t = 0.380$, Fail to reject $H_0$. There is not sufficient evidence that Arrangement A has a lower mean net annual income by at least \$10.

**11.** $H_0: \mu_1 - \mu_2 = 0$, $H_a: \mu_1 - \mu_2 \neq 0$,
Critical values $= -2.101$, $2.101$, $t = 8.771$, Reject $H_0$. There is sufficient evidence of a difference in the average weights.

**13. a.** $H_0: p = 0.10$, $H_a: p < 0.10$, Critical value $= -1.645$, 2009: $z = -16.29$, Reject $H_0$, 2010: $z = -5.84$, Reject $H_0$, 2011: $z = 0.53$, Fail to reject $H_0$.

**b.** $(-0.0316, -0.0133)$ We are 95% confident that the proportion of travelers using tablets or e-readers in 2011 is between 1.33% and 3.16% greater than the proportion of travelers using tablets or e-readers in 2010.

**c.** Yes. $H_0: p_1 - p_2 = 0$, $H_a: p_1 - p_2 < 0$,
Critical value $= -1.645$, $z = -4.77$, Reject $H_0$. There is sufficient evidence that the proportion of travelers using tablets or e-readers has increased between 2010 and 2011.

**15. a.** Yes. $H_0: p_1 - p_2 = 0$, $H_a: p_1 - p_2 \neq 0$,
Critical values $= -1.96$, $1.96$, $z = 4.78$, Reject $H_0$.

**b.** $(0.0450, 0.0750)$ We are 95% confident that the percentage of people with maxed out credit cards with IQs of 90 is between 4.55% and 7.5% greater than the percentage of people with maxed out credit cards with IQs greater than 125.

**c.** An independent experimental design was used.

**17.** $P$-value $= P(F \leq 0.5888) = 0.2211$. Fail to reject $H_0$

# Chapter 12

## Section 12.1

**15. a.** Probably not, the boxes overlap quite a bit. Variation appears to be large in comparison to the difference in the middle values. Answers will vary.

**b.** Probably, the variation in scores for juniors is larger than for freshmen but the middle value is greater for juniors than for freshmen. Answers will vary.

**c.** Probably not, the boxes overlap quite a bit. Variation appears to be large in comparison to the difference in the middle values. Answers will vary.

**17. a.** The experimental units are the items produced and the treatment is the shift.

**b.** First shift: mean $= 158$, median $= 161$; Second shift: mean $= 191.8571$, median $= 182$; Third shift: mean $= 111.5714$, median $= 111$

**c.** First shift: Min $= 127$, Max $= 181$, $Q_1 = 140$, $Q_3 = 173$,
Second shift: Min $= 162$, Max $= 224$, $Q_1 = 168$, $Q_3 = 219$,
Third shift: Min $= 77$, Max $= 147$, $Q_1 = 77$, $Q_3 = 145$,

**d.**

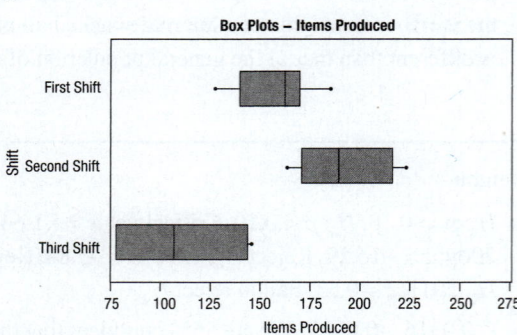

**e.** Probably, answers will vary.

**f.** Yes, answers will vary.

**g.** Probably, answers will vary.

**h.** Second shift, answers will vary.

**19. a.** 1.8464 is the variance of the sample means.

**b.** 2

**c.** 3.6928

**d.** 15

**e.** 14.1345

## Section 12.2

9.  Sample 1: No, the histogram represents data drawn from a population that has a negative exponential distribution. Middle histogram Sample 2: No, the histogram represents data drawn from a population that has an unknown distribution. Sample 3: Yes, the histogram represents data drawn from a population that has a normal distribution.

11. No, because the boxes are not basically the same width. The box plot at the bottom appears to be less than half the size of the box plot in the middle, indicating that the variation for this population is considerably smaller than that of the other two populations. Answers may vary.

13. Shapiro-Wilk Test: The $P$-values for each state in the test output are greater than 0.05 ( Iowa: 0.1417; Hawaii: 0.5628; California: 0.6889).

    Anderson-Darling Test: The $P$-values for each state in the test output are greater than 0.05 ( Iowa: 0.1445; Hawaii: 0.5122; California: 0.4609).

    Conclusion: The foot length measurements of all adult

males come from normally distributed populations for each of the three states: Iowa, Hawaii, and California. The assumption of normality is satisfied.

15. For this survey the three samples of consumers are surveyed from a shopping center close to the surveyor's residence. Thus, the samples are selected through a convenience sample as the participants of the sample were "conveniently" selected from shoppers at a nearby shopping center. The three samples cannot be considered random samples. One of the assumptions of the ANOVA test is that the samples are randomly selected.

    Therefore, the potential error of this ANOVA test is that the samples considered are not random samples. Thus, the results are not reliable.

17. It is possible that a student could be in more than one of these groups. In this case, the samples are not independent.

## Section 12.3

**11. a.** Yes. $H_0: \mu_1 = \mu_2 = \mu_3 = \mu_4$, $H_a$: at least one $\mu_i$ is different, $F = 16.8582$, $F_\alpha = 3.2389$, Reject $H_0$.

**b.** Hourly wages for employees are approximately normally distributed with equal variances. Observations were collected in an independent and random fashion. Answers will vary.

**13. a.** No. $H_0: \mu_1 = \mu_2 = \mu_3 = \mu_4$, $H_a$: at least one $\mu_i$ is different, $F = 3.4667$, $F_\alpha = 4.0662$, Fail to reject $H_0$.

**b.** Maximum heart rates for each workout are approximately normally distributed with equal variances. Observations were selected in an

independent and random fashion. Answers will vary.

**15. a.** Yes. $H_0: \mu_1 = \mu_2 = \mu_3$; $H_a$: at least one $m_i$ is different, $F = 5.1012$, $F_\alpha = 2.5893$, Reject $H_0$.

**b.** Dividends per share for each industry are approximately normally distributed with equal variances. Observations were selected in an independent and random fashion. Answers will vary.

**c.** It appears that the transportation industry pays lower dividends per share than the banking and energy industries, but the ANOVA test does not tell us which population mean(s) differ significantly. Answers will vary.

## Section 12.4

5.  There will be 6 pairwise comparisons that need to be tested individually.

$P(\text{at least 1 type I error}) = 1 - P(\text{no type I error})$
$$= 1 - (1-\alpha)^6 = 1 - (1-0.01)^6$$
$$= 0.05852 \approx 0.06$$

Thus, it can be seen that the probability of a Type I error increases from 0.01 to approximately 0.06 as the number of pairs for comparing means increases.

**7.** $df = n_T - k = 30 - 4 = 26$

$\dfrac{\alpha}{2} = \dfrac{0.05}{2} = 0.025$

Therefore, the critical value of $t$ corresponding to 26 degrees of freedom and a 0.025 level of significance is equal to 2.0555.

**9.** MSE from the ANOVA output is 102.2.

$df = n_T - k = 30 - 3 = 27$

$\dfrac{\alpha}{2} = \dfrac{0.1}{2} = 0.05$

Fisher's LSD = 7.700725

$|\bar{x}_A - \bar{x}_B| < 7.700725$
$|\bar{x}_A - \bar{x}_C| > 7.700725$
$|\bar{x}_B - \bar{x}_C| > 7.700725$

## Section 12.5

**13. a.** Yes, because the dealer believes that the average gas mileage of a particular car will vary depending on the person who is driving the car due to different driving styles. Blocking will reduce the variation in gas mileage which is not due to the type of car.

**b.** Yes. $H_0$: $\mu_1 = \mu_2 = \mu_3 = \mu_4$, $H_a$: at least one $m_i$ is different, $F = 696.8608$, $F_\alpha = 3.2874$, Reject $H_0$.

**c.** Yes. $H_0$: $\mu_1 = \mu_2 = \mu_3 = \mu_4 = \mu_5 = \mu_6$, $H_a$: at least one $\mu_i$ is different, $F = 101.7798$, $F_\alpha = 2.9013$, Reject $H_0$.

**15. a.** Yes, because the FAA believes that the number of on-time arrivals varies by airport. Blocking will reduce the variation in on-time arrivals which is not due to airline.

There is a significant difference in the mean fasting blood glucose levels between treatments A and C and between treatments B and C.

**11.** $df = (n_T - k) = (64 - 4) = 60$

The studentized range value corresponding to 60 degrees of freedom, four treatments, and a = 0.05 is 3.737.

**13. a.** $df = n_T - k = 15 - 3 = 12$, $a = 0.05$

The studentized range value corresponding to 12 degrees of freedom, three groups and 0.05 level of significance is 3.77.

The confidence interval is (2.50, 18.70). Since the interval does not include the value of 0, the null hypothesis is rejected. Thus, there is a significant difference between the mean test scores of students who took curriculum A versus those who took curriculum B.

**b.** The confidence interval is (4.10, 20.3). Since the interval does not include the value of 0, the null hypothesis is rejected. Thus, there is a significant difference between the mean test scores of students who took curriculum B versus those who took curriculum C.

**b.** Yes. $H_0$: $\mu_1 = \mu_2 = \mu_3 = \mu_4$, $H_a$: at least one $\mu_i$ is different, $F = 58.8261$, $F_\alpha = 6.9919$, Reject $H_0$.

**c.** Yes. $H_0$: $\mu_1 = \mu_2 = \mu_3 = \mu_4$, $H_a$: at least one $\mu_i$ is different, $F = 15.2609$, $F_\alpha = 6.9919$, Reject $H_0$.

**17. a.** So that any variation not due to the type of device used to measure systolic blood pressure can be reduced.

**b.** Yes. $H_0$: $\mu_1 = \mu_2 = \mu_3 = \mu_4$, $H_a$: at least one $m_i$ is different, $F = 9.9883$, $F_\alpha = 3.2874$, Reject $H_0$.

**c.** Yes. $H_0$: $\mu_1 = \mu_2 = \mu_3 = \mu_4 = \mu_5 = \mu_6$, $H_a$: at least one $m_i$ is different, $F = 135.8303$, $F_\alpha = 2.9013$, Reject $H_0$.

## Section 12.6

**13. a.** Yes, there appears to be interaction between airport location and major rental car company for all three cities.

**b.** Yes. $H_0$: There is no interaction, $H_a$: There is interaction, $F = 13.8127$, $F_\alpha = 2.9277$, Reject $H_0$.

**c.** We cannot test for effect of company on average daily rental rates because there is interaction.

**15. a.** There appears to be slight interaction between operator

and machine. If there was no interaction, the lines would be parallel. Answers may vary.

**b.** No, there is not significant interaction. This agrees with part **a.**, we only thought the interaction was slight. $H_0$: There is no interaction, $H_a$: There is interaction, $F = 0.0789$, $F_\alpha = 2.2858$, Fail to reject $H_0$.

**c.** Yes. $H_0$: $\mu_1 = \mu_2 = \mu_3$, $H_a$: at least one $\mu_i$ is different, $F = 4.5$, $F_\alpha = 2.6240$, Reject $H_0$.

**d.** Yes. $H_0$: $\mu_1 = \mu_2 = \mu_3$, $H_a$: at least one $\mu_i$ is different, $F = 14.2895$, $F_\alpha = 2.6240$, Reject $H_0$.

**17. a.**

| Source | SS | df | MS | F |
|---|---|---|---|---|
| Power | 1.270 | 1 | 1.270 | 14.7246 |
| Knowledge | 0.250 | 1 | 0.250 | 2.8986 |
| Interaction | 0.010 | 1 | 0.010 | 0.1159 |
| Error | 4.140 | 48 | 0.0863 | |
| Total | 5.670 | 51 | | |

**b.** No. $H_0$: There is no interaction, $H_a$: There is interaction, $F = 0.1159$, $F_\alpha = 2.8131$, Fail to reject $H_0$.

**c.** Yes. $H_0$: $\mu_1 = \mu_2$, $H_a$: at least one $\mu_i$ is different, $F = 14.7246$, $F_\alpha = 4.0427$, Reject $H_0$.

**d.** No. $H_0$: $\mu_1 = \mu_2$, $H_a$: at least one $\mu_i$ is different, $F = 2.8986$, $F_\alpha = 4.0427$, Fail to reject $H_0$.

## Chapter 12 Additional Exercises

**1. a.** Dividing the students of each class into blocks categorized as Below Average, Average, and Above Average. Answers will vary.

**b.** Dividing participants into blocks, categorized as <100 lb overweight, 50-100 lb overweight, and 0-50 lb overweight. Answers will vary.

**c.** Dividing the persons into blocks categorized as Low IQ, Average IQ, and High IQ. Answers will vary.

**3. a.** Yes. $H_0$: $\mu_1 = \mu_2 = \mu_3$, $H_a$: at least one $\mu_i$ is different, $F = 15.6429$, $F_\alpha = 6.3589$, Reject $H_0$.

**b.** Fat contents for each brand of margarine are approximately normally distributed with equal variances. Observations were collected in an independent and random fashion. Answers will vary.

**c.** The data are obtained from servings. There are no variables associated with the servings that we can block with. Answers will vary.

**5. a.** $\bar{\bar{x}} = 3.3064$, SST = 20.1246

**b.** $F = 35.3327$

**c.** Yes. $H_0$: $\mu_1 = \mu_2 = \mu_3$, $H_a$: at least one $\mu_i$ is different, $F = 35.3327$, $F_\alpha = 4.7623$, Reject $H_0$.

**d.** The distributions of all of the populations of interest are approximately normal with equal variances, each of the $k$ samples must be selected independently from each other and in a random fashion. They cannot be checked in this instance because the raw data are not available.

**7. a.** $H_0$: The average median starting salary for all four

types of majors is the same, $H_a$: At least one of the median starting salaries is different.

**b.** Yes, there is sufficient evidence that at least one of the starting salaries is different.

$H_0$: $\mu_1 = \mu_2 = \mu_3 = \mu_4$, $H_a$: at least one $\mu_i$ is different, $F = 17.1205$, $F_\alpha = 4.4156$, Reject $H_0$.

**c.** $H_0$: The average median mid-career salary for all four types of majors is the same, $H_a$: At least one of the median mid-career salaries is different.

**d.** Yes, there is sufficient evidence that at least one of the mid-career salaries is different.

$H_0$: $\mu_1 = \mu_2 = \mu_3 = \mu_4$, $H_a$: at least one $\mu_i$ is different, $F = 10.9840$, $F_\alpha = 4.4156$, Reject $H_0$.

**9. a.** The mixed question tests appear to result in the highest average test scores. It also appears that graduate students typically have higher test scores, on average. Answers will vary.

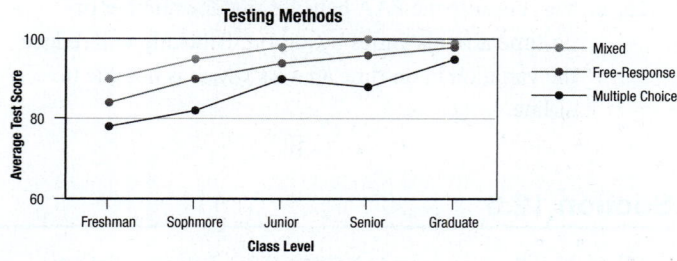

**b.** Yes. $H_0$: $\mu_1 = \mu_2 = \mu_3 = \mu_4 = \mu_5$, $H_a$: at least one $\mu_i$ is different, $F = 22.0786$, $F_\alpha = 3.8379$, Reject $H_0$.

**c.** Yes, because the blocking effects were significant. Answers will vary.

**11. a.**

| Source | SS | df | MS | F |
|---|---|---|---|---|
| Block | 2154.1333 | 9 | 239.3481 | 96.8876 |
| Treatment | 30.2 | 2 | 15.1 | 6.1124 |
| Error | 44.4667 | 18 | 2.4704 | |
| Total | 2228.8 | 29 | | |

**b.** Yes, $H_0: \mu_1 = \mu_2 = \mu_3 = \mu_4$, $H_a$: at least one $\mu_i$ is different, $F = 6.1124$, $F_\alpha = 3.5546$, Reject $H_0$.

**c.** $P = 0.0094$. The probability of a Type I error for this hypothesis test is 0.0094.

**13. a.** $H_0: \mu_1 = \mu_2 = \mu_3$, $H_a$: at least one $\mu_i$ is different.

**b.**

| Source | SS | df | MS | F |
|---|---|---|---|---|
| Block | 1103 | 3 | 367.6667 | 31.2908 |
| Treatment | 158.1667 | 2 | 79.0833 | 6.7305 |
| Error | 70.5 | 6 | 11.75 | |
| Total | 1331.6667 | 11 | | |

**c.** Yes, at least one of the revenues is significantly different. $H_0: \mu_1 = \mu_2 = \mu_3$, $H_a$: at least one $\mu_i$ is different, $F = 6.7305$, $F_\alpha = 5.1433$, Reject $H_0$.

**d.** Yes, $H_0: \mu_1 = \mu_2 = \mu_3 = \mu_4$, $H_a$: at least one $\mu_i$ is different, $F = 31.2908$, $F_\alpha = 4.7571$, Reject $H_0$.

# Chapter 13

## Section 13.1

**33.** $\hat{y}_i = b_0 + b_1 x_i$

**35. a.** Sales volume, since it is the value we want to predict.

**b.** Advertising expenditures, since we use it to predict sales volume.

**c.** $69,650

**d.** $99,510

**e.** Random error. This may cause the company to under or over-estimate sales volume, causing budgeting problems. Answers will vary.

**37. a.**

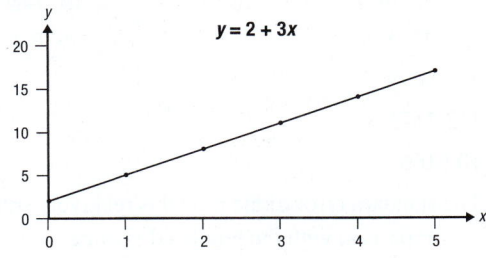

**b.**

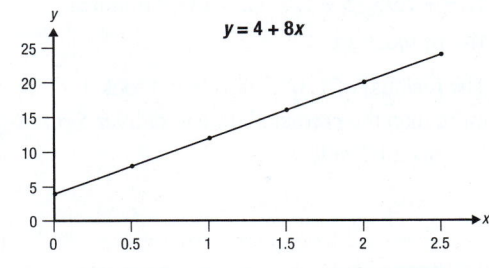

**c.**

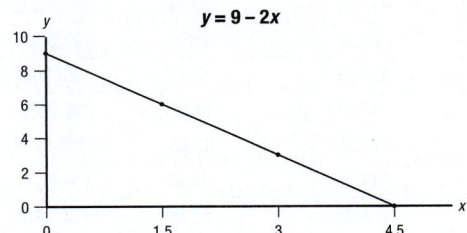

**d.**

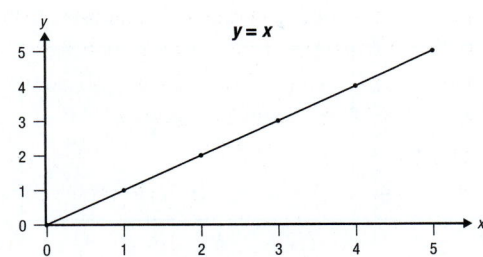

**39. a.**

| $\hat{y}$ |
|---|
| 15 |
| 45 |
| 65 |
| 85 |
| 95 |

**b.** Positive

**c.** We would expect $r$ to be positive, since the variables have a positive relationship.

**41. a.** Estimated Selling Price $= b_0 + b_1$(Square Footage)

**b.**

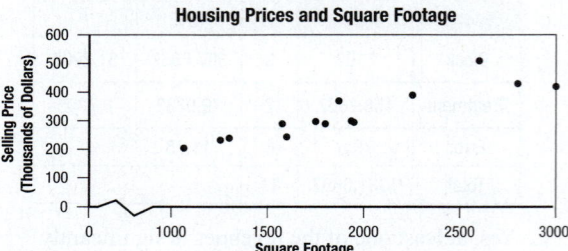

**c.**

| Predicted Selling Price (Thousands of Dollars) | Error | Squared Error |
|---|---|---|
| 201.45 | −1.55 | 2.4025 |
| 227.91 | 0.09 | 0.0081 |
| 234.35 | 0.65 | 0.4225 |
| 273.13 | 11.87 | 140.8969 |
| 276.35 | −37.35 | 1395.0225 |
| 297.35 | −4.35 | 18.9225 |
| 304.35 | −19.35 | 374.4225 |
| 314.15 | 50.85 | 2585.7225 |
| 323.25 | −28.25 | 798.0625 |
| 325.07 | −35.07 | 1229.9049 |
| 367.91 | 17.09 | 292.0681 |
| 416.35 | 88.65 | 7858.8225 |
| 444.35 | −19.35 | 374.4225 |
| 472.35 | −57.35 | 3289.0225 |

**d.** SSE = 18,360.123

**43. a.**

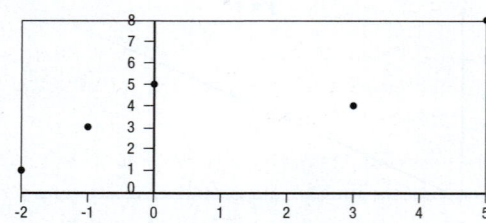

**b.** $\hat{y} = 3.4353 + 0.7647x$

**c.**

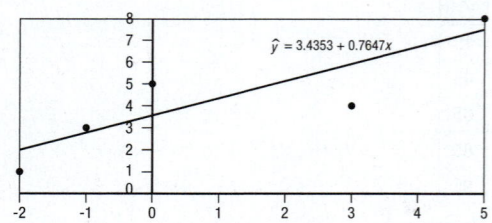

**d.** −0.9059, 0.3294, 1.5647, −1.7294, 0.7412

**45. a.** Current value of home

**b.** Annual salary

**c.** $112,811

**d.** $15,700

**e.** For each additional dollar earned in annual salary, the current value of home is predicted to increase by $3.14.

**f.** If someone is not earning any annual income, the predicted value of his or her home would be $12,331.

**g.** It is possible because if you have a greater annual salary you have more money to spend on a home. However, we cannot conclude that there is a causality from the estimated regression equation. Answers will vary.

**47. a.** $\hat{y} = 5.7333 + 0.6667x$

**b.** 16.4

**c.** No. $\hat{y} = 10.9732 + 0.3065x$

**d.** 15.8772

**e.** There is not likely a causal relationship between these two variables. Answers will vary.

**49. a.**

| Predicted $y$ | Error | Squared Error |
|---|---|---|
| 124.8372 | −14.8372 | 220.1425 |
| 131.1462 | 3.8538 | 14.8518 |
| 145.8672 | 4.1328 | 17.0800 |
| 143.7642 | 5.2358 | 27.4136 |
| 147.9702 | 10.0298 | 100.5969 |
| 177.4122 | −8.4122 | 70.7651 |

**b.** 450.8499

**c.** 112.7125

**d.** 10.6166

**e.** The standard error of the model is relatively small in comparison with the predicted $y$-values, so yes. Answers will vary.

**51. a.** $1.79 + 1.95 \cdot 5 + 1.57 \cdot 2 = 14.68$ minutes

**b.** 16−14.68=1.32

**c.** The residual of 1.32 means that it took 1.32 minutes more than the predicted time to deliver 5 pizzas at a distance of 2 miles.

## Section 13.2

**5.** The given scatterplot of $y$ on $x$ shows that the linearity assumption is violated.

**7.** The plot of residuals of the regression model against the experience shows that the constant variance assumption is violated.

**9.** No, this model is not appropriate for predicting the price of the car using the age of the car because the residuals of the model show a nonlinear relationship.

**11.** The histogram of the regression residuals is closely normal, so it can be argued that residuals are normally distributed, and the assumption of the regression model is satisfied. However, one should interpret the model with caution.

**13.** The normal probability plot of the regression residuals shows that the points form an approximately straight line, and hence, the assumption of the normality of the regression residuals is satisfied.

**15. a.** The scatterplot of the data, along with the least squares line, shows there is a linear trend that as the number of kilometers run in the 4 weeks prior increases, the marathon time decreases.

**b.** The scatterplot of Marathon Time vs. Km Run in 4 Weeks Prior in a. shows that the data follow a linear pattern, validating the linearity assumption.

The residual plot shows a random scatter about $y = 0$, indicating that the linear fit is appropriate. Also, since it doesn't show a pattern, it upholds the assumption that the errors are independent of each other. As $x$ increases the spread of the residuals remains fairly constant; therefore the assumption of equal variance is not violated.

By looking at the normal probability plot of the residuals, the residuals reasonably follow the diagonal line, indicating the errors are normally distributed.

## Section 13.3

**13. a.** Ratio

**b.** Savings

**c.** Income

**d.**

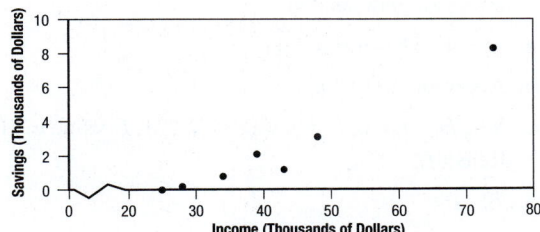

Yes, the data points in the scatterplot appear approximately linear with a positive slope trend.

**e.** $\hat{y} = -4.8658 + 0.171x$

**f.** $3684.20

**g.** Savings is estimated to increase by approximately $171 for each additional $1000 earned in annual income.

**h.** $R^2 = 0.9495$ Approximately 94.95% of the variation in savings is explained by the variation annual income.

**15. a.** Controlled experiment

**b.**

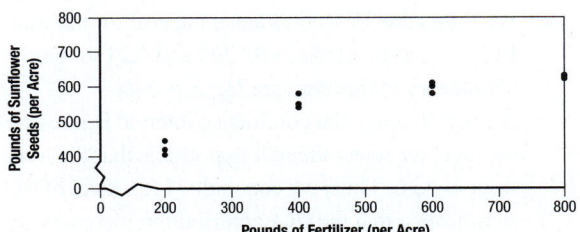

**c.** Pounds of sunflower seeds because the researchers are interested in the effect of fertilizer on the yield of sunflower seeds.

**d.** $\hat{y} = 389 + 0.323x$

**e.** As the amount of fertilizer per acre increases by one pound, the yield of sunflower seeds is predicted to increase by 0.323 of a pound.

**f.** $R^2 = 0.8456$. Approximately 84.56% of the variation in the sunflower yield can be explained by the variation in the amount of fertilizer used.

**g.** 550.5 seeds

**17.** Coefficient of determination $= r^2 = 0.64^2 = 0.4096$.

## Section 13.4

**7. a.** Independent variable: Month; Dependent variable: Sales

**b.** $\hat{y} = 403.7576 + 40.9091x$

**c.** MSE $= 2622.248$, $s_e = 51.2079$

**d.** $935,575.90

**e.** Approximately 90.12% of the variation in sales is explained by the time trend model, therefore the model seems to fit the data accurately.

## Section 13.5

**11. a.** (33.1498, 48.6684)

**b.** We are 90% confident that the true increase in sales for each additional month is between $33,149.80 and $48,668.40.

**13. a.** The variables appear to have a positive correlation.

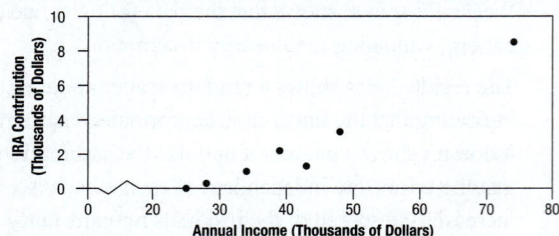

**b.** $\hat{y} = -4.8603 + 0.1740x$

**c.** (0.1293, 0.2186) We are 95% confident that the true increase in IRA contribution for each additional thousand dollars of income is between \$129.30 and \$218.60.

**d.** The error term is a normally distributed random variable, the expected value of the error term is zero, the variance of the error term is constant, and the errors are independent of each other.

**e.** We found the 95% confidence interval for the true slope $(\beta_1)$ to be between 0.1293 and 0.2186. Our hypotheses in this case are $H_0: \beta_1 = 0$ vs. $H_a: \beta_1 > 0$. Since the confidence interval is in entirely positive, we reject the null hypothesis that the true slope is zero. Thus, we conclude at 5% level of significance that the IRA contribution increases with the increase in income of the subject.

**15. a.** The two variables appear to have a positive correlation.

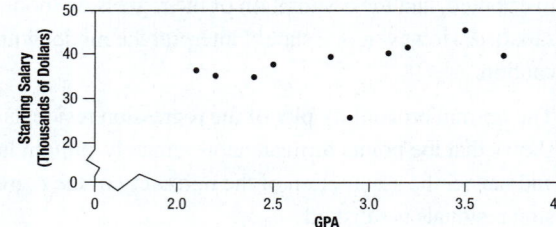

**b.** $\hat{y} = 24.9542 + 4.4175x$

**c.** No, $H_0: \beta_1 = 0$, $H_a: \beta_1 \neq 0$, $t = 1.471$, $P$-value = 0.1794, Fail to reject $H_0$.

**d.** \$35,997.95

**e.** For each additional one point increase in GPA, starting salary is expected to increase by approximately \$4417.50.

**f.** Approximately 21.3%

**g.** The error term is a normally distributed random variable, the expected value of the error term is zero, the variance of the error term is constant, and the errors are independent.

**17. a.** $\hat{y} = 42.5154 + 0.3914x$

**b.** Approximately 20.76%

**c.** Yes, $H_0: \beta_1 = 0$, $H_a: \beta_1 \neq 0$, $t = 2.231$, $P$-value = 0.0379, Reject $H_0$.

**d.** 72

## Section 13.6

**5. a.** There appears to be a weak linear relationship between actual spreads and betting spreads.

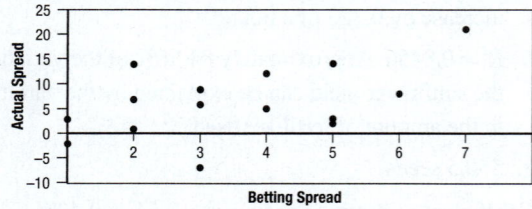

**b.** $\hat{y} = -0.3834 + 1.9198x$

**c.** No, $H_0: \beta_1 = 0$, $H_a: \beta_1 \neq 0$, $t = 2.043$, $P$-value = 0.0619, Fail to reject $H_0$.

**d.** Approximately 24.3%

**e.** The estimated increase in the actual spread for a one point increase in the betting spread is approximately 1.9198 points.

**f.** (−0.1104, 3.9500) We are 95% confident that the change in actual spread for a one point increase in betting spread is between −0.1104 and 3.9500 points.

**g.** 9.2156

**h.** (−5.27, 23.70) We are 95% confident that the actual spread when the betting spread is 5 is between −5.27 and 23.70.

**i.** (4.59, 13.85)

**7. a.** There appears to be a moderate negative linear relationship between age and sick days.

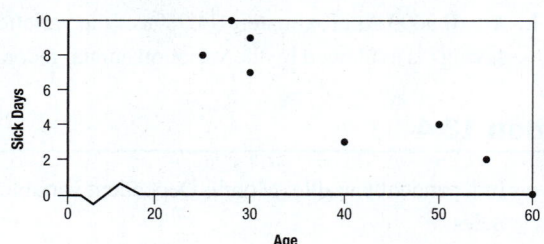

**b.** $s_e = 1.4765$. This is the estimated standard deviation of the errors associated with the model.

**c.** With each additional year of age, the number of sick says is expected to decrease by approximately 0.247.

**d.** Approximately 85.8%

**e.** Yes, $H_0: \beta_1 = 0$, $H_a: \beta_1 \neq 0$, $t = -6.01$, $P$-value = 0.001, Reject $H_0$.

**f.** (−0.3473, −0.1464) We are 95% confident that the true decrease in sick days for a one year increase in age is between 0.3473 and 0.1464 days.

**g.** Approximately 6.5 days

**h.** (5.184, 7.911) We are 95% confident that the average number of sick days for a 35-year-old employee is between 5.184 and 7.911.

**i.** (2.686, 10.409) We are 95% confident that a new 35-year-old employee will take between 2.686 and 10.409 sick days.

**j.** 4 days

**9.** We have found the 99% confidence interval for the slope to be (−1809.1701, −973.0039). Our hypotheses in this case are $H_0: \beta_1 = 0$ vs. $H_a: \beta_1 \neq 0$). Since the confidence interval does not include 0, we reject the null hypothesis that the true slope is zero. Thus, we conclude at 1% level of significance that there is a significant relation between the age of the car and its price.

## Chapter 13 Additional Exercises

**1. a.** There appears to be a moderate positive linear relationship between amount of drug and reaction time.

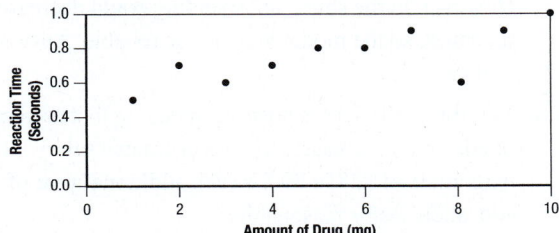

**b.** $s_e = 0.1101$. This is the estimated standard deviation of the errors associated with the model.

**c.** Reaction time increases by approximately 0.0394 seconds for each additional milligram of the drug.

**d.** Approximately 56.9%. Other factors could include age, weight, etc. Answers will vary.

**e.** $H_0: \beta_1 = 0$, $H_a: \beta_1 \neq 0$, $t = 3.25$, $P$-value = 0.012, Reject $H_0$ at the 0.05 level, Fail to reject $H_0$ at the 0.01 level.

**f.** (0.011, 0.067) We are 95% confident that the true change in reaction time for each additional milligram of the drug is between 0.011 and 0.067 seconds.

**g.** 0.6909 seconds

**h.** (0.600, 0.782) We are 95% confident that the average reaction time for an individual with 4 mg of the drug in the bloodstream is between 0.600 and 0.782 seconds.

**i.** (0.421, 0.961)

**3. a.** There appears to be a strong negative relationship between FICO score and interest rate.

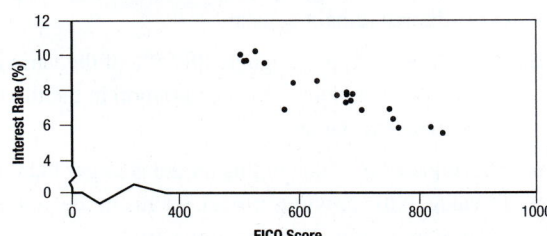

**b.** $\hat{y} = 16.2146 - 0.0129x$

**c.** 0.3202

**d.** Yes. $H_0: \beta_1 = 0$, $H_a: \beta_1 \neq 0$, $t = -10.281$, $P$-value = 0.0000, Reject $H_0$.

**e.** Interest rate decreases by approximately 0.0129 percent for each additional one point increase in FICO score.

**f.** (−0.0155, −0.0102) We are 95% confident that the true change in interest rate for each additional one point increase in FICO score is between −0.0155 and −0.0102 percentage points.

**g.** $R^2 = 0.8545$. Approximately 85.45% of the variation in interest rate is explained by the variation in FICO scores.

**h.** $r = -0.9244$. There is a strong negative linear relationship between FICO score and interest rate.

**i.** 6.8621%

**j.** (6.613, 7.148) We are 95% confident that the average interest rate for a FICO score of 725 is between 6.613 and 7.148 percent.

**k.** (5.864, 7.898) We are 95% confident that a person with a FICO score of 725 will receive an interest rate between 5.864 and 7.898 percent.

**5. a.** There appears to be a strong positive linear relationship between home size and annual energy usage.

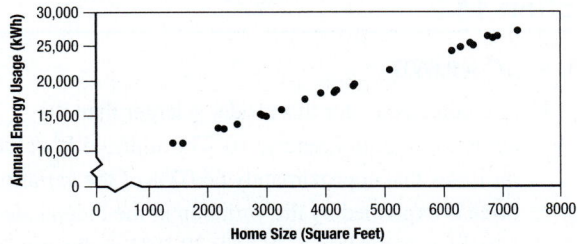

**b.** $\hat{y} = 6774.4571 + 2.8406x$

**c.** $R^2 = 0.9982$. Approximately 99.82% of the variation in annual energy usage is explained by the variation in home size.

**d.** Yes. $H_0: \beta_1 = 0$, $H_a: \beta_1 \neq 0$, $t = 112.114$, $P$-value = 0.0000, Reject $H_0$.

e. For each additional square foot of home size, annual energy usage is expected to increase by approximately 2.8406 kWh.

f. (2.7695, 2.9117) We are 99% confident that the true increase in annual energy usage for each additional square foot of home size is between 2.7695 and 2.9117 kWh.

g. $r = 0.9991$. There is a strong positive linear relationship between home size and annual energy usage.

h. 15,864.3771 kWh

i. (15,572.2, 15,976.5) We are 95% confident that the James family will use on average between 15,572.2 and 15,976.5 kWh in their first year in the home.

# Chapter 14

## Section 14.1

11. a. $b_0 = 6.1342$, $b_1 = 0.0108$, $b_2 = -0.0100$

b. $\hat{y} = 6.1342 + 0.0108x_1 - 0.0100x_2$

c. Yes, the coefficient of the number of pages is positive, indicating that more pages increases printing cost. The coefficient for the number of copies is negative, indicating that if you buy in bulk, the cost is less per book. The magnitudes of $0.01 per page and $0.01 per copy also seem reasonable. Answers may vary.

d. Type of paper, black & white vs. color printing, type of binding, etc. Answers will vary.

13. a. # of employees, average salary, advertising expenditures, research and development expenditures, charitable gifts to the community, etc. Answers will vary.

b. Revenue $= \beta_0 + \beta_1(R\&D) + \beta_2(Advertising) + \beta_3(Salary\ Paid) + \varepsilon_i$

c. Answers will vary. The coefficient could be positive because R&D expenditures may result in more and better products, meaning more sales, which would increase revenue. Alternatively, spending on R&D may reduce revenue if the resulting sales do not overcome the amount spent on development.

d. R&D expenditures and advertising expenditures are both costs that should increase revenue in the long run. However, in the short run, spending could decrease revenues, so the model may not be reliable. Answers will vary.

15. a. Yes, the coefficient is positive, meaning that as years of education increases, so does estimated salary. The magnitude of $2854.89 for each additional year of education seems reasonable.

b. Yes, the coefficient is positive, meaning that as years of experience increases, so does estimated salary. The magnitude of $839.64 for each additional year of experience seems reasonable.

c. For each additional year of experience, annual salary is expected to increase by $839.64, assuming years of education remains constant.

d. $38,251.52

e. His annual salary would be expected to increase by $839.64, assuming years of education remains constant.

f. The employee with the master's degree is expected to earn approximately $5709.78 more than the employee with the bachelor's degree.

## Section 14.2

11. a. $R_a^2 = 0.6607$

b. The adjusted $R^2$ for this model is larger than the adjusted $R^2$ from Exercise 10. The adjusted $R^2$ value indicates that approximately 66.07% of the variation in price is explained by the variation in the independent variables, compared with only 29.19% in the previous model. Answers will vary.

c. Yes, because the adjusted $R^2$ value is significantly larger for this model. Answers will vary.

13. a. Revenue $= \beta_0 + \beta_1(Television) + \beta_2(Newspaper) + \beta_3(Mail) + \varepsilon_i$

b. $\hat{y} = 73.9320 + 2.3830x_1 + 1.4544x_2 + 1.8160x_3$

c. For each additional $1000 spent on TV advertising, revenue is expected to increase by approximately $2383, assuming newspaper and mail advertising expenditures remain constant.

d. $R_a^2 = 0.8865$. Approximately 88.65% of the variation in revenue is explained by the variation in the three independent variables.

e. The adjusted $R^2$ value in this model is larger than the $R^2$ value in the previous model, so this model appears to be more useful. Answers may vary.

f. $R^2 = 0.9352$. The adjusted $R^2$ value should be used to compare this model to the simple model because additional independent variables have been added.

## Section 14.3

**17. a.** $H_0: \beta_1 = \beta_2 = 0$, $H_a$: At least one $\beta_i \neq 0$.

**b.** $F = 10.8947$

**c.** Yes. $P$–value $= 0.0001$, Reject $H_0$.

**d.** $(1466.6644, 4243.1181)$ We are 95% confident that the true increase in annual salary for each additional year of education is between \$1466.66 and \$4243.12.

**e.** $H_0: \beta_1 = 0$, $H_a: \beta_1 \neq 0$

**f.** Yes. $t = 4.140$, $P$–value $= 0.0001$, Reject $H_0$.

**19. a.** Rent $= \beta_0 + \beta_1$(Population) $+ \beta_2$(Income) $+ \varepsilon_i$

**b.** We would expect both coefficients to be positive since larger cities tend to have more expensive rental rates and more expensive rental rates would be expected in areas with greater incomes. Answers will vary.

**c.** $b_0 = 138.5023$; $b_1 = 0.1199$; $b_2 = 16.8207$; $\hat{y} = 138.5023 + 0.1199x_1 + 16.8207x_2$. A city with 0 population and 0 income would have an expected monthly rent of about \$138.50. For each additional 1000 people in the city, monthly rent is expected to increase by about \$0.12. For each additional \$1000 in average median income, monthly rent is expected to increase by about \$16.82.

**d.** Yes; $H_0: \beta_1 = \beta_2 = 0$; $H_a$: At least one $\beta_i \neq 0$; $F = 30.6224$; $P$–value $= 0.0000$; Reject $H_0$.

**e.** $(11.4624, 22.1789)$; We are 95% confident that the true increase in monthly rent for each additional \$1000 increase in median income is between \$11.46 and \$22.18.

**f.** Population: No; $H_0: \beta_1 = 0$; $H_a: \beta_1 \neq 0$; $t = 1.423$; $P$–value $= 0.1803$; Fail to reject $H_0$. Income: Yes; $\beta_2 = 0$; $H_a: \beta_2 \neq 0$; $t = 6.840$; $P$–value $= 0.0000$; Reject $H_0$.

**g.** Yes, the population variable should be removed, as it is not a significant predictor of monthly rent. Answers may vary.

## Section 14.4

**7. a.**

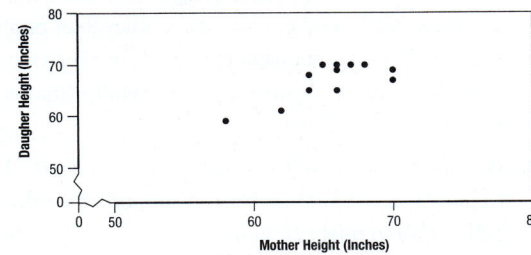

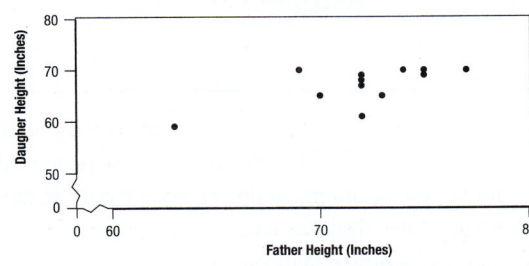

**a.** There appears to be a positive relationship between each parent's height and the child height. The mother–daughter plot appears more linear than the father–daughter plot. See plot on previous page.

**b.** $\hat{y} = -4.6456 + 0.5939x_1 + 0.4523x_2$

**c.** Yes; $H_0: \beta_1 = \beta_2 = 0$; $H_a$: At least one $\beta_i \neq 0$; $F = 11.2521$; $P$–value $= 0.0028$; Reject $H_0$.

**d.** No; $H_0: \beta_2 = 0$; $H_a: \beta_2 \neq 0$; $t = 2.176$; $P$–value $= 0.0546$; Fail to reject $H_0$.

**e.** No; $H_0: \beta_1 = 0$; $H_a: \beta_1 \neq 0$; $t = 2.628$; $P$–value $= 0.0253$; Fail to reject $H_0$.

**f.** For each additional inch in the mother's height, the daughter's height is expected to increase by approximately 0.5939 inch. For each additional inch in the father's height, the daughter's height is expected to increase by approximately 0.4523 inch.

**g.** Mother: $(0.0903, 1.0974)$; We are 95% confident that for each additional inch in the mother's height, the daughter will be between 0.0903 and 1.0974 inches taller. Father: $(-0.0108, 0.9153)$; We are 95% confident that for each additional inch in the father's height, the daughter will be between 0.0108 inch shorter and 0.9153 inch taller.

**h.** 66.8297 inches, or 5 feet 6.8297 inches

**i.** $(61.533, 72.126)$; We are 95% confident that a particular daughter whose mother is 5 foot 4 and father is 6 foot 2 will be between 61.533 and 72.126 inches tall.

**j.** $(64.785, 68.875)$

**9. a.** $\hat{y} = -15.2395 + 0.1319x_1 + 0.0869x_2$

**b.** Yes; $H_0: \beta_1 = \beta_2 = 0$; $H_a$: At least one $\beta_i \neq 0$; $F = 21.6521$; $P$–value $= 0.0000$; Reject $H_0$.

**c.** Approximately 59.89%. This is slightly lower than the percentage for the model with the 3 independent variables (59.99%). However, the difference is likely due to the additional independent variable in the previous model.

**d.** The model without the first downs variable is likely the better model. Though the $R^2$ value is lower, the adjusted $R^2$ value is larger, indicating that the difference is likely due to the addition of the first downs variable, which is not useful in predicting points scored. Answers will vary.

**e.** Approximately 4 (3.6890)

**f.** (−2.96, 11.16); We are 95% confident that the average points scored is between 0 and 11 points (since points

scored cannot be negative in football) when the offense has 102 rushing yards and 63 passing yards.

**g.** (−12.70, 20.91); We are 95% confident that in this particular game against Miami, Buffalo will score between 0 and 21 points (since points scored cannot be negative in football).

## Section 14.5

**9. a.** Yes, the overall model is significant.
$H_0: \beta_1 = \beta_2 = \beta_3 = 0$, $H_a$: At least one $\beta_i \neq 0$,
$F = 235.1310$, $P$–value $= 0.0000$, Reject $H_0$.

**b.** Approximately 51.06%

**c.** Yes. $H_0: \beta_3 = 0$, $H_a: \beta_3 \neq 0$, $t = -3.051$,
$P$–value $= 0.0024$, Reject $H_0$.

**d.** Yes, all of the independent variables appear to be significant in predicting GPA. However, the $R^2$ value is only about 51%. This indicates that there may be other factors influencing GPA that have not been accounted for. Answers will vary.

**e.** Answers will vary. Quantitative: hours of study time, number of credit hours, SAT score, etc. Qualitative: major, gender, extracurricular activities, etc.

**11. a.**

| School | Private | School | Private |
|--------|---------|--------|---------|
| 1 | 1 | 11 | 1 |
| 2 | 0 | 12 | 0 |
| 3 | 0 | 13 | 1 |
| 4 | 1 | 14 | 0 |
| 5 | 1 | 15 | 1 |
| 6 | 0 | 16 | 0 |

| 7 | 1 | 17 | 0 |
|---|---|----|---|
| 8 | 1 | 18 | 1 |
| 9 | 0 | 19 | 0 |
| 10 | 0 | 20 | 1 |

**b.** $\hat{y} = 105.9195 + 2.5203x_1 - 0.0033x_2 - 65.3808x_3$

**c.** Yes. $H_0: \beta_1 = \beta_2 = \beta_3 = 0$, $H_a$: At least one $\beta_i \neq 0$,
$F = 35.6335$, $P$–value $= 0.0000$, Reject $H_0$.

**d.** The coefficient for police is positive, indicating that as the number of police increases, so does the number of crimes. This is not what would be expected. The coefficient for enrollment is negative, indicating that as enrollment increases, the number of crimes decreases. This is also surprising, as one would think that a larger student body would result in increased crimes. The coefficient for private is negative, indicating that private schools tend to have less crimes than public schools. This is somewhat expected since private schools are more expensive and generally smaller in size. Answers will vary.

**e.** $H_0: \beta_3 = 0$, $H_a: \beta_3 \neq 0$, $t = -2.624$, $P$–value $= 0.0184$,
Reject $H_0$ at the 0.05 level, so it supports the officials' belief, Fail to reject $H_0$ at the 0.01 level, so yes, the decision would change.

## Section 14.6

**11. a.** Age and experience are likely correlated since the older you are, the more experience you tend to have. Answers will vary.

**b.** By using statistical software to find the correlation coefficient for each pair of variables. The closer the correlation coefficients are to 1 or −1, the more likely it is that collinearity exists. Answers will vary.

**c.** $58,921.91

**d.** Yes, because the overall model is not significant. None

of the independent variables appear to be useful in predicting salary. Also, the age and experience values for which we are predicting salary are outside the range of the observed data, so extrapolation could be an issue. Answers will vary.

**13.** There appears to be a strong positive correlation between $x_1$ and $x_3$. There also appears to be a moderate positive correlation between $x_1$ and $x_4$ and $x_3$ and $x_4$. Answers will vary.

## Chapter 14 Additional Exercises

**1. a.**

| Observation | Degree |
|-------------|--------|
| 1 | 0 |
| 2 | 1 |
| 3 | 0 |
| 4 | 0 |

| 5 | 1 |
|----|---|
| 6 | 1 |
| 7 | 1 |
| 8 | 1 |
| 9 | 0 |
| 10 | 1 |

| | |
|---|---|
| 11 | 0 |
| 12 | 1 |
| 13 | 0 |
| 14 | 0 |
| 15 | 1 |
| 16 | 0 |
| 17 | 1 |
| 18 | 1 |
| 19 | 0 |
| 20 | 0 |

**b.** $\hat{y} = 10419.8636 + 1501.2807x_1 + 13047.4495x_2$

**c.** $1501.28

**d.** Yes. $H_0: \beta_2 = 0$, $H_a: \beta_2 \neq 0$, $t = 3.845$, $P$–value = 0.0000, Reject $H_0$. This indicates that a master's degree is significant in predicting salary, and the estimated increase in annual salary for people with master's degrees is $13,047.45, which is greater than $10,000.

**e.** There are many other factors that influence annual salary. Answers will vary.

**3. a.** The relationship appears to be positive, but it does not look linear. Answers will vary.

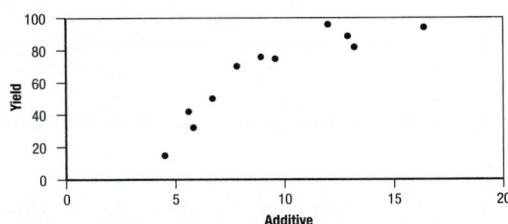

**b.** $\hat{y} = 5.9860 + 6.3361x_1$, $R^2 = 0.7971$, $s_e^2 = 163.5177$

**c.** Linear: 107.3638, Polynomial: 91.09. The observed value when the additive was 16.4 was a yield of 94, which is far closer to the polynomial result. Answers will vary.

**d.** The polynomial model has a higher $R^2$ value and a lower $s_e^2$ value, indicating that the polynomial model fits the data better than the linear model. Answers may vary.

**e.** The polynomial model because it has a higher $R^2$ value and a lower $s_e^2$ value. Answers will vary.

**5. a.** Answers may vary. GPA $= \beta_0 + \beta_1$(Xoom) $+ \beta_2$(Galaxy) $+ \beta_3$(iPad) $+ \varepsilon_i$

**b.** Answers may vary. $\hat{y} = 3.4867 - 0.2761$(Xoom) $- 0.2503$(Galaxy) $- 0.4182$(iPad)

**c.** No. $H_0: \beta_1 = \beta_2 = \beta_3 = 0$, $H_a$: At least one $\beta_i \neq 0$, $F = 0.4740$, $P$–value = 0.7031, Fail to reject $H_0$. Answers may vary.

**d.** Answers may vary. $\hat{y} = 3.5473 + 0.1115(\$30,000) - 0.6059(\$30,000–\$49,999) - 0.3448(\$50,000–\$74,999)$. This model is also not significant at the 0.05 level. There is not sufficient evidence that type of tablet or income is a significant predictor of GPA.

**e.** No, none of the tablet types are significant in predicting GPA. Answers will vary.

**7. a.** Benefit $= \beta_0 + \beta_1$(Family Size) $+ \beta_2$(Income) $+ \varepsilon_i$

**b.** Yes. $\hat{y} = 40.7903 + 3.6594x_1 + 0.1461x_2$, $H_0: \beta_1 = \beta_2 = 0$, $H_a$: At least one $\beta_i \neq 0$, $F = 397.9462$, $P$-value = 0.0000, Reject $H_0$.

**c.** Family size is significant at the 0.10 level ($t = 1.702$, $P$-value = 0.0972), and monthly income is significant at the 0.01 level ($t = 27.955$, $P$-value = 0.0000). Monthly income appears to be a much better predictor for benefits than family size. Answers may vary.

**d.** (410.70, 430.77), We are 95% confident that the average monthly benefit for a 4-person family with a monthly income of $2500 is between $410.70 and $420.77.

**e.** (338.93, 502.53), We are 99% confident that the monthly benefit for a particular 4-person family with a monthly income of $2500 is between $338.93 and $502.53.

**f.** The prediction interval is wider than the confidence interval by 143.53. This is due to the increased confidence level (99% vs. 95%) and the prediction interval accounts for individual variation. Answers may vary.

# Chapter 15

## Section 15.1

**9.** Yearly: Linear trend, big drop after 2013.

Monthly: More stationary. February is always the lowest point and August the highest.

**11.** There is a positive trend; however it is broken during 2008 (during the economy collapse) and looks like the sales numbers are reaching back to that level again.

**13.** The rate has been decreasing since 1981 and it seems to be stabilizing around 4%.

## Section 15.2

**7.** 192,686.67
Passenger: 582,790.80

**9.** Truck: 128,487.40,

5-month WMA: 188,083.05

**11.** 4-year SMA: 4.11%, 4-year WMA: 4.23%

## Section 15.3

**7.**

| Year | Truck Crossings | Forecast |
|------|-----------------|----------|
| 2011 | 1,474,775.00 | 1,474,775.00 |
| 2012 | 1,541,150.00 | 1,474,775.00 |
| 2013 | 1,533,049.00 | 1,494,687.50 |
| 2014 | 1,554,152.00 | 1,506,195.95 |
| 2015 | 1,544,702.00 | 1,520,582.77 |
| 2016 | 1,598,017.00 | 1,527,818.54 |
| 2017 | 1,574,771.00 | 1,548,878.07 |
| 2018 | 1,581,443.00 | 1,556,645.95 |
| 2019 | | **1,564,085.07** |

**9. a.** Simple exponential smoothing forecast for 2020: 4.35%

Adjusted exponential smoothing forecast for 2020: 3.45%

**b.** Until 1993, the adjusted exponential smoothing forecast is above the simple exponential smoothing forecast. After that, it's always below.

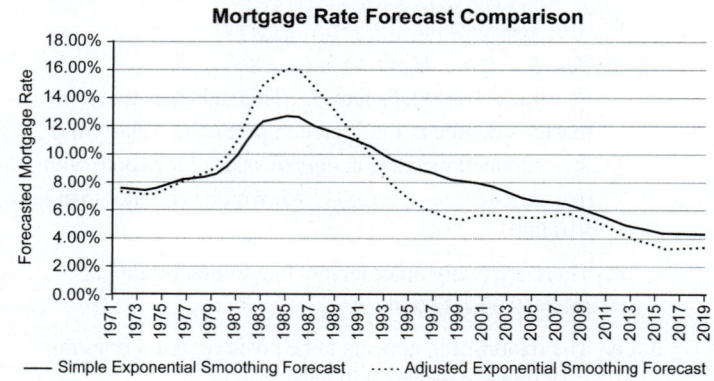

**Mortgage Rate Forecast Comparison**

— Simple Exponential Smoothing Forecast      ⋯⋯ Adjusted Exponential Smoothing Forecast

## Section 15.4

**11.** MAPD = 4.62%

From the tracking signal, we see it looks biased and out of control. Note in many places it is above 4. The forecast looks good from an MAPD point of view, but it is biased.

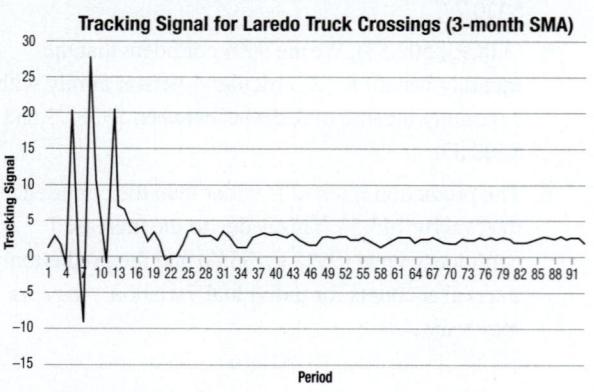

**Tracking Signal for Laredo Truck Crossings (3-month SMA)**

**13.** MAPD = 0.63%

No bias, all values of the *TS* are between 4 and −4. The forecast looks very accurate based on the MAPD since it's less than 1%. However, seasonality still needs to be

addressed.

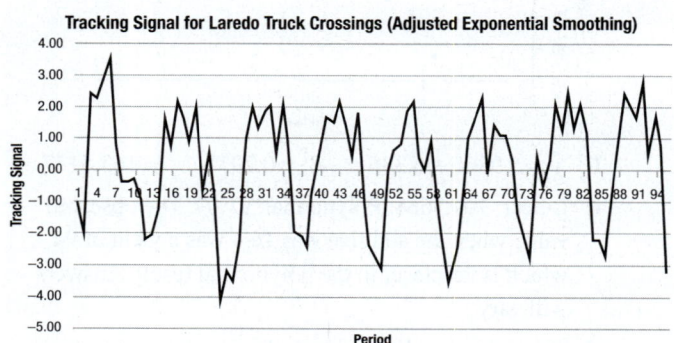

**Tracking Signal for Laredo Truck Crossings (Adjusted Exponential Smoothing)**

**15.** MAPD = 8.37%

**17.**

| Error Metrics |
|---------------|
| MSE |
| MAPD (best), MAPE and MAD |
| TS and E |
| MAPD |

## Section 15.5

**7.**

| Month | Seasonality Factor |
|-------|--------------------|
| Jan | 0.0788 |
| Feb | 0.0781 |
| Mar | 0.0863 |
| Apr | 0.0837 |

|  | |
|---|---|
| May | 0.0854 |
| Jun | 0.0851 |
| Jul | 0.0833 |
| Aug | 0.0864 |
| Sep | 0.0828 |
| Oct | 0.0893 |
| Nov | 0.0832 |
| Dec | 0.0775 |

| Month | Seasonality Factor |
|---|---|
| Jan | 0.0767 |
| Feb | 0.0781 |
| Mar | 0.0818 |
| Apr | 0.0849 |
| May | 0.0875 |
| Jun | 0.0878 |
| Jul | 0.0870 |
| Aug | 0.0866 |
| Sep | 0.0868 |
| Oct | 0.0839 |
| Nov | 0.0806 |
| Dec | 0.0782 |

9. Forecast $= 380010.875 + 5335.250\, Q_1 + 21527.000\, Q_2 + 3351.625\, Q_3$

11. The summer months, particularly May and June are the highest.

## Chapter 15 Additional Exercises

1. $MAD = 0.8333$
   $E(WMA) = -0.8$

3. $E(SMA) = -1.6667;$

5. $SF_1 = 16.28; SF_2 = 11.63; SF_3 = 8.73; SF_4 = 21.53$

# Chapter 16

## Section 16.1

9. a. 41.401
   b. 28.845
   c. 5.024
   d. 107.565
   e. 24.769

11. a. 66.766
    b. 27.488

   c. 7.378
   d. 33.196
   e. 63.167

13. a. 0.042982
    b. 30.0874
    c. 16.013

## Section 16.2

9. a. No.
   $H_0: p_1 = p_2 = \cdots = p_5 = \frac{1}{5}$
   $H_a$: Any possible difference.
   $\chi_\alpha^2 = 9.488, \chi^2 = 7.647$
   $P$-value $\approx 0.1054$
   Fail to reject $H_0$
   b. The distribution of the number of service calls has a multinomial probability distribution.

11. a. Yes.
    $H_0: p_1 = p_2 = \cdots = p_{12} = \frac{1}{12}$
    $H_a$: Any possible difference.
    $\chi_\alpha^2 = 24.725, \chi^2 = 26.376$
    $P$-value $\approx 0.0057$
    Reject $H_0$

   b. The distribution of fatal accidents has a multinomial probability distribution.

13. a. Yes.
    $H_0: p_1 = 0.15, p_2 = p_3 = 0.30, p_4 = 0.25$
    $H_a$: Any possible difference.
    $\chi_\alpha^2 = 11.345, \chi^2 = 28.000$
    $P$-value $\approx 0.0000036$
    Reject $H_0$
    b. The distribution of the survey responses has a multinomial probability distribution.

## Section 16.3

**13. a.** Yes.

$H_0$: Age and political affiliation are independent.

$H_a$: Age and political affiliation are dependent.

$\chi_\alpha^2 = 9.488, \chi^2 = 31.881$

$P$-value $\approx 0.00000202$

Reject $H_0$

**b.** Each variable satisfies the properties of a multinomial distribution.

**15. a.** No.

$H_0$: Type of abuse and gender are independent.

$H_a$: Type of abuse and gender are dependent.

$\chi_\alpha^2 = 7.815, \chi^2 = 3.575$

$P$-value $\approx 0.311164$

Fail to reject $H_0$

**b.** Each variable satisfies the properties of a multinomial distribution.

## Chapter 16 Additional Exercises

**1. a.** Yes.

$H_0: p_1 = 0.50, p_2 = p_3 = 0.25$

$H_a$: Any possible difference.

$\chi_\alpha^2 = 5.991, \chi^2 = 68.400$

$P$-value $\approx 0.0000$

Reject $H_0$

**b.** The underlying distribution is a multinomial probability distribution.

**3.** Yes.

$H_0: p_1 = 0.51, p_2 = 0.33, p_3 = 0.16$

$H_a$: Any possible difference.

$\chi_\alpha^2 = 5.991, \chi^2 = 10.861$

$P$-value $\approx 0.00438$

Reject $H_0$

**5.** No.

$H_0: p_1 = 0.226, p_2 = 0.611, p_3 = 0.075, p_4 = 0.088$

$H_a$: Any possible difference.

$\chi_\alpha^2 = 11.345, \chi^2 = 8.012$

$P$-value $\approx 0.045764$

Fail to reject $H_0$

**b.** Each variable satisfies the properties of a multinomial distribution.

**7.** Yes.

$H_0$: Opinion and region are independent.

$H_a$: Opinion and region are dependent.

$\chi_\alpha^2 = 16.812, \chi^2 = 96.416$

$P$-value $\approx 0.0000$

Reject $H_0$

# Chapter 17

## Section 17.1

**15.** $H_0$: Median = \$25,000, $H_a$: Median > \$25,000, Critical value = $-1.28, z = -3.25$, Reject $H_0$

**17. a.** $H_0$: Median = 14,400 hrs, $H_a$: Median < 14,400 hrs, Critical value = 2, $X = 6$, Fail to Reject $H_0$

**b.** The data are randomly selected.

**19.** $H_0$: # of Positive Signs = # of Negative Signs

$H_a$: # of Positive Signs > # of Negative Signs

Critical value = 6

$X = 9$

Fail to Reject $H_0$

**21.** $H_0$: # of Negative Signs = # of Positive Signs

$H_a$: # of Negative Signs ≠ # of Positive Signs

Critical value = 0

$X = 0$

Reject $H_0$

## Section 17.2

**11.**

| Mutual Fund | Price | Rank |
|---|---|---|
| American Funds | 24.4 | 7.5 |
| Columbia Management | 9.41 | 2.5 |
| Morgan Stanley | 23.74 | 6 |
| Fidelity Investments | 24.40 | 7.5 |
| John Hancock | 9.41 | 2.5 |
| DWS Investments | 15.57 | 5 |
| UBS | 12.15 | 4 |
| Prudential Investments | 9.23 | 1 |
| Value Line Funds | 32.82 | 9 |
| The Vanguard Group | 34.72 | 10 |

**13. a.** $H_0$: DHEA-S is the same
$H_a$: DHEA-S is increased
Critical value $= 0$
$X = 2$
Fail to Reject $H_0$

**b.** The data are randomly selected.

**c.** $H_0$: DHEA-S is the same
$H_a$: DHEA-S is increased
Critical value $= 4$
$T_+ = 5$
Fail to Reject $H_0$

**d.** Pairs of data have been randomly selected and are such that the absolute values of their differences can be ranked.

**e.** The signed-rank test because the magnitudes of the differences are not ignored. Answers will vary.

**15. a.** The paired differences have an approximately normal distribution.

**b.** $H_0$: Model A = Model B
$H_a$: Model A ≠ Model B
Critical value $= 1$
$T_+ = 0$
Reject $H_0$

**c.** Pairs of data have been randomly selected and are such that the absolute values of their differences can be ranked.

**d.** In all three tests the null hypothesis is rejected in favor of the alternative.

## Section 17.3

**11. a.** $H_0$: Mr. Ellis = Mr. Ford
$H_a$: Mr. Ellis ≠ Mr. Ford
Critical value $= 37, 68$
$T_x = 42, 63$
Fail to Reject $H_0$

**b.** The data are such that they can be ranked. The two samples are selected in an independent and random fashion.

**13. a.** $H_0$: New Battery = Old Battery
$H_a$: New Battery < Old Battery
Critical value $= 28$
$T_x = 35$
Fail to Reject $H_0$

**b.** The data are such that they can be ranked. The two samples are selected in an independent and random fashion.

**15. a.** $H_0$: Dramas = Comedies
$H_a$: Dramas > Comedies
Critical value $= 36$
$T_x = 32$
Fail to Reject $H_0$

**b.** The data are such that they can be ranked. The two samples are selected in an independent and random fashion.

**17. a.** $H_0$: Service City = Sunshine City
$H_a$: Service City < Sunshine City
Critical value $= 52$
$T_x = 47$
Fail to Reject $H_0$

**b.** The data are such that they can be ranked. The two samples are selected in an independent and random fashion.

## Section 17.4

**11. a.** $H_0: \rho_s = 0$
$H_a: \rho_s \neq 0$
$r_s = -0.8042$
Critical values $= -0.591$,
$0.591$
Reject $H_0$

**b.** Yes, but the assumption must be made that the relationship between the variables is linear. Answers will vary.

**13.** $H_0: \rho_s = 0$
$H_a: \rho_s \neq 0$
$r_s = 0.0182$
Critical values $= -0.648$,
$0.648$
Fail to Reject $H_0$

## Section 17.5

**9.** $H_0$: The sequence is random.
$H_a$: The sequence is not random.
$N = 26$, $m = 10$,
$n = 16$, $R = 14$
Critical values $= 8, 19$
Fail to Reject $H_0$

**11.** $H_0$: The sequence is random.
$H_a$: The sequence is not random.
$N = 31$, $m = 15$,
$n = 16$, $R = 19$
Critical values $= 10, 23$
Fail to Reject $H_0$

## Section 17.6

**9.** $H_0$: The milk production for all schedules is the same.
$H_a$: The milk production for at least one of the schedules is different.
$H = 7.0234$
Critical value $= 6.2514$
Reject $H_0$

## Chapter 17 Additional Exercises

**1. a.** The paired differences have an approximately normal distribution.

**b.** $H_0$: Vest Treatment = Traditional Treatment
$H_a$: Vest Treatment > Traditional Treatment
$X = 0$
Fail to Reject $H_0$

**c.** The data are randomly selected.

**d.** $H_0$: Vest Treatment = Traditional Treatment
$H_a$: Vest Treatment > Traditional Treatment
$T_+ = 0$
Fail to Reject $H_0$

**e.** Pairs of data have been randomly selected and are such that the absolute values of their differences can be ranked.

**f.** The signed-rank test because the magnitudes of the differences are not ignored. Answers will vary.

**g.** There is not sufficient evidence that the diameter of blood vessels in the lungs is significantly larger after using the vest treatment when performing a paired difference test. There is sufficient evidence using the nonparametric methods. Answers will vary.

$H_0$: Vest Treatment = Traditional Treatment
$H_a$: Vest Treatment > Traditional Treatment
$t = -2.75$
Critical value $= -3.747$
Fail to Reject $H_0$

**3. a.** The Wilcoxon rank-sum test because we are not dealing with paired data.

**b.** $H_0$: Diet B = Diet A
$H_a$: Diet B > Diet A

**c.** $H_0$: Diet B = Diet A
$H_a$: Diet B > Diet A
Critical value $= 131$
$T_x = 113$
Fail to Reject $H_0$

**d.** The data are such that they can be ranked. The two samples are selected in an independent and random fashion.

**5. a.** $r_s = 0.2727$ these two variables have a weak positive relationship.

**b.** $H_0: \rho_s = 0$
$\qquad H_a: \rho_s \neq 0$
$\qquad r_s = 0.2727$
$\qquad$ Critical values $= -0.648, 0.648$
$\qquad$ Fail to Reject $H_0$

**7.** $H_0$: The sequence is random.
$\qquad H_a$: The sequence is not random.
$\qquad N = 9$, $m = 5$, $n = 4$,
$\qquad R = 4$
$\qquad$ Critical values $= 2, 9$
$\qquad$ Fail to Reject $H_0$

**9.** $H_0$: The rankings are the same for the three conferences.
$\qquad H_a$: The ranking of at least one conference is different.
$\qquad H = 13.5755$
$\qquad$ Critical value $= 4.6052$
$\qquad$ Reject $H_0$

# Chapter 18

## Section 18.1

**11. a.** In order to assist customers in the best possible manner, the receptionist should understand how to handle various customer calls. Answers will vary.

**b.** The flowchart appears to be incomplete, since it does not tell you what to do in the case of something other than shipping questions, billing questions, problems with software function, and product information. The arrow following the "other" box goes nowhere. Answers will vary.

**13. a.**

| Complaint | Relative Frequency (%) |
|---|---|
| Not Enough Parking | 35.6 |
| Rude Personnel | 22.2 |
| Poor Lighting | 18.7 |
| Confusing Store Layout | 12.4 |
| Limited Sizes | 6.7 |
| Clothing Unattractive | 4.4 |

**b.**

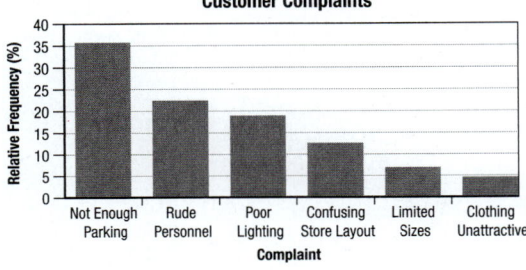

**c.** Not enough parking, rude personnel, poor lighting

**d.** No, 80% of the problems are explained by 50% of the causes. Answers will vary.

**e.** The store should implement more parking, better training for its employees, and better lighting. The parking issue should be addressed first because it is generating the largest percentage of complaints. Answers will vary.

**15. a.**

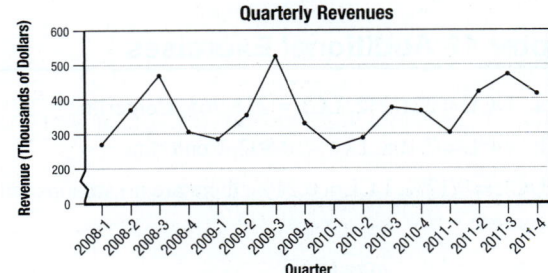

**b.** The revenues appear to be cyclical. Answers will vary.

**c.** Based on the run chart, yes. Answers will vary.

**d.** Families probably tend to eat out more when the weather is nicer. Answers will vary.

**e.** There is a cyclical pattern, so the process could be considered unstable. The restaurant could offer coupons or incentives in months where revenues tend to be lower. Answers will vary.

## Section 18.2

**9. a.** UCL $= 14.2$, LCL $= 5.8$, Centerline $= 10.0$

**b.** The average value of the measurements is 10. Measurements between 5.8 and 14.2 are considered in control, and can be attributed to normal process variation.

**c.** B, F, H, K, N

**11. a.** UCL = 36, LCL = 16, Centerline = 26

**b.** The average value of the measurements is 26. Measurements between 16 and 36 are considered in control, and can be attributed to normal process variation.

**c.** G, K, M

## Section 18.3

**11.** UCL = 31.125, LCL = 28.875

**13. a.** UCL = 4.3354, LCL = 3.6646, Centerline = 4

**b.** UCL = 4.6244, LCL = 0, Centerline = 2.1875

**c.** 1, 3, 4, 5, 8, 11, 13, 14, and 15

**d.** 13

**15. a.** UCL = 25.134, LCL = 23.046, samples 1, 4, 9, and 10 are out of control.

**b.** UCL = 9.1988, LCL = 2.4012

**17.** UCL = 0.235, LCL = 0.070, samples 5 and 9 are out of control.

## Section 18.4

**9.** UCL = 3.98%, LCL = 0%

**11. a.** UCL = 28%, LCL = 0%, Centerline = 10%

**b.**

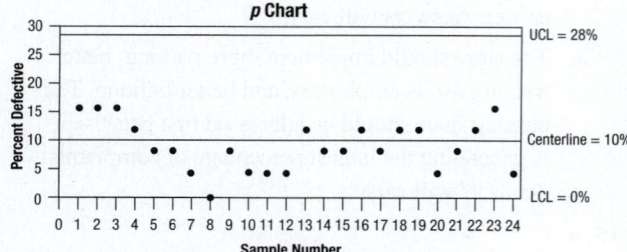

**c.** No, no samples are out of control.

**13.** UCL = 41.77%, LCL = 0%, Centerline = 11.5%, no samples are out of control.

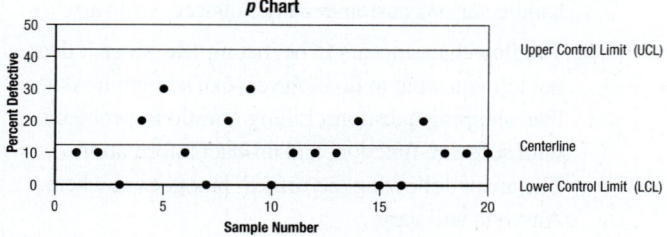

## Chapter 18 Additional Exercises

**1. a.** UCL = 221.232, LCL = 218.768, Centerline = 220

**b.** UCL = 7.108, LCL = 0.892, Centerline = 4

**3.** UCL = 9.17%, LCL = 0.21%, there are no seasons out of control.

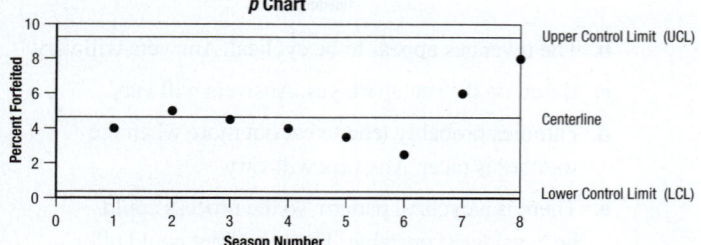

**5.** UCL = 54.03%, LCL = 0%, Centerline = 21.96%, there are no samples out of control, so the VP should not be concerned about the number of mistakes being made on financial statements.

**7.** $\bar{x}$ chart: UCL = 42.4908, LCL = 39.1800, Centerline = 40.8354, no samples are out of control.

R chart: UCL = 6.0651, LCL = 0, Centerline = 2.869, no samples are out of control.

# Index

## A

Achenwall, Gottfried 8
Addition rule 260
Adjusted exponential
  smoothing 925
Adjusted $R^2$ 854
Adrain, Robert 361
Alternative hypothesis
  definition 510
  one-sided 510
  two-sided 510
Analysis of variance.
  *See ANOVA*
Analytics
  business 43
  definition 43
  predictive 44
  prescriptive 44
Analyzing graphs 119–123
ANOVA
  assumptions 664
  definition 653
  degrees of freedom 655
  formulas 656
  masking 708
  table 656
  two-way 701
Anscombe, Francis 215
Aristotle 27
Arithmetic mean 152
Assignable variation 1084
Attribute sampling 1095
Average
  definition 152
  moving 161
    simple 917
    weighted 919

## B

Bar chart
  3-D 90
  definition 84
  histogram 101
  pictograph 123
  stacked 88
Bar graph 120–121
Base level variable 878
Bayes' theorem 273
Before and after study 32
Bell-shaped 159
Bias
  sample 416
  selection 416
Biased 416
Big data 41
Bimodal 159
Binomial
  distribution 315, 386

mean 386
  shape 320
  standard deviation 386
  experiment 316
  random variable 315
    expected value 321
    standard deviation 321
    variance 321
  table 319
Bivariate data 205
Black Swan event 255
Block(s)
  mean square (MSBL) 692
  randomized design 690
  sum of squares (SSBL) 691
Bohr, Niels 251
Box and whisker plot 182
Box plot 182
Brahe, Tycho 169
Bristol, Muriel 527
Business analytics 43

## C

Causal factors 35
Census 6
Centerline
  definition 1083
  mean and standard deviation
    known 1087
  mean and standard deviation
    unknown 1087
  process proportion known 1095
  process proportion
    unknown 1095
  process range 1088
Central Limit Theorem 425
Central tendency 150
Central value
  of the sample mean 421
  of the sample proportion 432
Chamberlain, Wilt 174
Chance variation 1084
Chart
  3-D bar 90
  3-D pie 92
  bar 84
  control 1083
  flow 1076
  $p$ 1095
  Pareto 1077
  pie 90
  $R$ 1086
  range 1086
  run 1079
  stacked bar 88
  $\bar{x}$ 1086
Chebyshev's Theorem 173
Chi-square
  critical values 968
  distribution 484, 556, 967
  hypotheses
    test for association 984
    test for goodness of fit 974
  statistic 556, 967
  table 968
  test for association 980

critical value 985
  hypotheses 984
  rejection region 984
  test statistic 984
  test for goodness of fit 971
  critical value 976
  hypotheses 974
  rejection region 974
  test statistic 974
Choropleth maps 111–113
Class
  boundaries 101
  midpoint 198
  number of 97
  width 97
Classical probability 250
Cluster sampling 441
Coefficient
  correlation 208, 784, 786
  of determination 783, 786
    adjusted 854
    multiple 854
  individual 865
  intercept 763
  slope 753, 763
    confidence interval 795
    critical value 802
    hypothesis 802
    $P$-value 803
    test statistic 802
  of variation 174
Combination 277
Common cause variation 1084
Common response 211
Complement 257
Complete factorial
  experiment 701
Completely randomized
  design 31, 579, 690
Compound event 256
Conditional probability 263
Confidence
  coefficient 464
  interval 464
    comparing two
      population means 580, 588
      population proportions 607
      population variances 617–
        618
    hypothesis test
      sigma ($\sigma$) known 544
      sigma ($\sigma$) unknown 544
    individual coefficient 865
    mean
      sigma ($\sigma$) known 544
      sigma ($\sigma$) unknown 544
    mean value of $y$ given $x$ 809,
      870
    multiple regression
      individual coefficient 865
    paired difference 600
    population mean
      sigma ($\sigma$) known 465
      sigma ($\sigma$) unknown 470
    population proportion 479
    population variance 485, 556,
      501

predicted value of $y$ given
  $x$ 812, 871
  sigma-squared 485, 556, 501
  slope coefficient 795
  to test a hypothesis 544
  unequal variances 593
  level 464
Confounding 212
Confounding variable 26
Contingency table 980
Continuity correction 387
Continuous
  data 46
  improvement 1083
  random variable 299, 357
  uniform distribution 357–358
Control
  chart 1083
    assignable variation 1084
    chance variation 1084
    common cause variation 1084
    normal process variation 1084
    $p$ 1095
    $R$ 1086
    range 1086
    special cause variation 1084
    $\bar{x}$ 1086
  group 5, 30
  limit
    centerline 1083
    in control 1083
    lower (LCL) 1083
    out of control 1083
    upper (UCL) 1083
Control group 5, 30
Controlled
  experiment 5, 30
Convenience sample 440
Correlated error 886
Correlation
  coefficient
    formula 208, 786
    properties 211
    Spearman rank 1036
  spurious 213
Critical value
  chi-square ($\chi^2$) 968
  chi-square test for
    association 985
  chi-square test for goodness of
    fit 976
  $F$ 615
  individual coefficient 863
  Kruskal-Wallis test 1047
  paired difference 603
  population
    mean
      comparing two 583, 591
      sigma ($\sigma$) known 521
      sigma ($\sigma$) unknown 533
    proportion 548
      comparing two 611
    variance 557
  rank correlation test 1037
  runs test for randomness 1042
  sign test 1013
  slope coefficient 802
  $t$ 470